STEPHENS' BOOK OF THE FARM

EDWARDIAN FARM EDITION

CROPS AND LIVESTOCK

by
James MacDonald, F.R.S.E.
Secretary of the Highland and Agricultural Society of
Scotland

Stephens' Book of the Farm was first published in 1844. This material is extracted from the Edwardian edition that was revised, reorganised and largely rewritten by James MacDonald in 1908.

Printed on acid free ANSI archival quality paper.

CROPS

CONTENTS OF VOLUME II.

THE MANAGEMENT OF PASTURES.

FORMATION OF PERMANENT PASTURE.

ENSILAGE.

LIST OF ILLUSTRATIONS IN VOLUME II.

THE

BOOK OF THE FARM.

———•———

EXPERIMENTS IN MANURING CROPS.

ROTHAMSTED EXPERIMENTS.

ROTHAMSTED has become a household word wherever science is applied to agriculture. In 1834 Sir (then Mr) John Bennet Lawes succeeded to the estate of Rothamsted, Hertfordshire, and soon after began to conduct experiments with different manuring substances, first with plants in pots, and afterwards in the field. In 1840 and 1841 somewhat extensive field trials were carried out, and in 1843 the experiments were begun upon the comprehensive and systematic form which they have ever since maintained. The foundation of the Rothamsted Experimental Station is therefore usually dated from 1843.

The experiments, the most elaborate and comprehensive of the kind ever attempted in any country, have from the first been maintained entirely by the late Sir John Bennet Lawes, Bart., LL.D., who gave in trust the handsome sum of £100,000, besides certain areas of land, to ensure to British agriculture the benefits and guidance derivable from the perpetual continuation of the Rothamsted experiments.

The Committee to whom the management of the Experimental Station is entrusted consists of representatives of the Royal Society, the Royal Agricultural Society of England, the Chemical and Linnean Societies, and the owner of Rothamsted.

From 1843 until the death of Sir John Lawes in 1900, Sir J. H. Gilbert was associated with the conduct of the experiments, and had the direction of the laboratory: the familiar names "Lawes and Gilbert" appear on the titles of about 60 papers published between 1851 and 1900. Lawes died in 1900, and Gilbert followed him a little more than a year later, in 1901.

During the lifetime of the founders of the Rothamsted experiments one or more trained chemists, besides the assistants engaged in routine work, were on the staff of the station: for many years also the late R. Warington, F.R.S., was at work in the Rothamsted laboratory, and there conducted his long series of investigations into nitrification and the nature and composition of drainage-waters with which his name is chiefly associated.

Since the death of Sir Henry Gilbert, Mr A. D. Hall has been Director of the Experimental Station, which, by the generosity of the Goldsmiths' Company and other individuals, has been enabled to enlarge its staff and bring its equipment more into line with that of a modern agricultural research station.

Main Features of the Work.—The special feature of the Rothamsted Experimental Station has always been the manurial experiments upon the principal farm crops, wheat, barley, roots, clover, hay ; field experiments which have been carried out without variation for more than fifty years, so that all fluctuations due to season or to inequalities of the soil have been swept away from the results, and the effect of the manure upon the yield, the quality of the crop, and again upon the soil, has become manifest in a fashion unrivalled in any other series of experiments. At the same time, investigations into soils and such collateral questions as rain and drainage-waters have always been pursued, while for many years feeding experiments upon cattle, sheep, and pigs were conducted, and led to the enunciation of some of the fundamental laws of animal nutrition.

The Soil.—The Rothamsted Estate adjoins the village of Harpenden. The land lies mostly about 400 feet above the sea. The average rainfall is about 28 inches. The surface-soil is a heavy loam, containing many flint stones; the subsoil is a pretty stiff clay, resting on chalk. The chalk is usually about 9 feet from the surface, and affords a good natural drainage. The land does not bear a high rent. The soil is a fair one for wheat, but would not be considered as specially suited for barley; it is still less suited for turnips. The land in the district is still largely under the plough, though much of it has been laid down to grass and affords fair summer grazing for bullocks and dairy-cattle, but is rather too heavy for sheep.

Scope of the Manurial Experiments.—Different fields on the farm have been set apart for the study of individual crops : thus one has been devoted to wheat, one to barley, one to roots, &c. In each of these fields the crop has, as a rule, been grown continuously for many years without the intervention of fallow or any other crop.

In the early years of the experiments trials were made with various miscellaneous manures, and the same plot of land did not each year receive the same manure, but since 1851 the present systematic treatment has been adopted, and the same manure applied to the same plot every year. In nearly every case farmyard manure has been annually applied to one portion of the experimental field, while another portion has been left entirely without manure. The other plots have received the various chemical constituents of manure, either singly or in mixture with each other. The substances applied have been generally—ammonium-salts, nitrate of soda, rape-cake, superphosphate of lime, sulphate of potash, sulphate of magnesia, and sulphate of soda. The object has been to supply the various constituents of plant-food (see p. 326, vol. i., Div. ii.) in their most soluble and active form, and thus obtain their greatest effect. By employing substances of known composition, it is also possible to calculate how much of each constituent has been applied to the land.

Each plot of land has, during the later systematic portion of the experiments, received each year, as a rule, the same manure. By this plan trustworthy averages of the amount of produce yielded under each condition of manuring are obtained, and also ample information as to the influence upon the produce of seasons of different character. The permanent or temporary effect of the manures is also shown.

By long-continued treatment of this kind the soil of the experimental field, which was at first practically the same throughout, has been altered, so that the different plots now represent extremely different conditions of food-supply. On certain plots the crop now grows in soil specially exhausted of nitrogen, or phosphates, or alkalies, to an extent which can very rarely occur in farm practice; while in the soil of other plots abundance of these constituents has accumulated.

The work has not been confined to a determination of the amount of produce obtained from each manure,—the crops have themselves been analysed at the Rothamsted laboratory. Information has thus been obtained as to the proportion of the manure that is recovered in the increase of the crop, and also respecting the alteration in the composition of the crop brought about by the differences in the composition of the soil and the character of the season. The effect of the manures upon the yield of the various crops may indeed be considered as settled by the fifty years of results that have accumulated : much, however, yet remains to be investigated as to their effect upon the quality of the crops and upon the soil.

Soil and Drainage-water Investigations.—The investigation has further extended to the soil. After applying the same manure to the same land for many years, it becomes possible to learn by soil analysis what accumulation or exhaustion has taken place, and the depth to which manure has penetrated. In one of the fields the drainage-waters are collected and examined : the nature and amount of the soluble matters lost by drainage, under various conditions of manuring, are thus indicated. The investigations relating to the soil are, from the difficulty of the subject, in a less advanced stage than those relating to the effect of manures on crops.

Scientific Character of the Trials. —It will be seen from the above sketch that the object of the investigations has been primarily scientific. It has not been the aim to demonstrate directly the most economical manuring for each crop. None of the experiments have been designed with a view to a money profit: on very few of them would there be any profit if conducted on a large scale. The whole investigation, therefore, might stand condemned by the so-called "practical" man as a mere scientific amusement, from which he has nothing to learn. He, indeed, may learn little, but if so it will be because he lacks the elementary knowledge which is necessary for an appreciation of the results.

The mode of investigation adopted is, however, one which must add largely to our true knowledge of crops, manures, and soil. This knowledge will be turned to practical account in a number of ways by a skilful farmer ; but to provide him with practical rules has not been the immediate object of the investigation. To have aimed directly at practical results would have cramped the whole inquiry, and defeated its highest purpose.

EXPERIMENTS ON THE CONTINUOUS GROWTH OF WHEAT.

The experiments on wheat are among the oldest of those at Rothamsted. Broadbalk field has been under arable culture for at least two or three centuries. It grew its last turnip crop in 1839 : this was followed by barley, peas, wheat, and oats. The last four crops were without manure.

The continuous growth of wheat commenced in 1843, and has since proceeded without interruption, so that the present crop (1907) is the 64th crop of wheat in succession upon this field. The cultivation of the land has been that usual in the district : there has been no deep ploughing. During the course of the experiments the variety of wheat sown has been changed three or four times; latterly Square Head's Master has been sown, a fresh stock of seed being procured each year. The area of the plots is in nearly all cases one half acre. Formerly the artificial manures were sown broadcast, screens being carried on each side of the sowers to prevent the manure falling on other plots ; latterly a machine has been used. The wheat is drilled in October, 2 bushels of seed being used. In the spring and early summer great care is taken to remove weeds, the most troublesome of which is the grass *Alopecurus agrestis*. The luxuriance of weeds, in the absence of fallow crops, will always prove a practical objection to the continuous growth of corn, and especially of winter corn.

Without Manure.

In Table I. (p. 4) is shown the average produce per acre on Plot 3, without manure, in the first eight years, and five succeeding periods of ten years.

If all the seasons had been perfectly alike, the produce of the unmanured land would doubtless have fallen throughout the experiment—rapidly at first, but very slowly after the first ten years. The very variable character of the seasons in our climate prevents any such regularity in the produce. Indeed, putting aside the fluctuations due to season, the crop during the last forty years would almost seem to have reached a stationary condition, a certain proportion of the original capital of plant-food in the soil becoming available each year. It is noteworthy that under such extreme conditions of soil-exhaustion as prevail on this plot the quality of the grain does not suffer: in 1906 it weighed 63 lb. per bushel, and when converted into flour made as good loaves as the grain from any other plot.

The average produce of fifty-eight years of continuous wheat-growing without manure is seen to be 13.7 bushels. It is interesting to note that this amount

TABLE I.—PRODUCE OF WHEAT WITHOUT MANURE, FIFTY-EIGHT YEARS (1844-1901).

| | Dressed Corn. | | Straw. |
	Quantity.	Weight per Bushel.	
	bush.	lb.	cwt.
Eight years, 1844-51 . .	17.2	60.0	15.5
Ten years, 1852-61 . .	15.9	55.8	15.2
,, 1862-71 . .	14.5	59.4	11.5
,, 1872-81 . .	10.4	57.9	8.5
,, 1882-91 . .	12.6	59.8	8.5
,, 1892-1901 . .	12.3	61.3	9.1
Mean of fifty-eight years .	13.7	59.0	11.2

is quite equal to the average yield of the principal wheat-producing countries of the world. Thus, the average yield of the United States is 13 bushels, of Russia 10 bushels, and of India 11 bushels.

With Farmyard Manure.

Ordinary yard manure, at the rate of 14 tons per acre, has been annually ploughed in in October on Plot 2; the produce is shown in Table II.

TABLE II.—PRODUCE OF WHEAT WITH FARMYARD MANURE, FIFTY-EIGHT YEARS (1844-1901).

| | Dressed Corn. | | Straw. |
	Quantity.	Weight per Bushel.	
	bush.	lb.	cwt.
Eight years, 1844-51 . .	28.0	61.1	26.6
Ten years, 1852-61 . .	34.2	58.8	33.9
,, 1862-71 . .	37.5	61.3	34.0
,, 1872-81 . .	28.7	59.8	28.0
,, 1882-91 . .	38.2	61.0	34.8
,, 1892-1901 . .	39.2	62.3	38.7
Mean of fifty-eight years	34.5	60.7	32.9

Plant-food in Dung.—The amount of plant-food supplied is much larger than on any other plot in the field. The fourteen tons of farmyard manure are estimated to contain 200 lb. nitrogen, 235 lb. potash, 35 lb. magnesia, 31 lb. lime, and 78 lb. phosphoric acid, with a number of other substances, including a large amount of silica, which is at present supplied to no other plot in the field. In consequence of this large supply there

has been a great accumulation of manurial matter in the soil, which is now far richer than that of any other plot in the field.

Limits to High Manuring.—The table shows a considerable rise in the produce during the earlier years of the experiment, owing to the accumulation of food in the soil. This rise afterwards ceases. Everything, indeed, in nature tends to come to an equilibrium. On the unmanured land the crop falls, till its demands equal the annual supply from soil and atmosphere. On the dunged plot the produce rises, till here, too, the crop equals the annual supply of assimilable food. With very high manuring we meet with another limit, that of season. A larger crop cannot be produced by manure than the character of the season will admit of.

The average produce with farmyard manure in fifty-eight years has been 34½ bushels: the highest produce was 48½ bushels in 1891.

Nitrogen in Dung.—Notwithstanding the richness of the soil, the farmyard manure plot very seldom yields the highest produce in the field, both nitrate of soda and ammonium-salts proving more effective. The nitrogen in farmyard manure is in fact principally combined with carbon, and exists as nitrogenous humic matter; only a limited portion of this is each season oxidised to nitrates, and thus becomes available to the crop.

Mechanical Influence of Dung.—Not a few of the advantages attending the use of farmyard manure are due to its improvement of the physical condition of the soil. In the present case the soil, while becoming less heavy, has also become more retentive of moisture, and the crop thus suffers less in time of drought (*Jour. Royal Agric. Soc.*, 1871, p. 91). The produce of this plot is more even, and less affected for good or evil by the vicissitudes of season, than on the other highly manured plots in the field.

With Ash Constituents.

When water has been removed, the constituents of a plant may be classed under two heads—the combustible and the incombustible.

The *incombustible* portion is very small: in wheat grain it is about 1.7 per cent, in wheat straw about 4.6 per cent. It consists of the phosphates, potash, lime, magnesia, silica, &c., derived from the soil.

The *combustible* part is made up of the carbon, oxygen, and hydrogen derived from the atmosphere and rain, and of the nitrogen derived from the soil. The quantity of the principal ash constituents, and of nitrogen, contained in a wheat crop of 30 bushels, has been already given on p. 326, vol. i., Div. ii.

Of the substances present in the ash, six—potash, lime, magnesia, iron, phosphoric and sulphuric acid—are quite indispensable for plant-growth.

Mineral Theory.—At the time when the Rothamsted experiments commenced, chemists had a very exaggerated notion of the amount of ammonia annually supplied by rain. Liebig, owing to this mistaken idea, taught in 1843 that the ashes of a manure contained its true active ingredients; that where the necessary ash constituents of a crop were supplied by manure, the crop would have no difficulty in obtaining all the nitrogen it required from the atmosphere. This view was known as the "mineral theory." The state of opinion at the time must be borne in mind in considering the Rothamsted field experiments, as they were planned to a considerable extent to test the truth of the mineral theory.

In the first season of the wheat experiments (1843-1844), one plot received 14 tons of farmyard manure, and a second plot the ashes from another lot of 14 tons, with the following result:—

	Dressed Corn.	Total Corn.	Total Produce.
	bush.	lb.	lb.
Farmyard manure, 14 tons .	20½	1276	2752
Ashes of ditto . .	14½	888	1992
Unmanured . .	15	923	2043

The plot receiving the ashes thus yielded no more produce than the plot entirely without manure.

Various systematic experiments have since been made with the ash constituents of wheat: these have been supplied in abundance, and the crop left to obtain its carbon and nitrogen from the natural resources of the soil and air.

Plot 5 has received superphosphate of lime and a mixture of the sulphates of potash, soda, and magnesia—generally termed by Lawes and Gilbert the "mixed mineral" manure, and representing the ash constituents of the crop without nitrogen. It consists of 3½ cwt. superphosphate, 200 lb. sulphate of potash, and 100 lb. each of sulphate of soda and magnesia, per acre.

This mixture (Plot 5) has given an increase of about 2 bushels of corn and 1¾ cwt. of straw over the produce of the unmanured land.

Nitrogen of the Soil and Atmosphere Insufficient.—As these manures have supplied all the ash constituents of the wheat crop (excepting silica, which we shall presently see to be non-essential), it is quite evident that the amount of the other necessary elements of plant-food supplied by the soil and atmosphere has been insufficient to produce a full crop of wheat. The crop grown with a full supply of ash constituents on Plot 5 has contained, on an average, about 20 lb. of nitrogen per acre per annum. This quantity represents the average amount furnished by the soil and atmosphere without the aid of nitrogenous manure.

We shall presently see that the growth of wheat on these plots was really limited by the small quantity of nitrogen at the disposal of the crop. When nitrogen is supplied, phosphates and potash become important elements in producing growth.

Ammonium-salts with Ash Constituents.

The ammonium-salts employed have been a mixture of equal parts sulphate and chloride: 200 lb. of this mixture are estimated to contain about 43 lb. of nitrogen. The systematic experiments with ammonium-salts did not begin, in several cases, till 1852. We shall therefore take the average produce after this date as the basis of our comparison:—

TABLE III.—PRODUCE OF WHEAT VARIOUSLY MANURED, AVERAGE OF FIFTY-FIVE YEARS.

Plot.	Manuring.	Average Produce, 55 Years, 1852-1906.				Average Produce last 10 Years, 1897-1906.		
		Dressed Corn.		Straw and Chaff.	Corn to 100 Straw.	Dressed Grain.	Weight per Bushel.	Straw.
		Quantity.	Weight per Bushel.					
		bush.	lb.	cwt.		bush.	lb.	cwt.
3	No manure . . .	13	59.0	10.5	69.5	11.6	61.2	9.7
5	Mixed ash constituents	14.9	59.6	12.3	68.6	13.9	61.3	12.1
6	Do., and ammonium-salts, 200 lb. . .	23.9	60.3	21.6	62.5	21.6	61.5	21.0
7	Do., do. 400 lb.	32.8	60.3	33.0	56.4	31.0	61.6	33.2
8	Do., do. 600 lb.	37.1	60.1	40.9	51.7	38.6	61.0	44.0
10	Ammonium-salts, 400 lb.	20.4	58.2	18.5	62.6	18.8	60.4	17.0
11	Superphosphate and ammonium-salts, 400 lb.	23.5	58.2	22.5	59.6	18.6	'59.5	19.0
2	Farmyard manure, 14 tons . . .	35.5	60.7	34.3	58.8	36.6	61.7	40.1

Table III. shows, that whereas the continued use of ash constituents alone increased the crop by less than 2 bushels, the addition of 200 lb. of ammonium-salts gave a further increase of 9 bushels, the addition of 400 lb. of ammonium-salts an increase of 17.9 bushels, and the addition of 600 lb. an increase of 22.2 bushels. The produce with ash constituents and 400 lb. ammonium-salts (Plot 7) nearly equals the produce from the annual application of 14 tons farmyard manure; while the produce with 600 lb. of ammonium-salts (Plot 8) exceeds both in corn and straw that yielded by the dung. The far greater effect produced by the nitrogen of the ammonia than by the nitrogen of the dung is very evident, *86 lb. of nitrogen as ammonia being on a long series of years nearly equal to 200 lb. applied as dung.*

Organic Manures Unnecessary.—

These results throw a flood of light on the conditions required for producing good wheat crops. The manure applied to these ammonia plots has been purely inorganic,—it has contained no carbon; yet the produce has been large, and in favourable seasons very large. In 1863 the yield of corn on Plot 7 amounted to 53½ bushels per acre. About 1 ton of ‘carbon is contained in the average crop of Plot 7, and still more in that of Plot 8. All the carbon assimilated by these crops has been derived from the atmosphere. *The atmospheric supply of carbon is apparently sufficient for the largest cereal crops.* Such crops may be obtained in favourable seasons by the use of purely inorganic manures.

Silica Unnecessary.—The results are equally conclusive as to the uselessness of applying silica in manure. The composition of cereal crops (p. 326, vol. i., Div. ii.) shows silica to be by far the largest constituent of the ash of straw, and to its presence the stiffness of the straw has been too hastily attributed. Many experiments have shown that silica is not an indispensable constituent of cereal crops, that fully developed plants can be obtained without it, and that in these plants the straw does not show any want of stiffness.

At Rothamsted, wheat crops, above the average produce of the country, have been continuously obtained for sixty years with manures supplying no silica. The produce with these manures has indeed been larger than that yielded by farmyard manure which supplies silica. To make the test still more complete, one-half of several of the plots received for four years an application of soluble silicates, and in the succeeding twelve years the straw of the crop was returned to the land. The half plots thus treated have not shown any increase of produce, save in those cases where the straw was helpful by supplying potash; nor had the wheat-straw so grown any greater power of standing in rough weather than that grown without silica in the manure.

Artificial Supply of Nitrogen essential for Wheat.—The evidence afforded by these experiments with ammonium-salts shows unmistakably the great need of the wheat crop for an artificial supply of nitrogen, if full crops

are to be continuously obtained. The assimilable nitrogen furnished by the air and rain is quite insufficient for the production of a full cereal crop. The annual application of 86 lb. of nitrogen per acre, in the form of ammonia, has raised the average produce from 14.9 bushels to 32.8 bushels per acre.

Manures best for Cereals.—The manures which experience has proved to be most effective for wheat, barley, or oats, are those which, like guano, nitrate of soda, and sulphate of ammonia, supply nitrogen in a form readily assimilated by plants. The enrichment of the surface-soil with nitrogen is also the main effect of a variety of agricultural methods commonly employed to render land fit to produce good crops of cereals.

Excessive Dressings Unprofitable.—It will be noticed that the application of 200 lb. of ammonium-salts per acre gave an average increase of 9 bushels of corn and 9¼ cwt. of straw. The addition of a second 200 lb. of ammonium-salts gives a further increase of nearly 9 bushels of corn and 11⅜ cwt. of straw. The 400 lb. of ammonium-salts was thus not an excessive dressing under the conditions of the experiment—*i.e.,* continuous cropping with wheat. When wheat is grown in a rotation, when the land is enriched by occasional use of farmyard manure, and by the growth of clover, &c., no such quantity of nitrogenous manure is needed. With a further addition of 200 lb. ammonium-salts, however, the return is greatly diminished, the increase only amounting to 4.3 bushels of corn and 7.9 cwt. of straw. It is plain, therefore, that 600 lb. was not an economical dressing.

For thirteen years, 1852-64, as much as 800 lb. of ammonium-salts were applied to one of the plots. The average produce of different amounts of ammonia during these thirteen years is shown in Table IV., p. 8.

We have here a successive increase of 10¼, 8⅝, 1¾, and ⅝ bushels of corn, and 10½, 11, 4⅝, 3⅞ cwts. of straw for each additional 200 lb. of ammonium-salts.

Corn and Straw from High Manuring.—It will be observed that there is a much larger increase of straw than of corn with the heavier dressings of ammonium-salts : the proportion of corn to straw diminishes, indeed, with each addition of ammonia.

The quality of the corn is improved by the use of 200 lb. and 400 lb. of ammonium-salts, but with further additions of ammonia the weight per bushel begins to decline.

TABLE IV.—PRODUCE OF WHEAT WITH VARIOUS QUANTITIES OF AMMONIUM-SALTS, AVERAGE OF THIRTEEN YEARS (1852-64).

| Plot. | Manuring. | Dressed Corn. | | Straw and Chaff. | Corn to 100 Straw. |
		Quantity.	Weight per Bushel.		
		bush.	lb.	cwt.	
5	Mixed ash constituents . . .	18¼	58⅛	16⅝	62.6
6	Do. with ammonium-salts, 200 lb. .	28½	58⅞	27⅛	58.8
7	,, ,, ,, 400 lb. .	37⅛	58¾	38⅛	54.6
8	,, ,, ,, 600 lb. .	38⅞	58¼	42¾	51.2
16	,, ,, ,, 800 lb. .	39½	58	46⅝	47.8

Ammonium-salts alone.

We come now to Plot 10, which has received annually 400 lb. of ammonium-salts, without any supply of phosphates, potash, magnesia, lime, or other ash constituents (saving the sulphuric acid and chlorine in the ammonium-salts). This treatment has dated from 1845. The average produce in fifty-five years has been 20.4 bushels and 18.5 cwt. of straw ; or 7.4 bushels and 8 cwt. of straw over that of the unmanured land.

Natural Supplies of Ash and Nitrogen.—While the crop on Plot 5 was entirely dependent upon natural sources of nitrogen, the crop on Plot 10 has been wholly dependent upon natural sources for its ash constituents. The supply of ash constituents from the soil has clearly been insufficient, for the same amount of ammonium-salts, when aided by a manuring of ash constituents (Plot 7), has produced a much larger crop than on Plot 10.

The natural supply of ash constituents, though insufficient, is, however, more effective than the natural supply of nitrogen ; for while, on Plot 5, the natural supply of nitrogen only produces 14.9 bushels, the natural supply of ash constituents is equal to the production of 20.4 bushels.

Soils better supplied with Ash than with Nitrogen.—The fact just stated is one that holds true in general agricultural experience. A purely nitrogenous manure will, in a vast majority of cases, produce a greater effect on wheat or other cereals than any manure supplying ash constituents,—not because the latter are less necessary for the growth of the crop, but because the soil is generally far better supplied with available ash constituents than it is with available nitrogen.

It must be remembered, also, that the average results obtained in these Rothamsted experiments with purely nitrogenous manures are by no means so good as would be obtained in ordinary practice. The soil on Plot 10 is now in fact exhausted of ash constituents by sixty-two successive wheat crops, removing at least 20 lb. of potash and 12 lb. of phosphoric acid per acre per annum. In the earlier years of the experiment the ammonium-salts applied alone gave a much better result than they do at present.

Importance of Ash Constituents.— The importance of ash constituents when nitrogen is supplied is strikingly shown by comparing the produce of the exhausted soil on Plot 10 with that of the soil of Plot 7, which has annually received an abundance of ash constituents, with the same amount of ammonia. The average produce with ash constituents and ammonia is 12.4 bushels greater than with the same quantity of ammonia applied alone. *As nitrogenous manures are by far the most costly that a farmer purchases, it is important to remember that economy in their use depends a great deal on there being a sufficient* supply of available phosphates and potash in the soil.

Ammonia with Individual Ash Constituents.

On Plot 11 the 400 lb. of ammonium-salts have been continuously applied with superphosphate. The average produce is 23.5 bushels and 22.5 cwt. of straw, or 3.1 bushels and 4.0 cwt. of straw more than that given by the ammonium-salts alone. Thus, on a phosphate-exhausted soil, superphosphate becomes a paying manure for wheat if nitrogen is not deficient.

The produce on this plot is, however, far below that on which all the necessary ash constituents are applied. The superphosphate has increased the produce of the ammonia by 3.1 bushels, but the mixture of ash constituents applied on Plot 7 increases the produce by 12.4 bushels. The mixed ash constituents include potash, soda, and magnesia.

A series of experiments has been made in which the sulphates of potash, soda, and magnesia have been used separately, each with ammonium-salts and superphosphate. In the earlier years of the experiments there was a great reserve of potash in the soil, partly natural and partly due to previous manurings, and this potash was so rendered available by the action of the soluble sulphates of magnesia and soda that all three plots gave much the same results. Table V., however, shows the results for the ten years, 1893-1902, when the original potash in the soil had become largely exhausted, and proves that neither magnesia nor soda can take the place of potash in a manure. That the action of the magnesia or soda is indirect may be seen from the fact that the sulphate of magnesia on Plot 14 does not increase the amount of magnesia in the ash, but does cause an increase in the quantity of potash taken up from the soil.

TABLE V.—PRODUCE OF WHEAT VARIOUSLY MANURED, AVERAGE OF TEN YEARS (1893-1902).

Plot.	Manured with Ammonium-salts and Superphosphate.	Total Corn.	Straw and Chaff.
		lb.	lb.
11	Alone	1299	2150
12	With soda	1760	2822
14	With magnesia	1641	2684
13	With potash	1938	3342
7	With soda, magnesia, and potash . . .	2076	3606

Relative Importance of the Ash Constituents. — Phosphoric acid and potash are the ash constituents of the greatest importance to the wheat crop, and indeed to every other crop. Magnesia is a less important ash constituent of wheat, and is usually found in sufficient abundance in the soil. Soda is found to a very small extent in the mature crop, but soda salts have some effect as manure: they act by liberating potash in the soil. Lime scarcely occurs in wheat grain, and to only a small extent in the straw; the natural supply is quite sufficient.

Effect of Autumn and Spring Applications of Ammonium-salts.

Up to the year 1872 the whole of the manures, with the exception of nitrate of soda, were applied to the land in au-

tumn at the time of wheat-sowing, and ploughed in.

With the season 1872-73 an experiment commenced on the comparative effect of autumn and spring applications of ammonium-salts. For five years (1873-77) Plot 15 received 400 lb. of ammonium-salts as a top-dressing at the end of March or beginning of April, while Plot 7 received the same amount when the wheat was put in in October. For the autumn of 1877 the manuring was reversed: Plot 15 now received the ammonium-salts in the autumn, and Plot 7 received them in the spring. Both plots had at all times a complete autumn manuring with ash constituents.

The comparative results in ten years of autumn and spring manuring are shown in Table VI.

TABLE VI.—COMPARATIVE EFFECT OF AUTUMN AND SPRING SOWING OF AMMONIUM-SALTS.

	Rainfall.		Drainage, 5-ft. Gauge.		Total Produce, Corn and Straw.		
	Autumn Manuring to Spring Manuring.	Spring Manuring to end of July.	Autumn Manuring to Spring Manuring.	Spring Manuring to end of July.	Autumn Manuring.	Spring Manuring.	Spring + or − Autumn.
	inches.	inches.	inches.	inches.	lb.	lb.	lb.
1872-73 . .	18.53	6.92	11.45	0.42	3344	5031	+ 1687
1873-74 . .	7.05	7.93	2.89	0.58	7094	4588	− 2506
1874-75 . .	10.55	13.55	5.21	3.86	5110	4915	− 195
1875-76 . .	12.17	7.58	10.14	1.94	3793	4083	+ 290
1876-77 . .	22.01	8.18	15.78	1.18	3048	4795	+ 1747
1877-78 . .	11.17	12.96	8.11	6.02	4486	7017	+ 2531
1878-79 . .	15.05	17.10	13.09	6.76	1275	4063	+ 2788
1879-80 . .	5.78	10.82	3.37	1.58	6309	6155	− 154
1880-81 . .	15.20	6.16	12.75	0.25	3489	3917	+ 428
1881-82 . .	10.34	14.73	7.62	4.48	5948	7981	+ 2033
Mean . .	12.79	10.59	9.04	2.71	4390	5255	+ 865

Spring Sowing preferable.—It appears that, out of the ten seasons, there was one (1874) in which the autumn sowing of the ammonium-salts gave decidedly the best results ; there were four in which the difference between autumn and spring sowing was very small ; there were five in which the spring sowing gave much the best result. The average result was thus decidedly in favour of spring sowing.

Rainfall and Time of Sowing Manure.—When we turn to the other columns in the table, it is plainly seen that the advantage or disadvantage of autumn sowing depends on the amount of the rainfall. The autumn application of ammonium-salts is advantageous only when a dry winter follows their application. *This is owing to the fact that ammonia is converted into nitrates in the soil ; and the soil having no power of retaining nitrates, they are liable to be washed into the subsoil by heavy rain, and to be carried in drainage-water beyond the reach of the roots.* This is what happens during a wet winter.

In the table, the quantity of rain, and the amount of drainage-water passing through 5 feet of uncropped soil (60-inch drain-gauge), in each season, are given.

It will be noticed that a wet winter, in some cases (1880-81), does little harm to the autumn-sown ammonium-salts. In these cases the wet winter is followed by a dry summer, and the crop is able to draw up from the soil the solution of nitrates which had passed downwards.

The worst results of autumn manuring are when a wet winter is followed by a wet summer (1877-78, 1878-79, 1881-82). In these cases the nitrates washed below are kept down by the subsequent spring and summer rainfall.

In consequence of these results, the time for applying the ammonium-salts to the experimental plots in the wheat-field has been altered. For 1878-83 the ammonium-salts were (save on Plot 15) applied entirely in the spring. Since then 100 lb. of ammonium-salts have been applied in autumn and the remainder in spring.

With Nitrate of Soda.

The trials with nitrate of soda commenced in 1852, but certain variations have been made from time to time in the quantities used, so that Table VII. shows the results of the ten years 1893-1902 only. As one object of the experiment was to compare the effect of nitrogen in the two forms of *ammonia* and *nitric acid*, the quantity of nitrate of soda employed was arranged to supply the same weight of nitrogen (43 and 86 lb.) as 200 lb. and 400 lb. of ammonium-salts. The nitrate of soda has always been applied as a top-dressing at the end of March or beginning of April.

TABLE VII.—PRODUCE OF WHEAT WITH NITRATE OF SODA AND AMMONIUM - SALTS, AVERAGE OF TEN YEARS (1893-1902).

Plot.	Manure.	Dressed Corn.		Straw.	Corn to 100 Straw.
		Quantity.	Weight per Bushel.		
		bush.	lb.	cwt.	
3	No manure	12.7	61.2	9.3	76.1
5	Mixed ash constituents . .	15.4	61.4	11.8	74.7
6	,, ammonium-salts, 200 lb. .	23.5	61.6	20.2	66.8
9	,, nitrate of soda, 275 lb. .	27.3	61.3	25.5	60.9
7	,, ammonium-salts, 400 lb. .	32.4	61.4	32.2	57.6
16	,, nitrate of soda, 550 lb. .	32.5	61.0	33.3	55.6

Nitrate of Soda excels Ammonium-salts.—From these results it will be seen that nitrate of soda is at Rothamsted a more effective source of nitrogen for wheat than the ammonium-salts, the straw being more particularly responsive to the nitrate of soda dressings. When 43 lb. of nitrogen are used the nitrate yields 16 per cent more grain and 26 per cent more straw than the equivalent amount of ammonium-salts: 86 lb., however, yields practically the same grain, and only about 1 cwt. more straw. It should be remembered that the soil of the Broadbalk field is well provided with carbonate of lime, hence the injurious effects arising from the acidity produced by the continuous use of ammonium-salts on some soils is not here apparent. The superiority of the nitrate with wheat is probably due to the fact that it remains soluble, thus diffusing deep into the soil and encouraging a greater range of roots.

Influence of Rainfall.—Ammonium-salts and nitrate of soda compare, however, very differently in different seasons: there are seasons in which the nitrate is immensely superior, and there are some seasons in which the ammonium-salts give an equal or better result. With a dry spring and summer the nitrate is generally much superior to a spring dressing of ammonium-salts, the nitrate being immediately available to the plant, while the ammonium-salts have to undergo the process of nitrification, which in dry weather is not speedy. On the other hand, in a wet spring the nitrate is subject to immediate loss by drainage, while the ammonium - salts are not lost till they

are nitrified, and thus for a few weeks partially escape the losses which the nitrate is undergoing. While this may be generally true, it is found that in exceptional seasons the relative position of the two manures is reversed. In an exceptionally dry season, such as 1893, the ammonium-salts give a better return than the corresponding nitrate of soda, whereas in continuously wet summers, such as 1879 and 1903, the nitrate of soda, despite its tendency to wash out of the soil, produces much the better crops. Doubtless this is due to the fact that the conversion of ammonium-salts into nitrates is greatly delayed by the low temperature and deficient aeration prevailing in the soil during a wet summer, so that the ready-formed nitrate is of greater advantage to the plant. The evidence is even in favour of the nitrate of soda generally in the wetter seasons, as may be seen from Table VIII., where thirty years' results have been divided into two groups according as the rainfall was above or below the average. In the dry seasons the yield from ammonium - salts was 86.6 per cent of that from nitrate of soda; in the group of wet seasons the ammonium - salts were less effective, and only yielded 78.8 per cent of the grain produced by the nitrate of soda. The question of the relative value of these two sources of nitrogen is complicated by several factors, which will vary from soil to soil: for example, the unfavourable action of nitrate of soda upon the tilth of the land will have its greatest influence in a dry season upon a comparatively heavy soil such as that of Rothamsted

TABLE VIII.—COMPARISON OF THE YIELD OF DRESSED GRAIN WITH NITROGEN AS AMMONIUM-SALTS OR NITRATE OF SODA, IN SEASONS WHEN THE RAINFALL WAS BELOW OR ABOVE THE AVERAGE, THIRTY YEARS (1873-1902).

	Rainfall.	Dressed Grain per acre.		Ratio of yield by Ammonium-salts to that of Nitrate of Soda = 100.
		Plot 9a. Nitrate of Soda.	Plots 6 & 7. Ammonium-salts.	
	inches.	bush.	bush.	
14 seasons *below* average rainfall .	24.23	30.6	26.5	86.6
16 seasons *above* average rainfall .	33.13	32.1	25.3	78.8

Proportion of Corn to Straw.

In Tables III. and VII. will be found the proportion of corn to 100 straw in the produce of the various manures we have considered. The proportion of corn is highest in the produce of the un-manured land, and on that receiving only the ash constituents of the wheat crop.

The addition of any manure producing luxuriance of growth increases the proportion of straw: thus, by the continuous application of farmyard manure, the proportion of corn to 100 straw falls from 69.5 to 58.8.

With increasing quantities of ammonium-salts, applied with ash constituents, the proportion of corn gradually falls, being 62.5, 56.4, and 51.7, with 200, 400, and 600 lb. of ammonium-salts. This considerable increase in the proportion of straw with the higher amounts of ammonium-salts is not, however, entirely due to the ammonium-salts, as on Plot 10, with 400 lb. of ammonium-salts alone, the proportion of corn is 62.6; and on Plot 11, with the same quantity of ammonium-salts with superphosphate, the proportion is 59.6 to 100. The increase in straw is clearly due in great part to the potash supplied on Plots 6, 7, and 8, which helps largely to form straw when the nitrogen necessary to nourish the crop is present.

The proportion of straw is much greater with nitrate of soda than with ammonium-salts (Table VII.)

Influence of Season.

The 64 successive wheat crops in Broadbalk field at Rothamsted, grown for the most part under the same conditions as to manuring every year, afford splendid material to the statistician for indicating the varying produce of the country in different seasons. We cannot in this place regard them in this wide aspect. The produce of each plot, and the character of each season, during 40 years, will be found in two papers by Messrs Lawes and Gilbert, in *Jour. Royal Agric. Soc.*, 1864, 93; 1884, 391. To these papers, and to a paper, "Our Climate and our Wheat Crops," *ibid.*, 1880, 173, we must refer for full details of the earlier years. Certain other effects of season upon yield are discussed in a paper by A. D. Hall in the *Journal of the Board of Agriculture*, 1905, ii. 716. We have here to regard the influence of season as a condition affecting the fertility of soil and the action of manures.

Every farmer knows that the effect of season is greater than the effect of manure. A season may be so bad that the best soil and manure may yield a miserable produce, and it may be so good that moderate manuring may nearly equal in result a liberal treatment. A suitable manuring will, however, assert itself in a large majority of cases, redeeming a bad season from utter loss, and securing from a good season the grand return which it is capable of yielding.

Influence of Light and Heat.—No large crop can be obtained without a sufficient amount of light and heat, as the assimilation of carbon from the atmosphere only occurs with suitable light and temperature. The formation of seed especially requires heat. A bulky crop in June will produce abundance of corn in July, if this month is warm and not too wet; but it will remain a crop of straw if July is cold and rainy. The corn produced in a cold wet summer is also imperfectly developed: it contains less starch, and a larger proportion of albuminoids and ash constituents, than well-ripened grain, and has a low weight per bushel. The same defect in the corn may be brought about by premature ripening, occasioned by sudden heat and drought; but this will seldom happen upon a clay soil like that at Rothamsted.

Autumn and Winter Weather.— The popular view of the character of a wheat season is confined to the meteorological conditions of spring and summer. Winter is taken into account only when frost or floods have injured the plant. We have already seen, however, when considering the very different results of the autumn and spring application of ammonium-salts, that the dryness or wetness of the autumn and winter is a most important factor in determining the character of the next summer's crop. In a wet winter, the nitrates produced in the soil since the last cropping, or resulting from autumn applications of nitrogenous manure, may be removed almost entirely in the drainage-water, and the soil reduced to an impoverished condition by the time the growth of wheat commences in the spring. A dry winter is thus essential if a full wheat crop is to be harvested throughout the country.

A detailed examination of the results would seem to show that the dryness of the late autumn and early winter months constitutes the most favourable factor towards a large wheat crop. It is at that time that the foundation is laid, and if the season is wet the crop can never recover from the adverse start: the value of a dry early winter may, however, be nullified by subsequent unfavourable conditions.

A dry period from October to January benefits the wheat plant not only by leaving in the soil the nitrates available for growth, but also by favouring the development of an extensive root-system. During the winter it is chiefly the root of the wheat plant that is forming, and should the soil be water-logged by repeated rain this growth of root becomes much restricted and unable to maintain an adequate development of leaf and stem later.

Conditions Favourable to Large Crops.— The years of greatest total produce during the Rothamsted experiments have been 1863 and 1854. These seasons had dry winters, and in the case of 1863 the winter was also mild. There was also during spring and summer a deficiency of rain, though enough fell at critical times to prevent any check to growth. The summers were not unusually hot,—indeed that of 1854 was decidedly cool; there was thus no premature ripening of the produce.

These are the conditions favourable to large produce on every description of soil, manured or unmanured. The dry weather between autumn and spring retains in the soil all the nitrates belonging to it; dry mild weather during winter and spring also occasions a maximum development of root; the plant is thus enabled to levy contributions from a considerable depth of soil. If moderately dry weather continue, the plant is afterwards fed with a concentrated solution of plant-food. The moderate warmth of the season allows full time for the collection of food from the soil. There is finally a somewhat late harvest, and a most abundant produce.

Ash Constituents and the Seasons. —During wet and cool seasons the supply of phosphoric acid has a potent effect in increasing the yield, and especially the proportion, of grain to straw. Phosphoric acid promotes the maturing processes in the plant, and has thus its greatest effect in seasons unfavourable to ripening. On the other hand, potash manures are most effective in dry, hot summers, since they tend to encourage growth and prolong the vegetative development of the plant.

Effects of Residues of Manures.

Residues of Previous Manuring.— The Rothamsted experiments supply numerous illustrations of the influence of the residues of previous manuring, and the wheat-field supplies one notable example of the relative duration of ammonium-salts and the ash constituents respectively.

The manures on Plots 17 and 18 have alternated each year since 1852. In each year one plot receives the usual full dressing of ash constituents, and the other plot 400 lb. oi ammonium-salts. In the following year the manuring is reversed: the plot that had received ash constituents now receives ammonium-salts, and the one which had received ammonium-salts now receives ash constituents. There is thus each year a crop manured by ammonium-salts plus a residue of ash constituents, and a crop manured by ash constituents plus the residue from ammonium-salts.

The average effect of these annual residues is shown in Table IX.

TABLE IX.—EFFECT OF ANNUAL RESIDUE OF ASH CONSTITUENTS, AVERAGE FIFTY-FIVE YEARS (1852-1906).

	Dressed Corn.	Straw.
	bush.	cwt.
Ammonium-salts and residue of ash constituents	30.4	29.6
Ammonium-salts alone, Plot 10 . . .	20.4	18.5
Excess, due to residue of ash constituents .	10.0	11.1

EFFECT OF ANNUAL RESIDUE FROM AMMONIUM-SALTS, AVERAGE FIFTY-FIVE YEARS (1852-1906).

	Dressed Corn.	Straw.
	bush.	cwt.
Ash constituents and residue of ammonium-salts	15.4	13.2
Ash constituents alone, Plot 5	14.9	12.3
Excess, due to residue of ammonium-salts .	0.5	0.9

The abundant residue of ash constituents remaining from the preceding year has proved its effectiveness, by raising the produce by 10 bushels per year.

We turn now to the result produced by the residue of the ammonium-salts. It has yielded, according to the table, an increase of but ½ bushel per year!

Of the 86 lb. of nitrogen contained in the ammonium-salts, not more than 43 lb. would be contained in the crop obtained by its use: what then has become of the remaining 43 lb.? It is quite clear that the missing ammonia is not present in the soil ready for use in the next season, for it produces no effect on the crop, and the evidence goes to show that the unused ammonia has been in great part lost as nitrates in the drainage-water.

Tracing the Fate of Manures.

The fate of the various manures applied to the wheat plots at Rothamsted has been studied in other ways, — by analysis of the crops that are removed and of the soil after the lapse of successive periods of years, and also by analysis of the waters flowing from the tile-drains which run below the experimental plots. The latter method gives us information as to the substances which are being lost to the soil by removal into the permanent store of subsoil-water below the reach of the plant's roots, though it is difficult to estimate the amounts so lost. The former method should enable us to strike a balance-sheet, and to distribute the manure applied between the crops, the soil, and wastage, were it not that, from the impossibility of drawing exactly representative samples of soil, only approximate estimates can be formed of the amount of any constituent present in the land at each period: furthermore, samples of soil were not taken until the experiments had been going on for some time.

Loss of Nitrates in Drainage-waters. — Considering the drainage-waters first, it has already been shown that they always contain considerable quantities of nitrates, derived from the oxidation of the nitrogen compounds, either originally in the soil or added as manure. Whatever the compound of nitrogen that is used, whether nitrate of soda, ammonium-salts, rape-cake, or dung, it only reaches the drainage-water in the form of nitrates. Even the freely soluble ammonium-salts are very rarely found in the drainage-water, and then only in small quantities, so quickly are they arrested by the soil. The nitrates, in fact, are the only compounds of nitrogen that are not held by the soil. The formation of nitrates is rapid when the conditions are favourable, as, for example, when the soil is warm and moist and has been recently stirred, but the rate of production will also depend upon the nature of the nitrogen compounds: many of those present in dung and in the original stock in the soil are extremely resistant to change. The point, therefore, which is brought out by the analyses of the drainage-water is that, even from the poorest soils, loss of nitrates can and does take place whenever there is percolation through the soil and no growing crop to utilise the nitrates as fast as they are formed.

On the rare occasions when the drains below the Rothamsted wheat plots have run in May, June, or July, the water was found to contain very little nitrate, because they had been utilised by the growing crop; but a heavy rainfall in October or November, after the land had been broken up and cultivated for the new crop, always gives rise to drainage-water rich in nitrates. The higher the condition of the land, as, for example, on the plot that receives farmyard manure every year, the greater will be the loss of nitrates. It has been shown that the rainfall of the autumn and early winter to a large extent determines the magnitude of the succeeding wheat crop: a heavy rainfall washes away the summer-formed nitrates, upon which the future crop must feed, and from the poor start thus occasioned the plant never recovers. This is well brought out at Rothamsted by a comparison of the wheat plot that is continuously unmanured with a pair of similarly unmanured plots which are alternately cropped and bare-fallowed, so that each only carries a crop once in two years. Taking the whole period, the wheat after fallow yields 16.9 bushels per acre, as against 12.6 bushels for wheat after wheat each year; but if the rainfall each year for the months of September to December are set out, and the crops are divided into two series according as they follow autumns of greater or less than average rainfall, it will be seen (Table X.) that after the wet autumns the average crop is no higher on the fallowed than on the continuously cropped land, whereas the dry autumns are succeeded by crops 50 per cent larger on the fallowed than on the continuous wheat plot. The summer fallowing results in the production of nitrates, which are washed out of the soil if the autumn is wet, but remain there for the benefit of the wheat crop if no percolation takes place.

TABLE X.—EFFECT OF WET OR DRY AUTUMNS ON THE INCREASE OF THE WHEAT CROP DUE TO FALLOWING (1870-1902).

	16 seasons *less* than average Rainfall.	16 seasons *more* than average Rainfall.
	inches.	inches.
Rainfall (Sept. to Dec. inclusive) . . .	8.88	13.66
Percolation through 60″ of soil (Sept. to Dec. inclusive)	4.03	8.92
	lb.	lb.
Total produce (wheat after wheat) . . .	1810	1627
Total produce (wheat after fallow)	2743	1757
Increase due to fallowing	933	130
Percentage increase due to fallowing . . .	51.5	7.9

Loss of other Manurial Elements in Rain-water.—Of the other elements of plant-food the analyses show little or no loss in the drainage-waters: on Plot 5, for example, 3½ cwt. per acre of super-phosphate, containing about 70 lb. of phosphoric acid, have been applied every year, yet the drainage-water from this plot contains only about 0.000091 per cent of *phosphoric acid*, so that the losses are trivial. The case is much the same for *potash*: despite the fact that this same Plot 5 receives sulphate of potash containing 100 lb. of potash annually, the drainage-water contains only about 0.00054 per cent, so that again the loss is trifling. In both phosphoric acid and potash, then, we may expect to find remaining in the soil all that has been applied as manure and not removed in the crops—an expectation that is justified by the analysis of the soil. Of *lime*, however, the drainage-water is always removing considerable quantities from the soil: from the unmanured plot it is removed as bi-carbonate, the solution of the calcuim carbonate being effected by the carbonic acid derived from the roots of the plants and the decay of the humus; but on the manured plots the loss of lime is much greater, because reactions take place between the salts applied in the manures and the compounds of calcuim in the soil, which result in the production of soluble salts of lime.

Evidence of Analyses. — Turning now to the evidence afforded by the analyses of the soils, Table XI. shows the nitrogen found in the surface-soil in 1893, calculated both as percentages and as lb. per acre: this is compared with the additions of nitrogen in manure and its removal in crops during the previous fifty years, so as to strike a rough balance-sheet of gain or loss to the land, taking the unmanured plot as a basis for comparison. The percentages of carbon are also given.

The unmanured plot has been parting with nitrogen to the crop at the rate of about 17 lb. per acre per annum. There are other unestimated losses through drainage and the removal of weeds, and these various withdrawals have resulted in the reduction of the percentage of nitrogen to 0.0992. Although the amount of nitrogen at starting is not known, yet analyses made in 1865 and 1881 show that the decline in the nitrogen has been continuous, the average fall being equivalent to about 11 lb. of soil-nitrogen per acre per annum between 1865 and 1893. This figure is subject to inevitable errors due to the impossibility of taking exactly equivalent samples of soil at the different dates, but yet it is so much less than the annual removal in the crop (about 16 lb. per acre per annum for the period in question) that we are forced to look for agencies recuperating the stock of nitrogen in the soil. In the first place there is the rain, which has been analysed month by month at Rothamsted for the last forty years, and is found to bring down nearly 5 lb. of combined nitrogen per acre per annum. This, added to the 11 lb. the soil is found to have lost, makes up the 16 lb. removed in the crop, but there are still the unknown losses by drainage and the removal of weeds to be accounted for. A little nitrogen is doubtless fixed by leguminous weeds growing on the plot, *medicago* being rather an abundant weed in certain seasons; but the bulk of the restoration of nitrogen is probably due to the bacterium *azotobacter*, an organism capable of fixing nitrogen without the intervention of a leguminous plant, and which has been found in all the Rothamsted soils. That it has not been more effective on the arable land is due to the complete removal of the crop there, thus leaving little or no organic material to serve as a source of energy for the *azotobacter*, which can only fix nitrogen when it is oxidising some carbonaceous material.

Natural Restoration of Nitrogen.— The possibility of recuperative agencies restoring nitrogen to arable soils is still obscure, but some light is thrown upon the question by the examination of a portion of the same field which was under arable cultivation until 1885, and was then left entirely to itself until it has become covered with self-sown grasses and weeds, which are never removed, but allowed to remain as they fall in the autumn. On examining the soil of this part of the field after twenty years of being left to itself, it was found to have gained nitrogen at a rate of 100 lb. per acre per annum,—far more than would be accounted for by the proportion of leguminous herbage present. It is, however, probable that the *azotobacter*, being in this case supplied with carbonaceous material from the *débris* of the vegetation, has been able to fix nitrogen in the large quantities indicated by the analyses. In the soil of the plot receiving ash constituents only (except for a little nitrogen in the first eight years of the experiments) there is a slightly larger proportion of nitrogen, representing the *débris* of the rather larger growth of crop on this plot, as may be seen from the fact that the percentage of carbon in the soil is also higher than on the unmanured plot. The differences, however, are small, nor is the amount of nitrogen very much greater on Plots 7 and 9, receiving a complete manure containing large amounts of nitrogen every year. The nitrogen is, however, supplied as ammonium-salts or as nitrate of soda : the latter is not retained by the soil, the former is also rapidly transformed to nitrates, hence both plots are subject to loss by drainage.

Residues of Nitrogen not Accounted for.—Of 4100 lb. of nitrogen supplied as ammonium-salts only 2360 have been recovered in the crop, the soil has been protected from loss to the extent of about 600 lb., but there is still 1990 lb. unaccounted for in either soil or crop, most of it having been washed out. The recovery of nitrogen from nitrate of soda by the crop is rather more complete, some 2700 lb. being harvested from 3570 lb. applied, but again the loss is great, 1210 lb., mostly by drainage. It is, however, on the dunged plot that the recovery of the applied nitrogen is most imperfect: something like 10,000 lb. have been put on during the fifty years of the experiment, of which only 2600 have been returned in the crop and 730 lb. stored up in the soil. Thus 5670 lb. of nitrogen—more than half of the whole that has been applied —has been lost entirely, the loss being in this case not so much due to the washing out of nitrates as to the action of bacteria, which destroy nitrogen compounds and liberate the nitrogen as a free gas.

TABLE XI.

Plot.	Manuring.	In soil 9″ deep, 1893.			Approximate supply of Nitrogen in Manure. 1844-93.	Approximate removal of Nitrogen in Crops. 1844-93.	Surplus of Nitrogen over Plot 3 unaccounted for in Soil or Crop.
		Carbon per cent.	Nitrogen.				
			Per cent.	Per acre.			
		%	%	lb.	lb.	lb.	lb.
3	Unmanured	0.888	0.0992	2,570	...	850	...
5	Ash constituents only . .	0.931	0.1013	2,630	590	1,180	200
7	Ash constituents and ammonium-salts . . .	1.101	0.1222	3,170	4,100	2,360	1,990
9	Ash constituents and nitrate of soda	1.162	0.1189	3,080	3,570	2,700	1,210
2	Farmyard manure . . .	2.230	0.2207	5,150	10,000	2,600	5,670

THE FATE OF MINERAL CONSTITUENTS.

Turning to the mineral constituents of the plants' food, the fate of the phosphoric acid applied to the Broadbalk soil has been studied in some detail. Dyer examined the samples taken in 1893, extracting the phosphoric acid in both soils and subsoils by means of strong hydrochloric acid as well as by a 1 per cent solution of citric acid. He was able to show that the subsoils of the various plots were very much alike as regards their phosphoric acid content, but that the surface-soil (0-9 in.) on the plots receiving phosphoric acid was richer in phosphoric acid by an amount practically equivalent to what had been supplied in the manure and not removed in the crop. His analyses went to show that the phosphoric acid supplied in the manure (superphosphate) had all been retained in the top nine inches of soil except as far as it had been taken up by the crop, but that none of it had been washed out or gone even lower down into the subsoil. This conclusion was confirmed by a later examination, when the soils were extracted repeatedly with fresh portions of citric acid solution, which process resulted in the recovery from the surface-soil of the whole of the phosphoric acid supplied since the beginning of the experiments, as shown in the following Table XII. :—

TABLE XII.

		Phosphoric acid, lb. per acre.			
		Supplied in Manure.	Removed in Crop.	Surplus.	Dissolved by 5 extractions with 1 % citric acid.
Broadbalk, Plot 3.	Unmanured . . .	0	550	...	565
,, Plot 5.	Ash constituents only .	3960	790	3170	3000
,, Plot 7.	Ash constituents + ammonium-salts . .	3810	1370	2440	2470
,, Plot 2b.	Farmyard manure .	4870	1650	3130	2060

From these and other results of the same kind, it may be concluded that when superphosphate (the most soluble of the phosphatic manures) is applied to the land, whatever phosphoric acid the crop does not immediately utilise is retained by the surface-soil, and retained also in a readily available form, as shown by its solubility in dilute citric acid solution. Dyer also determined the potash present in the soils and subsoils of the same plots, and found that it was washed down a little farther into the subsoil, so that after fifty years the second and even the third depth of nine inches had become enriched in potash by its application at the surface. Of actual loss by washing completely through the soil there was little evidence.

Behaviour of Lime in the Soil.

Though lime is hardly to be counted as a manure, yet its actions in the soil are so important that a brief summary may be given here of the very interesting evidence as to its duration in the soil that is afforded by the Rothamsted plots. On the Rothamsted estate, as has already been indicated, the lime, or more strictly calcium carbonate, in the soil is of artificial origin, having been dug from some 12 to 20 feet below, and spread upon the surface in large quantities at various times during the seventeenth and eighteenth centuries. On the adjacent common, on the woodlands, and even on many of the grass fields of the estate, the soil contains no appreciable quantity of calcium carbonate, nor does the subsoil of the arable fields. As the calcium carbonate applied is all located in the surface layer, an examination of the soil samples taken at different dates gives an opportunity of estimating the rate at which this valuable substance is being removed from the soil, either by natural causes or by the action of the manures also applied. In the first place, there is no evidence that the lime sinks in the soil, as it is popularly supposed to do: such sinking only takes place on pasture-land, where lime, stones, or anything else on the surface is steadily buried year by year by the action of earthworms. On the arable land the lime has never descended below the layer stirred by the plough, and none is found in the sample taken between 9 and 18 inches, though something like a hundred years must have elapsed since the lime was put on the land.

On the unmanured plot the loss of carbonate of lime by purely natural causes—*i.e.*, by solution in the rainwater—amounts to 800-1000 lb. per acre per annum. This loss is much increased by the use of ammonium-salts as manures. A dressing of sulphate of ammonia will reduce the chalk in the soil by nearly its own weight, and as may be seen on some of the Rothamsted grass plots and on the wheat and barley plots at Woburn, its repeated use will so remove the chalk as to render the soil eventually acid to test-paper and incapable of carrying crops. On the other hand, superphosphate, though itself an acid manure, occasions no great loss of carbonate of lime; while nitrate of soda and farmyard manure actually reduce or repair the losses which would otherwise be occasioned by the percolating rain-water.

CONTINUOUS GROWTH OF BARLEY.

The experiments at Rothamsted on the continuous growth of barley were begun in the Hoos field in 1852. The arrangement of the plots and the manures applied to each plot have practically been unchanged since, so that the plots to-day show the effects of more than fifty years' continuous growth of barley under the same treatment year after year.

Manures Applied.—The manures are sown in the spring, and ploughed in about a week or a fortnight before seeding. The plots do not run the whole length of this field, as in Broadbalk. Instead, there are four longitudinal strips receiving different combinations of the mineral manures; these are all crossed by four breadths receiving different nitrogenous manures. The mineral manuring on the strips is as follows: (1) none; (2) phosphoric acid only, no potash or alkali salts; (3) potash, magnesia, and soda, no phosphoric acid; and (4) complete mineral manure, supplying both phosphoric acid and the alkaline salts. Each of these is combined with the four different cross-dressings of nitrogenous manures—O. no nitrogen, A. ammonium-salts, N. nitrate of soda, and C. rape-cake. There are other plots, one of which has received farmyard manure each year, and a second which received farmyard manure for the first twenty years, but has since been unmanured. Table XIII. (p. 20) shows the average

production of grain and straw for the whole period, for the last ten years, and for the single year 1902.

Maintenance of Yield under the Continuous Growth of Barley on the same Land.—One of the plots, 1-0, has been without manure since the beginning of the experiments. Under the continuous barley-growing the decline in production has been much more marked than on the wheat plot similarly treated, the average crop having been only 10 bushels for the last ten years, against an average of more than 15 bushels for the whole period. The more limited root-range of the plant would bring about a complete exhaustion of the available soil much sooner with barley than with wheat, but there is evidence that the decline in the yield of these barley plots is to some extent due to other factors than soil exhaustion, since the continuously dunged plot, which must be gaining in fertility, also shows a steady decline in production for each of the last four decades.

Effect of Nitrogenous Manures.—The effect of nitrogenous manures upon the barley crop is best seen by comparing the yields of the various Plots 4, all of which receive the same mineral manures. Plot 4-0, receiving no nitrogen, has only given an average crop of 20.4 bushels per acre, and this has been more than doubled by the application of 43 lb. of nitrogen per acre to the other three plots. But little difference is seen in the return for this amount of nitrogen, whether it be applied as ammonium-salts, nitrate of soda, or rape-cake. Over the whole period the nitrate of soda gives the highest return by about 3 per cent, but during the last two decades the plot receiving ammonium-salts has been slightly the best of the three. In the straw, again, the differences are very small, though the superiority of nitrate of soda is rather more pronounced with the straw than with the grain. The fact that ammonium-salts answer better with barley than with wheat is due to their retention by the soil close to the surface: the comparatively shallow-rooted habit of barley and its growth during the warmer portion of the year when nitrification is active, renders such a surface accumulation of nitrogen as readily available to the plant as the nitrate of soda itself.

On the completely manured plots the rape cake, 4 C, is not quite so effective as the more active forms of nitrogen, giving over the whole period an average yield of 41 against 43.5 and 42.1 bushels of grain.

TABLE XIII.—EXPERIMENTS ON BARLEY, HOOS FIELD. PRODUCE OF GRAIN AND STRAW
PER ACRE. AVERAGES OVER FIFTY-ONE YEARS (1852-1902), AND OVER TEN YEARS
(1893-1902). ALSO PRODUCE IN 1902.

Plot.	Abbreviated Description of Manures.	Dressed Grain.			Straw.		
		Average, 51 years (1852-1902).	Average, 10 years (1893-1902).	Season 1902.	Average, 51 years (1852-1902).	Average, 10 years (1893-1902).	Season 1902.
		bush.	bush.	bush.	cwt.	cwt.	cwt.
1 O	No minerals, and no nitrogen .	15.3	10.1	14.3	8.8	6.4	5.6
2 O	Superphosphate only, do.	20.1	13.6	21.7	10.2	7.8	9.9
3 O	Alkali salts only, do.	16.1	8.9	13.3	8.9	5.9	6.3
4 O	Complete minerals, do.	20.4	12.4	18.2	10.8	8.0	10.9
1 A	Ammonium-salts alone .	26.5	16.2	21.7	14.9	10.5	9.7
2 A	Superphosphate and ammonium-salts	39.9	26.8	39.8	22.5	16.5	20.8
3 A	Alkali salts and do.	29.4	20.8	20.0	17.0	12.9	10.4
4 A	Complete minerals and do.	42.1	35.1	40.3	24.9	20.5	23.1
1 N	Nitrate of soda alone .	30.4	20.5	25.1	18.1	14.4	14.1
2 N	Superphosphate and nitrate of soda	44.0	35.9	45.9	26.2	23.0	26.1
3 N	Alkali salts and do.	31.5	23.4	22.0	19.7	15.3	10.0
4 N	Complete minerals and do.	43.5	34.9	36.5	27.4	22.6	23.4
1 C	Rape-cake alone .	39.2	31.0	38.7	22.4	18.4	20.5
2 C	Superphosphate and rape-cake	41.5	33.2	41.4	23.9	19.6	24.2
3 C	Alkali salts and do.	37.7	29.6	39.7	22.4	18.1	22.7
4 C	Complete minerals and do.	41.0	32.5	40.7	24.5	20.1	25.4
7-1	Unmanured (after dung 20 yrs, 1852-71)	27.0[1]	19.9	27.5	15.4[1]	12.8	14.3
7-2	Farmyard manure .	47.6	42.6	42.4	29.1	28.8	27.4

[1] Average, 31 years (1872-1902).

The plot receiving farmyard manure, 7-2, gives a higher crop than any other, but the amount of nitrogen supplied in this case is very high, being estimated at nearly five times as much as on any of the other plots.

Effect of Mineral Manures. — The great importance of phosphoric acid to the barley crop is seen on comparing Plots 3 and 4, which only differ from one another in the omission of phosphoric acid on Plots 3. It will be seen that Plots 3 give but little more crop than Plots 1, which receive nitrogen alone —only 32.9 bushels per acre against 32, taking the average of the three series A, N, and C,—but that a very marked

increase to 42.2 bushels per acre is found on Plots 4 for the addition of phosphoric acid. The straw shows just as marked an increase of crop brought about by phosphoric acid as does the grain, rising from 19.7 cwt. to 25.6 cwt. per acre. In the field the most striking effect is seen in the hastened maturity brought about by the phosphoric acid. Not only are Plots 2 and 4, which receive phosphoric acid, in the ear long before Plots 3 and (to a less extent) Plots 1, but they will have begun to yellow for harvest when Plots 3 still show only upright green ears.

Comparing Plots 2 and 4, we see that a manure supplying phosphoric acid and nitrogen is almost as effective as a complete manure containing also potash and the other alkaline salts. There is a great increase of crop caused by the superphosphate and nitrogen on Plots 2, over the nitrogen alone on Plots 1, and very little further increase for the further addition of potash and other alkaline salts on Plots 4. Where the nitrogenous manure is nitrate of soda or rape cake, the omission of the potash on Plots 2 compared with Plots 4, receiving a complete manure, shows no effect, whether we make the comparison over the whole period or for successive ten-year periods.

Potash plays a less important part than phosphoric acid in the manuring of barley. Very little increase of crop has resulted from its use on the Rothamsted soil, and the only indication of the supply in the soil giving out has been seen in the last twenty years on the plot receiving superphosphate and ammonium-salts. Of course the Rothamsted soil starts with a very large original store of potash.

Speaking generally, we find that barley is much more dependent on a supply of mineral manures than is wheat, a free supply of phosphoric acid in particular being essential to its proper development.

Relation of Manuring to Quality. —In many years the barleys grown on these plots have been submitted to valuation by an expert, and a comparison of the average results obtained year by year leads to the following conclusions. The first thing that becomes apparent is that it is impossible to grow high-class barley by simply starving the plant. It is found that the barley showing the highest average value, the best weight per bushel, the largest grains, and the smallest proportion of tail corn, is that grown on Plots 4, where a complete manure containing both nitrogen and minerals is supplied. It does not, however, follow that any kind of manure will improve the quality of the barley. The grain from the plot receiving farmyard manure every year, despite the high weight per bushel, and the bold berry indicated by the high weight of 100 grains, has yet a value considerably below the average. Again, the use of nitrogen alone on Plot 1 A or 1 N gives the lowest weight per bushel and the lowest valuations of the whole series. It has already been seen that the yield of the barley crop is very dependent on the supply of minerals, especially of phosphoric acid, and the same effect extends to the quality of the crop. The use of superphosphate on Plots 2 as compared with Plots 1 gives a better proportion of grain to straw, a higher weight per bushel, and a greatly increased value; similarly, the omission of superphosphate on Plots 3 as compared with Plots 4 results in a deterioration of all the qualities making for value in the barley. Comparing the barley from Plots 3 and Plots 1, in the absence of superphosphate the potash salts on Plots 3 do not effect much improvement, though their presence on Plots 4 as compared with Plots 2 results in an improved quality. The presence of potash in the manure increases the straw more than the grain. In all the series Plots 3 and 4, receiving potash, give a lower proportion of grain to straw than do Plots 1 and 2, without potash.

If we compare the series together, the rape-cake gives better barleys than either ammonium-salts or nitrate of soda, but the sample which on the average is the best is that grown with the full minerals and ammonium-salts.

The better quality of the barley grown with ammonium-salts as compared with nitrate of soda is doubtless due to the more shallow-rooting habit induced by the retention of the ammonium-salts near the surface. The more deeply rooted barley on the nitrate of soda plot continues to grow longer, and yields a later and more irregularly ripening crop.

Effects of Season upon Yield and Quality.—As to the effect of season, although it is noticed that in the very dry years the ammonium-salts often have the advantage, and that in very wet years the nitrate of soda is the more effective source of nitrogen, yet when averages are taken over the whole period, the seasons in which the ammonium-salts give a better crop than the nitrate of soda are wetter throughout than those in which the nitrate of soda is the more effective source of nitrogen. A wet March seems to be the most hurtful to the nitrate of soda plot. The comparative effect of the mineral manures in a wet and dry season are also similar to those noticed in the case of the wheat. In the wet season the crop is very dependent upon supplies of minerals in the manure, and especially on an abundance of phosphoric acid.

Doubtless in the wet season the ripening effect of the phosphoric acid is specially valuable, while in a dry season the potash, by inducing a longer period of growth, is more effective in increasing the crop. The ripening action of the phosphoric acid may also be seen in the way it increases the weight per bushel of the grain in a wet season, whereas in a dry season it has little or no effect.

TABLE XIV.—MANURING OF THE PERMANENT GRASS PLOTS PER ACRE PER ANNUM, 1856 AND SINCE.

Plot.	Abbreviated Description of Manures.	Nitrogenous Manures.		Mineral Manures.				
		Ammonium-salts.	Nitrate of Soda.	Superphosphate.	Sulphate of Potash.	Sulphate of Soda.	Sulphate of Magnesia.	Silicate of Soda.
		lb.	lb.	cwt.	lb.	lb.	lb.	lb.
3 12	Unmanured every year .	...	...	...	...	...	...	...
2 1	Unmanured ; following dung first 8 years . Ammonium-salts alone ; with dung also first 8 years .	... 200	... ...	... ...	... ...	... ...	... ...	... ...
4-1 8 7	Superphosphate of lime . Mineral manure without potash . Complete mineral manure .	... ...	... ...	3.5 3.5 3.5	 500	... ¹250 100	... 100 100	... ...
6 15	As Plot 7 ; ammonium-salts alone first 13 years . As Plot 7 ; nitrate of soda alone first 18 years .	... ...	... ...	3.5 3.5	500 500	100 100	100 100	... ...
5 17	Ammonium-salts alone (to 1897) . Nitrate of soda alone .	400 ...	... 275	... ...	... ...	... ...	... ...	... ...
4-2 10 9 13 11-1 11-2	Superphosphate and ammonium-salts . Mineral manure (without potash) and ammonium-salts . Complete mineral manure and ammonium-salts . As Plot 9, and chaffed wheat straw also to 1897 . Complete mineral manure and ammonium-salts . As Plot 11-1, and silicate of soda .	400 400 400 400 600 600	... ...	3.5 3.5 3.5 3.5 3.5 3.5	... ... 500 500 500 500	... ¹250 100 100 100 100	... 100 100 100 100 100	 400
16 14	Complete mineral manure and nitrate of soda . Complete mineral manure and nitrate of soda .	... ...	275 550	3.5 3.5	500 500	100 100	100 100	... ...

¹ Reduced in 1905 to 100 lb.

The results indicate that in wet and cool climates, and upon heavy soils, phosphates should always form part of the manure for barley, but that potash is only likely to be valuable on light soils and in the drier parts of the country.

In the dry season the weight per bushel is much higher than in the wet, and the grain is about equal in weight to the straw, whereas in the wet season the weight of grain only amounts to about 70 per cent of the straw.

Relative Effects of Weather and Manures.—Taking the results as a whole, it is seen that season has a much greater effect in bringing about changes in the composition of the barley grain than have variations in the manuring, but that the best barley will be grown with a fair but not large amount of nitrogenous manure combined with a free supply of phosphoric acid in some way or other.

It does not appear possible to establish any such critical periods for the rainfall in relation to the growth of barley as could be done for wheat.

TABLE XV.—PRODUCE OF HAY PER ACRE. AVERAGE OVER THE PERIOD OF FORTY-SEVEN YEARS (1856-1902), THE TEN YEARS (1893-1902), AND THE INDIVIDUAL YEAR 1905, ROTHAMSTED. TOTAL OF FIRST AND SECOND CROPS (IF ANY).

| Plot. | Abbreviated Description of Manures. | Averages over | | Season 1905. |
		47 years (1856-1902).	10 years (1893-1902).	
		cwt.	cwt.	cwt.
3	} Unmanured every year . . . . . {	21.9	15.9	19.4
12		24.5	18.5	24.7
2	Unmanured ; following farmyard dung for first 8 years	27.9[1]	17.4	23.2
1	Ammonium-salts alone (=43 lb. N.) ; with farmyard dung for first 8 years	35.4[2]	24.9	26.3
4·1	Superphosphate of lime	23.3[5]	17.8	22.3
8	Mineral manure without potash . . .	28.1	21.6	30.3
7	Complete mineral manure	38.8	36.5	52.9
6	Complete mineral manure as Plot 7 ; following ammonium-salts alone first 13 years . . .	37.4[3]	36.0	46.1
15	Complete mineral manure as Plot 7 ; following nitrate of soda alone first 18 years	37.0[4]	40.8	51.9
5	Ammonium-salts alone = 86 lb. nitrogen . .	(26.1)[6]	...	...
17	Nitrate of soda alone = 43 lb. nitrogen . .	35.3[7]	30.6	39.7
4·2	Superphosphate and ammonium-salts = 86 lb. N. .	35.5[5]	28.3	31.5
10	Mineral manure (without potash), and ammonium-salts = 86 lb. N.	49.3	38.1	37.3
9	Complete mineral manure and ammonium-salts = 86 lb. N.	54.1	46.8	48.6
13	As Plot 9, and chaffed wheat straw also to 1897 inclusive	62.5[6]		...
11·1	Complete mineral manure, and ammonium-salts = 129 lb. N.	65.5	64.6	59.0
11·2	As Plot 11-1, and silicate of soda . . .	72.0	68.0	74.5
16	Complete mineral manure and nitrate of soda = 43 lb. N.	48.0[7]	42.4	52.3
14	Complete mineral manure and nitrate of soda = 86 lb. N.	59.3[7]	53.4	57.6

1 After the change. Before the change, 42.9 cwt.
2 ,, ,, 49.5 cwt.
3 ,, ,, 30.6 cwt.
4 ,, ,, 35.4 cwt.

5 44 years only (1859-1902).
6 42 years (1856-1897).
7 45 years only (1858-1902).

GRASS FOR HAY.

The experiments upon grass at Rothamsted began in 1856, about 7 acres of the park close to the house being set aside for the purpose. The land has been in grass as long as any recorded history of it exists, for some centuries at least. It is not known that seed has ever been sown, and at the beginning of the experiments the herbage on all the plots was apparently uniform.

The plots, of which there are twenty in all, vary somewhat in size, from

one - half and one - eighth of an acre.
Up to 1874, inclusive, the grass was
only cut once, the aftermath being
fed off by sheep. Since that time there
has been no grazing, and the plots are
generally cut twice in the year. The
grass is made into hay in the usual way,
and the whole produce of each plot is
then weighed.

1. Effect of Manures upon the Nature of the Herbage.

In dealing, however, with the produce
of grass-land, which is a mixed herbage
consisting of many different species of
grasses, leguminous plants, and other
orders, it is not sufficient to consider
only the gross weight of produce. The
various species are differently stimulated
by particular manures : even among the
grasses themselves, such a difference of
habit as a deep or shallow root system
will determine to which manure the grass
will respond. The aspect of any meadow
represents the results of severe competition among the various species represented : the dominant species are those
most suited to their environment—i.e.,
to the amount and nature of the plant-
food in the soil, the water-supply, the
texture of the soil, and other factors. If
any of these factors be altered, as is done
in the case of the Rothamsted plots by
manuring in different fashions, the original equilibrium between the contending
species is disturbed ; some species are
favoured, and increase at the expense of
the others until a new equilibrium is
attained, and the general character of
the herbage from the botanical point of
view is completely altered. It thus becomes important to ascertain the nature
of the plants comprising the herbage
produced by a given manure, as well as
to determine its amount : from time to
time, therefore, at Rothamsted a carefully
selected fraction of the herbage from
each plot has been separated into its
constituent species, the relative proportions of which are determined by weighing. As this complete separation involves
a great amount of work, a partial separation only is made every year, in which
case the herbage is separated into three
groups — the grasses, the leguminous
plants, and the miscellaneous species
respectively.

2. The Unmanured Plots.

Two of the plots have remained without manure during the whole of the
experiment. They are situated near the
extremities of the field, and show a slight
but constant difference in crop. Taking
the average of the whole period, these
unmanured plots have produced rather
more than a ton of hay per acre per
annum. If we compare the successive
ten - year returns, there is no sign of
approaching exhaustion or great falling-
off in crop from year to year. The im-
poverishment of these unmanured plots
is more to be seen in the character of the
herbage than in the gross weight of produce. Weeds of all descriptions occupy
the land, and the relative proportion they
bear to the grasses and clovers has increased from year to year. A fair proportion of clovers, both red and white,
is found on these plots, but the weeds,
which amount to 26 per cent, taking the
average over the whole period, have of
late years constituted nearly one-half of
the herbage. The most prominent species
among the grasses are the quaking-grass,
so generally taken as a sign of poor land,
which constituted 20 per cent of the
whole herbage in 1903, and sheep's
fescue ; among leguminous plants the
bird's-foot trefoil ; and burnet, hawkbit,
and black knapweed among the weeds.

Speaking generally, these plots now
present the appearance, perhaps in a
rather exaggerated degree, of much of
the poor pasture and meadow - land in
this country, wherever milch cows and
wet flocks are habitually grazed and the
land occasionally hayed, without anything being restored in the shape of
artificial food or manure.

3. Use of Nitrogenous Manures alone.

Three of the plots—17, 5, and 1—
show the effect of the long-continued
use of nitrogenous without any mineral
manures. Plot 5 has been receiving 86
lb. of nitrogen as ammonium-salts, Plot
17 half the quantity of nitrogen in the
shape of nitrate of soda, and Plot 1 the
same half quantity of nitrogen as ammonium-salts, though on this plot dung
was applied in each of the first eight
years of the experiment. It is very evident that when a nitrogenous manure is

used alone for grass, nitrate of soda is far more effective than the ammonium-salts— e.g., on Plot 17 it has given an average crop of 35 cwt. against 26 cwt. produced by double the quantity of nitrogen in ammonium-salts on Plot 5.

TABLE XVI.—PERCENTAGES OF GRAMINEOUS, LEGUMINOUS, AND MISCELLANEOUS HERBAGE. AVERAGE OF FORTY-SEVEN YEARS (1856-1902, AND 1902 SEPARATELY). ROTHAMSTED. FIRST CROPS.

Plot.	Manures.	Average over 47 years (1856-1902).			Season 1902.		
		Gramineæ.	Leguminosæ.	Miscellaneæ.	Gramineæ.	Leguminosæ.	Miscellaneæ.
		per cent.	per cent.	per cent.	per cent.	per cent.	per cent.
3 12	} Unmanured every year {	64.8 64.9	8.9 9.0	26.3 26.1	34.3 38.1	7.5 16.1	58.2 45.8
2	Unmanured; following farm-yard dung for first 8 years	75.5	4.3	20.2	24.4	5.7	69.9
1	Ammonium-salts alone (= 43 lb. N.); with farmyard dung for first 8 years	87.8	0.7	11.5	77.6	1.4	21.0
4-1 8	Superphosphate of lime	68.0	5.8	26.2	54.4	15.4	30.2
	Mineral manure without potash [1]	70.6	6.8	22.6	28.8	22.1	49.1
7	Complete mineral manure	62.0	23.8	14.2	20.3	55.3	24.4
6	Complete mineral manure as Plot 7; following ammonium-salts alone first 13 years	...	...	..	18.4	61.0	20.6
15	Complete mineral manure as Plot 7; following nitrate of soda alone first 18 years	...	..	...	26.2	63.1	10.7
5	Ammonium-salts alone = 86 lb. N.	80.5	0.4	19.1	...	...	...
17	Nitrate of soda alone = 43 lb. N.	71.0	1.3	27.7	43.8	3.4	52.9
4-2	Superphosphate and ammonium-salts = 86 lb. N.	88.2	0.1	11.7	91.5	(0.01)	8.5
10	Mineral manure (without potash),[1] and ammonium-salts = 86 lb. N.	90.7	0.1	9.2	97.6	(0.01)	2.4
9	Complete mineral manure and ammonium-salts = 86 lb. N.	88.7	0.4	10.9	91.2	1.3	7.5
13	As Plot 9, and chaffed wheat straw also to 1897 inclusive	92.3	0.3	7.4	98.1	0.6	1.3
11-1	Complete mineral manure and ammonium-salts = 129 lb. N.	95.8	0.1	4.1	99.2	0	0.8
11-2	As 11-1, and silicate of soda	97.5	0	2.5	99.5	0	0.5
16	Complete mineral manure and nitrate of soda = 43 lb. N.	82.9	5.4	11.7	61.7	12.8	25.5
14	Complete mineral manure and nitrate of soda = 86 lb. N.	90.6	1.3	8.1	88.8	3.7	7.5

[1] Including potash first 6 years.

For this superiority of the nitrate of soda two reasons may be traced: being completely soluble it sinks deeply into the soil, and encourages grasses of a deeply rooting habit, which not only obtain more food from the soil, but also are better able to withstand the droughts of spring and early summer. On Plot 17 (nitrate) deep-rooting grasses like meadow foxtail and downy oat grass are prominent, whereas the plots receiving only ammonium-salts are almost wholly occupied by sheep's fescue and common bent, whose feeding-roots are close to the surface, where the ammonium-salts are caught and retained by the humus in the soil.

The continued use of large applications of ammonium-salts has also had an injurious effect upon the reaction of the soil, since it behaves as an acid, and continually removes carbonate of lime. A creeping surface vegetation tends to accumulate, and decays into a substance resembling peat ; at the same time the vegetation shrinks into tufts, between which are bare patches of black soil, showing an acid reaction to litmus paper. So pronounced had this effect become on Plot 5, which received the larger amount of ammonium-salts, that the application has been discontinued since 1897, lest the turf should be entirely killed. Another sign of the sourness caused by the use of ammonium-salts without minerals is seen in the prevalence of sorrel on this plot : it forms nearly 15 per cent of the whole herbage, and it is interesting to note that the only portion of the plot from which the sorrel is absent is a strip that was dressed with chalk in 1883 and 1887. The ill effect of continually manuring with a fertiliser containing only one of the constituents of a complete manure is seen in the very large proportion of weeds on Plot 17 (nitrate of soda alone), where the yield also has declined decade by decade. It is remarkable, however, how well the crop has been maintained, in spite of the fact that nitrate of soda is usually regarded as "exhausting."

4. *Mineral Manures used alone.*

On three of the plots no nitrogenous manures have been applied since the beginning of the experiments. On Plot 7 a complete mineral manure, supplying phosphoric acid, potash, magnesia, and soda, is used ; Plot 8 has received the same application, but without potash, since 1861, while Plot 4-1 receives superphosphate only. With the complete minerals a fair crop is grown, averaging over 1½ ton of hay for the first cut alone. The reason that the crop on this plot is maintained, although no nitrogen is supplied in the manure, lies in the free growth of leguminous plants.

It will be seen that, taking the average over the whole period the leguminous plants form 24 per cent of the herbage, and the proportion has increased from year to year. These leguminous plants are not only themselves independent of nitrogen in soil or manure, but by fixing the atmospheric nitrogen and leaving it behind in the residues of their dead roots, they provide a supply of combined nitrogen for the grasses and other plants which cannot of themselves feed on the free gas in the air. The predominant leguminous plant is *Lathyrus pratensis*, but red and white clover are also abundant.

The omission of potash on Plot 8 has caused a very striking difference both in the crop and in the character of the herbage. The average crop has been about one-quarter less over the whole period, and shows a progressive decline in fertility, until at the present time it is little more than half that of Plot 7. The poor results on this plot, as compared with Plot 7, must be put down to its poverty in leguminous herbage, the development of which depends on a free supply of potash. Of late years the proportion of leguminous plants on this plot has amounted to about one-half of that found on Plot 7 ; the grasses are about the same, the difference being made up by an increased amount of weed.

Plot 4-1, which each year has received superphosphate only, now presents a very impoverished appearance, and is giving no more crop than the unmanured plots. Indeed the aspect of this plot, where the most abundant grass is quaking-grass, and where weeds, chiefly hawkbit, burnet, and plantain, are unusually prominent, would seem to indicate that the land is more exhausted here than on the unmanured plot.

5. *Complete Manures—Nitrogen and Minerals.*

Among the plots which receive both nitrogenous and mineral manures, Plot 9, with a complete mineral manure and ammonium-salts, should be compared with Plot 14, which is exactly similar except that the nitrogen is applied in the form of nitrate of soda, and again with Plot 16, where only half the amount of nitrogen is applied, but again as nitrate of soda. The nitrate of soda gives the heavier yield, the herbage is also more diversified, and there is not the total absence of leguminous plants which marks the plots receiving ammonium-salts. Two characteristic plants, soft brome grass and beaked parsley (*Anthriscus sylvestris*), are found only on the plots receiving nitrate of soda, the corresponding umbelliferous plant where ammonium-salts are used being the earth-nut (*Conopodium denudatum*).

On Plot 11 the same mineral manures are applied with an extra amount of ammonium-salts, so that the nitrogenous manuring is excessive. As a result, the vegetation consists entirely of tufts of

three coarse grasses—meadow foxtail, Yorkshire fog, and tall oat grass. The soil has also become sour and unhealthy; the plant is dying in patches, except on the upper portion of the plot where lime has been applied, and on the half numbered 11-2 where the silicate of soda is used.

The effect of omitting potash from the complete manure is seen on Plot 10, and again on Plot 4-2, where superphosphate and ammonium-salts only are applied. It is noticeable that the grass on all the potash-starved plots is weak in the straw and liable to fungoid attacks.

TABLE XVII.—THE PARK, ROTHAMSTED. (FIRST CROPS ONLY.) PRODUCE PER ACRE.

Plots.	1903.		1904.		1905.		1906.		1907.	
	Limed.	Un-limed.	Limed.	Un-limed.	Limed.	Un-limed.	Limed.	Un-limed.	Limed.	Un-limed.
	cwt.	cwt.	cwt.	cwt.	cwt.	cwt.	cwt.	cwt.	cwt.	cwt.
3	16.34	10.61	30.20	22.46	18.78	15.79	11.88	12.18	23.08	19.44
7	51.91	49.46	61.83	61.87	47.15	44.34	41.40	34.38	60.02	54.35
9	60.49	50.07	69.76	63.69	52.18	36.87	49.95	39.01	66.81	63.14
11-1	80.84	70.20	88.40	85.42	50.97	24.71	51.62	42.89	58.44	34.89
16	45.68	48.68	52.12	53.34	41.97	46.19	38.47	39.25	47.24	53.56

6. Changes in the Herbage following changes in Manuring.

Plot 6 was up to 1868 manured with ammonium-salts alone, like the adjoining Plot 5; the ammonium-salts were then replaced by a complete mineral manure containing potash. The result is seen in the way leguminous plants have gradually invaded the plot, until they now are just as prominent as they are on Plot 7 where mineral manures have been used throughout. The southern half of Plot 5 has also been manured with minerals instead of ammonium-salts since 1898, and the gradual invasion of leguminous plants may now be seen in progress.

On Plot 15 nitrate of soda was applied up to 1875, when a change to a complete mineral manure was made, with the same result of the incoming of the leguminous plants.

7. Effect of Lime.

The southern halves of Plots 1 to 4-2, 7 to 11-2, 13, and 16, were dressed with ground quicklime, at the rate of 2000

lb. per acre, in January 1903, and the treatment was repeated in January 1907. Table XVII. shows the results each year (first crop only) on the limed and unlimed portions of some of the plots.

It will be seen that the lime has produced an increase of crop on the unmanured Plot 3; on 7, which receives minerals only; on 9 and 11-1, which receive a complete manure containing ammonium-salts; but not on Plot 16, where a complete manure is also used, containing nitrate of soda in place of the ammonium-salts. In many cases the benefit of the lime is more marked in the third and fourth crop after its application than in its first.

The greatest increase is to be found on Plots 9 and 11-1, where the soil had become acid through the long use of salts of ammonia, and where there was a considerable accumulation of organic matter; on Plot 16, which receives nitrate of soda, the soil is always slightly alkaline, and the lime produces no beneficial effect. These results confirm the opinion that the value of liming the soil consists in the neutralisation of acids, so that the organic matter can decay properly, and the liberation of reserves of potash in the soil.

ROOT CROPS.

The Barn Field at Rothamsted has been given up to experiments upon roots since 1843, and no change has been made in the system of manuring the different plots since 1856. At first the field was devoted to white turnips, then swedes, but as it proved impossible to grow these crops year after year upon the same land, mangel-wurzel have occupied the land since 1876, and no difficulty has been experienced in growing successive crops of this root, beyond those due to the bad effect of the repeated use of some of the manures upon the tilth of the soil.

1. *Continuous Growth of Mangel-wurzel.*

The following tables give the nature and amount of the manures applied to each plot, the average crop during the whole period of thirty-one years, and the crop, both root and leaf, of 1907.

TABLE XVIII.—EXPERIMENTS ON MANGEL-WURZEL, BARN FIELD, BEGINNING 1876.
QUANTITIES OF MANURES PER ACRE PER ANNUM.

| Strip. | Strip Manures. | | | | | Nitrogenous Manures running across all the Strips. | | | | | | |
	Farmyard Manure.	Superphosphate.	Sulphate of Potash.	Sulphate of Magnesia.	Chloride of Soda. (Salt.)	Series O. None.	N. Nitrate of Soda.	A. Ammonium-salts.[1]	AC. Rape-cake.	AC. Ammonium-salts.[1]	C. Rape-cake.
	tons.	cwt.	lb.	lb.	lb.	lb.	lb.	lb.	lb.	lb.	lb.
1	14	...	...	...	...	...	550	400	2000	400	2000
2	14	3.5	500[2]	...	...	...	550	400	2000	400	2000
4	...	3.5	500	200	200	...	550	400	2000	400	2000
5	...	3.5	...	...	...	...	550	400	2000	400	2000
6	...	3.5	500	...	...	...	550	400	2000	400	2000
7[3]	...	3.5	...	200	200	...	550	400	2000	400	2000
8	...	...	...	...	...	...	550	400	2000	400	2000

[1] Equal parts sulphate and muriate ammonia of commerce.
[2] The addition of potash to Plot 2 began in 1895.   [3] Commenced in 1903 only.

TABLE XIX.—BARN FIELD, MANGEL-WURZEL. AVERAGE PRODUCE OF ROOTS PER ACRE OVER TWENTY-SEVEN YEARS (1876-1902).

| Strip. | Strip Manures. | Cross-dressings. | | | | |
		O. None.	N. Nitrate of Soda.	A. Ammonium-salts.	AC. Rape-cake and Ammonium-salts.	C. Rape-cake.
		tons.	tons.	tons.	tons.	tons.
1	Dung only . . .	17.44	24.74	21.73	24.05	23.96
2	Dung, super., potash[1] .	17.95	25.19	22.35	24.91	24.43
4	Complete minerals .	5.36	18.01	14.86	25.49	21.33
5	Superphosphate only .	5.21	15.40	7.66	10.38	11.13
6	Super. and potash .	4.55	15.38	14.03	22.48	18.63
8	None . . .	3.91	10.24	5.89	9.84	10.00

[1] The addition of potash to Plot 2 only began in 1895.

TABLE XX.—BARN FIELD, MANGEL-WURZEL. PRODUCE OF ROOTS AND LEAVES
PER ACRE. SEASON 1907.

| Strip. | Strip Manures. | Cross-dressings. | | | | |
| | | O. | N. | A. | AC. | C. |
		None.	Nitrate of Soda.	Ammonium-salts.	Rape-cake and Ammonium-salts.	Rape-cake.
		tons.	tons.	tons.	tons.	tons.
1	Dung only . .	R 26.00	41.42	33.52	34.29	35.02
		L 3.64	4.64	5.27	4.90	5.17
2	Dung, super., potash .	R 26.52	42.13	41.68	43.52	40.74
		L 3.33	4.61	6.64	7.08	5.34
4	Complete minerals .	R 5.95	32.80	26.68	40.97	33.09
		L 1.09	4.47	3.42	5.25	4.11
5	Superphosphate only .	R 6.21	24.62	10.88	11.26	15.43
		L 1.17	3.42	2.86	2.18	2.18
6	Super. and potash .	R 5.78	25.05	25.22	35.88	28.15
		L 1.05	3.14	3.44	5.68	2.84
8	None . . .	R 5.15	18.60	9.87	10.90	13.24
		L 1.06	3.84	3.03	2.26	2.40

It will be seen that, considering the amount of manure that is used on many of the plots, the average returns are low, much below the weight a good farmer would expect to get in an average season with his land in good condition for man-golds. It is, however, to be remembered that these are average results from the actual weights removed from the land, with no allowances for blanks caused by insect attacks, drought, &c., and though the manuring appears excessive, the mangold crop grown year after year on the same land takes away a good deal more than does an ordinary rotation. There is no particular evidence that growing this crop continuously on the same land causes any progressive deterioration in the yield; in fact, some of the crops during the last few years have been higher than any during the earlier years of the experiment.

2. Value of Farmyard Manure for Mangel-wurzel.

On a first consideration of the results, the most notable thing is the value of farmyard manure in growing mangel-wurzel. This is due not to the plant-food the dung contains, but to its beneficial effect upon the texture of the soil. To secure a good crop of roots with a moderate rainfall on the stiff soil of Rothamsted, the chief difficulty lies in the preparation of a satisfactory seed-bed. In favourable seasons it is possible to obtain good crops on the plots receiving no organic manure, but in ordinary years the bad texture of the soil which results from such treatment, and its tendency to lose water because of the lack of humus, affect both the germination of the seed and the growth of the plant in its early stages. Thus the all-important start of the mangel-wurzel plant is promoted by the use of farmyard manure and, to a less degree, of rape-cake; but once the plant has been established the dung is best supplemented by the addition of some more active source of nitrogen.

Since the farmyard manure is applied every year at Rothamsted at the rate of 14 tons per acre, there is a great accumulation of fertility on the plots which receive it. Notwithstanding the store of nitrogen thus present, in years of abundant growth the crop cannot be supplied with nitrogen rapidly enough by the dung and its residues, so that the yield is considerably increased by the addition of more active fertilisers, especially of nitrate of soda. For example, in 1907 farmyard manure alone gave 26 tons of roots, which was increased to 41 tons by the further addition of nitrate of soda.

During the earlier years of the experiment, the plots numbered 2 received superphosphate in addition to the dung, while Plot 1 had dung only: no increase of crop resulted from the superphosphate. But for the last twelve years Plots 2 have received both superphosphate and sulphate of potash, and this mixture has given a fair return where ammonium-salts or rape-cake is used as an additional source of nitrogen. This increase must be attributed to the sulphate of potash, since the superphosphate alone made no difference to the yield of the dunged plots. It is somewhat remarkable that sulphate of potash should still be effective on a strong soil naturally rich in potash, and on plots to which farmyard manure in excess has been applied yearly for more than half a century, but the mangel-wurzel is, above all other crops, a potash-loving plant.

3. Effect of Nitrogenous Manures on Mangel-wurzel.

The experiments with artificial manures show the great dependence of the mangel-wurzel crop upon an abundant supply of nitrogen in the manure : up to as much as 180 lb. of nitrogen per acre the yield increases with each addition of nitrogenous manure ; beyond that point there is no increase in the crop from further nitrogen, and in practice probably 100 lb. of nitrogen per acre would represent the paying limit.

The experiments further bring out the fact that, as a nitrogenous manure for mangel-wurzel, nitrate of soda is much more effective than sulphate of ammonia containing the same amount of nitrogen. For this superiority there are two reasons : firstly, the soda base of the nitrate of soda renders more soluble the potash compounds in the soil or in the other manures used, thus economising the large amounts of potash the mangel-wurzel crop always requires. This is well seen on comparing the yields of Plots 1 and 2, Table XIX. When ammonium-salts or rape-cake are added

to the dung, as on Plots 1 and 2, A, AC, and C, there are much better crops on Plot 2 with sulphate of potash than on Plot 1 without sulphate of potash. But Plot 2 N gives practically the same crop as Plot 1 N. In this case the nitrogenous manure is nitrate of soda, and this can set free so much potash from the dung on Plot 1 that the use of sulphate of potash on Plot 2 produces no increase of crop.

Secondly, the deep-rooting habit induced by nitrate of soda gives the plant a better supply of water through the dry periods, and causes it to keep growing later in the autumn, thus increasing the weight of the crop. The leaves of the plants on the plots receiving ammonium-salts always turn yellow and begin to die off long before any change is to be seen on the plots receiving nitrate of soda.

4. Effect of Potash Salts upon Mangel-wurzel.

But the most striking fact which is demonstrated on these plots as to the nutrition of the mangel-wurzel is its dependence upon an ample supply of potash. We know that potash is necessary to the process in the plant by which carbohydrates like sugar and starch are manufactured out of the carbonic acid of the air, and the mangel-wurzel is essentially a sugar-making plant, since more than three-quarters of the dry matter in the root consists of cane-sugar ; hence the crop must have plenty of potash at its disposal. If the yield of Plots 4 and 6, especially in series A, C, or AC, be compared with the yield of the corresponding Plots 5 (no potash), the magnitude of the effect produced by potash will be plain. The following table will serve as an example from a particular year :—

TABLE XXI.—PRODUCE OF MANGEL-WURZEL, ROTHAMSTED, 1900.

Plot.	Manure.	Leaf per acre.	Roots per acre.	Sugar per acre.
		tons.	tons.	tons.
5 A	Sulphate of ammonia, superphosphate . . .	2.95	12.00	0.80
6 A	" " super., and sulphate of potash .	3.60	28.20	2.47

The only difference in the manuring of the two plots lay in the sulphate of potash applied to 6 A. This hardly affected the amount of leaf produced, but it increased the roots from 12 to 28 tons, and the sugar contained in these roots from 0.8 of a ton to 2.47 tons. Of course, Plot 5 A, by long cropping with nitrogenous and phosphatic manures without potash, has been impoverished as regards its potash in a way that would rarely or never occur in ordinary farming. Still, the example is instructive as showing how essential a part potash must be of a mangel-wurzel manure.

5. Proportion of Manure Recovered in Crop.

As the nitrogen in the crop is determined each year, it becomes possible to trace the return from each of the nitrogenous manures with which the mangel-wurzel are grown, so as to estimate the efficiency of the manure by the percentage recovered in the crop.

From Table XXII. it will be seen that when the nitrogenous manures were used in conjunction with phosphates and potash, the recovery ranged from 78 per cent with nitrate of soda down to 56 per cent with a mixture of rape-cake and ammonium-salts. This latter mixture contained, however, more than twice the amount of nitrogen supplied by the nitrate of soda, and the proportion recovered in the crop is always less with a heavy than with a light manuring, as is very evident from the second part of the table, in which the nitrogenous manures are used in conjunction with dung.

TABLE XXII.—MANGEL-WURZEL.—NITROGEN RECOVERED IN ROOTS FOR 100 IN MANURE.

Series.	Cross-dressings.	Average produce per acre of Roots.	Nitrogen.			
			Per cent in fresh Roots.	Per acre per annum in Roots.	Supplied in Manure per acre per annum.	Recovered in Roots for 100 in Manure.
	Plots 4.—Superphosphate, sulphates of potash and magnesia, and common salt.					
		tons.	per cent.	lb.	lb.	per cent.
N	Nitrate of soda . . .	17.95	0.164	67.2	86	78.1
A	Ammonium-salts . . .	15.12	0.145	49.3	86	57.3
AC	Ammonium-salts & rape-cake	24.91	0.184	103.0	184	56.0
C	Rape-cake	20.95	0.148	69.4	98	70.9
	Plots 1.—Farmyard dung, 14 tons.					
O	None	17.44	0.162	63.3	200	31.6
N	Nitrate of soda . . .	24.74	0.209	115.8	286	40.5
A	Ammonium-salts . . .	21.73	0.217	105.6	286	36.9
AC	Ammonium-salts & rape-cake	24.05	0.241	129.8	384	33.8
C	Rape-cake	23.96	0.207	111.1	298	37.3

On the plot receiving farmyard manure alone, less than one-third of the nitrogen supplied is ever recovered in the crop, and the recovery of the other nitrogenous manures added to the dung ranges from 40 per cent with nitrate of soda down to 37 per cent with sulphate of ammonia or rape-cake.

6. Effect of Manures upon the Tilth of the Soil.

Another secondary result of the manuring that is very manifest on these mangel-wurzel plots is the deleterious action of certain artificial fertilisers upon the tilth of the land. The soil becomes excessively sticky and tenacious after wet weather; on drying it sets very hard, with a tough, glazed surface that cannot be pierced by a young, freshly germinated seedling. These effects are seen where exclusively mineral manures are applied without dung or rape-cake; they are at their worst on the plots receiving nitrate of soda and sulphate of potash. Sulphate of ammonia and superphosphate leave the land in a friable state when dry, but where both nitrate of soda and sulphate of potash are applied the land becomes almost unworkable, and the plant is always defective and full of gaps,—in one or two seasons, indeed, it has failed entirely on this plot. The effect has been traced to the nitrate of soda. As the plant grows it takes up an excess of the nitric acid, leaving behind the soda as a carbonate. Like other soluble alkalies, the carbonate of soda causes the clay to divide into its finest particles: all the clay properties are intensified just as if the soil had been puddled in a wet condition. Sulphate of potash under certain conditions also gives rise to a little free alkali, and it is this small amount of alkali from the nitrate of soda and the sulphate of potash, alkali which can actually be extracted from the soil, that has destroyed the tilth.

7. Manures and the Incidence of Disease.

Another secondary result of manuring that is always to be seen on these mangel-wurzel plots is the variation in the susceptibility of the plant to fungoid disease induced by the different fertilisers. The leaves of the mangel-wurzel growing on the plots receiving an excessive amount of nitrogen are always attacked towards the end of the season by a leaf-spot fungus, *Uromyces betæ*, until in bad years the leaf is entirely shrivelled and burnt up before the crop is ready to lift.

The attack is chiefly to be seen on the high nitrogen plots, but even there it is kept off by the use of potash manures; so that although all the plots are close to one another, and equally open to infection, it is only the plots receiving an excessive amount of nitrogen or without potash which show any disease. The same thing is true of other plants and fungi—that an excess of nitrogen predisposes the plant to an attack, whereas potash strengthens it against the attack.

The *general conclusions* derived from a study of the mangel-wurzel field at Rothamsted — that a mangel-wurzel manure should be highly nitrogenous, nitrate of soda upon a basis of farmyard manure being the best combination, and that it should also contain an abundance of potash—does not, however, hold for all root crops. Swedes, cabbage, and other cruciferous crops are very specially de-

pendent upon phosphoric acid, and swedes at least do not require much nitrogenous manure.

OTHER EXPERIMENTS AT ROTHAMSTED.

Space does not allow us to give an account of the experiments upon crops grown in rotation in Agdell Field, and upon the leguminous crops in Hoos Field or the experiments in Little Hoos Field, designed to estimate the value of the residues left by manures after one, two, and three crops have been grown with them.

For such accounts the original papers issued from Rothamsted should be consulted (they numbered 205 down to the end of 1907, and are to be found in the *Philosophical Transactions and Proceedings of the Royal Society*, the *Transactions of the Chemical Society*, the *Journal of the Royal Agricultural Society*, &c., &c.); or the summary contained in the *Book of the Rothamsted Experiments*, published by J. Murray in 1905; or the lectures delivered in America by Sir J. H. Gilbert, published by the Highland and Agricultural Society in 1895.

The same sources must be consulted for accounts of the other scientific work on agricultural subjects that has gone on at Rothamsted during the last sixty years. The feeding experiments, though they have been largely superseded by the more refined methods of investigation perfected in Germany, established some of the fundamental facts in the nutrition of our domestic animals, and in one direction—the composition of the carcass of the whole animal—still form the only basis of our knowledge. Other investigations have dealt with the composition of the soil and the changes induced in it by the action of manures, with the bacteria causing nitrification, nitrogen fixation, and other actions beneficial or noxious to plants, with the composition of the plants themselves as affected by manuring and season. It is hardly too much to say that there is no branch of the chemistry of the growing plant which does not find elucidation and experimental illustration among the results that have been accumulated at the Rothamsted Experimental Station.

HIGHLAND AND AGRICULTURAL SOCIETY'S EXPERIMENTS.

At one time or another the Highland and Agricultural Society of Scotland has carried out a large number of useful experiments upon the manuring of crops, the feeding of live stock, and other matters relating to agriculture. The Society's most extensive schemes of experiments in manuring were those inaugurated in the year 1878 under the care of the late Dr A. P. Aitken, who was chemist to the Society from the year 1877 till the time of his death in 1904. The notes subjoined here contain the essence of Dr Aitken's official reports on these experiments.

Object of the Experiments.—The object of the experiments begun in 1878 was to test the accuracy of many views then prevalent regarding the efficacy of the various light manures in use among farmers, to discover what was the agricultural or crop-producing values of these substances, and to see how far these values corresponded with the prices at which the substances were being sold in the market.

It was believed by many advanced farmers that large sums of money were annually being spent in the purchase of manurial substances whose efficacy as manures was entirely out of harmony with their market prices, and that nothing short of an extended series of experiments, performed upon an agricultural scale over two rotations, would be capable of uprooting old prejudices, and of enlightening farmers regarding the true value of the substances in which so much of their capital was being invested. It was believed that such a series of experiments would not only determine, in a practical and reliable manner, what was the real value of manures, but would also supply much-needed information regarding the special utility of the various ingredients of manures, the forms in which they could be most profitably employed, and the most rational and economical methods in which to apply them.

The Stations.—For this purpose the Society rented two fields,—one at Harelaw, in East Lothian, and one at Pumpherston, in West Lothian. At each station 10 acres were set apart and divided into forty plots of one rood each. The soil of the former, a rich deep loam near the sea-level, in a dry early district; and that of the latter, a thin clayey loam, resting on the till or boulder clay, a somewhat wet and late district, 400 feet above the level of the sea.

No dung was applied to the stations during the course of the experiments, nor for four years previous to their commencement.

Manures tried.—The three classes of manures under experiment were phosphates, nitrogenous matters, and potash salts of the following kinds:—

Phosphatic Manures.

Carolina land phosphate. | Phosphatic guano.
Canadian apatite. | Coprolites.
Curaçoa phosphate. | Bones, in various forms.
Aruba phosphate, &c. | Bone-ash.

These were applied in a finely ground state, and also after having been dissolved in sulphuric acid.

Nitrogenous Manures.

Soluble { Nitrate of soda. / Sulphate of ammonia.

Insoluble { Meat-meal / Dried blood / Horn-dust / Keronikon / Shoddy or wool-waste } of animal origin. { Rape-cake dust / Cotton-cake dust } of vegetable origin.

Guanos, &c. { Peruvian guano. / Ichaboe guano. / Fish-manure. / Frey Bentos manure.

Potash Manures.

Sulphate of potash.
Muriate of potash.

These manures were so applied that each plot received the same quantity of *phosphoric acid, of nitrogen, and of potash,* whatever might be the form in which these were applied, and irrespective of the gross weights of the substances, or of their market prices.

Cropping.—The cropping consisted of a four-course rotation of turnips, barley, beans, and oats.

Manures for Turnips and Beans. —When the crop was turnips or beans, the manures applied to these plots contained—

		lb. per acre.
Phosphoric acid	. .	160
Nitrogen	. . .	40
Potash .	. . .	120

Manures for Cereals. — When the crop was barley or oats, the manure contained—

		lb. per acre.
Phosphoric acid	. .	80
Nitrogen	. . .	40
Potash .	. . .	60

The plots on which the various *phosphatic manures* were tested, received, in addition, their proper quantity of potash in the form of a mixture of muriate and sulphate, and their nitrogen in the form of nitrate of soda.

The plots on which the various *nitrogenous manures* were tested, received, in addition, their proper quantity of phosphoric acid in the form of superphosphate, and their potash as mixed sulphate and muriate.

The plots on which the two *potash salts* were tested, received their proper quantity of phosphoric acid as superphosphate, and their nitrogen as nitrate of soda.

The great majority of the plots on the stations were thus fully manured; and in so far as the essential ingredients—phosphoric acid, ammonia, and potash—were concerned, they all fared alike. It was only the outward and accidental form and fashion of these substances that differed.

In order to form a starting-point or basis of comparison for the whole station, three plots received no manure whatever.

In order to measure the specific effects of each of the three essential ingredients, three plots received one of each and nothing else, while from other three plots each of the three essential ingredients respectively was withheld.

In addition to the two series of experiments on the stations, there were annually carried out a selected number of experiments on farms in various parts of the country to test the accuracy of the results obtained, and to acquire additional information regarding the action of manures when applied to different soils and under different climatic conditions.

Full reports of the experiments were published annually in the Society's *Transactions*, and the following is a general statement of the chief results obtained and observations made.

I. Results with Phosphatic Manures.

Produce of Dry Matter from Pumpherston.—During the eight years comprised in the two rotations, the total amount of dry vegetable matter per acre, in the form of roots, grain, and straw, removed from the plots to which *complete manures* had been regularly applied on that section of the station at Pumpherston devoted to the study of phosphatic manures was as follows :—

	Tons of Dry Matter, per acre.	
	Undissolved.	Dissolved.
Bone-ash . .	12.69	12.66
Ground coprolites	11.80	13.22
Bone-meal . .	11.32	13.80
Phosphatic guano	12.47	14.11
Ground mineral phosphates . .	11.66	14.16
Average .	11.99	13.59

Conclusions. — The facts apparent from a mere glance at these figures are, that—

Soluble phosphates have produced about 13 per cent more actual fodder than insoluble phosphates.

Bone-meal, which is one of the dearest of the undissolved phosphates, has given the smallest return.

Dissolved mineral phosphate, which is just ordinary superphosphate, and made from the cheapest material, has given the largest return.

Among the insoluble phosphates, *phosphatic guano* and *bone-ash* are best.

Over a series of eight years, the amount of fodder raised by the application of different kinds of insoluble phosphates are not very different.

The following facts, although not apparent from a mere scrutiny of these figures, were attested from year to year during the course of the experiments :—

Insoluble Phosphates.—These vary in their efficacy far more than soluble phosphates. They are more dependent on moisture for their activity, and during dry seasons they are of very little use. Even during wet seasons they were found to be very capricious in their action. The phosphate which was the best one year might be the worst the next year.

Fineness of Grinding.—This uncertainty was found to be caused by the different *degrees of fineness* to which they happened to be ground. The finer they were ground, the more effective they were as manures.

A series of experiments made in 1886, on four plots of Pumpherston and on four Lowland farms, with the same mineral phosphate, in two slightly different degrees of fineness, showed uniformly a difference of about 11 per cent in favour of the more finely ground phosphate. The whole question of the efficacy of ground phosphates has been shown to turn on the point of the fineness to which they are ground.

Phosphatic Guano.—The reason why phosphatic guano is so effective a form of insoluble phosphate is presumably because it consists in great measure of very finely divided matter, and also because it contains from 5 to 10 per cent of precipitated or "reverted" phosphate which is in an infinitely fine state of division.

Bone - meal.—The reason why bone-meal is slowest in its action, is probably because it consists in large measure of very coarse particles.

Judged by the standard of fineness of division alone, bone-meal, which was enormously coarser than the other phosphates, should not have produced nearly so much vegetable matter. Its efficacy must therefore depend on other circumstances—notably its power of rotting in the soil, and of accumulating a store of phosphate, in no very long time becoming available as plant-food.

Soluble Phosphates.—Just as the undissolved phosphates differed from year to year in their fineness of grinding, so the dissolved phosphates differed from year to year in the fineness of their manufacture, or in their state of aggregation due to dampness, or the time during which they were kept in bags before being applied. Dissolved manures are liable to cohere into lumps from various causes, and the most careful riddling cannot restore the fine condition of a manure that has become lumpy.

Fine Powdery Condition essential.—Attention was early drawn to this circumstance during the course of the experiments, and observations made indicated that the efficacy of dissolved manures depends largely upon the more or less *powdery condition* in which they are applied.

Insoluble Phosphates for Mossy Land, &c.—A large number of experiments to determine the relative utility of soluble and insoluble phosphates were made on farms differing widely in their soil and climate, and it was found that insoluble phosphates produced their best results upon mossy land, and soils rich in organic matter in wet districts. In such circumstances they were a more economical manure than superphosphate.

Bones and Fineness of Grinding.—An extended series of experiments carried out on the stations, and on other farms, to test the relative manurial value of bone - meal of different degrees of fineness, showed that the finer ground bone-meals gave the best results during the season in which they were applied, and also during succeeding seasons where their after-effects were observed.

II. Nitrogenous Manures.

Produce of Dry Matter at Pumpherston. — The following are the amounts of dry vegetable matter removed from the plots at Pumpherston that were set apart to determine the relative efficacy of nitrogenous manures during the two rotations. The manures contained in each case the same amount of nitrogen, and there was given along with it a definite uniform amount of superphosphate and potash salts.

	Tons per acre.
Nitrate of soda . . .	12.22
Sulphate of ammonia .	11.62
Horn-dust, shoddy, &c.	9.28
Dried blood . . .	10.38
Rape-cake dust . . .	10.96

As in the case of phosphates, so also in the case of nitrogenous manures, the most soluble substances produced the largest return.

Nitrate of Soda.—This is the most active and efficient of all the nitrogenous manures, and its action has been studied under a variety of conditions at the stations, and on other soils of very different character.

Its chief peculiarity is that it acts almost immediately on the crop, and produces a marked effect whether ploughed in with the seed or applied as a top-dressing during the growth of the crop.

When applied to land in good condition, or when it forms part of a complete manure, it causes the crop to braird vigorously, and is sometimes the saving of a crop whose youth is precarious. It is especially valuable in seasons of drought, as it enables the young plant to root rapidly and become less dependent on surface-moisture.

When applied to cereals it causes a more abundant growth of straw than any other manure. When applied with the seed or to the young braird, it not only increases the bulk of the crop, but it hastens its development and causes it to ripen sooner. If applied at a later period, it causes the plant to grow too much to stem and leaf, and it unduly prolongs the period of growth. When applied late as a top-dressing to cereals, it causes a disproportionate growth of straw, retards the period of ripening, and favours the production of light grain.

When applied to a thin sharp soil during a wet season its effect is transient, showing that much of it has been washed down through the soil and out of reach of the roots of the crop.

When applied too liberally on good land, it causes a rapid growth of ill-matured vegetable matter, and produces a crop which is too abundant, unable to ripen, of poor feeding value, and liable to accidents.

When applied to plants grown for their seed, nitrate of soda must be used more sparingly; for increase of stem or straw, if overdone, is secured at the expense of the seed, both in quantity and quality.

It may therefore be used with greater impunity to crops which are grown for the sake of their stem and leaf—chiefly and notably to grass of one or two years' duration.

When applied liberally to grass, it increases the growth of the grasses proper, but diminishes the amount of clover and other leguminous plants; therefore, when a good crop of clover is desired, nitrate should be used very sparingly.

Sulphate of Ammonia.—Sulphate of ammonia is slower in its action than nitrate of soda. It is therefore to be preferred as a nitrogenous manure for crops which have a prolonged period of growth. When applied as a top-dressing to cereals, it retards the time of ripening. A similar effect is produced when applied with the seed in dry districts or during seasons of drought. It does not fail to benefit the crop even upon thin soils and during wet seasons. It is therefore more appropriate than nitrate of soda for application in these circumstances.

Sulphate of ammonia can do little for the germinating seed in dry weather, as it is not in an immediately available form. Even after rain comes, it is some time before the sulphate of ammonia comes into action.

Sulphate of ammonia has been found to check the growth of clover more effectively than nitrate of soda if applied in excess, but in moderate quantity it is an excellent manure for old grass. It is not suitable for application to leguminous crops, which are intolerant of strong nitrogenous manures, especially after the first period of their growth.

Insoluble Nitrogenous Manures.

Insoluble nitrogenous manures are substances containing albuminoid matter. They are very suitable for wet districts, but none of them can be considered a manure until it is finely ground, or rotted, or dissolved.

Rape - cake Dust. — Among the insoluble nitrogenous manures rape - cake dust has produced the greatest amount of vegetable matter. It is very probable that this is due in some measure to the large amount of carbonaceous organic matter contained in it. It was also noticed that the plot to which this manure was applied was singularly free from disease, and that the texture of the soil improved under its application.

Dried Blood, Horn - dust, &c. — Dried blood was found to be a good manure for roots, especially when applied early, but too slow in its action for cereals.

The same remark applies to *horn-dust*

and *keronikon*, which should be applied long before sowing. *Shoddy* was tried on only one occasion, and was found quite inoperative.

All these insoluble nitrogenous matters become, when dissolved in sulphuric acid, good quickly acting manures.

III. Potash Manures.

Potash salts are chiefly important on land that has not been dunged. On dunged land they frequently fail to produce any marked effect.

Sulphate and *muriate* of potash are nearly equal in their action. They are most effective when applied some months before sowing. The crops to which they are most beneficially applied are beans, clover, and leguminous crops generally. When applied to cereals, they increase the amount of grain to some extent, and they make the straw more elastic and less liable to lodge.

Manuring Turnips.

The results obtained in the experiments on the manuring of turnips were interesting and suggestive, but in some respects they have not been confirmed by more recent investigations.

In reference to his observations of the general effects of the different manures on turnips, Dr Aitken stated that an excess of potash manures was found to decrease the quantity of roots, and might even injure the crop. He considered that it is scarcely possible to overdo the application of phosphates to turnips, so far as the health and feeding quality of the roots are concerned; but too liberal an application of nitrogenous manure unduly increases the tops and retards the ripening of the bulbs, and also increases their liability to disease.

Dr Aitken stated that the nitrogenous matter in turnips is partly of a nutritive and partly of a non-nutritive kind. The former consists of albuminoid matter. The ratio of nutritive to non-nutritive nitrogenous matter varies extraordinarily in different turnips, and under different circumstances of weather and manuring.

Forced Turnips of Bad Quality.— Bulbs grown very rapidly, whether from excess of moisture or too liberal application of soluble nitrogenous manure, have a smaller proportion of their nitrogenous matter in the form of albumen.

Manures which unduly force the growth of turnips may increase the quantity of the crop; but the increase of quantity is got at the expense of quality, and the deterioration of quality is mainly expressed in the large percentage of water and the small percentage of albumen in the bulbs.

Manures for Rich Crops of Turnips.—In order to grow a large and at the same time a healthy and nutritious crop of turnips, such a system of manuring or treatment of the soil, by feeding or otherwise, should be practised as will result in the general enriching and raising of the condition of the land, so that the crop may grow naturally and gradually to maturity.

For that purpose a larger application of slowly acting manures, of which bonemeal may be taken as the type, is much better suited than smaller applications of the more quickly acting kind.

A certain amount of quickly acting manure is very beneficial to the crop, especially in its youth, but the great bulk of the nourishment which the crop requires should be of the slowly rotting or dissolving kind, as uniformly distributed through the soil as possible.

Manures for the Barley Crop.

The relative importance to the barley crop of the three manural ingredients may be seen from a comparison of the results obtained on the plots manured as under for five years:—

No. of Plot.	Grain per acre. lb.
22. Potash . . .	875
12. Phosphate (bone-ash) .	1175
17. Phosphate and potash .	1256
18. Nitrate . . .	1287
21. Nitrate and phosphate .	1706
11. Nitrate and potash . .	1814
13. Nitrate, potash, and phosphate	2596

Manures applied to the barley crop affect, in the first place, the quantity per acre both in grain and straw; in the second place, and to a much less extent, they affect the quality of both grain and straw, and they materially affect the time of ripening.

Nitrogenous Manure for Barley. —The most important constituent of a manure for the barley crop is nitrogen.

In ordinary circumstances, it is the quantity of nitrogen in the manure or in the soil which determines the bulk of the crop.

In an ordinary rotation of cropping, in which barley succeeds turnips, the *phosphate and potash* required by the crop are relatively abundant in the soil, and a good crop can be obtained if only some nitrogenous manure is applied in sufficient quantity to enable the plant to take up its mineral food.

The kinds of nitrogenous manure most suitable for barley are those which are soluble and rapid in their action, such as sulphate of ammonia and nitrate of soda. *Sulphate of ammonia*, if applied as a top-dressing, and *nitrate of soda*, if so applied, much later than three weeks after the date of sowing, may increase the quantity of the crop both in grain and straw, but the quality of the grain, as indicated by the weight per bushel, will be lowered, and the time of ripening will be retarded.

A difference of three weeks in the time of ripening occurred among the experimental crops. The earliest were those which were manured with soluble phosphate, and whose nitrogenous manure was nitrate of soda applied with the seed. The latest were those which received no nitrogenous manure, an overdose of it, or too late a top-dressing.

Slowly acting nitrogenous manures are of no use to the barley crop, unless applied some months before the time of sowing.

A deficiency in the amount of nitrogenous manure applied to barley not only diminished the total amount of the crop, but it also diminished the percentage of albuminoid matter contained in the grain.

Barley, top-dressed with nitrate of soda, contained somewhat more albuminoid matter than that which had the nitrate applied with the seed.

The amount of albuminoid matter varied from 8½ to 11½ per cent. The former amount was contained in barley, from whose manure all nitrogenous matter was withheld, and the latter from barley top-dressed with nitrate.

Phosphatic Manures for Barley.— *Phosphatic manures* are next in order of importance for barley. The more speedy their action the better; therefore *superphosphate* is the most reliable form of phosphate.

The plots to which soluble phosphates were applied came to maturity ten days before those with insoluble phosphates.

Potash for Barley.—Potash manures somewhat increased the quantity of grain on the station where no dung was applied, and they strengthened the straw. But it was noticed that the grain was somewhat darker in colour than that to which no potash was applied.

Manures for Oats.

The manures required for oats are quick-acting manures, to enable the crop to get a good hold of the soil before the nourishment contained in the seed is exhausted.

For this purpose superphosphate and nitrate of soda are peculiarly applicable.

Sulphate of ammonia, although a soluble manure, did not come into operation in time for the wants of the young plant during the dry season of 1885, and the crop which received that manure was a signal failure at both stations.

Potash manures, especially muriate of potash, had a very beneficial effect upon the oat crop, and considerably increased the yield of grain, and in a less degree the amount of straw.

The *general conclusions* to be drawn from the experiments with the oat crop are, that the treatment of the land should be such as to accumulate organic matter in it, to prevent too great a loss of moisture, and to provide the young plant with manures that come rapidly into operation.

When the young plant has safely passed the critical period of its growth it roots deeply, and lays hold of the moisture and nourishment contained in the subsoil.

Manures for the Bean Crop.

The usual practice in bean-growing districts is to apply dung to the bean break, and the opinion prevails that beans cannot be successfully grown without dung. But the experiments at Pumpherston station show that a full crop of beans may be grown with artificial manures upon land that has not been dunged for ten years.

The relative importance to the bean crop of the three chief constituents of a manure may be seen by comparing the produce of eight plots manured as follows for six years :—

No. of Plot.	Kind of Manure.	Bushels Dressed Grain per acre.
27.	No manure	2½
12.	Phosphate (bone-ash) . .	5⅙
18.	Nitrate	6¼
21.	Phosphate and nitrate . .	5⅓
22.	Potash	26½
17.	Potash and phosphate . .	42⅓
10.	Potash, phosphate, and nitrate	45½
38.	Potash, phosphate, nitrate, and gypsum . . .	51

The characteristic ingredient of a bean manure is potash.

Without potash in the manure, the other two ingredients are of very little use, unless, indeed, the land be very rich in potash.

Potash salts alone may be a sufficient manure on land in good condition, and may even produce a fair crop on land that is in poor condition.

Phosphate, when applied along with potash salts, or when applied to land rich in potash, has a marked effect upon the crop.

Nitrogenous manures, even when of the most favourable kind, have very little influence in increasing the bean crop.

Lime, in the form of gypsum (or sulphate of lime), has a beneficial effect upon the crop.

Dissolved phosphate acts far more powerfully on the bean crop than ordinary ground phosphate.

Phosphatic guano was more effective than ground mineral phosphate, presumably for the reason that a small proportion of it was in an easily dissolved form.

The nitrogenous manures that are most beneficial to the bean crop are those whose action is rapid and soon over. In this respect nitrates are preferable to all other nitrogenous manures.

Nitrogenous manures should either be applied in very small quantity or altogether withheld from the bean crop.

Nitrogenous manures that come into operation after the crop has made some growth have an injurious effect. Even sulphate of ammonia is too slow in its action, and retards the growth of the crop.

Nitrogenous manures should not be applied as a top-dressing to the bean crop.

Peruvian and other nitrogenous guanos are among the worst manures for the bean crop. They contain too much nitrogen and too little potash.

The muriate of potash has proved a more effective manure than the sulphate.

The beneficial effect of gypsum is to be ascribed, not to the sulphuric acid it contains, but to the lime, which, in combination with sulphuric acid, is a soluble manure, and has the power of liberating potash in the soil.

The general results of the experiments with different manures on the bean crop inform us that the bases potash and lime are the substances most required by the crop.

For land dunged in autumn—or for land in good condition—it would seem from the experiments at Pumpherston that the application of superphosphate, muriate of potash, and sulphate of lime, in equal parts, would be a very appropriate manure for the bean crop.

The composition of beans is very uniform whatever be the nature of the manures applied. It is the *quantity* of the crop, and not the *quality* of it, that is affected by the application of manures.

Lessons from Incomplete Manure Experiments.

The following are the amounts of dry vegetable matter yielded during eight years by those plots at Pumpherston from whose manures one or more of the three constituents—nitrogen, phosphoric acid, and potash—were withheld :—

	Tons per acre.
Nitrate and potash (no phosphate)	9.78
Nitrate and phosphate (no potash)	8.97
Potash and phosphate (no nitrogen)	7.65
Nitrate of soda alone . . .	8.68
Bone-ash alone . . .	6.50
Potash salts alone . . .	5.35
Unmanured . . .	5.40

From these figures it is evident that the manurial constituent most required for the production of the crops grown was *nitrogenous matter*, in the next place *phosphates*, and in the next *potash*.

Potash alone.—The plot to which potash salts alone were applied gave scarcely as much produce as the un-manured plot.

This plot went steadily from bad to worse, and was latterly the worst on the station, showing that the accumulation of potash was hurtful to most of the crops grown there.

There was one exceptional year, 1884, when the crop was beans, and then for the first time it threw up a crop five times as abundant as the neighbouring plot, to which no potash had been applied.

An Experiment for Farmers.—An experiment of the above kind—in which, along with a completely manured plot, there are arranged side by side a series of plots from which in turn one of the essential ingredients of a complete man-ure is withheld—forms a most instruc-tive lesson for farmers, and should be applied by them to all the fields on their farm. It serves to show what is the ingredient in the soil or in the manure that is most deficient for the production of a crop, and thus guides the farmer in the selection of the light manures that are most appropriate for his purposes.

Manures for different Crops.

A review of the manurial requirements of a rotation of crops, consisting of tur-nips, barley, beans, and oats, shows that while the three great constituents of a manure—nitrogen, phosphoric acid, and potash—are all required in order to raise full crops and to maintain the fertility of the soil, the predominance which should be given to one or other of these constit-uents varies with the crop. The pre-dominant constituent is—for

> Turnips—Phosphoric acid.
> Barley and oats—Nitrogen.
> Beans—Potash.

Forms of Manures for Turnips.—For turnips the phosphates should be applied either in a soluble form or in a state of very fine division—in the case of ground phosphates, they should be at least so finely ground as to pass through a sieve of 120 wires to the linear inch,—or they should be of a kind that rapidly rot in the soil (such as bone-meal), and at the same time so finely ground as to permit of their being rotted in great measure during the period of the crop's

growth. The nitrogenous manure should be partly of a quick-acting and partly of a slow-acting kind, so as to be of service to the crop during the whole period of its growth.

Forms of Manure for Cereals.—For cereals the nitrogenous manure should be very rapid in its action, so as not to re-tard the ripening of the crop. If applied as a top-dressing, it should consist of nitrate. The phosphate cannot be too rapid, and on that account superphos-phate is to be preferred to any other form of phosphate.

The importance of potash in a cereal manure will depend on whether grass and clover seeds are sown with the crop. If that is the case, potash salts take the second place, as the presence of potash in the manure is of importance for the nourishment of clover.

Forms of Manure for Beans.—For the bean crop, the form of potash salt that is most suitable is the muriate of potash. Superphosphate is preferable to other forms of phosphate, probably on account of the large amount of sulphate of lime contained in that manure; but if sulphate of lime is applied to the crop, any other good phosphatic manure may form part of the mixture. The only kind of nitrogenous manure that is to be re-commended for this crop is a soluble one, and that in small quantity, applied with the seed.

Dung for Turnips, Cereals, and Beans.—When farmyard manure is used for the turnip crop, potash salts should not be applied to it, and any nitrogenous manure added should be soluble.

The need which cereal crops have of nitrogen points strongly to the con-clusion that a part of the dung should be withheld from the root crop and applied to the white crop; and this is all the more to be recommended, as it is evident that a considerable loss of the nitrogen of the dung is inevitable when a heavy dunging is applied to the fallow break.

If dung is to be used for beans, it should be applied to the stubble, rather than put in with the seed.

Organic Matter.

While it has been stated that on ordi-nary soils the three constituents—phosphoric acid, nitrogen, and potash — are sufficient to form what is known as a complete manure, and that a manure containing two of these substances, or, it may happen, only one of them, is a sufficient manure to apply to certain crops in certain circumstances, it is of the utmost importance here to observe that, nevertheless, it must not be supposed that, in the manipulation of these three constituents, in reference to the crops they are producing, lies the whole question of manuring.

Consider Soil as well as Manure and Crop. — The rapidity with which light manures act upon the crops to which they are applied has tended to restrict our view too much to the two factors —manure and crop—and has caused us to think less of the *soil* than our forefathers did.

Before the days of light manures—a time comparatively recent — when the wants of a crop for phosphates, nitrates, and potash were unknown, farmers fixed their attention upon the soil, and used every means to raise its general fertility —to put it into what is called high "condition,"—and this they did by the use of heavy manures containing a large amount of organic matter.

Function of Organic Matter. — It has since been discovered that plants can grow to perfection without organic matter, but the circumstances in which that is possible for crops are not those which prevail in ordinary farming and in this climate.

It is to the organic matter in the soil that are due many of the changes going on there that are beneficial. to the roots of plants. The warmth and moisture of the soil are increased by the organic matter in it, and the acids formed by its decay have an important part to play in dissolving the mineral matter, which forms the food of plants. It is indeed the key to the treasures of the soil. But in the ordinary operations of agriculture —in the constant disturbing and working of the ground—organic matter is rapidly destroyed, so that if farmyard manure and organic composts or other substances rich in organic matter are not put into land under cultivation, or fed on it, it soon becomes unduly deprived of organic matter. And the soil is thus deteriorated as a medium for the growth of roots and for the retention of moisture, and as a store of fertility gradually becoming available for the nourishment of crops.

During very dry or cold seasons, and even during very wet ones, the want of organic matter in the soil is a source of danger to the crop. The fate of many plots at the stations during the recent drought showed how intimately the fertility of the land, and the health and safety of the crop, are concerned in the accumulation of organic matter in the soil.

Quick - acting Manures and Organic Matter. — However much, therefore, we may commend the application of quick-acting light manures—phosphates, nitrates, and potash salts—for the assistance of crops, it is quite evident that their proper position on most kinds of land is subordinate to that of the heavier manures and to the slowly acting manures rich in organic matter, which perform the important work of building up the fabric of the soil, and accumulating therein a reserve of fertility which is commonly known under the name of "condition," and which is also called "backbone" by those who are able to appreciate its importance.

OTHER RESEARCH WORK BY THE HIGHLAND AND AGRICULTURAL SOCIETY.

In more recent years the Highland and Agricultural Society has conducted a large amount of research work in various directions, part of it under the supervision of Mr James Hendrick, F.I.C., who succeeded Dr Aitken as chemist to the Society in 1904. Useful experiments have been carried out at different times on the fattening of cattle and sheep with different foods, on the effects of food upon the milk yield of cows, on the effects of various manures as top-dressing for grass-land of different character, on the produce of different varieties of oats, on the "boxing system" of preparing seed-potatoes, and on other questions of practical interest to farmers. The lessons to be derived from most of these investigations are made use of in different parts of this work.

EXPERIMENTS BY THE ROYAL AGRICULTURAL SOCIETY OF ENGLAND.

The Royal Agricultural Society of England has, from time to time, carried out a great amount of valuable experimental work. Most of this work has taken place at the Society's Experimental Farm, near Woburn, Beds, the use of which was offered, in the year 1876, to the Society by Hastings Russell, 9th Duke of Bedford. This Station has since been maintained almost entirely by the generosity of successive Dukes of Bedford, the superintendence, management, and expert advice being provided by the Royal Agricultural Society.

The Woburn Experimental Farm consists of about 140 acres, of which 100 are arable and the rest in grass. The soil is a light sandy loam, with subsoil of sand, and it belongs to the Lower Greensand formation. It is well suited for sheep - feeding and the growing of barley and potatoes. There are, in addition to a residence for the farm manager, suitable farm buildings, and specially arranged feeding-boxes for carrying out experiments on cattle, together with weighbridge, oil-engine, &c.

Pot-culture Station.—In 1897, and mainly as the outcome of a bequest from the late Mr E. H. Hills, a Pot-culture Station was added. This was built and equipped by the Royal Agricultural Society of England, at a cost of £1200, and is the first and the most complete of the kind erected in this country. It consists of a laboratory, a conservatory, and a spacious wired-in enclosure. It is in close proximity to the farm buildings.

The Staff.—The resident staff of the entire experimental station consists of a farm manager, with foreman, and six to seven farm hands; at the laboratory a chemist and a laboratory lad are also employed. The whole of the work of the Station is under the charge of the Chemical and Woburn Committee of the Royal Agricultural Society, the director of the entire experimental work being Dr J. A. Voelcker, the chemist to the Society.

Objects of the Woburn Experiments.—The inception—in 1876—of the farm was due to two main causes:

(1) The desire to put to the test, on a light soil, the results of the world-renowned experiments of Lawes and Gilbert on the continuous growing of wheat and barley, which had been conducted since 1843 on the heavy soil of Rothamsted (Herts).

(2) The desirability of testing in actual practice the accuracy of Lawes' and Gilbert's tables for fixing the compensation value of the unexhausted manure produced by the consumption, on the farm, of different purchased foods.

In regard to the former, it had often been asked whether the striking results obtained in Lawes' and Gilbert's experiments on a heavy soil would be borne out on a light one; while, in regard to the second question, the subject of compensation to an outgoing tenant for purchased foods consumed on the farm had derived special significance because of the passing of the Agricultural Holdings Act of 1875.

At the outset, the plan of the experimental inquiry was laid down by Sir John Lawes and the late Dr A. Voelcker, Dr Voelcker shortly afterwards taking sole charge and so continuing until his death in December 1884, after which he was succeeded in the Direction by his son, Dr J. A. Voelcker, the present chemist to the Society, to whom we are indebted for the subjoined notes.

Though the farm was originally intended for the elucidation of the above two questions, the work rapidly extended, and while the main experimental field (Stackyard Field) of 27 acres has been rigorously kept for these strictly scientific inquiries, a large number of other experiments, either supplemental to the foregoing or of general farming interest, have been, from time to time, added, so that the whole is now essentially an "experimental farm" in the wide sense.

Scope of the Work.—Not only have there been experiments on the manuring of different farm crops, but also on the introduction of new crops, and of new varieties, feeding experiments on cattle and sheep, and the testing of different agricultural questions, such as the value of ensilage, the losses in making and storing farmyard manure, green-manuring, the inoculation of leguminous crops, the prevention of "finger-and-toe" in turnips, the combating of "potato disease," &c.

At the Pot-culture Station, moreover, a very large and varied series of experimental work has been carried out, partly in conjunction with the field experiments, and partly on such inquiries as the influence on plants of some of the rarer constituents found in soils (the Hills' experiments), the prevention of fungoid diseases, the eradication of weeds, &c. The station is, further, a reporting one of the Meteorological Department, and observations are regularly taken and recorded of rainfall, temperature, dew-point, &c.

Stackyard Field.—The principal experimental field is Stackyard Field, 27 acres in extent. It is somewhat inconveniently situated, being one mile from the farm buildings. In other respects this disadvantage is more than made up for, inasmuch as the level situation, the even character of the soil throughout, and the ready response of the land to manurial agents applied to it, make the field an ideal one for experimental work. At the outset the soil of the field was sampled in different spots over the whole area, and analyses made of the samples thus taken showed the whole field to be of exceptionally uniform character. This has further been borne out by practical experience, duplicate plots similarly treated having given results in close agreement with one another. The following analyses represent the general composition of the soil of the field :—

(SOIL DRIED AT 100° C.)

	1st depth of 9 inches.	2nd depth of 9 inches.
[1] Organic matter and loss on heating .	4.132	2.432
Oxide of iron .	2.934	2.571
Alumina. . .	3.610	2.840
Lime . .	.308	.205
Magnesia .	.143	.162
Potash . .	.286	.235
Soda . .	.140	.217
Phosphoric acid	.156	.115
Sulphuric acid .	.027	.023
Insoluble siliceous matters	88.264	91.200
	100.000	100.000
[1] Containing nitrogen .	.166	.094

The top-soil is about 9 inches deep and of reddish colour, the subsoil being more yellow and sandy in character. The fact of the soil being poorly supplied in lime (as shown in the analysis) has brought out results of peculiar significance. But, apart from this, it may be said that the soil is one that is peculiarly responsive to any manurial treatment applied to it. Further, though light in nature, it does not suffer greatly, either in extremely dry or in very wet weather.

Continuous growing of Wheat and Barley.

These experiments are the complement — on a light soil — of the older Rothamsted experiments. The object is, primarily, by growing the corn crops year after year, and adopting the same manurial treatment each year, to ascertain what is the effect, on the crop, of the different manurial constituents of the various applications. Further, to ascertain what these will produce when used in differing amounts, and what may be learnt regarding the nature of their action.

It must be pointed out that these experiments had not, in the first instance, any direct bearing on the question as to whether it would "pay" to grow corn year after year on the same land, nor as to which of the applications would give the greatest money return; but they were distinctly scientific experiments for the purpose of ascertaining the action, both on crop and on soil, of the different applications. From them it is possible, certainly, to deduce conclusions as to what will be successful or otherwise in actual practice, but the main purpose was, as stated, to establish the principles of the action of different manurial constituents.

Five and a half acres of Stackyard Field are devoted to this work, 2¾ acres being every year in wheat and 2¾ acres in barley.

The varieties of wheat and barley grown on the respective areas are changed from time to time. The plots are, as a rule, ¼ acre each in extent, some, however, having been subdivided as the work progressed.

In each case there are two unmanured plots, one at either end of the field; there are plots with nitrogen supplied in the form of ammonium-salts (sulphate of ammonia and muriate of ammonia in equal quantities), of nitrate of soda, of farmyard manure, and of an organic manure (rape-dust). The nitrogen is applied in different amounts, and either alone or with mineral manures, these latter consisting of superphosphate of lime, and the sulphates of potash, soda, and magnesia. The effect of omitting the nitrogenous applications for a year, or for longer periods, is also studied, and, of later years, the effect of applying lime in addition to ammonium-salts has been observed, and important conclusions have been drawn from this part of the work.

The farmyard manure used is made by bullocks consuming in the feeding-boxes known quantities of decorticated cotton-cake, maize-meal, hay, straw-chaff, and roots, and being supplied with known amounts of straw as litter. The foods are analysed, and deduction is made for the constituents used in giving the increase in live-weight of the animals, so that the manure applied is of approximately known composition.

The harvest results are tabulated for each year, and as the first crop was taken in 1877, the year 1896 marks the conclusion of the first twenty years, and the year 1906 that of the next ten years (thirty years in all), of the continuous experiments both with wheat and barley. These results are summarised in Table I., opposite.

Continuous Growing of Wheat.

In the first place, it will be noticed that wheat has been grown year after year for thirty years without any manure being applied. For the first twenty years the average produce was 15.3 bushels per acre, this falling in the next ten years to 10.8 bushels (the results of plot 7 are taken, as plot 1 is somewhat affected now by trees near it). Mineral manures alone (plot 4) have not increased the crop at all, but when used with nitrate of soda (plots 6 and 9a) have given marked increases. The heavier applications of nitrate of soda, with the minerals, have not, however, given proportionate increase in crop. Nitrate of soda used alone (plot 3) continues to give fair yields even after thirty years, but

ammonium-salts have afforded exceptional and unexpected results.

While for twenty years ammonium-salts (plot 2a) did as well as nitrate of soda (plot 3), after this period the crop began to fail, getting worse each year; the soil was found to have become quite acid and to be capable of doing little more than grow the weed spurry (Spergula arvensis) in abundance. On the soil, too, a mossy green mould was found to spread. Lime was then applied, at the rate of two tons per acre, in December 1897 (plot 2b), when the spurry disappeared, and the wheat crop was again restored, the effect of the application still telling after an interval of ten years. Plots 8a, 8b, and 9a, 9b, show the result of applying (along with minerals) ammonium-salts and nitrate of soda respectively one year and leaving them off the next year. When nitrate of soda is left off for a single year the produce goes down nearly to the unmanured yield, but not to such an extent when ammonium-salts are left out, thus showing that nitrate of soda is exhausted in a single year, but ammonium-salts not so entirely.

Next, rape-dust, supplying 100 lb. of ammonia per acre, has done as well as farmyard manure giving twice that amount, and where either of these have been put on for several years and then left off, the effects of their application are still noticeable after an interval of twenty years, thus showing the "lasting" character of farmyard manure as compared with nitrate of soda and ammonium-salts.

Continuous Growing of Barley.

The results follow very much the lines of the wheat experiments. The unmanured produce was 21 bushels for the first twenty years and 12.4 bushels for the next ten. Mineral manures used alone have given a small increase, but much larger ones when in combination with nitrogenous salts. As in the case of wheat, there has been no failure with nitrate of soda used by itself, but the results with ammonium-salts have been of an even more striking character than with wheat, the failure coming earlier and being more marked. The recovery when lime was used was similarly very

striking and equally lasting. With rape-
dust and farmyard manure the results'
were much as with wheat, though the
farmyard manure proved rather the
superior.

Practical Conclusions from the Contin-
uous Growing of Wheat and Barley.

It follows from the above experiments
that both wheat and barley can be grown

TABLE. I.—CONTINUOUS GROWING OF WHEAT AND BARLEY.

(Stackyard Field, Woburn.)

Average results of first twenty years (1877-1896) and next ten years (1897-1906).

Plot.	Manures applied annually per Acre.	WHEAT.		BARLEY.	
		Average Produce of Corn per Acre.			
		First 20 years. (1877-1896).	Next 10 years. (1897-1906).	First 20 years. (1877-1896).	Next 10 years. (1897-1906).
		bush.	bush.	bush.	bush.
1	Unmanured every year	14.7	8.6	21.9	11.5
2a	Ammonium-salts containing 50 lb. ammonia	23.8	9.7	33.5	4.8
2b	Do. with 2 tons lime, December 1897	...	16.8	...	21.2
3	Nitrate of soda containing nitrogen equal to 50 lb. ammonia	23.6	17.0	35.6	23.6
4	Mineral manures	15.1	8.2	22.5	16.2
5a	Do. and ammonium-salts containing 50 lb. ammonia . . .	30.2	24.4	39.0	7.1
5b	Do. and ammonium-salts with 2 tons lime, December 1897 . .	...	...	...	33.8
6	Do. and nitrate of soda containing nitrogen equal to 50 lb. ammonia . .	31.2	23.6	43.5	35.3
7	Unmanured every year	15.9	10.8	20.5	13.3
8a	Mineral manures and, in alternate years, ammonium-salts containing 100 lb. ammonia	37.2	25.4	45.4	14.5
8b	Mineral manures, ammonium-salts omitted in alternate years	23.1	17.6	31.0	12.7
8aa	Same as 8a with 2 tons lime, December 1897	...	...	...	38.0
8bb	Same as 8b with 2 tons lime, December 1897	...	...	...	28.3
9a	Mineral manures and, in alternate years, nitrate of soda containing nitrogen equal 100 lb. ammonia	34.0	29.1	49.3	42.8
9b	Mineral manures, nitrate of soda omitted in alternate years	16.4	11.8	31.1	23.4
10a	No manure since 1890, after rape-dust for 1 year. Previously for 5 years farmyard manure equal to 100 lb. ammonia .	15.3	12.7	22.9	16.7
10b	Rape-dust (about 14 cwt.) equal to 100 lb. ammonia annually since 1890 . . .	27.2	26.6	37.4	33.4
11a	No manure since 1882. Previously for 5 years farmyard manure equal to 200 lb. ammonia	18.7	14.1	32.6	21.9
11b	Farmyard manure (about 7 tons) equal to 200 lb. ammonia annually . . .	27.2	24.0	39.9	36.6

NOTE.—Ammonium-salts are equal weights of sulphate of ammonia and muriate of ammonia. Mineral
manures are : 3½ cwt. superphosphate of lime, 200 lb. sulphate of potash, 100 lb. sulphate of soda,
100 lb. sulphate of magnesia, per acre.

perfectly well on the same land year after
year if the proper manures be used. If
the land be kept clean and free from
weeds it may go on producing moderate
crops even when no manure is used.
Mineral manures alone do not increase
the yield, but the proper manuring for
both crops is a combination of mineral
manures with nitrogenous salts such as
nitrate of soda or sulphate of ammonia,
or else organic manures such as farmyard
manure or rape-dust.

While the influence of nitrogenous salts lasts practically only for the year of their application, that of farmyard manure will not be entirely exhausted even in twenty years' time. Lastly, on land naturally poor in lime, sulphate of ammonia will, sooner or later, exhaust the land and render it acid. The remedy for this is to be found in the application of lime, the influence of a single dressing of two tons to the acre lasting for ten years subsequently.

Though it has not been possible to give, in the compass of a single table, more than an outline of the results, there have been recorded many other observations—e.g., those on the produce of straw and the quality of the grain. For these, reference must be made to the Journals of the Royal Agricultural Society of England.

It may be said, in general, that the results are in direct confirmation of those obtained at Rothamsted, and that to the latter have been added important contributions, mainly in regard to the influence of ammonium-salts on a soil poor in lime, the restoring effects of lime, and the duration of farmyard manure. The Woburn Experimental Farm is, indeed, the first one in any country where these striking results as regards the action of salts of ammonia and of lime have been brought out.

At the close of the thirtieth year (1906) further modifications of the original plan were introduced. These consist of (a) simplification of, and alterations in, the quantities of the artificial manures used, these being lowered to amounts commonly used in farming practice; (b) the use of lime in different quantities (5 cwt., 10 cwt., and 1 ton per acre); (c) the substitution of plots 10a and 11a by others in which the influence of potash and of phosphates is separately tested.

The Rotation Experiments.

The next series of experiments is that on rotation, the primary purpose of these being to put to the test, in actual practice, the accuracy of Lawes' and Gilbert's tables for the unexhausted value of manure produced by the consumption of feeding-stuffs by animals on the farm. The attempt was now made, by selecting two foods in Lawes' and Gilbert's tables —decorticated cotton-cake and maize-meal—and feeding them to animals on the farm, to see whether the manure left over after their consumption was capable of producing in increased crops results commensurate with the theoretical values assigned to them in the tables.

At the outset (1876) decorticated cotton-cake and maize-meal respectively were fed—with other foods—to two sets of bullocks in the feeding-boxes, and the resulting manure was used for growing a root crop; the roots were removed and a barley crop taken, clover was sown among the barley, the clover fed off by sheep, which received in addition decorticated cotton-cake in one case and maize-meal in the other; after this a wheat crop followed.

This course was pursued for two entire four-course rotations (1877-1884), at the close of which it was found that, taking everything together, the highly nitrogenous decorticated cotton-cake had done practically no better than the maize-meal, which latter was very poor in nitrogen. But, on examining the results more closely, it was found that the poorer maize-meal had produced crops which were quite up to the average of well-manured land on this class of soil, and hence it was not to be expected that the maximum produce could be exceeded by further manuring, so that the richer decorticated cotton-cake did not have a fair chance of showing its superiority. The land, in fact, had been over-manured, it was believed.

Hereupon a change was introduced in the manurial plan, and, from 1884 on, the crops of the rotation were only manured once in the four-course duration. The root crop, instead of being manured with decorticated cotton-cake dung and maize-meal dung respectively, was merely grown with a little superphosphate to give it a start; the root crop was fed off by sheep with decorticated cotton-cake on one portion and maize-meal on another; barley followed, then clover; but this latter was not fed off as before with cake and meal, but was cut and removed as hay, while wheat followed without further manure.

This plan was followed on one half of the original area, while on the other half the same four crops of the rotation were simultaneously grown, but without any manure, and they were all removed off the land. This was done with the intention of exhausting the believed over-fertility, and of seeing, if possible, whether by this method the previously applied decorticated cotton - cake would show its superiority over the maize-meal similarly applied. This plan was continued for another three rotations (12 years, 1885-1896), but, at the close, the results were very little more definite than in the former instance.

It was found, indeed, that the barley crop which immediately followed the cotton - cake manuring was in all cases superior to that after the maize - meal manuring, but the subsequent crops showed no differences, and, indeed, the wheat crop following the clover was practically as good on the half to which no manure whatever had been applied since 1884 as that on the manured half. This, it must be pointed out, however, was the case so long as clover was the preceding crop, and, with but few exceptions, it was found possible to grow a clover crop once in the four years.

Hellriegel's Discoveries.

During this interval the classic discoveries of Hellriegel came to light, and these supplied an explanation of the failure of the experiment to bring out the points expected. By showing that clover was able to obtain its supplies of nitrogen from the air, Hellriegel explained what had been long known in practice—how it was that wheat after clover needed no nitrogenous manuring. So here, at Woburn, as long as the clover crop could be grown, the land was being enriched with nitrogen, and further manuring could do no good, so that the differences between decorticated cotton - cake and maize - meal, due to the higher nitrogen of the former, could not be brought out, for, after the clover crop, the wheat was just as good whether decorticated cotton-cake, or maize-meal, or nothing at all was used.

Accordingly this long series of work, though seemingly failing in its ultimate object, afforded most striking proof of the practical application of Hellriegel's nitrogen theory.

A Change in Plan.

It was found hopeless, therefore, to attack the problem of unexhausted manure value in this way, and the next step was to devise a plan of rotation in which the disturbing influence of clover should not enter. But, as a preliminary, it was necessary to get the land—hitherto unevenly manured—into a uniform condition of fertility again, so that future work might begin on a fair basis. To effect this, crops of barley—all without manure—were grown year by year for the next 8 years (1897-1904), the produce of each plot being weighed as before, and this was continued until the land showed uniform results, after which (1905) the new series was begun.

In the new plan clover or other leguminous crop is omitted altogether, a green crop (mustard) taking its place. The experiment is in two parts—the one to represent the use of decorticated cotton-cake (or maize-meal) on light (sheep) land, where the cake is given to the sheep as they feed off the roots ; the other to represent the practice on heavy land, where the cake or meal is fed to bullocks in the yards, the manure being carted out and spread for growing the root crop. Barley in each case follows the root crop, then the green crop (mustard) is cut or ploughed in, and wheat follows. In this way it is hoped to ascertain, by the crops grown subsequent to the manurial application, what the difference in manurial value is between decorticated cotton-cake and maize-meal, whether fed direct on the land or consumed in the yard and carted out as manure.

Green-manuring Experiments.

When attention was called, by Hellriegel's discoveries, to the important rôle played by leguminous crops, it was resolved to try the practical application of the theory in the growing of cereal crops in alternation with leguminous ones. If it be the case that leguminous crops, by utilising the nitrogen of the air, enrich the soil in nitrogen to an extent that non-leguminous do not, then it may be reasonably argued that the cheapest manuring for a cereal crop is to grow a leguminous green crop before it ; and, similarly, if two green crops be grown, the one leguminous and the other non-leguminous, then the corn crop following the leguminous green crop ought to be much the better because of receiving the nitrogenous manuring stored up in the soil as the result of the growing of the leguminous crop. This latter point it was resolved to put practically to the test.

Lansome Field was selected, and three plots were set out on it. On one of these tares (leguminous) were grown, on a second rape, and on a third mustard (both the latter non-leguminous). Two crops of each were grown during the season and successively ploughed in green, barley following in the next spring. Moreover, that failure to ensure nitrogen-assimilation might not be due to need of mineral applications, superphosphate and sulphate of potash were given to one half of each plot, and lime to the other half.

After carrying on the experiment for three successive barley crops, it was invariably found that the barley after mustard was the best crop of the three, then that after rape, and that after tares —contrary to theory—the worst of the three.

Wheat was then substituted for barley as the corn crop, as it was thought that possibly the accumulated nitrogen might be removed by drainage in winter. Also winter tares were substituted for spring tares in order to give the crop a longer hold on the soil. But the same results were forthcoming whether barley or wheat was the crop, whether manured with minerals or not, whether tares were spring- or winter-sown : in all cases the best corn crop followed the growing of the non-leguminous green crop, mustard, and the worst was that after the leguminous crop, tares. Even when after a wheat crop a barley crop was next taken (in case the nitrogen had not become available in the first season), no different results were obtained. The experiments were too often repeated to admit of any doubt, and one is forced to the conclusion that, so far at least as concerns the growth of the tare crop on land of this light sandy character, the theory does not hold good in practice.

In order to see that more nitrogen was

actually being supplied by the tare crop, this and the mustard crop were, one year, cut, weighed, and analysed, and the tares were found to be supplying twice the amount of bulk, of organic matter, and of nitrogen, that the mustard did. The experiments are still in progress, and, side by side with them, investigations are going on at the Pot-culture Station close by with a view to supplying the explanation of these unexpected results. The work, so far, points in the direction of the differences being due to considerations of changes produced in the mechanical texture of the soil and of its water-holding capacity.

Other Experiments with Corn Crops.

From time to time different experiments have been conducted in order to ascertain the relative yielding powers of different varieties of wheat, barley, and oats. These, of course, have reference particularly to the soil of the farm, but it may be mentioned that foreign malting barleys, such as " Hauna," " Bohemian Chevalier," and Californian barleys, have been tried, also different kinds of Canadian wheat. With these it has been found universally the case that, while in the first year of their growth they exhibit their characteristic qualities, these disappear almost entirely upon the seed thus obtained being grown for a second season.

Other trials with corn crops have concerned the selection of seed, and experiments both with wheat and barley have shown that it is not the plump, well-grown seed which of necessity produces the largest crop, the same weight of smaller grains producing quite as good a return.

The kiln-drying of barley constituted a further inquiry, the result being to show that, unless in an exceptionally wet year, no benefit accrued from the kiln-drying.

Manurial experiments have been conducted with soot as a top-dressing for wheat, these showing that a dressing of 40 bushels per acre of good soot is quite as effective as one of 1½ cwt. per acre of nitrate of soda. At the same time, it was shown that applications of 60 and 80 bushels per acre of soot are needlessly large if the quality of the soot be good.

Experiments with Potatoes.

For several years in succession the spraying of potatoes with "Bordeaux mixture" (sulphate of copper and lime), to counteract "potato disease," has been carried on. In years when disease is prevalent the spraying undoubtedly wards off the disease, and gives profitable results; but when disease is not rampant, though the crop is slightly increased through its period of growth being prolonged, this does not "pay" for the application. An interesting series of manurial experiments on potatoes has also been carried out for several years in succession with the object of ascertaining whether nitrate of soda or sulphate of ammonia is the better nitrogenous manure for potatoes, and whether kainit or sulphate of potash is to be preferred as a source of potash.

The results have come out very uniformly, and have shown that while sulphate of ammonia has done rather better than nitrate of soda, there is a marked advantage possessed by sulphate of potash over kainit. A general artificial dressing which has proved most effective for potatoes on light land is—dung, 12 loads, with superphosphate 3 cwt., sulphate of potash 1 cwt., sulphate of ammonia 1 cwt., each per acre.

Different varieties of potatoes have further been tried in regard both to their cropping powers and their liability or otherwise to disease.

Experiments with Mangels.

The principal manurial experiment on the mangel crop has concerned the use of nitrate of soda as a top-dressing, with or without the addition of common salt. When dung—say 12 loads per acre—has been used, it has been found that a top-dressing of 1 cwt. per acre of nitrate of soda with 1 cwt. per acre of salt has been more effectual than the dung used with 1 cwt. of nitrate of soda, or even with 2 cwt. of nitrate, when salt is omitted altogether. This result has been confirmed by repetition in several successive years.

"Finger-and-Toe" in Turnips.

The application of different materials which have been suggested for the pre-

vention of "finger-and-toe" in turnips has been tried, among these being sulphate of iron, sulphate of copper, bleaching-powder, kainit, common salt, borax, &c. Of all the applications tried, only those which contained lime in quantity have been of any use, lime itself being the best, then gas lime. Basic slag did good for a time, but not permanently.

Experiments with Clover.

Experiments have been devised with a view to obtaining an explanation of the so-called "clover-sickness," but no manurial applications have been found to prevent this. Also trials with "nitragin" and similar inoculating materials have failed to reproduce the benefits attributed elsewhere to the use of these materials.

Different varieties of clover, including some received from Canada, Chili, and elsewhere, have been grown, but none have been found superior to the ordinary English red clover.

Experiments on Lucerne.

The influence of manurial ingredients in prolonging the growth of lucerne has been tried for many years. Though the soil is a light sandy one, very poor in lime, lucerne has been successfully grown and kept down for as many as thirteen years. Both nitrate of soda and sulphate of ammonia, when used alone, were found to be harmful, but sulphate of potash gave fair crops. The only really good crops, however, were those to which a mixture of artificial manures comprising superphosphate, sulphate of potash, and either nitrate of soda or sulphate of ammonia had been applied, so that potash would seem to be an essential element in the case.

In recent years different varieties of lucerne — viz., from Provence seed, American seed, and Canadian seed — have been grown, the Canadian seed giving decidedly the best return.

Pasture Experiments.

In 1886 a field (Great Hill Bottom) was laid down with different mixtures of grasses, and at different cost, some containing rye-grass and some not, and this with the object of seeing whether the rye-grass was permanent or not, and if its inclusion in a seed-mixture was to be recommended. The general result has been to show that the rye-grasses have remained constituent parts of the pasture, and have even invaded the plots on which they were not originally sown.

Manurial applications on this field also brought out the important result that, while basic slag (8 cwt. per acre) used alone exercised no benefit, this same quantity used along with 1 cwt. per acre of sulphate of potash caused a wonderful improvement, and increased the growth of clover remarkably.

Two plots laid down later (1901) in this field with Elliot's mixtures of deep-rooting plants (which include kidney-vetch, burnet, and chicory) have been very successful, and they still (1908) remain quite good.

Manurial experiments. in other fields (Broad Mead and Long Mead), in which farmyard manure, lime, basic slag and sulphate of potash, and superphosphate with sulphate of potash, are compared, have given the best yields with the two latter mixtures. Botanical examination of the hay has further shown that where the potash salts are used the percentage of leguminous plants is increased markedly.

Introduction of New Crops.

From time to time crops new to the land have been tried — e.g., *Lathyrus sylvestris*, maize, kohl-rabi, crimson trifolium, white trifolium, gorse. The two former were proved to be useless for the land, the lathyrus not being relished by stock; gorse, also, though readily enough taken by stock, failed to be practically useful because of the want of a proper machine to bruise or crush it adequately and economically; kohl-rabi and trifolium, however, though formerly not known to the neighbourhood, have now become regular crops of the farm.

Ensilage Experiments.

An extended inquiry was made into the subject of ensilage during the years 1884-7, both silos and silage-stacks being erected, and feeding experiments were conducted on bullocks with the produce. In this way the respective losses incurred in making silage and in making hay were ascertained, and a comparison was instituted between feeding bullocks on roots and hay and on silage respectively. The feeding with silage was not quite as good or as economical as that with roots and hay, and though it was established that silage was profitable whenever hay could not be made or roots were scarce, no further advantage could be claimed for it.

Feeding Experiments.

From the outset, feeding experiments both with bullocks and with sheep have been carried out, the latter in the open, the former in the eight specially erected feeding-boxes or "pits" at the farm. Besides the ensilage experiments already referred to, different foods have been compared in respect of their use for fattening cattle. Thus, the superiority of a mixture of decorticated cotton-cake and maize-meal over linseed-cake has been shown, and that of decorticated cotton-cake over undecorticated cake.

Losses with Farmyard Manure.

From the making of farmyard manure in the feeding-boxes, by bullocks consuming known quantities of food, important results have been obtained as to the losses incurred in the making and in the storage of farmyard manure. The manure has been analysed when made in the boxes, and again when removed from the heap after storing, and the result has been obtained that, even under the best conditions, farmyard manure loses about 15 per cent of its nitrogen during the making, and another 20 per cent during storage, so that one can say that under ordinary conditions only about one-half of the nitrogen originally passing out of the animal finds its way on to the land.

This has, of course, an important bearing on the question of compensation for unexhausted manure.

Pot-culture Station.

Last to be mentioned is the most recent development in experimental work, that of the Pot-culture Station, where a very large number of inquiries are being carried on. These are partly independent inquiries, such as those forming the Hills' Bequest work, in which the influence of the "rarer constituents" of plants are being studied. In this connection different salts of lithium, manganese, iodine, bromine, &c., have been experimented with, and the results show that these, even when used in minute quantities, have a marked influence, more particularly on germination.

Other matters, such as the "hardness" or "softness" of wheat, the prevention of smut in grain, the destruction of weeds, and the relative thickness of sowing of grain, are being investigated.

To a large extent, however, the Pot-culture Station is made use of in direct connection with the field experiments, and for the solution of problems that have arisen in the carrying on of these latter. Thus, the questions of acidity in the soil of Stackyard Field, produced by continued manuring with ammonium-salts, and its correction by the use of lime and other means, are being worked out, together with the collection of data regarding the removal of lime from the soil by nitrate of soda and different salts of ammonia. The field green-manuring experiments are also being more closely studied here, together with other soil problems, such as the physical changes produced by different salts, and the influence of magnesia in its relation to lime according as these two may exist in a soil.

By means of the Pot-culture Station a great deal of light is being continually thrown upon what has been noticed to take place in the field experiments, and further ways of developing inquiry are being constantly desired.

Within recent times the Pot-culture Station has also been utilised for conducting inquiries, on behalf of the Board of Agriculture, on the value of inoculating materials for leguminous crops, and for the Royal Commission on Sewage Disposal, on the value of different sewage sludges.

ABERDEENSHIRE EXPERIMENTS.

The following account of experiments carried on by the Aberdeenshire Agricultural Association was, at the request of the Editor, prepared for the Fourth Edition of *The Book of the Farm* by Mr Thomas Jamieson, F.I.C., chemist to the Association, who has all along had the active management of the experiments.

The experiments were commenced in 1875. They had been framed with great care, scrutinised and amended by several gentlemen conversant with the various aspects of the question, chiefly by the late Mr J. W. Barclay, who was familiar with the manure trade and with farming, and had given close attention to the scientific aspect of the question; by Mr John Milne, Mains of Laithers, farmer, manure manufacturer, and holder of the Highland Society's diploma; by the late Mr Ranald Macdonald, factor on the Cluny estates; and by the chemist to the Association.

The Experimental Stations.—Five different sites were fixed upon, at altitudes varying from 1 to 400 feet above sea-level; at distances from the sea varying from 2 to 30 miles; representing soils of different characters and different degrees of fertility, the depth of mould varying from 8 to 36 inches; while the subsoils represented crumbling granite, gravel, and sand, yellow clay, bluish clay, and stiff red clay.

Size of Plot.—Each site was about two acres in size, and was enclosed by a substantial fence. This area gave space for a large number of plots, of the size that had been so highly recommended by the late Professor Anderson, chemist to the Highland and Agricultural Society—viz., $\frac{1}{112}$th part of an acre.

It may be mentioned in passing that Professor Anderson arrived at this size after much experience with experiments on a larger scale. It may also be mentioned that the same experience was got in Aberdeenshire,—preliminary experiments on $\frac{1}{10}$th and on $\frac{1}{20}$th acre plots having been made, while along with the large number of $\frac{1}{112}$th acre plots a large field was divided into $\frac{1}{4}$th acre plots. This experience gradually led to a clearer discernment of the objectionable features of large plots, and to a distrust in their results: at the same time, Professor Anderson's opinion was abundantly confirmed, that the $\frac{1}{112}$th acre plot is a most suitable size for field experiments, while it is also very convenient for calculation, as every pound of manure applied, or of crop reaped, represents the same number of cwts. per acre.

Discussion as to Size of Plot.—It is only what is to be expected, that this subject of size of plot will crop up every now and again—familiarity with work on large areas engendering a leaning towards large experimental plots, while greater familiarity with actual experimenting leads to the small plot, as ensuring uniformity of soil, as well as identical cultivation under the same climatic conditions, and hence fair comparison. The $\frac{1}{112}$th acre plot is indeed too large; but it is probably as small as can be adopted, unless the soil is actually taken up, thoroughly mixed, and returned in equal quantities to the former position. Under such arrangement the $\frac{1}{1000}$th acre plot will be found in the highest degree satisfactory.

It is interesting to notice how steadily opinion grows in favour of small plots, and how constantly the above experience is repeated—namely, that every beginner, especially if associated or influenced, directly or indirectly, with practice on the large scale, begins with large plots, gradually works towards the smaller ones.

Duplicated Plots.—Especial care was taken to have each experiment duplicated, a feature too often neglected in experiments. It is indeed desirable that they should even be triplicated.

In the experiments having reference specially to phosphate applied with and without nitrogen, particular care was taken that there should be no hindrance to the action of these essentials by the absence of other materials understood to be essential. This was prevented by the application, all over the plots, of a mixture consisting of 3 cwt. potassic chloride, 1 cwt. magnesia sulphate, and ½ cwt. common salt. Each plot was surrounded by a deal-board 9 inches deep, driven edgewise into the soil.

Adjusting the Manures.—The soils were subjected both to chemical and mechanical analyses. The manures were also analysed, and care taken that equal quantities of the ingredients were used. In the earlier experiments, however, the proportion of insoluble phosphate was a half more than soluble phosphate,—an adjustment considered necessary in order that the two phosphates might be fairly

compared, assuming that the finer division or greater distribution of the soluble phosphate would give it undue advantage in a fair trial of the relative powers of the two substances. Possibly this adjustment was unnecessary: the probable effect was to provide a larger quantity of phosphorus in the case of the insoluble form than was requisite. In the later experiments, therefore, equal quantities were adopted, with about the same result as had previously been got in the crop. In the first instance, also, the soluble phosphates were exactly a half soluble (*i.e.*, in commercial terms, about 20 to 26 per cent superphosphate). In the later experiments, however, the highest practicable degree of solubility was sought—viz., about 35 per cent soluble.

On singling the plants (turnips) it was sought to have an equal number in each plot—namely, about 200; but that number, from various causes, which will be easily understood by those engaged in practice, was seldom maintained to the end of the season. Attacks by insects, weakly plants, frost, drought, &c., frequently reduced the number.

None of the operations on the plots were allowed to go on, nor weighing of the crop, except in the presence of the chemist who directed the experiments.

It may thus be seen that the most scrupulous care and attention were given to the whole work.

First Year's Conclusions.

At the end of the year the numerous and duplicated results of this large series of experiments were tabulated, and presented such a varied and confirmed series of results as probably had not previously been available. They were carefully considered by the individuals above mentioned, and others taking part in the direction, and finally the following conclusions were adopted:—

1. That *phosphates of lime* decidedly increase the turnip crop, but that farmers need not trouble themselves to know whether the phosphates are of animal or of mineral origin.

2. That soluble phosphate is not superior to insoluble phosphate to the extent that is generally supposed.

3. That nitrogenous manures have

little effect on turnips used alone, but when used along with *insoluble* phosphates increase the crop; that the addition of nitrogen to *soluble* phosphates does not seem to increase the solids or dry matter in crop; and that there is no material difference between the effects of equal quantities of *nitrogen* in nitrate of soda and in sulphate of ammonia.

Note.—Pure sulphate of ammonia contains about 5 or 6 per cent more nitrogen than nitrate of soda.

4. That fineness of division seems nearly as effective in assisting the braird and increasing the crop as the addition of nitrogenous manures. Hence the most economical phosphatic manure for turnips is probably insoluble phosphate of lime, from any source, ground down to an impalpable powder.

Condensed Results.—It would occupy too much space to give the results in detail. It may suffice to give a few condensed results—namely, a few results from the station that responded best to the action of phosphate, and therefore showed the relative action of the different forms most clearly; and also the results of the five stations averaged :—

		ABOYNE.	AVERAGE OF 5 STATIONS.
		Turnips. Tons per acre.	Turnips. Tons per acre.
GROUP I.	No phosphate given	5	10
	Insoluble phosphate (ground coprolite)	19	16
	Soluble phosphate (superphosphate)	22	18
GROUP II.	Insoluble phosphate and nitrate of soda	21	18
	Soluble phosphate and nitrate of soda	26	21
GROUP III.	Insoluble phosphate and sulphate of ammonia	23	20
	Soluble phosphate and sulphate of ammonia	24	20
GROUP IV.	Raw bone-meal	15	16
	Steamed bone-powder	23	20

Insoluble Phosphates as Plant-food.—From the point of view of new information, the first and last groups are by far the most important. Formerly coprolite was deemed of no manurial value until rendered soluble by sulphuric acid; and in placing a money value on a dissolved manure, no value was attached to the insoluble portion it contained. The above results indicated that this position was untenable. They led the Aberdeenshire Association to say decisively that insoluble phosphate in the form of ground coprolite was directly effective on plants, and to add the statement that the superiority of the soluble form is not so great as is generally supposed. It was thought well to limit expression to the latter general and tentative statement, reserving a definite statement till further results were obtained.

The fourth group indicates the excellent results got by using phosphate in a fine state of division, and led to the fourth conclusion stated above.

It may be remarked that these opinions are now generally accepted. No doubt there may constantly be heard dissentients from these doctrines. That is only what may be expected, when the subject concerns so large a body as the whole agriculturists of a kingdom. But no responsible person will now be found to take up an opposite position.

The bearing of the New Doctrine. —At this stage there ought to be prominently brought forward the real bearing of this new doctrine on agricultural practice.

What is the actual effect of the knowledge that the natural coprolite, merely ground, is able directly to feed the plant with phosphate? Being decidedly the cheapest form of phosphate, does it follow that it should be employed to the exclusion of all other phosphates? Assuredly not, when it is so clearly brought out that although it produces 16 tons per acre, other forms produce 18 tons, and others 20 tons per acre. Assuredly not again, when it is stated that greater assistance is given to the plant in the early stage, by more finely divided phosphate, or by soluble phosphate. So long as the latter two phosphates are not charged a higher price, as compared with coprolite, than is compensated by the larger crop, they should be used. So soon, however, as the price advances much beyond that point, the agriculturist can fall back on coprolite, which is found abundantly in many parts of the world, and requires no more manufacture than simple grinding.

It is thus wholly and solely a matter of price. And herein lies the important practical bearing of the new doctrine. It is well to grasp fully the significance of the knowledge that coprolite may be used directly. Put in few words, it is this— *that it provides a check to the undue raising of the price of manufactured phosphates.*

Experiments of Subsequent Years.

It would go beyond the limits of this article to explain the many points that engaged the Aberdeenshire Association during the following six years—viz., till 1882—during which the experiments of the first year were continued and repeated, providing altogether many hundreds of results. The proceedings of the Association, for that period of seven years, form a large volume, replete with tables, diagrams, and photographs, which provide the critic with full details, while at the same time the main points are clearly brought out for the general reader. It may suffice to say that the following points were very fully entered into :—

1. The *specific gravity of turnips*, which was found to give no reliable indication of their quality.

2. The *proportion of water in turnips*, which was found to be increased both by nitrogenous and, to some extent, by soluble phosphatic manures.

3. *"Finger-and-toe" disease* was investigated ; farmers' opinions regarding it widely ascertained ; many experiments conducted to ascertain the effect of manures in giving rise to the disease ; and other experiments with the view of finding a remedy. Speaking generally, it was found that whatever weakened the plant predisposed it to disease, and rendered it an easy prey to its natural fungoid enemy, which then produced the disease. But while many influences, both mechanical and climatic, caused weakness, it was found, in a remarkable and unmistakable manner, that soluble phosphate produced this effect in a very striking degree. Nor was this effect

confined to phosphate rendered soluble by sulphuric acid, but sulphur in various forms seemed more or less to have a similar effect. As to a remedy, the disease seemed lessened by whatever ensured healthy growth, or a condition of soil uncongenial to fungoid growth, as well as such lapse of time between the two turnip crops as would reduce the natural food of the fungus, while a heavy dose of lime markedly lessened the proportion of disease.

4. The *variation in weight on oat grain by storing ;* the solid nourishing matter in oats differently manured ; and the proportion of husk to kernel.

5. *Different methods of storing turnips* during winter were tried, and the method of storing in pits of two or three loads, and covered with three or four inches of earth, was found to answer best ; while the result was not greatly different whether or not the roots or leaves, or both, were cut off previous to storing.

The first series of experiments was, as mentioned, on turnips, and turnips were grown on the same ground successively for five years.

But in the second year of the experiments, the original experiments were repeated on new ground at each station, and the effect of the various manures ascertained over a rotation.

Relative Value of Phosphates and Nitrogen.

At the end of seven years it was considered that the subject which had been carefully avoided up to that time might then be approached—viz., to fix the relative agricultural value of phosphates and nitrogen. This was done, not by attaching a money value, which might vary every year, but by fixing on some large natural source of phosphate, and a similar source of nitrogen, and adopting these each as a standard, to be referred to by the figure 10. The standard adopted for phosphate was ground coprolite of the usual commercial degree of fineness, which was called 10 ; while the standard chosen for nitrogen was nitrate of soda, the value of which was also called 10.

It may be necessary later on to make these standards more definite, by specifying more distinctly the precise state of mechanical division ; and obviously the finer the division chosen for the standard, the less will be the difference between it and the forms standing above it. But for the immediate purpose the commercial forms were deemed sufficient.

The values thus carefully arrived at for phosphate were :—

Phosphate of iron 0
Phosphate of alumina (redonda) . . 3
Tribasic phosphate of lime in bone . . 10
Tribasic phosphate of lime in insoluble mineral 10
Monobasic phosphate of lime in soluble phosphate 12
Bibasic or tribasic phosphate of lime in precipitated form 13
Tribasic phosphate of lime in steamed bone-flour 14

While the values for nitrogen were as follows :—

Nitrate of soda 10
Sulphate of ammonia 10
Guano 10
Nitrogen (only) in bones (supplemented with dried blood) 8

At the same time the conclusions originally framed were more specifically drawn out, as follows :—

Final Conclusions.

1. Non-crystalline phosphate of lime, ground to a floury state, applied to soil deficient in phosphate, greatly increases the turnip crop, and also, though to a less extent, the cereal and grass crops, but always with equal effect, whether it be derived from animal or mineral matter.

2. Soluble phosphate is not superior in effect to insoluble phosphate if the latter be in finely disaggregated form— e.g., disaggregation effected by precipitation from solution, or by grinding bones after being steamed at high pressure. In such finely divided conditions, the difference is in favour of the insoluble form, in the proportion of about 12 for the soluble to 13 and 14 for the above insoluble forms respectively. In less finely divided form (such as mineral phosphate impalpable powder), insoluble phosphate is inferior to soluble phosphate in the relation of about 10 to 12.

3. Nitrogenous manures used alone have little effect on root crops, unless the soil is exceptionally poor in nitrogen and rich in available phosphate.

Nitrogenous manures used with phosphate on soils in fairly good condition give a visible increase of root crop, but this increase is due mostly, and often entirely, to excess of water in the bulbs.

Nitrogenous manures greatly increase cereal crops, and the increase in this case is not due to excess of water.

As to the relative efficacy of different forms of nitrogen : the ultimate effect of nitrogen in sulphate of ammonia, in guano, and steamed bone-flour, is nearly identical, whether used with soluble or insoluble phosphate. Nitrate of soda, when used with soluble phosphate, is also identical with the above forms, but is of less efficacy when used with insoluble phosphate.

4. Fine division (or perfect disaggregation) of phosphates assists the braird nearly as much as, and with more healthy results than, applications of nitrogenous manures.

The most economical phosphatic manure is probably non-crystalline, floury, insoluble phosphate of lime ; the cheapest form being mixed with an equal quantity of the form in which the highest degree of disaggregation is reached.

(At present these two forms are respectively, ground mineral phosphate (coprolite) and steamed bone-flour.)

Duplicate Trials in England.

It remains only to say that, it having been argued that while these results might apply to soil in Scotland, poor in lime, and not to soils in England, generally richer in lime, it was considered desirable to ascertain whether or not the results had only this limited application.

A station was therefore established in Huntingdon and another in Kent, while later on a large number of experiments were established in Sussex, and carried on by the Sussex Association for the Improvement of Agriculture, under the same chemical direction as the Aberdeenshire experiments. These experiments in England proved that while in soil actually on the chalk formation soluble phosphate showed to more advantage than on all the other soils tried, yet as regards Sussex and Huntingdon, where the soil was not so purely chalky, but yet contained the ordinary quantities of lime, the results were practically the same as those got in Aberdeenshire.

Outside Confirmation.

The value of these experiments in Aberdeenshire and Sussex would be uncertain unless confirmed not only in other places, but by other and independent experimenters. The importance of the question, however, was widely recognised; and after some time, both the Highland and Agricultural Society of Scotland and the Royal Agricultural Society of England established experiments on the same subject, as did also a number of private experimenters, all of whose results pointed more or less conclusively in the same direction.

Still the march has been slow, if we judge its progress by the amount of coprolite applied, or by the small effect on the superphosphate trade. But for this there are two obvious explanations: first, as already explained, that the effect is not to be looked for in the direction of the greater use of coprolite, but rather in the reduction of the prices of superphosphate and other phosphates—and this reduction has indeed taken place to a very marked extent; and, second, that the interest of the trade is more than able to cope with the agriculturist, who at the present day is hardly so skilled in the intricacies of manure as in a few years he is likely to become.

The Test of Time.

In response to the request of the Editor that he would revise the foregoing notes for the present edition of this work, Mr Jamieson writes as follows:—

"The preceding account is based on experiments made thirty-two years ago, but reperusal of it has not suggested any adjustment. It has therefore stood the test of time, and of numerous experiments since made over about one-third of a century.

Effects of the Aberdeenshire experiments.—" The effect of the discovery on the practice of agriculture has been in several directions: (1) partly in the direct use of insoluble phosphates by farmers; largely (2) in the introduction of an artificial insoluble phosphate (slag) which is extensively used in that form;

(3) in the general use by manure manufacturers of insoluble phosphate in their manure mixtures, justified as they are by its proved efficacy, and encouraged by the fact that in the valuation of manures, value is now given for insoluble phosphates which formerly had been regarded as valueless; and, finally, (4) by the cheap natural insoluble phosphate being found efficacious, having lowered the price of bones and other phosphates by at least £1 per ton. As about one million tons of insoluble phosphates are used yearly in agriculture, there is indicated, on this point alone, a saving to agriculture of one million pounds sterling annually.

"The only addition that may be made is, that in these experiments it was found that for the short-lived cereal crops, and for grasses whose rootlets are beyond easy reach, soluble phosphate was more efficacious than insoluble phosphate. But since that time, attention being so much directed to the need for fine division, manufacturers are now offering so finely divided insoluble phosphate that it is probable they would act on cereals and grasses equally as well as soluble phosphate, and experiments conducted in Germany appear to have proved that this is the case."

OTHER EXPERIMENTS.

A great variety of useful experiments have in recent years been conducted by other agricultural societies, by the Irish Department of Agriculture, by private individuals, and by Colleges of Agriculture. By Colleges of Agriculture in particular the volume of experimental work now being carried out every year is very large, and is constantly on the increase. The value of this work to practical farmers can hardly be overestimated, for it has a close practical bearing upon their daily duties.

For interesting accounts of numerous important experiments the Editor of this edition is indebted to the officials of the following institutions — viz.: The Agricultural Department, Cambridge University; the Agricultural Department, Leeds University; Armstrong College, Newcastle on-Tyne; South-Eastern Agricultural College, Wye, Kent; the University College, Reading; the West of Scotland Agricultural College, Glasgow; the Edinburgh and East of Scotland, Edinburgh; the Aberdeen and North of Scotland College of Agriculture, Aberdeen; the University College of North Wales, Bangor; the University College of Wales, Aberystwyth. Space unfortunately is not available for the production of these notes in their entirety, but most of the many useful lessons they afford to the farmer are incorporated in the different parts of the work dealing with the particular subjects to which the results specially relate.

THE SEASONS—WEATHER AND WORK.

The Agricultural Year.—The Agricultural Year practically begins immediately on the completion of the grain harvest, and the commencement of autumn wheat-sowing. And as this may be sooner or later, according to the lateness or earliness of the season, so the agricultural year may commence sooner or later in different years; just as it begins at different times in different climates when the seasons are normal. It may begin early in September in the south of England after an early harvest, or it may be postponed until near the end of November in less favoured districts when the harvest is very late. Here, however, we take the seasons in their natural sequence.

SPRING.

FIELD OPERATIONS AND SPRING WEATHER.

In the vegetable world winter is the season of repose, of passive existence, of dormancy, though not of death. Spring, on the contrary, is the season of returning life, of passing into active exertion, of hope, and of joy; of hope, as the world of life springs into view immediately after the industrious hand has scattered the seed upon the ground—and of joy, in contemplating with confidence the reproductions of the herds and flocks. It would be vain to attempt to describe the emotions to which this delightful season gives birth. It is better that the pupil of agriculture should enjoy the pleasure for himself; for "the chosen draught, of which every lover of nature may drink, can be had, in its freshness and purity, only at the living fountain of nature; and if we attempt to fetch it away in the clay pitchers of human description, it loses all its spirit, becomes insipid, and acquires an earthy taste from the clay."

Early Rising and the Joys of Spring.—To enjoy the beauties of spring in perfection, "it is necessary to take advantage of the morning, when the beams of the newly risen sun are nearly level with the surface of the earth; and this is the time when the morning birds are in their finest song, when the earth and the air are in their greatest freshness, and when all nature mingles in one common morning hymn of gratitude. There is something peculiarly arousing and strengthening, both to the body and the mind, in the early time of the morning; and were we always wise enough to avail ourselves of it, it is almost incredible with what ease and pleasure the labours of the most diligent life might be performed. When we take the day by the beginning, we can regulate the length of it according to our necessities; and whatever may be our professional avocations, we have time to perform them, to cultivate our minds, and to worship our Maker, without the one duty interfering with the other."[1]

Cares of Stock-owners in Spring.—Spring is the busiest of all seasons on the

[1] Mudie's *Spring.*

farm. The cattle-man, besides continuing his attendance on the feeding cattle, has now the more delicate task of waiting on the cows at calving, and providing comfortable lairs for new-dropped calves. The dairymaid commences her labours, not in the peculiar avocations of the dairy, but in rearing calves—the support of a future herd. The farrowing of pigs also claims a share of attention. The shepherd, too, has his painful watchings, day and night, on the lambing ewes; and his care of the tender lambs, until they are able to gambol upon the new grass, is a task of peculiar interest, and naturally leads to higher thoughts.

Field-work in Spring.—The condition of the fields demands attention as well as the reproduction of the stock. The day now affords as many hours for labour as are usually bestowed at any season in the field. The ploughmen, therefore, know no rest for at least ten hours every day, from the time the harrows are yoked for spring tillage until the turnips are sown. The turnip land, bared as the turnips are consumed by sheep, or removed to the steading, is now ploughed and prepared for spring wheat, barley, or oats — that is, should the weather be mild and the soil dry enough. The first sowing is the spring wheat; then the beans, the oats, and the barley. The fields intended for the root crops are then prepared. Grass seeds are then sown amongst the young autumnal wheat, as well as amongst the spring wheat and the barley or oats.

The hedger resumes his work of watertabling and scouring ditches, cutting down and breasting old hedges, and taking care to fence with paling the young quicks upon the hedge - bank, which he may have planted during fresh weather in winter, as also to make gawcuts in the sowed fields.

The steward is now on the alert, urges the progress of every operation, and intrusts the sowing of the crops to none but himself, or a tried hand. Thus every class of labourers have their work appropriated for them at this busy season; and as the work of every one is carefully defined, it is scarcely possible for so great a mistake to be committed as that any piece of work should be neglected by all.

The Farmer's Duties in Spring.—The farmer himself now feels that he must be "up and doing." His mind becomes stored with plans for future execution; and in order to see them executed at the proper time and in the best manner, he must now forego all visits, and remain at home for the season; or at most undertake an occasional and hasty journey to the market town to dispose of surplus corn and transact other pressing business. The work of the fields now requiring constant attendance, his mind as well as body becomes fatigued, and, on taking the fireside after the labours of the day are over, the farmer seeks for rest and relaxation rather than mental toil. He should at this season pay particular attention to the state of the weather, by observing the barometric and thermometric changes, and make it a point to observe every external phenomenon that has a bearing upon the changes of the atmosphere, and be guided accordingly in giving his instructions to his people.

Weather in Spring.—The weather in spring, in the zone we inhabit, is exceedingly variable, alternating, at short intervals, from frost to thaw, from rain to snow, from sunshine to cloud—very different from the steady character of the arctic spring, in which the snow melts without rain, and the meads are covered with vernal flowers ere the last traces of winter have disappeared. Possessing this variability in its atmospherical phenomena, spring presents few having peculiarities of their own, unless we except the cold unwholesome east wind which prevails from March to May, and the very heavy falls of snow which occasionally occur in February.

Spring Winds.—All the seasons have their peculiar influence on the winds. The character of the winds in spring is, that they are very sharp when coming from the N. or N.E. direction; and they are also frequent, blowing strongly sometimes from the E. and sometimes from the W. In the E. they are piercing, even though not inclining to frost; in the W. they are strong, boisterous, squally, and rising at times into tremendous hurricanes, in which trees escape being uprooted only in consequence of their leafless state.

Snow in Spring.—Very frequently snow covers the ground for a time in spring. The severest snowstorms and falls usually occur in February. It is a serious affliction to the sheep-farmer when a severe and protracted snowstorm occurs in spring.

Rain in Spring.—The character of rain in spring is sudden, violent, and cold, not unfrequently attended with hail.

Evaporation in Spring.—Evaporation is quick in spring, especially with an E. wind, the surface of the ground being as easily dried as wetted. Thus two or three days of drought will raise the dust in March, and hence the cold felt on such occasions.

Birds in Spring Storms.—During a snowstorm in spring, wild birds, becoming almost famished, resort to the haunts of man. The robin is a constant visitor, and helps himself with confidence to the crumbs placed for his use. The male partridge calls in the evening within sight of the house, in hopes of obtaining some support before collecting his covey together for the night to rest upon the snow.

Rooks now make desperate attacks upon the stacks, and will soon force their way through the thatch. Beginning their operations at the top, they seem to be aware of the exact place where the corn can be most easily reached. Sparrows burrow in the thatch; and even the diminutive tomtit, with a strength and perseverance one should suppose beyond his ability, pulls out whole straws from the side of the stacks, to procure the grain in the ear.

Cottage Gardening.—"By the time the season is fairly confirmed, the leisure hours of the cottagers," and of the ploughmen, who are cottagers of the best description, are spent, in the evening, "in the pleasing labour, not unaccompanied with amusement, of trimming their little gardens, and getting in their early crops. There is no sort of village occupation which men, women, and children set about with greater glee and animation than this; for, independently of the hope of the produce, there is a pleasure to the simple and unsophisticated heart in 'seeing things grow,' which, perhaps, they who feel the most are least able to explain.

"Certain it is, however, that it would be highly desirable that not only every country labourer, but every artisan in towns, where these are not so large as to prevent the possibility of it, should have a little bit of garden, and should fulfil the duty which devolved on man in a state of innocence, 'to keep it and to dress it.'"

The Farmer's Garden.—Farmers, as a rule, are bad gardeners. Not unfrequently the garden, or where the garden should be, is one of the most thoroughly neglected spots on the farm. This is much to be regretted, for the value of a good, well-stocked kitchen-garden to a household is very great. There should be a garden on every farm, and it may be kept in good order at trifling expense. The hedger, stableman, or some other of the farm-servants, should know as much of the art of gardening as to be able to keep the farmer's garden in decent order in the absence of a gardener, whose assistance may with advantage be called in to crop the ground in the respective seasons. A field-worker now and then could keep the weeds in subjection, and allow both sun and air free access to the growing plants. Besides carelessness about the garden, the same feeling is evinced by too many farmers in the slovenly state in which the shrubbery and little avenue attached to their dwelling are kept.

Fat Cattle.—In spring the farmer thinks of disposing of the remainder of his fat cattle. Should he not be offered the price he considers them worth, he may keep them on for a time—a few of them perhaps for a month or two on grass—for beef is usually plentiful and cheap in spring, and scarce and dear early in summer.

Grass Parks.—Spring is the season for letting grass parks. In the majority of cases the parks are held by landed proprietors. The ready demand for old grass induces the retention of pleasure-grounds in permanent pasture, and removes temptation from a landlord to speculate in cattle. It is not customary for farmers to let grass parks, except in the neighbourhood of large towns, where cowfeeders and butchers find them so convenient as to induce them to tempt farmers with high prices. Facility of obtaining grass parks in the country is useful to the farmer who raises grazing stock, when he can give them a better bite or warmer shelter than he can offer them himself, on the division of the farm which happens to be in grass at the time.

ADVANTAGES OF HAVING FIELD-WORK WELL ADVANCED.

The season — *early spring* — having arrived when the labouring and sowing of the land for the various crops cultivated on a farm of mixed husbandry are about to occupy all hands for several months to come, the injunction of old Tusser to undertake them in time, so that each may be finished in its proper season, should be regarded as sound advice. When field-labour is advanced ever so little at every opportunity of weather and leisure, no premature approach of the ensuing season can come unawares; and no delay beyond the usual period will find the farmer unprepared to proceed with the work. When work proceeds by degrees, there is time to do it effectually. If it is not so done, the farmer has himself to blame for not looking after it. When work is advancing by degrees, it should not be allowed to be done in a careless manner, but with due care and method, so as to impress the work-people with the importance of what they are doing. The advantage of doing even a little effectually is not to have it to do over again afterwards; and a small piece of work may be done as *well*, and in as short a time, in proportion, as a greater operation.

Keep the Plough going.—Even if only one man is kept constantly at the plough, he would turn over, in the course of a time considered short when looked back upon, an extent of ground almost incredible. He will turn over an imperial acre a-day—that is, 6 acres a-week, 24 acres in a month, and 72 acres in the course of the dark and short days of the winter quarter. All this he will accomplish on the supposition that he has been enabled to go at the plough every working day; but as that cannot probably happen in the winter quarter, suppose he turns over 50 acres in that time, these will still comprehend the whole extent

of ground allotted to be worked by every pair of horses in the year. Thus a large proportion of a whole year's work is done in a single, and that the shortest, quarter of the year. Now, a week or two may quickly pass, in winter, in doing things which, in fact, amount to time being thrown away.

Instances of misdirected labour are too apt to be regarded as trifles *in winter;* but they occupy as much time as the most important work—and at a season, too, when every operation of the field is directly preparatory to others to be executed in a more busy season.

Neglected Work inefficiently done.—The state of the work should be a subject for the farmer's frequent consideration, whether or not it is as far advanced as it should be; and should he find the work to be backward, he consoles his unsatisfied mind that when the season for active work really arrives, the people will make up for the lost time. Mere delusion—for if work can be made up, so can *time*, the two being inseparable; and yet, how can lost time be made up, when it requires every moment of the year to fulfil its duties, and which is usually found too short in which to do everything as it *ought to be done?* The result will always be that the neglected work is done in an inefficient manner.

Subdivision of Farm-work.—Field-labour should be perseveringly advanced in winter, whenever practicable. Some consider it a good plan, for this purpose, to apportion certain ploughmen to different departments of labour—some to general work on the farm, some constantly at the plough, others frequently at the cart. When the elder men and old horses, or mares in foal, are appointed especially to plough, that most important of all operations will be well and perseveringly executed, while the young men and horses are best suited for carting when not at the plough. Thus the benefits of the subdivision of labour may be extended to farm operations.

Advancing Field-work.—It is right to give familiar examples of what is meant by the advantage of having field-labour advanced whenever practicable. The chief work in spring is to sow the ensuing crops. It should therefore be the study of the farmer in winter to advance the work for spring sowing. When the weather is favourable for sowing spring wheat, a portion of the land, cleared of turnips by the sheep, may perhaps be ploughed for wheat instead of barley. If beans are cultivated, let the ploughing suited to their growth be executed; and in whatever mode beans are cultivated, care should be taken in winter to have the land particularly dry, by a few additional gaw - cuts where necessary, or clearing out those already existing. Where common oats are to be sown, they being sown earlier than the other sorts, the lea intended for them should be ploughed first, and the land kept dry; so that the worst weather in spring may not find the land in an unprepared state. The land intended for potatoes, for turnips, or tares, or bare fallow, should be prepared in their respective order; and when every one of all these objects has been prepared for, and little to do till the burst of spring work arrives, both horses and men may enjoy a day's rest now and then, without any risk of throwing work back.

Spring Preliminaries.—But besides field operations, other matters require attention ere spring work comes. The implements required for spring work, great and small, have to be repaired,—the plough-irons new laid; the harrow-tines new laid, sharpened, and firmly fastened; the harness tight and strong; the sacks patched and darned, that no seed - corn be spilt upon the road; the seed - corn threshed, measured up, and sacked, and what may be last wanted put into the granary; the horses new shod, that no casting or breaking of a single shoe may throw a pair of horses

out of work for even a single hour;—in short, to have everything ready to start for the work whenever the first notice of spring shall be heralded in the sky.

Evils of Procrastination.—But suppose all these things have been neglected until they are wanted—that the plough-irons and harrow-tines have to be laid and sharpened, when perhaps to-morrow they may be wanted in the field—a stack to be threshed for seed-corn or for horse's corn when the sowing of a field should be proceeded with; suppose that only a week's work has been lost, in winter, of a single pair of horses, 6 acres of land will have to be ploughed when they should have been sown,—that instead of having turnips in store for the cattle when the oat-seed is begun, the farmer is obliged to send part of the draughts to fetch turnips—which cannot then be stored—and the cattle will have to be supplied with them from the field during all the busy season.

In short, suppose that the season of incessant labour arrives and finds every one unprepared to go along with it, what must be the consequences? Every creature, man, woman, and beast, will then be toiled beyond endurance every day, not to *keep up* work, which is a lightsome task, but to *make up* work, which is a toilsome burden. Time was lost and idled away at a season considered of little value; thus exemplifying the maxim, that "procrastination is the thief of time"—and after all, the toil will be bestowed in vain, as it will be impossible to sow the crop *in due season*. Those implicated in procrastination may fancy this to be a highly coloured picture; but it is drawn from life.

SUMMER.

THE WEATHER.

As spring is the restoration of life to vegetation, and the season in which the works of the field are again in activity, so summer is the season of *progress* in vegetation and in the works of the field. This advancement involves no difference of practice, only impressing into its service many minor works for the first time,

in assistance to the greater. Many of these minor operations being manual, and performed in the most agreeable season of the year, they are regarded with peculiar interest and delight by the light-hearted farm-workers.

Atmospherical Complications in Summer.—The atmospherical phenomena of summer are not only varied, but of a complicated character, difficult of explanation, and apparently anomalous in occurrence. There is *dew*, with a great deposition of water, at a time when not a cloud is to be seen; there is a *thunderstorm*, which suddenly rages in the midst of a calm; and *hail*, which is the descent of ice and congealed snow, in the hottest days of the year.

Beneficial Influence of Dew.—The deposition of dew is a happy provision of nature. Often when the rainfall is insufficient, the wants of vegetation are supplied by dew. This is particularly the case in tropical regions where there may be little or no rain for months, and where, owing to the rapid radiation of heat at night, and the great evaporation of moisture from the soil into the surrounding atmosphere during the day, abundant dews are deposited.

Summer Rain.—The character of rain in summer is refreshing. Even in a rainy season, though we may feel displeased at being kept within doors on a summer day, we feel assured that it will in a great measure be absorbed by the varied mass of vegetation which is in constant activity during this season.

Light. — *Light* is a most important element in nature for the promotion of vegetation in summer. Its properties are most evidently manifested in this season, and have been shortly and forcibly enumerated by Lindley. "It is to the action of leaves," he observes,—"to the decomposition of their carbonic acid and of their water; to the separation of the aqueous particles of the sap from the solid parts that were dissolved in it; to the deposition thus effected of various earthy and other substances, either introduced into plants, as silex and metallic salts, or formed there, as vegetable alkaloids; to the extrication of nitrogen, and probably to other causes as yet unknown, —that the formation of the peculiar secretions of plants, of whatever kind, is owing. And this is brought about principally, if not exclusively, by the agency of *light*. Their green colour becomes intense, in proportion to their exposure to light within certain limits, and feeble, in proportion to their removal from it; till, in total and continued darkness, they are entirely destitute of green secretion, and become blanched and etiolated. The same result attends all their

other secretions; timber, gum, sugar, acids, starch, oil, resins, odours, flavours, and all the numberless narcotic, acrid, aromatic, pungent, astringent, and other principles derived from the vegetable kingdom, are equally influenced, as to quantity and quality, by the amount of light to which the plants producing them have been exposed."[1]

The advantage that summer possesses over the other seasons as regards light is seen in its comparative duration in the respective months. Summer indeed enjoys more than double the light of winter, a half more than spring, and a third more than autumn. Thus—

In Winter,

November has 8 hours 10 minutes of light a-day.
December || 7 || 8 || ||
January || 7 || 44 || ||
Making a ————
mean of 7 || 41 || ||

In Spring,

February has 9 hours 30 minutes of light a-day.
March || 11 || 49 || ||
April || 14 || 9 || ||
Making a ————
mean of 11 || 49 || ||

In Summer,

May has 16 hours 11 minutes of light a-day.
June || 17 || 16 || ||
July || 16 || 45 || ||
Making a ————
mean of 16 || 44 || ||

In Autumn,

August has 14 hours 34 minutes of light a-day.
September || 12 || 23 || ||
October || 10 || 17 || ||
Making a ————
mean of 12 || 25 || ||

Besides its existence for a greater number of hours each day, light is of greater intensity in summer than in the other seasons, because it is then transmitted through the atmosphere at a higher angle.

SUMMARY OF SUMMER FARM-WORK.

Calendar and Agricultural Seasons. —Practical farmers know well that farm-work cannot be sharply divided in accord-

[1] Lindley's *Theo. Horti.*, 52.

ance with the months of the calendar. Indeed the farming seasons, as commonly understood, differ considerably from the calendar seasons. For instance, there are the autumn and spring seed-times, which stretch respectively into winter and summer.

Root Sowing.—Early in summer the land for the root crops is worked, cleaned, drilled, dunged, and sown. The culture of roots is a most important and busy occupation, employing much labour in singling and hoeing the plants for the greater part of the summer.

Fat Cattle.—Feeding cattle not to be put to grass are now got rid of as soon as the state of the markets warrants. There is usually a deficiency of fat cattle for disposal early in summer, the winter supply becoming exhausted before grass-fed animals are fit for slaughter. It is thus a good plan, when it can be carried out conveniently, to have a few fat beasts for sale early in summer.

Fat Sheep.—The fat sheep are also sold, except when desired to have their fleece, in which case they are kept until the weather becomes warm enough for clipping.

Repairing Fences.—Before stock is put on grass, the hedger should mend every gap in the hedges and stone walls, and have the gates of the grass fields in repair.

Grazing Stock.—Young cattle, sheep, and cows are put on pasture, to remain all summer. Cattle and sheep graze well together, cattle biting the grass high, while sheep follow with a lower bite. For the same reason, horses and cattle graze well together. Horses and sheep, biting low, are not suitable companions on pasture. Horses, besides, often annoy sheep. Many successful stock farmers say better results are got by grazing cattle and sheep alternately than together.

Horses.—Horses now live a sort of idle life. They escape from the thral-dom of the stall-collar in the stable to the perfect liberty of the pasture-field, and there they do enjoy themselves. In the opinion of many farmers it is better for work-horses to have forage at the steading than to be grazed on the fields. The brood-mare brings forth her foal, and receives immunity from labour for a time.

Haymaking.—Haymaking is represented by poets as a labour accompanied with unalloyed pleasure. Lads and lasses are doubtless then as merry as chirping grasshoppers. But haymaking is in sober truth a labour of much toil and heat: wielding the hayrake and pitchfork in hot weather, for a livelong day, is no child's play.

Weaning Lambs.—The weaning of lambs from the ewes is now effected, and marks of age, sex, and ownership are stamped upon the flock.

Forage Crops.—The forage crops on farms in the neighbourhood of towns are now disposed of to cowfeeders and carters.

Dairying.—Butter and cheese are made on dairy-farms in quantities which the supplies of milk warrant.

Weeds.—Summer is the best of all seasons for making overwhelming attacks upon weeds, those spoilers of fields and contaminators of grain. Whether in pasture, on tilled ground, along drills of green crops, amongst growing corn, or in hedges, young and old, weeds should be day by day exterminated. And their extermination is, in many cases, most effectually accomplished by the minute and painstaking labour of field-workers; for which purpose they are provided with appropriate hand-implements.

Insect Attacks.—This is the season in which all manner of insects attack both crops and stock, much to their injury and annoyance. Reliable information for the combating of these plagues is provided in this work.

Top-dressing.—Top-dressings of specific manures upon growing crops are applied for the promotion of their growth and fecundity, at the fittest state of weather and crop.

Hours of Labour.—The hours of field-work in summer vary in different parts. In some parts it is the practice to go as early as four or five o'clock in the morning to the yoke, and the forenoon's work is over by nine or ten, time being given for rest in the heat of the day. The afternoon's yoking commences at one o'clock, and continues till six. Thus ten hours are spent in the fields. But in most parts of the country the morning yoking does not commence till six o'clock, and, on terminating at eleven, only two hours are allowed for rest and dinner till one o'clock, when the afternoon's yoking begins, terminating at six P.M. In some places the afternoon yoking does not commence till two o'clock, and, finishing at six, only nine hours are spent in the fields; or it is continued till seven o'clock. In other parts, only four hours are spent in the morning yoking, when the horses are let loose at ten o'clock, and, on yoking again from two till six in the afternoon, only eight hours are devoted to work in the fields, the men being employed elsewhere by themselves for two hours.

Many farmers maintain that the best division of time is to yoke at five o'clock in the morning, loose at ten, yoke again at one, and loose at six in the evening, giving three hours of rest to men and horses at the height of the day, and ten hours of work in the field. One drawback to this plan is that, without their night's rest being unduly curtailed, the horses cannot have had time to feed sufficiently before the day's work begins.

Day-labourers and field-workers, when not working along with horses, often work from seven till twelve, and from one till six o'clock in the evening, having one hour for rest and dinner. When labourers take their dinner to the field, this is a convenient division of time; but when they have to go home to dinner, one hour is too little for dinner and rest between the yokings—and rest is absolutely necessary, as neither men nor women are able to work ten hours without an interval of more than one hour. It is a better arrangement for field-workers to go to work at six instead of seven, and stop at eleven instead of twelve.

When field-workers labour in connection with the teams, they must conform with their hours.

Rest.—The long hours of a summer day, of which at least ten are spent in the fields—the high temperature of the air, which suffuses the body with perspiration—and the oft-varying character of field-work in summer, bearing hard both on mental and physical energies, cause the labourer to seek rest at an early hour of the evening. None but those who have experienced the fatigue of working in the fields, in hot weather, for long hours, can sufficiently appreciate the luxury of rest—a luxury truthfully defined in these beautiful lines:—

"Night is the time for rest.
　How sweet, when labours close,
To gather round the aching breast
　The curtain of repose—
Stretch the tired limbs, and lay the head
Upon one's own delightful bed!"
　　　　　　　JAMES MONTGOMERY.

The Farmer's Duties.—Every operation, at least early in summer, requires the constant attention of the farmer. Where natural agencies exert their most active influences on animal and vegetable creation, he requires to put forth his greatest energies to co-operate with the very rapid changes they produce. Should he have, besides his ordinary work, field experiments in hand, the demands upon his attention and time are the more urgent, and he must devote both assiduously if he expect to reap the greatest advantage derivable from experimental results.

The Farmer's Holiday.—Towards the end of summer is the only period in which the farmer has liberty to leave home without incurring the blame of neglecting his business. Even then the time he has to spare is very limited. Strictly speaking, he has only about two or three weeks before the commencement of harvest in which to have leisure for travel. A journey once a-year to witness the farm operations of other parts of the kingdom, or of foreign countries, enlightens him in many uncertain points of practice. He there sees mankind in various aspects, his mind becomes widened and raised above local prejudices, and a clearer understanding of places, manners, and customs is afforded him when reading the publications of the day. A month so spent may, in its experience, be worth a lease-length of local reading and of stay-at-home life.

AUTUMN.

AUTUMN WEATHER AND FIELD OPERATIONS.

Having passed through Spring and Summer, we now contemplate Autumn, the season of *fruition*, in which nature, bringing every plant to perfection, provides for the ensuing year sustenance for man and beast.

Rewards of Labour.—Autumn maturing its products, the toiling labours of the husbandman during the period of a year find their reward. In it, hope enjoys the possession of the thing hoped for; and the yield being plentiful, the husbandman is full of thankfulness. "It is this feeling which makes the principles of seasonal action thicken upon us as the year advances, and the autumn to become the harvest of knowledge as well as the fruits of the earth. Nor can one help admiring that bountiful and beautiful wisdom which has laid the elements of instruction most abundantly in the grand season of plenty and gratitude."

Rain in Autumn.—The greatest average quantity of rain falls in October. The heaviest rains come down in summer and early autumn months. In summer, over an inch will sometimes fall in a few hours in short and tempestuous torrents. In autumn the same quantity will occupy many hours in falling.

Autumn Work.

Harvest.—The great work of autumn is the harvesting of the corn crops. This necessarily engrosses most of the time and attention of the farmer and his assistants until these perishable crops are secured beyond danger in the stack-yard.

Weather and Harvest.—During this eventful period of a month at least, the farmer looks night and day at "the face of the sky," consults his glass, and fear or hope inspires him according to its indications. And no wonder, for the results of a whole year of labour are at stake on the exigency of the weather, and he feels that unless he exercise his best skill and judgment he will not be satisfied with himself.

Magnitude of Harvest-work.—The labour of harvesting a crop is almost incredible. Only conceive the entire cereal crop of such a nation as this reaped, carried, and stored in minute sheaves, in the short period of a few weeks! Then, besides the harvest of the cereal plants, the leguminous as well as portions of the root crops are stored in autumn.

Autumn Anomalies.—Some curious anomalies in farm labour occur in autumn. One is sowing a new crop of wheat, while the matured one of the same grain is being reaped; another, that while spring is the natural season for the reproduction of animals, autumn is that for the reproduction of sheep.

Field-sports.—The sports of the field commence in August. The gatherings on the hills in thousands on the historic "Twelfth" of August, in quest of the unique-flavoured red grouse (*Lagopus scoticus*), find a welcome home in shielings, which, at other seasons, would be contemned by their luxurious urban occupiers. Partridge-shooting comes in September, at times before the corn is cut down. Hare-hunting finds ample room by October. Last of all, the attractive "music" of the pack gathers around it, from hill and dale, all the active Nimrods of the country. At this season, however, most farmers are too much occupied in their important work to bestow time on field-sports.

Autumn Cultivation.—Autumn cultivation should be assiduously practised where the stubbles are in need of cleaning, when the weather immediately after harvest is dry and favourable. The weeds are then easier to destroy, and a little of this work in autumn saves an enormous amount of labour in spring, besides giving a great set forward to the agricultural work of the year. But autumn cultivation for the mere purpose of exposing an upturned furrow to the action of winter frosts cannot in all cases be accounted good husbandry. To preserve the soil fertility during the period of winter rains, the soil must be covered with a growing crop of some kind.

Autumn Crops.—Several crops sown in autumn succeed in England, but few can be sown in Scotland at that season with safety. Most of the forage plants sown in autumn in England, for affording early food in spring — as crimson clover, winter tares, &c.—would not withstand a Scottish winter; and some plants sown in England on stubble-ground—as the stone turnip and rape—could not with advantage be sown so late in Scotland.

The harvest in England is from two to three weeks earlier, the stubble is bare so much earlier, and the land is in a comparatively drier state, and may be worked to advantage before the wet weather sets in at the latter part of autumn. The later harvest in Scotland, and the earlier winter, would not, as a rule, permit working stubble after a grain crop; and there is but little done in Scotland in the way of growing winter forage crops.

WINTER.

THE WEATHER, AND FIELD OPERATIONS IN WINTER.

Work in the Steading.—The subjects which court attention in winter are of the most interesting description to the farmer. He directs his attention largely to work conducted in the steading, where the cattle and horses are collected, and this with the preparation of the grain for market affords pleasant employment within doors. The progress of live stock towards maturity is always a prominent object of the farmer's solicitude, and especially so in winter, when they are comfortably housed in the farm-steading, plentifully supplied with wholesome food, and so arranged in various classes, according to age and sex, as to be easily inspected at any time.

Field Work.—The labours of the field in winter are confined to a few great operations. These are chiefly ploughing the soil in preparation for future crops, and supplying food to live stock. The commencement of the ploughing for the year consists in turning over the ground which had borne a part of the grain crops, and which now bears their *stubble*—which is just the portion of the straw of the previous crops left uncut.

Water-channels in Ploughed Land.—The stubble land ploughed in the early part of winter, in each field in succession, is protected from injury from stagnant rain-water by cutting channels with the spade through hollow places, permitting the rain to run quickly off into the ditches, and leaving the soil in a dry state until spring.

Ploughing Lea.—Towards the latter part of winter the grass land, or *lea*, intended to bear a crop in spring, is ploughed; the oldest grass land being earliest ploughed, that its toughness may have time to be softened before spring, by exposure to the atmosphere. The latest ploughed is the youngest grass.

Best Season for Draining.—When the soil is naturally damp underneath, winter is selected for removing the water by draining. It is questioned by some farmers whether winter is the best season for draining, as the usually rainy and otherwise unsettled state of the weather renders the carriage of the requisite materials on the land too laborious. By others it is maintained that, as the quantity of water to be drained from the soil determines both the number and size of the drains, these are best ascertained in winter; and as the fields are then entirely free of crop, that season is the most convenient for draining. Truth may not absolutely acquiesce in either of these reasons, but, as a rule, draining may be successfully pursued at all seasons.

Planting Hedges.—Where fields are unenclosed, and are to be fenced with a quick thorn-hedge, winter is the season for performing the work. Hard frost, a fall of snow, or heavy rain, may put a stop to the work for a time, but in all other states of the weather it may be proceeded with in safety.

Water-meadows.—When water-meadows exist on a farm, winter is the season for carrying on the irrigation with water, that the fostered grass may be ready to be mown in the early part of the ensuing summer. It is a fact worth bearing in mind that *winter* irrigation produces more wholesome herbage for stock than summer irrigation. On the other hand, summer is the most proper season for forming water-meadows.

Feeding Stock.—The feeding of stock is so large and important a branch of farm business in winter, that *it* regulates the time for prosecuting several other operations. It determines the quantity of turnips that should be carried from the field in a given time, and causes the prudent farmer to take advantage of dry fresh days to store up a reserve for use in any storm that may ensue. All the cattle in the farmstead in winter are placed under the care of the *cattle-man.*

Threshing Grain. — The necessities of stock - feeding also determine the quantity of straw that should be provided from time to time; and upon this, again, depends largely the supply of grain that may be sent to market at any time. For although it is in the farmer's power to thresh as many stacks as he pleases at one time, and he may be tempted to do so when prices are high; yet, as new threshed straw is better than old, both as litter and fodder, its threshing usually depends to some extent on the wants of the stock.

Sheep on Turnips.—The feeding of sheep on turnips, in the field, is practised in winter. When put on turnips early in winter, sheep consuming only a proportion of the crop, a favourable opportunity occurs for storing the remaining portion for cattle in case of bad weather. The proportion of turnips used by cattle and sheep determines the quantity that should be taken from the field.

Attention to Ewes.—Ewes roaming at large over pastures require attention in winter in frost and snow, when they should be supplied with clover-hay, or with turnips when hay is scarce.

Marketing Grain.—The preparation of grain for sale is an important branch of winter farm business, and should be strictly superintended. A considerable proportion of the labour of horses and men is occupied in carrying grain to the market town or railway station—a species of work which used to jade farm-horses very much in bad weather, but railways have materially shortened the journeys of horses in winter.

Carting Manure.—In hard frost, when the plough is laid to rest, or the ground covered with snow, and as soon as

" By frequent hoof and wheel, the roads
A beaten path afford,"

farmyard manure is carried from the courts, and placed in large heaps on convenient spots near or on the fields which are to be manured in the ensuing spring. This work is continued as long as there is manure to carry away, or the weather is suitable.

Implements used in Winter.—Of the implements of husbandry, only a few are used in winter: the plough is constantly so when the weather will permit; the threshing - machine enjoys no sinecure; and the cart finds frequent and periodic employment. Most of the other implements are laid by for the winter.

Winter Recreation. — *Field - sports* have their full sway in winter, when the fields, bared of crop and stock, sustain little injury by being traversed. Although farmers bestow but a small portion of their time on field-sports— and many have no inclination for them at all—they might harmlessly enjoy these recreations at times. When duly qualified, why should not farmers join in a run with fox-hounds ? — or take a cast over the fields with a pointer ?—or shout a see-ho to greyhounds ? Should frost and snow prevent the pursuit of these sports, curling and skating afford healthful exercise both to body and mind.

Winter Hospitality.—Winter is the season for country people reciprocating the kindnesses of hospitality, and participating in the amusements of society. The farmer delights to send the best produce of his poultry-yard as Christmas presents to his friends in town, and in return to be invited into town to partake of its amusements. But there is no want of hospitality nearer home. Country people maintain intercourse with each other; while the annual county ball in the market-town, or an occasional one for charity, affords a seasonable treat; and the winter is often wound up by a meeting given by the Hunt to those who had shared in the sport during the hunting season.

Domestic Enjoyment. — Winter is the season of *domestic enjoyment*. The fatigues of the long summer day leave little leisure, and less inclination, to tax the mind with study; but the long winter evening, after a day of bracing exercise, affords a favourable opportunity for conversation, quiet reading, or music. In short, there is no class more capable of enjoying a winter's evening in a rational manner than the family of the country gentleman or of the farmer.

Weather in Winter.—The *weather* in winter, being very precarious, is a subject of intense interest, and puts the farmer's skill to anticipate its changes severely to the test. Seeing that every operation of the farm is to some extent dependent on weather, a familiar acquaintance with local prognostics which indicate a change for better or worse becomes a duty. In actual rain, snow, or hard frost, few outdoor occupations can be executed; but if the farmer have wisely "discerned the face of the sky," he may arrange his operations to continue for a length of time, if the storm is to endure—or be left in safety, should the strife of the elements quickly cease.

Winter Rain. — The character of *winter rain* has more of cold and discomfort than of quantity. When frost suddenly gives way in the morning about sunrise, rain may be looked for during the day. If it do not fall, a heavy cloudiness will continue all day, unless the wind change, when the sky may clear up. If a few drops of rain fall before mid-day after the frost has gone, and then ceases, a fair and most likely a fine day will ensue, with a pleasant breeze from the N. or W., or even E. When the moon shines brightly on very wet ground, the shadows of objects become very black, which is a sign of continuance of rain and unsettled state of the wind. Rain often falls with a rising barometer, which is usually followed by fine healthy weather, attended with feelings that indicate a strong positive state of electricity.

Cold and Frost.—*Frost* has been represented to exist only in the absence of heat; but, more than this, it also implies an absence of moisture. The most intense frost in this country rarely penetrates more than one foot into the ground, on account of the excessive dryness occasioned by the frost itself withdrawing the moisture.

Beneficial Influence of Frost.— Frost is a useful assistant to the farmer in pulverising the ground, and rendering the upper portion of the ploughed soil favourable to the vegetation of seeds. It acts in a mechanical manner on the soil, by freezing the moisture in it into ice, which, on expanding at the moment of its formation, disintegrates the indurated clods into fine tilth. Frost always produces a powerful evaporation of the pulverised soil, and renders it very dry on the surface; by the affinity of the soil for moisture putting its capillary power into action, the moisture from the lower part of the arable soil, or even from the subsoil, is drawn up to the surface and evaporated, and the whole soil is thus rendered dry. Hence, after a frosty winter, it is possible to have the ground in so fine and dry a state as to permit the sowing of spring wheat and beans, in the finest order, early in spring.

Plants uprooted by Frost.—On lands already planted, however, the effect of frost is not always so desirable, especially where the soil is at all wet or the drainage defective. Examine such a field in early spring and it will be found that many plants have been thrown out of the ground during the winter, or have their roots partially exposed. Wheat, grass, strawberry, and other plants are particularly liable to this summary ejectment, and considerable loss and disappointment are often sustained in farms and gardens from this cause. This uprooting of plants by frost only occurs where the soil is saturated with moisture. In each case the first act of frost is to solidify the surface layer. In doing this, the plants growing thereon are firmly caught and secured by the neck. If the frost continues, it next freezes the subjacent stratum, by virtue of the water it contains, which must enlarge, and consequently bear up on its shoulders the surface layer. Thus those plants which were secured to the surface more strongly than their roots adhered to the frozen soil below will be raised up, and to a certain extent eradicated, and when the thaw comes and the soil subsides, these plants are thrown out altogether.

Snow.—Rain falls at all seasons, but snow only in winter, or late in autumn, and early in spring. Snow is just frozen rain, so that whenever symptoms of rain occur, snow may be expected if the temperature of the air is sufficiently low to freeze vapour. Vapour is supposed to be frozen into snow at the moment it is collapsing into drops to form rain, for clouds of snow cannot float about the atmosphere any more than clouds of rain.

Snow keeps Land warm.—During the descent of snow the *thermometer sometimes rises*, and the *barometer usually falls*. Snow has the effect of retaining the temperature of the ground at what it was when snow fell. It is this property which maintains the warmer temperature of the ground, and sustains the life of plants during the severe rigours of winter in the arctic regions, where the snow falls suddenly after the warmth of summer; and it is the same property which supplies water to rivers in winter, from under the perpetual snows of the alpine mountains. "While air, above snow, may be 70° below the freezing-point, the ground below the snow is only at 32°."[1] Hence the fine healthy green colour of young wheat and grass after the snow has melted in spring.

Snow-water and Rain.—In melting, 27 inches of snow give 3 inches of water. Rain and snow-water are the *softest* natural waters for domestic purposes, and are also the purest that can be obtained from natural sources, provided they are caught before reaching the ground. Nevertheless, they are impregnated with oxygen, nitrogen, and carbonic acid, especially with oxygen; and rain-water and dew contain nearly as much air as they can absorb.

Uses and Drawbacks of Snow.— Snow renders important services to husbandry. If it fall shortly after a confirmed frost, it acts as a protective covering against its further cooling effects on soil, and in this way protects the young wheat and clover from destruction by intense frosts. On the other hand, frost and rain and snow may all retard the operations of the fields in winter very materially, by rendering ploughing and the cartage of turnips impracticable. Heavy snowstorms are also very detrimental to the interests of sheep-farmers.

Hoar-frost. — *Hoar-frost* is not always frozen dew, for dew is sometimes frozen in spring into globules of ice which do not at all resemble hoarfrost, which is beautifully and as regularly crystallised as snow. The formation of hoar-frost is always attended with a considerable degree of cold, because it is preceded by great radiation of heat and vapour from the earth, and the phenomenon is the more perfect the warmer the day and clearer the night. In the country, hoar-frost is of most frequent occurrence in autumn and winter, in such places as have little snow or continued frost.

In general, low and flat lands in the bottoms of valleys, and grounds that are land-locked hollows, suffer most from hoar-frost, while all sloping lands and open uplands escape injury. But it is not their relative elevation above the sea, independently of the freedom of their exposure, that is the source of safety to the uplands; for if they are enclosed by higher lands, without any wide open descent from them on one side or another, they suffer more than similar lands of lesser altitude.

Injury by Hoar-frost.—The severity of the injury by hoar-frost is much influenced by the wetness of the soil at the place; and this is exemplified in potatoes growing on haugh-lands by the sides of rivers. These lands are generally dry, but bars of clay sometimes intersect them, over which the land is comparatively damp. Hoarfrost will affect the crop growing upon these bars of clay, while that on the dry soil will escape injury; and the explanation of this is quite easy. The temperature of the damp land is lower than that of the dry, and on a diminution of the temperature during frost it sooner gets down to the freezing-point, as it has less to diminish before reaching it. Young potato-plants are exceedingly susceptible of being blackened by hoar-frost.

[1] Phillips's *Facts*, 440.

SEEDS OF THE FARM AND THEIR IDENTIFICATION.*

PART I.—GENERAL.

When examining farm seeds for purposes of identification the main points to be attended to are—
1. The nature of the seed. 2. Its size. 3. Its form and surface. 4. Its colour and gloss. 5. The impurities present. 6. Adulterants. 7. Parasites.

The Nature of Farm Seeds.—Under the name " seed " the farmer includes anything and everything which he sows to produce a crop. Sometimes this " seed " is a *tuber*, as for potato ; sometimes the *whole spikelet* of a grass, as for meadow foxtail ; sometimes only a *detached part of the spikelet*, as for oats and barley among the cereals, for rye-grasses, meadow-grass, cocksfoot, &c., among the forage grasses ; sometimes *a grain* (fruit), as, for example, wheat and rye, shelled timothy and shelled cocksfoot ; sometimes *a part split off from a fruit*, and composed of a seed enclosed in a case, as for carrot and parsnip ; and only sometimes the *real seed*, as for turnips and clovers. Occasionally, indeed, the part sown as seed is a whole cluster of fruits grown together into one body,—for example, mangel. The habit of calling things so various in their nature by the one name " seed " makes us only too ready to think that they really are seeds : we judge as if they were, and often pronounce them good or bad from the mere appearance of the coat.

Whenever we try to distinguish and identify, the very first thing to be done is to consider the component parts of the " seed," and this with the view of determining the real nature of the thing, and so of deciding the class of " seed " with which we are dealing.

* *The illustrations of seeds given in this section are reproduced by kind permission from an excellent series of photo-micrographs prepared by Messrs James Hunter, Limited, Seedsmen, Chester. The illustrations are the property of that firm, and may not be reproduced. The seeds in Figures 363 and 374 are magnified 4½ diameters ; those in Figures 371, 382, 387, 389, 394, 395, and 398, 10 diameters ; and the seeds in the other Figures 6 diameters.*

The Size of Seeds.—To determine the size of a seed is a very simple matter. We require for the purpose a sheet of paper ruled off into squares of one millimetre. A large sheet of such *millimetre paper* can be purchased for a few pence. We sprinkle a few seeds over the sheet and read off their size by the aid of the squares on the paper. If we use a pocket lens the size can easily be read to one quarter of a millimetre. When we are measuring a mangel seed, for example, we must remember that it is not the true seed we are measuring, but really the box that contains the seeds. In this case we are liable to think that the seed is a large one, but we must carefully notice that the *true seed* within the box is quite small, and when planted in the ground behaves as a small seed.

Form and Surface of " Seeds."—Certain forms are very characteristic. All common seeds of the chickweed family, for example, are approximately *kidney-shaped*, with small peg-like projections (tubercles) on the surface. Again, certain seeds are *globular*, and readily roll about on a sheet of paper ; such are the common cruciferous seeds and the " seeds " of timothy. Other " seeds " are marked by excessive *flatness*. Some of these flat seeds have a beak-like projection from one side of the apex (buttercups) ; some others a projection from the centre of the apex (red-shanks or persicarias). Seeds such as rib-grass are *two-faced* ; others, such as docks and sorrels, *three-faced*. Burnet has *four wing-like projections*, and so on. Details of surface, such as the honeycomb depressions on geranium seed, the tiny appendage at the narrow end of pansy seed, and so forth, are best seen by the aid of a pocket lens.

Examination of Seeds by Microscope and by Pocket Lens.—The microscopic examination is a very simple matter if we go about it properly. We require a compound microscope with a very low power, say a 2-inch objective, and an instrument perfect for such a purpose can be purchased for less than

£2. A suitable holder for the seeds to be examined is an ordinary match-box half filled with paraffin wax instead of matches. We soften the surface of the wax by heat, and sprinkle a few seeds upon the melted surface. In a minute or so the wax hardens and holds the seeds fast in various positions. We now examine the seeds under the microscope, and note the form and any peculiarities presented by the surface. Another match-box with a strip of millimetre paper pasted over the bottom enables us to determine sizes with accuracy sufficient for purposes of identification. Very small seeds, as poas, dodder, and clover-rape, require such microscopic examination for satisfactory determination. For most farm seeds, however, a pocket lens, which need not cost more than a sixpence, is all that is necessary.

Colour and Gloss.—Colour is best determined by examination of a bulk sample : for details the seeds should be spread out on a sheet of paper and examined by a pocket lens. Natural gloss on the surface of such seeds as clovers is an important indication of freshness : artificial gloss, however, which is imparted by oil to certain brands of "Welsh clovers," indicates age rather than freshness.

Impurities in Farm Seeds.—Taken generally, a sample of commercial seed is composed of the following classes of ingredients :—

1. Mature seeds true to kind, some capable of germinating, some incapable.
2. Immature seeds and husks true to kind.
3. Seeds of other plants.
4. Dust, stones, pieces of straw, &c.

Suppose we have a cocksfoot sample containing 50 per cent mature seeds and 50 per cent husks of cocksfoot, and nothing else, what is the purity of such a sample? At first, one might say that the sample is pure ; if, however, we consider that the germinating test does not apply to the empty husks, but only to those seeds that contain kernels, it is found better to agree to class the empty husks as impurity, and to say that the purity of such a sample is 50 per cent, thus regarding *all the chaff and husks as impurity.* We see, then, that

impurities may vary in their nature : some are immature seeds and husks, some are seeds of weed-plants, and some others are mere uncleanness.

It is not enough, therefore, to determine the amount of impurity by weight : the point of greatest import is the nature of the impurity and the effect consequent upon sowing it into our land. This nature is the very point which rises into prominence whenever we begin seriously to discriminate our seeds.

Suppose, for example, we seed our land with rye-grass seed containing Yorkshire fog. At hay-cutting the Yorkshire fog is partly ripe, the rye-grass unripe. The consequence is that the land is left foul with seed shed from the Yorkshire fog. On the hay-loft floor, too, further seeds of the same Yorkshire fog will form much of the sweepings, and these, if they get into the dung, will contaminate other fields. Turn now to the rye-grass field in pasture and see what is happening there. The stock depastures the rye-grass, but leaves the Yorkshire fog comparatively untouched. The weeds are thus allowed to bloom and seed freely, and as the rye-grass dies out the white uneaten grass takes its place, and a field chiefly composed of Yorkshire fog results. To avoid such dangerous consequences the seed used should be free from dangerous weeds.

Adulteration and Deterioration.—Sometimes samples of seed are sold containing impurities which cannot have been *naturally* present—that is to say, seeds of inferior value have been *intentionally* added : such additions are adulterants. Examples of adulteration are :—

1. Substitution of perennial rye-grass for meadow fescue.
2. Substitution of New Zealand tall fescue for Rhenish tall fescue.
3. Substitution of wavy hair-grass for golden oat-grass.
4. Substitution of slender foxtail and Yorkshire fog for meadow foxtail.

These cases of adulteration are easily detected by any one who is familiar with the genuine seeds, and with those points which distinguish them from their adulterants.

Sometimes a variety from one locality is sold as from another locality. Seed sold as Scottish timothy, for example, may not have been grown in Scotland at all. In the same way Chilian red clover may be substituted for English red, home Italian ryegrass for foreign, and so on.

Again, improved varieties are sometimes replaced by inferior strains, as in turnips. Although such deceptions are of the utmost importance, from a profit and loss point of view, it is often the case that no amount of skill in examining the seeds is of avail as a preventive. Sometimes, however, when the country in which the seed was grown is in dispute, it is possible to settle the point approximately by examining the weed seeds present, and noting to what country the weeds present naturally belong. Actual growth of the seeds on the land, too, may sometimes throw light on such deceptions; but there are often difficulties in bringing the deceptions home, even though the crop has been grown, since selected varieties are very liable to undergo change and deterioration under certain conditions of season and environment.

Parasites.—These are important, not from their amount but from their nature, their power of communicating disease to our crop, of spreading it from plant to plant, and of continuing it from year to year. Examples of such dangerous parasites are dodder in clover seeds, clover-rape in clover seeds, ergot in timothy, smut in corn, and bunt in wheat. Seeds free from such parasites should alone be sown.

PART II.—SPECIAL.

Classification of Farm Seeds.—For purposes of identification the common seeds of the farm, including their impurities, adulterants, and parasites, may be classified and arranged as follows:—

Class I. True Seeds.

The globular-seeded cruciferous plants. —1. Cabbage. 2. Swede. 3. Turnip. 4. Rape. 5. Mustard. 6. Charlock. 7. Runch. (P. 73.)
The kidney-shaped seeds of the chickweed family.—8. Corncockle. 9. Campion. 10. Chickweed. (P. 73.)
Leguminous seeds.—11. Red clover, its impurities and parasites. 12. Alsike

and its impurities. 13. White clover and its impurities. 14. Yellow suckling clover. 15. Lucerne and its adulterants. 16. Trefoil and its fruit. 17. Bird's-foot trefoil and its adulterants. 18. Kidney vetch. (P. 74.)
Rib-grass seeds.—19. Rib-grass. 20. Greater rib-grass or way-bread. (P. 77.)

Class II. Fruits.

The flat buttercup "seeds."—21. Upright buttercup. 22. Creeping buttercup. (P. 77.)
The three-faced "seeds" of the dock family. — 23. Buckwheat. 24. Dock. 25. Sheep's sorrel. (P. 78.)
The flat and beaked "seeds" of the dock family. — 26. Persicaria or redshank. (P. 78.)
"Seeds" of composites.—27. Yarrow. 28. Ox-eye daisy. 29. Field chamomile. 30. Cornflower. 31. Field thistle. 32. Chicory. 33. Dandelion. 34. Nipplewort. (P. 79.)

Class III. Mericarps.

Umbelliferous "seeds."—35. Parsnip. 36. Parsley. 37. Carrot. (P. 80.)
"Seeds" of the Cleavers family (*Rubiaceæ*). — 38. Sherardia. 39. Cleavers. (P. 81.)

Class IV. Nutlets.

40. Self-heal. 41. Myosotis. (P. 81.)

Class V. Multiple Fruits.

42. Mangel. (P. 82.)

Class VI. Grass "Seeds."

The whole spikelet.—43. Meadow foxtail. 44. Slender foxtail. 45. Yorkshire fog. (P. 82.)
The whole spikelet except the chaff (glumes).—46. Tall oat-grass or French rye-grass. 47. Timothy. (P. 84.)
Portion of spikelet with stalk on upper face (back rounded).—48. Italian ryegrass. 49. Perennial rye-grass. 50. Meadow fescue. 51. Rhenish tall fescue. 52. New Zealand tall fescue. 53*a*. Sheep's fescue. 53*b*. Hard fescue. 54. Crested dogstail. 55. Fescue hairgrass or hair-grass. 56. Soft brome. 57. Rye brome. (P. 85.)
Portion of spikelet—back V-shaped (keeled). — 58. Cocksfoot. 59. Poas. 60. Purple molinia. (P. 87.)

Portion of spikelet — back or base awned.— 61. Golden oat. 62. Wavy hair-grass. 63. Tufted hair-grass. (P. 88.)

Class VII. Seeds of Parasites.

64. Clover-dodder. 65. Clover-rape. 66. Ergot of timothy. (P. 89.)

CLASS I.—TRUE SEEDS.

In all plants with no exception the true seed is a body containing an embryo plant in its interior. Such a seed is always formed in a flower from an ovule by a process of fertilisation and ripening. This ovule in all the classes of farm plants is always formed and ripened within a closed case, and to have *true seed* this case must have been removed so as to leave the seed bare and detached from the enclosing parts.

THE GLOBULAR-SEEDED CRUCIFEROUS PLANTS.

The seed here is always a true seed, composed of an embryo plant so doubled up as to form a tiny sphere, which is completely enclosed within a seed-skin. The crops grown from such seeds are: Cabbage, swede, turnip, rape, and mustard. A common cruciferous weed, with globular seeds, is the well-known charlock.

The taste of the seed-contents becomes important when identification is the object. Ground mustard-seeds, for example, when mixed with water, become the well-known condiment mustard, with its pungent taste and odour. Ground turnip-seeds, if mixed with water, would have no pungency whatever, but only taste. Instead of mustard-oil, we have here the *sweet-oil* of commerce. The oily-tasting seeds are: Cabbage, swede, turnip, and rape; while mustards and mustard-seeds, such as charlock, can produce the pungent taste and odour of mustard-oil.

1. *Cabbage.*

The seed is dark-brown, with a greyish tinge, suggesting mouldiness. The size depends upon the variety, and varies between 1¼ and 2¾ millimetres.

2. *Swede.*

The seed is dark-purplish-brown. The size ranges from 2 to 2¼ mm.

3. *Turnip.*

The seed is very like that of the swede, but generally slightly paler in tint.

4. *Rapes.*

Rapes are non-bulbing varieties of swede and turnip. The seed is accordingly like that of the swede or like that of the turnip, but it affords no indication of the bulbing or non-bulbing character of the plant which it can produce.

5. *Mustards.*

There are two species of mustard— the black, with black-skinned seeds, and the white, with white-skinned seeds. The dimensions of the black seeds range from 1 to 1½ mm., those of the white from 2 to 2¼ mm. The seeds are distinguished from those of turnip and swede by their pungent taste when chewed.

6. *Charlock.*

The seeds of charlock and of turnip are practically the same in size, but the former are slightly darker in colour. Charlock is rightly called a mustard weed, for when the seeds are chewed mustard-oil forms; thus by the taste distinction is easy.

7. *Runch.*

(Although considered here for convenience, runch "seed" really belongs to Class III.)

Runch is another mustard weed, whose seeds can yield mustard-oil. Here, however, the "seed" is not only a true seed, but a seed completely enclosed in a case that looks like a harmless little piece of fine-ribbed straw. When this seed-case is cut open, the true seed is seen to be a light-brown globose body about 3 mm. long and 2 mm. broad.

THE KIDNEY-SHAPED SEEDS OF THE CHICKWEED FAMILY.

The seeds of the whole chickweed family are always true seeds, as a rule kidney-shaped, and having a skin with minute but well-marked pig-like projections. When cut into, *snow-white starchy contents* are conspicuous next the concavity, and a curved embryo is seen lying upon the white substance within the convex part of the skin. Many members of this family occur as impurities in farm seeds, the most frequent being the poisonous-seeded corncockle in wheat; campions, and common chickweed (*Stellaria media*), in grasses and clovers.

8. Corncockle.

The seed is specially large for this family—namely, $3 \times 2\frac{1}{2}$ mm. in typical seeds. The seed-skin is very dark, almost black, with distinct projections on its surface. The kidney-shape is irregular, for one end of the seed is narrow and beak-like instead of rounded off.

9. White Campion (fig. 347).

The seeds are grey and kidney-shaped, $2 \times 1\frac{1}{2}$ mm. The surface of the seed-skin has distinct projections arranged in curved lines.

Fig. 347.—White campion (*Lychnis vespertina*).

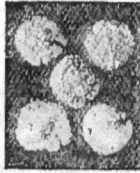

Fig. 348.—Chickweed (*Stellaria media*).

10. Common Chickweed (fig. 348).

This is a small-seeded form, 1 mm. × 1 mm. These dimensions show that the kidney form is very much rounded, the length and breadth being almost equal. The skin is yellowish-brown, becoming darker with age, and has the usual pig-like projections.

LEGUMINOUS SEEDS.

The seed of a leguminous plant is, in all the cases here dealt with, a true seed, composed of a seed-skin filled up with an embryo plant. For purposes of distinction, it is important to notice that the embryo is always doubled up in such a way that its radicle—the part whose tip becomes the root—lies along the edge of the cotyledons, the first leaves of the

plant. Thus is caused a ridge, very visible on the outside of the seed. The length, the thickness, the spread, and other characters of this external ridge correspond to the peculiarities of the radicle within, and such characters readily mark off the seed of one leguminous plant from that of another.

In the case of clover-seeds, and, indeed, in leguminous seeds generally, there is often a lack of swelling power, due to a peculiarity of the seed-coat. The outer cells of the coat are waterproof, from a glossy impregnation with silica and lime compounds. Such seeds are termed *hard*. All that is necessary to remedy this defect, and to allow the water to enter freely, is a mere scratch on the surface of the seed-skin. Accordingly, to improve the germinating power of commercial clover-seeds, they are sometimes shaken up in bags with sharp sand, and sometimes rolled about in specially constructed cylinders with a rough lining.

The clovers here considered are red, alsike, white, and yellow suckling. The other forage seeds noticed are lucerne, trefoil, bird's-foot trefoil, and kidney vetch.

11. Red Clover (fig. 349).

The seed varies in length from $1\frac{1}{2}$ to $2\frac{1}{4}$ mm. The shape is oval, with a stout, blunt-pointed, projecting ridge a little more than half the length of the

Fig. 349.—Perennial red clover (*Trifolium pratense perenne*).

seed. The glossy surface of the skin is coloured rich purple at the broad end with the ridge, and fleshy or yellowish at the narrower end. Unripe seeds are

The seeds illustrated on this page are magnified six diameters.

yellow all over, or of a greenish tinge. When the seed is old it loses its gloss, and the purple changes to a redder tint.

Sometimes the sample contains red clover fruit, composed of a thin seedcase with a rod-like projection from the apex, and one seed within. This seedcase readily opens in a transverse direction, and then the characteristic red clover-seed within is laid bare.

Common impurities are—

a. *True seeds.*—Campion (No. 9), ribgrass (No. 19), the commonest of all, and cut-leaved geranium.
b. *Three - sided fruits of the Dock family.* — Dock (No. 24) and sheep's sorrel (No. 25).
c. *Composite fruits.*—Ox-eye daisy (No. 28), field chamomile (No. 29), and nipplewort (No. 34).

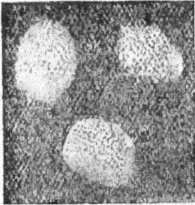

Fig. 350.—Cut-leaved geranium
(*Geranium dissectum*).

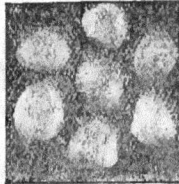

Fig. 351.—Alsike clover
(*Trifolium hybridum*).

d. *Cut-leaved Geranium.*—The broad oval seed 2 × 1½ mm. is slightly flattened at the broad end, and of a purplish-brown colour. The most characteristic feature is the *honeycomb appearance of the skin*, readily seen with the naked eye (fig. 350).
e. *Seeds of Parasites.*—Clover-dodder (No. 64) and clover-rape (No. 65).

12. *Alsike Clover* (fig. 351).

The seed is small—from 1 to 1¼ mm. long—with a glossy, mottled, dark-olive skin. The ridge on the surface is not quite so long as the body of the seed, and at its apex there is a gap appearing as a slight notch in the seed. Immature seeds are yellowish-green.

The impurities met with are: true seeds of soft geranium and field pansy; composite fruits: yarrow (No. 27),

nutlets of self-heal (No. 40); grass seeds: timothy (No. 47).

The seed of soft geranium (fig. 352) is oblong, 2 × 1 mm., reddish-brown,

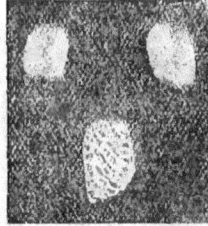

Fig. 352.—Dove's-foot geranium
(*Geranium molle*).

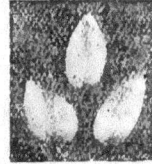

Fig. 353.—Pansy
(*Viola versicolor arvensis*).

and *smooth* on the surface, without the depression so marked in the cut-leaved species.

The seed of field pansy (fig. 353) is egg-shaped, 2 × 1 mm., glossy and yellow, or brown, with *a special appendage at the narrow end.* This appendage is tiny, shrivelled, and lighter in colour than the body of the seed.

13. *White Clover* (fig. 354).

The seed is pale-yellow, and practically of the same size as alsike—1 to 1¼ mm. The shape is like a heart, for the ridge is thick, *of the same length* as the body of the seed, and gapes away at the base. Immature seeds are greenish. Old seed loses its gloss, darkens, and may become quite red.

The *common impurities* are the same as in alsike and red clover-seeds. Yellow

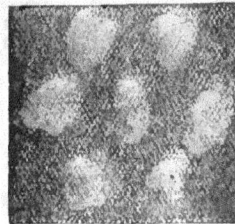

Fig. 354.—White clover
(*Trifolium repens*).

suckling clover (No. 14) is sometimes used as an adulterant, and is easily detected by its oval (not heart) shape.

The seeds illustrated on this page are magnified six diameters.

14. *Yellow Suckling Clover* (fig. 355).

The seed is specially glossy and of a warm yellow colour, somewhat like that of a small white clover, but oval rather

Fig. 355.—Suckling clover
(*Trifolium minus*).

than heart-shaped; for here the ridge formed by the radicle is scarcely noticeable.

15. *Lucerne* (fig. 356).

In typical examples the shape is kidney-like, and the length varies from 2 to 3 mm. The seed-skin is dull, destitute of gloss, and of a pale-yellow colour, inclining to brown when the seeds are old. The radicle of the embryo is a little more than half the length of the cotyledons, and the cotyledons have a characteristic curve to one side, so that the ridged part of the seed is in one plane and the remainder of the seed in another.

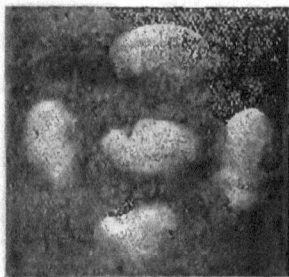

Fig. 356.—Lucerne or alfalfa
(*Medicago sativa*).

A common *impurity* is trefoil-seed (No. 16); a common *parasite* is dodder (No. 64); and an adulterant sometimes used is white melilot, otherwise called "Bokhara clover."

This adulterant, though closely resembling lucerne-seed, is easy to detect, because it contains an odoriferous principle—the same which gives the character-

istic odour to new-mown hay—namely, *cumarin*. By this characteristic odour an adulterated sample is detected.

16. *Trefoil* (fig. 357).

The seed is greenish-yellow, shaped like a bean, and smaller than lucerne—2½ mm. long and 1 mm. broad. The most useful feature for identifying this cheap seed is the very noticeable *projecting point of the radicle*, which is bent out from the middle of the concavity at right angles to the body of the seed.

In trefoil samples a small flattish black fruit, with raised curved lines on the

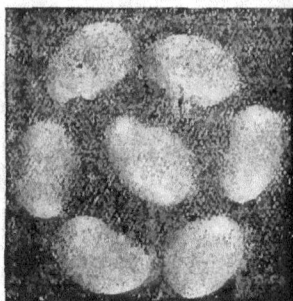

Fig. 357.—Trefoil or yellow clover
(*Medicago lupulina*).

surface, is often seen lying in the toothed calyx of its flower. This is the trefoil fruit, as we easily know, for on opening the black seed-case, a kidney-shaped seed is seen with the characteristic projecting point of trefoil.

17. *Bird's-foot Trefoil.*

The seed of the useful trefoil is almost globular, 1¾ mm. long, 1½ mm. broad, and 1 mm. thick. The colour is dark-brown, mottled with white spots. Marsh bird's-foot trefoil—the useless species—is often substituted. Its seeds, however, are only half as long, at most 1 mm., and the colour is greenish.

18. *Kidney Vetch* (fig. 358).

The seed is oval, 2½ mm. long, and easily distinguished by the colour—one end of the oval being green and the other yellow. The yellow end becomes brown with age.

The seeds illustrated on this page are magnified six diameters.

THE TWO-FACED RIB-GRASS SEEDS.

These seeds are either brown or at times quite black. When wetted the whole skin, like that of linseed, swells up into mucilage, and in the mouth this mucilag-

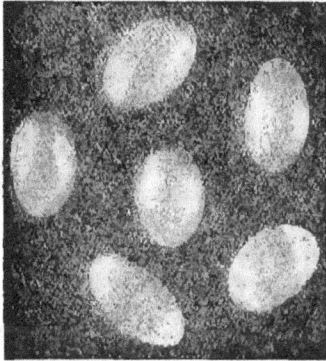

Fig. 358.—Kidney vetch (*Anthyllis vulneraria*).

inous character becomes immediately revealed. There are always two different faces on these seeds — an inner flat or grooved, and an outer convex. Tried by the knife the seed cuts like a piece of horn.

19. *Rib-grass (Plantago lanceolata)* (fig. 359).

The seed is smooth, brown, and of an oval outline, $2\frac{1}{2} \times 1\frac{1}{4}$ mm., like a miniature date-stone with a broad instead of a narrow furrow on the inner face. Sometimes the groove is hidden by a fragment from the seed-case, which remains ad-

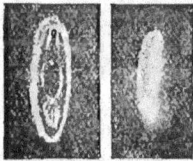

Fig. 359.—Rib-grass (*Plantago lanceolata*).

Fig 360.—Way-bread (*Plantago major*).

herent to the inner face. This is one of the commonest impurities in clovers, especially in red clover.

20. *Greater Rib-grass or Way-bread* (fig. 360).

The seed is rough and black, and the inner face is flat without the groove.

CLASS II.—FRUITS.

Fruit is always something more than a seed; there is a closed seed-case in addition. Take a common hazel-nut for example: here, the kernel is the seed, and the hard shell the seed-case. These two parts are always constituents of a fruit. It is easy, then, to identify these fruits: cut the fruit across, and the seed-case is seen enclosing a seed in its cavity.

THE FLAT BEAKED "SEEDS" OF BUTTERCUPS.

One buttercup flower produces many "seeds," but if we cut into one of these so-called seeds we find the two constituents which mark a fruit—namely, the seed-case and the seed within.

21. *Upright Buttercup* (fig. 361).

The "seed" is flat and lop-sided shaped, somewhat like the figure 6. The rounded

Fig. 361.—Upright crowfoot (*Ranunculus acris*).

Fig. 362.—Creeping crowfoot (*Ranunculus repens*).

part of the 6 corresponds to the body of the seed, and the stroke to the apical beak. The seed-case is brown and of leathery texture, about $3\frac{1}{2} \times 2$ mm.

22. *Creeping Buttercup* (fig. 362).

This species is larger than No. 21, and its beak is longer as well as more curved. Among the seeds of corn crops a common impurity is the fruit of the *corn buttercup*, readily distinguished from ordinary buttercups by its *spiny surface*. Compare the flat, black, symmetrical (not lop-sided) beaked "seeds" of red-shank, No. 26a.

The seeds illustrated on this page are magnified six diameters.

THE THREE-FACED "SEEDS" OF THE DOCK FAMILY.

In this family are included cultivated buck wheats, often called "grain crops," and many pestilent weeds, such as docks and sorrels. The invariable rule in the family is one flower, one fruit, never many as in buttercup. The fruit has a character which renders identification easy; it has three similar faces, or, in other words, the section is triangular. When cut with a knife the seed-case is distinguished, and the seed within the case is seen to be filled with snow-white starchy meal surrounding the embryo plant.

23. *Buckwheat* (fig. 363).

The three-faced fruit is of egg-shaped outline, pointed at both ends, 5 × 3 mm.,

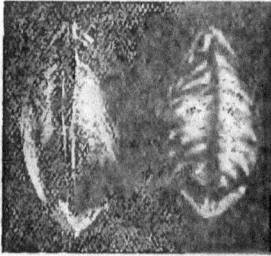

Fig. 363.—Buckwheat
(*Polygonum Fagopyrum*).

and girt at the base with the persistent calyx of the flower which made it. The colour is a glossy dark-brown with a greyish tinge. Old seeds lose their gloss, and split along the three edges.

In the Tartarian species of buckwheat the edges are not sharp and entire, but rounded off and slightly indented.

24. *Curled Dock.*

The three-sided fruit is like a small buckwheat, 2¼ × 1¼ mm., and of a glossy reddish-brown tint.

25. *Sheep's Sorrel* (fig. 364).

The three-sided fruit is like a small dock, *only* 1 *mm. long*, and glossy reddish-brown. This, like rib-grass, is one of the commonest impurities in clover-seeds.

THE FLAT AND BEAKED "SEEDS" OF THE DOCK FAMILY.

In certain members of the Dock family the "seed" is not three-faced, but a flat symmetrical disc with a beak projection from its apex. These

Fig. 364.—Sheep's sorrel
(*Rumex acetosella*).

seeds are the common impurities called persicarias or red-shanks. When cut across they resist the knife like tough horn, and are seen to be destitute of white meal.

26a. *Spotted Persicaria or Red-shank* (fig. 365).

The fruit is a round, glossy, black disc, 1½ to 2½ mm. in diameter, with a sharp point projecting from its apex. This disc is two-faced—the one face flat, the other slightly convex, and where

Fig. 365.—Red-shank
(*Polygonum Persicaria*)

the two faces blend they form a sharp edge round the disc. Red-shank is a common impurity in clover-seeds.

26b. *Pale-flowered Persicaria* (*Polygonum Lapathifolium*).

The seed is distinguished from that of spotted persicaria by the two faces being concave, and by the thick rounded (not sharp-edged) margin of the disc.

The seed illustrated in fig. 363 is magnified four-and-a-half diameters; the seeds in figs. 364 and 365 six diameters.

"SEEDS" OF COMPOSITES.

These are never true seeds, but always fruits, each composed of a seed completely enclosed in a hard, dry seed-case. The invariable rule is one flower, one fruit. It is specially noteworthy that this fruit is never made inside the flower, but always on the outside from what is technically called an inferior ovary. A composite "seed" is accordingly never found seated in a persistent calyx. The sign of external origin is found at the apex of the seed-case, where special scars or persistent relics of the flower-leaves are usually to be recognised.

The forage plants included among composites are yarrow and chicory, and the common weeds are ox-eye daisy, field chamomile, knapweed, field thistle, dandelion, and nipplewort.

Fig. 366.—Yarrow, or milfoil (*Achillea Millefolium*).

27. *Yarrow* (fig. 366).

The "seed" is flat and blunt at the apex, and tapers off gently to the rounded base. The colour is silvery grey, with narrow white wings like expansions along the two margins. The dimensions are 2 × 1 mm.

28. *Ox-eye Daisy* (fig. 367).

The "seed" is four-sided, and *strongly ribbed*. The apex is blunt, and tapers off gradually to the blunt base. The dimensions are 2 × 1 mm. Corresponding to the white florets of the daisy head there are the "seeds" with a membranous crown, and to the yellow flowers of the centre those other seeds destitute

of crown. These "seeds" occur as impurities in clovers.

29. *Field Chamomile* (fig. 368).

The "seed" closely resembles that of ox-eye. It is often bent, and none of

Fig. 367.—Ox-eye daisy (*Chrysanthemum leu-canthemum*).

Fig. 368.—Field chamomile (*Anthemis arvensis*).

the seeds have the membranous crown. These "seeds" also occur as impurities in clovers and cereals.

30. *Knapweed* (fig. 369).

The "seed" is stout, and the broad, blunt apex bears a crown composed of numerous fair hairs, inclining to red. The seed-case is quite smooth on the surface, and shines like white porcelain tinged with violet. At the side of the narrow base there is a well-marked scar, as if a morsel had been bitten out there. The dimensions are: for the seed-case, 3 × 1½ mm.; for the hairs, 1½ mm.

Fig. 369.—Knapweed (*Centaurea nigra*).

Fig. 370.—Field thistle (*Carduus arvensis*).

31. *Field Thistle* (fig. 370).

The "seed" is somewhat three-sided, and of egg-shaped outline, 3 × 1 mm. The glossy seed-case is whitish, and its broad apex bears a knob-like projection, surrounded by a ring-like swelling.

The seeds illustrated on this page are magnified six diameters.

32. *Chicory* (fig. 371).

The fruit is quadrangular, $2\frac{1}{2} \times 1\frac{1}{2}$ mm., with the broad apical end quite

Fig. 371.—Chicory (*Cichorium intybus*).

blunt, and *crowned with a multitude of minute blunt scales.* The colour is mottled fawn.

33. *Dandelion* (fig. 372).

The leather-like seed-case is 2 mm. long and $\frac{1}{2}$ mm. broad. Its surface is ribbed and spiny towards the apex, which is prolonged as a cylindrical beak, whose length depends upon where the break across occurred.

Fig. 372.—Dandelion   Fig. 373.—Nipplewort
(*Taraxacum officinale*).   (*Lapsana communis*).

34. *Nipplewort* (fig. 373).

The "seed" is flat and club-shaped, 4×1 mm. Under a lens the surface shows very fine longitudinal striations.

CLASS III.—MERICARPS.

The word is the Greek for *part of a fruit*. It signifies a part split off from a fruit, and is constructed of a seed-case enclosing one seed. In runch, for example, the fruit is jointed like a short string of beads, and each joint when it detaches is rightly called a mericarp, for it is a part

of a fruit, and it is constructed of a closed case containing one seed. (See Runch, No. 7.) In the mericarps dealt with here the fruit splits longitudinally into two pieces : each piece is a mericarp with two faces, the inner face being formed by splitting.

UMBELLIFEROUS "SEEDS."

The seed here is formed from a fruit which splits not as in runch transversely into many pieces, but lengthways into only two pieces or mericarps. Each mericarp is composed of a seed within a closed seed-case, with an inner surface (at which the split occurred) and an outer surface. A striking peculiarity of the seed-case is the presence of *canals containing coloured oil.* These appear on the outside as coloured lines. Among the plants included here are parsnip, parsley, and carrot.

35. *Parsnip.*

The seed is a *flat disc* about 7 mm. long and 5 mm. broad, with very low ridges on its outer face. The colour is like straw, with dark, well-marked oil-canals between the ridges.

Fig. 374.—Parsley (*Petroselinum sativum*).

36. *Parsley* (fig. 374).

The greenish-grey "seed" when lying on its flat inner split face shows an egg-shaped outline about $2\frac{1}{2} \times \frac{3}{4}$ mm., surmounted by a conical swelling at the

The seed illustrated in fig. 374 is magnified four-and-a-half diameters ; the seeds in figs. 372 and 373 six diameters ; the seed in 371 ten diameters.

narrow end. Three well-marked light-coloured ribs are seen on the outer face standing out clearly against the dark background between the ribs.

37. Carrot.

The grey seed has rows of *prickly ridges* on its surface, and if fresh, gives off a faint odour of aniseed. The length is about 2½ mm.

"SEEDS" OF THE CLEAVERS FAMILY.

The "seed" here is again a mericarp formed from a fruit which splits longitudinally into two mericarps. Each mericarp is constructed of a seed within a closed seed-case, which shows an inner surface (formed by splitting) and an outer surface. This seed-case, however, is destitute of ridges and oil-canals. The true seed is of horny texture, and when cut across, shows a section of the embryo embedded in the horny material. The "seeds" of two members of this family—namely, cleavers and sherardia —are ubiquitous impurities in seeds, the cleavers in corn and the sherardia in clovers.

38. Sherardia (fig. 375).

The seed-case, 3 × 1½ mm., is flat on the inner face, formed by splitting, but convex on the outer surface. The colour

Fig. 375.—Field madder (*Sherardia arvensis*).

is buff, and marked all over with white dots, which under the lens turn out to be little stiff hairs. The fruit, which splits into the two mericarps, was originally outside the flower, and the calyx leaves of this flower, five in number, persist at the apex as a crown of teeth. The five leaves of this calyx are shared

between the two parts into which the fruit splits, thus giving 2½ leaves per part. Hence each "seed" of sherardia is readily recognised, even with the naked eye, by this *crown of 2½ leaves which looks like three sharp teeth.* Sherardia "seed" frequently occurs as an impurity in clover-seeds, usually unsplit and crowned with the five teeth.

39. Cleavers.

The "seed" is purplish, and shaped like a very much curved kidney, the concavity representing the inner surface formed by splitting. The length varies up to 3 mm. The convex surface of the seed-case is, in typical examples, covered over *with hooked bristles.* The seed of cleavers is a common impurity in corn.

CLASS IV.—NUTLETS.

These occur in two distinct families of plants, represented by forget-me-not or scorpion-grass (*Myosotis*) and self-heal (*Prunella*). Each flower forms a cluster of four nutlets pressed close together in a ring thus. The nutlet accordingly shows three faces, two flat and an outer convex. The name is appropriate, for each nutlet is a nut in miniature, with a *thick hard shell,* enclosing one seed.

40. Self-heal (fig. 376).

The nutlet is egg-shaped, 2 × 1 mm., and of a glossy dark-brown colour. Its

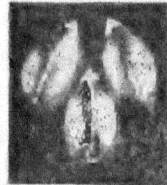

Fig. 376.—Self-heal (*Prunella vulgaris*).

narrow base bears a characteristic *white triangular flap.*

The "seeds" are remarkable for the following peculiarity which distinguishes

The seeds illustrated on this page are magnified six diameters.

them from all others. When in water for a minute, *four clear ridges of mucilage appear* and project, one along the middle line of the outer face, and three others, one along each of the three edges. This peculiar formation depends upon the presence of special cells located along the lines indicated and capable of swelling strongly in water.

41. *Forget-me-not or Scorpion-grass* [*Myosotis versicolor*] (fig. 377).

The nutlets here are black and egg-shaped, about half the size of the nutlets

Fig. 377.—Scorpion-grass
(*Myosotis versicolor*).

of self-heal. The shell of this nutlet has no mucilage.

CLASS V.—MULTIPLE FRUITS.

A multiple fruit is formed from a whole cluster of flowers, each of which becomes fruit, but the various fruits of the cluster are grown together to form one mass. The rule is, many flowers, one fruit. A large example is the pine-apple : each rhomboidal area seen on its surface corresponds to one flower of the cluster. In the same way each chamber of a mangel "seed" represents a mangel flower.

42. *Mangel.*

The "seed" is a multiple fruit formed of three or more very hard and thick-walled fruits, grown together, and each one of these may contain a true seed. When the "seed" is cut across it is seen to be a chambered body, and the snow-white meal contained in the true seed now becomes very noticeable. Here we have a case where the germination of the "seeds" may be 300 per cent. This is because each so-called seed is a multiple fruit containing three seeds, capable of producing three seedling plants.

CLASS VI.—GRASS "SEEDS."

The "seed" of a grass never is a true seed. It may be the grain fruit, as in shelled timothy; the detached part of a spikelet (fig. 384), as in rye-grass; the whole spikelet, wanting the chaff (glumes), as in the "double seed" of tall oat; or even the whole spikelet, as in meadow foxtail. In most cases, however, the "seed" is a detached part of the spikelet; and such a part is distinguished by the presence of a *stalk*, which lies next the *upper* valve of the husk. The *lower* valve of the husk is on the other face of the "seed," away from the stalk. For purposes of distinction, it is important to pay special attention to this lower valve of the husk. Sometimes the "seed" can lie on its back, and then, of course, the lower valve of the husk is rounded off thus ⌣. In certain other cases, where the seed lies on its side, and cannot lie on its back, this lower valve is ⋁-shaped. The seeds that lie on the side are : cocksfoot, all the *poas*, and purple molinia. In still other cases an awn springs sometimes from the back and sometimes from the base of this lower valve, as in golden oat, aira.

Grass "seeds" are accordingly—

a Complete spikelets.

b The whole of the spikelet except the chaff (glumes).

c Detached portions of the spikelet with rounded back.

d The same with ⋁-shaped back (keeled).

e The same with awn from back or base.

In exceptional cases, such as shelled timothy and shelled cocksfoot, the "seed" is the grain fruit.

a. THE "SEED" IS THE WHOLE SPIKELET.

This is the case in three grasses: meadow foxtail, slender foxtail, and Yorkshire fog, or unshelled holcus as it is sometimes called.

The seed illustrated in fig. 377 is magnified six diameters.

43. *Meadow Foxtail* (fig. 378).

Each "seed" of foxtail is a whole spikelet containing one grain. The length is 5 mm., and the breadth at the widest part—midway between apex and base—is 2½ mm. The spikelet is quite flat, and is covered all over with velvety hairs. Along each margin there is a conspicuous fringe of these hairs. At the apex there projects from the inside of the spikelet a long but fine awn. The colour of the "seed" is greyish-yellow, if ripe; if unripe, quite pale.

Fig. 378.—Meadow foxtail grass
(*Alopecurus pratensis*).

The adulterants used are spikelets of slender foxtail (No. 44) and Yorkshire fog (No. 45).

The impurities are portions of grass spikelets, usually tufted hair-grass (No. 63) and poas (No. 59), entangled in the hair fringes of the foxtail "seed."

44. *Slender Foxtail* (fig. 379).

This "seed," as compared with meadow foxtail, is longer (over 6 mm.), duller in colour, bald, and *destitute of the marginal fringes of hair*.

45*a*. *Yorkshire Fog* (fig. 380).

This spikelet, sometimes called "unshelled holcus" (on the left of fig. 380), is only 4 mm. long, and wants the projecting awn. When the spikelet is opened up it is found to contain two flowers (instead of one, as in foxtail), with husks which gleam like polished silver. The upper of the two flowers

is remarkable and very characteristic, for just behind the point of its husk it has a short awn, which at maturity is bent inwards *like a fish-hook*. The lower

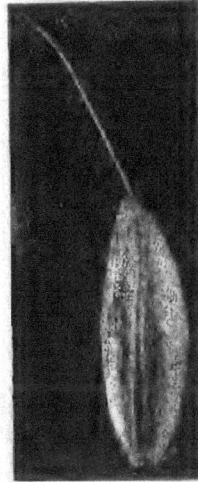

Fig. 379.—Black grass (*Alopecurus agrestis*).

flower with its husk is called "shelled holcus" when detached from the rest of the spikelet.

45*b*. *Yorkshire Fog* (*Holcus mollis*).

The creeping species of Yorkshire fog (*Holcus mollis*) is sometimes used as an adulterant for meadow foxtail. The spikelet of this species is like that of Yorkshire fog, but is immediately dis-

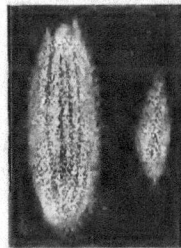

Fig. 380.—Yorkshire fog (*Holcus lanatus*).

tinguished by the presence of a long awn projecting from the apex. The awn here, instead of being bent in and hidden by the chaff (glumes), projects far beyond the chaff (glumes).

The seeds illustrated on this page are magnified six diameters.

45c. *Shelled Holcus.*

This "seed" is not the spikelet of Yorkshire fog, but the detached part of the spikelet (on the right of fig. 380) enclosing the grain in a husk which is much longer than the grain within. It is very like timothy seed, but *longer,* over 2 mm., and has a *stalk* on the flat inner face. To distinguish shelled holcus from timothy requires careful examination with a good pocket lens.

b. THE "SEED" IS THE WHOLE SPIKELET EXCEPT THE CHAFF (GLUMES).

This is the case with the two cultivated grasses, tall oat-grass or French rye-grass and timothy.

46. *Tall Oat-grass or French Rye-grass* (fig. 381).

The "seed" is 8 or 9 mm. long, and is called a "double seed," for it is composed of two flowers. These flowers are differently constructed. The lower flower is barren, and from the back of its husk there springs a long awn (12 mm.), coarse, dark, and twisted at the base, but fine, light coloured, and bent at the apex. The upper flower is fertile, containing the grain, and its awn is fine, short, and quite straight. The distinctive feature is the presence of *two awns*—one long and conspicuous, the other quite short and easily overlooked. An adulterant sometimes used is *rye brome* (No. 57). This "seed" is single, a mere part detached from a spikelet, and not a "double seed" with two awns.

47. *Timothy* (fig. 382).

The "seed" is the whole contents of the spikelet minus the chaff; but unlike tall oat, there is in this case only one flower to make the "seed." Timothy "seed" is composed of a grain enclosed in a thin, transparent, silvery husk. The dimensions are 1½ mm. to 2 mm. long and ¾ mm. broad. This husk readily allows the grain to escape, so that at times timothy seed is the mere grain, bare of husk, called *shelled timothy.* This grain is golden and easily distinguished; for, unlike the

grain of most other grasses, it is almost globular, without the flat face and groove familiar to us in the large wheat grain.

The common impurities are: (1) True seeds of rib-grass (No. 19); (2)

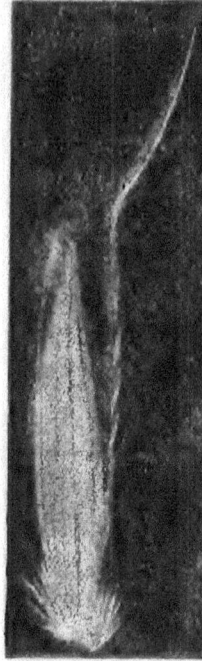

Fig. 381.—Tall oat-grass (*Avena elatior*).

fruits of sheep's sorrel (No. 25); (3) nutlets of self-heal (No. 40); (4) *shelled holcus* (No. 45c); (5) detached por-

Fig. 382.—Catstail or timothy (*Phleum pratense*).

tions of grass spikelets with awn from the base of the lower valve of the husk; tufted hair-grass (No. 63).

Seed of the parasite dodder (No. 64) may occur in timothy seed sieved off from clover. Ergot (No. 66) frequently occurs.

The seed illustrated in fig. 381 is magnified six diameters ; the seed in fig. 382 ten diameters.

c. THE "SEED" IS A PORTION OF THE SPIKELET WITH STALK ON UPPER FACE; BACK ROUNDED.

The "seed" in this group of grasses is composed of a grain enclosed within a two-valved husk. The lower valve of

Fig. 383.—Italian rye-grass (*Lolium italicum*). Fig. 384.—Perennial rye-grass (*Lolium perenne*).

the husk is rounded so that the "seed" can lie on its back. The valuable seeds are: rye-grasses, fescues, and crested dogstail; the worthless weed seeds: fescue hair-grass and bromes. Shelled holcus, which belongs here, is described under No. 45c.

48. *Italian Rye-grass* (fig. 383).

The "seed" closely resembles that of perennial rye-grass, but the presence of an apical beard or awn readily distinguishes. The common impurities are: (1) True seeds, rib-grass (No. 19); (2) fruits, composed of a seed-case enclosing one seed, buttercups (Nos. 21 and 22), docks, and sorrels (Nos. 24 and 25); (3) composite fruits, ox-eye daisy (No. 28), and nipplewort (No. 34); (4) detached portions of grass spikelets, the fescue hairgrass (*Festuca sciuroides* or *Festuca myurus*) (No. 55); (5) complete grass spikelets, Yorkshire fog (unshelled holcus) (No. 45*a*).

49. *Perennial Rye-grass* (fig. 384).

The "seed" is a detached portion of the spikelet, consisting of a grain enclosed in a two-valved husk. The length varies from 6 mm. to 7 mm., and the longer-seeded sorts, with the husk considerably longer than the enclosed grain, usually come from *annual* rye-grasses. There is no beard or awn on the lower valve of the husk as in Italian. The stalk is *flat, broader at the apex than at the base*, and has no apical flange.

The common impurities are: (1) True seeds, rib-grass (No. 19); (2) fruits, each composed of a case enclosing a seed, buttercups (Nos. 21 and 22); (3) detached portions of grass spikelets with rounded back, goose or brome grass (No. 56); (4) grass spikelets complete, Yorkshire fog (unshelled holcus) (No. 45*a*).

50. *Meadow Fescue* (fig. 385).

The "seed" is a detached portion of the spikelet. The husk is 6 mm. long, and ends in a broad blunt point, very frequently broken. The stalk is a

Fig. 385.—Meadow fescue-grass (*Festuca pratensis*). Fig. 386.—Tall fescue-grass (*Festuca elatior*).

narrow cylinder, 1¼ mm. long, with a flange at its apex.

The common impurities are: soft brome (No. 56) and rye brome (No. 57). An adulterant "seed" is perennial rye-grass (No. 49).

51. *Rhenish Tall Fescue* (fig. 386).

The "seed" closely resembles that of meadow fescue. It differs, however, in the following points: (1) *The colour is*

The seeds illustrated on this page are magnified six diameters.

darker; (2) the length is slightly greater; and (3) the whole "seed" is narrower, especially at the point, which is never broken. Colour alone suffices for distinction.

A common impurity is cocksfoot (No. 58), and a common substitute New Zealand tall fescue (No. 52).

52. *New Zealand Tall Fescue.*

The "seed" differs from that of Rhenish tall fescue in the following points : (1) *The colour is light,* as in cocksfoot; (2) the length is over 7 mm.; and (3) the point is hard and sharp, ending in a very short awn. Colour is the best distinction.

53a. *Sheep's Fescue* (fig. 387).

The "seed" is narrow, 4 *or* 5 *mm. long,* and only sometimes tipped with a very short awn (1 mm.) The husk is firm, and of a straw-yellow colour. The impurities are wavy hair-grass (No. 62) and purple molinia (No. 60), which lies on the side. Hard fescue is sometimes substituted.

Fig. 387.—Fine-leaved fescue (*Festuca ovina tenuifolia*).

Fig. 388.—Hard fescue-grass (*Festuca duriuscula*).

53b. *Hard Fescue* (fig. 388).

This is a longer "seed," its body 6 or 7 mm., and *its awn* 5 *mm.*

54. *Crested Dogstail* (fig. 389).

The "seed" is 4 mm. long, or including the awn, 4½ mm., and 1 mm. broad. The colour is bright yellow at the narrow end, shading to reddish-brown at the broad end. The bristly point of the

lower valve of the husk is bent slightly to the side, and ends in a rigid sharp point. The stalk is a very short cylinder, terminating in a disc-like flange.

Fig. 389.—Crested dogstail-grass (*Cynosurus cristatus*).

A common impurity is Yorkshire fog shelled (shelled holcus) (No. 45c).

55. *Fescue Hair-grass or Hair-grass [Festuca myurus]* (fig. 390).

The "seed" is excessively fine and dark, like a grey or brown hair, 15 mm. long. The base of this hair is slightly swollen for a length of 5 mm. The swollen base is the body of the seed, and the apical portion the awn. This occurs as an impurity in rye-grasses.

56. *Soft Brome* (fig. 391).

The "seed" is awned broad and flat, with each of its two edges forming an angle instead of a curve at the broadest part near the apex. The stalk is slightly conical, and broadest at the apex. The upper valve of the husk next the stalk is decidedly shorter than the other. The dimensions are 9 mm. long, and including the awn about 17 mm.

57. *Rye Brome* (fig. 392).

This is a narrow cylindrical seed about 7 mm. long, and including the short awn some mm. more. The cylindrical appearance and narrowness, as compared with soft brome, is due to the strong inrolling of the lower valve of the husk towards the upper face of the "seed."

The seeds illustrated in figs. 387 *and* 389 *are magnified ten diameters; the seed in fig.* 388 *six diameters.*

d. THE "SEED" IS A PORTION OF THE SPIKELET: BACK ∨-SHAPED.

This group of grass "seeds" includes cocksfoot and *poas* (meadow grasses), as well as the adulterant purple molinia or flying bent.

58. *Cocksfoot* (fig. 393).

The "seed" is a detached portion of the spikelet, consisting of a husk enclos-

Fig. 390.—Squirrel-tail fescue-grass (*Festuca sciuroides*).

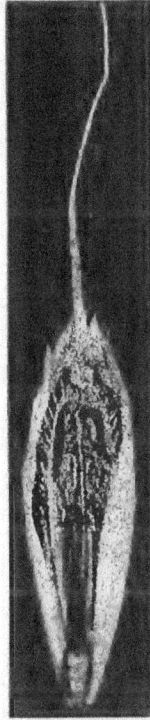

Fig. 391. — Soft brome-grass (*Bromus mollis*).

like that of crested dogstail, has a slight *bend to the side.*

Common impurities are: dock fruits (No. 24), grass spikelets of Yorkshire

Fig. 392.—Rye seeded brome-grass (*Bromus secalinus*).

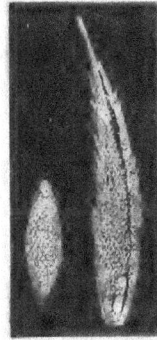

Fig. 393.—Cocksfoot-grass (*Dactylis glomerata*).

fog (No. 45*a*), and detached portions of grass spikelets rounded on the back—soft brome (No. 56).

Adulterants sometimes used are: detached portions of grass spikelets with rounded back, such as perennial rye-grass (No. 49) and hard fescue (No. 53*a*).

Fig. 394.—Smooth-stalked meadow grass (*Poa pratensis*).

Fig. 395.—Rough-stalked meadow grass (*Poa trivialis*).

59. *Poas or Meadow Grasses* (figs. 394 and 395).

The "seed" is a detached portion of the spikelet, consisting of a grain enclosed

ing a grain. Sometimes, in *immature* samples, the detachment is incomplete; then we have "double, often treble, seeds." These "double seeds," unlike those of tall oat, are not the whole contents of the spikelet, but only a detached portion. The length is 5 mm., or including the short awn 6 or 7 mm. The colour is very light, inclining to golden white. The point of the husk,

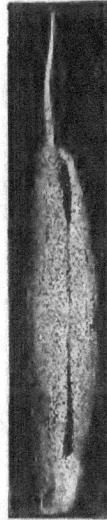

The seeds illustrated in figs. 390, 391, 392, and 393 are magnified six diameters; the seeds in figs. 394 and 395 ten diameters.

in a two-valved husk. It cannot lie on its back, for the lower valve of the husk is V-shaped. This husk is never transparent, and never has an awn. The "seeds" are always small, ranging in length from 2½ to 3½ mm. For distinguishing the various species microscopic examination is necessary, and so full consideration of the distinctive characters is unnecessary here.

By the aid of a good pocket lens, however, the "seed" of the *rough-stalked meadow grass* may be recognised if the *end "seed"* of the spikelet is selected for examination. The stalk of this seed is very slender, and has a *globular rudiment* at its end, whereas the corresponding seed of *smooth-stalked meadow grass* has a stalk twice as thick, and the *rudiment is long and pointed*. An adulterant of *poas* sometimes met with is purple molinia, a much larger seed (No. 60). Aira seeds (Nos. 62 and 63) are also used, but the presence of the basal awns at once detects such substitutes.

60. *Purple Molinia.*

The "seed" is a detached portion of the spikelet. The points of distinction from dogstail seed, for which this is sometimes substituted, are: (1) The molinia seed lies on its side; (2) it is often longer by about 1 mm.; (3) the colour is darker brown, with a purplish tip; (4) the stalk is comparatively long, from 1½ to 2 mm., and ends not in a flange, but *in a cleft knob*; (5) the husk often gapes at the point, like the open beak of a bird.

This gape is very characteristic, and is due to the excessive narrowness of the lower valve of the husk, which cannot surround and tuck in the point of the upper valve.

e. THE "SEED" IS A PORTION OF THE SPIKELET WITH THE BACK OR BASE AWNED.

The valuable grass here is golden oat, with the awn from the back; the worthless weeds, the *airas*—namely, wavy hair-grass and tufted hair-grass, with the awn from the base.

61. *Golden Oat* (fig. 396).

The "seed" is a detached portion of the spikelet, consisting of grain and husk. The length is about 4½ mm. The husk is very thin, and coloured like

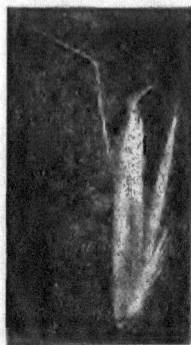

Fig. 396.—Golden oat-grass (*Avena flavescens*).

gold. From its back, *midway between apex and base*, there springs forth a bent awn. The stalk is very characteristic, and looks like a little white feather, for it is flanged with two rows of white hairs.

A common adulterant is wavy hair-grass "seed" (No. 62).

62. *Wavy Hair-grass* (fig. 397).

The "seed" is 5 mm. long, and most readily distinguished from that of golden oat, for which it is a common substitute,

Fig. 397.—Wavy mountain hair-grass (*Aira flexuosa*).

Fig. 398.—Tufted hair-grass (*Aira cæspitosa*).

(1) by the darker and browner colour; (2) by the straight awn springing from the *base*, not the back of the husk; and (3) by a conspicuous basal tuft of white hairs.

The seeds illustrated in figs. 396 and 397 are magnified six diameters; the seed in fig. 398 ten diameters.

63. *Tufted Hair-grass* (fig. 398).

The "seed" is a detached portion of the spikelet constructed of a thin, silvery, transparent husk containing a grain. The "seed" is only 2 mm. long, and very narrow. Round the base there is a conspicuous tuft of hairs, and from the base also there springs a fine awn which has the same length as the body of the "seed." This hair-grass "seed" occurs as an impurity in foxtail. When the foxtail is ripe the hair-grass is only in flower, with no grain as yet developed, and so impotent to reproduce the plant.

CLASS VII.—SEEDS OF PARASITES.

The parasitic flowering-plants, whose seeds are often found in clovers and in timothy seed sifted off from clovers, are dodder and clover-rape.

64. *Clover-dodder* (fig. 399).

The seeds look like little *pieces of grey earth*, but, unlike earth, they do not crumble when pressed beneath a knife blade. Under the microscope they are seen to be tiny globules more or less angular, ¾ mm. in diameter. To see the seed contents, which are peculiar, the

seeds are boiled in potash, and then squeezed between two glass plates. The cylindrical, yellow embryo, rolled in three

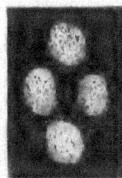

Fig. 399.—Clover-dodder (*Cuscuta Trifolii*).

turns round the transparent endosperm, now becomes apparent, and marks the dodder seed.

65. *Clover-rape* (*Orobanche minor*).

The seeds are very small — about ⅕ mm. long, like particles of *fine dark-brown or black sawdust*. Under the microscope each seed appears as an iridescent club with a netted surface.

66. *Ergot in Timothy "Seed."*

Ergot is not a seed, but the body of a fungus in a resting state. This ergot as it appears in a timothy is a black rod about 1 mm. broad, and from 2 to 4 mm. long. The rod when cut across is white inside. Such ergots when sown produce spores which spread the fungus to the timothy grass.

The seed represented in fig. 399 is magnified six diameters.

CORN CROPS.

CORN-GROWING.

The present position of corn-growing in the United Kingdom is highly interesting and instructive. Without doubt it is here and elsewhere the oldest of arts, and it is pretty certain to be the most enduring; for, notwithstanding the plenteous supplies of artificial foods promised us by the chemist, bread is likely to remain the staff of human sustenance.

The annual consumption of bread-corn in this country averages about 440 lb. per inhabitant, and far exceeds that of all

other foods put together, not excluding potatoes, which largely take the place of bread with meat. Its consumption increases not only with the increase of population, but also with every improvement in the social condition of the people as a result of higher wages, and a raising of the standard of living amongst the working classes.

But whilst we are much the largest consumers of corn per head of population, and ourselves barely raise a quarter of the amount needed for home consumption, we devote a smaller proportion of our cultivation to its production than any other corn-growing country. Under the unfair competition encouraged by our one-sided free-trade policy, most people in this country have come to regard corn as a crop not worth growing so long as it can be imported at a low price. So accustomed have we become to this view of the matter, that we unthinkingly take it for granted that it is cheaper to bring corn from Canada, Australasia, India, Russia, the Argentine, or any part of the world, than it is to grow it for ourselves. Yet, when we are being told that corn-growing here does not pay, and we are year by year reducing the area of our corn-fields, it is not a little interesting to note that other European countries, which already have a far larger proportion of their cultivated lands under corn than we have, are steadily increasing this proportion.

It is true that we have not the fine climate for wheat-cultivation that some other countries have, yet a large part of these islands is well adapted for wheat; while for oats and barley we need fear no competitors, and we have the immense advantage of the home market for all of them. There is, indeed, less difference than might be imagined in the value to the farmer of the different kinds of grain crops, although the market quotations would seem to indicate otherwise. Thus, according to the Board of Agriculture Returns, the grain prices in this country for 1906 were: Wheat, 31s. 6d.; barley, 26s.; oats, 19s. per qr.; and the average yields were 33 bushels, 40 bushels, and 52 bushels per acre respectively. From these figures we find that the average value of the grain crop last year, exclusive of the straw, was £6, 7s. 9d. per acre; but if we add the value of the straw, the average total value of the grain crop is found to have been £9, 2s. 1d., as seen in the following table:—

TABLE OF GRAIN CROP VALUES.

Crop.	Grain.						Straw.		Total value of crop per acre.
	Bushels per acre.	lb. per bushel.	lb. per acre.	Price per qr.	Price per lb.	Value per acre.	Value per qr.	Value per acre.	
				s. d.	Pence.	£ s. d.	s. d.	s. d.	£ s. d.
Wheat .	33	60	1980	31 6	0.79	6 9 11	14 0	57 9	9 7 11
Barley .	40	50	2000	26 0	0.78	6 10 0	8 0	40 0	8 10 0
Oats .	52	39	2028	19 0	0.73	6 3 6	10 0	65 0	9 8 6
Average	41⅔	49⅔	2002	25 6	0.76	6 7 9	10 8	54 3	9 2 1

But no good farmer is content to grow less than 40 bushels of wheat, 50 bushels of barley, or 60 bushels of oats per acre; nor to have the weights much less than 64 lb., 56 lb., and 42 lb. respectively; and 40 bushels of wheat per acre is often exceeded. But even 40 bushels, instead of 33, at the above prices, makes the wheat crop worth £11, 7s. 6d. per acre, which shows that corn-growing in this country is not played out yet, and that few, if any, of the other staple crops will pay better *for the same amount of labour.*

British Resources for Corn Production.—What will be the economic relation of British to Colonial and foreign wheat-growing in the next few years is a matter of doubt. But in the event of our being involved in a great European war, or any such calamity, there is much satisfaction in knowing that this country has an immense reserve for corn-growing in its thirty-three million acres of cultivated grass lands which could be put under corn if necessary, and one-fourth of which would suffice to grow all the grain we now need to import. That this is no exaggeration it would be easy to prove. It must here suffice to mention that we have this latent or potential power of corn-production in case of need, and that the production could be kept up year after year, as long as required, without the least disorganisation of the agriculture of the country.

VARIETIES OF CORN.

Wheat, like all true cereals and grasses, belongs to the natural order *Gramineæ*. In treating of the plants cultivated on a farm, systematic writers on agriculture describe their characters in minute botanical phraseology. This is right when different species of plants have to be distinguished from each other. When mere varieties and sub-varieties are numerous, they should be described in a more easily understood if less scientific method, so that others besides botanists may easily distinguish them.

Wheat.

Professor Low enumerated 11 different subdivisions of wheat[1] which were cultivated; while Mr Lawson described 83 varieties,[2] and Colonel le Couteur mentioned having in his possession, in 1836, no fewer than 150 varieties.[3]

Simple Classification.—In view of this large number of different varieties of wheat, it is much to be desired that a method should be established for easily recognising the different kinds of corn by the external characters of the ear and grain. Colonel le Couteur gave a classification of *wheat*, in which he divided all the varieties of wheat into two classes—namely, *beardless* and *bearded*. In so far he imitated the modern botanist, who divides the cultivated varieties of wheat into the two divisions of *barbatum* and *imberbe*, signifying the above conditions. But, unfortunately for the stability of this classification, that distinction is not immutable, for some bearded wheats lose their beards on cultivation, and some beardless ones are apt to become bearded when cultivated on poor soils and exposed situations.

Colonel le Couteur subdivided beardless wheat into white, red, yellow, and liver-coloured, smooth-chaffed, and vel-

[1] Low's *Ele. Prac. Agric.*, 229.
[2] Lawson's *Agric. Man.*, 29.
[3] Le Couteur *On Wheat*, ii., Dedi.; and 77.

vet-chaffed; and the bearded he divided under the same colours. Some varieties of wheat are, no doubt, decidedly downy on the chaff; but others, again, are so very little so, that it is difficult to distinguish them from some of the roughest varieties of smooth-chaffed; and it is known that the same wheat will be differently affected in this respect by the soil upon which it grows. A sharp soil renders the chaff and straw smoother and harder than a deaf one, and the deaf soil has a tendency to produce soft and downy chaff and straw. Downiness is thus not a more permanent character than the beard for establishing the denominations of the great divisions of wheat.

Conjoining the characters of the grain and ear of wheat seems unnecessary, inasmuch as the character of either separately cannot positively indicate the state of the other, and both characters are not required to indicate the superior properties of any variety of wheat for making bread. A miller at once distinguishes the *grain* which will afford the best bread; and neither he nor any farmer could indicate such a property from the *ear* of any wheat.

Colour of Wheat.—Colonel le Couteur assumed that a liver-coloured wheat was a distinctive colour. We never remember to have seen a wheat of a liver-brown colour. All the colours of wheat, we think, may be classed under two primary colours, *yellow* and *red*—for even the whitest has a tinge of yellow, and every dark colour is tinged with red; and as *white* and *red* are the terms by which the colours of wheat have been longest known, these should be retained. The sub-tints of yellow and red might be easily designated.

Classification by the Ear.—Were we to classify both the plant and grains of wheat by *natural marks*, we would make two classifications, one by the ear and the other by the grain, so that each might be known by its own characteristics. In this way confusion would be avoided in describing the ear and the grain. The farmer who grows the wheat plant, and sells it in the grain, should be acquainted with both; but the miller who purchases the grain need know nothing of the ear.

The ears of three classes of wheat are represented in fig. 400, which shows the ears half the natural size. The first, *a*, is a *close* or *compact* eared wheat, which is occasioned by the spikelets being set near each other on the rachis; and this

Fig. 400.—*Classification of wheat by the ear.*

construction makes *the chaff short and broad*. The second class of ears is *b*, the spikelets being of *medium* length and breadth, and placed just as close upon the rachis as to screen it from view; this ear is not so broad, but longer than *a*; the *chaff* is of *medium length and breadth*. The spikelets of *c* are set *open*, or as far asunder as to permit the rachis to be easily seen between them; this ear being about the same length as *b*, but much narrower, *the chaff long and narrow*. There is no chance of confounding these three structures of the ears of wheat.

These three classes of varieties constitute the *Triticum sativum imberbe* of botanists,—that is, all the beardless cultivated wheats. Formerly they were divided by botanists into *Triticum hybernum* or winter wheat, and *Triticum*

æstivum or summer wheat; but experience has proved that the summer wheat may be sown in winter, and the winter wheat sown in spring, and both come to perfection. Paxton says that Triticum is derived from "*tritum*, rubbed — in allusion to its being originally rubbed down to make it eatable." [1]

In *d*, fig. 400, is represented a bearded wheat, which shows the appearance the beard gives to the ear. The bearded wheats are generally distinguished by the *long shape* of the *chaff* and the open position of the spikelets, and therefore fall under the third class *c*. But cultivation has not only the effect of decreasing the strength of the beard, but of setting the spikelets closer together, as in the white Tuscany wheat. Bearded wheat constitutes the second division of cultivated wheat of the botanists, under the title of *Triticum sativum barbatum*. The term bearded has been used synonymously with spring wheat, but erroneously, as beardless wheat is as fit for sowing in spring as bearded, and the bearded for sowing in winter.

Classifying by the Grains.—Classified by the grain, wheat may again be grouped under three heads. The first class is shown in fig. 401, where the grains are *small, short, round*, and *plump*, with the median line distinctly marked and well filled up. Fine *white* wheat belongs to this class, and is enclosed in *short, round, thin*, and generally *white chaff*, which, when ripe, becomes so expanded as to endanger the grain falling out. Very few *red* wheats be-

Fig. 401.—*Short, round, plump form, and small size of wheat.*

Fig. 402.—*Long medium-sized form of wheat.*

long to this class. In reference to the ear, this class is found in *short-chaffed* and broad spikelets, which are generally compact, as *a*, fig. 400.

The second class is in fig. 402, where the grains are *long and of medium size*, longer and larger than the grains of fig.

[1] Paxton's *Bot. Dic. Tritic.*

401. The *chaff* is also *medium sized.* In reference to the ear, it is of the *medium* standard, in respect to breadth and closeness of spikelets, as *b*, fig. 400, though *medium-sized grain* is *not* confined to this sort of ear, and is found in the *compact* ear as well as in the *open* ear. Most *red* wheat belongs to this class of grain, though many of the *white medium-sized* also belong to it. This grain is the Caucasian red wheat, whose ear is bearded, and belongs to the open-spiked class *c*, fig. 400.

The median line is strongly marked, and the ends are sharp.

In fig. 403 is the third form of grain, which is *large and long.* Its *chaff* is *long*, and in reference to the ear, the spikelets are generally open. The median line is not distinctly marked. The ends of the grain are pointed but not sharp, and the skin is rather coarse.

Fig. 403.—*Large size and long form of wheat.*

The germ and radicle are boldly marked.

These three sorts of wheat are of the natural size, and indicate the forms of the principal varieties found in our markets.

Relation of Ear and Grain.—It will be seen from what has been stated, that no inevitable relation exists between the *ear* and *grain;* that the compact ear does not always produce round grain nor white wheat; that in the medium ear is not always found medium-sized grain; and that the open ear does not always produce large long grain.

Still, there exist coincidents which connect the *chaff* with the *grain.* For example, *length* of *chaff* indicates *length* of *grain*, upon whatever sort of ear it may be found; and, generally, the *colour* of the chaff determines *that* of the grain.

On desiring, therefore, to determine the sort of *grain* any number of *ears* of different kinds of wheat contain, the *form* and *colour* of the *chaff* determine the point, and not whether the ear carries compact, medium, or open, bearded or beardless, woolly or smooth spikelets.

Vilmorin's Classification.—M. Henry Vilmorin, in his beautiful work [1] on

wheat, adopts the following arrangement:—

Triticum sativum .	Soft wheat.
T. turgidum .	. Plump ,,
T. durum .	. Hard ,,
T. polonicum .	. Polish ,,
T. Spelta .	. Spelt.
T. amyleum .	. Starchy wheat.
T. monococcum .	. One-grained ,.

The most important of these, *T. sativum*, he subdivides according as the variety is awned or unawned, the ear white or red, smooth or downy, and the grain itself white or red. Familiar forms of *T. sativum* are Chidham, Hunter, Trump, Talavera, Hickling, Hallett, Dantzic, Shireff, Browick.

Judging Wheat.—But the classification is unimportant to the farmer, compared to the mode of *judging* wheat, to ascertain the external characters which best indicate the purposes for which the corn may be best employed, in the particular condition of the sample. The purposes are, for seed and for making into flour—whether the flour is to be employed in bread, in confections, or starch.

In its *best* condition, all wheat, whether red or white, small or large, long or round, should appear plump within its skin. The skin should be fine and smooth. The colour should be bright and uniform. The grains should be of the same size and form, and perfect. With all these properties wheat is fitted for every purpose.

Wheat for Flour.—When wheat is quite opaque, it is in the best state for yielding the finest flour. Such flour, from white wheat, confectioners use for pastry, and it contains the largest amount of starch, but it is too dear for the starch-maker. When wheat is translucent, hard, and flinty, it is suited to the baker, as affording flour that rises freely with yeast, having much gluten in it. For bread of finest quality a mixture of the two conditions of flour is best suited.

Some sorts of wheat naturally possess *both* these properties, and are great favourites with millers. Generally speaking, the purest-coloured white wheat indicates most opacity, and yields the finest flour; and red wheat is most flinty, and yields the strongest flour: translucent red

---

[1] *Les meilleurs blés.*

wheat will yield stronger flour than translucent white wheat, and yet red wheat rarely realises so high a price in the market as white—partly because it contains more bran, makes darker-coloured bread, and yields less starch.

Wheat varying with Soil.—Mr Powles says, in his translation of Kick's treatise : " Wheat varies very much according to the soil and country in which it is grown. Among the best kinds of wheat are the Hungarian and the Banater, though they frequently show a flinty appearance in cross-section. This is not the case with wheats grown in more northern countries. Their grain, which shows in cross-section a uniform white colour, gives better flour, and is called soft or white wheat, whereas that which shows a mottled or flinty cross-section takes the name of hard wheat, and gives less and inferior flour. A fine, clear, glistening exterior and oval form, are a sign of good quality with old wheat, which has been kept for many years in the granary ; the recovery of its original colour and lustre after washing and slow drying indicates good quality." [1]

Weight of Wheat.—The weight of wheat varies, according to the state of the season, from 55 lb. to 66 lb. per imperial bushel ; the 55 lb. being very light, and produced in a wet late season on inferior land—the heavy being very heavy, and produced in a hot season on the best soil. An average weight for wheat is 62 to 63 lb. per bushel. The average weight of all the wheat sold in the Edinburgh market in the thirteen years up to 1880 was 62.2 lb. per bushel.[2]

" A plump, rounded, white, smooth grain, without wrinkles, gives the heaviest weight per bushel. Wheat grain is heavier than water, its specific gravity ranging from 1.29 to 1.41." [3]

" High specific gravity is, above all, an indication of good quality. Wheat which weighs 50 to 60 lb. per imperial bushel is considered good—that is, rich in flour. The grains should be equal-sized, large, and full. In rare cases the weight rises to 66 lb. per imperial bushel." [4]

[1] Kick's Flour Manufacture.
[2] A Bushel of Corn, A. S. Wilson, p. 35.
[3] Church's Food.
[4] Powles's Kick's Flour Manufacture.

Number of Grains in a Bushel.—Of Chidham white wheat, weighing 65 lb. per bushel, 86 grains were found to weigh one drachm, so that the bushel should contain 715,520 grains. At 63 lb. to the bushel, and 87 grains to the drachm—the most common case—the bushel should contain 701,568 grains.

Wheat for Seed.—For seed, the root-end of the grain should be distinctly prominent, and the stem-end slightly hairy. When either end is rubbed off, the grain is deprived of its vitality. Kiln-drying also destroys vitality. Wheat unfit for seed may be detected in various ways. If it has been in sea-water, although not enlarged by moisture, it never loses the saline taste. When washed in fresh water and dried in a kiln, the washing gives it a bleached appearance, and the kiln-drying is detected by smell or taste. When shealed, the ends are rubbed down. When heated in the sack, it tastes bitter. When heated in the stack, it has a high colour. When long in the granary, it is dull and dirty, and has a musty smell. When attacked by weevils and other insects in the granary, which breed within its shell and eat the kernel, the shells are light, and have holes in them. Germinated, swollen, burst, bruised, smutted grains, and the presence of other kinds of corn and seeds, are easily detected by the eye.

Preserving Wheat in Granaries.—Difference of opinion exists in regard to the best mode of preserving wheat in granaries. The usual practice is to shovel the heap over from the bottom every few weeks, according to the dryness or dampness of the air, or heat or coldness of the atmosphere. In this mode of treatment a free ventilation of air is requisite in the granary, and the worst state of the atmosphere for the grain is when it is *moist* and *warm*. Extreme heat or extreme cold are preservatives of corn.

The practice of others is not to turn it over at all, but to keep it in the dark in thick masses, reaching from the floor to the ceiling. No doubt, if air could be excluded from a granary, the corn would be preserved in it without trouble ; and a good plan of excluding the air seems to be, to heap the grain as close together as possible. When kept long in heap

without turning, it retains its colour with
the fresh tint, which is also secured by
keeping it in the dark.

Ancient Practice in Storing.—The
ancients preserved grain many years, to
serve for food in years of famine. Joseph,
in Egypt, preserved wheat for seven
years in the stores; in Sicily, Spain, and
the northern parts of Africa, pits were
formed in the ground to preserve it; and
the Romans took great pains in constructing granaries, which kept wheat for 50
and millet for 100 years.[1]

Storing v. Immediate Selling.—The
practice of storing grain in farm granaries is not now pursued to so large an
extent as formerly; yet it is often found
necessary or desirable for the farmer to
store a moderate quantity for a limited
time. As to this point a cautious and
experienced farmer says:—

"As regards the farmer, the question
of preserving wheat in granaries should
little affect him, the best way of keeping
wheat being in the straw in the stack;
and when the stacks are threshed, that
the straw may be used, he should dispose
of his wheat immediately, and take the
current market prices. During the cur-
rency of a lease, this is the safest practice
for securing him an average price; and
it saves much trouble in looking after
the corn, much vexation when it becomes
injured, and much disappointment when
the price falls below its expected amount.
Loss is likely to be the fate of farmers
who speculate in corn of their own
growth; and when they become merchants besides, they are likely to become
involved in the intricacies of foreign
trade, and feel the effects of their thought-
lessness."

**Quantity of Ash in an Acre of
Wheat.**—Lawes and Gilbert found at
Rothamsted that the average quantities
of total mineral constituents (ash) yielded
per acre per annum, over sixteen years
on three plots, differently manured, on
which wheat was continuously grown,
were the following:—

	In grain. lb.	In straw. lb.	Total. lb.
By farmyard manure	36.3	201.1	237.4
Without manure	16.6	89.5	106.1
With ammonia salts alone	23.0	119.2	142.2

[1] Dickson's *Hus. Anc.*, ii. 426.

Kernel and Husk.—Mr A. S. Wilson
found that of the grain of wheat about
95.59 per cent consisted of kernel, and
4.41 per cent of husk.

Origin of Wheat.—"It is a very remarkable circumstance," observes Lindley, "that the native country of wheat,
oats, barley, and rye should be entirely
unknown; for although oats and barley
were found by General Chesney, apparently wild, on the banks of the Euphra-
tes, it is doubtful whether they were not
the remains of cultivation. This has led
to an opinion, on the part of some persons, that all our cereal plants are artificial productions, obtained accidentally,
but retaining their habits, which have
become fixed in the course of ages.[2]

Antiquity of Wheat Cultivation.—
A. de Candolle[3] observes that the cultivation of wheat is prehistoric in the Old
World. Very ancient Egyptian monu-
ments, older than the invasion of the
shepherds, and the Hebrew Scriptures,
show this cultivation already established;
and when the Egyptians or Greeks speak
of its origin, they attribute it to mythi-
cal personages—Isis, Ceres, Triptolemus.
The earliest lake-dwellers of Western
Switzerland cultivated a small-grained
wheat, which Heer has described under
the name of *Triticum vulgare antiquorum.*
The first lake-dwellings of Robenhausen
were at least contemporaneous with the
Trojan war, and perhaps earlier. The
Chinese grew wheat 2700 B.C.

Limits of Wheat Culture.—Only
the lower-lying parts of the United
Kingdom are well suited for wheat cultivation, yet at one time or other wheat
has been grown to a greater or lesser
extent in every county in England and
Wales, and also in most counties in
Scotland.

"Wheat is cultivated in Scotland to
the vicinity of Inverness (lat. 58°); in
Norway to Drontheim (lat. 64°); in
Sweden to the parallel of lat. 62°; in
western Russia to the environs of St
Petersburg (lat. 60° 15′); while in central
Russia the polar limits of cultivation
appear to coincide with the parallel of
59° or 60°. Wheat is here almost an
exclusive cultivation, especially in a zone

[2] Johnston's *Lect. Agric. Chem.*, 2nd ed., 928.
[3] Lindley's *Veget. King.*, 112.

which is limited between the latitude of Tchernigov, lat. 51°, and Ecaterinoslav, lat. 48°. In America the polar limits of wheat are not known, on account of the absence of cultivation in the northern regions. The physical conditions of these limits are, in the different countries where cultivation has been carried to the utmost extent, as follows :—

	Lat.	Mean temperature, Fahr. Year.	Winter.	Summer.
Scotland (Ross-shire)	58°	46°	35°	57°
Norway (Drontheim)	64	40	25	59
Sweden . . .	62	40	25	59
Russia (St Petersburg)	60 15'	38	16	61

This table shows how little influence winter cold has in arresting the progress of agriculture towards the north; and this is confirmed in the interior of Russia, where Moscow is much within the limits of wheat. The cultivation of wheat is very productive in Chili, and in the united state of Rio de la Plata. On the plateau of southern Peru, Meyer saw most luxurious crops of wheat at a height of 8500 feet, and at the foot of the volcano of Arequipo, at a height of 10,600 feet. Near the lake of Tabicaca (12,795 feet high), where a constant spring-heat prevails, wheat and rye do not ripen, because the necessary summer-heat is wanting; but Meyer saw oats ripen in the vicinity of the lake." [1]

Barley.

The botanical position of *barley* is the genus *Hordeum*, of the natural order of *Gramineæ*. Professor Low divides the cultivated barleys into two distinctions, —namely, the 2-rowed and the 6-rowed, and these comprehend the ordinary, the naked, and the sprat or battledore forms. [2] Lawson describes 20 varieties of barley. [3]

Classifying by the Ear.—The natural classification of barley by the ear is obviously into three kinds — 4-rowed, 6-rowed, and 2-rowed. Fig. 404 represents the three forms, where *a* is the 4-rowed, or bere or bigg, *c* the 6-rowed, and *b* the 2-rowed, which figures give the ear in half its natural size. Of these the bere or bigg was cultivated until a

recent period, but the 2-rowed has almost entirely supplanted it, and become the most commonly cultivated variety, the

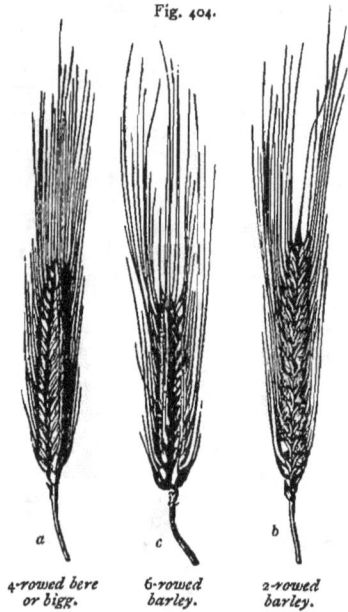

Fig. 404.

| 4-rowed bere or bigg. | 6-rowed barley. | 2-rowed barley. |

6-rowed being rather an object of curiosity than culture.

Classifying by the Grain.—In classifying barley by the *grain* we find there are just two kinds, *bere* or *bigg*, and *barley;* and though both are awned, they are sufficiently marked to constitute distinct varieties. In the bere, fig. 405,

Fig. 405.—*Scotch bere or bigg.*　　Fig. 406.—*English barley.*

the median line of the bosom is so traced as to give the grain a twisted form, by which one of its sides is larger than the other, and the lengthened point is from where the awn was broken off. The figure gives the grain of the natural size.

In barley, fig. 406, the median line passes straight, and divides the grain into two equal sides, short and plump,

[1] Johnston's *Phys. Atl.*—Phytol., Map No. 2.
[2] Low's *Ele. Prac. Agric.*, 244.
[3] Lawson's *Agric. Man.*, 33.

with a crenulated skin. The grain here is of the natural size. The bigg was long cultivated in Scotland, along with a 2-rowed variety named common or Scotch barley; but several English varieties are now cultivated which show a brighter and fairer colour, plumper and shorter grain, and are quicker in malting, though less hardy and prolific, than the common barley.

A variety known as *awnless* barley is now cultivated. When it becomes fully ripe the awns fall off, hence its name.

Judging Barley.—The crenulated skin is a good criterion of malting; and as most of the barley is converted into beer or spirits, both requiring malt to produce them of the finest quality, it is not surprising that English Chevalier barley should realise the highest price. In judging good barley it should break soft between the teeth, and show a white fracture, and be wrinkled in the bosom. When it breaks hard it is flinty, and will not malt well.

As to grinding barley, "the indications of good flour-producing qualities in barley are these: a fine pale-yellow colour, roundish rather than long form, and a high specific weight. Long, pointed, and flat grains yield less flour, and of a bluish tint. According to Neumann, barley one year old will yield flour whiter than fresh barley." [1]

Yield and Weight of Barley.—A good crop of barley yields a return of from 48 to 60 bushels per acre. Good barley weighs from 54 lb. to 59 lb. per bushel. A crop of 60 bushels per acre will yield of straw about 176 stones of 14 lb. to the stone, or $1\frac{1}{10}$ ton, and the weight of the grain of that crop, at 56 lb. per bushel, will be $1\frac{1}{2}$ ton. Mr A. S. Wilson states that the average weight of all the barley sold in the Edinburgh market in the thirteen years ending with 1880 was 54.93 lb. per bushel, the range being from 46 to 60 lb. [2]

Grains in a Bushel.—It takes of bigg 111 grains to weigh one drachm; of 6-rowed barley, 93; and of Chevalier barley, 75 grains; of which, with the weight per bushel of 57 lb., the number

of grains of Chevalier barley in a bushel will be 547,200.

About 90 per cent of the grain of barley consists of kernel, and 10 per cent of husk.

Utilisation of Barley.—By far the largest proportion of the best barley is converted into *malt* for making malt liquor and spirits. Barley is also used for distillation in the raw state.

Pot and pearl barley are made from barley for culinary purposes. Both meal and flour are manufactured from barley for making unleavened bread, which is eaten by the labouring class in some parts of the country.

Of the states of barley the soft is best adapted for making into malt and meal, and the flinty into pot barley. It was supposed that flinty barley contained the most gluten or nitrogen; but Professor Johnston showed that it contains less than the soft barley, in the proportion of 8.03 to 10.93.

Barley-meal.—"The meal so highly commended by the Greeks was prepared from barley. . . . It was not until after the Romans had learnt to cultivate wheat, and to make bread, that they gave barley to the cattle. They made barley-meal into balls, which they put down the throats of their horses and asses, after the manner of fattening fowls, which was said to make them strong and lusty. Barley continued to be the food of the poor, who were not able to procure better provision; and in the Roman camp, as Vegetius has informed us, soldiers who had been guilty of any offence were fed with barley instead of bread corn." [3]

Malting.—The malting of barley for use as food for live stock will be noticed in the section dealing with "Food and Feeding."

Limits of Barley Culture.—Barley is not found to be a profitable farm crop so far north as oats, but occasionally it is grown in small areas in Orkney and Shetland (lat. 61° N.) In Western Lapland the limit of barley is under lat. 70° near Cape North, the northern extremity of Europe. In Russia, on the shores of the White Sea, it is between the parallels of 67° and 68° on the western side, and

---

[1] Kick's *Flour Manufacture.*
[2] *A Bushel of Corn.*

[3] Phillips's *Hist. Culti. Veget.*, i. 50.

about 66° on the eastern side, beyond Archangel; in central Siberia, between lat. 58° and 59°.

Between the tropics barley does not succeed in the plains, because of its liability to suffer from heat.

Oats.

The oat-plant belongs to the natural order of *Gramineæ*, genus *Avena*. Its ordinary botanical name is *Avena sativa*, or cultivated oat. The term oat is of obscure origin. Paxton conjectures it to have been derived from the Celtic *etan*, to eat.[1]

There are a great number of varieties of this cereal cultivated in this country. Lawson describes thirty-eight.[2]

Classification by the Grain.—The natural classification of the oat by the grain consists only of two forms—one plump and short and beardless, as fig.

Fig. 407.—*Potato oat.*　　Fig. 408.—*White Siberian early oat.*

407, the potato oat, smooth-skinned, shining, having the base well marked, and the germ-end short and pointed.

The other form is in fig. 408, long and thin, and having a tendency to produce a beard, the white Siberian early oat. It is cultivated in the poorer soils and higher districts, resists the force of the wind, and yields a grain well adapted for the support of farm-horses.

The straw is fine and pliable, and makes an excellent dry fodder for cattle and horses, the saccharine matter in the joints being very sensible to the taste. It comes early to maturity, and hence its name.

Mr A. S. Wilson divided oats into three groups, which he designated as the *Oviform*, *Coniform*, and *Fusiform*. In the first he placed the short round oats approaching the form of an egg, the potato oat and Scots barley oat being types of

this class. The *Coniform* embraced the oats of medium length in proportion to their thickness, as the sandy oat. The long oats, such as the Tartarian and Arkangel oats, comprised the *Fusiform*.[3]

Classification by the Ear.—The natural classification of the oat by the ear is obvious. One kind, fig. 409, has its branches spreading equally on all sides, shortening gradually towards the top of the spike in a conical form, and the panicles are beardless. This is the potato oat. While the ear is yet recent, the branches are erect; but as the seeds advance towards maturity, and become full and heavy, they assume a dependent form. By this change, the air and light have free access to the ripening grain, while the rain washes off the eggs or larvæ of insects that would otherwise prey upon the young seed. This variety is extensively cultivated in Scotland on account of the fine and nourishing quality of its meal, which is largely consumed by the people—unfortunately not so largely now as in former times. It is cultivated in the richer soils of the low country.

Fig. 409.—*Spike of potato oat.*

The plant of the potato oat is tender, and the grain is apt to be shaken out by the wind. The straw is long and strong, inclining too much to reediness to make good fodder. It is late in coming to maturity. Its peculiar name of the potato oat is said by one writer to have been derived from the circumstance of the first

---

[1] Paxton's *Bot. Dict.*, art. Avena.
[2] Lawson's *Agric. Man.*, 44.

[3] *A Bushel of Corn.*

plants having been discovered growing accidentally on a heap of manure, in company with several potato-plants, the growth of which was equally accidental;[1] while another writer says plants of it were first found in 1789 in Cumberland, growing in a field of potatoes. The ear in the figure was taken from the stack, none being at the time available in the field, where it would have been more regular and beautiful.

The white Siberian oat, fig. 408, has an ear of this description.—The other kind of ear has its panicles shorter, nearly of equal length all on the same side of the rachis, and bearded.

Fig. 410, a head of Tartarian oat, taken from the stack, shows this form of ear. The seeds of this form also assume the pendant position. It is of such a hardy nature as to thrive in soils and climates where other oats could not be raised. This variety derives its name, most

Fig. 410.—*Spike of Tartarian oat.*

probably, from Tartary. It is much cultivated in England, and only to a limited extent in Scotland. It is a coarse grain, more suitable for animal food than for making into meal. The grain is dark-coloured, awny; the straw coarse, harsh, brittle, and rather short.

Yield and Weight of Oats.—The *crop* of oats varies from 30 to 80 bushels per imperial acre, according to kind, soil,

and situation, 40 to 48 being very general. Oats vary in weight from 33 lb. to 48 lb. per bushel. The average of all the oats sold in the Edinburgh market during a period of thirteen years was 42.22 lb. per bushel. Whiteness, of a silvery hue, and plumpness, are the criteria of a good sample. A crop of potato oats, yielding 60 bushels to the acre, at 47 lb. per bushel, weighs of grain 1 ton 5 cwt. 20 lb., and yields of straw 1 ton 5 cwt. 16 lb., in the neighbourhood of a large town; or, in other words, yields 8 kemples of 40 windlings each, and each windling 9 lb. in weight. A crop of Hopetoun oats, of no more than 60 bushels to the imperial acre, grown near Edinburgh, yielded 2 tons 18 cwt. 16 lb. of straw.

Grains in the Bushel.—The potato oat, 47 lb. per bushel, gave 134 grains to one drachm; the Siberian early oat of 46 lb. gave 109 grains; and the white Tartarian oat, 42 lb., gave 136 grains; so that these kinds respectively afford 806,144,651,792, and 731,136 grains of oats per bushel.

Kernel and Husk.—Mr A. S. Wilson gave the proportions of kernel and husk in the old varieties of oats as follows: *Oviform*—kernel 76.34, husk 23.66 per cent; *Coniform*—kernel 76.07, husk 23.93; and *Fusiform*—kernel 73.23, and husk 26.77 per cent. Average of all kinds—kernel 75.21, and husk 24.79 per cent.[2] In the new varieties of oats the percentage of husk is greater—often as much as 37 per cent.

Oatmeal.—For human food the oat is manufactured into *meal*, not into flour. Oats are always kiln-dried before being ground, in order the more readily to get quit of the thick husk in which the grain is enveloped. After the husk has been separated by a fanner, the grain, then called groats, is ground by the stones closer set, and yields the meal. The meal is then passed through sieves, to separate the thin husk from the meal. The meal is made into two states: one *fine*, which is the state best adapted for making into oat-cake or bannocks; and the other is coarser or *rounder* ground, which is best adapted for making the common food of the country people—

---

[1] Rhind's *Hist. Veget. King.*, 218.

[2] *A Bushel of Corn.*

porridge,—*Scottice*, parritch. A difference of custom prevails in respect to using these two different states of oatmeal, the fine meal being best liked for all purposes in the northern, and the round meal for porridge in the southern, counties.

There is, unfortunately, too good reason to fear that this wholesome article is losing its position as the " common food " of the country people of Scotland. Meat, fish, and milk food are now consumed much more largely by the rural classes of Scotland than in former times; and the " cheap loaf " is fast supplanting the more substantial oat-cake.

A sharp soil produces the finest cakemeal, and clay land the best meal for boiling. Of meal from the varieties of the oat cultivated, that of the common Angus oat is the most thrifty for a poor man, though its yield in meal is less in proportion to the bulk of corn.

Oatmeal was for long the principal food of the Scottish ploughman. In most parts it is still largely consumed by them, but it is now used to a smaller and gradually decreasing extent. It was considered a rather anomalous circumstance to find men thriving as well on oatmeal as on wheat bread and butcher-meat; but the anomaly was cleared up by the investigations of chemistry. The oat contains about 7 per cent of oil or fat, and 12 per cent of protein, making together nearly 20 per cent of really nutritive matter, capable of supporting the loss incurred by labour of the muscular portion of the body. All vegetables contain fat, and the largest proportion of vegetable fats contain the elaic and margaric acids, mixed with a small proportion of the stearic. The elaic is always in a fluid state, and the margaric and stearic in a solid; and of the latter two, the margaric is much less, and the stearic acid very much greater, in animal fat than in those of plants. It is by the dissipation of this oil or fat by heat, in baking, that the agreeable odour of the oat-cake is at once recognised on approaching the humble cottage of the labouring man.

Yield of Meal.—In regard to the *yield of meal* from any given quantity of oats, when they give half their weight of meal they are said to give *even meal.* Supposing a boll of oats of 6 bushels to weigh 16 stones, it should give 8 stones or 16 pecks of meal, and, of course, 8 stones of refuse, to yield even meal. But the finer class of oats give more meal in proportion to weight than this—some nearly 9 stones and others as much as 12 stones per boll. The market value of oats is therefore often estimated by the meal they are supposed to yield, and in discovering this property in the sample millers become very expert.

Composition of Oatmeal.—The following figures give an approximation of the percentage composition of fresh Scotch oatmeal:—

Water .		10.5
Albuminoids .		11.0
Carbohydrates	.	52.2
Fats	.	4.5
Fibre .	.	14.5
Ash	.	6.8

One hundred pounds of oats (weighing 45½ lb. to the bushel) commonly yield the following proportion of products:—

	lb.
Oatmeal	60
Husks	26
Water	12
Loss . . .	2

Kick,[1] quoting the mean of many analyses, gives for oats the following percentage composition:—

Starch .	.	56
Gluten .	.	12
Cellulose	.	12
Salts . .	.	3
Water .	.	17

Oats as Food for Stock.—Oats are now used much more extensively than formerly as food for horses, cattle, and sheep. Indeed this is now their chief function. For this purpose they are usually crushed flat.

Antiquity of Oat Culture.—" We find no mention made of oats in Scripture," says Phillips, " which expressly states that Solomon's horses and dromedaries were fed with barley;" but " the use of oats as a provender for horses appears to have been known in Rome as early as the Christian era, as we find

---

[1] *Flour Manufacture.*

that that capricious and profligate tyrant, Caligula, fed 'Incitatus,' his favourite horse, with *gilt oats* out of a golden cup." Oats are mixed with barley in the distillation of spirits from raw grain; and "the Muscovites make an ale or drink of oats, which is of so hot a nature, and so strong, that it intoxicates sooner than the richest wine." [1]

Origin of the Oat.—As all the varieties of oats are cultivated, and none have been discovered in a truly wild state, it is very probable that they are all derived from a single prehistoric form, a native of eastern temperate Europe and of Tartary (A. de Candolle).

Limits of Oat Culture.—"The oat (*Avena sativa*) is cultivated extensively in Scotland, to the extreme north point, in lat. 58° 40'. In Norway its culture extends to lat. 56°; in Sweden to lat. 63° 30'. In Russia, its polar limits appear to correspond with those of rye. Whilst, in general, oats are cultivated for the feeding of horses, in Scotland and in Lancashire they form a considerable portion of the usual food of the people. This is also the case in some countries of Germany, especially in the south of Westphalia, where the inhabitants of the 'Sauerlands' live on oaten bread. South of the parallel of Paris oats are little cultivated; in Spain and Portugal they are scarcely known; yet they are cultivated with considerable advantage in Bengal to the parallel of lat. 25° N." [2]

Rye.

Rye, botanically, occupies the genus *Secale* of the order *Gramineæ*. It is the *Secale cereale* of the botanists, so called, it is said, from *á secando*, to cut, as opposed to leguminous plants, whose fruits used to be gathered by the hand.

A spike of rye, fig. 411, is not unlike a hungry bearded wheat. There is only one known species of rye, which is said to be a native of Candia, and was known in Egypt 3300 years ago. But several varieties are raised as food, four of which are described by Lawson. [3]

A. de Candolle adduces historical and philological data to show that the species probably had its origin in the countries north of the Danube, and that its cultivation is hardly earlier than the Christian era in the Roman Empire, but perhaps more ancient in Russia and Tartary.

The grains of rye are long and narrow, not unlike shelled oats or groats, but more flinty in appearance. They are in fig. 412, of the natural size.

Rye is rarely cultivated in this country for its grain, but as a green forage crop it is grown to a considerable extent both in the south of England and the south of Ireland. In Scotland only small patches here and there are to be seen. It is extensively cultivated on the Continent, on all soils, and forms the principal article of food of the labouring classes.

"Closely resembling the wheat berry is that of rye. Its appearance is well known as naked, rather long, very tapering off at the lower end, curved or slightly keeled at the back, with a furrow at the front, and with hairs at the upper end. It is of a greyish-brown colour, and slightly wrinkled." [4]

Fig. 411.—*Ear of rye.*

Yield and Weight of Rye. — The produce of rye is about 25 bushels per acre, and the weight of the grain is from 52 to 57 lb. per bushel. The number of grains in 1 drachm being 165, at 55 lb., the bushel should contain 1,161,600 grains.

In a crop of 25 bushels to the acre, weighing 1300 lb., the nutritive matter derived from rye consists of 130 lb. to 260 lb. of husk or woody fibre; 780 lb. of starch, sugar, &c.; 130 to 230 lb. of gluten, &c.; 40 to 50 lb. of oil or fat; and 26 lb. of saline matter.

Fig. 412.—*Grains of rye.*

Limits of Rye Culture. — "Rye (*Secale cereale*) is cultivated in Scandinavia, on the western side to the parallel of 67° N., and on the eastern side to lat. 65° or 66° N. In Russia, the

[1] Phillips's *Hist. Culti. Veget.*, ii. 9.
[2] Johnston's *Phys. Atl.*—Phytol., Map No. 2.
[3] Lawson's *Agric. Man.*, 31.

[4] Kick's *Flour Manufacture.*

polar limit of rye is indicated by the parallel of the city of Jarensk, in the government of Wologda, lat. 62° 30'. . . . It is as common in Russia, Germany, and some parts of France, as it is rare in the British Islands. Rye-bread still forms the principal sustenance of at least one-third of the population of Europe; it is the characteristic grain of middle and northern Europe; in the southern countries it is seldom cultivated." [1]

Rye is much used in the distillation of gin in Holland. Rye-bread is heavy, dark-coloured, and sweet, but when allowed to ferment, becomes sour.

Rye-flour.—In Russia, 100 lb. of rye-flour, containing 16 per cent of water, yield from 150 lb. to 160 lb. of bread. There, horses get it on a journey, in lieu of corn. The following indicates the percentage proportion of water, albuminoids, fats, carbohydrates, fibre, and ash in rye-flour:—

Water	12.0
Albuminoids	13.6
Fats	2.9
Carbohydrates	63.2
Fibre	4.2
Ash	4.1
	100.0

STRAW.

The straw of the various farm crops presents marked differences in appearance and composition.

Wheat-straw.

Wheat-straw is generally long, often upwards of 6 feet in length, and most kinds are strong. Of the two sorts of wheat, white and red, the straw of the white is softer, more easily broken by the threshing-mill, and decomposed in the dunghill. Red wheat-straw is tough, and is used for stuffing horse-collars. The strength and length of wheat-straw render it useful in thatching, whether houses or stacks. It is yet much employed in England for thatching houses, and perhaps the most beautifully thatched roofs are in the county Devon, whilst excellent examples of this art may be seen in Wiltshire.

Since the general use of slates in Scotland, thatching houses with straw has

[1] Johnston's *Phys. Atl.*—Phytol., Map No. 2.

fallen into desuetude. Wheat-straw makes the best thatching for corn-stacks, its length and straightness ensuring safety, neatness, and despatch, which, in the busy period of securing the fruits of the earth, is valuable. It forms an admirable bottoming to the littering of every court and hammel of the steading. As litter, wheat-straw possesses superior qualities, and few gentlemen's stables are without it.

It is not so well suited for fodder, its hardness and length being unfavourable to mastication; yet farm-horses are fond of it when it is fresh.

Upholsterers use wheat-straw as stuffing in mattresses for beds, under the name of *paillasse*; but such a mattress is a miserable substitute for crisp, curled, elastic horse-hair.

Ash of Wheat-straw.—The ash of wheat-straw contains the following ingredients:—

	Berthier.	Boussingault.	Fromberg.
Potash	10.86	9.56	15.52
Soda	...	0.31	...
Lime	5.36	8.83	4.58
Magnesia	...	5.19	2.45
Oxide of iron	2.32	1.04	1.56
Phosphoric acid	1.12	3.22	2.92
Sulphuric acid	0.44	1.04	10.59
Chlorine	2.82	0.62	1.56
Silica	77.08	70.19	60.58
	100.00	100.00	99.76
Percentage of ash	4.40	7.00	

In Fromberg's analysis silica is deficient, and sulphuric acid abundant.

The following figures show the mean results of analyses of the ash of wheat-straw, grown under ten different conditions as to manuring, during two consecutive periods of ten years each, by Sir J. B. Lawes and Dr Gilbert, at Rothamsted:—

	10 years. 1852-61.	10 years. 1862-71.
Pure ash	55.6	55.6
Ferric oxide	0.32	0.22
Lime	2.86	3.50
Magnesia	0.81	1.03
Potash	11.19	10.46
Soda	0.23	0.34
Phosphoric acid	1.75	1.77
Sulphuric acid	2.42	2.25
Chlorine	1.95	2.17
Silica	34.48	34.28

These figures show the quantity of each ash-constituent per 1000 dry substance of straw.

Barley-straw.

Barley - straw is soft, has a clammy feel, and its odour, with its chaff, when newly threshed, is heavy and malt-like.

As will be shown in the section dealing with the different varieties of foods, barley-straw is not much relished by live stock, but cut into chaff it is used to a considerable extent in pulped mixtures for cattle. It does not make a good thatch for stacks, being too soft and difficult to assort in lengths, apt to let through the rain, and rot.

Ash.—The ash of barley-straw contains these ingredients :—

	Boussingault.	Sprengel.
Potash	9.20	3.43
Soda	0.30	0.92
Lime	8.50	10.57
Magnesia	5.00	1.45
Oxide of iron and a little oxide of manganese	1.00	0.65
Alumina	...	2.78
Phosphoric acid . .	3.10	3.06
Sulphuric acid . .	1.00	2.25
Chlorine . . .	0.60	1.33
Silica	67.60	73.56
	96.30	100.00
Percentage of ash . .	7.00	5.24

Strength of Straw.—"There exists a popular notion that strength of straw is dependent on a high percentage of silica ; but direct analytical results clearly show that the proportion of silica is, as a rule, lower, not higher, in the straw of the better-grown and better-ripened crop—a result quite inconsistent with the usually accepted view, that high quality and stiffness of straw depend on a high amount of silica. In fact, high proportion of silica means a relatively low proportion of organic substance produced. Nor can there be any doubt that strength of straw depends on the favourable development of the woody substance ; and the more this is attained the more will the accumulated silica be, so to speak, diluted — in other words, show a lower proportion to the organic substance." [1]

Oat-straw.

Oat-straw is used mostly as fodder, being too valuable for litter.

[1] Fream, *The Rothamsted Experiments on Wheat, Barley, &c.*

Ash.—The composition of the ash of oat-straw is as follows :—

	Levi. KURHESS.	Boussingault. ALSACE.
Potash . .	12.18	26.09
Soda . . .	14.69	4.69
Lime . . .	7.29	8.84
Magnesia . .	4.58	2.98
Oxide of iron .	1.41	2.24
Phosphoric acid .	1.94	3.19
Sulphuric acid .	2.15	4.37
Chlorine . .	1.50	5.00
Silica . . .	54.25	42.60
	99.99	100.00
Percentage of ash .		5.10

Chaff as a Foot-warmer.—The chaff of all the cereals is an admirable conserver of heat. Poachers in Scotland, when sitting out in winter nights in wait for ground-game, have effectually kept their feet from getting cold by letting them lie in a bag containing dry chaff. A bag of chaff may not be a convenient, but it is certainly a most effective, foot-warmer.

Rye-straw.

Rye-straw is small, hard, and wiry, quite unfit for fodder, and would be an unmanageable litter in a stable, though useful in a court, in laying a durable bottoming for the dunghill. It makes excellent thatch for stacks. It is much sought for by saddlers for stuffing collars of posting and coach horses. It is also in great request by brickmakers. Bottles of Rhine wine are packed in rye-straw.

Rye-straw is sometimes three or four times as heavy as the grain, which is a remarkable feature in this straw.

The plaiting of rye - straw into hats was practised as long ago as the time of the ancient Britons. Bee-hives and *ruskies*—baskets for supplying the sowers with seed—are beautifully and lightly made of rye-straw.

The ash of rye - straw contains these ingredients :—

	Will and Fresenius.
Potash . . .	17.36
Soda . .	0.31
Lime . .	9.06
Magnesia .	2.41
Oxide of iron .	1.36
Phosphoric acid .	3.82
Sulphuric acid	0.83
Chlorine .	0.46
Silica .	64.50
	100.11

Percentage of ash, about 4.

Ash of Straw.—100 lb. of the *ash* of the above sorts of straw gave the following weights of these constituents:—

Constituents.	Wheat-straw.	Barley-straw.	Oat-straw.	Rye-straw.	Bean-straw.	Pea-straw.
	lb.	lb.	lb.	lb.	lb.	lb.
Potash	0½	3½	15	1	53½	4¾
Soda	0¾	1	a trace	0½	1½	..
Lime	7	10½	2¾	6	20	54¾
Magnesia	1	1½	0½	0½	6½	6¾
Alumina	2¾	3	a trace	1	0¼	1¼
Oxide of Iron	..	..	..			0½
Oxide of manganese	..	0½	a trace	..	0¼	0¼
Sulphuric acid	1	2	1½	6	1	6¾
Phosphoric acid	5	3	0¼	2	7¼	4¾
Chlorine	1	1½	a trace	0¾	2¾	0¼
Silica	81	73½	80	82¼	7	20
	100	100	100	100	100	100

On comparing these numbers, one cannot fail to remark the large proportion of potash in bean-straw; the trace of soda in all the straws except the pea; the large proportion of lime in pea-straw compared with bean-straw; the large proportion of silica in wheat- and oat-straw compared with pea-straw and bean-straw; and the large proportion of phosphoric acid in bean-straw compared with oat-straw.

Yield of Straw.—The value of straw may be estimated from the quantity usually yielded by the acre, and the price which it realises. Arthur Young estimated the straw yielded by the different crops—but rejecting the weaker soils—at 1 ton 7 cwt., or 3024 lb. per English acre. Mr Middleton estimated the different crops in these proportions:—

	cwt.		lb.	
Wheat-straw	31	or	3472	per acre.
Barley "	20		2240	"
Oat "	25		2800	"
Bean "	25		2800	"
Pea "	25		2800	"
Average rather more than	25		2822	

or 1 ton 5 cwt. 22 lb. per English acre. In the immediate vicinity of Edinburgh, the produce, both in Scotch and imperial measures, per acre, has been found to be as follows:—

						Stones.		lb.		ton.	cwt.	lb.
Wheat-straw,	9	kemples of	16 st.	of	22 lb. =	144	or	3168	or	1	8	32
Barley "	7	"	"		" =	112		2464		1	2	0
Oat "	8	"	"		" =	128		2816		1	5	16
Average	8	"	"		" =	128		2816		1	5	16

or 1 ton 5 cwt. 16 lb. per Scotch, or 1 ton 0 cwt. 3 lb. per imperial, acre.

Ancient Uses of Straw.—The Romans used straw as litter, as well as fodder, for cattle and sheep. They considered millet-straw as the best for cattle, then barley-straw, then wheat-straw. This arrangement is rather against our ideas of the comparative qualities of barley and wheat-straw; but the hot climate of Italy may have rendered the quality of barley-straw better, by making it drier and more crisp, and the wheat-straw too hard and dry. The haulm of pulse was considered best for sheep. They sometimes bruised straw on stones before using it as litter, which is analogous to having it cut with the straw-cutter. Where straw is scarce, they recommend the gathering of fern, leaves, &c., which is a practice that may be beneficially followed in this country, where opportunity occurs. Varro says, "It is the opinion of some that straw is called *stramentum*, because it is strawed before the cattle."[1]

[1] Dickson's *Husb. Anc.*

Straw as food for stock will be dealt with in the section of this work entitled "Food and Feeding."

CROSS-FERTILISATION OF GRAIN.

The following notes on the cross-fertilisation of grain have been prepared for this work by Mr John Speir.

Degeneracy of Grain.—Most varieties of fixed types of plants appear to degenerate or become weakly after having been subjected, for a number of years, to the forcing influences of modern cultivation. Comparatively speaking, indeed, only a short time elapses between their introduction and the time when they commence to show signs of decay. With the grains this is in part averted by repeatedly and continuously using seed grown in some different locality, so that their rate of degeneration is slow in proportion to that of some other farm crops—potatoes, for instance.

As a rule, however, new varieties of grain, if otherwise good, are more vigorous in growth than most old ones, and in consequence their production is a matter of great importance to the arable farmer. The grains have not been improved to an equal extent with most other farm crops.

Mr Knight's Efforts.—Previous to the middle of the nineteenth century most of the new varieties of grain were natural crosses or sports, which were perpetuated and increased by selection. It appears that Mr Knight, a celebrated horticulturist who lived during the latter half of the eighteenth century, introduced a considerable number of new varieties of grain; but although he was aware how cross-breeding was done, it is unlikely that he obtained any of the varieties he introduced by directly crossing them. His method of procedure was to grow a number of varieties together, in the hope that a favourable natural cross might be produced. In this way he was able to introduce several new varieties, which were of such a strong constitution that, during the years 1795 and 1796, when most grain in this country was blighted, the varieties thus obtained are said to have more or less escaped.

Mr Raynbird's Experiments.—In 1851 Mr Raynbird and Mr Maund showed ears of cross-bred wheats at the great International Exhibition held in London in that year. These are supposed to be the first direct cross-bred grains which were ever offered to the public; and although many of them were considered more as curiosities than anything else, still one of them attained considerable popularity as Raynbird's Hybrid in after-years.

Mr P. Shirreff's Experiments.—About this date Mr Patrick Shirreff of Haddington commenced his experiments in cross-breeding and selection. In the twenty years or so during which he persevered in the work, he succeeded in introducing several new varieties; but although he may be considered the first methodical cross-breeder of grain, he still says he was as successful in getting new varieties from mixtures by natural crossing as from those directly fertilised.

Other Experiments. —About the year 1882 Mr Sharman, of the firm of Messrs James Carter & Sons, London, commenced experiments in the cross-breeding of wheats, which have been attended with a good deal of success. These experiments have been since carried on, and several new varieties have been offered to the public, most of which, as far as appearance of the grain is concerned, look well. All more or less differ in character, some having long straw, and some short. Others have slender straw, while many are stout; some are very early, while others ripen about the usual time. Messrs E. Webb & Sons, Wordsley, Stourbridge, also carried out extensive experiments upon the cross-breeding of grain, and here again considerable success has been attained.

Process of Cross-fertilisation.

In regard to the cross-breeding of grain it may be mentioned that in the vegetable world, as well as among animals, there is a male and female, and the process consists in fecundating the female of one variety with material called pollen taken from the male of another. The process, although a little delicate, is not by any means difficult, and to carry it out does not require any special training in, or knowledge of, botany.

Organs of Fructification.—The accompanying sketch, fig. 413 (for the use of which we are indebted to Messrs A. & C. Black), represents the organs of fructification, much enlarged, of a spikelet of wheat, the chaff-scales having been removed for the sake of convenience. The round part, o, is the ovary, and what ultimately becomes the grain; the feathery parts, s, are the two styles, or female portions of the flower;

Fig. 413.—*Organs of fructification in wheat.*

while e represents the three stamens, or male portions of the flower. The tops of stamens are called anthers, while the tops of the styles are called stigmas. In all the grains the organs of fructification are very much alike, so that what is said regarding one, as a rule will apply to all. For the purpose of effecting cross-fertilisation of any variety, the anthers, e, are cut away before they are old enough to have deposited any pollen on the stigmas. If such has happened, cross-fertilisation cannot be effected, and all labour in that direction will be lost.

Details of the Process.—As wheat is perhaps the easiest of all the grains to fertilise artificially, a description of the process, as applied to it, may be given. A variety having been selected, the stigmas of which it is desired to impregnate with the pollen-dust from the anthers of some other variety, the ear is taken as soon as it comes out of the sheath, and all the seed-vessels or spikelets are cut off except one, two, or three. This mutilation of the ear assists considerably the future operations, and if more than one seed-vessel is left on each ear, they should be left as far apart as possible. An ear is now procured of the variety which it is intended to use as a male parent, and which, if possible, should be from four to six days out of the sheath, while the ear which has been prepared, and on which it is intended to operate, should not be over two days out of the sheath, otherwise risk of self-fertilisation will be run.

For convenience in carrying out successfully the delicate process of fertilisation, the operator should provide himself with a very small pair of forceps, so as to be able readily to pluck out the anthers from the one flower, and lift up those of the other. These may be made of a strip of thin steel, brass, or tin, about a couple of inches long, and quarter of an inch wide. Both ends of this strip are narrowed to about one-sixteenth of an inch broad at the points, the strip being then carefully bent over a lead pencil placed at the middle, while the two points are brought together and held in position by the finger and thumb. The ear, which it is intended to make the male parent, is then taken, and the spikelet gently opened by pressing the point of one of the fingers on the tips of the glumes, B, and palea, A (chaff), fig. 414. The chaff-scales having been thus opened, the anthers, e, will be exposed to view. The

slender stems which support these are called filaments, which the operator now takes hold of with the forceps, and plucks out, laying each on a sheet of paper in order to be readily taken hold of again when required.

Enough anthers having been procured, the prepared ear, which it is intended to make the female parent, is taken, the chaff-scales *very carefully opened* as already described, and the anthers plucked out. If both ears have been taken at the proper stage, the anthers of the one which it is intended to make the female parent will present a decided greenish tint, while the others will be more of a cream colour.

In plucking the anthers from the female parent, care should be taken to

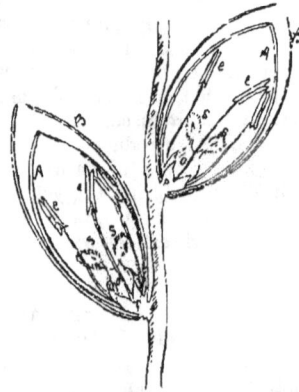

Fig. 414.—*Organs of fructification in wheat.*

catch them by the *filaments only*, otherwise, if caught by the anthers (if too ripe), a portion of the pollen might be shed on the stigmas, causing self-fertilisation. While the chaff-scales are being held open with the one hand, the anthers on the sheet of paper should be caught by the forceps, and dropped on the top of the stigmas, the chaff-scales or palea being then *very carefully closed.* In putting in the anthers, they are none the worse, but all the better, of being caught and pressed by the forceps, as if nearly ripe this forces out the pollen, there being no occasion to catch them by the filament, as when taken out.

In the most of cases it will be found fully as easy to convey the pollen from the one plant to the other by a small fine-haired brush. This is first drawn several times across a ripe anther, and in the process a considerable amount of pollen adheres to the brush. The brush is now gently passed over the stigmas, and in doing so sufficient pollen is generally left to fertilise them. If an anther is quite ripe, pressure by the forceps is sufficient to cause the case to burst and the pollen to be shed on the stigma ; and if everything is in proper condition, as good results will probably be obtained from this method as from any of the others.

Care should, however, be taken not to bruise the feathery stigmas, otherwise fertilisation will not proceed. If the flowering glumes or chaff-scales are *not most accurately closed*, damp gets in and rots the feathery portion of the stigma, thus preventing fertilisation.

The pollen-dust retains its fertilising properties for several days, so that, although the female parent is not ripe enough for fecundation when the operation is performed, it becomes so very soon after, and long before the pollen-dust becomes useless.

After fertilisation the ear should be securely tied to a stake and labelled with the names of both parents.

Time of Natural Fecundation.— It is a general belief among farmers that the grain is being fecundated when the anthers—or bloom, as it is called—appear on the outside of the ear. Such, however, is not the case, as fecundation has already been carried out, the expulsion of the anthers being an effort of nature to rid herself of what is now so much useless material, and the presence of which might interfere with the formation of the grain. The plant opens the chaff-scales and thrusts these out in good weather only, and as soon as they fall off by decay, or are broken off by the wind, the flowering glumes are again closed.

Good Weather Essential.— At this stage of the life of the plant, good weather appears to be necessary, not for the fertilisation of the plant, as has generally been supposed, but to prevent damp getting inside the glumes at the time they are partially opened to get clear of the anthers. The farmer's idea, that good weather is necessary at this stage to ensure a full crop, is thus well founded, although its effect is slightly different from what it is popularly supposed to be.

Period for Crossing.— In order to prolong the period during which crossing may be successfully carried on, a portion of the plants with which it is intended to operate should be cut over near to the ground before and after the stalks are formed, which has the effect of producing a late crop of ears. In this manner the period of crossing may at least be doubled.

First Year usually Unsatisfactory. —Seeds of grain which are produced by artificial crossing have a habit of always presenting themselves the first year in anything but a pleasing form. Whether or not this is brought about by injury to the stigmas or ovary during manipulation, or by the imperfect closing of the chaff-scales, it is difficult to say, so that inexperienced experimenters should not be discouraged when they are in the first year rewarded for their trouble with a badly formed or badly coloured grain, as the next season may quite change its character.

Time Required to Fix Type.— It is at least the second year, and often many years afterwards, before a variety can be obtained true to type. It is at this stage that the Mendelian law comes into force. The single grain which is the result of the first year's cross never produces grain similar to itself. In the following year it may produce from 50 to 100 grains, each of which, although on the one root or on the one stem, may be a different variety. One half of these will generally be permanent, and will produce grain true to their kind, but each of the other half will produce grain which will differ as much from each other in the following year as each seed in the ear of the first year did from its neighbours. Of those which have fixed characteristics, one-half, or quarter of the whole, will resemble the seed-bearing parent, while the other will take after the pollen-bearing one. A similar percentage of the remaining grains become fixed in the following year, and so on. It is only recently that Mendel's laws became generally understood, and with the aid of this knowledge the work of the hybridiser will be considerably simplified and less-ened.

Percentage of Success.—In a favour-able season, and in the hands of an ex-perienced operator, from 25 to 75 per cent of the spikelets operated on may produce grains, while, if the operation is clumsily done, none may be produced.

Protecting the Ears.—As soon, how-ever, as it is seen that the flowers have set, the ears should be encircled by fine wire gauze, or strong muslin, to prevent birds destroying the grain. The opera-tion of crossing, for the sake of con-venience, is generally performed near the side of a wheat plot; and the fixing of a stake to each plant is a necessity for identification, and this stake is almost sure to be made a resting-place by the sparrows and other small birds which infest the sides of wheat fields, so that if unprotected many grains are sure to be lost.

After Culture.—When the grains are ripened and thoroughly dried, they should at once be sown in 3- or 4-inch pots, one in each, in which they may be grown till late autumn or early spring, when they should be transferred to a piece of specially prepared land in the middle of an ordinary wheat field. Here they should be planted at least one foot asun-der each way, with a space a foot or two clear from the ordinary crop. By giving the plants so much room, each tillers to its full extent, while the grain when ripe runs little risk of being stolen by birds.

MESSRS GARTON'S WORK AS HYBRIDISERS.

Of all who have made attempts in the hybridisation of grain, no one else has attained anything like the success of the brothers Garton of Newton-le-Willows, Lancashire. Their business was flour-milling, and, of course, they had an in-terest in grain. As an amusement and hobby they made several attempts at crossing cereals between 1880 and 1883; but owing to the use of plants in which the organs of reproduction were too far advanced, their efforts were fruitless. After 1884 they used ears in a much earlier stage of development, and from that date forward their success may be traced. They very soon became expert hybridisers, and were not only successful with wheat,—as already said, perhaps the easiest of all the grains to fertilise, —but were soon almost equally successful with oats and barley.

At first Messrs Garton used only varieties common to the district or country, but by and-by they drew on the whole world for wild and culti-vated varieties from which to get some special quality which they desired to engraft on the new varieties they were about to produce. In a few years they had an immense number of varieties, amounting to many thousands, each of which they grew in short rows of 10 yards. By 1893 they had about 10 acres devoted to small patches of new varieties, and in order to let the outside public know something of their work, they decided to exhibit samples of their new sorts of grain at that year's show of the Highland and Agricultural Society of Scotland at Edinburgh. The directors of the Society were much impressed with the Messrs Garton's exhibit, and the botanist to the Society, Professor M'Alpine, and Mr Speir, Newton, were delegated to inspect their trial-grounds just before harvest, and report on them.[1]

So slow is the work of the hybridiser that, although the Messrs Garton had been steadily at the work from 1884, they were unable to make any public display of their work till 1893, and it was not till 1898 that they were in a position to put any new varieties in the hands of the public. Early in 1898 they presented to the Highland and Agricul-tural Society one bushel each of three varieties of oats, which were handed to Mr Speir of Newton, to be tested and reported on. These were unnamed at the time, but were ultimately designated Tartar King, Waverley, and Pioneer. These and others were grown at New-ton for two seasons, and reported on in the *Transactions* of the Society for 1900.

The most notable of the varieties of oats introduced by Messrs Garton has been *Abundance*, a cross between white August and white Swedish, which was first offered to the public in 1892. This is a white oat of good colour and milling

---

[1] *Trans. High. and Agric. Soc.*, 1894.

quality, which produces a large quantity of grain, and is now in general cultivation under various names all over the world.

Waverley is a longish white oat of heavy cropping quality. Its pedigree is as follows:—

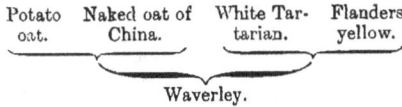

Potato oat.	Naked oat of China.	White Tartarian.	Flanders yellow.

Waverley.

This oat is very popular in Scotland. It was introduced in 1898.

Another oat of a different type—viz., *Goldfinder*, a yellow oat—was introduced in 1899. This oat has the following pedigree:—

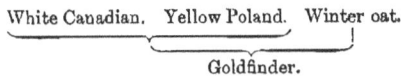

White Canadian.	Yellow Poland.	Winter oat.

Goldfinder.

Goldfinder has exceptionally well-flavoured grain and straw, and few sorts yield a higher proportion of meal to grain. Yet, on account of its yellow colour, it has never become a very great favourite with either farmers or merchants. It is a very heavy cropper, and, because of its colour, the grain is generally consumed on the farm. This oat is rather slow in ripening, but being a hardier oat than most varieties, it can be sown very early in the season where the circumstances permit, and where this is done it ripens about the ordinary period.

Bountiful is one of the most noted black oats which the Messrs Garton have introduced. It has the following pedigree:—

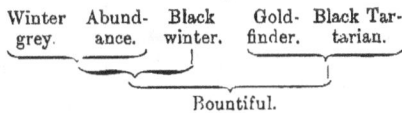

Winter grey.	Abundance.	Black winter.	Goldfinder.	Black Tartarian.

Bountiful.

Bountiful is thus descended from other varieties of exceptional merit, which have been in cultivation for only a few years. Being from such well-known heavy croppers as Abundance and Goldfinder, it is not to be wondered at that it is promising well. It is a black oat of exceptional dark colour.

An oat of quite a different kind, and suited for altogether different circumstances, is *Tartar King*. It was intro-

duced in 1898, and has the following pedigree:—

Black Tartarian.	White Tartarian.	White Canadian.

Tartar King.

Tartar King is a very early oat, with the grain all on one side of the ear. Its straw is of a remarkably stiff quality. It is specially adapted for lands which are in a high state of cultivation, or where the crop is inclined to lodge. The grain is coarse and husky, while the straw, like most other varieties having much of the Tartarian strain in them, is not very well suited for fodder purposes.

The Messrs Garton have also introduced several new varieties of barley, of which the following have given the best results: *Standwell, Maltster, Ideal,* and *Eclipse.* Each of these varieties has some good quality peculiar to itself. Eclipse is a six-rowed barley of the Chevalier type, which gives a very heavy yield of grain of good quality.

Messrs Garton also introduced several new varieties of wheat, but these do not seem to have met with the same success as their oats.

They have recently put on the market a perennial Italian rye-grass, which is a cross between perennial and Italian rye-grass. It seems as if it would be a useful addition to our grasses, but as yet has been too short a time in cultivation to be spoken of with great confidence. A perennial red clover has also been recently introduced by the firm. It is a cross between the ordinary red and wild red clover. Hitherto it has done well, but as yet has not had a very extended trial.

Progress of Hybridisation.

The possibilities of the cross-fertilisation of cereals have, since the work of the Messrs Garton was made known, been appreciated in a way they seem never to have been before. Not only are there now eminent scientists all over the world devoting attention to the subject, but there are also Government departments of various countries which have specialists set apart for this particular work. There are likewise various public companies which lay themselves out for the propagation and sale of new varieties.

It therefore seems probable that in the near future considerably more progress may be made in this department than has hitherto been the case. That this work is much needed is admitted by every one. The standard varieties of grain in every country are difficult to supersede, yet it is the case that their number is being gradually reduced.

MENDEL'S LAWS AND THE IM-PROVEMENT OF GRAIN.

The discovery of the operation of what are known as Mendel's laws of heredity has supplied a wonderful stimulus to those engaged in the production of new and improved varieties both of plants and animals. In a paper on "Heredity in Plants and Animals" in the *Transactions of the Highland and Agricultural Society of Scotland*,[1] by Professor Wood and Mr R. C. Punnett of Cambridge University, there is given an interesting *résumé* of experimental work with plants and animals, showing the operation of Mendel's principles. In reference to Mendel and his work the writers say: "Thanks to Mendel's remarkable discovery, we are at last beginning to recognise the true meaning of heredity, and to realise the great powers of control over living things with which this knowledge endows us. We begin to understand many of the mysterious things that happen when crosses are made among animals and plants—why a character often skips a generation, why the type is often broken to give rise to new forms, and what is the meaning of reversion. The foundations of this knowledge were securely laid by Gregor Mendel, an Austrian monk, in the garden of the monastery of which he afterwards became the head. Mendel has been dead for nearly thirty years, and it was as long ago as 1865 that his discovery was first given to the world. But his ideas were in advance of his time; they excited little interest and were soon forgotten. It was not until

[1] Fifth ser., vol. xx., 1908, p. 36.

1900 that his paper on the pea was unearthed, and scientific men began to realise what a far-reaching discovery this was that Mendel had made so many years ago. As a young man he had studied the natural sciences in Vienna, and had become interested in the problems of heredity. On returning to his monastery he devoted much of his leisure to carefully investigating the manner in which characters are transmitted in the common pea. From the results of his experiments he deduced certain principles which he found to hold for all the various characters he studied. During the past few years these principles have been confirmed and extended, not only for many plants but for animals as well."

WHEAT BREEDING ON MENDEL'S PRINCIPLES.

Experimenters have already attained encouraging success in the application of Mendel's laws to the raising of improved varieties of grain. A prominent investigator in this direction is Mr R. H. Biffen, Botanist to the Department of Agriculture in the University of Cambridge. In regard to work carried out by him we take the following from the paper above referred to :—

"On the rediscovery of Mendel's laws in 1900, Mr Biffen began to work out the inheritance of various characters in wheat and other farm crops. Some of his results are shown in fig. 415 A, the following description of which will best explain the method.

Inheritance of Beard and length of Ear.

"Two varieties, P, P, were crossed which differed only in two characters— viz., beard and length of ear. The first cross, F_1, is beardless, and we express this in Mendelian terms by saying that the beardless condition is dominant, the bearded condition recessive. The inheritance of length of ear shows neither dominance nor recessiveness, for in the first cross, F_1, the length of the ear is intermediate between that of the parents. The characters of the first cross are exactly the same, whichever way the cross is made, and all the individuals of which it is composed are practically identical.

Fig. 415.—A. *To illustrate the results of crossing two wheats differing in two pairs of characters.*
B. *To illustrate the results of crossing two barleys differing in two pairs of characters.*

Natural cross-fertilisation is so rare among the cereals that the first cross may be left to itself to produce seed by self-fertilisation. The seed so produced is sown, and the second generation, F_2, is found to be no longer uniform. In 1899 we should have said that the type had been broken, and that new variations had arisen without law or order. Now, in the light of Mendel's conception of unit characters, we realise that this is merely a rearrangement of existing characters in new combinations.

"These points are well illustrated by the six types shown in the figure.

"Type 1 has inherited the dense ear from one parent, the beardless condition from the other.

"Type 2 combines both the characters of the dense-eared bearded parent.

"Type 5 combines both the characters of the long-eared beardless parent.

"Type 6 has inherited the long ear of one parent, the beard of the other.

"In Types 3 and 4 the intermediate length of ear of the first cross is reproduced with and without the beard.

"If a numerous second generation is grown, and the plants showing the above six types sorted out and counted, it is found on the average that out of every sixteen plants there are three of Type 1, one of Type 2, six of Type 3, two of Type 4, three of Type 5, and one of Type 6.

New Types.

"It is at once evident that in the second generation there are two new types, Nos. 1 and 6, whose novelty consists in a recombination of characters existing in one or other parent. Thus the beard of the short-eared parent has been removed from No. 1 and transferred to No. 6, which thus starts a new bearded long-eared type. But No. 1 has thus been left with the short-eared character deprived of beard, and this combination is also new.

How to pick out fixed Types.

"Now in making a new variety we want not only to get a few specimens showing the desired characters, but to isolate a fixed type and grow enough to distribute for seed. Here, too, Mendel's work helps us. Mendel defined as recessive those characters which disappear in the first cross. In the case under consideration the beard is a recessive character, and from the definition a bearded individual of the second generation cannot contain the beardless character. For this is dominant over beard, and the individual would be beardless. In other

words, an individual of the second generation showing a beard or other recessive character must be pure as regards that character. We know, therefore, that seed of Types 2, 4, and 6 must, if grown, produce a pure bearded progeny.

"Again, an individual carrying both the long- and short-eared characters has, as in the first cross, ears of intermediate length. This intermediate length is, so to speak, the mark of the mongrel, and the seed of individuals having it will produce long, short, and intermediate ears as shown by the second generation. Types 3 and 4, therefore, will not breed true, and are useless. A long-eared type and a short-eared type must be pure, and will breed true. Types 2 and 6 are therefore pure as regards both characters, and can be at once taken as fixed.

"We are now left with Types 1 and 5. These must be pure as regards length of ear, but they contain the dominant beardless character, which may be masking the recessive bearded character. In order to pick out the pure beardless, we must resort once more to the breeding test. Seed must be collected from a number of plants separately and sown in separate plots. We shall then find that, on the average, one plant out of every three will breed true to the beardless condition. Further, Mendel indicates, and experience confirms, that if a plant begins to breed true it will continue to do so.

Inheritance of other Characters.

"We have now seen how to obtain new combinations by crossing two varieties of wheat differing in only two characters, and how to pick out those individuals of the new types which will breed true. Mr Biffen has worked out the inheritance, not only of the two characters quoted above, but of all the obvious characters of wheats, such as colour of grain and chaff, presence or absence of hairs, shape of glumes, and so on. Particulars are given in his papers in the *Journal of Agricultural Science*. He has also succeeded in working out the inheritance of the far more practically important characters of strength and disease-resistance, and his results in this direction are of such great interest that they must be given in some detail.

Definition of Strength.

" Strength is the term used by millers and bakers to sum up a number of properties which together make a flour valuable for baking. A wheat is said to be a strong wheat when it produces a flour which will make, when baked, large shapely loaves. The strongest wheats on the English market come from Canada and Russia. We may give a numerical value to strength by marking the very best samples of these foreign wheats at 100. On this scale ordinary samples of these wheats must be marked at 90-95, and good average samples of home-grown wheat at 60-65. The difference in market-price between a wheat marked at 95 and a wheat marked at 65 is about 5s. a quarter.

A Strong Parent.

"The best Canadian wheat, graded as No. 1 Hard Manitoba, is composed of at least 60 per cent of one variety known as Red Fife. This variety has been grown in England under its own name for seven years, and as Cook's Wonder for fifteen years, without the least sign of losing its strength, and we are therefore fairly justified in assuming that it breeds true to strength under English conditions. But it is not a suitable wheat for the British farmer, as it seldom yields a crop of more than about three quarters per acre. Now our common home-grown varieties are, as a rule, excellent croppers, but very deficient in strength.

Inheritance of Strength.

"We have, therefore, two kinds of wheat—Red Fife, which is strong but a bad cropper, and any of our home-grown sorts, which are weak but good croppers. If now strength and cropping power are characters which can be inherited according to Mendel's laws, like beard and length of ear, then all we have to do is make a cross between these two sorts, and we shall be able to pick out from the second generation fixed forms in which a recombination of strength from the Canadian parent and cropping power from the English parent have taken place. We shall thus obtain a new type of wheat which crops as well as any of our ordinary

varieties, and yields grain which, in virtue of its strength, is worth say 5s. per quarter above the market-price of present-day home-grown wheat.

New Types.

"Mr Biffen has tried the experiment outlined above, and proved that strength is a dominant character. He has picked out from the second generation a number of fixed types, the best of which have been grown for two years in field-plots, of which some in 1907 were an acre in extent. The soundness of Mendel's method of picking out the fixed types is shown by the fact that, even among the enormous number of plants growing on an acre plot, there was no sign of reversion to parent forms or any other departure from the type which had been picked out. Both in 1906 and 1907 grain from some of these new fixed types was sent to be tested in the mill and in the bakehouse. In 1906 they were marked at 84 to 88, in 1907 at 88 to 90. Remembering that ordinary Fife is 90 to 95 and ordinary English only 65, it is clear that these new types have to all intents and purposes inherited the strength of their Canadian parent and are breeding true. Up to the present the object has been to increase the stock of seed as rapidly as possible, and very thin sowing of under 1 bushel per acre has been practised. It is impossible to say, therefore, how well they will ultimately crop, but even with this excessively thin sowing a yield of 42 bushels per acre was produced this year by two of the new varieties. Adjacent plots of Square Head's Master and Browick, similarly sown and manured, yielded only 39 bushels per acre.

"With the seed of the best of these new types something like ten acres have been sown this autumn on the Cambridge University farm,—an area large enough to give a reliable verdict as to their cropping power, and if the verdict is favourable, to grow seed enough for extended tests throughout the country.

Damage due to Rust.

"A short time ago some striking figures were published showing the money value of the damage done annually to cereal crops by such fungoid diseases as rust.

According to the International Bureau of Plant Pathology, the loss of crop value in Germany alone amounted in 1891 to £20,000,000, or about one-third of the total value of the crop. It is estimated that the annual loss throughout the world is not less than £100,000,000. These diseases are so widely spread throughout the country that remedial or preventive measures are practically out of the question. The only hope of reducing the damage they cause is to produce immune varieties of crops.

A Rust-proof Parent.

"During the last seven years Mr Biffen has had under observation on the University farm plots of almost every known variety of wheat. These plots are crowded together in a cage of wire-netting to keep off birds, and it has been noticed every year that, whilst most of the varieties have suffered greatly from rust, one of them has almost kept entirely free. It is safe to assume, therefore, that this variety is immune to rust, or, in other words, rust-proof. This variety is a small-eared bearded variety without a single good character except its immunity to disease. It occurred to Mr Biffen that this immunity might be a character which could be inherited in accordance with Mendel's laws, in which case it could be transferred to any other variety just as the beard in the case already described.

Inheritance of Immunity.

"Another variety had been noted every year for its excessive susceptibility to rust, which was so great that in bad rust years it was unable to set seed. This susceptible variety was bearded like the immune one, but, when not destroyed by rust, formed long ears. These two varieties were crossed together, and the first cross came with ears of intermediate length, and was so susceptible to rust that it set seed with difficulty. How-

ever, enough seed was obtained to grow a fairly numerous second generation, when it was found that, on the average, one plant in four of this generation was rust-proof. This experiment has been repeated several times, and several thousand second-generation plants have in all been counted. The evidence is quite conclusive that immunity to rust is a recessive character. This being established, the transference of immunity from the original short-eared bearded variety, in which it was found, to the new types, in which strength and cropping power have been already combined, is merely a matter of routine, requiring one year to grow the first cross, a second year to grow the second generation, from which the first recessive rust-proof forms can be picked out at once, and finally two or three years to increase the stock. We shall then be possessed of wheats which will bake as well as the best Canadian, and, since they will be free from disease, crop better than any varieties at present on the market.

Possibilities of further Work.

"This demonstration of the possibility of transferring immunity to disease from one variety to another is an achievement of the greatest scientific and practical importance. It seems to offer the most hopeful method of checking disease yet suggested. If immune strains of any species of plant or animal can be found, no matter how useless they may be from other points of view, there seems to be no reason why their immunity should not be transferred by crossing to our present valuable varieties which are being ravaged by disease. In this way it may be found possible to produce high-class horses and cattle immune to the diseases which are the scourge of the South African farmers, sheep which will be proof against anthrax, black quarter, and other diseases that attack sheep at home, and pigs immune to swine fever."

SOWING CEREALS.

SEED-TIME.

For practically all his crops, excepting wheat, the farmer has his seed-time in the cheery months of spring. On all tillage farms the spring is a time of great stir and bustle. The prognostics and variations of the weather are watched with the keenest interest and anxiety, for not only the progress of the spring work but also the returns of the harvest are greatly influenced by the character of the weather during the seed-time.

Seasonable Working of Land.— Field-work will now be pushed on with all possible speed. Yet there are more points to be considered than the mere progress of the work. In particular, care must be exercised as to the condition in which the different kinds of soils are tilled and prepared for the crops. To stir stiff clay when it is soaked with wet would be ruinous. Better delay a little than commit the seed to a cold, unkindly, ill-prepared seed-bed. Better let the men and horses stand idle for a few days than run the risk of destroying the year's produce by working the land in an unseasonable condition. On the other hand, when the weather is favourable, and the land in good condition for tillage operations, let all hands do their very best, so that full advantage may be taken of every favourable spell of weather.

Selecting Seeds.

Farmers cannot be too careful in the selection of seeds. It matters not what the crop may be, the best possible seed should be secured. To ensure thoroughly reliable seeds of a high character, an extra outlay of a few shillings per acre may be entailed, but then these *few shillings* may add pounds to the value of the crop.

Improvement in Seeds.—In this matter of seeds, the farmers of the present day are well situated compared with their brethren in former times. The development of the Seed industry is indeed one of the most notable—one of the most beneficial—features in the progress of modern agriculture. The improvement of the animals of the farm has been accomplished on the farms by the stock-owners themselves. Equally important and equally great in its way has been the improvement of the plants of the farm. And this latter work has been carried out in the most thorough and energetic manner by a number of extensive and influential seed firms, who have for many years devoted great attention not only to the improvement of the old varieties of the farm crops, but also to the propagation and development of new varieties of increased producing power. There are many eminent firms who have in this way rendered good services to the country. Amongst the names most prominently associated with this great work of plant improvement are those of Sutton, Carter, Webb, Drummond, Dickson, Garton, and Hunter; but there are several other firms which have also been active in similar well-doing.

The part which these enterprising firms, who give us the improved, selected, and tested seeds, have played in the progress of modern agriculture, has been greater by far than is generally recognised. It has, of course, been a matter of business, not of philanthropy, with them; all the same, it is right to acknowledge the great power which the development of the seed trade has exercised in the advancement of agriculture.

Buy Seeds in Good Time.—Sowing is sometimes delayed by dilatoriness on the part of the farmer in providing the necessary supplies of seeds. Have these on the farm *before* they are required, so that they may be at hand when a suitable time arrives for sowing.

Change of Seed.—It is well known amongst practical farmers that great advantage may be derived by judicious change of seed. As a rule with roots, fresh seed is introduced every year, for it is only in exceptional cases where the farmer grows his own turnip-seed, although this latter practice is with advantage being pursued to an increasing extent. With grain, however, the rule is reversed. The home-grown seed is used for the most part; but it has been clearly shown that by an occasional change from one climate, one soil, and one system of farming to another, the vitality and producing power of a particular kind or "stock" of grain are substantially increased. When one considers the artificial influences by which our improved varieties of grain have been brought to their highly developed condition, one cannot be in the least surprised that such changes of scene and surroundings should often exercise a beneficial effect upon the crop.

But all changes are not successful. Neither are the conditions essential to success very fully known. In almost every change of seed, as in every change of a sire, there is something of the nature of an experiment. As a rule, a change of cereal seed from an early to a late district is followed by much benefit, notably in the earlier ripening of the crop, but also to some extent in the quantity and quality of the produce. The influence on the date of the harvest is most marked. For instance, by the habitual introduction of seed-oats from the south of Scotland every second or third year, the ripening of the crop on certain farms in the later districts of the north-east has been hastened by from six to ten days; and practical farmers acquainted with a late climate know that acceleration to that extent in harvest is a very important advantage—perhaps all the difference between a crop secured and a crop partially lost. The weight of the grain will also most likely be increased 2, 3, or more pounds per bushel. Then in taking seed from a late to an early district there may sometimes be an advantage—notably an increase in the bulk of the produce, though there are many exceptions to this experience.

A good plan in changing seed is to try the change on a small scale in the first year, and if the results are satisfactory, use the variety more extensively in subsequent years. Farmers should be experimenting in this way very frequently, for by introducing fresh varieties well suited to their land, the produce of their crops may be substantially increased. A change of seed from a clayey to a light loamy or sandy soil is generally beneficial.

New Varieties of Farm Plants.—Farmers also derive much benefit by taking advantage of the many new and improved varieties of grain and roots which are brought out by experimenting seedsmen. As already shown, our leading seedsmen are continually engaged in propagating fresh and improved varieties of farm crops, more particularly of grain, mangels, swedes, turnips, and potatoes, and by availing themselves of these new and vigorous sorts of proved excellence, farmers may to a marked extent enhance their produce.

At the same time, it is well to say that caution should be exercised in introducing new varieties. Let them be tried on a small scale at the outset, and adopted extensively only after their suitability and high qualities have been unmistakably established.

Testing Seed.—Farmers should carefully avoid using *weak* or *unreliable* seed. Seeds of all kinds may now be procured pure, and of certain germination. This should always be insisted upon, and farmers should themselves test the seeds when they take them home. Even home-grown seed, however well it may look, should never be sown without having been first carefully tested. This may be done very easily with grain or grass seeds, by placing say a hundred seeds between two folds of damp blotting-paper laid on a meat or soup plate, with another similar plate placed face downwards over that plate. No artificial heat need be used, and the plates may sit on an open shelf in the farmer's parlour. The blotting-paper should be damped every day by sprinkling a little water on it by the hand. The object of having the two plates placed face to face is to cause a current of air to pass over the seeds. In this way cereal seeds will germinate in about a week, and grass-seeds in about three weeks. An efficient testing apparatus may be purchased at a moderate cost.

Some consider it more reliable to test the germination of cereals in pots filled with soil.

Grain-seeds are often tested under a very thin damp turf in a well-exposed spot in the farmer's garden. We have also seen it done on damp turfs, placed on the rafters over the heads of cattle, where, of course, the temperature is considerably higher than outside early in spring, when testing is usually carried out.

Clover, turnip, or any other leguminous seeds may be tested in a more simple and expeditious manner. Count out say 100 seeds, roll them into a piece of flannel, and dip into boiling water for four or five minutes, and on opening the piece of flannel all the reliable seeds will be found much swollen, and actually germinated, with the elementary root shooting out. The seeds which do not present this swollen appearance cannot safely be reckoned upon, and the quantity of seed to be given per acre should be regulated by the percentage of the reliable germinating seeds.

SOWING WHEAT IN AUTUMN.

The autumn is the season in which the main portion of the wheat crop is sown. Indeed, comparatively little spring wheat is now grown in this country. Spring wheat is a more risky crop, and rarely yields so well as the autumn-sown; in fact, the earlier the sowing, and the longer the period of growth that wheat can be given, the better the crop, as a rule. A five-month crop is preferable to a nine-month one, in mere economy of time, of course, but only if the ground is under crop all the time: if the land is lying bare all winter it is losing nitrates which the autumn-sown crop appropriates to good purpose. And where wheat is to be the next crop, the rotation should be arranged to have the ground clear of its predecessor in September or October at latest, so that winter wheat may be sown. Where corn is to follow a root crop, and the latter cannot be cleared off the ground until spring, it is then better to sow barley or oats as the spring corn crop.

Land intended for autumn or winter wheat is ploughed as early in autumn as possible. Indeed, the plough is often at work for the next crop before the harvesting of the preceding crop is quite finished.

Fallow Wheat.—Bare fallow, as we have seen, has now almost disappeared, but any portion of land which has been fallowed during the summer is the first to be prepared for wheat. It has perhaps received a dressing of dung, which may have been drilled in. If so, the first operation is levelling the drills which cover the dung by harrowing them across with one double turn of the harrows. After the land has been harrowed down, any root-weeds brought to the surface should be removed, but the surface-weeds will soon wither.

Ploughing for Wheat.—The land is then feered, to be gathered up into ridges; and if thoroughly drained, or naturally dry, one gathering-up makes a good seed-bed; but wet land, to lie in a good state all winter, should be twice gathered up. The second gathering-up should not be immediately after the first, for a short interval should elapse to allow the land to subside, which rain will accelerate.

Should the fallow have had the dung spread upon the surface and ploughed in with feered ridges in gathering up, the feering left half-ridge at the sides of the field, now that the land is gathered up for the seed-furrow, is converted into one whole ridge.

Grubbing for Wheat.—But a practice came into use, with the introduction of the grubber, that possesses advantages on strong land in a dry state, which is, to put the seed-wheat into the ground with the grubber, upon gathered-up ridges that covered in the dung of the fallow, and to finish the work with one double-tine harrowing along the ridges. When the grubber is so used, the land is gathered up in finished ridges in the fallow, as grubbing cannot alter the form of ridges.

Advantages of Grubbing.—When a tough waxy clod arises on ploughing strong land, rather wet below, for a seed-furrow, or when unsettled weather threatens, the grubber will keep the dry ameliorated soil upon the surface, and accelerate the seed-time considerably.

New or Old Seed.—The land being thus prepared for the seed, it is quite possible for a part of the new crop to be thrashed out for seed in time for sowing in autumn; but those who sow early cannot procure new seed, and must use the old. But although the new crop were secured in good time to afford seed for sowing in autumn, it is better to sow old wheat than new. New wheat germinates quicker than old, but is more easily affected by bad weather and insects; and consequently its braird is neither so thick nor so strong as from old wheat — that is, from seed of the preceding year; for very old wheat may have been weakened in vitality even in the stack, or been much injured by the weevil in the granary.

Time for Sowing.—Some farmers sow wheat on fallow early in September; and where there is much fallow and strong land this is a proper season to begin. The objection is, that should late autumn and early winter prove mild, the plant will become too rank before cold weather sets in to check its growth. October is considered by many the best period for sowing winter wheat, as the risk of rank growth is avoided.

Quantity of Seed for Wheat.—The quantity of seed used for wheat varies from 1¼ to 3 bushels per acre. The smallest quantity is used for early autumn sowing, especially on good rich land, and the largest for sowing in spring. The questions of thick and thin sowing and different methods of sowing are discussed in the section on "The Germination of Seeds" in this volume.

Variety to Sow.—The varieties of white wheat well suited to be sown in autumn are now so numerous that it is impossible here to indicate which is the best for a particular locality. Upon inferior soils it is safest to sow a red wheat, which, although realising a lower price in the market, will yield a larger increase. In this matter the farmers in the different localities must exercise their own judgment, giving due consideration to the opinions of farmers in the neighbourhood as to the varieties best suited to the locality.

Methods of sowing are dealt with in connection with spring sowing.

Water-courses. — The finishing processes of harrowing and of water-furrowing are the same as in spring; but as water is more likely to stand upon the land in winter, gaw-cuts must be made with the spade in every hollow on the surface and across head-ridges, even on thoroughly drained land, to quickly carry off large falls of rain.

Harrowing.—As regards the harrowing, it is right to leave on wheat-land in winter a round clod upon the surface. Such clods afford shelter to the young plants from wind and frost, and, when gradually mouldered by frost, also increase the depth of the loose soil.

Frost throwing out Plants.—Wherever land is harrowed to a fine tilth in autumn, rain batters its surface into a crust, and frost heaves it up in spring like fermented dough, by which the plants are raised up with the soil, and, when the earth subsides in a thaw, left upon the surface almost drawn out by the roots. Thorough-draining is the only safeguard against rain and frost acting in this manner upon a fine surface in winter.

When land is naturally strong enough to grow wheat, and yet is somewhat soft and wet below, to make it probable that the plant will be thrown out, *ribbing* with the small plough will give a deeper hold to the plant than common ploughing. The wheat is sown broadcast over the ribs, and harrowed in with one double tine along them. Ribbing is not suitable on fresh-ploughed land, as even the small plough would go too deep, and make the drills too wide; nor is it advisable on land that has not been ridged up.

Another mode of preventing the throwing out of wheat on soft land, is first to feer the land into ridges, sow the seed broadcast between the feerings, plough the seed in with a light furrow with the common plough, and leave the surface unharrowed and rough all winter.

Deep sowing — from 1¼ to 1¾ inch deep—is regarded by many as the surest preventive. The sowing of the seed by the drill sower, which is now largely practised, greatly lessens the risk of loss by the plants being thrown out.

Rolling.—Rolling wheat in autumn is rarely practised, except to consolidate soft land. In reference to this point, a writer says: "We have for the last three seasons rolled nearly the whole of the autumn-sown wheat when a favourable opportunity could be obtained, making the surface perfectly smooth and fine. The old idea of rough lumps lying here and there being of advantage for shelter is neither more nor less than an antiquated fallacy. After wheat is sown, the land cannot be made too firm. It is good for all land where a roller can be used without what is called *poaching* the soil. We have tried repeated experiments, and the rolled portions were always the best. It is true the wheat is rolled in spring whenever the season will permit; but these clods are pressed down and for a time retard this, and check the tillering power of the plant. The autumn-rolled wheat requires only to be harrowed or hoed in spring. These observations are made from a three years' trial of the system." A firm seed-bed is unquestionably, as we have already pointed out, beneficial to. wheat. But so are surface clods in winter and spring. The clods lying on or in the surface cannot possibly prevent fresh rootlets pushing their way into the soil around them.

Wheat after Beans.—The bean-land is the next sown with wheat in autumn; and the land occupied by summer tares, or other summer forage crop, if in the same field with the beans, is sown with wheat at the same time. The land is feered and gathered up and sown when the soil has been allowed to subside for a few days.

Where bean-land is strong and the ridges sufficiently round, a four-horse grubber may be used to make the seed-furrow instead of the plough. The grubber succeeds in this case very well as far as the wheat is concerned, and it has the advantage in a late autumn of getting through the work expeditiously, and keeping the aerated soil upon the surface. But on strong soil, not thoroughly-drained, and in a comparatively flat state, grubbing is not the best preparation for wheat after beans, because the seed is apt to rot and the soil become sour.

Grubbed soil may require only one double turn of the harrows along the ridges.

Wheat after Potatoes.—The potato-land, having been harrowed after the potatoes have been raised, is feered and gathered up, and sown with wheat. It is better to let the soil subside a little, although the usual practice is to sow the wheat as soon as it is ploughed, the season getting late by the end of October. In many cases wheat after potatoes is sown broadcast on the un-ploughed surface, the seed in this case being either ploughed in with a light furrow less than two inches deep or covered with a cultivator. The same process is sometimes followed on strong land where the root crop has been consumed on the land by sheep.

Wheat after Grass.—To a large extent wheat follows grass and clover in the rotation. Indeed in Yorkshire and other parts of England wheat is now more largely grown after one year's hay than after any other crop, the seed being sown in the autumn. The skim-coulter plough is used to bury the sod, the Cambridge roller is then put on, the land is harrowed lightly so as not to bring turf to the surface, and the seed is sown with the seed drill.

SOWING WHEAT IN SPRING.

As already stated, only a small extent of the wheat crop is sown in spring, the main portion being sown in autumn.

Land for Spring Wheat.—To ensure a good crop of spring wheat, the land should be for. some time in good heart, otherwise the attempt will inevitably end in disappointment. Wheat cannot be sown in spring in every weather and upon every soil. Unless the soil has a certain degree of firmness from clay, it is not well adapted for the growth of wheat —it is more profitable to sow barley upon it; and unless the weather is dry, to allow strong soil to be ploughed in early spring, it is also more profitable to defer wheat, and sow barley in the proper season. The climate of a place affects the sowing of wheat in spring; and it seems a curious problem in climate why wheat sown in autumn should ripen satisfactorily at a place where spring wheat will not. Experience makes the northern farmers chary of sowing wheat in spring, unless the soil is in excellent condition, and the weather very favourable for the purpose.

Date of Spring Sowing.—In former times, even under the most favourable circumstances, wheat was seldom sown after the first week of March, but later varieties have been introduced which may be sown as late as April.

On farms possessing the advantages of favourable soil and climate, and on which it is customary to sow spring wheat every year, the root-land is usually ploughed with that view up to the beginning of March; and even where spring wheat is sown only when a favourable field comes in the course of rotation, or the weather proves tempting, the land should still be so ploughed that advantage may be taken to sow wheat. Should the weather take an unfavourable turn after the ploughing, the soil can afterwards be easily worked for barley.

Tillage for Spring Wheat.—The land should receive only one furrow—the seed-furrow—for spring wheat, because if ploughed oftener, it would be deprived of that firmness so essential to the growth of wheat.

It is probable that a whole field may not be obtained at once to be ploughed, and this often happens for spring wheat; but when it is determined to sow wheat, a few ridges should be ploughed as convenience offers, and then a number of acres may be sown at one time. In this way a large field may be sown by degrees, whereas to wait till a whole field can be sown at once, may prevent the sowing of spring wheat that season. Bad weather may set in, prevent sowing, and consolidate the land too much after it had been ploughed; still, a favourable week may come, and, even at the latter end of the season, the consolidated land can be ribbed with the small plough, which will move as much of the soil sufficiently as to bury the seed.

Double-furrow Plough.—To expedite the ploughing of the seed-furrow at a favourable moment, the double-furrow plough is used by some, though not so largely now as a few years ago. A double-furrow plough, made by Ransome, Sims, & Jefferies, is shown in fig. 416.

Advantages of the Double-furrow Plough.—The double-furrow plough is usually worked with 3 horses, and as to the question whether it effects a saving of draught as compared with two single furrow-ploughs, there has been much discussion. Experiments with the dynamometer have shown that there is little saving in this respect, and that the 3 horses have to exert about as much force as 4 horses, with 2 common ploughs doing

the same amount of work, with a slight difference in favour of the double-furrow plough. In a trial with the double-furrow plough and others, the common plough, with a furrow of from 6½ to 7½ inches deep, gave a draught from 4 to 5 cwt.; while 9 double-furrow ploughs, with an average depth of furrow of 5½ inches, gave an average draught of 7 cwt.

Latterly double-furrow ploughs have to some extent been losing favour in certain parts of the country where they obtained a footing. The modern Anglo-American plough is preferred by many for speedy

Fig. 416.—*Double-furrow plough.*

ploughing. Still, in some circumstances, the double-furrow plough may be employed with advantage.

Several improvements have lately been effected in the double-furrow ploughs, and now they are, as a rule, lighter in draught, and more easily manipulated, than in former times.

Sowing Operations.

Placing Sacks in the Field.—There is some art in setting down sacks of seed-corn on the field. The plan of placing the sacks of course depends on whether the seed is to be sown by the hand or by a machine. The sacks are set down across the field from the side at which the sowing commences. One row of sacks is sufficient, when the ridges are just long enough for the sower to carry as much seed as will bring him back again to the sack, and the sacks are then set in the centre of the ridge. When the ridges are short, the sacks are set upon a head-ridge; and when of such length as the sower cannot return to the sack by a considerable distance, two rows of sacks are set, dividing the length of the ridges equally between them, setting the two sacks on the same ridge. The sacks are placed upon the furrow-brow of the ridge, that the hollow of the open furrow may give advantage to the carrier of the seed to take it out easily as the sack becomes empty. In thus setting down the sacks of seed, it is intended to give the supply of seed more easily to the man who sows the seed by hand.

When a machine is employed to sow the seed, the sacks are set upon one of the head-ridges connected with the gate of the field, unless the field is so long that a row of sacks must be placed in the middle.

Where to begin Sowing. — If the surface is level, it matters not which side of the field is chosen for commencing the sowing; but if inclined, the side which lies to the left on looking down the incline should be the starting-point. The reason for this preference is, that the first stroke of the harrows along the ridge is most difficult for the horses to draw; and it is easiest for them to give the first stroke *downhill*. This first action of the harrows is called *breaking-in* the land. It is the same to the sower at which side he commences the sowing, but ease of work for the horses ought to be studied.

Seed Carrier. — In Scotland the carrier of the seed is usually a woman, and the instant the first sack of seed is set down, she unties and rolls down its mouth, and fills the *rusky*, basket, pail, or whatever she uses in conveying the seed, and carries it to the sower, who awaits her on the head-ridge from which he makes his start. Her endeavour should be to supply him with such a quantity of seed at a time as will bring him in a line with the sack where he gets a fresh supply; and as the sacks are placed half-way down the ridges when only one row is set down, this is easily managed; but with two rows of sacks, she must go from row to row and supply the sower, it being her special duty to attend to his wants, and not to consider her own convenience. Nothing can be more annoying to a sower than to have his sheet or sowing-basket served too full at one time and too stinted at another; as also to lose time in waiting the arrival of the seed-carrier, whereas she should be awaiting his arrival. When two rows are at a considerable distance, on long ridges, two carriers are required to serve one sower. Better that the

carriers have less to do than that the sower lose time and delay the harrows, which will likely occur when the carriers are overtaxed.

Seed-basket. — The basket or vessel in which the carrier conveys the seed is of various patterns — a deep or shallow basket, or ordinary pail, sometimes carried on the head, and in other cases in the hand or on the arm and haunches. The seed is most easily poured into the sowing-basket from the seed-basket on the head. It should be filled each time with just the quantity of seed the sower requires at a time.

The Seed-sacks. — The mouth of the sack should be kept rolled down, that the seed may be quickly taken out, for little time is usually at the disposal of the carrier. The carrier should be very careful not to spill any seed upon the ground on taking it out of the sack, otherwise a thick tuft of corn will unprofitably grow upon the spot. As one sack becomes empty, the carrier should take it to the nearest sack; and as the sacks accumulate, they should be put into one, and carried forward out of the way of the harrows. It is a careless habit which permits the sacks to lie upon the ground where they are emptied, to be flung aside as the harrows come to them.

One-hand Sowing. — In former times the sower by hand in Scotland was habited in a peculiar manner. He sowed by one hand only, and had a sowing-sheet wound round him, as shown in fig. 417. The most convenient sheet is of linen. It is made to have an opening large enough to admit the head and arm of the sower through it, and a portion of the sheet to rest upon his left shoulder. On distending the mouth of the doubled part with both hands, and receiving the seed into it, the loose part of the sheet is wound tight over the left hand, by which it is firmly held, while the load of corn is supported by the part of the sheet which crosses the breast and passes under the right arm behind the back to the left shoulder. A *basket* of wicker-work, such as fig. 418, was very common in England for sowing with one hand. It was suspended by a girth fastened to two loops on the rim of the basket, and passing round the back of the neck; the left hand holding the basket steady by the wooden stud on the other side of the rim.

Two-hand Sowing.—But the system of sowing with both hands is now more

Fig. 417.—*Sowing-sheet and hand-sowing corn.*

general than one-hand sowing. It should indeed be the universal method wherever hand-sowing is pursued. It is the most expeditious; and many people consider that the sowing can be done more evenly with two hands than with one.

For two-hand sowing a simple form of sowing-sheet is a linen semi-spheroidal bag, attached to a hoop of wood or of iron rod, formed to fit the sower's body, buckled round it, and suspended in front in the manner just described. Both hands are thus at liberty to cast the seed, one handful after the other.

Art of Sowing.—The following detailed description of the art of sowing by one hand is also so far applicable to sowing by both hands. Taking as much seed as he can grasp in his right hand, the sower stretches his arm out and a little back with the clenched fingers looking forward, and the left foot making an advance of a moderate step. When the arm has attained its most backward position, the seed is begun to be cast, with a quick and forcible thrust of the

hand forward. At the first instant of the forward motion the fore-finger and thumb are a little relaxed, by which some of the seeds drop upon the furrow-brow and in the open furrow; and while still further relaxing the fingers gradually, the back of the hand is turned upwards until the arm becomes stretched before the sower, by which time the fingers are all thrown open, with the back of the spread hand uppermost. The motion of the arm being always in full swing, the grain, as it leaves the hand, receives such an impetus as to be projected forward in the form of a figure corresponding to the sweep made by the hand. The forward motion of the hand is accompanied by a corresponding forward advance of the right foot, which is planted on the ground the moment the hand casts forward the bulk of the seed.

The action is well represented in fig. 417, except that some would consider the sower should give his hand a higher sweep, especially on a calm day. The curve which the seed describes on falling upon the ground is like the area of a portion of a very eccentric ellipse, one angle resting on the open furrow and the other stretching 2 or 3 feet beyond the crown of the ridge, the broadest part of the area being on the left hand of the sower.

Fig. 418.—*English sowing-basket.*

The moment the seed leaves it the hand is brought back to the sowing-sheet to be replenished, while the left foot is advanced and the right hand is stretched back for a fresh cast, and thrown forward again with the advance of the right foot.

The seed ought to be cast *equally over the ground.* If the hand and one foot alternately do not move simultaneously, the ground will not be equally covered, and a strip left between the casts. When the braird—that is, the young plants—comes up, these strips show themselves. This error is most apt to be committed by a sower with a stiff elbow, who casts the grain too high above the ground. The arm should be thrown well back and stretched out, though, in continuing the action, with the turning up the back of the hand, the inside of the elbow-joint becomes pained.

If the hand is opened too soon, too much of the seed falls upon the furrow-brow, and the crown receives less than its proportion. This fault young sowers are very apt to commit, from the apprehension that they may retain the seed too long in the hand. If the hand is brought too high in front, the seed is apt to be caught by the wind and carried in a different direction from that intended.

When the wind becomes strong, the sower is obliged to walk on the adjoining ridge to the windward to sow the one he wishes; and the sower should cast low in windy weather.

Some sowers take long steps, and make long casts, causing some of the seed to

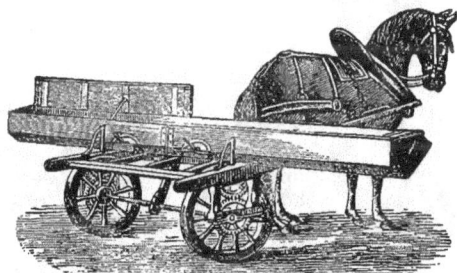

Fig. 419.—*Broadcast sower ready for work.*

reach across the ridge from furrow to furrow. Such a sower spills the seed behind the hand, and makes bad work in wind. The step should be short, the casts frequent, and the seed held firmly in the hand, then the whole work is under complete command. The sower should never bustle and try to hurry through his work; he should commence

with such a steady pace as to maintain it during the day's work.

A sower with *both hands* makes the casts alternate, the hand and foot of the same side moving simultaneously with regularity and grace.

Sowing - machines. — Hand - sowing has been to a large extent superseded by

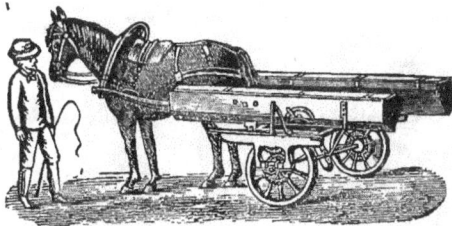

Fig. 420.—*Broadcast sower in transit.*

sowing-machines. These do the work better than it can possibly be done by hand, and their use is therefore to be commended. Of seed-sowing machines there are many patterns, some dropping the seed in drills, others scattering it broadcast. A material difference exists between these two classes of machines. The broadcast machine deposits the seed upon the surface of the ground, and is in fact a direct substitute for hand-sowing; and as it deposits the seed very regularly, this machine it now extensively used.

The *drill-machine* deposits the seed at once at a specific depth under ground in rows, and at such distances between the rows, and with such thickness in the rows, as the will of the farmer may decide.

The seed being left by the broadcast machine on the ground like hand-sowing, is buried in the soil more or less deeply as the harrows may chance to take it; whereas the drill-machine deposits the seed in the soil at any depth the farmer chooses, and all the seed at the same depth, thereby giving him such a command over the position of the seed in the soil as no broadcast machine or hand-sowing can possibly do.

Broadcast Sowers.—There are various forms of the *broadcast* sowing-machine. The one illustrated in figs. 419 and 420, made by Ben. Reid & Co., Aberdeen, exhibits the machine in the most perfect form. This machine not only does the work well, but it is so constructed that its long sowing-chest is divided into sections, the two end ones of which can be folded upon the central division, whereby it may pass through any field-gate

Fig. 421.—*Corn and seed drill.*

without the sowing-chest having to be removed. The *sowing-gear* of the broadcast machine is connected with the main axle of the carriage, as shown in the figure. The arrangements for regulating the quantity of seed per acre are very simple and effective, and altogether the machine is very easily worked and controlled. About 18 feet is the usual width sown at once by the machine.

Drill Sowers.—There are now many excellent seed-drills in the market, their general type being indicated by the representation of Garrett's drill in fig. 421. Ingenious and efficient devices are employed for regulating the quantity of seed per acre, the width of the drills,

Fig. 422.—*Horse-hoe.*

and the depth to which the seeds are deposited.

By the use of the drill-machine less seed will thus suffice, and another advantage is that the land between the rows may be hoed by the hand-hoe, or

by a horse-hoe, such as in fig. 422 (Martin, Stamford), thus tending to clean the land. Drilling is rightly enough in favour for good land in good heart, but on poor or medium land it does not give so much straw as broadcast sowing.

Width of Drill.—The width between the rows of wheat varies somewhat. On good land in high condition, 6 inches is a common width, but many consider that rather too great for ordinary land.

Hand Seed-drill.—There are useful small hand seed-drills both for grain and root crops, such as Boby's, which is represented in fig. 423.

Hand Broadcast Sowers.—Fig. 424 represents a very ingenious and most useful hand broadcast sower, the "Little Wonder," of American invention, and brought to this country by Mr J. H. Newton, West Derby, Liverpool. The illustration pretty well explains its appearance and action. A light box of thin wood is carried under

Fig. 423.—*Hand seed-drill.*

the left arm with a strap over the shoulder. To the top part of this is attached a canvas receptacle for the seed, while on front and below is fixed a little tinned iron wheel, or rather four crossed pieces revolving on a spindle. Round this spindle is passed a thong which forms the string of a bow, and by "see-sawing" this bow the wheel revolves in alternate directions. An eccentric on the spindle moves a little hopper which keeps a regular stream of seed falling on to the revolving "wheel," and this in its turn sends the grain spinning out all round. It will cover a width of about 30 feet, but some have found it best in practice to go up the centre of one rig and down another, thus taking 14 or 16 feet at a time. It is thus possible, if kept supplied with seed, to do four acres per hour, while three is easy of attainment. To ensure an even braird, the machine should be carried in a level position. It sows all kinds of grain ad-

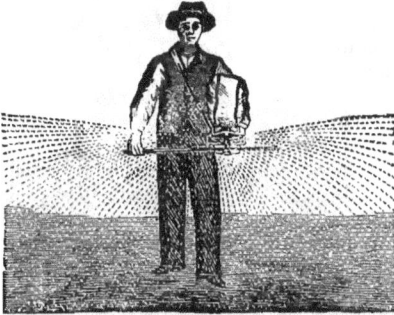

Fig. 424.—*Broadcast hand-sower.*

mirably, and is equally well adapted for sowing dry artificial manure. The quantity to be sown per acre is regulated by a little slide.

Strawson's ingenious air distributor may also be adapted for sowing grain broadcast.

Harrowing.

The land, whether sown by hand or with any sort of machine, must be harrowed. The order in time of using the harrows differs with the sort of machine used for sowing the grain. When the grain is sown by hand or with the broadcast machine, the harrow is used chiefly after the grain has been sown, although many consider it desirable to "break in" the surface by a single or double turn of the harrows before sow-

ing. But in sowing with drill-machines, the harrow is first used to put the land into the proper tilth for the machine.

Considering the operation the *harrow* has to perform in covering the seeds that have been cast upon the soil, and reducing the surface-soil to a fine tilth, it is an implement of no small importance; and yet its effects are apparently rude and uncertain, while its construction is of the simplest order. So simple indeed is this construction, that at a very remote period it appears to have taken that form which, in so far as the simple principles of its action are concerned, is almost incapable of further improvement.

Iron Harrows.—Fig. 425 represents Howard's set of iron harrows for a pair

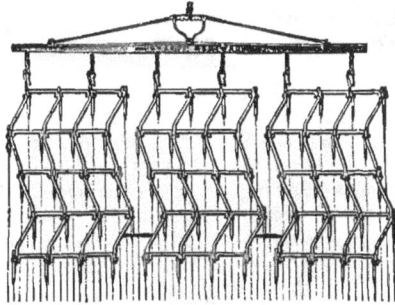

Fig. 425.—*English iron harrows.*

of horses. Sellar's harrows, suited for heavy land, are shown in fig. 426. Wooden harrows, once so common, are now out of date. Iron harrows are

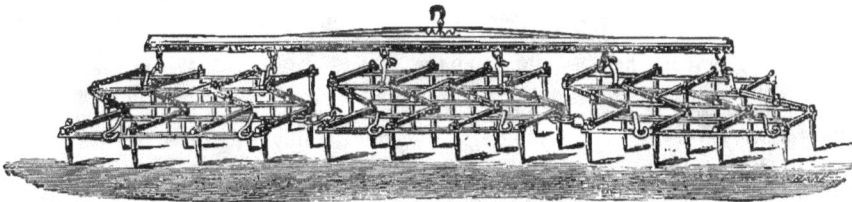

Fig. 426.—*Scottish iron harrows.*

made of many patterns. Most of them are wonderfully durable, light in draught, and very effective in reducing the soil to a fine condition. They are made heavy or light, according to the work intended to be done. In some the teeth or tines are held in by screw and nut,

and in others by being driven through holes of the required size.

Process of Harrowing.—Two pairs of harrows work best together. One pair takes the lead, by going usually on the near side of the ridge, while the other pair follows on the off side, but the leader takes the side of the ridge whichever is nearest the open field. Each pair of harrows should be provided with double reins, one rein from each horse; and the ploughmen should be made to walk and drive their horses with the reins from behind the harrows. If a strict injunction is not laid upon them in this respect, the two men may be found walking together, the leading one behind the harrows, the other at the head of his horses. The latter is thus unable to know whether his harrows cover the ground which they ought to cover, and the two are more engrossed in talk than in the work in hand.

To draw harrows as they should be drawn is really not so light work for horses as it seems to be. When the tines are newly sharpened and long, and take a deep hold of the ground, the labour is considerable. To harrow the ground well—that is, to stir the soil so as to allow the seed to descend into it, and bring to the surface and pulverise all the larger clods, as in the case of broadcast sowing—requires the horses to go at a smart pace; and for efficient working harrows should on all occasions be driven with a quick motion.

When the seed is sown by a drill-machine, it is deposited at a given depth; and in order that the harrows shall not disturb its position, the land is harrowed fine before the seed is sown, a single tine — that is, one turn of the harrows — along the drills covering the seed sufficiently.

Cross - harrowing. — After the appointed piece of ground, whether a whole field or part, has been broken-in and sown, the land is cross-harrowed a double tine—that is, at right angles to the former harrowing, and to the ridges. But as, for this operation, the ground is not confined within the breadth of ridges, the harrows cover the ground with their whole breadth, and get over the work in less time than in breaking-in.

Cross - harrowing is not easy for the horses, inasmuch as the stripes left in the ground by the breaking-in have to be cut through, and the irregular motion of the harrows, in jerking across the open furrows of the ridges, has a fatiguing effect upon the horses.

To finish the harrowing, another double tine along the ridges, as in the case of the breaking-in, may be necessary. This turn is easily and quickly performed, the soil having been so often moved; and should it seem uniform in texture, a single tine will suffice for a good finishing.

Efficient Harrowing.—To judge of the harrowing of land, the sense of feeling is required as well as that of sight. When well done, the soil seems uniformly smooth, and the small clods lie loosely upon the surface,—the ground feeling uniformly consistent under the tread of the foot. When not sufficiently harrowed, the surface appears rough, the clods are half hid in the soil, and the ground feels unequal under the foot,—in some parts resisting its pressure, in others giving way to it too easily.

The old saying that "good harrowing is half farming" has more wisdom in it than at first sight appears. The *efficient harrowing* of land is of more importance than seems generally to be imagined. Its object is not merely to cover the seeds, but to pulverise the ground, and render it of a uniform texture. Uniformity of texture maintains in the soil a more equable temperature, not absorbing rain so fast, or admitting drought too easily, as is the case when the soil is rough and kept open by clods.

Whenever the texture becomes sufficiently fine and uniform, the harrowing should cease, although the appointed number of double or single tines has not been given; for it is a fact, especially in light, soft soils, that over-harrowing brings part of the seed up again to the surface.

Water - furrows.—When the spring wheat was sown early in the season, in January or near the end of February, it was usually considered necessary in former times that the ridges should be *water-furrowed*, so that, in case of much rain falling or snow melting, it might run off the surface of the ground by the water-furrows. Where the spring wheat was sown late in the spring, in the end of February or beginning of March, the water-furrowing was not executed until after the sowing of the grass-seeds, if any were to be sown with the wheat crop.

Water - furrowing is making a slight plough-furrow in every open furrow, as a channel for rain-water to flow off the land. It may be executed lightly with a common plough and one horse, but better with a double mould-board plough and one horse; and as the single horse walks in the open furrow, the plough following obliterates his footmarks.

The better water-furrowing by the double mould-board plough consists in the channel having equal sides; and the furrow-slice on each side being small, compared with the one-furrow slice of the common plough on one side, the water can run more freely into the furrow. The plough simply goes up one open furrow and down another until the field is finished, the horse being *hied* at the turns into the open furrow. Water-furrowing finishes the work of the field.

Under-drainage v. Water-furrows. — On average soils there will be no necessity for water-furrows if the land is thoroughly under-drained. The importance of this latter is now universally acknowledged, and great benefit has been derived by the large extent to which drainage has been executed throughout the country. When the soil is exceptionally adhesive, and water apt to lie in pools on its surface, it is very desirable that water-furrows should be provided to prevent this.

Wheat after Grass. — The foregoing relates mainly to the sowing of wheat after a root crop. But a large extent of spring wheat is also sown after grass, chiefly in England, and some of the earlier and drier districts of Scotland. The success of spring wheat after grass in England attests the superiority of the English climate, which is too dry, and too warm in the southern counties, for the perfect growth of oats. A great obstacle to sowing wheat in Scotland in spring is the action of two classes of soil on the growth of that plant. Clay soils are too inert in the average climate of Scotland to mature the growth of wheat in a few months; and the light soils, though more favourable to quick vegetation, want stamina to support the wheat plant, and are, besides, too easily affected by drought in early spring — it being no uncommon occurrence in Scotland to experience a severe drought in March, and during the prevailing east wind.

Wheat cannot be safely sown in the autumn in Scotland after the end of October, which is the time for sowing after potatoes. Some sow it in November, to the risk of producing a thin crop. To plough up lea before October would be to partially sacrifice the aftermath. Many farmers do this rather than lose the advantage of sowing wheat in good time before the winter sets in. But with others the aftermath is of greater importance, and they accordingly defer the ploughing of the lea till winter, and the sowing of the wheat till spring. January is considered a good month for wheat-sowing, but it is only in exceptional seasons and in favoured districts that the weather permits of this. There is thus a considerable extent of spring wheat sown after grass.

Varieties of Spring Wheat.

As to the varieties of wheat which should be sown in spring in different localities, it would be imprudent to dogmatise. With the great attention now being given to the improvement of farm plants, and to the bringing out of new varieties and stocks of exceptional vigour and power of production, it is quite probable that the variety which is considered best to-day will be excelled in the near future. Farmers must therefore be constantly on the outlook for improved sorts, and be guided by the experience of the time as to which variety they should select.

It is this same consideration — the great ingenuity and enterprise employed in developing new sorts, and the rapidity with which one good sort is supplanted by a still better — which influenced us in deciding not to attempt in this work a detailed description of the different varieties or sorts of the respective farm crops now in use in this country.

For guidance as to the best varieties to use, no farmer need have any difficulty. By a careful study of the experience of other farmers, and due consideration of his own peculiar conditions as to soil and climate, he is not likely to be far wrong as to the selection of varieties.

Of course, care must be taken not to sow a *distinctly winter* variety of wheat in spring. As to a winter wheat no mistake can be made, for however early may be the habit of the variety sown, the very circumstance of its being sown in autumn, when sufficient time is not given to the plant to reach maturity before winter, will convert it for that season into a winter variety. The wheat plant is a true annual, but when sown late, and the progress of its growth is retarded by a depression of temperature, it is converted for the time into a biennial. It is therefore highly probable that, as the nature of wheat is to bring its seed to maturity in the course of one season, any variety sown in time in spring would mature its seed in the course of the ensuing summer or autumn. This is believed to be a fact; nevertheless, circumstances may occur to modify the fact *in this climate.* Under the most favourable circumstances, the wheat plant requires a considerable time to mature its seed; and a variety that has long been cultivated in winter, on being sown in spring in the same latitude, will not mature its seed that season should the temperature fall much below the average, or should it be cultivated on very inferior soil to that to which it had been accustomed. In practice, therefore, it is not safe—at least in so precarious a climate as that of Scotland—to sow *every variety* of wheat in spring.

Spring Wheat-seed from Early Districts.—Wheat taken from a warm to a cold climate will prove earlier there than the native varieties, and, in so far, better suited for sowing in spring; and if the same variety is an early one in the warm latitude—bringing its seed to maturity in a short period, perhaps not exceeding 4 months,—then it may safely be sown as a spring wheat, whether it be red or white, bearded or beardless.

The long experience of the late Mr Patrick Sheriff, East Lothian, led him to the conclusion that autumn wheats should not be sown in spring, as they will not produce a sufficient number of prolific ears.

Late Varieties of Wheat.—Special attention has been given in recent years to the bringing out of varieties of wheat suitable for sowing late in spring. Considerable success has been attained, and there are varieties now in use which in average years give fairly satisfactory results, although not sown till March or April.

As to the quantity of seed for wheat per acre, information is given on page 118, as also in the section on "The Germination of Seeds" in this volume.

Manuring Wheat.

In the description of the Rothamsted experiments in pages 1-32 of this volume, much useful and suggestive information as to the manuring of wheat will be found. Wheat is usually sown on land in good heart, for the most part after a potato or root crop, with which a heavy dressing of dung and artificial manure had been applied. In this case no special application of manure may be necessary for the wheat beyond perhaps a top-dressing with a little ammonia-salts or nitrate of soda in spring. The sulphate of ammonia may be sown at the same time as the seed for the spring wheat, or early in spring for winter wheat, but nitrate of soda should not be sown until the plants are able to immediately assimilate the manure. From 1 to 2 cwt. per acre are common quantities of these fertilisers for top-dressing wheat.

When the land has not been liberally manured with the preceding crop, a heavier dressing, including phosphatic and potassic manures, must be given to the wheat crop; or it may be manured with dung. See chapter on "Manures and Manuring."

Summer Culture of Wheat.

The amount of attention which the wheat fields demand in the summer months depends mainly upon the time they had been sown.

Autumn Wheat.—Autumn or winter sown wheat may be too far advanced in growth before the advent of summer to permit of any cultural work being given to it in that season. Such horse-hoeing or harrowing as it may require will therefore be performed in spring. The state of the autumn-sown wheat in summer depends on the weather in winter and spring, and the nature and condition of the soil upon which it was sown.

Over-luxuriance in Autumn Wheat.—Mild weather in winter will cause it to grow luxuriantly; and if the mildness continue till spring, the plants may, from over-luxuriance, lie down in spring, and become blanched and rotted at the roots.

In the early part of winter, if the ground is dry, sheep may eat down luxuriant wheat to a considerable degree. Even if not folded on it, sheep will do much good to luxuriant wheat by trampling upon it for a while every day, and eating off the tops of the plants.

But the winter luxuriance is frequently checked, and even the plants destroyed, by severe frosts at night and bright sunshine during the day in March. Should the winter luxuriance continue till spring, sheep cannot then crop it uniformly, and should not be allowed to attempt it. If luxuriance only commenced in spring, sheep can restrain it then as well as in winter.

Cropping Rank Wheat.—The winter luxuriance can be restrained in spring only by mechanical means—by cutting off the tops with the scythe. This may be done safely until the plant puts forth the shoot-blade, perhaps as late as the end of April. Before commencing cropping with the scythe, some of the most forward plants should be opened to ascertain the position and length of the ear, which should not be touched. The leaves cut off lie on the ground to decay. The advantage of cropping wheat when over-luxuriant is, that rain will no longer hang upon it, and air and light will have access to the stem to strengthen and support it. The risk of lodging is thereby greatly lessened. Spring wheat rarely becomes too luxuriant in summer, and requires no expedient to check its growth.

Soil and Over-luxuriance.—Of the classes of soil which produce over-luxuriance, dry deep clay loam is most apt to do it in a mild autumn and winter; and thin clay land, upon a retentive wet subsoil, is most liable to destroy wheat in March. Even when showing no luxuriance, and the crop promising, yet by the injurious effects of March weather the plants may not only be sickly and scanty, but too late to tiller.

Weeding.—The weeding of the cereal crops in summer where the land is foul is an indispensable work for their welfare. If the crop should be too far advanced to permit horse labour, the weeding must be done solely by the hand or with manual implements; if not, both manual and horse implements may be employed,—that is, where the

seed has been sown in drills. Among broadcast grain, weeding must be performed by the hand and with manual implements. An effective tool for this purpose is the simple weed-hook, fig. 427. It consists of an acute hook of iron, the two inner edges of which are flattened and thinned to cut like a knife, and which are as far asunder at one end as to embrace the stem of succulent herbaceous plants which are destined to be cut down. The cutting-hook is attached to a socket, which takes in the end of a light wooden shaft about 4 feet in length, which is fastened to it with a nail or screw, the hook having such a bend as that its under surface shall rest upon the ground, while the worker uses the shaft in a standing position. A sharp spud with a cross-head handle is the best instrument for cutting weeds with strong stems—as docks, thistles —with a push.

The best way for field-workers to arrange themselves, when weeding broadcast corn, is for two to take one ridge, each clearing one-half of the ridge from the open furrow to the crown. On weeding amongst corn, the point of the weed-hook is insinuated between the stems of corn toward the weed to be cut, and on its stem being taken into the sharp cleft of the hook at the ground, it is easily severed by a slanting cut upwards towards the worker. The weeds, cut over, are left on the ground to decay; but no weed should be allowed to grow *beyond* the time of its flowering. Docks should be pulled up by the root and carried away and burned.

Hoeing Drilled Wheat. — Wheat sown in rows may be weeded with the hand-hoe, or with horse-hoes. The hand-hoe is used by field-workers, who each take one row between the drills. To prevent jostling, the worker in the centre of the band takes the lead in advance

Fig. 427.—*Weed-hook.*

I

position, while the others follow on each side in echelon. Where drilled crops occupy much extent of ground, the ordinary number of hand-hoers are unable to clear the weeds before the crops advance too far to go amongst them. Hence the need of the more expeditious horse-hoe (fig. 422).

The horse-hoeing of corn should be intrusted to a careful man and a steady horse. A steady horse will not leave the row he walks in from end to end of the landing. A young horse is unsuited for this work. A careful man to steer the hoes is as requisite as a steady horse; otherwise the hoes may run through the rows of corn-plants, tearing them up as well as the weeds.

As already indicated, to wheat sown in the preceding autumn or winter this horse-hoeing has usually to be given in spring,—that is, if given at all.

Top-dressing Wheat.—If the crop is not making satisfactory progress, or if it is considered desirable for any reason to top-dress wheat late in the season, this may be done during the month of May. Mild moist weather is most suitable for the process. At this late period of the season a dressing of nitrate of soda, perhaps from 1 to 2 cwt. per acre, would likely give the best result. Some would add 2 cwt. of superphosphate. It is considered a good plan to delay a portion of the more quickly acting manures to be sown as a top-dressing in this way.

Flowering Season. — The flowering season is critical for wheat, since a difference in the weather of June may affect the yield to upwards of 50 per cent. Should the weather be rainy and windy in the flowering season, the produce will inevitably be scanty. Rain alone, unless of long duration, does not affect the produce as much as strong wind, which seriously injures the side of the ear exposed to it. Showers and gentle breezes do no harm; but sunshine, heat, and calm are the best securities for a full crop.

PICKLING WHEAT.

There is much to be said in favour of the pickling of seed-wheat—that is, subjecting it to a preparation in a certain kind of liquor—before it is sown. This treatment assists greatly in protecting

the crop against the attack of a fungoid disease in the ensuing summer called *smut*, which renders the grain comparatively worthless. Some farmers affect to despise this precaution, as originating in an unfounded reliance on an imaginary specific. But the existence of smut and its baneful effect upon the wheat crop are no imaginary evils; and when experience has proved, in numberless instances, that steeped seed protects the crop from this serious disease, the small trouble and expense which pickling imposes may surely be incurred, even although it should fail to secure the crop.

Various methods and materials for pickling are employed. A solution of blue vitriol is largely used, and the pro-

Fig. 428.—*Apparatus for pickling wheat.*

a Sackful of wheat.
b Basket to receive the wheat from the sack.
c Tub of pickle.
d Basket of pickled wheat.
e Drainer for basket.
f Tub to receive draining of pickle from the basket.
g Heap of pickled wheat.
h Sacks for the pickled wheat.

cess, as described in former editions of this work, is seen in fig. 428. Care must be taken not to have the "pickle" so strong as to retard or prevent germination. One pound of copper sulphate to a gallon of water would suffice for four bushels of seed.

The pickling may be done on the corn-barn floor. Two upright baskets are provided, each capable of holding easily about half a bushel of wheat, having upright handles above the rims. Pour the wheat into one basket from the sack, and dip the basketful of wheat into the tub of vitriol completely to cover the wheat, the upright handles protecting the hands from the vitriol. After it remains in the liquid for a few seconds, lift up the basket, so as to let the surplus liquid run from it into the tub again, and then place the basket upon the drainer on the empty tub, to drip still more liquid, until the other basket is filled with wheat and dipped in the vitriol tub. Then empty the dripped basket of its wheat on the floor, and as every basketful is emptied, let a person spread, by riddling it through a wheat-riddle, a little slaked caustic lime upon the wet wheat to dry it. Thus all the wheat wanted at the time is pickled and emptied on the floor in a heap.

Turning Pickled Wheat.—The pickled and limed heap of wheat is turned over and mixed in this way: Let two men be each provided with a square-mouthed shovel, one on each side of the heap, one having the helve of his shovel in his right hand, and the other in his left; and let both make their shovels meet upon the floor, under one end of the heap of wheat, turning each shovelful from the heap behind them, till the other end of the heap is reached. Let them return in a similar manner in the opposite direction, and continue, until the wheat is thoroughly mixed and dried with the lime. The pickled wheat is then sacked up, and carried to the field in carts.

SOWING BARLEY.

It may be laid down as an axiom that the seed-bed upon which barley is to be sown should be fine, moderately deep, and clean, with an abundant supply of all the ingredients necessary for the growth of the plant present in a soluble or readily available form. In Scotland land after turnips is the place in the rotation which is generally set aside for the growth of barley. In England barley is largely grown after either wheat or oats, especially on farms where the turnips are consumed on the ground by sheep to which cake or grain is given.

Tillage for Barley.—If the land is not of the heavy order of soils, all that is necessary is the ordinary ploughing, especially if it can be accomplished by the second week of February. The action of the weather and frost will break down and mellow the soil, rendering it friable, so that a double tine of the har-

rows before putting in the seed is all that is needed to obtain a seed-bed in good tilth. On the heavier class of soils, and where ploughing cannot be done until later, more especially where the turnip crop has been eaten off by sheep, two ploughings may be necessary as well as harrowing before seeding.

But the simplest and easiest mode of procedure is to plough the land with one of the new Anglo-American ploughs, which will break down the furrow, leave the land level, and in excellent tilth. By this plan the old method of cross-ploughing, scarifying, grubbing, ribbing, &c., may be obviated.

It is probable that some of the turnip-land which may have been ploughed for spring wheat may have to be sown with barley, on account of inclement weather preventing the sowing of wheat in seasonable time. In that case, whether the land had been gathered up from the flat or cast together, it should be seed-furrowed in the same form for the barley, to retain the uniform ridging of the field; for the ploughing for spring wheat being the seed-furrow, and the ridges made permanent, it would be impossible to reverse the ploughing with one furrow, without leaving one ridge on each side of the field half the width of the rest. The ridges would have to be ploughed *twice* to bring them back to their proper form, but for which there could not be time, so they must be stirred with the grubber, or ribbed with the small plough.

Another method which is being adopted by farmers is to plough the land after turnips, in breaks of six ridges, gathering four and splitting two. This has become advisable nowadays, owing to the advent of the reaper, for which the old open furrows were very unhandy, while the crop was uneven, as the growth on the crown of the ridge was heavier than that in the furrow which divided the ridges.

If the ridges have consolidated on being long ploughed, the grubber will make a suitable bed for the barley seed, and keep the dry surface uppermost. If the soil is dry and loose on the surface, and tilly below, it will be best preserved by ribbing with the small plough.

By putting such ridges thus into the best state for barley, there will be no difficulty in ploughing the rest of the land. The first furrow upon the trampled soil should be the *cross*-furrow.

Sowing.—Sowing barley upon a fine evenly pulverised surface requires strict attention, inasmuch as on whatever spot every seed falls, there it lies, the soft earth having no elasticity to make the seed rebound and settle on another spot. Hence, of all sorts of corn, barley is the most likely to be striped in sowing by hand, so every handful must be cast with great force. Walking on soft ground in sowing barley is attended with considerable fatigue. Short steps are best suited for walking upon soft ground, and small handfuls are best for grasping plump slippery barley.

The broadcast machine sows barley as well as oats on the ploughed surface, and so do the corn-drills across the ridges after the surface has been harrowed. The grubbed surface is best sown by a drill-machine, affording the seed a firm hold of the ground, while the surface ribbed with the small plough is best sown by hand, or with the broadcast machine, the seeds falling into the ribs, from which the young plants rise in rows, the ground being harrowed only a double tine along the ribs. Barley may be sown any time fit for spring wheat, and as late as the month of May. But the earlier crop will be of better quality and more uniform, though the straw may be shorter.

Quantity of Seed.—The quantity of seed sown broadcast is from 2½ to 4 bushels to the acre. When sown early, less suffices; when late, more is required, because less time is given to tiller and cover the ground. Sown with the drill, 2 bushels suffice.

Brown makes some sensible remarks on this subject. "Amongst the farmers," he says, "it seems a disputed point, whether the practice of giving so small a quantity of seed (3 bushels per acre) to the best lands is advantageous. That there is a saving of grain, there can be no doubt; and that the bulk may be as great as if more seed had been sown, there can be as little question. Little argument, however, is necessary to prove that thin sowing of barley must be attended with considerable disadvantage; for if the early part of the season be dry, the plants will not only be stinted in their

growth, but will not send out offsets; and if rain afterwards falls—an occurrence that must take place some time during the summer, often at a late period of it—the plants *then begin to stool*, and send out a number of young shoots. These young shoots, unless under very favourable circumstances, cannot be expected to arrive at maturity; or if their ripening is waited for, there will be great risk of losing the early part of the crop—a circumstance that frequently happens. In almost every instance an unequal sample is produced, and the grain is for the most part of inferior quality. By good judges it is thought preferable to sow a quantity of seed sufficient to ensure a full crop without depending on its sending out offsets. Indeed, when that is done, few offsets are produced—the crop grows and ripens equally, and the grain is uniformly good."[1]

Still, too thick sowing should be avoided, for it is liable to produce weak straw, and consequently a greater tendency to lodging.

Germination of Barley and the Weather.—No grain is so much affected by weather at seed-time as barley. A dash of rain on strong land is liable to cause the crop to be thin, many of the seeds failing to germinate. In moist, warm weather, the germination is certain and rapid. It has been observed that unless barley germinate quickly the crop will be thin.

Harrowing for Barley.—The harrowing required for barley land sown broadcast is generally less than for oat land, a double tine being given in breaking-in the seed, and a double tine across immediately after. When sown with the drill-machine, the harrowing is perhaps a double tine along and double tine across the ridges, before the seed is sown. When sown on ribbed land, the only harrowing may be a double tine along the ribs, just to cover the seed, as the ribs afford it a sufficient hold of the ground. Care, however, should be taken in all cases to ensure a fine even seed-bed for barley. The condition of the land will be the best guide as to the amount of harrowing required in individual cases.

When barley is grown on fairly strong land after roots consumed on the ground

[1] Brown's *Rur. Aff.*, ii. 45.

by sheep, extra harrowing may be required.

Finishing.—The grass seeds are then sown with the grass-seed sowing-machine; the land harrowed a single tine with the light grass-seed harrows, and thereupon finished by immediate rolling. On strong land, apt to be incrusted on the surface by drought after rain, rolling may *precede the sowing of grass seeds*, and the work is finished with the grass-seed harrows, and perhaps another turn of the roller. On all kindly soils rolling is usually the last operation.

Soil for Barley.—Medium and light loams of a calcareous and friable nature —such as are generally known as good turnip lands — are best adapted to barley.

Barley does not stand the winter in Scotland as it does in the warm calcareous soils of the south of England. Winter barley is early ripe, and prolific; but if the weather causes it to tiller in spring, it produces an unequal sample, containing a large proportion of light grain.

Varieties of Barley.—The varieties of barley are numerous. They are generally distinguished by the shape of the ear and the number of rows of grain which grow upon the ear.

Uses of Barley.—The great bulk of the better samples of barley is used for distillery purposes, a small proportion being employed for the manufacture of pot barley or barley-meal, chiefly confined for use in Scotland. The inferior or damaged barley is used as food for animals.

Manuring Barley.

When it follows a well-manured root crop, as it generally does in Scotland, barley seldom requires or receives any further manuring. Barley is a suitable crop for land on which a portion of the root crop has been consumed by sheep, and in this case the soil is usually in good heart, especially if the sheep have been allowed extra food, such as cake or grain along with the roots. The custom is to plough this land with a light or moderate furrow, and thus give the barley an abundance of readily available plant-food within the reach of its shallow roots.

But when the land has not, by previous treatment, become sufficiently stored with fertility for barley, this crop will, as a rule, respond satisfactorily to direct dressings of suitable manure. Being a rapid-growing shallow-rooted plant, barley should have plenty of readily available food within easy reach of the surface. Quickly acting artificial manures are thus specially suited for barley. Superphosphate and nitrate of soda, or sulphate of ammonia, in different quantities and proportions, according to the character and condition of the land, are extensively and advantageously used as top-dressing for barley. The first and last should be applied at seed-time; nitrate of soda, which acts more rapidly than sulphate of ammonia, may be applied in moist weather a few weeks later. Common quantities, per acre, are 2 to 3 cwt. superphosphate, and ½ to 1 cwt. of the nitrogenous manure. In many cases a light dressing of sulphate of ammonia or nitrate of soda is found to be very effective alone. In other cases a combined dressing of phosphatic, nitrogenous, and potassic manures gives the best results.

Rothamsted Barley Experiments.

The experiments on the manuring of barley at Rothamsted are full of interest to farmers. They have gone on continuously since 1852, and are capable of teaching some important lessons. Briefly summarised, the results are as follows:—

No Manure.—The plot which has had no manure of any kind since the beginning of the experiments gave an average of 17⅞ bushels for the thirty-two years up to 1883—4½ bushels less than the average of the first ten years.

Farmyard Dung.—Applied at the rate of 14 tons every year for thirty-two years, this gave for that period an average of 49½ bushels, or about 31½ over the unmanured plot.

Mineral Manures.—Mineral manures alone—that is, superphosphate of lime, and sulphates of potash, soda, and magnesia—gave very poor crops, both of grain and straw. Superphosphate alone, on an average of the thirty-two years, gave only about 5 bushels more than the plot with no manure; the increase from potash, soda, and magnesia over no manure was barely 2 bushels per acre, and from all these mineral manures combined scarcely 6 bushels.

Nitrogenous Manures.—These supplied in sulphate of ammonia or nitrate of soda gave more than double the increase produced by the mineral manures. Ammonia-salts, 200 lb. per acre (containing 43 lb. nitrogen), gave an average of 30¾ bushels for the thirty-two years—nearly 13 bushels over the unmanured plot. Nitrate of soda, 275 lb. per acre (containing 43 lb. nitrogen), gave nearly 4 bushels more per acre. *Rape-cake*, 1000 lb. per acre, calculated to yield 49 lb. of nitrogen, raised the produce to 43¼ bushels.

Nitrogenous and Mineral Manures combined.—These in combination produced excellent crops, more than the average of the country, continuously for thirty-two years. This result is very interesting, showing that barley responds admirably to the influence of readily acting artificial manures. Equal quantities of nitrogenous and mineral manures applied in the autumn to wheat, and in spring to barley, gave considerably more produce from the latter crop than the former.

Practical Conclusions.—From the results of the experiments with various manures for barley, it is inferred that in corn-growing the soil is most rapidly exhausted of its nitrogen, next of phosphates, and most slowly of potash. Nitrogenous manures are thus the first essential, but, especially for barley, phosphatic manures are also required, and give a good return. To most soils of a clayey tendency, dressings of potash will be unnecessary for cereals; but where it is deficient, a small allowance may be expected to exercise a wonderful influence on the crop. Here, as in general farm practice, it was found that superphosphate is more effective with the spring-sown than with the autumn-sown cereals.

Barley after Corn.—In reference to the practice of growing barley after a crop of wheat, Dr Fream says:[1] "It may be laid down as a general rule, applicable to the country at large, that, on the heavier soils, full crops of barley of good quality may be grown with great certainty after a preceding corn crop, under the follow-

ing conditions: The land should be got into good tilth. It should be ploughed up when dry, as soon as practicable after the removal of the preceding crop. In the spring it should be prepared for sowing by ploughing or scuffling, as early in March as possible, if sufficiently dry. The artificial manure employed should contain nitrogen, as ammonia or nitrate (or organic matter), and phosphates. From 40 lb. to 50 lb. of ammonia (or its equivalent of nitrogen as nitrate) should be applied per acre. These quantities would be supplied in 1½ cwt. to 2 cwt. of sulphate of ammonia, or 1¾ cwt. to 2¼ cwt. of nitrate of soda. With either of these there should be employed 2 cwt. to 3 cwt. mineral superphosphate of lime. Rape-cake is also a good manure for barley; from 6 cwt. to 8 cwt. would supply about as much nitrogen as would be equal to from 40 lb. to 50 lb. of ammonia. With this manure, as with guano, the addition of superphosphate is unnecessary. Whatever manure be used, it should be broken up, finely sifted, sown broadcast, and harrowed in with the seed."

Quality of Barley after Roots.—Interesting trials on the quality of barley grown after roots were conducted by the Agricultural College at Wye, Kent. The results indicated that a dressing of salt is detrimental to the value of the barley, and that sulphate of potash, though increasing the starch content of the barley, did not give any commercial return, whilst a dressing of 3 cwt. of superphosphate per acre produced a slight increase in yield and a marked increase in quality. Determinations of the chemical composition of English barleys led to the conclusion that high quality in barley is linked with a high starch and low proteid content, factors which are not readily attained when barley is grown after roots which have been consumed on the land by sheep.

Superphosphate for Barley.—Trials conducted by the Wye Agricultural College showed the effect of superphosphate on the barley crop to be substantial. Without increasing the straw a dressing of from 3 to 5 cwt. per acre produced a marked improvement in the grain, the gross yield being increased, and the quality — as indicated by the weight per bushel, the proportion of flinty corns,

---

[1] *Rothamsted Experiments*, 120.

the amount of tail corn, and the albuminoid content of the grain—beneficially modified. In a year of severe drought it was found that the heavy dressings of superphosphate ripened off the corn too soon, both quality and quantity suffering.

Yorkshire Trials in Manuring Barley.

In the East Riding of Yorkshire barley of very good malting quality is grown over a wide area on what are known as the wold soils. These soils overlie chalk and are of light character very suitable for the fattening of sheep and for the growth of corn. As a rule the rotation followed is the following, viz. :—

> Roots.
> Barley or oats.
> "Seeds" (clover mixture).
> Wheat.
> Barley or oats.

The land for roots as a rule receives no dung, but the crop is grown by means of artificial manures, and a part, or whole, of the crop is eaten on the land by sheep. The sheep receive some artificial food in the shape of cake and corn, and consequently the land is left in a condition that enables a satisfactory crop of corn to be grown. The "seeds," as a rule, are grazed by sheep, and during summer or early autumn dung is applied, and a good crop of autumn-sown wheat is secured in the following year. At this stage of the rotation, especially after a good crop of wheat, it is sometimes considered that for a crop of barley the land should receive some artificial manure. It therefore becomes a question as to what artificial manure might profitably be employed. To gain information on this point, experiments were planned and conducted by the Agricultural Department of Leeds University. Arrangements were also made for the barley grown by means of different artificial manures to be malted, so that it might be possible to determine if the manures had any effect upon malting quality. The results have shown that for the growth of barley as a second corn crop and on the soils indicated above, it is necessary that there be supplied a soluble nitrogenous manure, and further that nitrate of soda is a much more profit-

able source of nitrogen than sulphate of ammonia.

Although the nitrogenous is the important manure to use, still, profitable results followed when nitrate of soda was accompanied by superphosphate and kainit. The following may be taken as the mixture that, as a rule, could be depended upon to give profitable results, viz. :—

> 1 cwt. Nitrate of Soda ⎫
> 2 cwt. Superphosphate ⎬ per acre.
> 2 cwt. Kainit ⎭

The average profit per acre by the use of the above mixture in nine experiments was 17s. 10d.

As regards the effect of the manures upon the malting quality of the barley, it has been found that in the best barley season during the period in which the experiments were conducted, the finest malting barley was grown with the aid of the above-mentioned mixture.

Summer Culture of Barley.

Such weeding and hoeing as the barley may require is usually given as in the case of wheat.

Destroying Charlock.—This is one of the most troublesome weeds amongst corn crops. It is known variously as skellock, runch, karlock, yellows, and wild mustard (*Brassica Sinapis*, Vis., *B. Sinapistrum*, Boiss., *Sinapis arvensis*, L.). It is allied to the turnip, which it resembles in appearance. It produces an oily seed, which can lie dormant under grass for years and spring into life when the ground is again broken up.

It is now found that charlock can be effectively kept in check amongst corn by spraying with solutions of sulphate of copper. The dressing most generally employed is 50 gallons per acre of a 3 per cent solution of copper sulphate, not less than 98 per cent purity. For the making of this solution the following is a simple rule : A gallon of water weighs 10 lb., 10 gallons of water weigh 100 lb., and 3 lb. of copper sulphate added to this give approximately a 3 per cent solution. To make sufficient fluid for one acre, 15 lb. of copper should be mixed in 50 gallons of clean water. In some cases a stronger solution (4 or 5 per cent) is found more effective.

Spraying should be carried out when the plants are about three inches in height, in any case before flowering begins. The solution is sprayed over the plants on a calm day by a hand or horse power machine. The spraying is most effective in dull weather. Heavy rain immediately after the spraying lessens its effectiveness, and may necessitate a second spraying. In any case, a second spraying within about ten days of the first may be required to complete the cure.

Many wonder why it is that while this poisonous solution kills the charlock it does no harm to the corn. The reason is supposed to be that the leaves of the charlock, being rough in surface and horizontally disposed, catch and hold the poison, whereas the leaves of the corn, being smooth and erect, let the fluid run off quickly, thus escaping injury.

Top-dressing Barley.—Barley may be top-dressed like wheat. From 1 to 2 cwt. of nitrate of soda and 2 cwt. superphosphate would be a good late dressing.

SOWING OATS.

The oat is by far the most extensively grown of the cereal crops in Scotland and Ireland. In England oats are grown extensively after turnips or mangels. And in all northern and high-lying districts unfavourable for the ripening of wheat or barley, oats are the prevailing crop after turnips.

Oats are sown on all sorts of farms, from the strongest clay to the lightest sand, and from the highest point to which arable culture has reached on moorland soil to the bottom of the lowest valley on the richest deposit. The extensive breadth of its culture does not imply that the oat is naturally suited to all soils and situations, for its fibrous and spreading roots indicate a predilection for friable soils; but its use as food among the agricultural population generally, and its suitability to support the strength of horses, have induced its extensive cultivation.

Varieties of Oats.— The oat plant thrives best in a cold climate, and is grown in the chief countries lying in the temperate zone. It comes to its greatest perfection in Scotland. This is to a certain extent due to the climate, but the care which many Scottish farmers give to the oat crop also contributes to this result. The varieties which occupy the greatest breadth are the Common Improved or White oats, and to a lesser extent Black or Tartarian. Common oat is the name by which farmers designate the variety which is commonly grown in the respective districts in which they farm.

The following are the chief varieties: The Potato, Poland, Angus, Hopetoun, Sandy, Tartarian, Tam Finlay, Hamilton, Longhoughton, Canadian oats, Swiss oat, &c.

Sowing.—The sowing of the oat seed is begun with the common varieties of oats about the beginning of March. It is the custom in some parts to sow the improved varieties a fortnight after the common. The ploughed lea ground should be dry on the surface before it is sown, as otherwise it will not harrow kindly; but the colour of dryness should be distinguished from that arising from dry hard frost, a state improper to be sown upon. Every spot of the field need not be alike dry—even thorough draining will not ensure that, though spots of wet indicate where dampness in the subsoil exists.

Harrowing before Sowing.—Should the lea have been ploughed some time and from young grass, the furrow-slices will lie close together at seed-time; but when recently ploughed, or from old lea, or on clay land in a rather wet state, the furrow-slices may be as far asunder as to allow a good deal of the seed to drop down between them, and thus be lost, as oats will not vegetate beyond 6 or 7 inches deep in the soil. In such states the ground should receive a double tine or strip of the harrow before being sown. This should be done in every case unless the furrows are small and packed quite closely.

When oats are sown by hand upon dry lea ground, the grains rebound from the ground and dance about before depositing themselves in the hollows, in rows, accommodating themselves between the crests of the furrow-slices, and do not so readily show bad sowing as upon a smooth surface. Were the ground harrowed along the ridges, so as not to disturb the seed in the furrow-slices, the crop would come up as if sown by drill; but as the land is cross-harrowed, the braird comes up to some extent broadcast.

Quantity of Seed.—The quantity of common oats usually sown is from 4 to 5 bushels per acre. In deep friable land in good heart, and in early districts, from 3 to 4 bushels of the best tillering varieties is considered sufficient seed.

A man does a good day's work if he sows broadcast by one hand 16 imperial acres of ground in ten hours. Some men can sow 20 acres; and double-handed sowers will do even more than 20 acres.

Number of Seeds per Acre.—In experiments on different varieties of oats conducted by the West of Scotland Agricultural College, it was decided to regulate the quantity of seed to be sown by the number of seeds per acre instead of by measure. In the first year 2,500,000 seeds of each variety were sown per acre, but except on rich soils, the plants from this seeding were too thin, and in after years the number of seeds was increased to 3,000,000 per acre. The weights of seed required of the different varieties included in the trials to supply this number of seeds per acre were as follows:—

Variety.	Average Quantity of Seed required to supply 3,000,000 Seeds.
	lb.
Sandy	194
Potato	212
Banner	252½
Waverley	258
Wide Awake	267½

It is thus seen that approximately about ½ bushel per acre more Potato oat and 1½ bushel more of the Banner oat were required than of the Sandy oat in order to supply the 3,000,000 seeds per acre.

Harrowing after Sowing. — The tines of the harrows should be particularly sharp when covering in seed upon lea. After the land is broken in with a double tine, it is harrowed across with a double tine, which cuts across the furrow-

crests, and then along another double tine, and this quantity commonly suffices. At the last harrowing the tines should be kept clean from grassy tufts, and no stones should be allowed to be dragged along by the tines, to the injurious rubbing of the surface. On old lea, or hard land, another single tine across or angle-ways may be required to render the surface fine; and, on the other hand, on light soil a single tine along after the double one across may suffice. In short, the harrowing should be continued until the ground seems uniformly smooth and feels firm under the foot. The head-ridges are harrowed by themselves at the last.

Water-furrows.—If the land is liable to suffer from surface-water, water-furrows may be formed in the open furrow, after sowing. But with underground drainage now so general and thorough, this practice has become almost a thing of the past.

Machine-sowing. — Almost every farm with two or more pairs of horses, and even smaller holdings, has its broadcast or drill sowing-machine. Hand-sowing is thus being replaced by the machine. The practice in sowing oats with machines, whether broadcast or drill, is similar to that in sowing wheat and barley. To enable the drill to make good work in sowing on ploughed lea, the surface must be well broken up with the harrow. Where the surface is rough, and the furrows tough, the broadcast machine would be preferable.

Ploughing for Oats.—Difference of opinion exists as to the depth to which lea ground should be ploughed for oats. One opinion is that a depth of 4 inches is sufficient, with the furrow-slices laid down close; others contend that the land should be ploughed 9 inches in depth, and not laid over close. To determine which opinion is the more correct, it should be taken into account that the roots of oats are fibrous, and permeate through the soil to a greater depth than the roots of barley. This being their character, a good depth of furrow will be best for oats. Much of course will depend upon the depth and the character of the soil and of the subsoil; but as a rule, it is considered undesirable to plough lea shallower than 7 inches, to afford a considerable amount of pabulum to the roots of the plants.

Thick and Thin Sowing. — Uncertainty still exists in the minds of farmers whether thick or thin, drill or broadcast, sowing of oats is the better mode. Numerous experiments have been made on both these points in different parts of the country. The results generally have favoured thin or medium rather than thick sowing, 3½ to 4½ bushels of oats per imperial acre usually giving the best returns on good land and with good seed. As a rule, drill-sowing has been found to be the most economical method.

Sowing Mixed Varieties. — Experiments have shown that a mixture of varieties of oats sown together may produce a heavier crop than when sown singly. For example: J. Finnie of Swanston obtained, when sown singly, from potato oats 74 bushels, Hopetoun 65, early Angus 73, Sandy 56 to 61; whereas, when mixed, these results were obtained: Hopetoun 5 parts, and Kildrummie 1 part, produced 85 bushels; Hopetoun and Sandy, 80; Hopetoun and early Angus, 86; potato and early Angus, 66; and potato and Sandy, 66 bushels. It thus appears that potato oats alone produced 8 bushels more than when sown with either early Angus or Sandy oats; that Hopetoun, with Kildrummie, produced 20 bushels more than when alone, with Sandy 15 more, and with early Angus 11 more.

In these trials an average of 13 bushels more per acre was obtained by mixing seeds of oats of different varieties than when sown singly, and that from a space of ground which took 6 bushels of seed. Still, results of similar trials in other cases have been less favourable to the mixing of varieties.

It must be borne in mind that, in mixing varieties of oats, the varieties to be mixed should come to maturity at the same time.

Oats and Barley Mixed. — Another occasional practice in the north of Scotland is to sow a mixture of barley and oats in the proportion of 4 bushels of oats to 1 bushel of barley. Good results ensue, especially on land where oats, after brairding, become thin or die out. The gross produce is greatly increased,

and an excellent food for horses and cattle is obtained.

It is more than probable that the greater produce which is thus obtained from a mixture of oats and barley than from either alone, is that oats and barley search for their food in different layers of the soil—oats penetrating to a considerable depth, whilst barley confines its search mainly to the upper portion of the soil.

Manuring for Oats.

In manurial requirements oats are not much different from barley. They abstract a little more nitrogen and potash, and about the same quantity of phosphoric acid. Oats require more moisture than either wheat or barley, and delight in soils enriched by decayed vegetable matter. Thus oats give large yields on land which has been for a considerable time under grass.

Superphosphate of lime and nitrate of soda applied as a top-dressing give good results when the land requires manuring. The nitrate is specially useful when a bulky crop of straw is desired. Common dressings consist of from ½ to 1 cwt. of nitrate of soda, and from 1 to 2 cwt. of superphosphate. On light land a little potash is sometimes applied with advantage.

But the practice of top-dressing oats is not general. The oat crop, indeed, receives less manure in direct applications than any of the other ordinary farm crops — that is, when the oats follow either grass or roots. Of course, when the oats follow another corn crop some dressing is considered necessary.

Welsh Trials in Manuring Oats. — Trials conducted by the University College of North Wales at Aberystwyth showed that while it is the usual practice to grow the oat crop without the aid of any artificial manures, the application of dressings of complete manures proved highly remunerative, the dressing used consisting of $1\frac{1}{7}$ cwt. nitrate of soda, $2\frac{1}{3}$ cwt. superphosphate, and $1\frac{1}{2}$ cwt. kainit per acre. An increase of 20 per cent in this dressing gave only a very slight gain in the crop. A reduction of the dressing by 20 per cent lessened the crop by only a bushel per acre.

Summer Culture of Oats.

The weeding of oats is not often practised when the seed has been sown broadcast, except to remove docks or thistles. When the thistle flourishes amongst corn, it is extremely troublesome to reapers at harvest. This plant should not be cut down till it has attained 9 inches in height, otherwise it will spring from the root, and require another weeding; and by the time it has attained 9 or 10 inches, the oats will be about 1 foot high. In weeding oats in broadcast, the field-workers may be arranged in the manner described for wheat.

Charlock is also a troublesome weed amongst oats. It may be removed as described in the case of barley (p. 135).

A light top-dressing of from 1 to 1½ cwt. nitrate of soda and 2 cwt. superphosphate per acre, is sometimes given to the oat crop early in May.

Oats are as little affected by weather in the flowering season as is barley. Both are in flower about the same time, and the weather must be stormy for successive days to injure either.

Influence of Season and Soil on varieties of Oats.

In connection with the West of Scotland Agricultural College an extensive series of trials were conducted on the growth of different varieties of oats on a large number of farms in the five years 1902-1906. Numerous varieties were tried, but three varieties were taken as standards, the old Scottish Sandy and Potato oats, which have been grown extensively in Scotland for over a century, and the Banner oat, which was introduced from Canada in 1899, and is now largely grown in both England and Scotland. The Banner is known as a grain-rather than a straw-producing type; the Potato as a combined grain and straw producer; and the Sandy as a straw-producing variety.

For the five years these three typical varieties gave the following average yields per acre:—

	Dressed Grain, Bushels of 40 lb.	Light Grain. lb.	Straw. cwt.
Banner	66	225	37
Potato	57	219	41
Sandy	53	262	45

In reporting on the results of these trials for the five years, Principal Wright drew the following conclusions:—

1. That the yield of all varieties of oats varies greatly, according to the character of the season.

2. That the variation in the West of Scotland is less in the case of the acclimatised straw-producing varieties like the Sandy than in the case of new and imported grain-producing varieties like the Banner.

3. That seasons of low temperature, and especially late, cold, and wet spring and early summer weather, are very prejudicial to the yield of the oat crop, and that in such seasons the grain-producing varieties are liable to suffer special damage when grown on cold and unsuitable soils.

4. That, on suitable soils and in suitable districts, the grain-producing varieties of oats like the Banner, on the average of years, give larger and more profitable crops than the older Scottish varieties either of the Potato or of the Sandy type.

5. That, in favourable seasons and on suitable soils, the grain-producing varieties like the Banner will give much larger and more profitable crops than the Scottish varieties.

6. That the grain-producing varieties like the Banner are more liable to be seriously damaged or destroyed by grub than the common Scottish varieties, and that, if grown on land ploughed out of lea, special precautions should be taken to prevent a grub attack, or to assist the oat to resist it.

7. That the grain-producing varieties of oats are best adapted for cultivation either as a second corn crop or after a root crop, and that they should only be grown after lea on friable and open soils, or on land in high condition or liberally manured.

8. That the grain-producing varieties are not adapted for growth on cold or wet clays or mosses, or on poor, exposed, unproductive land, or on tough old lea land, and that on such soils, on the average of years, better crops will be got from the Sandy and similar varieties of the straw-producing type.

9. That the grain-producing varieties are best adapted for growth on good

soils or on soils well drained and in high condition, and in situations not too much exposed to cold north and east winds.

New and Old Varieties of Oats Compared.

In the years 1903-7 a large number of field trials with different varieties of oats were carried out by the Aberdeen and North of Scotland College of Agriculture.

In an extensive series of trials with the new varieties of oats in comparison with the sorts which have been long in use in Scotland, the results generally were strongly in favour of the former, except that in wet years the old varieties gave the better yields of straw.

From these trials it would seem that on the better class of soil, and in a climate not too late, Banner, Siberian, Thousand Dollar, Mounted Police, and Abundance are likely to prove more profitable than other strains. Abundance is the earliest, but the other three are quite as early as Potato or Hamilton. For very early districts the Wide Awake is a most prolific oat, and equal or superior to the Banner, but much later. The earliest oat yet experimented with is the Daubeney of Canadian origin. This variety is two or three weeks earlier than Sandy, but at present decidedly less prolific.

Cross-bred oats are not recommended for the North of Scotland, except for very good land in a good climate. There seems to be some foundation for the belief that they degenerate rapidly, and require to be constantly renewed by supplies from the breeder. Abundance seems to be an exception to this rule.

The Milling Properties of Oats.

In connection with the Aberdeen and North of Scotland College of Agriculture, numerous trials have been conducted on the milling properties of oats. With the object of discovering what foundation exists for the general prejudice entertained by millers against the new varieties of large-grained thick-skinned oats, over eighty milling tests with different varieties of oats from different classes of soils have been made. In 1904 a milling test was undertaken at five places.

The outstanding feature of the experiment is the comparatively poor result from Potato and Scots Birlie,—comparatively poor in the sense that the reputation of these oats for milling does not seem to be justified by the figures. The average produce from 168 lb. of oats is 100 lb. of meal, and the local kinds are capable of producing as much but apparently not more, whereas general opinion would credit them with a larger out-turn. The table below shows the meal from each of the five centres and the average from all, and there is nothing to indicate the superiority of Potato or Scots Birlie :—

MEAL OBTAINED FROM 168 LB. GRAIN.

	Echt.	Denwell.	Cairnbulg.	Tipperty.	Harestone.	Average.	
	lb.	lb.	lb.	lb.	lb.	lb.	Per cent.
Siberian . . .	101	101	103½	104½	101	102.2	60.7
Thousand dollar . .	100	100	100	105½	101	101.3	60.3
Waverley . . .	101	101	99	103	102	101.2	60.2
Scots Birlie . . .	98	104½	100½	105	96	100.8	60.0
Wide Awake . . .	96	104	103	98	103	100.8	60.0
Newmarket . . .	101	100	101	100	101	100.6	59.8
Banner . . .	100½	101	99½	98	103	100.4	59.7
Potato . . .	95	104	102	105	95	100.2	59.6

In 1906 a more complete test under the most careful supervision did nothing to enhance the reputation of Potato. Seed of Thousand Dollar, Potato, and Banner was obtained from one farm in Inverness-shire and sown at Berrymoss in Aberdeenshire, and Cowfords in Morayshire. Four quarters of dressed grain of each variety were ground at Berrymoss and five quarters at Cow- fords. The results obtained were as follows :—

Variety.	Per-centage of water.	Percentage of husk in		Percentage of dust in		Percentage of meal-seeds in		Percentage of oatmeal in	
		Raw grain.	Dried grain.	Raw grain.	Dried grain.	Raw grain.	Dried grain.	Raw grain.	Dried grain.
Berrymoss—									
Thousand dollar . .	18.82	13.61	16.68	4.98	6.05	3.05	3.76	59.52	73.33
Potato. . . .	19.50	14.06	17.47	5.57	6.93	2.82	3.51	58.03	72.09
Banner . . .	18.53	14.80	17.99	5.73	7.03	3.27	4.02	57.81	70.96
Cowfords—									
Thousand dollar . .	10.71	13.69	15.33	4.17	4.66	2.95	3.33	68.51	76.66
Potato. . . .	11.90	12.79	14.53	3.87	4.39	3.27	3.71	68.21	77.36
Banner . . .	13.69	13.69	15.80	4.76	5.53	4.17	4.83	63.69	73.79

The weight of evidence from these trials goes to show that the milling properties of the Potato are exaggerated in popular estimation, and that several cross-bred and foreign oats are quite equal, if not superior, to the common strains. When the total quantity of meal produced per acre is considered, the comparison is entirely favourable to the newer varieties.

Effect of Soil on Milling Property. —It is well known that millers have a preference for oats from certain soils, and even from particular farms. Two of the foregoing trials throw some light on this preference. Tipperty is on a stiff boulder clay not far from the sea ; Echt is on a light gravelly drift soil about twelve miles inland. The yields of meal from these two centres were as follows :—

	Boulder clay. Per cent of meal.	Gravelly drift. Per cent of meal.
Siberian . . .	62.2	60.1
Thousand Dollar .	62.8	59.5
Waverley . . .	61.3	60.1
Scots Birlie . .	62.5	58.3
Wide Awake . .	58.3	57.1
Potato . . .	62.5	56.5
Newmarket . .	59.5	60.1
Banner . . .	58.3	59.8
Average . .	60.9	58.9

Except Newmarket and Banner all the varieties gave more meal on the stronger soil.

SOWING RYE.

Rye is sown for its grain and straw, and also for forage. Where land is too light and sandy for wheat, after any green crop, rye may be sown in autumn ; and its culture is similar to that for wheat. One ploughing suffices, and two turns of the harrows finishes the surface, 2½ bushels of seed being sown per acre.

Rye will thrive in drifting sand, and will endure the hardest frost. It grows rapidly, and in Germany, where it is largely grown for bread, it is often harvested before the end of June.

As an ordinary cereal crop, however, it is now grown in this country only to a very limited extent. Rye is further dealt with in the section on "Forage Crops."

ROLLING LAND.

The common land-roller is an imple- ment of simple construction, the acting part of it being a cylinder of wood, of stone, or of metal. Simple, however, as this implement appears, there is hardly an article of the farm in which the farmer is more liable to fall into error in its selection.

From the nature of its action, and its intended effects on the soil, there are two elements that should be particularly kept in view—*weight* and *diameter* of the cylinder. By the weight alone can the desired effects be produced in the highest degree, but these will be always modified by the diameter. Thus, a cylinder of any given weight will produce a greater pulverising effect if its diameter is 1 foot, than the same weight would produce if the diameter were 2 feet; but then the one of lesser diameter will be much heavier to draw, hence it becomes necessary to choose a mean of those opposing principles. In doing this, the material of the cylinder comes to be considered.

Wood, which is still employed for the making of land-rollers, may be considered as least adapted of all materials for the purpose. Its deficiency of weight and liability to decay render it objectionable. *Stone*, though not defi-cient in weight, possesses the one marked disadvantage of liability to fracture. This of itself is sufficient to place stone rollers in a doubtful position as to fit-ness. *Iron and steel* are undoubtedly the most appropriate of all materials for this purpose.

Diameter and Weight of Rollers. —There has been much discussion from time to time as to the most advantage-ous diameter for a land-roller. The preponderance of practical evidence is to the effect that a diameter of 2 to $2\frac{1}{2}$ feet is, under every circumstance, the one that will produce the best effects with a minimum of labour from the ani-mals of draught. In many cases, how-ever, rollers of less as well as of greater diameter are in use. The weight is, of course, proportioned to the force usually

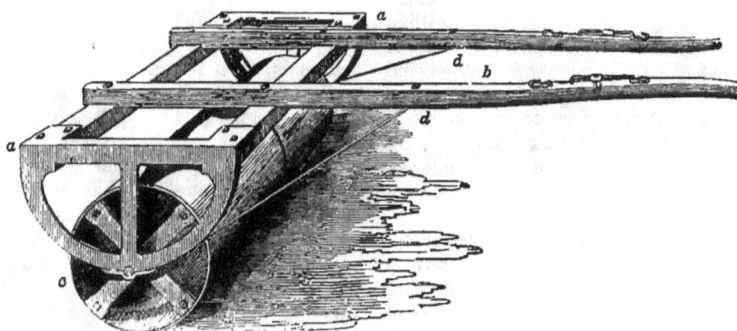

Fig. 429.—*Cast-iron land-roller.*

a a Carriage-frame.   *b* Horse-shafts.   *c* Cylinder.   *d d* Iron stays.

applied, generally 1 but often 2 horses. The weight of roller, including the frame corresponding to this, is from 10 to 15 cwt. But some think it better that the roller itself should be rather under these weights, and that the carriage be fitted up with a box in which a loading of stones can be stowed, to bring the ma-chine up to any desired weight. Such a box is, besides, useful in affording the means of carrying off from the surface of the ground any large stones that may have been brought to the surface by the previous operations.

Divided Roller. — In a large and heavy roller, in one entire cylinder, the inconvenience of turning at the headlands is very considerable, and has given rise to the improvement of having the cylinder in two lengths. This, with a properly constructed carriage, produces a very convenient form of land-roller. Fig. 429 is a perspective of the land-roller con-structed on the foregoing principles, with the carriage-frame crossed by the horse-shafts. The cylinder is in 2 lengths of 3 feet to 3 feet 3 inches each, and 2 feet in diameter; the thickness of the metal is according to the weight required. The axle, in consequence of the cylinder being in two lengths, requires to be of consider-able strength, and of malleable iron; upon this the two sections of the cylinder revolve freely, and the extremities of the axle are supported in bushes in the semi-circular end-frames. Two iron stay-rods pass from the end-frames to the shafts as an additional support to the shafts.

Excellent rollers are now made of steel sheets fixed on wrought- or cast-iron ends.

Water-ballast Roller.—A very convenient form of roller, made by Barford & Perkins, Peterborough, is represented in fig. 430. It is made in two enclosed cylinders of wrought iron, formed so that by filling or partially filling the cylinders with water, the weight of the roller may be varied as desired. These water-ballast rollers are made of many sizes for field and garden work, and are exceedingly convenient to work and move about. A water-ballast roller, 2 feet in diameter, weighs about 11 cwt. when empty, and 22 cwt. when quite full of water.

Process of Rolling.—The rolling is always effected across the line of ridges. Otherwise the open furrows would not receive any benefit from it. Although the dividing of the cylinder into two parts facilitates the turning of the implement, it is not advisable to attempt to turn the roller sharply round, as part of the ground turned upon may be rubbed hard by the cylinders, with the result that young plants may be injured or killed.

The rolling is sometimes executed in feers of 30 yards in width, *hieing* the horses one-half of the feering and *hupping* them in the other half. This, however, is unnecessary with care at the turning. When the ploughman becomes fatigued with walking, it may be allowable for him to sit on the front of the framing, where a space is either boarded or wrought with hard-twined straw-rope, as a seat from whence to drive the horses with double reins and whip. With this indulgence, an old ploughman, employed only in ploughing, could take the rolling when urgent work was employing the stronger horses in the cart.

Speed in Rolling.—Were a 6-feet roller to proceed uninterruptedly for ten hours, at the rate of $2\frac{1}{2}$ miles per hour, it would roll about 18 acres; but what with the time spent in the turning and the markings-off of feerings, 10 to 12 acres a-day may be considered a good day's work. When the weather is favourable, and a large extent of ground has to be rolled, it is a good plan to work the roller from dawn to nightfall, each horse or pair, as the case may be, working 4 hours at a time. In this way, 16 hours' constant rolling, from 4 in the morning till 8 at night, may be obtained in the course of 24 hours, and from 25 to 30 acres rolled with one roller.

Time for Rolling.—The usual time for rolling is immediately after the seed has been sown. But the condition of the land as to moisture must be considered.

Fig. 430.—*Water-ballast roller.*

The young braird on strong land is much retarded when the earth becomes encrusted by rain after rolling, so that such land in wet districts is in rainy seasons not rolled until the end of spring, when the plant has made some progress, and the weather continues dry. Light friable dry land should be rolled immediately after the seed is sown and harrowed, if there is time to do it. But the rolling of one field should not be allowed to cause delay in the sowing of others in dry weather. There will be plenty of time to roll the ground after the oat seed and other urgent operations at this season are finished.

On the other hand, the rolling is most effective in securing smoothness in the surface immediately after harrowing has been completed. And for the sake of the reaping-machine a smooth surface is of much importance.

In preparing land for grass and clover seeds the roller is not, as a rule, used so much as it should be. An even, firm seed-bed is of the utmost importance for these tiny seeds, just as it is for the seeds of the turnip crops.

GENERAL PRINCIPLES OF CORN CULTURE.

The sowing of the chief cereal crops has thus been dealt with. Much more might have been said on the subject, but there seems to be little necessity for describing at great length operations which are so simple as the cultivation of corn. Of all important work upon the farm this is, perhaps, the most simple and the most uniform in the methods of procedure.

The simplicity and the universality of the general principles of *corn cultivation* are well shown by Professor Wrightson in the following admirable epitome:—

"No business pursuit is easier than corn cultivation, and this is why we have such millions of bushels of corn thrown in upon us. It is a cheap cultivation. All we have to do is to plough the land, throw on the seed, and scratch it in. Of course we must do this at the right time of the year, and in the proper manner. When we take wheat or barley, or oats [to be sown in the autumn or winter], after a root cropped on the land, a very general method of cultivation is as follows: We first plough about 4 inches deep, then broadcast the seed upon the newly turned up fallow, and put the harrows on and give it a really good harrowing, so as to break the compact furrow and cover the seed thoroughly—that is all. Protect it from the ravages of the birds, and in the spring of the year roll and harrow it, and that is pretty much the cultivation of corn after roots. A great deal of corn is taken after grass and clover crops; and the cultivation of either oats or wheat, or barley after lea, is much the same thing. We plough and press, and often sow the seed upon the pressed furrow and harrow it in.

"Again, in other cases we plough, press, or heavily roll, harrow repeatedly, and drill. That again is the whole of the cultivation. Corn crops sometimes follow peas or beans, in which case the plan would be to dung the surface, and then proceed as before, ploughing in the dung, and either broadcasting or else producing a proper seed-bed with the use of the harrow, and drilling in the corn."[1]

[1] *Prin. of Ag. Practice*, 136.

Insect attacks upon corn and other crops are dealt with in a special chapter.

CROSS-PLOUGHING LAND.

The first preparation for cereal seed after turnips is ploughing the land across at right angles to the existing ridges. The surface of the ground where sheep consume turnips is left in a smooth state, trampled firm by the sheep, presenting no clods of earth but perhaps numbers of small round stones, which should be removed with carts before the cross-ploughing is begun. The small stones are useful for drains, or to repair farm roads, and the larger stones for dykes.

A plough then feers the ground for cross-ploughing. The reason that land is cross-ploughed for barley, and not for spring wheat, after turnips eaten off by sheep, is, that wheat thrives best when the soil is firm and not too much pulverised —whereas the land cannot be in too fine a condition for barley. Moreover, if the turnip-land were not cross-ploughed after the sheep left it, their manure would not be sufficiently intermixed with the soil, and in consequence the barley would grow irregularly in small rows, corresponding to the drills that had been manured for the turnip crop.

Harrowing before Cross-ploughing.—The portion of the stubble-land first to be cross-ploughed is for beans. Every winter-ploughed field for cross-ploughing in spring is freed from large clods by *harrowing*. The winter's frost may have reduced the clods of the most obdurate clay soil, and the mould-board of the plough may thus be able to pulverise them fine enough, while the lighter soils may have no clods upon them. In this case it would seem loss of time to harrow the ground before cross-ploughing, and some farmers do not then use the harrow; yet, in the majority of cases, the harrowing will be found beneficial. One cannot be sure that, in the strongest soil, all the clods have been reduced to the heart by frost; and should any be buried by the cross-furrow while still hard, they will not afterwards be so easily pulverised amongst the soft soil as when exposed upon the harder surface of the winter-furrow. Then in the lightest soils, the harrows not only make a smoother surface, but intermix the surface-dry, frost-pulverised soil with the moister and firmer soil below, as far as the tines of the harrows can reach.

There is not much time lost in harrowing before cross-ploughing; and although it should require a double tine to pulverise the clods, or equalise the texture of the ground, it should be *across* instead of along the ridges, to fill up the open furrows with soil, whether the land had been previously ploughed with gore-furrows or not.

If time presses, the feerings for cross-ploughings may be commenced by one

plough almost immediately after the harrows have started; and if the harrows cannot get away before the plough, the plough can take a bout or two round the first feering till the harrows have reached the second feering; or, still better, the harrows can go along each feering, preparing the ground for the plough, and then return and finish the harrowing between the feerings.

Thus, in fig. 431, after the first feering *e f* across the ridges has been ploughed, the plough can either take a bout or two round *e f*, till the harrows have passed the next feering *g h*, or the harrows can

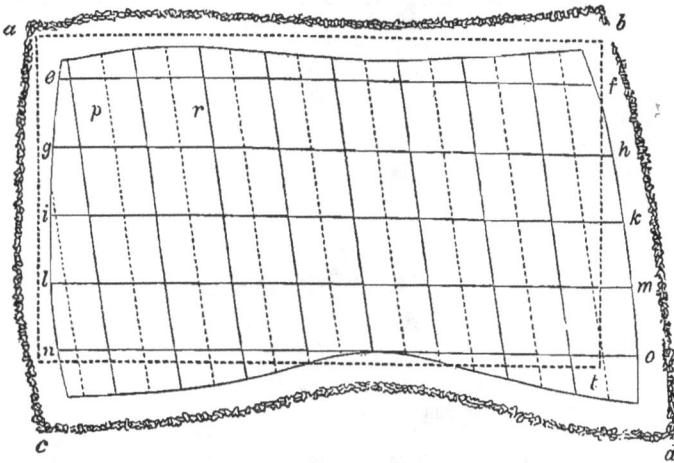

Fig. 431.—*Field feered for cross-ploughing.*

go along the line of each feering, at 30 yards' distance, first *e f*, then *g h*, then *i k*, and so along *l m* and *n o* in succession, and prepare the ground for feering, and then return and harrow out the ground between *e* and *g*, *g* and *i*, *i* and *l*, and *l* and *n*. In this way the harrowing and feerings, and ploughing the feerings, may go on at the same time.

System of Cross-ploughing.—But if time is not urgent, the systematic way is to feer the field across, at 30 yards' distance, from *e* to *n*, across the whole field, and the ploughs take up the feerings in succession. To illustrate this more fully, suppose that all or as much of the field to be cross-ploughed has been harrowed as will give room to a single plough to make the feerings without interruption. In choosing the side of the field at which

the feerings should commence, it is a good rule to begin at the side farthest from the gate and approach gradually towards it, because then the ends of the finished feerings will not be passed, and the trampling of the ploughed land be avoided. The convenience of this rule is felt not in cross-ploughing only, but in prosecuting every kind of field-work; for besides avoiding damage to finished work, it is gratifying to the mind that, as work proceeds, the approach is nearer home; while it conveys the idea of a well-laid plan to have the operations of a field commenced at the farthest end and finished at the gate, where all the implements meet, ready to be conveyed to another field. The gate is like home, and in most cases it is placed on the side or corner of a field nearest the steading.

Ploughing Ridges and Feerings.— Some farmers neglect the head-ridge in the cross-ploughing, and measure the feering from the open furrow which divides the head-ridge and the ends of the ridges. The head-ridges ought to be ploughed at this time along with the rest of the field, for, if neglected now, the busy seasons of spring and early summer will draw away attention from them, till, what with trampling in working the green crop and the drought of the weather, they will become too hard to plough, and will lose the ameliorating effects of sun and air in the best part of the year.

In cross-ploughing the ridges of the field, the head-ridges must be ploughed in length, for they can never be *cross*-ploughed.

Depth of Cross-furrow.—The depth of the cross-furrow varies with the character of the soil. It is often, in good soil, deeper than the winter-furrow. The deepness is easily effected by the plough passing under the winter-furrow and raising a portion of the fresh soil below it. If the under soil is suitable, the 2 inches of fresh subsoil mix well with the thicker winter-furrow.

Cross-ploughing the first furrow in spring is unsteady work for the ploughmen, the open furrows presenting little resistance to the plough compared with the crown of the ridge.

The depth of the cross-furrow may vary from 8 to 12 inches, 10 inches being quite common.

Grubbers or cultivators are now extensively employed in spring tillage. To these operations fuller reference will be found in the chapter dealing with sowing turnips.

BEANS AND PEAS.

Beans.

Beans are classed with a very different tribe of plants from the cereals which we have been considering. They belong to the natural order *Leguminosæ*, because they bear their fruit in legumes or pods. Their ordinary systematic name is *Faba vulgaris;* but the bean is also known as *Vicia Faba.*

The common bean is divided into two classes, according to the mode of culture to which it is subjected—that is, the field or the garden. Those cultivated in the field are called *Faba vulgaris arvensis,* or, as Loudon calls them, *Faba vulgaris equina,* because they are cultivated chiefly for the use of horses, and are usually termed horse-beans. Some farmers attempt to raise a few varieties of the garden-bean in the field, but without success. All beans have butterfly or papilionaceous flowers.

Field-bean.—Lawson has described 8 varieties of the field-bean. The variety in common field-culture is thus well described by him: "In length the seed is from a half to five-eighths of an inch, by three-eighths in breadth, generally slightly or rather irregularly compressed and wrinkled on the sides, and frequently a little hollowed or flattened at the end; of a whitish or light-brown colour, occasionally interspersed with darker blotches, particularly towards the extremities; colour of the eye black; straw from 3 to 5 feet in length. There is, perhaps, no other grain over the shape and colour of which the climate, soil, and culture exert so much influence as the bean. Thus, in a dry warm summer and harvest, the sample is always more plump and white in colour than in a wet and cold season; and these more so in a strong rich soil than in a light, and more so in a drilled crop than in one sown broadcast."[1]

Fig. 432 represents the horse-bean of its natural size.

Leguminous Plants.— "The leguminous order," observes Lindley, "is not only among the most extensive that are known, but also one of the most important to man, whether we consider the beauty of the numerous species, which are amongst the gayest-coloured and

[1] Lawson's *Agric. Man.*, 62.

most graceful plants of any region, or their applicability to a thousand useful purposes.

"The cercis, which renders the gardens of Turkey resplendent with its myriads of purple flowers; the acacia, not less valued for airy foliage and elegant blossoms than for its hard and durable wood; the braziletto, logwood, and rosewoods of commerce; the laburnum; the classical cytisus; the furze and the broom, both the pride of the otherwise dreary heaths of Europe; the bean, the pea, the vetch, the clover, the trefoil, the lucerne, all staple articles of culture by the farmer, are so many leguminous species. The gums, Arabic and Senegal, kino, senna, tragacanth, and various other drugs, not to mention indigo, the most useful of all dyes, are products of other species; and these may be taken as a general indication of the purposes to which leguminous plants may be applied. There is this, however, to be borne in mind, in regarding the qualities of the order in a general point of view— viz., that, upon the whole, it must be considered poisonous, and that those species which are used for food by man and animals are exceptions to the general rule; the deleterious juices of the order not being in such instances sufficiently concentrated to prove injurious, and being in fact replaced, to a considerable extent, by either sugar or starch." [1]

Fig. 432.—*Horse-beans.*

Beans containing Poison.—In various cases in Great Britain sudden deaths occurring amongst cows have been attributed to beans included in the food. Investigations conducted by Mr Hendrick, Chemist to the Highland and Agricultural Society of Scotland, showed that prussic acid existed in dangerous quantities in samples of beans imported from Java, and that this virulent poison was also found in Burma beans, though in smaller quantities. In purchasing imported beans, farmers should therefore exercise great care, requiring the sellers to guarantee that the beans are free from poison. It has been found that boiling twice over renders the poisonous beans harmless.

Yield and Weight of Beans.—The produce of the bean crop varies from 20 to 40 bushels per imperial acre, the prolificness of the crop palpably depending on the nature of the season. The average weight may be stated at 66 lb. per bushel. It requires only 5 beans to weigh 1 drachm, so that a bushel contains only 42,240 beans. Beans have been known to yield 2 tons of straw or haulm per acre.[2] A crop of 40 bushels, at 66 lb. per bushel, gives 1 ton 3 cwt. 64 lb. per acre.

Consumption of Beans.—Beans are given to the horse, whole, boiled, raw, or bruised. They are given to cattle in the state of meal, but can be ground into fine flour, which is used at times to adulterate the flour of wheat. Its presence is easily detected by the peculiar smell arising from the flour on warm water being poured upon it.

"There are several varieties of the bean in use as horse-corn, but I do not know that one is better than another. The small plump bean is preferred to the large shrivelled kind. Whichever be used, the bean should be old, sweet, and sound; not mouldy, nor eaten by insects. New beans are indigestible and flatulent; they produce colic and founder very readily. They should be at least a year old." [3]

Peas.

The *pea* occupies a similar position to the bean in both the natural and artificial systems of botany. The plant is cultivated both in the field and in the garden, and in the latter place to great extent and variety. The *natural* distinction betwixt the field- and the garden-pea is founded on the flower, the field-pea always having a red-coloured and the garden almost always a white one; at least, the exceptions to this mark of distinction are few.

The botanical name of the pea is *Pisum sativum*, the cultivated pea; and those varieties cultivated in the field are called in addition *arvense*, and those in the gar-

[1] Lindley's *Veget. King.*, 546, 547.

[2] *Brit. Husb.*, ii. 215.
[3] Stewart's *Stab. Eco.*, 205, 206.

den *hortense*. The name is said to have been given to it by the Greeks, from a town called Pisa, in Elis, in the neighbourhood of which this pulse was cultivated to a great extent: Paxton derives the name from the Celtic word *pis*, the pea, whence the Latin *pisum*.[1]

Lawson has described 9 varieties of the field-pea. Of these, a late and an early variety are cultivated. The late kind, called the *common grey field-pea* or *cold-seed*, is suited for strong land in low situations; and the early, the *partridge*, *grey maple*, or *Marlborough pea*, adapted to light soils and late situations, is superseding the old grey Hastings, or *hot-seed* pea.

The grey pea is described as having "its pod semicylindrical, long, and well filled, often containing from 6 to 8 peas. The partridge pea has its "pods broad, and occasionally in pairs, containing from 5 to 7 peas of a medium size, roundish, and yellowish-brown speckled, with light-coloured eyes. The ripe straw is thick and soft-like, leaves large and broad, and average height 4 feet."[2]

Fig. 433.—*Partridge field-pea.*

Fig. 433 is the partridge field-pea of the natural size.

Produce of Peas.—The produce of the pea crop varies greatly according to the season. In warm weather, with occasional showers, the crop may amount to 48 bushels, and in cold and wet it may not reach 12 bushels the acre. The grain weighs 64 lb. the bushel, and affords 13 grains to 1 drachm; consequently a bushel contains 106,496 peas.

Consumption of Peas.—The pea was formerly much cultivated in this country in the field, and even used as food, both in broth and in bread, *pease bannocks* having been a favourite food of the labouring class; but, since the extended culture of the potato, its general use has greatly diminished. It is now chiefly given to horses, and also split for domestic purposes, for making pea-soup—a favourite dish with families in winter.

[1] Paxton's *Bot. Dict.*, art. Pisum.
[2] Lawson's *Agric. Man.*, 70.

Its flour is used to adulterate that of the wheat, and is easily detected by the peculiar smell which it gives out with hot water. Peasemeal in brose is administered in some cases of dyspepsia. Pea-pudding is eaten as an excellent accompaniment to pickled pork. Pigeons are excessively fond of the pea, and it has been alleged that they can devour their own weight of them every day.

Bean- and Pea-straw.

Bean- and pea-straw, or haulm, are difficult in some seasons to preserve, but, when properly preserved, no kind of straw is so greatly relished as fodder by every kind of stock.

According to Sprengel, the ash of bean- and pea-straw contains the following ingredients:—

	Field-bean.	Field-pea.
Potash . . .	53.08	4.73
Soda . . .	1.60	...
Lime . .	19.99	54.91
Magnesia . .	6.69	6.88
Alumina . .	0.32	1.21
Oxide of iron . .	0.22	0.40
Oxide of manganese	0.16	0.15
Phosphoric acid .	7.24	4.83
Sulphuric acid .	1.09	6.77
Chlorine . .	2.56	0.09
Silica . . .	7.05	20.03
	100.00	100.00

Percentage of ash, from 4½ to 6.

Young cattle are very fond of bean-*chaff*, and, with turnips, thrive well upon it. Cows also relish it much.

SOWING BEANS.

On suitable soil the bean crop is usually a profitable one. As the bean takes about 7 months to come to maturity, the crop should be sown as early in spring as possible. It should be sown in February if the weather and the condition of the land permit; in no case later than March. A very favourable season may hasten the plant through its courses of vegetation in a shorter time; but a very unfavourable season will so retard it as almost to prevent the formation of the seed.

In Scotland the bean is not a reliable crop. It was never cultivated extensively there, and in recent years has lost ground slightly. Strong land is best suited for beans, which still hold an important place on good carse farms. The land must be in good heart, and is generally well manured with dung in the previous autumn or winter. Beans are sown on the flat surface, or in rows from 15 to 20 inches apart, or in raised drills from 25 to 30 inches wide. The bean crop occupies varying positions in the rotation. It may come in between two cereal crops, between two crops of wheat, between oats and wheat, or between wheat and barley.

The bean crop is valuable both for its straw and grain. Though the crop fail in seed, it seldom fails to produce good fodder provided it can be well secured. A dry season stints the growth of the haulm, but produces beans of fine quality; and a wet season prevents the growth of the bean, but affords a bulky crop of fodder.

The culture for beans is not dependent so much on the soil as on the peculiar growth of the plant. Bearing fruit-pods on its stem near the ground as well as near the top, it should have both light and air ; and its leaves being at the top, and its stem comparatively bare, weeds find room to grow. The plant should therefore be wide asunder in the row and between the rows, so that the crop may become luxuriant and the land cleaned.

Beans were long wont to be sown broadcast, and are so sown still in some cases. It is not a good plan, however, for it has a great tendency to leave the land full of weeds.

Varieties of Beans.—Several varieties are in cultivation. Those most largely sown are the common Scotch or horse-bean, and the common tick-bean. The former is the best suited for northern districts, and under favourable circumstances grows to a height of 4 or 5 feet, weighing from 62 to 65 lb. per bushel. The seed is large, flat, of a dingy whitish colour, with a black eye, and irregularly wrinkled on the sides. The tick-bean, which is shorter in the straw, and generally more prolific, is the variety most largely cultivated in England. The seed is smaller, plumper, a pound or two heavier per bushel than the seed of the horse-bean. Amongst the other best-

known varieties are the Russian or winter bean, the Mazagan, and the Heligoland bean.

Quantity of Seed.—From two to four bushels per acre are the most general quantities. In the north it is more frequently four, sometimes even five bushels. The seed is sown by machines of various patterns—barrow-shaped appliances, worked by hand or horse-power, and sowing usually one or three drills or rows at a time.

Manure for Beans.—Land intended for beans is usually well dunged in the autumn, or early in winter, with perhaps from 8 to 12 tons of farmyard dung, spread just before the land is ploughed. The dung will be all the better for this purpose if it is tolerably fresh, and it should be spread evenly on the land. In other cases, the dung is spread early in spring on the flat or in drills, as for turnips. When the dung is to be spread in drills, these are opened a little deeper than if the land were simply drilled to receive the seed.

Formerly it was thought that beans could not be grown satisfactorily without farmyard dung, but, as shown clearly by the Highland and Agricultural Society's experiments, that idea was not well founded. The artificial manures which gave the best results in these experiments are described at pp. 38, 39. Potash is the dominant ingredient. It is seen that, unaccompanied by potash, neither phosphates nor nitrate is of much use to the bean, whether applied separately or together ; but the addition of potash to either or both at once enormously increases the crop. The artificial manures were applied in March, three days before the seed was drilled in with the three-drill bean-barrow.

Beans and Nitrogenous Manure.—Seeing that a leguminous crop such as beans contains a great deal more nitrogen than cereal crops, it might be expected that nitrogenous manures would exercise a more beneficial effect upon beans than upon cereals. It has been found, however, that such is not the case. At Rothamsted extensive experiments have been carried out in the manuring of beans and other leguminous crops, but, curiously enough, the results have not been so clear or instructive as those obtained from the manuring experiments with most other crops. Sir J. B. Lawes says:—

"The general result of the experiments with beans has been, that mineral constituents used as manure (more particularly potash) increased the produce very much during the early years; and to a certain extent afterwards, whenever the season was favourable for the crop. Ammonia-salts, on the other hand, produced very little effect, notwithstanding that a leguminous crop contains two, three, or more times as much nitrogen as a cereal one grown under similar conditions as to soil, &c. Nitrate of soda has, however, produced more marked effects. But when the same description of leguminous crop is grown too frequently on the same land it seems to be peculiarly subject to disease, which no conditions of manuring that we have hitherto tried seem to obviate.

"Experiments with peas were soon abandoned, owing to the difficulty of keeping the land free from weeds, and an alternation of beans and wheat was substituted,—the beans being manured much as in the experiments with the same crop grown continuously.

"In alternating wheat with beans, the remarkable result was obtained, that nearly as much wheat, and nearly as much nitrogen, were yielded in eight crops of wheat in alternation with the highly nitrogenous beans, as in sixteen crops of wheat grown consecutively without manure in another field, and also nearly as much as were obtained in a third field in eight crops alternated with bare fallow."

Ploughing for Beans.—Strong land intended for beans is usually ploughed about the end of autumn or early in winter, so that it may have the benefit of the pulverising influences of winter. If the land is very heavy and liable to hold surface water, it will be useful to plough it in the direction of the greatest inclination or fall, so that there may be no cross-furrows to retain the water. But when the land can be ploughed across the inclination it will be well to do so, and then the drills, if the crop is to be grown in drills, will follow the inclination, thus crossing the autumn furrow.

Spring Tillage for Beans.—The amount and kind of tillage which bean land should receive in spring will depend upon the nature and condition of the land and the character of the season. If the land lying in the winter furrow is tolerably friable, harrowing may be sufficient. As a rule, however, a turn of the grubber or cultivator will be found highly beneficial.

Cultivators and Grubbers.—The improved grubbers or cultivators are excellent implements for pulverising surface-soil. They do their work well, and are very speedy—a consideration of special importance at this time of the year. They are extensively employed in almost all parts of the British Isles where tillage farming is pursued. Cultivators are now designed so that by changing fittings they can perform various tillage operations, thus economising outlay in the cost of implements.

The action of the grubber or cultivator in the soil is to stir it effectually as deep as the tines descend, and at the same time retain the surface-soil in its existing position. This advantage is especially appreciated in early spring, when it is precarious to turn over the soil with the plough, lest by a fresh fall of rain it should become wetter and worse to work than if it had not been ploughed at all. If the land be raw and not very clean, and the weather precarious, the grubber will prepare the soil for harrowing, of which it should receive one double tine along the ridges, the grubbing having been given across them. Should this not be sufficient to reduce the clod, another double tine should be given across the ridges, when the land will be ready for sowing.

If the weather in spring is favourable, and the beans are to be sown broadcast or in rows on the flat, ploughing across the winter furrow is by many considered desirable. The modern grubbers or cultivators, however, do their work so well that the necessity for the plough in spring is much lessened.

In preparing land in spring for beans, care should be taken not to grub or harrow more in one day than can be drilled up or sown on the same or the following. A fall of rain on this prepared ground before it is drilled for the seed would be detrimental to the crop.

Sowing Autumn-manured Beans.

—The process of sowing beans upon land which had been purposely dunged and ploughed in autumn or early winter is thus described by Mr F. Muirhead :—

" We will suppose the time has arrived for sowing the seed. The farmer should previously have had his bean-sowing machine examined, repaired if necessary, and well oiled. He should also have provided the requisite quantity of seed—say 4 bushels of common Scotch beans for every imperial acre; and he had better have an extra bag of beans for every twenty he intends to sow, in case he may need a little more to finish the field than he anticipated.

" He should visit the field a day beforehand, and ascertain the length of the proposed drills, and how many make an imperial acre; and the following table may assist him :—

Inches wide. Drills.	Yards long. Imperial acre.
26	6701
27	6453
28	6222

" The open furrows should be filled in with two or three bouts of a two-horse plough, and the ends or headlands marked off, say, to hold eight drills, which should be ample room to admit of horses and ploughs turning quickly without treading on the newly formed drills. If the land requires a double stroke of heavy harrows before being drilled, as much should be harrowed the afternoon previous to sowing (provided weather is somewhat settled) as to allow the ploughs to get to work *readily* the following morning, or the foreman had better be sent half a day beforehand to do this, and to open, say, ten or twelve drills; and care should be taken, if the field has much inclination from top to bottom, to begin at that side of it which will, in covering up the sown seeds, give the horses the heavy furrow *down* hill. The following morning fully as much seed is taken out to the field as will likely be needed during the forenoon, and the bags should be placed along the top headland, if drills are not too long to admit of the three-drill horse sowing-machine sowing a ' bout ' or six drills before it needed to be refilled, care being taken that the seed always covered the pinions for forcing out the beans.

" In placing the bags with the seed, suppose that it takes thirty drills to be an acre imperial, and we wish to sow 18 stones per acre, it will be more convenient to have the beans weighed up to that weight in each bag, and place the bags along the headland, one bag at the last drill of each acre; and in beginning to sow, it will be found of advantage to take out as much extra seed in a bag as cover the pinions of the sowing-machine, so that when the *first* bag is all sown, the person in charge knows at once whether the machine is sowing too quickly or too thinly. Perhaps if the first bag were accurately divided into two, and set down separately, at half an acre for each, the setting of the machine would be the sooner tested. The sowing-machine will now begin and sow the three outside drills, and the ploughs will commence and cover up the seed as they go *down hill*, and open fresh drills at the required width as they return. One sowing-machine will easily keep four or five pairs of horses at work." [1]

Sowing Spring-manured Beans.— When the dung has to be applied to the drills in spring, it is carted to the field, and thrown in graipfuls as the horse moves along the drills, just as in the dunging of roots or potatoes. The graipfuls are then spread evenly along the bottom of the drills, which, having received the seed, are thereupon closed.

If the dung has to be applied in spring, and it is intended to sow the beans *broadcast* or *in rows* on the flat, then the land receives a single or double turn of the harrow, the dung is spread evenly on the surface, and the land ploughed, the seed, perhaps, being dropped by the single bean-barrow into every third furrow. And as the furrows are about 9 inches in breadth, the three furrows will place the rows of beans at 27 inches apart. This ploughing finishes the operation.

When the land is manured in the spring, and the seed sown broadcast, the dung in the same state is spread broadcast upon the surface. The further part of the operation depends on the state of the weather. Should it promise well until the bean-sowing is finished, the dung may be ploughed in, the seed sown

[1] *Farming World Almanac,* 1888.

broadcast upon the ploughed surface, harrowed in with a double tine, and the ridges water-furrowed. Should the weather seem doubtful, a safer plan is to sow the seed broadcast upon the spread dung, and plough in both seed and dung together, and the surface will be secured from danger. In this case the plants will come up in rows of the breadth of the furrow—9 inches apart.

Harrowing Drills.—If it is considered desirable to harrow the drills, this may be done about a fortnight after the sowing, if the surface is at all dry. If the land is wet, the harrowing should be delayed, and the first dry state of the surface taken advantage of. The common harrow is sometimes used to harrow

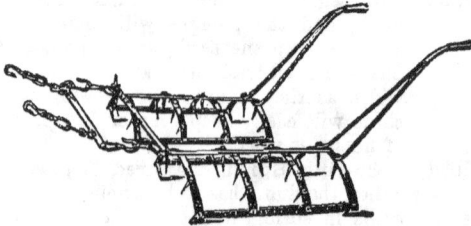

Fig. 434.—*Saddle drill-harrow.*

down drills; but a better implement is the *saddle drill-harrow* (Clay), represented in fig. 434. This harrow is worked in pairs; and, to render it applicable to its purpose, it is made of an arch form, partially embracing the curvature of the drill, and on this account is best fabricated of iron. The pair of harrows are drawn by one horse, walking between the drills. Chain-harrows are also useful for this purpose.

Summer Culture of Beans.

Beans require a good deal of labour and attention in summer.

Horse-hoeing.—As soon as the young plants growing on raised drills have attained 2 or 3 inches in height, the common drill-grubber or scuffler should remove the weeds that have appeared between the drills in the interval of time since the drill-harrowing. The grubbing will also reduce the clods and loosen the soil generally.

Hand-hoeing.—The field-workers follow the scuffler with the hand-hoe, and remove the weeds growing around the

plants, and displace clods that are seen to interfere with the plants. The workers should be careful in using the hoe amongst bean plants, which are very tender and easily cut and bruised.

After the plants have risen about 1 foot in height, which they will soon do in good growing weather, the blossom will begin to appear; and its appearance is with many the signal to finish the work amongst the crop. Time may be found to again drill-grub between the drills, and hoe the sides of the drills along the plants; but if not, the double mould-board plough should, as the last operation, set the earth up to the roots of the plants, to give them a firm footing on the top of the drill.

Rows on the Flat.—The summer culture of beans growing on flat ground in rows is the same, in as far as scuffling, hoeing, and drill-grubbing the ground are concerned, as on the raised drill. Almost the only difference is that the land is not set up with the double mould-board plough.

No Harbour to Weeds.—No amount of horse- and hand-hoeing should be grudged that may be necessary to make and keep the land free from weeds. It should be remembered that one of the objects in having beans sown in drills is to have the land well worked and cleaned.

Broadcast.—When beans are grown broadcast, no implement but the hand-hoe is of any avail in clearing the ground of weeds; and as hand-hoeing would require to be performed much oftener than time will allow, to keep the ground as clean as it should be, the consequence is that a crop of broadcast beans affords a harbour to weeds, unless growing weather pushes the bean plants forward to smother the weeds.

Cropping Beans.—After the bean plant has grown until all the pods are set, the practice of the garden indicates that, when the top of the plant is cut off in moist weather, at that period of its growth the crop will be sensibly increased. This is a probable result, it being a common observation that in moist weather the bean has a great tendency to grow in height long after the pods have ceased to form. As long as this tendency continues, the pods and beans do not enlarge; and the only mode of checking it is to cut off the top, when the vigour of the plants' growth will be solely devoted to the nourishment of the fruit.

Beans and Peas Mixed.—Beans and peas are often grown together, the seed being sown broadcast. The most general proportion is about one-third of peas to two-thirds of beans.

Botanical Character of Beans.—It was an observation of De Candolle, that "it is remarkable that the botanical character of the *Leguminosæ* should so strictly agree with the properties of their seed. The latter may be divided into two sections—namely, the first, *Sarcolobæ,* or those of which the cotyledons are thick, and filled with fecula, and destitute of cortical pores, and which, moreover, in germination do not undergo any change, but nourish the young plant by means of that supply of food which they already contain; second, the *Phyllolobæ,* or those of which the cotyledons are thin, with very little fecula, and furnished with cortical pores, which change at once into leaves at the time of germination, for the purpose of elaborating food for the young plant. All the seeds of the *sarcolobæ* are used as food in different countries, and none of those of *phyllolobæ* are ever so employed."

Ancient Notions regarding Beans.—The ancient Greeks had some strange notions regarding the bean. Thus Didymus the Alexandrian says: "Do not plant beans near the roots of a tree, lest the tree be dried. That they may boil well, sprinkle water with nitre over them. Physicians, indeed, say that beans make the persons that eat them heavy; they also think that they prevent night dreams, for they are flatulent. They likewise say that domestic fowls that always eat them become barren. Pythagoras also says that you must not eat beans, because there are found in the flour of the plant inauspicious letters. They also say that a bean that has been eroded becomes whole again at the increase of the moon; that it will by no means be boiled in salt water, nor, consequently, in sea-water," &c.[1]

[1] Owen's *Geoponika,* i. 82.

SOWING PEAS.

In recent years peas have been sown to a smaller extent than they were at one time in this country. They seldom take a prominent place as an ordinary rotation crop, but are largely grown near populous towns for sale in the green pod.

Peas give the best results on light and friable loamy soils of a calcareous character, or which had been recently dressed with lime or chalk. It is a general observation that annual weeds are encouraged in growth amongst peas; and the pea being a precarious crop, yielding a small return of grain, except in fine warm seasons, a mere good crop of straw is insufficient remuneration for a scanty crop of grain, accompanied with a foul state of land. Hence in many cases turnips have been substituted for peas.

Peas, for a long period, were invariably sown broadcast; but seeing their tendency to protect weeds, and that drill-culture rendered the land clean, the conclusion was obvious that peas sown in drills would admit of the land being cleansed. It was found that the straw by its rapid growth creeping along the ground soon prevents the use of the weeding instruments. To counteract this tendency, the practice was introduced of sowing peas and beans together, and while their seasons of growth coincide, the stems of the bean serve as stakes to support the bines of the pea. The proportion, as already stated, is about one-third of peas to two-thirds of beans.

Tillage for Peas.—It is somehow considered of little moment how the land shall be ploughed when the pea is to be sown by itself. Sometimes only one furrow after the stubble is given; and when the land is tender and pretty clean, a sufficient tilth may be raised in this manner to cover the seed, which requires neither a deep soil for its roots (which are fibrous and spreading near the surface), nor a deep covering of earth above them, 2 inches sufficing for the purpose. But a single furrow does not do justice to the land, whatever it may do for the crop. The land should be double drilled or grubbed after the spring ploughing.

Since the pea can be cultivated along

with the bean, it will grow on good strong soils; and its spreading roots enable it to grow on thin clays, where the bean does not thrive. But by itself, the pea, as has been indicated, thrives best on light soils. In clay, it produces a large bulk of straw, and the grain depends on the season being dry and warm; and as this is not the usual character of our climate, the yield is but indifferent.

Dung is seldom given to the pea when sown by itself, having the effect of forcing much straw with little grain.

When peas and beans are reaped together, they are separated when thrashed simply by riddling, the peas passing through the meshes of the riddle, while the beans are left upon the riddle.

Sowing Peas. — Peas are sown by hand when cultivated broadcast, and with the barrow when in rows, in every third, or in every furrow. With beans, they are sown by a barrow; on drilled land, broadcast by the hand; the seed falling to the bottom of the drills is covered by the harrows passing across the drills. Like beans, peas are sown on ploughed lea in some parts of England. On lea, the pea is dibbled in the harrowed surface, the holes being placed about 9 inches asunder. When varieties of the white garden-pea are cultivated in the field, as in the southern counties of England, these various modes of sowing them deserve attention; as also in the neighbourhood of large towns, where the garden-pea is cultivated and sent in a green state to the vegetable market.

The quantity of seed per acre varies, in drilling, from 2½ to 3 bushels per acre in the south, and sometimes as much as 4 in the north. The rows are usually from 12 to 15 inches apart. A little more seed is used in sowing broadcast.

The *varieties* of peas are very numerous.

Sowing Peas in Autumn.—Peas are sown in the field in autumn in some parts of England. Although manure is rarely given to peas sown in spring, it is given in moderate quantity to that sown in autumn. On clean oat-stubble the manure, 8 to 10 cart-loads to the acre, should be spread on the surface, and ploughed in with the common plough. In every third furrow the seed is sown with the bean drill-barrow. The ploughed surface should have two tines of the harrow, to close the openings in the ploughing and protect the seed from frost.

The crop ripens earlier than when sown in spring, and the land is worked, cleaned, and manured again for sowing wheat upon it in autumn. The after-culture and harvesting are the same as for peas sown in spring.

Summer Culture of Peas.

Although a common practice is to sow peas along with beans, yet, as they are also cultivated alone, it is necessary to bestow attention on them when so cultivated. When sown broadcast, the pea plant, growing quickly, especially in moist weather, soon overspreads the weeds growing along with it. But though it overspreads, it does not entirely destroy them. The consequence is, that the ground is left by the pea crop in a foul state.

When sown in rows, in every third furrow of the plough, or in raised drills, the ground is scuffled, hoed, and drill-grubbed, as are beans when sown in rows on the flat.

These operations require to be rapidly performed, the quick and straggling growth of pea-stems affording neither time nor room for dilatory work.

GERMINATION OF SEEDS.

At this stage it will be interesting to introduce the following notes on the germination of seeds, carefully revised for this edition by Professor A. N. M'Alpine :—

Though apparently lifeless to the sight and touch, the healthy seed of a plant is a living object, for it contains within it a dormant embryo plant capable of being excited to active life. What life is we do not know, perhaps never shall—it is a secret which Nature has hitherto kept to herself; but we do know the circumstances in which seeds must be placed in order that they may germinate and develop the embryo within into a seedling plant. The proof of the excitement is in their germination, which is the first movement towards the production of a plant.

Conditions essential for Germination. — Now, the circumstances which excite germination are the combined action of air, heat, and moisture. These conditions must be satisfied before the seed will germinate satisfactorily. They may all be supplied to the seed, and its germination secured in the air as certainly as in the soil; but on the development of a root, most plants would die if kept constantly in the air. The soil supplies all the requisites of air, heat, and moisture to the seed in a better state than the atmosphere could alone; and it continues to supply them not only when the seed is germinating and producing the seedling, but also during the entire life of the plant till maturity is reached.

A vital seed placed in the soil is affected by three agencies—1, physical; 2, chemical; and 3, physiological — before it can produce a plant.

Air and Germination. — When a seed is placed in pulverised ground, it is *physically* surrounded with air; for although the particles of soil may seem to the eye to be close together, on examination it has been found that the interstices between the particles occupy about $\frac{1}{4}$ of a given volume of soil. Hence, 100 cubic inches of pulverised soil contain about 25 cubic inches of air. Therefore, in a field the soil of which has been ploughed and pulverised, and cleared of large stones, to the depth of 8 inches, 1 acre of it may contain about 12,545,280 cubic inches of air; and hence also, as every additional inch of depth pulverised calls into activity some 260 tons of soil, at 1.48 of specific gravity, so the ploughing up of another inch of soil not before stirred and not hitherto containing any air, introduces into the workable soil an addition of perhaps nearly $1\frac{1}{2}$ million cubic inches of air. Thus, by increasing the depth of pulverised soil, we can provide a depot of air to any extent for the use of plants.

It should be noted that it is the oxygen of the air that is of chief importance in germination.

But this air must be above a certain temperature ere the seed will germinate —it must be above the freezing-point, else the seed will remain dormant and not display its vital powers. It is also

Fig. 435.—*Cloddy and stony soil.*

a The seed.　　b Hard clots.　　c A stone.

desirable that the soil should be well pulverised, and not as in fig. 435, where a seed is placed among hard clods on the one side, and near a stone on the other, conditions not likely to favour the development of strong regular plants.

Moisture and Germination. — Fig. 436 represents the seed placed in a pulverised soil, all the interstices of which are entirely occupied by water instead of

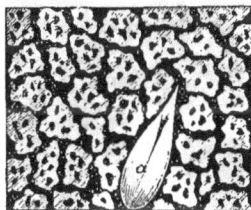

Fig. 436.—*Soil with water and without air.*

a The seed.

White spaces—pulverised soil.　Black spaces—water.

air. It is clear that, in this case too, the seed, being deprived of air, is not placed in the most favourable circumstances for germination. Besides the direct exclusion of the air, the water, on evaporation, renders the earth around each seed much colder than it would otherwise be.

But total want of moisture prevents germination as much as excess. Fig. 437 shows the seed placed in pulverised soil, and the interstices filled with air, with no moisture present between or in the particles of soil. In such a state of soil, heat will find an easy access to the seed, and as easy an escape from it.

Fig. 438 represents the seed in soil completely pulverised. Between every particle of the soil the air finds easy access to the seed, and in the heart of every particle of soil moisture is lodged,

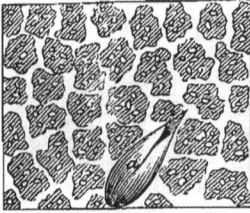

Fig. 437.—*Soil with air and without water.*
a The seed.
White spaces—air and heat.
Dark spaces—dry pulverised soil full of air.

which the seed can draw upon and use. All that is here required in addition is a favourable temperature, which the season supplies. Now all the conditions are fulfilled, everything is favourable, and germination proceeds smoothly without interruption.

Composition of Seeds.—The *chemical* components of seeds are organic and inorganic substances. The organic sub-

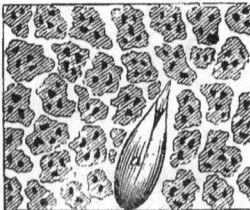

Fig. 438.—*Soil with water and with air.*
a The seed.
White spaces—air and heat.
Dark spaces—pulverised soil with darker water.

stances are of two classes, the nitrogenous and the non-nitrogenous; the inorganic, also of two classes, but comparatively small in amount, are the minerals from the soil and the water. The nitrogenous substances consist of albuminoid matter analogous to the caseine of milk, the albumen of the egg and of blood, and the fibrine of the flesh of animals; the non-nitrogenous consist of starch and

cellulose, with fatty and oily matters rich in carbon and hydrogen.

Changes incident to Germination. —When a seed is consigned to the ground, the first change which takes place in it is physical—it becomes increased in bulk by the absorption of moisture; and being also surrounded by air, it only requires the requisite degree of temperature to excite its vitality into action. If there is no moisture present, as in fig. 437, it will remain in a state of dormancy until moisture arrive, and in the meantime may either become the prey of the many animals which inhabit the soil, eager for food, or be scorched to death by heat. If it is placed in excess of moisture, as in fig. 436, its germination is prevented by the exclusion of the air, and its tissues are, by maceration in the water, destroyed by soil germs.

When the seed begins to germinate, a digestive juice is formed at the expense of its albumen, and the active ingredient in this juice is the same as in our own saliva, namely, *diastase*. The function of diastase is important. It is to convert the insoluble starch of the seed into soluble dextrine and sugar into a form fit to nourish the growing embryo plant within the seed. The digestive power which diastase has is most extraordinary. One part of diastase will convert into sugar no less than 2000 parts of starch. The diastase thus converts the starch which it finds into a state useful for the support of the embryo plant at its start in life.

The Embryo.—"Under fitting circumstances," says Lindley, "the embryo which the seed contains swells, and bursts through its integuments; it then lengthens, first in a direction downwards, next in an upright direction, thus forming a centre or axis round which other parts are ultimately formed. No known power can overcome this tendency, on the part of the embryo, to elevate one portion in the air, and to bury the other in the earth; but it is an inherent property with which nature has endowed seeds, in order to insure the young parts, when first called into life, each finding itself in the situation most suitable to its existence—that is to say, the root in the earth, the stem in the air."

The Young Plant or Seedling.—

When the germ has become a seedling plant, it is found to be possessed of a sweet taste, which is owing to the presence of grape-sugar in the sap which has already begun to circulate through its vessels. There is little doubt that the grape-sugar is formed subsequently to the appearance of diastase, and that diastase is merely a digestive agent for converting the insoluble starch into the soluble and diffusible form of sugar. The parts of this seedling plant are a root down in the ground for using the soil, and a leafy stem for using the light and air.

Seed dissected.—A seed always consists of a protective skin without and a germ or embryo within. Along with this embryo a mature seed always contains a comparatively large amount of food stored away sufficient to nurse the embryo till it has developed into an independent seedling plant. Fig. 439 represents a grain of wheat magnified, and so dissected as to show its component parts. It consists of two skins, an outer and an inner. Within the inner skin the nutritive matters, called the starch and albumen, are situate. There is also the little scale or sucker through which the nutritive matter passes in the sweet state, when the grain is germinating. The important part is *d* in the figure; its apex grows out as a leafy shoot, and a bunch of outgrowths from its base form the first roots.

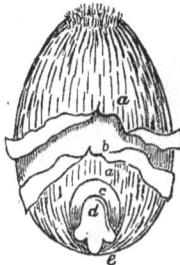

Fig. 439.—*Component parts of a grain of wheat.*

a a Outer skin.
b Inner skin.
c and *d* together, The germ or embryo.
c Scale or sucker.
d That part of the embryo which develops into roots and leafy stem.
e Base of grain.

Multiple Stems or "Tillering."—The embryo in the seeds of plants possesses such a structure that only 1 stem can proceed from them; but in many agricultural plants, particularly in the cereals, which yield human food, a remarkable departure from this structure is observed at a very early stage of life. In them the young plant is usually thickened towards its base, and so organised that, instead of 1 stem, 3 or 4 may spring from 1 grain.

The thickened base of the young plant has the habit of producing pimple-like projections, and each of these may be-

Fig. 440.—*Wheat plant in the state of germination.*

a First shoot of the embryo leaving the sheath.
b A tiller shoot just evolved.
c Tiller shoot yet unevolved.
d d Rootlets formed at the side.
e Main rootlet.

come an extra shoot. The peculiarity mentioned may be observed in fig. 440. The figure represents a grain in a state of germination. The shoot of the embryo has left its sheath, and two tiller shoots have been produced: the one to the left is as yet a mere pimple-like projection, whereas the other to the right is further advanced and partially evolved. The rootlets are seen extending downwards in a bunch. The root of the embryo has developed into a main rootlet, while 3 projections at the side of its base have also developed into rootlets. On other kinds of seedlings there would have been but 1 rootlet, since on these the basal projections do not form.

Different Methods of Sowing and Germination.

Disadvantage of Broadcast Sowing.—Of all the modes of sowing seeds, none requires so much seed as the *broadcast*. However regularly the land may have been ploughed, seed sown broadcast will braird irregularly—some falling into the lowest part, some upon the highest, some scarcely covered with earth by the harrows, some buried as deep as the ruts of tines have penetrated. To make the land smooth by harrowing, previous to sowing the seed, would not cure irregular covering, since it is impossible to cover a large seed as that of the cereals with tines without the assistance of a rough surface of mould. In fig. 441 the furrows are well and regularly ploughed; but while it is obvious that the seeds, when scat-

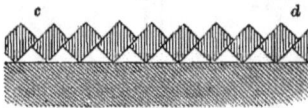

Fig. 441.—*Well-ploughed regular furrow-slices.*
c to d Regularly ploughed furrow-slices.

tered broadcast from the hand, will fall mostly in the hollows between the furrows, yet some will stick upon the points and sides of the furrow-slices. The seeds will thus lie in the ground, as in fig. 442, those which fell into the hollows of the

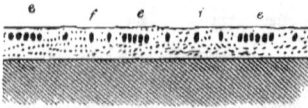

Fig. 442.—*Positions of seeds on regular furrows.*
e e e Seeds fallen in the hollows of the furrows.
f f Seeds scattered upon tops and sides of furrows.

furrows being thicker than the seeds which stuck upon their tops and sides. But it is not at all likely that the seeds will be so regular as represented. Some will be too deep and others too shallow in the soil, whilst some will be left on the sur-

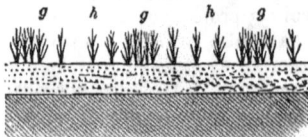

Fig. 443.—*Irregular braird upon regular furrows.*
g g g Plants growing in clumps.
h h Plants growing scattered.

face. From irregular deposition, plants will grow in irregular positions, as in fig. 443, where some are in clumps from the bottom of the furrows, and others are straggling too far asunder. Where the

seeds have been deposited at different depths, the plants will grow at more irregular heights than in the figure.

When the land is ill-ploughed, the case

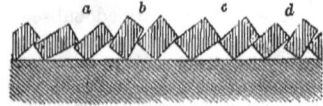

Fig. 444.—*Ill-ploughed irregular furrow-slices.*
a Furrow-slice too flat.   c Furrow-slices too wide.
b Furrow-slice too high.   d Furrow too deep.

is still worse. Fig. 444 shows the irregular furrows from bad ploughing. Bad ploughing entails bad consequences in any crop, but especially in cereal ones, inasmuch as irregularity of surface can-

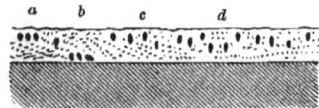

Fig. 445.—*Irregular positions of seed on ill-ploughed furrows.*
a Seed clustered and covered shallow.
b Seed clustered and buried deep.
c Seed scattered and covered shallow.
d Seed scattered and covered deep.

not be amended by a series of future operations, as in green crops. In the irregular furrow-slices of fig. 444, some are narrow and deep, some shallow, some too large, some of ordinary depth, and

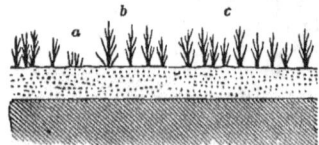

Fig. 446.—*Irregular braird on ill-ploughed furrow.*
a Late plants.   b Early plants.
c Regular growth of plants.

some too high and steep. The seed sown on these irregular furrows is shown in fig. 445, where some are clustered together with a shallow covering, others also clustered, but buried deeply, whilst many are scattered irregularly at different depths. Such a deposition of seed must make the braird come up irregularly; and the plants have not the chance of reaching maturity at the same time.

In fig. 446, where the seed was covered deeply, the plants will come up late; with shallow covering, they will come up early, and will push on in growth; while the remainder, coming up regularly, will form the best part of the crop. Where a crop of cereals does not mature at the same time, the grain cannot be equal in the sample.

Advantages of Drill Sowing.—One obvious advantage of sowing with a *drill* over a *broadcast* machine, is the deposition of seed at the same depth, whatever depth may he chosen. Fig. 447 shows the seed deposited at regular intervals.

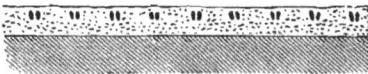

Fig. 447.—*Regular depth of seed by drill-sowing.*

The braird is shown at the same regular intervals in fig. 448, and its produce will reasonably be of the same quality. For drill sowing the land has previously received all the harrowing it requires for the crop, and by the coulter or tongue of the machine the seed is deposited regularly at a uniform depth and thickness.

Still there are many who prefer broadcast sowing, and, with careful preparation of the seed-bed, and skilful performance of the work of sowing, it will usually give satisfactory results.

Drill sowing leaves a blank between the rows of plants, which encourages the growth of weeds. On the other hand,

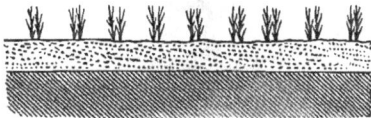

Fig. 448.—*Regular braird from drill-sown seed.*

this system permits of hoeing after the plants are advanced considerably, and if this operation is carefully performed by hand or horse-hoe it is usually found to be beneficial to the crop.

Dibbling.—*Dibbling* is distributing seed by means of a dibble at given distances, and at a given depth in the soil. The distribution by this system may either be in rows or broadcast. The difference betwixt dibbling and drilling is, that in drilling the seed is placed in lines, while dibbling places it at uniform distances in the line. The object of dib-

bling is to fill the ground with plants with the smallest quantity of seed. The seed planted in lines with the dibble appears as in fig. 447, and the plants like those in fig. 448. The depth of the seed and brairding of the plants are as uniform as in drilling, but the plants stand independent of each other in dibbling.

As would be readily understood, dibbling is not suitable where any considerable extent has to be sown, but it is very useful in filling up blanks.

Waste of Seed.—When sown in all these ways in equal quantities, the *waste of seed*, as determined by experiment, is surprising. *Wheat* at 63 lb. the bushel gives 87 seeds to 1 drachm, avoirdupois weight, or 865,170 to 1 bushel. Now, 3 bushels of seed sown broadcast on the acre gives a total of 2,595,510 seeds. Suppose that each seed produces 1 stem, and every stem hears 1 ear containing the ordinary number of 32 seeds, the produce of 1 acre would be 96 bushels. How far this exceeds the usual return need hardly he stated. Rarely, indeed, have we known the produce of wheat to exceed 64 bushels on 1 acre, so that in this case 32 bushels, or 33 per cent of the seed, would be lost, while in an ordinary crop of 40 bushels the loss of seed would be 58 per cent.

The waste in *barley* seed is estimated thus: Chevalier barley at 57 lb. the bushel, and 75 grains to 1 drachm, avoirdupois weight, gives 665,242 seeds; 4 bushels of seed sown on 1 acre gives 2,660,968 seeds; and allowing 1 stem from each seed, and 1 ear of 32 seeds, the produce would be 128 bushels! Even with an exceptional crop of 64 bushels there would be a loss of 50 per cent, while on the ordinary crop of 48 bushels the loss would be nearly 69 per cent.

In like manner the loss upon *oats* may be estimated, and will be found to be often more than one-half the quantity of seed sown.

In all these cases only 1 stem from 1 seed is reckoned, but many of the seeds produce 2 or 3 or more. The *actual* loss of produce sustained is thus not so great as of seed.

Another view of the waste of seed is this: 2,595,510 seeds of wheat on 1 acre give 536 seeds to 1 square yard; 2,660,968 seeds of barley give 550 seeds; and 5,879,808 seeds of oats give 1214. In wheat and barley the proportion of seed is in proportion to their respective weights, but in oats the seed is more than double in proportion to the weight, because of the thick husk of the oats.

Waste of Seeds by different Methods of Sowing.—P. M'Lagan of Pumpherston made experiments to ascertain the waste of seed in sowing oats in the three different ways of dibbling, drilling,

and broadcast. The oats weighed 42 lb. the bushel. The dibbled holes were made 6 inches apart, and 6 inches between the rows, making 36 holes in 1 square yard, and each hole was supplied with from 1 to 4 seeds, making the quantity sown from 1 peck to 4 pecks on 1 acre; and the seeds sown drilled and broadcast were in the same proportion. In drilling and dibbling, the seed was inserted 3½ inches into the ground. The results were as follows :—

		Dibbled.	Drilled.	Broadcast.
From 36 grains sown		26 plants	32 plants	19 plants came up.
" 72 "		49 "	53 "	52 "
" 108 "		75 "	78 "	68 "
" 144 "		120 "	94 "	87 "
360		270	257	226
Percentage		75	71	62

As might have been anticipated, there was not much difference in the brairding of seed dibbled and drilled, since the seeds were deposited much in the same position in the soil.

The *broadcast* involves a substantial loss—an anticipated result, since many of the seeds were unburied, or buried too deeply. The seeds were sown on the 19th March, and the thickest sown of the drilled and broadcast brairded first on the 16th April. Thick-sown seeds always braird earliest.

The experiments were extended by sowing 7 pecks of oats drilled, or 252 seeds to the square yard, and from these 208 plants came up, giving a percentage of 82. There were also sown 24 pecks to 1 acre broadcast, or 864 seeds to 1

square yard, which produced 570 plants, giving a percentage of 67, only a little more than in the former case of broadcasting. Thus, the smallest number of seeds gave the largest return of plants brairded.

G. W. Hay of Whiterigg, Roxburghshire, also made several experiments at the same time, by dibbling and drilling wheat, barley, and oats, and sowing oats broadcast. The *dibbled* seeds were put into holes within 3 inches square to the number of 1, 3, and 6 grains in each hole, which gave respectively 144, 432, and 864 grains to the square yard. The seeds were sown on the 16th March, and the plants counted on the 8th May. The results were these :—

DIBBLED.

	After 144 seeds.	After 432 seeds.	After 864 seeds.	
Of wheat . .	97	296	616	1009 plants came up.
Barley . .	95	335	687	1117 "
Hopetoun oats .	129	403	800	1332 "
Potato oats .	135	407	823	1365 "
Birley oats .	125	413	777	1315 "
Sheriff oats .	132	405	751	1288 "
Percentage of				
Wheat came up .	67	69	71 average 69	
Barley . .	66	79	79 " 75	
Oats . . .	90	94	91 " 92	

On the 25th March similar seeds were sown in *drills* at the same rates per

square yard, and the plants counted on the 8th May, when the results were :—

DRILLED.

	After 144 seeds.	After 432 seeds.	After 864 seeds.	
Of wheat . .	105	327	652	1084 plants came up.
Barley . .	86	318	747	1151 "
Hopetoun oats .	139	408	798	1345 "
Potato oats .	137	407	795	1339 "

Percentage of

				average	
Wheat came up .	73	73	75		74
Barley . . .	60	73	86	"	73
Oats . . .	96	94	92	"	94

On comparing the brairds of the drilled with the dibbled seeds in the barley and oats little difference is apparent, while the wheat incurs less loss of plants when drilled than when dibbled, in the ratio of 1009 to 1084. Comparing the results obtained by both experimenters with oats, we find that Mr Hay obtained a braird of $9/_{10}$ of the seed in dibbling and drilling; while Mr M'Lagan obtained only $7/_{10}$, and, in oats broadcast, $6/_{10}$.

Tillering.—After a lapse of ten days, on the 18th May, when rain had fallen in the interval, the plants after broadcast were counted, and were unexpectedly found greater in number than the seeds sown. The plants must have tillered after the rain, and the tillering was ascertained to be from—

Seeds.	Plants.	Tillering.
315 Barley	360 =	one-sixth.
325 "	405 =	one-fourth.
471 Sheriff oats	930 =	double.
520 " "	648 =	one-fourth.
666 Potato "	704 =	one-sixteenth.

The advanced state of the plants after the rain indicates that in spring oats tiller very strongly and rapidly.

Quantity of Seed.—Taking the respective quantities of seed sown on 1 square yard by both experimenters, they will be as follows on 1 acre:—

Seeds.		Seeds.	Per acre.
36 per square yard	=	174,240 =	1 peck.
72 "	=	348,480 =	2 "
108 "	=	522,720 =	3 "
144 "	=	696,960 =	1 bushel.
288 "	=	1,393,920 =	2 "
432 "	=	2,090,880 =	3 "
576 "	=	2,787,840 =	4 "
720 "	=	3,484,800 =	5 "
864 "	=	4,181,760 =	6 "

Produce from different Methods of Sowing.—Kenyon S. Parker made a comparative experiment between drilling, dibbling, and broadcasting wheat on

clover lea, and the results show that drilling produced more grain than dibbling; while the straw was longer and stronger, the ears larger, and the seeds heavier in the dibbled, thus:—

	1 acre. bush. peck.		1 acre. qr. bush. gal.		Weight. per bush. lb.
Broadcast	1 3	produced	3 7	1	62
Drilled, at 12 in.	1 2	"	4 3	1	63
Dibbled	1 0	"	4 3	0	63½

Importance of economising Seed. —The questions to which such results give rise are, What quantity is too thick and what too thin sowing? and, What is the least quantity of seed to yield the largest crop? The inquiry assumes much importance when we consider that from $1/_{10}$ to $1/_{14}$ of all the grain grown in the country is every year put into the ground as seed. A small fraction of either of these proportions saved would add a profit to the farmer to that extent. If 1 bushel of seed could be saved on each acre, a simple calculation would show that the gain to the farmer would amount to a vast sum of money.

Thick and Thin Sowing — Thick and thin sowing of seed is a subject of controversy among farmers. The saving of seed would be a sufficient argument in favour of thin sowing, provided the same return were received. But the results have been found to vary. There are many conditions to be considered in deciding as to the quantity of seed to be sown. The nature and condition of the soil, the climate, the quality of the seed itself, and even the character of the season, must all be kept in view.

Hewitt Davis, Spring Park, Croydon, who occupied 800 acres of high-rented poor soil, upon a warm subsoil of chalk, stated that "the practice throughout England is to sow 2 or 3 bushels of wheat to 1 acre, and the yield seldom reaches 40 bushels, and more commonly less than 20 bushels, so that $^1/_{10}$ at least of the crop grown is consumed as seed, whilst 1 single grain of wheat, planted where it has room to tiller out, will readily produce many 100-fold. The knowledge of these facts has induced me, in the course of years, to make a variety of experiments, the results of which have clearly shown me that, independent of the waste, a positive and serious injury of far more consequence is done to the crop from sowing so much seed. I bear in mind that, if so much be sown as to produce more plants than the space will allow to attain to maturity, the latter growth of the whole will be impeded, and a diseased state will commence as soon as the plants cover the ground, and continue till harvest." The quantities of seed Mr Davies determined on sowing, in accordance with these reasons, are, for—

Rye	.	.	.	1¼ bushel sown in	August and September.
Winter barley	.	.	2 " "		September.
Tares	.	.	.	1½ " "	3 sowings in Aug., Sept., and Oct.
Oats	.	.	.	6 pecks "	January, February, and March.
Barley	.	.	.	5 " "	January, February, March, and April.
Wheat	.	.	.	3 " "	September and October.
Peas	.	.	.	9 " "	December, January, and February.
Beans	.	.	.	9 " "	September and October.

The returns obtained by Mr Davis, after these scanty sowings, were 5 quarters of wheat, 13 quarters of oats, and 8 quarters of barley per acre on "very inferior land," from the manure available on the farm.[1]

Mr Barclay, Eastarch Farm, Surrey, *drilled* 2½ bushels of wheat at 9 inches apart, and obtained 37 bushels at 64¾ lb. per bushel, and 70 trusses of straw, value £16, 6s. He *dibbled* 1 bushel 3 pecks at 9 inches apart, and had 37 bushels at 64 lb. per bushel, and 72 trusses of straw, at a value of £15, 12s. 9d. He sowed *broadcast* 2½ bushels, and had 40 bushels at 65 lb. per bushel, and 84 trusses of straw, the value being £18, 1s. Here broadcast and thick sowing prevailed. Soil, deep loam on chalk subsoil.[2]

On the comparative merits of thick and thin sowing, it has been contended that experience has established that,—thick sowing is advisable on newly-broken-up land, containing a large amount of vegetable matter in an active state of decomposition, when it is beneficial in repressing, by its numerous roots and stems, that exuberance of growth which produces soft and succulent stems, which are easily lodged, and produce unfilled ears. Thin sowing has a tendency to make the roots descend deeply; and where a ferruginous subsoil exists, thick sowing keeps the roots nearer the surface, away from it. Thin sowing develops a large ear, grain, and stem, but delays maturity. Thick sowing on old land in high condition renders the plant diminutive and weak in straw, and hastens its maturity before the ear and grain have attained their proper size. Thin sowing in autumn affords room to plants to tiller and fill the ground in early spring, while thin sowing late in spring does not afford time to the plant to tiller. Thick sowing in autumn makes plants look best in winter, but gradually attenuates them in spring. Thin sowing makes plants look worst in winter, but to look better and fuller as the harvest approaches.

Different Methods of Sowing Compared.—On comparing the broadcast, drilled, and dibbled methods of sowing the cereal grains, it must be owned that the *broadcast* incurs a loss of seed by some being exposed on the surface, and others sent too deeply into the soil. Such effects are produced whether by hand or machine sowing, and cannot be avoided until a machine is contrived to sow corn broadcast at a uniform depth.

The *drill* does not work well in stony ground, which easily jolts the coulters to one side, or they displace small stones, or ride over large ones; while where land-fast stones or subjacent rocks are near the surface, they would be broken. Where there are many stones the drill should

[1] Davis's *Waste of Corn by Too Thick Sowing*, 6-12.
[2] *Jour. Eng. Agric. Soc.*, vi. 192.

not be used. Where the soil is fine, drilling has the advantage of having the land smooth before the seed is sown, and then seed escapes disturbance by cross-harrowing.

Dibbling may be done by a hand-dibble, or with an implement having pins attached to the bottom of a spar of wood, and which pins are thrust into the ground with a pressure of the foot. Another method is, to thrust small hand-dibbles through holes formed in a thin board of wood. In all these modes the seed is deposited in the holes at stated distances—perhaps 7 inches between the rows, 4 inches apart in the rows, and 2½ inches in depth. The earth is put over the holes with the foot. When a man uses a small dibbler, a convenient mode of keeping the lines straight is this : Take 2 long lines and stretch them along the side of the field, at a determinate distance between them ; *a b* and *c d* are the 2 lines at a distance between them of *a c* and *b d.*

```
a ————————————— b
c ————————————— d
e                f
g                h
```

Let him dibble in the seed along *a b*, and when at *b*, let him shift that end of the line from *b* to *f*, and then dibble the seed in from *d* to *c*, where let him shift the end of the line at *a* to *e*, which brings the line straight from *f* to *e*. Before starting with the dibbling from *e*, let him remove the end of the line at *c* to *g*, and then dibble the seed from *e* to *f*, where he shifts the end of the line from *d* to *h*, which brings the line straight from *g* to *h*. Shifting the line from *f* to *i*, he proceeds as he did at *b*, and so on alternately from one side to the other.

Dibbling-machines. — The dibbling-machine first brought into notice was invented by James Wilmot Newberry, Hook Norton, Chipping Norton, Oxfordshire. It is ingenious and elaborate in construction, and deposits every kind of corn at given distances, in any quantity, with the utmost precision. Fig. 449 is a view in perspective of a 1-rowed machine. It consists of a hollow flat disc, which contains the machinery that directs the corn from a hopper into hollow tubes, 18 of which are connected with and project

from the circumference of the disc like the spokes of a wheel from its nave, and their points pass through a large outer ring, which retains the hollow tubes or distributors of corn in their respective places, and prevents them sinking into the ground beyond the requisite depth. A fore-wheel, which is placed between the extremities of the stilts or handles, prevents the large outer ring being pressed closer to the ground than needful. A man pulls the machine forward by means of a rope attached to the fore part of the stilts, or, what is better, a bridle and shackle might be mounted there, for yoking a pony or horse to draw the machine. As the wheel is drawn forward by the horse, it turns round by contact with the

Fig. 449.—*Newberry's one-rowed dibbling-machine.*

a a Stilts.	*e* Projecting points or dibbles.
b Fore part of stilts.	*f* Large outer ring.
c Fore-wheel.	*g* Hopper.
d Hollow flat disc.	*i* Stay to support the machine.

ground, the projecting points of the hollow tubes acting as dibbles and making holes in the ground ; a portion of the dibbles, before leaving the ground, slides up upon the upper part, making an opening through which the corn is deposited in the holes. The corn descends of the requisite number from the hopper by means of feeding-rollers, moved by a pinion, which is set in motion by teeth placed on the circumference of the flat disc. The disc is supported in its centre by an axle revolving in its ends on plummer-blocks. In using this machine, a man holds by the two stilts, while a man or horse draws the machine in the given line. The line not being in the line of the body of the drill, a rigger is required for the horse to be yoked to. A stay supports the machine when at rest. This

1-rowed dibble is said to be well suited for sowing mangel seed on the top of the drill.

Another dibbling-machine, presented to public notice by Samuel Newington, of Knole Park, Frant, Kent, is shown in fig. 450—a view in perspective of one having 6 depositors. The box in front contains the corn, and the points of the depositors are seen to rest upon the ground, which has been harrowed smooth for the purpose. The depositors place the seeds at the desired depths, deeper or shallower, being kept in their places by pinching screws. The machine is worked by taking hold of the upper rail by both hands, and, on pressing upon it, the de-

Fig. 450.—*Newington's 6-rowed dibbling-machine.*

positors, when withdrawn, leave the requisite number of seeds in each hole the depositors have made, by the machinery in the interior of the machine. By pressing down the upper handle, the depositors press every seed firmly into a solid bed, which is so small as to preclude the fear of its containing water, and yet completely buries the seed. By changing the cups, the quantity of the corn is regulated, as well as the description of corn. With a machine having six depositors, 1 man can dibble 1 acre in 10 hours, so that the cost of dibbling may be easily ascertained by the rate of wages in the district.

In using the machine after the first line is laid off straight next the fence, the workman continues to keep the other lines straight at the stated distance by the mark left on the ground by the machine. The seeds are put in at 4 inches apart in the rows, and the quantity is varied by either altering the distance between the rows or increasing the number of seeds in each hole, but it is not desirable to exceed 3 seeds in 1 hole. The cups which contain the grains are of 4 sizes, and can be easily removed or replaced by means of screws.

As already indicated, dibbling is too slow a process for the modern necessities of farm practice, but on a small scale, and for filling up blanks, it may be pursued with advantage.

Deep and Shallow Sowing. — Another circumstance which affects the relation between the grains sown and the plants produced, is the *depth* to which the corn is buried in the ground. In ill-ploughed land, when the corn is sown broadcast, falling between ill-assorted furrows, some of it may sink to the bottom of the furrow-slice, where it will be buried, to become dormant or lose its vitality. Corn is differently affected by depth in soil, some sorts germinating at a considerable depth, whilst others become dormant or die if placed at a smaller depth below the surface of the ground. A stem of barley has been traced to a depth of 9 inches, while oat seed buried 7 inches cannot be depended on to braird. This accounts for oats which had slipped to the bottom of the furrow-slices of lea and perished. The risk of thus losing seed in fresh-ploughed lea induces us to recommend partial harrowing of ploughed old lea before the seed is sown.

Wheat and all cereals possess a peculiarity in the growth of the root. The grain will bear to be deep-sown— not so deep as barley, but deeper than oats. Most wheat seeds may germinate at a depth of 6 or 7 inches, but sowing at that depth is risky, for the crop will likely be thin. After the germ of wheat has developed its first leafy stem, it puts out another set of roots about 1 inch below the surface. The deeper may be called the *seminal*, and the upper the *coronal* root of the wheat plant. Fig. 451 shows the position of the roots under the surface, where *a* is the grain with its seminal roots *c*, and the stem *b* rising from it to the surface of the ground at *f*, above which is the stem with its leaves. About 1 inch below the surface *f*, at *d*, are formed the coronal roots, *e e*, and this same spot forms the site from which the tillers are sent forth. At whatever depth the seed may have been sown, the coronal roots are formed at 1 inch below the surface.

"As the increase and fructification of the plant depend upon the vigorous absorption of the coronal roots, it is no wonder that they should find themselves so near the surface where the soil is always the richest. I believe I do not err when I call this *vegetable instinct*. In the N. counties wheat is generally sown late. When the frost comes, the *coronal* roots, being young, are frequently chilled. This inconvenience may, however, be easily prevented by sowing more early, and burying the seed deeper. The seminal roots, being out of the reach of frost, will then be enabled to send up nourishment to the crown by means of the pipe of communication."

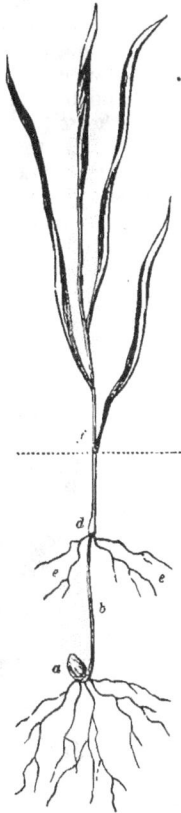

Fig. 451.—*Double roots of deep-sown wheat.*

Now the form which the plant assumes, when sown near the surface, is different, as in fig. 452, where *a* is the seed with its seminal roots; *b*, the stem or pipe of communication between them and the coronal roots *c c*, a little beneath the surface *d*. The coronal roots *c c* being at a short distance from the surface, the pipe of communication is at its shortest. "Hence it is obvious," continues the same writer, "that wheat sown superfi-cially must be exposed to the frost, from the shortness of the pipe of communication placing the seminal roots within reach of the frost. The plant, in that situation, has no benefit from its double set of roots. On the contrary, when the grain has been properly covered, it depends almost entirely on the coronal roots, which, if well nourished during the winter, will send up numerous stalks in spring; and on the tillering of the corn the goodness of the crop principally depends; but if not well nourished there will be no tillering. A field of wheat dibbled, or sown in equidistant rows by the drill, always makes a better appearance than one sown with the harrow. In the one the pipe of communication is regularly of the same length, but in the other it is irregular, being either too long or too short."[1] The con-clusions these statements would warrant in practice are: That wheat sown before winter should be deeply covered with earth, to be beyond the reach of ordinary frost; that in spring the coronal roots will set up abundance of tillers or stools; that wheat sown in spring should be lightly covered, the tillers being few; that autumn wheat should be drilled to secure the pipes of communication between the seminal and coronal roots being long and uniform; that spring wheat should be sown broadcast; and that autumnal wheat should have a smaller quantity of seed than spring wheat.

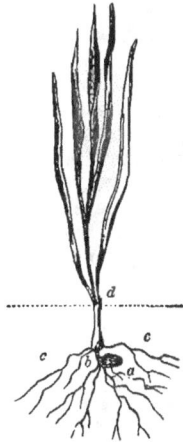

Fig. 452.—*Roots of shallow-sown wheat.*

Tillering.—The property of the cereal plants to *tiller* or *stool out*—that is, to send up a number of stems from the same root—is a valuable one in an economical point of view. But for this property, when the first shoot of the cereal happens to be destroyed by insects under

[1] *Georgic. Ess.*, i. 67-69.

ground, or by unfavourable environment, or by frost, or when young plants are injured by insects as they appear above the surface, the crop would be so scanty that it would be ploughed up by the farmer, and another substituted in its stead. The extent of tillering depends on the state of the soil and weather, and on the space allowed the plant to spread in. A loose soil, admitting the young shoots to penetrate easily, encourages tillering more than a stiff hard soil. Yet wheat tillers best on a moderately firm clay soil in good heart, for there the plant has manufactured a sufficient food-supply to give the tiller shoots a good set-off. If the plants have room enough, moist warm soil and sunny weather promote tillering to the utmost.

Unless plants have space for their roots and light for their shoots, they will not tiller and become strong plants, overcoming and killing their weaker fellows.

The question which such an occurrence gives rise to is, Whether it is better to allow few plants to fill the ground by tillering, or to fill the ground at once with the requisite number of plants? The answer to this question must be given conditionally. In naturally fertile soils, and in those rendered fertile by art, tillering will readily take place, and should be encouraged, inasmuch as the straw and ears of tillered plants are much stronger and larger than those of single plants. At the same time, the ears are less evenly ripe, for the tiller shoots are younger than the leading shoot that gave them birth. In favourable soils a small quantity of seed will suffice in early spring, and it is in that season that tillering takes place in a most marked degree; but the seed must not be sown so deeply nor so late as to deprive the plant of time for tillering, so as to occupy the ground fully.

The extent of tillering is sometimes remarkable. Le Couteur mentions a downy variety of wheat which tillers to the extent of 32 stems,[1] and from 5 to 10 stems are a common tillering for ordinary varieties of wheat. Barley also tillers, though late and thick sowing, with quick growth, overcomes that tendency. Oats indicate fully as strong a tendency

to tiller as wheat. In weak soils, and soils in low condition, the tendency to tiller is much checked, each single root being conscious of its inability to support more than its single stem. Hence the practice is to sow more seed in low than in high conditioned land, and yet the ability to support the larger number of plants is in an inverse ratio. Yet what can the farmer do but sow as many seeds as will produce as many plants as will occupy the soil? The best way for him to escape from the dilemma is to put the soil in high condition, and reap the advantages derivable from tillering.

If, however, uniform grain for selling is the chief desideratum, thick sowing to prevent tillering may then be advisable.

Destruction of Seed.—The great loss in plants compared with the numbers of seed sown may be accounted for by natural causes. Birds pick up seeds exposed on the surface after broadcast sowing. Many vermin, such as the rabbit, devour the young germ as it penetrates the soil, and many insects subsist on the stems and roots of young plants.

Frost and cold winds which come upon the young plant when specially tender and unprepared often account for much destruction.

Transplanting.—A mode of saving seed to a greater degree than dibbling and drilling is *transplantation*. This is done by sowing a small portion of ground with seed early in the season, taking up the plants as they grow, dividing them into single-rooted shoots, and transplanting these. By thus dividing the plants as they tiller, at four periods of the season, a very small quantity of seed will supply as many plants as would cover a large extent of ground. Though wheat no doubt bears transplanting, yet the amount of manual labour which the scheme entails would be so great as to render it impracticable upon any considerable scale. This method, however, has been pursued with a certain measure of success in the formation of permanent pastures.

When it is desired to propagate a new variety of grain quickly, this process of transplanting might perhaps be useful. It may therefore be interesting to preserve the following record of the details and costs of the operation: Suppose 440

---

[1] Le Couteur's *Wheat Plant*, 29.

grains of wheat are sown widely on the 1st of July, and that every seed germinates by the beginning of August, each seed will afford four plants, or in all,

1,760 plants

At the end of August
these will produce . 5,280 „
In September these again 14,080 „
And in November these
last will produce . 21,120 „

The time occupied in sowing the 440 grains, and dividing and transplanting their produce, stands thus :—

			Hours.	min.
July sowing,	. . .	440 grains,	0	20
August, beginning, taking up	.	440 plants,	0	20
„	dividing into .	1,760	1	10
„	planting .	1,760	3	30
August, end,	taking up .	1,760	1	28
„	dividing into .	5,280	3	30
„	planting . .	5,280	10	33
September,	taking up	5,280	4	24
„	dividing into .	14,080	9	23
„	planting .	14,080	28	9
November,	taking up .	14,080	11	44
„	dividing into .	21,120	14	4
„	planting .	21,120	42	14
			130	49

Equal to 13 days 49 minutes' work at 10 hours a-day. Of these 13 days, 5 days may be reckoned for women and boys occupied in taking up and dividing the plants, which, at 1s. 6d. per day, will cost 7s. 6d. The remaining 8 days are for men transplanting, at 14s. per week, which will cost 18s. 8d. more ; both 26s. 2d. per acre. The seed for the plants, ½ bushel at 48s. the quarter, or 6s. the bushel, would cost 3s. The entire cost would be £1, 9s. 2d. The saving of seed from the ordinary quantity sown would be the difference of cost between ½ bushel and 3 bushels, 15s. So that the loss on the transplanting over sowing would be 14s. 2d. Of course the cost of transplanting would vary with the rate of wages.

The best way of executing this plan is to dibble in the seed two grains in a hole, about 4 inches from each other, the plants to be taken up when in a proper state, and divided into five, which would be as many at that time as could be had, and then planted out at once, where they are to remain, thus getting rid of all the intermediate dividings.

GRAIN HARVEST.

The joy of the harvest has been extolled by emotional writers in all ages. The merry whirr of the modern reaper has drowned the dull hum of the primitive "shearing" of ancient times. By the genius of the inventor and the enterprise of the farmer the entire process of harvesting has been revolutionised. Yet with all those changes something of the glory of the harvest still survives. And who would have it otherwise !

Beginning of Harvest.—The nature of the weather during the season has of course much to do with the date of the harvest, as well as with the character of the crop. As a rule, reaping begins with the winter wheat in England in the third or fourth week of July, and about the middle of August in Scotland. Between a late and an early season there may be a difference of three, or even four weeks. The pulse crops follow the cereal crops in ripening.

Stage for Cutting.

The propriety of cutting wheat and oats before they are dead ripe has been so well established as to require no demonstration here. Not only is the yield of grain increased, but the quality of the straw is likewise improved by reaping at this stage.

Loss by too Early Cutting.— On the other hand, wheat, reaped one month before it was ripe, has been found to give an advantage of 22 per cent in weight of straw compared with the ripe, but suffered disadvantage in every other point.

Ripening in the Sheaf.—If there is not too much of the ripening process to be accomplished when the cutting takes place, it will be successfully completed in the sheaf. It is a nice point to decide how much of the ripening may be left till after the grain is severed from the roots. Experience alone can be relied upon as the guide. There seems little doubt that the cut grain can do more in the way of ripening itself in the sheaf than is generally believed.

Shedding or "Shaking." — One of the greatest risks of loss by allowing grain to become dead ripe before being cut is that of shedding or "shaking." Oats are particularly liable to loss in this way, both by wind and hail storms, before being cut, and by the shedding in the process of harvesting. From 10 to 30 per cent of the corn is often left on the ground in this manner, and the injury is all the greater since it happens that the plumpest and best matured grains are the most easily dislodged from the straw. With oats, therefore, it is especially desirable that early rather than late cutting should be the rule.

The greatest losses by shedding in oats usually take place when high wind follows rain and bright sunshine. After being swollen by the rain the chaff is thrown wide open by the sun, exposing the grain to the play of the wind.

Barley. — The stage at which barley should be cut has to be regulated mainly by the particular purpose for which the grain is intended. If it is to be used for feeding stock, or in the manufacturing of whisky, the crop should be cut just before it is fully ripe. If it is to be employed for brewing, especially for the production of the lighter coloured ales, the crop should be left uncut until it is dead ripe. For brewing, the value of barley depends greatly upon its colour, which, for this purpose, should be as bright as possible, not "steely-white" or gritty-looking, but a pure soft white, with a very slight golden tinge. The deep golden colour so common in barley grown in Scotland and the north of England renders it unfit for the manufacture of light-coloured ales.

To secure this bright colour the barley in the south of England is left uncut until it is dead ripe, when very little drying is sufficient to prepare it for stacking or threshing.

The main object of the farmer who grows brewing barley is to shorten the drying period. While on the root, barley, although drenched with rain, will regain its bright colour to a wonderful extent;

but after being cut it is liable to be permanently damaged in colour, even by one heavy shower of rain. The farmer, therefore, not only allows the grain to become dead ripe, but also lets the drying of the straw go on to a considerable extent before cutting the crop.

Happy Medium. — There is a happy medium in the time to begin cutting grain as in most things else. Much will depend upon the district, whether the season be late or early, and the weather good or bad. In late districts, in a backward season, it may be better to cut "on the green side," that is, to begin early, than to delay till the crop is more nearly ripe. In these cases there may be great danger of injury to the crop by inclement weather, perhaps by premature frost and snow, so that the prudent farmer will prefer to have a moderately ripened crop well preserved, than a well-ripened crop injured in the stook.

The exact time when cutting should begin is a matter which each individual farmer must, after careful contemplation, determine for himself every harvest as it comes round.

Ripening Process. — As a general rule, corn in a healthy state comes to maturity first in the ear, and then in the upper part of the straw downwards. When the straw becomes matured first at the root, the grain suffers premature decay; and when this is observed, the crop should be reaped, as it can derive no further benefit from the ground, and its grain will dry more speedily in the stook than on the root.

Judging Ripeness. — The most ready way of judging when the ear is ripe in wheat and oats is to note the state of the chaff in the ear, and two or three inches of the straw under the ear. If these parts are of a uniform straw-yellow colour, and feel hard in the ear of the oat, and prickly in the wheat, on being grasped, they are ripe. On examining the grain itself, it should feel firm under pressure between the finger and thumb, and the neck of the straw should yield no juice on being twisted with the fingers and thumbs.

Barley should be of a uniformly yellow colour in the grain and awns, and the rachis somewhat rigid; and as long as the head moves freely by a shake of the hand, neither the grain nor the straw is sufficiently ripe.

When very ripe, wheat bends its ear down, opening the chaff, and becomes stiff in the neck of the straw, all clearly indicating that nature intends that the grain shall fall out. Red wheat is less liable to be shaken than white; but any kind will shake out when too ripe, provided the plant is in good health and the grain of good quality,—for it is difficult to make immature grain leave the chaff even when hardened.

Degrees of Ripeness.—It might be supposed that when the ear and the entire straw are of uniformly yellow colour, the plant is no more than ripe; but by that time the straw has ripened to the root, and the ear becomes rigidly bent and ready to cast its seeds with the slightest wind. When the neck of the straw of barley is ripe, it is, as a rule, time to cut; and when too ripe the ear bends down, the awns, diverging, stand nearly at right angles with the rachis, and the whole head is easily snapped off by the wind. In oats, when over-ripe the chaff stands apart from the grain, which is easily shaken out by the wind.

It is not equally prudent to reap all sorts of grain at the same degree of maturity. When wheat is reaped too soon, it is apt to shrink, and have a bluish tint in the sample; and when too ripe, the chaff opens from the grain, which is apt to fall out on the least wind; and some sorts of white wheat are thus very subject to fall out, even before reaching the point of maturity. Barley, when reaped too soon, also shrinks, and assumes a bleached colour. Much less loss attends the reaping of oats too soon than the other grains.

Harvest Labour.

With the use of the self-binding reaper, fewer extra harvest hands are now required than in former times, yet in most cases some additional workers have to be employed in harvest. Farms near a large town may obtain the requisite number daily, and in these cases the labourers usually return to their own homes at night. These extra day-labourers are paid daily or weekly, according to arrangement. On farms at a distance from towns, no reliance can be

placed upon obtaining labourers at harvest. For them labourers are hired to remain all the harvest on or near the farm. Such labourers receive food daily, and their money wages are paid at the termination of the engagement.

REAPING APPLIANCES.

Of the many appliances designed for the reaping of the corn crops three alone —the sickle (or hook), the scythe, and the reaping-machine—have been extensively employed. These three came into use in the order named, the first having as yet had the longest lease of life. Except on very small holdings and for odd purposes, the reaping-machine has entirely superseded both the sickle and the scythe, and in its many forms and developments is performing the work in an expeditious, satisfactory, and economical manner.

SICKLE OR "HOOK."

As late as 1868, when the third edition of this work was being prepared for the press, the sickle was considered to be still employed so extensively as to warrant the retention of the detailed account given in former editions of the manner of reaping with the sickle. Since that time, however, the work of the harvest has been completely revolutionised. In most parts the scythe supplanted the sickle, and now the reaping-machine has driven both into disuse.

It is interesting to note that, in several parts of the country, the sickle survived until the reaping-machine was ready to take its place. In these parts the scythe never succeeded in obtaining a footing.

Form of Sickle.—Although the sickle has lost its position as an important farm tool, it will be of interest to reproduce the following notes and illustrations of the two forms, the toothed and the smooth-edged sickles, which were employed.

Toothed Sickle.—The toothed sickle was largely used in former times. It has a blade of iron, with an edging of steel, in which very small teeth are formed (fig. 453).

Smooth-edged Sickle.—The large, smooth-edged sickle, is represented in fig.

454. It has a curvature approaching very near to that which, in this implement, may be termed the *curve of least exer-*

Fig. 453.—*Hook or sickle.*

tion. From this circumstance it is an easier implement to cut with than the toothed sickle.

Sickle still used.—Upon very small holdings the sickle is still in use in many parts of the country. It has this advantage, that women can work it as well as men, while, for the former, the scythe is not a suitable implement. Upon large farms, too, the sickle is sometimes employed in reaping portions of the crop which have been laid and twisted by stormy weather. As late as 1889 large fields in the Lothians of Scotland were on this account reaped by the sickle. It is a tedious and costly process, however, and should be resorted to only where it

Fig. 454.—*Large smooth-edged sickle.*
a Centre of the handle of the sickle.

is impossible to work the reaping-machine or scythe with satisfaction.

Reaping by the sickle requires, of course, a larger supply of hand-labour than either scythe or machine reaping, the latter requiring least of all.

Thraving.—Reapers with the sickle were in most cases paid by piece-work—by so much per *thrave* cut. A thrave consists of two full stooks of corn, each stook of oats and barley consisting of 12 sheaves, and of wheat 14 sheaves, and each sheaf measuring 3 feet in circumference or 12 inches in diameter at the band. The proper size of sheaf was ascertained by a sheaf-gauge, shown in fig. 455. When used, the prong of the gauge embraces the sheaf when lying on the ground, along the band, and if the sides and top of the gauge slip easily down and touch the band, the sheaf is of the required size, the prongs being one foot long and one foot asunder inside.

Fig. 455.—*Sheaf-gauge.*
a b c d Prong of gauge.
a b Points of prong.
c d Upper part of prong.

SCYTHE.

The scythe is a more expeditious tool than the sickle. With a given number of men and women, it enabled the farmer to cut down his crop much more speedily than was possible with the sickle. The introduction of the scythe, therefore, led to the general abandonment of the sickle. In certain parts, where there was an abundant supply of cheap labour, the sickle maintained its hold until the overwhelming superiority of the reaping-machine drove it into the limbo of forgotten things.

It is therefore true, though it may seem strange, that in certain districts the scythe was first used to cut "roads" for the reaping-machine.

Hainault Scythe.—Many different forms of the scythe have been employed. The Hainault or Flemish scythe may be regarded as an intermediate implement between the sickle and the cradle-scythe. It is held in the right hand by a handle 14 inches long, supported by the fore-finger, in a leather loop. The blade, 2 feet 3 inches in length, is kept steady in a horizontal position by a flat and projecting part of the handle, 4½ inches long, acting as a shield against the lower part of the wrist. The point of the blade is a little raised, and the entire edge bevelled upwards to avoid striking the surface of the ground. By fig. 456 an idea of the form of the Hainault scythe and its hook, and of the mode of using them in reaping corn as described, may be formed.

In 1825, the author of this work accompanied the Flemish reapers, Jean B. Dupré and Louis Catteau, through Forfarshire, and drew up a report of their proceedings in that county for the Highland and Agricultural Society. The impression on the farmers present was, that a saving of about one-fourth might be effected by the Hainault scythe in comparison with the common sickle; but it was not equal in its work to our cradle-

Fig. 456.---*Reaping with the Hainault scythe.*

scythe, and therefore never came into general use in this country.

Common Scythe.—The scythes used in this country still exhibit different forms. The helve or sned is usually made of wood, in two short branching arms, as shown in fig. 457, or in one long piece, as in fig. 458.

Cradle.—The cradle-scythe, once very common in the north-east of Scotland, is represented in fig. 457. In this form the scythe-blade is 3 feet 4 inches to 3 feet 6 inches long.

The function of the cradle was to carry the cut corn round with the sweep of the scythe. Except for a very short crop, however, the cradle is really not necessary, and was latterly to a large extent dispensed with.

Setting a Scythe.—In setting the

blade, the following rule is observed : When the framed helves are laid flat on a level surface, the point of the blade should be from 18 to 20 inches above

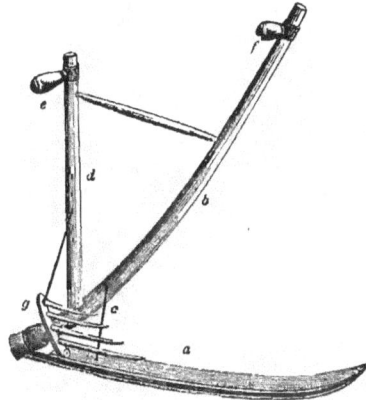

Fig. 457.—*Cradle-scythe for reaping.*

a Scythe-blade.	e Right-hand handle.
b Principal or left helve.	f Left-hand handle.
d Minor or right helve.	g Cradle or rake with
c Grass-nail.	its stay.

that surface, and measuring from a point on the left helve, 3 feet distant from the heel of the blade, in a straight line; the

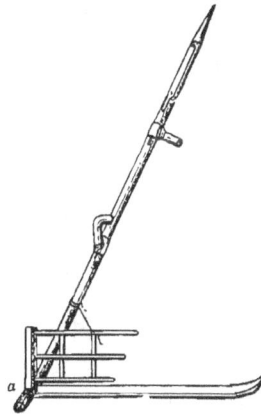

Fig. 458.—*Common reaping-scythe.*

a Cradle.

extremity of the blade should be also 3 feet distant from that point.

Iron Scythes. — Iron has, in many cases, been substituted for wood in the construction of the helves; but it is not by any means so well adapted to the purpose as the wooden helves.

Straight Sned. — The blade of the common scythe with the straight sned, fig. 458, is mounted on the same principle and the same manner as the blade of the bent sned.

Sharpening Scythes. — When any of the scythes are used in reaping, the strickle and the scythe-stone are much in requisition. They should be used only as often as to keep a keen edge on the blade. Experienced reapers keep a "long" rather than a "short" edge on their scythes, and thus require less of the sharpening tools. An edge put on at a short angle is easily and speedily blunted.

Method of Scythe-reaping. — Reaping with the scythe is best executed by the mowers being placed in *heads*— namely, one head of three scythemen, three gatherers, three bandsters, and one raker; or, as some would prefer, one head of two scythemen, two gatherers, two bandsters, and one raker.

A number of heads on the second arrangement may be employed on a large farm, while a small farm may employ one head on the first arrangement.

Speed with the Scythe. — The speed of the scythe is considerable. A good mower will cut an acre of wheat, or perhaps rather more, in one day; and from one and a half to two acres of oats and barley.

But the almost universal adoption of the reaping-machine has rendered it unnecessary for the present race of farmers to acquaint themselves with the working of either the sickle or the scythe, useful as these appliances have been in their day.

THE REAPING-MACHINE.

In all parts of the United Kingdom, and on almost all farms of any considerable size, the reaping-machine has superseded the slower and older appliances for cutting down the corn crops.

Historical.

Although it did not come into extensive use until near the middle of the nineteenth century, the reaping-machine is by no means a modern invention. It

is indeed much older than is generally believed.

Ancient Machine. — Both Pliny and Palladius describe a reaping-machine worked by oxen, which was much used in the extensive, level plains of the Gauls.[1] Pliny's words are: "In the extensive plains of Gaul large hollow machines are employed, with teeth fixed to the fore-part, and they are pushed forward on two wheels through the standing corn by an ox yoked to the hind-part; the corn cut off by the teeth falls into the hollow part of the machine."

Nineteenth-Century Machines. — It is known that before the advent of the nineteenth century several attempts had been made to devise a workable reaping-machine. No authentic information has come down to us as to the actual structure of these abortive machines.

But soon after the commencement of the nineteenth century, when agricultural improvements were making progress in every direction, and in particular by the extension of the use of improved machinery to the various branches of farming, active attention was successfully devoted to the invention of a reaping-machine. With the object of stimulating inventors, agricultural societies offered premiums, and we know that within the first twenty-five years of the century nearly a score of reaping-machines, less or more distinct in pattern, and invented by different men, were introduced into public notice in England and Scotland.

The principal of these machines were designed by Boyce, Plunket (London), Gladstone (1806, Castle-Douglas), Salmon (Woburn), Smith (1812, Deanston), Scott (1815, Ormiston, East Lothian), Mann (1820, Raby, Cumberland), and Ogle and Brown (1822, Alnwick).

First Effective Reaping-machine.

It is believed that not one of the early reapers mentioned was ever worked throughout a harvest. Even Smith's and Mann's machines, which were the most perfect, do not appear to have been worked beyond a few hours consecutively. Their actual capabilities, therefore, seem never to have been properly tested.

Bell's Reaping-machine. — The year

[1] *Dic. Gr. and Rom. Anti.*—art. "Agric."

1826 may be held as an era in the history of the reaping-machine, by the invention, and the perfecting as well, of the first really effective mechanical reaper. This invention is due to the Rev. Patrick Bell, minister of the parish of Carmylie in Forfarshire.

The principle on which its cutting operation acts is that of a series of clipping shears (fig. 459). When the machine had been completed, Mr Bell brought it before the Highland and Agricultural Society, who appointed a committee of its members to inspect its operation in the field, and to report. The trials and the report being favourable, the Society awarded the sum of £50 to Mr Bell for his invention, and a correct working-model of the machine was subsequently placed in the Society's Museum—the model, on the closing of that Museum, having been deposited in what is now the Royal Scottish Museum, Chambers Street, Edinburgh. The invention shortly worked its way to a considerable extent in Forfarshire; and in the harvest of 1834, the author of *The Book of the Farm*, in a short tour through that county, saw in operation several of these machines, which did their work in a very satisfactory manner. Dundee appears to have been the principal seat of their manufacture, and from thence they were sent to various parts of the country. It is known, also, that *four* of the machines were sent to the United States of America; and this circumstance renders it highly probable that they be-

Fig. 459.—*Bell's reaping-machine.*

came the models from which the numerous so-called inventions of the American reapers have since sprung. At the great fair or exhibition held at New York in 1851, not fewer than six reapers were exhibited, all by different hands, and each claiming to be a special invention; yet, in all of them, the principal feature—the *cutting* apparatus—bears the strongest evidence of having been copied from Bell's machine.

Construction. — The accompanying illustration, fig. 459, will enable readers to form a just conception of the construction and principles of Bell's machine, and to compare it with the modern implements now in use. The machine was worked by two horses pushing it before them by means of the pole to which they were yoked by the common draught-bar.

Work done.—In the process of work-ing this machine, Mr Bell's practice was to employ one man driving and conducting the machine; eight women to collect the corn into sheaves, and to make bands for these sheaves; four men to bind the sheaves, and two men to set the sheaves up in stooks—in all fourteen labourers, besides the driver of the horses. The work performed averaged 12 imperial acres per day. These data were obtained from fourteen years' experience of the machine, and are therefore reliable.

Cost of Reaping.—The expense in money for reaping by this machine about 1835 averaged 3s. 6d. an acre, including the expense of food to the workers. This, in round numbers, was a saving of one-half the usual expense of reaping by hand, at the lowest calculation; and the saving on a farm where there might be 100 acres of cereal and leguminous crop would do more than cover the price of a machine of the best quality in two years.

Slow Progress of Bell's Reaper.— It is difficult to account for the fact that Bell's machine was not more extensively adopted. For a period of nearly twenty years it was successfully used; and yet, with practical agriculturists, it did not seem to gain so high a reputation as its American rivals—the machines of Hussey and M'Cormick—yet to be described. It did its work really well, but its draught was exceptionally heavy, and the delivery web was liable to become disordered.

Subsequent makers improved on Bell's machine, and now it exists only as the groundwork of the modern reaper.

American Machines.

The two machines which, perhaps, did most to popularise the reaping-machine in this country were both introduced from America. They were known as Hussey's and M'Cormick's machines— Hussey's being manufactured by Messrs Dray & Co., engineers, of Swan Lane, London Bridge, London; M'Cormick's by Messrs Burgess & Key, Newgate Street, London. These firms introduced great improvements in the machines which they respectively manufactured, so much so, that there would be some difficulty in recognising in them the same machines, the appearance of which, at the Great Exhibition of 1851, created

Fig. 460.—Dray's Hussey reaping-machine in perspective.

such an interest in the agricultural world.

Dray's Hussey Machine.— In fig. 460 we give a perspective view of Hussey's reaper, improved by Messrs Dray. The cost of this machine was £25.

M'Cormick's Reaping-machine.— In M'Cormick's reaper, with the improvements introduced by Messrs Burgess & Key, Newgate Street, London, the cutting apparatus and driving-gear presented features somewhat similar to those of Hussey's machine. But while in Dray's machine the grain, after being cut, was delivered to a platform, the working of which required a special attendant, and the grain delivered to the ground in quantities sufficient to make a sheaf was required to be immediately bound up in order to clear the path for the return journey of the machine,—in Burgess &

Key's the cut grain was at once delivered to a screw platform, and passed to the ground at the side of the machine. A special attendant was therefore not required, and the grain, moreover, being delivered at the side, could be left till the whole could be conveniently bound up.

Modern Reaping-machines.

From these small beginnings in the invention and manufacture of reaping-machines a great industry has sprung up, from which the agriculture of this country has derived benefits of inestimable value. The firms in the United Kingdom who manufacture reaping-machines are now numbered by the hundred, and the larger firms send out several thousand machines every year.

Many improvements have been introduced with the view of simplifying the construction, reducing the draught, lessening the cost, and increasing the efficiency and general usefulness of the machines.

Varieties of Machines.—The reaping-machine is now produced in many forms, less or more distinct, suited for different purposes and different conditions of soil and climate. There are the simple mower, adapted merely for mowing hay and leaving it lying as it is cut; the combined mower and reaper, which may be arranged not only to cut the crop, but also to gather it into sheaves or swathes; the back-delivery, the side-delivery, the self-delivery, and the reaper in which the sheaves are turned off by the hand-rake. And last, and greatest of all, comes the combined reaper and binder, which is now an established success, performing its intricate and difficult work in a most admirable manner.

Draught.—Most reaping-machines are arranged for the draught of two horses. Some may be worked by one horse, and others occasionally require three horses.

Price.—The prices of the different reaping and mowing machines vary greatly, from £13 to £20, according to strength and other features. In recent years there has been a marked reduction in price, and this, accompanied by increased efficiency, has given a great impetus to the employment of machines in cutting the hay and corn crops. The combined reaper and binder costs from £25 to £35.

Perfect Workmanship.—The work accomplished by the leading reapers and mowers is now as nearly perfect as might be. Unless the crop is very seriously laid and twisted, the improved machine will pick it up and cut it from the ground in the most regular and tidy manner, leaving a short even stubble. Now and again a corn crop is laid and twisted by a storm so as to defeat the reaping-machine; but the possibilities of the modern machine are indeed wonderful.

Speed.—The speed of the reaping-machine varies considerably, according to the width and general make of the machine, the character of the ground and the crop, and the horses employed. The extent reaped in a day of ten to twelve hours would perhaps run from 8 to 14 acres, the greatest breadth of course being cut where the machine can work continuously along all sides of the field. This, however, is possible only when the crop stands tolerably erect. If a strong wind is blowing, or if the crop is bent to a considerable extent, it is advisable, to ensure good work, to cut only in one direction—against the "lie" of the crop. The machine in this case returns "empty" and out of gear; and it is not all lost time that is employed in the return journey—for the relaxation is appreciated by and is beneficial to the horses, which can thus go at a smarter pace when reaping and work longer without being fatigued.

Force employed.—The force of labourers required to keep a reaping-machine going varies chiefly with the rate of the reaping and weight of the crop, but partly also with the form of the machine, whether for self or manual delivery. With the self-delivery reaper one man to drive the horses is sufficient on or at the machine. The manual-delivery reaper requires an experienced and careful person to deliver the sheaves, and a man or lad to drive the horses.

To "lift" the sheaves, bind, and stook them, from six to ten persons will be required, according to the rate of reaping and the weight and bulk of the crop. It is the custom in some parts to have boys making bands, women lifting the sheaves on to the bands, and men to bind and stook. In other cases women make bands, lift and bind, the stooking being done by men. In many parts, chiefly in the south of England, men do all the manual harvest-work.

In many cases the raking is now accomplished by a rake attached to the rear of the reaper. When this is not provided the work has to be done by a horse-rake, or by a rake drawn by a man or lad.

Cost of Reaping.—The cost of reaping grain by the reaping-machine, including cutting, binding, stooking (or shocking), and delivery, will vary with the rate of wages, the nature of the ground and the crop, the rate of reaping, and the character of the machine, from about 8s. to 12s. per acre. Much depends upon the season—for it is obvious that when the crop is standing erect, so that it can be cut continuously around the field, the speed must be greater and the cost less. Then a heavy crop may require one couple more to lift and bind and stook than would suffice for a light crop.

It is well established that, by the introduction of the improved reaping-machine, the work of cutting the grain crops has been not only much accelerated, but also in most cases to some extent lessened in cost.

It is not necessary to describe in detail the mechanism of improved reaping-machines. They are now so simple and efficient that any man of average intelligence and care can work them perfectly. In their general construction there is much similarity. This is indicated by figs. 461 and 462, which represent self-

Fig. 461.—*Harrison, M'Gregor, & Co.'s self-delivery reaper.*

Fig. 462.—*Howard's self-delivery reaper*

delivery reapers made by Harrison, M'Gregor, & Co., and Howard respectively.

Manual and Self-delivery Machines.

Of the two main classes of reaping-machines, manual and self-delivery reapers, the former makes better work with heavy or tangled crops, but requires an extra man in working. In crops which are moderate in length and not much twisted, the self-delivery reaper, with the saving of one man's labour, is quite as efficient as the manual delivery.

Side-delivery Reapers.—Self-delivery reapers are of two classes, back and side delivery. In the former the sheaves are dropped behind the reaper in the same position as those left by the manual delivery, while in the latter they are deposited far enough to the side to permit of the machine passing, whether the sheaves are tied before it comes back again or not. With the side delivery a whole field may be cut without binding any sheaves, while they must be bound as the cutting proceeds if the back delivery or manual machine is used.

In the dry climate of the south and east of England it is often an advantage to cut a crop and let it lie a day or so before binding, hence side-delivery machines are those most in use there. In Scotland, however, the climate is so uncertain, that a crop cannot advantageously be left unbound even for one day, because should it once get wet when lying loose, the difficulty of drying it again is so great that it more than counterbalances any gain which might result from the method. In Scotland, therefore, the self-delivery reapers are mostly of the back-delivery pattern.

Another point in favour of the system of binding immediately behind the machine is that in this way the labourers work more expeditiously than when they are not pressed by the reaper.

Manual v. Self-delivery Reapers.— In districts of Scotland having a moderate rainfall, and where, consequently, the grain crops are moderate or short in the straw, or where labour is comparatively scarce in harvest-time, the back-delivery reaper is the kind most largely used. Indeed, in the eastern counties the back-delivery machine is found everywhere. In the western and south-western counties, however, a self-delivery reaper is seldom seen, and very seldom do they work well; because in these districts the rainfall is heavy, the straw long and soft, while the whole country is more exposed to wind than on the eastern side of the watershed. The consequence is, that grain crops are usually laid and often much twisted, and in reaping much better work can be done with the manual than the self-delivery reaper.

SELF-BINDER.

The most modern and most expeditious method of harvesting grain is by the automatic combined reaper and binder—one of the most useful agricultural inventions of the nineteenth century.

General Construction. — A sheaf-binding harvester has four separate operations to perform — viz., cutting, elevating, binding, and delivering the crop. In the binder the cutting apparatus differs only in details from the ordinary one-wheeled self-delivery reaper. The grain as cut falls across an endless web, which conveys it over the top of the driving-wheel to the knotter, where the straw falls into two arms called compressor-jaws, which keep it on the knotter-table until a sheaf of any specified size has accumulated. Whenever a sheaf of the desired size has been delivered to the compressors, these relieve the tripper, which sets in motion the needle (carrying the binding twine) and the knotting apparatus. The needle is circular, and in its course it passes the band (twine) round the sheaf, when the band is caught by the knotter, and almost instantaneously a firm and secure knot is tied, while the needle is drawn back ready to operate on a new sheaf. As soon as the knot is tied and the string cut, the sheaf is ejected from the machine in a horizontal position, dropping on the ground on its side, quite clear of the machine.

Efficiency of the Binder. — The binder, as now constructed, is admirably suited for cutting standing grain of any kind, more particularly where the straw is not very long. The land should be laid down with as flat a surface as circumstances will permit, otherwise a longer stubble will be left. Granted a moderate crop of standing grain and good weather,

these machines do their work in a way which cannot be equalled in any other manner.

When the binder was first introduced, wire was used in tying. As would be expected, there were strong objections to the wire, and the substitution of twine was a step of the greatest importance.

One drawback to twine is that it is easily cut by mice, and when these vermin get into stacks of twine-bound sheaves, much trouble may be caused by loose sheaves. The best method of prevention is of course to keep mice from getting into the stacks.

THE HORNSBY BINDER.

The binder made by Messrs R. Hornsby & Sons, Limited, Grantham, which won several prizes in important trials, is here described in some detail.

Cutting the Crop.

The cutting apparatus is, as already indicated, similar to that employed in the simple reapers, and need not be particularly described.

The illustration, fig. 463, shows Messrs Hornsby's arrangement of finger and its method of attachment to the framing of

Fig. 463.—*Finger arrangement.*

the machine, by which they obtain the lowest cut without the platform rubbing on the ground. This great advantage is gained by the use of a sheet steel platform, so that the fingers may be close to the ground whilst the platform is quite clear of it.

The platform canvas for carrying the cut crop to the foot of the elevator is kept as low as possible, so that even short crops fall readily upon it without any liability to choke the knife.

The new pattern reel and reel-support can be instantaneously adjusted, up or down, backwards or forwards, for dealing with laid, twisted, or heavy standing crops.

Both inside and outside dividers are made to suit all crops, making perfect division on the one side, and lifting up

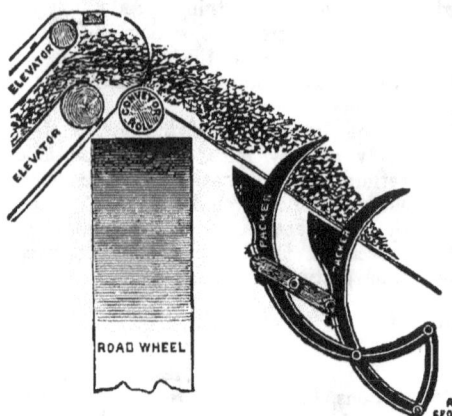

Fig. 464.—*Conveyor-roll.*

hanging ears and cutting every straggling straw on the other.

Elevating.

The cut grain is carried up to the knotter between canvas elevators. These elevating canvases are brought down below the level of the platform canvas, so that the one feeds the other evenly and regularly. This is done by means of a novel arrangement of strengthening plate enabling the canvas rollers to work lower than in any other. The canvases are also shorter, owing to the reduced height of the machine, which, with the increased diameter of the rolls, reduces draught and makes the canvases run perfectly with less frequent tightening and adjusting.

The laths are *riveted to the canvas*, and are of an improved shape, securing easy running over the rollers, and preventing any straws from sticking between them and the canvas.

The canvas rollers run in metallic bearings.

Elevator and Platform Rolls.—By a newly patented arrangement, the rolls for the platform and elevator are made the width of the machine, making the canvases more certain than ever in their action, the roll ends being recessed into the framing, so that loose straws cannot possibly wind round them. The vibrat-

ing buttor takes the place of the old canvas buttor, doing away with a considerable amount of friction and lessening the cost of repairs.

Conveyor-roll.—The height of the machine is considerably reduced by the use of a conveyor-roll, to pass the cut crop from the top of the elevators close over the top of the main road-wheel to the binder-table. See fig. 464.

Binding.

The "Hornsby" binding mechanism is exceedingly simple, perfectly automatic in its action, and perfectly reliable in operation.

To avoid wear and save power, the binding apparatus remains at rest, whilst the cut crop in a steady stream passes down from the elevator, and is pressed forward by the packers; but the moment enough has been accumulated to form a sheaf (of one of the five sizes before determined on, according to the crop), the binding mechanism is automatically started, the needle carries the string round the sheaf, the knot is tied, the string cut, the loose end retained for the following sheaf, and the operation is complete.

Needle.—A new patent needle is now used. It works with much less friction on the twine, and is also much easier to thread.

The Knotter.—The improved Hornsby knotter is shown in figs. 465, 466, 467; and fig. 468 shows the manner in which the knot is tied. Fig. 469 represents the tied knot. A bevel gear now actuates the knotter, replacing the old chain mechanism (fig. 470).

Fig. 465.—*Knotter.*

Delivery of Sheaf.—It is important that the sheaves should be delivered gently, for if they are subjected to rough usage a considerable quantity of grain might be knocked out, especially if the crop were over-ripe.

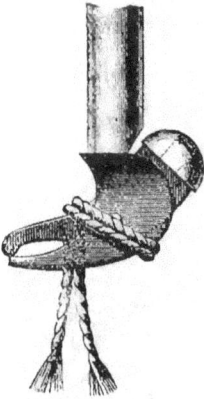

Fig. 466.—*Half turn.*

Fig. 467.—*Whole turn, jaws open and string entering.*

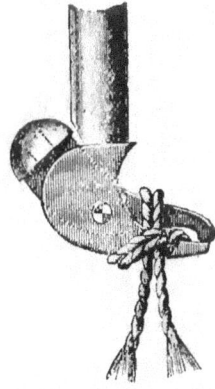

Fig. 468.—*Jaws closed on string; the string-knife then cuts the ends, and the lever draws the string off, completing the knot.*

Fig. 469.—*Knot tied by the Hornsby binder.*

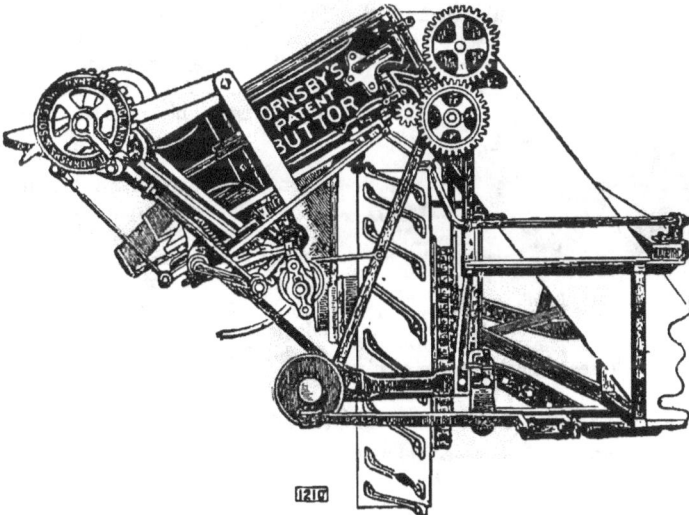

Fig. 470.—*Bevel gear and vibrating buttor.*

In the " Hornsby " binder the sheaf is firmly held whilst the knot is being tied ; the ejectors then coming into action press it forward, whilst the retaining boards fall, so as to slide it gently to the ground without liability to shedding.

It delivers the sheaf near to the ground, and thus avoids shaking.

At Work.—Roller bearings are freely used for the spindles, and materially lighten the draught. The Hornsby binder may be worked by two good

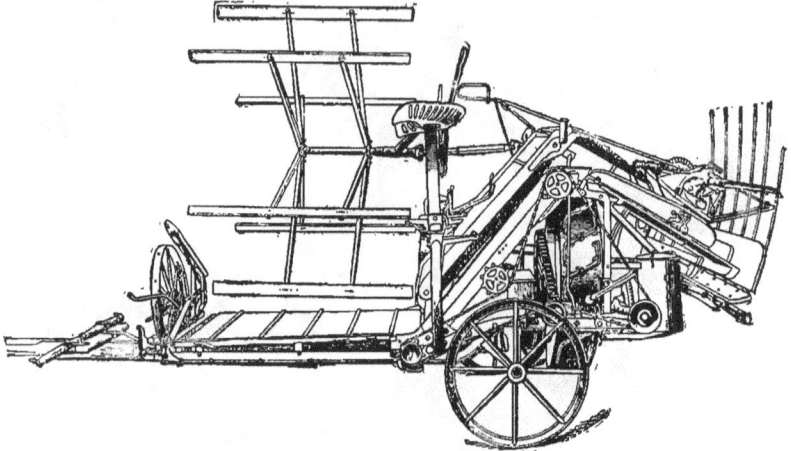

Fig. 471.—*Wood's New Century reaper and binder.*

horses, but many prefer to use three. The horses will perhaps travel about three miles per hour.

Speed. — At this speed and cutting around the field the binder may cut over an acre per hour. Its daily work is therefore a simple question of how many hours it is kept in action.

Fig. 472.—*Bisset's binder.*

By laying the sheaves in rows the sheaf-carrier lessens the labour in stooking.

Other Binders.

The manufacture of combined reapers and binders is now carried on extensively by many eminent firms—Canadian and American machines competing strongly against British-made ones in our own

country. Each machine has its own peculiar merits and special admirers, but, as with ordinary reapers and mowers, they are now all wonderfully efficient. The other binders best known in this country are the Massey-Harris (Canadian), the Walter A. Wood, the M'Cormick, the Deering, the Plano, the Milwaukee, the Osborne, and the Champion (American). The " New Century Binder," brought out by Walter A. Wood, is represented in fig. 471. Bisset's binder is shown in fig. 472.

Progress of the Binder.

On farms of average size the binder has made more rapid progress, and is a more serviceable machine, than the manual reaper was at its introduction. The experience gained in the manufacture of the ordinary reaper has been fully taken advantage of in the manufacture of the binder; and whereas the first reapers often failed, through the breakage of some more or less important part, the binder rarely does so. When it does fail to do its work, it will, as a rule, be found that the crop is too heavy or too much laid and twisted, or the land unsuitable.

With a light and serviceable binder,

which could be depended on for cutting all average crops, there need be little extra hurry or press of work at harvest more than at any other time of the year, while the whole might be accomplished without an extra hand being engaged.

Working the Binder.

Before beginning with a binder, a couple of swathes should be cut with the scythe round the whole field, or if more convenient, one swathe with a scythe and one with an ordinary manual or self-delivery machine. Two swathes are necessary, not entirely on account of the width of the machine, but to provide walking space for the three horses, often required to work the binder.

Where the circumstances permit, the easiest cut is round the field, but if the crop is bent in any particular direction it must be cut one way only.

Speed in cutting round.—Cutting round about, an acre an hour is easily accomplished, and if the horses travel freely an acre and a half may be done.

Hands required.—According to the weight and closeness of the crop, from two to four men will be required to stook, if going round about, while only half the number will be required if cutting is done one way only. Behind a binder going round the field, a man can stook more grain of an equal weight than he will do after any other method of cutting.

With good string, the knotting apparatus rarely gives trouble, and the whole machine is easily under the control of one man.

Raking.—Where the binder can work anything like satisfactorily, no raking is required, as very few straws are left.

Size of Sheaves.—With a moderately regular crop, the sheaves can be made much smaller than where the crop has to be tied by hand, without adding materially to the cost, the only increase being what extra will be required for binding-twine, and the little additional labour required in stooking a crop of small or moderately sized sheaves instead of large ones.

The small number of hands required in the harvest-field, where a binder is in use, would thus, on a moderate-sized farm, allow of reaping and stacking going on simultaneously—that is, if the varieties of crops were so regulated as to allow them to come forward in regular succession.

Cost of Cutting with Binder and Reaper.

The exact cost of cutting a certain area of corn will of course, as already pointed out, vary considerably in accordance with the kind, condition, and weight of the crop, the configuration of the ground, the rate of wages, and skill in management.

Saving in Labour.—It is now generally conceded that where circumstances are favourable — the fields moderately large and level, and the area under grain crops sufficiently large to warrant the somewhat heavy initial outlay in purchasing a binder,—the cutting can be accomplished at from 1s. to 3s. per acre cheaper by the self-binder than by the ordinary reaping-machine. Common estimates indicate a saving of about 1s. 6d. or 2s. per acre—the saving of course arising in manual labour.

Saving in Crop.—Another point in favour of the binder is that it gathers up the crop more cleanly than the reaper. It leaves fewer stray stalks of grain on the ground, and thus saves both straw and grain. The saving on this head alone has been variously estimated at from 1s. to 5s. per acre. In average circumstances from 1s. to 2s. would perhaps be tolerably near the mark.

Examples of Cost.—Mr John Prout, Sawbridgeworth, Essex, gave the cost of cutting with the binder and the reaper as follows, the extent cut per day being on an average 10 acres by the binder and 12 acres by the reaper :—

Binder.

						£	s.	d.
Six horses	.	.	.	.	.	1	10	0
Two men	.	.	.	.	.	0	14	0
Two boys	.	.	.	.	.	0	7	0
String, 2s. per acre	.	.	.	1	0	0
Oil	.	.	.	.	.	0	2	0
Total for 10 acres	.	.	.	£3	13	0
Per acre, 7s. 3d.								

Ordinary Reaper.

						£	s.	d.
Four horses	.	.	.	.	.	1	0	0
Tying by hand, 6s. per acre	.	3	12	0
Two men	.	.	.	.	.	0	14	0
One boy	.	.	.	.	.	0	3	6
Oil	.	.	.	.	.	0	2	0
Total for 12 acres	.	.	.	£5	11	6
Per acre, 9s. 3d.								

There is here a balance of 2s. per acre in favour of the binder. In many cases the costs with the binder would be less than above by about 1s. per acre.

HARVESTING BY MOTOR.

Only to a small extent as yet have motor tractors displaced horses in working the self-binder and other forms of harvesters. We are still labouring under the disadvantage of having to yoke the motor to machines built for horse draught, and this occasions an unnecessary waste of power, besides the employment of unnecessary labour.

Most likely when we get an actual motor-binder—that is, motor and harvester combined — the wheels of the motor-carriage will be the only travelling wheels, the cutting and binding mechanism being carried on the motor vehicle and doing their work without actually touching the ground.

THE STRIPPER HARVESTER.

In Australia and other large grain-growing countries, where a very small part of the straw is required for the stock-yard, the grain is stripped off and the straw left on the ground, as a rule. Mr M'Kay brought out the first Australian stripper in 1884, but it did not make much progress there until he remodelled his machine in 1894, and produced the first stripper harvester of the now well - known "Sunshine" type. These harvesters are made with varying sized combs, from 5 ft. to 11½ ft. in width, and are said to have reduced the cost of harvesting to less than 2s. 6d. an acre. Compared with the system of cutting, binding, and threshing formerly practised, the introduction of the Sunshine harvester has enabled the farmer to save 12s. or more per acre in harvesting. No wonder that it is now being largely used, not only in Australia, but also in the Argentine and other countries where large tracts of grain have to be harvested, and they can afford to leave the straw on the ground.

In its early form the machine used to strip the heads off, and thresh them, but did not clean the grain, and the winnow-

ing and bagging had to be done by manual labour. The modern harvester combines both the stripper and the winnower, and does the work of both machines at one operation. Drawn by three or four horses, and worked by only one man, it harvests at the rate of about 10 acres a-day, in a crop yielding 20 bushels to the acre; but it is profitable to employ a second man to sew the bags as they are filled, and drop them off the machine, and thereby avoid the stoppages which are necessary where the driver has to stop and lift off the bags.

PROCESS OF REAPING.

The detail-work of reaping corn with the machine varies considerably throughout the country. The process is not quite the same for the different varieties of grain even in any one district, and the climate and labour-customs are also responsible for differences in the methods of working.

Reaping Oats.

The main principles which should guide the farmer in arranging the practical work of reaping are applicable alike to wheat, barley, and oats; yet it will be convenient, in the first place, to describe the reaping of oats, and afterwards point out the distinctions relating to wheat and barley.

Preparing to Cut. — The prudent farmer will have the reaping-machines looked out and put into the pink of condition before the day arrives for the commencement of cutting. Any necessary repairs will have been effected at the end of the previous season. No judicious farmer would think of laying up a machine or implement of any kind for the idle season until the needed repairs have been attended to. It is very bad practice indeed to delay such matters until the time arrives for the active employment of the machine or implement. See that all preliminaries are attended to beforehand, so that when the work of cutting begins there may be no avoidable delay.

Sharp Knives.—Keep the knives as sharp as possible, as good work and light draught cannot be had without sharp knives. Where two or more reapers are kept going, it is advisable to keep one man sharpening knives, as then they are always in good repair, and cutting goes on more smoothly and rapidly than when the driver has to look after not only his horses but his mower and knives as well.

The most common method of sharpening the knives of reapers is with a fine file supplied for the purpose. Machines for sharpening are now in use to some extent, that made by Harrison, M'Gregor, & Co. being shown in fig. 473.

Cutting "Roads." — A day or two before reaping is to be begun, "roads" for the reaping-machine should be cut with the scythe all round the field or section about to be cut. One cut of the

Fig. 473.—*Reaper knife-sharpener.*

scythe is usually considered sufficiently wide, but at the ends where the machine is to be turned it is more convenient to have the "road" two cuts wide. The corn cut out of the "roads" is tied into sheaves, which are laid against the fence and stooked with the main crop when it is being cut.

The scythe is the most convenient appliance with which to form "roads" for the reaping-machine, but in some parts the hook is still employed for this purpose.

Cutting.—In working with either the manual or self-delivery reaper, cutting may be done in two ways—either along one side of a field or round about. If the crop be not laid in one direction, the weather is moderately calm, and the crop

mostly standing, the roundabout method is the best, as no time is lost returning.

Force of Labour. — If the crop is moderately ripe, and the straw dry and free from grass, so that fairly large sheaves may be made, each man binding may do from 3 to 5 roods per day, the extent depending very much on the size of the sheaves, the thickness of the crop, and the tidiness with which the work is done. In many districts of Scotland women do the bulk of the binding; and if the crop is light or the sheaves small, most women who know their work can do as much as the average of men. For heavy crops, however, where large sheaves are made, they are not so well suited, in which cases three women will be required for every two men, or it may even be two women for each man.

In reaping round about, the binders may be distributed in two ways: each may have a certain distance to do, or a certain number of sheaves to tie. If the crop is moderately regular, no better plan need be adopted than that of dividing the circumference of the field into equal divisions for each binder, and sticking a piece of wood into the ground where the one division ends and the other begins. After every half-dozen swathes or so of the reaper, the marking-posts may be moved nearer to the side of the uncut grain, so that disputes may be prevented from arising among the binders as to who should or should not tie certain sheaves. When a binder has finished his or her number of sheaves in a swathe, a stoppage is made until the reaper again comes past, when the work is resumed.

By so doing, all unnecessary travelling backward and forward is done away with, and each binder gets a regular share of the good and bad parts of the crop in the field. In this system of working every one gets an equal share to do, and cannot avoid doing it; yet good as the system is where the binders are all on an equality, it is not a suitable one to adopt where there are inefficient persons or learners, as one slow person in the lot keeps back the whole squad.

The same system can, of course, be pursued where two or three reapers follow each other, if a corresponding number of binders can be obtained. If two machines are working, the driver of the second should start as soon as the first is half-way round, and the drivers should endeavour to keep as near as possible the half circumference of the field apart. In this system of harvesting each binder should properly clean up the ground where each sheaf has been made, before leaving the spot.

Raking.—A drag-rake similar to a hay-rake, but smaller, is often attached to each machine, which rakes the ground that was cut the swathe previous. In order to allow the rake to work, the sheaves as tied are in some cases conveyed back from the standing grain fully two breadths of the reaper. The necessity for this is averted by each binder letting his sheaves lie just clear of the last swathe,—that is, where the binding and stooking are done by different men, —a plan which is largely followed.

Other Methods. — Various other methods are followed. By one method in which each binder has an equal number of sheaves, each begins near or where they ended the time previously, the number of sheaves, not the distance, regulating the place. This system is very well suited for comparatively small fields where the number of sheaves allotted to each binder is not very large, no larger than each binder can easily count his or her own share without moving backward to do so.

Detecting Bad Work.—Another advantage is that by this method each binder's sheaves are all together, so that the farmer can at once see if any one has been doing the work in a careless or slovenly manner; whereas by any of the other methods, it is impossible to say by whom the sheaves were bound and stooked after the first or second swathe.

Arranging the Force for Manual-delivery Reapers.—Where the cutting is round about, binding behind the manual-delivery reaper is the same as if a side or back-delivery machine were used. Most of the cutting with the manual-delivery machine is, however, done only along one side or end of a field. In this case the binders may have an equal or any number of sheaves each, or a certain distance.

If two manual-delivery machines are working together, the one should be just entering the swathe while the other is going out at the other end, so as to allow the binders an equal time to tie the sheaves behind each machine.

In some cases it has been the custom to have the band-making and lifting done by women and lads, and with men following to bind and stook. It is more expeditious, however, for each labourer to combine the lifting and binding, and leave the stooking or shocking to another. One man can stook to four or five lifting and binding, one known to be an expert at stooking being entrusted with this work.

Bands and Binding.

The corn-band, fig. 474, is made by taking a handful of corn, dividing it into two parts, laying the corn-ends of the straw across each other, and twisting them round so that the ears shall lie above the twist—the twist acting as a

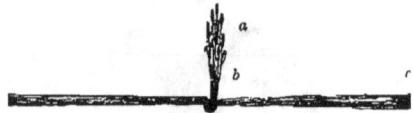

Fig. 474.—*Corn-band ready to receive the sheaf.*
a Corn-end of straw.   *b* Twisted knot.
c c Band stretched out.

knot, making the band firm. The lifter then lays the band stretched at length upon the ground, to receive the corn with the ears of the band and of the sheaf away from him.

The bands should always be made of two lengths of straw, as under no circumstances can a single length be used advantageously.

Method of Binding. — In approaching the sheaf the binder gathers the spread corn on either side into the middle of the band with both hands, and, taking a hold of the band in each hand near the ends, he crosses the ends of the band, pulls forcibly with the right hand close to the sheaf, and keeps the purchase thus obtained with the under side of the left hand, while he carries the end in the right hand, below and behind his left hand; and then, taking both ends in both hands, twists them firmly and thrusts the twist under the band with the right hand, as far as to keep a firm hold. In a bound sheaf, the corn-knot in the middle of the band is held firm by the pressure of the sheaf against the ears of corn and the twisted part of the band.

Position of the Band. — The band should always be put on as near the middle of the sheaf as possible, never below the middle, but if anything above it. If put on much below the middle the top of the sheaf spreads out, instead of keeping close together, and if rain comes on the water runs down the centre of the sheaf instead of the outside of it.

Size of Sheaves.

Although large sheaves add considerably to the speed with which a crop can be tied, they hinder materially the after-drying of the crop. In the end it will therefore be found the most profitable way to make the sheaves as small as possible, consistent with the length of the crop. In a short crop with grass among the straw, they should not be over 6 inches in diameter at the band, and in the longest and cleanest straw 10 inches will be quite sufficient.

Lifting broken Stalks. — If the crop is much twisted or tangled, a young lad is often employed with a fork to raise up the heads of the standing grain, should these be inclined to be broken down by the passage of the reaper. It is advisable to have this done, because if the ears are not lifted up they are liable to be cut off by the next passage of the machine, and left on the ground instead of being secured in the sheaf.

Raking.

With the use of reaping-machines there is much less of the crop to be gathered from the ground in the form of rakings than when the scythe was used, and this is one of the advantages of the changes in harvest operations.

A hand-rake of a type long in use is shown in fig. 475. Although this is commonly called a hand-rake, it is really pulled by a leather strap, attached to the shaft and passed over the shoulder of the raker.

If the machines are cutting only one way, in returning they must pass behind the second row of stooks, in order to give the raker time to rake the space between the first and second rows.

Drying Rakings. — The disposal of rakings during harvest operations has been a difficulty with many people. Some farmers maintain that the best, least costly, and easiest way is for the raker to thrust each lot as gathered into the very centre of a stook ; but a better plan is to tie them into sheaves, and set the sheaves together in threes and fours. They should be thrust as far in as to be clear of the end pair of sheaves. In this position they will be free from rain ; and owing to the current of air through the stook they dry quickly, and at stacking time they can either be brought in along with the sheaves or by themselves.

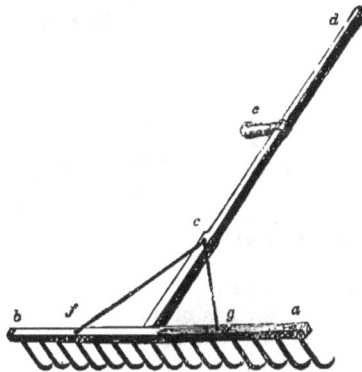

Fig. 475.—*Hand stubble-rake.*

a b Head of rake.	e Handle.
c d Helve.	f c g Iron braces.

Others, however, object to this plan, on the ground that the rakings placed in the centre of the stook interfere with the drying of the crop.

Stooking or Shocking.

Where the reaper is cutting one way only, two stookers or shockers will be required. If two machines are working together, three will be sufficient in a short-strawed crop, while four may be hard enough worked in a long crop. Shocking or stooking should always begin at the end of the swathe which is first cut, the second stooker beginning a new row as soon as he sees the first one started, and the others as soon as possible afterwards, each beginning as near as possible about the same distance from the standing grain. An easy guide is for all to follow some particular wheel-track of the reaper, as these in most circumstances are easily seen.

Stooking. — In setting a stook, the centre pair of sheaves should always be set up first. Each sheaf should get a good solid dump on its butt end, so as to give it a firm foundation; and the two sheaves should be firmly pressed together at the top, by putting a hand on the outside of each a little above the bands, and exerting considerable pressure on these parts. Each following pair of sheaves should be put at opposite ends of the stook, in such a position that they only very slightly incline their heads towards the centre of the stook.

It is sometimes said that in stooking, each pair of sheaves should stand perpendicularly or independently of the rest of the stook. Such instructions are decidedly wrong, and should not be followed in practice. If the first pair are perpendicular, and all the other pairs have a very slight lean towards the centre pair, a much more substantial stook is built than if all are set perpendicularly.

Direction of Stooks. — The direction to which the ends of stooks point is a very important one, to which, in many cases, too little attention is often paid. When finished, the stooks should always point as nearly as possible between south and south-west — to the one o'clock sun — as the prevailing winds then strike them on the end, and blow right through the stook. In this direction the sun dries each side of the stook about equally, which, in a wet or late harvest, is a matter of considerable importance.

Placing the Band-knots. — Were the corn-knots in the bands set outwards in the stook, the rain in a wet season might injure them; and as they bear a sensible proportion to the corn of the whole stook, the sample might thus be materially injured. By simply turning the corn-knots inwards and the root-knots outwards, the injury is prevented. In a fine season the corn-knots may be placed outside.

But in the hood-sheaves the corn-knots were generally placed uppermost, and exposed to the rain; because, were the other *side* of the sheaf exposed *upwards*, where a groove runs down the length of the sheaf, by the straw being gathered

into that form while making the root-knot of the band, the rain might penetrate by the groove through the body of the sheaf, lying in its horizontal position, to the corn in the standing sheaves below, and thereby inflict a much greater injury than merely spoiling the corn-knots.

Size of Stooks. — In no variety of harvesting work is there greater variation than in that of shocking or stooking, simple and plain as it may appear to be. In some districts the sheaves are entirely set up in stooks of four or six sheaves, while in others they will be found of all sizes, up to fourteen or sixteen pairs of sheaves. By making extremely long stooks, the risk of their being blown down by the wind is undoubtedly lessened, but at the same time so is the speed with which they are dried. In a large stook the end pair of sheaves will usually be found ready to stack several days before any of the centre pairs; and if any of the centre pairs get soaked with rain, they will scarcely dry at all, unless taken from the centre.

If the grain is not very tall, fine in the straw, and contains any rye-grass or other grass, the stooks should not, as a rule, be larger than eight sheaves, four on each side, and in sheltered situations the number may be advantageously reduced to six sheaves. Very short-strawed crops should also be set up in stooks of four or six sheaves, or in exposed situations eight sheaves may be used. Where the straw is full length, eight sheaves are most generally used, and if the crop is very long, the stooks may contain ten or twelve sheaves.

" Pirling." — A plan of stooking sometimes pursued in certain exposed districts of the west and south-west is to set up two pairs of sheaves, the one pair at right angles to the other instead of side by side, as in an ordinary stook. The butts of the sheaves are if anything kept a little wider apart than in ordinary stooking, and when set up, the tops of the four sheaves are tied together about 9 inches under the apex, by a few straws pulled out of the top. This system is called "pirling," and, unless in particular districts, was probably more common half a century ago than now.

Stooks of this class dry much quicker than those of any other, and withstand a gale which levels almost all other stooks. The time required to do the

Fig. 476.—*Barley or oat stook hooded.*

a First 2 sheaves set.   d d Fourth 2 sheaves set.
b Second 2 sheaves set.   e Fifth 2 sheaves set.
c Third 2 sheaves set.   f Hood-sheaves set.

extra tying, although a little, is not great, and need not deter any one from adopting it, where the circumstances call for such protection from wind and rain.

Hooding.—The use of hood-sheaves for oats, although at one time almost universally adopted, is now seldom resorted to. Owing to the earlier and shorter harvest of the present as compared with bygone times, some of the precautions once adopted are not now necessary.

A once common form of "hooding" is shown in fig. 476.

"Gaiting."—Another ancient method of setting up sheaves, which has now almost entirely been discarded, is "gaiting"—viz., setting up each sheaf singly, where the grain was wet when cut. The band of the sheaf is tied loosely round the straw, just under the corn, fig. 477, and the lower part of the sheaf is made to stand by spreading out the straws' end in a circular form. Gaitins are set by the bandster upon every ridge; the wind whistles and the rain passes through them. Gaiting is practised only in wet weather, and even then only when a ripe crop is endangered in standing by a shaking wind. It is used for oats and barley, wheat never being gaited, be-

cause when wheat gets dry, after being cut in a wet state, it is apt to shake out in binding the gaitins.

Reaping Barley.

In nearly all respects the harvesting of barley is much the same as that of oats. Barley, however, as has already been indicated, is rarely cut with a green tint like oats, but is allowed to stand till it is fully ripe, more particularly if it is intended for malting purposes. For malting it must all germinate at or near the one time, and if a portion of the grain is not fully ripe when cut, these grains will be more tardy in germinating.

Quick Drying.—As bright clear samples are of most value for malting purposes, those samples giving a price very much in excess of darkened ones, it is of great importance to a farmer to be able to shorten the period during which his barley crop runs risk of damage from the weather.

Small Sheaves.—Small sheaves, owing to their being quickly dried to the very centre, and a much larger proportion of the grain being exposed to the influ-

Fig. 477.—*Gaitin of oats.*

a Band loosely tied.   b to c Base of sheaf spread out.

ence of the sun and air, are much to be preferred to large ones, as the latter are apt to darken the grain in the centre of the sheaf. In an unsettled harvest large sheaves can scarcely be got dried through, whereas had they been small, they might have been at least so far dried as to be rickled, where the drying can be completed without much further risk.

A larger proportion of barley than any other grain is threshed from the stook, and small sheaves and stooks are as great an advantage for such in good bright weather as in times when it is dull and close.

Stooks of barley are occasionally "hooded."

Reaping Wheat.

Scythe unsuitable.—In the cutting of wheat the scythe has never been extensively used. The straw of wheat is so hard that the scythe does not readily cut it, and when cut by the scythe it is almost impossible to make a respectable sheaf of it. Scythe-cut sheaves of wheat are generally very long, a great many ears are in the butt of the sheaf, and the stooks rarely ever stand well, even when carefully put up.

Reaping - machines. — The ordinary self and manual delivery reapers all make excellent work in a wheat crop, and were suitable for harvesting wheat before they could be generally used for the softer-strawed grains.

Self-binders.—In a regular up-standing crop of wheat no class of machine can do work equal to the binder, and at so small a cost. In wheat of the proper class the binder's highest degree of perfection is attained, and harvesting is done by it with an ease, speed, and accuracy of which, before the days of binders, farmers could have formed no conception.

Time to Cut Wheat.—Wheat should not be so ripe as barley when cut, but riper than was suggested for oats. Whenever wheat becomes white or yellowish-white under the ear, it may be cut any time, as no more sap can then pass from the lower portions of the straw, much less from the roots, to the ear. If cut rather on the early side, the outer skin or brawn is generally thinner and clearer; while if the crop is allowed to become dead ripe, the colour is deadened or dulled, while the outer skin is much thickened. This thickening of the outer skin apparently is a provision of nature to prevent premature decay of the grain. Extra

ripe wheat also germinates freely if subjected to rough weather; and although early and strong germination is a good point in a seed sample, it is rather a bad one when it occurs in the stook.

Sheaves for Wheat.—Owing to the dryness, stiffness, and length of the straw of wheat, it is usually advisable to bind it in larger sheaves than any of the other classes of grain. A large sheaf of wheat dries about as easy as a small one of oats or barley; and whereas oats or barley are easily stooked if small sheaves are made, it is difficult to satisfactorily put up stooks of small sheaves of wheat. The straw of wheat is so hard and slippery that small sheaves easily slide past one another, and even in calm weather they are difficult to keep on end, and in stormy weather they are almost sure to go down entirely.

Wheat Stooks.—Stooks of wheat are most frequently built of ten or twelve sheaves. Small stooks of wheat, such as have been recommended for oats, are liable to be thrown over by wind.

Hooding.—Before the advent of the reaping-machine, it was customary to cover each stook of wheat with two hood-sheaves, as shown in fig. 476. For wheat these were tied as near as possible to the butt-end, and were laid along the top of the stook, the two butts meeting above the centre. Both hood - sheaves were laid on at a considerable angle, generally about half the slope of an ordinary roof.

When put on, the first hood - sheaf should cover fully one-half the stook, and when the other one is put on, the butt-end of the first one should be slightly pushed up. After fixing the second one similarly to the first, the workman steps to the side of the stook and carefully presses the two butts into each other.

The hood-sheaf on the east side should always be put on first, as the butt of the west one shelters it from the west wind, and prevents it from being thrown off.

Hooded in this manner, wheat stooks stand a considerable amount of either wind or rain; but if carelessly done it is worse than useless, as the first gust of wind knocks a large proportion of the hoods off, and if rain falls, both hood-sheaves and stooks are worse wetted than if they had not been hooded at all.

STACKING CEREALS.

The reaped corn must be allowed to remain some time in the stook in the field in order that it may become sufficiently dry to keep in the large quantity composing a stack.

Time for Drying.—The length of time required for drying depends largely on the weather, but partly also on the ripeness of the corn when reaped. If the air is dry, sharp, and windy, the corn will be ready in the shortest, while in close, misty, damp air, it will require the longest time. At least one week for wheat, and 10 days for barley and oats, will usually be required. Small sheaves of course dry more quickly than large sheaves. Corn having an admixture of grass in the ends of the sheaves is the most difficult to dry. In reaping with the machine the corn is more closely packed in the sheaf than when reaped with the scythe, and thus in the former case a day or two's longer drying may be required.

Judging of Dryness.—Mere dryness of straw in feeling does not constitute every requisite for making newly cut corn keep in the stack. The natural sap of the plant must not only be evaporated from the outside, but from the inside also. The outside may feel quite dry, whilst the interior may be moist with sap. The state of the internal condition, therefore, constitutes the whole difficulty of judging whether or not corn will keep in the stack.

Several criteria exist by which certainty is arrived at—namely, by the straws being loose in the sheaf; by easily yielding to the pressure of the fingers; by the entire sheaf feeling light when lifted off the ground, and dry when the hand is thrust in beyond the band; or by twisting a straw, and observing if any sap remains in it.

Weather and Drying.—Winning is effectual when the weather is dry. Wind is also winning, but the stooks are apt to be blown down, which incurs the trouble of setting them up again. Rain immediately following or accompanying wind injures stooks materially. When much rain falls, accompanied with cold, the corn becomes sooner ready than the straw for the stack; and, to win the straw, the bands may have to be loosened, and the sheaf spread out to dry in the wind and sun. In like manner, the sheaf may be spread out in dry weather, when a large proportion of young grass is mixed with the straw.

Corn wins in no way so quickly as when "gaitined" (fig. 477).

Sprouting.—When the air is calm, dull, damp, and warm, every species of corn is apt to sprout in the stook before it is ready for the stack. In this way the quality of the grain is often much injured.

Process of Stacking.

Temporary Stacking or "Rickling." —Oats which have had rye-grass, clovers, or other grasses sown with them are usually difficult to dry, and more particularly in a damp climate or a late season. With such crops "rickling," "coling," or "hooacking" is sometimes resorted to before the crop is dry enough for stacking. A "rickle," "cole," or "hooack" may contain from 6 to 8 or more stooks, according to the size of the stooks and length of the crop.

The centre of the "rickle" is composed of 4 or 6 sheaves, all set up together, with the bottoms slightly out and the heads close together. Around these are built another circle, the butts of which also rest on the ground, the next row being kept far enough up to just cover the straps or bands of the sheaves of the preceding one.

Advantages of "Rickling."—This manner of securing a crop allows of the butts of the sheaves being dried in a way attained by no other system.

Sheaves with grass in the butts very speedily kill off the young grasses under them, particularly in wet weather. "Rickling" allows of the crop being placed in a new position, and damage to the grasses avoided, while at the same time almost securing the crop, and putting it into such a position that it readily dries afterwards, and is seldom difficult to get dry enough for carting to the stack.

Preparing for Stacking.—Prior to harvest the stackyard should be put in order to receive the new crop by removing everything that ought not to be in it

—such as old decayed straw; and weeds, such as strong burdocks, thick common docks, tall nettles, rank grass, yellow weed; which in too many instances are allowed to grow and shed their seeds, and accumulate to a shameful degree

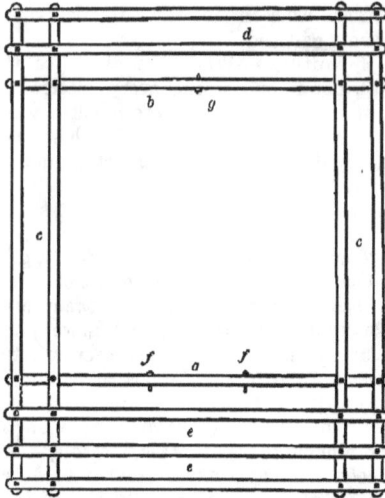

Fig. 478.—*Corn and hay cart tops or frame.*

a Foremost main bearer.   *e e* ₃ arched fore cross-rails.
b Hindmost main bearer.   *f f* Bolts through rail in
c c Pairs of slight side-   front of cart.
      rails.   *g* Bolt through rail on
d 2 hind cross-rails.   backboard of cart.

during summer. The larger classes of implements are often accommodated in the stackyard for want of sheds to keep them in, and these must now be removed.

Where stathels are used, they should be put in repair. Loose clean straw should be built in a small stack on one of the stathels, or other place, to be ready to make the bottomings of stacks as wanted. Drawn straw or other thatch material should be ready for thatching the stacks as they are built, in case of wet weather occurring—a little time being given for the stack to settle, else the ropes by which the thatch is held become loose, and require to be tightened.

Straw-ropes or coir yarn should be stored in the hay-house or elsewhere, ready to be used in thatching. New straw, being less brittle, is better suited than old straw for making into ropes. For this purpose a few loads of sheaves can be threshed early, even though only

partially dry. The tops or frames should be put on the tilt-carts; the corn-carts should be put on their wheels and the axles greased; and the ropes should be attached to the carts. The forks for pitching the corn in the field, and from the carts to the stacks, should be ready for use. Neglect and want of foresight in these particulars, small as they may appear, indicate mismanagement on the part of the farmer.

Cart Frames.—The tops or frames for placing upon tilt-carts are a light rectangular piece of framework, as shown in fig. 478. Two main bearers are fitted to lie across the shelvements of the cart; the foremost is slightly notched, and the hindmost rests against the backboard of the cart, the top sides of which are first taken off. One pair of slight side-rails is applied on each side, crossing the bearers, and notched upon and bolted to them with screw-bolts, these being crossed by two rails behind, and by three more in front; and as these last project over the back of the horse, they are made in arch form, fig. 479, to give freedom to the animal.

A simple and effective method of securing the frame to the cart is by means of the bolts in the bearers, the front ones passing through the head-rail of the front of the cart, and the hind one through the top-rail of the tail-board.

Harvest Cart.—But the common corn or hay cart is a more convenient and efficient vehicle for carrying the corn

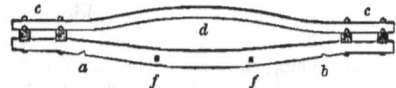

Fig. 479.—*Transverse section of the tops or frame.*

a b Notches in foremost main bearer.
c d c Arched rails over back of horse.
f f Bolt-holes in frame.

crops into the stackyard than the tilt-cart with the frame, inasmuch as the load is more on a level with the horse-draught, and the body being dormant, the load is not liable to shake with the motion of the horse. Fig. 480 gives a perspective view of such a cart. Lightness is a special object in its construction, so that it is made of light strong wood.

Corn-carts are not in all cases furnished

with wheels of their own. The body may be set upon those belonging to the tilt-carts. The cart weighs about 8 cwt.

It is easily converted into a *dray-cart* by simply removing the framework, which should then have the standards based upon two longitudinal rails, instead of being mortised into the shafts. In such a form it is eminently useful in carrying large timber.

Farm Waggon.—The English farm waggon is often preferred in carting corn, especially where the distance from the field to the stackyard is considerable.

Harvest Forks.—Forks used in the loading of corn require to have long shafts, not less than 6 feet, and small prongs. Such a length of shaft is required to lift the sheaf from the ground to the top of a loaded cart, or from the cart to the top of a stack. The fork used in the field should have a strong stiff shaft, as the load on the cart is at no great elevation. That for unloading the cart to the stack should be slender and elastic, as many of the sheaves have to be thrown a considerable height above the head.

The prongs, being small (about half the length of the prongs of the hay-fork), just retain hold of the sheaf, without being so deeply pierced into the band as to be withdrawn from it with difficulty. A deep and firm hold with long prongs renders the pitching of a

Fig. 480.—*Common corn and hay cart.*

a a Shafts of Baltic fir.	c c Oak standards.	e Broad load-tree.
b b Cross-heads.	d d, d d Inner top-rails.	f f, f f Outer rails, front and rear.

sheaf a difficult matter; and if one of the prongs happens to be bent, or a little turned up at the point, the difficulty is much increased.

The prongs of the forks are now made of steel, and are therefore much lighter, more durable, and far superior in every way to the old-fashioned iron fork.

The best fork for the person on the top of the stack to use, in assisting the builder, is the short stable-fork.

Cart-ropes.—The loads of corn and hay on the carts are fastened with ropes, which should be made of the best hemp, soft and pliable. Ropes are either single or double, and both are required on the farm. Double cart-ropes are from 20 to 24 yards long, and single ones half those lengths.

Fig. 481 shows the rope coiled and suspended when not in use.

Care of Ropes.—Cart-ropes last according to the care bestowed on them. When used with the corn-cart, they should never be allowed to touch the ground, as earthy matter, of whatever kind, soon causes them to rot. When wetted by rain they should be hung out in the air to dry. On being loosened when the load of corn is to be delivered to the stacker, they should be coiled up before the load is disposed of, and not allowed to lie on the ground till the cart is unloaded.

A soft rope holds more firmly, is more easily handled, and far less apt to crack, than a hard one.

Forking.—The carts, forks, straw, and

ropes being in readiness at the steading, and the corn fit for carrying to the stack-yard, the first thing is to provide an efficient person to fork the corn in the field to the carts. That man is the best for this work who is able to wield the sheaves from each stook with ease, and has dexterity to place them in positions most convenient for the carter to build them on the cart. Throwing the sheaves in an indiscriminate manner, or too quickly upon the cart, makes the work less easy for the carter; for he has the trouble of turning the sheaves to arrange

Fig. 481.—*Coiled-up cart-rope.*

them aright, while his footing upon the load is insecure. A delay of two or three minutes thus occasioned in loading each cart makes a considerable loss of time upon the day's work.

Injury to Young Grass.—In carrying the crop off the ground, care should be taken to do as little injury as possible to the land with the cart-wheels, particularly to young grass. If the track is frequently changed little damage will be done.

Order in Forking Sheaves.—In forking a hooded stook from the ground, the hood-sheaves are first taken, then the sheaves from the body of the stook, from one end, sheaf by sheaf, in pairs, to the other end.

When stooks have stood long upon the ground, they may require considerable force to remove them.

Carting "Gaitins."—On removing gaitins from the field, they must first be

bound into sheaves, which is done by loosening the slack band from its tying and slipping it down the body of the gaitin to the proper place, and binding it in the manner of a sheaf when reaped. They are not stooked when bound, nor left scattered on the ridges as they stood before, but laid in heaps on alternate ridges with the corn-ends away from the cart, as near the furrow-brow as most convenient for the forker and the carter. A number of hands are required to bind gaitins as fast as they are carted off.

Loading a Cart.—A corn-cart is loaded with sheaves in this way: The body is first filled with sheaves, their butt-ends to the shaft-horse, and to the back-end of the cart. When these come to the level of the frame, other sheaves are placed across them in a row along both sides and both ends of the frame, with the butt-ends projecting as far beyond the outer rail as the band. Another row of sheaves is placed upon these. Sheaves are then placed along the middle of the cart with the butt-ends like those in the body upon the corn-ends of the side sheaves to fill up the hollow of the load.

Thus row after row of sheaves is placed, and the hollow in the middle filled well up at last, 12 full stooks making a good load upon an ordinary cart.

Before finishing, it should be seen that the load is neither too light nor too heavy upon the horse's back.

A load thus built will have all the butt-ends of the sheaves on the outside, and the corn-ends in the inside.

Roping Loaded Carts.—The ropes keep the load from jolting off the cart upon the field and the road. They are thrown across the load diagonally from the hind part of the cart to the opposite shaft at the front, and one end is made fast to each shaft, the forker holding on the slack, while the carter on the load tightens the rope by pulling from behind, and tramping on the sheaves to make them firm. The crossing of the ropes at the centre prevents the load splitting asunder over each side of the cart.

Hours of Carting.—Carrying often is continued from break of day to twilight. From a little after sunrise to a little after sunset, corn may be taken in with safety. Morning and evening dew may occasionally interrupt the carrying.

Commencing Stack - building. — While the first cart has gone to the field, the builder of the stacks, or stacker, collects his forks, ladders, and trimmer; and his assistant, who pitches the sheaves conveniently for him on the stack, fetches a few straw-ropes and a hand-rake into the stackyard.

The first stacks are built on the stathels, which are arranged along the fence of the stackyard, and which require no preparation for the reception of the stacks.

When more than one stacker is required, each should have one head of carts leading to him; and the number of carts in one head depends on the distance the corn has to be brought. There cannot be fewer than two carts to one head, to come and go. The same forker and carts should serve the same stacker, because the same workers together understand each other better in their work.

Arranging a Stackyard.—In filling a stackyard, the barley being the first crop threshed—being the first in demand in the market—their stacks should be placed nearest the barn; and wheat being last threshed, their stacks are placed upon the stathels. Oats being required at all seasons, their stacks may be placed anywhere.

Stacks of peas and beans fill up the heart of the stackyard when there is room, or are placed on the outside.

Foundation for Stacks.—When stacks are built upon the ground, stools of loose straw or other material are made, to prevent the sheaves at the bottom receiving injury from the dampness of the ground. A stool for a stack is made in this manner: Stick a fork in the ground, on the spot where the centre of the stack is desired to stand. Put a quantity of dry straw round the fork, shake it up with a fork and spread it equally thick over the area the stack shall occupy. Then take a long fork, with the radius of the stack notched upon its shaft, 7½ feet; embrace the shaft of the upright fork between its prongs, and in walking round push in or pull out the straw with a foot, so as to form a circle having a

diameter twice the radius notched upon the shaft of the fork (fig. 482).

Process of Stack - building. — In setting a loaded cart to a stack, the carter should take advantage of the wind in forking the sheaves from the cart. The stack should be built in this way: Set up a couple of sheaves leaning on each other in the centre of the stathel, and another couple against their sides. Place other sheaves against these in rows round the centre, with a slope towards the circumference of the stathel, each row being placed half the length of the sheaf beyond the inner one, till the circumference is completed, when it should be examined; and where any sheaf presses too hard upon another, it should be relieved, and where a slackness is

Fig. 482.—*Making a stool for a corn-stack.*

a Fork stuck into the ground. *b* Fork 7½ feet long.
d Man making the circle of stool with his feet.
e c d Circle of stool 15 feet in diameter.

found, a sheaf should be introduced. Keeping the circumference of the stack on the left hand, the stacker lays the sheaves upon the outside row round the stack, placing each sheaf with his hands upon the hollow or intermediate space between two of the sheaves laid in the preceding row, close to the last one, and pressing it with both his knees (fig. 483)

When the outside row is thus laid, an inside row is made with sheaves whose butt-ends rest on the bands of the outside row, thereby securing the outside sheaves in their places, and at the same time filling up the body of the stack firmly with sheaves. A few more sheaves may be required as an inmost row, to raise the heart of the stack at its highest part.

It is of immense benefit to a stack to have its centre hardened with sheaves.

It is the heart sheaves which retain the outside ones in their places in the circle, with an inclination from the centre to the circumference; and it is this incline of the outside sheaves that prevents the

Fig. 483.—*Building a stack of corn.*

e Loaded cart of corn alongside a stack.	*h* Stacker kneeling on the outside row of sheaves.
f g Sheaves of corn with their butt-ends outwards.	*i* Sheaves of the inside row.
m Carter forking up a sheaf.	*l* Sheaf placed most conveniently by the field-worker for the stacker.
k Field-worker receiving the sheaf with a fork.	

rain passing along the straw into the heart of the stack, where it would soon spoil the corn.

Size of Stack.—The number of rows of sheaves required to fill the body of a stack depends on the length of the straw and the diameter of the stack. For crops of ordinary length of straw, such as from 4¼ to 5 feet, a stack of 15 feet diameter is well adapted. In such a stack one inside row, along the bands of the outside one, with a few sheaves crossing one another in the centre, form sufficient hearting. Where wheat grows long, from 5 to 6 feet, the stack should be 18 feet in diameter, to give room to a few sheaves for the hearting.

Second Forker.—The stacker should receive the sheaves within easy reach, as he cannot reach far on his knees to take them without loss of time, and risk

of making bad work. To facilitate the building, a second forker may be employed to receive the sheaves on a short fork from the carter, and to throw them to the stacker in the position he wants them, to save him the trouble of turning them.

For regularity of work, the carter should pitch the sheaves just as fast as the builder can place them, and no faster, having only one sheaf in reserve on the stack in advance of the builder: any more can be of no service to him, and may be a hindrance.

It is necessary for the second forker to use the fork equally with the right hand and the left, so as to avoid having to swing the sheaves across the body for half the round of the building of the stack.

By another plan which many prefer, the second forker becomes unnecessary. When a stack gets near completion, and the cart at the stack nearly empty, another stack is begun until a cart arrives with a full load, when from the top of it one forker is easily able to send up sheaves for the completion of the former stack.

Trimming Stacks.—As each cart is unloaded, the stacker descends to the ground by means of a ladder, and trims the stack by pushing in with a fork the end of any sheaf that projects farther than the rest, and by pulling out any that may have been placed too far in. As the stack rises above the stacker he cannot trim it with a fork. He uses a *trimmer*, consisting of quarter-inch thick flat board, about 20 inches in length and 10 inches broad, nailed firmly to a long shaft, fig. 484, with which he beats in the projecting ends of sheaves, giving the body of the stack a uniform roundness. An improved trimmer has its edges formed into thick strong teeth.

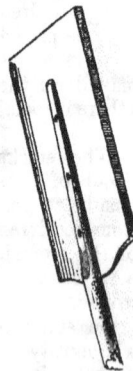

Fig. 484.—*Stack-trimmer.*

Form of Stack.—Many stackers make the stack swell out as it proceeds in height, but this is not necessary for throwing off the drops of rain from the eave, as the eave itself, on the stack subsiding after being built a few days, or the thatching, projects sufficiently to throw off the drops. The body of the stack should be carried up perpendicularly.

Height of Stack.—As a stack of 15 feet in diameter should ultimately stand 12 feet high in the body to maintain a due proportion, an allowance of about one foot for subsidence, before making the top, is generally given. The height is measured with the ladder, and allowing two feet for the height of the stathel, a 15-feet ladder will just give the desired height of the body before the top is built up. Fig. 491 is a stack built upon a stathel.

Eave.—The eave of a stack is formed according to the mode in which it is to be thatched. If the ropes be placed lozenge-shaped, the eave-row of sheaves is placed just within the topmost row of the body. If the ropes are to run from the crown of the stack to the eave, the eave sheaves project two or three inches beyond the topmost row of sheaves.

Topping Stacks.—In building the top of a stack, every successive row of sheaves is taken as much in as to give the slope the same angle as a common roof—one foot below the square. The bevelled bottom of the sheaves, acquired by standing in the stook, answers the slope of the top pretty nearly. The hearting of the top of a stack should be particularly attended to, as on rain obtaining admission at the top it cannot be prevented descending to the heart. After the area of the top has contracted to a space on which 4 sheaves only can stand upright, they are placed with their butt-ends spread a little out, and the tops pressed together, so as to complete the apex of a cone. The top sheaves are held in their position against the wind by means of a straw-rope wound round them and fastened to the stack.

Process of Thatching.

Seldom is leisure found to thatch stacks as long as there is corn to carry in. The finer the weather the less the leisure. A damp day, however, which prevents carrying, answers well for thatching, as thatch-straw is none the worse of being a little damp; but in heavy rain it is improper to thatch and cover up the wet ends of sheaves. The materials for thatching should all be at hand before commencing —drawn bunches of straw, coils of straw-ropes or coir-yarn, ladders, forks, hand-rakes, and graips. To get on with the business quickly, one man and two assistants are required for each stack — the most thrifty assistants being field-workers, to supply the thatcher with straw and ropes, and tie the ends of the ropes.

Stack-ropes.—For tying down thatch or holding firm the tops of stacks, straw-ropes, once universally used, are now being supplanted by coir-ropes or yarn. This latter material is cheap, durable, and convenient to use. If well cared for, it should last three or more years ; and many farmers contend that, especially on large farms, or where straw and labour are both scarce and dear, the coir-rope is cheaper than the straw-rope.

Straw-rope making.—Nevertheless, straw-ropes are still largely employed, and where they can be made without any

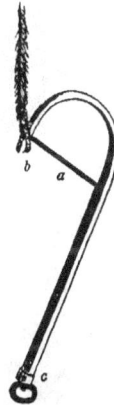

Fig. 485.—*Old throw-crook.*

a Iron stay.
b Protection for end of rope.
c Ferule and swivel ring.
c b Line of direction of rope.

Fig. 486.—*Best throw-crook.*

a Hook, and
a e d Curved spindle of iron.
b Perforated cylindrical handle of wood.
c Swivel-ring.
c b d Curved part of iron spindle.

appreciable addition to the labour bill, they will likely continue to be used. It will thus be useful to repeat here the information in former editions of this work as to the making of straw-ropes.

Straw-ropes are made by means of the

implement named the *throw-crook*. Various forms of this instrument are in use, and one of the simplest is fig. 485.

A better form of throw-crook is fig. 486, where the strain of the straw-rope is in a straight line from the hook, along the spindle to the handle. The left hand holds a swivel-ring, and the right hand causes the curved part to revolve by means of a perforated cylindrical handle of wood, the rest of the instrument being made of iron.

An improved form of spinner consists of a simple contrivance by which one person is enabled to spin two or three ropes at one time. The contrivance hangs from the shoulders of the spinner, who, by turning one handle, gives motion to two or three spindles, to each of which a rope is attached, the spinner moving backwards as the ropes increase in length.

The once common method of twisting straw-ropes by a throw-crook is shown in fig. 487. The left hand of the twister, a field-worker, holds by the swivel-ring, fig. 486.

Straw for Ropes.—The best sort of straw for making into ropes is that of the common or Angus oat, which, being

Fig. 487.—*Making a straw-rope with a throw-crook.*

soft and pliable, makes a firm, smooth, small, tough rope.

The ordinary length of a straw-rope for a large stack is about 30 feet. Counting every interruption, a straw-rope of this length may take five minutes in the making—that is, 120 ropes in ten hours.

Winding Straw-ropes.—After the rope has been let out to the desired length, the man winds it firmly in oblique strands on his left hand and arm into an oval ball, the twister advancing towards him as fast as he coils the rope, which is finished and made firm by passing the end of it below one of the strands.

Fig. 488 represents a straw-rope *coiled* up in this form. With the ends smaller than the middle, the rope can be easily taken hold of and carried; and in the oval form instead of the spherical the

Fig. 488.—*Coil of straw-rope.*

coil can be more easily thrown upwards to the top of a stack. Still many prefer large circular coils, except for use in forming a network of ropes over a stack, for which small oval bundles are most convenient. Straw-ropes should be spun of such lengths as are suitable to the size of the tops of the stacks.

Thatching.—The material to be used in thatching the stacks must also be in a state of readiness before the crop is brought into the stackyard.

Material for Thatch.—The material most largely used in thatching stacks is straw, which has been previously drawn parallel by the hand and tied into sheaves. The roughest and rankest straw is generally used for thatch. In some cases rushes or other coarse herbage is employed instead of straw.

Drawn Straw.—A common method of drawing straw for thatch is as follows: In commencing to draw straw in the straw-barn, the man takes a wisp from the mow, and, placing it across his body, takes hold of each end of the wisp, and spreading out his arms, separates the wisp into two portions. Holding the ends of both portions in one hand, he takes hold of the other ends with the other hand, and spreading out his arms, draws the straws parallel and straight;

and he does this until he finds the straws parallel and straight, when he lays down the drawn wisp carefully upon the floor of the barn.

The state of the straw, and the kind, render the drawing more or less easy and

Fig. 489.—*Bunch of drawn straw.*

expeditious. When straw is much broken in thrashing, it requires the more drawing to make it straight; and of all the kinds wheat-straw, being long and strong, is most easily and quickly drawn, barley-straw being shortest and most difficult to draw. Oat-straw is the most pleasant of any to draw.

After as much has been drawn and laid down as to make a bunch of about 15 inches in diameter, the man makes a *thumb-rope* by twisting a little of un-drawn straw round the thumb of his right hand, drawing it out with his left and twisting it with his right alternately, until a short coil is made, one end of which he places on the floor by the side of the drawn straw, and puts his foot upon it; and, keeping hold of the other end in his left hand, puts the drawn straw into the rope with his right; and then, holding both ends of the rope, binds the straw into a bunch as firmly, and in the same manner, as a bandster does a sheaf of corn. A bunch of drawn straw is represented in fig. 489.

Thatch-making Machine.— The genius of the inventor has now come to the aid of the farmer in the making of thatch, as in most other of his operations. A thatch-making machine is represented in fig. 490 (Barnard & Lake). The drawn straw is fed into the machine by hand, and the form of the thatch when completed is well shown in the figure. It is found that this machine economises straw, and saves time in thatching.

This machine is also employed in making straw matting to protect pits or clamps of trees, as well as race-courses from frost. It likewise produces excellent material for providing shelter in sheepfolds.

Method of Thatching.—The thatching of a stack with drawn straw is done in this manner: On the thatcher ascending to the top of the stack by means of a ladder, which is immediately taken away by an assistant, one bundle or two of drawn straw are forked up to him by the other assistant, and kept beside him behind a graip stuck into the top of the stack. The straw is first laid and spread upon the eave, beyond which it projects a few inches, and then handful after handful is laid in an overlapping manner to the top. Where a butt-end of a sheaf projects, it should be beaten in; and where a hollow occurs, a sheaf should be drawn out a little, or the hollow should be filled up with additional straw. In this manner the straw is evenly laid all round the top of the stack, to the spot where the thatcher began.

Forming the Apex.—After putting

Fig. 490.—*Thatch-making machine.*

the covering on the top of the stack, fig. 491, he makes up the apex with a small bundle of well-drawn long straw, tied firmly near one end with a piece of cord, and the tied end is cut square with a knife; the loose end being spread upon the covering, and giving the finish to the thatching. To secure the apex in its place, a straw-rope is thrown down by the thatcher, the end of which his assistant on the ground fastens to the side of the stack. After passing the rope round the apex, he throws it down in the same direction, where it is also fastened to the

Fig. 491.—*Lozenge mode of roping the covering of a corn-stack.*

a Apex or ornamental top.
c f First rope for securing the apex in its position.
f g Second rope for farther securing the apex in its position.
h Last rope on that side of the stack.
i Last rope on the opposite side of the stack.
l Eave of thatch.
l k Eave-rope.
c d Diameter of the stack, 15 feet.

stack. In like manner he throws down a rope round the opposite side of the apex, and their ends are also fastened by the assistant.

Roping Stacks.—Having thus secured the ornamental *top*, the thatcher comes down the thatching, closing up the covering in the descent of his track, and descending by the ladder placed to let him down. Taking a longer ladder, he inclines its upper part nearly parallel to the covering of the stack, and secures its lower end from slipping by a graip thrust against it into the ground. He then stands upon the ladder at a requisite height above the eave, where he receives a number of coils of ropes from his assistant, which he keeps before him between the steps of the ladder.

The thatch-straw is made smooth by being stroked down with a supple rod of willow, before the ropes are successively put on. Holding on by the loosened end of a coil of rope, he throws the coil from where he stands on the ladder down to the right hand to his assistant, who, holding it in the hand, allows the thatcher to coil it up again upon his hand without ruffling the covering of the stack, till as much of it is left as to allow the assistant to fasten it to the side of the stack, while the thatcher adjusts its position parallel to the rope he had placed round the apex. The thatcher then throws the other end of the coil to the right hand to his assistant, who takes hold of its end, while he retains the rope in his hands, and places its double parallel with the rope round the apex, and the assistant pulls it tightly down, and makes it fast to the stack, or perhaps to a brick or other weight.

He then takes the ladder to the opposite side of the stack, and puts on each rope on that side, as he had done on the other side.

Lozenge Roping.—Ropes thus placed parallel from opposite sides of the stack, crossing each other, make the lozenge-shape in fig. 491.

Number of Ropes.—On a stack 15 feet in diameter at the base, 16 feet diameter at the eave, 12 feet high in the body, and 6½ feet high in the top, 10 ropes on each side will secure the thatch.

Tying Ropes.—The ends of the ropes are fastened to the stack by pulling out a little straw from a sheaf, twisting the rope and straw together, and pushing through the twisted end between the rope and the stack.

Windy gusty weather is unfavourable for the thatching of stacks.

Another Method of Roping.—Another method of roping the thatching of a stack is fig. 492. The thatching of straw is put on in the same manner as described above. The ropes cross over the crown of the stack, and subdivide the top into equal triangles, their ends being fastened to the side of the stack. The ropes, at their crossing over the crown, are fastened together by a straw-rope, which is tied above them with cord, and cut off in the form of a rosette. The cross-ropes are either put on spirally round the top till they terminate at the eave, or in separate bands, parallel to the eave. In either case the cross-ropes are twisted round each crown-rope, at equal intervals, from the top to the eave.

This mode of roping requires more ropes than the last, but it secures the

Fig. 492.—*Net-mesh mode of roping the covering of a corn-stack.*

a Top or rosette.
a b, a c, a d Form of triangles on the thatch.
a to d is the spiral rope round the top, from the top to the eave.
g h are ropes round the top parallel to the eave.

thatch against any force of wind, and is therefore well adapted for exposed situations.

Round Tops.—A third mode of roping the covering of a stack is applicable to where the eave is formed of sheaves projecting beyond the body. It is shown in fig. 493, and is common on the Borders. The first thing is to put a strong eave-rope round the stack, below the projecting row of sheaves. The covering straw is then put on in a similar manner to that described, but rather thicker, and it projects farther down than the line of the eave - rope. The tops of the finishing sheaves of the stack are then pressed down, and a hard bundle of short straw is placed upon them, to serve as a cushion for the ropes to rest upon. Upon this the thatcher perches himself, where he receives the ropes as thrown up to him on the prongs of a long fork.

An English Custom.—A mode of thatching stacks, common in England, is the insertion of handfuls of well-drawn wheat-straw into the butts of the sheaves, which are kept down with stobs of willows, or sewed on with tarred twine, being an imitation of thatching cottages. No straw-ropes are used, and, finished by an experienced thatcher, it gives the stacks a neat and permanent appearance. This method would not resist much wind, but its smooth surface would detain the snow a much less time than any of the ropings described above.

Thatching with machine-made thatch is much more expeditious and more simple than any of the methods described above.

Finishing Stacks.—Seldom is the thatching of a stack finished when the straw and ropes are put on,—the object of these being to place, in the shortest time, stacks beyond danger of rain. Besides, stacks subside in bulk after covering. Stacks to be early threshed, as barley, seldom receive finishing; and many farmers only finish the outside row of stacks. It is slovenly management to leave stacks unfinished in the thatching, as wind readily strips them of their thatching.

The finishing of the thatching in fig. 491 is done in this manner: A rope is spun long and strong enough to go round the stack as an eave-rope. Wherever two ropes from opposite directions cross the eave-rope, they are passed round it, and, on being cut short with a knife, are fastened to the stack. After all the ends of the crossed ropes are thus fastened to the stack, the projecting part of the

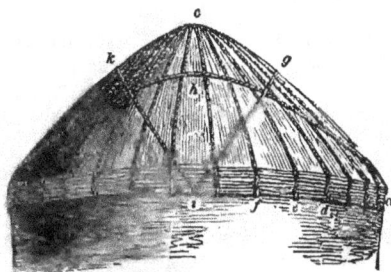

Fig. 493.—*Border mode of roping the covering of a corn-stack.*

c Crown of stack, upon which the thatcher stands.
c b to c a Ropes passing over the crown of the stack.
f e d Ends of ropes fastened to the eave-rope.
a b Band of rope-ends round the stack.
h, i k, i g Strap-ropes quartering the top of the stack.

thatch at the eave is cut of equal length with a knife round the stack. Of all the modes of thatching, there is none more efficient or better-looking than the lozenge-shaped.

The finishing of thatching, fig. 492, is as follows: An eave-rope is first put round the stack. The crown-ropes are passed at each end round the eave-rope, and fastened to the stack. The projecting straw at the eave is cut with a knife at equal length.

The difficult part of roping, fig. 493, is in finishing the eave, which, if well done, looks neat; if not, slovenly. The eave is finished in this way: The eave-rope having been put up at the thatching, the ends of the ropes are loosened from the stack, and passed between the eave-rope and the stack, and each successive end is so passed and carried horizontally along its length; and thus every rope all the way round the stack at both ends is treated; and in carrying the ends of the ropes round the eave, the band of ropes should be of the same breadth round the stack. From four to eight ropes, according to the exposure of the situation, are strapped across the crown-ropes, quartering the top of the stack, and fastened to the eave-rope.

Cutting Thatch.

Where rough grass grows on a farm, as on a bog which is partially dry in summer, it should be mown and sheafed, for thatching stacks. One or two days given to mowing such grass, after the harvest is over, are well spent, even at the rate of wages and food of ordinary harvest-work. Such vegetable materials save the drawing of clean straw when it is scarce, and form good covering for stacks soon to be threshed; and when it has served the purpose of thatch, it is suitable for littering courts. Bog-reeds (*Arundo phragmites*) might be used in the same way where they do not find a profitable market as thatch for cottages. Such materials add many tons to the manure-heap.

Stack-heating.

Barley.—Of all kinds of corn, barley is most liable to heat in the stack, partly owing to the soft and moist character of the straw, and partly because clover is usually mixed with it. On this account it is advisable, in most seasons, to make barley-stacks smaller than the others, both in diameter and height, and to build them upon bosses. Much care should be bestowed on building barley-stacks to heart them properly, which is the best expedient to prevent heating.

Injury from Heating.—The least heat spoils barley for malting, and it should be remembered that malting barley always fetches the highest price in the market. Besides injuring the grain, heating compresses barley-straw very firmly, and soon rots it.

Remedy.—When a stack is seen to heat, it should be instantly carried into the barn and threshed, to cool both grain and straw. If this should be inconvenient, the stack might be " turned "—that is, forked down and rebuilt, the hotter sheaves being kept to the exterior.

Symptoms of Heating.—When a stack begins to lean to one side about twenty-four hours after being built, or shows a depression in the top a little above the eave, you may suspect heating to have proceeded to a serious degree. Incipient symptoms of heating are moisture on the top of a stack early in the morning—indicated by cobwebs—before the sun evaporates it, as also when heated air is felt, or steam is seen to rise.

Heated barley lubricates the threshing-machine with a gummy matter.

Oats.—Oats are less apt to heat than barley, though their heating is stronger. If sap remain in the joints of the straw, oats will be sure to heat in the stack. Heating gives to oat-straw and grain a reddish tinge, may render the straw unfit for fodder, and give the corn a bitter taste.

Wheat.—Wheat seldom heats, but when it does the heat is most violent. Heated wheat is bitter to the taste.

Partial Heating.—Partial heating is induced in a compressed part of a stack caused by bad building, and it is indicated by the stack leaning over.

Propping Stacks.—To prevent a stack from leaning to one side, props of weedings of plantations are loosely set around it to guide subsidence, especially if it has been rapidly built; but in using props caution is required not to push one harder in than others. Stacks often sway whenever their top is finished, when props should be set to keep them upright. To push a prop firmly into a stack much swayed requires the strength of two men —one to push up backwards between the stack and the prop, with both hands clasped upon the outside of the prop, the other to push forward with the shoulder planted against the prop below the other man's hands.

Stack Ventilators.

Various contrivances have been introduced as safeguards against heating in stacks. These are most generally wooden structures, in Scotland named *bosses*, which signify hollows. The mode of using them is to occupy the space which would have been filled with the heads of the sheaves of corn, with a void into which the air shall find access. When stacks are built on bosses erected on stathels, the air finds access through the stathel; but when built upon the ground, a trestle of woodwork may be connected with the boss, by which the air is led into the interior of the stack. When such a trestle is placed at both sides of a boss, a ventilation is maintained through the body of the stack.

Common Boss.—The most common form of boss is a three-sided pyramid, formed of three small trees, of larch or Scots fir, tied together at the small ends,

Fig. 494.—*Pyramidal boss and trestle.*
a Tying of 3 sticks together.   b b Fillets of wood.
c Trestle.

and the thick ends placed at equal distances upon the stathel or ground. A common boss is shown in fig. 494, where three trees are fixed at the top, standing about 8 feet in height and 3 feet asunder. They are fixed together by rows of fillets of wood nailed on, stiffening the

pyramid and preventing the sheaves passing into the boss. A trestle, about 2 feet high, is placed on one side to conduct the air into the boss.

An objection to this form of boss is, that as the stack subsides, the sharp apex penetrates through the sheaves and disfigures the upper part of the stack.

Prismatic Boss.—A form of boss which many prefer is shown in fig. 495. It consists of three stems of small trees, 7 feet long, held together in the form of a prism, 3 feet in width, by fillets of wood nailed to them. The prism is set on end, and upon a stathel, which is nailed to it; but as a further means of stability, a spur from each tree might be nailed to the stathel within the prism.

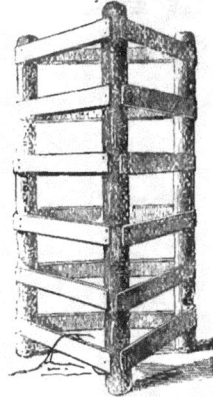

Fig. 495.—*Prismatic boss.*

Upon the ground it requires a trestle for the conveyance of air.

The advantage of this boss over the other is, that it supports the top of the stack evenly when it subsides, thereby relieving the body of the stack of the weight of its top.

Other Methods.—Other means are employed to form a hollow in the heart of a stack, such as by setting the upright sheaves which form the foundation of the stack around a long cylindrical bundle of straw, firmly wound with straw-rope, or sack filled with hay, and as the stack rises in height, the bundle or sack is drawn up through its centre to the top, where it is removed, leaving a hole through the height of the stack. This hole creates a current of air through the stack, allowing the heated air to escape, while the cool air enters from below by means of a trestle or stathel.

Measuring Heat in Stacks. — The degree of heating in a stack may be found by the stack-thermometer. (Fig. 5, vol. i., p. 26.)

Improved Ventilator.—An excellent

stack ventilator (Taylor's) is represented in fig. 496. It is an effectual ventilator, alike for stacks of grain and hay.

Field Stacks.—It is a common practice with some farmers to build a portion of the crop in the field. This is not commendable in good weather, as, besides the trouble of carrying thatch to the field, much waste is experienced in carrying corn to the steading in winter, when the stacks are wanted, perhaps in bad weather or through deep snow. The stacks there are beyond protection, and subject to depredation. A scheme may be justifiable under peculiar circumstances which would not be in ordinary practice, and the building of stacks in the field is one of them.

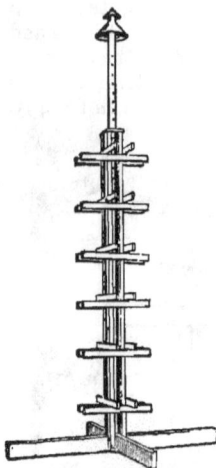

Fig. 496.—*Taylor's stack ventilator.*

ARTIFICIAL DRYING.

Many attempts have been made to introduce some practical method of drying the cereal crops by artificial means. So far, however, little success has been attained.

Hot - air Drying. — The system of drying by a hot-air blast, referred to in the section on Haymaking in this volume, was at one time looked to with considerable hope. Unfortunately, however, it has proved to be impracticable.

Nelson System.—This method, also described in the section on Haymaking, has been tried with fairly good results by some farmers; but it has nowhere come into general or extensive use.

Drying Racks.—The most successful and useful efforts have been those directed to the devising of what are known as "drying racks." These racks consist of contrivances by means of

which the sheaves of damp corn may be stored in thin layers in such form as to induce the play of currents of air, and expose the sheaves to the drying influence thus created or encouraged.

In late wet districts it has long been the custom, in cases of slow drying, to build imperfectly dried corn on hurdles or other wooden erections in what is called the "sow" form. The sheaves are built, perhaps only one sheaf deep, on both sides, and on the top of a long wooden frame. The hollow centre is left open at both ends, and a current of air is thus kept playing upon the thin layers of sheaves. The sheaves are built with the head towards the hollow centre of the rack, so that the grain is not only protected from wet, but directly exposed to the internal current of air.

By resorting to some such plan as this, corn which would otherwise be seriously damaged is often tolerably well preserved.

THE RICHMOMD GRAIN-DRYING RACK.

An ingenious and efficient rack, invented and brought into use by Mr John Richmond, Dron, Bridge of Earn, Perthshire, is worthy of special notice.

Its construction is described by Mr George W. Constable in the *Transactions of the Highland and Agricultural Society of Scotland* for 1897. The appearance of the rack when empty is shown in fig. 497, and when filled with corn in fig. 498. As to the details of construction, Mr Constable writes:—

The two sides are 5 feet apart, and can be formed of any length, from 100 to 130 yards long being considered amply sufficient for a moderately sized farm.

The usual height is 16 feet, and the whole erection is covered with a roof of corrugated iron. There are four large straining-posts, one at each corner, while intermediate standards (*X X*, fig. 499) are erected 4 yards apart, on which the wires, after being secured to the straining-posts, are supported. The straining-posts, which should not be less than 10 inches in diameter at the small end and 22 feet long, must be sunk into the ground at least 6 feet. A beam of wood 9 feet long and 6 inches in diameter at

Fig. 497.—*Richmond drying-rack - Empty.*

Fig. 498.—*Richmond drying-rack—Full of grain.*

the small end ought to be put across the foot of both posts to act as a heel, and the strainers should be embedded in concrete from the foot of the post to about

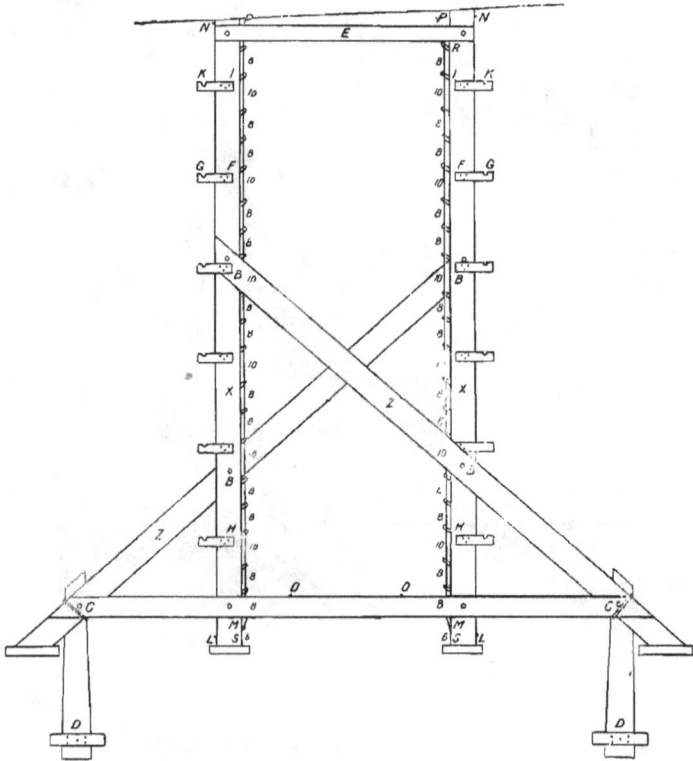

Fig. 499.—*Richmond drying-rack—section.*

2 inches above the level of the ground (fig. 500), and thus secure the post from yielding a hair's-breadth. Strong stays are also required, firmly founded and

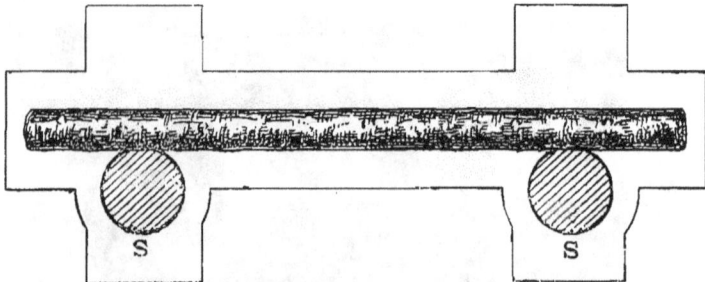

Fig. 500.—*Cross-beam fixing strainers.*

secured, to assist the strain of so many wires.

The uprights (*X X*, fig. 499) and the struts or rances (*Z Z*, fig. 499) should rest on square fireclay bricks or flat stones, raised so as to have the top about 2 inches above the ground.

Staples are used to support the wires

on the intermediate standards. The staples should be driven in the standards at a steep angle, and so driven that the wire will rest on both legs of the staple, and just sufficient left undriven as will support the wires, besides allowing room for a wire to pass down *through* the staples, so as to keep the wires resting on the staples confined between the wood of the standard and this upright wire passing down through the staples. About 1 inch undriven will serve the purpose. Staples 2½ inches long, made of No. 4 galvanised wire, will suit. The distances between the staples on the *inside* of standards should be 8 inches, 8 inches, and 10 inches respectively from top to bottom, or according to the size of sheaves made—the first staple beginning 6 inches from the foot of the inside of the standard (*N*, fig. 499). Each standard should be ranced both ways, and the rances should pass through the 10-inch or wide spaces on both uprights (*B*, fig. 499).

Bolts should be used for fixing the rances and crossbars to the intermediate standards.

A strong larch post should be pitted, not driven, at least 3½ feet into the ground at the foot of each rance, and bolted or strongly nailed thereto, also strongly secured by twisting and stapling wire several times round post and rance (*C*, fig. 499). The thick end of the post should be sunk in the ground, and a crossbar 1 foot long should be notched and nailed to foot of post (*D*, fig. 499). The foot crossbars of standards (*E*, fig. 499) should be about 2½ inches wide by 1½ inch thick. The top crossbar (*F*, fig. 499) should be 3½ inches by 2 inches and 9½ feet long.

The uprights of intermediate standards should be made of battens 16 feet long, 6½ inches by 2½ inches, and the rances should be 6 inches by 2 inches.

A batten 2 inches by 3 inches should be fixed by bolts to top and bottom crossbars, and also to the rances (*G*, fig. 499). This batten should be 2½ inches distant from the standard; a staple should be driven into the batten opposite *each wire-space of 10 inches* and 2 inches below the level of the staple at the top of the 10-inch space on the standard (*H*, fig. 499).

The staples in the batten should be driven in the same way as those in the standard, so as to support the outside wires, and a wire should be passed down through the staples, as in the standards. The outside wires will thus be 1 foot distant from the inside wires.

Wires. — No. 6 wire (galvanised annealed) should be used, and a hand ratchet should be employed for stretching each wire. The wires are stretched between the end straining-posts and rest on the staples on the inside of the standards, also on the staples on the outside of the batten (see *K* for inside wires and *L* for outside wires, fig. 499).

An additional wire should be firmly stapled on the *outside* of the standards 1 inch below the level of the lowest *inside* wire, to keep the posts from moving off the bricks or stones (*M*, fig. 499). The lowest *inside* wire should also be firmly stapled for the same purpose (*N*, fig. 499).

A wire ought also to be stretched between the straining-posts and affixed to the top of each intermediate standard (*O*, fig. 499), so as to keep the standards in an upright position. A couple of wires, or more, should be stretched from beams fixed to the outside of the straining-posts, the said beams being on a level with the foot crossbars of the standards, the wires resting on those foot crossbars (*P*, fig. 499). Nails can be used to keep these wires in their proper place, enough being left undriven for that purpose. These wires form a floor to keep the heads of the lowest sheaves off the ground.

Roof.—The most suitable roof is corrugated iron. If the inside wires are 5 feet apart, the roof should be 10 feet wide. An incline of half an inch to the foot will be sufficient.

The four runners (*R*, fig. 499) supporting the roof should be 5-inch battens, in lengths sufficient to cover two spaces between intermediate standards, and should be notched 2 inches into the top of intermediate standards and top crossbars (*R R*, fig. 499).

The runners, besides being nailed to the top of the standards and crossbars, should be strongly stapled down with wire. The corrugated iron should be bolted or screwed with washers to the runners.

The runners must not be nailed to the ends of the strainers, but should be kept in their places by being passed loosely through an iron socket, so as to allow for the slight yielding of the strainer brought about by the wire pressure. If this is not done the roof may bulge. Top crossbars should be attached to the strainers, the same as those on the intermediate standards,

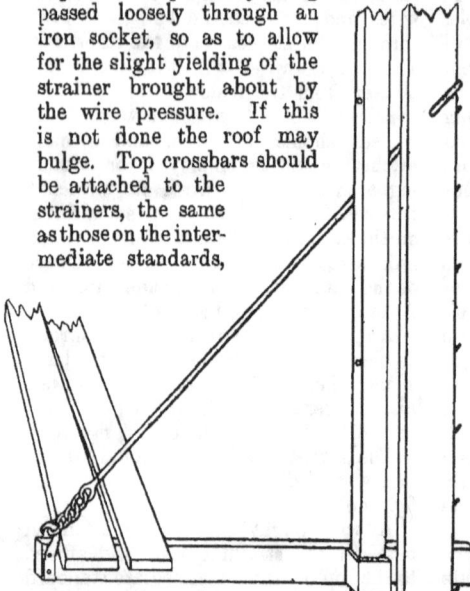

Fig. 501.—*Platform for filling.*

but the runners should pass through an iron socket and not be nailed to the crossbar.

Filling the Rack.

Three buildings or layers of 5 feet each will be found most convenient. One building from the ground and two from platforms are thus required.

The platforms are formed of two planks supported on two brackets, a sketch of which is seen in fig. 501. The hanging rod (fig. 502) of the bracket is hooked on to the standard above any wire at the height necessary, and a few links of a chain at the other end of the rod affix it to the bracket. The bracket (fig. 503) is made of hard wood, 4 inches by 2 inches, with a strong iron clasp.

In filling, raise the second lowest wire off its supporting staple and hang it by an "S" hook to the wire immediately above, so as to allow room for the first wire to be filled with sheaves. The sheaves should slope downwards by being overbalanced *outwards*, the stubble end of the first row of sheaves resting on

straw, branches, or stones laid on the ground, the heads of grain being kept off the ground by the wires (*P P*, fig. 499), on the foot-ties or crossbars.

When the lowest wire is filled, let down the wire above to its original place, and lift the next one above to the wire above it; then lay the sheaves across the one wire, the stubble end of the sheaf resting on the stubble end of the sheaf below. The next row should be done in the same way.

When the outside wires are reached, a handful of the stubble end of every third sheaf (or if the crop is a short one, every second or each sheaf) should be passed underneath the outside wire, and the sheaf drawn outwards till the wire nearly touches the band of the sheaf. The double wires at these regular 10-inch spaces give a downward slope to the sheaf,

Fig. 502.—*Hanging rod for platform.*

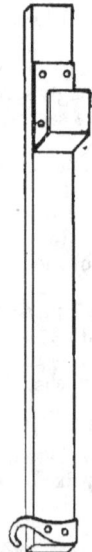

Fig. 503.—*Bracket for platform.*

as in a well-built stack; they also prevent the body of sheaves from slipping outwards, acting like a through-band in a wall, and they further admit ventilation between this row of sheaves and the set below. Air-holes may be left at intervals if required, according to the condition of the grain.

Newly cut grain can be packed more closely than grain which has been soaked with rain after being cut.

When filled, a space exists within the rack from end to end between the heads of the sheaves, allowing air to play on the loose heads of the sheaves, and thus obviating the smallest chance of heating.

To prevent a half side-wind from blowing sheaves outwards, both ends of the rack should be sparred to about half-way down: the spars should be about 1 inch apart. Put wire-netting on the lower half in order to keep out birds.

The end of the rack should if possible face the prevailing winds, and it should not be put in a sheltered situation. The most convenient site for it is alongside a farm-road.

The ground does not require by any means to be level. The only precaution necessary is to see that there is not a hollow between the extremities of the rack, as such would have a tendency to loosen the wires when tension is applied.

Contents.

A rack 100 yards long, 16 feet high, with wires distant from each other 8 inches, 8 inches, and 10 inches respectively from top to bottom, will provide room for about 20,000 sheaves.

About 800 sheaves, 8 inches in diameter, may be taken as an average crop per imperial acre; and thus a rack of the above dimensions will hold the produce of fully 25 acres.

It must, however, be clearly understood that it is quite possible to fill and empty the rack at least twice in one harvest. Thus if a crop, say, of barley, be put in dry immediately after cutting, should the weather be at all favourable, in a short time—a few days at most—it will be ready to thresh, and the rack can be again filled. Of course, if the crop be put in *very* wet, it necessarily will require a longer time to get into condition, but it cannot heat or go wrong.

Advantages of the Rack.

The advantages claimed for the Rich-

mond rack are not confined to the saving of the grain and straw in a bad harvest. When the rack is used the crop can be carted straight from the reaper or binder without being stooked, and each day's cutting might be secured by night. Then the rack saves thatch, ropes, kilns, bosses, and props, and the setting up and moving stooks in bad weather. No skilled hands are required for building; any ordinary farm hand can build. It dispenses with keeping on extra hands at a high wage in a late protracted harvest, and the owner of a rack can have all his hands employed filling the rack when other harvest work is at a standstill. It enables the farmer to clear the ground of grain much sooner, so that stubble-ploughing or other work may be proceeded with. The grain gets into condition for threshing much sooner than in a stack; and invariably the weight per bushel is increased and the quality of the grain is superior.

In a good season all crop round woods, hedgerows, and in sheltered positions can, with the aid of the rack, be at once secured after cutting; and thus what on most farms is the worst conditioned grain—no matter how good the season may be—is by the rack made equal to the best.

Cost.

The cost of erecting a rack of 100 yards long amounts to about £58, with a patent fee of from £2, 2s. to £5, 5s. for each farm.

Convertible into a Shed.

If the season be exceptionally fine, the rack, by removing the wires (a simple process), can be converted into a shed for filling with grain, when in this way it will hold four times as much as with the wires. Thus a rack 100 yards long, turned into a shed, will hold 100 acres of average crop, and no thatch, ropes, bosses, or props are required.

GRAIN-DRYING SHED.

An improvement upon a grain-drying rack alone is a combination of a rack and a shed. A structure of this kind, erected by Mr M. G. Thorburn of Glenormiston, in the county of Peebles, is described in the *Journal of the Board of Agriculture and Fisheries* for December 1907, full information being given for the building of the shed and rack. By the kind permission of the Board of Agriculture and Fisheries a representation of this combined shed and rack is given here in fig. 504.

Fig. 504.—*Combined grain-drying rack and shed.*

OTHER RACKS.

Several other forms of drying racks are in use over the country, most of them quite simple in construction. A simple and serviceable rack was invented in Perthshire in the wet harvest of 1907 by two neighbours, the Rev. Mr John M'Ainsh and Mr C. Robertson. Grain racks need not be elaborate or costly, and any handy man can erect one for himself without the aid of a trained joiner.

HARVESTING BEANS, PEAS, AND TARES.

The leguminous crops, having stiff or trailing stems, are more difficult to reap than the cereals.

Beans.

If sown tolerably early, beans will likely in an average season be ready for reaping towards the end of September. They should not be so ripe as that the pods will open and allow the beans to escape. Examine the crop carefully as it approaches maturity, and cut it whenever the eye of the bean is black and the skin has acquired a yellowish colour and leather-like appearance.

The reaping is generally done by the common sickle, and the produce of four drills is laid in handfuls in one row.

In some cases the beans are left lying in this form for a few days, and then turned, and shortly afterwards lifted and bound into sheaves with short straw-ropes, previously provided, and the sheaves are then stooked. In other cases they are bound and stooked as reaped.

In many instances the reaping-machine is now employed in cutting beans.

When peas are sown with beans, the haulm of the peas makes excellent bands. In some cases beans are bound by their own straw.

A bean-stook, which consists of 4 or more sheaves, is never hooded. Bean-sheaves should always be kept on end, as they then resist most rain. If allowed to remain on their side after being blown over by the wind, little rain soaks them, and a succeeding drought causes the pods to burst and spill the beans upon the ground.

Sometimes to admit of preparations for a succeeding wheat crop, the beans as soon as cut are removed from the field and stooked elsewhere.

Peas.

Whenever the straw and pods of peas become brown they are fit for reaping. In seasons when the straw grows luxuriantly, it is cut down whilst retaining much of its greenness. On account of their trailing stems, peas are difficult to cut, and the best work is perhaps made by the hook. The reaping-machine and scythe are, however, both used for the purpose.

Peas as a rule are not bound at first, but laid in loose bundles on the ground, where, after drying for some time, according to the state of the weather, the bundles are rolled into an oblong form and made firm by a wisp of their own straw. The bundles may be set together in pairs to form a sort of stook, or left singly over the surface of the field. Many consider that it is a better plan not to tie the bundles at all, but to turn them over once a-day until sufficiently dry, and then carry direct to the stack in the loose form. A man or boy, with a pitchfork, will turn the pea-bundles at the rate of an acre an hour. The turning over of the bundles should be done in the cool of the morning, and with as little shaking as possible, not to open the pods.

Tares or Vetches.

Tares are most easily and quickly reaped by the reaping-machine or scythe. To win they are placed in bundles and treated in the same way as the pea crop. Tares are sometimes seen spread over strong hedges as they are cut, and there they dry quickly.

In dry and warm seasons, peas and tares may be harvested as early as the cereal grains; but beans are always long in winning, and sometimes are not harvested until three weeks after the other crops.

Peas.—Pea-straw is very apt to compress in the stack, and to heat, and should therefore be built with bosses, either in round stacks or oblong ones like a haystack. When peas become very dry in the field, the pods are apt to open and spill the corn in sunny weather — to avoid which, they should be carried and built on bosses.

Beans.—Beans are a long time of winning in the field in calm weather. If the land they grow on is desired for wheat, they should be carried to a lea-field and stooked till ready to be stacked. Being hard and open in the straw, they keep well in small stacks, though not quite dry; and there is risk of keeping them in the field in dry weather after much rain, when the pods are apt to burst and spill the corn. In building both pea and bean stacks, the sheaves are laid with their corn-end inwards, and tramped with the feet. The stacks receive but little trimming, the peas none at all, the beans with the back of a shovel.

Thatching Peas and Beans.—The thatching of pea and bean stacks is done in the manner described for grain, but less pains are bestowed in finishing it. As, however, a good deal of corn is exposed on the outside of pea and bean stacks, their bodies are also thatched with straw, kept on with straw-ropes.

The buckwheat is a plant remarkably affected by the weather. It requires dry weather on being sown, and it springs in the greatest drought. But after putting forth its third leaf, it must have rain for the development of its flowers. During the long time it continues in flower, alternate rain and sunshine are requisite to set the flower. The flower drops off in thunderstorms, and it withers in cold east winds. After flowering, dry weather brings the seed to maturity.

Buckwheat ripens very unequally, for the plant is continually flowering and setting, and therefore the crop is cut at the time the greatest quantity of grain is ripe.

In the south of England a period of

hot and dry weather is necessary in autumn to harvest buckwheat.

Cutting.—Buckwheat may be reaped with the sickle, scythe, or machine, or it may be pulled up by the roots, which last method is recommended by some as less likely to shed the seed when fully ripe. In dry weather it should be reaped early in the morning, or late in the evening, when the dew is upon it, and should not be moved too much in the day.

Drying.—Buckwheat may be tied up in sheaves, or made into bundles like peas; but, in either way, it should be protected from birds, which are very fond of the seed. Owing to the thick knotty stems of the straw, the green state in which it is cut, and the late period it comes to harvest, a succession of fourteen or fifteen fine days is requisite to dry it sufficiently for stacking. It should be turned and moved several times in preparing it for the stack, and these acts should be done gently and in the dew, the least to disturb the seed; but the plant does not easily spoil when lying on the ground. To be early carried, it should be built in small stacks with bosses.

CORN AT THE STEADING.

In the first volume of this edition (pp. 175-184) some information is given on the providing of suitable means for the stacking and threshing of corn, the storing of straw, and the dressing and storing of grain. Here these matters will be dealt with in fuller detail.

THRESHING-MACHINES.

An Ancient Threshing-machine.— The following quaint "Advertisement anent the threshing-machine" appeared in the *Caledonian Mercury* of August 26, 1735: "Whereas many have wrote from the country to their friends in town about the price of the threshing-machines, the following prices are here inserted, for which the machines will be furnished (with the privilege of using them during the patent) by Andrew Good, wright in Edinburgh, whose house and shop are in the College Wynd—viz., to those who have water-mills already, one which will thresh as much as 4 men, costs £30 sterling. . . . One which threshes as much as 6 men, £45; 8 men, £60; and so on, reckoning £7, 10s. for each man's labour that the machine does, which is but about the expense of a servant for one year, whereas the patent is for 14 years. One man is sufficient to put in the corn to any one of 'em and take away the straw. . . . About 6 per cent of the grain which is lost by the ordinary method of threshing may be saved by this machine. . . . One of the machines may be seen in said wright's yard in the College Wynd."

The old-fashioned forms of built-in threshing-machines, at one time so extensively used throughout this country, and some of which were illustrated and described in former editions of *The Book of the Farm*, are now rarely met with. Machines of a much more efficient character have taken their place. The portable threshing-machine is now largely used, and is growing in favour. Still many farmers, especially in Scotland, prefer to have a good modern threshing-machine, built permanently in their steadings, to be always at their hand for use when desired.

These built-in machines now, as a rule, accomplish their work in a most admirable manner, threshing the grain at a rate formerly undreamt of, and in many cases not only at the same time dressing the grain so as to be fit for market, but conveying it into the granaries, which may be some considerable distance from the threshing-machine, and also carrying the straw to the remotest end of a long straw-barn — all this being done automatically, no human hand touching either grain or straw, after being fed into the drum, until each is deposited in its appointed quarters. These modern built - in threshing-machines are of many patterns, several of which may be said to be equally efficient.

Scottish Threshing - machine. — A section of the threshing-mill wing of a modern Scottish steading is represented in fig. 505. This shows at a glance not only the position of the threshing - machine, but also the courses of the grain and the straw until the former is dressed and carried by elevators and oscillating spout into the granary, and the latter by shakers and a travelling web to the extreme end of the straw-barn.

The following is a working description of this machine, as erected by the late Mr R. G. Morton, Errol, Perthshire :—

The sheaves to be threshed are fed through the hopper A, the grain being driven from the husk by the drum B in its grated concave cc', which is regulated for the different kinds and conditions of grain, by an instant and parallel acting set-gear Dd'. A large portion of grain and chaff fall through c, while the remainder is discharged, at a tangent, amongst the straw by the centrifugal force of the drum against the reflecting board D', then dropping upon the shakers E. The straw is tossed forward to the straw-carriers by the action of cranks on shaft F, and the patent balance throw-

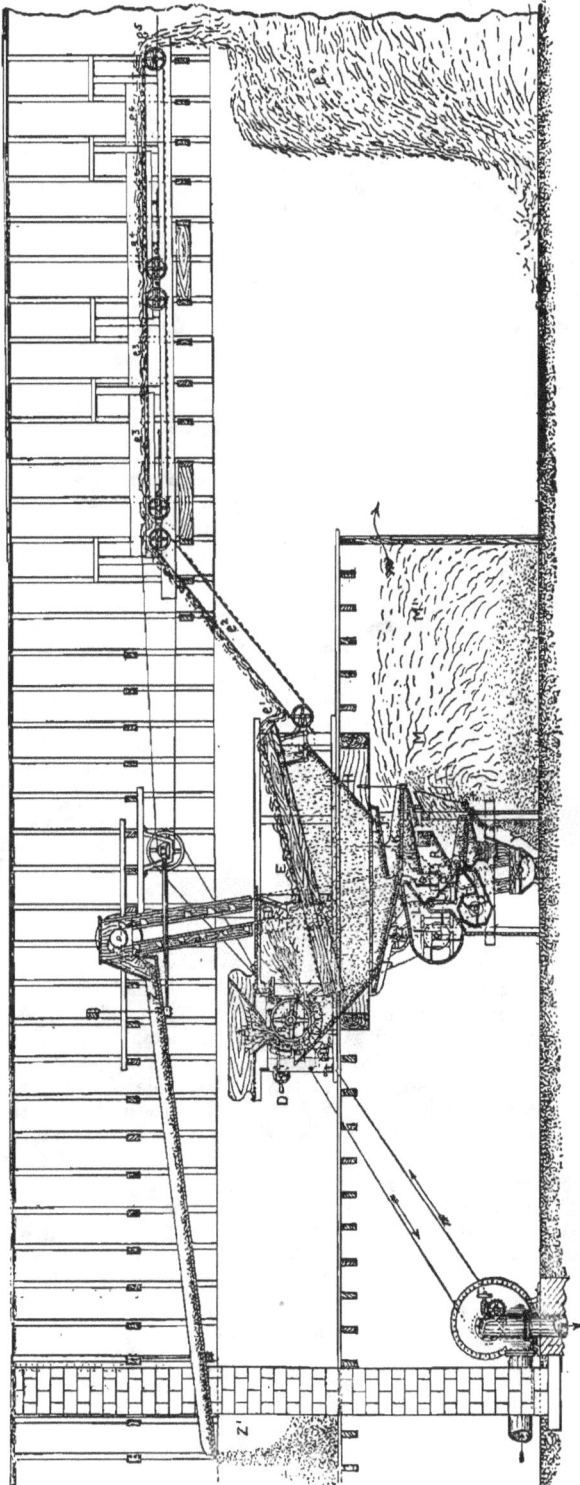

Fig. 505.—Modern Scottish threshing and dressing machine.

gear G. The grain falling through the shaker and concave gratings is gathered by the inclined planes H*h*, and oscillating planes JJ' of first riddle K, which, by the current of air from first blast L, carries the chaff, short straws, &c., to chaff-room MM', while the good grain falls through K to plane N, from which it slides down to cross-spout o. The light grain, &c., blown over N, falls on plain P, over which it slides and falls into the current of air from second blast T, to be further cleaned as it falls into the light-grain compartment Y; while the good grain falls through a trap-door in spout o, from whence it slides down the inclined planes Q and R to oscillating plane s, and receives the current of air from second blast T, as it falls from s to second riddle v, the wind carrying all the lights over v into y. The remaining husks and dust are blown into chaff-room M. The good grain falls through v into oscillating spout w, which, with a perforated bottom for extracting the small seeds and sand, delivers the good grain to elevator ark X, to be carried up to the roof and discharged into vibrating inclined plane Z, passing over a series of sand, seed, and small grain extractors, and being delivered through the gable into the granary z' for storage or bagging.

When the grain requires awning the trap-door in cross-spout o is closed, the grain passing over same to the patent pneumatic awner (shown by dotted lines at back of machine) and discharged into vertical tube, the air carrying the dust, &c., up same, while the grain falls down to the bottom of the tube and enters the machine by the port aa, and slides down the plane R to the oscillating plane s, to be winnowed and riddled as above.

The threshed straw as it leaves the shakers E at e 1 falls on the first section of straw-carriers e 2, to be delivered above the balks on to second section e 3, which in like manner delivers on e 4, which drops it into straw-barn at e 5, to be stored as at e 6 until full, when pinion 7 is caused to turn in rack 8. The frame and pulley 9 are drawn along with it, making an opening at 10, where the straw drops to the barn-floor, the operation being repeated at 11.

The whole parts of this machine are driven off drum-coupling, which is driven by "Morton's" direct-acting poncelet turbine, set in corn-room, designed for 34 feet fall and 26 feet suction, with 130 cubic feet of water per minute, making 509 revolutions, and developing 11½ horse-power. The power required is 9 horse-power and 100 cubic feet water per minute.

The best of these modern machines often thresh and dress from 12 to 16 quarters of ordinary oats per hour. Much, of course, depends on the length of the straw and the wealth of grain in the crop. From 6 to 8 quarters per hour are common quantities.

By ingeniously constructed blasts worked in conjunction with the threshing-machine, the newly threshed and dressed grain is in some cases conveyed, or rather blown, into granaries situated at awkward angles from the threshing-machine, where one would scarcely consider it possible to have such work accomplished.

By extension or contraction of the travelling-web the straw may be carried to the extreme end of the longest straw-barn, or, as already indicated, dropped at intermediate points as desired.

Saving of Labour.— A remarkable saving of labour is effected by these mechanical contrivances, and this, of course, is a point of great importance. The dressing-machines attached to the most improved threshing-mills are so effective as to dress the grain sufficiently well for ordinary purposes; and thus, when the grain is to be immediately sent to market, it may also be bagged —and not only bagged but also weighed automatically, as it issues from the spout at z'. To accomplish this automatic bagging and weighing, means are provided for hanging a bag upon or underneath the mouth of the spout, so as to catch the grain. The bag rests upon a portable weighing-apparatus, upon which are placed weights equal to the weight which it is desired to have in each bag. As soon as the bag receives the proper quantity of grain, it of course presses down its side of the weighing-machine, and in the act of thus descending it disengages a sluice, which thereupon shuts up the mouth of the spout. The attendant instantly removes the full bag, hangs on an empty one, and lifts the sluice, and the operation goes on with admirable speed and precision.

It is only in exceptional cases that the automatic work is carried on to this extent, but the practicability as well as the advantages of the plan are obvious enough.

Portable Threshing-machines.—The portable form of threshing-machines prevails in England. As a rule, there is no threshing-machine of any kind in English farm-steadings. The threshing is done by travelling machines owned by companies or individuals, who may have several machines at work in different parts of the country at one time. This system is now also pursued to a large extent in Scotland. Fig. 506 shows Clayton & Shuttleworth's portable threshing-machine at work in a stack-yard.

Several leading firms of implement-makers have given much attention to the manufacture of portable threshing-machines, and now the farmer has ample choice of machines of the highest efficiency. These portable threshing-machines are usually worked by steam traction-engines, which also draw them from one place to another. In some

Fig. 506.—*Portable threshing-machine at work.*

cases portable steam-engines are employed in working the machines, but then horses have to be used in taking the machine from farm to farm.

In fig. 507—a longitudinal section of a modern portable threshing-machine, made by Marshall & Sons, Gainsborough—the operations of threshing, dressing, and bagging, all going on simultaneously, are shown clearly. The working is seen so distinctly in the sketch, that no detailed description of the process is necessary. The machine is supposed to be working in the stackyard. The sacks of grain as they get filled have to be conveyed to the granary—but that is easily done.

The disposal of the straw entails more labour. It is usually formed into a large stack at the rear of the threshing-machine, and the conveyance of the straw from the shakers to this stack is, in most cases, accomplished by means of elevators, which can be lengthened and raised in the pitch as the stack increases in height. Fig. 508 represents the Hayes patent elevator (Clayton & Shuttleworth).

Hands required for Threshing-machines.—The number of persons required to work these portable threshing-machines varies according to the operations performed and the speed of the machine. Ransomes, Sims, & Jefferies, whose portable threshing-machine is represented in fig. 509, point out that the economy of threshing must depend in a great measure on the proper distribution

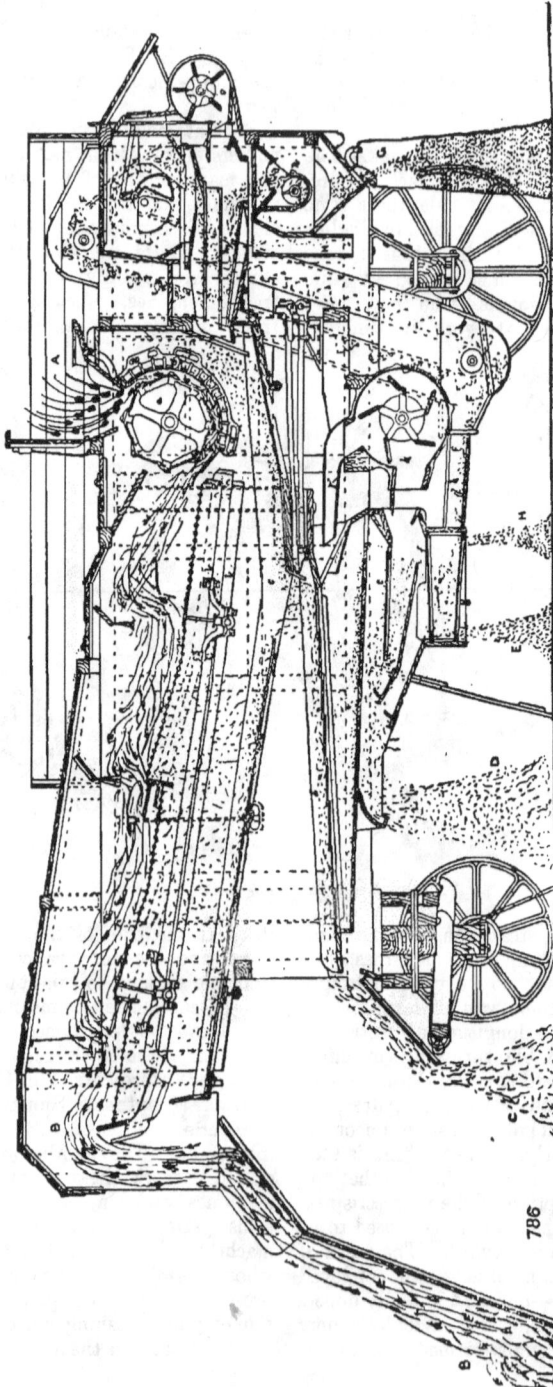

Fig. 507.—*Section of portable threshing-machine.*

A Unthrashed corn.
B Straw.
C Cavings.
D Chaff.

E Cobs.
F Corn.
G Finished grain.
H Dust.

a Drum.
b Shakers.
c Top shoe.
d Caving riddle.

e Chaff riddle.
f Caving riddle.
g Screen.
h Corn spout.

j Elevator tins.
k Main blower.
l Smutter.

m Finishing riddles.
n Separating screen.
o Back-end blower.

786

of the hands employed, and state that the force, when straw-elevators are not used, should consist of eleven men and boys, to be engaged as follows: "One to feed the machine; two to untie and hand the sheaves to the feeder; two on the corn-stack to pitch the sheaves on to the stage of the threshing-machine;

Fig. 508.—*Straw and hay elevator.*

one to clear the straw away as it falls from the straw-shaker; two to stack the straw; one to clear away the chaff from underneath the machine, and occasionally to carry the chobs which fall from the chob-spout up to the stage, to be threshed again; one to remove the sacks at the back of the machine as they are filled; and one to drive the engine. The feeder, on whom very much depends, should be an active man, and should have the control of the men stationed near the machine. He should endeavour to feed the drum as nearly as possible in a continuous stream, keeping the corn uniformly spread over the whole width. The two men or boys who untie the sheaves should stand on the stage of the threshing-machine, so that either is in a position to hand the feeder a sheaf with ease, but without obstructing the other. The men on the stack must keep the boys or men on the stage constantly and plentifully supplied with sheaves, which must be pitched on to the stage, so that the boys can reach them without leaving their position. The man who removes the straw from the end of the shaker should never allow it to accumulate so that it cannot fall freely. The man whose duty it is to clear away the chaff and cavings from underneath the machine must not allow these to accumulate so as to obstruct the free motion of the shoes; he

must watch the basket under the chob-spout, and as soon as it is full, empty its contents on to the stage, in a convenient position for the feeder to sweep the same, a little at a time, into the drum to be threshed over again. The man who attends to the sacks must remove them before they get so full as to obstruct the free passage of the corn from the spouts, otherwise the clean corn may be thrown out at the screenings-spout.

When a large quantity is being threshed at one time, additional hands may be required to take away and stack the straw. It is better to cart the sheaves to the threshing-machine than to shift its position in the stackyard. The engine-driver, during threshing, should be as prompt as possible in attending to the signals for stopping and starting, and he should carefully attend to the bearings of the drum-spindle and other spindles of the threshing-machine.

Safety-drums.—The frequency of serious accidents to those engaged in feeding threshing-machines led to the passing of an Act of Parliament providing that the drum and feeding-mouth of every threshing-machine must be sufficiently and securely fenced so far as practicable. Great ingenuity has been displayed by

Fig. 509.—*Portable threshing-machine.*

leading manufacturers in devising means for preventing these accidents, and now there are several patent safety-drums or drum-guards, most of which seem to render accidents by contact with the drum, if not absolutely impossible, at least extremely improbable.

Straw-trusser.—The straw-trusser is a most useful contrivance. It is attached to, and worked in conjunction with, a threshing-machine. The straw as it leaves the shakers of the threshing-machine is caught by the trusser, securely tied in convenient bundles—which may at will be varied in size — with stout

Fig. 510.—*Straw-trusser.*

twine, and thrown on the ground behind the machine, ready to be forked on to a stack or cart. This excellent machine, represented in fig. 510, made by J. & F. Howard, Bedford, will tie up the straw as fast as it leaves the most speedy threshing - machine, thus performing the work of five or six men. As indicated in the illustration, the threshing - machine and trusser attached takes up very little more space than the threshing - machine by itself.

Hand Threshing-machines. — Several tiny threshing-machines are made for hand - power, and there are machines somewhat larger, but still, of course, of a comparatively small size, for one-horse or pony-gear. Some of these thresh-ing - machines, adapted for hand, pony, or horse power, are cap-able of threshing from 10 to 25 bushels of grain per hour. They are extensively used on small holdings, where they are supplanting the "flail," which is now almost a thing of the past.

A very useful little threshing-machine, arranged for driving by hand and foot, or by other power (Ben Reid & Co.), is illustrated in fig. 511.

MOTIVE POWER FOR THRESHING-MACHINES.

Steam- or oil-engines are fast taking the place of horse - power in working threshing-machines. Where the supply is plentiful, water still holds its own, and will continue to do so, for it is the cheapest of all motors for the purpose. But the horse-wheel is gradually disap-pearing, and, for threshing purposes, the windmill may be said to have gone.

Steam-power.

The steam-engine in its various forms, suitable for farm-work, has already been fully explained (vol. i. pp. 404-428), so that nothing more may be said in regard to it here. Steam-power possesses two im-portant advantages : it is always at com-mand and can be completely controlled. By the use of steam the threshing may proceed continuously as long as may be desired ; while, except in the rare cases in which the force of running

Fig. 511.—*Small threshing-machine.*

water is sufficient to drive the mill-wheel, the threshing for the time ceases with emptying of the "mill-dam." Ex-perience has abundantly proved that threshing-machines dependent on water derived chiefly from the drainage of the surface of the ground, frequently suffer from a short supply in autumn, and late in spring or early in summer, thereby creating inconvenience for the want of straw in the end of autumn, and the want of seed or horse-corn in the end of spring. Wherever such casualties are likely to happen, it is better to adopt a steam-engine or oil-engine at once.

The other advantage is also important. Water- or horse-power cannot be so nicely governed as steam or oil, and, as a consequence with these powers, irregularities in feeding-in the grain or variations in the length of the straw are apt to make the motion of the corn-dressing appliances irregular, which, of course, causes imperfect dressing.

Water-power.

But wherever there is a sufficient fall and a reliable supply of water, it is desirable, for the sake of economy, that the latter should be utilised for threshing purposes. There are various methods by which water-power is made available for driving the threshing-machine, pulper, chaff-cutter, grist-mill, &c. The turbine is a comparatively modern invention, and a valuable contrivance it is. The water-wheels are usually of two kinds—*undershot* and *bucket* wheels.

Undershot Water-wheels.—The *undershot* or *open float-board* wheel can be advantageously employed only where the supply of water is considerable and the fall low. It therefore rarely answers for farm purposes, and need not be discussed.

Bucket Water-wheel.—A much more useful kind is the *bucket*-wheel, which may be *overshot* or *breast*, according to the height of the fall. It is this wheel that is adopted in all cases where water is scarce or valuable, and the fall amounts to 6 or 7 feet or more, though it is sometimes employed with even less fall than 6 feet.

Measuring the Water-supply.—When it is proposed to employ a stream of water for the purpose of power, the first step is to determine the *quantity delivered by the stream in a given time*. This, if the stream is not large, is easily accomplished by an actual measurement of the discharge, and is done by damming up the stream to a small height, say 1 or 2 feet, giving time to collect, so as to send the full discharge through a shoot, from which it is received into a vessel of any known capacity, the precise time that is required to fill it being carefully noted. This will give a correct measure

of the water that could be delivered constantly for any purpose. If the water be in too small a quantity to be serviceable at all times, the result may be found by a calculation of the time required to fill a dam of such dimensions as might serve to drive a threshing-machine for any required number of hours.

If the discharge of the stream is more than could be received into any moderately-sized vessel, a near approximation may be made to the amount of discharge by the following method : Select a part of its course, where the bottom and sides are tolerably even, for a distance of 50 or 100 feet; ascertain the velocity with which it runs through this space, or any measured portion of it, by floating light substances on its surface in a calm day, noting the time required for the substance to pass over the length of the space. A section of the stream is then to be taken, to determine the number of superficial feet or inches of sectional area that is flowing along the channel, and this, multiplied into five-sixths of the velocity of the stream, will give a tolerable approximation to the true quantity of discharge—five-sixths of the surface velocity, at the middle of the stream, being very nearly the mean velocity of the entire section.

Supposing the substance floated upon the surface of the stream passed over a distance of 100 feet in 20 seconds, and that the stream is 3 feet broad, with an average depth of 4 inches—here the area of the section is exactly 1 foot, and the velocity being 100 feet in 20 seconds, gives 300 feet per minute, less one-sixth = 250 feet, and this multiplied by the sectional area in feet, or 1 foot, is 250 cubic feet per minute for the discharge.

It is to be borne in mind that this is only an approximation, but it is simple, and from repeated experiments we have found it to come near the truth.

The next step is to ascertain the fall, by levelling, from the most convenient point at which the stream can be taken off, to the site where the water-wheel can be set down, and to that point in the continuation of the stream where the water can be discharged from the wheel, or what is called the outfall of the tail-race. If the water has to be conveyed to any considerable distance from the point where it is diverted from the stream to the wheel, a lade must be formed for it, which should have a fall of not less than 1½ inch in 100 feet; and this is to be deducted from the entire fall. Suppose, after this deduction, the clear fall be 12 feet, and that the water is to be received on a bucket-wheel whose power shall be equal to 4 horses.

Horse-power in a Stream.—It may be useful to know the rule for calculating the number of horse-power any stream may exert if employed as a motive power. It is this: multiply the specific gravity of a cubic foot of water, 62½ lb., by the number of cubic feet flowing in the stream per minute, as ascertained by the preceding process, and this product by the number of feet in the fall, and divide by 33,000 (the number of pounds raised 1 ft. high in 1 min. by a "horse-power"); the product is the answer.

Thus, — Multiply the number of
cubic feet flowing per minute in
the stream—suppose . . . 350
By the weight of a cubic foot of
water 62½ lb. 62½

175
700
2100

21,875

And then multiply the product
by the number of feet of fall
available—suppose . . . 12

Divide the remainder by 33,000)262,500(7.9 horse-power.
And the quotient, 7.9, gives the number of horse-power.

This is, of course, the theoretical horse-power, of which only a proportion can really be utilised, varying from 35 per cent in undershot wheels to 75 per cent in turbines.

Mill-dam.—If the stream does not produce this quantity, a dam must be formed by embanking or otherwise, to contain such quantity as will supply the wheel for three or six hours, or such period as may be thought necessary.

Forming Mill-dam.—The dam may be formed either upon the course of the stream, by a stone weir thrown across it, and proper sluices formed at one side to lead off the water when required; or, what is much better, the stream may be diverted by a low weir into an intermediate dam, which may be formed by digging and embankments of earth, furnished with sluice and waste-weir, and from this the lade to

the wheel should be formed. The small weir on the stream, while serving to divert the water, when required, through a sluice to the dam, would, in time of floods, pass the water over the weir, the regulating sluice being shut to prevent the flooding of the dam. This last method of forming the dam is generally the most economical and convenient, besides avoiding the risk which attends a heavy weir upon a stream that may be subject to floods. When water is collected from drains or springs, it is received into a dam formed in any convenient situation, which must also be furnished with a waste-weir, besides the ordinary sluice, to pass off flood-waters.

The position of the sluice in the dam should be so fixed as to prevent the *wrack* floating on the surface of the water finding its way into the sluice, and thence to the water-wheel. To avoid this inconvenience, the sluice should not be placed at the lowest point of the dam, where it most commonly is, but at one side, at which the water will pass into the lade, while the rubbish will float past to the lowest point.

Dimensions of the Bucket-wheel.— The water-wheel should be on the *bucket* principle, and, for a fall such as we have supposed, 12 feet, should not be less than 14 feet diameter; the water, therefore, would be received on the breast of the wheel. Its circumference, with a diameter of 14 feet, will be 3.1416 × 14 = 44 feet; its velocity, at 5 feet per second, is 44 × 5 = 220 feet a minute; and 234 cubic feet per minute of water spread over this gives a sectional area for the water laid upon the wheel of $\frac{234}{220} = 1.06$ feet; but as the bucket should not be more than half filled, this area is to be doubled = 2.12 feet; and as the breadth of the wheel may be restricted to 3 feet, then $\frac{2.12}{3} = .704$ foot, the depth of the shrouding, equal to 8½ inches nearly; and if the wheel is to have wooden soling, 1 inch should be added to this depth already found, making 9½ inches.

The Arc.—The *arc* in which the wheel is to be placed must have a width sufficient to receive the wheel with the toothed segments attached to the side of the shrouding. For a bucket-wheel it is not necessary that it be built in the arc of a circle, but simply a square chamber —one side of it being formed by the wall of the barn, the opposite side by a wall of solid masonry, at least 2½ feet thick: one end also is built up solid, while the opposite end, towards the tail-race, is either left entirely open, or, if the water is to be carried away by a tunnel, the water-way is arched over and the space above levelled in with earth. It is requisite that the walls of the wheel-arc should be built of square-dressed stone, having a breadth of bed not less than 12 inches, laid flush in mortar, and pointed with Roman cement.

Construction of the Wheel.—Fig. 512 is a *sectional elevation* of the wheel. The barn-wall, and the sole of the arc

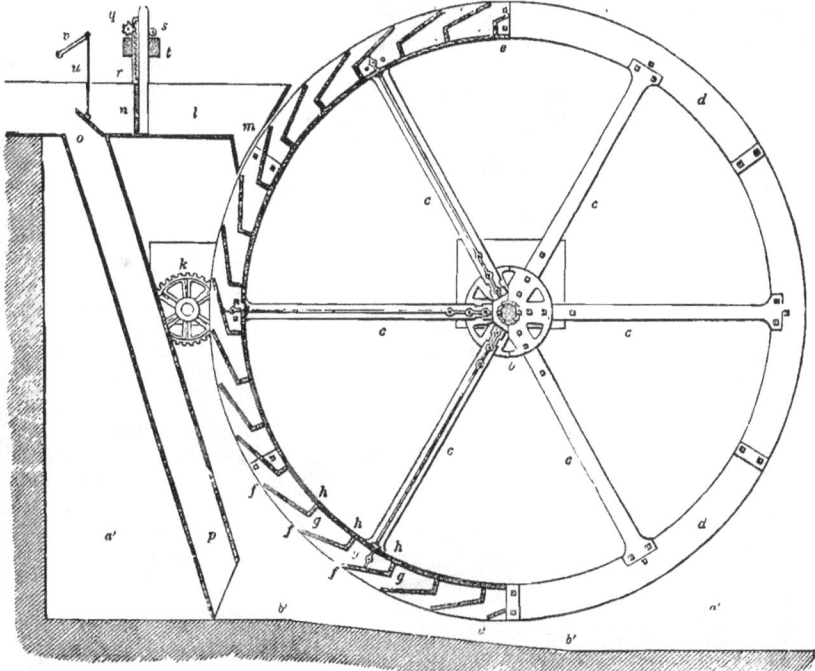

Fig. 512.—*Section and elevation of a bucket water-wheel.*

a' a' Barn-wall.	f g Front of bucket.	r Sluice-stem.
b' b' Sole of arc.	g h Bottom of bucket.	s Friction-roller.
b Eye-flanges.	k Pinion.	t Cross-head.
c Arms.	l Trough.	o Trap-sluice.
d d Shrouding.	m Spout.	o p Spout.
e Groove for securing the buckets.	n Regulating-sluice.	u Connecting rod.
f f Pitch of the buckets.	q Pinion.	v Crank-lever.

or chamber, are formed of solid ashler, having an increased slope immediately under the wheel, to clear it speedily of water. The shaft, the arms, and shrouding are of cast-iron, the buckets and sole being of wood; and to prevent risk of fracture, the arms are cast separately from the shrouding. The width of the wheel being 3 feet, the toothed segments 4 inches broad, and they being 1 inch clear of the shrouding, gives a breadth over all of 3 feet 5 inches, and when in the arc there should be at least 1 inch of clear space on each side, free of the wall. The length of the shaft depends upon how the motion is to be taken from the water-wheel. In the case of the wheel illustrated in the sketch it is taken off by the pinion shown on the left hand, in a line horizontal to the axis of the water-wheel.

The eye-flanges, 2 feet diameter, are

separate castings, to which the arms are bolted; the flanges being first keyed firmly upon the shaft. The shrouding is cast in segments, and bolted to the arms and to each other at their joinings. On the inside of the shroud-plates are formed the grooves for securing the ends of the buckets and of the sole-boarding.

The form of the buckets should be such as to afford the greatest possible space for water at the greatest possible distance from the centre of the wheel, with sufficient space for the entrance of the water and displacement of the air. In discharging the water from the wheel also, the buckets should retain the water to the lowest possible point. These conditions are attained by making the pitch of the buckets, or their distance from lip to lip, 1½ times the depth of the shrouding; the depth of the front of the bucket inside, equal to the pitch; and the breadth of the bottom as great as can be attained consistently with free access of the water to the bucket immediately preceding: this breadth, inside, should not exceed two-fifths of the depth of the shrouding.

If there is the least danger of backwater — that is, of interruption to the discharge in the tail-race—it is a good plan to keep the bottom of the arc high at the up-water side. This gives the water discharging at the higher points a velocity greater than 5 feet per second, and assists in driving the water away from below the lowest discharging buckets.

In the illustration one-half of the shrouding-plates are removed, the better to exhibit the position of the buckets. The *shrouding-plates* are bolted upon the buckets and soling by bolts passing from side to side; and in order to prevent resilience in the wheel, the arms are supported with diagonal braces. The toothed segments which operate on the pinion are bolted to the side of the shrouding through palms cast upon them for that purpose, and the true position of these segments requires that their pitch-lines should coincide with the circle of gyration of the wheel: when so placed, the resistance to the wheel's action is made to bear upon its parts, without any undue tendency to cross strains. For that reason it is improper

to place the pitch-line beyond the circle of gyration, which is frequently done, even upon the periphery of the water-wheel. The determination of the true place of the circle of gyration is too abstruse to be introduced here, nor is it necessary to be so minute in the small wheels, to which our attention is chiefly directed : suffice it to say, that the pitch-line of the segment wheel should fall between one-half and two-fifths of the breadth of the shrouding, from the extreme edge of the wheel.

Overshot or Breast ?—An important point to decide is whether the wheel is to be worked on the overshot or breast method. Where the fall is ample but the supply of water small, or moderate, the overshot is the best; where the water is fairly plentiful and the fall not so great, the breast may be most suitable. But whether the water be delivered over the top of the wheel or on the breast, the water should be allowed to fall through such a space as will give it a velocity equal to that of the periphery of the wheel when in full work. Thus, if the wheel move at the rate of 5 feet per second, the water must fall upon it through a space of not less than .4 foot; for, by the laws of falling bodies, the velocities acquired are as the times and whole spaces fallen through to the squares of the time. Thus the velocity acquired in 1″ being 32 feet, a velocity of 5 feet will be acquired by falling .156″; for $32 : 1'' :: 5 : .156''$, and $1''^2 : 16 :: .156''^2 : .4$ foot, the fall to produce a velocity of 5 feet. But this being the minimum, the fall from the trough to the wheel may be made double this result, or about 10 inches.

So as to secure the proper filling of the buckets, the breast-wheel at 5 feet per second should have the pen-trough, with two or three guide-vanes set to turn the direction of the water into the bucket at a velocity of about 6 feet per second. This will also prevent the force of the water from opposing the wheel. For an overshot wheel at 6 feet per second at the periphery, it would be well to have the water entering the bucket at about 7 feet per second.

Trough and Sluices. — The trough which delivers the water upon the wheel should be at least 6 inches less in breadth than the wheel, to give space for the air escaping from the buckets, and to prevent the water dashing over at the sides: the trough and spout convey the water to the wheel. It is convenient to have a regulating sluice, that serves to give more or less water to the wheel; and this is worked by a small shaft passing to the inside of the upper barn. The shaft carries a pinion working the rack of the sluice-stem, a small friction-roller being placed in proper bearings on the cross-head of the sluice-frame; and this apparatus is worked inside the barn by means of a lever handle upon the shaft of the pinion. As a waste-sluice, the most convenient and simple, in a mill of this kind, is the trap-sluice, which is simply a board hinged in the sole of the trough, which in opening turns up towards the wheel. It is made to shut close down to the level of the sole, and when so shut the water passes freely over it to the wheel. The lifting of this sluice is effected by means of the connecting-rod and crank-lever, the latter being fixed upon another small shaft, which passes through the wall to the interior of the barn, where it is worked in the same manner as the lade-sluice. When it is found necessary to stop the wheel, the trap is lifted, and the whole supply of water falls through the shoot, leading it to the bottom of the wheel-arc, by which it runs off, until the sluice at the dam can be shut, which stops further supply.

Speed of the Wheel. — The wheel here described, if it moves at the rate of 5 feet per second, will make 6¾ revolutions per minute. The pinion-shaft will carry a spur-wheel, by which all the other parts of the machine can be put in motion. The rate of the spur-wheel depends on the relation of the water-wheel and its pinion. In the present case they are in the proportion of 8 to 1, and as the water-wheel takes 6¾ revolutions per minute, this, multiplied by 8, will give 54 to the spur-wheel.

The Turbine.—The turbine is much superior to the ordinary vertical wheels for utilising water-power, and it is rapidly taking their place. It is an ingenious and powerful water-engine, one of the many useful inventions we owe to the development of science. It is suitable for high or low falls, and, as a rule, can be fitted in at much less cost than the common vertical water-wheel. The power which the turbine generates can be applied very easily, and the "engine" can be worked at different degrees of its capacity, so that it may be adapted either to the working of the chaff-cutter, root-pulper, or grist-mill alone, or to the threshing-machine and all the smaller machines combined. The turbine makes the most both of the water and the fall. As its action is not impeded by back-water, the turbine may be placed on a level with the tail-race, and thus give the water before entering the turbine the full benefit of the entire available fall. Its small size is another advantage, and a small bed of masonry is all that is required for its foundation. Turbines revolve with such velocity that the motion for driving machinery may be obtained direct from the wheel-shaft, thus saving intermediate gearing.

Turbines of various types are in use in this country and elsewhere.

Horse-power.

As already indicated, horse-power for threshing purposes is gradually giving place to other motors. Still it is in use on many farms, and it demands brief notice.

Formerly there were two leading types of horse-wheels, known as *under-foot* and *over-head*. The under-foot was used chiefly where small powers are required, and the over-head on large farms where four horses and upwards were employed. But on nearly all large farms either the steam-engine, oil-engine, or turbine water-wheel has taken the place of horse-power for threshing, so that the over-head horse-wheel is now rarely seen in use. It is therefore the under-foot horse-wheel that now prevails, and with it the horses draw by means of trace-chains and swing-tree. The horses usually worked singly, one at each lever or beam; but sometimes they are yoked in pairs, two horses at each lever. It is often found that horses accustomed to go together in the plough work most willingly in the horse-wheel when yoked side by side; and in this way also a greater force may, if desired at any time, be employed than with one horse to each lever or beam.

Horse-gear for one or two horses is now provided in great variety, and, as a rule, of a very convenient and serviceable description, easily fitted up or removed from one place to another. Only thresh-ing-machines of small proportions can be worked by this form of horse-gear. Its most general function on the farm is to drive the chaff-cutter, turnip-cutter, cake-breaker, and grist-mill.

THRESHING AND WINNOWING CORN.

In bygone days the first preparation for threshing corn—that is, separating the grain from the straw by the thresh-ing-machine—was usually taking in the stack to be threshed, and placing it in the upper or threshing barn. Now the almost universal plan is to cart the un-threshed corn from the stack to the threshing-mill or machine. The corn is usually carted to the sheaf-barn as the threshing proceeds; but in many places where there is sufficient sheaf-barn ac-commodation, a stack is stored there at some convenient time and threshed out at another time, or at intervals, according to circumstances.

In many cases where the threshing-machine is fed from the ground-floor, or where a cart-way can be made up to the level of the sheaf-barn on the first floor, the sheaf-door is made wide enough to admit the load of straw, which is de-posited there by a tip-cart without any further handling. Where the sheaf-door is not wide enough to admit the load, the sheaves are usually forked off the cart into the barn. Where the floor of the sheaf-barn is level with the ground out-side, the load of sheaves may be tipped at the door and carried or forked in. This is an expeditious plan when only one cart is employed in taking in the stack; but it has this drawback, that the tipping is apt to shake out grain from the straw.

Ladders.—Ladders are most useful about a farm-steading. They are best formed of tapering Norway pine spars, sawn up the middle. A useful form of ladder for farm purposes is in fig. 513, where the rounded form of the Nor-way spar, divided in two, is placed out-

most, though it is as often placed inmost. Those spars are connected together by steps of clean ash, pushed through auger-made holes in the spars, and rendered firm by means of wedges driven into the outside ends of the steps. The steps are 9 inches apart, and 16 inches long at the bottom and 13 inches at the top, in a ladder of 15 feet in length, which

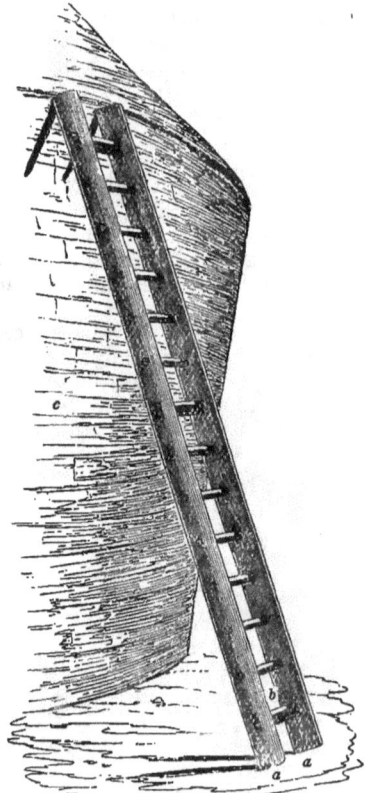

Fig. 513.—*Ladder, 15 feet long.*
a a Spars of ladder.   *c* Stack.
b Steps of ladder.

is the most useful size for use in a stack-yard. To prevent the ladder from falling to pieces in consequence of the shrinking of the round steps, a small rod of iron is passed through both spars, having a head at one end and a screw and nut at the other, *under* the upper, middle, and lower steps, the head end keeping its hold firmly while the screw end is rendered tight by the nut. When well finished and painted, such a ladder will last many years. A couple of ladders 10 feet, a couple of 15 feet, and one of 24 feet long, will suffice for all the purposes of a farm, as also for the repairs of the steading and houses.

Preparing for Threshing.—Before setting on the threshing-machine, its several parts require to be *oiled*. Fine machinery-oil should be employed for this purpose, though too often a coarse dirty oil is used. It should be put for use into a small tin flask, having a long narrow spout (fig. 514) to reach any gudgeon behind a wheel. It is important that the machine should be thoroughly oiled, and it should therefore

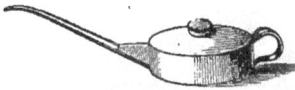

Fig. 514.—*Oil-can.*

be carried out with great care, and by one acquainted with the construction of the machine.

When steam is employed as the motive power, the fire should be kindled by the engine-man in time to get up the steam by the moment it is wanted. From half an hour to an hour may be required for this purpose, according to the state of the atmosphere. Less time will suffice to start an oil-engine.

When water is the power, the sluice of the supply-dam should be drawn up to the proper height, to allow the water time to reach the mill-wheel sluice when it is wanted.

When the power is of horses, the horses are yoked in the wheel by their respective drivers, immediately after leaving the stable at the appointed hour of yoking; and while one of the men is left in charge of driving the horses, the other men go to the straw-barn to take away the straw from the shakers of the mill with straw-forks, and fork it in mows across the breadth of the barn, which mows may be tramped down by a woman in narrow breadths—that is, where the straw is not carried away automatically, which is done with most modern threshing-machines, as in fig. 505.

Every preparation ought to be completed before the machine is started by the order of the person who is to feed the

machine, and who should be a careful man of experience. The power should be applied gently at first, and no sheaf should be presented until the machine has acquired its proper momentum—the *threshing motion,* as it is termed.

Care in Feeding.—The capacity of the modern threshing-machines compels the feeder to be active at his work. The efficiency of the threshing, however, is not now so much dependent upon the care and skill of the feeder as was the case with the old-fashioned machines formerly in use. With the improved high-speed drums, the best modern machines make a perfect separation of the grain and the straw even with unskilled feeding, yet it is desirable that this important piece of work should be executed carefully.

Irregular Driving.—There are certain circumstances which greatly affect the action of the machine in the *foulness* of its threshing. One depends—where horse-power is used—on the *driving of the horses,* in which a considerable difference is felt by the feeder when one man keeps the horses at a regular pace, whilst another drives them by fits and starts. The regular motion is most easily attained by the driver walking round the course in *the contrary direction to the horses,* in which he meets every horse twice in the course of a revolution, and which keeps the horses upon their mettle, every horse expecting to be spoken to when he meets the driver.

Removing Straw.—The straw, as it is threshed, is either carried away to the straw-barn automatically, as shown in fig. 505, or trussed as in fig. 510, or is removed by forkers, who place it in the part of the straw-barn where it is intended to be stored. One method of storing straw in the barn is to spread it in a line, across the end or along one side of the straw-barn, in breadths or mows of 5 or 6 feet, and tramp it firmly with the feet; and when this breadth has reached such a height as the roofing of the barn will easily allow, another one is made upon the floor beside it, and so on in succession, one breadth after another, in parallel order, until the stack is threshed or the barn filled. When stored in this form the straw can be more easily taken away in forkfuls as required.

Dressing Corn.

In former times the threshing and dressing of grain were distinct operations performed at different times. Now they may be said to be but two parts of one operation. The modern threshing-machine of the most improved type is so admirably equipped as to efficiently clean and dress the grain, as well as separate it from the straw; also "hummelling" or "beating" the barley, and, as has already been explained, conveying the grain to the granary, and the straw to the extreme end of the straw-barn—all this in one continuous operation.

Still there are many farms on which the threshing-machines only partially dress the grain, and not a few indeed, mostly of small size, where the threshing-machines do little or nothing except separate the grain and the straw. It is therefore necessary to have at least one detached corn dressing-machine at every farm to give the grain such finishing touches as may be required.

Modern Winnowers. — The modern corn-dressers, worked in conjunction with

Fig. 515.—*Combined winnower and bagging-machine.*

the threshing-machine, are, as a rule, built upon similar principles to those on which the old-fashioned detached winnowers so long in use in this country were constructed, but many improvements have been introduced which enhance their efficiency. The blowing, finishing, and dusting—that is, the blowing away of the chaff, the separating of the light grain from the heavy, and the removal of sand and dust—are now all performed by one machine,—a machine of which there are numerous patterns, made by various firms, many of them very ingenious, and nearly all very efficient.

Very thoroughly most of these modern dressers do their work, yet many farmers still adhere to the custom of putting the grain once through a separate winnower after leaving the threshing-machine. In some cases arrangements are made whereby the grain as it leaves the elevator, raising it from the threshing-machine as at z' in fig. 505, falls into the hopper of a winnower, which is worked by hand, and which remains in the granary to give the grain its finishing touches.

In fig. 515 a representation is given of an improved corn-dressing machine, with apparatus attached, for bagging and weighing the grain automatically, made by T. Corbett, Shrewsbury. The elevator is worked by a strap from the driving-wheel spindle of winnower, and raises the grain by conducting tins, as it passes from the machine, into a hopper, sufficiently high to apply a bag at full length. Under the spout is an ordinary weighing-machine, to which a cord passes from a catch applied in the elevator spout, so that when the bag (which is placed on the weighing-machine) has its proper weight, the descent of the machine disengages the catch, and the slide falls instantly, and thereby prevents a further discharge. The hopper is sufficiently large to receive the grain while the attendant is moving the bag and applying another, by which arrangement the laborious work of filling the bag from the machine is dispensed with, and at the same time a saving of two men is effected. The grain may be elevated at the rate of from 60 to 80 bushels per hour, with wonderfully little difference in the power required to work the winnower.

Many of the improved modern winnowing-machines are so constructed that by change of riddles, screens, &c., all kinds of grass as well as grain seeds can be cleaned by them. Notwithstanding their more varied accomplishments, the modern corn-dressing machines are easier to work than the simple and less efficient but clumsier machines of former times.

Corn-screens. — Screens of various patterns are often used in addition to winnowers in dressing and finishing grain. By these screens sand and dust are thoroughly removed, and small seeds are separated from those of the proper size. Screening is specially serviceable in dressing barley for malting purposes, uniformity of the size of the grain being an element of some importance in the manufacture of malt. Some of these screens are made on the flat, and others on the rotary, principle.

Barn Implements.—There is less use now than formerly for such appliances as riddles, wechts or maunds, and shovels, in the corn-barn. In former times hand-riddles played an important part in the cleaning of grain, but they have been almost entirely supplanted by improvements in the threshing and dressing machines. The riddling of the corn was heavy and tedious work; and altogether

there are few branches of farm-work in which there has been greater saving of

Fig. 516.—*Old wooden wheat-riddle.*

manual labour through the introduction of improved mechanical contrivances

Fig. 517.—*Old wooden barley-riddle.*

than in the threshing, dressing, and handling of grain.

Fig. 518.—*Old wooden oat-riddle.*

Riddles. — Although they are now little used, it may be interesting to preserve here some of the illustrations of the barn implements in use in former times, and illustrated in previous editions of this work. (See figs. 516 to 520.) In earlier times riddles were, as a rule, made of wood, but latterly wire came to be extensively used. In the illustrations the meshes are shown at full size, and the diameter of the riddle was usually

Fig. 519.—*Old wooden bean-riddle.*

about 23 or 24 inches. The mesh for wheat was ¼ inch square, for barley and beans ⁵⁄₁₆ inch, and for oats ⅜ inch;

Fig. 520.—*Old wooden riddle for the roughs of wheat and corn.*

while for riddling the roughs of wheat and oats a riddle with meshes of 1 inch square was employed.

Sieves.—The use of the sieves was to sift out dust, earth, and small seeds from corn. The wooden sieve, fig. 521, had meshes of ⅛ inch square, and the iron-wire sieve, fig. 522, 64 meshes to the square inch, including the thickness of

the wire. Fig. 523 is a triangular-meshed iron-wire sieve, with an oak rim.

Barn Wechts or Baskets.—A form of *wecht* or *maund* long in use for taking

Fig. 521.—*Wooden sieve.*

up corn from the bin or floor is represented in fig. 524, made either of withes or of skin, attached to a rim of wood. A young calf's skin with the hair on, or sheep's skin without the wool, tacked to the rim in a wet state, after becoming dry and hard, makes a better and more durable wecht than wood.

Fig. 522.—*Iron-wire sieve.*

Baskets of close and beautiful wicker-work, such as fig. 525, have been used in barns in parts of England instead of wechts. The articles for lifting the loose corn, either for pouring into the bushel, the bag, or the winnower, are now usually made of wood, almost square, with a deep frame at three sides; and these are much more expeditious than the older forms.

Barn-hoe.—A *wooden hoe*, fig. 526, 7 inches long and 4 inches deep in the

Fig. 523.—*Triangular-meshed iron-wire sieve.*

blade, fixed to a shaft 9 inches long, made of plane-tree, is better than the hands to fill wechts with corn from the floor.

Corn-shovels.—A couple of wooden *scoops* or shovels, such as fig. 527, to shovel up the corn in heaps, and to turn it over, are indispensable implements in a corn-barn.

Barn-brooms.—Excellent brooms for

Fig. 524.—*Wecht of skin.*

the corn-barn and granaries are made of stems of the broom plant, about 3 feet in length, simply tied together with twine at one end, and used without a handle. The broom is also in the best

Fig. 525.—*Corn-basket of wicker-work.*

state when fresh, and becomes too brittle when dry. When long straight stems of the common ling (*Calluna vulgaris*) can be procured, they make both good and durable brooms. A hard birch-broom is required to clear the dirt from between the stones of a causeway, while the softer broom answers best for the barn-floor.

Fig. 526.— *Barn wooden hoe.*

Measuring Grain. —Corn is now invariably measured by the imperial bushel, fig. 528. It is of cooper-work, made of oak and hooped with iron; and, according to the Weights and Measures Act, must be stamped by competent authority before it can be legally used. Having been declared the standard measure of capacity in the country for dry

measure, it forms the basis of all contracts dependent on measures of capacity when otherwise indefinitely expressed. The bushel must contain just 2218.19 cubic inches, though its form may vary. The form represented in this figure is somewhat broader at the base than at the top, and furnished with two fixed handles. It is not too broad for the mouth of an ordinary half-quarter sack, nor too deep to compress the grain too much; and its two handles are placed pretty high, so that it may be carried full without the risk of over-turning.

Fig. 527.—*Wooden corn-scoop.*

In connection with the bushel is the *strike* for sweeping off the superfluous corn above the edge of the bushel. It is made

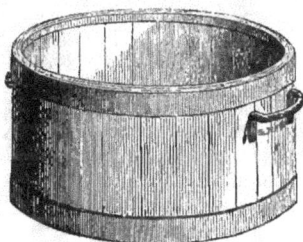

Fig. 528.—*Imperial bushel of a convenient form.*

of two forms—one a flat piece of wood, the other a roller (fig. 529). The

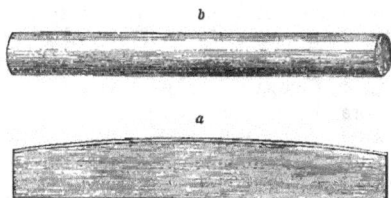

Fig. 529.—*Corn-strikes.*

a Flat corn-strike.　　b Cylindrical corn-strike.

Weights and Measures Act prescribes that the strike shall be round, of light wood, 2 inches in diameter; but many maintain that the flat strike is best fitted for the purpose.

Bagging Grain.— As already indicated, a great deal of grain is now run into bags or sacks right from the threshing and finishing machine, or from a detached dressing-machine; but in former times the universal practice was to fill sacks by the use of hand wechts or baskets.

Some care is required in measuring corn. The basketfuls should be poured into the bushel from a small height, the higher fall compressing more grains into the bushel. The bushel should be striked immediately after it is filled. The corn raised in the centre of the bushel by the pouring should be levelled lightly with a wave of the fingers of the left hand, not lower than the edge of the bushel farthest from the heap, and sweeping the edge clear of corn, the strike is applied to make the superfluous corn fall off near the heap.

Formerly, the almost universal custom was to put exactly four bushels into each bag and ascertain the weight of the grain by weighing two or three bushels taken from the body of the heap of grain. Now, grain is more frequently bagged and sold by weight than by measurement, the quantity sometimes reduced or measured, according to whether it is unusually heavy or exceptionally light, to bring it to the standard weight of four bushels. The standard weights of the different kinds of grain vary throughout the country—oats from 38 to 42, barley about 54, and wheat 60 or 62 lb., per bushel.

Hummellers.— Wheat and oats are dressed clean by the winnower; but it is otherwise, at times, with barley. When barley has not been thoroughly ripened, the awns are broken off at a distance from the grain by the threshing-machine; and as the part left must be got rid of before the corn can be clean dressed, a *hummeller* is used for the purpose. Improved modern threshing-machines are provided with hummellers, so that barley as well as oats and wheat is threshed and dressed at the one operation. The hummellers in use are of various patterns; and besides those worked in conjunction with threshing-machines, there are hummellers which work separately.

The hand-hummellers are now seldom used, except on small holdings. A very simple form of hand-hummeller is shown in fig. 530. It consists of a square frame of iron, 12 inches each way, 2 inches in depth, and ⅛ inch thick. Bars of similar dimensions are riveted into the sides of the frame, and crossing each other, form compartments of from 1½ to 2 inches square. Such hummellers are also made with parallel bars only, in which case they are less expensive but much less effective. The hummeller is used with a mincing motion on a thin layer of barley on the floor.

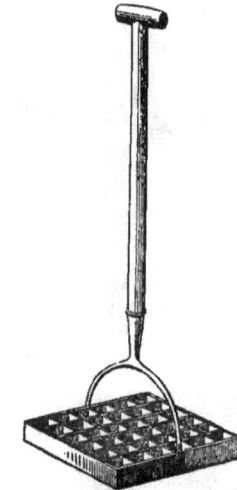

Fig. 530.—*Simple hand-hummeller.*

Tying Sacks of Corn.— Filled sacks wheeled aside should have their mouths flat-folded. On tying sacks, which they must be when intended to be sent away by cart, the tying should be made as near the corn as possible, so as to keep the sack firm.

Lifting Sacks of Corn.—There are four modes of lifting a sack to a man's back. One is, for the man to bow his head low down in front of the sack, bending his left arm behind his back, across his loins, and his right hand upon his right knee. Two persons assist in raising the sack, by standing face to face, one on each side of it, bowing down and clasping hands across the sack near its bottom, below the carrier's head, and thrusting the fingers of the other hands into its corners. Each lifter then presses his shoulder against the edge of the sack, and with a combined action upwards, which the carrier seconds by raising his body up, the bottom of the sack is placed uppermost, and the tied mouth downmost, the sack resting upon the back of the carrier. The lifters leaving hold, the carrier keeps the sack steady on his back, with his left arm across its mouth.

Another plan is, for the carrier to lay hold of the top of the shoulder of the sack with both hands and crossed arms. His two assistants do as directed before; and while they lift the sack between them, the carrier quickly turns his back

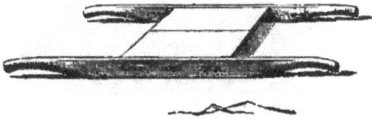

Fig. 531.—*Sack-lifter.*

round to the sack and receives it there, retaining a firm hold of all the parts he had at first.

A third plan is for the assistants to lift the sack upon another one, and the carrier lowers his back down against the side of the sack, laying hold of its shoulders over his own shoulders, when he is assisted in rising up straight with it on his back. A fourth plan, and by far the easiest, is with a sack-lifter, of which there are different patterns. A simple form of sack-lifter is shown in fig. 531. On using it, the sack is lifted upon the board; two assistants, taking hold of the handles, lift it up simultaneously, while

Fig. 532.—*Combined sack-barrow and lifter.*

the carrier turns his back to the load to receive it.

An improved and very convenient sack-lifter is shown in fig. 532. This is a combined sack-barrow and sack-lifter, made by Clayton & Shuttleworth. The barrow is pushed below the sack in the usual way, and then, by the handle shown, the bag is screwed up sufficiently high to enable a man to take it easily on to his back.

The more upright a man walks with a loaded sack on his back, with a short firm step, the less will he feel the weight of the load.

Loading a Cart with Sacks.—In regard to loading a cart with filled sacks, the general principle is to place all the mouths of the sacks within the body of the cart, that should any of the tyings give way, the corn will not be spilled upon the ground. Two sacks are laid flat on the bottom of the cart, with the mouths next the horse. Two are placed with their bottoms on the front; two on the tail-board, with their bottoms in the rear; other two on edge above

Fig. 533.—*Sack-barrow.*

a Handle.   b Shelf.
c Shields over wheels.

these four; and one behind, on its side, with the mouths of all three pointing inwards.

Corn-sacks.—The sacks for corn require attention to keep them serviceable. They are usually made of tow yarn, manufactured, tweeled, or plain. The ties for tying the mouth of the sack when it is full should be fastened to the seam of the sack. Every sack should be marked with the initials of the owner's name, or the name of the farm, or with both. The letters are best painted on with a brush, rubbing the paint upon open letters cut through a plate of zinc.

When sacks become wetted with rain they should be shaken and hung up to dry; and if dirtied with mud, they should be washed and dried. If the air in winter, when sacks are most used, cannot dry them in time to prevent mouldiness, they should be dried before a fire. Where steam is used for threshing, sacks may be dried in the boiler-house. An airy place

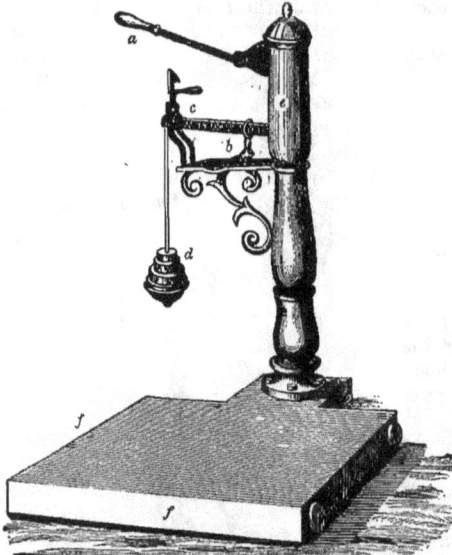

Fig. 534.—*Portable weighing-machine.*

| a Lever. | d Weight-plate. | f f Platform. |
| c Balance. | b Sliding-weight. | e Standard. |

to keep sacks is across the granary, over ropes suspended between the legs of the couples.

The best thread for darning even canvas sacks is strong worsted. When a sack is much torn, it may be used to patch others with. The person who has the charge of threshing and cleaning the corn has also the charge of the sacks, and must be accountable for their number.

Sack-barrow.—Sacks, filled, are most easily moved about by a sack-barrow, fig. 533—see also fig. 532. The height of the barrow in fig. 533 is 3½ feet, and breadth across the wheels 1½ foot.

Weighing-machines.—A weighing-machine is an important article of barn furniture, and various forms of it are resorted to. The common beam and scales is the most correct of all the instruments of the class; but it is defective, as being less convenient for the purposes of the barn than several others that are partially employed. Steelyards of various forms are also used.

Fig. 534 is a perspective view of a portable lever weighing-machine, extensively used. When the lever of the standard is up, as in the figure, any weight placed on the platform does not affect the balance-beam; but on pulling it down, it puts the platform in connection with the beam. Weights are then put on the weight-plate connected by a rod with the extremity of the balance-beam, these weights representing cwts. and imperial stones. The balance-beam is divided into parts, each showing a pound, from 1 to 14 lb. After bringing this beam nearly to the balance-level, the sliding-weight is moved along it till the balance is accurately obtained. The weights on the weight-plate show the number of cwts. and stones, and the sliding-weight indicates on the scale of the beam the number of pounds beyond the cwts. and stones. The article to be weighed is placed on the platform.

GRASSES AND CLOVERS.

VARIETIES OF GRASSES.

The grasses all belong to the natural order *Gramineæ*. Those varieties which are principally used in agriculture are detailed below. For the descriptions of them we are indebted to Mr Martin J. Sutton, the author of the well-known work on *Permanent and Temporary Pastures*.

Agrostis alba—*var.* stolonifera.

(Fiorin, or Creeping Bent Grass.)

Fr. *Agrostide blanche stonolifère.* Ger. *Fioringras.*

Roots creeping, rootstock perennial and stoloniferous. Stems 6 inches to 3 feet. Leaves numerous, flat, and usually scabrid; sheath rough; ligule long and acute. Panicle spreading, with whorled branches. Spikelets one-flowered, small. Empty glumes larger than flowering glumes, unequal, smooth, and awnless. Flowering glumes slightly hairy at the base, with occasionally a minute awn. Palea minute and cloven at the point. Flowers from July to September. Grows in pastures and damp places throughout Europe, Siberia, North Africa, and North America (fig. 535).

Although none of the creeping bent grasses are considered particularly nutritious for cattle, yet this variety is sometimes desirable in permanent mixtures, in consequence of its value in affording herbage early in spring and late in autumn, before and after other grasses

Fig. 535.—Fiorin, or creeping bent grass (*Agrostis alba,* var. *stolonifera*).

have commenced or left off growing. Its long fibrous roots and creeping habit are naturally adapted for moist situations.

Alopecurus pratensis.

(Meadow Foxtail.)

Fr. *Vulpin de prés.* Ger. *Wiesen Fuchsschwanz.*

Roots fibrous, rootstock perennial. Stems 1 to 3 feet, erect and smooth. Leaves flat and broad; sheath smooth and longer than its leaf; ligule large and truncate. Panicle spike-like, cylindrical, and obtuse. Spikelets one-flowered, and laterally compressed. Empty glumes larger than flowering glumes, hairy on the keel, awnless. Flowering glumes with straight awn inserted at the middle of

the back. Palea none. Flowers from the middle of April to June. Grows in meadows and pastures throughout Europe, North Africa, Siberia, and North-western India (fig. 536).

Meadow foxtail is one of the earliest and best grasses for permanent meadows and pastures, and may also with advan-

Fig. 536.—Meadow foxtail (*Alopecurus pratensis*).

tage be included in mixtures for 3 or 4 years' lea. It furnishes a large quantity of nutritive herbage, produces an abundant aftermath, and is eagerly eaten by all kinds of stock. The leaves are broad and of dark-green colour. The habit is somewhat coarse, hence it is unfit for lawns or bowling-greens, but its very early growth recommends it as eminently suitable for ornamental park purposes. It succeeds best on well-drained, rich, loamy, and clay soils, makes excellent hay, and should be included in a larger or smaller proportion in most mixtures for permanent pasture. Meadow foxtail is admirably adapted for irrigation. It also flourishes under trees, and should be sown plentifully in orchards and shaded pastures.

Anthoxanthum odoratum.

(Sweet-scented Vernal.)

Fr. *Flouve odorante.* Ger. *Gemeines Ruchgras.*

Roots fibrous, rootstock perennial. Stems 1 to 2 feet, tufted, erect, glabrous, and with few joints. Leaves hairy, flat, and pointed;

sheath ribbed and slightly hairy; ligule hairy. Panicle spike-like, pointed at summit, uneven below. Spikelets one-flowered, lanceolate. Empty glumes in two pairs; outer two much larger than the flowering glumes, unequal, hairy at the keels and pointed at the ends, awnless; second pair shorter and narrower than first pair, equal; also hairy and both awned, one with short straight awn inserted at the back near the summit, the other with long bent awn inserted at the centre of the back. Flowering glumes small, glabrous, and awnless. Palea adherent to the seed. Stamens two. Anthers large. Flowers April and May. Grows in fields, woods, and on banks throughout Europe, Siberia, and North Africa (fig. 537).

To the presence of this grass our summer hay-fields owe so much of their fragrance that it should be included in all

Fig. 537.—Sweet-scented vernal
(*Anthoxanthum odoratum*).

mixtures for permanent meadow or hay. The scent is less distinguishable in a fresh than in a dried state, but its very pleasant taste, somewhat resembling highly flavoured tea, is discernible at all stages of its growth. In point of productiveness this grass is inferior to foxtail, cocksfoot, and other strong-growing varieties; but the quality is excellent, the growth very early, and the plant continues to throw up flowering stalks till quite late in the autumn. On account of the broad foliage, this grass is ill adapted for grounds where short grass is indispensable; but for parks and pleasure-grounds it is especially suitable, on account of its bright green colour. Pastures in which this grass abounds naturally produce the finest mutton; and, both in a young state and when mixed with other varieties, it is much relished

by cattle and horses. It is valuable in hay, as its flavour enhances the price, and it also yields a good quantity of feed after the hay crop is cut. It constitutes a part of the herbage on almost every kind of soil, particularly on such as are deep and moist.

Avena flavescens.

(Yellow Oat-grass.)

Fr. *Avoine jaune*.   Ger. *Goldhafer*.

Rootstock perennial, creeping, and somewhat stoloniferous. Stems 1 to 2 feet, erect, glabrous, and striated. Leaves flat; radical leaves and sheaths hairy; ligule truncate and ciliate. Panicle spreading, with many branches, broad at the base and pointed at the summit. Spikelets three- or four-flowered, small, shining, and of a bright yellow colour. Empty glumes unequal, keeled, and rough. Flowering glumes hairy at the base and toothed at summit, with slender twisted awn springing from below the middle of the back. Palea narrow, short, and blunt. Flowers June, July, and August. Grows in pastures throughout Europe, North Africa, and Asia (fig. 538).

This grass may easily be discerned in July by its bright golden cluster of flowers, and is among the latest varieties in coming to maturity. The leaves are of a pale-green colour, hairy, and al-

Fig. 538.—Yellow oat-grass
(*Avena flavescens*).

though they are not produced in great abundance, are much relished by cattle. It affords sweet hay, and yields a considerable bulk of fine herbage. After the crop is cut for hay, a large aftermath is produced. This grass thrives on calcareous land, and in the Thames Valley it contributes no inconsiderable portion of the herbage of the water meadows.

Avena elatior.

(Holcus avenaceus ; Arrhenatherum avenaceum.) (Tall Oat-grass.)

Fr. *Arrhénathère élevée.*   Ger. *Hoher Wiesenhafer.*

Rootstock perennial, widely creeping. Stems 2 to 4 feet, erect and smooth; leaves scabrid and flat ; sheath smooth ; ligule short and truncate. Panicle erect and sometimes slightly nodding at the apex, widely spreading during flowering, closed before and after. Spikelets two-flowered. Empty glumes unequal and pointed. Flowering glumes two, the lower with long twisted awn, the upper with short straight awn. Flowers June and July. Grows in meadows and pastures throughout Europe, Africa, Asia, and America.

A strong-growing and rather coarse grass of good feeding quality. The flavour is slightly bitter, and on this account cattle do not at first manifest a liking for it, but when mingled with other grasses the objectionable characteristic is imperceptible. Although this plant is classed among perennials, it cannot be relied on as strictly permanent, and therefore we do not advise its employment for a longer period than three or four years. For alternate husbandry, however, it may be freely sown among other grasses, and its presence will augment the weight of the crop. On poor thin land tall oat-grass is useless, but on drained clays and rich soils generally it grows luxuriantly. The plant is a gross feeder, and must be liberally treated to bring it to perfection. The seed needs to be buried more deeply than is safe with other grasses.

Cynosurus cristatus.

(Crested Dogstail.)

Fr. *Cynosure crételle.*   Ger. *Kammgras.*

Rootstock perennial, stoloniferous. Stems 1 to 2 feet, tufted, erect, smooth, and wiry. Leaves very narrow, ribbed, slightly hairy; sheath smooth ; ligule short and bifid. Panicle spike-like, secund. Spikelets many-flowered, ovate, flat, with a barren spikelet consisting of empty glumes arranged in a pectinate manner at the base. Empty glumes sharply pointed, shorter than flowering glumes, unequal, with prominent rough keels. Flowering glumes lanceolate, with short awn at summit. Palea very thin, slightly ciliated. Flowers July and August. Grows in dry hilly pastures throughout Europe, Western Asia, and North Africa (fig. 539).

Crested dogstail is a fine short grass, and constitutes a considerable portion of the herbage of sheep-walks and deer-parks. It is found in most meadows and pastures used for grazing. Sinclair describes it as forming "a close dense turf of grateful nutritive herbage, and is little affected by extremes of weather." From our own experience

Fig. 539.—Crested dogstail (*Cynosurus cristatus*).

and observation, we can fully indorse the opinion of this eminent authority, and recommend its being included in all best permanent mixtures. We have especially noticed the beneficial results obtained by its use with other grasses in sheep-pastures ; and it is generally believed that sheep fed on pastures containing dogstail are less liable to foot-rot than when fed on pastures composed of the more soft-leaved varieties. On account of its close-growing habit and evergreen foliage, it is particularly valuable for lawns, pleasure-grounds, and other places kept under by the scythe.

Dactylis glomerata.

(Rough Cocksfoot.)

Fr. *Dactyle gloméré.*   Ger. *Gemeines Knaulgras.*

Roots fibrous, rootstock perennial. Stems 2 to 3 feet, erect, stout, and smooth. Leaves broad, keeled, and rough; sheath scabrid; ligule long. Panicle secund, spreading below, close and pointed above. Spikelets three- to five-flowered, laterally compressed, and closely clustered at the end of the branches. Empty glumes smaller than flowering glumes, unequal, keeled, and hairy on upper part of the keel, pointed at the summit. Flowering glumes with hairy keel, pointed and ending in a short awn. Palea bifid at summit, and fringed at base. Flowers June and July. Grows in pastures, woods, orchards, and waste places throughout Europe, North Africa, North India, and Siberia (fig. 540).

This well-known grass grows luxuri- antly in deep rich soils and low-lying meadows. For the enormous quantity

Fig. 540.—Rough cocksfoot
(*Dactylis glomerata*).

of produce it yields, the rapidity with which it shoots forth again after having been eaten or cut, and also for the im- portant fact of its being so much relished by horses and cattle, it is eminently suitable for sowing with other quick- growing grasses for alternate husbandry. It should be included in permanent mix- tures for tenacious soils and damp situ- ations; but in parks and ornamental grounds its tufty habit of growth renders it inadmissible. It withstands drought well, and succeeds under trees, &c. It is very useful for sowing in covers, if allowed to grow without checking.

Festuca pratensis.

(Meadow Fescue.)

Fr. *Fétuque de prés.*　　　Ger. *Wiesen Schwingel.*

Rootstock perennial. Stems 18 inches to 3 feet, tufted, erect, and smooth. Leaves flat and smooth; sheath smooth; ligule short. Panicle spreading, but closer and narrower than in *F. elatior*, with fewer branches. Spikelets many-flowered, lanceolate. Empty glumes shorter than flowering glumes, un- equal and acute. Flowering glumes rough, and slightly awned. Palea acute and ribbed, with hairy nerves. Flowers June and July. Grows on good pastures throughout Europe and Northern Asia (fig. 541).

One of the earliest, most nutritious, and productive of our natural grasses. Both in its green and dried state it is eagerly eaten by all kinds of stock. It is useful for 3 or 4 years' leas, but is especially suitable for permanent pasture purposes. It is more adapted for moist than dry soils; still it constitutes a con- siderable portion of the herbage of all high-class pastures. Meadow fescue is thus referred to by Commander Mayne, in his *Four Years in British Columbia and Vancouver's Island:* "Cattle and horses are very fond of *F. pratensis*, or sweet grass, and it has a wonderful effect in fattening them. I have seen horses on Vancouver's Island, where the same grass grows, which had been turned out

Fig. 541.—Meadow fescue
(*Festuca pratensis*).

in the autumn, brought in in April in splendid condition, and as fresh as if they had been most carefully treated all the time." Although particularly robust in habit, it never grows in large tufts, as is the case with some coarse-growing grasses. The hay from it is plentiful, and of excellent quality.

Festuca elatior.

(Tall Fescue.)

Fr. *Fétuque élevée.*   Ger. *Hoher Schwingel.*

Rootstock perennial, somewhat stoloniferous. Stems 3 to 6 feet, erect and smooth. Leaves broad, flat, and scaberulous; sheath smooth; ligule short. Panicle diffuse and nodding.

Fig. 542.—Tall fescue
(*Festuca elatior*, var. *fertilis*).

Spikelets many-flowered, half an inch long or more, lanceolate. Empty glumes shorter than flowering glumes, acute and unequal. Flowering glumes broad, rough, and toothed at the apex. Palea acute and ribbed, with hairy nerves. Flowers June and July. Grows in damp pastures and wet places throughout Europe, North Africa, and North America (fig. 542).

Some botanists consider the *F. elatior* and the *F. pratensis* to be identical, and these grasses are consequently to be found in many botanical works bracketed together as synonymous. There is, however, a decided difference, which is clearly manifest not only in the seed, but in the growth of the two varieties. The seed of the true *F. elatior* is broader and longer than that of *F. pratensis*. The growth, too, is more robust, of much greater size in every respect, and it will consequently produce a heavier bulk of hay or feed. The panicles also of the *F. elatior* are quite distinct from those of the *F. pratensis*, being branched, bent, and drooping, and composed of large clusters. Those of the *F. pratensis*, on the contrary, are decidedly upright in their early stages of growth, becoming slightly bent as the flower approaches maturity. On account of its luxuriant habit, we do not recommend the use of *F. elatior* where a fine turf is required; yet as a productive grass, and one which is greedily eaten by stock, it may form a part of permanent mixtures for moist and strong soils where the crop is intended for grazing, and also for irrigation purposes. It is admirably adapted for covers, in which its large seeds are useful as food.

Festuca heterophylla.

(Various-leaved Fescue.)

Fr. *Fétuque feuilles variées.*
Ger. *Wechselblätteriger Schwingel.*

Roots fibrous, rootstock perennial, tufted. Stems 2 to 2½ feet, numerous, erect, and smooth. Leaves various, dark green, lower ones folded triangular, upper ones flat. Ligule almost obsolete. Panicle diffuse. Spikelets many-flowered. Empty glumes unequal, shorter than flowering glumes, with prominent midrib and long awn. Flowers June and July. Grows in meadows and pastures throughout Central Europe; introduced into Great Britain for cultivation in permanent pastures.

This species is a native of France, where it is extensively grown, and was introduced to England in 1814. It is well adapted to our climate, and is valuable for parks and ornamental grounds, for its beautiful dark-green foliage. It is also particularly suited to pastures, on account of its large bulk of herbage; but it produces little feed the same season after mowing.

Festuca ovina tenuifolia.

(Fine-leaved Sheep's Fescue.)

Fr. *Fétuque des brebis.*　　　Ger. *Schaf Schwingel.*

Rootstock perennial, tufted. Stems 6 to 12 inches, erect, and densely tufted, rough at the upper part and smooth below. Leaves very slender, chiefly radical, upper ones rolled; sheath smooth; ligule long and bilobed. Panicle small, erect, contracted, and subsecund. Spikelets many-flowered, small, upright. Empty glumes shorter than flowering glumes, unequal, and acute. Flowering glumes small, with minute awn. Palea toothed, with hairy nerves. Flowers June and July. Grows in dry, hilly pastures throughout Europe, Siberia, North Africa, North America, and Australia (fig. 543).

This grass is supposed to have received its specific name from Linnæus, on account of its being so much relished by

Fig. 543.—Sheep's fescue
(*Festuca ovina*).

sheep; and Gmelin, the eminent Russian botanist, says that the Tartars generally pitch their tents during the summer months in close proximity to it, on account of its value to their herds. There is no question but that on good upland pastures, especially if used for sheep grazing, this grass should form a large proportion of the herbage. In produce it is inferior to some others, but deficiency in quantity is more than counterbalanced by its excellent nutritive qualities. From its remarkably fine foliage it is particularly suited for lawns and pleasure-grounds, which are constantly mown.

Festuca duriuscula.

(Hard Fescue.)

Fr. *Fétuque durette.*　　　Ger. *Harter Schwingel.*

Rootstock perennial, slightly creeping. Stems 1 to 2 feet, erect, and tufted, but less so

than in *F. ovina tenuifolia.* Stem-leaves flat, lanceolate, and striated; sheath downy; ligule almost or entirely obsolete. Panicle erect and spreading when in flower. Spikelets many-flowered, and larger than in *F. ovina tenuifolia.* Empty glumes lanceolate and unequal. Flowering glumes narrow, with a short awn. Palea toothed, with hairy nerves. Flowers June and July. Grows in hilly places throughout Europe, North Africa, Siberia, North America, and Australia (fig. 544).

This is one of the most valuable and important of the smaller fescues, and its presence in hay is generally indicative of superior quality. It comes very early, retains its verdure during long-continued drought in a remarkable manner, and is one of the best of pasture grasses. All kinds of stock eat it with avidity, but especially sheep, which always thrive well on the succulent herbage it produces.

Fig. 544.—Hard fescue
(*Festuca duriuscula*).

From the fineness of its foliage, and the fact of its resisting the drought of summer and cold in winter, it is eminently adapted for sowing in parks and ornamental grounds. A large quantity of food is produced after the grass is cut for hay.

Festuca rubra.

(Red Fescue.)

Fr. *Fétuque rouge.*　　　Ger. *Rother Schwingel.*

Rootstock perennial, with long creeping stolons. Stems erect, 2 to 3 feet. Leaves flat and rolled; sheath hairy; ligule long. Panicle spreading, and slightly drooping at apex. Spikelets many-flowered, of a reddish colour. Empty glumes unequal. Flowering glumes lanceolate, with a short awn. Flowers June and July. Grows in dry low-lying places near the sea, throughout Europe, North Africa, Siberia, and North America.

Although this grass is considered by some to be merely a variety of *F. durius-cula*, altered in habit by frequent cultivation on dry soil, yet to the careful observer there will appear an appreciable difference between the two varieties. The leaves are broader, of darker colour than the *F. duriuscula*, while the growth is more robust. The principal difference, however, is in the creeping habit of *F. rubra*, which enables it to live on loose, light, dry soils, where most other grasses fail. Its creeping roots penetrate so deeply into the soil, as to enable the plant to maintain a fresh and green appearance when other varieties are burnt up. It is particularly adapted for pastures by the seaside. The leaves and stems are more nutritious, and of superior bulk, at the time of ripening seed than earlier in the season.

Glyceria fluitans.

(Floating Sweet Grass.)

Syns.—POA FLUITANS and FESTUCA FLUITANS.

Fr. *Glycérie flottante.*   Ger. *Schwimmgras.*

Rootstock perennial, stoloniferous. Stems branched, floating or creeping, stout and smooth. Leaves short, flat, and broad ; ligule long, broad and pointed at apex. Panicle erect and branching. Spikelets oblong and many-flowered. Empty glumes unequal, flowering glumes scabrid, and blunt at apex. Palea with ciliated nerves. Flowers July and August. Grows in damp places throughout Europe, Siberia, North Africa, and North America.

This grass is found growing naturally by the sides of ditches, pools, lakes, and rivers, and is perhaps the only water-grass which is eaten with avidity by both sheep and cattle. The leaves are narrow, of a pale green colour, and succulent. It is valuable for moist situations, and thrives especially in the Fen districts.

Lolium perenne.

(Perennial Rye-grass.)

Fr. *Ivraie vivace.*   Ger. *Englisches Raygras.*

Roots fibrous, rootstock perennial, sometimes stoloniferous. Stems 1 to 2 feet, bent at the base, ascending, smooth, and slightly compressed. Leaves flat, narrow, and obtuse ; edges and upper surface scabrid ; sheath smooth and compressed ; ligule short and blunt. Panicle spiked. Spikelets many-flowered, solitary, sessile, distichous. Empty glumes, only an outer one to each spikelet, except in the case of the upper spikelet, which has two, lanceolate, smooth, distinctly ribbed, and shorter than the spikelets. Flowering glumes obtuse, ribbed, and with sometimes a minute awn. Flowers May and June (fig. 545).

In the year 1882 a warm discussion arose as to the character and value of rye-grass, and the part which it should play in the formation of permanent and temporary pastures, the former in particular. In that year the late Mr C. D. L. Faunce

Fig. 545.—Perennial rye-grass
(*Lolium perenne*).

de Laune of Sharsted Court, Sittingbourne, contributed a paper to the *Journal of the Royal Agricultural Society of England* (vol. xviii., sec. ser., part 1) "On Laying Down Land to Permanent Grass." There he condemned rye-grass, and urged "the necessity of eliminating" it from all mixtures of seeds to be sown in the formation of permanent pastures. In the same publication and through other channels he continued his denunciation of rye-grass, stating that— "My observations lead me to believe that rye-grass is detrimental to the formation of new pasture, not only because it is a short-lived grass, but because, owing to the shortness of its roots, it exhausts the surface of the soil; and when it dies, the bare space left is so impoverished that, though grass seeds may germinate upon it, they fail to live unless highly manured by accident or on purpose."[1]

Mr de Laune certainly formed excellent permanent pastures without the assistance of rye-grass, and it cannot be denied that some good resulted from the discussion which he aroused. It was well known that farmers did not then, as a rule, give sufficient attention to the selection of seeds for pastures, and it is also more than probable that rye-grasses sometimes bulked more largely in seed-mixtures than was desirable.

Mr W. Carruthers, consulting botanist to the Royal Agricultural Society, joined with Mr de Laune in the controversy, in so far as to contend that rye-grass is no more perennial than the wheat plant; that it would die out in two years unless kept free from seeding; and that it should therefore be excluded from permanent pastures. But he recommended rye-grass for temporary pastures, and admits that if it were eaten close down and not allowed to seed, "they might keep it alive as long as they like."

But the attack upon rye-grass did not long prevail. It was successfully repelled by the late Dr Fream, the late Sir John B. Lawes, and others, who demonstrated the important and significant fact that rye-grass with white clover form the dominant constituents of many of the finest old pastures in the country, including the celebrated feeding-pastures of Leicestershire. The results of Dr Fream's investigations are recorded in the *Journal of the Royal Agricultural Society*, vol. xlviii., sec. ser., part 2.

There is no thought now of removing rye-grass from its wonted place in grass-seed mixtures, whether for permanent or temporary pastures. As to the relative quantity of rye-grass and other grasses, hard-and-fast rules should not be insisted upon. The quantities we have stated will not suit equally well in all circumstances; and while some may think it well to use still larger quantities of rye-grass, others may perhaps find smaller give better results.

[1] *Jour. Royal Agric. Soc. Eng.*, xviii., sec. ser., part 2.

Lolium italicum.
(Italian Rye-grass.)

Fr. *Ivraie d'Italie.*   Ger. *Italienisches Raygras.*

Annual or biennial. Root fibrous. Stems 2 to 4 feet, erect, stout, smooth. Leaves long, broad, glabrous, and succulent; sheaths slightly rough; ligule short and obtuse. Spikelets many-flowered, sessile, distichous on a long rachis. Upper empty glume only present in the terminal spikelet; lower empty glume persistent, lanceolate, obtuse, scarcely reaching to middle of spikelet. Flowering glumes lanceolate. Awn almost as long as glume. Palea ciliate at base. Flowers June and July. Not known in a wild state (fig. 546).

The Italian rye-grass was introduced into this country in 1831 by the late Charles Lawson. It is very distinct in its character and seed from ordinary rye-

Fig. 546.—Italian rye-grass
(*Lolium italicum*).

grass, and as it is not perennial, it is only suitable for alternate husbandry, and producing early feed in the spring for sheep and cattle; but in permanent pastures it is to be avoided entirely. For sewage cultivation it stands in the first rank of all forage plants.

It has produced extraordinary crops at various sewage farms. On account of its rapid growth, and for its succulent herbage, it is invaluable for early sheep feed. It may be sown with safety any time between the months of February and October. If alone, 3 bushels per acre is the quantity required; but if sown on a corn crop with clovers, a much smaller quantity will suffice. In the latter case, it should not be sown until the crop is up. The mode of cultivation is exceedingly simple—harrowing the ground before and after sowing, and rolling subsequently, being all that is required. If the land is in good condition, three or four heavy cuttings per annum may be obtained, even without liquid manure; but undoubtedly, the more manure applied, especially in liquid form, the more abundant the crop; and it is important that the liquid should be applied immediately after cutting.

It is a common notion that wheat will not answer after Italian rye-grass. The following opinion of the late Mr William Dickinson on this point is worth consideration: "Thirty sheep may be kept upon Italian rye-grass, fed through hurdles, upon as little land as ten can be kept upon the common system upon common grass; and the finest crops of wheat, barley, oats, and beans may be grown after the Italian rye-grass has been fed off the two years of its existence. *Wheat invariably follows the Italian—splendid crops are grown where wheat had not been grown before.*"

it attains a great height, and forms the bulk of the grass hay of that country. In England it is largely cultivated in conjunction with other strong-growing grasses. For early feeding timothy is superior to cocksfoot. It may be pastured for some time through the spring

Fig. 547.—Timothy (*Phleum pratense*).

without damage to the hay crop. It succeeds well on soils of a moist and retentive nature, and is keenly relished by all kinds of stock, whether in a green state or made into hay. In addition to its usefulness for permanent pasture, it possesses a high value for alternate husbandry.

Phleum pratense.

(Timothy Grass, or Meadow Catstail.)

Fr. *Fléol des prés.*   Ger. *Timothygras.*

Rootstock perennial, somewhat creeping. Stems 1 to 3 feet, erect and smooth. Leaves short, flat, and soft; sheath smooth; ligule oblong. Panicle spike-like, cylindrical, elongate, and compact. Spikelets one-flowered, laterally compressed. Empty glumes larger than flowering glumes, equal, each with stiff hairs on the keel and a short scabrid terminal awn. Palea minute and pointed. Flowering glumes much smaller than empty glumes, toothed and awnless. Flowers end of June to August. Grows in meadows and pastures throughout Europe, North Africa, Siberia, and Western Asia (fig. 547).

One of the most common of our meadow plants. In some parts of America

Poa pratensis.

(Smooth-stalked Meadow-grass.)

Fr. *Paturin des prés.*   Ger. *Wiesen Rispengras.*

Rootstock perennial, creeping and stoloniferous. Stems 1 to 2 feet, erect, smooth, and rather stout. Leaves flat, rather broad and slightly concave at the tip; sheath smooth and longer than its leaf; ligule short and blunt. Panicle loose, spreading and pyramidal in shape. Spikelets three- to five-flowered, compressed. Empty glumes much webbed, lanceolate, almost equal. Flowering glumes larger, webbed, keeled, and acute. Palea short. Flowers June and early in July. Grows in meadows and pastures throughout Europe, Siberia, North Africa, and North America (fig. 548).

This variety in early spring presents a beautiful green appearance, and is easily distinguished from *Poa trivialis* by its smooth culms and leaves. Being of a more creeping habit than other Poas, it is sometimes condemned as exhausting the soil. On account of its unusual

Fig. 548.—Smooth-stalked meadow-grass
(*Poa pratensis*).

earliness and great productiveness at a period of the season when other grasses are comparatively dormant, it should be included in permanent pasture mixtures where early feed is of importance. *Poa pratensis* flourishes in dry soil, makes excellent hay and aftermath, and is valuable for garden lawns and ornamental grounds.

Poa trivialis.

(Rough-stalked Meadow-grass.)

Fr. *Paturin commun.*   Ger. *Gemeines Rispengras.*

Rootstock perennial, somewhat creeping, but not stoloniferous. Stems 1 to 2 feet, erect, rough and slender. Leaves flat, narrow, acute, and rough; sheath rough and equal to its leaf; ligule long and pointed. Panicle loose, spreading and pyramidal in shape. Spikelets three- to five-flowered, compressed. Empty glumes webbed, lanceolate, and nearly equal. Flowering glumes keeled and acute. Palea short and slightly fringed. Flowers June to end of July. Grows in meadows and pastures throughout Europe, Siberia, North Africa, and North America (fig. 549).

This grass is somewhat similar in appearance to *P. pratensis*, but the two varieties differ materially in habit and general properties. It will be seen, on referring to the illustrations, that the

flower-stems of the *P. pratensis* are slightly drooping in habit, while those of the *P. trivialis* are more erect; that the ligule (or small tongue) of the leaf in the latter is pointed, while in the former it is blunt. *P. trivialis* is adapted for good deep rich moist loams, stiff heavy clays, and irrigated meadows. It is unsuited for dry upland pastures, and if sown in such positions will soon disappear. Opinions differ as to the merits of this grass, some botanists declaring it to be only a second-rate variety. Our own experiments quite confirm Sinclair, who thus refers to it : "The superior produce of this Poa over many other species of grass, its highly nutritive properties, the season at which it arrives at perfection, and the

Fig. 549.—Rough-stalked meadow-grass
(*Poa trivialis*).

marked partiality which horses, oxen, and sheep have for it, are merits which distinguish it as one of the most valuable of those grasses which affect rich soil and sheltered situations."

Poa nemoralis sempervirens.

(Hudson's Bay, or Evergreen Meadow-grass.)

Fr. *Paturin des Bois à feuilles persistantes.*
Ger. *Wintergrünes Hain Rispengras.*

Rootstock perennial, slightly creeping, but not stoloniferous. Stems 1 to 3 feet, erect, smooth. Leaves narrow, pointed, rough on the surface and outer edges; sheath smooth; ligule none or very minute. Panicle diffuse, slender, and nodding; spikelets lanceolate, compressed. Empty glumes acute, nearly equal, sometimes slightly webbed. Flowering glumes rather large, lanceolate, with three hairy ribs. Palea with nerves slightly fringed. Flowers June and July. Grows in woods and shady places throughout Europe and Northern Asia (fig. 550).

The great recommendations of this grass are its perpetual greenness, and dwarf, close-growing habit. These qualities, as well as its reproductiveness, ren-

Fig. 550.—Evergreen meadow-grass
(*Poa nemoralis sempervirens*).

der it one of the very best varieties for lawns or pleasure-grounds, and the fact that it thrives under the shade of trees considerably enhances its value. It yields a good bulk of herbage, endures drought, and starts growth early in spring.

Poa aquatica.
(Water Meadow-grass.)

Fr. *Paturin aquatique.*   Ger. *Wasser Rispengras.*

Rootstock perennial, creeping and stoloniferous. Stems erect, smooth, and very stout. Leaves broad, rough, and with prominent ribs ; ligule short and truncate ; sheath smooth. Panicle spreading, with many branches. Spikelets many-flowered, oblong and compressed. Empty glumes unequal and short. Flowering glumes short, broad, and with prominent nerves. Flowers July and August. Grows in wet places throughout Europe, Siberia, and North America.

Poa aquatica grows luxuriantly in the Fen counties, where it forms a rich pasturage in the summer, and constitutes the chief winter fodder. In districts which are wholly or partially flooded, it is entitled to increased attention. It may be cut three or four times a-year, and produces an immense quantity of herbage on soils which will not grow other grasses. The seed is generally scarce.

VARIETIES OF CLOVERS.

The clovers belong to the natural order *Leguminosæ*, genus *Trifolium*. The generic name is evidently derived from the triple leaves of the plants.

The following are the usually cultivated forms of *Trifolium* :—

Systematic Name.		Common Name.			Colour of Flower-head.
T. incarnatum	.	" Trifolium "	.	.	Crimson.
T. pratense .	.	Meadow clover .	.	.	Red or purple.
T. hybridum	.	Alsike	.	.	Pink and white.
T. repens	.	Dutch clover	.	.	White.
T. minus	.	Suckling clover	.	.	Yellow.

Importance of the Clovers.—This tribe includes, therefore, the most valuable herbage plants adapted to European agriculture—the white and red clovers. Notwithstanding what has been said of the superiority of *lucerne*, and of the excellence of *sainfoin* in forage and hay, the red clover for mowing, and the white for pasturage, excel, and probably ever will, all other plants.

Soils and Climate for Clovers.—The *soil* best adapted for red clover, *Trifolium pratense*, is deep sandy loam, which is favourable to its roots ; but it will grow in any soil, provided it be dry. Marl, lime, or chalk promotes its growth. The *climate* most congenial to it is neither hot, dry, nor cold. Clover produces most

seed in a dry soil and warm temperature ; but as the production of seed is only in some situations an object of the farmer's attention, a season rather moist, provided it be warm, affords the most bulky crop of herbage.

Clover Seed.—Red-clover seed is imported into Britain from America, Germany, Holland, France, and even Italy, where it is raised as an article of commerce. What has been obtained from the last two countries has been found often too tender to stand an English winter. In Switzerland, clover seed is prepared for sowing by steeping in water or oil, and mixing it with powdered gypsum, as a preventive to the attacks of insects.

Perennial Red Clover.—The perennial red variety — *Trifolium pratense perenne*, or cow-grass (fig. 551)—bears a great resemblance to the biennial in its general habits and appearance, and is

Fig. 551.—Perennial red clover (*Trifolium pratense perenne*).

thus accurately described in *Permanent and Temporary Pastures*, Sutton, sixth edition, p. 78 :—

"*Trifolium pratense perenne* differs from broad clover (fig. 552) in having a somewhat taller, smoother, and, except in its very young state, a less hairy stem, and a stronger, less fibrous, and more penetrating root. It carries its flowers some way above the foliage, surpasses broad clover in succulence and weight of crop, and stands frosts much better.

"The root of perennial red clover reaches down into the subsoil, enabling it to obtain moisture and nourishment in the hottest weather, when red clover gives up from drought. This penetrating habit also affords a means of sustenance to the plant on land which is too poor to grow broad clover, and frequently makes it desirable to increase the proportion of this seed for pastures on thin uplands.

"Perennial red clover has two characteristics which greatly augment its value : flowering does not begin until at least ten days later than broad clover, and the more robust and solid stems remain succulent and eatable by stock long after broad clover has become pithy and withered. Perennial red clover fills up the gap between the first and second

cuttings of broad clover, and comes into use at a time when no other green food is available for the horses of the farm, but it rarely gives a second crop of any consequence.

"Cow - grass produces comparatively little seed from its single crop ; whereas red clover yields a good crop of seed from the second cutting, after the first has been taken as fodder. Consequently seed of the perennial variety is necessarily high in price."

Sinclair says, in his *Hortus Gramineus Woburnensis :* "In the fertile grazing lands between Wainfleet and Skegness in Lincolnshire, this true perennial red clover (*Trifolium pratense perenne*) is abundant. . . . Last summer, when examining the rich grazing lands in Lincolnshire, I found this plant to be more prevalent than any other species of clover. . . . The natural appearance of this plant in these celebrated pastures is such as to recommend it strongly for cultivation. It being strictly perennial, and the root only slightly creeping, it may be used for the alternate husbandry, for which the *Trifolium medium* is inadmissible on account of its creeping roots,

Fig. 552.—Red or broad clover (*Trifolium pratense*).

constituting what, in arable lands, is termed *twitch*. . . . The nutritive powers of this species are superior to those of the *Trifolium medium*. . . . It thrives better when combined with other grasses than when cultivated by itself ; but this, indeed, is also the case with all the valuable grasses. . . . The slightly creeping root remains permanent in the experimental garden, while the roots of the common broad-leaved clover have almost disappeared in the third season from sowing. For permanent pasture, therefore, this variety (*Trifolium pratense perenne*) is the only proper one to cultivate."

Meadow Trefoil.—*Trifolium medium* —meadow trefoil—is often confounded with perennial red clover, otherwise so worthless a weed would never have been recommended as a valuable constituent for our permanent pastures on light soils, where it never fails, by its obtrusive character, to destroy the more valuable pasture-plants around it. Sinclair owns that "the *Trifolium medium* is inadmissible in alternate husbandry, on account of its creeping roots, constituting what, in arable lands, is termed *twitch*"; and the twitch is most abundant, and therefore most troublesome, in light soils, not only in arable fields, but in pasture, where it usurps the place of better plants.

Creeping Trefoil.—*Trifolium repens* —creeping trefoil, Dutch white, or sheep's

Fig. 553.—Perennial white clover (*Trifolium repens perenne*).

clover (fig. 553)—is indispensable for low-lying pastures, and is, indeed, better adapted to pastures than to meadows. Curtis affirms that a single seedling covered more than a square yard of ground in one summer.

White Clover.—White clover is sometimes called shamrock, but it is not the true Irish shamrock. In the eastern counties it is called white suckling, which fact causes it to be confounded

Fig. 554.—Alsike clover (*Trifolium hybridum*).

with *Trifolium minus*—yellow suckling, which latter plant in Norfolk and Suffolk, singularly enough, is invariably called red suckling.

Alsike Clover.—*Trifolium hybridum* —hybrid trefoil, Alsike clover (fig. 554) —is a species possessing the properties of the red and white clovers, and was considered by Linnæus a hybrid between them. It is a native of the south of Europe, but has been introduced into the agriculture of Germany and Sweden, where it is cultivated to considerable extent in the district of Alsike. Its average duration is three years, it resists cold well, it thrives in moist lands and under irrigation, but is susceptible to drought.

Trifolium incarnatum. — *Trifolium incarnatum*, a most beautiful dark crimson-flowered clover, makes good food for cattle, and grown with winter barley, or sown alone on wheat stubbles in August, it makes excellent fodder for sheep in the month of May. It is strictly an annual, and can never be sown without risk north of the Humber. There are now in cultivation four distinct varieties— *T. incarnatum, T. incarnatum tardum, T. incarnatum tardissimum Suttoni*, and *T. tardissimum album.*

By sowing all these varieties at the

same time in the autumn, the period during which Trifolium can be fed or cut the following summer will be extended to at least a month; whereas when the early Trifolium is sown alone, it has to be all consumed in about a week, to prevent its getting pithy and worthless.

Trifolium minus—yellow suckling—is often confounded with *Medicago lupulina*, yellow or hop trefoil. Suckling, however, is much harder, and more wiry in the stem, darker in the foliage, and has paler flowers than the *Medicago*. Although an annual or biennial, it is much more suited to permanent pastures than trefoil is, and is equally at home on dry soils and strong land.

Medicago lupulina (fig. 555).—Although not a *Trifolium*, no account of

Fig. 555.—Common yellow clover or trefoil
(*Medicago lupulina*).

the agricultural clovers would be complete without reference to this plant, commonly known under the names of trefoil, black medic, or hop clover. This is the earliest of all the clovers to come to maturity in spring. On calcareous soils it is invaluable.

These are all the species of clover that seem to deserve special notice, out of 166 described by botanists.[1]

Impurities in Clover.—The most frequently occurring impurities in samples of clover seed are the seeds of dodder, plantain, sorrel, dock, cranesbill, wild carrot, self-heal, corn bluebottle, chickweed, chamomile, and scorpion grass.

[1] Don's *Gen. Sys. Garden. Bot.*, ii.—"Legumen."

VARIETIES OF GRASSES SOWN.

For one year's lea it has been usual for them to consist only of red clover, *Trifolium pratense;* white clover, *Trifolium repens;* rye-grass, *Lolium perenne;* Italian rye-grass, *Lolium Italicum;* and, on light soils, the yellow clover, *Medicago lupulina.* These, in common parlance, are called the *artificial grasses*, because they are sown every year like any other crop of the farm, and are of temporary existence.

But of late it has been found very desirable to include other strong-growing perennial varieties, such as cocksfoot and timothy, even where the mixture is to remain down but one season, and they are still more indispensable for 2, 3, 4, or 6 years' leas. The quantities sown vary but little over the country. The seeds are proportioned according as the grasses are to remain for one year or longer.

SEEDS FOR ROTATION GRASSES.

It is not advisable to attempt to prescribe definite mixtures of seeds for extensive areas of land in different parts of the country. Every county and district has peculiarities of climate and soil, which should be taken into consideration when deciding upon the exact varieties and proportions of the grasses and clovers sown. But the following mixtures will generally be found a useful standard to work by:—

For *One Year's Lea.*—Where clovers are to be sown alone, 16 lb. should be sown per acre, in the following proportions:—

	lb.		lb.
Trefoil . .	5	Red clover .	6½
White clover .	1½	Suckling . .	1
Alsike . .	2		

Cost about 12s. per acre.

Where rye-grass is the only grass used, 20 lb. in all should be sown, and the following will be found a desirable prescription:—

	lb.		lb.
Rye-grass .	8	White clover .	½
Red clover .	8	Suckling . .	½
Trefoil . .	3		

Cost about 12s. 6d. per acre.

But a far better prescription (20 lb. in all), and one costing no more, is the following :—

	lb.		lb.
Cocksfoot	½	Suckling	½
Rye-grass	4½	Alsike	1
Italian rye-grass	3	Trefoil	4
White clover	1	Timothy	1
Red clover	4½		

Two Years' Lea.—When a lea has to remain down for two seasons, a slightly heavier seeding is required, and 24 lb. in all should be sown. The following is an extremely useful prescription :—

	lb.		lb.
Cocksfoot	2	Red clover	2½
Rye-grass	6	Alsike	3
Italian rye-grass	4	Trefoil	2½
Timothy	3	Suckling	1

This will cost about 14s. 6d. per acre, but must not be depended upon for more than two years.

For 3 *or* 4 *Years' Lea* other valuable grasses, like foxtail, meadow fescue, and lucerne, may be included with advantage : 32 lb. should be sown per acre, made up as follows :—

	lb.		lb.
Foxtail	1	White clover	2
Cocksfoot	2	Cow-grass	3
Meadow fescue	1	Alsike	1
Rye-grass	12	Suckling	1
Italian rye-grass	4	Lucerne	1½
Timothy	2½	Trefoil	1

Cost about 20s. per acre.

For 5, 6, *or* 7 *Years' Lea*, from 36 lb. to 40 lb. of seed should be sown per acre, and may consist of the following :—

	lb.		lb.
Perennial rye-grass	12	Cocksfoot	2
		Timothy	2
Italian rye-grass	8	Cow-grass	1½
Foxtail	1	White clover	1½
Meadow fescue	2	Suckling	1
Hard fescue	3	Lucerne	½
Smooth-stalked		Trefoil	2½
meadow-grass	2	Alsike	1

The cost of this need not exceed that of the foregoing mixture.

The process of sowing these temporary mixtures is so identical with that practised in the sowing of permanent grasses, that the whole subject may be treated under one head.

GRASSES AND CLOVERS FOR PERMANENT PASTURE.

With the decline in the price of wheat in this country there has come a great extension in the area under permanent pasture, and there is every reason to believe that this area will continue to increase. Still, soil and climatic influences must determine in a great measure the extent of arable land that can with profit be converted into permanent pasture. Districts like the eastern and southern parts of England, being dry, are better adapted for corn than grass, and a glance at the returns for the various counties will show that the proportion of land under grass is smallest where the rainfall is lightest. In the western and northern districts, where the rainfall is heavy and strong lands abound, the summer is colder, and thus grass preponderates.

Permanent seeds like lea mixtures are generally sown in corn, and a wheat plant is perhaps best for this purpose, though oats and barley are much more commonly chosen.

Grasses for different Soils.—It is impossible to give exact advice as to the kinds and quantities of grasses and clovers required, in consequence of the extreme diversity of the soils of the country, but the following table will help greatly to determine which varieties are most suitable for any particular soil under consideration. An ample seeding per acre is 28 lb. of the larger grasses and 12 lb. of clovers, &c. ; and nearly all prescriptions include the following varieties :—

Grasses.		Especially suitable for—
Agrostis stolonifera (fiorin)	. . .	Heavy and alluvial soils.
Alopecurus pratensis (meadow foxtail)	. .	Rich deep soils.
Anthoxanthum odoratum (sweet vernal)	. .	Medium and light soils.
Avena elatior (tall oat-grass)	. .	All soils.
Avena flavescens (yellow oat-grass)	. .	Dry and calcareous soils.
Cynosurus cristatus (crested dogstail)	. .	Medium and light soils.
Dactylis glomerata (rough cocksfoot)	. .	All soils.
Festuca duriuscula (hard fescue)	. .	Medium, light, and thin soils.
Festuca elatior (tall fescue)	. . .	Deep heavy soils and clays.
Festuca ovina (sheep's fescue)	. .	Calcareous and thin soils.

Grasses.	Especially suitable for—
Festuca pratensis (meadow fescue) . . .	Medium and heavy soils.
Lolium perenne (perennial rye-grass) . . .	All soils.
Phleum pratense (timothy grass) . . .	Deep heavy soils, clays, and alluvial.
Poa nemoralis (wood meadow-grass) . . .	Rich medium soils.
Poa pratensis (smooth meadow-grass). . .	Light thin soils.
Poa trivialis (rough meadow-grass) . . .	Rich, heavy, and alluvial soils.

Standard Seed Mixtures.—The following prescriptions may be considered very safe standards :—

Good Loamy Soil.

	lb.
Foxtail	2½
Sweet vernal . . .	½
Cocksfoot . . .	4
Meadow fescue . . .	3½
Sheep's fescue . . .	1½
Hard fescue . . .	3
Red fescue	2
Perennial rye-grass . .	9
Smooth-stalked meadow-grass	½
Rough-stalked meadow-grass	1
Wood meadow-grass . .	½
Dogstail	½
Timothy	2½
Lucerne	1
White clover . . .	2½
Cow-grass . . .	2
Alsike	1½
Suckling	½
Yarrow	¼

Costing about 35s. per acre.

Gravelly Soil.

	lb.
Fiorin	½
Golden oat-grass . . .	½
Sweet vernal . . .	½
Cocksfoot . . .	2
Meadow fescue . . .	2
Sheep's fescue . . .	1½
Red fescue	3
Hard fescue . . .	3½
Perennial rye-grass . .	9
Smooth-stalked meadow-grass	3½
Wood meadow-grass . .	1½
Dogstail	1½
Timothy	1
Lucerne	1
White clover . . .	2
Cow-grass . . .	2
Trefoil	1
Suckling	3
Yarrow	¼
Lotus corniculatus . .	¼

Costing about 32s. per acre.

Clay Soil.

	lb.
Fiorin	2
Foxtail	4
Cocksfoot . . .	4
Meadow fescue . . .	3
Tall fescue . . .	1
Hard fescue . . .	1½
Perennial rye-grass . .	9
Rough-stalked meadow-grass	1½
Timothy	4
White clover . . .	1

	lb.
Cow-grass	2½
Alsike	3
Trefoil	1½

Costing about 36s. per acre.

Peaty Soil.

	lb.
Foxtail	2
Agrostis	4
Cocksfoot . . .	2½
Tall fescue . . .	1
Meadow fescue . . .	4½
Water meadow-grass . .	1
Smooth-stalked meadow-grass	2½
Rough-stalked meadow-grass	1½
Timothy	3½
Perennial rye-grass . .	9
Trefoil	3½
Alsike	1½
White clover . . .	1½
Cow-grass . . .	2

Costing about 34s. per acre.

Mr De Laune's Mixtures.

Reference has been made to the objections raised by the late Mr Faunce de Laune to the inclusion of rye-grass in seed mixtures for permanent pastures (p. 237). Although, as indicated there, good reason has been shown why farmers should still put faith in rye-grass, it may nevertheless be of interest to produce here the particular mixtures of seeds recommended by Mr De Laune for the formation of permanent pastures on different soils. They are as follows :—

	Good or Medium Soils.	Wet Soils.	Chalky Soil.
	lb. per acre.	lb. per acre.	lb. per acre.
Foxtail . . .	10	4	...
Cocksfoot . . .	7	10	14
Catstail . . .	3	3	3
Meadow fescue . .	6	3	2
Tall fescue . . .	3	8	...
Crested dogstail . .	2	2	5
Rough meadow-grass .	1½	2	...
Hard fescue . . .	1	1	4
Sheep's fescue . .	1	...	4
Fiorin	1½	2	...
Yarrow	1	1	2
Golden oat-grass . .	...	...	1
Perennial red clover .	1	1	1
Cow-grass . . .	1	1	...
Alsike	1	1	1
Dutch clover . . .	1	1	1
Total lb. .	41	40	38

SOWING GRASS SEEDS.

It is important that the most careful attention should be given to the process of sowing grass seeds, for much depends on the manner in which the sowing is carried out.

Time of Sowing.—The best time for sowing depends much upon the weather, and no hard-and-fast period can be named. April may be properly regarded as a safe and favourable month in which to sow; but if the seed-bed is ready, and the land in working order by the beginning of March, there need be no scruple as to putting in the seed. On heavy soils grass seeds are frequently sown in August or September. Grass seeds are usually sown along with a cereal crop.

Sowing before is better than immediately after a shower, even supposing the land can be worked soon after rainfall. The seeds sown before rain gradually absorb moisture from the soil and dew until wet weather sets in, and then the plants spring up with great rapidity.

Depth for Grass Seeds.—Depth of sowing affects no plants so sensibly as the grasses. Some experiments were made at Glenbervie, Falkirk, to ascertain the depth at which the common grass and clover seeds should be set, to produce the greatest number of plants. The same weight of seed was sown of each kind, and as different seeds differ in bulk and weight, the numbers of each kind differed materially. A better plan would have been to have sown the same number of seeds of each kind whatever their weight, and the proportion which came up of the plants would have been more easily ascertained than by the method adopted. Each kind of seed was covered from ¼ of an inch to 3 inches of depth in the soil. They were sown on the 1st of July, and counted on the 1st of August, and the results are shown in the following table :—

KINDS OF SEED EXPERIMENTED ON.	No. of seeds sown altogether.	COVERED AT												No. of plants that came up.	Proportion of plants that came up.
		¼ in.	½ in.	¾ in.	1 in.	1¼ in.	1½ in.	1¾ in.	2 in.	2¼ in.	2½ in.	2¾ in.	3 in.		
Perennial rye-grass (*Lolium perenne*)	348	29	30	27	19	16	19	14	21	11	9	8	4	198	·57
Italian rye-grass (*Lolium Italicum*)	276	24	21	20	13	13	10	11	8	9	6	5	5	145	·51
Cocksfoot (*Dactylis glomerata*)	300	30	22	15	15	10	9	7	5	2	..	..	..	115	·38
Large fescue (*Festuca elatior*)	312	20	24	20	16	13	13	11	9	4	2	1	..	142	·42
Meadow fescue (*Festuca pratensis*)	324	28	28	16	12	10	6	9	4	2	2	..	..	117	·36
Varied-leaved fescue (*Festuca heterophylla*)	348	31	23	20	18	12	9	6	4	1	..	..	..	124	·35
Hard fescue (*Festuca duriuscula*)	300	30	23	10	15	10	8	5	3	1	..	..	..	114	·38
Meadow foxtail (*Alopecurus pratensis*)	192	17	17	16	15	12	7	6	3	1	..	..	..	94	·49
Timothy grass or meadow cats-tail (*Phleum pratense major*)	528	52	39	37	19	16	15	7	5	..	..	..	..	190	·36
Evergreen wood meadow-grass (*Poa nemoralis sempervirens*)	228	24	14	4	1	..	..	..	..	..	..	..	..	43	·18
Rib-grass (*Plantago lanceolata*)	252	22	25	19	17	14	11	10	8	6	2	..	..	134	·53
Red clover (*Trifolium pratense*)	192	17	16	14	11	11	8	4	4	..	.	..	..	85	·44
White clover (*Trifolium repens*)	144	13	11	6	4	3	1	..	..	..	..	..	..	38	·26
Yellow clover (*Medicago lupulina*)	96	12	10	8	6	4	2	..	..	..	..	..	..	42	·43
	3840	349	303	232	181	144	118	90	74	37	21	14	9	1581	·46

In only 3 cases did the number of plants exceed ½ the seed sown, those being perennial and Italian ryegrass and large fescue—the average of the whole being under ½—viz., .46. The clovers came up in small proportion, particularly the white, which is considered a hardy plant in this climate. Of the depths, the ¼-inch covering gave the largest return of plants, and 16 per cent more than ½ inch.

Mr John Speir, Newton Farm, Newton, Glasgow, states that, in a series of trials with grass and clover seeds sown at different depths up to 1½ inch, he obtained results which do not agree with these recorded at Glenbervie. Mr Speir remarks that his experience does not favour so shallow a covering as is likely to be got by first rolling, then sowing, then harrowing with a light-toothed or chain harrow, and finally rolling. He is thus opposed to rolling prior to sowing grass seeds.

In such soil as prevails on Mr Speir's farm, which is not difficult to reduce to a fine tilth, there will rarely be any necessity for rolling before sowing. Rolling is unquestionably beneficial when by the harrows a fine smooth surface cannot be prepared for the grass seeds. If chain or light-toothed grass-seed harrows will not provide a sufficiently deep covering for the seeds after rolling, then ranker-toothed harrows may be used. The object aimed at in using the roller before sowing is to secure for the small seeds a firm level bed, where their regular brairding will not be interfered with by clods and heights and hollows. Where this can be obtained without prior rolling, there is no need to occupy time with this operation.

Methods of Sowing. — Grass seeds are sown by hand and with machines. The hand-sowing is confined mainly to small farms, while on moderate and large farms the machine is almost universally used.

Hand-sowing.—Sowing grass seeds by hand is a simple process, although it requires dexterity to do it well.

Clover and rye-grass seeds are so different in form and weight, that they should never be sown at one cast. The sower has little control over the grass seed, the least breath of wind taking it wherever it may. His sole object is to cast the seeds equally over the surface, and, as they cannot be seen to alight on the ground, he must preserve the strictest regularity in his motions. Being small and heavy, the clovers, even in windy weather, may be cast with tolerable precision. It is pleasant work to sow grass seeds by the hand. The load is comparatively light, and the ground having been harrowed fine, and perhaps rolled smooth, the walking is easy.

Machine-sowing. — But now the grass-seed broadcast sowing-machine, fig. 419, has superseded the necessity of hand-sowing on most farms. This is a most perfect machine for sowing grass seeds, distributing them with the utmost precision, and to any amount, and so near the ground that the wind affects but little even the lightest grass seed. Its management is easy when the ground is ploughed in ordinary ridges. The horse starts from one head-ridge, and walks in the open furrow to the other, while the machine is sowing half the ridge on each side, the driver walking in the furrow behind the machine, using double reins. On reaching the other head-ridge, the gearing is put out of action till the horse, on being *hied*, enters the next open furrow from the head-ridge; and on the gearing being again put on, the half of a former ridge is sown, completing it with the half of a new one by the time the horse reaches the head-ridge he started from. Thus 2 half-ridges after 2 half-ridges are sown until the field is all covered.

The seed is supplied from the head-ridge, upon which the sacks containing it were set down when brought from the steading.

The head-ridges are sown by themselves. But the half of the ridge next the fence on each side of the field cannot be reached by the machine, and must be sown by hand.

When ridges are coupled together, the horse walks along the middle between the crown and open furrow, the furrow-brow being the guide for one end of the machine, and 2 ridges are thus sown at every bout. Where ridges are ploughed in breaks of 4 ridges in width, the furrow-brow is the guide in going and the crown in returning, while sowing 2 of the ridges; and the crown in going and the furrow-brow in returning, while sowing the other 2 ridges.

Speed of the Sowing-machine. — Were this machine to sow without interruption for 10 hours, at the rate of 2½ miles per hour, it would sow about 45 acres of ground; but the turnings at the landings, and the time spent in filling the seed-box with seed, cause a large deduction from that extent.

Grass-seed Harrows. — After the grass seeds are sown, the ground is harrowed to cover them in. For this purpose lighter harrows are better than the ordinary, which would bury clover seeds too deeply in the ground. These light harrows are arranged (with wings) to cover a large breadth at a time, so that the sowing of grass seeds is a speedy process. Fig. 556 is grass-seed harrows, with wings, covering a ridge of 15 feet wide at one stretch. The harrows have a set of iron swing-trees. Modern harrows well suited for covering grass seeds are shown in fig. 557.

Working wide Harrows. — Some dexterity is required to drive these wide grass-seed harrows. They should not be moved from one ridge to the adjoining, as part of the implement would then have to turn upon a pivot, which might

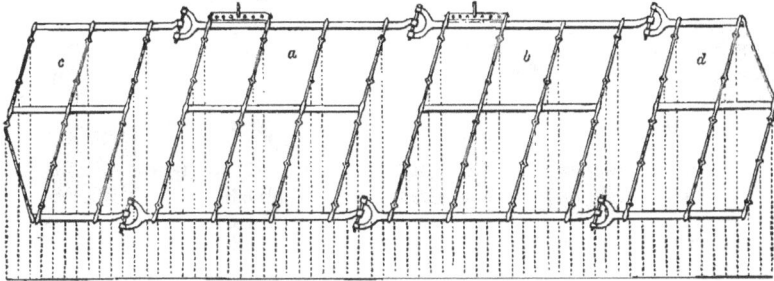

Fig. 556.—*Grass-seed iron harrows, with wings and swing-trees.*

a b Main leaves of the harrows.   *c d* The 2 wings.

wrench off a wing. Besides, it is inconvenient to *hup* the horses with these harrows. To avoid the inconvenience is to *hie* the horses at the end of the landings, round an intermediate unharrowed ridge.

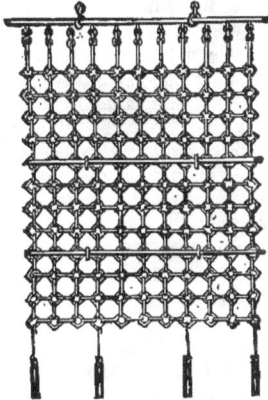

Fig. 557.—*Chain harrows.*

Harrow Carriage. — Fig. 558 is a convenient and safe form of carriage for conveying harrows. This is much better for the purpose than the ordinary cart. It consists of a frame of wood sparred in length to take on a pair of harrows coupled with their master-tree, and in breadth 3½ feet. The hind part of the frame rests on crutches supported upon the axle of 2 wheels, the upper part of the rim of which is below the top part of the frame; and the fore part rests upon

a castor, which allows the carriage to be turned when desired. A horse, to draw the carriage, is yoked to 2 eyes of the fore-bar of the frame by the hooks of the plough-chains. The harrows are piled one above the other on the framing. Such a carriage may convey other articles to and from the fields.

Rolling for Grass Seeds.—The importance of thorough rolling in sowing grass seeds is not fully realised by the general body of farmers. It is of great moment that the small seeds should have an even firm bed, and this can best be secured by rolling, which also helps to retain the moisture in the soil, a matter of great importance in dry soils.

Rough land, if dry enough, should therefore be rolled before the grass seeds

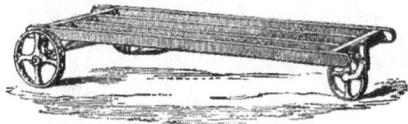

Fig. 558.—*Carriage for conveying harrows, &c.*

are sown. The rolling will reduce the clods before they become hard, and give a kindly bed to the small seeds. If the land is naturally dry, the roller is the more required to consolidate it after the winter's frosts. On light loams and turnip soils, the roller is often used with advantage both before and after sowing, the ground getting a turn of light harrows after receiving the seed.

When strong land is in a waxy state, between wet and dry, the rolling had better be deferred, but the sowing of the grass seeds may proceed, if the season, or state of the crop amongst which the grass seeds are to be sown, is already sufficiently advanced.

Crops accompanying Grass Seeds. —The cereal crops, amongst which grass seeds are sown, are winter wheat, spring wheat, oats, and barley. Winter wheat on bare-fallow clay sometimes grows so strong as to injure the young grass plants, but on lighter soils the grasses are always safely sown amongst it. There is little fear of spring wheat attaining to such growth as will injure the grasses amongst it. Oats are the usual vehicle by which to introduce grass seeds to the ground. Remaining but a short time on the ground, they permit young grass plants to grow considerably before winter, and become able to withstand the vicissitudes of that season. Barley, in some seasons, grows rank and thick, so as to endanger the existence of the grasses. Barley receives grass seeds in the same way as oats, but for some reason or other, probably because of the horizontal spread of the barley roots in the surface-soil, grasses do not thrive so well with barley as with oats.

Unless the winter wheat is too forward, the latter end of March will be the best time to put the grasses in. If the plant is strong, the common harrows will be required to obtain a hold of the ground ; if weak, and the ground tender, the grass-seed harrows will be better.

Harrowing the Wheat-braird.— Winter wheat will be all the better for a harrowing in spring, even although some of the plants should be torn up by the tines, as it loosens the ground compressed by the rains, and admits the air to the roots of the plants. After such a harrowing, rolling will press the weak plants into fresh earth, and induce an immediate formation of new shoots by tillering ; but should the plants have grown rank, the rolling should be dispensed with, in case of bruising the stems. The difference between bruising

and bending the stems of wheat by rolling should be considered, so that rolling be done or left undone. A cereal crop, on a rolled surface, affords great facility for being reaped at harvest.

Many farmers sow grass seeds without harrowing them in, trusting that they may find their way into the soil amongst the clods, and be covered by their mouldering. But the safe and correct practice is to cover every kind of seed when sown.

Sowing with Spring Crops.—Although double-harrowing across prepares the land on which spring wheat is sown for the grass seeds, these are not sown at the same time as the wheat. The wheat may be sown during winter or early spring at any time when the state of the weather and soil permit. But when wheat is sown at the latest period, the grass seeds should not only be sown then, but also amongst the spring wheat previously sown ; as also amongst the winter wheat, should there be any in the same field.

In fields in which wheat has been sown at different times, the grass seeds should be sown first on the latest sown wheat, then on the next latest, and last of all on the winter wheat. The reason for this is that it is desirable to finish the land most recently worked, in case the weather should change, and so prevent the finishing of the grass seeds over the whole field.

Frost Injuring Clover Seeds.—Frost injures clover seeds, and will even kill them when exposed to it, so they cannot safely be sown very early in spring, nor left unprotected without harrowing. But they run little risk of damage from frost in March when harrowed in, which is best done with the grass-seed harrows, the roller of course following.

If rolling the grass seeds amongst the corn cannot be done at the time of sowing on account of the raw state of the land, it should be done as soon as the state of the ground will permit, as it is of vast importance to have a firm bed for the grass seeds and a smooth surface in reaping the crop.

THE HAY CROP.

The importance of the hay crop to the agriculture of the United Kingdom is indicated by the fact that its annual yield reaches something like fifteen million tons. It occupies a larger area than all the cereal crops put together,— more than twice the area of the oat crop, which comes next to hay in regard to acreage. The treatment of the hay crop, therefore, deserves careful attention in any work dealing with the agriculture of the United Kingdom.

Varieties of Hay.—There are three main classes of hay : (1) "rotation hay," made from grasses and clovers sown in the preceding year ; (2) hay made from sown grasses which have endured for several years ; and (3) "meadow hay," or that made from natural meadows, the herbage of which consists of plants which have grown up at nature's own sweet will, or which had been sown many years back. The first is the class of hay most general where arable farming prevails, notably in Scotland and the north of England. The last abounds chiefly in the south of England and in certain districts in Ireland. The second variety is found here and there throughout the United Kingdom.

Tillage for Hay.—Nothing need be said here regarding tillage for hay. "Rotation-hay" usually follows a cereal crop ; in some cases the seeds of grasses and clovers are sown without an accompanying crop, and then the tillage is in the main the same as for the cereal crops. For the reception of the seeds of grasses and clovers a fine tilth is of especial importance, and it is highly advantageous to have the seeds firmly packed in the soil by rolling.

MANURING THE HAY CROP.

The manuring of the hay crop is dealt with in the report on the Rothamsted Experiments in this volume, pp. 23-27. What is said there should be consulted at this stage ; it would also be well to refer to the section on "The Fertility of Soil,"

vol. i. p. 317, as well as to the section on "Manures and Manuring," vol. i. pp. 446-520. It is of special importance to note the effects of the different manures on the herbage at Rothamsted —the encouragement which is given to deep-rooted plants by nitrate of soda, to shallow-rooted plants by ammonium-salts, and to leguminous plants by potash. Observe also the peculiar effects of lime upon the herbage at Rothamsted—effects which confirm the opinion that the value of liming the soil consists largely "in the neutralisation of acids, so that the organic matter can decay properly, and the liberation of reserves of potash in the soil " (vol. ii. p. 27).

Manuring Rotation Hay. — As a rule, rotation hay does not get any manures beyond what is applied to the preceding root crop. In a good many instances—and the practice is increasing —a portion of the manures is withheld from the root crop and spread on the " seeds " soon after the removal of the corn crop. Some farmers consider it a good plan to apply a part of the farm-yard dung in this way—that is, on the stubble where a hay crop is to be grown in the following season. About 8 or 10 tons is a good dressing for this purpose. Instead of dung, others give dressings of artificial manures, perhaps from 3 to 5 cwt. of basic slag or superphosphate, and on the lighter soils about 2 cwt. of kainit. For nitrogen, when it is needed, nitrate of soda in most cases does better than sulphate of ammonia. It is seldom that more than about $1\frac{1}{2}$ cwt. of either is required for hay on well-managed land.

Manuring Meadow Hay. — The systems of manuring meadow hay vary greatly, according to the character of the soil, the situation, and the rainfall of the district. For land capable of giving a good response to liberal treatment, farm-yard dung, applied in the autumn at the rate of from 8 to 12 tons per acre, usually does well. Such a dressing as this may be repeated at intervals of three, four, or more years. It is not often profitable when applied every year.

The most general custom is to dress meadow-land with mixtures of artificial manures, known as " complete manures " —mixtures containing the three main essentials of plant - food which farmers have to supply — viz., nitrogen, phosphoric acid, and potash. These dressings are given as may be required, in some cases annually, and in others at varying intervals.

English Trials in Manuring Meadows.—Trials in the manuring of old grass-land for annual crops of hay, conducted over many years by the Armstrong College, Newcastle-on-Tyne, in Northumberland and Cumberland, gave interesting and useful results. Phosphatic manures were, as a rule, the most profitable, but on the lighter and peaty soils potash also did well. Nitrogenous manures, either in mixtures or alone, were seldom profitable, especially as they discouraged clover. Basic slag was in most cases the most effective and most economic phosphatic manure. Bone-meal also gave fairly good results. Neither dissolved bones nor superphosphate did so well, due, it is believed, to the soils being deficient in lime. It is advised that the basic slag and potash should be applied in early winter and be well harrowed in. Where the herbage is coarse a good plan is to run strong harrows over it before sowing the manure. For soils in poor condition up to 8 or 10 cwt. of slag and 2 cwt. of muriate of potash, or 6 cwt. of kainit per acre may be given.

About half this dressing may be applied every three or four years thereafter. Good results have been got from 10 tons of dung and 10 cwt. of basic slag per acre, with the repetition of half the dressing of both dung and slag every three or four years, — the combination of dung and slag being, on the whole, the most profitable dressing tried.

In poor land it is recommended that the aftermath be grazed instead of being cut and carried away.

Trials in Yorkshire.—Trials conducted in Yorkshire by the Agricultural Department of the Leeds University showed that 10 tons of dung given annually increased meadow hay by 16¾ cwt. per acre, but that a more profitable increase was got from 10 tons of dung the one year and a dressing in the next year of 2 cwt. superphosphate, 1½ cwt. of nitrate of soda, and 3 cwt. of kainit per acre. That mixture of artificial manures applied yearly without any dung gave results which " left a fair profit." In these Yorkshire meadows the aftermath is almost invariably grazed.

Reading Trials.—Extensive trials in the manuring of old grass-land for hay have been carried out by the University College of Agriculture at Reading. A dressing of 1 cwt. nitrate of soda, 5 cwt. basic slag, and 3 cwt. kainit per acre had a good influence on the hay crop for a period of five years. In a comparison of basic slag and superphosphate it was found that 5 cwt. of the former gave a better return than 3½ cwt. of the latter, though the percentage of phosphates was the same in the two dressings. The addition of 1 cwt. of nitrate of soda to the slag slightly improved the yield for the first two years. Even in clay soils kainit had a good effect. All over, the manurial dressings were beneficial, not only as regards the annual yield of hay but also in its quality, as well as in the quantity and quality of the aftermath.

Trials in Wales.—In trials on the manuring of meadows, carried out by the Agricultural Department of the University College of Wales, Aberystwyth, the best results were obtained from a " complete " dressing of artificial manures — 1 1/7 cwt. nitrate of soda, 2⅓ cwt. superphosphate, and 1½ cwt. kainit per acre. It was noticed that in all cases nitrate of soda had a tendency to produce rank herbage and to promote the growth of weeds and worthless grasses. Phosphates and potash, on the other hand, had the effect of improving the herbage, the potash plots in particular containing a much higher proportion of clover and finer grasses.

In North Wales experiments conducted by the Bangor College of Agriculture tended to show that the practice of dressing meadows with dung every year is not profitable. It was found that equal crops of better quality are grown at less expense by replacing the farmyard manure with artificials in alternate years. On land in fair condition, a dressing of farmyard manure every other year even may be too frequent to give good financial results. For use alternately with farmyard manure the best combination of artificial manures has been basic slag and nitrate of soda, about 4 cwt. of the former and 1 cwt. of the latter per acre. Though kainit has occasionally given good results, as a rule it has not been needed. Excellent results have been obtained by the use of artificials alone applied every year, and for this purpose the above combination (4 cwt. basic slag and 1 cwt. nitrate of soda) has proved the best, with the addition in some cases of a little potash manure.

Yield of Hay. — The yield of hay varies greatly throughout the country. The general average of rotation hay for the United Kingdom is a little over 3 tons per acre, the yield in Ireland being over, and in Wales under, the average. The average yield from meadow hay is slightly under 3 tons per acre.

Aftermath.—The general practice is to graze the "aftermath" following meadow hay. It is only in exceptional cases that the meadow-land would long withstand the cutting and carrying away of more than one crop each season. In many cases the "aftermath" following "rotation hay" is cut at least once.

HAYMAKING.

There are few agricultural operations that are of greater importance to the British farmer than the converting of fresh grasses and clovers into hay by the drying influences of sun and wind.

Object of Haymaking.—Haymaking is the handmaid of stock-rearing. As stock-rearing increases or diminishes, so in all probability will haymaking as a branch of agriculture. Haymaking is the means by which the farmer endeavours to preserve for the winter feeding of his stock the class of food which they pick up for themselves on the fields in summer. The quality and feeding value of this preserved grass much depend upon the manner in which it has been transformed from the green to the dry condition. It is thus of the utmost importance that the process of haymaking should be conducted on the best method known and attainable. We say "attain-

able," because in our precarious climate the best-laid schemes of farmers are often upset by tantalising outbreaks of unfavourable weather.

Weather and Haymaking. — Haymaking is peculiarly subservient to climatic conditions. It goes without saying that hay cannot be made in wet weather. Even the proverbial injunction to "make hay while the sun shines" is not without limitation. By the too industrious "making" of hay, under clear scorching sunshine, the quality of the food may be considerably impaired. To expose the fresh grass to such drying influences as will preserve it with the least possible loss in its bulk and nutriment necessitates the exercise of the utmost skill and care.

It is therefore desirable that the various methods of haymaking pursued in different parts of the country should be discussed fully.

The chief principles of haymaking are alike applicable to all classes of hay. Several modifications desirable for particular varieties of grasses, and for certain localities, will be noticed presently.

The process of haymaking may, for the purposes of description and study, be conveniently divided into three sections—(1) cutting, (2) treatment between cutting and carrying, and (3) carrying and stacking.

CUTTING HAY.

Time for Cutting Hay. — An important preliminary is to decide when the crop is ready for cutting. For the moment the probability of unfavourable weather will be left out of consideration, and the weather will be assumed to be all the most ardent haymaker could desire. The precise time at which it is most desirable to cut down the crop will depend upon the object in view.

Hay for Seed.—If it is intended to obtain seed from the crop, then of course the plants must be left until the seeds have matured. This is easily determined. A few heads may be rubbed lightly in the hand, and the seeds and the heads examined. A common plan is for the farmer to sweep his hat smartly along the heads of the plants and note the seeds it catches.

Hay for Feeding.—But if the object is to raise good hay for feeding, and not to procure seed, then the cutting should take place at an earlier stage. It is well established, though the fact has not in all cases its due consideration in practice, that the production of seed and the securing of the maximum feeding value in the hay are incompatible. This is not due to the mere loss of the seed in the food, but rather to the fact that the seed is matured at the expense of the nutrient juices of the plants. It is the soluble ingredients of the hay — those soluble in water—which are chiefly valuable for feeding. Nearly all grasses and clovers contain the greatest quantity of these soluble ingredients when they are in full flower, and before the seed has been formed. The formation of the seed and/ the general ripening of the plants have a strong tendency to increase the proportion of woody fibre, and thus lessen the nutritive properties of the hay.

Best Stage for Cutting.—The stage, therefore, at which hay (from which seed is not to be taken) should be cut is when the plants are in full bloom, or at latest within a few days after the bloom or flower has disappeared. Many farmers delay cutting in the belief that any little loss by the maturing of certain of the plants will be more than made up by the increase which they imagine they obtain by growth of the under or bottom grass. This, however, is often fallacious.

Aftermath. — Moreover, the subsequent cutting or aftermath should be kept in view. The longer the first crop is left on the ground the poorer, as a rule, will be the after-growth. If the plants are left uncut until their stems become withered at the bottom, the plants may be so much impaired as to seriously lessen after-growth.

To ensure the greatest possible quantity of feeding matter in the course of the year—and this in the great majority of cases will be the object of the farmer—the best plan is to cut the first crop early rather than late. All things considered, it will be admitted that it is far more common for farmers to lose by delaying cutting too long than by cutting too early.

Premature Cutting.—Yet it is well to bear in mind the fact that the cry for early cutting may be carried too far. The agricultural chemist has shown clearly that the nutritive ingredients of grasses are not fully elaborated until the plants have reached the flowering stage, —in a few cases not indeed until the seed has been nearly ripened. Water is the principal constituent of young grasses, and it is not until they have reached the full stature of the flowering stage that the feeding properties are fully developed.

Thus the farmer should wait for the bloom before putting the mower into the hay-field. But when the flower appears he should have everything in readiness, and begin operations as soon as the weather justifies him in doing so.

Study the Weather.—In the foregoing remarks upon the time to begin cutting, it was assumed that the weather was favourable. Unfortunately, however, inclement weather has often to be contended with by the haymaking farmers of this country.

In the hay harvest the farmer must study the weather indications with unceasing care. In this he would do well to procure the aid of an efficient barometer, which can now be purchased for a very small sum.

Notwithstanding all that has been said as to the advantages of cutting the crop before the seed is formed, it will, as a rule, in the case of wet unsettled weather, be safer to delay the mower for a few days, until more settled weather sets in, than to cut down the hay and get it spoiled in the swathe by drenching rains. In contingencies of this kind, which are of frequent occurrence, the prudent farmer will choose the least of the evils which afflict him. There is no operation on the farm which demands more constant and careful attention or better judgment than haymaking. At best, in unfavourable seasons, it will often be a matter of compromise, involving not a little of the experimental element. Yet there are certain known conditions and influences which the haymaking farmer should bear in mind. The object here is to set forth these, leaving the farmer to apply them to his own individual circumstances.

Hay Injured by Wet. — One consideration which the farmer should bear in mind is that rain is much more injurious to cut than to uncut hay. No nutriment is washed out of the stalk or blade of a grass while it remains in life, no matter how heavy the rainfall may be. When the plant is dead, however, every shower of rain to which it is exposed is liable to dissolve and wash away a certain portion of its most valuable feeding ingredients.

The warmer the weather the greater the loss from the wasting of hay by rain—this for the reason that warm water is, as a rule, a more powerful solvent than cold water.

Hot and Cold Rain.—The difference in the influence of hot and cold rain upon half-made hay is very noticeable in September, when a second cutting of grass is made into hay. At that season it is observed that the half-made hay will bear with impunity double the quantity of rain it would stand in June or July.

When wet weather sets in at the beginning of the hay harvest, it is not wise to go on mowing, in the expectation that the hay will be easily made safe when dry weather returns. Before the return of dry weather the cut grass may be seriously damaged by the drenching rains to which it is subjected.

It is better policy to delay cutting until the weather has become favourable. If there are indications of more rain at hand, the mowing should be prosecuted slowly, and then, when there is reason to believe that a spell of dry weather has set in, the order to all hands should be "full speed ahead."

Cutting Rotation Hay.—The two first varieties of hay mentioned above, those made from sown grasses and clovers, are roughly classed as rotation hay, as distinguished from hay made from natural meadows that lie permanently in grass. First years' hay, that grown from seeds sown in the previous season, as a rule consists chiefly of perennial or Italian rye-grass and clovers, or it may be all three. Perennial rye-grass and clovers are most largely used. If the weather is favourable the mowing of this hay should be begun when the rye-grass has been in flower for a day or two. If the breadth of hay to cut is great in comparison with the available

force of labour, begin early, so that the main bulk of the crop may be cut down at the right time. In case of wet weather, delay a little as advised above.

Cutting Early and Late Grasses.—In hay from subsequent years' growth (as in meadow-hay), several of what are known as natural or permanent grasses are included in varying proportions. Of these permanent grasses cocksfoot and foxtail are among the earliest, and when these are plentiful the crop should be cut as soon as they go out of bloom. In a piece of meadow-land having a variety of grasses it is bad policy to lose the substance of the earlier grasses in waiting for the flowering of the later plants, more particularly if it should happen that the early varieties predominate. Here again it is erring on the side of safety to begin cutting early.

Ill-suited Mixtures.—Early and late grasses, so advantageous for grazing purposes, are not well suited for companionship in the hay crop. A certain amount of variation in this respect is practically unavoidable. It would be well, however, to guard against the association of extremes. For instance, it is imprudent to sow timothy and cocksfoot together for a hay crop. When the latter is ready for cutting, the timothy is not nearly at its best; while, if cutting were delayed till the timothy attained its greatest value, the cocksfoot would be deteriorated by over-ripeness. When timothy is sown for hay, which is extensively done, it is best sown by itself, as none of the other plants principally grown for hay ripen at the same time as it.

Clover, Sainfoin, and Lucerne.—Many experienced farmers consider it desirable to cut clover and sainfoin as soon as the first traces of the flower appear. Lucerne is often cut even earlier. In dry hot seasons its growth seems to cease before the flowering stage is reached, and in that case it is the practice with many to cut it down at once.

Preparing to Cut.

The prudent farmer will have the mowing-machines looked out and put into the pink of condition before the day arrives for the commencement of cutting. Any necessary repairs will have been effected at the end of the previous season. No judicious farmer would think of laying up a machine or implement of any kind for the idle season until the needed repairs have been attended to. It is bad practice indeed to delay such matters until the time arrives for the active employment of the machine or implement.

These general remarks apply with special force to preliminary arrangements for hay-cutting. See that all preliminaries are attended to beforehand, so that when the work of cutting begins there may be no avoidable delay.

Methods of Cutting.

Mowing-machine.—To ensure satisfactory work, the mowing-machine must be in good order. Have the knives well sharpened, and see that they work smoothly and close to the face of the fingers.

It is advisable to use a good set of knives for cutting the hay crop, more particularly if the crop is heavy or contains a large quantity of soft grass in the bottom. Half-worn knives, although good enough for cutting oats or wheat, often make very unsatisfactory work in soft grass. Whenever cutting has begun, see that the cutter-bar is as near level as possible. In many cases the outside end is by far the closest—in fact, so close that the knives are often considerably damaged—while the inside end is so high that far too much of the crop is left on the ground.

Close Cutting.—Moderately close cutting is no doubt advisable on account of the greater weight of produce obtained than by higher cutting. Very close shaving, however, is doubtful policy. Indeed most good farmers regard it as decidedly undesirable. It incurs greater risk of delays and breakages in cutting. Then it is also observed that when the plants are cut excessively close to the ground, they are, as a rule (rye-grass and clover especially), unusually long in springing up again. In very hot dry weather the roots may be injured by undue exposure to the sun.

Sharp Knives.—Keep the knives as sharp as possible, as low-cutting and easy-drawn mowers cannot be had without sharp knives. Where two or more mowers are kept going, it is advisable to keep one man sharpening knives, as then they are always in good repair, and cutting goes on more smoothly and rapidly than when the driver has to look after not only his horses but his mower and knives as well.

The most common method of sharpening the knives of reapers is with a fine file supplied for the purpose. Machines such as is shown in fig. 473, vol. ii., for sharpening are now in use.

Mower *v.* **Scythe.** — The mowing-machine is now almost universally employed in cutting hay. Except in the case of holdings too small to employ horse power, the advantages which the mowing-machine possesses over the scythe are so decided and great that the time-honoured scythe has been relegated to quite a secondary position.

Types of Mowers.—In the section on the harvesting of the grain crops, the introduction of the mowing and reaping machine has been referred to fully. Here it will suffice to say that this most useful appliance has reached a very high state of efficiency, and that the improvement in the working of the machines has been accompanied by the further advantage of a reduction in price. A very large number of firms are now extensively employed in manufacturing mowing (and reaping) machines, and farmers have the privilege

Fig. 559.—*Howard's mower.*

of selecting from a very ample collection of different "makes" and patterns, nearly all efficient, durable, and cheap, some of course better suited than others for certain localities and other conditions.

Excellent mowers are represented in figs. 559 and 560, made respectively by Howard, Bedford, and Harrison, M'Gregor, & Co., Leigh, Lancashire. The combined reaper and mower is a popular and most useful machine. Some of the best forms of these are mentioned and illustrated in the section on the harvesting of grain. Fig. 561 represents a well-known machine made by Jack & Sons, Maybole, Ayrshire.

All-round or Side Cutting.—The greatest speed is of course, as a rule, attained by working the machine in a

Fig. 560.—*The "Albion" mower.*

continuous course all round the field. This, however, is not always practicable or advantageous.

If the crop is moderate, and mostly standing, it may be cut round about where the field is no larger than can be cut in one day. Should the field be large, however, it will be advisable, even in a moderately standing crop, to

Fig. 561.—*Jack's reaper and mower.*

cut two ways only, after having thoroughly opened up the field by taking four or five swathes round the outside. The field may then be cut into breaks about 50 or 60 yards wide, and the horses driven through the crop where a beginning is desired to be made. The crop, flattened by the horses' feet and the wheels of the machine, should be cut on the return journey, as the fingers then get easily in below it, and cut it

clean. Unless the crop is very light, a narrow track should, before coming back, be cleared by the rake for the inside wheel, otherwise it is liable to get blocked up by loose hay. The trouble is very little, and much neater work can be done. The cutting may then be proceeded with till the breadth cut is equal to what remains between the first beginning and the side of the field, after which the mower should go round the remaining part.

By cutting in this manner little unnecessary time is lost at the turnings. Many fields, if inclined to be laid, or if heeled over by the wind, can be easily enough cut in this way, while they would be anything but pleasant to cut round about. Moreover, by the cutting being done from one side or end of a field right forward, the crop can be much readier coiled and ricked afterwards, more particularly if the work is interrupted by bad weather.

Laid Patches.—Where for any reason it is desired to cut a field round about, and patches here and there are lying in the wrong direction, it is the custom in some districts to turn such back by hauling a heavy plank broadside over the crop in the opposite direction to that in which the mower is moving, and a swathe or so in advance of it. A horse is yoked by a pair of plough-chains to the centre of the plank, a boy then gets on the horse's back and drives it where required.

Cutting Laid Crops.—If the crop is very heavy and laid, it, as a rule, can be cut only in one direction. In this case a beginning is generally made at that side of the field which admits of the mowing-machine going almost right against the direction in which the crop is laid.

Direction to Cut.—In choosing the direction in which to cut, it is always advisable to let the crop lie, if anything, against the divider instead of falling towards the horses' feet, as by doing so the face of the standing crop remains more erect and clean than where the opposite course is followed.

Clearing Swathe Ends.—Where the crop is moderately heavy, it is a great convenience to the person in charge of the mower to have a boy or girl along with each machine, or one for two machines, who, with a rake, can clear a small space at the entry to each swathe, and draw back the cut hay after the first swathe and before the second; and when the finishes are not parallel, rake the cut crop out of the way, so as to allow a free passage out and in.

Head-ridges.—If the head-ridges are much laid or twisted, it may be impossible to cut them satisfactorily by the mowing-machine. In this case the scythe will have to be resorted to.

With the exceptions here mentioned, no other hand-labour need be used during the cutting process.

One-way Cutting.—If mowing is done one way only, the horses should always travel back in the clear spaces between the swathes, and the wheels of the mower should straddle the swathe of cut-grass. By so doing the grass lies much more open, and dries quicker and more regularly than where it is carelessly trampled on and pressed close to the ground. Another method is to form return roads at intervals by putting two swathes together.

Loosening Lumps.—The boys or girls who are keeping the ends of the swathes clear, or the man who sharpens the knives, should also come behind, and regularly throw out any unusually thick pieces of grass which have been pulled together by the bar of the mower, raked up by the boys, or from any cause whatever are gathered into a thicker part than usual.

Difficulties with Heavy Crops.—With very heavy crops, it often happens that the shedder of the machine is unable to turn back the cut grass. In that case boys or women should go along and turn it back with a rake handle. In a heavy crop which is tangled the growing plants are often pulled over by the shedder-board. In that case, unless the side of the standing crop is straightened up, the wheel of the inside shoe runs on the top of it, and the result is very indifferent work.

"MAKING" THE HAY.

The operations between the cutting and the carting or stacking of the hay may be conveniently described under the above heading. Indeed these operations may be said to constitute the "making" process.

Variations in Practice.—In this, which is really the chief part of the work of the hay harvest, practice varies greatly throughout the country. To a large extent, no doubt, these variations have no other grounds for their existence than the peculiar tastes and notions of the farmers themselves, who, it may be frankly confessed, own a full share of the contrarieties and perversities of human nature.

Grounds for Variations.—In most cases, however, the differences in the methods of haymaking are accounted for by variations in the soil, climate, system of farming, and the purposes for which the hay is intended. In particular, the making process must be varied with wet and dry seasons, heavy and light crops, and with the particular class of hay.

There is thus good reason for variety in the practice of haymaking; and while notes are appended descriptive of methods known to be pursued with success on widely separated parts of the country, and in different conditions of soil, climate, and system of farming, it is deemed right to say that it is not presumed that these are positively the best methods for all circumstances.

In haymaking, as in most farm practices, each individual farmer must think for himself. The prudent farmer is eager to know the methods which are pursued with success by others. Having acquainted himself with these, he must carefully consider their adaptability to his own peculiar circumstances. He will not hesitate to adopt such features of these methods as seem to improve upon his practice hitherto, yet he rightly deems it wise to introduce radical changes in a tentative way.

Haymaking controlled by Weather.—Haymaking, beyond almost every other farm operation, is incapable of being conducted with success upon any definite or hard-and-fast system. It is so thoroughly within the control of the "clerk of the weather," and that important "functionary" is so fickle that every season, nay, even every week, may demand treatment peculiar to itself. The farmer must watch closely these uncertain and shifty conditions, and be prepared at any moment to vary his practice to suit them.

This very fact renders it all the more important that the farmer should acquaint himself as fully as possible with the various methods of haymaking pursued with success throughout the country, so that he may have the greater resource in battling with untoward circumstances as they arise.

English Methods.

In England, speaking generally, the prevailing methods of haymaking are somewhat different from those most largely pursued in Scotland and Ireland. As a body, Scottish and Irish farmers are not so highly accomplished in the art of haymaking as are their English brethren. Less experience and less encouragement are mainly accountable for this. There is only a very small extent of Scotland really well suited for hay-culture, while in many parts of England the hay crop plays quite a leading part in the economy of the farm.

We have many a time observed and contemplated with delight the care, intelligence, and methodical precision exhibited on well-conducted English farms in the harvesting of hay. The practice would seem to be reduced almost to the nicety of a fine art, and it is conducted with the enterprise and forethought happily characteristic of British agriculture.

Meadow Hay.—The making of natural or meadow hay—hay grown from permanent grasses—is, as a rule, slightly different from the making of rotation or clover hay. The former abounds largely in England and Ireland.

Haymaking Machines. — Tedding, swathe-turning, and other improved hay-

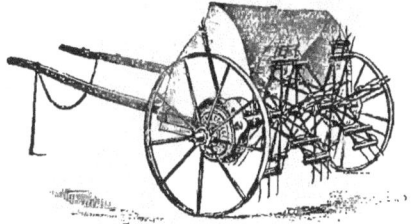

Fig. 562.—*Howard's haymaker.*

making machines are now largely used in all parts of the country. Types of these useful machines are represented in fig. 562, a double-action tedder or hay-

Fig. 563.—*Swathe-turner.*

maker, made by Howard, Bedford, and fig. 563, a modern swathe-turner, made by Blackstone, Stamford.

Swathe-Turning and Tedding.— Opinions differ somewhat as to the

tedding or turning of hay. In favourable weather and with a good crop cut by the mower, most farmers set the tedder to work to scatter the swathes as soon as a few acres have been cut down. Others think it better to leave the swathes to wither for a day or two before being disturbed. Obviously grass which lies in a thick swathe gets withered and dried on the upper side, while the portions on the lower side remain fresh and damp. If the weather should be favourable it is therefore inadvisable to delay stirring the swathe.

On many farms the crop is immediately turned a second time, and by this the "making" process is greatly expedited in dry weather.

Forward and Backward Tedding.— Tedders are made with both a forward and a backward action. By the first the grass is carried forward below the machine right over the top, and then scattered behind. In the backward action, the tedder merely picks up the grass and gives it a less or more vigorous "kick" backwards. The former movement is of course much more violent than the latter, and many farmers are opposed to it in the belief that the hay is thereby injured, seeds dashed out, and the leaves and stems bruised and broken.

Experiments in Tedding.—An important point in working the haymaking machine is the speed at which it revolves. In a series of experiments made by Mr Howard of Bedford, it was found that the barrels with the lowest speed for the back action made the best work, the crop being left looser and more hollow; the higher-speeded machines, owing to the greater violence of the throw, left the crop flatter on the ground.

Speedy Haymaking.—With a light crop the back action will be quite sufficient for spreading the swathes. Indeed, in some cases, in favourable weather, crops of a ton or more per acre are cut the one day, the swathes drawn into "windrows" by the horse-rake next forenoon, shaken out by the haymaker in the back action in the afternoon, then raked together, and carted in capital condition towards evening.

Tedding with High Wind.—High wind may be troublesome in haymaking. During a strong wind it is desirable to arrange for working the haymaker, when used in the forward action, sidewise to the wind: this may often be done by working obliquely across the swathes. It is, however, desirable to avoid using the

forward action when the wind is troublesome, inasmuch as the crop becomes very unevenly spread.

Tedders injuring Clover Hay.—Hay in which clover forms a considerable part is liable to injury by the use of certain haymaking machines. The leaves of the clover become so brittle that the violent motion of the machine breaks them in pieces, and thus causes loss in the crop. Easy back action may be employed in fresh clover hay with impunity, if it is done carefully and at a slow pace. Turning with the hand-rake or modern swathe-turner is safer for clover-hay. Indeed, the swathe-turner, such as Blackstone's shown in fig. 563, is a most efficient machine for making clover, ryegrass, or timothy hay.

Collecting Hay.—In collecting the hay after it has been scattered for drying, manual labour has to a large extent given place to horse-labour and mechanical appliances. The horse-rake, such as that in fig. 564 (Ransome, Sims, & Jefferies), is

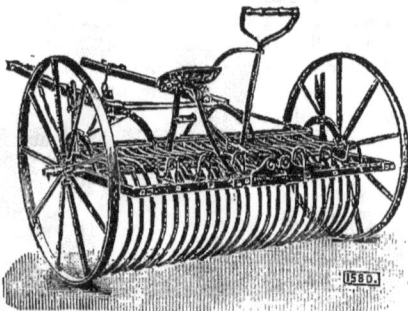

Fig. 564.—*Horse-rake.*

an excellent labour-saving machine, and is now universally employed. By it the partially withered grass is drawn into "windrows." The rake is started at the side of the field first cut, and emptied at intervals, regulated according to the carrying capacity of the rake and the weight of the crop. The rows of gathered hay thus formed are called "windrows." In these rows the hay lies loosely, and in this condition, with favourable weather, it dries speedily.

Where a modern swathe-turner is available, the rows may be turned once or twice during the day, in such a way that four swathes are put into one.

When this is done, gathering together by the horse-rake may be dispensed with prior to cocking.

Unless the weather is dry and quite settled, no more hay should be gathered into windrows than can be cocked in the same day. As will be readily understood, a fall of rain upon the windrows would cause more delay than a similar fall of rain upon hay in the swathe or thinly spread as by the tedder, for in this case the windrows would again have to be "spread out to dry." In dry, settled weather, more freedom may, of course, be exercised in these all-important operations.

"Cocking" Hay.—As soon as the hay is in a fit condition for putting into cocks, the horse-rake is run along the windrows, drawing the material into heaps, which by the hand-fork are speedily formed into cocks.

The practice in the cocking of hay varies greatly in different parts of the country, and is likewise modified to suit the weather at the time.

Large and Small "Cocks."—In the greater part of England the half-made hay is usually collected into very small cocks, often containing no more hay than a man could lift by the hand-fork at two or three turns, sometimes even less. In Scotland, Ireland, and the north of England the more general practice is first to put the hay into very small and then into larger heaps. In a damp climate, or in cases where the hay has to be carted a long distance to the homestead, big cocks may be desirable; but for moderately dry districts, or where expedition is the order of the day, there is much less advantage in the method. The practice of putting hay first into small and then into larger heaps is no doubt attended with waste, owing to so much hay being so long exposed. Still it has certain advantages, especially in the north, where, through the clashing of haymaking and turnip thinning, the former operation has often to be carried out by fits and starts. If large cocks are properly formed, a large amount of rain must fall before they will suffer much damage.   •

Differences in Methods.—Although it is convenient to treat of English, Scottish, and Irish customs separately, it should be stated that no very distinct line can be drawn between the practices of the three countries. In each country nearly all the known methods of hay-making are pursued to a lesser or greater extent. Certain methods are more prevalent in one district than in others, but the differences lie in degree more than in principle.

The English custom of taking the hay right from the first small cock or coil into large stacks or hay-sheds has the effect of saving time, and of inducing in the hay in the stack or hay-shed a certain amount of fermentation, which is regarded as rendering the hay more palatable to stock.

Scottish Methods.

In Scotland and the north of England first year's hay—that grown from grasses and clovers sown in the previous year—is the variety most general. Hay of this kind, especially when it contains a considerable proportion of clover, must be handled more tenderly in making than hay from permanent grasses, as the former is more liable to injury by breaking and bruising. This circumstance is in a large measure responsible for the distinctions between the methods of haymaking most prevalent in England and Scotland respectively.

Rotation Hay.—A light crop of ryegrass and clover, if cut before mid-day in clear dry weather, may be in condition to "coil" or "cock" by the afternoon of the day following. A heavy crop, however, even in good weather, will require a clear day between cutting and coiling; while an extra heavy crop, or one including a large amount of clover, will most likely require two clear days. If the weather be dull, damp, or wet, the process of coiling may be of necessity delayed an indefinite time.

Turning Clover Hay.—It has been explained that the turning of the swathes of hay grown from permanent grasses is usually performed by the haymaking or tedding machine. Rotation or clover hay, on the other hand, is generally turned by the swathe-turner or by a small hand-rake or the hand-fork, such as is shown in fig. 565 (Spear and Jackson). The reason for this difference in practice is, as already explained, that the rotation or clover hay is more easily bruised and broken than the softer and tougher produce of the permanent grasses.

Process of Turning.—In the forenoon of the day on which it is considered that the grass should be ready to coil, the swathes should be turned over by rakes, so as to expose the under surface to the sun, and allow the wind to easily play through the whole mass. This it cannot do if allowed to lie one or more days in the condition in which it was left by the mower, as the longer it is left untouched the closer it lies to the ground, and although the upper surface may be hard and dry, the underlying stalks and leaves will be wet and damp.

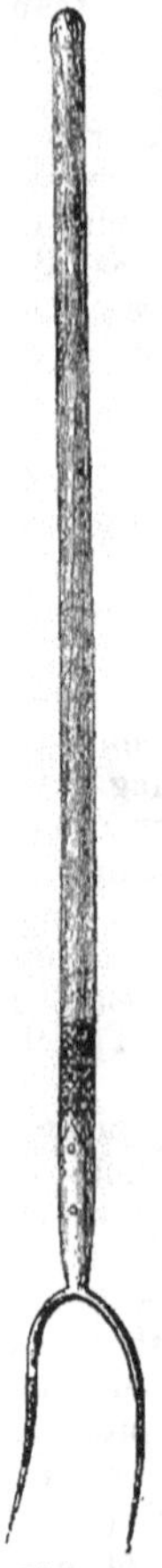

The turning-over process should, if at all practicable, be done early in the forenoon, so that as long time as possible may elapse between the turning and coiling, without allowing it to lie exposed to the dews of another evening, and the risk of rain on the day following.

The operation of turning by hand is best performed by the person walking in the cleared space left by the mower on the side of the swathe farthest from the turned-over part, and then by catching the folded-over part of the swathe with the teeth of the fork or hand-rake, and sharply pulling it towards him, the whole under surface is brought to the top. The work is more easily performed when the operator walks in the direction in which the mowing-machine was drawn. Several persons generally go together when turning-over is being done, the one following at a yard or two behind the other.

Care should be taken to turn the whole swathe upside-down, because if the work is slovenly done, the near side of the swathe will not be turned over at all, and, of course, when the hay is gathered together, those parts may in

Fig. 565. Hand hay-fork.

that case be quite damp, while the bulk of the hay may be in good condition.

Fit for Coiling.—If after mid-day the grass is withered and free of all positive damp or wet, even although it should have a raw feel, it will be in condition for coiling. Should the crop have lain on the ground for several days, and there be damp or wet parts in it which have not been exposed to the sun and air, it is a good plan to shake out the rows by hand or machine, or to rake it together an hour or two before coiling begins. By doing so the rake on being relieved turns the hay upside-down, so that any damp parts are exposed to the sun, and have a chance of getting dried before being covered up in the coil.

Again, it may be remarked, that so much depends on the crop, season, and weather, that it is impossible for any one to properly describe what are the requisite conditions of dryness suitable for coiling. It is not difficult to learn, but can be learned only by practice and experience.

Methods of Collecting the Hay.—It having been considered that the requisite dryness has been obtained, the crop may be gathered together by the horse-rake preparatory to coiling. The best rakes are the self-relieving ones, as the man can then devote all his attention to his horse and machine, and put off the rows more nearly straight and regular.

Coils or Cocks.—Coils or cocks are small conical heaps of half-dried hay, put together of such a size and shape as to admit of the hay continuing the drying process, and yet preserving it from serious deterioration by rainfall. According to the class of crop, the climate, and dryness of the hay at the time, coils range upwards in size from a yard in diameter and a yard in height.

Forming Coils.—Many used to contend that coils could best be made by the hands, unassisted by graips or forks, but this is rarely done nowadays. A forkful or an armful of hay is taken as left by the horse-rake, and roughly shaken and allowed to drop on itself as regularly as possible. A second forkful is then put on the first, being, however, a little more carefully shaken or spread, and kept within as little space as possible, so as to leave the top narrow. A third and smaller quantity should now be taken and, more carefully than the previous two, shaken over the top. The coil will now present the appearance of a blunt-pointed cone widened at the base. The base is narrowed by pulling all the loose hay out and spreading it carefully over the top. The handfuls of hay, when pulled out, are comparatively straight, and when spread over the top of the coil assist·very materially in throwing off wet.

Well-made Coils.—The protection of the hay from damage by rain in the coil much depends on the carefulness with which the bottoms or bases are pulled, and how the pullings are spread on the top. The operation is comparatively simple, yet many farm-hands do it badly; and while a properly-made coil is proof against any moderate rainfall, a badly-made one may be much damaged by a single heavy shower.

Badly-made Coils.—The great faults of badly-made coils are portions of the hay being put into the heap in a doubled-up condition, which holds the rain instead of throwing it off,—the one forkful or armful being put on the top of the other without being properly shaken out, which allows the rain to go down the division between the two, and admits of the top half being easily blown off; and neglect of pulling the bases causes the loose hay round the bottom to get easily wet, and when once wet it is difficult to dry — a well-pulled coil drying at the base in half the time required by a badly-pulled one. Neglect of pulling at the base also leaves the top of the coil unprotected, for the straight hay carefully spread over the top acts very much as thatch to an ordinary stack. It leads the rain from the top to the side, down which it readily flows, while without it the rain would run right down the centre of the coil.

Small Coils.—If the hay is not in thorough condition for coiling, and the weather looks as if rain were about to fall, the crop may be secured in smaller coils than usual, which will act as a partial protection against the rain, and yet be small enough to allow the wind to blow through them, and in part complete the drying.

Remaking Coils.—As soon as the weather has become bright again, the coils should be remade; and if the crop is now in moderately dry condition, two heaps may be put into one. Should the crop have been put together too green, the coils may be shaken out in the sun a few hours before being remade; but on no account should coils be shaken out to any great extent, unless the weather is such as to give a reasonable assurance of their being rebuilt before rain again falls.

Avoid Over-working.—At this stage it may be mentioned that rye-grasses and clovers are rarely improved by much shaking out and remaking. Hay from these plants, indeed, as formerly indicated, should always be secured with as little knocking about as possible.

Timothy Hay.—Timothy, like natural meadow hay, will stand a good deal of shaking. Timothy is the easiest secured of all the grasses, owing to the small proportion of leaves which it contains, and the length and strength of its seed-stalks. A heavy crop of timothy can often be cut the one forenoon and coiled the afternoon of the day following, whereas a similarly heavy crop of rye-grass and clover would require two or three days to get ready for coiling. The great preponderance of stalks which this crop contains over all others keeps it so open, that, in the coil, it dries much quicker and more efficiently than any other grass. Owing to the length of the stalk, timothy coils have generally to be made much larger than coils of other classes of hay. But its openness in texture, although an advantage in drying, is also a disadvantage in case of heavy rains. On account of its exceptional length and strength, timothy is probably the most difficult to coil properly of all the grasses, and in consequence it often is the worst coiled, the result being that heavy rains damage the crop very badly by running down through it.

Thunderstorms and Haymaking.—Coils, although a fair protection against moderate showers, are often little protection against heavy thunderstorms. Where thunderstorms are frequent and heavy, the system is therefore little resorted to for open crops such as clover.

Securing Wet Coils. — During a

heavy thunder-shower, or light continuous rain for one or more days, as occasionally happens, the rain often runs under the coils, and so wets them that only in exceptional cases could they ever be expected to dry if allowed to remain in their original position. After such rains the probability is that in a few days it will be found that, while the very top of the coil is dry enough, three or four inches farther down it is quite wet. If the weather is settled, and the centre of the coil is sufficiently dry to enable it to be put into the field-rick, or tramp coil or cock, the tops (which will likely be damp with the morning dew) may be taken off and laid to one side, while the body of the coil is taken away and secured in the rick, the bottoms also being left alongside the tops. These, if carefully spread out in the sun for an hour or two, very soon dry, and can ultimately be gathered up with the rakings.

If, however, the body of the coil is not dry enough to admit of being secured in the field-rick, the top should be taken off to well under the damp portion, and a new coil made, the tops and bottoms being loosely spread over the top. In this way the damp material very soon dries, unless the quantity of it has been all the greater, while at the same time the operation can be performed without exposing the crop to further damage from the elements.

Hours for Coiling.—As a rule, the dew prevents coiling early in the morning. In most cases coiling is done between eleven o'clock in the forenoon and evening, the bulk being done after midday.

Raking.—As soon as possible after the coiling is finished, the land should be raked clean between the rows of the coils before it gets any rain, if at all possible. These rakings are in some cases carefully spread over the tops of the nearest coils, and in others coiled by themselves. The rakings become soaked easily, and if once wet are difficult to dry, more particularly if they have been put on the top of the coil without being methodically shaken out. Great care should therefore be taken to see that only a few are put on each coil—if put there at all,—and that these are thoroughly shaken out.

Coiling in High Wind.—If a good breeze should be blowing during the operation of coiling, considerable annoyance is often caused by the tops being carried off. Under such circumstances the hay should be built well to the windy side of the coil, the operator always standing on that side, with his or her back to the wind. In this manner the coils can be built so that they are less liable to be blown over. After the field or plot is finished, the whole should be again gone over, and the damaged ones repaired, as when rain falls on them in this state they are liable to be seriously injured.

Time for Field-stacking.—The time hay should stand in the coil before being transferred to the stack is regulated solely by the dryness of the crop and the weather at the time. Where the crop is light, and has been well dried before being put into the coil, it occasionally may be stacked the day following, should circumstances and the weather permit. In fact, during dry and settled weather it often is not coiled at all, although in most parts of Scotland, unless under exceptional circumstances, the practice is not considered a good one.

The stage at which the crop will keep in the stack without loss of colour or excess of fermentation is one which must be seen to be learned, as it cannot be described in words. When, however, the hay is considered in condition to stack, no time should be lost in making it secure if the weather is at all favourable, for hay can never be considered anything like secure until it is in the stack or shed.

Temporary Stacks. — In districts where it is customary to sell hay and cart it direct from the field to the consumers' premises, field-stacks are usually made about one ton in weight, more as a matter of convenience for loading the carts than for any other reason. In other districts, however (and they are the most numerous), where the hay is consumed at home, the hay may not be stacked on the fields but in the stackyard near the steading, or it may be put into hay-sheds. If put into the smaller size of field-stack, there is much less risk of damaging the hay by stacking it too soon. As a rule, hay can safely be put into a 12-cwt. or 14-cwt. stack a day or two earlier than it would be judicious to put it into one weighing a ton. This alone is no mean consideration, for every additional hour the hay stands in the coils the greater will be its risk of damage.

Stack "Kilns" or "Bosses."—For the purpose of saving one or two days' exposure in the coil, it is customary in some districts to use triangles, "kilns," or "bossings" for the centres of the stacks. According to the district, these are usually from 6 to 10 feet high, and are made of thinnings of plantations or other suitable wood. Some farmers have a supply of such permanently bolted together at the top and nailed by spars at the sides, while others form them as required out of the ordinary supply of stackyard props. In the latter case they are usually tied at the top with a piece of stack-rope, which is made tight, no side-spars being used at all. Used in this way, stack-props serve a double purpose, and as single props they are more easily handled and stored away when not in use than are permanently made triangles. The labour of setting temporary triangles up is also very little, as a man can tie the three props together and set them up in a minute or two; and if there is likely to be little time for such while ricking is going on, they can often be made and erected in the morning before the dew is off the hay. Still the permanent "kilns" are in some respects more convenient.

Situation of Field-stacks.—In some localities it is the custom to build the stacks anywhere over the field, wherever hay can be got, while in others the usual method is to build a row of stacks across the rows of coils. If the rows of coils are short, say under 150 yards long, the stacks are built at the middle, and the crop brought in from both sides. If, however, the rows are longer, two or more rows of ricks may be required.

Hay-collector.—For the purpose of hauling the hay from each end of the row to the stack, several methods are in common use. An efficient method is by the hay-collector shown in fig. 566 (East

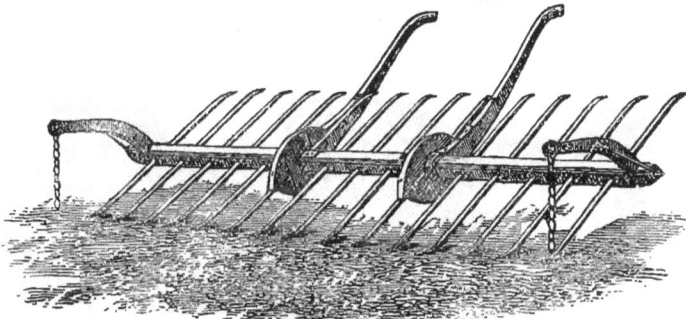

Fig. 566.—*Hay-collector.*

Yorkshire Cart and Waggon Co.) Appliances of this type are extensively used, and they do their work speedily and well.

Hay - sledge. — Another excellent method of collecting hay is by the hay-sledge shown in figs. 567, 568, and 569 (John Wallace & Sons). This sledge has two runners, which are usually straight or nearly so on the upper edge, and more or less curved on the under one. Across these are fixed four cross-bars, on which are nailed thin boards running the whole length of the sledge, which may either be fitted closely together or have spaces between each strip. The sledges may in size be from 7 feet to 9 feet wide, and say from 8 to 11 feet long. They are inconvenient if made wider than can easily pass through an ordinary field-gate. For the smaller sizes one horse will be sufficient to draw all that can conveniently be put on, while for the larger sizes two horses will be required.

In working, the sledge goes to the end of the row farthest from where the rick

Fig. 567.—*Wallace's hay-sledge, empty.*

Fig. 568.—*Wallace's hay-sledge, in the act of loading.*

Fig. 569.—*Wallace's hay-sledge, with load.*

is to be, the coils being loaded on it as it moves nearer the rick. If the crop is light and the rows short, the whole row between the end and the rick will be cleared by one sledgeful. If longer and heavier, two loads may be required.

Loading the Sledge.—One person, usually a boy, will be required to lead the horse in the sledge, while one, two, or more boys, women, or men lift on the coils with light steel graips or forks, the former being preferable, as by the graip or spade handle the workman prevents the load from turning round in his hand, an accident which is frequently happening with the ordinary fork. Two persons, particularly women or boys, work best together, as they have then quite sufficient strength to lift between them a full coil each time. In this way fewer rakings are left, the hay is not unnecessarily tossed about, and is in consequence easier forked at the rick, while more can be put on the sledge than if each coil has been lifted at two or three times.

Unloading the Sledge.—On arriving at the rick two or more persons put their forks into the front part of the load, against which they throw their whole weight, when the lad moves forward the horse, thus pulling the sledge out from under the load. If the sun is bright and the hay dry, the load of hay is shoved off with little exertion. In fact, in going up even a slight incline the load may sometimes come off of its own accord; but if the sledge has been wetted by rain or dew, or by dampness from the bottoms of the coils, it is sometimes difficult to get the hay removed.

Other Methods of Hauling.—Besides the methods of bringing in the hay here described, there are several other ways in which the same operation is performed. The most primitive is to carry it in by forks, or as occasionally has to be done on soft meadows, by thrusting two poles under each coil and then carrying it off by two persons, after the manner of a double-handed barrow. This system is of course adopted only where the land is too soft to safely carry a horse, or where labour is abundant and wages are small.

Another method is to join the chains from two horses by a pair of stout ropes, travel a horse along each side of a row of

coils, catch the coils by the rope connecting the horses, and sweep all that can be caught into the rick.

American Hay-collector.—A hay-collector invented in America is now much used in this country. This machine as now made is capable of collecting hay from the swathe, the coil, or shaked-out coils. One man with a pair of horses will bring in sufficient to keep from two to four men building stacks, and will leave much less to be gathered by the rake than by any other method yet introduced. This machine is about 12 feet wide, and rests on three light wheels, one on each side and another a few feet behind. To the under side of the framing of the machine twelve teeth 5 feet long are bolted. These teeth are of wood and pointed with iron, and are fixed about a foot above the ground. The driver, sitting on a seat above the hind wheel, can raise or lower the points of the teeth, whether loaded or unloaded, until they are a foot above the ground. On the outside of the apparatus, at each side of the teeth, a pole is placed, outside of each of which a horse is yoked to a plough-tree at one end of the pole and by a breast-strap or chain at the other. The points of the teeth being lowered in front of a windrow of loose hay or coils, and the horses guided down each side, the teeth slip under the hay until a load is gathered, when the driver lifts the points of the teeth from the ground, turns round his horses if need be, and carries off his load to the rick. Here he drops the teeth of the apparatus on the ground, backs his team, when the breast-straps of the horses pull the apparatus from under the load of hay, the points of the teeth are again raised, and the team move off for another load.

This apparatus is very light, being built almost entirely of wood, and very strong. It is said to be able to gather and carry from 4 cwt. to 7 cwt. of hay. It was introduced into Scotland by Mr John Speir of Newton in 1889, and since its introduction it has been more largely adopted in the south of England than in Scotland. In the former it is used principally for bringing the hay to the large stack, which in these circumstances is usually built at one side or corner of the field. When combined with the elevator, the hay is brought from the windrow or coil, and delivered on the top of the stack, with remarkably little handling.

Preparing for Stacking.—If there is any dampness worth speaking of on the bottom of the coils, they should be turned upside-down an hour or so before stacking is commenced. In turning them over they should be turned so that the bottom faces the sun, and if the day is cloudy the bottom should be turned to the wind. As it is never advisable, even in good weather, to turn up a great many coils at one time, no more should be done than will just allow them to dry before being removed, an odd person going in front turning the coils over at or about the same speed as the crop is being stacked.

It often happens that mere turning of the coils upside-down is not sufficient to make the crop dry enough for stacking. In this case the rows of coils must be thoroughly shaken out, the space occupied by the spread hay depending very much on whether or not the crop requires much or little drying.

If the day is dull, or the crop more than usually damp, the coils not only require to be shaken out, but to be turned over with rakes or forks, in much the same way as the swathes are turned upside-down, before the hay is brought into the stack. Soft grown meadow hay often requires to be treated in this way, although it is very unusual to do so with timothy or clover and rye-grass. Where hay has been thus spread out, it should be gathered into narrower rows by hand-rakes or forks, or by the horse-rake, before beginning to stack, as the time thus spent is saved when stacking is begun, while the crop is drying all the time. If one or more persons can be spared to collect the scattered hay, this may be done as the stacking proceeds.

Ground for Stacks.—For the stacks, as dry and level a portion of the field should be chosen as possible. If the foundation be damp, the hay in the bottom of the stack is often considerably damaged, and if the land is not level it is difficult to build the stack so that it will not ultimately lean over to the low side.

Owing to soft places being in many meadows, it is also advisable to see that the stacks are placed in such a position that no difficulty will be experienced in getting the load out of the field. The position of the gates and water-courses must also be considered in selecting the positions of the stacks. These different obstacles, in the case of water meadows particularly, often necessitate departure from the rule already laid down as to building the stacks in a line across the rows of coils, and an equal distance from either end.

Work with the Wind.—Unless there is some reason for doing otherwise, a beginning with stacking should always be made at that side of the field from which the wind is blowing, as what is left after each stack is completed, and the rakings and dressings from it, can much more easily be conveyed to the next stack, when going with the wind, than when they have to be taken against it. Again, by this plan, hay which blows off the stack in course of being built is carried towards the next stack, whereas if building were continued in the opposite direction, the wind would always be blowing the loose hay on ground which had quite recently been raked, thus causing unnecessary work.

Building Field-stacks.—Little direction is required as to this. Some think it well to lay a foundation of straw, but this is not a general practice. When a few courses have been built, the stack should be tightly pulled at the very base, either by some one there for that special purpose or by the forker. Little taper should be given to the stack until four-fifths or five-sixths of the whole quantity intended to be put in the stack has been built. The stack may now be rapidly tapered, the forker or some other handy person in the meantime roughly smoothing down the stack with a rake. The rake should now be given to the builder, who, during the operation of building the head, should keep constantly raking the hay down. As the apex is reached, he should watch and put the hay only under his feet, and not at the sides, so as to have the centre as high and firm as possible. If the stack is intended to stand for some time, the top should be finished carefully, and the stack should be well roped.

In some cases, where the hay is not in good condition, the stacks are built without any one being on them—all the hay being placed in position by forks alone. To build stacks in this manner, two men working together make better work than where each works alone.

In early and dry localities the field-stacks, as a rule, are made much wider in proportion to their height than in higher and later districts. In the latter it is difficult to get the crop dry enough to keep well even in a narrow and very high stack. The shapes of stacks, like many other points in haymaking, vary in different localities.

Premature Stacking.—Hay stacked in too green a condition readily heats, occasionally moulds, and in many cases it loses its aromatic flavour and green colour, getting changed to a light-brown colour, with a musty sickening smell. The aromatic smell of hay may be preserved even where much heating occurs (as under the English system of haymaking), but in that case it must not have been stacked and dried in the field-stack—it must be taken direct from the coil or windrow. The same aromatic smell is often preserved in well-made silage under the heated or "sweet" silage process; but in this case it has a sweet flavour added to the original aromatic one. By many this flavour is supposed to be caused by sweet vernal grass; but well-made hay generally has it, although in many cases it may not contain a single blade of vernal.

Early v. Late Stacking.—There is now a pretty general consensus of opinion to the effect that it is better to be on the early than the late side with the stacking—that is, that of the two evils of early and late stacking the latter is the greater.

Straw and Hay Mixed.—The practice of interspersing imperfectly dried hay with layers of dried straw is pursued with good results in some parts, chiefly in wet districts. There is a twofold advantage in this plan. The excess of moisture in the hay is absorbed by the straw, and the interchange is beneficial to both the hay and the straw. The former is thus prevented from injury by heating, while as food for stock the straw is rendered more juicy and palat-able. The proportions of this mixture may be 1 ton of straw to from 3 to 6 tons of hay, according to the condition of the hay as to moisture.

This plan is often adopted with advantage in saving aftermath.

Haymaking by Stages.

Whenever the first portion of hay has been secured in the stack, an opportunity is offered, if the weather is favourable, for advancing one stage further the different sections of the remaining work. To prevent hay being any longer exposed to the elements than is necessary, only as much should be in each stage as the available force of the farm can advance a stage further in a full day's work. Thus, if the weather were moderately favourable, and the crop from its nature could in the most of cases be coiled after exposure to the sun for from two to two and a half days, we would have the hay cut, say, on Monday, lying undisturbed in the swathe on Tuesday, turned on Wednesday morning as soon as the dew disappears, and coiled during the afternoon. On Wednesday night there would thus be a portion of the crop in three different stages of manufacture—viz., one portion cut that day, another partially dried which had been cut the day before, and a third in coil.

Unless the hay has been in all the better condition when coiled, it is rarely it can be put into the stack without standing at least one day and a half; and if the weather is dull or showery, it is quite uncertain how long it may have to stand.

Presuming, however, for the sake of illustration, that the weather has been favourable, this section of the hay crop which was coiled on Wednesday afternoon might be put into the stack on Friday afternoon or Saturday forenoon. In the meantime an equal area should be cut and another coiled each day, and on Friday afternoon or Saturday three different sections will be advanced a stage —one will be cut, another coiled, and a third put in the stack; while of the other three remaining sections which have been cut during the week, and which are not touched to-day, one will be drying in the swathe, and two in the coil.

As soon as any one stage is ready to move on to the next, no time should be lost in advancing it, as every stage the crop is moved forward it is the more secure. In a wet day the crop is always less damaged in the coil than had the same crop been in the swathe; and in the stack it may, practically speaking, be considered safe, although absolutely not so, as isolated persons occasionally find to their cost.

When the stacking stage of the process of haymaking has been attained, no more should be cut than there is a reasonable prospect of getting coiled on an early date. It is well to have little more in coils than can reasonably be stacked in from two to three days, as should unsettled weather set in, the intervals during which hay can be handled are often so brief, that where a small piece can be easily secured, a large area may be completely spoiled.

Hay-barns.

In many parts, particularly in dairying districts, it is the custom to put the hay crop, in whole or in part, into barns or sheds. These are usually buildings without sides, the roofs of which are supported on pillars, set wide enough apart to allow a cart loaded with hay to readily pass between them. Where the hay is to be used for the daily food of the stock of the farm, as it must be on the majority of farms, it is always advisable to have a considerable portion of the crop stored in sheds. For convenience, these sheds should be situated as close as possible to the byres, feeding-boxes, or other buildings in which the stock are kept. Stored in this way, the crop is always accessible in all kinds of weather, and is safe from injury from storms. The floor of the shed should be raised a few inches above the surrounding surface, and if possible filled with some dry material such as furnace-cinders, small stones, or gravel; and before any hay is laid down the bottom should be covered with some old dry straw, fern, or other rubbish. Failing this, rough boughs and branches of trees do well for keeping the hay off the damp ground.

The value of these hay-sheds, where large quantities of hay are grown, can hardly be overestimated. They have been provided very extensively in Ireland, where, upon medium and large holdings, the greater portion of the hay crop is preserved in this manner.

The hay-barns are now, as a rule, constructed entirely of iron, the roof consist-

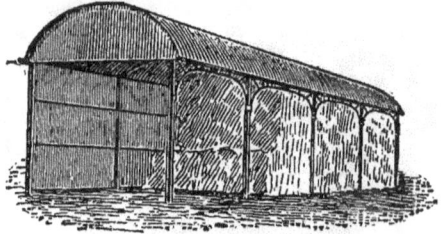

Fig. 570.—*Hay-shed.*

ing of sheet-iron, corrugated and galvanised. They are remarkably durable, and, in view of the great storing capacity afforded, the outlay is indeed very moderate. An excellent sample of the improved hay-barn (erected by A. and J. Main & Co.) is represented in fig. 570.

Permanent Hay-stacks.

Where hay-sheds have not been provided, the hay may be finally stored in round or oblong stacks either in the field or at the farm steading.

Foundation of Stacks.—A good, dry, level foundation for the hay-stack is a point of considerable importance. It is a good plan to have the ground upon which these stand permanently marked off by being raised a foot or so above the surrounding surface. In mining districts furnace-cinders are the best and cheapest material available for the bed of the hay-stack. Gravel and road-scrapings, or burned clay, are also suitable, and where neither of these can be easily obtained, the foundation may be raised by digging a small gutter round the base, and throwing the cleanings into the centre. Over this may be placed old pieces of wood, which, if covered with some old straw, prevent the hay from being spoiled to any appreciable extent. Branches of trees in the bottoms of stacks of grain are often objected to, on the score of their forming a harbour for vermin, such as mice or rats. These, however, rarely do any damage in a hay-stack, and tree branches are surpassed by few other materials as stack foundations.

Bosses or Flues.—In permanent as well as temporary stacks, bosses, kilns, or flues are occasionally used for hay which is in a very indifferent condition.

Size of Stacks. — On small farms, where the available power for forking is generally not over-plentiful, round stacks of 12 feet in diameter will often be found large enough. If the hay is quite dry, it undoubtedly remains more palatable in a larger stack; but then the large stack entails extra labour and delay in forking.

The larger size of round stacks are usually made from 15 to 18 feet or even 20 feet in diameter, which, if of a corresponding height, are of twice or thrice the capacity of the smaller size of stacks, with a corresponding decrease in outside waste.

The oblong form of stack is very general. The dimensions vary in the main, in accordance with the extent of the holding, or rather with the area under hay. A common and convenient width is from 5 to 6 yards, and height from 8 to 16 feet to the eaves when settled down, while the length may range from 10 to 20 yards or more.

It is a common practice in many parts of the country, in England especially, to increase the width or diameter of the stack as it rises in height. One advantage of this is that wet dropping from the eaves falls on the ground clear of the sides of the stack.

Force in Building Stacks.—The size of the stack having been arranged in accordance with the power at the farmer's command, a beginning may be made in building. To keep the builders and others at the stack constantly employed, at least two hay-waggons or carts with harvest frames must be employed. Hay-waggons are much to be preferred to carts. Being longer and lower they are much easier loaded. Where the field is close to the stackyard, or other place of stacking the hay (it occasionally being stacked in the field it which it grows), one forker in the field will be able to keep both horses going, for while the one is loading the other is unloading. One builder will be required on the stack, who may have the assistance of three or four men, women, or boys to tramp or pack the hay, and throw it to him from the side on which the forker deposits it. Where another man can be spared, one should be put on to act as guide to the builder, who, while keeping the stack in proper shape as it is being built, throws up any loose hay, and may or may not put on the top on one stack while the builder is making the bottom of another.

Process of Building. — In building the stack the hay is first regularly laid in fork-loads over the whole bottom; the builder then puts on a ring of forkfuls all round the outside, taking care to throw the ends of the hay well out. While he is doing so, his assistant may put in other forkfuls of hay behind those he lays down, so as to form an inner circle, the forkfuls of the inner circle lying partially on the outer one, and holding them in position. The second row being finished, the centre should now be filled level with the outside circle, and on no account should it be kept hollow. The second and third courses should always project well over the first, after which the others should only project a very little. By following this plan less has to be pulled off the base to make it firm.

The trampers scatter the forkfuls of hay evenly over the surface of the stack, and the more carefully and thoroughly this scattering is done the more substantial and symmetrical will be the stack.

In this way the building continues until the eaves of the stack are reached, the person acting as guide all the while pulling off by his hand or the rake any hay which is loose or too far out. When the top is being formed, the person acting as guide will in all probability now require to go on the ladder and fork, as the top of the stack will be beyond the power of the carter to pitch the hay on to it.

High-forking by Hand.—When the stack gets too high for the forkers to be able to put up the hay, and no horse-fork or other similar appliance is available, a forking-stage of some sort must be provided. The following simple plan is pursued in some parts. The extra man mounts a ladder set against the stack, and taking the loaded fork out of the hands of the carter, at as great a height as the latter can hand it to him, he is enabled to fork it several feet higher than can be done by the man on the cart.

To enable the second forker to get a good footing on the ladder, and yet to keep the latter as nearly straight up as possible, a stout plank about 8 or 9 inches wide and 4 feet or so long should be thrust through the spars of the ladder, above the spar where the man is intended to stand. By lowering the end of the plank nearest the stack, thrusting it through until it meets the hay, and then putting downward pressure on the outward end, it is brought into an almost horizontal position, which is quite safe and easy to stand on. Standing on the plank, with his back to the ladder, a man can hoist an ordinary forkful of hay without any danger of falling.

If the plank is not used, the inclination of the ladder must be much flatter so as to allow the man to stand on it easily and safely, and, in this case, his weight against a stack of any moderate size would almost certainly cause it to lean to the opposite side.

The use of the horse-fork or elevator, to be noticed presently, obviates all this difficulty.

Forking from Different Parts.—To prevent the stack leaning to one side, the forking should be done from as many positions as possible. By the continual dropping of the hay on one side, and the weight of the persons removing it always pressing down the hay at that place, the stack becomes more solid there than at the other side, with the result that when the stack settles down, no amount of propping will keep it from leaning over in the opposite direction.

Heading Stacks.—Before the top is begun to be formed, the whole outside ring should, as a rule, be made as level as possible. If the ground is slightly sloping (a position which should be chosen as seldom as possible), or if the forking has of necessity been all done from one side, then the eaves-ring on the opposite side should be left just a trifle higher than the other. In all other circumstances the eaves-ring should be level. The first ring of the top should be only slightly drawn within the eaves-ring, the second a trifle more, and the third a little more

still, after which the decrease in width should be regular. As soon as the third ring and hearting have been completed, the builder should go round the whole top with a rake (fig. 571), and, with its head downwards, rake the top of the

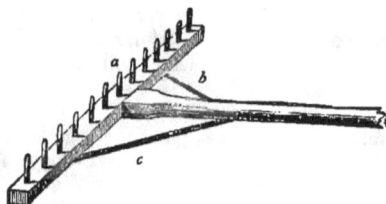

Fig. 571.—*Hay hand-rake.*
a Wooden teeth.   b c Iron stays.

stack into proper shape. This done, the finishing of the top may be continued.

As the apex is approached, all should leave the top of the stack except the builder, who even alone will have barely room to stand. After laying on every couple of courses or so, he should carefully rake the roof. Towards the end the forker should be careful to put up only very small forkfuls, and at the finishing touches only mere handfuls. It is very desirable that this should be attended to, as in the building of a stack few things are more harassing or provoking to the builder than to have large forkfuls sent up, when there is not sufficient room for them.

The apex having been reached, the builder should carefully clear away any loose hay lying on the roof.

Roping Hay-stacks.—A couple of ropes should now be passed at right angles to each other over the top, and a half-brick hung to each end. Weights of any kind, hung to the end of the ropes, are much to be preferred to tying the end to the side of the stack. When the stack settles down, the tied rope becomes slack and does little good, whereas by the other method the strain on the rope is always the same.

Form of Stackhead. — When the head of a stack is built as already described, it should have the outline of an open umbrella: it should be flattest from the apex two-thirds down, and then more steep, because from the apex until half-way down the slope the thatch has little more rain to carry off than falls on that particular part, whereas farther down the thatch has not only the rain which falls on itself to convey, but likewise the rain that has fallen on the part above.

By making the top of this form another gain is obtained. There is much less hay, proportionately speaking, in the narrow part under the apex, and as this part is often more injured by drying than any other portion, the form recommended reduces the loss to the minimum.

Large Stacks.—The building of large stacks is, of course, very similar to that of smaller stacks. Where more forkers than one are employed in the field, of course more carters will be required as well as more persons on the stack to receive it from them. As a rule, however, there is only one main builder, who puts on the outside courses, while there may or may not be another, who attends to the second or inside rings, while others spread and throw the hay across.

Propping Stacks.—In building either round or oblong stacks, particularly those small in diameter, short props should be put in whenever a height of 6 or 7 feet has been reached, and another set as soon as 10 or 12 feet is attained.

Height of Stacks.—The height of the haystack may be arranged to suit the taste and convenience of the farmer. It is contended by some that, as a rule, the eaves when built should be in height at least twice the diameter of the stack at the ground. Most farmers make their stacks much lower than this, because they are not possessed of sufficient power to put them higher; but when it is remembered that a high stack may contain twice as much hay as a low one, while the thatching is almost the same, the gain of having moderately high stacks will be obvious. This is more particularly the case when the extra height can be obtained at little or no extra cost, as where the horse fork is in use.

Improved Hay-stacking Appliances.

The stacking of hay has been greatly facilitated by the introduction of various ingenious appliances, such as stack-lifters and horse-forks.

Stack-lifters.—These appliances, designed for lifting stacks in the field and bringing them entire into the stackyard, have been in use for many years. They consist of a flat body, with shafts and tipping arrangements (similar to those in use on a tip-cart), mounted on two wheels 2 feet in diameter. Across the front of this low, broad, and flat cart is placed an iron roller, with lever and ratchet-wheel attached to one end. To the centre of this roller the ends of two stout ropes or light chains are attached, at the outer end of which are hooks for joining the one to the other.

When it is desired to put on a stack on this apparatus, the tipping arrangement is unlocked, the back end dropped on the ground, and the horse then backs it close into the stack. Meanwhile the attendant raises the hay forming the base of the stack from the ground, to permit of the edge getting better under it : the ropes are now unwound from the roller and passed round behind the stack, where they are joined and made firm, about 9 inches above the ground. The lever and ratchet are then brought into play, and the rope coiled on the roller. As the strain becomes greater on the roller, the edge of the stack-lifter is gradually pressed in under the stack. At about a foot or 18 inches under, it remains stationary, and as the attendant continues to wind in the rope, the stack is seen to move, and from that forward to glide gently up the incline until the front of the roller is reached, by which time the weight is generally heavy enough in front of the wheels to cause the body of the apparatus to drop down on the shafts, where it is locked.

This apparatus, if carefully worked, generally lifts all the stack without any trouble, brings it to the stackyard in its original position, and there, by unlocking the tipping arrangement, drops it in its former upright position. With this apparatus the bottom of a stack could be much more easily built than by first building the hay on carts, because from five to seven minutes is all that is required to draw a field-stack on to the lifter.

Defect in the System.—When, however, the stack became so high that one man could with difficulty pitch the hay on to it from the ground, it was found that little advantage was gained, because two were required to fork in the stack-yard ; whereas with carts, and one man in the field and the carter in the stack-yard, an equal height could be attained. For this reason, therefore, the original stack-lifters made little progress, and were to be found only on isolated farms.

Horse-forks.—In the spring of 1886 it occurred to Mr John Speir, Newton Farm, Glasgow, that this defect might be removed if this apparatus were strength-ened and improved and used in connec-tion with one of the many classes of horse-forks or elevators so largely used in America, and also employed on many farms in England and elsewhere in this country. He believed that double the value would be got out of each apparatus when combined that was possible when used separately, because the horse-fork could only do the forking in the yard, leaving that in the field to be done as formerly. With the stack-lifter, the field-forking was done away with, and with the horse-fork that in the stackyard was reduced to a minimum. Arrangements were at once made for procuring a couple

of stack-lifters and a horse-fork of the most improved pattern, and after thorough trial, the results fully justified the antici-pations which had been formed.

Types of Horse-forks.—Horse-forks are of various patterns. Most of them fairly well answer the purpose for which they were designed, but they are often found unsuitable where the circumstances are altered.

Clip-fork.—For lifting hay from solid stacks, preference is usually given to the clip-fork, the principle of which is the same as a mason or quarryman's shears for lifting stones by a crane, except that while the mason's shears have only one prong on either side, the hay-fork has two or three. When dropped open on to a portion of hay, the mere matter of the horse raising it forces the points together, a firm hold of the hay being thus secured. An arrangement exists by which the prongs can be pulled asunder at any instant by a person standing on the ground, so that the load of hay can be instantaneously dropped anywhere and at any time.

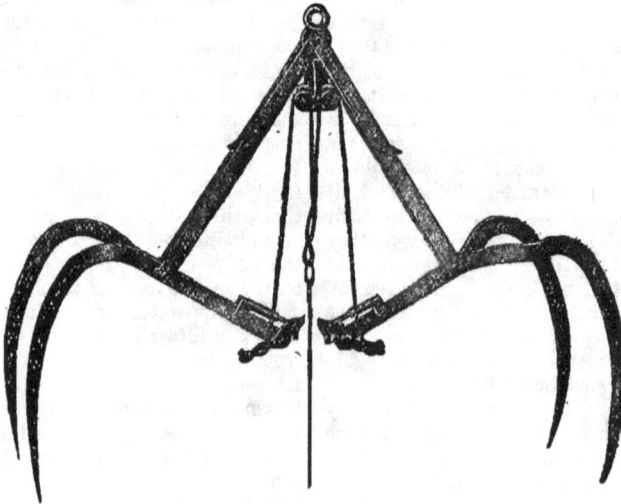

Fig. 572.—*Wallace's hay clip-fork (open)*.

The working of Wallace's hay clip-fork is shown in figs. 572 and 573 (John Wallace & Sons, Glasgow).

Elevating Apparatus.—A simple and convenient arrangement for hoist-ing the hay is to have a pole about 35 feet high, which is held in upright posi-

tion by three or four guy-ropes from the top of the pole to iron pins driven into the ground. A short "jib" or "gaff," 10 feet long or so, is arranged to slide up and down the pole, being worked by pulleys from the ground. The fork is attached to an inch hemp rope or ½ inch steel strand rope, which passes over a pulley at the point of the jib or gaff, thence down the upper surface of the gaff to its lower end, where it passes over another pulley, from which it runs down the side of the pole to about 3 feet from the ground, where it passes through

Fig. 573.—*Wallace's hay clip-fork (closed).*

the pole and under a pulley fixed in it, where it is attached to the tree or chains by which a horse draws up the forkful.

The pole is set up with a slight lean to the stack which is being built, so that as soon as the ascending load has been raised above the portion already built, the gaff or jib with its load always swings round over the top of the stack, where it can be dropped on almost any part of even a large stack.

Hay Elevator. — Many prefer the hay-elevator as shown in fig. 508, vol. ii., to the horse-fork in the stacking of hay. In stacking hay this elevator may be worked by a single-horse gear. The hay is thrown on to the bottom of the elevator and carried in a continuous stream to the desired height, the elevator being constructed so that its height can be varied.

Advantages of the Horse-fork. — The great speed in forking is not the only merit of the horse-fork. Stacks built by the horse-fork are more easily kept perpendicular than those built from hand-forking. With the horse-fork the forkfuls are regularly discharged only where they are wanted, and where the builder is working for the time being, so that the stack has not the same tendency to settle down at one side more than another, as it would have if the hay were forked by hand. A certain number of persons can also distribute regularly over the stack a much greater quantity of hay when forked by the horse-fork than by hand, because it is always dropped just where wanted, and they do not require to spend their time carrying it from one part to another.

A Good Day's Work. — The horse-fork can quite as easily lift hay from an ordinary hay-waggon or cart as from the ground; and where there are not as many stack-lifters as will keep it going, a portion of the hay, if desired, may be brought in by carts. On a moderate-sized farm each stack-lifter will bring in from 10 to 15 field-stacks per day, even where each may weigh from 15 cwt. to 20 cwt., the load being pulled on all the newer patterns by the horse instead of the man. Going a moderate distance, therefore, each carter may be supposed to be able to bring in about 10 tons of hay daily; and three stack-lifters, although they will be sufficient to bring in the hay of most farms, will not keep the fork going if there are sufficient people on the stack to build it. If three stack-lifters are in use, at least one man building, and three women or men spreading, will be required on the stack, one man will be necessary to act as guide, another will be required at the fork, and a woman or boy to lead the horse, while an extra woman will be required in the field to rake up the bottoms and assist in putting on the ropes. There are thus engaged six men and six women, who may bring from the field and build in the stack from 25 to 30 tons or over of hay daily—this, too, with very little exertion to any of them compared with what would be necessary were the hay lifted by hand-forks. A single horse-fork has been known to put up 50 tons of hay in one day.

Horse-fork Working in Sheds.— Where the hay is stored in sheds, and it is desired to use the horse-fork, a different arrangement from that already described must be adopted. A stout wire, one half-inch in diameter, or double track of angle iron, is stretched tightly about a foot under the apex of the roof, from the one end of the building to the other. On this wire or other track runs a small carriage on two or four wheels, under which is a pulley. The double track and four-wheeled carriage is now in almost universal use, all other designs being gradually abandoned. The rope from the horse comes in at one end of the shed, passes up between guide-pulleys to the pulley under the carriage, over which it passes. In this case a pulley of a peculiar shape must be attached to the top of the fork, and the rope from the carriage passes under this pulley and back to the carriage, to which it is knotted. When the fork has been loaded, and the horse moves forward, the load is raised perpendicularly until the top of the pulley over the fork strikes a spring on the carriage, when a catch which has hitherto held it in position is released, and the carriage, with its load hanging from the under side, is pulled along the track till any particular division of the shed has been reached, whereupon the attendant drops the load and pulls back the fork. The arrangement is very neat, and admirable in working.

Although some modification of it is in use in almost every hay-barn in America, it for long gained little favour in this country. But after the combination of the stack-lifter and horse-fork was adopted in 1886 by Mr Speir, Newton, Glasgow, its use extended rapidly.

Horse-forks for Small Farmers.— In this country the capabilities of the horse-fork are as yet but partially understood. It has only been on the largest farms where, as a rule, they have been introduced; yet as a matter of fact they are a much greater gain to the small farmer than to the larger one. The small farmer could bring in the hay with one stack-lifter, unyoke his horse, and fork it up with the assistance of another person to lead the horse or build, as the person who leads the horse can easily enough work the fork also. Or, if more convenient, the small farmer may bring in two or three stacks, place them down as conveniently as possible, and then fork them afterwards. With the assistance of four persons in all, he may keep two stack-lifters going, have one person building and another at the fork, and neither of them need be able-bodied. The builder, indeed, is the only person requiring to exert much strength, as the horses load the stack-lifters as well as fork the hay.

Dressing Stacks.

After a hay-stack, or the hay stored in a shed, has been standing for three or four days, the sides should be carefully hand-pulled. Before beginning, all props should be taken away, so that the work may be more accurately done. The bottom or base of the stack should be pulled as firmly as possible, as the hay gets damaged very quickly if it is allowed to rest on the damp ground. Any parts of the sides which have a greater projection than the rest should be pulled down until the whole is uniform. If the stack has been well built at first, and has not been allowed to lean over to one side, the pulling is neither a difficult nor a tedious process. If, however, the stack has been badly built, is too far out at one part and too far in at another, or leans over in one direction, almost no amount of pulling will make it a good one. At most all that can be made of it is to give it a fairly respectable appearance, and reduce its outside waste as far as possible.

If stacks are not pulled after they have consolidated, there is usually much loose hay on the outside exposed to damage by the weather. On the other hand, if the stack is well built and carefully pulled, the loss is reduced to a minimum. After pulling, such a number of props should be returned as are necessary for the stack, considering its width, height, and accuracy of build. No more should be used, however, than are absolutely indispensable.

Irish Methods.

Besides the methods of harvesting hay just described, many others, differing from these in a lesser or greater degree, are pursued throughout the country.

In Ireland, where a very large area is occupied by the hay crop, there is great variety in the systems of haymaking.

The moist climate prevailing there compels the Irish farmer to resort to many devices to secure his hay crops in safety. In the natural meadows the tedder is employed freely, and as soon as at all possible this hay is put into tolerably large cocks or coils, and thence into the hay-barn—a valuable institution nowhere in this country so fully appreciated or so largely utilised as in the Emerald Isle. Rye-grass and clover hay is made, for the most part, as already described..

Often in Ireland the hay has, on account of the prevalence of drizzling rain and the absence of good drying weather, to lie long on the fields after it has been cut and before it can be stacked or stored in sheds. In this way more hay is unavoidably lost or injured in Ireland than in England and Scotland put together. By the employment of more expeditious and improved methods of handling the hay, and the introduction of the hay-barn for the storing of it, the loss in the fields has been very considerably curtailed. The hay harvest, however, is still a time of great anxiety to Irish farmers.

Rick-cloths.

The use of rick-cloths when stacking hay in the open field or yard is to be

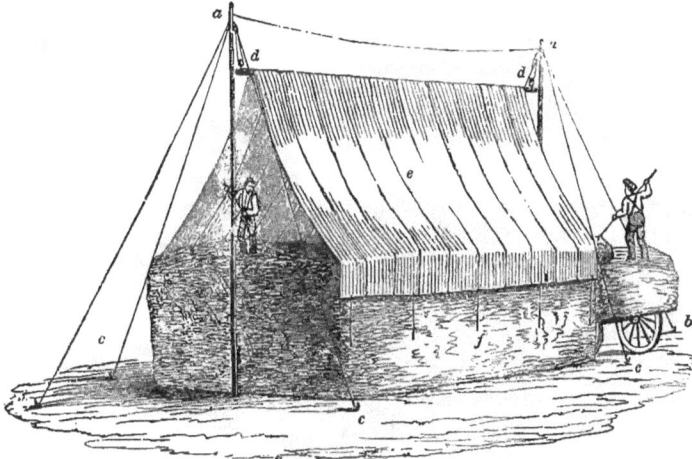

Fig. 574.—*Mode of erecting a rick-cloth over a hay-stack when being built.*

a a Wooden spars.	of a stack.	*d d* Blocks and tackle.
a to a Top-rope.	*b b* Guy-ropes from end to end	*e* Rick-cloth.
c c c Guy-ropes from side to side	of stack.	*f* Reef-points of cord.

recommended, especially in unsettled seasons and wet districts. One form of erecting these cloths, long pursued, is shown in fig. 574. Modern ingenuity has devised various methods of erection, more or less convenient. These cloths are of great service in protecting a partially built stack from showers of rain.

When a rick-cloth has not been provided, large waterproof tarpaulins should be at hand, to be drawn over unfinished stacks to protect them from a sudden downpour of rain.

Salting and Spicing Hay.

It is a common practice—and in many cases a good one—to scatter a little common salt amongst hay while it is being stacked. When hay is in a damp or badly made condition, in consequence of bad weather, salt is an excellent remedy against mouldiness. It is sown by hand by the builders upon every portion tramped down. The quantity used should correspond to the state of the hay, and must be left to discretion. Perhaps ¼ bushel to a ton is enough. Salt renders the ill-made hay more palatable to stock. "Hay spice," admirably suited for the sweetening of musty hay, is now offered by several firms. It is extensively used instead of salt, and is sown in stacking in the same way.

GROWING RYE-GRASS SEED.

In some parts of the country rye-grass is grown primarily for the production of seed. This is particularly the case in the province of Ulster in Ireland and the Scottish county of Ayr, where the bulk of the rye-grass seed produced in the British Isles is grown.

Methods of Growing and Handling Rye-grass Seed.—The methods pursued in the growing of the seed and in handling the seed are similar in Ulster and Ayrshire. A crop with much clover in it is not desired for the production of rye-grass seed, as in such cases the seed is generally small in quantity and less perfect in quality than where the crop is composed of rye-grass only. The crop is cut by the manual reaper, tied and stooked the same as cereals, only the sheaves and stooks are both smaller.

After the sheaves have stood a few days in the stook, they are put into small cocks, variously called "rickles" or "huts." In this condition the crop stands for from two to three weeks, when it is threshed or stacked. If for threshing, the crop is carted from the field to the farm or travelling threshing-machine, and after threshing, the hay is stacked in the ordinary way.

Instead of being threshed from the field, the crop is occasionally stacked in much the same way as cereals. This, however, is a wasteful practice, as the seed has such a tender attachment to the straw that a considerable proportion of it may be lost in the handling. "Rickling" or "hutting" is almost a necessity, as it is very difficult to get the seed sufficiently dry to keep unless the crop has been put up in this way for several weeks.

A certain amount of dressing is done to the seed by the growers, but it is all put through various machines which take out several weed seeds before it is offered for sale by the wholesale seed merchants.

Yield of Rye-grass Seed.—A fair yield of rye-grass seed would be from 26 to 30 bushels per acre. The weight of the seed per bushel usually ranges from 24 to 28 lb. There should be nearly 1 bushel of seed from each cwt. of hay.

The Flail for Threshing Rye-grass. —In some cases the flail is still used in threshing rye-grass, but it has disappeared from the greater part of the country. This antiquated and once most serviceable tool consists of two parts, the hand-staff or helve, and the supple or beater, fig. 575. The hand-staff is a light rod of ash about 5 feet in length, slightly increased in breadth at the farther extremity, where it is perforated for the passage of the thongs that bind the beater to it. The beater is a rod of from 30 to 36 inches in length, made of ash, though a more compact wood, as thorn, is less likely to split;

Fig. 575.—*Hand-flail.*

and to prevent this disintegration of the wood the beater should be constructed to fall upon the *edge* of the segmental portions of the *reed* of the wood. The usual form of the beater is cylindrical, the diameter being from $1\frac{1}{4}$ to $1\frac{1}{2}$ inch. For the most part it is attached to the hand-staff by a thong of hide untanned: eel-skins make a strong durable thong.

ARTIFICIAL HAY-DRYING.

Soon after the disastrously wet, sunless summer of 1879, a considerable amount of attention was given to the devising and testing of appliances for the artificial drying of hay.

Hot Air and Neilson Systems.— Two systems attained wide notoriety. These were Gibbs's hot-air method and Neilson's exhaust-fan system. The former, the invention of Mr Gibbs, of Gillwell Park, Essex, involved the use of a huge, unhandy apparatus, costing about £350. For the Neilson system, the inventor of which was Mr Neilson of Halewood Farm, near Liverpool, fans of numerous patterns were brought out, costing in most cases about £14 or £15. As is often the case with new devices, great hopes were raised by these inventions. They excited much interest and attracted many enthusiastic advocates, who assured farmers that at last their trials and losses of harvest were at an end. Alas! it was little better than a dream. Both systems failed under the crucial test of practical work. They never came into general use, and have long since been given up.

PRESSING HAY.

The bulky character of hay renders it desirable that means should be provided for compressing it tightly into convenient trusses or bundles. In these trusses it can be more easily and more cheaply carried alike by road, rail, and steamer, than in its natural bulk.

Much ingenuity and enterprise have therefore been exerted in the devising of hay-presses — additional impetus being given to these efforts by the railway companies offering a reduced rate for carriage when 50 cwt. or more is packed on to an ordinary railway waggon. For this purpose, such pressure as will pack

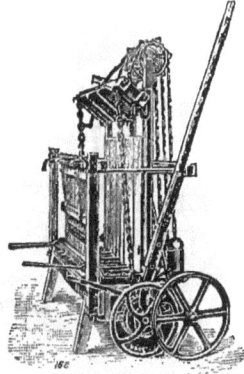

Fig. 576.—*Hay and straw hand-power press.*

nearly 8 lb. of hay or straw into a cubic foot is sufficient.

At various times trials of hay-presses have been conducted throughout the

Fig. 577.—*Morgan's hay- and straw-press.*

country, and in this way several efficient appliances for the purpose have been brought into notice. Large presses for steam-power have been introduced, but smaller presses for horse- or hand-power are more widely used. A combined hay- and straw-press for hand-power, made by Barford & Perkins, Peterborough, is shown in fig. 576. Morgan's hay- and straw-press is represented in fig. 577.

THE MANAGEMENT OF PASTURES.

Insufficient Attention to Pastures. —In view of the falling off in corn culture, greater importance is nowadays attached to the grazing division of the farm than in former times. It is therefore especially unfortunate that the majority of the farmers of the United Kingdom do not give sufficient attention to the management of land under grass. Often, indeed, it would seem to be thought by those who have not adequately considered the subject that pastures may be largely left to take care of themselves. A greater mistake could hardly be made. Not only does the management of pastures require capable and careful attention, but few departments of the farm will, as a rule, better repay proper care when it is bestowed. The difference between good and bad farming may be seen very strikingly in the condition of grazing land ; and it has to be confessed that in many parts of this country appearances in this respect are far from what they ought to be.

Too frequently one sees pasture-land strewn with destructive weeds which are rarely timely or effectually cut down. Lack of attention to surface-water runs or underground drains often greatly lessens the value of pastures. More skilful treatment in manuring is also required, while, generally speaking, there is room for improvement in the methods of grazing by different classes of stock.

Pastures on Different Soils.—In the management of pastures the character of the soil has to be carefully considered. Pastures lying on good soils usually tend to improve, whereas those lying on poor soils, as a rule, deteriorate, unless maintained by manuring or other remedial treatment. The best pastures carry a feeding stock which return valuable manurial ingredients in the droppings, whereas the poorer pastures are usually grazed by either a growing or a breeding stock, by which much more of the elements of fertility is removed from the land.

Generally speaking, the stronger classes of soils make the best pastures. A strong clay soil, when under cultivation, is often either somewhat baked or puddled, whereas, when the same soil is under pasture, the better conditions of tilth which it obtains, as a result of successive winter frosts not being interfered with by implements, are retained by the soil. It is therefore advisable that, as far as this can be controlled, the permanent pasture of the farm should be on the stronger soils, and that the lighter class of soils should be under cultivation, assuming, of course, that the quality of the latter is good enough for the required purposes.

Effect of Climate on Pastures.— Climate is a most important factor in connection with pasture. The best grazing districts in Britain are to be found where the rainfall is fairly abundant and well distributed, where the sunshine is not excessive, and where prolonged frosts or droughts are rare. The western counties of Britain have more rainfall, cooler summers, and milder winters than the eastern counties, and as a result the former are much better suited for grazing than the latter. The pasture advantages of the western counties are shared to a still greater extent by nearly the whole of Ireland, a result which gives that country the name of the Emerald Isle. On the other hand, the smaller rainfall of the south-eastern counties of England and of the counties around London, combined with the warmer summers and colder winters of these counties, give unfavourable conditions for pastures, especially when they are lying on a subsoil of sand or chalk. It is under these latter conditions that the greatest difficulty is experienced in laying down land to pasture.

Pastures should not be Mown.— A pasture, like a lawn, should be regularly rolled and kept close down, but the rollers and mowers of a pasture should be the grazing stock. A pasture that carries a heavy stock has, therefore, a great advantage over that which supports a light stock. The custom in the midland and southern counties of England to occasionally take a hay crop from pasture-land tends to make the herbage coarse, as the growth of the hay injures the close bottom herbage, which is so essential to a good pasture. For this reason also a pasture should never be allowed to become coarse and benty, and it is very advisable that it should be grazed bare at least twice every season. Permanent pasture, therefore, if it is to be made the most of for grazing purposes, should never be mown for hay. In many cases when pasture becomes unusually rank and coarse, benefit is derived by running a high-set mower over it in the autumn, but this is very different from taking a full crop of hay from the land.

The Prospects of Profit from Pastures.—A question that must always be considered is a profitable return from the outlay in improving pasture. Before land became depreciated in value, operations like draining and liming were considered to be essential in commencing the improvement of poor clay soil. It is now possible, however, to purchase clay land in Essex and other counties at a considerably less price per acre for the freehold than would cover the cost of draining and liming such land. Under present conditions, therefore, these operations should be carried out only if they are likely to give a profitable return, and full consideration should be given as to whether either of these operations is actually necessary.

Drainage of Pastures.—When a pasture is wet and in a marshy condition, drainage in some form or another is essential. At the same time, care should be taken not to *overdrain*, as there is no doubt that large areas of pasture-land have been injured by doing so. In a dry season a fair supply of subsoil water is of great importance.

Manuring of Pastures.

The manurial ingredients usually required by soils which are naturally poor, or have been reduced in fertility, are four—namely, nitrogen, phosphates, potash, and lime. It has long been known that lime not only acts as a plant-food, but that it has a great value in making other plant-food in the soil suitable for the growth of plants. Recent and extensive manurial trials, however, on all kinds of pasture have shown that lime in a great many cases not only has little effect in improving pasture, but that when applied with certain other manures it may retard their action. It has also been claimed for lime that it is a most effective agent for removing moss in pastures, but this has been proved by experiments not to be so. Moss is most easily got rid of by encouraging good pasture-plants, and therefore the system of manuring that will improve a pasture to the greatest extent is the best means of removing moss.

The manurial requirements of pastures may thus be shortly summed up :—

(1) Manures containing phosphates— *i.e.*, superphosphate, bones, basic slag— are of the first importance, as phosphates are most likely to be exhausted to the greatest extent by the grazing stock.

(2) As pastures improve they tend to become richer in nitrogen, and to enable them to do this the encouragement of clovers and other leguminous plants in the pasture is all important. This can be most effectually done by the judicious use of phosphatic manures with the addition of manures containing potash and lime if necessary.

(3) A potash manure is usually not needed on the heavier classes of soils, but is likely to be an essential adjunct to phosphates on the lighter soils, especially if recently laid down. The grazing stock return practically all the potash to the land in their droppings, and for the same reason dung is rich in potash. Old pastures, therefore, which have a rich covering of black organic matter, or pastures on which dung has been freely applied, are not likely to respond to a potash manure, even when lying on a somewhat light soil.

(4) Lime will in most cases give its best results when applied in small quantities per acre, and in a condition that it can be thoroughly distributed. This is undoubtedly one of the reasons that basic slag is so effective on many of our old pastures, for the large percentage of lime that slag contains is invariably well spread over the land. Ordinary dressings of lime have been tried with unfavourable results on poor clay in Northumberland and Cumberland, on poor sandy soils in Northumberland

and Dorset, and on both classes of soils elsewhere, and yet in all cases the soils were poor in lime. Even on peaty soils in Cumberland ordinary lime has not been very effective. In practically all these cases, however, basic slag has given excellent results by itself on the clay soils, and when accompanied with a potash manure on the sandy and peaty soils. There can be no doubt that basic slag acts not only because of the phosphates, but because of the lime that it contains. Further, the use of basic slag makes the application of either common lime or ground lime unnecessary on the great bulk of pastures. It is only soils of a peaty character, or those with a good deal of rough, matty herbage, that are likely to give a good return from these forms of lime. The results of recent experiments in Germany, especially on peaty soils, fully corroborate the foregoing conclusions. In fact, 'for pasture as well as for most other purposes, lime may be effectively applied in quantities not greatly exceeding that of the other manurial ingredients, provided it is in a fine condition, so that it is thoroughly distributed in the soil.

Great care must be exercised in applying either nitrate of soda or sulphate of ammonia to pastures. These manures diminish clover herbage, make the grass much more benty in character, and ultimately, as a rule, tend to reduce the feeding value of the herbage.

Feeding Cake on Pastures.—The cakes usually fed to stock grazing on pastures are decorticated cotton cake, undecorticated cotton cake, and linseed cake. Voelcker and Hall (1902) give the residual manurial value of a ton of each of these as 56s. 5d., 33s. 9d., and 38s. 7d. respectively. The values attached to the nitrogen in each are 41s. 5d., 21s. 3d., and 28s. 6d. per ton respectively, so that, roughly speaking, nitrogen accounts for three-fourths of the residual value of decorticated cake and linseed cake, and for nearly two-thirds of that of undecorticated cotton cake. There can be no doubt that the feeding of cake to grazing stock, if judiciously done and in moderation, is a very effective method of improving pastures. As, however, cakes contribute nitrogen to the soil principally, and only a small amount of phosphates, pastures can-

not be effectively improved by this means only, as the main requirement, phosphoric acid, is not in this way applied in anything like sufficient quantities.

The feeding of cake improves a pasture, not only because of its manurial residue, but because a heavier grazing stock can be kept, and this has a most valuable result in keeping the pasture finer and closer at the bottom. When cake, however, is fed beyond moderate quantities on pasture, the herbage becomes more benty in character and clover decreases, no doubt owing to the nitrogen contained in the cake residue.

Caution in using Cakes on Pastures.—But while the consumption of cake upon the land may usually be an effective, it does not always follow that it is also an economical, method of maintaining pasture. It is both, as a rule, where the conditions as to soil and climate are favourable; but the matter is very different when both soil and climate are backward, where the soil is so bad as to yield produce deficient alike in quantity and quality, and where the rainfall is so large that the loss of plant food in drainage is exceptionally heavy. In these latter circumstances cakes have to be used on pastures with the utmost care to ensure profit.

Suggested Dressings of Manures.—A poor pasture lying on a poor clay soil will probably be effectively improved by the application of 10 cwt. basic slag per acre. This should contain not less than 37 per cent phosphate of lime, and if so, 10 cwt. slag will contain about 170 lb. phosphoric acid per acre. If the slag is poorer in quality, the dressing should be correspondingly increased. Eight cwt. of finely ground bone-meal might be applied instead of slag, but this is likely to be slower in its action, and has not the advantage of the free lime contained in the slag.

A dressing of half the quantity of slag might be repeated every three years thereafter, and the feeding of cake to the grazing stock will probably assist the improvement in the earlier years. There is now ample evidence to show that the good effects of slag will continue after repeated dressings. If the soil is of a sandy or a peaty character, 2 cwt. an acre of muriate of potash (50 per cent potash), or 6 to 8 cwt. kainit, will probably be useful in addition to the first dressing of slag, along with half of either of these quantities with the later dressings.

For soils rich in lime, superphosphate will probably be more effective than basic slag, but it is not advisable to apply more than 5 cwt. per acre of this manure in one dressing.

Application of Manures. — Basic slag, bone-meal, and the potash manures are best applied in the autumn or early winter, while superphosphate is probably best applied about February. As slag is ground to a very fine state of division, it should always be applied by itself and not mixed with other manures, as the same perfect distribution will not be secured as when the slag is sown alone. Superphosphate should be in a fine and friable, and not in a pasty, condition. Bone-meal should be finely ground, and manures like kainit and muriate of potash should have all lumps larger than shot broken into a fine condition. A pasture should always be closely eaten down if possible before the manures are applied. If rough and in a benty condition it should be subjected to a heavy harrowing, and, if necessary, it should be mown, otherwise the effects of the manures will be retarded. The sowing of the manures, preferably by a machine, should be efficiently and thoroughly carried out.

Stocking of Pastures.

The overstocking of pastures greatly reduces the total amount of herbage produced, especially in a dry season, as when the sward is bare the bad effects of drought are greatly increased. Understocking, however, does even more harm, as the herbage becomes coarse and the fine grazing plants are greatly reduced. Wherever possible, pasture fields should be grazed in rotation, so that each field is grazed for a few weeks and rested for about a fortnight alternately. At least twice each season pastures should be closely eaten down. When pastures are saved for winter foggage, care should be taken that all the rough herbage has been eaten off by the beginning of May.

Horses are the worst stock for pastures. They deposit their droppings only on

certain parts of the fields, and on these they never graze. On the other parts they graze very closely, which they can do, as they have incisor teeth on both jaws. A pasture, therefore, grazed by horses only, soon becomes very unsightly —coarse and tufty where their droppings are deposited, and poor and bare where they do graze. On this account horses should never be grazed for long on the same field, but should be kept moving round the different fields. Cattle have no incisors on the upper jaw, and cannot graze nearly so closely as horses. Their manure is dropped all over the fields they graze, although each dropping is not so well spread as the droppings of sheep. Cattle are the best stock for the grazing of a pasture.

Sheep, like cattle, have no incisors on their upper jaws, but they bite more closely, partly because their lower lips are cleft. By means of their small mouths they select the fine bottom herbage and reject that of a stemmy character, so that a pasture grazed by sheep only soon becomes of a benty character. Sheep do best when grazing as a mixed stock with cattle, but as the latter do not bite so closely they lose much of the finer and more nutritious herbage when sheep are grazing with them, and cattle, therefore, do better when grazing by themselves than with sheep.

Sheep droppings are distributed much more evenly over the land than either those of cattle or horses. The spreading of the droppings of horses and cattle by harrowing is a very desirable operation, and where convenient should be done both in autumn and in spring. Where horses are the principal stock, their droppings should be collected and spread on the bare parts of the field or on other pastures.

Pasture recently laid down should be closely eaten in the early summer, especially if rye-grass has been largely sown, otherwise this grass will "run to seed." Young pasture is always earlier in spring than old pasture.

The best quality pastures are usually reserved for feeding animals, while secondary or inferior pastures are grazed by young or growing, breeding, and milking stock.

As feeding animals are usually full

grown, they exhaust pastures to a very small extent, whereas young or breeding or milking stock all considerably exhaust the pastures, especially in phosphates. It is therefore clear that a really good pasture has a much better chance of retaining its character than a poorer pasture, and this partly explains why such a pasture may not require any manurial treatment when a secondary or inferior pasture may frequently do so.

Cows in milk rapidly impoverish a pasture, and still more so if they graze on it during the day only, and are housed at night, as, when this is so, considerably less than half of their droppings is left on the pasture.

Water - supply. — Cattle and horses must always have access to water when they are grazing. This is necessary also for sheep in a dry season, but they seldom drink water in showery weather. A running stream is by far the best supply, but if this is not available, the ponds should be of sufficient size to contain water to last throughout a dry season. These should always be approached by a hard bottom, as otherwise the water soon becomes dirty. Where running water is not available a gravitation supply should be secured if at all possible. The progress of grazing stock is very much retarded by a bad water - supply, and now that so much attention is being given to a proper water-supply for milch cows by sanitary authorities, this has become an important question to dairy farmers.

Weeds. — Thistles and docks are the most troublesome of the larger weeds. These should be cut or hoed out twice in each season. When the soil is fairly moist docks should be pulled out by the roots. The best time to cut thistles is when the flower is beginning to form, as the roots are now weaker than when the plants are only a few inches above the ground, and they are not yet mature enough to produce seed. The common thistle spreads principally by underground stems, which make this plant much more difficult to eradicate, while the Scotch thistle is produced from seed only, and is therefore more easily dealt with. The smaller weeds, such as daisies, buttercups, sorrel, and plantains, can be effectively checked by judicious manurial

treatment and the proper grazing of pastures.

Drains and Fences. — All ditches should be regularly cleared once a-year, and the outlets of field drains entering into them should be examined to see that they are clear and properly trapped to prevent the ingress of vermin. Fences need regular attention, and should be at once repaired when they are injured by the grazing stock, or have become deficient from decay. There should be a clear understanding as to the upkeep of tenant's fences, landlord's fences, and march or boundary fences.

On many estates a large extent of pasture is let annually. In these cases the landlord usually provides for the fencing, the manurial treatment when necessary, and also for a certain amount of supervision of the grazing stock, and he pays all rates and taxes. The annual rent paid for such pasture is therefore an inclusive one, and is considerably higher than is paid under ordinary farming conditions. It is also fairly common to allow for the manurial value of cake fed to the grazing stock.

Grazing after a Hay Crop. — The "aftermath," or foggage, which follows a crop of first year's hay may have a considerable value for grazing, especially if clover is abundant. Foggage of this character is frequently let for eating off by lambs in Scotland and the North of England, and usually makes the equivalent of from 3d. to 4d. per lamb per week. A good aftermath should carry five lambs an acre for eight weeks in the autumn, and would thus be worth from 10s. to 12s. 6d. an acre. If very good it would be worth more, and if poor, considerably less. The aftermath of a second and third year's hay has a considerably less value, as well as that of old land hay.

Recent experiments have shown that the aftermath of old land hay has a greater feeding value when neither nitrate of soda nor sulphate of ammonia is used for the hay crop.

Value of Pastures. — The annual value of a pasture varies from a few shillings to over £5 per acre, and its value can be enhanced or reduced to a much greater extent by good or bad management than land under cultivation for ordinary farm crops. A first-class pasture will feed a good bullock on each acre, and may even feed two in succession in one season. On secondary pastures one and a half acres, and even more, may be required per beast, with cake in addition if fattening is the object. On the poorer pastures the amount of stock carried is considerably less, and these will be kept in store condition only.

In vol. i. pp. 340 - 346, information is given as to methods of laying poor land down to pasture, and of renewing and improving hill pasture.

FORMATION OF PERMANENT PASTURE.

At pp. 245-247 information is given as to grasses and clovers and standard mixtures of seeds suitable for permanent pastures. As already indicated, the low prices for cereal crops and high rates of wages now prevailing in this country have led to great importance being attached to the maintaining of land under grass, and on this account it is thought well to introduce here some additional information regarding the laying of land down to permanent pasture. In the first place, some supplementary notes may be given as to special seed mixtures which have been experimented with.

Most important Grasses and Clovers for Permanent Pasture. — The most important of the grasses for per-manent pasture are meadow fescue, timothy or catstail grass, cocksfoot, meadow foxtail, and perennial rye-grass. While the clover best suited for this purpose is wild white clover, which has a more perennial character than the common white clover of commerce, a certain proportion of other clovers should be included to give good herbage in the first few years.

Trials in Northumberland. — From trials that have been made in connection with Armstrong College, at Cockle Park, Northumberland, it has been found that the following mixtures per acre have given promising results, both as to hay in the first year and as to pasture after-wards. :—

	Cost of Seed.	Mixture.		
		No. 1. lb.	No. 2. lb.	No. 3. lb.
Perennial rye-grass . . .	4s. 6d. per bush.	6	6	...
Italian rye-grass	5s. 0d. "	6	6	...
Cocksfoot	9d. per lb.	6	6	12
Timothy	4d. "	3	3	3
Meadow fescue	6d. "	8	8	6
Tall fescue	1s. 3d. "	...	...	4
Tall oat-grass	1s. 0d. "	...	...	4
Red clover	8d. "	4	4	4
Alsike clover	9d. "	2	2	...
White clover	9d. "	4	...	...
Wild white clover . . .	1s. 8d. "	...	4	4
		39	39	37
Cost of seeds per acre		19s.	22s. 8d.	31s. 4d.

Trials in Cumberland. — At Whitehall, Cumberland, the following mixture, at a cost of 29s. an acre, gave excellent crops of hay in the first two years after sowing, and resulted in an excellent pasture : perennial rye-grass, 9½ lb.; cocksfoot, 5 lb.; timothy, 3 lb.; meadow fescue, 13 lb.; rough-stalked meadow grass, 1 lb.; cow-grass clover, 2½ lb.; alsike clover, 2½ lb.; white clover, 5½ lb.; yarrow, ⅓ lb. Although in the third year cocksfoot was present to the extent of 40 per cent, the pasture was closely grazed, and was not at all coarse. It was found that rough-stalked meadow grass had not been of value, and at any rate was not present at that time. It is likely that meadow foxtail might have been added to this mixture with advantage. This grass, however, will probably develop satisfactorily on rich loam soils only.

The three Cockle Park mixtures and the Whitehall mixture are good examples of different types of mixtures at a reasonable cost. Cockle Park No. 1 mixture has proved excellent so far as grasses are concerned, but not one of the clovers has proved to be permanent. Cockle Park No. 2 mixture is much more promising in its results, as the 4 lb. of wild white clover included has given a close bottom herbage of this plant, which promises to be perennial, or at least to be so until natural plants of white clover have developed themselves. In both these mixtures Italian rye-grass has been included to give bulk of hay in the first year. Cockle Park No. 3 mixture excluded the rye-grasses, and although the hay crop was not as heavy in the first year, it was a fairly satisfactory one, while the promise of the herbage for hay or pasture was encouraging. The Whitehall mixture has resulted in an excellent pasture in six years after sowing, while it also gave excellent hay crops in the first two years, and grazed well immediately afterwards.

These are good inexpensive mixtures which may form a useful guide to farmers who wish to lay down their land so as to take one or two crops of hay and then follow with pasture. The soil at Cockle Park is a poor boulder-clay in an exposed position, while that at Whitehall is a strong loam of a good character.

Wild White Clover.—The wild white clover is obtained from meadows in the south of England, where it grows as a naturally wild plant. Care should be taken that the seed has been obtained from true wild plants.

Trials in South of England.—In the southern counties of England the following mixture of seeds was found, in trials conducted by the University College of Agriculture at Reading, to give excellent results:—

	lb. per acre.		lb. per acre.
Perennial rye-grass	10	Cow-grass clover	1
Cocksfoot grass	2	Alsike clover	2
Timothy grass	2	Lucerne	10
Tall fescue grass	2	Sainfoin (seed in	
Trefoil	1	the husk)	10

Cost about 21s. per acre.

This gave good results as a mixture for a few years' lea on gravelly, sandy, and loamy soils, but was not so suitable for stiff clays. The trials showed that, under suitable conditions, sainfoin, and especially lucerne, make excellent plants for including in such a mixture. This mixture was tested alongside several other mixtures, and gave the best results under the foregoing conditions.

Local Circumstances to be considered.—But, as already pointed out more than once, farmers must use their own judgment in the making up of seed mixtures, as in the choice of manures. The local circumstances as to soil and climate are variable, and these must be carefully and intelligently considered if the best possible results are to be secured.

Sowing Permanent Pastures.

Sowing Grass Seeds without another Crop.—When grass and clover seeds are sown without a crop, there is no risk of their being killed out by a "laid" or lodged cereal crop, and there is a better chance of the grass and clover plants being well established in the year of sowing. When this is done it is usual to sow a small amount of rape or mustard with the seeds. It is the exception, however, to sow grass and clover seeds in this way.

Sowing Grass with another Crop.—The prevailing practice, as has been seen, is to sow grass and clover seeds with a corn crop, now most commonly barley or oats, in most cases which ever happens to be the crop grown after roots or potatoes. This corn or clover crop acts as an excellent check for weeds in the first year, and gives a good return to the farmer, which, if partly expended on the manuring of the young seeds, will give an excellent result if the manures are properly chosen. Such a corn crop should never be top-dressed with an active nitrogenous manure, and if in spite of this it is likely to lodge, it should be cut early so as to prevent damage to the young seeds. Land should always be clean and in good tilth on which young seeds are sown, so that the proper time for sowing them is with the corn crop which succeeds a fallow or fallow (root or potato) crop.

When the corn crop has been harvested, a dressing of 5 cwt. basic slag an acre, or an equivalent of some other phosphatic manure, will be useful ; and to this should be added a potash manure (say 1½ cwt. muriate of potash an acre) if the soil is of a light sandy character.

Hay Crop from Permanent Grasses.
—In the following year many prefer to graze their seeds and not to mow them as hay, but the roots of the young plants will be better developed if a crop of hay is taken. This, however, should be cut about ten days earlier than seed hay is usually cut, as by doing so the roots of the grass and clover plants retain considerably more vitality when the plants are cut before they begin to mature their seeds.

Early Grazing of Permanent Grass.
—The field should now be regularly and judiciously grazed, and care should be taken that it is always well eaten down in the early summer. This is especially important if perennial rye-grass has been included in the mixture, owing to the tendency of this grass to run to seed. The hay crop must not be top-dressed with nitrogenous manures like nitrate of soda or sulphate of ammonia, as if so clovers will be checked and the grass will not make as good grazing herbage afterwards.

In laying down wet stiff clay soils to pasture, it may be advisable to mow the field for the first two years and not to pasture it, so as to give time for the soil to become firmer, and so stand better the treading of the grazing stock.

Manures for Permanent Pasture.
—Another dressing of about 5 cwt. of basic slag or some equivalent may be repeated in the autumn three years later. If the soil is of a light character, the formation of a sward of pasture will be greatly assisted by a dressing of 10 to 12 tons dung an acre in the autumn after the first crop of hay has been removed ; in fact, such a dressing will be found very useful for any young pasture. Active nitrogenous manures will not be required if clover plants are produced and maintained, as the roots of these collect nitrogen from the air. Instead of slag, finely ground bone-meal may be used as the phosphatic manure, 4 cwt. of this being used instead of 5 cwt. slag. It must be remembered, however, that slag contains a considerable amount of free lime, and this, with its fineness of grinding, gives it two distinct advantages over bone-meal. But on certain soils, especially on those which have a fair proportion of lime, bone-meal may give better results. There is no doubt also that bone-meal can always be relied on in the long-run.

Fairy Rings in Pastures.—What are known as "fairy rings" in permanent pastures are so called because of the supposition prevalent in olden times that the circles were formed by fairies in their nocturnal dances. The rings are caused by fungi, which spring up at random. The fungus grows outwards from a centre, and spreads wider year by year, the decay of each season's growth of fungus bringing up a gradually widening ring of grass exceptionally fresh and green. The rings do little or no harm in pastures, but disfigure ornamental grounds. In the latter the fungus should be dug out immediately it is observed at a centre, and not allowed to spread. By breaking and forking-out the fungus two or three times a-year, just when it begins to show above-ground, the rings may be prevented from becoming very prominent in appearance.

Clifton Park System of Growing Grass.

The system of growing grass introduced by Mr Robert H. Elliot, owner of the estate of Clifton Park, in the Scottish county of Roxburgh, has attracted much attention. The main object aimed at by Mr Elliot was to sow such deep-rooting plants as would increase the amount of vegetable matter in the soil, and thus raise its fertility. His practice in laying his land to grass has been to sow for this purpose chicory, burnet, kidney vetch, and yarrow, along with natural grasses and clovers. Mr Elliot has not only obtained excellent pasture from these seeds in average seasons, but also in exceptionally dry years, when the deep-rooted plants show wonderful drought-resisting properties. Moreover, when broken up, the soil is found to be greatly increased in productiveness. Mr Elliot has published a full account of his system in the form of a small volume, which is well worthy of perusal.

In a new edition of this volume, published in 1907, Mr Elliot recommends the following as an "improved mixture" of seeds, adjusted in accordance with his "experience of the last twelve years" :—

Improved Kaimrig Selection.	Quantity of Seed per acre.	Guaranteed germination.	Number of germinating Seeds per statute acre.	Price.	
	lb.	per cent.		s.	d.
Cocksfoot (*Dactylis glomerata*) . . .	10	95	4,047,000	12	6
Meadow fescue (*Festuca pratensis*) . .	5	97	1,144,600	4	7
Tall fescue (*Festuca elatior*) . . .	4	97	954,480	6	0
Tall oat-like grass (*Avena elatior*) . .	3	94	389,160	3	0
Hard fescue (*Festuca duriuscula*) . .	1	95	549,100	0	8
Rough-stalked meadow grass (*Poa trivialis*) .	½	96	1,072,800	0	10
Smooth-stalked meadow grass (*Poa pratensis*)	1	75	1,395,000	1	4
Golden oat-grass (*Avena flavescens*) . .	½	80	560,000	1	9
Italian rye-grass (*Lolium italicum*) . .	3	98	793,800	1	0
White clover (*Trifolium repens*) . . .	2	98	1,434,720	2	0
Alsike clover (*Trifolium hybridum*) . .	1	98	703,640	1	1
Late-flowering red clover (*Tri. prat. per var.*)	2	98	427,280	2	8
Kidney vetch (*Anthyllis vulneraria*) . .	2½	98	472,850	2	6
Chicory (*Cichorium intybus*) . . .	3	90	904,500	3	6
Burnet (*Poterium sanguisorba*) . . .	8	130	561,600	4	8
Sheep's parsley (*Petroselinum sativum*) .	1	85	195,500	0	6
Yarrow (*Achillea Millefolium*) . . .	½	90	1,579,500	3	0
	48	...	17,185,530	51	7

Permanent Pasture on Poor Clay Soil.

In the *Journal of the Board of Agriculture* for October and November 1905 (Nos. 7 and 8, vol. xii., New Series), the "Formation of Permanent Pastures" is discussed by Professor T. H. Middleton.

As an adaptation of the seed mixture used by Mr Elliot of Clifton Park, with the cost of the seed restricted to about £1 per acre, Professor Middleton suggests the following mixture for the formation of permanent pasture on poor clay land:—

Seeds Mixture for Laying Down Poor Clay Soil to Permanent Pasture per Acre.

Plant.	Weight.	Number of Seeds in Thousands.	Cost.	
	lb.		s.	d.
Italian rye-grass	4	1,069	1	1
Perennial rye-grass	3	635	0	7
Timothy	1	1,307	0	6
Cocksfoot	2	809	1	10
Meadow fescue	2	467	1	2
Tall fescue	½	120	0	5
Hard fescue	1	555	0	6
Meadow foxtail	1	441	1	2
Tall oat-grass	½	62	0	6
Golden oat-grass	¼	280	0	8
Rough-stalked poa	¾	1,626	1	2
Smooth-stalked poa	¾	1,085	0	6
Crested dogstail	¼	210	0	9
Perennial red clover	1½	320	1	5
Alsike clover	1½	1,055	1	6
White clover	2	1,434	1	10
Lucerne	1	219	0	10
Sainfoin	5	110	0	6
Burnet	4	259	1	7
Chicory	1	284	0	11
Yarrow	⅛	417	0	9
Total . . .	33⅛	12,764	20	2

Professor Middleton recommends that burnet and sainfoin, which may be purchased mixed, as also lucerne, chicory, and yarrow, should be included in the mixture, as "these plants are either useful in themselves or are indirectly useful in opening up the soil by means of their strong roots ; and, further, in the case of burnet and chicory, by keeping the hay crop erect and open, thus allowing light to reach the white clover and young grasses."

Further, Professor Middleton suggests that, as poor clay soils are frequently deficient in phosphates, a dressing of 3 to 4 cwt. of superphosphate per acre should be applied before the seeds are sown, and a dressing of 5 to 7 cwt. of basic slag per acre in the autumn, after the first hay crop has been cut. "Basic slag will greatly encourage the growth of the clovers, and if subsequently the stock grazing the pastures receive oilcakes, the permanent grass will benefit. The above quantity of seed should be enough for clean land in good tilth. On a rough surface increase the rye-grass and red clover to ensure a cover ; and to counteract the effects of rye-grass on the other pasture plants, manure liberally about the third or fourth season." [1]

The treatment of permanent pastures on high-lying and rough land is dealt with in vol. i. pp. 340-346.

ENSILAGE.

The practice of Ensilage may be defined as the preservation of green food by the exclusion of air. In a modified sense the system is an ancient one, but in the United Kingdom it was not pursued to any considerable extent prior to 1882.

History of Ensilage. — From time immemorial the storage of grain in underground pits for preservation has been practised in Eastern countries. Pliny speaks approvingly of this method as being adopted, in his time, in Thrace, Cappadocia, Barbary, and Spain. Varro endorsed his opinion of its merits, and asserted that wheat could be thus kept sweet and entire for fifty years, and millet for a century. The main object, especially among nomadic tribes, was to prevent marauders or victorious enemies from obtaining their stores of food.

In later days the practice was adopted —in Spain for example—for commercial reasons, as by its means the surplus in years of plenty and low prices could be kept for disposal in times of scarcity and high prices.

It appears to have been in Germany that the system of ensilage was first applied to the preservation of fodder crops, as distinguished from grain. In an article in the *Transactions of the Highland and Agricultural Society* so far back as 1843, Professor Johnston gave a detailed description of the German system of making "sour hay."

In 1874 Professor Wrightson, in his "Report on the Agriculture of the Austro-Hungarian Empire," published in the *Royal Agricultural Society's Journal*, remarked that "the system of making 'sour hay' is also well worth the attention of English agriculturists. It is done by digging graves or trenches, 4 feet by 6 or 8 feet in depth and breadth, and cramming the green grass or green Indian corn tightly down into them, covering the whole up with a foot of earth. The preservation is complete, and the wetter the fodder goes together the better. . . . This sour hay affords a capital winter fodder, and when cut out with hay spades it is found to be rich brown in colour, and very palatable to stock."

Introduction into Great Britain.— In 1882 the practice of ensilage began in the United Kingdom in real earnest, and in the course of the next few years the progress was rapid. In 1883 the Ensilage Commission—a private but highly influential body, whose labours were endorsed by the Government, and embodied in official Blue-books—sat and collected a mass of invaluable evidence.

[1] *Jour. of the Board of Agric.*, New Series, No. 8, vol. xii. p. 462.

The Agricultural Returns first included
ensilage in their survey in 1884, and
enumerated 610 silos as being in existence in Great Britain.

Silos.

A silo was originally a pit—the word
being derived from the Greek σιρὸς,
which, according to Liddell and Scott, is
"a pit or hole sunk in the ground for
keeping corn in." The word came to us
through the Spanish and French, in which
languages the r was naturally changed
to l. Soon, in practice, "a pit or hole"
was found to be adaptable only to special
soils and situations, and a large variety
of receptacles for ensilage, both below
and above ground, have been constructed,
which have widely extended the original
term silo.

It would be impossible even to enumerate the different descriptions of silos
which have been tried. From the most
elaborately designed and expensively
constructed buildings to the simplest
and cheapest "converted" structure, the
variety of methods resorted to has been
remarkable. Some idea of their diversity may be gathered from the fact that
the cost per ton capacity has ranged
from 8s. up to 30s. or 40s., and even 50s.

Silos, Above or Below Ground?—
In making a silo, the first question is
obviously whether it shall be dug out
or erected. Sunk silos were thought
in some respects preferable, but the
cost of excavation has to be considered.
Latterly, where the system is still pursued, silos above-ground have almost
entirely superseded sunk silos.

Concrete Silos.—In many cases silos
have been formed largely of concrete.
On Lord Ashburton's estate a silo with a
total capacity of about 96 tons was substantially built of concrete at a cost of
£113. It was formed in three compartments, and roofed with corrugated iron.

Silo with Lever-pressure.—A silo
erected by Mr C. G. Johnson with
special lever-pressure is worthy of mention. It was built of brick, with slated
roof. It was 18 feet long, 10 feet wide,
and 28 feet high up to the eaves, but
6 feet of this height was left for working
the machinery, so that the total capacity,
at 50 cubic feet to the ton, would be
about 80 tons. But Mr Johnson's

silage weighed very much more than
usual, as it reached 60 lb. per cubic
foot; and at this rate, if the whole space
were occupied, fully 100 tons could be
put in. The total cost of the silo and
apparatus was about £150, of which
£65 was for masonry, £40 for pressing
apparatus, and the remainder for roof,
&c.; but Mr Johnson was his own en-
gineer, and the cost would have been
higher had a professional man been employed to superintend the work. Deducting £40 for pressing apparatus, the cost,
at the same rate as in other cases, would
be about 22s. per ton.

Wooden Silos.—Wooden silos also
came into favour when it was discovered
that lateral pressure was practically non-
existent in making silage.

American and Canadian Silos.—
In the United States of America and
Canada, most farms which have milch
cows as part of their stock are provided
with silos. These silos are most generally made of concrete, and are variously
shaped, in different parts of the country.
They also vary greatly in capacity.
In most cases they are built at the
end or side of the cow-house. They
are generally filled with green maize
stalks cut into half-inch lengths, the
fodder being conveyed from the cutter
to the silo by a mechanical conveyer.
For the production of winter milk, cut
maize in the shape of silage is to the
American or Canadian dairymen what
roots are to dairymen in the United
Kingdom.

Methods of Pressure.

The methods of obtaining pressure in
silos are as varied and numerous as the
forms of the silos themselves. Dead
weights—earth, stones, bricks, iron, &c.
—were the elementary forms, and are
still largely used. The labour of putting
on and taking off the weight is, however,
obviously great, and this led to the introduction of a great variety of mechanical
appliances for the purpose.

Stack Ensilage.

Soon after it was discovered in this
country that there was practically no
lateral pressure in the silo, experiments
were tried in the making of silage in
stacks. The results were favourable,

and this practice extended rapidly. It was found that there was usually more waste in the stack than in the silo, through exposure of the food material at the sides of the former; but, on the other hand, the saving of the cost of erecting silos was an important consideration.

Sweet and Sour Silage.

"Sweet" and "sour" are arbitrary terms which have perhaps been somewhat abused in the ensilage controversy. Perfect silage—that at which all makers should aim—is neither the one nor the other. But it was found easier to make "sweet" silage in a stack than in a silo, and easier to make "sour" silage in a silo than in a stack. That is to say, in a stack the temperature rises very rapidly, and the difficulty lies often in preventing too great heat. In a silo it may be necessary to wait at intervals for the temperature to rise, and the work of filling has thus to be interrupted.

Making Sweet Silage.—The general practice at the outset in this country was to apply pressure directly the silo was filled, and the product was sour silage. It was discovered that the sourness could be got rid of, at least to a large extent, by deferring the weighting of the mass for two or three days, until the temperature of the silage rose to about 120° or 140° Fahr. The theory was that this temperature, about 120° Fahr., is sufficiently high to kill the bacteria which produce acid fermentation; and if the bacteria be thus killed, and the silo then covered

and weighted, the enclosed mass of green fodder will remain sweet, and be practically preserved under the same conditions as fruits, vegetables, or meats are preserved when canned.

In practice the theory was found to be a sound one, but to ensure good sweet silage careful attention has to be given to the temperature during the making process. With the aid of a stack-thermometer, it is easy to ascertain exactly the rise and fall of temperature in either a silo or silage stack.

If a tube 1 inch in diameter be built in a perpendicular position in the centre of the silo, a thermometer attached to a string may be dropped into it at any time and the temperature found. As the building of the contents proceeds, this tube may be pulled up, so as always to have the bottom of it 6 to 10 feet from the surface. After a temperature of 125° to 140° Fahr. has been reached by loose building, a further addition of material to the top will so compress what is 10 feet lower that the temperature will begin to fall. Silage so made will invariably turn out sweet.

Analyses of Sweet and Sour Silage. —As would naturally be expected, the stock prefer the sweet to the sour silage. By analysis it was found that there is little difference in the feeding properties of the two kinds of silage, sweet silage having usually a slight advantage. The following are analyses of samples of sweet and sour silage made by the late Dr A. P. Aitken :—

| | EARLYPIER. | | HARCUS. | |
	Sweet.	Sour.	Sweet.	Sour.
Water	75.09	76.08	69.39	77.77
Solids	24.91	23.92	30.61	22.23
	100.00	100.00	100.00	100.00
Solids (dried at 212° Fahr.)—				
Albumen	6.52	6.33	6.71	6.33
Non-albuminoid nitrogenous matter reckoned as albumen	4.43	3.64	2.02	2.28
Carbohydrates	44.55	46.18	46.05	47.87
Ether extract	6.20	5.95	6.85	6.35
Woody fibre	28.85	25.15	30.20	28.70
Ash	9.45	12.75	8.17	8.47
	100.00	100.00	100.00	100.00

Pressure for Ensilage Stacks.—A good method of applying pressure to ensilage stacks is shown in fig. 578.

Another system of pressure in use is that known as Blunt's patent (fig. 579). It combines the two principles of the screw and lever, and one main advantage claimed for it is that by its means "continuous pressure" is secured.

Crops for Silage.

There are many kinds of crops which are suitable for making into silage. It is well to bear in mind that the quality of the silage is directly dependent upon the quality of the material from which it is made. This may appear to be a simple truism, but it is by no means unnecessary to insist upon it. In the early days of the system there seemed

Fig. 578.—*Johnson's ensilage press.*

to be a common idea that silage was silage—so to speak—whatever it might be made from, some people apparently

Fig. 579.—*Blunt's ensilage press.*

thinking that coarse, useless grass, or waste substances, might be ensiled and transformed into valuable food. That such materials may be, and are, made into silage with advantage, is no doubt true; but it is essential for those who use them to remember that the process of ensilage does not give them any higher feeding value, other than possibly to make them more palatable or digestible to stock. No formation of food constituents goes on in the silo or stack, and practically that which a farmer puts in, that will he—if the silage be well made—take out. It follows, therefore, that those who wish for valuable silage must make it of a valuable crop.

The crop which has been most used in this country for silage has been meadow grass, but sewage-grown Italian rye-grass, tares, green grain of any kind, clover, lucerne, sainfoin, &c., may each be made into silage. All the grass crops should be put in the silo without any chopping;

if green and succulent, tares, grain, and lucerne may be treated in the same way, but if the stalks of either are the least hard the crop should be run through a straw-cutter and cut into one-inch lengths. If this is not done, air remains in the stalks, and the material does not pack sufficiently firm to exclude air. Such material rarely keeps well, and almost invariably turns out mouldy.

Such soft materials as mangold, turnip, and carrot leaves are not suited for turning into silage. They have too little fibre in them, and in the silo they go into pulp.

Feeding Value of Silage.

From time to time there has been much discussion as to the feeding value of silage. Advocates of the system claim that in numerous experiments which have been carried out the results have in the main favoured silage. **Hay** *v.* **Silage.**—Feeding trials conducted by the Royal Agricultural Society of England with hay and silage made from equal areas of the same field finished with a very slight advantage on the side of the silage. Other trials by the same Society with oat-silage as against roots and chaffed straw ended more decidedly in favour of the silage.

Advantages claimed for Ensilage.

Reference has already been made to the English Ensilage Commission, which inquired into the subject in 1883 and prepared a Report, which was published by the Government as a Blue-Book. The general advantages claimed for ensilage are set forth fully and clearly in that Report. The advantages of the system are classified by the Commission under these three heads: " I. In rendering the farmer independent of weather in saving his crops. 2. In increasing the productive capabilities of farms: (a) in greater weight of forage saved; (b) in greater available variety and rotation of crops; (c) in increased facility for storage. 3. In connection with feeding: (a) dairy stock; (b) breeding stock; (c) store stock; (d) fattening stock; (e) farmhorses." Taking each of these points in order, the Commission remark upon them as follows:—

" I. *Independence of Weather in sav-*ing *Crops.*—In this respect it has been abundantly proved to us that ensilage is of great economic value. In Scotland, in Ireland, and in the north and west of England, few seasons occur in which more or less difficulty is not experienced in reducing green fodder crops to a sufficiently dry condition for stacking in the ordinary way. This is especially the case with second crops of clover and aftermath. The loss occurring through ineffectual attempts to dry such crops, or through their inferior condition when carried, is often very considerable; and it is obvious that any system which enables a farmer to store these in good condition for future use must be a great saving of expense and anxiety.

" 2. *Advantages in increasing the Productive Capabilities of Farms: (a) In greater Weight of Forage saved.*—It is obvious that unless the forage in a weighty condition be of more feeding value per acre than when saved in a less weighty form, there can be no gain to the farmer. It has been contended that the loss of weight in the process of drying is simply loss of water by evaporation, and that by avoiding this nothing is saved. If such were truly the case, dry forage should give the same feeding results per acre as green forage. No practical farmer would contend that it does so, and the difference is especially noticeable in the case of dairy stock. So far as we have been able to ascertain the opinion of competent men on this subject, we estimate the value of green forage well preserved in a silo at somewhat more than one-third, weight for weight, of the value of the same material made into hay under favourable conditions. The very wide difference of value between good and bad silage cannot be too strongly insisted upon. It is found that grass well preserved in a silo, after deduction for loss, will yield approximately five times the weight of the same grass made into hay. We have therefore, say, five tons of silage, which, taken at one-third the value of hay per ton, yields a profit of over 60 per cent as compared with one ton of hay. If we take it at one-fourth, it still leaves a profit of 25 per cent. Any waste that may occur to reduce the weight of nutritious forage, whether by evaporation or by excess of chemical change, must necessarily affect this calculation, which is based upon the highest degree of perfect preservation so far known to be attainable.

"(b) *In Available Variety and Rotation of Crops.*—By the process of ensilage many crops can be preserved which would not otherwise be found profitable if used in the form of green forage. Rye, oats, millet, maize, barley, and even wheat, if cut about the time of attaining their full development, but before the seed begins to harden, have been successfully used as food for cattle through the medium of the silo. Such of these crops as are found to reach the required condition before the middle of June, if cut before that time, will leave the land free for a second sowing, and thus increase its capabilities of annual production, while maintaining the fertility of the soil. Where land is well treated, maize, buckwheat, or, in some parts of England, also turnips, can be sown after green rye or oats are cut and carried, and thus a second crop may be secured for preservation in the silo, or for consumption by sheep on the land.

"(c) *In Increased Facility for Storage.* —This advantage has been forcibly impressed upon us. It enables farmers to guard themselves against emergencies, such as frequently arise in our climate through prolonged cold in February, March, and April, causing great scarcity of food for cattle and sheep, where the supply of roots is inadequate.

"3. *Advantages connected with Feeding:* (a) *Dairy Stock.*—We have received the strongest evidence of the undoubted advantage of the system for the feeding of dairy stock. The effect of dry winter food given to such stock has always been to reduce in quantity and to deteriorate in quality milk, cream, and butter, as compared with the same products resulting from green summer food. Although the degree of perfection attainable in summer has not been reached, it has been at least much more nearly approached by ensilage than by the use of hay and other dry foods, while at the same time the objections inseparable from the employment of roots for this purpose have been overcome. A sensible improvement in the colour of butter has been especially noticed.

"(b) *Breeding Stock.*—Green fodder preserved by ensilage has been successfully employed in feeding sheep and cattle at the time of breeding; and as it has been shown to increase the flow of milk, it will undoubtedly be found useful for this purpose, although the proportion of its admixture with other kinds of food must always require care and judgment.

"(c) *Store Stock.*—It forms a complete and wholesome food for store stock.

"(d) *Fattening Stock.* —The value of this process for the purpose of forming flesh and fat has not yet perhaps been so widely demonstrated as in the case of dairy produce. At the same time the results attained show that it compares favourably with the use of roots, and, if given in proper proportions with other food, it affords a cheap substitute for the same bulk, which would otherwise be required in some different form. The advantage of its use is most apparent in the degree to which it enables a farmer profitably to consume straw-chaff, rough hay-chaff, and other dry materials, which, without admixture with some kind of moist food, would not be palatable or advantageous to the growth of stock.

"(e) *Farm-horses.*—Strong as the evidence has been of the advantage of ensilage for keeping all stock in healthy condition, farm-horses have by no means been excepted. We have received highly satisfactory accounts from several quarters of the health of working teams when given a limited proportion of silage mixed with other food."

In conclusion, the Commissioners state that they endeavoured to discount all exaggerated estimates, as well as to make allowance for a considerable amount of prejudice and incredulity which they met with, and they add: "After summing up the mass of evidence which has reached us, we can without hesitation affirm that it has been abundantly and conclusively proved to our satisfaction that this system of preserving green fodder crops promises great advantages to the practical farmer, and, if carried out with a reasonable amount of care and efficiency, should not only provide him with the means of insuring himself to a great extent against unfavourable seasons, and of materially improving the quantity and quality of his dairy produce, but should also enable him to increase appreciably the number of live stock that can be profitably kept upon any given acreage, whether of pasture or arable land, and proportionately the amount of manure available to fertilise it."

Ensilage Losing in Favour.

For a number of years after the issue of the Report of the Ensilage Commission the system was largely pursued in different parts of the British Isles. On the whole, the results continued to be fairly satisfactory. Gradually, however, it became apparent that the system had not obtained an extensive permanent footing in this country. Early in the last decade of the nineteenth century the practice of ensilage began to wane, and in the course of a few years it almost altogether disappeared from several districts in which it was for a time pursued. It may now be said that ensilage is practised in this country only to a very limited extent.

It is no doubt the case that the occurrence of better seasons in the matter of weather had something to do with the change of feeling towards ensilage. There are some who still continue the system, and there are many who look upon it as a valuable agent in the hands of the agriculturists of the British Isles, especially in seasons of inclement weather. In view of these considerations, the system seemed still deserving of the notice it has received in this edition of *The Book of the Farm.*

THE POTATO CROP.

The potato crop is one of increasing importance to the farmers of the United Kingdom. It is a costly crop to grow and prepare for the market, and it involves a good deal of risk, but in average seasons it is moderately profitable. Not unfrequently it is the best paying crop on the farm.

Varieties of Potatoes.

The potato belongs to the class and order *Pentandria Monogynia* of Linnæus; the family *Solaneæ* of Jussieu; and to class iii. *Perigynous exogens;* alliance 46, *Solanales;* order 238, *Solanaceæ;* tribe 2, *Curvembyræ;* genus *Solanum* of the natural system of Lindley.—Lindley observes that this family of plants are "natives of most parts of the world without the arctic and antarctic circles, especially within the tropics, in which the mass of the order exists in the form of the genera *Solanum* and *Physalis.* The number of species of the former genus is very great in tropical America."

The *Solanaceæ,* or Nightshades, comprise 900 species, of which we have only five in Britain. The genus *Solanum* has only two British representatives—*Solanum dulcamara,* a pretty climbing shrub, found occasionally in hedges ; and *Solanum nigrum,* with a herbaceous stem. Both these plants, like the rest of the tribe, are strongly narcotic. The *Solanum dulcamara,* bitter-sweet, or woody nightshade, has a purple flower and bears red berries ; the *Solanum nigrum,* or garden nightshade, bears white flowers and black berries. These plants can be identified botanically only by an examination of the leaves and berries. The active principle in both is an alkaloid, *Solanine,* which is itself a poison, although not very energetic: two grains of the sulphate killed a rabbit in a few hours. According to Liebig, this poisonous alkaloid is formed in and around the shoot of the common potato when it germinates in darkness; but there is no evidence that the potatoes are thereby rendered injurious. Their noxious qualities are probably due to other causes.

Introduction into Europe.—It is asserted that Sir Francis Drake introduced the potato into Europe in 1573, but this is very doubtful, since it has also been ascribed to Sir John Hawkins in 1563 : it is, however, certain that Raleigh brought it from Virginia to England in 1586; and it is believed that the Spaniards had established its cultivation in Europe before this time. It was first cultivated extensively in Belgium in 1590, in Ireland in 1610, and in Lancashire in 1684. Between 1714 and 1724 it was introduced into Swabia, Alsace, and the Palatinate; in 1717 it was brought to Saxony; it was first cultivated in Scotland in 1728; in Switzerland, in the canton of Berne, in 1730; it reached Prussia in 1738, and Tuscany in 1767. It spread slowly in France till Parmentier, in the middle of the eighteenth century, gave it so great an impulse that it was contemplated to give his name to the plant; the famine in 1793 did still more to extend its cultivation.

Distribution of the Potato.—According to Humboldt, the potato is generally cultivated in the Andes, at an elevation from 9800 to 13,000 feet, which is nearly the same elevation to which barley attains, and about 9800 feet higher than wheat. In the Swiss Alps of the canton of Berne the potato reaches, according to Katsoffer, an elevation of 4800 feet.

In the north of Europe the potato is grown farther north than barley, and it is cultivated in the mountainous regions of India.

Varieties in Use.—The varieties of potatoes now in use are very numerous. Several hundreds indeed there are, and every year adds to the number.

The principal kinds planted vary from time to time, new sorts taking the place of older kinds which ultimately degenerate. Great interest is now taken in rearing new varieties, and new sorts of considerable promise are offered to farmers almost every season.

The influences of soil and climate introduce variations in the different sorts, but the multiplicity of varieties is due mainly to the raising of new sorts from the seed. It has been found that an occasional new variety successfully resists disease for a few years, and this, of course, has given a great stimulus to propagation from seed.

Good Potatoes.—A good potato is neither large nor small, but of medium size; of round shape, or elongated spheroid; the skin of rough and netted appearance, and homogeneous; and the eyes neither numerous nor deep-seated. Smooth potatoes are almost always watery and deficient in starch.

Some kinds of potatoes are fit to use when lifted, but other kinds improve with keeping, and are best in spring.

The *intrinsic* value of a potato, as an article of commerce, is estimated by the quantity of starch it yields on analysis; but, as an article of domestic consumption, the *flavour* of the starchy matter is of as great importance as its quantity. Almost every person prefers a mealy potato to a waxy one, and the more mealy the better flavoured it usually is. The mealiness consists of a layer of mucilage immediately under the skin, covering the starch or farina, which is held together by fibres.

Light soil yields a potato more mealy than a strong soil; and a light soil produces a potato of the same variety of better flavour than a clay soil. Thus soil has an influence on the flavour, as well as on culture; and the culture which raises potatoes from soil to which dung had been applied some time before planting, imparts to them a higher flavour than is obtained when the tubers are grown iu immediate contact with dung.

Potato Disease.—The fungoid diseases which attack potatoes will be dealt with in connection with other "Fungoid Diseases" of crops, as will also the spraying of potatoes, now so largely resorted to for the checking of disease.

SOIL AND TILLAGE FOR POTATOES.

In most cases the potato crop is grown after a crop of oats, which, in turn, had succeeded pasture. In some districts potatoes come after lea, and this many consider the best place in the rotation for the crop.

Land for Potatoes.—Potatoes thrive best on light dry, friable, or sandy loams. They also do well upon virgin soils and mossy or turfy land, but seldom give good results on strong, tenacious clays, with retentive subsoils.

Tillage for Potatoes.—The stubble land intended for potatoes is ploughed early in the autumn, so that it may have the full benefit of the ameliorating influences of winter. Potato land should be tilled early in spring, and cleaned as well as possible. The time for cleaning land is usually limited in spring, so that the cleanest portion of the fallow-break should be chosen for the potatoes to occupy.

After cross-ploughing or grubbing, the land is thoroughly harrowed with a double tine along the line of the furrow, and a double tine across it, and any weeds brought to the surface and gathered off. If the land be clean, it is then ready for drilling ; if not, it should receive a strip of the grubber in the opposite direction, and again be harrowed, and any weeds gathered off.

The cross-ploughing of potato land in spring is not so extensively practised now as formerly. With deep autumn or winter ploughing little spring stirring suffices, and if the land is clean, and the soil fine and friable, rank harrows may do all that is necessary. Most likely, however, a strip of the grubber will be beneficial ; and as grubbing is a speedy operation it need not long delay the planting.

In cultivating land for potatoes, it is important to remember that the roots and tubers should have free scope to ramify in the soil.

MANURING POTATOES.

It is well known that no farm crop is more variable in its yield than the potato crop. In two successive seasons, when the land and the treatment are the same, the one crop may be twice as great as the other. The chief element of success is the weather,—not the weather of the whole season, but the weather at certain critical periods—the favourable distribution of shower and sunshine. In this respect potatoes are far more dependent on weather than are most other farm crops. Whether the season prove such as to favour a big crop or not, the manuring for potatoes must be liberal if the best yield is to be gathered ; but in some years the same manuring will give a far greater increase than in others.

Dung for Potatoes.—Farmyard dung is the staple manure for potatoes. Without a certain amount of dung they are seldom grown ; and heavier dressings of dung are employed for potatoes than for any of the other ordinary crops of the farm. From 15 to 20 tons per acre are common quantities, and the dressing is often as much as 25 tons or more per acre.

Mechanical influence of Dung on Potatoes.—It is obvious that the potato crop cannot make immediate use of more than a small portion of the plant-food contained in these heavy dressings of dung. Farmyard manure, however, would seem to be far more to the potatoes than a source of nutrition. Its mechanical influence upon the soil has evidently a peculiarly beneficial effect on this crop. It not only opens up the soil and renders it more friable, but by the decay of the organic matter the temperature of the soil is raised, thus surrounding the potato with more kindly conditions than would have existed in the absence of dung. These heating and pulverising influences of dung are undoubtedly of great importance ; some, indeed, are inclined to think that a good deal more of the value of dung as a manure lies in these functions or influences than has usually been associated with them.

Typical Dressings.—In the Lothians of Scotland, where potato-growing is a prominent feature in the system of farming, the dressing of dung ranges from 20 to 30 tons per acre—even as much as 35 tons or more per acre being occasionally used for an early crop. In addition to these allowances of dung, from 5 to 10 cwt. of artificial manure is applied, the latter consisting of guano, dissolved bones, superphosphate, and perhaps a little potash, or instead of guano, nitrate of soda or sulphate of ammonia. A dressing of 6 cwt. of artificial manure sometimes used with about 20 tons of dung, consists of 2 cwt. bone meal, 1 cwt. vitriolised bones, 1½ cwt. of mineral superphosphate, 1 cwt. sulphate of potash, and ½ cwt. sulphate of ammonia. On light lands in Ayrshire, where potatoes are successfully grown for early consumption, very heavy manuring is practised. In some cases here as much as 30 tons of dung, and 12 to 15 cwt. of artificial manure, are applied per acre—the artificials consisting of 4 or 5 cwt. of kainit, and 8 or 10 cwt. of a mixture containing from 8 to 10 per cent of ammonia and 20 to 30 per cent of phosphate.

Quickly acting Manures for Potatoes.—For potatoes, manure should be supplied in a readily available form near at hand, as it is a moderately rapid growing, feebly-rooted plant. It is thus desirable that, however much dung may be applied, a certain quantity of more quickly acting manure should also be given. It may be that the dressing of dung will contain far more nitrogen, phosphoric acid, and potash than the crop of potatoes will require. Experience, however, has clearly shown, notably in the case of the Rothamsted experiments, that only a very small portion of the plant food in the dung can be utilised by the first crop of potatoes. Thus, while the heavy dressing of dung is beneficial, partly as a source of manure and partly on account of its mechanical influence on the soil, it is necessary to apply phosphates, nitrogen, and potash in forms in which they will be immediately available to the crop. For this purpose, mineral superphosphate, guano, sulphate of ammonia or nitrate of soda, and kainit, are most largely used. Some prefer sulphate or muriate of potash because of the excess of salt in kainit.

For early potatoes, which grow rapidly and are only a short time in the ground, quickly acting manures are most suitable.

Dr Aitken on Manuring Potatoes.

With special reference to the experiments conducted under the auspices of the Highland and Agricultural Society, the late Dr A. P. Aitken wrote as follows as to the manuring of potatoes:—

"The potato is a feeble rooter, and requires its manurial food to be closely within reach. In order that the tubers may be able to expand, the soil about them must be loose. Manures which keep the roots free are therefore very appropriate. There is nothing so suitable as dung for that purpose, but any very bulky manure is also good, and especially if it has a large proportion of organic matter to keep the soil warm and make a soft compressible seed-bed.

Form of Manures for Potatoes.— "The potato is not well adapted for utilising insoluble materials, and therefore any artificial manures applied with the dung should be of a soluble, or, at least, not very insoluble kind. Superphosphate is better than ground mineral phosphates or bone meal; and even dissolved bones is rather a slow manure for this crop.

Nitrogen for Potatoes.—"The most important ingredient of a potato manure is nitrogen, and the most of it should, as has been said, be of a soluble quick-acting kind. Insoluble nitrogenous matters do not come into activity quickly enough for the wants of the crop. When dung is used, there is no better way of increasing the nitrogenous manure than by giving some nitrate of soda or sulphate of ammonia along with the dung.

Potash for Potatoes.—"As regards potash the wants of the potato crop are peculiar. The potato plant takes away a great deal of potash; and fields on which potatoes are frequently grown very soon become exhausted of potash. This must be made good, for the potato is very dependent on potash manures. Where much dung is used there is little need of applying extra potash manure, seeing that dung is so rich in potash; yet much of the potash in dung is not very readily available, and an addition of potash salts is therefore to be recommended.

Too much Potash Injurious. — "Where no dung is used, potash forms an exceedingly important ingredient. The limit of potash manure required for potatoes is nevertheless very soon reached; and no good but rather harm is done by overdoing the application of potash.

"It is a common practice to apply very large doses of light manures to the potato crop. Much extravagance may occur in that way, and sometimes more harm than good result. It is important in such cases that the manure should not be placed in direct contact with the sets. It should rather be applied some time before planting, and should be well incorporated with the upper layer of the soil as a general fertilising application.

Proportion of Manurial Ingredients.—"When no dung is used, the proportion of the manurial ingredients in a well-balanced potato manure will be just about equal parts potash, ammonia, and phosphoric acid. When applied along with dung the potash may be diminished by half, and the nitrogen slightly increased." On the most of soils good results have been obtained from the following mixture: 1 cwt. sulphate of ammonia, 1½ cwt. sulphate of potash, and 2 to 4 cwt. superphosphate, given along with about 10 tons of dung, the dung being applied in the drills at the time of planting.

Rothamsted Experiments with Potatoes.

An interesting series of experiments upon the manuring of potatoes was conducted at Rothamsted. The results were fully explained by Sir Henry Gilbert in his lectures at Oxford and Cirencester, and several of them are of considerable practical value to farmers.

Farmyard manure was tried by itself and in conjunction with nitrogenous, phosphatic, and potassic manures. The potatoes were grown on the same land every year, and on this account the results cannot be unreservedly applied to potato culture under ordinary rotation farming. Still some lessons of importance may be learned.

Farmyard Dung.—This was applied at the rate of 14 tons per acre every year, and the average yield for six years was 5¼ tons per acre—just over 3 tons more than the plot which had no manure of any kind in those six years.

Dung and Superphosphate.—The addition of 3½ cwt. superphosphate of lime to the dung had very little influence on the crop. The yield rose to 5 tons 12 cwt., or an increase of 7 cwt. over the dung alone.

Dung, Superphosphate, and Nitrate of Soda.—But when the dung and superphosphate were supplemented by some nitrate of soda, supplying 86 lb. of nitrogen per acre, a marked difference upon the crop became apparent. The produce rose to 7 tons 2 cwt.—an increase of 1½ ton, due to the 86 lb. of rapidly acting nitrogen.

Artificial Manures.—Artificial manures were also tried by themselves, separately, and in different combinations.

Superphosphate of lime (3½ cwt. per acre) applied alone gave an average of 3 tons 13⅜ cwt. for twelve years—nearly 2 tons 6 cwt. more than the no-manure plot.

Mixed mineral manure (consisting of 3½ cwt. per acre of superphosphate, 300 lb. sulphate of potash, 100 lb. sulphate of soda, and 100 lb. sulphate of magnesia) gave only 2 cwt. per acre more than the superphosphate alone.

Salts of ammonia (459 lb.) alone gave a poor result—only 2 tons 5¾ cwt. per

acre, or 6 cwt. more than the unmanured plot.

Nitrate of soda (550 lb.) alone did little better. It exceeded the salts of ammonia by 7 cwt. per acre.

Nitrogenous and mineral manures mixed produced very different results. Applied together to the same plot they raised the produce to an average of over 6½ tons—6 tons 14½ cwt. for salts of ammonia and mixed mineral manure, and 6 tons 13 cwt. for nitrate of soda and mixed mineral manure.

Conclusions.—In contrasting these experiments with various kinds and dressings of manure, some noteworthy results are observed. As to *artificial manures* it is shown (1) that the exhaustion of phosphoric acid by the potatoes was greater than that of potash; (2) that in the continuous growth of potatoes here it was the available supply of mineral constituents within the root-range of the plant, more than that of nitrogen, which became deficient—hence the greater produce from mineral manures alone than from nitrogenous manures alone; (3) that it is only when all the essential elements of manure are present in sufficient quantity that the full benefit of any kind of dressing can be derived; and (4), that when thus applied together in a well-balanced dressing, artificial (nitrogenous and mineral) manures produced a crop which for twelve successive years exceeded the average yield of the United Kingdom—decidedly greater indeed than the yield from farmyard manure alone, and only about 8 cwt. per acre behind the produce from a combined dressing of dung, superphosphate, and nitrate of soda.

The efficacy of well-proportioned artificial manures for potatoes thus demonstrated at Rothamsted is a consideration of great importance to farmers. Equally valuable to the practical farmer is the unquestionable conclusion that, in efficient and profitable manuring, an essential condition is that the dressing shall be properly balanced—that is, contain all the necessary elements of plant-food in due proportion.

Slow Exhaustion of Dung.—In the Rothamsted experiments with potatoes some interesting information has been brought out as to the behaviour of farmyard manure in the soil. The most striking point in these results is the slow action of the dung, particularly of the nitrogen it contained. The dressing of dung applied annually to the potato crop contained, per acre, about 200 lb. of nitrogen—besides, of course, an abundance of mineral matters, &c.; yet from 86 lb. of nitrogen, supplied in the form of nitrate of soda or salts of ammonia, along with an artificial mixture of mineral manures, the average produce was considerably greater than from the dung. Thus it is observed that, while the dung supplied far more nitrogen than the crop required, it did not contain enough in such a readily available condition as could be at once seized by the crop.

Further striking evidence of the slow action of nitrogen in dung was furnished by the fact that by supplementing the dung with some quickly acting nitrogen —86 lb. of nitrogen per acre in nitrate of soda—the produce of the tubers was increased by over 1½ ton per acre.

Residue of Dung.—Then as to the residue of dung, the results showed that it acted very slowly. Of the nitrogen supplied in the annual dressing of dung only about 6.4 per cent had been recovered in the crop of potatoes in the first six years. In the succeeding six years potatoes were grown every year on the same plot without any further application of dung or other manure, and in that time only 5.2 per cent of the unrecovered nitrogen was taken up in the crop. Thus in twelve years only 11.6 per cent of the nitrogen supplied in the dung during the first six years had been recovered in the crop.

Sir Henry Gilbert on Dung for Potatoes.—Referring to this point in his Cirencester lectures, Sir Henry Gilbert said: "In the case of other crops it has been found that only a small proportion of the nitrogen of farmyard manure was taken up in the year of application. But these results seem to indicate that the potato is able to avail itself of a less proportion of the nitrogen of the manure than any other farm crop. Yet in ordinary practice farmyard manure is not only largely relied on for potatoes, but is often applied in larger quantity for them than for any other crop. It is probable that, independently of its liberal supply

of all necessary constituents, its beneficial effects are in a considerable degree due to its influence on the mechanical condition of the soil, rendering it more porous and easily permeable to the surface-roots, upon the development of which the success of the crop so much depends. Then, again, something may be due to an increased temperature of the surface-soil, engendered by the decomposition of so large an amount of organic matter within it; whilst the carbonic acid evolved in the decomposition will, with the aid of moisture, serve to render the mineral resources of the soil more soluble."

In considering these results obtained at Rothamsted it should, of course, be borne in mind that the system of cropping pursued on the experimental plots there—the same crop on the same plot every year—differs greatly from that followed in ordinary farming. In all probability, with the more thorough and varied tillage, and the cropping with plants of different depth of roots and different powers of assimilating food which obtain in ordinary rotation farming, the residue of dung would be more speedily recovered by crops than was the case at Rothamsted. Still it can hardly be gainsaid, that the Rothamsted experiments have proved that the beneficial influence of dung upon potatoes is due in a larger measure to its mechanical effect, and in a less degree to it as a source of plant-food, than was before generally believed.

A Practical Lesson.—The chief lesson which the practical farmer is to draw from these conclusions, as to the action of dung, is that, while a large dressing of dung may with advantage be applied for potatoes, it must not be relied upon as the sole source of plant-food for the crop—that the dung must be supplemented with a substantial allowance of quick-acting nitrogenous manure, such as nitrate of soda, and with a smaller application of phosphates and potash.

Potash for Potatoes.—The undoubted value of potash for potatoes is not very clearly shown in the Rothamsted experiments. In soils deficient in available potash the application of about 2 or 3 cwt. of kainit per acre, or, better, of 1 cwt. of sulphate of potash, will be found to have quite a marked effect on the produce. Good results have been obtained from kainit when sown as a top-dressing just before the drills of potatoes are earthed up for the first time, especially when rain happened to fall soon after, thus carrying down the potash to the roots of the crop, which is then ready to absorb it. Mr John Speir says he prefers to sow the kainit on the ploughed land in autumn or early winter, which indeed is generally believed to be the best plan.

General Observations on Manuring Potatoes.

In some general notes prepared for this work on the Manuring of Potatoes Dr Bernard Dyer says :—

"While dung should be used with fair liberality, excessive dressings should be avoided. Dung moderately, rather for the sake of mechanical texture and warmth than with the view of fully nourishing the crop. Nourish the crop by adding artificials. Sour or peaty soils should be occasionally well limed. Potash should always be used, unless experience has proved it locally useless. It is best to sow 1 cwt. sulphate of potash per acre in winter or early spring. Failing earlier application, let 1 cwt. sulphate of potash be mixed with the phosphatic manure. On soils containing a fair quantity of lime 3 to 6 cwt. per acre of superphosphate or dissolved bones should be given ; on soils poor in lime a like quantity of phosphatic Peruvian guano, or half as much again of basic slag, fine bone meal, or 'basic superphosphate' — well mixed with the soil. For early potatoes nitrogen should be given, either as sulphate of ammonia or nitrate of soda (or both), or dissolved Peruvian guano. For late yielding varieties, planted early, a few cwt. of rape dust or fish guano may supply part of the nitrogen. Sulphate of ammonia must not be mixed with slag. If slag is used, nitrate is more appropriate. With all other forms of phosphate 1 cwt. to 2 cwt. sulphate of ammonia (per acre) may be mixed. This will only be if sufficient nitrogen in dung has been used. If no dung has been used, one to two later dressings should be given of 1 cwt. each time of nitrate of soda or sulphate of ammonia. The former commends itself for top-dressing, except in wet districts. The nitrogenous manure must be given discreetly. In some seasons more is wanted than in others, and the need is best gauged by watching the haulm. Clearly, if the total application is to be controlled, this can only be done by withholding part of it. Hence only 1 or 2 cwt. of sulphate or nitrate (according to soil and locality) should be sown at planting time—the rest being given at intervals as required in top-dressings. In recent experiments 4 cwt. of nitrate per acre, used with phosphates and potash, was found to be a remunerative dressing, even when 12½ tons of stable manure were used. When no dung is used, or in some seasons with dung, as much as 6 cwt. of nitrate per acre has been used with remunerative results."

College Trials in Manuring Potatoes.

A large number of trials in the manuring of potatoes have been conducted in connection with the leading Colleges of Agriculture in this country. As a rule, in these trials a dressing of 10 tons of farmyard manure has done better than twice that quantity, the increase from the larger quantity not being sufficient to cover the extra cost, while the tendency to disease was greater, and the quality and keeping properties of the tubers usually inferior with the heavier dressing of dung. In most cases, the most profitable dressing of artificial manures to accompany dung was found to be a mixture of about 1 cwt. sulphate of ammonia, 2 cwt. superphosphate, and 1 cwt. sulphate of potash per acre. This, or some other similar mixture, along with 10 tons of dung, usually gave a moderately profitable return. The increased yield from any substantial increase in the quantity of artificial manure was usually too small to be profitable. In some cases where no dung was applied good crops were got from artificial manure alone, a dressing of double the quantity mentioned above. Some of the College trials indicated that large allowances of dung, nitrates, and ammonia-salts tended to lower, and phosphates and potash to improve, quality in potatoes.

Application of Manure for Potatoes.

Autumn Dunging.—If the land is strong and a supply of dung happen to be then on hand (which, however, is not often the case), it is considered a good plan to spread the dung for potatoes on the stubble just before ploughing in the end of autumn or early in winter. This, no doubt, tends to the better preparation of strong land for potatoes ; and the carting and spreading of the dung in autumn or winter lightens the pressure of work in spring. One great hindrance to this system is the fact that a sufficient supply of dung is not usually available in the autumn or beginning of winter. Where summer-house feeding is practised, there is generally an ample supply of dung in good time for this purpose; and these two systems, which, in suitable circumstances, are both to be commended, fit well into each other. On dry soils and in dry seasons dunging before ploughing usually gives the best results. In the opposite conditions spring dunging will be found to excel.

Spring Tillage with Autumn Dunging.—When the potato land has been dunged and deeply ploughed in autumn or winter, the spring tillage is simple and soon finished. If the land is clean no further ploughing may be necessary. A single or double stripe with the grubber and moderate harrowing will most likely suffice, and then shallow drills are opened from 28 to 30 inches wide. The seed is planted in these drills with from 10 to 14 inches between the sets, and an allowance of artificial manure is sown, and the drills are closed by splitting each ridgelet in two with the drill plough, such as that made by Newlands, shown in fig. 580.

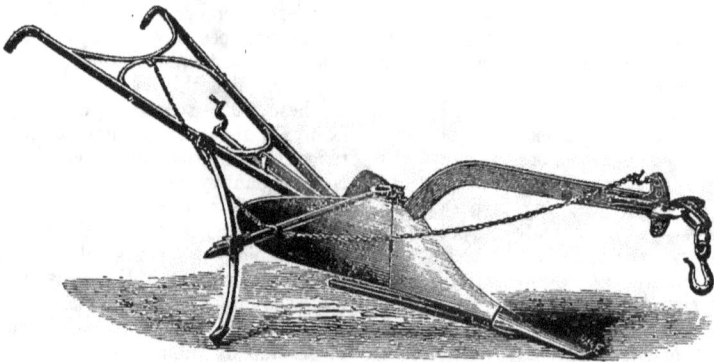

Fig. 580.—*Scottish drill-plough.*

Spring Dunging.—A system largely followed is to apply both the dung and the artificial manure in the drills at the time of planting in spring. If the land is clean and not too rough, a small amount of spring tillage will suffice. When they have to hold dung, as well as the seed and artificial manure, the drills have to be a little deeper than is necessary with previous dunging on the surface.

Carting Dung for Potatoes.—The more expeditious plan is to have the dung carted in a heap on the field before the rush of spring work sets in. In this case the carting of the dung into the drills is speedy work. But it is very often found that better crops of potatoes are obtained from dung carted right from the cattle-court to the drills than from exactly similar dung which had some time before been carted into a heap on the field. Where the lessening of spring work is of special importance, as in late districts, with their long winter and short growing season, it will perhaps still be best to pursue the practice of carting the dung to heaps on the field in winter. In cases, however, where there is no such excessive rush of spring work, and where the potato field is within easy distance of the homestead, the dung had better be left in the courts till required for the drills in spring, or until it is to be spread upon the land in one form or another.

Filling Dung.—To avoid delay in the field, and keep the horses as active as possible, one or two men may be employed at the dung-heap in assisting the drivers of the carts in throwing the dung into the carts, which is done with ordinary four-pronged steel graips. The movements of the carts are so arranged that only one, or at most two carts are at the dung-heap getting filled at one time.

Distributing Dung from Carts.— The dung is thrown from the cart into the drill in graipfuls as the horses move on at a moderate pace. The quantity of dung intended to be given to the land is evenly apportioned by the farmer or overseer, fixing the length of drill which loads of certain size should cover, and seeing that the man throws out the dung in uniform graipfuls at regular distances. Intelligent horsemen very quickly become expert at this practice, which is far more expeditious and satisfactory than the antiquated method of dragging the dung out of the cart into heaps in the drills.

Three ordinary drills are just about the width of a farm-cart. The dung is often thrown into the drill in which the horse is walking, and one wheel of the second cart thus runs over the dung thrown out from the first cart. This packing of the dung by the cart-wheel should be avoided by throwing the dung, not into the centre drill, but into the drill on the side of the cart next to where the dunging was begun.

Spreading Dung.—The spreading of the graipfuls of dung in the drills is done by men, lads, and women. On many farms this work is now done by mechanical means. An excellent machine for spreading dung in drills is shown in fig. 344 (vol. i. p. 513).

In England women are rarely seen at this work, and in Scotland also the custom of employing them at it is by some strongly condemned as unbefitting modern civilisation. Certainly much less female labour is now engaged in outdoor farm-work than in former times, and the tendency is still towards diminution.

A long-shafted steel fork or graip, with three or four prongs, is best suited for spreading dung. It is very important that the dung should be finely broken and evenly spread in the drill. Lumps of dung should be thoroughly broken, and rolls of straw or other litter undone, so that the dung may not only be evenly distributed over the land, but be so exposed to the surrounding soil as that it may speedily and regularly decompose.

Four or five workers will, in average circumstances, spread as fast as one drill-plough can cover in.

Sowing Artificial Manure.—Whatever artificial manure is to be given at the time of planting is sown broadcast by hand or machine along the drills, after the dung is spread, and either before or after the seed is planted. It is considered preferable to sow the artificial manure before the planting of the seed, so that the manure may not lodge in the "eyes" of the seed. A man sowing with two hands will sow as fast as three drill-ploughs can cover in. Machines to sow two or more drills at a time are now in use on many farms.

PLANTING POTATOES.

The planting of potatoes demands attention early in spring.

Potato-seed.—While the land is being prepared for the potatoes—and it will not be possible to prepare it continuously, as the sowing of grain has to be attended to—the *potato-seed* should be prepared by the field-workers. When preparing potato-seed a great saving of time will be effected if the seconds (*i.e.*, after the ware has been taken out) are dressed over 1¼-inch riddle, and then over 1¾-inch riddle. The tubers above the 1¾ inch should be taken to an outhouse and cut, while the smaller ones can be covered up again and planted whole. In selecting tubers to cut into sets, the middle-sized, that have not sprouted at all, or have merely sprouted buds, will be found the soundest; and wherever the least softness is felt, or rottenness seen, or any suspicion as regards colour or other peculiarity is indicated, the tuber should be entirely rejected, and not even its firm portion be used for seed. The very small potatoes should be picked out and put aside to boil for poultry and pigs.

Potatoes intended for seed should always be turned in the pits between February and March, in order to prevent sprouting.

Potatoes are planted whole or cut into parts or *sets*. Large whole potatoes should not, as a rule, be planted, as it is a waste of seed. Some kinds of potatoes, however, such as Magnum Bonum, are best planted whole. With the Magnum Bonum, growth begins so late in the season that at cutting time it is impossible to say whether or not any eye will grow. They have few "eyes," which are nearly all at one end, and when cut the part containing the eye is thin. Very small sets, or very small whole potatoes, should not be used as seed, as they are liable to produce a light crop of puny tubers. Moderately small tubers, if they have not too many eyes, make good seed.

The usual practice is to cut a middle-sized potato into two or three sets, according to the number of eyes it contains. It is well to leave two eyes in each set, lest one of them may have lost its vitality. The sets should be cut with a sharp knife. When fresh, the tubers cut crisp, and exude a good deal of moisture, which soon evaporates, and leaves the incised parts dry.

A common practice was to heap the cut sets in a corner of the barn until they were planted. Had they been exposed previously to drought they might remain uninjured, but if heaped immediately on being cut, in a moist state, they will probably *heat*, and heated potatoes, whether whole or cut, rarely vegetate. Much of the injudicious treatment which sets of potatoes receive arises from want of room to spread them out thin. The straw-barn is most generally used, but in many cases it cannot be spared, as the cattle-man and ploughman must have daily access to it. The corn-barn may be occupied with grain. The implement-house has too little room, on account of the many small articles which it contains. The only alternative is an outhouse; a large one should be in every steading.

A considerable quantity of seed should be prepared before planting is begun, and the rest can be made ready when the horses are engaged with the barley or oat seed, or during any broken weather. A considerable quantity of seed can usually be stored about the steading for cutting in wet or stormy weather, while the rest can generally be prepared at the pit-side.

Preserving Sets.—When cut indoors the sets should be spread out thin and dusted with lime. This forms a crust on the incised surface, and prevents the sap from exuding. Potatoes are best cut a day or two before planting, as they keep longer fresh in the ground when there is a crust on the incised surface of the set.

Quantity of Seed per Acre.—The sets required to plant an acre of land vary very much according to size of sets and kinds of potatoes. From 10 cwt. to 15 cwt. of most of the late varieties of potatoes will be sufficient to plant an acre, but many use up to 20 cwt.

Since disease became prevalent among potatoes in all soils and situations, numerous expedients have been devised to prepare the seed, with the view of warding off disease, but without much practical effect.

Potatoes not required for seed, firm and of good size, whether intended for sale or for use in the farmhouse, should remain in the outhouse until disposed of or used, kept in the dark, with access to air, and examined as to soundness when the sprouts, if any, are taken off.

Planting on the Flat.—In many parts of England, where the climate is dry, potatoes are grown extensively on the flat. There autumn or winter dunging is often practised, and by grubbing, harrowing, and ploughing, the land is prepared for the seed in spring. As already stated, the seed in this case is dropped into every third furrow, the seeds being from 9 to 14 inches apart in the furrow. The spaces between the rows of potatoes are hoed by horse and hand hoes just as in the case of drills, and when the plants are well grown they are earthed up by passing a double mould plough between the rows, thus forming ordinary drills.

The Drill System.—But the growing of potatoes in drills is far better than planting on the flat. The drill system is all-prevailing in Scotland (except in some parts of the Highlands and Western Islands) and the northern counties of England, while it greatly predominates in Ireland, except on small holdings where the "lazy-bed" method is still pursued extensively.

Single v. Drill Plough.—Mr Speir, Newton Farm, Glasgow, says: "On any land, fine or firm, where one or more ploughs can be kept in constant work, a single mould-board plough with a specially narrow mould-board on, makes *much better* work and is easier held than the double one. With it the soil is lifted and turned right over on the dung and sets—not *shoved* over as with the other. Here no double mould-boards are used at planting time." It is important that the double mould plough should be formed so as to turn over rather than press the soil outwards. On "cloody" soil the single plough does the better work. The double plough is apt to *enclose* the clods, whereas with the single plough they roll over the crest into the bottom of the next drill and get crushed.

Upon a large farm, where a considerable area is devoted to potatoes, the operations of opening drills, carting dung, spreading dung, planting seed, sowing artificial manure, and closing in drills, all proceed simultaneously. There is no more active scene upon a farm in the course of the whole year than this; and few operations afford greater opportunities for the exercise of skill and forethought in arranging and controlling farm labour.

Planting the Seed.—The spreaders of dung are followed immediately by a similar force planting the seed. Women make the best planters. Five or six planters with the seed regularly supplied to them will plant as fast as the four or five workers will spread the dung; and this force of spreaders and planters, with one man to sow the artificial manure, will keep one drill-plough at full work in

Fig. 581.—*Potato-planter.*

covering in. It is perhaps better that each planter carry her own sets, as a relief to the stooping posture is thus obtained, and each planter should have a separate drill, otherwise parts may be missed. The sets are dropped into the drill upon the top of the dung, at from 9 to 14 inches apart.

Potato-planting Machines. — Machines have been invented for planting potatoes, and some of them do excellent work. Richmond's ingenious potato-planter is shown in fig. 581.

Conveying Seed to the Planters.— The sets are shovelled either into sacks like corn, or into the body of close carts, and placed, in most cases, on one or both head-ridges or middle of the field, according to the length of the ridges. When the drills are short, the most convenient way to get at the sets is from a cart; but when drills are long, sacks are best placed along the centre of the field. A still better plan, if a horse can be spared, is for a boy to drive the potatoes alongside the planters. The cart can go in the drills that are covered, and the boy will carry the sets to the workers.

In some cases a small round willow basket, with a bow-handle, fig. 582, is provided for each person who plants the sets. Others prefer aprons of stout sacking. As a considerable number of hands are required, boys and girls may be employed beyond the ordinary field-workers. The frying-pan shovel, fig. 342 (vol. i. p. 512), with its sharp point, is a convenient instrument for taking the sets out of the cart into the baskets.

Covering in.—The drill-plough should at once follow the planters, as both dung and potatoes suffer by being exposed to

Fig. 582.—*Potato hand-basket.*

the sun. The drills are split in the same way as they are set up—that is, the plough splits the drill, throwing one-half of the land on one row of sets and the other half on the other, both of the drills being completely covered in two rounds of the plough. The whole of the drills dunged and planted should be covered every day—the man who has been opening helping the one who is covering, after he has opened enough to serve for the day, with a few for a start next morning.

Where there is only one drill-plough, it is employed alternatively in opening and covering in, or opens one way and covers the other. Or the single board plough is used as described on page 304.

Complete Planting as it proceeds.—It is undesirable to open many more drills than can be planted and covered in before nightfall, lest inclement weather should set in, and so render the opened drills too stale before the planting can be resumed. The work of planting is done most satisfactorily where, as far as it goes, it is begun and completed in the same day.

Danger of leaving Dung and Seed uncovered.'—Most farmers are specially careful as to the completion of the work of potato-planting as it goes on. In many cases it is insisted that, even at loosing from the forenoon yoking, every drill should be covered in, although the plough-man should work a little longer than the

rest of the work-people; for which detention he would delay as long in yoking in the afternoon. In dry hot weather he should make it a point to cover in the drills at the end of the forenoon yoking in a complete manner, as dung soon becomes scorched by the mid-day sun, and in that state is not in good condition—not on account of evaporation of valuable materials, as what would thus be lost would be chiefly water, but because dry dung does not incorporate with the soil for a long time, and still longer when the soil is also rendered dry. If all the ploughs cannot cover in the drills at the hour of stopping at night, give up dunging the land and planting the sets a little sooner, rather than run the risk of leaving any dung and sets uncovered.

An Ayrshire Practice in Potato Planting.—On the earliest farms along the coast of Ayrshire, principally around Girvan, Maybole, and Ayr, where the area of potatoes grown is excessively large in proportion to the size of the farms, a different class of plough is used from that which is commonly met with throughout the country. There the area planted is so great in proportion to the power at the command of the farmer, that a speedier method must be adopted than that in general use. The plough (one of which, made by T. Hunter, Maybole, is shown in fig. 583) has very much the appearance of an ordinary 3-horse grubber, at least so far as the frame is concerned, while it has also two similar side-wheels, a fore-wheel, and lifting lever.

Fig. 583.—*Triple drill-plough.*

Instead, however, of from five to seven tines, it has only three, all set abreast, each tine being in fact a double mould-board plough hung from the frame. The mould-boards are a little less in size than those in use in the ordinary mould-board plough, but otherwise they are the same. By this plough, with three horses, three drills are opened or covered at one passage of the plough. For its proper working the land must of necessity be well prepared beforehand, and any farmyard manure used has generally been ploughed in some time previously. This plough, indeed, is not well suited for covering in dung.

In this district, where much of the best potato-land is very sandy, a peculiar class of double mould-board plough is also used to earth up the potatoes. In it the mould-boards are solid and continuous from the sole of the plough to the top of the drill, so that in working in dry weather the whole weight of the plough is exerted in pressing the sandy earth on the side of the drill, and in the driest weather slipping down rarely happens. The same class of plough is in use on the early potato-lands of Cheshire, from which district the Ayrshire men adopted it.

Width of Potato-drills.—Abundance of air is of great importance to the potato plant. Where the earlier varieties are grown drills are usually not more than from 26 to 27 inches wide, but for the main crop varieties the most general width of drill is from 28 to 30 inches.

Depth of Sets and Distance apart. —The distance between the sets in the drill varies from 9 to 14 inches, according to the width of the drill, the variety of potato, whether the stems are tall, medium, or short, and the character and condition of the soil and climate, whether likely to favour a heavy or light yield, and the size of the sets. In light soils sets are usually placed about 6 inches under the surface, and in heavy soils slightly less.

Time for Planting Potatoes.—The time for planting potatoes varies with different districts and sorts of potatoes. Early sorts are planted in February and March, and late varieties in April, or in late cold districts in May.

Experiments with Late Planting. —It has been suggested that in certain circumstances potato-planting might be more remunerative if the tubers were not planted until June. To throw light on this point Mr John Speir, Newton Farm, Glasgow, conducted experiments with plantings on different dates in June and July. Early varieties were planted— whole seed which had been strongly sprouted before being set. Farmyard

dung alone was used. In the first year the planting on June 30 gave about one-third more produce than the planting ten days later. The July plantings were still more unsuccessful in the second year, but in both years the June plantings gave satisfactory results. In the second year three plantings gave the following results :—

Planted.		Produce.
June 10		7 tons per acre.
" 20	. .	5½ "
" 29	. .	4 "

Mr Speir is quite convinced that the system is capable of great expansion, more particularly on market-garden farms or in late districts. But he adds, that to be attended with any measure of success at all, *the seed must be sprouted, and the sprouts must never have been broken off, but be the first ones which come.*

Culture after Planting.—Potatoes require a considerable amount of horse-work both before and after brairding. As soon as convenient after planting, the drills should be harrowed down either with a set of light zigzag harrows or chain-harrows, or, better still, with a saddle-drill harrow, such as is illustrated in fig. 434 (vol. ii. p. 152). Immediately after, the drills are again set up with the double-moulded plough. When the plants are well sprung, but before they are too far advanced, the drills should be again harrowed down. This makes a fine surface for the young plants, and helps to keep down weeds. A very suitable implement to crush clods on strong land is a fluted roller to embrace two drills, such as that made by T. Hunter & Sons, fig. 584.

Fig. 584.—*Drill roller.*

This crushes the clods, and leaves the drill in good condition for being harrowed by the saddle-harrows.

Hand-hoeing.—The drills are then hand-hoed, loosening the soil around the young plants and removing weeds. The hollows of the drills are stirred with the drill-harrow or horse-hoe, and then the drills are set up with the double mould plough. Unless weeds are so abundant and strong as to necessitate another hoeing, no further tillage may be required.

Planting Potatoes in Autumn.

Amongst the expedients suggested for evading the potato disease was planting the sets in autumn. The plan, however, has been only occasionally tried, and even where the circumstances were favourable the success has been very partial.

Autumn Planting unsuitable. — Planting potatoes in autumn cannot be practised everywhere nor extensively anywhere. Potatoes are not only a green but a fallow or cleaning crop, and a green crop being taken after a crop of corn, the stubble of the corn crop is not in a fit state to receive manure before undergoing the process of cleansing by the plough, the harrow, and the grubber, as the land for a fallow crop ought to be; and, in Scotland, too short time intervenes between the harvesting of the corn crops and bad weather in winter to permit the land to be sufficiently cleaned. Hence very few cases occur in which the stubble can be manured in October for potatoes, which occupy so important a position as a green crop. This is perhaps to be regretted, for in the majority of cases where autumn planting was tried it seemed to check the disease.

Method of Autumn Planting. — Potato-planting in autumn is the same as in spring; but there will not be time to stir the land as much. The stubble should be cross-ploughed or grubbed to a considerable depth. Harrowing with a double tine brings up any weeds there are, and these should be gathered off. If the land is first cross-ploughed, the grubber may follow, to cut the furrows in pieces, if there be time for that efficient operation; but if not, the land should be drilled up in the double way, in preparation for the dung.

The after operations are the same as in spring.

Seed for Autumn Planting. — Many think it advisable to use whole potatoes for seed instead of cut sets in autumn.

Small potatoes answer well, and save time in cutting. The whole potatoes should be planted in the drill at from 10 to 12 inches asunder.

The Boxing System of preparing Potato-sets.

Mr John Speir, Newton Farm, Glasgow, thus describes the system of preparing the potato-sets in boxes:—

This system was introduced for the purpose of maturing the potato crop sooner than could be attained by the ordinary manner of planting. It is said to have been first introduced in Jersey, where it is extensively practised. Along the whole of the Firth of Clyde it is more or less in use on all the earlier farms, and more particularly in the neighbourhood of Girvan it has been carried to such an extent that several farmers there have upwards of a hundred acres of potatoes all planted from boxes.

Boxes.—The boxes may be of any convenient size or shape, provided they are not too deep, the size in most common use being about 2 feet long, 18 inches broad, and from 3 to 4 inches deep. Each box generally holds from 3 to 4 stones of potatoes, the former being about the average. The boxes are made of ½-inch deal, and have pins 1 inch square and 6 inches high nailed in each corner. The top of these pins therefore projects from 2 to 3 inches above the edge of the box. These pins are strengthened in their position by having another bar, 1 inch square, nailed across the ends, and reaching from the top of the one corner pin to the top of the other. These cross-bars also serve as handles for carrying the boxes, besides being in other ways useful. In Jersey, and in many districts of Britain and Ireland, boxes about one-third smaller than above are preferred. These smaller boxes are much lighter to carry about, and the sets are planted direct from them.

Tubers Boxed.—The potatoes used may be of any variety, but where early maturity is the main object, only the earliest varieties are boxed. Only small or medium-sized potatoes are used, all over 1¼ inch and under 2 inches in diameter being considered suitable.

Cut Seed Unsuitable. — Cut seed cannot so satisfactorily be used, because the sets remain so long in the boxes, and such a quantity of the moisture evaporates from the sets that they ultimately shrivel up, and become so dry that the bud never starts into life.

Boxing the Seed.—The seed may be placed in the boxes any time between the end of July and the New Year, the most suitable time being probably September or October. At the latter end of July, all potatoes which are at that time dug and of too small size for table use may at once be put into boxes, and thus preserved for seed to the following spring. In the boxes they keep with very little loss, even although quite soft and green when put in, whereas if stored in the ordinary manner all would be lost.

Storing Boxes. — During autumn the boxes may be stored in any unused barn, byre, shed, or other house which is rain-proof. The boxes are placed in tiers one above the other to any convenient height, the corner pins and cross-bars of the one box supporting the weight of those above, the extra height of the pins over the depth of the box giving sufficient room for the ventilation of the tubers and growth of the sprouts.

When cold weather sets in, the boxes should be removed to some position where they will be free from the effects of frost. Very many are stored on the joists of byres, bullock-houses, &c., where the heat from the animals is always sufficient to start germination and keep out most frosts. Others, again, are stored in empty cheese-rooms and other houses specially built for the purpose, which are provided with artificial heat in the shape of a stove or other heating apparatus. It is not often that the heating apparatus requires to be called into use, but it is almost a necessity against occasional extreme frosts, and it comes in handy for pushing on late boxed or tardy germinating tubers.

Planting Boxed Seed.—Planting is generally begun about the first of March, and in the most favoured localities a little earlier. Before this system was adopted, the localities which now use it generally began to plant in January or

February, but now there is nothing to be gained by beginning so early, and much may be lost by frost cutting off the haulms of the plants after they have come through the ground. Previous to planting, the boxes with the potatoes in them are removed to the field in carts, and distributed along the side of the land, being placed in much the same way as sacks of cut potato-sets are put down before planting begins.

The *sprouts* at this time may be from 2 to 4 or more inches long, but instead of being white and brittle like those seen on potatoes in an ordinary pit, they are blue or pink according to variety, are tough, and not at all readily broken off. The tubers are generally planted direct from the boxes. If the large-sized box is used, two planters carry one between them ; while if the smaller one is adopted, each planter has a separate box. If the sprouts are comparatively short, the sets may be transferred from the boxes to the planters' aprons in the usual way of carrying cut potato-sets ; but by doing so the plants get much rougher handling, and a few are always more or less damaged.

Seed per Acre.—The seed required to plant an acre on this system varies very considerably according to the size of the potatoes used. Where the smallest size of potatoes are planted, 30 boxes containing from 3 to $3\frac{1}{2}$ stones will be found amply sufficient, even where 26-inch drills are made, and close planting in the drill is followed. If, however, the potatoes are larger, say about $1\frac{3}{4}$ inch in diameter, 50 boxes of the same capacity may not be more than sufficient. In the former case, therefore, 12 cwt. or so will be sufficient for an acre of land, while a ton may be required in the latter.

Advantages of the System. — The reasons of the success of this system appear to be : 1st, the gain in time by the sets being sprouted before being planted ; and 2nd, the long period of drying to which the seed is subjected in the boxes seems to enable the plant to mature its tubers in the following season more quickly than if it had been stored and planted in the usual way.

A crop from seed which has been boxed is usually ready to lift three weeks earlier than one grown from similar seed which has not been boxed.

In the forcing of rhubarb, hyacinths, narcissi, spiræa, &c., the plants or bulbs must all be rested a certain time before growth will begin, no matter what heat and moisture are used; and in the case of the potato, the dry-keeping in the boxes instead of the damp-keeping in pits appears to have a somewhat similar effect, as more time is gained than is accounted for simply by sprouting. An unsprouted crop may indeed look as far forward as a sprouted one, and yet not yield the same weight of tubers.

The Lazy-bed System.

Another mode of field-culture for potatoes is in *lazy-beds*, very common in Ireland. This system is becoming less general on arable land, though on lea-ground it gives very good results, and seems indeed to be best suited to certain circumstances. In the island of Lewis drill-sowing was at one time adopted, but found unsuitable for the soil and climate, and the lazy-bed system had again to be resorted to. The usual method is to remove a line of turf along the margin of the proposed lazy-bed, after which a slight covering of dung is given, and the next line of turf turned over green side under upon the top of the sets. The next line is turned over without sets, then dung, &c.,—this proceeding being adopted along the whole bed, until finished, after which a trench is cut or formed round the edges to carry off any surplus moisture.

In reference to Ireland, Martin Doyle says: "In bogs and mountains, where the plough cannot penetrate through strong soil, beds are the most convenient for the petty farmer, who digs the sod with his long narrow spade, and either lays the sets on the inverted sod,—the manure being previously spread,—covering them from the furrows by the shovel, or, as in parts of Connaught and Munster, he stabs the ground with his *loy*,—a long narrow spade peculiar to the labourers of Connaught,—jerks a cut set into the fissure when he draws out the tool, and afterwards closes the set with the back of the same instrument, covering the surface, as in the case of lazy-beds, from the furrows.

"The general Irish mode of culture on old rich arable lea is to plough the fields in ridges, to level them perfectly with the spade, then to lay the potato-sets upon the surface, and to cover them with or without manure by the inverted sods from the furrows. The potatoes are afterwards earthed once or twice with whatever mould can be obtained from the furrows by means of spade and shovel. And after these earthings, the furrows, becoming deep trenches, form easy means for water to flow away, and leave the planted ground on each side of them comparatively dry.

"The practice in the south of Ireland is to grow potatoes on grass-land from one to three years old, and turnips afterwards, manuring each time moderately, as the best preparation for corn, and as a prevention of the disease called fingers-and-toes in turnips. In wet bog-land, ridges and furrows are the safest, as the furrow acts as a complete drain for surface-water; but wherever drilling is practicable, it is decidedly preferable, the produce being greater in drills than in what may be termed, comparatively, a broadcast method." [1]

RAISING NEW VARIETIES OF POTATOES.

For the success of the potato-growing industry the raising of new varieties of tubers is of the utmost importance.

New Varieties Resisting Disease. —That baneful fungoid disease, known as *Peronospera infestans*, played such havoc with the older varieties in use when it made its earlier attacks in the country, that a great cry arose for new sorts, which it was believed would be less liable to the malady. Experience showed that there is foundation for this belief, yet varieties which seem to be almost disease-proof when first introduced gradually lose vitality, and have ultimately to be abandoned because of their liability to disease. Most new sorts begin to show signs of this deterioration in ten or twelve years—many, indeed, in a shorter time. It is thus of the greatest importance that the creation

[1] Doyle's *Cy. Prac. Husb.*—art. "Potato."

of new varieties of potatoes should be encouraged, and with this object in view we produce here the following description of the process, prepared for this work by Mr John Speir, Newton Farm, Glasgow. The illustrations used in describing this process are taken (by the kind permission of the publishers, Messrs A. & C. Black) from Balfour's *Elements of Botany*.

Potato-seeds.—As most people know, new varieties of potatoes are raised from the plum, as it is popularly called. The plum holds the same relation to the potato-plant as the apple does to the apple-tree. It is the fruit, and, within, the fruit contains the seeds. The seeds of the apple or orange every one is familiar with. The potato also, like them, has its seeds contained in a mass of pulp, which, however, unlike the apple or orange, is not of such a pleasant taste. Hence the seed of the potato is not so well known.

Seedless Varieties.—Some varieties of potatoes do not throw up flowers, and therefore cannot have plums. Seed must thus be looked for only on those varieties which have flowers. Again, all varieties which have blossoms do not have plums, as some appear unable to set a single bloom, unless on very rare occasions. With plants, as with animals, in-breeding, if I may so express it, although not at first very hurtful in its effects, is very liable if persisted in to have a deleterious influence on either plant or animal—the stamina of both evidently becoming so reduced, that they fall a ready prey to disease.

Cross-fertilisation.— The methods which nature has adopted in plants, not exactly to prevent self-fertilisation, but to favour cross-fertilisation, are numerous, curious, and very interesting. Darwin proved beyond doubt that certain plants if self-fertilised would attain a moderate size ; if cross-fertilised from plants growing alongside of them, they would attain a much greater size; and if fertilised from plants of the same variety grown on different soil, some miles away, their size would be still further increased.

These facts are of very great importance to the raiser of new varieties of potatoes. In fact, it is principally on cross-fertilisation that he relies for success. A new and quite good enough

potato may be raised from seed where no intentional cross-fertilisation has been done, but the chances are that such a plant has been self-fertilised, or cross-fertilised, by a plant of its own variety. The consequence will be that a much smaller proportion of the seeds sown will produce plants having vigorous constitutions, than if a different and improved variety from a dissimilar class of soil had been used in fertilisation.

Male and Female Organs.—Among plants, as among animals, there are male and female organs, which in most plants are situated in the same flower. On some, however, the male blossoms are on one part of the plant and the female ones on another, while in others the males and females are on separate plants.

Fig. 585.—*Section of a flower.*
c, Calyx. p, Petal. s, Stamen.

In fig. 585 is shown a section of a flower, in which c represents the calyx or short green hard leaves at the base of most flowers, p is the petal or flower proper, s is a stamen, or male part, of which there are two shown, one on either side of the central figure ; while the pistil, or female part, is seen in the centre.

Fig. 586 is a horizontal section showing the organs of fructification of the potato, where the calyx or outer scales are five in number, and the blossom proper contains five petals, the one overlapping the other. Inside the circle of petals are shown five stamens, and inside

Fig. 586.—*Horizontal section.*

that again the pistil, with seed-pod at its base.

In fig. 587 is shown a vertical section of a potato-blossom, in which c represents the calyx ; p, the petals, or bloom proper; e, the stamens ; s, the pistil; and o, the ovary or seed-vessel. At a certain age the stamens throw off from their top a very fine powder, which is called pollen.

Fig. 588 represents a stamen in the act of discharging its pollen, which in

Fig. 587.—*Vertical section of a potato blossom.*

c, Calyx.   p, Petals.
e, Stamens.   s, Pistil.
o, Ovary.

Fig. 588.—*Stamen discharging pollen.*

a, Slits in the anther.
p, Pollen.

some cases is thrown out through slits in the anther, *a*, or top part, while in others, like the potato, it comes out through holes or tubes.

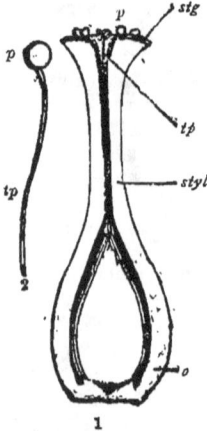

Fig. 589.—*Pistil with pollen-grains on top.*

(1) stg, Stigma.
    p, Pollen-grains.
    tp, Tubes.
    styl, Style.
    o, Ovule.
(2) p, Pollen-grain.
    tp, Its tube.

Fig. 589 is a pistil with pollen-grains on the top. The uppermost part, *stg*, is called the stigma, with pollen grains, *p*, adherent to it, sending tubes, *tp*, down the conducting tissues of the style, *styl*; the ovule is *o*; while in (2), *p* is a pollen-grain separated, and *tp* its tube.

Fig. 590 represents a pollen-grain very much magnified, showing three points where the tubes come out, one of which is considerably elongated.

Fig. 591 is a very much magnified vertical section of the style and stigma of the pistil, showing two pollen-grains on the top, throwing out their protruding tubes which descend to the ovules.

Process of Cross-fertilising.—When

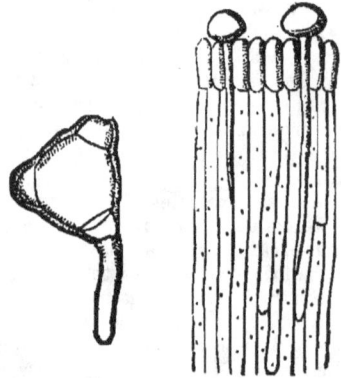

it is wished to cross-fertilise a potato-blossom, the flower is held steadily in the left hand, while with the right the stamens, or male parts of the flower, are cut away with a pair of fine-pointed scissors, or a sharp and fine-pointed knife. These are the parts marked *s* in fig. 585 and *e* in fig. 587, all of which must be destroyed soon after the bloom has expanded. Three or four days afterwards, on a bright clear day, the bloom of some plant which it is intended to cross with the one on which we have operated, is taken, and the pollen scattered on the stigma of the mutilated

Fig. 590.—*Pollen-grain magnified.*

Fig. 591.—*Vertical section of style and stigma magnified.*

plant. If the anthers are ripe, this can be very readily done by bending the stamens back with the tip of one of the fingers, then letting it spring forward again, when the pollen will be thrown off. Another way is to brush them with a dry feather or small camel-hair brush, which takes on a certain amount of the pollen-grains, which, by drawing across the stigma, are in part conveyed to it. The top of the stigma always contains more or less glutinous matter, on which the pollen-grains readily stick.

If it is desired to make the cross-fertilisation very accurate, and to be certain that no other pollen-grains are conveyed to the stigma by insects or the wind, the bloom may be tied to a stake and covered with a small fine canvas bag, or a glass globe.

For the purpose of raising seedling potatoes these precautions are, however, unnecessary. It may be here mentioned that when a potato-bloom has been only a day or so opened, the organs of fructification then have a more or less greenish tint, the colour of the stamens and pistil being as yet only partially developed;— that is *the time to cut away the stamens.* At first the stamens are much shorter than the pistil, but as they approach maturity they become more of one length.

Fig. 592 represents the ripened plum, while fig. 593 is one cut across the centre, showing the seeds inside. According to

Fig. 592.—*Potato-plum.* Fig. 593.—*Plum cut, showing seeds inside.*

variety the size of the plum may vary from that of a cherry to the size of a damson plum. Fig. 594 represents a magnified seed.

Marking Fertilised Plum. — In order that the plum of the flower on which cross-fertilisation has been practised may not be mistaken for some self-fertilised one, each bloom as operated on should be tied to a stake, to which a label is affixed, giving the name of parent, date of cutting the stamens away, date of fertilisation, and name of the variety used for crossing. Flowers thus labelled are easily found, as the white stakes and labels are good guides, and worth all the labour for that alone.

Fig. 594.—*Potato-seed magnified.*

Ripe Plums.—When the plums are thoroughly ripened, they should be gathered and the seeds separated from the pulp. The ripening stage is easily known, because as maturity is approached the stalk bearing the plum first withers, then gradually shrivels up, ultimately becoming so dry that it breaks, when the plum drops on the ground. If left to themselves the plums soon rot, the hard seeds alone remaining fresh, and if these are kept moderately dry and out of the reach of birds they remain dormant till spring, when they begin life anew.

Securing and Storing Seeds.—For experimental purposes, however, the plums should be cut up when ripe, and the seeds picked out and dried under cover, preferably on a window-sill, or in a dry greenhouse; and when dried sufficiently to keep during the winter, they may be stored away in any dry situation. Instead of thoroughly drying the seeds, they may be mixed with dry earth or sand, and thus stored during the winter, the whole (the earth or sand and seeds) being sown in a seed-bed in spring.

The plums may even be treated in this way, by surrounding them with dry earth and letting them so remain till spring, by which time the pulp will have rotted or dried up, leaving the seeds more or less mixed up with the soil. Either plan may be adopted successfully enough, but personally I prefer the first.

Sowing the Seed. — In spring the seeds may be sown under glass any time during February, March, or April, the young plants being kept under glass, particularly at nights, until all risk of frost is gone. If no glass is at hand, the sowing of the seed should be deferred till April or May, when it may be sown thinly on any garden soil in a small bed by itself.

The Young Seedlings. — In the month of May the young seedlings should be planted out, in rows not less than 20 inches apart, with 1 foot between the plants. If the seeds have been raised from strong-growing varieties. the seedlings will be all the better of more space; while if they are from smaller-stemmed varieties, thay can do with less space. If the ground is dry-bottomed, they should be planted in the bottom of the drills, so as to give a suitable opportunity for thoroughly earthing them up. But if the young plants are likely to run any risk of being soured at the root by heavy rain they will be better planted on the flat, the earthing up in either case being done as the plant grows. If moderately manured, and kept in good clean order, the plants will soon cover the ground, the time they will take to do so being not very much longer than if ordinary potato sets had been used.

Lifting and Selecting.—At the end of October, or beginning of November, storing should commence, when the experimenter's real difficulties begin. When storing, all varieties of a very bad shape, coloured or partly coloured skins, or bad colour of the flesh, should at once be rejected, as their preservation will likely only lead to trouble and expense, with very little chance of any corresponding gain.

The first year all plants not positively bad should be preserved, and a note kept of any peculiarities of growth, shape, colour, size, or productiveness of each.

Storing Seedling Potatoes.—With many experimenters, the separation and preservation of, it may be, several hundred varieties has been a serious drawback to their continuing the search for improved kinds. This, however, may be easily overcome in the following manner: A number of ordinary drain-pipes should be procured, and, for the first year's crop, the smaller the bore the better. One end of the pipe having been closed by a small wisp of hay or straw, the tubers of each variety are put in along with their number, when another small portion of straw, or piece of turf from an old pasture, is put on the top, then another variety, and so on till the tiles are all filled. The first year each tile may hold several varieties, whereas the second year one variety may be more than enough for one narrow-bored tile, in which case two may be used, or larger-sized ones procured. Small strips of wood, coated with white lead, and marked with an ordinary lead pencil, serve for numbering each lot; or pieces of tin, with the figures stamped on, may be used, if a set of figure stamps can be procured. After packing, the tiles may be built up in a heap, and covered as if it were an ordinary potato-pit, when they will require no further attention until planting-time in the following spring.

Period of Development.—It is a common belief that seedling potatoes require several years to form ordinary-sized tubers. Such, however, is not the case, as even the very first year many of the varieties may yield potatoes of a medium size and upwards, while the second year all worth preserving should have one or more full-sized potatoes.

Second Year.—When the potatoes are taken out of the pit or clamp the following spring, they may be planted in the usual way, and at the usual time, no particular advantages of soil, situation, or manure being given, in order to facilitate the elimination of the worthless varieties as soon as possible.

At the end of the second year they should again be stored in tiles as formerly. This time each variety will require one or more tiles for itself.

The selection must this year be much more searching than the former one, and instead of rejecting only what appeared to be positively bad varieties, those only should be kept which show really good points, or some noted peculiarity.

If notes have been taken during both the growing seasons of the robustness, earliness, lateness, liability to disease, or other peculiarity of the plants, the grower will be greatly aided in his selection.

Third Year.—The third year, only tubers should be kept which show positively some good points, all others being laid aside for consumption. In this way the list will be gradually reduced each autumn and spring, as many good cropping varieties will be found to be bad keepers. A large portion of the plants will thus be rejected every autumn, with a smaller portion in winter and spring, as some which have stood the test of several years may be found not to keep well, or to cook badly.

Retain only Superior Varieties.—After the fourth year, no variety should be kept which the *grower does not consider better than those already in cultivation,* because to propagate any that are not superior to those already in use, undoubtedly *in the end will be sure to bring pecuniary loss on the grower.* It is on this rock that most raisers of new varieties wreck themselves. They find it difficult to cast away seedlings on which they have expended much time and care: to do so appears to them like sacrificing their own children. For want of courage to apply the pruning-knife severely enough, they continue to propagate and keep in existence a large number of varieties which are of little merit. The consequence is, that in the end they become overwhelmed with varieties which are of no commercial value, and most likely give up the work.

Need for New Varieties.—At present there are more than enough varieties of potatoes in commerce or cultivation. Unfortunately, however, there are only a few good ones, so that the field for research is not only a wide but a very varied one. The qualities requisite to make a first-class early, medium, or late potato are so many, and even good varieties retain these for such a short time, that there is likely always to be a demand for really first-class varieties. Although the personal attention requisite may be too heavy a drag on the ordinary farmer, who already has as much to do as he can well accomplish, the raising of new varieties of potatoes might well form a very suitable and interesting pastime for a proportion of our farmers' sons and daughters, as well as older farmers and gardeners, who have the time to spare.

RAISING POTATOES.

The raising of potatoes is a part of the work of the farm which demands careful attention. It requires a great deal of manual labour, even when machines are used in digging the tubers.

Time to Lift Potatoes.—Potatoes indicate fitness for being lifted by decay of the haulms. As long as these are green, the tubers have not arrived at maturity. In an early season some potatoes ripen before October; and although the weather should continue fine, it is best to let them remain in the ground until all the corn crops have been carried in. But in ordinary seasons the corn is cut down and carried before the potatoes are ready for lifting. In this case the potatoes should be lifted just when the haulm has almost completely died down, but before it has become bleached. There is the minimum of disease at this stage, and the longer the tubers are left in the ground after this the greater the risk of injury by disease.

Methods of Potato-raising.—Potatoes are harvested in three different ways —viz., by the digging graip or fork, by the plough, and by the potato-digger.

Raising Early Potatoes.

The practice of consuming potatoes before they are fully matured is steadily increasing. These potatoes are usually raised by the graip, because at that time the tubers have a firm hold on the roots, and are themselves so tender in the skin that, even if the tubers could be separated from the roots, they would be so bruised by the plough or digger that the damage done might be more than the total cost of raising them by the graip. By careful working, the ordinary steel graip or dung-fork may be used in digging potatoes without any serious injury being done to the tubers. There are, however, special kinds of potato graips or forks which are often used, one of these being shown in fig. 595. A graip of an improved pattern has triangular prongs, about an inch broad at the top and tapering to a fine point.

The digging of early potatoes begins in June —even earlier than that in the Channel Islands.

Raising by the Graip. —In raising potatoes by the graip, each person, as a rule, digs one or two drills at a time. In some districts digging is done by men, women or boys doing the gathering; but in many cases, in Scotland, where the drills are narrow and the land light, as in the districts where early potatoes are grown, women occasionally do the digging, but even there digging may be said to be generally done by men, and the gathering by women or boys.

Barrels for Early Potatoes.—At this season of the year potatoes are sent to market in barrels holding 1½ cwt., barrels being used because in them the potatoes do not rub so much the one against the other in handling as they do in bags, in which their skins get ruffled and torn.

Fig 595.—Potato-graip.

While the potatoes are being gathered, the tubers over 1½ inch in diameter or thereby are put into one basket and those under that size into another. The large ones are put into barrels on the spot without weighing, all barrels used being about the one size.

Preparing Barrels.—As a rule, the barrels have been used previously for bringing flour from America or from the home mills, and before being employed for potatoes they are strengthened by an extra hoop or two, and strings put on the top. For this purpose six or eight half-inch holes are bored in the staves of the barrel an inch or more from the top. Cord, three-eighths of an inch thick, is then taken and passed through one hole forward to the centre of the barrel, and then back through the next hole. A knot is now put on each end of the cord on the outside of the barrel, and large enough for the cord knot not to pass through the hole. The other holes are then roped in the same way, when the barrel is ready for use.

Filling Barrels.—When the barrel is being filled, a very few of the best-shaped potatoes are dropped into a basket at the side of the barrel, and when about filled, the barrel is completed by putting as many as are necessary of those in the basket on the top. A few green potato-stems are now taken and packed firmly over the top, and slightly above the rim of the barrel. A stout string, a foot or so in length, is then put through each of the loops of rope fixed on the top of the barrel, which, when tightly drawn and knotted, securely holds down the top.

Handling Barrels.—In letting the barrels down from a man's shoulders, or from a cart, they are always dropped on the mouth. If dropped in any other manner the barrel would run a great risk of being damaged.

Speedy Marketing.—Where the circumstances permit, the usual custom is to send as many of the potatoes which have been dug on any particular day to the railway station that evening, so that they may be carried during the cool of the night to the place where they are to be consumed. Early potatoes being soft and immature, very soon lose their flavour, and deteriorate in quality, if allowed to lie about.

Small Potatoes.—As soon as the crop has begun to harden and ripen a little, the small potatoes are usually carried to a bare smooth place in the field, where they are redressed, the tubers over 1¼ inch or so being set aside for sale at a low price early in the season, and for seed purposes. Those tubers under an inch or inch and quarter in diameter are used for feeding cattle or pigs.

Medium - sized Potatoes.—The medium-sized ones, or seconds as they are commonly called, when they are to be preserved for seed, are at once packed in the sprouting boxes referred to on p. 308, and carefully stored in any well-ventilated shed or empty house. Even although very immature, they can be preserved in this way with comparatively little loss, whereas were they stored in the usual way in pits till the end of the year, before putting them in the boxes, a large proportion of them would be spoiled.

Disease.—Disease rarely makes its appearance in time to do harm to early potatoes, but later on, when it is plentiful on the foliage, although scarcely perceptible on the tubers, the small seed, even although fairly well matured, keeps very much worse, and is often almost entirely lost. The reason is supposed to be that in digging, the disease spores, owing to the shaking the plant receives to separate the tubers from the roots, are shed from the leaves over the potatoes, where they vegetate and live after the potatoes are stored. Any seriously diseased potatoes should at once be separated, but tubers only slightly affected may be thrown into the basket containing the small ones, to be afterwards rejected if the small ones are dressed for seed purposes. If the small tubers are not to be used for seed, they, along with the slightly diseased tubers, may be used as food for cattle or pigs.

Raising Late or Main-crop Potatoes.

As already indicated, the later varieties which form the main bulk of the potato crop are not usually raised till after the grain crops have been harvested. Indeed, the raising and storing of late potatoes are seldom begun before the advent of October. When the potato leaves have been dead for a couple of weeks or so storing may begin in earnest. The work now to be accomplished in a limited time is too great for such a slow tool as the graip, and the potato-digger is therefore called into use.

Potato-digger.—The modern potato-digger is an ingenious contrivance which does its work well and is extensively used all over the country. The general type is indicated in the illustration of Jack's digger in fig. 596.

Speed of Potato-diggers.—With three good sharp-moving horses, and in a fairly large field, where everything is in readiness for it, and a sufficiency of gatherers are at hand, the digger will accomplish from 4 to 5 acres per day. If digging is done only one way, as is

Fig. 596.—*Potato-digger.*

the usual custom on small farms or on steep land, from 2 to 2½ acres will be a good day's work.

Preparing for the Digger.—When the digging is to be done one way only, one drill, and in many cases two drills, should be dug by the graip or plough along the side at which a beginning is desired to be made before the digger is brought on the field. Besides, if the head-ridges have been planted with potatoes, which they occasionally are, one drill should be dug next to the hedge all round the field, and the potatoes from it gathered. The digger may then commence at any convenient corner and go round the field, taking a new drill each time till the whole of the head-ridge drills, and an equal number at both sides, are dug.

Gatherers Required.—If there is no disease in the crop, and if the crop is moderate, and all the potatoes large and small are thrown into one basket, from twenty to twenty-four persons will keep a digger constantly digging, both ways.

If there are diseased tubers in the crop, or if the small and diseased are to be in any way separated, thirty to thirty-two persons may be required to keep a digger going, according to the weight of the crop and amount of diseased tubers to be taken out.

Arranging the Gatherers.—The gatherers should be arranged in pairs, or singly as is thought most desirable, each plan according to circumstances having its advantages. If there is no disease, and the small tubers are not taken out, two baskets will be sufficient for each gatherer; but if the diseased or small tubers are to be separated, three baskets at least must be given to each.

Before beginning, the land should be measured off into equal lengths for each gatherer or pair of gatherers, a lath, twig, pole, or other mark being put in where the one's lot ends and the other's begins.

The gatherers begin to collect from the end of their division nearest to where the digger starts, and work towards the other end.

Assorting Potatoes.—A general practice now is to gather all sizes of sound tubers into one basket, and to have them assorted afterwards by one of several good machines designed for the purpose.

Harrows following Digger.—As soon as the first space is gathered, a boy should follow with a pony or other light quiet horse, yoked to a single division of a set of zigzag harrows, or, better still, to a half of a set of the old pattern of hinged harrows. The latter is about the proper breadth, while the former is a little too narrow. The harrow should be of moderate weight, but should have good long tines, so that it may fully search the loose earth thrown out by the digger, and bring the buried potatoes to the surface.

As soon as the harrow has passed the first division, the gatherer of that piece of ground picks up the exposed potatoes, and works towards the end where a beginning was made. By the time the gatherer has collected the potatoes exposed by the harrow, the digger will again be round at the place where a beginning was made—that is, if the number of gatherers has been properly proportioned to the work which the machine can do. In the next and succeeding drills the same routine is of course followed.

Collecting Carts.—When the digger has made a circuit of the field, there should follow one or more carts, into which the baskets should be emptied when full. The ordinary farm-cart (fig. 597) is used in carting potatoes. The emptying should be the work of the man in charge of the cart, and not the gatherers.

Diseased Tubers.—If many diseased or small tubers are taken out, a separate cart will be required to carry these away as soon as the baskets holding them are full, or nearly so. With late varieties it is specially important that no tubers showing the slightest symptoms of disease should be stored with the sound potatoes. Potatoes that are only slightly diseased may be of some value at digging time; later they are not only of no value, but they become positively hurtful by conveying the disease to those healthy tubers with which they come in contact.

Tubers free from Earth.—The gatherers should take care to put the full baskets far enough to the side to prevent the machine throwing the earth on them when passing, as the more free potatoes are from earth, the less they are liable to sprout in the pit.

While the headlands and drills along the side of the field are being dug in above

Fig. 597.—*Jack's farm-cart.*

manner, a single drill should be dug by hand from one end of the field to the other—say sixty drills from the side of the field. If there are not sufficient persons to dig and gather such a drill, it may be done by the double mould-board or other plough, if the land is thoroughly clean and the shaws quite rotted. The harrow must of course follow the plough in the same way as it follows the digger, and in many cases it will be advisable to turn the drill to one side with the single plough, and as soon as the potatoes are gathered to turn it back again to the other side, after which the harrow should level the land before the digger comes along the next drill.

Process of Digging.—One drill having been cleared by any of the above methods, the regular digging of the field should begin. According to the side of the field from which the drills were counted, the digger should either go along the outside drill, and back the one next to the hand-dug one, or else along it first and then back the one next the fence. Digging is proceeded with in this way until the space dug at the side next the hand-dug drill is at least 15 or 20 feet wide, after which the digger may go down one side of the hand-dug drill and back the other.

This course is necessary, or at least advisable, to allow space for the digger to throw out the potatoes to each side, and still leave enough space between the two rows of ungathered potatoes for the carts to go along without bruising the potatoes.

The carts should not pass each other in this narrow space, but come in empty at the end farthest from where they are to be stored, and fill towards the other end. After a space has been dug quite wide enough to allow the carts sufficient space, and when it becomes inconvenient for the carters to lift the potatoes from both sides at once, owing to their increasing width, the carts should follow the direction of the digger, and empty the baskets along the one side, and then return along the other.

When the space so dug equals the number of drills remaining next the side of the field, the digger should then go round about them until they are finished. By proceeding in this manner, the greatest amount of land is dug with the least travelling across the ends, and fewest shifts of the gatherers.

It is seldom the headlands are sufficiently wide to allow the digger to get easily out and in without missing some of the potatoes. To prevent loss here, it is advisable to have a person with a graip at one or both ends, digging as much as gives the horses quite space enough to turn easily.

One-way Digging. — If the land is steep, or the number of gatherers limited, it may be advisable to dig only along one side of the field. In this case digging is always done down-hill, the machine returning up-hill out of gear. As in the other system, the gatherers have each their own division; but digging drills at intervals through the field by hand is done away with after it has been properly opened up.

Horses Required. — While three horses are required to work the digger when digging both ways, two may do where only one side is dug. In the latter case the number of gatherers is of course correspondingly small.

There are few operations on the farm which require greater personal supervision than potato-digging.

Adjusting the Digger.—Before beginning, the farmer or other person in charge should examine the digger carefully and see that it is in good working order. They should in particular see that the graips or forks are the proper distance behind the sock, and the sock sufficiently below the graips to allow a considerable proportion of earth to pass through, without compelling the forks to throw it to the side, or making it so wide that the potatoes pass through without being brought to the surface by the forks. If the forks are about three-fourths of an inch behind the sock, and the sock fully an inch below their points, it will be found that the digger is easier drawn, and scatters the potatoes better than if either are kept wider or narrower.

Depth of Digging. — Whenever a commencement has been made in digging, particular attention should be directed to the depth at which the digger is working. If too shallow, some of the potatoes will be cut, and others left in the ground below the reach of the harrows; and in both cases they are, practically speaking, lost. A digger working badly in this manner may cause as many potatoes to be damaged and left in the soil as would nearly pay for properly digging the whole crop. If, again, the digger is set too deep, owing to the extra quantity of earth which the forks have to displace, a considerable number more of the potatoes are covered than if it is set a little shallower, while the difference in draught will be not a little.

Adjusting the Sock.—The several causes which influence the depth at which the sock moves should be well understood by the person working the digger. The principal are the depth at which it is set, the depth of the drills in which the digger is for the time working, and the condition of the sock itself. After the sock has been set at any particular depth by the lever controlling it, the height at which the wheels stand thereafter regulates the level at which it moves. If the crop has been high-ploughed up, the space between the drills will be deeper; and the digger-wheels getting deeper compel the sock to go also deeper, so that high-ploughed drills require the sock set at less depth than where the drills are more flat.

Again, a new sock, broad and thin on the face, goes to a much greater depth with the lever set at any particular place than one badly worn, short and thick on the face. A worn sock may work satisfactorily enough on soft easy land, such as moss or sand, while it would be of no use in a drier, harder, and firmer soil.

The most of the potatoes which have been raised by the digger and buried will, as a rule, be found a little out from the point of the sock. The centre of the harrow should therefore move along there. Contrary to expectation, the finer and more sandy a soil is, the more potatoes will be left in it after the digger. The reason for this is that the unseen potatoes are covered with a very thin coating of soil, which in great part would have rolled off and exposed the tubers had the soil been rougher.

Adjusting Force of Labour.—Where the labour of one person depends so much on that of some other, as in raising potatoes by the digger, it is of the utmost importance that each class of labour should bear a certain due proportion to all the others. As many gatherers should be provided as will keep the digger moving at a moderate, steady pace; and as many carts should be employed as will easily remove the potatoes when collected; while great care should be taken that each gatherer is provided with a sufficiency of baskets. Plenty baskets help considerably to steady working, for if a cart may at a time be a little longer in coming than another, the whole work is brought to a standstill if the baskets are full; whereas had a few more been provided, no interruption would have occurred.

Second Harrowing.—It is customary to harrow the land and gather all remaining potatoes as soon as possible after being dug. But if the work is at first carefully and methodically performed, this second harrowing and gathering will not pay expenses. If, however, a second harrowing and gathering are to take place, they should be done every evening, either by the horses from the digger and the whole company of gatherers, or by a separate pair of horses and another company of gatherers. In the potato-digging season the weather is so uncertain that it is impossible to say what a day or night may bring forth; and potatoes in dug land which have been subjected to rain or frost are rarely afterwards worth the cost of collecting. As potatoes collected after the harrows

usually contain a large proportion of small, diseased, and damaged ones, they should invariably be put into a pit by themselves. They rarely keep so well as the ordinary crop.

Weather and Digging.—Potato-digging should not be persevered with during very showery weather, nor when the land is wet. A large proportion of the potatoes are left unseen in the land when the soil is damp and adhesive; and those tubers which are collected in a wet condition seldom keep well.

Plough Digging.

On farms where the potato-digger has not been introduced, or on fields very steep and otherwise unsuitable for the digger, the aid of a plough of one kind or other is usually invoked in lifting potatoes. In some cases the double mould-board plough is used; in others, it is the ordinary single furrow swing-plough; in others, the American chill plough; while not a few employ specially fitted "potato-ploughs."

Potato-plough.—The "potato-plough" may be an entirely distinct implement, or an ordinary plough fitted with a potato-raiser—a series of iron or steel fingers running out from one or both sides of the body of the plough.

The specially fitted ploughs are certainly superior to the ordinary ploughs for lifting potatoes, yet it is generally agreed that if a separate implement is to be procured for this work, that implement should be the potato-digger proper.

The original potato-raiser attached to an ordinary plough, designed by the late Mr J. Lawson, Elgin, is shown in fig. 598; but numerous modifications have been introduced by different makers. The cutter of the plough should be removed, so as to avoid injury to the potatoes.

Digging by Drill-ploughs.—If it is desired to raise potatoes by the double mould-board plough, every second drill should be split by it, the mould-boards being set much wider than for drilling, or removed altogether, and the potato-prongs put on. Each gatherer has a section to collect, in the same way as behind the digger. When nearly a half-day's work has been done by splitting every second drill, the ploughman then begins and splits the remaining drills. All drills which have been split during the day should get a double stroke of the harrows before night, the gatherers all the time keeping close up to the harrows.

Harrowing.—As soon as the gatherers have collected their potatoes, a harrow should pass along in the same manner as was described in connection with the digger; and any tubers exposed by the harrow should be collected before another drill is turned over by the plough.

The harrowing is a most important part of the operation of potato-raising, and one which should not be neglected, either when the plough or the digger is used.

Forking after Ploughs.—In many cases it is the practice to further scatter with the graip the drills which have been split by the plough. This makes more thorough work, leaving less to be done by the harrows, and making it easier work for the gatherers. But this forking, although much more speedily done than the forking of drills which have

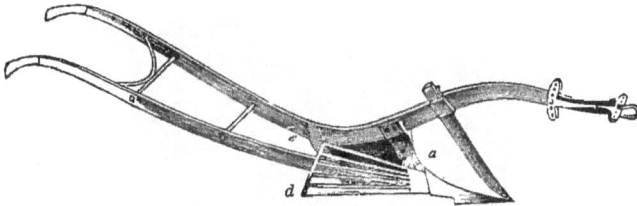

Fig. 598.—*Potato-raiser attached to a plough.*

a Narrow end of brander.　　　c Upper angle of brander.　　　d Lower angle of brander.

not been split, entails a good deal of extra labour.

Improved Digger best.— Where a proper digger is not available, a plough of some sort should certainly be used to split the drills. But while lifting with the plough is a decided step in advance of lifting with the graip or fork, both methods are much inferior to lifting with the modern digger.

STORING POTATOES.

Supplies of Early Potatoes.—Formerly it was necessary to store a larger quantity of potatoes for use in winter, spring, and early summer than is now required. Early in spring large quantities of new potatoes are now imported from Malta and elsewhere; and as soon as these are exhausted, supplies come in from Jersey and early districts along the south coast of England. Then about the middle of June the first Ayrshire and Irish crops are generally ready, and this is at least a month, if not six weeks, earlier than potatoes could be depended on prior to 1860. Moreover, the nation as a whole would appear to be using fewer potatoes and more bread than for-

merly. And when new potatoes can be obtained at anything like moderate prices, they are preferred to old ones which have been stored.

Potatoes more difficult to Preserve. —For some reason or reasons not very well known, potatoes are now more difficult to store successfully than they were prior to the middle of last century. Then they could be stored safely in pits at least double the capacity which can be trusted now; and in what were then called potato - houses they were easily preserved many feet deep, while now they would spoil if kept two or three feet deep.

Field Pits.—Where large quantities have to be handled, storing in pits in the open field is the easiest, cheapest, and most satisfactory plan pursued at the present day. A situation is selected near the farmhouse, if possible, having a dry bottom and free working soil. There shallow trenches, from $2\frac{1}{2}$ to 3 feet wide are made, running as nearly south by west and north by east as the lie of the land will permit. If the land is not thoroughly dry in the bottom, or is very clayey, no trench should be made, the potatoes being placed on the surface and piled up into long conical heaps, the sides of the heaps being kept as steep as possible. Where this is done the heap of potatoes may be from 3½ to 4 feet wide at the ground. If the subsoil is naturally dry and open, a trench from 6 to 12 or 14 inches deep may be dug out between the two guiding lines. The earth taken out should be neatly packed on the sides, a considerable slope being given to the sides of the trench.

Prismatic Pit.—Into this trench the potatoes are emptied from the carts as they come from the field, the heap carefully and neatly built as high as possible by hand. This pit is known as the long or prismatic pit, and is shown in fig. 599.

Conical Pit.—A conical pit is often preferred for storing small quantities of potatoes. It is well adapted for small farms and cottars—the prismatic being used for storing larger quantities.

A common method of forming a conical pit is as follows : if the soil is of ordinary tenacity, and not very dry, let a small spot of its surface be made smooth with the spade. Upon this let the potatoes, as taken out of the cart, be built up by hand in a cone not exceeding two feet in height, which height will give the diameter of the cone at its base 6 feet. The potatoes are then covered thick with dry clean-drawn straw. Earth is then dug with a spade from the ground as a trench around the pit, fig. 599, the inner edge of which being dug as far from the pile

Fig. 599.—*Conical and prismatic potato-pits.*

a b c Conical pit of potatoes.   e d h g Prismatic potato-pit.
  b Apex of cone.   d h Side of pit.
  c a Diameter of cone and inner   d e End of pit.
      edge of trench round the   f g Crest of pit.
      cone.   i k Straw chimneys.

of potatoes as to allow the covering of straw and earth to be put upon it, about one foot. The first spadeful is laid upon the lower edge of the straw, round the circle of the heap—the earth being clapped down with the spade, to form a smooth outward surface. Spadeful after spadeful of earth is thus taken from the trench and heaped on the straw, until the entire cone is formed, which is then beaten smooth and round with the back of the spade.

The apex of the cone is about 3 feet 3 inches in height, and the diameter about 8 feet. The trench round the pit should be cleared of earth, and an open cut made at its lowest side to allow surface-water to run freely away.

When the soil is naturally dry, the pit may be dug out of the solid ground a spade-depth, and the height of the heap will be proportionately less. But unless the soil is as dry as sand or gravel, potatoes should be piled upon the surface of the ground.

A Yorkshire Method.—The following method of covering potato-pits is practised in Yorkshire and many other parts of the country: The prismatic pit is covered with a layer of straw 6 inches thick. The straw is practically in a "drawn" condition, having come from the threshing machine in *trusses*. A plank 1 foot broad and 8 to 10 feet long is then placed along the top of the pit, and the sides to the length of the plank are covered with an inch or two of soil. The plank is then moved on, and another length is covered with soil. In this way the top is kept free from soil, and even when more soil is added to the sides later in the year, it is left untouched. In a time of severe frost potato haulms may be put over the top.

Ventilating Pits.—Small sheaves of straw are placed along the gutter left by the plank after its removal, on each of which a spadeful of earth is put to keep it in position. This straw provides thorough ventilation, and it, at the same time, keeps out rain and frost until the whole crop is secured, or opportunity is afforded of covering them more securely.

Instead of leaving a gutter the whole way along the ridge, it is the custom in some localities to only put in wisps of straw here and there, the rest of the ridge being covered with earth. Where the crop is thoroughly matured and firmly grown, such a provision against heating is probably ample enough; but in many seasons and in many circumstances, particularly with late varieties, it is not sufficient.

Heating.—As a rule, even the best ripened potatoes, after being put in the pit and closed on the sides, generate more or less heat, and give off moisture which passes out at the top, and through the earth on the sides. For this reason it is never advisable to heavily cover up potatoes when first brought in from the field, no matter how well ripened they may appear to be; for should a mild autumn and winter follow, premature growing, if not worse, will be almost sure to follow.

In order to check heating many farmers adopt the simple plan of only partially covering the layer of straw with earth at the outset. A covering of earth is put on the straw at the base of the pit, and again about its middle, leaving the straw uncovered with earth not only on the top but also along a section 8 or 10 inches wide, about 15 to 18 inches from the ground. When the pit is left in this way for a few days the potatoes rarely suffer from heating.

Seed Pits.—If the potatoes are intended for seed, the pits should be made specially narrow, in order to reduce premature sprouting to the lowest limit possible.

Storing Wet Potatoes.—Any potatoes which come in wet from passing showers, which are common at this season of the year, should be put in a pit by themselves, or along with those gathered behind the harrows. Potatoes stored wet rarely keep well. Where it can be carried out, the best plan is to take all carts which get caught in a heavy shower to the farm steading, and leave them there in an open shed till the potatoes are quite dry. Treated in this way, they are usually little the worse of getting wet, whereas they would almost certainly not keep well if put into the pit in a wet state.

It is advisable that any potatoes that may not be dry when stored should be cleared off as soon as possible after the whole crop has been safely secured.

Potato-pits are occasionally thatched in much the same manner as an ordinary stack, the ropes running lengthways and crossways, being fixed to wooden pins driven into the sides. All this trouble is, however, quite unnecessary, as the other plans are equally as secure, and cost much less. Many persons also put on a second covering of earth before putting on the thatch. This also is unnecessary, unless where thatch is very scarce and labour plentiful. Six inches of straw, with 6 inches of dry earth under, will keep out stronger frost than 18 inches of bare earth unthatched.

It is of great importance that the straw should be put on as early as possible after the crop is stored, and before the earth on the sides has become soaked with rain, as in the damp stage it takes in frost much more readily than when dry.

Potato-shaws as Thatch.—Instead of putting on ropes or earth over the top of the thatch, some farmers spread a thick covering of potato-shaws. This practice became common when the "Champion" potato with its rank shaws was introduced. It has not much to recommend it, however, as the cost of putting on and taking off the shaws is considerable.

Frosted Potatoes.—Potatoes are subject to damage by frost. When they have been exposed to frost they become soft and develop a sweet taste. The sweet taste gives place to sourness, and soon putrefaction follows.

THE TURNIP CROP.

It is not easy to over-estimate the importance of the part which turnips have long played in the agriculture of the United Kingdom. Their introduction as a field crop completely revolutionised the methods of farm practice.

Advantages of the Turnip Crop.— For the light land in the northern districts of the British Isles, where the climate is too cold for the sugar-beet or even the mangel crop, the turnip crop is of primary importance. It enables the farmer to clean and fallow his land, and at the same time to grow, even from poor light soil, an immense quantity of nutritious food for cattle and sheep. It has been said that the greatest improvement in arable farming during the last hundred years was due to the introduction of the turnip crop into the rotation.

The turnip crop has, to a large extent, given to Scottish agriculture the eminence it has attained, and it has made the eastern half of Great Britain the greatest cattle-breeding region in the world. If properly managed, the crop is a moderately reliable and valuable one on the lightest and shallowest of soils; its introduction has been of immense advantage, and its place in the rotation cannot be filled so well by any substitute which has as yet been tried.

On stiff clay soils its cultivation is not of so much advantage. The cost of reducing these to a proper tilth for the seed is great, and if the weather is either too wet or too dry, the crop is precarious and uncertain. Then clay land is liable to be injured either by carting the roots off the land or by the treading of sheep in consuming them upon it.

Unlike the potato crop, the turnip crop is usually consumed on the farm, and the unappropriated matter returned to the soil. With properly constructed dung-pits there should, therefore, by the growth and consumption of roots, be comparatively little waste of manurial elements, and consequently little exhaustion of the land.

Turnip-growing may be Overdone. —The serious injury which the turnip crop has so frequently sustained from insect and fungoid plagues, together with the heavy costs involved in its cultivation, have somewhat weakened the hold which it obtained on the affections of the British farmer. The decline in the price of grain has also tended, indirectly, to lessen the area under turnips. It has been contended, with a good show of reason, that the unfavourable experience with the crop has been in a large measure due to an attempt to grow roots upon the same land too frequently—that is, with too short an interval between the successive crops of roots. It has been clearly proved that the growing of roots, like most other things, can be easily overdone, and that the results of indiscretion with this tendency may be almost disastrous.

With this qualification, there are few who would not endorse what is said above as to the advantages of the turnip crop and the part it has played in building up the fabric of British agriculture.

INTRODUCTION OF TURNIPS.

Like that of many other cultivated plants, the history of the turnip is obscure. According to the name given to the swede in this country, it is a native of Sweden; the Italian name *Navoni de Laponia* intimates an origin in Lapland; and the French names *Chou de Lapone*, *Chou de Suède*, indicate different origins. Swedes, it is believed, were first cultivated in Scotland in 1777 by a Mr Airth, who then farmed in Forfarshire, and who obtained the seed from a son who was settled in Gothenburg. Mr Airth sowed the first portion of seed he received in beds in the garden, and transplanted the plants in rows in the field. In this way he succeeded in raising good crops for some years, before sowing the seed directly in the fields.

It is probable that the yellow turnip originated, as supposed by Professor Low, in a cross between a white and the swede, and, as its name implies, the cross may have been effected in Aberdeenshire. The origin of the yellow turnip must therefore, on this supposition, have been subsequent to the introduction of the swede.

It is remarkable that no turnips should have been raised in this country in the fields until the end of the 17th century, when they were lauded as field-roots as long ago as the time of Columella, and even then the Gauls fed their cattle on them in winter. The Romans were so well acquainted with turnips, that Pliny mentions having raised them 40 lb. weight. Turnips were cultivated in the gardens in England in the time of Henry VIII.

VARIETIES OF TURNIPS.

The varieties of turnips now in use are very numerous. Of the Swedish turnip (*Brassica campestris, rutabaga*—smooth-leaved summer rape) there are over 20 field varieties, more or less widely cultivated; and of the common turnip (*Brassica campestris, rapa*—rough-leaved summer rape) and hybrids there are more than 50 varieties in cultivation.

Swedes.—The Swedish turnip has a blue-green smooth foliage. It is a comparatively slow-growing plant, and therefore requires to be sown earlier than the common turnip. It requires for its successful growth, and will resist without injury, a greater degree of heat; is less watery; of harder texture; will stand several degrees of greater cold without injury; and will keep longer than the common turnip. The bulbs of some of the varieties are green-topped; some are purple or bronze-topped. The purple-topped varieties are usually more or less tankard in shape (*b*, fig. 600), and thus stand farther out of the soil. In consequence, they are more apt to be injured by severe frosts, and should be lifted and

Fig. 600.—*White globe.*   *Purple-top swede.*   *Green-top yellow.*

stored early. From their habit of growing well out of the ground, they are thought to be better suited for shallow soils than the green-topped varieties, the general shape of which is globular. The bulbs of the latter are more deeply seated in the ground, and are thus better protected from winter frosts.

Common Turnips. — The common turnips have rough foliage of a more decided green colour. The yellow-bottomed varieties are looked upon as a cross between the swede and the white turnip. They grow more rapidly than the swede, and come to maturity sooner. They may therefore be sown successfully much later. They will grow on a poorer soil, and in a colder climate. The bulbs contain less solid matter, and are more easily injured by hard frosts. They should therefore be used or pitted sooner than is necessary for swedes.

There are numerous hybrid varieties which are usually soft in flesh, tankard in shape, and most of them ill adapted for resisting hard frosts. The white-bottomed varieties are even more rapid in growth, more soft in texture, more easily injured, and more watery than the yellow-bottomed varieties.

Some varieties of the yellow-fleshed are green, some are purple-topped. The white-fleshed varieties are white, green, grey, purple, or red-topped. The green-top yellow turnip is shown in *c*, fig. 600. The white globe, shown in *a*, fig. 600, is an excellent turnip for early use, but is readily injured by frost.

New Varieties. — In recent times much has been done by enterprising firms engaged in the seed trade in the introduction of improved varieties of both swedes and common turnips. Several of these new varieties show marked improvement over most of the older sorts, alike in yield per acre, quality, and keeping properties.

Produce of different Varieties. —

Experiments purposely conducted to test the point, and general experience in turnip culture, have shown clearly that there is a very wide range in the productive powers, not only of the various kinds of roots, but it also of each individual variety, propagated and grown under different conditions. In the midland and southern counties of England the crop of swedes generally runs from 12 to 18 tons per acre; in Scotland, Ireland, and north of England, from 18 to 30 tons. Common turnips may give from 1 to 4 or 5 tons more per acre. Often, indeed, the extremes are still greater.

Much of course depends upon soil and climatic conditions, which are beyond the control of the farmer, and still more perhaps upon the system of culture, which is almost entirely within his direction; yet it is unquestionable that, by selecting sorts which have been distinguished for abundant production upon the different classes of soils, the yield of the crop may be sensibly increased. In regard to the feeding and keeping properties of roots the same remark holds good. With turnips, as with all farm plants and animals, the selection of the sorts best adapted for the surrounding conditions and the purposes in view is a point which demands, and will repay, the most careful attention from the farmer. Indeed it is a point which the farmer who would be successful cannot afford to overlook or disregard.

Climatic Influences on Turnips.—The turnip has a moderate range of temperature. A summer isotherm of about 56°, with a moderately moist atmosphere, is the most favourable. Before getting into the rough-leaf stage, it is easily adversely affected with night frosts. These, with hot scorching days, such as are frequently experienced in the end of May and first half of June, are very inimical to the young turnip in its cotyledon stage, and often cause its destruction, and necessitate resowing.

Insect Attacks.—This condition is generally aggravated by the attacks of insects that puncture and nibble at the seed-leaves, which injury in dry weather tends to kill the plant from bleeding or drying up. But insects seldom do much harm at this stage if the weather should be damp and the nights free from frosts.

Proportion of Leaf and Root.—Regarding the question of the proportion of top to root, interesting experiments were conducted at Rothamsted. It was found that common turnips yield a much higher proportion of leaf to root than swedes; and if the leaf be unduly developed, there may even be more nitrogen, and more total mineral matter, remaining in the leaf to serve only as manure again, than accumulated in the root to be used as food. In the case of swedes, however, not only is the proportion of leaf to root very much less under equal conditions of growth, but the amount of dry matter, of nitrogen, and of mineral matter, remaining in the leaf, is very much less than in the root. In one case, with a highly nitrogenous manure, whilst there was, with an average of $10\frac{1}{4}$ tons of white turnip root, nearly $6\frac{1}{4}$ tons of leaves, there was with swedes, with more than 12 tons of roots, not quite 1 ton of leaf. In a series of experiments, moreover, with different manures, whilst white turnips gave from 300 to 600 parts of leaf to 1000 of root, the highest proportion by weight of leaf to root in the case of swedes was $78\frac{1}{2}$ to 1000. Whilst in yellow or white turnips a very large amount of the matter grown is accumulated in the leaf which is not always consumed as food, in swedes a comparatively small amount of the produce is useless as food for stock. Generally the proportion of top to root in swedes may be stated at from 10 to 14 per cent, and in common turnips from 16 to 20 per cent.

Order of Using Turnips.—White varieties come earliest into use, and will always be esteemed on account of their rapid growth and early maturity; and though unable to withstand severe frost, their abundance of leaf serves greatly to protect the roots from the effects of cold. Being ready for use as soon as pasture fails, they afford the earliest support to both cattle and sheep.

Yellows then follow, and usually last for about two or three months. Swedes come into use after yellows, and with care may be in good condition for consumption till the month of June.

Ill-shaped Turnips.—In *a*, fig. 601, is shown an ill-formed turnip, as also one, *b*, which stands so much out of the ground, represented by the dotted line, as to be liable to injury from frost. The turnip a is ill formed, inasmuch as the part around the top is hollow, where rain or snow may lodge, and find its way into the heart, and corrupt it.

Number of Turnips per Acre.—It may be useful to give a tabular view of the number of turnips there should be on an imperial acre at given distances between the drills, and between the plants in the drills, and of the weight of the crop at specified weights of each turnip, to compare actual receipts with defined data, and to ascertain whether differences in the crop arise from deficiency of weight in the turnip itself, or in the plants being too much thinned out. The distance between the drills is the usual 27 inches; the distances between the plants as stated. As the imperial acre contains 6,272,640 square inches, it is easy to calculate what the crop should be at wider and narrower intervals between the drills:—

figures enables one to realise how important careful thinning is, and how serious may be the loss from carelessness in this operation. For example, 5-lb. turnips, at 9 inches asunder, give a crop of 57 tons 12½ cwt.; whereas

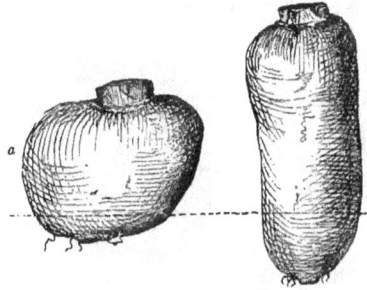

Fig. 601.—*Ill-shaped turnip.*   *Tankard turnip.*

the same weight of turnip at 11 inches apart gives only a little more than 47 tons. Every farmer knows how easy it is for careless workers to thin out the plants to 11 instead of 9 inches, and yet, by so doing, 10½ tons of turnips per acre are sacrificed. The figures as to weight are also worth looking at. A difference of only 1 lb. on the turnip—from 5 lb. to 4 lb.—at 9 inches asunder, makes a difference of 11½ tons per acre.

Width of drills.	Distances between the plants.	Area occupied by each plant.	Number of turnips there should be per imperial acre.	Weight of each turnip.	Weight which the crop should be per imperial acre.
Inches.	Inches.	Square inches.		lb.	tons. cwt.
27	9 between the plants of white turnips.	243	25,813	1	11 10½
				2	23 1
				3	34 11½
				4	46 2
				5	57 12½
				6	69 3
				7	80 13½
				8	92 4
27	10 between the plants of yellow turnips.	270	23,232	1	10 7
				2	20 14
				3	31 1
				4	41 8
				5	51 15
				6	62 2
				7	72 9
				8	82 16
27	11	297	21,120	1	9 8
				2	18 18½
				3	28 5
				4	37 14½
				5	47 2
				6	55 11½
				7	65 19
				8	75 8½

Careful and Careless Thinning of Roots. — A close inspection of these

Specific Gravity of Turnips.—All turnips, except swedes, are lighter than water. This is remarkable, because all the ingredients composing turnips—sugar, gum, proteine compounds, fibre, &c.—are heavier than water: the conclusion is, that all turnips contain a larger proportion of air than swedes.

Distribution of the Turnip.—The turnip is a plant whose constitution is eminently suited to the damp and comparatively cold climate of the British Isles. The crop indeed reaches its most certain and highest development in the northern parts of the islands and in the moist climate of Ireland. The cool climate of Caithness, Orkney, and even Shetland favours its bulb growth. In the Hebrides it grows well, the damp air causing increased luxuriance of top.

In the south of England the turnip is often a failure in dry seasons. The hot dry winds occasionally experienced

there are liable to kill the plants in the
early stages, and to cause stunting, and
sometimes mildew, if the growth is
farther advanced. There, in some
seasons the plant has a struggle for
days and weeks with dry warm winds
and a parched soil, and makes little
progress until the shorter days and cool
nights of autumn set in.

The turnip thrives in a temperature
too cold for the profitable cultivation of
cabbage, kohl-rabi, or mangels. These
in the British Islands do best in moderately dry warm seasons. The oat
luxuriates in a climate similar to what is
required for the growth of turnips, and
wherever heavy, well-filled oats can be
grown, there the cultivation of the
turnip will succeed.

Soils for Turnips.—The soils most
suitable for turnip cultivation are those
of a light friable description. The fine
state of division to which these can be
readily reduced favours the germination
of the small seeds. On such soils cultivation is easy, and they also suit the
habits of the plant, which spreads its
roots like a network into every part of
the soil. Alluvial and sandy soils are
of all the best for the turnip plant.
Next come the lighter soils formed from
trap or volcanic rocks, and the lighter
soils resting on Silurian, Cambrian,
Devonian, granitic, and New Red Sandstone rocks.

Clay Soils Unsuitable for Turnips.
—The soils least suitable are the clays,
from whatever derived. The London,
Oxford, and Kimmeridge clays being
especially stiff, are not well suited for
turnip cultivation, partly from the great
difficulty in securing a braird among the
rough particles in dry seasons, from the
hardness of such soils preventing the
free spreading of the roots, and from the
absorbed and retained water injuring the
roots in wet seasons. On the stiffer
clays, which occupy a large area in the
southern part of England, the cultivation of the crop is so precarious that it
cannot be profitable in the average of
seasons, although good crops are occasionally grown. Were it not that the
working of the land for the crop acts
upon the soil similarly to a bare fallow,
its cultivation on these soils would not
be attempted to any considerable extent.

TILLAGE OF LAND FOR TURNIPS.

Variety of Systems.—The system of
tilling land so as to prepare it for the
turnip crop necessarily varies greatly
upon different classes of soils, and in
the different parts of the country. The
condition of the land as to foulness or
freedom from weeds has likewise to be
considered in deciding upon the system
of tillage likely to be most effective and
economical.

It has also to be remembered that in
many farming operations there are variations in local customs, for which there
is no apparent or sufficient explanation
beyond the simple influence of long-continued usage. In regard to most branches
of farm-work, it is assuredly true that
there are several ways of doing the same
thing,—several methods by which the
same piece of work may be accomplished, and this, too, with almost equal
efficiency, and with little difference in
outlay.

That teaching which would seek to in-
culcate the idea that any one way is the
right way and the best way, and all other
methods wrong or inferior, is essentially
narrow and unsafe, arising most likely
from limited experience or a dogmatic
spirit—or from both ; for there is a close
kinship between dogmatism and limited
knowledge. The more one sees of the
detail-work of farming in the various
divisions of our own country and in
foreign lands, the less inclined one is to
dogmatise,—the more indeed is one impressed with the almost infinite variety
of methods and practices which farmers
may, with prudence and good results,
pursue in the prosecution of their calling.

The introduction of these remarks at
this particular point has been suggested
by the fact that in his observations as
to the methods of root culture in nearly
every corner of the British Isles, and in
foreign countries as well, the editor of
this work has noted with special interest
the almost endless variety in the details
of practice. In perusing the remarks
which follow as to the system of preparing turnip land, and in contrasting the
practices described and recommended
with different practices which may prevail in certain localities, it should there-

fore be borne in mind that it is not presumed that the methods described here are the only methods worthy of description and commendation. Indeed one may go further, and suggest that any farmer who has been moderately successful with methods different from those described here should think well before introducing a change, doing so at first only to a small extent, and in an experimental way. To describe all the good systems of root culture is out of the question. The details set forth here are those of certain methods known to be pursued with success in different parts of the country.

Soil, Climate, and System of Tillage.—The character and condition of the soil are of course the main considerations in determining the system of tillage. Stiff clay land requires very different treatment from light friable soil. The former must not be touched in wet weather, or while it is in a very wet condition. The latter is much less liable to injury from unseasonable working.

The climate is also answerable for variations in systems of tillage. The comparatively mild open winter of the southern and lower-lying parts favours autumn and winter tillage. In the higher-lying and colder districts, with their severer winter, much of the tillage work must be delayed till spring.

Prevailing System.—The system which prevails most largely in the principal turnip-growing district of this country is to plough the land with a strong furrow in the autumn or winter, allow it to lie in this condition under the disintegrating influences of winter, and in spring clear it of weeds and reduce it to the desired condition for the reception of the manure and the seed. Unless the land happen to be exceptionally foul, or is of a strong clayey nature—in which cases other methods to be explained presently may be adopted—this system of autumn or winter ploughing and spring cleaning and manuring answers admirably for the turnip crop.

Normal Conditions.

In the first place, there is described the process of preparing land for turnips, under what may for convenience be called *average or normal conditions.* By this term is meant land well, or at least moderately well, suited for turnips— heavy clays excluded; in average condition as to weeds, fertility, and drainage, and with average weather.

Exceptional circumstances will receive treatment subsequently.

Autumn and Winter Ploughing.— Turnips almost invariably follow a grain crop. As soon as practicable after the completion of the grain harvest, the stubble land intended for roots next year is—unless very foul—ploughed with a deep strong furrow, varying in depth according to the character and depth of the total surface-soil from perhaps 10 to 14 inches—rarely over 12 inches. In deep ploughing care has to be taken not to bring to the surface more than a very small quantity (if any) of the subsoil at one time. Many subsoils contain matter which is positively injurious to vegetation, and which, if mixed freely with the surface-soil, may for a considerable time have a deleterious influence on the crops. If the land be strong loam, it may be advisable to yoke three horses in the plough. When the land is very steep, and it is desired to run the furrow up and down the incline, the plan of going up-hill empty and taking a strong furrow down-hill is often resorted to. With this method no feerings are required after the first side furrow, as all the ploughs follow each other at convenient intervals in the one furrow.

A good deal of time, however, is unavoidably wasted by this plan, and farmers generally contrive to get a furrow each way by running the plough so as to avoid the direct line of the incline. For ploughing with a strong furrow in steep land, the one-way ploughs described and illustrated in vol. i. pp. 372, 373 are very useful. With the one-way plough the furrow can always be thrown down-hill, which of course lightens draught greatly.

In this strong furrow the land lies over winter, deriving much benefit from the frost and snow to which it is thus freely exposed.

Spring Tillage.—In average seasons the land intended for turnips, which has been ploughed in autumn or winter as just described, may probably not be touched again until the sowing of the grain crops has been completed. The spring working of the turnip land is usually begun in April, but the greater portion of it will most likely have to be gone through in May, some of it perhaps even later.

The extent and nature of the spring tillage will depend upon the character and condition of the land and the state of the weather. For even in what may still be called normal or average conditions, there are many variations which demand the careful consideration of the farmer.

Ploughing or Grubbing. — Most likely one spring ploughing will be sufficient, this time with a moderate furrow, perhaps from 6 to 9 inches deep. Many farmers now prefer to stir the land with some kind of strong iron-toothed implement of the grubber kind ; or it may be a half-plough, half-grubber, usually spoken of as a digger.

Whether it is advisable to plough the land, or only to drag it with a grubber, cultivator, or digger, will depend upon the kind of soil and the weather at the time. If the subsoil is very hard, the plough will be the best for breaking it up. Again, if the season is wet, the plough is preferred by many, in the belief that less injury is inflicted by the treading of the horses in ploughing, unless the grubbing is done by steam-power, which is very effective if not done too deeply.

The cultivator gets over the land much more quickly than the plough. But if the soil is stiff or full of weeds, a second turn after harrowing and gathering will be necessary. By the use of the cultivator the fine surface-mould produced by the winter frosts will be kept nearer the surface, and will make the germination of the turnip seed more certain. In stiff soils fewer of the large clods will be brought to the surface ; in dry weather less evaporation from the surface will take place, and the success of the crop will be more assured.

Some of the modern diggers are admirably adapted for their work. Used on stubble land the digger pulverises the soil to a good depth, and turns only the upper few inches, thus exposing the roots of weeds which lie near the surface to the winter's frost, and leaving the soil thoroughly broken up. This also facilitates the removing of surface weeds from the land. Amongst land after turnips it also does good work. While pulverising the soil it does not turn it over, and expose the dung, as is often done by the common plough.

The Tennant grubber (Hunter) is shown in fig. 602 ; Clay's grubber in fig. 603 ;

Fig. 602.—*The Tennant grubber.*

and Martin's spring-tined cultivator in fig. 604.

Pulverising Ploughs. — By the attachment of revolving prongs, ploughs are made which, at the one operation, plough the land and pulverise it, throwing most of the weeds on to the surface.

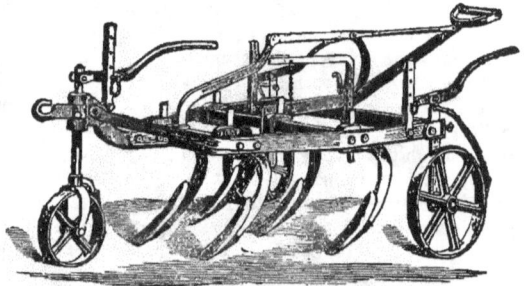

Fig. 603.—*Clay's grubber.*

Harrowing Turnip Land.—Whether ploughed or stirred, the land must be afterwards harrowed, the weeds picked off, and the large stones, if any, gathered. If the weather permit, harrowing and rolling must be continued until the clods are reduced, and a fine mould formed.

Removing Weeds.—The harrowing brings the weeds loosely to the surface. Chain-harrows are frequently used to collect the weeds into heaps, and so are horse-rakes, the work being concluded by hand-rakes and forks, or graips. Hand-picking is preferred by many farmers, and is of course the most thorough system. The weeds may be burned in heaps on the field and the ashes scattered around, or carted to some convenient corner to be united with lime to form a compost-heap. In the latter case care must

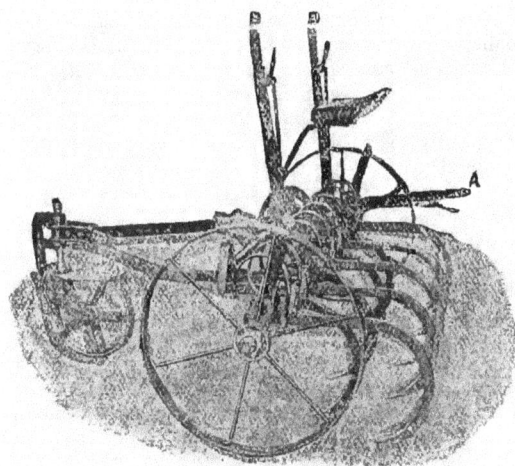

Fig. 604.—*Martin's cultivator.*

be taken not to spread the compost on the land until the vegetable matter in it has been thoroughly decomposed.

Exceptional Conditions.

In soils well suited to turnips, and kept in good heart and condition as to cleanliness, the foregoing process of tilling and cleaning will most likely be sufficient to prepare the land for the sowing or laying down of turnips, as it is often termed.

But there are many circumstances which render deviations from the prevailing system necessary or advisable. For instance, stiff clayey land, land which is excessively foul, and land unusually free from weeds, all receive peculiar methods of treatment. Again, the land may be both stiff and foul, and in this case still another plan will be adopted. The questions as to whether the roots are to be sown in drills or on the flat, and at what time the farmyard dung is to be applied—whether in the autumn or winter, on the flat in spring, or in drills at sowing-time—are also re-

sponsible for variations in the preparatory work.

Preparing Foul Clay Land.

This is often a serious undertaking. No progress can be made with it in wet weather. Indeed, any attempt to cultivate or clean clay land when it is in a wet condition must inevitably result in failure. Far better let men and horses remain idle than allow them to work stiff clayey land unseasonably. In this condition the more it is worked the greater is the injury inflicted.

When, therefore, the farmer has before him the unenviable task of having to clean stiff clay land which is in very foul condition, he must watch the weather carefully and seize every suitable day for the purpose.

Autumn Cleaning.—For cleaning land of this kind the autumn is the best time — that is, if the weather should be favourable. Begin the work as soon as the grain crops are secured. The first operation will either be the cultivating (or grubbing) or the ploughing of the land with a shallow furrow,—a furrow just deep enough to turn over, but *not to bury*, the weeds. The depth of the first furrow is indeed regulated mainly by the character of the weeds, whether they are deep-rooted, creeping, or surface weeds. Some of the surface weeds may be killed by being buried with a deep furrow; but couch-grass, docks, thistle, knapweed, and other well-known troublesome weeds, require more drastic treatment. Grubbing or dragging and harrowing follow ploughing, and if necessary to break clods holding weeds, the land is then rolled, again harrowed, and the weeds collected and burned or carted away.

A Second Crop of Weeds.—An examination of the land may reveal the fact that it is still far from clean. In this case the whole process should, if the weather permit, be at once again gone over. The ploughing may perhaps be omitted. The grubber or cultivator, followed by the harrows, will take the remaining weeds to the surface, and this time, in particular, it will be advisable to hand-pick the weeds, so as to ensure that all the little particles of couch-grass roots may be removed.

Do not Break Weeds. — Excessive tillage is liable to break these weed-roots into small pieces, each of which, if left, will form a centre of filth. It is therefore important to have the weeds brought to the surface with as little knocking about as possible.

Cross-cultivation.—The subsequent ploughing and grubbing are usually given at right angles to or in a slightly different line from the preceding. The object of this cross-cultivation is of course to ensure that all portions of the soil may be stirred.

Steam-power for Cleaning Clay Land.—For the cultivating and cleaning of strong land steam-power is very suitable. The steam-cultivators go over the ground quickly, and they can be as easily regulated as to depth as implements for horse-labour.

Half-ploughing.—Land which is *excessively foul* is sometimes cleaned by another process, — a piecemeal method known as break-furrowing, raftering, or half-ploughing. Only half the surface is at once disturbed, each furrow being thrown on to its own breadth of ploughed land. Harrowing and weed-collecting follow, and this fleece of weeds being removed, the strips of the land formerly undisturbed are then turned over by the plough, harrowing and weed-collecting completing the process. This is a tedious and costly process, which is not often adopted, and need not be resorted to except in such rare cases as where the land is so excessively foul that the entire mass of weeds in it could not be conveniently dealt with at one time.

This system is more frequently adopted for the purpose of killing surface-weeds. In this case the land lies over winter in the ridged-up appearance which the half-ploughing gives to it.

Autumn Dunging and Ploughing. —Assuming that the weather has been sufficiently dry and free from frost to enable the farmer to complete in autumn and early winter the cleaning processes described above, the next step—with strong land intended for roots—will perhaps be to spread its allowance of farmyard dung and plough in this with a shallow furrow. This is the usual practice, and by far the best plan in stiff land of this kind, where the turnips are to be sown on the flat, and where there is a sufficient supply of dung ready in time for application before the last ploughing in autumn or early winter. The advantages of the autumn instead of the spring dunging of heavy land will be mentioned in dealing with the manuring of turnips. If the dung is not to be applied at this time, the land is turned over in a strong furrow before the rigours of winter fairly set in.

Spring Tillage of Strong Land.— The spring tillage of stiff clays intended for roots has to be carried out with the utmost care and caution. Clay is stubborn material, in the working of which the farmer, who has not before had practical experience of it, is liable to unwittingly commit errors, which may seem trifling at the time, but which may result in serious injury to the crop. If the land has been cleaned and dunged in the autumn, the spring work is thereby greatly simplified. Lying over winter in a strong furrow, the land becomes pulverised and more easily prepared for the seed. In southern parts, where the winters are open, the spring tillage of this land is begun as early as possible—as early as January or February if the weather is sufficiently dry. It is then cross-ploughed at least once. Often, indeed, strong land is ploughed two or three times in spring, in the attempt to reduce it to that fine tilth which is so advantageous to the root-crop.

Grubbing or Cultivating in Spring. — Grubbing or cultivating is preferable to repeated ploughing in spring, for while the former leaves the finely pulverised soil on the surface, the plough turns this underneath. An excellent implement for this purpose is the modern cultivator, such as that in fig. 604, which has spring tines set in such a manner as to enter the hardest soil.

By repeated harrowing, rolling, and grubbing or dragging, the rough strong land is reduced as finely as possible, and is thus prepared for the reception of the seed.

Preparing Clean Land.

When the land intended for turnips is in a cleanly condition, the preparatory tillage operations may be considerably lessened. As early as possible in the autumn or winter, the land, whether light or strong, is ploughed with a deep furrow or cultivated with a rank grubber. On strong clay land the dung—as much of it as is then made—is spread on the stubble just before ploughing.

In spring, strong clean land will require similar treatment to that just described in speaking of the preparation of foul clay land, which had been cleaned in the autumn.

Spring Tillage of Light Clean Land. —But in the case of light land free from weeds, very little spring tillage may suffice. Indeed, such land may be allowed to lie in the winter furrow till the work of grain sowing is finished. It may then receive a strip of the harrows across the winter furrow, be turned over once with the plough, or stirred by the grubber or cultivator, and again harrowed two or three times. In many cases this will be found sufficient; but if the tilth is not reduced as finely as desired, another turn of the grubber or cultivator and harrows may be prescribed, and with this the preparation will be completed. Where injury from lack of moisture is feared, early harrowing to check evaporation is sometimes resorted to.

Overworking Injurious.—Not unfrequently injury is inflicted upon the turnip crop by the overworking of the land in spring. Turnips delight in a fine moist soil. The finer the soil is the better, but it must also be damp. In preparing turnip land, therefore, the farmer must strive not only to break down the soil but also keep in the moisture. This, as will be at once understood, is not so easy to accomplish. Repeated ploughing and opening up the land, late in spring and early in summer, encourages the escape of moisture. It is thus important that in dry districts the deep turning and stirring of the land should be done in autumn, winter, and early in spring, so that when the dry season has set in, shallow stirring and surface-scratching may be sufficient to provide the desired tilth.

The dissipation of moisture during spring may be to some extent lessened by immediately following the ploughing or grubbing by harrowing.

Even in moderately moist climates this matter is deserving of more attention than farmers, as a rule, bestow upon it. Indeed it may be described as one of the cardinal points in successful turnip culture. The importance of retaining moist soil around the young turnip plant is perhaps the consideration which has been most powerful in maintaining the system of growing turnips on the flat in England.

Forking out Weeds.—A practice much pursued in England, with land not so foul as to require a special course of tillage to clean, is to send several workers over the stubbles in the autumn with graips or forks to dig out couch and other visible weeds. This is a good plan, likely to save after-labour in removing weeds.

It is the habit, indeed, of some particularly careful farmers, to send two or three labourers over the entire farm in this way, forking out any weeds to be seen, and giving special attention to head-ridges and sides of fences, which often form perfect nurseries for weeds.

Turnips on very Strong Clays.— In some cases in England, on very strong clays, which are by nature ill adapted for turnip culture, crops of swedes, which would delight the heart of any farmer, are occasionally grown. The main secret of success, in these instances, has nearly always been the studied and careful preparation of the land. Such deep tillage and cleaning as the land receives are done in dry weather in autumn, when the dung is also put in. Then in some cases which we have known to be successful in an eminent degree, no further stirring of any kind is given to the soil till sowing-time, when, after a turn of the harrows, the seed is sown in rows on the flat.

This plan, of course, would not succeed in land containing many weeds; but on some of the strongest clays in England we have seen it carried out with the most gratifying results—upon land so strongly adhesive that it would sometimes exhibit in spring with little effacement the footprints made upon it by labourers five months before.

The chief difficulty in turnip culture, on strong clay land in a dry climate, is to obtain a strong regular plant. This is most effectually promoted by retaining the winter moisture in the soil. And the best method of conserving the moisture is to clean, dung, and plough the land in autumn, and stir it as slightly as possible after the advent of warm weather in the following season.

Still, when the farmer has done his very best, turnip growing upon very strong clayey land will often fail. And while it is interesting and may be useful to record these instances of exceptional success, one cannot with confidence recommend the extensive culture of turnips upon such land.

SOWING TURNIPS.

In the early days of turnip growing in this country the seed was sown broadcast on the flat surface of the land. At one time, indeed, that was the universal custom with all farm crops.

Introduction of Drill Sowing.—For the introduction of that most serviceable system of drill sowing we are indebted to Jethro Tull, whose writings, during the first generation of the eighteenth century, did much to promote the improvement of farm practice. In his book on *Horse-hoeing Husbandry*, published in 1731, he advocated the system of drill-sowing wheat in narrow ridges. The success of the method attracted much attention, and it was soon after tried for other crops. For turnips it was found specially suitable, and as early as 1745 the drilling of turnips was practised in Dumfriesshire by Mr Craig of Abbeyland. The system rapidly won many converts, and soon after the middle of the eighteenth century, turnip culture in drills or rows was being pursued successfully in various parts of the country—notably, besides Dumfriesshire, in Cumberland, Northumberland, Roxburgh, Berwick, and Norfolk. Indeed, to the last-named county, still noted for turnip culture, an improved system of turnip cultivation was as early as 1730 introduced from the Netherlands by Charles, Viscount Townshend of Rainham.

Turnips in Raised Drills.—In Scotland, Ireland, and the north of England, turnips are now almost universally grown in raised drills. This method, it is said, dates from about 1760, when it was begun by Mr Dawson of Harperton, Kelso. For districts with a moist or moderately moist climate, it has long ago proved itself to be superior to all other methods of root culture.

Disadvantage of Raised Drills.—The one drawback to raised drills is that throwing up the land in this form encourages evaporation, and thus intensifies the effects of drought. Mainly for this reason, the system of sowing in rows on the flat is preferred in the greater part of England.

Advantages of Raised Drills.—The system of raised drills possesses several advantages of the highest importance. In the first place, the gathering of the finely pulverised soil together in the raised drill gives the roots the benefit of a deeper and freer soil than they would obtain on the same soil in the flat system. The stores of plant-food in the surface-soil, and the manure applied at the time, are brought into closer proximity to the young plants, whose growth in the early and most critical stages is thus effectually stimulated. The thinning and hand-hoeing of the crop are more easily and expeditiously accomplished in the raised drill than on the level surface, while the subsequent hand-hoeing and horse-hoeing, or drill harrowing, bring back the land to a nearly level condition by the time the crop is throwing out its spreading root-fibres.

Width of Drills.—This varies from 24 to 30 inches, the most general width being 27 inches. In narrow drills there is difficulty in covering rank dung thoroughly, and there is less facility for horse-hoeing. On the other hand, the yield of the crop per acre will be lessened by having the drills much wider than about 27 inches.

Drill-plough.—The raised drills are now most generally made by the drill-plough, the construction of which is well shown in fig. 580. The breast and mould-boards of this, as of all other improved drill-ploughs, are formed so as to throw up the soil loosely rather than to squeeze it together, as would be done by a wedge-shaped plough. The width of the drill can be easily regulated by the screw shown between the shafts. The "marker" is adjusted to the corresponding width, and with these improved ploughs a skilful ploughman makes drills that are pleasing to the eye of a tasteful farmer—straight in line, and uniform in depth and width.

The *depth* of the drill must be sufficient to thoroughly cover the dung. Where there is no dung to cover, the drill may be shallower, yet deep enough to make the ridge complete on the top.

In many districts the drilling is done by the ordinary single plough. In drilling with the single plough, the tail of it is purposely kept high, which leaves the bottom of the drill narrow. One passage of the plough is quite sufficient for either opening or closing, and many consider it preferable to the double mould - board plough, unless for earthing up potatoes. One point in favour of the single plough for drilling is, that by it the clods are thrown over the drill and fall into the bottom of the previous furrow, instead of being thrown, as by the drill-plough, into the centre of the drill, above the dung and under the seed.

Raised Drills on Strong Clays.— The system of raised drills is not so suitable for strong adhesive clays as for more friable soils. Still, even in very stiff land it is often practised with success. On land of this kind—which, as we have seen, may be seriously injured by much tillage late in spring—perhaps the best plan is to form the drills in autumn or winter, after the land has been cleaned, dunged, and ploughed. In these open drills the land lies till the time of sowing, when a light harrow, chain-harrow preferably, is drawn over the land in the direction of the drills. Any artificial manure to be given is then sown broadcast, the drills are set up by the drill-plough, and the seed at once sown.

This provides a moderately fine tilth for the seed on the top of the drill, and yet it does not unduly promote evaporation. But it leaves the soil hard a few inches below the surface, sometimes checking slightly the development of the crop.

Drilling on the Flat.—In the midland and southern counties of England, the prevailing system is to sow the turnip seed on the flat surface in rows from 15 to 22 inches apart. As already explained, the main object in pursuing this plan is to avoid the dissipation of moisture, which is to a considerable extent unavoidable in raising the soil into loose ridges.

As a sort of general rule, it is recommended that, in districts with an average rainfall of less than 24 inches per annum, the flat system should be the prevailing one. A maximum crop is not likely to be obtained by this method, but in dry climates it is the safest, and is therefore extensively pursued in the south.

Width of Rows.—The rows on the flat are invariably narrower than raised drills. The most general width between the flat rows is from 18 to 20 inches, occasionally more and frequently less. With a greater width in the midland and southern counties of England, where the roots seldom attain the weights that are common farther north and in Ireland, the crop would fall off in yield per acre; yet it will be readily understood that the comparatively little space thus left between the rows on the flat does not permit of satisfactory horse-hoeing while the plants are growing. Moreover, the horse-hoeing cannot be begun so soon — not until the plants are sufficiently far up to ensure that they may not be unwittingly buried.

Broadcast Sowing of Turnips.— The broadcast sowing of turnips is now rarely practised. Where turnips are grown for the development of root it is quite unsuitable.

Still, in certain cases, when a crop of turnips cannot be got in in time to grow roots satisfactorily, a useful supply of green food in spring may be provided, only in a good climate of course, by sowing in August, with the broadcast barrow, from 1¼ to 2 lb. per acre of turnip seed. For this purpose the ground is harrowed before and after sowing, and then rolled (vol. i. pp. 341, 342).

When not to be systematically thinned, turnips do better sown broadcast than in rows.

In some parts of the south, where an abundant supply of field food for sheep is a matter of great importance, land planted with beans is occasionally thinly broadcasted with turnip seed. In a mild autumn, after the harvesting of the beans, the turnips develop a wonderful bulk of very useful food.

The actual details of the process of sowing turnips depend upon whether the raised-drill or flat-row system is pursued, and what manure has to be applied at the time of sowing.

Dunging and Sowing in Raised Drills.

Taking first the system which prevails in Scotland, Ireland, and the northern counties of England, it is found that the detail work of sowing—assuming the land to be already cleaned of weeds, and sufficiently pulverised—consists, in succession, of opening the drills with the drill-plough, carting the dung and spreading it in the drills, perhaps drawing a light harrow along the drills, sowing artificial manures most likely broadcast, covering in the drills with the drill-plough, and sowing the seed with the drill-sower.

Simultaneous Drilling and Sowing. —Upon large holdings possessed of a sufficient force of horse and manual labour, all these processes go on at one time. The result in the crop is generally most satisfactory when there is no appreciable delay between the opening of the drills and the completion of the operation by the sowing of the seed. It is bad practice to open many more drills in one day than can be manured, closed, and sown before nightfall of the same day.

Stale Seed-bed Undesirable.—Turnip seed does not take kindly to a "stale" seed-bed. It comes away most satisfactorily when sown upon a freshly turned-up mould, fine in the texture, and tolerably moist — about two to four hours after the drills are closed in. When, therefore, it does happen that a portion of land has lain for a few days in finished drills unsown, perhaps on account of wet weather, some farmers consider it advisable to draw a light harrow along the drills, and set them up afresh with the drill-plough. This, however, takes time and labour when these can ill be spared, and unless the surface of the drills has become firmly packed or caked by heavy rains, the harrowing down and drilling up again may be dispensed with.

The Force Employed.—The arrange-

ment of the force of horses and workers in sowing turnips on a large farm, so that there may be no delay and no collisions or interruptions to any of the force, requires considerable skill and forethought. We will assume that there are two drill-ploughs at work, and that the force for carting and spreading dung and sowing manure will be sufficient to keep these fully employed opening and closing drills during the entire day. The number of carts, and men to fill them, required to keep the two drill-ploughs busy, will depend upon the proximity of the dung-heap to the drills, and the quantity of dung to be applied per acre. With the dung in heaps on or near the field, and not more than about 15 tons of dung per acre, four carts, with one or two men to assist the carters in filling, should be amply sufficient. Assuming that the two drill-ploughs would open and close about four acres per day, the four carts would thus convey to the drills about 60 tons of dung per day, perhaps from 18 to 20 loads each, in the full working day of ten hours.

Four or five workers — men, lads, or women—will be required to spread the dung, one man will sow the artificial manure, and another will follow all with the two-drill turnip-sower, drawn, perhaps, by a good-sized cob or farmer's pony. The steward, bailiff, or grieve (as the farm manager is variously called), or the farmer himself, usually sows the turnip seed, and as the turnip-drill takes two drills at a time, and the draught is very light, it will usually go over the whole day's opening and closing in rather less than a half-day. There will thus be employed in the "laying down" of about four acres of turnips eight horses, at least eight or nine men, and five or six lads and women for a whole day, and an additional man and horse for four or five hours. The cost, per acre, involved by the employment of this force would vary with the rate of wages, price of horses, and cost of horses' food in different districts and seasons.

Arranging the Force. — So as to avoid interruptions and ensure the maximum amount of work done in an efficient and satisfactory manner, it is important to have the duties of each person clearly and intelligently defined and understood beforehand. About a dozen drills or so should be opened the night before, so that the full force may at once get to work in the morning.

Opening and Closing Drills.—As to the two teams with the drill-ploughs, the better plan is for the one to open and the other to close in the drills. In some cases the practice is for the two ploughs to follow each other, and open in the one direction and close in the other.

In cases where there is only one drill-plough at work, the best plan is to open in the one direction and close in the other. This plan is followed where the single plough is used to open and close the drill with one furrow.

Another Method.—The following is another method of arranging the force, which some would prefer. In the evening before, 20 drills or so are opened, so that an immediate start may be made with the dunging operations in the morning. Three ploughs being used, they open the drills up-hill, and close them down the slope, if the field is not level. Three persons are placed at the dung-heap to load the four carts employed in dragging the dung. Each man throws the dung out of his own pair of carts, which come to the drills in rotation. This plan gives an interval of leisure to each man, so that he is not constantly kept in one position. A boy is sometimes employed in driving the carts between the dung-heap and the drills. When the first drill receives its dung, 4 spreaders are placed in divisions of equal length along the drill. This enables the manager to check the work of any spreader, which can be readily known. A machine for sowing manure, and another for sowing turnips, complete the operations. There is no waiting in any division of the work, but the whole proceeds in a regular manner. In this way 12 horses, and 15 men, women, and boys, can lay down from 6 to 8 acres of turnips per day without any undue pressure.

Turnip Seed Drill. — Various have been the forms of turnip-sowing machines, and modes of distributing the seed. The old heavy square wooden-framed machine with its revolving seed-barrel, once so common, is now seldom seen. Its weight was useful in heavy soils, but it was cumbrous, and the seed-barrel required great care to give an equal delivery. The improved modern turnip-drill sowing-machine is light, elegant, and easily managed. It consists of a simple iron frame, with shafts, handles, two rollers, seed-boxes, spouts, and coulters. The arrangements for working the seed-boxes, and for regulating the quantity of seed deposited, vary considerably, but the better known drills are all thoroughly efficient and reliable in working. The general formation of the modern turnip-drill is shown in fig. 605, which represents an excellent machine, made by

Fig. 605.—*Turnip drill-sower.*

James Gordon, Castle Douglas. Most of these modern drills can also be arranged with larger boxes for the sowing of mangels. Rollers can be attached to the rear of the machine, but these are only sometimes used.

Drilling Manure and Seed.—Manures are occasionally drilled along with the seed. In the raised drill system this is not often practised, as it is tedious and not of much practical advantage over broadcasting, unless where very small quantities are used. When large quantities of manures are used, they should be distributed over and mixed through the soil.

Water Drill. — The water drill, so common in the flat-sowing system of England, has been used with advantage in the north in dry seasons. It will sometimes secure a braird which would otherwise have failed. A stream of water, in which superphosphate may be dissolved, is run into the seed-rut. It acts as a moistener of the soil, and stimulant to the young plants.

Consolidating the Drill-top.—If the weather is very dry and the soil open, it is found advantageous to go over the drills a second time with the turnip machine, although no seed is sown. The rollers consolidate the drills, and make a braird more certain. A drill-roller such as that shown in fig. 584 is admirably suited for this purpose.

The braird seldom comes well if the soil is so damp that the rollers clog with earth.

Drill for Sowing on the Flat.—A machine of a different description is employed for sowing turnip seed in rows on the flat surface. It is in general form similar to the Suffolk drill, shown in fig. 421, vol. ii. p. 124, but is provided with means for sowing either dry artificial manure or water or liquid manure along with the turnip seed.

Water and Dry Drills compared. —Both the water and the dry drills are used in the south of England. There is much difference of opinion as to their respective advantages and disadvantages. By the use of the water drill, the seed of course is provided with more moisture in the seed-bed. In average seasons this would most likely be an advantage, yet it is maintained by many that it may often prove to be the reverse, or at any rate a doubtful advantage. In very dry seasons, for instance, it has been found in numerous cases that a stronger and more regular plant has been obtained from the dry drill than from the water drill. The reason assigned for this is that the superabundance of moisture at the very outset caused the seeds to germinate too rapidly, and set up a rate of growth which could not be maintained when the artificial supply of moisture began to fail.

Manure Injuring Seeds.—One important drawback to the system of applying artificial manures in close contact with the seed, is that the vitality of the seed is thereby apt to be injured or destroyed. An obvious remedy for this is not to sow the manure along with the seed, but to incorporate it with the soil before, as is done in the northern system.

A machine is now made which at the same time sows artificial manures in two rows and forms drills over them.

Time of Sowing.

The period for sowing the Swedish turnip in Scotland usually extends from the 1st May to the 1st June; for yellows, from the 20th May to the 20th June. In the south of England, and most parts of Ireland, the sowing may be almost a month later. But there, as in Scotland, turnips sown early have, as a rule, the best chance of becoming a full crop. Notwithstanding an occasional season in which mildew attacks the earlier-sown crops, it is an undoubted advantage, if the land can be properly prepared, to sow the seed as early as possible after the 1st of May.

In the south of England late turnips are frequently sown up till the end of August, and occasionally even in September. In these cases, however, heavy crops of solid roots are not looked for.

Quantity of Turnip Seed.

The quantity of seed required varies with the season and soil. The seeds of the Swedish variety are about one-fifth larger and heavier than those of the common turnip, and a correspondingly larger quantity must be used. Swedes sown in May require from 3 to 5 lb. of seed per acre. The latter quantity of good seed should be used in the earlier part of the season, if the soil is rough or of a stiff nature. If the season is somewhat advanced, and the soil finely moulded, from 2½ to 3 lb. may be sown. For yellows sown in May the quantity may be from 2 to 4 lb.; if sown in June, from 1½ to 2 lb. may be sufficient

Thick and Thin Sowing.—While thick seeding is expensive and injurious, in some seasons producing a rush of spindly plants, it is not prudent to sow too thinly. Not a few crops have been lost where a little more seed would have saved them. Moderately thick sowing sometimes helps to overcome an attack of the fly. As the season approaches midsummer, the weather gets warmer, the risk of fly diminishes, and a smaller quantity of seed is sufficient.

If the seeds could be evenly distributed, and if all could be depended upon to germinate and grow, 3 ounces of average yellow turnip seed, and 3½ ounces of swede seed per acre, would give a plant for every six inches of drill.

Selection of Turnip Seed.—Great care should be exercised in the selection and purchase of turnip seed. By unscrupulous sellers of seeds, turnip seed is sometimes mixed with old stock, or even with wild-mustard seed killed by immersion for one minute in boiling water. Home-grown seed, if fresh, is usually most reliable. Fine plump seed is better than that which is small and immature, and will produce a stronger plant and heavier turnip.

Depth for Turnip Seed.—The depth to which turnip seed should be put into the soil varies with the state of the weather and the condition of the soil. In dry weather, with the soil moderately dry, the seed should be put fully an inch under the surface. In wet weather, with plenty of moisture in the soil, from a quarter to a half inch will be sufficient.

MANURING TURNIPS.

The manuring of the turnip crop is a subject that demands, and will repay, careful attention from the farmer.

Dependence upon Manure.—It is quite essential for root crops of all kinds that a dressing of manure, usually very liberal, shall be given for their own special benefit. Turnips are gross feeders: they produce a great weight of material in a comparatively short space of time, and must therefore, if their success is assured, have within easy reach an abundant supply of readily available plant-food.

It is a characteristic of the turnip crops that they fail entirely upon impoverished soil. Upon a deteriorating unmanured soil grain will continue to produce some considerable yield long after turnips have failed upon it completely.

This peculiarity has been well shown at Rothamsted. There Norfolk white turnips grown for three successive years on two plots—one with no manure, the other with 12 tons of farmyard dung every year—gave in roots (omitting tops or leaves) the following results per acre:—

	No manure.		12 tons dung.	
	tons.	cwt.	tons.	cwt.
1st year	4	3¾	9	9½
2nd ,,	2	4¼	10	15¼
3rd ,,	0	13¾	17	0¾
Average	2	7¼	12	8½

On another piece of land an unmanured plot was cropped continuously on the Norfolk four-course system,—roots, barley, clover (or beans or fallow), wheat,—and while the turnips coming at intervals of four years fell from 3 tons 5½ cwt. in the first year to 1 ton 6 cwt. in the second crop of roots, and to 5 cwt. in the tenth crop, the barley following after these miserable crops of roots, without any manure whatever, gave the respectable average of 31⅝ bushels per acre for the whole of the eleven crops grown in this way at intervals of four years.

It is thus evident that turnips readily exhaust the soil of the available supply of plant-food suitable for them, and that as foragers in poor soil they are not equal to the grain crops.

In quite an exceptional degree, therefore, turnips are dependent upon dressings of manure applied for their own special benefit. No farmer attempts to grow turnips without an allowance of manure, in that or the previous season, no matter how fertile naturally or how high in condition the land may be.

An Exhausting Crop. — Assuredly the turnip crop is an *exhausting* crop. The fact that, in prevailing farm practice, it generally leaves the land better than it found it, is due, not to the influence of the roots, but entirely to the tillage and cleaning the land received in preparation for the roots, and to the surplusage in the dressing of manure.

It was at one time supposed and contended that turnips enriched the land by their large extent of leaf-surface absorbing nitrogen from the atmosphere, and leaving it in the soil for the benefit of succeeding crops. Careful investigations have shown that this idea is not well founded, and that the root crop, if wholly removed from the land, is the most exhausting of all the ordinary farm crops grown in this country.

These considerations all tend to emphasise the importance of the question of "manuring for turnips." Information bearing on the subject will be found in various parts of this work,—in the articles on "The Fertility of Soil," vol. i. pp. 317-334, the "Highland and Agricultural Society's Experiments," vol. ii. pp. 37, 40, and in the large section on "Manures and Manuring," vol. i. p. 446.

Elements Absorbed by Roots. — First, let us see what are the elements and the quantities of these elements absorbed by an acre of turnips. Reverting to the table on p. 326, vol. i., giving the weight and average composition of ordinary crops in pounds per acre, we find that the figures relating to common turnips and swedes are as follows :—

	TURNIPS.			SWEDES.		
	Roots, 17 tous.	Leaf.	Total crop.	Roots, 14 tons.	Leaf.	Total crop.
	lb.	lb.	lb.	lb.	lb.	lb.
Dry matter . .	3126	1531	4657	3349	706	4055
Total pure ash . .	218	146	364	163	75	238
Nitrogen . . .	63	49	112	74	28	102
Sulphur . . .	15.2	5.7	20.9	14.6	3.2	17.8
Potash . . .	108.6	40.2	148.8	63.3	16.4	79.7
Soda . . .	17.0	7.5	24.5	22.8	9.2	32.0
Lime . . .	25.5	48.5	74.0	19.7	22.7	42.4
Magnesia . .	5.7	3.8	9.5	6.8	2.4	9.2
Phosphoric acid . .	22.4	10.7	33.1	16.9	4.8	21.7
Chlorine . . .	10.9	11.2	22.1	6.8	8.3	15.1
Silica . . .	2.6	5.1	7.7	3.1	3.6	6.7

Elements to be Supplied in Manure. — Now the next and all-important question is, What proportion of these elements has to be supplied in manure? In ordinary farm practice the only essentials of manure are nitrogen, phosphoric acid, and potash. To most soils not by nature calcareous, lime has to be applied at intervals; but the functions of lime in the soil are well known to be so different from those of what are generally understood as manures, that we will not here embrace the question of liming, but will assume that the soil is sufficiently provided with it. Of the other elements mentioned in the above table the natural supplies will almost invariably be ample enough for the wants of the crop.

Subordinate Elements. — On some soils the application of magnesium, calcium, and sulphur has produced a considerable increase in the weight of the turnip crop. But the good effects seem limited to certain soils, and are probably due more to chemical and mechanical agency than to the supplying of direct food to the plant. Thus caustic and carbonate of lime act upon soils by disintegration, and by causing a more rapid decay of the organic matter, liberate nitrogen, which acts as a plant-food to all crops, and in the turnip give an increase of shaw equal to that obtained by a small application of sulphate of ammonia. The addition of sulphuric acid, especially in a free state, must act upon and change some of the soil constituents.

Uncertainties in the Manuring Question. — Confining attention, therefore, to those three important constituents of plant-food—nitrogen, phosphoric acid, and potash—we have to consider what quantities of each of these should be applied to the different kinds of turnips in different conditions as to soil and climate. This, unfortunately, is not a simple mathematical question. There are so many uncertainties as to the character and contents of the soil, and so many disturbing influences in climatic variations, that the farmer, however scientific, careful, and capable generally, must always be to some extent working by chance. Moreover, the farmer has to keep in view the important considerations of profit and loss as well as the perfection of the crop. He is not content to discover merely what quantities of nitrogen, phosphoric acid, and potash would be likely to ensure a full crop of turnips. His great object is to learn what quantities of these constituents should be applied in order to secure the greatest possible return for the outlay involved. The prudent farmer, like all prudent business men, works for profit. He wants not merely a *big* crop but a *paying* one as well.

Now in practice it is found that to apply to the land the exact quantities of essential manurial elements which analysis shows that the particular crop removes, would not be efficient and economical manuring. The reason for this is twofold. In the first place, there are the stores of fertility already in the land, which may be sufficient to provide much of all, and all or the greater portion of some, of the elements. On the other hand, the whole of the plant-food in the manure applied may not, in an available form, come within the range of the roots of the crop for which it was intended.

For guidance in manuring, therefore, the farmer has to rely largely upon practical experience as well as upon scientific formulæ. For instance, the general system of cropping pursued on the farm has to be considered,—whether the manure to be applied to the root crop has to serve for future crops, for what other

crops, and for how many years the manuring is intended to last. This, indeed, is a most important point in arranging the allowance of manure for turnips,—a point which has been fully discussed in the chapter on "Manures and Manuring." Another important consideration is the manner of utilising the crop of roots— whether they are to be in whole or in part consumed on the ground by sheep, or entirely removed.

Turnip-tops are now seldom removed from the land : they are either consumed on it, along with the roots, by sheep, or they are cut off and ploughed in when the roots are pulled. In considering the after fertility of the land, the elements absorbed by the tops would therefore not have to be taken into account. In manuring for the roots, however, the entire contents of the crop must be kept in view.

Nitrogen, Potash, and Phosphoric Acid for Turnips.—It is found, then, that crops of common turnips and swedes absorb about the following quantities of nitrogen, potash, and phosphoric acid per acre :—

	Common Turnips, 17 tons (of bulbs).		Swedes, 14 tons (of bulbs).	
	Bulbs and tops.	Per ton (of bulbs).	Bulbs and tops.	Per ton (of bulbs).
	lb.	lb.	lb.	lb.
Nitrogen . . .	112.0	6.88	102.0	7.28
Potash . . .	148.8	8.58	79.7	5.69
Phosphoric Acid . .	33.1	1.95	21.7	1.55

These yields per acre are above the average for England, and below what would be reckoned good crops in Scotland, Ireland, and the best turnip districts of the north of England. From these figures, however, it will be easy for any farmer by a simple mathematical question to form a useful *estimate* as to the quantities of these constituents of plant-food which his crops of turnips are likely to absorb. We use the word estimate advisedly, because it should be remembered that such figures as these, giving the average composition of turnips, cannot be held to represent the composition in all cases with precise accuracy. Yet by multiplying the number of tons he expects to grow by the quantities per ton

shown above, the farmer will come sufficiently near the actual facts to afford him a useful guide as to the supplies of the important constituents of plant-food which should be available to his crops.

Chief Manure for Turnips. — To judge by these analyses of turnips, one would conclude that potash and nitrogen should bulk more largely than phosphoric acid in manures for turnips. In practice, however, it is found that such is not the case. The dominant element in all special manures for turnips is phosphoric acid. It must in some form or other be applied to all soils, and in many cases constitutes the sole application for the turnip crop.

Nitrogen for Turnips.

In farm practice it has not usually been found that any considerable direct application of nitrogen has been repaid by an increase in the turnip crop. Yet it has been proved that the presence in the soil of readily available nitrogen is essential for the healthy growth of the crop.

Atmospheric Nitrogen for Turnips. —With their broad leaf-surface, turnips have by some been credited with the ability to draw a considerable quantity of nitrogen from the atmosphere. As to this question, there has long been differences of opinion among scientists, and these differences have been accentuated by reports issued by Mr Thomas Jamieson, Aberdeen, regarding his researches. The views expressed by Mr Jamieson, referred to in vol. i. p. 332, are strongly controverted by other agricultural chemists, and on the whole, for the present at any rate, one does not feel disposed to advise farmers to rely entirely upon atmospheric nitrogen for turnips.

Nitrogen in the Soil.—Practical experience has tended to show that, in most soils in good average condition as to cultivation and fertility, the turnip will find as much nitrogen as it can profitably take up. Certainly wherever a reasonable quantity of short or well-rotted farmyard manure is applied, there will be little or no need for any further application of nitrogen. On the other hand, where no dung can be spared, and where it is known or suspected that the soil is deficient in available nitrogen, the application of a small quantity may be expected to produce an increase in the crop.

Rothamsted Trials.—In summarising the results of the Rothamsted experiments upon different manures for turnips, the late Sir Henry Gilbert gave the following conclusions in reference to *nitrogenous manures* :—

1. It is entirely fallacious to suppose that root crops gain a large amount of nitrogen from atmospheric sources by means of their extended leaf - surface. No crop is more dependent on nitrogen in an available condition within the soil ; and if a good crop of turnips is grown by superphosphate of lime alone, it is

a proof that the soil contained the necessary nitrogen. In fact, provided the season be favourable, the *condition* of the land, as far as nitrogen is concerned, may be more rapidly exhausted by the growth of turnips by superphosphate than by any other crop.

2. If nitrogenous manures are used in excess—that is, in such an amount as to force luxuriance, that the roots do not properly mature within the season— there will be, not only a restricted production of root, but an undue amount and proportion of leaf.

3. Excess of nitrogenous manure tended to lower the percentage of dry matter and increase the percentage of nitrogen in the roots.

English Practice.—Notwithstanding the importance which the Rothamsted experiments place upon nitrogen for the turnip crop, it is not the rule in English practice to apply nitrogenous manures directly to turnips. As to this point, Professor Wrightson remarks : " Ammonia-salts and nitrate of soda, although producing an increase of leaves, do not greatly increase the yield of bulbs. Their effect, when applied alone on exhausted soils, is trifling ; but where there is an abundance of available mineral food, an increase is no doubt effected by their application. This increase is, however, not commensurate with the expense, and the wiser system is to employ superphosphates in root cultivation, and hold back the ammonia-salts and nitrate of soda for application on the cereals or grasses."

Experiments with Nitrogen at Carbeth.—Mr David Wilson, D.Sc., of Carbeth, Killearn, Stirlingshire, conducted various experiments on the manuring of turnips, and with regard to nitrogen arrived at the conclusion that while turnips are not, upon soils in average condition, nearly so dependent upon supplies of soluble nitrogenous manures as the cereals, it will pay in most soils, when growing them without dung, to use a little nitrate of soda, or sulphate of ammonia, say fully 1 cwt. per acre. Nitrogen, he believes, is most likely to be required for roots when the soil is deficient in organic matter, and where the climate is warm and dry.

In a review of a large number of experiments on the use of nitrogenous manures for turnips, Principal R. P. Wright states that the most effective method of applying nitrate of soda to the turnip crop is to give half the allowance in the drills at time of sowing, and the remainder as a top-dressing after the thinning of the young plants. In these trials the best results were got from nitrogenous manures by giving one-half in the form of sulphate of ammonia in the drills at the time of sowing, and the other half in the form of nitrate of soda as a top-dressing later on.[1]

Potash for Turnips.

Although potash bulks largely in the analysis of the root crop, the application of potash in the form of manure would not in all cases be followed with advantage. In most soils there are great natural supplies of potash, and, as a rule, all the additional potash required will be provided in a moderate dressing of farmyard dung. But in certain soils, notably those of a light sandy and gravelly nature, and in cases in which little or no dung is applied, it is more than probable that the addition of a small quantity of potash to the dressing of manure would be profitable.

Many instances have been observed of quite a remarkable increase in the crop from a moderate allowance of potash. These of course have taken place where all the elements and conditions necessary for the production of a large crop of roots are present excepting available potash. In manuring, the farmer should never forget the significance of the law of minimum—that law whereby the produce is limited, not by the combined quantity of all the elements present in the soil, but by the producing power of the supply of the essential element present in the smallest proportion.

Thus when potash is deficient, the application of it is followed by a marked increase in the crop.

Potash is usually most deficient in light gravelly soils in poor condition. Still it is the exception rather than the rule for land to be in need of potash for turnips. The conclusion which the majority of experimenters and observing

farmers have arrived at is, that unless there is good reason to suspect that the particular field is deficient in available potash, it need not be included in the manure, at all events when a moderate quantity of dung is used.

An Excess of Potash Injurious.— Indeed it has been found in several cases that an excess of potash has injuriously affected the yield of roots, as in the Highland and Agricultural Society's experiment referred to on p. 37 of vol. ii. At Carbeth, Stirlingshire, Dr Wilson had similar experience. Potash salts equal to 2 cwt. of kainit per acre were tried on 22 plots, alongside 22 similar plots without potash, but dressed also with dung like the other plots. The results were—

	Average of 22 plots, per acre.	
	tons.	cwt.
Dung alone . . .	21	9
Dung with 2 cwt. kainit .	20	13
Decrease due to the kainit, 16 cwt. per acre.		

Tried at Carbeth without dung on four different soils, potash gave a profitable increase in roots in only one soil. In the other cases the supply of potash already in the soil was sufficient. As to excess of potash, Mr Wilson remarks that "the mineral acids combined in these salts seem to be set free, and to do mischief to the crop." Mr Wilson therefore advises the withholding of potash, unless it is believed from actual experiment or observation that there is a deficiency of it in the particular field.

On some fields where kainit had not benefited the turnip crop, increased produce has been obtained from small allowances of potash in the form of muriate or sulphate of potash.

Test the Soil.— Here again let us urge the farmer to watch closely and *test* every year the condition of his land as to its supplies of the leading constituents in plant-food (vol. i. p. 326).

Phosphates for Turnips.

In all manures specially adapted to turnips, the dominant ingredient should be phosphoric acid. Under all circumstances, in all soils and situations, with dung and without dung, it is the almost invariable practice to furnish turnips with a phosphatic dressing in some form or other.

[1] *Trans. High. and Agric. Soc.*, 1906.

Too much Reliance on Phosphates.
—There is a tendency in some parts of
the country to place too much reliance
upon phosphatic manures alone for the
turnip crop. This should be guarded
against, for with imperfectly balanced
manuring the results cannot be fully
satisfactory. It is more than probable
that in many cases where phosphatic
manures alone are applied, the addition
of a small allowance of nitrogenous and
potassic manures would very substantially
increase the produce of the crop.

This would not likely be the case in
land which is naturally fertile and in
good heart from liberal manuring with
dung and other lasting manures in previous years. But in land in poor or
medium condition, it would be advantageous to add small quantities of nitrogenous manures and potash to the phosphates.

**Superphosphate Manuring Ex-
hausts the Soil.** — In the economical
manuring of any particular farm crop, it
is important to keep in view the after
condition of the soil—that is, the effect
which, under the dressing of manure
now applied, the crop is likely to exercise upon the general fertility of the soil.

In the manuring of turnips this consideration demands more attention than
many farmers have been in the habit of
giving to it. For it is tolerably well
authenticated that by the injudicious
— the excessive or exclusive — use of
superphosphates for turnips, the standard fertility of the soil has in many
cases been appreciably lowered.

Recouping the Soil.—The exhaustion of the lime of the soil which thus
takes place by the growth of turnips
from exclusive or excessive dressings
of superphosphates may be prevented,
or rather recouped, by the consumption
by sheep on the ground, not only of the
root crop but also of some other food,
such as cake or grain. This is extensively done in many parts of the
country, and is especially commendable
where dung cannot be spared for the
root crop. Indeed it is the rule in
many districts to consume on the land
by sheep the whole or greater part of
any section of the turnip crop which
had not received farmyard dung and was
grown solely by artificial manure.

Phosphates with Dung.—In four
years' experiments at Carbeth, Stirlingshire, the only artificial manure sown
along with dung which repaid its cost in
an increased crop of roots was superphosphate, applied at the rate of from
3 to 5 cwt. per acre. From a large
number of plots in different fields and in
different years, dressed with from 10 to
13 tons of rich covered-court dung, the
addition of 5 cwt. of superphosphate
gave an average increase of 2 tons 10
cwt. per acre in bulbs. The dung given
here was probably sufficient to supply all
the phosphoric acid required by the roots.
The increase from the addition of superphosphate is therefore attributed mainly
to its assisting the plant with easily
assimilated phosphoric acid before it
could lay hold of the more slowly acting
dung. Another advantage, often one of
great importance, is that the quickly
acting superphosphates force the plants
more rapidly past the stage in which
they are attacked by the fly.

Phosphates without Dung. — Dr
Wilson also experimented with phosphates without dung, and with and without the aid of other artificial manures.
On land at Carbeth which is evidently
above average fertility, the average produce of roots without dung or phosphates
was 7 tons 17 cwt. per acre. With
8 cwt. 25 per cent superphosphate, the
average produce rose to 17 tons 19 cwt.
—an increase of 10 tons per acre.

Cheapest Phosphate for Turnips.—
An important question, as to which there
is a good deal of difference of opinion,
is that of the most economical form of
phosphate for the turnip crop.

From 1840 to 1870, Peruvian guano
and roughly crushed bones, with the
occasional addition of dissolved bones
and superphosphate, were the manures
chiefly employed to supply the nitrogen
and phosphates to the turnip crop. Since
the Chincha Island deposit of guano
became exhausted, the other deposits,
being inferior in ammonia and high in
price, are comparatively little used.
Crushed bones, more finely ground than
formerly, are still in much repute for
turnip manure in all light soil districts;
while on the heavier soils dissolved bones
and, still more, superphosphate, have become the general manures.

Mineral Phosphates.—Notwithstanding some opinions to the contrary, carefully conducted experiments have shown that phosphates from mineral sources, such as rock guano, coprolites, Carolina phosphates, and the phosphate from basic slag, when finely ground, act on the turnip crop more quickly than the phosphate in finely crushed bones, while the mineral phosphates can usually be bought at a much less price per unit. So long as mineral phosphates are cheaper the farmer in favourable circumstances as to soil and climate may do well to use them, in part at least, in place of crushed bones—taking care that *the grinding is as fine as possible*, and avoiding all the phosphates of alumina, and the crystalline apatite, which latter should always be dissolved before application.

Discrimination in use of Mineral Phosphates.—To use undissolved mineral phosphate successfully as a turnip manure, the farmer must exercise not a little discrimination. Some mineral phosphates will give excellent results in one soil, while in another soil, not very different in appearance, the effects will be disappointing. An exception is phosphatic slag, which is almost invariably effective on light and medium soils. The various forms of phosphate are fully described in the section on "Manures and Manuring," vol. i. p. 493.

Superphosphates.—Superphosphate of lime, the characteristics of which are fully discussed at p. 501, vol. i., is extensively used as the source of phosphoric acid for turnips. In great parts of England, where the soil and climate are dry, it is indeed almost the only form of phosphates now used for turnips along with dung. In many cases it has been found the most economical form of phosphatic manure for this crop, producing a heavier yield than the same value of crushed or dissolved bones.

At Carbeth, Dr Wilson compared superphosphates with equal money's worth of ground Charleston phosphate. He obtained in four years an average of 10 per cent more weight of bulbs from the superphosphate than from the ground mineral phosphate. Dr Wilson also contrasted the superphosphate with Thomas slag. The results again were in favour of the former, at the existing prices of the two articles. Dr Wilson likewise considered the phosphates in guano dearer than those in superphosphate; but in contrast with the same value of *steamed bone-flour*, the superphosphate failed at Carbeth to sustain its supremacy. Steamed bone-flour mixed with superphosphate produced 13 cwt. more per acre than an equal money value of superphosphate alone. Dr Wilson adds that, making allowance for the nitrogen contained in steamed bone-flour, more phosphoric acid is got for the same money in this form than in superphosphate.

Along with dung, Dr Wilson prefers superphosphate (mainly for its quick action) to all other forms of phosphates. Without dung, he would provide the phosphates in a mixture of steamed bone-flour and superphosphates.

The Aberdeenshire experiments, described on pp. 51 to 56, vol. ii., have a very direct bearing upon this point. Note in particular what is said (p. 54) as to the influence of phosphates rendered soluble by sulphuric acid upon the tendency to "finger-and-toe," and as to fineness of grinding or perfect disaggregation (pp. 53 and 55) being as effective as dissolving in sulphuric acid.

Climate and Soil to be Considered.—In deciding as to the form of manure used, the characteristics of the climate and soil must be carefully considered. As to this point, Mr John Milne, Mains of Laithers, Aberdeenshire, who is a practical chemist as well as an extensive experimenter and successful farmer, remarks:—

"In cold wet districts, or if the crop is late in being sown, the quantity of soluble phosphate should be increased, as its effect is to force the crop to early maturity. In these circumstances, if farmyard manure is applied, little or no nitrogenous manure should be used, as its tendency is to keep the crop growing longer, and thus retard its maturity. Undissolved mineral phosphates always act best in warm early seasons, and do not show quite so well as soluble phosphates in cold wet years.

"In manuring, the farmer should be guided by the quality of his soil, the period of sowing, and probable character of the weather. If his soil is rough or stiff, the sowing late, or the climate cold or wet, a pretty large proportion of soluble, precipitated, or very finely ground phosphate is advisable. If the soil is soft, the season early, and the climate dry, the phosphate need not be so finely divided, and a larger proportion of nitrogen may be beneficially used."

Basic Slag for Turnips.—This important fertiliser (described in vol. i. p. 499) is used with excellent results as a source of phosphates for turnips. It is specially suitable for soils deficient in lime.

Farmyard Manure for Turnips.

Throughout the country generally farmyard dung is the standard manure for turnips. It is the rule—which, however, has a good many exceptions—to apply the whole or the greater portion of the farmyard dung to the potato, turnip, and mangel crops. The prevalence of the practice is a tolerably sure indication that a dressing of dung is well suited to the turnip crop.

Supplementing Dung.—But while a dressing of dung is highly beneficial to the turnip crop, it may be found advisable to supplement it with some more quickly acting fertilisers, such as superphosphate or slag, nitrate of soda, and potash. Much will, of course, depend upon the condition and quality as well as the quantity of the dung. Well-rotted dung acts more quickly than fresh dung, while if it has been enriched by the consumption of concentrated foods, it will be still more efficacious. It is highly important that the plants be pushed forward rapidly in their first few weeks, so that they may get beyond the ravages of insects. For this purpose a dressing of some quickly acting phosphatic manure will be a valuable supplement to the more substantial but slower farmyard dung. As we have seen, superphosphate or a mixture of very finely ground mineral phosphate and steamed bone-flour will likely be most suitable. When dung is applied to soils in good condition, only a small quantity of any readily acting phosphate is required, but when quick growth is wanted superphosphate will serve the purpose very well.

Is Dung Essential in Turnip Culture?—This question has been much discussed. It is still the subject of difference of opinion. Many noted agriculturists, including Professor Wrightson, contend that good crops of swedes cannot be grown without dung. Others hold that it is not by any means essential, and that better results will be obtained by applying the dung to other crops, such as potatoes, or on pasture or meadowland, and growing the turnips entirely or mainly with substantial artificial manure. It is going too far, we think, to hold that swedes cannot be grown advantageously without dung. As a matter of fact, good crops of swedes *are* grown without dung ; and the feeling is gaining ground that some proportion of the excessive dressings of dung which are often applied to swedes might be more advantageously utilised for other purposes.

Assuredly it is most desirable that a substantial dressing of good farmyard dung should be available for swedes. It is the best foundation of all for a successful crop ; and, as a rule, it will be found the safest practice to devote the main portion of the dung to the swedes. But while dung is probably necessary to ensure a maximum crop of swedes, it is not absolutely essential for the production of a profitable crop. In some cases it may be desirable to grow a greater breadth of swedes than the available supply of dung will cover ; and this may be done by the use of artificial manures. Generally, however, it is deemed prudent to substitute yellow turnips when the dung becomes exhausted.

It would be unnecessarily restricting the operations of the educated and skilful farmer to tell him that he must not attempt to grow swedes without farmyard dung.

The softer varieties of turnips are grown very extensively, and with great success, without the slightest particle of dung,—great care, skill, and liberality being of course necessary in these cases in the use of artificial manure, so as to maintain the fertility of the land. Some farmers consider it imprudent to apply dung in drills for common turnips in light open soils, believing it to be better in this case to use artificial manures only, and to eat off the roots wholly or partially with sheep. This is a specially good plan for light land in outlying parts of a farm. Unless the turnips are consumed on the land by sheep, it will most likely be necessary to top-dress some of the other crops which follow upon the land which received no dung for the roots. In particular, an autumn top-dressing to young grass would be advisable.

Quantities of Manures for Turnips.

The quantities of manure applied to the root crop vary greatly throughout the country. The ruling influences are the climate, the natural character of the soil, its condition as to accumulated fertility or exhaustion, the purposes for which the roots are intended, and the general system of farming pursued.

Yield and Quantity of Dung.—The consideration which most largely regulates the amount of manure—that is, where the objectionable practice of applying all the manure for the rotation with the root crop has been abandoned—is the suitability of the district and the field for the production of a heavy or light crop of roots. Where a crop of 25 to 30 tons per acre is to be looked for, the allowance of manure must, as a matter of course, be much larger than where the yield is not likely to exceed 12 to 15 tons. These figures roughly represent the respective yields of the best turnip-growing districts of Scotland, Ireland, and the north of England, and in the midland and southern counties of England, and thus in the latter the prevailing quantities of manure applied are much less than in the Green Isle and north of the Humber.

The general questions to be considered in deciding as to the quantities of manure for the various crops have already been fully discussed in the chapter on "Manures and Manuring," vol. i. pp. 446-520. See in particular pp. 486-520. Here, therefore, a very few notes as to the prevailing customs will suffice.

Scottish Dressings.—In Scotland, in the north of England, and in Ireland, the allowances of dung vary from 5 to 20 tons per acre, and the accompanying dressings of manure from 3 to 8 or 10 cwt. of phosphatic manures, ½ to 3 cwt. of nitrogenous manures, and ½ to 3 cwt. of potash salts. More general quantities of dung run from 8 to 15 tons. Along

with from 10 to 12 tons of dung, from 3 to 5 cwt. of phosphatic manures, 1 cwt. of nitrate of soda or sulphate of ammonia, and 1 to 1½ cwt. of kainit, would be a liberal dressing. For swedes some farmers give as much as 12 to 14 tons of good dung, 4 cwt. of mineral superphosphate, or an equivalent of basic slag, 2 to 3 cwt. crushed or dissolved bones, 1 cwt. of nitrate of soda, and 1 cwt. of kainit. Others curtail the artificial manure to about 3 or 4 cwt. superphosphate, or an equivalent of basic slag, ½ cwt. of nitrate of soda, and ½ cwt. of kainit. Often the two latter are omitted altogether; still more often the potassic manure is omitted, and the small allowance of nitrogenous manure included.

Advantage of Heavy Dressings Questionable.—Several of these dressings of artificial manures along with dung are assuredly very heavy. Many careful and successful farmers are doubtful as to the economy of such liberal and costly additions to the supplies of dung. By his carefully conducted experiments at Carbeth, Stirlingshire, Dr Wilson was led to the conclusion that the usual practice in many turnip-growing districts of expending from 30s. to £2 per acre upon artificial manure, to apply along with dung, is not a profitable one, and that in many of these cases half the rent of the land might be saved by reducing this outlay.

Certainly the once practised method of applying manure—dung, bones, and guano—to the turnip crop to serve for the entire rotation, has been exploded as thoroughly unsound. The allowance of dung for the rotation may of course be, and is still, applied to the roots, and with good effect; but with the artificial manure the case is entirely different. In regard to these, it is a safe rule to apply no more at any one time than you expect the first crop will profitably utilise or repay. A reasonable exception to this rule would be a dressing of crushed bones, particularly for grass land.

Moderate Dressings of Dung.—When the supply of dung is not sufficient to go over the entire root break, it is a good plan to lessen the allowance per acre, and make the dung go as far as possible, increasing the quantity of artificial manure in proportion. Better far give 8 tons to the entire break than 12 tons to a certain portion, and none to the remainder—better especially for the after fertility of the land.

Artificial Manures alone.—When no dung can be spared, the allowance of artificial manures has to be very liberal. In some cases the allowance is as high as from 5 to 6 cwt. superphosphate, 2 to 3 cwt. steamed bone-flour or crushed or dissolved bones, 1 to 2 cwt. of nitrate of soda, and 2 to 3 cwt. of kainit, or an equivalent of muriate or sulphate of potash. In other cases, again, from one-half to two-thirds of these quantities are supplied, the potash often being omitted altogether. In many cases superphosphate at the rate of 8 to 10 cwt., and 1 to 2 cwt. of nitrate of soda, constitute the sole dressing. Others use a portion of finely ground mineral phosphate, and basic slag is now largely used as the phosphatic manure.

But the variations in the individual dressings are so numerous that it would be impossible to fairly represent them here.

Southern Dressings.—The most general dressing in England, where a crop of from 12 to 18 tons is expected, is from 8 to 12 tons of dung and 3 cwt. of superphosphate per acre. A small allowance of guano or nitrate of soda, from ½ to ¾ cwt per acre, is often drilled along with the superphosphate and the turnip seed, but this plan is regarded by many leading authorities as unprofitable.

Necessity for Individual Judgment.—In arranging the quantities of manure for turnips, as in most other farm operations, the circumstances of each individual case must be carefully considered. General rules are subject to many variations, which each farmer must decide upon for himself. A careful study (aided by a few experiments, which should always be going on) of the condition of the soil and its capabilities under favourable circumstances as to fertility will be the safest guide as to the most profitable quantities of manure to apply. It is a point in farm management which demands the very best attention from the farmer.

Application of Manure for Turnips.

The general methods of applying manures, and the principles upon which these should be regulated, have already been dealt with (vol. i. pp. 511-520). What is said there should be carefully studied in connection with the culture of turnips.

Dung.—As to the merits and demerits of the various practices of applying dung in the autumn, and on the flat surface, and in the drills in spring, enough has been said in the pages just referred to.

Upon heavy lands where the dung is available in time, the best and most general practice is to plough down the dung with a shallow furrow in the autumn or early in winter.

Where this has not been done, and where the turnips are to be sown on the flat surface, the dung is spread on the flat surface and ploughed down with a moderate furrow early in spring. Late dunging in this case is not to be commended, as the rank dung would be liable to unduly encourage the escape of moisture by keeping the surface-soil open.

The general practice where the turnips are grown in raised drills, is to spread the dung in the bottom of the drills at the time of sowing the seed; yet, as just explained, if the land is stiff and the dung available, autumn dunging, even with sowing in raised drills, is in many cases a beneficial method. It lessens work at sowing-time, and the dung helps to disintegrate the adhesive soil.

Carting Dung into Drills.—The old-fashioned method of emptying the dung from the carts in small heaps in every third drill is still in vogue in some parts. As a rule, however, it has long since given place to the much more expeditious and economical plan of throwing the dung in graipfuls from the cart into the drill as the horse moves along. A careful workman distributes the dung in this manner with admirable precision as to quantity, and it is left so as to make the work of the spreaders comparatively easy. The spreading of the dung is rendered still easier if the carter throws the graipfuls into the side drill (next to the drills already dunged), so that the wheel of the next cart may not go over the graipfuls, which would be the case if the dung were, as is often the case, thrown into the drill in the centre of the cart. With short well-made dung thrown out in this way we have often seen two smart women spread as fast as one team with a drill-plough could open and close in.

Cart for Steep Land. — Ordinary farm carts are employed in carting out dung. In vol. i. p. 455, information

Fig. 606.—*Farm tip-cart.*

which should be consulted at this stage is given as to the position of dunghills and carting out dung. In steep land, when a load has to be conveyed down-hill, a cart similar to that shown in fig. 606, made by the Bristol Waggon Works Company, will be found useful. It is a

Fig. 607.—*Tip-cart going down-hill.*

tip-cart, with screw arrangement, whereby the load may, as shown in fig. 607, be raised off the horse's back.

Dung - spreading Apparatus.—Appliances have been invented for spreading the dung as it is thrown from the cart. A very useful apparatus of this kind is that shown in fig. 344, vol. i. p. 513, by which well-made dung is spread in drills even better than is possible by hand. This machine is attached to the rear of the cart, and is fed with the dung by the carter. It scatters the dung by its revolving prongs. Still, in most cases, dung is spread by the hand-fork or graip.

Sowing Artificial Manure. — The artificial manures are often sown either by hand or by drill or broadcast machines just before the drills are closed in. Manure-sowing machines are shown in figs. 345 and 346, vol. i. p. 520. It is a good plan to run a light harrow along the drills after the dung is spread, and before the artificial manure is sown. In some cases, instead of harrows, a long heavy pole is drawn over the drills. This helps to keep the quickly acting artificial manure nearer the rootlets of the young plants, and likewise still further pulverises the seed-bed. When there are many clods, some roll the drills before sowing the artificial manure. The manure is sown along the drill rather than broadcast, and may be done so quickly by a two-hand sower that one man will keep two drill-ploughs going, and supply himself with manure from the bags or carts deposited at the ends of the drills.

Southern Customs.—In England, wherever the turnips are sown in rows on the flat surface, the artificial manure is generally drilled in along with the seed with the dry or water drill, as already explained. A better method is the use of a machine, which at one operation sows the manure in two rows on the flat and raises drills over the rows, thus avoiding the bringing of the seed into direct contact with the manure. In many cases in England all the artificial manures are sown broadcast just before the seed is sown, although with a light application sown in this way the crops grown in rows on the flat are often disappointing. Broadcasting artificial manures is more satisfactory with raised drills, as in this case the scattered particles of the manure are gathered towards the plants by the operation of the drill-plough.

For the flat-row system the best plan perhaps is, where the artificial dressing consists entirely of superphosphate, to drill the whole of it along with the seed, and where other manures as well as superphosphates are given, to drill the greater portion of the superphosphate along with the seed, and sow the remainder with the other manures broadcast, and harrow in, following with the roller.

Kainit is in many cases found to give the best results when sown in the preceding autumn.

Top-dressing Turnips.—The practice of top-dressing turnips is rarely pursued. If nitrogenous manure is required, some consider it a good plan, especially in wet climates, to hold it back till the plants are about ready for singling, and then apply it in the form of a top-dressing of nitrate of soda.

Experiments in the North of Scotland.

Aberdeen and North of Scotland College of Agriculture has carried out a considerable number of experiments on the manuring of the turnip crop. As has been found in most other Scottish experiments on this crop, under ordinary conditions of practice available, phosphate is by far the most important artificial manure for the turnip crop. Under average conditions, and especially where dung is used, nitrogenous manures have comparatively little effect, while potassic manures come intermediate. On the average they produce a greater increase of crop than nitrogenous manures, but their effect depends much upon the nature of the soil, and is very variable.

Fine Grinding in Manures.—A series of experiments upon the effect of fine grinding in the case of bones and other insoluble phosphatic manures showed similar results to those obtained in the experiments of the Highland and Agricultural Society on the same subject—namely, that in order to obtain a rapid and remunerative return for bones used as manure for the turnip crop, the bones must be finely ground, and that the more finely they are ground the better is the return they give. In particular, the experiments brought out that steamed bone-flour when used so as to give an equal weight of phosphate gave a better return than even the finest commercial bone-meal. Steamed bone-flour is more finely ground than even the finest bone-meals, and its superior action could only have been due to its superior fineness of grinding. Otherwise it was at a disadvantage compared with bone-meal, as it contains both less organic matter and less nitrogen. The dressings of bone-flour had the further advantage of costing less than the bone-meal.

Basic Slag *versus* Superphosphate.—In two series of experiments the composition and feeding value of turnips manured with basic slag were compared with those of similar turnips manured with superphosphate. The feeding value was determined by feeding experiments with cattle. Very little difference in quality was found between the two different lots of turnips. Certainly the slag-manured turnips were not found to be of any poorer quality than those which received superphosphate.

Experiments in the North of England.

Experiments conducted by the Durham College of Science, Newcastle-on-Tyne, have given better returns from dung when applied for roots in spring than when applied in autumn or winter. They have also shown that large dressings of dung are not so profitable for roots as moderate dressings of about 10 to 12 tons per acre.

A striking result in these trials has been that dung alone has frequently given nearly as good results in the crop of swedes as dung and accompanied by artificial manure, though it has been noticed that when the effects upon the other crops in the rotation are taken into account the combination is distinctly profitable. A good average dressing for swedes and common turnips was found to be from 10 to 12 tons of farmyard dung, 4 cwt. basic slag, 1½ cwt. superphosphate, and ¾ cwt. sulphate of ammonia per acre.

When artificial manure was used alone the following dressing did well for turnips of all kinds: 4 cwt. basic slag, 1½ cwt. superphosphate, ¾ cwt. nitrate of soda, 1 cwt. sulphate of ammonia, and 1 cwt. of muriate of potash. All these manures were usually applied in the drill just before the seed was sown. In some cases it is thought well to apply the nitrate of soda as a top-dressing in moist weather two or three weeks after hoeing.

Experiments in the South of England.

At different centres in the southern counties of England trials have been made in the manuring of the turnip crop. An exhaustive series of experiments, conducted on about thirty farms by the University College, Reading, brought out suggestive results.

It was found that the omission of phosphatic manures made it impossible to obtain remunerative crops of swedes. A small allowance of nitrogen was always beneficial, and so was potash, at about half the centres. A general conclusion was that moderate dressings were the most profitable.

Manures Preventing Blanks. — A striking result in these trials was the influence which the manures had upon the regularity of the plants. Where the land was unmanured, or only received small dressings, many gaps were met with in the rows. On the other hand, where the dressing of manures was fairly complete very few plants failed, the crops being remarkably even in growth. The presence of readily available manure close to the germinating seed greatly assists the seedlings in their early growth, and helps to overcome the attacks of fungi and insects as well as the influence of drought.

Welsh Trials.

A large number of experiments in the manuring of swedes have been conducted by the Agricultural Department of the North Wales University College at Bangor. The objects were to ascertain—

(*a*) The comparative returns from moderate and large dressings of farmyard manure.

(*b*) The extent to which it is profitable to supplement farmyard manure with artificial manures.

(*c*) The best forms and quantities of artificial manures for use: (1) alone, (2) along with farmyard manure.

The results of the trials showed that, in average circumstances, in North Wales a fair dressing of farmyard manure, say, 10 or 12 tons an acre, will give a profitable return. As a rule, if more than this be given, the extra manure will not pay for use. With such a dressing (10 or 12 tons an acre) artificial manures will only pay if used in small quantities, not more than 3 or 4 cwt. an acre, most or all of which should be superphosphate, or, in special cases, basic slag. This

combination, if climatic and other conditions are favourable, can be relied on to produce a full crop, and in several cases crops of over 40 tons an acre have been grown with it.

Artificial Manure alone.—The most profitable crops have, as a rule, been those grown with artificial manures alone, and not only have they been the most profitable, but, on the whole, have been little inferior in actual yield per acre to those grown with farmyard manure, either alone or along with artificials.

In this connection it is necessary to point out that—particularly in Anglesey and Carnarvonshire, where the majority of the root experiments have been carried out—the rotation usually followed includes grass left down for five or six years, so that, apart from any dressing of farmyard manure, there is always a fair amount of organic matter left in the soil for the succeeding root crop.

The most profitable mixture of artificial manures for use without farmyard manure has generally been a moderate quantity (4 or 5 cwt.) of superphosphate, together with 2 or 3 cwt. of kainit per acre.

Nitrogenous manures, as a rule, have barely paid for use even when no farmyard manure has been given.

In some cases, chiefly on the heavier soils, basic slag has given better results than superphosphate, and in most cases it has been only slightly, if at all, inferior to an application of superphosphate of equal money value, even when applied in the drills in late spring.

Dissolved bones, fish-meal, basic superphosphate, and bone-meal, have in most cases given fairly good returns, but not sufficiently so to justify the use of these manures in preference to superphosphate or basic slag at current market prices.

Similar trials in the manuring of swedes were conducted by the Agricultural Department of the University College of Wales at Aberystwyth, and, on the whole, similar results were obtained. In this case the manuring was suited to soil of medium fertility on the Silurian formation. Here 15 tons of dung without any artificial manure did better than 10 tons, the 5 additional tons of dung raising the yield by about

2 tons per acre. With a small dressing of artificial manures (3 to 6 cwt. of superphosphate, ½ to ¾ cwt. sulphate of ammonia, and about 1½ cwt. potash), 10 tons of dung per acre gave a profitable return in swedes. Basic slag did almost as well as superphosphate. An application of potash manure seemed necessary when no dung was applied. Without dung the best yield was got from 6¼ cwt. superphosphate, 3 cwt. kainit, and ⅞ cwt. of sulphate of ammonia per acre.

SINGLING AND HOEING TURNIPS.

The turnip crop requires prompt attention at the time of the singling of the plants and the hoeing of the drills.

Influence of Weather.—The seed-leaves usually appear in from three to seven days after sowing. The plants grow rapidly in fine dry weather, if the nights are free from frost. Until the plants are of considerable size, heat and dryness favour their growth, while at this stage much rain is not favourable.

Turnips should be singled when the leaves measure about an inch across.

Drill-harrowing or Horse-hoeing.—But before singling or hand-hoeing is commenced, operations may be performed which will make the labour of hoeing more easily performed, and by further loosening the soil tend to promote the growth of the plants. If the weather is dry, the drills should be run between by a drill-cultivator or scuffler, let in as

Fig. 608.—*Drill cultivator.*

deeply as possible. But the width stirred should not exceed twelve inches, for if set wider the land will be too much drawn away from the plants before the process of singling is finished, and the raised drill too much reduced.

Fig. 608 represents a type of a drill-cultivator made by T. Hunter, Maybole, and fig. 609 Martin's drill horse-hoe,

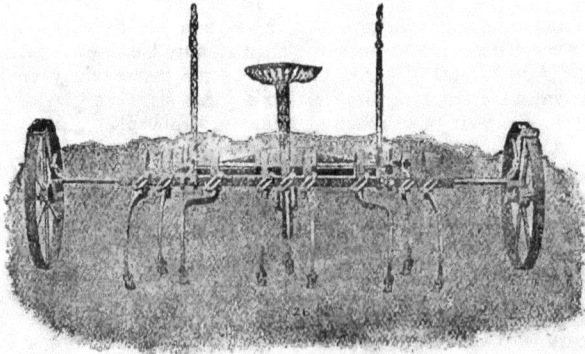

Fig. 609.—*Drill horse-hoe.*

which can be adjusted to work either in drills or in rows on the flat.

Harrowing across Flat Rows.—In the south a sort of drag-harrow, in some cases similar to a light Scottish drill-harrow, is drawn right across the flat rows before the first horse or hand hoe-ing, the object being to loosen the sur-face-soil, pull out surface weeds, and

thin out the plants a little. Careful turnip-growers in the north do not ap-prove of disturbing the plants thus early and in such an irregular fashion.

Drill-Scarifier.—A drill-scarifier such as that made by T. Hunter, Maybole, shown in fig. 610, is used largely and successfully for paring away the sides of the drills, thus destroying weeds, and bringing the drills into the intended form, leaving less work for the hand-hoe.

Thinning - machines.—The singling of turnips by machine has been found to be a very difficult task, and complete success has not yet been attained. Still,

Fig. 610.—*Disc drill scarifier.*

machines have been brought out which save a certain amount of labour by partially thinning the plants, rendering easier the perfect thinning by hand.

Hand-hoes.—The hand-hoe used in thinning turnips is a simple instrument. Yet even in it improvements have been introduced in recent times. Instead of the shaft or handle being closely

Fig. 611.—*Improved hand-hoe.*

attached to the blade, it is now often made with a bow-shaped attachment, as shown in fig. 611. A hoe of this pattern works more lightly and cleanly than the hoe of the old shape, shown in fig. 612. The length of the blade of the hoe varies

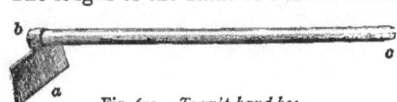

Fig. 612.—*Turnip hand-hoe.*

a Thin iron plate.   b Eye of plate.   c Wood shaft.

from 5 to 8 inches, according to the width usually left between the plants. Excellent hoes can be made cheaply from disused scythe blades.

Process of Hand-hoeing on Raised Drills.—As already mentioned, the turnips should be thinned when they measure about an inch across, when the tops are well into the rough or second leaf.

The hoer ought to be taught to draw the hoe towards himself or herself in pulling out the spare plants, and to work as lightly as possible. If the plants are pushed away from the hoer, a deeper hold of the soil must be taken, a greater quantity of soil will be removed from the remaining turnips, the drills will be more pitted and levelled, and the plants thus too much denuded of support. Hoers generally take pride in their work, striving to leave the drills as high and symmetrical and as smooth in the surface as possible, with all weeds thoroughly removed—uprooted, not cut—and the plants thinned to precise distances as arranged, care being taken to leave strong well-formed plants, and never two together.

Hoeing-matches.—In some parts of the country hoeing-matches in the evening are quite an institution, and there is often great enthusiasm amongst the rival hoers—the farmers' families and servants of the surrounding district. These friendly contests are very properly encouraged by farmers, for they stimulate tasteful and careful hoeing, which in turn has a considerable influence upon the yield of the crop—far greater than would at first thought be imagined.

Good and Bad Hoeing.—It is quite within reason to say that the difference in the yield between a carefully hoed piece of ground—hoed as we have indicated above—and another hoed carelessly, with irregular intervals between the plants, weak plants left instead of strong, two plants sometimes left together, and the drills cut deeply into, and weeds only partially removed,—in short, between good and bad hoeing,—may very easily amount to from 2 to 4 tons per acre !

Hand-hoeing in Flat Rows.—Here, also, the plants should be drawn towards the hoer. Indeed, as will be readily understood, the great part of the hand-hoeing on the flat must be done in this way, as it is more difficult to push out weeds in the flat row than on the raised drills.

Speed of Hoers. — The amount of work done by hoers varies according to the soil, the width of the drills or rows, the intervals left between the plants, the thickness of the seeding, and the stage at which the hoeing is done. If the soil is clean in raised drills, the plants not too thick, and taken at the proper size, an average hoer should overtake an imperial acre in from twenty-five to twenty-seven hours. If circumstances are very favourable, it may be done even in twenty hours; and if very unfavourable, it may take forty to forty-five hours to single an acre. If the drills are well scarified, the work is much lighter.

Expert men-hoers often go over the ground almost as quickly in the flat-row system of the south, where the rows may be only from 18 to 20 inches apart. But it would be all the better for the crop if a little more pains were taken with the hand-hoeing than is often the case.

In Scotland, Ireland, and the north of England, women do a large portion of the hoeing; but in the midland and southern counties of England it is performed almost entirely by men and lads.

Thinning by Hand.—In some cases when a greater breadth of plants comes forward at one time than can be gone over with the hand-hoe as quickly as may be considered desirable for the sake of the crop, thinning by hand is resorted to. This is an expeditious method of averting injury by the overcrowding of the young plants. In the long-run, however, it increases the cost of thinning and hoeing.

The better system of management, therefore, is to have a sufficient force of hoers to overtake the thinning as the plants become ready for the process. In average seasons the sowing is done so that the plants come forward to the hoe in breaks; but irregularities in the weather may upset this arrangement, and result in a pressure of work at certain times in the hoeing season, perhaps justifying recourse to hand-thinning if an extra force of hoers cannot be obtained. In any case, some farmers, who are particularly careful of their turnip crop, would give the preference to the hand-thinning, because by it a little more care can be exercised in leaving the strongest plants. At the same time weeds are usually more thoroughly dealt with by the hoe.

In some parts the thinning is done partly by the hoe and partly by hand. The hoers go on before, taking gaps out of the row of plants, leaving little bunches of perhaps three to half a dozen plants, while lads and women follow, and single these bunches by the hand, taking care to leave in the strongest and most promising-like plant in each bunch.

Transplanting Turnip Plants.— Common turnip plants cannot be transplanted with success. With swedes, however, transplanting is often done, to fill up blanks in the drills. The results are fairly satisfactory, sometimes yielding nearly half the weight of an average bulb.

Distance between Plants.

There has been much discussion, and there is still wide difference of opinion, as to the distances which should be left between turnip plants. The prevailing practice in this matter has undergone many modifications and alterations since the introduction of turnips as a regular field crop throughout the country generally.

Results from Short Intervals.—The late Mr Stephen Wilson, North Kinmundy, Aberdeenshire, conducted exhaustive trials with turnips thinned to different widths on seven separate farms, and in reference to the results he stated that such uniformity was shown as hardly left any doubt that in general 6-inch intervals will ensure a heavier crop of swedes or of common turnips than either 8- or 9-inch intervals. Indeed he satisfied himself, after all his trials with many sorts of turnips, in favourable and unfavourable seasons, under ordinary rotation of cropping, that 6-inch intervals will give a heavier crop than any wider interval.

Prevailing Intervals.—These experiments by Mr Wilson, and other more limited trials, no doubt tended to shorten the intervals left between turnip plants in certain districts. The prevailing intervals are still, however, considerably wider than Mr Wilson advised. Where the system of raised drills obtains, the intervals most general are from 9 to 10 inches in the cases of swedes, and from 7 to 9 inches for common turnips. A good deal depends upon the known habit of the particular variety of roots, whether it is inclined to develop large or medium bulbs. The soil and climate must also be considered, for under conditions which favour the growth of large roots the intervals should be longer than where small roots are expected.

The space between the plants should of course vary with the width of the drill, or between the rows of plants if the crop is grown on the flat surface. The most general width of the raised drill is 27 inches, and, as will be readily understood, the plants may be left nearer each other in these wide drills than in the much narrower flat rows which abound in the midland and southern counties of England. These flat rows are usually only from 16 to 20 inches apart, and so the intervals between the plants there most frequently vary from 13 up to 16 inches for swedes, and about 2 or 3 inches less in the case of other varieties of turnips.

Growing Roots in Squares.—There is little doubt that the maximum weight per acre of roots would be obtained by growing them at equal distances apart in all directions, in squares of one foot or 14 inches for instance. Indeed it was found by experiments in Canada that a better crop resulted from placing the plants in the centre of a square unit than in the middle of an oblong unit, as in the case of common drilling.

Advantages of Drills.—But there is a practical advantage in the drill and row systems which far outweighs any loss in the produce of roots. The cleaning and tilling of the land are facilitated, and thus by growing them in tolerably wide rows or drills the root crops take the place of the costly "fallows" of olden times.

Medium and Large Roots.— One important point which should be kept in view in discussing and deciding as to the best intervals to be left between turnips is the ascertained fact that, as a rule, medium-sized bulbs show a higher specific gravity and contain a greater percentage of useful feeding material than exceptionally large-sized roots. This is the case in a marked way with the common varieties of turnips; it is slightly less marked with most kinds of swedes. "Large" and "small" are comparative terms. It is claimed for some of the improved varieties of swedes that large

bulbs are more nutritious than small ones. What is meant in this case of course is, not roots of abnormal dimensions, but what the practical farmer would regard as large roots grown under normal conditions.

The object of every farmer should certainly be to grow a big—that is, a heavy—root in relation to the space allotted to it. What has been taught by investigations as to the nutritive properties of roots of different sizes is not that small varieties of roots should be cultivated, but that the maximum quantity of good feeding material per acre is more likely to be obtained by growing (at shorter intervals) a greater number of medium-sized roots than a smaller number (at longer intervals) of abnormally large roots—this, too, even although in both cases the gross weight of the produce may be equal. In other words, three medium roots—"big-little" roots—weighing each 3 lb., and grown in, say, 30 inches of an ordinary drill, will, as a rule, contain less water and more solid nutritive matter than two bulbs of $4\frac{1}{2}$ lb. each, grown in the same area of ground.

Moderate Intervals.—The teaching of modern investigation is therefore decidedly in favour of shortening the intervals between the turnip plants. For common turnips from 7 to 9 inches should perhaps be the range in drills from 26 to 28 inches wide, and for swedes 1 or 2 inches more. In flat rows from 16 to 20 inches wide, suitable intervals would be from 9 to 11 inches for common turnips, and from 11 to 13 inches for swedes. In dry seasons favourable to mildew the wider intervals will likely give the best results. The free exposure of the plants to the atmosphere, as in wide singling, has a tendency to check the development of mildew.

Irregularity in Growth of Turnips.—Notwithstanding every care taken to single the plants at equal distances, it will usually be found after the crop has made some progress in growth that irregularities appear both in the distances apart and in the size of plants. Unless the seeding is very liberal, plants are apt to appear and grow somewhat irregularly, especially in dry weather. A good hoer will strive to leave a strong plant, even if an extra inch or two beyond the distance intended. After hoeing, plants are occasionally pulled up by crows and wood-pigeons, and cut across by wireworm and grub-worm. Some of the plants receive other injuries which prevent growth. The smaller and more backward get shaded and overtopped. The available supply of manure is appropriated by their more vigorous neighbours. Manure, especially dung, is not always so evenly spread as it ought to be, and it often happens that an average field will show not a little irregularity both in size of bulbs and distance apart. Indeed it is only on the most fertile and easily pulverised soils, and under the most favourable circumstances of soil and climate, that the bulbs approach equality of size and regularity of distance apart.

After Cultivation.

The cultivation required by turnips after the singling has been completed consists of hand-hoeing once or twice, and horse-hoeing between the rows of plants two, three, or more times. The season and condition of the land as to weeds and tilth will regulate the number of hoeings.

About ten or fourteen days after singling, the horse-hoe or drill-harrow is run along the drills or between the rows of plants, to stir up the soil and eradicate weeds. The second hand-hoeing may follow in a few days, the hoers removing all weeds left by the horse-hoe or drill-harrow, and loosening, but not displacing, the earth around the plants. If in any case two plants have been left together, in singling one should now be carefully pulled by hand.

Care in Hoeing Strong Plants.— It is no doubt beneficial to stir the soil around the plants even after they have grown almost to cover the drill with their tops. In this operation, however, the greatest care must be exercised not to cut the rootlets, which are now spreading like net-work in all directions, and which cannot be cut or seriously disturbed without less or more injury to the crop. For this reason the third hand-hoeing is often abandoned.

Earthing-up Turnips.— It is sometimes found beneficial, chiefly on wet soils, to earth-up turnips immediately after the second hand-hoeing. The main advantage of this is, that surplus surface-water is carried away more freely. In dry soils, however, the earthing-up may do more harm than good. Some of the rootlets may be cut or injured by the plough, and their development thus impaired. Then the sharp, deep furrows are troublesome, even dangerous, in case of sheep feeding on the roots, as sheep may get upon their backs in the ruts, and perish if not released in time.

If earthing-up is to be done at all, it should be carried out as soon as possible after the second hoeing. The younger the plants the less will be the injury or disturbance to the rootlets. But the earthing-up of turnips is neither a general, nor, as a rule, a commendable practice, except for the protecting of the roots from frost, and then it is done after growth has ceased.

Turnip Pests.

The pests of various kinds which prey upon the turnip crop are dealt with separately in this volume.

STORING TURNIPS.

The turnip crop is not only an expensive one to grow, but the amount of attention it requires in its different stages to secure a satisfactory yield is very great. It is therefore desirable that when it has been grown every effort should be made to turn the produce to the best possible account. The first step towards this end is the timely storing of the roots.

Advantages of Storing Roots.— The advantages which arise from the storing of roots are manifold. Chief amongst these are, the preservation of the crop from the effect of the frosts and thaws of winter, the procuring of a regular supply of fresh and clean food for the animals upon the farm, the prevention of the growth of the tops in spring, and keeping the land free from carting and consequent poaching in unsuitable weather. Roots, like fruit, ought to be stored before they become over-ripe; the months of October and November are therefore the most suitable for the work. The other operations of the farm allow time for it at this season; and the crop is generally in a fit state of maturity, as well as the land being dry.

Keeping Properties of Turnips.— Yellow turnip will continue fresh in the store until late in spring, but the swede has a superiority in this respect over all others. A remarkable instance of the swede keeping in the store, in a fresh state, was observed in Berwickshire, where a field of 25 acres was pulled, rooted, and topped, and stored in the manner shown in fig. 613, in fine dry weather in November. The store was opened in February, and the cattle continued on the roots until the middle of June, when they were sold fat, the turnips being then only a little sprouted, and somewhat shrivelled, but sweet to the taste.

Time for Storing Swedes.—It has been contended by some that the best time for storing swedes is before vegetation makes its appearance, in March or April, when they are heaviest. By experiments made in England, it was found in weighing swedes on the 16th of November, and again on the 16th of February, from the same field, that the crop had increased in weight in that time no less than 2½ tons per acre. This experiment corroborates the belief largely held that swedes gain weight until vegetation recommences in April. If there were no danger of damage by frost, it would therefore be prudent to

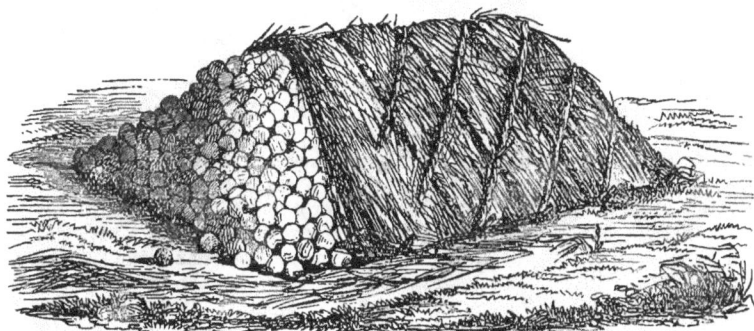

Fig. 613.—*Triangular turnip-store.*

delay storing swedes until the end of February. There is, however, great risk of damage from frost, and it is safer to store the root early in winter before severe frosts set in.

When to Store White Turnips.— All white turnips, when allowed to remain on the ground after they have attained maturity, become soft, spongy, and susceptible of rapid putrefaction, which reduces them to a saponaceous pulp. This affords a good motive to store white turnips when they come to maturity, which is indicated by the leaves losing their green colour.

Turnips Consumed on the Ground by Sheep.—When different sorts of live stock are supported on the same farm, as is the case in mixed husbandry, the sheep are usually provided with the turnips they consume upon the ground on which they grow, which saves the trouble of carrying off a large propor-

tion of the crop. The proportions carried off are not taken from the ground at random, but according to a systematic method, which requires attention.

One object in leaving turnips on the ground for sheep is, to afford a greater quantity of manure to the soil than it received in its preparation for the turnip crop; and as sheep can withstand winter weather in the fields, and are not too heavy for the ground, they are selected to consume them on it. This is a convenient method of feeding sheep, giving them their food on the spot, and returning great part of the food to the land in the form of manure.

Quantity of Roots to be left for Sheep.—The quantity of roots left upon the field to be consumed by sheep depends upon the weight of the crop, and whether the land is in a high state of fertility or not.

In ordinary practice on a mixed farm, worked on the five-shift rotation, at least one-half the crop will be required to be consumed by cattle to convert all the straw into manure. The other half may be consumed where grown by sheep, which are left in the manner described in the paragraphs which follow. When a small crop is the result of the growth of the season, the foregoing plan must be modified, and two-thirds or even a larger portion of the crop may have to be left for the sheep; but this will depend upon the soil, whether fertile or otherwise. However arranged, it must be always kept in view that cattle will thrive much better on artificial feeding than sheep, and that it is sound economy to give a certain portion of dry food, along with roots, to cattle and sheep, so that the proper ratio of nutrients be established, and every constituent of the food be economised and waste prevented.

Stripping Turnips.—When one-half of the turnips is to be left for sheep, the other half can be pulled in various ways, but not all alike beneficial to the land. It can be done by leaving 2 drills and taking away 2 drills; by taking away 3 drills and leaving 3 drills; by taking away 6 drills and leaving 6 drills; or by taking 1 drill and leaving 1 drill. In ordinary farm practice, where half the crop is to be eaten off by sheep, the plan of taking 6 or 8 drills and leaving 6 or 8 drills is largely adopted, as in the other methods there is not sufficient space left clear of roots to allow a cart and horse to turn without damaging the roots.

The first break of turnip given to the sheep ought to be as large as possible, and therefore 8 drills should be taken and 4 drills left. This plan should also be adopted when it is desirable to leave a smaller quantity for consumption by sheep than one-half, which may be either due to a high state of fertility of the soil or to a short crop of turnips.

Whatever the proportion removed, the rule of having 2 or more empty drills for the horses and carts to pass along when taking away the pulled turnips, without injury to those left, should never be violated.

Turnip-tops as Food.—The tops of turnips possess greater value as manure than as food, and should therefore, as a rule, be left to be ploughed in on the ground. But when food for sheep is scarce, the tops may be given to ewes up to the latter end of December. This practice good farmers have for years pursued with excellent results, especially when given in conjunction with some concentrated food of a costive tendency, to counteract the laxative tendency of the turnip-tops. Cotton-cake is pre-eminently suitable for this purpose. Sheep are not so easily injured by turnip-tops as cattle, on account, perhaps, of their costive habit; but in the spring it is dangerous to let sheep consume them freely, as fatal results have often followed. Then many farmers have the idea that turnip-tops make good feeding for young beasts or calves at the beginning of the season—not from the knowledge that the tops really contain a larger proportion of bone-producing matter than the bulbs, as chemical analysis informs us, but from a desire to keep the turnips for the larger beasts, and to rear the young ones in any way. But such a notion is a mistaken one. No doubt the large quantity of watery juice the tops contain at this season makes young cattle devour them with eagerness on coming off perhaps a bare pasture; and indeed any cattle will eat the tops before the turnips, when both are presented. But experience favours the condition that the time in consuming turnip-tops is worse than thrown away, inasmuch as tops, in their cleanest state, are apt to produce looseness in the bowels of cattle as well as sheep, partly, perhaps, from the sudden change of food from grass to a very succulent vegetable, and partly from the dirty, wetted, or frosty state in which tops are often given to beasts.

Turnip-tops as Manure.—Tops are not thrown away when spread upon the ground—indeed, as already stated, they are more valuable as manure than as food, and should therefore, as a rule, be left to be ploughed in on the field. In systematic trials it has been found that from 2 to 3 bushels per acre more of corn was obtained when the turnip-shaws were ploughed down than when they were carried off the field.

Turnip-Lifting Appliances.—The

tops and tails of turnips are easily removed by means of very simple implements. Figs. 614 and 615 represent these
in their simplest form, fig. 614 being an
old scythe reaping-hook, with the point

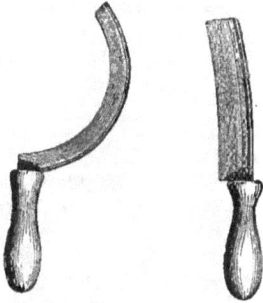

Figs. 614, 615.—*Implements for topping and
tailing turnips.*

broken off. This makes a light instrument and answers the purpose pretty
well ; but fig. 615 is better. It is made
of a worn-out patent scythe, the point
being broken off, and the iron back to
which the blade is
riveted driven into
a helve protected
by a ferrule. This
is rather heavier
than the other, and
on that account re-
moves the top more
easily. Some prefer
the implement seen
in fig. 616. If the
turnip requires any
effort to draw it, the
claw c is inserted
gently *under'* the
bulb, and the lift-
ing is easily effected
with certainty.

Mode of Pull-
ing Turnips.—The
mode of using these
implements in re-
moving tops and
tails from turnips
is this : When 2 drills are pulled and 2
left, the field-worker moves along between the 2 drills of turnips to be pulled,
and pulling a turnip with the left hand
by the top from either drill, holds the
bulb in a horizontal direction, as in fig.
617, over and between the right hand

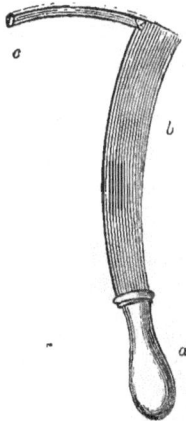

Fig. 616.—*Turnip-trim-
ming-knife.*

a Handle.
b Cutting edge.
c Claw welded to the ex-
tremity of the back.

drills, and with the knife first takes off
the root with a smart stroke, and then
cuts off the top between the turnip and
the hand with a sharper one, on which
the turnip falls into the row, the tops
being thrown down on the cleared ground.
Thus, pulling one or two turnips from
one drill, and then as many from the
other, the two drills are cleared from
end to end.

Checking Turnip-growth in Spring.
—It frequently happens, especially in
spring, when the second growth of the
turnips requires to be checked, that the
ordinary method of pulling and cleaning
the turnips cannot be quickly enough
performed to prevent the crop becoming
useless. More speedy means must therefore be adopted. The old style of slash-

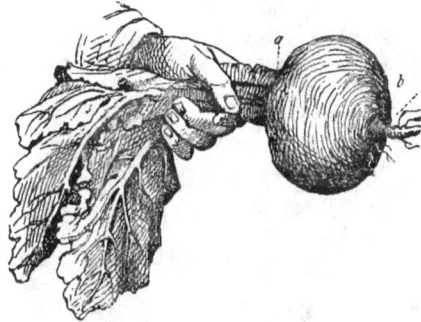

Fig. 617.—*Mode of topping and tailing turnips.*
b Root, first cut off. *a* Top, where cut off.

ing the tops off with a scythe or hook
does not overcome the difficulty. The
following method is pursued on some
farms with a fair amount of success. A
common scuffler is taken, and after the
cutting part of the hoes is extended to
about 12 inches, the hoes are reversed
—that is, change the side, so that the
cutting part is turned out instead of
inwards. Operations may then be commenced, after fixing the body of the
implement to the required breadth between the drills. The hoes cut the tap-
root beneath the surface without disturbing the bulb, which remains in the
position it grew. The growth is thus
completely checked, and the bulb will
remain fresh, as there is a sufficient
number of the small roots left to provide the moisture lost by evaporation,
but not enough for continued growth.

Many farmers run the chain-harrows across the rows, which leaves the crop lying on the surface, ready to store; but the bulbs require to be cleaned and partly trimmed before being used.

Turnip-lifters.—Implements known as "Turnip-lifters" are used to a considerable extent for topping and tailing turnips, but they are not as yet quite satisfactory in their working. The turnip-lifter would come into more general use if it were made to do the work throughout the season. When the shaws are strong and plentiful, these seem to clog the parts of the machine which top the turnips, and the coulter can seldom be set to cut away cleanly the spreading roots. With yellow and white turnips these machines often make excellent work, but as a rule they are not so successful with swedes.

Further Hints to Turnip-lifters.—Due care should be taken, on removing tops and tails, that none of the bulb be cut, as the juice of the turnip will exude through the incision. When turnips are consumed immediately, an incision does little harm; but slicing the bulb does much injury when the roots are to be stored for a considerable time.

Carting Turnips.—When the field is to be entirely cleared of turnips, the clearance is begun at the side nearest the gate; and if the workers move abreast, the carting on the land will be made as easy as possible. On removing prepared turnips from the ground, the carts are filled by the field-workers, as many being employed as will keep the carts agoing,—that is, to have one cart filled by the time another approaches the place of work in the field. If there are more field-workers than are required to do this, they should be employed in topping and tailing. The topped and tailed turnips are thrown into the cart by the hand, and not with ordinary forks or graips, which would puncture them. The cart is driven between the rows or lines of turnips, fillers being placed on each side. The carter manages the horses and assists in the filling, until the turnips rise as high in the cart as to require trimming, to prevent falling off in the journey.

Lifters one Yoking ahead of Carters.—It implies bad management to make horses wait longer in the field than the time occupied in filling a cart. It is well therefore to let lifters be one yoking ahead of the carters. The driving away should not commence at all until a sufficient quantity of turnips is prepared to employ the available carts one yoking; nor should more turnips than will employ the available carts for that time be allowed to lie upon the ground before being carried away.

Dry Weather best for Turnip-storing.—Dry weather should be chosen for pulling turnips, not merely for preserving them clean and dry, but that the land may not be poached. When so poached, sheep have an uncomfortable lair, ruts forming receptacles for water not soon emptied; for let land be ever so well drained, its nature cannot be entirely changed—clay will always have a tendency to retain water on its surface, and loam will rise in large masses with the wheels. Unless absolutely necessary, therefore, no turnips should be led off fields during or just after rain; nor should they be pulled at all until the ground has become consolidated. They should not be pulled in frost, and, as a rule, if they are urgently required from the field in frost or rain, a want of foresight is manifested either by the farmer or his manager, or by both.

On the weather proving unfavourable at the commencement of stripping, or an important operation intervening—as wheat-sowing,—no more turnips should be pulled and carried off than will suffice for the daily consumption of the cattle in the steading; but whenever the ground is dry at top and firm, and the air fresh, no opportunity should be neglected of storing a large quantity.

Importance of Storing Roots.—To store turnips in the best state should be regarded as a work of the first importance in late autumn or early winter; and it can be done only by storing a considerable quantity in good weather, to be used when bad weather comes. When a large quantity is stored, the mind remains at ease as to the state of the weather, and having a store does not prevent taking supplies from the field as long as the weather permits the ground to be carted upon with impunity, to be immediately consumed, or to augment the store. No farmer would dissent from this truth ; yet many violate it in practice.

Methods of Storing Turnips.—The storing of turnips is well done in this manner. Choose a piece of lea ground, convenient to access of carts, near the steading, on a 15-feet ridge, running N. and S., for the site of the store. Fig. 613 gives the form of a turnip-store or turnip-pit. The cart with topped and tailed turnips is backed to the spot of the ridge chosen to begin the store, and there emptied of its contents. The ridge being 15 feet wide, the store should not exceed 10 feet in width at the bottom, to allow a space of at least 2½ feet on each side towards the open furrow of the ridge, to carry off surplus water. The turnips are piled by hand up to the height of 4 feet, but will not pile to 5 feet on that width of base. The store may thus be formed of any length.

Thatching Turnip-store. — There are various ways of thatching turnips. In some cases straw drawn out length-wise is put 6 inches thick above the turnips and kept down by means of straw ropes arranged lozenge-shaped, and fastened to pegs driven in a slanting direction into the ground, along the base of the straw. Or a spading of earth, taken from the furrow, may be placed upon the ends of the ropes to keep them down. The straw is not intended to keep out either rain or air—for both preserve turnips fresh—but to protect them from frost, which causes rottenness, and from drought, which shrivels them. Another method is merely to cover the roots with a layer of earth about 8 inches deep, and if care is exercised to see that the roots are quite dry before being covered, this system suits very well. To avoid frost, the end and not the side of the store should be presented to the N., which is generally the quarter for frost. If the ground is flat, and the open furrows nearly on a level with the ridges, so that a fall of rain might overrun the bottom of the store, a furrow-slice should be taken out of the open furrows by the plough, and laid over to keep down the ropes, and the furrow cleared out as a gaw-cut with the spade.

Turnips may be heaped about 3 feet in height, flat on the top, and covered with loose straw ; and though rain pass through them readily, they will keep very fresh.

Turnip Pits in the Field.—In many cases turnips are speedily and effectually stored by being thrown into small heaps on the land, with from one to ten loads in each heap, and covered with earth. This method is called pitting, and is sometimes done without the aid of horses and carts. It is useful to place a tuft of straw in the apex of each pit as a ventilator.

Taking Roots from the Store.— When turnips are to be used from the store in hard frost, the straw on the S. end is removed, as seen in fig. 613, and a cart, or the cattleman's capacious light wheelbarrow, backed to it. After the requisite quantity for the day has been removed, the straw is replaced over the turnips.

Storing in Furrows.—One plan of storing is to pull the roots from one drill in which they have grown, and set them upright in the neighbouring drill, covering the bulbs with furrows. This plan, with slight alterations, has been followed in many parts of Scotland. Instead of leaving the tops, the bulbs are both topped and tailed. The plough returns in the furrow, opening up both sides as deeply as possible, and into this furrow the turnips from the drills at each side are thrown, the plough then covering up the whole. Turnips stored in this manner come out very fresh in spring, but can only be got out in dry weather, when a broad sock on a potato-lifting plough effectively brings them to the top, ready for cleaning by hand. It is a speedy and very effective method. But if there is any indication of an inclination of the bulbs to run to seed, this plan should not be employed.

Temporary Storing on Lea.—Another still more temporary method of storing is to pull the turnips and carry them to a bare or lea field, and set them upright beside one another, as close as they can stand, with tops and roots on. An area of 1 acre will thus contain the growth of 4 or 5 acres of the field. A turnip-field can be quickly cleared in this way for a succeeding crop. But turnips cannot be so secure from frost here as in a pit or store; and after the trouble of lifting and carrying them has been incurred, it is much better to take them to a store at once, where they would always be at hand.

Storing in Houses Objectionable. —Defective as these temporary plans are, compared to triangular or flat-topped stores, they are better than storing turnips in houses, where they engender heat and sprout on the top, and seldom fail to rot. Roots should not lie in the turnip-house at the farm-steading more than two or three days at most.

Storing in Hurdle Enclosures.— The following method is frequently adopted for a temporary store. Ordinary hurdles are taken, and the spaces between the bars wattled up with the old straw ropes that have been used for thatching. These hurdles, when thus finished, are set with stays 9 or 10 feet apart, one of the ends being closed by a hurdle placed across. Into the enclosed space the turnips are backed in the carts and tilted, after which they are trimmed until about 3 feet high. The store may be made any length by adding hurdles; the whole being finished by throwing over the top old thatch, straw ropes, &c. Rain and air

which permeate through the mass do no injury, but rather the opposite, as their tendency is to keep the turnips fresh and sappy.

Storing Turnips for Ewes.—A quick but rough-and-ready plan of storing turnips, when they are intended for ewes on the grass fields in spring, is coming more and more into favour in some districts. The turnips are pulled roughly, shaws and roots, and carted to a convenient corner in or near the grass field where they will be required, where they are simply tipped out of the cart as closely as they will lie. It is not possible to make the heap more than 3 feet deep by this plan, and therein seems to lie the success of the preservation. Hundreds of loads are often laid together in this way, and though they sprout a little, they usually continue in good condition, and the expense is not great.

Earthing-up Turnips.—The double mould-board plough is frequently employed to place earth upon the turnips in the drills, as a mode of temporary storing. The extra time required in pulling turnips after this process involves some loss, but the earthing-up protects the roots from damage by game and frost.

THE MANGEL CROP.

The mangel-wurzel, known more commonly as mangel, also as mangold, is embraced in the general term of "root crops." It belongs, however, to a race of plants quite distinct from the *Cruciferæ*, to which turnips and cabbages belong. The mangel cultivated on farms is the *Beta vulgaris* of the natural order *Chenopodiaceæ*. It is really a cultivated form of the wild sea-shore beet found in countries of the temperate zone. It was first grown as a garden plant, and it is understood that the field mangel was raised by crossing the red and white varieties of garden beet (*Beta hortensis*), the great development of root and distinctive features being obtained by persistent careful cultivation and selection.

It is believed that the field mangel was introduced into this country in 1786 by Thomas Booth Parkins, who obtained the seed in Metz. It is grown in the United Kingdom solely as food for stock. It is cultivated largely in France, Germany, and other countries for the production of sugar.

Varieties of Mangels. —There are many sub-varieties of mangels in use. The principal sorts are the long red, red globe, and orange and yellow globes. The last two are the hardiest, of excellent quality, with good keeping properties, and suitable to most soils in which mangels grow satisfactorily. The long red mangel is extensively grown on heavy soils, and produces great crops in favourable circumstances. They stand high out of the ground, and are therefore exposed to damage from early frosts. The red globe is better suited for lighter soils.

Climate for Mangels.—Mangels require different climatic conditions from those most favourable to turnips. Dry, hot summers are best suited to mangels. They thrive admirably, and yield a great weight per acre, in the southern counties of England and in the warmest parts of Ireland; but even in the best favoured districts of Scotland they are unreliable, and north of the Tweed are grown only to a very limited extent. Mangels stand drought much better than turnips.

Soils for Mangels.—Mangels need good soils. Thin poor soils, and the bleak, cold, high-lying lands upon which turnips luxuriate, are quite unsuited for mangels. Rich alluvial loams in high condition and well cultivated are best adapted for mangels, and they also grow well on strong lands in a warm climate, if these are carefully prepared and liberally manured. For the strong lands of the south of England they are better suited than turnips.

Cultivation for Mangels.—The preparation of land for mangels is in the main similar to that for turnips. And having already discussed so fully the various methods of autumn and spring tilling, cleaning, and manuring land for turnips, it will be unnecessary to do more here than point out wherein these practices should be varied to suit the mangel crop.

Autumn Tillage.—The great object to be aimed at in preparing land for mangels is to have it cleaned, dunged, and deeply ploughed in autumn. When the land is stiff these should be done as early in autumn as possible, generally before the end of October. Deep autumn ploughing is especially beneficial for mangels, and where the subsoil is inclined to form into a "pan," it should be broken up by subsoil ploughing.

It is a good plan, after the land has been thoroughly cleaned and deeply ploughed in September or October, to at once open drills, spread the dung, and cover in the drills just as at seed-time for turnips. Some recommend that before spreading the dung a drill-grubber

should be run along to loosen the bottom of the drills. In these ridges the land lies throughout the winter, admirably exposed to the disintegrating influences of the season, and is found easily prepared for the seed next spring.

Spring Tillage.—When the land has been cleaned, ploughed, and dunged in drills in autumn, as just described, little has to be done in the way of tillage in spring. A light harrow is drawn along the drills (not across them), the drills are again set up by the drill-plough, and the seed thereupon sown. Such artificial manure as is to be given may be sown broadcast either before the harrowing or before the setting up of the drills.

When the dung has been simply ploughed in with an ordinary furrow in autumn, the land has to be grubbed and harrowed in spring just sufficiently to secure as fine a tilth as possible. Deep spring ploughing when the land has been dunged in autumn is not to be commended.

It often, perhaps generally, happens that a sufficient supply of dung is not available till well into the winter or early in spring. In this case the land is cleaned in the autumn and left in a strong furrow till early spring, when it is grubbed or ploughed, or both, then harrowed, drills opened—if the raised drill system is pursued—the dung spread, artificial manure sown, the drills closed, and the seed sown.

Strong land should be stirred as little as possible in spring. Mangels, like turnips, delight in a fine moist seed-bed, and it is difficult to obtain this with much stirring of strong land late in spring.

Drills and Flat Rows.—Mangels are sown both in rows on the flat and in raised drills. The latter is the better plan, as it affords greater facilities for the after tillage and cleaning of the land. The rows on the flat usually vary from 20 to 26 inches wide, and the raised drills from 25 to 29 inches. In dibbling, each hole should be about 1 foot apart, two or three seeds being placed in each.

Mangel Seed.—The seed of mangels is encased in a rough woody capsule which makes germination very slow, unless special means are taken to hasten it. For this purpose the seed is steeped, for from 12 to 36 hours, before sowing —by some in warm water, by others in cold water, and by others again in liquid manure. If warm water is used, 12 to 14 hours should be sufficient. The seeds, when removed from the steep, are spread on a wooden floor, or on canvas cloth or sieve, and allowed to attain such a state of dryness as will prevent adhesion. In some cases the saturated seed is coated with a quantity of finely powdered charcoal, which is freely mixed with it.

The seed is then sown either by the flat-row drill or raised-drill machine, as the case may be. The peculiarities of the mangel seed necessitate the attachment of specially devised seed-boxes. Water or ashes may be sown along with the mangel seed, as in the case of turnips. The seed must not be put deeper in the soil than from ½ to 1 inch.

Quantity of Seed.—The quantity of mangel seed sown per acre is usually about 6 to 10 lb. If dibbled in, 3 lb. of seed per acre will suffice.

Time of Sowing.—Mangels have to be sown earlier than turnips. April is, as a rule, the best month for mangel sowing, but portions of the crop are usually sown earlier, sometimes even as early as February. When, owing to a crop of winter rye or some other catch crop occupying the land, or when, from some other cause, it cannot be prepared sooner, sowing may be done in May. After the middle of that month it would be very risky.

MANURES FOR MANGELS.

Dependency on Manure.—Mangels require, and will under favourable conditions repay, liberal manuring. It would be useless to attempt to grow them upon scanty fare.

They produce an extraordinary yield in a comparatively short space of time. To enable them to realise their full capabilities in this respect, an ample supply of the kinds of plant-food best suited to them must be furnished in a readily available condition. And the farmer must discriminate as to the kinds of plant-food to be supplied.

Ingredients absorbed by Mangels.—The following table, compiled from that on page 326, vol. i., shows the quantities of nitrogen, potash, and phosphoric acid—the three chief manurial ingredients—taken out of the soil by a crop of mangels weighing 22 tons of roots per acre:—

	Roots. lb.	Leaves. lb.	Total. lb. per acre.
Nitrogen	96	51	147
Potash	222.8	77.9	300.7
Phosphoric acid	36.4	16.5	52.9

Ingredients of Manure for Mangels.—These figures, compared with the corresponding analyses of a crop of turnips, show at a glance that the manurial wants of the mangel differ considerably from those of the turnip crop. Phosphates are essential for both, but do not by themselves exercise such a marked effect on mangels as on turnips. On the other hand, nitrogen, so little required for turnips, must be freely given to mangels. Then mangels will turn heavy dressings of good farmyard dung to better account than turnips can, while the palate of the mangel would seem to delight in having its food seasoned with a pinch of common salt.

Rothamsted Experiments—In Manuring Mangels.—Exhaustive trials in the manuring of mangels have been carried out and are still going on at Rothamsted. As to the results obtained, full information is given in the report on Rothamsted Experiments in this volume, pp. 28 - 32. The effects of different manures on the mangel crop there are shown clearly.

Salt for Mangels.—It is not in the least surprising that the application of common salt has been found in general farm practice to substantially increase the yield of mangels. The plant, we have seen, is indigenous to the sea-coast, and its ash is found to contain from 25 to 50 per cent of common salt. The late Dr A. Voelcker applied salt to mangels on deep sandy soil, and obtained an increase of 2 tons 6 cwt. per acre from 3 cwt. of salt; 5 tons 11 cwt. from 5 cwt. salt; and 4 tons 1 cwt. from 7 cwt. salt. Still, at Rothamsted heavy dressings of salt, over 5 cwt., were found to be hurtful rather than beneficial.

Useful Dressings.—In general practice farmyard manure is almost always applied for mangels. To obtain a maximum crop, and yet maintain the land in a high state of fertility, a fairly liberal allowance of dung may be regarded as essential — say from 8 to 12 tons per acre, though many farmers apply as much as 15 tons or more.

Still it has been shown in various experiments that in certain circumstances a mixture of artificial manures, consisting of from 3 to 5 cwt. of superphosphate per acre, 3 to 4 cwt. of sulphate of potash, 2 to 3 cwt. of common salt, and 2 to 3 cwt. of nitrate of soda, and perhaps a little rape-cake, might produce a fairly satisfactory crop of mangels without any farmyard manure. Some give kainit instead of sulphate of potash, and lessen the allowance of common salt.

Along with from 10 to 12 tons or more of dung a liberal allowance of artificial manure would be from 2 to 3½ cwt. of superphosphate, 2 or 3 cwt. of common salt, and 1 to 2 cwt. nitrate of soda or sulphate of ammonia. In many cases even larger quantities of artificial manures are applied; but with land in good average condition as to fertility these doses should, as a rule, be sufficient.

Basic slag has not, as a rule, done so well with mangels as superphosphate.

With a full allowance of dung there will seldom be much necessity for the application of special potash manure. If there is any reason, however, to suspect that there is a deficiency of potash in the soil, from 1 cwt. to 2 cwt. of kainit per acre should be applied.

As to whether it should be nitrate of soda or sulphate of ammonia which should be used along with the dung, the farmer must think for himself. He will especially consider the market price of the two commodities at the time, and buy whichever happens to be the cheaper. In a rainy climate and wet seasons sulphate of ammonia will most likely give better returns than nitrate of soda, and even in dry districts similar results are sometimes obtained.

The condition of the dung as to rottenness should also be taken into account in deciding whether to sow nitrate of soda or sulphate of ammonia. In well-rotted dung there is more readily available nitrogen than in fresh dung. With fresh dung, therefore, nitrate of soda would, as a rule, be preferable to sulphate of ammonia.

Dr Bernard Dyer, who has experimented largely on the manuring of mangels, places a high value on nitrate of soda as a manure for this crop. With less than 10 tons of dung he would give 3 or 4 cwt. of nitrate of soda per acre— one cwt. being sown with the seed, one cwt. as a top-dressing just after singling, and the remainder as a top-dressing some weeks later.[1]

APPLICATION OF MANURE FOR MANGELS.

It would be well here to refer to what is said in the special section on "Manures and Manuring" as to the general principles to be observed in applying the various manures to land—pp. 511-520, vol. i.

Dung.—If dung is available it should be applied in the autumn, and ploughed down or spread in drills, and covered in. If it cannot be applied in the autumn or winter, and if the mangels are to be sown in rows on the flat surface, the dung should be spread and ploughed in as early as possible in the spring. Where the mangels are to be sown in raised drills, and the dung cannot be applied till spring, it is spread in the bottom of the drills at sowing-time, as in the case of turnips — carted out and spread as described for turnips.

Artificial Manures. — Perhaps the most general plan is to sow these by the hand or machine in the drill or row at the time of sowing the seed, as for turnips. As a rule, however, it will be found advantageous to reserve the whole or part of the nitrogenous manure, especially nitrate of soda, and apply it as a top-dressing some time in July. The allowance of common salt is also by many held back till July, when some careful farmers apply a mixture of from 1 to 1½ cwt. of nitrate of soda, and 2 to 4 cwt. of common salt in two sowings.

There is no denying the advantage of such a top-dressing for mangels. It has been well established in extensive practice. By holding back the nitrate

[1] *Fertilisers and Feeding Stuffs,* p. 60.

of soda till the plants are ready to make use of its nitrate, loss by washing into the subsoil and drains is minimised.

Theoretically, one would expect that the slower acting sulphate of ammonia should give better results by being applied at the time of sowing, and this is the practice on many farms. Nevertheless, some farmers prefer to use it as a top-dressing—prefer it to nitrate of soda for this purpose also. These are points as to which hard-and-fast lines cannot in all cases be followed.

THINNING AND AFTER CULTIVATION.

Preliminary Cleaning. — Mangel plants are slower in growth at the very outset than are those of turnips. To keep down weeds, therefore, it may be necessary to horse or hand hoe the rows or drills before the plants are ready for thinning. This preliminary hand-hoeing need be resorted to only in narrow rows on the flat, or where weeds are encroaching injuriously upon the plants. The horse-hoe or drill-scarifier will suffice, as a rule, in raised drills.

Thinning. — As soon as the plants show a fairly strong leaf, they should be thinned and hand-hoed as in the case of turnips. From 12 to 14 inches are common intervals between the plants. The narrower the drills, the greater should be the interval between the plants in the rows.

As with turnips, it has been found that mangels of medium size usually contain more solid nutritious matter than mangels of excessive size. And by moderate, rather than large, intervals between the plants, the maximum yields of good food per acre are likely to be obtained.

After Hoeing. — The treatment of mangels after thinning is, in regard to hoeing by hand and horse-power, very similar to that of turnips. The horse-hoe or scarifier should be kept at work as long as the leaves of the roots will permit.

Transplanting Mangels.—The young mangel plant may be successfully transplanted. Blanks in the rows should be filled up by transplanting. This should be done with care, so that the tap-root may be dibbled right down into the soil. Unless the weather is showery at the time or the soil moist, the transplanted plants should receive a spray of water.

Mangels of exceptional weights have been grown experimentally from plants raised in a seed-bed (sown in January), and planted out in February. How far this system could, with advantage, be extended into farm practice is uncertain.

Injuring Mangel Plants. — Mangels are peculiarly liable to suffer from injuries to the leaves of the plants. Cuts or bruises to the leaves, even if inflicted when the plants are very young, do not heal up as would be the case in turnips —they remain as open, "bleeding" sores, robbing the plant of not a little of its life-juice, and rendering it liable to ready attacks of frost and decay. In the thinning of mangels, therefore, the plants should be guarded with the greatest care.

Yield of Mangels. — The yield of mangels per acre varies greatly. A good average crop should yield about 18 to 20 tons per acre. Sometimes it is as low as 12 tons, and occasionally over 40 tons per acre. In favourable years crops of from 25 to 35 tons are not uncommon.

Mangel Plagues. — Mangels suffer less than turnips from fungoid and insect plagues. See chapters in this volume on crop plagues.

STORING MANGELS.

The storing of mangels requires greater care than the storing of turnips. Pulling usually begins in the third week of October, and it is well not to pull more at a time than can be put into "clamps" (heaps) in one day.

Cover with Dry Straw.—Opposite opinions are expressed as to covering. Upon the whole, it might be safest to cover with dry straw or bracken. Special care, however, must be taken to see that the material used for covering is perfectly dry.

Pulling Mangels.—In pulling mangel-wurzel, care should be taken to do no injury to the roots. Cleansing with the knife should on no account be permitted: rather leave some earth on the root. The drier the weather is, the better for storing the crop. The roots are best prepared for the store by twisting off the top with the hand, as a mode of preventing every risk of injuring the root, though little damage may be done if the knife is only used to cut off the top, and that not too closely. Mangels not being able to withstand severe frost, should be entirely cleared from the field before its occurrence.

The best way of pulling mangels is where two drills are pulled by one worker and the adjoining two drills by another, and the prepared roots placed in rows in the hollow intermediate to the four drills, the leaves being also thrown into rows between the roots. The leaves thus treated, when intended to be fed either by sheep folded on land or carted off and thrown on pastures for cattle or sheep, are always clean and fit food for stock, which they are not when thrown over the land and trampled on. Mangels not pulled, and protected by the broad leaves, will stand frost (if not very severe) without injury ; but a very slight frost will damage those roots which are pulled, therefore it is wise to store the roots as soon as pulled. If the leaves are not desired to be used as food, they may be scattered over the ground. Mangel-leaves can be given to cattle with greater freedom than turnip-leaves.

Carting Roots.—On removing any kind of roots, the cart goes up between two rows of pulled roots, and thereby clears a space at once of the breadth of eight drills. In this manner the work proceeds expeditiously, and with little injury to the land by trampling.

To save the land still further, and also to lessen the draught to the horses, the carts should be driven up and down the drills and not across them, whether going with a load or returning empty.

Method of Storing Mangels.—A general practice is to form heaps of the roots on a base varying from 6 to 8 feet wide, the roots on the surface being packed with their crowns outwards. The heap is drawn to a narrow ridge which may be from 5 to 8 feet high. To guard against fermentation, a thick handful of straw should be inserted in the ridge every 6 feet : this affords ventilation and keeps the roots from heating. It is usual to thatch the heap with about 4 to 6 inches of straw, covered with 6 inches or more of earth, care being taken to leave the ventilators uncovered. Some farmers defer earthing the uppermost two feet of the heap for a month or two, so as to lessen the risk of injury by fermentation.

Mangels Stored in Houses.—Mangels can be stored in houses with greater safety than turnips. Many farmers store several hundred tons of mangels in a store-house at a time, and if the top of the heap is only loosely covered with straw no harm will be done by fermentation.

A Suffolk farmer says he stores his mangels in a barn formed of wood, the inside of which is first lined with barley-straw 18 inches thick, to protect the roots from frost, the heap of roots being 12 feet deep and 18 feet wide, and left uncovered on the top. He has pursued this plan for several years, and it has preserved them admirably up to March, or even longer.

FORAGE CROPS.

Forage crops may be defined as those which are grown for the sake of their leaves and stems, as distinct from crops grown for seeds and roots. Chief amongst the forage crops are the grasses and clovers. These have already been described, and here will be given some information regarding several other for-age crops which may be grown to provide wholesome green food for farm live-stock. These are vetches, lucerne, sainfoin, rye, cabbages, rape, mustard, crimson clover, kidney-vetch, gorse or whin, buckwheat, maize, sorghum, and prickly comfrey. Sainfoin, lucerne, buckwheat, maize, and sorghum are confined to southern parts, where the climate is mild; the others may be grown in almost any part of the kingdom.

Importance of Forage Crops.—The growing of forage crops, particularly of crops to be cut and used as green food, has not yet received from British farmers so much attention as it deserves. Our acquaintance with forage crops is still imperfect, and the extent to which they are capable of contributing to the saleable produce of the farm is not fully understood or appreciated. In those districts of Britain where pasture does not succeed very well, or which are subject to severe summer droughts, the providing of a plentiful supply of green succulent food coming into use in succession all through the year is one of the greatest objects of the stock-owner. The forage crops at present in use, as they are now known and cultivated, are far from adequate for this purpose, and assuredly no subject could more worthily engage the attention or employ the resources of our great agricultural and experimental bodies than furnishing to farmers the knowledge and the means which would enable them to grow a more abundant supply of green food for stock throughout the year.

VETCHES.

The vetch or tare belongs to the natural order of *Leguminosæ*, and the cultivated tare or vetch is named *Vicia sativa*. There are numerous wild varieties.

The vetch is a most valuable forage crop. It is hardy and prolific, and affords palatable and wholesome food for stock. There are two varieties, the winter and the spring vetch. The former, through repeated sowing in winter, has acquired a hardiness that is quite remarkable.

Winter Vetches.—The winter vetch is sown to provide green food in spring before a full supply of grass is available. It is sown at various intervals from September till February where the climate is variable; in cold northern districts there is little use in sowing after severe winter frosts set in. It is a good plan to sow rye or oats along with vetches, so as to keep them from lodging and rotting at the root. Winter vetches

should provide a good supply of food from April till June, according to the district.

The importance of having a supply of fresh succulent food at this season of the year, when roots are wholly or nearly exhausted, and before the pasture fields can sustain the animals, will be readily acknowledged by all farmers, and it is surprising that winter vetches are not sown much more extensively than they are, especially when it is remembered that they can be off the ground in time for a root or potato crop in the following season.

Spring Vetches.—Vetches should be sown at different times in spring, so as to afford a succession of cuttings when green food is likely to be most urgently required. If the weather and the state of the land permit, the first sowing may be made in February, and successive sowings may take place every second or third week up till towards the end of June. It is advisable to sow small breadths at a time, so as to have a succession of cuttings when the crop is in full bloom. By judicious sowings at different times in autumn, winter, and spring, supplies of fresh-cut tares may be had from the end of April till October.

Utilising Vetches.—In the colder northern districts vetches are for the most part cut as house food, chiefly for cattle, but partly also for horses, and likewise for sheep where the house feeding of sheep is pursued. In other parts the great bulk of the crop is consumed on the ground by sheep. In most cases sheep or lambs being pushed on for sale get the first run of vetches, store and breeding sheep following after and picking up what is left. Sheep are folded on vetches just as they are on rape or roots.

Vetches for Horses.—Horses eat vetches with a keen relish, and thrive well upon them. They should be provided for horses during the harvest work, and given in moderate quantities along with dry food. It is considered by many that on strong land there is no better or cheaper way of keeping farm-horses in summer than by feeding them in the stable or yards with vetches and a little dry food.

Land for Vetches.—Vetches usually

follow a grain crop. They thrive best on strong loams and tenacious clays, just the sorts of soil upon which turnip culture is most difficult. But they also afford a good return on lighter soils. In some cases vetches are sown upon strong land, which is fallowed in summer as a preparation for wheat. In other cases turnips or potatoes succeed winter vetches, so that the latter come in as a sort of "catch crop"—and a most useful one it is. Land for vetches does not require much tillage. If clean, ploughing, harrowing, and rolling will suffice. If the land should be foul, it may be grubbed and cleaned before being ploughed for the seed.

Seed.—The seed of vetches is usually sown broadcast, but often in rows about 8 inches apart. The quantity of seed varies from $2\frac{1}{2}$ to $3\frac{1}{2}$ bushels per acre, supplemented, perhaps, with from a half to a whole bushel per acre of oats or rye. Vetches are also sometimes sown along with rape. Vetch seed more than two years old should not be used. For growing the seed of vetches it is a good plan to sow vetches and beans together, at the rate of about half a bushel of vetches to $2\frac{1}{2}$ bushels of beans per acre. The seed is harrowed in the same way as a corn crop.

Cutting Vetches.—Vetches are most valuable for feeding when cut just in full bloom, and before the seed has begun to form. It is thus important to sow small quantities at a time, so as to be able to use the crop as it comes into bloom. When vetches are grown for seed they are, of course, allowed to ripen, and are cut and harvested in the same way as peas.

Manuring Vetches. — Land for vetches may be easily and cheaply manured. If the land is in moderately good condition it may receive a light dressing of farmyard manure,—from 6 to 10 tons per acre,—which it is preferable to let lie on the surface for a few weeks previous to ploughing in. Along with the dung, or at any suitable time before or after dunging, 2 to 3 cwt. of kainit, and half that weight of superphosphate, should be sown over the unploughed land, and the same or more on the surface, as soon as the land is ploughed and before it is harrowed. Where the land is in good heart substantial crops of vetches may be grown with artificial manures alone, say 3 to 5 cwt. of kainit and 3 to 4 cwt. of superphosphate per acre. Like other leguminous crops, vetches can do much in the way of providing nitrogen for themselves.

Vetches and cleaning Land.—With a fairly liberal system of manuring, vetches seldom fail to do well in moderate seasons, and with a full crop they smother root-weeds well, while owing to the early cutting or eating of the crop, seed-weeds have no time to ripen their seeds. The land being bare comparatively early may be bastard fallowed and cleaned. Vetches, therefore, if well done to, offer an excellent opportunity of keeping down weeds, and of cleaning the land after the removal of the crop, thus leaving it in good condition for what is to follow.

LUCERNE.

In warm climates, notably in the southern counties of England, lucerne is a prolific forage crop. It is the *Medicago sativa* of botanists (Nat. Order *Leguminosæ*); root sub-fusiform, stem erect, flowers large and violet-coloured. Lucerne is said to have been brought to Greece from Asia. The Romans were well acquainted with its properties as a forage plant, particularly for horses. Hartlib endeavoured to introduce its culture into England in the time of the Commonwealth, but did not succeed.

Lucerne, though sometimes included in seed mixtures for rotation grass, is usually grown alone. When well laid down in suitable soil — deep calcareous loam, clean and in good heart—it affords every year several cuttings of excellent green food, which is relished by both cattle and horses. If kept free from weeds, the crop may remain productive for six or seven years. Weeds, however, are liable to disturb it, and may cause it to be ploughed up earlier. Land should therefore be prepared with great care for lucerne. It should be well cultivated, and as thoroughly as possible cleared of weeds of all kinds, and the crop responds to moderate dressings of dung and superphosphate. Occasionally the year's produce amounts to 30 tons per acre, and 20 tons are by no means rare.

The seed is sown in April, in rows from 6 to 10 inches apart, at the rate of 20 to 30 lb. per acre. One cutting will be obtained in the autumn of the same year, but it is advisable to leave a rank growth to protect the roots from the winter's frosts.

The Crop for Dry Seasons.— Lucerne withstands drought wonderfully. It thrives best in a dry climate, and is therefore cultivated extensively on the continent of Europe. It is an exceptionally deep-rooted plant, and is thus comparatively independent of rain. The late Sir John Bennett Lawes found it the best of all the forage crops for a drought.

Lucerne is not well suited for extended cultivation in a wet climate.

SAINFOIN.

Upon the calcareous soils of the southern counties of England, sainfoin has proved a most useful and reliable forage crop. Belonging to the Natural Order *Leguminosæ*, it is the *Onobrychis sativa*, the cultivated sainfoin, of botanists.

Sainfoin Hay.—The sainfoin yields the finest quality of hay when cut before the blossom comes out. Jethro Tull declared this sanfoin hay, cut before blossoming, "kept a team of working store-horses, round the year, fat without corn, and when tried with beans and oats, mixed with chaff, refused it for the hay. The same fatted some sheep in the winter in a pen, with only it and water; they throve faster than other sheep at the same time fed with peas and oats."

Sainfoin, like lucerne, is a deep-rooted plant, and thrives best on dry soils in a dry warm climate. It is grown extensively, and with great success, on the chalky soils of the south of England. It is useful as an ingredient in mixtures for temporary grass and hay, but is perhaps still more valuable as a forage crop grown by itself.

If well laid down in clean suitable land, it will endure, and yield liberally, for six or seven years. It should not be resown upon the same land for some twenty or more years. Indeed it is a common saying that land will not successfully carry sainfoin more than once in a lifetime. Sainfoin is both cut and pastured, and especially for sheep a run of old sainfoin is much esteemed.

It is not a reliable crop on strong lands or in wet climates.

Land intended for sainfoin should be thoroughly clean and in good heart. The seed is best sown with barley or oats, and it may be mixed and drilled with the grain seed. In other cases it is drilled separately at the same time across the rows of the grain seed. The quantity of sainfoin seed used per acre is usually about four bushels of unmilled seed,—rough seed in the pod. Sainfoin does not develop fully until the second year, and it is therefore considered a good plan to sow from 6 to 8 lb. trefoil (*Medicago lupulina*) per acre along with it.

It would be well to defer grazing the sainfoin until after the first cutting has been removed. Young sainfoin is liable to be damaged by being grazed too soon by sheep.

RYE.

Rye makes a very useful forage crop. It is wonderfully hardy, and may be sown in autumn or winter for spring use as forage in northern parts, where even vetches cannot be depended upon. It throws up a rank growth, and although it is not so succulent as the vetch, it is, nevertheless, a valuable forage plant, affording, as it does, the earliest green food for sheep or cattle in spring. As already mentioned, it is often sown along with winter vetches.

For spring forage, rye should be sown in autumn immediately after the removal of a grain crop, at the rate of about 3 or 4 bushels per acre. If the land is in good heart, or the crop well manured with dung or superphosphate, and nitrate of soda or sulphate of ammonia, the rye will afford a large produce in the following April, when it may be consumed on the land by sheep, or cut and fed to cattle in the house.

In market-gardens at Biggleswade and other parts, strips of rye about 3 feet wide and 18 feet apart are grown to provide shelter for spring vegetables.

CABBAGES.

The cabbage (*Brassica oleracea capitata*, Nat. Order *Cruciferæ*) is a most suitable plant for field culture. It is not grown so extensively as might be expected, when one considers the vast amount of wholesome food which it is capable of producing.

The cabbage succeeds best on deep good loams, with porous or well-drained subsoil, and it also does well on well-farmed strong clays. It is a gross feeder, and requires liberal manuring and careful tillage.

There are three main groups of cabbage plants—(1) what are known as *early cabbages*, planted in autumn and ready for use next spring; (2) *summer cabbages*, planted in spring and consumed early in summer; and (3) *late cabbages*, planted in spring or early summer, ready for use in autumn and winter. The late cabbages are grown the most largely on farms: before dealing with the culture of these a few notes may be given regarding the raising of the earlier sorts, which are the varieties most extensively cultivated in family and market gardens.

Early Cabbages.

For Spring Use.—For the varieties fitted to withstand the winter and be ready for earliest use in spring the seed is sown in a carefully prepared seed-bed about two months before the permanent planting is to take place in autumn. The land for this variety must be in good heart, free from weeds, and fairly firm, a very loose soil not being well suited for holding the plants during winter. In the way of tillage, therefore, a shallow ploughing and slight harrowing will suffice.

The earliest cabbages do best on land well dunged for the preceding crop, the direct manurial dressing consisting of from 3 to 6 cwt. per acre of nitrate of soda, or partly nitrate of soda and partly sulphate of ammonia. This allowance is given as a top-dressing after the plants begin to grow in spring, and part of it is spread round each plant, not nearer the root than about 2 inches, and not farther away than about 5 inches.

For the planting of early cabbages the land is marked off in lines about 15 to 18 inches apart on the flat with a plough or other implement. In these lines the plants are put in with the dibble at intervals of from 12 to 15 inches. To keep the land clean a good deal of hand-hoeing has to take place.

The earliest cabbages are largely used for table consumption, and the main object of the grower is to have the crop as early in the market as possible.

For Summer Use.—The varieties of cabbages to be planted in spring for use early in summer are for the most part grown on land which has been well cultivated in the preceding autumn and winter and formed into drills. A fairly liberal allowance of dung—10 to 15 tons —is given in the drills, and this is supplemented by from 1½ to 3 cwt. each of kainit and superphosphate per acre, applied on the top of the dung.

The plants are dibbled on the top of the drills, and in order to get a firm hold for the plants the drills should be rolled with a drill-roller, such as that shown in fig. 584, p. 307 of this vol.

When these plants have started into vigorous growth they should get a dressing of nitrate of soda at the rate of about 2 cwt. per acre applied round the plants as already indicated. Three or four weeks later another similar dressing of nitrate of soda should be given.

By repeated horse-hoeing weeds must be thoroughly kept down.

Crops of cabbages thus treated should be ready for use early in July.

Late Cabbages.

As already stated, it is the late cabbages that are most largely grown as a field crop. For the information which follows on the culture and utilisation of field-cabbages we are mainly indebted to Mr John Speir.

Probably no ordinary farm green crop admits of growth in a moderately successful way, in a greater variety of soils or climates, than Drumhead cabbages. With suitable manuring they may be grown on sand, loam, or clay, and on the sea-shore, or well up the mountain-side.

Sowing and Planting.—The seed should be sown in a seed-bed early in autumn, and the plants transplanted from it to the field in spring. Spring planting may be done in any suitable weather during March or April, the best crops being usually obtained from the earliest plantings, all other things being equal. By planting moderately early few plants fail to catch root, and as they are rarely hurt by frost after being planted out, they have thus a much longer season in which to mature a full crop.

The drills should not be less than 28 or 30 inches in width, and the plants about 1½ ft. apart in the drill. Planting is best done by the dibble, although some people prefer the spade.

Cabbage Plants.—About 1½ to 2 lb. of seed sown on 2 square roods of land should yield sufficient plants for an acre of the crop. The seed should be sown from six to eight weeks before the plants are to be planted out. For the most part, farmers purchase cabbage plants from market-gardeners or others who make a speciality of the growing of them. From 8000 to 10,000 cabbage plants are required to plant an acre.

Produce.—The produce of cabbages on good land under liberal and skilful treatment may reach from 30 to 50 tons per acre.

Manuring.—Where possible, cabbages should always have their farmyard manure applied to them in the drill. The cabbage is such a gross feeder that it is almost impossible to spoil it by excessive manuring. Any available quantity of farmyard manure, from 20 tons per acre upwards, may therefore be applied, and whatever assistance the crop afterwards requires can be made up by surface manuring with artificials.

As soon as the plants have thoroughly taken with the ground, and have begun to spread their leaves across the drill, they should receive from 1 cwt. to 2 cwt. of nitrate of soda or sulphate of ammonia per acre. For a first manuring this is best applied by dropping a little at the root of each plant, 1 cwt. doing as much good at this date, applied in this manner, as 2 cwt. applied broadcast, the plants being so far asunder that a large proportion of such a soluble manure sown broadcast runs to waste.

Before the crop is earthed up for the last time, it is always advisable to apply 1 cwt. or 2 cwt. of nitrate of soda or sulphate of ammonia, no matter how much manure may have been applied in the drill. This may either be dropped near each plant or sown broadcast, after the drills are grubbed and before the crop is earthed up. By this means the whole nitrate is turned over on the top of the roots of the plants and under their wide-spreading leaves, so that it is protected from washing, no matter whether the season prove wet or dry. Manured in this way, an enormous crop of cabbages can be grown almost any year, on nearly any kind of land.

Few who have not seen a crop thus manured can form any idea of the weight which may be produced, even under unfavourable circumstances; and certainly for autumn consumption no other crop will produce anything like the same weight of leaves and of an equal feeding value.

Dr Bernard Dyer recommends from 4 to 6 cwt. of nitrate of soda per acre for cabbages along with dung—the nitrate to be sown in successive doses at intervals of a few weeks.

Utilising Cabbages.—Cabbages are well suited for consumption by any kind of farm stock, but for dairy cows, lambs, or older sheep, they are particularly valuable. They are usually given to the cows raw, although a few people give them boiled or steamed; this, however, is generally considered to be unnecessary. In ordinary seasons the Drumhead cabbage will be ready to use from the beginning of October till the New Year.

In consuming the crop it is always best to begin by using the largest and ripest cabbages first, as these are the ones to suffer most by frost. In the interval the smaller and greener ones increase considerably in size, and the labour so spent is doubly repaid by the better preservation of the crop, as the small green cabbages suffer little from even severe and protracted frost.

Cabbages and Italian Rye-grass. —Where early cabbages are grown for table use, a crop of considerable value is got in autumn from the second growths, and from Italian rye-grass sown broadcast among the plants after the last hand-weeding and before the crop is earthed up for the last time. If this course is followed, it is better to earth up the crop with a drill harrow and plough, which will not only mix the seed with the soil, but also cover the bulk of it. This crop makes little progress during the period that the main crop is in full growth, but as soon as it is removed the Italian rye-grass comes up quickly and soon covers the ground. This class of a cabbage crop is generally consumed during July and August, and after the main crop is removed the waste cabbages, second growths, and Italian make excellent feeding for sheep or lambs. So utilised, this catch crop may, according to season and situation, be worth from 40s. to 80s. per acre. It has also the further advantage of helping to keep down annual weeds, and conserving in its roots the excess of nitrogen which it is necessary to apply to the cabbage crop. Along the sea-shore of the southern counties thousand-headed cabbages may be grown after early potatoes. Those come in very handy in spring for feeding ewes and lambs, when other green food is extremely scarce.

Cabbages are usually regarded as an exhausting crop. This, however, is only partially true. Certainly, as already stated, they are gross feeders, and require heavy manuring; but if they are consumed on the farm the exhaustion does not arise.

Storing Cabbages. — Cabbages are generally consumed direct from the ground in a green state. They are not so easily stored for future use as are turnips or mangels; still there are some methods by which they may be safely preserved for several months. The mistaken idea that cabbages cannot be successfully stored or protected from frost except in a barn or other building specially prepared for them has, no doubt, prevented the more extensive cultivation of this most useful crop.

Amongst the various methods of storing cabbages which have been practised and recommended are the following: Taking them up and replanting them in a sloping manner, and covering them with straw; pitting them; hanging them up in a barn; turning them head downwards, and covering them with earth, leaving the roots sticking up in the air. But every one of these plans

is attended with great labour, and some of them forbid the hope of being able to preserve any considerable quantity.

A plan which has proved successful is this: Throw up a sort of ridge with the plough, and make it pretty hard on top. Upon this ridge lay some straw. Then take the cabbages, turn them upside down, and, after taking off any decayed leaves, place them, about six abreast, upon the straw. Then cover them, not very thickly, with straw, or leaves raked up in the woods, throwing here and there a spadeful of earth on the top, to keep the covering from being blown off by the wind. Only put on enough of straw or leaves to hide all the green, leaving the cabbage-roots sticking up through the covering.

Stored in this way, cabbages of all sorts will be found to keep well through the winter, and they are at all times ready for use. They are never locked up by frost, as often happens with those pitted in the earth; and they are never found rotting, as is often the case with those stored with their heads upwards and their roots in the ground.

The bulk of this crop is so great, that storing in buildings of any sort is not to be thought of. Besides, the cabbages so put together in large masses would heat and rot quickly. In some gardens, indeed, cabbages are put into houses, where they are hung up by the roots; but they wither in this state, or soon putrefy.

By adopting the mode of storing recommended above, however, all these inconveniences are avoided. Any quantity may be so stored, in the field or elsewhere, at a very trifling expense compared with the bulk and value of the crop.

Lifting Cabbages. — Some recommend the cabbages to be pulled up by the roots; others prefer cutting the stem close to the ground and leaving the root in the ground, which will throw up a fresh growth of leaf early next spring. These sprouts are most valuable for ewes in early spring. Where the sprouts are not desired for food, the better plan is to pull up the roots along with the cabbages, as the spring growth tends to exhaust the soil.

Utilising Cabbage-stalks.—In regard to cabbage-stalks after cutting off the cabbages, a farmer says: "I do not get these 'out of the way as quickly as possible,' by shooting them into a ditch to rot and wash away, as is too often the case. I lay them thin to dry, and then clamp and char them for absorbing urine, and drilling with a compost drill, or broad-casting, where most likely to be advantageous."

Thousand-headed Kale.

This is another Brassica of the cabbage sort, which is much esteemed as food for sheep. This variety may be sown in rows on the flat about the end of April, on rich well-prepared land, at the rate of 4 or 5 lb. of seed per acre. It will produce a bountiful yield of excellent food for sheep in the autumn.

Transplanting Kale. — Thousand-headed kale gives the best return when the plants are raised in a seed-bed, and planted out like ordinary cabbage. For transplanting, the seed is sown in a seed-bed, early in August, and the plants are dibbled into well-prepared, well-dunged land in October and November. This should afford an abundant growth for folding in the following summer. If required, a moderate dressing of nitrate of soda would force on the growth of the plants. But this is a troublesome plan, and in many parts of the country it has been entirely given up, and the seed is sown in rows on the flat.

Consuming Kale.—Thousand-headed kale thus grown, may either be consumed by sheep being folded upon it, or by the heads being cut off and consumed by sheep on pasture-land. If by the first method, the stems are not too closely eaten or peeled by the sheep, the plants will throw out new leaves, and afford a supply of delicious green food in the following spring. In cutting off the heads the bottom leaves should be left, and by taking care not to injure the stocks by either eating or cutting, and not allowing them to run to seed, the plants will endure, and supply useful fodder for several seasons.

RAPE.

Rape (*Brassica napus*, Natural Order *Cruciferæ*) is grown to a considerable extent as autumn food for sheep in the fold.

The main crop is usually sown in June, but small patches may be sown as early as April, to afford successive folds of green food as they may be required. Rape is usually ready for consumption about three months after being sown. If sown by the grain-drill, from 8 to 12 lb. an acre may be used, but if sown broadcast, from 10 to 16 lb. may be given. If it is intended to eat the crop when very young, half a bushel of Italian rye-grass may with advantage be mixed with the seed. The land should be well dunged, and a dressing of from 2 to 4 cwt. per acre of superphosphate along with the seed will be useful.

Rape delights in fen or peaty soils rich in vegetable mould. It is sometimes sown upon newly reclaimed peaty land, and consumed by sheep, thus helping to reduce the rough soil to a useful condition. In some cases rape is sown after an early crop of potatoes, and consumed early in winter.

Rape should be hand-hoed like turnips, but it is not so carefully thinned, although it undoubtedly affords the largest yield when the plants are thinned out to from 12 to 14 inches apart. Between the rows weeds must be kept down by the horse-hoe or drill-harrow.

Rape is sometimes sown along with vetches, the vetches being sown broad-cast over the rows of rape. This mixed crop affords admirable green food for sheep. Rape is also, in some cases, sown in seed-beds, and planted out like cabbages. It is well suited for clay lands when it is sown early and consumed in summer and early autumn, when these lands will bear sheep without injury. Then the early removal of the crop admits of the land being prepared in good time for wheat.

In upland districts, where a grain crop is less desired than a good green crop for summer and autumn feeding, the rape is best sown broadcast along with the grass-seeds in spring. Broad-leaved summer rape is the best variety, and it should be sown at the rate of 16 lb. an acre, in addition to the usual grass-seeds and clovers. The rape and young seeds will be ready to feed by sheep in July, and will provide good feeding between then and the end of November. Moreover, the young seeds will be much better than they would have been if sown down with a corn crop, while the land will be greatly enriched by feeding off the rape. This is an excellent plan to pursue in laying land down to permanent pasture. The sheep should not be folded on the young seeds, but the rape and seeds stocked at the rate of 8 to 12 sheep per acre.

Rape is known to possess high fattening properties. It is also highly stimulating, and is the best of all food for flushing ewes at tupping time. To lambs, and also to ewes in lamb, it should be fed in moderation, and along with hay or other dry food.

MUSTARD.

Mustard (*Sinapis alba*, Natural Order *Cruciferæ*) makes a very useful catch crop. It grows up very rapidly, being ready for consumption on the land by sheep in about eight or nine weeks after being sown. The white mustard may be sown in southern counties after an early corn crop, about a peck of seed being sown broadcast. It is sometimes also sown in spring before a late crop.

Mustard is very sensitive to frost, and should therefore be consumed before the end of October.

In many cases it is sown to be ploughed under as a green manure. For this purpose it is also very useful. Besides affording useful manure itself, it helps to prevent the waste of nitrates, which, instead of being washed away in drainage-water —which would probably happen if the soil were bare — are stored up in the growing plant.

SOWING CRIMSON CLOVER.

The crimson clover (*Trifolium incarnatum*) is one of the most beautiful plants cultivated in the field. Its stem rises up to 2 feet or more in height, with spikes of tapering, nodding, brilliant scarlet-coloured flowers. It was long cultivated in the garden as a border annual. It is an excellent forage plant, and when sown in autumn, so quick is its growth that it affords the earliest cutting in

spring of any plant of its age. In Scotland it is useless to attempt its cultivation except as a garden plant, damp autumns and winter frosts never permitting it to come to a good result. It is successfully grown in some parts of Ireland.

Culture in England.—Trifolium is grown largely in the south of England. It produces the best results on soils of a loamy nature; but on thin, poor, highlying land the plant appears quite out of place.

Of all known plants it is best suited for the stubble of a cereal crop in England. It may either be drilled in summer in rows of from 8 inches to 1 foot distant, or sown in autumn broadcast on stubble, and covered by harrowing. From 20 lb. to 28 lb. of seed are sown per acre broadcast; and the quantity is increased or lessened according to the nature of the climate and soil. From the beginning of August till the first week of September is the best time to sow.

Crimson clover ripens its seed easily in England; and English seed of the first year after importation is the best, being heavier and more free of the seeds of weeds than the foreign seed.

As a Forage Crop.—When sown in autumn, the entire crop may be grown, cut down, and cleared off by June following, allowing the ground to be worked up for late turnips, to be consumed in autumn. When cut in full flower, it makes hay much relished by horses, and its entire yield may be nearly twice the weight of a crop of red clover. It is better suited to sow on stubbles than even the white turnip. It is more rapid in its growth than winter tares. On light land a crop of buckwheat may be readily obtained after it. Italian ryegrass may be sown with it, and will grow as rapidly. After the crimson clover has been cut, the rye-grass will continue to grow and afford an excellent second crop.

The crimson clover, having the property of smothering early weeds, is not well suited for sowing among a corn crop.

Late Variety.—A variety of crimson clover named by the French *tardif*, or late-flowering, was introduced to notice in France about 1836. If sown at the same time as the common variety in autumn, it will flower next season after that has yielded its crop, and thus form a valuable successor. Its characteristics are lateness of flowering and tallness, with vigour of growth.

Extra late Variety.—A few years ago Messrs Sutton & Sons, of Reading, introduced another variety, called Extra Late Red (*Trifolium incarnatum tardissimum*), which usually comes in from ten days to a fortnight after *T. tardif*, thus prolonging the supply of this valuable forage crop up to midsummer. The Extra Late Red has a special value on account of its suitability for filling up deficient clover leys, as if sown after the corn is off it is ready for cutting with the clover sown in the previous spring.

White Variety.—A fourth variety, with white flowers (Late White Trifolium), which comes in about the same time as *T. tardif*, is also spoken of favourably by some growers.

ITALIAN RYE-GRASS.

Italian rye-grass (*Lolium italicum*), growing rank and quick, is not so well suited for sowing among a corn crop as by itself when used as a forage crop. Its nature certainly indicates that it is better adapted for a forage than a pasture plant.

Sowing for Forage. — As a forage crop it should be sown by itself in a portion of the dunged fallow land in August, or the middle of September at latest, that it may acquire sufficient strength to stand the winter. It may be sown broadcast, there being no use of drilling it, since it will grow as early in spring as any weed, and will outstrip it in growth.

Seed.—On account of its natural tendency to produce many stalks from the same root, and its upright habit of growth, not forming a close turf, the seed should be sown at the rate of from 1½ to 2 bushels per acre. If sown in August or early in September, the seed should be well harrowed in, as the plants stand the winter better when the roots are moder-

ately well covered. The seed germinates well at 1 inch to 1¼ inch deep.

If the ground and weather are dry in spring, the roller should smoothen the surface.

Produce.—The crop will be ready for cutting in May, and yield from 3 to 5 tons of forage per acre.

Irrigation Crop.—Italian rye-grass is an extremely valuable crop for irrigated land. It is in general use on sewage farms throughout the country. For this purpose one seeding will generally last two years, one-half of the land being ploughed up and reseeded every year. Under sewage cultivation the same land may be resown with Italian rye-grass as often as is thought desirable, as when one crop follows another in that way, the crop does not seem to suffer in any degree from disease or other cause.

OTHER FORAGE PLANTS.

Furze, gorse, or *whin* (*Ulex europæus*, Natural Order *Leguminosæ*) as a forage crop will be referred to in the section Food and Feeding in vol. iii.

The kidney-vetch is regarded by some as a useful forage plant. Professor Wrightson thinks it worthy of a trial, and says that it ought to form an ingredient in mixtures of permanent pasture seeds intended for light and thin soils, in which this plant finds its most suitable position.

Maize and *sorghum* are both recommended as forage plants for southern counties, but they have not been largely grown. They are unsuited to northern districts.

Regarding the merits of *prickly comfrey* (*Symphytum asperrimum*, Natural Order *Boragineæ*) as a forage crop there is much difference of opinion. For odd corners it is of some use. It requires heavy manuring. It is perennial, and the plants are dibbled in 18 inches apart, in rows from 18 inches to 2 feet apart.

Buckwheat has also a certain value as a forage crop. It is very susceptible of injury from frost, and can seldom be sown with safety earlier than May.

HOP CULTURE.

The hop is the most speculative of all the farm crops grown in the United Kingdom. Its produce varies from little more than 2 to 20, or perhaps even 25 cwt. per acre, worth from less than the cost of picking to upwards of £20 per cwt. The more usual prices run from 50s. to £5. The area under hops in England has varied greatly. At one time it exceeded 70,000 acres, but with lower prices, due mainly to the importation of hops, the growth of the crop has diminished to a large extent. Many fortunes have been made and lost in the growing of this crop, around which has gathered a halo of romance which hop-farmers delight to contemplate and talk of.

The hop requires a fine climate and good land. Its cultivation in this country is confined mainly to the English counties of Kent, Surrey, Sussex, Hampshire, Worcestershire, and Herefordshire.

One feature of the agriculture of the principal hop-growing districts is, that the hop may be almost said to monopolise the attention of the farmer, with the result that the other crops of the farm occasionally suffer.

The hop (*Humulus lupulus*) belongs to the class and order *Diœcia Pentandria* of Linnæus, Natural Order *Urticaceœ*. It is generally believed that the hop was introduced into this country in 1524.

Varieties of Hops.—No fewer than about 160 varieties of hops are said to be in culture throughout the world. In this country only a small number are in regular cultivation, the principal of these varieties being — Goldings, Bramlings, Grapes, Jones, Farnham Whitebines, Mathons, Cooper's Whites, Fuggles, and Colegates. The Golding is generally acknowledged as the best variety in this country, but for certain localities other sorts are more suitable.

The selection of the most suitable variety for a given locality and soil requires considerable experience and good judgment. Whichever kind is chosen, it is desirable either to have only one variety within one hop-ground, or the varieties separated in the same ground. Different varieties require to be pulled at different times. It is desirable, in choosing different varieties, to have them to ripen in succession, in order that the hops may not all be ready for picking at the same time.

Male and Female Hops.—The male and female being on separate plants, there has been a good deal of discussion as to the necessity or desirability of planting male plants so as to ensure fertilisation. It is contended by many that there should be at least one male plant on each acre of hops. But in practice no attention is paid to this. The male plants are generally grubbed up, and the fertilisation of the female plants left to chance.

Soil for Hops.—The soil for the hop plant should be deep and mellow, and if resting on a fissured rock, so much the better. An old meadow forms the best site for a hop-ground. In every case the ground should be dry—not subject to stagnant water, and, if not naturally dry, it should be made so by thorough drainage. To afford sufficient room for the roots of the plants, the drains should be not less than 4 feet deep, and the distances between them from 15 to 35 feet, according to the tenacity of the subsoil.

Preparing Land — Land which is about to be planted with hops is either trench-ploughed or trenched by hand in the autumn before planting. In the former practice the land is ploughed deeply with an ordinary plough, followed by a subsoiling apparatus, to break up the hard bottom as shown in figs. 323 and 324, pp. 401 and 402, vol. i. If this plan is not thought advisable or practicable, then the land is dug by hand labour to the depth of two spades. It is considered a good preparation to fold sheep on the land and feed them well before ploughing or digging; and at the time of ploughing or digging a heavy dressing of farmyard manure is given.

Rearing Hop Plants.—Hop plants are raised from cuttings taken from the "hills," or plant-centres, when these are being dressed early in spring. The cuttings are reared in a nursery until about the end of the autumn of the same year, by which time they have formed a strong root. These sets, or young plants, may be purchased from those who give special attention to their culture, and good judgment is required in selecting the kind best suited for each particular locality. Local experience is the best guide as to this, as well as in regard to many other points in farm practice. Attempts to raise hop plants from seed have not been successful, owing to the strong tendency of the plant to revert to its wild type.

Planting Hops.—The planting takes place either just before winter sets in or early in spring. The cuttings or shoots are planted in "hills," two or three to each "hill." Changes in the systems of training have involved modifications in the standard distances between the "hills." When poles were in vogue the "hills" were usually from 5 feet 6 inches to 6 feet 6 inches apart each way, but now the alleys are wider and the plants closer together in the rows. The common width between the rows under modern conditions is 6 feet 6 inches, while the "hills" are from 4 to 5 feet apart in the rows.

Since the fourth edition of this work was published, poles as a medium of cultivation have been very largely abandoned, their place being taken by a system of wire and string or coir yarn. In different districts variations of the wire and string arrangement are met with,—what is known as the Butcher system obtaining in some localities, while the umbrella and cross-over designs find favour in other parts. The principle is the same whatever the precise method adopted, and the net result is that the hops are trained on coir yarn instead of poles. The young shoots of the plant make much more rapid progress on string than on poles, and as a rule there is a stronger growth of bine, the main shoots reaching a higher point before throwing out laterals than when poles were used. The laterals as well as the upright bine are stronger, and spread over a larger surface when string is employed, and in this way the plants

have a better exposure to sunshine and air, which conduces to the production of a larger crop of bigger and better developed hops than it is possible to grow on poles. It has been contended that the depreciation in the hop markets is largely due to increased production per acre, and if this is the case there is no doubt that the adoption of the wire and string system of training is chiefly responsible, for it has enabled growers to obtain a substantial increase in the average return of hops.

Cost of Wiring.—The cost of wiring an acre of hop land will run from £30 to £45 per acre, the price being influenced largely by the price of the posts or poles required for standards. If poles are plentiful and can be obtained within easy carting distance of the hop farm, the lower figure may cover the expenditure, but in a district where poles are difficult to procure and have to be brought from a distance, the outlay may easily reach the higher sum. At the lowest estimate 250 poles are required per acre. The outside or straining poles are usually 5 or 6 inches in diameter at the bottom, and the intermediate poles from 3½ to 4 inches in diameter. Larch fir poles are most sought after for the purpose, as they are generally straighter and last longer than other kinds. After larch, chestnut is perhaps the most suitable. The coir yarn used in training costs from £19 to £23 per ton, and about 2 cwt. is required per acre.

Where poles are used it is necessary that they be sufficiently strong and long to support the plants effectively. The best poles are of larch, ash, chestnut, willow, oak cut in winter, alder, beech, in the order enumerated.

The crop is not expected to give any produce the first year. In that season the ground between the hills may be utilised in growing potatoes, cabbages, or some such crop, though it is better not to do this, as the hop plants require a great quantity of manurial substance.

The hop land has to be thoroughly dug or ploughed every autumn or winter, the "spud"—a three-pronged fork or "graip" with broad points—being used for the former purpose.

Dressing Hop Plants.—Early in spring the adult hop plants are dressed as soon as the soil is sufficiently dry to be worked satisfactorily. The old bines and fibrous growth of the previous year are cut away, and some fine earth is thrown over the "hills."

These are nice operations, and require an experienced hand to execute them, otherwise the success of the future will be rendered doubtful. Mr Rutley writes thus particularly on this subject—after stating that one boy or woman opens around the stock of the hill, with a small narrow hoe, a little below the crown of the hill—"one man follows with a pruning-knife and a small hand-hoe, with which he clears out the earth on the crown of the hill between the sets or shoots of last year that were tied to the poles; and which, from having earth put on them the preceding summer, swell out to four or five times their original size, and form what we call sets or cuttings; and it is the cutting them off at the right part that should be particularly attended to, or great injury may be done. It is therefore necessary that the .person cutting them should ascertain exactly where the crown of the hill is, that he may not cut them too low or too high; and the place where they should be cut off is between the crown of the hill and the first joint, for it is around the set close to the crown where the best and most fruitful bine comes. If the set is pared off down too close to the stock or crown, it takes away the part from where that bine comes, as little buds are seen ready to shoot forth at the time of cutting, which, if cut off, the bines come weakly and few. On the other hand, if the set is cut off above the first joint, which sometimes will be the case if the man in cutting does not pay the attention to it he ought, the bines which come from that or any other joint higher up the set grow fast, but are coarse, hollow, or what we call pipy, and unproductive: all such should be discarded at the time of tying. Consequently the operation of cutting or dressing, on which the future well-doing of the plant so much depends, is not left so much to the judgment or skill of the operator as to his care and attention. Many planters have their hops dressed by the day, paying extra wages to persons in whom they can confide to do it with care. After all the old bine and runners,

as the roots and small rootlets near the surface are called, are cut and trimmed off clean, some fine earth is pulled over the crown, and a circle made round with the hand-picker, to intimate where the hill is before the young shoots appear."

The dressing should be finished before the bines begin to show. Such of the sets as have two or more joints are selected to put into a nursery, or sold for that purpose. But the cuttings should be taken only from the most healthy bines.

The stringing operation follows. The strings are tied to the wires, three or four strings being placed to each "hill," and after the hop shoots have been served, two or three to each string, the remaining shoots are removed by cutting or by pulling out.

After Culture. — The vacant spaces between the "hills" must be well cultivated, kept free from weeds, and heavily manured. For the cultivation of the land a sort of horse-hoe, or "nidget" as it is called in Kent, is used.

In the month of June the "hills" are earthed up by the spade, and in some cases by the plough. After this, until picking, attention is confined to the cultivation of the vacant ground—that is, unless fungoid or insect foes attack the plants and demand serious treatment.

Manuring. — Hops are greedy for manure. The annual produce of a hop-ground consisting of hops and bines is very considerable, and as the perennial nature of the plant does not permit it to be placed in the category of those plants of the farm which follow each other in any given rotation, it is necessary to manure the ground at least once, if not twice, every year. The first manuring after the crop may be given in autumn or spring; and if in spring, the time is before the digging of the ground commences. The best plan is to apply the manure twice a-year: in the autumn or spring, with farmyard manure and woollen rags, and during the summer with some such manure as guano, rape-cake dust, and super-phosphate of lime. Of farmyard dung, from 25 to 30 cubic yards should be given to an acre. Black mould is an excellent application about the crown of the roots. The dung should be carted on to the ground before winter ploughing or digging, and if applied afterwards, is drawn on to the land between the rows of "hills" in long narrow carts called "dollys" in Kent. Of woollen rags from 12 to 20 cwt. per acre; woollen waste or shoddy from 20 to 30 cwt. per acre; and guano rape-cake dust, and superphosphate of lime, 6 or 7 cwt. per acre, are convenient applications, in June and July, generally dug in closely around the "hills," and sometimes spread over the surface, and hoed in with horse-hoes or "nidgets." Mustard-cake makes a good manure for the hop plant.

There is difference of opinion as to the advisability of applying large dressings of nitrate of soda to hops. Many growers of wide experience are opposed to it, and at most do not give more than 2 cwt. per acre—1 cwt. in spring to push on the plants, and another cwt. about the time the bine is coming into burr to keep the hops in colour. With some growers it is a custom to apply 1 cwt. per acre any time before June sets in.

Experiments in Manuring Hops.— Dr Bernard Dyer and Mr F. W. E. Shrivell conducted continuous experiments on the manuring of hops for a number of years on the latter's farm at Golden Green, Tonbridge. They consider that, in the past, most growers have paid too much attention to supplying nitrogen and too little to the use of phosphatic manures. They recommend the use of superphosphate at the rate of 8 to 10 cwt. per acre annually for calcareous soils, or basic slag at the rate of 10 cwt. per acre annually for soils very poor in lime; or an alternation of the two manures for soils moderately poor in lime, or equivalent mixtures of superphosphate and bone-meal or guano. They also recommend 2 cwt. per acre per annum of sulphate of potash, except on soils naturally rich in available potash, or when a dressing of dung is used. In the presence of abundance of phosphates and potash, they find that nitrate of soda may be used to the extent of as much as 8 cwt. per acre year after year, in the absence of other nitrogenous manure, with advantage to the yield of hops and without detriment to quality. But they recommend that a variety of nitrogenous manure is also good, and that, therefore, a portion should be supplied as dung or

as rape-cake, fish manure, or shoddy, in accordance with custom, but with the addition of nitrate of soda. Their experience, confirmed by observation on other farms, shows that even when the soil is otherwise liberally manured by autumn or winter dressings of dung or of rape-dust, fish guano, &c., 4 cwt. of nitrate of soda per acre, applied early in the season, may be regarded as a thoroughly safe dressing for hops, even in a season which turns out to be wet. This, however, is a dressing which, under these circumstances, it would be best not to exceed in the case of the more delicate varieties of hops. If neither dung nor any of the other ordinary nitrogenous fertilisers has been recently applied, there appears to be no reason to anticipate that 6 cwt. of nitrate of soda per acre would be otherwise than safe; while for freely-growing and heavily-cropping varieties as much as 8 cwt. per acre might be used. But a more general manuring, including a smaller quantity of nitrate of soda than this, will probably in the end commend itself to most growers.

Messrs Dyer and Shrivell advise that large dressings of nitrate of soda should be divided into separate applications of not more than 2 cwt. per acre each, with an interval of some weeks between the different dressings. The most favourable time for application on soils of medium consistency is probably April or May. For stiff or not readily permeable soils they regard this as the latest time at which, under normal conditions of weather, nitrate of soda should be applied, and they are inclined to prefer April to May for the final application.

But once again they urge that it should be borne in mind that neither nitrate of soda nor any other nitrogenous manure can be relied upon to produce a healthy growth and a heavy crop unless at the same time the soil is kept abundantly supplied with some form of readily available phosphatic manure, and also of potash, except on those soils on which it may have been experimentally ascertained that the latter is superfluous.

Creosoting Poles.— Hop-poles or standards are now universally treated with creosote at the ends, and this preparation makes them last about twice as long as before the practice was introduced. The creasote, purchased at about 2½d. or 3d. per gallon, is poured into a tank into which the poles are set on end and kept there, sunk about 18 inches in the creasote for fully twelve hours. By this treatment the end of the pole which is stuck into the ground is rendered quite impervious to wet. There are three standard lengths of poles, 12, 14, and 16 feet.

Poling.—In most cases there are three poles to each "hill," but often only two, and in other cases the one "hill" has three poles, and the next only two. The two or three poles are set around each "hill" at equal distances apart. A hop-pitcher makes a hole deep enough to give the end of the pole a firm hold of the ground, which should be about as many inches in depth as the pole is feet in height. The pole is pushed down to the very bottom of the hole, and if it have any crook or set at the lower end, that is placed inwards, to be out of the way of the horses in "nidgetting" the ground; and the top should have a slight lean outwards, to give room to the bines to branch, and let in air and light.

Tying up the Bines.—Whenever the bines shoot to a length to be fastened, they are tied to the poles. In some seasons when the bine comes very early the coarser bines are pulled out. Three of the best bines are selected to be tied to each pole, and the rest are cut away. Withered rushes are used for tying; and the tie is made with a slip-knot, so that the tying may give way as the bines enlarge in diameter. The tyings are done from near the ground up to 5 feet above it, and when above that height ladders are used, which stand independently upon the ground. In the string system two or three bines are placed to each string. The bines do not require tying, as they cling to the coir-yarn. The tying begins about the end of April or beginning of May. From 18 inches to 2 feet of the lower end of the bines may be stripped of their leaves, to allow air to get to the crown of the roots.

Longevity of the Hop.—The power of some hop-grounds to produce a great crop year after year, when external circumstances are favourable, is extraordin-ary. Many grounds have borne crops for upwards of half a century, and some exceed in age an entire century. It must not be supposed from this, however, that any plant which had been planted at the formation of the ground remains alive such a length of time. Whenever a plant or an entire "hill" indicates symptoms of decay, it is removed and another substituted, care being taken to plant the same kind of hop as that cultivated in the ground.

Insect and Fungoid Attacks.

The hop is unfortunately subject to serious injury from various insects and fungi. As to these attacks, see the sections on Fungus and Insect Plagues on the subject in this vol.

Harvesting Hops.

Picking.—The harvesting of hops is really autumn work, yet the brief description to be given of the process may be conveniently introduced here.

Hop-picking usually begins about the last days of August or the first week in September. The picking is a tedious process, demanding the employment of a great number of hands. As a rule, the picking is done by bands of immigrants, men, women, and children, who wander to the hop-growing districts from large towns and villages. The process generally extends over three weeks, and the immigrant pickers live in extemporised villages, in huts, hopper-houses, or tents provided for the purposes.

The process of picking on the poling system was thus described by Sir Charles Whitehead, who did much by his writings to improve the practice of hop culture:[1]—

"The pickers are divided into companies of eight or ten, each of which is under the charge of a ganger or 'binsman,' who pulls up the poles for the pickers with wooden levers having iron teeth, called 'dogs,' and holds the 'pokes,' or sacks, or 'sarpliers,' for the measurer when he comes round to measure the hops that have been picked. In most cases the bines are cut about 2 feet from the

[1] *Jour. Bath and West of Eng. Agric. Soc.*, 1881, 208.

ground, and the poles are pulled by means of 'dogs,' or wooden levers with iron teeth, and carried to the pickers, who pick the hops from them into the bins or baskets. Occasionally when the hops are not quite ripe, or when the plants are weak, the poles are not pulled, but left standing. The bines are cut 4 or 5 feet high, and the bines with hops upon them are pushed up and over the poles with forked sticks, as bines cut high and kept to the poles in the upright position do not 'bleed' so much, or lose so much sap, as when cut short and left lying on the ground. The hop-grounds are marked out into as many portions or 'sets,' containing 100 hills, as there are companies, for which lots are drawn by each binsman, so that there may be no wrangling about good or bad sets. The hops are picked into bins — long light wooden frames with sacking bottoms. There is one of these for every two adult pickers. In Mid and West Kent and the Weald of Kent and Sussex, and in Worcestershire and Herefordshire, these bins are used. In East Kent large baskets are used for picking into, holding 15 to 20 bushels. In Hampshire and Surrey the hops are picked into baskets holding 7 bushels, which are emptied into long bags, called 'sarpliers,' holding 14 bushels, in which they are taken to the kilns. In Kent, Sussex, Worcestershire, and Hereford-shire the hops are measured into pokes —sacks holding 10 bushels—in which they are taken to be dried. The measurer, who generally takes from six to eight companies, is accompanied by a boy, who enters the number of bushels picked into a book kept by each picker, and into a book retained by himself.

"The *price* of picking hops ranges from 1¼d. to 3d. per bushel. The average price is 2d. per bushel. Binsmen are paid from 2s. 4d. to 3s. per day. Measurers get from 4s. to 5s. per day. Driers, who work night and day, earn from £2, 10s. to £3, 15s. per week. Before picking commences, the planter generally fixes a price for picking. Sometimes it is not fixed until after a day or two, that it may be better ascertained how the hops come down."

The alteration in the system of training the hops has necessitated changes in the methods of picking. Whereas it was the usual practice when poles were in use to have the hops brought to the pickers at a given point, the latter now move from hill to hill until the whole plantation is completed, just as they would do in the pulling of any kind of fruit. There is the additional difference in the method of harvesting hops grown on strings from those grown on poles, that the former are not cut at the bottom, but are merely detached from the top wires and fall down within convenient reach of the pickers. In harvesting, as in other respects, the wire and string system is more convenient than that which it superseded.

Drying Hops.—Immediately on being picked, hops are artificially dried. They are dried in square or circular kilns, 16 or 18 feet square or in diameter, on hair-cloth, and heated by Welsh coal, coke, or charcoal. The kiln-floor is situate at 10 to 13 feet above the fire, and the height of the kiln is 18 to 20 feet above the kiln-floor, surmounted with a cap-cowl 7 or 8 feet in height and 3 or 4 feet diameter in the bottom, a free circulation of air being kept up through the fire and hops to the top of the kiln. The hops require to be rapidly dried to keep the pickers in operation. The kilns ought to take 1 bushel of green hops on 1 square foot of flooring, and be filled twice a-day, giving from 9 to 12 hours to each kilnful, so that from 200 to 250 bushels may be drying on one kiln at a time.

Artificial draught by means of fans, either exhaust or forced, is now largely used, and by its aid the foregoing quantities are easily more than doubled. Fans for exhaust, used in conjunction with the open fires, can be installed for a moderate outlay, and in all cases they should be placed as high up in the roof of the kiln as possible, as when they are high the heat is more evenly distributed over the floor than when they are placed near the hair-cloth or drying floor. Several systems of hot-air drying are in use. In most cases they are on the forced-draught principle— viz., a fan propelling air through hot-air furnaces or between coils of pipes heated by steam or hot water. This is claimed to be the ideal system for hop drying, as it allows only heated air to pass through the hops, and thus prevents impurities that might arise from bad or inferior fuel coming in contact with the hops. It is only fair to say, however, that if the best anthracite coal is used in drying in the open-fire system the impurities are practically nil. The chief drawback to the use of hot air is the original cost. The average grower has difficulty in persuading his landlord, or himself, to expend a sum which may be roughly calculated at 10s. per square foot of drying space when hop-growing is a declining industry.

For two kilns of these dimensions, a cooling-room of 20 feet in width and 40 feet long is required. This should be on a level with the kiln-floor. And there should be another room of similar dimensions, under the cooling-room, for stowing and weighing the hops in the pockets.

Great caution is required to regulate the fires of the kilns. If too strong at first, when the hops are naturally moist, they will be drawn down to the haircloth and be much deteriorated in quality. The fire may be increased as the drying proceeds, until a temperature not exceeding 130° Fahr. is reached, and the drying process finished at a temperature gradually decreasing to 115° Fahr. About 13 cwt. of coal, with a little charcoal, will dry one ton of hops.

Sulphur is also used in drying hops, from ¼ to 1 cwt. to 1 ton of hops. The object of using sulphur is to improve the colour of the hops. It is of importance to the seller to present his hops in the market with a light-coloured delicate primrose hue.

When taken from the kiln, the hops are laid in heaps on the cooling-floor, not only to cool but to acquire a state of adhesiveness, which, though dry, causes them to lump together when squeezed in the hand, and yet not so much as to lose elasticity.

The drying will cause a loss equal to about three-fourths of the weight in a green state—giving 1 lb. of prepared for 4 lb. of green hops.

Pocketing. — Hops are put into pockets in the stowing-room, through an opening in the floor of the drying-room, under which the pockets are suspended.

A *pocket* is 3 feet wide and 7½ feet long, consisting of 5 yards of cloth, weighing 5 to 6 lb., and when filled contains 1 cwt. 2 qrs. to 1 cwt. 3 qrs. and a few pounds gross weight of hops.

Hops are packed in the pocket by powerful screw-pressing machines devised for the purpose. These presses, which are formed upon one principle, differing only in detail, were thus described by Sir Charles Whitehead :—

"A wooden circular foot, just large enough to go into a pocket 3 feet in diameter, is fitted to a ratchet lever, which is worked up and down by handles. This is fixed immediately over the 'pocket hole' cut in the floor. The empty pocket is fastened to a movable frame or collar, so as to keep its mouth firm to the floor while it is being filled, suspended in mid-air. There usually are two posts set up below, into which two rods, connected with a wooden stand, run up to hold the pocket up and to keep it straight. In place of these guiding rods, some pressers have circular iron cases to surround the pocket and keep it from bulging. Pressers cost from £14 to £27." [1]

The pocket is neatly sewn up, leaving a lug, or ear, projecting from each side of the sewn mouth.

The produce is then ready for the market.

Stacking Poles.—When the bines are cleared of the hops, they are taken off the poles, which are then put up in small conical stacks at equal distances apart on the hop ground, with the sharped ends on the earth, having four equidistant divisions striding over the "hills." Each division of the stack should be bound round with three bines, deprived of their leaves and twisted into a rope, which binds the division close and compact, and prevents the poles being stolen, or makes a theft more easily detected. The small refuse poles are bound together, separating those which may be used for the young bines of the first year from those which may be burned into charcoal, or used as firewood.

Analysis of Hops.—The quantity of mineral matter removed from the soil per acre by the different parts of the

---

[1] *Jour. Bath and West of Eng. Agric. Soc.,* 1881, 214.

Golding hop plant is, according to Way and Ogstone :—

	Flowers.	Leaves.	Bine.	Whole crop.
	lb.	lb.	lb.	lb.
Silica	32.6	97.3	12.9	142.8
Phosphoric acid	29.5	40.6	15.1	85.2
Sulphuric acid	8.7	8.2	3.0	19.9
Carbonic acid	3.4	52.4	15.4	71.2
Lime	16.3	134.0	31.0	181.3
Magnesia	8.2	21.1	4.9	34.2
Peroxide of iron	1.1	0.8	1.0	2.9
Potash	54.0	57.0	22.9	133.9
Soda	...	...	...	...
Chloride of potassium	15.3	10.0	19.9	45.2
Chloride of sodium	1.3	13.6	3.4	18.3
Total	170.4	435.0	129.5	734.9

The hop plant is peculiar in the quantity of phosphoric acid required for all its different parts, as is seen above.

Spent Hops as Manure. — Spent hops are used largely for manure. The analysis of their ash, by Nesbit, is as follows :—

Potash	.	.	.	1.45
Lime	.	.	.	23.70
Magnesia	.	.	.	2.75
Phosphate of iron	.	.	2.50
Sulphuric acid	.	.	.	3.05
Phosphoric acid	.	.	.	4.10
Carbonic acid	.	.	.	9.00
Chloride of sodium	.	.	2.95
Chloride of potassium	.	.	0.70
Silica (soluble)	.	.	.	27.10
Sand and charcoal	.	.	21.80
				99.10
Percentage of ash	.	.	10.40

Cost of Hop-planting and Cultivation.—The planting of hops is very costly—generally about £24 to £25 per acre. The yearly outlay on the cultivation and handling of the crop is still more expensive. It usually runs from £35 to £45 per acre.

Produce of Hops.—The yield of hops varies greatly with the seasons, both in quantity and quality. It usually averages about 8 to 11 cwt. per acre.

FLAX CULTURE.

Flax (*Linum usitatissimum*, Nat. Order *Lineæ*) is cultivated for fibre or for seed, or for both. In the United Kingdom it is cultivated most largely in the north of Ireland, where it is grown with great success to supply fibre to the extensive linen-mills of Ulster.

With the exception of the scutching of the flax—*i.e.*, the separation of the fibre —the Irish farmers undertake all the processes connected with the crop, from the preparation of the land and the sowing of the seed to the selling of the scutched flax to the spinners.

In many other parts, notably on the continent of Europe, the farmer merely grows the crop and leaves its manipulation to others. In order to obtain the finest fibre, the flax has to be pulled before the bolls or seed are ripe. In Ireland the crop is grown for fibre alone, on the Continent for seed and fibre, and elsewhere almost solely for seed.

The flax plant is stated to be a native of Britain ; and yet flax seed was not sown in England until A.D. 1533, when it was directed to be sown for the making of fishing-nets.[1]

Ure says of the flax plant : "In it two principal parts are to be distinguished— the woody heart or boon, and the *harl* (covered outwardly with a fine cuticle), which encloses the former like a tube, consisting of parallel lines. In the natural state, the fibres of the harl are attached firmly not only to the boon but to each other, by means of a green or yellow substance. The rough stems of the flax, after being stripped of their seeds, lose in moisture, by drying in warm air, from 55 to 65 per cent of their weight, but somewhat less when they are quite ripe and woody. In this dry state they consist, in 100 parts, from 20 to 23 per cent of harl, and from 80 to 77 per cent of boon. The latter is composed, upon the average, of 69 per cent of a peculiar woody substance ; 12 per cent of a matter soluble in water ;

[1] Haydn's *Dict. of Dates*—art. "Hemp and Flax."

and 19 per cent of a body not soluble in water, but in alkaline lyes. The harl contains, at a mean, 58 per cent of pure flaxen fibre, 25 parts soluble in water (apparently extractive and albumen), and 17 parts insoluble in water, being chiefly gluten. By breaking the harl with either hot or cold water, the latter substance is dyed brown by the soluble matter, while the fibres retain their coherence to one another. Alkaline lyes, and also, though less readily, soap-water, dissolve the gluten, which seems to be the cement of the textile fibres, and thus set them free. The cohesion of the fibres in the rough harl is so considerable, that by mechanical means—as by breaking, rubbing, &c.—a complete separation of them cannot be effected, unless with great loss of time and rupture of the filaments. This circumstance shows the necessity of having recourse to some chemical method of decomposing the gluten. The process employed with this view is a species of fermentation, to which the flax stalks are exposed. It is called *retting*, a corruption of rotting, since a certain degree of putrefaction takes place." [1]

Flax is manufactured into twine, rope, and thread, and into fabrics, varying in texture from coarse bagging, employed to pack cotton or hops, to canvas, linen, cambric, and finest lawn.

In former times the seed of the flax was occasionally used with corn to make bread, but it was found to be ill suited for the purpose, being difficult to digest.

As is well known, linseed, the seed of flax, possesses great value as an article of food for cattle.

Soil for Flax.—The flax plant thrives best on a rich kindly loam. On sandy soils the crop will not, as a rule, give a good yield of fibre. Good crops are often grown on clay land. Any soil in too high condition causes flax to be rank, branching, and coarse, and if the crop lodges the fibre will be deteriorated.

Rotation for Flax.—The finest flax is perhaps best obtained after corn. In the north of Ireland, where flax cultivation is pursued extensively and with great success, it is not considered good practice to grow flax after lea. It is

difficult to get lea-land into a sufficiently fine tilth, and insect attacks are more frequent after lea than after oats. Flax does well after potatoes. It should not be repeated on the same land at shorter intervals than about seven years; some say nine years would be better still. Flat land is preferable to undulating, hilly, uneven land rarely producing flax of a uniform reed. The following is a general rotation in some districts in Ireland: 1st year, oats after lea; 2nd year, roots; 3rd year, oats; 4th year, flax with clover and grass seeds; 5th year, seeds hay; 6th and 7th years, pasture. In another rotation flax is taken between lea oats and roots. In both these cases, however, the flax is liable to be weedy, and some farmers prefer, therefore, to grow flax directly after the cleaning root crop.

Tillage for Flax.—After cereals, the land for flax should be ploughed early in winter, to receive the full effects of the frost. It cannot be in too fine a state of pulverisation when the seed is sown provided a firm seed-bed is secured. To promote this fine state of the soil, cross-ploughing should be executed early in spring, taking care to avoid wet weather, or the soil in a waxy state, as dry weather following renders the soil difficult to be pulverised. Clods left on the surface, after a double turn of the harrows, should be reduced by a clod-crushing implement.

The cross-ploughing in spring should be done about two months before sowing. Medium land after potatoes will do with one ploughing from four to six weeks before sowing. Heavy land after potatoes should be ploughed as early in the year as possible. Plough shallow, about 4 inches deep, after potatoes. It is recommended that as far as possible weeds should be removed by forks and graips in the preparation of the seed-bed. Flax delights in a firm, even seed-bed, and if the land is naturally dry or well drained, it thrives best broadcast or in rows on the flat. In drills it is more apt to be uneven in length, and it is very important that flax should be as uniform in length as possible. Light land should not be too much stirred, but heavy land cannot be too much pulverised.

Clod - crushers. — Crosskill's clod-

[1] Ure's *Dict. of the Arts* —art. "Flax."

crusher, shown in perspective in fig. 618, is a most efficient implement. The roller consists of a number of toothed wheels, supported on feathered arms, and

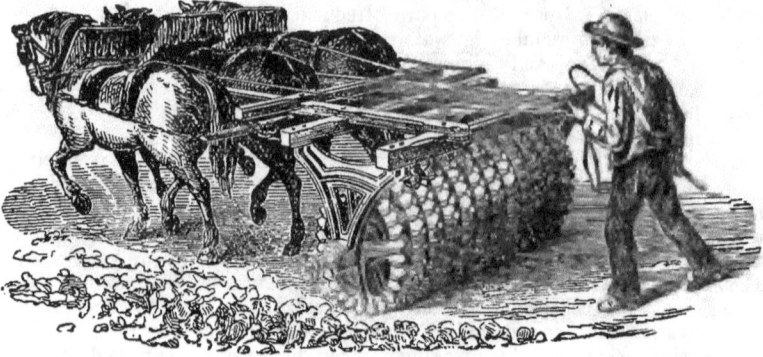

Fig. 618.—*Crosskill's clod-crusher.*

an eye formed in the centre fitted to move easily on the axle of the roller.

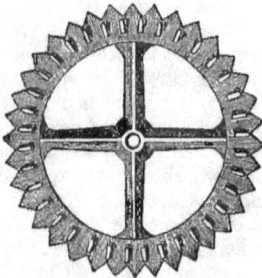

Fig. 619.—*Side view of one wheel of the clod-crusher.*

Fig. 619 shows a side view of one of those wheels, by which its action upon the soil may be easily understood.

This clod-crusher has been only slightly used by Scottish farmers, but it is extensively used in England to break down clods on strong clay soils.

Norwegian Harrow.—Another very useful pulverising implement is the Norwegian harrow (Clay), shown in fig. 620. The action of this machine is to reduce large clods into very small ones, by the insertion of the points of the rays into them, and to split them into pieces by their reiterated action. The larger clods are split into smaller pieces by the first row of rays, the second row splits these into smaller ones, and the third row splits those smallest pieces into still smaller ones; so that, by the time the clods have undergone those various splittings, they are probably sufficiently pulverised.

Spring - Tined Cultivator. — The

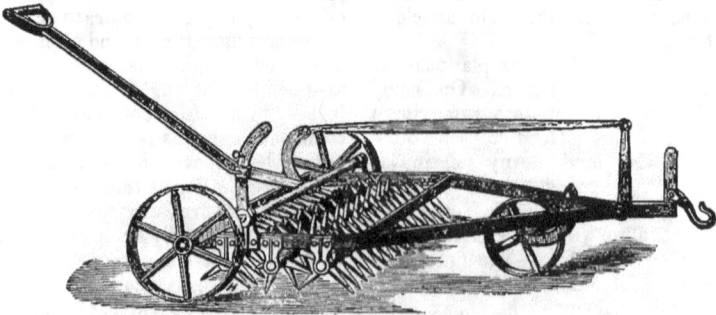

Fig. 620.—*Norwegian harrow.*

spring-tined cultivator, such as is shown in fig. 604 in this vol., is very useful in the preparation of a seed-bed. It is, in the long-run, quite as efficient as the Norwegian harrow, being, of course, of much lighter draught.

Sowing Flax.—The time for sowing flax will of course depend partly on the climate of the district and partly on the character of the particular seasons. As a rule, from the last week in March till the third week in April is the flax seed-time. The young flax plants, if the sowing has been done too early, are liable to injury from frost, which causes the plant to branch, thereby greatly lessening the value of the crop.

Prepare a fine, smooth, firm seed-bed with the harrow and roller. Some harrow, roll, and then sow; others harrow, roll, and harrow again, and then sow. The land may be marked off on the flat for the casts of seed with poles or footprints.

Seed. — Russian (Riga) and Dutch seed are most commonly used. Recent experiments have demonstrated that there is no foundation for the opinion held in many flax-growing districts, that Riga seed suits light soils and Dutch seed heavy soils. Good seed gives uniformly good crops on all classes of soil. Care should therefore be taken to secure pure seed, of high germinating power and a relatively heavy sample. A good average sample will be 98 per cent pure, show at least 95 per cent germination, and a thousand seeds or "pickles" will weigh 4½ grams. Riga seed is not usually so pure as Dutch, and at times, owing to the adverse weather conditions which have prevailed in the Continental flax-growing districts during the seed-harvest, very inferior samples of sowing seed are placed on the market. No pains should, however, be spared to secure good seed for sowing purposes.

As to the quantity of seed, if the crop is grown for fibre, 7 pecks (7 stones) of a sample conforming to the above figures will be required per acre; if for seed, a thinner sowing is advisable. Many samples of Riga seed do not, however, possess a germinating power of even 90 per cent, and are not 98 per cent pure. More seed will in such cases be therefore required. Flax-seed may be sown by a broadcaster, a seed-barrow as used for clover and grain, or by hand. If sown by hand, it must be done by a skilful person as the seed is very slippery. A strip with a light harrow will suffice to cover the seed.

Flax seed is oblong, lenticular in shape, having a smooth oily surface, feels heavy, and should be plump and fresh. A most useful hand-broadcaster is that named "The Little Wonder," commonly called the fiddle sower, illustrated in fig. 424, p. 125, of this vol. Since its introduction this machine has become most popular in flax-growing districts of Ireland.

Grass Seeds with Flax. — Land is frequently sown out into grass with the flax crop. Italian rye-grass is injurious to the flax on account of its vigorous growth, but perennial rye-grass, clovers, and natural grasses may be sown. In a wet season, however, this is an objectionable practice, for the grasses deprive the flax of its nourishment. Such flax when scutched presents a brown "root"—i.e., the fibre is browned at the base of the stem and its value accordingly decreased. The grass and clover seeds should be sown immediately after the flax seed, and the two harrowed in together with a light harrow. If it is desired to have Italian rye-grass after flax, the seed may be sown as soon as the flax is pulled, late in July or early in August.

As a catch-crop for districts where the climate is suitable, scarlet clover (*Trifolium incarnatum*) may be sown when the flax is pulled, and this will provide a useful cutting in the following May, after which the land may be prepared and sown with turnips. Others sow rape or winter vetches and rye after flax for spring food for stock.

Manuring Flax.—The flax crop does not bear being sown upon dung. The land should be in high condition from previous manuring. Recent experiments show that a potassic manure is to be recommended for this crop, particularly on lighter soils. A dressing of 4 to 5 cwt. kainit, or 1 to 1¼ cwt. muriate of potash per acre, applied preferably some weeks before the flax is sown, will increase the yield of fibre, improve the quality, and effectually prevent a diseased appearance known as "yellowing" in the early stages of growth. "Yellowing" is often attributed to frost, but the disease is caused in most instances by the presence of a fungus. The Belgians profusely top-dress their flax-ground with liquid manure with which have been mixed both rape-cake and nightsoil.

Weeding Flax.—The only attention which the flax crop requires in summer is to keep it free from weeds. These will appear as soon as the crop itself; and when the crop can be identified from them the ground should be weeded. Being in broadcast, and thickly sown, the only practicable way of weeding flax ground is by the hand. As the plant is firm and elastic, the stem is not injured by the weeders treading on it, if they are careful. The weeders should not wear shoes, so as to avoid injuring the flax. The crop should be weeded twice if the land is not clean before the plants are 7 inches high. Close hand-weeding is costly, but the increase of crop will more than repay the outlay.

Careful observation shows that, with the exception of a little corn-cockle, practically none of the weed seeds present as impurities in the sowing seed, as now imported from the Baltic and Rotterdam, germinate and grow with the flax crop. One of the dodders attacks the flax plant, but this is fortunately extremely rare. "The flax-dodder, *Cuscuta europæa*, is a plant," says a writer, "which germinates in the ground, and sends up a slender threadlike stem, which, twisting itself about, soon touches one of the stems of the flax amongst which it is growing. As soon as this takes place, the dodder twists itself round the flax, and throws out from the side next to its victim several small processes, which penetrate the outer coat or cuticle of the flax, and act as suckers, by which *the parasitical dodder appropriates to its own use the sap which has been prepared in the flax, upon which the growth of the flax depends.* The dodder then separates itself from the ground, and relies solely upon the flax for its nourishment, producing long slender leafless stems, which attach themselves to each stem of flax that comes in their way. Thus large masses of crop are matted together, and so much weakened as to become almost useless."[1] A thorough weeding will remove this pest from the soil before it has the power of injuring the flax plant.

[1] *Garden. Chron.*, Feb. 10, 1844, 189.

From the time the weeding is finished the crop needs no further attention till the pulling time approaches, usually in July or early August.

Pulling, Steeping, and Drying Flax.

Pulling.—The flax plant is pulled up by the root. The pulling is done after the plant has flowered and the seed attained a certain degree of maturity in the capsule or boll which contains it. As to the proper time to begin pulling, great care and judgment must be exercised. If pulled too soon the fibre will be weak; if allowed to ripen too much, the fibre will be dry and coarse, and deficient in spinning quality.

Flax is considered to be ready for pulling when all the leaves from the base up to half the height of the plants have fallen from the stalks. A more technical method of determining the fitness for pulling is to cut the ripest seed boll of an average plant transversely with a sharp penknife. The crop is ready when the seeds in the boll have lost all traces of milkiness, are firm and of a brownish colour. To defer pulling until too late a stage is a more usual fault than to pull the crop in too green condition. If allowed to become too ripe, the stems, particularly on light soils and in bright sunny weather, "fire"—*i.e.*, turn brown at the lower end —with disastrous results to the fibre.

Method of Pulling.—The crop is pulled by hand, and it is usual for the workers to pull a flat or ridge each. If the flat is narrow, he will catch the flax and pull with the right and left hands alternately until he has as much as he can hold. The flax is then tied in sheaves, or *beets* as they are called in Ireland. The sheaves are often tied with rush bands and made 15 to 18 inches in circumference, and in dull weather may be set up in stooks for a day or so before being carted to the retting pond. It is most essential to keep the flax *even* at the root-end, and this cannot be done without time and care; but it can be done, and *should always be done.* The *beets* or sheaves should always be equal-sized, straight and even.

Rippling.—In the flax-growing districts of Ulster rippling, or the removal of the seed bolls, is seldom if ever practised, the flax being steeped as soon as practicable after pulling. Recent experiments prove that rippling green flax is not remunerative, unless the crop is much branched and has produced an abundance of seed.

The process is carried out by pulling the seed heads of handfuls of the flax through an iron comb, a foot or so in length, fixed in an upright position across a plank which is carried on two supports about 18 inches high. The two men engaged in rippling the flax sit astride the plank facing the comb and each other, and strike their handfuls alternately through the comb. The seed bolls fall on to a sheet placed on the ground.

The arrangement of labour should be such that the rippling goes on simultaneously with the pulling. The rippled plants should be tied in sheaves, to be taken to the watering-pool to be steeped.

The Bolls.—The green bolls containing the seed should be thinly spread on the floor of an airy loft, and frequently turned to prevent heating. When thoroughly dry the bolls may be crushed and the linseed separated by a winnower or sieves. Seed obtained by this method is not suited for sowing purposes, as it is not well developed and matured. It can, however, be usefully employed as a food for cattle.

Another Method of Harvesting.—In Yorkshire and in certain districts of Holland and Belgium the crop when pulled is dried on the field. One plan is to stand up the flax in loose small sheaves, with the root end well spread out, so that the air can circulate freely between the stalks. If heating or fermentation takes place in the sheaf, the fibre is at once spoilt. When dry, the flax is taken to a central rettery, and there rippled before being steeped. Seed thus obtained is better developed and more mature than that from green bolls.

Steeping or Retting.—Next comes the steeping, a most important process. The object of steeping the flax plant is to separate the outer fibre of the stem from the inner core to which it is attached by adhesive mucilage. By retting this mucilage is changed, and

the process of separation of the fibre is completed in the scutch mill. Wherever possible, a pond with a clay bottom should be selected for the retting.

The question of water for the steeping pond is most important, for the quality of the fibre is largely dependent upon the water used. The most suitable is soft, and waters containing lime or iron compounds should be carefully avoided. Care should be taken to prevent flood water getting into the pond whilst retting is in progress. A very gentle even flow or trickle of water through the flax may, however, be beneficial. The finest of the Continental flax is retted in sheeted crates in the slow-flowing Lys at Courtrai. Good results are often obtained by retting flax in a large volume of water.

The sheaves or beets of flax are placed in rows, root or boll end down, in a somewhat sloping position in the pond, which should be narrow enough to enable the sheaves to be placed in position with a pitchfork from the bank. It is a good practice to allow some water to stand in the pond for a few days before the flax is put in. The sheaves are then weighted with stones or sods and the flax covered with about 3 inches of water. If during fermentation the sheaves rise, more stones are placed on them. The length of time required for the retting operation depends upon the water, its temperature, and the nature of the straw. Suitable water, warm weather, and coarse straw will conduce to quick retting. The average time taken for retting in the north of Ireland with water at 64° Fahr. is about nine days.

To determine when flax is properly retted demands much skill. Of all the operations performed by the farmer in connection with the crop, this is the most technical. The most practical and simplest test to employ is to draw straws from various parts of the retting pond, and to break each stalk at two places about 5 inches apart. Should the core be easily drawn—i.e., should it not adhere to the fibre—the flax is sufficiently retted and may be taken from the pond.

Under-retted flax requires severe scutching to get rid of the woody core, and loss of fibre ensues. If the flax be over-retted, the fibre is soft and breaks away in the scutch-mill, and a large quantity of tows results.

Drying. — When taken from the retting pond the sheaves may be heaped for a short time, three or four hours, in order to allow the water to drain away. They should not be heaped for long, otherwise they will heat. The flax is then taken to a rather bare grass field, where the sheaves are opened and the flax spread evenly and lightly in rows to dry, being well shaken out in order that the stalks do not adhere to one another. A slight shower will not be deleterious to the flax when spread, but with continued rain retting will proceed, and there is danger of the flax being over-retted.

In unsettled weather the flax may be dried by setting it up in handfuls with the root or butt end on the ground in the form of a hollow cone, the boll ends being slightly twisted together. But in the case of high winds following this practice, the flax is liable to injury by being blown about.

Lifting. — When dry, the flax is carefully gathered, tied as evenly as possible with the rush bands, stooked for a week or so, and then stacked or stored for a little time before being scutched. Care should, however, be taken that the rush bands are properly dried, or otherwise "streaky" flax will result.

In all processes involving the handling of the flax—*i.e.*, pulling, drying, and lifting—the utmost care should be taken to keep the stalks as even as possible and square at the root end.

"The proper culture and preparation of flax require more care, exertion, and expense than the old slovenly method; and those who will not give those requisites would do wisely to abstain from growing flax altogether. Any other crop will abide more negligence. Flax is proverbially either the very best or the very worst crop a farmer can grow." [1]

Scutching. — This is more distinctly a part of the manufacturing process, and therefore does not come within the scope of this work.

Growing and Saving Seed. — The saving of mature seed by drying the flax

[1] Henderson's *Cult. of Flax*, I.

straw on the field entails much labour, particularly in a rainy season. Furthermore, to ensure the seed being fully mature the flax must be allowed to ripen to a stage beyond that desired for the yields of fibre of highest quality. For these reasons Irish flax-growers do not save the seed of their crop, but are content to import their seed for sowing purposes.

Yield. — In Ulster 4½ cwt. of scutched flax, or 36 to 40 cwt. of dried straw, is regarded as a fair crop, though these quantities are often exceeded. When the crop is grown specially for seed, a yield of 8 to 10 cwt. of linseed per acre may be expected.

Selecting Seed for Sowing. — Good sowing seed should be of bright colour, not too dark, but plump and free from broken, immature, or otherwise imperfect grains and weed seeds. A good sample should conform to the standard abovementioned as regards purity, germination, and relative weight.

Old Seed. — Seed two years old is sometimes sown, but only when in any season the weather during the seed harvest on the Continent has been particularly bad, and when in consequence a good sample of new seed is extremely difficult to procure, is this plan to be followed; and should a grower propose to use old seed, he should take steps to ascertain the germinating power of a sample, by sowing a little in moist sand placed in a shallow earthenware dish, which may be kept for a few days in a living-room.

Flax-pond Water. — The water left in the pond after retting flax is extremely poisonous to salmon and trout, and should not be allowed into a stream stocked with these fish, unless in time of flood, when the pond water will become well diluted with that of the stream. The smell from retting pond water is very objectionable. Contrary to the opinion sometimes expressed, such water has no manurial properties.

FLAX-GROWING IN GREAT BRITAIN AND IRELAND.

The flax-growing industry has almost died out in Great Britain. A few decades ago the crop was largely grown in certain districts of Yorkshire and Lincolnshire, and also in Scotland. There has been a most serious decline likewise in Ireland, but during the last few years more land has again been devoted to flax in Ulster. The reasons for these changes are: (1) the fall in price of flax fibre consequent upon increased imports from the Continent, particularly from Russia; (2) the relative large amount of manual labour required in the cultivation, harvesting, and after-treatment of the crop, and also, in regard to Ireland, (3) the importation and use of inferior sowing seed.

Former Uses.—For domestic purposes very small patches of flax have been grown in some parts of England and Scotland for a very long time. But except in certain districts it has never taken rank in these countries as an ordinary farm crop. Indeed, even the small patches of flax or lint for domestic use have in most cases become things of the past. The "lint-pools," once dotted pretty freely over Scotland, have nearly all disappeared, and many of the Scottish and English farmers of the present day have never seen a stem of flax growing.

Recent Trials.—Between 1880 and 1887, and again in the early nineties, there was a good deal of discussion as to the propriety of introducing flax as a farm crop in England and Scotland. In several parts of the country the crop has been tried upon small patches, but the results financially have not been sufficiently good to warrant any great extension of the enterprise.

Outlet for Flax Straw.—In the north of Ireland flax-growing is pursued because, in the extensive linen-mills of that industrious province of the Green Isle, there is a sure and ample demand for the fibre. There, indeed, flax is grown for the sole purpose of supplying flax fibre to these linen-mills. The two industries go hand in hand, although the one is not now essential to the success of the other. Little attention is given to the seed in Ulster, for the reason that, by allowing the seed to ripen, the fibre of the flax, the main concern of the Irish flax-grower, would be somewhat injured in quality.

It has been well shown in the various trials conducted that the climate and soil in most parts of Great Britain are well suited for the successful growth of both flax straw and flax seed; but even in the neighbourhood of flax spinning-mills, for instance near Dundee, no flax is grown as a farm crop, and the spinners depend entirely on fibre imported from the continent of Europe.

Uses of Flax Seed.—For flax seed, or linseed as it is more commonly called, there will always be a reliable and satisfactory market. Its high feeding properties are well known, and it is an article which is easy of transport. From linseed is prepared linseed-oil, a valuable article of commerce.

Uses of Flax Straw.—With the flax straw the case is different. It makes admirable thatch, but the demand for this purpose would never be worth reckoning. Its most remunerative use is the manufacture of linen. Unfortunately in England there is no such general demand for it for this purpose as there is in Ulster and Scotland. Whether or not the demand may arise, or could be raised up by any concerted action, is a very doubtful point.

Flax Straw for Paper-making.— Flax straw is also adapted to the manufacture of paper. Its value for this purpose, however, is kept severely in check by the abundance of other commodities which are more suitable, and which— by the processes of manufacture now known — can be manufactured at less outlay. It is just possible that methods of paper-making may yet be discovered which will provide a remunerative outlet for flax straw in that ever-growing industry.

Experiments on the growing of flax straw for paper manufacturers [1] were carried out by Mr Richard Stratton, The Duffryn, Newport, Monmouth, in 1880 and subsequent years, the seed being also saved for marketing for feeding purposes. These experiments showed that the cultivation of the crop for these purposes would be remunerative if for the dried straw a price of £4, 10s. per ton, and for the linseed 16s. per cwt., could be obtained. Consequent, however, on the introduction of new pro-

---

[1] *Journal Royal Agric. Soc.*, Eng., sec. ser., xviii. 461.

cesses of manufacture and the utilisation, or containing water of regulated
tion of other and cheaper raw materials, temperature. It is claimed that by these
the paper manufacturers could not con- means the conditions are rendered uni-
tinue to guarantee this price for the form and the retting is more under
straw. control. Though a relatively high yield

Other Methods of Retting. — Ret- of fibre of good quality, strength, and
ting in ponds as carried out by flax colour has been obtained in some of
growers is a putrefactive or rotting pro- these processes, the extra cost incurred
cess, and is brought about through the in apparatus and labour has rendered
agency of bacteria, which are naturally them unremunerative. Other systems
present in pond water. Attempts have of artificial retting consist in treating
therefore been made to use pure cultures the flax straw, usually in the dried state,
of certain putrefactive bacteria obtained with solutions of certain chemicals to
from good retting ponds for this purpose. dissolve the mucilage which unites the
Up to the present, however, these trials fibre to the woody core of the stems.
have not met with success, for the result- The fibre obtained by these systems is
ing fibre has been of a harsh nature and coarse and weak, and hackles badly in
wanting in spinning quality. Various the spinning-mills. None of these pro-
systems of so-called artificial retting have cesses has yet been found as satisfactory
been patented. Some such systems com- as pond retting, but new processes are
prise retting in tanks of special construc- constantly being devised.

SUBSIDIARY FARM CROPS.

KOHL-RABI.

Properly speaking, the kohl-rabi (*Brassica oleracea*) is not a root crop. Its bulb
is formed by an enlargement of the stem
or stalk, and it is thus grown for its stem
and not for its root. Nevertheless it
falls into the rotation with the root
crops, and is cultivated for the same
purpose, namely, to provide winter food
for farm stock.

Kohl-rabi was cultivated in this country as far back as 1734, but it was not
generally known till about 1837. Two
varieties are grown in this country, the
one purple, the other green, and of sub-
varieties some are round and others
oblong.

Advantages of Kohl-rabi. — Kohl-
rabi undoubtedly possesses high merits
as a field crop, and it is surprising that
in England especially its cultivation has
not extended much more widely than
has been the case. In Scotland, and
other parts well suited for turnips, there
may be little necessity for it, but on the
stiff clay and soft fen lands of England,
which are well suited for kohl-rabi and

badly fitted for turnips, it ought to be
grown more largely.

The advantages of kohl-rabi as a field
crop are thus forcibly stated by Professor
Wrightson : [1]—

"It is subject to no diseases and few
insect attacks. Like the turnip, the
young plants are liable to the depredations of the turnip flea-beetle, but in a
less degree to swedes and turnips. It
thrives on two classes of soils upon
which turnip cultivation cannot be very
successfully carried out—namely, upon
the stiffest classes of clays, and the fen-
lands of East Anglia. It possesses great
powers of resistance to drought, and in
fact thrives best in hot and dry seasons.
It is exceedingly hardy, and resists frosts
successfully. The crop may therefore be
left over till the spring of the year. The
leaves are of the same quality for feeding
purposes as the stems, and resemble rape
or kale leaves in nutrient properties. It
is well suited for cow-feeding, as it does
not impart an unpleasant flavour to
milk. It is well adapted for sheep-
feeding on the ground, because the bulb

---

[1] *Fallow and Fodder Crops*, 190.

being supported upon a footstalk it can all be eaten without the waste which is inevitable when turnips are fed. It is an excellent feed for ewes and lambs in the spring, as it supplies leafy herbage as well as more solid food."

Uncertain Crop.—Perhaps the influence which has been most instrumental in restricting the cultivation of kohl-rabi is the belief that it is rather uncertain in its growth—liable to grow to a mass of leaves with insufficient development of bulb. This drawback is being gradually removed by the raising of improved varieties which are more reliable in their development.

Soil for Kohl-rabi.—Kohl-rabi grows well on all soils adapted to swedes, and may also, as we have seen, be cultivated with success on stiff clays and fen lands.

Tillage and Manuring.—These should be very similar to what are best suited for mangels—deep autumn tillage and dunging, grubbing or cultivating in spring, with a liberal dressing of nitrogenous manures.

Planting or Sowing.—Kohl-rabi may be sown either on the flat or in raised drills exactly like turnips; or the plants may be raised in seed-beds, and transplanted into rows 25 to 27 inches apart, with from 10 to 16 inches between the plants. The seed should be sown in the seed-bed fully two months before the time for transplanting, as the plants should be about 8 inches high before being transplanted. From 10 ounces to 1 lb. of seed sown thinly in rows a foot apart, in a well-prepared seed-bed, should produce sufficient plants to cover one acre.

Some particularly careful farmers always raise a few kohl-rabi plants with which to fill up blanks and odd corners in the root field.

Time for Sowing.—The seed may be sown in the drills in March or April. Transplanting may take place from the first of May till the middle of August.

Thinning and Hoeing.—When sown directly in the field the plants are thinned like turnips, with wider intervals between the plants—from 10 to 15 inches. The after tillage and hand-hoeing is exactly the same as for other root crops.

Produce.—From 20 to 25 tons per acre are common crops. Occasionally the produce reaches from 30 to 35 tons or more.

Storing Kohl-rabi.—Kohl-rabi may be stored in the same way as swedes.

CABBAGES.

The growing of cabbages is dealt with under the section on Forage Crops, p. 371 of this vol.

CARROTS.

Carrots (*Daucus carota*, Natural Order *Umbelliferæ*) are more of a garden crop than a crop for general field culture. Yet on most farms with suitable soil a small patch of them may be grown with advantage. Carrots are, in limited quantity, excellent food for horses; and with capital results the carrot-tops may be used as food for cows in milk.

Carrot-tops as Food for Stock.—The high feeding value of carrot-tops is not generally known or acknowledged among farmers, for the tops are, as a rule, left on the ground, just as in the case of turnip-tops. Mr John Speir, Newton Farm, who grows carrots extensively for the Glasgow market, says that carrots form an excellent class of food for any kind of farm stock. They are relished extremely by both sheep and cattle, dairy cows doing particularly well on them. A good few tons of leaves are yielded by each acre of well-grown carrots, and a ton of carrot-leaves appears to be as valuable as the same weight of turnips.

Mr Speir has given carrot leaves to dairy cows for many years, and he prizes them highly for this purpose. He says :—

"I have repeatedly been laughed at for the opinions I held regarding the value of carrot-leaves, by those who had little or no experience of their use, my friends saying they were not worth the cartage, and that I was impoverishing the land by giving them to the cows. I grant I was decreasing the fertility of the land, but I was increasing my milk production, and there was no greater occasion why these carrot-leaves should not first pass through the bodies of animals before being applied to the land, than should a second crop of hay or clover, diseased potatoes, or, for that matter of it, any palatable farm crop, be it a first or second one.

"**Analysis.**—In support of this view are appended, for the sake of comparison, the analyses of the digestible constituents of a few similar plants, as given by Professor Stewart :—

	Carrot-leaves.	Cabbage.	Turnips.	Potatoes.	Pasture Grass.
Water	82.2	84.7	92.	75.	80.
Albuminoids	2.2	1.8	1.1	2.1	2.5
Carbohydrates	7.0	8.2	6.1	21.8	9.9
Fat	.5	.4	.1	.2	.4

"From this it will be seen that carrot-leaves compare favourably with any of the other articles of food, and although considerable latitude be allowed for variation in the analyses of the samples here given, there is still room enough for much to be said in favour of carrot-leaves. If we allow the stock themselves to be the judges, the point will be easily settled, as the carrot-leaves will be taken in preference to almost any of the other foods named.

"I may here mention in support of this view that it is well known that hares and rabbits travel long distances to feed on carrot-leaves, and they are as dainty in regard to their food as any animals on the farm. In my own case I have for very many years used the produce of from 15 to 20 acres annually with the most satisfactory results, while throughout the country generally it is only the few here and there who do use them anything like extensively. Those farmers who allow them to lie and rot would be considering they were extremely careless if they allowed even a tithe of the same weight of turnips to lie rotting in the fields, while they pay no attention to what is an equally if not more valuable crop, although at the same time a more perishable one."

It is a difficulty of carrot-tops that they cannot be stored like roots. Still, in small heaps they can be kept fresh for a few weeks in the cold season.

Soil for Carrots.—Carrots, having a strong deep root, require soil of considerable depth. A good sandy loam is best adapted for the crop. Excellent crops of carrots are grown on well-manured land with a mossy tendency.

Tillage and Manuring.—The land must be deeply cultivated in the autumn. If the subsoil is stiff, it should be loos-ened by the subsoiler. Autumn dunging is generally preferred, but where this is not convenient the dunging may be done in spring as for other root crops. Carrots will take as heavy dunging as any of the root crops. The dressing of artificial manure applied may consist of from 3 to 5 cwt. superphosphate, 1 cwt. of potash salts, and 1 to 2 cwt. of nitrate of soda, sown as a top-dressing just after singling, or in two portions, one before and the other after singling.

If the land has been dunged and well tilled in the autumn, little cultivation in spring will suffice. A fine tilth and a loose range for the searching root are very desirable. Yet harm may be done by overworking the land in spring.

Cleaning for Carrots.—It is specially important to have the land for carrots as thoroughly cleaned as possible. Land should only be selected which is in good order and clean, as no amount of cleaning of dirty land just before the crop is sown is likely to turn out satisfactorily. Land inclined to grow chickweed or any other quick-growing weed with fibrous roots should be avoided if possible, for the keeping of such weeds in subjection among a crop of carrots is not only costly, but the disturbance to the growth of the crop is often so great that the crop is rarely a profitable one. The leaves of the carrot are small, compared with those of mangels and turnips, and the growth of weeds is thus encouraged by the large amount of space which remains uncovered.

Sowing Carrots.—Carrots are usually sown about the end of March and beginning of April. They may be sown in rows on the flat, from 15 to 18 inches apart, or in raised drills from 27 to 30 inches in width. In the latter case, two rows of seed are sown on the one drill.

The wide raised drill is preferred by many, because of the greater facility it affords for cultivating and cleaning the land by horse-tillage.

Preparing Carrot Seed.—About 6 to 8 lb. of seed will sow an acre. The hairy covering of the seeds makes their separation rather difficult, and to overcome this difficulty it is a good plan to mix the seed with fine sand, perhaps at the rate of 1½ to 2 bushels of sand per acre. The seed and sand should be thoroughly intermixed by rubbing with the hands; moisten the mixture with water, spread it out on a dry floor, turn daily, sprinkle with water if it become dry, and when it has lain from a week to ten days in this form, it should be sown, just before the seed germinates. When this preparatory process is gone through with proper care, the plants will come up more quickly than if sown without the week or ten days of incubation. This is an important point, because the carrot plants are so tiny in their earliest stages that, when sown on the flat surface, they are liable to be covered and overcome by weeds.

In many cases a little oats, barley, or turnip seed is sown along with the carrot seed for the purpose of indicating the rows, and thus enabling the hoe to be used with freedom before the carrot plants are very clearly visible.

It is unsafe to use old carrot seed. It should be the produce of the previous year's crop.

Thinning Carrots. — The plants should be thinned when the leaves are from 1 to 3 inches high. Intervals of from 4 to 8 inches, according to the variety grown, are left between the plants. When grown on raised drills, with two rows on each drill, the plants are singled so that they alternate rather than sit directly opposite or abreast in the two rows. Horse and hand hoeing should be pursued as with other root crops.

Carrots and Rye.—The late Professor Wilson described a practice followed by some enterprising Continental farmers, in which crops of rye and carrots are grown upon the same land, so as to overlap each other in rather an ingenious way. He said: "In the light-soil districts of Belgium and Holland, where carrots are cultivated to a far greater extent than with

us, it is a common practice to grow them mixed with a crop of rye or flax. In the former case the rye is sown early in the autumn, so as to root well before the winter sets in, and thus come early to harvest the following year. In the spring the carrot seed is sown broadcast as late as the growth of the rye will admit of the harrows being used to cover the seed. This germinates and continues its growth until the rye is ready for cutting, which usually takes place about the second or third week in June. It is then mown with a cradled scythe, care being taken not to cut it so close as to injure the top of the root of the young carrot plants, which by this time have acquired a size about the thickness of one's finger. The field is cleared as quickly as possible of the stooks, the harrows are sent over the ground to disturb the surface, and to drag up the roots and stubble that are left, while the remaining weeds are carefully removed by the hand. The liquid-manure cart follows with a supply of rape-cake mixed up with 'purin,' and in a few days the young plants begin to show themselves again, and by the end of the autumn are in a condition to yield a weighty crop of roots, which, when forked up in the usual manner, leaves the land in excellent condition, both chemically and mechanically, for the succeeding crop of corn."

Varieties.—There are many varieties of carrots in use. The best known are Stump-rooted, James's Intermediate, the Altringham long red, white Belgian, large red, and short red.

Produce.—The produce in average seasons should reach from 12 to 20 tons per acre. It is sometimes more, often less.

Storing Carrots.

Liability to Rot.—Of all the root crops of the farm or garden, carrots are the most difficult to preserve. Unless under exceptional circumstances, they are too tender to be allowed to stand out all winter; and in pits (more particularly in wet or mild winters) they are often extremely difficult to keep till May or June. A little mould may be all that is noticed, with an odd soft carrot here and there, and in a couple of weeks after the whole pit may be a mass of pulp. Where, therefore, carrots are grown, their proper storing is a matter requiring the most careful consideration.

Time of Storing.—Unless in the very earliest districts, carrots are rarely ripe enough to be stored before the last few days of October or the first week in November. In every case the latter is to be preferred to the former.

Pulling.—One person should pull the roots by catching them tightly and close by the bottom of the leaves, laying them out in rows flat along the top of the drill in which they grew; while another should be intrusted with the selecting and cutting off the leaves. Each pair of pullers should be provided with a potato-graip, with which to dig out any roots the leaves of which may happen to break. Between each pair of those cutting off the leaves, one basket should be provided for holding all split, forked, or otherwise deformed roots; while one or two baskets should be provided for each cutter, into which to put the selected roots. A cart generally goes behind, and each basket when full is emptied into it, the extra baskets being filled while the cart is away.

Carrot-leaves.—The leaves should be kept clean and thrown into heaps behind the baskets, in readiness to be carted to the cattle and sheep at the farm. All classes of cattle are exceptionally fond of these leaves, and thrive well upon them.

Growing and Heating in Pits.—It is of the utmost importance that the leaves of carrots be cut off as near to the crown as possible. If such is not done, the roots grow much quicker in the pits, heat is thereby generated, and the whole in a very short time becomes a mass of putrefaction.

Carrot-pits.—The pit in which carrots are stored should never be so large as potato-pits, from 2 feet to 2½ feet wide being quite large enough. The sides should be as steep as possible, so as to keep them thoroughly dry, and at first they should not be very heavily covered. In other respects the pitting is much the same as has been recommended for potatoes. Carrots, however, require more ventilation, and must on no account be pitted wet.

In a winter in which there is very little frost, a carrot will keep quite sound and fresh lying on the surface of the ground exposed to wind and rain; but shut it up in an unventilated and warm pit, and it will become soft in a week. As a protection against frost, there is not the same necessity to thatch carrot-pits as potato-pits; but as a protection against rain, thatch is about as needful for the one as for the other.

Hares and rabbits are especially fond of carrots, and will do much damage to the crop if they are not shut out by wire netting.

Carrot Pests.—Insect pests which prey upon carrots are dealt with in the sections on Fungus and Insect Plagues in this vol.

PARSNIPS.

The parsnip is the *Pastinaca sativa* of the same Natural Order as the carrot, *Umbelliferæ*. Indeed the two plants are so very similar in their habits of growth that the remarks as to soil, tillage, and manuring for carrots may be held as applying also to parsnips.

Parsnips go still deeper into the soil than carrots, and grow to their best in a heavier loam than that which carrots specially delight in. In the south the parsnip seed—6 or 8 lb. per acre, with a little oat, barley, or turnip seed, as with carrots—may be sown as early as February.

Parsnips are usually grown in rows on the flat surface about 14 or 15 inches wide, and the plants are thinned to from 6 to 8 inches apart.

Among the varieties of parsnips most largely grown are the long-rooted parsnip, the Student, the long Jersey (or hollow crowned), large Guernsey, and Cattle parsnips.

Both parsnips and carrots are found growing wild as weeds in this country. No doubt the cultivated varieties have been raised from these.

The parsnip is possessed of very high fattening properties.

Parsnips may be treated similarly to carrots in regard to tillage and manuring

Storing Parsnips.—Parsnips may be stored in precisely the same manner as carrots. They are not much affected by frost, and will keep fresh in the store till April. Care, however, should be taken that none of the leaves remain attached to the roots.

Parsnip-Leaves as Food for Cows. —The leaves of parsnips make good food for cows. As they begin to decay they should be cut off and given to the cows. In October the leaves come in as a convenient auxiliary to grass ; and, if given *moderately*, a good armful per day to each cow will impart almost as much richness to the milk as the parsnip itself.

MARKET-GARDENING.

The decline in the prices of ordinary farm products has induced farmers, in many parts of the country, who formerly devoted their land and means almost entirely to the production of grain and beef or mutton, to give a share of their attention to the growing of vegetables, fruit, and in some cases flowers, for the use of the dwellers in towns, who have no facilities for the growth of such, but are able and willing to pay good prices for them. The area under these crops is on the increase in this country, and under judicious management there is room for further extension.

Vegetable culture, as a rule, is extensively carried on only in the vicinity of some populous centre, where large quantities of town dung can be procured. Fruit is easier of transit than vegetables, and may be cultivated anywhere, where the soil, climate, railway facilities, and supply of manure are suitable ; while the same may be said of flowers.

Dairying and Market - gardening. —Often the growth of vegetables, fruit, and flowers is combined in one establishment, and to this system of farming a profitable adjunct is a dairy. The cows composing the dairy can get almost all the year round a supply of green stuff very suitable as food, which would scarcely pay for its cartage to a more distant dairy, and which is too valuable for manure, and as such is often difficult to plough under.

For this purpose imperfectly grown kale, cabbages, cauliflower, broccoli, and Brussels sprouts are available almost through the entire year. Then carrots and parsnips, unfit for table use, make good food for cows, and their leaves, which are very bulky where a heavy crop is grown, are much relished by cows, and weight for weight appear to be as valuable as turnips. In the culture of turnips for table use, as well as of such other plants as beet, peas, and beans, there is also always a considerable quantity of what would otherwise be waste food, which can be turned to good account through the medium of milch-cows. Then in the growth of such crops there are plots of land which are bare at a season of the year when it is unsuitable to put these under any of the regular crops, and which may be profitably utilised in growing tares for the cows. Tares, with the winter and spring varieties, form a very accommodating and valuable crop for such circumstances. They do not necessarily occupy the ground for any great length of time, and they generally do well on heavily manured land.

For these and many other reasons the growth of market-garden crops should, wherever practicable, be combined with dairying. And unless there are circumstances of an exceptional character, the class of dairying most suitable will be the production of milk, and the sale of it fresh from the cow.

This system of combined dairying and market-gardening has been pursued extensively and with success by Mr John Speir, Newton Farm, Glasgow, and to him we are indebted for the following notes on the culture of subsidiary farm crops.

Savoys.

The culture of the savoy is much the same as of the cabbage, and equally as easy. Plantations are usually made at intervals from spring to midsummer, the latter plantings generally following some early cleared spring or winter crop.

Around Dublin they are grown to an enormous extent. The Dublin population are probably the heaviest cabbage-eating people in Britain ; while large quantities are also annually exported to Glasgow, Liverpool, Manchester, and Bristol.

The farmers around Dublin grow first a crop of early potatoes, which is cleared off in June and July. The land is then immediately planted with savoys, which form good heads by the New Year, thus allowing the land to be cleared in time for any spring-planted crop.

Greens, Cauliflowers, &c.

Instead of savoys, greens or kale may be planted after an early potato crop is cleared off. This custom is followed very largely around Belfast, and on some of the earliest farms along the Ayrshire coast, from Girvan northwards. Greens are always in fair demand, and often give a good return per acre.

Brussels sprouts, cauliflower, and broccoli resemble the savoy very much in culture; and in manure they all have much the same requirements. General dressings are 10 to 12 tons town dung, 4 to 6 cwt. of superphosphate, 2 to 4 cwt. of nitrate of soda, and perhaps 4 cwt. kainit per acre.

Carrots, Parsnips, Beetroot, and Onions.

Though nowadays grown on farms to a moderate extent, these are standard crops in market-gardens. Their culture is very similar on the field and in the garden. Carrots succeed best on sandy or mossy soil; parsnips on loam or clay; and beetroot on any free soil, which, however, must be in the very highest state of cultivation and fertility. Beetroot requires a good climate.

Onions also thrive best on good open soil well manured and kept free from weeds.

A portion of the crop is often sown in early autumn. This is cleared off before the spring-sown onions are ready to pull, while they in their turn are removed by the beginning of September to make room for cabbages to stand the winter.

Leeks.

Leeks are constantly in demand from September till April, when new autumn-sown onions again come in. There is thus a sale for the one or the other throughout the whole year.

Leeks, owing to their excessive demand for manure, are not so easily grown as a field crop as onions. There is, however, no great difficulty in cultivating them successfully where the land is clean and in moderate condition.

Manuring. — Probably no crop can stand so much forcing with nitrogenous manures as leeks, so that, whenever they are not growing to one's mind, a thorough weeding and heavy dressing of any soluble nitrogenous manure will be sure to bring them away rapidly.

Culture.—Leeks are generally sown in beds or frames, from which they are afterwards transplanted in the field in rows about a foot wide, with about four inches between each plant in the row. In dibbling in the young plants, they are put as deep in as the size of the plant will permit, so that when full grown they shall have as much of the stalk blanched as possible. It is also found that the plants grow much better when only a minute quantity of earth is put into the dibble hole after the plant has been dropped in. When the plants are grown so shallow as to whiten only an inch or two of the stem, they are generally earthed up, to increase as much as possible the length of white stem.

Turnips for Table Use.

Turnips for table use are also a very suitable crop for growing either in market-gardens or on farms near large towns. Their culture is in no way different from the ordinary field culture of turnips, except that two rows are generally sown along one drill. This is necessitated by the smallness of the bulbs and the tops, which, if sown on one drill only, would return a very small crop per acre. Moreover, with only one row of these small roots, a considerable portion of the land would be left bare, and would be immediately monopolised by the weeds. Such turnips are usually taken to market with their tops on, a dozen or fourteen being tied in a bundle by passing a straw or other rope round the tops.

Turnip-tops.—Farmers in the counties around London and some other large towns derive a substantial return by selling turnip-tops, particularly the young sprouts in spring, for use as "greens" in soups and with meat.

Beans and Peas.

Beans and peas can also be profitably grown where there is a market for them in a green state. In most large towns they are bought up in considerable quantities. Both are pulled green and sent to market in hampers or sacks, and sold at so much per stone or cwt. The return per acre is not generally very high, but the crop does not require much manure, and the ground is early cleared for something else.

Beans do best on moderately firm clayland thoroughly drained, while peas are more suitable for free-working lands. When grown on a large scale for green pulling for market, peas are seldom staked. The drills are made from 3 to 4 feet apart, and the haulm is allowed to grow along the ground, but not to catch the next row.

Rhubarb.

There are few crops of the market-garden class so easily grown by the ordinary farmer as that of rhubarb. The cost of the roots is considerable, ranging according to size, variety, and date of sale from £20 to £80 per acre. The roots are usually planted in rows 3 feet or less apart, and about 30 to 33 inches between the plants. One, two, or occasionally three pullings may be taken each year. Summer rhubarb does not, however, sell high, from 20s. to 30s. per ton being a common price, with 40s. early and late in the year.

Glasgow is the principal centre of the growth of rhubarb in Scotland, and Leeds in England. Around the former city many farmers have from 20 to 30 acres under rhubarb. Most of these have forcing-houses, in which it is grown during winter, several of them being able to force the roots on from 5 to 10 acres in a single winter. Roots which are intended to be forced in winter have few of the leaves taken off during the preceding summer. After forcing the roots are of little value. If again planted they yield no crop the first year, but give a moderate crop the second one. They are, however, not ready to force again until four or five years old.

Rhubarb can make profitable use of liberal manuring. Moderate dressings of dung and liberal additions of artificial manures usually give the best results, common allowances of the latter being from 4 to 5 cwt. superphosphate, 2 to

3 cwt. nitrate of soda, and about 4 cwt. kainit per acre.

In all cases where lime is supposed to be deficient, basic slag should be substituted for superphosphate.

Strawberries.

Strawberries are the fruit easiest brought to a bearing condition and requiring least skill for their culture. They therefore form the fruit most suitable for the ordinary farmer to begin with. In the valley of the Clyde, between Lanark and Hamilton, the growing of this fruit has been largely pursued by the farmers. The farms there are all more or less devoted to dairying, yet for a distance of six or seven miles along both banks of the river every one has made a trial of strawberries.

Irregular Produce.—A year occurs every now and again when they give a comparatively poor return; but on the other hand there are years when an enormous production is obtained.

To such an extent has their culture been carried in this locality, that as many as fifteen railway waggon-loads of strawberries have been known to leave a single station in one day, and yet there are three stations which are all more or less fed from this district.

Near Crieff, Dumbarton, and Aberdeen, in Scotland, and in different parts of England, strawberry culture has likewise made rapid progress in recent years.

Planting.—The plants are generally dibbled in beds in spring,—three rows, about 15 inches apart, being allowed to each bed.

The Fruit.—The finest of the fruit is sent to market each morning for dessert purposes, the remainder being pulled during the day for making preserves.

Price.—The preserve-makers generally arrange at the opening of the season for so many tons from each grower, the ordinary price of recent years being from £16 to £28 per ton. Few have ever been sold at less than £12 per ton, and occasionally the price exceeds £30 per ton.

Duration of Plants.—The plants yield no fruit worth speaking of the first year, and are at their best the second and third years, after which they deteriorate quickly. Most growers do not crop them more than four years, after which the plantations are ploughed down, and a grain or green crop taken.

The best plants for making a new plantation are yielded by those which have been put down the year previous.

Labour and Soil. — To enable the cultivation of strawberries to be successfully carried on, a plentiful and cheap supply of labour must be at hand. The soil should lean to the heavy side rather than the light. They rarely do well on very light soil.

Bush Fruit.

The cultivation of bush fruit, be it black, red, or white currants, or gooseberries, is not so easy of attainment as that of strawberries, and is not so well suited for a farmer holding land under a short lease.

Cost of Planting. — The purchase of the young bushes is a rather costly business, and instead of only one year being lost before fruit-bearing begins, as with strawberries, two, three, or even four may be said to elapse before a very large return is obtained, even where the climate is good and the bushes fairly well grown when put in.

Catch Cropping. — As the rows of bushes are, however, generally from 5 to 6 feet apart, a good deal can be made out of the spaces between the rows by growing vegetables. In these spaces turnips, cabbages, cauliflowers, beet, parsley, leeks, and onions may advantageously be grown; and if the culture of flowers is attempted, these spaces form very suitable places for the growth of wallflowers, narcissi, snowdrops, and annuals generally.

Disease in Black Currants. — Black currants, when they do yield well, usually bring a large money return to the fortunate grower — not because they bear a heavier crop than other fruit, but because of their extreme liability to disease. To such an extent has this pest prevailed in the fruit-growing districts all over the country, that the larger plantations are being rapidly grubbed up. The disease, which is dealt with in another part of this vol., seldom attacks the red or the white currant.

Gooseberries. — Gooseberries are occasionally seriously attacked by the caterpillar, but it does not carry with it any of the destructive effects of the black-currant mite. In recent years gooseberries have not been bringing anything like the same price in the market as formerly, and in consequence growers are rather chary of making new plantations. For good gooseberries from £8 to £12 per ton are common prices. The attacks of the fungus and insect enemies of gooseberries are noticed in other parts of this vol.

Manuring for fruit may be similar to that for most vegetables, with, in all cases, a liberal allowance of a potash manure.

Raspberries.

Raspberries are now largely planted in many districts, particularly in the neighbourhood of Blairgowrie, in Scotland, where they have succeeded very well. They are usually planted in rows about 5 feet apart. One or, it may be, two wires are stretched along the rows on posts, and to the wires the canes are tied. This prevents them rubbing against each other when knocked about by the wind, and admits of the fruit being more easily gathered.

Orchards.

In many parts of England, especially in the southern counties, the cultivation of tree-fruit, such as pears, apples, plums, and damsons, is carried on to a large extent and with good results.

It is only on a very limited area of Scotland that the cultivation of large or tree fruit for the purpose of sale has been attempted. The largest breadth in one lot probably centres round the village of Crossford, on the Clyde below Lanark. Both sides of the river there are for several miles devoted to the cultivation of fruit-trees, which has been carried on for a very long time. During the days of the old stage-coaches, it was a comomon remark that a handful of plums could always be gathered in the season from the trees on the roadside, while the coach passed underneath them. There is still a considerable extent in plums; but apples hold the largest share, pears being grown to a less extent.

New orchards are being continually

planted and old ones uprooted, the new ones being, as a rule, planted with small fruit in the intervening spaces. Owing to the shelter which the deep and narrow valley affords, this locality is extremely well suited for fruit, and all the farmers in the valley have more or less of their land under it. Considerable orchards at one time existed in the Carses of Stirling and Gowrie, but lately these have not been well attended to, and are fast disappearing. Fruit importations have considerably checked home planting all over Britain, and it is very questionable if there are many districts in Scotland where it can be carried out with much prospect of profit.

In the British Isles tree-fruit has never received such careful cultivation as is given to it in the United States and Canada. There the best orchards are regularly cultivated the whole year over, and the growth of the trees so treated is almost double what it is where the land is sown with grass or is left uncultivated. The usual method in Canada is to either summer fallow the land between the trees or to do so up to the end of June, and then sow a cover crop such as buckwheat or clover, which is ploughed under in autumn.

FLOWERS.

The demand for pot-plants for house decoration, and cut flowers for the house, personal decoration, marriages, and funerals, has so much increased among the dwellers in towns, that there is ample room for the growth of these by those farmers who have a taste for flowers, are conversant with their growth, and suitably situated in regard to climate, soil, and disposal.

In sheltered and early situations, wallflowers may be grown by the acre, and if early or late enough, are sure to give a fairly good return. In the middle of the season it may often be difficult to sell the wallflowers even at the cost of carriage; but at the beginning and end of the season they generally do well. The same may be said regarding mignonette, forget-me-nots, and many other flowers of a similar class, which are likely to be grown by the farmer attempting flower-cultivation as an adjunct to his farm.

Bulbs. — Amongst bulbous plants which may be grown for cut blooms may be mentioned the whole narcissi family as plants which are easy of growth, and the blooms of which sell well. The great drawback to the culture of such on a more extended scale than most persons have yet attempted, used to be the enormous cost of planting even one acre with these bulbs. They are, however, much less costly now than they used to be, so that there is more inducement for beginners to invest in them. Two farmers in the south-west of Scotland grow something like 10 acres each of these bulbs alone, and are generally presumed to be doing well.

For sale as plants, the farmer might also grow wallflowers, daisies, pinks, primroses, and other flowering-plants, a limited quantity of which can profitably be sold in most large towns.

Indoor plants cannot be grown without the aid of glass, and therefore need not be noticed here.

RED-CLOVER SEED.

The growing of the seed of red-clover (*Trifolium pratense*) is pursued to some extent in various parts of England.

Where grown for its seed, red-clover is sown alone. Were it to stand for seed at the first cutting, when the blossoms do not appear simultaneously, the seed of one plant would be matured, while that of another would be scarcely formed. At the second cutting the flowers blossom at one time, and the plants attain the same height, the crop appearing one of the richest description in our fields. The first cutting in ordinary practice is delayed until the plant is in full bloom, and sometimes till the bloom has begun to decay; so no surprise need be felt when a full second cutting is not obtained after such treatment.

From some of the imported seed it is impossible to get a good second cutting; but if the seed is obtained from a reputable British firm of seed-growers, there should be little difficulty about this, whether the seed is of home or foreign growth.

To secure a good second cutting, the first crop should be cut before the plant comes into bloom; or sheep might eat down the crop by the end of May or beginning of June, when the second growth will come away thick and vigorously.

The red clover is injured by insects when in bloom.

Cutting and Drying. — When the blooms of the plants become withered, the crop should be cut by reaper or scythe in August or September. If put together in heaps, a slight degree of heating will cause the seed to leave the husk more readily on being threshed; and on the heated heaps being spread out to the sun, the crop will soon be dry enough to lead home to the steading, to be threshed with the flail or threshing-machine.

When the weather is good this plan may be adopted, but should it prove damp the crop should be sheafed, and set into stooks to dry, and afterwards carried to the stackyard and built into stacks, to be threshed at a convenient time. There is little danger of the seed falling from its husk, as it is difficult to thresh out; but the heating recommended renders the husk brittle, and easily broken by threshing.

Threshing.—In most districts where red-clover seed is grown, special machines are taken from farm to farm to do the threshing. In other cases the threshing is done by the ordinary threshing-machine or by the flail.

HEMP CULTURE.

Hemp (*Cannabis sativa*, Natural Order *Urticaceæ*) is grown to a very limited extent in this country, chiefly in the counties of Lincoln and Dorset. The climate of Scotland does not suit it. It grows best in deep, rich, moist, alluvial soil. Its mode of culture is in several respects similar to that of flax. Hemp responds well to a heavy dose of dung, the finest fibre being grown after a dressing of about 20 tons of dung, applied in the autumn before sowing.

Hemp is sown towards the end of April, in rows about 18 inches apart, with 3 to 5 pecks of good seed per acre.

The plants are thinned out in the rows to nearly a foot apart. The plants throw up a rapid and bulky growth, so that little weeding early in the season is sufficient to keep the land clean. The crop is pulled, stacked, and steeped similarly to flax. The object of the steeping in water is, of course, to rot away the woody part of the stem and separate the fibre.

When the crop is growing, the ground should be watched after sowing until the plants are in leaf, to keep off birds of the finch tribe, which are very fond of hemp seed. Even the young plants are injured by them, — the capsules of the seed, being brought above ground by the embryo, are greedily devoured by those birds. Care should be taken in weeding not to break the young plants, as, if broken, they will never rise again.

A good crop of hemp yields about 16 bushels of seed, and from 6 to 8 cwt. of fibre per acre.

The hemp plant has the male and female flowers on different plants. The male plants are recognisable by the difference of their inflorescence, and in thinning, a number of them must be left in order to the formation of the seed. The male plants ripen long before the female plants, and should be pulled first, so as to promote the formation of a good crop of seed.

The stem of hemp is upright, from 5 to 8 feet high, and is strong and branching. Its valuable fibre makes the cordage of our ships.

An oil is expressed from the seed of hemp, which is used in various industries. The proportion of oil from the seed varies from 14 to 25 per cent. The seed is used for feeding cage-birds.

LAVENDER.

Lavender (*Lavandula*, a genus of *Labiatæ*) has for many years been grown to a small extent in England, chiefly in the counties of Surrey and Herts. It is grown mainly for the distillation of its essential oil, and, according to whether the season is dry and bright or dull and wet, the yield of oil varies from 12 to 30 lb. per acre. The plant does best on sandy loam lying on a calcareous subsoil, with a sunny exposure, not liable to fog or early frost. The plants are propagated by slips or by dividing the roots. The flowers are collected in August and at once taken to be distilled. The finest oil is got from the flowers. The growing of lavender is not extending to any great extent.

WOAD FARMING.

Though an old industry, woad farming has never been pursued except to a very limited extent in this country. It is now mainly confined to a few centres in the Fen district of England.

The woad plant is used as a colour for fixing the dye, which it seems to have the property of making perfectly fast. Natural dyes are now so easily procured that woad is not much used except when stipulated for, as it is by Government for cloth for the police force, whose blue uniforms rarely lose colour, thanks to the influence of woad.

The woad plant yields three crops annually, and grows from 6 to 7 inches high, when it is plucked in a green state.

EXCISE RESTRICTIONS ON CROPPING.

Seeing that for all our staple agricultural products, corn, hay, potatoes, &c., our markets are free to the produce of the whole world, it may seem rather hard that our farmers should not be at liberty to grow any crop whatever which might yield them good returns.

Tobacco is the only crop the growth of which has hitherto been absolutely forbidden in this country, but for all practical purposes chicory might be classed with tobacco; while the cultivation of flowers and aromatic herbs, for the manufacture of perfumery, is also hampered by excise restrictions. If the sugar-beet industry could be established in the United Kingdom, it would also be chargeable with an excise duty.

Sugar-beet.

Experiments on a fairly commercial scale (¼ acre and over) carried on in Britain clearly proved that not only can a heavier crop of beet be grown in Britain than on the Continent, but also beet equally as rich in saccharine material. Therefore if the possibility of bounty-fed sugar entering British ports were completely done away with, it is possible that sugar factories would be erected here, and would succeed in producing sugar as cheaply as anywhere else. Farmers are much in need of a new crop, particularly such a one as the sugar-beet, which not only produces a large quantity of sugar without taking much fertility from the land, but at the same time leaves a substantial amount of residue which is very useful for feeding milk cows or fattening cattle.

Root-alcohol for Industrial Purposes.

It is no doubt the case that several of our staple crops offer large possibilities for industrial uses, if they could be dealt with without hampering excise restrictions. In particular, the production of industrial alcohol for motor-fuel and other purposes might well engage the attention of our farmers. It promises to become one of the largest industries of the day, and in the production of root-alcohol, from potatoes, turnips, sugar-beet, and other green crops, British farmers could compete, on advantageous terms, in home markets, against petroleum or any other kind of imported motor-fuel. There would have to be denaturation, of course, but if the revenue charges were limited to that and the bare cost of supervision, the charge would not hinder the growth of the industry.

Tobacco.

Whether tobacco could be made a paying crop in this country is doubtful. Tobacco is one of the few articles on which we levy import duty, and the £10,000,000 or so which foreign-grown tobacco annually contributes to our customs receipts is a very strong argument in other than farming circles for not taking the duty off the home-grown article, especially if, as has been officially stated, a £50 per acre tax on home-grown tobacco would only produce half the present revenue from the imported article.

No doubt our farmers could easily grow all the tobacco needed for home consumption. Experiments in different parts of Ireland as well as of Great Britain have shown that the production of tobacco in quantity in this country would be an easy matter; but in this moist and comparatively sunless climate the fine - flavoured weed, which is the natural product of tropical and semi-tropical regions, cannot be equalled. It is doubtful, therefore, if much will come out of the facilities the Government is now offering for trials with tobacco culture in Great Britain.

Aromatic Herbs.

There is no duty chargeable on distilled waters made from home - grown herbs, but a 10s. licence is necessary to authorise the keeping of a still for this purpose. It is right and needful that there should be some safeguard in this direction, but this tax brings in nothing really to the national exchequer, and only serves to deter small cultivators who might find it profitable to make the growth of flowers and aromatic herbs an addition to other branches of small farming.

Chicory.

Likewise, although there is no official interference with the *growth* of chicory, there practically is, since the only value of chicory is as a commercial crop, and a *drying* of the root is subject to excise regulations and a duty of 12s. 1d. per cwt. of the dried chicory delivered for consumption. A drier of chicory is required to make entry, with the proper officer of excise, of his premises, kilns, and utensils, and to provide to the satisfaction of the Commissioners of Inland Revenue (1) a warehouse for depositing chicory when dried, (2) proper accommodation on the premises for the officers, and (3) scales and weights and assistance to the officers in taking the necessary accounts therewith. The grower is also required to give notice to the officer of the times respectively when he will begin to dry chicory and remove it from the kiln to the warehouse. Dry chicory to be forthwith secured in the approved warehouse, and it may only be removed therefrom in the day-time, on a four hours' notice. Apart from the 12s. 1d. per cwt. duty on dried chicory, these unnecessarily vexatious and expensive regulations do not encourage farmers to grow chicory, even when it might be done with considerable profit to themselves.

ELECTRICITY IN AGRICULTURE.

Before leaving the subject of crops we may refer to attempts which have been made to stimulate the production of the soil by the influence of electricity. At one of the meetings of the British Association at Dublin in September 1908, Sir Oliver Lodge read a paper on this subject, giving the results of experiments conducted in the south of England. He stated that in these experiments it had been found that a discharge of positive electricity into the air above growing plants stimulated and increased their growth. The effect of the electricity was greatest when the sunshine was not very strong and the soil not too dry. In strong sunshine and very dry soil the electricity seemed to over-stimulate the plants. The trials took place on various farms, and all over grain crops gave a substantial increase in yield from the application of electricity.

The results of these trials are quite in keeping with results obtained in similar but cruder experiments conducted many years ago in this and other countries, and it is probable that ere long electricity may come to be recognised as an important factor in the growing of field and garden crops.

THE FUNGUS DISEASES OF PLANTS.

It is unfortunately the case that farmers have nowadays a greater number of fungus, insect, and other plagues to contend against than in bygone times. In a work such as this, therefore, it is important that information should be provided which is fitted to assist farmers in the combating of these plagues. For much of this information the Editor is indebted to Dr R. S. Macdougall.

Fungi belong to that lowest group of plants known as the Thallophytes, or the plants which in their vegetative parts show no distinction into root, stem, and leaf. The plant in the case of a fungus consists typically of a series of much-branched threads known as the mycelium.

These mycelial threads, which may be septate or non-septate, have walls enclosing protoplasm. In feeding, absorption may take place over the whole surface of the mycelium, or special absorbing processes may be developed known as suckers. The fructification of a fungus may be a large structure, and may not appear thread-like, but the seemingly solid fructification is made up of a mass of interwoven threads: thus in the field-mushroom the real plant is in the soil, and consists of numerous threads, while the familiar stalk and cup make up the fruit of the plant.

An extremely important fact to note in connection with Fungi is that none of them possess chlorophyll — the green colouring matter characteristic of most plants, and the possession of which makes green plants, speaking generally, independent. Plants that lack chlorophyll are not able to do for themselves, and Fungi, therefore, have to obtain the carbonaceous food-material, which they cannot manufacture for themselves, either by stealing it from a live host or by living on decaying organic matter. The latter class of Fungi is known as the Saprophytes, of which the mushroom and many moulds are good examples. The Fungi which live off live hosts are known as Parasitic Fungi. The parasite may live on an animal, as in the cases of ringworm and the Fungi which infest salmon and kill flies and locusts; or the Fungus may be parasitic on a plant, and it is with such Fungi that we have to deal here.

Multiplication of Fungi may take place either asexually or sexually. In asexual multiplication pieces of the mycelium broken away and placed in suitable conditions feed, grow, and develop into new plants—e.g., in cultivating mushrooms from the so-called "spawn"; or in asexual multiplication special cells may be produced known as conidia or spores, and any one of these in suitable conditions can alone and without fusing with another develop and give rise to a new mycelium. In sexual multiplication we have special parts produced—male and female—a fusion of male and female elements takes places, and a new individual is the result.

There are interesting examples of Fungi living in helpful partnership with the higher plants. For example, many woodland plants in soil rich in humus are indebted for supplies of carbonaceous and nitrogenous food-material to Fungi which can make use of the carbon and nitrogen in the humus.

THE PHYCOMYCETES.

The Phycomycetes is a family of fungi characterised by a branching mycelium without partitions or septa. Sexual and asexual multiplication are met with.

The "Damping off" of Seedlings (Pythium de Baryanum).

When seedlings—e.g., of cress, or mustard, or clover—are overcrowded and kept in very moist conditions, they fall over, the weak place showing just above the surface of the soil, become pale, then brown, and ultimately rot. This is the result of the attack of the fungus Pythium, the much-branched mycelium of which invades the tissues of the host plant, passing between the cells or eating its way into the cells. The mycelium of this fungus passes from one seedling to another till at last the seed-bed may be ruined. Pythium multiplies both asexually and sexually. In the asexual multiplication, the ends of some of the mycelial threads swell and get cut off by partitions. These cut-off pieces behave in either of two ways. In drier conditions the round or oval cut-off piece, known as a conidium, gives rise directly to a delicate thread which enters a new host plant and results in a new mycelium. In the presence of water the behaviour is different : the protoplasm of the cut-off pieces divides up into zoospores, which escaping, move through the water by means of cilia—two delicate cilia for each zoospore ; these zoospores come to rest, and each gives out a delicate thread or germ tube, which piercing a host plant results in a new mycelium.

In the sexual multiplication of this fungus, the male reproductive organ is known as an antheridium, which is a cell cut off by a partition from one of the threads of the mycelium : a portion of the contents of this antheridium forms the male gamete. The female sexual organ is known as the oogonium : it is a rounded swelling of one of the hyphæ or threads, and a portion of its protoplasm forms the egg-cell. The antheridium comes in contact with the oogonium, a fertilisation process pierces the oogonium wall and the male gamete is carried to the egg. The fertilised egg becomes surrounded by a thick coat and is known as the oospore. The oospore is a resting-spore which may remain dormant in the soil until new seedlings are being grown, when the oospore germinates directly, or gives rise to zoospores, and the new seedlings are infected. An important feature in the life of this Pythium is that though a parasite it is capable of living for a time as a saprophyte.

Treatment.— 1. Do not overcrowd. 2. The seed-bed should not be too moist. 3. There must be a supply of fresh air, and access for fresh air and for sunlight. 4. The burning of infected plants with the soil surrounding them. 5. The burying of the oospores by deep ploughing. 6. The fact that the oospores may remain dormant for a time, and that the fungus can live for a time as a saprophyte, should prevent a seed-bed which has been affected from being immediately used again for the same purpose.

Potato Disease (Phytophthora infestans, D.B.)

This fungus first shows itself in Britain towards the end of July and on into August. It is first seen on the leaves, the yellow spots or patches soon turning brown and black. The fungus generally appears during close weather, with a humid atmosphere, especially when mists hang over the fields in the evenings and mornings, and the days are hot and damp. These are conditions well known to favour fungus growth. In dry weather the diseased patches on the leaf do not increase much, but in the humid conditions described, the diseased patches spread over the leaves very quickly, and the above-ground parts of the plant shrivel and rot with an attendant foul odour. Examination of the under surface of a diseased leaf would show round each affected patch a mealy- or floury-looking rim, this consisting of mycelial branches and conidia. The disease may pass from leaf to shaw, and may also infect the tubers either by the mycelium travelling down the stem, or by spores or conidia tumbling away on to the soil and getting washed down to the tubers, where, after infection, brown, dead, sunken patches show on the surface.

Description and Life - history. — The following is part of the description given by Worthington G. Smith in *Diseases of Field and Garden Crops* (Macmillan). This book is no longer completely up to date, but it contains useful figures :—

" For an exact examination of *Phytophthora infestans*, a very minute and extremely thin and transparent slice must be cut from a diseased leaf at a spot where the white bloom caused by the presence of the fungus is visible underneath. A good plan is to cut a diseased leaf in two through a disease spot, and then with a sharp lancet cut an extremely thin slice off from one of the exposed cut surfaces. If the slice last cut is somewhat longitudinally wedge - shaped, it will often best show the structure of the leaf and the contained fungus at the thinner end of the section. The atom to be examined should be placed on a glass slide in a drop of glycerine (this is preferable to water, as the latter often dries too quickly), and then covered with a clean thin cover-glass.

" The magnification given by an ordinary lens is useless for the observation of the minute fungus now before us, so we must at once place it under the higher powers of the microscope. If the slicing

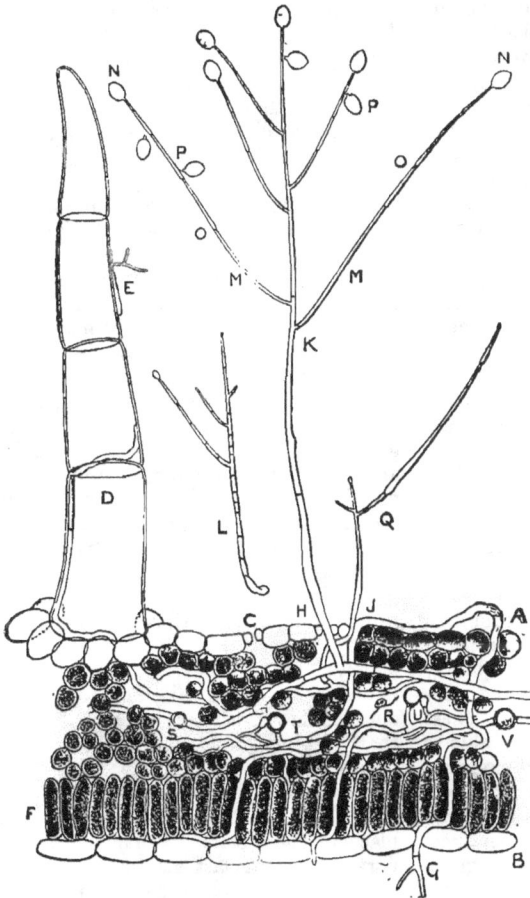

Fig. 621.—*Section through a fragment of a potato-leaf, with the potato fungus,* Phytophthora infestans, *growing within its substance, and emerging through the epidermis. Enlarged 100 diameters.*

through a disease spot is successful, we shall probably see the atom when magnified 100 diameters (as at fig. 621). The thickness of the lamina of the leaf is shown at A, B; the under side of the leaf is represented at A, from which surface the fungus almost invariably springs. The true upper surface is shown at B.

If we confine our attention for the present to the section of the leaf, we shall note that it is made up of minute cells, loosely packed together; and that the cells at top and bottom, representing the lower and upper epidermis of the leaf, are devoid of the shading, which is meant to indicate the green colouring matter or chlorophyll within. An opening into the interior of the leaf will be seen at C; this is one of the stomata or organs for gaseous interchange, sugar-making, and respiration, and in the giving off of water vapour. At D may be seen a hair built up of four transparent cells, the two lower being traversed by a mycelial thread of the potato fungus. On the upper part of this hair, attached to the outside at E, may be seen one of the small branches of the fungus; this branch has burst and thrown out a mycelial thread from its side. Every fragment of the potato fungus is capable of growth, and of ultimately reproducing the parent fungus. The cells immediately under the true upper epidermis of the leaf at F are termed palisade cells; and their disposition in the manner illustrated serves to give the necessary firmness to the exposed upper surface of the leaf.

"If we now look within the fragment of the leaf we see transparent threads running between the small spherical leaf-cells: these are the spawn-threads or mycelium of the fungus. It should be especially noticed that wherever the spawn touches the cells it discolours them (as indicated by the darker shading), and causes putrescence by contact. If we again look at the palisade cells near G, we observe that a spawn-thread has pushed itself between them and between the cells of the upper cuticle, and is emerging into the air. If we trace the spawn-threads to the stoma at H, we notice that a thread in its passage from the body of the leaf has blocked it up. This choking prevents the transpiration of water-vapour, and hastens putrescence. Two other threads have pushed themselves between the leaf-cells at G and A. When the larger of the emerged threads is traced upwards to K, a tree-like growth is noticed. The whole fungus is perfectly transparent, like colourless glass, and extremely fine and thin. If we now look at the branches, M M, we observe that each is surmounted by a transparent conidium, as at N N. It must also be noticed that all the branches are more or less constricted or jointed in a peculiar manner, as at O O; and that each joint has at one time carried a conidium, the lower conidia having been pushed off as the branches have continued their growth, as at P P. Sometimes a weakly impoverished thread, if grown in dry air, will quickly become strong and robust in growth if transferred to warm moist air, as in the thread illustrated at Q.

"If ripe conidia [N N] are placed in water, it will be noted that a differentiation of the contained protoplasm takes place; and that the interior mass of each conidium becomes divided into portions. These differentiated portions speedily emerge from the top of the conidium when placed on any moist surface; and each portion, now free, becomes quickly furnished with two extremely fine hair-like cilia or vibrating hairs. These secondary spores or zoospores are able to sail about in the slightest film of moisture. After a brief time the little motile zoo-spores rest and take a globular form. After a short rest the now quiescent zoo-spores germinate, each producing a germ-thread which can give rise to a new mycelium.

"Sometimes the conidium, which, when it bears zoospores, is really a sort of spore-case, sporangium, or zoospor-angium, does not differentiate within, but germinates directly, giving a mycelial thread capable (like the thread from the zoospore) of carrying on the life of the potato fungus. It must be specially noted that water or moist air is essential for the existence of the fungus, for nearly every part speedily perishes in dry air, heat, or frost. When the conidia burst and set free the minute zoospores, the latter move over the damp surfaces of leaves. A zoospore swimming in an intercellular space is shown at R.

"One has only to imagine a large field of potatoes, with all the leaves moist and swaying backwards and forwards with the wind, to perceive that such a field, say on a warm misty morning or evening, would form a sort of continuous lake of moisture on which the zoospores could float from one plant to another. The conidia, with the contained zoospores, are also carried through the air in millions by the wind; they are so lightly attached to their supporting stems, and so extremely small and light, that the faintest breath of air wafts them away. Insects and other creatures also carry the conidia from place to place. The germ-threads from conidia or from zoospores may enter by a stoma, or may directly eat their way into the potato leaf.

"The fungus possesses such wonderful powers of spore production and rapid growth, especially when the air is moist and the temperature ranges from 60° to 70° Fahr., that in a few days one fungus growth will become ten thousand, so that the disease seems to almost suddenly cover the potato fields."

Smith believes that this potato-disease fungus can also multiply sexually, producing oospores like Pythium de Bary-anum, mentioned above. There is reason, however, to doubt this, and instead of dormant hibernating oospores, it is likely that the mycelium may hibernate in refuse from diseased potato fields. Experiments at Kew have shown that hibernation of the mycelium can take place in the tubers. The practical importance of the last statement is manifest. A sudden outbreak of potato disease will not then be due necessarily to a great infection by conidia or spores, but may be due to the presence of latent mycelium in the tuber, called into active life by weather and other conditions that favour its growth. The importance of planting clean "seed" is further emphasised.

Phytophthora infestans attacks other plants in the Natural Order Solan-aceæ.

Preventive and Remedial Measures.
—1. Plant clean tubers, the tubers not having been saved from diseased fields. The tubers should have been stored in the best conditions. 2. The earthing-up of the potato drills with a deep covering of earth has been recommended, with a view to preventing the spores of the fungus from reaching the young tubers. 3. Growing the potato crop under such general sanitary and manurial conditions as will ensure to the fullest extent possible the healthy and vigorous development of the crop.

4. *Planting Varieties that have been known to be successful in resisting the Disease.*—It is known that certain varieties are, for the time being, exceptionally successful in resisting the attacks of the fungus. This valuable property is most generally found in some comparatively new variety—a variety recently raised from the seed—whose constitution and vitality of growth are unusually robust.

It is obviously advantageous, therefore, as a means of guarding against loss from the disease, to plant for the main crop such varieties as are at the time known to be the most successful in resisting the onslaught of the fungus.

Unfortunately there is a tendency in all the cultivated varieties of potatoes to lose vitality with long-continued culture. The "Champion," for instance, which for many years was almost disease-proof, at last fell an easy prey to the fungus. It is therefore desirable that the propagating of new and robust varieties should be liberally encouraged by potato-growers.

5. *Spraying with Bordeaux Mixture.*—The germination of the spores is prevented, and the spread of the disease kept in check. Further, the spray has an invigorating and stimulating effect on the leaves of the plant. The plants should be sprayed before the disease shows itself. If rain follows within some hours, the spraying must be repeated. The first spraying might be about the end of June or the beginning of July, followed by a second when the plants are well grown. In wet seasons there ought to be repeti-

Fig. 622.—*The Strawsoniser at work.*

tions of the spraying. Bordeaux Mixture consists of copper-sulphate and lime in water. In Leaflet No. 23 of the Board of Agriculture the proportions of sulphate of copper and lime are given, and various modifications pointed out in detail. Recently Mr Pickering has given the following as the best recipe for 100 gallons of Bordeaux mixture : [1]—

"Dissolve 6 lb. 6½ oz. of crystallised copper sulphate by suspending it in a

[1] *Woburn Experimental Fruit Farm—Eighth Report.* By the Duke of Bedford and Mr Spencer Pickering. 1908.

piece of sacking in two or three gallons of water. This must be done in a wooden or earthenware pail : zinc or iron must not be used. Take 2 or 3 lb. (not less) of fresh quicklime, slake it with a little water, and put it into a tub with some 120 gallons of soft water ; stir several times and allow to settle till the solution is quite clear. Then run off 86 gallons of the clear lime-water and mix it with the copper sulphate, bringing the whole up to 100 gallons by the addition of soft water.

"This Bordeaux Mixture should be

tested to make certain that all the copper has been thrown down. The way to apply the test is to put a few drops of a solution of potassium ferro-cyanide into a white saucer with some water, and to drop into this some of the clear liquid

Fig. 623.—*Warty disease of potatoes; different stages of infestation.*

obtained after the Bordeaux mixture has been allowed to settle: any brown or red coloration indicates that there is copper in solution, and a little more lime-water must be then added to the mixture and the test repeated."

An excellent machine for spraying crops is the Strawsoniser, shown in fig. 622.

Warty Disease of Potatoes (Chrysophlyctis endobiotica, Schilb.)

This fungus does not consist of branching threads, but is a single-celled structure, a droplet of protoplasm, which enters the tuber, probably at an eye, and lives at the expense of the contents of the invaded cell. A yellow-brown discoloration accompanies attack, and externally the disease is recognisable by the warty outgrowth and distorted tubers (see figs. 623 and 624).

The fungus has swarm-sporangia and resting-sporangia. In the first case, a wall forms round the naked protoplasm of the fungus; within the wall the protoplasm breaks up into a number of swarm-spores, and these spores issue and spread the infection. The other sporangium is a resting-stage which tides the fungus over the winter, or till the conditions for its activity become favourable again. In this stage the proto-

plasm is protected by a stout wall (see figures). The disease may attack the young sprouts, or later, the tubers.

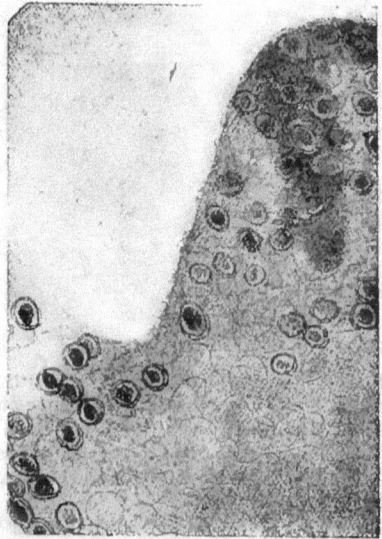

Fig. 624.—*Warty disease in potatoes; section through malformation to show its resting sporangia.*

The warty outgrowth ceases to grow when the potatoes are lifted.

Treatment.—Potatoes should not be planted for at least two years on land that has shown the disease. Crucifer or grass or Umbelliferous crop plants are not attacked. Diseased tubers should not be used for seed, but infested tubers should be collected and burned. Leaflet No. 105 of the Board of Agriculture treats of the Warty Disease of Potatoes, and this disease is scheduled in the Destructive Insects and Pests Act of 1908.

Onion Mildew (Peronospora schleidenii, Unger).

Yellowish patches show on the leaves, which ultimately coalesce into greyish-lilac patches. An early symptom of attack is the marked increase in length of the neck of the onion. The mycelium living in the leaf sends threads out through the stomata; these mycelial threads branch and give off the conidia or spores. If the disease show early, then on account of destruction of leaf the bulbs fail to develop. This fungus also produces oospores, which lie dormant for a time, and can, on germination, set up a new infection.

Treatment. — Remove and burn infested plants. If allowed to lie, the oospores will infect the soil. Dust the crop when the plants are covered with dew with 1 part of powdered quicklime to 2 of sulphur.

White Rust of Crucifers (Cystopus candidus, Pers.)

This Phycomycete attacks cabbage, radish, and wild crucifers such as shepherd's purse, whose leaves, stems, and fruits often show bad attack and accompanying distortion. Glossy white patches show on the parts attacked. The skin of the plant ruptures, and chains of conidia show on the outside in the form of a white powder. These give rise to zoospores, which infect new plants. Oospores are formed later (not in the case of shepherd's purse), and oospores attack seedlings in the next season.

Treatment. — (1) Keep down shepherd's purse. (2) Remove and burn diseased leaves, and burn distorted stems so as to destroy the oospores.

THE ASCOMYCETES.

This is a family of Fungi with a septate mycelium. The name is given

from the spores being enclosed in sacs called asci. The family contains both saprophytic and parasitic members.

The Mildews.

The term mildew is popularly applied to many fungi that may not have any definite relationship to one another. Botanically the term is applied to the Erysipheæ, a group of Ascomycetes parasitic in habit. A great characteristic, too, is that the mycelium of the fungus is external, and the fungus would therefore be easy to reach with treatment, only the worker is baffled often by the widespreadness of the disease. The mildew fungi send suckers from the external mycelium into the host plant, tapping it for food-material. Typically there are two modes of multiplication. First, conidia are produced during the summer, which conidia being scattered spread the infection to other plants. Secondly, in autumn small dark spots show on the mycelium; these are fruits which contain one or more asci, the spores from which are resistant to conditions that generally would kill the mycelium and the conidia; after resting during the winter, the ascospores germinate and set up the disease. In some of the Erysipheæ both the conidial and the ascospore stage have not yet been demonstrated, and this is the case with

Turnip Mildew (*Oidium balsamii*, Mont.)—This fungus is only known in one reproductive stage, the conidial. The host plants are crucifers, like turnip and swede; but plants outside Cruciferæ are also attacked. The mycelium is found all over the leaves, covering them often so completely as to choke up the stomata; respiration, carbon-assimilation, and transpiration are interfered with, and the tubers fail to fill. The disease begins towards the end of July, and by the end of August the plants may be white. The whiteness is due to a mass of mycelium, from which threads are sent up that end in barrel-shaped conidia. The further stages in the life-history are not known.

Treatment.—Stimulating dressings to tide the plant over the early stages of attack. Spraying with Bordeaux Mixture, care being taken that the under sides of the leaves are reached by the spray. Cucumbers and vegetable-marrows also fall before this fungus. Should Bordeaux Mixture be used in their cases, care should be taken that the delicate blossom is not harmed, and, further, the spraying should not be done when the fruit is ripening. Powdered sulphur or washes containing sulphur are also useful against mildews.

Mildew of Grasses (*Erysiphe graminis*, DC.)—The lower leaves of cereals and grasses—specially where the growth is rank—sometimes show irregular brown-white spots, due to the mycelium of this fungus. During the summer conidia are produced, which are scattered and spread the infection. Later in the year brown somewhat flask - shaped structures (the perithecia of the books) appear : each contains asci, whose spores, passing the winter in leaf and straw, germinate in the next spring.

Mildew of Pea (*Erysiphe martii*, Lév.) — This enemy of leguminous plants, worst in dry seasons and on late varieties, is sometimes so destructive to the leaves that pods fail.

Mildew of Rose (*Sphœrotheca pannosa*, Lév.)—Leaves, shoots, and flower-buds may be covered with a white or a greyish-white felt or film of the mycelium of this fungus. The leaves as a result pucker and fall early. In the summer, spread is by conidia. Later, round dark-coloured flasks appear, each with one ascus. The ascospores, after winter, give rise to a new mycelium. The disease is kept in check by spraying with a solution in water of liver of sulphur, ½ an ounce in a gallon of water, stronger than this if the leaves are not tender.

Gooseberry Mildew (*Microsphœria grossulariœ*, Lév.)—The mycelium is found on the leaf, giving rise to white patches. On magnification the powdery material on the leaf is found to consist of conidia. An ascospore stage is also found. The leaves fall early, and the fruit is small. The bushes, after several attacks in successive years, may die.

Treatment.—Burn all fallen infested leaves to destroy the ascospores. Deep digging about the bushes will bury the spores on the ground. Against the mycelium the bushes should be sprayed with 1½ lb. of sulphate of potassium to 50 gallons of water.

American Gooseberry Mildew (*Sphœrotheca mors-ursœ*, Berk.)—This is a far more harmful fungus than the last, for it attacks leaf, shoot, and fruit. There have been outbreaks of this fungus in several parts of England, and as a result the disease has been scheduled in the Destructive Insects and Pest Order, 1908. Full details as to life-history and instructions as regards treatment are given in the newly issued Board of Agriculture Leaflet, No. 195. Red currants have been attacked in Ireland. It is possible that black currants and raspberries may also be attacked.

The two forms of Gooseberry Mildew are shown in fig. 625.

Winter Rot of Potatoes (Nectria solani, Pers.)

This disease is one which attacks stored tubers. Tubers are stored with the mycelium present, the result of infection by spores in the soil ; but whether or no the disease develops, so as to give rise to sunken shrivelled portions on the outside of the tuber, depends on the conditions. The conditions favouring the fungus are the storing of the potatoes before being well dried, and then the absence of aeration.

As the disease makes progress, white mycelium is seen to be present on the shrivelling areas of the tuber. Spores produced by this mycelium spread the disease from tuber to tuber until the potato is ruined, the decay being hastened by secondary parasites. A second kind of spore is produced in the next season from little red warts on the skin of the potato, and the spores lying in the soil infect the next crop.

Treatment. — Have well - ventilated pits and store in dry condition. Powdered sulphur, 2 lb. to the ton, dusted over the tubers, kills the fungus.

In Leaflet No. 193 of the Board of Agriculture it is advised that kainit be applied to infected land. "When land is infected this manure should be used in preference to sulphate or muriate of potash ; but the quantity should not exceed 5-6 cwt. per acre, or the quality of the potatoes may be injured. Kainit may be applied in the drills before planting ; but in this case, where it is required both as a manure and a fungicide, it would probably be better to apply it as a

Fig. 625.—1, *Gooseberry mildew;* 2 and 3, *American gooseberry mildew.*

top-dressing before the horse-hoe is used for the last time.

"If the land needs potash, and especially if the potato crop is to be followed by a crop likely to be benefited by potash, *e.g.,* barley and mangolds, a dressing of kainit may be applied to the infected land as soon as the potatoes

have been lifted. If potash is not required, and if the land is likely to be
benefited by lime, then it would be desirable to dress the affected field with
1 to 3 tons of lime per acre. See also
Leaflet 170 on Liming of Land."

Ergot (Claviceps purpurea, Tul.)

This fungus is parasitic in the flowers
of cereals and grasses, where as a result
of its presence the ovary is destroyed,
there being of course no fruit or grain.
The ergot grains seen in autumn projecting from the grass heads are black or
black-blue in colour and ribbed. Each
grain is the result of a compacting of
mycelial threads, and is a hard structure
fitted to act as the dormant winter stage
of the fungus. The grains fall away
from the host plants and lie till the
next season, when, in the presence of
favourable conditions, they send out
branches that end in swollen heads.
Each head has all round it internally,
but with openings to the outside, a series
of flask-shaped structures. These contain asci with ascospores. The ascospores escape and reach the flower of a
grass; here they germinate, and a mycelium is formed which penetrates the
ovary and lives at the expense of the
host. During the summer, conidia are
produced in great numbers. These are
carried by insects to other grass flowers,
and so the disease spreads. At this time
a quantity of "honey-dew" is secreted,
and this may be attractive to insects,
which carry away conidia attached to
their bodies. As the season goes on the
mycelial threads compact into the hard
grain with which we started the life-
history.

The ergot grain contains a poisonous
principle, and cases of ergot poisoning
are found on the Continent where bread
has been made of flour that contained in
it ground-up ergot. Where pasture
grasses are ergoted the flowering heads
should be cut, raked together, and
burned. Only clean samples of grain
or grass should be sown. It is not at
all uncommon to find the wild grasses
along roadsides ergoted.

Phoma.

Potter has described a species of Phoma
which attacks the roots of swede. The
fungus is probably a stage in a more
complex life-cycle, part of the life-
history being spent away from the
root.

The disease shows at the upper part of
the swede; depressions appear from the
drying up of attacked cells. Black spots
appear at the outside, and from these
points the spores are given off.

The Uredineæ or Rust Fungi.

These are parasitic forms in which
there are several kinds of spore produced
in the same life-history: the mycelium is
generally intercellular.

Black Rust of Wheat (*Puccinia
graminis*, Pers.)—A wheat plant suffering
from this parasite would present early in
the summer a somewhat yellowish appearance, and soon, on careful looking,
long rusty spots would be seen on leaf
and stem. Under the microscope these
are seen to be places where the epidermis has been broken, and where
numerous orange-coloured spores are
hanging out (fig. 626, p. 420). These
spores, known as uredospores, pass to
other wheat plants and send out germ-
tubes or threads which enter the new
host by a stoma, or by directly boring
through the epidermis. This freshly
infected plant will soon harbour the
developed mycelium of the rust fungus,
from which uredospores will be given off
in turn. Towards the end of the summer
this mycelium on the wheat produces a
different kind of spore—the teleutospore.
This teleutospore is a two-celled spore,
and is dark-brown to black in colour. The
teleutospores hibernate over winter, and in
the next spring germinate by sending out
two germ tubes, the upper ends of which
become partitioned. Each jointed portion gives rise to a conidium or spore,
which in order to germinate must reach
not a wheat plant but a barberry. In
the barberry as a result a mycelium is
developed, which gives out, through
the ruptured upper skin of the barberry leaf, conidia, and from the broken
lower skin spores called æcidiospores
(fig. 627, p. 421). The destiny of the
conidia is unknown, but the æcidiospores on reaching the wheat plant
germinate and give rise to the mycelium
in wheat with which we started the
life-history.

In this life-history it will have been noted that two hosts were necessary for its completion. It is possible, however, that the barberry is not absolutely necessary, and Eriksson has suggested that this parasite in wheat can be transmitted to the grain, where as a "plasm" it lies dormant in the embryo, and that in the presence of favouring conditions the uredo-mycelium can be developed after the germination of the seed and the growth of the plant.

A fact of great interest, proved by the researches of Eriksson and Henning, is that while Black Rust is parasitic on wheat, barley, and oats, the disease cannot pass from one to the other—i.e., the uredospores of wheat cannot set up the infection in barley or oat, and the uredospores from barley cannot set up the disease in wheat, and so on. Under the microscope these varieties cannot be distinguished in form or appearance, but they are "biological varieties." The practical importance of this is further seen when we know that while Black Rust occurs on various wild grasses, the uredospores from these cannot infect wheat.

Yellow Rust of Wheat (*Puccinia glumarum*, Schm.)—This rust is also known as Spring Rust, as it appears earlier than the last. According to Biffen,[1] and opposed to the popularly accepted view, this is the most common and important of the rusts in this country, spoiling leaf and shoot and flower-envelopes and grain.

The life-history of the wheat plant is the same as given for Black Rust, uredospores and teleutospores being produced. No second host, however, has been found in the case of this rust corresponding to the barberry in the other, and it is known that the teleutospores are not necessary; the uredospores can hibernate, and can set up infection directly in wheat in the next season. We have here also "biological varieties," one being found on wheat, another on barley, a third on rye.

Treatment. — Immense loss results from rusted crops in other parts of the world. In Britain we have no estimate,

[1] *Journal of Board of Agriculture*, July 1908.

but the lessening of feeding and assimilative area represented in the destruction of leaf and shoot means poorer harvest and shrivelled grain. Apart from the advantage of good cultural operations, practically nothing can be advised as remedial. It is felt that the rust problem throughout the world is most likely of solution by the breeding of rust-resistant varieties. Experimenters have been active with this end in view for a good many years now. More recently, the interest in, and the wide acceptance of, the principles underlying Mendel's Law —by which plant-breeding is no longer the lottery that it seemed to be, but a thing of definite system — have stimulated experiment, and have raised hopes that at no distant date there will be on the market varieties of cereals immune to rust.

Other Rusts.—The rust of beans (*Uromyces fabæ*, Pers.) attacks the leaves and stems. Uredospores and teleutospores are produced, but there is no intermediate host. The Uromyces of the pea has two hosts, one the pea and the other a wild spurge.

Two troublesome rusts in gardens are Chrysanthemum Rust (*Puccinia hieracii*, Mart.) and Hollyhock Rust (*Puccinia malvacearum*, Mont.) In the first its whole life-history is completed on the one species of host, there being no second host with æcidiospores. The disease is spread in the summer by uredospores, and later the teleutospores are produced. The disease also affects wild composites — e.g., hawkweed and burdock. Spray with sulphide of potassium, half an ounce to a gallon of water. The leaves with teleutospores should be collected and burned to prevent infection in the next year.

In Hollyhock Rust, leaves, stems, sepals, and ovary are all attacked. There is no second host, and only one kind of spore is produced—viz., teleutospores. The little wart-like swellings which bear the clusters of teleutospores are grey-pink at first, but later dark-brown or black. If fading diseased plants be pulled and burned, there will be no disease in the next year. As a preventive measure spray with Bordeaux Mixture. Do not use seed from plants whose fruits have been rusted.

The Ustilagneæ.

This is a set of parasites interesting to the agriculturist from their attacks on cereals and grasses.

Smut of Oats (*Ustilago avenœ*, Jens.) —This parasite attacks the ears and invades the ovary, feeding off the reserve material which would normally have been laid up in the grain for the use of the embryo plant. In the ovary immense quantities of minute spores are produced, these giving the characteristic dark-brown sooty appearance characteristic of the spoilt oat-ears in July or earlier: chaff and grain are both destroyed. The spores are carried about, but infection of other plants does not take place at this stage. Under the microscope the spores are seen to have two coats, a somewhat rough dark-brown outer coat and a more delicate inner one. Infection of the oat plant can only take place in the seedling stage. When the oats are sown, spores attached to the grain germinate, and the germ-tube of the fungus invades the seedling. The fungus mycelium grows up with and branches in the host plant, ultimately passing into the ear, and giving rise to the spores which break through the epidermis of chaff and ovary and show on the outside.

Treatment.—Soak the seed oats in warm water thus: Place the grain in a sack and steep in a tub of water, with the temperature at 121° Fahr. to warm the seed; then soak for five minutes in water at a temperature of 131° Fahr. Remove and plunge the seed into cold water. Dry and then sow. Or dissolve 1 lb. of copper sulphate in a gallon of boiling water in a *wooden bucket* or *a copper pan.* Allow to cool. Empty a sack (4 bushels) of corn on to the barn floor and spread out. Pour the solution of copper sulphate over this, turning the seed over to make sure that all be wetted. Dry and sow. Or make a solution of 1 pint of 40 per cent formalin in 36 gallons of water. Place the seed in a bag, and allow to dip into the formalin solution for ten minutes. Dry and sow. A pint of the formalin costs 2s., and would serve for between 40 and 50 bushels of grain.

Wheat Smut (*Ustilago tritici*, Jens.) —This smut attacks the grain and the chaff. The spores are shed before the crop is harvested. A very important difference in life-history as compared with the smut of oat has to be noticed. Infection does not take place in the seedling stage in the next season, but Brefeld's experiments show that in the season in which the spores of wheat smut are shed the spores are carried by wind, &c., to wheat *flowers* and there germinate, the mycelium reaching the ovary. The mycelium hibernates inside the seed and develops when the seed is sown. If Brefeld's experimental work be accepted, then the "steep" treatment or "pickling" will not be efficacious as applied to wheat seed.

Naked Barley Smut (*Ustilago nuda*, Jens.) and **Covered Barley Smut** (*Ustilago tecta*, Jens.)—These two kinds of smut are found on barley: the spores of naked smut come to the outside and are spread before harvest. Infection is, as in the case of wheat smut, through the flower, the mycelium hibernating in the grain. In covered smut the spores remain within the chaff until the crop is harvested.

Bunt of Wheat (*Tilletia tritici*, Bjerk.) —This fungus enters the young wheat plant by germ tubes from conidia in the soil. The spores of the fungus are found in immense numbers in the ovaries of the infested plant. These spores do not show on the outside, but on crushing a diseased grain the olive-black spores escape. They have a disagreeable odour, and this, as well as the dark colour, is imparted to the flour when diseased grains are present in a sample of wheat and are ground up.

When a bunt spore germinates, it sends out an elongated mycelial thread, from the apex of which arises a tuft of secondary spores. These secondary spores become connected one with the other; those that join in pairs germinate by giving out a delicate germ-tube which enters the wheat seedling and grows up with it, giving rise to the fungus mycelium, whose spores are formed in the ovary.

Wheat plants attacked by the fungus of bunt are generally in their early stage a deeper green or blue-green, and appear more luxuriant; later, the ears are erect and rigid, the chaff is pale and bleached looking, and the grains are swollen and out of shape. Inside the grain is the mass of black spores.

Treatment. — The treatment is the same as that recommended for smut of oats.

Hyphomycetaceæ.

A number of fungi, probably representing for the most part stages in the life-history of higher fungi, come under the above title. The following are examples met with on agricultural plants.

Early Blight or Potato Leaf Curl (*Macrosporium solani*, M. and E.)—This fungus may attack the potato when the plants are young (5 or 6 inches high), and especially in dry weather. The disease begins at the base of the stem, and the mycelium spreads to the leaf. The leaves curl, light-brown patches of dead tissue show, and the rest of the blade is sickly yellow; in time the stem collapses. Spindle-shaped conidia or spores are produced which break through the tissue: these spores, which are brown in colour, are partitioned. The mycelium may spread to the tuber. Tubers of badly infested plants remain small from the destruction for feeding purposes of the leaves above. Massee holds that the above-ground parts cannot be infected by the spores, in which case the spraying with Bordeaux Mixture would not directly affect the fungus.

Treatment.—As the disease appears from the presence of the mycelium in the tuber or from spores in the soil, burn diseased shaws; do not plant tubers from a diseased area.

Potato Scab (*Oosporascabies*, Thaxter). —This is a disease of the tubers: the mycelium is superficial, and gives rise to scabs or patches, which in time may run together and the potatoes be cracked and spoiled for the market. The potatoes, however, may be eaten. In the multiplication of the fungus, conidia are produced at the apices of delicate threads.

Treatment.—Do not follow potatoes where the disease has been present with another crop of potatoes, nor with such crops as swede, cabbage, carrot, beet, as these may also suffer from scab. A cereal crop may be taken in safely.

Do not use "scabby" potatoes for "seed" unless treated by steeping them for two hours in a solution made by mixing 1 pint of commercial formalin (=formaldehyde, 40 per cent) in 36 gallons of water. The tubers are allowed to dry and are then planted.

The Slime Fungi or Myxomycetes.

These are chiefly saprophytes, but the one causing club-root in cruciferous plants—viz., *Plasmodiophora brassicæ*— is a parasite. They are characterised by their protoplasm being naked, not enclosed, as it generally is in plants, within a wall or walls: the naked protoplasm is known as a plasmodium, and is capable of an irregular amœboid creeping movement.

Finger-and-toe or Club-root (*Plasmodiophora brassicæ*, Wor.)—This enemy (fig. 628, p. 422) attacks turnip, swede, cabbage, especially, but also cauliflower, brussels sprouts, broccoli, rape, mustard, kohl-rabi, charlock, wallflower.

Attack is in the young stage, and is marked by sickly foliage; the tops of the attacked turnips become yellow and droop; as the disease advances the bulb rots, and with its destruction an offensive odour is given off. If a thin slice be cut with a razor through one of the swellings and placed under the microscope, the cells would be found to contain a turbid brown slime which would either fill the cells or stretch across them in bands. This is the plasmodium of the fungus, which feeds at the expense of the cell contents and causes the tissue of the root to break down. As the season goes on a change comes over the masses of brown plasm; these break up into numbers of small spores, each surrounded by a thick wall. These spores are resting-spores, and as such they are resistant to outside influences. The spores are colourless if examined individually, but are yellowish in the mass. They may remain all the winter in the affected turnips, or lying in the ground or on the dung-heap. On germinating after the winter the wall ruptures, and the protoplasm issues as a tiny droplet provided with a cilium or lasher, by which movement is possible. This moves about in search of a host, gaining entrance by a young rootlet, or by a root-hair, or through a wound. Inside the root the plasmodia move from cell to cell, and the characteristic swellings appear. Other enemies may assist in the destruction of the bulbs, which become the happy hunting-ground of insects and their larvæ, and millepedes and eelworms.

Treatment. — General principles: Keep down cruciferous weeds. Try to prevent diseased material from being

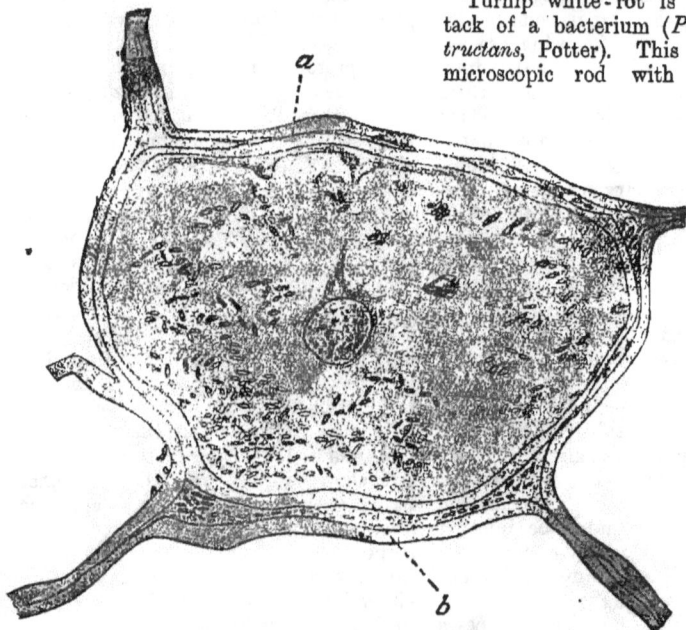

Fig. 629.—*Turnip white rot.*

A cell from a turnip infested by *Pseudomonas destructans*. The bacteria are seen in the cell-cavity and along the track of the middle lamella. At *a* the cell wall is beginning to separate along the middle lamella. At *b* the dissociation is more strongly marked. (After Potter.)

carried to uninfected ground. Specially do not feed the diseased roots to stock at the homestead, else the spores of the fungus will get into the dung, which later may be carried to the fields where turnips will be grown. Finger-and-toe does not attack grasses, hence infected dung may be used on a permanent grass-field. Potatoes, cereals, and mangolds can be grown on finger-and-toe infected ground. The spores of the fungus can lie dormant for a long time, therefore lengthen out the rotation where the disease is prevalent, so that the turnips be separated by, say, eight years. Let

the headlands be watched, so that the disease may be stamped out on its appearance. Apply lime, unslaked or recently slaked, some time before the turnip crop is taken.

BACTERIA.

The diseases of plants caused by bacteria are, as compared with those due to fungi, very few.

Turnip white-rot is due to the attack of a bacterium (*Pseudomonas destructans*, Potter). This bacterium is a microscopic rod with rounded ends (fig. 629). The rot induced is the work of a cytase ferment produced by the bacterial activity. The ferment acts on the walls of the cells of the turnip, causing them to swell and soften; the central part of the common walls of the cells is dissolved, and so the cells fall apart and the whole tissue becomes disorganised. The bacteria gain access to the root by a wound.

Black-rot of Cabbages (Pseudomonas campestris, Smith).

This disease is also bacterial in origin. Cultivated crucifers are the plants attacked. Leaflet No. 200 of the Board of Agriculture describes the disease and treatment.

The bacteria enter the leaves by water-stomata or by wounds, or enter the roots through wounds. The bacteria spread by the vascular bundles to the stem, and thence to other leaves. A characteristic symptom is the blackening of the veins.

DESCRIPTION OF FIGS. 626, 627, 628, pp. 420, 421, 422.

Fig. 626.—*Rust in wheat.*

1 and 2, The æcidia or cluster-cups on barberry leaf.

3 and 4, The spermagonia on the barberry leaf.

5, A section through an attacked barberry leaf.

A, An æcidium with æcidiospores escaping.

B, A young cluster-cup or æcidium.

C, Vein of leaf.

D, Spermagonia with escaping conidia.

E, Younger stage.

In 5, the threads of the fungus mycelium can be seen between the cells of the leaf.

Fig. 627.—*Rust in wheat.*

1, 2, 3, Fragments of leaf of wheat invaded by uredo mycelium.

4, The same under a higher magnification to show the mycelium and the numerous uredospores at the ruptured skin of the wheat leaf.

5, Germinating uredospore.

6, 7, 8, 9, Wheat showing later stage of attack: in 9 the mycelium is seen and many teleutospores.

10, Magnified teleutospores as seen in autumn and winter.

Fig. 628.—*Finger-and-toe in turnips.*

1, 2, 3, Young plants distorted by the attack of the finger-and-toe parasite.

4 and 5, Sections through a piece of turnip: in 5 the slime is seen stretching across the cells.

6, The plasmodium has divided up into spores.

7, Spores under high magnification.

8, Spores germinating.

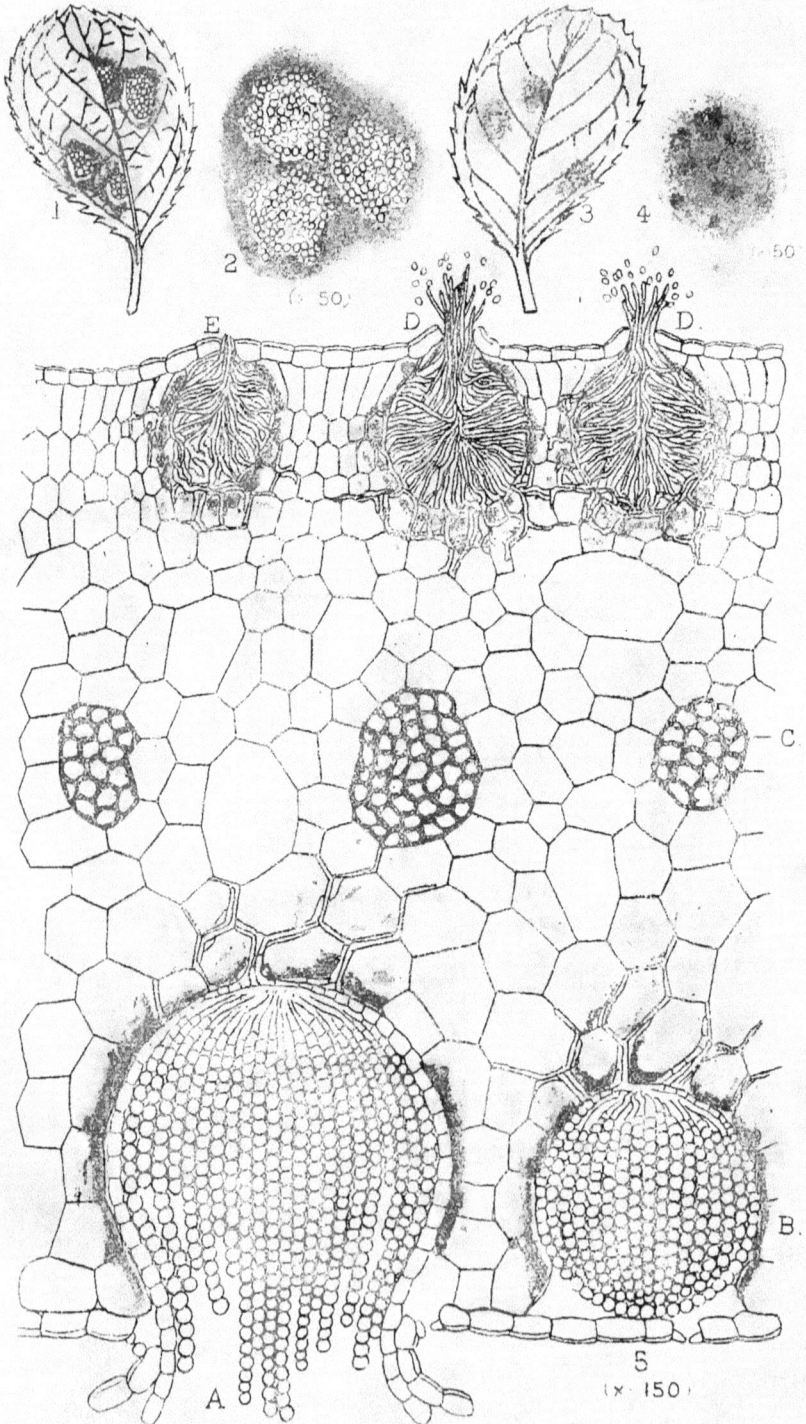

Fig. 626.—*Rust in wheat*—see p. 414.

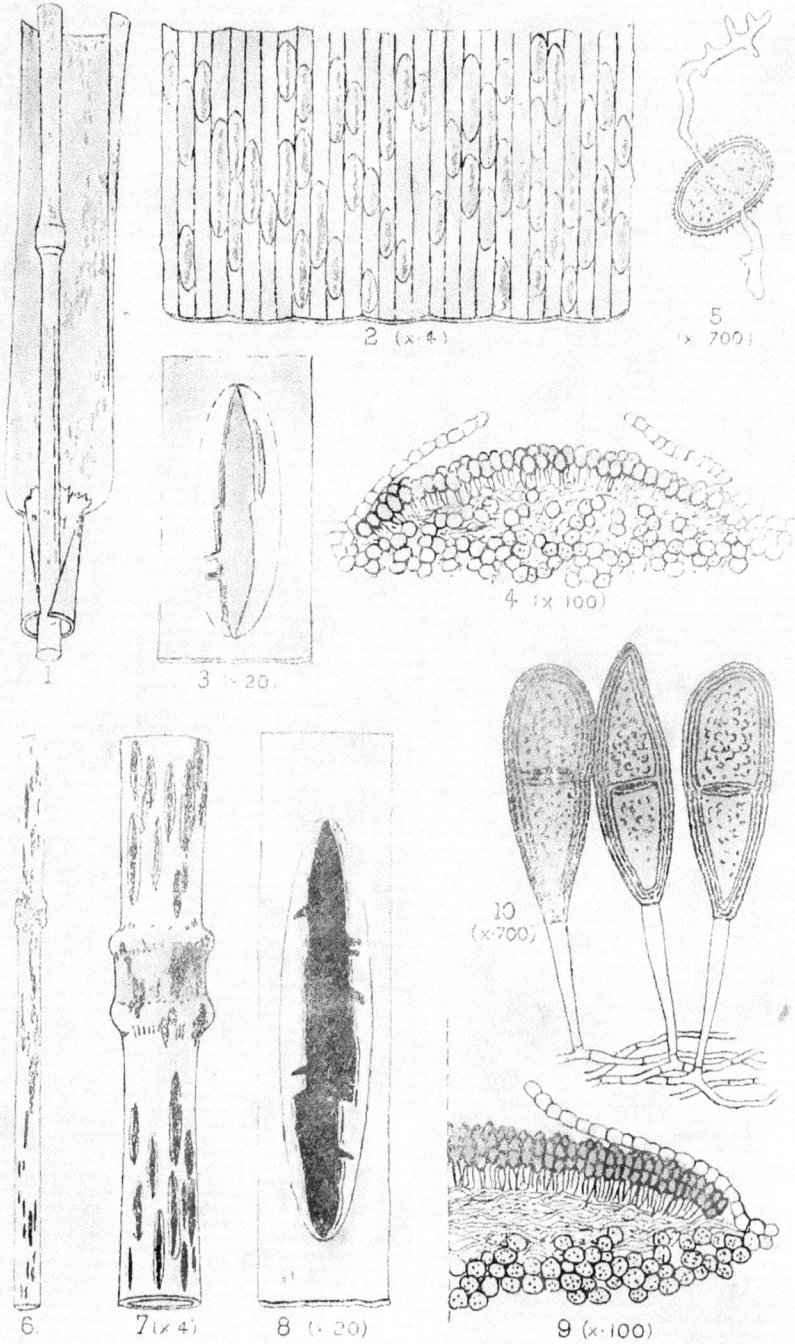

Fig. 627.—*Rust in wheat*—see p. 414.

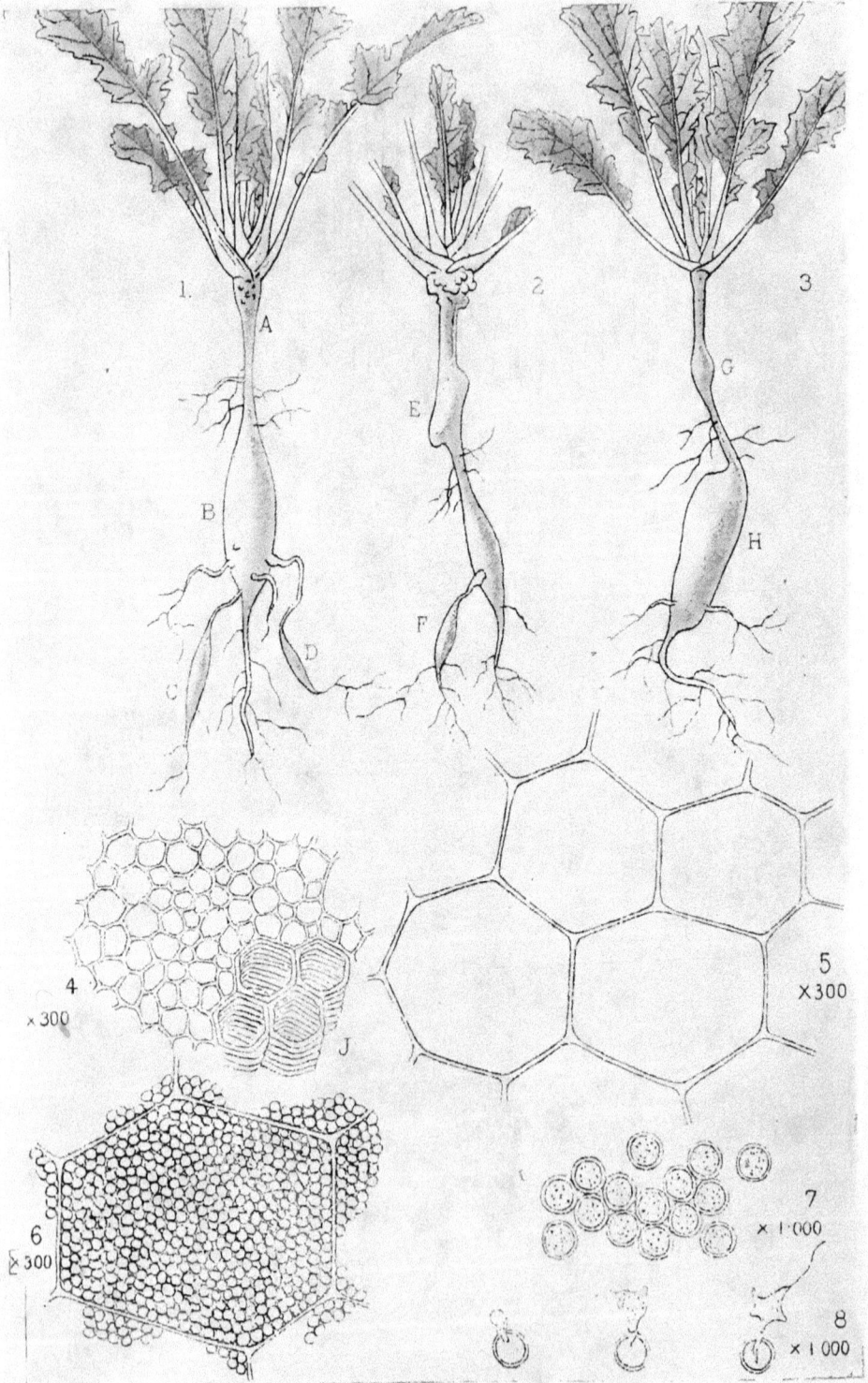

Fig. 628.—*Finger-and-toe in turnips*—see p. 417.

DODDER—A PARASITIC FLOWERING-PLANT.

Dodder or Cuscuta is a genus of flowering-plants belonging to the Natural Order *Convolvulaceæ*. Dodder is harmful in a crop because of its parasitic habits, the host plant perishing owing to the drain upon its strength from loss of sap drawn away by the parasite.

Typically, an independent flowering-plant feeds by its roots and its green leaves. By the roots water is taken in and mineral matter in solution; by the leaves CO_2 is taken in, and in the presence of certain necessary conditions, one of which is the possession of chlorophyll or leaf-green, the CO_2 taken in is broken up and made use of in the manufacture of carbonaceous food-material. A dodder plant has no roots, and, once it has attached itself to a host, no connection at all with the soil; neither have the dodders any chlorophyll—or at most the merest trace. The dodder plants therefore are complete parasites, fixing themselves to, and feeding entirely at the expense of, the host plant. Attachment to the host is secured by means of haustoria or suckers, which are sent into the sap-conducting tissues of the plant (the vascular bundles), and by them the host is tapped for food-supplies.

There are four British species of Cuscuta, but at least other three species have been found, introduced in seed from foreign countries. The British species are as follows:—

Clover-Dodder (*Cuscuta trifolii*, Bab.) —It has very delicate thread-like stems, brownish-red in colour; it produces clusters of small whitish flowers. The fruit of each flower is a two-chambered box, each chamber holding two seeds. The seed of dodder is in almost all cases smaller than the seed of the crop plants. The seed of *C. trifolii* (fig. 399, p. 89, vol. ii.) measures less than one millimetre in diameter; it is dim, not glossy like the clover seed, and if examined with a lens will be seen to be irregularly pitted and somewhat flattened and angular, owing to pressure in the seed-box. Dodder seeds contain, in addition to the embryo plant, a store of

reserve. In this reserve the embryo lies coiled in a spiral. *C. trifolii* attacks specially red clover and lucerne and other leguminous plants; but in other European countries there are records of attack on beet, potato, and carrot.

Lesser Dodder (*C. epithymum*, Murr.) —This dodder is very close to the last. It flowers from July onwards; the flowers vary somewhat in colour. Its host plants are thyme (hence the specific name), whin, and ling.

Greater Dodder (*C. europæa*).—This is a larger plant, with thicker yellow-red stems. Its host plants are the stinging-nettle, hops, and some leguminous plants; it has proved harmful, too, to young osiers in plantations. It is found from near the north of England southwards, but is not common.

Flax Dodder (*C. epilinum*, Weihe).— The stems of this species have a trace of green; the flowers are whitish. The host plant is flax. Although as big as the last, it is not such a strong plant.

Life-history.

Clover-seed may contain in the sample sown some seeds of dodder. When the Cuscuta seed germinates, the embryo grows out; the lower radicle end of it is thickened, and gives support to the whip-like stem that is growing into the air. This aërial portion twines round a clover plant and sends suckers into it. Once attachment is secured the dodder loses all connection with the soil. Unless the dodder succeeds in reaching a host it dies soon after the reserve laid up for it in the seed has been exhausted. If it fix itself, however, after feeding for a short time, it sends out branches which reach and twine round other plants, and so the parasite spreads. Some species of dodder have a greater capacity for branching and spreading in this way than others. It has been recorded that spreading so, a single Cuscuta plant can in a growth period of three months spread to, twine round, and destroy the host plants over an area equal to about 30 square yards.

In addition to this mode of spread, dodder has another mode of multiplication, as any pieces of cut or broken-through stems can in favourable conditions give rise to a new plant. Reproduction is also by seed which may be harvested with that of the crop plant so that the seed sample is impure, or may fall away from the ripe dodder fruit and lie dormant in the soil.

Treatment.

Although dodder is not the pest in Britain that it is in other European countries, yet botanical records and reports show that samples of seed contain, and that not rarely, dodder seed as an impurity. Seed in which dodder is found should be sieved—*i.e.*, the impure sample should be shaken in a sieve or set of sieves whose meshwork is fine enough to keep back the seed of the crop while allowing the seed of the parasite to pass through. The dodder seed thus separated out should be burned. The *Journal of the Board of Agriculture* some time ago quoted a recommendation from the *American Agriculturist* as to the making of a suitable sieve. "Make a light wooden frame, about 12 inches square and 3 inches deep, and tack over the bottom of it a 20-mesh wire screen (mode of No. 32 English gauge, round wire). In this one quarter to one half pound of seed should be placed and vigorously shaken for a half a minute. A man should be able to deal with from 5 to 10 bushels of seed per day." Samples of clover or other seed containing dodder should never be sown.

If dodder be present in a field, it should be annihilated with the plant round which it has twined. The destruction of the dodder should take place at once and in the spot where it is found; the best means is by fire.

Instead of firing the infested plants, some prefer to kill the dodder by putting over it some material which will kill the dodder. Several different materials are used, and a French worker, quoted by the author of an article on dodder, in the *Journal of the Board of Agriculture* for September 1906, praises above all calcium sulphide. It should also be kept in mind that feeding plants infested with dodder to stock is attended with danger, for it has been shown that dodder seeds can pass through the alimentary canal of stock without losing its powers of germination. In cases of very bad infestation the surface-soil should be buried deeply, and no leguminous crop taken for two years.

THE SMALLER MAMMALS IN RELATION TO FARM LIFE.

The Mammals which enter at all into the life of the farm are embraced in the following orders :—

Ungulata, or hoofed mammals — *e.g.*, horse, pig, ox, sheep, goat, deer.

Carnivora—*e.g.*, dog, fox, cat, weasel, stoat, polecat, badger, otter.

Rodentia, or gnawers — *e.g.*, rabbit, hare, rats, mice, voles, squirrel.

Insectivora — *e.g.*, hedgehog, mole, shrew.

We are concerned here with the smaller and non-domesticated mammals.

CARNIVORA.

These are typically flesh-eating forms which prey on other animals. They have small incisor teeth, but the canine teeth are strong, and some of the cheek teeth are modified for cutting; the toes are always clawed.

Fox.—The dog-like carnivores—*e.g.*, the fox—have the claws non-retractile. The fox (*Canis vulpes*) is in colour well suited to its environment. It may live above ground in hollow trees, or under old tree roots, or amongst heather, but, typically, it shelters in holes below ground — holes made by other animals — *e.g.*, badger and rabbit, or "earths" shovelled out by its clawed fore-limbs. To facilitate escape there may be several passages from the shelter place of the fox. There is one litter in the year, in the spring, 3 to 6 or 7 cubs being produced.

The senses of the fox are keen, especially sight and smell, and his "intellect" generally has been sharpened as a result of the struggle for existence. "More elegant than his relatives in mien and bearing; sharper, more prudent, calculating, and adaptive; of strong memory and sense of locality, resourceful, patient, resolute; equally skilled in jumping, slinking, crawling, and swimming, he seems to unite in himself all the qualifications of a perfect highwayman, and when his lively humour is also taken into account, produces the impression of a highly-educated artist in his own line." The food consists of game-birds, poultry that are not protected, hares, rabbits, and to a very great extent field-mice; insect grubs are also eaten.

Wild Cat. — The cat-like carnivores have retractile claws, and have rounder skulls than the dogs. Characteristic also are the rough tongue and the flexible backbone. The wild cat (*Felis catus*) is still to be found in some places in Scotland. In colour it is grey or yellow-grey with black stripes. It is larger than the domestic cat, while its tail, unlike that of the domestic cat, does not become thinner to the tip. In relation to the farmer the wild cat may be said to be useful in its destruction of mice, but it can be harmful among young deer and in game preserves.

Other Small Carnivores. — The family Mustelidæ, embracing the remaining small carnivores, shows the following features: body somewhat elongated and slender; short legs; small flat elongated head; on each side in upper and lower jaw is a tuberculate molar tooth; glands near the anus that give out a strong odour.

Stoats. — The stoat or ermine (*Putorius erminea*) measures—including the tail—15 or 16 inches in length. In its summer dress it is yellow-brown above and white below; the tail has a black tip. In winter, especially in the north, the colour of the fur is white, save for the tip of the tail, which retains its black colour (the weasel never has the tip of the tail black). The stoat hunts at night, its food consisting of rabbits, hares, rats, mice, birds. There is no doubt, however, that it can do much harm in poultry-yards and among game-birds. In packs, stoats may attack man, and there is a recorded case of a single stoat attacking, at short intervals, a girl and then a man.

Weasel (*Putorius vulgaris*).—This is the smallest of the British Mustelidæ. The length, including the short tail, is about 10 inches. The female is smaller than the male. The elongated body is red-brown above and white below. The slender snake-like body enables the weasel to pass into the holes and runs of mice and rats and voles, all of which are favourite food. The weasel also takes rabbits, young hares, birds building not high up, eggs, and, it may be, poultry. The weasel is a most courageous animal, and, though bloodthirsty and capable of harm, does much good in the destruction of mice and voles. The weasel's nest is in a bank or in a hollow tree; the litter is produced in spring, and numbers 4 to 5 or 6. The weasel, unlike the stoat, does not assume a white coat in the winter time.

Polecat. — The polecat or foumart (*Putorius fœtidus*). Foumart means foul marten, the name being given from the strong scent given off from the anal glands. The colour is chestnut brown-black with yellow soft woolly hairs. The tapering tail is shorter in comparison to the size of the animal than in the case of the stoat, and is bushy. In summer it lives in the open, in hollow trees and in rabbit burrows; in winter the polecat comes nearer to human habitations. The young are produced in summer 4 to 6 at a birth. The food consists of mice, voles, rats, fowls, ducks, game, frogs, eggs. The eggs are sucked without being broken. From its size the polecat is certainly able to do great harm amongst poultry, but this is somewhat atoned for by the destruction of rodents.

The *ferret* is an albino domesticated variety of the polecat.

Badger.—The badger (*Meles taxus*). This is a clumsier and much less active form than any of the preceding. The body is plump and heavy. The head is long, with the muzzle pointed, white in colour, with a broad black stripe on each side. The hair is short and close, and on the back earthy-grey in colour; the belly and feet are black. The tail is short. The badger measures in length 2½ feet and over. Shy and nocturnal, it lives in a burrow lined with leaves and moss and having long passages leading from it, the openings of which may be far apart. In diet the badger is omnivorous,—roots, bulbs, fruits, acorns, young ground-birds, eggs, mice, rats, young rabbits, insects and their larvæ which are dug up by its sharp broad claws, worms, snails. On the whole, the badger is certainly more useful than harmful, and the animal should not be killed out.

Otter.—The otter (*Lutra vulgaris*) measures 2½ feet in length ; the fur is dark brown in colour, and smooth and shining brown. It is well fitted in structure for its aquatic habit. The thick fur with spaces containing entangled air aid in keeping the animal warm ; fat glands in the skin keep the skin oiled and so protect it from wet ; the flexible body, oiled and slippery, enables the creature to move quickly through the water, while the tail, compressed at the sides, plays the part of a rudder. The toes have swimming membranes on them, and the legs are used as oars ; the nostrils can be closed ; touch and sight are acute, and the sharp teeth enable it to seize and tear the prey. Otters live in the banks of streams in a chamber, the entrance to which is below the water-level, ventilated by an opening higher up. The food consists of frogs, insects, water-rats, fish. The toll it levies on fish is great.

RODENTS.

These are herbivorous mammals characterised by their incisor teeth being chisel-edged and adapted for cutting or gnawing. The hard enamel of these incisor teeth is typically restricted to the front, and so the hind-part wears away more quickly and leaves the chisel-edge. There are no canine teeth, and the skin projects into the mouth as a hairy pad into the space between the incisors and the cheek teeth. They may be divided into the *Simplicidentata*, where there is a single pair of upper incisor teeth—*e.g.*, rat, mouse, vole, squirrel ; and the *Duplicidentata*, with two pairs of incisor teeth in the upper jaw (the extra pair are too small to be of use)—*e.g.*, rabbit and hare.

Rats, Mice, and Voles.

The true rats and mice belong to the family *Muridæ*, and the voles to the family *Arvicolidæ*. These two families may be distinguished as follows :—

Rats and Mice.	Voles.
Muzzle pointed.	Muzzle rounded.
Ears large and prominent.	Ears short, almost hidden in the fur.
Tail long, naked, scaly.	Tail short and may be hairy.

The Black Rat (*Mus rattus*).—This rat has long been decreasing before the superior strength of the brown rat, until now it is becoming extinct in Britain. The black rat is smaller and darker than the other. Its food is varied, with a preference for vegetable matter.

The Brown Rat (*Mus decumanus*).— This rat has also such names as the house rat, the Norway rat, the Hanoverian rat, the barn rat. It has a very wide distribution, having been spread over the world in ships. The brown rat has lighter fur than the black rat, and a dusky grey belly ; its muzzle is broader, and the ears and tail shorter.

In diet the rat is omnivorous, eating grain of all kinds, peas, beans, potatoes, carrots, turnips, truffles, fruit, the bark of trees, insects, mice, young rabbits, eggs, chickens, pigeons, ducklings, and even one another. In the obtaining of its food the rat often shows great ingenuity. This list of varied foods is one example of how the rat is favoured in the struggle for existence, but there are other factors favouring the rat—namely, its prolific nature,—eight at a litter, and this several times in a year,—and the rapidity with which the rat attains sexual maturity, and the fact that the animal can run, leap, climb, and swim.

One way of fighting rats is by trapping, but in such work one must keep in mind the ingenuity and the cunning of the rat. Another means of destruction is by ferrets. Harvie-Brown recommends, in order to get rid of rats in a poultry-yard, the making of some chloride of lime into a thick paste and smearing the rats' runs with it ; or chloride of lime mixed with water in a large watering-can and poured into their holes ; or pouring into the holes a strong solution of carbolic acid.

A correspondent of *The Scotsman* recommends the following as a certain way of ridding houses and buildings of rats: "Get from a drysalter a supply of cream caustic soda. It is in a solid state and is cheap. Break the soda up into small pieces; melt some in an iron or stoneware vessel and pour it into the rat holes so that the ground may be saturated with it; then jam one or two pieces into the holes so that the rats may undermine or scrape it away. When the rats come to the mouth of the hole and smell the soda they will begin to scratch under it to remove it; but the fluid soda has wet the soil or stones around and their feet will get blistered, and they cannot remove the solid pieces. Exposure to the air keeps the surface of the soda always damp; but long before all the pieces are entirely melted the rats will have forsaken that hole. As to dogs or poultry suffering by its use, care should be taken to keep them from touching it. Where the ground is undermined by a series of holes, pieces of wood covered by caustic soda should be inserted into the holes and a quantity of melted soda slowly poured on the ground around, giving it time to dry in. Rats are exceedingly cunning, and if they find themselves constantly liable to get severely burned when running about their favourite haunts they will entirely forsake the premises. As to handling caustic soda, it should not be touched with the ungloved hand, and care should be taken in breaking it not to let it spark on the face or eyes."

Very successful results in the war against rats have been got recently from the use of a virus which inoculates the animals, setting up in them a disease which kills them. Of the preparations in the market some are solid and some liquid. The material used contains in it the bacteria or organisms which, introduced into the body of the rat, cause a disease which is allied to typhus fever among human beings. Dry bread is taken, and after being cut into pieces, is soaked with the virus and spread where the rats may find and eat. The virus does not affect any of the domesticated animals. In a case recorded in the *Journal of the Board of Agriculture*, the Agricultural Chamber of Saxony had

an experiment on seven selected farms. On six of these the rats were practically exterminated, while on the seventh there was little or no effect, perhaps due to the possibility that the rats on this farm were survivors of the attack of an allied disease and had thereby been rendered immune.

If such experiments be undertaken in Scotland the wisdom of joint action is clear, else a farm rid of the pests by virus treatment might easily receive a new infection from an untreated place. Any treatment, then, with virus should be as the result of organisation and arrangement.

Some of the increase and spread of rats is due to the killing down of such of their natural enemies as owls, kestrels, and weasels.

Mice.

The Common House-mouse (*Mus musculus*).—This mouse is well known; it varies somewhat in colour but typically is grey or grey-brown. Like the rat it is very prolific, and can be a great scourge in wheat-ricks, in corn-fields, and in farm buildings.

The Long-tailed Field-mouse or Wood-mouse (*Mus sylvaticus*). — The burrows of this mouse may be found in banks, in corn-fields, in gardens, and the mouse may also infest stacks. Deserted nests in trees may be made use of for the young. It is a very harmful form, eating grain and all kinds of seeds, beans, peas, fruit, nuts, carrot. This mouse does not go to sleep for the winter, but lives off stores of varying food materials— *e.g.*, it may store great quantities of grain. The wood-mouse is yellowish brown, with a greyish tinge on the back, while below it is white; the breast is white, with a dark or fawn patch; feet white; the tail is brown on the upper surface and white on the under.

Here is a table quoted in *The Zoologist* for April 1881 to show how prolific the wood-mouse is:—

					Interval since last litter.
March 7 or 8.	A	3	young .	.	...
" 19.	B	5	"	.	...
" 31.	A	3	"	.	24 days.
April 18.	B	5	"	.	29 "
" 24.	A	3	"	.	24 "
May 11.	B	5	"	.	23 "
" 17.	A	4	"	.	23 "
June 12.	A ?	4	"	.	26 "
July 9.	A ?	4	"	.	27 "

This mouse does some good by feeding on insects, moths and their larvæ, and beetles. The mouse itself is dug out and eaten by crows and rooks, and foxes also hunt for them.

The Harvest-mouse (*Mus messorius* or *minutus*).—This is the smallest member of the family, and, save for the lesser shrew-mouse, is the smallest British quadruped. It measures 2½ inches. It lives in stacks and in the open. In colour this mouse is red-brown or sandy-yellow, with the belly white. In climbing among grasses and shrubs it makes use of its prehensile tail. It builds a neat nest of dry grasses twined cleverly round the stems of the plants. The food consists of grain and insects.

Voles.

The Water-rat or Water-vole (*Microtus* or *Arvicola amphibius*).—

This vole is 6 inches in length. It is dark-brown or black above, and grey to greyish-black or brown below. The cusps on the back teeth are angular, whereas in the true rat they are rounded. Commonly found along the banks of streams, it also burrows in meadows and fields at some distance from the water. The passages it makes in the river-banks and in canal-banks may cause the destruction of the banks. This vole is a good diver and swimmer. Its chief food is of vegetable nature,—aquatic plants, roots like turnip, carrot, and mangel, and the bark of root and stem of shrubs and trees, osier beds for example being sometimes much harmed.

The Short-tailed Field-vole (*Microtus agrestis*).—This (fig. 630) is one of the greatest enemies in pasture and other land. The back is dark-brown and the under surface grey. It has great powers

Fig. 630.—*The field-vole.*

of multiplication. The field-vole is found not only at low levels but in upland and hill pastures, and there are records both in Scotland and elsewhere of plagues of this species. In 1890 these voles had been noticed to increase, and in 1891 the hill pastures of Roxburgh and Dumfries were overrun with their burrows, and 90,000 acres of land were more or less affected. In addition to the loss entailed by this pest on the farmer, the forester may also suffer severely, both young broad-leaved species and conifers having their roots gnawed through, or being stripped of their side-shoots, or being barked. In 1814 practically all the oak saplings in the Forest of Dean were destroyed owing to their roots being eaten through by

this species. Of the field-vole and the wood-mouse in this plague of 1814, no fewer than 100,000 tails were brought in for reward. This vole nests below ground, there being several litters, especially in the warmer part of the year, four to five or more in a litter; but breeding may take place from the spring to the late autumn.

The Bank-vole (*Evotomys glareolus*) is also harmful in field and forest, the edges of woods and sheltered banks being characteristic places for it. In length it measures up to 4 inches; the colour is red or red-brown above and white below; the feet are also white. The tail is more hairy and the ears are larger than in the last species. It is harmful to trees, gnawing various broad-leaved and conifer species, and eating also the buds of the pine.

Measures against Mice and Voles.

The turning out of cats has been tried. The ground, too, should be cleared, so as to better expose the pests to their natural enemies—*e.g.*, owls and kestrels; the weasel is worthy of mention, too, though it has been admitted as possibly harmful in another connection. The grazing of cattle keeps down the undergrowth, and so prevents shelter and destroys breeding-places. Several times in Scotland and England, in bad infestations, much good has resulted from digging trenches 30 yards apart, 18 to 20 inches broad at the bottom, 9 inches broad at the top, 1½ foot deep, and 2 feet long: mice and voles falling in were unable to climb the inward-sloping sides, and so were trapped. In a devastation at the Forest of Dean 30,000 field-voles were caught by this means.

Injury to trees can be prevented by smearing them with asphalt-tar or one of the patent tars. A further method of preventing injury consists in spreading here and there, beside the young trees to be protected, a few branches of soft-wooded trees. These are used as food, the voles preferring to bark or nibble what is lying on the ground rather than the trees. These branches should be changed at intervals, for as soon as they are dry the rodents will cease to use them for food but will utilise them for shelter. Should poison be resorted to, such collections of branches would be good positions in which to place the bait. In poisoning, the oatmeal or flour mixed with arsenic or other poison should not be left exposed, but should be placed in glazed pipes 1·½ inch in diameter: the mice then get access to the bait, while larger animals cannot reach the poison.

Hares and Rabbits.

The **Rabbit** (*Lepus cuniculus*) and the **Hare** (*Lepus timidus*) can be harmful both in agriculture and in forestry. They eat such plants as young grass, cabbages, turnips, vetches, carrots, grain, and the bark of a large number of species of trees. Not only does the rabbit multiply more quickly than the hare — 5 or 6 litters in a year, with 4 to 6 young at a time, and reproduction capable at 6 months old — but its burrowing habit makes it more difficult to keep down. Shooting, fencing with wire, netting, ferreting, are the measures practised against these animals.

The Squirrel.

The **Squirrel** (*Sciurus vulgaris*) has relation more to the forester than the agriculturist. While the squirrel by its grace and liveliness is an ornament to our woods, if the numbers be great immense damage can be done. Its chief food consists of acorns, beech-nuts, hazel-nuts, and the seeds of conifers. Buds of trees and young shoots are destroyed, and the bark of pine and larch and other trees is stripped or gnawed off in large patches. Insects may be taken, and sometimes eggs and young birds.

INSECTIVORA.

These are small plantigrade mammals with five strongly-clawed toes. The teeth vary, but the molars have pointed prominences on them. The nose is sensitive, being prolonged into a proboscis, useful as a tactile organ. Examples are the hedgehog, mole, shrew.

The Hedgehog.

The **Hedgehog** (*Erinaceus europæus*), while not fitted in structure for a burrowing habit, has managed to survive in the struggle for existence, to a great extent by means of its protective spiny-armour and the resemblance of its colour to its surroundings; it is also nocturnal in habit. It hides in the day-time in its dwelling by ditch or hedge sides, and under bushes or brushwood, or in hollow trees. It has a varied bill of fare—slugs, snails, insects, worms, mice and voles, frogs, snakes, eggs of game-birds and poultry, chickens and ducklings. Fruit in the shape of windfalls may be taken.

The Mole.

The **Mole** (*Talpa europæa*) is specially fitted by its structure for a life in the earth. The fore-limbs are excellently fitted for digging and throwing aside the soil; they are short and broad, and have strong claws, and are worked by powerful muscles; the collar-bones are strong and give attachment to the fore-limbs; the wedge-shaped head and round body aid in the movement through the soil; the hind-limbs are used in propelling the animal; powerful muscles give such strength that the mole can bury itself very quickly in a firm soil; there is no flap-like external ear to get filled with earth; the nostrils by being directed downwards do not get choked up with sand or fine soil; the mouth can be closed tight by a skin-fold on the upper lip; the fur standing erect and always smooth does not interfere with a backward movement should this be necessary; while the shortness and closeness of it

keep out water and dirt; the sharp canine teeth and the projecting tubercles on the molars are fitted for seizing, holding, and crushing the prey.

The most acute senses are smell and touch. Eyes are present, small and bead-like, and buried in the fur; they are not of service underground but are of use when the mole comes to the surface.

The mole-hills (fig. 631), which are formed of soil thrown up from shafts leading from the runs, must not be mistaken for the nest. The nest of the mole is under a heap of earth larger than the ordinary mole-hill; it is lined with vegetable matter, such as leaves, dry grass, or moss, and consists of a central chamber

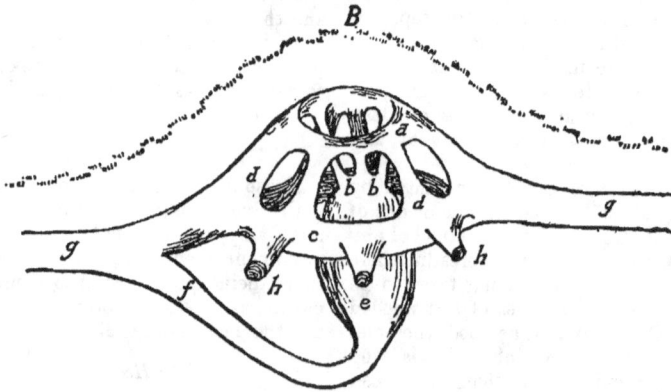

Fig. 631.—*Mole-heap.* (Schlich and Fisher.)

B, Surface of ground.
a, Upper gallery.
b, Descending passages.
c, Lower gallery.
d, Ascending passages.
e, Central chamber.
f, Passage to chamber.
g, Moles' run.
h, Diverging runs from lower gallery.

with passages varying in position; the passages are in association with the runs. The depth of the runs varies with the weather and with the season.

The young are born in May, two to six being a litter (the female has six teats). They are born blind and naked, and begin to run about in five weeks or so.

The food of the mole consists chiefly of insect larvæ and earth-worms, and its appetite is voracious; slugs are also eaten. In a nursery or a garden a mole may be a nuisance by throwing up the heaps of soil, so also in a pasture; the mole-hills, too, interfere with the mowing of the grass, and later with the cutting of the corn, but it should be remembered that

the diet can be mainly an insectivorous one.

Owls, buzzards, and the weasel are natural enemies of the mole. The animal, if too numerous, can be kept down by trapping; and in a garden it is said to be kept off by putting in thorns, these injuring the sensitive snout of the mole.

The Shrew.

The Shrew-mice are small insectivores with soft fur, a somewhat pointed snout, external ears, and hairy tails. There are three British species.

The Common Shrew (*Sorex vulgaris*) lives in runs below the soil but near the surface. The fur is velvet-like, the eyes are small, and the proboscis is sensitive.

The colour is reddish or brown-black above and grey below. Like other species, it gives off a strong musk-like odour, which may serve as a protection against enemies, but Harting has found numerous skulls of shrews in the "castings" of the barn-owl, and once found two little shrews in the stomach of a curlew. The shrew feeds on insects, worms, slugs.

The Water Shrew (*Sorex fodiens*), a larger species, black above and white below, and measuring up to 3½ inches, feeds on insects, including the aquatic caddis grubs, and on fresh-water molluscs and crustacea: it is said to eat the fry of fish. This shrew swims well, long hairs on the hind-feet being spread out when swimming, and acting as oars

The Lesser Shrew (*Sorex pygmœus* or *minutus*), about 2 inches long, is the smallest British mammal. It feeds on insects.

In spite of several unkind superstitions, shrews are practically harmless animals.

BIRDS IN RELATION TO THE FARM.

Many birds are of economic importance to the farmer, the gardener, and the fruit-grower: some because they are distinctly harmful to crop plants and their seeds; some because they are useful, inasmuch as a great part of their food consists of injurious insects; and still others, with a mixed diet, whose position as useful or harmful cannot be stated with definite certainty. Part of the difficulty in deciding into which set, economically, a bird must be put consists in the absence of organised investigation, and of direct examination, of the contents of the crops and stomachs of shot birds. Such an examination, too, should extend over all months of the year where the species is a resident one. There is the further difficulty that there may be a certain adaptability on the part of birds to the different kinds of food material available in different districts, and therefore, as the result of inquiries in a single district only, too sweeping generalisation as to food taken may be fallacious.

The most important birds in relation to the farm are briefly reviewed here.

BIRDS OF PREY.

The Kestrel (*Falco tinnunculus*).— This bird may be recognised from its habit of hovering motionless for long periods in the air, hence one of its names, the "Windhover." In colour it is reddish-brown, the male having dark spots on the back, and the female dark bars; the under parts are buff, streaked and spotted with black; the legs and the feet are yellow; the beak is blue, but the cere at the base of the bill is yellow. The length is 14 inches. Although the kestrel has been blamed for taking game birds, the consensus of opinion is certainly in favour of the usefulness of this bird. Its food consists, to a very great extent, of mice and voles; insects are also taken. On the Continent it is regarded as most useful to the cultivator.

The Sparrow-hawk (*Accipiter nisus*). —From the farmer's standpoint the sparrow-hawk is useful and not harmful. It is chiefly a bird of the woodlands. Its food consists of small birds and game birds, and also poultry; it is an enemy of the wood-pigeon. Insects also, and some of them very injurious, are taken. It has not the long wings of the kestrel, and the wings are rounded; it is thus not a very quick flier in a long chase, hence its cunning when hunting for prey.

The Barn Owl (*Strix flammea*).— This owl, known by its white breast and buff-coloured upper parts, is also called the White and the Screech Owl. In spite of the way it has been persecuted, it is a most useful bird. Like other birds of prey, the barn owl has the habit of disgorging, by way of the mouth, indigestible particles in the form of pellets. Dissection and examination of great numbers of these by competent and unprejudiced observers has proved that fur and not feather is the chief food. Seebohm states that in 700 pellets from this owl, there were found the remains of 16 bats, 2513 mice, 1 mole, and 22 birds of which 19 were sparrows. Mr J. E. Adams, in 1124 pellets examined by him, found the remains of 2397 mice and rats and 97 sparrows. This owl, along with the brown owl and the short-eared owl, have proved useful enemies of the voles in the plagues on hill farms.

The **Brown or Tawny Owl** (*Syrnium aluco*).—This owl has the upper parts ash-grey, mottled with brown; the wing-covert feathers have large white spots at their outer webs; the tail is tipped with white; the under parts yellowish-white, and mottled with brown. The male measures 15 inches in length; the female is larger, and often more red-brown. From its note this owl is called the "hooter." It is, like the last, a useful bird; its food consists of rats, mice, voles, small birds, *e.g.*, finches, and injurious insects.

The **Long-eared Owl** (*Asio otus*) and the **Short-eared Owl** (*Asio accipitrinus*).—These should also receive protection, their food being chiefly rats, mice, and voles. The short-eared owl is typically a winter visitor with us, leaving in spring; but in the vole plague of 1892-93, in the south of Scotland, food, in the shape of voles, was so abundant that the short-eared owls stayed, and bred and reared more than one brood. Some fifty or more pairs could have been counted on a single farm. This owl has a somewhat hawk-like appearance as seen by daylight on a dull day—for it hunts by day as well as in the evening. The long-eared owl is resident in wooded parts of Britain all the year round, the residents having their numbers added to in autumn by migrants.

The Crow Family (Corvidæ).

In the family we have the raven (*Corvus corax*), hooded crow or hoodie (*C. cornix*), carrion crow (*C. corone*), rook (*C. frugilegus*), jackdaw (*C. monedula*), chough (*Pyrrhocorax graculus*), magpie (*Pica caudata*), and the jay (*Garrulus glandarius*).

Raven.—The raven is the largest of the family, and is black, with a purple-blue sheen on the upper parts. It is omnivorous in diet, and is looked upon as an enemy by sheep-farmers, where the bird is found.

The **Hooded Crow** is recognisable by its ash-grey back, the head, throat, wings, tail, and thighs being black; there are varieties.

The **Carrion Crow** is entirely black, the male being glossier than the female; this crow is distinguished from the rook by the presence of a tuft of feathers at the base of the bill, instead of the grey patch; it has also a harsher cry, and in habit is not gregarious. In food habits the carrion crow and the hoodie practically agree: carrion, eggs of game and other birds, poultry, moles, rats, fish, insects and their grubs. Weakly ewes and lambs may be attacked.

Rook.—The rook—or the crow as it is called in Scotland—is black, with a grey warty patch at the base of the bill. Concerning this bird there is always great controversy. There is no doubt that it does much harm, but one must hurry to add that it also does good. Many people, with regard to the rook, find it difficult to say guilty or not guilty, and fall back on the Scottish verdict of not proven. Sir John Gilmour's[1] verdict, as a result of his investigation with rooks shot from January to December, and dissected for determination of the food contents, is that this bird is "a cunning rogue." Of 355 rooks shot, 336 had food contents. Of the contents, 58 per cent consisted of the grains of wheat, barley, oat, and rye; 23 per cent of insects and grubs, 12 per cent miscellaneous, 7 per cent roots (potatoes), no leaves, flowers, fruit, or seed being found. Dr Hollrung of Halle has given, in a German agricultural journal,[2] the result of post-mortem examinations made on 4030 rooks, over a period of years from 1895-1905. In this report there appears—

GRAIN, &c., TAKEN BY 4030 ROOKS.

	Grains.		Grains.
Sprouted—		*Sprouted—*	
Wheat	15,578	Potatoes	587
Barley	10,465	*Unsprouted—*	
Oats	12,787	Barley	247
Maize	987	Maize	40
Buckwheat	1,777	Rye	358

[1] *Transactions of Highland and Agricultural Society*, 1896.

[2] *Landwirtschaftliche Jahrbucher*, 1906.

INSECTS TAKEN BY 4030 ROOKS.

Cockchafer beetles	. .	2,222
,, grubs	. .	2,264
Click beetles	. . .	2,307
Wireworms .	. .	1,589
Weevils	. . .	14,710
Caterpillars .	. .	9,126
Leather-jackets .	. .	3,411
Tortoise beetle	. .	2,062
Leaf beetles	. . .	2,133
Garden-chafer and other small chafers	. . .	1,717
Burying beetles .	. .	984
Maggots of Bibio flies .	.	406
Corn Ground-beetle	. .	86

Mr Cecil Hooper,[1] writing from south-east England, points out that rooks damage fruit, eating "green strawberries, cherries, gooseberries, especially in dry weather, and sometimes apples and pears; it dearly loves walnuts, also cob and filbert nuts." As to the fruit-eating habit, Phil Robinson has written: "The poor farmer never suspected the rooks to have eaten his cherries. No one did. They said it was the black-birds, and the thrushes, and the sparrows, and those boys of Chinnery's. So they caught blackbirds in traps, and harried the sparrows' nests, and had a row with the Chinnerys. And yet is it not written in every book that the rook is *Corvus frugilegus*, the fruit-eating crow?" As far back as the time of Henry VIII. we find enactments against the rooks as destructive to corn and grain, thus, that "every one should do his best to destroy rooks, crows, and choughs, upon pain of amercement, and that every hamlet should provide and maintain crow-nets for ten years, and that the taker of the crows should have after the rate of 2d. per dozen."

On the other hand, the rook's insect-eating habit must be stated. In Sir John Gilmour's investigation the insects found in the rook were wireworms, leather-jackets, turnip-moth caterpillars, and a number of ground-beetles, adult and larva. Most of the ground-beetles are predaceous and useful. Rooks certainly are of service in the destruction of the leather-jackets' or daddy-long-legs' larvæ and of wireworms; we have found them gorging themselves on the grubs of the garden-chafer in a hill pasture; they

[1] *Journal of Society of Arts*, vol. iv., Dec. 14, 1906.

take winter-moth caterpillars, and have been serviceable in destroying the caterpillars of the green tortrix moth of the oak.

It cannot be denied that farmers may suffer much loss from rooks. Hollrung, whose tables were quoted above, concludes from his examinations that while the extermination of rooks is under no circumstances justifiable, it being clearly proved that they destroy harmful insects, some of which are with difficulty reached by artificial means, yet in the neighbourhood of rookeries the harm done easily outweighs the good, and that where rooks occur in large flocks excess of damage is to be feared, especially at times of the year when insects are scarce: where rooks are few in number or widely dispersed, and where harmful insects abound, rooks are undoubtedly useful.

The Highland and Agricultural Society has issued warnings, and the Board of Agriculture has suggested ways in which rooks in arable districts might be kept down. Several methods have been quoted by the Board in its *Journal*.[2] Thus a plan reported from Scotland as adopted with success: "A frosty night was selected, just after the rooks had laid their eggs, but before they were sitting closely, and a man was placed, with a gun, in every clump of trees where the rooks were for an area extending over two or three miles of country. At a fixed time shooting was begun, not so much with the idea of killing the birds, but more particularly for the purpose of keeping them off the nests. Firing was kept up for about three hours from just before dusk, with the result that the eggs were frosted and became infertile. This has now been done regularly for four years, with the result that the rooks are considered to have decreased by 80 per cent, and their numbers are now within limits."

Another method practised to keep the numbers in control is to shoot the young birds. Mr H. W. Slater[3] says: "My own practice is to kill two young birds to each nest, and to leave the rest, which appears just enough to counterbalance the waste of accident and old age."

[2] *Journal of Board of Agriculture*, July 1906.

[3] *Journal of the Farmers' Club*, April 1895.

Two mixtures[1] have been given for protection of seed against rooks—2½ oz. of coal tar, 2½ oz. of petroleum, 1 quart of water; this suffices to dress a bushel of seed. And again, 11½ pints of coal tar, 5¾ pints of petroleum, 1¹¹/₁₂ pints of carbolic acid to 5 quarters of seed. The instructions given are: "The proportions named should be strictly adhered to in order to obtain successful results. Commercial carbolic acid of full strength must be used. The mixture should be made as follows: place the coal tar in a pot on a slow fire until it is quite hot and shows signs of boiling, remove it from the fire and add the petroleum while stirring, and finally the carbolic acid. The mixture should be thoroughly well stirred, and it will remain quite liquid after cooling.

"In order to treat the grain, about 4 bushels of seed should be spread out on a water-tight floor, and a tenth part of the mixture, which will be a little less than 2 pints, poured on it. This must then be stirred up quickly until each grain is blackened, and the whole is about the colour of roasted coffee. The seed, however, cannot be sown in this condition, as it would stick to the cups of the drill, and in order to dry the seed add afterwards to each 4 bushels of seed about 2 pints of phosphate of lime. When mixed the grain will be quite dry, and can be sown in the ordinary way. Seed treated in this way will not be touched by crows, though the growth may be retarded by some two or three days. The cost per bushel is insignificant."

Jackdaw.—Has rounded wings and grey nape; it is generally distributed. It is an egg stealer, and kills young birds, and does some harm to cherries, but it takes insects and their larvæ, and also keds from the sheep.

Chough.—A bird of the west coast and islands. A little bigger than a jackdaw, it is black in colour, with the bill, legs, and feet red.

Magpie.—This handsome bird has head, neck, back, and breast black, glossed with green; belly and scapulars white; the secondary feathers are black,

the primary feathers black, with white on their inner parts; the tail is black and has a greenish-bronze sheen. In diet it is omnivorous—insects, slugs, mice, birds, and eggs.

Jay.—This is a bird of the woods, a beautiful bird with a crest of white, tipped with black; nape and back light brown; the rump white; tail black; the wing coverts barred with black, white, and blue.

Pigeons.

Wood-pigeon, or Cushat, or Ring-dove (from the white feathers that encircle the neck) (*Columba palumbus*).—A resident in Britain, and generally distributed in wooded districts. It is increasing in numbers; the native wood-pigeons are added to in winter by great numbers which arrive on the east coast from the Continent. Two or three broods are reared annually. By common consent this bird is one of the great farm enemies. In many places—*e.g.*, the north-east of Scotland—the wood-pigeon is proving at present a very severe enemy of the farmer. Its powers of destruction may be estimated by the wholesale levy it makes on the products of the fields and of the woods, as thus enumerated by Macgillivray:[2] From its roost in the larger branches of trees "it issues at sunrise to search the open fields for its food, which consists of seeds of the cultivated cereal grasses—wheat, barley, and oats; as well as of leguminous plants—beans and peas, and of the field-mustard and charlock. In spring it also feeds on the leaves of the turnips and picks the young blades of the red and white clovers. At this season I have several times found its crop distended with the farinaceous roots of *Potentilla anserina*, obtained in the ploughed fields. This root is highly nutritious, and formerly, in seasons of scarcity, was collected in the West Highlands and Hebrides as an article of food, and eaten either boiled or roasted in the peat-ashes (the plant is a troublesome weed). In summer the wood-pigeons eat grass and other vegetable substances; in autumn grain, beech-mast, acorns, and leguminous seeds. The beech-masts and acorns they swallow entire, their bill not

---

[1] *Journal of Board of Agriculture*, December 1904.

[2] Macgillivray's *Brit. Birds*, i. 115, 263.

being sufficiently strong to break them up."

The wood-pigeon destroys the growing crop in the manner described by an eye-witness watching near:[1] "The wood-pigeon has a weak bill, but nature has provided her with very strong wings. When the flock, therefore, settle upon the lying portion of a wheat-field, instead of breaking off the heads and carrying them away, they lay themselves down upon their breasts upon the grain, and, using their wings as flails, they beat out the pickles from the heads, and then proceed to eat them. The consequence is, that the pickles having been thrashed out upon a matting of straw, a great proportion of them fall down through it to the ground, and are lost even to the wood-pigeon; in short, they do not eat one pickle for twenty which

they thrash from the stalk. I have repeatedly watched this process from behind the trunk of a large willow-tree, growing in a thick-set hedge on the edge of a wheatfield, and seen the operation go on within two yards of me. The pigeons descend first singly; but, having left a watcher upon the highest tree in the neighbourhood, the whole flock are soon at work on the same spot, and the loss of corn to the farmer is very great. They are also gluttons in quantity."

Sir John Gilmour, in the report of the investigation referred to above, gives a post-mortem record of what MacAlpine found in the case of 245 cushats shot from January till December. The record is a most interesting one, showing, alongside of the heavy toll levied on the farmers' crops, the weeds which are also taken as food.

Kinds of Food.	Crops.	Weeds.
Cereal grains.	Barley.	
	Oat.	
	Wheat.	
	Rye.	
Leaves.	Red clover.	Charlock.
	White clover.	Runch.
	Alsike clover.	Creeping buttercup.
	Turnips.	Oak.
	Swedes.	
Fruits and seeds other than cereals.	Beans.	Charlock.
	Peas.	Runch.
	Tares.	Goose-grass.
	Clovers.	Spurrey.
	Turnips.	Mouse-ear chickweed.
	Swedes.	Common chickweed.
	Rye-grass.	Field speedwells.
	Seeds of mixtures.	Docks.
	Maize (from artificial foods).	
	Elm fruits.	
	Beech-nuts.	
	Wild cherries.	
	Haws.	
Roots and underground stems.	Turnips.	
	Swedes.	Lesser celandine.
	Potatoes.	
Flowers.	Beech.	Charlock.
	Elm.	Creeping buttercup.
		Annual meadow-grass.

A few earthworms were taken for the young birds. It is pointed out by Gilmour that it is the best of the crops which are attacked — the grains of cereals, the leaves of clovers and turnips, the seeds of beans, peas, &c.

[1] Burn Murdoch's *Observations on Game*, p. 11.

That cereal grains and leaves are the foods freely used by wood-pigeons is seen from this additional table:—

Kinds of food.	Number of times taken during the year.	Per cent.
Cereal grains	123	33
Leaves	103	27.5
Other fruits and seeds	88	23
Roots	31	8.5
Flowers	29	8

The crop or store-chamber of the pigeon is large, and T. H. Nelson [1] gives various records to show its capacity—viz., a cupful of turnip tops (December 1884); 61 acorns in another; 76 acorns and a quantity of swede tops (18th December 1883); 73 hazel-nuts (January 1884); another containing 838 grains of corn.

The task of fighting the wood-pigeon is rendered difficult by the bird's wariness, while its nesting habits do not bring the bird into the farmer's power in the way that the habits of the crow or rook do.

Stock-dove.—The Stock-dove (*Columba œnas*) can be told from the wood-pigeon by its being smaller and by the absence of the white feathers on the sides of the neck. This bird has been increasing of late years in England, and in some parts of the north of Scotland is now fairly plentiful. In food habits it resembles the wood-pigeon.

The Starling (Sturnus vulgaris).

This is a common resident in Britain, where it has increased enormously in late years. Additional hosts of migrants add to the native starling population in autumn and winter. Its long strong legs with their blunt nails fit it for running rapidly about for hours. Its long beak helps the bird in its careful search of grass-tuft and crevice. It is, from the standpoint of the farmer, a most useful bird. Insects form the great bulk of its food, wireworms, leather-jackets, caterpillars of grass-moth, surface caterpillars, diamond-back moth, &c. Further, the starlings perch on the backs of sheep and help to clean off the keds, although it has been stated that the excrement of the starling attracts the sheep-maggot flies. The fruit-grower, however, looks upon the starling as levying too severe a toll upon fruit, cherries, raspberries, strawberries, damsons, plums, pears, apples. Two broods may be reared in a year; the birds are hardy and able to hold their own.

In the investigation conducted by Gilmour, the starling was included. [2]

The number of starlings shot from January to December, in which post-mortem examinations were made, was 175. As a result, it is concluded that while the starling can take and use many and varied articles of food, the range of staple food-stuffs is very narrow. Three-fourths of starling food was found to be insect, and one-fifth grain. The general estimate based on the above examinations is, that the "starling is a bird to be fostered rather than destroyed, a benefactor rather than a foe." The table added indicates the relative importance of the foods:—

Food-crop.	Number of times taken during the year.	Per cent.
Grub . . .	30 } 182	70
Adult insect . .	152 }	
Grain . . .	58	22
Miscellaneous . .	21	8
Roots . . .	0	0

Of the insects found, there were rove-beetles and some ground-beetles, both useful in agriculture; and of injurious insects, Pterostichus (a ground-beetle injurious to strawberries); adult click-beetles, and weevils, and the larvæ of click beetles (wireworms), daddy-long-legs (leather-jackets), the grass-moth, and a species of Hepialus.

Thrush Family (Turdidæ).

Of this family may be mentioned the song-thrush, the missel-thrush, and the blackbird.

Song-thrush (*Turdus musicus*) or Mavis or Throstle.—The mavis is a resident in Britain, the places of those which migrate in the autumn being taken by others from the Continent. The food consists of insects, slugs, snails, fruits. As to fruit, it can be extremely troublesome. This bird rears two or three broods in a season.

Missel-thrush (*Turdus viscivorus*).—This is the largest of the British thrushes, and is also a resident. From its singing out defiance to the storm it is known as the storm-cock. Its food consists of the berries of such plants as holly and mountain ash, fruits, snails, worms, insects. In the south it is very destructive to fruit. The bird is rather a bully, charging away and scattering smaller birds.

Blackbird (*Turdus merula*). — This

---

[1] *Ornithology in Relation to Agriculture and Horticulture*, p. 90.

[2] In connection with these post-mortem examinations of rooks, wood-pigeons, and starlings, the birds were shot on Sir John Gilmour's estates in Fifeshire.

bird is also a resident—*i.e.*, Britain is not without blackbirds at any time of the year; the places of such as fly south in autumn are taken by numerous winter visitors. The male is black, with the beak yellow; the female black - brown, and the bill is brown. There are several broods in the year. The food consists of fruit, insects and their larvæ, worms. It is looked upon by fruit-growers as a great scourge.

For keeping off the blackbird and thrush from fruit it is suggested in Bulletin No. 11 of the Yorkshire College, Leeds, that an effective method is "to attach a cat by a ring and swivel to a long cord fastened securely at each end, so that the cat can walk up and down the length of its tether."

Swallows and Martins (Hirudinidæ).

The Swallow (*Hirundo rustica*), the House-martin (*Chelidon urbica*), and the Sand-martin (*Hirundo riparia*) are entirely useful birds, their diet being an insectivorous one, and the fact that these birds are decreasing in numbers is a loss to the agriculturist. The birds are often confused; they may be distinguished thus—

SWALLOW.	MARTIN.
Forehead and throat chestnut brown.	
Upper surface steel blue, including the rump.	Upper surface steel blue, except the rump, which is white.
	Lower surface white.
Lower surface dusky reddish white (the female has the lower surface white).	
Tail markedly forked (the female has the outer feathers of the tail shorter).	Tail not markedly forked, and the wings shorter. Feathers on feet and toes.
Eggs white, but speckled with brown or dark red spots.	Eggs pure white.
Nest open at the top.	Nest with a small hole only for entrance.

The sand - martin is smaller than either of these two birds, and its upper surface is brown or mouse-coloured. It nests in steep river - banks, sand - pits, gravel quarries, and railway cuttings, the tunnels made by the birds being from 18 inches to 6 feet in length, with the nest in a chamber at the end.

The Spotted Fly-Catcher (Muscicapa grisola).

The Fly-Catcher comes to Britain in April, leaving again for the South in autumn. It is rare in Scotland. Saunders [1] describes this bird thus:—

"The adult has the crown light brown, with dark streaks down the centre of the feathers; upper parts hair-brown, slightly darker on the wings and tail, and paler on the margins of the wing-covers; chin and under parts dull white, with brown streaks on throat, breast, and flanks. Length, 5.8 inches. The young are very much spotted."

The bird is entirely useful, being insectivorous in diet; all kinds of insects are taken.

[1] *Manual of British Birds.* By Howard Saunders. P. 158.

The Water-Wagtails (Motacillidæ).

There are five British species, of which the three commonest are the pied wagtail (*Motacilla lugubris*), the grey wagtail (*M. melanope*), and the yellow wagtail (*M. campestris*). All the wagtails deserve protection, as they are entirely useful. They feed greedily on insects and small molluscs.

The Titmice (Paridæ).

This is another useful family. Perhaps the most useful of all the tits is the blue tit (*Parus cœruleus*), a bird about 4½ inches long, and with blue, yellow, and green coloured plumage. Insects are its chief food, but Mr Hooper accuses it of sometimes "doing serious damage by pecking cherries and the sweeter-flavoured apples and pears."

The Great Tit. — The great tit, or tomtit (*Parus major*), known also as the ox-eye, is a resident species. In length it measures about 6 inches. The upper side is yellow-green, the under surface light yellow; the top of the head, the throat, and a stripe on the breast are black; under each eye is a white patch. The legs are short and strong, and have curved claws

that fit the bird for climbing and clinging. The short conical beak is well fitted
for picking up insects and their eggs.
Besides destroying the eggs of numerous
harmful moths and insects, it takes seeds,
beechmast, and nuts; like the blue tit it
pecks holes in apples and pears. The
blue tit and the great tit destroy bees.
The planting of sunflowers—of the seed
of which these two tits are fond—lures
them away from the fruit.

Buntings (Emberizinæ).

Yellowhammer. — The yellowhammer (*Emberiza citrinella*) is a common
resident in Britain; there are two broods
in the year. It takes insects, but prefers
for its own eating corn and seeds, particularly oats; and in new-sown fields of
oats, as well as wheat, it may be seen
busily picking up the grain from the
moment it is sown to the period of its
brairding; the seed of such weeds as
plantain, dock, thistle, chickweed, and
pod-grass are taken. By autumn, when
the broods are reared and the corn crops
begin to ripen, the yellowhammers assemble with sparrows and corn-buntings,
and leave little alongside the hedges but
empty husks on the standing straw.
When feeding in the stubble-fields, they
advance by short leaps, with their breasts
near the ground; when danger approaches
they crouch down motionless; and when
alarmed, give out their ordinary short
note, *yite, yite*. They are more shy than
chaffinches, but less so than the corn-
buntings.

Reed-Bunting.—The reed-bunting, or
black-bonnet(*Emberiza schœniculus*), lives
mostly on weed seeds, though small
patches of oats on the crofts in the upland districts attract its notice. It also
takes insects and their larvæ. Not being
shy, it is not easily scared away. It is a
British resident, but may be migratory
in parts of Scotland, departing in October, and reappearing in the beginning of
April.

Corn - bunting.—The corn - bunting
(*Emberiza miliaria*) feeds on corn, and in
spring, together with the yellowhammer
and others, devours considerable quantities of seed-corn of oats and barley.
After the breeding season it feeds on
beans, peas, wheat, oats, or barley, while
during autumn it feeds on the stubble-
lands, sits close, and is shy. It visits the
new-sown fallow and potato-wheat. In
winter it is remarkably fat, and superior
as an article of food to most of our small
birds.

"It could hardly be supposed," observes Knapp, "that this bird, not larger
than a lark, is capable of doing serious
injury; yet I this morning witnessed a
rick of barley, standing in a detached
field, entirely stripped of its thatching,
which this bunting effected by seizing
the end of the straw, and deliberately
drawing it out to search for any grain
the ear might yet contain—the base of
the rick being entirely surrounded by the
straw, one end resting on the ground, and
the other against the snow, as it slid
down from the summit, and regularly
placed as if by the hand; and so completely was the thatching pulled off, that
the immediate removal of the corn became necessary. The sparrow and other
birds burrow into the stack, and pilfer
the corn, but the deliberate operation of
unroofing the edifice appears to be the
habit of the bunting alone."

The Finches (Fringillinæ).

The Chaffinch (*Fringilla cœlebs*), or
Shilfa, is a generally distributed species
in Britain in cultivated and wooded parts.
The male is a handsome bird: forehead
black, back of neck slate - blue, back
brown, rump greenish, the breast red-
dish-pink, wings black, with a white and
a yellow transverse band. The female
is duller. A beautiful nest is made of
moss, wool, and lichens felted together
and lined with hair and feathers. Two
broods are generally reared in a year.
After the breeding season they collect in
flocks, roving about, males and females
separately. There is a migration southward at the end of the autumn, but
one can always find examples, males
chiefly, in the winter time. The chaffinch takes insect food, especially at the
breeding season, but its diet to a great
extent consists of seeds, coniferous seed,
beech-mast, groundsel, chickweed, charlock, knot-grass. The short, thick, conical beak, with sharp edges, is well suited
for removing the husks of seeds. On
recently - sown beds it is troublesome,
taking out the sprouting carrot, lettuce,
radish, cabbage, turnip, onion. It is

also a disbudder of gooseberries, currants, and plums.

The Greenfinch or Green Linnet (*Ligurinus chloris*).—Like the last, it is a resident in cultivated and wooded districts. Yellowish-green is the prevailing colour. There is a golden-yellow stripe over each eye; bill dull flesh colour. Two broods are often reared in a season. It takes some insect food, but in the seed season, accompanied by the young brood, will attack almost every sort of seed that is ripe or ripening—*e.g.*, turnip, mustard, sainfoin; oat-fields and even wheat-fields near woods and hedges may suffer considerably; the seeds of weed plants are also taken; it pulls hop flowers to pieces, and may do the same to fruit blossom. Where present in numbers, its seed-eating propensity renders it capable of considerable harm.

Linnet.—The grey or brown linnet (rose lintie) (*Linota cannabina*) does more damage than is generally supposed. It visits patches of turnips left for seed, and frequents the newly sown turnip-fields. When the young families begin to wander in small companies as the corn becomes ripe, they devour large quantities of the standing corn, voraciously living upon it from the moment it begins to whiten until led to the stackyard. After this period the smaller families associate in larger flocks, frequently combining with the greenfinch, and subsist on the stubbles. It frequents newly sown wheat-fields, and thins the seed-corn in detached patches so much, that the scantiness of the braird is ascribed to the attacks of a grub. It is easily scared. There is a marked seasonal change of plumage. In spring and summer the feathers of the head and breast are red-brown tipped with bright red, but on the approach of winter the latter tint almost wholly disappears.

Goldfinch.—The goldfinch (*Carduelis carduelis*), local and rarer in Scotland, is generally distributed in the summer in England and Ireland. It takes insects at the breeding season, but its chief food is weed seeds, *e.g.*, thistle—its moderately long-pointed bill is well suited for picking the thistle heads,—knapweed, and dock. It is a pity the bird is not in greater numbers.

Bullfinch.—Bullfinch (*Pyrrhula euro-*

pœa)—known as "the canon" in Germany from the black patch on the top of the head resembling a priest's skull-cap—takes numbers of weed seeds, but in gardens and fruit plantations can be most destructive by destroying the buds; indeed, fruit-growers rank the bullfinch with the blackbird and the sparrow as their worst enemies.

Protective Measures against Disbudding.

Dusting the buds with quicklime while the plants are still wet after winter washing is a practised method. A suggested spray for protective purposes is described by Mr W. E. Bear,[1]—"60 lb. of quicklime, 30 lb. flowers of sulphur, 12 lb. caustic soda, 10 lb. soft-soap, 100 gallons water. The method of mixing is important. Mix the sulphur into a paste, beating it up well while somewhat stiff, and gradually thinning it, and pour it over the lime. Stir the ingredients thoroughly until the lime is slaked, adding only as much water as is necessary to allow of stirring. Then add the caustic soda, and stir it in until the renewed boiling action which it sets up is finished. Dissolve the soft-soap separately by boiling it in two or three gallons of water, and stir it well in with the other ingredients of the wash, afterwards adding enough water to make up 100 gallons. Pass the mixture through a strainer of fine brass-wire gauze. The lime should be of the best quality and freshly burnt.

"The period for spraying varies with the kind of fruit, the season, and the situation. Gooseberries are generally attacked sooner than plums, and nearer the homestead sooner than farther away. In a mild winter little or no damage is done before the buds begin to swell, but in severe weather birds, from lack of other food, may begin the attack prematurely." January and February may be taken as suitable times for the spraying.

The Sparrow.

The House-Sparrow (*Passer domesticus*).—This bird is a scourge to farmer, gardener, and fruit-grower. The charges against the sparrow may be summarised thus: (*a*) Causes great loss by eating

[1] *Journal of Board of Agriculture*, February 1907, p. 668.

cultivated grain. Especially before harvest time flocks of the birds, young and old, are found in the fields gorging themselves on grain; (b) harmful by destruction of the blossom of garden and fruit plants; (c) harmful also by its destruction of buds, e.g., gooseberry, cherry, red currant, and later to fruit; (d) harmful to such plants as pears, lettuce, cauliflowers, &c. ; (e) drives away soft-billed birds without in turn doing their useful work; (f) chokes up the rhones of houses. Against these charges it is to be admitted that weed-seeds are taken and insects, some of these most injurious ones. As a result of examination [1] of about 1000 house-sparrows taken at different times over a period of 15 years, the following estimate was arrived at the average of the food taken throughout the year :—

Grain	.	.	.	75 per cent.
Seeds of weeds		.		10 " "
Green peas		.	.	4 " "
Insects		.	.	6 " "
Other matters		.		5 " "

Abroad, in the United States, in Australia, and in New Zealand, the sparrow has proved a plague. Just as at home here, when it was permissible, poisoned wheat was used against the sparrows, so

in Australia wholesale poisoning was resorted to; and in an Adelaide newspaper the sufferers were addressed and exhorted as follows :—

" What means this sadly plaintive wail,
    Ye men of spades and ploughs and harrows?
Why are your faces wan and pale?
    It is the everlasting sparrows !

No more your wasted fruits bewail,
    Your crops destroyed of peas and marrows,
A cure there is that cannot fail
    To rid you of the hateful sparrows !

The remedy is at your feet :
    Slay them, and wheel them out in barrows,
Poisoned by Faulding's Phœnix wheat,
    The one great antidote to sparrows ! "

The Board of Agriculture has issued suggestions in connection with sparrow clubs, and evidence generally is overwhelming as to the bird's destructive work and the great necessity for an organised campaign to reduce its numbers.

Tree-Sparrow. — The tree-sparrow (Passer montanus) is a rarer and more local species. It is not parasitic on civilisation like the house-sparrow, but building in trees it comes into towns and villages in winter when food fails. It can be distinguished from the house-sparrow thus :—

Top of head ash-grey ; white patch on side of head in the region of the ear ; a band of white across the wings.

Top of head bright brown-red ; the white patch on the side of the head has a black spot on it ; two bands of white across the wings.

The Larks (Alaudidæ).

The Skylark or Laverock (Alauda arvensis) is a resident species. Being a ground feeder, it is of all our song-birds the best runner, the hind toe or spur being unusually long. Omnivorous in diet, it does good by feeding on insects and weed-seeds ; it also takes seed-corn and the young sprouting plants; in severe seasons it can do harm by feeding on the leaves of such plants as swede, and thousand-headed kale.

The Plovers.

The Lapwing, Peewit, or Green Plover (Vanellus cristatus) is generally

distributed throughout Britain. It is a bird of marsh, moorland, and meadows, and is well known from its cry and flight, and the tuft of feathers at the back of the head. This is one of the most useful birds in the country; it destroys quantities of harmful insects, taking both adults and larvæ ; it feeds also on snails and slugs, taking, among the rest, the small snail that is the intermediate host of the liver-fluke.

The Cuckoo (Cuculus canorus).

The Cuckoo, from the cultivator's standpoint, is an entirely useful bird. In addition to other insects, it eats the looper caterpillars of the magpie moth which are not eaten by other birds, and it also takes hairy caterpillars like those

[1] The House-Sparrow. By J. H. Gurney and Colonel C. Russell.

of the brown-tail moth and the lackey moth which are destructive to the leafage of a number of trees.

Gulls (Laridæ).

The gulls of the genus *Larus* have three toes webbed, and the fourth high up on the metatarsus. The bill is hooked at the tip of the upper mandible. While normally fish-eaters, gulls are often met with inland, where they are distinctly useful by eating insects. The characters given in the subjoined table will aid identification.

Head.	Back.	Legs and Feet.	Species.
White	black	flesh colour . . .	The Giant Black-backed Gull (*L. marinus*).
		yellow	Lesser Black-backed Gull (*L. fuscus*).
	light-grey	flesh colour . . .	Herring Gull (*L. argentatus*).
		greenish-yellow . .	Common Gull (*L. canus*).
		crimson	Black-headed Gull (*L. ridibundus*).
Black	light-grey	crimson	Black-headed Gull in summer plumage.

Of these, three may be chosen for reference as found in fields and useful to the farmer.

The Black-headed Gull.—This gull is very common in Britain along the flat shores; it breeds inland in marshes and by rivers. It follows the plough diligently, taking the grubs that are turned up: in addition to its insect diet, the bird, like other gulls, is a scavenger.

The Common Gull (*L. canus*).—This gull is not so common as the last, in spite of its name. In addition to frequenting the sea-coast it breeds also inland by lochs. It is found, like the last, following the plough, and in company with the rook and lapwing. This gull and the black-headed gull resemble one another in plumage in the winter time.

Herring Gull.—The herring gull is widely distributed round the shores of Britain. By the shore it feeds on garbage and carrion; it takes the eggs of cliff birds; out at sea it is said to follow the herring shoals; it is also found inland searching for insects and grubs and worms.

Game Birds.

The Pheasant (*Phasianus colchicus*).—The pheasant is accused of committing havoc amongst corn crops. Its true habits are thus described by Macgillivray: "Its favourite places of resort are thick plantations or tangled woods by streams, where, among the long grasses, brambles, and other shrubs, it passes the night, sleeping on the ground in summer and autumn, but commonly roosting on the trees in winter. Early in the morning it betakes itself to the open fields to search for its food, which consists of the tender shoots of various plants, grasses, bulbous roots of grasses, and *Potentilla anserina*, turnip-tops, as well as acorns and insects. In autumn, and the early part of winter, it obtains a plentiful supply of grain, acorns, beech-mast, and small fruits. In severe weather, however, especially where great numbers are kept, the pheasants require to be fed with grain, when they learn to attend to the call of the keeper.

In the natural state, and in small numbers, pheasants prefer insects and the young shoots of plants to corn, of which they pick at a time only a few grains; but when semi-domesticated, and congregated in large numbers, they assume the habits of the domestic fowl, and will eat and trample down extensive patches of the growing corn in the immediate vicinity of their preserves—and this they do between the ripening and the reaping of the crop. Pheasants devour quantities of wireworm.

Partridge.—The common partridge (*Perdix cinerea*) is troublesome in the corn-fields, but more than compensates by the destruction of insects and their grubs.

INSECTS INJURIOUS TO FARM CROPS.[1]

Insects are Invertebrate animals belonging to the Phylum, Arthropoda or Jointed-footed animals. They have the following characteristics that mark them out from related classes: A body divided into three regions—head, thorax, abdomen; one pair of antennæ; six legs when adult; respiration is by tracheæ; the majority have wings.

The body is covered by a protecting horny material, known as chitin. The head carries the antennæ, the eyes, the mouth parts. The antennæ are sensory and exploring in function; they function very importantly in the sense of smell—a sense by which insects, to a great extent, find their food. The eyes are of two kinds—simple and compound. The simple eyes are borne on the top of the head; "they are confined to the perception of very near objects," and are useful in dark places. The compound eyes, two in number—one at each side of the head—are very complex in structure, having often thousands of facets, each with a lens in association. It is believed, in spite of this, that by these eyes form is perceived very imperfectly, and that the perception of movement and light sensations are their chief uses. Probably no insect can see an object farther away than six feet.

As regards mouth-parts, insects are divided into two sets: the mandibulate insects, which can take solid food by biting jaws, and the haustellate insects, those which take liquid food. The mouth-parts of haustellate insects are modified in various ways, as piercing and sucking, sawing and sucking, and so on.

The thorax of the insect shows three divisions—a front or prothorax, a middle or mesothorax, and a hind or metathorax. Each region of the thorax bears a pair of legs, and the two hind regions of the thorax very commonly carry each a pair of wings. The legs show the following parts: (*a*) the coxa, jointing the leg to the thorax; (*b*) a small trochanter; (*c*) a femur, often stout and strong; (*d*) a tibia, often bristly; (*e*) a tarsus, with a varying number of joints, and ending in claws. The abdomen has in it the bulk of the internal organs, but seldom has external appendages.

Nervous System.—At the head end, on the upper surface, a collection of nerve matter—the supracœsophageal ganglia—form the brain, which supplies nerves to the eyes, antennæ and mouth-parts. From the brain a nerve collar passes round the gullet, on the under side of which is a ganglion, from which a double nerve chord runs backwards along the under surface of the insect; this chord has ganglia along it. These ganglia, present in thorax and in abdomen, control to a great extent the segment in which they are found, and to this extent are independent of the brain. Each ganglion acts as a motor centre for that segment; thus, if an insect lose its head, it can yet fly, walk, breathe. As to sense organs, insects have the sense of sight, smell, taste, touch, hearing; the senses of touch and smell are highly developed.

Respiration.—The air is taken in by means of openings down the sides (as a rule) of the body; the openings are called spiracles, and they lead into tubes known as tracheæ. In many insects the tracheæ are swollen into air-sacs. The air taken in at the spiracles passes through the tracheæ by diffusion, and the free passage of air is aided by the muscular movements of the insect. Advantage is taken of this mode of respiration to kill insects by some contact-insecticide, which chokes up the spiracles, and so suffocates the insect. Birds, while bathing themselves on the highway in sand and fine dried mud and dust, are ridding themselves of insect parasites, whose spiracles get clogged.

Reproduction.—Insects are male and female. Fertilised eggs, typically, are laid to the outside, but in most insects development is not direct; there is a

[1] *The following Figures in this section and other sections dealing with farm pests are, by permission kindly granted, reproduced from Leaflets issued by the Board of Agriculture and Fisheries—viz.*, 633, 637, 638, 643, *and* 655.

metamorphosis. Thus the adult butter-fly, known as the imago, lays eggs, from each of which there hatches a larva totally unlike the parent; after a time the full - fed larva passes into a resting condition known as the pupa, which gradually assumes the adult condition. Insects with all these stages in their life-history are known as Holometabolic. With some insects the metamorphosis is incomplete; the imago lays eggs from which come tiny young forms resembling the parent externally; this young form feeds and grows and moults until after the last moult maturity is reached with-out any resting pupal stage. Such insects are known as Hemimetabolic.

In the larval condition feeding is vora-cious, many larvæ eating more than their own weight in a day. To different kinds of larvæ different names are applied: thus a larva with numerous legs is known as a caterpillar, one with six legs and biting jaws as a grub, and one without legs a maggot. Once the adult mature state has been attained, there is no further growth, although the adult insect may live for a long time.

CLASSIFICATION OF INSECTS.

Insects may be divided into Orders, the wing characters being made the basis of classification. The Orders are—

ORDER.	METAMORPHOSIS.	MOUTH-PARTS.	EXAMPLES.
Thysanura and Collembola.	No metamorphosis.	Fitted for biting, but withdrawn into the head.	Springtails.
Coleoptera.	Complete.	Mandibulate.	Beetles.
Hymenoptera.	"	Bite, and may also suck.	Bees, wasps, ants.
Orthoptera.	Incomplete.	Biting.	Cockroaches and earwigs.
Neuroptera.	Complete or incomplete.	Biting and sucking.	Dragon-flies, May flies, caddis flies.
Lepidoptera.	Complete.	Haustellate.	Butterflies and moths.
Diptera.	Complete.	"	Flies.
Thysanoptera.	A slight metamorphosis or none.	"	Thrips.
Hemiptera.	Incomplete.	"	Aphides and scale insects.

Orders containing Insects of importance in Agriculture.

Collembola.—These are wingless in-sects with no metamorphosis; the jaws are pushed back into the head. Carried below the body, with its tip pointing forwards, is a process known as the spring; as this spring is straightened out the insect is sent with a leaping movement into the air.

Coleoptera or Sheath-winged In-sects.—Wings 4, the front pair horny and serving as wing covers or elytra; hind pair used in flight. The larva is a grub with 6 legs and biting jaws, or, as in the weevils, legless, and with a curled, wrinkled body.

Hymenoptera or Membrane-winged Insects.—Wings 4, the front pair larger; all 4 wings used in flight. The larva may be a legless grub—*e.g.*, bee, wasp, ant, ichneumon flies; or a caterpillar with more legs than 16—*e.g.*, sawflies; or a grub with 6 very short legs, and living in the wood of trees—*e.g.*, wood-wasp. The females have an egg-laying tube at the end of the body, which is sometimes modified as a sting.

Lepidoptera or Scale-winged In-sects.—Wings 4, all used in flight, and covered with scales. The larva is a 16-legged caterpillar (only the front 6 are true legs, the others are skin projections), or some of the abdominal legs may be absent, in which case the larva is known as a looper or geometer.

Diptera or Two-winged Insects.—Wings 2; the hind pair of wings replaced by two balancers. The larva is a legless maggot with pointed head end and blunt hind end.

Thysanoptera or Fringe - winged Insects.—Wings 4 and alike; they are narrow and have fringes; mouth-parts pierce and suck.

Hemiptera.—Wings 4. The four wings may be alike and membranous, and are held over the back when the insects are at rest — *e.g.*, aphides; or the wings may be dissimilar, the front pair stiff and horny, except at the tip, and the hind pair membranous, and when the insect is at rest the wings are folded flat on the back—*e.g.*, the bugs.

CROP PESTS.

Bean Aphis or *Collier* (Aphis rumicis, Linn.)

The attack by this aphis (fig. 632) is begun by wingless females, which

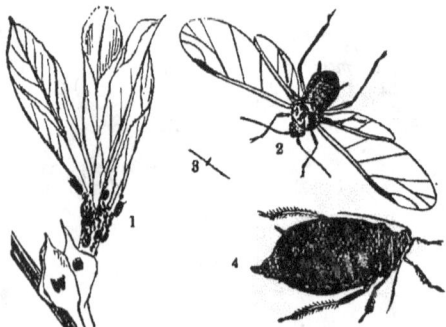

Fig. 632.—Bean Aphis (*Aphis rumicis*, Linn.)
1, Bean-shoot, with aphides; 2, Male, magnified; 3, Natural size; 4, Wingless female, magnified.

establish themselves near the top of the bean stalks at flowering time. These produce living young, and as one generation succeeds another the upper part of the plant soon becomes coated with the colliers. When this overcrowding takes place winged viviparous females appear, which spread the infection to other bean plants and to such wild plants as dock. A sexual generation is produced towards the end of the year, the eggs being laid on wild plants, and from these in the next year infection comes to the beans. In addition to the loss of sap from the feeding of the aphides, the plants are spoiled by the sticky honey dew.

Treatment.—The infested tops of the beans should be cut off as soon as "the colliers" appear. This treatment is very successful, if care is taken to carry off the infested tops and to destroy them. Spray with soft-soap, quassia, and water.

General Treatment for Aphides or Plant Lice.—In dealing with the attack of aphides, treatment should be immediate, because in favourable weather-conditions multiplication is exceedingly rapid, and as attacked leaves curl over, the aphides receive protection and the spray does not reach them. A soft-soap wash is an excellent means of getting rid of aphides, 6 to 10 lb. of soft-soap in 100 gallons of water. Against such as give off quantities of honey dew quassia should be added. Boil 6 to 8 lb. of quassia chips to extract the bitter principle, and then pour into the 100 gallons of soft-soap wash. One of the very best sprays against aphides is paraffin emulsion.

How to make Paraffin Emulsion.—Dissolve ½ lb. of soft-soap in 1 gallon of boiling water, and whilst still hot add 2 gallons of paraffin; churn violently till a creamy liquid is formed. This is the stock, which for use should be diluted with 5 to 50 times its bulk of water, according to the time of the year and the nature of the plant —tender or hardy.

Mr Pickering, in the *Eighth Woburn Report*,[1] recommends as a formula for paraffin emulsion: iron sulphate, 10 ounces; lime, 5 ounces; paraffin (solar distillate), 16 to 24 ounces; water to make up to 10 gallons. Dissolve the iron sulphate in about 9 gallons of water; slake the caustic lime with a little water, making it into a milk, and run it into the iron sulphate solution through a piece of sacking; then churn the paraffin into the mixture. This should be tested, before use, for any unprecipitated iron, by adding to some of the material a few drops of potassium ferricyanide; if there be iron in solution a blue coloration results (Prussian blue), and a little more lime-water should be added and the test repeated.

[1] *Eighth Report, Woburn Experimental Fruit Farm*, 1908, pp. 29, 93, 127.

BEAN AND PEA BEETLES.

Bean Beetle (Bruchus rufimanus).

This beetle (fig. 633) measures about ⅙ inch. The ground colour is black, with a pubescence of brown hairs. The pea beetle (*Bruchus pisi*) resembles it, but the bean beetle has the thighs of the front legs red and the exposed tip of the abdomen nearly covered with white grey pubescence, while the tip of the abdomen of pisi has marked dark spots, and its thighs are black. The larva in

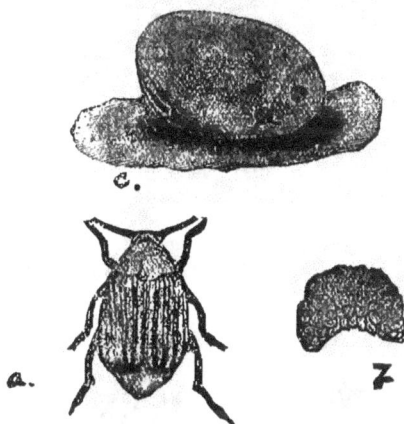

Fig. 633.—The Bean Beetle.
a Beetle, and *b* Grub, magnified.
c Exit holes of beetle.

both species is a fleshy wrinkled whitish-yellow grub; it has a horny head and biting jaws.

The eggs are laid in each case on the pods when these are very young; the grub, on hatching, passes into the seed and completes its growth there; pupation also takes place in the seed, the position being marked by a little round depression in the skin. A round hole marks the place of the exit of the adult.

Treatment.—(1) Do not sow peas or beans containing the beetles. (2) The pests can be killed in the peas or beans by fumigating these with bisulphide of carbon. The seed to be treated should be placed in an air-tight receptacle and some bisulphide of carbon placed in a saucer, the saucer being then laid *on the top* of the peas or beans, and the receptacle closed and kept closed for twenty-four

to forty-eight hours. The heavy poisonous fumes pass down through the material and kill the pests.

One lb. of bisulphide of carbon will do for 100 bushels, or in smaller quantities 1 oz. for 100 lb.

Bisulphide of carbon fumes are poisonous to human beings, and explosive if a naked light be brought near.

Bean Weevils (Sitones)—see p. 459.

Beet Carrion Beetle (Silpha opaca).

Beet and mangold crops are sometimes attacked by the beet carrion beetle, which begins to prey upon the leaves as soon as they appear above ground, giving great trouble to the sugar-beet growers in France. The beetle measures about ²/₅ in.; it is black, with yellow hairs; the wing-covers have three raised lines on each.

The grubs are much like wood-lice in shape, black, and about ¾ of an inch long when full grown.

Prevention.—As the eggs are laid in putrid matter, it is advisable (to avoid repeated attacks) to put the manure on in autumn, and only use artificial at the time of sowing. Stronger manures, such as offal and sea-weed, or shore refuse, may bring it; and as it winters in decayed leaves, they should be removed. Spray with paraffin emulsion (see p. 444).

When the mangolds are swept off in the seed-leaf stage, it is advisable to put in immediately another kind of crop. Turnips, carrots, parsnips, potatoes, peas, beans, and cabbage have been recorded as succeeding perfectly on land where the mangolds had been destroyed.

The Carrot Fly (Psila rosæ).

This fly measures ¹/₅ inch in length; the wings are iridescent and have yellowish-brown veins; the colour of the body is black or dark green; the head and legs are yellow. The larva is legless and smooth; at the pointed head end are two curved hooks used in feeding; the hind end is blunt, and has two small spiracular plates. The puparium is light-brown and wrinkled, and has two small black spiracular plates.

The first brood of flies issues in spring from puparia that have lain over winter in the soil. The eggs are laid on the carrots a little below the surface of the ground. The maggots bore into the root, making, especially in the under parts, winding tunnels which get filled with excrement, and soon the rotting which follows is accompanied by a nasty smell, and, if the carrot be eaten, it has a disagreeable taste. The tunnels show a rusty colour. The larva on being full fed leaves the carrot and pupates in the soil. There is more than one generation in the year.

Treatment.—An excellent preventive measure is to spray with paraffin emulsion (see p. 444). The carrot-bed should be sprayed after sowing, again after germination, and a third time after thinning. Or sand and paraffin can be sprinkled over the plants (see "Onion Fly," p. 459).

Carrots seen to be attacked — the withering and yellow colour of the leaves will be a guide—should be removed from the ground ; the removal, however, should be done in good time and carefully. The careless wounding of the roots may be a means of promoting attack, as the odour from the wounded carrots gets into the air and acts as a guide and an attraction to the carrot flies. This is partly why the pest is worst at thinning-time, as thinning means a good deal of bruising of the plants. One should thin early or thin late, and after thinning leave the soil well compacted, so as to make it difficult for the fly to reach the roots for egg-laying. In connection with the last suggestion, we received some time ago a communication which read : " For the past two years I have left a portion of the carrots in my garden unthinned, only pulling them from time to time for use. I found that portion of the crop untouched by maggots, while the portion thinned was. Of course the unthinned carrots were not so big as they might have been." A turning up of the soil, after infestation, in the winter time would expose any over-wintering pupæ ; while the burying of the surface soil would leave any puparia that might be present at such a depth that even if the pupa-cases gave out their flies these would not be able to make their way above ground.

For Wireworm in carrot, see p. 453.

Beet and Mangold.

The Pigmy Mangold Beetle (Atomaria linearis).—This beetle now and again does immense damage to the mangold crop, sweeping off the whole crop when it is in the seedling and young stage. The beetle is so small that it is very often never seen, and the damage ascribed to something else. The young plants are gnawed in the parts below the soil, and the leaves are also eaten. The life-history is uncertain, but the grubs probably live in the soil, while the beetles live in the soil round about the plants, but also come above ground.

Treatment.—If the crop be hopelessly infested, plough it in deeply. On the Continent thick seeding is practised where this insect is a pest. The young plants should be dressed with soot, and then the soil round the plants hoed in. As an experiment, we should suggest steeping the seeds for a short time, before sowing them, in paraffin.

Mangold - leaf Maggot. — Another well - known beet and mangold pest is the maggot of the Pegomyia betæ, Curtis (fig. 634). This damages the crops by

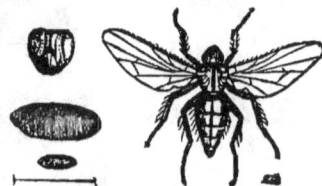

Fig. 634.—Beet-fly (Pegomyia betæ, Curtis).

Female, magnified ; Line showing spread of wings, natural size ; Head, magnified ; Pupa, natural size and magnified.

feeding, between the upper and lower epidermis, on the tissue of the leaf. The white legless maggots are about $\frac{1}{3}$ of an inch long, and of a yellowish-white colour, and as soon as they are hatched bore through the skin of the leaf by the aid of two mouth hooks. There are two broods in the year.

Treatment. — Pull and burn badly attacked plants. Spray with paraffin emulsion (see page 444) as a preventive. Keep down Composite and Chenopodiaceous weeds, as these may harbour the enemy.

Cabbage Butterflies.

Cabbage very often suffer from cabbage-butterfly attack. We give illustrations of the two principal offenders, — the large white cabbage butterfly (*Pieris brassicœ*, Latreille) (fig. 635) and the small white

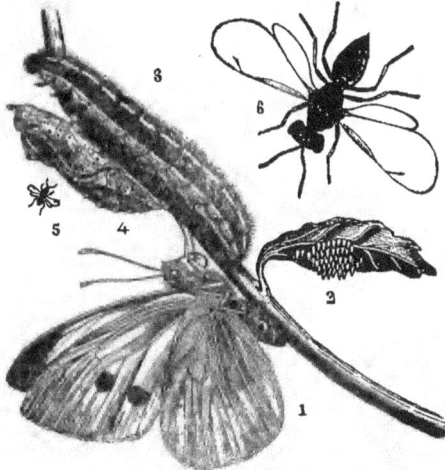

Fig. 635.—Large White Cabbage Butterfly (*Pieris brassicœ*, Latreille).

1, Female butterfly; 2, Eggs; 3, Caterpillar; 4, Chrysalis; 5 and 6, Parasite Chalcid-fly (*Pteromalus brassicœ*), natural size and magnified.

cabbage butterfly (*Pieris rapœ*, Latreille), (fig. 636). Leaves on which the eggs are laid should be picked off, and the caterpillars searched for and destroyed. The caterpillars are more common in gardens, where they find congenial shelters, than in large open fields. Measures to promote a healthy growth should be adopted.

1. The Ichneumon fly (*Microgaster glomeratus*) comes to our aid as a natural enemy to these caterpillars, in which it lays its eggs. The maggots from these eggs feed inside the caterpillar, and as a result the metamorphosis is not completed. The small yellowish cases, collected in bunches, often seen on cabbages, are those of the ichneumon pupæ. They should not be destroyed.

2. The great *measure of prevention* is searching for the chrysalids, which may be sometimes collected in handfuls from shelter-places under eaves, boards, &c., in the neighbourhood of gardens.

3. Sending boys to hand-pick the caterpillars has been found useful as a remedy.

4. Destroy the egg clusters.

5. A good drenching with ordinary water is often useful, as it causes the caterpillars to suffer in health. Some drench with soap-suds.

The Cabbage Moth (Mamestra brassicæ).

The destructive caterpillars of this moth are general feeders, found not only on cabbage, which is the commonest host plant, but on other cruciferous plants and on fruit plants and garden flowers. The caterpillars have 16 legs; they measure 1¼ inch when full grown, and have a habit of rolling themselves into a ring when they are touched : the colour varies, green or grey-green above and yellowish below, with a dark line down the back. The pupa, shining chestnut brown, is in the soil. The caterpillars eat into the heart of the cabbages, and the plants are made disgusting by the excrement.

Treatment.—1. Hand-pick the caterpillars before they get into the heart. 2. Dust the plants with lime. 3. Destroy the pupæ when the ground is dug in winter. Poultry turned on to the land at this time would be helpful.

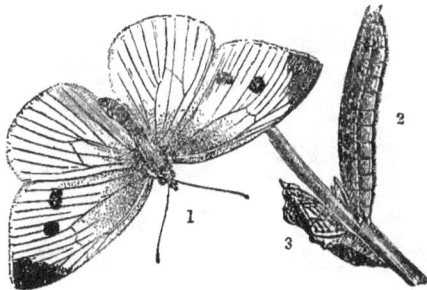

Fig. 636.—Small White Cabbage Butterfly (*Pieris rapœ*, Latreille).

1, Female butterfly; 2, Caterpillar; 3, Chrysalis.

Cabbage Root Fly (Phorbia brassicæ).

This (fig. 637) is one of the most troublesome enemies of the cabbage : not only are the maggots found infesting cabbages, but turnip, swede, cauliflower, brussels-sprouts, radish, are all attacked.

2 F

The plants are checked in their growth, the leaves discolour, wither, wilt, and the infected parts rot and the plant falls away.

The maggot, which does the harm, is whitish-yellow and legless; the head end is pointed, and has two dark curved mouth hooks; the hind end is truncate, the last segment having two dark spiracles; all round the edge of this last segment are 12 little projections (seen with a lens),—the two lowest are the largest and are forked. The full-grown maggot measures ¼ inch. The puparium is brown in colour and oval.

The females lay their eggs close to the plant; the maggots bore into the plants, making galleries in the roots; they may be found, too, tunnelling in swede leaf-stalks. Pupation takes place in the soil or in the infested plant. There is more than one brood in the year, the first flies showing about the end of April.

Treatment.—Successful treatment is

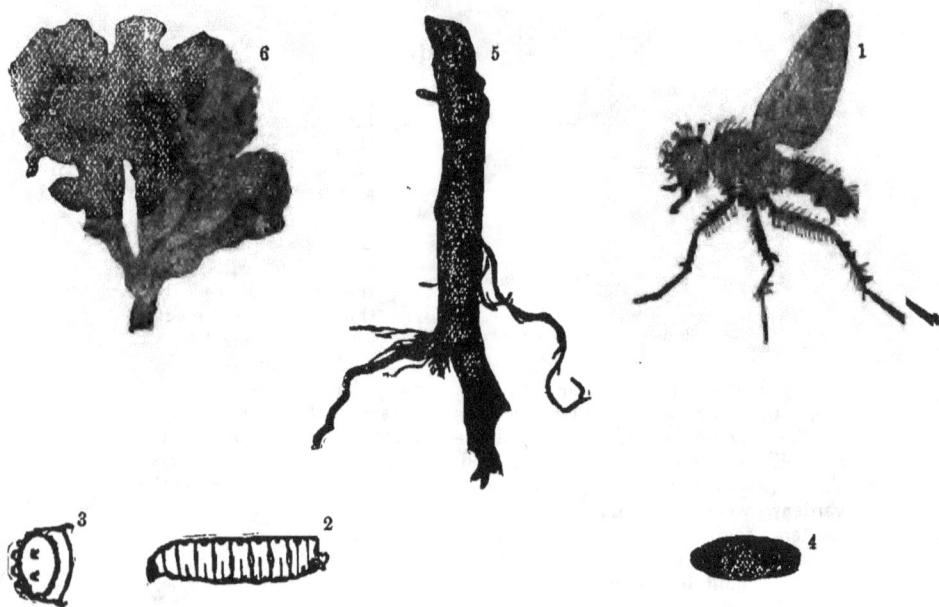

Fig. 637.—The Cabbage Root Fly.

1. Male fly, greatly magnified.
2. Maggot, magnified.
3. Last segment of maggot, enlarged, showing tubercles.
4. Puparium, magnified.
5. Attacked cabbage with larva *in situ*.
6. Withered leaf of attacked cabbage.
(Figs. 1 and 3 after Slingerland.)

extremely difficult. In America the cabbages and cauliflowers are protected by having tarred paper or cards placed round them; the flies are not able to get near enough to lay their eggs. This treatment has proved too expensive in Britain.

In garden cultivation a cupful of paraffin should be added to a pailful of sand, and the sand sprinkled once a-week round the stems of the cabbages, as a deterrent from egg-laying. Remove and burn badly infested plants. After bad attack do not follow with a crucifer crop.

Very early sown plants are noticed largely to escape.

Cabbage Root Gall Weevil (see Turnip Gall Weevil).

The **Cabbage Flea Beetle** (*Haltica oleracea*) is glossy bright blue-green in colour; it measures from $1/7$ to $1/6$ in. in length. The adult beetles feed on the leaves not only of seedling but of older plants. The grubs feed externally on the leaves. Turnip is attacked as well as cabbage. There are a number of broods in the year.

Cabbage Aphis (Aphis brassicæ).

This aphis, which has a white mealy coat, is a pest on cabbages. The infested leaves pucker and blister, and the plants are spoiled by their being covered with the honey dew excreted by the aphides.

Treatment (*see* Treatment for Aphides, page 444).—There are natural enemies (*see* Useful Insects, page 469) of the aphis —viz., hover-fly larvæ, which eat the aphides, and Ichneumonidæ, whose larvæ are parasitic inside them.

The Celery Fly (Acidia heraclei).

The maggots of this fly (fig. 638) mine the leaves of celery and parsnips, causing blister-like patches, pale at first but brown later. The eggs are laid by the female fly on the upper surface of the leaves; the pale-green larvæ hatch in a week, and are full grown in another fortnight, the puparium being found in the leaf, or in the soil if the full-fed maggot has fallen away. There are several broods in the year.

Treatment. — (1) As a preventive measure the plants should be sprayed with paraffin emulsion, to deter the fly from egg-laying. (2) Spoiled plants should be burned with the enclosed larvæ or pupæ. (3) When the crop is removed, the surface-soil should be buried deeply, so that the flies from buried puparia will be unable to reach the surface.

The Cereals.

Daddy-long-legs or **Crane-flies** (*Tipula*).—The larvæ of the crane-fly, known as leather-jackets, are often the cause of great destruction to cereal and grass crops, and to plants of other Natural Orders as well. The Tipulidæ or Crane-flies are the large, sprawling, awkward flies seen in greatest numbers from August; the name Crane-fly is given to them from the snout that projects somewhat from their head, and from the long legs.

There are several species of Tipula, the best-known three being *T. oleracea* (fig.

639), *T. paludosa*, and *T. maculosa*. All three may be present, or two of them, at the same time, or infestation in a district may be due to only one of them.

The general life-history is that the females lay numerous eggs on the ground, pasture and coarse rank herbage being favourite places. From the eggs hatch larvæ which live in the soil, gnawing the roots of plants; they may come to the surface at night and destroy leaves, *e.g.*, the

Fig. 638.—The Celery Fly.

1. Fly, magnified.
2. Larva, magnified.
3. Pupa, natural size.
   Lines showing natural size of fly and larva.

young leaves of corn. Among the many different plants attacked are corn, grass, turnip, pea, bean, clover, mangold. The larvæ are earth-coloured or rust-coloured, and have small hard heads, with short antennæ; the head carries a strong pair

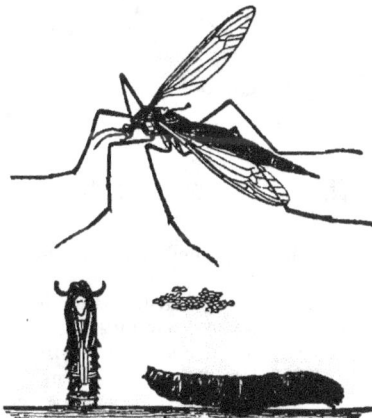

Fig. 639.—Daddy-long-legs (*Tipula oleracea*, Linn.) Fly (after Taschenberg); Pupa and larva (after Curtis).

of mandibles, which bite against the teeth of another pair of jaws; this head can be retracted within the segments that follow it. Round the margin of the hind end of the leather-jacket are little projections, and on the hind surface of the last segment are the two spiracles;

the tough, strong skin is very resistant to external influences.

When this larva is full fed it moults its last skin and becomes a pupa. After a short resting-stage—it may in the summer be only a fortnight—the pupa, by the aid of a series of little hooks on the abdominal segments, wriggles to the surface, where it may be seen standing erect with quite half of the body projecting above the surface. The head end carries two little tracheal horns by which respiration is accomplished. At length the skin cracks on the upper sur-

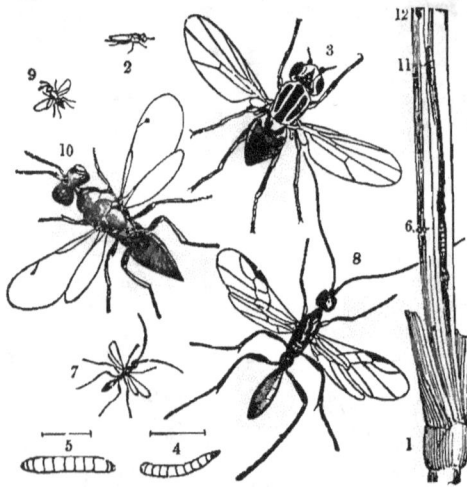

Fig. 640.—Ribbon-footed Corn-fly ; Gout (*Chlorops tæniopus,* Curtis).

2–6, Larva, pupa, and fly of *Chlorops tæniopus*, natural size and magnified parasite flies ; 7 and 8, *Cœlinius niger ;* 9 and 10, *Pteromalus micans*, natural size and magnified ; 1, 11, and 12, Infested corn-stem.

face at the front end, and the mature crane-fly issues.

Treatment. — Close grazing in the autumn in order to discourage egg-laying. The absence of a rank growth will tend to reduce egg-laying. "Grub-on-corn," as the pest is often called, is worst on corn after grass, and the females should be deprived of their chance of laying on pasture by having the land ploughed, if possible, before the flies in large numbers have proceeded to their egg-laying in August and September. Failing the possibility of a ploughing early in the autumn, it is recommended to dress grass and clover leys with gas-lime, three to four tons to the acre ; this is fatal to eggs and larvæ.

In the north there is in some years great loss of crop from the leather-jackets, and some practise late sowing to give the leather-jackets a chance to have done a great part of their feeding on the roots and leaves of the ploughed-up pasture before the young oats are ready for them. As a remedial measure, when the pest has got to work in the young oats the crop should receive a dressing of nitrate of soda, 1 to 2 cwt. to the acre. When the crop is evidently beyond saving, it should be ploughed up and some other crop taken. The crop planted will differ according to the locality and the conditions, but white turnip, rape, and mustard have been recommended. Harrowing and rolling kill some of the grubs, and turn others up for the birds ; where rolling is practised (with a heavy roller) it should be done at night, or very early, as the larvæ will be found at or near the surface. Professor Carpenter,[1] in a laboratory experiment, found that if powdered naphthalene were mixed with the soil the leather-jackets succumbed. Fumigants, containing naphthalene as an ingredient, may prove helpful as a mode of treatment against soil insects.

Gout - fly or The Ribbon-footed Corn-fly (*Chlorops tæniopus*).—This fly (fig. 640), 1/6 inch long and 1/4 inch in spread of wings, is, in general colour, yellow, with the antennæ black ; there are three black longitudinal stripes on the thorax. The flies lay their eggs in May and June on young barley plants —other cereals and grasses may be used, —and the maggot, on hatching, tunnels its way down the stem, from the base of the ear to the first knot in the stem. The maggot is legless, whitish-yellow in colour, and with two tubercles at the hind end. The full-grown maggot pupates at the bottom of the burrow, the flies appearing in autumn. This brood of flies lays on wild grasses, and it is from these that the first brood of flies of the year comes in the next season. As a result of attack on barley, the ears

[1] *Economic Proceedings of the Royal Dublin Society*, August 1905.

may be unable to break through the enveloping sheaths, this giving a characteristic swollen appearance.

Treatment.—Sow early, in order that the plants may be well on before the appearance of the flies; plants whose ears have broken through before the flies appear are safe. The refuse from a threshed infested crop should be burned.

Frit-fly (*Oscinis frit*).—This fly (fig. 641) measures less than ⅛ inch in length. The body is shining black; the legs also are black, except the feet, which are yellow-brown. The maggots of the fly are injurious to cereals and pasture grasses. In Britain the chief damage is to oats, although barley may also suffer. Symptoms of attack are pale spots on upper surface of leaves that are still green, a reddening or browning of the leaves, stunted growth, and at last failure : tillering may take place, but the shoots are twisted and swollen.

The first flies of the year appear in April and May to lay their eggs on the young leaves of the oat. The maggot is round, fleshy, and legless; it has two mouth-hooks at the head end, and at the blunt hind end two projecting spiracles. The maggots eat to the heart of the plant by means of their mouth hooks. The full-fed maggots, ⅛ inch in length, pupate in the infested plant, and the flies of the new brood appear in July. These new flies lay on pasture and wild grasses, or in the ears of oats and barley, if these be in a suitable stage with young grains. The result of attack on the ears is a light sample, with shrunken, shrivelled grains. By August and September a third brood of flies may appear, and these lay on grasses from which infestation of the oat will follow in the next April and May.

Treatment.—Sow early, so that the plants may get a start before the fly. The late Miss Ormerod quotes cases thus : "All early spring fields seem to have escaped, in some others sown late 90 per

cent of the crop is gone"; and again, "One field of oats sown on 29th March enjoyed almost complete immunity; and in another field, sown on 29th April, over 70 per cent of the first stems were destroyed." A stimulating dressing may save the plant if the attack is noticed early, but badly infested plants are doomed, and should be ploughed in deeply. Frit-flies found swarming in granaries, stores, &c., that have issued from puparia in harvested corn should

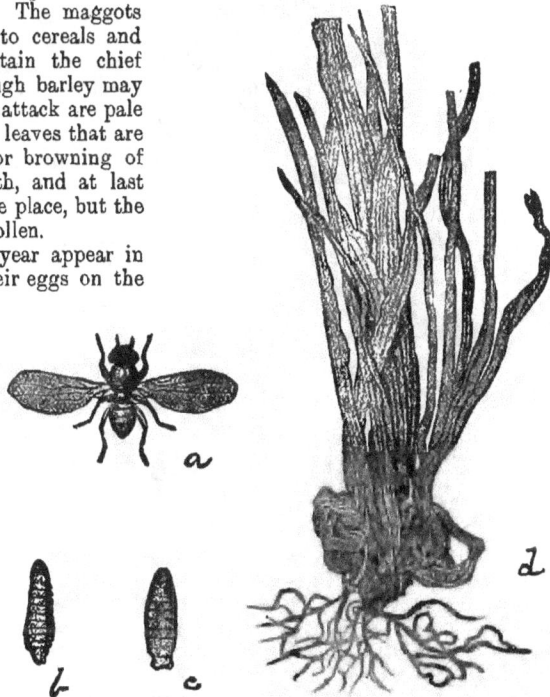

Fig. 641.—The Frit Fly.

a The Frit Fly (*Oscinis frit*). b Larva. c Puparium. d Young infested plant.
a b c Much enlarged.

be destroyed. Destroy such wild grasses, in the winter, as are known to be infested.

Wheat Bulb-fly (*Hylemyia coarctata*).—This fly, which measures ¼ inch in length, lays eggs on young wheat plants, and the maggots feed inside the stems. The puparia are found both in the soil and in the spoiled plants. At present we have no known remedial measure against this insect. The other cereals, oat and barley, are not attacked.

Hessian Fly.

In 1886 a new pest was discovered in England. The Hessian fly (*Cecidomyia destructor*, Say) (fig. 642) was found to be present in some barley-fields near Hertford.

The following abstract, from a German source, gives its life-history: "The larvæ live in the haulm of wheat, rye, and barley. The female flies usually lay their eggs on the young leaves twice in the year—in May and September,—out of which eggs the maggots hatch in fourteen days. These work themselves in between the leaf-sheath and the stem, and fix themselves near the three lowest joints, often near the root, and suck the juices of the stem, so that later on, the ear, which only produces small or few grains, falls down at a sharp angle. Six or eight maggots

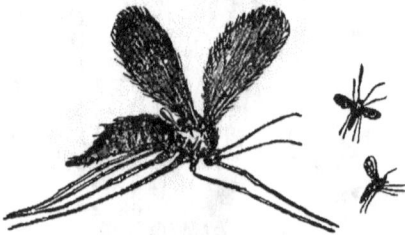

Fig. 642.—Hessian Fly (*Cecidomyia destructor*, Say). Natural size and magnified.

may be found together, which turn to pupæ in spring or about the end of July, from which the flies develop in ten days."—(*Stett. Ent. Zeit.*, xxi. p. 320.)

Miss Ormerod (from whose pamphlet on the subject this information is taken) found, on visiting the infested fields, the stems doubled sharply down a little above the joint; and between this double and the joint there lay, closely pressed to the stem and covered by the sheathing-leaf, the flax-seed-like puparia. The injury is caused by the fly-maggots, lying at the same spot, sucking the juices from the stem, which is thus weakened, and falls.

The Hessian fly has commonly two broods in the course of the year—the winter attack to the young plants, and the summer attack to the growing straw. The flies which come out in August or September from the "flax-seed" chrysalis-cases (sheltered above the second joint of the straw from the ground) lay their eggs, we are informed by various observers—the late Professor Riley amongst the number,—in the grooves on the surface of the leaves, or between the stalk and sheath where loose, and as soon as the footless larva or maggot hatches, it makes its way down the leaf to the base of the sheath, which in the young winter wheat is at the crown of the root.

This form of attack has not yet been reported in England. The summer attack with us is started chiefly from "flax-seeds" or chrysalids which have survived the winter. The flies from these "flax-seeds" come out in spring, or about the beginning of May, and as, where the corn is running up to stem the tender ground leaves are no longer to be found, which are used for autumn egg-laying, the flies have no choice, but they lay them instead, as we know, so that the maggot, when hatched, shelters itself between the stem and sheath, just above the first or second joint from the ground, and there it turns to the flax-seed chrysalis, from which the autumn brood presently comes out.

Prevention.—A chief method of prevention is in *late sowing, so that the young wheat will not be up until the autumn brood is over*: this is a most important precaution. All measures to secure hearty good growth are very desirable; so is rotation of crop, and it should be borne in mind that strong-stemmed corn is less liable to attack than the kinds of which the outside is more readily injured by the maggot.

One most important measure to prevent recurrence of attack from infestation present in any locality is *destruction of siftings*, in which the flax-seeds, as they are called, are thrown by the threshing-machines. These chrysalids are often present in great numbers, and would, if left, be the origin of next year's attack; and if burnt together with the rubbish in which they lie, great danger will be spared. Where there has been infestation the stubble should be ploughed in deeply.

The Wheat Midge (*Diplosis tritici*). —The adults measure $1/25$ to $1/16$ of an inch; they are yellow in colour, with antennæ black; the two wings are hairy. The female midge (fig. 643), recognised by her hair-like projecting egg-laying tube, lays the eggs in summer time in the flowers of grasses, wheat, and rye, rarely

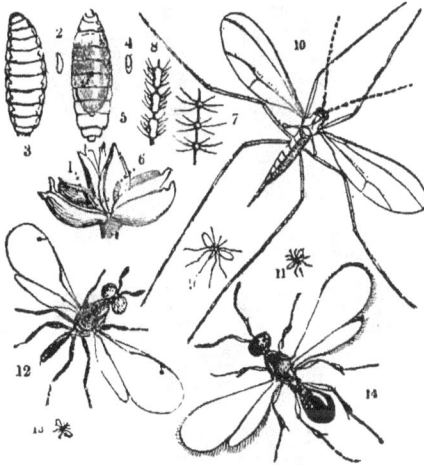

Fig. 643.—Wheat Midge
(*Cecidomyia* (*Diplosis*) *tritici*, Kirby).

1, Infested floret; 2-6, Larva and cased-larva (? pupa), natural size and magnified; 7 and 8, Joints of antennæ, magnified; 9 aud 10, *C. tritici*, natural size and magnified. Parasite flies: 11 and 14, *Platygaster tipulæ*; 12 and 13, *Macroglenes penetrans*,—natural size and magnified.

oats and barley. Several eggs are laid in each flower, and the maggots live, inside the flowers, on the grain ; they are legless, and in colour orange ; on the under sur-face at the front end there will be seen under the microscope a process bifid at the apex, and known as the breast-bone or anchor process. The full-fed maggots pass into the soil for pupation, or they may be harvested with the grain. Ears which have been infested are light.

Treatment.—Deep ploughing of the stubble of an infested crop to bury the puparia. The refuse and chaff after threshing should be burned, as where there has been an attack maggots and puparia may have been carried in in the crop. From any collection of screenings and refuse containing puparia the flies will issue in the next season, and may pass to the crop for their egg-laying.

The Corn Sawfly (*Cephus pygmæus*). —The sawflies are Hymenopterous, not Dipterous, insects. The corn sawfly (fig. 644) is a four-winged insect, black in colour, spotted with yellow ; it measures ⅓ of an inch in length. The female makes a hole in the stem—wheat especially is attacked—whilst the stem is young and

soft. The larva which hatches is legless, and yellowish-white in colour, with the head yellow-brown ; it tapers to the hind end. The larva tunnels the stem, piercing the knots in its pass-age ; when full fed it passes to the base of the stem, to the part which will be left as stubble, and passes the winter in a cocoon of silk. Before spinning this cocoon the larva gnaws the stem through in a circle, above the place where it will lie—about the ground-level. As a result of this, the stem falls over. In the next spring the larva which has hibernated as such pupates, and the adult issues about May. Loss in grain from the twisted state of the straw follows attack.

Treatment.—Burn the stubble, or bury it deeply, so as to destroy the hibernating larvæ and prevent a new infestation.

Wireworms.—These are among the very worst pests of the farm. Some insects can be fought by alter-ing the rotation, but the wireworms destroy all kinds of crop irrespective of plant relationship—cereals and grass, potatoes, turnips, hop, mangold, carrot

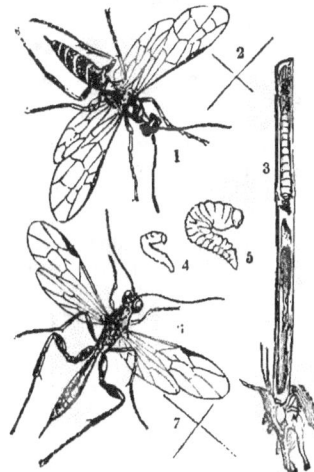

Fig. 644.—Corn Sawfly (*Cephus pygmæus*, Curtis).

1 and 2, Sawfly, magnified and natural size ; 3, Stem containing larva ; 4 and 5, Larva, natural size and mag-nified ; 6 and 7, Parasite fly (*Pachymerus calcitrator*), magnified and natural size.

(fig. 645). These wireworms are the larvæ of the beetles known as Click

beetles; they are elongated worm-like forms, with biting jaws, three pairs of thoracic legs, and a fleshy sucker at the

Fig. 645.—Wireworm on carrot.

downwardly bent tail end, which aids in progression; the joints have horny shields on upper and lower surface for protection. In colour these larvæ are yellow-brown.

Click Beetles.—There are several species of click beetle—*e.g.*, *Agriotes* (*Elater*) *lineatus* (fig. 646), which is ⅜ inch long and ½ inch in spread of wings; the wing-cases are brown and have yellow-brown stripes; the legs are rusty red.

A. obscurus, black, but covered with brown hairs; the upper parts of the legs are black, the rest rusty red.

A. sputator is rather smaller; it is black, with red legs.

Athöus (*Elater*) *hæmorrhoidalis* is larger than the other species; it is pitchy-black or brown and hairy, the wing-cases are lighter; characteristic are the red-brown tip of the abdomen and edges of the wing-cases.

The dingy colours of the beetles and sluggish habits prevent them from being noticed on plants. If touched they drop to the ground. If the beetle fall on its back, and there is nothing within reach

that it can catch on to by means of its short legs, a spring apparatus on the under side of the body, consisting of a curved spine which fits into a groove on the mesothorax, is brought into play, and the beetle rights itself. It raises the middle region of the body so that the spine comes out of the groove, then the body is suddenly straightened, the spine being forced back into the groove with a click — hence the name click beetles,—and the cases of the wing-covers striking the ground smartly, the beetle is sent into the air, and when it falls it lands on its feet.

The beetles lay their eggs close to the roots of plants, and the grubs on hatching feed by tearing away the soft parts and tunnelling into the tissue. Several years are spent in the larval condition, the wireworms moving from plant to plant. In the winter they go deeper, but on the return of mild weather come nearer the surface again to renew their feeding. When full fed the wireworm goes deeper into the soil and becomes a pupa under cover of a cocoon of soil particles: in due course the mature beetle makes its way to the surface.

Treatment.—Attack is worst in clover-

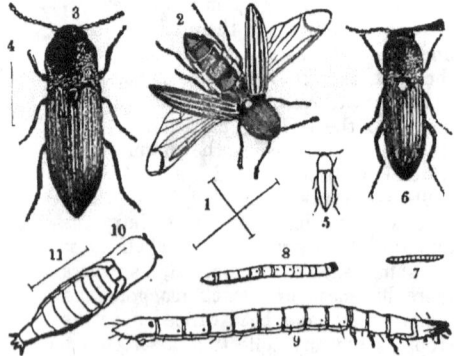

Fig. 646.—Click Beetles.

1 and 2, *Agriotes* (*Elater*) *lineatus;* 3 and 4, *A.* (*E.*) *obscurus;* 5 and 6, *A.* (*E.*) *sputator,* natural size and magnified; 7, Larva of *A.* (*E.*) *sputator*(?); 8 and 9, Larva of *A.* (*E.*) *lineatus,* natural size and magnified; 10 and 11, Pupa of wireworm magnified—the straight lines show natural length.

leys or broken-up pasture. Wireworms should be anticipated under these circumstances, and preparation made by feeding the pasture as bare as possible before ploughing and spreading gas-lime on the surface—the gas-lime must be allowed to lie some weeks before it is ploughed in. Some attempts have been made to combat the pest by a fumigant named vaporite (Strawson). For arable land the vaporite is sown broadcast before the plough, and some time in advance of seed-sowing: the vaporite is mixed with the soil to the depth of 4 to 6 inches at the rate of 2 to 3 cwt. per acre. In gardens the vaporite is dug in. The action of vaporite is said to be slow, and therefore the material should be applied early.

The beetles may be trapped in gardens by being lured to little bundles of the crop plant that have been steeped in Paris-green, or little heaps of clover are laid here and there to be used for shelter and egg-laying: the trap material in each is covered with a board.

Many birds do good by devouring wireworms—e.g., peewit, gulls, starling, rook (see Birds and their Food Habits).

The Cockchafer (*Melolontha vulgaris*).—The large white grubs of this beetle feed at the roots of cereals and grasses as well as on nursery forest-trees; the grubs have six legs, a horny head, and strong biting jaws; they lie in a curved position, and their bodies are swollen and dark in colour from food contents at the hind end.

The Garden Chafer or Bracken Clock (*Phyllopertha horticola*). — The chafer or beetle is about 1/3 inch in length; it is shining greenish-black in colour, with blackish or grey hairs; the head and thorax are glossy blue-green; the wing-covers yellow-brown. The grub is whitish, fleshy, wrinkled, with a brown head and gnawing jaws; it has three pairs of short legs on the thorax. The adult beetles eat the leaves of orchard trees and also damage fruit; the females lay their eggs in garden soil, and plentifully in grass land; and the grubs which hatch gnaw the roots through, and the plants die off. In hill-pastures much harm is done sometimes by these grubs, strips of pasture looking as if they had been burned, owing to the withering away of the above-ground parts on account of the destruction of the roots. Parts of golf links have been destroyed in the same way.

In working against this enemy the beetles should be shaken down, in the early morning, off the plants on which they are settled, on to cloths spread for the purpose, and then collected. The grubs are not easy to reach. Birds are, in cases of infestation, most useful—crows, starlings, plovers, pheasants; the common hen takes the grubs greedily if it gets the chance.

Corn Ground-beetle (*Zabrus gibbus*). —The ground-beetles are mostly useful, being carnivorous in diet (see Useful Insects, p. 469), but this beetle is harmful to cereals, both in the adult and in the larval stages; the adults have been found crawling up the stems and destroying the grain; the grubs destroy the plants at or near the surface. The beetle is half an inch long and black in colour, with a dash of purple; legs and antennæ are brown. The grubs are pale-brown; their head is large, and carries a pair of strong sickle-shaped jaws; they are active, moving about by means of six thoracic legs; in common with the ground-beetle larvæ they have two projections at the tail end, with a smaller one between.

Corn and Rice Weevil.—The corn weevil (*Calandra granaria*) (fig. 647) and the rice weevil (*C. oryzæ*) are both very troublesome to stored grain—wheat, barley, oats, and the rice weevil to rice, and also to cargoes in ships. *C. gran-*

Fig. 647.—Corn Weevil (*Calandra granaria*).
1. Beetle, magnified.   2. Infested wheat grain.

aria is 1/8 of an inch long, and has the long beak characteristic of weevils; the colour is brown-black; the antennæ, which spring from the beak, are reddish in colour and bent; the legs are also red; the wing covers are striated; the beetle is wingless. *C. oryzæ*, which has wings, is smaller, smoother, paler, and has two orange-coloured patches at the apex and base of the wing-covers.

The beetles lay their eggs—an egg to a grain—in a hole made in the grain, and the larva on hatching feeds on the contents of the grain, and when full fed pupates in the spoilt and hollowed-out grain, the mature weevil issuing in due course. The adult beetle, too, is harmful by puncturing and feeding off the grain. The length of the cycle from the laying of the egg to the development and appearance of the adult beetle varies much with the conditions. At a temperature of 80° Fahr., and with the other conditions favourable, the whole life-cycle can be completed in a month. The adult beetles have a long life.

Treatment.—Infested grain from cargoes is run through a sieve or down a screen whose mesh-work is fine enough to keep the grains back and yet let the beetles pass through, these being caught in a vessel containing paraffin arranged below for the purpose.

The best method to rid the grain of the pests is fumigation, the fumigant being bisulphide of carbon. This liquid vaporises readily, and the vapour being heavier than air sinks or falls. Place the grain to be treated in a bin or airtight receptacle, pour into a saucer or shallow vessel some bisulphide of carbon, and then lay the saucer *on the top* of the grain. Close the receptacle tightly, and leave for from twenty-four to forty-eight hours. The fumes kill all the insects. Two pounds of bisulphide of carbon are sufficient to fumigate one ton of grain. Bisulphide of carbon has a disagreeable odour, and as the fumes are poisonous they should not be breathed more than is necessary by human beings. A naked light must not be brought near the fumes, else an explosion takes place.

The most favourable environment for the development of these grain weevils in cargo is heat, a certain amount of moisture, and a confined atmosphere.

Flour Beetles.—Two little beetles (*Tribolium confusum* and *T. ferrugineum*) are sometimes very troublesome in flour. *T. confusum* is reddish-brown in colour, and measures about ⅙ inch in length; the joints of the antennæ get larger towards the apex. *T. ferrugineum* is very like *confusum* in colour and size, but the antennæ end is a distinct three-jointed club.

The beetles lay their eggs on and in the neighbourhood of the flour, meal, or other cereal product. The small six-legged hairy grubs which hatch may be found crawling about amongst the material. In a favourable temperature there can be a number of broods in a year, so that infested material can quickly get worse. In addition to the presence of the pests in the flour, the sample gives off a disagreeable odour.

Treatment.—Fumigate with bisulphide of carbon (see p. 445), five ounces of bisulphide of carbon to a barrel of flour. The flour after fumigation and before being used should be exposed to the air.

Corn Thrips (*Thrips cerealium*).— The insect is ¹/₁₂ inch long. The male is wingless; the female has four wings; eggs are laid in the ears of cereals; the larvæ which develop are deep yellow. The result of attack is a shrivelling of the grain.

When at work in the ears the insect cannot be reached, but hibernation takes place in the stubble, and in decaying roots deep ploughing would destroy many.

The Order Thripidæ.

This is a family of elongated insects with four long narrow wings that have long fringes. As regards metamorphosis, the thrip insects make a connecting link between complete metamorphosis and the absence of a resting stage. From the egg hatches a young form which feeds for a varying length of time. Then follows a non-feeding stage, when the full-grown larva is sluggish, the rudiments of the wings being encased in a sheath and the limbs hidden by a film. The next is the mature stage. The mouth-parts are modified for piercing and sucking. The legs have a small bladder or sucker at the tip of each tarsus.

Hops.

Hop Aphis (*Phorodon humuli*).—This aphis (fig. 648) is the worst enemy of the hop, its damage being in some years very great. The hop aphis appears upon the hop plants generally about June, and if the conditions of temperature and of the plants are favourable, it multiplies with astonishing rapidity. The swarms live upon the sap of the plants, which is drained away by the long, piercing and sucking beak of the insect.

They attack first the youngest and smallest leaves of the leading shoots, which are more succulent than the older leaves. After a week or two the growth of the plants is checked, and the plants struggle in vain to reach the top of the poles. Their juices are exhausted by the continuous suckings of the insects; the respiratory and sugar-making and transpiring functions of the leaves are interfered with by the honey dew — a peculiar glutinous sweet secretion ejected from the bodies of the aphides—falling on their upper surface, and by the aphides congregating and feeding on the lower surface. In three weeks to a month after the appearance of the aphides—if these are in number—the plants shrivel, the leaves turn black and fall off, and all chances of a crop are lost. If the numbers have not been overwhelming and the plants survive, the aphides may go into the hop cones, which, as a result, are deteriorated in value.

There are two possible cycles in the life-history. The wingless females that may have hibernated pass to the plants in the spring, and by these living young are produced. This mode of multiplication continues during the season. In the course of the summer winged females appear, and, flying to other hop plants, spread the infection. As the season ends males appear for the first time in the year's life-cycle, and the females after fertilisation hibernate.

The alternative life-cycle is: The fertilised females desert the hop and fly to sloe and damson and plum, and on the shoots of these lay their eggs. These eggs hatch in the next spring, and the young develop into winged females, which migrate to the hop from May onwards.

Treatment. — (1) Spray the plants with soft-soap and quassia (see p. 444). (2) The plum and the damson should be washed in winter, to destroy the eggs laid on them.

Winter Washing of Fruit-trees.

The advantage of the winter washing of fruit-trees is that mosses and lichens and such vegetable growths that cover the bark are removed. These provide shelter and hibernating places for a number of injurious insects, which are destroyed by the wash; eggs also, which would hatch in the next year, are destroyed. In Leaflet No. 70 of the Board of Agriculture full details are given, and the Woburn Winter Wash quoted.[1] Mr Pickering has suggested the following modification: Sulphate of iron, $\frac{1}{2}$ lb.; lime, $\frac{1}{4}$ lb.; caustic soda, 2 lb.; paraffin (solar distillate), 5 pints; water to make 10 gallons.

Dissolve the iron-sulphate in about 9 gallons of water. Slake the lime in a little water, and then add a little more water to make it into a milk; run this into the first through a piece of coarse sacking, to keep back grit. Next, churn

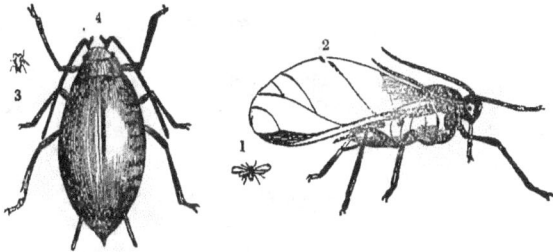

Fig. 648.—Hop Aphis; Green Fly (*Phorodon humuli*, Schrank). 1 and 2, Female aphis, natural size and magnified; 3 and 4, Larvæ, natural size and magnified.

the paraffin into the mixture and stir. Just before using add the caustic soda in the powdered condition.

It should be noted that this wash is a burning one, and the hands and face of the worker require protection during its use. February would be a suitable month for the treatment. The principle is that this wash, with its caustic properties, may only be used when the plant is in a dormant condition.

Natural Enemies of the Hop Aphis. —Many aphides are devoured by the larva of the lady-bird and by the larva of the lace-wing fly; others are parasitised by the chalcid flies (see p. 469).

The Hop Frog-fly (*Euacanthus interruptus*) is now and again found troublesome. The insect (fig. 649) in the adult

[1] *Eighth Report of the Woburn Experimental Fruit Farm.* 1908. By The Duke of Bedford and Spencer U. Pickering.

stage can fly, and is also capable of making long jumps. In the case recorded by the late Miss Ormerod, the insect was found "principally infesting the tops and therefore most succulent portions of the bines; when once the plant becomes a prey to them, it is visibly checked in

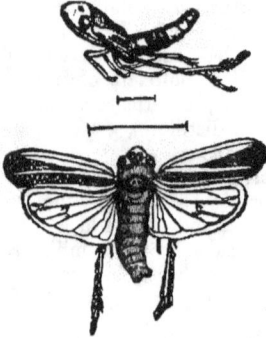

Fig. 649.—Hop Frog-fly (*Euacanthus interruptus*, Linn.)

growth, the leaves assume a deformed shape, and curl at the edges." The only measure so far attended with success is to jar the insects off the plants on to tarred boards or sacking.

Fever - fly (*Dilophus febrilis*).—This black fly, which sometimes appears in swarms, has been blamed as an enemy to the hops, the larvæ being accused of spoiling the roots: the larvæ are also found in cow and horse manure, and it is possible that in this way they may be conveyed to plants.

The Hop - flea or Brassy - flea Beetle (*Phyllotreta concinna*). — This tiny beetle measures only between 1½ and 2 millimetres; yet, after the hop aphis, it is the worst enemy of the hop. It is oval in shape, and greenish in colour, with a brassy tint. The beetles are excellent jumpers. The harm is done both by the adult beetle and the grub. The beetles bite holes in the leaves and shoots when these are quite young, and when in numbers ruin the plants. The eggs are laid by the beetles on the under side of the leaves, and the grubs on hatching bore into the leaf, which they proceed to mine. The full-fed grub— the growth is completed in 7 or 8 days —falls to the ground, and in the soil the pupation stage is passed. There are several broods in the year.

Treatment. — The beetles should be jarred off the plants on to tarred boards or sacking (see Turnip-fly, p. 463).

Mr Fred. V. Theobald[1] has described the hop cone-flea (*Phyllotreta attenuata*). It is about the same size as the last, and is deep shiny green, with the wing-covers slightly hairy at the apex. Mr Theobald finds that its damage is done not in spring but late in the year. The harm in the late part of the year lies in the attack on the cones, the beetles "riddling the bracts until the crop is spoiled." The eggs may also be laid in the cones, the grubs mining in the bracts.

Spiders.

Hop Red - Spider (*Tetranychus*). — Red-spiders are not insects, but belong to the Mite section of the class Arachnida or Spinners. They can be told from insects by the absence of the marked division into head, thorax, and abdomen, and by the adults having 8 legs and not 6 (fig. 650). The red-spiders spin webs on the under surfaces of the leaves; here eggs are laid, from which hatch small 6-legged forms, which feed and moult, attaining with their last moult a fourth pair of legs. All stages of the mite from egg to adult may be found in the webs. In hot dry weather multiplication is very rapid, and the web-covered leaves, deprived of their sap, become sickly and yellow, and wither. Hibernation of the mites takes place in cracks in the hop-poles and in other shelter places.

Fig. 650.—Red-Spider (*Tetranychus telarius*, Linn.)

Treatment. — The pests are difficult to fight when their numbers are great, and when their webs have been formed in mass. Early treatment then should be practised, this consisting of spraying with paraffin emulsion (see p. 444), to which it is advisable to add 2½ lb. of liver of sulphur (sulphide of potassium) for every 100 gallons of the wash.

[1] *Notes on Economic Zoology*, p. 14. By Fred. V. Theobald, M.A.

The Onion-fly (Phorbia cepetorum).

This fly (fig. 651) measures ¼ to ⅓ of an inch. The colour is grey, the males being distinguished by having a dark line down their back; the compound eyes of the male, too, are closer together than they are in the female, a sexual distinction very common in the Diptera. The wings are iridescent, and have brown veins. The maggots, which are the direct cause of the damage, are legless, smooth, and fleshy, with a pointed head, provided with two rasping hooks, and a blunt posterior end, in the centre of which are two brown spiracles, while round the margin of this hind segment are a number of little projections. The puparium is oval and red-brown.

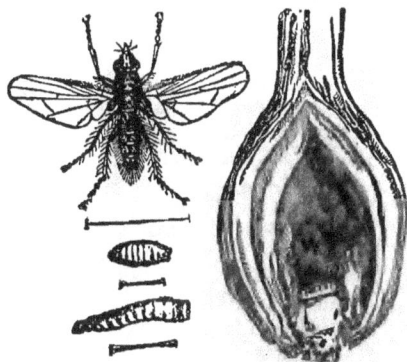

Fig. 651.—Onion-fly (*Phorbia cepetorum*, Meade).

The signs of infestation are a yellowing and drooping of the leaves, slimy decaying misshapen bulbs in which the maggots will be found at work, or the red-brown puparia. In April and May the flies proceed to lay their oval white eggs close to the leaves at the surface of the ground, or on the neck of the plant—6 to 8 eggs on each plant. The maggots on hatching descend between the leaves into the bulb. Here the maggot feeds for a fortnight before pupating, and in 10 to 20 days from the puparia being formed the new brood of flies may come away. There are several broods in the year, the first flies of the year being seen in April, and the last ones as late as November in an open season.

Treatment.—As a deterrent to egg-laying, the young plants should be sprayed with paraffin emulsion (see p. 444), this spray to be repeated three times, at intervals. Another protective measure is to sprinkle sand mixed with paraffin-oil—a cupful of paraffin-oil to a pailful of sand—at the base of the onion plants. Earthing up the onions to the neck and broadcasting soot over them is a good measure. All infested plants should be carefully removed and destroyed before the maggots have had time to develop themselves: in this removal pieces of attacked bulb should not be left behind.

A stimulating dressing of nitrate of soda, 1½ to 2 cwt. per acre, on infested land may help the plant in resisting the attack. A plan practised in the United States and quoted by Smith is, "Keep a close look-out for the first sign of the maggot. Turn away the earth from the rows with a hand-plough so as to expose the root-system in part, then apply broadcast 600 lb. kainit and 200 lb. nitrate of soda per acre; turn back the earth to the plants, and this will stop the injury. The application is best made just before or during a rain, or immediately after a good shower. The object is to get the salty fertilisers dissolved rapidly and brought into direct contact with the roots of the plant, and, of course, with the insects as well. This plan has proved entirely satisfactory on light lands, but it has not been tested on heavy land."

Onions should not be taken on the same land in the year following an attack.

Parsnip-fly (*Acidia heraclei*). (See p. 449.)

Peas.

The Pea Beetle (*Bruchus pisi*). (See p. 445.)

Pea and Bean Weevils.—The pea and bean weevils (*Sitones lineatus* and *Sitones crinitus*) (fig. 652) are harmful—the first especially—in the adult and in the larval stages, the adults eating pieces out of the leaves of young plants and the grubs feeding on the roots. The attack is known by the leaves being scooped out at the edge. The beetles begin their ravages at the outside of the leaves, and often eat all except the central rib. The striped pea weevil (*Sitones lineatus*) is of an ochreous or light clay colour, the antennæ and legs are reddish; it measures up to $\frac{1}{4}$ inch long. The spotted pea weevil (*Sitones crinitus*) is rather smaller and more of a grey colour; the wing-cases have short bristly hairs down the furrows, and are spotted with black.

The grubs have been found, by the observations made in the last few years, to feed at the roots of peas and clover and other leguminous plants, and may be found in large numbers at clover roots during the winter. The weevils until lately were supposed to feed by day, and shelter themselves in the ground under clots or rubbish at night, but more recently they have been observed to be night-feeders also.

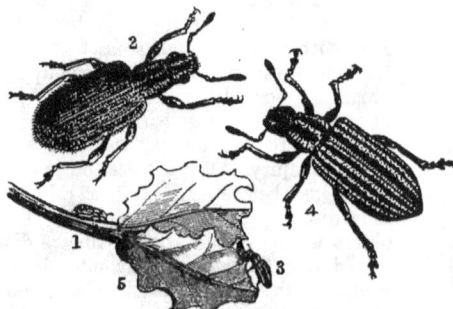

Fig. 652.—Pea and Bean Weevils (*Sitones lineatus*, Linn.; *Sitones crinitus*, Olivier).

1 and 2, *S. crinitus*, natural size and magnified; 3 and 4, *S. lineatus*, natural size and magnified; 5, Leaf notched by weevils.

The weevils come from their winter quarters in spring, and in May lay their eggs on and under the soil beside their food-plants. The grubs are white and fleshy, with brown heads and gnawing jaws. A second brood of beetles may develop from these by July. Hibernation may be in the adult in the larval stage. In a communication received from Yorkshire last October, we were told that "beans standing in the field in stooks are covered with countless thousands of Sitones, which cause a rattling sound, as one walks through the lines of stooks, by their falling and feigning death. They are eating through the bean pods and attacking the beans themselves, but only those with green pods apparently. The adjoining field is young clover (oat stubble), and this

the Sitones are now attacking—hardly a clover leaf being free from attack, and many Sitones clearly visible at their work." When the weevils fall to the ground on the plants being touched, and lie motionless, feigning death, they are very difficult to see.

Treatment.—As pea crops suffer most from weevil attacks in the early stages of their growth, it is most important that the soil should be well pulverised, and an available supply of manure given to push on the growth of the plant. Dressings of lime and soot (applied when the peas are wet) are good. Starlings and insectivorous birds are very fond of these weevils.

Care should be taken to have the weevils swept out from the bottom of waggons and carts when the crop is being carted home, and also to remove them from the platforms of the threshers. The collected beetles can be destroyed by dropping them into paraffin or into boiling water.

Pea Moth (*Grapholitha pisana*). —The caterpillar of this little moth hatches from the egg laid in the pea blossom; it eats through the young pod and into the seed. The caterpillar has 16 legs, measures about $\frac{1}{2}$ an inch long, and is pale-green in colour, the head and joint behind the head being brown; tubercles and hairs show here and there. This larva, on being full fed, leaves the pea and enters the soil where a cocoon is made, inside which the caterpillar, after sheltering for the winter, passes on to pupation stage.

Treatment.—The peas should not be left, without being harvested, long enough for the caterpillars to complete their growth. If the surface-soil of an infested area be buried deeply the moths from cocoons so buried will not appear in the next season.

Pea Midge (*Diplosis pisi*). — The maggots of this midge, present sometimes in large numbers in the pea-pods, cause malformation of the pod, and may spoil the crop. A number of attacked pea plants reached us in the beginning of August 1906, and in the accompanying letter it was stated, "As far as outward appearances go I never saw a more robust lot of plants; but I noticed that the blooms did not open properly, and some not at all, and now my crop is being ruined." Many maggots—whose springing habit was noticeable—and pupæ were found in the material. Deep ploughing was recommended, and the clearing away of attacked plants and burning them after the collection of the pods. Mr Fred. V. Theobald recorded the fly in his *Report on Economic Entomology*, 1906, and described it in his report for 1907. The larvæ measure 3 mm., and are yellowish in colour, with a marked anchor process: when full grown they reach the soil by the opening of the pods.

Pea Thrips (*Thrips pisivora*).—This insect attacks the blossom of pea and scarlet-runner beans, spoiling the flowers by their feeding, so that the pods are misshapen or may fail. Mr Theobald, in his *Report on Economic Zoology* for 1906, pp. 84 and 85, describes an attack on the pods of the pea thus: "The adults laid their eggs on the growing pods. The little yellow thrips collected in groups on the outside of the pods, and in one case ruined several rows of garden peas. The damage was very noticeable, even in the beginning of the attack. The tenderer pods were soon contracted and deformed, slightly older ones ceased to grow and remained thin, and no peas were produced. At first the skin around the attacked areas became of a silvery hue, then brown rusty patches appeared, and by degrees the pods, even those nearly mature and ready to pick, dropped off."

The female thrips, which has the characteristically fringed wings (see p. 456), is very small, measuring $1/16$ to $1/12$ inch, and in colour is black-brown, with pale bands on the abdomen; the larvæ from the eggs are orange-yellow in colour.

Treatment.—In the case just quoted from Mr Theobald's report successful results were got from two different sprays—viz., 1 ounce pyrethrum powder and 1 ounce soft-soap, in 2 gallons of water; or boil 6 ounces of coarse tobacco in 2 gallons of water with 2 ounces of soft-soap.

To reduce attack in the next year, it is recommended that the old sticks used for the peas and beans and the haulms which are used as places of hibernation by the adult strips should be burned.

Potatoes.

The Colorado Beetle (*Doryphoru decemlineata*).—This beetle (fig. 653), which had done much damage in the United States to potatoes, was found at Liverpool in 1877. The possibility of its becoming acclimatised in this country led, in view of its bad record in the United States, to the Colorado Beetle Order, 1877, legislation extended by the passing of "The Destructive Insects and Pests Order, 1908," an order which gives greater powers to the Board of Agriculture. Its native food-plant was a wild Solanaceous plant, but about 1850 it was found feeding on the potato-

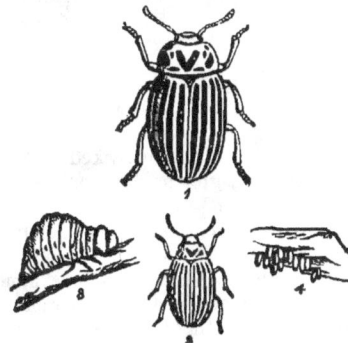

Fig. 653.—Colorado Beetle (*Doryphoru decemlincata*, Say).

patches of early settlers in the West United States. Gradually it spread east, until by 1874 it had reached the Atlantic coast. It has since been recorded from the continent of Europe; and, as mentioned above, it appeared in England in 1877; it appeared again in 1901 and 1902 at Tilbury Docks. Exterminating measures applied rigorously at this time were successful.

The Colorado beetle measures about ½ inch in length; it is oval in shape, robust in build, and yellowish or cream coloured, with five longitudinal black lines down each wing-cover. The oval orange-coloured eggs are conspicuous when seen fixed to the leaves of the potato. The larva is a six-legged grub, red-brown in colour, with the head and the legs black, and two rows of black spots on each side : the body is arched,

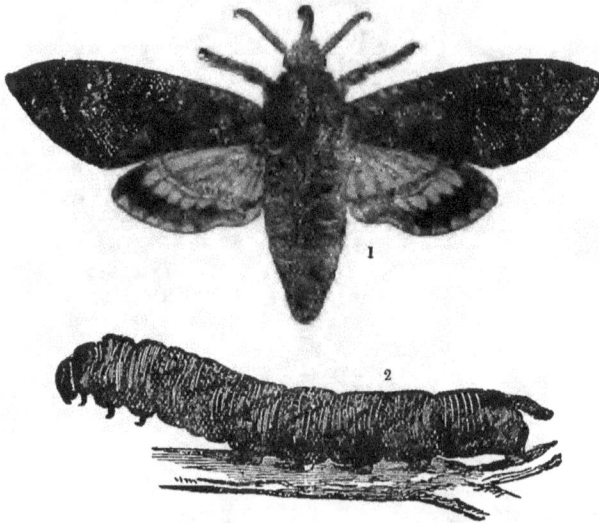

Fig. 654.—Death's-head Moth (*Acherontia atropos*).
1, Moth; 2, Caterpillar.

giving the grub a humpbacked appearance. The full-grown larva enters the soil for pupation.

In the United States the Colorado beetle was greatly feared as long as the only practised treatment was hand-picking, but since the introduction of arsenical sprays the insect is not considered to have its old importance, control being so much easier.

The Death's-head Moth (*Acherontia atropos*). — This is the largest of the British moths (fig. 654). Its caterpillar is sometimes found feeding on potato-leaves. It usually feeds by night ; and when it is noticed as doing great damage, it is advisable to resort to hand-picking in the twilight or by moonlight.

Turnips.

The Turnip Mud-beetle (*Helophorus rugosus*).—Although this beetle (fig. 655) has a wide distribution in Britain, yet complaints of damage by it are practically limited to Aberdeenshire. The leaves may be eaten ; the leaf-stalks may be tunnelled ; the root tubers may be gnawed and tunnelled on the outer surface. The curled-up leaves draw attention to the attack. The crown of

the tuber is a favourite place for the pests, which, sheltered in the leaf bases, destroy the young leaves. Both adult beetle and grub are destructive, and the worst damage is done when the plants are small. The beetle measures ¼ inch in length ; it is dark-red in colour, but the redness is often obscured by mud ; the thorax is ridged and knobbed ; the wing-covers have longitudinal ridges, between which are rows of punctures. The grub is six-legged ; the head is dark coloured and the jaws brown ; a transverse curved line marks the upper surface of the three thoracic segments ; down the back there are two rows of

Fig. 655.—The Turnip Mud-beetle.
1, Beetle, magnified; 2, Larva, magnified;
3, Turnip showing gnawings of grub.

square spots and smaller spots at the sides ; the body ends in two processes.

The complete life-history is not known, and so far treatment has been confined to helping the plants by stimulating dressings.

The Turnip and Cabbage Gall-weevil (*Ceuthorhynchus sulcicollis*).—This weevil is ⅛ inch long, and is dull-black in colour, with greyish scales which show in fresh specimens; a proboscis or beak is present from which the bent antennæ spring. The eggs may be laid on or in the root of turnip or cabbage. The grubs are yellowish-white, with brown heads and gnawing jaws; they are legless, and their body is wrinkled. From their irritating presence galls form on the roots, and inside these the feeding grubs are found. Pupation takes place in the soil under cover of a small earthen cocoon. Such cruciferous weeds as charlock act as breeding and feeding places for the weevil early in the year, and later on the turnip is attacked. Infested plants should be removed early. The cabbage stocks and removed plants should be burned, as, if merely left to die or thrown into a heap, the insects may complete their development; badly infested turnips could be fed off by stock, and then deep ploughing should follow.

C. assimilis, a smaller dark-grey species, is harmful as adult in the flower-heads of various crucifers—*e.g.*, turnip, mustard, rape; while the grubs feed on, and destroy, the seeds in the pods. Harvested seed containing the larvæ or pupæ should be fumigated with bisulphide of carbon (see p. 445).

A third species, *C. contractus*, is a troublesome enemy to young turnip and mustard. This beetle is smaller than the two preceding species, measuring only up to ¹/₁₆ inch; it is shining black in colour, and has a slender proboscis; if examined with a lens the front part of its body is seen to be deeply grooved. Attack may be on the leaves above ground, or in the stem above ground, or on the seed-leaves just above ground, or the newly sprouted seedling may be attacked and destroyed before the plant reaches the surface. Mr R. B. Greig pointed out, in the *Journal of the Board of Agriculture* for April 1905, that the seed-leaves were attacked just as they emerged from the seed, and showed light-brown spots sometimes ascribed to frost or the turnip-fly. Mr Greig, along with Professor Trail, made a series of observations on this weevil, and they found that the insect is kept off if the seed be steeped for an hour or two in paraffin before sowing.

The Flea-beetles (*Halticæ*).—This family of Leaf-beetles is characterised by a leaping habit, hence the name flea-beetles; the femora of their hind-legs are much thickened: Mr Fred. V. Theobald has been at much pains to make clear that the most harmful forms belong to the genera *Phyllotreta, Haltica, Psylliodes*, and in his *Notes on Eco-*

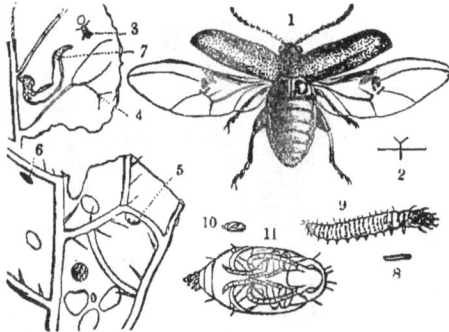

Fig. 656.—Turnip-fly (*Phyllotreta nemorum*, Chevrolat).
1-3, *H. nemorum*; 4 and 5, Eggs; 6-9, Maggot; 10 and 11, Pupa: all natural size and magnified.

nomic Zoology has given a full and extremely useful description of the various species of economic importance.

Turnip-fly or Flea-beetle (*Phyllotreta nemorum*).—This (fig. 656) is a small shining beetle, measuring from ¹/₁₂ to ⅛ of an inch. It is black, blue-black, or green-black, and has a yellow longitudinal stripe down each wing-cover. The adult beetles come out of their winter quarters in spring, and lay their eggs, and feed on charlock and such other wild crucifers as Jack-by-the-hedge and shepherd's purse. The grubs feed internally, mining in the soft tissue of the leaf. When full grown they drop to the ground, pass into the soil, and there pupate. The whole life-cycle, from egg to adult, may be completed in a month; hence there may be several generations in a year. The adult beetles especially are harmful from their destruction of the seed - leaves, but they also feed on the rough leaves.

Phyllotreta nemorum is one of a large number of related species—*e.g., P. consobrina* and *P. undulata* and *Haltica oleracea;* and in his exhaustive paper in the *Notes on Economic Zoology* mentioned above, Mr Theobald notes this most practical point, that in the genus *Phyllotreta* the larva feeds internally, and that, therefore, treatment has to be directed chiefly against the adult, whereas in the genus *Haltica* the eggs are laid on the outside of the food-plant, and the larvæ on hatching continue to feed externally, so that the *Haltica* species can be fought both in the larval and adult stages.

Treatment.—As the fly will appear before turnips are sown, it is most important to clear away charlock and other weeds suitable for its food.

One spring, a field in good tilth, ready for turnips, suddenly became a mass of charlock. This was entirely cleared away by the fly, which again appeared and preyed on the turnips when sown. Had the precaution of harrowing up the charlock as soon as it appeared been taken, the turnips would in all probability have been saved.

In coping with the turnip - fly the following important principles should be observed: 1st, cleaning the ground; 2nd, destroying rubbish round the fields which might serve as shelters to the flea-beetles; 3rd, so preparing the ground by good cultivation and plenty of manure that the growth of the turnips may be pushed on vigorously past the first leaves, in which stage they are most vulnerable to the attack of the fly: where it is possible, autumn cultivation is desirable, so that at turnip-sowing the upper surface will only require slight disturbance, and thus the moisture beneath, which is a great desideratum for the young turnips, will remain to aid the growth.

As a deterrent to the fly, the seeds should be steeped before sowing in paraffin or turpentine, and the young plants might be sprayed with paraffin put on by a Strawsoniser (fig. 622), by which one gallon of paraffin to the acre can be distributed.

Mr Fisher Hobbs's remedy for the fly was as follows:—

"1 bushel of white gas-ashes" (gas-lime) "fresh from the gas-house, 1 bushel of fresh lime from the kiln, 6 lb. of sulphur, and 10 lb. of soot, well mixed together and got to as fine a powder as possible, so that it may adhere to the young plant. The above is sufficient for two acres, when drilled at 27 inches. It should be applied very early in the morning *when the dew is on the leaf,* a broadcast machine being the most expeditious mode of distributing it; or it may be sprinkled with the hand carefully over the rows."

Boards tarred on the under surface, placed on wheels and run through the fields, will trap many of the beetles, as these in leaping stick to the tar.

In order to poison the larvæ and the feeding beetles, the plants should be sprayed with an arsenical spray put on with a Strawsoniser.

The Diamond-back Moth (*Plutella maculipennis*).—The caterpillars of this moth (fig. 657) are harmful to cruciferous crops generally. The insect, while

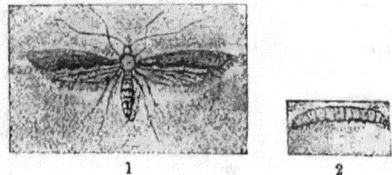

Fig. 657.—Diamond-back Moth.
1, The moth, magnified; 2, The caterpillar of the moth.

not markedly prevalent every year, now and again, when circumstances favour it, works great havoc. In favourable conditions for the moth there are two and more broods in the year. The front wings of the moth are grey-brown with dark markings, with a whitish waved stripe along the hind margin of each: when the moth is at rest and the wings lying on the back the two hind edges of the front wings meet, and the wavy stripes give the appearance of diamond-shaped areas, hence the common name of the moth. The hind wings are pale ash - grey, and are fringed. When the moth is at rest the wings are somewhat tilted up at the ends. From the yellow-white head the antennæ project straight forwards. The caterpillar is faint-green in colour, with a darker head; it has 16 legs, measures about ½ an inch when full grown, and tapers to both ends. A light cocoon is spun on the leaf by the full-fed caterpillar.

Treatment.—Apart from the destruction of crucifer weeds, so that the food-plants on which the earliest moths of the year can lay are lessened, the remedial measures are directed against the caterpillars.

To dislodge the caterpillars drag furze or other branches across the turnips, taking care that the under sides of the leaves are reached, then follow with the scuffler so as to bury the fallen caterpillars. Dress the plants with a mixture of soot and lime in the proportion of 1 part of lime to 3 of soot. Spray with paraffin emulsion (p. 444).

The caterpillars are poisoned if the leafage be sprayed with arsenate of lead. This can now be bought in the paste form as Swift's Arsenate of Lead from Strawson's, Queen Victoria Street, London. The proportions recommended are 2 lb. thoroughly mixed up in 50 gallons of water.

Surface Caterpillars.

These are also known as "cut-worms," from their biting the stems through. The name is applied to such caterpillars as hide below the surface of the soil and attack plants, generally at night, at or just below the surface. Three very troublesome surface caterpillars are those

HEART-AND-DART MOTH.

About 1½ inch long, brownish in colour; spiracles black, and larger than the spots before and behind them.
A pear-shaped blotch on the first 5 segments.

The caterpillars of the above three moths may pass the winter in the larval stage, renewing their feeding in the next spring.

The pupæ are red-brown.

Treatment.—If the attack be on a small scale the caterpillars may be hand-picked; the worker provided with a lantern makes a round of the plants, and turns up the cut-worms by means of a blunt knife or a pointed piece of wood. Disturbance and destruction of the

of the turnip moth (*Agrotis segetum*) (fig. 658), the heart-and-dart moth (*Agrotis exclamationis*), and the yellow underwing moth (*Tryphæna pronuba*). Turnips, swedes, potatoes, mangolds, carrot, lettuce, wheat, grass, and many other field and garden plants are injured by these caterpillars.

The yellow underwing caterpillars are 1¾ inch long when full grown; the

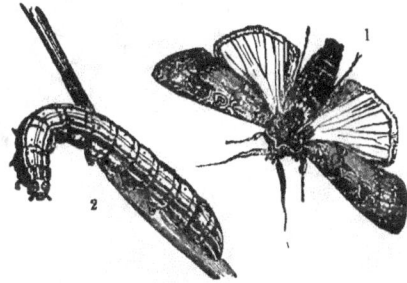

Fig. 658.—Turnip Moth (*Agrotis segetum*, Westwood).
1, Moth; 2, Caterpillar.

head is ochreous, and has two black stripes; there is considerable variation in shade of colour, greenish-brown being the prevailing tint, with a line down the back, and black spots or streaks along each side; the colour below is pale green.

On being disturbed the caterpillars coil themselves into a ring.

The caterpillar of the turnip moth and of the heart-and-dart moth may be contrasted thus:—

TURNIP MOTH.

When full grown about 1½ inch long, greyish in colour.

No pear-shaped blotches.

caterpillars by horse and hand hoes; caterpillars or pupæ are killed or brought up for the birds.

In America there are records of successful treatment by poisoning the caterpillars. This may be done by spraying clover or grass with Paris-green—1 lb. of Paris-green to 50 gallons of water, the mixture to be kept well stirred,—afterwards tying these sprayed plants into bundles and distributing them at intervals through the crop.

The Turnip Sawfly (Athalia spinarum).

The damage done by the larvæ of this sawfly (fig. 659) may be very great, partly because of their numbers and their vorac-

Fig. 659.—Turnip Sawfly grub at work.

ity and the rapid succession of broods. The caterpillars are known as "blacks," "black palmers," "niggers," &c. They will sometimes clear off the leafage of a whole field. The figure will give an idea of their destructiveness.

One sawfly (fig. 660) will lay from 200 to 300 eggs, and these will hatch in five days, or less in warm weather. The larva is a caterpillar with 22 legs; it is greenish-white at first, but at last it is black; full grown it measures ¾ inch. When full grown, in about three weeks the larvæ go down to the earth and spin a silken cocoon; from these cocoons the sawflies emerge in about three weeks. It should be remembered that, when they are moulting (that is, every six or seven days during the time they continue in the grub form), if they are disturbed they die; for if they lose hold with the pair of feet at the tip of the tail during the operation, they cannot fix themselves so as to pull themselves out of the old skin, and therefore they perish in it.

Prevention.—In order to effect dislodgment of the caterpillars a very light bush harrow should be drawn over the turnip leaves. Spray the plants with arsenate of lead (see p. 465). A field that has been badly infested should be deeply ploughed to bury cocoons.

Spring-tails or Collembola.

These small wingless insects live in concealed places, under stones, bark, moss, and in damp soil. They feed on decaying organic matter; but evidence is accumulating to show that they are capable of doing damage, first hand, bean seeds and seedling turnips and cabbages being mentioned in the literature as suffering. Their respiration is

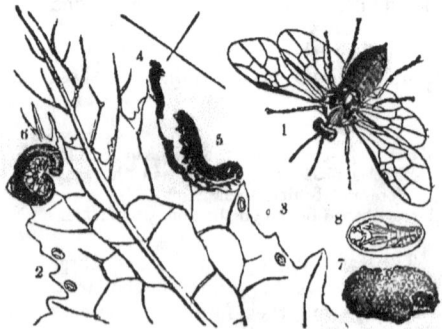

Fig. 660.—Turnip Sawfly (*Athalia spinarum*, Fabricius). Caterpillars, pupa, and pupa-case; Sawfly, magnified, with lines showing natural size.

typically through the skin, tracheæ being wanting, and therefore moisture is a necessity. Dressings of lime or of soot and lime are effective against them.

FRUIT.

In view of the development which has taken place in fruit-growing, some notes are added on the insects which attack raspberry, strawberry, gooseberry, and black currant.

Raspberries.

Raspberry Weevils.—The clay-coloured weevil (*Otiorrhynchus picipes*) and the red-legged weevil (*O. tenebricosus*) are destructive not only to raspberry but also to a number of other fruit plants.

Harm is done both by adult weevil and grub, the grubs living in the soil and destroying the roots, the weevils feeding on above-ground parts. The clay-coloured weevil measures ¼ inch in length, and the red-legged weevil up to ½ an inch. These weevils feed at night and shelter in the day-time; they cannot fly. The grubs are yellow-white in colour, and have brown heads and biting jaws.

Treatment.—Shake the beetles, at night, down off the plants on to sheets or tarred boards. They may also be trapped by twining bands of hay among the plants, for the weevils will use these to shelter in during the day-time.

Raspberry Beetle (*Byturus tomentosus*).—This is a small but very harmful beetle, measuring ⅙ inch in length; it is brown in colour, and, examined with a lens, is seen to be covered with a greyish-yellow pubescence. The full-grown grubs reach about ¼ inch in length; they have biting jaws, 6 legs on the thorax, and 2 processes at the hind end, with a smaller projection between them. The beetles appear about the middle of May, and lay their eggs in the blossom; the grubs on hatching pass into the fruit. When full fed, towards the end of the season, the grubs leave the fruit and pupate under cover of a cocoon in the soil below the plants or in cracks in the bark.

Treatment.—Shake the beetles off the plants on to tarred boards or into a vessel containing paraffin. Burn prunings and old canes likely to be used as places for the pupation stage. Bury the surface-soil with the cocoons.

Raspberry Moth (*Lampronia rubiella*).—The caterpillars of this moth are very harmful to the raspberry canes. The life of the caterpillar extends into two seasons; in the first year the caterpillars, which come from eggs laid in the blossom, do not complete their growth but leave the plants and enter the soil for wintering; in the next spring the caterpillars climb up the plants, bore into the buds, and tunnel the shoot below the buds; in this hollowed shoot they pupate. The full-grown caterpillar measures ¼ of an inch; it has 16 legs, and is red in colour, with the head black, and a black patch on the joint behind the head.

Treatment.—Cut off and burn infested canes while the caterpillars or pupæ are present; this can be done from the end of April till the middle of May. The over-wintering caterpillars should be disturbed in their shelter-places by the surface-soil being buried. If the bases of the canes be coated in March with a sticky composition, the caterpillars in coming from their hibernating quarters will be prevented from ascending the canes.

Strawberries.

The Black Vine Weevil (*Otiorrhynchus sulcatus*).—This is a close ally of the two weevils mentioned under raspberry, and is an enemy of the strawberry. The adult weevils gnaw the runners and the shoots, and their grubs attack the roots and crown of the plant. The beetles are found in summer when egg-laying takes place; the grub stage lasts over the winter till the next spring, when pupation takes place.

For treatment reference should be made to the raspberry weevils.

Ground-Beetles (*Carabidæ*).—The ground-beetles as a family are useful, being predaceous both as adult and larva; but there are several species which as adult have often proved most destructive to the strawberry fruit.

The beetles are: *Pterostichus* (*Omaseus*) *vulgaris*, a shining black beetle, with two longitudinal depressions on the thorax and with striated elytra; there are no flying wings; it measures 6½ to 7½ lines in length; *Pterostichus* (*Steropus*) *madidus*, which measures up to 9 lines long; it is black, has no flying wings, and shows three deep punctures on the elytra; *Harpalus ruficornis*, common in gardens, which is 6 to 7 lines long, deep black in colour, with the legs and antennæ red; the wing-cases are striated, and are covered with a greyish-yellow pubescence, the antennæ also have a golden pubescence; flying wings are present; and *Calathus cisteloides*, which measures 3 to 6 lines in length; it is black, with antennæ and legs red-brown; there are no functional flying wings.

These four beetles shelter in the daytime in the soil about the plants or among the straw of the beds; they come out to feed at night on the fruits.

Treatment. — Go over the beds, removing the straw, section after section; turn over the exposed soil with a trowel and collect the beetles. Miss Ormerod in one of her reports quotes the treatment adopted on a large scale by the Laxton Brothers of Bedford: " Let into the soil level with the surface a number of cheap pudding-basins, at distances of a few yards apart, keep these baited with pieces of lights and sugar-water. When the weather was dry we often caught half a basinful of a night until the number gradually diminished. The process is laborious, but well worth the trouble, as we have lost no fruit this season."

Gooseberries and Currants.

The **Magpie Moth** (*Abraxas grossulariata*).—The caterpillars of this moth are very harmful to the leafage of gooseberry and currant. The caterpillar is easy to recognise; it is a looper with 10 legs, has a black head and a yellowish-white body with black spots. The moth flies in late summer and the eggs are laid then; the caterpillars do not complete their growth in the autumn, but hibernate in some shelter-place. Warning is therefore given by the appearance of the caterpillars in autumn, when they don't do much harm, that infestation may be expected in the next spring.

Preventive Measures. — Therefore, when the pests have been noticed in autumn the ground below the plants should be covered with quicklime and dug deeply in winter. In pruning the bushes care should be taken to cut away and destroy any caterpillars that should have found shelter-places on the bushes.

Against the feeding caterpillars in spring and summer nothing is better than arsenical washes—viz., arsenate of lead —bought as Swift's arsenate of lead in the paste form—2 lb. to 50 gallons of water; or Paris-green, 1 lb. to 250 gallons of water.

Gooseberry and Currant Sawfly.— The caterpillars of the gooseberry and currant sawfly (*Nematus ribesii*) are sometimes confused with those of the magpie moth, but they are not the least like the moth caterpillars in colour, and they have 20 legs. There are more broods than one of this pest in the year.

Treatment.—The treatment is to bury surface-soil, below bushes that have been infested, in winter along with the cocoons; and against the caterpillars to spray with 1 ounce of hellebore to 3 gallons of water. The hellebore, however, should not be used within six weeks of collecting the fruit.

The Black Currant Gall Mite (*Eriophyes ribis*).—This pest of currants feeds in numbers in the buds and gives rise to what is known as "big bud." Buds that contain large numbers of mites never open but wither away. The mites pass the winter in the buds. Migration of the mites, from dried-up buds that cannot provide food, and from buds that have opened somewhat and so deprived the mites of shelter, takes place from April onwards. The migrating mites seek new buds, and having entered these proceed to lay their eggs. The eggs hatch in due course, and by the end of August "big buds" are found swollen with the new generation. The most favourable time for spray or dusting treatment is while the mites are engaged in moving from the old buds to the new ones. The dusting during this migration period with lime and sulphur—1 part of unslaked lime to 2 parts of flowers of sulphur—is well worthy of trial. The dusting should be repeated three times at intervals of a fortnight. The grower should cultivate from clean stock only.

Aphides. — Currant and gooseberry plants are often badly infested with aphides. If attack has been severe the bushes should be pruned hard in autumn, and the prunings, on which are the eggs, burned. In February the plants should be sprayed with emulsion - soda wash (see page 457 and Board of Agriculture Leaflet 70), and then careful outlook should be kept for any signs of aphides in later spring, these being sprayed, if seen, with dilute paraffin emulsion before the leaves have time to curl over and so provide shelter for the aphides from the spray.

USEFUL INSECTS.

Certain insects deserve protection from the cultivator, either because as predaceous they prey on other harmful insects, or as parasites they destroy them.

Useful Beetles.—The Tiger beetles, both as adult and as larva, feed upon other insects, the adult chasing them and killing in the open, the larva trapping them in its burrow previous to tearing them to pieces.

Carabidæ or Ground - Beetles.— Although it contains a few harmful species, this family decidedly is useful. The adult beetles are good runners, and prey upon insects and insect larvæ; the Carabus grub is also carnivorous; such grubs are known by their elongated body, 6 legs fitted for running, biting jaws, and two projecting bristle - like processes at their hind end, with a tube between them.

One of the ground beetles (Zabrus gibbus) is harmful to corn, while four of them are pests in strawberry-beds—viz., Harpalus ruficornis, Pterostichus vulgaris, Pterostichus madidus, and Calathus cisteloides.

Rove Beetles.—The Rove Beetles— family Staphylinidæ — are also carnivorous. They can in the adult stage be easily recognised by their very short wing-covers, which do not nearly reach the tip of the abdomen. One of the family is the devil's coach-horse (Ocypus olens), whose habit of erecting its tail-end is well known.

The Lampyridæ or Glow-worms, and the Telephoridæ or soldiers and sailors —so called from their red and blue colours—are predaceous beetles.

Ladybirds.—Very useful beetles are the ladybirds of the family of Coccinellidæ. These small rounded beetles vary in colour,—red with black spots, black with red spots, yellow with black spots. The larvæ are elongated grubs, broadest towards the front end and tapering towards the hind end; they have 6 legs and biting jaws; they come from eggs laid on the leaves of plants, and when full grown they attach themselves by the hind end, gluing themselves by it to leaf or other object, becoming pupæ, which pass on to the adult stage in a fortnight or so in favourable weather. The adult beetles are predaceous, and their active larvæ feed greedily on aphides and scale insects.

Useful Flies.

The Asilidæ or Robber-flies are predaceous, the adult chasing on the wing other insects, which are impaled on the rostrum of the robber-fly. Dr Sharp, in mentioning their insatiable appetite, records how a single individual killed eight moths in twenty minutes.

The Syrphidæ or Hover - flies.—This is a large family, of which the genus Syrphus is made up of black or green flies banded, like wasps, with yellow or orange. The larvæ are legless, their body soft, tapering in front and thicker behind; the mouth is circular, and has two mouth-hooks. These hover-flies lay their eggs amidst a colony of aphides, and the maggots on hatching feed on the aphides, catching them up by their mouth-hooks and sucking them dry.

The family Tachinidæ consists of robust, stout, bristly flies. Their larvæ are parasitic on the caterpillars of moths and butterflies. The flies lay their eggs on the caterpillar near the hind end, and the maggots, on hatching, bore into the caterpillar, feeding at first on store of reserve and muscle; later they eat other parts. When full grown they pupate in barrel-shaped puparia inside the spoilt host, or sometimes in the soil.

Useful Hymenoptera.

Apart from the hive or honey bee, and useful for a different reason, we have three families—the Chalcididæ, the Ichneumonidæ, and the Braconidæ. The larvæ of the species of these three families are almost without exception parasitic. Eggs may be parasitised and pupæ, but most commonly larvæ. The range of insect host parasitised is a very wide one, scarcely any insect Order escaping.

The Neuroptera or Nerve-winged Insects.

The members of this Order may none of them be described as harmful. The ant lions are insectivorous, and so also are the larvæ of the lace-wing fly (*Chrysopa*). The female lace-wing fly in her egg-laying touches a leaf with the tip of her abdo-

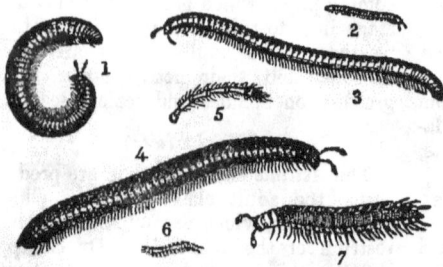

Fig. 661.—False Wireworms.

Snake millepedes: 1, *Julus Londinensis;* 2 and 3, *J. guttatus* natural size and magnified; 4, *J. terrestris;* 5, Horn; 6 and 7, Flattened millepede, *Polydesmus complanatus*, natural size and magnified.

men, and elevates the abdomen, giving out at the same time a delicate thread, at the end of which the egg is found. The eggs thus stand out of the reach of other insects. The larva is 6-legged, and has sickle-shaped mandibles; aphides form its chief food.

Centipedes and Millepedes.

The *Myriapoda* form a class of jointed-footed animals that have a relation to insects. They have a head with a single pair of antennæ, followed by a number of resembling joints, each carrying legs. There are two sub-classes, the Centipedes and the Millepedes. The centipedes have a flattened body and a single pair of legs to each joint. They are useful, being carnivorous, preying upon slugs and insect larvæ.

The millepedes have a rounded body and two pairs of legs to every seeming joint. They are vegetarian in diet, and are often troublesome in the fields to crop plants with roots and underground stems.

A very common millepede is *Julus guttatus*, which has a thin body, pale-pink in colour, with rows of purple spots; it measures about ½ an inch. *Julus terrestris* is another; it is black, with a pointed tail. Some millepedes

are flat—*e.g.*, *Polydesmus complanatus* of the figure.

Millepedes (fig. 661) are often distributed in leaf-mould, and this should therefore be examined before being used. They can also be trapped by placing pieces of hollowed-out mangold or potato beside the food-plants, and slightly covering these with soil. Such traps must be regularly visited. Pieces of mangold and potato soaked in Paris-green poison the millepedes. Another suggested measure is to dig here and there small holes, like the hole on a putting-green, but not so deep; fill with bran and cover with a flower-pot. After some days pour boiling water on to the bran, to kill the millepedes that have been attracted.

Eelworms.

The eelworms belong to the great class of Round Worms and the family *Anguillulidæ*. One of the characteristics of the *Tylenchus* genus is the presence in the gullet of a sharp process, by which plant tissue can be bored and the sap extracted.

Ear-cockle of Wheat. — This is a disease of the ears or flowers in which

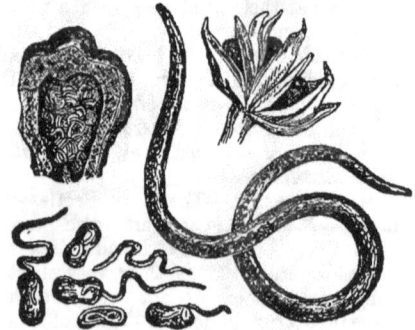

Fig. 662.—Ear Cockles (*Tylenchus scandens*). Wheat cockle-gall; Eelworm.

the grains are replaced by dark-coloured structures. The galls characteristic of the infestation are the result of attack by the minute worm *Tylenchus scandens* (fig. 662). The developing parts of the flower are attacked while in rudimentary condition, and the increased growth which results in the worms being enclosed in the swelling or gall. If such a gall be broken through, eelworms in all stages, egg, young, larger worms, full-grown worms, will be found. A symptom of attack is the puckering of the leaves at their edges.

Treatment.—(1) Do not sow galls along with the grain, as the worms will issue and make their way into the young growing plants, and then make their way to the flower. (2) If the suspected grain be placed in water, the galls rise to the surface and can be removed. (3) Bos recommends steeping infested grain in weak sulphuric acid, 1 pint of sulphuric acid to 33 gallons of water: this kills the worms.

Tulip Root of Oat or Segging (*Tylenchus devastatrix*).—This species of eelworm is the cause of failure in oats, and a contributing cause of clover sickness. In addition, many other plants are attacked—*e.g.*, wheat, onion, lucerne, hop, &c. The roots of infested plants get twisted and swollen (fig. 663), and on their being teased out with a needle, in a drop of water, a more or less spongy appearance is found under the microscope, with decayed matter and the pests. The worms measure up to $1/24$ inch, and are like miniature eels, with both ends of the body pointed. Eggs and young worms getting into the soil are able to endure desiccation for a long time.

Treatment.—(1) Lengthen out the rotation between crops that are infested —*e.g.*, if the land be clover-sick, clover must not be taken once every four years. (2) As the eelworms leave the plants and get into the soil, plough so deeply that they will be buried and unable to return to the surface. (3) Dressings of various kinds have experimentally proved helpful, one of the most helpful being sulphate of potash.

In experiments at Kew[1] with clover plants it was shown that—(1) Eelworms can infect and kill otherwise healthy clover. (2) That infection can only be effected during the seedling or quite young stage of clover. (3) In the case of infected land, if the eelworms are buried to a depth of 5 inches in the ground, no infection can take place. Hence the advantage of deep ploughing.

[1] *Journal of Board of Agriculture*, July 1907.

(4) If a diseased crop is treated with sulphate of potash at the rate of 4 cwt. per acre, the eelworms are destroyed.

Tomato Root-rot (*Heterodera radicicola*).—This eelworm causes knots or swellings on the roots of tomatoes and cucumber. Attack is followed by the stems becoming limp and the plant falling away. The swellings, if dissected, will reveal under the microscope the eelworms. From the eggs come tiny worms, shaped like eels. Later there is a swollen or cyst stage from which the male emerges, still with the shape of an eel, but the female, swollen with eggs, is pear-shaped. If the soil be saturated with 1 part of carbolic acid in 20 parts of water, the eelworms are killed ; such soil cannot be used for a month afterwards. The mixing of naphthalene with the soil as a fumigant has some success.

Fig. 663.—Tulip-rooted Oat plant infested by Eelworm (*Tylenchus devastatrix*).

Other plants are attacked by this same eelworm—*e.g.*, beet, carrot, clover, black medick,—and this increases the risk of spread.

LITERATURE.

The literature of Economic Entomology is an ever-increasing one. Never before has there been so much activity. The Board of Agriculture extends a general invitation to farmers to avail themselves of its aid as regards insects of the farm. The Royal Agricultural Society of England and the Highland and Agricultural Society of Scotland offer facilities to their members, while agricultural colleges and societies up and down the country are doing useful entomological work. The Reports published by these various societies, and the series of leaflets distributed free by the Board of Agriculture, form a yearly increasing part of the literature. The Annual Reports of Mr Fred. V. Theobald, Vice-Principal of Wye, of Mr Cecil Warburton of Cambridge, of Mr W. E. Collinge, late of Birmingham University, of Professor Carpenter of Dublin, and of Dr Stewart MacDougall, University of Edinburgh, should be consulted, as every year they contain something new. The old Annual Reports of the late Miss Ormerod and of Mr Charles

Whitehead, accessible only in libraries, will be found very useful. The two volumes on Insects by Dr Sharp, in the Cambridge Natural History Series, form a magnificent contribution to the literature of Entomology on its scientific side. The late Miss Ormerod's *Manual of Injurious Insects*, now out of print, is being rewritten, and the new edition will not be long delayed.

In the United States there is in the Annual Reports of the various States a mine of useful information on crop insects, and the literature of Canadian Entomology cannot be neglected. Finally, the farmer who may chance to settle abroad will find in the various British Colonies not named an Official Consulting Entomologist whose reports give the salient features in the Economic Entomology of the year.

SNAILS AND SLUGS.

The *Helicidæ* (Snails) and the *Limacidæ* (Slugs) are two families of the class *Gasteropoda*, a class belonging to the Phylum *Mollusca*. This Phylum of invertebrate animals numbers in it mussels, cockle, oyster, snail, slug, whelks, limpet, and the cuttlefishes. As is only too well known, snails and slugs do a good deal of harm to many garden and field crops.

Snails and slugs have a fairly well-developed head region, and a muscular process known as the foot by which the animal moves. They are further characterised by the possession of a rasping ribbon known as the radula; this radula has on it rows of horny teeth, and under the microscope resembles a file; this file is pulled backwards and forwards by the working of muscles, and by means of it particles of food are scraped off. On the roof of the mouth, and seen on separating the lips of the snail, is a horny, crescent-shaped jaw, which acts with the rasper, and so leaves and leaf-stalks are cut. Traces of rasper and jaw can be found on leaves, and if there be silence, the working of the mouth-parts of a grazing snail can be heard. Eyes are present, but the most delicate and accurate sense is the sense of smell.

Both snails and slugs are hermaphrodite. The eggs, which have a parchment-like shell, are laid on or in the soil, or under heaps of leaves, or under stones. There is no metamorphosis in the life-history, a tiny snail, resembling the adult externally, hatching from the egg.

Air is breathed directly, being taken in by a respiratory pore to a pulmonary chamber or lung.

In habit, snails and slugs are crepuscular or nocturnal, but in damp weather or after heavy rain they may be found out and about, feeding, in the daytime. In dry sunshiny weather shelter is sought in the soil or under stones.

SNAILS (*Helicidæ*).

These have a spiral shell into which the whole animal can be withdrawn. In habit they are almost exclusively vegetable feeders.

The Garden Snail (Helix aspersa, Müll.)

This snail measures up to $1\,^2/_5$ inch; the spiral shell has $4\frac{1}{2}$ turns, is brown in colour, and has spiral bands or lines crossed by transverse patches. White globe-shaped eggs are laid in batches in the soil; these hatch in about a month. The snails remain dormant in a place of shelter during the winter. The length of life may be five years. The species is harmful in gardens to vegetables and to the leaves and blossoms of fruit-trees.

The Wood Snail (Helix nemoralis, Müll.)

This snail measures an inch in length; the shell has $5\frac{1}{2}$ turns; the colour of the shell varies much; the lips of this snail are brown or black.

Helix hortensis, Müll.

Measures an inch; has spiral bands, but no transverse patches; the lips are white.

The Strawberry Snail (Helix or Hygromia rufescens, Penn.)

Measures up to $\frac{1}{2}$ an inch long; 6 to 7 turns to the shell; the shell is compressed above, is reddish-brown in colour, and often has a white band on the last turn; the snail itself has brown stripes on neck and tentacles, and the foot is narrow and pale. This snail is sometimes very troublesome in strawberry beds, and vegetables and garden flowers are also destroyed.

The Hairy Snail (Helix or Hygromia hispida, Linn.)

Measures ⅓ of an inch; six to seven turns to the shell; colour grey-brown. Short hairs are present, white on the epidermis and red on the first part of the shell to be formed. Found in gardens and in osier beds.

The Small Banded Snail (Helix or Helicella virgata, Da Costa).

Measures ½ an inch; colour white, with purple-brown stripes. Its typical habitat is dry downs with short turf, but besides its destruction of grass there are records of harm to wheat and root crops.

The Allied Banded Snail (Helix or Helicella caperata, Mont.)

Measures ⅓ of an inch or less; the shell has keels or ribs on the turns, and has coloured stripes. Found on dry downs, gardens near the sea, in cornfields and woods.

SLUGS (Limacidæ).

Except the genus Testacella, none have an external shell; most, however, have a rudimentary plate-like shell hidden below the skin of the back, or at least a few limy granules; the region in the slugs which is over the rudimentary shell is known as the shield. Some have a large slime pore at the end of the foot.

The Large Black Slug (Arion ater, Linn.)

The species of the genus Arion have the respiratory pore half-way along the shield; the shield is shagreened, the skin is wrinkled, and there is a slime gland at the hind end of the body.

In Arion ater the prevailing colour is black, but it may be brown, yellowish-green, &c.; the edges of the foot yellow-white; the shell is made up of loose limy granules, irregular in size. A large number of eggs is laid; one in-

dividual laid 577 from 13th October to 30th November; the eggs hatch in forty days or more, according to the temperature. The slug is full grown in a year and a half, and measures 4 inches in length. The large black slug is found in gardens, woods, and along hedge-sides. It is omnivorous in diet—vegetarian and carnivorous—and can be harmful both in field and garden.

The Garden Slug or Small Arion (Arion hortensis, Fer.)

The colour is very variable; the shield shows three stripes, and there are also grey stripes on the back and sides; the oval shell is granular. This slug is found both in field and garden, and may do much damage.

The Black Striped or Mottled Slug (Limax maximus, Linn.)

In the genus Limax the respiratory opening is at the hind end of the shield or in the posterior half; the shield is also concentrically wrinkled.

Limax maximus measures up to 6 inches and over in length; it is grey in colour, with black spots, but the colour is variable; the tentacles are purple; the body is tubercled; the foot is edged with white. In diet it is vegetarian and carnivorous.

The Grey Field Slug (Limax or Agrolimax agrestis, Linn.)

Measures when full grown 1½ inch; the colour is mottled ash-grey, but, as with other slugs, it varies much; the body is elongated and spindle-shaped. The shell is small and oval. A large number of eggs is laid from May onwards, these being laid in small batches up to fifteen in each. This is an exceedingly destructive species, probably the worst of all, levying toll on many plants, peas, beans, clover, cabbage, young turnips, wheat, strawberries, pansies, carnations, and many other garden plants.

The Household or Cellar Slug (Limax flavus, Linn.)

Measures up to 4 inches in length; colour variable, but typically is yellowish with brown-black spots; the head and tentacles are grey or blue-grey; the under surface of the foot white, and the edges of the foot yellow-white. It gives out abundant slime. This slug is found in the open, but it is more commonly found · indoors in cellars, sculleries, pantries, &c., where it feeds on bread, meal, flour, &c.

The Root- or Bulb-eating Slug (Milax Sowerbii, Fer.)

In the genus *Milax* the respiratory opening is on the right side in the posterior half of the shield ; the shield is shagreened ; there is also a well-marked keel along the back of the body.

Milax Sowerbii is yellow in colour, but may be grey or brown, or even on occasion black. It is a destructive garden pest to such strange organs as stem and root tubers and bulbs, and to many garden plants whose leaves are pulled into the ground ; it is also carnivorous, taking caterpillars and small slugs.

PREVENTIVE AND REMEDIAL MEASURES.

1. Snails and slugs have numerous animal enemies—*e.g.*, rats, voles, hedgehogs, and shrew-mice.

There are several bird enemies of these pests. The thrush is well known for its destruction of snails, while blackbirds and starlings and ducks greedily take slugs.

Among insect enemies are ground beetles and beetles of the family *Staphylinidæ*.

2. *Hand-picking* should be practised in gardens. In some parts of the Continent the people have a most effective way of keeping *Helix pomatia* in check —viz., by collecting and eating it. In some parts of the west of England the garden snail is said to be collected for use in chest complaints, and the same slug has been reported elsewhere as of use in preparing a substitute for cream, particularly in rearing calves.

3. *Trapping*—By means of lettuce or cabbage leaves (in the case of the strawberry snail) or by little heaps of bran, or by means of boards laid along garden beds ; these boards are sometimes greased or larded on the under side, and here the pests collect.

4. *Irritant dressings*—In the application of these, two points must be kept in mind : firstly, that the time for a dressing will not be in the middle of a hot day,

but at such times as the pests may be out—*e.g.*, in early morning or late in the day, or after rainy weather ; secondly, it should be remembered that a single dressing is not enough, for these molluscs have the power of secreting much slime, and in this way they may receive protection, or the dressing may be moulted off : after a third outpouring of slime, however, the animal has exhausted for the time its power of slime-secretion, and then the dressing accomplishes its result. Irritant dressings are soot and lime, or soot and salt, or salt and lime. Theobald records success with a mixture of fine lime with 4 per cent of caustic soda, 12 bushels to the acre, or with salt with 4 per cent of caustic soda, 4 to 5 bushels to the acre. These were broadcasted on the field early in the morning or late at night. Plants, too, round which the material was spread, received protection.

5. On land that has been very badly infested with slugs, gas-lime should be spread late in autumn, and after being allowed to lie for six weeks should be ploughed or trenched in ; this is destructive to the slugs and their young.

Slugs that devour Worms and Insect larvæ.

Slugs of the genus *Testacella* are, from the grower's standpoint, useful, and should be protected. They are predaceous, living in the soil and preying on worms, insect larvæ, smaller snails, and slugs— *e.g.*, *Limax agrestis*. The teeth on the radula are long. Slugs of the genus *Testacella* can be recognised by their having a rudimentary external shell at the tail end of the foot. There are three species of *Testacella* found in Britain— viz., *T. haliotidea*, Drap., *T. scutulum*, Sby., and *T. mangei*, Fer. The first is the longest, measuring from 3½ to 4 inches. The eggs are laid singly in the soil.

LITERATURE.

British Conchology, by T. Gwynn Jeffreys, F.R.S.

Injurious Mollusca, by Fred. V. Theobald, M.A., in *Zoologist*, June 1895.

Injurious and Beneficial Slugs and Snails, by Fred. V. Theobald, M.A., in the *Journal of the Board of Agriculture*, January and February 1905.

Slugs, by Scharf, in *Scientific Transactions of the Royal Dublin Society*, iv. (2), p. 520.

DAIRY WORK.

Along with other matters relating to the live stock of the farm, the breeding, feeding, housing, and general management of cows and their live produce are fully discussed and described at convenient points in vol. iii. of this work. The manipulation and utilisation of milk and its products are dealt with here.

The cow—the best friend to man of the entire animal creation — gives a bountiful yield of a delicious fluid, which, although universally familiar to the eye and welcome to the palate, is neither so well understood nor so skilfully manipulated and utilised as might be reasonably expected in the enlightenment of the present day. What, then, is this fluid? and what shall we do with it?

The notes in the following pages, intended to elucidate these two comprehensive questions, are not addressed in any exclusive sense to those who are usually described as "dairy farmers"—those who make dairying their sole object, or the one great, all-absorbing feature in their farming. The information is designed for all who have milk to manipulate, in large or in small quantities. Whether it happen to be the milk of the crofter's one cow, the stinted produce of the breeding herd, or the fuller flow of the heavy milkers on the large dairy farm, the object all through should be the same—to turn the milk to the best possible account. Whether this will be as food for the residents on the farm, for calf rearing, for sale as wholemilk, skim-milk, cream, butter, or cheese, will depend upon circumstances too numerous, involved, and variable to be discussed here with advantage. Whatever the destiny of the milk may be, it is important, in order to ensure the best possible results in its utilisation, that the operator should be acquainted with the characteristics, the inherent properties, the weak points and the strong, of the commodity which he or she is handling. It is thus desirable that all who keep cows, whether few or many, should make themselves familiar,

not only with the characteristics and properties of milk, but also with the best methods of preparing it for the various purposes for which it may be used.

THE DAIRY.

In the section on Farm Buildings information is given as to the erection of buildings for dairy farms (vol. i. p. 147). It is not necessary, therefore, to say much here regarding the construction of dairies. In capacity and equipment the special apartments designed as the dairy will be regulated in accordance with the extent and nature of the dairying operations carried on upon the farm.

Upon mixed husbandry farms, where dairying is quite a subsidiary interest, or where, indeed, only as many cows are kept as will supply the wants of the farm itself, the dairy or "milkhouse" is often merely a small compartment easy of access from the dwelling-house. Formerly on such farms the "milk-house" was usually in direct communication with the kitchen, but now the sanitary regulations of almost all counties prohibit direct access from any part of the dwelling-house to an apartment in which milk is kept.

Upon holdings where dairying bulks largely, the manipulation of the milk and its products may require a distinct set of apartments of considerable dimensions.

Be the extent of the dairy what it may, there are some important conditions which should be common to all. Leaving individual farmers to secure the dairy capacity required for their respective holdings, and also allowing them the fullest freedom in what may be called the embellishment of their dairy buildings, we would press for general adoption only such conditions and arrangements as are known to be essential for the hygienic handling and successful manipulation of milk, butter, and cheese.

Situation of the Dairy. — In the first place, the milk compartment or dairy should be so situated as to be free from strong or unpleasant odours. Unless it is kept perfectly sweet, airy, and wholesome, successful dairying is out of the question. Hence the importance of avoiding direct communication with the dwelling-house. Where the milk-room has direct access from the dwelling-house, odours from the kitchen, scullery, and pantry are liable to find their way into this compartment, and play havoc with the milk, cream, and butter. If the dairying interests of the farm are too small to justify the erection of a separate dairy, make a point of having the milk compartment as far removed from the kitchen, scullery, and pantry as possible. Let it be in a cool, airy position, on the north side of the house if possible. Keep nothing in the milk compartment except milk or butter — above all, nothing that gives off a strong or unwholesome smell.

A Medley in the Milk-room.—An arrangement by no means uncommon upon farms where little attention is given to dairying, is to have the milk-house and pantry combined in one compartment. Here, in close proximity, perhaps on one shelf, are milk, butter, cheese, old and new; cold meat from the table, dripping, fish, fresh and cured, and such odorous, savoury and unsavoury articles. A worse arrangement for the milk and butter could hardly be conceived. Those who desire to have first-class, good-keeping dairy produce, must protect it from all such contaminations.

Separate Dairy.—A convenient position for the separate dairy is right back from the kitchen on the north side of the house. It is a good plan to have the kitchen and dairy connected by a covered passage, leading through a yard perhaps from 5 to 10 yards wide. It is desirable, of course, that the dairy shall be within easy access from the cow-house, yet not so close as to endanger the tainting of the milk with smells from the byre, dungstead, or piggery.

Compartments in the Dairy.—In all cases it is desirable to have at least two compartments in the dairy,—one where the milk may be kept as cool as possible, and where the cream may be

prepared for churning and the butter stored previous to despatch or use; the other being reserved as a washing-up room or scullery. Particular attention must be given to ventilation, so that the steam from the boiler and wash-tubs may be speedily carried off. There should not be a drain within the building, but an open channel provided to carry away all waste. This open channel should have a run towards one end into a drain-trap outside the building. Where cheese is made one room is also necessary for the making of cheese and another for the ripening of it, as the temperature and moisture in the air of the dairy may be unsuitable for the ripening of cheese.

Verandah.—It is a good plan to have a covered way or verandah along the south front of the dairy. This provides a shade from the noonday sun, and permits the dairy utensils being dried and aerated in rainy weather.

Finishings of the Dairy.—There is no need for elaborate or costly buildings for dairy work. They should be fairly roomy, not less than 10 feet in height, well ventilated and thoroughly dry, with a subdued rather than bright light. The ceiling should be lathed and plastered, and the flooring formed of some material which will be hard, have no crevices, be proof against damp, and easily cleaned. Encaustic or enamelled tiles and polished pavement are often used, but there is nothing better than well-formed concrete with a smooth surface. The concrete floor is rounded at the edges and corners, and declining towards the door or other exit, so that with a hose-pipe and a good supply of water it can, with little trouble, be thoroughly flushed.

Dampness to be avoided.—Dampness is very injurious in the dairy. If the situation is of a damp nature, special precautions must be taken in constructing the dairy to ensure that its floor and walls shall be proof against damp.

Milk Shelves.—Since the introduction of the cream-separator we have seen many milk shelves taken out of the dairy and the setting-pans discarded. Yet there are many farms where the milk is still stored on flat pans where it is intended for sale, or is "set" for the hand-skimmer. Such shelves should be of some non - absorbent material such as slate or flagstone. These may, with advantage, be erected with a space of about 3 inches clear from the wall, so that cleansing may be thorough, and no crevices for the accumulation of objectionable matter exist when this plan is adopted.

Temperature of the Dairy.—It is a matter of great importance to have the dairy kept cool, sweet, and fresh. To secure, as far as possible, a cool, even temperature, the dairy is sometimes sunk partly into the ground on a hillside, great care being given to the drainage; while walls and roofs are made double, with an air-space between. In summer the windows and doors are well shaded from the sun, and it is considered that a subdued is preferable to a bright light in the dairy. Specky or streaky butter is sometimes attributed to exposure to strong rays of light.

Precise limits need hardly be laid down as to the temperature of the dairy. The object to be aimed at is to have the atmospheric temperature of the milk-room lower than the temperature of the milk itself.

Professor Sheldon remarks that "milk that has been cooled by water or ice should not be exposed to an atmosphere 10° or 20° warmer, for it then becomes a facile condenser and absorbent. While the air is seldom pure enough not to injure milk that is 10° colder, it is seldom so impure as to vitiate milk that is 10° warmer." [1]

There is thus good reason for keeping the atmosphere of the milk-room cool, fresh, pure, and dry.

Importance of Temperature. — At every step in dairy work, no matter what branch of dairying may be pursued, the guidance of the thermometer must be constantly resorted to. Temperature is a controlling influence in all the operations of cream-raising, milk-ripening, butter-making, and cheese-making; and without due attention to this influence success cannot be reckoned upon. Guess-work is quite unreliable, both as to the temperature of the room and of the commodities under manipulation. But there is a simple and efficient guide at hand—

[1] *The Farm and Dairy*, 62.

the common thermometer—which may be had specially adapted for dairy work at from 1s. to 3s. each.

Thermometers.—Glass or porcelain thermometers are required for inserting in and marking the temperature of milk, curd, &c. There should also be a wall-thermometer hanging in every compartment of the dairy, so that the temperature in each may be seen at a glance. The use of a thermometer provided with a case of wood, to prevent breakage under rough handling, is objectionable for insertion amongst dairy produce, for it is sometimes the cause of much trouble in butter - making. It is liable to be imperfectly cleansed after use, and day by day the cream may thus be inoculated with undesirable organisms from this thermometer.

Dairy Utensils.

In regard to the kind of dairy appliances to use, it would be imprudent to dogmatise. In recent years vast ingenuity and enterprise have been employed in the bringing out of "new and improved" dairy appliances. We heartily acknowledge the benefits which have thus been conferred upon the dairy interest, for it has been established beyond question that many of these modern contrivances for use in the dairy are possessed of merits of the highest order.

Still it is not necessary that in order to ensure first - class dairy produce the dairy farmer should discard all his old appliances and adopt new ones. He will do so only as far as his experience, observation, and means seem to dictate and justify. This is pre-eminently one of those points as to which the dairy farmer may, with ample justice to the produce of his dairy, give considerable scope to his purse and his fancy.

Important points to look for in all dairy appliances are simplicity and economy in working, facility in cleaning, and durability. Cheapness is, of course, also to be kept in view. The greatest consideration of all is efficiency.

Power for the Dairy.

There are many cases where all the separating of milk, churning of cream, and even pumping of water are now done by hand, while at little expense these could be driven by the existing power on the farm. Where there is a water - wheel or turbine for driving the farm machinery, the power might be conveyed to the dairy at very slight expense, and for dairy work this power might be had early in the morning without the necessary delay where steam has to be raised. Where a separate installation is to be made for the dairy, a small vertical cross-tube boiler and a steam-engine serve the purpose well. So much cleansing of utensils has to be done, that the greatest benefit is derived from the use of live steam. The water may be boiled by introducing a jet of live steam into the cleansing-tank. For the cream - separator an electric motor is an excellent source of power, as both are high-speed machines.

MILK.

Milk possesses characteristics which should be carefully studied by those who have to handle it and manufacture its products.

Composition.

The composition of cow's milk varies greatly. The following may be taken as fairly representing (1) the analysis of an average sample of milk; and (2) the extremes in milk-analyses :—

	Average Analyses.	Extremes of Ingredients.
Water .	87.40	81.00 to 91.00
Casein . . .	3.30	3.00 ,, 4.10
Butter-fat . .	3.40	1.85 ,, 9.50
Milk-sugar . .	4.55	3.00 ,, 5.00
Albumen . .	0.60	0.30 ,, 1.20
Ash . . .	0.75	0.70 ,, 0.80

The range of solid matter in milk is as great as from 9 to over 19 per cent. A good sample should contain over 12 per cent. Different breeds vary greatly in the standard percentage of solids in their milk. Dutch cows rank lowest, Jersey cows highest, shorthorn cows being about the average. Even with the same cows there will be marked variations under good and bad treatment, and in accordance with the period of lactation.

Butter-fat.—It will be observed from the figures given above that the greatest variation occurs in the butter - fat. In many samples of milk from Jersey cows,

analysed at the London Dairy Show by Mr F. J. Lloyd, the percentage of total solids ranged from 13 to over 19 per cent, and this variation arose almost entirely in the butter-fat. The percentage of butter-fat ranged from about 4.10 to about 9.50 per cent, while total "solids other than fat" showed only very slight variation, at most considerably under 1 per cent. With the milk of shorthorn cows analysed on the same occasion, exactly similar results were obtained—that is, in regard to the variability in the percentage of butter-fat, and the comparative fixity in that of the other solids. The total solids in the shorthorn milk ranged from 11 to 15 per cent, while the solids *other than butter-fat* did not vary more than about one-half per cent.

Yield of Butter.—The yield of butter varies considerably, not only as between breeds but also amongst cows of the same breed. From about 1⅓ to 1⅔ lb. of butter per day are general yields. In regard to butter ratio—that is, the number of pounds of milk to one pound of butter—the Jersey breed takes the highest place with from about 18 to 20, Guernseys coming next with 19 to 21, and Kerries and Dexters following with about 21 to 22. The following table gives the highest and lowest yields of butter from over 600 cows of the breeds mentioned, at the London Dairy Show, during the twelve years from 1895 to 1907 :—

Breed.	Average No. of days in Milk.	Average weight of Butter.	Average Butter Ratio.
		lb. oz.	
Shorthorns . .	44	2 0½	26.69
"	58	1 6¾	32.87
Jerseys . .	117	1 13½	19.62
"	141	1 9½	17.80
Guernseys . .	82	1 12½	18.90
"	138	1 3¼	27.00
Red Polls . .	80	1 8⅝	25.50
"	76	15	39.15
Ayrshires . .	52	1 13¼	26.35
"	77	1 2½	28.07
Kerries and Dexters	33	1 13	.22.4
"	72	14¾	21.31

Composition of Milk from Different Breeds.—The following table gives the average daily yield of milk in pounds, with the fat and total solids in the milk, calculated out for each breed from the results of trials with 482 cows at the London Dairy Shows in the years stated :—

Breed.	Years.	No. of Cows.	Lb. of Milk.	Fat. %	Total Solids. %
Shorthorns . . .	1900-1904	123	48.8	3.72	12.61
" . . .	1907	17	47.9	3.56	12.51
Lincoln Red Shorthorns .	1907	7	51.8	3.41	12.36
Jerseys . . .	1900-1904	111	31.3	5.20	14.40
" . . .	1907	12	34.9	5.10	14.27
Guernseys . . .	1900-1904	36	31.5	4.58	13.65
" . . .	1907	5	35.7	4.59	13.93
Red Polls . . .	1900-1904	33	40.5	3.70	12.70
" . . .	1907	8	41.5	3.62	12.49
Ayrshires . . .	1906	3	42.5	3.56	12.58
" . . .	1907	3	33.5	3.22	12.07
South Devons . .	1906	5	48.3	3.72	12.93
Kerries . . .	1900-1904	43	30.9	4.12	13.26
" . . .	1907	5	40.3	4.30	13.28
Dexters . . .	1907	8	31.0	3.66	12.72
Crosses	1900-1904	50	45.1	3.92	12.86
"	1906	13	46.4	3.54	12.68

A comparison of these figures corroborates many points with which we are already familiar. Thus Shorthorn milk is not very rich in total solids or fat; Jerseys and Guernseys are very high in fat and total solids; Kerry cows above the average; and the others medium. If the solids other than fat are worked out, it will be found that they vary only within a half per cent, not only within the limits of the breed, but between the averages of all—that is, they exist in the proportion of from 8.85 to 9.34 per cent, as against a variation in the total solids of from 12.07 to 14.40 per cent.

It is thus established beyond question that by far the most variable ingredient in milk is fat. It is the commodity which is most within the control of the farmer, and which will be chiefly influenced by the breeding of the cow. The fat, of course, is the ingredient from which the butter is derived.

Testing Percentage of Cream. — The ordinary *test-tube*—a glass tube with graduated lines at the top to mark the percentage of cream as it rises—is useful, but not quite reliable. The milk is put into this tube as it is drawn from the cow, and when the cream has risen, it shows on the graduated scale the percentage of the bulk of the cream. But the cream of different cows varies so much in the size of the butter-fat globules, and therefore in specific gravity, that this test will not always show the entire and exact comparative quantities of butter-fat in samples of different milk.

The *Lactocribe* is a useful invention

for ascertaining the exact amount of butter-fat in milk. It is worked similarly to the De Laval separator, and tests the milk by centrifugal force — making as many as twelve tests at one time. It is specially useful in dairy factories or creameries, where cream is purchased from the farmers.

The "Gerber" butter-fat tester is largely used in private herds. It does its work well, and by it each cow's milk can be tested at a cost of about ¼d. per test.

Fat-globules.—The butter-fat may be seen by the microscope to be in suspension in the milk in tiny globules. These globules vary in diameter in the milk of different breeds and different cows from $\frac{1}{2000}$ to $\frac{1}{20000}$ of an inch, and it has been estimated that there are over forty thousand millions of these fat-globules in a pint of milk containing 4 per cent of fat.

Casein.—This useful ingredient of milk, so important in the manufacture of cheese, is described as existing in the milk in the form of an extremely attenuated jelly, owing to lavish absorption of water. There has been much discussion as to how far the percentage of casein can be influenced by the food given to the cow. Mr F. J. Lloyd says he is sure every chemist who has analysed milk will confirm his statement, that "we cannot by feeding perceptibly increase the casein contents of milk"; and he adds, that therefore the object in cheese-making should be to feed so as to increase the flow of milk and keep down the fat —that is, unless a rich cheese is desired, when more albuminoids are given in the food to increase the fat. Some practical dairy farmers contend that, by changes in feeding, they have been able to alter the casein contents of the milk, but the general experience is in accordance with the views expressed by Mr Lloyd.

Milk-sugar.—This is usually present in a larger quantity than any of the other solid ingredients. It is the most active agent in the decay of milk, as by the action of germs, *Bacterium lactis,* it is transformed into lactic acid, producing sour coagulated milk.

Albumen. — This nitrogenous substance is very similar to casein. Yet the two are so different that, while the rennet precipitates casein in the form of curd, the albumen passes off with the whey. It is this albumen and the sugar of milk that give to whey the feeding value it has been shown to possess (see section on Food and Feeding, vol. iii.) This albumen coagulates on boiling the whey *after* the removal of the casein. It forms the skin which appears on boiled milk.

Weight and Specific Gravity.—A gallon of whole-milk weighs as near as might be from 10.25 to 10.35 lb. For simple calculation, it is a common practice to reckon 10. lb of milk to the gallon. The specific gravity of whole-milk would vary from about 1.025 to 1.032, as compared with 1.000 for water as the standard. The higher the percentage of fat the lower the specific gravity of the whole-milk. The specific gravity of cream itself is about .90.

Milk Statistics.—Taking the stock of cows in this country as a whole, the average yield of milk would probably be somewhere between 500 and 550 gallons of milk each per annum. Good average dairy cows of the heavier milking breeds —shorthorn, cross-bred, Ayrshire, Dutch, and red polled—should give from 700 to 900, some of them even more than 1000 gallons each in the twelve months. Fuller information as to the milk yield of cows of the different breeds will be found in the section on Breeds of Cattle in vol. iii. The produce of butter from a given quantity of milk varies greatly. The choicest butter cows, such as Channel Island cows, often give 1 lb. of butter from rather less than 2 gallons of milk,

and average dairy cows in a butter dairy should give a pound of butter from 25 to 30 lb. of milk. Cheese-makers expect to get about 1 lb. of hard cheese, such as Cheddar, from each gallon of milk, and a little less in Stiltons.

Milk Records.—It is very desirable that the product per day of every cow in milk should be tested by weight, and for butter-fat at least one evening and one morning every two or three weeks. Many, indeed, insist upon this being done every day. Useful information will thus be obtained as to the return each individual cow is giving for the food she consumes. In the absence of precise facts as to the yield, unprofitable cows may be occasionally kept on longer than would otherwise be the case. The great value to the farmer of records of the milk yield of cows has been well demonstrated by the movement set on foot by the Highland and Agricultural Society of Scotland in the year 1904, when it introduced a scheme for the systematic recording of the produce of milk given by cows in herds of Ayrshire cattle. Farmers are being greatly aided by these records in raising the milk production of their herds. Not only do the records guide farmers in buying and selling cows, but also in the selection of animals for breeding purposes.

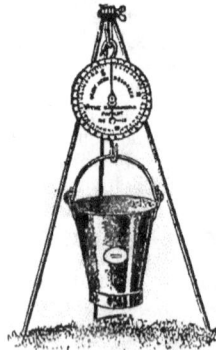

Fig. 664.—*Sandringham dairy herd recorder.*

Appliances for Weighing Milk. — A convenient appliance for weighing and measuring milk is represented in fig. 664.[1] The dial shows the weight of the milk in pounds and ounces, and the measure or quantity in gallons and pints.

Along with this useful instrument the Gerber butter-fat tester may be used, and each cow's milk tested at a cost of about ¼ d. per test.

[1] *For a number of the illustrations presented in this section we are indebted to the Dairy Supply Co., Ltd.*

PURIFYING AND PRESERVING MILK.

It has been the custom for ages to treat milk for the purpose of keeping it pure, fresh, and sweet as long as possible. In olden times the chief means used for this purpose was the mixing with the milk such preservatives as boracic-acid or borax, formaline, and saltpetre. In small quantities these preservatives may not be very harmful to health, but strong objections are now urged to the use of any such agents in milk. Fortunately the more enlightened methods of Refrigeration, Pasteurisation, and Sterilisation are at our service nowadays.

There are several processes by which milk may be to some extent purified and preserved sweet and wholesome for a time. As soon as milk is exposed to the atmosphere it is liable to absorb not only bad odours, but also living organisms (minute vegetable growths), which are continually floating about, and which accelerate the souring and decaying of milk. The action of these organisms is very much impaired by cooling the milk to about 50° or 55°

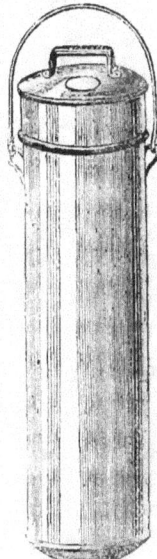

Fig. 665.—*Pasteuriser tin can.*

Fahr., and this is frequently done as soon as the milk is taken to the dairy. It is also believed that by heating the milk to at least 170° the most of these living germs are killed. In hot weather, therefore, when it is difficult, mainly on account of the activity of these organisms, to keep the milk sweet for any length of time, it is the custom in some dairies to heat the milk up to from 150° to 170° Fahr., and then rapidly cool it to below 50°. The boiled taste in milk begins to be developed about 156°. Fahr., and care should be exercised in heating milk not to go much beyond this temperature

when it is intended to be used in the fresh state as either food or drink.

Pasteurisation of Milk.

The method adopted in heating milk depends upon the quantity to be dealt with and the form of heat available. If only a small quantity is to be heated, a pasteurising can (as in fig. 665) may be used. This is made very deep so as to afford as much heating surface as possible. It is fitted with a plunger, which works through an opening in the lid. The tin can is immersed in hot water, and by means of the plunger the milk is agitated and kept from singeing while heating. Such an appliance is very well suited for dealing with small quantities of milk or cream.

Where large quantities of milk have to be dealt with, other methods have to be adopted. The practice of raising the temperature of milk by means of a jet of live steam introduced into it is to be condemned. The water used for raising steam may contain many impurities, which decompose and give off obnoxious gases at a high temperature. These gases may be transmitted to the milk with disastrous results. There is also a considerable amount of water added to the milk which is so heated.

The Turbine-driven Pasteuriser.— This machine (fig. 666) consists of a double-jacketed cylinder. A jet of steam drives a turbine wheel and exhausts into the jacket, which surrounds the cylinder through which the milk passes. Here it is condensed, and passes out through a siphon situated beneath the machine. On the steam inlet pipe a pressure gauge is fitted, so that

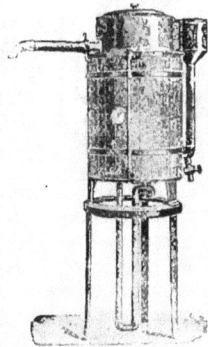

Fig. 666.—*Turbine-driven pasteuriser.*

the speed of the stirrer driven by the turbine wheel may be regulated. The receiving-funnel communicates with the bottom of the inner cylinder, and the milk passing through this opening is caught by the wings of the stirrer and revolved round the hot wall. In flowing out at the top of the machine the milk passes a thermometer, which indicates the temperature. If, however, the inflow of milk is kept constant, and the pressure of steam constant, then little trouble will be experienced in variation of temperature. This machine is particularly well suited for dairies and creameries where there is no steam-engine or other similar source of power.

Belt - driven Pasteurisers. — This machine differs from the turbine pasteuriser in the manner in which the stirrer is driven. A stirrer is necessary to prevent the milk-sugar burning and imparting a cooked flavour to the milk, and also to prevent the casein from being deposited on the walls of the machine. Where there has been a deposit of casein the efficiency of the pasteuriser is greatly impaired, and much steam may pass out at the exhaust siphon before it has an opportunity to condense. This deposit must be removed with a brush; and a solution of washing-soda left overnight in the machine will greatly aid in its removal. This pattern of pasteuriser may be fitted with a tight-fitting lid and the stirrer driven at such a speed as will elevate the milk 6 to 10 feet above the machine. In creameries this may be a great advantage, for in many cases the milk has to be raised to a high level in order to flow by gravitation over the milk coolers. Where there is exhaust steam from an engine this may be used for heating purposes in the pasteuriser, care being taken to have a safety-valve fitted to the jacket to prevent injury to the thin copper cylinder into which the milk passes.

Motor-driven Pasteuriser. — A small engine, with oscillating cylinder fitted on to the frame of the pasteuriser, may be employed in place of the turbine wheel. The exhaust steam from this cylinder enters the heating jacket, which may also be fitted with a steam inlet for live steam to ensure sufficient heating power independent of the speed at which the motor may run.

Sterilisation of Milk.

Many people consider pasteurisation and sterilisation to be synonymous terms.

This, however, is not the case. The object of pasteurisation is to destroy the bacteria which are present in the milk, and in this process a temperature between 160° Fahr. and 175° Fahr. serves the purpose. Milk so treated in sultry trying weather will keep sweet for 24 hours longer than milk which has not been raised to that temperature, but it must not be assumed that it has been rendered sterile or that germ life has been entirely destroyed. The object of sterilisation, on the other hand, is not only to render the milk free from bacteria, but also free from spores. The spore or resting stage in the life of the bacillus is much more resistant than the active bacillus, and must be subjected to a temperature about 230° Fahr. for 20 minutes. This temperature cannot be attained under atmospheric pressure, and so the design of a steriliser differs from a pasteuriser in being made to withstand a pressure of 40 lb. to 50 lb. per square inch.

Fig. 667.—*Bottle for sterilised milk.*

Sterilised Milk in Bottles. — It is usual in practice to sell all sterilised milk in bottles (such as that in fig. 667). These bottles are filled with raw milk and then set into the machine. When the bottle is taken from the machine, the milk, under atmospheric pressure, is boiling vigorously, and consequently a stream of vapour passes from the mouth of the bottle, which prevents contamination from the atmosphere gaining access until the bottles are sealed up. In the more modern sterilisers the bottles are corked while still under steam-pressure.

A steriliser which is known as a "Sterlicon," and is designed to deal with large quantities of milk, is fitted with an automatic arrangement for closing the bottles as they pass out of the machine. The "Simplex" steriliser is arranged for small quantities. The bottles have to be closed by hand, the operator wearing stout leather gloves. After the bottles have been closed they are removed and set in a bath of cold water. The "Simplex" steriliser is fitted with water connections, so that a second cooling chamber is unnecessary. Great care must be taken to admit the cold water gradually, otherwise sudden contraction of the glass will result, and the bottle and its contents will be lost.

Homogenised Milk. — Sterilisation does not prevent the butter-fat from rising. The natural flavour and character of the milk has been changed, but all the fatty constituents are still present, and in a very short time the cream will rise in the neck of the bottle and become quite solid. Where, therefore, this bottling process has been adopted with a view to supplying pure sweet milk, intended to be used, it may be, months after the sterilising, means must be taken to so incorporate the globules of fat amongst the other constituents of the milk as to prevent the cream rising, no matter how long the milk may stand in the bottles. This is accomplished by means of the homogeniser, which consists of a strong force-pump by which the milk is forced through the pores of a porcelain baffle-plate, and the fat globules are so broken up into fine particles that they do not again become detached or rise to the surface.

Cooling of Milk.

In bygone times, long before dairy farmers had become aware of the existence of bacteria, or knew anything definite as to the cause of sourness in milk, they pursued the practice of cooling milk because they found from experience that it tended to keep milk longer fresh and sweet. In our day the cooling of milk is pursued because we know that bacteria, the active agents in the souring of milk, are hindered in their development by low temperature. It thus happened that in the custom of cooling milk to a low temperature, as in many other details of agricultural practice, intelligent practical experience had long anticipated the teaching of the skilled scientist.

Milk Coolers. — An appliance for cooling milk consists of a corrugated copper plate, built so as to form a water

chamber between the plates. Water passes in at the bottom, fills the chamber, and exhausts at the top. The milk from a receiving-pan flows over the outside of the corrugations and is cooled by the water within.

The *flat* cooler (fig. 668) is well suited for farm use. It may be erected on a stand, or simply hung from two brackets on the wall.

The *cylindrical* cooler (fig. 669) has lately come into use in this country, and deserves the popularity it has met with.

The *conical* cooler claims as its special feature that all the milk must pass over its entire surface—that is, a jet of milk

Fig. 668.—*Lawrence flat cooler.*   Fig. 669.—*Laval cylindrical cooler.*

cannot fall from the receiving-tray above to that below without passing over the corrugations.

Tubular Coolers.—Where the temperature of milk has to be reduced below the temperature of the water-supply in the dairy, artificial methods must be adopted in order to obtain the desired result. Refrigerating machines are installed in all up-to-date creameries, and the saving they effect during warm weather fully justifies the expense incurred.

The "Tubular" cooler is largely used for this purpose. Brine made from chloride of calcium, or a mixture of salt and rice, is reduced to a temperature below freezing-point, and circulated through the tubes of this cooler by means of a pump. The milk run over the face of the cooler can thus be reduced to any temperature between that of the water-supply and freezing-point.

Cold Store.—Where a refrigerating-machine has been installed a cold store is almost an essential adjunct. The walls of the store, along with the ceiling and floor, may be insulated with any non-conducting material, one of the best of which is silicate of cotton. The store may be cooled by air forced over a coil in which brine is circulating, but for dairy use, brine coils, drums, or a tank, erected in the ceiling of the store, will be found to give the best results, as the cooling process continues as long as there is cold brine overhead, while with the air system the cooling process does not continue after the machine is stopped.

Refrigerating Machine.—A refrigerating machine consists essentially of a cylinder in which a gas is compressed, a coil over which water flows or which is exposed to the atmosphere, and in which the compressed gas is condensed, and a second coil in which the liquid gas is evaporated. Heat is necessary to expand the gas. This heat is obtained from whatever surrounds the coil in which the gas expands. Usually this coil is surrounded by brine, which is thus cooled in giving out heat to expand the gas. This brine is then circulated through the brine - cooler for cooling milk, or through the brine storage in the cold chamber.

The power of a refrigerating machine should be stated in the number of British

Fig. 670.—*Hall's refrigerating machine.*

thermal units which it is capable of eliminating per hour, and before installing such a plant it is well to know the quantity of water consumed in the condenser and the horse - power necessary

to drive the machine. Different gases are used in working refrigerators. In Hall's refrigerating machine (fig. 670), carbonic anhydride is employed, and in Enoch's (Telford, Grier, & Mackay), shown in fig. 671, ammonia is used.

Fig. 671.—*Enoch's refrigerator.*

Condensed Milk. — The preparation of condensed milk is now quite an important industry. The process consists of evaporating the water of the milk and preserving the solids in sealed tins. The nourishing elements in the milk may be thus preserved for any length of time, but the natural flavour has been to a large extent dissipated.

DESTINATION OF THE MILK.

The treatment of the milk, from the moment it leaves the cow-house, will to some extent vary in accordance with the purposes for which the milk is to be employed—whether (1) for consumption as milk upon the farm—as human food, and for calves, (2) for selling as whole-milk, (3) selling as cream and skim-milk, (4) butter-making, or (5) for cheese-making.

It is not intended to discuss here the relative advantages or disadvantages of these various methods of utilising milk and its products. Circumstances as to supply and demand, and other conditions, may so vary in a comparatively short space of time as to completely upset former reasoning, and warrant farmers in altering their plans. Farmers are left to decide for themselves, after due consideration of the circumstances of the particular time and locality, as to the method of utilisation that will afford them the best return for the milk. Here attention is devoted solely to describing the details to be gone through in each of the methods of utilisation pursued.

CONSUMPTION AND SELLING OF WHOLE-MILK.

When the milk is to be consumed on the farm, or sold as new milk, little manipulation in the dairy is required. In the former case it may be measured out to the consumer just as it leaves the cow-house.

In the latter case, the milk should be run over the refrigerator as soon as it is taken to the dairy. In view of a journey by road or rail, the immediate cooling process, down to 50° or 55°, is very desirable. The refrigerator should be erected so as to admit of the railway churn standing under it. A receiving-pan is erected above the cooler, into which the milk is poured from the milking-pail. Where this pan is fitted with a strainer and a circular sieve placed in the mouth of the railway churn, the milk is ready for despatch in first-class order as soon as it has been run into the churn.

Milk-selling Trade.—The selling of whole-milk for consumption in towns and villages has grown into a business of vast proportions. And as the taste for milk-food grows amongst townspeople —it is growing fast, and will surely enough continue to do so for many years —the milk-selling trade will go on increasing. In some cases the milk is conveyed from house to house by the farmer. The general custom, however, has been to consign the milk and deliver it by road or rail to extensive milk-sellers, who contract with farmers for a certain supply during the year or season.

Co-operative Milk Depots.—The central milk depot or creamery on co-operative lines has become a new feature of the dairy industry, and already several of these have been erected in important milk-producing districts, and others are in course of construction. The creameries or depots have been built and equipped by landed proprietors, and the farmers who are members of the respective co-operative associations at these centres pay rent in the shape of interest on the outlay, and have the option of acquiring the buildings and equipment on easy terms of repayment.

The milk produced on the farms of the members is brought to the depot, and as much as can be profitably sold as whole-milk is passed over the refrigerating plant with which these institutions are equipped, and thus reduced to a very low temperature to secure that it will keep in good condition during transit to the distant centre to which it may be consigned. The balance, for which there may not be a profitable sale as whole-milk, is either passed through the separator and disposed of as cream and separated milk, or partly made into butter, or alternatively made by an expert cheese-maker into a uniform high-class cheese, for which in recent years there has been a capital demand at remunerative prices, ranging from 58s. to 62s. per cwt. By this system all waste both to the producer and middleman who distributes is obviated. It is also claimed that under the system the necessity for excessive early rising on dairy farms will be obviated, as evening's milk properly cooled will suit for next morning's sales.

Methods of Milk Distribution.— In recent years marked changes have arisen in the methods of distributing milk. In cities, and even in certain villages, the milk-bottle has to some extent taken the place of the pitcher.

Milk Bottles.—Bottles for the distribution of milk are usually made of flint glass, can be sterilised, and stand a good deal of rough usage. They may be had in various sizes, from one-eighth of a pint to a quarter of a gallon. One pattern of bottle (fig. 672) is sealed with a wooden disc which fits into a groove in the neck. This disc must be destroyed in order to pour out the milk. If the discs are kept under strict charge at the dairy the milk cannot be tampered with by the messengers employed in delivery. This system of delivery is gaining in favour daily, and as it is cleanly and sanitary it should be encouraged.

Railway Milk Churn or Can.—For

transit by road or rail the steel railway-churn or can (fig. 673) is to be preferred to the oak-butts, which are still in use in some parts of the country. Cleanliness is the watchword in dairying, and in respect of being more easily cleansed the tinned-steel churns or cans have many obvious points in their favour.

Cleaning Dairy Utensils. — With this system of disposal no other dairy work is involved, excepting, of course, the cleaning of the vessels used in conveying the milk. This latter is a most important matter, which should be attended to with the greatest care. Wash and scald the utensils thoroughly as soon as they come into the dairy empty. Upon no account have any dairy utensil

Fig. 672.—*Milk bottle and wooden disc.*

dirty over-night. Where live steam is available all utensils should be scalded on a steaming-block (fig. 674) or table after being thoroughly washed. This method is particularly convenient for sterilising milk-cans and railway-churns.

Milk for Calves. — The system of feeding whole- and skim-milk to calves is dealt with in the section on Calf-rearing in vol. iii.

CREAM-RAISING.

An important piece of dairy work is the separating of the cream from the milk. The manner in which this process is carried out has much to do with the success of dairying,—more perhaps than is generally recognised.

Principles of Cream-raising.—The term cream-*raising*, which is extensively used, affords in itself an indication of the theory of the process of separating cream from milk. The factors involved are specific gravity and temperature. Cream is the lightest ingredient of milk, and therefore rises to the surface. The period of time which the cream requires to make its way to the surface depends largely on the difference in the size of the fat globules in the milk and the difference between the temperature of the milk and that of the air or water surrounding it.

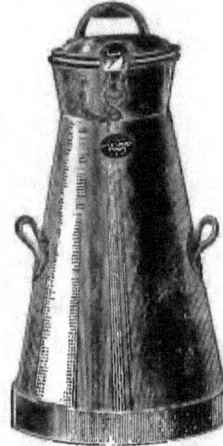

Fig. 673.—*Railway milk churn.*

Water is the largest element in milk, fat the chief ingredient of cream. Water is a better conductor of heat than fat—the former expanding with heat and contracting with cold rather more quickly than fat. Thus it happens that with a falling temperature, and the water in the milk cooling and contracting—increasing in specific gravity — more rapidly than the fat in the cream, the latter is more quickly forced to the surface. On the other hand, when the temperature of the mass is rising, the difference in specific gravity between the milk and cream becomes less, and collecting of the cream on the surface therefore slower.

The discovery of these facts has been of great service to dairy farmers, for it has enabled them to so manipulate the forces of

Fig. 674.—*Steaming block.*

nature as to raise the cream much more speedily than was attainable in former times.

In practice it is found that the sudden cooling of milk, as soon as it is drawn from the cow, retards the rising of the cream, while the setting of the milk, while it is warm, hastens the process.

Methods of Raising Cream. — At one time the setting of milk in shallow pans was almost universal in this country. Now, however, several other methods of raising cream are in use, and some of them have unquestionable advantages in their favour. The deep-pan system has many advocates, and so likewise have the centrifugal separator, the "Dorset" system, the Devonshire scalding system, and the Jersey water-cooling system.

Shallow-pan System.

This system, the oldest of all, is still pursued by many successful farmers. The theory of this plan is that by setting the warm milk in pans from 2 to 4 inches deep in a cool milk-room, the temperature of the milk will rapidly fall, and thus accelerate the rising of the cream. With a steady temperature of about 58° to 60° in the milk-room, this shallow setting gives satisfactory results, raising almost the whole of the cream within from 24 to 30 hours.

Airing Cream. — It is believed that the butter made from cream raised on these shallow pans is rendered superior to what it would otherwise be by the cream being brought freely into contact with a pure cool atmosphere in the process of rising. This was confirmed by the late Professor Arnold, who stated that "cream makes better butter if raised in cold air than in cold water. . . . The deeper milk is set, the less airing the cream gets while rising."

Disadvantages of the Shallow-pan System. — The chief disadvantages of this system are — (1) that it is liable to be rendered unsatisfactory by changes of temperature in the milk-room, or if the atmosphere gets contaminated through the milk-house being in close proximity to anything giving forth obnoxious odours, such as a cattle court or dung midden, in which case the considerable exposed surface is a source of danger; (2) that it requires a great deal of shelving space for the setting of the milk; (3) that it also involves much time and labour; and (4) that the loss of fat by imperfect skimming compared with centrifugal separation is very great.

Temperature and Shallow Pans. —

In the first place, if the temperature of the milk-room rise to unusual height, to anything over 60°, the milk is liable to become sour very rapidly, perhaps before all the cream has risen. Then by exposure to a temperature warmer than itself the cream is liable to absorb impurities. The importance of this latter point was enforced by the late Professor Arnold, who wrote : "While milk is standing for cream to rise, the purity of the cream, and consequently the fine flavour and keeping of the butter, will be injured if the surface of the cream is exposed freely to air much warmer than the cream. When the cream is colder than the surrounding air, it takes up moisture and impurities from the air. When the air is colder than the cream, it takes up moisture and whatever escapes from the cream. In the former case the cream purifies the surrounding air; in the latter the air helps to purify the cream."

The depth of setting, Professor Arnold added, "should vary with the temperature : the lower it is the deeper the milk may be set; the higher, the shallower it should be. Milk should never be set shallow in a low temperature nor deep in a high one."

Then if it should happen that the milk-room is unusually cold, under 50°, the milk may have to stand for 48 hours, and even then the whole of the cream may not have risen. The loss of a certain percentage of cream is not the only result of this slow rising of the cream. A great deal of shelving space must be provided for the milk, and this on a large dairy farm might involve considerable expense. Then the skim-milk will not keep so long sweet as if it had been separated sooner, while the labour in skimming and cleaning so many pans is also an item worthy of consideration.

Shallow Pans. — The pans in which milk is set in the shallow system consist of either stoneware, tinned iron, or wood. Common stoneware is the least durable of the materials employed, and is not now so extensively used as in former times. The harder and better finished varieties of stoneware are preferable.

Fig. 675 represents an excellent milk-pan made of best quality white porcelain, oval in shape, 16 inches long and 3 inches deep inside measure. Milk-dishes of this material are wonderfully durable, nice-looking, and easily kept clean.

Fig. 675.—*White porcelain milk dish.*

The form of milk-pan now most common, and perhaps on the whole the best, is shown in fig. 676. The former has a mouth to facilitate pouring. This pan is made of tinned iron, and similar tins are made of block tin stamped in one piece, or of iron with enamelled interior, and with or without a lip to pour out the milk by. This material admits

Fig. 676.—*Iron milk-pan.*

of perfect cleanliness, while it is practically unbreakable.

Zinc Unsuitable.—Zinc or galvanising should never be used on dairy utensils, except perhaps on outside parts, where the milk or its products do not come into contact with the metal. Milk always tends to sour, the souring being due, as we have seen, to the formation of lactic acid from the milk-sugar by the fermentive action of a particular

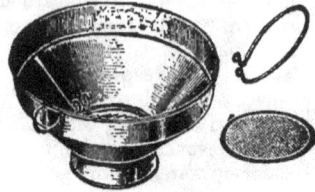

Fig. 677.—*Milk-sieve.*

germ—the *Bacterium lactis*—which is always present. The acid so formed has a great affinity for zinc, forming zinc lactate, a substance which is highly poisonous, giving rise to nausea and vomiting.

Milk-sieve.—Another utensil required in a dairy is a milk-sieve (fig. 677), which consists of a bowl of tinware, 9 inches in diameter, having an orifice covered with wire gauze in the bottom for the milk to pass through, and to detain the hairs that may have fallen into the milking-pails from the cows in the act of milking. The gauze is of brass wire, and, when kept bright, is safe enough; but silver wire is less likely to become corroded. The wire gauze is set in a tin ring,

Ready for use.   Fig. 678.—*Milk filter.*   Dismantled for cleaning.

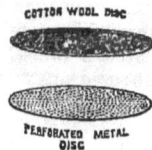

so that it may be easily removed for cleansing.

The straining of the milk through a sieve such as this should in all cases be the very first operation after the milk is drawn from the cow. A very useful strainer is made for attaching to the side of the milk-pail. The use of cotton wool for straining milk is gaining favour, and an illustration of such a filter is given in fig. 678, showing how the wool may be inserted.

The creamery strainer (fig. 679) must be of suitable design to run a large quantity of milk. If the wire gauze is inserted where the milk falls, any dirt which may collect on the surface is forced through the gauze by the next milk poured on to

Fig. 679.—*Creamery strainer.*

its surface, so that it is better to have the gauze inserted in the sides.

Skimmer. — The creaming-dish (fig. 680), also of tin-ware, skims the cream off the milk. It is thin, circular, broad, and shallow, having on the near side a sharp edge to pass easily between the cream and milk, and a mouth is formed for pouring the cream into any vessel.

At the bottom are a number of small holes for milk to pass through.

Cream-jar. — In small dairies the cream, until churned, is usually kept in a jar of stoneware or white porcelain. This should be covered by moist muslin to prevent foreign material entering.

Shelves. — The shelves in dairies should be made of materials easily and

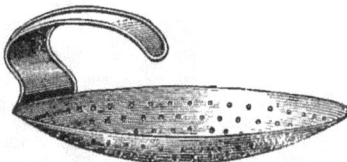

Fig. 680.—*Cream skimmer.*

quickly cleaned. Wooden shelves are easily cleaned, but are too porous and warm in summer. Stone ones are better, but must be *polished*, otherwise they cannot be cleaned without being rubbed with sandstone. Marble or slate shelving is the

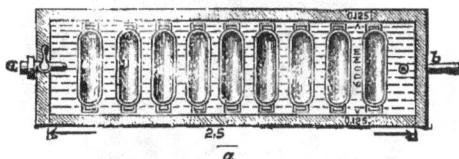

b

Fig. 681.—*Swartz system.*

a Trough with pans immersed in water.   b Empty pan.

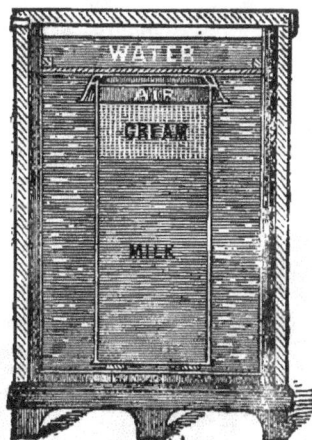

Fig. 682.—*Cooley system.*

best for coolness and cleanliness combined, and now neither is expensive.

Deep Setting.

The earliest departure from the old-fashioned shallow-pan system was the setting of the milk in deep pans.

The Swartz System.—In the Swartz system, represented in fig. 681, deep cans of milk are set in a trough filled with

cold water, or water and ice, the water being kept continuously running through the trough. By these means the milk, set at blood-heat, or even warmed up higher, is rapidly reduced in temperature, and, as already explained, this falling temperature hastens the rising of the cream to the surface.

The Cooley System.—In the Cooley system, somewhat similar to, but in most respects an improvement upon, the Swartz plan, a lid is fitted to each can on the principle of the diving-bell, so that the cold water is allowed to rise over the top. Slips of glass are fixed into the sides of the cans to show the depth of the cream, and taps are provided to run off the milk. Fig. 682 indicates the arrangement of the Cooley system.

Swartz and Cooley Systems Compared. — The main principle in the working of these two systems is the same—the accelerating of cream-raising by a falling temperature in the milk. The pans are about 20 inches deep, and in both the cream will have risen in about 12 hours. The main difference is in regard to the exposure and enclosing of the cream and milk, as to which there is some difference of opinion.

Atmospheric Influence on Cream.—In the Swartz system, as we have seen, the pans are open at the top. Some regard this as an advantage, holding that the exposing of the cream to the air has an influence which improves the butter. Others, again, prefer the Cooley system, mainly for the very reason that in it the pans are closed and submerged, so that atmospheric impurities and changes are entirely prevented from coming into contact with the milk and cream. The condition of the atmosphere immediately around is the regulating influence. Exposure to a *pure*, cool atmosphere is beneficial to cream; contact with impure, hot air is distinctly the opposite. It is therefore claimed that the Cooley system is more to be relied upon in securing uniformly good results, in spite of impurities and changes in the atmosphere.

Ice used in Summer. — For these two systems, especially in the open Swartz trough, ice has to be employed in summer unless cold spring water is available.

Advantages of Deep Setting.—The setting of warm milk in deep pans in cold water economises time, labour, and space, and lessens the risk of injury to the cream from impurities and changes in the atmosphere.

Disadvantages of Deep Setting.— The appliances are more costly than for shallow setting, and the providing of the necessary supplies of water (and ice in summer) may be troublesome and costly. The improvement imparted to butter by the free exposure of the cream, when rising to a pure, cool atmosphere, cannot be so fully obtained in the deep as in the shallow pans.

Devonshire Scalding System.

The Devonshire system of raising cream by scalding is of long standing. Fig. 683 represents the appliances employed in this system.

Method of Working.—The milk is first set in the ordinary way in pans in

Fig. 683.—*Devonshire cream stove.*

a cool dairy (temperature about 60°), and at the end of about twelve hours the pans are placed on a stove, as shown in the figure, and the milk scalded to a temperature of about 180°—until the surface of the cream becomes wrinkled — when the pans are removed. The milk and cream are allowed to cool, when the cream is removed and put into crocks or jars, in which it becomes thick and clotted.

Merits of the Scalding System.— This method of scalding raises more cream than would be obtained in the ordinary setting system. The butter is very easily made, and the scalding has the effect of purifying the cream and making it keep longer sweet.

Other Methods of Raising Cream.

There are some other useful appliances for raising cream rapidly. The "Dorset," the "Richmond," and the "Speedwell" cream-raisers are all well spoken of; the "Dorset," in particular, has been largely used with excellent results. Before the centrifugal separator was introduced the "Jersey Creamer," a very ingenious contrivance, was largely employed in some parts of the country, and is still used to some extent.

Centrifugal Separator.

The most remarkable and most useful of all the modern contrivances for separating cream from milk is unquestionably the "centrifugal separator." By the use of this admirable invention the cream and milk can be separated immediately upon leaving the cow. In recent years great improvements have been effected with the object of increasing the capacity of the machines and diminishing the already small percentage of butter-fat left in the skim-milk. Of these machines there are several patterns, all working upon similar principles. The first of them were manufactured on the continent of Europe, but several firms in this country are now turning out machines for each of which superiority is claimed.

Working of the Separator. — The milk passing into the machine is received in a bowl or cylinder. This cylinder revolves at a speed varying from 5000 to 10,000 revolutions per minute, according to the design of the machine. The specific gravity of the fat in milk is less than that of the other ingredients, consequently we find that the fat takes up a position near the centre of the cylinder with which it is revolving, and the water and solids forming the skim-milk are found close to the wall of the bowl. Tubes from this surface convey the skim-milk to the skim-milk cover, the spout of which may be turned round to a suitable position to command the can, cooler, or whatever receptacle may be provided. The cream passes up a special tube opening nearer the centre of the bowl. The design of this varies. It may take the form of a disc or other

bell-mouthed structure. The essential point is that it will catch the contents of the bowl which passes upwards nearer the centre. A screw on this outlet is the usual means of regulating the quality of the cream. If the cream outlet is partly closed by the screw less milk passes out with the fat, giving a richer cream; if thin churning cream is desired, the screw may be adjusted to allow a larger outflow to pass out by the cream opening. The cream leaving the bowl is caught by a circular cover which is provided with a spout. Practically all separators are provided with a float, which checks the inflow of milk.

The less complicated gearing there is about a separator the better, provided a sufficiently uniform high speed can be maintained to effectually separate the fat from the milk. The most essential points to look for in purchasing a separator are, first, clean skimming,—not more than .10 to .15 of 1 per cent of butter-fat should be left in the skim-milk if the machine is working properly; secondly, the design

Fig. 684.—*Hand separator.*

of the machine, so that a maximum of speed may be obtained with a minimum of power and consequent tear and wear. Facilities for taking asunder and cleaning are also points of great importance.

The capacity of hand-power separators varies from 10 to 135 gallons per hour. The farmer who keeps only one or two cows for family use may have a small-sized, inexpensive machine (such as is shown in fig. 684) capable of dealing with his limited milk-supply. Many of these small machines are now being used regularly in the private dairies of country gentlemen. In the case of the larger size of hand-power machines it is sometimes desirable to have these fitted with pulleys so that they may be driven by either hand or power, if such be available.

The best machine for creamery use is undoubtedly the belt-driven separator (fig. 685). The high speed is attained by an intermediate spindle, which is complete on its own frame. This spindle is driven from the main shafting, and communicates with the separator by means of a rope drive.

The steam turbine plant answers well where a steam boiler is used. This machine is driven by a jet of steam which impinges against the buckets of the turbine wheel. This does away with the expense of an engine, but the

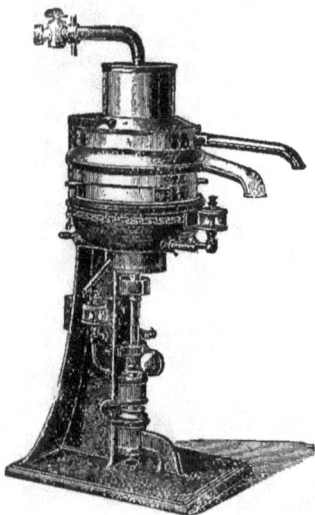

Fig. 685.—*Belt-driven separator.*

wear and tear with this style of power is greater than with the belt-driven separator.

Advantages of the Separator.— The advantages which may be derived from the use of the centrifugal separator are of great importance. In the first place, the work of the dairy is facilitated and simplified, for the setting and skimming are done away with. The cream and skim-milk are obtained separately in a perfectly sweet and fresh condition, and therefore more suitable for marketing than if the slower system of setting and skimming had been followed. All foreign matter or sediment is effectually detached from both milk and cream.

One of the chief merits of the centrifugal separator therefore is, that by its

use the maximum quantity and highest quality of butter may be obtained.

Selling Cream and Skim-milk.

Since the introduction of the centrifugal separator a large and growing trade has arisen in the selling of sweet cream and sweet skim-milk.

Separated Cream.—The inhabitants of towns and villages have a keen relish for sweet cream for tea, fruit, and puddings, and the separated cream supplies this demand admirably. The cream removed by the centrifugal separator is much fresher, more wholesome, and will keep longer sweet, than cream from any of the old setting systems.

Preserving Cream.—One difficulty in the sweet cream trade is that of preventing the cream from getting sour. Various methods are tried with the view of overcoming this difficulty. The introduction of a little boracic acid into the cream has the effect of keeping it fresh longer than would otherwise be the case. But, as in the case of milk, enlightened scientific opinion is now against all forms of preservative, and dairymen and farmers must now deal with their cream just as they deal with their sweet-milk, by pasteurising and then cooling it to a low temperature, thus obviating the necessity for using preservatives.

Some dairymen adopt the more troublesome expedient of enclosing the cream in hermetically sealed tins of various sizes, containing quantities suitable for family and hotel use.

Devonshire Clotted Cream. — The system of scalding the whole-milk, which has been so long associated with the county of Devonshire, tends to strengthen the keeping properties of cream (see p. 490). The scalding destroys or impairs ferments in the whole-milk, and this followed by the cooling of the cream to a low temperature, perhaps 40° to 50°, tends to preserve the cream. This scalded cream becomes unusually thick and clotted, and is largely sold for family use in London and elsewhere. It is retailed in sealed tins, or small or large jars, which should be kept, in the shops, in suitable refrigerators.

Separated Milk.—The milk which is deprived of its cream by the centrifugal separator is no doubt poorer in a sense than milk skimmed in the old way, for, as a rule, by the latter system, about one per cent of fat is left in the skim-milk, whereas in separated milk, as already indicated, there is rarely more than one-tenth of that quantity. But if the separated milk loses in fat it gains in freshness. It is in a better condition for selling, and for consumption both by man and beast, than if it had sat perhaps till acidity had begun.

But it should be remembered that, after all, the most nourishing and strength-giving ingredients of the milk, as it leaves the cow, remain in the milk after the cream has been removed. What it has lost in the butter-fat can be easily made up by other articles of food, and assuredly there is no more healthy or muscle-making food than plenty of fresh skim-milk. Happily its consumption amongst townspeople, to drink by itself, and for use in puddings and other food, is on the increase, though hitherto the rate of increase has not been so rapid as could be wished.

BUTTER-MAKING.

After the separating of the cream and milk, attention is given to the ripening of the cream and the making of butter. In some parts of the country, as will be afterwards mentioned, the whole mass of milk and cream is churned, but the pre-vailing custom is to churn the cream only.

Ripening Cream.—The importance of the ripening of cream for butter-making is not realised as it should be. Much of the bad butter made to-day in certain districts is due, not to the makers' inability to churn and make up the butter, but to lack of knowledge as to how the cream or milk should be pre-pared for butter-making.

In Continental dairies famed for the choicest brand of butter, the proper ripening of the cream is regarded as a matter of primary importance. If the ripened cream is as it ought to be in flavour and consistency, there need be no anxiety as to the resulting product.

To set cream aside in a foul or damp atmosphere, or in a vessel which has not been thoroughly cleansed, is to court disaster. The ripening process is effected by the action of living organisms. Foul air is laden with injurious bac-teria, which will produce undesirable results, and as milk or cream forms an excellent nutritive medium for these germs, they speedily convert good sweet cream into that which will yield rancid, bitter, spongy, or discoloured butter. Hence the absolute necessity for scrupul-ous cleanliness and pure air.

The Use of a "Starter" or Pure Cul-ture.—The practice of using a "starter" in ripening cream has been criticised as a "newfangled method," but it is by no means new. Butter-makers and cheese-makers in time past added buttermilk or whey to their milk in order to make it ready for churning or for the rennet. This was nothing more nor less than adding a "starter." The modern "star-ter" is, however, handled upon scientific principles, since it has been confirmed that bacteria are the active agents in-volved in producing ripe milk or cream.

In many cases buttermilk may with advantage be used; but the more reliable method is to cultivate, from day to day, a "starter," by inoculating sterilised or at least pasteurised milk with such bac-teria as produce a pleasant flavour. It is usual to speak of this bacterium as *lactis acidi*, but it has been demonstrated that there are other bacteria—bacteria which at least are of different form when examined by the microscope, and which cause sourness in milk, imparting to the butter that full flavour and keep-ing qualities so much desiderated by the butter-maker.

These ripening agents, with instruc-tions as to their preparation and use, should be obtained from sources that can be thoroughly relied upon.

Starter Jelly.—A suitable culture for the ripening of milk and cream, prepared in sterilised milk, so that it is ready for use, is obtainable from the Dairy Supply Company, and also from several of the leading Agricultural Colleges and Dairy Schools. The advantages which are claimed by those who advocate the use of a special culture, in the form of a "starter," are that the flavour is superior, that the keeping quality of the butter is improved, that the colour is clearer, and that the quality may be kept uniform.

Uniform Ripening of Cream.—It is very important that the mass of cream to be churned at any time should be as uniformly ripened as possible. This is most easily secured, of course, where the churning takes place daily or frequently. It can be fairly well obtained, however, by care in the mixing of the cream as it is removed from the milk. Each "creaming" should not have a separate vessel to itself unless it is to be churned by itself. The better plan is to have a cream-holder sufficiently large to hold all the cream to be churned at one time, and as each quantity of fresh cream is added, the whole should be thoroughly stirred, the stirring being perhaps repeated once or twice between the times of creaming.

As has already been explained, temperature plays a very important part in connection with the activity of bacteria, and consequently it must be kept in control by the successful butter-maker. It is now generally admitted that cream ripened at a comparatively low temperature, about 56°, gives superior butter to that made from cream ripened between 60° and 70° Fahr. In creameries where large quantities have to be dealt with daily, this may be accomplished by the use of a submerged cream-cooler, which consists of a coil of pipes through which chilled water or brine circulates. When low temperatures are adopted, larger quantities of "starter" have to be added to procure the desired acidity within twenty-four hours—a period that seems to give better results than any other in the flavour and keeping property of the butter.

The uniform ripening of all the cream in one churning is essential to obtain both the greatest quantity and the choicest quality of butter. The reason is not far to seek. Ripe cream passes into butter more quickly than fresh cream. Then with a quantity of ripe and a quantity of fresh cream in one churning the former would be over-churned before the whole of the butter-fat in the other would be transformed into butter. The result is usually a compromise, a little over-churning of the ripe cream and a slight under-churning of the fresh cream. Avoid the evils of this compromise by attending to the proper mixing and uniform ripening of the cream.

Influence of Salt in Ripening.—Salt may be added if the ripening process appears to be too rapid, where the temperature cannot be further reduced. Being a preservative, the salt retards the action of the bacteria, but it renders the buttermilk practically useless.

Sweet-cream Butter.—For immediate consumption, butter made from perfectly sweet cream is by many preferred to butter from sour or well-ripened cream. But it does not keep so well as the latter, and the weight of butter from a given quantity of sweet cream will be less by from 3 to 6 (perhaps even more) per cent than from the same quantity of sour cream.

Times of Churning.—It is a common practice to churn only once a-week. Others think it preferable to churn twice, and many do so three or four times a-week, or even daily. In the majority of cases of churning once or twice a-week, the whole of the cream then in the dairy, excepting that taken off on the previous day and day of churning, is well mixed together and churned. The fresh cream is usually held over till the next churning. The times of churning must of course be regulated by local circumstances, such as the quantity of cream to be handled and the demand for butter.

Temperature of Cream for Churning.—There is not a little difference of opinion, amongst both theoretical and practical butter-makers, as to what should be the temperature of cream when put into the churn. Fortunately, it would seem that upon this point some latitude may be allowed without seriously injuring the produce. Much, of course, depends upon the temperature of the churning-room. From 55° to 58° in summer, and from 58° to 63° in winter, are common ranges of temperature for the cream just on being put into the churn—56° to 58° in summer, and 60° to 62° in winter, are perhaps most general. Some prefer to keep the dairy at the same temperature—about 58° to 60°—in summer and winter, and so churn the cream at the same temperature all the year round.

A high temperature hastens the churning—the "coming" of the butter—but this gives pale, soft, and often spongy butter.

In the best creameries it is now found that the introduction of refrigerating machinery, facilitating experiments in churning at low temperatures, has led to the adoption of what at one time was considered a ridiculously low temperature—viz., between 45° and 50° Fahr. But still, where the churn is driven by the hand the higher temperatures are retained.

Effect of Food on the Churning Temperature of Cream.—In his experiments on the effect of food on milk and butter in the three years 1895-97, Mr Speir, Newton, found that if various foods were freely used it was necessary to churn the cream at a higher temperature than if the cows had been fed on pasture. Unless this was done, the percentage of fat left in the buttermilk was very much increased, and the time required to do the churning was considerably lengthened. The following table gives a rough idea of the range of temperature found necessary to give the best results :—

Food used.	Degrees Fahr.
Young pasture . . .	56 to 58
Dried grains, bran, and treacle .	57 " 59
Paisley (gluten) meal and grains	58 " 60
Pea-meal and grains . .	59 " 61
Sugar-meal . .	60 " 62
Sugar-meal and malt coombs .	61 " 63
Bean-meal .	62 " 64
Meat-meal and decorticated cotton-cake . . .	64 " 66
Decorticated cotton-cake .	66 " 68
Young heather . .	66 " 68

The above figures apply where the cream is sufficiently ripened, and the temperature of the air ranges from 58° Fahr. to 62° Fahr., and the cows are neither recently calved nor approaching the end of their lactation.

Churning Whole-milk.

Strange as it may seem to some, the old-fashioned system of churning the whole-milk after it is properly ripened or lappered is still practised in several parts of the country.

Advantages of Churning Whole-milk. — The chief advantages claimed for the churning of the whole-milk are,

that less dairy space and milk-setting appliances are required, that in certain districts more money can be obtained for the buttermilk than for skim-milk, and that more butter is obtained than where the milk and cream are separated by skimming, and only the latter churned. There is, no doubt, a saving in outlay for buildings and utensils, and in a district where a good market can be got for the butter and buttermilk, and both can be sold from the cart, the system works well.

In all probability a little more butter may be obtained by churning the whole-milk than when the cream is skimmed off by hand, as a portion of the butter-fat is left in the milk. But with the more effective method of separating the cream by means of the "centrifugal separator," the churning of the whole-milk will not likely compare favourably, especially in regard to weight of butter.

The improved contrivances for more speedily and effectually separating the cream from the milk have removed the strongest argument advanced in favour of the churning of the whole-milk.

Disadvantages of Churning Whole-milk. — Amongst the reasons urged against the churning of the whole-milk are, that it involves a great deal of labour in churning such a large quantity of fluid, and that the resulting buttermilk, excepting in industrial districts, meets with a poor demand, and is not so suitable for calf-rearing as new separated milk. The butter made in this way is more difficult to work, and considerable skill is required to effectually remove the water and the casein, with which it has been in close contact in the churn.

The buttermilk finds a ready sale in large towns for human food, but in the country districts the demand for this purpose is of course very limited. It is useful for feeding pigs, but, as already indicated, not so suitable for calves.

Methods of Preparing Whole-milk for Churning.—In preparing the whole-milk for churning it is necessary that it should be well soured. If churned while sweet a good deal of the butter-fat may remain in the buttermilk. A little buttermilk is often poured in amongst the fresh whole-milk to hasten its souring, but this is not a good plan, as the butter-milk is liable to contain organisms that would be detrimental to the butter.

The system which is coming largely into use is to add a "starter" as soon as the milk is set. The lapper is greatly improved by stirring it occasionally while ripening, but the milk must be allowed to coagulate. Large oak butts may be used for this purpose, as tin is apt to impart a flavour to milk which sits for some time, especially where the milk is set at milking temperature.

Continental Method.—By carefully regulating the temperature of the dairy and the depth of the milk in the butts, Continental dairymen, who churn the whole-milk, secure the proper degree of ripeness without introducing any ferment. For this purpose they keep the temperature of the milk-room somewhere between 45° and 59° Fahr. If the temperature is low, say between 45° and 50° Fahr., the milk should be filled into the butt to a depth of about 24 to 28 inches. If the temperature is higher the milk should be set shallower, so that when the maximum temperature of 59° Fahr. occurs the depth of the milk should not be more than from 12 to 16 inches.

The milk should be put into the butt just as it comes from the cow. No previous cooling is necessary, nor is it advantageous, as it retards the ripening too much. Should the butt not be big enough to hold an entire milking, the milk should be divided between two butts, but quite equally, so that the ripening may go on at the same pace in both, for unequally ripened milk makes bad butter. On no account should warm milk be added to partially ripened milk, for if that is done, bad-keeping flavours are sure to develop in the butter. If the temperature of the room rises higher than 59° Fahr. the milk should be cooled, and if it should be too cool the milk must be heated, otherwise there will be imperfect ripening and consequent loss of butter. In about thirty-six hours the milk will likely have attained the proper degree of ripeness, and then, before being put into the churn, it is thoroughly mixed, so as to be rendered quite homogeneous.

CHURNS.

How many first-class churns there are in the market at the present day one would not venture to say. Not only is the number considerable, but it goes on increasing. It is well, indeed, for the dairy farmer that his wants are thus so admirably provided for.

No attempt will be made to draw up a list of the first-class churns in order of merit, but it may be useful and interesting to illustrate and indicate the working of two or three of the well-known churns.

Types of Churns.—In general use throughout the country there are three types of churns less or more distinct: (1) those in which the fluid and the containing vessel with its agitators (if it has any) are in rotative motion; (2) those in which the containing vessel is at rest and the agitators are in rotative motion; and (3) those in which the agitators rotate in one direction while the containing vessel rotates in another.

Plunge-churn.—The old-fashioned *plunge-churn*, in which the agitator is worked by hand upwards and downwards in a stationary cylinder of cooper-work, is never seen now in a well-equipped dairy. It is still employed on some farms where

Fig. 686.—*End-over-end churn.*

dairying receives little attention, and where few dairy improvements have been introduced. It is heavier to work, and altogether inferior to the modern barrel churns.

Barrel Churns.

End-over-end Churn.—The end-over-end churn (fig. 686) is widely popular with up-to-date butter-makers. The mouth is large, admitting of easy cleansing, while there is no inconvenience with a leaking joint. The grain of the butter made in the end-over-end churn is as near perfection as it may be obtained in any other pattern. The construction is simple, and consequently a real serviceable churn may be had of this type at a reasonable price.

It is important to notice that churning operations are greatly retarded if an end-over-end barrel churn is more than half full; indeed it is advisable not to fill the churn more than one-third full.

Diaphragm Churn. — Messrs Bradford & Co. have fitted a special appliance in the form of beaters or breakers inside their churn, so that the diaphragm churn of to-day (fig. 687) is simply an "end-over-end churn" with breakers fitted longitudinally in the interior.

It is claimed for this churn that the time of churning is shortened while the grain of the butter is not inferior to that obtained in the simple barrel churn.

Other Forms.

Holstein Churn. — The Holstein churn is an example of the third class of churn mentioned. It is an upright barrel, with agitators which revolve horizontally while the churn is at rest. The Holstein churn is extensively employed in large factories and creameries, where a number of churns require to be driven separately from one shaft.

Streamlet Churn. — This churn, made of fire-clay enamelled, is extensively used for churning the whole-milk in the west of Scotland. It is usually made in large sizes, with dashers. It is rather difficult to clean out, and for this purpose steaming is most effective.

Fig. 687.—*Diaphragm churn.*

"Speedwell" Crystal Churn. — This churn consists of one or more glass jars or cells, mounted in a revolving frame —a frame adapted for sitting upon a table (fig. 688). This churn is very con-

venient for churning small quantities of cream in private dairies, or for experimental purposes.

Swing Churn. — This churn is in the form of a box or child's cot, and effects the churning by oscillation.

High-speed Churns. — From time to time various designs of high-speed churns, such as the Disc and Fishback, have been introduced, claiming to reduce the

Fig. 688.—*Speedwell crystal churn.*

time of churning from thirty to three minutes, and at the same time to produce the choicest butter. These churns, however, have not met with the general approval of butter-makers, and it may therefore be doubted whether the high-speed principle is adapted for so delicate a product as butter. The high-speed churns are liable to leave an exceptionally large percentage of fat in the buttermilk.

Factory Churns. — With the development of butter-factories in this country the demand for churns to deal effectually with large quantities of cream has increased. In order to meet this demand makers of dairy appliances have departed from the design of hand churns. The factory churn (fig. 689) most commonly used is simply a large cylinder built of wood, with a spindle passing through the centre from end to end. To this spindle is attached two or more beaters. The churn is erected in a horizontal position and the beaters driven by a belt from the dairy shaft. In one end there is fitted a large door through which the buttermilk with the butter is drawn off into a tank in which the butter may be washed, and from which it may be easily transferred to the butter-worker.

Combined Churn and Butter-worker.—The latest and perhaps best

Fig. 689.—*Factory churn.*

device is in the form of a churn and butter-worker combined (fig. 690). Here, after churning is completed, rollers are placed in position within the churn. The churn is then set in gearing, which causes the barrel to revolve ; this carries round the butter, which drops between the rollers, and is pressed together while the water is worked out. The advantage of this arrangement is that the loss of butter is reduced to a minimum, no grains being lost on the way from the churn to the butter-worker, or carried away with the buttermilk.

Important Features in a Churn.— While the farmer may exercise consider-

Fig. 690.—*Combined churn and butter-worker.*

able freedom in the choice of the pattern of churn, there are a few important features which he should look for and insist upon. Amongst these are, that the churn should be easily cleaned, with no crevices wherein dirt may lodge and escape observation ; that it should afford ample facility for removing the

butter ; that the churn may be easily ventilated ; and that means should be provided for seeing the cream and ascertaining its temperature during churning. Light working as well as efficiency should, of course, also have due consideration.

It is a good plan to have a small pane of glass in the churn through which to note the progress of the churning. And to permit the escape of gases evolved in the process of churning, there should be a ventilation valve in the lid of all churns.

Churning.

The details of churning have now to be dealt with. These require little explanation.

Preparing the Churn.—The preparation of the churn for the reception of the cream requires careful attention. It may be assumed that, after the previous churning, it had been thoroughly cleaned— first rinsed with cold water, then well scalded with boiling water, and again rinsed with cold. If it has not been in use for a few days, the churn may be scalded with hot water the day before churning. Some heat the churn with hot water just before putting in the cream. Others consider this a bad plan, and prefer to rinse out with water about the same temperature as the cream, or perhaps even two or three degrees lower than the cream. In cold weather the churn is heated, and in hot weather cooled.

Upon the whole, perhaps the safest plan is to have the temperature of the churn just the same as that of the cream to be churned. The temperature of the cream rises a little, about 3° or 4°, with the friction in churning.

Some sprinkle a little salt in the churn before the cream is put into it, and others put salt into the water used in rinsing. The object of this is to counteract any taint that may possibly be present.

Straining Cream.—To prevent, as far as possible, impurities getting into the milk, the cream is run into the churn through a strainer, perhaps a coarse linen cloth, well known as cheese-cloth. This cloth is dipped in clean water, and held over the mouth of the churn while the cream is poured into it. The thickest of the clotted cream will be held back, and impurities, such as dust and flies, will be prevented from getting into the churn.

The straining-cloth is washed without soap, and kept sweet by exposure to air.

Speed of the Churn.—The speed at which the churn and agitators, if there are such, should be driven depends entirely upon the design of the churn. The end-over-end pattern should be driven about 60 revolutions per minute.

With very slow churning the butter is long in coming. With rapid churning the butter is liable to be soft and oily. With every individual churn, and in the varying circumstances of temperature and condition of the cream, the operator must exercise careful judgment as to the rate of speed in the churning.

Ventilation.—This must be carefully attended to in the first 8 or 10 minutes' churning. In the stirring of the cream at the outset, some gas is evolved, and the ventilator in the churn should be opened frequently during the first 10 minutes, to provide the desired ventilation.

Breaking Water.—When the butter "begins to come"—i.e., when the butter is just forming a grain—it is advisable to add a small quantity of water. When added, the temperature of the water should be at about 10° below that of the cream. This addition of water assists in securing a granular condition in the butter, and restores the temperature to nearly the original temperature of the cream.

As soon as the granules become about the size of shot, or about ⅛ inch diameter, churning should be stopped, the buttermilk drawn off, and the washing water added. This is a point when care should be taken to remove the last traces of buttermilk. Change the washing water. until it is drawn off quite clear. A solution of ordinary salt may be added to assist in removing the buttermilk.

Time Churning.—The churning will probably have occupied from 30 to 40 minutes. With less time, there is a liability to softness in the butter, and great risk of an excessive amount of fat being left in the buttermilk, while with much more time the flavour is apt to be injured. During the time of churning, the agitation will raise the temperature by perhaps 3° or 4°.

Sleepy Cream.—Occasionally the complaint is heard from the dairy that "the butter won't come," "the cream is sleeping." Most probably the cause will be that the cream is too cold. Test the temperature with the thermometer, and if it is below 55° in summer, or 58° in winter, raise it to slightly over these points by immersing a vessel filled with hot water (fig. 665).

But the temperature may be high enough and still the butter, or a portion of it, may refuse to come. In this case also, scalding the cream may be effective. If not, the use of a little churning powder, which it is well to have at hand, will most likely make the sleepy, frothy cream give up its butter. Dr Aitken considered a little bicarbonate of soda (baking soda) as efficacious as any butter-powder.

This difficulty is most liable to occur in the cold months of the year, and may be due to various causes besides cold cream, such as the feeding of the cows, a sickly cow, dirty milk-vessels, or to cream from cows that have been long constantly giving milk.

Butter-working.

The working of the butter is part of the operation which demands the most careful attention in its minutest detail.

Object of Working.—The object of working the butter is the complete removal of the superfluous water, the working-in of the salt, and the consolidation of the butter into a solid mass. This should be done by pressure, not by rubbing, in order to avoid injuring the "grain" of the butter. If any portion of casein is left in the butter, it will speedily ferment and spoil the butter. Good keeping butter must be free from casein; and, to obtain this, butter-makers cannot be too careful.

Hand-working Objectionable.—Many eminent authorities state emphatically that in all the process of working, butter should never once be touched by the bare hand. The temperature of the hand is usually so high as to have a tendency to make the butter soft, while there is also some risk of the flavour of the butter being slightly injured by contact with the hand. It is no doubt true that a great deal of first-class butter is worked by the bare hands of the operator, yet the safest plan, unquestionably, is to avoid this practice, and use some of the modern butter-workers. With one of these and the deft use of the "Scotch hand" (fig. 691) there is no need to let the bare hands touch the butter.

Fig. 691.—
Scotch hand.

If in any case the bare hands are to come into contact with the butter, they should be first washed with warm water and oatmeal, and then rinsed in cold water, and this rinsing should be done frequently while the hand-working proceeds. A person with hot clammy hands is not suited for dairy work.

The Butter-worker.—In all well-equipped dairies the butter is now worked by a "butter-worker," such as in fig. 692. With this appliance the butter is never touched by the hands. The butter, taken from the churn in a granular state, is placed in the trough of the worker and the roller passed over it. The pressure removes the water and forces the grains into a solid mass. The butter

Fig. 692.—Butter-worker.

is then rolled up by reversing the motion of the roller, and placed at right angles in the trough to its former position. The roller is again passed over the butter, and this process is repeated until the butter is dry enough to be made up. Over-working at this stage must be guarded against, or the result may be greasy butter.

Circular butter-workers fitted for power are in use in large dairies.

Salting Butter.—The method adopted most generally in dairies and factories is that of dry-salting. After the roller has been passed over the granular butter once or twice the dry salt is sprinkled over the surface of the butter as it lies spread out over the worker. Working then proceeds as usual; but it is an advantage to stop before all the water has been removed, and allow the butter to lie for several hours in a tub of cold water. This allows the salt to dissolve before the final working, which of course facilitates the thorough incorporation of the salt with the mass of butter. Here, again, care must be taken not to overwork the butter, or the grain and texture will be impaired and it will have a greasy appearance.

The quantity of dry salt used for incorporating with the butter is rarely more than $\frac{1}{4}$ to $\frac{1}{2}$ oz. to the pound, according to taste and length of time it is to be kept. Formerly, cured butter was much more heavily salted, but the taste nowadays is all in favour of milder salting. Even when it is to be used as fresh butter, a very little salt will improve the flavour of the butter.

Centrifugal Butter-drier.—This is an ingenious invention in which centrifugal force is employed to remove superfluous moisture from the butter. It is named the "Normandy Délaiteuse." The butter, after leaving the churn, is, while still in a granular state, placed—about 16 lb. at a time—in a canvas bag. This bag is then put into a metal cylinder, perforated with holes, like a colander, which, from motion communicated by the horizontal spindle, is made to revolve rapidly —700 to 800 turns per minute. The buttermilk, and any other moisture the butter may contain, is driven off to the circumference, and thence through the holes into the outer case, whence it passes out by the pipe into a receptacle underneath, the butter remaining in a perfectly dry condition, in immediate readiness for being worked up into pats of whatever shape may be required. The whole operation only takes four minutes, and directly one lot of butter is dealt with another may be put in.

The machine is similar to the hydro-extractor of laundry fame; but it has not proved a success in dairying.

Scotch Hands. — Scotch hands are used for making up the butter into bricks or prints; and where quantities of over 1 lb. are to be made up, it may be done conveniently by butter-beaters. For rolls a board is more serviceable. A "Scotch hand" is represented in fig. 691.

Packing into Crocks. — If the butter is to be kept for a considerable time, it is packed into crocks. And the packing process requires both skill and care. The object is to thoroughly exclude the air, and this will be effectually secured by packing the butter in shallow layers, not much over an inch in thickness. It is a good plan, after placing the first layer in the bottom of the crock, to line the sides with a similar layer as high up as it is intended to fill the vessel. Then proceed to press in one layer after another. Over the butter place a muslin cloth, and cover this with fine salt to the depth of about 1 inch.

To this covering some prefer to have about an inch deep or more of brine floating on the top of the butter. By this method, always taking care to keep the surface of the butter covered with brine, the writer has kept butter, which had been given merely a trace of salt in the working, quite fresh, from the beginning of October in the one year till into May of the following. Even at the very last the butter was perfectly free from any rancid or undesirable flavour, and was so slightly salt to taste as to almost pass for fresh recently made butter. But this butter was made by a skilled hand, who was careful to leave in it the least possible traces of casein, which is so destructive to the keeping properties of butter.

Fresh Butter. — If butter is properly made from pasteurised well-ripened cream, well washed in the churn, and worked so as to have the surplus moisture removed, it may be kept sweet and fresh for eight or ten days without any salt whatever. Care should be taken to keep fresh butter in a cool temperature. In warm weather the farmer should have the butter made and conveyed to market at night or early in the morning; and in retail shops refrigerators should be provided for holding the fresh butter in summer.

Butter Boxes for Transit. — Attractive and convenient boxes are supplied by vendors of dairy appliances for conveying fresh butter to market or to private consumers by post or rail. In some of these boxes there are ingenious contrivances for keeping the butter cool.

If the housekeeper discovers that her supply of fresh butter is likely to become rancid before being used, she will find it a good plan to pack the butter firmly into some fine glazed stoneware vessel and pour some strong brine over it.

Colouring Butter. — A rich golden colour is most esteemed in butter. When it is naturally pale or not sufficiently "gilt-edged" it is a common practice to colour it artificially. This may be done by introducing a little liquid annatto into the cream just before churning is commenced. Experience is the best guide as to the quantity needed to give the required tint to the particular make of butter.

But artificial colouring, like the introducing of preservatives, is an objectionable practice, and where high-coloured butter is desired, the better plan is to have on the farm one or two cows known to produce high-coloured butter. Jersey and Guernsey cows are noted for this property, and one of these will most likely give sufficient "colouring" to the butter of ten or twelve other cows.

CHEESE-MAKING.

The systems of cheese-making pursued in this country are numerous. It is a more intricate process than butter-making, affording scope for the exercise of greater skill in manipulation, and of more ingenuity in producing differences in the manufactured article.

In making the hard cheese of this country the whole-milk as it comes from the cow is dealt with. In making Stilton cheese a little extra cream is usually, and ought always to be, added. The cheese-maker has thus a bulky article to handle, and one which requires to be handled with the utmost skill and care if uniformly good results are to be obtained.

Apartments for Cheese-making.— In well-equipped dairies there are at least three separate compartments for cheese-making — (1) the making-room, (2) the press-room, (3) the curing- or ripening-room. A convenient arrangement is to have the curing-room over the other compartments, or over the press- and making-rooms only. Some prefer to have the curing-room on the ground-floor. An important point is to have this room, and indeed all the compartments, protected as much as possible from variation in temperature, and so arranged that the temperature can be controlled summer and winter. For this reason all curing-rooms, whether on the ground-floor or higher, should have a ceiling as some protection against rapid changes of temperature. The floors of all the rooms, other than the curing-room, are usually of some material which is impervious to water, preferably of good concrete with a smooth surface. The walls may be plastered, cemented, or lined with enamelled tiles. The ceilings, unless where there is considerable vibration, are plastered. Ceilings lined with wood are sometimes a cause of trouble in dairying, as they allow particles of matter to pass through. All drains inside the buildings must be open or surface drains, and where these reach the outside of the walls they must be well trapped, the traps being covered by a grating only.

As in butter-making, the apartments and all appliances and utensils must be kept perfectly clean and fresh. Bad odours and impurities in the milk are fatal to successful cheese-making.

Utensils.—The utensils required in cheese-making are numerous, though few of them are costly. They include, besides the ordinary milk-pails, &c., a milk-vat or tub, strainers, dipper, curd-knives, rake, cooler, draining-racks, curd-mill, curd-shovel, cheese-moulds, hoops or chessets, and cheese-presses. The curing-room is usually fitted with turning shelves, so that two or three shelves, with their contents of cheese, turn round on an axle.

The Vat.—The vessel in which the milk is collected to be coagulated by rennet is called the cheese-vat, or cheese-tub. It is usually oblong, as shown in fig. 693, and mounted on wheels, so as to be easily moved about, and from one room to another. Vats are made of many sizes to suit different dairies. Modern vats have a double casing, so as to admit between the two cases cold water for cooling, and hot water or steam for heating the milk or curd. The inner case should be of the best tinned steel. The vat is provided with a large brass tap for running off the whey, and with smaller taps at different levels for running off the cooling and waste water. Formerly the milk-vat took the form of a circular tub without the double casing, and the milk and curd could be heated only by removing a quantity of the milk or whey, scalding it to a high temperature, and pouring it back into the vat. This was a troublesome operation, and had to be frequently

Fig. 693.—*Milk-vat.*

repeated. In the modern vats with double casing the contents can be heated to any desired temperature by circulating steam or hot water between the two cases. In the same way the evening's milk can be cooled in the vat by the use of cold water. With the round tubs cooling of the milk could be facilitated only by stirring the milk in the vat, or by exposing the bulk of it in shallow pans. The almost perfect control of the temperature of the milk and curd which the modern cheese-vat gives, is of great importance in the making of uniform cheese.

Curd-mill.—The frame of the curd-mill (fig. 694, Pollock's, Mauchline) is of wood or iron, and consists essentially of two horizontal bars supported firmly on four legs. Firmly held between the two bars is a metal grating. Running along the middle of the grating is a regular open space, into which is fitted a revolving iron axle with two or three rows of pins or teeth fixed in it spirally. A wheel with a handle drives the toothed axle, and one tooth of each row passes through each bar of the grating. On the top is fastened a hopper with hinges. The curd is put into the hopper in slices, is cut and broken by the toothed axle in passing through the grating, and falls into the receiver below. The object is to tear the curd into pieces to receive the salt, and facilitate the removal of free moisture in pressing, but not to reduce the curd to pulp and render it greasy, as with many of the older machines.

Fig. 694.—*Curd-mill.*

Presses.—The varieties of cheese-presses are numerous. The more common forms at the present day are fitted with levers, from the simple lever to the combination of lever and toothed wheel, and the double levers (figs. 695 and 696). The essential feature of a good cheese-press is that the load descends automatically after the cheese, which sinks as the whey is expelled. Each press should be capable of giving a pressure of at least 30 cwt.

Rennet. — Rennet is the agent universally employed in cheese-making to coagulate the milk. It contains as its chief constituents the enzymes "rennin" and pepsin. The ferment rennin is found in the gastric juice of all mammals, and specially in the young while still suckling. Rennet is commonly prepared from the mucousmembrane of the fourth stomach of the

Fig. 695.—*Single cheese-press.*

calf. Healthy stomachs are selected, dried, and kept for some time. They are next cut into small pieces, and put into a 5 per cent solution of common salt containing a little boracic acid. After some days a further 5 per cent of salt is added, and the liquid filt-

ered. The clear liquid is extract of rennet. The general practice is to use one of the commercial extracts of rennet, which are now to be obtained of uniform strength and at moderate prices. The strength of any sample of rennet extract can readily be compared with rennet of known strength by adding the same measured quantity of each to two equal quantities of milk taken from the same bulk under the same conditions, and comparing the time that elapses in each case before coagulation becomes evident.

Action of Rennet.—The action of the rennet enzyme on milk is to produce a fermentation, whereby the caseine

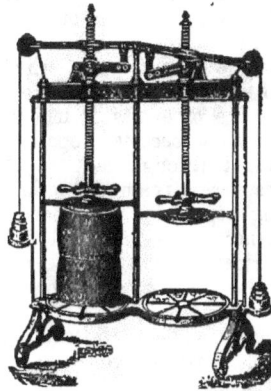

Fig. 696.—*Double cheese-press.*

which is present in partial solution in combination with lime becomes in large measure precipitated in the form of a clot. This clot mechanically encloses and holds most of the fat globules, and forms the curd. The pepsin of the rennet, aided by acidity, plays a most important part in the ripening of the cheese.

Rennet acts only in neutral or acid solutions, and its action is affected by the acidity in milk. The greater the acidity the more rapid its action. Like other enzymes, its action is also dependent on temperature. It acts best at about 105° Fahr. Near this temperature the curd produced is firm, at low temperatures soft and flocculent. Heating rennet to 140° or over causes it to rapidly lose its properties. Milk which has been heated to this temperature and cooled does not coagulate properly with rennet, owing to the rendering insoluble of some of the lime salts. The addition of soluble lime salts to such milk causes it to curdle with rennet in the usual way.

Acidity. — Acidity has a most important and far-reaching effect in the making of cheese. The manufacture of quantities of uniformly fine cheese necessarily implies the successful control of the acid development in the milk and curd at the different stages of making. In the Cheddar system a certain degree of acidity is absolutely essential to success. Unless the milk is allowed to become sufficiently acid before adding the rennet, and the curd before salting, coagulation will be imperfect, the curd obtained will be soft and flocculent from lack of acid to assist in the cooking of the curd, the cheese will develop a bitter flavour for want of the acid, which tends to purify the milk and curd, and from the presence of excess of moisture, the cheese will not become properly ripened or digested through lack of the acid which is necessary to the action of the digesting pepsin, but will remain tough and leathery in texture, and the cheese during ripening will become distended by the development of strong-smelling gases.

Too much acidity is also injurious in a cheese. It will injure the texture and make the flavour too sharp. Such a cheese will be dry and crumbly and not sufficiently mellow. In poor milk less acidity should be developed than in rich milk. For this reason the acidity is kept more moderate in spring than in summer or autumn. Acidity has much to do with the rate of ripening and the keeping properties of a cheese. If too much acidity has been developed the cheese will ripen quickly but will not keep well. With little acidity a cheese ripens slowly, usually develops undesirable qualities, and cannot be said to keep well. The best keeping cheese is one which has the proper degree of acidity developed. For the best results about 1 per cent of acid is necessary in the curd when salted.

Ripening. — The acid which develops naturally in milk and curd is lactic acid, and is the product of bacteria in lactic fermentation. This fermentation is brought about by the important class of organisms termed lactic bacteria. Part of the milk sugar becomes converted into lactic acid, which gives a sour flavour to milk, and in time coagulates it. The development of lactic acid, being a bacterial change, can be controlled by regulating the temperature of the medium; so that temperature is one of the most important factors in successful dairying. Cooling the evening's milk and holding it at a low temperature overnight retards acid development. Heating milk to 80° or 100° Fahr. greatly hastens this change. The development of acidity in milk can also be greatly hastened by the addition of a "starter." Adding a starter is a method of inoculating the milk with large numbers of bacteria, and ensuring that the milk is thickly seeded with the germs necessary for lactic acid production.

"Starters." — Starters used in this way are either home-made or culture-starters. The former include sour whey, sour cream, buttermilk, and milk which has been allowed to sour under natural conditions. In most cases such starters are more or less impure, containing some species of bacteria which bring about undesirable changes in the milk and curd, and give rise to various defects in cheese. The method, once common, of ripening milk by mixing with it a little sour whey is not to be recommended. Mr Drummond, of the Kilmarnock Dairy School, preferred to this system that of developing acidity naturally in the milk by regulating the temperature, by having the evening's milk cooled, so that sufficient heat is left in it to develop by morning the degree of acidity required. The warmer the milk overnight the more rapidly it becomes sour. It was generally found that sufficient acid was developed when the evening's milk was at a temperature of from 66° to 70° in the morning. No doubt a little pure sour whey, judiciously used, gave better results than making cheese with too little acidity in the curd. Since the introduction of culture-starters, however, better and more uniform quality can be obtained from their use in cheese-making than from any of the older methods of ripening milk.

Culture - starters. — It is usually a matter of chance as to what kinds of bacteria a home-made starter will contain, and of late years the use of pure culture or commercial starters has become almost general. A pure culture is a culture of a single species of bacteria in an artificial medium. The species is selected on account of its ability to produce a desirable ripening in milk and curd, and is cultivated and sold commercially by bacteriologists in the form of a liquid or a powder. The powder forms are to be preferred. The principles of pure culture ripening are—(1) the elimination as far as possible of the bacteria usually present, by strict attention to cleanliness so as to prevent unnecessary contamination of the milk, and by cooling the milk as quickly as possible after it is obtained, to check the growth of the bacteria present ; (2) the addition of a large number of the desirable germs to start the proper fermentation, and to enable these bacteria to gain the ascendancy over the other species present.

Important advantages have accrued during recent years from the use of culture-starters in cheese-making in the south-west of Scotland. The quality of the cheese is more uniform than formerly, and of a higher standard of merit, due largely to the more regular control of the ripening processes.

The credit of having introduced these culture-starters for cheese-making into this country belongs to Mr Drummond. He was of opinion that many of the defects then common in Cheddar cheese— such, for example, as discoloration, a trouble then very prevalent, and causing great loss in the south-west of Scotland —were due to the development of undesirable organisms in the ripening milk overnight, and that the keeping of this milk at lower temperatures, followed, in the morning, by the addition of quantities of culture - starter containing the desired organism for ripening, would produce a purer fermentation in the ripening curd. The use of culture-starters in this way was subsequently greatly extended through the influence of the itinerant instructors in the counties of Ayr, Kirkcudbright, Wigtown, and Dumfries.

Propagation of a "Culture-starter." —The culture when received should be fresh and the seal unbroken. In old cultures the germ may have died out. The whole of the powder is added to about half a gallon of freshly pasteurised skim-milk at a temperature of 80° Fahr. The vessel is laid aside, covered with a clean muslin or cheese-cloth, in a suitable part of the dairy, until the milk has become partially coagulated, which, under ordinary conditions, will be in about 18 hours. This process is repeated daily, some of the previous day's starter being used to inoculate the next lot of milk. The quantity of starter used for propagating in this way is about 10 per cent, or 1 lb. of starter to a gallon of pasteurised milk. The vigour of the starter for the first few days daily increases, and all starters should be thus " built up " for a few days before being used in the cheese-vat. A starter should at no time become completely coagulated, but only partially so. Too much acidity diminishes the vigour of a starter. To prevent over-development at any time, dilute with freshly pasteurised skim - milk. Commercial starters may be propagated as long as they retain the pleasant acid flavour and smooth appearance, and give no indication of the presence of gassy or other unpleasant odours. For those who have little experience of starters a safe rule is to obtain a fresh culture every two weeks.

Controlling the Acidity.—The controlling of the acidity is unquestionably one of the most important and most difficult points in the entire process of making. It is important that exact knowledge be acquired, not only of the part which acidity plays in the making and maturing of cheese, and the means by which it may be developed, but also of how it may be measured and controlled. The most rapid development of acid in cheese-making takes place in the milk just before coagulation. After coagulation the acid formation is considerably checked by removal of moisture from the curd, as the more moisture contained in curd the more rapid the acid development. Early renneting of the milk and cooking of the curd both tend to retard ripening. After renneting the most rapid ripening takes place just before drawing off the whey. The cooling of the evening's milk, the adding of the starter, the renneting, the drawing of the whey, are all important methods of controlling the rate of acid development. Other important factors influencing the ripening are the salting and pressing of the curd, and the temperature of the curing-room.

Measuring the Acidity.—The acidity in milk or whey can be measured fairly accurately by the acidimeter. This is a simple and reliable guide in competent hands as to when to draw the whey. Unfortunately, the testing of milk for renneting is not such a simple matter, as the comparatively large and variable quantity of carbonic acid (carbon dioxide) present in the milk at this stage affects the reading, causing it to vary considerably and be unreliable. The subsequent heating of the whey drives off most of this acid gas before the whey is drawn. The most reliable guide at present as to the time of renneting is the rennet test. This is a comparative test only for comparing the acidity in the milk one day with another. It is based on the principle that, other conditions being equal, the greater the acidity the more quickly will milk coagulate with rennet. Four ounces of milk at 84° and one drachm of rennet extract are employed, and the time that elapses before coagulation becomes evident noted.

Perhaps the most reliable and simple test for acidity in curd is the hot-iron test. It is found that acidity has the effect of partially digesting the curd, so that it will string or draw out in fine threads from a hot iron. Within certain limits, the longer and finer the threads the greater the acidity in the curd. This test is a convenient one for all cheese-makers, and in the hands of an experienced person reliable results are obtained.

The methods pursued in making different varieties of cheese will now be dealt with in detail.

CHEDDAR CHEESE.

The Cheddar variety of cheese, which takes its name from the village of Cheddar in the county of Somerset, has been well known for centuries. It was intro-duced into the south-west of Scotland by the late Mr Joseph Harding, Marksbury, Bristol, who is said to have been the first to establish the practice of Cheddar cheese-making upon a regular system. It is now extensively made in that part of Scotland, as well as in Somersetshire and other districts of England. It is also manufactured in very large quantities in Canada and the United States.

The making of Cheddar cheese has received very special study in Canada, and it is remarkable that the employment of Canadian specialists as instructors led to great improvement in the system of Cheddar cheese-making in the south-west of Scotland. Subsequently the Scottish Dairy Institute was established at Kilmarnock, under the management of Mr Drummond, and this school of dairying has had a remarkable influence on the cheese-making not only of Scotland but of the United Kingdom. In more recent years the extension work done by itinerant instructors in the south-west of Scotland, under the joint auspices of the West of Scotland Agricultural College, the various County Dairy Associations, and County Councils, has also been the means of greatly raising the quality of the cheese of these districts to a higher and more uniform standard. The position which Scots Cheddar now occupies on British markets is well exemplified by the fact that cheese from Scotland has in several instances gained the bulk of the premiums at the London Dairy Shows, besides securing valuable prizes at other important English dairying centres.

Character and Composition of Cheddar Cheese. — Before giving the details of the operation of Cheddar cheese-making, it will be well to define as clearly as possible what a typical Cheddar cheese is. Cow's milk, with which we have to deal in cheese-making, usually contains about 13 per cent of solids and 87 per cent of water. Cheese-making is the process of preserving the valuable food solids of milk in the best possible form as human food. A Cheddar cheese, speaking generally, is composed of about equal parts of butter-fat, caseine, and water, with a small proportion of sugar and mineral matter. A good cheese has a quality rich and mellow, a characteristic pleasing flavour, a uniformly close smooth texture, a bright even colour, and a prepossessing attractive appearance. "Quality" in cheese may be defined as the nature of the inherent properties, considered relatively. Thus a rich mellow cheese is of good quality, a hard dry cheese of poor quality. Texture refers to the arrangement of the matter composing the cheese. Thus a cheese is either smooth or rough in texture, according as the compound particles are combined to form a smooth or grainy body of cheese, and open when the pieces of curd, after pressing, do not form a completely solid mass in the cheese. The colour of a cheese may be defined as the quality that affects our senses with regard to its hue or tint. The colour of a cheese is true when the body of the cheese after being cut appears of exactly the same colour throughout, and untrue when the body of the cheese has a mottled or streaky appearance. The appearance or finish is seen on the outer surface. In a cheese this should be considered good or bad according as it is symmetrical in shape, with a smooth clean skin, or unshapely, and with a surface dirty, cracked, and open.

Scale of Points.—The following scale of points of merit is that adopted by the Ayrshire Agricultural Association at their great annual Cheese Show at Kilmarnock: Flavour 40, body and texture 40, colour 12, appearance and finish 8.

Character of Milk.—As has already been shown, milk has a natural tendency to decay, as it offers very favourable conditions for the rapid growth and multiplication of many different species of bacteria. These readily gain access to it in large numbers from the surroundings generally, even when special precautions are taken in regard to cleanliness. And when strict cleanliness is not observed, the introduction of impurities in this way is sure to affect the flavour and other valuable properties of cheese prejudicially. Thus too much care cannot be taken to protect the milk as far as possible from all contamination.

Treatment of the Milk.—In Cheddar cheese-making the evening's milk is kept in the vat, to be mixed with the warm milk in the morning. Besides keeping

it where the surroundings are fresh and pure, it is important that the evening's milk be properly cooled. The aim, where a starter is employed in ripening the milk, is to keep the evening's milk as fresh and free from all germ action as possible, till it can be inoculated with the starter in the, morning and mixed with the morning's milk. This is accomplished by rapid and thorough cooling of the milk in the double-jacketed vat. It should be remembered that early cooling is much more effective in checking germ action than later cooling even to a lower temperature. The aim should be to reduce the milk to about 72° Fahr. as soon as possible after it leaves the cow. The method when no starter was used was to cause the milk to cool gradually from its natural heat to about 68° in the morning. It was then usually in proper condition to be mixed and ripened with the morning's milk without the addition of any form of starter. But when kept warmer than this the lactic acid was in most cases too far developed in the morning to allow sufficient time for the proper cooling and working of the curd.

If the milk has been properly cooled the previous evening, it will be at from 62° to 66° Fahr. in the morning, and the mixed morning and evening milk will require for proper ripening about ¼ or ½ per cent of starter. This is added early in the morning, as soon as the temperature of the milk has been noted and the cream which has collected on the surface removed and again mixed with the milk. The quantity of starter is carefully measured, strained through a clean cloth, and thoroughly stirred with the milk. Heat is applied, and the whole contents of the vat raised to 84° or 86° for ripening, which should occupy from 30 to 60 minutes after the whole of the milk has been collected in the vat.

Testing Acidity.—The most reliable test at present is that founded on the rennet coagulation principle, and with great care in measuring and in thermometer readings, fairly reliable results are obtained. Useful information as to time of renneting can also be obtained by consideration of the previous treatment of the milk, such as the rate of cooling, the temperature in the morning, the quantity of starter added, and the time of ripening. No standard for acidity at time of renneting can be given. This must be determined by each maker according to previous experience in his own particular case. The rennet test is useful only for comparing the acidity of the milk with that earlier or later, or in some other body of milk, such as the milk of the previous day.

Colouring.—When colouring is to be used, it should be added as soon as the whole quantity of milk is together in the vat. About 1½ ounce of annatto to 100 gallons of milk gives a medium bright colour. From 1 to 2 ounces per 100 gallons are commonly used, according to brightness of colour desired. The colouring should be diluted in not less than five times its bulk of pure water, to facilitate its thorough incorporation with the milk. The practice of using artificial colouring is being very properly discouraged.

Renneting.—Pure rennet of known strength should be used, and that also should be well diluted with pure water to ensure its rapid and equal distribution throughout the milk. In spring months sufficient rennet should be used to thicken the curd ready for cutting in 30 minutes, and in the summer months in 45 minutes. When the rennet is added, stirring of the milk should not exceed 5 minutes, as the milk should be quite still when coagulation begins. The surface may, with advantage, be slowly agitated to prevent the cream from rising till coagulation has commenced. From 4 to 5 ounces of rennet extract per 100 gallons of milk is usually sufficient in the spring months, and about 4 ounces in the summer months.

Cutting the Curd. — The curd is ready for cutting when it splits clean before the finger when inserted at an angle of about 45°. Horizontal and perpendicular curd-knives (fig. 697) are used, and the cutting is done very gently to avoid undue loss of fat in the whey. The object of cutting is to facilitate escape of the whey and the cooking or firming of the curd. In some cases a curd-breaker such as is shown in fig. 698 is used.

Cooking.—The cooking of the curd implies the change from the bulky, soft, flocculent mass when cut, to the drier, firmer, and tougher condition observed before the whey is run off. The process is a gradual removal of whey from the curd particles, and is brought about by

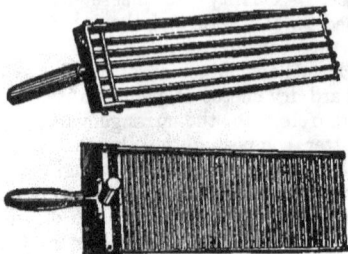

Fig. 697.—*Curd-knives*

the effects of heat and acid. Each of these agencies causes the curd to contract and whey to be pressed out. So that within limits the higher the temperature of scalding, or the greater the acidity, the firmer and more thoroughly cooked the curd. Cooking is facilitated by fine division of the curd in cutting.

After the curd has been reduced to small uniform cubes, heat is applied very slowly, and the temperature raised at the rate of 1° every four or five minutes. Rapid application of the heat at this time forms a skin on the pieces of curd, and thus prevents the proper expulsion of the whey from the curd that the heating is intended to accomplish. The heat in the modern vat is applied, as we have seen, by circulating hot water or steam between the two casings of the vat; and in the older circular tub by heating whey and pouring it over the curd. The temperature of scalding is that which has been found sufficient for the proper cooking of the curd, and will vary considerably in different dairies, and in the same dairies at different seasons. Curd which shows a tendency to drain freely will require

Fig. 698.—*Curd-breaker.*

lower temperatures of scald than curd which is difficult to drain. The adoption of the proper temperature of cooking is an important factor in the making of uniform cheese. In early summer a lower temperature is employed than in late summer and autumn, otherwise the earlier cheese would be too hard. For summer and autumn common temperatures are from 98° to 102° Fahr. The temperature for cooking is reached in about one hour after cutting, gentle stirring of the curd being continued during the process of heating, and for 20 or 30 minutes after. The curd may then be allowed to settle until the necessary degree of firmness and acidity has been attained. It is most important in the making of a good keeping cheese that the curd is properly cooked and the acidity developed to exactly the proper extent before running off the whey.

Testing the Acidity.—A good test for acidity in the curd at this stage is the hot-iron test already referred to. A handful of curd is pressed in the hand until it is as dry and closely matted as possible, and applied to a hot iron just warm enough to melt it without charring. When applied lightly and removed very gradually, fine silky threads are observed to draw out between the curd and the iron. The length and fineness of these threads are reliable indications of the degree of acidity in the curd. When fine threads about a quarter of an inch in length are obtained, the whey, in general, may be removed at once, but the length of threads for the best results will vary considerably in different cases, and can only be arrived at from experience in a given dairy. With proper manipulation, the time from renneting to drawing off the whey will be from 3 to 3¼ hours.

Draining.—As soon as the whey is drained off, the curd is removed from the vat and placed on a cooler over racks covered with a cloth, through which the whey may readily escape. Unless the curd is very firm and dry, it should be stirred for a few minutes to prevent it matting at once, thus facilitating the escape of whey. The curd is then packed to a depth of 5 or 6 inches, and well covered to maintain the heat. In half

an hour it is cut into pieces about 10 inches square and turned. It is subsequently turned and doubled up every half-hour until ready for milling.

Milling.—When the curd has developed so that fine threads from 1½ to 2 inches draw out on the curd when applied to the hot iron, it is put through the curd-mill (fig. 694). The curd at this stage of the process should have a smooth velvety feel, with a flavour like well-ripened cream. The length of time from placing the curd on the racks till milling varies from 1½ to 2 hours.

Salting.—After being milled or broken, the curd is stirred by hand for about 10 minutes to liberate accumulated gases and allow of aeration. It is then piled up and covered for another half hour to further mature before salting. Curd which is salted too soon does not ripen properly, but remains somewhat tough and open in texture and bitter in flavour. In salting, the curd is mixed with salt at the rate of about 2 lb. to 100 lb. of curd, the quantity varying with the time of year and the time to be allowed for ripening. After salting, the curd is left for 15 or 20 minutes to allow the salt to dissolve and to penetrate the curd before pressing.

Pressing.—Unless the curd is specially soft it should not be allowed to cool below 78° or 80° before packing into the chesset (figs. 699 and 700). Pressure is applied to the curd in the chesset very lightly at first, just sufficient to start the whey to run, and gradually increased to 30 cwt. by the end of three hours, so that the curd may become a completely solid mass. The curd is generally pressed in the chesset for three days, and turned and covered with a fresh dry cloth each day. To secure a smooth skin, the cheese is turned in the press on the evening of the day on which it is made, and next morning it is wholly immersed in water at 140° Fahr. for one minute. The bathing gives a tough thin rind to the cheese, and tends to prevent cracking. When sufficiently pressed, the cheese is well greased with pure lard, the ends protected with cotton cloth, and the sides well bandaged to maintain the shape. When the surface is exposed to the atmosphere it dries and tends to crack.

Curing.—The curing-room should be kept dry and well ventilated, and at a regular temperature of from 58° to 64°

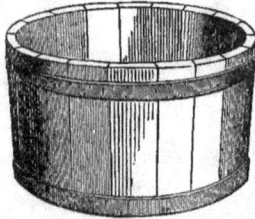

Fig. 699.—*Wooden chesset or cheese-mould.*

Fahr. A warmer temperature ripens the cheese quicker and a lower one slower. The cheeses are turned each day in the curing-room. A Cheddar cheese made and kept in this way should be ready

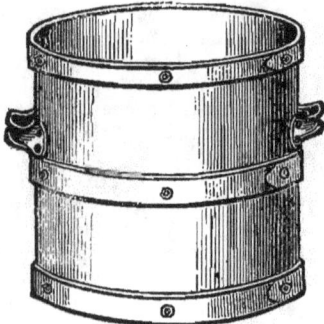

Fig. 700.—*Metal chesset or cheese-mould.*

for use in from two to three months, and should keep well for months afterwards if required.

SKIM-MILK CHEESE.

Skim-milk cheeses are made in several parts of the country, chiefly in Scotland, but without the addition of some portion of cream the cheese is dry and rather tasteless.

Attempts have been made in different countries abroad to replace the fat removed in the cream by the introduction of lard or other animal fat into the skim-milk. But the oleomargarine cheese thus produced is an inferior article, which has been very properly classed as a "dairy abomination."

STILTON CHEESE.

The making of cheese of one kind or another is an important industry in various parts of England. For the notes which follow upon the better known English cheeses the editor is indebted to Professor J. P. Sheldon. In appearance and character quite distinct from any other kind of British cheese, save the Yorkshire Cotherstone, which resembles it more in looks than anything else, Stilton cheese is at once one of the most modern and perhaps the most famous of all the many different kinds that are produced in the British Islands. Late in the eighteenth century it had a local reputation in the district around Melton - Mowbray, chiefly because the well-known Cooper Thornhill, who kept the Bell Inn at Stilton, on the Great North Road between London and Edinburgh, had it always at hand to regale travellers in the old coaching-days. It was first made by Mrs Paulet of Wymondham, a relative of Thornhill's, whose customers were sometimes "gratified" with it "at the expense of half-a-crown a pound,"—so we are told in Marshall's *Rural Economy*, which was published in 1790. It thus received the name of "Stilton cheese"; and the place where, as well as the method by which, it was made, were kept secret for some time. At length, however, the place and the method became known, and it was then made at various farms in the counties of Leicester and Rutland, while in modern days it has been produced in many parts of England, in the United States of America, in Canada, and elsewhere.

Characteristics of Stilton Cheese. —The distinguishing feature in the old-time Stilton method of cheese-making was the presence in the milk of a double quantity of cream—that is, the cream of the evening's milk was added to the morning's milk, which was then made into cheese. Hence, indeed, its superior quality, and the price it used to command. True Stilton cheese is still a double-cream cheese, wherever it may be made; but it is doubtful if more than a small proportion of the Stilton now made is a double-cream cheese. Modern science has modified the method so far as to produce fine Stilton from new milk only, without the addition of cream. Well-made cheese, indeed, ought to be rich enough when yielded by good milk that is clean and fresh, without additional cream. And so, indeed, it is, except on the tongue of an epicure. But we must admit that Stilton is an epicurean cheese after all, and that the highest qualities of Stilton cheese have still extraneous cream in them.

Climate and Soil.—It is said that no other county can produce Stilton cheese equal to that of Leicester and Rutland, soil and herbage having so much to do with the result. It is probable, however, that this claim cannot be sustained, and that the finest qualities of Stilton can be produced elsewhere, on the same method, with a double quantity of cream, and from rich old pasture-land. It is also said that really fine Stiltons can only be made in the five months beginning with May and ending with September. This is probably correct, but the statement is equally applicable to other kinds of cheese.

Method of Making.—The Stilton method is as follows: The evening's milk is put into shallow "leads," or pans, and is skimmed next morning, the cream being mixed with the fresh milk of the morning. The rennet is added when the milk has been raised to a temperature of 83° Fahr., and coagulation is perfected in an hour afterwards. The coagulum is then broken a little and very gently, after which it remains at rest for a quarter of an hour: it is then put into the "leads," over which cloth strainers have been spread to receive it, and the whey drains slowly out of the curd. As the draining proceeds, the corners of the cloth are drawn together and tied, and this is repeated time after time until the curd has become fairly dry and firm. The curd is then put into a draining vat and broken up into pieces, during which time, exposed to the atmosphere and its germs, it becomes impregnated not only with the lactic acid ferment, but also with the spores of the mould, which is so marked a feature in all well-ripened Stilton cheese. A layer of curd is then put into the hoop, and on it a sprinkling of salt, care being taken not to let the salt get too near the outside; then another layer of curd, and on

it salt as before, and so on until the hoop is full, when the mass of curd is lightly pressed down in the hoop.

The hoop is a cylinder of perforated tin, but without bottom or top, and it is placed on a shelf over which a cloth has been spread, and where the whey may still drain away. The hoop is turned "other end down" two or three times a-day, until the cheese is firm enough to be taken out of it, the time required being from five to ten days, or even longer, according to the temperature. The cheese is then bound tightly round with a cloth, which is repeatedly changed for a dry one, until the crust of the cheese has firmed, and the shape can be maintained without the aid of cloths. The cheese is then placed on a shelf in the cheese-room, where it ripens, and the blue mould so highly prized is developed—a process, as a rule, occupying a good many weeks.

No curd-mill is used in the Stilton method, and the cheese is not put into press. Grinding the curd, indeed, would liberate the cream, a portion of which would be lost in the draining, and pressure would cause more of it to escape. In a double-cream cheese the danger is obvious; and even in single-cream cheese there is always a loss of butter-fat through grinding and pressing. In the Stilton method the curd is a good deal exposed to the air. It oxidises it and admits of inoculation by the pollen of the well-known blue mould or fungus—by name, *Penicillium glaucum*—whose influence is predominant in developing the characteristic flavour, and to no small degree the mellow consistency, of the thoroughly matured Stilton cheese. Young persons aspiring to make Stilton cheese will woo success more favourably if they buy a thoroughly ripe, old, mellow Stilton cheese, break a portion of it up into pieces like raisins, and keep it on plates in the dairy. The rest of the cheese should also be in the dairy, helping to impregnate the air with the spores of the mould.

CHESHIRE CHEESE.

Made in the County Palatine of Chester, whose soil in some parts is presumably more or less influenced by saline deposits beneath, this grand type of British cheese was famous, not in England only, but in various Continental countries, long enough before Stilton cheese had been heard of outside its native county of Leicester.

A West Indian planter, some generations ago, was boasting in a Cheshire dairy farmer's house of the fruits they grew in Jamaica, two crops in one year. The farmer went out, and came back bearing a huge Cheshire cheese. "This," said he, placing the cheese on the table, "is the fruit we grow here once, or even twice a-day!" Legend leaves it at that.

Cheshire cheese is no longer the only famous dairy "fruit" we grow in the British Islands to-day. Its reputation has not been lessened, save in a relative sense. Once it stood "first," and the rest were "quite out of it," if one may use a current expression. It still stands where it long ago did, not any longer as first, but on its own lofty pedestal. More famous still are the Stilton and Cheddar, though their fame is modern in comparison. It is still true that a fine old Cheshire cheese, a year and a half old, and mellow, and blue-moulded throughout, is simply unbeatable in its own special way as an epicurean treat.

The method of making Cheshire cheese is an embodiment of simplicity, which time elevated into a science in all the oldest dairies and not a few of the newer ones. Yet for all that, in many dairies given to experiment, various modifications have been introduced, not always with pleasing results. Some have aimed at making an early ripening cheese, the profit of which would not be so long in coming to hand. And the modification in that consists in using more rennet in the milk, and in developing lactic acid in the curd. This is done in some of the best establishments, but only to a safe and reasonable extent, by keeping the loose curd in its vat, in an oven for a day or a night, not pressing it beforehand. Where slow-ripening cheese is made, lactic acid is not intentionally developed in the curd.

It is not unreasonable to assume that the saline element which presumably impregnates the soil of large portions of the county has some direct influence in causing Cheshire cheese to ripen slowly, when lactic acid is not intentionally developed. And a slow-ripening is a long-keeping cheese, as a rule, whilst its blue-mould and its mellow consistency have not been hurried in their natural course of development.

Much less Cheshire cheese is being made nowadays than was the case a few score years ago, because the demands of the milk trade have been so imperative in recent times, and also because the importation into this country of huge supplies of cheese from Canada and the United States, since the beginning of the last decade of the nineteenth century, caused cheese-making to become much less remunerative in these islands than it was before. All this, fortunately for cheese-making dairymen, was collateral with the rapid expansion of the urban demand for country milk in the period denoted. And hence it is that much less cheese is made in Cheshire to-day than was the case a quarter of a century ago. Yet for all that, there is Cheshire cheese to be had, now, quite as excellent in all respects as that of fifty years ago.

LEICESTERSHIRE AND DERBYSHIRE CHEESE.

Between the types of cheese produced in these two adjoining counties there is a good deal of dissimilarity, alike in appearance and in character, though there is much in common between the methods under which they are produced. A first-rate specimen of either of them is, from a gastronomical point of view, a really excellent and attractive article of food. They are of open, flaky texture, and quite uncommonly mellow, whilst their flavour is remarkably pleasant and mild. At the same time, their quality is of the best, because the rich loams of the two counties produce milk which has long borne a high reputation. The marly soils of Leicestershire old turf land are admitted to have a marked influence on the flavour of the cheese made thereupon—a flavour which seems to be incapable of plenary imitation elsewhere. The method of making the cheese of either county is quite as simple as any other. The carboniferous limestone soils of northern Derbyshire produce cheese of good quality and pleasant flavour, from milk which, coming from sound, dry land, is in brisk request for the milk trade amongst great centres of population, as in Lancashire and even in London.

LANCASHIRE CHEESE.

In the district known as "The Fylde," lying to the north of the Ribble, and bordering along the coast of the Irish Sea, a type of cheese is made—rich in quality and mellow in texture, like the land from whose herbage it is produced —which has a good deal of character all its own. To some extent, as in flavour and texture, it resembles the best of southern Derbyshire cheese, than which nothing simpler or nicer, or more appetising, can be found in any country. But the best of the Fylde cheese may be regarded as equal to the best of any other county, not of special manufacture, unless we except that of Cheshire and of Leicestershire. The finest qualities of it have been produced by the special employment of the lactic acid bacillus, developed in a few pounds of curd, not salted, which is held over to become slightly acid and mixed with the fresh curd of the following day. For the rest, the method of making it differs not at all, or differs very slightly, from the old-time method generally employed in the county of Derby. Local demand absorbs all of it, and therefore it is seldom seen away from its own county, within whose borders, indeed, it enjoys close relationship to a vast and excellent market.

YORKSHIRE CHEESE.

In times long gone by the North Riding of Yorkshire evolved two local specialities in the line of cheese—the "Cotherstone" and the "Wensleydale." The latter, made in any ordinary way of cheese-making that suggests itself to the dairymaid, and done up in different shapes and sizes, is the produce of small dairies in many instances, and is consequently without the uniformity of type which prevails in more extensive districts and larger prevalent herds of cows. The former has points of resemblance to the Stilton, alike in outward shape and in-ternal appearance. Whether it is a copy of the Stilton or just a spontaneous and purely local evolution is perhaps conjectural. But in any case it is more like the Stilton than is any other type of cheese that has been evolved—so far as we know—in this or in any other dairying country. It is also popular, though not widely so in the sense that distinguishes the Stilton, and some of its admirers prefer it to Stilton.

OTHER WEST OF ENGLAND CHEESE.

The "blue veiny truckles" of Wilts have an ancient popularity, though the cheese has never attained a very large scale of manufacture. Not uncommonly called "loaves," they are usually about 9 inches deep, and of diameter corresponding thereto. The method of making is not different in any essential point from that employed in many other counties, and the cheese is of average merit.

"Single" and "double" Gloucesters —the difference being for the most part of thickness—are flat cheeses, like those of Derbyshire. They had at one time, long ago, a sort of cosmopolitan popularity. This is now mainly a matter of the past, and one can hardly see why it grew up, except on the ground that Gloucester is practically a seaport, though not on the sea, and that the local cheese, from small beginnings that could hardly be dignified by the name of business, gradually became an article of export to various countries whence came articles of import to the city of Gloucester.

ENGLISH SOFT CHEESES.

For good or for ill, English people have never taken kindly to soft cheese of any kind save to a very limited extent. Cream cheese we eat, at times, but more as a relish and a luxury than as a practical everyday article of diet. Cream cheese, indeed,—the easiest and simplest of all sorts in make,—is produced more or less in most dairying counties, though there is hardly anywhere a regular trade in it, save on a very slender scale. Any one with milk at command can make cream cheese: a piece of muslin and a perforated box comprise the apparatus; in some cases the box is dispensed with. The process, indeed, consists in the cream automatically draining away its own superfluous moisture and becoming about as firm as fresh butter. This takes three or four days, during which time the cheesy little mass of cream becomes—at all events in warm weather—ripe enough for eating.

A genuine soft cheese of English origin is made in the smallest of English counties, at a place called Wissenden. It is named "Slipcote" cheese, because, when ripe, its skin becomes loose and seems to be slipping off. It is made of milk, just fresh milk, which is coagulated with rennet. The coagulum is put into a strainer to allow the whey to drain off. When dry enough, curd that will form a small cheese (of about 6 inches diameter and 2 inches thick) is put into the concavity of a plate, where it drains still more. When firm enough, the little cheese is placed between cabbage-leaves that are daily changed, and is ripe in a week or so.

BATH CHEESE.

In the days when Bath was one of the most fashionable resorts of pleasure, there was a cheese made in the vicinity and named after the town. The cheese, we are informed, was popular with visitors, who make a point of duty to patronise local specialities. It is no doubt true that the cheese had sufficient intrinsic merit as an article of food, quite apart from its claim as a local product.

This speciality in the domain of the dairy had its day, and passed, some fifteen to twenty years ago, into the limbo of things that are being rapidly forgotten. It is, however, now being revived in a form which, based on the science of dairying greatly improved in our time, may haply win back, it is to be hoped, all the ancient popularity of its prototype, with something added to it. Its modern exemplar is Mrs Mabel Loxton, of the Bath Creamery, and its location is promising.

The cheese is to all intents and purposes a "soft cheese," and ripens inside a week. In colour it is white, and in texture fairly firm; it cuts close and

creamy in grain, and in taste it is piquant with the acid of the lactic acid bacillus. The cheese is small, varying in weight from one to two or three pounds, in accordance with the size of the moulds and the requirements of different customers. At present the demand for these soft cheeses is, we may hope, merely in its infancy.

CAERPHILLY CHEESE.

Caerphilly is by origin a Welsh cheese, but since about 1890 it has been making its way into different parts of the south-west of England—notably into the county of Somerset. It is a cheese that is much in demand amongst mining and other industrial communities in South Wales, and is gaining in popularity. Large quantities of the cheese now made in the counties of Somerset, Wilts, Dorset, and Devon are sent to Welsh towns at prices which range from 45s. to as high as 74s. per cwt. Each cheese weighs about 8 or 9 lb. It is lightly pressed, and is sold in a "green" state, usually when not more than ten to fourteen days old.

Method of Manufacture.—The method of the manufacture of Caerphilly cheese is thus described by Miss Jessie Stubbs, itinerant Instructress in cheese-making to the Somerset County Council:[1]—

"The method of making Caerphilly cheese varies with the length of time they are kept before being marketed. Sour or acid cheese take longer to mature, hence sweet cheese are more profitable, as they show quality at a much earlier date, and so do not shrink to the same extent as sour ones. In the summer months the milk is made up daily, but during the cold weather only once or twice each week, as all farmers did, and may do still, find great difficulty in obtaining the desired amount of acidity during the winter period unless the milk is very stale. The difficulty of obtaining acidity or sourness has been overcome in a number of Caerphilly dairies by the introduction of a 'pure starter,' so if the quantity of milk is large enough it may be made up every day with as

[1] *Jour. Brit. Dairy Farm. Assoc.*, 1908.

much ease as if it were the middle of summer.

"The evening's milk is strained into the cheese-tub and stirred occasionally to prevent the cream from rising; or if the day has been close and hot, it should be cooled to 70° Fahr. or under as soon as it is brought into the dairy. This prevents the milk from developing acidity during the night, and also makes the cheese more regular. Next morning starter is added at the rate of 1 quart per 100 gallons of milk, and the night's milk heated to 90° Fahr. It is then left for the acidity to develop, and the addition of the morning's milk. When all the milk is in, take a test with the acidimeter, and when the acidity is .01 higher than on the previous evening, put in the rennet. The amount of rennet varies on different farms, but, usually speaking, the quantity is 1 oz. to 50 gallons of milk. This will, if there is sufficient acidity present, thicken or coagulate the milk in three-quarters of an hour. The temperature for renneting depends on the time of year. As a rule, it should be from 86° to 90° Fahr. When firm, the curd is cut with American knives into pieces about the size of broad beans, and then gently stirred with the shovel breaker for twenty minutes, or until the curd is firm. The maker must use his or her own judgment at this stage, as the degree of firmness varies with the quality of the milk (the curd from rich milk requiring more firming than that from poor). After the right degree of firmness has been obtained the stirring ceases, and a test is taken to enable the maker to calculate the length of time before drawing off the whey. If the test shows the acidity to be .15 the curd is allowed to 'pitch' or settle for half an hour. It is then pushed back with the hands from the tap and the whey started off. If the curd is well pushed back from the tap the whey drains off, leaving the curd comparatively dry. This is now cut round the sides of the tub, piled in the centre, and allowed to drain for fifteen or twenty minutes. Then cut into oblong pieces, throw the curd from outside into the centre, again pile, and test for acidity. If the acidity is .2 to .25, leave for

twenty minutes, and the curd is then ready to vat. No salt should be added. Take 9 to 11 lb. of curd for each cheese, according to the size of the vat, and carefully break it into the vat by hand without squeezing it, as unless care is taken over the breaking a loss of fat will occur. The cheese are now placed in the press, with about 4 cwt. pressure applied. Then take a test: the drainings from the press should be .3 to .35. In three-quarters of an hour the cheese are turned, and a little salt is rubbed on the outside to prevent the cloths from sticking. The cheese are then returned to the press, increasing the pressure to 6 cwt. Leave in the press until the following day. A brine strong enough to float an egg is prepared, and next morning the cheese are taken from the press and placed in the brine, with a small handful of salt on the top of each. At night they are turned, and the other side salted. The following day the cheese are taken from the brine and the moisture wiped off with a damp cloth, after which they are taken to the curing-room. There should be a certain amount of draught through the curing-room, as this will make the cheese coat faster, and they should be turned daily until ready to market. Some makers sell out weekly, keeping only about eight days' cheese in hand."

FOREIGN CHEESES SUITABLE FOR BRITAIN.

For the following notes regarding foreign varieties of cheese capable of manufacture in this country the editor is indebted to Professor James Long :—

The variety of cheese made upon the continent of Europe is much greater than can be realised by those who have not examined the subject. Those which are suitable, however, to British trade and taste are becoming numerous, although the importers confine themselves to Gruyère, Roquefort, Parmesan, Gorgonzola, Edam, and Gouda among the pressed, and to Camembert, Pommel, Port du Salut, and an occasional Brie among the soft varieties, most of which are included in the list, details of the manufacture of which are given below. France claims the longest record, after which come Italy, Switzerland, and Germany. In Germany there is no specially fine variety, such as is usually recognised as a leading cheese, as in the case of the Gruyère of France and Switzerland. Nor do we find any important cheeses in such well-known dairy countries as Sweden, Norway, or Denmark.

Gruyère.

This cheese, which is made chiefly in Switzerland, and in those departments in France bordering upon that country, is well known in England. It is of great size, weighing from 100 to 150 lb., and often being more than 2 feet in diameter and 6 inches in thickness. It is a cheese which, at its best, is mellow, melting on the tongue, homogeneous, a light yellow in colour, without cracks on the crust, with a number of small holes which should not exceed three-eighths of an inch in diameter. The interior of these holes is moist, and the walls glazed, and they usually contain a little brine. The flavour is at once rich and nutty, somewhat resembling the very best Cheddar.

Gruyères are made in three qualities, —fat, half-fat, and lean,—or from full milk, half skim-milk, and skim-milk respectively. Most of the cheeses are made at factories or *fruitières*, to which the milk is delivered by the small producers.

It is warmed to 93° Fahr., and the curd brought by means of rennet in from 25 to 35 minutes. It is then cut with a long wooden knife, and subsequently stirred until the pieces of curd are no larger than peas.

The operation takes place in a handsomely made vat, or kettle of copper, frequently 5 feet in diameter. This kettle hangs upon a crane, and is swung over a wood-fire in the floor. Sometimes it is fixed, and the fire, made in a movable grate upon wheels, is run on a pair of rails from kettle to kettle.

The curd is next heated up to 135° Fahr., the stirring continuing until it has reached a proper consistence, which can only be ascertained by experience. The whole is then allowed to settle, and the cheesemaker skilfully passes a cloth beneath the curd, which has settled at the bottom of the vessel, brings up the ends on the other side, attaches the four corners to a hook hanging from a pulley, and in a few moments the curd is swung over a table and dropped into a mould waiting to receive it.

This mould is open at the side, and can be tightened at will. When once within it, the curd is carefully wrapped up with cloths, and after standing for a short time it is put under a press for the removal of the whey.

It is salted the next day, the salting continuing from day to day for a considerable period, two men being required to move the cheeses, which are placed upon shelves in the ripening-room.

Here three temperatures are, if possible, introduced, at the lowest, middle, and top shelves. These temperatures vary between 52° and 60° Fahr. Poor, or skim-milk, is, however, set at a lower temperature. What art there is in making Gruyère is chiefly displayed in removing the curd from the vat at the right moment, and in efficient pressing, salting, and ripening.

The other cheeses made in Switzerland, but all of which are unknown in the ordinary markets of this country, are the Spalen, the Bellelay, the Battelmatt, the Vacherin, the poor man's cheese, and the Schabzieger, in which the sugar of milk plays an important part. This cheese resembles the Myseost of the Scandinavian countries, and is not likely to become an important article of commerce.

Dutch Cheeses.

The two important cheeses made in Holland, both of which are sold in the English markets in large quantities, although much less popular than was formerly the case, are known as round or *Edam*, and flat or *Gouda* Cheese. We have seen Dutch cheeses made in various parts of the province of North Holland, and have found that, although the systems of most makers differ, it is only in minute details—which, however, are sufficient to improve the flavour of the cheese.

Edam.—A round Dutch cheese weighs about 5 lb. The milk is sometimes partially skimmed, but the best makers remove no cream from it. The cows are milked in the meadows, and the milk—

placed in round wooden tubs, which are taken to the cows by boat along the dykes which divide each farm—is renneted before starting for the dairy, at from 85° to 90° Fahr.

The curd usually forms in from 15 to 30 minutes, in accordance with the custom on the farm, and it is slowly cut with a wire cutter during 10 minutes, when the whey commences to separate. ·

The colouring is added with the rennet, if it is used.

After manipulation with the hands, the whey is baled out with a ladle, and the curd gathered together and again worked. As no mill is used, it is broken in the tub, and more whey removed, after which it is gathered together in a round mould and pressed for a short time. This pressing continues until sufficient whey is removed, when the curd is placed in a mould which resembles an egg-cup with a lid, which gives it its circular form. In this it is placed under a unique lever press, which is common in Dutch dairies.

Next day some salt is placed upon the top, but the cheese is reversed from time to time, while always being salted on the top. This continues for from 8 to 10 days, when it is put into a vat of thick brine for from 12 to 24 hours, being subsequently washed and removed to the ripening-room, where it stands upon a shelf, as near as possible at 70° Fahr. The cheese is turned daily until it is fit to sell.

It is well rubbed with linseed-oil and coloured yellow or red, in accordance with the market to which it is destined, the surface being scraped smooth and fine.

This cheese is made to an enormous extent in the province of North Holland, the chief markets being Hoorn, Edam, and Purmurend, all of which are within a convenient distance of Amsterdam.

Gouda.—The *Gouda*, or flat Dutch cheese, when at its best closely resembles the fine flavour of English Cheddar. It is much larger and heavier than the Edam, and although flat, has rounded sides. It is generally possible to purchase cheese of prime quality in Amsterdam, although its very high price prevents any considerable sale in this country. The milk is set at 92° Fahr.,

sufficient rennet being added to bring the curd in 25 minutes. It is then cut either with a knife or a lyre-like implement common among Dutch cheesemakers.

As the whey exudes, it is removed from the tub, and the curd carefully and gradually broken up into fine pieces with the hands. It is subsequently pressed and squeezed in a large perforated basin-like mould, in which it is again pressed for the removal of the whey.

The cheese afterwards goes into the mould which gives it its shape, and in 24 hours is salted, salting continuing from day to day until it is fit for the brine-vat, where it sometimes happens that hot water is added to the curd after the withdrawal of the whey, in order to harden it. This is a rougher plan of heating up than the operation as performed in England.

The Dutch cheeses are undoubtedly a boon to our working classes, many of whom prefer them to inferior home-made cheese at similar prices.

Parmesan.

Parmesan cheese is manufactured in Italy, chiefly in Parma and Emilia. It is generally known as Grana, on account of the fine grain into which the curd is brought during manufacture. In size and shape it resembles Gruyère, but often weighs more than 150 lb.

Parmesan requires keeping for a considerable period, sometimes three years, until it is fit for the market, and for this reason the export trade is in few hands, the makers being obliged to sell to the dealers while the cheese is new, for they complete the process of ripening in the marvellous caves which are built beneath their premises.

The true Parmesan is full of minute holes, and when cut in halves emits a sticky sweet substance, which has caused the term "honeyed" to be applied to cheeses of the finest quality. The flesh of the cheese is a pale straw colour, but the crust is often almost black from its age and the colour which has been applied to it. Like the Gruyère, the Parmesan is made in factories, where the milk is carried by small farmers, as in Switzerland. In one of these establishments, where we were enabled to learn the process, as many as eighty persons brought in their milk, varying from 4 to 60 litres apiece. Two men only were required to conduct the work.

The milk is put into a kettle of solid brass, and resembling in shape an inverted bell. This hangs from a crane over a fire in the floor. The milk is heated to 92° Fahr., when the cheese-maker takes a piece of solid rennet, the size of a walnut, which he places in a cloth, and dipping this into the milk, wrings it for some minutes, until its virtue has passed into the milk. The strong-smelling animal matter is then thrown away.

The curd is sometimes brought in fifteen minutes. It is then roughly cut, and subsequently broken up with two implements — one called the *rotilla*, a long staff with wire-work bound around its head, and the other a rod with a disc at the end. Stirring is continued until the grain is almost as fine as large shot; some cold water is then sprinkled over the surface, the kettle is swung over the fire a second time, and the milk heated to from 104° to 110° Fahr., stirring being continued the while.

When the Grana, which is continually tested, is fit, the whey is dipped out, and the curd, which has been pressed into the bottom of the kettle, is removed into a cloth by two men, and placed in a large vessel for half an hour, after which it is removed into the mould. Here it is wetted with whey two or three times, in order to keep it sufficiently flexible; but it is also pressed by lying between two boards, and having weights placed upon the top.

The cloths are removed from time to time, when the cheese is covered with buckram, which gives an imprint to the skin. The buckram is subsequently cut, and the cheese is salted and again pressed. This process continues every other day for a fortnight, when the cheese is cleaned and scraped and taken to the ripening room, where it is greased and turned from time to time at suitable temperatures until it is ripe. In the ordinary way, however, it is sold to the dealer while it is yet young and green, very few of the makers venturing to complete the ripening process.

Gorgonzola.

This blue-moulded cheese, which somewhat resembles Stilton, is made chiefly in Lombardy, in moulds which are 12 inches in diameter by 12 inches high.

The curd is chiefly prepared by owners or drivers of cattle, and sold to the merchants when it has become solid, and formed into a cheese to ripen. The practice is to add the rennet to the evening's milk while it is from 85° to 95° Fahr., so as to bring the curd in fifteen minutes. It is then cut and broken up and ladled into cloths, which are hung up to drain in a cool apartment until the following morning.

The milk of the morning is served in a similar manner, except that the cloth holding the curd is placed into a bucket or vat to drain for some ten to fifteen minutes. At the end of this time the curd of the evening, which is cold, and the warm curd of the morning are placed in the cheese-mould, care being taken that the top and bottom, as well as the sides, are composed of the warm curd. The middle of the cheese is built up of alternate layers of cold curd and warm, the maker plunging his fingers occasionally into the mass to amalgamate them.

When filled, the cloth which envelops the curd is folded over the surface, and the cheese is allowed to settle until it has sunk into the lower half of the mould—for it is divided into two pieces: the top portion is then removed and the lower one reversed, that the cheese may drain and present a better face.

At the end of twelve hours it is again turned, and the mould tightened. Next day the cloth is removed, and the cheese begins to take its form.

It is then removed into an apartment of 65° Fahr., where it remains for three or four days, at the end of which time the mould is removed altogether, and salting commences.

One-half of the cheese, and that always the top, is daily sprinkled and rubbed with salt, being reversed the following morning. This salting continues until, in the judgment of the maker, sufficient has been given; brining is then practised for a few days, and the cheese is next taken to the cave, which must be cool and moist, and is preferable if a damp draught is passed through it.

By this time a red mould has commenced to grow over the surface, and the cheese now requires great care in management and frequent turning. In from four to five months it will be ripe for the market, and will be veined throughout the interior with green mould.

The following are analyses of Gorgonzolas, some of which, it may be mentioned, are made without any mould, for the higher Italian classes, many of whom prefer it, as the green fungus has been at times produced by artificial means, which are objectionable :—

German Analyses.

					Soxhlet.
Water	.	.	.	36.72	43.56
Fat	.	.	.	33.69	27.95
Casein	.	.	.	25.67	24.17
Salt	.	.	.	3.71	4.32

Professor Kinch's Analyses.

				White.	Blue.
Water	.	.	.	48.99	24.96
Fat	.	.	.	26.50	26.10
Casein	.	.	.	21.11	43.46
Salt	.	.	.	3.40	5.22
Sugar	.	.	.	..	.26

The analyses by Professor Kinch were of cheeses made under the direction of the writer, at the Royal Agricultural College, Cirencester—10 gallons of milk made from 14 to 15 lb. of cheese.

Roquefort.

This cheese, also somewhat popular in this country, is properly made in the Aveyron, in France, from the milk of the ewe, some half-million of these animals being kept in one district alone for the purpose; but its large sale has now induced makers to use the milk of the cow.

The Roquefort is a small, round, flat cheese, weighing about 5 lb., and, like Gorgonzola and Stilton, it is veined with blue mould. This, however, is obtained in a different way, as will be seen.

The evening's and morning's milks are mixed together, and brought to a temperature of about 90° Fahr.; the rennet, which is made from the stomach of the lamb, is added, and the curd brought in a short time.

It is then cut and broken down, and much of the whey removed. The curd is afterwards conveyed into the mould in three layers, between each of which a quantity of specially prepared mouldy bread crumbs are sprinkled, the bread being made from a mixture of wheat and barley flour.

After pressure, and when the cheese has attained a distinct form, it is removed to the drying-room for two or three days, when it is carried to the celebrated caves which have made the district so famous, and which are extremely humid, the temperature being about 46° Fahr. Here it is from time to time scraped, as mould grows upon it, salted, and ripened. Machines, however, are now used in some instances for brushing the rind instead of scraping it, and also for piercing the cheeses with needles, in order to encourage the growth of the fungus within.

Cantal.

The Cantal cheese, which is an extremely important one upon the Continent, and which is probably destined to make its appearance in this country, is chiefly made in the Auvergne, and varies in weight from 40 to 100 lb. It is of piquant flavour, has a solid consistence, and may be termed a hard cheese.

Cantal is made from milk at a temperature of 75° Fahr.; the curd is broken up in an hour, the whey removed, and the solid remnants gathered together in fifteen minutes, when they are kneaded and further drained. The curd is then put into a vessel pierced with holes, and again pressed with the hands, and indeed with the body, the maker frequently getting on to the top of the mould and pressing with his knees. The mass is then reversed, and left under heavy pressure for twelve hours, being kept warm the while. Each lot of curd manipulated in this manner is called a tome—a full-size cheese requiring from three to four tomes in its manufacture.

When the real cheese-mould is about to be filled, the masses of now solid curd tomes are broken up with the fingers into small pieces, the whole salted, and finally put into the moulds in cloths, and sent to the press. Here the cheese obtains its final form, and when sufficiently solid to be removed from the mould, it is taken to the cave to ripen.

The Cantal is ripened in about two months, and when made of full rich milk is of very fine quality.

Camembert.

This is the most popular of the small cheeses of the Continent which are sent to this country. It is made upon one principle, under various forms, in the department of Calvados and in the neighbouring districts of Normandy.

In a general way the evening's milk is skimmed and added to that of the morning, and heated to from 80° to 85° Fahr., sometimes higher. There are makers, however, who make three batches daily from three several milkings, thus preventing the necessity for heating the milk.

The curd is brought by the use of rennet in from one and a half to four hours, according to the custom on the farm. While still warm it is ladled into cylindrical moulds, placed upon mats made of rush or reed, upon benches of cement or galvanised metal. Each mould is nearly filled with each batch of curd, and by the time the next curd is ready, the first will have sunk in the mould by reason of drainage, when it is filled again.

When the second curd is sufficiently low the mould is skilfully reversed, and kept upon the cheese until it is firm enough to handle; it is then salted upon one side, and left until the following day to be salted on the other.

After salting it rests upon shelves for a few days, when it is carried to the *séchoir*, or drying-room, an apartment through which currents of air are induced to travel in all directions. Here a white mould appears; and when the velvety pile is at its best the cheeses are conveyed to the cellar, which is usually dark, damp, and free from draught.

It is then turned daily until covered with a green mould. During the growth of this fungus the flesh of the cheese will have gradually changed its condition, and in from five to six weeks it will be fit for market.

Camembert and Brie Bacteria.

M. Georges Roger, Director of the Laboratory of La Ferté, France, has isolated bacteria, which he claims are responsible for the maturation of Brie and Camembert. He has studied the mode of cultivation of these types, is acquainted with both the technique of the cheese dairy and the scientific causes of the changes which occur, and has adopted a system with the object of securing the best results. He cultivates the pure organisms, employs absolutely pure milk and sterilised rennet, giving special attention to the question of the purity of the dairy. It has been noticed that in successful French cheese dairies certain moulds are practically in possession. In some cases the results are only second-rate, while in others, and these the best, they are practically always good. The walls of the apartment and its shelves and plant are sterilised by the aid of boiling-water. Lime-washing follows, and this is succeeded by the process of inoculation with the spores and cells of the mould and the bacteria.

Brie.

This cheese, which is the most popular in France, is chiefly made in the department of Marne, not far from Paris, and sent to the Paris markets, where it obtains high prices. It varies from an inch to an inch and a half in thickness, and from 9 to 12 inches in diameter. Its character very much resembles that of Camembert, although it is differently made.

The new milk set at 83° Fahr. is brought to a curd in from three to four hours, although details differ upon various farms. The mould is made in two parts, the top portion fitting into the bottom. This is placed upon a mat and a beech board, and the curd is laid within it in large, thin, unbroken slices until it is full. It remains to drain until the top portion of the mould can be removed; the cheese is then reversed by the aid of a clean mat and board, and in time becomes firm, when the mould is removed altogether.

It is then salted, as in the case of the Camembert, and finally taken, first to the drying-room, and subsequently to the cellar, an apartment which, as we found in the Brie district, was not only extremely dirty, but positively reeked with fungoid growth upon the walls and shelves.

The cheese is speedily covered with white mould, with specks of blue here and there; it then goes to the cellar, and is soon covered with blue mould, upon which patches of a vermilion mould commence to grow.

There is no more delicious cheese than the Brie, not even excepting the Camembert.

The Neufchatel.

This cheese, sometimes called the Bondon, is largely made in the department of Seine Inférieure. It is a small loaf-shaped cheese, about 3 inches high, and 1½ to 2 inches in diameter, and is properly made from new milk, although the majority of makers, many of whom we have visited, in this department use milk which has been partially skimmed. For such cheeses the makers obtain only a penny apiece in the Paris markets.

The majority of the makers are farmers of the smallest class, who have not sufficient milk to make large cheeses.

The milk is set directly it comes from the cow, and sufficient rennet is added to bring the curd in twenty-four hours. It is then ladled into a cloth, stretched by the four corners over a draining-tub, and left to drain for twelve hours; the partially solid curd is then removed to a press, in which it remains for some hours, until in the judgment of the maker it is fit for moulding, At this moment it is worked up with the hand, and each cheese is moulded separately in a small brass cylinder, and placed upon a straw-covered shelf to dry. Here it becomes covered with white mould, which subsequently changes to blue, the apartment being maintained at 60° Fahr., or a little less.

The cheeses are turned daily; and when a second lot of white fungus has covered the blue it is ready for the market, and will keep a long time.

This cheese is salted after it has been dried upon the shelves for a day.

It should be added that although the Neufchatel is sometimes sold in a white or fresh condition, yet if ripened it becomes veined with blue mould as in varieties like the Stilton and Wensleydale, and in this stage it is one of the choicest varieties in the French market.

The Bondon, which is made in a similar form to the Neufchatel, is an unripened cheese made of new milk, and it is often produced in London by East End purchasers of sour or stale milk.

Pont l'Evêque.

This cheese, which is fairly well known in English dairy schools, although it is not upon the market, is one of a tasty and admirable character. It is made in different qualities, but the true cheese, which was formerly known as the Augelot, is produced from whole-milk. We have seen the process of manufacture in the dairy of one of the best makers near the village of Pont l'Evêque. The milk is brought to a temperature of 100° Fahr., and, the rennet being added, it is mixed with the milk by stirring or by the hand of the maker, the curd being produced in about fifteen minutes. The curd is subsequently cut with a knife, when a shallow vessel is placed upon it in order to express the whey, the whole being covered with a clean cloth. At the end of about ten minutes the whey is removed by baling, while the comparatively dry curd is placed upon rush mats, where it continues to part with its whey. When the curd is ready it is placed in square wooden moulds, which are turned several times during the succeeding half-hour; it is then placed upon a second mat, where it is turned some half a dozen times during the day. On the following morning the young cheese is firm and is salted with the finest salt—which has been well dried in the oven,—one side being salted in the morning and the other during the evening following. The cheeses are then taken to the first ripening- or drying-room and placed on the shelves upon rye straw, the apartment being well ventilated, and here it remains for three days, being turned each day. When quite dry the Pont l'Evêques are carried to the "cave," or final ripening-room, which is not of necessity a basement. During the process of ripening in the "cave" the cheeses are turned every other day, care being taken to prevent the entrance of flies. The best class of Pont l'Evêque remain in the "cave" from a fortnight to three weeks, depending on their thickness, cheeses of the larger size requiring a still longer time before they are fit for the table.

The best class of Pont l'Evêque is made only in the autumn, whereas the summer cheeses are made from May until autumn commences. These cheeses are made of the mixed milk of the morning and the evening, a portion of the cream being removed. The chief features of the process of manufacture are the rapid separation of the whey by early breaking of the curd and the prevention of the development of mould upon the surface, so that the process of ripening depends entirely upon the work of bacteria within the cheese.

ASSOCIATED BUTTER AND CHEESE MAKING.

An important development of the dairy industry is the organising of establishments in which the milk produce of the cows on several different farms can be collected together for united manipulation. These establishments are of different kinds, with different yet similar aims and objects.

Creameries.—The original conception of the creamery was, as a rule, to receive only the cream from the farmer, for which he was paid either by measure or weight, or, as is more preferable, according to percentage of butter-fat, or actual butter yield: in the latter case each farmer's consignment requires to be tested or churned separately. The former is the more convenient to the creamery staff, but the latter is more satisfactory, as a rule, to the consigner. A third plan, which has certain points to commend it, is for the farmer to bring the whole-milk to the creamery twice a-day—just after it is milked—and get it at once separated by the centrifugal separator. In summer he will likely take back with him the fresh skim-milk, for in that season he can usually turn it to better account on his farm than could be done by the creamery; but during the months of October to March inclusive there is usually a capital demand for separated milk in the large cities, at fairly remunerative prices, ranging from 4d. to 5½d. per gallon. If there are means of properly cooling the milk at the farm, and care be taken to keep the milk of each milking separate, it is not usually necessary to convey the milk to the creamery twice daily excepting during the period of warm weather.

In the creamery the cream is made into butter in large quantities at a time, thus securing a product of uniform character and appearance, which is so important in the sale of butter. In some cases creameries also do a considerable trade in selling fresh cream for table use in towns, this latter, indeed, being the more profitable method of disposal.

Butter-blending House.—This is a sort of modified butter factory, in which butter is collected in small quantities from farmers, graded according to quality, and submitted to a certain amount of remaking. The object here is to rectify the home defects in the "working" of the butter, to grade, blend, and remake it, so that it may be presented in the market in large quantities of uniform character and attractive appearance. This system has been highly successful in France, especially in Normandy, and is being carried out at various centres in this country with results which seem to be fairly satisfactory. In Ireland a considerable trade of this description was developed, but the great spread of the co-operative creamery system has tended to circumscribe the operations of the blending factory.

Dairy Factory.—Then there is the dairy factory, which, known perhaps by different names, embraces all the branches of dairying (excepting, most likely, the keeping of cows)—buying in new milk from many farmers, selling some of it as fresh whole-milk, making cheese of various kinds, separating the cream from the milk by the centrifugal separator, making butter, selling fresh cream, selling the skim-milk back to the farmers or through towns and villages, and perhaps feeding part of it to calves and pigs.

These latter are generally establishments of considerable size, and, like the creameries and butter-blending houses, are conducted in some cases as distinct businesses, and in others in co-operation with the farmers who produce and supply the raw material.

Co-operative Milk Depot. — The formation of co-operative milk depots by proprietors and tenant farmers in milk-producing centres in Scotland is referred to on p. 485.

LIVESTOCK

LIVESTOCK

CONTENTS OF VOLUME III.

LIST OF ILLUSTRATIONS IN VOLUME III.

ANIMAL PORTRAITS.

NOTE.—*The portraits in the plates are in all cases reproduced from photographs of the living animals, the great majority of the photographs having been taken by those highly successful and widely known live-stock photographers, Mr C. Reid, Wishaw; Mr G. H. Parsons, Alsayer, Cheshire; Mr F. Babbage, London; Messrs Brown & Co., Lanark; and the Sport and General Illustrations Co., London. Excellent photographs for the work were also received from Messrs Lafayette, Dublin; Messrs Chancellor, Dublin; Mr D. M'George, Coupar-Angus; Mr G. Wickens, Bangor; Mr Hayworth, Knighton, Radnorshire; Mr Marshall, Henley-on-Thames; Mr Gay, South Brent, South Devon; Mr Abernethy, Belfast; and others. Messrs Hislop & Day, Edinburgh, gave the utmost care to the preparation of the plates.*

GENERAL ILLUSTRATIONS.

THE

BOOK OF THE FARM.

———•———

FARM LIVE STOCK.

In the number, variety, and character of its races of farm live stock, the United Kingdom possesses a source of wealth that is practically inexhaustible, and that may for many years go on increasing. Not only are the numbers of animals that are maintained exceptionally large for the extent of territory, but by the skill and enterprise of the owners and occupiers of land in this country the many breeds and varieties of British farm stock have been raised to a general standard of merit that is universally acknowledged to be unique.

It was vastly to the advantage of early improvers of live stock in the British Isles that in the native races they had the very choicest of material to work upon. In each of the different classes of stock there was a variety of type that seemed almost endless, and not only this, but the dominant characteristics throughout all were those represented in symmetrical formation and high value for practical purposes.

It is interesting, indeed, to note that all through the ages of recorded history this country has been strangely free from animals, either tame or wild, with "humps" or with bodies otherwise prominently ill-proportioned. True it certainly is that, long before scientific breeders got their hands upon them, British live stock were distinguished for the symmetry of their formation as well as for the robustness of their constitution. To seek for an explanation of these characteristics would be difficult and of little avail. Not so difficult is it to account for the almost endless variety of type represented in the native races of British live stock. It is by nature's own bountiful design that the *fauna* as well as the *flora* of a country become modified by environment. Great as was the variety in the type of British native live stock, it was no greater than the infinite variation in the climatic conditions, geological formation, and general natural phenomena of the British Isles would lead one to expect.

In itself this wonderful variety of type has been an element of inestimable value in the hands of skilful breeders. In the raising up of races of stock specially adapted for peculiar surroundings and for different purposes, it has enabled them to reach a higher degree of success than could have otherwise been attained. But it is also true that the value of that element of variety has been vastly enhanced by the inherent regularity of structure so characteristic of almost all the numerous native types of stock.

In nearly all civilised countries British live stock are being employed iu the improvement of the native races. The results obtained are everywhere striking.

In no respect are they more so than in an unvarying tendency towards a truer harmony in formation. Under the influence of British stock the coarse, ungainly irregularities of native foreign races disappear with unfailing certainty, often with a rapidity that is amazing. To students of heredity this result is not in any sense surprising. Of all the useful characteristics possessed by British breeds of live stock, none are older or more strongly established than their unique symmetry of structure; and so, in obedience to the beneficent laws of nature, this valuable hereditary force exercises a paramount influence wherever it is employed.

Not only have live stock improvers in this country guarded with jealous care the finely proportioned structure and other valuable properties of our native races of animals, but have, as the outcome of long years of skilful breeding and general management, so developed the general utilitarian qualities as to vastly increase their value for the various practical purposes for which farm live stock are bred and reared. And of all this the net result is that the United Kingdom has come to be looked upon as the parental stud-farm for enterprising stock-owners in all advancing countries.

It is not surprising, therefore, that, with the greatly depreciated values of grain in spite of a growing population, live stock interests bulk more largely now than ever before in the fabric of British agriculture. This important development naturally demanded, and has received, due consideration in the preparation of the Fifth Edition of *The Book of the Farm*. It was decided that this, the Third Volume of the work, should be devoted exclusively to the Live Stock branch of agriculture, and in order to ensure, as far as practicable, that the volume shall be worthy of its great purpose—the promotion of British live stock interests—the matter for it has been almost entirely rewritten. In this important work valuable assistance has been willingly afforded not only by many leading writers on live stock matters, but also by a large number of men who have attained distinction as breeders of different classes of farm animals, and to all these the grateful thanks of the Publishers and Editor are heartily accorded.

BREEDS OF HORSES.

There are differences of opinion as to whether the horse had a single or a multiple origin. The former is the prevailing view, but Professor Cossar Ewart and others maintain the latter.[1] Low gives the following as the six species of the "one genus of the tribe—namely, Equus": (1) *Equus asinus*—the ass; (2) *Equus zebra*—the zebra; (3) *Equus quagga*—the quagga; (4) *Equus Burchellii*—the striped quagga or zebra of the plains; (5) *Equus hemionus*—the dziggetai; and (6) *Equus caballus*—the common horse.[2] To this classification Professor Cossar Ewart, in the article just quoted, adds *Equus caballus celticus*—the Celtic pony.

It is, of course, with the common horse that we are mainly concerned. Low thinks it natural to refer the origin of the horse to the countries of Western Asia to the southward of the Euxine and Caspian Seas, but he acknowledges the probability that the species may also have been diffused from Africa and Eastern Asia.

When the horse was introduced into Great Britain is not known. Julius Cæsar found it here in large numbers when he invaded the country fifty-four years before the Christian era. Whether these early British horses were brought from the East in a state of domestication or reclaimed from the wild horses roaming in the wastes of Europe is uncertain. Little is known of the character of the horses in Great Britain at the time of the Roman invasion, but it is believed that they were of a somewhat coarse, draught-horse type, strongly built, but lacking in quality and action. It is further believed that for several centuries little change was effected upon British horses, beyond such variation as would be induced by the differences in their environment throughout the country. It was apparently not till some time after the Norman Conquest that the systematic improvement of British horses had made

[1] "The Multiple Origin of Horses and Ponies." *Trans. High. and Agric. Soc. of Scotland*, 1904.
[2] Low's *Domesticated Animals of the British Islands*, 1842.

substantial progress. Superior breeds of horses were introduced by the Normans from the continent of Europe, including the great black horse of Flanders and Germany; and gradually better classes of horses were bred in this country.

More marked still were the changes for the better which were brought about in British horses by the establishing of the race of English Thoroughbreds in the time of James I. In the building up of that noble breed the choicest of material was drawn from the countries of the Mediterranean, and still more effectually from the distant deserts of Syria and Arabia. Of the influence exerted upon British horses by the Thoroughbred, Low writes: "The effect has been that a breed of horses has been formed of peculiar lineage and characters, and been mingled in blood with the native varieties in every degree. Iu this manner the property of blood, as it is technically termed, has been communicated to the inferior races, and varieties have been multiplied without limits. Not only does there exist the diversity of what may be termed natural breeds, but those further differences by the greater or less degree of breeding communicated to individuals. Many remain with little or no admixture of the blood of the race-horse, and so may be regarded as native breeds or families; but others are so mixed with the superior horses, or with one another, that they cannot be treated as breeds, but must be regarded as classes suited to particular uses."[1]

It is of the varied material thus described that the present-day breeds of British horses and ponies have been built up. Some of the mixed "classes" referred to by Low have disappeared, but others have been cultivated with such skill and enterprise that they have developed into well-established breeds of great value.

It is known that the ass existed in this country in the time of the Anglo-Saxon kings, but it did not become numerous for several centuries afterwards. As the poor man's horse the ass has long had an established position in the British Isles, and especially in Ireland it is now kept in large numbers.

The faithful and patient mule is a follower of the ass, and so useful and thrifty is it that one wonders it is not reared to a much larger extent than is the case in this country.

The Thoroughbred, with as free will as ever, still stands at the head of British horses. It is followed by a large number of races of riding and driving horses and ponies, including the Hunter, Cleveland Bay, Yorkshire Coach-horse, Hackney, Hackney Pony, Polo Pony, Welsh Ponies, Fell Ponies, the Connemara Pony, Exmoor and Dartmoor Ponies, New Forest Ponies, Highland Ponies, and Shetland Ponies. Several of these varieties, notably the Cleveland Bay, Yorkshire Coach-horse, and the heavier of the Connemara, Fell, and Highland Ponies, are used largely for draught purposes, as well as for riding and driving.

Of draught-horses in the stricter sense there are now only three recognised breeds in the British Isles. These are the Shire, the Clydesdale, and the Suffolk breeds. Outside the limits of these distinct breeds there are numerous varieties of draught-horses of a generally useful character. They are of mixed breeding, differing greatly in type and weight, but, on the whole, well fitted for their respective spheres of usefulness.

BREEDS OF CATTLE.

All varieties of cattle, whether wild or domesticated, belong to the Taurine group of the Bovine race, the other groups being the Bisontine—the bison tribe, and the Bubaline—the buffalo tribe. As to the early history of the British varieties of cattle there has from time to time been much discussion. Even yet their true origin is to some extent shrouded in mystery, and it is unlikely that all uncertainties will ever be fully cleared up.

Amongst leading authorities the prevailing view is that the breeds of cattle to be found in the United Kingdom at the present day all trace their descent from those two types of the sub-genus *Bos taurus*,—the *Bos urus* and *Bos longifrons*.

[1] Low's *Domesticated Animals of the British Islands*, 1842.

The *Bos urus*, known also as *Bos primigenius*, was a type that attained to gigantic dimensions, far in excess of any living variety of cattle. It is recorded that in specimens of the type the length of the body, including the head, was about 11 feet, the height at the mane 6 feet 6 inches, the span of the horns 2 feet 2 inches, and the girth of the horns at the base 14 inches. The *Bos longifrons* type was much smaller—often smaller, it seems, than some of the existing varieties of cattle.

Except in the matter of size, there was little if any difference between the two types, and thus there are those who regard the *Bos urus* and the *Bos longifrons* as belonging to the same species. Low says: "We can, by all the evidence which the question admits of, trace existing races to the ancient uri which, long posterior to the historical era, inhabited the forests of Germany, Gaul, Britain, and other countries. It is a question involving an entirely different series of considerations whether these uri were themselves descended from an anterior race, surpassing them in magnitude, and inhabiting the globe at the same time with other extinct species. While there is nothing that can directly support this hypothesis, there is nothing certainly founded on analogy that can enable us to invalidate it. There is nothing more incredible in the supposition that animals should diminish in size, with changes in the condition of the earth, than that they should be extinguished altogether and supplanted by new species. The fossil urus inhabited Europe when a very different condition existed with regard to temperature, the supplies of vegetable food, and the consequent development of animal forms. Why should not the urus, under these conditions, have been a far larger animal than he subsequently became ? We know by experience the effects of food in increasing or diminishing the size of this very race of animals. The great ox of the Lincolnshire fens exceeds in size the little ox of Barbary or the Highland hills, as much as the fossil urus exceeded the larger oxen of Germany and England ; and we cannot consider it as incredible, that animals which inhabited Europe when elephants found food and a climate suited to their natures, should have greatly surpassed in magnitude the same species under the present conditions of the same countries."[1]

It is believed that the *Bos longifrons* was the only type of domesticated cattle in Britain at the time of the Roman invasion. Many of them, it is said, were driven with their owners into the remote regions of the country, where they remained in purity for ages. It has been suggested by high authorities that the purest descendants of these cattle are to be found in the horned breeds of the Highlands of Scotland, of Wales, and of Ireland. It is further recorded that by the Romans, the Danes, and others, improved varieties of large cattle were imported into Britain and crossed with the cattle of the *longifrons* type, but little is known as to the particular varieties thus introduced. It has been held by some ancient writers that the *Bos urus* never was domesticated in the British Isles, or, at any rate, only to a limited extent. That it was tamed on the European continent is well established, and it is supposed that, amongst other sorts, the Romans had taken animals of the *urus* type with them to Britain. Thus, while there is much that is obscure in the ancient history of British cattle, it may safely be assumed that the breeds of cattle which now exist in this country can claim descent from different branches of the Taurine group.

WILD WHITE CATTLE.

There still exist in this country some singularly interesting remnants of the wild cattle which at one time roamed in freedom through British forests. The most notable herds of these old-world cattle are the Duke of Hamilton's herd in Cadzow Park, in the county of Lanark, and the Earl of Tankerville's herd in Chillingham Park, Belford, Northumberland.

Cadzow Park Wild White Cattle.

Cadzow Park formed part of the great Caledonian Forest, and it is believed that the herd of semi-wild cattle now enclosed

[1] Low's *Domesticated Animals of the British Islands*, 1842.

there are direct descendants of the wild cattle which, as late as the sixteenth century, roamed through that vast preserve of wood and moor. Low states that all the characters of the Cadzow Park cattle show them "indubitably to be the descendants of the ancient race." He adds: "They are of the size of the cattle of the West Highlands; they are of a dun-white colour; the muzzle, the inside of the ears, the tongue, and the hoofs are black. They are very wild, and cautious of being approached; when suddenly come upon they scamper off, turn round, as if to smell and examine the intruder, and generally gallop in circles, as if meditating an attack. They are not, however, vicious, though some of the bulls have manifested the savage and dogged temper of their race. Some persons have been pursued to trees. . . . The females conceal their calves amongst thickets or long grass, returning to them cautiously twice or thrice a-day to suckle them. The little creatures exhibit the instincts of their race: when suddenly approached they manifest extreme trepidation, throwing their ears close back upon their necks and squatting upon the ground. The only method of killing the older animals is by shooting them."[1]

In the main, the Cadzow Park cattle have been bred within themselves, but the influence of excessive in-breeding has gradually impaired their constitutional strength. In the hope of correcting this tendency, a bull from the Chillingham Park Wild White herd was introduced in 1886 and mated with a number of selected cows. Most of the earlier crosses were unsatisfactory in their colours and were not used for breeding, but in 1888 two bull calves, true in colour to the Cadzow type, were obtained, and through the use of these and their progeny a marked improvement was effected in the stock.

Again, in 1896, a bull was introduced from the Wild White herd at Vaynol Park in Wales, and as the result of these two infusions of kindred yet fresh blood the Cadzow Wild herd has obtained a new lease of life.

The Cadzow Park cattle have main-

[1] Low's *Domesticated Animals of the British Islands*, 1842.

tained in a wonderful manner the old-time features of their race. They are less timid, but in colour and form there is little change.

In Plate 49 there are reproduced photographs of Cadzow Park Wild White cattle, and of the Chillingham Park Wild White bull introduced in 1886.

Chillingham Park Wild White Cattle.

The Wild White cattle at Chillingham Park, Northumberland, have been declared by various high authorities to be the purest and most characteristic representatives extant of the aboriginal wild cattle of this country. They are more timid than the Cadzow Park cattle, and they are wonderfully robust in constitution considering the closeness with which they have been bred for hundreds of years. There are good grounds for believing that towards the close of the eighteenth century a portion at least of a herd of Wild White cattle, long kept at Drumlanrig in Dumfriesshire, found its way to Chillingham Park, and it may be assumed that the Chillingham Park herd had gained in constitutional strength by that infusion.

The Chillingham Park cattle are wonderfully uniform in their main features. At birth the colour is almost pure white, but gradually it changes into a creamy white. The upper surface of the tongue is slate-coloured, and the under side reddish brown; the horns white, with black tips; the ears red inside and partly red outside; the eyes fringed with long eyelashes; the hoofs and noses black. Their general formation is well proportioned, and it has been said of them that they have such finely set shoulders that they can trot briskly and with the gaiety of race-horses.

As ruling monarch of the herd there is always a "King Bull," the same animal holding this high office usually for two or three years, when, after a fierce fight, he is deposed by a younger and stronger sire. Here, as in Cadzow Park, the calves are secreted by the mothers when born. When killed, the bulls weigh from about 500 to close on 600 lb.; the cows about 50 or 60 lb. less, and the steers 10 to 20 lb. more, than the bulls.

In 1875 Lord Tankerville began experiments in the crossing of the Wild cattle with pure-bred Shorthorns. The trials of a Wild bull with Shorthorn females did not succeed on account of the resulting female crosses failing to breed. The mating of a Shorthorn bull with Wild females was quite successful, and a useful variety of cattle has thus been established. Steers bred in this way have won third prizes in the Smithfield Fat Stock Show. One, three years and eight months old, reached 18½ cwt. live-weight, and yielded a carcase of 96 stones. The outward features as well as the fattening properties have been improved by the Shorthorn influence: there has been no loss in hardiness, little change in colour, except that the dark colouring of the nose has nearly disappeared. Shorthorn bulls continue to be used, as the sires.

As already mentioned, the Chillingham Park Wild bull taken to Cadzow Park for crossing with that herd is represented in Plate 49.

Other Wild White Herds.

Another interesting herd of Wild White cattle similar to the two already mentioned was long maintained with success in Chartley Park, near Uttoxeter, in the county of Stafford. The Chartley Park herd, which traced back to early in the thirteenth century, ultimately became seriously affected with tuberculosis and other ailments, which, on account of their long-continued in-and-in-breeding, the cattle were not well able to withstand. Partly through deaths, and partly by the sale of a number of animals to the Duke of Bedford in 1905, the herd became reduced to very small numbers. The Chartley Park cattle were similar to those at Cadzow and Chillingham, but black calves occasionally appeared amongst them.

At Vaynol Park, near Bangor, a herd of Wild White cattle has been maintained since 1872. In that year the herd was founded by Mr G. W. Duff Assheton-Smith by the purchase of twenty-two cattle from Sir John Orde of Kilmory, Argyllshire, the remainder of the Kilmory herd being taken to Vaynol Park fourteen years later. The Kilmory herd was founded by stock tracing from a Wild White herd which was kept for a time by the Duke of Atholl, at Blair-Atholl

in Perthshire, and which in 1834 was purchased partly by the Duke of Buccleuch, Dalkeith Park, Edinburgh, and partly by the Marquis of Breadalbane, Taymouth Castle, Perthshire. Crosses of white West Highland cattle were introduced at Kilmory, and the beneficial influence of the West Highland blood can easily be traced in the thick, well-fleshed, handsome cows of the Vaynol Park herd shown in Plate 49.

It was from the Vaynol Park herd that, in 1890, a Wild White heifer was sent to the Zoological Gardens, London, where she bred successfully to a Wild White bull introduced from the herd at Chartley Park.

MODERN BRITISH CATTLE.

Not for a long period of time has foreign blood been infused into British breeds of cattle to any considerable extent. Such changes and improvements as recent generations of breeders have effected—and they have assuredly been remarkable alike in character and value —have been brought about by skilful handling of native material. It is doubtful if in the annals of Agriculture there is to be found a more striking feature than the very marked improvements effected upon British breeds of cattle since the middle of the nineteenth century. The nature and extent of these improvements will be indicated more fully in the appended descriptions of the different breeds. It suffices here to observe that, with the wealth and variety of material which has long existed in the British races of cattle, the skilful breeder has no need to resort to infusions of foreign blood.

Breeds of cattle are commonly divided into horned and hornless varieties. Of the hornless cattle there are now only three recognised breeds in the British Isles—the Aberdeen-Angus, the Galloway, and the Red Polled breed of Norfolk. Of the horned cattle there are many breeds and varieties, the more important being the Shorthorn, Red Lincoln Shorthorn, Hereford, Devon, South Devon, Sussex, Long-horn, Ayrshire, Highland, Welsh, Kerry, Dexter, Jersey, Guernsey, and. Dexter-Shorthorn.

In addition to these established breeds

of cattle there are numerous sub-varieties and types of crosses which are bred extensively in different parts of the country. Some of these might well be developed into distinctive breeds, but it happens that the tendency is rather in the other direction, to allow these mixed types to lose such individuality as they possess. In this way several types of cattle that at one time or other commanded attention in various parts of the country have disappeared either wholly or partially. Amongst these may be mentioned the black horned cattle of the North-east of Scotland, the black horned breed of the county of Fife, the Glamorgan and White Pembroke cattle of Wales, and varieties long associated with the county of Gloucester.

BREEDS OF SHEEP.

It is generally agreed amongst naturalists that the domesticated races of sheep trace descent from certain wild species of the genus Ovis. The numerous varieties now existing throughout Europe are believed to be descended from the Argali or Wild sheep of Asia and the Wild musmon of Southern Europe, the latter being a species almost identical with the Rocky Mountain sheep of America. Similarly the domestic sheep of Africa seem to have been raised from wild species native to that great continent.

In ancient history, both sacred and profane, there is ample evidence of the useful part played by the sheep in the life and affairs of the human race even in its earliest days. Man has always been keen to appreciate the benefits derivable from the cultivation of the Ovine tribes, and it is safe to say that at the opening of the twentieth century varied flocks of domesticated sheep made up a larger proportion of the great fabric of agriculture than was the case at any previous time in the history of the world.

The varieties of sheep established in Europe at one time or other have been very numerous. They have also shown much diversity in form and character. Broadly speaking, all varieties were divided into two classes — long-tailed sheep and short-tailed sheep. The former were by much the more numerous, the greater proportion of the flocks in the West of Europe being of the long-tailed sorts. In most cases there was a strange development of fat on the tail, but, except in a few varieties, this peculiarity has to a large extent disappeared. Short-tailed sheep, which were favoured by Slavonic nations, made their way to northern parts of the British Isles through the agency of Scandinavian invaders, but they were not of sufficient utility to secure for them an enduring position amongst the more profitable races of long-tailed sheep which hold sway throughout this country.

In the British Islands at the present day there are to be found a greater number of races of high-class rent-paying sheep than are known to exist in any other country. Alike in size, form, and outstanding features generally, they present variation that is quite remarkable. To some extent this diversity may be due to a difference of descent. In large measure it has arisen from the long-sustained influence of environment—the influence of soil, climate, and food. To a still greater extent the diversity, as well as the general high standard of merit displayed by the numerous types, has been brought about by the skill and the enterprise of British flock-owners in pursuing the science and the art of stock breeding.

It is a common practice to classify sheep according to whether the staple of their wool is long or short. In the case of some races there are differences of opinion as to the class in which they should be placed. It may be as well, therefore, to arrange the different breeds and types into three groups—(1) Long-wools, (2) Short-wools and Downs, and (3) Mountain and Moorland sheep.

Of long-wooled sheep the recognised existing varieties are—Leicester, Border Leicester, Lincoln, Cotswold, Devon Long-wools, South Devon, Wensleydale, Kent and Romney Marsh, Roscommon, and Half-bred.

The short-wooled and Down races are —Southdown, Shropshire, Hampshire, Oxford Down, Suffolk, Ryeland, Dorset Down, Dorset and Somerset Horn, Radnor and Norfolk.

The Mountain and Moorland sheep comprise the Blackface, Cheviot, Exmoor, Dartmoor, Lonk, Herdwick, Welsh, Kerry Hill (Wales), Derbyshire Gritstone, Shetland, &c.

A number of other varieties are kept to some extent in certain districts, but the more important breeds and types are enumerated above.

GOATS.

The goat is so closely allied to the sheep that naturalists have not been in complete agreement in distinguishing between the two. The domesticated goat is generally regarded as being descended from one or more of the caprine group—most largely, it is believed, from *Capra œgagrus*, but partly also from *Capra ibex*, the Alpine ibex, and probably some of the other varieties of the tribe.

The goat would seem to have made its way to the British Isles from the continent of Europe. The best variety of goats in this country resemble pretty closely the more highly-prized goats in the countries of the Mediterranean. Less attention is now given to the rearing of goats in the British Isles than prior to about 1830; but in Ireland and in some other parts they are still cultivated,— this, too, with much advantage to their owners.

BREEDS OF SWINE.

The Wild Hog, *Sus aper*, is universally regarded as the progenitor of the many existing domesticated races of swine. That species was widely distributed throughout the old continent in early times, and still roams in a wild state through woods and wastes in the European continent and in countries farther east. Wild swine existed in this country prior to the Norman Conquest, but that their numbers were gradually declining is indicated by the fact that William the First passed a law providing that any one

found guilty of killing a Wild Boar should have his eyes put out. Various writers refer to wild swine as existing in the English and Caledonian forests in the twelfth century, but in course of time, at a date not definitely known, the species ceased to exist in this country.

Not only were the native domesticated swine of this country derived from the Wild Hog, *Sus aper*, but at various times fresh draughts of the pure wild blood were infused into the tame varieties, which were thereby reinvigorated to a marked extent.

British races of swine for long consisted mainly of two varieties. The one was of large size, somewhat ungainly in form, with long drooping ears, and slow in fattening. The other variety was smaller, with short erect ears, more easily fattened than the larger sort, but rather coarse and fibrous in flesh. While these were the only races that were distinctive enough to be regarded as breeds, many other varieties of greatly diversified characters existed throughout the country.

Happily the British Islands are now in possession of numerous excellent well-established breeds and types of swine. Broadly speaking, almost all these varieties have been built up from the native breeds by the admixture of strains introduced from the East—chiefly from China and the Mediterranean countries. For the production of high-class ham and bacon at an early age the British swine have been vastly improved by the influence' of these imported races, and from this excellent material modern breeders and feeders have attained great success in the swine-rearing industry.

The leading varieties of swine now kept in the British Islands are the Large White, Middle White, Berkshire, Large Black, Lincoln Curly, Tamworth, and the Ulster Large White. Several other sorts are still kept to a certain extent, the most important of these being the Small Blacks, Small Whites, and local varieties reared in Gloucester, Dorset, Hampshire, and Sussex.

THE SHIRE HORSE.

For a long period of time the Shire has been the leading variety of draught-horses in England. The breed is widely distributed over the country, and is a valuable source of wealth and power.

Origin.—The Shire horse of to-day is the lineal descendant of the Old English War-horse, which, alike for its strength and courage, excited the surprise and admiration of the Romans when they first invaded England. It may not be the only surviving descendant of that noble race of horses, but it is now generally regarded as the purest living representative of that earlier type.

In an interesting brochure entitled *A Short History, tracing the Shire Horse to the Old English Great Horse*, Sir Walter Gilbey states that investigations appear to establish that the Shire horse is the closest representative of—the purest in descent from—the oldest form of horse in the island. "A thousand years ago," says this authority, "this form was written of as 'The Great Horse'; and nearly a thousand years before that we have evidence which goes to prove that the same stamp of horse then existed in Britain, and that it was admitted by those who saw it here to be something different from—and something better of its kind than—what any of the witnesses (of that day) had seen before : and they had seen most of the horses of those times."

Name of the Breed.—For a long time prior to the advent of the nineteenth century, and for many years thereafter, the breed was widely known as the Large Black Old English Horse. It is now universally recognised by the title of "Shire," derived from "the Shire counties in the heart of England," in which, according to Arthur Young, who wrote near the end of the eighteenth century, the Old English Horse was principally produced.

Shire Horse Society. — The desirability of taking steps to encourage the improvement of the old English breed of cart-horses was brought into public notice by a paper read by Mr Frederick Street at the Farmers' Club, London, in 1878. The result was the establishment of the Shire Horse Society, under whose fostering care the breed has been vastly improved,—made more uniform in type and character, and much sounder in wind and limb. The Shire Horse Society issue a volume of the *Shire Stud-Book* every year. The first volume, published in 1880, contains the pedigrees of 2380 stallions, many of which were foaled in the eighteenth century. The Society likewise holds a great Shire Horse Show in London every spring, and this show has done much to further the interests of the breed. The first show was held in 1880, and as many as 862 entries have been recorded, and as much as £2220 in prizes offered at a single show.

The Society's schemes for the improvement of the breed include distribution of medals, of which the winners may obtain their equivalent in money if preferred.

Veterinary Inspection.—The Shire Horse Society adopted veterinary inspection at its shows, and only awards prizes and medals to animals passed as sound by recognised veterinary inspectors. As indicative of the condition to which the breed has been brought by attention to this matter, it may be stated that in 1908, of 248 horses examined 237 were passed sound. The 11 which were rejected were cast as follows : 2 roarers, 1 whistler, 1 wind, 2 shiverers, 1 side-bone, 2 ringbone, 1 cataract, and 1 lame. This is a small percentage, and clearly proves the wisdom of making it a condition of showing that hereditary unsoundness is a disqualification.

Distribution. — The Shire horse is found in every part of England and Wales, from Northumberland in the north to Cornwall in the south, and from Lincolnshire in the east to Carmarthen in the west. It is to be seen at its best in Lincolnshire and the fen country generally, in the midlands, in Derbyshire and Lancashire.

Land and Water for Breeding.—The heaviest lands are the best for breeding heavy horses. Limestone land, such as exists in Derbyshire, is also greatly favoured. The two most famous and historic parts of England in the breeding of draught-horses are the fens of Lincolnshire and the valleys and flats of Derbyshire. Shire horses are bred on comparatively light soils in the south of England, but difficulty is experienced in growing them big enough without forcing feeding. In summer, when the land bakes and cracks, it is almost impossible to keep young foals on their joints. Consequently many stud owners have established the practice of hiring grazing in a more suitable neighbourhood for the summer season.

Breeders are not agreed as to what constitutes growing qualities in the land. Many appear to think that the water-supply has much to do with it. The writer has had confirmation of this theory from many sources. Fields on a particular farm were known to grow stock much better than others, and the water for these issued from a different source. The contention has been put forth that soft water is an important element in producing big draught-horses.

Horse - breeding Societies.—Horse-breeding societies have rapidly increased in number, and the hiring of stud-horses of good breeding and free from disease has done much to drive off the road the "guinea" horse of doubtful parent-age and poor character, and afflicted with many of the ailments which horse-flesh is heir to. There are, however, so many studs scattered over England where the use of horses can be conveniently obtained at reduced fees to tenant-farmers (frequently involving an option on the foal), that the establishment of horse-breeding societies is less necessary than would otherwise be the case. In the south and in Wales, however, the movement has made admirable progress.

Fees and "Retainers."—Good horses can be used at prices varying from 3 guineas to 15 guineas. For tenant-far-mers the abatement of a 10 guinea fee to 7 guineas is freely made, so that the smaller owner of pedigree mares has every encouragement. The "retainers" offered by societies vary according to the class of mares in the neighbourhood, the wealth of the society, and the enterprise of the farmer. The Welshpool Society has frequently hired at 1000 guineas.

Others, again, give a small "retainer," usually about £50 with a guarantee of so many mares, say about 80, at a fixed fee, generally about 3 guineas.

Value of Pedigree Mares.—It is probably a correct estimate to assume that about 85 per cent of the mares served by pedigree stallions are without pedigrees. Farmers are realising, however, that the pedigree mare is a valuable asset. Thus, assuming that two foals are dropped to the same horse—one from a pedigreed mare, and the other from an unpedigreed mare—the one may be worth 100 guineas, and the other from 20 guineas to 30 guineas, while the fee for the service is the same in both cases. In this way a saying has come about that the small farmers of Derbyshire are accustomed to sell a foal to pay the rent.

Public and Private Prices.—Many high prices have been given for Shire horses in the public sale-ring. Those reported to have been offered privately and accepted are: 2000 guineas for the London champion two-year-old, "Bear-wardcote Blaze"; 2500 guineas for "Bury Victor Chief," another London champion. In the public sale-ring the highest price has been 1550 guineas for the stallion "Hendre Champion," bought by Mr Leopold Salomons at the late Mr Fred. Crisp's sale. The Premier horse "Chancellor" was sold at a Calwich sale for 1100 guineas. The highest priced mare was his Majesty's three-year-old "Seabreeze," which made 1150 guineas at a Sandringham sale, the late Sir J. Blundell Maple being the buyer. The well-known mare "Hendre Crown Princess" drew 1100 guineas at one of the late Lord Wantage's sales, Mr Smith Carrington being the purchaser.

Amongst averages obtained at public sales the highest was Lord Rothschild's, £266, 14s. for 35 head in 1908. More remarkable still is this figure when it is remembered that all the stock offered were home - bred. The top price was 900 guineas — the highest figure ever given in public for a yearling colt. Lord Llangattock's fine average of £226 for 44 head, and his Majesty's (then Prince of Wales) of £224, 7s. 9d. for 54 head, are likewise landmarks in the history of the breed. Lord Llangattock's sale was held in 1900, while the Sandringham fixture took place in 1898. In 1899 Mr (now Sir) Alexander Henderson sold 39 head at an average of £209, 3s. 10d.

Mating.—In the mating of mares the skill of the horse-breeder is tested. To some extent it is a matter of luck rather than skill—the use of the nearest horse mayhap. Otherwise how can the fact be accounted for that the smallest breeders, with least pretensions to an extensive and accurate knowledge of breeding, occasionally breed a champion? The using of a heavy stallion to a mare of quality is a safe rule. With a big mare almost any type of horse may be used. The short-legged, lengthy mare is usually associated with the "brood" mare type —*i.e.*, she is that class of mare which in the majority of cases throws a first-class foal. When pedigree is a consideration, back breeding requires to be studied. It is also invaluable to have a knowledge of the leading strains of blood.

Peculiarities of Stallions.—Eminent stud-horses have been known to show marked peculiarities. Thus "Premier" was known chiefly for the high character of mares he left; so was "Royal Albert." "Lincolnshire Lad II." produced stallions which exercised an overmastering influence on the breed: "Harold" was one of them. His mares had a name for lasting well, and his stallions were London champions. "Hitchin Conqueror," again, stamped his produce with wonderful quality and joints, and imparted activity to his stock.

Stallions may be foal-getters—*i.e.*, the young animals are at their best as foals. Others may produce foals that do not do well till they attain maturity. Yet, again, one meets with that class of horse which brings coarse stock; others breed them too fine. And so the catalogue might be continued, but enough has been said to emphasise the importance of a knowledge of the sire and his breeding.

Shire breeders are accustomed to study very closely not only the pedigree but the appearance of an animal. Nowadays the fact that the dam of a good young colt or filly is a good sound mare is sufficient to increase the value of the progeny materially. It is the custom with careful breeders to acquaint themselves with the character of the dam before buying a stud horse.

Foaling.—Foals are sometimes dropped in January, but they are too troublesome to favour the general adoption of a February service. It is found, if they are good enough for the autumn foal shows, that they are seldom so big as those dropped in March and April, which can go on to grass immediately and suffer no check in growth. Late spring and early summer foaling is the rule on most farms, and it is the most convenient.

Forcing Young Stock.—The system of forcing young stock for shows obtains generally among breeders of pedigree Shire horses. It is one of those practices almost unanimously condemned in theory but encouraged in practice. It is impossible to win prizes unless a young animal is very big and weighty. It undoubtedly shortens the period of an animal's usefulness, and may perhaps in some remote degree affect its soundness. The cynic has declared the dominating principle of Shire horse showing has been "soon ripe, soon rotten." While far from approving the contention, the forceful feeding of young stock must be condemned as laying the foundation of future trouble.

In few breeds can two-year-olds be put to service. As a rule, it stunts the growth, but through the method of forcing young stock to great height and weight, it is frequently practised by breeders without ulterior effect upon the subsequent stature of the mare.

Characteristics.

Colour.—As would be inferred from the use of the title the Old English Black Horse, black was no doubt at one time the prevailing colour of the breed. A large number are still black, but bay and brown of varying shades predominate. Many are grey, roan, or chestnut, but light colours are not, as a rule, in favour in the market.

Size.—It is undisputed that the Shire horse is the largest of all the varieties of draught-horses which exist in this country, or indeed in any other country. About 17 hands is a common height amongst the stallions of the breed, although many attain to 17.1 hands and 17.2 hands.

Dimensions and Weight.—The dimensions of the Shire horse form an interesting study. The American system of weighing heavy horses has not yet been adopted in this country, but in the course of time that may come about. Horses over a ton weight are far from being uncommon. The well-known horse "Tatton Friar," owned by the Earl of Egerton, scaled 25 cwt.

Mr Walter Crosland, agent to Sir Alexander Henderson at Buscot Park, Farringdon, has supplied the following measurements of two Shire horses, "Markeaton Royal Harold" and "Buscot Harold." They are both champion winners at London, and "Buscot Harold" was sired by the former out of the London champion mare "Aurea," so that he has a double dose of champion blood. Particulars of these measurements are as follows, those of "Markeaton Royal Harold" being made when he was five years old, and of "Buscot Harold" when three years of age, immediately after each had won his championship at the London Shire Horse Show:—

	Markeaton Royal Harold.	Buscot Harold.
Height .	17 hands ½ in.	17 hands.
Girth . .	8 ft. 1 in.	7 ft. 11 in.
Knee . .	1 ft. 8 in.	1 ft. 6½ in.
Below knee .	1 ft. 1¼ in.	1 ft. ½ in.
Round coronet	1 ft. 8¾ in.	1 ft. 8¾ in.
Hock . .	1 ft. 11 in.	1 ft. 10 in.
Below hock .	1 ft. 2 in.	1 ft. 1¾ in.
Across foot (fore)	7½ in.	7¾ in.
Length of head	2 ft. 6 in.	2 ft. 9 in.
Middle of knee to ground	1 ft. 9 in.	1 ft. 8¾ in.

Doubtless when he reached maturity "Buscot Harold" had considerably improved upon these figures.

"Birdsall Menestrel," Lord Rothschild's champion stallion at the Shire Horse Show (represented in Plate 12), has been measured, and Mr Richardson Carr, in response to our inquiry, supplies the following dimensions:—

BIRDSALL MENESTREL.

Height	17 hands.
Width across sole of fore foot (without shoe) . . .	8 in.
Round fore coronet . . .	21½ in.
Below knee	13 in.
Round knee	19½ in.
Round fore arm . . .	28 in.
Round hock	23 in.
Below hock	15 in.
Girth	8 ft.
Weight at end of 1907 .	19½ cwt.

The following are the measurements at ten years old of the 1550 guinea horse "Hendre Champion," owned by Mr Leopold Salomons of Norbury Park, Dorking:—

HENDRE CHAMPION.

Height . . .	17 hands 1 in.
Width of foot . . .	8 in.
Round coronet . . .	20½ in.
Bone below knee . .	14 in.
Round knee . . .	18½ in.
Round arm . . .	23 in.
Bone below hock-joint .	16 in.
Round hock . . .	23¼ in.
Girth	8 ft. 6 in.
Weight	21 cwt.

Form.—When the Shire Horse Society began its good work, the rank and file of the breed presented defects which materially impaired the value of the horses for hard work. Chief of these were short upright pasterns, wide hock action, unsoundness, and sluggish movement. In all these points a marked improvement has been effected. To one who, like the writer, has been regularly attending the London Shire Horse Shows, the contrast between the general characteristics of the animals exhibited at the earlier and later shows is most striking. Nowadays symmetry is as essential in a draught-horse as it is in a Smithfield champion. It is of little use showing the fine muscular development of loin, depth of rib, and beautiful fore end of a draught-horse if he has not thighs to fill the breeching. Symmetry is a great thing, but it does not constitute the alpha and omega of a Shire horse breeder's catechism. The old type of Shire was heavy and cumbersome, set on short legs, but disproportionate in his weight of top. To-day he is a beautifully balanced animal, with better feet and action than he had in bygone times.

Type.—Type may be regarded as the governing force in horse-breeding. A good gelding might be a very poor type of Shire. As a rule, type is shown as much in the character of the head as in the formation of limbs and body. The gaunt, leggy type of horse is of little use to any one who wants equine power concentrated. There are, roughly speaking, two types—the quality and the rougher kind of Shire. In-breeding and the use of pedigree make for quality : without it stock-breeding could not prosper. Through its aid the type of Shire produced in modern times is more lasting, and will keep on its joints longer with fewer limb ailments than at one time were common.

Quality is found most fully shown in the mares. England is to-day full of breeding stock of high quality. Occasionally there is a tendency to fine down the stallions too much. They lack masculine appearance, and begin to lose that cresty sex-like boldness so characteristic of the older type. The rougher sort of horse has his uses. We do not refer to the round-boned horse whose spongy legs are the happy hunting-ground of grease and divers ailments. But the stallions whose hair has a tendency to curl, and whose bone is not of superfine quality,—in fact, whose appearance is impressive but will not bear close inspection, — that type of horse has his uses, to correct the effeminacy which comes from concentrating attention too much on quality.

Legs and Pasterns.—It is now universally acknowledged that a short upright pastern is an objectionable feature in a horse, whether for draught or other purposes. With such a pastern the shock to the system, in walking, trotting, or galloping under a burden, must obviously be much greater than with the "springy" action of a moderately long sloping pastern. This point is more keenly appreciated in England now than in former times. A long pastern may be a source of weakness, and the Shire breeder aims at a happy medium.

Similar remarks would apply to the general conformation of the legs. The angle of the hind leg has been very carefully studied, more so since horses like "Royal Albert" set the fashion. A common fault is "back at the knee." It indicates muscular weakness of the fore limb. The most common of all faults, however, is lightness of bone below the knee.

Action.—The unduly wide hind action, so prevalent at one time in English draught-horses, is happily becoming much rarer. A bent hind leg, set outside the body, so to speak, is undoubtedly a source of weakness in a draught-horse. An animal with limbs of this sort can have little endurance under hard work. Wide hind action was one of the most notable defects in the earlier London shows of Shire horses. It is rapidly becoming the exception to find wide movers. Of course it must be remembered that the Shire horse is much wider in frame than most other breeds, consequently the closeness of the Clydesdale's action need not be expected, otherwise we should have a race of "cowhocked" animals.

The demand nowadays runs on very active geldings, and the tendency of the times is altogether favourable to the further development of speedy movement in the draught-horse.

Feet.—The foot of the horse is a point of the utmost importance. It is there very often that, under hard labour, the animal first gives way. Flat soft hoofs cannot be durable, and with the persistent striving for large sound feet with deep strong walls, which has been fostered by the show system, the feet of the rank and file of Shire horses have greatly improved. The breeder looks for a wide and deep heel, and is averse to a too wide coronet, but more so to a narrow one. The most wearing type of hoof is blue, but the advantage of white limbs for show purposes is too apparent to enable the breeder long to resist the incursion of the white and more brittle hoof.

A Typical Shire.—What is a typical Shire? The description given by Mr Frederick Street, in a paper read before the London Farmers' Club in 1878, holds true to-day as it did then. Mr Street said : "The feet should be firm, deep and wide at the heel, not too long or straight in pastern, flat bone, short between fetlock and knee. A stallion should not measure less than 11 inches below knee, and girth from 7 feet 9 inches to 8 feet 3 inches; should not stand more than 17 hands; should have wide chest, shoulders well thrown back, head big and masculine, without coarseness ; full flowing mane, short back, large muscular development of the loin, long quarters with tail well set on, good second thighs (this is a point where so many fail), large flat clean hocks; plenty of long silky hair on legs,—or, to sum up in few words, a horse should be long, low, and wide, and thoroughly free from all hereditary disease. A main point is action : he should be a good mover in the cart-horse pace, walking ; and, if required to trot, should have action like a Norfolk cob."

Hair.—Mr Street's reference to silky hair touches upon a remarkable development in the breed. The hair denotes the quality of an animal as accurately as anything else. The suspicion of a curl in the hair of the limbs, or "feather," as it is technically termed, is not looked on with favour. The older type of Shire was a much befeathered animal. The introduction of fine silky hair was coincident with the diffusion of flat bone and the supersession of such stable troubles as grease and Monday morning leg. The American buyer likes as little hair on the limbs as possible. The English buyer objects to its absence. Hence we have warring elements. Whether there is any truth in the contention that, as with Samson of old, the hair denotes strength we do not pretend to assert, but it is a fact that hair goes a long way to obscure defects of knee, cannon-bone, pastern, and hoof.

English and Scottish Notions.—It used to be a trite saying that in judging a horse a Scotsman began with the feet and legs of the animal, an Englishman with its top. By this it is meant that the chief consideration with the Scottish judge is the feet and legs, and with the English the body of the horse. Of both judges there was truth in the statement, and in these habits both were mistaken. It is true, no doubt, as the Scotsman argued, that without good, sound, well-formed, well-set legs and feet to carry and propel it, the best body one could conceive would be of little value. It is equally true, as the Englishman contended, that a horse with a big well-formed body will usually fetch more money in the market than one with a small weak body. Fortunately they have differed and agreed—at least the English breeder has absorbed the whole creed. If a horse lacks foot he has a poor chance of recognition in an English showyard. A short pastern is quickly detected and condemned. But one thing the Shire breeder will never part with is weight. He wants the avoirdupois in

the collar to start a load. The Scottish breeder pits against this the superior activity of the Clydesdale. Perhaps the reason for this difference of opinion is found in the fact that London is paved with wood, while the large towns of the north mainly employ stone setts. Weight is especially necessary where there is a bad foothold.

FEEDING AND MANAGEMENT.

Bringing out Shires for Show.—The bringing out of Shire horses for show is now so much a question of the expertness of the individual, that directions, while forming a useful guide, must be applied with all the skill of an expert to be successful. In the choice of suitable young animals for show the "expert" eye is the most necessary agent. In a foal, a big quantity of hair right from the back of the knee should be looked for. The foal that comes out in a naked condition has very little chance to win : therefore cultivate hair. The next things to be looked for are weight, the naturalness of the joints, the openness of the hoof heads and the heels, the quality and set of the limbs. Bad action can sometimes be corrected by judicious shoeing ; but when the indifferent action is due to a physical defect, it is impossible by artificial means to correct it, although it may be improved. It is sometimes difficult to determine when a fault in the action is due to weakness and when it is inherent. The writer remembers many occasions when foals have been penalised because of their indifferent action, yet coming out in maturer years with few signs of their earlier defects. The defects sometimes vanish when the young animal gathers strength, when the bones and muscles become set.

Rearing Foals.—In brief, the first thing to do is to choose a type and stand by it. The foal that gives early promise will do well on mother's milk if the dam is doing her duty. The longer that corn-feeding can be staved off the better, as it has a tendency to affect the limbs and joints. If the mare, however, does not nurse the foal well enough, a feed consisting of crushed oats, bran, and chaff once or twice a-day should be given. The foal will soon eat along with the mare. The quantity of artificial food may be increased as weaning time approaches, usually when the foal is about five months old. There are foals one sees at the foal shows which are dropped in January and February still looking for mothers' milk. It is a strain upon the mare to have her foal suckling too long, and usually those early foals are difficult to keep correct on their joints. They do not have a chance for some months of enjoying a run at pasture, consequently they are very troublesome to the owner. Moreover, they are almost certain to lose their hair, and have a stale appearance ere the September foal shows come round.

Some breeders trust to the milk-pail, sweetening cow's milk and occasionally diluting it. It is a penny-wise and pound-foolish policy. It is never very difficult to detect a foal that has had too much of this kind of treatment. The foal grows by it, it is true, but it gets very shaky about the limbs, goes off its joints, and lacks that fresh brisk appearance characteristic of younger foals with harder feeding.

It is well to get the young foal to start the winter well. A check in the earlier months is a bad preparation for the future. It means at least two to three months lost, and that is a serious handicap when judges are so pronouncedly in favour of big young colts and fillies. Give the weaned foal a companion in the paddock night and day, with a shed for shelter. As a rule, they only use it as a feeding-box. Night and morning a mixture of ground oats, bran, and hay chaff may be given with the addition of linseed-cake and a few carrots. It is better to damp the mixture with warm water prior to feeding. This combination of feeding-stuffs grows bone, assists the coat, and stimulates the appetite. At no time should more be fed than the foal will clean up readily.

Young Horses.—If the young stock are intended for spring exhibition they may, in December, be brought into a loose-box and receive a quantity of long hay as well as the chop mixture, which should be gradually increased in oats and bran. The object is to have the colts handled as much as possible and to accustom them to confinement, which they have to undergo at such shows. It also prevents them taking cold so easily as they otherwise would.

Attention to Hair.—Particular attention should be directed to keeping the hair on. What with rubbing and clogging with mud it is apt to become worn and fragmentary. A good dressing to use is equal parts of sulphur, paraffin, and train-oil, which should be well rubbed in to get to the roots of the hair and also to prevent it running off. The day after the application a sawdust dusting should be given, otherwise the hair will mat and even rot with the adhesions of mud and dirt which young animals delight to walk in. To walk in mud no doubt is cooling to the feet, an important consideration when the food is heating.

Before the animals are brought out, the hair should be carefully cut away above the joints with a sharp knife, and a wet cloth tied round the limb for some time before exhibition to make the hair lie naturally. The "feather" should be carefully washed, and dried with sawdust, and brushed freely. The old practices of soaping and resining are now not recognised by the Shire Horse Society, so that the hair must be naturally straight and silky to do the animal justice.

Attention to Feet.—Another point that must be carefully looked to is the paring of the feet. The hoofs have a tendency to wear down on the outside, and rasping must be the remedy, otherwise the action of the colt may be interfered with.

Grooming and Handling.—Grooming should be regular some time before showing, and great care should be taken to train a young animal to the halter. Find out his best pace at the trot, and keep him to it. An indifferent and slouching walker may need the sharpening of the whip. Again, the action may be improved by the shoeing, the calkins being raised and lowered as desired.

Adult Animals.—In bringing out older animals, the advantages derivable from a run at pasture with shoes off should not be forgotten. A cooling mash is also freely given. Some exhibitors medicate their horses very freely to stimulate the appetite, but the practice is reprehensible. To get a good coat, sleek and glossy, linseed or linseed-cake should be used in the ration. Maize, boiled and flaked, and saccharine preparations are likewise used for conditioning.

As already stated, the Shire stallion "Birdsall Menestrel" is represented in Plate 12. A portrait of the Shire mare "Pailton Sorais" is given in Plate 13.

CLYDESDALE HORSES.

The Clydesdale is the native Scottish breed of draught-horses. The history of the breed has been often written, and little that is fresh can be said by any writer on that subject.

Origin.—All attempts to demonstrate that the breed sprang from one sire are destined to failure. No breed owes its existence to such a cause. Before the influence of one sire could be regarded as alone responsible for the success of a breed it would require to be proved that the breed had no existence prior to the advent of the sire.

The broad facts connected with horse-breeding in Scotland are not difficult to state. As early as the fifteenth century Scotland was famous for its horses. It did a large export trade with the Continent then, and from time to time Royal edicts were issued regulating that trade. Sometimes exportation was prohibited; sometimes it was carried on subject to a heavy export duty; sometimes it was unrestricted. But however conducted, the fact stands out clear that Scottish-bred horses were coveted by Continental buyers during the long period of the Stuart dynasty. The kings were each after his own manner patrons of horse-breeding, but it cannot be said that this patronage did much to improve the native breed for draught purposes.

The Clydesdale in Peace and War.—The Clydesdale as a draught-horse came into being after wars had ceased, so that men could without molestation pursue the arts of peace. Therefore the history of the draught-horse is pre-eminently associated with the rest which the land enjoyed after the Revolution Settlement in 1690.

This is specially true of the district from which the Scottish breed takes its name. Clydesdale is the old name for Lanarkshire, just as Angus is the old name for Forfarshire, and Tweeddale the old name for Peeblesshire. The name indicates that the home of the breed, in its modern draught type, is to be found in the valley of the Clyde. There the internecine warfare of the later Stuart era was waged with relentless fury, and the very places and parishes associated with the early history of the improved breed are those which formed the theatre of many of the most stirring incidents in the Covenanting struggle. The arts of peace required a horse of a different type from that called for by the exigencies of war. The era of road-making and the era of industrial development in Lanarkshire were contemporary. The industrial development demanded a horse that could pull as well as carry, and the formation of roads on which carts could be pulled created the demand for a heavier horse than the sure-footed nag on which the Upper Ward and Avondale farmer had hitherto relied.

Early Improvement.

Various traditions point to Flemish stallions having been the instruments employed in the work of improvement. There are three traditions of this nature. One credits the sixth Duke of Hamilton (1742-1758) with having kept a dark-brown Flemish stallion at Strath-aven Castle for the use of his tenantry. Another speaks of a Duke of Hamilton a century earlier who kept "six fine black stallions from Flanders" there; and a third gives one John Paterson, of Lochlyoch, on the slopes of Tinto, the credit of introducing about the year 1720 a black stallion from England named "Blaze," which became the founder of the celebrated breed of Clydesdale horses. So the chronicler of these things avers. It may be accepted, therefore, as truth, that outside influences so enhanced the weight and substance of the native breed in Lanarkshire that about the beginning of the nineteenth century they became noted for their properties as draughthorses.

Their reputation extended far afield, and the markets held at Rutherglen, Lanark, and Biggar were frequented by dealers from all quarters, south of the Border as well as north. Mobs of young colts and fillies were drafted from the Lanarkshire breeding-grounds into other areas. The general influence of these importations was towards the extended breeding of animals possessing the Lanarkshire type, so that "Clydesdale" became the trade-mark of a type of horse bred in areas far apart. The name occurs in literature as early as 1823, and could not have been applied in such widespread fashion had the type which it represented not been generally recognised. That type was indigenous to Lanarkshire or Clydesdale, and was not imposed upon the horses of the Clyde valley by any external influence. Such influences imparted properties which enhanced the value of the existing type for draught purposes, but it would not be true to say that the influence of one sire, or of six, created a new type within the Clydesdale area.

An Unfounded Theory.—Attempts have been made to connect the entire modern breed of Clydesdales with one Lanarkshire tribe in a definite and direct way. The theory is that a filly directly descended from the Lochlyoch mares, improved by John Paterson's black stallion from England, became the dam of Thompson's black horse "Glancer" 335, and that the whole Clydesdale race can, through this one link, be connected with the historic tribe. That the whole modern race of Clydesdales is connected with the Lanarkshire race does not admit of doubt, but a close examination of the facts, and especially a comparison of dates, does not warrant the theory that Thompson's black horse was a son of the Lochlyoch-descended Lampits mare—

VOL. III.

otherwise the Shotts Hill Mill filly. This filly was bought at the displenishing sale at Shotts Hill Mill in 1808. It is assumed in the Retrospective Volume of the *Clydesdale Stud-Book* (published in December 1878) that "Glancer" 335, Thompson's black horse, was her son, foaled in 1810. But Thompson's black horse ("Glancer" 335), on the authority of one who knew him, was the sire of Paton's horse of Bankhead, Yoker, Renfrewshire. This Paton's horse won second prize at the Highland and Agricultural Society's Show at Edinburgh in 1842, when he was six years old. He was therefore foaled in 1836, and, if the entry in the *Stud-Book* is correct, his sire, "Glancer" 335, must have been twenty-five years old when he was got. The travelling-card of Thompson's black horse, which has been reproduced in the second volume of the *Stud-Book* (published in February 1880), is unfortunately not dated, and it tells nothing about his pedigree. It is, however, stated in the Introductory Essay in which that card is embodied (no doubt on good authority) that "Glancer" 335 died when ten years old. Consequently the theory that he was the son of a filly sold at Shotts Hill Mill in 1808 may be dismissed, along with the fabric of pedigree which has been built upon that theory.

Thompson's Black Horse and his Descendants.—Thompson's black horse was a sufficiently noted horse in his time. His service fee was one guinea, with one shilling additional to the leader. He was the progenitor in direct line of "Broomfield Champion" 95, and that horse did quite notable work in fixing for many generations the type of Clydesdale horse. His most noted son was "Clyde," *alias* "Glancer" 153, known as "Fulton's ruptured horse," and through him he may be said to have made the modern Clydesdale breed.

"Clyde," *alias* "Glancer" 153, was a "mickle strong horse." He had seven sons, which made Clydesdale history. These are "Baasay" 21, "Clyde," *alias* "Prince of Wales" 155, "Farmer," *alias* "Sproulston" 290, Erskine's "Farmer's Fancy" 298, "Muircock" 550, "Prince Charlie" 625, and Barr's "Prince Royal" 647. "Clyde," *alias* "Prince

B

of Wales" 155, "Farmer's Fancy" 298, and "Prince Royal" 647, made their mark in the showyards of the Highland and Agricultural Society, and all of the seven were successful in leaving an indelible impression on the breed as a whole.

If the descendants of these seven sons of "the ruptured horse" were eliminated from the breed, it would be weak indeed. They were not all regular in their breeding as sires, and "Prince Royal" 647 had the gift of breeding some of the best as well as some of the least satisfactory of stock. It is possible his dam was of southern extraction. He bred some stock chestnut in colour, and others having a tendency to roundness of bone. Generally, however, the stock descended from Fulton's "ruptured horse" conformed to the standard set by his owner, William Fulton, and were of the "razor-legged" type. By these seven sires and their direct descendants, in a marked degree, was the Clydesdale type which dominated the show-ring for about thirty years, from 1850-1880, determined.

Other influences began to make themselves felt from about 1880 and onwards, and in the end a type, especially in respect of formation of feet, obliqueness of pastern, and hardness of bone, was evolved which controls the Clydesdale world to-day.

Spreading of the Breed.

Other horse-breeding districts were so influenced by these sires and their descendants, and by other sires imported from Clydesdale, that the stock reared within these areas bore the same name. The Clydesdale influence can be directly traced in districts so widely separated as Galloway, Cumberland, Kintyre, and Aberdeenshire.

Galloway Horses.—The native breed of horses in Galloway had a character of its own—which has obtained renown in history, poetry, and romance. To this day the town of Inverness is the scene of races in which horses called "Galloways" are included. Writers of parish records in the province of Galloway have left descriptions of horses bred in Galloway, prior to the introduction of the Clydesdale, about the beginning of the nineteenth century. One of the

most noted of these writers is the Rev. Samuel Smith of Borgue, whose *Survey of Galloway* was published in 1810. He writes eloquently concerning the merits of the old Galloway nag, and shows how the demands of an improved agriculture led, by judicious selection, to the improvement of the native breed. They were, he says, deservedly held in high estimation for the purposes of husbandry. They were "round in the body, short in the back, broad and deep in the chest, broad over the loins, level along the back to the shoulder, not long in the legs, nor very fine in the head and neck. Their whole appearance indicated vigour and durability, and their eye commonly a sufficient degree of spirit." Mr Smith admits that they were inferior in size to the dray horses of many other districts, but were not inferior in respect of capacity to perform labour or endure fatigue. These horses were, according to Mr Smith, improved by the use of sires from England, Ayrshire, and Ireland, and the Clydesdale influence from Lanarkshire was imposed upon the product of this union.

Improving the Breed in the Stewartry.—The Stewartry of Kirkcudbright began early to improve the breed of native horses by hiring stallions from Lanarkshire. Two of the earliest recorded sires so hired were "Samson" 1288, foaled in 1827 or 1828, and his grandsire, "Smiler," which must, therefore, have been foaled early in the nineteenth century. Since that date, in unbroken succession, Clydesdale sires were hired for service in the Stewartry, and the breeding of Clydesdales there received an additional impetus when representatives of the Muir family migrated from Sornfallo, on the slopes of Tinto, in Lanarkshire, one of them to Maidland, Wigtown, the other a few years later to Banks, Kirkcudbright. The blending of blood taken into Galloway by them produced in a later day "Lochfergus Champion" 449. A later blending of the blood of that horse with that of native mares in the parish of Twynholm gave the Clydesdale world "Conqueror" 199, the sire of "Darnley" 222.

Early Improvement in Wigtownshire.—Wigtownshire was early engaged in importing and hiring stallions from Lanarkshire. The horse "Clydeside," credited with being the sire of Agnew's "Farmer" 292, could hardly have borne that name had he not been of Clydesdale origin. The Dumfries Highland Show of 1830 marked an era in the history of the breed. At it "Farmer" 292 gained a premium of £30 from the Highland and Agricultural Society. The late Colonel M'Douall of Logan was a spirited exhibitor of Clydesdales at that period. He had a formidable rival in Mr Robert Anderson, Drumore, Kirkmaiden, who, in 1835, made a historic tour into Lanarkshire and Renfrewshire, the history of which is given in the Retrospective and second volumes of the *Clydesdale Stud-Book*. The Clydesdale of the latter half of the nineteenth century was the direct result of the blending of these two streams of breeding in Galloway. Horses, the product and descendants of the Lanarkshire importations of 1835, travelled in the Kirkcudbright area for many seasons, and the strong family likeness and fidelity to one type of their produce ensured the success of almost any sire with which they might be mated. When the selected sires happened to be bred very much on the same lines as the native mares, as in the case of "Darnley" 222 and his descendants, the results went to the making of Clydesdale history.

The Breed in Kintyre.—Kintyre enjoys all the advantages of an insular position without actually being subject to its disadvantages. The history of the importation of the Lanarkshire breed into the peninsula is fortunately very clear. The native breed would no doubt be of Highland origin. The first operating source of improvement was the importation of Lanarkshire stock by the laird of Lee in Carnwath parish, Lanarkshire, who also owned Largie estate in Kintyre. One horse in particular, bearing the local title of Lockhart of Lee's black horse, had quite a good reputation, and the tradition connected with his name survived up to the time when tradition gave place to record by the establishing of the *Clydesdale Stud-Book*.

On account of the geographical formation of the district the record of improvement can be clearly traced. It is prior to 1878 associated with a succession of horses, the principal of which are "Farmer's Fancy" 298, "Rob Roy" 714, "Largs Jock" 444, "General Williams" 326, and "Lorne" 499. The influence of these horses for good can be traced with considerable clearness, as can also their defects. "Farmer's Fancy" 298 had "boxy" feet and upright pasterns, and that defect long persisted in the Clydesdales of Kintyre. "Rob Roy" 714 had very good feet and legs, but his back was hollow, and he was locally known as the "laigh-backit horse." "Largs Jock" 444 was a good horse with excellent feet, but his hind legs were too straight. He was locally known as the "straight-legged horse." "Lorne" 499 was a horse with a splendid top and well-sprung ribs, but he lacked spring and length of pasterns.

Other sires came and went to Kintyre for a season, but the aforementioned travelled in the peninsula for several years in succession. Consequently they, and not the premium horses imported in later days, dominated the type produced in Kintyre. As far as prizewinners are concerned, "Rob Roy" and "Largs Jock" made the best mark.

The Clydesdale in Cumberland.—Cumberland, from its geographical position, is a county in which a struggle for the mastery between the northern and the southern breeds of draught-horses might be looked for. To a certain extent this took place, but when the matter is examined closely it is found that there is a much stronger admixture of Clydesdale blood there than of any other. Good Shire horses have from time to time travelled in Cumberland, but the records of the Clydesdale Horse Society show clearly that the Clydesdale element predominated in the native horses.

The links between Lanarkshire and Cumberland are clearly established. The first can be traced back to "Old Bay Wallace" 572, bred in Ayrshire and foaled in 1827; "Old Stitcher" 577, bred in Dumfriesshire prior to 1815, probably about 1810; and Pringle's "Young Clyde" 949, the most impressive of all the old Cumberland sires. He was bred at Hyndford Bridge, Lanark, in 1826, and was a horse of great size and strength. He lacked depth of rib, but was big and well coloured, and as he lived long he moulded the Cumberland type along Clydesdale lines.

Of Shire horses that travelled in Cumberland one deserves special mention because of his having also for two seasons travelled in Ayrshire. This is "Farmer's Glory," owned in Ayr by Andrew Hendrie, horse-dealer, and in Cumberland by John Robinson, Wallacefield. He won a £50 premium at Ayr in 1857, and left good stock, his female progeny far excelling his males. In this respect he resembled another excellent Shire horse, Mr Alexander Galbraith's "Tintock," which won second prize at the Highland and Agricultural Society's Show at Glasgow in 1867, and the Strathendrick premium. Both of these horses did good service in improving the breed, but curiously enough their influence did not continue into their second season in Scotland.

The Clydesdales in Aberdeenshire. —Aberdeenshire, like Cumberland, had to some extent a mixed breed of mares to begin with. Clydesdale sires from Lanarkshire were taken north as early as 1823. One of them was "Young Glancer," supposed to have been a son of Thompson's black horse, but it is doubtful whether a rigid application of the age test would support this theory. Other horses bearing names suggestive of Clydesdale lineage were "Young Champion of Clyde," foaled in 1840; "Farmer's Fancy," foaled in 1847; and "Justice" 420. The Earl of Kintore, Inglismaldie, and the well-known Captain Barclay of Ury, were owners of horses whose reputations survive. The horses which live, however, and fairly put the Clydesdale mark on the draught-horses of Aberdeenshire, were "Grey Comet" 192 and "Lord Haddo" 486. They were both prize-winners at the Highland and Agricultural Society's shows, and being horses of sound constitution, as well as of true Clydesdale character, they stamped their own image on the horses of the northern counties.

The Clydesdale Horse Society.

The Clydesdale Horse Society was formed in June 1877, and the first volume of the *Clydesdale Stud-Book* was issued in December 1878. These institutions owe their existence chiefly to the enterprise of the late Earl of Dunmore and Mr John M. Martin, now residing at Lasswade, Mid-Lothian. The Earl of Dunmore was instrumental in securing the support of 100 Life Governors, who subscribed £10, 10s. each to the funds of the Society at the outset, thus giving it a unique start. Mr Martin was at that time tenant of Auchendennan Home Farm, and Hawthornhill, Dumbartonshire, and owned several of the best Clydesdales of their time.

The initial work connected with the Society was carried through by these gentlemen and a Council and Editing Committee, on which were the late Sir William Stirling Maxwell of Keir and Pollok, Bart., the late Sir Michael R. Shaw Stewart of Greenock and Blackhall, Bart., and other gentlemen keenly interested in Clydesdales. They had as their secretary, from 1877 to 1880 inclusive, Mr Thomas Dykes, who was also Agricultural Correspondent for the *Glasgow News*. In the latter year the office of secretary was filled by the appointment of Mr Archibald M'Neilage, who has rendered valuable services to breeders of Clydesdale horses, and to whom we are indebted for this sketch of the breed.

Infusion of Shire Blood.—A primary difficulty had to be encountered. For many years prior to 1877 an occasional Shire horse or mare had been imported from England, and in several cases, as has already been indicated, good results had followed from blending the two races. A small but influential body of breeders, led by the late Lawrence Drew, of Merryton Home Farm, Hamilton, indulged the idea that there should be but one Stud-Book for the English and Scots breeds, and refused to join the Clydesdale Horse Society. They maintained an attitude of opposition until the lamented death of Mr Drew in March 1884.

Standards for Admission to Stud-Book.—The difficulty was to fix a standard of admission into the *Clydesdale Stud-Book*, which while conserving the distinctive character of the Clydesdale breed, would not disqualify a large number of animals, true to Clydesdale type, in which there was admittedly a strain of Shire blood.

The standard fixed for the Retrospective volume, which included stallions foaled prior to 1st January 1875, was to recognise all stallions reputed to be Clydesdales, which were dead when the Stud-Book movement commenced, as such, and to recognise as Clydesdales all stallions foaled before the date named and then living, if got by a recognised Clydesdale sire or out of a mare got by such a sire. This disqualified a number of horses, such as Hendrie's "Farmer's Glory" and Galbraith's "Tintock," already named, because their pedigrees were given and known as Shire.

For horses foaled on or after 1st January 1875, the standard was made, registered sire and dam got by registered sire, and a similar standard was set for mares foaled on or after 1st January 1877. The policy then outlined of proceeding cautiously, and not making rules more stringent than Nature admits of in her reproductive functions, has been stedfastly adhered to.

For several years the standard has been registered sire and registered dam, or registered sire and dam having herself two registered crosses.

Practically the whole of the Clydesdale breeding interest now supports the Stud-Book movement. The membership of the Society in 1908 numbered about 1500. In the first thirty volumes of the *Stud-Book* there are 14,432 entries of stallions, and 20,650 entries of mares.

Characteristics.

The Clydesdale has undergone modification in type at the hands of man during the past century.

Ancient Types.—Portraits of a mare named "Meg" and a stallion named "Young Clydesdale," which won at the Highland and Agricultural Society's Show in 1826, represent the mare to have been a big handsome animal with clean hard bones, good round hoofs, and well-sprung pasterns. She looks a big mare, standing possibly 16.2 hands, with a well-set-on head and neck and high withers. The feature which distinguishes her from the Clydesdale mare of the present day is her somewhat "gyp" appearance. "Young Clydesdale" appears to be a much lighter animal, shown in plough harness. He would be regarded now as rather much of a "van" horse.

About the year 1840 and onwards the demand was for a thick, wide, low-set horse, with strong forearms and thighs, broad bones, plenty "feather" on his legs, and not too much spring of pasterns. The feet were always an essential point in the Clydesdale, and at no time can it have been a matter of indifference with breeders as to the wearing quality and openness of hoof-head of the Clydesdale horse.

Action about the year 1860 reached its highest illustration in the stallion "Sir Walter Scott" 797, which won first in that year at the Highland and Agricultural Society's Show at Dumfries, and first at the International Show at Battersea in 1862. In 1870 "Rantin' Robin" 685 beat "Prince of Wales" 673 at the next Highland Show at Dumfries. His pasterns were short and upright, but he had a clear advantage over his rival in respect of the breadth and openness of his face, and his grandly rounded barrel, with deep ribs.

Favourite Type in 1850-1880.—The Clydesdale stallion of the period from 1850 to 1880 was generally a horse standing from 16.2 hands to 17 hands, with good open-hoofed feet; pasterns not too oblique; broad flat bones, fringed with plenty of hair; broad hocks, not too straight; well-developed forearms and big knees, broad in front; good walking action and moderate trotting action. Colours were mostly browns, bays, or blacks, with an occasional grey among the mares, but chestnuts were anathema. Only one really good chestnut horse was seen during that period, "Topsman" 886, and while his breeding on the sire's side is undoubtedly Clydesdale, dispute was keen as to his dam. His granddam was bought in Glasgow market in foal to a horse called "Samson" (so it was said), but the "Samson" was never identified. The foal was "Topsman's" dam.

Prince of Wales.—The sire which modified this type was "Prince of Wales" 673, foaled in 1866. He lived until the autumn of 1888, when he died in the possession of his first owner, Mr David Riddell, Blackhall, Paisley, who bought him at Mr Drew's dispersion sale at Merryton in April 1884 for 900 guineas. "Prince of Wales" was an upstanding, tall horse, with rather a hard "Roman nose," and somewhat straight hocks. He was marvellously healthy and sound. He could trot like a Roadster, and imparted much greater style to the Clydesdale than the breed had possessed up to this time.

Darnley.—His great rival, and in the end his stable companion, was "Darnley" 222 (1872-1886). He was a more regular and impressive sire than "Prince of Wales" 673. If the head of the latter was rather long and narrow, the head of "Darnley" was rather small and pony like. He had slightly drooping quarters, but otherwise he was the ideal Clydesdale. He was a magnificent walker, but lacked the dash and vim of the "Prince of Wales" strain. When on his season in the year 1882 he weighed a ton (2240 lb.)

Stock of Prince of Wales and Darnley.—The "Prince of Wales" mated with mares by "Darnley" produced some of the highest priced Clydesdales on record, including "Prince of Albion" 6178, sold when two years old for £3000 to Sir John Gilmour of Montrave, Bart.; his own brother, "Prince of Kyle" 7155, sold when rising two years old to the late Mr James Kilpatrick, Craigie Mains, Kilmarnock, for £1700; "Prince Alexander" 8899, sold when a foal, not twelve months old, to the late Mr James Lockhart, Mains of Airies, Stranraer, for £1200. Sons of "Darnley" also made high prices, and many of them were the best breeding horses of their time. As sires they bred with greater uniformity than did the sons of "Prince of Wales" 673. "Topgallant" 1850 was sold when rising four years old to the late Sir Michael R. Shaw Stewart, Bart., for £1600, and "Flashwood" 3604 was sold when one year old to Mr John Pollock, Langside, for £900.

A New Era—Measurements of Clydesdales.

Sir Everard.—A new era in Clydesdale breeding began with "Sir Everard" 5353, a son of "Topgallant" 1850, and out of a mare by a son of "Prince of Wales" 673. He was foaled in 1885 and died in August 1898. He stood fully 17.1 hands, girthed, when in low condition, 8 feet round the heart, and weighed, in June 1890, 20¾ cwt. He measured round the forearm, above the knee, 26 in.; 17 in. round the knee; 11 in. bone immediately below the knee; 12 in. bone immediately below the hock; 11½ in. from the centre of the knee to the centre of the fetlock joint; 21½ in. from the stifle joint to the point of the hock; and 18½ in. from the point of the hock to the hind fetlock. He mated very successfully with mares got by "Prince of Wales" 673, or mares by sons of "Darnley" 222.

Baron's Pride. — "Sir Everard's" most celebrated son is "Baron's Pride" 9122, foaled in 1890, and still alive (1908). Without cavil, this is the greatest breeding horse the Clydesdale race has known. He stands 17.2 hands, and in show bloom, in 1894, when he was champion of the show of the Highland and Agricultural Society at Aberdeen, he girthed 8 ft. 2 in. "Baron's Pride" is represented in Plate 10. His son "Silver Cup" 11,184, with a prize record exceeding that of his sire, and still alive, stands 17 hands. In November 1905 "Silver Cup" girthed 8 ft. 1 in. He measured 17 in. round the forearm, and 19 in. round the gaskin. He had then 10½ in. bone below the knee, and 12 in. below the hock. At the date named he weighed 2156 lb.

"Baron of Bucklyvie" 11,263, the first prize aged stallion at the Highland Show at Aberdeen in 1908, was foaled in 1900. When five years old he stood 17.2 hands, and girthed (in November 1905) 7 ft. 2 in. Around the forearm, 1½ in. above the knee, he measured 15½ in., and 18½ in. round the gaskin, about 1½ in. above the hock. He measured 10½ in. bone below the knee, 11 in. bone below the hock, and at the date named, in low condition, weighed 1876 lb. The corresponding measurements for "Sir Hugo" 10,924, when seven years old, and in lean winter condition, were: height 17.1 hands, girth 7 ft. 3 in., 15½ in. round the forearm, and 18 in. round the gaskin, above the hock; 10½ in. bone below the knee, 11¾ in. bone below the hock; weight when on his season travelling in 1905, 1950 lb.

Hiawatha.—Passing from the "Sir Everard" tribe, to which all of these horses whose measurements have been given belong, the most notable show horse of modern times is "Hiawatha" 10,067 (Plate 9). Four times he won the Cawdor Cup for the best Clydesdale stallion at the Glasgow Show, and he was awarded the supreme championship of the breed at the Highland and Agricultural Society's Show at Edinburgh in 1899. He belongs to the most modern type of Clydesdale, and has more of "Prince of Wales" 673 character in him than any other horse of his time.

In November 1905 "Hiawatha" stood 17.2 hands, girthed 7 ft. 6 in. in lean condition, and then weighed 2128 lb. His bone, below the knee, measured 10½ in., and below the hock, 11 in. His son, "Hiawatha Godolphin" 12,602, was foaled in 1902, and when three years and four months old stood 17.2½ hands, girthed 7 ft. 6½ in., and weighed 1960 lb. Below the knee his bone measured 10½ in., and below the hock 11¾ in. He measured 17 in. round the forearm, and 19½ in. round the gaskin above the hock.

"Marcellus" 11,110, another son of "Hiawatha," was foaled in 1898, and in November 1905 he stood 17.1½ hands, girthed 7 ft. 8 in., and weighed 1988 lb. He had 16½ in. muscle above the knee, and 19½ in. round the gaskin, above the hock. He had 11 in. bone below the knee, and 12¼ in. bone below the hock.

Royal Favourite.—Of a different type and of another line of breeding, but still combining "Prince of Wales" and "Darnley" blood, is "Royal Favourite" 10,630. He was foaled in May 1897, and in November 1905 stood 16.2½ hands, girthed 7 ft. 3 in., and weighed 1960 lb. He has 16 in. above the knee, 18 in. above the hock. Bone below the knee 10 in., and below the hock 12 in.

Measurement of Mares.—Mares may be taken as a rule to measure about 2 in. less in height, from 15.2 to 16.2 hands being the average. And in respect of weight of bone and other measurements, these are in proportion. "Chester Prin-

cess" 16,371, the champion mare at the Highland Show for two years, is considerably over these measurements, and in proportion in every respect.

Features of the Modern Clydesdale.

A Clydesdale, whether male or female, must walk close behind—that is, the points of the hock must be turned towards each other, and they must not be too open in the thighs. In front, their legs should be planted well under the shoulders and chest, and not at all on the outside, like those of a bulldog. It is a very bad fault for a Clydesdale to stand "easy" on its forelegs, so that its knees are shaky.

A true Clydesdale gives the ideas of strength, spirit, and soundness. Activity is essential, along with soundness of wind and limb.

Markets for Clydesdales.

Export Trade.—The Clydesdale has long been in great demand for foreign export. As early as the second quarter of the nineteenth century stallions and mares were being exported to the Australian colonies and to Canada. During the next quarter of the century Australia and New Zealand bought many of the choicest specimens of the breed, and prices over £1000 were recorded for horses like "Time o' Day" 875; "Pride of Scotland" 602, the best two-year-old colt of 1874 went to Australia at £750, and his son "Bonnie Scotland" 1076 followed in 1878 at £900.

Many Highland Society first-prize winners were exported from 1850 to 1880, and although the volume of trade in any single year might not have gone much, if any, over a score, the value of each animal was high.

In 1880 a totally different trade was developed with the United States and Canada. Numbers rather than quality were its characteristic, although this rule did not universally apply. In 1881 a large number of the best females at the shows of that year went to the United States and founded studs there from which valuable animals have since been brought back to Great Britain.

Export Certificates.—In 1884 the Clydesdale Horse Society began to keep accurate records of the Export Certificates issued, and the following list indicates what these were :—

Year.				No. of Certificates issued.
1884	.	.	.	500
1885	.	.	.	514
1886	.	.	.	600
1887	.	.	.	920
1888	.	.	.	1149
1889	.	.	.	1040
1890	.	.	.	554
1891	.	.	.	349
1892	.	.	.	158
1893	.	.	.	110
1894	.	.	.	21
1895	.	.	.	15
1896	.	.	.	56
1897	.	.	.	57
1898	.	.	.	132
1899	.	.	.	250
1900	.	.	.	178
1901	.	.	.	167
1902	.	.	.	266
1903	.	.	.	411
1904	.	.	.	536
1905	.	.	.	653
1906	.	.	.	1317
1907	.	.	.	1100

Home Market. — With respect to values in the home market the following tables of averages show the rise and fall in prices over the period from 1876 until 1908. It will be observed that the average made at the Knockdon sale of 1876 has only once been surpassed during the generation that has passed away since it was held :—

Clydesdale Sales from 1876 to 1908.

The following is a list of the average prices obtained at the leading public sales of Clydesdale horses from 1876 to 1908 :—

PLACE.		Nos.	Average.		
1876.					
Knockdon	. . .	22	£209	15	2
1878.					
Merryton (partly Shire and Crosses)	. . .	50	168	11	0
1879.					
Merryton (partly Shire and Crosses)	. . .	55	112	11	0
Auchendennan	. .	13	114	5	9
1884.					
Auchendennan	. . .	14	161	14	6
Merryton (Dispersion, partly Shire and Crosses)	.	63	152	3	0
1887.					
Whitehill, Sanquhar (Mares only)	. .	11	138	18	2

PLACE.	Nos.	Average.		
1892.				
Montrave (highest, Queen of the Roses, two-year-old filly, 1000 gs.) . . .	29	£149	15	0
1893.				
Croy-Cunningham . .	19	88	12	10
Blairtummock . . .	31	51	1	6
Kippendavie . . .	25	48	2	0
1894.				
Mains of Airies . . .	32	80	9	1
Edengrove and Robgill (Joint)	23	77	15	10
Craigie . . .	20	26	11	3
Seaham Harbour (Draft) .	42	36	2	6
1895.				
Eastfield (Stallions only) .	38	168	8	10
Glasgow Cattle Market (Mares only) . . .	10	68	19	8
Earnock . . .	27	79	5	10
Polmont (Joint) . . .	39	60	15	10
1896.				
Sinclair Scott's (Glasgow) .	29	63	5	0
Lochburn . . .	12	43	5	5
Edengrove (Carlisle) .	16	116	19	8
Blairtummock . .	19	52	3	4
Keir (Dispersion) . .	53	67	11	6
Seaham Harbour (Draft) .	52	41	12	1
1897.				
Moncreiffe . . .	10	44	10	0
Seaham Harbour (Draft) .	41	49	5	8
1898.				
Kippendavie . . .	30	48	16	2
South Acomb . .	38	45	13	0
Seaham Harbour (Draft) .	47	48	11	1
1899.				
Overdale (Mares) . .	7	83	8	0
Morton Grange . . .	38	10	8	0
Seaham Harbour (Draft) .	67	48	11	5
Mertoun . . .	18	45	0	8
1900.				
Kippendavie . . .	11	83	2	10
Balmedie (Dispersion) .	18	110	6	2
Seaham Harbour (25 Foals)	61	52	18	5
South Acomb . . .	37	40	17	7
1901.				
Milton Ardlethen (Females only) . . .	11	45	5	8
Mains of Airies (Dispersion)	19	141	7	9
Morton Grange . . .	36	74	18	7
Seaham Harbour (Draft) .	53	42	3	11
Perth—Montrave .	23	64	16	6
Blairtummock .	5	61	6	5
Rosehaugh .	10	47	2	11
Orchardmains .	10	44	4	1
Glamis . .	4	45	8	3
Sands (Fillies) .	2	51	9	0
Cavens . . .	22	44	2	0

PLACE.	Nos.	Average.		
1902.				
Bellsfield	6	£66	10	0
A. B. Matthews (Draft) .	4	73	4	9
Seaham Harbour (Draft) .	38	53	16	0
Millfield	16	77	6	1
Seaham Harbour (Draft) .	32	39	15	8
Lambton	36	47	15	2
1903.				
Swinton House . .	18	46	10	6
Orchardmains (Dispersion) .	24	76	9	6
Milton Ardlethen . .	19	50	1	11
Morton Grange . .	20	62	4	3
Seaham Harbour (Draft) .	43	44	5	7
Drumflower . . .	9	134	12	8
Garthland . . .	11	45	11	8
1904.				
Seaham Harbour (Draft) .	37	45	5	4
Perth—Glamis . .	9	56	0	0
Rosehaugh . .	9	42	0	0
Mertoun .	13	38	0	0
Mersehead (Pilkington's) .	19	47	4	0
1905.				
Blacon Point (at Lanark) .	30	152	3	7
Charleston . · .	20	38	18	0
Seaham Harbour (Draft) .	45	35	1	0
1906.				
Perth	43	60	16	4
Seaham Harbour (Draft) .	39	50	15	0
Blacon Point (at Lanark, Dispersion) . . .	14	216	10	6
1907.				
Scotstoun (A. B. Matthews)	5	84	17	0
Challoch (Newton-Stewart) .	12	73	2	1
Perth (Gross) . .	74	83	5	5
Same Detailed—				
Harviestoun . .	15	149	17	5
Bullion . . .	3	107	2	0
Mertoun . . .	12	67	12	9
Nether Bogside .	5	69	1	9
Mains of Edzell .	5	55	10	0
Arnprior (Yearling Fillies) . . .	5	56	18	2
Lochlane . .	2	93	19	6
Seaham Harbour (Draft) .	57	38	1	6
Lambton . . .	40	39	8	8
1908.				
Ardimersay (at Ayr) .	6	40	15	6
Perth (Gross) . .	84	54	5	0
Same Detailed—				
Mains of Edzell .	3	55	13	0
Jordanstone . .	2	136	10	0
Ledlanet . .	12	74	16	3
Dunure Mains .	5	38	8	7
Harviestoun .	9	120	19	8
Crieff . .	5	26	13	5
Mertoun . .	5	24	15	7
Bullion . .	4	71	18	6
Seaham Harbour (Draft) .	53	45	10	6

MANAGEMENT OF CLYDESDALE STUDS.

The system of management pursued in Clydesdale studs varies to some extent according to the district. In the extreme south-west of Scotland, for example, where the temperature is comparatively high even in winter, it is possible to have the animals out in fields practically all the year round. In such circumstances all that is required is a shelter-shed or some such structure for providing cover in a specially cold or stormy day. In other parts, on the other hand, winter-housing to a greater or less extent is almost a necessity, especially in the case of mares. Young colts running rough may do with a warm shed for use at nights and on cold days, but mares require comfortable housing at that season of the year if colds and other troubles are to be avoided, and the best foaling and other results obtained.

Apart from this, there is no great difference in the methods of treatment between north and south, although naturally where the winter is open and grass comes early in the spring, it is possible to bring out young stock earlier in the year for show purposes. Open-air rearing in winter, especially where the climate is moist, is also a great aid in the growth of hair. This circumstance explains to a considerable extent why young horses reared, say, in Wigtownshire, have usually a greater profusion of hair in the spring than those reared in Lanarkshire, Perthshire, or Aberdeenshire. Later in the year the two classes compete on fairly equal terms in this respect.

Brood Mares. — The treatment of brood mares is perhaps the most important element in stud management. Where mares are kept exclusively for breeding purposes, the artificial feeding should be of the lightest possible description consistent with keeping them in fresh breeding condition. In one very eminent stud in Central Scotland the custom a few years ago was to give the mares of this class only one bushel of oats per head per week mixed with chopped oat-straw, a few swedish turnips in the forenoon, a pailful of boiled food—turnips, cut hay, and bran—in the afternoon, and long oat-straw *ad lib.* Very excellent results were obtained on this feeding for a considerable number of years, the mares going out in sheltered fields all day, summer and winter. Of late years there has been a tendency to restrict the boiled food and to give everything cold, as being less liable to set up colic. In an equally well-known Fifeshire stud, where as many as fifteen mares were wont to be kept for breeding purposes alone, the custom was, and still largely is, to give the mares twice daily in winter bruised oats and chopped hay along with a few raw swedish turnips. The daily allowance of the combined oat and hay mixture was about 14 lb. per head, one half being given in the morning and the other half in the evening during the period from the end of September until the grass came in the spring. In summer the animals got nothing but grass.

In a Wigtownshire stud, where seven mares were formerly reserved for breeding purposes, the regulation winter diet was 6· lb. of crushed oats and 1½ lb. bruised linseed-cake each per day, with as much timothy hay as the animals cared to eat,—this being in addition to what they picked up in a rough pasture. In this stud rock-salt was always kept within reach of the mares, and this is a course that should be generally followed. Salt not only sharpens the palate of animals but helps to keep them in good general health. An Aberdeenshire breeder gives his breeding mares in winter 1 lb. of oats daily along with cut hay, a boiled mash, and a few turnips; while a Forfarshire breeder gives from the end of October onwards one feed of oats in the morning and a feed of boiled barley and beans at night, in addition to hay and straw and what the mares pick up on the field during the day.

These instances afford an indication of the lines followed generally in the feeding of brood mares kept exclusively for breeding purposes. The important point is to keep the animals in as fresh natural condition as possible, so that they may breed regularly and produce strong healthy foals.

Possibly the greater number of mares are kept for working as well as breeding, and in these cases somewhat different methods of feeding and management have to be adopted. A mare that is working will always require more liberal treatment than one that is doing nothing. Up to about a month before foaling the working mare can be fed on pretty much the ordinary horse rations of the farm—that is, 10 to 14 lb. of grain can be given along with a few swedish turnips, and possibly a bran mash on Saturday nights, with, of course, straw or hay as required. Some people also give an occasional feed of linseed, while others give a two-ounce dose of Epsom salts at least once a-week. About a month before foaling it is usual to reduce the oats and increase the bran or linseed ration. Mares in foal must at all times be worked with special care. They should be backed as little as possible, and for two months at least before foaling they should not be carted. In chains, however, they can be wrought quite safely, up practically to foaling time, provided that they are not hurried and get their own time in turnings and other awkward positions.

Mares worked fairly regularly make the most reliable and satisfactory breeders. They have also, as a rule, the easiest foaling time, and produce the strongest, most thriving, and healthy foals. Idle mares are very apt to get fat no· matter what is done to prevent it, and when such is the case foaling risks are greatly increased. Working mares, of course, require careful handling, but where this is given there are fewer losses, as a rule, with mares of this class than with those that go idle. Mares that are hard worked and mares that are highly fed are usually the most difficult to settle in foal. Stallion owners on this account prefer, where they have a choice, districts away from towns, where neither the work nor the feeding is heavy.

Foaling.—In Scotland, at any rate, the bulk of the foaling has necessarily to be done under cover. Several experienced stud-owners never allow their mares to foal in the open before the middle of June in any case. For inside foaling a roomy loose-box is almost a necessity. Mares should be sheeted for a day or two after foaling, and fed on soft, sloppy food, such as boiled barley, pulped turnips, or cut hay mixed with moistened or boiled meal. Bean-meal is regarded as being about the best for this purpose.

Care of Foals.—Perhaps no young animal on the farm is more precarious to handle for the first few days of its existence than a foal. Immediately a newly-born foal gets to its legs it should be taught to suckle its dam. Many mares, even at best, are poor nurses, and it will be the business of the attendant to see that the foal is getting the nourishment it requires. Mares that foal early and have little nourishment for their offspring should be fed as much as possible on sloppy food, and given occasionally a meal-drink. On the other hand, there are mares which have milk so plentiful and so strong as to cause diarrhœa in the young foal. This is an evil to be guarded against. In such cases the mare should be at once put upon dry concentrated food and straw fodder; if at grass, she should be put on the oldest and driest pasture available, and kept there until the flow of milk ceases somewhat in volume. In obstinate or rather over-milky conditions it may be necessary to systematically drain off some of the milk by hand.

Great care should be taken in the choice of a day for turning out the mare and foal for the first time. The selected day should be dry and the grass free from white frost. For the first eight or ten days, should rain begin to fall, the mare should be at once sheltered. A young foal, if exposed in such weather, may contract joint-ill and other diseases, and the mare herself may become chilled. Even after the foal gets older, it is sound policy never to let it out when there is hoar-frost on the ground until at any rate it has had a meal in the house.

Foals are usually weaned when they are from four to six months old. In special cases they may be allowed to go the full six months or even a little longer with their dams, but that involves a good deal of strain on the mare if she is again to have a foal next year, and most breeders are content when they get foals eighteen weeks with the mare. Foals just taken from their mothers should be put in a field or paddock by themselves and given a little artificial food,—nothing is better for keeping the flesh on a foal than a chop mixture of oats, cut hay, and beans, fed twice a-day. Most breeders feed both the mare and the foal before they are separated, for by October, when most of the weaning takes place, the pastures are beginning to get dry and bare. The foal in such cases is accordingly accustomed to eat out of a box before it is weaned, and thus it takes readily to the new conditions, and does not lose condition. If it be important to keep on the "calf" flesh, it is equally important, if ground is not to be lost, to keep on the "foal" flesh. A foal neglected at weaning time or earlier never grows to the size that it otherwise would.

Putting Mares Dry.—As a rule, no difficulty is experienced in getting the mare dry, but where such is the case a doze of 4 drachms of aloes, along with a pint of linseed-oil, has been found beneficial. In extra difficult cases part of the milk should be drawn off at extending intervals and the udder bathed with vinegar.

Other Classes.—The "other classes" in a Clydesdale stud will include colts and fillies of various ages up to three years, by which time the fillies should be qualifying for the brood-mare stage. It is desirable in rearing young horses of this class to run the colts separately from the fillies. In each case the fields or pasture outruns should be equipped with warm wooden sheds fitted with both feed-boxes and hay-racks. Where the animals come into the steading at night warm sheds can be dispensed with, although shelter-sheds, with their faces to the south, may still be desirable. "Chop," as already noted for foals, forms an excellent winter food for young horses of either sex, and this should be accompanied by a few swedish turnips and what long hay or straw they will take. In some studs as much as 10 lb. each animal of chop is allowed per day, but this will depend on the nature of the weather and the character of the pasture. For animals that are being pushed on for showing in the ensuing season, a daily mash of bran, pulped turnips, and barley or beans, with a few raw carrots, are very useful, as are likewise a few green tares, the young animals eating the latter even more readily than hay.

Colts that are not good enough for breeding purposes should be castrated, at the very latest, before they are two years old. Preferably, this operation should not be delayed beyond twelve to fifteen months. Colts left too long entire become coarse about their heads and necks, and too rough about their legs, to make first-class geldings. Moreover, they never settle or thrive so well as colts castrated earlier. Colts intended to be left entire require to be boxed at two years old. By this time, however, they are usually in the hands of the regular stallion owners, who have premises specially suitable for this class of stock.

Although a few breeders put their fillies to the horse at two years old, the general custom is to leave them until they are a year older. Fillies served at two years old, unless they are extra big in size, are apt to become stunted in their growth, and rarely make such big mares at four or five years old as they would otherwise do. At the same time, many people hold that a filly served at two years old, like a heifer started to breed at fifteen months old, breeds more regularly and with greater certainty in subsequent years than one that is not served until she is three years old.

Management of Show Stock.

Showing is to a considerable extent a business by itself, and in its highest form at any rate not for the amateur. To win a prize in important shows, an animal must, in the first place, be fairly perfect in shape and correct according to the ideas of the time in its detailed points. It must, for instance, be of fair size, as clean as possible in its legs and ankles, and sound in its feet. Moreover, it ought to be a good, straight mover, both behind and in front. More or less solid reasons can be advanced for each of these requirements, and in addition it must be brought out in what is called show form. The latter, assuming that the fundamental groundwork is right, is mainly a matter of physical labour and judicious feeding.

Where the science comes in is in regard to detailed points. Thus blistering is pretty extensively resorted to for the growth of hair over the hoof heads. In the same way many animals are made to stand on soft, prepared stances to encourage the growth of their feet. While most of these devices are perfectly legitimate and harmless so far as the general public are concerned, there is another practice that has crept in of late years that cannot be so well defended,—that is, to force out young animals, colts especially, with cow's milk. By this means it is possible to have big lustrous-looking yearlings with great hair; but animals so forced rarely do much good in succeeding years when the milk-supply is not forthcoming, and the custom of using milk in this way has been blamed for encouraging the objectionable disease known as wind-sucking. The rush for prizes is, however, so keen that expert show men usually use all means open to them to present their animals in the best possible prize-winning form.

The great majority of breeders are wise enough to avoid all these dubious methods, and to rely upon more natural means of embellishment, which, after all, pay best, and are in every respect most satisfactory in the end.

Clydesdale stallions are represented in Plates 9 and 10, and a Clydesdale mare in Plate 11.

THE SUFFOLK HORSE.

The Suffolk Punch is a distinct type of horse. It has its headquarters in the English county of Suffolk; but although it has long been held in high esteem there, it has never obtained an extensive footing beyond the south-eastern counties of England.

Historical.—As to the origin of the Suffolk Punch, various accounts have been given. Low says: "The colour distinctive of this variety connects it with the race widely diffused throughout the north of Europe and Asia, from the Scandinavian Alps to the plains of Tartary, in which the dun colour prevails. It is believed to have been carried to the eastern counties of England from Normandy, which yet possesses many fine horses of this variety, introduced, it may be believed, by the Scandinavian invaders." [1]

Arthur Young was a native of Suffolk, and in his report on the Agriculture of this county, compiled about the end of the eighteenth century, he speaks of "the old breed" of horses as if it had been specially associated with the district long prior to that date. Writing in 1878, Mr Herman Biddell says: "Two hundred years ago there were draught-horses peculiar to the county, and of standing enough as a distinct breed to maintain their prevailing characteristics through generations of descendants, long after the original type had been considerably modified by repeated selection, and the introduction of incidental crosses. How long prior to Young's time the breed had existed we have no evidence to show." [2]

Continuing, Mr Biddell says: "It clearly appears that there is scarcely a Suffolk stallion in the county, of any note whatever, whose pedigree is not clearly to be traced in a direct male line for seventy years. The records in the possession of the association, which relate to a period between 1790 and 1810, throw some light on the matter, and point to the introduction of materials not ill calculated to bring about the transformation that has taken place. Infusion of the Thoroughbred, Flemish, and heavier blood of native horses, has tended to exert upon the 'old breed' the influence such elements would be likely to produce; but as far as a careful search through the lineage of the horses now extant in the county will show, not one seems to have inherited the alloy in the male line, all of which terminate in an ancestry in all probability tracing back to the old breed mentioned by Arthur Young."

Since then opinion has been modified. In a lecture before the Framlingham Farmers' Club in 1907, Mr Herman Biddell remarked—

"I have seen over and over again the statement made that the Suffolk horse was the result of a cross from the

[1] Low's *Domesticated Animals of the British Islands*, 1842, 619.
[2] *Live Stock Journal Alk.*, 1878.

Continent. It is said they are descendants of some Flemish ancestry. I have searched every available source from which this statement was likely to have emanated. I have been unable to find the slightest foundation for this mythological origin of the Suffolk sire. . . . There was a strain of Suffolks fifty years ago which came from a mare imported into the country from abroad. But the horse that had this strain in his pedigree could only have had an eighth of his parentage of this doubtful origin, but this is the only instance of authentic introduction of foreign blood that I could discover."

The most salutary influence on the modern Suffolk was exercised by a trotting stamp of horse from Lincolnshire, brought down about the middle of the eighteenth century by a Mr Blake of St Margaret's, Ipswich. This blending of "Blake's strain" with the old county type of horse produced, in the opinion of qualified authorities, animals as handsome, or nearly so, as horses of our own day.

Characteristics.

Colour.—The colour is the most distinctive feature in the Suffolk breed. It is a chestnut of varying hue, with lighter coloured mane and tail. The bright chestnuts are the favoured colour. It is a peculiar fact that no breed reproduces truer to colour than the Suffolk. Silver-haired horses are met with, but so long as they are not roans they are not objected to. The mealy colour is one of the worst.

Form.—Arthur Young apparently had not a very high opinion of the breed. He cuts it off with this sarcastic touch: "Sorrel colour; very low in the fore end; a large ill-shaped head, with slouching, heavy ears; a great carcass, and short legs: an uglier horse could hardly be viewed." Now, however, the breed is gainly, although it is still a thick, chubby, or punchy animal, with a body disproportionately large for the length of its limbs. Its legs are stout, full of substance, and now flatter in the bone than was at one time the case. The charge that the Suffolk is a round-boned horse need not seriously be considered, for there is a tendency to run to the extreme of quality with bone of "razor"-like character.

A great improvement is discernible at the ground. Up till near the end of the last century bad feet were comparatively common, and while it cannot be urged that the attainment of perfection has yet been reached, still the casual observer who remembers the earlier appearance of the breed must admit that there are plenty of open coronets and good thick feet to be found in modern Suffolk studs. It is not the case that the Punch is unpopular for the London dray trade because of bad feet. The fact of the matter is, that he is hardly big or weighty enough to compete for this traffic with the strong and massive Shire, which all but monopolises the market. But for heavy van work there is no better horse than the Suffolk.

Measurements.—The limbs of the Suffolk horse have a naked appearance, for they carry little long hair. A good measurement of bone below the knee for a stallion is 10½ in. It is mentioned that Mr Smith's famous sire "Wedgewood" girthed 7 ft. 11 in., and measured as much as 10¾ in. below the knee, which must be considered good, as there was no hair to include. In point of height, 16.2 hands is about the limit. Mr Biddell goes as far as to say that "unless extremely well put together, anything over 16.1 hands should be viewed with suspicion."

Action and Handiness.—The action of the Suffolk is not the least potent recommendation of the breed. A fine agile walker, he uses his joints well, snapping his knees sharply. For farm work his speedy and willing service is much appreciated. When in the early days of Clydesdale "horse-breeding by the book," when that breed was making new supporters in the north of Scotland, not a little of the opposition it encountered in Aberdeenshire was from the Suffolk mares. At the end of a drill they are speedy and handy to turn, and give equal satisfaction between the shafts.

This fine spirit with which the Suffolk horse goes to work is no doubt inherited. In the olden days it was the custom in Suffolk to enter horses in pulling matches.

Prior to the institution of agricultural shows these pulling matches appeared to be of the nature of sporting institutions. Sir Thomas Cullum writes of them thus:—

"A trial is made with a waggon loaded with sand, the wheels sunk a little in the ground, with blocks of wood laid before them to increase the difficulty. The first efforts were made, as usual, with reins fastened to the collar, but the animals cannot when so confined put forth their full strength; . . . that they may not break their knees in the operation, the area on which they draw is strewn with soft sand."

It requires no stretch of the imagination to connect the willing service of the modern Suffolk with the exercise of his talents in this peculiar way. Although capable of growing to a ton weight, his powers of haulage are not entirely to be measured by the avoirdupois he can put into the collar. Activity and sustained muscular effort count for much.

Docility and Longevity.—Docility and longevity are two points which can be claimed for the breed. The value of the former need not be emphasised: it is apparent.

Many cases of wonderful longevity are on record. One of these is worthy of mention. At one of the earlier exhibitions of the Suffolk Agricultural Society there was a mare shown which had entered her thirty-seventh year. More remarkable still, she was suckling a foal.

The Suffolk Horse Society.

The Suffolk Horse Society was established in 1876, and had in 1908 239 members. The chief scheme of the Society for the advancement of the breed was that of assisting farmers to acquire mares, which, by partial payment and subsequently by realising the foals, became their virtual property in three years. This scheme was inaugurated in 1897, when the Society was empowered to purchase thirty nominations to approved sires. The owner from whom a nomination was bought was asked to restrict the service of mares to the number of eighty to the stallion in that year. Tenant-farmers whose holdings did not exceed 200 acres were to apply under the scheme, in response to an advertisement, and if chosen they were required to sign an agreement to deliver the foal, unweaned and free of cost, on a sale day to be fixed, and to accept £15 from the Society as purchase price, the Society taking the risk of getting a higher or lower price. Thirty nominations were taken up, at a cost of £2 each, and fifteen foals followed in September 1899, which realised, after deducting the auctioneer's charges, £17 each. Each live foal cost the Society £3, 6s. 8d., so that there was a deficit of £1, 6s. 8d. on each of the fifteen foals.

The main difficulty was to find suitable mares, so the Society went one step further and bought mares for farmers, who, on getting possession, paid 25 per cent of the purchase money, the remainder being loaned at 4 per cent. If the mare proved barren for two years, or in case of accident, the Society had the right to dispose of the mare, dividing the proceeds with the tenant-farmer in proportion to his indebtedness. The price agreed on for the delivery of the foal was raised to £16, 10s.

Foreign Trade.—The foreign trade for the breed is considerable. On the Continent it is popular as a useful cross for producing a heavy class of artillery horse. Many go south of the Equator, but the demand is chiefly from Russia and other Continental countries. There is a possible field for the breed in the United States, where the draught-horse type most popular is very much like the Suffolk. With their indifferent roads, the heavy horse with hirsute heels does not seem to enjoy the same popularity as the clean-legged type in certain parts of America.

Leading Shows.—The chief shows at which the Suffolk is exhibited are Woodbridge, Suffolk, Essex, and Norfolk county meetings, and the Royal Agricultural Society's Show. Suffolks are also included in the Cart-horse Parade in London, and fine teams have been exhibited at the International Horse Show at Olympia.

MANAGEMENT.

The Suffolk stud does not differ in respect of management from studs of other heavy breeds. The same routine is followed, the same foaling dates arranged, the same feeding carried out.

Half the management of the stud lies in the right treatment and mating of the brood mare. The selection of a stallion, with the travelling facilities afforded by railway companies, is not limited to the horses in the immediate neighbourhood, for those farther afield may be visited.

Brood Mares.—During summer the brood mare has her foal at pasture and requires little attention. If she is a valuable brood mare only the lightest work sufficient to exercise her should be given. If she has no particular claims in the way of pedigree she may take her share in the work of the farm. Many are opposed to this working of mares suckling foals. It is not productive of evil effects, however, if the mare is not over-driven and heated and the foal not allowed to suckle when she is heated. Mares with very young foals should certainly not be worked.

The Foal. — Early weaning is favoured by many Suffolk breeders. On the farm it is an advantage if the foal can be weaned before the mare is wanted for the stress of harvest work. It is, however, a disadvantage from the point of view of producing a big growthy foal. Some mares have been allowed to suckle their foals from seven to eight months, but that is too heavy a drain upon the system to be advised. In show studs the temptation in this way is great when a foal is doing well, but for not more than six months, and preferably five months, should the mare be asked to rear her offspring.

Foaling time is a more or less anxious period. There are mares that like the human presence when foaling, while others dislike it greatly. The usual indications of approaching foaling are the uneasiness of the mare and the waxiness of the teats. For the first few days after the arrival of a foal it should be kept indoors. Then a sheltered paddock should be chosen, and in the course of a fortnight it may be turned out to grass in the ordinary way. Foaling in the open is not favoured, because it is more convenient to have the mare at hand. It would, however, avoid the danger of contamination arising from the ordinary foaling - box, which manifests its evil effects in navel-ill and other troubles. If the foal is intended for show it should be kept in good condition by other means than mother's milk when the supply is scanty or the quality inferior. Cow's milk sweetened may be given, but only as a last resort. Concentrated food, like corn, should also be avoided as long as possible. The immediate benefit of extra food of this description may be apparent, but it only lays the foundation of future trouble.

Service.—Service is usually offered the mare the ninth or tenth day after foaling. If she fails to respond thus early the mare is tried again at three weeks, and thence every fortnight.

Food and Care in Winter.—Extra rations are allowed in winter, usually beginning about November. A little corn, and perhaps hay in rough weather, will carry the brood mare well through winter. Work in the chains is beneficial to the breeding mare. Carting may also be done up to within a short time of foaling, provided the mare is not backed. During the hard wintry weather a more varied ration is given than at ordinary times. There should be no frosty roots fed, and the allowance of turnips in any case ought to be small. Mares and draught-horses at work in winter will do well on a ration of crushed oats, bran, and a little bean-meal. A peck of maize (soaked) and an equal quantity of bran

with pulped roots night and morning is another ration that is not uncommonly fed in Suffolk studs when the price of maize permits. Flaked maize is good for conditioning. The prejudice against maize for horses is well founded, unless the feeding is carried out with discretion. It will speedily find out the bad-legged horses when they are not sufficiently worked to throw off the surplus of the fat-producing material.

Grooming regularly, watering before meals, and periodical exercise in frosty weather, are important matters in successful stud management. It is a good plan to give draught-horses, young and old, the run of an open court for a time. They will get the exercise they want in this way.

Stallions.—The management of stallions has nowadays been reduced to the simplest of methods. In many studs they run out summer and winter, being taken up to the service - shed when wanted. Shelter-boxes in the paddocks are necessary as feeding-places, but they are not much used by horses for the purpose for which they are erected. It is a good plan to encourage "constitution" by allowing stallions to "rough it" during the winter, taking them under cover about three weeks prior to the show at which they are intended to be exhibited.

A portrait of a Suffolk mare is given in Plate 14, and a portrait of a Suffolk stallion in Plate 15.

THE CLEVELAND BAY.

Cleveland Bay horses have played an important part both in road and farm work. This is only what would be expected of a breed of horses of such size, action, power, and hardy constitution as the Cleveland Bays can claim. They are not now so widely used as in the pre-railway days, but they are still recognised as a very useful class of horses. At one time the variety bore the name of the Chapman horse, but it is now known by the name of the Yorkshire district with which it is mainly associated.

Origin.—The origin of the Cleveland Bay has exercised the ingenuity of several writers, who have puzzled themselves and their readers in vain efforts to account for the existence of the Cleveland Bay by promulgating elaborate theories of crossing between the Thoroughbred stallion and the cart-mare. It is unnecessary to enter into minute detail respecting these theories. The very conformation of the Cleveland Bay clearly points out that he cannot be descended from the cart-horse, the elegance of his quarters especially showing that there can be no kinship between them ; whilst the way in which, as a rule, the Cleveland Bay breeds to type, both in colour and conformation, precludes the possibility of his being the result of an elaborate system of crossing between the Thoroughbred and the cart-horse.

Mr W. Scarth Dixon, to whom we are indebted for notes on the breed, considers it very probable that the Cleveland Bay derives a certain proportion of his courage and endurance from a pretty large infusion of Eastern blood, which doubtless did take place in the earlier years of the Christian era.

It is also possible that the Cleveland Bay may have been crossed with the Scandinavian horse during the time that the Danes effected a settlement on the north-east coast of Yorkshire.

Characteristics.—The Cleveland Bay is a short-legged horse, standing from 16 hands to 16 hands 3 inches, seldom being found under the one, and only a few specimens being met with that exceed, or even attain to, the other. His head is rather plain, but is well set on, his neck is well placed, and his shoulders generally lie well back. His back is rather long, from the standpoint of a riding man, but it is strong and muscular ; his quarters are long, level, and elegant ; and his tail is well put on and well carried. He is remarkable for the quality of bone, which is as clean and flat as that of a race-horse, and his legs are almost clear of hair. His action is of a high standard of excellence, both in a walk and a trot ; and although he has none of that knee action so much admired by the lover of the hackney, he moves his shoulders and hocks in rare style, and in a manner highly suggestive of getting over the ground. In modern times the complaint is sometimes heard that substance is being sacrificed to quality.

It is recorded of the famous stallion "Cleveland" that he measured 16 hands 1½ in. high, 9⅜ in. round the pastern, 10 in. round below the knee, 21 in. round the arm, 15⅝ in. round the knee, and 6 ft. 10 in. round the girth. Cleveland Bays are excellent workers on farms, especially on the lighter classes of land. They are hardy, active, and endurable.

Value for Crossing.—The value of

the Cleveland Bay for crossing with other breeds is difficult to estimate, and to this very fact is to be attributed in no small measure that falling off in the numbers of the pure breed which a few years ago nearly led to its extinction. It was used in Scotland in the early part of the present century to improve the breed of agricultural horses in that country, and the results were, as a rule, satisfactory. Valuable riding and driving horses have been bred by crossing a short-legged Hackney sire with a Cleveland mare ; and Cleveland mares crossed with a Thoroughbred horse have bred some of the best hunters that ever went out of Yorkshire.

Great care, however, is required in the selection of a stallion. The latter should be of an active, wiry character, with good shoulders, and a short strong back, and rather under than over 15 hands 3 inches. Especial care should be taken to select a horse with short legs, this being a far more important matter than size, for great size is to be avoided, even if the horse is ever so well put together. The second cross from a Cleveland mare makes the best hunter as a rule, retaining the size and substance of the Cleveland, and naturally possessing more quality and pace ; but after the second cross the tendency is for the breed to lose size and degenerate. As an instance of the value of the Cleveland Bay as a foundation for breeding hunters may be cited the fact that some of the best hunters bred by Lord Middleton at Birdsall came third in direct descent from a Cleveland mare.

THE YORKSHIRE COACH-HORSE.

The Yorkshire coach-horse is an offshoot of the Cleveland Bay. It originated in the demand which sprang up in the earlier years of last century for big flash carriage - horses. The short-legged compact Cleveland mare was crossed with a big, lengthy, and flash Thoroughbred horse ; the produce, whether horse or mare, was bred from, and eventually the Yorkshire coach-horse, or—as he was sometimes called, from the locality in which he was principally bred — the Howdenshire Cleveland, became recognised as a distinct breed.

Characteristics. — Possessing the length and fine level quarters of the Cleveland Bay, as well as others of his good properties, the Coach-horse also has much of the elegance of the Thoroughbred. He is apt, however, to grow leggy in the course of a few generations; what is gained in quality is lost in bone; and recourse has to be had to the old breed to restore that substance which is so essential in a good coach-horse.

From the Cleveland Bay and the Coach-horse are bred a large proportion of what are known in the trade as London carriage-horses. The larger and stronger animals are bred from mares of the former breed, and sired by either Thoroughbreds or Coach-horses; whilst the lighter and lesser horses are bred from mares of the latter breed, and sired either by Coach-horses of high quality or by Thoroughbred horses of a coaching type. Efforts made to unite the two strains were not attended with success.

MANAGEMENT.

The management of the Cleveland Bay and Yorkshire coach-horse does not differ very materially from that of other similar breeds. The method pursued by Mr Frank H. Stericker of Westgate House, Pickering, may be thus described: Brood mares four or five days after foaling are turned out into a paddock a few hours at midday, being put into well-ventilated boxes at night. The time out is increased as the foal gets older and stronger. The mare has a liberal diet of bran, oats, and chopped hay or straw, with the addition of a few carrots. The foal is haltered in its early youth, as it saves time and patience later on. The mare will "do" the foal better if later on she has the run of a bigger pasture or seeds and clover. A good supply of water is necessary, and so also is the provision of shelter. The foal is weaned usually about the first or second week in October.

Mr Stericker holds that brood mares are all the better of a little light work. When near foaling a feed of corn and bran with hay twice daily along with a few carrots or swedes is recommended, the idea being to get her into good though not high condition.

Foals benefit by running out during the day and having the shelter of a covered yard at night. With bran, oats, and a little boiled food, chopped hay, and carrots, the youngster will go on thriving, and gain bone and muscle. Do not house yearlings too much, but keep them on pasture as long as possible.

Regularity in feeding is one of the secrets of successful rearing. When bringing animals out for show, it is important to give no more food than will be eaten up. Exercise should be regular, and when breaking, a fortnight must be given with the bit in the mouth, leading about, and sending the animal round in a circle. The girth is then put on, and later the "dumb jockey." Driving about in strings should not be neglected, the mouthing being particularly attended to. Many accidents are caused and good horses spoiled by the groom being in too great a hurry in the early handling.

A typical Cleveland Bay mare, "Woodland Briar" 1318, is represented in Plate 18.

THE THOROUGHBRED HORSE.

Properly speaking, the Thoroughbred is not an agricultural horse. He does not come within the province of farm live stock, for the main object for which he is bred is speed. The only measure in which the "blood" horse trenches upon farm stock-raising is in the production of hunters, military horses, or, it may be, according to the cross, harness horses. It is therefore from that standpoint that he must needs be treated here.

It is unnecessary even to glance at the history of English horse-breeding as represented by the rise and development of the Thoroughbred. It will suffice here to say that it is evident that the modern Thoroughbred is chiefly indebted to three great sires for his present position. These are: the Byerly Turk, the Darley Arabian, and the Godolphin Arab (which, by the way, many writers contend was a Barb).

It is natural that the customs of the race-course have placed their imprint upon the character of our Thoroughbred horses. Thus we find it freely contended that the Thoroughbred of to-day has not the stamina of his ancestors, which may or may not be true. But we do not now ride such punishing races as formerly, and doubtless, if the occasion demanded it, the necessary training would speedily vindicate the constitution of the modern Thoroughbred. It is also contended that breeders have sacrificed speed, which, if not quite accurate, we may assume to arise from the fact that the stop-watch is a more accurate method of checking a performance than was in existence in the earlier days of racing. One thing at least can be claimed for the modern Thoroughbred in which he is superior to his ancestors—he is a bigger horse. His descent may be assumed to be largely but not entirely Eastern. The pedigrees of some of the more famous horses in the olden time lend colour to the belief strongly held by several writers that the stamina of the English native mares, highly spoken of, was in some degree responsible for the "blood" horse as we know him to-day.

Thoroughbreds for Hunter Breeding.

The breeding of the pure Thoroughbred is not pursued to any extent by the ordinary farming classes. It is a business or hobby by itself. The fact, however, is undisputed that the most valuable animal in the equine world is a good Thoroughbred stallion.

The farmer has his uses for "blood." He finds it of inestimable service in the production of hunters endowed with pluck and stamina. The Royal Commission on Horse Breeding annually presents 28 King's Premiums, of the value of £150 each, for Thoroughbred stallions which are allotted districts to trave in, the classes in which they compete. In later years the selection of stallions has been partly influenced by their period in training and their racing performances. These King's Premiums used to be offered as the King's Plates for racing, but for many years have been diverted to the much more useful purpose of assisting to stock the country with horses of a military type.

In the production of the hunter the Thoroughbred plays a prominent part. For harness purposes, too, an oblique dash of blood has been found to give character, colour, and courage to the stamp of animal produced; but it is obvious that a first cross is more likely to be productive of a saddle than a harness animal.

Forcing Young Stock.—It is the custom to force young stock, which is not likely to help the constitutional vigour of the breed.

Character.—Most people are familiar with the Thoroughbred type. Centuries of careful breeding have imparted an aristocratic air and carriage such as no other breed possesses. The fine sweeping arch of the neck, the thin nostrils, the prominent eye and short head, are familiar features. The shoulders slope, and are thin at the withers, which rise high. The deep rather than rounded rib, powerful loin, and graceful sweep of the limbs at the trot, no less than the perfect motion of every joint at the canter, are as well known to all admirers of equine style and symmetry as they are to those most deeply versed in the points and lore of the breed.

Plate 16 represents the Thoroughbred stallion "Diamond Jubilee."

THE HUNTER.

Strictly speaking, the hunter is a type and not a breed. It may in the course of time. attain to the latter status, but for the present it is bred in so many different ways that it cannot even be considered as the product of a first cross—the dam usually being a half-bred or nondescript. There is a great field, however, for the production of horses of the hunter type. That field may indeed be considered limitless, in view of the fact that the hunter misfits are frequently suited for military purposes.

Type.—The typical hunter is a class of horse by itself. Usually the classification provided at shows divides them into light weights, carrying up to 12 st.; middle weights, from 12 to 14 st.; and heavy weights, from 14 st. upwards. The Hunters' Improvement Society divides them into horses for weights not exceeding 13 st. 7 lb.; over 13 st. 7 lb. and under 15 st.; 15 st. and upwards.

The most valuable horse, as a rule, is a weight-carrier. Here the difference between a blood-like weight-carrier and a heavy-boned hunter without "blood" characteristics is apparent. The former is invariably the more courageous and the faster type of horse, and most favoured by the *cognoscenti*. The points of a hunter are good forehand, deep sloping shoulders, short back, muscular and flat limbs, a strong loin, and well-developed quarters.

Method of Breeding.—The Hunters' Improvement Society hopes ultimately by the registration of foundation stock to build up a breed of hunters in the same way as the Hackney and Cleveland Bay breeds have been built and maintained. There is no reason why in the course of time this scheme should not be successful; but ere the breeder will cross the Rubicon and burn his boats much prejudice must be conquered, and a plain, practical demonstration of the fundamental truths of the Society's scheme be afforded to all and sundry. The half-bred registered sire has still to win his spurs against the Thoroughbred, and in the opinion of those who have grown grey in the study of the many problems involved, that sire must be phenomenally successful to do so. Not the least of the difficulties to be encountered is to prevail upon breeders to keep on for breeding purposes entire animals whose appearance indicates that they might sell well as hunters.

Scarcity of Mares.—The chief difficulty at the present time is not altogether that of securing blood sires up to sufficient weight and with speed and a racing record behind them. The average man possessing a half-bred mare expects with a Thoroughbred union to have a full-fledged hunter type of offspring. He expects too much of the sire, in spite of the fact that the Thoroughbred is the most impressive of all equine breeds. The mares themselves must first be bred ere the breeding of hunters can become universally profitable.

All too frequently the mares used in the production of hunters are themselves the result of a happy-go-lucky cross, so that the breeding of hunters becomes under such circumstances more a game of chance than skill. Cart blood is often traceable, and not a few of the mares mated with the Thoroughbred sire are either of the light runner type or simply active cart mares. Under these conditions the breeder may by chance breed an animal of the hunter type, but he is just as likely to obtain a nondescript. The ideal mare to mate to a weighty blood sire should herself have blood characteristics. Perhaps the heavy-weight hunter is not always bred in this way, but it is the surest and the simplest way of preventing the sportive tendency of cross-breeding.

Irish Hunters.—Hunting horses bred and "made" in Ireland have earned a wide reputation alike for their build, quality, stamina, and manners. On the female side they were mainly descended from the old varieties of Irish draught-horses, but they are deeply saturated with the blood of the Thoroughbred, and are entirely deserving of their good name.

The Cleveland Bay mare, as already stated, is frequently used in the production of hunters, the blood stallion giving the results that are the most satisfactory.

MANAGEMENT OF HUNTERS.

The management of a hunter stud may be said to begin with the mare. Having found a sire that "nicks" well with the mares in the stud, it is provident management to stand by him.

The mare should not be turned on to poor pasture when carrying her foal, as the brood state entails a considerable physical strain. The difference between a high priced and a moderately priced hunter is often only the difference between a very light-weight and a weight-carrier; therefore treat the brood mare well. When the foal arrives it is not necessary to artificially feed the mare and her progeny if there is good grass available. A little corn will do the dam no harm if the foaling occurs early in the year. Towards weaning-time it sometimes happens that the mare's milk is not sufficient, and neither mother nor offspring is thriving. A bite of corn and a fresh pasture are good correctives. It is as well to teach the foal to eat a little corn before it leaves the mother, so that possible loss of condition due to the severance may be the more easily averted.

Winter Treatment.—In the winter time a very good ration is 5 lb. of crushed oats and 1½ lb. of white pea-meal, divided into two feeds, and fed morning and night. This, with sweet meadow-hay and a handful or two of bran, will bring the young hunter fresh through the winter. A field provided with a shelter is all that is necessary, thus giving the youngster constitution as well as stamina — which latter is the great thing aimed at in successful hunter-breeding. Select a growing rather than a fattening pasture—land rich in lime being preferred. If the animals have a good stretch of land in which they can exercise there will be little attention wanted.

"Making" Hunters.—The education of the hunter really begins about three years old, although the elementary duties of teaching the young horse to be readily handled and accustoming it to the use of the halter are earlier attended to. Mounting is most important in a saddle horse — equally as important as in a horse prepared for harness. This being accomplished, saddling should be taken in hand and gentle riding exercise given for about six weeks. They may then be left to run at pasture till the autumn, and again put through their exercises. It is not wise to jump them till later— say in spring, when the horse attains his fourth year. The bones are better to set first ere they are put to the strain which leaping entails. They can then be gently ridden to hounds, but on no account should they be tired out on the initial journeys.

The work begins in earnest in the following winter, and they are ready to market as five-year-olds. The aim of the breeder should be a heavy horse capable of carrying at least 14 stone.

A portrait of a famous hunter is given on Plate 17.

THE HACKNEY HORSE.

The Hackney horse as it exists to-day is a breed possessing distinctive type and distinctive uses. It was originally associated with the old Norfolk Trotter, and in past days made many notable performances against time. Farmers used to employ them as cobs and hacks, their constitutional vigour and muscular power enabling them on occasions to carry to market, not only the farmer himself, but also his spouse on a pillion behind. So far as the history of the breed is concerned, Mr H. F. Euren's admirable essay in the first volume of *The Hackney Stud-Book* still stands as the best epitome of what is known of the Hackney in the early part of last century.

Historical.

The name Hackney, writes Mr Euren, came in with the Normans, but the old Danish name Nag held its own. Hackney was applicable only to a pacing or trotting horse, while nag was and is used as a name for any riding-horse.

Hackneys and Trotters are frequently mentioned in old farm accounts from the year 1331 to 1518 (Thorold Rogers's *History of Agriculture and Prices*). In 1340, by 14 Edward III., s. 1, c. 19, one of three Acts passed to regulate purveyance and to make illegal the practice of sending the "king's great horses" on to farmers' lands; but there was reserved to the king's Master of the Horse privilege of purveyance for "a Hakeney," which he might have: in the Paston Letters, under date 1470: in Acts of Henry VIII.—1535-36, 1540, 1542—the last named providing that cart-horses or sumpter-horses were not to be reckoned as trotting horses: by Blundeville, the Norfolk parson, who was the first English writer on horses (A.D. 1558): by Thomas de Grey, *The Phœnix of our Times* (A.D. 1624), who spoke of the trotting horse as the English breed of horse, the troop-horse of his day.

The Hackneys of the eighteenth and nineteenth centuries trace back, almost without exception, to one horse, named "Shales," foaled about the year 1755. His sire was "Blaze." The sire of "Blaze" was "Flying Childers," which horse was a mixture of Barb and Arab blood. The dam of "Blaze," known as "Confederate Filly," had Barb or Turk blood in equal proportions with English blood of unknown breeding. The dam of "Shales," as of "Hopeful," another son of "Blaze," was a trotting mare.

From 1750 to 1780 Barb blood was freely used in Norfolk on trotting mares. The horse "Shales" is said in an old advertisement to have been "the fastest horse of his day." Through his get, "Scot Shales" and "Driver," came all the famous "Shales" and "Fireaway" stock of the end of the eighteenth century. Many of the good ones were bred in the Long Sutton district. The "Driver" stock first won popularity in Yorkshire—the "Shales" stock in Norfolk; but there was a regular interchange of the two strains from the outset. Their descendants, Burgess's "Fireaway," Wroot's "Pretender," and his son, Ramsdale's "Performer," Bond's (two) "Norfolk Phenomenon," Chamberlain's "Marshland Shales," and the "Norfolk Cob" family, are a few of the horses existing between the years 1788 and 1850, whose names occur often in the full pedigree of the horses which have won the Society's champion honours.

Notwithstanding that examination of an extended pedigree shows that the modern Hackney is frequently an inbred horse, it is claimed for the breed that it retains its old-time characteristics—good action, high courage, and great powers of endurance. M. de Thannberg, who for nearly forty years was connected with the Government studs in France, declared in 1873 that the Norfolk Trotter had transmitted these very qualities to the French horses, and thus established what is now known as the French coach-horse. The old custom of trotting against time and in matches, which prevailed in England in the early years of this century, having been discontinued, the qualities which won for the Hackney its old reputation are not now so plainly in evidence; but those who have a knowledge of back-breeding have no difficulty in selecting horses which shall transmit the old-time powers to the progeny.

Practice of Breeding.—The practice of the breeders of the Hackney, as shown by records from 1780 to 1820, was that of using the Hackney stallion on half-bred mares, the produce of Thoroughbred stallions and trotting mares. This has continued to be an almost universal practice in Yorkshire. In Norfolk there have been experiments made of using Thoroughbred stallions on trotting mares, and the result has not been so satisfactory as is the breeding in Yorkshire, as regards form, endurance, or action. The most experienced breeders are agreed that the truest mode of breeding Hackney stallions, so as to get a certain result, is to put the necessary Thoroughbred blood into the breed through the mare, and, better still, through her dam. The examination of hundreds of pedigrees received from Yorkshire has shown me that in a very small proportion of cases—certainly not more than two per cent—Yorkshire breeders have followed this plan of using Hackney stallions—putting Thoroughbred blood into the breed through the mares only.

Characteristics.

Type.—When we come to the consideration of the modern Hackney from the point of view of type, we are confronted not with one but with several. The truest Hackney character is expressed in the animal that does not exceed 15.2 hands. In passing, it may be mentioned that the Royal Agricultural Society used to stipulate that the maximum height of a Hackney admissible under their classification was 15.2 hands. Beyond that height it is rare, if not impossible, to find a Hackney with that sweetness of character so freely seen in animals of smaller stature. The true Hackney is a beautifully built horse. He stands very squarely on good feet. His limbs are hard and flinty. He should possess short cannon-bones, but frequently one sees them longer than is desirable. There should be as much substance as the limbs can conveniently carry without losing quality. The back should, in the stallion, be short, the rib round, and the loins beautifully filled, the quarters more rounded than in the blood horse, and the tail carried like a bedecked spike, almost on a level with the top. The shoulders should be well laid, thin at the top but not too sloping. The forehand should be long, and the crest pronounced in the stallion. The eye should be bright, and the head express intelligence where no vice can lurk. The carriage is everything in a ride and drive horse. There should be a perfect blend of style and form, no angles being perceptible. The "blood" type was at one time more prevalent than now. Many breeders would welcome more "blood," as they agree that a bigger type of Hackney of better colour and more style could then be bred.

Height.—Hackneys are bred to a greater height than formerly. It is quite common to meet with 15.3 hands and 16 hands mares and stallions, but, as before remarked, the smallest type is preserved by animals 15.2 hands and under.

Colour.—Colour is an important matter in a harness breed. If the hues are difficult to match it is only natural that the value of an animal is reduced. The most prevalent colour is chestnut, the shades varying from a light chestnut to a dark and liver colour. White markings are very prevalent, being handed down from stallions which have won important prizes. They are an undoubted defect from a harness point of view. The soundest colour is bay with dark points. Browns and blacks are also good. Roans and skewbalds are sometimes met with.

Action.—The commercial value of the Hackney is determined by one thing—action. With action the most indifferent horse will meet a ready market. Without it the most perfectly formed horse will be neglected. The Hackney clearly excels all other breeds in brilliant use of its limbs. High, free, and rhythmic movement is most of all encouraged. At one time it was more important to move high, after the style of the funeral horse, than to exhibit that liberty of shoulder nowadays demanded. The knee must be snapped to give style and sharpness to the movement, while the hocks should be closely carried, the more nearly parallel to the belly the better. A first-class Hackney showing his paces as nearly represents "the poetry of motion" as it is possible to conceive.

For Harness and Saddle.—The question might be asked, Is the Hackney a harness or saddle horse, or both? To some extent the types conflict, especially when extravagant action is demanded of a harness breed. It may be readily conceived that if the ordinary hack, which represents the acme of comfort in saddle, derives much of its popularity from its unattractive action, the free use of the shoulders, and the propulsive power of the hocks and loins brought into play in the type of movement demanded of the Hackney, would not conduce to a comfortable seat. This has led to the abolition of saddle classes for Hackneys at many shows, because animals were winning on action which was more suitable for leather than pigskin. The Hackney Horse Society provides no class for Hackneys in saddle, which implies in a negative way that the real vocation of the breed is to supply the harness horse market. The old Yorkshire type of Hackney, now rapidly altering, was essentially bred to meet the two markets, but the tendency of the times is undoubtedly in the direction of breeding for the carriage and harness market generally.

Soundness.—The Hackney is one of the soundest, if not the soundest, breed of horses we possess. The charge is sometimes unjustly made that the modern type lacks stamina. This may be dismissed as an idle tale. Some showyard animals may not be in a state of training. for hard work, and to judge a breed by the artificially pampered specimens trained to show their paces for a brief period only is to do an injustice to the breed. No better proof of the soundness of the Hackney can be adduced than the following table, extracted from *The Hackney Stud-Book,* showing the number of rejections under the veterinary examination at the Hackney Show at Islington :—

Year.		Number Examined.	Passed.	Rejected.
1890	.	170	161	9
1891	.	116	112	4
1892	.	186	179	7
1893	.	249	241	8
1894	.	217	204	13
1895	.	223	219	4
1896	.	396	379	17
1897	.	438	415	23
1898	.	436	407	21
1899	.	437	379	30
1900	.	400	382	21
1901	.	406	418	16
1902	.	434	401	21
1903	.	422	392	24
1904	.	416	385	23
		4946	4674	241

Alterations in Form.—The appearance of the modern Hackney is vastly altered from what it used to be. It has lost some of that depth of frame and shoulder so common to the old Norfolk strains. Indeed, the blood cross appears to have been the dominating influence in moulding the modern Hackney. The head is clean cut but sweeter than the Thoroughbred. The stallion character is perhaps less marked than in the blood horse, but head and neck in the true Hackney should be beautifully proportioned. The muzzle is not so sharp, the appearance suggesting more docility than is associated with the Thoroughbred.

The shoulders are sloping more so than in the riding-horse. In the older type there was a tendency to loaded shoulder points, but these have been fined down in the modern representatives of the breed.

MANAGEMENT OF HACKNEYS.

In the matter of managing a Hackney stud there is no specific axiom which can be laid down as the basis of success. Successful Hackney breeding is confined largely to the tenant-farmers, and save in cases where there are outstanding horses at the head of studs, it not infrequently happens that the breeding of this type of horse becomes an expensive hobby. The Hackney is kept very generally throughout the country. It is mostly in the hands of farmers, and except where studs of considerable size are maintained the ordinary farm buildings are made to suit the purposes of the breeder.

Buildings.—Those who equip their farms with expensive buildings are indulging their fancy. To the majority of Hackney breeders the inexpensive pile will answer the purpose quite well. The main buildings should be roomy, light, and sunny, free from draughts, and with adequate ventilation. At Mr A. W. Hickling's stud, at Adbolton, near Nottingham, the stud buildings are composed of wood, strong and serviceable, forming three sides of a large yard, facing south. They consist of a range of foaling-boxes, 20 ft. by 20 ft., 12 ft. high to the eaves, match-boarded under corrugated roof, well lighted, with brick on edge or rough finished concrete floors. Another series of boxes measures 15 ft. by 12 ft., and is similar in construction and ventilation to the foaling-boxes. There are ten roomy, airy yards, with open board roofs, which Mr Hickling says form capital half-way houses between field and stable for rough horses. The usual fodder and saddle rooms complete the buildings.

Close to the stud farm, in the fields around, are several large sheds with corrugated circular roofs and boarded ends, having long half-drain-pipe mangers down the centre, which, being boarded to the roof the entire length, afford protection from whichever quarter the wind blows. For the purpose of training show horses an enclosed level exercise-ground, oblong in shape, and equipped with a good sound track, is a great acquisition.

Many people erect riding-schools which are useful but costly. Young animals trained in the open do as well as, if not better than, those which receive their education indoors.

The Brood Mare.—The management of the brood mare is of first importance, particularly when the object in view is a full-sized Hackney. Mr A. W. Hickling gives his experience as follows :—

"Brood mares running out the year round produce better and stronger foals, and with less risk of accident, than those kept in yards or mares that have been going the round of the summer shows. When within a few days of foaling the mares are brought into their boxes and watched at night; then out again by day, often foaling in the field, with no bad results. Prompt attention to the newly born foal's navel is imperative. It should be tied and thoroughly disinfected, then for three days the mare should be fed sparingly on oat and bran mashes and chilled water, the foal haltered and handled. If all be satisfactory both mare and foal may go out every day except when wet.

"A frequent change of pasture during the summer keeps the foal in a forward growing condition, and until September no corn is needed. Then a mixture of crushed oats, bran, and chopped hay is given to mares and foals in tumbrils placed in the fields preparatory to weaning in early October."

Weaning.—Mr Hickling's method of weaning is to take mares straight from the foals to an outlying pasture beyond

call, being fed during the winter months on pulped kohl-rabi and swedes, crushed oats, and chopped hay. The mare's udder should receive constant attention. The foals are better not confined in yards, but in the open, provided there is a good shed in the field, under which, however, they seldom go. Fed on similar lines to the mares, they usually grow into big and strong if not fat yearlings. Rock-salt is recommended within reach of all stock.

Young Stock.—With young stock Mr Hickling's plan is to arrange for all young mares and geldings not required for show to be "boarded" out with farmers having few horses of their own, and, where possible, run out not more than two together on dry sheltered land, and fed only on good hay or seeds. Even in winter they can remain out. By this method a change of pasture is provided for the mares, and it assists the vigorous development of the foals.

Stallions.—"Stallions," says Mr Hickling, "require when standing at home a good-sized airy box, the larger the better for their health's sake. Adjoining this should be a covering yard, enclosed and roofed. From experience, I am opposed to concentrated heating food for stallions, even with the prospect of a full season, finding on good oats, bran, hay, and green food a greater percentage of foals next year. Exercise must be regular and not less than two hours daily; in summer, early morning and late afternoon will be found the best times.

"I advocate plain open-air treatment, with plain living, for all breeding stock."

A Hackney stallion is represented in Plate 19.

ENGLISH AND IRISH PONIES.

Active public interest in pony breeding was greatly stimulated by the South African War, while the earlier establishment of the Polo Pony Society—afterwards styled the Polo and Riding Pony Society—has been of immeasurable service in stimulating and directing the improvement of the native races of ponies in this country. There are many

kinds of ponies—ranging from the tiny Shetland to good-sized polo and harness ponies. Before briefly considering the distinctive races, some information may be given as to what a pony really is.

Pony Type.

The most natural answer to the question propounded above is that a pony is a small-sized horse. This is true to some extent; but the bantam horse — say of the hackney type — can never, by any stretch of imagination, be called a pony. If an undergrown or dwarf hackney constitutes a pony, it makes an immediate departure from recognised pony type. Compare, for instance, a harness pony—say a hackney cross on a Welsh strain — with a pure-bred hackney that has remained of pony stature, and the reason for assuming that a pony is not determined by its inches immediately becomes apparent. Therefore let it be conceded that the pony is a separate and distinctive type of animal from the horse.

Lord Arthur Cecil, who has taken so much interest in the rescue of British and Irish native pony races, speaks with authority when he thus describes the true pony type :—

TRUE PONY TYPE.

Head.—Somewhat small, *ears* small, pointed and extremely sensitive.

Eye.—Bright, prominent, and with quite a distinctive look of intelligent determination. Very often of a light brown or hazel colour.

Mane.—Very often thick and coarse, and often lying on the near side.

Shoulders.—Thick and somewhat wanting at the withers, but generally fairly deep, which gives a look of being loaded at the point, but they are well laid. This is specially noticeable in ponies which have never been under cover, and have had to stand for hours, or even days, huddled up under a bush or rock for shelter with very little to eat.

Knees.—Generally big and strong, but apt to be rather close together from same cause as above. Cannon bone very short.

Feet.—Almost invariably excellent, but apt in action to be lady toed. This is almost an universal fault in mountain and moorland ponies, but of very great service in feeling their way over bad or soft ground. Nearly every deer-stalking pony goes so.

Croup.—Low and goose rumps. Faulty from same cause as shoulders. Disappears with first cross of good breeding.

Hocks.—Always of good shape and sound, but apt to be turned in at the point, especially if the pony is in weak condition.

Colour.—A rich brown is a colour which all the varieties seem to incline to, with a mealy or tan muzzle. Highlanders and Fell incline strongly to black, and Highlanders to dun and mouse colour,

which I have reason to suppose may also be a very old Welsh colour too. A bright chestnut is rare, and generally means a cross of other blood somewhere.[1]

Points in Pony Breeding.

There is a great field for the extension of pony breeding in this country. Large tracts of moorland and forest—some of it Government land—are devoted to the raising of ponies which lead a wild untamed existence. The first point that the breeder has to consider is how to keep the stature within limits. It may be assumed that for harness purposes a pony should not exceed 14 hands, and for riding and polo 14.3 hands. The breeder is naturally confronted with the problem how to rear his ponies so that they shall not exceed these limits. In the polo pony the difficulty is more pronounced than in other breeds, for the pony to be valuable should reach not less than 14.1 hands and not more than 14.3 hands. Poor land is necessary. A bare existence is found to be the most practical method to adopt in early years and when fitting for show, a few weeks' preparation prior to the event, until the animal's stature is fixed by maturity.

HACKNEY AND HARNESS PONIES.

The hackney pony, which is the most brilliant pony for harness purposes, is either a variety of the true Hackney into which out-crosses have crept, and which have gradually worked into the Stud-Book, or merely a pocket edition of the larger breed. It is of two types—the essentially hackney type or little hackney and the pony type. It is safe to say that for pure adroit use of the limbs the hackney pony excels its big brother the Hackney. It is probably the most profitable form of pony breeding.

Not a little of the success of the hackney pony has been due to Mr Christopher Wilson of Rigmaden, Kirkby Lonsdale, whose daring feats in in-and-in-breeding are now a matter of history. The modern hackney pony owes much to Mr Wilson's brilliant sire "Sir George,"

[1] *Farmer and Stock-Breeder Year Book,* 1906.

which on eight occasions won first prize at the Royal English Show. He possessed a strong dash of Norfolk blood.

It is undoubtedly the case that the hackney pony has inherited his chief merits from the hackney, which has exercised a masterful influence in such crosses as have been made. Thus we find that Mr Christopher Wilson's success was founded partly on his selection of Cumberland mares and crossing them with the hackney pony stallion. His experiments in in-breeding were too bold to be universally adopted, but the breeder knowing the material with which he works, there is no room for doubt that the best means of fixing a type is by judicious close breeding. In Mr Wilson's stud the result of the first union with "Sir George" was again mated with him, this process being repeated successfully a third time. The most curious fact is that neither constitution nor substance were lost in this daring experiment in breeding, which seems to imply that the initial course must have been more or less violent, though apparently not sportive.

There is less to be said concerning the hackney pony than almost any other type, for it is so closely associated with the hackney in blood and, shall we say, in form, that an extended description would savour of repetition. One thing, however, may be remarked concerning the hackney pony, and that is, that it is usually of a sounder colour than the hackney. There are fewer mis-marked animals, more bays, browns, and blacks. This is probably due to the fact that the successful show ponies which have in later years been most fortunate at stud were themselves of sound colour.

The hackney pony stallion "Bantam King" is represented in Plate 20.

THE POLO PONY.

The progress of polo in England gave birth to the Polo and Riding Pony Society, which has done much to foster the systematic breeding of ponies suitable for the game. The polo pony is a product of no one recognised cross, but the aims of the Society if carried to fruition are likely to provide an accepted type to breed to, and to guide breeders in the selection of stallions and mares which are likely in the future to furnish the right class of animal.

So far breeders have been feeling their way in a somewhat perplexing manner. The claims of mountain and moorland ponies as the progenitors on one side,—chiefly Welsh,—and the Thoroughbred, Barb, and Arab stallions on the other, have been advocated with vigour, if not with warmth. From the chaos of conflicting opinions certain facts concerning the breeding of polo ponies emerge. In the first place, the limit height according to the Hurlingham standard is 14 hands 3 inches. Hurlingham and the Polo Pony Society now see eye to eye. The type of pony most in demand is the hunter type with pony character.

With these two salient facts before them, breeders are asked, so to speak, to produce the material which will enable the polo player to discard the imported horse and the pure Thoroughbred in favour of home industries.

From the breeders' point of view, it is unfortunate that the margin of height is liable to create so many misfits in what is, after all, the progeny of a cross. Again, unless the breeder is competent to mouth and train his own stock, he may produce the best-looking ponies in the world and be badly recompensed for the trouble. The value of a polo pony is dependent upon its stamina, appearance, speed, and training.

It has been estimated that the number of polo ponies in connection with Ranelagh, Hurlingham, and Roehampton is 3500. It is stated in the records of the Roehampton Club that in one year no fewer than 6000 ponies passed through its gates.

The best ponies are bred from mares which themselves have done real hard work. For reproductive purposes, probably, there is no better cross than the small Thoroughbred or the Welsh pony. There are acknowledged polo pony sires which have won prizes at shows in London, and their reputation is justified by the stock they produce. For the average breeder, however, the first direct cross is still the only means of production.

A portrait of the characteristic polo pony mare "Ruby" is given in Plate 20.

THE DARTMOOR PONY.

Dartmoor, with its rough range of 20,000 acres of moorland, has long nurtured a breed of pony which has distinctive features. Not a little portion of this land belongs to the Duchy of Cornwall, the rights of common being let out. Although at one time the regulation was promulgated that no sire over 12 hands high was allowed to run on the moor, that regulation did not long exist. The one temptation against which the breeders of the Dartmoor, in the eyes of the best informed judges, should fight is increase of stature.

Various efforts have been made to improve these ponies by the use of the small Thoroughbred and the Arab. No doubt numerous crosses, including the old Devonshire Pack Horse, have crept in, but there are still pure-bred animals,— at least in type they approximate to it,— which show that expressive countenance which usually stamps the true Dartmoor.

Early improvers of the breed included a well-known farmer named Eliot, Lord of the Manor of Brent, Mr John King, whose herd ran on Buckfastleigh moors, and Mr Hamblin of Buckfastleigh.

Of the capacity of the Dartmoor pony to carry weight there can be no doubt. Its conformation approximates to the hunter type. Not a few of them, however, exhibit indifferent heads and shoulders, no doubt due partly to the efforts of "improvers," by the introduction of unhappy alien crosses. First crosses on these ponies, as a rule, are not very successful.

Nature has made the Dartmoor pony strong in constitution, and like many of our hill races, lean fare and inclement weather have not formed them at the rumps and hocks as well as might be wished. The pony never stands with his head to a storm. The height of the Dartmoor pony should not be more than 13 hands.

A portrait of a Dartmoor pony is given in Plate 24.

THE EXMOOR PONY.

The Exmoor pony, which is the near neighbour to the Dartmoor, is equal to him in stature. History accords to this pony a lengthy tenure of that great stretch of moorland. Near the beginning of the nineteenth century Sir John Knight acquired some 20,000 acres of the moorland with the main object of raising ponies thereon. He subsequently extended his proprietorship by taking in part of the land owned by Sir Thomas Acland, and at the same time purchased the famous herd of ponies which Sir Thomas had reared. About this time various crosses were tried, including the Arab and the Thoroughbred, which had the effect of raising the stature. Sir Thomas Acland took a prominent part in the improvement of the Exmoor pony, his strains being highly valued by breeders generally.

The Exmoor is a very hardy, surefooted pony, with a rare constitution. The head is cleaner cut than that of the average Dartmoor pony, the ears being sharp and intelligently carried. The shoulders are finer than in most pony breeds, the back short and powerful, and the legs and feet good. The typical Exmoor pony is very active, as might be expected of the denizen of a rough moorland tract.

Reference to this breed would be incomplete without mention of the fiction-famed Katerfelto, whose appearance amongst the native ponies is supposed to have exercised a wonderful influence on their character. This dun stallion, it is asseverated, was no creation of the imagination, his mysterious appearance being assumed to be due to a wreck on the adjacent coast, whence he escaped inland.

In Plate 24 a portrait is given of the famous Exmoor pony stallion "Twilight," the property of Mr H. Dyson, Priory Farm, Pamber, Basingstoke.

THE NEW FOREST PONY.

The New Forest pony is stated to have held the "field," by which we mean the Forest, since the times of King Canute. There are some 70,000 acres of Crown property in the Forest, so that the indigenous race of ponies has had every opportunity to lead a wild and roving existence. A society was formed for the purpose of protecting and improving this race of ponies, but its work is not much known to the public.

The Arab stallion has undoubtedly exercised a considerable influence on these ponies, the late Prince Consort taking an interest in their welfare. Perhaps to the influence of Arab blood is due the large number of grey ponies to be found in this particular race.

The New Forest pony is in some ways not comparable with the other ponies of the south-west. In the New Forest there are many typical ponies, if we accept the head as exemplifying pony type, but the original character of the race has been to a great extent submerged in the multifarious crosses to which it was subjected. The main attempt on the part of improvers was to keep up the size, which has gradually been dwindling since the plantings and enclosings began in 1834.

A portrait of a New Forest pony is given in Plate 24.

THE FELL PONY.

The Westmoreland and Fell ponies have in all probability much in common, and a very useful class of ponies they are. This type of pony has been used to a considerable extent in the production of the well-known "Galloway," and a certain element has percolated through the famous stud of Mr Christopher Wilson, Rigmaden, Kirkby Lonsdale, into the harness pony.

The Fell or Dale pony is very hardy, and withal has an appearance of breeding. It is nothing if not full of stamina. It is a larger race than the southern moorland type, but has the same bright eye, alert head and ears. The winter coat is usually exceptionally heavy, for the snowstorms are severe. A degree of sure-footedness is acquired to which southern breeds, if not altogether alien, are at least not called upon so freely to exercise.

The trotter and the roadster blood have so altered the original Fell pony that the type has undergone an undesirable change on the mountains surrounding the Lake District. The aggressive sheep, too, has been responsible for much of the neglect surrounding the maintenance and improvement of the Fell pony.

Lord Arthur Cecil mentions as indicative of the stamina of this race, that one pony which he knew carried eighteen stone on parade with mounted infantry every day for a month, "sometimes doing her twenty miles a-day when she was only three weeks off grass, in the month of March, having been out of doors all winter without a bite except what she picked up."

The same authority continues: "They are the kind that carried Kinmont Willie and Jock Elliot in their Border frays, and are probably therefore identical, or at any rate freely crossed, with the old 'Galloway,' probably now quite extinct."

A portrait of a Fell pony is printed in Plate 23.

THE WELSH PONY.

The Welsh pony is probably the most serviceable type of hill pony that we possess. There is no cob to equal him either for rough saddle or harness work. In past times he has been bred without due regard to the future, and the variety of types that nowadays masquerade under the guise of a Welsh cob or pony is truly bewildering.

Much good was done by establishing a stud-book, which has led to greater care and skill being exercised in the improvement of the breed. The chief modern influence on the Welsh pony is the Hackney cross, which is largely permeating Wales.

The points of a Welsh mountain pony are thus officially set forth :—

General character. — Hardy, spirited, and pony-like.
Height.—Not exceeding 12 hands 2 in.
Colour.—Any.
Head.—Small, clean cut ; well set on, wide between the eyes, and tapering to the muzzle.
Ears.—Well placed, small, and pointed.
Eyes.—Full, bright, and sensible.
Nostrils.—Prominent and open.
Throat and Jaws.—Finely cut.
Neck.—Fairly lengthy, and moderately lean, with a stronger crest in the case of a stallion.
Shoulders. — Long, and sloping well back ; fine at the points, with a deep girth.

Back and Loins. — Muscular, strong, and short-coupled.

Hind Quarters. — Lengthy and fine ; tail well set on, and carried gaily, undocked and long preferable, but the reverse not a disqualification.

Fore Legs. — Well placed, free at the elbow ; long, strong fore-arm, well-developed knee, short, flat bone below knee, pasterns of proportionate slope and length ; feet well-shaped, and hoof dense.

Hocks. — Wide, large, and clean, parallel with body, and well let down ; shank flat and vertical.

Action. — Quiet, free and straight from the shoulder ; knees and hocks well flexed, with straight and powerful leverage, well under the body.

The type of Welsh cob suitable for remount work is thus described :—

Head. — Small and flat, showing pony character, with fine silky hair under the jaws when rough.

Neck. — Well defined where it joins the shoulder, giving the cob a good "look-out."

Shoulders. — Well laid and strong.

Back and Loins. — Back not too long ; loins muscular and strong ; tail well set on and not goose-rumped.

Second Thigh. — Well developed, not too long from stifle to hock or from hock to the ground.

Fore Legs. — Should stand well outside the body and placed well forward ; big knees, flat bone, moderately sloping pasterns, feet round, well-formed, not "boxed" or too big. When in the rough there should be a moderate quantity of fine silky "feather" on the back of the legs. Hard wear and tear fetlock joints are absolutely essential.

Action. — Free, true, and forcible, and they should bend their knees and hocks as much as is compatible with pace and staying powers.

A portrait of the famous Welsh pony stallion "Greylight" is given in Plate 21.

THE CONNEMARA PONY.

The Connemara pony is a thoroughly useful type of a small horse. For light farm work it is well suited ; in most respects it is almost an ideal horse for the small holder.

A peculiar interest attaches to the history of the Connemara pony. Low, writing of it in 1842, says—

"The horses of Spain have been re-ferred to as having contributed to form the mixed races of the British Islands ; but it is not generally known that a race of horses of Spanish descent, nearly if not altogether pure, exists in this country in considerable numbers. They inhabit the Connemara district of the county of Galway. The tradition is that from the wreck of some ships of the Spanish Armada on the western coast of Ireland, in the year 1588, several horses and mares were saved, which continued to breed in the rugged and desolate country to which they were thus brought. But the aid of tradition is in no degree necessary to prove the origin of these horses, since all their characters are essentially Spanish. They are from 12 to 14 hands high, generally of the prevailing chestnut colour of the Andalusian horses, delicate in their limbs, and possessed of the form of head characteristic of the Spanish race. They are suffered to run wild and neglected in the country of mixed rock and bog which they inhabit, and where they are to be seen galloping in troops amongst the rugged rocks of limestone of which the country consists. When they are to be captured, which is usually when they are three or four years old, they are driven into the bogs and haltered. They are hardy, active, sure-footed in a remarkable degree, and retain the peculiar amble of the Spanish Jennet. Any selection may be made from the wild troops, after being hunted into the bogs ; and individuals are obtained at a trifling expense.

"It must be regarded as remarkable that these horses should retain the characters of their race for so long a period in a country so different from that whence they are derived. They have merely become smaller than the original race, are somewhat rounder in the croup, and are covered in their natural state with shaggy hair, the necessary effect of a climate the most humid in Europe. From mere neglect of the selection of the parents in breeding, many of these little horses are extremely ugly, yet still conforming to the original type."[1]

To Low's interesting sketch of the

---

[1] Low's *Domesticated Animals of the British Islands*, 1842.

Connemara pony little need be added. The general features of the breed are still fairly well maintained, though the variety have lost some of their value through lack of care or method in breeding and rearing. In later years, mainly at the instigation of the Irish Department of Agriculture, interest in the breed has been revived to some extent, and there is reason to believe that Connemara ponies have a highly useful future before them.

Chiefly as a result of indiscriminate crossing, the draught-horses of Ireland have unfortunately lost all claim to recognition as a distinct breed, and practically, therefore, the only distinctive race of Irish horses are the ponies of Connemara. This consideration should in itself act as a strong stimulus to Irishmen in their revived efforts to improve these ponies and extend their use throughout the country.

By the kind permission of the Irish Department of Agriculture, a portrait of a typical Connemara pony is given in Plate 23.

MANAGEMENT OF PONIES.

The management of ponies naturally varies with the class of pony kept, the character of the farm, and the objects the breeder has in view.

Size. — Generally speaking, the first thing to be looked to is to treat the individual so that he or she will not exceed what is the recognised limit of size. Taking Hackney ponies as an example, it is a mistake to keep them on rich land. They are apt to overgrow, and no class of stock is harder to sell than an animal that is too big to be a pony and too small to be a horse. Hard fare up to a certain age—say, three or four years—may therefore be considered good management on the part of the pony-breeder.

There is a great temptation to run ponies thickly on the land to keep the height down, but more than one breeder has found to his cost that horse-sick land is a greater evil than over-sized ponies. Hay in winter is the usual feed.

Wintering Hill Ponies. — In the mountains and moorlands ponies cost next to nothing to keep, the chief expense being wintering. It is the custom in Wales to bring ponies from the hills in November and graze them in the lowlands until March. When snow is on the ground they usually get a little hay, but often they have to fend for themselves. After the bare living of the hills, the ponies usually return from the low winter quarters quite fat and sleek without having even a handful of hay. For show purposes special feeding is necessary, but the average hill pony mare is all the better if kept in natural condition.

Young Cobs.—In the rearing of cobs from 13 hands to 14.2 hands a little hay and corn may be fed during the winter. When it reaches three years old a cob can begin to earn its living, and it is a good thing to put it to light work in chains. This makes the cob more tractable and easy to handle.

Foals.—In the treatment of the foal the mother's milk comes first. The youngsters should be handled early, and at the period of breaking, which should not be delayed too long, it is advisable to undertake the work thoroughly. Many a good cob misses a market because this precaution has been neglected.

Ponies on Rough Pasture. — One aspect of the keeping of hill ponies should not be overlooked. They are invaluable for eating the rank grass in a pasture field which bullocks and other kinds of stock would altogether neglect. Moreover, they tear up the mossy herbage which makes it possible to manure effectively. One good authority declares that ponies and sheep, followed by a good strong chain-harrow with a little basic slag, will convert many acres of useless hill land into good mountain pasture.

Mr Tom Mitchell of Eccleshill relates, as indicative of the value of the pony on barren hill-land and the all but costless system of the keep, that he turned out twenty mares, chiefly Welsh, to forage for their living on hill-land in Ireland. He mated them with a pony stallion by "Sir Horace." He had 18 foals at the first foaling-time, and 14 on the second occasion. When sold as yearlings the produce averaged £20. This, he pleads, is more profitable than sheep would be on such ground.

The moral of it all is—keep ponies on bare living.

Training for Shows.—With regard to the Hackney pony and its training for show purposes, condition in the older ages is necessary, and it is well to allow as much corn as will give a little extra stamina to those of younger years. The chief thing sought after is to get style and action. The former is inherent ; the latter is cultivated in two or three different ways. The clay-box makes the pony lift its limbs and develops muscle. Shoeing with heavy shoes has the same effect, the pony when exhibited being very lightly shod. Action developers are also used.

Limitation of Weight of Shoes.—The abuse to which the shoeing of show horses and ponies has given rise has compelled the Hackney Horse Society to take action. It has passed a resolution refusing to sanction a shoe heavier than 2 lb. on an animal exceeding 14 hands, and 1½ lb. on animals under 14 hands, as well as for yearling colts and fillies.

HIGHLAND PONIES.

The origin of the Highland pony is lost in the mists of antiquity. There is little doubt, however, that, like the Galloway pony in the ancient province of Galloway, and the Fell pony in the hilly parts of Cumberland and northern Yorkshire, the Highland pony was the original general purpose horse of the Highlands and Islands of Scotland. The three types partook to a large extent of the same characteristics. They were all equally strong and sturdy in the make, equally sure on the foot, and equally docile. These characteristics would be bred into them through a long process of selection, for in the days before the country was opened up by roads, the ponies, apart from walking, would be the only means of locomotion. The other qualities of hardiness and endurance, for which all three classes of ponies were equally distinguished, would be developed in the same way from the circumstances of their existence and the nature of their environment.

The Galloway Ponies.

A good many years ago, as the result of crossing and the invasion of their country by other and heavier breeds, the Galloway ponies ceased practically to exist. At the present time scarcely a real Galloway pony is known to be left ; indeed, their old ground now forms part of the head-quarters of the Clydesdale draught-horse breed.

Resuscitation of Highland Ponies.

For a time it looked as if the Highland pony would also be allowed to pass out of existence. It had been greatly crossed and degenerated by other breeds, and no one seemed disposed to lift even a little finger to save it. But the outbreak of the South African War, and the demand for mounted infantry, caused a fresh view to be taken of the utility of these hardy medium-sized animals, and since then Highland pony breeding has become quite popular, and numerous men of public spirit have taken up the work of improvement on systematic lines.

A "Highland" Committee has been added to the Polo and Riding Pony Society, while classes for the variety have again been introduced into the annual shows of the Highland and Agricultural Society. Fortunately there is still a fair amount of the old original material left, so that the work of resuscitation may now be expected to proceed on well-ordered lines, to the benefit of all associated interests.

Points of Highland Ponies.

The points of a model Highland pony have not been officially defined, but details which will have to be kept steadily in view are the size, stamina, and strength of the animals. In the past, Highland ponies have not, as a rule, exceeded 14 hands in height. Considerable numbers, and these not the least useful, have been under that limit. The head in the better class of ponies is small and neat rather than large and coarse. The neck is deep and strong without being abnormal, and it should run gracefully into the shoulders, the ridge in the case of the stallion being slightly arched.

In ponies of the old and unimproved breeds the shoulders have always been a weak part. They are inclined to be too heavy and upright for the tastes of the modern hunting- or saddle-horse enthusiast. This heaviness and uprightness of shoulder rather detracts from the appearance of the animals when walking or trotting, and does not permit of the fore-legs being thrown out, as is done in the case of breeds fitted with more oblique shoulders. But, of course, pace and action were only secondary considerations with the old Highland pony breeders. Much more important matters in their ponies were surety of foot and the provision of a reliable seat for their owners when they set out to make a journey. All the same, some little modification in this respect will probably have to be effected to bring many of the ponies into line with modern ideas and requirements. Whether this can be done by selection within the breed itself, or by the introduction of fresh blood as from the Arab or other source, is a matter that time only can prove. But whatever is done, care will have to be taken that the useful and rather special points of the ponies are not sacrificed to a showy daintiness which can quite well be obtained in other existing breeds.

The back of the best ponies should be short and muscular and the ribs well arched. The quarters should be deep and muscular and carried well down to the hocks, which should be slightly bent as in a Clydesdale horse. The legs should be hard and clean, and well covered, in winter at least, with a warm coat of hair. The pasterns should be of fair length, and the feet fairly wide and deep. Narrow feet are as objectionable in a pony as they would be in a horse of large size: such feet are apt to develop side-bones and other forms of disease.

As to colours, there is little doubt that

the majority of the ponies in the early days of recorded history were greys, blacks, chestnuts, and duns, the duns having nearly always an eel-stripe along the back and down the quarters to the tail. Indeed there are those who maintain that the dun is the oldest colour of the four — the greys and blacks being of later date, and probably introduced through alien blood in the days when cross-Channel traffic with the Continent in horses was active. But however that may be, there is no hard and fast sticking to colours nowadays, although the majority of the ponies still to be seen in Highland markets and fairs are of one or other of the already mentioned colours. Should Highland ponies ever be required largely for mounted infantry work—as they will very probably be—a further modification in colour is by no means unlikely to come about.

Highland ponies have the reputation of living to great ages, and remain comparatively active and useful up to the last.

Early Studs.

Although the early origin of the breed is to all intents and purposes unknown, the history of several of the existing or bygone studs can be traced back for a great many years. Thus, as shown by Mr Thos. Dykes, in an article which he contributed to the *Transactions of the Highland and Agricultural Society* for 1905, the district of Glenorchy—or, as it was formerly spelt, Glenorquhey,—in Perthshire, was a great centre of Highland pony-breeding as early as 1600. In 1609 Lord David Murray, then Private Secretary to James I. of England and VI. of Scotland, writing from Whitehall, London, to the Laird of Glenorquhey (Glenorchy), says: "The Prince received a pair of eagles very thankfullie, and we hade good sport with theme, and according to his promiss he hath sent you a horse to be a stallion, one of the best in his stable for that purpose, and commendis him kyndlie to you and says that seven yeers hence when he comes to Scotland that he hopes to gett some of his breed."

The Royal Mews at that time are known to have contained many varieties of horses, Barb, Arab, Turk, &c.,

all of which were considered to be superior to our own horses as regards pace, style, and symmetry. It would, no doubt, have been a horse of one or other of these breeds that was sent north, and so kindly commended to the recipient.

Even earlier than this, however, mention is made of the ponies of this district. Writing of the great snow-storm of 1554, the chronicler of Finlarig says: "There was no thaw till 17th January. It was the greatest snow-storm that was seen in memory of man living. Many wyld horses and mares, kye, sheep, and goats perished and died for want of food in the mountains and other parts." These "wyld horses," as Mr Dykes says, were, no doubt, the ponies indigenous to the district.

That the Glenorchy stud at this time had a considerable reputation even outside of the district is proved by the records of Mr Cosmo Innes, who, writing of the Thanes of Cawdor, in Nairnshire, in his interesting work, *Scotland during the Middle Ages*, says: "Somewhat more care is shown of the breed of horses. Long before this time the lairds of Glenorchy had introduced English and foreign horses for their great stud in Perthshire, and the example was followed at Cawdor."

As early as 1638, Duncan Campbell, writing from Islay to his brother Colin of Galcantray, says: "I wishe if you may Cromarties old Spanish horse, provyding he be of a reasonable prys." In these days the ponies seem to have been kept on the hills in droves like sheep. The following entry, applying to fully a century later, appears in *The Black Book of Taymount* :—

"John, Earl of Breadalbane, lets to John M'Nab for five years the grazing hills of Bentechie and Elraig, with the full accustomed places where his Lordship and his predecessors' horses were wont to pasture in Glenorchy, delivering to him thirty stud mares either with foal or having foals at their feet, the one-half worth 30 merks apiece, as also 100 merks Scots to buy a sufficient stallion not exceeding five years of age, to be kept with mares on the said grass; and the said John M'Nab is to keep the mares and stallion on his own peril, and to be an-

swerable for them in all cases, excepting only the case of daylight depredations and public harrying in a hostile manner, and to keep the stallion from labour. To pay the Earl the sum of ten pounds Scots for each of the lands yearly in name of tack duty, and at the expiry of his tack to re-deliver to the Earl the same number of mares and foals and a stallion of equal value with these he received, or to pay the foresaid prices, for the mares and the stallions which are awanting. And in like manner ten pounds for every foal which shall be short of the number of thirty as above mentioned, delivering also the Earl's burning-iron, which he received for marking the horses.—FINLARIG, 11*th June* 1702."

The Atholl Ponies.

Another very old stud appears to have been that at Atholl owned by the Dukes of Atholl. This stud fortunately is still, in part at any rate, in existence, and representatives of it have been seen of late years at the annual shows of the Highland and Agricultural Society. In 1904, at the Perth show of that Society, one of the Atholl ponies, "Bonnie Laddie," a three-year-old dun-coloured colt, carried off the president's medal as the best Highland pony. This pony (which is represented in Plate 22) was got by "Herd Laddie," also a very successful prize-winner. "Herd Laddie" in turn was got by "Highland Laddie," while his dam was "Jeannie," by "Campbell Lofty." "Bonnie Laddie" was an exceedingly purpose-like pony. He stood fully 14 hands high, had deep, well-filled thighs and quarters, a short, nicely coupled back, and a handsomely set on head and neck. His legs were very clean and strong, with just that little tuft of hair at the back of the fetlock joint so characteristic of the old equine breeds. He looked like a pony that could travel a long way and do a big day's work without much trouble. Unlike many Highland ponies, "Bonnie Laddie" had comparatively good sloping shoulders.

Although there are records of mares which existed before that time, the first recorded stallion of the Atholl stud was "Glentilt," a grey-coloured pony which was foaled in 1862. This pony was bought from Mr Donald Cameron, Glengarry, Inverness, for £13, 10s., and sold to the Earl of Southesk for £60. According to a statement supplied by the Marquis of Tullibardine to Mr Dykes, he was the sire of several of the best hill ponies at Atholl, notably "Lady Jean" in 1867, afterwards used as a brood mare. This mare's dam was "Polly," a "garron" mare bought from Mr Halford, the tenant of Foss, who bought her in a Muir of Ord market.

In reference to the term "garron," it is well to remember that although in recent years it has been used as descriptive of the heavy mainland type of ponies, its real meaning is gelding, and in the early premium lists of the Highland and Agricultural Society it was so applied, there being separate classes for stallions and mares as well as for "garrons."

At Atholl the ponies are principally used for hill purposes in the shooting season. They travel long distances over the roughest ground, and are invaluable either for saddle or game-carrying purposes. A number of the ponies also formed mounts for the Atholl detachment of the Scottish Horse which did duty in Edinburgh on the occasion of the visit of his Majesty the King in 1903, and they created a very favourable impression amongst those who saw and recognised them, their sturdy make and hardy-like appearance being novel features in a great military display.

Inverness-shire Ponies.

A number of first-class studs are known to have been owned in different parts of the mainland of the county of Inverness. At least the remnants of a few of these still exist.

One of the best of the old Inverness-shire studs was that at Corriechuille, in Lochaber, which flourished some time prior to 1833. It was from this stud that the Gaick strains of blood so extensively used by Lord Arthur Cecil and Professor Cossar Ewart originally came. The Corriechuille ponies were of all colours — bays, browns, duns, yellow-creams, and piebalds. Little is known of their early history, but judging from the fact that some of them were taken

to Gaick, and that they have been maintained practically pure there ever since, they must have been of a good class.

One of the best mares in the Gaick collection was "Gaick Calliag," a black by "Glentilt," which latter was bred in the near neighbourhood, and afterwards became principal stud-horse at Atholl. "Calliag" was purchased by Lord Arthur Cecil when sixteen years of age, and carrying her ninth foal, for £64, and afterwards passed into the New Forest, in Hampshire. At the same time her son was sold for £75 to Mr Forsyth of Quinish for the Congested Districts Board, and was forwarded to Professor Cossar Ewart, who used him, under the name of "Atholl," in some of his experiments at Penicuik. Several descendants of "Calliag" are still at Gaick, and the outstanding feature of the stud is its great hardiness. "Calliag" herself until she went south was never under a roof, this, too, notwithstanding the severe winters which were frequently experienced in Lochaber. At Gaick, as at Atholl, the ponies are mainly used for hill-carrying purposes, although they also do any carting that is required, being hand-fed only when at the latter class of work.

To some extent allied to the Gaick ponies were the Guisachan ponies owned by Lord Tweedmouth. This stud was descended from old Highland blood, but in late years the ponies had been crossed with outside blood with the view of getting more quality and style. The outside ponies used were "Seaham," by "Lord Derby II.," and "Guisachan Miracle," by the famous "Little Wonder II." Both were of Hackney pony blood to some extent, but that the cross was successful in producing at least a saleable pony was proved by the fact that at a draft sale in 1903 a pair sold at 130 guineas, a single pony at 68 guineas, and others as high as 50 guineas.

Ross-shire Ponies.

In regard to Lord Middleton's stud at Applecross, the following note from Lord Middleton, in the article in the *Transactions of the Highland and Agricultural Society* already referred to, may be quoted :—

"The present Applecross stud of ponies

was formed about the year 1878, though previous to that time my father, the eighth Lord Middleton, kept and bred ponies at Applecross. About that time he came into the possession of a grey mare, 'Kitty,' which he bought with the property from the Duke of Leeds in 1861. This mare had been bred by the M'Kenzies of Applecross, who had ponies at the time on the place, which was brought from Skye.

"The mare 'Kitty' was a good type of the Highland pony. In 1878 I bought a bay mare in foal from Mr Macrae of Glenbaragait in Skye. He (Macrae of Glenvarait) was of the same family as the Macraes of Camsunary, near Coruisk, in the Isle of Skye. This mare was a beautiful type of the Highland pony, small, strong, full of mettle. At that time she was in foal to a pony which took first prize at the Highland and Agricultural Society's Show. She dropped a bay filly, and both go respectively now by the names of the Old Skye mare and the Young Skye mare. From these two mares many of my ponies have been bred.

"In 1882 I bought a beautiful grey mare, 'Molly' (foaled 1872 or 1873), at the sale of Lord Dacre's ponies at Garve, Lord Dacre having then given up his forest. She was his favourite hill pony. I bought another, which did not breed. This mare 'Molly' was larger than the two Skye mares, about 14 hands, and strong. She had a family of three colts and two fillies to 'Glen.' 'Glen's' sire used to travel in Skye, and was a chestnut with a white mane. The eldest colt, foaled in 1884, was a chestnut with silver mane and tail. I have ridden him for the last fifteen years, and have always taken him with me to Scotland. He is a wonderful pony, very strong, up to 16 stone, can walk five miles an hour, is exceedingly wise and clever, and never makes a mistake. A sister (grey) was a carriage pony, and is breeding now. Another sister travels at Birdsall with the stallions. A brother goes in harness. The other colt I sold.

"In regard to types and colours, all my ponies are thick-set, strong, short-legged, and bred especially for carrying weights (deer) and for riding on the hill. Their colours are black, chestnut, grey, and bay. The chestnut probably comes in from 'Glen,' as I hold that chestnut and black are akin.

"In the spring they plough, cart, and execute the general work of the foresters' crofts. In the autumn they of course do the work required of them in the forests. Some I use as carriage ponies, some also I use at Birdsall for going messages—post-office communication and the like—or travelling as groom's mounts with the Shire or Thoroughbred stallions. All are brought to Birdsall to be broken. They usually arrive in a truck with the Highland cattle. They are then broken at the Hunter Stud Farm, and used for the different classes of work alluded to in order to make them quiet and tractable. Those required at Applecross are returned for work there.

"Some I have successfully bred from here (Birdsall) to the Arab stallion—beautiful hardy ponies, fit for polo or hacks, and I should think just the sort for mounted infantry. I have all through tried to keep up the Highland pony hardihood. Here and at Applecross they only get hay or silage during the snowtimes. Of course during the stalking season they get a feed of corn daily. Except those used for carriage purposes, they are never under cover, and the latter are only kept up during the period they are used for carriage work, being turned out for the winter."

Fell and Arab Crosses.

The first recorded sire used at Applecross was "Glen," a black or brown, bred by Mr M'Leod, Coulmore, who was a noted Skye breeder for several years. Afterwards "Fitz George," a son of Mr C. W. Wilson's famous show and breeding pony "Sir George," was secured. The dam of "Fitz George" was a well-known Cumberland Fell pony. "Fitz George" himself was a 14-hands grey, very stout in the make, and with good action.

The stock of "Fitz George," from the Highland mares, showed great improvement in appearance and quality, and it was no doubt some of these that did so well at Birdsall, Lord Middleton's Yorkshire seat. But whether this cross or others available would be best for breeding military ponies is a matter that experience in the work alone could settle. The Arab crosses undoubtedly have sweeter heads, but they are apt to lose bone and constitution, and it may be hardiness as well.

Island Ponies.

The ponies of the Islands are legion. There are the ponies of the Inner and Outer Hebrides, the Barra ponies, the Benbecula ponies, the Mull ponies, the Rum ponies, and the Skye ponies, differing all more or less in points of detail. At one time there was a widely prevalent tradition that the general excellence of the Island ponies, and, indeed, of many of the mainland ponies as well, was entirely attributable to the comparatively latter-day misfortunes of the Spanish Armada. While that particular theory is nowadays greatly discounted, there seems to be little doubt that Spanish blood, which of course was Asiatic blood in some form or other, found its way on to these Island shores. But the probability is that the greater part of the improvement was effected earlier, and in the ways otherwise mentioned.

Mull Ponies.

None of the Island ponies enjoyed perhaps a greater reputation in the old days than the Mull ponies. They were keen rivals even to the Galloway pony in its best days. No doubt the accessibility of the Island, and the fact that large droves were taken annually to the Falkirk and other trysts, had something to do with the preference, but all the same they were very useful general purpose small horses. They rarely exceeded 14 hands in height, but were so thick-set and strongly built that they could do as much general work as horses considerably larger, while they could exist on very moderate fare.

An excellent type of the modern Mull pony was the stallion "Islesman" (253 Polo Pony Stud Book), the property of Mr J. H. Munro Mackenzie of Calgary, Mull. With the view of improving the backs and shoulders of the Highland ponies Mr Mackenzie crossed the mares with "Syrian," an Arab hack brought from Algiers. The progeny is being put to "Islesman," and in this way Mr Mackenzie is hopeful that he may be able to effect some improvement in the form of the animals without deteriorating them in other respects. The owner of the Calgary stud is a firm believer in the theory that it was from the Arab that the good points of the Mull pony, and, indeed, of most of the Highland ponies, originally came.

Skye Ponies.

Skye ponies also had a good reputation for a great many years. Indeed the M'Leods of Coulmore, the Macraes of Glenvarait, and one or two others, kept their strains pure for a very long time, and bred many animals which were taken on to the mainland for breeding purposes. But in later years, especially in the southern parts of Skye, there has been a demand for rather bigger equine stock, and many of the ponies have been crossed with horses intermixed with Clydesdale blood. The result has not been very satisfactory.

Uist Ponies.

The Uist ponies, being farther removed from the mainland, have not fallen under the same adverse influences, and here many fine representatives of the old breed are still to be found. It was from Uist that the late Mr D. Stewart, Drumchorry, Perthshire, got his noted prize-winning stallion "Mosscrop," as well as the mare "Heather." Both were bred at Balranald, and both had the old Highland characteristics in a marked degree.

Most of the outer Hebridean ponies are believed to have been of old Norse stock. Even yet some of the Uist strains of ponies have white or silver manes, this being also a characteristic of the Faroe Island ponies.

Rum Ponies.

Lord Arthur Cecil, who has long been a champion of pony breeding, has through various channels made known much interesting information regarding numerous varieties of ponies. In the article in the *Transactions of the Highland and Agricultural Society* for 1905 before referred to, his Lordship tells of "nine very good black ponies coming to Hatfield, which were said to have been running quite wild in the Island of Rum." All of them were too wild to be broken except two, which "we hunted and drove till they were twenty-eight or twenty-nine years old." In 1888 his lordship bought eight Rum ponies with which he continued the breed. It was for use in this stud that Lord Arthur purchased the famous stallion "Highland Laddie," which became the sire of the Duke of Atholl's "Herd Laddie," and which his lordship says was identical with the Rum ponies in appearance.

Recent Experiments with Highland Ponies.

Numerous experiments have in recent years been made by the Congested Districts Board in the use of sires of various breeds in crossing with Highland pony mares. Strains of the Thoroughbred, the Arab, the Hackney, the Connemara, the Fell, and other breeds have all been tried. The results have varied greatly, and opinions regarding them differ somewhat. The Arab cross has gained a good deal of favour, but amongst experienced breeders there is a growing belief that in the main the wisest course is to seek for improvement by the skilful mating of selected animals of the native types.

Professor Cossar Ewart's Experiments.

In connection with the work of the Congested Districts Board an interesting feature was formed by the pony-breeding experiments conducted by Professor Cossar Ewart of Edinburgh University. For these experiments animals of practically all the noted Highland and Island pony breeds, as well as of such kindred varieties as the Connemara, Iceland, and Norwegian ponies, were obtained, and amongst the out-bred sires used were a Thoroughbred and an Arab.

The Celtic Pony.

Professor Cossar Ewart has made a careful study of the native races of Highland and Island ponies, as well as of other varieties of horses, and has come to the conclusion that what he classifies as the Celtic pony (*Equus caballus celticus*) is one of the most specialised of all the members of the Equidæ family. The typical Celtic pony of the present day he regards not as a product of artificial selection, "but as an almost pure representative of a once widely distributed wild species." The considerations which led the Professor to this conclusion are stated fully in an interesting paper on the "Multiple Origin of Horses and Ponies" which he contributed to the *Transactions of the Highland and Agricultural Society of Scotland* in 1904.[1]

Management.

To a large extent Highland ponies forage for themselves. It is the usual custom to turn them on to the hills or rough pasture out-runs and let them gather their food. As a rule, it is only when they are being hard worked that the ponies get corn, and it is the exception to allow them even hay when idle. In many cases little or no house accommodation is provided for them. What is provided is usually only partially enclosed.

A Highland pony stallion is represented in Plate 22.

SHETLAND PONIES.

The Shetland pony is unique in at least one respect—it is, so far as known, the very smallest of the equine races of the world. In some of the islands off the Swedish coast and in Norway there are ponies that are not greatly dissimilar in appearance from Shetlanders, and at one time there was a belief that the "Shelties," as they are frequently called, were introduced into the Shetland Islands by the Norsemen between 1300 and 1400. But these Swedish and Norwegian ponies are larger, as a rule, than the Shetland ponies, and the generally accepted belief now is that the latter were originally of practically the same stock as were to be found in the northern districts of the mainland of Scotland in

[1] Vol. xvi., Fifth Ser.

very early times, and that they were "ferried" across to the islands at a date much anterior to the period mentioned.

The long-continued maintenance of the Shetland ponies at such a small size is regarded as being due to the hard conditions under which they have been reared and the struggle which successive generations, even yet, have to engage in to obtain a bare existence.

Purity of Shetland Ponies.

The Shetland pony is probably also unique in respect of purity of blood. While most other breeds, large and small, have been more or less the subject of experimental crossing, the Shetlander has probably in most parts remained uncontaminated for hundreds of years.

A Norwegian Cross not Successful. —In the eighteenth century, according to Goudie, who writes on the early history of the breed in the first volume of the *Shetland Stud-Book*, published in 1891, an attempt was made to increase the size of the breed by crossing with a Norwegian pony of the stamp probably already referred to. But it was never carried to any great length, for the reason that the progeny would not have stood the rigorous conditions under which the "Sheltie" had to exist. There are still a few of the crosses in the Dunrossness district, where the land is comparatively fertile, and where it is the custom to stable the horses, but they range in size from 12 to 13½ hands, and are not Shetlanders at all in the generally accepted sense of the term. Moreover, they cannot be registered in the Stud Book, which is confined to animals 10.2 hands or under.

A Mustang Stallion Tried.—The same writer (Goudie) mentions that about the middle of the last century, on the island of Fetlar—there are nearly one hundred islands, great or small, in the Shetland group—Sir Arthur Nicolson introduced a mustang stallion among the ponies there. A remarkably fine stock of ponies was, it is stated, the result; but again they got too large, their size ranging from 12 to 13½ hands. A number of them also inherited the excitable temper of their feral ancestors and were difficult to tame, although very

useful in many cases where a little size over the ordinary Shetlander was desired. These ponies are still spoken of as Fetlar ponies, as distinct from Shetland ponies.

With these exceptions, the Shetland pony has remained undisturbed by attempts at crossing, and even in the cases where crossing was attempted it never made great headway or touched the real heart of the breed.

Early Description of Breed.

Although several of the early historical writers make reference to the small horses of Orkney and Shetland, the first really good description of the Shetland breed was given by Brand, who visited the islands in 1700. In his book, *A Brief Description of Orkney, Zetland, Pightland Firth, and Caithness,* published at Edinburgh in 1701, this author says:—

"They have [in Shetland] a sort of little horses, called shelties, than which no other are to be had, if not brought hither and from other places; they are of a less size than the Orkney Horses, for some will be but 9, others 10, Nives or Hand-breadths high, and they will be thought big Horses there if 11, and, although so small, yet are they full of vigour and life, and some, not so high as others, often prove to be the strongest. Yea, there are some whom an able man can lift up in his arms, yet will they carry him, and a woman behind him, 8 miles forward and as many back! Summer or Winter they never come into an House, but run upon the Mountains, in some places in flocks; and if at any time in Winter the storm be so great that they are straitned for food, they will come down from the Hills, when the Ebb is in the Sea, and eat the Sea-ware (as likewise do the sheep), which Winter storms and scarcity of fodder puts them out of ease, and bringeth them so very low that they recover not their strength till about St John's Mass-day, the 24th of June, when they are at their best. They will live to a Considerable Age, as 26, 28, or 30 years, and they will be good riding Horses in 24, especially they'le be the more vigorous, and live the longer, if they be 4 Years old before they be put to Work. Those of a black Colour are Judged to be the most durable, and the pyeds often prove not so good; they have been more numerous than now they are; the best of them are to be had in *Sanston* and *Eston*, also they are good in *Waes* and *Yell*, those of the least size are in the Northern Isles of *Yell* and *Unst*.

"The Coldness of the Air, the Barrenness of the Mountains on which they feed, and their hard usage, may occasion them to keep so little, for if bigger Horses be brought into the Country, their kind within a little time will degenerate; and, indeed, in the present case we may see the Wisdome of Providence, for, their way being deep and Mossie in Many places, these lighter horses come through when the greater and heavier would sink down; and they leap over ditches very nimbly, yea, up and down rugged, Mossy braes or hillocks, with heavy riders upon them, which I could not look but with Admiration. Yea, I have seen them climb up braes upon their Knees, when otherwise they could not get the height overcome, so that our horses would be little, if at all, serviceable there."

The Modern Type.

With the exception that the height of the ponies is even less now than then, and that the colours are not so much mixed, thanks in large measure to careful breeding, Brand's 1700 description remains true down to the present day. The ponies are still exceedingly hardy, nimble on their legs, and docile and tractable. Pretty much the same aims in the matter of the form of the ponies are also pursued by breeders. It would still be quite true to say that "some not so high as others often prove to be the strongest." The object of the breeders for many years has been to get as much power as possible on the shortest legs possible. This, of course, can only be got by strengthening the bone of the leg and widening the bodies and ribs of the animals.

Some years ago, when there was a keen demand from America for children's ponies, an idea got abroad that the very wide ponies were not the best for this purpose, as it was said that they led to cases of rupture. A more slim and narrowly got-up pony was accordingly in

fashion for a time, but the theory was never greatly credited on this side of the Atlantic, and now next to nothing is heard of it. At the present time the wide sturdy-made ponies are almost exclusively the class that are in demand. The wider the pony is, provided his legs be strong and he is not too far from the ground, the more valuable he is considered.

Ponies in the Mines.

The preference for this class of pony is no mere fancy. By far the largest market for Shetland ponies is found in the coal-fields of Northumberland, Durham, and the southern districts of Scotland. There they are used for underground haulage—principally running the little waggons of coal to the pit-shafts. In some cases the ponies have to pass through workings little higher than themselves, and in these the gradients are often fairly stiff. It can easily be realised, therefore, that a low-sized pony, and at the same time a powerful one, is a real necessity.

Shetland ponies were first introduced into the coal-pits of the north of England about 1850. In 1851, according to Mr Robert Brydon, Mr Hunting, of South Hetton, than whom there was no better authority on animals for work in mines, bought thirty Shetland male ponies—all three, four, and five years old—at £4, 10s. per head, delivered at the collieries. Since then the prices have increased to a great extent. Average yearlings when the Stud Book was published in 1891 were worth in the north of England £15 per head; two-year-olds fetched about £18, and older ponies considerably more. These prices continue to be well maintained. Indeed, very small ponies are dearer now probably than ever they were, this partly being due to fancy, but to some extent also to the fact that they are suitable for working in thin seams where large ponies cannot enter.

The Bressay Stud.

Being an extensive coal-mine owner in Durham, Lord Londonderry would no doubt have had his attention early directed to the question of a good supply of ponies of the proper stamp and height for use in pits. Early in the 'seventies of last century his lordship acquired the grazings on the islands of Bressay and Noss, and at once began to found a stud from the best animals that were to be found in the islands. The most careful selection was pursued in breeding, with the result that the Bressay stud soon attained to a distinguished position. Indeed for many years it was the acknowledged fountain-head of the breed, and the annual sales which were held at Seaham Harbour attracted buyers from great distances. The stud was dispersed some years ago, but even to the present day the blood of Lord Londonderry's ponies dominates the showyards through their descendants in the hands of other breeders.

Pony Management on Crofts.—The great aim of those who were in charge at Bressay was to produce a low-set sturdy animal which would have great power on short legs. Accordingly the stud was managed on somewhat different lines from those that generally prevailed among the other breeders. The average crofter — the Shetland Islands are composed very largely of small holdings—simply turned his ponies into the "scatholds," or common grazings attached to the crofts, and left them to mate themselves with whatever stallions happened to be in the vicinity. The result of this, and the great privations which the animals suffered in winter, coupled with the fact that the foals were allowed to suckle the mares for a year or more, was that foals were only produced, as a rule, once in two years and often at longer intervals.

Management in the Bressay Stud. —In the Bressay stud a more scientific system of selection and mating was adopted. About the end of May in each year the mares were divided into lots of from a dozen to fifteen and put into separate enclosures along with a stallion specially suited to each lot. A very perceptible difference soon appeared in both the quality and numbers of the progeny, the mares managed under this system seldom missing a foal every year. The foals were weaned in November and put on good pasture which had been saved for the purpose. They were taught to eat hay as soon as the state of the weather rendered it necessary, and in this way they stood the winter better, and reached the spring much stronger, than foals that were allowed to suckle their dams all through the year. The mares also got a very desirable rest, and were in good condition again by the end of May.

Points of the Breed.

As will have been gathered, the first and principal point in the formation of the Shetland pony is its height. According to the rules of the Stud Book, no Shetland pony can be registered that is over 10.2 hands high. As a matter of fact, the great majority of the island-bred ponies run from 9 to 10 hands. The smaller they can be got without loss of other essentials the better.

In the best-bred ponies the head is small, the countenance pleasant and even intelligent looking, and the neck short, with a fine tapering in to the throttle. The back should be short, the quarters expanded and powerful, the legs flat and fine, with, however, a comparatively large measurement of bone below the knee, and the feet round.

The ribs should be laid on till within two inches of the hip bone. There should be great depth and width over the heart and lungs; the shoulders should be well sloped back from the brisket, and the fore arms and the thighs strong and muscular. Colours can pretty well be anything, but the most popular are blacks, with a considerable number mouse-coloured and a few dark-brown.

The pyeds or piebalds of the olden times are not numerous nowadays.

A point of great importance in connection with the breed, and which makes them extremely valuable as children's ponies, is that they are practically free from vice. A naturally vicious Shetland pony hardly exists. Of course, like other horses, they can be taught tricks, but properly treated they become companions and pets, equally willing, as Mr. Brydon says, to draw a carriage, carry panniers or saddle, or be led by a rein.

Distribution.

Since the founding of the Stud-Book in 1891, a good many studs of the breed have been established in different parts of Scotland and England.

For some time back considerable numbers of Shetland ponies have annually gone to the United States of America, principally for the use of children, a few also going from time to time to the Continent for the same purpose. This demand from America and the Continent has been very welcome to breeders, for it is mare ponies that these outside customers usually prefer, and this is the class which the native breeder has most difficulty, as a rule, in turning into cash, male ponies only being used in the mines.

Male Ponies for Mines.—The preference of the mine manager for male ponies implies no reflection on the capabilities of the mares, which are quite as strong, hardy, and quiet as horses. It is due simply to the fact that in such a limited space as a coalmine it is practically impossible to have mares and stallions working alongside of each other without trouble and loss of work. Mare ponies going to the United States command readily from £10 to about £15.

"Sheltie."—The term "Sheltie" applied to the ponies is, according to Mr Goudie, derived from the old form of the name Hjaltland given to the Shetland Islands by the Norsemen. But the word "shalt" or "shelt" has for many years been used in Scotland to describe a saddle or other lighter class horse.

MANAGEMENT.

As already noted, the Shetland Islands are for the most part colonies of small holders, who devote part of their time to fishing, the cultivation of the land being mostly done by the women-folks. Spade labour is still the prevailing method of tillage, ponies being used only to a very limited extent for work on the crofts. Their principal function is to carry their owners and their families about as required, and bring home the peats towards the end of the summer. At that season of the year it is no uncommon sight to see strings of ponies coming home from the hill with loads of peats on their backs, either carried in baskets, pannier form, or built on to a shaped frame of wood. These ponies are usually in charge of youths and maidens, who gaily mount the ponies' backs on the return journey.

Most of the small holdings are situated in townships, and while only three to four acres may be cultivated, each township has usually a common hill grazing, or "scathold" as it is called. This may extend to a hundred or more acres. The ponies are usually turned into these common grazings about the month of April, and remain there practically all the summer without further attention.

In the olden times, the owners being generally very poor and not over enterprising, it was the custom to leave the serving of the mares to chance—to any stallion or stallions that might happen to be in the "scathold." But now that ponies have become valuable, the thrifty owner rather turns his stallion pony into cash than allow it to be used at random without payment, and in many cases without thanks. The result is that there is now a shortage of stallions, and many mares pass several years without breeding at all. The more enterprising crofters, however, arrange for a stallion or stallions on co-operative lines, and this custom is becoming general.

The mares, except those that have been brought in and are tethered on the crofts to foal, remain on the "scathold" until the crops are off the ground, when they are brought in and have the general run of the fields. The fresh clean bites of grass which have been preserved round the patches of cultivated land afford the ponies, and the sheep which accompany them, an agreeable and much relished change. The ponies remain there in the open practically the whole of the winter, very often in severe weather being reduced for sustenance mainly to the seaweed which grows on the rocks or is cast up on the beaches of their storm-beaten islands. This the pony eats and thrives upon to a certain extent.

All this time the mares are probably suckling their foals, as the whole of the ponies go together in droves. The result is that, even when the mare is got in foal next season, the strain is too much, and nature intervenes in the form of abortion. Mares kept in this way produce foals only once in two years, and sometimes the interval is longer. The crofters, as a whole, are so poor that they cannot afford to wean the foals earlier or to keep the mares better in winter. Such spare winter food as they have has to be given to the cattle and sheep, without which the holders could not exist. While the school is an excellent one for ensuring hardiness, it has not led to any great increase in the numbers bred, at any rate by the crofters. In the studs on the larger holdings matters have improved considerably.

Nearly all the ponies exported from Shetland are conveyed by boat to Aberdeen and Leith, and from these centres are distributed over the kingdom.

The Pit Ponies.

The bulk of the male ponies are taken by dealers to the north of England, where they are sold to the various collieries. Mine work is no doubt a hard life for the ponies, but it is not so dreadful as might be supposed. The ponies are well fed and cared for, and they live under it to old ages. Many pathetic tales are told of the attachment of the mine boys to their dumb charges, and *vice versâ*.

A portrait of a typical Shetland pony stallion is given in Plate 21.

THE ASS AND THE MULE.

By high authorities the origin of the varieties of the ass in this country is assigned to *Asinus tœniopus*, a wild species which existed in Abyssinia and other parts in the north-east of Africa. The ass, it is believed, was domesticated before the horse, and this belief is supported by the fact that in sacred history it is referred to much more frequently than the latter.

Varieties of the Ass.—Many varieties of asses are known to exist. While they have all the leading characteristics in common, they vary greatly not only in size but in strength and stamina. Perhaps the best of the modern day asses are to be found in Spain, Italy, Greece, and the old French province of Poitou. The French or Poitou ass is a brown breed, with long shaggy coat, powerful limbs, great bone and feet, standing from 13.3 to 14.3 hands high. Hardly less famous than the Poitou ass are the Catalonian and Andalusian breeds of Spain, which are of great merit.

The Andalusian asses are exceptionally powerful animals. One prize jack of this breed imported from Spain by that enthusiastic patron of the ass, Mr H. Sessions, Wooton Manor, Henley-on-Thames, measured 15 hands in height when four years old, and had great bone and substance in addition. The donkey stallion, belonging to Mr Sessions, which is represented in Plate 25, though only two years old when photographed for this plate, was then 14.2 hands high, his girth being 5 feet 5 inches, while the leg-bone under the knee measured 8 inches.

The Egyptian donkeys, which are practically all grey in colour, have neither the strength nor the stamina of the French or Spanish kinds, and having little value for mule-breeding, are not much used out of their own country. Sometimes a few are imported into Great Britain by visitors who have been struck by their fine appearance as compared with our native donkeys, and they do very well for children and for light classes of work.

A donkey mare of this type with its foal is represented in fig. 701.

In this country, and particularly in Ireland, a large number of donkeys are kept. No systematic attempts have been made to form distinct breeds, and the animals are accordingly simply donkeys and nothing else. They are mainly of the small kinds found in Eastern countries, but long since acclimatised to the conditions as existing here.

In recent times considerable numbers of both French and Spanish jacks have been imported by such enterprising private owners as Mr Sessions, but their influence has not yet reached the common stock.

Uses of Donkeys.—The donkey, in all parts of the world where it is found, is a most useful animal, especially to the poor man. There is no kind of work to which horses are put that cannot be more or less successfully performed by the ass. In Ireland it is almost a *sine quâ non* to the small cultivator. Not only is it employed to turn over the small patch, but it is usually the only power available for carrying produce to market. In parts of the west of Ireland especially, the spectacle of the farmers' wives driving along in their little donkey-cart is a characteristic feature of the landscape.

The donkey is also largely used for market-garden work, and for hawking vegetables and other produce through the larger towns. The London coster would hardly know himself without his donkey.

Asses are also very largely used by children, both for riding and driving. Asses do not appear to know fatigue, are very easily kept, and, given time, will get through a great amount of work.

Longevity of the Ass.—Donkeys live to great ages. In Brettell's Account of the Isle of Wight, it is mentioned that an ass for the space of fifty-two years drew up the water daily from the deep well at

Fig. 701.—*Donkey mare and foal.*

Carisbrooke Castle. The animal might have continued at the operation for considerably longer had it not fallen over the ramparts and been killed. It is stated that up to the hour of its accidental death it was "in perfect health and strength."

The period of gestation in the ass varies from 360 to 375 days, being thus nearly a month longer than in the horse.

MULE BREEDING.

The fact that the horse and the ass breed together is proof of the close affinity that exists between the two. The offspring of this union is a sterile animal known as a mule. Even did the donkey perform no other useful function, it would still be entitled to consideration as one of the agents in the production of the highly useful mule.

In this country the mule is not so well known as it deserves to be. In Spain it is *par excellence* the beast of burden. The larger kinds perform all agricultural and general draught work, while the more slender and finer-boned varieties are extensively used for saddle purposes, being preferred by the rich, in many cases, even to horses.

Mules are also used very largely for a variety of purposes in the United States of America. Before the days of electricity they did nearly all the tram-haulage, besides being extensively employed in agricultural and commercial operations generally. Many of the American mules are as big and powerful as horses. In addition to being powerful, mules are very hardy and tough, and give less trouble, as a rule, than horses, with their legs on hard causewayed streets.

Donkeys and Mule Breeding.—For mule-breeding only the bigger size of donkey jacks are of much use. To serve a mare and get produce of any value the animal must be fairly upstanding and have a large amount of bone. In Spain, Italy, and other countries where mule-breeding is extensively carried on, the best mules are considered to be bred from the jack put to the mare, the produce appearing to follow the mother in the external form. Those bred from female asses are said to be longer in the ears, of less comely form, and duller in temperament. Occasionally trouble is experienced in getting a donkey to serve a mare, but the difficulty is not insuperable, as a rule.

A pair of Poitou mules, which be-

Fig. 702.—*Pair of mules.*

longed to Lord Arthur Cecil, are shown in fig. 702. These mules took a full share of farm work, day by day, alongside average Clydesdale horses, for a period of twelve years, and their food rations were only two-thirds of what had to be provided to the horses.

FOREIGN BREEDS OF HORSES.

Comparatively few foreign breeds of horses are known even by name to British agriculturists of the present day. The Arab and Barb races are, of course, familiar to us, and are deservedly held in high repute, for they have played a useful part in the formation of the best varieties of our saddle-horses.

Amongst draught-horses the best known are the French Percheron, now the most highly valued heavy draught breed in the United States of America, the Boulonnais breed of France, and the Flemish breed, which has its home in Belginm. The Flemish breed has contributed its quota to the improvement of British draught-horses.

ABERDEEN-ANGUS CATTLE.

This breed of black hornless cattle is native to the north-eastern counties of Scotland, although within comparatively recent years it has spread largely throughout the different parts of the United Kingdom, and has also secured a firm footing in many of the cattle-raising countries abroad. The outstanding feature in the history of the breed is the remarkable rapidity which has characterised its development. It may indeed be safely said that no other breed of cattle has spread so rapidly to new homes as has the Aberdeen-Angus since its existence as an improved race began. Its rise, development, and progress form a most interesting chapter of British cattle history.

Origin.—Although the origin of our different races of domesticated cattle can only be to a greater or less extent matter of speculation, there is abundant evidence to show that the Aberdeen-Angus breed is of great antiquity. The earliest writings dealing with the agriculture of those districts chiefly recognised as the homeland of the breed, and in which any attempt is made to characterise the different varieties of cattle, show the existence of a black polled race. There is existing legal documentary evidence showing that in the early part of the sixteenth century black hornless cattle constituted in Aberdeenshire an important commercial commodity. Not only in that county, but also in Forfarshire, Kincardineshire, and Banffshire, where the breed was also retained in more or less purity in the early days before its establishment as an improved race, records of the eighteenth century contain numerous references to the "hummel" and "hornless" cattle in these parts. Many of the present-day herds can trace their direct descent for considerably over a century.

Early Improvement.

Mr Hugh Watson.—The first great improver of the breed was Mr Hugh Watson, Keillor, Forfarshire. His father had been a breeder of black polled cattle

as early as 1735, but the systematic improvement of the breed may be dated from the year 1808, when, as a young man of eighteen years of age, Hugh Watson entered the farm of Keillor in the old territory of Angus. There can be little doubt that the wonderful success which was attending the efforts of the Brothers Colling, especially those of Charles Colling at Ketton, in the improvement of the Shorthorn breed, spurred on the young Forfarshire farmer, who indeed lived for a time as a student with Charles Colling. General Simson of Pitcorthie in Fifeshire was then buying at great prices some of the products of the Ketton herd, while the tidings of the sale of the Shorthorn bull "Comet" at Mr Colling's sale in 1810 for a thousand guineas must also have proved an incentive to Mr Watson to persevere in the improvement of the native cattle of his county. He had many co-workers, such as Mr Bowie, Mains of Kelly, who was born in 1809; Mr Fullerton, Mains of Ardovie; Lord Panmure, Sir James Carnegie, and the late Mr Ferguson, Kinochtry. But it is especially by Mr Watson's persistent efforts that the greatest services were done to the interests of the breed while yet in an embryo state.

Mr William M'Combie.—In the north of Scotland an outstanding name in Aberdeen-Angus history is that of Mr William M'Combie, Tillyfour, who, along with Mr Watson, took a great part in the establishment and early development of the breed. Mr M'Combie was only three years of age when Mr Watson began his work at Keillor, but by 1830 he owned a breeding herd, and about 1848 he gave himself up entirely to the cause of Aberdeen-Angus breeding, or what was then styled Polled Aberdeen cattle as distinguished from Polled Angus, although the necessity for this differentiation soon passed away.

About that time, near the middle of the nineteenth century, the black polled breed was threatened with complete extinction, as the result of the crossing craze which followed upon the introduction of the Shorthorn to the north of Scotland. As a matter of fact, one northern race of cattle, the Aberdeenshire Horned breed, entirely disappeared as the direct result of this new system of breeding. Mr M'Combie is the recognised rescuer of the polled breed at this juncture, and by setting himself to bring out the great feeding capabilities of the breed, he undoubtedly gave it a new lease of life. From this point the history of the breed has been one of continuous and unbroken progress.

Sir George Macpherson Grant.—In more recent years the central figure in the improvement of the breed was the late Sir George Macpherson Grant, Bart., of Ballindalloch, who vastly advanced the cause of Aberdeen-Angus breeding and perfected the type which had been evolved by those who went before him.

Early Show Successes.—Although as early as 1867 Mr M'Combie, after repeated trials, managed to secure the blue ribbon of the Smithfield Fat Stock Show, thereby greatly advancing the interests of the breed, there is little doubt but that a most important agency in the spread of the breed was the French International Exhibitions. These exhibitions, from 1856 up to 1878, were taken part in by a number of leading breeders, such as Mr Bowie, Sir George Macpherson Grant, Mr Walker, Portlethen, and Mr M'Combie; and great successes were won both for feeding and breeding stock.

It was in the year last mentioned (1878) that the greatest victory of all was won by the Aberdeen-Angus breed, and there was no doubt a very direct connection between this success and the great and remarkable demand which was about that time being experienced from America for cattle of the breed. A prize given by the French Government for the best animals for breeding purposes, in the sections other than French, was won by Mr M'Combie, and Sir George Macpherson Grant was reserve. But the greatest trophy of the show was in the competition for the best group of beef-producing animals, when all varieties of European cattle competed together. The bench of judges, by twenty-four votes to seven, decided in

favour of the representatives of the Tillyfour herd, and thus both the fat stock championship and the championship of the breeding classes went to the Aberdeen-Angus cattle. This proclamation to the world of the superiority of the Aberdeen-Angus breed in the realm of beef-production gave a great impetus to the growing popularity of the cattle not only in this country but likewise in America.

Characteristics of the Breed.

Record as Beef-Producers.—The breed possesses valuable dairy qualities, which are capable of greater development. Many strains of the breed are found to be exceptionally heavy milkers, and the milk, in the various tests that have been made, has been found to be very rich. The breed has, however, been all along cultivated primarily for its beef-producing properties. The aim which Mr M'Combie ever kept in the forefront was the production of size, symmetry, fineness of bone, strength of constitution, and disposition to accumulate flesh. Keeping these objects in view, Aberdeen-Angus breeders have been able to evolve a type of animal which holds an unrivalled position in the estimation of feeders and butchers. The remarkable success of the breed at the leading fat stock shows of the country has also tended to greatly increase the admirers of the breed, and to enhance the reputation of the cattle as grazers.

Graziers on a large scale have borne testimony to the fact that cattle of the breed give a better return for the same amount of keep than any other kind of cattle, and the statistics of the Board of Agriculture and Fisheries show that both in Scotland and England enhanced prices are paid by butchers for Aberdeen-Angus cattle and their crosses as compared with other breeds, while on the London cattle market it is a generally recognised fact that this class of cattle sell first and sell dearest. The reason of this is that the Aberdeen-Angus produce beef of the finest quality, and have the best cover of meat on the most valuable parts. It may be that the beautifully rounded form set on short legs may be deceptive to the eye, and may cause the cattle to bulk less largely in appearance than some other breeds, but the well-filled rump and loins, the thick cover along the back, and the long well-filled-out quarters appeal at once to the butcher, and constitute them his primest favourites. Cattle of the breed are found to feed very smoothly, unlike some other breeds which are much more apt to run into lumps and bumps of fat, which are absolute waste. The breed holds a record of 76¾ per cent of dead- to live-weight, and in addition to great returns at the block, butchers find that the flesh of cattle of the breed is admirably mixed and beautifully marbled throughout.

A breeder of extensive experience has put the following on record, speaking of Aberdeen-Angus cattle: "I may state that my resolve to keep this particular breed is the result of having carefully watched the breeding, feeding, and general health of cattle for some years. Having for many years been engaged in a large veterinary practice with special opportunities for forming an opinion on the merits of the different breeds of cattle from a professional point of view, and having for a number of years been a farmer and feeder of stock, I have had not only my own farming experiences to guide me, but also the cattle market, and the health of the large and varied cattle population of this district — the result being that I believe this breed of cattle stands pre-eminently forward both to the farmer and the butcher as being hardy and healthy, good milkers both in quantity and quality, easily fed, good beef-producers, coming early to maturity, and highly prized by butchers."

Reputation in America.—In America this same characteristic has been brought out, and has led to the phrase "market-toppers" being applied to the Aberdeen-Angus cattle. It is a rather striking fact that for a space of about twenty years the top price in the Chicago Meat Market has been made each year by cattle of Aberdeen-Angus breeding. Here, too, it has been found that no other class of cattle put on flesh so quickly in proportion to what they eat as Aberdeen-Angus cattle do, and few breeds can stand the cold winters so well. They give most satisfactory returns both when rustling on the scanty

herbage of the ranch and when foraging on the luxuriant pastures of the fertile farms. A representative of one of the largest packing firms in America stated that "in buying cattle for our trade in the United States, and especially for export, we give the preference to Aberdeen-Angus steers. These well fattened will dress from one to two pounds more per hundred pounds of live weight than either Shorthorns, Galloways, or Herefords. Although the Aberdeen-Angus may appear very fat, they will show more lean meat and be less wasteful for the retail butcher than animals of any of the other breeds above mentioned, and the meat itself will show a better and richer grain, and is more juicy."

Records in Fat Stock Shows.—As regards the fat stock show record, it will be sufficient to deal with the two greatest shows of the world—the London Smithfield Show and the International Fat Stock Show of America, held annually at Chicago. At the London Show, where eleven different breeds of cattle compete, Aberdeen-Angus cattle during the fifteen years, 1894-1908, won the championship upon eight occasions, while a cross showing Aberdeen-Angus lines of breeding has won it once. On the occasion of the other six shows, the Aberdeen-Angus breed has provided three reserve champions, and crosses of the breed have also produced four reserve champions. In other words, at these fifteen shows pure or cross Aberdeen-Angus cattle each year provided the champion or the reserve champion. This constitutes a record that is quite unique. In the case of the Chicago Show, which was begun in 1899, the championship in the first nine years was won upon five occasions by Aberdeen-Angus cattle, twice by Herefords, once by a cross, and once by a Shorthorn. This, coupled with the equally successful record of the breed at the other fat stock shows throughout Britain and America, demonstrates clearly that the great popularity attained by the breed has been built on a sure foundation.

Carcase Competitions.—The breed has also won many of the higher honours in the carcase competitions at fat stock shows, the great return given by cattle with a dash of Aberdeen-Angus b'ood in them leading to their being largely represented in this department of the show.

Weights.—Animals of this breed attain heavy weights at an early age. At the Smithfield Show in 1908 the steers of the breed under two years old weighed alive from 11 cwt. to 14 cwt., the exact age of the heaviest animal being 1 year and 11 months. In the class for steers between two and three years old the live-weights ranged from 15 cwt. 1 qr. 14 lb. for a steer 2 years and 9 months old to 16 cwt. 1 qr. 27 lb. for a steer one month younger. The live-weight of a heifer of the breed at the age of 2 years and 11 months was 16 cwt. 1 qr. 20 lb.

Prepotency of the Breed.—An outstanding characteristic of the breed is its remarkable prepotency in imparting its properties to its offspring. This is seen in the demand for polled cross oxen for feeding purposes, and by the extent to which Aberdeen-Angus blood is represented in the cross sections at the fat stock shows. So prepotent are bulls of the breed that it is found that quite 75 per cent of the calves come black and hornless, even when the cows belong to a pronounced horned breed.

A breeder in Ireland writing of the breed says: "This breed of beef-producing cattle has made rapid progress in the Sister Isle, and its crosses, whether made with the Shorthorn, the Hereford, or the native Kerry and Dexter cattle, are amongst the most useful stores for the feeder to buy that can be produced."

In the case of a large dairy farm where Ayrshire cows were kept, the owner, to improve his calf stock, introduced Aberdeen-Angus bulls. The result was that 90 per cent of the calves were black and hornless, and fetched greatly enhanced prices when sold. Again, an American experiment showed that where an Aberdeen-Angus bull was used on fifty horned cows there was not a single horned calf, while 95 per cent of them were black. Even when used on the long-horned Texan cows, bulls of the breed produce a very large percentage of black and hornless calves. On the great ranches of America the breed has proved to be most prolific.

Influence of the Breed in England.—A writer in *The Times*, in November 1908, in commenting upon the character of the cattle exhibited at English fat stock shows, referred thus to Scottish polled breeds: "The Norwich [Fat Stock] Show of last week provided an instructive illustration of the popularity of the hornless black breeds of the north, especially for crossing with the English varieties. Of the 110 head of cattle stalled at Norwich, 12 were red polls and 15 shorthorns, and deducting these 27, which, of course, were exempt from the influences of the black breeds, 83 remain, and of this number 53 were either black or blue-grey. Thus, nearly 64 per cent of the exhibits, other than red poll and shorthorn, revealed the characteristics of the Scottish black poll breeds, the Aberdeen-Angus greatly predominating. The latter influence was as marked in the county and butchers' classes as in the others, and it was the general opinion among graziers present that the change, as compared with past years, is beneficial to the eastern counties, the compact, short-legged, thick-fleshed bullocks of the present time being much more economical feeders and more popular with butchers than the leggy, plain steers they have displaced."

Early Maturity.—The property of the breed to mature early has already been indicated. In the early days Mr M'Combie brought out this feature, his champion group of six at the French Exhibition being with one exception only two-year-olds. In later times, as showing that there has been no falling off in this respect, it may be recalled that the Aberdeen-Angus is the only breed that has produced at any of the leading British fat stock shows a champion animal at one year old. It is also a rather interesting fact, in view of the chief aim of the Smithfield Club to encourage early, maturity, that the first occasion in the history of the Club upon which the championship was taken by a two-year-old, the successful animal was of the Aberdeen-Angus breed.

Prices.—During the twenty-five years, 1882-1907, average prices for the breed have ranged from £23 to £25, although early in the 'eighties, when the American "boom" was being experienced, the average was from £45 to £55. The highest single price in this country at a pbulic sale is £504 for a bull-calf from Ballindalloch. In America, a bull bred by Sir George Macpherson Grant sold for £1820.

Points of the Breed.

In the formation of Aberdeen-Angus cattle, well-defined points are kept in view. In the case of the bull, there should be sought both size and quality. The head should be neatly put on, and the throat clean. The distance between the eye and the nose should not be over long, and the eyes should be bright and prominent, with a good breadth between them and surmounted by a good, high poll. The neck should be of good length, and clean—a little but not over full on top; chest full and deep; legs short, but not so as to give the animal a dumpy appearance; bone clean and free from coarseness; shoulders not too full, and top free from sharpness, but not over broad; back level and straight; ribs well sprung; deep barrel; well ribbed down towards hook; full behind shoulder; hooks level, but not too broad for other proportions; and well and evenly fleshed to tail; twist full and long and well fleshed down, but not protruding behind; tail of moderate thickness and hanging straight; hair soft and plentiful; skin of moderate thickness and mellow to the touch; body fully developed, and the animal when in motion to have a blood-like look and style about him.

A cow of the breed should differ from a bull in the head in having, instead of a broad masculine-looking head, a neat feminine-looking one. The ear should also be of good size, with plenty of hair in it; the neck well put on, clean and straight, and without any prominence on the top or abrupt hollow where it joins the shoulder; the top of the shoulder sharper than in the bulls, and the shoulders themselves thinner.

Present Position of the Breed.

The leading position which the breed has taken at the fat stock shows, both when shown pure and in the form of crosses, has led to a marvellous growth in the numbers of the breed both at home and abroad. As indicating the progress which has been made, the fol-lowing facts may be mentioned. In the first volume of the *Polled Herd-Book*, published in 1862, there were eighty-three owners of animals, and in the early volumes the names of only two English breeders and two Irish breeders appear. In 1879 a meeting was held for the formation of The Polled Cattle Society, —changed in 1908 to The Aberdeen-Angus Cattle Society,—and at the first annual meeting in the following year the membership totalled 56. By 1908 the Society reached a membership of about 530, of which about 120 resided in England and 70 in Ireland. In volume xxxii. of the Herd-Book,—which brought the registered numbers up to 27,662 bulls and 43,173 cows,—there are 2837 entries.

The Breed in Canada and United States. — Although the first breeding herd was established in Canada only in 1876, and in the United States about a couple of years later, the breed has in the course of the thirty intervening years spread widely in these countries. As an indication of the demand for the breed in the early 'eighties, it may be remarked that in 1882 there were landed on North American soil 104 Shorthorns, 173 Herefords, 222 Galloways, and 586 Aberdeen-Angus. Within a space of two or three years over two thousand head of cattle of the breed had been introduced into America.

In the first seventeen volumes of the *American Aberdeen-Angus Herd-Book* 112,500 animals were registered, and the 120,000 entries in the seventeenth volume represent over 2000 breeders. The American Aberdeen-Angus Breeders Association was instituted in 1883, and when the first volume of the Herd-Book was issued in 1886 the membership was only 112. It is thus seen that the progress of the breed there has been very rapid.

The Breed in other Countries.— In several other countries the breed has secured a firm footing. In the case of Argentina there is a steadily growing demand, for this breed is found to be pre-eminently suitable in the northern districts on account of the thrifty and hardy properties which characterise it. Aberdeen-Angus cattle were first introduced to the Argentine about 1876, and there are now a good few herds of the breed in that country, though for the most part the bulls of the breed have been used for the grading up of the native cattle of the country. In the pastures of Argentina, cattle of the breed are found to thrive excellently.

In several of the Australian Colonies the breed is also largely represented. As early as 1863 Aberdeen-Angus cattle were introduced to New Zealand, and the Aberdeen-Angus is now the second most numerous breed in that country. During recent years large numbers of the cattle have been imported to South Africa, and recent advices state that bulls imported into that country, and especially into Rhodesia and the Transvaal, are giving excellent results. To various other countries, such as France, Spain, Germany, Sweden, Russia, and even to Demerara, India, and China, representatives of the breed in small numbers have been introduced, mostly for crossing purposes.

MANAGEMENT OF ABERDEEN-ANGUS HERDS.

Systems of management in Aberdeen-Angus herds vary considerably. They are influenced to some extent by locality and climate, by the accommodation afforded by the farm-steading, and also by the consideration whether the chief end in view is the rearing of bulls for sale. Ages of Aberdeen-Angus cattle are reckoned from 1st December, and the principal calving months accordingly are December, January, February, and March, the object being to get the calves as early in the year as possible. Especially is this of importance in the case of bull-calves.

Calving.—At the calving time each cow is placed in a loose-box, or given a whole stall to herself, all depending on the accommodation that is available. A week or two prior to calving time the quantity of turnips fed to the cows is reduced, and they are allowed a soft feed, such as bran, once a-day. The decreased ration of turnips is continued for about a week after calving, the bran mash being also continued, when the cows are generally placed again on their full feed of turnips and straw.

Calf-rearing.—The calves suckle their dams, and in many herds a couple of calves are put to one cow, the cow sthus relieved being hand-milked to supply the ordinary requirements of the farm. The calves are, as a rule, tied up beside the dam, but in several large herds they are allowed to wander about in the open area behind the cows, being allowed access to suckle four times a-day. The breeding byre, unlike the byre for the feeding cattle, should always be a single one.

When about three months old the calves are allowed a small supply of hay and sliced turnips with a little linseed cake, although in a great many well-managed herds they get nothing beyond the dam's milk until they are put out to grass about the month of May. The bull- and heifer-calves are put into separate fields, and where the pasture is poor, and it is wished to keep them going on, the bull-calves receive about 2 lb. of cake per day. When they are from seven to eight months old the calves are gradually weaned, and thereafter put out to grass again so long as the weather permits.

In any case they are allowed plenty of room for exercise. In a few cases the heifer-calves get a little cake each morning, which is found to be of great value in helping them to retain the calf flesh. In the great majority of cases, however, nothing is given them beyond turnips and straw.

The bull-calves, where the necessary accommodation can be got, are put up two by two in loose-boxes. In large herds this cannot always be done, and the practice then is to put them into a court. Their principal diet consists of turnips and straw, with about 3 lb. of cake per day, and a hot mash of bran and barley twice a-day. But on this point treatment varies considerably, and in many cases the allowances are much less liberal. The bull-calves are sold off in the spring when about a year old.

Young Heifers. — The heifers are kept out all summer, being again, when the weather becomes severe, housed up in the open courts or in the byre, according to available room. The only feed is turnips and straw, so as to keep them in natural condition for breeding.

The best mating season is about the month of March, and, as a rule, heifers are not served until they are two-year-olds. Breeders prefer to have all their females settled in calf before they are put out to grass.

Stock Bulls.—Stock bulls should be kept in healthy condition by avoiding too heating or heavy feeding. They should be given plenty of exercise, and it has been found that the most beneficial form of exercise is to walk the bulls along the hard road for about an hour each day. Turnips and straw or hay form the principal foods during the winter, and as the mating season approaches an allowance of dry crushed oats is frequently added.

Ballindalloch Herd.

In the course of his history of the Ballindalloch herd the late Mr Campbell Macpherson Grant gave the following notes on the system of management :—

"The principal calving months are December, January, February, and March, although calves are dropped all the year round. When due to calve each cow is allowed a double stall, and the calf when dropped is tied at the opposite side, while a strong bar, angled lengthwise down the stall, prevents any risk of accidental injury to it.

Calf-rearing.—"When strong enough and able to take all the milk the calves are allowed to move at will through the byre, their beds being made up for them behind their dams. A trough with cake and sliced turnips, as also a rack with good, sweet hay, is always within their reach. The bull-calves when at grass are kept separate from the cow-calves, and have an allowance of cake daily. They are gradually weaned when six to seven months old, and are then, so far as accommodation permits, placed two together into loose-boxes with an outside court for exercise. They are liberally fed on yellow turnips and hay or oat-straw, with an allowance of cake, care of course being taken not to overfeed.

"Heifer-calves are treated in much the same way, but get no cake on the grass; and they run in the covered courts during the winter, getting a fair allowance of yellow turnips, good oat-straw, and 2 lb. bruised cake each day.

Winter Treatment in the Herd.— "As soon as the nights begin to turn cold, all the cattle are housed at night and turned out during the day. When finally brought up for the winter, at a date determined by the character of the season, the cows get a fair quantity of turnips twice a-day, with plenty of oat-straw, but get no artificial food except for a fortnight before and after calving, when they get 2 lb. of cake daily; and during the fortnight after calving, in addition to the cake, a bran mash daily, which twice a-week contains a little nitre. The two-year-old heifers have nothing but turnips and straw. Except in quite an exceptional case heifers are not served until they are two-year-olds."

Pictstonhill Herd.

Mr W. S. Ferguson, Pictstonhill, Perth, writes: "I aim at having the cows in fairly fresh condition at calving. This is done by giving them straw and turnips in limited quantity, in covered courts, in autumn and winter after the grass is done. The cows are tied in stalls when they show signs of calving, and when the calf comes it is tied not far from the dam. It is let to her four to six times a-day to begin with, and afterwards three times a-day, when the calf takes all the milk freely. Great care is exercised at the beginning to take all the milk from the cow. Some calves cannot take nearly all the dams can give, and if not milked dry nature seems to meet the case by drying up the cow to suit the requirements of the calf, and the cow will not then come back to milk when the large, grown calf requires more. One of the most important matters the cattlemen have to attend to between calving and grazing time is to keep the cows in full milk. Every cow requires different treatment: some more food and some less. When a calf is becoming too fat, as sometimes may happen, it is not permitted to take all the milk from its dam. After the calves are two months old they get some cake, meal, and pulped turnips, but not much, as we rely mostly on the milk.

Aberdeen-Angus Cows as Milkers. —"It is a mistake to suppose that Aberdeen-Angus cows are not good milkers. They give milk, as a rule, according to the treatment they receive; and I find that when passed on to the dairy, as I sometimes do, they give as good an account of themselves as any excepting Ayrshires.

"The cows after calving and up till grass time get mostly turnips and straw, along with a drink of bran and meal once a-day, while if an individual beast begins to look thin and dry she may get a bit of cake extra.

Calves.—"The calves go with their dams at grass. The heifer-calves get nothing but their mother's milk and what they pick up on the field, but after a short time the bull-calves are trained to eat cakes and meals. The calves are weaned when about seven months old, the heifers going anywhere at little expense, and the bulls to folds and boxes to be trained and fed for sale in the following spring. According to modern ideas it is not easy to overdo a bull-calf to sell him as a yearling, but care must be exercised to keep his feet and appetite always in good order. This is where the expert cattleman comes in, for fixed rules are of little use.

Objections to Forcing.—"But the modern system of forcing young stock for showing and selling is a mistaken one. By it many young animals are impaired in growth and health, and are not in the end as useful as are animals that are kept in moderate growing condition. I never put too much flesh on a calf intended for breeding purposes, and if sometimes I am constrained to put a good, young bull in prime show order I always grudge it. I seldom do it till after he is two years old, and then he can stand it better. My efforts—as were those of my father before me—have been to keep a good, healthy, presentable herd at as little expense as possible, and to make the cattle leave a profit.

"The heifer-calves after being weaned get a small allowance of cake or meal for the first winter, along with turnips and straw. After that nothing in the way of short concentrated food is given them until they reach the cow stage. Of course this does not apply to a few females now and then put into training for show purposes. With these few it is

a case of feed as hard as you can without making them patchy."

Mulben Herd.

Mr John Macpherson, Mains of Mulben, Banffshire, states that his cows, except an occasional animal for showing, receive very little artificial food,—turnips and straw during winter, and grass in the fields during summer, being all that is necessary to keep them in good healthy breeding condition. For a week or two after calving, or if at any time any animal seems to be down in condition, a little linseed-cake is given.

Calves.—The calves are all suckled, bull calves singly, and heifer-calves in pairs on good milking cows. With the exception of a little cake for a few weeks at weaning time, the heifer-calves get no extra keep.

The bull-calves, being intended for early sale, require more attention. After the grass begins to fail they are taken into a court overnight, and get some tares and a small allowance of linseed-cake. Bran and feeding-meals mixed together, and scalded with boiling water, are fed to them in boxes, the food being thinly scattered on the bottom until they begin to eat it. During the day they go to the field, but they soon learn to gather about the gate to get in. The cows are left in the field, and cows and calves are thus accustomed to be separated, so that when the final weaning time comes there is far less noise and trouble than there would otherwise be.

Heifers.—The earliest and strongest heifers are served when fifteen or sixteen months old, so as to get them to calve when about two years old. Mr Macpherson has found that when the animals are strong and fairly well kept, although they may take a little longer time to mature, the ultimate growth and size of the heifers thus served are not very much affected, while their milking qualities are improved.

The stock bulls are well kept and regularly exercised, and during the mating season a little extra grain is added to their feed.

The whole steading is thoroughly cleaned and disinfected every summer, and the byres and courts are frequently sprayed with a solution of Jeyes' fluid.

Spott and Inverquharity Herds.

Writing of the management of the herds at Spott and Inverquharity, Mr Archd. Whyte states that all cows in calf go out every day to rough pasture till the calving time comes on. After calving they are kept in till early spring, when the weather becomes favourable. During winter the cows are fed very moderately on turnips and straw.

Calves.—As soon as practicable, the cows and calves are turned out to pasture, the bull-calves being weaned in August and the heifer-calves a little later. After weaning time the cows remain outside till the end of November, and then only get shelter overnight. Bull-calves, after weaning, go out to clover during the day, being taken in at night to a bite of hay and cake. Thereafter they are put gradually on to turnips, &c., getting out for an hour every day for exercise. Heifer-calves get moderate keep. They are out every day, and are allowed a few turnips, straw, and a little cake night and morning.

Heifers and Cows.—Yearling and two-year-old heifers get very ordinary fare—when on grass only what they can gather. This applies also to cows with calves at foot. When on grass they get nothing extra, and when weaning time comes round they are always in fine condition.

Stock bulls get ordinary fare all the year round, a little cake being added if other keep be scarce.

The farms being situated in a very high-lying district, winter keep is never plentiful, but cattle keep themselves in wonderfully good condition on very small rations.

Dr Clement Stephenson's Herd.

Dr Clement Stephenson, Balliol College Farm, Newcastle-on-Tyne, writing on the subject of herd management, states that up to a few days before calving the cows may remain in their stalls, and for calving should be isolated in a box or stall. In no case should a cow be allowed to calve in a byre beside other in-calf cows.

Calves.—Calves should suckle their dams, and when in the byre should be tied up in such a way as to allow

them to get a fair amount of exercise. A large piece of rock-salt, and sometimes chalk also, are kept in the racks, so that old and young may lick them when they choose.

As soon as weather permits cows and calves are put to grass, the bull and heifer calves being put into separate fields. Should the dams of the bull calves begin to fail in their milk-supply, the calves should be given a small allowance of cake. Great care should be taken in the breaking-in and training of the calves. From weaning time till turning-out time in the following spring the calves should be well attended to, and their food must be of good quality. They do not want coddling up in warm places, but should be kept in covered folds which are well lighted and ventilated, and in which they have plenty of room to move about. Twice a-day they should be let out into a yard to scamper and play about. They should be accustomed to being handled, and kept clean with brush and comb.

Heifers.—After being turned out at May-day the heifers need not be brought into the house again until next spring, and then only for service. A shed in the field into which they can go if inclined, and oat-straw in the winter, are all they require, but if it be thought advisable to give them cake it should be linseed-cake.

Bulls.—Stock bulls should be well fed —not made fat, but kept in vigorous condition. When in free use, their ordinary diet should be supplemented by stronger, more nitrogenous food, such as bean-meal or crushed oats. The bull-house should be well ventilated, and have a walled exercise-yard adjoining.

When in the house a sloppy mash, sweet hay, and a few turnips are all the cows require. Cake is not necessary, nor is it advisable to give it, at any rate until the cows are again safely settled in calf.

Preston Hall Herd.

Rev. C. Bolden, Preston Bisset, Buckingham, writes: "I endeavour to get cows and heifers to calve in December and January. They lie out in the fields until within a week or a fortnight of calving, when they are housed. A week after calving they go out for three or four hours daily in all kinds of weather. In ordinary seasons yearling heifers are left out all winter, getting hay when there is snow on the ground or during hard frost. In some seasons I am obliged to put them in open yards to prevent damage to pastures, as my land in Bucks, being heavy clay, treads into holes in very wet weather. I find that yearling heifers do best lying out all the year, and I generally manage to keep a field fresh with plenty of grass for them during winter.

"Calves are gradually weaned in October, and put into covered yards in November, the heifer-calves getting hay and roots and 2 lb. of cake daily. The bull-calves get more cake with meal, and are fed on as well as possible with a view to sale in February, but I object to any free use of condiments or forcing them into overfed condition, as this, I believe, shortens the period of their usefulness as sires, and in some few cases may render them uncertain, or possibly useless, as stock-getters during their first year of service. My covered yards have a hard level bottom, either paved or solid gravel. They are frequently cleaned out, and no accumulation of muck is allowed in them."

An Irish Herd.

Mr H. Bland, Kilquade, Greystones, County Wicklow, Ireland, writes: "Owing to our exceptional climate it is possible to keep our cattle under the most natural conditions. The cows are out at grass all the year, and only come in, say, a week before calving. We keep them tied up after calving, with their calves behind them, the calves going out daily in a sunny court. About the first of May all get to grass. We take up the calves about the last week of October, and feed the bulls and such females as we decide to exhibit. Stock bulls we keep out all the year unless in very bad weather.

"Tuberculosis is unknown, and the veterinary surgeon seldom visits us. We keep the byres and boxes in a very sanitary condition. The cattle always have access to salt. In hot weather we spray them with dip to keep off the warble-fly."

A portrait of a noted Aberdeen-Angus bull is given in Plate 36, and of a characteristic cow of the breed in Plate 37.

GALLOWAY CATTLE.

Early History.—This breed took its name from the province of Galloway, which at the present time includes only the counties of Kirkcudbright and Wigtown—at one time known respectively as the Stewartry and the Shire of Galloway. At a very early date the term Galloway was applied to almost the whole southwest of Scotland lying south of the Clyde, and the only cattle then kept in that extensive area were of this polled breed. Indeed they were often termed "Carrick cattle," from the title of the southern division of the county of Ayr. Ortelius, the celebrated geographer, says: "In Carrick are cattle of large size, whose flesh is tender and sweet and juicy." In very ancient times Cumberland was under the same rule as Galloway, and over the northern counties of

England adjoining the Border Galloways were long the native breed.

Even in the area comprised in the present restricted province of Galloway the breed has been to a great extent supplanted by the Ayrshire dairy breed, and in the north of England the cosmopolitan Shorthorns have made a serious inroad on their territory.

There was a time in the distant past when sheep and not cattle were the principal live stock kept in Galloway. The breed of sheep peculiar to Galloway were celebrated for the fineness and superior quality of their wool. There is an adage of unknown antiquity—

"Kyle for a man,
Carrick for a cóo,
Cunningham for butter and cheese,
And Galloway for woo'."

Early Export to England.—What led to the very early improvement of the breed of Galloway cattle, and to a great increase in their numbers, was a demand for them which sprang up from Norfolk and other south-eastern counties of England. Before this southern trade for lean cattle developed there was little demand for beef from a province so far removed from any great centre of population. This outlet for the native cattle had been opened up by the middle of the seventeenth century; for the Rev. Andrew Symson, Episcopal minister at Kirkinner in Wigtownshire, in his work entitled *A Large Description of Galloway*, published in 1682, states that "the bestials are vented in England." He also mentions that Sir David Dunbar of Baldone kept in his park, extending to about 2½ miles in length, both summer and winter, about 1000 head of Galloways of different ages, and that he was in the habit of selling from eighteen to twenty score of the four-year-olds annually to dealers who took them to the English fairs. This trade in lean cattle led to a great increase in the breed, for through it breeders received large sums — a new experience, compared with the times when little money was received for that class of live stock from any outside quarter. It is said that there was an old proverb in Galloway that a good farmer would rather kill his son than a calf, which is a strong form of expressing the value which those engaged in the cattle industry put on their bovine stock.

This trade had become so large a century ago that from 20,000 to 30,000 three- and four-year-old Galloways were annually sent in late summer and in autumn from Dumfriesshire and Galloway to England—principally to the counties of Norfolk and Suffolk. They were taken on foot in droves, iron plates being put on the hoofs of such as proved tenderfooted during the long journey. They were finished on the rich pastures in these counties, and disposed of in the London market.

The Norfolk purchasers, tiring of paying so much money to Scottish farmers for lean cattle, adopted the plan of extending their own breeding herds; and as they wished to have them after the type of the Galloways, they took South Galloway bulls of a colour similar to that of their own native red polled cattle. In this way the present excellent breed of Norfolk Red Polled cattle claim descent from the Galloways on the one side.

While this extensive and lucrative trade led to a great increase in breeding in the south-west of Scotland, it also gave a powerful stimulus to the improvement of the breed. In fact, the Galloway was among the first, if not the very first, breed which was actively and systematically improved in Great Britain. The quickened demand and the greatly enhanced prices naturally induced the breeders to strive energetically to supply their southern customers with an improved type of beast which would respond to the richer and more generous keep they got in the south.

Origin.—The origin of the breed is lost in the mists of antiquity. But no authority of any weight has ever thrown a doubt on the claim that it is a pure breed, and that the improvement was not brought about by the introduction of alien blood from any quarter. Aiton, in his *View of Ayrshire*, written for the Board of Agriculture in 1810, says that "the breed was brought to its present improved state by the unremitting attention of the inhabitants in breeding from the best and handsomest of both sexes, and by feeding and management."

Improvement of the Breed.

Early Improvers.—No man stands out conspicuously among his fellows as having been chiefly instrumental in improving the Galloways at the early period of their history. Smith, in his *Survey of Galloway*, written in 1810, says: "Among Galloway farmers have arisen no enthusiasts in the profession, none who have studied it scientifically, or dedicated their talents almost exclusively to this one object. No Bakewells, no Culleys, no Collings have yet appeared in Galloway, who with a skill, the result of long study and experience, have united sufficient capital, and by the success of their experiments have made great fortunes and transmitted their names to the most distant parts of the kingdom."

That the production of the same ideal

type of Galloway was aimed at a century ago as at the present day is proved by comparing the points or characteristics of a typical animal of the breed given in Aiton's work, published in 1810, and the statement of characteristics which was drawn up by the Council of the Galloway Cattle Society in 1883 and which is given below. It is somewhat remarkable that there is a very close resemblance between the two descriptions.

Later Improvement.—The improvement effected since the commencement of the second quarter of the nineteenth century has been great, and it was the result of much enterprise and skill. Landowners and tenant-farmers vied with each other in this commendable work, and the latter received great encouragement and assistance from the former. In many instances on both sides of the Border proprietors purchased the best bulls which could be got and gave the use of them to their tenants. Sir James Graham, Bart., of Netherby, the celebrated statesman, had a novel but influential method of encouraging and assisting his tenants in their efforts after improvement. Instead of money or medals, bull-calves from his own very superior select herd were given as prizes to the tenant who showed the best lot of five yearling Galloways and as many two years old, the choice of the prizes in kind being given to the winners according to their order in the prize list. This was recognising past and contributing to future success in an admirable manner.

There is one man who stands out as having bred a number of bulls by one sire from which are descended almost all the best Galloways in the Herd-Book—namely, Mr George Graham, a tenant-farmer at Riggfoot in Cumberland, who has been called by "The Druid" in *Field and Farm* the "Black Booth of Cumberland and the Border Counties," from his having done for Galloways what Booth did for Shorthorns. The sire above alluded to was "Cumberland Willie" 160, bred by Mr Sproat, Borness, in Kirkcudbright. There were bull sales by auction established at Lockerbie and Castle-Douglas at the middle of last century which were the means of diffusing the best blood in all districts where pure Galloways were bred. Males of the choicest lineage and of the greatest individual merit were entered for these sales, and the introduction of railways provided a ready means of getting them conveyed to their respective destinations.

About this time the rapid extension of dairy farming and the great increase in Ayrshire cattle threatened, if not to completely supplant the breed in Galloway, at all events to restrict its numbers as well as to endanger its purity. What has been termed "a dairy wave" swept over the south-west of Scotland, to the detriment in various ways of the native polled breed.

Herd-Books.

The improvement of the breed has been greatly promoted by the establishment of the *Galloway Herd-Book*. From the outset the editor of the Herd-Book has been the Very Rev. John Gillespie, LL.D., Mouswald, to whom we are indebted for information on the breed, and who has rendered to its breeders services of the highest value. The first four volumes of the *Polled Herd-Book*, published by Dr Ramsay of Banff, included pedigrees of both Aberdeen-Angus and Galloway cattle. But in 1877 a Galloway Cattle Society was established which purchased the copyright of the Galloway portion of the *Polled Herd-Book* and published it as the first volume of the *Galloway Herd-Book*, twenty-eight volumes of which had been issued in 1908. About 20,000 females and one-half of that number of males have been registered in it.

Owing to a misunderstanding a substantial section of breeders in the north of England hived off from the parent society, and, joined by a number of breeders of pure-bred Galloway cattle, who had not registered their animals, formed a new organisation called the English Galloway Cattle Society. They had issued four volumes when, in 1908, negotiations took place between the two societies, which resulted in each of these being dissolved and a new body, called the Galloway Cattle Society of Great Britain and Ireland, being formed and registered under the Companies Acts. This new organisation includes in its membership breeders in all parts of the United Kingdom, and it bids fair to conduce to the extension and prosperity of the breed.

Galloways have been exported in large numbers to North America, and in the *American Galloway Herd-Book* there have been registered at least as many of the breed as in the Herd-Book of Great Britain and Ireland.

Characteristics of Galloways.

Milking Properties. — It is not claimed for Galloway cows that in general they are deep milkers, although there have always been individuals which have been good at filling the pail. Their milk, however, ranks very high in respect of richness in butter-fat.

Galloway Beef. — It is as beef-producers that Galloway cattle are most highly esteemed. The quality of Galloway beef is exceptionally high. This fact has long been acknowledged, but it has been strikingly demonstrated in connection with the carcase competitions at the Smithfield Fat Stock Show. For years after these carcase classes were instituted the Galloways regularly, year by year, carried off the lion's share of the prizes against all other breeds. "The Druid," the well-known H. H. Dixon, author of the Royal Agricultural Society of England's Prize Essay on Shorthorns, published in 1865, says, "There is no better or finer mottled beef in the world than the Galloway and the Angus, and so the Smithfield prices show." Mr William M'Combie, the celebrated Aberdeen-Angus breeder, testifies that "there is no other breed worth more by the pound weight than a first-class Galloway."

A Natural Breed. — Galloways arrive at maturity at different ages, according to the way they are kept when young. They are essentially a natural breed, and have been kept as such, never having been pampered in any way. In the lowlands they come to maturity early, though it is not claimed for them that in an exceptional degree they are early maturers. In the uplands, where many of them are bred and reared, the climate is cold, the fare scanty, and little or no artificial food is given; the progress they make is, as might be expected, not rapid, although when Galloways so reared are taken to the lowlands they come away amazingly, after being put on more generous keep.

Weights. — Where there is so much diversity in the way they are kept and fed, only an approximation can be made of the average weight of Galloway cattle at different ages. The following may be taken as a fair estimate of the live- and dead-weights respectively of good well-fed cattle of this breed:—

Age.	Live-weight.	Dead-weight.
1 year 3 months, . .	900 lb.	540 lb.
2 years 3 months, .	1400 „	840 „

But far heavier weights are reached where the diet has been fairly generous all along, and where an effort has been made to force forward individual animals. At the Smithfield Fat Stock Show in 1883, a pure-bred Galloway steer, when 2 years 10 months 3 weeks old, weighed 19 cwt. 0 qrs. 20 lb.—that is, he turned the scales at 2148 lb. when 1055 days old, which makes an average daily increase of 2 lb. in live-weight.

At the Smithfield Show in 1908, a Galloway steer 1 year and 9 months old gave a live-weight of 11 cwt. 2 qrs. 18 lb., and a steer 2 years and 11 months old a live-weight of 15 cwt.

A Hornless Breed. — The Galloway has always been a hornless breed. If a member of the breed shows the slightest trace of a horn or even a scur, there is reason to suspect its purity. The prepotency of the breed is remarkable when crossed with other breeds, but in no respect so much so as in the matter of obliterating horns. Even when mated with the majestic horned West Highland variety of cattle, it is very rarely indeed that the produce has any trace of horns, and certainly it is no mean achievement to get quit of any trace of such horns as it were at one single stroke.

Hardiness. — With the exception of the shaggy picturesque West Highlander, the Galloway is admitted on all hands to be the most hardy among British breeds of cattle, and the difference between the two breeds in the possession of this characteristic is very slight, if it exist at all. This outstanding quality is highly prized, and is sedulously sought to be preserved. For this end the class of skin and coat is regarded as of no little importance. A moderately thick but mellow skin is preferred, and a typical Galloway should have two coats of hair—an outer coat and an inner— the former moderately long, but soft and not curly, and the under coat should be thick, mossy, or woolly. It is the latter which is the more valuable in retaining the heat and keeping out the cold. The manner in which the cattle are reared conduces to their exceptional hardiness. The young ones generally pass the winter in the open air. The Druid says, "The sky and the hills of the glen are their only winter shelter, and however deep the snow may be they are kept out in the field." As the same authority puts it, "Unsheltered bullocks come to hand quicker in the spring than if they have the shed option." It is claimed for Galloways that, as a result of their being kept so much in the open air, they are in a special degree free from tuberculosis. A few years ago 80 were exported to the United States in one lot, and when the tuberculin test was applied to them by the Republic's veterinary inspector, every one of them passed the ordeal successfully.

Colour.—Until about a century ago there was much variety in the colour of Galloways. While the great majority were then black, some were brindled and dun, while a few were belted—that is, white round the middle, as if a white sheet had been fastened round them. During the last one hundred years almost all of them have been black—those of that colour being reckoned the most hardy. "Black and all black" is what is insisted on, but a very few belted and dun ones are still to be met with.

For Crossing.—Galloways have long enjoyed the highest reputation for crossing with other breeds. Their remarkable prepotency makes them valuable for this purpose, and while crosses bred from them are superior beef animals, they have the invaluable quality of hardiness to an extent which is a strong recommendation of them in this severe and variable climate. Pure Galloways have been crossed with Ayrshires, Herefords, and representatives of other breeds with success. In the south-west of Scotland Galloway bulls are mated extensively with Ayrshire cows in the dairy herds, and the produce are well thought of for the production of both beef and milk.

Blue-greys.

One of the most fashionable and highly prized class of beef cattle in this country is a first cross between the Galloway and the Shorthorn—these being widely and favourably known under the name of "blue-greys." They have that appearance in respect of colour from the coat having an almost equal admixture over the entire frame of black and white hairs. The districts where these are most extensively bred are the northern counties of England—especially Cumberland, Northumberland, and Westmoreland. They are larger in frame, come very early to maturity, and their beef is as choice as any put on the market. Some breeders mate the Shorthorn bull with the Galloway cow, while others follow the plan of using the Galloway bull and the Shorthorn cow. It is impossible to say which of these systems of mating produces the better animal.

The use of the Galloway cow is preferred by many on the well-founded ground that she can be kept decidedly more cheaply than the Shorthorn, and indeed the pure black female will thrive on poor land and in high altitudes where the more tender and dainty-feeding Shorthorn might experience difficulty in living. White Shorthorn bulls are chosen, preferably those of a white family, because they leave produce of more uniform colour than where the sire is a coloured Shorthorn. Blue-greys are almost invariably hornless in whatever way they are bred.

Many specimens of this cross have been prominent prize-takers at fat stock shows. In 1892 a steer out of a Galloway cow by a Shorthorn bull was supreme champion at Smithfield in the hands of Sir John Swinburne. At three years and five months old he weighed 2276 lb. In 1897 a steer by a Galloway bull out of a Shorthorn cow was champion at the same show after being champion at Norwich and Birmingham. At two years and ten months old he weighed 1800 lb. He was bred by Mr Parkin-Moore of Whitehall, Cumberland. In 1907 the champion at York Christmas Show, a blue-grey, turned the scales at 2310 lb., and was sold for £72.

Great auction sales of these "blue-greys" are held at Carlisle in the early summer and autumn, as many as 3000 head of them being sold at the two auction marts on a single day at each season of the year. The estimation in which they are held may be judged by the fact that they often realise up to 5s. per cwt. live-weight more than animals of equal weight of any other pure or cross breeding. By far the largest number of blue-greys are first crosses. Galloway bulls have been successfully used in Ireland for producing blue-greys.

Points of the Breed.

The following statement of the points of a typical animal of the Galloway breed was drawn up by the Council of the Galloway Cattle Society of Great Britain and Ireland in 1883 :—

Colour.—Black, with a brownish tinge.

Head.—Short and wide, with broad forehead and wide nostrils ; without the slightest symptoms of horns or scurs.

Eye.—Large and prominent.

Ear.—Moderate in length and broad, pointing forwards and upwards, with fringe of long hairs.

Neck.—Moderate in length, clean, and filling well into the shoulders ; the top in a line with the back in a female, and in a male naturally rising with age.

Body.—Deep, rounded, and symmetrical.

Shoulders.—Fine and straight, moderately wide above ; coarse shoulder-points and sharp or high shoulders are objectionable.

Breast.—Full and deep.

Back and Rump.—Straight.

Ribs.—Deep and well sprung.

Loin and Sirloin.—Well filled.

Hook Bones.—Not prominent.

Hind Quarters.—Long, moderately wide, and well filled.

Flank.—Deep and full.

Thighs.—Broad, straight, and well let down to hock ; rounded buttocks are very objectionable.

Legs.—Short and clean, with fine bone.

Tail.—Well set on, and moderately thick.

Skin.—Mellow, and moderately thick.

Hair.—Soft and wavy, with mossy undercoat ; wiry or curly hair is very objectionable.

MANAGEMENT IN GALLOWAY HERDS.

For the most part the system of management pursued in herds of Galloway cattle is natural and simple. The cattle are so hardy that they spend a great deal of their time in the open fields, even throughout the winter months.

Chapelton Herd.

The following system prevails in the choice herd of Messrs Biggar & Sons, at Chapelton, Dalbeattie :—

Calves.—The calves are dropped as soon after 1st December as can be secured. Each calf is put to its dam three times a - day until grass time. Calves then go out to pasture with their dams, where they remain until September, when they are weaned. After being weaned the calves get a mixture of about 2 lb. of oats, maize-meal, and linseed-cake. This is increased later on. The calves are wintered out in fields.

Cows.—Cows lie outside until calving time. After calving they get roots and fodder—3 lb. of mixed oats, bran, and bean-meal, with chaff— until the grass comes. After the grass is sufficiently forward they get no artificial food.

The heifers are never in a house. In summer they have to depend on the pastures alone. In winter they get from 3 to 4 lb. of concentrated food, with a few roots and hay during the first year, and straw afterwards. Heifers are put to the bull at two years old.

Bulls. — Young bulls, after being weaned in September, are put on to clover-grass till about the 1st of November, getting 2 lb. per day of cakes and meals. After 1st November they are shifted on to old pasture, and the artificial food is gradually increased to 4 lb. per day, with, in addition, roots and a little hay.

The stock bulls run with the cows in summer. In winter they go out and in, getting cake, bruised oats, maize, and bran, with a liberal supply of roots and hay.

Messrs Biggar do not believe in pampering their cattle, and keep them out of doors as much as possible. They find that if they look after the young stock pretty well during the first year (after weaning), the animals can look after themselves thereafter. No yeld cow or heifer in the house ever gets any concentrated food.

Castlemilk Herd.

In Sir Robert Buchanan-Jardine's herd at Castlemilk calves are dropped from December to April. It is found, however, that those dropped in February and March generally do best.

Calves.—The cows are allowed to calve in a box, and the calf is left with the dam for a week. Afterwards the calves are taken from their dams and led out to suckle three times a-day. When the calves get "the cud" (at about three weeks or a month after birth) they receive a small quantity of hay with an allowance of pure linseed-cake. The cake is broken very small, and given in a trough immediately after the calf has finished sucking. This prevents them sucking each others' ears, &c.

Cows.—Cows that are extra good milkers suckle two calves. Cross-bred calves are got for this purpose, the cow's own calf being put on first to receive the largest share. In May the cross calves are weaned, and the cow and the pure-bred calf are turned out to pasture, and are allowed to run together until August.

When the cows go dry they are fed on straw and turnips. After calving they receive an allowance of ground oats, bean-meal, and bran, with roots and meadow-hay. As the cows suckle their calves there is no record of the yield of milk which they give. One, having lost her calf, was milked by hand, and gave 18 quarts daily. This, however, was exceptional, and cannot be taken as an average yield for the breed. Probably about 13 quarts may be set down as a fair average when the cows are in full milk.

Heifers.—Heifers after being weaned are kept all winter in a small field, with an open shed for shelter, and are fed with hay and turnips and a daily allowance of 2 lb. each of linseed-cake. During the second winter they lie outside with no shelter, and are fed on hay and turnips alone. Heifers are put to the bull at two years old.

Bulls.—Young bulls have the same treatment as heifers when suckling. During the first winter they are kept in field, with hedge or plantation shelter. The 2 lb. of cake allowed to the heifers is supplemented by from 2 to 4 lb. of meal (bean and Indian in equal parts). Young bulls are generally sold at from 12 to 15 months old. Any kept over this age receive the same treatment as stock bulls.

From the beginning of November, or a month before service is expected to begin, stock bulls are allowed about 4 lb. of bruised oats daily. After the season is finished they get grass during summer and straw and roots in winter.

Broomfield Herd.

In Mr F. N. M. Gourlay's herd at Broomfield, Moniaive, the undernoted system is followed :—

Calves.—The calves are dropped in January and February. All calves are suckled by their dams morning and evening until the grass comes, which is generally (in this neighbourhood) about the middle of May, when they are turned out with their dams. All calves get meadow-hay and about ¾ lb. of small linseed-cake daily. While in the byres the calves are taken to the cows on halters.

Heifer-calves when weaned are put on meadow stubble, or young grass if available, and given 1 lb. of linseed-cake. When grass fails they are put into a well-sheltered field and wintered there on good bog-hay, cut swedes, and linseed-cake. When grass comes the heifers run on the hill among the sheep until November, and are again wintered out on hay and roots. Heifers are put to the bull at two years old.

Bull-calves are treated in the same manner as heifer-calves until about the November term, when the allowance of concentrated food is gradually increased. Crushed oats, bran, and Indian meal mixed with cut hay, are given, in addition to linseed-cake, cut swedes, and bog-hay. Stock bulls run out in a quiet, well-sheltered field, and are given hay and roots in winter. No cake or meals of any description are given.

Cows.—The cows run along with their calves on hill-land from the middle of May until October, when the calves are weaned. The cows are housed at night about the middle of November, and run out in a rough field every day, except for a week after calving. Good meadow-hay and a few turnips are given, but no meals or cakes except in special circumstances.

Preparation of Animals for Shows.—Animals for exhibition are generously fed, great care being taken never to surfeit. Special attention is paid to punctuality in feeding. All young stock are washed as often as may be necessary to keep the skin clean and the hair in good order, and are regularly haltered and led on these occasions. Cows are served before being turned out to grass with their calves, as if not settled before leaving the winter quarters they are very apt to miss the bull altogether.

Mr Gourlay keeps his Galloways on a sheep farm where he has good meadows, but practically no arable land.

A portrait of a handsome two-year-old Galloway bull is given in Plate 38, and a portrait of a typical Galloway cow in Plate 39.

RED POLLED CATTLE.

The Red Poll breed of cattle is native to East Anglia. The counties in which the breed was cradled are Norfolk and Suffolk.

Origin.—The cattle of this breed bear a close resemblance to the polled cattle of Scotland, and from the fact that in former times Scottish cattle were in large numbers transported to Norfolk for fattening, it is assumed that this likeness in form arises in some part from kinship in blood. Be this as it may, the Red Polled breed can be traced as a distinct and well-defined variety far back into the eighteenth century. In his *Review of Norfolk*, published in 1782, Marshall states that the native cattle of the county were "a small hardy thriving race, fattening as freely at 3 years old as cattle in general do at 4 or 5. They are small-boned, short-legged, round-barrelled, well-loined, the favourite colour a blood-red, with a white and mottled face."

Writing twelve years later, Arthur Young says the Suffolk breed of cattle "is universally polled—that is, without horns; the size small, few rise when fattened to above 50 stones (14 lb.); the milk-veins remarkably large; cows upon good land give a great quantity of rich milk."

The Improved Red Poll.

The improvement of this breed may be said to date from the year 1846, when the Norfolk and Suffolk types became merged. In that year the East Norfolk Agricultural Association established separate classes for Norfolk Polled cattle. Descendants of the winning animals at that show, exhibited by Mr G. B. George of Eaton and Mr T. Edwards of Hatton, were registered in the first volume of the *Red Polled Herd-Book* in 1874. The amalgamation of the Eastern and Western Division Societies in Norfolk and Suffolk gave a strong impetus to the improvement of Red Polled stock. Then in 1847 Mr T. Crisp of Butley Abbey, Suffolk, won at the Norfolk Show with his two bulls. The struggle between the two counties continued with varying success.

In 1868 the late Mr Clare Sewell Read, M.P., before the British Association at Norwich, declared that "as a set-off against the loss of the Devons we have to commemorate a grand revival of the Polled Norfolks as a numerous and distinct breed; ... horns and slugs are studiously avoided, and milking properties well cared for. They possess a uniformity of character, style, and make that would do credit to many of our established herds."

In July 1862, Mr Ellis at an agricultural meeting declared that "there is much in your native breed which is deserving of your notice, and which your forefathers knew was valuable. . . . I have never heard in Norfolk of the existence of a herd-book of stock; . . . there is a great deal in a herd-book. . . . I can only express my astonishment that as you have animals of such a class and of so good a stock you have not done more."

Establishing a Herd-Book.—Eleven years after Mr Ellis urged the establishment of a herd-book, a meeting was held at Norwich and a Society formed. The late Mr C. S. Read was president. It is a fact worthy of note that those responsible for the Herd-Book instituted the system of recording tribes. Thus the "A" group consisted of Elmham stock. Here the cows known to be of the old Elmham stock were registered ; secondly, the cows for a long period in the possession of the tenantry ; and, thirdly, recent additions. Tribes were thus associated with groups, and the system is still in operation.

Standard Description.

What is a Red Poll ? The answer to this question was settled when the Herd-Book was instituted. The standard description is as follows :—

ESSENTIALS.

Colour.—Red. The tip of the tail and the udder may be white. The extension of the white of the udder a few inches along the inside of the flank, or a small white spot or mark on the under part of the belly by the milk-veins, shall not be held to disqualify an animal whose sire and dam form part of an established herd of the breed or answer all the other essentials.

Form.—There should be no horns, slugs, or abortive horns.

POINTS OF A SUPERIOR ANIMAL.

Colour.—A deep red, with udder of the same colour, but the tip of the tail may be white. Nose not dark or cloudy.

Form.—A neat head and throat. A full eye. A tuft or crest of hair should hang over the forehead. The frontal bones should begin to contract a little above the eyes, and should terminate in a comparatively narrow prominence at the summit of the head.

This interesting description has the merit of brevity and terseness. But that, in one sense, may be regarded as its weakness. It may be advisable, therefore, to supplement it with a pen-picture of a modern Red Poll.

The Modern Types.—It is necessary to refer in the plural to the types of Red Poll. A breed which has won its way largely by reason of its milking qualities must necessarily be also a good feeder if it is to find support amongst East Anglian farmers. As might be expected, the Norfolk—or larger type of Poll—is the class of animal we find in the open courts laying on flesh for Christmas markets. Norfolk is the paradise of the feeder, there being a virtue in Norfolk roots

unequalled by the produce of any other county out of Scotland. So much so, indeed, is this the case that animals can grow fat on roots and hay.

The larger type of Red Poll is usually the showyard favourite. It is hardly necessary to remark that when a milking type meets a beef-producing type in the show-ring the odds are in favour of flesh. A big, well-grown, level-fleshed animal is the Norfolk Poll. Occasionally we find a lack of sweetness in the females, but a typical Norfolk cow may be thus described : A neat head with a befringed poll, which distinguishes it from the Aberdeen-Angus type, is well set on a clean-cut neck. The eye is prominent but not bold. The muzzle broad and free from specks. The shoulders should be well laid, and not pointed. The dewlap is square, but the width of breeds such as the Shorthorn is absent. The ribs spring well from the back, and carry a good covering of flesh, a large proportion of lean to fat. The back is level, the pin-hooks are smoothed over with flesh in the male, but prominent in the female. The tail should be moderately thick, and fall at right angles to the back. The hind quarters are not square, but slightly rounded. The underline is lengthy, and fills the hand at the flanks. The vessel in the cow is not exceptionally capacious in appearance, but the teats are well placed and large.

In the milking or Suffolk type there is less flesh, smaller stature, and a larger vessel.

Colour.—The colour should be an attractive red not too dark, without suspicion of yellow, and not too bright. The yellow shade is very insidious, and hard to breed out. It is frequently accompanied by white markings, which are a distinct objection. It is an old fancy, probably founded on fable, that cows of a yellow shade are specially good milkers. The truth of the statement is not borne out by observation.

Red Polls in the Showyard.—The showyard has a tendency to run a dual purpose breed to flesh. This tendency is sometimes observed in the Red Poll. There can be no doubt that the fattening qualities have been greatly improved, but the primary aim of a Red Poll is to produce milk. That, at all events, is the chief reason why the breed has found a home in counties far removed from those of its birth.

Locality. — The breed flourishes in Norfolk and Suffolk. There are also herds in Essex and the neighbouring county of Herts. In Shropshire there is one very thriving herd, while in far-away Ireland there is a colony of admirers. The foundation of the Irish demand was laid about the middle of the nineteenth century. Lord Dartrey was one of the pioneer importers into Co. Monaghan as far back as 1861. Animals have also been sent into Wales, and there is a considerable export trade to North and South America.

Weights.—The breed has for many years been classified at Smithfield and East Anglian fat stock shows. As indicative of the weights to which good bullocks will grow, it may be stated that a prize steer, aged 1 year 11 months 2 weeks, scaled alive 12 cwt. 1 qr. 16 lb., which would be a very good average for prize fat stock. A two-year-old steer in 1908 reached a live-weight of 16 cwt. 2 qrs. 23 lb. at 2 years 10½ months old,—rather a greater weight than the average, which may be stated at about 1 cwt. less.

Milk Yields.

Illustrative of the excellent milk-yielding capacity of the Red Poll, reference may be made to the very complete records which are kept in Lord Rothschild's herd at Tring Park. In 1907 there were 40 cows in the herd throughout the whole year. These produced a total of 262,859 lb. of milk, averaging 6571½ lb. per cow. The highest individual yield was obtained from "Clarissa," whose 12,005 lb. was spread over 303 days, making the very high average of 39.61 lb. daily while in lactation. The best yield of the first-calf cows drafted into the herd was obtained from "Parody," calved on March 5, 1904. She gave 7150 lb. in 332 days,—a remarkably good result for a three-year-old.

In the Hon. A. E. Fellowes' herd at Honingham, near Norwich, the average for twelve cows and heifers, averaging 284 days in milk, was 6300 lb., one cow giving 11,833 lb. in 329 days.

The average milk yields recorded in Sir Walter O. Corbet's small herd at Acton Reynold, near Shrewsbury, are as follows :—

Year.	Number of cows.	Average yield in lb.
1903	9	6434.33
1904	9	7236.72
1905	12	7753.45
1906	8	8073.75
1907	9	7363.77

MANAGEMENT OF RED POLLS.

Feeding occupies a prominent place in the management of a Red Polled herd of dairy cattle. Each owner has his own method.

Eldo House Herd.

Mr A. H. Cobbald of Eldo House, Bury St Edmunds, who has kept upwards of sixty cows for several years, has tried numerous rations with varying results. He believes in the cheapest kind of corn ground into fine meal, allowing from one to two gallons daily, with chaffed hay and straw mixed, and about half a bushel of pulped mangels. A shredder, preferably to a mincer or pulper, is used in preparing the food. In addition to this, the cows get about 10 lb. each daily of long hay in racks over their mangers, and an iron pan filled with clean water is provided between two cows.

Cows.—From 1st May to 1st November the cows lie out in the open, and grass is the chief food. Only about half a gallon of meal mixed with chaffed hay is given daily while they are being milked. From 1st November to 1st May the cows are tied up in a shed, with stalls about 7 feet wide to hold two cows. The cows are turned out for exercise every day except on very cold wet days. During these months their food is increased to the rations first mentioned.

The Honingham Herd.

In the herd of the Hon. A. E. Fellowes, the cows are turned out to pasture in the ordinary course in the summer months. They have lucerne in the open yards when they come up to be milked. During the winter months they are turned out upon pastures daily for some time to exercise, no matter how cold. If wet, or snow is on the ground, they go out for a little time. The remainder of the time they are kept in partially covered yards made of corrugated iron. Their food chiefly consists of chopped oat-straw and hay mixed, with a little long hay. At milking they have one bushel of kohl-rabi, with bran and oats, and a small quantity of cake daily when in full milk.

Calves.—The calves are taken off the cows when a week old and placed in calf pens. They are then fed on new and separated milk to which linseed and oatmeal are added. When old enough they get crushed oats, bran, linseed, and cake in small quantities. In particular cases the calves are kept upon the cows for some weeks to bring them up fit for exhibition. When the calves are able to take their food freely they are turned into open yards during winter and summer, it not being the custom to turn them out to pasture until they are from ten to twelve months old. Calves are not allowed to remain out in damp, cold nights, nor in the heat of the day if the flies are troublesome. At the same time they receive hay, lucerne, and artificial food.

Heifers.—The heifers are not put into service at Honingham until they are about a year and ten months old. They are generally allowed some time before being put to service after the first calf, as otherwise they would not develop sufficiently.

The steers are readily bought by butchers, who give 6d. more per stone for them than for other local breeds. Their weights as dressed carcases are surprising, having little offal. They can be kept in courts without danger, as they are peaceable feeders.

Acton Reynold Herd.

In Sir Walter Corbet's herd at Acton Reynold, near Shrewsbury, the plan of putting two calves on a foster mother is adopted. Mr Reginald Astley, the agent, says: "The foster mothers generally cost about £14 or £15. After the calf that is on them when bought has been sold, they are always sent to some bull in the neighbourhood, and they sell very well when they calve." The ordinary cows in the Acton Reynold herd calve in January, February, or March.

Young Stock.—In the following September the training and education of the young show stock commences. Only the best bull-calves are kept, the others being made into steers and fed. The heifers not intended for exhibition are kept in roomy yards where they get exercise, and are fed on hay *ad lib.*, two feeds of roots, and about 1½ lb. to 2 lb. of cake and meal daily. In summer they are turned out to pasture, remaining there the following winter, and getting from 3 to 4 lb. of cake and meal daily, three feeds of roots, and hay *ad lib.*

The heifers go to the bull at from sixteen to eighteen months old, but the exhibition heifers go to the bull in November or December, so that they may not be too heavy in calf when exhibited. Mr Astley says: "It is most important that all heifers should be got in calf before being exhibited as two-year-olds, as otherwise there is a very great probability that after being fed up, which is absolutely necessary if they are to have a proper chance of winning, they will not hold to the bull and be ruined for breeding purposes."

Combination of Beef and Milk.—With regard to the combination of beef and milk, an instance is furnished by an Acton Reynold cow, which averaged for six years 10,039 lb. of milk, and was the dam of two Smithfield Breed cup winners, the weights and ages of which were—

Heifer, 2 years 5 months 27 days; 16 cwt. 2 qrs. 14 lb.
Steer, 2 years 5 months 9 days; 15 cwt. 3 qrs. 18 lb.

Mr Astley believes that Red Polled cattle will, better than any other breed, fulfil the condition of producing both milk and beef in the same animal.

A portrait of a noted Red Polled cow is given in Plate 34.

SHORTHORN CATTLE.

It is acknowledged by all that the Shorthorn has abundantly earned the right to the premier position amongst British breeds of cattle. It is by far the most numerous, as it is the most widely diffused. . More wealth is bound up in it than in any other variety of the bovine race. In the development of the live-stock industry of the United Kingdom it has played a great part, far exceeding that of any other distinct class of animals. And the breed has done more than de-velop wealth at home. It has gone in vast numbers to foreign countries, bring-ing in exchange foreign gold to British farmers, and creating wealth, and pro-moting agricultural prosperity wherever it has been given a habitation. The breed which has done all this—and is as busy at work as ever, widening its field of operations from time to time—well merits a word of homage from the live-stock historian.

Origin of the Breed. — Extremely little is known of the foundation ele-ments of the Shorthorn breed, and next to nothing of the moulding influences exerted by breeders during the seven-teenth century. Even for the period between 1700 and 1750 there does not exist much of a practically useful char-acter in the form of breeding records. The breed was probably in more or less complete possession of Durham and North Yorkshire for two or three hun-dred years before it began to attract the attention of outsiders. Some writers have associated the early history of the breed with Holland, but there is now a general agreement that it is not of Dutch origin. Further, it is fairly well established that the occasional importa-tions of Dutch stock referred to by Culley, William Ellis, and others, had comparatively little influence on the Durham or Teeswater breed during the first forty or fifty years of the seven-teenth century. Later alloys of Gal-loway and Highland blood were rather incidentals than disturbers of the breed's course. The main elements were power-ful enough to assimilate such factors without betraying outward signs of the blending.

Mr James Cameron, to whom we are indebted for information regarding the breed, states that in the early decades of the seventeenth century the Teeswater cattle were mostly large-framed, yellow-ish-red, red-and-white, and white stock, odd specimens being of a "mealy-roan" hue. Old Northumberland traditions also had it that numbers of the cattle showed dark noses and patches of blue on the skin, such markings being no doubt due to previous crossings with the native black cattle of surrounding dis-tricts. Persistence of "unfashionable" noses and a dull blue slatey-roan may thus be accounted for, but to what ex-tent the occasional blue or blackish tips in horns are due to very old out-crosses it is impossible to say.

With reference to the blood-red colour which is now so much prized, there is no evidence to show that it was common in the early part of last century. It is to all intents a relatively modern evolution, the result of careful and persistent selec-tion.

Early Improvers.

Among Shorthorn improvers of the earlier part of the eighteenth century, high positions must be given to Smith-son of Stanwick; the brothers George and Matthew Culley of Winton; John Maynard of Eryholme; Waistell of Great Burdon; John Hunter of Hurworth—breeder of the remarkable bull "Hub-back" 319; Stephenson of Ketton; John Charge of Newton Morrell, well known as a friend of Bakewell; Jolly of Wor-sall; and Michael Jackson, who bred the sire of Maynard's cow, "Favourite." Those men, and large numbers of their contemporaries, were of untold benefit to the Shorthorn interest. They pre-pared admirable materials for the great breeders, the brothers Charles Colling of Ketton (1750-1836) and Robert Col-ling of Barmpton (1749-1820), both frank admirers of Bakewell, and ready appreciators of his selective and moulding methods,—Charles having, indeed, lived with Bakewell as a pupil.

The Brothers Colling. — With intuitive knowledge of animals while such were still "in the making," the brothers Colling purchased all round their own neighbourhood, and then proceeded to fix their ideal type by means of in-breeding. They were, at the same time, judicious advertisers of their own cattle. The "Durham Ox" of the one and the "White Heifer" of the other—both by "Favourite" 252—turned the attention of a larger public to the merits of the improved Shorthorn. In short course, the Collings had high prices for bulls and cows. On the wonderful cattle bred by the brothers or owned by them, the best being full of "Hubback" blood, there is no need to dilate here. The bull "Foljambe" 263, grandson of "Hubback" and sire of "Phœnix"; "Old Cherry," "Old Daisy," "Duchess" by "Favourite," and dam of Bates' "Duchess 1st"; "Red Rose," "Favourite," or "Lady Maynard," the 216 guineas "Lady," and many others, are easily called to remembrance by students of Shorthorn history.

Charles Colling's sale at Ketton in 1810 was the first great event of its order. At that disposal 29 cows and heifers averaged £140, 4s. 7d., while 18 bulls reached an average of £169, 8s. The marvel of the time was the sale of the light roan bull "Comet" 155, at 1000 guineas, to Messrs Wetherell, Trotter, Wright, & Charge. Eight years later Robert Colling had the astonishing average of £128, 9s. 10d. at Barmpton for 61 head, although agriculture was then in a depressed condition. Looking back, it is practically impossible for any student of Shorthorn affairs to over-estimate the importance of the work done by the brothers Colling.

Among the many gifted men who took up Shorthorn breeding at the end of the eighteenth or beginning of the nineteenth century were Christopher Mason of Chilton, Robertson of Ladykirk, Thomas Booth of Killerby and Warlaby, and Thomas Bates, whose name will always be associated with Kirklevington, which he purchased in 1811. For his foundation materials Mason went to Maynard and Charles Colling. On the male side he relied upon Colling bulls, and so great was his success as a breeder that the Booth family and Thomas Bates, while at the opening of what proved to be lifelong forms of animosity, agreed that Mason had blood to suit the two "rival houses."

Captain Barclay's Pioneer Work. —In the north of Scotland the pioneer breeder of Shorthorns was that remarkable man, Captain Barclay of Ury, who at the dispersion of the Chilton herd in 1829 acquired the grand three-year-old cow "Lady Sarah" for 150 guineas. She was then said to be in calf to Mason's "Monarch" 2324, and at Ury she produced a bull-calf, which was named "Barclay's Monarch" 4495. She bred, further, the notable "Mahomed" 6170, "Pedestrian" 7321, and "Sovereign" 7539, and the females "Julia," "Cicely," and "Helen."

Booth Cattle. — Reverting to the Booth family, the steadfastness of purpose shown by that race of breeders for well over a hundred years is probably without parallel in the whole world of stock-breeding. Thomas Booth and his two sons, John and Richard Booth, were of one mind in regard to type, but the remarkable matter is that their tastes should hold such overpowering dominance over strong minds of the third generation. The Booths fixed on a type which in the main showed a pronounced tendency towards beef-production. Milking powers were cultivated to a reasonably full extent, and when capabilities at the pail came easily and naturally they were always welcomed. Still, the beef-carrying carcase was the family ideal.

In the time of John and Richard Booth it was wont to be said that the Warlaby, Killerby, and Studley cattle lacked gaiety and style. The representative bulls had frequently round, strong, forward-staring or slightly high-set horns, big curly heads, wide crops, very deep fore-quarters, arching ribs, and usually fairly long and deep hind-quarters, but they did not walk with the easy dash of animals showing something of the "racehorse shoulders" and less compact knitting of frame. To a very great extent the old criticism on the Booth cattle retained force until the end of the nineteenth century.

Bates Cattle.—Bates, with his more artistic nature, was captivated by style, while, on the practical side, his leanings were strongly towards milking powers. He was a great admirer of a beautiful head. His bulls, with their flat and generally well-set horns, broad foreheads, large staring eyes, nicely chiselled faces, expansive nostrils, long, clean, arching necks, high and rather narrow shoulders, and general length and "liberty" of frames, cut a dash while on parade. Opponents of the Bates' cult were not loth, as a rule, to note such defects as bare shoulder-blades and flat fore-ribs, nor did they hesitate at times, even during the life of the old man of Kirklevington, to hint that constitutions were in danger, and that milk was departing from one or two ultra-fashionable families.

The Booth and Bates' partisanships lasted for well over thirty years, and during the period of faction the Shorthorn breed lost many friends in the English tenant-farmer ranks. A complete break-down of the unfortunate petty divisions did not come until the early 'eighties of the past century. Before that time Lord Dunmore had two great sales and fortunate "escapes" with cattle, mainly of Kirklevington descent—first in 1875, when 30 cows and heifers averaged £576, 5s. 6d., and 9 bulls and bull-calves £992, 16s. 8d.; and, finally, in 1879, when 54 head realised £13,118, 14s., an average of £241, 14s. 3d. The sensation of the 1875 sale was the disposal of the two-year-old bull "Duke of Connaught" 33,604 to Lord Fitzhardinge at 4500 guineas. At the 1879 sale, "Duchess 114th," her yearling daughter, and her bull-calf, "Second Duke of Cornwall," made a total of £7507, 10s. "Duchess 117th" and "Duchess 114th" passed to Sir Henry Alsopp at 3200 guineas and 2700 guineas respectively.

Later Improvers.

While the Booth and Bates' fashions were running their course, work of great excellence on behalf of the Shorthorn was overtaken in England by such men as Sir Charles Knightley, Colonel Towneley, and Wilkinson of Lenton; in Ireland by W. T. Talbot Crosbie and others; and in Scotland by Captain Barclay—who never really experienced the bitterness of the contentions—and after him by Amos Cruickshank, Sittyton; Wm. Hay, Shethin; Sylvester Campbell, Kinellar; Wm. S. Marr, Uppermill; the Duthies, and others.

Cruickshank Shorthorns.—Amos Cruickshank, to whose memory world-wide homage is now paid, purchased his first heifer in Durham. That was in 1837. In the following year he went south to Nottingham, and returned home with about a dozen heifers. From that stage onwards for many years he and his brother Anthony were constantly on the outlook for good animals at reasonable prices. The first sires used were of Ury blood, and these were followed for about a quarter of a century by bulls of high repute from many herds, such as those of Torr, Wiley, Richard Chaloner, Colonel Towneley, Smith of West Rasen, Wilkinson of Lenton, Foljambe, Pawlett, Willis, Sir William Stirling Maxwell, and the Duke of Montrose. Looking backwards, the existing race of breeders are struck by the apparent want of system in the Sittyton selections. One is forced to the conclusion, however, that Amos Cruickshank was never really able during those years to reach his ideal. Booth blood preponderated in the sires which he selected, but his search was for a good animal. He paid little regard to pedigree.

A turning-point in the history of the Sittyton herd was reached at the fall of 1858. Cruickshank was in need of a young red bull for use during the following spring. He applied to Wilkinson of Lenton, and was strongly recommended to take "Lancaster Comet" 11,663, a fleshy short-legged roan over eight years old, and in-bred to the remarkably prepotent sire "Will Honeycomb." This "Comet" was not liked by some of Cruickshank's neighbours, on account of his long "Highland-looking" horns. After limited use, he left about a dozen calves, two of these being "Champion of England" 17,526, which was used in the herd for nearly twelve seasons, and "Moonshade" 18,419, which passed into the Inverquhomery herd. When the merits of "Champion of England" as a sire were clearly seen, his blood was gradually worked through the whole stock by means of sons, grandsons, and other descendants. This concentration of blood gave the Sittyton cattle great uniformity of character and singular impressiveness.

In the year 1889, when Amos Cruickshank was in his eighty-second year, the whole herd was purchased by Mr Robert Bruce for Messrs James Nelson & Sons, the aim being exportation to the Argentine. The great South American country, however, was then passing through financial trouble, and most of the cattle had to be disposed of privately in this country. Mr William Duthie, Collynie, and Mr J. Deane Willis, Bapton Manor, fortunately for the interests of the breed in general, secured a large number of the best animals.

Cruickshank Cattle in England.— In the 'seventies and 'eighties of the past century Amos Cruickshank had excellent customers for his best bulls and spare heifers in North America, but the prejudice against his cattle was still strong in England. The hiring of the Sittyton-bred "Field-Marshal" in 1884 for the Windsor herd was considered a rash step. From about 1890, however, a gradual change in favour of the Sittyton type began to set in all over the United Kingdom. South America also began to patronise Cruickshank cattle strongly, and since that time stock of Sittyton descent have gone everywhere in pure form, and have blended admirably with Booth, Bates, Knightley, and other strains. Many of the finest cattle to be seen in the English showyards of the present day are of Booth, Bates, or other old southern descent, with two or three Cruickshank top crosses. History is prone to repeat itself. The existing danger is that good cattle may be neglected because they are not quite in the fashion.

The Ideal Shorthorn.

The type of perfection in Shorthorns, as in other stock, has varied slightly from time to time. There is now a tendency towards the breeding of a rather smaller and more closely-knit Shorthorn than was common prior to the closing decade of the past century. In the main, also, the beefy type wins most prizes at the open shows, but sensible attention is paid to milking properties in heifers and cows.

In the ideal Shorthorn bull of to-day the horns should be flattish, with a wide space between the roots, rich in colour, and free from black or blue at the tips; the forehead should be broad, the eyes prominent and gentle, with expressive chiselling under them; the length from eyes to nostrils should be moderately short, and the nose should be perfectly free from black spots or even faint bluish stains. Free from throatiness, yet robust looking and with a fair crest, the neck should taper gradually into fairly wide well-covered shoulders and crops, and the brisket should perfectly fill the space between the fore legs. A broad chine or back, arching ribs, great heart girth, strong well-covered loins, neatly turned hooks or hips, long, smooth, and deep quarters, squarely set-on tail, straight hind legs, flat bone, mellow hide well clad with mossy hair, and jaunty easy carriage complete the picture.

In a female, more refinement of face, neck, and shoulders are of course looked for, and the hips are more on the square, but still they ought not to be unduly prominent. Some representative Cruickshank bulls were rather plain in horn, and although they had grandly covered backs, their quarters were relatively short or wanting in finished appearance. Their thick shoulders and general compactness of build were also to some extent against liberty of movement.

In most parts of the United Kingdom fairly strong efforts have been put forth to breed out light "washy" roans and gaudy reds-and-whites, because the exporting demand has been much set on rich roans and blood-reds. Representative herds are consequently a good deal darker in colour than they were a quarter of a century ago. In practice it is found advisable to make occasional use of white bulls for the purpose of preserving a balance of mellow roans.

Mr John Thornton's Ideal Shorthorn.—The typical characteristics of the breed were thus described by the late Mr John Thornton, the celebrated Shorthorn auctioneer:[1] "The breed is distinguished

[1] *Cattle, Sheep, and Pigs of Great Britain.* By John Coleman. Horace Cox, *Field* Office, Breams Buildings, London.

by its symmetrical proportions and by its great bulk on a comparatively small frame; the offal being very light and the limbs small and fine. The head is expressive, being rather broad across the forehead, tapering gracefully below the eyes to an open nostril and fine flesh-coloured muzzle. The eyes are bright, prominent, and of a particularly placid, sweet expression; the countenance being remarkably gentle. The horns (whence comes the name) are, by comparison with other breeds, unusually short. They spring well from the head with a graceful downward curl, and are of a creamy white or yellowish colour; the ears being fine, erect, and hairy. The neck should be moderately thick (muscular in the male), and set straight and well into the shoulders. These, when viewed in front, are wide, showing thickness through the heart; the breast coming well forward, and the fore legs standing short and wide apart. The back, among the higher-bred animals, is remarkably broad and flat; the ribs, barrel like, spring well out of it, and with little space between them and the hip-bones, which should be well covered with flesh. The hind quarters are long and well filled in, the tail being set square upon them; the thighs meet low down, forming the full and deep twist; the flank should be deep so as partially to cover the udder, which should be not too large, but placed forward, the teats being well formed and square set, and of a medium size; the hind legs should be very short, and stand wide and quite straight to the ground. The general appearance should show even outlines. The whole body is well covered with long, soft hair, there frequently being a fine undercoat; and this hair is of the most pleasing variety of colour, from a soft, creamy white, to a full, deep red. Occasionally the animal is red and white, the white being found principally on the forehead, underneath the belly, and a few spots on the hind quarters and legs; in another group the body is nearly white, with the neck and head partially covered with roan; whilst in a third type the entire body is most beautifully variegated, of a rich, deep purple° or plum-coloured hue. On touching the beef points, the skin is found to be soft and mellow, as if lying on a soft cushion.

In animals thin in condition a kind of inner skin is felt, which is the 'quality' or 'handling,' indicative of the great fattening propensities for which the breed is famous."

Attributes of the Breed.

Enough has already been said to indicate that the Shorthorn can claim attributes of the very highest order. It is universally acknowledged that in the production of beef and in general utility combined the Shorthorn is unsurpassed. It may be excelled by some other varieties in special aptitude for peculiar purposes or for certain limited districts; but for a combination of all the more useful properties of domestic cattle, as well as adaptability to varying conditions of soil, climate, and treatment, there is no other breed of cattle that can equal the Shorthorn.

Beef-Production.

It was perhaps most largely by its remarkable beef-producing properties that the Shorthorn gained its early fame. "From the very outset the improved Shorthorn took up a position of pre-eminence as a beef-producer, which it has ever since maintained. Its fame was won by its rapid feeding properties at a time when there was a keen struggle between various breeds to supply an improved type that would meet the growing requirements of the public. No doubt, at first size was the main consideration, though in sending round the country the 'Durham Ox' and 'The White Heifer that Travelled,' the object of the Collings was to arrest attention to these as specimens of what the breed was capable of accomplishing, rather than as the sort of animals which they wished farmers to keep and breed. Shorthorns of less imposing size and fatness were more suitable for ordinary purposes, but for years before and after the Collings the various breeds were recommended by the abnormal specimens they could produce. Very soon quickness of growth and ripening, reduction of waste, and finer bones and choicer quality were required, and the Shorthorns were found not only to supply these requisites themselves, but to stamp them on the inferior races with which they were crossed. In converting the herbage of the farm into wholesome nutritious food for the increasing population of the country the Shorthorn was unsurpassed, and when to this was added the good milking properties of the cows, which soon made up when dry into a thoroughly good carcass of beef, the claims of the breed received wide recognition."[1]

In the annals of the breed there are instances of great weights attained by individual animals. A twin heifer slaughtered at three years of age weighed in carcase 980 lb. A three-year-old ox, slaughtered off the pasture, yielded a dead-weight of 1330 lb. Many cows give from 1000 to 1200 lb. of carcase. For two-year-old Shorthorn steers dead-weights of from over 800 lb. are now by no means rare. It has become quite common to fatten off steers of the breed at from eighteen to twenty months old, and by that age they attain wonderful weights and show well-matured carcases.

The live-weights recorded in the shows of the Smithfield Club are worthy of note. Average live-weights for Shorthorns there are—steers under two years old, 1400 lb.; steers over two and under three years old, 1830 lb.; oxen over three years, 2250 lb.; heifers under three years, 1730 lb.; and cows over three years old, 1900 lb. The average daily gain in live-weight for Shorthorn steers under two years at the Smithfield Show has been about 1.93 lb., for steers over two and under three years 1.67 lb., and for heifers under three years 1.58 lb. At the Smithfield Show of 1908 twelve Shorthorn steers, whose average age was 22½ months, gave an average live-weight of 12 cwt. 2 qrs. 14 lb., the lightest being 10½ cwt. and the heaviest 14 cwt. 19 lb.

Shorthorns have taken creditable positions in the competitions at the Smithfield Show for carcases. An animal of the breed has yielded no less than 73.75 per cent of live-weight in carcase.

For Crossing Purposes.

Another outstanding attribute of the Shorthorn is its unequalled value for

[1] *History of Shorthorn Cattle.* Edited by James Sinclair. Vinton & Co., Limited.

crossing purposes. "No variety of cattle fits itself more easily and readily to varying conditions of life than the improved Shorthorn. This undoubtedly is one of the most valuable attributes of the breed. Without it Shorthorns would have made but little headway in foreign countries, where they are now doing good work. In both Scotland and Ireland they have thriven admirably,—nearly as well, indeed, in the cold dry climate of the north-east of Scotland, with close house winter feeding, as in the mild, moist climate of the south of Ireland, with daily field exercise all the year round. In both countries there are numerous pure-bred herds of high individual merit, a few of them ranking among the finest in the kingdom. The Aberdeenshire Shorthorn has attained a well-recognised type—somewhat deficient in high-class Shorthorn character perhaps, but, at the same time, broad, deep, well-fleshed, and thoroughly useful. Then as to well-bred and well-cared-for Irish Shorthorns, who has not been struck by their rich, soft, natural touch and beautiful, rank, glossy coats of hair, as well as by their attractive character generally?

"But while the breed reared in its purity has maintained a high character in these countries, it cannot be doubted that in crossing with other varieties of cattle it has achieved still more remarkable results. Shorthorns have been crossed freely with all the local races and sorts of cattle, and have everywhere and upon every sort effected marked improvement. In all that adds value to cattle, improvement has followed in the wake of the Shorthorn—in size, form, quality, rapidity of growth, and aptitude to fatten at an early age. Among the small, scraggy, old-fashioned Irish cows, Shorthorn bulls have produced results truly wonderful. Stock from an ordinary Irish cow and a good Shorthorn bull will, it is estimated, reach maturity at least a year sooner than unimproved cattle—at two and a half or three, instead of from three and a half to four years old. Moreover, the cross, besides being far superior in quality, will also show an increase in weight of from 1 to 1½ cwt. per head. It is certainly within the mark to place the increase in the value of one-year-old Irish cattle due to the use of Shorthorn bulls at from £2 to £3 a-head on an average. In many instances it has risen as high as £5, and in few cases has it failed to reach £2—that is, above the value of the corresponding class got by native or cross-bred bulls.

"In Scotland the experience of the breed has been equally satisfactory. The stock of native cows in Scotland are, as a rule, larger and finer than those of Ireland, and therefore the contrast between the native cattle and the improved crosses has generally been less marked in the former country than in the latter. In some parts of Scotland, however, where the native cattle were small and slow in growth, the transformation effected by Shorthorn bulls has been quite as remarkable as in Ireland." [1]

Milking Properties.

The milking properties of the Shorthorn are of a high order. Sure evidence of this is found in the great predominance of Shorthorn features in cross-bred dairy herds throughout the country. In many of the pure-bred herds large yields of milk are recorded, and this too from cows which produce fattening stock of the highest merit. "The late Mr E. C. Tisdall, of the Holland Park Dairy, Kensington, who long cultivated a Shorthorn dairy herd, reported, among the records of many years' experience, an average of 885 gallons apiece yielded by twenty-five cows of this breed during the ten or eleven months of the year when they were in milk, and ten selected cows had yielded as much as 1200 gallons apiece in the same time. The yield of butter by the Shorthorns has been exceeded by other breeds, but the returns of milk and butter together have not. There is a record of a cow having produced 1650 gallons of milk between May 20, 1888, and April 7, 1889, which is, of course, an exceptional quantity. The cows in the Duke of Westminster's dairy herd gave an average yield, in 1890, of 714 gallons each. Others report yields over the whole herd of 885 gallons, and single cows have given 1050 gallons

annually for several consecutive years. At the London Dairy Show, for ten years the milk produced by Shorthorns averaged 43.13 lb. per day, and the total solids showed a percentage of 12.87, of which 3.73 was fat, and 9.14 other solids. Taking a later period of five years the averages are—age 6 years and 1 month, days in milk 42.8, daily milk yields 49.2 lb., fat 3.91 per cent, solids other than fat 9.08 per cent, total solids 12.99 per cent. The breed standard of the British Dairy Farmers' Association for Shorthorns is 8500 lb. of milk, and pure butter fat per diem 1.25 lb.; and with respect to other fat, the Shorthorn has the same weight assigned as the Jersey and Guernsey, the Dutch being put at 1.00 lb.

"In recent years, careful records have been kept of the milk yields in a number of Shorthorn herds. Lord Rothschild's herd, at Tring Park, is a noteworthy example, the statistics being published annually by Mr Richardson Carr, the agent. Several cows in the herd have records of over 10,000 lb. of milk per annum. The average yield in a herd of thirty-eight cows for the year ending September 30, 1905, was 7031 lb. per annum. 'Decentia 24th,' 371 days in milk, gave 10,069 lb. For fifty-seven cows in the year ending September 29, 1906, the average was 6706 lb. per annum. 'Wild Queen 10th,' 364 days in milk, gave 10,044 lb. 'Darlington Cranford 3rd,' in the herd eight years, gave a total of 60,524 lb., or an average of 7565½ lb. per annum. 'Darlington Cranford 5th,' in the herd six years, gave a total of 59,921 lb., or an average of 9986⅚ lb. per annum. 'Lady Rosedale,' in the herd eight years, gave a total of 69,018 lb., or an average of 8627¼ lb. per annum." [2]

Shorthorn cows in the herd of Mr C. R. W. Adeane, Babraham, Cambridge, have for some years given an average of over 7500 lb. of milk per annum, one cow yielding 8507 in one year. In the herd of Bates Shorthorns owned by Mr George Taylor, Cranford, Middlesex, the average yield of several cows is from five to six gallons per day when in full

---

[1] From a Paper by James Macdonald in the *Jour. of the Roy. Agric. Soc. of Eng.*, 1883.

[2] *History of Shorthorn Cattle.* Edited by James Sinclair. Vinton & Co., Limited.

milk. A number of cows have exceeded 10,000 lb. in a year, and one reached a total of 12,320 lb.

Shorthorn Society and Herd-Books.

The interests of the breed are well looked after by the Shorthorn Society of Great Britain and Ireland. The society was founded in 1875, and in 1908 had over 1600 members. The *Shorthorn Herd-Book* was established in 1822 by George Coates, and the work still bears the name of its founder. Volume xxxvii. of Coates' Herd-Book, containing the lists of births for 1890, had 1834 entries for bulls and 3920 for cows with produce. Volume liii., published in 1907, and consequently containing a record of the births for 1906, shows a registry of 3800 bulls and 6760 cows with produce.

In the United States of America there is an enterprising Shorthorn Society which issues a Herd-Book for the breed. In the seventieth volume there are entries of 8299 bulls and 12,000 females. Yet there is no evidence to show that the States had even one Shorthorn previous to 1811. Canada's first importation was in 1832, when some animals of the breed were introduced from the States. Then the fourth volume of the Argentine Herd records 1084 bulls and 1173 female animals.

Exports of Shorthorns.

The trade in the exportation of Shorthorns continues to be large. From 1882 to 1890 the Shorthorn Society of Great Britain and Ireland issued certificates for the exportation of 3131 animals of the breed, while from 1891 to 1908 the certificates issued numbered close on 16,000.

For a quarter of a century the Argentine has been by far the best foreign customer for our high-class Shorthorns, but many fine animals have also been taken by Chili.

MANAGEMENT IN SHORTHORN HERDS.

No very hard and fast rules as to the management of Shorthorns can be laid down. Much depends on the district in which the herd is situated, on the object the particular owner has in view, and the outlet there may be for young stock. In Scotland, generally speaking, no one need attempt to raise Shorthorns successfully who has not comfortable buildings for winter. The same holds good to a considerable extent also in England and Ireland, although in the southern districts of both these countries it is possible to winter young cattle at any rate almost wholly in the open. Shorthorns also require a fairly liberal dietary all through the year. In most districts they will do quite well on grass alone during summer, especially where the calves suckle their dams, but in winter they must be liberally hand-fed, even when running on the pastures. In dairying districts where Shorthorns are used for the production of milk they have to be fed like ordinary dairy cows —on cake, bean-meal, or other material, in addition to grass, at least during the latter part of the grazing season.

North of Scotland Methods.

The northern counties of Scotland have achieved notable distinction in connection with Shorthorns. The management here is on rather special lines so far as other parts of the country are concerned, although it does not differ greatly from that of other classes of cattle kept in the same district. In Aberdeenshire, and the north of Scotland generally, it is necessary owing to the severity of the winter to house cattle for five if not six out of the twelve months of the year.

During this time the cows are tied up in byres and have everything brought to them. In former days it was quite a common thing for the animals to stand there right through the winter without once being turned out. But of late years, since the tuberculin test was discovered and the prevalence of tuberculosis has been more fully recognised, most breeders try to give their cows a turn out every day, if it is only into the yard. In justice to the old plan, it should be stated that the byres, as a rule, are airy and comfortable, and give a fair amount of cubic air space per cow.

Heifers and young bulls are usually accommodated separately in partially covered courts. Stock bulls frequently stand in the end of byres alongside the cows, although in the larger herds they are usually housed in loose-boxes.

Feeding Methods.—Feeding in the north of Scotland follows the general custom in consisting for the most part of turnips and straw. North - country turnips have a feeding value of their own, and are fed in quantities which may surprise those who are not familiar with the local conditions. Thus, when the crop is a good one, they are fed three times a-day—morning, noon, and evening,—cows consuming from 25 to 35 lb. per head at each feeding-time. When the crop is a short one, the quantity is either reduced over the three periods or otherwise turnips are wholly omitted at mid - day, and a meal of cake, bruised grain, bran, or other artificial food substituted, with, of course, what straw the animals require. But there is no food that cattle, in this part of the country at any rate, do better on than turnips and straw, and all breeders grow a regular quantity of turnips every year for their cattle.

The young animals in the boxes are fed on pretty much the same lines as the cows, except that they usually have fewer turnips and more artificial food. A common allowance of turnips in the case of young growing heifers and bulls is 50 to 60 lb. per day. Linseed-cake is a good deal used for young stock, although cotton-cake also has its patrons. In the case of both, they are almost invariably fed before the turnips, as this is thought to prevent "hoven" and troubles of that kind.

Yellow turnips in Aberdeenshire and adjoining counties keep perfectly fresh up to March and April, when they are succeeded by swedes until the grass comes. In the case of swedes, it is usual to slice them, but to all except to animals rising two years old yellows are fed whole without much risk of choking or bolting.

While turnips are very wholesome as a rule, they should never be fed when in a frosted state, especially to in-calf cows. Cattle fed largely on turnips will not usually drink much water, but all the same, it is customary where water is not always available to give them the offer of it at least once a-day.

Calving.—Not having dairy exigencies to contend with, northern breeders usually aim at having calving-time arranged for the months of January, February, and March. Odd calves will come at other times, but the bulk of the calves have these months as their birth-dates. One advantage of this is, that when the cows go out to grass in May they have comparatively strong calves at foot. Calving usually takes place where the cow stands during the winter, although many breeders aim at having special accommodation for this purpose. When the cow calves, the calf is usually tied up beside the cow in a double stall. Until the calf is able to take all her milk, the cow is regularly milked by hand, the calf sucking at the same time, so as to encourage the cow to let down her milk.

Feeding Calves.—Many of the frequent and discouraging losses among young calves are believed to be caused by the allowance of too much milk at a tender age. It is better to keep the calves hungry than to allow them to gorge themselves for, at any rate, the first three weeks of their existence. Scouring, indigestion, the formation of wool balls in the stomach, and other evils, arise from too liberal or irregular feeding. When the cows go to grass the milk generally increases, and sometimes it is again necessary to resort to hand-milking to take away the surplus. After the calves are weaned, such of the cows as require it are also regularly milked. But this is only necessary, as a rule, in the case of extra heavy milking cows. Cows bred on beef lines, as they generally are in the north of Scotland, do not usually have more milk, unless shortly after calving, than the calf is able to utilise. With an extra heavy milking cow the expedient is sometimes adopted of putting on a second calf to suckle her along with her own calf.

In this part of the country suckling is the almost universal method of rearing Shorthorn calves. The calves are trained to eat oil-cake and sliced turnips as soon as possible, and are weaned at seven to eight months old.

The young bulls which are to be sold in autumn or spring get some oil-cake in the fields during the latter part of summer; but heifer calves, as a rule, depend entirely on their mothers and the grass.

Age for Breeding.—Heifers are generally put to the bull so as to calve at from 20 to 26 months old. This early breeding tends to reduce size, but this can usually be counteracted by a little extra feeding. The danger of putting off breeding until another season is that permanent infertility may ensue. A year's rest at three or four years' old generally enables an early-bred heifer to come to her full size.

Treatment of Bulls.—In the late Mr Cruickshank's herd at Sittyton, when the cows had calved about six or seven weeks, they were turned out with the bull every day, and in summer the bull grazed regularly with the cows. Running pretty constantly with the bull, it was thought that the cows came into use sooner than they would if separated from him, and were in no danger of being missed. On the other hand, one bull under this system does not usually beget so many calves as if kept alone and used sparingly. On this account many breeders adopt the alternative plan of keeping their bulls wholly in the house, and only bringing them out as required. When kept in this way bulls are fed on green tares or cut grass, with the addition of a feed of bruised oats, linseed-cake, or other artificial foods two or three times a-day. Exercise in such a case is given by the attendants taking the animals out for an hour or so each day.

Methods in South of Scotland and North of England.

In the central and southern districts of Scotland management is on pretty much the same lines as farther north, except that turnips are not quite so extensively fed. In a few cases hay is also substituted for part of the straw, the straw farther south not always being so valuable from a feeding point of view as it is in the north.

Alnwick Park System.—The system followed in the south of Scotland and the northern districts of England, outside Cumberland, is well illustrated by what is done in the Duke of Northumberland's extensive herd at Alnwick Park. Here the method of management is substantially as follows: Roots, hay, and straw constitute the principal winter food, with the daily addition of not more than 3 lb. of linseed- or cotton-cake, for each breeding cow or heifer. When turnips are scarce or not available at all, mashes of ground oats, barley, beans, and maize and bran, are given, or a liberal supply of linseed- and cotton-cake is used along with the hay and straw. The food of the stock bulls in winter is usually turnips and hay, ground oats, and about 3 lb. of linseed-cake per day. In summer they get grass and tares in lieu of hay, with the same quantity of ground oats and linseed-cake. Bulls kept in the house get exercise every day.

The majority of the calves suckle their dams. As soon as the 12th of May comes round, and the weather is favourable, all the breeding animals are turned out to pasture, their calves with them, until late autumn. The cows get no artificial feeding in summer, but a corner is railed off somewhere to which only the calves can have access, and here they get a little cake once or twice a-day.

As the season advances, the cows with early bull-calves are separated from those having heifer-calves. An effort is always made to have a good aftermath field for the bull-calves and their dams. This not only gives a nutritious feed to the calf, but increases the flow of milk in the cow.

Housing time depends on the weather. October, however, is the general month. When housing does take place, most of the cows are tied in byres, and the strongest calves—bulls and heifers—put in batches into separate folds. Such calves are allowed to suckle their dams twice a-day up to weaning time. Cows with very young calves are put into boxes together. By the time the weaning of the calves begins they have been taught to eat cake and possibly cut turnips. The check from the milk is therefore scarcely felt.

At Alnwick Park, any more than in the north, it is not found that the suckling system prevents the cows from coming early into use after calving, though occasionally heifers which have had their first calf while still very young are long in taking the bull in the same season. Indeed they often take a considerable rest before having a second calf. Cows in the Alnwick Park herd breed regularly up to twelve or thirteen years old;

a few will go on even to sixteen or seventeen.

Cumberland Methods.—In Cumberland and Westmorland a somewhat different system of management prevails. There, the cattle are mainly in the hands of tenant-farmers, and are kept chiefly for milking purposes. The management is more economical than it usually is in the case of costly herds in the hands of wealthy owners. The Cumberland and Westmorland farmer hand-milks all his cows, and feeds his calves by pail. The calves get a small allowance of new milk for a time, but gradually they are turned on to skim-milk, to which is added porridge made of linseed and maize-meal when the animals are old enough to take such food with safety. When they begin to nibble, dry food, consisting of broken cake, bruised corn, or bran, is placed within their reach.

In these parts cows lie out all summer and autumn. Their winter food consists, as a rule, of turnips and straw, although some breeders are rather more liberal, and give a moderate allowance of crushed oats and decorticated cotton-cake along with the pulped roots and oat-straw.

Young bulls which are being fed for sale receive extra keep in the shape of linseed-cake and bruised oats. They are usually kept in well-ventilated sheds, so that they have abundant coats of hair. In spite of their economical system of feeding, the Cumberland and Westmorland farmers often turn out remarkably good Shorthorns, and this from farms ranging in height from 700 to 800 feet above sea-level.

South and West of England Systems.

Different systems of management prevail in the southern and western districts of England. Cattle here can be kept much more in the open, and they do with less substantially built houses than farther north. At Morgenau, South Wales, for instance, Mr Morgan Richardson's cattle are sometimes in the field as late as the middle of December, and return to them again as early as the middle of March.

In these districts, as elsewhere, management depends on the particular object of the owner. Should the herd be a specially valuable one, and devoted to bull-breeding and beef-production, the cows, as in the north, are timed to calve, as far as possible, in the three first months of the year. For most of the leading shows, as well as for the Herd-Book, ages are reckoned from the 1st of January in each year, and if the calves are born much outside the first three months they are apt to be out-classed for the first season, if not for succeeding seasons also. The aim of the breeder is, therefore, to have the calves as early in the year as the climatic and other conditions of his district will permit.

Morgenau Herd.—A good example of the system in a bull-breeding and showing herd is that followed in the Morgenau herd already referred to. In this herd no corn or cake is given to the breeding cows. In winter they get nothing but hay and chopped straw, with roots and cabbage. Mr Richardson says that at one time he tried milking his cows by hand and feeding calves by pail, but he found it unsatisfactory, and now his cows suckle their calves. Under this system there is sometimes a difficulty with a young bull that has been suckled for six or eight months, and whose dam is getting well forward in calf again. But in such a case the calf can usually be induced to draw from another cow and allow his own dam to go dry. Mr Richardson, like most breeders who bring out young animals for show purposes, is a great believer in the virtue, in such cases, of milk, and plenty of it. Nothing, he affirms, will grow bone, muscle, and hair like milk, preferably suckled by the young animals as they require it.

Those who give attention to the question of sustained progress in young Shorthorns will be interested to know that at Morgenau a system prevails of taking the girth of calves every fortnight. Every bull-calf is expected to girth not less than 2 feet 6 inches at birth, and to make an average increase of 1 inch a-week until he is six months old, and about 1½ inch per fortnight between the ages of six and twelve months. A bull-calf, in Mr Richardson's experience, should measure no less than 4 feet 6 inches at six months old, and 6 feet at twelve months old.

Buscot Park Herd.—Similar methods are pursued in Sir Alex. Henderson's herd at Buscot Park. The cows are out at grass all through the summer. Some of the best milkers, and especially those that are rearing calves, have a small allowance of feeding cake, crushed oats, and mangolds, but otherwise they have to provide for themselves in the fields. The cows are brought up twice a-day for milking, or to suckle their calves. The winter feeding consists of an allowance of about 7 lb. of meadow-hay twice a-day, mangel pulp, and oat-straw chaff *ad lib.*, with, in addition, 5 to 6 lb. of linseed or other cakes, crushed oats, and patent foods. The bulls are fed in a similar way. The best bull-calves are allowed to run with their dams until they are five or six months old. As soon as they will eat they have some sweet meadow-hay given them, and some finely-ground linseed-cake, crushed oats, and a little bran and hay-chaff.

Methods in Ireland.

Except that the animals can be kept out of doors longer than even in the south of England, the management of Shorthorns in Ireland does not differ materially from what is practised on the English side of the Channel. If the rainfall is heavier, the general conditions otherwise are not unfavourable. In the case of bull-breeding herds calves arrive, as in Scotland and England, during the first three months of the year. Dairying herds, on the other hand, have their calves arriving all through the year, to suit the requirements of the milk trade. Where turnips are not largely grown their place is taken by cabbages, hay, mangels, or artificial food.

Calves, as a rule, are pail-fed here, unless in the case of heifers with their first calves, these being allowed to suckle their calves. Some years ago breeders, in order to save new milk in rearing calves, adopted the plan of boiling down whole flax-seed into a mucilage and adding it to milk. The flax seemed to do well enough for a time, but ultimately it was found to set up disease of the kidneys, and is not now used to any large extent. Linseed and maize-meal is now the general partial substi-

tute for milk, although various kinds of calf-foods are also used. Young stock in the south of Ireland especially can go out practically all the year round.

In Irish herds the usual practice is for heifers to drop their first calf when they are about two or two and a half years old.

Management in Dairy Herds.

Where dairying is the principal object different times of calving have necessarily to be adopted. In some cases it takes place all over the year; in others, mainly in the autumn.

In the Shorthorn dairy herd kept at Kelmscott, Lechlade (by Mr R. W. Hobbs), cows go to grass all summer, those giving 20 lb. of milk daily being allowed 4 lb. of cotton-cake. In winter they are tied up in sheds and fed with one meal of hay and chopped straw, with about 56 lb. of mangels or cabbages, about 8 lb. meal and cake (mixed dried grains, soaked maize, germ meal, and decorticated cotton-cake). The stock bulls are kept loose in boxes as much as possible, having cut grass in summer and hay and straw chaff with pulped mangels in winter. Young bulls for sale have, in addition, linseed-cake, crushed oats, and bran.

Calves go with their dams until they are fourteen days old, when they are taken away, taught to drink, and given milk for a few days. As soon as possible they are turned on to some cream equivalent. This is continued for twelve or thirteen weeks, after which they are allowed 2 lb. linseed-cake, hay, and a few roots. The linseed-cake is continued when they are turned out to grass in May. By September they are taken into the yards and given one meal of hay, straw-chaff, mangels, and 2 lb. meal, in addition to straw at nights. The following spring they run on grass with no additional feeding, and most of them run out all the succeeding winter, coming into the yards for hay, which, if short, is given sparingly, and 3 lb. of cotton-cake added.

Bulling is begun in December, so as to ensure a winter supply of milk in the following year. The ordinary cows, however, calve from 1st September to 1st June.

Milk is also the principal consideration in Mr C. R. W. Adeane's herd at Babraham Hall, Cambridgeshire. Here the cows are kept in sheds during winter, but go out three or four hours a-day on the grass, while in summer they stay out the whole time on the pastures. When grass is short, mangels, kohl-rabi, swedes, and oat-chaff with a little hay are given. Cotton-cake and crushed oats are the principal artificial foods in winter. Bull-calves in this case are taken from their dams when three days old and brought up by pail, having milk for about six weeks to two months. Young stock, as soon as they eat, have crushed oats, linseed-cake, bean-meal, and bran. The cows, not including heifers with their first calf, will average from 650 to 700 gallons of milk per annum.

A portrait of a characteristic bull of the Shorthorn breed is reproduced in Plate 26. A noted Shorthorn cow is represented in Plate 27.

THE LINCOLNSHIRE RED SHORTHORN.

Origin.—The Lincolnshire Red Shorthorn has since 1890 attained to the status of a distinct type, if not a distinct breed. It is sometimes referred to as a "sub-variety" of the Shorthorn. Yet, while the Shorthorn has been used successfully in forming the modern type of the Lincolnshire Red cattle, it is known that for over a hundred years red shorthorned cattle have been associated with the county of Lincoln. The cattle were then of enormous size but of slow growth. The growth has been accelerated by modern improvements.

Early Improvement.—The date of the improvement of the Lincoln Reds is first traceable to the year 1810, when three bulls were sent into Lincolnshire from Charles Collings' sale.

Mr Thomas Turnell's Herd.—Probably the most potent factor in producing the breed as now known was the herd owned by Mr Thomas Turnell at Reasby, near Wragby, towards the close of the eighteenth century. Arthur Young says that "Mr Turnell has a breed of cattle which are not surpassed by any in the county for points highly valuable, or their disposition at any age to fatten rapidly. His bull covers at a guinea and has many cows sent to him. This breed originally came from the neighbourhood of Darlington." He further describes these cattle as of medium size, but he preferred the larger ones.

There are no minute records available, but the fact that the Reasby herd attained to considerable eminence is made clear by the acknowledged influence which the "Turnell Reds" exercised. The fine rich cherry-red colour which has been the fashion in all ages was one of the special features of these cattle. The scale Mr Turnell reduced, aiming at more flesh and quality than they apparently then possessed.

Later Improvement.—At a later time, approaching the middle of the nineteenth century, herds owned by Mr Coulam of Withern, Mr Baumber of Somersby, and Mr Oliver of Eresby did much to extend the county reputation of the Lincoln Reds. Mr Cartwright of Tathwell had likewise a celebrated herd whose dispersion in 1844 scattered good blood throughout Lincolnshire. Again the name of Chatterton stands high in its association with the breed, and by the use of the Coates' Shorthorn was partly responsible for altering the character. The "Old Welbourn Reds," too, had a fine reputation, Messrs Burtt of Welbourn being amongst the oldest supporters of the race.

Records of herds exist for a period of 100 years, the type of cattle gradually conforming to one colour.

Herd-Book.—Volume i. of the Lincolnshire Red Shorthorn Association was issued in 1895, and contains, besides herd histories in brief, a record of 293 bulls. The Association has been conspicuously successful in bringing the Lincolnshire Red Shorthorn to the front, by offering prizes at leading shows and in other ways serving the best interests of the breed, which stands under a separate classification at the Royal English Show.

Characteristics.

Description.—It is unnecessary to elaborate a description of the breed. There is no official standard, save that the cherry red is the acknowledged colour, and white markings are no disqualification, although looked upon with disfavour. By taking a good type of Shorthorn with a little more than average size and robustness we have the model for the Lincoln Red.

Aims of Breeders.—To thoroughly understand and sympathise with the objects Lincolnshire breeders have in view, it is necessary to remember that Lincoln is a county where the ideal of the breeder is to produce big stock. The land is capable of carrying large-sized animals, therefore why not make the most of it? Perhaps this point may be presented with greater emphasis if it is borne in mind that the Lincoln sheep is amongst the weightiest and sturdiest of the ovine race; the Shire horse associated with Fenland is the weightiest type of that breed; the curly-haired pig, one of the latest recruits to pedigree, is deemed to be about the largest and heaviest of the porcine tribe in this country. Moreover, Lincolnshire markets can assimilate heavy stock.

Robustness of Constitution.—The Lincolnshire Red Shorthorn cattle owe much of their popularity to the robustness of their constitution. Breeders declare that while they have to house their Coates' Shorthorns, they can leave their Lincoln Reds on the fields to look after themselves. Any one with a knowledge of the flat lands where they are wintered in the southern parts of the county will readily grant that only animals of great constitution could "rough it" as the native Reds do there during an inclement season.

Size, therefore, is one of the chief distinctions between the Lincoln Red and the Coates' Shorthorn. The second point is that they have superior constitutions.

Flesh-bearing Qualities.—No doubt as long as there is a north and a south, Lincolnshire breeders in extremes of the county will never quite agree as to the correct type. There will be large cattle and medium-sized cattle—the latter still larger than the average Shorthorn. The use of Coates' Herd-Book bulls has done much to increase the flesh-carrying qualities of the modern type.

At one time it was commonly noticed that many of the show cattle lacked finish and wealth of flesh. To-day, however, breeders are more experienced, and show their stock with as great a wealth of flesh as almost any other breed. Flesh-bearing properties can be bred into stock as well as cultivated by skilful feeding. The fact that Coates' Herd-Book bulls have been freely used with success, and that the best cattle are now well got up for show, will undoubtedly affect the flesh-bearing character of the produce.

Type.—There is less divergence in type to-day than at any time in the previous history of the breed. Gradually the thick, short-legged, wealthily-fleshed Red Shorthorn type is prevailing. There is a greater size of frame than is noticeable in the Scottish stamp of Shorthorn, and breeders, in their efforts to keep to a type denoting quality, are not likely to forget that if they do not maintain the scale they are losing a potential characteristic of the breed.

Colour.—The colour favoured is a cherry red. Faded reds and reds of yellow shade are often met with, but they are rapidly disappearing from the best herds. Bulls of incorrect shade find few buyers, and the prices obtained speedily impress upon the breeder the necessity of keeping the rich cherry red in view. A few white marks on the vessel or underline are not a disqualification, although if they get as far as the dewlap they are a distinct objection.

Weights.—The weights to which the breed will grow are remarkable. Bulls scaling over 23 cwt. alive have been known. Stall-fed show cattle will weigh up to 24 cwt., while grass-fed three-year-old bullocks average from 8 to 10 cwt., scaling much more when fat. Lincolnshire is essentially a grazing county, and a large number of cattle are fattened there as three-year-olds. Good root crops and rich pasture are obtained in return for caking animals on the land —a system locally known as "begging keep."

Milking Qualities.—As a rule, the system of rearing in Lincolnshire herds is to allow the calves to suckle the cows. This does not encourage a high yield of milk. Yet the reputation which the breed has established outside of the confines of the county is to all intents and purposes that of a fine milking race.

The Burton Herd.—The eminence of the Burton herd, owned by Mr John Evens, and situated close to the county town of Lincoln, has provided another feather in the cap of the breed. For over twenty-three years Mr Evens has kept exhaustive milk records. His aim he tersely describes as "milk combined with size, quality, and constitution." He began showing at the London Dairy Show in 1887, and since then he has had one long record of success. The following comparative statement of the annual yield of milk by cows in his herd is interesting :—

No. of Cows.	Year.	Average yield per Cow. Gallons.
31	1890	740
35	1891	720
34	1892	795
38	1893	732
39	1894	834
43	1895	867
43	1896	889
36	1897	881
38	1898	824
34	1899	860
36	1900	785
48	1901	758
40	1902	776
42	1903	780
43	1904	842
54	1905	816
48	1906	802

Individual yields have been very large. Thus in 1906 ten cows out of forty-eight gave over 1000 galls., the highest being 1602 galls., an average daily yield while in milk of 32 lb. One of his cows holds the record for the largest yield in twenty-four hours at the famous Tring milking trials.. She gave 7½ galls.

MANAGEMENT.

The management of herds varies according to the aims of the breeder—whether the primary object is milk or bull-breeding. Mr John Evens believes that "like produces like," and has saved his bulls from deep milking cows. The female calves are kept in natural condition, the best being always retained in the herd. Mr Evens is of opinion that the bull has more influence in transmitting dairy qualities than the dam. He buys one or two of the best pure-bred dairy cows in order to breed his own stud bulls, thus procuring a change of blood.

Treatment of Cows.—The methods of cow-feeding pursued by Mr Evens are well planned and are carefully carried out. In May or June, if the grass is plentiful, the cows are given 2 lb. cotton-cake, and later, if the grass is scarce or dried up, about 3 or 4 lb. of mixed meal or bran per day with it, and either cabbages or lucerne thrown in the fields. Towards the autumn a change of pasture is provided if possible, usually grass "eddish."[1] The winter daily rations are 4 lb. cotton-cake, 2 lb. malt coombs, 2 lb. dried grains, 2 lb. bran, and 3 lb. mixed meal, generally oats and wheat. In autumn, 40 to 50 lb. cabbages, and later, 40 to 50 lb. swedes, are provided; after Christmas, 40 lb. mangels, when ripe, good oat-straw, long hay once a-day, water always before them, a trough between two cows.

In his method of preparing the foods Mr Evens steeps the dried grains and malt coombs for twenty-four hours. Then these wet grains, coombs, bran, and meals, with a very few pulped roots, are mixed with good oat-straw about twenty hours before using. A few handfuls of salt are thrown in. The mixture must not be allowed to ferment, otherwise it will taint the milk. Cows receive two feeds per day, and one feed of long hay at night. This latter is necessary to enable them to raise the cud. The cake is given dry—roots and cabbages being fed twice, morning and afternoon.

Mr Evens milks his best cows, two or three calves being suckled on cows not intended for use in the herd. The heifers are calved just under three years old. His land is not capable of growing them big enough to admit of a calf being taken earlier.

General Methods.—Cattle in Lincolnshire are usually housed from the middle of October to the end of April to protect them from the cold east winds and to tread down a large quantity of straw.

---

[1] Aftermath.

The usual method of managing a herd in the county is to suckle one or two calves on a heifer and sometimes a third on the cows. The cows are either fed off or sold lean after the third calf. The young stock are allowed to grow in store condition. The heifers are put to the bull at two years old. The steers are brought out fat from two to three years old, and if well done from birth will finish about 60 st. (14 lb.) beef from two to two and a quarter years.

The cattle are wintered out of doors. The wind-swept, bleak countryside is no nursery for the delicate constitution, but the cattle do fairly well with a little hay. Lincolnshire is a corn-growing county, and manure must be made and trampled. Open yards are usually provided on Lincolnshire farms, which while they may not improve the quality of the manure, at least ensure healthy stock. Large numbers of these bullocks go in spring at about two to two and a half years old to the better pasture lands of the county to fatten off during summer. These will kill about 60 st., and if kept on to the following autumn will "die" up to 80 st.

A typical Red Lincoln Shorthorn cow is represented in Plate 28.

HEREFORD CATTLE.

There is no other breed in this country comparable with the Hereford in its happy combination of commercial beef-making qualities and picturesque appearance in the field. It is unsurpassed as a grazier's beast; indeed, grass-fed Herefords sell better than any other class of cattle in the fat markets of the midlands of England.

Origin.—The generally accepted opinion as to the origin of improved Hereford cattle is, that they trace directly from the aboriginal cattle of the county of Hereford and adjoining districts. The improvement was begun far back in the eighteenth century, by the Tomkins family. There is abundance of evidence to show that, as early as 1766, it was taken up in a systematic manner by Benjamin Tomkins, who continued the work with great energy and success until his death in the year 1815. For four years after, his herd was maintained by his daughters, and when it was dispersed by public auction in 1819, one year after the famous Barmpton sale of Shorthorns, twenty-eight breeding animals realised an average of £149 per head —four adult bulls bringing £267, 15s. each, and two bull-calves £181, 2s. 6d. each.

Other early breeders of skill and enterprise took up with commendable spirit the work which had been so well begun by Tomkins, and to their successful efforts the Hereford farmers of to-day are in-

VOL. III.

debted for a valuable race of rent-paying cattle.

It is generally considered that infusions of foreign blood have contributed to some extent to the building up of the improved Hereford. In the history of this breed,[1] it is mentioned that in the seventeenth century cattle had been imported into Hereford from France by Lord Scudamore, and that in later times there have been introductions of stock into Hereford from various parts of England and from Wales. Undoubtedly, however, the dominant ingredient in the improved Hereford is the aboriginal race of the county—the same race of cattle which under different conditions of soil, climate, and management, have given us such breeds as the Devon and Sussex.

The white face has been well described as the "tribal badge" of the Hereford, and we are told that this distinctive mark is traceable to the infusion of foreign blood referred to.

Many animals of the breed were at one time grey or spotted in the face, and even yet there exists a strain of Herefords known as "Smoky-faced Montgomerys."

Characteristics.

Uniformity of Type. — No other breed has more clearly defined character-

[1] *History of Hereford Cattle*, by Macdonald and Sinclair. Vinton & Co., Limited, London.

istics than the Hereford. It is certainly a unique tribute to its wonderful constancy in breeding—and thereby one of the strongest proofs of the purity of its ancestry—that the markings should be so clearly and persistently maintained in successive generations.

Colour of Herefords.—The colour of the Hereford is the first thing that strikes the observer. The white clean face, the white shoulder tops, the white dewlap, the rich red hue, all go to form a striking picture. In the matter of colour it is worthy of note that dark-reds are not favoured, neither are light-coloured cattle. The red that does not contain even the suspicion of a black hair, nor the suggestion of a yellow one, has always been associated with the best animals in the showyard. The rich curly coat is as sure a sign of a truly bred Hereford as the white face and clean muzzle.

General Appearance.—The typical Hereford is a fine massive animal. Its broad back, deep ribs and well-lined flanks, square quarters and well-built-out rumps, undeniably indicate the prime butcher's animal. No other breed possesses such a rare wealth of dewlap, such conspicuous spread over the top, nor in the average such well-rounded ribs. The typical Hereford is level in flesh, bulky in form, and built nearer to the ground than almost any other breed.

Standard Description.—In 1905 the Hereford Cattle Breeders' Association issued a standard description of the breed. It is pointed out that there are difficulties surrounding a scale of points for the breed, as breeders' ideas are so much at variance. The circular remarks: "It is a common saying that beef does not grow on horns, yet a breeder who aims to produce fine breeding stock would fail in his purpose if he neglected to place full value upon the shape and colour of the horns." The description is as follows:—

"The bull should have a moderately short head, broad forehead, and horns nearly resembling the colour of wax, springing straight out from the side of the forehead, and slightly drooping; those with black tips or turning upwards are not regarded with favour. The eye should be full and prominent, the nose should be broad and clear. A black nose is objectionable. The body should be massive and cylindrical, on short legs, the outline straight; chest full and deep, shoulder sloping but lying well open at the top between the blades; neck thick and arched from the head to the shoulders, ribs well sprung, flanks deep, buttocks broad and well let down to the hocks; the tail neatly set and evenly filled between the setting of the tail and the hip bones, which should not be prominent. The whole carcass should be evenly covered with firm flesh; the skin should be thick and mellow to the touch, with soft curly hair of a red colour; but the face, top of neck, and under parts of the body should be white.

"The same description should apply to the cow, excepting that she should be grown upon more feminine and refined lines, the head and neck being less massive, and the eyes should show a quiet disposition."

The circular embodying the above description, which, curiously enough, does not refer to colour, concludes as follows: "The Hereford is essentially a beef breed, and reaches maturity at an earlier age and at less cost than any other breed; the steers readily fatten at two years old on grass alone, and in the summer months they command the top price in the London market."

Constitution.—At one time the Hereford was used as a beast of burden, in the sense that it bore its share in the tillage work of the farm. To this is no doubt attributable the strength of its frame and its constitutional vigour. Nowadays it is employed in a more peaceable and equally useful manner, turning a profit from the fine grazing lands in the midlands of England.

Freedom from Tuberculosis.—A noteworthy feature in the Hereford breed is its freedom from tubercular disease. Shipments of cattle to the number of one hundred have been sent abroad, not one of which reacted to the tuberculin test. This advantage has not been purchased at the expense of aptitude to fatten. It is attributable, in the first place, to the constitution built up in the early days at the plough. That vigour has not been assailed by a pampered system of rearing. The Hereford is a grass fattener, and the open air is the finest antidote to tuberculosis that we have yet discovered. Fattening at grass or finishing in the open court, the Hereford has access to the open air. In the case of stall-fed bullocks the confinement implies less fresh air, and providing disease with a lodgment where it can be communicated.

Milking Qualities. — The Hereford has won so great a reputation as a beef-producer that probably few people outside of the circle of breeders would associate it with milk-production. There are milking strains, however, which give no mean results. The majority of Hereford breeders do not wish to breed for beef alone, without recognising the importance of the cow's ability to rear her own calf. This is the prevailing practice in Hereford herds. Admittedly, this is not the means to be used if milk is to be encouraged as a commercial asset, but no one is likely to take the Hereford cow for milk-production when he can do better with breeds like the Shorthorn. The assertion is sometimes made, however, that the show Hereford is purely an animated block of beef. Milk secretion is deficient. Doubtless it is in many cases, but the fact should be borne in mind that Mr William Tudge of Summer Court, Kington, has bred cows that have won prizes at dairy shows.

A Milking Herd. — In the herd owned by Mr White of Zeals, Wilts, attention is particularly devoted to the cultivation of milking qualities. The calves are allowed to suckle the dams for a week, and are then reared by hand, too much condition not being favoured. Mr White, from eighty cows, sent in two months, May and June—this being an ordinary extract from his records—5400 gals. of milk to the factory, from which 5444 lb. of cheese was made. This is no mean performance, considering that it was only part of the milk. In 1905 the eighty cows at Zeals Park produced 38,500 gals. of milk. Although this does not seem an exceptionally high average, it must be borne in mind that the cows had no artificial food during the year except in the spring and after calving. A fair sample of May milk was submitted for examination, and it averaged 4.3 per cent of fat.

Weights. — Good grass-fed Hereford steers weigh alive from 10 cwt. to 12 cwt., handy weights, which are much appreciated by Midland butchers. At Smithfield Show a class of nine steers under two years averaged over 13 cwt., and in the class between two and three years old the weights averaged close on 17 cwt., which is clear proof of the breed's aptitude to fatten rapidly.

For Crossing. — The Hereford is perhaps, considering its fine beef-producing qualities, not so much used for cross-breeding as could be desired. Probably it is thought best to keep its grazing qualities unalloyed. Abroad on the prairies of the new hemisphere, on the bare lands of the veldt, and on the sun-burned pastures of the antipodes the Hereford flourishes. It is no mere trite observation to say that it thrives under these conditions better than any other breed. One of its chief claims to the support of the foreign buyer is that it is the best of all foragers when circumstances compel it to seek for its living. A large export trade is carried on to North and South America, to South Africa, and Australia.

In the Showyard.

Hereford cattle do well in the showyard. There is no lack of herds in the bull-breeding business, and that being so, there are numerous exhibitors. For a time breeders seemed to attach almost undue importance to quality, sometimes at the expense of scale and weight, favouring very short, compact, thick animals. There is, however, a greater disposition now prevailing to give substance and size their due, recognising that in breeding it is easier to lose weight than to regain it.

MANAGEMENT.

The management of Hereford breeding herds does not vary much. In Mr Allen E. Hughes' herd at Wintercott, Leominster, the practice is to run the cows at pasture with the heifer-calves during the summer months. The bull-calves are separated from the dams, being suckled night and morning.

Treatment of Cows. — When the cows come into the yards in the autumn they get out straw and "rowings" (the chaff and riddlings from the straw when threshing) until they produce their calves. Then they have pulped roots and chaff once a-day. The cows are allowed to run in a meadow for a few hours daily, 'and later on in the spring they have hay until turning out to grass. The natural plan of keeping the cows out in the open yards all the winter is followed. When they are ready for calving they are put into loose-boxes, and are in them for a few weeks, and then turned out in the open yard, coming in to suckle their calves night and morning.

Treatment of Calves.—Mr Hughes tries to get his cows to calve after 1st January. He takes the calves from the cows when about eight months old. The heifer - calves receive about 1½ lb. of oat-flour in the morning with chaff, finger-sliced roots twice a-day and hay. The bull - calves during the summer have mixed flour and cake, and run out at grass. After they are weaned they are put in boxes and have flour, roots, and hay.

Management of a Milking Herd. —As an example of management under different conditions — *i.e.*, with milk as a prime object—Mr White's system at Zeals Park in Wilts may be summarised. The herd has a run of 180 acres grass-land, of which one - third is mown for winter consumption. The cows are kept throughout winter on oat-straw. They are allowed the run of pasture until a fortnight prior to calving, when hay and straw - chaff with roots are given them, with the addition of 4 lb. of cotton-cake per day. The calving season is in April and May, when the grass begins to be plentiful. The calves suckle the cows for a week, and are then hand-reared, most of them being sold for veal, which in Mr White's opinion is more profitable than keeping them on for beef. Several bull-calves are saved, however, and are sold for use in dairy herds as far south as Cornwall. The heifer-calves to be kept in the herd are reared by hand, receiving about a gallon of milk daily, till they can eat a little cake and other artificial food and hay. They run on the grass throughout the summer, receiving a little linseed-cake. In winter they are trans-ferred to a sheltered field and subsist on hay only.

The bull is put to the yearlings about the end of July, so that they may breed at two years old in April. This early breeding is encouraged in the belief that the udders developing early do better before the natural beef-making qualities of the breed begin to assert themselves.

The cows are generally at their best as milkers with the third calf. The cows are milked up to the day of calving. Mr White's experience is that they are difficult to dry off, while the heavy milk-ing does not in any way mitigate their natural aptitude to fatten. They pro-duce the milk more cheaply in this way, and cost nothing in artificial food to fatten them.

Management in the Montford Herd. —Mr T. S. Minton, Montford, Shrews-bury, believes that in rearing bulls it pays to be liberal with them, provided the breeder is careful not to surfeit. Discussing the question when to take a calf he says: "Many breeders differ in opinion as to the time a heifer should have her first calf. I think it is at two years two months, which would be March if she was calved in January. Her calf would then be ready to take all the milk by grass time. The dam would not have lost her milk by calving too long before grass is ready, which is often the case with heifers. The plan of heifers not having their first calf till three years old is very prejudicial to their milking pro-perties. If you want a good-looking herd have your first calf at two years two months, and then rest your cow, having your second calf at four years."

In Plates 29 and 30 portraits are given of a noted bull and cow of the Hereford breed.

DEVON CATTLE.

The Devon breed has played an important part in the history of beef-production in England. It is kept in a circumscribed area in the south-west from Dorset to Cornwall, and on the north bounded by the Bristol Channel. There are a few herds to be found farther afield, but if we except the royal herd at Windsor, they have not played a prominent part either in the public ring or in racial improvement. The history of the Devon, or "Ruby" as it is sometimes called, goes back far into the past. The red lands of the south-west have reared for generations red-coated cattle, and a singularly brilliant, active, and useful agent the breed has been in the agricultural evolution of Devonshire and the surrounding counties.

Early History.—Of early breeders of note information is provided by Arthur Young's Chronicles. In his famous report of 1776 the agricultural historian of two centuries ago makes prominent mention of the old Quartly race of cattle. Mr Quartly of Molland was the most celebrated of breeders in North Devon in the time of Arthur Young. The objects which Quartly and other breeders had in view were thus described by Young:—

"The points they have aimed at in breeding have chiefly been to gain as great a width as possible between the hips; to have the hip-bones round and not pointed; that the space from the catch to the hips should be as long as possible; the catch full, but not square; that the tail should fall plumb, without a projection of catch and rump; to have the tail not set on high—not to rise—but be snug, and the line to be straight with the backbone—no pillow just below the cross-line from pin to pin; to be thick through the heart under the chine; that the shoulder-point be not seen—no projection of bone, but to bevel off to the neck, all elbowing out being very bad. All the bones to be as small as possible; the rib-bones round, not flat; the leg as small as possible under the knee; not an atom of the side to have any flatness.

In respect to size, if other points be the same, he prefers a small cow rather than a large one for breeding a bull, because it is very rare to see any very large one handsome; but to breed oxen, a large cow. To have them sharp and thin from the throat to the nose; in the throat the cleanest have small variations from the perfect snake; though fat there, it should not bag. To be thin under the eyes and tapering to the nose, which should be white, but the original breed was yellow. Between the eyes to be rather wide; eyes themselves to be very prominent, like those of a blood-horse, and no change of colour round them. The horns to be white, with yellow tips; thin at root and long, spreading at the points. The breast or bosom should project as much as possible before the shoulder and legs; and the wider between the fore legs the better. To have the line of the neck from the horns to the withers straight with that of the backbone. The belly to be light and rather tucked up; if fat before the udder, it is a sign of a good milker."

The Quartly Herd.—Any one who knows the Devons of to-day would readily recognise in the ideal cattle thus portrayed the true progenitors of the improved breed. Intelligent breeders with so clear an ideal, so well-defined a model, and the relation of all important points so well reasoned out, could not fail to leave an almost indelible stamp upon the race on which they operated. No wonder that the fame of the Quartly Devons still lives, for the efforts of Mr Quartly must have done much to mould the breed into the strongly set type which it now displays. Young tells us that the points which he describes so fully are the points which these gentlemen considered desirable to breed for in Devon cattle, which "they consider as the best in England"; and he adds, "Of their fattening qualities they observed that the Somerset graziers are the judges, who are known to prefer them. For working none can excel them. As milkers they are represented as possessing much merit. They had two cows that gave each 17 pints at a meal, and would make in general 10½ lb. of butter per week in the flow of the season." The systematic improvement of Devon cattle as a breed began with this Mr Quartly. He stated to Young that his father had begun breeding Devons about twenty years earlier—twenty years prior to 1776—and that he thought the breed there or elsewhere was no better then, or at any rate "two years ago," than it was when his father commenced, so little progress had there been made by any one in improving the breed. About this time, however, the demand for Devon cattle began to increase, giving a stimulus to the good work so systematically taken in hand by Mr Quartly and a few other men of "light and leading."

Down to this day the Quartly strains are held in high esteem by all the leading breeders.

The Modern Devon.

The modern Devon embodies two different types—the North Devon and the Somerset Devon. This distinction has gradually come to be recognised. The former is the smaller of the two, and, on the whole, we may aptly term it the sweeter. Of late years a disposition has been manifest to increase the size without sacrifice of quality. It is felt by breeders that a little more weight would not come amiss, provided the "waste" of the carcase could be reduced. It has been amply demonstrated that to carry beef, bone and muscle are necessary, and in Devon cattle there may have been a tendency to run to the extreme of quality. At all events, there was a great temptation to produce a pretty little beast which, on some of the strong lands which carry Devon cattle, could not be fed so profitably as Devons of larger scale. It has come about in the showyard that the larger type is winning most of the honours, and what the showyard says to-day all Devon breeders must agree to to-morrow.

The old type of mottled Devon is still met with, but there are fewer of the old drooping-horned cattle than were at one time seen. Indeed, the tendency in this respect is rather in the other direction, and who shall say that the horn gracefully curving upwards does not make as neat and pretty a head as any? At the same time, there is often a lack of sweetness and femininity in the heads and horns seen in the modern showyard. The Shorthorn type of head is not infrequently met with, the "form" being as shapely and symmetrical as ever.

The complaint is all too frequently heard, "We can't win with the little things nowadays." But there is room for the two types, although one of them must become the predominant partner, if we may judge from the manner in which events are shaping themselves. It should be borne in mind that on the higher-lying lands the Devon has to find a living, and very often the smaller cattle are, under these conditions, the more profitable to rear.

Appearance. — There is no official description of the Devon breed. In appearance it is of a rich uniformly red colour. Symmetry is a strong point. The frame should be well balanced, the flesh being carried right down to the hock joint. The typical Devon is built near to the ground. His head should be well set on a clean-cut neck. The horns curve outwards, then upwards, and should be fairly open. The head is wide at the base, the eye prominent and kindly, the nose short, and the muzzle broad and free from dark spots. The neck should fit into the shoulders, which should be free from coarseness at the points. The top of the shoulder should be broad, the chine of the Devon being essentially built for beef-carrying. The dewlap should be well developed and hang at a right angle. The ribs should be well hooped, so that a deep roast can be carried. They should also be deep, the flank forming the lower side of a parallelogram. The quarters should be long from the pin to the tail head, and any tendency to shortness of second thigh, although a somewhat common fault, should be condemned. The legs should be well placed outside the body. No white patches should be noticed on the skin, which should be mellow and thick under hand. Level flesh is a feature of all fattening stock of pronounced merit, and it is hardly necessary to say that it is as important in the Devon as in any other breed.

As a Show Beast.—As a show animal there are few equally attractive breeds. It is true that there is a tendency to uneven fleshing at the summer shows, perhaps still more apparent at the fat stock meetings, but this is due to the feeding as much as to anything else. A really well ripened Devon is remarkable for two things—plumpness and good killing qualities. In the smaller types so ripe are they that on parade they have a tendency to waddle like a well-fleshed duck, but that must be regarded as a tribute to their wonderful width and flesh-bearing qualities.

On the Farm.—On the farm Devons are kept either as grazers or feeders. Many of the south country feeders replenish their courts with Devon steers, which move off rapidly, enabling three batches a-year to be fitted for the butcher. At one time they were popular with eastern feeders in Norfolk and elsewhere, but as a rule the Norfolk beef-grower prefers an animal of greater scale.

The Milking Type.

Devon cattle are not devoted to beef production alone. There are milking strains which find much appreciation in Dorsetshire, where herds are kept for the express purpose of supply-milk for town consumption. It may be asked why keep Devons of a milking type when the Shorthorn is available? The reason is that when a breed becomes acclimatised and retains its ability to put on flesh when dry, its capacity to adapt itself to the locality is worth a good deal in size and substance. The milking type of Devon is a beautiful animal, with a good vessel and all the breed characteristics in form, character, and quality. Perhaps in some cases they are not quite such a deep red in colour, but their form and qualities are essentially dual purpose—milk and beef. Few particulars can be obtained of herd yields.

Devons in a Yorkshire Dairy.—Mr G. J. B. Chetwynd, who has established a herd near Doncaster, has a very high opinion of them for the purposes of a private dairy. He picked the best blood in Dorsetshire, and has some beautiful cows as his foundation stock. The chief points to be remembered in choosing a breed such as this are that it is capable of milking well and giving richly in quality. They have been so long bred on uniform lines that they throw their young stock very true, and when the calves are not wanted for milk purposes they fatten rapidly either for veal or young beef.

Mr Chetwynd considers the milking Devon one of *the* dairy breeds of the country. They are moderate eaters, and in return are rich milkers. In Mr Chetwynd's belief, if the breed had been run on milk records, classes would have been provided at the dairy shows. The milk testing in the Wyndthorpe herd is for butter-fat in carefully mixed samples. Each cow's milk is tested monthly, beginning one month from the date of calving. An extract from the results recorded in these tests is interesting, as the following will show:—

Date calving.	Date tested.	Quantity of Milk per day.	Per cent fat.
16th Dec.	16th Jan.	32.25 lb.	4.4
19th „	19th „	30 lb.	4.4
26th „	26th „	30.75 lb.	4

Another cow was giving 35 lb. of milk daily only three weeks off calving.

Classes have recently been established for milking Devons at the Bath and West shows, and doubtless as they become better known they will extend their radius of influence.

Antiquity of Milking Type.—The milking type is of course no product of the new century. It goes back more than a hundred years. In the year 1808 Vancouver mentions a cow which, three weeks after calving, yielded in seven successive days 17½ lb. of butter, averaging 14½ pints of milk daily. Another cow, Mr J. G. Davis's "Cherry," gave 2 lb. 5 oz. of butter from 33 pints of milk. Instances are on record of 2½ lb. of butter per cow being made daily from the rich milk of this breed.

Weights.—The popular London handy weight Devon is probably from 1 to 2 cwt. lighter than most of the larger breeds. It is a popular fallacy to assume that because the breed's reputation has chiefly been built up by the Devon of smaller scale, animals that weigh well cannot be found. At Smithfield Show in 1907 the heaviest Devon steer was

18 cwt. 12 lb., the age being 2 years 10 months 3 weeks. Steers under two years averaged 10 cwt. 3 qrs. 9 lb., while the class of major age varied in weight from 12 cwt. 24 lb. to 18 cwt. 12 lb. A good weight for fat show heifers is about 13 cwt., although from 1 to 1½ cwt. less is often recorded.

MANAGEMENT OF DEVONS.

The management of Devon herds may best be described in the words of breeders themselves.

Mr Chick's Herd.

Mr W. J. Chick of Stratton, Dorchester, who is an advocate of the milking Devon, says that the "dairies" in Dorset are let to dairymen at £11 to £12 per cow annually, the owner finding house and premises for the dairyman's use, and food for the cow.

When let, the calving season is during the months of January to April. "As a rule," says Mr Chick, "the cows first fat their calves, but those required for breeding purposes are taken from their dams at a week old and reared by hand on skim-milk. During the winter and spring the cows in milk stay in at night, and are fed on hay with some corn or cake, and by day run out on the pastures. During April or May the cows go on fresh pastures, and stay out at night, the cake and hay then being stopped.

"By this time most of the calves are fattened, there being a good supply of milk if the separator is not used. Best cheese or butter and blue cheese are made. The buttermilk and whey are given to the pigs. The cows are dried off about 22 weeks after service, are fed on straw in the yards, with a run out at pasture until they calve again.

"The best cows give from 40 to 50 lb. of milk per day, but not many in a herd will keep up this quantity. When managed as a letting-dairy, I have known a herd that made 180 lb. to 220 lb. butter per cow, the average being taken from 12 to 14 consecutive years. When the herds are managed by the owner, the milk, as a rule, is sold. Then the cows are calved from September to April."

Mr Huxtable's Herd.

Mr J. L. Huxtable of Overton, Bishops Tawton, writes: "The cow is generally dried off from six to eight weeks before being due to calve. She is put on not too rich pasture in summer. In winter she is fed on roots and oat-straw or hay. After calving, the food is usually light and digestible, such as bran and a few mangolds and a little hay for two or three days. A plentiful supply of water is at hand, the chill first being taken off. It has been my practice for 27 years to give a cleansing and cooling drench when necessary, and I have had only two bad cases of milk-fever, and both speedily recovered.

"Methods of rearing calves vary. Some run with the dams, and others are kept indoors and allowed to suckle the cow morning and evening. More generally, however, the calf is taken from the cow at from two to three weeks old, and fed on skim or separated milk with a little calf-meal mixed with it. Bruised oats, cake, hay, and roots are also given.

"In summer fattening, the cattle are put on pasture with an allowance of cake. In winter, they are either yarded or tied in stalls with an allowance of roots, hay, and cake or meal. Cattle not kept for breeding are fattened at from two to three years old, their weights varying according to age from 30 to 40 scores (of 20 lb. dead). I have sold one two years and eight months old for £25, about 44 score."

Mr Tribble's Herd.

Mr Abram Tribble of Halsdon, Holsworthy, North Devon, says that the heifers in calf generally run during the summer months on the moor-ground or common pasture-land attached to most farms, and are wintered in an open shed, where they are given hay night and morning, with free access to some old grass by day. The in-calf cows are generally milked to within six weeks to two months of calving, and during that time have ordinary rations; in winter, hay twice, roots twice, out during day, and perhaps a little crushed oats or pulped roots with chaff once.

"The method of rearing calves for the

commercial market is to wean after ten days old, and feed twice daily on separated milk, to which should be added a calf-meal or boiled linseed. They are given hay from the rick, and a little cake, crushed oats, and cut roots. When old enough to eat more the allowance of food is gradually increased. Wean off the separated milk at from three to four months. To rear bull calves for show purposes let them suck from three to five months old. The steers are usually sold at two years old for fattening and grazing up-country. Many go to Chichester from here. These usually make £15 at two years old. The cows and calves go chiefly to Exeter Cattle Market to supply the dairies around there, and the calves come back again and are reared here.

"Devon cattle are practically free from tuberculosis. I have never had an animal which has failed under the tuberculin test."

The portrait of a characteristic Devon bull is given in Plate 31.

SOUTH DEVON CATTLE.

The South Devon breed of cattle may best be described as the antithesis of the Devon. The "South Hams" or "Red Devons," as they have been variously called, have a lengthy and historical connection with the county.

Early History.—The early history of the breed has not been brought together into concrete form, but here and there in early literature of a purely agricultural character and otherwise it is referred to. Thus, in 1700, Prince in his *Worthies of Devon* singles out the breed as distinct from any other in England. It was then located between the Teign and the Tamar, being described as of great size, and peculiarly adapted for tilling the steep hills that are found in that neighbourhood. In these early days the largest calves were kept for stock purposes, the thick-backed, fleshy youngsters being sent into the veal market. The reason given for sending these latter calves to market so early was that they would not in all probability grow tall enough for the "collar" work, which was then an important part of the duties of the breed. It thus came about that size, which is to-day a distinguishing feature of the South Devon, was one of the earliest points cultivated by the breeder.

Old breeders refer to a famous ox, owned by Mr Toms of Coyton, Ivybridge, about the middle of the nineteenth century, which, when killed, weighed 16 cwt. He is said to have shown a fine carcase, thickly fleshed, with no waste. That was, of course, in the time of work-oxen. Mr William Treneman of Burraton also brought an ox to the block which weighed over 16 cwt. dressed. Other cases are recorded of oxen, without artificial food, scaling 14 cwt. in the carcase. It was only in the latter half of the nineteenth century, when the work of the farm was transferred to the horse, that the early maturing qualities of the breed were considered, and amongst the pioneers in this direction Mr John Wroth of Knowle deserves honourable mention.

The Herd-Book Society.—In the olden days it was the custom for farmers to breed from the same class of stock for generations. The stock-bulls were kept at farms perhaps two or three miles apart, and breeders drove their cows to these "custom places," as they were then termed. All this was altered by a period of depression and by the inroads of the Shorthorn in the "'fifties" of the last century. Consequently, to maintain purity in the best and smallest breeding-herds, a herd-book became a matter of necessity.

Locality.—The breed has not penetrated beyond the area of the south-western counties of England—Devon and Cornwall—but with the establishment of a herd-book it has certainly very effectually captured the farmers of the south-west, who, having tried Shorthorns, found them less satisfactory from a rent-paying point of view than the

South Devons. From Exeter to Lands End may therefore be considered the home of the breed.

Description.

No standard of points has been fixed by the Herd-Book Society. Those who have seen this breed at home on the fine pasture-lands of the south-west or in the showyard are never likely to forget their characteristics. The recognised colour is described by breeders somewhat indefinitely as a "medium red." The North Devon red and the type of curly coat found in that breed are two things which are strictly avoided, for there is not a little rivalry between the two breeds.

The "points of excellence" are thus described :—

"Rich medium red in colour, hide of moderate thickness, loose and mellow, well covered with soft curly hair, straight over the back and rump, deep and full in girth and full at the chest, shoulders covered at the points and flat on the top, bone of moderate size, tail commencing from line of back and hanging below the hock with a good brush, pins fairly wide but not very prominent, flanks deep, forming straight underline, full and deep in rounds, rump well filled and straight from peg to pin, ribs wide, deep, and well back to the pin, nose white and wide.

"Bulls. — The head massive and broad in the forehead, but not coarse, long from eyes to nose and well covered with curly hair, eyes wide apart, nose white and wide, horns white or yellow, wide at base, and tapering with downward tendency.

, "Females.—The head broad and of medium length, eyes full, horns white or yellow, wide at base, tapering, and fairly straight, the udder well forward and projecting behind, not too fleshy, teats of fair size, regular, and well distributed."

Recent Improvement in the Breed. —These "points of excellence" somewhat inadequately describe the appearance of the South Devon. In the first place, it may be explained that the "South Ham" cattle are without doubt the largest and heaviest of our bovine races. The improvement that breeders have effected

since about 1890 is marvellous. In the early days they were hard-fleshed cattle, if we judge them by the standards set up by other breeds. To-day they are big, wealthy, symmetrical animals of an eminently rent-paying kind. "No more sneering at symmetry," said a well-known breeder, and he was right. In the bulls there is immense sap and very thick flesh, with an evenness of fleshing which is altogether remarkable in animals of such great weight. Occasionally we meet with a lack of style, and sometimes an absence of quality, but the preference as exhibited in the showyard is undoubtedly towards quality, although those breeders whose ideas were nurtured on the older type of cattle needlessly deplore a distinct preference in that direction. There is sometimes a lack of second thigh and weakness of fore rib, but these defects are found in every breed. Certainly the most remarkable development in modern times has been in the way of thicker fleshing, greater symmetry, and more quality.

The cows are fine milkers, and a thousand gallons are not infrequently yielded in the course of the year.

The South Devons Abroad.—The breed has enlisted many supporters abroad. The fact that it is capable of imparting great size to the progeny is a strong point in its favour. South Africa, the Antipodes, Jamaica, and many other countries have been purchasers, and there is every appearance of a great development in this direction in the future.

Weights. —When dealing with the weights of this breed, reference must first be made to Mr W. J. Crossing's champion bull "Good Gift," which as a six-year-old turned the scale at 27½ cwt. Bulls of mature age not infrequently scale from 23 cwt. to 26 cwt. A good fat steer was "Jumbo," which won a fat show championship in the south-west of England for Mr W. M. Roberts of St Germans. He turned the beam at 22 cwt. 1 qr. 27 lb. As an illustration of rapid maturity and great weight for age, reference may be made to a young steer shown at Smithfield in 1894 by Mr J. Sparrow Wroth of Coombe, Aveton Gifford. This steer, aged 668 days, scaled 1833 lb., giving an average daily gain of 2.74 lb. The weight of the dressed carcase was 1190 lb., and the average daily gain of carcase 1.78 lb. There were 120 lb. of loose fat and 112 lb. of hide. The butcher's report was altogether favourable to the quality of the flesh and the profitable nature of the carcase. The kidneys were remarkably fine ; one of the "nobs" weighed 17 lb. cut straight across. Similarly, in the competition for the best carcase at the London Smithfield Show, the South Devon breed has on occasion won highest honours.

Milking Qualities.—In South Devon there are many cows which give 5 and 6 gallons daily in the full flow of milk. At the London Dairy Show in 1906 the cow "Iris," 150 days in milk, shown by Mr Cundy, yielded in twenty-four hours 63.15 lb. milk, from which 2 lb. ½ oz. of butter were made. The second cow, "Primrose," yielded 50.4 lb., 153 days in milk ; and "Sally II." gave 48.1 lb., 136 days in milk.

MANAGEMENT OF SOUTH DEVONS.

The systems of management pursued in herds of South Devon cattle vary to some extent. They are usually natural and fairly liberal.

Mr W. J. Crossing's Herd.

In the herd of Mr W. J. Crossing of Woodford, Plympton, the calves are weaned from the cow at a week old and then reared by hand. No heifer is timed to calve under three years old, the object being to prevent a check in growth which seriously affects perfection in the cow. The herd numbers from 25 to 30 cows, and some of them yield from 20 to 24 quarts of milk daily, the average being about 12 quarts.

Each cow is kept in the herd till she produces about four calves, although some of them produce six or seven. For fattening cattle linseed-cake and a mixed cake are used at the rate of about 10 lb. daily, and in winter hay and roots are given in addition. The carcase weights would be from 7½ to 10 cwt., and in some cases more.

Food for the milking cows varies. Crushed oats, dairy meals, maize-meal, bran and cake mixed, are used. Mr Crossing adds, "The quality of the milk obtained is all that can be desired."

Messrs Whitley's Herd.

Messrs Whitley of Primley Farm, Paignton, Devon, keep their herd at grass throughout the summer, giving a small allowance of corn or cake once a-day when the cows are milking heavily. They supply a large quantity of milk and cream to the town of Paignton. The cattle are kept out in winter unless the weather is exceptionally severe. Thus, in the winter of 1907-8, the cows were taken in only on six nights.

The food consists of hay, roots, corn, or cake. The best cows will average about 20 quarts of milk daily. The calves are taken from the cows at a week old and fed on scalded milk and linseed until they are old enough to pick up a living for themselves, when they go out to graze.

The steers are fattened for beef at about two to two and a half years old, and generally realise from £18 to £28, 10s., the live-weight being from 12 to 16 cwt. Messrs Whitley add, "We have had a steer at two years and eleven months old weigh 19 cwt. and make £41, its dead-weight in beef being 12 cwt. 3 qrs. 6 lb."

The best heifers at Primley are kept for breeding purposes. At seven or eight years old the cows are fattened off, selling at from £20 to £30.

The bulls grow to an enormous size, some weighing up to 30 cwt. "We measured one of ours the other day in reference to an inquiry we received," write Messrs Whitley, "and the following are the particulars :—

Height at shoulders . .	5 ft. 1 in.
Height at croup, . . .	5 ,, 1 ,,
Length of body from top of shoulder to root of tail . .	5 ,, 4 ,,
Length of neck from top of shoulder to centre of horns .	2 ,, 7 ,,
Girth behind shoulder . .	7 ,, 6 ,,

Mr B. Luscombe's Herd.

Mr B. Luscombe of South Langston, Kingston, Kingsbridge, does not sell milk, but makes use of the separated milk on his farm for feeding calves. He adds a little cake and crushed corn for this purpose as soon as the calves are old enough to take it. His cows have yielded up to 7 gallons daily. Animals intended for exhibition are allowed the use of a courtyard with a shed. They are fed in the shed on roots, hay, corn, and cake. Mr Luscombe adds: "The live-weight of beasts in this neighbourhood runs from 14 to 18 cwt. by the time they are ripe for slaughter; but in many cases they turn the scale at a ton. Some of the heaviest are from 24 to 26 cwt."

Messrs Butland's Herd.

Messrs Butland Brothers of Leigham, Plympton, milk about 50 cows and send the milk to Plymouth. They average from 2 to 3 gallons per head daily, and that includes cows that are getting on in calf. "The largest quantity," Mr B.

Butland writes, "I ever knew a South Devon to give in one day was 32 quarts, but we have several that will give from 20 to 25 quarts per day." The cattle are out by day in winter, and have turnips in the field. By night they are under cover, and have mangels, crushed oats, dairy meal, maize meal, bran, and a little linseed-cake. In summer a little cotton-cake with linseed and bran is fed. The general weight of fat beasts is from 8 to 12 cwt. "We rear our own calves by the stall cows," adds Mr Butland. "We cannot allow them more than 2 to 3 quarts of milk per day, but we get them to take a little linseed-cake as soon as possible."

In Plate 32 a portrait is given of a typical South Devon bull.

SUSSEX CATTLE.

The Sussex breed is one of the old indigenous varieties of cattle. It has remained, so far as England is concerned, a purely local breed.

History.—The history of the breed has been comparatively uneventful. It has come through no sensational periods, either in the show or sale-ring, but breeders can point to a record of solid useful work. As far back as 1795, Arthur Young, writing in the *Annals of Agriculture*, declared that the Sussex cattle were recognised as a well-established breed of high repute. He refers to an experiment in feeding which proves the cattle at that time to have shown a tendency to very early development. He sums up their merits thus: "Sussex oxen are as remarkable for the fineness of their hides as they are for the closeness and delicacy of their flesh."

Early Aims.—One of the chief objects of breeders in the early days was to breed Sussex oxen for the plough. They were largely used for multifarious draught purposes. They were able, owing to their wonderful size and weight, to move heavy loads, and on account of their steady pull they made few breakages. Probably the local demand for oxen for this purpose did not

suggest to breeders a wider market than the mere slavish work of the farm and estate. But it is not too much to say that they thereby laid the foundation of a magnificent constitution, which has been inherited by the modern representatives of the breed.

Locality.—To-day the red-coated Sussex cattle are found on the wealden clays of Sussex, Kent, and Surrey. These counties comprise all qualities of grazing land, poor, thin soil, and stiff clays. On the poor lands the breed was at one time raised, and finished off for the butcher on the stronger lands. There are few herds outside of the counties of its birth.

Standard Description.

It was as far back as 1855 that the *Sussex Herd-book* was established. During the period of its existence it has witnessed a radical alteration in the type of the breed and the objects which the breeder has in view. The work-ox has all but gone, and in his place has come the early maturing steer. The following is the standard of excellence as adopted by a general meeting of members of the Sussex Herd-book Society in 1907:—

Bulls.

Registered pedigree.

Head.—Masculine and fairly long.

Forehead.—Broad.

Eyes.—Bold.

Ears.—Of medium size and thickness, fringed with fine hair, and clear flesh-coloured inside.

Nose.—Broad, flesh-coloured, and free from dark spots.

Horns.—Clear, not coarse, starting at right angles to the head and slightly curved inwards, with dark tips.

Throat.—Clean.

Neck.—Muscular and of medium length, spreading out to meet the shoulders, which should not be coarse but neat; sloping, well covered, and showing no projection at the point when looked at from the front.

Chest.—Broad and deep.

Back.—Straight, not rising above the top of the shoulders, and level thence to the setting on of the tail.

Loins.—Broad and full.

Hips.—Moderately wide and on a level with the back.

Ribs.—Well sprung and nicely arched.

Rumps.—Full and level.

Hind Quarters.—Deep, thick, and square.

Tail.—Set in the back, level with the top line, and hanging at right angles to the back; to be of medium thickness, showing strength but no coarseness.

Underline.—To be as nearly as possible parallel with the top line.

Arms and Thighs.—Muscular.

Legs.—Short, good quality bone, with flat, strong, clean hocks, and to be squarely placed when viewed behind.

Flesh.—Even.

Skin.—Moderately thick, mellow to the touch, and covered with an abundant coat of rich, soft, red hair (preferably dark); a little white in front of the purse is admissible but not desirable, and must not extend beyond the navel or appear on any other part of the animal, but a few grey hairs are no disqualification.

General Appearance.—Masculine and active.

Cows.

Registered pedigree.

Head.—Feminine character, moderately long.

Forehead.—Broad.

Eyes.—Bright and prominent.

Ears.—Thin, fringed with fine hair, and clear flesh-coloured inside.

Nose.—Broad, flesh-coloured, and free from dark spots.

Horns.—Clear, not coarse, starting at right angles to the head, well balanced and spreading, with an even, graceful curve slightly upwards, with dark tips.

Throat.—Clean.

Neck.—Of medium length.

Shoulders.—Not coarse but neat and sloping and well covered, showing no projection at the point when looked at from the front.

Chest.—Broad and deep.

Back.—Straight, not rising above the top of the shoulders, and level thence to the setting on of the tail.

Loins.—Broad and full.

Hips.—Moderately wide and on a level with the back.

Ribs.—Well sprung and nicely arched.

Rumps.—Full and level.

Hind Quarters.—Deep, thick, and square.

Udder.—Square, not fleshy, teats set evenly apart.

Tail.—Set in the back, level with the top line, hanging at right angles to the back, and to be of medium thickness.

Underline.—To be as nearly as possible parallel with top line.

Legs.—Short, good quality bone with flat, strong, clean hocks, and to be squarely placed when viewed behind.

Flesh.—Even.

Skin.—Moderately thin, and mellow to the touch, and covered with an abundant coat of rich, soft, red hair (preferably dark); a little white about the udder is admissible but not desirable, and must not extend beyond the navel or appear on any other part of the animal, but a few grey hairs are not a disqualification.

General Appearance.—Smart and gay.

There is nothing to add to the above description, which accurately paints the type of Sussex animal which breeders are aiming to produce. The most common fault of the breed is a shortness of second thigh, more accurately described as "hamminess." The breed is very even in flesh, and is deeply fleshed over the back and ribs.

Weights and Early Maturity.—The weights to which the breed grow are clear and convincing evidence of successful breeding for early maturity. Perhaps this is best exemplified by an examination of the Smithfield Show cattle for a number of years. Over a consecutive period of eight years the class of steers under two years of age has averaged 678 days old and 1422 lb. weight, equivalent to an average daily gain of 2 lb. 1.55 oz. Taking the heaviest beast each year, they averaged over eight years a daily gain of 2 lb. 8¾ oz., while the average of the lowest in the class was 1 lb. 12.90 oz. In the two-year-old class over eight years the average age was 1024 days, and the average daily gain 1 lb. 11.93 oz. The best steer each year averaged a daily gain of 2 lb. 2.28 oz., which must be considered a remarkable tribute to the early maturing propensities of the Sussex cattle. The heifers averaged 2 lb. 2.02 oz. for the best in eight successive years, while the class average was 1 lb. 10.72 oz. daily. It is worthy of special note that the fine young steer which made the highest gain in 1902— viz., 2 lb. 8.34 oz.—killed 68.02 per cent carcase. The highest known yield of carcase to live-weight was recorded in 1888, when a steer weighing 1422 lb. killed 71.67 per cent.

Sussex Bulls for Crossing. — In many ways the Sussex bull might with profit be more widely used for crossbreeding purposes. In 1900 two crosses were exhibited at the Smithfield Show, but a third prize was the highest prize won, although the carcases were very meaty. In 1899 a first prize was won with a heifer.

Sussex Cattle Abroad.—The breed should be very useful in those countries where a draught type of ox is required. A number have been taken to Egypt. In America the breed is appreciated. One purchaser in Tennessee, who has supported the Sussex breed for many years, declares that "the introduction of the Sussex breed has proved an unqualified success, and the breed has stood the crucial test of yielding a reasonable return over the cost of production. . . . The domiciliation of the Sussex in this country is an accomplished fact, and in Tennessee, its peculiar *habitat*, is doing its full share in the improvement of native cattle."

MANAGEMENT OF SUSSEX CATTLE.

There is little specially to record as peculiar to the management of Sussex herds. Mr A. Heasman says that the most successful way of breeding is to calve the cows down in October and November, to let them have their own calf through the winter, weaning in spring, and thereupon putting another calf to the cow. By this method one cow rears two calves. It may be added that the Sussex cow is only a moderate milker, the chief use of the breed being beef-production.

Lord Winterton's Herd.

Mr W. Massie, agent to Earl Winterton, says: "We find the best results are obtained from calves dropped as early as possible in January. The calves are allowed to run with their mothers during the summer months, one cow sometimes bringing up two calves. The average time of suckling is nine months. The cattle are usually wintered in open or covered courts, their feeding consisting of crushed oats mixed with roots and hay. In the following summer they are fattened off the grass with the assistance of linseed- and cotton-cake, and are all sold to the butcher before they are twenty months old. The best average price I have been able to make under these conditions is £24, 8s. for fourteen head sold during summer."

With regard to weights of commercial animals, Mr Massie adds: "We frequently get them to weigh from 100 to 110 stones (14 lb.) per bullock at twenty months old, which at the present price of 4s. 10d. in Guildford market would be £24, 3s. 4d. for the 100 stones. I find that I can generally get from 2d. to 4d. per stone more for them than for cross-breds."

Mr Steven Agate's Herd.

Mr Steven Agate of Horsham usually lets the calves run with the cows until October. They are then weaned and put on rations so as not to lose the calf-flesh. They are fed according to their requirements, some doing so much better than others. Mr Agate generally makes his beasts fit for the butcher at two years old, weighing then between 90 and 100 stones.

The cows are run in the yards during winter, before calving, and have a little hay but no roots. After calving they get what meal and roots they can clear up, as the better the cow does the more satisfactory is the calf's progress. Mr Agate believes in plenty of exercise for all stock, keeping them as clean and comfortable as possible, and, in the matter of feeding, giving salt with their chaff and roots.

Mr Hubble's Herd.

Mr H. T. Hubble of Maidstone owns a half Sussex, half Aberdeen-Angus herd. His method of rearing is to allow the calves to run with the cows in the meadow until the autumn, and then separate the steer calves from the heifers, giving them meal, cake, and hay, and, following the old plan, giving a few swedes in the mid-day ration. Mr Hubble says: "I find a cross with an Aberdeen-Angus bull on the Sussex cow produces good quality of beef. I have this year taken some first and second prizes with this cross, and sold them for £27, £28, and £30 at 20 and 21 months old."

In Plate 33 a portrait is given of a well-known Sussex cow.

AYRSHIRE CATTLE.

The Ayrshire breed of cattle affords a striking example of how the farmers of Great Britain have in years past gradually developed classes of live stock to suit the physical features of each distinctive district of the country. The breed is native to the poorer arable land of Ayrshire, Renfrew, and Lanark. There the farms have always been small, and the surrounding circumstances were such as to point to dairying in preference to the raising of store stock or the production of beef. Of necessity, the cows had to be small in size and hardy of constitution. Before draining became general and artificial manure was available, the Ayrshire had no land of Goshen to enjoy life in, and clearly the breed was evolved long before the advent of either of these aids to the advancement of agriculture.

Historical. — The breed must have originated in the local cattle of the district referred to. What these were no one can tell with any degree of confidence. There is, however, fairly conclusive evidence that for at least a hundred years the Ayrshire cow has not varied much in character.

Early Ideals. — In the minute-book of the Kilmarnock Farmers' Club there is a report, dated 7th August 1795, of a discussion opened by Gilbert Burns (brother of Robert) on "What may be further done to improve the cattle in this country." The summing up of that paper ran thus: "That although much has been done of late in this country in selecting proper individuals of the species to breed from, yet much remains to be done. That particular attention ought to be given to the whole form of the animal as well as to its colour and horns. That much attention also ought to be given in the selection of the cow as well as of the bull. That young cattle, while in a growing state, ought to be more liberally fed than they too generally are in this country; and that as great a proportion of succulent food as possible ought to be given them in winter while they are calves, and thereafter plenty of rye-grass hay each spring."

Something more specific than these generalities was forthcoming at a discussion held shortly thereafter on "The particular form of cattle the Ayrshire farmer ought to select to breed from." The result of this discussion is thus summed up in the minute: "Long and small in the snout, small horns, small neck, clean and light in the chops and dewlap, short-legged, large in the hind quarters, straight and full in the back, broad above the kidneys and at the knuckle bones, broad and wide in the thigh, but not thick-hipped, a thin soft skin of the fashionable colours, whatever these be, and the mother carrying her milk pretty high and well forward on the belly." It would seem that this description is intended to apply to both the male and female form.

Points of the Breed.

In 1884 it was submitted to a committee of the Ayrshire Agricultural Association "to revise the points indicating excellence in the Ayrshire breed of cattle, and to consider other matters connected therewith." The following is the finding of this committee:—

Points.

1. Head short, forehead wide, nose fine between the muzzle and eyes, muzzle large, eyes full and lively, horns wide set on, inclining upwards   10

2. Neck moderately long, and straight from the head to the top of the shoulder, free from loose skin on the under side, fine at its junction with the head, and enlarging symmetrically towards the shoulders   5

3. Fore quarters—shoulders sloping, withers fine, chest sufficiently broad and deep to ensure constitution, brisket and whole fore quarters light, the cow gradually increasing in depth and width backwards   5

4. Back short and straight, spine well defined, especially at the shoulders; short ribs arched, the body deep at the flanks   10

5. Hind quarters long, broad, and straight; hook bones wide apart, and not overlaid with fat; thighs deep and broad; tail long, slender, and set on level with the back   8

6. Udder capacious and not fleshy, hinder part broad and firmly attached to the body, the sole nearly level and extending well forward, milk veins about udder and abdomen well developed. The teats from 2 to 2½ inches in length, equal in thickness, the thickness being in proportion to the length, hanging perpendicularly; their distance apart, *at the sides*, should be equal to about one-third of the length of the vessel, and across to about one-half of the breadth   33

7. Legs short in proportion to size, the bones fine, the joints firm   3

8. Skin soft and elastic, and covered with soft, close, woolly hair   5

9. Colour, red,—of any shade, brown or white, or a *mixture* of these, each colour being distinctly defined. Brindle or black and white is not in favour   3

10. Average live-weight, in full milk, about 10½ cwt.   8

11. General appearance, including style and movement   10

————

Perfection . 100

There is a later pronouncement than this on the part of a committee appointed by the Ayrshire Cattle Herd-Book Society and adopted by the latter in 1906. Though much the same in effect as the Agricultural Association's schedule, it is hardly so much to the point. The Herd-Book Society devote a separate schedule to the bull. The Society's schedule allows for the udder maximum marks of 20 compared with the 33 set aside by the Association; and the desirable size for teats is "2½ to 3½ inches, and not less than 2 inches." Under head of colour brindle alone is stamped as undesirable. "Black and white" passes. Escutcheon gets 1 (in the case of the bull this is stretched to 4). Weight gets 4 only, but the newer schedule standardises the weight of the cow at maturity from 800 to 1000 lb. This we suspect is nearer the mark than the 10½ cwt. stated above. And the increase in length of the teats in the later schedule indicates the recent breaking away from the smallness of teat which fashion unfortunately upheld in the closing quarter of last century.

Type Similar for 100 Years.——There seems, as we have said, to be little difference between the ideal Ayrshire of to-day and that of over a hundred years past. In little, indeed, except as regards horns is there any real difference. A hundred years ago the Ayrshire cow, as old engravings show, had smaller horns than now, and they curled inwards on the forehead instead of standing out wide apart and pointing upwards as on present-day cows. Colour does not seem to carry many marks with it. Even the Kilmarnock experts of a century ago allowed this point to be ruled by fleeting fashion. At present, judging from the cattle one sees at shows, it would appear that the breed will ere long be colourless. The majority of these are white, with a splash here and there of red.

In the fields, however, the rank and file of the breed are considerably diversified in this respect, as one would expect from the terms of the foregoing schedule. Black, though slightly unpopular in the judging ring, is by no means uncommon in the field. But in common with the recognised red and brown overlays the patches thereof are unmixed. Black is indeed understood to be the original ground of the Ayrshire's coat, and at Ayr show there has been of late a special class for animals with black markings.

In this connection it may be noted that the nondescript cattle of the district (those which, unlike the Ayrshire, the West Highlander, and the Galloway, had never been differentiated from the common herd, but had drifted on in accordance with casual circumstances) are spoken of in the minute-book above referred to as black cattle, to distinguish them from the Ayrshires proper.

Infusion of Strange Blood.—There are instances on record towards the end of the eighteenth century and the beginning of the nineteenth of Shorthorn blood having been introduced into the Ayrshire strain. Dutch or Holstein blood is said by some of the writers on agricultural subjects about that time to have been used too. But it would appear that in both cases this new streamlet soon became toned down and lost in the general current. As regards the Shorthorn breed, however, this may have brought the change of colour which distinguishes the modern from the old Ayrshire. Otherwise, the Ayrshire has little in common with the Shorthorn. And, luckily, the Dutch or Holstein dash of blood has not interfered with the established graceful lines of the Scottish breed of dairy cattle, however it may have helped to improve their milking powers.

Useful Properties.

Milking Properties All-important. —The regrettable custom of recent years of judging Ayrshire cows solely on their physical points, giving as much as 33 per cent of total points for a well-turned udder, without the slightest reference to the most important matter of all in the why and wherefore of the existence of the cow, has begun to be understood by the practical farmer, and he is looking around for a remedy. This is not so easily found, however, although the Highland and Agricultural Society is seeking to give a lead in this respect which is now being accepted by the Herd-Book Society of the breed.

Milk Records.—In 1903 the Highland and Agricultural Society began a movement with the object of inducing owners of dairy herds to keep careful records of the quantity and quality of milk given by each cow, it being recognised by that Society that only by such means can even the most experienced man obtain reliable knowledge of the milking qualities of his stock. The National Society instituted local societies

in several districts to conduct the keeping of records, and the work gained so rapidly in favour with dairy farmers that in 1908 the direction of the movement was taken over by the Ayrshire Cattle Herd-Book Society. In the year 1908 records were kept in thirteen centres, embracing over 8000 cows.

Value of Milk Records.—Although the points of the Ayrshire cow are on the lines clearly indicative of high milking power, an animal may approach very closely upon excellence, as defined by the schedule, and yet be poor in contributing to the milk-pail. But with the advent of record-keeping there is less chance of this state of matters occurring often. The symmetrical cow is at any time of more value than the comparatively unsightly one. When, however, in addition, the former can show a good record that can be relied upon, her value is increased considerably. And the same, of course, applies to a cow less pleasant to look upon. A good-service testimonial of the kind will always make such a cow easy to sell.

Care will, however, need to be taken that the Ayrshire cow is not unduly pressed in this connection. There is a limit of her milk-yielding powers beyond which it is unwise to seek to press her if we wish to maintain her present desirable stamina. There is room and to spare in raising the average by equalising (levelling up rather) as far as we can the individuals of the herd. It would not be wise to endeavour to build up a herd of "freak" milkers altogether; but the average will stand considerable augmentation without approaching the "freak" stage.

Milk Yields. — In many Ayrshire herds the annual average yield of milk does not much exceed 550 gallons. But gradually, as a result of selection in breeding and greater liberality in feeding, the yield is being increased, and there are now a large number of herds that substantially surpass 600 gallons on an average. Individual cows often yield from 750 to 1000 gallons, and in the milk records conducted by the Highland and Agricultural Society there are several yields between 1000 and 1200 gallons. Ayrshire milk is of medium quality.

Beef-Production.—Ayrshires are not famed as beef-producers. Animals of the breed have not the right kind of frame for the accumulation of beef. Irrespective of this, however, individuals of the breed seldom get the chance of showing what they can do in this respect. Generally speaking, only the bulls are rounded off in body in readiness for the butcher. The bulls have served their turn earlier in life than the cows; after three years' service the best of them are face to face with their own offspring, and a change must be made. A well-fed bull of the age this implies affords at the best but third-rate beef. The younger representatives are hardly so coarse. As a rule, moreover, the cast bulls are fed off in a somewhat haphazard manner.

The cows are seldom specially prepared for the butcher. They are kept pretty scraggy as milk-suppliers, and when they show signs of permanent failure in this respect they are passed on for slaughter as they stand. Youngish ones may pass muster in a way in the shambles; but the sinewy matrons can hardly be otherwise disposed of than in the mincing-machine. And when we add that the calves over and above the number required for stock purpose are slaughtered almost as soon as dropped, it will be allowed that the Ayrshire breed of cattle does not directly contribute much to the meat-supply of the country, and what it does contribute is, on the whole, of a poor description.

The Ayrshire is in her sphere matchless as a milk-supplier; she may be excused, therefore, in failing to do much as a beef-producer. Now and again it is suggested that the type so admirable of its kind should be modified so as to increase the bulk and square up the frame of the Ayrshire, and in this way get more beef on the carcase. It is doubtful, however, if it would be prudent to attempt much in this direction.

Weights.—Cattle of the Ayrshire breed attain moderate weights. For cows, when full-grown and in good milking condition, common live-weights are 8 or 9 cwt., some exceeding 10 cwt. Bulls run from 11½ to 14 cwt., fat bullocks about 11 to 12½ cwt., and heifers about 1 cwt. less.

MANAGEMENT OF AYRSHIRE HERDS.

The management of herds of Ayrshire cattle varies not only with the character of the districts in which they are kept, but also according to the purposes for which the milk of the cows is utilised. In many herds the main object is the selling of fresh milk in Glasgow and other centres of consumption; in others, cheese-making is the mainstay; in others, again, it is partly butter-making and partly cheese-making; while some dairy-farmers engage to a certain extent in all three methods. Obviously, the seasons for calving, methods of feeding, and general treatment must be varied to suit the peculiarities of these different systems of turning milk into hard cash.

Housing Cattle.—The general custom is to tie up cows in houses throughout the winter, taking them in when the grass fails and the weather gets cold in the autumn, and letting them out again in the late spring when grass is available. In most cases the cows get out for a time every day when the weather permits. All through the grazing season cows are usually kept on pasture.

Calving Season.—Where the main object is the selling of fresh milk, the great aim is to have calves dropped so that the supply of milk may be as nearly as possible equal in quantity throughout the year. In cheese-making and butter-making herds calves are mostly dropped early in spring, so that the cows may be in full milk throughout the grazing season.

Feeding in Milk-selling Herds.—In milk-selling herds the system of feeding must be liberal, and costly feeding-stuffs must be used to a considerable extent. In the grazing season the cows, as a rule, go to the pasture fields daily, and as additional food they may get moist mixtures of distillers' grains and meals. In summer the allowance of concentrated food varies from about 4 to 7 lb. each per day, the quantity varying with the supply of grass, and is generally increased as autumn approaches. In many cases the cows get a small quantity of freshly cut grass—about 20 to 25 lb—per day during the summer, in addition to what they eat in the fields.

In winter the allowance of grains and

meals is substantially increased, in some cases reaching about 10 or 12 lb. of grains and 6 or 8 lb. of meals per day. The meals usually consist most largely of the meal of beans, peas, or Indian corn, but they vary according to market prices. In addition, the cows get hay, oat-straw, and turnips or cabbages, and in some cases mangels in succession. The quantities of these allowed varies greatly in different herds, common allowances being 8 to 10 lb. of hay, 5 to 7 lb. of straw, and 25 to 35 lb. of roots.

In other cases in winter less is given in the way of grains, and the daily quantity of meals (including bruised oats) is raised somewhat, probably to 10 or 12 lb. per head. Bran is also used, and so is treacle, the latter at the rate of .1 or 1½ lb. per day.

On a good many farms the practice of giving turnips or other roots to dairy cows has been abandoned. In these cases quite the maximum quantities of concentrated foods mentioned are given.

As would be expected, these systems of feeding pursued on milk-selling farms entail heavy outlays. For the full year's keep the cost per cow cannot be much under £11 to £12. For the winter months it may cost about 1s. per day.

Feeding in Cheese- and Butter-making Herds.—In herds where the main objects are the making of cheese and butter the cows can usually be kept through the winter at less expense than in milk-selling herds. In the former herds most of the cows are wholly or partially dry in part of the winter season, and in that condition they are fed quite sparingly. If giving milk in winter, these cows are fed similarly to those in milk-selling herds, the daily allowance of concentrated food being perhaps rather smaller.

In the grazing season cows in cheese- and butter-making herds often get little in the way of extra food unless grass is deficient, but as autumn advances concentrated food is given in gradually increasing quantities, beginning with perhaps not more than 1 lb. per day.

Stewartry Customs.—In the counties of Kirkcudbright and Wigtown, where large quantities of turnips are grown, and where cheese-making is the main object with dairy farmers, the system of feed-ing pursued has its own peculiarities. There turnips enter largely into the dietary of the cows. Indeed it is in many cases calculated that there ought to be from 5 to 7 tons of turnips available for every cow in the herd.

In the early part of the winter season the cows in these Stewartry cheese-making herds get little or nothing but turnips and straw. In spring, after calving, the cows get a mixture of dried grains and cotton-seed meal or some other meal, at the rate of from 4 to 6 lb. per day. This mixture is most likely made into a sloppy condition by hot water, and it is given to the cows in a warm state. Oat-straw is given in moderate quantity—about 12 to 16 lb. —also turnips at the rate of from 50 to 80 lb. per day. In summer in these herds cows generally live entirely on pasture. The cost of keeping a cow for a year in cheese-making herds is estimated at from £8 to about £10, 10s.

Calves.—A large proportion of Ayrshire calves are sold soon after birth to be consumed as veal. Those that are reared usually get fresh whole-milk for a few weeks, skimmed-milk or butter-milk being gradually substituted. When the calves are about 8 to 10 weeks old gruel made of oatmeal and linseed is added to the milk, and they are taught to eat broken linseed or other cake. If still in the house, the calves get a little hay, and when the grass is ready and the weather favourable they go to the pastures.

Heifers.—During the grazing season heifers are kept entirely on pasture. In winter, in the colder parts, they are housed most of the time, getting turnips and straw, or instead of roots hay, and a small allowance—not over 2 lb. per day—of such foods as meals, cake, and bruised oats.

Boiled Food.—The practice of giving boiled food to cows was at one time largely followed in Ayrshire herds, but it has lost favour in most parts. The more general custom is to have the concentrated foods scaled with hot water, and fed when slightly warm.

The "Bowing" System. — Under what is known as the "Bowing" system the farmer lets his cows to a "Bower" at a rent of from about £12 to £15 a-year per cow, the farmer replacing any cow whose yield of milk falls below a standard agreed upon. The farmer provides house accommodation and all food for the cows, certain quantities of meals, &c., being stipulated for. The "Bower"

milks the cows, and disposes of the milk to his own advantage. It is mainly Ayrshire cows that are kept under this system.

A characteristic cow of the Ayrshire breed is represented in Plate 42.

HIGHLAND CATTLE.

This singularly handsome breed of cattle, often spoken of as the "Kyloe," has its headquarters in the Western Islands of Scotland and on the high-lying grazing farms of the counties of Argyll, Perth, and Inverness. Amongst all the varieties of British cattle there is none more striking in appearance than the typical Highlander. It is quite as handsome in form as the most improved Shorthorn, is almost as large in size, and with its long shaggy coat of hair, wide-spreading, gracefully-turned horns, hardy muscular appearance, and defiant gait, throws all other varieties of cattle into the shade in picturesque beauty.

Origin.—The prevailing opinion as to the origin of Highland cattle is that they are descended, and that in a direct line, from the ancient native cattle of the districts still regarded as the home of the breed. Indeed it is generally considered that the wild white cattle of Chillingham, the wild cattle of Hamilton Park, the useful little Kerry of Ireland, and the Highland cattle of Scotland, are the purest representatives that we now have of the ancient cattle of the British Isles.

In the early days of the breed—and to this is no doubt due the fact that many people still put West before their name—there appears to have been two classes of Highland cattle. There was the West Highlander, which was largely an Island race, and the ordinary Highlander, which was more a mainland breed. The latter being kept, as a rule, on better fare than the Island cattle, were usually bigger in frame, although the Island cattle had the advantage in the matter of hair. But although there was this distinction at one time, it has long since disappeared. So far as both the

name and general position of the breed are concerned, Highland cattle have for many years been Highland cattle wherever they have been bred. The breed society is called the Highland Cattle Society, and all cattle coming within its purview are judged by one standard of points.

The term "Kyloe," as applied to the breed, is believed by many to be simply an adaptation of the word kyloes or ferries which separate the Western Isles from the mainland of Scotland. This fits in with the theory that the name was first applied to Island cattle. On the other hand, some maintain that the name is a corruption of the Gaelic word which signifies "Highland," and if this be its proper derivation the term would, of course, lose significance.

Characteristics.

The outstanding characteristics of Highland cattle are their wonderful hardiness and their ability to adapt themselves to varying conditions. Mr Andrew S. Grant, to whom we are indebted for information on this and other breeds, states that Highland cattle can live and do well in altitudes and climatic conditions in which few other varieties of cattle would survive. Taken, on the other hand, to the richer pastures of the low grounds, they will yield almost as good a return for their keep as any other class of stock. Nothing sells better than a well-fleshed steer of the Highland breed, and for many years they have been in excellent demand for stocking gentlemen's parks and purposes of that sort.

Coat and Colour.—Owing to the comparatively hard conditions under which they have to exist in their native districts, a first essential in a

Highlander is a good thick although soft skin, and a good coat of hair. Hair is of importance, not only for keeping the animal warm but for throwing off moisture, of which it is hardly necessary to say the Western Highlands have always a full share. The coat, as in the case of a Blackface sheep, should be of the jacket and vest order—that is, there should be long but not too dense outergrowth and a short close covering next the skin. The outer hair should always be of fair strength, but free from harshness and dryness. The colour of the coat is not of so much importance, although, as a rule, it is easier to get the strength of hair required in a dark-coloured animal than in a very light-coloured one. In the early days of the breed wholly black animals were much run upon as being believed to be hardiest. Indeed, there is a tradition that almost the whole of the Island cattle were originally black, and that the lighter-coloured sorts came for most part from the mainland, and Perthshire in particular. Even yet, notwithstanding a slight tendency to harshness of hair on the part of many of them, some people still have a favour for blacks, although the majority of the cattle that one sees nowadays are either brindled, red, yellow, or dun. Very light yellows are rather apt to be associated with soft woolly hair; but apart from that no objection can be taken to them, as they frequently make good feeders and the cows capital milkers. In the case of bullocks, red is possibly the most popular colour, while as to bulls there seems to be a preference for brindles, with here and there a few reds and occasionally yellows.

Points of the Breed.

It cannot be said that the points of Highland cattle, apart from hair and horns, differ greatly from those of most other beef-breeds. The back should be straight and wide and the quarters wel. carried down to the hock. This latter is, and always has been, a point of great importance with Highlanders. It is believed to have been their fine quarters, combined with their excellent coats of hair, that first attracted the attention of the late Mr Thomas Bates to them as a possible source of improv-

ing material when he was building up his noted strains of Shorthorns. In late years some people have seen a tendency to neglect length and depth of quarter in favour of some other points, although it can be said with truth that good quarters are still a strong feature of the Highland breed.

The legs, both before and behind, should be short and strong, the bones strong, broad, and straight, the hoofs well set in and large, and the legs well feathered with hair. The importance of a good hoof, especially in a bull, should be emphasised; many people will not buy an animal that is not well equipped in this respect, as they hold that it is one of the surest indications of stamina and constitution. The hind legs should be hooked a little rather than straight up and down. The underline should be as straight with the back as possible, and both this and the thighs should be well covered with hair.

In olden days Highland breeders liked their cows to be wide between the hookbones. This is still a sign of a good, robust animal, although it should not be overdone, extra wide hooks being frequently associated with bareness over the loins. The head should be short and broad in front rather than narrow, the brow being covered with a long fringe of hair hanging over the eyes and without curls in it. The eyes should be prominent and clear—even to the extent of having a slight "staring" appearance when the animal is at attention.

In both the bull and the cow the horns are important, not only as enhancing the appearance of the animal but as an indication of breeding and constitution. They should be wide apart at the roots and show mellowness and "sap" to the points. A clear, hard, "shiny" horn is apt to be associated with slow-feeding qualities. Black-tipped horns are not liked except in black and dun animals. In the case of a brindle it is always a recommendation to have a waxy-yellow tinge right out to the end of the horns. As to the carriage of the horns, they should in the case of the bull come level out of the head, slightly inclining forwards, and also slightly rising towards the points. Some do not care for this rise, though any drooping is considered to be a very bad fault when between the crown and the commencement of the curve. On the other hand, when the horns rise directly from the crown they detract from the appearance of the animal.

Two styles of horns are commonly seen in cows. In the first case they come squarer out from the head than in the male, rise sooner, and are somewhat longer, though they preserve their substance and reddish-yellow appearance to the very tips. The other taste is for horns coming more level from the head, with a peculiar back-set curve and very wide sweep. A large number of breeders prefer the latter style, which gives possibly the more graceful appearance.

An old fault in Highlanders is a tendency to flatness on the fore rib. The ribs should spring well out from the backbone, and should not only be rounded but deep. In the bull the neck should be of fair length and nicely arched, with a fringe of long hair hanging from the top.

Finally, the animal should be wide-set between the fore legs, and should move with great dignity and style, the movement of a Highlander being of great importance as an index of true and careful breeding.

Early Improvement.

Although the Highland breed never had its Bakewell or its Watson or M'Combie, it still had its own crop of distinguished early improvers. Amongst the earliest that are known to history were the Macneils of Barra and the Macdonalds of Balranald, families that are known to have had large folds[1] in the Long Island from time immemorial. Although the former fold was dispersed when the last Macneil sold his property, the produce of many of his cattle were, up to a comparatively recent date, to be found in the Long Island. The Balranald fold is still as extensive as ever, and has had a widespread influence for good, especially in the Islands.

Other great early improvers were the brothers Donald and Archibald Stewart, who went from Garth, in Perthshire,

[1] Herds of Highland cattle are spoken and written of as "folds," the word "fold" being a legacy from the far bygone days of cattle-lifting, when, for protection, cattle had to be folded at night.

about the beginning of last century, to the farm of Luskentyre, in Harris. These gentlemen started a fold which, from the care bestowed upon it, and the skilful selection of bulls from the leading folds in Perthshire, soon became known as one of the most famous in Scotland. Mr Donald Stewart was the father of the late Mr John Stewart of Ensay, and through that distinguished breeder, in the later years of the century, the cattle became even better known at the Highland Society and other shows, Mr Stewart being almost invincible, especially with bulls.

On the mainland, the Duke of Sutherland in the extreme north, the Earl of Seafield in the county of Moray, the brothers Stewart of Auch, Cashlie and Chesthill in Argyll and Perth, and the Marquis of Breadalbane likewise in the latter county, also rendered great service to the breed. Indeed, the Taymouth sale of the Marquis of Breadalbane is one of the landmarks in Highland cattle history. It was held in 1863 and excited extraordinary interest. One of the bulls sold—viz., "Donull Ruadh," a two-year-old—made £136. Cows sold up to £57; three-year-old heifers to £125; two-year-old heifers to £71; yearling heifers to £46; and three- and four-year-old bullocks to £43. These, of course, were remarkable prices for the time, and they held the record for Highland cattle up to the dispersion of the Earl of Southesk's Kinnaird fold in 1905, when 78 head of breeding animals made the splendid average (for Highland cattle) of £48, 12s.

Several of the finest of the Taymouth cattle of 1863 were bought by the Duke of Atholl, and they became the founders of the famous Atholl fold, which has produced so many noted animals in late years. Other breeders, in addition to the Duke of Atholl and the Earl of Southesk, who have rendered special service to the Highland cattle cause in recent years, have been the late Mr Smith of Ardtornish; the late Lord Malcolm of Poltalloch; Sir Wm. Ogilvy Dalgleish, Bart.; Mr Turner of Kilchamaig; the Countess Dowager of Seafield; and the brothers, Mr Duncan M'Diarmid, Camusericht, and Mr Robert M'Diarmid, Castles, Loch Awe.

Size and Early Maturity.

A much debated question of late years has been whether Highland cattle have been kept up to the size that they used to display in the old days of the Taymouth fold. A good many people hold that they have not, although it would not be difficult to show that they have been improved in other respects, notably in the quality of their hair and in their ability to fatten quickly. In the olden days steers were not considered ripe until they had reached four or five years old. Now three years old is the most common age at which to fatten them off, while many are made fit for the butcher at two and a half years old.

The cattle have also been greatly improved in their flesh-carrying capacity. At Smithfield Show in 1907 the younger steers averaged 12 cwt. 62 lb. live-weight at practically thirty-two months old, while the older steers averaged 16 cwt. 20 lb. at forty-four months old, individual specimens of course scaling considerably more. These are weights that could never have been attained at the age in the olden days. All the same, it has been becoming increasingly difficult of late years for the Highlander to hold his own, in face of the difficulty of paying rent, and the advance of the more quickly maturing breeds into various of the straths and glens once occupied almost exclusively by Highland cattle.

For Crossing Purposes.—Cattle of the Highland breed are well adapted for crossing with other races. They in particular produce excellent results when crossed with Shorthorns. The most general practice is to mate the Shorthorn male with the Highland female. The resulting crosses are usually animals of handsome proportions and excellent beef-producers.

Temper of Cattle.—Notwithstanding the rather ferocious-like appearance which Highland cattle sometimes present, they are not naturally vicious or evil-disposed. Indeed, the majority of Highland bulls are as easily managed as the bulls of any other breed. A Highland cow will certainly defend her young with vigour, but left to herself she is usually much more docile than her appearance would warrant. The general good character of Highland cattle in this respect is believed to be due to the weeding out of ill-tempered animals, which had of necessity to be done in the old days when there were few marches and fewer fences, and when it was dangerous to the owner's pocket, if not to his person, to have a doubtful animal roaming about. When high-strung cows or bulls are met with they should not be bred from, the former being very apt to become troublesome at calving-times. For the same reason it is always well to know that the bull in use is of as placid and evenly-tempered a disposition as possible.

Herd-Book Society.

A Herd-Book Society in connection with the breed has been in existence since 1884. The Society has been very successful, and has published some fourteen volumes of the *Highland Herd-Book*, the last one bringing the pedigrees of bulls up to 2217, and of females up to 7142. In addition to keeping the pedigrees of the animals the Society holds a sale of pedigree animals twice a-year at Oban—that for bulls taking place towards the end of February, and for cows and heifers in October.

The Breed Abroad.—In recent years Highland cattle have been exported to both North and South America, as well as to New Zealand and one or two other countries. But so far the outside demand has not been so great as the decided merit and adaptability of the cattle for special situations would have led one to expect.

MANAGEMENT OF THE BREED.

The management of the Highlander is simplicity itself compared with what it is in the case of some other breeds of cattle. Even on the mainland many Highland cattle go out practically all the year round, having only a shed or suchlike protection in winter. On the islands it is not even customary, unless in the case of specially valuable animals, to take in cows at calving-time. This is done, as a rule, on the mainland, and the custom is quite a good one, for in addition to the saving of casualties at this important period, the housing of cows at calving-time enables both cows and calves to be handled more or less. Calves treated in this way are not so apt to become wild or untractable as those born and reared wholly in the open.

As to feeding in winter, a few turnips and a little straw or hay are about the most that the average stock requires. Even in the ease of housed cattle this is usually found sufficient, although a little cake or corn can, as a rule, be fed with profit where it is desired to bring the animals specially well forward for spring.

Newly weaned calves must be kept on good fresh pastures so as not to let them lose their first flesh.

It is a good plan to hand-strip the cows after the calves are taken from them. Many breeders do this regularly, and believe that cows so treated do better in the following year. Although exceptional Highland cows will rear two calves at a time, the general plan is to have only one calf suckling a cow. Considering the rough pasture on which many of them exist, they pay their way well enough when they rear one calf per season and do it well.

Young stock, after the weaning stage has been got over, usually give little or no trouble, and with a little hay or straw and a few turnips on the pasture in winter will make a wonderful existence.

It is, however, a mistake to suppose that Highland cattle will exist on next to nothing. They will certainly live on poor fare compared with most other breeds, but starvation can only have one result.

In most regular folds the surplus young stock are sold in lots at the Oban, Inverness, Stirling, Perth, or other markets in the back end, usually October. In 1908, at Oban, three-year-old bullocks were making up to £17, 10s. per head, two-year-old bullocks to £14, 15s., and stirks to £12, one specially fine lot selling at £15, 10s. per head.

The Atholl Fold.

The following notes, kindly supplied, in regard to management in the Atholl fold, will be of interest :—

The fold of Highland cattle kept by the Duke of Atholl at Blair Castle is limited in number. The cows, of which there are not at any time more than fifteen, begin calving about 1st January. The calf is left with the cow for about twenty-four hours, when it is placed in a loose-box at the end of the byre, and thereafter let with the cow only twice daily—morning and evening. As soon as the calf is able to eat it gets turnips and meadow-hay.

The cows are all kept in their natural condition, and are not taken to shelter until within about three weeks of calving, unless during severe storms, when they get oat-straw. They get nothing but what they are able to gather when on rough pasture.

The erection they are housed in when near calving is in the form of loose-boxes with outside pens or runs. The loose-boxes are each 10 by 8 feet, and the outside runs 25 by 8 feet. There they are fed twice daily on turnips and meadow-hay, with a plentiful supply of water.

In spring, when turned out on low-lying pasture the cows and calves get together, and there they are allowed to remain until the hill-pasture, which is a month later, is able to support them. They are sent to the hill (part of the Atholl Forest) about 1st June, depending upon the season, and are left there until the middle of October, when the calves are weaned.

Heifers drop their first calf at four years old. The young stock are never housed. They are fed once a-day on meadow-hay and turnips, getting the hay in haiks in an open shed or shelter, the turnips spread on the pasture. In winter the newly-weaned calves are kept by themselves, and the two- and three-year-olds in other enclosures, getting the same feeding as the younger animals.

All the spare stock are sold in autumn.

Castle-Grant Fold.

Owing to the very severe winters in Strathspey the cattle in the fold kept by the Dowager Countess of Seafield at Castle-Grant, Grantown-on-Spey, are housed from the last week of November till the first week of April. The cows are tied two in a stall and do not get out during the day. The young cattle are all out together every day from daylight till dark, and are housed at night in loose-folds according to age and size. All are fed on turnips and oat-straw only, the cows getting one feed per day of bruised oats from the time of calving till they go out in April. When they go out they get turnips and ensilage till they have grass, which usually keeps them by the second or third week of May. Summer feeding consists of permanent grass, at 750 feet above sea-level.

The bull runs with the cows and heifers at 1st of April when they go out. Heifers are served at three years old. The feed of bruised oats has been found to bring the cows earlier to the bull, as they are in better heart when put out. The early calves are suckled till the end of August; the late ones, a month or two longer.

The bull-calves when taken•from the dams are housed in separate boxes with open pens in front, and they get a run to grass (separately) for several hours every day till sold.

For the summer shows only the cow or cows to be exhibited get cake. The heifers get nothing but the grass. For the fat shows the steers are housed all the winter and summer previous to the show, and are sold rising three years old. A heifer or two, usually the worst of the age (three years old), are kept from the bull and fed also.

The home farm at Castle - Grant is worked as an ordinary arable farm, the housing of all cattle for four months and the feeding of steers and heifers providing a large supply of manure. Sheep are grazed and fed on the rotation pasture, and the cattle grazed on the permanent pasture.

Plate 40 represents a handsome Highland bull, and Plate 41 an equally typical cow of the breed.

WELSH BLACK CATTLE.

Welsh Black cattle as they exist to-day have perhaps departed as little from the original type as any of our British breeds. During the early part of last century there were several distinct varieties, all of which appear to have had a common origin. Most of these have now practically disappeared, having been replaced by other breeds from across the English border. For example, the old Glamorgan breed, which was at one time much esteemed both for the production of beef and milk, is now almost extinct, and the White Pembrokes exist in such small numbers as to be negligible.

Historical.

Of these old varieties the Castle Martins, or Pembrokes, and the Anglesey breed are the only ones which remain. For the last century or more these have been described by all writers as similar in character, and there can be no doubt that they, at any rate, sprung from the same stock. Such differences as do exist, or have existed, are mainly due to environment, and are not greater than are to be found in Shorthorns bred and reared under varying conditions.

In describing the Castle Martins

Youatt said : " Great Britain does not afford a more useful animal. . . . They combine to a considerable degree, and as far perhaps as they can be combined, the two opposite qualities of being very fair milkers with a propensity to fatten. The meat is generally beautifully marbled. It is equal to that of the Scotch cattle, and some epicures prefer it. They thrive in every situation. They will live where others starve, and they will rapidly outstrip most others when they have plenty of good pasture. . . . Great numbers of them are brought to the London market. They stand their journey well and find a ready sale, for they rarely disappoint the butcher, but, on the contrary, prove better than appearance and touch indicate." Further, Youatt said that Castle Martin cattle were essentially the same as those of North Wales, but finer in the neck, head, and breast than the Anglesey beasts.

For a good many years the North and South Wales breeders had each their own herd-books, but in 1904 the two Societies amalgamated, so that now the Welsh Black Cattle Society represents the interests of all breeders of Welsh cattle.

Characteristics.

In appearance Welsh have a greater resemblance to the Sussex than to any of the other English breeds of cattle. The only recognised colour is black, but a little white on the udder or scrotum is not objected to. The horns, which are of a creamy colour with dark tips, are long, and are bent slightly forward in the males, in the females they usually incline forward and upward. There are frequent complaints that the high-pitched horn, so characteristic of the breed, is giving place to a shorter horn, which in many cases would seem to indicate doubtful parentage. As a rule, the cattle are short in the leg, with long bodies, are occasionally inclined to be rough on the shoulders, and are not always straight in the top line. They are deep in the ribs, which are better sprung than formerly. They are not so wide across the hips as the Shorthorn, and are more rounded in the hind quarters. The skin should be of moderate thickness, as experience has shown that animals with thin skins are not hardy .enough to withstand the varying conditions of climate which are experienced in many parts of Wales. In the best Welsh herds most of the cattle have lost the high-set tails which were universal a generation ago. They are also stronger at the heart girth, and, generally speaking, are more regular in outline than prior to about 1880.

Hardiness.—One of the main characteristics of the breed is its hardiness. In this respect they are equal to the Highland (Scotch). They will live under conditions in which most of our breeds could not exist. In the lowland districts many cattle are wintered out and receive no food except the grass they pull, unless the ground be covered with snow, when they usually get a little hay. In the mountain districts the shelter provided is not usually of the best, and as the stock of hay, which is always poor in quality, is limited, it is only given to the cattle when the weather is very severe. The hardy character of the breed has always been recognised by the large graziers in the Midland counties of England, who buy the best of the store cattle, which, though forward in condition, will live out even in March and continue to improve.

Rate of Maturity and Weights.—Welsh cattle have been subjected to criticism on the ground that they are slow feeders. This is certainly true of some of the cattle which have been reared so hard as to become stunted in their growth, but that it is true of Welsh cattle as a whole there is no evidence to show. The contrary, indeed, can easily be proved. *The Live Stock Journal* publishes annually particulars of a large number of the cattle exhibited at the Smithfield Show and afterwards slaughtered. The following figures for the years 1900 to 1905 relating to cattle under two years old, indicate the position which Welsh occupy with regard to other breeds :—

Breed.	No. of Entries.	Average age (days).	Average daily gain (lb.)
Aberdeen-Angus .	37	678	2.12
Devon . . .	33	679	1.81
Galloway . .	25	663	1.65
Hereford . .	30	672	2.14
Red Poll . .	20	644	1.87
Shorthorn . .	33	677	2.17
Welsh . . .	36	705	2.08

From these figures it would appear that Welsh cattle are little, if at all, behind the Aberdeen-Angus, Hereford, and Shorthorn, when considered from the standpoint of early maturity.

Grazing Properties. — The figures quoted above refer solely to stall-fed animals, but it is as grazing stock that the Welsh show to greatest advantage. They are for the most part in the hands of small farmers, who have, unfortunately, in order to secure a little more money, been induced to sell to English graziers many of their best animals, with the result that many inferior animals have been kept at home for breeding.

Quality of Beef. — As producers of beef of high quality the Welsh occupy a position very near to that of the Aberdeen-Angus. They furnish beef in which the fat and lean are well mixed, and which generally is of that marbled character so much sought after by those who can afford to pay the best prices. The carcase competitions at the Smithfield Show provide a remarkable testimonial to its quality. During the five years 1901 to 1906 Welsh Black cattle won in these classes, in addition to minor prizes, 6 first prizes and 1 Championship prize. They are looked upon by Scottish breeders, who have always been successful exhibitors in these classes, as their greatest rivals. As they also yield a high proportion of carcase to liveweight, they are always regarded with favour by butchers. This is nowhere more evident than at the Smithfield Show, where all the Welsh cattle are invariably sold on the first day of the show.

Milking Properties. — Although chiefly known outside Wales as beef-producers, Welsh cattle are able to hold their own as dairy stock. It is open to question if the returns in the form of dairy produce obtained from Welsh cows by the small holders of Wales are excelled in any part of Great Britain. Dairy records are almost unknown in the country, and the only figures at hand relate to the Madryn herd of 30 cows, the property of the University College of North Wales, which are not above the average as dairy cattle. In this herd the milk and butter sold from each cow have, during the five years 1903-8, produced on the average from £17 to £20 per year. In addition to this, all the cows have, with one or two exceptions, reared their own calves.

As may be inferred, the conditions under which the produce is disposed of are not unfavourable, but any advantage in this respect is possibly more than outweighed by the fact that many of the cows are essentially beef-producers and do not give much milk. While it cannot perhaps be said that, taken all round, Welsh are equal to Shorthorns as dairy cattle, it is no doubt true that on many Welsh farms the milk yield from the dairy cows averages from 500 to 600 gallons a-year.

On stock-raising farms two, and frequently three, calves are reared for every cow kept. The returns of the Board of Agriculture show that in no part of the country are there more cattle on a similar area than in the county of Anglesey, where from forty to fifty young cattle may often be found on the farm on which they have been reared, and where not more than six or seven cows are kept.

Need for Pioneer Improvers.

It is evident from what has been said that Welsh cattle appear to have retained the characteristics for which they have long been famous. No less an authority than Bakewell thought that, next to the breed with which he commenced his operations, the Welsh approached his ideal more nearly than any other. It is perhaps surprising that a breed which occupied such a prominent position in Bakewell's time should have made so little progress. It must be remembered, however, that they have known no improvers in the ordinary sense of the term, and would have been unable to hold their ground but for the good qualities which are inherent in the breed. A more active policy must be pursued if the breed is to keep its place in the front rank, not to speak of extending its boundaries both at home and abroad.

Where Improvement is Required.

In these days no breed can hope for wide popularity if it is lacking in symmetry, and in this respect, says Professor Winter, Welsh cattle are apt to be wanting. They are also frequently rough in the shoulders and flat on the ribs, while in many of them there is more than a tendency to bareness on the rumps.

If Welsh breeders can improve their cattle in these particulars without materially changing the character of the breed, they will accomplish work which cannot fail to be of service to the country, and which, at the same time, is certain to bring profit to themselves. They must see to it, however, that the breed maintains its hardy character, and continues to produce beef of the quality which is now so much prized. Further, it is absolutely essential in the districts in which Black cattle are found that the cows should be at least fair milkers.

More Enterprise Required.—It is feared that these improvements will take long to accomplish unless more enterprise is shown. In purchasing bulls farmers must set up a high standard, and make up their minds that they will not be satisfied with anything which falls short of it. It may be urged that a small farmer cannot afford to pay much for a bull to serve half a dozen cows.

This must be admitted, but where a number of farmers join together the difficulty disappears.

There are many circumstances which seem to indicate that there is a great future in store for Welsh cattle, and that those who register their cattle in the volumes of the Welsh Black Cattle Society and continue to breed on systematic lines will have no cause to regret their action.

Risk in Introducing Out-crosses. —It has been urged that, in order to secure well-sprung ribs and highly developed hind quarters, resort should be had to a Galloway or even a Highland cross. Such a course is open to the gravest objections, and there is reason to expect that its adoption would be more likely to produce deterioration than improvement. The crosses suggested would most probably produce good animals in the first generation, but there would be a great likelihood that for many years afterwards the country would be flooded with mongrels. There are no worse cattle in the whole of Wales than in the regions which lie between the Welsh Black cattle areas and the districts which are monopolised by the English breeds. These cattle are the results of indiscriminate crossing and changing over a long series of years.

Selection Preferable to Crossing. — A good breed is never formed by casual crossing, but by long perseverance in breeding from similar animals until a uniform class of characters is acquired and rendered permanent. For this reason it is generally better to adopt a good breed already formed than to attempt to produce a new one by a mixture of the blood of dissimilar animals. We are confident that by careful selection and the exercise of good judgment the defects which now exist in Welsh cattle can be bred out. The late Mr William Housman, in his report on cattle exhibited at the Royal Show, Windsor, said: "Welsh cattle have unquestionably vast capabilities of both milk and beef production, and their rude health is an important recommendation. Where hardy, active cattle are required—cattle which can live roughly, yet answer to keep and care, grow beef of the first quality and, under favouring conditions,

great in quantity — the Welsh breed should claim a trial, and they would doubtless prove ready to adapt themselves to districts and countries to which hitherto they have been strangers."

A High Standard to be Aimed at. —In order to realise these vast capabilities breeders of Welsh stock must take a wider view of the situation than they have done in the past, and be satisfied with nothing but the best for breeding purposes. At the present time nearly all the bull-calves in many of the leading herds in Wales are steered, as breeders find that good young steers are worth as much money as, and can be reared at much less cost than, young bulls. There can be little doubt, however, that as the demand for good stock increases the breeders of pedigree cattle will be prepared to meet it by supplying in increasing numbers such animals as may be required.

MANAGEMENT OF WELSH CATTLE.

In a country in which the climatic conditions are so varied the systems of management must necessarily differ widely. The following particulars relating to the herd of the University College of North Wales indicate broadly the methods pursued in the management of most of the herds on the best lowland farms. The College herd in point of numbers is one of the largest in the principality. It is kept at the College farm, Madryn, near Bangor, and consists of about thirty cows and their offspring.

Calf-rearing. — All the calves are reared; a few of the males are kept for bulls, but the majority are castrated. Those that are intended for show purposes suckle their dams, the rest are fed on new milk for about three weeks, after which separated milk and boiled linseed are gradually substituted, until at five or six weeks old the new milk has been entirely replaced. All calves, except those on their dams, are fed from the bucket until they are at least four months old. The feed usually consists of separated milk and boiled linseed; but oatmeal and, to a certain extent, wheat-flour are also used.

As soon as the calves will eat they are given small quantities of crushed oats and linseed-cake, with a little sweet hay, and during the autumn and winter a few sliced swedes. The quantities of these are gradually increased, so that by the time the calf is six months old it is receiving from a pound to a pound and a half of concentrated food per day. All calves six months old or over are turned out to grass about the middle of May. The concentrated food is continued for a time, but if grass is plentiful no concentrated food is given after the first fortnight. Shelters from the heat are always provided where possible.

About the middle of August the calves are brought in in the evenings, but continue to run out during the day until the end of October, and often throughout the winter when the weather is favourable. From October onward they are given hay and a small quantity of crushed oats and linseed-cake, and as soon as the grass begins to fail, pulped roots and chaffed hay and straw in addition, the quantities of prepared foods being gradually increased as the winter advances.

Young Stock.—During their second summer the young stock get nothing except what they pull from the pastures. If the weather is favourable and there is plenty of grass, they are kept out until December, when they are brought in at nights and given a feed of pulped roots and chaffed hay and straw night and morning, with long straw in their racks.

Fattening Stock.—The bullocks are fed off for the butcher at from two to two and a half years old.

Breeding Stock.—Heifers are put to the bull so as to calve at from two and a half to three years old.

The cows run out all the year round, but are housed at nights from November to April. During the winter they receive a moderate supply of roots and long hay. Those in milk get, on the average, about 4 lb. of concentrated food (the bulk of which is cotton-cake) daily throughout the year. Some of the milk is sold, and the rest is made into butter.

A portrait of a Welsh cow is given in Plate 43.

KERRY AND DEXTER CATTLE.

Of the different native varieties of cattle that have from time to time been reared in Ireland, the Kerry and its sub-variety the Dexter-Kerry alone remain. The Kerry and the Dexter-Kerry are now, by most people, looked upon as practically separate breeds; but there are those who still think it more accurate and preferable to regard the Dexter as a sub-variety of the Kerry. Both varieties have earned good names for themselves in England as well as in their native country.

Origin of the Kerry.

It is generally acknowledged that nowhere in the British Isles is there a purer representation of the smaller varieties of the aboriginal cattle of Northern Europe than is provided in the Kerry cattle of Ireland. They are the smallest of the many varieties of British cattle, and none possess more distinctive features or more certain marks of purity of descent. Their individuality is indeed very striking, and although small in size and slow in maturing, they are most useful cattle in their own proper sphere.

Headquarters.—The breed has its headquarters in the bleak steep hills of county Kerry, where it has had to subsist upon scanty fare, exposed to wind and rain, with little artificial shelter or attention of any kind. This untoward treatment has of course told its inevitable tale. It has cramped the stature of the cattle, and made them slow in maturing; but it has also endowed them with a sound constitution and exceptional hardiness, as well as the rare and useful faculty of existing and feeling happy on small fare.

Kerries as Emigrants.—Proverbially, Irishmen make the best of emigrants. A similar property can be claimed for the Irish national breed of cattle; for the little Kerry "adapts itself readily and agreeably to change of scene, and seems as much at ease in the wooded parks of England as on the rocky heights of its Irish home." As already indicated, Kerry cattle have been largely introduced into England, where they are found specially useful on poor land and in small family dairies.

The Typical Kerry.

In outward form Kerry cattle are somewhat similar to the cattle of Brittany, fully as high, but rather more slender and deer-like. The Kerry is active and graceful, long and light in the limb, head small and fine, throat and neck slight and clean, eyes prominent and keen, horns turned upwards, and white with black tip; shoulders thin and sloping, and sharp on the top; chest fairly wide, back straight but rather narrow; ribs fairly well sprung, barrel not deep, hooks wide, quarters long but often uneven; tail-head sometimes rather high, tail slight and long, thighs thin; udder large in size and well shaped, being full behind and carried well forward; milk-veins full and well defined, skin moderately thick and mellow, colour usually black, but some have white spots underneath, and now and again a red Kerry is seen.

The bull is thick, straight, fine in the skin, with good masculine head and neck. Many of the animals have curiously "cocked" horns, first projecting forwards and then taking a peculiar turn backwards—caused, says an Irish wag, by the strong winds the cattle have to face in mounting the Kerry hills!

Origin of the Dexter.

The origin of the Dexter variety, so distinct from the typical Kerry, is enveloped in uncertainty. There is general agreement in regarding the Kerry breed as the main parental stem. As to the development of the sub-variety different theories are put forward. One theory is that the variety was established by the interbreeding of carefully selected Kerries of a low-set thick type, without the aid of any extraneous blood. A more general belief is that' the Dexter sort was originated by mating thick, short-legged Kerry cows with bulls of a similar

type belonging to another breed, most probably the Devon.

This latter is the view put forward by Professor James Wilson in an exhaustive paper on the subject submitted to the Royal Dublin Society in November 1908.[1] There are authentic records of the introduction of Devon cattle into the south of Ireland early in the nineteenth century, and, on the whole, Professor Wilson seems to be well justified in stating that the probability "that Dexter cattle are descended from black Kerries and red cattle of Devon type is very high." He adds: "If further proof were wanted, it can be found by setting a red Dexter cow side by side with a red Devon. The only difference between them is that the Devon cow is now slightly larger—a matter that can be accounted for by the Devon having been much better cared for and increased in size during the last hundred years."

Origin of the name of Dexter.

The origin of the name as well as of the variety itself has long been the subject of speculation. Less or more directly it would seem to be associated with a Mr Dexter who, about the middle of the eighteenth century, went from the south-west of England to act as land agent to Lord Hawarden in Tipperary. Arthur Young in his *Tour in Ireland*, published in 1780, refers to Mr Dexter as a stock breeder, remarking that "there have been many English bulls introduced for improving the cattle of the country at a considerable expense, and great exertions in the breed of sheep: some persons, Mr Dexter chiefly, have brought English rams, which they let out at seventeen guineas a season, and also at 10s. 6d. a ewe, which indicates a spirited attention."

To that Mr Dexter the credit was given by Low of having founded the Dexter variety of cattle.[2] Others, including Professor Wilson, doubt whether Mr Dexter really founded the variety. They believe that the Dexter type was developed in Co. Kerry, not so far from the home of the Kerry breed as Co. Tip-

[1] "The Origin of the Dexter-Kerry Breed of Cattle." Royal Dublin Soc., 1908.
[2] Low's *Domesticated Animals of the British Islands*, 1842.

perary, and they account in another way for the thick, squat Kerries getting to be known as "Dexters." As Arthur Young indicates, Mr Dexter was a distinguished breeder of sheep. Sheep bred by him became known in the south of Ireland as "Dexters"; they were thick and short-legged, and it is said that in course of time the habit grew of applying the term "Dexter" to all animals, and even to men abnormally short in stature and thick in the body.

Type of the Dexter.

The Dexter is an animal of a very different type from its parent the Kerry—much shorter in the leg, thick and plump in the body; indeed, in all respects excepting that of size, an admirable sample of a beef - producing animal. If less elegant and "milky" looking than the typical Kerry, the Dexter is decidedly more symmetrical; and if increased in size, as by selection in breeding and liberal treatment it might soon be, it would be found to be a very profitable class of stock. The Dexter is very hardy and docile, easily fattened, and produces beef of the very choicest quality.

Deformed Dexter Calves.—A tendency in the Dexter variety of cattle to occasionally produce deformed calves has long been a cause of perplexity to breeders. Sometimes the calf is dead born, and in other cases it may be alive but is so misshapen as to be of no value. The occurrence of this misfortune became more frequent after herd-book registration required the Dexter variety to be bred within itself, no longer permitting the use of Kerry bulls. To thoughtful breeders this tendency in the "Dexter" has been the subject of interesting and earnest study. It is generally looked upon as "proof positive" of the "hybrid" origin of the "Dexter." That the defective strain can and will be bred out is not doubted, but skill and perseverance will be required. In the meantime it is suggested that breeders of Dexters should be permitted to make occasional fresh infusions of Kerry.

Characteristics of Kerries and Dexters.

The general body of farmers in Ireland have never given anything like so much attention as they might, with advantage to themselves bestow upon the breeding and rearing of Kerry and Dexter cattle as pure breeds. The breeds certainly possess characteristics which, with careful cultivation, would earn for them an excellent reputation as rent-paying stock. They are specially adapted for high, cold situations where food is not too plentiful. As dairy cattle they have gained a good name. Indeed it may be doubted whether there is any breed of cattle in this country which will beat the Kerry in the return in milk and butter from a given quantity of food. Youatt gave the Kerry cow a high character, and yet he was well justified in what he said: "Truly the poor man's cow, living everywhere, hardy, yielding for her size abundance of milk of a good quality, and fattening rapidly when required."

Improvement.—Early in the closing quarter of the nineteenth century a number of enterprising breeders, prominent amongst which were the late Mr James Robertson, La Mancha, Malahide, Co. Dublin; Mr Pierce Mahony of Kilmorna; and Mr Richard Barker, St Ann's Hill, Co. Cork, organised efforts for the systematic improvement of Kerry and Dexter cattle, and from these efforts great benefit has been derived. A Herd-Book for Kerries and Dexters was published by the *Irish Farmers' Gazette* in 1887, and it was soon after taken in hand by the Royal Dublin Society, which has done much to improve and popularise the breeds. An *English Kerry and Dexter Herd-Book* was established in 1900 by the English Kerry and Dexter Cattle Society, which was founded in 1892.

Weights and Measurements.—Since the systematic improvement of the breed set in the weights of Kerry cattle have somewhat increased. The following are the live-weights of animals in Mr Mahony's herd at Kilmorna: Bull, 2 years 8 months and 2 weeks old, 8 cwt. 1 qr.; bull, 15 months old, 5 cwt. 2 qrs. 7 lb.; bull, 12 months old, 4 cwt. 1 qr. 15 lb.; bull, 8½ months old, 4 cwt. 2 qrs.; cow, aged 4 years, 6 cwt. 2 qrs. 7 lb.; heifer, 2 years and 2 months old, 6 cwt. 21 lb.; heifer, 14 months old, 4 cwt.

Mr James Robertson's famous prize cow "Rosemary" was one of the most handsome Dexters ever seen. Her height at the shoulder was only 3 feet 4 inches, and yet behind the shoulders she girthed 5 feet 7 inches, her length from the shoulder-top to the tail-head being 3 feet 9 inches; and what is more remarkable* still, her udder when in full milk girthed no less than 34 inches.

Milking Properties.— In regard to milking properties, Kerry cows occupy a high position. They often excel both the Jersey and Guernsey breeds in quantity, and are not far behind either in the quality of their milk. In the milking trials at the London Dairy Show Kerry cows have in different years averaged from about 28 to 40 lb. of milk per day, with percentages of butter fat ranging from about 3.50 to 4.50. Cows of the breed have exceeded 1000 gallons in a year. Dexter cows are only to a slight extent inferior to the Kerry in milking properties. Many Dexter cows give from 25 to 35 lb. of milk per day, and the butter fat usually reaches 3.30 to close on 4 per cent.

Kerries and Dexters as Beef-Producers. — Though more valuable for dairy purposes than for beef-production, Kerry cattle are not deficient in fattening properties. When well treated with food they take on condition speedily, and the quality of the carcase is well spoken of by butchers. But as beef-producers Dexters stand decidedly higher. They not only mature early but they take on a remarkably thick, firm, level cover of meat of the choicest quality, the cut being exceptionally deep on the parts where the most valuable meat is carried. For its size there are few better or handsomer butchers' beasts than a well-finished Dexter bullock.

For Crossing Purposes.—Both Kerries and Dexters are well adapted for crossing with other breeds. For dairy purposes they are often crossed with Channel Island and Ayrshire breeds, and for general purposes with other breeds— Shorthorns in particular. By skilful crossing with the Shorthorn a valuable breed of Dexter-Shorthorns, described in page 135, has been built up at Straffan, Co. Kildare.

Colour.—Black is the predominating colour of both breeds. Kerry bulls should be pure black, but a small amount of white on the organs of generation and a few white hairs in the tail are permissible. Cows and heifers of the Kerry breed should be pure black, but there may be a small extent of white on the udder and in the tail. Dexters, both male and female, may be either black or red with little streaks or patches of white.

MANAGEMENT.

Little need be said here regarding the management of herds of Kerries and Dexters. As a rule, the methods pursued are simple in the extreme. The hardy character of the cattle enables them to live largely in the open air, and they are singularly frugal in their fare. They can subsist on poorer pastures than any other of the pure breeds of this country. On moderate pastures they need little and seldom get any extra food.

When kept in houses, Kerry and Dexter cows are fed similarly to other dairy cattle, but the allowances are smaller than for animals of larger size. The statement is often made, and it is well founded, that no other variety of cattle in this country can be depended on for a better return for a given quantity of food in either milk or beef or both combined than is usually obtained from a good class of Kerries and Dexters. Better testimony than that need not be asked on behalf of any race of stock.

A portrait of a characteristic Kerry cow is given in Plate 46; an equally typical Dexter cow is represented in Plate 47.

JERSEY CATTLE.

Jersey cattle, which are the most numerous of the Channel Island breeds, have long been recognised for their beauty of form and excellent dairy qualities, especially in the economic production of butter.

Origin.—The origin of Jersey cattle is somewhat obscure, but some writers are inclined to think the breed analogous to the small cattle which abound in Brittany and Normandy. In vol. i. of the *English Jersey Herd-Book*, the late Mr John Thornton directed attention to this view.

As far back as 1763 measures were taken on the Island to keep the breed pure, an Act being then passed by the States of Jersey prohibiting the importation of cattle from France.

Introduction to England.—At what period Channel Island cattle were introduced into England it is hard to say, but at the close of the eighteenth century they were shipped in small numbers to this country. In 1794, in *A General View of the County of Kent*, by John Boys, farmer, of Betteshanger, a description is given of an experimental test between a "home-bred cow," probably a Suffolk, and a small "Alderney" (as Channel Island cattle were then generally termed), in which the Alderney cow produced twice the quantity of butter per gallon of milk yielded.

The year 1811, however, marks the opening of the English trade. In that year Mr Michael Fowler visited Jersey and commenced exporting. His practice was to take the animals to the various fairs in England for the purpose of sale, and in that way he was greatly the means of popularising the breed.

Improvement of the Breed.—The original type of Jersey was not all that could be desired in point of beauty, but by dint of careful breeding and management the animals at the present day have been brought to a high standard of perfection both in their form and produce. The late Colonel Le Couteur did much for the breed on the Island, and was followed by Colonel Le Cornu.

In 1834 the Island Society published detailed scales of points for judging bulls and cows; in 1852 farmers' clubs were started in Jersey, and in 1871 separate classes were established for Jersey cattle at the show of the Royal Agricultural Society of England at Wolverhampton.

Characteristics.

Jersey cattle are the smallest in size of the Channel Island varieties. They are mostly fawn or greyish fawn in colour, a few having patches of white, the majority being whole-coloured—that is, without any white on them.

The males are generally of a darker hue, and in both sexes extreme dark colours are occasionally found, a few being nearly black. They are generally described, for registration in this country, as whole- or broken-coloured, the former being more sought after for exhibition purposes.

Their attractive appearance and gentle character render them an ornament for the park, and their abundant and long-sustained supply of rich milk is a valuable asset for the dairy.

Scale of Points.—The scale of points adopted by the Royal Jersey Agricultural Society is as follows :—

Cows.

	Points.
1. Head fine, face dished, cheek fine, throat clean	4
2. Nostrils high and open, muzzle encircled by a light colour . .	2
3. Horns small and incurving, eye full and placid	2
4. Neck straight, thin and long, and lightly placed on shoulders . .	5
5. Lung capacity as indicated by width and depth through body immediately behind the shoulders . .	3
6. Barrel deep, broad and long, denoting large capacity; ribs rounding in shape	10
7. Back straight from withers to setting of tail ; croup and setting on not coarse	6
8. Withers fine and not coarse at point of shoulders	4
9. Hips wide apart, rather prominent and fine in the bone	2

10. Hind legs squarely placed when viewed from behind and not to cross or sweep in walking 2
11. Tail thin, reaching the hocks, good switch 2
12. Udder large, not fleshy, and well balanced 10
13. Fore udder full and running well forward 10
14. Rear udder well up, protruding behind and not rounding abruptly at the top 8
15. Teats of good uniform length and size, wide apart and squarely placed . 7
16. Milk veins large and prominent . 3
17. Richness as indicated by a yellow colour on horns, escutcheon and inside of ears 3
18. Skin thin, loose and mellow . . 4
19. Growth 3
20. General appearance : denoting a high-class and economical dairy cow . 10

Perfection . . 100

Bulls.

Points.
1. Head broad, fine ; horns small and incurving ; eye full and lively . . 5
2. Muzzle broad, encircled by a light colour ; nostrils high and open ; cheek small 5
3. Neck arched, powerful and clean at the throat 7
4. Withers fine ; shoulders flat and sloping 5
5. Lung capacity as indicated by depth and breadth immediately behind the shoulders 8
6. Barrel deep, broad and long, denoting large capacity ; ribs rounding in shape 12
7. Back straight from withers to setting of tail ; croup and setting on not coarse 10
8. Hips wide apart, rather prominent and fine in the bone . . . 5
9. Loins broad and strong . . . 5
10. Legs rather short, fine in the bone, squarely placed and not to cross or sweep in walking 5
11. Rudimentary teats squarely placed and wide apart 5
12. Tail thin, reaching the hocks, with good switch 2
13. Well grown according to age . . 3
14. Hide thin, loose and mellow . . 5
15. Showing a yellow colour on skin and horns 3
16. General appearance : denoting a high-class male animal, typical, and of a class suitable for reproduction . 15

Perfection . . 100

Weight. — The average weight of Jerseys is well under 900 lb. At the English Jersey Cattle Society's Show at Kempton Park in 1890, the only exhibition solely for Jersey cattle ever held in England, all the animals were weighed. Twenty-four English-bred animals 5 years 8 months old weighed on an average 826 lb. live-weight, while thirty Island-bred, each 5 years 4 months, averaged 735 lb. live-weight. These figures show that the Island-bred animals are about one-ninth less in weight than those bred in England.

English Improvers.

The efforts of Colonel Le Couteur and Colonel Le Cornu on the Island to improve the condition and milking qualities of the breed were seconded in this country by several breeders. Mr Philip Dauncey founded his celebrated herd at Horwood, Winslow, Buckinghamshire, about 1825. The herd usually numbered about fifty cows, and 14 lb. of butter weekly were often obtained from one cow ; in June 1867 the fifty cows gave an average of over 10½ lb. of butter in one week. Mr W. G. Duncan of Bradwell, whose herd was started in 1849, and Lord Chesham, who began his at Latimer, Chesham, in 1850 ; Mr Palmer of Stewkley, Buckinghamshire ; Mr Edward Marjoribanks, Watford ; Mr Selby Lowndes ; and Sir Walter Gilbey, Bart., were among the earliest English breeders ; also Lord Braybrooke at Audley End, and Mr Archer Houblon at Great Hallingbury, Essex.

English Jersey Cattle Society.

At Mr George Simpson's sale in 1878, at Wray Park, Reigate, a Committee was formed to establish a Herd-Book for Jerseys in England. The late Mr John Thornton was appointed Honorary Secretary, and the first volume of the *English Herd-Book* was issued in 1880, an exhaustive history of the breed being published in that volume. From that date a steady improvement proceeded in England.

In vol. xix. of the *English Jersey Herd-Book*, published in 1908, there are 346 entries of bulls, and 1252 entries of females.

Milk and Butter Tests.—Milk and

butter records were kept by some of the leading herds and printed in the Herd-Books and Supplements. In 1886 the late Mr John Frederick Hall, then living at Erleigh Court, Reading, suggested that public butter tests should be carried out under the auspices of the Society at the various agricultural shows, and he conducted the first one personally at the London Dairy Show in that year. From that time till 1908 no fewer than 2476 Jersey cows were tested at the leading agricultural shows, and the results give an average of 1 lb. 10½ oz. of butter from 31 lb. 13 oz. of milk 112 days after calving, —a record that reflects the highest credit on the breed.

These tests have had great influence on the improvement of the dairy qualities of Jersey cattle.

Records of Dr Herbert Watney's herd at Pangbourne, Reading, one of the most successful in these competitions, have been published in the Herd-Books since 1899. The average weight of butter per head in 1898 was 373 lb., whereas in 1907 it was 404 lb.

Accurate accounts of milk yields have also been kept of Lord Rothschild's herd at Tring Park, Herts. In 1907, eighteen cows that had been in the herd during the whole year averaged 7455 lb. of milk.

Merits of Jersey Cows.—Mr Ernest Mathews, in a paper on "The Jersey Cow," read before the British Dairy Farmers' Association, in Jersey, 1905, stated : "A Jersey cow will give as good a return to her owner, where milk is sold, as the larger breeds of English cattle, since her live-weight, which has something to do with the quantity of food she consumes, is considerably less than theirs, her period of lactation is much more prolonged, while the quality of her milk is so much richer that not only a higher price can be obtained for it, but there need never be any fear of legal proceedings on the ground that 3 per cent of fat and 8.5 per cent of solids other than fat are not present in the milk."

The Jersey cow is to be found in all parts of the world—large numbers being exported annually to Denmark and the United States.

MANAGEMENT OF JERSEY CATTLE.

The feeding and management of Jerseys vary somewhat according to the constitution of the animals and the locality in which they are situated. "The majority of English breeders regard home-bred Jerseys as stronger in constitution than those bred on the Island, while the minority consider that Island-bred cattle if taken care of the first two winters eventually become as hardy and profitable as the English-bred animal." [1]

With regard to bulls, breeders agree that they should have had good dairy ancestors for two or three generations, and that individual appearance should be closely studied.

In the Island of Jersey a masculine-looking bull is preferred, but in England by some breeders a bull of feminine appearance is selected.

It is the general practice to turn the cows out during the day excepting in very inclement weather.

A breeder in Hertfordshire writes : "If a herd of Jerseys is to be kept solely for dairy purposes, regardless of showing and appearance, I should turn the cows out during the day from about the middle of May if the weather is suitable—as the nights get warm." [2] A breeder in Kent "turns out all the animals daily, and considers it a matter of great importance if you desire a healthy herd."

A shed should always be provided where the animals can take shelter in wet or stormy weather, — the object being to keep the animals in a healthy comfortable state.

In the feeding of Jerseys the peculiarities of the animal have to be considered. In order to keep up her flow of milk the cow should receive just as much food as she can properly digest, care being taken to avoid a diet too rich in albuminoids. The best and most natural food is the early grass in the spring of the year, which generally lasts till June, when lucerne, rye, trifolium, sainfoin, and other rotation crops can be ready. The flow of milk can be maintained by such

---

[1] *Jersey Cattle : Their Feeding and Management.* Published for the English Jersey Cattle Society by Vinton & Co., London.

[2] *Jersey Cattle : Their Feeding and Management.*

feeding until later in the year. When this food becomes coarse, roots may be substituted.

For milk-production cabbages, carrots, swedes, mangels, grains, both wet and dried, crushed oats, bran, cotton- and linseed - cake are recommended. For butter - production swedes and turnips are to be avoided, and carrots, parsnips, cabbage, and kohl-rabi are recommended.

Example Rations. — The following four examples of feeding rations are given in the English Jersey Cattle Society's Handbook on the Feeding and Management of Jersey Cattle :—

EXAMPLE I.	Lb. per day.
Carrots	12
Chaff, oat-straw . . .	5
Chaff, good hay . . .	5
Decorticated cotton-cake . .	2
Crushed oats	2
Coarse wheat-bran . . .	2
Hay, good	7

EXAMPLE II.	
Drumhead cabbage, inner leaves .	12
Chaff, as in Example I. . .	10
Linseed-cake	2
Crushed oats	2
Bran	2
Good meadow-hay . . .	7

EXAMPLE III.	
Mangels	14
Chaff, as above . . .	10
Decorticated cotton-cake . .	3
Maize-meal	3
Hay, good , . . .	7

EXAMPLE IV.	Lb. per day.
Mangels	14
Chaff, as above . . .	10
Decorticated cotton-cake . .	2
Maize-meal	2
Malt, sprouted . . .	2
Hay, good	7

The number of feeds are naturally fewer when the animals are turned out.

In the Channel Islands it is the custom to tether the cows on the pasture fields, the animals being moved every two hours or so. They are in some cases milked three times daily.

Care of Bulls.—Bulls should be kept in good hard condition ; an addition of crushed oats and linseed-cake with the ordinary feed of roots and chaff is good. In summer they should have grass, lucerne, and cabbages, but vetches are not recommended. Generally speaking, the bulls are in service from one to four years old.

Calving.—The cows are usually dried off from four to six weeks before calving. As soon as the calf is born, if left with its mother, it should be rubbed over with a little salt, which induces the cow to lick the calf. The cow should be kept warm and free from draughts, and milked frequently, a little being drawn off at a time.

Heifers, as a rule, produce their first calf when two years old.

A portrait of a beautiful Jersey cow is given in Plate 44.

GUERNSEY CATTLE.

The origin of Guernseys, like that of other Channel Island cattle, is not definitely known, but it may be assumed that they are a branch of the Normandy breed. In his work on Domestic Animals (1845) Professor Low remarks that Guernseys deviate from the ordinary type of Channel Islands cattle, and present a greater affinity to the races of Normandy.

In common with other Channel Islands cattle, which were at one time generally termed " Alderneys," they have long been valued for their dairy qualities. Being of a larger frame than Jerseys,

bigger in bone and more prone to fatten, they may be considered more of a dual purpose type. Their flesh makes excellent beef, but being of a high colour like Jerseys, with yellow fat, is not popular with butchers in this country.

Early in the last century it was proposed in Guernsey to allow cattle to be imported from France and other neighbouring countries with the object of reducing the price of butchers' meat and increasing the export trade, but this was negatived by an act of the Royal Court, and all cattle now landed are slaughtered, the only exception being Guernseys that have been sent to England for exhibition.

In the year 1811 Mr Michael Fowler began importing from the Channel Islands, and from that period a steady trade has increased for Guernseys in England.

Characteristics.

Guernseys generally range from about 900 lb. live-weight to 1200 lb. and upwards. Their colours vary from a bright fawn to a reddish fawn, with more or less white about the body. The head is long and well formed, with well-shaped horns; eyes large and prominent, and general appearance and character docile.

The present scale of points, adopted in October 1905 by the English Guernsey Cattle Society, is as follows:—

Cows.

		Points.
Size, Symmetry, and Constitution, 48.	1. Head fine and long; muzzle expanded, with wide open nostrils; eyes large, with quiet and gentle expression; forehead broad; horns curved, not coarse . . .	5
	2. Long thin neck; clean throat; backbone rising well between shoulder blades; chine fine .	5
	3. Back level to setting on of tail; broad and level across loins and hip; rump long; thighs long and thin; tail fine, reaching to hocks; good switch .	5
	4. Ribs amply and fully sprung, and wide apart; barrel large and deep, with strong muscular and navel development .	15
	5. Hide mellow and flexible to the touch, well and closely covered with fine hair . .	5
	6. Hair a shade of fawn, with or without white markings; cream-coloured nose .	3
	7. Size—Cows four years old and over, about 1000 lb. .	10
Indication of Milk yield, 10.	8. Escutcheon wide on thighs, high and broad, with thigh ovals	2
	9. Milk veins prominent, long and tortuous, with large and deep fountains . . .	8
Udder formation, 32.	10. Udder full in front . .	8
	11. Udder full and well up behind	8
	12. Udder of large size and capacity, elastic, silky, and not fleshy . . .	8
	13. Teats well apart, squarely placed, and of good and even size	8

		Points.
Indicating Colour of Milk, 10.	14. Skin yellow in ear, on end of tail, at base of horns, on udder teats, and body generally; hoofs amber-coloured .	10

Total . 100

Bulls.

		Points.
Size, Symmetry, and Constitution, 65.	1. Head fine and long; muzzle expanded, with wide open nostrils; eyes large, with quiet and gentle expression; forehead broad; horns curved, not coarse . . .	5
	2. Long masculine neck; clean throat; backbone rising well between shoulder blades; chine fine . . .	10
	3. Back level to setting on of tail; broad and level across loins and hip; rump long; thighs long and thin; tail fine, reaching to hocks; good switch .	15
	4. Ribs amply and fully sprung, and wide apart; barrel large and deep, with strong muscular and navel development .	15
	5. Hide mellow and flexible to the touch, well and closely covered with fine hair .	5
	6. Hair a shade of fawn, with or without white markings; cream-coloured nose . .	3
	7. Size—Bulls four years old or over, about 1500 lb. .	12
Appearance and Style, 15.	8. General appearance: vigour, style, alertness, and carriage; hind legs not to cross or sweep in walking . .	15
Rudimentaries, 10.	9. Rudimentaries squarely and broadly placed in front, and free from scrotum .	10
Indicating Colour of Milk in Offspring, 10.	10. Skin yellow in ear, on end of tail, at base of horns, and body generally; hoofs amber-coloured . .	10

Total . 100

The heifers generally drop their first calf when about two years old, and frequently continue breeding and milking to the age of twelve years and over.

The bulls become very heavy when old, and readily fatten for the butcher.

Milking Properties.

Guernseys are excellent dairy cattle, yielding a large quantity of milk rich in butter-fat. The butter produced is high in colour and excellent in quality. In vol. iii. of the *Royal Guernsey Society's Herd-Book* an official test, dated May 28, 1885, states that the cow "Vesta 6th" (625), born November 20, 1881, yielded 13 lb. 15¾ oz. of butter in one week.

Since the general inauguration of butter tests in England in 1866 there have been numerous examples of the value of Guernseys for the dairy, and in competition with other breeds in the open butter tests they have obtained many awards. The following records of Guernseys, tested at the London Dairy Show from 1895 to 1907, appeared in the *Journal of the British Dairy Farmers' Association* for 1907 :—

Year.	No. tested.	Average days in Milk.	Average Butter yield.	Average Ratio = lb. of Milk per lb. Butter.
			lb. oz.	
1895 to 1900 }	23	71¾	1 9½	21.86
1901	8	81	1 8¾	21.43
1902	1	17	1 3¾	21.46
1903	5	52	1 1	27.77
1904	3	98¼	1 10	20.65
1905	3	165⅔	1 6¾	19.66
1906	2	138	1 3¼	27.00
1907	2	82	1 12½	18.90

The following are the average results of tests carried out during the five years 1904-8, which have appeared in the *Journal of the Royal Agricultural Society of England :—*

Year.	No. tested.	Average yields.		Average days in Milk.	Ratio.
		Milk. lb. oz.	Butter. lb. oz.		
1904	1	46 4	2 7¾	51	18.61
1905	4	38 5½	1 8¼	55	25.29
1906	2	35 3	1 11¾	41	20.28
1907	4	35 4½	1 9¼	123	22.35
1908	1	35 12	1 12¾	135	19.89

MANAGEMENT OF GUERNSEYS.

The systems of management of Guernseys are necessarily those specially adapted for dairy cattle. Tethering is generally practised on the Island, but not in this country. The cattle are considered fairly hardy after becoming acclimatised. Like other milch breeds, they should not be exposed to inclement weather, although they should have plenty of air and exercise, even in winter, care being taken to provide a shed where they can take shelter when necessary.

From May to about the third week in October they are generally allowed out in the fields night and day if the weather is fine, but it is considered that the flow of milk is retarded by exposure to cold and damp. The comfort of the animals should therefore be assured.

In Alderney, where cows and heifers have been in past years imported from Guernsey, and from whence cows and heifers are now exported to Guernsey, mixing with the breed and being entered in the Herd-Book, the custom is to allow the animals out in all weathers. This, no doubt, tends to harden them, but in the general treatment of milch cows warmth and comfort are essential. The supply of food should be ample and of a kind that promotes good butter-yielding qualities, too sloppy food having a tendency to affect the constitution although increasing the flow of milk. The individual digestive capacities of the animal should be studied.

In spring and summer the early grass, followed by such succulent crops as clover, lucerne, sainfoin, &c. (and maize where it can be grown), form a very good diet.

During autumn and winter, cabbage as well as the root crops—carrots, turnips, mangels, kohl-rabi, and parsnips—may be given, but for butter production turnips are usually avoided.

An old breeder in the south of England considers that "the quantity to be given depends on the cows; a good herdsman finds out far better than following any set rule." He recommends the following daily ration :—

Good meadow-hay chaff, according to what the cow will clear up.
Bran, 1½ lb.
Crushed oats, 3 to 4 lb.
Linseed-cake, 1 lb.

Half the quantity to be given in the morning and half in the evening. For roots he recommends carrots, parsnips, a small allowance of mangels, not more than about a pailful in all.

Guernsey Herd-Book.

The first volume of the English Guernsey Cattle Society's Herd-Book was issued in 1885. The Society has done much to improve the dairy qualities of the breed by giving prizes for dairy tests at the various shows in this country. In the twenty-fourth volume of its Herd-Book, issued in 1908, there are 99 entries of bulls and 376 entries of females.

Guernseys are frequently exported abroad, especially to the United States of America and the Colonies.

A portrait of a representative Guernsey cow is given in Plate 45.

THE DEXTER-SHORTHORN.

This is a new variety of cattle of quite a distinctive type. About 1880 it was founded by Major Barton of Straffan, Co. Kildare, Ireland, by the mating of a Shorthorn bull with a red-coloured Dexter heifer. It was believed that, by the judicious mating of Shorthorn sires with the female descendants of this cross, a useful class of cattle could be established, and the results attained have even more than fulfilled early expectations, high as these were.

For a period of about thirty-five years the only sires used on the Dexter-Shorthorn females were registered Shorthorn bulls of a thick, compact, well-fleshed, short-legged type, chiefly red in colour. All through that period the male calves were castrated, and no heifers were bred from that showed objectionable colours or did not conform to the desired thick, short-legged, well-fleshed type.

By the end of the thirty-five years, during which Shorthorn sires alone were used—that is, in the closing decade of the nineteenth century—the progeny were eligible for entry in Coates's *Shorthorn Herd-Book*. The object of the owner, however, being to establish a distinctive race of cattle, the use of Shorthorn sires was discontinued, and from that time onwards the new variety of cattle has been bred strictly within itself.

A more complete or harmonious blend than is the Dexter-Shorthorn of the two parent strains could hardly be conceived. How long the blood of the Shorthorn and the blood of the Dexter have been running in separate channels no man can tell. Whether the two races had a common origin in the *Bos longifrons*, or whether the Dexter has come down to us from that species, and the Shorthorn from the more gigantic *Bos urus*, must ever remain a mystery. The fact, however, is well established that the Dexter and the Shorthorn breeds have had a distinctly separate existence for many hundreds of years. That the two breeds should blend well is by no means a far-fetched idea. Many is the time the writer has heard typical modern Dexters described as miniature Shorthorns. A good red Dexter cow seemed to want little but size to enable her to rank as a model Shorthorn.

It was a happy idea which led to the two breeds being blended as they have been at Straffan. By Mr Thomas Milne, manager at Straffan, the new variety has been tended from an early period in its existence with a parental care that is pathetic, and in association with his public-spirited employers he has attained marked success in the building up of what is already entitled to rank as an established race of cattle.

The Dexter-Shorthorns differ from the parent breeds only in that they are larger than the one and smaller than the other, and that in colour the black of the Dexter is never repeated, while the whole white of the Shorthorn rarely shows itself. The pure Dexter cow is, for its size, one of the best milking cows in this country. The cow of the new breed retains that characteristic to the fullest extent, giving usually from 18 to 22 quarts of exceptionally rich milk per day. In the production of high-class beef at an early age, the new breed comes quite up to the highest Shorthorn level in proportion to size. In constitutional stamina the Dexter-Shorthorn is all that could be desired. Outstanding features are the thickness of the body and the shortness of the leg. For the most part the cattle are red or dark roan in colour.

No females of the new breed have been sold, but for bulls there is an active demand. They are specially adapted for mating with cows on small holdings, and have been used with excellent results amongst the native cattle in Shetland. Numerous direct and indirect descendants of the Straffan Dexter-Shorthorns have won prizes in fat stock shows.

At the London Smithfield Show in 1908 the cup for the best animal in the classes for "Small Cross-bred Cattle" was awarded to Sir Walter Gilbey for a Dexter-Shorthorn steer bred at Straffan. At the age of two years and eight months this steer reached a live-weight of 1496 lb., showing a daily gain of 1.53 lb.

Plate 48 represents a group of Dexter-Shorthorn cattle. The following are the ages, weights, and measurements of the three heifers named in that group:—

	DAISY.	DORA.	TIDY BELL 3RD.
Age	5 years.	4 years.	5 years.
Live-weight . . .	10 cwt. 2 qrs.	10 cwt.	10 cwt. 1 qr.
Height . . .	3 ft. 10 in.	3 ft. 11 in.	3 ft. 10 in.
Girth [1] . . .	7 , 0 ,,	6 ,, 8½ in.	6 ,, 8 ,,
Length [2] . .	6 ,, 3 ,,	6 ,, 6 ,,	6 ,, 2 ,,
Fore leg [3] . . .	1 ,, 6 ,,	1 ,, 10 ,,	1 ,, 10 ,,
From dewlap to ground .	1 ,, 1 ,,	1 ., 3 ,,	1 ,, 4 ,,

[1] Behind shoulder.   [2] From root of horn to square of tail.   [3] From arm-pit to ground.

OTHER BREEDS OF CATTLE.

LONG-HORNED CATTLE.

This ancient and characteristic breed of cattle, once numerous and widespread in England, has become reduced to very narrow limits and to quite a few herds. It was the first breed upon which Bakewell, the great pioneer breeder of farm live stock, began his experiments in the improving of cattle. Those experiments were begun as early as 1755, and from that year dates the breeding of farm live stock in this country upon scientific principles.

The Long-horned cattle at one time existed in large numbers throughout England, chiefly in the Midland counties. They were also introduced into Ireland, but long ago they have been supplanted at one point after another by the Shorthorns or some other variety, and now the last few remnants of the breed are to be found in the Midlands of England.

The Long-horns are big, long-bodied, rather ungainly cattle, with long drooping horns, which are often so shaped as to make it difficult for the animals to graze short pasture. The cows are fair milkers, and the bullocks attain great weight. In the general properties of rent - paying stock, they are surpassed by most other improved breeds. Yet, partly on account of the unique historic interest attached to the breed, one delights to know that it is still being maintained in all its purity and antique character by a few devoted breeders.

A portrait of a typical Long-horn bull is represented in Plate 35.

ORKNEY AND SHETLAND CATTLE.

The native cattle of the Shetland and Orkney Islands are quite distinct in character from the races in the mainland. They show a considerable resemblance to the Kerry cattle of Ireland, and, like these hardy animals, are well adapted to their surroundings.

They are small in size, and, as a rule, not of a high character. The true Shetland cow, however, is a fairly handsome animal of a dairy type, with fine brown mellow skin and silky hair. On the poor scanty feeding which she generally obtains she gives a wonderfully rich yield of milk.

Fig. 703 represents a Shetland cow.

FOREIGN BREEDS OF CATTLE.

The foreign breeds of cattle best known in this country are those which occupy prominent positions on the continent of Europe. In these countries there are breeds and varieties of cattle almost without number. At the Paris International Exhibition of 1878 there was held the largest and most widely representative show of farm live stock that has ever taken place. In that display there were sixty-five distinctive races and sub-races of cattle, besides thirty different crosses. The United Kingdom claimed eleven of the breeds, winning the Champion Prize for the best group of beef-producing cattle with a group of the Aberdeen-Angus breed. The other breeds and crosses came from continental countries. France itself contributed close on thirty varieties, the most noted being the Norman, Flemish, Charolais, Limousin, and Brittany breeds.

Dutch or "Holstein-Friesian" cattle,

Fig. 703.—*Shetland cow.*

favourably known in this country and in America for their deep milking properties, were well represented at the Paris Exhibition, and so also were the cattle of Belgium, Denmark, Switzerland, Portugal, and Italy.

Polled Durhams in the United States.

A peculiarly interesting class of cattle in the United States of America is known as the "Polled Durham" breed. The cattle are in reality Shorthorns *minus* the horns. Some of the strains are pure-bred Shorthorns, descended directly from British-bred stock, the loss of the horns in these cases having apparently arisen originally through the exercise of nature's inherent power to give forth variety. Animals belonging to these strains are eligible for both the American Shorthorn Herd-Book and the American Polled Durham Herd-Book. Other strains of the polled breed trace from American native hornless cows, Shorthorn bulls being the sires continually used.

Polled Herefords.

Equally interesting is a tribe of polled Hereford cattle which has been established in Canada. Originating no doubt in a "spontaneous variation," the "hornless whitefaces" have been cultivated so skilfully that they now breed to the polled type with wonderful regularity.

LONG-WOOLLED BREEDS OF SHEEP.

THE LEICESTER SHEEP.

The Leicester sheep has been described as the parent long-wool. At one time it was commonly known as the Dishley sheep, and has probably occupied a larger space in ovine history than any other single breed. This is due to Bakewell's association with it.

Bakewell's Influence.

Bakewell's great influence in the history of live-stock improvement of this country first asserted itself towards the close of the eighteenth century. He began his life's work in the year 1755, and in the height of his success, some thirty years later, rented three Leicester rams for 1200 guineas. In the year 1789 he let seven rams for 2000 guineas, and in the same year the Dishley Society hired the remainder of his rams for 3000 guineas.

It is difficult to follow Bakewell's methods, for he was careful to let the public know little about them; but by his selection of big sheep, and having the eye of a genius for form and proportion, he undoubtedly brought the Leicester to its highest pinnacle of fame.

According to Culley, Bakewell began by making a tour of selection amongst neighbouring flocks. In Lincolnshire, which was not far removed from his farm, he had the assistance of Mr Stow of Long Broughton, who was the purchaser of many of his sires. Even in these early days, about 150 years ago, they were noted for their fine sheep in the Fen country; and that they were jealous of that reputation, the hire of one of Bakewell's tups for 1000 guineas by four Lincolnshire breeders for a season is more than *prima facie* evidence.

Culley, the authority already mentioned, expresses the opinion that prior to Bakewell's time there was no criterion in sheep but size. Before Bakewell's improvements, the description of the sheep then generally found is interesting. "They had," says Culley, "a large hollow behind the shoulders, upon the top as well as the side, now known by the technical name fore flank, which in a fat sheep now not only fills up the former defect, but even projects beyond the shoulder and gives a great roundness to the form of the carcass."

There was a great air of mystery about Bakewell's improvement of the Leicester. In these days of Flock-Books and public registration, no doubt many of his methods would be condemned, though the results of his handiwork were undeniably successful. He was supposed to have a black ram in the background for one thing. Sir John Sebright was moved to protest in the *Farmer's Magazine* of 1827 against "the mystery with which he [Bakewell] is well known to have carried on his business. The various ways which he employed to mislead the public induce me not to give that weight to his assertion which I should do to his real opinion could it have been ascertained."

Then the Bakewell Ram Club consisted of twelve members pledged to absolute secrecy. One of the conditions was that "the much dreaded members of the Lincolnshire Society should not have a ram unless four joined and paid 200 guineas for him."

Youatt, in his well-known volume, says that Bakewell selected sheep "of the most perfect symmetry with the greatest aptitude to fatten, and rather smaller in size than the sheep then generally bred." He did not object to use "near relatives." Referring to the supposition that Bakewell created the new Leicester by crossing different "sorts of sheep," Youatt remarks, "There does not appear to be any reason for believing this, and the circumstances of the new Leicesters varying in their appearance and qualities so much as they do from the other varieties of long-woolled sheep can by no means be considered as proving that such was the system which he adopted."

Locality.

The Leicester is chiefly found in the more northerly parts of England, on the bleak wolds of Yorkshire. There they feed very quickly, and have the constitutional stamina to resist the inclement weather to which they are so freely exposed.

Characteristics.

In appearance the Leicester sheep has a bold head and the rams are slightly Roman-nosed. The head is broad at the poll, which is sometimes covered with a forelock and sometimes not. The lips and nose are black. The back is broad and level, the breast deep and wide, carrying a full bosom. A good sheep must be deep through the heart, the ribs being well-sprung, the loins wide, and the dock carried level with the spine. The fleece, which is a valuable portion of the sheep, should be free from black hairs. The sheep should stand squarely, with a leg at every corner.

In the Flock-Book Mr Joseph Crust writes as follows: "The Leicester has during the last few years made rapid strides towards perfection. . . . By continuous and judicious crossing with other sires of large size and heavy fleeces, a class of sheep has been produced of corresponding proportions, with a fulness of wool, yet retaining the original propensity to fatten. They are hardy and well adapted for any climate and soil, during the severe winter months being folded on turnips in the open fields on the bleak wolds of Yorkshire, where they feed quicker than any other class of sheep that have been wintered in the same situation, requiring less artificial food, and suffering a minimum proportion of loss; they are also remarkably sound in their feet, being seldom attacked with foot-rot."

Clip and Weight.

The Leicester is known to clip and weigh well. It is not surprising to learn that fleeces of 21 lb. to 28 lb. of washed wool from rams are not uncommon. A good flock average would be about 12 lb. The breed attains to heavy weights up to 240 lb. As long ago as 1793 a ewe at Mr Paget's sale, in Leicestershire,

scaled 36 lb. per quarter. She had 16½ lb. rough fat, and including the offal weighed 177½ lb.

For Crossing.—The Leicester has in the past been particularly favoured by colonial and foreign buyers for crossing purposes. Its aptitude to fatten is a strong point in its favour. In New Zealand the rams are highly esteemed in the production of freezers. They are also much used as ordinary commercial sheep in the north of England, where their freedom from foot-rot enables them to do well in rough country.

It would be ungracious not to recognise the part played by the Leicester in the improvement of other breeds. From the time that Bakewell gave it a preponderating influence in the work of English sheep-breeding, it has been used in the evolution of the present-day type of the following breeds: Lincoln, Wensleydale, Kent or Romney Marsh, Border Leicester, South Devon, Devon Longwool, and Cotswold.

MANAGEMENT OF LEICESTERS.

In Leicester flocks the system of management is exceptionally systematic and skilful. The method of feeding is liberal, and this is well justified by the yields of wool and mutton obtained from the breed. The general practice in the leading flocks is well indicated by the following notes received from Mr T. H. Hutchinson, Manor House, Catterick, Yorkshire, who has long maintained a very celebrated flock of the genuine old English Leicesters:—

"I keep a flock of pure-bred Leicesters, which I find to answer my purpose better than any other breed. My aim is to produce as much wool and mutton as possible from the produce of my farm, and to keep the land in a very high state of cultivation.

"I annually put 200 ewes to the ram, and generally average about 1½ lambs to a ewe. The ewes are put to the ram in the last week in September.

"Besides the lambs I breed, I buy from 150 to 250 to 'turnip' during the winter. As I cannot buy pure Leicesters, I generally buy 'north' lambs—that is, lambs bred from Cheviot ewes with three crosses of the Border Leicester. These do remarkably well on turnips, and go off fat in February and March, weighing from 16 lb. to 22 lb. per quarter.

Feeding of Ewes.—"The ewes run on the grass in autumn, and have roots with cut oat-sheaves given in addition before lambing, also hay if I can spare it. After lambing, the ewes get roots with a mixture of malt-combs, linseed-cake, bran, oats, and cut hay, until the pastures are good enough to keep them going.

Feeding Lambs.—"The lambs are weaned in July either on to some after-grass or good old pastures, until cabbages or thousand-headed kale are ready. After that they go on to Fosterton Hybrid turnips, then finish on the swedes. As soon as the lambs go upon cabbage, &c., they are allowed a mixture of crushed tail corn, linseed-cake, malt-combs, bran, &c., made into a kind of lamb-food. I prefer a mixture to cake alone. When put upon turnips the roots are all cut, the turnips all being stored in October and early part of November. Hay and straw are also given. I find nothing like plenty of dry food for sheep on turnips.

"A piece of rock-salt should always be kept in a trough, for the sheep to go to when they like.

"The lambs and ewes are all dipped after clipping, and again in autumn.

"The rams for show purposes are kept as well as possible, and get the best of everything likely to do them good.

"You ask me what quantity of turnips or other food should be consumed per day. I am sorry to say I cannot tell you. I always let the sheep have plenty to go to, and fancy they are better judges than I am as to the quantity they require; at any rate, I leave it to them to decide."

Gainford Hall Flock.

In Mr George Harrison's well-known flock of Leicesters at Gainford Hall, Darlington, lambs are dropped in March and weaned in July. For some time before weaning the lambs get a mixture of cake and oats in a pen from which the ewes are excluded. After weaning this mixture is continued to the lambs on clover or other foggage. A number of fat lambs are sold for killing in June, July, and August, the prices ranging

from 35s. to 40s. each. Fat hoggets are sold in the following January at from 50s. to 60s. each.

Ram breeding is a special feature in Mr Harrison's flock. He sells ram lambs in September and October for the getting of lambs for early fattening. He also sells a number of shearling rams for stud purposes. Young rams are put on clover, thousand-headed kale, cabbages, turnips, and swedes in succession, and get a mixture of cake and corn in addition. This flock has taken a leading position in showyards, and rams for it find their way not only to all parts of this country where Leicesters are kept, but also to many foreign lands, including New Zealand, Australia, Tasmania, France, Denmark, and South America.

During the winter the ewes in this flock get plenty of good hay, a run on grass, and a few roots after Christmas; after lambing, for about a month or six weeks, they get a mixture of cake and oats with plenty of roots. During the other parts of the year the ewes depend entirely on grass. They drop their first lamb when they are one year old.

A portrait of a Leicester ram is given in Plate 50.

BORDER LEICESTERS.

It is only since about the year 1860 that the Border Leicester breed of sheep has been known by this name. Prior to that, although its distinct characteristics had become well established, it was classed along with the English sheep descended from the same source, and termed the Leicester, or the Improved Leicester. The Border and the English Leicesters were so widely different in their form and wool, that it became impossible they could compete satisfactorily in the same classes, as the judges in the showyards, however carefully chosen, could not be otherwise than biassed. When the majority happened to be breeders of the English variety, the premiums went very naturally to the type they favoured, and *vice versâ*. If the judges were solely on one side, then a grave injustice followed. The position became the more acute as in the course of time the differences apparent in the two varieties widened; and at length it was found necessary to have distinct classes for the variety under notice, which has since been known as the Border Leicester. Down to 1868, all Leicesters, whether from the Midlands of England or the Border counties, were forced to compete in the same classes at the shows of the Royal Agricultural Society of England; but next year, at the Manchester meeting, they were divided as in Scotland.

Origin—Bakewell's Improvements.

Both varieties had their origin at Dishley, near Loughborough, where Mr Robert Bakewell began to improve the sheep he found around him in or about 1755. The precise method adopted by Bakewell is unknown, as a certainty. It is believed by some that he crossed the sheep of his shire of Leicester—"said to have been large coarse animals, with an abundance of fleece and a fair disposition to fatten"—with other long-woolled breeds, probably possessing smaller frames and more symmetrical proportions. Another and highly probable theory is, that without going beyond the sheep at his hand, he boldly adopted this material, and by breeding from selected animals of close affinities, and continuing this system as far as was advisable, he managed to establish a distinct breed, the main characteristics of which were large yet symmetrical frames, carrying heavy flesh upon fine bone; decided aptitude to fatten upon a moderate quantity of food; and capable of being brought early to maturity, while bearing a fleece of large weight and superior quality.

It is evident that the materials which Bakewell used must have been very plastic, since his improvements were quick in displaying themselves. So early as 1760 he commenced letting his rams for a guinea for the season's use. The reputation of the Dishley flock increased by "leaps and bounds," rising to such a pitch that twenty years after he commenced to let rams, Bakewell received no less than £3000 in hire fees in a single season. In 1789 it is stated that he netted £6000 by the letting of his tups. So general was the rush for

improvement in sheep stock about this time that it was computed no less than £100,000 were annually spent by Midland agriculturists in procuring sires. Large although this sum is, it is not altogether improbable, considering that in 1789 Bakewell received from £1000 downwards for the season's use of a single ram.

Such being the condition of sheep-breeding in the Midlands, it need not be marvelled at that agriculturists in far off shires, even in those days of slow, tedious, and imperfect communication, should have desired to share in the results which others had accomplished.

Messrs Culley's Flocks.

When Bakewell commenced his improvements, two brothers, George and Matthew Culley, were farming at Denton, not very far from Darlington. In 1762 and 1763 the brothers visited and became intimate with Bakewell, and from Dishley they brought rams with which they crossed the native Teeswater sheep, which then stood in high favour as a long-woolled breed. Proceeding in this manner, they were not long in forming a flock of their own, which was transferred to North Northumberland in 1767, and they took farm after farm until they paid an aggregate rental of about £6000 a-year. After having bred sheep in North Northumberland for nearly forty years the Culleys retired in 1806, when their sheep were sold off.

Other Early Improvers.

Mr Robert Thomson, who, like one of the Culleys, had been a pupil with Bakewell, also took a leading part in the introduction of the breed. He farmed at Lilburn, and afterwards at Chillingham Barns; and his flock, long known as one of the very best on the Borders, was bred directly from Bakewell's. It was at Lilburn that the first sale by auction of Bakewell sheep took place in the north. At Chillingham Barns Mr Thomson held annual lettings, and there, in May 1814, his entire flock was dispersed.

A part of Mr R. Thomson's flock passed into the possession of Mr James Thomson, Bogend, Duns, who had also formerly obtained rams from the Culleys and from Chillingham Barns; and it may be stated that this flock was until comparatively recently still in existence, being owned by Mr James Thomson, Mungo's Walls, a grandson of the tenant of Bogend. Having been bred by the Thomsons for upwards of a century, the Mungo's Walls flock formed the most perfect connection between the time of Bakewell and the present day in the history of the breed. It seems there were in Bakewell's flock two types of sheep, known as "blue-caps" and "red-legs," the latter being much the hardier of the two; and from what the writer has been able to learn Mr Thomson's sheep were of this hardier sort.

Rams from the flocks of the Culleys and Mr Thomson must have been very early and very generally used in the district around them as well as north of the Tweed, since flocks had multiplied to a wonderful extent at the time of the dispersion of the Culley sheep. Whether these other breeders obtained ewes of the improved breed or "bred in" from the ordinary country stock with Dishley rams it is difficult to determine satisfactorily at the present day. Yet this question affects the purity of the breed in later times to a grave extent, and certainly gives weight to the prevalent impression that there is much Cheviot blood in the Border Leicester. It is next to an impossibility that all of the Tweedside, Glendale, and "Barmshire" breeders could have obtained their stock ewes and gimmers from Thomson and the Culleys so quickly and to the extent necessary to account for the size and number of the flocks in 1806. Besides, these pioneer breeders were chary of parting with females.

The subject is a difficult one to treat of satisfactorily, and is referred to here only in a suggestive way. One thing is most evident: that flocks of "improved Leicesters," whether pure — relative though the term may be—or not, sprang up, like the proverbial mushrooms, in North Northumberland, from which they quickly spread into Berwickshire and Roxburghshire.

In later years the breed has spread widely over Scotland, and although the counties of Roxburgh and Berwick may still be said to be the headquarters of the breed, Border Leicesters are bred as successfully in the north as in the south. Indeed, the late Mr David Hume, whose flock was located in Forfarshire, was invincible for several years for shearling rams at the annual shows of the Highland and Agricultural Society. There are also very good flocks of the breed as far north as Ross-shire, and even in the county of Caithness.

The Mertoun Flock.

To a large extent the history of the Mertoun flock is the history of the breed, in Scotland at any rate. It was founded by Mr Hugh Scott of Harden, grandfather of the present Lord Polwarth, in 1802, and for over forty years was entirely a self-supporting flock, not a single animal, male or female, having been introduced from any other flock during all that time. The system on which Lord Polwarth worked was to carefully select animals from the different strains of blood existing in his flock and mate them together on prearranged lines, and in this way it was possible, not only to minimise the risks of in-and-in-breeding, but at the same time to secure a uniformity of type and character in the flock that could hardly have been obtained in so large a measure by any other method of breeding.

Lord Polwarth was careful to note the breeding qualities of rams which were bought from the flock, and never hesitated, whatever the price asked, to buy back for use in the parental flock any ram which bred specially well and which seemed likely to be of advantage in the flock. Sheep brought back in this way, it was thought, answered to a certain extent the same purpose as introducing new blood, for their stay for some time under different conditions as to soil and climate was believed to have an effect upon them which enabled them to exercise a reinvigorating influence upon the parent stock.

There is no doubt this system was very successful for a long period of time. Towards the close of the last century few breeders of Border Leicesters considered their flock properly equipped without a "Polwarth" ram as principal sire. Many first-class breeders would hardly buy anything else for ram-breeding purposes. At that time Lord Polwarth was accustomed to get very high prices for his annual draft of tups sold at Kelso, thirty sold in 1890 realising within a few pence of £54 per head, while one sold at £155. A few years later one very fine ram was sold to Mr Lee of Congalton for £275, which up to 1907 was the record price for a ram of the breed.

Since the advent of the present century other breeders have been getting a larger share of patronage from buyers of high-class rams for stud purposes. The Mertoun flock continues to show the remarkable gaiety and strong family likeness which so long characterised it, but there are those who incline to the belief that it would be benefited by the infusion of fresh blood.

Other Noted Breeders.

The rising popularity of a considerable number of other flocks of Border Leicesters has been a gratifying feature of the Kelso ram sales in recent years. At these sales in 1907, Mr Matthew Templeton, who is a tenant on the Mertoun estate, obtained an average of £40, 5s. for each of his best "cut" of twenty-five shearling rams, one ram bringing £160. On the same occasion Messrs Smith, Leaston, Upper Keith, received an average of £35, 13s. for each of eight shearling rams, one of these rams breaking all previous "records" by realising the handsome sum of £280, the buyer being a New Zealand flockmaster. Another ram from the same flock was sold for £200 in 1908. It is much to the advantage of all interested in the breed that so many flocks of it of the highest character and quality exist throughout the country.

Characteristics.

As already stated, there were two families in Bakewell's flock, the "blue-caps" and "red-legs." Formerly "blue-caps" were pretty common on the Border, but for a long period the "red-legs," owing to their greater hardiness, have held possession. Their representatives of the present day are admirably described by Mr John Usher, in *The Border Breeds of Sheep*, thus: "The head of fair size, with profile slightly aquiline, tapering to the muzzle, but with strength of jaw, and wide nostril; the eyes full and bright, showing both docility and courage; the ears of fair size and well set; the neck thick at the base, with good neck vein, and tapering gracefully to where it joins the head, which should stand well up; the chest broad, deep, and well forward, descending from the neck in a perpendicular line; the shoulders broad and open, but showing no coarse points; from where the neck and shoulders join to the rump, should describe a straight line, the rump being fully developed; in both arms and thighs the flesh well let down to the knees and hocks; the ribs well sprung from the backbone in a fine circular arch, and more distinguished by width than depth, showing a tendency to carry the mutton high, and with belly straight, significant of small offal; the legs straight with a fair amount of bone, clean and fine, free from any tuftiness of wool, and of a uniform whiteness with the face and ears. The sheep ought to be well clad all over, the belly not excepted, with wool of a medium texture, with an open pirl, as it is called, towards the end. In handling, the bones should be all covered; and particularly along the back and quarters (which should be lengthy) there should be a uniform covering of flesh, not pulpy, but firm and muscular. The wool, especially on the ribs, should fill the hand well."

Mr Usher's description is still generally accepted as on the whole accurate. No doubt many of the best sheep in most flocks are occasionally flesh-coloured in the muzzle, but it is regarded as an evidence of hardiness to have it perfectly black; therefore in any general description this point ought to be emphasised. The ears should be of moderate length, and boldly set, but thickness, even at the base, in shearling rams as well as in ewes should be a disqualification. Again, the ear should neither be too much flesh-coloured, nor blue nor purple. A darkish —not a brownish—skin, covered with fine white hair, is most to be preferred, although black spots, when distinct and separated from each other, are not at all objectionable. Then the face should be covered with short white hair; and on no account should there be any blueness. Corded or scored faces are specially to be avoided in females and shearling rams. These are considerations of primary importance. The legs, from where the wool ceases to grow, should be covered with short white hair; the "cluits" should be black, and the pasterns as upright as possible.

The "blue-caps" of by-past times have been described by the late Mr John Grey, Dilston, as having "blue faces, generally bare on the scalp, and red when lambed; and when mature, easily broken by flies; on which account they were not favourites with the shepherds. They were good feeders, but light of wool." It is evident that what are now termed English Leicesters are the representatives of this Bakewell family, although time has both modified and accentuated their former characteristics. The English sheep is not such a large-framed sheep as the Border Leicester, and is considerably shorter in both neck and legs; but it is much more compact, wider through the heart, and heavier in front in proportion to its size. It is also a much deeper sheep, and carries a heavier fleece. The bone, too, is finer, and the amount of mutton yielded, size considered, is greater than that usually carried by the Border Leicester.

Value for Crossing.

Except as ram-breeding flocks, Border Leicesters are not now kept to the same extent as they formerly were, the mutton being too coarse in its texture, and the fat too much of the consistence of tallow to be put to a profitable use, or please the palate. These are drawbacks to the general utility of a breed which has few rivals in reaching early maturity, and which produces a great weight of mutton and wool in a given time. Its outstanding merit lies in its pre-eminent suitability for crossing with the Cheviot, Blackface, and other varieties. The cross with the Cheviot is the most popular, the produce being the Half-bred variety now largely reared. On most turnip and grain farms a stock of Half-bred ewes are kept, which being again crossed with the Border Leicester ram, produce three-parts-bred lambs: These quickly develop, and being fed at high pressure, are generally in the fat market before they are a year old.

MANAGEMENT IN BORDER LEICESTER FLOCKS.

Leaston Flock.

The management of Border Leicester flocks follows fairly general lines. In the well-known flock of Messrs A. & J. K. Smith at Leaston, Upper Keith, lambing takes place, as a rule, from the 12th of March to the end of April. Weaning takes place about the beginning of August. Single lambs before weaning get no extra feeding; twins get about ¼ lb. of a mixture of compound cake, bruised oats, and locust-meal. After weaning ram lambs are put on clean pasture or foggage, and the extra food continued. In addition they get cut tares daily to begin with, followed by cabbages, white turnips, and yellow turnips, as they are in season. Ewe lambs are also put on clean grass after being weaned. No box food is given in the case of the ewe lambs, although they may get a few cabbages and then turnips.

Rams are sold as shearlings in September and October at from £4, 10s. upwards, one, as already stated, having brought £280 in 1907. Surplus gimmers are sold in September at from £3 to £10 per head, and cast ewes in October at from £4 to £5 also per head.

Ewes after being settled to the tup are kept on rough pasture until the middle of December—sometimes till the end of the month. Then they get five hours daily on turnips, with a run off on a grass field, and as much hay as they can eat. If the frost should be hard fresh turnips are laid out daily on the pasture. Nothing in the owner's experience is worse for in-lamb ewes than frosted turnips. At Leaston box-feeding is begun ten days before lambing, the practice being to give 1 lb. each of a mixture consisting of bran, oats, and compound cake. Ewes with single lambs get no extra feeding after lambing. Ewes with twins are fed up to the end of May with the mixture already mentioned.

Lambs about this time are shed off from the ewes every morning and get their extra feed. As a rule, a corner of the field is railed off for this purpose. With this exception grass is the staple food all through the summer. Good breeding ewes are sometimes retained in the flock until they are ten years of age.

The ram lambs are fed on yellow turnips, and get the concentrated food already mentioned up to about the middle of February, when they get swedes. The concentrated food is also changed at this time to a mixture of linseed-cake, bran, bruised oats, and locust-meal. All through the winter the young animals get the best of hay, and have access to rock-salt, getting also a little common salt in their concentrated food. As soon as young grass is ready they are put upon it. Tares are begun about the middle of June, and cabbages in August. Sometimes thousand-headed kale is used. The extra food is increased daily, bran being stopped on ·grass, and peas and Bombay cotton-cake substituted.

A Border Leicester flock requires very careful attention, and involves much hard work, especially where considerable numbers of rams are turned out annually. The annual output of these from Leaston is about two hundred. A close study has to be made of the mating of the ewes to suit the different rams. When a ram is bought he is not given many ewes the first year, but his produce is carefully watched, and ewes drawn to suit him for the following year. Messrs Smith like to run their pure-bred sheep thinly over the pastures. No class of sheep, in their experience, do well heavily stocked.

Galalaw Flock.

Lambing in the Galalaw flock, belonging to Messrs J. & J. R. C. Smith, takes place in March and on to the middle of April, the lambs being weaned in the latter half of July. During summer the lambs get a little box-feeding—lamb-food not more than ¼ lb. daily, the lambs being run off from their mother for this purpose. After weaning they go for three weeks on clean old grass, and have the box-feeding continued. The ram lambs are carried on to the shearling stage and then disposed of at Kelso, prices running from £5 upwards, the highest being £150 per head. Ewe

lambs are drawn in the spring, when they weigh from 18 to 20 lb. per quarter. Those not required for breeding purposes are sold at about 8d. per lb. as a rule.

The ewes in this flock get ten weeks turnips before lambing, and a little box-feeding and hay as lambing time approaches. The box-feeding is continued until June. Ewes with twin lambs get young grass; those with singles the two-year-old grass. After weaning ewes are all the better of a change off the farm to higher-lying land if possible. They return to their own pastures a few weeks before the tups go out. This change helps to ensure a good crop of lambs, and admits of the cleaning up of the home ground.

Ewes have their first lamb at two years old, and may either go on for four or five years or be cast at two and a half, as the appearance of the ewe or her produce warrants.

Ram lambs are brought through the autumn on seeds, with tares (if available) or cabbage. Hay foggage is valuable, but not always easy to get. In November they go on to turnips, and get cut turnips, hay, and a little box-feeding during winter. The sooner they get young grass in the spring the better; and they depend upon it, with tares in addition as the sales draw near. Any change of food is made gradually, and surfeiting is avoided. Stock rams are kept among the ewes, except for two months before tupping time, when they are kept by themselves, and get a little extra keep in the way of box-feeding.

Pictstonhill Flock.

In Mr W. S. Ferguson's flock at Pictstonhill ewes are kept from the end of May until August with their lambs on nothing but grass. The lambs are weaned in August, and ewes go on the grass fields until the following March, when they are taken in to lamb. In an open winter the ewes require little more than they pick up, but when frost and snow prevail they get a small quantity of turnips daily (not ad lib.) and hay. If the snow continues long 1 lb. of oats and cake per day is given. When turnips are scarce, oats, cake, and hay bring them through the winter quite as well as, if not better than, roots, but at greater cost.

At Pictstonhill the ewes lamb in March and April. A large lambing shed is available, but is used as little as possible; sheltered fields and dyke-sides are better if the weather be at all favourable. The ewes after lambing are liberally hand-fed to keep them in milk, the quantity of food varying with the weather. Cut turnips, mostly yellows, are given. The time turnips are given and the quantities depend entirely on the spring and the grass. Nothing in Mr Ferguson's experience will put ewes and lambs into sound health and thriving condition like the first flush of young grass.

The ewes and lambs get a little hand-feeding until about the middle of May; after that, grass and nothing else. When lambs are weaned they get the best of the grass, cabbages, and ½ lb. of hand-feeding. When ram and ewe lambs are separated in September the rams are continued on cabbages, turnips, and grass, along with the ½ lb. of hand-feeding, while the ewe lambs get the run of grass, and if turnips are added the cake is taken away. About 1st December the rams are generally folded on turnip land for the winter and cut turnips fed to them in boxes, and hay given. The hand-feeding may be increased to 1 lb. a-day, but more is seldom given at any time except for a month before the September sales, when the allowance is doubled if the rams will take it. They get some cut tares on the grass in July and August, but only if the grass is insufficient.

Rams in summer have always to gather their food, and so make them muscular and active. If all the food is taken to a Border Leicester ram so that he can lie and feed and sleep at grass in summer, Mr Ferguson thinks that it is all the worse for the man who buys him. At Pictstonhill rams are never housed except in rough days in early spring after clipping. The owner is strongly of opinion that the modern artificial bloom put on rams for sale through house-feeding is doing harm to the breed. It was never done in the old days. This, however, does not apply to the few animals drawn out for show-training, but even with these the practice at Pictstonhill has generally been to walk the sheep on a hard road for at least a quarter of a mile to a bit of pasture every good day.

Oldhamstocks Flock.

In Messrs Clark's old-established flock at Oldhamstocks, Cockburnspath, lambing begins, as a rule, about the 1st of March, and the lambs are weaned in the beginning of August. From the beginning of December the ewes in this flock are allowed every day a cart-load of white turnips to every seventy, and about ½ lb. each per day of cake and oats. As soon as the turnips are finished ewes with double lambs get an extra allowance of the artificial food. After weaning the ewes are put for the first fortnight on to the poorest pasture on the farm. A few weeks before the tups are let out they are put on to better pasture.

Messrs Clark feed a considerable number of half-bred lambs which they buy in August. In addition to turnips the lambs are allowed daily about ¼ lb. of cake and oats. The lambs are generally sold about the end of May in the following year as soon as they are clipped. The better half of them by this time will run from 80 to 100 lb. of mutton per carcase, and they realise about 50s.— fully 10s. per head less than they fetched a few years ago.

Females of the pure-bred flock have their first lamb when about two years old. They are usually cast about five years old. When the tup lambs are weaned in the beginning of August they are put on to foggage if there be any on the farm. If not, they get tares on a clean pasture. Whenever the turnips are ready they get a few of these, bringing them gradually on to as many as they will eat, with an allowance of cake. Stock rams on the farm receive no extra feeding except when at service.

Deuchrie Flock.

Mr Jeffrey at Deuchrie, Prestonkirk, has the lambs arriving from the second week of March onwards. They are weaned about the beginning of August. Twin lambs get a little lamb-food before weaning; single lambs nothing but the milk and grass. After being weaned the tup lambs are put on foggage if there be any available. Ewe lambs are kept on first-year's grass.

Shotts of the tup lambs are sold in spring in the fat market, making, in 1908, 53s. The others are kept on until shearlings and sold at the Kelso, Edinburgh, and Lockerbie sales. Ewe lambs are drawn at the same time as the tup lambs, and the best only kept for breeding purposes.

Ewes before lambing are generally on turnips for several hours every day, and run off on to heather at night. When nursing they are grazed in the fields, getting about ¾ lb. each of some feeding mixture, not always the same. When the lambs are weaned the ewes are turned on to the hill or poor pasture, and are brought in a week or two before the tups are put out. Ewes, as a rule, have their first lamb at two years old, and they are cast after having four crops.

Young rams are managed as ordinary feeding sheep until clipping time, after which they are allowed some extras in the way of green tares and cabbages in preparation for the ram sales. Stook rams are generally wintered among the ewes, and do not call for special care, as they generally get fat enough without any extra feeding.

Whittingehame Flock.

In the Whittingehame flock of Border Leicesters, belonging to the Right Hon. A. J. Balfour, M.P., lambs are dropped between 15th February and 1st April. No extra food is given to the lambs before weaning. Before lambing the ewes get yellow turnips and straw; after lambing, swedes, with a mixture of cake, bran, and oats until pasture grass is ready.

After weaning, about the end of July, the ram lambs to be kept for breeding purposes are put on to young grass, where they get cake, or a mixture of cake, Indian corn, and bran. This mixture is continued until September of the following year,—about 1 lb. per day is the full quanity, but less is given at first. In winter, when the pasture is done, the young rams get turnips and hay.

PREPARING RAMS FOR SALE.

Messrs Smith, Leaston, have been good enough to supply the following information regarding their method of preparing Border Leicester rams for sale. They begin clipping about the 1st of April. All new wool is left on, but on no account do they leave old wool. Sheep are never washed when newly clipped. The grease that there is in the wool makes the sheep less liable to catch cold. The animals are kept in a well-ventilated dry shed, and are let out for a run daily. When warmer weather comes, say about the middle of May, the sheep are washed with soap and water and made to swim through cold water. When dry and coats thoroughly set, the sheep are dressed all over. In a week's time they are dipped with fairly strong dip, which helps the growth of the wool.

Nothing more is done until the second week in July, when the rams are again washed. Great care must be taken not to rub out the curl in the wool; indeed, they should not be rubbed at all—just clapped with the hand. When the coat is set (a full week is usually given) dressing is again done. This is the most important dressing of the year. Backs are well taken down, coats squared out at the rump, tails dressed to set off quarters, wool left full between hind legs, necks close taken in and tapered off to front of breast. The sheep are then dipped again in strong dip. Care is taken not to let the sheep out in strong sun for a few days after.

As the sales come on generally in the second week in September, the next washing, &c., is done two weeks before. This time dipping is done before washing, this being to give the sheep a nice rich bloom after they are washed. They are again washed a week before the sales, and carefully gone over with a pair of shears. Care is taken to show the sheep in first-class bloom, but it is very difficult to do, especially if the weather be wet and when such large numbers have to be dealt with.

The rams' heads are capped during the later part of summer. This helps to prevent them fighting, and also keeps away the flies. In very hot weather they are sometimes kept in the house during the day and let out at nights— the select lot at any rate. The Leaston sheep are carefully classed to suit the different markets. The best breeders' sheep are sold at Kelso—viz., those with nice curly wool, well covered white heads, well-set ears, strong loins, stylish walkers, with good flat bones. For Edinburgh the rams are of the same style, but closer in coats, as most are sold for crossing with Cheviot ewes for breeding half-bred lambs. The rams for the Perth sales must have very strong curly coats; the character of the head is not so important, as the Perth rams are used for crossing with Blackface ewes. Sheep used for this purpose must be active on their legs. The remainder are sold privately at home.

A portrait of a Border Leicester ram is given in Plate 50.

THE LINCOLN LONG-WOOL.

The Lincoln Long-wool sheep is native to the county from which it takes its name. It is one of the oldest of our breeds, being known in Lincolnshire for upwards of 150 years. Many of the chief flocks are, so to speak, family heirlooms handed down from one generation to another.

Lincolnshire has always been noted for the size of its stock. It produced the Fen type of Shire horse, the red variety of Shorthorn cattle, the Long-wool sheep, and the curly-coated pig—every one of which stands out in respect of size and substance.

Doubtless, in the earlier days, the Leicester was employed to get fleece and form; but not a few hold that the Leicester is much indebted to the Lincoln, yet perhaps not so much as the Lincoln to the Leicester.

Noted Early Flocks.

There are records in existence tracing the descent of flocks in the present day as far back as 160 years. Thus Mr J. E. Casswell's flock at Laughton, Folkingham has been in the hands of the family since 1740, and that of Mr Tom Casswell at Pointon since 1755. Other well-known flocks can claim almost as ancient a record—notably that of Mr Henry Dudding at Riby Grove, which for so many years was maintained by his father at Panton. The names of Clarke, Kirkham, Need-

ham, Smith, Cartwright, Howard, and Wright, are prominently identified with the development of the Lincoln sheep. It is interesting to note that Mr Henry Dudding's grandfather was one of four who hired a Bakewell ram for a season at the record price of 1000 guineas.

In the olden days ram-lettings were great institutions, but in modern times they have been superseded by auction sales. As far back as 1837 Mr G. Casswell, the grandfather of Mr J. E. Casswell, let a ram for £90. The averages made in the middle of last century indicate that the very high prices paid in recent years are founded on a long period of high figures. In the old Biscathorpe flock (Mr Kirkham), for instance, the average of 150 sheep in 1864 was £22, 12s. 4d. This flock also averaged in 1872 £25, 11s. 6d. for 120 head; while in 1873 the average was £35, 17s. 7d. for 70 sheep.

Modern Records.

All previous records, of course, sink into insignificance before the 1450 guinea sheep at Mr Dudding's sale in 1906. That sheep was purchased for the Argentine. Mr Dudding has on two other occasions realised 1000 guineas for a single ram, and in the year 1907 the top figure at his auction was 900 guineas. In 1906 Messrs Wright of Nocton Heath, Lincoln, sold their flock to be exported to the Argentine. The buyer was Señor Cobo, and the total sum paid is said to have been in the neighbourhood of £40,000.

Characteristics.

The Lincoln is a big, bold type of Long-wool. One of its chief sources of value is its matchless fleece. A sheep of fine symmetry, it carries a strong head on a thick neck. The ears should be wide set and carried forward. The eye should be bold, the nostrils broad in the ram, and the muzzle shapely. There should be no spots on a white face. The sheep should stand squarely, be broad in the back, with no weakness of rib to be detected under hand. The rump and loin should be well filled, and the dock carried high. A low-set dock is a sign of weakness. The leg of mutton should be full—a point in which defect is most frequently seen. The limbs should be white—a grey leg being a bad fault.

The fleece should be as nearly as possible of one quality all over the body, and extend down to the hoofs behind. The staple is very broad and wavy. A shearling will sometimes grow locks of close on 2 feet long. The cod should be well covered, otherwise there is likely to be a scarcity of belly wool. The forelock is a strong point in a show sheep.

Clip and Weight. — There are instances on record of exceptional clips, such as 32 lb. of washed wool. Well-grown rams commonly clip 25 lb. of washed wool. Probably the best flocks will average 14 lb. of wool, though 12 to 13 lb. is a good figure.

Mr Henry Smith, jun., of Cropwell Butler, Nottingham, says that "a flock of well-bred ewe hoggets will yield a stone of wool each; rams much more. I believe that the 350 guinea ram sold by Mr John Pears in 1896 to Messrs Kirkham of Biscathorpe and Cartwright of Keddington clipped 31 lb. This was an unusual weight certainly, but many go over 20 lb."

The breed is the heaviest in England. It has produced phenomenal weights at Smithfield. Mention might be made of the winning pen of three wethers in 1907, which weighed 10 cwt. 6 lb at 21 months 3 weeks and 4 days. An analysis of Smithfield weights shows that the average gain of lambs in live-weight is about 11¼ ounces daily, though it is sometimes as high as 12¾ ounces. The wethers gain about 8¼ ounces, the best turning 9 ounces daily.

Mr Henry Smith, jun., writing of his champion pen in 1896, says that they averaged 219 lb. when dressed, and the butcher reported that they were "very good fleshed sheep."

Early Maturity.—The breed is very free from foot-rot and matures early. Sheep of the Lincoln type are kept largely for crossing purposes, the Down cross being much favoured.

Constitution. — A strong point is made of the constitution of the Lincoln sheep. It is kept in a natural state, although the show specimens are brought out in a condition of obesity. In winter the flock makes its living on turnips.

Foreign Trade.—Without the foreign demand the Lincoln flockmaster could not boast of high prices. Several breeders are their own exporters, but most of the business in high-priced sheep is done through agents. At the present time the Argentine is the chief market, although in former years New Zealand and Australia were splendid customers. As indicative of the strength of the demand, it may be mentioned that the Lincoln Long-wool Sheep Breeders' Association, founded as late as 1892, issued in 1905 4855 export certificates. In 1906 the number was 6928, and in 1907 3566.

MANAGEMENT OF LINCOLN FLOCKS.

The management of Lincoln flocks does not vary greatly in the leading flocks. As befitting sheep of large size, carrying exceptionally heavy fleeces of wool, the system of feeding is liberal in all successful flocks.

Cropwell Butler Flock.

In Mr Smith's celebrated flock of Lincoln sheep at Cropwell Butler, near Nottingham, the majority of the lambs are dropped in the month of February. The ewes are kept on pastures till about Christmas. On the approach of bad weather they are given in troughs a mixture of different foods, such as cotton-cake, oats, offal peas, and barley, with as much chopped straw as they care to eat, additional fodder being supplied in racks. Between Christmas and lambing time the ewes get white turnips or kohl-rabi, with plenty dry food. After lambing they go on to good pasture, and get mangels with some extra food if required. The ewes drop their first lamb when two years old, and are kept on as long as they continue to breed well, some of them till they are nine or ten years old.

The earlier ram lambs are weaned early in June, the others after midsummer. The ewes and lambs are penned with "creeps," which admit the lambs to boxes containing mixtures of crushed linseed-cake, cotton-cake, and oats. After weaning the lambs go on to vetches till the clover "eddish"[1] has grown. Towards August the lambs get Enfield

---

[1] Aftermath.

Market cabbages spread to them, and afterwards they are penned, first on ox cabbage, then on kohl-rabi, and at a later stage on swedes. None of the lambs in the Cropwell Butler flock are castrated. The best of them are sold for breeding purposes, many of them being exported to various countries. The "culls" are shorn in March or April when about 13 or 14 months old, and are sold at the Nottingham fat stock market. In April 1907 a clipped hogget from this flock brought 61s. for slaughter, while its fleece of 20 lb. realised 1s. per lb.—in all, £4, 1s. for a hogget under 14 months old.

The female hoggets are grazed on pastures without extra food during the summer. The best of them are added to the home flock, and the others are sold for breeding purposes elsewhere.

Young rams are clipped in April. They are kept on "seeds" and "clovers," with swedes, mangels, Enfield Market cabbage, and vetches in succession, supplemented by concentrated food, such as cake and corn, as may be required.

Riby Grove Flock.

In Mr Henry Dudding's famous flock at Riby Grove, Stallingboro', Lincolnshire, lambs are dropped during February, March, and April, and are weaned in May or June. For some time before being weaned the lambs get a mixture of oats and cake and a few cut swedes. When taken from the ewes the lambs are put on to new "seeds," where cabbages and mangels are thrown out to them. As soon as turnips are ready the lambs are folded on them.

Ewes are kept on pasture till turnips are ready, when they are folded on the roots, getting cut straw and a mixture of cake, oats, and peas. After lambing the ewes go on to grass, where they get roots and the same dry food. Ewes are two years old when the first lambs are dropped, and they are cast when their teeth give way.

Young rams are treated similarly to lambs after weaning, the allowances of concentrated food being gradually increased. They get vetches in racks. Stock rams are kept on pastures till they go amongst the ewes.

Phenomenal prices have been obtained by Mr Dudding for the produce of his flock—up to 1450 guineas for a shearling ram, and 200 guineas for a ram lamb; shearling ewes, 15 to 30 guineas, and 10 to 15 guineas for ewe lambs. Ram lambs weigh about 25 lb. per quarter.

A portrait of a Lincoln ram is printed in Plate 51.

COTSWOLD SHEEP.

The Cotswold breed of sheep is to be found chiefly in Gloucestershire and the neighbourhood of the Cotswold hills.

Origin. — The early history of the breed ascribes the name Cotswold to "the range of oolite hills running from north-east to south-west, and occupying the eastern division of Gloucestershire." That point, however, is in dispute—the other suggested derivation of the name of the breed being "cotes," buildings, and "wold," the wild open country. The manufacture of cloth in the Cotswold neighbourhood by the Romans implies the presence of sheep, so that there is some ground for the assumption that the Cotswold is one of the oldest breeds of which we have record.

Improvement.

The improvement of the breed since the early times of last century has been very marked. From a large slab-sided, long-limbed, and heavily coated animal the modern well-ribbed, clean-cut type of sheep has been evolved. In the days of Bakewell, no doubt, the Leicester was used for grading up; and amongst the names conspicuous in the advancement of the breed in the early days are Garne, Hewer, Large, Lane, Barton, Gillett, Fletcher, and others. About the middle of last century, when agricultural shows began to play a strong part in live-stock breeding, the excellence of the breed attracted widespread attention. Ram sales were established, and the Cotswold was dispersed all over the British Isles — particularly to the southern and midland counties. They were certainly very adaptable sheep, and were capable of making themselves at home under every condition of soil and climate. In 1847 Mr R. Smith, in the course of a prize essay, mentions that

Cotswold rams were "much sought after for crossing with short-woolled breeds, and with good effect."

The breed has always been a tenant-farmer's sheep, and its earlier successes are therefore all the more creditable. The Oxford Down is perhaps the most pronounced example of the value of the Cotswold for cross-breeding. The old name of this type was Down-Cotswold, it having been directly descended from the Cotswold on the one hand and the Hampshire Down on the other,

About fifty or sixty years ago in Gloucestershire it was estimated that 5000 rams were sold and let in a season at a total revenue of £50,000. A good export trade prevailed to America, Australia, and the Continent.

Characteristics.

In appearance the modern Cotswold is a noble sheep. The head is a fine index of a sire. In the ram it should be masculine, wide between the eyes, the eye full and prominent but kindly. The nostrils should be well expanded and somewhat broader than the face, the colour of the nose being dark. The cheek should be full and covered with white hair, a slightly blue tinge on the cheek and round the eye being an attraction rather than otherwise. The ear should be fairly long, not too thick, and well covered with hair. They should be well carried, and a dark spot or two on the tips is not an objection. The forelock of wool should be plentiful and full from the top of the head, which should be free from coarseness. Grey faces are, of course, not fancied, although difficult to breed out entirely. In the ram the neck should be big and muscular, and should be long enough to enable the sheep to carry his head with gaiety. The neck should fit into the shoulders, which should lie well back. The point of the shoulder should have a good covering of flesh, which should be well spread over the chine. The ribs should be well sprung, the hips broad and well covered, the fleshing deep. The frame should be square, the legs set on straight and well outside the body. Long lustrous wool is looked for, the wool being regarded, as in all breeds, as an indication of the character of the flesh.

Mutton and Wool Production.—The Cotswold is a ready mutton and wool producer. It can be brought to market at from 9 to 12 months old, with ordinary feeding, at from 90 to 112 lb. deadweight, and not infrequently the best flocks will turn out sheep from 120 to 130 lb. at that age. It is on record that a Mr Cother of Middle Aston killed a sheep aged 3 years and 9 months, weighing 336 lb. or 84 lb. per quarter, one of the legs of mutton weighing 54 lb. Good Smithfield live-weights for pens of three lambs, 9 months and 3 weeks old, are 5 cwt. 14 lb., and for three wethers, 20 months and 3 weeks old, 7 cwt. 3 qrs. 1 lb. Another good pen of three scaled 8 cwt. 2 qrs. 20 lb., at 21 months and 1½ weeks. The show of Cotswolds at Smithfield has fallen to very small dimensions, however.

In good flocks, from 9 to 11 lb. of washed wool will be clipped.

Prices.—The old prices and averages obtained at Cotswold ram sales are merely memories nowadays. There is a restricted home demand due to the advance of other breeds, although the foreigner takes a number, chiefly to North America, where there is a big trade for the Cotswold type of sheep. As a matter of history, it may be interesting to mention that in 1861 Mr R. Lane's average at his ram sale was £34, 10s. 8d.; in 1873 Mr R. Garne averaged £28, 16s. 4d. In 1864 Mr W. Lane of Broadfield bought one of Mr W. Hewer's rams for 230 guineas. Prices are very much lower to-day, the best figures being made privately.

MANAGEMENT OF COTSWOLDS.

The following plan of management is pursued by one of the largest, most prominent, and successful breeders. He mates his ewes in August so as to get as many lambs as possible in January and February, but the mating is continued so long as the lambs will be born by April. The ewes are put on grass or mixed seeds after mating, until the middle of November if the weather keeps open, and then on roots, which are fed sparingly, with a liberal allowance of hay. If the weather is very wet the flock ewes are run on pasture with hay only. In January the first ewes are brought to the lambing-pen. They go on pasture during the .day, and have roots carted to them. At nights, in the pen, they have as much hay as they can eat.

As fast as the lambs are born the twins are separated from the singles, the ewes with doubles being allowed a supply of Egyptian cotton-seed-cake and oats. The lambs are encouraged to eat oats and linseed-cake. The single lambs are treated in the way of feeding like the doubles, but the ewes are not given cake. The show lambs are selected when a fortnight old, and put by themselves with their dams. They receive oats, old split beans, linseed-cake, and a proprietary cake. The ewes are given a liberal allowance of cotton-cake and oats, with plenty of roots and the best hay. The lambs have pulped roots and hay. Pasture by day and the shed at night is the rule.

About the 1st of May the lambs with their dams are put on pasture, and the roots are carted to them. Weaning takes place about the middle of May, but the show lambs are not weaned till they go to the first exhibition, usually the Bath and West. All ewes- and lambs are brought into the sheds at night, till the end of March or thereabouts. There are about two-fifths twins.

Ewes are only discarded when their breeding days are over. One ewe, for instance, was breeding until she was fourteen years old. Rarely are any marketed but old culled ewes. The yearling rams and ewes, and ram and ewe lambs, are sold principally for breeding purposes, a number going to Canada and the United States,—the remainder going for crossing purposes, only a few of the best being sold to home breeders.

When the lambs are weaned they go on young mixed seeds, sainfoin or vetches (tares), till roots are ready in autumn. About fifty of the best yearling ewes come into the flock every year.

The system of management described above is typical of the Cotswold ram-breeding flock. The principle upon which flockmasters go is to get lambs early, so that they will be well grown by the autumn, to feed them well, using the lamb-creep to enable the lambs to have the freshest bite. Culling is done

in the summer, young sheep not up to the standard being dispensed with, and the good breeders kept as long as possible. Sainfoin is the popular legume for Cotswold sheep. The store lambs have turnips and hay in September, and come on to swedes about Christmas. The culls are sold fat to the butcher. Those fit to be kept for shearlings are retained. About two-thirds of the ewe-lamb crop are drafted into the flock.

A portrait of a Cotswold ram is represented in Plate 51.

THE DEVON LONG-WOOL.

The Devon Long-wool is one of four ovine tribes found within the confines of the county of Devon. It is a very ancient breed, although in point of character there is a great resemblance between three of the types common to Devonshire.

Early History.—The early history of the Devon Long-wool is somewhat obscure. It is maintained that it is descended from the old Bampton Nott sheep which were marketed in the town of Bampton in large numbers. *Bell's Gazeteer* in 1836 refers to these sheep as "of large size and an uncommonly fine quality from the excellence of the pastures." A little later Professor Wilson, writing of the Bampton Nott, remarked that "it is very difficult to find a pure Bampton unmixed with other blood, a few only remaining in Devonshire and West Somerset."

There is no doubt that in the time of Bakewell and since, Dishley Leicester blood was used to improve the fleeces of Devon Long-wools, and there is equally little doubt that Lincoln blood was likewise introduced. The South Hams rams from Totnes district were also used, so that flockmasters freely borrowed from the best sources in evolving the modern type of sheep.

Characteristics.

It is apparent that at the present day the Devon Long-wool has much in common with other Long-wool breeds of the Lincoln and Leicester type. It is a big framed sheep, with a plenitude of bone and substance. It is rather bolder in the face than the Leicester, being larger in the head. It is wide at the base of the skull, and the nostrils in the ram are full and well developed. The ears are a good length, and a good tuft of wool should grow on the forehead. In appearance the Devon Long-wool is a bulky sheep, with a broad back, good loin, and strong dock. The leg of mutton is sometimes deficient. The skin is a nice pink. The coat should be uniform, the fleece being one of the important recommendations.

It may often happen that Long-wool sheep show great variety in the class of wool in a flock, and any tendency in the individual sheep to have coarse breech wool should at once be noted and that ram discarded for breeding purposes. When the writer inspected several of the leading Devon Long-wool flocks about the time the Flock Book was established, he was most struck by the lack of uniformity in the fleeces. That defect, however, is rapidly improving under the critical eye of the showyard judge and the flockmaster. The coat is often curly, in which respect it more resembles the Leicester than the Lincoln.

The flesh of the breed is of excellent quality, and should touch well under hand. Breeders have very carefully and successfully bred for "form," and the fact that so much success in the production of the fat lamb is attained in Devon, Somerset, and Cornwall from this breed, is independent testimony to the high character of its flesh.

Clip and Weights.—The breed clips and weighs well. The clip of a shearling ram would be from 18 to 24 lb., and perhaps exceed that figure. The ewes will produce to 12 or 13 lb. of wool, but 8 or 9 lb. is the average, and the lambs when shorn about 3 lb. and over. The breed is largely used for crossing with the Dorset Horn, the Dartmoor, and the Exmoor. It develops rapidly. In from 10 to 12 weeks fat lamb will dress to 10 lb. a quarter. The wethers are mostly sold as yearlings, dressing from 22 to 24 lb. a quarter.

MANAGEMENT OF DEVON LONG-WOOLS.

Rams are usually put with the ewes about the middle of September. When the tups are sound on their feet and vigorous, about fifty ewes may be allotted to each of them. In Mr E. R. Berry Torr's flock at Instow, North Devon, it is the custom after the rams have started work to take them in each morning and give them a few white peas and oats. The ewes are run on the best pastures, and a little cake and corn are given to them during the critical time. In Mr Berry Torr's flock rather over 50 per cent of twins are thrown by this treatment. The lambs come from the end of January to the middle of February. In the above-mentioned flock roots are avoided as far as possible for the ewes, the grass run being assisted by hay and straw chaffed and a few pulped roots. Just before lambing a little crushed oats or dried grains are given with the chaff and pulp.

The lambs from the best ewes are selected for rams, and the ewe lambs retained to keep up the flock. Of the remainder the fattest are sold when they reach about 9 or 10 lb. per quarter. Those not getting fat Mr Torr keeps on for turnips and sells them with others bought in about Christmas, when they scale from 18 to 20 lb. per quarter.

As soon as the lambs are born the ewes are dotted about in small lots on old pasture, and given a few roots with cake and corn; and when the lambs begin to pick up they are put on seeds, with the usual lamb creep, the youngsters having access to linseed-cake and lamb food, or home-grown oats and a few crushed beans or peas. For later consumption cabbage, rape, and kale are grown, and the flock maintained in a thoroughly healthy condition.

A portrait of a Devon Long-wool ram appears in Plate 58.

THE SOUTH DEVON.

The South Devon sheep is bred chiefly in South and Mid Devon. In Cornwall it may be termed the leading ovine breed kept by farmers. There is a great similarity amongst the long-wool breeds of the south-west if we except the sheep that roam on the moors.

Origin. — The origin of the South Devon is difficult to discover, but there can be no doubt that the modern type of sheep has been produced by the aid of Leicester and other Long-wools. They are supposed to have originated in the vale of Honiton, and descended from the South Hams Nott sheep, whose origin is wrapped in obscurity.

Characteristics.

In the early days they were described as having been inferior and badly shaped sheep, with heavy and coarse fleeces, but like most of the stock in the south-west of England, coming to great weight. They had brown faces and legs, which seemed to suggest a Devon affinity. The characteristics, however, were very materially altered by union with the Leicester.

Description. — The South Devon should carry a well-balanced head, broad, and rather long, and well covered on the upper portion with wool. The nostrils should be open and of a dark colour. The muzzle should be broad. The ears should be fairly long and of medium thickness, covered with hair, and are often spotted. The neck is strong and of medium length. A straight and level back from the withers to the setting on of the tail gives a symmetrical turn to the sheep. The shoulders should be flat and well covered, and ribs well sprung. The loins should be broad and the bosom deep. The sheep should stand squarely, with the legs well on the outside. The tail should be thick and fill the hand, the hind quarters being well filled and square. The skin should be pink and mellow. The fleece should be thick and even, of great length of staple, curly, and free from kemp or hair.

The South Devon is a sheep of nice symmetry, well grown, with plenty of bone and muscle. It thrives well on poor land, and responds very rapidly to generous treatment. Like most of the lustre long-wools it can grow fat when desired, but its strength of bone ensures that there is a good percentage of lean meat.

Weight and Clip. — It is on record that a seven months' old lamb of the breed weighed 224 lb., which is exceptional for any breed. The fleece on the average would weigh about 9 lb.

MANAGEMENT OF SOUTH DEVON SHEEP.

In the present day the South Devon sheep has attained a wonderful degree of perfection in symmetry and the weight of mutton carried. The efforts of the flockmaster in management are therefore to a considerable extent concentrated on the improvement of the fleece. The importance of the fleece can be realised when, even with wool on the down grade, the better class of flocks were able to realise from ½d. to 1d. per lb. more on their clips than was paid for ordinary wool in the district. In Messrs Tippett & Sons' flock at The Barton, North Petherwin, Egloskerry, the ewes average about 14 lb. of wool, whilst the rams clip from 26 to 33 lb. unwashed wool.

In the winter months hay, chaff, and roots are given on the grass, most of the flocks being kept in a natural state. Fattening is generally accomplished on roots and rape with artificial food. Fat lamb is one of the objects for which the breed is kept, and they attain good weights by Christmas. The ewes are very good mothers.

Mating takes place in September and October, but in the earlier flocks they put the rams to the ewes in August. Lambs arrive as early as the first few days in January, but February and March are the usual lambing months.

When the lambs are eight weeks old they can command from 40s. to 42s., making about 10d. to 1s. per lb. When sold by weight at a little over three months old, lambs scale about 68 lb., and later in the season they weigh up to 81 lb. They make the highest prices, as they carry a lot of flesh.

Ewes are drafted after the fourth lambing, but in the Barton flock favourites have been kept until ten years old.

A South Devon ram is represented in Plate 58.

WENSLEYDALE SHEEP.

The Wensleydale sheep is a product of Yorkshire. It is descended from an old breed called Mugs which were introduced into Wensleydale about the middle of the eighteenth century, and which were apparently a variety of the old Teeswater sheep. The Wensleydale doubtless resulted from a cross of the Leicester on this breed, and has taken on a distinctive character.

The dark countenance of the breed is in a good measure due to the use of a celebrated ram called "Blue Cap," whose sire was a Leicester ram. This sheep was shown at the Royal Agricultural Society's Show at Liverpool in the year 1841.

At a later period the Lincoln ram was used, but the success of this cross is doubted by breeders, who had the ram trade, rather than the grazier, in view.

Characteristics.

Appearance.—The Wensleydale ram is strong boned, with great length of side, and a big proportion of lean flesh. A scale of points has been drawn up by the Wensleydale Long-wool Sheep Breeders' Society, as follows :—

	Points.
Head.—Face dark; ears dark and well set on; head broad and flat between the ears; muzzle strong in rams; a tuft of wool on the forehead; eyes bright and full; head gaily carried	20
Neck.—Moderate length, strong, and well set on shoulders	10
Shoulder.—Broad and oblique	5
Chest.—Wide and deep	10
Wool.—Bright lustre, curled all over body, all alike in staple	10
Back and Loins.—Ribs well sprung and deep; loin broad and covered with meat; tail broad; flank full	20
Legs and Feet.—Straight, and a little fine wool below the hock; fore legs well set apart; hind legs well filled with mutton	20
Skin.—Blue, soft, and fine	5
	100

For Crossing. — The Wensleydale sheep depends to a large extent for its prosperity upon the demand for rams for crossing purposes. As far back as 1847 the tup breeders of the Dale presented Mr Macqueen of Crofts, in the south of Scotland, with a silver snuff-box "as a token of esteem for his encouragement of the breed." The rams are chiefly used on the Scotch Blackface ewe, on which they have been particularly successful, producing what is known in Yorkshire as the Masham sheep. One of the reasons of their success is that the Wens-

leydale mutton, unlike the mutton of many other long-wool breeds, is hard and firm to the hand. As a hill sheep, too, it is active, and the Wensleydale ram never fails to keep up with its quarry, be it a mountain ewe or one of the larger breeds.

Interesting Crosses. — In certain trials carried out at Newton Rigg in Cumberland, the Wensleydale ram cross on the Border half-bred ewe (Cheviot Border Leicester) came out very satisfactorily. The experiments of 1904-5 enabled the verdict to be passed on this cross that it produced the fastest growing lambs, although less capable of fattening as they grew: "It would appear that these lambs fatten easiest when near mature growth. The lambs were not allowed to arrive until the herbage came, and they were not weaned till four months old. They are run at grass and fattened on turnips in the early part of the year, being killed and sent to the London market.

Clip and Weight.—A good Wensleydale flock of ewes will clip from 9 to 10 lb. of wool. The rams will produce from 14 to 21 lb. A celebrated ram, "Royal Darlington," clipped 20 lb. The breed is kept at an altitude of from 700 to 1400 feet above sea-level, and such weights as 30-stone rams are not unknown, although the general run of shearlings is from 18 to 25 imperial stones.

MANAGEMENT.

In the management of a Wensleydale flock well defined lines are followed. In the choice of a sire most breeders have a leaning towards a twin ram,—some, indeed, will not use a single in the belief that precocity and prolificacy can thus be bred into the flock. The ewes are excellent nurses, and it is therefore not surprising that as many as two lambs to the ewe should occasionally be a flock average. The ewes themselves are capable of rearing, and do sometimes rear, as many as three lambs in a season.

Previous to turning the ewes to the ram a gentle system of flushing by change of pasture is adopted. Not only do the sheep take the ram earlier in consequence, but a better crop of lambs is believed to result.

The ewes, owing to the lateness of the district, are not put to the rams until October. Early lambing has no ascribable advantages. On the contrary, to face even a month of short-keep with a big crop of lambs does not appeal to the average flockmaster on the uplands of the north of England.

Little hand-feeding is done in a mild and open winter, but when necessity compels, as the severity of the weather frequently does, the simplest extra fare suffices. A little oats and hay will easily pull the flock through. At lambing, oats or cake are provided with dry fodder in the form of hay. When the turnips last they are also given, but not every farmer has a large enough breadth of arable land to grow them in sufficient quantities.

Shelter is provided for the new-born lambs, which are drafted out into the fields as the accommodation becomes limited.

A portrait of a Wensleydale ram appears in Plate 61.

KENT OR ROMNEY MARSH SHEEP.

The Kent or Romney Marsh sheep belongs to a race that is not of yesterday's creation. It is peculiar to the Romney Marsh district, where it thrives as no other breed could.

Sir Charles Whitehead has declared through the *Journal of the Royal Agricultural Society* that some one had suggested "that the aboriginal Kent sheep posed as the model of the cube upon four legs representing sheep in toy Noah's arks, and as toy manufacturing has long been carried on in the low countries, perhaps the breed, like hops and other good things, was fetched from Flanders."

There is a certain similarity between the sheep of the Netherlands and this breed. Mr Arthur Finn has recorded, in a lecture delivered before the Rye Farmers' Club, the formation of a town flock at Lydd as long ago as 1572. This flock was founded in return for certain people giving up rights of common land.

No doubt, about Bakewell's time, the Improved Leicesters were extensively used in the Marsh, and the type of sheep grazing there was materially changed in consequence.

Characteristics.

The breed has a very hardy constitution. This can be readily understood from the nature of the land on which it thrives without the assistance of artificial food. In their native county reclaimed pastures are not uncommonly found side by side with the poorest and barest lands sparsely covered with vegetation. On the one the breed rapidly fattens, and on the other it can find sustenance.

Points.—The chief points of the breed may be considered as follows: The head should be wide; the ears should be thick; there should be no dark hair on the poll, on which a covering of wool is looked for. The head is white and the nose black. In form the typical Kent sheep is very thick, and shows great width of chest. It stands on very short legs, with thighs, loin, and rump well developed. The fleece should be of one kind, without coarse breech wool, the staple being good and thick on the pelt. The breed is essentially a mutton one, the favourite cross being the Hampshire or Southdown.

A good crop of lambs would be $1\frac{1}{3}$ per ewe, although Mr Arthur Finn, in his flock at Westbroke, Lydd, records a fall of 519 lambs from 300 ewes on one grazing occupation.

Clip.—The clip of good ewes would be from 8 to 10 lb., the former figure being about the average. A good flock, in which the ram lambs are shorn, would average from $6\frac{1}{2}$ to 7 lb. per fleece. These weights are for washed wool, in some flocks washing being performed twice.

The Kent sheep is wonderfully immune from foot-rot, and is inured to the fluke trouble which visits most marshy lands.

A foreign trade of considerable dimensions has sprung up since the Flock Book was established.

Weights.—An average weight for fat wethers fed on grass would be from 10 to 11 stone. Taking the Smithfield Show weights, $10\frac{1}{2}$ oz. daily is a very high gain for lambs, the average being

9.8 oz. per day. The wethers average 6.9 oz. daily increase.

MANAGEMENT.

Breeders of Kent or Romney Marsh sheep are to some extent divided in opinion as to the type of ram to use. Some of them endeavour to grade the flock to a level, and thereby obviate the necessity of using a strong or coarse tup to correct the fault of too much quality. Mr J. B. Palmer of New Shelve Manor, Lenham, does not believe in having coarse or fine rams to mate with ewes of opposite character, but to fix the type and draft all ewes that do not conform to it. His plan is to flush the ewes for about a week before admitting the rams, as by so doing he considers that he gets a greater crop of lambs. When the rams are taken from the ewes he keeps the latter in fair condition. It is important, however, that they should be in good condition when they drop their lambs. Last year his ewes had quite 50 per cent of twins.

The general management of a flock on the Marsh is not an elaborate matter, as sheep can live there without extra food except in very severe weather. Early maturity is not a strong point with Kent sheep-breeders. In some of the flocks the policy carried out is to mate the old rams with the young ewes, and the yearling rams with the ewes of more mature age. The matrons showing symptoms of a weak constitution are drafted out after weaning. The limit age in the ordinary flock is four years. At times the best of the old ewes are retained for a special reason, and are kept perhaps a year longer. It is not advisable, however, to keep ewes too long in the flock, for grazing on good pastures and coarse grass makes long and therefore loose teeth. When that happens the ewes are likely to come to weaning-time in very poor condition.

Mr F. Baker of Manor Farm, Frindsbury, Rochester, does not think that the crop of lambs is so large as formerly. "This," he says (1908), "I attribute to the fashion of putting up the yearling ewes to such a useless extent. Some thirty years since the increase of 25 to 30 per cent of lambs to ewes tupped was not unusual, but now it only amounts to 5 or 10 per cent, and in the starvation years of the 'Nineties one lamb to each ewe was scarcely weaned."

The fleeces are becoming more uniform and better in quality, from 7 to 7½ lb. being a good flock clip.

Mr Baker estimates that the average weight which the ewe flock attains is 9 stone in the first year, 10 stone in the second, and 11 stone of 8 lb. in the third.

A ram of the Kent or Romney Marsh breed is represented in Plate 62.

ROSCOMMON SHEEP.

Of several native varieties of sheep which at one time existed in Ireland the only breed now surviving is the Roscommon Long-wool. The breed is believed to have been reared in the province of Connaught for centuries, though it is doubtful if it was distinguished for either good looks or high merits till wellnigh the middle of the nineteenth century.

It appears that strains of the race kept on the higher and poorer lands were of an inferior character, but that the bulk of the breed kept on the lower and richer parts were big useful sheep, though lacking in symmetry.

Improvement of the Breed.—In due time the improvement of the breed was taken in hand by the more enterprising of its supporters, and partly by the moderate infusion of the blood of English long-woolled breeds, notably of the Leicester, and partly by skilful selection within the breed itself, a marked change for the better was introduced. To a large extent this improvement was effected during the third quarter of the nineteenth century; but since then, by careful selection and liberal and judicious treatment generally, much has been done not only to enhance the appearance of the sheep, but also to raise to a higher level their characteristics from a rent-paying point of view.

Characteristics.

The Roscommon sheep of the present day where well kept are large-sized, handsome sheep, hardy in constitution, and excellent grazers. They do not mature quite so rapidly as some of the other long-woolled breeds, the explanation of this being the fact that Roscommon sheep have from time immemorial had to pick up their living from pasture-lands, and have only in quite exceptional cases had the forcing feeding applied to most other breeds. With moderate time to mature, Roscommon sheep attain heavy weights. Rams three to four years old have yielded from 300 to 380 lb. of carcase. Mr Matthew Flanagan, Tomona, Tulsk, Co. Roscommon, usually sells his wedder hoggets for killing in November and December, when about eighteen months old, their carcase weights running from 27 to 32 lb. per quarter. The price obtained is always the highest rate in the markets at the time. Indeed, the Roscommon mutton is superior in quality to that of most of the other long-woolled breeds.

The Roscommon is a hornless breed, carrying a long, lustrous fleece; the head is well shaped and well posed; face long and white, sometimes with and sometimes without a tuft of wool on the forehead; the ears fine, white, and of medium length, with perhaps a pinky tinge; the muzzle strong in the ram; the tail well hung and broad; and the legs strong.

Fleece.—The Roscommon wool has a good reputation amongst wool-buyers. The fleece is very white and bright in colour, and lustrous. From sheep kept entirely on pastures the fleece weighs from 8 to 11 lb., and from sheep that are partly hand-fed and generally well cared for the weights will rise to from 12 to 16 lb.

MANAGEMENT.

The management of Roscommon flocks is usually simple in the extreme. As already indicated, the sheep are, in the main, left to forage for themselves both in winter and summer. It is only in a few flocks where ram-breeding or feeding for early maturity is pursued that any hand-feeding is resorted to.

March and April are the lambing months, and the lambs are weaned about the second week in June.

Mr Flanagan, already mentioned, gives his ewes about 1 lb. each per day of a mixture of cake and oats for a short time before lambing, and for a similar time after lambing about 2 lb. daily of linseed-cake and crushed oats. Throughout the rest of the year there is no hand-feeding. Mr Flanagan sells a number of young rams for breeding purposes, getting from £7 to £12 each when they are about eighteen months old. The ram lambs are taught to eat cake along with their mothers in spring, and they get a small allowance of this food up till the selling time.

The portrait of a Roscommon ram is given in Plate 64.

HALF-BRED SHEEP.

This is the name usually given in Scotland and the northern districts of England to the first cross between the Border Leicester ram and the Cheviot ewe, and the produce of these crosses when mated together. Strictly speaking, the Half-bred is not a breed at all, but a variety or type. Yet the name has, through use and wont, come to be specially identified with this particular cross, and nowadays Half-breds are looked upon almost in the light of a breed. Half-bred sheep have had separate classes at the shows of the Highland and Agricultural Society, and at other leading shows in Scotland, for many years, and although they have no flock book or breed society, they are as carefully bred, and have as clearly marked characteristics, as most of our registered breeds.

Founding of the Breed.

Northumberland is entitled to the credit of having been the county where Border Leicester rams were first systematically put to Cheviot ewes, the pioneers of the cross being generally believed to have been Mr John Borthwick of West Newton, his son Mr Charles Borthwick, also of West Newton and Mindrum, and the late Mr Elliot of Lamberton. Each of these gentlemen is known to have bred Half-breds from a Border Leicester ram and a Cheviot ewe many years ago. Mr John Borthwick, indeed, had a regular flock of Half-breds early in, if not before, the opening of the Victorian era. At that time it was customary to breed Half-breds through the medium only of the first cross. But as the merit and great value of the sheep for general purposes became more widely known and appreciated, and as they began to spread over the Border districts, breeders took to mating half-bred to half-bred, in the first instance at any rate, as a means of getting up numbers quickly and cheaply. Good Border Leicester rams in these days were not so numerous as they are now, and the half-bred to half-bred system enabled their influence to be carried further in a short period of time.

The practice of putting half-bred to half-bred is still pursued to a considerable extent, and there is a good deal of difference of opinion among the followers of the two systems as to which is the better. Those who give attention to showing and tup-breeding adhere almost exclusively to the first cross. They maintain that a sharper-headed and finer-boned animal can be got in this way than through the second generation of the cross. They also hold that the white hair on the face is purer, as a rule, in the case of a first cross than in the produce of subsequent crosses, the wool being also usually closer and denser on the body. On the other hand, the females of the second and subsequent crosses usually grow to bigger sizes than first crosses of the same class, while they feed fully as quickly.

Two Classes of Half-breds.

Writing some years ago on the difference between the two classes of Half-breds, Mr Andrew Elliot, Newhall, Galashiels, who has been a prominent breeder of half-breds for many years, said—

"In some minds there is a prejudice against the half-bred and half-bred breed, but in every instance where the rams are selected with judgment and care, they can be bred in this way for any length of time without deteriorating in size, style, or value. In this part of the country we have many instances of flocks that have been bred in this way for the last twenty-five years, and have not only been successful but are growing daily more in favour. Although it is usual to have the rams of the first cross, I am quite convinced that it is perfectly practicable to breed them pure half-bred and

have even better sheep if done with the skill of a judge. There might be a prejudice against them for a time, but I feel confident that the result would be a success. As show animals the ewes got by half-bred rams will always beat those of the first cross (that is to say, if they be bred with care and skill), as they show so much greater weight, which is always an advantage if you have quality along with it."

Practically the same views are held to the present day, and it is not very wide of the mark to say that nearly one-half of the Half-breds, in Northumberland in particular, are of the half-bred to half-bred cross. Although the one class—unless for special purposes—sells as readily as the other, it is usual at sales, especially in the case of breeding gimmers, to intimate whether they are of the first or the second cross.

Distribution of Breed.

For a good many years Half-breds were confined to Northumberland and the arable districts of the south of Scotland adjoining the Border. But in course of time they spread widely over the country, and large numbers are now bred as far north as the counties of Sutherland and Caithness. Indeed, Sutherland and Caithness Half-breds, like Cheviots from the same localities, have a special place in the market, and are very popular in the feeding districts of the Lothians and elsewhere. Although many fairly high-lying farms in Berwickshire, and a good part of the lower slopes of the Lammermoors, are under Half-breds, the breed does not attain its best results on very high grounds. Half-breds are essentially a low-ground sheep; they require plenty of food of a good quality, and do best in association with turnip husbandry. Properly managed, no sheep have paid better in recent years than Half-breds.

Early Lambs from Half-bred Ewes.

Half-bred ewes are very prolific, producing usually on the average from one and a half to two lambs apiece per season. They are also good mothers, milking excellently as a rule. In addition to their value for ordinary Half-bred breeding, Half-bred ewes have a special value for crossing with other breeds. Thus, in late years they have been extensively crossed with Oxford and Suffolk rams for the production of fat lambs. Lambs of these two crosses grow to big sizes very early; indeed, lambs from Half-bred ewes and Down rams now constitute fully one-half of the early lambs bred in Scotland.

Three-parts-bred Lambs.—Half-bred ewes are also to a large extent used for the production of what are called three-parts-breds—that is, sheep having three parts of Border Leicester blood to one of Cheviot, the Border Leicester being again the ram used. This was a very popular animal in East Lothian and one or two other districts before the Down crosses became so popular, and it is still bred by many in preference to all others, especially where the animals are intended to be fed off as lambs or in the hogget stage.

Increasing Popularity.

Since crossing with Down rams for fat lambs became general, Half-breds have increased still further in popular favour, and may now be said to be used in one or other of their forms from one end of Scotland to the other. They have also greatly increased in numbers in Northumberland, where, owing to their suitability for being fattened on turnips, they are now the prevailing low-ground sheep.

Characteristics.

Appearance.— From the way it is bred it is hardly necessary to say that the Half-bred is a white-faced breed. It is also hornless. The head of a well-bred sheep should be well covered with pure white hair. The ears should be erect and mobile, with a slight inclination forward, and also well covered with white hair. The eye should be bold, bright, and prominent; the forehead should be wide and open; and the muzzle black, like a Border Leicester, and fairly wide, with good open nostrils. The neck should be strong and well set on the shoulder; the chest should be wide, and the ribs well arched. It is a strong point in favour of a sheep to be thick through the heart. Indeed, with many judges sheep that are not

thick through the heart stand little chance of getting notice in the show-ring.

The back should be straight and well carried out to the rump, with quarters wide and deep. The wool should incline more to the Cheviot than the Border-Leicester in closeness, and should be very fine in staple and uniform all over the body. Finally, the animal should be well set on fine flat-boned legs, should carry its head well, and be a good walker. The last is a point of great importance, and is never overlooked by a careful capable judge.

Weights and Feeding Qualities.—From a commercial point of view there is nothing to excel a good class of Half-breds. They grow to big sizes, come early to maturity, and, whether as hoggets or hoggs, make excellent butchers' sheep. The weights to which Half-breds can be brought may be judged from the facts that at the Scottish National Fat Stock Show in Edinburgh in 1907, a pen of three wedders of the breed under two years old scaled alive 865 lb.—an average of 288 lb., and a pen of three ewes 708 lb., an average of 236 lb. Cheviots on the same occasion scaled an average of 244 lb. for wedders and 217 lb. for ewes—these weights, however, being rather exceptional for Cheviots.

Clip.—Half-breds are also very good wool-producing sheep. A ewe flock should clip from 5¾ to 6 lb. of wool per sheep, and where hoggs are included a little more. Half-bred wool realises practically as much as Cheviot wool when the sheep have been well fed and are of a good class.

Sale Centres.

The great sale centre of half-bred ewes, gimmers, and lambs is St Boswells, although finely bred half-breds can now also be bought at Rothbury, Perth, Inverness, and other centres in Scotland and the north of England. Rams in the same way are mainly sold at Kelso, although sales are also held at Lockerbie, Edinburgh, and other places.

Being purely commercial sheep, half-breds have not the aristocratic support which is frequently extended to other breeds. Both rams and females, however, sell very well, and occasionally realise comparatively high prices. A half-bred ewe stock will usually realise from 40s. to 75s. per head according to age, while rams make anything up to £40, specially choice ones occasionally going as high as £50. In 1906 ten specially fine Half-bred rams from Mr Jeffrey's flock at Deuchrie, Prestonkirk, averaged as much as £19, 13s. per head. The highest price in 1907 was £40, again for a Deuchrie ram.

MANAGEMENT OF HALF-BRED FLOCKS.

The general management of Half-bred flocks does not differ materially from that of Border Leicesters. Both are essentially low-ground sheep, and if they are to give the best results they must be liberally treated. No one, for example, who has not a fair supply of young grass in the spring need hope to breed Half-breds very successfully. Ewes of the breed rarely average under 1½ lambs per head. A good supply of milk in the spring is therefore a first necessity, and in no way can it be got or kept on ewes more easily than through a good supply of succulent young grass.

With either Border Leicesters or Half-breds it is also very desirable to have at call a fair quantity of turnips. Nothing makes better winter food, and supplemented with a little cake, corn, or hay, the roots will bring the ewes on to lambing in first-rate form, and carry them on to the grass. With Half-bred ewes, particularly when they are crossed with rams of the Down breeds, lambing begins earlier than it does in the case of Border Leicesters. Many aim at having the lambs arriving as soon after the New Year as possible. In such cases it is possible to have the lambs ready for the market by the end of April or the 1st of May. Lambs bred in

this way usually fetch from 35s. to 44s. per head.

On regular Half-bred farms, where breeding for the ordinary breeding and feeding market is the object aimed at, lambs arrive from March onwards. Such lambs are usually drawn and sold during the month of August. Ewe lambs suitable for breeding purposes will then realise quite readily 40s. per head, and occasionally a little more. Wedder lambs usually realise slightly lower figures, and are bought either for feeding off on turnips during the ensuing winter, or for keeping on to the shearling stage.

Cast ewes are usually drafted out after they have nursed their fourth crop of lambs. Ewes of this age are sold towards the end of September or early in October, and are largely bought for putting to a Down ram, the ewe and the lamb going away together, fat, as early as possible in the ensuing spring. Rams are sold in September, and go amongst the ewes early in the following month. In special cases where exceptionally early lambs are wanted the rams are turned out earlier.

Half-bred ewes do not, as a rule, give much trouble at lambing time, being hardier than Border Leicesters. All the same, they require close attention on the part of the shepherd at this time if the best results are to be obtained. Lambing, as in the case of the Border Leicester and other low-ground sheep, takes place, as a rule, in specially prepared pens, near the steading, the ewes being kept in adjoining paddocks for some days both before and after they lamb.

With many of the outlying parts of farms being laid down to grass, Half-breds have the prospect of having an even extended sphere of usefulness.

A portrait of a Half-bred ram is given in Plate 56.

SHORT-WOOL AND DOWN BREEDS OF SHEEP.

THE SOUTHDOWN SHEEP.

The *doyen* of the short-wool breeds of sheep is undoubtedly the Southdown. It holds amongst these the same estimable position that the Leicester does amongst long-wools. It is native to the range of hills which runs through Sussex. There can be no doubt that it has been largely used in the building up of other breeds, such as the Shropshire, Oxford Down, and Suffolk.

Early Improvers.

In the early times of Arthur Young speckle faces were common, but to-day the demand runs on a nice mouse-brown colour. Ellman of Glynde was one of the earliest improvers of the breed. He brought his flock to a high pitch of excellence. We know little or nothing of his methods, but it has been suggested that he may have introduced a dash of Leicester blood.

Arthur Young states that the "Ellman flock of sheep is unquestionably the first in the country. . . . He has raised the merit of the breed by his unremitting attention, and it now stands unrivalled."

According to Youatt, the Ellman type of sheep, as exemplified in the head, was as follows: "The head small and hornless; the face speckled or grey, and neither too long nor too short; the lips thin, and the space between the nose and the ears narrow; the under jaw or chops fine and thin; the ears tolerably wide and well covered with wool, and the forehead also; and the whole space between the ears well protected by it as a defence against the fly; the eye full and bright but not prominent."

When Ellman sold out, in 1829, his ewe flock of 770 head averaged £13, 1s. 6d.; 320 lambs averaged 36s.; 32 ram lambs 110s.; 360 rams of mixed ages 125s.; and 241 wethers 21s. These were big prices in those days. Francis Duke of Bedford gave Ellman 300 guineas for the hire of a tup for the two seasons of 1802 and 1803.

Subsequent improvement was brought about by Jonas Webb of Babraham, whose ram-lettings were famous.

Characteristics.

The characteristics of the Southdown are first flesh, second wool. The breed is recognised as the finest mutton-producer, the great aim being to make it the sheep of the epicure. The Southdown Sheep Society, an amalgamation of two pre-existing societies, has approved the following descriptive scale of points:—

Description and Scale of Points.

	Points.
Characters.—General character and appearance	10
Head.—Wide, level between the ears, with no sign of slug or dark poll	8
Face.—Full, not too long from the eyes to nose, and of one even mouse colour, not approaching black or speckled; under jaw light	4
Eyes.—Large, bright, and prominent	2
Ears.—Of medium size, and covered with short wool	2
Neck.—Wide at the base, strong, and well set on to the shoulders; throat clean	5
Shoulders.—Well set, the top level with the back	7
Chest.—Wide and deep	5
Back.—Level, with a wide flat loin	10
Ribs.—Well sprung, and well ribbed up, thick through the heart, with fore and hind flanks fully developed	7
Rump.—Wide and long, and well turned	4
Tail.—Large, and set on almost level with the chine	4
Legs of Mutton. — Including thighs, which should be full, well let down, with a deep wide twist	10
Wool.—Of fine texture, great density, and of sufficient length of staple, covering the whole of the body down to the hocks and knees and right up to the cheeks, with a full foretop, but not round the eyes or across the bridge of the nose	10
Skin.—Of a delicate bright pink	5
Carriage.—Corky, legs short, straight, and of one even mouse colour, and set on outside the body	7
	100

Disqualifications.

Judges at Breeding Stock Shows are advised not to award a prize to otherwise good sheep on which are to be seen — (a) horns, or evidence of their presence; (b) dark poll; (c) blue skin; (d) speckled face, ears, and legs; or (e) bad wool.

Types.—It would perhaps be erroneous to say that there are two types of Southdown—one the original small, compact hill type, and the other a larger and weightier sheep. The fact is that, when the Southdown is taken on to very good mutton-producing land, it has a tendency to reach greater weight, which can be counteracted only by the use of small sires. Grey faces and muzzles are frequently met with, some of the best types otherwise having that lightness of countenance which breeders profess to avoid.

Mr Ellis of Summersbury, Shalford, had a famous flock which won many honours in the showyard. On the question of type in the Southdown sheep he declared his opinion as follows: "When I first began breeding (and looking at the judgments passed, especially at the fat stock shows), it may be said that Lord Walsingham's sheep were greatly in favour. They were large, well fleshed, but somewhat coarse. They were not of the type of the Ellman flock, nor had they the symmetry of the Duke of Richmond's or the Throgmorten sheep. At that time there was nothing like the disparity in price which now exists between the coarser and the finer carcases of mutton, and small joints were not so much in request. . . . I have always stuck to the finer type whatever the judgments of the year may have seemed to favour." Speaking of the different types he says: "Some may be delicate and too refined, with extremely small bone, but generally with good wool; others, again, high on the leg, with poor legs of mutton and narrow in the chest; while others, without being in any way coarse, are of the square, blocky, short-legged type. I think there is no doubt that soil and climate do affect and alter the type of sheep as of other animals. Without wishing to dogmatise on the matter, I think that Southdowns removed from the south of England, especially if on rich land, tend to increase in size, and at the same time lose some of the especial characteristics of the breed. I am bound, however, to say that there are exceptions. I have always aimed at a sheep very low on the legs and very square, with the legs well outside of it and width between the fore legs, giving plenty of room for the vital organs. I have never finished judging a sheep until I have turned him up. Then the wool should be short, close, and hard as a board. Such fleeces always weigh well, besides being splendid non-conductors of heat and cold."

Dead Weight.—A good shearling wether will kill about 20 lb. a-quarter, and lambs well done will reach about 15 lb. The smaller type of lamb kept in the hill district will probably dress a 50-lb. carcase if well fattened. It may be mentioned that the Southdown kills very light of offal, as much as 65 per cent of dead to live weight being common.

Weight and Value of Fleece.—The clip on the Downs is probably in the neighbourhood of 4 to 4½ lb. In the eastern counties, where there are many good flocks, from 5 to 6 lb. is shorn. The wool is exceptionally fine, and easily earns the top price in the market—next to Merino.

For Crossing.—The Southdown has been more used as a parent cross in the production of other breeds than for crossing purposes in ordinary commercial flocks. It is very popular abroad, particularly in the United States, France, and the Antipodes. By its use good carcases for freezing are produced. The Southdown has impressed experimenters more by the quality than by the quantity of its produce.

MANAGEMENT.

In the course of a lecture which he delivered in 1865 before the Royal Agricultural Society, Ellman mentioned that the one great point to bear in mind was that the Southdown should be made to graze pastures closely and thus prevent the growing up of coarse herbage. The supplementary forage crops he used included rape sown in the early part of hay and vetches, while sainfoin was considered particularly suitable for fat lambs. These views are practically those of the flock-masters to-day.

The management of a Southdown flock may be divided into two classes—hill flocks and those occupying the lower and more fertile lands. As a typical instance of the latter we may take the method pursued in Mr C. Adeane's noted flock at Babraham, near Cambridge. The breeding ewes have the run of grass as their sole food from October till the end of November. Their night fold is on the arable land. In the early part of December a little clover or grass-hay is given every evening in addition to what they graze. Should the weather be very cold, the rations are further supplemented and varied by folding on a small portion of white turnips. The belief, however, prevails that when carrying their lambs it is better for the ewes to have as few roots as possible. About ten days prior to the time when the lambs are expected the ewes receive a little cake or other artificial food.

The lambs usually begin to arrive early in February, the breeding season extending over two months. After the lambs arrive the ewes are allowed as many turnips as they can consume on grass. If a grass field does not lie convenient to the temporary lambing-pen, the lambs when three or four days old go with the ewes to the fold on a turnip field.

Mr Webb, the agent at Babraham, is convinced that it is preferable before lambing to give long hay, a run at grass, with very few turnips, to feeding oat-straw chaff and a liberal allowance of roots.

A good crop of lambs to rear would be about 125 or 130 to each 100 ewes. The ewes are culled in the autumn, the retention of the flock ewes being largely determined by a system of recording the pedigree and the produce. The peculiarities of ewes, some breeding females better than male lambs, and *vice versâ*, can by this means be accurately studied. Mr Webb also finds it a great help when deciding which lambs to save for rams.

About half the males are saved for rams, about one-third of these finding buyers as ram lambs. In the autumn the number of ram lambs is reduced to 70, the culls being killed for mutton.

The ewe lambs are wintered as stores, and about 80 of them are drafted into the flock when nearly sixteen months old, the remainder being disposed of as yearling ewes for breeding purposes.

On the hills the lambing date is later than on the lowlands, the end of March being a favourite time, although the tendency is towards an earlier period. Prior to lambing a little cake and hay are given. Running the newly lambed ewes on rape sown in August, and later on vetches, is a common practice. Successive sowings of rape are made, so that at weaning in July the lambs may pass on to an April-sown green crop. Other green foods popular in the south are sainfoin and a mixture of white clover, trefoil, and Italian rye-grass.

Drafting takes place before tupping, most of the flocks being in three ages. The usual practice is to use shearling and two-shear rams, but ram lambs are more frequently brought into service now than they at one time were.

A portrait of a Southdown ram is given in Plate 52.

THE SHROPSHIRE.

The Shropshire breed is common to the county from which it derives its name. In stature and weight it fills a place midway between the Southdown and the Hampshire.

Origin.—The origin of the breed is a much-disputed question. Some contend that it is the result of a cross on the Morfe Common sheep which led an untamed existence on that stretch of land near Bridgnorth. Others, again, believe it to be a cross on the original Longmynd or old Shropshire sheep. Yet a third party holds to the belief that its foundation was laid on a breed known as the Whittington Heath sheep. From conflicting views, it is difficult at this late period to arrive at an accurate judgment. Those who assert that it is a cross-bred mention the Leicester, the Cotswold, and the Southdown as probable crosses. Possibly a dash of the Merino was also infused.

Early Breeders.—Two of the earliest and foundation breeders were Mr Samuel Meire and Mr George Adney. In 1858 Meire stated at a farmers' meeting that it was not his intention to deny that the Shropshire was a cross-bred sheep, and that the Southdown had been used to get rid of horns.

Early Types.—When the Shropshire was first afforded separate classification at the Royal Show at Gloucester in

1853, the description then given of the breed mentioned faces and legs of grey or spotted colour. The head was well carried on a thick neck. The back was straight, the breast deep and broad, though the hind quarters were hardly as wide as the Southdown's. The dead-weight of the tegs would average from 80 to 100 lb. each. The fleece was described as more glossy and longer than that of other short-wools, the weight of it being about 7 lb.

Modern Types.—It is a far cry to 1853. Now the Shropshire is a beautifully formed sheep with a soft thick fleece, well covered head muffled to the nostrils. It stands on short legs, is very lengthy in frame, and kindly to the hand. The skin must be pink—a strong point in breeding—and there must be no suspicion of black hairs in the wool, or incipient horns at the poll.

There are two types of sheep—the breeders' and the farmers'. For convenience the latter are usually termed pasture-rangers. The farmer requires a larger, and what the pedigree breeder would probably call a coarser, type of sheep than would be used in the production of a Royal Show winner.

Merits of the Breed.

Mr Alfred Mansell of Shrewsbury thus epitomises the good points of the Shropshire sheep :—

Prolific Character.— 150 to 175 lambs per 100 ewes is the usual crop. In 1896, 11,666 ewes reared 168 lambs per 100 ewes.

Shropshire ewes are excellent nurses. Nature has endowed them with great milk-yielding properties.

The Shropshire sheep cuts a heavy fleece of the most marketable description, being of good staple, fine in texture and dense, with small loss in scour.

The Shropshire sheep is ubiquitous, being found in the Highlands of Scotland, the humid climate of Ireland, the mountainous districts of Wales, and is frequently found at an altitude of 1000 feet over sea-level.

If well cared for, wethers are fit for the butcher at ten to twelve months old, and that on a moderate consumption of food. Shropshire lambs mature very early as fat lambs.

The breed is notoriously sound in constitution, and capable of withstanding extreme variations of heat and cold. A Shropshire ewe nineteen years old, still hale and hearty, had reared 33 lambs, and enjoyed immunity from foot-rot during the whole of that period.

The quality of the mutton is rich in flavour, contains a large proportion of lean flesh, and commands the highest price in the London, Manchester, Liverpool, and other markets of Great Britain.

The Shropshire is placid and contented, not given to roaming and trampling down pasture.

The Shropshire-Merino is preferred by many who have tried it to any other cross. The half-bred is a deep square sheep, well covered with a fine close fleece, which gives a high percentage of clean scoured wool. The sheep are hardy, and fatten to nice handy weights at a very early age.

Lambs from Clun ewes by a Shropshire ram have realised 49s. each at the Shrewsbury Easter market.

Progress of the Breed.—Some evidence of the progress of the breed may be obtained from the great displays it has made in leading showyards. In 1860, when the Royal Show was held at Canterbury, there were no fewer than 192 entries. All records, however, were beaten when the Shrewsbury Royal Show took place in 1884. No fewer than 875 Shropshires were exhibited by sixty breeders hailing from fifteen counties. The breed has continued to hold its own, having a remarkable export trade to the United States and the Antipodes.

Weights. — Shearling wethers kill from 22 to 24 lb. per quarter, and the clip will vary from 8 to 10 lb.

For Crossing. — The Shropshire is largely used in the Midlands for crossing with white-faced sheep. It is also extensively employed for crossing with different native breeds in Scotland and Wales. Its most signal triumphs, however, have been recorded abroad — in Australasia in particular.

MANAGEMENT OF SHROPSHIRE FLOCKS.

Shropshire sheep are capable of repaying liberal treatment, and they usually receive it. A niggardly system in respect to food would be unwise with sheep that yield so well as the Shropshires do in both wool and mutton.

The methods of management pursued generally in Shropshire flocks in England are fairly well indicated by information which, in response to our request, Mr T. S. Minton, Montford, Shrewsbury, has been good enough to supply as to the system followed in his own flock.

Lambs.

The lambs are dropped in February and March. They are weaned near the end of June. For a time before weaning the lambs are allowed to run on clover ahead of the ewes, through hurdles that let lambs pass but hold back ewes, and there they receive 2 to 3 oz. each daily of a mixture of split-peas, linseed-cake, and bran. The "lamb-hurdles" are moved every three or four days. After weaning, the lambs go on to thousand-headed kale for two or three hours daily, and receive mangels on clover aftermath.

Young Rams.

Most of the ram lambs are kept for breeding purposes, and are sold when shearlings. The majority are bought by home breeders, but many of them are exported to the United States, Canada, South America, Russia, Japan, &c. The cast ram lambs are fattened and sold to the butcher, yielding from 76 to 80 lb. dead-weight when about twelve months old.

In the rearing of young rams a careful system is pursued to ensure steady growth and vigorous constitution. When thousand-headed kale and mangels are finished, which usually happens about the end of August, the young rams are hurdled on root-land and receive white turnips cut into finger-pieces: here they remain till Christmas, when they get cut swedes, at the same time receiving clover-hay ad lib. in racks. They also get ½ lb. per day of a mixture of corn and cake, this allowance being gradually increased till it reaches 1 lb. by the month of April.

The best of the rams, which may be intended for showing, are clipped in March, the others being clipped later. After being clipped they are housed at night for two or three weeks, but as soon

as the weather permits they are turned on to "seeds," with plenty of roots, being housed in very wet weather.

Ewes.

A number of the best of the ewe lambs are every year added to the flock, and they drop their first lamb when they are two years old. Ewes that are specially good breeders are often retained in the flock till they are seven or eight years old. Before being put to the ram, ewes are "flushed" by feeding on reserved clover leas, and they remain on these leas till near lambing time. As soon as grass begins to fail, or frosty nights set in, the ewes receive a good feed of clover-hay in racks, care being taken to have plenty of racks to prevent crushing.

A week or two before lambing the forward ewes are drawn out in turn, and receive about 1 lb. per day of a mixture of bran, oats, and clover-chaff. After lambing the ewes receive a very few roots, either swedes or mangels, on grass-land. Ewes and lambs are not put on to "seeds" until the lambs have begun to graze.

Mr Alfred Mansell, who has done much to promote the interests of Shropshire breeding, dealt exhaustively with the management of breeding flocks in a paper read at the Ninth International Conference of Sheep-Breeders at New-castle-on-Tyne in June 1908. Young breeders would do well to peruse that interesting paper.

Mr T. A. Buttar's Flock.

Mr T. A. Buttar, Corston, Coupar-Angus, Forfarshire, has at our desire furnished the following description of his methods of management:—

I keep a flock of about 260 pure-bred Shropshires, fully pedigreed and registered in the Shropshire Flock Book.

I find them a very hardy, thrifty breed; they can be run thickly on the ground, and they produce the best class of mutton and wool.

The flock was started in 1870 by my father, and the pedigree of each individual has been carefully kept.

System of Ear-marking.

Each ewe in the flock has a separate and distinct ear number, and her lambs, when one day old, are ear-marked, so that there is no chance of making mistakes.

I adopt a cipher system of ear-notching, as shown in fig. 704. Metal ear-tags are

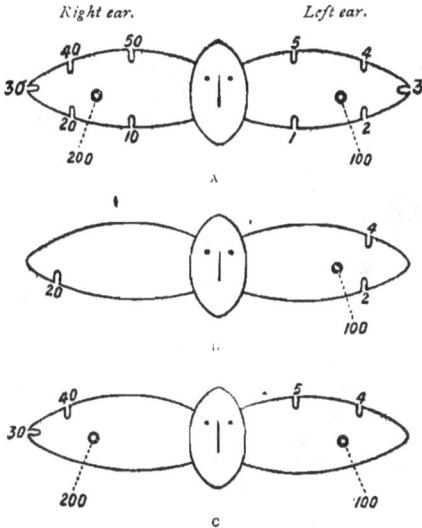

Fig. 704.—*System of ear-marking sheep.*

Diagram A shows the system whereby the numbers are marked on the ears, the units being on the left and the tens on the right ear. Numbering up to other 200 could be obtained by forming another hole near the middle of the right ear. Diagram B shows the marking for No. 126, and Diagram C for No. 379.

not satisfactory; they often cause festering, and are apt to be torn out, when of course the pedigree of the sheep cannot be traced. It is not necessary to use large ear-notches, as these disfigure the ear. Small notches, $\frac{1}{8}$ inch wide, are never noticed, and yet suffice for the purpose.

Mating Rams and Ewes.

I consider this one of the most important duties of the careful ram-breeder. Having culled all old and indifferent breeding ewes during the early autumn, and their places in the flock being now taken by about 50 of my best shearling ewes, I proceed to mate about 1st October, so that the bulk of the lambs will arrive in March. For 260 breeding ewes I generally use about 8 stud-rams.

I erect a pen, with a division for every stud-ram, as is shown in fig. 705.

About 30 ewes at a time are driven into the central pen; each ewe is caught in turn, her pedigree, general type, and form examined, and she is put to the stud-ram which is strongest in her weak points, and which we consider will make the best match.

A robust ram will easily serve 60 ewes: some of my rams get 60 and others only 20—according to the suitability of the mating. It is only by careful mating that a uniform flock—all of one type—can be bred.

It is also important to adhere to the same line of blood, which can be done without in-breeding. Violent out-crosses are dangerous, and rams, the produce of

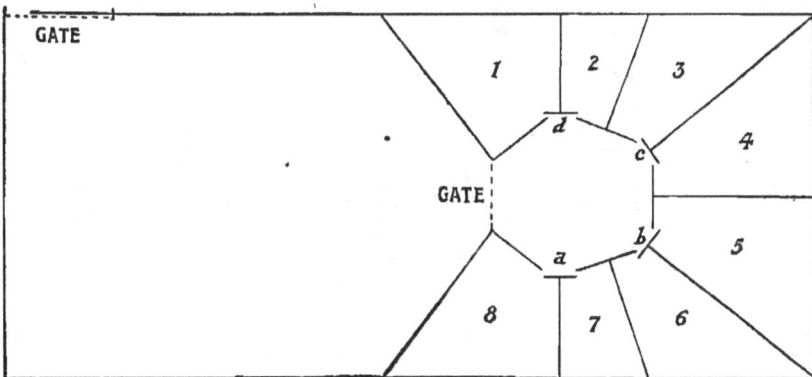

Fig. 705.—*Sheep-drawing pen.*

such, are not likely to be impressive sires.

After mating, each ram with his selected group of ewes is sent to a separate pasture-field.

All the rams have their briskets

smeared first with "yellow" paint, so that the ewes will be marked on the rump when served; when a third of the total number of ewes are marked yellow, "red" paint is used, and when two-thirds are served, "blue" paint is substituted.

This changing of colours serves a double purpose: it is only necessary to take in one-third of the ewes at a time to the lambing-fold in the order in which they were served, and it also shows if the rams are settling their ewes. If the ewes turn twice they must be given to another ram that is a sure stock-getter.

All the ewes ought to be settled in lamb in four or five weeks, but I leave the rams with them till about 1st December in case of any late ones turning.

Treatment of Ewes.

The ewes have the run of the pastures all winter. About 1st January, or earlier if the weather is severe, I begin to give them a few fresh Aberdeen-yellow turnips on the pasture. The turnips are driven out and cut into finger-pieces with Allan's turnip-cutting cart. I find when they are thus cut that fewer turnips are required; the ewes thrive much better, and live longer, as their teeth become badly broken with whole turnips, especially in frosty weather.

Each ewe also gets from ½ to ¾ lb. of a mixture of distillers' dried grains, bruised oats, and linseed-cake, also clover-hay in racks.

It is a great mistake to let ewes in lamb have too many turnips; they should rather be encouraged to eat a larger proportion of fodder or dry food. They should never get more than 20 lb. each per day, or say 1 ton to 130 ewes.

In-lamb ewes ought to be treated so that they will come to the lambing-fold in fine, healthy, robust condition,—neither too fat nor too poor,—and it is important for the flockmaster to watch the general condition of his ewes as the lambing season draws near, because in some cold, changeable, wet winters ewes require more extra keep, whereas in fine, dry winters they are apt to get too fat.

It is by constant care and observation that success is attained, and by lack of it that so-called "bad luck" during the lambing season occurs.

Lambing Season.

Ewes carry their lambs on the average 21 weeks, and a day or two before the first ewes are due to lamb, I draw out all those marked "yellow" and put them in a clean pasture as close to the lambing-shed as possible. The lambing-shed is large enough to hold 100 ewes comfortably at night, when they can be conveniently and thoroughly attended to by the shepherd. The ewes run out all day, and are only housed at dark,—getting all their feed outside.

When a ewe lambs, she and her lambs are shut up in a small pen, 6 feet square, for a day or so, till they are seen to be going on all right, and the lambs getting plenty of milk, when the lambs are ear-marked, as already explained, and they are turned out to another field with natural shelter if possible. I do not believe in housing them again if the weather is at all moderate; if lambs are getting plenty of milk they will stand severe cold.

I keep the ewes with twin lambs in separate fields from those with single lambs, when the doubles can be better done to. When I get about 30 doubles out, they are sent on to a more distant field to make room for a younger lot, and so on. Lambs thrive much better when in small lots.

Strict attention should be paid to cleanliness in the lambing-shed, and plenty of disinfectants and antiseptics used.

After lambing, the ewes get as many cut turnips as they will eat, and their concentrated food is also increased to 1 lb. per ewe. This treatment is continued till there is plenty of grass, when the trough-food is considerably reduced, as the ewes get too fat.

Lambs are weaned about 1st July; the ewes are put on the worst pasture in order to reduce them somewhat, till about 1st September, when they again get better keep to bring them into proper condition for the rams.

Feeding of Lambs.

The lambs, on the other hand, are put on the cleanest and best pastures, and get about ¼ lb. each of a mixture of linseed-cake, bruised oats, and bran. It is very important to keep lambs at this time from getting affected with stomach and lung worms, and there is no better preventive than changing their pastures frequently and keeping them as much as possible on young pasture. Old pastures should be avoided. Every endeavour must be made to keep lambs growing and improving.

About 1st August the ram lambs, having by this time been separated from the ewe lambs, are folded on vetches or early cabbage for part of the day, running on clover aftermath at night.

When the pasture fails, they are kept folded on cabbage, and later on thousand-headed kale, with an allowance of concentrated feeding-stuff and hay.

About 1st November, before the kale is finished, they get a feed of pulped roots and chopped hay, and they are gradually worn on to a full feed of pulp as the kale becomes exhausted.

They are fed entirely on pulp during the winter and spring.

The most convenient and economical mode of consuming vetches, clover, cabbage, thousand-headed kale, &c., by sheep is by using folding hurdles (fig. 121, vol. i. p. 117). These hurdles are placed close up against a row of cabbage, &c., and the sheep eat through the bars of the hurdles, thus getting their feed clean and not being able to trample on and soil it. One row is eaten at a time, and a man will easily move 50 of these hurdles in 10 minutes.

The ewe lambs get the run of the best pastures till about 1st November, when they also are fed on pulped roots and chopped hay, with an allowance of ¼ lb. each concentrated food mixed in the pulp.

Pulped Food for Sheep.

One very important advantage gained by pulping food for sheep is that the sheep always get a clean, fresh feed instead of a bellyful of cold, watery turnips, which are often dirty and frozen in the ordinary way of folding. Turnips, which are a most expensive crop to grow, are economised, and a larger proportion of fodder is consumed, thereby making the ration more natural and richer in feeding value.

Nothing is wasted, and more sheep can be kept on the same quantity of turnips. The percentage of deaths is very much less.

Feeding-boxes.

Feeding-boxes should be regularly shifted a few yards every day, so that the whole ground is equally manured. The best feeding-box for sheep is made of a pentagonal shape, as in fig. 706. At one of these boxes 10 large sheep or 15 hoggets can feed comfortably; the food is not thrown out and wasted, as it often is with long, narrow troughs; and

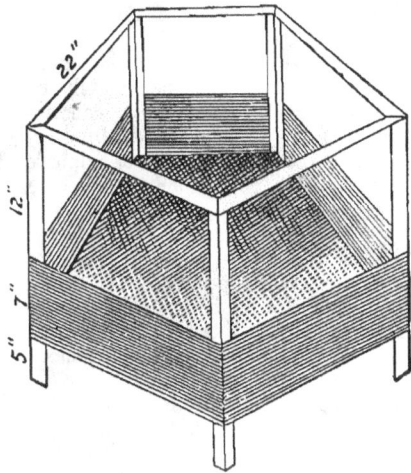

Fig. 706.—Feeding-box for sheep.

the sheep cannot crush each other, which is an important consideration in the case of ewes heavy with lamb.

Young Rams.

About 1st March I commence to shear my young rams; they are then kept in large, airy sheds till the wool grows sufficiently so that they can be turned out to grass about 1st May.

The swedes being by this time nearly exhausted, the young rams get a feed of pulped mangels instead, and when young clover and vetches are ready to cut they are gradually turned on to them for summer feeding.

Ewe Lambs.

About 50 of the best ewe lambs are selected to be put into the flock, and these are not shorn till 1st May. The remainder are treated in much the same way as the rams, and are sold throughout the summer and autumn. Many of them go to foreign and colonial buyers; whilst a large number of the rams are sold for crossing with Border Leicester, Cheviot, Half-bred, Cross-bred, and other ewes, with which they produce the best quality of fat lamb and butchers' sheep.

Prevention of Foot-rot.

In the prevention of foot-rot much depends on the shepherd. On seeing a sheep go lame he should at once examine and carefully dress the affected feet to keep the disease from spreading, and if several show signs of lameness, the whole flock should immediately be passed through a shallow trough containing a solution of arsenic — 1 lb. to 3 gallons of water, or a solution of sulphate of copper—1 lb. to 1 gallon.

Solution for Foot-rot.

Boil 2 lb. of arsenic with 2 lb. of potash (pearl-ash) in 1 gallon of water over a *slow* fire for half an hour; keep stirring, and when like to boil over pour in a little cold water; then add 5 gallons of cold water.

Put this solution to the depth of 1 to 1¼ inch, just sufficient to cover the hoofs of the sheep, in a trough 12 feet

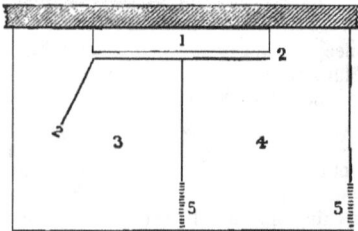

Fig. 707.—*Trough and pens for foot-rot dressing.*

| 1 Trough. | 3 First pen. | 5 Gates. |
| 2 Fence. | 4 Second pen. | |

long, by 18 inches wide, and about 6 inches deep—the trough to be set *perfectly level* along the side of a wall or other fence in some place out of the way, with a good waterproof lid on it, and secured by a padlock to prevent danger from the poison which might be left in it. A convenient arrangement for this trough is shown in fig. 707. There

should also be a wooden fence on the other side of the trough, carried out a little at one end to conduct the sheep into the trough as indicated in the figure.

Before the sheep are passed through the trough their feet should be well pared; then walk them quietly through, and let them remain in the second pen twenty minutes or so before taking them back to their pastures.

THE HAMPSHIRE DOWN.

Amongst contemporary breeds there is no more striking evidence of progress recorded than in the Hampshire Down. This sheep is for the most part quartered in Wiltshire and Hampshire, although it exercises influence over a wide area beyond these counties. The first step forward made by breeders collectively was in 1861, when they induced the Royal Agricultural Society and the Smithfield Club to provide the breed with a separate classification. Prior to that date Southdowns were the only breed thus honoured, the other Downs being shown in an inclusive class.

The Hampshire Down is largely reared on the high-lying and barren uplands of chalk in the south-western counties, where the flocks, as a rule, are large, numbering from 1000 upwards. Where the custom of the district is to keep smaller flocks than the figure named, it will generally be found that the Hampshire Down flocks are in excess of the other breeds in point of numbers.

Characteristics.

Early Maturity.—The great claim which breeders make, and have rightly established, on behalf of the breed is that it matures early. Indeed there is no Down or other breed which has so much advanced the cause of speedy maturity, and therefore of quick turnover. The pioneer work of the late Mr A. de Mornay must be remembered in this connection.

Weight of Hampshire Lambs.— A well-bred Hampshire lamb on good keep will grow at the rate of ¾ lb. daily, and will weigh 113 lb. on May 31. Calculating the carcase-weight at

60 per cent of the live-weight, we get an average of 17 lb. per quarter. That figure is very frequently exceeded, and 20 lb. at the time of sale a little later in the season is not uncommon. The chief claim made on behalf of the breed is that it progresses with amazing rapidity. The fact that the lambs come to such heavy weights in July and August is striking testimony to the progressive policy of breeders.

Examples of Precocity in Breeding.—The late Mr A. de Mornay, in the course of an article in the *Farmer and Stockbreeder Year-Book*, gave the following instances of the precocious instinct in the Hampshire Down: "Three ewes, each having two lambs by their side, were tupped by one of the lambs in the flock, which could not have been more than three months old. They gave birth to six more lambs in August, one having three lambs.

"Another example of this precocious and prolific instinct may be mentioned in the case of a ewe which gave birth to two lambs in January. She lambed again early in July, when she gave birth to two more lambs, and in January following had again two lambs, making in all six lambs in twelve months. The first two were ram lambs, and were sold at Oxford Fair for 14 guineas. The two young lambs were sold at Wallingford market for £4, and the lamb ram of the last couple was also sold at Oxford, and brought 6 guineas, making £25 for five out of six lambs. The sixth, being a ewe lamb, was saved for stock."

Constitution.—No doubt need be entertained concerning the constitution of the Hampshire Down. *Prima facie* evidence of capacity to endure hardship is afforded by the bare and somewhat bleak downs which they have made their home. Reverting to the very severe winter of 1894-95, it may be pointed out that the tegs from nine to twelve months old lived through that time on partially rotted turnips and hay without the aid of supplementary feeding of any kind. Flocks are frequently brought through the winter without loss by death, and save at the troublesome time of lambing, losses are seldom encountered.

It is the usual custom to sell ewes at four and a half years old, or in the early autumn when they have borne their third set of lambs. There are favourite ewes in most flocks, however, and they continue fruitful up to fourteen years old, cases of the latter age being on record.

Breeding from Lambs.

Mating.—Lambs of six or seven months old are preferred by flockmasters as sires, and ewe lambs may be put to the ram to produce lambs as yearlings. This is one of the means adopted of breeding early maturity into the flock. At the same time, it involves a certain amount of risk. As a rule, the lambing is more difficult, and the ewe's growth is stunted.

One method favoured by many breeders is to breed from a twin lamb. They have the reputation of being more fruitful, and unlike the custom with some other breeds, a large percentage of twin lambs is encouraged. Probably a correct estimate of the lamb-producing capacity of the Hampshire Down would be a lamb and a quarter.

Early and Rapid Breeding.—With regard to the possibility of getting lambs from ewes in the first year of their existence, and the possibility of getting two crops of lambs in the year from the whole flock of ewes, the late Mr de Mornay's views are interesting. "It may," he writes, "in general terms be said that on the same area of land a saving would accrue in the reduction of the flock of ewes, the ewe tegs being productive the first year and the ewes producing a second crop of lambs. A saving would be effected in consequence of the rapid growth and feeding of the second crop of lambs, which would be reared in the summer on the succulent green crops and fed with little cake and corn. On the other hand, account would have to be taken of the extra amount of food required to nourish the tegs during the period of their gestation.

"It is difficult to get at the exact amount of artificial food given to the different flocks on the farm; but, as near as I could ascertain it, in regard to the ewe lambs it amounted, for the eight or nine months from their birth to the time the ram lamb was introduced, to about 28s. per lamb, and during the period of gestation from 5s. to 6s. per lamb; and in regard to the wether lambs until they were fat, about 35s. per lamb, according to the quality of the hay. With good hay less artificial food is required."

For Crossing.—The Hampshire Down is one of the parents of the Oxford Down. It has been singularly successful when used for crossing. The ram trade is to all intents and purposes a lamb trade, large numbers of ram lambs being sold in the Midlands and the eastern counties of England to beget stock for supplying an immense business in fat lamb. One of the first to demonstrate the possibility of the Hampshire Down for cross-breeding was Mr Thomas Rush, whose series of successes with lambs and wethers of the Hampshire-Oxford Down cross at Smithfield and other fat stock shows did much to popularise the use of both breeds.

Mutton.—The quality of the mutton is of the very best. Nothing handles more kindly than a well-nurtured lamb fatted for the fat stock shows. Dark mutton is always in request.

Fleece.—The wool of the Hampshire is of medium length. It is dense, and fills the hand well. Tegs will clip from 12 to 14 lb. of unwashed wool, the ewes, of course, yielding a smaller return.

Description.

A well-set Hampshire Down is a smart, even gay sheep. It carries a dark strong head, free from horns or "slugs." Speckle faces are not recognised. The poll is well covered with wool, which should intrude upon the forehead. The neck must fill the hand in the case of a sire. Many breeders insist on two strong points in the Hampshire—a big neck and a strong dock, the latter indicative of well-sustained vertebræ. The carcase is symmetrical and square, not cylindrical. The ribs must be well arched, and the loin flat and well packed. The rump should be wide, and the legs of mutton well carried down. The skin should be pink. The following is a scale of points drawn up by the Hampshire Down Sheep Breeders' Association :—

Scale of Points.

	Points.
Head.—Free from horns or snigs; face and ears of a rich dark brown—approaching to black—absolutely free from white specks and well covered with wool over the poll and forehead; intelligent bright full eye; ears well set on, not drooping, fairly long and slightly curved towards tip. In rams, a bold masculine head is an essential feature	20
Neck and Shoulders.—Neck of strong muscular growth, not too long, and well placed on gradually sloping and closely fitting shoulders	20
Carcase.—Deep and symmetrical, with the ribs well sprung, broad straight back, flat loins, full dock, wide rump, deep and heavily developed legs of mutton and breast	30
Legs and Feet.—Strongly jointed and powerful legs of the same colour as face, set well apart, the hocks and knees not bending towards each other; feet sound and short in the hoof	15
Wool.—Of moderate length, close and fine texture, extending over the forehead and belly, the scrotum of rams being well covered	10
Skin.—Of a delicate pink and flexible	5
Total	100

Shepherds' Competitions.

One of the contests inaugurated is that for shepherds. Prizes are offered to those shepherds rearing the largest number of lambs. In 1906 thirty-two entries were received, involving a total of 15,248 ewes and 17,742 lambs. The gross number of lambs reared was 116.35 per 100 ewes. The gross average loss of ewes (including barren or other ewes sold to be killed) was 1.77 per cent. The highest percentage of lambs reared was 132.25.

In another competition twenty-seven shepherds reared their flocks without loss of tegs and shared the prizes. The entries numbered fifty, the ewe tegs aggregating 9180 and the total loss 37, equivalent to a percentage of .40.

Flock-Book.—The Hampshire Down Sheep Breeders' Association was established in 1899, when it issued its first Flock-Book.

Foreign Trade.—A foreign trade has been established, and from several parts there is a growing demand.

MANAGEMENT OF HAMPSHIRE FLOCKS.

The system of management pursued in flocks of Hampshire Downs is fairly well indicated by the following notes relating to Mr H. C. Stephens' famous flock at Cholderton, Salisbury, kindly supplied by the manager, Mr James G. Kerr.

The ewes begin dropping their lambs about the 1st of January, and by the end of the month the bulk of the ewes have lambed, a few late lambs coming in the first and second weeks of February. The lambs are weaned on the 12th of May, or as near that date as possible. The flock being a ram-breeding one, the feeding of the lambs is commenced as soon as ever they will eat out of a trough, and by the time they are weaned they are able to eat ¾ lb. per day of a mixture of feeding-stuff consisting of linseed-cake, peas, and pea-chaff. After weaning beans and locust-beans are added to the above mixture, the quantity being gradually increased until sale time, when they will be consuming 2½ lb. feeding-stuffs per day.

The ewe lambs, after weaning, get ½ lb. each per day of a mixture of linseed-cake and peas. At Michaelmas this is changed to ½ lb. cotton-cake, which they have all through the winter and spring, and is discontinued after shearing in May. After this they get no more feeding till they lamb down in the following year.

With the exception of a few that are sent to the butcher, all the ram lambs are sold for breeding purposes. Breeders of Hampshires prefer to use ram lambs, and at Cholderton all the ram lambs to be sold are sold before they become shearlings. The average price realised in 1908 for all male animals sold, including those sent to the butcher, was £8, 4s. 11d. each, the number sold being 339, the male produce of 612 ewes.

The ewe lambs are all wintered, and a great number are sold for exportation as shearling ewes, at prices ranging from £5 to £15 each, according to the selection of the purchaser. The ewes up to the first of December are penned on arable land, generally on a piece of cabbage, and running on the down for exercise during the day. During December they are removed to a grass lea where cabbages are carted to them, and they still go to the down by day for exercise.

About Christmas Day the early lamb-ing ewes get ½ lb. linseed-cake each per day, which produces a nice flow of milk and helps wonderfully in lambing. As the ewes lamb they are divided into three flocks, consisting of single ram lambs, single ewe lambs, and the twin lambs. The mothers of the single ram lambs receive ¾ lb. decorticated cotton-cake and ¾ lb. bran each per day. The mothers of the single ewe lambs receive ½ lb. decorticated cotton-cake each per day. The mothers of the twin lambs receive ¾ lb. decorticated cotton-cake, ¾ lb. linseed-cake, and ½ lb. bran each per day. In addition to the artificial feeding-stuffs, the ewes have hay, mangels, cabbage, kale, rape, vetches, winter barley and rye, each in its season. After weaning the ewes go to the downs during the day, and at night they are put into pens after the lambs to clear up anything the lambs have left.

The ewes drop the first lamb when they are two years old, and are cast when they have reared their fourth lamb. This is not, however, the general custom. Most breeders only take three lambs, as by this system they get a better price for their cast ewes. But on the Cholderton flock it is found that the old ewes produce the best lambs, hence an extra crop of lambs is taken from them.

Young rams kept for breeding in the Cholderton flock get 1 lb. of cotton-cake each per day, with roots and straw chaff, and as soon as they are shorn they are turned out into a pasture and receive no further feeding. The old stock rams when they come from the ewes are turned out into a pasture, and only receive a little hay in bad weather. If they were given extra food they would get too fat and heavy, and be useless for stock purposes. Only such old rams are kept as have proved exceptionally good stock-getters.

A portrait of a Hampshire ram is given in Plate 53.

THE OXFORD DOWN.

The Oxford Down, like most of our other breeds of farm live stock, is of a composite type. Its origin is not wrapped in obscurity. It is the result of a direct cross between the Cotswold and the Hampshire Down. A few breeders may have used the Southdown, but the dominant force in the cross was admittedly the Hampshire sheep. After a long series of years of pure breeding, it preserves to this day the characteristics of both parents. The carriage and form of the Cotswold are apparent, whilst the influence of the Hampshire is seen more in the mutton-producing properties.

Early Efforts. — Early last century the possibilities of the Cotswold cross on the Hampshire ewe first impressed Mr Twyman of Whitchurch, Hants. He was undoubtedly the chief of an able band of pioneer breeders, which included such names as Hobbs, Treadwell, Bryan, Stilgoe, and others familiar at the present time. The constitution of the Cotswold sheep has been a particularly valuable asset to the breeder of Oxford Downs.

Characteristics.

The characteristics of the Oxford Down sheep may be thus briefly described. In the ram a bold, masculine head is looked for, with slight inclination to a Roman nose. The neck should be strong and the poll well woolled, with a prominent top-knot. The face should be uniformly dark-brown, the deeper colour being more and more favoured by breeders. There should not be any black wool behind the ears. The eyes should be prominent and the ears a good length. The shoulders should be wide set, the back level, the dock strong. The ribs ought to be well sprung, the barrel thick and lengthy. The underline must be well clad. The legs ought to be short and dark in colour. Spotted legs are objectionable. The sheep should stand squarely on his limbs, which should be, so to speak, at every corner, with twist well developed. The skin should be a healthy pink in colour.

Fleece. — The wool should be dense and of good texture and free from openness, and without spot or patches of black. Short wool should extend down the legs. Rams will clip 14 to 15 lb. and ewes about 8 lb.

Changes in Type. — That the present day Oxford Down is of a different type from that prevailing thirty or forty years ago is evident from the impressions of one of the oldest breeders, Mr John

Treadwell. He recollects an old breeder saying that "the Oxford should have the Cotswold fleece and the Down mutton." That, however, soon got out of date. The close fleece was then favoured, and has continued to be one of the primary objects of the breeder. In the olden days the Oxford Down was quite as big a sheep as it is now, but it was "fatter natured." Breeders nowadays look for sheep with more bone than they used to possess, this being probably the most effective antidote to the formation of excessive fat.

In the 'thirties and 'forties of the nineteenth century the common name for the breed was the Cotswold Downs. This was changed to New Oxfords; and finally, when the Breed Society was established in 1888, the modern designation was formally adopted.

Points in Breeding. — The modern tendency is to dispense as far as possible with black wool. As this is usually associated with dark-skinned sheep, the importance of the colour of the skin can be readily appreciated.

Location. — The Oxford Down is true to the county which gave it birth. Flocks are to be chiefly found in Oxfordshire and Gloucestershire. It has, however, gone wide afield. Its most valuable market is the south of Scotland, where it is a prime favourite with owners of whitefaced sheep for crossing purposes. The cross has been unusually successful. Germany takes a number, but the trade with the Continent is fitful.

For Crossing. — The Oxford Down ram lamb is a favourite in the Midlands of England for crossing purposes. It produces not a little of the fat lamb that finds its way to the chief centres of population. Some experiments were carried out in the north of England by Mr W. T. Lawrence of Newton Rigg. The produce of the Oxford Down on the Scotch half-bred ewe (Cheviot-Border Leicester) lambed in March weighed in thirteen weeks 70 lb. liveweight. This weight was attained by double lambs, the singles turning the scales at a similar weight in ten weeks. In 1904 and 1905 further comparative trials were instituted, the competing breeds being Oxford Down, Wensleydale, and Border Leicester. The outstanding feature of the Oxford cross was that the lambs grew so quickly from birth.

Show Classification. — The Royal Agricultural Society and the Smithfield Club afforded separate classification for the breed in 1862.

MANAGEMENT.

In Oxfordshire rams are put to the ewes on grass-land in August, so as to get early lambs, and it is believed that early lambs are less subject to scour than later arrivals. Clovers are avoided, as they have a tendency to cause ewes to return to the ram. Towards the last month of the year grass-lands by day and root by night, or the reverse, is the rule. Prior to lambing the flock spend the night in the yards. The lambs are drafted on to grass-lands, and are given a few oats, bran, and easily digested foods. Weaning takes place in June, when rye and vetches are ready. The flock is folded on forage crops, the ewes following the lambs from fold to fold.

Mr Treadwell's Flock.

Mr John Treadwell, Upper Winchendon, Aylesbury, Bucks, favours us with the following notes as to the management of his famous flock of Oxford Downs: "This flock being entirely devoted to ram-breeding, is in many respects managed differently from an ordinary flock kept for mutton-producing.

Management of Ewes.—"About the middle of August the ewes are separated into lots, according to their suitability to the different rams to be used; and as many of the sires used are home-bred ones, care has to be taken as to the different pedigrees, as well as to size, wool, and symmetry. This adapting the rams to the different ewes is considered the most important factor in the whole matter of breeding.

"This farm containing a large proportion of grass-land—two-thirds—enables the ewes to be placed in lots as they are drawn in the different pastures.

"About the beginning of November when the ewes are all served they are put together, and clear up mangel-tops, stubbles, seeds, or anything there is for them. When this is done they are again drafted into smaller lots about the pastures, until they come up to the lambing-pen for lambing.

"Rather a large number of rams are used, as some have only a very few ewes and others have a fair number, varying from 10 to 70 to a ram.

"When the ewes come up to the lambing-pen they get a little hay or straw, according to the weather and their condition; and they run on pastures by day. As soon as they have lambed they return to the pastures, and have about 2 pints of cake each, and hay if they require it. The oats are continued until April, when they are gradually taken off, as the grass comes on.

"They are shorn about the end of May, and the lambs are generally weaned in June—the ewes being put to vetches or clover, or a rough pasture, or anywhere where they can be kept cheaply until tupping-time.

"The draft ewes get better treatment at this time. They are fed on the pastures, sometimes getting some cake and corn until they are sold off fat or put to roots or cabbage to finish. These get to very heavy weights if put on roots and brought out in January. They will average about 16 to 18 stone when well finished. Sometimes some of the best of them are sold to breeders in the autumn to keep on another year or two.

Treatment of Lambs and Rams.— "The lambs when weaned are separated, the ram lambs getting a little cake and corn at once. The ewe lambs do not get anything with the grass, as a rule.

"The ram lambs have their cake and corn increased slightly as the season advances, but do not get much attention until after the shearling rams are sold in August, when they are put on to the arable land as soon as some rape or turnips or something can be got for them. They then follow on to swedes and mangels until about the beginning of April, when, if the weather permits, they are shorn, kept in for a few nights, and out in the day, but left out entirely as soon as possible. They get on to rye, and then to vetches, with which they receive mangels until the cabbages come, when these take their place.

"These rams grow very fast and get big by the first Wednesday in August, when about 60 of the best of them are annually sold by auction at home, when buyers from almost every county in England and from many distant countries attend. A few of the rams are sold privately to foreign buyers, chiefly Germans. Then every year a number of rams are sent to the Scotch sales at Edinburgh and Kelso, where there is a great demand for them for crossing purposes—the Oxford ram on the Half-bred ewe answering better than anything else.

"The ewe lambs generally go off the pastures on to rape in October, and then on to turnips, with which they get a little cotton-cake. In the spring about half are selected for the flock, and they are fed on vetches or seeds or pasture until turned into the ewe flock, when the rams are put amongst them. The draft ones are put into the pastures, and sold during the summer for stock or to the butchers, the majority now going to Germany and other countries for breeding purposes. The stock rams are not highly fed."

For many years Mr Treadwell was the leading prize-winner in the Oxford Down classes at National and other shows, but soon after the advent of the new century he discontinued exhibiting. Since then the demand for Mr Treadwell's rams has increased, and so also has the run of prices for them. At the Jubilee Sale in 1907 the average for 58 shearling rams was £23 — with a top price of 150 guineas.

Maisey Hampton Flock.

In the well-known prize-winning flock of Oxford Downs belonging to Mr James T. Hobbs, Maisey Hampton, Gloucestershire, lambs are dropped between the 1st of January and the middle of March. For some time before lambing the ewes get a limited supply of roots and plenty chopped hay and straw. After lambing they get a liberal allowance of roots and good hay, with 1 lb. of corn each per day. After the lambs are weaned the ewes are kept on grass, and they clear up behind the lambs.

The lambs for some time before being weaned are allowed to run in front of their mothers, where they get a little linseed-cake and crushed oats and bran in boxes, sliced roots and hay being also given. After weaning the lambs are usually started on young "seeds" until vetches are ready for them, the concentrated food being continued, with the addition of a little split peas, the quantity allowed being about 1 lb. each per day.

Ewes drop their first lamb when two years old, and are usually cast when they have reared four crops of lambs.

Young rams in winter get roots and hay, with about 1 lb. of corn each per day. They are put on to rye and vetches in spring and summer, their allowance of corn being gradually increased till it reaches 2 lb. each per day. They are generally sold in August.

An Oxford Down ram is represented in Plate 53.

THE SUFFOLK.

The Suffolk breed of sheep has come to the front very much during the closing years of the nineteenth and opening of the twentieth centuries. It is kept in its native county, a few flocks being found in Essex, Norfolk, and Cambridge. It is chiefly in the hands of tenant-farmers.

Origin.—The origin of the breed is not difficult to trace. It was evolved by a cross of the Southdown on the Norfolk horned sheep. The horns were in course of a few generations eliminated. It is curious to note how the predominant features of the old Norfolk breed have asserted themselves. The Suffolk has all its leanness of flesh and darkness of limb and face. It has kept the size of its Norfolk progenitor, on which it has grafted the quality of the Southdown. The cross was made early last century, in the middle of which the breed was commonly known as the Southdown-Norfolks. It was in 1859 that the breed was finally christened the Suffolk.

Characteristics.

The Suffolk is a bare polled sheep, with greater length of limb than most of the other short-wools. It is very dark in the face and on the limbs, jet-black in fact, a characteristic inherited from its Norfolk ancestry. The eye is bold, the nose fairly long, and the muzzle square. The ears come forward parallel to the poll, and should not droop. In the ram the neck should be very full, and fill the hand when gripped. The back should be broad, and touch kindly under hand. Length of frame is necessary to carry flesh. The whole appearance of the sheep differs from other Down breeds in its bareness of limb and poll. It suggests activity.

Scale of Points.

The following scale of points has been adopted by the Suffolk Sheep Society:—

	Points.
Head.—Hornless; face black and long, and muzzle moderately fine—especially in ewes (a small quantity of clean white wool on the forehead not objected to); ears a medium length, black and fine texture; eyes bright and full . .	25
Neck.—Moderate length and well set (in rams stronger, with a good crest) .	5
Shoulder.—Broad and oblique . . .	5
Chest.—Deep and wide	5
Back and Loin.—Long, level, and well covered with meat and muscle; tail broad and well set up; the ribs long and well sprung, with a full flank .	20
Legs and Feet.—Straight and black, with fine and flat bone; woolled to knees and hocks, clean below; fore legs well set apart; hind legs well filled with mutton	20
Belly (also Scrotum of Rams).—Well covered with wool . . .	5
Fleece.—Moderately short; close fine fibre without tendency to mat or felt together, and well defined—*i.e.*, not shading off into dark wool or hair .	10
Skin.—Fine, soft, and pink colour . .	5
Total	100

Prolificacy.—The Suffolk is a prolific breed. It is on record that one ewe dropped no less than eight healthy lambs in the brief space of 12½ months. It is interesting to note, as indicative of the prolificacy of the breed, that since 1887, when returns were first made to the Suffolk Sheep Society by the owners of registered flocks, the number of lambs reared has been 132.25 per 100 ewes. Roughly speaking, therefore, one may conclude that the breed is capable of producing a lamb and a third a-year.

Lean Mutton.—The Suffolk more than any other breed has distinguished itself since the carcase contests were instituted at Smithfield Show. No doubt the quality already referred to—the large proportion of lean to fat—has enabled it to excel when the block is the objective. As a show sheep the Suffolk has not quite the width, depth, and wealth of some of the others, hence its absence from representative honours in inter-breed contests.

Produce of Mutton.—Experiments carried out at the Hollesley Bay College with a Suffolk on Merino ewes resulted in a lamb and a half per ewe. This lamb, slaughtered at 15 months, weighed 94 lb. live-weight, and gave a dressed carcase of 54 lb.—equal to 60.64 per cent. The washed fleece weighed 6.65 lb. The winning carcase in the short-wool wether sheep class at Smithfield Show in 1907 was a Suffolk, and so was the second. Weighing 208 lb. on arrival and 640 days old, the carcase-weight was 133 lb., this showing the highest daily gain in the class. The first, second, fourth, and fifth prizes in the short-wool lamb class were also won by Suffolks. The winner scaled 144 lb. 265 days old, killing 92 lb. Still further triumphs, including the championship in the carcase competition, fell to the breed at the Smithfield Show of 1908.

For Crossing.—The breed has been exploited for crossing purposes, particularly in the south of Scotland, where it finds patronage for mating with the whitefaced ewe.

MANAGEMENT OF SUFFOLK FLOCKS.

In the best of the Suffolk flocks a liberal and thoroughly up-to-date system of management is pursued. That this is the case is clearly shown by the rapid progress which the breed has made in regard to early maturity and mutton-producing properties generally.

The majority of the Suffolk flocks are kept on land of poor quality, and in these flocks March is the principal lambing month. The general system of management here is less expensive than in ram-breeding flocks.

Mr Herbert E. Smith's Flock.

In the well-known Suffolk flock owned by Mr Herbert E. Smith, The Grange,

Walton, the lambs are dropped in January and February, and they are weaned about the first of June. Before lambing the ewes run on grass during the day, and are folded on turnips at night, getting also a little hay. After lambing they are folded on turnips, cabbages, &c., and run out on rye; later on they go on to mixed grasses, and get a small allowance of mangels.

After weaning the lambs get about ½ lb. per day of mixed cake and oats, are folded on tares and rape, and have a daily run on clover or sainfoin. The draft ewe and wedder lambs are sold about the second week in July, realising about 50s. each. The ram lambs are sold in August and September, and bring about £20. Young rams are fed well on cabbages, rape, and sainfoin, getting in addition about ¾ lb. per day of a mixture of corn and cake.

The Playford Flock.

In Mr S. R. Sherwood's valuable flock at Playford, Ipswich, the ewes for about a month before lambing get ¾ lb. each per day of linseed-cake and crushed oats and bran, mixed in equal proportions. The lambs are dropped in January and February, and are weaned in April and May. For a time before weaning the lambs run through "creeps" in front of their mothers, and get as much as they care to eat of the same mixture, with cracked peas and beans. Lambs run on turnips, rye, savoys, swedes, and trifolium in succession.

The culled ewe lambs are sold in July at about 50s. to 55s. each, the best being retained for breeding. Ram lambs are sold in August, September, and October, at an average of about £12 each. Young rams are pushed on from the start, getting swedes and savoys mixed, and as much cake, crushed oats, bran, and cracked peas as they will eat.

After weaning ewes are kept for a time on moderate food, but they are gradually put into good condition for tupping in August. Just before tupping they are "flushed" on cole-seed or good grass and stubble. Mr Sherwood does not breed from ewe lambs.

A portrait of a Suffolk ram is given in Plate 57.

THE RYELAND.

The Ryeland breed is one of the oldest English breeds, although perhaps it has not contributed much to the ovine history of the country. It is found chiefly in Herefordshire and Worcestershire. Originally it had an extensive run on the Welsh Borderland, being prized for its wool.

The modern Ryeland is a vastly improved sheep. It can hold its own with any breed for symmetry, closeness of fleece, and firmness of flesh. Breeders freely advertise its suitability for fat lamb production.

Appearance and Weight.—In appearance the Ryeland has something in common with the Shropshire in quality and symmetry, although of course its colouring is a dull white, and it is not so severely muffled on the face. It carries a close, thick fleece of excellent quality. In weight it scarcely attains the scale of the Shropshire, but 10-month-old lambs will turn out as high as 18 lb. per quarter, and wethers at 16 or 17 months will kill 22 lb. per quarter. The old Ryeland breed was a sheep of much smaller frame, and did not fatten so readily as the modern type, which has been increased in weight to the extent of 6 to 8 lb. per quarter, age for age.

Fleece.—A still greater improvement is noticeable in the weight of the fleece, which has been advanced from about 3 lb. to close on 8 lb. in a well-bred flock. The wool of the Ryeland is said to be the best for carding purposes produced in England, and doubtless the competition of foreign wools has affected the popularity of the breed in England.

Management.—There is little that is exceptional in the management of Ryeland flocks. They are treated with enterprise and care.

A Ryeland ram is represented in Plate 57.

THE DORSET DOWN.

This breed, which supports a flock book established in 1906, is native to the south of England. Its origin was a cross between the Southdown and the Berkshire, Hampshire, and Wiltshire ewes.

Early Improvement. — The earliest exponent of this cross was Mr Thomas Homer Saunders of Watercombe, near Dorchester, who created a type of sheep known as the "Watercombe Breed of Improved Hampshire Downs." He and his son, Mr T. Chapman Saunders, were closely identified with it.

Contemporaneously with the work of Messrs Saunders was that of Mr Humfrey of Chaddleworth, near Newbury. His method was to procure a Webb Southdown ram and cross with the Hampshire and Wiltshire ewes. These sheep were known as "West Country Downs," and were exhibited at the Royal shows at Chester in 1858 and Warwick 1859.

Characteristics.

The Dorset Down is closely related to the Hampshire · Down, but is of finer bone and often of lighter colour. A good Dorset Down should be free from coarseness, have a long, full, clean face and under jaw, a bold eye and full muzzle. The ears should be thin, fairly long, pointed, and whole-coloured, being carried well above the level of the eyes. The bone should be fine. The fleece should be dense, growing well down to hocks and knees, round the cheeks, between the ears, and on the forehead. Wool under the eyes or across the bridge of the nose, on the ears, or below the hocks and knees, should be avoided. The face and legs should be of a brown colour. There should be no tendency to legginess.

Early Maturity and Weight.—The breed matures early. The ewes ' are capable of producing sucking lambs weighing from 40 to 48 lb. at 10 to 12 weeks old, or a well-finished carcase at from 8 to 9 months of from 66 to 72 lb. mutton.

MANAGEMENT.

Flock management in the south of England implies early lambing. The average Dorset Down flockmaster is well content if he rears just over a lamb to the ewe. In the Forston flock Mr Cecil

Boatswain writes that, from 400 breeding ewes, in 1908, he reared 385 lambs. Mr G. Wood Homer of Bardolf Manor, Dorchester, reckons that his flock of 580 ewes rear rather more than a lamb apiece.

The mating in Dorset Down flocks takes place early in July, and the lambs are dropped from December onwards. The lambs run with the ewes until not later than the beginning of May. A Dorset Down ram lamb will serve from 70 to 100 ewes. Mr Wood Homer estimates that not more than 8 per cent require second service, and 2½ per cent a third service. There should not be more than 1 per cent of barren ewes.

Prior to lambing, ewes fed on grassland get a few turnips and hay. The increased acreage of land laid down enables flockmasters to keep their flocks on grass. The hay-cribs should be out early in October, and about ½ lb. of hay given to ewes forward in lamb. ', The quantity is gradually increased, being given in two portions—morning and evening.

When the lambs are a week to ten days old they are put on turnips. The best lambs are pushed forward with cake, and are ready for the first draft early in May, when about four months old. They realise up to about 36s. per head. The second draft comes on in July, making about 33s. The off-going ewes are fit for market in May, making over 50s., and weighing as much as 100 lb. dead-weight.

It is of the highest importance to provide adequate shelter, otherwise the cold winds cause heavy losses. Shelter-hurdles are commonly used for this purpose.

In the ram-breeding flocks selection of the rams takes place about March, and those chosen are pushed forward with extra food. The ewes should be carefully drafted about August.

Mr Wood Homer considers that his couples, Chilver hoggs and fattening sheep, run to about 2½ sheep to the acre on light hill-land. This, however, is possible only by the liberal use of artificial food.

A Dorset Down ram is represented in Plate 59.

THE DORSET OR SOMERSET HORN SHEEP.

The Dorset Horn sheep appears amongst the earliest records of pastoral husbandry in the south of England. As far back as 1757, in his *Observations in Husbandry*, Edward Lisle records that in the course of his journeys into Dorsetshire between 1693 and 1772 he was struck with the fecundity of the native horn sheep. He remarks "that his tenant, Farmer Stephens, had ewes which brought him lambs at Christmas, which he sold fat to the butcher at Lady Day, anno 1707; and at the beginning of June, thinking his ewes to be mutton, they looked so big, he went to sell them to the butcher, who handled them, and found their udders springing with milk and near lambing, and they accordingly did lamb the first week in June."

Again, William Ellis, in his *Shepherds' Guide*, published in 1749, describes the west country sheep as whitefaced, with white and short legs, broad loins, and fine curled wool, "the Dorsetshire variety being especially more careful of their young than any other."

There is probably no better or more continuous record of a breed being associated for a long period with a county than this.

Another name for this breed is the Somerset Horn sheep.

Characteristics.

This is a whitefaced horned breed. It is essentially a meat sheep, in some respects not unlike the Cheviot in form, but longer in frame.

The head should be broad, the nostril full and open, the poll well woolled to the brow, the face white, the nose and lips pink. The ears are of medium size and thin. The teeth are flat, chisel-shaped. The neck is short and round, well sprung from the shoulders, and in the ram strong and muscular. The chest is well forward, full, and deep. The fore flank is full, with no depression behind the shoulder. The shoulders must be well laid and compact.

The back and loin should be broad, long, and straight, with deep well-sprung ribs. The quarters must be full, broad, and deep, and fleshed to the hocks. The tail should be well set in a line with the back, wide, firm, and fleshy. The legs must be well planted at the four corners, with plenty of bone, and well woolled to or below the knees and hocks.

The fleece should be compact and firm to the touch, of good quality and staple.

The rams should have a bold masculine appearance, carrying a handsome head, with strong and long horns well apart at the crown, springing out in a straight line with each other, and coming downwards and forwards in graceful curves as close to the face as may be without involving the necessity of having to be cut.

The ewes should have feminine characteristics and a more delicate set of horns.

It is a distinct objection to have a spotted skin or fleece. Markings on the horns are also disliked, while the tendency to grow the horns back is viewed with strong disfavour. The legs should be free from coarse hair.

In the Showyard.—The breed was first afforded separate classification at the Battersea meeting of the Royal Agricultural Society in 1862, the judges reporting limited competition but superior quality. The breed was again exhibited three years later at the Plymouth Royal Show, and subsequently at the Oxford and Cardiff meetings in 1870 and 1872.

Flock Book.—The Flock-Book was established in 1892. The volume for 1907 contains entries of 69,577 sheep.

Fecundity.— As already indicated, the outstanding characteristic of the Dorset Horn breed is its fecundity. The ewes receive the male as early as April or May, and the lambs are born in September, October, and November, the Royal Agricultural Society classifying them to be born 1st November. The lambs are produced early for the Christmas trade. The produce of a flock varies from 130 to 180 per cent of lambs, and in warmer countries two sets of lambs a-year have been bred. Occasionally this is done in this country, but the practice is not favoured.

Early Maturity.—About a ewe and a half are kept to the acre, varying with the quality of the land. The lambs remain with the ewes until May. The general lambing time is about two months in advance of other breeds, the flock ewes dropping about Christmas. The earlier lambs receive good feeding, the object being to fatten them as quickly as possible. October or November lambs, well nurtured, will be ready for the butcher at from ten to twelve weeks old, averaging from 10 to 14 lb. per quarter. They find a market in London at prices reaching up to 50s.

Dorset ewe lambs have been bred from under twelve months old, the rams being used on them in November and December. Their produce is fit for the butcher by midsummer.

For Crossing.—The Dorset Horn has not been used extensively for crossing. The most general cross is the Horn ewe and a Down ram, producing a very good grazing sheep, which may be fattened off pasture at eighteen months to kill from 20 to 25 lb. per quarter.

Where the Breed Thrives. — The breed is of course native to Dorset. It flourishes on the chalk farms of the Isle of Wight and Isle of Purbeck, and from Dorchester to Bridport, Crewkerne, and into the richer lands of Somerset and Devon. In the west of England it produces the early "house" lamb. Morton's *Cyclopædia of Agriculture* mentions the Horn sheep of the west of England as one of the oldest and best of the upland short-woolled Horn races. The breed has also, on a small scale, been tried in Scotland and Ireland.

The Somerset Horn Sheep.—This sheep was at one time bred on divergent lines to the Dorset, although they are of common parentage. Somerset breeders claim to have introduced the pink nostril as opposed to the dark. The Somerset sheep in the earlier times was lankier than the Dorset variety, but by judicious crossing greater plumpness and better form have been gained. Spooner says: "The Somerset sheep is a variety of the Dorset, possessing the same peculiarities and differing from it in being larger and taller, and having more arched profiles and heavy pink noses instead of black and white."

Clip.—The lambs clip from 2½ to 3 lb. of wool; the ewes from 5 to 7

lb., and the shearling rams from 10 to 14 lb. The particular virtue of the wool is its whiteness and the fine point · it possesses.

MANAGEMENT.

The management of a Dorset Horn flock is naturally determined to some extent by the period when the lambs are marketed. If very early lambing is the case, say in October and November, naturally Christmas lamb is the chief object. In the main, however, fat lamb is turned off from the month of April up till Christmas. A general lambing time is in November and December. This necessitates early ram sales, which take place in summer. One of the objects of the Dorset Horn flockmaster is to get his lambs forward to the London market before the Down breeder is ready with his consignments.

The wintering of the flock is very much like that of flocks of other breeds in the south. The root crops—mangels and turnips — play an important part, with plenty of hay to counteract the watery character of the roots. The twin ewes are specially fed, as they have a larger family to bring up, cake and corn being the chief ingredients of the artificial food mixture. Peas and old beans are also used. Mr James Attrill, who has a flock in the Isle of Wight, declares that "nothing fattens a lamb so quickly as plenty of milk." It pays, therefore, to look well after the ewes.

Mr Samuel Kidner's System.

In Somersetshire the system prevailing may be described in the words of Mr Samuel Kidner of Bickley, Milverton: "The breeding flock," he says, "consists chiefly of three ages, but a few of the best are retained for the fourth crop. The percentage of twins dropped would be about 66 per cent, with a few triplets last season, 3 per cent. The tupping begins about the first week in July, a few lambs being born in the last week of November, but the chief crop through December. None of the lambs are fattened, but are kept in a healthy growing state. The twins are kept separate, with more liberal treatment.

"Weaning takes place in about three

months from birth, when those to be kept for rams are selected. The lambs are then kept on cut swedes until we have green food for them, some linseed-cake being given. The over-age ewes are put forward as early as possible, being usually fit for the butcher when their lambs are weaned, there being always a demand for this class of sheep up to Lady Day.

"The wether lambs are maintained in store condition through the summer, in early autumn kept better, and sold at from twelve to thirteen months old, fat. The ewe lambs are selected for the flock in the autumn, there generally being a demand for the draft lots for breeding purposes. The rams are sold in their wool as yearlings about the second week in May."

Mr F. J. Merson's Flock.

Mr Frank J. Merson of North Petherton, Bridgewater, mates his ewes twice, as two tooths and four tooths. Thereafter a few of the best are retained in the flock as six-tooth ewes. The latter are put to the ram about the end of May to bring fat lamb, the progeny being fattened along with the ewes. The Chilver lambs from the younger ewes go into the flock, and a few of the best ram lambs are kept as tups. After lambing the flock is kept on grass for about six weeks, with cake, corn, and hay; and then on roots, rape, kale, and cabbage; finally, white turnips and cut swedes. The ram lambs run forward through "creeps." Fat lambs generally make from 35s. to 40s., fat ewes from 50s. to 55s., and fat hoggs up to 60s. There are about 50 per cent of twins. The ewes clip about 6 lb. and the ewe hoggs 7 lb., lambs 3 lb.

A Dorset Horn ram is represented in Plate 59.

RADNOR SHEEP.

This breed is associated with the county after which it is named. It has extended farther afield than that, however, being found on the Montgomery and Merioneth hills. The type has not been constant, being subject to extraneous influences which have altered it considerably.

Characteristics.

In point of colour some of the Radnor sheep are tan, some grey, and some speckled in the face. At one time their faces were yellow or, as they prefer to call it locally, tanned. Their fleeces were short and close, and they were built on short legs. They were well suited to resist the rough climate of the hills.

When the Radnorshire hills were fenced off, and the plough invaded what was hitherto the domain of the sheep, an effort was made to increase the size of the breed, Shropshire blood being introduced. This produced a hardy, clean-limbed, somewhat long-faced sheep, rather darker in visage. Latterly the Kerry Hill ram has been used extensively, and the time does not seem far removed when it will be difficult to distinguish between the two.

In appearance the modern Radnor is black of countenance, though some are tanned or grey. The rams are horned and the ewes should be polled. They are short-legged sheep, somewhat slow feeders, but their mutton is of excellent quality. When three or four years old the wethers will weigh from 14 to 15 lb. dead-weight per quarter, and clip from 4 to 5 lb. of wool.

The ewes are good nurses, and are largely used in the rearing of fat lamb.

MANAGEMENT.

At one time it was the custom to sell off the wethers when three or four years old, the wool paying for the sheep's keep. The cost of feeding was small. Nowadays the wethers are sold off at a year and a half to go on to Midland pastures, where they rapidly fatten and command a good price. The drafting of the ewe flock is done annually, and two- and three-year-old ewes are much in demand in September for the production of fat lamb.

The ewe flocks kept are much larger than formerly, owing to the disposal of wethers at an earlier age.

MOUNTAIN AND MOORLAND BREEDS OF SHEEP.

BLACKFACE SHEEP.

The early history of the Blackface sheep is pretty much a matter of conjecture. One eminent writer, Dr Walker, supposes that it is of foreign origin, and that the forest of Ettrick was selected as its first locality in Scotland. He mentions that a flock of 5000 sheep was imported by one of the Scottish kings, and from that stock the whole of the Blackface race, it is supposed, succeeded.

Other writers maintain that it originated among the mountains of Cumberland, Westmoreland, and Lancashire. Some people hold, on the other hand, that the Blackface had its rise among the mountains of southern Scotland. One Hector Boethius, writing about 1460, and speaking of sheep in the vale of Esk, says: "Until the introduction of the Cheviots the rough-woolled blackfaced sheep alone were to be found."

It is therefore pretty certain that from time immemorial it has held undisputed possession of the hills of southern Scotland and north of England.

The introduction of the breed to the Highlands of Scotland, which took place about the middle of the eighteenth century—when black cattle began to give way to sheep—was not altogether welcomed. In the Highlands at that time was a small white breed carrying a fine fleece, and its admirers felt sadly grieved over the inroads of the hardy Blackface. A Dr James Anderson, writing regarding the improvement of wool in the northern counties, says: "The coarse-woolled sheep" (meaning the Blackface) "have been debasing the old breed under the name of improving it, so that I am inclined to believe that in the mainland of Scotland the true unmixed breed is irretrievably lost." Since the beginning of last century, when flockmasters began to direct attention to the improvement of the breed, many defects have been removed. In modern times a healthy emulation and enthusiasm have taken possession of sheep-farmers to raise the value of their flocks, and of recent years a marked improvement in the character of the Blackface has been accomplished.

In many parts of the country the Blackface has been supplanted by the Cheviot, owing to the better price obtained for the wool of the latter.

Distribution of Breed.

The localities most noted for this breed are Lanarkshire, Ayrshire, Mid-Lothian, Perthshire, and Stirlingshire. Lanarkshire may be said to be the nursery of the Blackfaces, thousands of lambs being transported annually from this county to be reared upon the extensive pastures of the more elevated districts.

The southern districts of Scotland, as a rule, raise the best stock, being the districts in which the spirit of improvement has been longest and most actively at work. In the counties of Lanark, Ayr, Dumfries, and Mid-Lothian great pains and attention have been bestowed on the breeding process for a long period. The northern counties, though at one time behind, have been rapidly coming to the front during recent years.

In the more northern districts of Scotland extensive tracts abounded unconnected with any breeding farms, upon which the stock of wethers were maintained by buying in lambs.

Towards the end of last century, a demand having arisen for younger mutton, the grazing of three years became unprofitable, and the land had to be devoted to other purposes.

In the southern districts a ewe or breeding stock prevails; while in central and northern Scotland a mixed stock, ewe and wether, is the general rule.

Characteristics.

Strongly defined and distinctive characteristics and peculiarities distinguish the hardy Blackface. The general form is robust, muscular limbs with wide chest, body short and well barrelled, face and legs black and white or entirely black in colour. Endowed with great animation, the slightest alarm rouses them to action. Both sexes have horns,—large and spirally-twisted in the male, small and flattish and standing more out from the head in the female.

The wool is long and wavy, somewhat coarse, inclining to hairy.

Wild and restless in their habits, the nature of the sheep is to climb the highest hills. Remarkably hardy of constitution, they endure hunger and cold to a wonderful degree, boldly wintering it out where other breeds would succumb, and working with their feet among the snow for a bare subsistence with an energy and determination truly surprising. Their powers of endurance under the most trying circumstances is marvellous, instances being on record where some of the breed after being buried under snow-drifts for three or four weeks came out alive and apparently wonderfully well.

Strong in maternal or "homing" instinct, with a special attachment to a certain locality, ewes have been known to travel long distances so as to produce their offspring at the favoured spot.

Their mutton is so delicate and finely flavoured that it is preferred to every other.

An important property of this breed is its adaptation to heath lands; and it is this property that has rendered it so suitable to the extensive tracts of heath-covered hills throughout the country where it is acclimatised. There are many extensive Blackface sheep-runs, ten to fifteen thousand acres not being uncommon, with flocks of from five to eight thousand.

A Typical Blackface Sheep.

The following points are considered essential in a good specimen of the breed : Broad muzzle with strong aquiline nose and wide nostrils; forehead wide and full; the colour of the face to be either entirely black or black and white distinctly defined; both face and legs to be clean and free from all dunness or tuftiness; horns hard and free from blood-red, inclined to be wide set and not rising high on the crown, but coming out level with the top of the head, assuming a spiral formation; shoulder broad, with wide chest; straight broad back, not drooping behind; erect on hind legs, which should be well apart. The flow of the wool should almost reach to the ground.

The Blackface ewe is in good demand for crossing purposes — that with the Border Leicester proving very successful. The lambs, the result of this crossing, are excellent feeders, coming quickly to maturity, and yielding mutton of a high character and fine flavour.

Weights. — A well-known breeder gave the following as the average deadweight of the various classes of Blackface sheep taken off the hill :—

3-year old wethers from	14	to	16	lb. per qr.
2-year old do.	"	12½	"	14	"	"
Yeld ewes	"	13	"	15	"	"
Gimmers	"	12	"	13½	"	"
Cast ewes	"	10	"	12½	"	"

Prices of Blackface Sheep.

The following are the general prices for Blackface wethers and cast ewes in each of the years 1893-1907 :—

	Wethers.				Cast ewes.			
	s.	d.	s.	d.	s.	d.	s.	d.
1893 .	21	0 to	37	0	12	0 to	24	0
1894 .	20	0 "	37	6	14	6 "	26	6
1895 .	23	0 "	41	0	16	0 "	28	6
1896 .	19	0 "	35	4	13	0 "	24	0
1897 .	21	0 "	36	6	15	0 "	25	6
1898 .	22	0 "	37	0	16	0 "	26	6
1899 .	20	0 "	33	6	13	0 "	24	0
1900 .	23	0 "	36	0	16	0 "	26	0
1901 .	20	0 "	35	0	14	0 "	25	6
1902 .	18	6 "	34	0	12	0 "	24	0
1903 .	21	0 "	36	0	15	0 "	28	0
1904 .	23	0 "	38	6	18	0 "	30	0
1905 .	21	6 "	37	0	19	0 "	31	0
1906 .	23	0 "	38	0	20	0 "	33	0
1907 .	21	0 "	33	6	17	0 "	28	0

Prices of Wool.

The following are the prices per stone of 24 lb. of unsmeared wool of Blackface sheep for the years 1893-1907 :—

		s.	d.	s.	d.
1893	from	10	0 to	12	0
1894	"	10	0 "	12	0
1895	"	10	0 "	11	6
1896	"	10	0 "	11	6
1897	"	10	6 "	12	0
1898	"	10	0 "	11	6
1899	"	8	6 "	9	6
1900	"	8	0 "	9	6
1901	"	8	0 "	9	0
1902	"	8	6 "	9	6
1903	"	11	6 "	12	6
1904	"	14	0 "	15	0
1905	"	15	0 "	16	0
1906	"	16	0 "	17	6
1907	"	16	0 "	17	0

MANAGEMENT.

The management of the Blackface is, generally speaking, pretty much the same all over, varying little from north to south. On most farms the flocks are allowed to roam at free will. There are some farms, however, on which the flocks are divided into what are termed *hirsels*, each hirsel being confined to a certain portion of the farm.

The ewes have their first lambs at two years old. The rams are put to the ewes between 20th and 30th November, and the lambs are dropped towards the end of April.

During winter these Blackface sheep live on rather scanty fare,—auxiliary feeding being resorted to only when the ground gets covered with frozen snow to such a depth that they are unable to get at the herbage by scraping with their feet. Flockmasters in high exposed districts consider it necessary to keep a supply of hay in reserve against a protracted storm, as judicious feeding at such a time becomes indispensable.

The male lambs are castrated when about eight or ten weeks old, the best being left uncut for sires.

The fleece is removed in the months of June and July, the male and yeld portion of the flock coming to clipping condition earlier than the breeding ewes.

It is the custom on many farms to wash the sheep before clipping them. In the shearing operations mutual assistance is frequently given. Neighbouring shepherds help each other during the clipping. The sheep are generally branded or marked with tar after the fleece is removed.

The fleeces are rolled up and packed ready for sending to market. The average weight of the fleece is between 4 and 5 lb. The wool being inferior in quality to that of other breeds is chiefly used in the manufacture of carpets and the coarser fabrics. The clip is consigned to wool-brokers in the large towns, who dispose of it by auction, at prices ranging over a series of years, from 4d. to 8d. per pound. America is a good customer for this class of wool.

Within recent years there has been a tendency to favour the production of large, heavy fleeces of strong wool, although some breeders lean to the opinion that the advantage to the animal lies with the thick-set soft wool evenly distributed.

The lambs are weaned about the second week in August. The ewe lambs, with the exception of what have to be retained to keep up the numbers of the stock on the farm, are sold for breeding purposes. The wether lambs are disposed of according to the nature of the farm. Where a mixed stock is kept the best of the wether lambs are retained till two or three years old. Only the inferior class, or what are called *shotts*, are sold. Where only a ewe stock prevails the whole of the wether lambs are sold. They pass into the hands of low country and arable farmers, who, after feeding them for a few months, generally dispose of them at remunerative prices.

Hoggs on the majority of grazings are sent sometimes long distances to the country for wintering. This proves an expensive item in the economy of sheep-farming, the cost averaging from 7s. to 8s. a-head.

The old or cast ewes—that is, all above five years or so—are drafted in October, and sold for rearing a crop of cross lambs, after which they are fattened for the butcher.

Dipping.—The process of dipping hill and other sheep is universally practised, being for a time made compulsory by legislative enactment. It consists of a bath composed of certain ingredients, administered twice a-year. This is for the purpose of destroying parasites and the prevention of skin diseases, promoting the general health and comfort of the animal, as well as enhancing the quality of the wool.

Markets.—The principal markets for the sale of the Blackface are the various auction marts throughout the country, the once famous Falkirk Trysts now being a thing of the past. A sheep and wool fair is held at Inverness in the month of July. This market is unique of its kind, there being neither a sheep nor a fleece on view, all purchases being based on previously proved character.

Qualifications of a Shepherd.—Farmers place their flocks under the care of trustworthy and capable shepherds. At all seasons interested shepherds can by care and judgment do a great deal in improving the condition of flocks. Mr Little, a. writer on the subject, gives the following qualifications of a mountain shepherd : " The shepherd should be honest, active, careful, and, above all, calm-tempered. A shepherd who at any time gets into a passion with his sheep not only occasionally injures them, but acts at a great disadvantage both in herding them and working among them. A good-tempered man and a close-mouthed dog will effect the desired object with half the time and trouble that it gives to the hasty, passionate man. The qualifications of a shepherd are not to train his dog to running and hounding, but to direct the sheep according to the nature of the soil and climate, and the situation of the farm, in such a manner as to obtain the greatest quantity of safe and nutritious foods at all seasons of the year. Those shepherds who dog and force their flocks I take to be bad herdsmen for their masters and bad herdsmen for the neighbouring farmers."

Glenbuck Blackfaces.

Mr Howatson of Glenbuck has been good enough to supply information regarding the management of his famous flock of Blackfaces.

Age of Draft Ewes.—Mr Howatson takes only four or five crops of lambs from his ewes before parting with them, as he finds that better and stronger lambs are bred from robust young ewes than from exhausted old ewes, and that, as a matter of course, five-year-old draft ewes sell better than ewes a year older. The draft ewes are sold early in October, and the whole remaining flock is then dipped, the dipping being repeated as weather permits to meet the wants of the Board of Agriculture's Regulations.

Early Lambs.—Mr Howatson lets his rams to the ewes in the second week of November, which is about a week earlier than the general custom. The best lot of rams go first, and then in about three weeks the remainder of the rams are put amongst the ewes so as to pick up those not already served.

Ram Lambs.—Mr Howatson has so much improved his flock that he finds a ready demand for his ram lambs for breeding purposes, so that few of them

are castrated. He retains a few of the choicest of the ram lambs to bring out for shearlings, from which the best are again selected for home stud purposes, and the remainder, with the spare ewe lambs, are sold at sales in August, September, and October. The system of selling ram lambs, so successfully inaugurated by Mr Howatson about 1870, is growing in favour, as thereby the purchaser gets possession of the young sire which he can feed and treat as may seem best to suit his purposes.

Mr Howatson is opposed to the early clipping of rams for sale or breeding purposes.

Ewe Lambs. — The Glenbuck ewe lambs are weaned in August. The ewe lambs selected to be retained in the flock are dipped and sent back to the hill till the second week in October, when they are despatched to the low country, where they are wintered at a cost of from 8s. to 8s. 6d. per head.

Clipping. — Clipping begins in the second week of June with the ewe hoggs. At this time care is taken to mark for sale any of the ewe hoggs which may not in every respect be satisfactory for breeding purposes, special attention being given to the fleece, in the improvement of which Mr Howatson has been very successful. Mr Howatson thinks it advantageous to delay clipping ewes until the new wool is well raised, and the clipping of them is therefore postponed till the latter part of July.

BLACKFACE RAM-BREEDING.

The breeding of rams for sale to other flock-owners has become an important industry with many of the leading owners of the Blackface breed. With skilful and careful management the returns are usually substantial, the prices obtained for young rams of choice quality and character generally reaching high figures. For single shearling rams as much as from £150 to £200 has been realised at auction sales.

Information on the systems of management pursued in the breeding and rearing of Blackface rams has been kindly given by a number of owners of well-known flocks, including Mr Howatson of Glenbuck; Mr Archibald, Overshiels, Stow;

Messrs Cadzow Brothers, Borland and Stoneyhill, Carstairs; Mr Hamilton, Woolfords, Cobbinshaw; Mr Fraser, Rankinston, Ayr, and others.

Mating.—Special care is taken in the mating of ewes and rams so as to secure stock of the highest merit. The best ewes in the flock are naturally chosen for ram-breeding, but however good a ewe may be her lamb is not selected for stud purposes unless it is itself satisfactory in every way. In all judiciously managed flocks the breeding character of every strain is well known, and this knowledge assists greatly not only in the mating of ewes and rams, but also in the selecting of lambs, both male and female, to be retained for breeding purposes.

The few selected stud ewes are, as a rule, kept by themselves in fields where the pasture is good, and for the most part it is from these ewes that the successful show sheep are obtained. Still, in many cases the rams sold for stud purposes are bred from ewes that run with the general flock excepting at the time of tupping, when each tup is isolated with the ewes allotted to him.

The Overshiels System.

Feeding Young Rams.—Mr Archibald, Overshiels, writes: "The lambs to be kept as rams are weaned about the middle of August, when they are put on clover-foggage or cabbage, and taught as soon as possible to eat artificial food, such as linseed-cake. A good plan is to confine the lambs in a small enclosure where they can get nothing but cabbage, which they will eat greedily in a few days; then give them access to no cabbage except what are cut into troughs, and on the cut cabbage sprinkle linseed-cake and locust-meal. In a day or two the lambs will eat this food readily, and thereafter they will feed out of troughs and eat cabbages off the ground like older sheep.

"The ram lambs are put into the house not later than the first of October. There in some flocks they get a feed in the morning of a mixture of boiled barley and bran, with a pinch of salt. At midday and again at night they get a dry feed, consisting mostly of linseed-cake. Care must be taken not to give too much. At first ¾ lb. is ample, the quantity being gradually increased as the lambs get bigger, and it is found they can eat it with safety. Always have a rack filled with natural hay and a trough of fresh water within their reach. In course of time the lambs will come to eat over 2 lb. each per day of the concentrated food. In some cases a little cod-liver oil mixed with treacle is given in each boiled feed. Young rams intended for exhibition are by some considered the better of getting new milk twice a-day. Few of the animals can be got to drink the milk, so it has to be poured down their throats from a bottle. An ordinary cow will give enough milk for five or six shearling rams.

"After the grass comes, usually about the middle of May, the young rams should be put out a short time during each day, and put back to the house overnight and fed on green food, such as grass and tares, till the cabbages are ready. They should be well treated in this way up to the show or sale.

"The young rams should be clipped along the bellies and half-way up the ribs as early as possible in November, and the rest of the body should be clipped in December.

"The wool often gets so long that the animal is apt to pull it out of its breast by its feet and knees when rising; to avert this some tie the wool with tape in tassels about the thickness of four fingers. It is also a good plan to sew a sheet along the back to prevent the sheep from rubbing and spoiling the fleece.

"Rams that are out-wintered get the same treatment as the ordinary hogs. These out-wintered rams should be clipped if possible about the beginning of April, and if the farm is high and exposed they require to be housed for about six weeks, or until the weather gets favourable."

There has from time to time been much discussion over the question of the high feeding of rams. In theory high feeding is almost universally condemned, yet it is the practice of flock-owners to give the preference to highly fed rams in the sale-ring.

Messrs Cadzow's System.

Messrs Cadzow Brothers write: "In entering upon the breeding of rams, we in the first place made up our minds as to the ideal type to produce for all practical purposes, and have kept that ideal before us all the time without the slightest deviation. Our ideal is a sheep wide in the back and ribs, walking freely and straight on not long but strong well-planted legs, and carrying a thick coat of wool, not hair. Our system of breeding is to mate our females with sires of a masculine type embracing all the characteristics of our ideal, strictly avoiding in-breeding, and purchasing fresh blood whenever we see suitable animals for sale, more especially when we can get animals which may excel in those points which need correcting in our own flock. In mating, we at all times see that the males are strong in the points in which the females may be lacking.

"The lambs are dropped from the middle of April till the middle of May. The ewes get nothing but pasture during summer, the lambs being weaned about the middle of August.

"The ram lambs when weaned are put on hay and stubble or foggage till about the middle of October. They are then housed. Their winter food consists of from an eighth of a pound to half a pound of boiled barley mixed with good bran, and from an eighth to a pound of a mixture of linseed-cake, Indian corn, and oats once a-day, with as much meadow-hay as they can eat, and plenty of good water. For showing we clip a few of the rams in January, but most of them are clipped in February and March. They are put to grass in spring, and get from half a pound to one and a half pounds of the raw mixed feed till the time of the sales in September."

The Woolfords System.

Mr Hamilton, Woolfords, writes :—
"Before the ram sales, in fact all the year round, I try to find out the weakest points in the breeding ewes, and if possible keep and buy rams strong in these points. At about the 15th November the ewes are all hand drawn to the rams, and each lot put into different fields for about 34 days.

"The ewes here have to be carefully drawn, with regard to pedigree as well as points, as there are always some home-bred rams used, and they have half-

sisters and other near relatives in the stock. I have never gone in for close breeding, but I like a little of the same blood when practicable ; of course when a ewe has done well with a sire one year she is put back to the same ram again.

"Each ram's lot of ewes are keeled differently, and when dropped in the spring the lambs are ear-marked with a different mark for each individual sire. Thus the sire of each ram and ewe on the farm is known.

"The ewes during winter are all kept on the hill pasture, and get nothing extra in the way of feeding except in time of heavy snow, when they get hay, on which they do very well. It never pays to let them get lean, as with the lot of twins here the loss in lambs and ewes would be great.

"When the twin lambs are able to walk they are driven down into fields, but do not get any extra feeding until weaned, unless when they are on very old grass ; in that case the mothers get a little hand feeding during April and May. The single lambs get no extra feeding until weaned. If the forcing is commenced before weaning the lambs are apt to get coarse, and it is not good for the stock ewes.

"In the first place, as to the lambs that are to be sold as ram lambs at the ram sales in September and October, when weaned generally in the first week of August, they are put on to the best foggage on the farm, and get in addition lamb food and cabbages, as much as they will eat until sold.

"The ram lambs that are to be wintered and sold next year as shearlings are weaned at the same time and put on to clean grass, sometimes foggage has to be taken from home, the one object being to keep them growing steadily. They are put into the house about the middle of October, and are commenced with a little boiled barley mixed with bran, treacle, and salt for one feed, and lamb food or other mixed grains for the other meal. In about a fortnight they are getting 3/8 of a lb. of barley in a boiled condition, and 3/8 of a lb. of lamb food, this feeding being gradually increased until they get up to exactly double the quantity by the month of March.

"The rams are clipped in January or beginning of February, and are put out whenever there is grass for them in April. They are kept thriving steadily until the beginning of August, when they are put on to cabbage and as much corn as they will eat, to give them a flush for the sales. The important thing is to keep them steadily thriving from the day they are born until sold, with an extra flush in the last six weeks."

Mr M. P. Fraser's System.

Mr Fraser, Rankinston, Ayr, writes:—
"The ram lambs are weaned in the beginning of August, and put into fog-gage fields on the farm, where they are taught to take a feed of oats, Indian corn, and cake. About the middle of October they are put into houses (15 to each house or division), and in a few days they will have taken kindly to their winter rations.

It is very advisable at this time to carefully examine the sheep and see that they are free from foot-rot, because when once they commence to thrive, any back-set from the above cause, or from an overdose of feeding, may lead to a mal-formed turn of the horns. The lambs are fed at 6 A.M. on oats, Indian corn, and cake; at 1 P.M. on boiled barley and Indian corn, with oil-cake and beans; and at 7.30 P.M. on oats, Indian corn, and cake. They receive a fresh supply of hay twice each day, and water is always before them. The amount of feeding is gradually increased till by December each lamb will be eating 1 lb. of raw food and ½ lb. boiled food per day.

"With the exception of a few show rams that are clipped in the middle of December, all are clipped after the New Year, and their feeding is altered to a boiled feed night and morning and a raw feed in the middle of the day.

"About the beginning of May the shearlings are gradually accustomed to the grass, the boiled feed is stopped and the raw feed increased, till by June they will be eating 2 lb. of oats, Indian corn, and cake. There is not the same danger of giving them an over-feed on the grass as there was in the house. Towards the end of July the cabbages will be ready and may be given freely to the shear-lings. From now to the September sales it is just a steady plodding on upon these lines. With this feeding I have practically no losses from deaths between weaning and selling."

In Plate 54 portraits are given of a group of rams bred by Mr Howatson of Glenbuck, the group being arranged to represent the development effected in the type of Blackface rams between the years 1869 and 1894.

Portraits of a Blackface ram and ewe are produced in Plate 55.

CHEVIOT SHEEP.

What the Blackface is to the heathery hills of Scotland and the extreme northern districts of England, the Cheviot is to the grassy hills and uplands of the same range of country. The Cheviot at one time, indeed, was a serious rival to even the Blackface on what are known as the black hill sheep-runs.

In the early 'twenties of last century, and perhaps a little earlier, when the finer wools were rising in value, many heather-clad hill farms in Dumfriesshire and Ayrshire, and even as far north as Perthshire, Argyllshire, and Inverness-shire, were denuded of their Blackfaces to make way for Cheviots. But some years after this a number of very severe winters were experienced, and the new-comers were not found to stand the stress so well as the Blackfaces did, and, indeed, many were killed out. There was, accordingly, a reversion on most of these farms to the original stock, and since then Cheviots have for the most part been confined to the Cheviot range on both sides of the Border, to Dumfries, Selkirk, and Roxburgh shires close by, and to the more luxuriant of the grassy slopes of Inverness, Ross, Sutherland, and Caithness shires in the far north.

Origin.

There is little doubt that Cheviots are natives of the Cheviot range, still to a large extent the headquarters of the breed. How long the breed has occupied these towering grassy heights it is impossible to say; but it was there, and apparently flourishing, when, in the interests of the British Wool Society,

Sir John Sinclair visited the locality in 1791. Not only did Sir John report very favourably upon the breed from a wool-growing point of view, but he was so much impressed with the merits of the sheep that he introduced them into his own county of Caithness, where they have ever since remained. After a time they also got a firm hold in the neighbouring county of Sutherland, which they have likewise succeeded in retaining. Indeed Caithness and Sutherland shire Cheviots have long enjoyed quite a fame of their own. No doubt, owing to the deeper and heavier land on which they are kept, they grow larger than the South-country Cheviots, and on this account are very popular for feeding purposes, especially on turnips. Caithness or Sutherland Cheviot wedders nearly always realise a shilling or two more per head than South-country bred Cheviots of the same class bring.

But, on the other hand, there seems to be something in either the soil or the climate of the south-east country which produces a finer type of bone and wool than the north does. As a consequence, nearly all the most noted flocks of the breed are in the south, and even the north country breeders have to come there from time to time for fresh supplies of rams to maintain their stocks.

Early Improvement.

In the early improvement of the Cheviot breed Lincoln blood seems to have been used in smaller or greater quantity. One specific statement is that "Mr John Edminstoun, late of Mindrum, Mr James Robson, then at Philhope, and Mr Charles Kerr, then at Ricaltoun, went to Lincolnshire about the year 1756, and bought fourteen rams with which they crossed their sheep with great success."[1] Substantially the same statement is made in the *Farmers' Magazine*, published some considerable number of years before. There it is stated that these Lincoln tups so improved Mr Robson's stock as to give his sheep a decided superiority over those of his neighbours, and for many years after making this cross "he sold

[1] Douglas's *Survey of Roxburghshire*, published in 1876.

more tups than one-half of the hill farmers put together."

All this happened a good many years prior to Sir John Sinclair's visit to the Borders. The introduction of the Lincoln blood would, no doubt, have had an important effect in improving the quality of the wool remarked upon by Sir John Sinclair, but of what other advantage it could have been to such a sheep as the Cheviot—much smaller as a rule than the Lincoln—it is not easy to see.

It has also been stated that Cheviots were crossed with the Border-Leicester type of the Dishley Leicester shortly after this breed was introduced into the Border districts from Leicestershire, but of that infusion such definite records do not seem to exist. Still, one can readily imagine that a dash of the improved Leicester blood would have been advantageous to the Cheviots of that period, when in many cases sheep of the breed were lacking in symmetry, and were inclined to be brownish in hair in parts and not nearly so white generally as at the present time.

Characteristics.

The Cheviot sheep as it exists to-day is one of the most handsome and vigorous-looking animals of the whole ovine race. Entirely white in appearance, it is very active on its legs, carries itself with great dignity and courage, and when put into a tight corner will make a bold dash for liberty even against considerable odds.

Appearance of Rams. — According to the first volume of the breed Flock-Book, which was published in 1893, the Cheviot tup should weigh alive at maturity when fat 200 lb. His head should be of medium length, broad between the eyes, and well covered with short fine hair. His ears should be nicely rounded and not too long; they should be well up from the eye and rise erect from the head. Low-set or drooping ears are a decided fault. At the same time, they should not be what are called "hare-lugged"—that is, too near to each other. This indicates a narrow face, which generally denotes a narrow body. The neck should be short and strong, and in the ram well arched. The nose should be arched and broad, and the nostrils black, full, and open, and the ribs well sprung and carried well back towards the hook bones. Though occasionally a ram will appear that has rudimentary horns, the breed on both the male and female side is a hornless one.

A long weak back is about the worst fault a Cheviot can have. The back should be broad and well covered with mutton, the hind quarters full, straight, and square, and the tail well hung and nicely fringed with wool. The legs must stand squarely from the body; bent hocks, either out or in (the latter especially), are looked upon as a weakness. The bone should be broad and flat, and must be covered with short, hard, white hair. The wool should meet the hair at the ears and cheeks in a decided ruffle. Bareness there or at the throat is inadmissible, and the wool should grow nicely down to the hocks and knees. The belly and breast ought also to be well covered.

Appearance of Ewes. — The same description suitably modified will also apply to ewes, which usually weigh alive from 100 to 150 lb.

Wool. — The fleece of the Cheviot ram should weigh about 10 to 12 lb., of the ewes about 4½ lb., and of the wethers about 5 lb. Although the Cheviot is an excellent mutton sheep, its outstanding feature is the high quality of its wool. Cheviot wool is of a close, dense, beautifully fibred type, and has always been in great demand for the production of the best class of tweeds. Indeed, it was Cheviot wool very largely that made the name and fame of the Hawick, Galashiels, and other Border district tweed manufactures.

Crossing Purposes. — In addition to its other merits the Cheviot is of great value for crossing with the Border Leicester. Cheviot ewes put to Border Leicester rams give the popular half-bred —one of the most valuable commercial sheep that is to be found in Scotland. Half-breds are extensively used on arable farms all over the south of Scotland, and they make not only excellent grazing sheep but first-class stock for fattening on roots. It has been stated that half-breds pay more rent in the arable parts of the south of Scotland than any other breed or class of sheep, and the claim is believed to be well founded. They

are also most extensively used in Northumberland, and are found as far north as Aberdeen. From the half-bred, again, by the use also of a Border Leicester ram, is bred the very plump three-parts-bred—one of the quickest maturing of the sheep tribe, and greatly run upon for feeding rapidly off foggage or turnips.

Improvers of Cheviots.

One of the first and most noteworthy improvers of Cheviot sheep in comparatively modern times was Mr James Brydon of Moodlaw and Kennelhead, in the county of Dumfries, who held biennial sales of rams at Beattock from 1851 to 1881. Mr Brydon favoured what was known at the time as the west-country type of Cheviot—that was, a sheep with more length and substance than the original east-Border kind, but neither so stylish nor so dense in the character of its wool. It has been said that Mr Brydon introduced Border Leicester blood, and that he got the extra length in this way; but however this may be, his sheep had a great run of success for many years, both in the showyard and at his biennial sales. At the latter he was accustomed to average from £15 to £17 per head for from 150 to 180 rams—figures which could hardly be excelled even at the present day.

Individual prices were much higher. In 1867 Mr John Miller of Scrabster, Caithness, gave no less than 185 guineas for one specially good ram, "Craigphadrig" by name. This, it is noteworthy, was the record price in Scotland for rams of any breed for several years; indeed it was not exceeded until 1873, when Messrs Clark gave £195 for one of Lord Polwarth's Border Leicester rams from Mertoun. Successful as they were for many years, Mr Brydon's sheep latterly gave way to softness, and to a considerable extent lost their pre-eminent position.

For this result some people blamed the introduction of Border Leicester blood, while others alleged that the softness was due to the winter house-feeding of rams which was introduced in Mr Brydon's day. While both may have been predisposing causes, some part of the trouble may also have been due to the fact that attempts to raise mountain breeds of stock above their natural size have practically always ended in failure.

At any rate, where Mr Brydon met with failure, success was attained by Mr Thomas Elliot, Hindhope, Jedburgh, who had been working almost contemporaneously with the east-country and smaller type of sheep. Mr Elliot took the place which was gradually vacated by Mr Brydon, and his type of sheep as represented by the Hindhope flock—which is now carried on with great success by his son, Mr John Elliot—is still the dominant type of the breed.

Flock-Book.

In 1891 the Cheviot Sheep Society was formed and flock-books with a register of rams have been published annually since 1893. The secretary is Mr John Robson, Newton, Bellingham, Northumberland, himself a noted breeder of Cheviot sheep.

MANAGEMENT IN CHEVIOT FLOCKS.

The management of Cheviot flocks is comparatively simple. Except in the case of rams intended for sale for breeding purposes, little housing or special feeding is resorted to.

Newton and other Flocks.

Mr John Robson, Newton, Bellingham, whose valuable and old-established flock of Cheviots has for several years taken a leading position in the showyards, has favoured us with some notes relating to the management of his own and other similar flocks. His flock is entirely home bred. He casts ewes 6 years old. West of the Carter Fell ewes are sold at 6 years old, north of it generally at 5.

Selling Young.—Wether lambs used to be hogged on the farm, and kept till 3 or 4 years old, then sold fat—or in plentiful turnip years, for turniping. Now, on account of bad seasons, increase of sickness, and low price of wool, they are mostly sold as lambs, to go to better land to be fed off as shearlings; or if kept on hill farms, they are sold at 2 years old.

Weights.—Ewes weigh when sold probably 60 lb., wethers, 72 lb.; but, of course, when very fat they greatly exceed these weights.

Hirsels.—On the Cheviot Hills a farm is generally divided into two hirsels. On large farms the number of hirsels is of course multiplied indefinitely. But take a sixty-score farm—the ewe hirsel will contain three ages of twelve scores of ewes each, 3, 4, and 5 years old; the hogg hirsels, two ages of about twelve scores each of 1- and 2-year-old sheep. At clipping time the 2-year-old ewes or "young ewes" are brought from their "hogging" and put amongst the ewes, their ground being hained till the end of July, when the ewe lambs are weaned and taken to it.

Land "tired of Hogging."—Thus lambs never follow lambs, the ground always getting a year's rest from lambs, as they are allowed to remain till 2 years old. If lambs follow lambs too often, the land is apt to get "tired of hogging," which, if continued, means that the hoggs either die freely of sickness or of poverty.

Age for Breeding.—When farms are managed on this system, the gimmers are not, except on the very best low-lying farms, expected to bring lambs; only a few of the strongest are put to the tup.

West-country System.—The other or West-country system is to allow the ewe lambs to follow their mothers—none but those on the draft ewes being weaned, and those only for ten days, when they are put back to their mothers. Here the gimmers in good seasons are expected to bring lambs; all but a few of the worst get the chance of the tup, and the ewes are generally sold at 6 years old.

On land addicted to louping-ill this is much the best way, as there is less change; but on the healthy and stormy Cheviot Hills the former plan has this advantage, that it provides a stock for the harder and higher ground which would not keep ewes, and also allows of the hoggs being better looked after in a storm.

Feeding in a Snowstorm.—The only difference between winter feeding and summer is, that if a snowstorm comes which blocks up the ground so thoroughly that little or no natural food can be got, the sheep are given hay. About 1 lb. each is the usual quantity once a-day, as early in the morning as possible. Great care should be taken to keep sheep in as small "cuts" as possible — 100 is about the best number, and every farm should have a stell for every cut of sheep.

Hand-feed judiciously.—Hay should only be given to prevent hunger, as on some land sheep which have been heavily hayed do not thrive next summer so satisfactorily as those which have not been so much pampered. Corn or cake has also the same tendency, and ewes which have been hand-fed one winter always look for the same indulgence afterwards.

Wethers on Turnips.—Wethers are mostly kept on turnips about 20 weeks the first winter, and 6 or 8 weeks the next.

Extra Food with Turnips.—As a rule, no additional food is given to sheep on turnips, but sometimes when turnips are taken by the week sheep get hay or straw; feeding-stuffs are rarely given. If a hill-farmer has turnips of his own, he is generally a generous feeder, giving cake or corn and hay to fattening sheep, and hay or straw to hoggs. In a storm all sheep get hay, but seldom corn or cake.

Rams.—The rams are usually kept amongst the other sheep during summer. In winter they get turnips, and when being prepared for sale a little cotton-cake.

Price and Quantity of Turnips.—Turnips for wethers cost about 5d. or 6d. per week; for hoggs, 3d. And as an acre of fair turnips is said to winter a score of hoggs, it may be supposed that the same quantity will keep 20 wethers ten weeks. Probably an acre and a half will be required to feed 20 wethers.

There is now a greater tendency to treat Cheviots as park sheep than there was prior to 1890. Owing to so much of the worn land being now stocked with blackfaces it is possible to give Cheviot ewes more indulgence in the spring than they used to get, seeing that they have good land to return to. This change in management probably accounts for the greater demand for larger sheep than was the case formerly. And this was also helped by a cycle of good seasons which Border farmers have experienced after the disastrous 'eighties. Now practically all the wedder lambs are sold to feeders, none being left for breeding farms. A few wedder flocks are still left in Sutherland, and some shearling

wedders are fed off in parks on turnips, but none are now left on the hill pastures of the Borders.

Mowhaugh Flock.

In Mr J. R. C. Smith's flock at Mow-haugh, Yetholm, lambing begins usually about the 20th of April and extends on until about the end of May. Ewe lambs are weaned about the 20th of July, and wedder lambs from the 12th of August onwards. After being weaned ewe lambs get three weeks' change to a freestone country, and then go on to their winter hirsel. They do not, however, follow their dams. Cheviot lambs sold in August realise from 14s. to 20s. apiece. In the spring the same lambs should weigh from 48 to 56 lb., and be worth from 36s. to 45s.

Cheviot ewes in the flock depend almost entirely on their hill grazing, getting hay in very stormy weather. Ewes of this breed bring their first lambs, as a rule, at 3 years old, and are drafted out at from 5½ to 6½ years old.

Rams are sold at 2 years old, and are lightly fed the first year, getting a limited allowance of turnips but plenty of hay and ¼ lb. box-feeding per day. In the second winter they require better feeding in preparation for the sale-ring.

Alton Flock.

Lambing in Mr Michael Johnstone's flock at Alton, Moffat, begins on 18th April, and weaning takes place about the beginning of August. When running with their mothers the lambs may get a chance of a little oats or Indian corn, but they get nothing but grass after weaning. All lambs are sold at Locker-bie Auction Mart. In 1908 top wedder lambs realised 15s. 3d., and mid ewe lambs 15s. 6d. Ewes on the hill get nothing but what they gather. Any lean ones are brought in to the fields. Ewes to be mated with Border Leicester rams are kept in the fields, and get turnips, oats, and hay, beginning in the month of February.

Ewes drop their first lamb at 2 years old and are cast at 6. Young rams are run on "seeds" after being weaned, and are wintered on cut turnips and corn. Stock rams are summered on the hills;

in winter they are brought down to a field, and get hay and cut turnips.

Mr Johnstone brings his Cheviot ewes from the hill when they are 6 years old. They are run in the fields all winter, and a half-bred lamb taken off them. The following summer they are sold in the market, generally to go to Ireland. Half-bred lambs bred in this way usually begin to arrive about the 26th of March. These lambs are also sold at Lockerbie, realising for the best, in 1908, 25s. Ewes, after nursing half-bred lambs, fetch up to 23s. apiece.

Dalchork, Lairg.

A good example of the management of Cheviot flocks in the North is afforded by the system which prevails in Messrs W. and C. Mundell's flock at Dalchork, Lairg, Sutherlandshire. Here the lambs arrive from the 20th of April until the 28th of May. They are weaned about the 8th of August. No extra food is given to the lambs before weaning, but after weaning the wedder lambs are sold, and they usually get extra food almost as soon as they arrive at their destinations. No extra food is given to the ewe lambs until October, when they go to Ross-shire to wintering, and the worst of them get turnips in the spring.

Lambs in this flock, like those in most other flocks in the North, are sold at the Inverness wool market and are delivered about the 8th of August. The price realised in 1908 was about £1 per head for the tops. "Shott" lambs are put on to foggage after being weaned, and are sold about a month later at Inverness. In 1908 they realised 16s. per head. Ewes are disposed of at Lairg sale in the end of September, averaging in 1908 32s. 6d. Shearling tups are sold at Dingwall, the average price in 1908 being £7.

The ewes of the flock receive no artificial feeding of any kind except about sixty of the worst, which get, for about a month before lambing and until the grass comes on the hill, about 1 lb. of whole oats and bran and the run of a good park. In very bad winters all the ewes get hay, but only when they cannot have sufficient natural food. Ewes bring their first lamb at 2 years old, and are cast at 5 years old.

Tup lambs after being weaned are sent to a farm in Ross-shire, and remain there until the end of April. They are grass wintered up to the 1st of January, when they are put on to turnips for about a month. After that they get cut Swedish turnips and good hay.

Stock rams are sent to Ross-shire also when they come from the ewes, and are put on to turnips and get good clover hay.

In addition to the other classes mentioned, Messrs Mundell sell every year about two hundred gimmers (shearling ewes), those disposed of in 1908 making 40s. to 55s. per head. These gimmers are a little more liberally fed than the gimmers that are kept for stock purposes.

A Cheviot ram is represented in Plate 56.

THE EXMOOR HORN SHEEP.

Sir T. D. Acland, writing in the *Journal of the Royal Agricultural Society* in 1850, describes the horned flocks which run on the Somersetshire hills. He mentions that the ordinary sheep of the country when fat do not weigh above 10 or 11 lb. per quarter. "Where pains have been taken to improve a flock, they may reach on the average 16 to 18 lb. per quarter, and some are brought up to 24 lb. per quarter, fed on Bridgewater marshes."

The Exmoor Horn sheep is stated by some authorities to have a common origin with the Dorset Horn—a belief which may not be far wide of the mark, seeing that there is a similarity in appearance.

Characteristics.

A fine open curly horn decorates a white head of pleasing appearance.

The fleece is close, and the wool comes right up to the cheeks. The appearance of the breed is not unlike the Cheviot in formation of top, loin, and quarters. The wool is of medium length, superior in quality, and the fleece is so dense as to defy the storms which so frequently cover them over in winter for days at a time.

The ewes are prolific, producing from 30 to 50 per cent of doubles. Record is made of one ewe, owned by Mr Tom Elworthy of Simonsbath, which had 25 lambs, having reared 24, and was then nursing twins.

As indicative of the hardy character of the breed, a writer chronicles that lambs reared on the Wiltshire Downs from Exmoor ewes at three months old, without artificial feeding, realised 38s., the land being so poor in quality that its rent was only 1s. per acre. Breeders aim to produce a wether which at sixteen to eighteen months old will give a carcase, matured at small cost, of from 16 to 18 lb. per quarter. Such sheep, carried on for Christmas, would kill 30 lb. a quarter.

MANAGEMENT.

The management of Exmoor flocks, as a rule, is of the simplest. The ordinary grazing is at times supplemented with artificial food. In the best flocks green food is specially grown to keep the young sheep thriving. In the flock of Mr D. J. Tapp of Highercombe, Dulverton, weaning takes place about the middle of June—the lambs being turned on to the best ' grasses, pasture and clover. Water is available. If any appear to pine or do not thrive well, they are removed to vetches and mustard, which are grown expressly for the purpose. This is continued till they go on roots, when they get a little hay, and the wethers a little cake and oats. They are grazed the following summer on rape. When fat, they vary from 60 to 72 lb. per carcase.

After weaning, the ewes are drafted—the drafts being kept on poor enclosed land till they are sold in August. The breeding ewes are turned out on poor common land, where they stay till about the middle of September, when the rams are put to them. They have to subsist on grass up to Christmas, when the yearling ewes and weaker ones have hay and a few roots carted to them. The stronger ewes come after the hoggs on roots, and get a run on grass till the middle of February, when the lambing ewes are selected to get a few mangels and go on the best pasture. At that time there is usually plenty of rough grass.

After lambing, the ewes with single lambs are put on the worst meadows, and a few oats and perhaps cake are given. This is continued till May, when they go on to clover.

The number of lambs reared is about four lambs to every three ewes. If the season is fine, there is a larger crop—the number depending to a considerable extent on the weather. This can be understood when the altitude at which they are reared is remembered.

An Exmoor ram is represented in Plate 60.

THE DARTMOOR SHEEP.

This picturesque breed of sheep is named after the fine open tract of country in Devon and Somerset in which it is reared. It is one of the old local breeds of England, dating far back. In late years the hand of the improver can be traced. Like all breeds which have the open moorland or the hill for their home, it thrives amazingly on wild herbage.

Characteristics.

Description.—No doubt the Lincoln and the Leicester have been used to get substance as well as strength and weight of fleece. The old hardy character of the breed, however, is still maintained. To live on the bare expanse of Dartmoor a sheep of great constitution is necessary, and this the native breed possesses. When the additional fact is mentioned that the rainfall is excessive, averaging over 60 inches in the year, the importance of having a breed of sheep sound in hoof and liver will become apparent.

In size the Dartmoor of to-day is different from the little Moor-dag of olden times. The fact that in the best flocks a fleece of close on 14 lb. (in the grease) is clipped, implies a sheep of some substance and stature. The fleece is thick, strong, glossy, and curly, growing long, after the moorland type. It is the custom to shear the lambs.

Appearance.—In form, symmetry is much looked for, and lean flesh has not been bartered for fat. Good sheep should carry themselves well, and gaiety of carriage comes from good vertebræ and a strong neck. The head is bold, the face broad and somewhat coloured, the eyes full and bright, and the nostrils black (in the ram prominent). The ears should be thick and well covered with clean smooth hair. A small horn is not objected to, as it is supposed to indicate strong constitution.

As kept on the moorlands the Dartmoor was a whitefaced sheep, horned, and somewhat coarse in the fleece. The wethers were kept on the moor all the year round, and in the olden times were expected to yield a profit out of their wool. They were then hand-feeders.

The other type of Dartmoor, the greyface, mottled with black spots on a grey face, the legs being similarly marked, is found only on the moor during summer. They are very ready fatteners, and respond well to a cross for fat lamb.

Lambs and Wethers.—The ewes are good mothers, giving abundance of milk even on inferior pasturage. The ram is usually put to the ewes towards the end of September. The Down cross is frequently resorted to for the production of lambs suitable for fattening, the South Devon ram also being used for this purpose. The wethers are usually fed from one to three years old, and at the latter age they come to from 80 to 100 lb., the weights respectively representing the old-fashioned whiteface and the modern greyface. The ewes are prolific. Mr J. R. T. Kingwell of Great Aish, S. Brent, records a crop of 166 lambs from 112 ewes.

Clip.—Good fleeces are borne by the Dartmoor. Ewes in good condition will clip from 10 to 11 lb. each, and wethers from 12 to 14 lb.; rams sometimes up to 30 lb. It is recorded that Mr F. Ward of Burnville, Tavistock, once clipped 33 lb. of wool from a ram—the wool, of course, being in the yolk.

MANAGEMENT.

Dartmoors are generally fed on grass and turnips, to which a little corn or cake is added as the sheep draw near marketing. The wethers are advantageously used to graze bullock pastures in the autumn and winter months in Somerset and elsewhere. The change is highly beneficial, as they grow very rapidly. They also resist fluke better on those pastures than most breeds.

The moorland sheep have never a very rich pasture. For the most part they find their own living, but when hard pressed in winter are supplied with hay made from coarse moor herbage. Occasionally they may have a few turnips, but the heavy rainfall and wet soil often prevent the carriage of roots when most wanted.

Chief Markets.—Amongst the chief markets are Tavistock, Brent, Plympton, Okehampton, and Mortonhampstead.

A portrait of a Dartmoor ram is given in Plate 60.

THE LONK SHEEP.

The Lonk is a breed of sheep of a type peculiar to itself. It is found in Yorkshire, Lancashire, and Cumberland. It is a hill-breed with a fine presence, particularly when arrayed in full fleece.

Origin of the Name.—The derivation of the name Lonk is somewhat obscure. Probably it is an obsolete provincial term. According to *Holloway's Dictionary of Provincialisms* (1839) Lonk means Lancashire sheep. From another source we derive the information that Lonk means a Lancashire man, also a Lancashire sheep. In Lowland Scotch Lonker means a hole in the dyke through which sheep pass. Then, again, Lonk is another word for lank or leggy.

Locality.—The Lonk exists at a great altitude. It lives on poor land which is valuable mainly for shooting. The main force of the breed is found in the hill districts of Lancashire and the West Riding of Yorkshire — on Longridge Fells, Clitheroe, Whiterwell, Pendle Hill, Craven, and other districts, besides on the hills of the county Palatine.

The breed is chiefly in the hands of small farmers, and is largely used for crossing purposes, chiefly with the Scotch Blackface sheep, resulting in a heavier weight of mutton and a better class of wool.

Weight. — The usual age at which Lonk sheep are fattened for the butcher is three years. A good four-year-old would average about 65 lb., and a top weight probably 80 lb.

They are a very hardy breed, and have some affinity with the Scotch Blackface sheep.

Characteristics.

Seen in full fleece, the Lonk sheep has a very commanding appearance. Breeders look for size. The body is long, thick, and deep. The tail must be long for protection, stout, and straight. The colour of the legs and face is clear black and white streaked, making a dark face. The legs should not be as black as the head. The horns should be waxy in colour, strong and curled, very much like those of the Scotch Blackface mountain breed. They should be equally set in the head, not too close.

The head should be large, a good strong face being a point aimed at. The nose should be thick, deep, and heavy, the eyes full and large, and the ears long. On the forehead a tuft of wool is cultivated. The legs should be thick and full of bone, although a trifle "shanky." They should be wide set, and rather short from the knee to the pastern. The hoof should be sound. The chest must be wide and deep, the back long and rather narrow at the lumbar region. A thick, full fleece is cultivated with a long staple. The fleece should be carried down to the knee and hock, and should be free from kempiness.

Clip.—In a Lonk flock the average clip is from 9 to 10 lb., although a shearling will sometimes produce as much as 17 lb.

MANAGEMENT.

The management of a Lonk flock may be said to pursue an even course. There is a great similarity in the methods adopted in all hill breeds. The average Lonk flock will drop from 1⅓ to, in the case of the smaller flocks, 2 lambs per ewe. For instance, in the flock of Mr David Hague of Copynook, Bolton by Bowland, in the year 1908, 79 lambs were born from a total of 40 ewes.

The ewes are turned to the ram about the end of September, lambing in March and April. They winter on grass, except in very rough weather, when they have the assistance of hay. The practice of giving roots before lambing is not favoured, but after they have lambed a little corn-and-root ration is an advantage.

Towards the end of April the show stock are separated from the others, which are turned out to pasture. The ram and ewe lambs intended for show are housed in October and fed on cake, corn, and roots.

Mr Hague sells his draft ewes at home. The ram lambs go into the Fells to cross with the Scotch Blackface ewes. The draft ewes are sold to farmers, who cross them with other breeds, of which the Wensleydale is as popular as any. The half-bred sheep-raising business engages much attention in the north of England, and the size, substance, and springy coat of the Lonk are favoured, as they give the progeny a fine bulky appearance.

A Lonk ram is represented in Plate 61.

HERDWICK SHEEP.

Probably the hardiest of all British breeds of sheep is the Herdwick, whose ancestral home is the cragland of Cumberland and Westmorland. These sheep lead a roving life, exposed often to very inclement weather, and living on what they can pick up on the mountain-tops even in winter. Like other breeds, it is reputed to be a descendant of a number of sheep which came ashore from Spain's Grand Armada. Be that as it may, it is a useful breed, living where others would starve. It is said to be a cherished tradition with the best breeders that sheep of the breed refuse even hay in winter.

The flocks are usually taken over from the landlords at valuation, succeeding tenants keeping the same blood.

Characteristics.

In appearance the breed is small, the head is light in colour, open horns springing from the base of the skull. The fleece is very strong.

The breed has a reputation for the quality of mutton, which has that epicurean flavour associated with mutton raised on the lean fare of the mountains.

One peculiarity of the breed is that the lambs are born with black heads and shanks, the ears, however, being tipped with white. The colour gradually lightens, until as three-year-olds they are either white or hoary in appearance.

In the words of Mr James Bowstead, a Herdwick sheep should have "a heavy fleece of fairly fine wool, disposed to be hairy on the top of the shoulder and growing down to the knees and hocks; poll and belly well covered; a broad, bushy tail, and a well defined topping; head broad; nose arched or Roman; nostrils and mouth wide; teeth broad and short; jaws deep, showing strength of constitution and determination; eye prominent and lively, and in the male defiant; ears white, fine, erect, and always moving, as has been said, 'like a butterfly's wing.' There should be no spots or speckles, nor any token of brown on the face, as these are considered sure tokens of a cross. Horns in the ram are desirable but not essential. They should rise out well at the back of the head, be smooth and well curled. White hoofs are much preferred. The females are polled."

MANAGEMENT.

The breed is unique in its "late" maturity. At four and a half to five years old they are ready for the butcher, and when fattened on the mountains they kill from 10 to 12 lb. per quarter. They do not take kindly to rich food. The ewes are put to the ram when from two and a half to three years old.

May is the usual lambing time, and the time for mating the ewes is regulated to suit the lambing period. The tups are in some parts turned to the ewes on the Fells in order that lambs may fall early in May. The gimmer shearlings are bratted, or "clouted" as it is called—i.e., a piece of cloth is tied over their tails to keep them from service. When the ewes are kept on bare fare the percentage of twins is negligible, but on slightly better pasture the doubles may be reckoned up to 20 per cent.

Mr James Todd of Rougholme remembers showing a number of draft ewes at Ambleside Fair, 13 of which were sold to a farmer in the Ulverston district. These 13 ewes dropped 27 lambs in the following spring.

The ewes are not drafted at any particular age, that process being determined as much by constitution as any-thing else. On the average, from 4 to 6 lambs will be taken from the ewes before being drafted. The ewes disposed of usually go for crossing.

The wethers are now usually sold off, either as lambs or one or two years old. At one time they were kept until full-mouthed or four times clipped. They are usually turnip-fed, and have been known to bring over 30s. direct from the Fells. The hoggs are put out to winter in October on better land than they occupy in summer, costing from 5s. to 6s. each till they are returned in April.

A Herdwick ram is represented in Plate 62.

WELSH SHEEP.

This breed is widely distributed throughout Wales. It is one of the oldest types in the country. It is, too, a well-defined type, although the efforts of improvers and the variation in the quality of pasture are liable to alter the old-fashioned Welsh sheep and present it in different sizes. Thus we find that the eastern slope of the Berwyn, Merioneth Hills, and Plynlimmon is decidedly superior to the western in pasturage, and the sheep grown thereon are larger and possess finer wool.

The mountainous portion of Wales is divided into sheep-walks, and flocks vary in size from 200 to 4000. Here the thorough acclimatisation of a flock is said to be worth to the owners from 5s. to 8s. per head over the market value.

Characteristics.

Type.—The Welsh Flock-Book Society has determined the type of sheep that it wishes to encourage. The head of the ram should be wedge-shaped and tapering towards the nose. A broad forehead, black muzzle, face slightly tanned or white; horns strong and well curved, but not too close at the roots; eyes prominent; ears small, thin, and obliquely set; scrag strong and thick; brisket prominent; back straight; loins strong; tail long, strong, and bushy; legs short, white, and slightly tanned; skin pink; wool short and thick; handling firm ; a small proportion of kemp permissible practically completes the qualities of the Welsh mountain sheep.

Infusion of Alien Blood.—Efforts have been made by the introduction of Cheviot, Dorset Horn, and Kerry Hill blood to breed a bigger sheep, but the results have not been wholly satisfactory, although in Breconshire the Cheviot cross is favourably spoken of. It is worthy of note that the Cheviot cross has made itself pronounced in succeeding generations in the character of the fleece and the shape and colour of the head. The influence of the Dorset Horn, too, is noticeable in a big collection of show sheep such as one witnesses at the Welsh National Show at Aberystwyth. The writer remembers a prize-winning ram which had almost every characteristic of a pure Cheviot, and yet had only a twelfth of Cheviot blood in him. Probably the most satisfactory results will be obtained by such a mild cross as the exchange of rams from different localities—such, for instance, as Cader Idris and Plynlimmon.

Dead-weight.—Welsh sheep have a deservedly high reputation in the London market. Wethers at from three to four years old kill from 9 to 11 lb., but greater weights are got on good pasture, although the hill breeder protests that the name of Welsh mutton must in the future be maintained by small sheep. A real typical Welsh leg of mutton should run to about 5 lb. in weight.

Wethers were at one time kept till four years old, but the lamb trade has developed much of late years. Wethers off the poorest pastures will kill when ripe up to 35 lb. October-sold sheep, caked and corned in spring and summer, weigh in carcase up to 45 lb., and exceptionally well-wintered sheep up to 55 lb.

Crossing Experiments. — Experiments in crossing have been conducted at several centres. At the University College of North Wales, Madryn Farm, Wiltshire and Southdown rams proved very successful. Contrasting the Wiltshire and Southdown cross, one dealer remarks that "the difference between the Wiltshire and the Southdown cross is that for Salford market and for overhead sale I prefer the Wiltshire, as they look bigger in the pens, but for selling

and retailing in the shop on the coast Southdown crosses give most satisfaction."

In some experiments conducted by Mr D. D. Williams of the Aberystwyth College with Welsh ewes the average weight of the Shropshire cross lambs was 56 lb., the Kerry Hill crosses 68 lb., and the pure Welsh 46 lb. Taking the weight of lamb per ewe—*i.e.*, including twins—the Shropshire averaged 68 lb., the Kerries 96 lb., and the Welsh 61 lb.

MANAGEMENT.

The management of Welsh mountain flocks has not varied much for generations. The same strains of sheep have been kept on the different sheep-walks for many decades, the incoming tenant, as a rule, taking the flock over at valuation. During the severity of winter the flocks are removed from the uplands, those inhabiting the higher altitudes usually leaving their summer habitations from October till April.

It is the custom to sell the wethers at three or four years old, when they are either disposed of in their coats in June, or are, in the month of October, sold to be caked on roots. The change from the bare mountain fare to the rich lowland lands effects a wonderful transformation, and they fatten very rapidly. The tendency of the breed, however, is to grow naturally on the hillside, and forced feeding generally results in a somewhat fat carcase.

Latterly farmers have endeavoured to get their lambs fit for the market early, but obviously this must be accomplished on the lowlands, the youngsters being immediately after birth transferred to the more hospitable pastures. When failure to fatten early has resulted, it has been due to inability to appreciate the fact that the ordinary hill grazings are not the most suitable lands to push young stock forward. The most common practice is to run lambs on good pastures through winter and spring, enabling them to be fattened in the following summer or early autumn. They then command the top market price. In the poorer districts, and where the full severity of the climate is felt, instead of being fattened the young sheep are sold in May or early June for grazing during the summer months.

Wool.—Welsh mountain ewes will clip from 1½ to 2½ lb., and rams up to 6 lb. On most Welsh sheep farms mutual help is provided at shearing time.

Plate 63 contains a group of Welsh shearling ewes.

KERRY HILL (WALES) SHEEP.

Many people have a confused notion that this breed hails from the Emerald Isle. It has nothing whatever to do with Co. Kerry, being named after the range of hills in Montgomeryshire. It has latterly come into prominence, as a result no doubt to the fact that many hill-sheep farmers have been revising their notions concerning the size and weight of mountain sheep. Greater weight is now being aimed at.

Characteristics.

The Kerry Hill breed is speckle-faced (black and white), not too dark. The head is broad at the base and tapering to the muzzle. Wool should cover the poll, and a tuft of wool should decorate the forehead. The cheeks should be clean, but the jaw-bones are covered with wool. The ears are short, thick, and speckled. The symmetry of the sheep should be preserved, the points aimed at being the production of a mutton sheep with broad back, full brisket, well-packed loin, and full thighs. The tail should be fleshy and well set on, the legs squarely planted, speckled, and free from wool below the knee. The skin should be pink, although a red skin is not objected to. A tinge of blue is, however, a bad fault.

Official Description.

The following is the official description of a Kerry Hill sheep :—

Head.—Fairly long, not too broad, tapering to nose, well covered with wool on top between ears, brown or black objectionable, with bunch or tuft of wool on forehead.

Face.—A good speckled face, black and white—the colours clearly defined and not mixed—the black not too dark, but inclined to dark grey; clean cheeks, well woolled to jawbone.

Eyes.—Prominent, bright, and bold looking.

Ears.—Fairly short, thick, well set, and speckled.

Scrag.—Strong and muscular, and well set into shoulders.

Throat.—Well woolled, free from loose or hanging skin, well sloped to brisket.

Brisket.—Should be very wide, deep, and well covered with wool.

Shoulders.—Blades wide and flat, blending with neck ; shoulders full of flesh down to arms.

Ribs.—Well sprung and deep, giving a straight underline from arm to thigh, with plenty of heart-girth.

Back.—Strong, level, with plenty of length from hip to tail.

Loins.—Wide and strong.

Hind Quarters.—Wide and deep, well covered with flesh to hocks.

Tail.—A long tail well set on fleshy, large dock, with plenty of wool to point.

Legs.—Four good short legs, set four-square, with large bone, speckled, and free from wool below the knees and hocks.

Under parts.—Well covered with wool.

Skin.—A nice pink or red skin free from black or blue spots—a blue-tinged skin is objectionable.

Wool.—A tight, close fleece of good length and pure white wool, showing a little fledge on face, coarser on breech and tail.

Size.—This should be kept within reasonable bounds—large sheep are apt to lose hardiness and activity and become less fitted for living on the hills ; smaller-sized sheep are more saleable.

MANAGEMENT.

In the drafting of ewes greater care is now exercised than was at one time common. When ewe lambs are numerous they are culled more rigorously. The stock ewes are generally brought down from the hills about September and the early part of October, mating taking place in the latter month, or earlier if early lamb is desired. The flocks are kept at an altitude of from 500 to 1500 feet in summer. When the majority of the ewes have been served, they are sent to the hills again with a ram in case any should return for second service. They are kept there as long as the weather permits. When too severe they are brought back to the lowlands and have hay given to them.

Hill breeders are not dissatisfied if, when the season has finished, they can count a lamb to the ewe. On the lower lands, however, there would be about a lamb and a half to the ewe.

Clip.—Shearing usually takes place in June. Ewes will clip from 5 to 7 lb., yearling wethers from 6 to 8 lb., rams from 10 to 14 lb., and lambs from 1 to 1½ lb.

Weights. — The wethers are chiefly fed off as shearlings, being sold from May to October. They will average about 14 lb. per quarter, though many will weigh from 16 to 20 lb. per quarter; and fat lambs will average from 10 to 12 lb. dead-weight.

The ewes are fine mothers, and if given cake and corn, are fit for the butcher simultaneously with the lambs.

The breed is very largely used for cross-breeding, and has a great future before it.

A Kerry Hill ram is represented in Plate 63.

DERBYSHIRE GRITSTONE SHEEP.

It may be taken for granted that these sheep, indigenous to the mountainous district which forms what is known as the "Peak of Derbyshire," are as old a breed as can be found in Great Britain. Documentary evidence to prove all this has not yet come to light, but traditions amongst the hills aver that from time immemorial these sheep have existed where they flourish now. The breed has been preserved in yeoman families whose antiquity rivals that of the sheep themselves, and preserved without intrusion of alien blood. The mountainous chain locally known as "Axe Edge" —the home of the Gritstone sheep— extends from Cheshire, through the Peak of Derbyshire, and away into Yorkshire; but the Peak is recognised as the central home of the breed.

Improvement. — "The Derbyshire Gritstone Sheep Breeders' Society," founded on October 15, 1905, and now an influential body, has set itself most commendably to the task of securing pedigree for the breed, as an addition to the local habitation and name of which these sheep have long been in possession. The distinguishing appellation, "Gritstone," is appropriately derived from the "millstone grit" which forms the geological basis of a large portion of the district to which the sheep belong. Similarly, in the southern portion of the Peak country, where carboniferous limestone prevails, the sheep—of the Leicester type—are locally and generically called "Limestone" sheep.

The secretary of the Society, Mr W.'J. Clark, says that in some localities a considerable amount of alien blood has been introduced, "and still the Gritstone character of such crosses strongly predominates"; and he aptly adds that "the sheep have been for many generations bred pure, or otherwise their characteristics would have almost disappeared" from the districts in which the crossing has taken place. This prepotency may be taken to indicate not only antiquity of breed but also vigour of constitution.

The alien blood introduced in recent years has been that of the "Lonk," the Scotch "Blackface," and the "Limestone" sheep. The hoped-for improvements do not appear to have been realised, and as the infusions of alien blood have had but small apparent influence in modifying the type of the Gritstones, so it is doubtful if they will effect any marked improvement.

Flocks of Antiquity.—The breed is of high antiquity in the valley of the Goyt, near Buxton. A well-known breeder there, Mr W. Truman, can trace back to the middle of the eighteenth century the possession of these sheep by members of his own family, during which long period the breed has been kept pure and undefiled against alien blood.

Characteristics.

Hardiness.—There can be no doubt, indeed, in the mind of him who has seen these sheep in the wilder parts of the Peak country, amidst the furze and the ling, the rocks and the boulders, that they are exceptionally wiry and sound, possessing immunity from certain ills that lowland sheep are heir to and the energy that is characteristic of denizens of the hills. Hence their physical prepotency when crossed with other breeds of sheep, — prepotency, it will be noted, that is exercised wholly by the ewes of the Gritstone breed, to which rams of other breeds have been introduced by way of experiment and in hope of good results.

It must be understood,-however, that there are many pure - bred flocks of Gritstone sheep in the great districts over which the Society's scope extends. The Society's object is not only to establish pedigree on a sound and readily ascertainable basis, but to secure the identity of pure - blooded animals alive to-day, and to encourage and systematise the propagation of pure blood throughout the wild and mountainous district whose short commons and rigorous climate have made these sheep what they are.

Face-colour, Wool, and Weight.— The Gritstone is not a white-faced, or black-faced, or even brown-faced breed, but " mottled," with irregular patches of black on a white ground, on faces, ears, and legs alike. Their fleeces, however, are free from black spots ; free also from hairs and from roughness of " skirt." The wool is fairly long and dense, and of texture that is considered fine. Fleeces of ewes average about 4 lb., of yearlings 6 or 7 lb., and of rams up to 9 or 10 lb. The mutton is said to be of the best quality, and the dressed carcasses average 14 or 15 lb., but sometimes running up to 20 lb. a quarter or more in exceptional cases.

From the parasitic disease known as " liver rot" the Gritstones enjoy enviable immunity, though the land on the mountains. where they roam is in many places water - logged. These sheep, indeed, have thriven and multiplied for centuries, unimpaired, where white-faced breeds of the lowlands would perish in a year.

Scale of Points.

The first volume of the Flock-Book was published in 1907, and has entries of 67 rams and 1306 ewes.

The following is the standard type and points for the breed adjusted by the Society of its breeders :—

Points.

Face.—Black and white mottled . . 10
Head.—Fairly long, polled, free from wool, and wedge-shaped . . . 10
Eyes.—Bright and prominent, and set wide apart 5
Ears.—Black and white mottled, and carried slightly forward, but not pricked or drooping 5

Neck.—Medium length, well set on and nicely fleshed, and woolled nearly to the head 10
Body. — Rather long, with well - placed shoulders, good quarters, well-sprung ribs, good top and bottom outlines, well and evenly covered with flesh and wool 20
Wool.—Fairly dense, of medium length and fine texture, free from black spots and hairs, and not rough in the skirt . 20
Skin.—Bright and clear pink and free from spots 5
Legs.—Mottled black and white, free from wool, with good bone, joints, and feet, well placed at each corner of the body and set wide apart 10
Tail.—Fairly high and well set on ; in the rams long, in the ewes docked . . 5

MANAGEMENT.

The Derbyshire Gritstone sheep belong to the still existing grass - land breeds. They are yet, to all intents and purposes, gramineous sheep, even in the lower foot-hills of the range. It follows, therefore, that their feeding and management are characterised by simplicity and economy to a degree that cannot be surpassed elsewhere. Grass—commonly enough of the coarsest—is the natural food, year in and year out, of these sheep ; and it suits them exactly, for they are proof against flukes and foot-rot, and make a good living on bleak and water - logged soils of which even rabbits fight shy.

The less domesticated flocks of the Gritstone tribe still inhabit the wild moors for the most part, picking up a livelihood where any breed of sheep to the south of them would perish. These are in the semi-wild and wholly natural state which has been the lot of the breed for centuries. The chief trouble with them is to persuade the " roving blades " to keep within reasonable limits of distance from the respective homesteads down below to which they belong.

There are, however, many domesticated flocks of Gritstone sheep away down in the valleys. Some of these have been trained into a fair degree of docility and contentment within boundary fences, but still without extraneous feeding. Grass is the staple food everywhere, with hay in the bitter snowstorms of winter when grass is buried out of reach beneath the snow. They know not the taste of corn, or even of turnips, and still they breed and thrive to all satisfaction.

The Gritstones are independent of lambing-sheds, however severe the storms of spring may be. Young lambs just born are sometimes taken, with their dams, into sheltered spots, or perchance under an open shed, until they get well on their feet. But commonly enough the lambs are born out in the snow, and are "all alive and kicking" when the shepherd comes on his round in the night. But even then the ewes do not receive any extra food except hay.

It is probable, however, that on some of the valley farms ewes are gradually and experimentally being trained to the taste of concentrated foods as a preparation for coming parturition. This is so indeed, if anywhere, a practice where cross-bred flocks are kept. So far, however, as pure-bred flocks are concerned, all sorts of stimulating foods are considered unnecessary.

A ram of the Derbyshire Gritstone breed is represented in Plate 64.

THE CLUN SHEEP.

The Clun or Clun Forest sheep is chiefly at home in South Shropshire, Radnorshire, and Montgomeryshire. There is a similarity in type of the various races of sheep found on the Welsh Borderland.

Characteristics.

The Clun sheep may be fawn-coloured or mottled, and black in feature. At one time it was a small breed, like most of the other ovine inhabitants of the hills, producing a 3-lb. fleece and killing a dressed weight of 12 lb. per quarter. Bigger sheep were demanded, however, and through the influence of the Ryeland ram the modern type was probably evolved.

The ewes are much in demand for crossing purposes, the large sales established at various centres in Shropshire being attended from all quarters. The Shropshire cross is one of the most popular. The lambs mature early, and produce mutton of first-class quality.

Properly speaking, the Clun sheep is a type rather than a breed, originating with the intermingling of the Ryeland, the Shropshire, and Welsh breeds. Little information of a definite character is available concerning the earlier history of the Clun.

Prof. W. J. Malden, writing on Clun Forest sheep in the *Journal of the Royal Agricultural Society* (vol. iii., 1892), says that the Clun "perhaps does not show the effect of the skill of the breed-maker as do some older established breeds, yet there is undoubtedly in it those characteristics which can be moulded by skilful hands into a sheep which would be hard to beat. The excellence of the meat and wool cannot be denied; while the shapely well-covered head, with slightly Roman nose, the bold scrag, and the free imperious step, denote a robustness with which the breeder may take liberties in order to produce a more rapid maturity without being afraid of rendering it effeminate or weakly. The horns are being bred out." In a good flock the clip will average about $4\frac{1}{2}$ to 5 lb.

MANAGEMENT.

In Clun flocks the rams are put to the ewes from about September 20 to the middle of October, producing on the average about a lamb and a quarter. The ewes are generally drafted out of the flock after two crops of lambs have been taken. They are then sold to go to the lowlands usually to breed lambs for the fat market. The reason why no more than two crops of lambs are taken on the hills is that the mutton value of the ewe depreciates after the second lamb. The wethers and ewes are generally sold when three or four years old, but earlier drafts are made as yearlings and two-year-olds. They vary in price from 35s. to 50s.

During the winter months the flock subsists chiefly on grass, with the addition of hay and clover in bad weather.

OTHER BREEDS OF SHEEP.

The Norfolk.

The Norfolk breed of sheep, one of the most ancient and a parent of the Suffolk, is nowadays in few hands. The

Earl of Leicester and the Executors of the late Colonel M'Calmont both own flocks.

In appearance the Norfolk breed is of coal-black visage and horned. It is a very active sheep, but a slow maturer. The hoggs will clip from 8 to 9 lb., and the ewes from 5½ to 6 lb. wool. As a rule, the wethers are not mature till two years old, when they kill about 30 lb. a quarter.

The breed is prolific, the flock at Cheveley averaging about a lamb and a half to each ewe. Owing to the difficulty of obtaining fresh blood, the breeding of these sheep is naturally close.

Wiltshire Sheep.

The old Wiltshire horned breed of sheep was at one time more kept than now in its native and adjoining counties. They are not, perhaps, such ready feeders as some of the more improved breeds.

Fig. 708.—*Shetland sheep.*

The breed is horned, with white face and legs. For crossing purposes there is a demand for rams from Wales.

Masham Sheep.

A variety of sheep known as the Masham is freely encountered in Yorkshire. It is the product of a cross of the Wensleydale ram on the Scotch Blackface ewe. The Yorkshire Society at one time provided classes for this eminently thrifty type of sheep.

The lambs run with the dams on the high moorlands, the ewes only coming to

a lower altitude to lamb, and staying till the young lambs find their feet. They are then sent back to the moors, where they remain, with the exception of dipping and clipping times, till weaned.

The wether lambs find their way to lowland farmers for feeding on turnips. The best of the ewe lambs are drafted out and again crossed with the Wensleydale, producing a three-parts-bred sheep. With this double cross of Wensleydale the best feeding sheep are produced. They are good mutton sheep, and their clip weighs almost as well as the pure breed's. York is a big market centre. No doubt the name Masham arose from the fact that at one time it was the great centre for the disposal of this type of sheep.

Penistone Sheep.

This type is found on the borders of Yorkshire, Lancashire, and Derbyshire. It is, however, dying out. In appearance the sheep are white-faced, with wool of medium length and rather harsh, clipping about 4 to 5 lb. a fleece. No doubt the name is derived from the town of Penistone.

Shetland Sheep.

Amongst the Island varieties of sheep one of the most useful is the Shetland breed. It is a small sheep, not weighing much more when fat than 30 lb. The colour varies greatly, some being black, some white, some brown, and many strangely mixed, as in fig. 708. The body is thick and well set upon short clean legs, the head attractive, and eyes prominent; tail short and fine at the point.

The rams usually have horns; the ewes, as a rule, are hornless, and are excellent mothers. Exceptionally hardy, the sheep thrive well on poor pasture and exposed situations. The wool of the Shetland sheep is of remarkably fine quality, and is turned to admirable account by the natives in the celebrated Shetland shawls and other similar fabrics. The fleece weighs only about 2 lb. The sheep are not clipped, the wool being pulled off by hand.

The breed crosses well with rams of improved breeds.

Other Types.

The sheep of *Iceland* are well suited for the conditions under which they are

Fig. 709.—*Iceland sheep.*

reared, but are not of great value for any part of the mainland. They are small-sized, hardy sheep, some of them with strangely shaped horns, as seen in fig. 709.

The *St Kilda* breed of sheep is a characteristic one, very hardy, with dark-coloured mutton. The wool is fine in texture. Some of the sheep have four or even six horns, growing out from the head with fantastic irregularity.

In different parts of the British Isles there are numerous other types of sheep which are bred to a lesser or greater extent. Amongst these may be mentioned the *Swaledale* sheep of Yorkshire, &c.

FOREIGN BREEDS OF SHEEP.

Of all the foreign and colonial breeds of sheep the best known in this country is the Merino, a Spanish breed that has played a great part in improving the wool-production of sheep in many parts of the world. The outstanding feature of the breed is its remarkable fleece. Every inch of the Merino, from its nose to its hoof, is densely coated with wool so fine as to number up to almost 50,000

Fig. 710.—*Merino ram.*

fibres to the square inch. And as if to increase the number of square inches, the skin develops into great folds and wrinkles, giving the animal quite the strangely unique appearance shown in fig. 710, reproduced here by permission from the *Live Stock Journal Almanac,* 1909.[1]

A large quantity of white greasy oil gathers in the Merino fleece. From ewes the fleece weighs 15 lb. or more, and from rams 20 to 25 lb., exceptional animals yielding considerably heavier fleeces.

Merinoes were introduced into England from Spain by King George III. in 1792, and during the first quarter of the nineteenth century the breed was tried to a considerable extent in crossing with several English breeds, including the Southdowns. An improvement was observed in the wool, but the quality of the mutton was deteriorated, and gradually the Merino lost the moderate hold it gained in this country.

In the multitude of other varieties of foreign sheep there are scarcely any whose reputation has extended to this country.

At the great show of live stock in connection with the Paris International Exhibition of 1878 already referred to, there were in all fifty different races and sub-races of sheep, about forty of them being from European countries. Not one of the foreign breeds other than the Merino showed merits that would attract the attention of British flock-owners. Amongst a large number of interesting crosses the best from a British point of view were those bred from the Leicester and Merino races.

GOATS.

The goat has not unfittingly been called the "poor man's cow." In wide districts of Central Europe, in the northern regions of Africa, and in other parts of the world, the peasantry have little else to depend upon for their daily supply of fresh milk and cheese.

Habitat.—Goats are natives of the mountainous countries of the East, notably Asia and Africa. Few domestic animals have so wide a range as the goat. While seeming to thrive best under an ardent sun, they are nevertheless to be found in considerable numbers as far north in Europe as Norway.

At one time goodly numbers of goats were kept in this country, the majority of them being run on the hills like sheep. But this form of rearing, except in Ireland and in some parts of Wales, has now all but ceased. The few goats that are now to be seen in the country districts are kept for most part in ones and twos for milking purposes, and are treated pretty much as a small cow would be. They suit this purpose exceedingly well, and the wonder is that more of them are not kept by cottagers and others having small patches of pasture land.

[1] Vinton & Co., London.

Goats as Milkers.—The improved class of goats are excellent milkers: indeed there is no class of animal of its size that will give a better return in milk for the food consumed than a well-bred goat. Mr Woodiwiss, an English fancier, had a Swiss goat which gave daily for several days in succession 10 lb. 5 oz. of milk, or more than a gallon per day. At the time of the test the little animal had been in milk for more than five months. In another case a herd of five goats, owned by another English breeder, Mr C. A. Gates of Guildford, gave over 3 tons of milk in a year, equal to about 140 gallons each. These goats were also bred from Swiss stock. No doubt yields like these are exceptional, but there are said to be several breeds of goats in the Alpine regions of Switzerland which give regularly during their milking period 3 and up to 5 pints of milk in a day.

In Switzerland the goat is such an important animal that the Government gives a subsidy to selected and approved "Billies," pretty much as in other parts of the Continent and in Ireland premiums are given for bulls. This policy, combined with the skill and enthusiasm of the small owners, has had a most gratifying result, and nowadays most of the milch goats which are to be seen in this country—at the dairy shows in London and elsewhere — are bred from stock which has been imported from Switzerland or other parts of the European continent.

On account of the restrictive legislation on the importation of live animals into this country it is not easy to import goats, but a few selected specimens for stud purposes can usually still be passed in through the agency of the British Goat Society. In any case, most of the well-known strains are already represented in this country in herds established prior to the practical shutting up of the ports.

Goats' Milk.—Not only is the milk yield of goats surprisingly large in quantity, but it is exceptionally rich in quality. It is not usual for it to fall below 3.50 per cent in fat, and very frequently it reaches 6 or 7 per cent. In 1879 Dr Voelcker, F.R.S., reported on samples of goats' and cows' milk to the effect that they contained respectively 7.02 and 3.43 per cent of pure butter-fat, and 5.27 and 5.12 per cent of sugar. In a later comparative analysis (the cow in this case having won the champion milking prize at the London Dairy Show) the figures stand as follows :—

	Goat's milk.	Cow's milk.
Water	83.21	87.56
Butter-fat . .	7.30	3.63
Casein	4.18	
Milk-sugar . .	4.10	8.81
Ash	1.21	
	100.00	100.00

There is still in many minds a slight prejudice against goats' milk on the ground that it has an unpleasantly strong flavour. That prejudice in nearly every case has arisen through drinking the milk of goats kept in a semi-wild state. Where they are kept in captivity and fed on grass, hay, or other low-ground foods, goats' milk has no unpleasant flavour whatever ; indeed, were it not for its exceptional sweetness and richness, it would hardly be possible, under these conditions, to distinguish it by taste from cows' milk.

The milk of the goat has the important advantage, that it can be guaranteed practically free from the tubercle bacilli. While as a breed goats are not believed to be entirely immune from the fell disease of tuberculosis, cases of the trouble have occurred so rarely amongst them that it may be said to be practically non-existent. On account of this consideration many people have in late years taken to keeping goats for the supply of milk for children.

Mr Bryan Hook states, in his book on goats,[1] that he adopted the plan of taking a couple of goats with his family to the seaside on the occasion of their annual holiday. The goats were given the run of a little yard behind the house. Their breakfast, given while they were being milked, consisted of a good half-pint of oats or scalded maize, with a double handful of coarse bran, to which was added any available kitchen-refuse. At mid-day they received each an armful of weeds or grass cut from a disused piece of garden, and in the evening they followed members of the family to the beach, where they ranged the neighbouring waste lands for what they could pick up. Their supper consisted, like their breakfast, of corn and bran. The goats did very well with this treatment. At the end of the holiday of six weeks one of them was giving 6 lb. 14 oz. of milk, or nearly 5¾ pints per day. On the basis of the cost of cows' milk this goat gave during the six weeks produce to the value of £2, 7s.

Varieties of Goats.

There are a great many varieties of goats throughout the world. In Switzerland alone there are said to be sixteen practically distinct kinds. There are also the huge shaggy-haired Pyrenean goats, the pigmy goat of Sumatra, the Surats of India, the short-haired reds of Southern Spain, the Nubian goat, and several others besides. While several of these have leading characteristics in common, they vary a good deal in size and colour, as well as in being horned and minus horns.

The common goat which one sees up and down the country, on railway em-

[1] Vinton & Co., London.

bankments and the like, is mainly of Irish origin, or a cross between the Irish goat and one or other of the imported Continental breeds. Occasional Irish goats prove good milkers, but the majority have little to recommend them except their comparatively small price. An objection to the Irish goat from the point of view of those who keep goats for milk is, that they can rarely be induced to breed except as their half-wild nature prompts them. They are therefore of little use for winter milk. Irish goats are nearly always small in size, with long shaggy coats and large horns.

Of the crossed British goats the most successful have been bred from Nubian or Abyssinian strains. The Nubian goats are hornless, and black and tan in colour. The females are, as a rule, good milkers.

Closely resembling the Nubian goats are some of the Indian varieties that have occasionally been brought to this country. While some of these have also proved good milkers, they were not found to stand the climate so well, and have not been largely used.

Swiss Goats.—For milking purposes in this country probably no other kind of goats surpass the Toggenburg and Alpenzell varieties of Switzerland. It was from goats of these breeds that were obtained the large yields of milk already referred to. Both are big-sized, handsome varieties, the Toggenburg especially giving a large yield of milk. In their native districts these goats are taken out and in for milking just as a herd of dairy cows would be in this country. They respond readily to liberal treatment, and cheese is freely made from their milk. Both of these goats are hornless, and both are white in colour, except that the Toggenburger has usually markings on the head.

One of the most beautiful varieties of Swiss goat is the Schwartzhals, which runs for most part at large in the mountains. This breed has short horns, and is black and white in colour. Its flesh is much appreciated by those who like goat meat, but it is not such a good milker as the other two varieties named.

Selection of Goats.

Whatever variety of goat one fancies, it is wise to be careful in making the proper selection. A milch goat should be large, and her udder should correspond to her size. It is found from experience that a large-sized udder means plenty of milk; indeed, in good milkers the udder usually reaches far back between the thighs, and causes the goat to walk with an awkward gait. A good milking goat, too, has prominent eyes, and ears which are rather large; while the horns in the horned breeds should be short and fairly upright in the females, and longish and gracefully turned back in the case of the males. But most of the best milking goats are altogether without horns.

The coat in the case of the native stocks is usually shaggy and rough; in most of the finer imported breeds it is close and short, with a glossy appearance on the surface.

Like the sheep, the goat has no incisors on the upper jaw. There are, however, light incisors on the lower jaw, and these assist the buyer in selecting a young animal. The first pair of incisors fall after the goat has reached a year old; the second, third, and fourth pair after each succeeding year has passed. These young teeth are followed by permanent incisors, which fall out one by one when from seven to eight years has been reached, a good deal depending upon the nature of the food which the animal has been consuming. A goat may be said to have a full mouth at five years old.

In selecting females for breeding, care should be taken to have them high at the shoulder, wide across the loins, and well sprung in the ribs.

MANAGEMENT OF GOATS.

The normal period of the year for mating in the goat is the end of September to the 1st of March, but it may be possible, where the animals are housefed in winter and have in them a dash of Oriental blood, to get them to breed out of the ordinary season. In this way, where numbers are kept, kidding can be done at different periods of the year and a continuous supply of milk kept up.

The period of gestation in the goat is 149 to 154 days, two kids, as a rule, being dropped at each parturition.

Goats live well on grass or other rough pasturage in summer. In winter the best foods are hay, oats, maize, crushed wheat and barley, bran, and occasionally a few ground peas or beans. Turnips are quite suitable where available, and acorns also make a very acceptable food, but they are not always cheaply and easily obtained.

Rearing Kids.—When kids are reared by hand they should be allowed to take milk from the udder during the first three days: thereafter it will be found better to draw it off by hand and teach the kid to drink. If the milk should be too rich, it may be slightly diluted with skim-milk or water. At the end of ten to fourteen days a little well-cooked linseed gruel may be added to the milk, the quantity of which should be reduced. In two or three days further the quantity of gruel may be increased and the milk again diminished, and so on until at twenty days the young animals commence to feed. A female should not be allowed to breed until she has reached the age of eighteen or, better still, twenty months.

Liberal Feeding Required.—Goats, like cows, if they are to milk well, should

Fig. 711.—*Swiss horned goat and kid.*

be liberally fed. Their food should be varied as much as possible, and only hay of the soundest kind used. In the hilly districts, where they are run like sheep, the animals have to depend mainly on what they can gather, getting only a little hay or straw in winter.

Goats' Hair and Skins.—The hair of goats has a considerable value for upholstery work, and goat-skin rugs are also very useful for carriage and household purposes.

Objections to Goats.—One objection to keeping goats, at any rate in confined quarters, is that the male goat usually has a rather pronounced and penetrating smell, especially in autumn and winter.

This is the quality which commends "Billy" to owners of pedigree cattle for running with their cows as a preventive against abortion. Whether it has this effect or not is uncertain, but many people still believe it has, and a "Billy" goat is still part of the equipment of several well-known pedigree herds.

When kept on grass the smell of the "Billy" is not so offensive as when he runs wild. In any case, it does not apply to "Nanny" goats, as the females are generally called. These can be kept under confinement all the year round without the least trouble or objection. Owing to the pugnacious inclination of many of the animals, it is usually advisable to keep them tied up by the neck in little stalls. . A goat will live in about as much accommodation as will suit a medium-sized St Bernard or other large dog.

The country with the largest goat population is India, which has over 24,000,000, Caucasian Russia coming next with over 6,000,000.

A Swiss horned goat and kid are shown in fig. 711.

SWINE AND THEIR MANAGEMENT.

LARGE WHITE PIGS.

The most universally kept and the most popular of English breeds of pigs is admittedly the Large White. Other breeds have been exported, and have assisted to build up the marvellous porcine resources of such countries as the United States, but in Europe, wherever pig - breeding has received prominent attention, the Large or Middle White breed has formed the basis of improvement. Germany, Denmark, Scandinavia, Russia, Austria - Hungary, and other countries have freely imported White pigs from England, and on the strength of the improvement effected have they built up a wonderful bacon trade chiefly with Britain.

Historical.—There are few points in pig-breeding so obscure as the origin of some of our best-known breeds. The improvement of the White pig of England, and indeed the basis of the modern white breeds, is universally credited to Yorkshire. The Neapolitan and the Chinese crosses are spoken of as effecting a partial transformation of the race. Suffice it to say that the Windsor Royal Show of 1851 first set the seal of excellence on the Large White or Improved Yorkshire.

There were several breeders who exercised their skill in this process of evolution, none more prominent than the weaver, Joseph Tuley, and Mr Wainman of Carhead, Yorks. In their days the local shows in the counties of the Rose drew a magnificent entry from small pig-keepers. The extremes of the Large White and the Small White were freely met with in the north of England, and

it was not surprising that out of the chaos of conflicting fancy the Middle White should appear.

The Large White pig was then a monster of great excellence, and so long as the public taste was ripe for heavy sides of bacon, breeders continued to supply them.

Type and Characteristics.

The type of pig in demand is regulated by two important factors—the commercial market or bacon factory and the show-ring. We rarely see a pig weighing up to 90 stones nowadays. The tendency is to clear them off at handy weights, for it is more profitable to feed to 8 score than to 16. The Large White as we know it to-day is a different type from that prevailing thirty or forty years ago. All coarseness has been eliminated. The thickness of shoulder has been fined down. The capacity to feed to big weights is dormant, not discarded, for substance is too important in any breed to be lightly dispensed with. The general idea which the breeder has kept in view has been to reduce the cost of feeding the pig by refining those parts where the cheapest pork is grown, and steadily aiming at an early maturing pig of a good bacon type. The following is a description of the breed as approved by the National Pig-Breeders' Association :—

LARGE WHITE.

Colour.—White, free from black hairs, and as far as possible from blue spots on the skin.

Head.—Moderately long, face slightly dished, snout broad, not too much turned up, jowl not too heavy, wide between ears.

Ears.—Long, thin, slightly inclined forward, and fringed with fine hair.

Neck.—Long, and proportionately full to shoulders.

Chest.—Wide and deep.

Shoulders.—Level across the top, not too wide, free from coarseness.

Legs.—Straight and well set, level with the outside of the body, with flat bone.

Pasterns.—Short and springy.

Feet.—Strong, even, and wide.

Back.—Long, level, and wide from neck to rump.

Loin.—Broad.

Tail.—Set high, stout and long, but not coarse, with tassel of fine hair.

Sides.—Deep.

Ribs.—Well sprung.

Belly.—Full, but not flabby, with straight underline.

Flank.—Thick, and well let down.

Quarters.—Long and wide.

Hams.—Broad, full, and deep to hocks.

Coat.—Long and moderately fine.

Action.—Firm and free.

Skin. — Not too thick, quite free from wrinkles.

Large-bred pigs do not fully develop their points until some months old, a pig often proving at a year or fifteen months old a much better animal than could have been anticipated from its appearance at five months, and *vice versâ;* but size and quality are most important.

Objections.—Black hairs, black spots, a curly coat, a coarse mane, short snout, in-bent knees, hollowness at back of shoulders.

Blue spots have not been entirely obliterated, but they are more infrequently met with than used to be the case.

Weights.—The carcase contests at Smithfield Show afford the clearest evidence of the killing qualities of different breeds, assuming of course that the ancient prejudice arising from the colour of a breed is discarded. In these contests the preparation of the animal approximates very closely to feeding for an ordinary market. Pigs are shown of three weights. The youngest age is for pigs not exceeding 100 lb. live-weight, equivalent to about four-score dead. Generally they rather exceed this proportion, killing about 81 or 82 per cent. In the middle age from 100 to 220 lb. live-weight; in 1907 one Large White weighed 193 lb. alive and dressed 153 lb. The daily gain in live-weight was 12.2 oz. In the class for big pigs between 220 lb. and 300 lb. live-weight, the Large White and Large Black cross was successful, making the very rapid daily gain in

live-weight of 1 lb. 3 oz., and dressing a 236 lb. carcase from an arrival-weight of 288 lb.

Prolificacy.—The Large White is a prolific breed, and the sows are good mothers. Litters of twelve to sixteen are not uncommon.

A boar and a sow of the Large White breed are represented in Plate 65.

MIDDLE WHITE PIGS.

Many breeders whose views carry weight unhesitatingly affirm that they have found the Middle White the most profitable type of pig.

Origin.—The Middle White is undoubtedly a compound of the joint excellencies of the Large and Small White breeds, both of which were commonly kept and shown in Yorkshire and Lancashire many years ago.

The Middle White has come to be regarded as a distinct type. It is occasionally found creeping out in the Large White, particularly at fat stock shows. To the Yorkshire breeder is attributed the originating of the Middle White, and at a Yorkshire show it first found separate classification. The breed is not so well diffused as the Large White, and fewer opportunities are afforded by agricultural societies for its exhibition.

Characteristics.

The Middle White pig occupies a position that is difficult to maintain. In the first place, we frequently meet with rather large-framed pigs with decided Middle characteristics of countenance and type. Conversely we are more frequently confronted with under-sized pigs, the chief difficulty being not to strike but to maintain the happy medium which justifies the breed's existence. In some ways the Middle is a White Berkshire. They have points in common, save that a little more size than is common in the White breed is favoured in the Blacks.

Scale of Points.

The approved points of the Middle White pig are thus indicated by the National Pig-Breeders' Association :—

Colour.—White, free from black hairs or blue spots on the skin.

Head.—Moderately short face dished, snout broad and turned up, jowl full, wide between ears.

Ears.—Fairly large, carried erect and fringed with fine hair.

Neck.—Medium length, proportionately full to the shoulders.

Chest.—Wide and deep.

Shoulders.—Level across the top, moderately wide, free from coarseness.

Legs.—Straight and well set, level with the outside of body with fine bone.

Pasterns.—Short and springy.

Feet.—Strong, even, and wide.

Back.—Long, level, and wide from neck to rump.

Loin.—Broad.

Tail.—Set high, moderately long, but not coarse, with tassel of fine hair.

Sides.—Deep.

Ribs.—Well sprung.

Belly.—Full, but not flabby, with straight underline.

Flank.—Thick and well let down.

Quarters.—Long and wide.

Hams.—Broad, full, and deep to hocks.

Coat.—Long, fine, and silky.

Action.—Firm and free.

Skin.—Fine, and quite free from wrinkles.

Objections.—Black hairs, black or blue spots, a coarse main, in-bent knees, hollowness at back of shoulders, wrinkled skin.

Weights.—The chief merits of the Middle White are its capacity to fatten readily, its docility, and prolificacy. It is particularly well suited to produce the 8-score pig now so much in demand. Breeders are seeking a lengthy pig, as the middle piece with its wealthy cut of streaky meat is the most valuable portion of the pig.

Prolificacy.—The prolificacy of the Middle White is a strong point in its favour. Litters run from ten to thirteen in number, and will average double figures. No doubt the reason for the superior prolificacy of the White breeds is that at one time an extra pair of teats was cultivated as being a strong point in a sow.

A Middle White sow is represented in Plate 66.

LARGE WHITE ULSTER PIG.

Of the multiplication of breeds, like the making of books, there seems to be no end. The Royal Ulster Agricultural Society has established a Register of the native breed of pigs in Ulster known as the Large White Ulster. Classes are provided for the breed at the Belfast Spring Show.

Scale of Points.

The following is the official scale of points of the breed :—

	Points.
Head.—Moderately long, wide between the ears	5
Ears.—Long, thin, and inclined well over the face	6
Jowl.—Light	5
Neck.—Fairly long and muscular	2
Chest.—Wide and deep	3
Shoulders.—Not coarse, oblique, narrow plate	8
Legs.—Short, straight, and well set, level with the outside of the body, with flat bone not coarse } *Pasterns.*—Straight	5
Back.—Long and level (rising a little to centre of back not objected to)	12
Sides.—Very deep	10
Ribs.—Well sprung	5
Loin.—Broad	3
Quarters.—Long, wide, and not drooping	8
Hams.—Large and well filled to hocks	12
Belly and Flank.—Thick and well filled	5
Tail.—Well set and not coarse	1
Skin.—Fine and soft } *Coat.*—Small quantity of fine silky hair	10
Total	100

Objections.

Head.—Narrow forehead.

Ears.—Thick, coarse, or pricked.

Coat.—Coarse or curly; bristly mane.

Colour.—Any other colour than white is a disqualification.

The breed has for many years been reared with success in the north of Ireland, and in recent years a good deal of attention has been given to its improvement. In form and characteristics generally it resembles the Large White English breed, which has been used to a considerable extent in its development.

An outstanding difference between the Ulster and Large White breeds is in the length and formation of the ear, the Ulster pig having exceptionally long ears.

A portrait of a Large White Ulster boar is given in Plate 66.

THE BERKSHIRE PIG.

The Berkshire pig has greatly extended its sphere of influence since the nineteenth century entered upon its closing quarter. The origin of this, as of most of our other breeds of pigs, is a matter of conjecture, and it is immaterial whether or not the Neapolitan Black pig was used in its production. Certain it is that the Berkshire pig, as it is known to-day, is a very different animal from the Berkshire of the early half of last century.

Characteristics.

It is sometimes a fault of the showyard that it is liable to emphasise minor and fancy points to the detriment of commercial qualities. If it has not altogether succeeded in doing so with the present-day Berkshire, it has at least exercised an influence that has not always been for good.

Many breeders deplore the extent to which the markings of the breed hold sway in the minds of show judges. Any one acquainted with leading herds of Berkshires knows that many of the very best pigs have practically to be discarded because they lack a white hair in the tail, or because a few white hairs appear on the tip of the ear.

Scale of Points.

The British Berkshire Society has drawn up the following revised standard of excellence :—

Colour.—Black, with white on face, feet, and tip of tail.

Skin.—Fine, and free from wrinkles.

Hair.—Long, fine, and plentiful.

Head.—Moderately short, face dished, snout broad, and wide between the eyes and ears.

Ears.—Fairly large, carried erect or slightly inclined forward, and fringed with fine hair.

Neck.—Medium length, evenly set on shoulders ; jowl full and not heavy.

Shoulders.—Fine and well sloped backwards, free from coarseness.

Back.—Long and straight, ribs well sprung, sides deep.

Hams.—Wide, and deep to hocks.

Tail.—Set high, and fairly large.

Flank.—Thick and well let down, making straight underline.

Legs and Feet.—Short, straight, and strong, set wide apart, and hoofs nearly erect.

Objections.—A perfectly black face, foot, or tail ; a rose back ; white or sandy spots on the body ; a white ear ; a very coarse mane ; or in-bent knees.

Size.—There is no doubt that the Berkshire pig has deteriorated in size. In the days of the old Berkshire, when sandy spots were not uncommon, pigs grew to greater weights than they do nowadays. Breeders, however, affirm that the trade for very heavy pigs is merely local, and that medium weights find the readiest markets. The pig that kills 8 score under nine months old can command a good price. At the Smithfield Show of 1907 the champion in the carcase section was a Berkshire which, at 255 days old, weighed 190 lb. alive and 158 lb. dead, equivalent to a daily gain of close on $\frac{3}{4}$ lb. If the Smithfield carcase contests teach anything, it is that the Berkshire can mature quite as rapidly as, if not more rapidly than, other breeds.

Distribution of Berkshires.—The Berkshire pig is in full strength in the county from which it takes its name. It is found all over the south of England, where a black pig seems mostly favoured. The counties south of the Thames afford it most encouragement. No doubt the fact that a black pig is less liable to blister than a white pig has something to do with its popularity. The Berkshire is not quite so hardy as the Large White.

Changes in Type.—Changes in the type of pig favoured are not infrequent. They vary according to the accepted notions of breeders. Most of them object to a very pug face and prominent jowl, the chief difference of opinion arising over the length of snout. The necessity of maintaining the dish face is not disputed, as it is characteristic. A longer type of pig is more favoured than was the case some years ago. Breeders recognise that to have a bacon pig of the highest standing in the market length of side is necessary.

For Crossing.—The Berkshire is one of the most valuable breeds for crossing. A point that should be noted, however, in connection with this breed is the danger which some believe exists in using in a pure herd a boar which has been employed for crossing with white pigs. They say it will almost inevitably result in badly marked litters.

Prolificacy.—Although not so prolific as the Large White, the Berkshire rears a good litter, averaging about eight pigs reared. Breeders reckon that ten is a very good litter for a mature sow to rear.

A portrait of a Berkshire sow is presented in Plate 67.

LARGE BLACK PIGS.

The Large Black pig has risen from comparative obscurity to rank as one of our most useful registered breeds.

Progress. — The Large Black Pig Society was established as recently as 1899, but during its brief existence it has contrived to bring the breed very much under notice of the public.

At one time an excellent farm-scavenger, the breed has risen to a higher point of excellence than merely grubbing for a living. In the showyard nothing has been more remarkable than the progress made by breeders in bringing out their stock. Experience has enabled them to bring out their exhibits in condition more in keeping with the standard adopted in other breeds.

Characteristics.

The Large Black is designed as a bacon pig. It has been conclusively shown that in point of flesh-making, attested by the weighbridge, this breed can hold its own. Perhaps it provides most profit for the feeder as a 10 to 11 score carcase pig. In the past some great weights have been achieved, as much as 190 lb. per side dead. At the present time the breed is used more for the production of heavy than early and handy weights, but as early maturity becomes more recognised as the best and cheapest form of bacon production, we may expect the feeder to turn over more capital by keeping more sows and shortening the store period in a pig's life. The proportion of lean to fat is considerable, and the prolificacy of the breed one of its strong features.

Scale of Points.

The following is the scale of points drawn up by the Breed Society:—

	Points.
Head.—Medium length, and wide between the ears	5
Ears.—Long, thin, and inclined well over the face	6
Jowl.—Medium size	3
Neck.—Fairly long and muscular . .	3
Chest.—Wide and deep . . .	3
Shoulders.—Oblique, with narrow plate .	6
Back.—Long and level (rising a little to centre of back not objected to) . .	12
Sides.—Very deep	10
Ribs.—Well sprung	5
Loin.—Broad	5
Quarters.—Long, wide, and not drooping	8
Hams.—Large, and well filled to hocks .	10
Tail.—Set high, and not coarse . .	3
Legs.—Short and straight . . .	5
Belly and Flank.—Thick and well filled .	8
Skin.—Fine and soft	4
Coat.—Moderate quantity of straight, silky hair	4

Total   . 100

Disqualification.

Colour.—Any other colour than black is a disqualification.

Objections.

Head.—Narrow forehead or "dished nose."
Ears.—Thick, coarse, or pricked.
Coat.—Coarse or curly; bristly mane.

Weights.—If evidence were required of the great weights to which this breed can and does grow, the reader might be referred to the figures of the Smithfield Show catalogues.

Location. — The breed is located chiefly in Devon and Cornwall in the west, in Suffolk and Essex in the east, and in Sussex in the south. A number of pigs have been sent abroad, and the demand for them continues to expand.

A Large Black sow is represented in Plate 67.

THE TAMWORTH PIG.

The Tamworth is one of the old breeds of pigs handed down to the present generation from the time of forests and unenclosed lands. It is distinct from every other breed of pig that we possess —distinct in colour, form, and character.

Origin and Progress. — The Tamworth pig is a native of the Midland counties of England, where it is frequently seen running at pasture and about homesteads. Nature designed the Tamworth to be its own forager. It is remarkably active, and during the past twenty or thirty years has undergone some change, doubtless chiefly owing to careful selection and mating.

It is under the fostering care of a special Breed Society, although for many years, along with the White breeds, its interests were looked after by the National Pig-Breeders' Association. The colour favoured is a beautiful golden russet. It is not an easy matter keeping to the correct hue, and sometimes equally difficult to discard the spotted skin.

Scale of Points.

The standard of excellence adopted on behalf of the breed is as follows :—

Colour.—Golden red hair on a flesh-coloured skin, free from black.

Head.—Fairly long, snout moderately long and quite straight, face slightly dished, wide between ears.

Ears.—Rather large, with fine fringe, carried rigid and inclined slightly forward.

Neck.—Fairly long and muscular, especially in boar.

Chest.—Wide and deep.

Shoulders.—Fine, slanting, and well set.

Legs.—Strong and shapely, with plenty of bone, and set well outside body.

Pasterns.—Strong and sloping.

Feet.—Strong, and of fair size.

Back.—Long and straight.

Loin.—Strong and broad.

Tail.—Set on high and well tasselled.

Sides.—Long and deep.

Ribs.—Well sprung, and extending well up to flank.

Belly.—Deep, with straight underline.

Flank.—Full and well let down.

Quarters.—Long, wide, and straight from hip to tail.

Hams.—Broad and full, well let down to hocks.

Coat.—Abundant, long, straight, and fine.

Action.—Firm and free.

Objections.—Black hair, very light or ginger hair, curly coat, coarse mane, black spots on skin, slouch or drooping ears, short or turned-up snout, heavy shoulders, wrinkled skin, in-bent knees, hollowness at back of shoulders.

Form and Fattening Properties.— Great progress has been made in grading up the fleshing qualities of the breed. The best Tamworths of to-day are deeply fleshed, with a greater width of top than was at one time discernible. It is eminently a bacon pig, and for a judicious mixture of flesh and fat no breed can show a finer side of bacon.

Fresh Blood Wanted.—One of the leading breeders has declared that unless fresh blood can be imported from America the progress of the pure-bred Tamworth is impossible. Undoubtedly breeders work under great disadvantages. Those in the front rank who stand high in the show-ring are very few, and the difficulty of securing an out-cross of blood is a serious matter.

Character.— As a farmer's pig the Tamworth perhaps lacks depth, but it is a good farm-scavenger. It is in all probability not the sweetest-tempered of our breeds, and is given to rooting; but those who have had most experience of it declare that it grows to weight well, finds a ready market for bacon purposes, and crosses well with the Berkshire.

A portrait of a Tamworth sow is given in Plate 68.

LINCOLNSHIRE CURLY-COATED PIGS.

Lincolnshire has its own breed of pigs which have attained to a separate and corporate existence.

Characteristics.

The Lincolnshire Curly-coated pig has some points in common with the Large White, from which, however, it is essentially different. It is a quick-growing variety, with more capacity to turn out prime fat pork than bacon. Those who have had most experience of it declare that it has no rival in the Fen county for early maturity.

To understand the Lincolnshire farmer's point of view, it must be remembered that the native live stock of all descriptions are of exceptional scale. The Shire horse, the Red Shorthorn, and the Lincoln sheep are all of remarkable stature. The Curly-coated pig harmonises with accepted local ideas in livestock breeding. It is descended from earlier times when the yeoman families in the county were more numerous than now.

On the fen lands and marshes pigs are largely kept, frequently mustering herds to the number of 100 head and over. They run in the open, thus acquiring constitutional vigour and strength of frame. The latter is doubtless attained from the soil and climate. It is a custom of the county to allow the labourers a measure of pork in lieu of wages, consequently there is a strong demand for fat pork locally.

Appearance.—In appearance the Lincoln Curly-coated pig is white, with curly or wavy hair, with blue spots not infrequently found on the skin. The head should not be too long, the nose must be straight, without the suspicion of a dish, the ears thick and pendent but not obscuring the eyes. The body should be square and symmetrical, the shoulders wide set and deep, the belly parts thick and close to the ground, the legs straight, and the weight of bone pronounced.

It is only natural in these days, when pedigree is the great directing force in stock-breeding, that a breed or distinct variety with which Youatt was familiar should be placed on a registered basis. A society was formed in 1906 at Boston, and the first Herd Book issued in 1907.

Scale of Points.

This society drew up a scale of points as follows :—

	Points.
Colour.—White.	
Face and Neck.—Medium length and wide between eyes and ears . . .	5
Ears.—Medium length and not too much over face	10
Jowl.—Heavy	3
Chest.—Wide and deep	3
Shoulders.—Wide	15
Back.—Long and level	10
Sides.—Very deep and ribs well sprung .	10
Loin.—Broad	5
Quarters.—Long, wide, and not drooping	5
Hams.—Large and well filled to hocks .	15
Tail.—Set high and thick . . .	3
Legs.—Short and straight . . .	5
Belly and Flank.—Thick and well filled .	3
Coat.—Fair quantity of curly or wavy hair	8
Total . .	100

It is objectionable to have a narrow forehead and thin ears. If the ears are pricked, the nose dished or long, the coat coarse, strong, or bristly, or the colour of the hair other than white, the pig would be practically disqualified.

Weights.—At from 9 to 12 months pigs weigh up to 30 imperial stones. The sows are stated to be good mothers, and are usually fed after producing one litter. At 20 months old they weigh from 40 stones upwards. As indicative of the capacity of this breed to grow weighty pigs, an interesting contrast is made of the two winning gelts at the Lincoln County Show at Gainsborough in 1906 and the weight of the champion cup winners at Smithfield in the same year. The former at 10 months 2 weeks 2 days old weighed 8 cwt. 15 lb.; and the latter, a cross-bred pen, at 11 months 2 weeks 2 days old scaled 7 cwt. 2 qrs. 27 lb.

A Lincoln Curly boar is represented in Plate 68.

SMALL BREEDS OF PIGS.

The star of the small pig breeds has set. There is not now that demand for very fat small pigs that at one time existed, consequently the Small White and the Small Black breeds as commercial assets on the farm are all but non-existent.

The Small White.

The Small White variety is still kept as a "Fancy" pig. It has been brought to a wonderful state of perfection. It is a pure white in colour, with a dished head and broad turned-up snout. It is very full about the jowl, and breadth between the small erect ears is a characteristic feature. Its shoulders are wide, chest full, back broad, and sides deep. It is set on short legs, is small in stature, and ought to be free from wrinkles.

Small Black Pigs.

The Small Black is closely allied to the Black Suffolk, the black pigs of the neighbouring counties of Essex and Suffolk having much in common both in form and character.

The Small Black is a very straight symmetrical pig, set on short legs, very fine in bone. The snout is short and slightly dished, but essentially different in point of character from the full squat face of the Small White. The coat of the Small Black is somewhat strong.

This breed is an easy and rapid fattener, and this property, coupled with greater size than is apparent in the Small White, makes the Suffolk cross appreciated by farmers. The Small Black is decidedly prolific, the litter usually reaching double figures. Its chief defect, apart from lack of size, is a tendency to produce too great a proportion of fat to lean in the carcase.

OTHER TYPES OF PIGS.

Apart from the recognised and registered breeds of pigs there are many porcine types associated with different counties.

The **Black Dorset**, for instance, has a long-established local reputation. It is credited with a good character for ordinary farm purposes.

The **Improved Dorset**, as it was known in later years, was probably a cross on the native breed.

In Sussex there is frequently found on farms a black pig, which enjoys a good reputation locally. It is almost slate-coloured. It has length of body but is lacking in quality. This type is largely used in the production of " four-score " pigs for the neighbouring markets.

The **Hampshire Pig** has points in common with those kept in the neighbouring counties.

The **Gloucestershire Spotted Pig** is largely reared in that county. In the Midlands black and white spotted pigs are also to be found.

MANAGEMENT OF PIGS.

Farrowing.

There is as much diversity of opinion as to the best system to adopt with a sow at the time of farrowing as there appears to be on most other points connected with the management of pigs. Some persons advise that the sow should be left entirely to herself whilst she is farrowing, and others just as strongly urge that the sow ought to have some one in attendance on her.

There is much to be said in favour of both systems,—everything depending on the temperament of the sow and the manner in which she has been previously treated.

Many of the common "anyway-bred " country sows, whose time is spent in a strenuous search for the bare necessaries of life, and whose aim is to give as wide a berth as possible to every human being lest they should meet with the punishment they have already deserved (or most likely will, at some future time, deserve) for their predatory habits, resent the presence of an attendant when they are farrowing. At such a time sows of this class are naturally in a somewhat excited condition.

On the other hand, the well-bred, carefully tended sow, whose experience of man is of an exactly opposite nature, appears to like rather than dislike the attendance of the person who is in the habit of feeding and looking after her. It would, of course, be most unwise to have a stranger to attend to the sow at such a time. In most of the leading piggeries it is the custom for the pigman to be with sows at the time of farrowing, and it is only in exceptional cases that sows give serious trouble with their tempers if they are kindly and carefully treated.

Occasionally a sow, when farrowing her first litter, becomes rather excited, especially when the newly-born pigs happen to come near her head in struggling on to their legs in search of the teat. The wisest course is to gently remove the pigs as farrowing proceeds, and thereafter return them to the sow, when the excitement will most probably have passed away.

Preparation for Farrowing.—It is a good plan to have the sow placed in the sty or house where it is intended that she should farrow, at least a fortnight before her time is up.

Period of Gestation.—The period of gestation with sows is as nearly as possible sixteen weeks. Some aged sows, and yelts with their first litters, will often farrow a day or two before the four months have elapsed ; whilst the more robust sows will as frequently carry their pigs one hundred and fifteen or eighteen days, and in a few cases even a little longer.

Symptoms of Farrowing.—The pigman will easily foretell the arrival of the litter. The sow will be restless, her udder will become swollen and heated, and on the teats being drawn, moisture of a sticky glutinous nature, and sometimes milk, will be found at least twelve hours before the little pigs arrive on the scene; the vulva will become enlarged, and the muscles on either side of the tail will give way.

Bedding for Young Pigs.—It is not advisable to allow the sow to have much long straw for bedding during the first few days after she has pigged, or the little pigs may become entangled in it, and get lain upon by the sow. Some persons give their sows at this time long cut chaff for bedding, but the best material for the purpose is the wheat screenings or "cavings" from the riddles of the threshing-machine. This is both short and soft, and has no sharp ends such as are found in cut chaff.

Treatment of the Sow and Produce in Farrowing.—When the sow commences to farrow, the attendant should have ready a three-dozen size hamper, three-parts filled with wheat-straw, and as the little pigs come into the world they should be wiped with a cloth, placed to a teat so that they obtain a few drops of milk, and then put into the hamper, where they will rest contented and warm until the sow has finished farrowing—unless it be a very prolonged case. In the latter event the piglings should be taken out of the hamper and placed near the udder of the sow, when they will soon begin to forage about for that which nature almost invariably provides for them.

After the sow has suckled the pigs it will be advisable to again place them in the hamper and to give the sow a little slop composed of bran and sharps stirred with tepid water or skim-milk. The sow will then soon lie down again, when the pigs may be placed with her, and the family party will generally rest comfortably until the return of feeding-time. In cold weather it is better to cover the hamper with a sack or cloth, as the little pigs are easily chilled before they have become dry.

The After-birth.—In some cases the sow is allowed to eat the placenta or after-birth. This should be carefully avoided. The placenta should be re-moved from the sty as soon as it is clear of the sow.

It will be found advisable to walk the sow out of the sty the day after she has farrowed. The little exercise will generally cause her to relieve the bowels and the bladder.

Assistance in Farrowing.—It is not often that the sow requires any assistance in farrowing, but it will occasionally be necessary to give her help. Sometimes the little pig will present itself crosswise. At other times there may be a double presentation, or the fœtus be abnormally large. There is seldom any great difficulty in relieving the sow. The great essentials are patience, care, and a plentiful supply of lard. The hand and arm of the operator should be small and well smeared with grease. After farrowing, 2 oz. of sulphur and $\frac{1}{4}$ oz. of nitre should be given to her in a pint of skim-milk or thin gruel. She will readily drink this, and generally it will be all the medicine needed.

Pigs Biting Sow's Udder.—It will sometimes be found that when the newly born pigs are placed with the sow, they will fight for the teats to such an extent as to bite the udder of the sow, which at the time is especially sensitive. The sow will jump up in a hurry, and should no steps be taken to prevent the youngsters injuring her, she will often lie flat on her body and refuse to suckle the little pigs. This occurs more frequently when the sow carries her pigs beyond the usual period of sixteen weeks. The eight tusk-like teeth of the piglings will be found abnormally long, and generally of a dark colour at the root. Old-fashioned pigmen were wont to say that "these black-teethed pigs are never any good, and are sure to pine away and die." In this they were doubtless correct, unless the simple remedy of breaking off these offending teeth was applied. If this were not done the pigs would naturally become more hungry, and consequently more combative, whilst the sow's udder would become more sensitive and inflamed owing to the milk not being extracted. The usual result would be that the pigs would be starved to death from want of their natural food, and the sow would suffer from inflammation of the udder.

The remedy, a most simple and efficacious one, is to remove the pigs out of hearing of the sow, and to cut off the teeth of the piglings well into the gums with a small pair of cutting-pliers. If the pigs are then placed with the sow no further trouble will be experienced. Each pig will soon settle down to its selected teat, which it will make its headquarters for obtaining lacteal nutriment until it is weaned.

Weaning Pigs. — This should take place when the pigs are about six weeks old, if in summer, and about eight weeks old in the colder months. The weaning should be done gradually, by extending the time during the last eight or ten days of keeping the sow from the pigs.

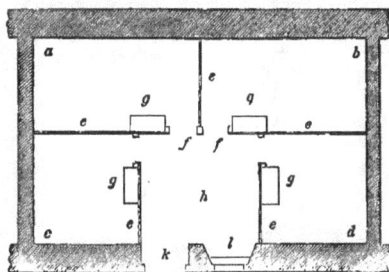

Fig. 712.—*Sties for brood-sows under one roof.*

a b Two sties, 7½ by 12 feet.
c d Two sties, 7½ by 8 feet.
e e e e Wooden partitions.
f f Four doors of sties.
g g g g A feeding-trough in each sty.
h Area from which to overlook the sties
    and to fill the troughs.
k Outer door of sties.
l Window for the sties.

Housing Brood-sows.—In the section on Farm Buildings in vol. i. information is given as to the construction of house accommodation for pigs (see vol. i. p. 184). Fig. 712 represents an arrangement of four sties or compartments for brood-sows, all under one roof, and communicating with a compartment in which the attendant may provide a bed for himself. It is a great advantage to have stout battens fixed along the sides of that part of the sty on which the bedding is laid. The battens require to be from 1½ to 1¾ inch thick, and from 4 to 6 inches broad, depending somewhat on the strength and nature of the wood. They should be firmly fixed with their under surface from 8 to 9 inches above the level of the floor, and should be at least 4 inches distant from the wall.

Galvanised iron tubing 2½ inches in diameter may be used instead of the battens, and is considered better from a sanitary point of view, but the iron is cold. The wood is much more comfortable for the pigs.

This arrangement is a useful protection to the young pigs, as they can creep in between the mother and the wall and obtain a share of the maternal warmth without running the risk of being overlaid. The expense incurred will soon be repaid in the saving of the lives of the young pigs.

Drains proceed from all the sties to the nearest liquid-manure drain; and the apartment is rendered comfortable by having the ceiling and walls plastered, a ventilator placed on the roof in connection with the ceiling, and the floor of brick. When two sows only are kept, the other two sties may be occupied by the weaned pigs.

Prolificacy in Swine.—In the different varieties, and even in the different strains or families of each breed of pigs, there is a marked difference in the prolific powers. This is most noticeable in those strains which have been bred for a number of years for showyard points alone, without due regard to those more useful and general-purpose qualities which are the only really valuable ones for the pig-breeder to study and cultivate. We would not for one moment wish to be understood as expressing the opinion that prolificacy, utility, and ability to win prizes are not to be found combined in several families or tribes of the different kinds of pigs. There are, indeed, numerous instances of such a happy blending, but it is undeniable that the rule is "the other way about."

Sows are capable of breeding—that is. of conceiving—when about seven months old; but it is imprudent to begin at such an early age. About the eighth month is quite soon enough to mate a sow with the boar.

A good breeding-sow will produce and nurse two litters in a year.

Seasons for Farrowing.

In former times it was the prevailing custom for farmers to fatten pigs during autumn and winter only rather than through the year. This was a mistaken practice, for it is well established that a feeding-pig will make considerably greater increase in condition from a given quantity of food fed to it in cool quarters during the summer months than in cold weather. Moreover, the average price of pork in the months of July, August, and September is higher than in the winter months.

These considerations, together with changes in methods of bacon - curing and in the tastes of the consuming public, have led to the abandonment of the old custom, and to the introduction of the practice of carrying on the fattening of pigs throughout almost the whole year.

An inevitable accompaniment of these changes has been the extension of the farrowing season over at least ten of the twelve months; and the greater difficulties to be encountered in the rearing of very young pigs in the cold season of the year render it more important now than ever that pig-men should be well trained for their duties.

Early Maturity in Pigs.—In no other class of stock does "early maturity" pay the feeder better than with pigs. Young pork commands a readier sale and higher price than old. Then the saving of food is important. It is generally considered that a pig of 100 lb. weight requires about 3 lb. of corn per day simply to keep the animal machinery going—merely to supply animal heat and repair the natural waste in the body. It therefore follows that if, by judicious feeding and attention, a pig can be made to realise as much at seven months old as one managed after the old-fashioned plan would at the age of twelve months, the gain in food alone must be substantial. And, in addition to this, there would be a saving in the cost of attendance and risk.

Attention to Pig-rearing.—There are thus several important circumstances which favour the feeding of pigs in summer and autumn rather than in winter. Economy in pig-feeding should have as careful consideration as economy in any of the more important operations of the farm, yet it is well known that, as a rule, farmers give but little thought to the management of pigs. Too often pigs are looked upon as little else than the scavengers of the farm. This is a great and unfortunate error, for with proper management pigs generally pay well. Indeed it may be doubted if any other variety of stock will give a better or quicker return for kind and judicious treatment and liberal feeding than may be obtained from a good class of pigs.

The pig assuredly deserves more attention from the general body of farmers than it has hitherto received. An important point, we have seen, in the profitable management of pigs is the season of the year in which the fattening is mainly carried out. Swine are more susceptible of cold than either cattle or sheep; and, upon the whole, it is desirable that farmers should aim at fattening the majority of their pigs (except porkers for home consumption) between March and October.

Winter Farrowing Risky.—Litters of young pigs are troublesome and risky in winter, and are to be avoided except where the delicacy of roast sucking-pig is desired at the Christmas dinner. But although the feeding of pigs should be carried out mainly in the warmer months, there will always be less or more pig-feeding in winter—perhaps a few pigs of late litters to finish off, or it may be only two or three young porkers for home consumption during winter and spring. For information on the feeding and general management of swine we are mainly indebted to Mr Sanders Spencer, Holywell Croft, St Ives, who has made the profitable breeding and rearing of pigs a life-study.

Rearing and Feeding Pigs.

In the methods of pig-feeding pursued throughout the country there is great variation, much depending upon the foods most economically available, and the purposes for which the animals are being prepared.

Feeding the Sow and her Litter.—It may be assumed that six is a fair number for a young sow or yilt, and ten to twelve for an aged sow, to rear at each litter. These numbers may be larger in the summer months, but it will be found most profitable not to attempt too much in pig-breeding any more than in most other things. From the time the piglets are three days to about four weeks old, the sow should be fed twice a-day with just about as much as she will clear up at once of thoroughly stirred slop, composed of seven-eighth sharps, thirds, or randan, and one-eighth broad bran. By this time, or even before, the little pigs will begin to lick round the trough, and show signs of a desire to become less dependent on their mother for the necessaries of life. This natural want must be satisfied either by allowing the sow to have a run on the grass field or in the straw-yard for an hour or two, or, if the weather is too rough and cold, letting the little pigs into an adjoining place, and there feeding them with a little sharps, or oatmeal stirred with milk; or a small quantity of oats, peas, or wheat will be thankfully received and turned to good account by the now hungry "squeakers."

This system of feeding may be continued until the pigs are weaned, the only variations being a gradual addition to the food given to both sow and pigs, and the warming of the milk or water with which the food for the little pigs is mixed during the cold weather.

Weaning Pigs.—The little pigs will be best left on the sow in the summer months until they are seven or eight weeks old, and in the winter months a week or two longer. The weaning should be effected gradually, by letting the sow remain away from the pigs a little longer time each day until the flow of milk gradually ceases, and the pigs think more of the arrival of the pail than of their mother. By adopting this plan the sow's milk will be no trouble, and the sow will desire to receive the attentions of the boar within two or three days after the pigs are weaned.

Castrating Pigs.—Those little pigs which are not required for breeding purposes should be attended to when they are about five or six weeks old. This is by no means a difficult operation, but it is better to employ a competent castrator, especially with the sow pigs, or, as they are variously termed, hilts, elts, yilts, yelts, gilts, or gelts.

Feeding Young Pigs.—After the pigs are weaned, their food should be very similar to that on which they had been previously fed, with the addition of a few more peas. As the pigs reach the age of three months, a proportion, amounting to one-sixth, of barley-meal may be added. This may be gradually increased until it becomes the principal food of a five-months-old pig.

Cocoa-nut Meal for Pigs.—We have of late years used a considerable quantity of cocoa-nut meal, and have found it a most economical food to use with the barley-meal. From experiments carried out at our wish, it was proved that not only was pork made at a less expense by the introduction of cocoa-nut meal to the extent of about one-eighth of the whole allowance of food, but the quality of the flesh was superior, and the appearance of the carcase much improved.

Cod-liver Oil for Pigs.—Owing to the high price charged until recently for cod-liver oil, its use for stock has been very slight; but it may now be procured at such a reasonable price as to come within the limit of profitable foods for young growing pigs, if not for those in the fattening stage. The flavour of the pork is affected if the oil be used within a month of the pig being killed, but we can recommend it with every confidence for newly weaned pigs and young stores. During one winter we have given it to some two or three hundred young boars and gilts which were being reared for the spring trade, and the result was most satisfactory.

A Golden Rule in Pig-feeding.—If it be desired to rear and fatten pigs at a profit, one "golden rule" must not be lost sight of—*never allow the pigs to become poor.* Keep them ever in a progressive state, and if this is done properly, they will be fit for the butcher a month or two earlier than is the rule, while the pork will be of better quality, and the loss from disease will be reduced to a minimum. Should illness attack any of the pigs, they will thus be always fit for the knife, and realise pretty nearly their full value.

Variety of Food.—Variety of food is as beneficial and as welcome to pigs as to human beings. It may not be practicable to change the course of feeding to any great extent, but it will certainly be beneficial to give the fattening and even the growing pigs a *mixture* of meals.

Meals for Pigs.—Barley-meal has been proved to be the best single food for fattening pigs, and to a great extent it is necessary for the manufacture of a high quality of meat. Maize-meal may be used somewhat largely at the commencement of the fattening, but if used extensively at the latter stage, the pork is not so saleable. Instead of maize a small quantity of bean-meal, or even better still, pea-meal, may be given with great advantage. Upon this the older pigs will thrive well, and the pork prove firm and sweet in flavour. Oatmeal will generally be found too expensive for pig-feeding. It may, however, be profitably used if the pigs are required to be made ripe at an early age, and exceptionally high quality of London porket-pig desired. The use of some condiment with fattening pigs of a restless disposition will be found of great benefit.

Condimental Food for Pigs.—Some object to the use of condimental food for pigs; but the experience of others is that for fattening-pigs, and for pigs that are newly weaned, some good well-manufactured stimulating food is of very great benefit, and is withal most profitable.

Cooked Food for Pigs.—There has been considerable discussion as to whether or not the cooking or steaming of meal as food for pigs is an advantage. Some writers on pig management strongly recommend the practice; but Mr Sanders Spencer states that his experience is decidedly against it. He has given it fair trials, and in every case where the experiment has been fairly and thoroughly carried out, it has been found unprofitable to cook or steam the meal for the pigs. In very cold weather it is advisable to mix the meal with tepid water, so that the food is given to the pigs at about the temperature of new milk. But a better plan even than this is to feed the pigs on dry meal, and to give the water to them in a separate trough. The pigs may be much longer in eating their food in this way, but it will be more thoroughly masticated and mixed with saliva, so that it is more fully digested; and the pigs will then only consume as much water as nature and the weather render needful. There

is certainly no need to warm the food in summer; but in winter there is an undoubted benefit in having the food warmed.

Experiments on this question have also been carefully carried out at different agricultural colleges in the United States of America, and in almost every case it was proved that the cooking of the food resulted in a considerable loss.

Upon many farms potatoes form a large part of the food of. pigs. The potatoes should be steamed or boiled.

Kitchen "Slops" for Pigs.—The "slops" of the kitchen are turned to good purpose as food for swine; but great care should be taken not to give pigs any liquid in which salt meat has been boiled or to which soda has been added. We have heard of several cases of death amongst pigs owing to their having been fed on such "slops" or boilings. The safest system to use house or hotel slops is to steam it, let it cool, and remove the fat which rises to the surface. The soup will in this form be far more valuable for pig-keeping, especially for young pigs. The "pig's-pail" should always be at hand to receive food-refuse from the kitchen.

Skim-milk, buttermilk, and whey are extensively used as food for pigs. These, of course, do not require cooking.

Feeding Old Pigs Unprofitable.—The fattening of old boars is, as a rule, unprofitable. One cannot afford to convert good food into pork which sells at from 1½d. to 3d. per lb., and even this only when not made very fat. The importation of low-priced foreign meat, and the great reduction in the price of lard, have rendered the manufacture of inferior, or very fat, meat a losing game. And a word of caution here may not be out of place as to the making of the bacon pigs too heavy and too fat. The well-fed, meaty pigs of 160 lb. dead-weight will realise much more per lb. than can ever be obtained for the over-fat pig of double the weight.

Green Food for Pigs.—Many pig-keepers seem to forget that the pig is naturally a graminivorous animal, and that in a state of nature it lives for a great portion of the year on grass, or the roots of certain plants, which it unearths by the use of its long snout; whilst its chief food during the remainder of the season consists of beech-mast, acorns, chestnuts, or similar tree-seeds. Those who are generally most successful in the feeding of our domesticated animals are those who study most carefully the natural habits of the animals in their charge.

To make pig-feeding a complete success, it is imperative that a certain amount of green food should be supplied to those pigs which are confined in close quarters. It does not appear to matter much what this vegetable food consists of, whether it be grass, clover, lucerne, beet, mangels, swedes, turnips, cabbages, or kohl-rabi. All seem to have a beneficial effect on the health and progress of the pigs; whilst great numbers of pigs are fattened on cooked potatoes, and a little meal stirred with buttermilk or whey.

Pigs which are not allowed their liberty should also have an occasional supply of small coal, cinders, or even a lump of earth or mould. This will greatly tend to keep the pigs in health, and cause them to settle and thrive much better.

Exercise for Feeding-pigs.—It is sometimes found necessary to allow highly bred pigs a certain amount of exercise during the short time they are shut up in close quarters at the latter part of the fattening period. This difficulty, if it may be so termed, is not often experienced with the common-bred pig, whose spirit of unrest forces it to take a sufficient amount of exercise to keep the

Fig. 713.—*Ring pigs' trough, to stand in a court.*
a b Hollow hemispherical trough, 30 inches diameter.
c Eight subdivisions within it, 9 inches high, converging and meeting at a central pillar.

various organs of the body in good working order, and for the formation of that lean meat and muscle which is the natural result of a free use of the locomotive powers.

Keep Pigs Clean. — Pigs are accused of dirty habits, but the fact is otherwise. The accusation really applies more to their caretakers, who oblige them to be dirty, than to the animals themselves. When constrained to lie amongst dirt, and eat food fit only for the dunghill, and dealt out with a grudging hand, they can be in no other than a dirty state. Let them have room, choice of clean litter, and plenty of food, and they will keep their litter clean, place their droppings in one corner of the court, and preserve their bodies in a wholesome state. The pig-house or pig-yard should be cleaned as regularly as the cow-house, and kept in a fresh wholesome condition.

It is the duty of the cattle-man to supply the store-pigs with food, and clean out their court-yard; and this part of his duty should be conducted with as much regularity as feeding the cattle. Whatever food or drink is obtained from the farmhouse is usually brought to their court by the dairymaid.

Pigs in Cattle-courts.—Pigs often get the liberty of the large courts, amongst the cattle, where they make their bed in the open court when the weather is mild, and in the shed when cold. Though thus left at liberty, they should not be neglected of food, as is too often the case. They should be fed regularly, and in addition to other food many give them sliced turnips in troughs. Pigs, when not supplied with a sufficiency of food, will leap into the cattle-troughs and help themselves to turnips; but this dirty practice should not be tolerated, and it can arise only from their keeper neglecting to give them food.

A convenient pigs' trough, adapted for standing in the middle of a court, is represented in fig. 713. The divisions have a convexity on the upper edge, to prevent food being dashed from one compartment into the other. This trough stands upon the top of the litter, is not easily overturned—the cattle cannot hurt themselves upon it, while it is easily pushed about to the most convenient spot.

Rest for Feeding-pigs.—When pigs are fattening, they lie and rest and sleep a great deal, no other creature showing "love of ease" so strongly in all their doings; and, in truth, it is this indolence which is the best sign of their thriving condition. The opposite effects of activity and indolence on the condition of animals are thus graphically contrasted by Liebig. "Excess of carbon," says he, ".in the form of fat, is never seen in the Bedouin or in the Arab of the desert, who exhibits with pride to the traveller his lean, muscular, sinewy limbs, altogether free from fat. But in prisons and jails it appears as a puffiness in the inmates, fed as they are on a poor and scanty diet; it appears in the sedentary females of oriental countries; and, finally, it is produced under the well-known conditions of the fattening of domestic animals;"[1] and amongst these last the pig may be instanced as the most illustrative.

Bedding for Pigs.—Wheat-straw is best suited for this, especially for the breeding-sow and her litter of young ones. In the cattle-courts, the pigs, of course, make litter of whatever is used for the cattle.

Nomenclature of Pigs.

The denominations of pigs are the following: When new-born, they are called *sucking pigs, piglings, piglets,* or simply *pigs;* and the male is a *boar pig,* the female *sow pig, hilt, elt, yilt, yelt,* or *gilt.* A castrated male, after it is weaned, is a *shot* or *hog.* Hog is the name mostly used by naturalists, and very frequently by writers on agriculture; but to avoid confusion with the name given to young sheep (hogg), it is convenient to use the terms pig and swine for the sake of distinction. The term *hog* is derived from a Hebrew noun signifying "to have narrow eyes," a feature which is characteristic of the pig. A spayed female is a *cut sow pig* or *gelt.* As long as both sorts of cut pigs are fat and young, they are *porkers, porklings,* or *London porket-pigs.* A female that has not been spayed, and before it bears young, is an *open sow* or *hilt, elt, yilt, yelt,* or *gilt;* and an entire male, after being weaned, is always a *boar* or *brawn.* A cut boar is a *brawner.* A female that has taken the boar is said to be *served* or *lined;* when bearing young she is an *in-pig* or *brood-sow;* and when she has brought forth pigs she has *littered* or *farrowed,* and her family of pigs at one birth form a *litter* or *farrow* of pigs.

THE PRINCIPLES OF STOCK-BREEDING.

The breeding of farm live stock is pursued with varying degrees of method or with no method at all. Far too many still mate their stock in a haphazard manner, availing themselves of the cheapest sire within reach, and practically leaving everything to chance. It is amazing that, at this time of day, there

should be this lack of care in the breeding of stock, for to all who keep their eyes open the advantages obtained by giving due regard to the underlying principles of systematic stock-breeding must be clearly apparent.

Few men have risen to recognition as great breeders. The essential gifts and opportunities are not widely spread. But while there may not be many who

----

[1] Liebig's *Ani. Chem.,* 89.

can attain fame as breeders, it is quite within the reach even of men of average intelligence to accomplish good work in the production of improved farm stock if only they will give careful heed to plain lessons taught by the experience of others. To set forth some of these lessons clearly, and in as few words as possible, is the object of these notes.

Heredity.

The subject of heredity in animal and plant life has engaged the minds of many of the ablest naturalists and scientists who have ever lived, and yet some of its problems still await solution. Enough, however, has been made known regarding the laws of heredity in animals to afford valuable guidance to the intelligent breeder of farm live stock. In the old familiar saying that "like produces like," there is a simple interpretation of hereditary force in plants and animals. This "hereditary force" may be for good or it may be for evil, according to the character of the parental stock. The object of the breeder is to select as parents, stock or plants which he has reason to believe are likely to possess hereditary tendencies in the direction of the characters desired in the produce.

The universality of its application is a valuable property in hereditary force. It is not merely in conformation and outward appearance generally that heredity makes its influence felt, although it is in these features that its effects are most familiar to casual observers. The influence of heredity applies to the physiological, pathological, and other conditions of animals—to every one, indeed, of the parts and properties in animals which the breeder desires to develop or control for the good of mankind. Thus, whatever may be the particular object of the breeder, the careful study of the mysteries of heredity is to him a matter of the highest importance.

Over and over again it has been found in experience that by the skilful manipulation of hereditary forces possessed by individual strains or families, or even by individual animals within families, certain features can be "bred out" and others developed if not actually created.

Unpopular colours in breeds of horses and cattle have been obliterated or lessened in the frequency of their occurrence. Tendencies to constitutional weakness or certain forms of disease in particular families may be partially or entirely removed. In like manner, desirable qualities or characteristics can be fixed and strengthened, and thus through the influence of heredity transmitted to the family or tribe generally.

For the breeder it is well to bear in mind that, as already pointed out, heredity applies to the psychological as well as to the physiological characters. Of this fact there is proof in the transmission of the wonderful instincts possessed by some animals. Not infrequently it has been found that vicious tempers can be weakened in certain strains, just as in other strains different characters and instincts have been developed.

It is equally important for breeders to keep in view the significant fact that pathological conditions are likewise affected by hereditary forces, and that unless care be exercised, strains of stock hitherto quite healthy may become tainted with or rendered predisposed to diseases the occurrence of which had originally been merely accidental. The safe course is to avoid breeding from animals known to be either actually affected by, or to be predisposed to, disease of any kind.

Variations in Breeding Results.

Fundamental and powerful as are the laws of heredity in the raising of both plants and animals, it is well known that they are by no means absolute or unfailing in their application. To the surprise of the breeder—it may be to his gratification or it may be to his disappointment—they are now and again found to have been quite unavailing, to have been for the moment pushed aside, as it were, by some other mysterious force, which displayed its influence in the production of a "variation" or a "sport," as it is differently called. Sometimes this "variation" may be merely a "reversion" to a type at one time characteristic of the ancestors on either side or both. Just as likely it may be à true "sport" displaying features entirely strange to the family and the tribe. It accords with the experience of breeders to say that the tendency to variation is contributed to by change of environment—by change in habits, in the uses to which animals are put, in the climatic conditions under which they, live, and, in particular, in the methods of feeding, which are nowadays much more of a forcing character than in olden times. It is probably true that the more highly artificial the conditions of animals have become, the greater is the liability to unexpected " variations " in type. The tendency to variation is also increased by indiscriminate crossing different strains.

But while these no doubt are the prevailing views regarding unlooked-for results in breeding, there are those who believe that their occurrences are just as surely the product of laws of nature as are the typical progeny of related parents. To give forth variety, it is claimed, is an inherent power in nature, a provision not really antithetical but rather beneficently complementary to those other natural laws which lead men to look for like begetting like as the normal condition of things.

" All. the organs and tissues of which an individual is compounded possess the power of independent variation. Every single cell may possess this power. . . . Every variation, when once it has started, may be looked upon as a structure capable of independent variations in an almost infinite number of directions, regressive and progressive. . . . Two forces are constantly at work in nature— Natural Selection and Reversion. The former causes progressive evolution, the latter regressive evolution. They are opposed, but one would be inadequate without the other. They are warring forces, but their resultant is a near approach to perfection." [1]

But be the causes what they may, it is well that " variations " do occur with moderate frequency. They have played a useful part in the development of stock-breeding. Indeed, it is by the skilful cultivation of adventitious " variations " that some of the most valuable improvements in British live stock have been brought about.

Transmission of Acquired Characters.

There has been much discussion and sharp differences of opinion as to the extent to which abnormal and acquired characters may be transmitted to future generations. Prior to the 'eighties of the nineteenth century it was the belief of many eminent biologists that " sporting variations," as well as modifications induced by sustained treatment, or arising as the result of accident, might be so " bred into " strains of stock as to ensure transmission to future generations like hereditary characters in a family current. Herbert Spencer wrote that " change of function produces change of structure. It is a tenable hypothesis that changes of structure so produced are inherited." That doctrine, however, was to a large extent set aside by the publication of Weismann's elaboration of Galton's Germ Plasm theory of heredity,[2] which in course of time claimed the support of most of the leading biologists.

It was declared by Weismann that the germ cells concerned in reproduction are distinct from, and quite independent of, the body or soma cells; that while the germ or reproductive cells are "housed" and nourished in the body, they do not absorb.transmissible characters from the body, but reproduce only those characters conveyed to the germ cells from the two parents in the act of fertilisation.

The continuity of the germ plasm may be admitted, but it does not necessarily follow that it is not subject to any modification by its successive hosts in its progress from generation to generation. Indeed, it is admitted by some of the foremost living biologists that the vitality and stamina of the germ cells are affected by the nourishing body for the time being, and with that admitted, and having also in mind the acknowledged inherent power of " independent variation " possessed by " all the organs and tissues of which an animal is compounded " (Reid), it is permissible for the breeder to assume that he is not so absolutely devoid of the power of initiative as a strict interpretation of the Weismann doctrine would suggest.

---

[1] *The Principles of Heredity.* By G. Archdall Reid.

[2] *The Germ Plasm : A Theory of Heredity.* Walter Scott, Ltd., London.

Is there not reason to believe that the inherent power and tendency in organs and tissues to give out variety may be usefully stimulated by "the play of forces from the environment"? How far the form or tendency of the "variation" may be guided by the breeder is matter of doubt. That he has exercised substantial guiding influence in the progressive evolution of the cultivated races of stock does not admit of denial, attested as it is by living testimony that is unmistakable.

"The discriminating sense of the fox-hound as he distinguishes on the moist earth the fresh track of the fox, or of the bird-dog that is insensible to the fox tracks, but becomes immediately excited in the proximity of birds, is an interesting phenomenon. The Scotch collie seems, as a result of long-continued breeding and training, instinctively to know how to assist in the handling of domestic animals, but is utterly foolish in its attempts to catch rats. Most terriers, on the other hand, are tremendously in earnest in their frantic efforts to tear up wooden floors or undermine buildings for the sake of securing a rat, but as stock-dogs are utterly useless. The wonderful productive capacity of the modern dairy cow, producing ten thousand or even twenty thousand pounds of milk in one year, and the transmitting of these qualities to her offspring, are recognised facts among dairymen. Families of horses have acquired speed at the trot and transmitted this quality with considerable certainty."[1]

There is no reason for breeders being in any way discouraged by the spread of the Weismann theories of heredity. Whatever the governing scientific principles may be, the fact remains that the useful features and properties of plants and animals are constantly undergoing important modification and development at the hand of man,—developments and modifications many of them indubitably influenced by the play of environment.

Other Breeding Problems.

Telegony. — Amongst many knotty problems which have troubled breeders

[1] F. B. Mumford in *Cyclopædia of American Agriculture.* The Macmillan Co., London.

of high-bred stock, what is known as Telegony demands mention. It is believed by not a few breeders that occasionally a calf, a foal, a pup, or other animal resembles or "takes after" neither its mother nor its own sire, but another sire mated with its mother at some former time. By Darwin it was stated that "the influence of the first male by which a female produces young may frequently be seen in her future offspring by different sires," and numerous instances have been mentioned which it was believed supported the idea that an early sire had so "infected" a female as to influence her future progeny by other sires. In later times the belief in "telegony" has lost ground, and few breeders now pay any attention to it. Professor Cossar Ewart, who has conducted many experiments on the subject, gives it as his opinion that the doctrine is not well founded. Be the facts as they may, the wise course to pursue in the breeding of valuable stock is to avoid even the occasional use of any sire whose "infectious influence" could be to any extent or in any way detrimental to the strain.

Mental Impression. — Another disputed question is the part which mental impression on the part of the dam is supposed to play in determining the colour or other character of the progeny. Ever since Jacob peeled wands and stuck them up before Laban's stock and his own, in order to increase the proportion of spotted and speckled produce which fell to him, this doctrine has continued to receive some little attention in stock-breeding. It has been the practice of certain breeders of black cattle to avoid keeping light-coloured animals within sight of their black cows when the latter are conceiving or are in the early days of pregnancy. At the same time, it is generally held by scientists and naturalists that mental impression is not a factor of any significance in the breeding of stock. It is of course known that unborn young may be seriously affected by extreme nervous shock sustained by the mother, and it is desirable that pregnant animals should be as far as possible protected from the risk of such occurrences.

Controlling Sex. — Much attention

has from time to time been given to the controlling of sex in stock. Various theories have been propounded. It has been held by some that if service takes place early in heat the produce will be a female, if late in heat, a male. Another theory is that ova are alternately male and female, and that if an animal has produced a male, and a male is wanted again, the female should be served not in her first but in her second heat. Yet another idea is that the sex of the produce will correspond to that of the parent that preponderates in stamina and general vigour at the time of mating. Some, again, believe that sex can be regulated by food. Little success has attended the prosecution of any of these or of other theories that have been advanced for the same purpose. Fortunately, it would seem that the controlling of the sex is one of nature's secrets not to be brought within the ken of man.

SYSTEMS OF BREEDING.

Four main systems are pursued in the breeding of live stock. These are generally known as (1) Cross-breeding, (2) Grading, (3) Line-breeding, and (4) In-and-in-breeding.

Cross-Breeding.

This term is applied to breeding from animals of different species, breeds, or varieties—to a mixing of strains as distinguished from systems of breeding in which the main purpose is the concentration of breed or tribal currents. Sometimes the word crossing is applied also to the interbreeding of different families of the same breed, but this is not in accordance with the general understanding of the term.

The general experience of breeders is that judicious crossing has a stimulating effect on the more useful properties of animals. There is often an increase in size, in vigour of constitution, in fecundity, and in rate of maturing, as well as in improved fattening properties in meat-producing stock. It has thus come about that by cross-breeding the rent-paying qualities of farm live stock have been enhanced substantially.

But there are some effects of crossing to which breeders must give careful heed. Crossing, as already stated, has a tendency to break up family currents, and unless great care is exercised in the introduction of an out-cross into pure or well-established strains, much injury may be done to their breeding properties, particularly in respect to regularity of type. As already indicated, the tendency to "variation" is increased by cross-breeding.

In the selecting of animals for cross-breeding, there is ample room for the exercise of care and judgment. Care is required in choosing varieties or breeds that blend well together, and also in selecting sires well adapted for mating with the females in the stock. The most general practice is to use on mixed-bred females pure-bred sires of well-established character. It is in this way that the quality and value of the produce can be most speedily and economically improved and maintained.

The first cross between two pure breeds is usually the most successful of all kinds of crosses. It is not very often that first or subsequent crosses give satisfactory results when thoughtlessly bred together. There would, as a rule, in these cases seem to be such a breaking up of the forces of heredity that the character of the progeny becomes a mere matter of chance. Family currents of any considerable strength cannot be established by indiscriminate breeding such as this.

Students of Mendel's laws of heredity are of opinion that a full acquaintance with the operation of these laws would enable breeders to make use of the principle of cross-breeding in the improvement of their pure-bred stock without incurring the risks hitherto believed to be involved. To this important question fuller reference will be made presently.

Grading.

This is an American term which very aptly describes the practice of raising improved races of stock by mating pure-bred sires with females of mixed breeding and secondary or inferior character. Pure-bred sires from the same breeds respectively continue to be put to the progeny for a greater or lesser number of generations, and in this way there may be established different types of animals of high utility, and possessed of fairly reliable breeding properties. In the building up of several of the existing pure-bred races this method has been largely pursued.

Line-Breeding.

This term implies the mating of animals that are related to each other. It differs from "in-and-in-breeding" in that the mating is not restricted to near relations, but, as a rule, applies rather to animals not closely related though mostly claiming some measure of blood-relation with the same family. The tendency of line-breeding is to concentrate and strengthen hereditary force. Here lies its advantage over crossing, the influence of which is in the opposite direction. It is further claimed for line-breeding that it is safer than in-and-in-breeding, in that it is not so liable as the latter is alleged to be to lead to an impairing of the fecundity and constitutional vigour of families. In the vast majority of pure-bred stocks line-breeding is pursued to a lesser or greater extent, and it is unquestionable that the judicious use of this method has done much to establish the high character of British pure-bred stock, alike in regard to stability of type and practical utility.

In-and-In-Breeding.

This system is the mating of closely related animals. It embraces the breeding together of animals of various degrees of relationship, no very distinct line of demarcation being drawn between it and line-breeding.

Over the merits and demerits of in-and-in-breeding there has been endless discussion. It is undeniable that its power for good is great. It is the surest and speediest of all methods for establishing character and fixing family type. The forces of heredity are more intensely concentrated by this system than by any other. As would be expected, therefore, in-and-in-breeding has been a predominating influence in the building up of most of our many distinctive breeds and tribes of live stock.

On the other hand, it is known that persistent close in-and-in-breeding tends to loss of size, fecundity, and constitutional vigour. Weaknesses and other defects are just as surely intensified by

it as are good points, and unless conducted with consummate skill and care it is not likely to be long followed with impunity.

The Value of In-bred Families.

In discussing the systems of breeding pursued amongst herds of Aberdeen-Angus cattle, the authors of the History of that breed expressed the opinion that it was very desirable more attention should be given to the building up of distinct well-defined families of as pure line-breeding as might be found practicable. "It seems to us," they continued (and the remarks have a general application to all breeds), "that it would be well for the interests of the breed if there existed several herds or strains which could be regarded as refined and reliable fountains of that mysteriously beneficial influence which may be generated by skilfully concentrating and assimilating the ever-present forces of heredity. Without entering upon a discussion of the question of in-and-in-breeding, we may remark that we believe it to be a most powerful agent either for good or evil. In competent hands it is perhaps the surest and shortest pathway to the highest pinnacle of a breeder's success. Unwisely employed, it becomes simply the broad road to ruin. We would not, therefore, desire that in-and-in-breeding should be pursued by the general body of breeders. We would, however, rejoice to see a few of those best able, intellectually and financially, to undertake the work, following the example of Thomas Bates, the Booths, and other noted Shorthorn breeders, and establishing distinct line-bred families. We should like to see a few families reared in such a way that they would not only be uniform in shape and character, but would also be possessed of one strong, unbroken, unadulterated, unvarying family current. We believe in the doctrine that 'like begets like'; but if we breed from composite animals—animals containing several conflicting family currents, perhaps the living influence of dead ancestors—we can have little confidence in the result. We cannot know which *likeness* may be produced—that of the immediate or of more remote ancestors. Practical experience and scientific reasoning both teach that no animal is so likely to reproduce an exact copy of itself as one that has been in-bred, or, in other words, one that contains one dominant, all-prevailing family current. We therefore think that the existence of a few well-defined in-bred families of really high individual merit would help greatly to maintain, and even still further improve, the high character of the breed generally. These families would be, as it were, strong springs of rich, pure blood, from which fresh draughts might be drawn from time to time for the refining and ameliorating of mixed herds." [1]

In-and-In-Breeding in Pioneer Herds.

To Professor James Wilson, Royal College of Science, Dublin, we are indebted for the notes which follow on the use which pioneer breeders of cattle made of in-and-in breeding in the establishing of their herds. It is a striking and remarkable fact, he says, that the operations of our greatest stock-breeders have always been accompanied by the same three phenomena in every case. The

breeders themselves have been unparalleled judges; they took enormous pains to secure the highest quality of stock for their herds, and, having done so, they bred from remarkably close relations. There is also strong presumptive evidence that they have all been masters of the art of culling or elimination. And these phenomena were to be observed, not only among the pioneers to whom breeds were indebted for their start in life, but also among subsequent workers.

It is well known how Bakewell, in the middle of the eighteenth century, scoured the country, going as far afield as Westmorland and Warwickshire for cattle, Yorkshire and Lincoln for sheep, and Holland for horses. Then, having secured the best stock, and afterward finding none so good as the progeny which he had bred himself, he put his own stock to his own for successive generations.

The following diagram showing the pedigree of Bakewell's bulls "Twopenny" and "D," and of "D's" son "Shakespeare," bred by Mr Fowler, of Rollright, in Oxfordshire, will show Bakewell's method:— [2]

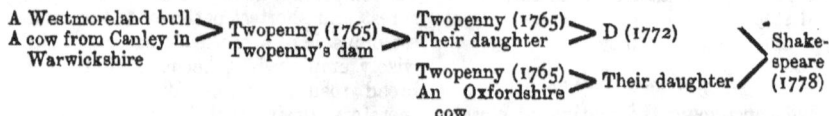

A Westmoreland bull A cow from Canley in Warwickshire	>	Twopenny (1765) Twopenny's dam	>	Twopenny (1765) Their daughter	>	D (1772)	>	Shake- speare (1778)
				Twopenny (1765) An Oxfordshire cow	>	Their daughter		

Similarly we know how the brothers Colling and Hugh Watson, with far less trouble, became possessed of the best cattle in Durham and Forfarshire respectively, how they mated their cattle as Bakewell had mated his, and how from their efforts the Shorthorn and the Aberdeen-Angus breeds of cattle were set upon the track

which they have since pursued. That these great breeders should have followed the example of Bakewell was not astonishing, since one of the Collings (Charles) had visited Bakewell, and Watson was intimately acquainted with the Collings' successors. A Colling and a Watson pedigree will suffice to show their methods:—

A COLLING SHORTHORN PEDIGREE.

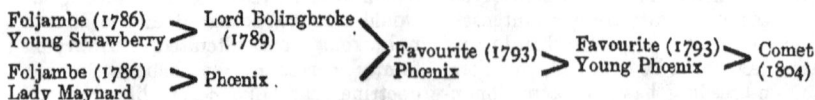

Foljambe (1786) Young Strawberry	>	Lord Bolingbroke (1789)	>	Favourite (1793) Phœnix	>	Favourite (1793) Young Phœnix	>	Comet (1804)
Foljambe (1786) Lady Maynard	>	Phœnix						

A WATSON ABERDEEN-ANGUS PEDIGREE.

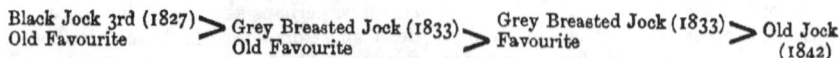

Black Jock 3rd (1827) Old Favourite	>	Grey Breasted Jock (1833) Old Favourite	>	Grey Breasted Jock (1833) Favourite	>	Old Jock (1842)

[1] *History of Polled Aberdeen-Angus Cattle.* By James Macdonald and James Sinclair. Vinton, London.

[2] The figures attached to bulls in these diagrams indicate, approximately in the case of pre-herd-book animals, the dates of their birth.

If the latter of these pedigrees were traced still farther back it would still show close breeding, although less close than in the diagram.

The greatest of all the non-pioneer breeders, and perhaps the breeder whose work is of most educative value, was Amos Cruickshank, who in 1837, about fifty years after their type was first established, began to breed Shorthorns. Of all the great breeders, Bakewell alone perhaps excepted, Cruickshank took the greatest trouble to secure for his herd the kind of stock that approached his ideals and to reject those that did not. Indeed, the story is almost pathetic. For more than twenty years he travelled up and down the country, securing occasionally a cow or a bull good enough in character and pedigree, but especially in character, to add to his herd. These were put upon trial, as it were, and retained or discarded according to the stock they produced. But in 1859 Cruickshank had the fortune to secure a bull, "Lancaster Comet," which produced him a bull-calf, "Champion of England," which approached so near to his ideals that only one or two more bulls were bought in. Then he used "Champion of England," "his sons, grandsons, and great-grandsons, until at the time the herd was sold (in 1889) every pedigree was saturated with 'Champion of England' blood." The following diagraphic pedigree of Cruickshank's great bull "Cumberland" will show this :—

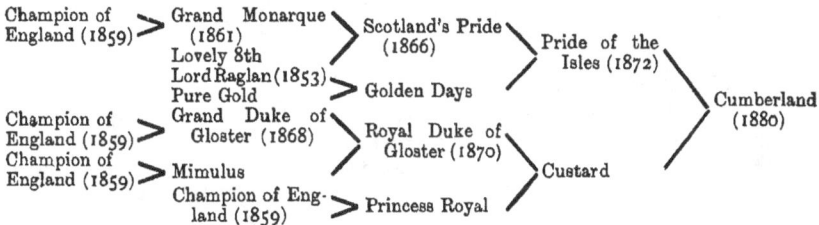

Between the intense in-breeding of Bakewell and Hugh Watson and the milder form pursued by Amos Cruickshank there is a large gap; between Cruickshank's system of mating and that of the ordinary breeder of high-class stock there is another gap which is often by no means large. In many cases, indeed, because of the incomplete manner in which pedigrees are usually set forth, animals are much more closely related than is generally supposed.

MENDEL'S LAWS IN STOCK-BREEDING.

It is believed by many that a new era is to be opened up for breeders of both plants and animals by the application of what are known as Mendel's Laws of Heredity. In vol. ii. (pp. 110-114) information is given as to experiments with these laws in the improvement of grain. That information is transcribed from a paper by Professors Wood and Punnet of Cambridge in the *Transac-*

tions of the Highland and Agricultural Society of Scotland for 1908.

Mendel and his Work.

From the paper just mentioned the following note is taken regarding Mendel and his work : "We begin to understand many of the mysterious things that happen when crosses are made among animals and plants — why a character often skips a generation, why the type is often broken to give rise to new forms, and what is the meaning of reversion. The foundations of this knowledge were securely laid by Gregor Mendel, an Austrian monk, in the garden of the monastery of which he afterwards became the head. Mendel has (1908) been dead for nearly thirty years, and it was as long ago as 1865 that his discovery was first given to the world. But his ideas were in advance of his time; they excited little interest and were soon forgotten. It was not until 1900 that his paper on the pea was unearthed, and scientific men began to realise what a far-reaching discovery this was that Mendel had made so many years ago. As a young man he had studied the natural sciences in Vienna, and had become interested in the problems of heredity. On returning to his monastery he devoted much of his leisure to carefully investigating the manner in which characters are transmitted in the common pea. From the results of his experiments he deduced certain principles which he found to hold for all the various characters he studied. During the past few years these principles have been confirmed and extended, not only for many plants but for animals as well." [1]

Mendelism Explained.

The following notes on the application of Mendel's laws to the breeding of live stock are from the pen of Professor James Wilson, Royal College of Science, Dublin :—

In explaining Mendelism we shall take our examples chiefly from cattle, because more is known in regard to them than in regard to other farm stock.

When red cattle are bred together their progeny are red, and when white cattle are bred together their progeny are white. But when red cattle are crossed with white their progeny are roan ; and

(a) When these roan crosses are bred together their progeny are 25 per cent red, 50 per cent roan, and 25 per cent white ;

(b) When they are crossed back again with red cattle their progeny are 50 per cent red and 50 per cent roan ; and

(c) When they are crossed back again with white cattle their progeny are 50 per cent roan and 50 per cent white.

A similar series of phenomena occurs when absolutely pure-bred black cattle are crossed with white, excepting that in this case the crosses are blue roans instead of red.

Mendel's conception [2] which explains these phenomena is that an animal, at its very start, receives from its parents the determinants of its future character-

[1] *Trans. High. and Agric. Soc. of Scotland,* 1908.

[2] Mendel worked with plants, but his theories are applicable to animals also.

istics in respect of colour, size, length of limb, length of horn, presence or absence of horns, mental powers, and so on ; that these determinants are made up of two halves ; and that they are passed on to the next generation through the sperms of the male and the ova of the female. But through each parent passing on a determinant and the young requiring only one, a half of each parental determinant is dropped in the melting-pot of fertilisation, and the young starts off with one only, the two halves of which are derived one from each parent.

This can be made clear by a diagram. A red Shorthorn carries a determinant for redness which may be represented by two small filled circles, thus ● ; a white Shorthorn carries a determinant for whiteness which may be represented by two small unfilled circles, thus ○.

When a red Shorthorn is bred to a white, either of the two halves of the red determinant may meet either of the two halves of the white, thus :—

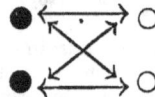

and the young starts off with a determinant which is half white, half red, thus ● ; and, as we know, its colour is roan, a mixture of white and red.

When two roan animals are bred together, either half determinant of each parent may meet with either half of the other, thus

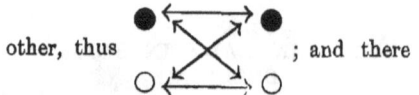

; and there are four chances : one that a red will meet a red, two that a red will meet a white, and one that a white will meet a white. Thus, over a sufficient number of calves from roan parents 25 per cent are red, 50 per cent roan, and 25 per cent white.

When a roan is bred back to a red or to a white, the chances are that half the young will be roan and the other half red in the one case, and half the number roan and the other half white in the other case, thus :   (red by roan) gives either ● (red) or ○ (roan),

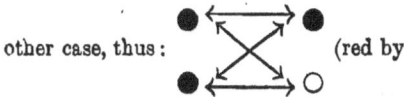

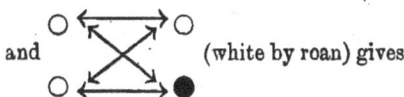

and   (white by roan) gives either ○ (white) or ● (roan).

In cases like the above, although it is possible eventually to change the colour of a breed from one colour to another by the continued infusion of that other colour, it is not possible to change the shade. There is no chance of gradually turning a white breed black by breeding each successive generation of a darker and darker shade.

But there are cases in which the first crosses are not intermediates with regard to one or more determinants, but are all like one of the parents. This happens, for instance, when absolutely pure black breeds are bred with red breeds. The first crosses are all black; and when they are bred together some of their progeny are black and others red; while when they are bred back to either parent race their progeny are all black in the one case, and some are black and some red in the other. Mendel's explanation is that these first crosses are not pure but impure blacks : they carry both determinants, but the black has its way and dominates or hides the red.

Let us put it graphically, using letters instead of circles, with capitals for the dominant and small letters for the hidden or recessive [1] colour.

Black crossed by red gives an impure black cross, thus :—

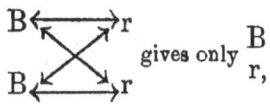

$$B \lessgtr r$$
$$B \lessgtr r$$

gives only $\dfrac{B}{r,}$

_____

[1] Mendel called the one kind "dominant," the other "recessive." "Subdued" would be a better word than "recessive."

in which the animal carries both determinants, but black hides the red.

When these crosses are bred together, 75 per cent of their progeny are black and 25 per cent are red; but of the black ones only one in three is pure black, the other two being impure, thus :—

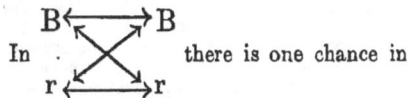

In $\begin{matrix} B \lessgtr B \\ r \lessgtr r \end{matrix}$ there is one chance in four of the young being $\dfrac{B}{B}$, two chances of them being $\dfrac{B}{r}$, and one chance of them being $\dfrac{r}{r}$; and those that are $\dfrac{B}{r}$ are impure black like their parents.

When these crosses are crossed back to pure black cattle all their progeny are black, but only half of them are pure, the other half being impure blacks, thus :—

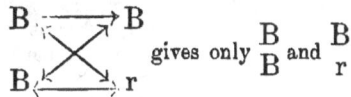

$$\begin{matrix} B \longrightarrow B \\ B \longrightarrow r \end{matrix}$$ gives only $\dfrac{B}{B}$ and $\dfrac{B}{r}$

in equal proportions.

But when these same crosses are bred back again to red cattle, half the young are impure blacks and the other half reds, thus :—

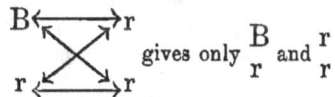

$$\begin{matrix} B \lessgtr r \\ r \lessgtr r \end{matrix}$$ gives only $\dfrac{B}{r}$ and $\dfrac{r}{r}$

in equal proportions.

Because of these phenomena it is possible, by crossing black and red breeds, to turn the red breed black and the black breed red, the latter being easier.

It will be noticed that the red cattle produced from the above crosses are always pure. Animals carrying recessive characters are always pure for that character. Thus to turn a black breed red it is only necessary to cross them with a red breed, breed from the first crosses, and keep the red calves they produce. If it were too expensive to sacrifice all the black calves, then by always putting red ones to black ones, the black ones would gradually become so few that their sacrifice would be comparatively inexpensive.

The following table shows the percentage of both colours that might be expected if this method were followed :—

	Black calves. per cent.	Red calves. per cent.
(1) 100 absolutely pure black cows crossed by red bulls would give . .	100	0
(2) 100 black first cross cows, *e.g.*, crossed by first crosses, would give	75	25
(3) 100 black (75) and red (25) of the second cross generation crossed by red bulls would give	50	50
(4) 100 cows of the third generation crossed by red bulls would give .	25	75
(5) 100 cows of the fourth generation crossed by red bulls would give	12½	87½
(6) 100 cows of the fifth generation crossed by red bulls would give . .	6¼	93¾

and so on.

And this process has actually been employed to turn the old black Highland breed red. The result is masked by the presence of other colours—brindle, dun, and yellow,—but when these other colours are eliminated, a breed that less than a hundred years ago was nearly all black is now nearly all red.

In the second volume of the *Highland Herd-Book* — the first in which cows and their progeny are entered—(published in 1887), the proportion of red calves registered as compared with black ones was 1.63 to 1, whereas in the fifteenth volume the proportion is as 7.8 to 1.

The process of turning a red breed black—that is, from a recessive to a dominant colour—is only slightly different, the added difficulty being that, unlike the red ones, the black cattle are not all pure for their own colour, and thus, although the continued use of the black colour will eventually eliminate the red, the process may take longer, and will be accompanied by the appearance of red calves—"reversions,"—the number of which, however, will gradually decrease. But the process could be hastened by testing the black cattle for purity and making use of those that come through the test: which is to breed the black ones to red ones. Those whose calves are all black are themselves pure for blackness.

And just as the Highland breed is an example of turning a black breed into a red one, so there are other breeds which, if they are not examples of turning red into black, can be quoted as examples which show the intrusion of red and the difficulty of its elimination, unless systematically taken in hand.

The Aberdeen-Angus is one of the breeds in question. Like all the other black breeds, it absorbed some red blood at some time in the past, and a red calf still appears occasionally. These red calves are really "reversions," and they appear in this way : The intrusion of the red cattle produced a number of impure black cattle $\left(\dfrac{B}{r}\right)$, and although these have grown gradually fewer, there are still some in the breed, and when two meet their progeny have one chance in four of being red, thus :—

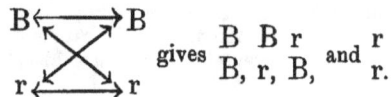

$$\begin{array}{c} B \longleftrightarrow B \\ \times \\ r \longleftrightarrow r \end{array} \quad \text{gives} \quad \begin{array}{cc} B & B & r & r \\ B, & r, & B, & r. \end{array}$$

A very famous Aberdeen-Angus cow completely lost her character by giving birth to a red calf. It will be seen from the above that the bull was equally to blame.

Besides the above cases of colour, a few more instances in which a breed or a race of cattle has been similarly affected by another might be quoted :—

(*a*) The long-legged, light-bodied, black Kerry cattle were crossed a century or more ago by short-legged, stout-bodied, red cattle of Devon type from the south of England, and there was produced the short-legged, stout-bodied Dexter Kerry, which is sometimes red, but more often black, the breed not being fixed as yet for one colour or the other. This is a case of shortness of leg, stoutness of body, and blackness all being dominant.

(*b*) A number of white-faced and finch-backed cattle were brought to England from Holland and the neighbouring countries in the seventeenth and eighteenth centuries, and through them these markings were handed on to many cattle in England, Scotland, and Ireland, and they still occasionally occur as "reversions." The Herefords still retain the white face, which is dominant over other face colours, and the Long-horns the finch-back.

(c) Hornlessness, which in all proba-bility came to Britain from Scandin-avia, and is now common to several breeds, is dominant to hornedness, and can be handed on to horned cattle. There are hornless Shorthorns and Herefords in America. By the reverse process horns could be put upon polled cattle.

(d) Nearly two thousand years ago the Romans brought cattle to Britain, whose long, wavy horns were handed on to many English and Scots cattle.

It is clear, therefore, that where a character is found dominant to another, either can be transferred from family to family and from breed to breed, and in this way a new variety can be produced. The importance of this will be realised if we mention a few pairs of characters which we should like to have under con-trol, and which we should like to be able, as the case may be, to impart to or eliminate from our stock. The char-acters we will mention are such as there is hope to believe may be Mendelian, viz. :

High-milking and low-milking qual-ities.
A high and a low power of producing fat in milk.
Fatness and leanness.
Straight and tilted horns.
Black noses and white noses in cattle.
Short legs and long legs.
Hairy and non-hairy legs in cart-horses.
Long wool and short wool in sheep.
Stiff wool and soft wool.

And, to take only a single case, if much milk and fatness are found to be Mendelian characters, it will at once be possible to combine the two characters in any breed of cattle: not, however, in some breeds without crossing with others.

The light which is thrown upon the methods of various breeders by Mendel-ism is already possessed of considerable illuminative capacity, which will in-crease as our observations, which have been confined so far almost to colour alone, pass on to other less obvious but more important characteristics. Mean-time, let us make use of the knowledge now at our command.

The extreme methods of breeders are crossing and in-breeding. It is well known that stock-breeders can usually tell what to expect when two breeds are crossed, but that when first crosses are bred together or to some strange breed their progeny are very irregular, some being like their parents, some like their grand-parents, others like breeds now extinct, others like no animal ever known. The explanation is that when breeds are mixed up the determinants for colour and other things are also mixed up, and shake themselves down in any possible manner. A good example may be taken from Highland cattle. Four races have gone to the making of this breed,—a black race, a red, a light dun, and a brown or donn ; and through the interbreeding of these, five new hybrid colours have been produced—namely, yellow and dun and black brin-dle, red brindle and dun brindle ; and if three brindle bulls—a black brindle, a red brindle, and a dun brindle—were put to a large herd of brindle cows, every one of all the nine colours would appear in their progeny.

And if this kind of thing happens with colours, and similar things may happen with other characteristics, we can readily understand why careful stock-breeders are so very chary of cross-bred animals. In-breeding, on the other hand, brings together fewer determinants, eliminates the unexpected, and produces a breed which is more and more regular in all its characteristics—those, at any rate, that are dominant and recessive—the longer it is persisted in. Most of our breeds of stock have been built up from mixed foundations, and it was only by in-breeding that regularity and some part of what is vaguely called "pre-potency" was achieved.

In line-breeding, again, the phenomena usually attendant upon crossing are avoided, although steadiness to type is not got in this way as it is got by in-breeding—a matter that is of less moment in a breed that has already been steadied by in-breeding.

These points could all be illustrated fully if only the work of breeders who are still alive, and of others recently deceased, could be referred to, but every stock-breeder knows how very tentatively and tenderly the greatest of his own colleagues proceed in the introduction of "fresh blood" or "out-crosses," and how very frequently, unless they can be graded up by being always mated to the breeder's own type, the descendants of these out-crosses have to be eliminated from the herd.

PERIODS OF GESTATION.

The periods over which the females of the various classes of live stock carry their young are as follows:—

Mare . . .	48	weeks.
Cow . . .	40	"
Ewe and goat .	21	"
Sow . . .	16	"
Bitch . . .	9	"

The egg of the goose hatches in 30 days, of the turkey, duck, and pea-fowl in 28 days, of the pheasant and partridge in 24 days, and of the barn-door fowl in 21 days.

POULTRY.

Poultry-rearing as a rural industry has not yet taken the position it ought to occupy in this country. In particular, as an adjunct to other branches of agriculture it should receive a great deal more attention than has hitherto been devoted to it. To realise that there is much room for extension in the raising of eggs and table poultry, one has but to look at the official returns showing the vast sums of money sent over the seas for eggs and table fowl to meet the demand in this country for these choice and popular articles of food.

In the belief that an extension in poultry-rearing is much to be desired amongst agriculturists of almost all classes, whether their holdings be small or large, it has been thought well that in this edition of *The Book of the Farm* the section dealing with Poultry should be entirely rewritten with that important object prominently in view. For this new matter the editor is indebted to Mr Alex. M. Prain, who has had much successful experience in the rearing of poultry.

New-laid Eggs.

A glance at the monthly and yearly returns of the imports of commodities for the food of the people will indicate that, especially in the production of new-laid eggs, there is room for great development in this country. For these there will always be a ready demand at prices far higher than can be offered for foreign eggs, which at the best cannot be placed in our markets under ten days' old—at which age, though they may be perfectly fresh, yet they cannot be regarded as *new-laid*.

A very considerable proportion of the imported eggs have been preserved in lime, and these are used for cooking and for confectionery purposes. With such an unlimited demand from our large cities, the British farmer has every advantage over his foreign rivals, and it should be his privilege to supply the demand for the top quality of new-laid eggs, leaving the foreigner to supply the second and third qualities if he likes. The benefit is a mutual one, shared equally by buyer and seller. The aim of poultry-rearers in this country should be to put a large supply of newly laid eggs on the market *all the year round*.

Table Poultry.

As with eggs so it is with fat poultry. But the average farmer has yet to learn that the surplus cockerels and old hens require to go through a process of fattening before being put on the market. The fattening is to a large extent in the hands of a few large firms, who buy the young birds at from three to four months old, put them through a three weeks' process of fattening, during the latter half of which they are crammed, and then sell them at very high prices.

PURE BREEDS.

In the breeding of poultry, farmers as a rule seem to have very hazy notions. No definite system could possibly be traced from the appearance of an average flock of farm fowls, unless the mixing up of as many breeds as possible in the composition of the flock could be called a system. Considering that certain well-defined rules guide the breeding of other classes of farm stock, such as horses, cattle, and sheep, it is surprising that totally different ideas should prevail in regard to poultry. Apart from the breeding of exhibition stock, every owner of a flock of fowls should have a definite object to strive for,—either the production of the greatest possible value in eggs or the best table fowls.

Now, to mix up indiscriminately laying breeds and table breeds in one flock is to court failure in both purposes, for it is recognised by all authorities on poultry-rearing that mongrels are economically unprofitable. Seeing that there are now available so many pure breeds having certain characteristics clearly established, it is a matter for surprise as well as regret that mongrels are still so common.

Classification of Pure Breeds.— These pure breeds, which are the result of the most careful breeding and selection, may be divided into four main groups, as follows : (1) Laying Breeds, (2) Table Breeds, (3) General Purpose Breeds, and (4) Fancy Breeds.

Among the best known and the most useful in each section are—

Laying or Non-sitting.

Minorcas.	Anconas.
Leghorns.	Campines.
Andalusians.	Hamburgs.
Houdans.	Scotch Greys.

Table Breeds.

Dorkings.	Old English Game.
Indian Game.	Sussex.

General Purpose Fowls.

Orpingtons.	Faverolles.
Wyandottes.	Langshans.
Plymouth Rocks.	

Fancy Breeds.

Modern Game Bantams.
(some varieties).

Laying Varieties.

A short description of each of these breeds is given here, with a note of their main characteristics.

Minorcas.—There are two recognised varieties of the Minorca breed, the Black and the White, but the latter are very rarely seen. The breed is one of the Mediterranean family, believed to have been imported into this country from the island of Minorca. It has all along been a favourite, more especially in certain districts of England.

The plumage is beetle-green black, with brilliant red comb and wattles, and smooth white ear-lobes. The carriage should be sprightly and stylish, with nice long body carried on legs of medium length. For size and quality of eggs no breed can beat them, and, if kept from severe frost and cold winds, they will lay well all the year round except during the moulting period. It is a general characteristic of all the non-sitting or laying varieties that they will not lay well during winter in very exposed situations or in periods of severe frost—that is, of course, unless suitable shelter is provided for them.

The Minorca is justly regarded as one of the oldest and most reliable of all the breeds of poultry. Fig. 714 represents a Minorca cock, and fig. 715 a Minorca hen.

Leghorns.—The Leghorns are also of Mediterranean origin, and are now divided up into a great many subvarieties. The best known of these are the Whites, Browns, Buffs, Duckwings, Piles, Cuckoos, Blacks, and Blues.

The Whites are the largest in body, but all have the same main points— namely, a very graceful body carriage, with bright, clean, yellow legs, and a very active foraging disposition, which makes them economical to feed and easy to rear. All varieties are capital layers of good-sized white eggs. A White Leghorn cock is shown in fig. 716, and a White Leghorn hen in fig. 717.

Andalusians.—The Andalusians are another of the Mediterranean group. They are slate-blue in ground-colour, with a purple-black lacing round each feather in the hen, and the same colour on the neck, hackle, and back of the cock. The body is rather slim, with fairly long legs, and the head-points less fully developed than in either the Minorca or the Leghorn.

Houdans.—The Houdan is a French breed once very popular, but not so common now. It is of large size, broad and massive, mottled black-and-white plumage, and a full round head-crest. The legs are short, pale in colour, free from feathers, and carry a fifth toe.

Anconas.—This is a comparatively new variety. It has brilliant beetle-green plumage, each feather being tipped with white. In style it resembles closely some of the smaller Leghorn varieties, the hens being excellent layers, inclined to be small in body, but very hardy and precocious.

Campines.—This is considered the great egg-producing breed of Belgium. Except in comb, it resembles our Pencilled Hamburgs. The body is small, but they are fairly hardy, and excellent layers.

Hamburgs.—There are two distinct varieties of these lovely fowls, the Pencilled and the Spangled. The Pencilled were probably imported from Holland, where they were known as the Everlasting Layers. The Spangled and Black varieties belong to this country.

The plumage of some of the varieties is truly magnificent, and the well-shaped bodies show it to every advantage. Though all varieties are good layers, they are not so strong in constitution as some of the other breeds mentioned, and the eggs are rather too small to realise the highest price.

Scotch Greys.—This is a very old typical breed, very hardy, and capital layers of large-sized eggs.

Table Breeds.

Dorkings.—The Dorking is one of the oldest and best known of our truly English breeds. There are several varieties—Darks, Silver Greys, Cuckoos, Whites, and Reds. The first two are by far the most popular. In fig. 718 a coloured Dorking cock and hen are represented.

The Dorking is essentially a table breed, the flesh being pure white in colour and very delicate in texture. The body is large and deep, and, looked at sideways, should appear almost square. The legs are short, pure white in colour, and carry the characteristic of the breed—the fifth toe. They attain a very large size on favourable soils, but some breeders think the chickens are delicate to rear.

Indian Game.—This is truly a valuable breed for table purposes. Though somewhat heavy in bone, they carry a large amount of flesh on the breast. The head is broad and massive, neck arched, the body very broad across the shoulders and wide in chest, legs rich orange colour, medium in length, and set well apart. For crossing with other breeds the Indian Game is even more valuable than as a pure breed. Fig. 719 represents an Indian Game cock and hen.

Old English Game.—Though smaller than the Indian Game, the Old English Game are of superior quality of flesh, which is close in texture and pure white in colour. There are a good many sub-varieties, but the white-legged ones are preferred. The body is medium in size, broad in breast, close, compact, and hard in feather. It is a very hardy breed, suitable for almost any climate. Fig. 720 shows an Old English Game cock and hen.

Sussex.—The Sussex is a very large square-bodied fowl, resembling the Dorking in type though not in colour, and without the fifth toe. The flesh is of excellent flavour and very white in colour.

General Purpose Breeds.

Orpingtons.—This is one of the best, if not the very best, of the general purpose fowls ever introduced. There are now a few varieties of this deservedly popular breed—Blacks, Buffs, Whites, Jubilees, and Spangled, the best known being the first three. All are of the same blocky type,—full round breasts, very deep in body, short in back, and short on leg,—perfect models of symmetry and shape. The chickens grow rapidly, and are extremely hardy, being suitable for any climate. As winter layers no breed can excel them, and the eggs are of fine shape and brown in colour.

For general farm fowls the Buffs and Whites are hard to beat. A Black Orpington hen is shown in fig. 721. Fig. 722 represents a pen of White Orpingtons.

Wyandottes. — This is a breed of American production, and is a credit to our cousins across the ocean. It has taken an extraordinary hold on poultry-breeders in this country, and has been a source of great profit to them. There is no need to describe all the varieties of the breed. It will be sufficient to enumerate the best known of them. The Silver, Golden, Blue-laced, Buff-laced, White, Partridge, Silver-pencilled, and Black are all as distinct in colour as their names imply, though in shape and general characteristics they are much the same. They stand on longer legs than the Orpington, and are not so heavy, but they have much of the same compact cobby build of body. All have rosecombs, and rich yellow legs.

For egg-production they are equal to the Orpington, and the egg is about the same shade of colour, though rounder in shape. An excellent all-round fowl it is. Fig. 723 represents a White Wyandotte cock, and fig. 724 a White Wyandotte hen.

Plymouth Rocks. — The Plymouth Rock is another American production, and the favourite breed of that country. The barred variety is the most popular, though there are Buffs, Whites, and Blacks as well. The Rocks are a very large, rather heavy-boned breed, with clean, rich yellow legs, and a constitution so vigorous that they can stand the most exposed situation. They lay a rich brown egg of good size, and come earlier to maturity than some of the other breeds. A Plymouth Rock cock is shown in fig. 725, and a Plymouth Rock hen in fig. 726.

Langshans. — These are jet black fowls of Chinese origin, with long, slightly feathered legs. In recent years the modern type has developed such length of limb as to make it ungainly, though the original fowls imported from China were a most useful breed, and grand winter layers of large deep-brown eggs.

Fancy Breeds.

The " fancy " breeds — those kept mainly for showing — need not be described in detail here, as they are not suitable for farmers, though a great interest is taken in their production by other classes. The breeds of poultry mentioned and briefly described above are the best known, and probably the most profitable from the utility point of view ; and, speaking generally, it will be found much more advantageous to keep one or other of them only, or a first cross between two of them, than to keep a mongrel stock. Keepers · of poultry should study and settle definitely what they mean to breed for. If eggs are likely to give most profit, then by all means keep a variety, or varieties, suited for that purpose, and not such a breed as Indian Game. If, on the other hand, table fowls are desired, then choose one or other of the table breeds. Where good egg-production combined with good table qualities is wanted, then one of the general-purpose fowls will suit best.

Cross-bred Poultry.

Regarding the raising of cross-bred poultry, excellent results in eggs will be got by crossing Brown Leghorns with Buff Orpingtons, or White Leghorns with White Orpingtons, or indeed any of the Leghorn or Minorca breeds with any of the Orpington, Wyandotte, or Plymouth Rock breeds. For table purposes nothing can be much better for quality than a cross between the white-legged Old English Game cock and the Dorking, Sussex, or Buff or White Orpington hen, though by using the Indian Game cock with the same hens or with Faverolles bigger chickens will be got. The latter cross is, however, rather coarser in bone and bigger in thigh. The chickens of either cross will be found extremely hardy, will grow rapidly, and will be ready for the table, weighing from 3 to 4½ lb. each at from thirteen to fifteen weeks old, at which period they are at their best. If allowed to grow beyond this age, the frame, especially in the cockerels, begins to rush up, the first real feathers begin to come in, and the birds have to be kept till full grown, and moulted before they will fatten properly. If they are kept, however, till from eight to ten months old, they will be grand specimens, with plenty of beautiful breast meat, and weighing from 7 to 10 lb. each.

Advantages of Pure Breeds.

While crossing can be strongly recommended for certain purposes, there are still a great many advantages in keeping the breeds pure. The first is, that a better price can be got for any surplus stock which has to be disposed of. The best of the pure cockerels can usually, with a little judicious advertising, be sold for breeding purposes at from 4s. to 10s. each, and the pullets, particularly of winter-laying breeds, can be easily sold in the autumn at from 3s. to 4s. 6d. each. These prices are by no means overstated, and no account is taken of any birds which might be good enough for the show pen. For these any price may be got, according to their quality.

Another advantage of pure breeds is the uniformity of the eggs as regards shape and colour, and even this point tells in the marketing. Still another benefit is that in the spring of the year, when eggs get cheap commercially, a fair trade can be done in selling sittings for hatching purposes at from 2s. 6d. to 10s. per sitting, according to the quality of the stock birds. Nothing of this can possibly be done with mongrels, and all the time the pure birds are eating no more, neither are they costing any more to manage.

Numerous instances could be given where the fowls kept on the farm or at the cottage are of one pure breed, and where trade of the kind indicated is profitably carried on. Large sums of money frequently pass from the big exhibitors to the small careful breeders for the pick of their season's chickens. For small crofts or holdings, or even cottages, the greatest profit will undoubtedly come from a carefully selected stock of a single pure breed.

DUCKS.

The principal breed of ducks is the Aylesbury, which are of large size, with long, deep, straight keel, pure white in colour, and of pronounceedly rapid growth. The name is taken from the Vale of Aylesbury in Buckinghamshire, where the breed flourishes remarkably well, and from which a very large business is done with London in the duckling trade. This, then, is the variety for early maturity, the ducklings coming up to 4 lb. weight at eight to ten weeks old. Rouen ducks are in plumage almost identical with the Mallard or Wild-duck. This variety grows to a larger size than any other variety, but it matures slowly, and so is more suited for winter fattening. When fully matured, some specimens attain from 9 to 11 lb. each. They lay well, the flesh is of fine quality, and they are extremely hardy.

Pekin ducks are of Chinese origin, and have been largely used for crossing purposes to give stamina to our home breeds. Though as a pure breed it does not equal in usefulness the Aylesbury or the Rouen, unless in that it is a slightly better layer, still the progeny of the cross between it and either of these breeds will mature with greater rapidity, and attain greater weight than the Aylesbury or Rouen, and it is for the purpose of crossing that the Pekin is most largely used. The colour is a very pale shade of canary, and the carriage is upright, somewhat resembling that of a penguin. The legs and bill are a deep orange, and the body is profusely feathered.

Indian Runner ducks, the great egg-producing variety, are noted also for their great foraging habits. They are in many respects an ideal farmer's breed, for, though small in size, they yet make a fair appearance on the table, and they may be said to be, in suitable places, everlasting layers. The colour is most attractive, being a mixture of fawn and white, and they very seldom go broody.

A first cross between the Indian Runner and Aylesbury makes an excellent all-round duck, combining both laying and table qualities.

Fig. 714.—*Minorca cock.*

Fig. 715.—*Minorca hen.*

Fig. 716.—*White Leghorn cock.*

Fig. 717. *White Leghorn hen.*

Fig. 718.—Coloured Dorking cock and hen.

Fig. 719.—Indian Game cock and hen.

Fig. 720.—Old English Game cock and hen.

Fig. 721.—Black Orpington hen.

Fig. 722.—White Orpingtons.

Fig. 723.—White Wyandotte cock.

Fig. 724.—White Wyandotte hen.

Fig. 725.—Plymouth Rock cock.

Fig. 726.—Plymouth Rock hen.

Fig. 733.—*Incubator.*

Fig. 731.—*Shelter coop.*

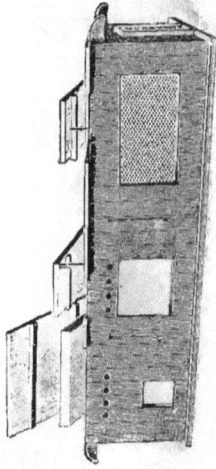

Fig. 730.—*Coop and run.*

Fig. 734.—*Foster-mother for rearing chickens.*

Fig. 732.—*A colony poultry-farm.*

Fig. 737.—*Poultry-house on wheels.*

Fig. 738.—*Movable poultry-house.*

Fig. 739.—*Movable poultry-house.*

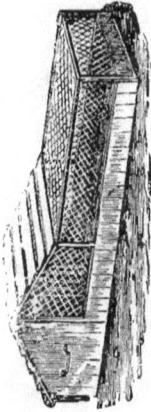

GEESE.

Geese might well be more numerously kept than they are. They forage so well for themselves that the cost of keeping is not large. Whether they will be profitable or not depends on the situation of the farm. On waste or marshy ground they will practically require no feeding. Geese always command a ready sale at Christmas time.

There are two main varieties of geese, —the Toulouse or Grey, and the Embden or White.

Toulouse Geese.—The Toulouse is the more common variety, and has a most solid substantial appearance, being short in leg and very square and massive in body. In colour it strongly resembles the grey-lag wild goose, with bill and feet a dark orange. It lays wonderfully well, and, as a rule, is a non-sitter. This variety is slow in maturing, though it finally attains a great weight. When growing the frame develops rapidly, but very little flesh is put on till the body is full grown. This variety is thus not suitable for killing as green or Michaelmas geese, its special use being for the Christmas trade.

Embden Geese.— This variety is white in plumage, and of more upright carriage than the Toulouse. It also matures much earlier, and is thus ready for the autumn demand. The quality of flesh is about equal in both varieties, and very often they are crossed for general purposes. The white feathers of the Embden are of value, so it has this advantage over its rival. The Embden is an excellent sitter and mother.

TURKEYS.

It is now generally admitted that our domestic turkey is descended from the wild species of North America.

American Bronze Turkeys.—The American Bronze is the most common variety, as well as the largest and handsomest. The colour is a dazzling lustrous bronze on the back, neck, and tail, with black breast and body, which is pencilled with white. The flavour of the flesh is said not to be so delicate as that of our English breeds, but this is compensated for by a much greater size,

some specimens weighing up to 50 lb., though 35 lb. is a very good weight for a cock and 20 lb. for a hen. An additional advantage of large size is that more is given per lb. for the weightier birds.

English Turkeys.—The Cambridge variety is common in some parts of England, and so is the Norfolk or Black turkey; but both are smaller than the American Bronze, and also somewhat more delicate to rear.

Pure and Cross Stocks of Ducks, Geese, and Turkeys.

To ducks, geese, and turkeys the same general remarks apply in regard to pure breeds as apply to poultry. In every case much of the ultimate success depends on the judicious choice of a breed or breeds. For that no hard and fast rule can be laid down, so much regard must be paid to soil, climate, situation, distance from markets, &c. Each individual breeder must therefore choose for himself: first, whether eggs or table fowls shall be made the first consideration; and secondly, which breed or cross will suit his special circumstances best.

Some are reluctant to take up pure breeds because of the initial expense and trouble incurred, but these objections are very easily overcome.

One very cheap and easy method of changing a stock is to buy a sitting or two of eggs from a reliable breeder of the new breed selected, the following spring buy a few more sittings of the *same breed,* meantime selling the old stock off gradually, and in a few years a complete change will thus be effected.

Another easy plan is to buy a cockerel and five or six pullets, and hatch only the eggs from this pen. This means separating them from the rest of the stock, but that can be very cheaply and easily done with some wire-netting and a movable house.

In any case, no initial trouble should be spared to get a start with the best varieties—that is, the varieties which will be the most profitable.

HOUSING POULTRY.

In no department of poultry-keeping has so much change taken place as in that of the housing of the birds. Old ideas of warmth for the fowls, which usually meant overcrowding and no ventilation, have been entirely given up, and more hygienic methods have been introduced. There is still, however, far too little attention paid to this important matter. Hen-houses at farm-steadings are too often in the very worst position possible. It is not uncommon to see a cart-shed, implement-shed, or tool-house with a nice sunny southern exposure, while the hen-house is facing the north. In looking broadly at the subject, housing may be considered under two heads—"fixed houses" and "movable houses."

Fixed Houses.

As regards fixed houses, it will be better to indicate a few general principles which should apply to them rather than lay down hard and fast rules.

First, then, all houses should have light. There is nothing which can purify or warm the air of the house like the light of the sun, and this should be admitted freely by a large window set in the wall so that the light can reach the floor and walls. So much the better if the window is fitted on the inside of the wall, and made in two halves to slide fully open each way. With wire-netting over the outside to keep out unwelcome intruders, the window can be left open night and day in summer.

The perches should be all on one level, about 2 feet from the ground, 18 inches apart, and easily movable. Each perch should be about 2 inches broad, and rounded at the edges. Nest-boxes should also be easy to move, and set quite low, about a foot from the floor, and not made fixed in tiers right up to the top of the wall. The reason for movable perches and nest-boxes is to make the process of cleaning out as easy as possible. By removing everything to the door it is a simple matter to go over walls and floor thoroughly. This should be done, and fresh chaff put in, 4 to 6 inches deep, at least once a-week. Where the floor is of cement a hose-pipe can be used to scour the whole place out properly at intervals.

Cleanliness. —Cleanliness is of the utmost importance. The house should be brushed out once a-week or so, and the walls should, at least once a-year, get whitewashed with hot lime, to which a little carbolic acid has been added. Cleanliness applies to more than the house: it applies to all drinking-vessels and food-troughs, and to the birds themselves. Very few would believe the number of insects which may be found on a hen of any average flock. This can be remedied by providing a good dust-bath, roofed over, but quite open to the front, with a board nailed up about 8 inches to keep the material in. Good sharp sand and ashes mixed make a capital dust-bath, and should be always available.

Ventilation. —About ventilation in poultry houses some curious ideas are entertained. Some people cannot distinguish the difference between a current of air being allowed to blow straight in on the birds and proper ventilation. Ventilation means the proper regulation of a current of fresh air getting into the house, with equal means for the bad air to get out. This can be secured in several ways which are well known, and which need not be detailed here. It is sufficient to state that abundance of fresh air should be provided, for there is no more frequent cause of disease than vitiated air.

These, then, are the main principles of housing—Light, Cleanliness, and Ventilation. They are not mere details, as some think, but matters of the very utmost importance, because on their observance depends the health of the birds, and it is folly to expect good laying results unless the fowls are in perfect health.

Movable Houses.—Movable houses are becoming more and more in evidence every year. The "colony system," as it is called, of dividing up the fowls into small flocks, of from 15 to 30 or 40 birds in each, has practically revolutionised poultry-keeping. The houses used with this system are usually made in sections to bolt together, so that they can be readily taken down and put up again. For convenience in moving from field to field or for changing to a fresh piece of ground, a great many of the houses are on wheels (as in fig. 727), or on slides (as in figs. 728 and 729).

Each house has a shelter of some kind for bad weather, either under the raised floor of the house or as part of the house itself. Shelter-coops such as are shown in figs. 730 and 731 are also used largely. Fig. 732 gives a general view of a colony poultry-farm, photographed with the camera looking northwards.

In every case there is light and ample ventilation. Some of the newest designs are almost entirely open-fronted, being only boarded up about 2 feet, the whole of the rest of the front being lined with wire-netting. With both sides and back solid, all fear of draught is avoided, and the birds seem to do excellently.

Such a system as this has everything to recommend it. The fowls are in a natural state, living in healthy surroundings, and picking up a large share of their own food. After harvest it is usually a profitable plan to stock the stubble-fields with groups of young birds. The grain is there in plenty,—grain which would otherwise go to feed the multitude of wild-fowl, and small birds of all kinds, and it is noteworthy how plump pheasants and partridges usually become from just this kind of feeding. Besides gathering the grain, which would otherwise be lost, and turning it into profit, the fowls consume a very large number of insects. This, with the open free life, builds up such a constitution that disease is almost unknown in well-managed "colonies." The cost of attendance is also reduced to a minimum.

Not only on stubbles can this system be practised, but on pastures as well. Oftentimes the houses are put beside some natural shelter, such as a clump of trees, a hedgerow, or dyke; and no doubt some natural shelter is desirable from bad weather, and from the sun as well. Were it not for foxes this system would be much more widely adopted than it is.

Feeding Poultry.

It is undesirable to prescribe very definite rules regarding the feeding of poultry, because ideas are always changing, and there is still a good deal to learn. Chemistry has been of the greatest value in determining the component parts of the various foods, but experience only can teach the action of the different foods on the body. By combining the knowledge chemistry has put at our disposal with the experience gained from observation of the suitability of certain foods, we are able to compose a properly balanced food.

In the feeding of ordinary laying stock, the point to be aimed at is to keep the hens up to full laying limit and yet keep their bodies properly nourished without running to fat.

Much, naturally, depends on the conditions under which the birds are kept, and the quantity and nature of the food which they can find for themselves. Birds kept in confined runs must have their bill of fare much more carefully selected than those running out on pasture or stubble-fields, where worms, slugs, and snails can be picked up freely.

Hand-feeding for Laying.

The usual plan of feeding ordinary laying stock is to give a hot meal of soft food in the morning, and grain for the evening meal. For the preparation of the soft food it is very convenient to have a stock-pot, into which are put all the house scraps, such as beef bones, meat or fish scraps of any kind, crusts of bread, potato or vegetable leavings,—in fact, anything of the food kind left over from the table. Cover over with water and boil the whole at night. In the morning it has only to be heated to be ready to mix with the meals. In mixing, a handful of common salt should be added, and once a-week, or once a-fortnight, Epsom salts should be substituted for the common salt. The meals to be used should vary with the season of the year, the heat-giving and fat-forming meals being discontinued or reduced in quantity in the warmer months.

Meat in some form is now considered imperative. Where large quantities of fowls are kept, raw horse-flesh is the cheapest and best form in which it can be supplied. In the case of cattle and sheep that have died, it is better to boil the flesh for fear of disease, and this also holds with butcher's offal, which should be cooked till it is soft.

Fresh-cut bone will take the place of meat to some extent, but it is expensive to buy, and though there are hand-machines for cutting it, the work is rather stiff. When neither flesh nor bone can be got cheaply or easily, meat-meal should be used. This should contain 70 per cent of albuminoids, and for small stocks of poultry it is cheaper and involves less trouble than other kinds of flesh.

Twice a-week is often enough to supply a flesh diet if the fowls get as much as they can eat. The price of meat-meal is about 14s. per cwt. The principal meals are sharps, oatmeal, barley-meal, Indian meal, pea-meal, and bran. For summer feeding to mix with the contents of the stock-pot, assuming that meat-meal is used instead of flesh, a fair ratio would be—

2 parts sharps.
2 parts meat-meal.
1 part oatmeal.

For the afternoon feed of grain, 2 parts oats to 1 of wheat.

For winter feeding—

3 parts sharps.
2 parts meat-meal.
1 part Indian or pea-meal.
1 part oatmeal.

For the afternoon feed—

3 parts maize.
3 parts wheat.
4 parts oats.

These, of course, may be altered to give variety, as, for instance, rice boiled in milk for summer, and boiled wheat or maize in the winter time. Vegetables ought to be freely used; swedes, mangels cooked or raw, also chopped clover and cabbages. In summer, clovers, green pea-haulm, lettuce, or any garden vegetable may be given. Cooked turnips are excellent for mixing with the meals, and so are potatoes occasionally, but not regularly.

Grit. — Grit is so essential that it might almost be considered a food. A considerable variety of this material should be constantly available, such as road scrapings, broken brick, coal-ash, lime or mortar, broken crockery, oyster and other shells. From the grit fowls get mineral matter, so that it is really something more than a mere aid to digestion.

Fattening Poultry. — The fattening of poultry is now a specialised industry. In this case the feeding adopted is largely Sussex ground oats, with milk and fat added. The birds are usually finished by a period of "cramming," which leaves the flesh very white in texture and delicate in flavour. "Hopper feeding" is now extensively adopted in America and some parts of this country. The idea is to have constantly before the birds a supply of food which they can eat at will. These hoppers, which are made of wood, consist of a reservoir with sloping lid, and a tray below into which the food falls. As the birds eat more comes down, so that the action is automatic.

Biscuit meals of various kinds are also extensively used in feeding, either alone or in combination with other meals.

Feeding Chickens.—Chicken-feeding may be said to be an industry by itself. Each system has its group of adherents. The system of "Dry Feeding," which originated in America, has many advocates, though its opponents say that the chickens so reared never attain the same size of frame as those which have had soft food supplied to them.

The dry chick feed consists of small seeds such as the seeds of dari, lint, and hemp, with wheat, groats, and rice. The commonest feeding for chickens is usually dry stale bread-crumbs, oatmeal, and hard-boiled eggs.

Another kind of feeding is a custard made with eggs and milk. As the unfertile eggs can be used in this way, it is cheap and certainly gives good results.

Another system is to give nothing but fine grit and water for the first two days, and then begin with stale bread-crumbs soaked in skim-milk and squeezed fairly dry. After this the dry chick feed is partly adopted, along with soft food cooked with milk.

Many specially prepared chicken meals are also most successfully used, either by themselves or in combination with other meals.

The greatest care has to be taken for the first fortnight or three weeks, as the heaviest losses occur during the first week through over and improper feeding. Tainted ground, body lice, and dirty drinking-vessels are also frequent causes of mortality. Cleanliness is absolutely essential, and milk given freely to drink is a splendid source of nourishment.

General Points in Poultry - feeding.—The whole question of poultry-feeding is of absorbing interest, and demands careful study. Different breeds require different treatment. Mediterranean non-sitting varieties can stand a richer diet than the heavier, less active sitting varieties. It is a good plan for poultry-feeders to make frequent experiments with different materials and rations.

INCUBATION.

The first essentials for the securing of good hatching results, whether by natural or artificial means, are the health and stamina of the stock birds from which the eggs are gathered. Eggs from birds properly mated, and enjoying their liberty, as they do when the "colony" system is pursued, are very little trouble to hatch, and the means used for hatching are of secondary importance. The natural and artificial methods have each their followers. Often both systems are used together with excellent results. With the spread of non-sitting varieties, broody hens are getting every year more difficult to obtain when wanted, and so the manufacture of incubators has gone up by leaps and bounds.

When eggs are being kept for hatching it is wise to turn them every other day, and not to set any over one week old if possible. Both the very large and very small eggs should be discarded for hatching purposes, and only the well-shaped ones free from all blemish selected.

Hatching Nest.—When hens are to be used the nest should be formed on a turf or sod, cut about 18 inches square, and from 4 to 6 inches thick. From the under side of the sod scrape away a little of the earth and then turn it back, green side up, and press it down in the centre to form a hollow big enough to hold the eggs. Cover this with chopped hay, straw, or chaff, and the nest is ready for the eggs.

A coop of some kind should be used in the early months — one made with the front hinged near the bottom, so that it will fold down, is very convenient for the hen leaving or returning to her nest.

Little more need be said about the care of the broody hen, except that her food should be of hard grain, that she should be taken off once a-day and have fresh water to drink and a dust-bath to clean herself in. Before the chickens are due, both the hen and the nest should be dusted over thoroughly with insect powder.

Use of Incubators.

The use of incubators is now very general. They are practically a necessity in order to get chickens when they are wanted. The principle of artificial incubation harks back to the ancients of Egypt and China. Large ovens were used for the purpose in those days, and it is really remarkable under what circumstances a strongly fertilised egg will hatch.

Many years of careful study and experiment have brought artificial incubation to a very high standard of perfection. In the machines now most popular there are two methods of supplying the heat. In the one the heat is supplied from a hot-water tank, and in the other by means of hot air. Both have their advocates, the hot-water machines being more common in this country, and the hot-air machines in America. Fig. 733 shows an incubator made by Phipps.

If the temperature of the room in which the machine is working be liable to great variations, then the hot-water principle will probably work the better, but so much really depends on the operator that it is unwise to discriminate too closely.

Moisture in Incubators.—The regulation of moisture to the eggs during hatching is one of the problems which is not yet finally settled. The quantity of moisture in the air is constantly changing, and this complicates the problem.

In hot-air machines there is no direct supply of moisture, the theory being that the ingoing air is raised in temperature in the heater, and gains moisture as it gains warmth, till the degree of humidity of the warm air is relatively equal to that of the outside air. After the air is heated in the heater, it is passed into the top of the machine, whence it travels by diffusion through a felt diaphragm to the egg-chamber, and finally, still travelling downwards, it is ejected into the fresh-air inlet of the heater.

With hot-water machines the moisture is supplied from a water-tray placed immediately under the egg-drawer. The moisture-tray is covered with canvas, and the heat of the tank draws the air up by way of the ventilation holes in the bottom of the machine through the moist canvas to the eggs.

Much of the success of hatching depends on the regulation of the moisture. Some operators believe in dispensing with the water-tray altogether, or in putting it in about the eleventh day.

The room in which the incubators are to be worked should be thoroughly well ventilated, as the air in the egg-drawers must be constantly renewed for the proper development of the embryo in the egg.

Temperature in Incubators. — The temperature in incubators should vary as little as possible. Cellars are very often utilised as incubating-rooms, but it is usually difficult to get such places properly ventilated. Probably the safest temperature for the incubating-room is about 60° Fahr., and, as has been said, it should remain as uniform as possible.

The incubator must be set level, and on such a solid foundation that vibration will be avoided. The usual temperature recommended for the egg-drawer is 103° or 104°, but many operators now keep the drawer at 102° for the first week, 103° the second week, and 104° the third week, putting in the moisture-tray at about the eleventh day.

All well-made incubators are perfectly simple to work, regulating their heat quite automatically, so that no possible objection can be taken to them on that score.

The lamp must of course have attention. It must be kept perfectly clean and free from smell, but that is really a detail.

Necessity for Incubators. — The development of the poultry industry to meet modern demands can only be possible by an extended use of appliances. Early pullets are a necessity to supply the demand for fresh winter eggs, and spring chickens and ducklings must be hatched before the natural brooding time of hens arrives. Incubators are, therefore, bound to be more and more

required, and improvements in their construction may even yet be possible.

Testing Eggs. — The testing of the eggs is a matter of economy as well as of necessity. This can be easily done after the fifth day of incubation, and the sooner it is then done the better. Testing lamps are simple and cheap, so that even the novice can, with a few lessons, detect the germ in a fertile egg. The removal of the infertile eggs leaves more room in the drawer, which can be filled up if desired, so long as the fresh eggs are not allowed to touch the older ones till they have been heated up, and this is easily avoided by putting a strip of cardboard in a piece of flannel between them. The infertile eggs can also, if removed before they have been too long in the machine, be used as food for chickens. One method of utilising them, as has already been mentioned, is to boil them into a custard with milk, this being really a capital food for newly hatched chickens.

When the chickens are hatching, the machine should only be interfered with occasionally to remove the chickens to the drying-box. The less disturbance the better. After each hatch, the water-tray, egg-drawer, and canvases should be thoroughly washed and disinfected before another lot of eggs is put in.

Rearing Chickens Artificially. — Artificial rearing is the natural sequence to artificial incubation, and there are now a very large number of rearers and foster-mothers for this purpose, such as is represented in fig. 734. The rearers are again worked on the two principles of hot-air and hot-water heating. There is a sleeping chamber, well ventilated, and warmed by a hot-water tank or hot air. This sleeping chamber usually occupies about one-third of the whole rearer, the rest being without floor and wire-netted in the front. They can be used outside in all weathers, and are of simple design and easy to manage.

Artificial and Natural Rearing compared. — Chickens artificially hatched and reared do quite as well as those reared by the hen, and are not in the least more delicate, although a prejudice still prevails against that practice. Many claim that having no contact with the hen keeps the chickens free from vermin, which is in itself a great consideration, and also that the chickens get the full benefit of the food provided for them, and not the hen as often happens. Then with a machine there are no broken eggs or cases of desertion at a critical period. For convenience and economy, also, the balance is in favour of artificial methods.

MARKETING POULTRY.

There is often a great deal of waste through the want of a proper system of marketing. In the usual stock of farm fowls there are generally two kinds, the profitable and the unprofitable.

Unprofitable Hens. — It may be taken for granted that the young hens are paying their way, but too often there are a good many old hens which are not only unprofitable in themselves but which are eating away the profit the others are making. The question to decide, therefore, is: When does a hen cease to be profitable? Generally speaking, the answer is, after her second laying season is completed.

Assuming that a hen is hatched in March, she should in ordinary circumstances be marketed before the beginning of the August after she completes her second twelve months. The proper time is just after she has completed her period of laying, probably during June or July, and just before she begins to go into moult. To make sure a system of this kind is carried out, it is wise to mark each year's chickens with a ring on the leg. A brass, copper, or india-rubber ring does quite well, rings being made for the purpose.

The 1st of August sees the wild-fowl in season, and by the middle of the month the shooting is in full swing, so that fat hens are at a discount. The London markets, which really rule the prices all over the country, also invariably fall after August, so that there is nothing to be gained by keeping the hens over till Christmas. If this is done, the hens moult and fall into poor condition; they have to be fed up again while meantime laying no eggs, and Christmas markets are always glutted with foreign frozen poultry of all kinds. Moreover, these moulting hens are taking up the room of the younger birds,

and, owing to cold and other causes, a good many of them contract disease, which is easily spread, and a few always succumb.

Chickens for Christmas. — For the Christmas markets it is an excellent plan to bring out a batch of chickens about July and August, which can be put out to the stubbles to grow. These chickens pick up a large part of their food in the corn-yards during and long after the stacking of the corn crops. With a little extra food before Christmas, the young birds always command a very good price, and leave a handsome profit. For this purpose the Game-Orpington cross can hardly be surpassed.

Winter Eggs.

Just as there is a right time to market the old hens, so there is a correct period for hatching. Winter eggs are a sure source of profit if they can be got, and that is now largely a matter which can be controlled. Taking advantage of the winter-laying varieties of poultry we now possess, beginning hatching operations about the middle of February, and continuing till the end of April, there will be no difficulty, under proper management, in securing a good supply of winter eggs.

Grown under ordinary conditions, without forcing in any way, pullets will naturally begin to lay at from six to eight months old. Pullets hatched in February, March, and April will therefore, as a matter of course, begin to lay in September, October, and November, and they should continue laying till spring, when they will have earned a rest.

Early Moulting. — Another distinct advantage accruing from the hatching of pullets in the months mentioned is that they will moult early, probably in July and August, and so be ready for laying again in the winter months. It is quite possible to induce the moulting process by keeping the birds on short rations for two or three weeks, then shutting them up in an open-fronted shed, and supplying them with heat-giving food, such as hemp and linseed. The period of moulting is also shortened by such special treatment.

Laying Competitions. — A great deal of good has been done by the laying competitions which have been carried on from year to year. It is most satisfactory that the period of competition is now extended over a whole year. What is equally satisfactory is that a grant has been obtained from the Board of Agriculture towards the expenses of these competitions, thus for the first time giving them the advantage of official recognition.

These laying competitions have not only established facts regarding winter laying, but have indirectly yielded a vast amount of information on the different methods of housing and feeding.

Co-operative Marketing.

In the marketing of eggs this country still lags far behind some of her Continental neighbours, more particularly Denmark. Individual marketing, with all its inconveniences and losses, is still unfortunately the rule. Co-operation is slow to spread, even though our markets are practically controlled by eggs from other countries marketed on that system.

A great awakening must take place in this country before long if we are ever to attempt to supply our own markets with home-grown eggs. Were this country organised as Denmark is organised, the consumers who are willing to pay for them—and there are plenty of such—could depend on having on their tables every day guaranteed new-laid eggs not over three days old. With individual marketing, and the eggs passing through so many middlemen's hands as they do without organisation, the consumers do not know what they are buying. It is not to the credit of the British farmers that they allow their own markets, the best in the world, to be so largely at the mercy of the foreign producer when, by agreeing to combine, they could greatly improve this state of matters.

Co-operation in marketing poultry is not something new which has to be experimented with and tried with caution. It is already an established principle, ruling and guiding purchase and sale with manifest advantage to all poultry-rearers who have availed themselves of it. Under co-operation the eggs are collected, frequently tested for freshness, graded into sizes and colours, and marketed direct to the consumer. There

can be no comparison between the individual and co-operative methods in dealing with the distribution of eggs.

In poultry-keeping British farmers have an ideal industry awaiting development. It is not from large farms devoted entirely to poultry-raising that our egg-supplies are likely to be obtained. It is from small flocks at every farm and croft in the country. The industry is eminently suited for cottagers with a small piece of ground and for small holdings, and every one of these should have poultry as part of their regular stock. There is no soil so poor, no climate so bad, no situation so exposed, as to render impossible the keeping of hens, ducks, geese, or turkeys; and, as has been previously pointed out, there are plenty of each of these classes of poultry to choose from. Poultry, however, must get attention, and it is urged that the same intelligence and forethought which are devoted to other kinds of farm stock should be given to them. Old ideas must cease to dominate this branch of agriculture, just as they have been superseded in other branches.

PRESERVING EGGS.

Use of Waterglass.—The advent of waterglass has rendered the process of preserving eggs so simple and cheap that it is now adopted in very many households. Waterglass is an alkaline silicate which effectually closes the pores of the shell, rendering it perfectly air-tight. There can thus be no evaporation, and the contents of the egg are preserved for months in a fresh state. When the eggs are taken out of the preservative and wiped with a clean cloth, they look as fresh and marketable as new-laid eggs.

It is advisable that all the water which is to be used to dilute the waterglass should first be boiled to kill the germs. Almost any kind of vessels are suitable for storing the eggs, but probably wooden barrels or earthenware jars are the best. The liquid must fully cover all the eggs, and a cool place is best for storage. Full particulars for mixing the liquid are printed on each tin, and it is universally sold.

Lime-water.—Lime-water used to be the common preservative for eggs, and it is still very largely used on the Continent and in this country too where large quantities are dealt with. A useful recipe for the lime-water is: 2 lb. lime, 1 lb. salt, 2 oz. cream of tartar, and 6 quarts of water.

The lime-preserved eggs are almost exclusively used for kitchen and cooking purposes, but it is claimed that by the waterglass method the eggs can be kept in a state fit for use on the table as boiled eggs. To prevent the shell of eggs thus preserved from cracking when being boiled, it is usual to prick the thick end of each egg with a needle.

Cold Storage.—Cold storage is also well adapted for preserving eggs, though evaporation is not prevented by the process. This method is, however, suitable only where very large quantities are handled. Both in the cooling down of the eggs and in the returning to the natural temperature a good many of the shells are apt to get broken.

Essential Conditions in Storing Eggs.—With all methods of preserving eggs the observation of the following rules is essential to success:—

1. The eggs must be perfectly fresh when put in.

2. Only eggs infertile and without flaw of any kind should be selected. Thin-shelled eggs should never be preserved.

3. Store in a cool place, as free from vibration as possible.

DISEASES.

In dealing with diseases of any kind, it cannot be too strongly emphasised or too often reiterated that "prevention is better than cure." Particularly is this so with poultry, because very often the disease is too far gone for cure before it is found out.

At farms where the fowls have their liberty and plenty of scope to roam, with healthy surroundings and a good supply of natural food, disease should practically be non-existent. Almost every outbreak that occurs can be traced to bad management in some form or other. Common causes are—

(a) Cold, damp, badly ventilated, and dirty houses;

(b) Over-feeding on too nutritious or fat-forming foods, such as maize and potatoes, and impure water;

(c) Overcrowding in a bad atmosphere;

(d) Want of healthy exercise, due to an improper system of feeding;

(e) Injudicious in-breeding.

Infectious Diseases.—In an ordinary farm stock, where the birds are not worth more than a few shillings each, by far the cheapest and most effectual plan is to kill off any bird which shows the slightest signs of having contracted an infectious disease, and one of the first precautions against disease breaking out is to see that every bird that is being bred from is in sound health. Further, if the principles laid down in the foregoing pages relative to the cleanliness and ventilation of the houses and the feeding of the fowls are adhered to, the chances of disease breaking out are remote in the extreme—more particularly if the colony system of housing is adopted.

Vermin.—Some of the worst plagues of the poultry-yard can scarcely be described as diseases, and one of the commonest of these is vermin. Unless fowls are minutely examined, particularly round the rump and under the wings, it is impossible to believe how badly infested they may be with insects. The presence of insects is a serious cause of loss both directly and indirectly. The constant irritation to the skin set up by these active workers is very often the cause of broody hens breaking their eggs; and further, the growth of the young stock is much retarded and the system so reduced that the way is paved for disease.

Bird-Lice.—Probably the most prejudicial kind of poultry parasites are the bird-lice. Eight species of these are found on the fowl, four on the duck, five on the goose, and three on the turkey. They may be said to spend most of their lives on their hosts, though certain species may live in the nests part of the time. These lice do not suck the blood, as is sometimes supposed; but they have a true biting mouth by which they gnaw away at the roots of the feathers, the scales, and the skin itself. On chickens these lice have a most injurious effect, and naturally the tender skin of the chicken is chosen for their attentions rather than that of the adult hen.

The simplest way to check the ravages of these insect pests is to have a dust-bath, as already described, always handy, with some strong insect powder mixed in it. Finely divided gypsum mixed with a small quantity of paraffin or carbolic acid is very effective for these dust-baths, and soon gets rid of any insects which the birds cannot reach.

Broody hens should always be treated before being put on the eggs, and also before hatching, and so should each individual member of the flock occasionally.

A strong insect-powder such as Keating's, or two parts of that to one of powdered sulphur, well dusted into the feathers—more especially round the tail and under the wings — will effectually kill all insects. This treatment, however, has to be repeated, because the eggs of the insects are laid mostly round the roots of the downy feathers, to which they are attached by numerous fine threads, and in six to ten days the eggs develop into young lice.

Mites.—Mites are another source of trouble, but their haunts are the cracks and fissures of the perches, nest-boxes, &c. Hence the necessity for lime-washing all the wood-work regularly, and painting over with kerosene or spraying with dilute carbolic acid.

A very small tick-like mite also attacks the heads of chickens, and this attack can be met by a very small dressing of mercurial ointment or white precipitate, or by dressing with olive-oil to which a few drops of paraffin have been added.

Gapes.—Gapes is perhaps the worst scourge in the poultry world. It is due to the presence in the windpipe of a number of very small worms, which kill the affected fowl either by wasting or actual suffocation. The symptoms are yawning and stretching of the neck, a wheezing cough, and a frothy saliva oozing from the mouth. The surest preventive of this fatal disease is to use fresh ground every year for the rearing of the chickens, and to colour the water two or three times a-week with permanganate of potash.

When the disease does occur, the ground very soon becomes contamin-ated and the whole flock may be affected. One remedy is to put the affected chickens in a box and fumigate them with the fumes from carbolic acid. The vapour from burnt sulphur is also fairly effective. A little camphor added to the drinking-water is also a safeguard.

The disease is frequently connected with a large insect found on the heads of newly hatched chickens; and it seems to be established that gapes will not break out if the chickens' heads are anointed with the following ointment: mercurial ointment, 1 ounce; pure lard, 1 ounce; flowers of sulphur, ½ ounce; crude petroleum, ½ ounce. The ointment is gently rubbed in after being warmed to semi-fluidity. On clean dry ground, however, the disease seldom appears.

Scaly Leg.—Scaly leg is another common disease for which there is little excuse, as it is so easily cured. It is caused by an insect burrowing under the scales of the leg. The treatment is to wash the legs thoroughly in warm water, using carbolic soap and a hard nail-brush to get well under the scales. After drying thoroughly, rub well in sulphur ointment or creosote and lard ointment (1 to 20). Another cure is to boil equal parts of paraffin and water, and add a little soft-soap, rubbing this in under the scales after washing as described.

White Comb.—White comb, or favus, is another noxious disease, often caused by overcrowding in a dark, damp house. It attacks the comb, wattles, head, and neck, which appear crusted with a whitish-like growth. The method of treatment is to bathe the infested parts with warm water and soft-soap; then apply either red oxide of mercury ointment (1 part of mercury to 8 of lard) or sulphur ointment with a few drops of benzine, just enough to moisten the sulphur before mixing it with the lard. Iodine is also said to be successful.

Roup.—Roup and diphtheric roup are the most troublesome and loathsome diseases with which the poultry-keeper has to contend. The symptoms are easily detected, as there is an offensive smelling discharge of white cheesy like matter from the nostrils and mouth, and the bird is highly fevered. There is also often a swelling round the eyes.

As this disease is highly contagious, the first thing to do is to isolate at once any bird affected, and disinfect the drinking-troughs, &c., it has been using.

In *diphtheric roup*, which is really distinct from common roup, the inside of the mouth and round the tongue will have small patches of matter growing which have to be scraped off with a quill or blunt knife, and the place anointed with an antiseptic such as salicylic acid. The mouth must be washed out by using cotton wadding attached to a small stick of wood dipped in peroxide of hydrogen. Sometimes hard white spots are found, to remove which lunar caustic will have to be used. To reduce the swelling round the eye, foment with hot water and drop into the eye a little powdered borax. Sometimes in bad cases the swelling has to be opened and the cheesy matter extracted.

The general treatment is to keep the bird in a warm, dry, airy room, giving a laxative in the form of half a teaspoonful of Epsom salts or castor-oil. Give the soft food seasoned with a little cayenne pepper, and administer a copaiba capsule a few hours after the laxative.

Specially prepared roup powders are now sold by most poultry chemists, and if these are given as directed on the first symptoms appearing, further trouble is often avoided. A few days' quarantine after cure is essential.

Liver Disease.—Liver disease, though more often associated with the larger breeds of fowls, is yet common to all, and is generally brought on by injudicious feeding on such heavy foods as Indian corn and potatoes, with insufficient exercise. The symptoms are moping, and a dark purple colour about the head. If the bird is handled it feels heavy, and if held head down for a minute or two, it will turn almost black, sometimes collapsing altogether.

A simple cure for a hen affected by this disease is to give her a sitting of eggs to hatch and let her rear the chickens. The fat in the body gets reduced, and in a manner the whole system renewed.

The treatment is to provide as much space and exercise as possible for the affected birds. A good dose of Epsom salts should be given. If given dry, in crystal form, which is the best way, a piece about the size of a marble to each bird is a good dose.

The following recipe by a well-known authority can be strongly recommended: "Get one pennyworth of gentian root, ditto of powdered rhubarb, ditto of bitter aloes, ditto of black Spanish, ditto of best cayenne. Add the above to one quart of water, and simmer down to a gill. Then strain through a fine sieve and let it cool. Boil till the nature is out of the herbs, when it is ready for use. Give eight to ten drops in a tablespoonful of water three times a-day for a week. Give also plenty of green food and grit, and a few cod-liver oil or chemical capsules."

A stock of birds affected with liver disease should not be bred from. It is far better to kill them and have a fresh lot put in. This is a disease for which there is no excuse, as it is so easily avoided by proper feeding.

Tuberculosis of Poultry.

In regard to tuberculosis, which is one of the most common diseases of fowls, turkeys, pheasants, and other birds, the following useful information is given in Leaflet No. 78, issued by the Board of Agriculture:—

Symptoms.—Affected fowls become anæmic, thin, emaciated, and they lose weight. Their appetite is impaired, and erratic feeding is noticeable. The comb and wattles and mucous membranes become pale, and there is usually persistent diarrhœa. As a result of extreme emaciation, which is the most noticeable symptom, the bones become very prominent.

Post-mortem Appearances.—The flesh is scanty and the muscles pallid. The liver is dotted all over with small pale spots, or larger patches of a white, grey, or yellow colour. The spleen is usually enlarged and beset with small or large tubercles. The intestines and the lymphatic glands of the mesenteries may be also the seats of tubercular deposits. Tubercles may likewise occur on the skin. There are very rarely small tubercles in the lungs.

Cause.—The exciting cause of the

disease is a bacillus which may be considered a variety of the bacillus of mammalian tuberculosis. It gains entrance with the food, fouled by means of droppings of affected birds.

Prevention and Remedy.— 1. The most frequent source of infection is the poultry - house or yard, which receives the droppings of the affected birds, these droppings containing bacilli. Damp, dirt, and absence of sunlight greatly favour the spread of the disease. It is necessary that there should be good ventilation and strict cleanliness in the runs and sheds.

2. All diseased birds should be killed and buried in lime. The house where they have been should receive several applications of disinfectant, and the tainted run should be dug over and heavily dressed with quicklime.

3. Many months should elapse before birds are put back in old quarters that have been cleaned. It is best to clear off all stock where this disease breaks out, and make a fresh start with new stock later. Strong and healthy birds should be carefully selected and put into a new house and run, and if any show indications of disease, they should be removed at once and the house disinfected with chloride of lime ($\frac{1}{4}$ lb. to 1 gallon of water). In this way a disease-free stock may be obtained, and until this is accomplished all that can be done is to observe all possible sanitary precautions.

Vices in Poultry.

Poultry have, unfortunately, a few vices which are as troublesome as the diseases.

Egg - Eating. — The habit of eating eggs is a common vice not easily detected or stopped. The habit is usually acquired from the devouring of a broken egg, thus creating an appetite for more. If the criminal (for there is usually just one real culprit) can be caught, the best cure is to twist its neck. The absence of grit, oyster-shell, and lime is given as a cause of egg-eating, and certainly these should be supplied in plenty. But the vice will appear even where there is no want of these substances.

A simple preventive of egg-eating is to have a good many nest-eggs lying about, so that these may get the attention of the culprits and disgust them. Another plan is to blow the contents out of an egg and fill it up with mustard, alum, and cayenne-pepper, so as to give a lesson to the hen which breaks it. Nests are also constructed so that the egg when laid rolls out of sight, but with big flocks the surest and best way is to execute the criminal.

Feather-Eating.—Feather-eating is a much commoner vice, though more prone to occur where the birds are cooped up. Probably the habit is caused in the first place by insects, but other causes are usually at work as well. A feverish state of body, through want of a plentiful supply of green food, or a craving for animal food, are undoubtedly predisposing causes. The cock often suffers, too, through the hens pecking at his comb and wattles till he is a pitiable object.

The treatment for feather-eating is to isolate any bird attacked, and see that the flock gets a regular supply of green food; also twice a-week, at least, some animal food, either raw or cooked flesh or green bone. The affected parts of the birds attacked should have carbolised vaseline well rubbed into them. This will cure the wounds, and at the same time prevent any more feathers being pulled out.

There are, no doubt, other simple and complex troubles which arise in the path of the poultry-keeper, but there is now an ample supply of literature available on almost any specific subject. No better medium of information can be wished than the weekly penny journals specially devoted to this subject, through whose columns information on any particular matter affecting poultry can be had for the asking.

BEE-KEEPING.

The keeping of bees is not only, as a rule, a profitable industry where it is conducted with skill, but is also one of absorbing interest and fascination. Originally the following notes on the subject were prepared for this work by the late Mr William Raitt, Beecroft, Blairgowrie. By another capable bee-keeper they have been revised for this edition.

Bee-keeping as a Farm Industry.— It is undoubtedly the case that bee-keeping ought to receive more attention as a farm industry than has hitherto been devoted to it. In many instances it has been cultivated as such with the best results. It is an industry peculiarly adapted for a place on the farm, as is indicated by the ancient and sacred association of "milk and honey." The same pastures yield both—though, alas! the latter is too often left to waste its sweetness on the air.

In America and many Continental countries bee-keeping already occupies a prominent place among rural industries, and is generally most successful when associated with farming. A few regions, like San Diego County in California, the Basswood tracts in other States, and, to a degree, our own heath-clad hills, afford unlimited natural honey-yielding bloom.

Clover for Bees.—But more generally success depends on the neighbourhood of clover-fields. Than these there are no better pastures for bees, as every farmer must perceive when he hears the joyous hum of *other people's* bees rollicking amongst *his* clover heads.

These "small cattle" are so independent of fences, that in a notice of the sale of an apiary there was added after the inventory of hives the words, "with unlimited right of pasturage." But just because these cattle are so small, they are often neglected. One forgets, however, that what they lack in bulk they compensate for in energy and in strength of numbers, so that the results of their united labours are, under proper conditions, out of all proportion to their "stature."

Bees v. Shorthorns.—Some years ago the writer was at tea in the company of several farmers, who chaffed him not a little on having a "bee in his bonnet." *Their* talk was of shorthorns. "I'll tell you what it is," said I, "I have a single bee at home that has this year put more money into my purse than the best shorthorn cow you have has put into yours." I of course referred to the queen-bee of one of my hives, the mother of all its inhabitants. It so happened that I had that season taken from that stock no less than 130 lb. of first-class honey, in such splendid condition that I sold it to a dealer, after winning a handsome prize besides, for £10, 16s.

Produce of Hives.—It is but fair to say, however, that that result was exceptional, though I have several times greatly exceeded it in quantity since. For instance, I had in one season from a single hive 204 lb. of bottled honey of first-class quality, and an almost equal amount from a hive the year before, and all without killing the bees or interfering with their necessary winter stores. These figures indicate the possibilities that lie in bee-keeping—though, taking one season with another, I should estimate the average produce of a well-managed apiary at from 30s. to 40s. per hive.

Commencing.

The times are propitious for commencing this industry.

Improved Practice.—A great revolution has taken place in the practical management of bees since the "seventies" of last century. The old straw skep and brimstone system have been improved away, and the new humane and profitable movable comb system has taken its place.

After many years' experiments with mixed success, the best form of hive and system of management became pretty well fixed. The era of experiment is past, at least to a large extent, and everything has been greatly simplified.

Cheap and Improved Appliances.— Not only so, but while in former years new hives and appliances were rather expensive articles, they are now very moderate. I remember when no hive was considered good for anything under £1 or 30s. Now they can be had for half the amount, and simpler forms for a good deal less—so simple, that with one as a pattern any handy man can make his own hives.

Marketing Honey.—Moreover, the chief initial difficulties connected with making a market for honey are overcome. It has become a staple article of trade in the best shops of all our large towns. To be sure the price, like that of all other sweets, has come down in late years; but even yet it has not fallen to the price that used to be considered a fair one for old-fashioned skep honey, and it is not likely to come lower.

Bee Information.—And lastly, information is now more easily attainable than ever it was before. Besides weekly and monthly journals entirely devoted to bees, most agricultural and horticultural weeklies have columns devoted to the industry and to the queries of correspondents. And special handbooks and more elaborate volumes are easily obtained.

Exhibitions illustrative of the whole art and mystery are held annually in connection with the shows of most of the leading agricultural societies in the three kingdoms, and at many local shows besides. Then almost everywhere a handy man can be picked up who will be delighted to tell all he knows, and give all the help he can to intending beginners.

Knowledge necessary.—Bee-keeping as much as sheep-farming and other rural employments requires the application of a good deal of acquired information. One may, however, commence practice and the study of principles at the same time—that is, commencing on a small scale, and increasing one's stocks as one's knowledge and ability advance. The limits of space here forbid anything more than a digest of the knowledge any one may easily acquire more fully from books and experience.

In regard to books, beginners should be careful to get only the latest editions of the latest published works. The be-

ginners should on no account allow themselves to become enraptured over any particular form of hive recommended by the maker. Study the latest information obtained from a disinterested quarter, and then judge for yourself what would best suit the object you have in view in the way of system and appliances.

After having thus formed a decided plan of operations, there need be no objections to reading any good works on bees, with a view to obtaining more scientific knowledge than most handy manuals can afford to give. Much also may be at the same time learned, and more especially in the art of handling bees, by a visit to some successful bee-keeper.

Principles of Bee-keeping.—As some guide towards judging as to the suitableness of any reading that may be undertaken, we give the following condensed summary of what we consider ought to be learnt from it: that modern bee-keeping is an art founded on strict scientific principles; that it can be depended upon, weather alone permitting, for yielding certain fixed results, as surely as can any other industry about a farm; and that to enable one to use his scientific knowledge to advantage, hives must be adopted that give every facility for controlling all the operations of the bees, and for assisting them by the use of comb-foundations and other modern aids.

Hives.—Such hives are variously called bar-frame or movable comb-hives, and the tendency is towards great simplicity in these. The books and dealers' lists may, with great plausibility, recommend costly hives with elaborate fittings and adjuncts; but for profit and convenience none excel those that consist of simple box bodies fitted with plain frames with roof and floorboard. To allow of tiering up, with a view to the production of either comb or extracted honey, the bodies should all be exactly alike, and so fitted as to sit accurately one over another. That is, one may have any number of bodies or stories in use as a hive or stock, though with only one roof and floorboard. Hives with fixed legs should specially be avoided, any plain stand being substituted.

Appliances for Special Conditions.
—The student ought also to learn that, in certain localities and under certain circumstances, it may be better to adopt appliances specially with a view to producing comb-honey, this especially where heather is plentiful; or that it may be better to work for extracted honey, as may be in most demand; or to work for both — say for clover-honey to be extracted, and for heather - honey in the comb.

Study Surroundings.—At the same time he ought to have his observing powers at work, more especially noticing the favoured bee-flowers peculiar to his neighbourhood, and their period of bloom. This knowledge will greatly aid him in forming his plan, for one of the great secrets of success is in having one's stocks in the very best condition, just when the prevailing honey-flow comes on, and not either still weak from spring neglect, or what is almost as bad, weakened by swarming after having been strong. The peculiarities of his location as to climate and exposure also merit attention. And as a result of all, he must make up his mind whether he can afford to give his bees the necessary time and attention, and in what particular direction he shall go to work.

Caution in Practice.

Obtaining Stocks.—Should such preliminaries chance to occupy him during the winter or early spring months, he may at once look out for the needful stocks. If these are already on hand, even though domiciled in ancient straw-skeps, so much the better; otherwise he may easily obtain by purchase one or more such. These are usually to be had so much cheaper than stocks in modern hives, and the experience gained in the course of working them into the new system is so valuable, all the more so because it compels him to "go slowly," that on the whole we generally advise beginners to commence with such.

By exceptional diligence in gathering information, and with that knack of managing live stock that many have as a peculiar gift, it might be safe enough to embark boldly in a wholesale fashion at first, but generally we recommend caution.

"Bee - fever." — Few become really successful bee-keepers until they have at least one whole year's experience, and it is better to try and control the "bee-fever" than to let it run riot, to the imminent danger of collapse and misfortune.

Appliances.

The needful appliances are by no means so numerous or costly as some of the many large and finely illustrated price-lists issued by dealers may suggest.

To begin with, at any rate, one's wants may be sufficiently met by the possession of a hat-veil, a smoker, a supply of hives, with the necessary frames, crates, and sections, and a stock of comb-foundations.

Hat-veil.—The veil is simply a yard and a half of black hexagon net, sewed up one seam with an elastic band, to go round a broad-brimmed hat, the lower edge to be tucked away inside the vest.

Smoker.—The smoker is a bellows contrivance for burning rags, brown paper, or touchwood, in such a way as to permit of directing a stream of smoke upon the bees when they are to be handled. A loosely tied roll of rag (corduroy or moleskin is best) may serve a turn instead, or the fumes of tobacco may be utilised by those who can use the pipe. This frightens and quiets the bees.

Hives.—The hives, as already hinted, should be of simple construction, each body made to hold not more than eleven frames.

The frames should be of the standard size used in the neighbourhood, hung in the hives, so that ten of them occupy a space of $14\frac{1}{2}$ inches, that being also the dimension of the hive the other way. We prefer eleven frames, so that our hives inside measure $14\frac{1}{2} \times 16$ inches, and are deep enough to hold the frames suspended, with the necessary bee-space below and around.

This size of hive is just about right for permitting ordinary-sized crates of sections to be piled up inside the upper storeys.

Sections. — Sections are those neat dovetailed boxes to hold one or two pounds of honeycomb, and are generally imported from America, and sold by dealers very cheaply.

Crates.—Crates are the bottomless boxes or trays in which the sections are arranged in groups of 21 or less, accord- to their size.

Comb - foundations.—Comb-founda- tions are sheets of bees'-wax impressed with the exact form of the cells as made by the bees. These are turned out by special machinery, and are a great help both in supplying the bees with material of which to build combs, and in com- pelling them to build them straight in the frames or sections where wanted, at the same time putting it in the power of the bee-keeper to limit the production of useless drones.

Other Appliances. — A few other minor appliances might be found useful, though not absolutely necessary, such as a queen cage or two, some queen-exclud- ing zinc, bottle - feeders, and a honey- knife. The cast carpets or blankets about the house will supply all the quilts needed for a commencement.

Honey Extractor.—The question of having the rather expensive machine for emptying combs without breaking them —called the honey extractor—may be deferred till experience warrants the expense.

Management—Preliminary.

Driving Bees.—The first concern of those commencing should be, as soon as may be best, to get their bees domiciled in the new frame-hives. It is quite easy for experts to transfer both bees and combs from the one to the other at al- most any season. The bees are "driven" into an empty skep, according to direc- tions in the book referred to ; the combs are then cut out, and 'pieced and tied into the new frames ; these, with the bees, are then placed in the new hive, when they soon fix all nicely up.

But we advise rather to await the natural swarming season, when either swarms may be allowed to come off or the plan afterwards described adopted.

New Swarms.—If natural swarms be got, they should be treated thus : the first that comes off should be placed in the new hive on the stool where the skep stood, the latter being removed to a new location. This causes many more bees, accustomed to the old place, to join the swarm and strengthen it. The likelihood

is that the skep will not swarm again. Should it do so, the swarm should be returned, and more ventilation given, as a preventive, till the 21st day from first swarming, when all brood will have been hatched out.

A second good bar-frame stock can now be had by driving all the bees and transferring any combs found straight and sweet. On no account would we advise more than two stocks to be made from one.

Another Plan.—The other plan is to set the skep when crowded with bees on top of a new hive fitted with comb- foundation, compelling the bees to work downwards through a 6-inch hole in the quilt, by closing their old entrance. If it be done at the right time, the bees will generally have some combs worked out below within a week, when an ex- amination should be made of these to see whether the queen has gone below. The presence of eggs in the cells may generally be accepted as proof sufficient, but we should prefer in all cases to see her majesty. This being so, the skep may be lifted off and set in a new loca- tion, to be afterwards treated as if it had swarmed naturally, as before described.

Rapid Increase of Stocks.—To those anxious to increase their stocks as much as possible, it is a good plan to rear or purchase spare queens, so as to be able to introduce one into each skep as soon as it has been removed from its old place and queen. In that case the same pro- cess of stocking new hives may be carried out at the rate of one every fortnight or three weeks during the honey season.

In backward and ungenial seasons less must be expected, and, indeed, it is com- mon to leave the skep in place on the first hive until all its brood is hatched out, when it is taken and treated as a honey super.

Purchasing Swarms. — Some may prefer, or have no alternative but to make a start by purchasing swarms wherewith to stock the new hives. These should be secured as early as possible, say by the first week of June in the south of Scotland, and a fortnight later in the north. They ought to weigh not less than 4 lb., an ordinary top skep swarm, though 6 or 7 lb. are usually had in a swarm from a good frame stock.

Collecting Driven Bees.—Still others may adopt the more economical though more troublesome plan of gathering up driven bees in the autumn, and by joining these into large colonies, and feeding rapidly with bottle syrup, get them into good shape before winter. Any one having learned the art of " driving," and having the soft side of cottagers who are going to brimstone their bees, may generally have them for the trouble of driving, though in most localities the cottagers are getting too knowing to give away what they may as well learn to use to their own benefit.

In whatever way obtained, let us suppose the reader to have in the autumn several good stocks of bees in modern hives. We would now indicate in the order of the seasons the system and treatment we consider best for him to adopt.

Wintering.

Secret of Success.—The great secret of successful wintering is in keeping the bees in as quiet a state and as constant a temperature as possible. Of course abundant supplies are the first consideration to this end, the next is careful packing and ventilation, and the third is to let them rest free from the least disturbance till the first of spring.

Preparing for Winter.—A warm day late in November is our chosen time for arranging hives for the winter. If made very comfortable long before this, the bees incline to fly too much and to dwindle. But left just as they were after the honey harvest, they have free ventilation and plenty of room, never get too warm, and stay more at home. As steady cold weather approaches we need not be so afraid, and so we choose such a day as mentioned to make all trim and comfortable.

Armed with smoker or other quieting agent, a bag of chaff, a quantity of extra pieces of carpet or other quilt materials, and some flat cakes of "bee-candy," we set to work. Hives still containing bees on every comb, or nearly so, we do not disturb further than to lay a cake on top of frames, cover closely with several thicknesses of quilt, and over all, if the make of the hive permits, pour a few inches of loose chaff, or stuff in a chaff cushion. The doorway is left full

width, or at any rate not under six inches long.

The candy is given not solely to increase the supply of food, but because it supports the coverings, so that when eaten away there is a nice warm cavity left that forms the best kind of winter passage from one frame space to another.

Weak Hives.—Weaker hives, containing bees on six frames only, or under, are contracted by removing all the outside beeless combs, inserting division-boards next the remaining combs, and filling the spaces with chaff. Otherwise they are treated as before.

Very small stocks are united two and two, though this should have been done in autumn.

For the rest, no further attention is required till spring, unless one chooses to keep the snow well cleared away from the ground in front, and to watch on sunny days when the snow is soft, keeping the bees at home by heaping soft snow over the entrances. This shades and cools the hive, and affords the necessary water to the bees that are trying to get out to find it.

Bees not shut in.—On no account should bees be actually shut in, as they often get into such a state as to suffocate. Only *tempt* them to stay at home when it is dangerous for them to be out.

Experiments.—Quite probably the experiments we are conducting in the line of cellar-wintering, or by burying the hives in pits or clamps, may result in an improved system in that direction, which is so much in favour in America.

Spring Treatment.

Provided all goes well in wintering, there is really no necessity for disturbing the bees during early spring.

Breeding resumed.—They naturally recommence breeding about the New Year, and their stores thereafter more rapidly diminish; but they ought to have sufficient left them in autumn to carry them through till the first new honey is to be got, or till gooseberry and fruit-trees are in bloom.

Supplementing the Winter Food.—Wherever there is any doubt as to the supply of food, it is our custom to take a peep into all stocks on the first fine day when bees are flying. We are loath to disturb the winter packing, which is of most value when the bees are breeding with diminished numbers in spring. We therefore simply raise the packing and quilts along the back edge of the combs, when it is possible to see whether there remains still a quantity of sealed comb in at least the most of the frames. If so, all is well so far as food is concerned, and it is too soon to inquire into other matters.

Where there is an evident deficiency in food, there must be a more thorough examination, and any want supplied, either by giving back any combs of honey reserved for this purpose, or by laying a cake of candy under the quilt.

Liquid Food.—Liquid food should not be given unless in desperate cases, when it may be poured into empty combs and hung in the hive.

Stimulating Stocks.—Later on, say when willows are in bloom, it will be of advantage to contract the brood-nest by removing all beeless combs and closing in the division boards, though many think it better to leave them alone. All depends on whether the district is one for very early honey, making it necessary to stimulate the bees by every means, so as to come to full strength before the honey season opens. With us the clover is the main harvest, commencing on an average about the 15th June, and our average stocks usually come to swarming strength by that time without any special stimulation, and thus the energies of the queen are conserved for keeping up the population till the close of the harvest.

Stocks stimulated to undue exertions early in the season are more apt to swarm excessively, and thus to imperil the honey returns.

Continuous Treatment. — As the bee-keeper's summer may be considered as commencing with the swarming season, or say from June 1st, we may add that whatever style of treatment may be adopted, in view of getting hives filled with bees and brood, should be continued without intermission till that period arrives. That is, care must be taken to see that once the bees have got started in earnest to brood-rearing there should be suffered no check from want of food or room. Both should be given in moderation, yet continuously; when plenty of natural stores are coming in, leave well

alone, but supplement these either by bottle-feeding whenever the weather is unsuitable for outdoor work, or by uncapping portions of their sealed stores every day or two.

Pea-meal may be given as an equivalent or supplement to natural pollen when that is deficient, the meal being sprinkled on shavings in an old skep set to face the sun in a sheltered corner. Room need only be given where combs have previously been removed, by adding one at a time in the centre of the brood-nest, as the bees are able to cover all closely. So soon as the hive is full of bees from side to side, with brood in every frame, the summer treatment should begin.

Summer Treatment.

It should previously be matter for consideration and decision whether the various stocks are to be worked for (1) increase, or (2) honey.

Working for Honey.—If the latter, it has to be decided whether it is for extracted or comb honey. Every preparation should be made accordingly. New hives, ready fitted to receive swarms, should be prepared beforehand, upper storeys filled with spare combs or foundation for extracting purposes, and crates ready fitted with guided sections for comb honey.

Working for Increase of Stocks.—If increase be wanted, some such plan should be followed as indicated on preliminary management.

Extracted or Comb Honey.—As to whether one should aim at getting extracted or comb honey, each must discover for himself which is likely to be more saleable in his district. We may, however, indicate our opinion that, generally, extracted honey is likely to be more in demand than comb. They are rapidly approaching each other in price, the former being obtained with more ease and certainty, and in perhaps a third greater quantity. It is in demand all the year round, while comb unfortunately has its "season."

The Writer's Practice.—Our own practice, adopted after many years' experience, is as follows: We work for honey, but allow a moderate natural increase, partly to ensure our having old queens replaced by young ones, partly to keep up our stock, so as to permit of doubling up weak colonies, and partly to allow the bees a little of their own way, which seems to keep them in better heart for work. That is, we do all we can towards getting honey, and in doing so to *prevent* swarming; but as occasional swarms will come off in spite of us, we do not try to thwart the bees by returning these, but make the best of them, by giving them a good start on combs ready built, or on combs of brood and foundation.

If second swarms issue, we cut out all royal cells and return the swarm.

By placing first swarms on the old stool, they are made stronger by the old bees returning to their accustomed place, and the removed stock is so weakened that it does not often swarm a second time. Sometimes we break up the latter, giving nearly all the bees to the new swarm, and dividing the combs of brood amongst those not yet at full strength. Of course we cut out royal cells, in case they may tempt the other stocks to swarm.

Controlling Swarming.—To prevent swarming, or at least reduce it to the lowest as a natural impulse, we find it generally enough to see that the bees have plenty of doorway and plenty of room for storage and for clustering inside.

This room we give them by tiering on upper storeys of combs for extracting, or of crates of sections, and this as long as the honey season seems to warrant. That is, from experience we know about what date the honey - flow, say from clover, usually ceases, and we take care not to give more accommodation than is likely to be made use of.

This is important when finished comb honey is wanted, though of little consequence if extracted honey is the object. The latter can be taken at the close of the season, whether in full-finished combs or not.

Securing well-ripened Honey.—To get either extracted or comb honey well ripened and sealed, we require at least two upper storeys or two crates of sections to each hive. As soon as the first put on is well forward, and the bees need more room, we raise it, placing the empty one between. If the latter have

foundation only, the bees are compelled to store all their honey for a day or more in the upper story, which generally ensures its being well finished.

Produce.—Towards the close of the season we place the empty tier uppermost, as the other has more chance of being finished off when left next the brood-nest. By careful calculation, and with favourable weather, we thus get from good stocks from 50 lb. to 100 lb., and often more, of nice comb honey each, and from others 150 lb. to 200 lb. of extracted honey.

For details of how to manipulate the bees and combs when harvesting the honey, or of using the extractor, and preparing the honey for show or market, and for other minute matters, the reader must seek in books and journals specially dealing with bees.

Autumn Management.

In many districts the autumn treatment includes part of the honey harvest —viz., the heather.

Heather Honey.—Usually a week or ten days intervene between the close of the clover season and the time that heather yields. Where this most magnificent of all honey is to be had, special pains must be taken to secure it.

The secret is, barring the weather, to have only strong stocks, and to make them warmer by soft coverings than during the earlier season. Where swarming has been allowed *ad libitum*, neither swarms nor old stocks are fit to do much in the way of surplus. Stocks previously worked for extracting are best of all. They have always more bees left than those which have been worked for comb.

There should be some change in the plan of working these — that is, comb honey only should be sought from heather. Heather honey will not leave the combs in the extractor, but has to be broken up and pressed; nor does it sell so well as in the comb.

There should be no more room given than the bees can crowd comfortably into, as the nights are chilly, causing them sometimes to desert the supers.

After Honey Harvest.—The general autumn treatment for stocks after the honey harvest consists mainly in doing all one can to keep the bees quiet, and so prevent robbing.

Bees Plundering. — Not a drop of honey or bit of comb should be left anywhere within their reach, for if once started, the bees get on at once for plunder; and so vicious do they then become, that the apiary is a place to be dreaded by man and beast. As soon as all surplus honey is taken, and that under every precaution, all hives should be closely though not warmly covered, doorways contracted a little, and left alone till early winter.

Necessary operations should be done towards evening, when flying bees have all gone home. If food be needed, either as a result of a poor season, or of the honey having nearly all been stored in supers, it should be given rapidly as soon as the supers are taken away, and before the time of dearth and robbery has come.

Queenless stocks should be attended to, weak hives united till strong, and all left to settle till the time for winter treatment arrives.

Food for Bees.

Liquid Food for Bees. — Boil together 5 lb. white sugar and 1 quart of water; a few minutes' boiling will suffice. It is improved by boiling with it a pinch of cream of tartar. This is the proper food for autumn. Spring food may have a half more water, and the tartar omitted.

Sugar-cake for Bees in Winter. — Boil together 5 lb. white sugar, less than a pint of water, and a pinch of cream of tartar, until a drop cooled on a plate stiffens so as to draw out as a thread. Take off the fire and set in a cool place, or in cold water, stirring briskly until the mass begins to cool and turns white and thick. Then pour out on thin sheets of paper laid in flat dinner-plates. When cold, the cakes should be white and firm, yet not hard.

Spring Food. — For early spring food, a handful of flour for each pound of sugar may be stirred in shortly before pouring out. These cakes should be slipped under the quilts, paper side up.

SHEEP-DOGS.

The collie dog is well entitled to mention amongst the live stock of the farm. He is a faithful and worthy servant, absolutely essential upon sheep farms.

Origin of Collies. — The origin of collies is not very clear. Darwin has stated that the type approximates more closely to the old feral type than does any other of the domesticated varieties of dogs. But it is extremely probable that the collie as we know it to-day is a created race, although the work of moulding the different types must have taken place very early — before probably some of our other breeds of dogs were much known, or even in existence.

The name Collie is believed by many to have been derived from the association of the dogs with certain Highland sheep which were known at one time as colleys on account of the black colour of their faces and legs. Others have held that Collie is simply a variation of the words Cooly, Colley, or Coley, signifying "black." Webster, in his dictionary, gives the derivation as from the Gaelic *cuilean*, a whelp, puppy, or dog.

Whatever may be the exact significance of the name, there is little doubt that collies in the early days of their history were specially associated with Scotland. Even yet in many parts of England it is customary to hear collies spoken of as Scotch collies, in contradistinction to the Old English sheep-dog, sometimes also called the hob-tailed dog, on account of his short, stumpy tail. Collies now hold, with fox-terriers, the distinction of being the most widely distributed breed of dogs that we have. In addition to the large numbers that are kept and used on farms, and by herds and drovers, many collies are now kept for fancy purposes and as pets.

Broadly speaking, there are three varieties of collies—rough-coated dogs, smooth-coated dogs, and bearded dogs. If one included the Old English dog already referred to, which is well entitled to be included amongst sheep dogs, there would be four. Practically all of the different kinds of collies or sheep-dogs which one sees up and down the country, both in England and Scotland, as well as in Ireland, are bred from one or other of these types, or a mixture of them.

Bearded Collies.—Of the three first-mentioned varieties the beardie is perhaps as distinct a type as any. This class of dog is a sort of combination of the ordinary collie and the Old English sheep-dog. Beardies are nearly always dark or hazel grey in colour, roughly haired over the upper part of the face and eyes, and rather pronounced in the hook of the hind leg. This class of dog is very intelligent, but some years ago, on account of their generally bigger size and heavier weight, they became less popular with hill shepherds than the smaller class of collies. This, however, was followed by efforts for their reintroduction which have been attended with a considerable amount of success, and one now sees more of them than formerly. It is a tribute to the beardie that he is often seen in the hands of drovers—a class of men who waste little sentiment, as a rule, on their dogs, but usually put points of utility and usefulness in the forefront. From the point of view of the hill shepherd, however, the smaller collie has its advantages. Being lighter in weight, its feet are not so apt to get torn or frayed by rough heather roots or stumps.

A good specimen of a Bearded Collie is represented in fig. 735.

Smooth- and Rough-coated Collies. —Although they differ in their coats, the other two classes of collies mentioned have many points in common. They are made pretty much after the same model, are equally varied in colour, and have the same general cast of features. In recent times the rough-coated class have been to a large extent spoiled for work through

the crossing which has taken place for fancy showyard points. A long sharp nose and a narrow contracted forehead has been practically the be-all and end-all with fancy breeders, the result being that much of the old intelligence of the dogs has been lost.

Fig. 735 —*Bearded collie.*

On this account, for practical purposes, many prefer the smooth-coated breed which has not been crossed to the same extent, or a cross between the smooth- and the rough-coated varieties. Many of the smooth-coated dogs are exceedingly valuable either for hill or field work. They are usually much more cautious than the rough-coated dogs and

Fig. 736.—*Smooth-coated collie.*

are easier trained, but are not, as a rule, so swift when a special spurt is required. In fig. 736 a portrait is given of Mr Robert Chapman's famous smooth-coated bitch "Young Trim." A modern representative of the rough-coated type is shown in fig. 737.

Rough-coated dogs, like dogs of the smooth-coated type, may be of almost any colour or combination of colours, although, as a rule, they are black and white, black and tan, and black and tan and white, or variations of these. Sable has for several years been a popular colour in the case of "fancy" (show) collies, but few of these are to be met with in the possession of shepherds or farmers who keep collies for working purposes.

Old English Sheep-Dogs.—The bobtailed dog, as he is frequently called, is not the least handsome of the four varieties. He makes a first-class companion, and by many is even preferred to the ordinary collie for working purposes. Indeed, on account of his sagacity and utility, he is often spoken of as the Smithfield or Drover's dog. Some people hold that the bearded collie of Scotland is a cross between the Old

Fig. 737.—*Rough-coated collie.*

English sheep-dog and the ordinary collie. In both the head is squarish, in place of being long and narrow, as in the case of the ordinary collie. The two types are certainly different, in respect that the one has a long tail and the other a short tail, but in most other respects they are not greatly dissimilar.

The colour most sought after in the English dog is some shade of blue or grey, with white markings. In many specimens the white predominates, but grey, grizzly blue, or blue merle, with or without white markings, are typical colours. The coat in both this and the bearded variety must be abundant, hard in texture, and shaggy, without, however, any great tendency to curliness. The under coat—this, however, applies to collies of all classes—must be very dense

and waterproof. The usual height of a bob-tailed dog at the shoulder is about 22 inches; bitches measure, as a rule, about 20 inches. The short tail and shaggy coat of an English sheep-dog gives him a distinctly bear-like appearance, and makes him easily identified wherever seen.

Other Kinds of Collies.—In late years two or three more or less distinct types of collies have been evolved by selection from particular specimens. One of the best known of these is the marled or marbled collie of Wales and different counties of England. This is a bluish-coloured dog, very much after the type of an ordinary smooth-coated collie, but much more mixed in colour. The best specimens are very good workers, and Welsh shepherds use them largely in their daily avocations.

In the same way in late years, in the south-east of Scotland and northern districts of England, shepherds have produced what is practically a distinct breed. This is a medium-sized black and white dog, with sharp, pricked ears, and a rough rather than smooth coat. These dogs are specially bred and trained for sheep-herding purposes, and are greatly valued by those who own them.

Training Dogs.

Bad Training of Dogs.—The natural temper of the shepherd may be learned from the way in which he works his dog among the sheep. When an aged dog is observed making a great noise, bustling about in an impatient manner, running fiercely at a sheep and turning it quickly, and biting at its ears or legs, it may safely be assumed that the shepherd who owns it is a man of hasty temper. Most young dogs exhibit these characteristics naturally, but it is the business of a competent man to curb them and not allow the dog to do as he pleases. A man who allows his dog to deal with the sheep in the manner described is culpably careless of his flock. If, on the other hand, a shepherd be observed allowing his dog, whether old or young, to take a range round the fences of a field, driving the sheep as if to gather them, it may be concluded that he is a lazy fellow, more ready to make his dog bring the sheep to him than to walk his rounds to see them.

Great harm may accrue to sheep by working dogs in these ways. Whenever sheep hear a dog bark that is accustomed to hound them every day, they will instantly start from their grazing, gather together, and run to the farthest fence, and a good while may elapse ere they settle again. And even when sheep are gathered, a dog of high travel, and allowed to run out, will drive them hither and thither, without an apparent object. This is a trick practised by lazy herds every morning when they first see their flock, and every evening before they take up their quarters for the night, in order to count them with what they deem to be the least trouble to themselves.

When an imperfectly trained dog is allowed to run far out, it gets beyond the control of the shepherd; and such a style of working among sheep of any class puts them past their feeding for a time: with ewes it is very apt to cause abortion; and with lambs, after they are weaned, it is apt to overheat them and induce palpitation and high breathing. Whenever a sorting takes place among sheep, with such a dog they will be moved about far more than is necessary; and intimidated sheep, when run into a corner, are far more liable to break off than those treated in a gentle manner.

Judicious Training.—A judicious herd works his dog in quite a different manner. He never disturbs the sheep when he takes his rounds amongst them at morning, noon, and night—his dog following at his heel as if he had nothing to do, but ready to fulfil its duty should any untoward circumstance arise, such as breaking out of one field into another. When he gathers sheep for sorting, or catching a particular one, the gathering is made in a corner, to gain which he will give the sheep plenty of time, making the dog wear to the right and left, to direct the sheep quietly to the spot; and after they are gathered, he makes the dog watch, and, with an occasional movement, prevent any sheep breaking away. When a sheep does break away, and must be turned, he does not allow the dog to bite it, or even to bark, but to circle well in front of it and thus turn it back. Some single sheep are very obstinate to turn, and in such a case a snap by the dog at the animal's

ear may be justified, but unless in extreme cases "teething" of the sheep ought to be forbidden.

A thoroughly good shepherd only lets his dog work when its services are actually required, he bestowing his own labours ungrudgingly, and only demanding assistance from his dog when he cannot do it so well by himself. At no time will he allow his dog to go beyond the reach of his immediate control.

Well-trained Dogs. — Dogs, thus gently and cautiously trained, become very sagacious, and will diligently visit every part of a field where sheep are most apt to stray, and where danger is most to be apprehended—such as a weak part of a fence, water-runs, deep ditches, or deep furrows into which sheep may possibly fall and lie *awalt* or *awkward*—on the broad of their back, unable to get up. Many dogs are so sagacious as to assist in raising up sheep lying awalt by seizing the wool at one side and pulling with all their power till the sheep get upon their feet.

Experienced dogs also know when foxes are on the move, and give evident symptoms of uneasiness on their approach to the lambing-ground. They also hear footsteps of strange persons and animals at a considerable distance at night, and announce their approach by unequivocal signs of uneasiness. A shepherd's dog when at active work is incorruptible, cannot be bribed with a bite of food, and will not permit even a known friend to touch it or its charge when intrusted with an act of duty.

Skill in Training.—Most shepherds profess to train young collies. In this delicate work many shepherds display little knowledge of the nature of the breed, and of the aptitude of the particular individual for its peculiar work. Hence many dogs are rendered unfit for useful service. Every collie-pup has a natural instinct for work amongst sheep; nevertheless, they should be trained with an old dog. Their ardent temperament requires subduing, and there is no more effectual way of doing this than by keeping them in company with an experienced dog. A long string attached to the pup's neck, in the hands of the shepherd, is necessary to make it become acquainted with the language of the various evolutions connected with work. With this contrivance it may learn to "*hold away out by*," "*come in*," "*come in behind*," "*lie down*," "*be quiet*," "*bark*," "*get over the dyke*," "*wear*," "*heel*," "*kep*." It will learn all these terms, and others, in a short time. It is said that the bitch is more acute in learning than the dog, though the dog will bear the greater fatigue. Of the two, the quietly disposed shepherd prefers the bitch as a rule, and is chary of working her when in pup.

Sagacity of the Collie.

Much may be said of the sagacity and faithfulness of the collie. "If he be but with his master," observes Youatt, "he lies content, indifferent to any surrounding object, seemingly half asleep and half awake, rarely mingling with his kind, rarely courting, and generally shrinking from, the notice of a stranger. But the moment duty calls, his sleepy listless eye becomes brightened, he eagerly gazes on his master, inquires and comprehends all he has to do, and, springing up, gives himself to the discharge of his duty with a sagacity and fidelity and devotion too rarely equalled even by man himself."

"If we consider," says Buffon, "that this animal is superior in instinct to all others; that he has a decided character, in which education has comparatively little share; that he is the only animal born perfectly trained for the service of others; that, guided by natural powers alone, he applies himself to the care of our flocks — a duty which he executes with singular assiduity, vigilance, and fidelity; that he conducts them with an admirable intelligence, which is a part and portion of himself; that his sagacity astonishes at the same time that it gives repose to his master, while it requires great time and trouble to instruct other dogs for the purposes to which they are destined,—if we reflect on these facts, we shall be confirmed in the opinion that the shepherd's dog is the true dog of nature, the stock and model of his species."

The Ettrick Shepherd truly says that "a single shepherd and his dog will accomplish more, in gathering a flock of sheep from a Highland farm, than seventy shepherds could do without dogs; in fact, that, without this docile animal, the pastoral life would be a blank. It would require more hands to manage a flock of sheep, gather them from the hills, force them into houses and folds, and drive them to markets, than the profits of the whole flock would be capable of maintaining. Well may the shepherd feel an interest in his dog: he it is indeed that earns the family bread, of which he is himself, for the smallest morsel, always grateful and always ready to exert his utmost abilities in his master's interests. Neither hunger, fatigue, nor the worst of treatment will drive him from his side, and he will follow him through every hardship without murmur or repining."

Dog Trials.

Notably since the advent of the twentieth century, competitive trials for working collies have become an interesting feature in many rural districts. Prizes are given for the best working dogs over a stated course. Three or more sheep are usually penned in the distance, but in sight of the shepherd competitor. The sheep are liberated when the competitor takes his stand at the appointed place, and the dog has then to be run out and bring them to his master. Stakes are frequently erected through which the sheep have to be passed in a stated way. One or two of the sheep have usually to be separated from the others and held close at hand for a stated time by the dog. Finally the whole of the sheep have to be penned, the shepherd being permitted to assist the dog in this operation. Otherwise he is supposed to direct the dog only by words, signs, or whistles. The prizes are awarded not only on a basis of time, but on the exactitude with which the different operations are performed and the general behaviour of the dog, rough usage of the sheep being an almost fatal fault.

These trials are objected to by many sheep-farmers, on the grounds that the operations performed at the trials are not such as are met with in ordinary sheep-farming practice, and that a good deal of harm is inflicted upon considerable numbers of sheep by excessive driving in the process of training the dogs for the competitions.

VARIETIES OF FOOD.

The farmers of the United Kingdom have ample choice of materials for the feeding of their different classes of stock. A fairly substantial home supply is augmented by ever-increasing imports of moderately priced foods of good quality from colonial and foreign countries, and in order that farmers may be assisted in deciding from time to time as to the kinds of food which, at the current prices, can be most economically employed, full information is here presented as to the composition and character of the feeding-stuffs available in this country.

Brief notes regarding the different materials used as food for farm livestock are given here. Detailed analyses will be found on page 290.

Milk.

Milk has a good right to rank first amongst foods. It is the most perfect and most natural of all foods for young animals. As already observed, there must be a proper mixture of the nitrogenous constituents or albuminoids along with the non-nitrogenous (carbo-hydrates and fat), to form a perfect food. A perfect illustration of this mixture is found in milk, the first food upon which the young animal is expected to subsist. It contains, 1st, casein or curd, which is a substance of the same class as the *fibrin* or lean part of the flesh; 2nd, fat in the shape of butter; 3rd, sugar, the most easily digested of all carbohydrates; and 4th, certain substances which are converted into the earthy part of the bones, and the saline matter of the blood. The saline or earthy portion of milk consists of the phosphates of lime, magnesia, and iron, chloride of potassium, and common salt.

In its ordinary state the milk of the cow consists on the average of about 3½ per cent of casein or flesh-forming matter, 3½ per cent of butter-fat; 4½ per cent of sugar; ¾ per cent of saline matter; and 87¾ per cent of water. Everything, therefore, which is required to promote the development of the growing animal is contained in the milk, blended together in proportions suited for the purpose.

Wheat.

Wheat is a very starchy food. In the form of flour it is not suitable for stock; but as it leaves the straw with the bran and other coats, it is a fairly well-balanced food, coming pretty near to the albuminoid ratio of 1 to 7.

Damaged Wheat for Stock.—Wheat which has been damaged by wet in harvesting is sometimes turned to good account in feeding stock. It should be first kiln-dried and then mixed with chaffed hay or straw.

Feeding Value of Wheat.—When wheat was selling at from 40s. upwards per quarter, it was too expensive to be used in feeding stock; but when it sells at not more than about 30s. per quarter, it may in some cases be employed for this purpose with advantage. Mr John Speir, Newton Farm, Newton, Glasgow, has used wheat with very satisfactory results in the feeding of dairy-cows. He points out, however, that to be a successful feeding-stuff by itself it would require much more oil than it possesses, and considers that the addition of one-fourth of linseed or one-third of linseed-cake would much enhance its feeding value. He says that to cattle—mixed with an equal proportion of decorticated cotton-cake and peas or beans, all ground into rough meal (not flour)—it has given excellent results. It is better boiled and given whole than ground into flour, but as rough meal it is better than either, as then it never gets into the doughy state, and it mixes freely with chaff and pressed or sliced turnips.

Wheat for Sheep.—Experiments conducted by the Royal Agricultural Society of England at Woburn showed wheat in a favourable light. It was tried along with linseed-cake, decorticated cotton-cake, and barley. The best results were got from decorticated cotton-cake, wheat coming next. The wheat was given whole.

Bran.

Bran, which in milling wheat for use as human food is usually separated from the flour, is much used as food for livestock. It is sometimes given in the form of mashes, and at other times mixed with other kinds of foods. When used by itself, or mixed with cold water, it has a slightly laxative effect, which renders it useful in preparing horses for physic, and in some cases may so act as to obviate the necessity of giving purgative medicine. The ash of bran contains a large proportion of phosphates, much larger than the ash of barley or oats. Hence it is particularly useful as part of the food given to milch cows, when such are "in profit," or full milk—milk being rich in phosphatic constituents.

Bran acts beneficially in counteracting the heating properties of maize and other similar meals.

Barley.

Barley is exceedingly rich in the fattening constituents of food. It is seldom—and never should be—given in its dry whole state as food for stock; but in the form of rough meal, or cooked, it is fed very extensively. Like wheat, it has fallen in price, and its home consumption has increased proportionately.

Cooked Barley. — When barley is being cooked, it must be allowed to simmer slowly twelve hours, until the whole forms a mass of rich pulpy matter, perfectly free from whole grains. The greatest care must be taken to prevent the barley from becoming burned, by adhering to the boiler in which it is prepared. When thoroughly cooked, it becomes a most valuable ingredient in the food of fattening animals. Horses thrive remarkably well upon it—so much so, that a course of boiled barley given at least once a-day will very soon renovate horses that have been worn out with hard work.

Boiled barley is used by some of the most successful exhibitors of Shorthorns in the preparation of their cattle for the showyards. Along with a little oilcake, it gives that finish—brings out that mellowness in handling—which is so much desired in such cases.

Steeping Barley. — Whole barley should be steeped in water at least twenty-four hours before being given to stock; but the more common practice now is to grind it or to crush it into rough meal. Some think it advisable to steep the ground barley in water.

Malt.

Barley is converted into *malt* by being first steeped and then allowed to germinate, the original object of this process being to prepare the barley for distillers and brewers. As to the simple question of the relative feeding merits of *malted* and *unmalted* barley, there was a lively and long-continued controversy. Formerly the duty now levied directly upon manufactured spirits, ales, and porters was imposed upon malt, and then farmers could not malt barley for feeding stock without paying the malt-duty. This was a momentous grievance to farmers, on whose behalf it was urged that malt was much more valuable as food for stock than unmalted barley. Human nature is a little curious in some of its moods, and it is just possible that the barrier which formerly existed to the use of malt as food for stock may have had something to do with the high opinion then expressed as to its value for that purpose. Be that as it may, the duty was removed from the malt, and now that farmers can make malt for their stock as freely as they desire, much less is heard of its alleged special feeding virtues than when they had no such liberty. Indeed, malt has almost entirely ceased to be used as food.

That malt is a valuable and palatable food there is no doubt whatever. The contention that it is superior food to unmalted barley has not been borne out by practical experience.

Rothamsted Experiments with Malt.—Sir John Bennett Lawes carried out an elaborate series of experiments upon the use of malt in feeding various kinds of stock. In all these experiments he compared a certain weight of barley with the same weight of barley converted into malt. Given to cows, he found that the same quantity of milk was produced, but the quality was better with unmalted barley. In a feeding experiment with twenty cattle, the ten getting unmalted barley increased more in weight and were more even in condition than the ten which got malt. In his experiments on sheep and pigs, the results were also rather in favour of the barley unmalted.

Special Properties of Malt.—It has, however, been proved that malt does possess certain useful properties in the feeding of stock which are not possessed to the same extent by unmalted barley. The late Mr Richard Booth, of Warlaby, considered that malt was superior to any other article for feeding cattle up to the very "tip-top" condition to which they require to be brought when they are intended for the showyard. Malt has been used with good results in rearing young pure-bred bulls.

The truth probably is, that such special value as malt possesses is to a great extent a *condimentary* value. Just as cattle-spices are valuable for imparting a relish to diets in which straw-chaff or poor hay predominates, so malt, owing to its sweet and appetising flavour, may impart a relish to food that may be of value.

But it by no means follows that a food which best puts the finishing touches on an abnormally fat animal (which is rarely produced at a profit) is to be regarded as, on that account, an economical article of diet for profitable meat-production. As a matter of fact, the balance of evidence is in the opposite direction, and is confirmed by so little being heard of the use of malt for commercial animals in recent years.

It usually costs close on 2s. per quarter to convert barley into malt.

Malt-combs.

When barley is converted into malt, the effect of the steeping process is to cause the grain to throw out young shoots, just as the seed does when put in the soil. These young shoots are afterwards separated from the malt, and are known as "*malt-combs*," or "*cummins*," or "*malt-dust*." The combs are used as feeding-stuff, and have been found useful, along with other articles, as food for milch cows. Sir Charles Cameron says that the composition of this food indicates a high nutritive power, but adds that it is probable that its nitrogenous matters are partly in a low degree of elaboration, which greatly detracts from its alimental value.

Malt-combs for Cows.—The late Dr A. Voelcker considered that malt-combs possessed high milk-producing qualities, and that the food might be given with great benefit to dairy-cows.[1]

Malt-combs as Manure.—Malt-combs are also used as manure, but the late Dr A. Voelcker considered it wasteful to apply them directly to the land; they should first be passed through the animal's body.

Bere and Rye.

In feeding value these are very similar, but slightly inferior, to barley. Rye is generally used in this country in a green state when given to cattle. The grain is useful for feeding purposes, although somewhat inferior to barley.

Rye-meal is given with advantage to milch cows.

Brewers' and Distillers' Grains.

Brewers' grains, or "draff" as the article is called in some parts, consists of the refuse malt after it has undergone mashing. The grains left in the distillation of spirits are usually slightly richer than those left in brewing ale or porter. Both are now very extensively used as food for different classes of stock, and the results are, on the whole, very satisfactory.

Dried Grains.—A process of preparing grains by drying and other modes of manipulation has been invented, and the article so prepared is sold under the name of "Dried Grains." The grains in this state are more concentrated than they are in the ordinary state, and may be given to all kinds of live-stock. For horses it is a frequent custom to substitute at first 3 lb. of grains for 3 lb. of oats, and increase the proportion until half the feed is composed of grains. For cattle the grains may be mixed with other food, and should be damped where oilcake is used. The animals should be supplied with water when equal parts of grain and cake are used. For cows it is usual to damp the grains with boiling water, and allow them to swell; 8 to 12 lb. per day may be given. To sheep the grains may be given alone, or with an equal weight of corn or cake. For pigs,

[1] *Jour. Royal Agric. Soc. Eng.*, xiv. 248.

damp well with boiling water as much as will be required for a day's use.

Dried v. Wet Grains.—A ton of dried grains would be equal to between three and four times its weight of wet grains. The drying chiefly effects economy in carriage. But when a brewery or distillery is within easy reach, it is, of course, more economical to use wet grains than the artificially dried, and therefore somewhat more costly, article.

Grains for Dairy-cows.—Grains are a particularly favourite food with cow-keepers, as they produce a large flow of milk—more remarkable, however, for its abundance than its richness, that is, where grains are the preponderating food.` When mixed with a fair proportion of other richer concentrated food, such as cake or grain, the grains form an admirable article of diet for cows in milk.

Difference in Composition.—The average of the analyses of a large number of samples of the two varieties, conducted in the Edinburgh College of Agriculture (1908), gave the percentage of oil in distillery grains as 5.96 and in brewers' grains as 4.99, the percentage of nitrogenous matter as 18.75 and 20.61 respectively, and the percentage of soluble carbohydrates as 54.37 and 48.85 respectively. In price the distillery grains are usually higher in comparison with brewers' grains than the difference in analyses would seem to justify.

Other Distillery Food.—Useful feeding material is found in other distillery by-products, such as the "wash" or "burnt ale" of malt distilleries and the "dreg" of the raw grain distilleries.

Oats.

No other variety of grain is so extensively used in this country as food for live-stock as are oats. And in the form of meal it is a very wholesome food for man, still used very largely—but not so extensively as in former times—in Scotland. It was Dr Johnson who described oats as "the food of men in Scotland, and horses in England." It was probably a Scotchman who retorted,—"Ay; and where will you find such men and such horses?"

Oats are highly favourable to the formation of muscle. Their nutritive value,

however, is by no means regular, some varieties being one-third more nutritive than other kinds.

Bruising Oats.—Oats ought generally to be bruised before being given to animals, as the food then becomes not only more thoroughly masticated, but also much less liable to produce inflammatory action, which sometimes arises from the over-liberal or inconsiderate use of the whole grain.

Nutriment in Oatmeal.—In the form of meal it is seldom used as cattle-food, except as nourishing drinks or gruel; but when ground into meal, the more thoroughly it is sifted the more nutritious it becomes. This is exactly the reverse of what takes place in the case of wheat-flour, because a large proportion of the flesh-forming and also of the fat-forming substances contained in wheat is removed in the bran. In fine oatmeal there is much more oil than in fine wheat-flour, and in the former one and a half times as much albuminoids as in the latter.

There is no need to enlarge here upon the merits of a food which is so generally esteemed for this purpose as oats are.

Indian Corn.

The prevailing cheapness and high nutritive properties of Indian corn or maize have brought it into extensive use as food for farm live stock.

This food is very rich in starchy matters. Given by itself, or in large proportions, it has a heating and binding tendency; but it does well with other foods, such as linseed-cake. On the whole, maize is usually about the cheapest form in which starch for feeding purposes can be purchased in this country, and it is therefore one of the most extensively employed articles of food for all kinds of farm live-stock, including poultry. For all stock except poultry maize should be bruised or kibbled.

Buckwheat.

This plant is comparatively little grown in this country, being easily susceptible of injury from frost, especially if the seed is sown earlier than the middle of May. The crop is sometimes cut green, and used for soiling. The grain is used chiefly for feeding game or poultry.

In Ireland the term "buckwheat" is sometimes locally applied to some of the varieties of common wheat, with which the true buckwheat has no connection.

Rice.

Rice is sometimes used as food for poultry, and is of a very fattening nature. It is exceptionally high in starchy matter.

Rice-meal.

Much more important than rice, as a feeding-stuff, is the so-called "rice-meal," which consists of the ground refuse left after dressing or trimming rice for human food. This rice-meal consists mainly of the coating of the reed (or bran), with more or less of the adherent starchy matter.

Rice-meal contains a fair quantity of albuminoids, and is rich in oil, and is in much request for pig-feeding. It is also used as food for cattle, and the experience of it has, on the whole, been satisfactory. Mr Garrett Taylor, Trowse House, Norwich, has used it largely both for dairy-cows and young store-cattle, and he speaks of it very favourably.

Care should be taken to obtain the genuine article, as this food is sometimes adulterated with ground rice shudes—the outer husks of the rice—which have very little nutritive value, but consist mainly of a silicious woody fibre.

Dari or Durra.

This is the seed of the plant called Indian millet or Guinea corn, which is largely cultivated in India, China, Africa, Italy, the West Indies, &c., where it is used for feeding horses, pigs, and poultry. It weighs upwards of 60 lb. a bushel, is of the size of a large millet-seed, is covered with a husk or envelope, and gives, when crushed, a beautiful white flour.

Ground into meal, this grain is an excellent fattening food for cattle. Dr Voelcker remarked: "It contains an appreciable amount of ready-made fat, and a large proportion of starch, which is with ease transformed into fat in the animal economy; but it is rather deficient in albuminoids, and for this reason Dari meal should be given to stock in conjunction with cake, beans, or peas, or,

speaking generally, with food rich in albuminous compounds." [1]

Dari grain is also good food for poultry.

Beans, Peas, and Lentils.

These leguminous plants closely resemble each other in their composition. From their nature they are better suited to be used as a portion of the food of working or growing animals or milch cows, than of those which are being fattened for the butcher. At the same time, when used along with other kinds of food, particularly such as are of an oily nature, they may be given with much advantage to fattening stock. Lentils are chiefly imported, but they may be profitably grown in this country on light, dry, sandy, or calcareous soils.

Vegetable Casein.—It is worthy of note that the albuminoids in these three seeds (and also in other leguminous seeds) are in a form somewhat similar with the casein of milk, and hence termed "vegetable casein." It is on this account that meals made from these seeds form useful ingredients in mixtures for calves.

Beans for Dairy-cows.—Bean-meal is by many recognised authorities assigned the very highest position as an article of diet for dairy-cows. Mr John Speir, Newton Farm, Newton, Glasgow, gave it as his opinion that "for the production of butter or cheese of the best quality, no other feeding-stuff ever gained or so long maintained so high a reputation as beans"; and he adds, "They are also very palatable to all stock of the horse, sheep, and cow kind, although swine are not so fond of them."

Beans, like the other leading leguminous foods, have a high albuminoid ratio, and, therefore, are well suited for mixing with other foods rich in carbohydrates, such as turnips, potatoes, oats, rice, straw, and hay.

Mr Primrose M'Connell says that "beans have made a name for themselves as food for dairy-cows, but prices and handiness make it more desirable to use something else." He adds that he gives his cows a mixture of crushed beans, oats, and bran.

---

[1] *Jour. Royal Agric. Soc. Eng.*, xiv. 247.

Preparing Beans as Food.—Beans should invariably be ground into rough meal before being given to stock, but should not, as is sometimes done, be steeped in water before being mixed with the other foods, as then, on account of its highly albuminous nature, the meal is apt to get into a doughy, indigestible mass. Bean-meal holds the premier place as a milk-producer; but being so highly albuminous, it requires to be mixed with some more bulky food in order to keep its particles apart, and allow the juices of the stomach and intestines to dissolve them. Mixed with cut hay or straw, the meal becomes one homogeneous mass of such a porous nature that each atom of its constituents can separately be attacked by the juices of the digestive organs; whereas if mixed in water alone, the bulk of it is voided undigested, if it does not also produce indigestion by the way.

Soy Beans.—The Soy bean (*Soya hispida*) is a leguminous plant extensively cultivated in China, Japan, and Manchuria, where it is an important article of human food. Large quantities of the bean are now coming to this country, where it is used as an oil seed, most of the oil being expressed, and the remaining cake is a valuable addition to our cattle foods. The Soy cake is a highly concentrated food, equalled in richness in albuminoids only by decorticated cotton-cake and decorticated earthnut-cake. There is little starch in Soy beans. Being so exceptionally rich in albuminoids, Soy beans are best suited for use in mixed foods.

Peas for Sheep.—Peas are capital food for sheep—along with linseed-cake there is perhaps no better as a concentrated food. In a series of experiments conducted at Woburn by the Royal Agricultural Society of England, pea-meal was contrasted with barley-meal and malt as food for sheep, each along with turnips, hay and straw, chaff and linseed-cake. In summing up the results, which were favourable to the pea-meal, Dr A. Voelcker stated that "linseed-cake and pea-meal in equal proportions, and used at the rate of ¼ lb. each per head per day, in conjunction with some hay and straw, chaff and swedes, given *ad libitum*, is a better food for young sheep than either a mixture of linseed-cake and barley-meal, or linseed-cake and malt." [1]

Poisonous Beans.—In the years 1905 and 1906 deaths occurring amongst dairy-cows in Scotland were attributed to eating Java beans (*Phaseolus lunatus*), which were found to contain prussic acid. The Java beans were ascertained to be the most dangerous, but traces of the poison were also observed in the Burma variety. [2]

There is a large variety of beans in existence, and great caution is necessary in using kinds not thoroughly well known.

Lupin.

The seeds of the lupin contain a larger proportion of flesh-forming substances than either beans or peas or lentils.

The cultivation of the plant is quite simple, and it grows well on poor, sandy, and gravelly soils. It is cultivated extensively in the northern parts of Germany, and it is grown to a small extent in England. The stems make excellent hay, and the seeds are found to be very superior food for sheep. They are also given to horses and cattle, mixed with oats or beans; and lupin-meal is given with milk to calves.

Linseed.

Linseed has not inaptly been described as the sheet-anchor of the stock-feeder. He is now less dependent upon it than when he first began to use concentrated foods extensively. But although many other useful articles of food for farm live-stock have been brought into notice in recent years—thanks in a large measure to the application of science to the question of economical stock-feeding—it is still true that for almost all classes of stock linseed is a feeding material of the highest value.

There are a great many varieties of linseed, some kinds being much richer than others. Linseed of fine quality, weighing 52 lb. per bushel, readily yields from 11 to 12 gallons of oil per quarter of 8 bushels, weighing 9 lb. per gallon, or about 25 per cent of its weight.

[1] *Jour. Royal Agric. Soc. Eng.*, xix. 430.
[2] *Trans. High. and Agric. Soc.*, 1907.

Preparing and using Linseed as Food.—Linseed is an exceedingly rich food, especially in oil. It is therefore not advisable to use it in its natural state, as, when so used, a considerable proportion of the seeds will be found to pass undigested. Being of a laxative nature, it requires to be used with caution, and in combination with other articles which have a counteracting effect.

The seed is sometimes boiled in order to prepare it for use as the food of animals, but a better mode of preparation is to grind it. When this is done, chaff, or the husks which are separated from oats.in the process of milling, should be passed through along with the linseed, as either of these articles helps to prevent the linseed from "clogging" the millstones; and besides, they absorb a portion of the oil which exudes from the seed in the grinding, and thus they become useful articles of food, although they are of little value in their natural state.

Linseed chaff also makes excellent food.

Boiling Linseed.—Meal made of pure linseed may be given in combination with other kinds of food, such as bean-meal, barley-meal, Indian-meal, &c., but it is also frequently prepared for use by boiling. When the seed is prepared in this way, it is generally steeped for some hours in hot water and then boiled, but it is very apt to burn during the process of boiling unless it is carefully watched. In order to prevent burning, it has been found better to raise the water to the boiling-point before putting in the linseed, instead of putting the linseed into cold water and then boiling it. When the linseed is put into boiling water, add a little cold water, and then let it again come to the boil, and allow it to remain boiling for twenty minutes, stirring it occasionally.

Linseed for Calves. — Linseed-oil commands a price for other uses, which renders it too costly for fattening commercial stock. The cheaper residue of the crushers' trade—linseed-cake—contains sufficient oil for most feeding purposes, but for calf-rearing no kind of food surpasses linseed, and in conjunction with cheaper commodities, it is very largely employed for this purpose.

Growing Flax for Fibre and Seed.—As has been shown in another part of this work (vol. ii. p. 391), flax is grown extensively in the north of Ireland for the production of a fine class of fibre, which is incompatible with a large crop of seed. If the value of the seed as a fattening material were inculcated more strongly than it has been by those who advocate the extension of flax-cultivation, it is probable that flax-growing would be more largely practised throughout the kingdom than it is at present. As it is, this country has to depend upon foreign sources of supply for the linseed required for various purposes.

Linseed-cake.

Linseed-cake consists of what is left of the seed in the process of extracting linseed-oil. Formerly from 12 to 14 per cent of the oil usually remained in the cake; but the means now employed in extracting the oil do the work so efficiently that less oil is left, although linseed-cake with high percentages of oil can still be got.

High Reputation of Linseed-cake.—Enough oil still remains to impart great value to linseed-cake as feeding material. It is suited for all kinds of farm live-stock. For fattening cattle no other food has such a high reputation as the best home or Russian linseed-cake. In fact, owing to the consensus of practical opinion, it almost always commands a higher price than its analysis seems to justify. Horses become extremely fond of linseed-cake, and 3 lb. per day has been given to farm-horses with good effect.

Linseed-cake is crushed into small pieces before being given to stock.

The dung of cattle fed on linseed-cake is very rich, nearly half the weight of the ash of linseed-cake consisting of phosphate of lime; and one result of giving cake to cattle or sheep feeding on grass land during summer and autumn is to improve the pasture, besides hastening the fattening of the animals.

Storing Linseed-cake.—The late Dr A. Voelcker remarked — and this is a point of the very greatest importance —that "the nutritive value of feeding-cakes depends not merely upon their proximate composition, but likewise upon their physical condition. Like all other perishable articles of food, linseed-cake, when kept in a damp or badly ventilated place, rapidly turns mouldy, and after some time becomes unfit for feeding purposes." [1]

Linseed-cake should be stored in as dry a place as can be found. The floor should be a wooden one if possible. If it be of plaster or concrete it is advisable to lay some old timbers on the floor, forming a stool, and pile the cakes in stacks thereon, about 8 or 10 inches from the walls, so that a current of air can get round. The cakes should be packed in such a way that the air can get through the pile and come in contact with all the edges of the cake.

The ventilation of the store should be good, and as much air as possible allowed to get inside when the atmosphere is dry, but the doors and windows should be closed when it is damp.

The same remarks are applicable to the storing of cotton-cakes, but these cakes do not keep well beyond a month or six weeks.

When good linseed-cakes, manufactured without the use of water, are stored in the manner thus described, they have been known to keep for 12 months without any appreciable deterioration.

Adulteration of Cakes.—Unfortunately, it has become so much the practice to adulterate cakes of all kinds in the process of manufacture, that the greatest caution is necessary in purchasing any article of the kind. Impurities also exist in the seed, varying from 1¾ per cent to 70 per cent; and these impurities are sometimes added artificially. Farmers are now well protected against loss in this way by the Fertilisers and Feeding Stuffs Acts, the provisions of which all purchasers of feeding-stuffs should study carefully.

Rape-cake.

Rape-cake, when pure, is a valuable food for cattle. In albuminoids it is richer than even the best of linseed-cakes.

It is not much relished by cattle at first, but if care is taken to prevent it from getting damp and mouldy they will take to it by degrees. If the animals refuse to eat it in its fresh state by itself, the difficulty may be got over by covering the cake for some time with sawdust, chaffed straw, or any substance that will prevent it from becoming damp or moulded.

Preparing Rape-cake for Cattle.—The cake is of course crushed, and it is of advantage to pour boiling water over the crushed cake, and allow the mixture to stand for a time before it is used. Steaming the cake along with chaffed straw is also a good mode of preparing it for cattle; and in so preparing it bean-meal or bran is added, in the proportion of 4 lb. of cake to 2 lb. of bran or 1 lb. of bean-meal. With these articles, 16 lb. of chaffed straw should be blended before steaming.

Impurity of Indian Rape-cake.—On account of the amount of wild mustard or charlock (*Sinapis arvensis*) it usually contains, it is not safe to use Indian rape-cake as food for stock. Dr A. Voelcker stated that from ¼ lb. of Indian rape-cake he obtained enough essential oil of mustard to convince him that half a cake of it, if not a smaller quantity, might kill a bullock.

Even the best rape, when mixed with linseed-cake, imparts a turnip-like flavour to the latter, which of course reduces its value.

Cotton-cake.

This cake is made from the seeds of the cotton-plant. There are two varieties of it in use—the decorticated, from which a portion of the husks has been removed, and the undecorticated, which contains the whole of the dark-brown husks of the seed.

As in the case of linseed-cake, the improvements in the crushing machinery have greatly reduced the percentage of oil in decorticated cotton-cake.

The decorticated variety is of a uniform yellow colour, while the presence of the dark-brown husks in the undecorticated at once indicates its nature.

Caution in using Undecorticated Cotton-cake. — The undecorticated variety is not only less valuable than the other, but it is also apt to produce inflammatory symptoms in the animals

[1] *Jour. Royal Agric. Soc. Eng.*, ix. 3.

fed upon it, and death has frequently been the result. This arises from the quantity of cotton which adheres to the seed, and from the harsh nature of the husk. Although undecorticated cotton-cake may be employed as food for cattle when used cautiously, it is generally safer to use the decorticated variety, which, as will be gathered from the analysis, is also superior as a feeding material.

In the eyes of graziers, the unde-corticated cake has a special value owing to its astringent properties, which render it useful to obviate any scouring tendency amongst cattle or sheep when on young or luxuriant pasture.

Two varieties of undecorticated cake very extensively used are Egyptian and Bombay cake, each of which has well-known characters.

Uses of Decorticated Cotton-cake. —Decorticated cotton-cake is very rich in flesh-forming matters, as well as in phosphates, or "bone-formers," and is therefore specially adapted for growing stock and cows giving milk. For calves and lambs it is not so suitable—indeed, for these young animals it is somewhat dangerous—on account of its aptitude to give rise to digestive trouble. It should be given to them only in very small quantities.

Value of Undecorticated Cotton-cake.—Undecorticated cotton-cake has gone on gaining in public favour as an ingredient in feeding mixtures. Numerous experiments by Mr John Speir and others have shown its high value as a food for cows in butter-making herds, and for other classes of stock. In almost all fattening experiments it is put on trial, and almost invariably it gives a good account of itself.

Preparing Cotton-cake for Feeding.—Some think it better to have the cake ground into meal than merely broken by the usual cake-crusher. If the cake is to be merely crushed, it would be useful to have this done some time, perhaps ten or fourteen days, before giving it to cattle, so that it might absorb moisture, and thereby get softened and more easily digested.

It is important in buying undecorticated cotton-cake to avoid hard rock-like samples, and accept only cake that is oily and soft.

Manurial Value of Cotton-cake.— Cotton-cake imparts an exceptionally high value to the manure of the animals fed upon it. As will be shown presently, in dealing with the "Manurial Value of Foods," it stands above most other foods in this respect.

Palm-nut Meal.

Palm-nut meal is made from the cake which remains in pressing the oil out of the palm-nut. There are some very inferior kinds of palm-nut cake or meal in the market, and these have been used in adulterating linseed-cake — not so much lately, on account of the vigorous and commendable efforts that have been made to put down the adulteration of all feeding-stuffs as well as of manures.

Palm-nut meal has been found an admirable food for calves, but it is not extensively used. For calves it is prepared by being steeped in hot water. It is also well suited for cows in milk— increasing both the quantity and quality of the milk. It gives good results in feeding pigs, if used with such other foods as barley-meal, Indian meal, beans or peas. An equal mixture of palm-nut meal and decorticated cotton-cake is equivalent in feeding properties to linseed-cake, and considerably less in cost.

Cocoa-nut Cake.

Cocoa-nut cake is manufactured from the cocoa-nut palm. It is a wholesome food, and cattle take to it readily. It is, however, inferior to pure linseed-cake.

The ordinary cocoa-nibs, as sold by grocers, are occasionally employed as a feeding material for young calves. The nibs are boiled over a slow fire for two or three hours—6 or 8 quarts of water to 1 lb. of nibs. They are then strained out, and the liquid is mixed with milk and given to calves when it is milk-warm.

Minor Oilcakes.

Other cakes prepared from oil seeds of which a considerable amount is used are niger-cake, earthnut-cake, sunflower-cake, and hemp-cake. All of these are residues left after the corresponding seed is crushed for oil. Earthnut-cake is specially rich in albuminoids. When it is made from the decorticated seed it surpasses even decorticated cotton-cake in this constituent. A considerable amount of it is made, however, with the husk only partially removed or not removed at all. When the husk or pod is crushed with the seed, the cake, like undecorticated cotton-cake, contains much fibre.

Cotton-cake, rape-cake, and all the minor oilcakes like earthnut-cake, niger-cake, cocoa-nut cake, poppy-seed cake, are much used in the preparation of compound or mixed cakes, which are now very common. Many cakes which are comparatively seldom used by farmers under their own names are largely used when made up into special compound cakes and meals.

Poppy-cake.

Poppy-cake must be used when quite fresh, as the oil is apt to become soon rancid. When fresh, it is a useful feeding material.

There are two varieties of poppy-cake, —one a light-coloured or whitish cake, made from white poppy; the other a dark or brownish cake, made from common poppy-seed.

Locust or Carob Beans.

These are the beans or pods of the locust-tree (*Ceratonia siliqua*). The locust-meal contains fully half its weight of sugar, but it is deficient in albuminous compounds or flesh-formers; consequently it should be given to stock in combination with peas or bean-meal, or with decorticated cotton-cake. The pods are either roughly crushed or ground into meal, and either way are much liked by cattle and sheep.

Molasses or Treacle.

Used with skill and care, treacle is an exceeding useful feeding commodity.

Treacle risky for Breeding Stock. —It is pretty generally believed that food excessively rich in saccharine matter, while highly valuable in the fattening of stock, is unsuitable for breeding animals, in that it tends to impair their procreative powers. About 60 per cent of the weight of molasses or treacle consist of sugar, so that it lies very specially under the above condemnation.

In his instructive paper on "The Reproductive Powers of Domesticated Animals" contributed to the *Journal of the Royal Agricultural Society of England*, the late Professor Tanner laid great stress upon the detrimental influence of "sugary" foods upon the reproductive powers, and considered it very doubtful if any stock which have been fed for a length of time upon food largely mixed with molasses ever regain their breeding powers.

These extreme views have not been universally accepted, but there is no doubt that the general drift of the contentions emphasised by Professor Tanner are well founded.

Useful Properties of Treacle.— Treacle possesses special properties of considerable value. Diluted with water, and sprinkled over layers of food-mixtures, it renders the material more palatable to the stock, and its laxative characteristic makes it a useful ingredient in many mixtures with an opposite tendency. Then its own intrinsic properties in laying on fat are very high; and therefore, properly and carefully employed, treacle is of considerable service to stock-owners.

Professor Tanner acknowledges the high fattening properties of treacle, and remarks that it has "the effect of suppressing these periodical returns of restlessness which prevent heifers feeding as well as steers"; and adds that, "whilst avoiding it for breeding animals, we may encourage its employment when cows or heifers have to be fattened."[1]

Treacle for Young Bulls.—There is no doubt that, in spite of all the warnings that have been given, treacle is still used extensively for breeding stock, notably in forcing young bulls and heifers into blooming condition for show or sale. It is included in the feeding mixture for young bulls in several of the leading herds of the day; but in almost all these cases there is a skilful hand at the helm, and the tasty but dangerous treacle is given sparingly and judiciously, so as to benefit the animals without impairing

---

[1] *Jour. Royal Agric. Soc. Eng.*, sec. ser., i. 267.

their fecundity. Unfortunately, there is too good reason to believe that in some cases harm is really being done to the breeding properties of young animals, bulls more largely than heifers, by the too liberal and imprudent use of treacle.

How Treacle is used.—Treacle is often given to sickly animals mixed with bran or gruel, and it is sometimes put amongst milk for calves. Owing to its highly laxative nature, from 2 lb. to 3 lb. per day is the most that can be given with advantage even to full-grown beasts, and from ¼ lb. to 1 lb. to a calf, according to the age of the animal. The late Dr R. Thomson of Glasgow found that about 3 lb. of molasses mixed with 9 lb. of barley-meal, and given along with 25 lb. to 30 lb. of hay, kept milch cows in full milk, and did nearly as well as 12 lb. of either linseed-cake or bean-meal. A few ounces per day, diluted with hot water, and sprinkled over the dry food of horses or of fattening sheep, will be found beneficial.

There are now on the market some prepared foods which contain a good deal of treacle, and for which it is claimed that they overcome its laxative and depressing qualities. These foods are well spoken of by many stock-owners, and are not so highly appreciated by others. Most experienced feeders prefer to mix their own foods, just as they think it best to mix their manures for themselves.

Turnips.

In the winter food of cattle and sheep roots bulk largely. To a smaller extent they are also used for horses and pigs. Turnips, with the swede as the chief variety, supply the largest proportion of this description of food.

Variation in Nutritive Value of Roots.—The nutritive value of turnips varies with the variety, the climate, soil, and also the manures used in their cultivation, so that any description of their constituent elements can be regarded as only an approximation to the truth, even in the case of the same kind of turnips if grown under different circumstances. All the varieties of the turnip contain a large percentage of water—namely, from 86 to 94 per cent, leaving only from 6 to 14 per cent of dry matter. Turnips grown in some parts of the kingdom, particularly in the north of Scotland, will, with the aid merely of fresh oat-straw, be found to fatten cattle without using much artificial food of any kind; whereas large quantities of cake and hay must be given along with the same kind of turnips to effect that object, when such turnips are grown in some other districts. This is more especially the case with turnips grown in the south and east of England.

Advantages of Storing Turnips.—Turnips become more nutritious after they have been stored for some time than they are when taken fresh from the field. By storing they lose a proportion of the water which they naturally contain; and there are also some chemical changes which take place in them tending to render them more nutritious.

When turnips are allowed to remain in the field until the leaves begin to put forth a fresh growth, as they will be found to do early in spring, a decided deterioration in their quality is the result, owing to certain of their elements becoming changed into indigestible woody fibre. Hence the necessity for storing turnips at the proper season, say in November and December.

Composition of Turnips.—The following table gives the average composition of five varieties of turnips, as deduced by Cameron from the results of the analyses of Anderson and Voelcker:—

	Swedes.	White Globe.	Aberdeen Yellow.	Purple-top Yellow.	Norfolk Turnip.
Water	89.460	90.430	90.578	91.200	92.280
Albuminoids	1.443	1.143	1.802	1.117	1.737
Sugar, &c.	5.932	5.457	4.622	4.436	2.962
Fibre	2.542	2.342	2.349	2.607	2.000
Ash	0.623	0.628	0.649	0.640	1.021
	100.000	100.000	100.000	100.000	100.000

The means of the analyses of 66 differently grown lots of roots, numbering in all nearly 3000 bulbs of Fosterton hybrid turnips, gave the late Dr Aitken[1] the following results on soils at Pumpherston and Harelaw respectively :—

	Pumpherston.	Harelaw.
Water . .	91.3	92.6
Dry matter .	8.7	7.4
	100.0	100.0

Composition of dry matter :—

		Pumpherston.	Harelaw.
Albumen	. .	7.7	7.5
Fibre .	. .	10.8	11.7
Ash .	. .	5.8	6.4
Carbohydrates (sugar), &c.	. .	75.7	74.4
		100.0	100.0

The mean results of 27 somewhat more detailed analyses of Aberdeen yellow turnips, comprising about 800 roots, grown with a great variety of manures at Carbeth, Stirlingshire, gave Mr David Wilson, jun., the following figures :[2]—

	In fresh roots.	In dry matter.
Water . .	91.09	...
Sugar . .	4.72	52.94
Fibre . .	1.03	11.54
Albuminoids .	0.54	6.06
Non-albuminoid nitrogen × 6.25 .	0.60	6.76
Extractive matter free of nitrogen .	1.36	15.23
Ash . .	0.66	7.47
	100.00	100.00

Variation in Composition of Turnips.—The quantity of nutritive matter in the same variety of the turnip varies greatly. In white turnips it may vary from 7 to 11 per cent, and in the yellow turnip from 8 to 13 per cent.

In an interesting paper contributed to the *Transactions of the Highland and Agricultural Society* for 1906 by the Society's chemist, Mr James Hendrick, much useful information is given regarding the variation in the composition of turnips, and also dealing with the increasing of the sugary contents of roots.

Sugar in Roots.—Inasmuch as feeding-roots are essentially *sugar crops*, the sugar they contain is very valuable for meeting the respiratory requirements of sheep and cattle, also for fat-forming and for milk-production. The following table, based on the experiments of Sir John Bennett Lawes and Sir Henry Gilbert at Rothamsted, records estimates of the approximate average percentages of dry matter, and of sugar, in the kinds of roots mentioned :—

	Dry matter.	Sugar per cent.	
		In fresh roots.	In dry matter.
	per cent.	per cent.	per cent.
White turnips .	8.0	3.5 to 4.5	44 to 56
Yellow turnips .	9.0	4.0 to 5.0	44 to 56
Swedish turnips	11.0	6.0 to 7.0	55 to 64
Mangel . . .	12.5	7.5 to 8.5	60 to 68

A bushel of turnips weighs from 42 lb. to 45 lb.

Excess of Water in Roots.—In feeding with roots farmers are sometimes apt to forget or overlook with how much water the feeding matter in the roots is associated. Unless an animal gets some dry food as well as roots, it is forced, in order to obtain sufficient solid nutriment, to consume a very large quantity of water — very much more, in cold weather, than is necessary for it. This water when swallowed has to become warmed at the expense of the heat of the animal, which has simultaneously to be replaced by fresh heat—so that part of the sugar, &c., of the roots, instead of going to fatten the animal, is wasted in furnishing fuel to warm the superfluous water swallowed in the root-substance.

Desirability of Economising Turnips.—Over and over again, in various parts of this work, prominent reference is made to the great and avoidable waste which thus takes place in the old-fashioned and time-honoured system of turnip-and-straw feeding. It is needless here to reason out the point at great length. The above statement as to the deleterious influence of the excess of cold water the animal has to swallow in a full meal of roots will suffice here for that part of the subject. It is also unnecessary to enter into any lengthened arguments to show that other reasons exist which make it

[1] *Trans. High. Agric. Soc.*, xvi. 1884.
[2] Ibid., xviii. 1886.

very desirable that the more economical use of roots in the rearing and feeding of stock should be practised. The root crop is a very costly one to grow, and unfortunately its cultivation is attended with great risks of loss from unfavourable weather, and fungoid and insect attacks. In dealing with the practical work of feeding the different kinds of stock, special attention is given to the question of how turnips may be most effectually and satisfactorily economised.

The avoidance of this waste is the great plea in favour of giving sheep in the turnip-fields a small daily allowance of cake or corn. They will then eat less of the roots, but will turn what they do eat to much better account.

Turnip-tops.—As a rule, it is better to leave turnip-tops on the field, for they possess considerable manurial value, and, except when other food is scarce, will give a better return in that way than used as food. Still, they contain more nutritive matter than some would imagine, and are useful when scattered on a green field for the use of young cattle or sheep. They should, however, be given with caution, for when eaten too freely they are apt to produce scour. The ash of turnip-tops contains a large quantity of phosphate of lime and potash.

Mangel-wurzel.

This is a most valuable root, grown extensively and with great success in England and Ireland. It needs a warm climate, and is grown in Scotland only to a very limited extent. The orange globe and long yellow kinds have been found to contain a larger amount of the respiratory or fat-forming elements than the long red variety, which agrees with the practical results obtained by the use of those varieties in feeding cattle.

Advantages of Storing Mangels.— The tendency in fresh mangels to produce scour when these are given to cattle is well known to all who have used them, and so also is the fact that this property disappears after the roots have been stored for two or three months. Like good wine, mangels improve by keeping, and it is desirable, as a rule, to delay the consumption of them till spring.

In comparison with turnips, it has been considered that 75 lb. of mangel are equivalent in feeding value to 100 lb. of turnips; but the two varieties vary so much in nutritive value that these proportions cannot be relied upon. The leaves of the mangel are also useful, especially for milch cows, but have a scouring tendency.

The solid matter in mangels ranges from 10 to 16 per cent, but about 12 per cent is general.

Medium v. Large Roots.—It is to be borne in mind, with reference to both turnips and mangels, that moderate-sized roots are commonly more nutritious than very large ones. The huge over-sized roots often seen at root-shows are commonly watery, and such dry matter as they do contain is intrinsically less valuable than in normal roots.

Sugar-beet.

Sugar-beet has given excellent results in the feeding of dairy-cows, but as food for stock it is cultivated only to a very limited extent.

Considerable attention has been given to the cultivation of sugar-beet for the production of sugar, and the late Dr A. Voelcker published the results of some very elaborate investigations made by him as to the composition of sugar-beets grown under different circumstances.[1]

Nutritive Value of Sugar-beet.— Dr A. Voelcker considered that the farmer "will run very little risk in trying the experiment to grow sugar-beets instead of common mangels; for although he may not get so heavy a crop as he does when he plants common mangels, it has to be borne in mind that 1 ton of sugar-beet is equivalent, in nutritive qualities as cattle-food, to at least 1½ ton of good common mangel."

Beet-root Pulp.

Beet-root pulp is the refuse left in extracting the sugary juice from the beet-root. It is much esteemed on the Continent for its fattening properties. It is, however, deficient in flesh-forming compounds, and requires the addition of some cake or meal to supply this deficiency. Cotton-cake is recommended

[1] *Jour. Royal Agric. Soc. Eng.*, vii., sec. ser.

for this purpose. Milch cows fed on beet-root pulp and a fair allowance of bean-meal or cotton-cake produce abundance of milk of good quality. Pigs also thrive on the pulp if they get some bean-meal or pea-meal mixed with it. Dr A. Voelcker considered beet-root pulp at 12s. a ton a cheap and valuable food.

Carrots.

The carrot does not contain any appreciable quantity of starch, but this deficiency is counterbalanced by its having about 6½ per cent of sugar. Carrots are excellent food for horses, and are greatly relished by them.

Carrot-tops are admirable food for cows giving milk.

Parsnips.

Parsnips contain more starch, but less sugar, than carrots. The starch in parsnips exists only in the external layers of the root, none whatever being found in the heart. There is nearly double the quantity of solid matter in parsnips of that in turnips; so that 1 ton of parsnips ought to go as far, as a fattening material, as 2 tons of white turnips.

Kohl-rabi.

Kohl-rabi is a valuable food, especially for milch cows. It increases the yield of milk, but does not impart to it any particular flavour of a disagreeable kind, such as is produced by turnips. The leaves of kohl-rabi form an excellent description of food for cattle and sheep.

Cabbages.

Cabbages are not cultivated anything like so extensively as they deserve to be. Cabbages are excellent food for sheep and other stock, and few other crops will give as good a return per acre.

Thousand-headed kale of the cabbage variety is most valuable as a green food for sheep or cows in autumn, early winter, or spring. Sprouting broccoli and winter greens are also cultivated for similar purposes.

Potatoes.

The demand which exists for potatoes as human food, generally renders them too expensive to be employed largely in feeding animals, although as food for most kinds of stock they are valuable.

Value of Potatoes for Cattle.—It has been stated that when potatoes can be purchased for £1, 10s. or £2, 10s. per ton, they will pay to be employed in feeding cattle. This, however, will depend upon circumstances which are liable to variation, such as the market price of other foods, and the selling price of beef. Second and small-sized potatoes are as useful for feeding purposes as larger tubers; and as the potato is a bulky and therefore an expensive article to send a long distance to market, those who grow potatoes to some extent in remote districts will be able to turn their crops to better account by converting the small tubers into meat than by selling the entire crop in its natural state. It is necessary to give potatoes to stock with caution, as the excess of starchy matter, unless counteracted by other foods, may injure the health of the animals.

There are many farmers who have an objection to potatoes as food for cattle. The late Mr M'Combie of Tillyfour said: "I would rather throw potatoes to the dunghill than give them to a store bullock, though I would give them to my fattening bullocks." He would never give them to animals intended to be afterwards grazed.[1]

The albuminoids, or flesh-forming matters, in potatoes are deficient, but there is an exceptionally high percentage of the respiratory or fat-forming elements, which constitute the largest part of the dry matter. For cattle, therefore, potatoes should be used in combination with such other foods as cotton-cake, bean-meal, or pea-meal.

Utilising Diseased Potatoes.—When the potato is attacked with disease, it is the albuminoids, or flesh-forming constituents, that are destroyed: these are partially converted into ammonia and other volatile matters, and hence the offensive smell which is emitted from diseased potatoes. The starch, &c., remains intact, and advantage is taken of this by employing diseased potatoes in the manufacture of starch.

---

[1] *Cattle and Cattle-Breeders*, p. 13.

Diseased potatoes may, for the same reason, be turned to account for feeding animals, particularly swine. In order to do this, it is necessary to thoroughly cook the potatoes either by boiling or steaming—the latter, when practicable, being the better way—and then pack the cooked potatoes into flour-barrels or casks, ramming them well down, and sprinkling some salt occasionally through the mass. When the barrel or cask is filled to the top, it must be closed from the air, and the potatoes will keep for some time fit for use.

Potatoes for Horses.—Potatoes are often fed to horses, but when freely given in a raw condition, they are liable to produce colic.

Water with Potatoes.—Water should not be given to animals fed on either raw or cooked potatoes, for some time after the meal.

Green Rape.

Rape in a green or growing state is usually fed off with sheep, or cut and used as soiling food for house-fed cattle. It is a nutritious and valuable plant for these purposes, and for spring and autumn food it should be grown much more extensively than it is.

Furze, Whins, or Gorse.

Like many other useful and beautiful plants indigenous to this country, furze —in some parts called whins, in others gorse—is not so highly esteemed as it ought to be, perhaps on account of its being so common, and of its tendency to grow where it has not been sown and is not wanted. Nevertheless, as food for cattle, sheep, and horses, it possesses very considerable value, and for this purpose it may be grown in any part of the country with success, financially and otherwise.

Furze as Winter Food.—The chief value of furze is as a green food for the winter months. It should be cut at least once every year, so that the plants may not be allowed to become too woody and hard. When sown thickly on fairly good land the shoots come up fine and juicy, growing to a length of from 2 to 2½ feet. The crop may be cut with the scythe, or with a strong mower past its best for regular harvest work —generally with the scythe.

Preparing Furze as Food.—Before being given to stock the furze should be cut into short pieces by a strong chaff-cutter, or, better still, bruised and cut by a machine which has been specially designed for the purpose, aptly named the "Masticator," and made by Mackenzie & Sons, Cork. This useful machine may be driven by horse, water, or steam power.

Some think it desirable to chop or masticate the furze daily as required; but others perform this work twice a-week, and find that the chop keeps well enough.

How fed to Stock.—An agriculturist, who had thirty years' experience of furze as food for stock, said: "Cut up the furze with hay for milking cows, and you will make first-quality butter, but pale—with hay for horses, but do not feed too heavily; add 3 or 4 lb. mangels to counteract a resin the furze contains. Young stock thrive amazingly upon it. Furze-fed cattle are hard to be fattened on other food; oaten straw, with cut furze for them."

Mr R. O. Pringle stated that horses may be kept through the winter on furze without hay, and only a moderate allowance of oats; and the furze gives the horses a fine coat of hair. An acre of well-grown young furze, which is regularly cut, will keep four or five horses or cows during the winter and early spring months with very little assistance in the shape of hay or roots. For hard-working horses it should be accompanied by a liberal allowance of bruised oats or other concentrated food. Both horses and cattle take to it readily, but sheep do not eat it willingly except when there is snow on the ground. When grown as food for sheep, the crop is not cut, and in a snowstorm a few acres of young juicy furze are most valuable for sheep.

Composition of Furze.—That furze should, in practice, prove to be a useful food, will not surprise any one when it is considered that it contains about 3.20 per cent of albuminoids and about 9.40 of sugar, digestible fibre, &c. Its proportion of water is about 72 per cent, and fibre 13.33 per cent.

Grasses and Clovers.

It may seem strange, but it is nevertheless true, that farmers possess less reliable knowledge as to the relative cropping and feeding value of the various grasses which cover their pastures, than as to that of any of the other leading crops of the farm. A good deal of fresh knowledge as to the habits and characters of our grasses has, no doubt, been gained in recent years, but investigations and experiments must be prosecuted much further before it can be said that we know our pastures and our hay crops as well as we do our crops of roots and grain.

There are special circumstances which render the investigation of this subject very difficult. Root and grain crops are usually matured, or almost so, before they are fed to stock, and thus it has been comparatively easy to obtain reliable information as to the average composition of food-mixtures consisting of these commodities. With grasses, however, especially those which are constantly grazed, the case is different. They are eaten at all stages of their growth, in extreme youth, full bloom, old age, and all the intervening stages. It is well known that the nutritive properties of plants vary at different stages of the development of the plants; and it does not follow that all grasses which show a useful composition when fully grown, are equally useful and suitable for grazing by stock in the earlier periods of their growth. Again, plants which would not stand well in an analysis of hay, may be extremely serviceable as an element in pastures to be regularly grazed.

The determining of the amount of nutrition—the grazing value—possessed by the different plants which compose our pastures, sown and natural, temporary and permanent, is thus at once a matter of the greatest difficulty and the utmost importance. No subject could more worthily engage the attention or employ the resources of the leading agricultural societies; and it is a matter, too, in which a great deal of good might be done by private experiment and investigation by farmers who have opportunities of studying their pastures, and watching the progress of the animals which feed upon them.

Composition of Grasses and Clovers at Different Stages of Growth.—In the 'Eighties of the last century the most extensive investigation ever carried out in this country regarding the nutritive value and produce of British-grown grasses was conducted by Mr (now Dr) David Wilson at Carbeth, Stirlingshire. Much valuable information on the subject is afforded in lengthy reports on these investigations which appear in the *Transactions of the Highland and Agricultural Society* for the years 1886 and 1889. From the later of the reports are taken the tables on page 285, giving (1) the Average Composition of Grasses at Different Stages of Growth, and (2) the Average Composition of Clovers at Different Stages of Growth.

Feeding Value of Clovers.—As to the clovers, which occupy a prominent place in pastures as well as in hay, they are well known to possess high feeding properties.

Alsike clover, which is much esteemed for damp soils, and is specially suited for meadows, shows a rather higher nutritive value than either of the other clovers.

In composition, lucerne and sainfoin closely resemble the clovers.

Composition of Grasses.—Mr Martin John Sutton's valuable work, *Permanent and Temporary Pastures* (the first edition of which was issued in 1886), contains a series of analyses of the principal agricultural grasses, made specially for the purpose by Dr John A. Voelcker. A precise and clear description of each grass accompanies the analysis, and this, with the beautifully coloured illustrations of grasses, and Mr Sutton's practical directions as to the formation and treatment of temporary and permanent pastures, renders the work one of remarkable value.

Dr J. A. Voelcker explains that each variety thus analysed was grown separately and was perfectly pure—the sample being taken, in every instance, as nearly as possible at the time when it would have been cut for hay. And the analysis of each grass is shown in its natural state and dried at 212° Fahr.—that is, until nothing but the solid or dry matter remained.

(1) Average Composition of Grasses at Different Stages of Growth.

| | Number of Analyses Averaged. | Percentage Water in Green Grass. | Composition of 100 Parts Dry Matter. | | | | | | Percentage of total Nitrogen Non-Albuminoid. | Percentage of Albuminoids Digestible. | Comparative Value of Dry Matter. Cocksfoot, 1st cut=100. |
			Total true Albuminoids.	Non-Albuminoid Nitrogen ×6.25.	Fat, Wax, and Chlorophyll.	Extractive Matter free of Nitrogen.	Ash (including Sand).	Fibre.			
Young undergrowth .	21 complete [1]	74.50	13.66	3.13	4.64	46.23	11.43	20.91	18.6	60.1	137
Longer undergrowth, before shooting	6 complete 11 partial [2]	74.92	9.08	2.80	2.55	47.71	10.31	27.55	23.5	60.1	110
Shot, but before bloom, one or two beginning to show bloom	11 complete 12 partial [3]	76.26	7.93	2.84	3.02	47.65	9.07	29.49	26.3	52.6	102
Coming into bloom and till full bloom	9 complete 6 partial [4]	72.59	8.30	2.56	2.34	47.53	8.47	30.80	23.5	55.8	103
After bloom till seed full-sized	2 complete 5 partial [5]	67.33	5.77	1.24	1.49	52.78	6.85	31.87	17.7	..	88
Seed full-sized till ripe	13 complete 7 partial [6]	65.66	5.29	1.22	2.27	48.51	7.95	34.76	18.7	44.6	83

The analyses of the following cuttings of grasses were averaged to obtain the above mean composition for each stage of growth :—

[1] 3d cut of cocksfoot, foxtail, tall fescue, meadow fescue, sweet vernal, golden oat, timothy, perennial rye-grass, wood meadow, rough-stalked meadow, crested dogstail, hard fescue, fine-leaved sheep's fescue, various-leaved fescue, smooth-stalked meadow, Hungarian forage grass, floating sweet grass, reed canary, wood fescue, and 1st cut of florin and floating sweet grass.

[2] 1st cut of timothy, Hungarian forage, reed canary, and wood fescue. 2d cut of various-leaved fescue, smooth-stalked meadow, Hungarian forage, floating sweet, reed canary, and wood fescue. 3d cut florin. 1st cut floating sweet. 2d cut florin, various-leaved fescue, smooth-stalked meadow, floating sweet, and reed canary.

[3] 1st cut of cocksfoot, tall fescue, meadow fescue, golden oat, P. rye-grass, rough-stalked meadow, crested dogstail, tall oat, and various-leaved fescue. 2d cut of hard fescue and florin. 3d cut of tall oat. 1st cut of wood fescue. 2d cut of cocksfoot, tall fescue, meadow fescue, sweet vernal, timothy, P. rye-grass, rough-stalked meadow, crested dogstail, hard fescue, and Hungarian forage.

[4] 1st cut of foxtail, sweet vernal, wood meadow, hard fescue, fine-leaved fescue, and smooth-stalked meadow. 2d cut of rough-stalked meadow and tall oat. 1st cut of florin, Hungarian forage, and reed canary. 2nd cut of foxtail, golden oat, tall oat, and wood fescue.

[5] 2d cut of cocksfoot, tall fescue, meadow fescue, and golden oat. 1st cut of timothy and various-leaved fescue. 2d cut of wood meadow.

[6] 2d cut of foxtail, sweet vernal, timothy, P. rye-grass, wood meadow, crested dogstail, fine-leaved sheep's fescue. 1st cut of cocksfoot, foxtail, tall fescue, meadow fescue, sweet vernal, golden oat, P. rye-grass, wood meadow, rough-stalked meadow, crested dogstail, hard fescue, tall oat, and smooth-stalked meadow.

(2) Average Composition of Clovers at Different Stages of Growth.

| | Number of Analyses Averaged. | Percentage Water in Green Clover. | Composition of 100 Parts Dry Matter. | | | | | | Percentage of total Nitrogen Non-Albuminoid. | Percentage of Albuminoids Digestible. | Comparative Value of Dry Matter. Cocksfoot, 1st cut=100. |
			Total true Albuminoids.	Non-Albuminoid Nitrogen ×6.25.	Fat, Wax, and Chlorophyll.	Extractive Matter free of Nitrogen.	Ash (including Sand).	Fibre.			
Very young leafy growth . . .	1 complete [1]	86.78	23.75	4.02	5.12	40.16	11.07	15.88	14.5	66.5	201
Young growth before bloom . . .	6 complete [2]	80.92	15.70	4.36	3.74	47.60	11.41	17.19	21.7	67.3	156
In bloom . . .	7 complete 6 partial [3]	79.71	13.46	3.87	2.96	47.18	10.93	21.60	22.3	63.2	137
After bloom . .	2 partial [4]	81.51	10.66	..	..		..	29.86	..	..	116

The analyses of the following cuttings of clovers were averaged to obtain the above mean composition for each stage of growth :—

[1] 1st cut of perennial red clover 1885.

[2] 3d cut of perennial red clover 1885 ; 1st and 3d cuts perennial red (1887), 3d cut Dutch (1887), 1st and 3d cuts alsike (1887).

[3] 1st cut perennial red 1885, 2d cut and 1st and 2d cut perennial red (1887), 1st cut Dutch (1885), 1st and 2d cut and 1st and 2d cut Dutch (1887), 1st cut alsike (1885), 2d cut and 1st and 2d cut alsike (1887).

[4] 2d cut and 2d cut perennial red (1885).

The following analyses of four of the grasses, taken from this volume, will indicate the great value of the work to practical farmers :—

	COCKSFOOT.		MEADOW FOXTAIL.		RYE-GRASS.		MEADOW FESCUE.	
	Grass in natural state.	Dried at 212° Fahr.	Grass in natural state.	Dried at 212° Fahr.	Grass in natural state.	Dried at 212° Fahr.	Grass in natural state.	Dried at 212° Fahr.
Water . . .	60.74	...	55.58	...	62.01	...	71.04	...
Soluble albuminoids [1]	.25	.62	.50	1.13	.38	1.00	...	...
Insoluble albuminoids [2]	1.50	3.81	2.56	5.75	2.06	5.38	1.13	3.88
Digestible fibre . .	11.30	28.78	14.22	32.01	7.98	21.01	8.91	30.77
Woody fibre . .	16.24	41.36	16.42	36.96	17.71	46.62	12.51	43.19
Soluble mineral matter [3]	2.04	5.19	2.58	5.81	2.90	7.64	1.05	3.62
Insoluble mineral matter [4]	.91	2.32	.94	2.11	.78	2.05	.64	2.21
Chlorophyll, soluble carbohydrates, &c. . .	7.02	17.92	7.20	16.23	6.18	16.30	4.72	16.33
	100.00	100.00	100.00	100.00	100.00	100.00	100.00	100.00
[1] Containing nitrogen .	.04	.10	.08	.18	.06	.16	...	...
[2] Containing nitrogen .	.24	.61	.41	.92	.33	.86	.18	.62
Albuminoid nitrogen .	.28	.71	.49	1.10	.39	1.02	...	...
Non-albuminoid nitrogen	.18	.46	.30	.67	.38	1.00	.18	.62
Total nitrogen .	.46	1.17	.79	1.77	.77	2.02	.36	1.24
[3] Containing silica . .	.35	.89	.37	.83	.05	.13	...	...
[4] Containing silica . .	.51	1.29	.52	1.17	.32	.84	.39	1.35

Hay.

Loss in Hay-making.—In considering the feeding value of hay it must be borne in mind that the analyses of fresh grasses cannot be relied upon as a key to the nutritive properties and value of hay made from these grasses. If hay were simply *dried* grasses and clovers, then there would be no loss of nutriment and no deterioration in feeding value—water only would have passed away. But in farm practice it is impossible to secure this. In hay-making, even in the best of weather and under the most careful management, there always will be some loss of feeding material.

Damaged Hay.—Here it will suffice to state that in the process of making, hay may be so much spoiled as to be almost worthless as food for stock. In experiments at Rothamsted it was found that sheep would increase in weight on well-made hay alone; but in experiments conducted by Dr A. Voelcker, and lasting three months, it was shown that, fed wholly upon hay which had been damaged by wet weather in making, sheep actually lost in weight. This result proves at once the great importance of exercising skill and care in hay-making, and the imprudence of attempting to maintain stock solely on damaged hay.

It thus becomes apparent that, in giving hay to stock, the physical condition as well as the original quality of the hay must be carefully considered, and the quantities of it and accompanying other foods regulated accordingly.

Feeding Value of Hay.—In average years the price which good hay commands for use as food for horses is usually too high to admit of its extensive employment in the feeding of cattle. If the better qualities of hay are used for cattle at all, they are generally given sparingly. Secondary qualities, especially of meadow-hay, are fed largely to both cattle and sheep.

Straw.

Even yet the value of straw as food for stock is not fully understood and appreciated by the general body of farmers. In recent years more prominent attention has been given to the subject, and much good will be done if farmers are induced to exercise greater care in the utilisation of straw. A large quantity of straw must no doubt be used as litter for cattle and horses; but in many cases, especially when hay is scarce and dear, it will be found advantageous to substitute, say, peat-moss as litter, and utilise the straw, at least oat-straw, which is the most easily digested, for feeding purposes. In any case there should be no waste — no more straw under the cattle than they can effectually convert into manure, and—what is still more unsightly—no loose bundles or handfuls of straw lying about the steading where no straw should be.

Feeding Value of Straw.—The value of straw as a feeding material depends not only on the kind of grain to which it belongs, but also on its condition as regards ripeness when it is harvested, and on the land and climate where it is grown. The straw of grain which is cut just as the grain is ripe, while there still remains a tinge of green in the straw, is much more nutritious than that which has been allowed to become over-ripe. Strong, coarse straw is, of course, not so much relished by cattle as that which is finer in the growth.

The preference which is given to oat-straw as food for stock is fully justified. Fine oat-straw, cut before the crop has become quite ripe, is relished alike by cattle, sheep, and horses, and is given extensively to all, especially in the northern parts where the oat is the prevailing cereal. Indeed in many parts of Scotland good oat-straw (given with a small allowance of roots and perhaps a little cake) forms the main portion of the winter food of young store cattle and dry cows. In pea- and bean-straw it will be seen there are also high feeding properties; while wheat- and barley-straw, although less nutritive, likewise possess considerable value as food for stock.

Straw-chaff.—When straw is used as feeding material, it is given either in its natural state, as it comes from the threshing-machine, or it is cut into very short lengths by a machine constructed for the purpose, in which state it is known as straw-chaff or chaffed straw. The latter method is by far the more economical, as by it the amount of waste may be reduced to a minimum. In order to render straw-chaff more palatable to animals, it is either moistened with treacle mixed with water, or it is employed to absorb a quantity of linseed-meal gruel. The dry chaff is also mixed with the oats given to horses, as such admixture has the effect of causing the horses to masticate their oats more fully than they might otherwise do. Then in pulped mixtures straw-chaff is very extensively employed; and it has been clearly proved by experience that by the economical and careful use of cut straw and concentrated foods a greatly increased stock of cattle and sheep may be kept.

Preparing and Storing Straw-chaff.—There is perhaps no better way of turning straw to good account than by cutting it into chaff and storing it for a few months in large quantities with a slight admixture of chaffed green food, salt, and perhaps ground cake. By this system straw remaining over from the previous winter is cut into chaff in spring or summer and stored in barns till the following winter, when it is consumed; thus enabling the farmer to hold over a quantity of the fresh straw for similar treatment next spring or summer. The system is pursued with great success by many farmers, and particularly when roots are scarce the straw-chaff thus prepared will be found most valuable.

Some now use pulped mangels instead of vetches or rye for mixing with the dry straw-chaff as it is being stored, and get better results therefrom.

Compound Cakes and Meals.

Compound or mixed cakes and meals are, as already mentioned, very largely used. Nearly all manufacturers now make a number of such cakes and meals for various purposes, just as manure manufacturers make up special mixed manures for various crops. Thus we have compounded cakes and meals for fattening animals, for dairy cows, for young stock, for pigs, and so forth.

In the manufacture of such cakes, oil-cakes, like cotton-seed cakes, rape-cake, earthnut-cake, and the minor oilcakes are largely used. These are mixed with various cereal by-products like rice meal, dried dreg, and various wheat and maize by-products, and with sweetening materials like locust bean and treacle. Strongly-flavoured spice seeds, like aniseed, fenugreek, dill seed, &c., are also generally introduced in small quantity to give an appetising odour and flavour.

Such mixed cakes are of various qualities. Unfortunately it is easy to mix inferior materials with such good food substances as those mentioned above, and it is difficult to detect admixtures of such valueless substances as ground oat-husks, sweepings, ground screenings of cereals, and such other substances. Purchasers should always be careful to buy mixed cakes and meals on analysis, and they should in particular look to it that the percentages of albuminoids and oil are high, and that the percentage of fibre is low. They should also see that the analysis does not show the presence of more than a very small proportion of sandy matters.

Condimental Foods.

In modern times remarkable ingenuity and enterprise have been directed to the production of "condimental cattle foods" and appetising spices for all kinds of farm live stock. No attempt will be made to describe the composition and individual characteristics of these condimental foods. It may be well, however, to advise farmers never on any account to purchase any of these foods without receiving therewith a warranty as to its freedom from adulteration, and a guaranteed analysis of its chemical composition, in full accord with the provisions of The Fertilisers and Feeding Stuffs Acts. With this analysis before him, and a reference to what is said in this work as to the elements of nutrition in cattle foods, the farmer will be able to form a tolerably correct idea as to the value of the food. A sample of the food may be analysed for a mere trifle, and if it should fall short of the guaranteed analysis, the vendor is fully responsible for the deficiency.

Vetches.

Of all green forage crops, the vetch is the most extensively grown in the United Kingdom. In almost all kinds of soils it can be grown easily, and at comparatively little expense.

Vetches cut when in full bloom, and before seeding, are much relished by all kinds of stock, and it is desirable that this crop should be grown much more extensively than it is.

How Vetches are Fed to Stock.— Vetches contain an exceptionally high proportion of albuminoids, and they are thus very suitable for giving to stock along with starchy foods, such as rice-meal and Indian corn-meal. For this same reason it is not desirable to give highly nitrogenous foods, such as decorticated cotton-cake and beans, along with vetches, for then the food would be badly balanced—there would be an unprofitable, even a dangerous, excess of albuminoids. Vetches should be sown in successive patches, so as to afford a continuous supply of fresh food. When cut just before becoming fully ripe, vetches make excellent hay or silage. Vetches are also largely consumed on the land by sheep.

Green Maize.

Maize has never been grown to any great extent as a forage crop in this country. It would be a gain to British farmers if it could be successfully acclimatised, for maize is capable of producing an enormous yield of succulent food, which is much relished by cattle, and which is well adapted for feeding in a fresh condition along with other foods, such as chopped hay or straw and decorticated cotton-cake, or for converting into silage for winter feeding.

For use as a forage crop, maize is cut green, and before the cobs have formed. Its chemical composition in this form has been found to vary greatly.

Green maize is not a rich food. Its merit lies in the great quantity of palatable succulent food it produces per acre. It is deficient in nitrogen, but along with highly nitrogenous foods such as decorticated cotton-cake it is most suitable and acceptable to all kinds of stock. It has been found that 120 lb. of green maize and 2 lb. of decorticated cotton-cake made an excellent food for dairy-cows.

In a dry summer, when grass is scarce, green maize, chopped and mixed with chaffed straw or hay, will be found to be a valuable food for cows or young cattle. A sprinkling of crushed decorticated cotton-cake—1½ or 2 lb. per head—would make this a nutritive mixture for cows giving milk.

Sorghum.

Sorghum saccharatum is a tall-growing plant, similar in appearance to maize, but finer in the stem. In warm climates it grows with great luxuriance, and when cut green, forms excellent forage for stock. It is hoped that hardy varieties of it may be raised, so that it may be successfully cultivated in this country. As yet experiments have been confined to the south and centre of England, and so far the experience has been variable.

Sorghum is exceptionally rich in sugar, and is therefore specially suitable for feeding along with decorticated cotton-cake. Fed alone to cows it has a tendency to cause looseness; but 2 lb. of decorticated cotton-cake to 100 lb. of green sorghum corrects this, and makes an excellent daily ration for cows in milk.

Lucerne.

Lucerne is exceptionally rich in albuminous matters, and is even more nutritious than red clover. It affords a large yield, under favourable circumstances sometimes close on 20 tons per acre; and is most useful when sown in a small patch near the steading, to be cut as required for consumption.

Young lucerne given alone, or as the principal food, has a tendency to cause the animals to become blown; but this danger is avoided by giving it along with straw, the two being chaffed together—an excellent method of turning straw to good account as food.

Sainfoin.

Sainfoin is peculiarly valuable in sheep-farming districts, and seems to sustain little or no permanent injury by being grazed by sheep. Unlike lucerne, it has no tendency to cause blowing in cattle.

Sainfoin is less nutritious than lucerne, and does not give nearly so large a yield per acre.

Prickly Comfrey.

There is much difference of opinion as to the value of prickly comfrey as a forage crop. It is a hardy and prolific plant; and in good soil, well manured, will afford a large yield. There is probably no forage-plant that has made warmer friends or more bitter enemies than prickly comfrey. It is a somewhat coarse watery food, not much relished by cattle at the outset, but useful as a green food for dairy-cows.

Dr Augustus Voelcker considered that prickly comfrey "has about the same feeding value as green mustard, or mangels, or turnip-tops, or Italian rye-grass grown on irrigated lands."

Sugar.

In animal economy sugar and starch perform similar functions, and experiments conducted by the late Sir John Bennett Lawes [1] showed that, "whether for the purpose of supporting the functional actions of the body, or of ministering to the formation of increase, . . . starch and sugar have, weight for weight, values almost identical. . . . Starch and sugar, therefore, as foods, appear to be equivalent; or, in other words, a pound of one, properly used, can produce no more increase in our stock than a pound of the other."

Remarking upon the exaggerated value which had been placed upon sugar as a food for stock, Sir John Bennett Lawes stated that it is nevertheless an excellent food; and that the only question is, what price is sugar worth (in comparison with other foods) for feeding purposes?

Sir John considered that it would not be advisable to use sugar with such foods as cereal grains, maize, rice, roots, or even meadow-hay, as all these are somewhat low in nitrogen; and to dilute the nitrogen that exists still more, by the use of sugar, would tend to waste it. On the other hand, foods containing a large amount of nitrogenous substance, such as leguminous seeds—especially lentils, tares, and beans—as well as linseed-cake,

[1] *Jour. Royal. Agric. Soc. Eng.*, vii. 388.

cotton-cake, and clover-hay, might be safely diluted with sugar.

Fish Products as Animal Food.

The frugal farmers of Norway turn fish-offal to use as food for cattle. Their custom has been to boil down the heads of cod-fish into a kind of soup, which they mix with straw or other fodder and give to cattle, and by the means of this cheap and nutritious food many Norwegian farmers have been able to maintain a much larger stock of cattle than would have been otherwise practicable. From the refuse of herring a cattle-

feeding meal of a useful kind is also made in Christiania. Various trials with it were made in this country, but it did not come largely into use.

Fish guano has useful feeding properties, but it has never been used as food to any appreciable extent.

COMPOSITION OF FOODS.

The composition of the common foods used for farm live stock is shown in the following table taken from Warington's unique little volume entitled *Chemistry of the Farm* :[1]—

PERCENTAGE COMPOSITION OF ORDINARY FOODS.

Food.	Water.	Nitrogenous substance. Albuminoids.	Amides, &c.	Fat.	Soluble carbohydrates.	Fibre.	Ash.
Cotton-cake (decorticated) .	8.2	43.2	1.8	13.5	20.8	5.5	7.0
" " (undecorticated)	12.5	20.7	1.3	5.5	34.8	20.0	5.2
Linseed-cake . . .	11.7	26.9	1.1	11.4	33.2	9.0	6.7
Rape-cake	10.4	28.1	4.6	9.8	29.1	10.3	7.7
Earthnut-cake . . .	11.5	45.1	1.9	8.3	23.1	5.2	4.9
Beans	14.3	22.6	2.8	1.5	48.5	7.1	3.2
Peas	14.0	20.0	2.5	1.6	53.7	5.4	2.8
Wheat	13.4	10.7	1.3	1.9	69.0	1.9	1.8
Rye	13.4	10.5	1.0	1.7	69.5	1.9	2.0
Oats	13.0	10.6	0.7	5.4	57.3	10.0	3.0
Barley	14.3	10.2	0.4	2.1	66.0	4.5	2.5
Maize	11.0	9.8	0.6	5.1	70.0	2.0	1.5
Malt sprouts . . .	10.0	16.6	7.1	2.2	44.1	12.5	7.5
Wheat bran . . .	13.2	12.1	2.0	3.7	56.0	7.2	5.8
Brewers' grains . .	76.2	4.9	0.2	1.7	10.7	5.1	1.2
" " (dried) .	9.5	19.8	0.8	7.0	42.3	15.9	4.7
Rice meal . . .	10.3	11.3	1.0	12.0	47.8	8.6	9.0
Oat-straw . . .	14.5	3.5	0.5	2.0	37.0	36.8	5.7
Barley-straw . . .	14.2	3.2	0.3	1.5	39.1	36.0	5.7
Wheat-straw . . .	13.6	3.3		1.3	39.4	37.1	5.3
Pea-straw . . .	13.6	9.0		1.6	33.7	35.5	6.6
Bean-straw . . .	18.4	8.1		1.1	31.0	36.0	5.4
Pasture grass . .	76.7	2.9	1.1	0.9	10.9	5.2	2.3
Clover (bloom beginning) .	81.0	2.6	0.8	0.7	8.0	5.2	1.6
Clover-hay (medium) .	16.0	10.5	2.5	2.5	37.2	25.0	6.3
Meadow-hay (best) .	15.0	10.2	1.8	2.3	39.5	24.0	7.2
" " (medium)	15.0	8.0	1.2	2.2	42.0	25.4	6.2
" " (poor)	14.0	6.3	0.5	2.0	41.1	31.0	5.1
Grass silage (stack) .	67.0	3.3	1.5	1.5	13.2	9.7	3.8
Clover silage (stack) .	67.0	3.3	2.7	2.2	10.5	11.9	2.4
Maize silage . .	79.1	1.0	0.7	0.8	11.0	6.0	1.4
Potatoes . . .	75.0	1.2	0.9	0.2	21.0	0.7	1.0
Cabbage	85.7	1.7	0.8	0.7	7.1	2.4	1.6
Carrots	87.0	0.7	0.5	0.2	9.3	1.3	1.0
Mangels (large) . .	89.0	0.4	0.8	0.1	7.7	1.0	1.0
" (small) . .	87.0	0.4	0.6	0.1	10.2	0.8	0.9
Swedes	89.3	0.7	0.7	0.2	7.2	1.1	0.8
Turnips	91.5	0.5	0.5	0.2	5.7	0.9	0.7

[1] Vinton & Co., Ltd., London.

ANIMAL NUTRITION.

In order to understand the value of foods to the animal it is not sufficient to know their chemical composition, some account of which is given in the immediately preceding pages. The use of the food to the animal, and the part which its different constituents play in animal nutrition, have also to be taken into account. For the appended notes on these subjects the editor is indebted to Mr James Hendrick, B.Sc., F.I.C., chemist to the Highland and Agricultural Society of Scotland.

Constituents of Foods.

The chief constituents of the dry matter of foods may be divided into four classes: (1) proteins, (2) fats, (3) carbohydrates, and (4) ash constituents. In addition to these, foods contain more or less water, and a number of minor constituents such as amide bodies, organic acids, &c.

The essential substances which all animals require to consume in their diet are: (1) water, (2) proteins, (3) fats and carbohydrates, and (4) ash constituents.

In the analysis of feeding-stuffs it is usual to state (1) the moisture or water, (2) the albuminoids, (3) the oil, (4) the soluble carbohydrates, also often called the non-nitrogenous extract, or the starch, sugar, gum, &c., (5) the fibre, and (6) the ash. In this statement the term albuminoids generally means the protein substances, together with a greater or less percentage of other nitrogenous bodies, such as amide substances. In concentrated foods the so-called albuminoids are composed almost entirely of true protein bodies, but in some of the ·bulby foods, such as turnips and mangels, a large part of what is commonly called albuminoids consists of the much less valuable amides.

The term oil in a food analysis indicates, or should indicate, fatty oils or fats. The fats in vegetable foods are generally liquid fats, and hence the rather vague term oil is commonly applied to them. In a feeding-stuff the term oil should be synonymous with fat.

The substances classed as soluble carbohydrates, and as fibre in a food analysis, are practically all carbohydrates. Some of the carbohydrates, such as starches and sugars, are easily dissolved, and these form the soluble carbohydrates; while other carbohydrates, like the celluloses, are very difficult to dissolve, and these form the main part of what is called fibre, or sometimes, quite wrongly, "indigestible fibre," or "woody fibre," in a food analysis.

The ash constituents consist of a variety of more or less valuable mineral substances.

The Functions of Food.

The functions of the food of an animal are (1) to build up the body itself, (2) to supply the body with a source of energy, and (3) to form fat in the body.

The bodies of animals are composed of (1) water, (2) nitrogenous matters, (3) fat, and (4) ash or mineral constituents. Water is, except in the case of very fat animals, the constituent which is present in greatest quantity. It usually forms over half the weight of the body and, except in the case of fat animals, over 60 per cent of the weight.

The nitrogenous matters of the body consist mainly of proteins, but various nitrogenous bases are also present in smaller amount. The muscle and nerve fibres, the blood cells, the skin and, generally speaking, the working mechanism of the body, are built up mainly of protein substances.

The fat of the body is essentially a reserve material, which is stored up from excess of food as a reserve supply of energy-producing material, to be drawn upon when the food-supply is deficient. The store of fat is comparatively small in the working animal. When an animal is starved for some time his fat store may be almost or entirely exhausted. On the other hand, highly fed animals which have not any great amount of work to do lay up large stores of fat, especially in the case of those animals in which the tendency to store up fat has been specially cultivated in order that they may be used for human food.

The ash constituents of the body form a comparatively small proportion of the total weight. They include potash, soda, lime, phosphoric acid, chlorine, sulphuric acid, and other constituents in lesser quantities. They are found in all parts of the body. The bones are specially rich in ash, which consists mainly of lime and phosphoric acid; but the blood, nerves, muscles, skin, &c., also contain mineral constituents.

Carbohydrates are also found in the bodies of animals, but only to a small extent. They are not built up in any large quantity into the tissues, but merely form a small floating reserve of material, which can be immediately drawn upon for the supply of energy.

Proteins and ash constituents must be present in the food in order to supply material for the building of the body. Every animal requires a certain minimum of these two constituents in its food. It has been pointed out above that the essential parts of the body are built up of protein substances and ash constituents, together with water; only protein substances appear to be capable of building up the nitrogenous tissues, and a supply of potash, lime, phosphates, &c., in suitable forms of combination, is essential for building up the bones, and contributing the necessary ash constituents to the other tissues.

A supply of common salt is also essential to the carrying on of certain functions of the body, such as digestion.

Fats and carbohydrates cannot build up the essential parts of the body, but are useful as sources of energy. It is only the smaller portion of the food which is necessary for building up the body and repairing its waste; the greater portion is used as a source of energy. Energy is the power or capacity to do work. The body is constantly engaged in doing work. Even where no external work is done, internal work is constantly being done by the heart, respiratory system, digestive system, &c., and some source of energy is required to enable this work to be done. In the food the body obtains supplies of potential energy, which are capable of being turned to kinetic energy, or energy of motion, when the food undergoes oxidation in the animal. The use of carbohydrates and fats to the animal is to act, so to speak, as fuel which can be consumed in the body to supply power of doing work. Any excess of these materials which is not required for the immediate production of energy may be stored in the body as fat, which is reserve fuel.

The organism is constantly expending energy on internal and external work, and in maintaining the body temperature. It may be said that to supply energy is the function of food. In the last resort most of what has been used in building up the body itself will be consumed as a source of energy. If the organism is not supplied with food it soon exhausts its stores of energy and runs down, and food has constantly to be supplied to maintain the store of energy.

Digestion.

It is only that portion of the food which is digested, that is taken into the blood stream, which is really of use to the animal. Strictly speaking, it is only what is digested that is really food.

The digestive system is a somewhat complicated apparatus for grinding and dissolving the food so as to get it into a fit state for absorption. The food undergoes both mechanical and chemical processes during digestion. It is ground up by the teeth. In ruminants it is soaked, reground, and sifted through the ruminating apparatus before it is fit to proceed to the true stomach, which is commonly called the fourth stomach.

It is acted upon chemically by the saliva, the gastric juice, the pancreatic juice, and the bile. All of these contain substances which assist in breaking up the constituents of the food and rendering them soluble. These dissolving substances are chiefly what are called enzymes or unorganised ferments. Thus the saliva contains an enzyme, ptyalin, which, like the diastase of malted grain, dissolves starch and changes it into sugar.

Much of the material which composes food is in an insoluble state. In the case of fibrous foods like hay and straw the greater part of the carbohydrates is in forms which are very difficult to break down and dissolve. The digestive system of animals which naturally live on a tough fibrous food is large and elaborate, so as to enable them to deal with such food. This is particularly the case with ruminants, which are able to live upon a much more fibrous food than the pig or even than the horse.

The carbohydrates of the food are changed into sugar before they are resorbed. Whatever is the form in which the carbohydrate is taken into the mouth of the animal, whether as sugar, starch, cellulose, or mucilage, it is taken into the blood as sugar. The protein substances are also broken up into the more soluble and simpler proteoses and peptones before they are resorbed. To a certain extent they are broken up still further into amides during the digestive processes. The fat which is digested is first changed into an emulsion, and to a certain extent saponified, that is, turned into soap, by the bile.

In addition to the processes already mentioned, the food undergoes a certain amount of fermentation due to bacteria. Such action takes place especially in the long and complicated digestive systems of animals which feed on fibrous food. Owing to these fermentative actions much cellulose is broken up and some of it rendered available for digestion. At the same time a large amount of gas is produced, as well as other substances such as butyric and acetic acids. The gas which is produced in the intestines is excreted and is really an addition to the matter excreted in the dung. It contains much methane. This subject will be referred to again later.

The portion of the food which is not digested is excreted in the dung. The dung of herbivorous animals consists almost entirely of the undigested part of the food. The digestibility of food is determined by weighing the food and subtracting from it the weight of what is excreted as dung. This is not quite accurate, (1) as part of the digestive secretions of the animal are contained in the dung, and (2) on account of the gas which is excreted from the intestine, which is not weighed with the dung. In the case of herbivorous animals this gas may cause a very serious error, and therefore special means of measuring it have been devised.

In Warington's table,[1] given on p. 294, the digestibility for ruminants of a number of the most common foods is given. These figures express the percentages of the total organic matter, and of the chief constituents, nitrogenous matters (proteins and amides), fat, soluble carbohydrates, and fibre which are digested. Such figures are sometimes called digestion coefficients. They are obtained by weighing and analysing the food eaten by animals over a period, and weighing and analysing the dung excreted over the same period, and calculating the digestibilities of the total organic matter and its different constituents from the difference between the amount eaten and the amount excreted in the dung.

The figures in the table are averages taken from German experiments. Very little investigation into the nutrition of farm animals and the digestibility of their food has been done in this country. For information on such subjects we are almost entirely dependent on foreign, and especially on German, work. In recent years much work on the nutrition of farm animals has been done in the United States also.

It is to be remembered that some of the digestion coefficients, given in tables derived from foreign investigations, may not apply strictly to foods grown in our climate. For instance, straw grown in the cool moist climate of Scotland may differ materially from straw grown in the drier and warmer continental climate of Germany or the United States.

As the table (p. 294) shows, the fibre of a food is by no means indigestible to ruminants; and, on the other hand, it shows that the soluble carbohydrates, fat, and nitrogenous matters are far from being entirely digestible. Generally speaking, the more fibrous a food the lower is the digestibility. But the lowering of the digestibility is not confined to the fibre, but is exhibited by the soluble carbohydrates, fats, and nitrogenous matters. The more fibrous a food the

[1] *The Chemistry of the Farm.* Vinton & Co., Ltd., London.

lower is the digestibility of the total organic matter, including the proteins, fats, and soluble carbohydrates.

In the natural fibrous foods of ruminants, such as hay and straw, the fibre is, generally speaking, almost as well digested as the other constituents. For instance, in a sample of hay of medium quality, a ruminant digests about 60 per cent of the total organic matter, and also about 60 per cent of the fibre. So in a sample of oat-straw a ruminant digests about 50 per cent of the total organic matter, including quite 50 per cent of the fibre.

Another illustration of the effect of fibre is seen by comparing the digestibilities of decorticated and undecorticated cotton - cakes. Undecorticated cotton-cake is much more fibrous than the decorticated cake, and while 76 per cent

EXPERIMENTS WITH CATTLE, SHEEP, AND GOATS. (Warington.)

Food.	Digested for 100 of each constituent supplied.				
	Total organic matter.	Nitrogenous substance.	Fat.	Soluble carbo- hydrates.	Fibre.
Pasture-grass	74	74	64	77	69
Meadow-hay (best)	67	65	57	68	63
" " (medium)	61	57	53	64	60
" " (poor)	56	50	49	59	56
Clover-hay (best)	61	62	60	70	47
" " (medium)	57	55	51	65	45
Lucerne-hay (bloom beginning) . .	62	77	39	70	43
" " (full bloom) . . .	56	70	39	63	42
Maize silage	62	48	85	68	56
Oat-straw	48	20	33	44	54
*Barley-straw	53	20	42	54	56
*Wheat-straw	43	11	31	38	52
*Bean-straw	55	49	57	68	43
*Cotton-cake (decorticated) . . .	81	87	95	76	?
* " " (undecorticated) . .	54	74	90	51	16
*Linseed-cake	80	86	90	80	50
*Peas	90	89	75	93	66 ?
Beans	89	88	82	92	72 ?
Oats	71	78	83	77	26
*Barley	86	70	89	92	?
*Maize	91	76	86	93	58
Rice meal	75	63	85	86	26
Wheat bran	71	78	72	76	30
Malt sprouts	81	78	50	86	85
Brewers' grains	62	70	82	63	39
Potatoes	88	66	?	93	?
*Mangels	88	77	?	96	?
*Turnips	88	62	?	99	?

* These results are derived from a few experiments.

of the soluble carbohydrates of the decorticated cake are digestible, only 51 per cent of the same constituents in the undecorticated cake are digestible.

Generally speaking, concentrated foods like grains and cakes, and succulent foods like roots, are more digestible than coarse fibrous foods like hay and straw. Of the cakes and grains about 80 to 90 per cent is generally digestible. Similarly nearly 90 per cent of the organic matter of roots has been found digestible. On the other hand, in straws generally from 40 to 55 per cent of the total organic matter is digestible, and in hays generally from 55 to 65 per cent is digestible. In succulent green herbage from about 60 to 75 per cent of the total organic matter is generally digestible.

The digestibility of any food will vary according to the kind of animal which eats it. Thus the horse has not so powerful a digestive apparatus as a ruminant, and therefore the digestibility of foods, and particularly of fibrous foods, is lower for the horse than for the ruminant.

In the case of concentrated foods there is less difference in digestibility. Thus the digestibility of the organic matter of medium meadow-hay for the sheep was found to be 59 per cent, but for the horse

only 48 per cent. In the case of oats, however, the sheep digested 71 per cent and the horse 68 per cent, while the digestibility of maize was found to be 89 per cent for both sheep and horse.

The following table, from Warington's work,[1] shows approximately the amount of digestible matter obtained by ruminants from 1000 lb. of material in the case of a number of the best known feeding-stuffs.

DIGESTIBLE MATTER IN 1000 LB. OF VARIOUS FOODS.

	Total organic matter.	Nitrogenous substance.		Fat.	Soluble carbohydrates.	Fibre.
		Albuminoids.	Amides, &c.			
Cotton-cake (decorticated) . .	691	374	18	128	158	13
" " (undecorticated) . .	422	150	13	50	177	32
Linseed-cake	655	230	11	103	266	45
Peas	747	175	25	12	499	36
Beans	733	196	28	12	446	51
*Wheat	786	92	13	15	656	10
Oats	600	81	7	45	441	26
Barley	715	70	4	19	607	15
Maize	786	73	6	44	651	12
Rice meal	612	67	10	102	411	22
Wheat bran	585	90	20	27	426	22
Malt sprouts	681	114	71	11	379	106
Brewers' grains . . .	137	34	2	14	67	20
" " (dried) . .	529	136	8	57	266	62
Pasture grass	156	19	11	6	84	36
Clover (bloom beginning) . .	123	17	8	5	63	30
Clover-hay (medium) . .	440	47	25	13	242	113
Meadow-hay (best) . .	511	60	18	13	269	151
" " (medium) . .	485	40	12	12	269	152
" " (poor) . .	460	29	5	10	242	174
Maize silage	124	1	7	7	75	34
Bean-straw	412	40		6	211	155
Oat-straw	381	7	5	7	163	199
Barley-straw	426	4	3	6	211	202
Wheat-straw	351	4		4	150	193
Potatoes	213	5	9	1	195	3
Mangels (large) . . .	89	1	8	½	74	6
" (small) . . .	109	2	6	½	96	5
Swedes	87	2	7	1	71	6
Turnips	68	1	5	1	56	5

* In the absence of experiments, it is assumed that wheat is digested like other foods of the same class.

The actual amounts digested will vary somewhat with the quality of the food, and with the personal peculiarities of the animal. The figures shown in the table are averages calculated from a number of experiments upon sheep and oxen with foods of different qualities, and with different quantities of foods.

The figures in this table are as important as those in tables of composition in indicating the true nutritive value of foods. The nutritive value depends upon digestibility as well as upon composition.

[1] *The Chemistry of The Farm.* Vinton & Co., Ltd., London.

Metabolism.

The blood nourishes all the tissues of the body. It carries to them the digested food from the constituents, of which the tissues are built up. It also carries to them the oxygen of the air taken in through the lungs. This oxygen combines with and oxydises materials in the tissues, producing oxydised waste products, and at the same time setting free kinetic energy, which enables the muscular tissues to do work, or which appears in the form of heat. These changes by which the tissues are built up by materials from the blood, and subsequently undergo oxidation and degradation, are known as changes of metabolism.

The oxydised waste products, which are the resultants of metabolism, are chiefly carbonic acid gas, water, urea, and salts. These are collected by the blood, and excreted through their several channels. The carbonic acid gas is carried by the venous blood to the lungs, and there given off with the expired breath. The waste water is partly removed in the moisture contained in the expired air, partly evaporates from the surface of the skin, and partly escapes in the urine. The nitrogenous waste of the body is chiefly urea, though other substances, such as uric acid and hippuric acid, are also excreted.

Herbivorous animals excrete a considerable proportion of their nitrogenous waste in the form of hippuric acid. The nitrogenous waste is collected from the blood, together with the waste salts, by the kidneys in solution in water, and is excreted in the urine. The amount of nitrogen in the urine therefore measures the nitrogenous waste of the body. The amount of carbonic acid gas expired by the lungs measures the waste of carbonaceous matter in the body.

Fats and carbohydrates consumed in the body are completely oxydised, and yield carbonic acid gas and water just as if they had been burned in oxygen outside the body. On the other hand, nitrogenous substances, whether proteins or amides, are not completely oxydised. They are only partially oxydised, and part of their carbonaceous material is excreted as carbonic acid gas, but their nitrogen is excreted in the incompletely oxydised forms of urea, hippuric acid, uric acid, &c. These are capable of undergoing further oxidation, and they remove from the body a certain amount of potential energy which has not been utilised in doing work or producing heat.

Water Requirements of Animals.

Along with their food animals always require a large proportion of water. Water is required since in digestion the great part of the food is absorbed in solution, since the food is carried by the blood mainly in the form of a solution, since the waste products are collected in solution, and in the urine and perspiration excreted in solution, and since not only do the tissues contain a large percentage of water, but the changes which take place in them demand the presence of much water. Much water is also excreted in the dung of animals, especially in the case of animals like the ox, which consume much fibrous food, and yield a bulby wet excrement.

Generally speaking, animals require 2 to 5 parts of water by weight for 1 part of dry food. Sheep require only about 2 parts of water to 1 of dry food; horses, 2 to 3 parts to 1; and cattle, 3 to 4 parts to 1. The proportion of water required is increased by a very fibrous diet, or by a diet rich in protein matters. It is also increased by hot weather, which increases evaporation from the body.

Excess of water is wasteful. Unless the water is given warm, heat is consumed in raising it to body temperature and, as the excess is excreted at body temperature, this heat is wasted. This means waste of food used up as fuel to heat the water. Increased consumption of water also leads to increased waste of nitrogenous matter in the body. The excess of water cannot be passed through the body and excreted without causing increased oxidation of protein matter, the nitrogenous waste of which is excreted by the urine.

In the case of sheep fed upon turnips alone the excess of water consumed is very great, and in consequence quite a large proportion of the dry matter consumed is wasted in heating this excess of water and pumping it through the system. It therefore tends to economy of food to give sheep dry food along with turnips.

Energy Value of Foods.

It has been shown above that foods are mainly valuable to the animal as sources of energy. One method of comparing the values of foods, therefore, is to compare their energy values—that is, their values as means of producing heat and work. To do this we measure the fuel value of the food or the amount of heat which it produces when burnt in oxygen. Later, the amount of this fuel value which is recoverable in the animal will be considered.

The heat value of fuels is measured by the calorimeter. In this instrument a given weight of the substance under experiment is burnt in oxygen and the amount of heat produced accurately measured by suitable means. The fuel values of coals, oils, and other combustibles are constantly measured in this way for industrial purposes, and similarly the fuel values of all ordinary foods and food constituents have been measured. It is especially to German and American investigators that we are indebted for our knowledge of food calorimetry.

The unit of heat employed is called the "calorie," and it represents the amount of heat required to raise 1 gram of water 1° Centigrade in temperature. For ordinary purposes the "Great Calorie" or "Calorie," which is a thousand times as great as the "calorie," is employed. It represent the heat necessary to raise 1 kilogram of water 1° C.

The fuel value of 1 gram of fat is about 9.4 calories, of 1 gram of protein about 5¾ calories, of 1 gram of carbohydrate about 4 calories, and of 1 gram of urea about 2½ calories. It will thus be seen that the fuel value of fat is much greater than that of proteins, more than twice as great as that of carbohydrates, and nearly four times as great as that of urea.

Since fats and carbohydrates are useful to the animal only for fuel purposes their comparative values entirely depend on their fuel values. In the case of proteins it is otherwise. The fuel value which is recovered in the calorimeter is not entirely recovered in the animal, for the nitrogenous matter is excreted as urea, which has a fuel value of 2½ calories. The fuel value of the urea excreted has therefore to be deducted from the fuel value of the protein. As a matter of fact, a greater fuel value than is represented by the equivalent of urea has to be deducted, for a portion of the nitrogen is excreted in forms such as hippuric acid which have a greater fuel value than urea.

On the other hand, the proteins have a special value to the animal which fats and carbohydrates have not, for they are essential for the formation of the nitrogenous tissues. They have also a special value to the farmer apart altogether from their use in the nutrition of the animal, for their nitrogenous waste which is excreted in the urine has a high manure value. On the other hand, nothing of manure value is derived from fats or carbohydrates. The actual value of nitrogenous substances, therefore, does not by any means depend entirely on the fuel value.

Energy Value of Foods to the Animal.

As has been shown, only digested food is of any value to the animal. The energy value of this has been measured in a large number of cases with much accuracy by Continental investigators. For this work the respiration calorimeter is needed. It is only in the German and American experiment stations that respiration calorimeters are to be found.

It has been shown that the law of conservation of energy holds true for the animal as for less complicated machines. Fats, carbohydrates, and proteins produce the same amount of energy in the animal as they do when oxydised to the same extent outside the animal. So that if we deduct from the digestible food the energy value of the incompletely oxydised nitrogenous substances excreted in the urine, and the energy value of the methane and other gases excreted from the intestine, we obtain its true energy value to the animal.

The table on p. 298 [1] shows the heat values found by actual experiment for the digestible organic matter of a num-

[1] *The Chemistry of the Farm.* Vinton & Co., Ltd., London.

ber of foods by Professor Kellner in the case of the ox.

It will be seen from the table that though straw has as great a total calorific value as hay, owing to the greater loss of gas from the intestine during its digestion, the actual calorific value recovered by the animal is less for wheat straw than for hay.

Energy Consumed in Digestion.

The total energy value of the digested food which can be recovered in the animal is not available for productive purposes. In order to digest the food a certain amount of energy has to be consumed, and it is only the balance which is left over after this is supplied that is available for other purposes. The energy consumed in digestion may be very great. It depends mainly on the mechanical condition of, and on the amount of fibre in, the food. The harder, coarser, and more intractable a food is, the more energy must be expended on chewing it and reducing it mechanically to a fine state of division.

In the case of coarse fibrous food such as wheat-straw, the amount of energy spent on its digestion is so great that little is left over for any other purpose. In the case of a horse, which extracts

Food.	Fuel Value of 1 gram Digested Organic Substance.	Losses of Combustible Matter.			Actual Heat Value to the Ox.
		In the Urine.	As Methane.	Total.	
	Cals.	Per cent.	Per cent.	Per cent.	Cals.
Earthnut oil . .	8.8	...	...	...	8.8
Wheat gluten . .	5.8	18.7	...	18.7	4.7
Starch . . .	4.1	...	10.1	10.1	3.7
Meadow-hay . .	4.5	8.5	10.3	18.8	3.6
Oat-straw . . .	4.5	4.7	12.2	16.9	3.7
Wheat-straw . .	4.5	5.6	20.0	25.6	3.3

less from such a food than a ruminant, and does it with greater difficulty, it has been shown by experiment that the amount of nutriment extracted from the straw may not be sufficient to supply energy for the digestion of the food. The results of experiments show that in no case can a horse extract sufficient nutriment from wheat-straw to supply energy to carry on the internal work of the body, and therefore a horse if fed on wheat-straw will starve even if it has no external work to do. On the other hand, an ox can live on wheat-straw and have a balance of energy over after performing the work of digestion and the internal work of the body.

Generally speaking, only a small proportion of the available energy of the digestible matter of such concentrated foods as cakes and grains is required to do the work of digestion itself: the proportion required is larger in foods such as hay, and still larger in straw.

Other Internal Work.

In addition to the energy consumed in the digestion of the food itself, there is a continual consumption of energy for such necessary internal work as that of the heart and of the respiratory system. The circulation and respiration and other functions of the body go on continuously whether the animal is doing any useful work or not. Even when an animal is apparently at rest it is constantly at work internally. A certain amount of food has constantly to be consumed to supply energy for this internal work. This non-productive work has constantly to be done before the animal can utilise any of its food for a productive purpose, such as the performance of external work, or the production of increase of flesh and fat, or the production of milk.

All the internal work, whether mechanical or chemical, appears ultimately as heat, and therefore helps to maintain the body temperature. The bodies of farm animals have to be maintained at a temperature considerably above that which is normal to their surroundings. So long as the heat produced in internal work is not in excess of that necessary to maintain the body temperature it is not wasted. If sufficient heat is not developed by internal work to maintain the body temperature at its normal, food has to be oxydised in the system merely to maintain that temperature. On the other hand, if the heat produced in the body by internal work is in excess of that required to maintain the normal temperature of the body, the excess is merely run to waste.

It will thus be seen that the energy used for digestion and other internal work, though it is useless for any productive purpose, is not valueless for maintenance purposes, as it serves to maintain the body heat, and if the body heat were not maintained in this way, food-materials would have to be oxydised in the body to maintain it.

Values of Foods for Maintenance and for Production.

Foods may have very different relative values for maintenance and for productive purposes. Thus Kellner showed that in the case of a fattening ox the heat value of the increase of flesh and fat obtained in the animal was about 42 per cent of the total available heat value of the digested food in the case of meadow-hay, but only about 38 per cent in the case of oat-straw, and only about 18 per cent in the case of wheat-straw. The remainder was consumed in non-productive internal work, physical and chemical.

The table [1] (p. 300) shows approximately the comparative values of ordinary foods for ruminants.

All the foods are calculated to their equivalents as starch, on the assumption that the energy value to the animal of protein is 1.25 times that of starch, of

[1] *The Chemistry of the Farm.* Vinton & Co., Ltd., London.

amides 0.6 times that of starch, of fat 2.3 times that of starch, and of fibre and other carbohydrates the same as starch. Of course only digestible constituents are taken into account, and it has been shown by experiment that these comparative values approximately represent their real energy values to the animal.

The values of all the foods are calculated both for maintenance and for production. The valuation of foods for maintenance does not involve special difficulties, but the valuation for productive purposes is more uncertain, as the basis of accurate experiment on which the calculations are founded is as yet very incomplete.

It is to be remembered that the value of starch itself is not the same for production as for maintenance. Kellner found that for the fattening ox digestible starch had a value of 3.7 calories per gram for maintenance, but only of 2.2 calories per gram for production.

The table (p. 300) shows that while concentrated foods and succulent foods have, in terms of starch, nearly as great values for production as for maintenance, fibrous foods, and especially straw, have very much lower values for production than for maintenance.

The calculation of the comparative values of foods by the methods here adopted is very much more accurate and valuable than the crude method often adopted in this country of calculating what are called "food units." In calculating these food units digestibility is, as a rule, left out of account, and the fibre is treated as if it had no food value at all. The number of food units is commonly calculated by multiplying the sum of the albuminoids and oil by 2½ and adding the soluble carbohydrates. Such a calculation is of little real value.

The comparative food values given in the table apply only to ruminants. Many of the foods, and especially the fibrous foods, have very different values if calculated for horses.

Albuminoid Ratios.

As the protein constituents of a food have a peculiar value as building constituents for the body of the animal, and as a certain amount of protein matter is necessary in the diet of every animal, the ratio of protein to non-protein material in a food is of some importance. To obtain such a ratio it is necessary to reduce all the non-protein organic matters to a common value. They are, therefore, generally reduced to their equivalent in starch. The ratio then obtained between the protein and non-protein nutrients is called the nutritive ratio or the albuminoid ratio.

The old rough method of calculating the albuminoid ratio was to multiply the total fat or oil by $2\frac{1}{2}$ and add this to the total soluble carbohydrates, and then get the ratio between the total nitrogenous or albuminoid matter and the sum so obtained from the fat and soluble carbohydrates. This method is still often used in this country. It is very inaccurate, and the ratio so obtained is of little or no value. It is only digestible

COMPARATIVE VALUES OF ORDINARY FOODS FOR OXEN AND SHEEP.

	For Maintenance.		For Production.	
	Value of 1000 lb. expressed as starch.	Quantities equivalent to 1 lb. of starch.	Value of 1000 lb. expressed as starch.	Quantities equivalent to 1 lb. of starch.
	lb.	lb.	lb.	lb.
Cotton-cake (decorticated) . .	944	1.06	826	1.21
Maize	859	1.16	825	1.21
Wheat	823	1.21	783	1.28
Linseed-cake	842	1.18	733	1.36
Barley	755	1.32	721	1.39
Rice meal	758	1.32	713	1.40
Peas	796	1.25	702	1.42
Beans	786	1.27	670	1.49
Oats	676	1.48	626	1.60
Wheat bran	635	1.57	578	1.73
Brewers' grains (dried) . .	634	1.58	533	1.88
Malt sprouts	695	1.44	518	1.93
Cotton-cake (undecorticated) . .	519	1.93	442	2.26
Meadow-hay (best) . . .	536	1.87	359	2.79
" " (medium) . .	506	1.98	337	2.97
Clover-hay (medium) . . .	459	2.18	319	3.13
Meadow-hay (poor) . . .	479	2.09	294	3.40
Bean-straw	421	2.38	252	3.97
Oat- and barley-straw . .	412	2.43	207	4.83
Potatoes	212	4.72	202	4.95
Mangels (small) . . .	108	9.26	99	10.10
Wheat-straw	357	2.80	96*	10.41*
Maize silage	131	7.63	92	10.87
Clover (bloom beginning) . .	131	7.63	92	10.87
Mangels (large) . . .	87	11.49	76	13.16
Swedes	86	11.63	75	13.33
Turnips	68	14.71	59	16.95

* These figures are the production values actually obtained in Kellner's experiments.

nutrients which are of value to the animal as food. Only digestible nutrients, therefore, should be considered. Further, since digestible fibre has a real food value to the animal, especially for maintenance purposes, it should be included in calculating nutritive ratios.

In the table given on p. 301 [1] the nutritive or albuminoid ratios are calculated from digestible constituents for ruminating animals, and the digestible fibre is taken into account. The non-nitrogenous matters are calculated into their equivalent in starch by multiplying the fat by 2.3 and adding this to the sum of the digestible carbohydrates and fibre. The first column of figures gives the ratio of the total digestible nitrogenous matters to this sum of non-nitrogenous matters.

It has been clearly shown in numerous

[1] *The Chemistry of the Farm.* Vinton & Co., Ltd., London.

investigations that the amide substances which occur in food have not the same value to the animal as true albuminoid or protein substances. Though the amides may to a certain extent save protein in the organism, it cannot take its place. Therefore, to calculate a true nutritive ratio between the real flesh-building substances or proteins, and the constituents which are only of use to the animal for fuel purposes, the amides should be reduced to their equivalent in starch and added on to the non-nitrogenous matters. This has been done in the second column of figures in the table below by multiplying the amides by o.6 and adding the result to the sum of the fat, carbohydrates, and fibre reduced to their starch equivalent. This column gives the nutritive or albuminoid ratio in its best and most accurate form.

	Total Nitrogenous Substance to Non-Nitrogenous.	Albuminoids to Non-Albuminoids.
Cotton-cake (decor.) .	I : 1.2	I : 1.3
„ (undecor.)	I : 2.0	I : 2.2
Linseed-cake . .	I : 2.3	I : 2.4
Beans . . .	I : 2.3	I : 2.8
Peas . . .	I : 2.8	I : 3.3
Brewers' grains . .	I : 3.3	I : 3.5
Malt sprouts . .	I : 2.8	I : 4.9
Wheat bran . .	I : 4.6	I : 5.8
Red clover (bloom beginning) . . .	I : 4.2	I : 6.4
Oats	I : 6.5	I : 7.1
Pasture grass .	I : 4.5	I : 7.4
Meadow-hay (best) .	I : 5.8	I : 7.7
Wheat . .	I : 6.7	I : 7.7
Clover-hay (medium) .	I : 5.3	I : 8.5
Barley . . .	I : 9.0	I : 9.5
Bean-straw . .	I : 9.5	—
Rice-meal . .	I : 8.7	I : 10.1
Maize	I : 9.7	I : 10.5
Meadow-hay (medium)	I : 8.6	I : 11.4
„ (poor) .	I : 13.2	I : 15.2
Potatoes . . .	I : 14.3	I : 41.1
Swedes . . .	I : 8.8	I : 41.8
Mangels (small) . .	I : 12.8	I : 52.9
Oat-straw . .	I : 31.5	I : 54.4
Turnips . .	I : 10.6	I : 66.3
Mangels (large) . .	I : 9.0	I : 86.0
Barley-straw . .	I : 61.0	I : 107.2
Maize silage . .	I : 15.6	I : 129.3
Wheat-straw . .	I : 88.1	—

A far greater value was formerly assigned to the albuminoid ratios of foods than is warranted by the results of experience and of experiment. Precise albuminoid ratios were laid down for working animals, for fattening animals, for milking animals, and so on, and it was supposed that close adherence to these ratios was necessary in order to secure the best results. The modern view is that albuminoid ratios can be treated with much more freedom than was formerly supposed. While they have a certain value in indicating, within certain limits, whether a diet is likely to be suitable to the animals we are feeding and for producing the results we wish to obtain from them, the ratios may have in most cases a considerable amount of elasticity, and may be varied through fairly wide limits without necessarily interfering with the results which are desired.

Thus, in the case of fattening cattle, it was at one time supposed to be necessary to provide a ratio of 1 : 5 or 1 : 6, but it has been shown that cattle may be rapidly fattened with a much wider ratio —say 1 : 16, provided a sufficient excess of digestible food be supplied to build up fat in the body of the animal.

On the other hand, it is still recognised that young animals which are rapidly building up their nitrogenous tissues, and milking cows which are yielding a large quantity of the highly nitrogenous secretion, milk, must have a diet with a comparatively narrow albuminoid ratio, as well as a liberal diet, if the best results are to be obtained.

It is to be remembered that the true albuminoid or nutritive ratio of a food is not the same for different animals with different powers of digestion. Thus meadow-hay of medium quality may have an albuminoid ratio of 1 : 9 for an ox and of 1 : 12 for a horse. This is because the ox is able to digest more of the fibre and carbohydrates of the food than the horse.

FOOD REQUIREMENTS OF ANIMALS.

The food constituents required by animals vary according to the age of the animal and to the use which the animal is to make of its food.

The Young Animal.

The young animal requires a diet rich in those constituents which build up the nitrogenous tissues and bone. It follows that it requires a diet containing a considerable proportion of protein, and therefore of narrow albuminoid ratio. It also requires a sufficiency of suitable mineral constituents, like phosphate and lime, in its diet. Such a diet is provided by nature in the milk of the mother. Cow's milk has a nutritive ratio of about 1 : 3.7.

As the animal becomes older and more active it requires more food for merely fuel purposes, and the nutritive ratio may therefore be gradually widened. But a young growing animal should always receive a considerable proportion of protein substance in its food, as well as a liberal and digestible diet, if the best results are to be obtained.

The young animal requires more food per 1000 lb. live-weight than it does at a later stage. As the animal increases in size the amount of food consumed increases, but the amount of food consumed per 1000 lb. live-weight diminishes.

The Adult Animal.

The adult animal which is merely being maintained, and which is neither working nor increasing in weight, requires food to do internal work, to maintain the body temperature, and to make good the waste of the body and form hair, horn, hoof, &c. For these purposes a poor diet, containing comparatively little nitrogenous matter, is sufficient. The nitrogenous matter is required only to form hair, hoof, &c., and to repair the small amount of waste of nitrogenous tissue in the body. Carbohydrates and fat can be used for all fuel purposes, to maintain the body heat and to supply energy for internal work. A wide albuminoid ratio is therefore sufficient.

It has been shown, for instance, that oxen can be maintained with a diet having a ratio as wide as 1 : 15. Experiment has shown that to maintain an ox of 1000 lb. live-weight about 6.5 lb. of digestible organic matter, reckoned as starch, is the minimum required. This should contain about 0.6 lb. of digestible

protein. These requirements would be met by a diet of about 13 lb. of meadow-hay of medium quality.

The Working Animal.

If external work has to be done the quantity of food required increases rapidly with the amount of work. A working animal therefore requires much more food than a mere maintenance diet.

Work is not done at the expense of the nitrogenous tissues of muscle as was at one time supposed, but at the expense of fuel materials consumed in the muscle. Such fuel may be supplied by any of the organic constituents of food, whether proteins, fats, or carbohydrates, and therefore a highly nitrogenous diet is not required in order that work may be done. What is required is sufficient digestible food in excess of that required for mere maintenance, to supply fuel materials for consumption in the working tissues. If that be not supplied the work will be carried on at the expense of the tissues themselves, and the animal will rapidly lose in weight and condition.

While it is not essential to supply food of narrow albuminoid ratio for a labour diet, it is generally advisable to supply working animals with a considerable amount of digestible protein. It is found that a diet fairly rich in protein causes the animal to be more active and to work with more spirit. Large horse users generally provide working horses with a diet having an albuminoid ratio of about 1 : 7.

The Fattening Animal.

Fat may be formed from any excess of organic nutrients over what is required to repair the body and to produce heat and work. Fat is not necessarily formed from the fat or oil in the food. The fat stored up in the body is not the same as the fat given in the food. Thus, if an animal is fed with linseed-cake it does not store up linseed-oil in its fatty tissues.

In the case of our farm animals fat may be formed from carbohydrates or from proteins, and probably any excess of food may go to form fat. The diet of fattening animals does not therefore need to be a highly nitrogenous one.

At one time it was usual in laying down standard rations for fattening animals to insist on narrow albuminoid ratios. These are by no means necessary. Indeed, a fattening diet for an adult animal may have a wider albuminoid ratio than a maintenance diet.

So long as there is excess of digestible food the animal will fatten. It has been shown in Continental experiments that adult animals may fatten rapidly on a diet having an albuminoid ratio wider than 1 : 20 provided plenty of digestible food is given. Similarly in feeding experiments carried on in this country animals have been equally well fattened on diets having very narrow and very wide albuminoid ratios.

At the same time, it is generally convenient and economical in farm practice to feed fattening animals on a diet of moderate albuminoid ratio. There are various reasons for this. Most of the nitrogen of the food of all animals is excreted and returned to the farmer in the urine and dung of the animals. If foods rich in nitrogen, and therefore of narrow albuminoid ratio, are used, the farmyard manure is made richer and more valuable, and the farmer recovers a considerable part of the value of his foods in the manure. Further, a diet containing a moderate amount of proteins and of oil is better digested and utilised by the animal than one consisting mainly of carbohydrates, and for this reason also a certain amount of protein and oil in the food of fattening animals is generally economical. Still, there is no necessity to have the albuminoid ratio of the food of adult fattening animals narrower than about 1 : 10.

The Milking Cow.

Milk is a fluid rich in nitrogenous matter, and if the excess of nitrogenous matters necessary to produce it be not supplied in the food, the animal will take them from her own nitrogenous tissues and will rapidly lose in condition. A cow giving the quantity of milk which is expected from a good dairy animal is returning far more organic matter in the milk than a fattening animal produces in his increase of weight. Thus a cow giving 3 gallons of milk per day gives about 26 lb. of dry matter per week, of which about 7.6 lb. consists of proteins. On the other hand, a fattening ox increasing 2 lb. per day produces about 10.6 lb. of dry matter per week, of which only about 1.1 lb. is protein.

The milking cow therefore requires a plentiful and digestible diet of narrow albuminoid ratio. If properly fed, the milking cow gives a much bigger return for the food consumed than a fattening animal. A milking cow giving a large yield of milk, say 3 gallons per day, does not require more digestible food per 1000 lb. live-weight than a fattening ox. But it is advisable to supply her with food of narrower albuminoid ratio. It is also advisable to feed her largely on such foods as are easily digested, and which do not require a great consumption of energy in their digestion.

The amount of food given to a milking cow should be proportional to the quantity of milk yielded. The albuminoid ratio may safely vary from about 1 : 6 to 1 : 8, provided plenty of food be given.

THE MANURIAL VALUE OF FOODS.

The value of animal excreta as a manure has been recognised perhaps as long as we have any records of agriculture.

It is learned from chemistry that—apart from mere mechanical effects on the texture of the soil—this value is due to the presence of nitrogenous and mineral compounds, of which latter the most important are the compounds of potash and of phosphoric acid. Seeing that, directly or indirectly, all the constituents of animal excreta are derived from the food consumed, it is at once reasonable to suppose that the composition of the food must influence that of the excreta derived from it—that food, rich in nitrogen and in phosphates, for example, should produce manure rich in these materials; and that food comparatively poor in these respects should produce manure comparatively poor in them.

Urine.—The urine is the richest part of animal manure, for it is the means whereby is eliminated from the animal system the waste nitrogenous materials which have undergone digestion and served their purpose physiologically. It is rich also in soluble salts of potash and phosphoric acid.

Solid Excreta. — The solid excreta consist only of those materials which have passed through the animal undigested; and if an animal could be fed on theoretically perfect principles, they would possess hardly any manurial value. But practically, an animal always consumes more nitrogenous and phosphatic food than it really digests, and the excess thus consumed gives value to the solid excreta, although this value is generally small compared with that of the urine if the animal is properly fed.

Proportion of Food assimilated and voided.—An animal in order to gain a given increase in live-weight has to consume an enormously greater quantity of food than would be required to produce that increase alone, for the mere sustenance of life involves a large daily consumption of food-material. What becomes of the carbon, hydrogen, and oxygen that is consumed by the animal beyond what it stores up as increase, does not here concern us; but it is essential that we should know what is the destination of the nitrogen and of the mineral matters in the food consumed.

Much attention has been devoted to this point at Rothamsted, and careful inquiry has shown that, of every 10 lb. of nitrogen consumed by an animal in its food, not more than about 1 lb. will be stored up as increase of live-weight,

except by a very young animal, the remaining 9 lb. or so being voided in the manure, partly as undigested matter, partly as soluble nitrogenous compounds, which readily become converted, first into ammonia, and then into nitrates, when applied to the soil. Similarly, only from about one-fifth to about one-tenth of the phosphates in food are stored up by the animal, and a still smaller proportion of potash salts, the great bulk of both going to enrich the manure.

The actual proportion of any of these fertilising ingredients retained in any given case will depend upon how liberally the animal is fed, and also upon whether it is a growing animal, having to build up its bony frame and muscles, or whether it is an already adult animal adding little but fat to its carcase-weight; or whether again it may be a cow having to produce its calf, and yield a flow of milk, which will make a heavier demand on the food than will the mere fat-forming processes going on in the case of a stall-fed ox.

Lawes and Gilbert's Manurial Tables.—Taking one case with another, however, it is possible to arrive at an average which shall in no case err very widely; and the careful experiments and calculations made at Rothamsted have furnished us with the following classical tables, indicative both of the original proportions of the chief fertilising ingredients contained in the various foods, and the proportions of these which will, on the average, be voided by animals consuming, say, a ton of any of them.

These tables were first published by Sir John Bennett Lawes and Sir Henry Gilbert in the *Journal of the Royal Agricultural Society of England* in 1885, and were subsequently revised by them and brought up to date in 1897 and 1898.

[TABLES.

LAWES AND GILBERT'S TABLES (1897) OF COMPOSITION AND MANURIAL VALUE OF FOODS.

TABLE I.—Average Composition, Per Cent and per Ton, of Cattle Foods.

No.	Foods.	PER CENT.					PER TON.		
		Dry Matter.	Nitrogen.	Mineral Matter (Ash).	Phosphoric Acid.	Potash.	Nitrogen.	Phosphoric Acid.	Potash.
		per cent.	per cent.	per cent.	per cent.	per cent.	lb.	lb.	lb.
1	Linseed . .	90.00	3.60	4.00	1.54	1.37	80.64	34.50	30.69
2	Linseed-cake .	88.50	4.75	6.50	2.00	1.40	106.40	44.80	31.36
3	Decorticated cotton-cake	90.00	6.60	7.00	3.10	2.00	147.84	69.44	44.80
4	Palm-nut-cake	91.00	2.50	3.60	1.20	0.50	56.00	26.88	11.20
5	Undecorticated cotton-cake	87.00	3.75	6.00	2.00	2.00	84.00	44.80	44.80
6	Cocoa-nut-cake	90.00	3.40	6.00	1.40	2.00	76.16	31.36	44.80
7	Rape-cake .	89.00	4.90	7.50	2.50	1.50	109.76	56.00	33.60
8	Peas . .	85.00	3.60	2.50	0.85	0.96	80.64	19.04	21.50
9	Beans . .	85.00	4.00	3.00	1.10	1.30	89.60	24.64	29.12
10	Lentils . .	88.00	4.20	4.00	0.75	0.70	94.08	16.80	15.68
11	Tares (seed) .	84.00	4.20	2.50	0.80	0.80	94.08	17.92	17.92
12	Indian corn .	88.00	1.70	1.40	0.60	0.37	38.08	13.44	8.29
13	Wheat . .	85.00	1.80	1.70	0.85	0.53	40.32	19.04	11.87
14	Malt . .	94.00	1.70	2.50	0.80	0.50	38.08	17.92	11.20
15	Barley . .	84.00	1.65	2.20	0.75	0.55	36.96	16.80	12.32
16	Oats . .	86.00	2.00	2.80	0.60	0.50	44.80	13.44	11.20
17	Rice-meal [1] .	90.00	1.90	7.50	(0.60)	(0.37)	42.56	(13.44)	(8.29)
18	Locust-beans [1]	85.00	1.20	2.50	...	...	26.88	...	...
19	Malt culms .	90.00	3.90	8.00	2.00	2.00	87.36	44.80	44.80
20	Fine pollard .	86.00	2.45	5.50	2.90	1.46	54.88	64.96	32.70
21	Coarse pollard	86.00	2.50	6.40	3.50	1.50	56.00	78.40	33.60
22	Bran . .	86.00	2.50	6.50	3.60	1.45	56.00	80.64	32.48
23	Clover-hay .	83.00	2.40	7.00	0.57	1.50	53.76	12.77	33.60
24	Meadow-hay .	84.00	1.50	6.50	0.40	1.60	33.60	8.96	35.84
25	Pea-straw .	82.50	1.00	5.50	0.35	1.00	22.40	7.84	22.40
26	Oat-straw .	83.00	0.50	5.50	0.24	1.00	11.20	5.38	22.40
27	Wheat-straw .	84.00	0.45	5.00	0.24	0.80	10.08	5.38	17.92
28	Barley-straw .	85.00	0.40	4.50	0.18	1.00	8.96	4.03	22.40
29	Bean-straw .	82.50	0.90	5.00	0.30	1.00	20.16	6.72	22.40
30	Potatoes .	25.00	0.25	1.00	0.15	0.55	5.60	3.36	12.32
31	Carrots . .	14.00	0.20	0.90	0.09	0.28	4.48	2.02	6.27
32	Parsnips .	16.00	0.22	1.00	0.19	0.36	4.93	4.26	8.06
33	Mangel-wurzels .	12.50	0.22	1.00	0.07	0.40	4.93	1.57	8.96
34	Swedish turnips	11.00	0.25	0.60	0.06	0.22	5.60	1.34	4.93
35	Yellow turnips [1]	9.00	0.20	0.65	(0.06)	(0.22)	4.48	(1.34)	(4.93)
36	White turnips .	8.00	0.18	0.68	0.05	0.30	4.03	1.12	6.72

[1] In the case of neither rice-meal, locust-beans, nor yellow turnips, have records of ash analyses been found. For rice-meal the same percentages of phosphoric acid and potash as in Indian corn, and for yellow turnips the same as in swedes, are provisionally adopted; but in all the Tables the assumed results are given in parentheses. For locust-beans no figure has been assumed, and the columns are left blank.

TABLE II.—(1897) SHOWING THE DATA, THE METHOD, AND THE RE-
OF CATTLE FOODS

No.	DESCRIPTION OF FOOD.	Fattening Increase in Live-weight (Oxen or Sheep).		NITROGEN.						
				In Food.		In Fattening Increase (at 1.27 per cent).		In Manure.		
		Food to 1 Increase.	Increase per ton of food.	Per cent.	Per ton.	From 1 ton of Food.	Per cent of total consumed.	Total remaining for Manure.	Nitrogen equal Ammonia.	Value of Ammonia at 4d. per lb.
			lb.	%	lb.	lb.	%	lb.	lb.	£ s. d.
1	Linseed	5.0	448.0	3.60	80.64	5.69	7.06	74.95	91.0	1 10 4
2	Linseed-cake	6.0	373.3	4.75	106.40	4.74	4.45	101.66	123.4	2 1 2
3	Decorticated cotton-cake	6.5	344.6	6.60	147.84	4.38	2.96	143.46	174.2	2 18 1
4	Palm-nut cake	7.0	320.0	2.50	56.00	4.06	7.25	51.94	63.1	1 1 0
5	Undecorticated cotton-cake	8.0	280.0	3.75	84.00	3.56	4.24	80.44	97.7	1 12 7
6	Cocoa-nut cake	8.0	280.0	3.40	76.16	3.56	4.67	72.60	88.2	1 9 5
7	Rape-cake	(10)	(224)	4.90	109.76	2.84	2.59	106.92	129.8	2 3 3
8	Peas	7.0	320.0	3.60	80.64	4.06	5.03	76.58	93.0	1 11 0
9	Beans	7.0	320.0	4.00	89.60	4.06	4.53	85.54	103.9	1 14 8
10	Lentils	7.0	320.0	4.20	94.08	4.06	4.32	90.02	109.3	1 16 5
11	Tares (seed)	7.0	320.0	4.20	94.08	4.06	4.32	90.02	109.3	1 16 5
12	Indian corn	7.2	311.1	1.70	38.08	3.95	10.37	34.13	41.4	0 13 10
13	Wheat	7.2	311.1	1.80	40.32	3.95	9.80	36.37	44.2	0 14 9
14	Malt	7.0	320.0	1.70	38.08	4.06	10.66	34.02	41.3	0 13 9
15	Barley	7.2	311.1	1.65	36.96	3.95	10.69	33.01	40.1	0 13 4
16	Oats	7.5	298.7	2.00	44.80	3.79	8.46	41.01	49.8	0 16 7
17	Rice-meal	7.5	298.7	1.90	42.56	3.79	8.91	38.77	47.1	0 15 8
18	Locust-beans	9.0	248.9	1.20	26.88	3.16	11.76	23.72	28.8	0 9 7
19	Malt culms	8.0	280.0	3.90	87.36	3.56	4.08	83.80	101.8	1 13 11
20	Fine pollard	7.5	298.7	2.45	54.88	3.79	6.91	51.09	62.0	1 0 8
21	Coarse pollard	8.0	280.0	2.50	56.00	3.56	6.35	52.44	63.7	1 1 3
22	Bran	9.0	248.9	2.50	56.00	3.16	5.64	52.84	64.2	1 1 5
23	Clover-hay	14.0	160.0	2.40	53.76	2.03	3.78	51.73	62.8	1 0 11
24	Meadow-hay	15.0	149.3	1.50	33.60	1.90	5.65	31.70	38.5	1 12 10
25	Pea-straw	16.0	140.0	1.00	22.40	1.78	7.95	20.62	25.0	0 8 4
26	Oat-straw	18.0	124.4	0.50	11.20	1.58	14.11	9.62	11.7	0 3 11
27	Wheat-straw	21.0	106.7	0.45	10.08	1.36	13.49	8.72	10.6	0 3 6
28	Barley-straw	23.0	97.4	0.40	8.96	1.24	13.84	7.72	9.4	0 3 2
29	Bean-straw	22.0	101.8	0.90	20.16	1.29	6.39	18.87	22.9	0 7 8
30	Potatoes	60.0	37.3	0.25	5.60	0.47	8.39	5.13	6.2	0 2 1
31	Carrots	85.7	26.1	0.20	4.48	0.33	7.37	4.15	5.0	0 1 8
32	Parsnips	75.0	29.9	0.22	4.93	0.38	7.71	4.55	5.5	0 1 10
33	Mangel-wurzels	96.0	23.3	0.22	4.93	0.30	6.09	4.63	5.6	0 2 2
34	Swedish turnips	109.1	20.5	0.25	5.60	0.26	4.64	5.34	6.5	0 1 10
35	Yellow turnips	133.3	16.8	0.20	4.48	0.21	4.69	4.27	5.2	0 1 9
36	White turnips	150.0	14.9	0.18	4.03	0.19	4.71	3.84	4.7	0 1 7

SULTS OF THE ESTIMATION OF THE ORIGINAL MANURE VALUE AFTER CONSUMPTION.

PHOSPHORIC ACID.						POTASH.						Total original Manure value per ton of Food consumed.
In Food.		In Fattening Increase at (0.86 per cent).		In Manure.		In Food.		In Fattening Increase at (0.11 per cent).		In Manure.		
Per cent.	Per ton.	From 1 ton of Food.	Per cent of total consumed.	Total remaining for Manure.	Value at 2d. per lb.	Per cent.	Per ton.	From 1 ton of Food.	Per cent of total consumed.	Total remaining for Manure.	Value at 1½d. per lb.	
%	lb.	lb.	%	lb.	s. d.	%	lb.	lb.	%	lb.	s. d.	£ s. d.
1.54	34.50	3.85	11.16	30.65	5 1	1.37	30.69	0.49	1.60	30.20	3 9	1 19 2
2.00	44.80	3.21	7.17	41.59	6 11	1.40	31.36	0.41	1.31	30.95	3 10	2 11 11
3.10	69.44	2.96	4.26	66.48	11 1	2.00	44.80	0.38	0.85	44.42	5 7	3 14 9
1.20	26.88	2.75	10.23	24.13	4 0	0.50	11.20	0.35	3.13	10.85	1 4	1 6 4
2.00	44.80	2.41	5.38	42.39	7 1	2.00	44.80	0.31	0.69	44.49	5 7	2 5 3
1.40	31.36	2.41	7.68	28.95	4 10	2.00	44.80	0.31	0.69	44.49	5 7	1 19 10
2.50	56.00	1.93	3.45	54.07	9 0	1.50	33.60	0.25	0.74	33.35	4 2	2 16 5
0.85	19.04	2.75	14.44	16.29	2 9	0.96	21.50	0.35	1.63	21.15	2 8	1 16 5
1.10	24.64	2.75	11.16	21.89	3 8	1.30	29.12	0.35	1.20	28.77	3 7	2 1 11
0.75	16.80	2.75	16.37	14.05	2 4	0.70	15.68	0.35	2.23	15.33	1 11	2 0 8
0.80	17.92	2.75	15.35	15.17	2 6	0.80	17.92	0.35	1.95	17.57	2 2	2 1 1
0.60	13.44	2.68	19.94	10.76	1 9	0.37	8.29	0.34	4.10	7.95	1 0	0 16 7
0.85	19.04	2.68	14.08	16.36	2 9	0.53	11.87	0.34	2.86	11.53	1 5	0 18 11
0.80	17.92	2.75	15.35	15.17	2 6	0.50	11.20	0.35	3.13	10.85	1 4	0 17 7
0.75	16.80	2.68	15.95	14.12	2 4	0.55	12.32	0.34	2.76	11.98	1 6	0 17 2
0.60	13.44	2.57	19.12	10.87	1 10	0.50	11.20	0.33	2.94	10.87	1 4	0 19 9
(0.60)	(13.44)	2.57	(19.12)	(10.87)	(1 10)	(0.37)	(8.29)	0.33	(4.00)	(7.96)	(1 0)	(0 18 6)
...	...	2.14	...	...	...	...	...	0.27	...	...	...	...
2.00	44.80	2.41	5.38	42.39	7 1	2.00	44.80	0.31	0.69	44.49	5 7	2 6 7
2.90	64.96	2.57	3.96	62.39	10 5	1.46	32.70	0.33	1.01	32.37	4 1	1 15 2
3.50	78.40	2.41	3.07	75.99	12 8	1.50	33.60	0.31	0.92	33.29	4 2	1 18 1
3.60	80.64	2.14	2.65	78.50	13 1	1.45	32.48	0.27	0.83	32.21	4 0	1 18 6
0.57	12.77	1.38	10.81	11.39	1 11	1.50	33.60	0.18	0.54	33.42	4 2	1 7 0
0.40	8.96	1.28	14.28	7.68	1 3	1.60	35.84	0.16	0.45	35.68	4 6	0 18 7
0.35	7.84	1.20	15.31	6.64	1 1	1.00	22.40	0.15	0.67	22.25	2 9	0 12 2
0.24	5.38	1.07	19.89	4.31	0 9	1.00	22.40	0.14	0.63	22.26	2 9	0 7 5
0.24	5.38	0.92	17.10	4.46	0 9	0.80	17.92	0.12	0.67	17.80	2 3	0 6 6
0.18	4.03	0.84	20.84	3.19	0 6	1.00	22.40	0.11	0.49	22.29	2 9	0 6 5
0.30	6.72	0.88	13.10	5.84	1 0	1.00	22.40	0.11	0.49	22.29	2 9	0 11 5
0.15	3.36	0.32	9.52	3.04	0 6	0.55	12.32	0.04	0.32	12.28	1 6	0 4 1
0.09	2.02	0.22	10.89	1.80	0 4	0.28	6.27	0.03	0.48	6.24	0 9	0 2 9
0.19	4.26	0.26	6.10	4.00	0 8	0.36	8.06	0.03	0.37	8.03	1 0	0 3 6
0.07	1.57	0.20	12.74	1.37	0 2	0.40	8.96	0.03	0.34	8.93	1 1	0 3 2
0.06	1.34	0.18	13.43	1.16	0 3	0.22	4.93	0.02	0.41	4.91	0 7	0 2 11
(0.06)	(1.34)	0.14	(10.78)	(1.20)	(0 2)	(0.22)	(4.93)	0.02	(0.34)	(4.91)	(0 7)	(0 2 6)
0.05	1.12	0.13	11.61	0.99	0 2	0.30	6.72	0.02	0.30	6.70	0 10	0 2 7

TABLE III.—(1897) PLAN AND RESULTS OF ESTIMATIONS OF THE COMPENSATION VALUE OF UNEXHAUSTED MANURE, STARTING FROM THE ORIGINAL MANURE VALUE, THAT IS THE VALUE, DEDUCTING THE CONSTITUENTS OF INCREASE IN FATTENING LIVE-WEIGHT ONLY.

Foods.	Original Manure value, deducting increase in live-weight only.	Compensation Value of unexhausted manure.								Total.
		Last year.	Second year.	Third year.	Fourth year.	Fifth year.	Sixth year.	Seventh year.	Eighth year.	
	£ s. d.	£ s. d.	£ s. d.	£ s. d.	s. d.	s. d.	s. d.	s. d.	s. d.	£ s. d.
DEDUCT ½ OF ORIGINAL MANURE VALUE THE LAST YEAR, AND ⅛ FROM YEAR TO YEAR.										
One Ton										
1. Linseed	1 19 2	0 19 7	0 13 1	8 9	5 10	3 11	2 7	1 9	1 2	2 16 8
2. Linseed-cake	2 11 11	1 6 0	0 17 4	11 7	7 9	5 2	3 5	2 3	1 6	3 15 0
3. { Decorticated cotton-cake }	3 14 9	1 17 4	1 4 11	16 7	11 1	7 5	4 11	3 3	2 2	5 7 8
4. Palm-nut cake	1 6 4	0 13 2	0 8 9	5 10	3 11	2 7	1 9	1 2	0 9	1 17 11
5. { Undecorticated cotton-cake }	2 5 3	1 2 7	0 15 1	10 1	6 9	4 6	3 0	2 0	1 4	3 5 4
6. Cocoa-nut cake	1 19 10	0 19 11	0 13 3	8 10	5 11	3 11	2 7	1 9	1 2	2 17 3
7. Rape-cake	2 16 5	1 8 3	0 18 10	12 7	8 5	5 7	3 9	2 6	1 8	4 1 7
8. Peas	1 16 5	0 18 3	0 12 2	8 1	5 5	3 7	2 5	1 7	1 1	2 12 7
9. Beans	2 1 11	1 0 11	0 13 11	9 3	6 2	4 1	2 9	1 10	1 3	3 0 2
10. Lentils	2 0 8	1 0 4	0 13 7	9 1	6 1	4 1	2 9	1 10	1 3	2 19 0
11. Tares (seed)	2 1 1	1 0 7	0 13 9	9 2	6 1	4 1	2 9	1 10	1 3	2 19 6
12. Indian corn	0 16 7	0 8 4	0 5 7	3 9	2 6	1 8	1 1	0 9	0 6	1 4 2
13. Wheat	0 18 11	0 9 6	0 6 4	4 3	2 10	1 11	1 3	0 10	0 7	1 7 6
14. Malt	0 17 7	0 8 10	0 5 11	3 11	2 7	1 9	1 2	0 9	0 6	1 5 5
15. Barley	0 17 2	0 8 7	0 5 9	3 10	2 7	1 9	1 2	0 9	0 6	1 4 11
16. Oats	0 19 9	0 9 11	0 6 7	4 5	2 11	1 11	1 3	0 10	0 7	1 8 5
17. Rice-meal	(0 18 6)	(0 9 3)	(0 6 2)	(4 1)	(2 9)	(1 10)	(1 3)	(0 10)	(0 7)	(1 6 9)
18. Locust-beans	...	...	...	...	...	...	...	...	...	...
19. Malt culms	2 6 7	1 3 3	0 15 6	10 4	6 11	4 7	3 1	2 1	1 5	3 7 2
20. Fine pollard	1 15 2	0 17 7	0 11 9	7 10	5 3	3 6	2 4	1 7	1 1	2 10 11
21. Coarse pollard	1 18 1	0 19 1	0 12 9	8 6	5 8	3 9	2 6	1 8	1 1	2 15 0
22. Bran	1 18 6	0 19 3	0 12 10	8 7	5 9	3 10	2 7	1 9	1 2	2 15 9
DEDUCT ⅔ OF ORIGINAL MANURE VALUE THE LAST YEAR, AND ⅙ FROM YEAR TO YEAR.										
One Ton										
23. Clover-hay	1 7 0	0 9 0	0 7 2	5 9	4 7	3 8	2 11	2 4	1 10	1 17 3
24. Meadow-hay	0 18 7	0 6 2	0 4 11	3 11	3 2	2 6	2 0	1 7	1 3	1 5 6
25. Pea-straw	0 12 5	0 4 1	0 3 3	2 7	2 1	1 8	1 4	1 1	0 10	0 16 11
26. Oat-straw	0 7 5	0 2 6	0 2 0	1 7	1 3	1 0	0 10	0 8	0 6	0 10 4
27. Wheat-straw	0 6 6	0 2 2	0 1 9	1 5	1 2	0 11	0 9	0 7	0 6	0 9 3
28. Barley-straw	0 6 5	0 2 2	0 1 9	1 5	1 2	0 11	0 9	0 7	0 6	0 9 3
29. Bean-straw	0 11 5	0 3 10	0 3 1	2 6	2 0	1 7	1 3	1 0	0 10	0 16 1
DEDUCT ½ OF ORIGINAL MANURE VALUE THE LAST YEAR, AND ⅛ FROM YEAR TO YEAR.										
Ten Tons										
30. Potatoes	2 0 10	1 0 5	0 13 7	9 1	6 1	4 1	2 9	1 10	1 3	2 19 1
31. Carrots	1 7 6	0 13 9	0 9 2	6 1	4 1	2 9	1 10	1 3	0 10	1 19 9
32. Parsnips	1 15 0	0 17 6	0 11 8	7 9	5 2	3 5	2 3	1 6	1 0	2 10 3
33. Mangel-wurzels	1 11 8	0 15 10	0 10 7	7 1	4 9	3 2	2 1	1 5	0 11	2 5 10
34. Swedish turnips	1 9 2	0 14 7	0 9 9	6 6	4 4	2 11	1 11	1 3	0 10	2 2 1
35. Yellow turnips	(1 5 0)	(0 12 6)	(0 8 4)	(5 7)	(3 9)	(2 6)	(1 8)	(1 1)	(0 9)	(1 16 2)
36. White turnips	1 5 10	0 12 11	0 8 7	5 9	3 10	2 7	1 9	1 2	0 9	1 17 4

On the basis of the figures set out in Tables I. and II. (1897), Lawes and Gilbert constructed a further table in which, beginning with the allowance of one-half the "original manure value" for food consumed during the last year of the tenancy, to the subsequent years, up to the eighth, were assigned compensation values in a regular descending scale, one-third being deducted each year. Thus, with linseed-cake, the "original manure value" being £2, 11s. 11d. per ton (1897), the compensation value for the unexhausted manure from one ton consumed was:—

Last year.	Second year.	Third year.	Fourth year.	Fifth year.	Sixth year.	Seventh year.	Eighth year.
£ s. d.	s. d.	s. d.	s. d.	s. d.	s. d.	s. d.	s. d.
1 6 0	17 4	11 7	7 9	5 2	3 5	2 3	1 6
(being one-half of the original manure value, £2, 11s. 11d.)	(being the previous year's value less one-third.)	(being the previous year's value less one-third.)	for each year one-third of previous year's value deducted.				

An outgoing tenant would, from this, be able to claim an allowance of 1s. 6d. for each ton of linseed-cake he had consumed seven years previously, there being assumed to be still some slight benefit accruing from it for the eighth crop grown.

In the first of these tables we have the total quantities of ingredients capable of contributing to the fertility of the land contained in the principal varieties of foods in use on the farm, stated both as percentages and as pounds per ton. These figures represent the manurial matter that would reach the land, supposing that the foods were simply ground up and applied directly to the soil, without the intervention of the stock that consumes them.

In Table II. we have indicated to us the average destination of this fertilising matter—how much of it, that is to say, may be assumed to be retained by the animal in increasing its weight, and how much will find its way into the manure. Then we have the theoretical money value of this latter portion calculated for each fertilising constituent; and finally, we have stated what would be the total value of the manure from a ton of the food, supposing its value to be completely realised.

To make the matter clearer, we will select an instance—say that of linseed-cake. From Table I. we learn that linseed-cake contains 88.5 per cent of dry matter, which includes 4.75 per cent of nitrogen, 2.00 per cent of phosphoric acid, and 1.40 per cent of potash; or otherwise stated, one ton of linseed-cake contains 106.40 lb. of nitrogen, 44.80 lb. of phosphoric acid, and 31.36 lb. of potash. From Table II. we learn that 6 lb. of linseed-cake go to make 1 lb. of increase in live-weight, so that 1 ton of cake yields 373.3 lb. of increase in live-weight. We also learn that of the 106.40 lb. of nitrogen in the ton of cake, 4.74 lb. are retained by the animal, while 101.66 lb. pass into the manure. This quantity of nitrogen is equal to 123.4 lb. of ammonia, which, at 4d. per lb., is equal to £2, 1s. 2d. per ton. In like manner we find that of 44.8 lb. of phosphoric acid in the ton of cake, 3.21 lb. are retained by the animal, while 41.59 lb. pass into the manure, which, at 2d. per lb., would be worth 6s. 11d. Of 31.36 lb. of potash in the ton of cake, 0.41 lb. is retained, 30.95 lb. passing into the manure, giving at 1½d. per lb., 3s. 10d. The three money figures added together give £2, 11s. 11d. as the "total original manure value" of one ton of linseed-cake. This value in the case of decorticated cotton-cake is as high as £3, 14s. 9d., while for maize it is but 16s. 7d., or for barley, 17s. 2d., and for swedes it is less than 2s. 11d.

There can be no doubt that the proportions which these "original manure values" bear to one another, correctly represent the proportions borne to one another by the actual manurial values realisable in the field, provided that the circumstances are favourable for their comparative realisation; though it has happened, as in the Woburn experiments, that practical trial has occasionally shown that manure made by the use of a food like decorticated cotton-cake has done no more immediate good than manure made from a like quantity of maize. But this has no doubt been caused partly by the land being in such good heart that the maize manure was in itself sufficient to bring out its maximum fertility, and that the richer manure supplied by the decorticated cotton-cake was of the nature of a superfluity; and partly by the influence of the clover crop in the rotation, which, by taking up nitrogen from the air, tended to equalise the manure residues.

It is of course to be borne in mind that the values calculated in each case are average ones, and any given ton of lin-seed-cake, for example, may differ a good deal from another ton ; but it is only on the *average* quality of each kind of food that a table for general reference could well be based without becoming bewilderingly cumbersome.

Theoretical and Realised Manure Values.—But even putting aside this consideration, there are obviously a vast number of circumstances affecting the question of how far the theoretical value given in the tables is capable of actual realisation in the field. The nearest approach to the perfect application of the whole of the manure to the crops is found in the consumption of food on the land itself, as when grazing cattle or sheep consume cake in the field. Their excreta go directly on to the land, and so the whole of the manurial matter at least reaches the soil.

The other extreme is found where the food is consumed in the farmyard, and the manure badly cared for—as when it is left to lie about in the open, exposed to the free and prolonged action of rain, in such a way as to allow the drainage from it to be lost. Wherever the rich drainings from dung are allowed to run to waste, there is a serious loss of fertilising matter,—for the most valuable part of manure is the soluble salts of ammonia and potash which it contains.

What proportion of the manurial value originally contributed to the dung really finds its way on to the land from the farmyard depends, therefore, upon individual care and management, of which no exact account can be taken in tables.

Furthermore, a herd of dairy cows will rob the food of much more nitrogen and phosphoric acid than a herd of fattening oxen, since oxen, while fattening, store up but little of these materials compared with that which is required by the cows to produce a flow of milk, and to build up the bodies of the young calves which they have yearly to produce. The same applies to the case of young growing stock as compared with fattening stock, the former storing up more of the food-materials, and producing consequently the poorer manure. There are obviously, then, difficulties to be surmounted in forming an estimate of the manurial value that may fairly be assumed to be realisable in any given case.

To meet in some way these difficulties, Lawes and Gilbert published in the *Journal of the Royal Agricultural Society of England* for 1898 a revision of their already quoted tables, in which these are made specially applicable to the case of cows producing milk.

As a guide to the farmer in regard to the value of the respective foods, and as to the best foods to use in order to at once fatten his stock and best fertilise his land, the mere " original manure values " supply sufficient information ; but when the question at issue is the realisable unexhausted value of manure from food consumed, such complexities as we have glanced at arise and give serious trouble.

Unexhausted Value of Consumed Food.—The " county customs " which are often brought in to assess, under the provisions of the Agricultural Holdings Act, the compensation due to an outgoing tenant for unexhausted manurial value for foods consumed, are in most cases absurdly fallacious, being too often based on the *cost* of the foods used, this having really no relation whatever to their manurial value. The difficulty which the valuer who proceeds on rational principles has to face is to decide on how much of the " original manure value " is to be assumed to be still left on the farm—the " compensation value," as Sir John Bennett Lawes has called it.

With a view to putting the matter on a broad general basis for practical purposes, Lawes and Gilbert drew up a further table (Table III.), in which they suggested that in the case of an outgoing tenant claiming compensation for the unexhausted value of consumed food, the "original manure value" of each ton of food (as shown in Table II.) should be discounted to the extent of 50 per cent for the food consumed within the last year. This deduction of 50 per cent was in order to allow for all the losses to which the manure was subject before it came to be actually applied to the land. The compensation was further spread over a period of eight years, for which period it was assumed to exercise an influence. In the case of food consumed in the last year but one, a deduction was made of one-third of the allowance for the previous year—while for food consumed three years previously a further deduction of one-third was made ; and so on, for any particular number of years, down to the eighth year.

Let us, as an instance, take again the case of linseed-cake, the "original manurial value" of which is £2, 11s. 11d. For each ton of this cake consumed in the last year of tenancy, it would be assumed that a practical unexhausted value of £1, 6s. remained on the farm, realisable by the new tenant. For a ton of cake consumed in the last year but one, this sum would be reduced by one-third, making 17s. 4d. If consumed a year previously, it would be still further reduced by a third, making 11s. 7d., and so on. In the eighth year back, the compensation would be only 1s. 6d.

As a matter of fact, most farmers would, no doubt, object to paying "compensation values" for food used more than two or three years previously; but the principle of compensation suggested —taking it as far back as may be deemed judicious—is a perfectly sound one. In applying it, the valuer, if he knows his business, will be influenced by his observations taken on the farm as to the mode in which manure is treated, and as to the information available in regard to the consumption of the food. Sir John Bennett Lawes and Sir Henry Gilbert, in the paper already quoted from, very rightly observe : "It is pretty certain indeed that every claim for compensation will have to be settled on its own merits; that the character of the soil, the cropping, the state of the land as to cleanliness, and many other points, will be taken into consideration both for and against any claim."

Voelcker and Hall's Tables.

What Lawes and Gilbert's Tables really effected was, to put the whole matter of the valuation of unexhausted manurial residues on a sound basis —viz., that of the value of the actual constituents supplied and not the mere cost of the foods producing them, this latter, as explained, having often no relation whatever to the manurial value. Previously to the issue of these tables the assessing of "unexhausted value" had been ruled entirely by "local custom," this varying very widely in different districts. Each system, moreover, was of a purely arbitrary character, and in almost every instance the actual *cost* of the respective foods was taken as the starting - point. Lawes and Gilbert's Tables rightly superseded these "local systems," and supplied, at least, a reasonable ground on which to frame a valuation.

At the same time, it was generally felt that, in practice, the period of eight years, over which Lawes and Gilbert spread the application of their system, was an unduly long one. Undoubtedly it could be shown from the Rothamsted and the Woburn experiments that manure made from purchased foods would exercise an influence for as long a period as eight years on crops subsequently grown with it, but there was the fact to consider that no one would, in practice, apply farmyard manure only at such long intervals, nor would any one expect it to have any practical bearing after so long a period as eight years after its first application.

Such considerations as these militated much against the general adoption of Lawes and Gilbert's Tables in the practical business of valuation, and, though the general principles of the tables were in a measure accepted, and though certain modifications were introduced into "local systems" consequent on these, the latter were not by any means wholly superseded. There was a general feeling, in short, that the tables, though perhaps they were right in principle, could not be fully applied in practice.

These facts led Dr J. A. Voelcker and Mr A. D. Hall—the then directors respectively of the Woburn and Rothamsted Experimental Stations—to review the whole position, and in 1902 they put forward a revision of Lawes and Gilbert's Tables, and this was published in the *Journal of the Royal Agricultural Society of England* for 1902. This table is here set out at length.

TABLE IV.—Voelcker and Hall's Tables (1902) of the Composition, Manurial and Compensation Values of Feeding-Stuffs (Revised from Lawes and Gilbert's Tables of 1897).

No.	Foods.	Nitrogen. Per cent in food.	Nitrogen. Value at 12s. per unit.	Nitrogen. Half of value to manure.	Phosphoric acid. Per cent in food.	Phosphoric acid. Value at 3d. per unit.	Phosphoric acid. Three quarters of value to manure.	Potash. Per cent in food.	Potash. Value at 4s. per unit, all to manure.	Last year.	Second year.	Third year.	Fourth year.
		%	s. d.	s. d.	%	s. d.	s. d.	%	s. d.	s. d.	s. d.	s. d.	s. d.
1	Decorticated cotton-cake	6.90	82 10	41 5	3.70	9 4	7 0	2.00	8 0	56 5	28 2	14 1	7 0
2	Undecorticated cotton-cake	3.54	42 6	21 3	2.00	6 0	4 6	2.00	8 0	33 9	16 10	8 5	4 2
3	Linseed-cake	4.75	57 0	28 6	2.00	6 0	4 6	1.40	5 7	38 7	19 3	9 7	4 9
4	Linseed	3.60	43 2	21 7	1.54	4 7	3 5	1.37	5 6	30 6	15 3	7 7	3 9
5	Palm-nut cake	2.50	30 0	15 0	1.20	3 7	2 8	0.50	2 0	19 8	9 10	4 11	2 5
6	Cocoa-nut cake	3.40	40 10	20 5	1.40	4 2	3 1	2.00	8 0	31 6	15 9	7 10	3 11
7	Rape-cake	4.90	58 10	29 5	2.50	7 6	5 8	1.50	6 0	41 1	20 6	10 3	5 1
8	Beans	4.00	48 0	24 0	1.10	3 4	2 6	1.30	5 2	31 8	15 10	7 11	3 11
9	Peas	3.60	43 2	21 7	0.85	2 7	1 11	0.96	3 10	27 4	13 8	6 10	3 5
10	Wheat	1.80	21 7	10 9	0.85	2 7	2 0	0.53	2 1	14 10	7 5	3 8	1 10
11	Barley	1.65	19 10	9 11	0.75	2 3	1 8	0.55	2 2	13 9	6 10	3 5	1 8
12	Oats	2.00	24 0	12 0	0.60	1 10	1 5	0.50	2 0	15 5	7 8	3 10	1 11
13	Maize	1.70	20 5	10 2	0.60	1 9	1 4	0.37	1 6	13 0	6 6	3 3	1 7
14	Rice-meal	1.90	22 10	11 5	0.60	1 9	1 4	0.37	1 6	14 3	7 1	3 6	1 9
15	Locust-beans	1.20	14 5	7 2	0.80	2 5	1 10	0.80	3 2	12 2	6 1	3 0	1 6
16	Malt	1.82	21 10	10 11	0.80	2 5	1 10	0.60	2 5	15 2	7 8	3 10	1 11
17	Malt culms	3.90	46 10	23 5	2.00	6 0	4 6	2.00	8 0	35 11	17 11	8 11	4 5
18	Bran	2.50	30 0	15 0	3.60	10 10	8 2	1.45	5 9	28 11	14 5	7 2	3 7
19	Brewers' grains (dried)	3.30	39 7	19 9	1.61	4 10	3 8	0.20	0 10	24 3	12 1	6 0	3 0
20	Brewers' grains (wet)	0.81	9 9	4 11	0.42	1 3	0 11	0.05	0 2	6 0	3 0	1 6	0 9
21	Clover-hay	2.40	28 10	14 5	0.57	1 9	1 4	1.50	6 0	21 9	10 10	5 5	2 8
22	Meadow-hay	1.50	18 0	9 0	0.40	1 2	0 11	1.60	6 5	16 4	8 2	4 1	2 0
23	Wheat-straw	0.45	5 5	2 8	0.24	0 9	0 7	0.80	3 2	6 5	3 2	1 7	0 9
24	Barley-straw	0.40	4 10	2 5	0.18	0 6	0 4	1.00	4 0	6 9	3 4	1 8	0 10
25	Oat-straw	0.50	6 0	3 0	0.24	0 9	0 7	1.00	4 0	7 7	3 9	1 10	0 11
26	Mangels	0.22	2 8	1 4	0.07	0 3	0 2	0.40	1 7	3 1	1 6	0 9	0 4
27	Swedes	0.25	3 0	1 6	0.06	0 2	0 1	0.22	0 11	2 6	1 3	0 7	0 3
28	Turnips	0.18	2 2	1 1	0.05	0 2	0 1	0.30	1 2	2 4	1 2	0 7	0 3

In drawing up this revised table, Voelcker and Hall were able to utilise information obtained from further experiments conducted at Woburn and Rothamsted as well as on the Continent, and these were specially useful as affording more definite knowledge as to the actual losses incurred in making and storing farmyard manure. Up to that time these losses had been merely assumed, but now they were able to be more nearly defined.

Further, the tables were revised in two important respects: firstly, in respect of the money values of the manurial constituents, these having undergone some modification since the earlier tables were issued ; and, secondly, in regard to the period over which compensation was spread. Voelcker and Hall limited this period to that of the ordinary farm rotation—viz., four years—substituting this for the eight years adopted by Lawes and Gilbert, and adjusting the tables in this sense. In addition to these alterations, they simplified the tables by the exclusion of certain foods which only very occasionally came under consideration, and, in place of the three tables of Lawes and Gilbert, they put forward a single table (Table IV.), which comprised practically all the details that were necessary for showing the basis of calculation and the final valuation of the unexhausted residue for each year of the rotation.

In arriving at their conclusions, Voelcker and Hall, it must be understood, closely followed the principles laid down by Lawes and Gilbert, adopt-ing, in great measure, the figures of these investigators as to the composition of foods, the constituents stored up in live-weight increase, &c. The main differences were in regard to the losses in making and storing the manure (these being now based on actual experiment), and in respect of the period over which compensation was to be spread.

Simultaneously with this revision by Voelcker and Hall, the Central Chamber of Agriculture, which had appointed a Committee for the purpose, and had received evidence from valuers, farmers, and others, issued a table for the assessment of unexhausted value. This Table, it may be said, differs but slightly from that of Voelcker and Hall. The tables put forward by the latter have been very favourably received, and may be said to have practically superseded the "local systems" formerly in use ; they have now secured general adoption, and may be taken as ruling the question of "compensation for unexhausted manure value of foods."

TREATMENT OF FARM-HORSES.

As would be expected, the management of the work-horses of the farm differs in many details from the system pursued in pure-bred studs. In the former case the methods are more simple and less varied.

In the greater part of England, horses when not at work are tended by lads or men employed specially for the purpose. In Scotland and Ireland it is the prevailing custom for the men who work the horses to attend at all times to their feeding and general treatment. The latter system is the better one for both men and horses.

Watering Horses.—The first attention to horses in the morning is to take them out to water—that is, if there is not a regular supply of water within their reach in the stable.

When horses are allowed to drink water freely immediately after feeding, they are liable to suffer from colic, as the water is apt to carry some of the undigested food into the intestines. Water should therefore always be given to horses before, and not after, feeding.

The quantity of water drunk by horses varies greatly, some drinking more than others. If allowed frequent access to fresh pure water, horses will not, as a rule, drink more than necessary. When an animal is very hot, or chilled, or exhausted, or has been long without water, only a small quantity of water should be allowed at first. In such cases, a safe drink is water thickened with a handful or two of oatmeal, or, better still, oatmeal gruel. Very cold water should be given in small quantities at a time. Keep watering-troughs scrupulously clean, and see that the water in them is changed frequently.

When the horses have received their morning feed, the men, before going to breakfast, remove the dung and soiled litter to the dung-pit.

Morning Feed.—Immediately after getting water, the horses receive their first allowance of bruised oats or other food, with long hay in the rack or manger, the latter being usually preferred. Horses should be given peace at feeding-time. Harness can be quickly enough put on after the feed is eaten, and time should be taken to groom the horses very carefully. An allowance of a little time between eating and going to work is of advantage to all horses.

Mid-day Care of Horses.—When the horses come in from the morning work they get a drink of water, a feed of bruised corn, and chaffed hay or oat-straw, and the men get their dinner. Some keep the harness on during this interval, but it should be taken off, to allow both horses and harness to cool. After dinner the men return to the stable, when the horses will have finished their feed, and a small ration of fresh straw or hay will be well relished. The men have a few minutes to spare, when they should wisp down their horses, put on the harness, comb out the tails and manes, and be ready to put on the bridles the moment one o'clock strikes.

When work is in a distant field, rather than take them home between yokings, some farmers feed the horses in the field out of nose-bags, and make the men take their dinners with them, or it is brought to them. This, however, is not a good plan.

Hours of Work.—The hours of work vary in different parts of the country, and, of course, also with the season. The most general rule is ten hours per day—from six to eleven A.M. and one to six P.M. In Scotland this method is strictly adhered to, when daylight admits, but in England there is less regularity in working hours.

Work expected of Horses.—No definite rules can be laid down as to the amount of work which should be accomplished by horses. The local circumstances, such as the character and fitness of the horses, the nature of the work, the exigencies of the time, and the supply of food, must always be duly considered, and the farmer must at the time decide for himself how much work of any particular kind he is to expect from each horse or pair of horses.

One general principle may be laid down—one not so fully observed as is desirable—and that is, that in working horses long days are preferable to quick pace. It will be much easier for a pair of horses to plough a certain equal extent of land in six days of ten hours than in six days of nine hours each, easier still than in six days of eight hours. With the farm-horse, as with the roadster and hunter, "it is the pace that kills."

Evening Care of Horses.—When the horses come in from their day's work they are well rubbed down, and receive hay or straw and bruised oats. The stable has had but half litter all day, since its cleansing out in the morning, and the horses have stood on the stones at mid-day. This is a good plan for purifying the stable during the day, and is not so much attended to as it deserves. Fresh straw is brought by the men from the straw-barn, and shaken up with the old litter to make the stalls comfortable for the horses to lie down upon for the night. In most cases horses receive food again about eight o'clock, when the litter is once more shaken up and adjusted for the night.

Grooming Horses.—The grooming consists first in currying the horse with the curry-comb b, fig. 738, to

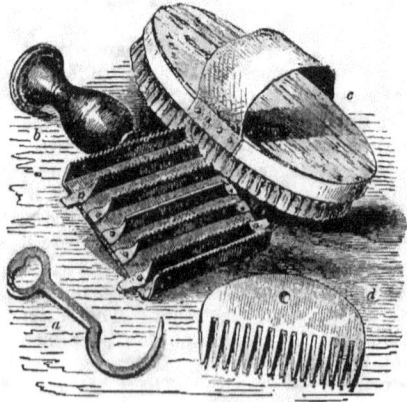

Fig. 738.—*Curry-comb, brush, foot-picker, and mane-comb.*

free him of the dirt adhering to the hair, and which, being now dry, is easily removed. A wisping of straw removes the roughest of the dirt loosened by the curry-comb. The legs ought to be thoroughly wisped—not only to make them clean, but to dry up any moisture that may have been left in the evening. At this time the feet should be picked clean, by the foot-picker *a*, of any dirt adhering between the shoe and foot. The brush *c* is then used to remove remaining and finer portions of dust from the hair, dust being cleared from the brush by a few rasps along the curry-comb. The wisping and brushing, if done with some force and dexterity, with a combing of the tail and mane with the comb *d*, should render the horse pretty clean. But there are more ways than one of grooming a horse, as may be witnessed by the skimming and careless way in which some ploughmen do it. It is the duty of the farmer himself, or his steward or manager, to see that the horses are well attended to.

Brushing preferable to Combing. —The use of the iron curry-comb is disapproved by many. Dr Fleming says it "should never, as a rule, be applied to the skin of horses." For long rough coats, he considers nothing is better than a good dandy-brush to remove dandruff, dust, and dirt; for finer-coated horses a good bristle-brush, supplemented by the wisp and rubber, will suffice. He regards the brush as the best appliance for cleaning the skin thoroughly, and he points out how essential it is, for the health of the animal, that the skin be kept clean, so that it may at all times be in a fit condition to perform its important functions.

Rubbing Wet or Heated Horses. — If a horse comes into the stable heated or wet, it should at once be well rubbed down with a handful of straw. If it has been excessively warm, it may be well to throw a rug over it till it has regained its normal temperature. It may perhaps, after two or three hours, or sooner, break out into a cold perspiration, and if so it should again be well dried.

Water-brush.—For washing the legs and heels of a horse, a water-brush, fig. 739, is very useful.

Exercise for Horses.—When idle, work-horses should be taken out for exercise every day, and groomed as carefully as when at work. Exercise is necessary to prevent thickening of the heels, a "shot of grease," or a common

cold. Fat horses, unaccustomed to exercise, are liable to molten grease.

Breeding Horses.—It is advisable for most farmers to breed their own horses. On a farm which employs three, four, or more pairs, two mares might easily bear foals every year, and perform a share of the work at the same time, without injury to themselves.

Wintering Young Horses.—In the south of England young horses are kept out on the fields all the year round, and in many cases do not even have a shed in which to lie overnight. Most people consider it necessary to have field-sheds for shelter, but in some cases nothing of the kind is provided.

Housing Necessary in Cold Districts.—But in the colder districts the young as well as the adult horses have to be housed in winter, at any rate overnight. When the weather is not wet or very cold the young animals should have

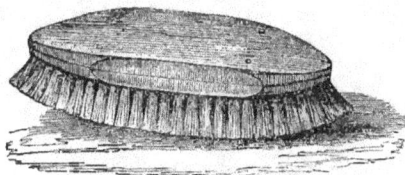

Fig. 739.—*Water-brush.*

a run out daily, and be brought back to a dry but airy bed at night. Covered sheds afford excellent shelter for young horses in winter nights, and where these are not available, loose-boxes or hammels are preferable to stalls.

Handling Young Horses.—Young horses should be frequently handled by their attendant, who by his kindly handling should make himself welcome and familiar amongst them. Young horses are not regularly groomed, but they will be all the better of a turn of the brush now and again. They should be rubbed with straw, if wet, and any clay or earth adhering to their hair should be removed.

Colts and fillies may be kept together their first winter, but afterwards entire colts should be kept by themselves.

Intelligence of Horses.—The horse is an intelligent animal, and seems to delight in the society of man. It is remarked by those who have much to do with blood-horses, that when at liberty, and seeing two or more persons standing conversing together, they will approach, and seem to wish to listen to the conversation. The farm-horse will not do this; but he is quite obedient to call, and recognises his name readily from that of his companion, and will not stir when desired to move until *his own name* is pronounced. He discriminates between the various sorts of work he has to do, and will apply his strength or skill in the best way, whether in the threshing-mill, the cart, or the plough. He will walk very steadily towards a feering-pole, and halt when he has reached it. He seems also to have an idea of time. We have heard a horse neigh daily about 10 minutes before the time of loosening from work in the evening, whether in summer or winter. He is capable of distinguishing the tones of the voice, whether spoken in anger or otherwise.

Horses are fond of nearly all kinds of music. Work-horses have been known, even when at their corn, to desist eating, and listen attentively, with pricked and moving ears and steady eyes, to music on various instruments. We have seen a kilted Highlander playing the bagpipes riding on the back of a farm-horse, which showed every sign of pleasure. The recognition of the sound of the bugle by a trooper, and the excitement occasioned in the hunter when the pack give tongue, are familiar instances of the power of particular sounds on horses, in recalling old associations to their memory. The horse's memory is very tenacious, as is evinced in the recognition of a stable in which he had at times been well treated. He is very susceptible of fear, and will refuse to pass into a road or a particular locality in which he had received a fright.

FEEDING HORSES.

In its way, the feeding of horses is quite as important as the feeding of cattle. The one is fed to perform work, the other to produce meat. In each case the performance will depend mainly upon how the matter of feeding has been attended to. He who would feed his horses perfectly must know and con-

sider not only the duties, powers, condition, and consequent food requirements of each animal, but also the composition and character of the available articles of food. It is only by properly adapting the one to the other that he can ensure the best possible results. Perfection may be beyond our reach. Let us get as near to it as possible.

Articles of Food for Horses.—The articles of food most largely used for horses are hay, straw of various kinds, oats, wheat, barley, beans, Indian corn, bran, linseed, linseed-cake, turnips, mangels, carrots, parsnips, potatoes, furze (or gorse), silage, vetches, fresh grass, clover, &c. Information regarding the composition and character of these and other feeding materials is given in the section on "Varieties of Food" (vol. iii. p. 269), in this volume. Before attempting to arrange mixtures of food for horses, farmers should give careful consideration not only to that information, but also to what is said in the succeeding section on "Animal Nutrition" (vol. iii. p. 291).

Food Requirements of Horses.

It is far from easy properly to understand and determine the food requirements of different horses—horses of various kinds, ages, conditions, and sizes, and performing different kinds of work.

Ration for Idle Horses.—For a horse doing no work, the food, to properly maintain its bodily functions for twenty-four hours, should contain over 12 lb. dry matter, made up as follows:—

Albuminoids .	. .	8.36 oz.
Fats	. .	3.19 "
Carbohydrates	. .	11.4 lb.
Salts	. .	0.5 oz.

Total food, free from water, 12.15 lb.

It is calculated that this amount of food, so composed, is capable of producing force equal to 27,855 foot-tons. "And if the weight of a horse," says Dr Fleming, "is estimated at 1000 lb., he would require 87.3 grains for each pound of body weight; or the whole body would require about 1-80th part of its weight in food every twenty-four hours, the animal undergoing no toil of any kind. A pony weighing 440 lb. requires 46 grains of nitrogenous matter for each 2 lb. 3¼ oz.

of weight. This essential diet is supposed to be theoretically totally devoid of water, but in reality it would contain from 15 to 20 per cent of that fluid; so that, to allow for it, something like 1.87 lb. to 2.49 lb. must be added to the 12.472 lb." [1]

But this is merely a ration for the bare subsistence of a horse. To enable the horse to perform work, additional food is necessary.

Additional Food for Work.—The amount of additional food required to enable a horse to perform work and maintain its condition will depend upon several circumstances, such as the nature and amount of work to be done, the season of the year, condition and size and powers of the horse, &c. The mere weight of the animal is not so reliable as a guide to the quantity of food required by a horse as it is in the case of cattle. The food requirements of small horses are relatively greater than those of larger ones.

Quick Pace and Food Requirement. —A point of some importance is this, that there is less waste of energy and tissue—and therefore less food requirement — when the labour performed is slow and prolonged than when it is brief and severe. Dr Fleming says it has been calculated that the useful work of a horse, which would be represented by 100, with a velocity of 2 miles per hour, would not be more than 51 with a velocity of 7½ miles, or more than 7 with a speed of 11½ miles an hour. In practice it has been found that the amount of food sufficient for slow work for ten hours will not suffice for more than five hours' exertion at a trot. Increased speed in work increases the demand for albuminous food.

A horse working at walking-pace requires from 6 to 9 grains of albuminoids for each 7233 foot-pounds of work performed; while for work at a trot the requirements of albuminoids would be as much as from 15 to 24 grains for the same number of foot-pounds of energy expended.

Force exerted by Horses.—In order to know how to properly adjust the quantity and composition of food, it is

necessary to ascertain as nearly as possible the amount of force exerted by horses in performing work, be it pulling a load or carrying a rider. With regard to this Dr Fleming says: "It may be mentioned that a one-horse engine, working ten hours per day, raises 19,799,360 pounds 1 foot high—this being the calculated amount of energy expended in ten hours if it could be all at once exercised. But this is probably much more than a horse could exert; a very hard day's work would in all likelihood not be more than 16,400,000 foot-pounds, which would be exercised by a horse pulling a load along at a walk for eight hours. Eight hours' slow walking, with a traction force of 100 lb., is equal to 8,436,571 foot-pounds per day. Slow farm-work is equal to 11,211,000 foot-pounds per day. With regard to fast work, the amount of foot-pounds raised is less, for the effort required is sudden, and the waste of tissue or force is consequently greater. The actual amount of work done is less, for the reason that the animal cannot sustain the effort, and owing to the greater waste incurred, more food is needed."

The amount of energy expended at work both at fast and slow pace must vary considerably, but Dr Fleming gives the following estimate as "fairly correct":—

	Foot-pounds.
A hard day's work for a horse at a walk would be	11,500,000
A moderate day's work, ditto .	8,500,000
A hard day's work for a horse at a trot of fast pace would be	7,233,000
A moderate day's work, ditto .	3,500,000

Rations for Degrees of Work.— The following table, showing the amount of food required by a horse under different conditions of labour—the proximate principles of the diet being stated —is given by Dr Fleming:—

Proximate Principles.	Moderate Work.		Active Work.		Severe Work.	
	lb.	oz.	lb.	oz.	lb.	oz.
Albuminoids .	1	4	1	8	2	0
Fats .	0	8½	0	10	0	12½
Carbohydrates	6	13	6	0	10	0
Salts .	1	5	1	7	1	9
Total .	9	14½	9	9	14	5½

---

[1] *The Practical Horse-Keeper.*

X

It is necessary to explain that these are merely approximate quantities, and must not be followed blindly. In each individual case carefulness and judgment must be exercised; and the appetite, health, condition, and working powers of each animal duly considered.

Winter Feeding of Horses.

There is almost as much variety in the systems of feeding horses in winter as in the methods of the winter feeding of cattle.

As to the methods of feeding different classes of horses so much information has already been given in describing the systems of management pursued in pure-bred studs of the various breeds of horses and ponies, that comparatively little need be added here.

Preparing Food for Horses.—On the best-managed farms all kinds of grain are bruised, and the larger portion of the hay and straw cut into chaff before being given to horses. As to the propriety of bruising grain there can be no question. Not an ounce of grain of any kind should be given to horses without being ground; for when given whole, a portion of the grain is liable to pass through the animals undigested. The husk of grain is so dense and difficult to dissolve, that if it should be given whole and escape being ground by the animal's teeth, the gastric juice acts feebly and slowly upon it, and will most likely be unable to dissolve it, so that a portion of the whole grain will pass through the animal unaltered.

As to the chaffing of hay and straw, there is some difference of opinion. But there is no doubt the chaffing both economises fodder and is advantageous to the horses, by assisting them to masticate their food. It should therefore be encouraged, for both these points are important.

Many who regularly pursue chaffing give their horses in addition small allowances of long hay or straw, which may be relished by the horses when they are not hard worked and have plenty of time to eat their food. The bruised grain and chaffed fodder are usually given together, and are of course mixed in varying proportions according to the work being performed at the time.

Beans and peas should be merely cracked or split, and not ground into flour. Care should be taken to mix the various ingredients thoroughly, so that each animal may receive its due proportions of all of them. The chaffed fodder and bruised grain may be conveniently mixed in a large iron vat or box, or in a wooden box lined with sheet-iron.

Mashes for Horses.—Farmers now, as a rule, prefer dry food to mashes for horses, but in many cases in winter mash is still given once or twice a-week. The mash generally consists of boiled barley, oats, or beans, mixed at times with bran and seasoned with salt, and an ounce each of sulphur and nitre is sometimes added. Raw potatoes or swedes are given one time and mash another, or the potatoes and swedes are boiled with either barley or oats. The articles are prepared in the stable boiler-house in the afternoon, and when given to the horses at night should not exceed milk-warmth. The corn put into the boiler is as much as when given raw, and in its preparation swells out to a considerable bulk. The horses are exceedingly fond of mash. The ingredients should be well mixed and well cooked.

For horses at light work, cooking food may be commended on the score of economy, for a small allowance of cooked grain will render a large quantity of chaffed fodder palatable. For horses, old or young, whose teeth and digestive systems are weak or defective, cooked food is highly advantageous. Mouldy hay is made safer and more palatable by being steamed, and damaged grain should in all cases be cooked. Horses will relish a sprinkling of salt in their cooked food. Be careful to give the cooked food to horses before it begins to ferment.

Oats for Horses.—The quantity of raw bruised oats given to farm-horses, when at moderate work, is usually from about 7 to 12 lb. per day in two or three feeds. Some give less when the horses are not at full work. Others give still larger quantities in the busy season.

Substitutes for Oats.—Some farmers withdraw the corn altogether from their horses in the depth of winter, giving them mashes of some sort instead; whilst others give them only one feed of oats a-day, divided at morning and noon, and a mash or raw turnips or potatoes at night. In many cases the mashes used when horses are at light work consist too largely of chaffed straw, upon which horses soon lose condition and strength. A fair proportion of grain of some kind should always be included.

Both turnips and potatoes are good food for horses, but they should be given in moderation. Large quantities of soft food do not prepare horses well for hard work, and so mashes should be given to them sparingly.

A Group of Rations.—The following are food mixtures used by leading farmers in different parts of the country when horses are at full work, the quantities mentioned being for one day :—

(1) 10 lb. of cut straw ; 10 lb. of oats ; 16 lb. of turnips.

(2) 16 lb. of hay ; 5 lb. of oats ; 16 lb. of turnips.

In these two cases the turnips are pulped and mixed with the cut fodder twelve hours before being used.

(3) 10 lb. of bruised oats ; 20 lb. of hay ; 12 lb. of cut straw.

(4) Hay, maize, oats, and beans, mixed in the proportions of 4 cwt. hay, 3 cwt. maize, 2 cwt. oats, and 1 cwt. beans—the hay being chaffed and the grain bruised. Heavy farm and cart horses, doing full work, get as much of this mixture as they will eat, which is about 24 lb. each daily, with a little long hay twice a-day.

(5) 2 bushels of oats, ½ bushel split peas, with 2 trusses of hay and straw chaffed per week per head when in full active work.

(6) As much as they can eat of long straw and uncut swedes, with 1½ bushel oats per week.

(7) 18 lb. crushed oats and 2 lb. linseed-cake, with hay *ad lib.*

(8) 12 to 14 lb. crushed oats ; a mash of bran, with a gill of linseed-oil and some saltpetre every Saturday night ; an allowance of hay and oat straw *ad lib.*

(9) 10½ lb. crushed oats, 5½ lb. crushed Indian corn, and 7½ lb. cut hay, with long hay *ad lib.*, and rock-salt to lick.

(10) 13 lb. of crushed oats, 3 lb. bran, 6 lb. raw swedes (sliced), and 14 lb. cut chaff (two-thirds hay and one-third straw)

—the oats, bran, and chaff being mixed together and macerated with water, and prepared twelve hours before being used.

In most cases these rations, which apply to the winter and spring, cost from 10s. to 12s. per week ; in some cases more. In the grazing season the rations are varied and greatly lessened, especially in grain.

Roots for Horses.— Swedes, either raw or cooked, are given largely to draught - horses. When the roots are cooked alone, from 50 to 60 lb. are put into the boiler or steaming-vat for each horse, and this gives about 35 to 45 lb. of cooked food, which should be prepared in time to allow it to become cool, but not cold, before being given to the horses in the evening. A little chaffed hay, perhaps not more than 1 lb. for each horse, is mixed with cooked roots, and some add about 1 lb. of ground oilcake, while others have about half a pound linseed for each animal cooked along with the roots. This warm food is given either in two meals—one when the horses come in from work in the evening, and the other at 8 P.M.—or in one meal at the latter hour.

But the most general plan is to give the turnips to horses raw and uncut, as the last meal for the night. Mangels are given in a similar way. Roots of all kinds should be thoroughly cleaned before being given to horses.

Carrots and Parsnips.—There is no kind of root equal to carrots for horses. They are especially suitable for hunting and other horses which are hard-worked. They are given raw and usually sliced. For ordinary farm-horses, however, they are generally beyond reach on account of their cost. Parsnips are also given to horses. Both carrots and parsnips are supposed to be injurious to the eyes of horses.

Furze for Horses.—Furze (whin or gorse) is relished by horses, and makes useful winter food for them. It is the young shoots of furze that are fed to horses, and they are best when bruised by a furze "masticator." In the absence of a "masticator," the furze is cut as fine as possible by a chaff-cutter, but this does not cut and bruise it so fine as is desirable. A hand furze-bruiser, which does its work wonderfully well, is represented in fig. 740. When fresh furze is crushed, it throws off a fine aromatic odour, which is much relished by horses. The furze is bruised every second or third day. It should not be allowed either to heat to any extent or to become dry. If it should get dry before being used, it would be well to sprinkle water over it by a garden watering-can.

Draught - horses will eat from 20 to 25 lb. of crushed furze per day, but it will be as well to give smaller quantities, mixed with chaffed hay or straw and bruised oats.

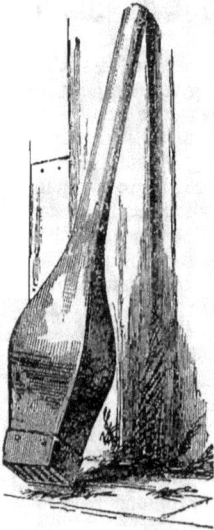

Fig. 740.—*Hand whin-bruiser.*

Feeding Young Horses. — Young growing horses are often stinted in food. No greater mistake could be made. They ought to be fed liberally and with as much care and punctuality as the hardest-worked horse on the farm. Let them have as much good hay or oat-straw as they can eat two or three times a-day. In addition to this, on many farms one-year-olds get 3 lb., and two-year-olds 4 lb., gradually increasing as they get bigger to 5 lb.,

of the following mixture : 3 parts crushed oats, 1 part beans, and 1 part linseed,—this food being given when the animals are housed at night, and before being put out in the morning. In wet stormy days, when they are out only half an hour or so for exercise, they should have their food thrice, instead of twice a-day.

This is liberal feeding, and less of the concentrated food, perhaps from 4 to 6 lb. per day, may suffice to keep the youngsters growing and in good condition. Many think it is desirable to give young horses once or twice a-week a warm mash, consisting of boiled roots, boiled linseed or linseed-meal, mixed with bran. Young horses will thrive admirably on 3 to 4 lb. of crushed oats, and 1 lb. of crushed linseed-cake per day, mixed with chaffed hay or straw, and raw swedes.

Young Horses not to be Pampered. —But while young horses should be fed liberally, they should not be forced in feeding, or pampered in any way. Keep them in good growing condition, full of natural flesh; and, without exposing them to excessive cold or wet, let them have plenty of exercise and fresh air, so that, as far as possible, their muscular and constitutional strength and hardiness may be developed.

Rations for Town Horses. — The rations given to horses for town haulage vary greatly, the following indicating usual allowances for light and medium van and lorry horses in the towns named :—

	Glasgow.	Edinburgh.	Birmingham.	London, South.	London, Street.	Liverpool.	Dublin.
	lb.	lb.	lb.	lb.	lb.	lb.	lb.
Oats	6	8	10	7	3	...	3
Maize	11	4	6	7	12	12	14
Beans or peas	...	4	4	1	1	4	...
Hay	8½	14	12	11	11	14	12
Straw	1	2	...	3	...	...	...
Bran	0½	...	...	...	1	1	0½
Total lb.	27	32	32	29	28	31	29½

Riding and Driving Horses.—Carriage-horses are often fed more highly than is necessary or is really beneficial for their health and usefulness. If their work is light, from 8 to 10 lb. of bruised grain and 12 to 14 lb. of chaffed hay per day will be sufficient. When the work is heavier the grain must be increased, perhaps 2 or 3 or 4 lb. per day. Hunting-horses, and all riding-horses which are kept at hard work, should be liberally fed—horses over 15 hands, perhaps from 15 to 16 lb. of bruised oats per day, with 10 or 12 lb. of chaffed hay; the allowance of oats for smaller horses being reduced by 2 or 3 lb. per day. A few pounds of carrots—not more than 3 lb. per day—may occasionally be given with advantage. For hunters, Dr Fleming recommends the substitution of 2 lb. of split beans for 2 lb. of the allowance of oats.

Army horses usually receive 10 lb. of oats per day with 12 lb. of hay, and—for litter—8 lb. of straw. When on severe duty, or in camp, the allowance of oats is increased to from 10 to 14 lb. per day. The hay is given uncut.

Quantity of Food.—It is considered that, as a rule, an average-sized draught-horse will require about 29 lb. of food per day. Much less than that, even although it should be highly nutritious, will not be sufficient to maintain the animal in a healthy and vigorous condition. Reynolds states that such a horse, when moderately worked and well housed, will consume from 29 to 34 lb. per day, of which the hay and straw should constitute about two-fifths.

Bulk of Food.—It is undesirable, in ordinary cases, to attempt to feed horses mainly upon highly concentrated food. In order to enable the digestive organs to properly perform their functions, a certain considerable degree of bulk in the food is necessary. When horses are hard-worked, the morning and mid-day meals may advantageously be small in bulk—a feed of oats can be speedily eaten, and does not interfere with the breathing organs as does a bulky feed of hay or straw. But at night, in these cases, bulky food should be given.

Frequency of Feeding.—Horses should be fed at least three times a-day—before 6 in the morning, about mid-day (as soon as brought in from work), and in the evening. The exact hours will depend upon local circumstances as to the work being carried out. But it is very important that precise feeding hours should be arranged, and that these should be rigidly adhered to. Punctuality in feeding is a most important consideration.

Long fasts are detrimental to horses. The standard hours of farm-work seldom permit of more than three meals per day; but it would be far better for the horses if they could be fed four times a-day, at intervals of not more than four hours. Let the evening meal be the largest and bulkiest, as the horses have then plenty of time for thorough mastication. Long fasts and rapid and heavy feeding often give rise to disorder of the digestive organs, and care should be taken to give the animals ample time to consume their food in comfort. Improper mastication, often caused by too hurried feeding, renders the process of digestion more difficult. At long spells of work, a feed of grain, even if it should be very small, given in a nose-bag, will be found very beneficial. Do not give more food of any kind at a time than the animal is likely to consume, as if any were left it would become stale and unpalatable, and probably be wasted.

Care of Horses in Summer.

The care of horses in summer causes little trouble or anxiety.

Pasturing Work-horses.—On many farms, especially in Scotland, the rule is to graze horses. As soon as the warm weather of summer has fully set in, the horses lie out in a pasture field overnight. Between the yokings they either get cut grass in the stable or are put on pasture, the latter plan saving the trouble of cutting grass. Work-horses are liable to suffer much from chilly nights, cold often laying the foundation of diseases—such as rheumatism, costiveness, stiffness of the limbs. The aftermath is good pasture in the interval of work at noon, and the second cutting of clover may last for suppers until the time to betake to the stable altogether.

Soiling Horses.—Many farmers disapprove of pasturing farm-horses, and support them at the steading upon forage. Where there are hammels or courts which could be easily divided, we would adopt this plan at once, but we are doubtful of its advantage in a stable. The heat of a stable in summer—and the doors cannot be left open—with the evaporation of the increased issue of urine due to the green food, cannot fail to vitiate the air. The cattle-courts are more open; and if they can be divided so that each pair of horses may have a compartment to themselves, they will thrive admirably here.

In the tillage districts of England this system of summering horses in the cattle-courts is extensively pursued. Many farmers, indeed, maintain that there is no better or cheaper method of keeping draught-horses in summer than in the courts, fed with green vetches or other similar succulent food, and dry hay, with perhaps a little bruised oats. Very often the grain is omitted.

Still it is a good plan to give the horses a week or two of the fresh air in an open pasture field.

Pasturing Young Horses.—Young horses are put to pasture during the day as soon as they can obtain a bite. They should be brought at night into their hammels until the grass has passed through them; after which they should lie out all night in a field which offers them the protection of a shed or other shelter. Work-horses do not care for a shed on pasture, being too much occupied with eating during night to mind it. In rainy weather young horses should be kept in the hammel on cut grass, and not exposed to rain in the field overnight.

The farmer's saddle-horse should usually have grass in summer, as it is the most wholesome food it can have. But it is more convenient to give it cut grass in a court or hammel than to send it to pasture, where it may be difficult to catch when wanted.

Peculiarities of the Horse in Grazing.—It is surprising with what constancy a work-horse will eat at pasture. His stomach being small in proportion to the bulk of his body, the food requires to be well masticated before it is swallowed; and as long as that process is proceeded with while the grass is cropped, no large quantity can pass into the stomach at a time.

The horse, like all herbivorous animals, grazes with a progressive motion onwards, and smells the grass before he crops it. His mobile lips seize and gather the stems and leaves of the grass, which the incisors in both jaws bite through with the assistance of a lateral twitch of the head. When grass is rank, he crops the upper part first; and when short, bites very close to the ground. Horses should not graze amongst sheep, as both bite close to the ground; and work-horses often injure sheep that come in their way, either by a sly kick or by seizing the wool with their teeth.

It is proverbial that horses do not graze well upon many of the very best bullock pastures. Horses often do better on rough pasture than on land which has been altered in its herbage by thorough drainage.

Horses Injured by Green Food.—Care must be exercised in beginning horses with green food every year. If allowed to gorge themselves too freely at the outset, serious illness may follow. Begin them sparingly with it, and if it should be wet or very succulent at any time during the season, it will be all the better to be accompanied or mixed with a little dry food such as hay.

Littering Horses.

Straw as Litter.—Straw is the most largely used, and is the best of all kinds of litter for horses. Wheat-straw, being stronger and tougher, is preferable to oat, or any other variety of straw, but in many parts of the country wheat-straw is not available. The stall should be thoroughly cleared out every morning, the wetter portions of the litter sent to the manure-pit with the dung; and the drier parts, which may be fit to be used for another night's bedding, retained in some convenient corner, or if the weather is dry, spread out near the stable, and taken in again in the evening.

Litter which has been used should never, as is sometimes the case, be stored beneath or in front of the manger, as the ammonia is apt to rise and injure the eyes of the horse, as well as taint its food.

From 8 to 14 lb. of straw is generally used as litter for each horse per day. With care, 8 to 10 lb. should be quite sufficient.

Peat-moss Litter.—The best substitute for straw as litter for horses yet introduced is "peat-moss litter"—peat-moss which has been broken and compressed by machinery till most of the moisture has passed away, leaving soft, spongy, fibry-looking vegetable matter. It makes cheap and comfortable bedding, absorbs and conserves the urine, and is a powerful deodoriser, keeping the stalls sweet and wholesome. It has a highly beneficial effect upon the feet of horses, keeping them cool, and encouraging the growth of strong tough hoofs. For animals with tender feet it is most beneficial. It is also valuable as manure, and its use should be commended on account of the saving of straw thus effected.

Peat-moss litter is now a regular article of commerce.

Other Varieties of Litter. — Many other substances are used as substitutes for straw in littering horses, such as sawdust, fine sand, spent tan, leaves of trees, and ferns. Sawdust is often used, but by itself it does not make comfortable or desirable litter. As a padding beneath a thin layer of straw it is very useful, comfortable, and economical, and may be resorted to where peat-moss litter cannot be procured at reasonable cost. Sawdust should be spread in a layer 2 or 3 inches deep, and raked daily. At least once every week the stall should be thoroughly cleaned out, and an entirely fresh layer of sawdust laid down. Where ferns are plentiful, they may be cut and stored for use as litter in winter. Spent tan, about 6 inches deep, makes durable and useful litter. If the surface is carefully cleaned of the dung every morning, and the tan raked by an iron garden-rake, one layer will last over a month. Scatter a little gypsum over it now and again.

General Hints.

Exercise.—Horses that are not regularly at work should be exercised punctually every day, say, just after breakfast. In very cold weather in winter towards mid-day may be preferable. Unless idle horses have plenty of exercise given to them methodically, they are liable to contract "grease" in the legs, and become soft, flabby, and unfitted for active work. Horses that are entirely idle should have two hours' daily exercise. It is specially important for young horses to have plenty of exercise.

Rest.—Farmers are often not so careful as they ought to be in providing quietness and comfort for hard-worked horses during hours of rest. In particular, during the two hours of mid-day rest horses should have as little disturbance as possible. As soon as they have been made comfortable in their stalls, and been fed, they should be left in perfect quietness. The stable-door should be shut, and no one let in to disturb the repose of the animals till their own attendants return to prepare for the work of the afternoon. Again, when the horses come in at night fatigued by a hard day's work, they should as soon as possible, after being fed and rubbed down, be left for quiet rest till supper-time.

Washing Horses' Legs. — Horses working on wet land are apt to have their legs so besmeared with mud that nothing but washing will clean them. In that case the legs should be washed at night, great care being taken to dry the legs thoroughly. Washing is itself undesirable, and should be resorted to only when absolutely necessary.

Cracked and greasy heels are often caused by imperfect drying after washing or after exposure to wet and mud. Referring to this point, Dr Fleming gives a few words of warning which are well worthy of careful attention. He says: "It must, however, be regarded as essential to proper management, that under no pretext is a horse to be left for the night until all his legs have been thoroughly dried. Nor is this precept very difficult of execution; a handful or two of light wood sawdust, rubbed for a few minutes well into the hair, will absorb all the moisture from the most hirsute legs, affording not only a sense of comfort to the animal, but preventing those undesirable consequences engendered by continued application of cold and wet to the extremities." [1]

Shoeing Horses. — Highly satisfactory methods of shoeing the various classes of horses are now pursued in almost all parts of the country. Skilled shoers are everywhere to be found, and farmers should be careful to see that the feet of their horses are well shod and kept in good condition.

Clipping and Singeing.—For horses which have rank coats of hair and have fast trotting work to do, clipping or singeing is found advantageous. This is seldom practised with farm-horses; and if they are well groomed there will be little need for any interference with the length of the coat. Clipping is most generally pursued with the ranker coats, and this is done speedily and efficiently by a clipper such as those (Clarke's) represented in fig. 741.

Shorter coats are singed, either oil or

[1] *The Practical Horse-Keeper*, 93.

gas being used in the singeing-lamp, the latter being preferable.

Injurious to Clip Legs of Horses. —But while this system of clipping or singeing has its advantages, there is one practice often resorted to which is entirely mischievous and should be strictly forbidden, and that is clipping the hair from the legs of draught-horses. In condemnation of this practice we cannot do better than quote the words of Dr Fleming, who regards it as "highly pernicious," and adds: "Hair is the natural protector of the cuticle, and is especially required to warm and shield the delicate skin of the heels; its removal from these situations is certain to induce a predisposition to *grease*, and other equally serious consequences. If the legs are muddy on return from labour, they should be dried as far as practicable, and the adherent clay subsequently removed

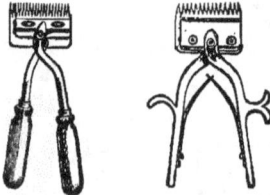

Fig. 741.—*Horse-clippers.*

with a hard brush. The application of the thinnest possible film of pure neat's-foot oil to the surface of the hair of the legs will prevent the adhesion of clay, but it should only be used when absolutely necessary. . . .

Protection to Skin from Wet.—"A predisposition to cracked heels is engendered by clipping the legs and pasterns in winter: this should never be done, if possible; but if necessary, then the skin should be protected from the action of wet and dirt by rubbing into it, before the horse leaves the stable, hard vaseline or zinc ointment. A very good protection against the action of icy cold water, or the salt slush which is so common on tramway lines in winter, is a mixture of one part of white-lead and three parts common oil, rubbed around the pasterns and the coronets by means of a brush."

Method in Stable Management.— Method and punctuality contribute as much to successful stable management

as to success in business. Let the stable rules be arranged on a well-thought-out, workable plan, such as will, in the most effective manner possible, contribute to the comfort and usefulness of the horses. And when the rules are laid down, see that they are rigidly adhered to. Irregularity in the feeding and general treatment of horses is most detrimental to their wellbeing. Among horses let all things be done quietly, kindly, and in order. Horses appreciate kindly treatment, and will repay such behaviour by confiding obedience. Good horsemen and good horses get warmly attached to each other. There is more of the "social element" about the horse, the greatest of all our quadruped friends, than the casual observer would be inclined to give him credit for.

The Foaling Season.

The foaling season is an anxious time for the owners of brood mares. The risks in foaling are greater than the calving risks, for the bovine race is hardier than the equine. With moderate skill and timely attention, however, serious losses in foaling are not likely to be of frequent occurrence.

Insurance against Foaling Risks. —Several insurance companies provide special facilities for insurance against losses in foaling, and farmers are prudently taking advantage of this provision of safety. The cost of insurance is comparatively small, and the sense of security it affords to the farmer is very comforting.

Abortion in Mares. — Abortion in mares, as in other animals, is often difficult to account for. In the large majority of cases, however, it may be attributed to injury inflicted in one form or other. A fright, chasing, running away, hurried driving, a kick from another horse, over-exertion at work, being driven in too heavy a grubber or cultivator, ploughing hard beaten headlands, being bogged in soft land, a shake between the shafts of a heavily loaded cart or waggon, or being upset or cast in shafts, are amongst the more violent actions liable to cause abortion. But it may also be induced by serious illness, improper feeding, especially with forcing food, exposure to wet stormy weather, eating poisonous plants, consuming frosted food, drinking an excess of cold water, &c.

When abortion does occur, the mare should be kept apart from other mares in foal until they have produced their young. And these other mares should not be allowed access to the spot where the unfortunate mare aborted.

Care of Brood Mares.—The greatest possible care should be exercised all through the period of pregnancy, alike in feeding and working the mare. She should be fed liberally but not excessively, for overfeeding may itself cause abortion. It is a well-known fact that overfed mares are liable to produce small foals, and the tendency to this is still greater when the overfed mare is an idle animal, kept perhaps solely for showing and breeding purposes.

Little need be said here in regard to the care of brood mares in and near the foaling season. The subject has already been fully dealt with in earlier parts of this volume in connection with the management of the leading breeds of pure-bred horses.

There is considerable difference of opinion and practice amongst farmers as to the working of mares up to foaling-time. Mares accustomed to steady farm-work may safely enough be kept at the lighter kinds of work up to within a few days, or at most a week, of the expected date of foaling. Carting, however, is dangerous, and should be avoided.

About ten days before the date upon which the foal is expected, the foaling compartment should be prepared. This should be free from draughts, comfortable in every way, and large enough to allow the mare to turn herself with ease at any part of it without incurring the risk of crushing the foal in so doing.

Watching Mares at Foaling.—It is very desirable that an eye should be kept on the mare night and day at foaling-time. Mares carry their foals from 330 to 360 days, eleven months being the time most generally "reckoned." They are by no means punctual, however, and very often a mare has to be watched for a week or ten days, occasionally even longer.

Symptoms of Foaling.—One of the surest signs of the approach of foaling is afforded by the udder. It of course becomes larger, and a waxy material appears like a bean at the tip of each teat. After this is present, in three cases out of four the mare will foal within twenty-four hours, and should not be left till the event has taken place.

Less definite indications of the completion of the period of pregnancy are the drooping of the belly, the enlargement and relaxation of the external organs of generation, and the flanks sinking inwards. The mare becomes dull and disinclined for exercise, while the movements of the foal may be seen to grow more distinct and active.

Assistance in Foaling.—Mares seldom need assistance in foaling. When aid is required, great skill and care must be exercised in rendering it. In cases which threaten to be protracted, or show any unusual and dangerous symptoms, the veterinary surgeon should at once be called in. Rarely, indeed, is a case of difficult foaling carried through successfully by any except an experienced and specially trained man in obstetrical work amongst farm animals.

Difficult Foaling.—If the mare has gone the full time of pregnancy, any exceptional difficulty in foaling is more than likely to arise from the foal lying in an abnormal position. The head and fore-feet should come first, the head resting upon the two fore-legs, just as in the case of a calf. If the labour pains are protracted without any apparent or sufficient progress, the hand and arm should be well lubricated with soft warm water and an antiseptic soap, and gently inserted to discover the position of the foal. If it is in its natural position as indicated, a little time will likely complete the process. If the foal is not yet in the passage, give the mare more time, and if necessary make another examination. If the foal is not presenting itself in the usual position mentioned, it may be necessary to adjust it, or at any rate to make some alteration in its position before birth can take place.

But this delicate work requires so much skill that, as already stated, it cannot be safely intrusted to any but a well-trained veterinary surgeon. If at all possible, have the veterinary surgeon at hand in such cases. If this is impossible, obtain the advice and assistance of the most experienced person within reach. Do not be too hurried in assisting the mare. Watch carefully, and assist nature when assistance seems likely to be useful. The mare needs more skilful and more careful operating than the cow in difficult parturition, and constant attention may be required to prevent her injuring herself should she become violent.

Such a case as this, however, is quite exceptional. As a rule, all that need be provided for the mare is a comfortable and cleanly compartment, with just a little less than the usual amount of food given to her when at work. The rest will, in most cases, be accomplished by nature.

Reviving an Exhausted Mare.—If the mare should seem to be weak or exhausted she may be revived by a drink of milk-warm oatmeal gruel, with the addition of a quart bottle of good ale.

Support to Mare's Belly.—Brood mares which have produced several foals are liable, when well up in years, to show a large extension of belly. For the sake of appearance as well as comfort to the mare, it would be well in extreme cases to support the belly for a time after foaling with wide, strong bandages wrapped several times round the body.

Mare's Udder.—Inflammation sometimes occurs in the udder of a mare being sucked. The udder is found to be hard and hot to the touch, and evidently painful to the mare. Foment the udder with warm bran-water, rub gently, and draw away a little milk at frequent intervals. It may be necessary to remove the foal for a few days and give the mare a dose of physic. Do not give medicine unless the foal is taken away from the mare for the time. A change of diet and low feeding for a few days may give relief. In a bad case, lose no time in calling in the veterinary surgeon.

After Foaling.—When it is seen that the foaling has been completed successfully, and the mare and foal are on their feet, a drink of warm gruel, made of oatmeal and water, or oatmeal, bran, and water, with a little salt in it, should be given to the mare, some sweet hay being placed in the rack. The two should then be left alone for a little

time, but carefully watched. As a rule, they speedily become accustomed to each other's society, and only in exceptional cases is any further interference required, either on behalf of the foal or the mare.

Cleansing.—In ordinary circumstances the "after-birth" will come away of its own accord very shortly after delivery. If it has not done so within at most ten or twelve hours, it will very likely have to be removed by the hand. This must be done gently and carefully; and if the after-birth has begun to decompose, the passage and uterus should be cleansed and disinfected by plentiful injections of some mild antiseptic fluid.

After-straining.—If the mare should continue to strain heavily for some time after birth, it may be assumed that all is not well with her, and that the advice of the veterinary surgeon would be useful.

Attention to the Foal.—The foal needs attention the moment it is born. First see that it has broken through and freed itself from the enveloping membranes, so that it has freedom to breathe. Then examine the umbilical cord, or navel-string, and see that it has been severed, and that there is no serious bleeding. The navel-string may be snapped in the act of foaling, but it is much safer to tie it. The attendant should tie a piece of thoroughly clean cord that has been soaked in an antiseptic solution around the navel-string about three-quarters of an inch from the skin; tie again an inch and a half farther down, and divide between with a clean knife. The stump of the cord should then be dressed with a strong solution of carbolised glycerine up to and over the edge of the skin.

Reviving Weak Foals.—It occasionally happens that a foal, although still living, is to all appearance dead when born. In this case, efforts should at once be made to induce respiration. A moment's delay may result in the extinction of the vital spark, which, with prompt action, might be fanned into active life.

Weakly foals will be all the better of a little extra attention at the outset, in the way of rubbing and drying with a woollen cloth. The limbs as well as the body should be well rubbed. It helps to promote circulation and give strength to the young creature.

Rearing Foals.

The feeding and general treatment of foals in pure-bred studs have been fully dealt with earlier in this volume, and therefore little detailed information will be required here.

Foals are not so robust as calves, and are more subject to injury from cold and wet. In the great majority of cases, the foal is reared almost entirely on its mother's milk for a period ranging from four to six months.

Troublesome Mares. — Unless exceptional circumstances have arisen—unless from some cause or other the mare becomes an inefficient or unkindly nurse —it will rarely happen that the mother and offspring require any special aid or interference until weaning-time arrives.

Occasionally it happens that a mare requires coaxing to admit the youngster to the udder, but with patience, tact, and kindliness success is generally attained. If sterner measures should be necessary, it is a good plan to put a net-muzzle on the mare's mouth and allow her to reach the foal with her mouth, but in a position that she cannot get at it with her feet. This should only be tried when the mare and foal can have sufficient attention. A bridle with blinkers may be required, and even a fore-foot held up, but do not use a twitch or strike the mare. Sometimes it is found that the bringing of a strange horse within sight of the mare a few days after foaling induces her to abandon her indifference and take the foal under her protection.

Beginning the Foal to Suck.—The foal will often be very awkward in its first efforts to suck. Do not attempt to assist or direct it except towards the proper quarter. Keep the mare quiet, and let the youngster feel its way itself. The instincts of nature will be its best teacher, and it will soon learn how to proceed. The mare's udder may be hard, and the teats dry. If so, rub the udder with the hand, and draw away a little milk, leaving the teats moist, so as to lead on the foal in its first attempt to suck.

Nursing Motherless Foals.—When a mare dies and leaves a living foal, or when a mare is unable to rear twin foals, or even to rear one, the best course for the sake of the foal is undoubtedly to procure a nurse-mother. No system of hand-rearing is quite equal to the mare's udder; and especially in the case of an exceptionally valuable foal an effort should certainly be made to procure a nurse-mother. This, however, is usually difficult to obtain, and, as a rule, foals that cannot be suckled by their own mothers have to be reared by the hand.

Rearing Foals by Hand.—For the young foal cow's milk is the next best food to the mare's milk. If the foal is newly born, the milk must at the outset be poured gently into its mouth out of an old teapot or kettle. By the time the foal is a week or ten days old it may be taught to drink the milk out of a pail, just as the hand-fed calf drinks milk. Give the foal your fingers to suck, and gently lead its head into the pail until it draws up milk between the fingers. In this manner it will readily learn to drink of its own accord when the pail is placed before it.

Cow's Milk for Foals.—Cow's milk, as we have said, is the best food on which to rear a foal for which mare's milk cannot be obtained. For some time at the outset, at any rate, the milk should be new and warm as it comes from the cow. Many experienced breeders think it desirable to dilute the milk with warm water and a little sugar. The foal should get little at a time, and be fed four or five times a-day. It may not be convenient to milk a cow so often as five times a-day, and therefore, at least for two of the meals to the foal, the cow's milk may have to be kept for two or three hours. In this case the milk should be heated to about the temperature of new milk by the admixture of a little hot water in which a very little sugar has been dissolved. When it is desired to give the milk undiluted, the best way of heating it is to insert the tin vessel holding it into another vessel containing hot water.

Bean - milk and Cow's Milk for Foals.—It sometimes happens that foals do not thrive satisfactorily on cow's milk alone. In this case the substitution of bean-milk for perhaps about one-half of the cow's milk may be tried. The bean-milk is prepared by boiling the beans almost to a pulp, removing the shells, and pressing the pulp through a fine hair-sieve. The result is a thick creamy fluid or paste. Sprinkle a pinch of salt over it, add the cow's milk entire or diluted, and the compound is ready for the foal. This system of feeding is highly spoken of by breeders of great experience.

Linseed, Bean-meal, and Milk for Foals.—Another liquid mixture used successfully in rearing foals consists of skimmed milk, linseed, and bean-meal. One formula for preparing the daily food of a foal from these substances is as follows: 12 pints sweet skimmed milk, 1 quart of linseed, which has been previously boiled for three or four hours, and 3 lb. of fine bean-meal added in a dry state. In some cases where the mares are hard-worked on the farm, the foals are weaned when only a few weeks old, and reared by the hand in some way similar to the above.

Health of the Foal.—During the nursing period the health of the foal must be watched carefully, so that its progress may not be interrupted by any derangement of the system that might be avoided or remedied. Young foals are liable to suffer from constipation, especially if they have not been able to suck the *first milk* from the mare's udder. The first milk is by nature provided with a moderate purgative tendency which is very beneficial to the offspring; but if the slightest symptoms of constipation appear in the foal it should at once receive a light dose of castor-oil.

Diarrhœa must also be carefully guarded against. Fresh air, exercise, protection from inclement weather, and good sound food to the mare, are the surest preventives.

Housing Mares and Foals.—The best treatment is to remove the cause, and if that cannot be done, call in a veterinary surgeon. Unless the weather is dry and genial, it will be prudent to keep the mare and foal under cover for a week or more. At the end of that time they will both be able to go out to the field for a short time. Every change should be gradual, whether it be a change from one kind of food to another, from a cosy box to an open field, or from idleness to work.

Working Nurse-mares.—Draught mares are often returned to work in a week or ten days after foaling. If circumstances permit, it would be better to delay till the beginning of the third week—better for the mare and the foal too. In any case, the work for a time should be light, and for several weeks the mare should not be kept longer from the foal than two or three hours on end. With good feeding the mare will be able for two yokings, of three hours' duration each, at light work, in six or eight weeks after foaling. As long as the foal is depending mainly upon the mare for its sustenance, it will be better, in a pressure of work, to take three yokings of three hours each daily from the mare, with intervals of not less than an hour, than to keep her longer in work at one time. Two short yokings daily, however, are as much as any nursing-mare should have to accomplish.

Some recommend that the foal should accompany the mare to the work, and be allowed to suck her at frequent intervals. It is safer to keep the foal in more comfortable quarters, and bring the mare to it at intervals of from two and a half to three hours, according to the stage in the nursing period.

It is injudicious, dangerous indeed to both mare and foal, to keep the mare away from the foal until her udder is much engorged and distended. Inflammation may arise in the udder, and unless it be at once checked, the life of the mare may be endangered. Then it will be risky for the foal to allow it to suck the milk from the inflamed udder. If there is any reason to suspect that inflammation has begun, a portion of the milk should be drawn away by the hand and the udder bathed with cold water before the foal is admitted.

Nursing-mares should never on any account be overheated at work.

Brood-mares while nursing their young should be liberally fed, especially so when working hard at the same time.

Weaning Foals.

The weaning-time is a critical period in the existence of a young horse. It is usually the first great trial of its life, and if the animal is not properly cared for at the time, its progress may be seriously impaired. In ordinary circumstances foals are weaned when they are from four to six months old.

As already indicated, the foal should be trained to eat other food some time before weaning. As the time for weaning approaches, the intervals during which the foal is withdrawn from the mare will be lengthened, and the extra food increased. And if the foal takes kindly to its other food, this process need not be long continued.

Whether the weaning process is to be short or protracted will depend mainly upon (1) the manner in which the foal takes to and thrives upon the other food; (2) the condition of the mare's udder; and (3) the necessities of the time as to the working of the mare. If the foal is weakly, and does not seem to thrive satisfactorily upon the other food, it may be well to continue a little of its mother's milk for some time : better submit to some inconvenience in this way than spoil a good foal. Then the mare may have such an abundant flow of milk that the sudden withdrawal of the food would be undesirable for her sake. On the other hand, the pressure of work may require that the weaning shall be completed as quickly as possible. Thus, in weaning, there is need for experience and careful consideration.

Feeding Foals at Weaning-time.— As to feeding, the foal should be well attended to at weaning-time. Feed it liberally but not to excess, taking care to keep its bowels and general health in as good order as possible. Bruised oats, bran, and beans make a capital mash for foals ; and some add boiled linseed.

Attention to the Mare at Weaning-time.—At weaning-time the feeding of the mare also needs careful attention, so that the flow of milk may be stopped. Hard work and spare feeding will diminish the secretion of milk. Let the food be dry and lessened somewhat in quantity. Even the allowance of water may be slightly restricted. Draw some milk from the udder once or twice a-day, or oftener if it becomes very full, but do not empty it at any time. If the secretion of milk is not diminishing satisfactorily, it may be well to give a light dose of phy-

sic. This is sometimes necessary with mares maintained solely for breeding, but rarely with mares kept hard at work.

In the event of a mare having to be dried soon after foaling, by the death of the foal or other cause, the flow of milk will usually be stopped by drawing away a little milk by the hand once or twice daily for a few days, and by giving the mare some purgative medicine, a short allowance of dry food and little water, and plenty of work or other exercise.

The Mating Season.

In regard to the mating of horses little need be said here. The information already given in this volume on this subject in reference to pure-bred stock is equally applicable to the breeding of ordinary farm-horses.

The latter end of spring and early summer is the *mating season* for horses. Both mares and stallions are in the best form for breeding when in robust health, in good natural condition—just such condition as should be shown by hard-worked well-cared-for horses. Overfeeding should be avoided; it is as injurious as insufficient feeding.

A mare will usually come into *season* about nine or ten days after foaling, but occasionally not in less than twice that period. It is generally quite apparent when a mare desires to receive a stallion ; but if there is any doubt, the point may easily be settled by *trying* her with the stallion.

It is advisable to serve the mare in the first heat of the season. As a rule, with healthy animals one service will be sufficient. About twenty days after the first service the mare should again be shown to the stallion, and if the usual symptoms of desire are not then exhibited by her, it may be assumed that she is pregnant. Still she may "come round" again in about three weeks, and the attendant should watch carefully for the symptoms. Some breeders think it desirable to have the mare served twice at one time, with an interval of ten to twenty-four hours ; but this is not the rule.

Number of Mares to one Stallion. —The number of mares allotted to one stallion in a season varies considerably with circumstances, such as the age, condition, and value of the horse. An adult horse in robust active condition may have from 60 to 70 mares. The number often exceeds 80, but it is highly imprudent to overdo a stallion, and it may incur the risk of many *blanks* amongst his mares.

Nomenclature of Horses.

The names given to the horse are these: the new-born is called a *foal*; the male being a *colt foal*, the female a *filly foal*. After being weaned, foals are called simply *colt* or *filly*, according to the sex. The colt when broken into work becomes a *horse*, and remains so all his life; and the filly is changed into *mare*. When the colt is not castrated he is an *entire colt*, which he retains until he is fully grown or serves mares, when he is a *stallion* or *entire horse*; when castrated he is a *gelding*, and it is in this state that he is a draught-horse. A mare, when served, is said to be *covered by* or *stinted to* a particular stallion; and after she has borne a foal she is a *brood mare* until she ceases to bear, when she is a *barren mare* or *eill mare*; and when dry of milk she is *yeld*. A mare, while with young, is *in foal*.

Names Suitable for Farm-Horses.

Names for horses should be *short and emphatic*, not exceeding two syllables, for long words are difficult to pronounce when quick action is required. For geldings, Tom, Brisk, Jolly, Tinker, Dragon, Dobbin, Mason, Farmer, Captain; for mares, Peg, Rose, Jess, Molly, Beauty, Mettle, Lily, seem good names. For stallions, they should be important, as Lofty, Matchem, Diamond, Blaze, Samson, Champion, Bold Briton, &c.

The language spoken to horses by their drivers is referred to in vol. i. p. 381.

MANAGEMENT OF COWS AND CALVES.

In the notes on the breeds of pure-bred cattle in this volume a good deal of information is given regarding the feeding and general management of cows and the rearing of calves. What little need be added here will apply mainly to ordinary commercial cattle, though, as a rule, it is only in small details that the treatment of cows and calves in mixed-bred stocks differs from that in pure-bred herds.

CALVING SEASON.

In exceptional cases, mostly in milk-selling herds, calving takes place all the year round. In the vast majority of stocks, however, the great bulk of the calving occurs in the months of January, February, March, and April, the spring months being most in favour in all excepting pedigree herds.

The risks of the calving season are considerable, and at this time breeding stocks require the most careful daily attention from their owners and attendants.

Symptoms of Pregnancy. — Cows may be ascertained to be in calf between the fifth and sixth months of their gestation. The calf quickens at between four and five months, and it may be felt by thrusting the half-closed hand, in which the point of the thumb projects over the curved index finger, against the right flank of the cow, when the calf should be felt as a distinct hard lump. All the flank should be explored, and strong, deep, but not violent, punches given before failure to detect it is acknowledged. Or when a pailful of *cold* water is drunk by the cow, the calf moves, when a convulsive sort of motion may be observed in the flank, by looking at it from behind, and if the open hand is then laid upon the space between the flank and udder, this motion may be distinctly felt. It is not in every case that the calf can be felt at so early a period of its existence, for lying then in its natural position in the interior of the womb, it may not be felt at all; and when it lies near the left side of the cow, it is not so easily felt as on the opposite one. Therefore, although the calf may not be *felt* at that early stage, it is no proof that the cow is not in calf.

When a resinous-looking substance can be drawn from the teats by stripping them firmly, many consider it a sure sign of pregnancy. After five or six months, the flank in the right side fills up, and the general enlargement of the under part of the abdomen affords considerable evidence of pregnancy.

But there is seldom any necessity for thus trying whether a cow is in calf, for if she has not sought the bull for some months, it is almost certain to be because she is pregnant.

Reckoning Time of Calving.—The exact time of a cow's calving should be known by the cattle-man as well as by the farmer himself, for the time when she was served by the bull should be registered.

Gestation.—A cow is reckoned to go just over 9 months with calf, although the calving is not certain to a day. The late Earl Spencer found from records of the calving of 764 cows that 314 cows calved before the 284th day, and 310 calved after the 285th; so he considered that the probable period of gestation ought to be regarded as 284 or 285 days, and not 270, as generally believed. In those observations the shortest period of gestation when a live calf was produced was 220 days, and the longest 313.

Prolapse of the Vagina.—Cows are most liable to this complaint when near the period of calving, about the eighth and ninth months, and, from whatever cause it may originate, the position of the cow, as she lies in her stall, should be amended by raising her hind quarters as high as the fore by means of the litter. No great danger need be apprehended from the prolapse, but it is better to use means to prevent its recurrence than to incur bad consequences by indifference or neglect.

Feeding In-calf Cows.—Much more care should be bestowed in administering food to cows near the time of their reckoning than is generally done. The care should be proportioned to the state of the animal's condition. When in high condition, there is risk of inflammatory action at the time of parturition. It is therefore the farmer's interest to check every tendency to obesity in time. Moder-

ate quantities of turnips suit well, so also do barley mashes and small quantities of oil-cake, the laxative tendency of the oil-cake being a special advantage for in-calf cows.

Critical Period in Pregnancy.—The eighth and ninth months constitute the most critical period of a cow in calf. The bulk and weight of the foetus cause disagreeable sensations to the cow, and frequently produce feverish symptoms, the consequence of which is costiveness. The treatment is laxative medicine and emollient drinks, such as a dose of 1 lb. of Epsom salts with some cordial admixture of ginger and caraway-seed and treacle, in a quart each of warm gruel and sound ale.

Calving.

Symptoms of Calving.—Symptoms of calving indicate themselves in the cow about fourteen days before the time of reckoning. The loose skinny space between the vagina and udder becomes florid; the vulva becomes loose and flabby; the udder becomes larger, firmer, hotter to the feel, and more tender-looking; the milk-veins along the lower part of the abdomen become larger, and the coupling on each side of the rump-bones looser; and when the couplings feel as if a separation had taken place of the parts there, the cow should be watched day and night, for at any hour afterwards the pains of calving may come upon her. In some cases these premonitory symptoms succeed each other rapidly, in others they follow slowly. With heifers in first calf these symptoms are often slow.

Attendance in Calving.—Different practices exist in attending on cows at calving. In most cases the cattle-man attends on the occasion, assisted sometimes by the shepherd, and other men if required, but in some districts in Scotland the calving is left to women to manage. The large and valuable breeds of cows almost always receive assistance in calving. The cows of the smaller varieties frequently calve without assistance.

In cases of difficult calving a veterinary surgeon should be summoned.

Preparation for Calving.—A few preparatory requisites should be at hand when a cow is about to calve. Flat *soft* ropes should be provided for the purpose of attaching to the calf. The cattle-man should have the calf's crib well littered, and pare the nails of his hands close, in case he should have occasion to introduce his arm into the cow to adjust the calf; and he should have some anti-septic oil or ointment or antiseptic soap and soft warm water with which to lubricate his hands and arms, although the glairy discharge from the vagina will usually be sufficient for this purpose. It may be necessary to have bundles of straw to put under the cow to elevate her hind-quarters, and even to have block and tackle to hoist her up in order to adjust the calf in the womb. These last articles should be ready at hand if wanted. Straw should be spread thickly on the floor of the byre, to place the new-dropped calf upon. All being pre-pared, and the byre-door closed for quietness, the cow should be carefully watched.

The Calf.—On the extrusion of the calf, it should be laid on its side upon the clean straw on the floor. The calf should never be allowed to fall with its full weight on the floor. The breathing is assisted if the viscid fluid is removed by the hand from the mouth and nostrils. The calf is then carried by two men, suspended by the legs, with the back downwards, and the head held up be-tween the fore-legs, to its comfortably littered crib.

Navel-string.—The state of the navel-string is the first thing that should be examined in a new-dropped calf, that no blood be dropping from it, and that it is not in too raw a state. The bleed-ing can be stayed by a ligature on the string, but not close to the belly. In-attention to the navel-string may over-look the cause of the navel-ill; and, in-significant as this complaint is usually regarded, it carries off more calves than most breeders are aware of. The navel-string should be dressed two or three times daily until dried up with a dress-ing consisting of one part of pure car-bolic-oil to twenty parts of olive-oil.

Inflammation of the navel is often caused by one calf sucking another.

Reviving Calves.—Some calves, though extracted with apparent ease, appear as if dead when laid upon the straw, but they may only be in a condi-tion of suspended animation. A power-ful attendant should seize the calf by the hind-legs above the hock, swing it verti-cally clear of the ground, while another strips all viscid material from its mouth and nostrils. It should then be laid flat on its side at all its length, with head, neck, and legs extended. An intelligent operator should then use artificial respira-tion in the same way it is used in the apparently drowned, by elevating and depressing a fore-leg. The leg should be pulled upwards and forwards until it is evidently pulling at the chest-wall by its attachments, then pressed gently down-wards and backwards over the lower part of the chest, the weight of the operator's arm going with it. This should be repeated from six to ten times for less than half a minute. It should then be lifted vertically by the hind-legs again, its mouth and nose stripped as rapidly as possible, and laid on its other side a little roughly, and the process repeated with the other fore-leg. It should be turned this way after every eight or ten movements of the leg, verti-cally over its long axis, not horizontally over its back, its mouth and nostrils kept clear, and assistants applying friction to the skin and drying it at the same time with handfuls of hay or straw. The first sign of life may be a slight cough, after which care should be taken that the movements of the operator should har-monise with the efforts of the animal at natural respiration. Success has resulted by the use of this method after fifteen and even twenty minutes' steady persist-ent work.

Extracting a Dead Calf.—When the symptoms of calving have continued for a time, and there is no appearance of a presentation, the operator should in-troduce his arm to ascertain the cause, and the probability will be that the calf has been dead in the womb some time. A dead calf is easily recognised by the hand of an experienced cowman. It should be extracted in the easiest man-ner; but should the body be in a state of decay, it may not bear being pulled out whole, but may require to be taken away piecemeal.

Mistaken Idea.—A notion exists in some parts that a cow, when seized with the pains of labour, should be made to move about, and not allowed to lie still, although inclined to be quiet. As a rule, she should not be interfered with.

Refreshing the Cow.—When a cow seems exhausted in a protracted case of calving, she should be supported with a warm drink of gruel, containing a bottle of sound ale. Should she be too sick to drink, it should be given her with the drinking-horn.

After the byre has been cleansed of the impurities of calving, and fresh litter strewed, the cow naturally feels thirsty after the exertion, and should receive a warm drink. There is nothing better than warm water, with a few handfuls of oatmeal stirred in it for a time, and seasoned with a small handful of salt. This she will drink up greedily. A pailful is enough at a time, and it may be renewed when she indicates a desire for more. This drink should be given to her for two or three days after calving in lieu of cold water, and mashes of boiled barley and gruel in lieu of cold turnips. At this critical period oil-cake is specially suitable, as it acts as an excellent laxative and febrifuge. Nothing should be given at this time of an astringent nature. The food should rather have a laxative tendency.

Immediate Milking.—It used to be considered desirable to milk the newly calved cow dry as soon as possible after calving. This is most unnatural. Her own calf would not take all her milk for days, and it is now recognised that milking dry soon after calving tends to induce milk fever. A little milk should be drawn from each quarter, but only sufficient to relieve the tension, and although this is done several times a-day the udder of a good milker should not be emptied for some days after calving.

Licking and Rubbing Calves beneficial.—Many skilled breeders systematically let the newly dropped calf be licked by the cow. There is more in this apparently small matter than is generally supposed. "The bloomy appearance of suckled calves is partly due to this motherly attention; and the licking along the calf's spine, which the cow, with her rasp of a tongue, gives her calf immediately after birth, has evidently an important meaning. All careful

managers, when the calves are not reared by the cow, take care to imitate this process, rubbing well over the spine with a wisp of straw. This not only dries the calf and prevents its taking cold, but evidently strengthens it; and the calf, if a healthy one, responds to the rubbing by vigorous efforts, soon successful, to gain its feet."[1] It is, moreover, held by experienced breeders that the licking of the calf has a beneficial effect on the cow, and in the case of breeds liable to milk-fever this is especially so. It is good practice in such cases to leave the calf beside its mother for at least two days.

Bulling.

Coming in "Season."—A cow will desire the bull in four or five weeks after calving. The symptoms of a cow being in season need not be described.

Too Early Bulling Unwise.—There is good reason to believe that many cases of cows not holding in calf with the first serving after calving arises from the want of consideration on the part of breeders as to whether the cow is in that recovered state from the effects of calving which may be expected to afford a reasonable hope that she will conceive. The state of the body, as well as the length of time, should be taken into consideration in determining whether or not the cow should receive the bull when she first comes into "season."

Leading Cows. — A cow is generally easily led to the bull by a halter round the head. If she is known to have a fractious temper, it is better to put a holder in her nose than to allow her to run on the road and have to stop or turn her every short distance. A simple form of holder is shown in fig. 742.

Fig. 742.
Bullock-holder.

a Joint.
b Knobbed points, meeting.
c Screw-nut.
e Ring for rein-rope.

[1] *Jour. Royal Agric. Soc. Eng.*, sec. ser., xvi. 428.

ABORTION.

It is now recognised that there are at least two forms of abortion. ⸱ The one, known as *Sporadic Abortion*, arises from many different causes, including accidents; the other, known as *Contagious* or *Epizootic Abortion*, is caused by a specific living organism. Heavy losses are often sustained by stock-owners from abortion, especially in herds of pure-bred cattle. Ewes abort frequently and mares occasionally.

SPORADIC ABORTION.

Causes. — Most frequently the direct causes of sporadic abortion are violent exercise, frights, bruises, careless attendance, diseased bulls, unwholesome food, impure water, and hay affected with ergot.

Ergot causing Abortion. — As to the part which ergot has played in causing abortion there is difference of opinion. Ergot is a fungus which attacks the ear or panicle of grasses and cereals, rye particularly, and is recognised as a black spur, seen in fig. 743. Farmers should certainly regard ergot as a dangerous enemy, and should burn any portions of hay in which it is seen to exist to any considerable extent.

Prevention. — Immediately a cow shows symptoms of aborting, she should be separated from her companions and watched carefully. She should be kept perfectly quiet, and should get laxative food such as

Fig. 743.—Head of timothy with numerous ergots.

oil-cake and mashes, and if there is straining, frequent doses of opium, belladonna, or anti-spasmodics.

After abortion cows must be carefully attended, in order to get them back into a healthy natural condition, and to pre-vent abortion spreading. In cases of slow cleansing it may be well to give a dose of laxative medicine, such as 1 lb. of Epsom salts, 1 oz. powdered ginger, and 1 oz. caraway seeds.

Preventing Recurrence of Abortion. — There is great risk of recurrence of abortion amongst cows that have once aborted, and, as a rule, the wisest course is to fatten off aborted cows. When abortion occurs the byre should be thoroughly cleaned and disinfected, and every possible precaution taken to get the animals and premises into a clean healthy condition.

EPIZOOTIC ABORTION.

For the following notes we are indebted to Principal Dewar, Edinburgh: Early in the closing quarter of the twentieth century acute observers began to think that a form of abortion was not uncommon which manifested contagious properties. It was not till 1896, however, that Professor Bang of Copenhagen published his article on "Infectious Abortion," showing that abortion in cows was caused by a micro-organism, and that he could communicate the disease to pregnant animals by cultures of that organism. For most of our subsequent knowledge of the subject obtained up till about 1908 we are indebted to Professor Bang.

Causes of Abortion.

The causal organism, isolated by Professor Bang, is a fine short bacillus, and is found in an almost pure state in a slimy poultaceous exudate met with in animals that have just aborted, as well as in pregnant animals affected with the disease, between the uterus and the placental membranes.

For years after the contagious nature of the disease was accepted it was generally believed that it was not communicated like other contagious diseases, but only through the vulva and genital canal,—that it was mostly communicated in the byre owing to the discharges from affected animals passing into the gutter, and that each cow infected herself, and it might be her neighbour, by means of her tail, which became soiled with the contaminated fluids in the gutter. It has now been proved, however, that animals may contract the disease by the injestion of food or water soiled with these virulent discharges, and it is very probable that the virus frequently gains access to the system in this way.

It has not yet been proved that the disease can be communicated by inhalation, by means of the respiratory organs, although the possibility of infection by this portal should not be lost sight of in dealing with the disease.

It is generally believed that a common method of infection is by means of the bull. Should a bull serve a cow that has aborted and that has not been properly treated for it, or any cow the genital passage of which contains abortion bacilli, unless he is carefully and thoroughly disinfected after service, there is a risk of him communicating the disease to every cow he serves for some time. That this means of infection has not been sooner and more generally recognised is due to the fact that the disease is of a very insidious nature, that the incubative period is very irregular and often very prolonged. A cow may contract the disease at the time of service from an infected bull, and may not abort until the seventh or eighth month of pregnancy. In such a case the real cause is likely to be overlooked and a less remote cause suspected.

Treatment.

It is seldom that contagious abortion is suspected in a herd until one or more cases occur, and by that time it is probable that a large number, in fact the great majority, of the cows and heifers may be affected, the bacillus proliferating in the uterus and setting up those chronic inflammatory changes which ultimately lead to abortion. Although we can hardly hope to prevent the accident in cases in which the disease is far advanced, still no one can be sure of the stage the disease has reached, and it is well to treat as affected all the pregnant animals that may have been exposed to the contagion.

For this purpose it has been recommended to wash or sponge the tail, anus, vulva, and perinæal region of each cow every morning with a reliable antiseptic wash. In addition to this a large syringe-ful of antiseptic wash, which should not be quite so strong as that used externally, should be injected into the vagina of each cow once a-week. The flooring of the byre should also be thoroughly scraped, cleaned, and disinfected every week.

Should some of the cows be giving milk, care should be taken not to use as antiseptic and disinfecting agents medicines which have strong penetrating odours, as the odour is apt to be communicated to the milk, and render it useless for human consumption.

Bräuer — on the Continent — recommended medicinal treatment with the view of getting at the organisms through the blood stream. He started with a subcutaneous injection of from half an ounce to an ounce of a 2 per cent solution of carbolic acid once a-fortnight, in addition to the external cleansing and washing out of the genital passage already mentioned, but he subsequently used double the quantity.

For a number of years the administration of carbolic acid by the mouth has been strongly recommended in this country for the purpose of destroying or hindering the proliferation of the organisms in the uterus. It is found that considerable quantities of the acid can be given in this way without causing any untoward symptoms. One well-known authority, writing in an agricultural paper, recommends half-ounce doses of a somewhat crude carbolic acid to be given to each cow three times a-week in bran-mashes. Some animals, he says, may be unwilling to eat mashes containing the acid.

As the purpose is to get at the organisms through the blood, it is necessary that the acid should be absorbed: it is therefore better to give it in smaller doses—quarter-ounce doses—and repeat oftener if necessary; and there is no doubt that a purer acid is less pungent, less irritating, much more readily taken by the cows, and quite as useful.

Preventive Treatment.

As the discharge from the uterus of aborting cows is the chief source of the contagion, it is necessary to use means to prevent its being spread in the byre or scattered in the field. Any cow, therefore, that shows the slightest symptom of abortion should at once be removed from the others and put in a byre or building by herself. And as the byre in which the abortion takes place requires to be properly cleaned and disinfected afterwards, one with a cemented smooth floor and no underground drains is to be preferred.

It should be remembered that the aborted calf and placenta are both fertile sources of infection, and these, as well as all discharges, should be buried, burned, or otherwise destroyed. Although the calf may be born alive, it is not a very desirable addition to the stock. Aborted calves often die within a few days of birth. In any case, it is necessary to remember that it is as fertile a source of contagion as if it had been dead, and even more so, as it is capable of moving about, and should be effectually removed from contact with other animals until old enough to be more than once disinfected. It is highly probable that for a short time the fæces of the calf may be contagious.

After abortion the uterus of the cow should be thoroughly cleansed and disinfected. If the placenta does not come away — which is quite common after abortion — it should be removed by the veterinary surgeon before twenty-four hours have elapsed. The uterus should then be flushed out with some reliable antiseptic wash several times daily for a few days, or as long as easy access to it can be obtained. The antiseptic should not be used in a strong condition, but a large quantity of fluid should be run through the uterus. The thorough cleansing and disinfection of the womb not only destroys the contagium and serves to prevent the spread of the disease, but tends to prevent barrenness in the cow and the recurrence of abortion during the next pregnancy.

It should also be remembered that the cattle-man may easily convey the disease on his boots or clothes, and by contaminating fodder or food of any kind quite unsuspectingly spread the disease.

Should any suspicion attach to the bull, or should he have been serving suspected cows, his prepuce should be carefully and gently but thoroughly syringed out with an antiseptic twice daily for several days. Some breeders who have had contagious abortion in their herds have regularly had the bull syringed out—disinfected—each time before and after service, and with the best results. Should there be a lot of strong hair about the orifice of the prepuce, likely to harbour dirt and germs, it should be clipped off and the skin around the opening disinfected.

There is no reason to believe that the organism of contagious abortion multiplies outside the animal body, but as it is possessed of a great amount of vitality (Bang found living bacilli in uterine exudate that had been kept seven months), the greatest care should be taken in disinfecting buildings and everything that could by any possibility have come in contact with the discharges.

Breeders, as a rule, have hitherto been inclined to dispose of their cows after abortion. There cannot be any harm in this if they are fattened and sent to the butcher, but to dispose of a cow that may be carrying the organisms of abortion in her system, as a breeding animal, to whomsoever cares to buy her, is, if not legally a criminal act, undoubtedly one morally, and should never be done. It is now considered a better policy to keep the cows, if they are good ones, disinfect them thoroughly as well as everything they could have been in contact with, and try to get rid of the disease. This has been done successfully, in some cases even during the first season. There is no doubt that replacing the cows which have aborted by purchasing fresh ones is a risky business, as the imported cows frequently abort, and thus serve to maintain the contagion.

It is well to bear in mind the possibility of the disease being conveyed between neighbouring farms by individuals, or the interchange of any commodity that has been in contact with the virus. We have known of cases where this seemed the only possible way by which the disease could have been communicated.

In this country contagious abortion has not yet been scheduled as a contagious disease, but in Norway this has been done since 1894, owners there being required to notify the existence of the disease in their herds. The restrictions in Norway are not severe, but the notification serves to warn probable buyers against the risk incurred by purchase.

. Professor Bang has been experimenting with a view to finding a means of conferring immunity on animals by inoculation, and not without obtaining considerable encouragement, but up to 1908 had not found any practicable method which could be of general application.

BOARD OF AGRICULTURE INQUIRY INTO EPIZOOTIC ABORTION.

In 1905 the President of the Board of Agriculture and Fisheries appointed a Departmental Committee to "inquire, by means of experimental investigation and otherwise, into the pathology and etiology of Epizootic Abortion, and to consider whether any, and if so, what, preventive and remedial measures may with advantage be adopted with respect to that disease." The investigations were begun on temporary premises, and were continued later on a small experimental farm which has been leased by the Board and equipped as a veterinary laboratory. A very considerable amount of experimental and bacteriological work was performed for the Committee by Sir John M'Fadyean of the Royal Veterinary College, London, and Mr Stockman, chief veterinary officer of the Board of Agriculture and Fisheries, and the results are embodied in the Committee's Report. The first part dealing with the disease in bovine animals was issued along with an appendix giving the work in detail in June 1909.

Microbe of Cattle Abortion.

The most definite and important announcement in the first Report is that of the discovery of the microbe of abortion in cattle in Great Britain. Abortion was experimentally induced in cows, ewes, goats, bitches, and guinea-pigs, by introducing into their bodies the microbe found in the uterine exudate of cows that had aborted; but although other species may be experimentally infected in the laboratory, the Committee do not think that they are likely to become infected

with bovine abortion in practice except as the result of gross carelessness in the disposal of infected material, and so the conclusion is that "bovine abortion is primarily a disease of cattle."

The bacilli of cattle abortion which proved to be the cause of abortion in cows in these researches are small oval rods, differing in several respects from the abortion bacillus found by Bang in his Danish experiments. But, while the bacillus isolated in England differs in so many material ways from that described in Denmark, the investigations established by exhaustive and laborious studies that the apparent differences arose from a faulty and incomplete study of the biological characters of the bacillus in Denmark, and instead of coining a new name for the bacillus isolated in England the Committee magnanimously suggest that it should be known as "Bang's Bacillus of Cattle Abortion." The English bacillus is non-motile, and is an aerobe —that is, it requires oxygen for its development. It can be cultivated on various substances, such as agar-gelatine-broth-serum, agar, potato, milk, &c. It grows best at temperatures between 30° and 37° C.

Tests made as to the temperature necessary to destroy the vitality of the bacillus of cattle abortion showed that it was not destroyed at a temperature of 55° C. maintained for an hour in the stove, but that after two hours at the same temperature its vitality was destroyed. When it was kept 10 minutes in water at a temperature of 55° C. it retained its vitality, but when kept in water for 10 minutes at a temperature between 59° and 61° C. its vitality was destroyed. The comparatively low temperature at which its vitality is destroyed promises well for the disinfecting effects of the homely bucket of boiling-water.

Microbe of Sheep Abortion. — In regard to abortion in sheep, the Report states that while the bacillus of cattle abortion can experimentally cause abortion in ewes, it was never found in the membranes of ewes aborting in the field: "a totally different microbe—a vibrio—has repeatedly been isolated from outbreaks of abortion in ewes, and has been successfully employed at the laboratory to experimentally infect other ewes pregnant for the first time. Pregnant cows, however, cannot be infected with this vibrionic abortion of ewes."

Methods of Infection.

The Report states that for experimental purposes the most certain method of infecting an animal with abortion is to inoculate natural virulent material or active cultures into the blood stream. As to natural methods of infection, the virulent material may gain access to the pregnant uterus by the vagina and by the mouth. The Committee do not regard infection by the vagina as likely to very often happen, but are inclined to believe that the disease is more frequently contracted by the mouth than in any other way.

The risks of infection being carried by the bull from one cow to another are regarded as comparatively slight, the Committee stating that "without denying that the disease may be spread by coition, we think that nothing more than a quite subsidiary *rôle* in the spread of epizootic abortion can now be assigned to the bull."

One of the most insidious ways of spreading abortion is the introduction into clean herds of in-calf cows affected with the disease, and it is difficult to guard against this risk, for it is impossible for the ordinary individual to say whether a pregnant animal is affected or not, but in the section dealing with diagnosis the Committee hold out hope that one or other of the new methods they have elaborated for diagnosing the disease in cows before abortion occurs may eventually solve this difficulty.

Cows which have aborted are, of course, a dangerous source of infection. The materials expelled from the uterus of an infected cow in the act of abortion are all virulent, for they contain the microbe, and so also will the discharge from the genital organs for a varying time after abortion. It is still uncertain how long virulent material may remain infective after leaving the animal, but if kept fluid and free from putrefaction it may remain virulent for seven months or even more. This significant consideration increases the importance of careful attention being given to the thorough disinfection or destruction of all materials in connection with aborting cows.

The Committee are not of opinion that many cases of abortion arise from any other cause than infection. They add: "We do not deny that odd cases of abortion may arise from accident or poisoning by such substances as lead, but we have no hesitation in stating that we believe 99 per cent at least of the outbreaks of cattle abortion which assume epizootic characters are due to infection by the bacillus of cattle abortion, and that the fact of a cow having aborted on premises formerly believed to be clean is a sufficient reason for suspecting that the disease has been introduced." They add to this the important statement that there is no difficulty in diagnosing the bacterial disease once an animal has aborted, if an examination of the fœtal membranes be made at an early date after abortion.

Immunisation of Animals.

The investigations have not brought out any evidence that could be regarded as showing that natural immunity from the abortion bacillus is possessed by any individuals of the bovine species. On the other hand, it has been found that there are serious obstacles in the way to a practical success by the use of a protective serum. The protection derivable even from potent serum cannot be depended upon to last more than 2 or 3 weeks, and as the period of the risk of infection extends over at least $7\frac{1}{2}$ months of pregnancy in cows, it is at once seen that it would neither be practicable nor economically possible to give the ordinary cow a sufficient number of doses of a rather expensive serum to protect her from infection during that long period. The idea of hyper-immunising animals for the production of serum was therefore abandoned.

Inoculation with Pure Cultures for the Production of Immunity.— The Report proceeds: "The most hopeful line of inquiry seemed to be the production of immunity by inoculation of large doses of pure culture. One of the great objections to the protective inoculation methods in practice is the number of operations necessary to ensure protection. But owing to the harmlessness of large quantities of pure cultures of the abortion bacillus when injected into non-pregnant animals, it seemed possible that whatever degree of immunity could be established by a practicable number of small doses might be conveyed by inoculating one large dose." Trials with pure cultures of the bovine abortion bacillus were therefore made with ewes and heifers, the animals being inoculated about 60 to 148 days before becoming pregnant. The results with sheep were so irregular as to be regarded as of little or no practical value, and these trials were discontinued. With heifers the results were more encouraging. Two heifers were inoculated with a rich liquid culture of the bacillus, the one 148 days and the other 106 days before becoming pregnant. The former heifer, 40 days after becoming pregnant, was inoculated intravenously with 10 c.c. of a dense emulsion of virulent uterine exudate, yet when killed 112 days thereafter she was found free from infection. The immunity of the other heifer "was tested by giving her enormous doses of virulent exudate both by the mouth and the vagina 36 days after becoming pregnant and 142 days after immunisation, and 16 days later she received 10 c.c. of a dense emulsion of a virulent exudate into the jugular vein. She was killed and found free from infection 122 days after receiving the first infecting dose."

These results with heifers, says the Report, "are all the more encouraging when one remembers that not a single negative result followed the intravenous inoculation of unprotected heifers with uterine exudate, and it should be noted also that the tests applied were in point of severity far beyond anything likely to be met with in practice." In addition to the above experiments, they show by infecting experiments on animals which have aborted that these may be absolutely immune to the disease at their next pregnancy.

Curative Measures.

The Report discusses the various methods which have hitherto been most largely used for the prevention and eradication of abortion, and which have already been described, the comment

being that it cannot be said "that either singly or collectively they have brought about any material improvement in the general condition of our herds in relation to abortion."

The spraying of the external genital organs and hind quarters of cows with disinfectant solutions is regarded as useless "so long as the animals remain in an infected byre."

As to the isolation of animals as soon as they show signs of abortion, it is remarked that the necessity for this measure is obvious, and cannot be too much insisted upon. "Isolation of the affected animals, however, must be complete immediately before and after the act to be of any real value," and the Report indicates possible methods of accomplishing this.

Carbolic acid and other antiseptics are regarded as useless as curative agents, and "as a preventive agent by internal administration we believe carbolic acid to be equally useless," an opinion which is supported by direct experiment.

The irrigation of the genital passages of animals which have aborted with antiseptic solutions is recommended, "but not on the grounds that the injections will disinfect the uterus. We are of opinion that it will seldom be necessary to continue the injections for more than a month, and that after three months there should be small risk in putting the cow to the bull, provided she is afterwards protected against fresh infection."

Cows which have aborted should not be sold except for slaughter till they have ceased to discharge. The Report states that cows which have once aborted are, as a rule, less liable to infection during a subsequent pregnancy than if they had not before aborted,—are indeed often absolutely immune, though it is known that a considerable number of cows abort twice in succession. The Committee consider "that on infected premises the animals which have already aborted are to be looked upon as valuable assets for purposes of eradication, —much more valuable than new and susceptible animals brought in. We find, however, that a small proportion of those which have aborted will not hold to the bull for an indefinite period after abortion, and it may be found better to fatten off such animals, unless they are of high value."

The keeping of a goat amongst cows as a preventive against abortion is stigmatised in the Report as the product of "ignorant superstition."

Very properly the Committee speak with reserve as to the part which preventive inoculation is likely to play in the combat with abortion in the field, but the Report would seem to hold out good hope for the future in this direction.

The Board of Agriculture and Fisheries is to be congratulated upon the success of the initial stage of this important and interesting investigation.

MILKING COWS.

The milking of cows is a process that demands greater care and skill than most people realise. The peculiar variations in the milk-yield of cows is due more frequently to imperfect milking than is generally believed. Too much care cannot be given to the operation.

The Udder. — The udder should be capacious, though not too large for the size of the cow. It should be nearly spherical in form. The skin should be thin, loose, and free from lumps, filled up in the fore part of the udder, but hanging in folds in the hind part. Each quarter should contain about equal quantities of milk, though sometimes the hind ones yield the most.

The teats should be at equal distances every way, neither too long nor too short, but of moderate size, and equal in thickness from the udder to the point. When the teat is too long and inclined to taper at the point, it is invariably tough to milk. A medium-sized teat, from $2\frac{1}{4}$ to $2\frac{1}{2}$ inches long, is considered the most desirable and most easily milked. On the other hand, nothing is more objectionable than too small teats. The teats should be smooth, and feel like velvet, firm yet soft to handle, not hard or leathery. They should yield the milk freely, and not require to be forcibly pulled.

When the milk is first to be taken from the cow after calving, the points of the teats will be found plugged up with a resinous substance, which, in some instances, requires the exertion of some force before it will yield.

Milking Period.—Cows differ much in the time they continue to milk without again bearing a calf, some not continuing to yield it more than 9 months, others for years. The usual time for cows that bear calves to give milk is 10 months. Many remarkable instances of cows giving milk for a long time are on record.

Hours of Milking.—The hours of milking vary in different parts of the country. On small farms, where the milk produced is required for consumption on the holding, the cows are often milked three times daily—morning, noon, and evening. In the great majority of dairy herds the milking takes place twice daily—early in the morning, and in the evening or late in the afternoon. The precise hours vary according to local habit, which is regulated mainly by the use made of the milk. Where the milk has to be sent long distances to market the morning milking takes place from 3 A.M. onwards, and the afternoon milking from 4 P.M. onwards. More general hours are from 5 to 6 A.M. and 5 to 6 P.M.

Hours of Milking and Percentage of Butter-fat.—Careful observation has shown that the tendency of the evening's milk to be richer than the morning's milk in butter-fat is partly due to the fact that the interval between the evening and morning hours of milking is usually longer than the interval between the morning and evening hours of milking. The discovery of this has led to the intervals being more nearly equalised. Why the differences in the intervals should have this effect is a problem that still awaits solution.

Milk-pails.—The vessel used for receiving the milk from the cow was at one time mostly made of thin oak staves bound together with thin galvanised hoops, but the pail now most generally in use on all properly conducted dairy-farms is made of tinned iron or tin, and is preferable for cleanliness and lightness. This pail should be 3 to 4 inches wider at the mouth than the bottom, and when placed between the milker's knees ehould be deeper at the under side in order to prevent spilling of the milk when held in a slanting position.* The pail should be large enough to contain all the milk that a cow will give at a milking without becoming quite full. It is undesirable to annoy the cow by rising from her before the milking is finished, or by exchanging one pail for another.

The milking-stool, as in fig. 744, is made of wood, to stand 9 inches in height, or any other height to suit the convenience of the milker, with the top 9 inches in diameter, and the legs a little spread out below to give the stool stability. Some milkers do not care to have a stool, and prefer sitting on their

Fig. 744.—*Milking-stool.*

haunches; but a stool keeps the body steady, and the arms have more freedom to act, particularly to prevent accidents to the milk in case of disturbance by the cow.

Cows holding back Milk.—The holding back of milk is a curious property which cows possess. How it is effected is not very well understood, but there is no doubt of the fact occurring when a cow becomes irritated or frightened by any cause. Cows should therefore at all times be treated gently, and neither struck nor shouted at. The cow will yield more milk to the skilled milker than to an unskilled person, who may tug and pull the teats instead of gently squeezing them. Not all are affected to the same degree; but, as a proof of their extreme sensitiveness in this respect, it may be mentioned that very few can be milked so freely by a stranger as by one to whom they have been accustomed.

The Milking Side.—Usually, the near side of the cow is taken for milking, and it is called the *milking side.*

Some think cows should always be milked on the same side, but in many dairies where the cows are stalled in pairs the milker steps up between the cows and milks the one from the left side and then turns round and milks the other from the right side. This practice is to be commended. It is rare to see a cow milked in Scotland by a man, and women as rarely do the milking in England.

The Operation of Milking.—Milking is performed in two ways, stripping and nievling. *Stripping* consists of seizing the teat firmly near the root between the front of the thumb and the side of the forefinger, the length of the teat lying along the other fingers, and of pressing the finger and thumb while passing them down the entire length of the teat, and causing the milk to flow out of its point in a forcible stream. The action is renewed by again quickly elevating the hand to the root of the teat. Both hands are employed at the operation, each having hold of a different teat, and moving alternately. The two nearest teats, the fore and hind, are first milked, and then the two farthest. In the case of cows with properly sized teats stripping should be resorted to only at the finish of milking in order to draw out the last drops.

Nievling is done by grasping the teat with the whole hand, or *fist*, making the sides of the forefinger and thumb press upon the teat more strongly than the other fingers, when the milk flows by the pressure. Both hands are employed, and are made to press alternately, but so quickly in succession that the alternate streams of milk sound on the ear like one forcibly continued stream; and although stripping also causes a continued flow, the nievling, not requiring the hands to change their position, as stripping does, draws away a large quantity of milk in the same time.

Thus stripping is performed by pressing and passing certain fingers along the teat, and nievling by the doubled *fist* pressing the teat steadily at one place.

Of the two modes the *nievling* is preferable, because it is more like the sucking of a calf. When a calf takes a teat into its mouth, it seizes it with the tongue against the palate, causing them to play upon the teat by alternate pressures or pulsations, while retaining it in the same position. Nievling does this: the action of stripping is quite different.

Milking should be done *fast*, to draw away the milk as quickly as possible; and it should be continued as long as there is a drop of milk to bring away.

An Improved System of Milking.

An improved system of milking was introduced into Denmark by Mr Hagelund, an eminent veterinary surgeon. It is claimed for this system that not only is an increased flow of milk obtained but a slight increase of butter-fat as well. The process consists of manipulating or massaging the udder in a special way, first by rubbing gently with a dry cloth, this process not only cleaning the udder but tending to bring down the milk into the teats; next, by milking slowly at first the two front teats and then the two rear ones alternately until all is drawn that will come in this way. The udder is then manipulated in the following manner:—

First.—The right quarters of the udder are pressed together by placing the left hand on the hind quarter and the right hand in front of the fore quarter, the thumbs being placed on the outside of the udder and the four fingers between the two divisions of the udder. The hands are now pressed towards each other, and at the same time lifted towards the body of the cow. The pressing and lifting are repeated three times, the milk collected in the milk-ducts is then drawn out, and the manipulation repeated until no more milk is obtained, when the left quarters are treated in similar manner.

Second.— The glands are pressed together from the side. The fore quarters are milked each by itself by placing one hand with the fingers spread on the outside of the quarter, and the other hand in the division between the right and left fore quarters; the hands are pressed against each other, and the teat then milked. When no more milk is obtained by this manipulation the hind quarters are milked by placing a hand on the outside of each quarter, likewise with

fingers spread and turned upward, but with the thumb just in front of the hind quarter. The hands are lifted and pressed into the gland from behind and from the side, after which they are lowered to draw the milk. This manipulation is repeated till no more milk is obtained.

Third.—The fore teats are grasped with partly closed hands and lifted with a push towards the body of the cow, both at the same time, by which method the glands are pressed between the hands and the cow's body. This is repeated three times, and the teats are then stripped dry. When the fore teats are emptied the hind ones are treated in a similar manner.

The process thus described may seem elaborate and intricate, but in actual practice it is quite simple, and cows in full milk can be milked by an expert milker in from six to eight minutes. It had long been known by observant dairymen that the flow of milk from a cow may be increased by gentle manipulation of the udder: indeed all skilled milkers have in the past been in the habit of bringing the hand gently round the udder before commencing to milk, but no systematised method had been adopted or published till it was adopted in Denmark. Now it is being largely practised at Scandinavian dairy farms and in the United States of America.

Milking-Machines.

During the closing decade of the nineteenth and the opening decade of the twentieth century quite a number of milking-machines have been put on the market, the inventors claiming that each in turn had solved the milking problem.

In many cases where dairy farmers had fitted up expensive machines it was found that whilst they milked fairly satisfactorily when the cows were in full milk, the operation was less efficient when the cows were drying off, and in many cases hand-milking was again resorted to. In other cases, the machine was dispensed with on account of the keeping properties of the milk being impaired.

Two Scottish milking-machines—the Lawrence-Kennedy and the Wallace—have stood the test of practice better than the earlier inventions, and a good many farmers both in this country and abroad are using them with a fair measure of success.

These two machines are similar in their main features, both working on the suction principle. By means of an exhaust pump a vacuum is created in a system of piping which is attached to the cow's teats by rubber cup and which leads into sealed milk-pails: ingenious contrivances impart to the teat-cups a pulsating movement which closely resembles the sucking action of the calf, and in this way the milk is drawn from the cow. Fig. 745 represents the teat-cups and milk-pail of the machine made by J. & R. Wallace, Castle-Douglas.

A simpler appliance is the self-acting milker. In this system the milk flows by gravitation through perforated siphons inserted into the teats. This method, however, is rarely used except in the case of sore teats or udder.

Spaying Cows.

The spaying of cows has sometimes been practised to secure the permanency of milk without continued calf-bearing. The operation of spaying a cow, which is

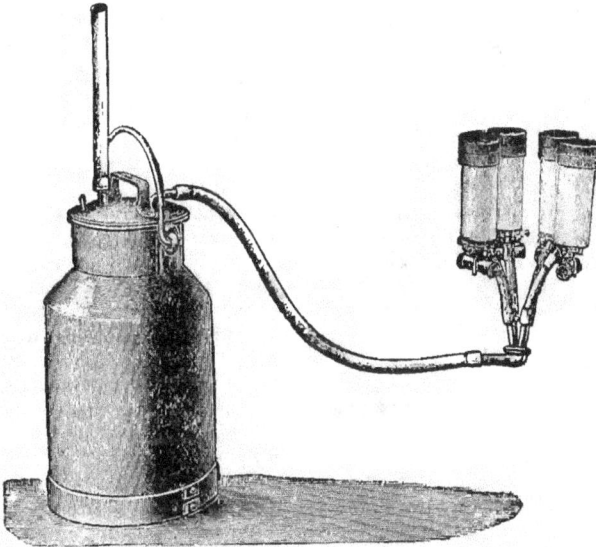

Fig. 745.—*Wallace's milking-machine—Teat-cups and milk-pail.*

performed some time after calving, consists in cutting into the flank of the cow, and, by the introduction of the hand, destroying the ovaries of the womb. The cow must have acquired her full stature, so that it may be performed at any age after 4 years. She should be at the flush of her milk, as the future quantity yielded depends on that which is afforded by her at the time of the operation. The operation may be performed in ten days after calving, but the best time appears to be 3 or 4 weeks after. The cow should be in robust health, otherwise the operation may kill her or dry up the milk. The only preparation required for safety in the operation is, that the cow should fast 12 or 14 hours, and the milk be taken away immediately before the operation.

The wound heals in a fortnight or three weeks. For two or three days after the operation the milk may diminish in quantity; but it regains its measure in about a week, and continues in full flow for the remainder of the animal's life, or as long as the age of the animal permits the secretion of the fluid, unless from some accidental circumstance — such as an attack of a severe disease — it is stopped. But even then the animal may easily be fattened.

Advantages of Spaying.—The advantages of spaying are: "1. Rendering permanent the secretion of milk, and having a much greater quantity within the given time of every year. 2. The quality of the milk being improved. 3. The uncertainty of, and the dangers incidental to, breeding, being to a great extent avoided. 4. The increased disposition to fatten, even when giving milk, or when, from excess of age, or from accidental circumstances, the secretion of milk is checked; also the very short time required for the attainment of marketable condition. 5. The meat of spayed cattle being of a quality superior to that of ordinary cattle."[1] With these advantages breeders of stock can have nothing to do; but since the operation is said to be quite safe in its results, it may attract the notice of cowfeeders in town.

FEEDING OF COWS.

In the feeding as in the general treatment of cows, practice varies greatly. The conditions which most largely regulate these variations are, the class or breed of cows, the purposes for which they are kept, the locality, and general systems of farming pursued.

Dairy Herds.

As would be expected, where *dairying* is the sole or dominant feature in the system of farming, the cows are fed and managed differently from what they are in *mixed farming*, where cows are kept chiefly to breed and rear calves, and provide milk and butter to the farmer's household. Again, even within the limits of dairying itself, there are distinctive conditions which induce different methods of feeding. Where the main object is the production of milk for disposal as milk, the feeding differs—unfortunately, sometimes differs too much for the quality of the milk—from that considered best for butter-production. Then surrounding circumstances, such as the varieties of food which may be most easily and most cheaply grown or procured, also tend to regulate and modify the systems of feeding; while it is well known that food which does well with one lot of cows is often less acceptable

and profitable as food for others. Thus it becomes manifest that there are good reasons for great variations in the systems of feeding cows.

Regarding the details of the systems of feeding cows pursued in dairy herds throughout the country little need be said here. Information on the subject is already given in the sections of this volume dealing with the management of pure breeds. The exceptionally bountiful methods of feeding pursued in the herds of Ayrshire cows are described in pages 114 and 115 of this volume. Of the detailed systems followed in herds of Jersey and Guernsey cattle, particulars will be found in pages 132 and 134. A typical system in dairy herds of Shorthorn and Shorthorn crosses is indicated at page 93, and the highly successful method of feeding pursued by Mr John Evens in his famous milking herd of Lincolnshire Red Shorthorns is described at page 96.

Regulating Food by Yield of Milk.—There are few points of greater importance in connection with the management of cows than that of maintaining the proper relation between the allowance of food and the production of milk. Fortunately a good deal of attention has been given to the investigation of this aspect of the question in recent years, and, generally speaking, the feeding of dairy cows is now carried on upon much more economic lines than till wellnigh the close of the nineteenth century.

Typical Rations.—The typical rations noted below are arranged in relation both to the weight of the cows and the quantity of milk they are yielding. They are based on experience gained in trials conducted in connection with the Durham College of Science, Newcastle-on-Tyne.[1]

No. 1.—*Ration for cows giving* 18⅓ *lb. of milk* (*roughly* 1⅘ *gallon*) *per day.*

Quantities for cows 9 cwt. live-weight and giving 16½ lb. of milk daily are given within brackets.

39 lb. swedes or 52 lb. yellow turnips (35 lb. or 47 lb.)
19 lb. oat-straw (17 lb.)
4¾ lb. decorticated cotton-cake (4¼ lb.)

Roughly speaking, 1 lb. less of decorticated cotton-cake might be given if the yield is 12½ lb. of milk daily instead of 18⅓ lb. daily.

---

[1] Ferguson's *Distem. among Cat.*, 29-36.

[1] *Jour. Board of Agric.*, March 1909.

No. 2.—Ration for cows giving 30½ lb. of milk (roughly 3 gallons) per day.

Quantities for cows 9 cwt. live-weight and giving 27½ lb. of milk daily are given within brackets.

46½ lb. swedes or 62 lb. yellow turnips (42 lb. or 56 lb.)

19 lb. oat-straw (17 lb.)

6⅔ lb. decorticated cotton-cake (6 lb.)

4½ lb. undecorticated cotton-cake (4 lb.)

No. 3.—As for No. 2, with hay instead of oat straw.

Quantities for cows 9 cwt. live-weight and giving 27½ lb. of milk daily are given within brackets.

46½ lb. swedes or 62 lb. yellow turnips (42 lb. or 56 lb.)

19 lb. meadow-hay (17 lb.)

5 lb. decorticated cotton-cake (4½ lb.)

3⅞ lb. Indian cotton-cake (3¼ lb.)

Heavy milkers, giving about 4 gallons of milk daily, should have all the foods of the best quality possible, and might be given either of the following additions to Ration No. 3 :—

2 lb. seeds hay	2 lb. linseed-cake.
2 lb. linseed-cake or	2½ lb. maize-meal.
1 lb. maize-meal	

Ration No. 1 might also have hay substituted for oat-straw, and if so the decorticated cotton-cake could be reduced by about 2 lb. daily. This, however, is not quite an exact equivalent. In any of these rations the roots can be considerably reduced if desired and a substitute used.

Rations for Dry Cows.

Cows giving reduced quantities of milk as the lactation period progresses should have the concentrated food given to them lessened ; but cows that are heavy milkers and have become lowered in condition, owing to their heavy milk yields, must not have the food restricted too greatly, but must be allowed to regain condition before coming to the next calving.

For cows that are to be fattened off at the close of their milking periods, the ration should not be reduced as indicated above, but should be gradually altered as the flow of milk decreases to that suitable for fattening animals. Cows of

10 cwt. live-weight, dried off previous to calving, would probably do well with either of the following rations. The quantities for cows of 9 cwt. live-weight are given within brackets.

No. 1.—39 lb. swedes or 52 lb. yellow turnips (35 lb. or 47 lb.)

19 lb. oat-straw (17 lb.)

2¾ lb. maize-meal (2½ lb.)

2½ lb. decorticated cotton-cake (2¼ lb.)

No. 2.—39 lb. swedes or 52 lb. yellow turnips (35 lb. or 47 lb.)

19 lb. meadow-hay (17 lb.)

2¼ lb. maize-meal (2 lb.)

2/7 lb. decorticated cotton-cake (¼ lb.)

The following are the winter rations given to cows in four dairy herds in Scotland, the cows in herd No. 4 being Shorthorns or Shorthorn crosses, and in the others mostly Ayrshires :[1]—

HERD No. 1.	Per cow per day.	
Turnips (yellow) or mangels .	43	lb.
Straw (oat) . . .	9½	,,
Hay (Italian and clover) . .	6½	,,
Meals (bean, pea, rice and dec. cotton-cake)	8	,,
Bran	1½	,,
Treacle	½	,,
Distillers' grains or draff (wet) .	10	,,

The roots are fed whole, mid-forenoon and afternoon, and the meals three times daily. A portion of the straw is cut long and mixed in the cooler with the meals and draff for each lot of animals. This is done immediately the cows have been fed. Hot water is then run on the mass in sufficient quantity to thoroughly wet it, the whole being left lying in this condition till next feeding-time, when it is fed at about blood-heat. There is always plenty of the coarsest of the straw left uneaten to sufficiently litter the animals.

HERD No. 2.	Per cow per day.	
Turnips or mangolds (raw) .	40	lb.
(Some of these are steamed.)		
Hay	11	,,
Corn, chaff, or cut straw . .	3½	,,
Bean-meal	3½	,,
Undec. cotton-cake . . .	3	,,

[1] *Trans. High. and Agric. Soc.*, 1909.

HERD No. 3.				Per cow per day.
Turnips	.	.	.	28 lb.
Straw	.	.	.	7 ,,
Hay	.	.	.	4 ,,
Bean-meal	.	.	.	4 ,,
Bibby dairy meal	.	.	.	3 ,,
Bran	.	.	.	1½ ,,
Treacle	.	.	.	1 ,,

HERD No. 4.				Per cow per day.
Turnips	.	.	.	78 lb.
Straw	.	.	.	24 ,,
Pease-meal	.	.	.	3 ,,
Compound cake	.	.	.	2½ ,,
Dried brewers' grains	.	.	2 ,,
Cummins (barley sprouts)	.	2½ ,,

Dairy Cows in Summer. — Little need be said as to the feeding of cows in summer. They are kept mainly on pasture, sometimes getting allowances of concentrated food according to the supply and quality of the pasture, the condition of the cows, and the quantity of milk they are giving. Recent trials have indicated that on reasonably good pastures cows rarely give a sufficiently increased yield of milk to pay for extra food.

Feeding Dry and Breeding Cows. — This point is also dealt with in the sections relating to breeds of pure-bred cattle. The foregoing notes relate mainly to the feeding of cows where the production of milk is the chief, or at any rate a specially important, consideration, and where, on this account, the cows are fed with such quantities and qualities of food as are calculated to stimulate and maintain a bountiful flow of milk. In herds in which the yield of milk is a secondary consideration, the systems of feeding are somewhat different, and, as a rule, the rations are arranged upon a more moderate scale.

Then, in all cases, cows are fed more sparingly when not giving milk. By far the most general practice is to feed dry cows upon oat-straw or hay and turnips or mangels. Formerly turnips were given to cows much too freely. Large meals of cold watery turnips are positively injurious to cows that are heavy in calf; and in all respects it is better practice to feed roots sparingly to cows. About 50 or 60 lb. of roots per day, given in two meals, are now very general quantities in well-managed herds, and with plenty of good sound fodder, either oat-straw or hay, or both, the cows should thrive well and sustain no harm. Many still give larger quantities of turnips, but dry cows may be kept in good condition with even less than 40 lb. of roots, as is often the case where the pulping system is pursued, or where recourse is had to warm mashes composed of cheap food, largely of chopped hay, straw, chaff, and perhaps a few roots.

In England dry cows are usually kept on hay, straw, and turnips or mangels, and in many cases they receive no roots of any kind. With plenty of good hay, a run out daily—in fine weather, of course,—free access to water, and perhaps a small allowance of bran or some other cheap food, they thrive fairly well.

It is not a good plan, however, to let cows get low in condition, and this is sometimes allowed to happen by too poor feeding when they are wholly or partially dry.

EFFECTS OF VENTILATION AND TEMPERATURE ON MILK-YIELD.

The extent to which milk-yield may be affected by variations in the temperature in which cows are kept has long been an open question. The opinion has been widely held that in the winter months in this country it was only by keeping cows moderately warm that the maximum yield of milk would be obtained. It has also been extensively believed that if the temperature of a byre were allowed to fall to say 40° F., or lower, there would in consequence be a marked decline in milk-yield from cows kept in that byre. But while these have hitherto been the prevailing views, a few dairy farmers of an inquiring turn of mind began, towards the end of the last century, to doubt whether those views were well founded.

Experiments with Cows.

In 1907 the subject was brought before the Highland and Agricultural Society of Scotland by Mr John Speir, Newton, Glasgow, and it was resolved to conduct a series of experiments in the hope of solving the problem. In the winter and spring of 1908-9 two similar lots of cows at five farms in different parts of Scotland were fed and housed alike, except that the byre containing one lot was freely ventilated in all weathers, so that its air, which was relatively pure, was kept comparatively cool, and that in the other byre the ventilation was so restricted that the temperature was maintained at about summer temperature.

It was intended to have a difference of about 10° F. between the temperatures of the two byres. This was very nearly attained, the general average for the whole period of the experiments — 18 weeks—being 49.82° in the freely-ventilated byres and 59.40° in the byres with restricted ventilation.

The results obtained, which are extremely interesting, are reported fully in the *Transactions* of the Society for 1909.[1] The following table gives a summary of the yield by the two lots of cows :—

YIELD OF MILK IN FREE *VERSUS* RESTRICTED VENTILATION.

For 18 Weeks—From 22nd November 1908 to 27th March 1909.

Farm.	No. of cows in each lot in "A" and "B."	"A"—Free Ventilation.				"B"—Restricted Ventilation.			
		Total milk in lb.	Milk per cow per day in lb.	Average per cent of fat.	Average temperature of the byre.	Total milk in lb.	Milk per cow per day in lb.	Average per cent of fat.	Average temperature of the byre.
Newton	18	60,302.5	26.6	3.65	49.35	59,453.6	26.1	3.59	60.81
Woodilee	10	29,242.7	24.6	3.38	52.24	29,011.5	24.4	3.27	60.57
Crichton	8	25,811.0	27.0	3.66	50.50	26,055.5	27.4	4.58	59.40
Hartwood	8	30,500.0	32.5	3.33	47.87	31,627.0	33.7	3.44	56.00
*Rosslynlee	6	12,466.0	30.2	3.82	48.92	11,185.5	27.1	3.43	57.53
Total	100	158,322.2	...	...	...	157,333.1	...	...	...
Average milk per cow per day for the whole period		27.5	3.55	...			27.3	3.49	...
Difference			...			989.1	.2	.06	...
Average temperature in proportion to the cows		49.82				...	...	...	59.40
Difference		9.58				...	...	...	

* For 10 weeks only.

It is thus seen that the popular belief in the advantage for milk production of a warm temperature as attained by restricted ventilation was not supported by these experiments. In the total yields of milk of the two lots of 50 each in eighteen weeks there was a difference of only about 100 gallons—less than ¼ lb. of milk per cow per day. In other words, cows kept in an average temperature of 49.82° F. gave slightly more milk per day over a period of eighteen weeks

than cows kept in an average temperature of 59.40° F.

It will be observed that in percentage of butter-fat in the milk, the advantage lies also on the side of free ventilation, the averages being 3.55 and 3.49 per cent.

Even more remarkable than the results, as seen in the general averages for the whole period, are the records of the

[1] *Trans. High. and Agric. Soc. of Scotland,* fifth ser., vol. xi., 1909.

yields of milk obtained in two periods of exceptionally cold weather that occurred in the course of the experiments. In the first cold period (four days in December) the average temperature of the cold byres was 41.2° F., and the average yield of milk per cow, 29.0 lb. per day. In the same byres the average temperature for the four days before and the four days after the cold period was 53.76° F., and the average yield of milk per cow per day precisely the same as in the four cold days, with a lower temperature of 12.56° F. Another cold period occurred in March, and the results obtained in it agree entirely with those of the first cold period.

General Conclusions.

The most important general conclusions drawn from these experiments are :—

1. That fresh air is a much more important factor in the production of milk in mid-winter than it is generally considered to be by milk-producers in this country. While most people agree to the need of fresh air in regard to the health of the animals, it seems almost as desirable in mid-winter if a full supply of healthy milk is to be produced.

2. In order that the greatest advantage may be derived from the fresh air, the animals should at no time have the ventilation restricted in autumn, but should be kept as cool as possible, so that they may not only retain all their hair, but if necessary increase it.

3. There is no difficulty, much less impossibility, in producing milk in freely ventilated byres in the coldest weather likely to be met with in this country, if the cows are kept sufficiently cool in early autumn.

4. While the present experiment shows that rather more milk has been produced under conditions of free ventilation than where ventilation was restricted, it would be injudicious, till these results have been corroborated by other trials, to consider that this will invariably happen. It is unquestionable that the general health of

the cows would be better under free than under restricted ventilation.

5. Milk produced in a building kept at a high temperature by restricted ventilation, or during a warm period, does not seem to be any richer in fat than that produced at a low temperature or during cold weather.

6. It seems hopeless to expect to be able to keep the air of any byre, no matter how constructed, at from 60° F. to 63° F. during the ordinary weather of an average winter without excessive pollution of the air.

7. Any saving in food which is effected by keeping the animals at a higher temperature seems to be equalled, if not exceeded, by improved digestion when they have plenty of fresh air but a lower temperature.

8. There is reason for believing that those great scourges of the dairyman, mammitis or weeds and tuberculosis, may be considerably reduced if cows are kept in freely ventilated byres in winter.

CALF-REARING.

It is only too true that calf-rearing, the root and the rise of the cattle-breeding industry, has not received from the general body of farmers such full and careful attention as it deserves, or as it is capable of repaying. It is undeniable that the live-stock resources of the United Kingdom might advantageously be developed to a much greater extent. The growing importance of live-stock interests in British agriculture is manifest to all. In this expansion calf-rearing must play a leading part. Breeding is of course the starting-point, and the rearing of the calf is the first great step in the progress of the industry.

Aversion of Farmers to Calf-rearing. — With many farmers calf-rearing finds little favour—often, one may venture to say, for no better reason than that it is a troublesome business, demanding constant and careful attention. With skilful and careful management, calf-rearing, where circumstances are at all favourable, is almost invariably remunerative. This much, however, it must have, and it rarely succeeds where not well conducted. The young animals must be fed with skill and regularity, and their health and comfort carefully attended to in every way. When this responsible work is left entirely to hired servants, it may be imperfectly or irregularly performed, with the result that the calves make unsatisfactory progress, or perhaps become impaired in health. The farmer thus loses faith in the benefits of calf-rearing. He has, perhaps, at last learned that the cause of the mischief is improper treatment; but personal supervision, or supervision by some member of his family or employees in whom confidence could be placed, may be found irksome or inconvenient, and thus again the industry of calf-rearing loses in favour.

Calf-rearing on Large Farms.— This demand which calf-rearing makes upon the careful personal supervision of the farmer or some member of his family, is undeniably the main reason why upon many large farms well suited for breeding, so few calves are brought up. A little of the blame for this may be laid at the door of modern social fashion. Upon a large farm the farmer himself has many other duties which draw him away from superintending the feeding and treatment of calves; and it is not the fashion for sons and daughters of large farmers to give their attention to such matters. This conception of social life upon the farm may easily be carried too far. It is not suggested that the sons and daughters of men of capital should be expected to put their hands to the manual work of calf-rearing. There is a difference between this, however, and the superintending of work done by hired servants. The daughters and sons of farmers will be none the less ladies and gentlemen if they make themselves acquainted with certain details of their father's business, and assist him in seeing that these details are carried out with due care and regularity.

Deficiency of Store Cattle.— The growth in the breeding of cattle has not kept pace with the increase in the consumption of beef. The supply of home-

bred store cattle has not been equal to the demands of the feeders. Farmers have been complaining of unsatisfactory financial results from fattening cattle, and the main difficulty has been the fact that, on account of deficient supply, store cattle have been dearer than fat animals —that feeders have had to pay more for the lean cattle than the price of beef would warrant.

Home-breeding, not Importation, the Remedy.— The proper remedy for this state of matters is the extension of home-breeding — assuredly not the importation of foreign lean cattle. Let that be resorted to only when our own resources in cattle-breeding have been developed to the fullest advantageous extent. We are far short of that limit yet; and one would fain hope that until it is reached the best efforts of our leaders of agriculture may be directed to the encouragement of home-breeding rather than to the devising or providing of means of increasing the embarrassments of home-breeders by importing foreign-bred lean stock.

Rear more Calves.— In any scheme for increasing the supply of home-bred store cattle, calf-rearing must play an important part. We must not only breed more calves, but we must also rear more. We should rear all we breed, or nearly so, and rear them well, too; for let it ever be kept in view that what an animal loses with bad treatment as a calf, it can hardly ever fully recover. But by rearing well, one does not mean any sort of extravagant treatment. As a matter of fact, there is in many cases room for much greater economy in the rearing of calves. In connection with calf-rearing on dairy farms, or wherever milk can be turned to good account, this point is of special importance.

Breed longer from Cows. — Cows that prove to be good breeders should be bred from to a greater age than is the general rule at present. A custom by no means uncommon is to buy a cow for a temporary supply of milk, and fatten her off when she gets dry. Now this is a serious loss. Breed from all suitable cows as long as practicable.

Breeding from Heifers.— From all heifers that are suitable, whether intended for cows or not, take one, two, or perhaps even a third calf. Keep them well all the while, letting the calves suckle; and if the heifer is not to be kept for a cow, she may be fattened off and sold as heifer-beef. The calf or two will have done her little or no harm in the butcher's eye, if only she does not show the udder of a cow. This will not often arise when the calves suckle. This question was put to an extensive salesman in the north of England, who replied that his experience was that two calves or so in no way spoiled the sale of the young heifer, if only there were no display of udder, and if she were plump, level, and well fattened. He added that a lot of young heifers never came before him for sale but he regretted that so much valuable material was being wasted. Premature fatting of heifers is really killing the goose that lays the golden eggs. In these times farmers cannot afford such waste as that.

Are Calves Nuisances?—Unfortunately not a few dairy farmers look upon calves as little else than nuisances—as necessary evils,—something which they would never wish to have if only they could without them get cows in milk. This is a great misfortune, and shows clearly that while the cry is for more store stock, there must be something radically wrong somewhere. The fact is, calf-rearing is very imperfectly understood.

It is undeniable that dairy farmers, as well as other farmers in all parts suited for breeding, would find, in well-conducted calf-rearing, returns which would amply repay careful treatment and judicious and liberal feeding. The dairy farmer may dislike the calf because he has found it a greedy and bad-paying customer for its mother's milk. But if he has done so, he has had himself to blame. A good calf will well repay a moderate allowance of its mother's milk for a short time; and one would emphasise this point, that it is only for a very short time at the outset that there is any necessity to give milk—at any rate, new milk—to calves.

Milk Substitutes.—Scientific research and commercial enterprise have placed us in possession of many advantages unknown to our forefathers. In the simple matter of calf-rearing much has been gained in this way. Why, the market is teeming with cheap milk substitutes; and, without going the length of affirming that these foods are worthy of all their energetic vendors say of them, yet one may unhesitatingly say that, with substantial advantage to themselves and the general public, farmers might draw upon them much more largely than they have done heretofore. Undoubtedly the use of these prepared foods is on the increase; and by a judicious use of them and other simple natural foods, calf-rearing might be increased to a very great extent, both on dairy and mixed husbandry farms.

Rearing or Selling Calves.—It is not suggested that all farmers should rear their calves. It may suit some better to sell the calves when one, two, or three weeks old. If the calves are of a good class they will sell readily at handsome prices. While it may suit some to breed calves and sell them young, it will undoubtedly pay others to adapt their arrangements specially for rearing. Instead of keeping large stocks of cows, they may buy in young calves, and rear them partly on milk and other suitable food. In certain cases these bought-in stock may be carried on and fattened when about two years old or less. In others they may be simply reared, and sold as lean stock when from ten to eighteen months old.

Housing Calves.

The comfortable and economical housing of calves is a matter that demands careful attention. Calves are either suckled by their mothers, or brought up by the hand on milk and other substances. When they are suckled, if the byre be roomy enough—say, about 18 feet in width—calves may be tied up to the wall behind the cows; or, what is a less restrictive plan, they may be put together in large loose-boxes at the ends of the byre, or in an adjoining apartment, and let out at stated times to be suckled.

When brought up by the hand, calves are put into a suitable apartment, preferably each in a crib to itself, where the milk is given to them. The advantage of having calves separate is, that it prevents them, after having had their allowance of milk, sucking one another, by the ears, teats, scrotum, or navel, by which malpractice ugly blemishes are at times produced. When a number of calves are kept together, they should all be muzzled to prevent this sucking.

Calf - crib. — The crib for each calf should be 4 feet square and 4 feet in height, sparred with slips of tile-lath, and have a small wooden wicket to afford access to the calf. The floor of the cribs, and the passages between them, should be paved with stone, or laid with asphalt or concrete, though asphalt and concrete make cold floors which should be well covered with litter. Abundance of light should be admitted, either by windows in the walls or sky-lights in the roof; and fresh air is essential to the health of calves, so that ventilation should be carefully attended to. So also should the cleaning of the calf-cribs. The cribs should be regularly cleaned out; and it is a good plan to sprinkle the floors daily with some disinfectant, such as diluted carbolic acid—one part of acid to twenty of water. This will keep the atmosphere pure and wholesome, which is very desirable for the young animals.

The crib should be fitted up with a manger to contain cut turnips or carrots, and a high rack for hay, the top of which should be as much elevated above the litter as to preclude the possibility of the calf getting its feet over it.

The general fault in the construction of calves' houses is the want of light and air—both great essentials; light being cheerful to animals in confinement, and air essential to the good health of calves. When desired, both may be excluded. Calf-houses are often also too cold. The walls of the house should be plastered, to be neat and clean, and should be lime-washed at least once every year.

In some cases the cribs are so constructed that the calf has access, either at will or when the door of the crib is opened, to a larger enclosure in which the young animal can exercise its limbs.

Care in letting out Calves.—When the calves are fit to be put out in the open air, after it becomes mild, they should be put into a shed for some nights before being turned out to grass, and also for some nights when at grass.

When put right out to the open from the crib they are apt to run about so much as to get chills, but this risk is lessened by the calves being loose in a shed for a little time before being put out. The shed should be fitted up with mangers for turnips, racks for hay, and a trough of water.

Calf's First Food. — The first food the calf receives is the biestings—the first milk taken from the cow after calving. It is of the consistence of the yolk of an egg, and is an appropriate food for a young calf. By the time it gets its first feed, the calf may have risen to its feet. If not, let it remain lying, and pour a little of the biestings into its mouth, introducing a finger or two with it for the calf to suck, when it will swallow the liquid. Let it get as much as it is inclined to take. When it refuses to take more, its mouth should be cleaned of the biesting that may have run over.

Composition of Biestings. — The biestings or first milk after calving differs considerably in composition from ordinary milk. It contains an exceptionally large proportion of casein or cheesy matter, as the following analysis of ordinary milk and biestings will show:—

	Ordinary Milk.	Biestings.
Casein (cheese) .	3.30	4.83
Butter fat . .	3.40	3.37
Milk-sugar . .	4.55	2.48
Albumen . .	0.60	15.85
Ash . . .	0.75	1.78
Water . .	87.4	71.69
	100.00	100.00

Feeding Calves.

Reform in Calf - feeding.—In the method of feeding calves during the first few months of their existence, there has been almost as great a revolution as in any other branch of farm practice. The old notion, that at least three months of feeding upon whole milk as it comes from the cow was necessary for successful calf-rearing, has been exploded. In many cases, almost entirely in herds of pure-bred cattle, the calves still suckle their dams. But beyond these herds comparatively little new milk is now employed in rearing calves, reliance being more largely placed upon skim-milk and milk-substitutes.

The introduction of the cream-separator led to important improvements in the system of calf-rearing. Although bereft of nearly all the butter-fat, separated milk is usually more wholesome for calf-rearing than skimmed milk. Separated milk is fresh and sweet, while in the case of skimmed milk a certain amount of change may have taken place which more than counteracts the advantage of the additional percentage of butter-fat.

Calf-feeding in Pure-bred Herds.—The methods of feeding calves pursued in herds of pure-bred cattle are detailed in the sections of this volume dealing with these breeds. Nothing need be added here in regard to pure-bred calves. The methods, it will be seen, vary considerably, yet there is a general agreement in the main features.

Calf-feeding in Ordinary Mixed-bred Herds.—The feeding of calves in ordinary mixed-bred stocks does not differ fundamentally from that in pure-bred herds. The general principles are the same in both cases. The main difference comes in on the score of economy. In pure-bred herds the main purpose aimed at is often the fullest possible development of the animal regardless of a little extra cost in the process of feeding. In ordinary commercial stocks strict attention must be given to economy from the very outset. Thus, as a rule, in the latter case the cheaper feeding materials are more largely used than in pure-bred herds.

Suckling and Hand-rearing.—Suckling, of course, is nature's method of calf-rearing. As has been seen, it is followed largely in pure-bred herds. For ordinary fattening stock it is too expensive, and in this case is rarely pursued, except with cows that have just had their first calves, or where two calves are put to one cow. Hand-rearing is by far the most widely prevalent system.

Prevalent Methods.—Perhaps the most widely prevalent method of rearing calves is to feed them entirely on new milk for a short period at the outset—that period varying from two to six weeks,—and afterwards partly on new milk, separated milk, and artificial food;

or upon separated milk and artificial food, without any of the rich milk as it comes from the cow. It is, no doubt, a good plan to let the calf have all the new milk it can readily consume for at least two or three weeks at the outset. By degrees separated or skimmed milk may be substituted for new milk, and when the new milk is wholly, or almost wholly, withdrawn, the separated or skimmed milk must be supplemented by some other richer food.

Separated Milk for Calves.—Separated milk alone is not a well-balanced food for calves. As the butter-fat has been almost wholly removed from it, the remaining constituents are not sufficient for the healthy development of the young animal. Skim-milk, left by an efficient system of creaming, will, on an average, contain the following per 100 lb. :—

Casein	.	3.5 lb.
Albumen	.	.7 "
Fat	.	.5 "
Sugar	.	4.0 "
Ash	.	.8 "
		9.5 lb.

The skim-milk thus retains almost all the casein and sugar in the new milk; but so effective are most of the modern processes of separating the cream from the milk, that only the merest traces of butter-fat may remain in the separated milk. About one-sixth of the casein and albumen consists of nitrogen, and as far as it goes, skim-milk is undoubtedly a valuable food, and may be used with great advantage in conjunction with other feeding material.

Separated or skimmed milk should not be fed largely by itself to calves, for calves so fed are liable to scour, indigestion, and other bowel-complaints. The withdrawal of the new milk should take place gradually, and other substances should be introduced in corresponding ratio to make up for the deficiencies of the separated or skimmed milk.

Artificial Food for Calves.—The other substances most largely used either in supplement of or as substitutes for milk in rearing calves are linseed, linseed-cake, oatmeal, Indian corn-meal, palm-nut meal, malt, pea-meal, barley-meal, or some specially prepared food.

Tho characteristics and composition of these articles are described in the chapter on " Foods," which should be referred to and consulted carefully in arranging the dietary of animals.

Preparing Foods for Calves.—These articles of food are given to calves in the form of gruel, and they can hardly be too well steeped or boiled. It is desirable to have the linseed and linseed-cake ground into meal before boiling. Gruel from linseed-cake is often prepared by adding four parts of boiling-water to one part of the meal derived by grinding the cake, and allowing the mass to remain covered up for twelve hours. Palm-nut meal may be prepared in a similar manner. In making linseed-gruel, water should be added so as to give almost a gallon and a half of gruel for every pound of linseed. If the gruel is found to purge the calf, add a little more water, and for a day or two give rather less of the gruel and more of the skim - milk. A little wheat-flour, mixed with gruel, is also a useful and simple remedy in cases of purging. Mixtures of these meals are often made into gruel for calves, and the selection of the particular articles to be used will be regulated mainly by their market prices at the time.

Quantities of Milk for Calves.—In the majority of cases where calves are raised by hand-feeding, they get about two quarts of new milk twice or three times a-day—four to five or six quarts in all—during the first two, three, four, or six weeks of their existence. At these various periods, according to custom or to the supply of new milk and the other demands for it at the time, a beginning is made with the substitution of separated or skimmed milk for new milk. A very small proportion of the latter is given at first, by degrees it is increased, and soon the new milk is wholly withdrawn. Some, indeed, give new milk only for about two weeks, and others continue it for six weeks or two months, perhaps even longer. The new milk and separated or skimmed milk are given together. Some feed calves three times a-day in the first few weeks, and others only twice: it is advisable that they should be fed often.

Allowances of other Foods.—Supplementary foods should be begun soon, as soon indeed as the curtailing of the

new milk has commenced. The artificial food, made into gruel, is given along with the milk, and at the outset the gruel should be given in very small quantities. Sudden changes of food may inflict serious injury upon the health of the tender young animal. Some begin to give gruel to calves before they are a month old, others delay till the animal is in its sixth or seventh week. The daily allowance of gruel will of course vary with the age of the calf, and the quantity of milk it is receiving. No fixed "bill of fare" can be prescribed with safety. The appetite of the young animals must be watched closely, and special care taken to keep the bowels in good order. Feed calves liberally, but never overdo them. Let them have just as much as they can readily consume at the time, keeping on the scrimp rather than the abundant side.

North of England Rations.—The following table of rations was long in use by an experienced breeder in the North of England for calves of the large breeds :—

1st week—4 quarts of new milk at three meals.
2nd week—4 quarts of new milk and 2 quarts boiled skim-milk at three meals.
3rd week—2 quarts of new milk and 4 quarts boiled skim-milk at two meals, and ½ lb. boiled linseed.
4th week—6 quarts boiled skim-milk and ⅔ lb. boiled linseed at two meals.
5th week—6 quarts boiled skim-milk and 1 lb. boiled linseed at two meals.

General Notes.

Feeding Calves for Veal.—A large number of calves are slaughtered for veal, and these are of course forced with rich food from the very outset. New milk is the best of all foods for this purpose, although it may be to some extent supplemented by rich gruel, made perhaps from barley-meal or Indian - corn meal. The new milk is given in three meals. The daily quantities of new milk may be a gallon and a half by the end of the first week, two and a third gallons by the end of the second week, rising gradually to three gallons by the end of the fourth week. Milk turned into veal is not likely to realise more than 6d. per gallon.

Some give raw fresh eggs to veal-calves, which are generally allowed to suck the cow at will, or at least three times a-day.

The usual period of fattening for veal is from six to ten weeks, and with the view of improving the colour of the flesh the calves are frequently bled. In fattening veal-calves, most careful attention must be given to cleanliness, ventilation, and regularity of feeding.

Danger of gorging Calves.—Great care should be exercised in the feeding of calves in their tender days, especially during the first three weeks. At this time they should be fed sparingly rather than liberally. Many calves are lost by sucking or drinking more milk when they are quite young than their weak digestive system can readily dispose of. Whether the calf is fed by the hand or suckled by its dam, take care that it does not over-feed itself. Never let it suck or drink till it is quite satisfied—at any rate during its first three weeks. If the cow has too much milk for the calf, take away a little by the hand.

Many calves are killed by gorging with milk after a long fast—perhaps after a journey. When a purchased calf is taken to its new home it should be fed very sparingly for at least two days.

Weaning Calves.—Weaning is usually a critical event in calf-life. In dairy and ordinary stocks, where only a small portion of the milk is given to the calves, the youngsters are weaned when very young. The process may be said to begin in some cases at the end of the second week, when some skim-milk or gruel is substituted for so much of the new milk. In pure-bred herds, and wherever calves are reared largely on milk, weaning, as has been seen, is generally completed in the sixth, seventh, or eighth month.

In the weaning of calves there is scope for the exercise of the utmost skill and care. If success is to be attained, both skill and care are essential. Prepare the young animal for the weaning—the complete withdrawal of its mother's milk—by feeding it partially for some time before with such food as will form its main support after it has been weaned. Let the milk be lessened, and the other food gradually increased in quantity, so that the transition may be effected almost imperceptibly. The more carefully and intelligently this is done, the more satisfactory will be the result in the calf. The amount of milk allowed to a suckled calf may be regulated by drawing away as much of the cow's milk by hand as may be desired, and at last, just before final weaning, the calf may have access to the cow only once a-day.

There is perhaps no better food for calves at weaning-time than good linseed-cake—from 1 to 2 lb. per day, and a few sliced turnips or mangels, and fresh well-made hay. If accustomed to this fare before being entirely deprived of their mother's milk, they will be found to pass through the ordeal of weaning without any loss in condition or delay in progress.

Setoning.—A seton is a piece of string or tape passed through a certain part of the body, with the object of either drawing an abscess, or acting as a counter-irritant, or for the purpose of inoculation. As a prevention against black-leg, or quarter-ill, it is a useful custom to insert a seton in the calf's brisket in the spring. It is considered desirable to soak the seton in some irritant such as the following embrocation—viz., hartshorn, 1 ounce; turpentine, 2 ounces; spirit of camphor, 2 ounces; laudanum, ½ ounce; olive-oil, 6 ounces.

Castrating.—The male calves can be most easily castrated when a few weeks old. They can then be cut standing, by twisting the tail round one hind leg. Stand behind the calf, cut through the bag, twist the stone several times, and scrape the cord closely through with a blunt knife. When the calves are several months old they must be cast. This may be done by tying the hind legs together with a rope, placing a halter round the neck, taking the shank end of the halter and running it through the rope that unites the hind legs, tying it back, passing it through the portion that is around the neck, and drawing the legs tight, then fastening the rope. The fore legs can be held by a man. The stones may then be removed by the clams and hot iron, as in the case of the horse—place the stone in the clams, and with a red-hot iron saw the cord slowly through close to the clams.

MANAGEMENT OF STORE AND FATTENING CATTLE.

The subjects to be dealt with under this heading bulk largely in the agricultural economy of the United Kingdom. The importation of fat stock and dead meat has grown to great dimensions, yet a substantial proportion of the agricultural community of this country derive a large part of their living from the rearing and fattening of cattle. In this work, therefore, these branches of the live-stock industry demand careful attention.

PREPARATION OF FOOD FOR CATTLE.

In order to ensure the best possible results in the progress of the animals, careful attention should be given to the methods of preparing food for the different classes of cattle. In this, as in most other farming matters, it is impossible to lay down hard and fast rules which would be equally applicable to all cases. This much, however, is applicable to all—let the food be prepared and presented to the animals in as cleanly and palatable condition as possible. Depend upon it, the animals, be they mere calves or adult cattle, will amply repay in increased progress any extra care required in presenting their food to them in a cleanly, inviting, and wholesome condition.

Washing Roots.—Dirty roots should never be placed before cattle, either cut or uncut. If turnips should become very wet and muddy, they should, by some means or other, be washed before being given to cattle. Several machines have been made for washing roots, but in a brook or pond they can be washed satisfactorily by hand.

Frozen Roots.—It is very unwise to give frozen turnips to cattle. A speedy way of thawing turnips is to steep them in a pond or tank of cold water. But here, as in many other cases, prevention is better than cure. Timely storing prevents the necessity of having to use frozen roots.

Cutting Turnips.—Young cattle and sheep, with tender, imperfectly developed teeth, cannot comfortably consume uncut roots, and should never be expected to do so. Fully grown cattle can quite well eat whole roots; yet even with these it is desirable, in all cases where practicable, to have the roots cut before being given to them. The slicing is the most common method of cutting turnips for cattle. The slices, as a rule, vary from a half to three-fourths of an inch in thickness. It is bad practice to slice more turnips at one time than can be used immediately.

Turnip-cutting machines are almost innumerable, and most of them do excellent work. There are large turnip-slicers, which are driven by horse, steam, or water power; and in very many cases the old-fashioned hand-lever slicers, with some modern improvements, are still in use.

Pulping.

Where the pulping system is pursued, the roots are cut by machines into pulp or small chips, and mixed with cut straw, chaff, or other fodder, and this mixture is given to cattle either with or without the addition of crushed cake, meal, or other concentrated food, according to the class and condition of animals receiving it.

Economy of Pulping.—The pulping system economises food of all kinds, especially roots. To be sure it increases the cost of labour somewhat, but the question to determine is not merely whether pulping increases the labour bill or outlays of any kind, but whether it enables the farmer to turn his roots, straw, and chaff to better account—in short, whether it is more profitable than the older method of giving the roots by themselves whole or sliced. For the pulping system may be more costly and yet more profitable. Experience has proved it to be both in most cases; and, as would therefore be expected, it is practised extensively throughout the country. A common expression amongst farmers who have pursued the pulping system is that it makes their roots "go a great deal further" than under the old method. Greater advantage can be derived from pulping in the rearing than in the fattening of cattle. It is also admirably adapted for sheep.

Preparing Pulped Mixtures.—The pulping process is very simple. The pulped mixture should be prepared every day, and allowed to lie from 12 to 24 hours before being given to the animals. The fermentation which takes place in this time is entirely beneficial. It softens the fodder and cake or meal, or whatever else there may be of dry food, sweetens the whole mass, and renders it not only more pleasant to the palate of the animal, but also more easily digested and assimilated than if the roots and dry food had been given separately. Never on any account allow the pulped mixture to lie so long as to become mouldy or sour.

The roots must be cut or pulped, the grain either bruised or ground into meal, the cake broken, and the straw and hay cut into chaff. For bruising and grinding grain, breaking cake, and cutting fodder into chaff, there are numerous machines of the highest efficiency.

Food-preparing Compartment.— Where pulping or any of the other modern systems of feeding are extensively pursued, it is found convenient to have a food-preparing compartment adjoining, or part of, the turnip-store. Adjoining this also, or in the same house practically, should be the cake and meal

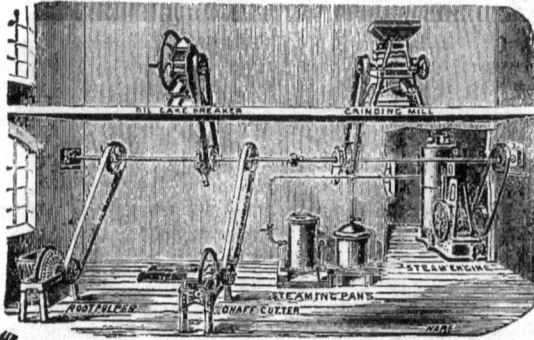

Fig. 746.—*Steam food-preparing machinery.*

compartments. A handy arrangement is to have the cake and meal stores on a floor right over the food-preparing compartment. In this floor the cake-breaker and grinding or bruising mill are situated, as also the chaff-cutter; and the broken cake, cut fodder, and bruised grain are dropped through hoppers into the apartment below, where the mixing of the food takes place.

This system is, of course, subject to many variations in detail, in accordance with the peculiarities of different steadings and the extent of the holding. The chief points to be aimed at are convenience and the saving of labour, these two terms being, in this connection mainly, but not entirely, synonymous. Fig. 746 represents one of many excellent and convenient food-preparing sets erected in farm-steadings, by Barford & Perkins, Peterborough. Provision is also made in this set for steaming the food. The small vertical engine is fixed in an outhouse or lean-to, and in addition to driving the grinding-mill, oilcake-breaker, root-pulper, and chaff-cutter, &c., it supplies steam to the two steaming-pans, one of which is used for roots, chaff, &c., and the other for boiling milk or compounds.

Cooking or Steaming.

The cooking or steaming system of preparing food for cattle has lost in favour. It was at one time practised to a considerable extent for cows and fattening cattle, but in most cases it has been abandoned wholly or partially. As a rule, food for cows receives nothing more in the way of cooking than scalding with hot water.

Bruising Grain.

The importance of having all kinds of grain bruised flat or ground into meal before being given as food to stock is now very generally acknowledged. Still, it is only too true that even yet farmers not unfrequently permit the feeding of whole grain, especially to horses. It is a wasteful practice, and should not be pursued on any account.

WINTER HOUSING OF STORE CATTLE.

The influence of locality is very great, and must be carefully considered by the successful stock-owner. In the cold regions of the north, even the young store cattle have to be housed throughout the entire winter. In the greater part of Ireland, and in the southern and milder parts of Great Britain, young growing cattle spend a good deal of the winter, when the weather is dry and favourable, on the pasture-fields. Between these two extremes of *in* all winter and *out* all or the greater part of it, there are many gradations, which individual farmers must judiciously and carefully arrange for themselves. So much depends upon local circumstances as to climate, house and field shelter, class of cattle, supply of outdoor and indoor food, &c., that to lay down hard and fast rules would be worse than useless.

Err on the Side of Shelter. — This one rule, however, one would lay down with all the emphasis and firmness that can be given to it. It is better to err on the side of caution—better to have the animals *inside when you think they might perhaps suffer little harm by being out, than outside when they would have been better in.* How often is it the case that even a reputedly careful farmer allows his cattle to remain out on the fields when he *thinks* they *might* be as well in? "As well in." Depend upon it, that means that the animals ought to be inside. The *thought* may or may not be expressed—when there is *thinking* in the play, be it ever so little, always let the animals have the benefit of the doubt —and the shelter too!

Fresh Air for Cattle. — Not for a moment would one depreciate the value of fresh air for cattle. Fresh air is most essential, particularly for young growing cattle. But it is easy to provide this without exposing the cattle to excessive cold, and drenching, chilling sleet, and winter rains. Cattle certainly cannot thrive well in close, stuffy, ill-ventilated houses. But while a few farmers are so careless as to let their cattle suffer in health and be retarded in progress by want of proper ventilation or fresh air, the prevailing error is entirely the other way.

Loss from Exposure to Bad Weather.—It is not in the least overstating the case to say that for every twenty shillings lost by want of ventilation in cattle-houses, there are hundreds of pounds sterling sacrificed by the exposure of cattle to inclement weather. If the value of property, in the shape of raw material for producing meat and dairy produce, which is lost every year through the imprudent and avoidable exposure of cattle to inclement weather, could be accurately stated in plain figures, the vastness of the sum would astonish everybody, no one perhaps more so than the defaulting stock-owners themselves. It would certainly run into millions of pounds sterling per annum!

For be it remembered that exposure to bad weather does more than retard the progress of cattle. It likewise incurs great waste of feeding material. While the animals are thus exposed more food is required to maintain the animal heat, not to speak of increase either in size or condition. It is a proverbial saying amongst observant if not always painstaking farmers, that cattle will thrive better upon moderate feeding with sufficient shelter, than with all the food they can eat in exposure to cold and wet.

Economical Rearing of Cattle. — The proper housing of cattle has much to do with their economical feeding. It is perhaps not overreaching the mark very far to say that the thriving of store cattle in winter is regulated almost as much by how they are housed or sheltered as by the system of feeding. This statement will suffice to show the young farmer that, if he wishes his cattle to make satisfactory progress, if he desires to secure in his store cattle the greatest possible progress, at the lowest possible outlay of time and money, he must give as careful attention to shelter as to feeding. Unfortunately this is not always done.

In very many cases, farmers who are known to be liberal and careful feeders are lamentably negligent in providing proper shelter for the stock.   More particularly does this remark apply to England and to Ireland—still more notably to those very districts in which comparatively little house or shed accommodation would supply all the shelter that is required.

Houses for Cattle in Cold Districts.—Where the winter is long and usually severe, as in the greater part of Scotland and colder parts of England and north of Ireland, substantial houses have to be provided for all kinds of cattle in winter; but where the winter is usually mild and open, very cheap erections are quite sufficient for store cattle. In cases where close houses or courts are required, care should be taken to have them well ventilated.

Cattle-sheds in Southern Districts. —Going at once from the one extreme to the other, from where the winter is severest to where it is mildest, one finds in the latter parts simple forms of winter shelter for store cattle used with satisfactory results.   Very often it is a large open court, with access to a roofed compartment where the animals can take shelter from rain or snow, eat their food, and lie over night.   Perhaps a roof is thrown over a portion of the court—a roof of sheet-iron or wood resting upon the wall of the court at one side and upon pillars at the other.   The roofed compartment may be merely a "'lean-to" on another building.   It matters little how it is provided, and in these mild districts it need not be costly, substantial, or elaborate.

The main object is to make sure that there is plenty of roofed space to protect the cattle from rain, to enable them to eat their food in comfort, and have a dry warm bed.   Store cattle need not be kept in such a *warm* temperature as milking cows and fattening cattle.   Keep them dry and *comfortable*, and so long as *comfort* is secured, the young growing animals will be all the better of some open space to move about in when the weather is favourable.

Cattle - courts. — Between the close byre and open court and shed there are many forms of winter shelter for store cattle.   The most general is the partially covered court, which is perhaps, upon the whole, the most serviceable and advantageous of all.   With surrounding buildings and boundary walls the court is usually well sheltered from "'a' the airts the win' can blaw"; and with a half, two-thirds, or three-fourths of it roofed, there is ample protection from rain and snow.

The equipment of houses for cattle is dealt with in vol. i. pp. 151-167.

WINTER FEEDING OF STORE CATTLE.

There are endless variations in the systems of feeding young store cattle in winter.   These variations are regulated mainly by (1) the locality and methods of cropping and general farming pursued; (2) the condition and time at which the animals are to be sold; and (3) the class and character of the stock.

Apportioning Home-grown Foods. —The farmer will have to consider and arrange at the beginning of winter what proportions of his supply of home-grown winter food, such as roots, straw, hay, silage, and grain, he is to allocate to the various kinds of stock.   The proper allocation of the home supply of food amongst the various kinds of stock, and the careful distribution of that supply so as to make it extend evenly throughout the entire season, are points of the very greatest importance in farm management.   For instance, too free use of roots or fodder at the beginning of the winter may cut short the supply before the next grass season comes round, and the blank thus created through want of forethought may have to be filled up at disproportionate outlay by the purchase of expensive foods.

At this particular time the farmer will take special note of the quantity of roots available for the young store cattle, so that he may be able to decide and explain to the cattle-man not only what daily allowance of roots is to be given to these store cattle, but also what kinds and proportions of other food will have to be provided for them. Probably the supply of roots available for the store cattle may decide whether or not the pulping system is to be pursued. If the supply of roots is very abundant, possibly the farmer may think it better to give the store cattle a liberal quantity of roots in the ordinary way by themselves, than to give a larger proportion of the roots to other kinds of stock or to buy in more store cattle. Circumstances alter cases ; and the farmer must, at the beginning of every winter, consider carefully how he can turn the produce of his farm to the best possible account.

Economise Turnips.—Now that the turnip-break is being curtailed, it is more probable that the supply will be scrimp than abundant. In any case, it may prudently be urged as a general principle that farmers should endeavour to economise the turnip crop. It is the most costly and most risky crop in the ordinary rotation ; and, all things considered, it is not by any means cheap food. As a rule, therefore, farmers should be encouraged to adopt methods which would advantageously economise the supply of roots, and render them less dependent upon the turnip-break than they have been in the past.

What Foods to be Bought and what Sold.—When it has been ascertained what quantity of roots can be had for the store cattle, the farmer will next consider what kinds and quantities of other foods are to be given to them. Whether these other foods are to be home-grown or bought, or part of both, will depend upon the supply of such home-grown foods as straw, hay, silage, and grain, and the current market prices of these and other commodities used as food for cattle. For instance, hay may be worth more in the market than as food for store cattle, so that it may be advantageous to sell hay, and — if the home supply of straw be deficient—buy oat-straw or some other food. Again, "ups" and "downs" in market prices may enable the farmer to derive profit by selling grain and buying maize, cake, or other food ; or the home-grown grain may be selling so badly, and the cattle so well, that he may find it beneficial to use the grain in pushing on the live stock, instead of sending it to market.

Advantage in Using Home-grown Food.—There is a growing tendency to use more and more of the home-grown produce as food for cattle and sheep, the low range of prices of grain being the chief influence in bringing this about. Other things being equal, there is an advantage in consuming instead of selling farm produce. It is true economy to make the produce of the farm "walk itself" to market, in the bodies of well-conditioned cattle, sheep, and swine.

No Hard and Fast Rules.—Yet farmers must not be tied by rules. They should sell their farm produce, and buy food whenever it is advantageous to do so. Thus it will be seen that if the farmer is to turn his produce to the best possible account, and rear his cattle as economically and efficiently as may be, he must be able to watch the condition and tendency of market prices, as well as the quality and quantity of his own crops, with keen intelligent perception, and sound, ready, and careful judgment.

Ages of Store Cattle.—Formerly there were two generations of *store cattle* to receive attention at the beginning of winter—namely, the calves of this and those of the previous year. Latterly, however, the adoption of the "early maturity" movement, of which more anon, has advanced the calves of the previous year, now from eighteen to twenty months old, into the ranks of *fattening cattle.*

Now, therefore, the winter feeding of store cattle begins with mere calves, some of them eight or ten months old, others considerably younger. Late calves may be either sucking their dams or receiving milk in other ways at the beginning of winter ; but, as a rule, the calves will have been weaned from two to several months before then, and have become well accustomed to eat such foods as grass, hay, cake, and meal.

Care in beginning Winter Feeding.—In the rearing of calves, the importance of keeping them progressing steadily from birth should be constantly kept in mind. "Never let your cattle lose the calf-flesh," is sound advice to give to farmers ; and it is one which the farm-student should store up carefully in his mind. In this particular section of the work we take up the care of these young cattle at the threshold of winter. They are, as indicated, of various ages, mostly from six to nine months, and in good thriving condition. As the supply of grass diminished and the evenings became chilly, the calves had been receiving indoor food, such as cake, meal, vetches, grass, or hay. By degrees they are worked into their winter rations. It is well to avoid sudden changes in the feeding and treatment of cattle. Give small quantities of the new food at the outset, increasing the new and lessening the old, until almost imperceptibly the complete substitution has been effected.

Turnips and Straw for Store Cattle.—In the colder districts the young store cattle, which may now be said to have emerged from calfhood, will be entirely dependent upon house-feeding by the time the winter has fairly set in. In the turnip-growing districts the food throughout the winter will consist mainly of turnips and oat-straw. Very many farmers still give the young cattle all the turnips they can eat comfortably; but, as has already been indicated sufficiently, the allowance of roots is being lessened with advantage.

Study the Animal's Appetite.—Where it is intended to feed the young store cattle solely with turnips and straw, and where there is an abundance of both, the cattle-man may decide for himself, from time to time, by carefully watching the appetite and progress of the individual animals, what quantity of each kind of food is to be given to each animal. He will be careful not to gorge the young beasts with cold roots, for in all probability some of them, of a greedier disposition than others, would eat more turnips than would be good for them. Keep within the limit of sufficiency rather than overstep it. Do not on any account give more roots at one meal than will be eaten up cleanly without delay at that time. It is a bad, wasteful practice to have roots lying for hours before cattle. Valuable food is thus destroyed, and the animals thrive best when they have their stated meals at fixed hours, getting no more roots at each time than will be at once consumed. The same remark applies to meals and cake, but with straw and hay the case is different.

Feed Sparingly and Frequently.—The long fodder is usually, and ought always to be, supplied in a rack sufficiently high to be just within easy reach of the animal's head. Many good farmers think it beneficial to have a little fodder always in the rack, so that the animals can take a mouthful when they feel the desire for it. There is something to be said for this, and the fodder in the rack is not so liable to get spoiled by the animal's breath as are roots or other food lying in a box or crib lower down. Still, it will be found more advantageous to supply the fodder sparingly and frequently than in large quantities at a time. The fresher and sweeter it is, the more keenly will it be relished by the animals; and if too much is given at a time, the cattle are apt to pull out more than they eat and waste it amongst their feet.

Feeding Hours.—The most general custom where the turnip and straw system prevails is to give the roots in two meals, one in the forenoon, between 8 and 10 o'clock, and another between 2 and 3 in the afternoon; and the fodder in three meals, between 5 and 6 in the morning, between 11 and 12 in the forenoon, and between 3 and 4 in the afternoon. In some cases a fourth meal of straw is given between 6 and 8 o'clock at night.

In many instances the daily allowance of turnips is divided into three meals, given at 6 A.M., 10 A.M., and 3 P.M.; and the young animals will be more contented and most likely thrive better with three small or moderate meals of roots than with the same quantity in two meals.

Different Kinds of Roots for Store Cattle.—At the outset, perhaps for two or three weeks, soft white turnips are given whole, "tops and all," but if the tops are very wet and muddy, they should be given very sparingly, or, better still, not at all, as in that condition they will be apt to cause scour. The white turnips are succeeded by yellows, and where a large proportion of swedes is grown, these take the place of the yellow turnips perhaps as early as the second or third week in November, probably not for several weeks later, according to the proportionate supplies of the two kinds of roots.

It is not often that the soft white turnips need to be cut; but in every instance yellow turnips and swedes should be cut for young cattle—for all kinds of cattle, indeed, whose teeth are not fully developed and in good order.

Roots, Cake, Meals, and Fodder for Store Cattle.—Partly from choice and partly from necessity store cattle are now being reared with much smaller allowances of turnips than in former times—say, prior to 1875. The advantages of this change have already been noticed. In certain cases the curtailment of the root-supply has been moderate, and little or nothing introduced in place of that withheld, excepting an increased quantity of straw or hay, and an offering of fresh pure water.

The more general plan, however, has been to give, along with the lessened allowance of roots, small quantities of other more concentrated foods, such as cake, bruised grain, or Indian corn meal, and the usual full supply of long fodder. With two small rations of roots, from 35 to 50 lb. altogether, plenty of good oat-straw or hay, and from 1½ to 3 lb. of cake or meal per day, young store cattle will be found to thrive admirably. The allowance of meal or cake is usually given early in the morning, perhaps about 6 A.M., and the roots at from 9 to 10, and about 3 P.M.; the fodder as already stated. It is considered undesirable to give a large feed of cold roots upon an empty stomach in the morning.

In other cases where still fewer roots are allowed, these are given at one time, perhaps about 10 or 11 A.M., the concentrated food being given early in the morning and afternoon, the former meal smaller than the latter. Again, in some farms the whole of the cake or meal is given in the morning, and the roots reserved till the afternoon. It cannot be said that any one plan is best for all cases; but as a rule, at any rate where the animals run out daily, it is considered most suitable to give the turnips in the forenoon.

Where the animals are able to pick up a little grass outside, they will relish a feed of cake or meal as soon as they come in, and an allowance of fodder may be reserved till later in the afternoon. Where no food is to be had outside, the animals, after a run in the fresh air and a drink of cold water, will welcome a substantial ration of oat-straw or hay.

Southern Systems of Feeding Store Cattle.—In the principal grazing districts of England and Ireland, and also in the south-west of Scotland, where the climate is mild, and the winters comparatively free from frost and snow, the young store cattle are out on the pastures almost daily throughout the winter—out many a day when they ought to be in. Where there is a good deal of rough pasture, and where care is taken to have the animals comfortably housed at night and in wet or exceptionally cold weather, the young cattle thrive wonderfully well under this system, with but very little extra food of any kind. Most likely no roots are given, perhaps nothing but long oat-straw, or a little hay or silage, once or twice a-day. In other cases a small allowance of cake or meal, from 1 to 2 lb. per day, is given.

Occasionally in these parts the extra food is given in racks and boxes outside. This, however, is not a good plan. Let the animals have it under a roof, with a dry place to stand upon, where they will have plenty of fresh air, but be free from draughts and wet.

It is not uncommon, indeed, to see turnips given to cattle on fields even in cold days in winter. In an exceptionally mild dry day there may be little harm in this, but, generally speaking, the practice is to be condemned. The animals will turn the cold roots to better account if allowed to consume them in comfortable quarters.

Pulped Food for Store Cattle.—As already indicated, the pulping system is specially serviceable in the feeding of store cattle. It enables the farmer to turn his straw and chaff to better account as food for stock than could be done otherwise. The straw of wheat and barley are not much relished by cattle when given by themselves, and cattle will not willingly eat chaff. Yet there is considerable feeding value in all these, and in a judiciously prepared pulped mixture cattle will eat them with appreciation. There is not the same advantage in pulping good oat-straw and hay, for if given in a fresh condition, and in small quantities at a time, cattle will consume these in the long form with exceedingly little waste. But the utilisation of the less palatable kinds of fodder is an important consideration, and this, together with its great influence in economising roots, commends the pulping system very strongly as a most useful agent in the rearing of store cattle.

Proportions of Pulped Mixtures.—Already some information has been given as to the manner of preparing pulped mixtures (p. 356). The proportions of roots to other foods will, of course, depend largely upon the supply available for the store cattle. Some mix equal quantities, bushel by bushel, of pulped roots and chopped fodder; but a much smaller proportion of roots is more general. One bushel of pulped roots is often made to serve for two, three, or even more bushels of chopped fodder, and when the allowance of roots is very small, it is desirable to add to the mixture a little crushed cake, meal, or bruised grain, perhaps from 1 to 2½ lb. for each beast per day. Decorticated cotton-cake is most largely used for store cattle, but many give a mixture of this and linseed-cake or linseed-meal. The market prices should be watched carefully, and the kind of cake or other food bought which is comparatively cheapest at the time. Many careful feeders sprinkle a little common salt over the pulped mixture, and still a larger number sweeten it with dissolved treacle.

When it is intended to push the animals from their youth, and have them fattened at an exceptionally early age, the richer and more concentrated foods are increased in quantity.

Store Cattle on Pastures.

Store cattle go to the fields as soon as the grazing season begins. If the supply of pasture is fairly ample the growing cattle may get no extra food. If the pastures are poor, and if it is desired, as it ought to be, to keep the animals progressing, concentrated food of some kind should be given on the fields. The extra food may consist of whichever of the ordinary cattle foods may be cheapest at the time, and the quantities may vary from 1 to 3 or 4 lb. daily.

Keep Stock Progressing.—There is one point which demands most careful attention about the end of spring and beginning of summer. It is this—to see that the animals are carried from the one season to the other in a steadily progressing condition. Do not on any account let the animals fall off towards the end of the house-feeding season. If the supply of turnips and other home-grown food become scarce, buy in food, or reduce the stock by selling. Then if the supply of grass should be deficient at the outset, supplement with other food—with purchased corn and cake, if need be. In the period of transition from one season to another, cattle are often allowed to fall back in condition. This is very detrimental to the interests of the stock-owner, and should be avoided by hook or by crook.

Give the Pasture a Good Start.—Do not be impatient to turn the cattle from the winter quarters to the summer grazing. Let cattle of all ages remain in the steading until the grass is quite ready to receive them, and able to maintain them in a satisfactory condition. In late seasons, when the turnips and other winter food are exhausted before the grass can afford them a bite, the animals should be partly supported upon extraneous food—as oilcake, beans, oats; or those in fairly good condition should be disposed of, to leave some turnips for the young cattle and cows until the grass grows up.

The cattle are let out in relays as the grass progresses. It is a good plan at the first of the grazing season to take up the cattle at night, and give them dry fodder. This tends to counteract the laxative influence of the fresh grass.

Overgrowth of Pastures Injurious.—An important point in the successful grazing of land is to keep the pastures from growing too rank. In the earlier part of the season, in particular, they should be well eaten down, cropped frequently, but not so as to injure the plants. Pasture-grasses should never be allowed to mature and produce seed, for both the land and the plants will be thereby impaired in their productive powers. Pastures do best when grazed for about two weeks, and rested for a similar period all through the season.

All kinds of stock thrive best on moderately short pasture. Rough bunches of grass should be regularly cut down by the scythe.

In some cases, in a good growing season, it may be advisable to buy in more stock to keep down the pasture. In other cases, especially when cattle are dear, it may be better to save a portion for hay, and thus curtail the grazing area.

On many farms the droppings of the cattle are daily collected into heaps, and in the autumn spread upon the inferior parts of the field. On others they merely scatter the droppings over the field where they are found, once or twice a-week.

Changing Stock on Pastures.— Grass-land requires skilful management if it is to yield the maximum amount of pasture in every sort of season. The circumstances under our own control which most injure grass are *overstocking* and *continual stocking*. There should be no more stock upon the farm than its grass will maintain in good condition; and the stock should not be allowed to remain too long in the same field.

The safest way to treat each grazing-field is to stock it fully at once, in order to eat it bare enough in a short time, and then to leave it unstocked for two weeks or so, that the grass may grow up to a fresh bite. One advantage of this plan is, that it provides new-grown grass; and another is, that the grass does not become foul by being constantly trodden upon. Stock delight to have fresh-grown grass; and they loathe grass which has been trampled and dunged upon, times out of number.

To facilitate the frequent changing of stock to fresh grass, many farmers run a temporary wire-fence across a pasture field, letting the animals crop first one division and then the other.

Mixed Stock on Pastures.—Another principle affecting the treatment of pasture-land is the different way in which different animals crop grass: cattle crop high, sheep nibble low, while horses bite both high and low. This is a wise distinction between the two classes of ruminants, sheep being suited to short mountain-pasture, which their mobile lips hold firmly while it is severed from the ground with the incisors of the lower jaw with a twitch of the head aside; whereas the ox is as well suited to the plains and valleys, where grass grows long, and which it crops with the scythe-like operation of its tongue and teeth.

From these different modes of cropping grass, it is inferred that the horse or sheep should follow the ox in grazing, or accompany him, but not precede him. On pasture eaten bare by horses or sheep, the ox cannot follow; and when all are in company, the horse and sheep will eat where the ox has eaten before, or the horse will top the grass before the ox, the horse being fond of seizing the tops of plants by his mobile lips, and pinching them off between the upper and lower incisors. The accompaniment of them all in the early part of the season is a good arrangement, because all have the choice of long and short grass; but the horse should be separated from the sheep in the latter part of the season, as both bite close.

Water and salt should always be within the reach of cattle on pastures.

FATTENING CATTLE IN WINTER.

In the study of the scientific aspects of cattle-feeding we have not kept pace with some other countries, yet we do know a great deal more about early maturity, and the economical production of beef, than was known in this country prior to 1870.

Early Maturity.

In the rearing and fattening of their stock the farmers of the present day are now turning both time and food to better account than their forefathers did. The progress that has been made in the matter of " early maturity "— in the rearing of stock at a more rapid rate, and fattening them in less time and at an earlier age—has been very marked and gratifying. Along with this movement—as an essential element in it, in fact—has come a great saving of cattle food. Apart from the question as to the influence which this early " forcing " of stock may exercise upon the constitutional stamina of the bovine race, in regard to which some misgivings are entertained by eminent authorities, there can be no doubt that substantial immediate benefit has resulted from it to feeders of cattle. In feeding cattle, as in most other industries, time means money. It is important, therefore, that time as well as food should be economised. Indeed, the economical use of the one involves the thrifty use of the other, and by a careful study of these considerations farmers have raised their system of "meat manufacture" to a decidedly better footing.

Age for Fattening.—As a rule, cattle are now fattened during the second year of their existence. Large numbers are slaughtered before that year is completed, when about twenty or two-and-twenty months old. It is the exception now to find three-year-old English or Scottish bred bullocks on British farms. Many farmers practically keep on fattening their cattle from their very birth, never stinting them in food, thus not only maintaining a rapid rate of growth, but also a steady increase in the accumulation of fat and muscle.

Cheaper Meat from Young than from Old Animals.—There is no longer room for doubt that meat can be produced at a lower cost per pound on young than on old animals. To throw light on the question of the most profitable age at which to fatten animals, many interesting experiments have been carried out in this and other countries. At Rothamsted, in particular, the trials bearing on this point were numerous and instructive. In most of these trials it was found that the older and fatter an animal became the more costly it was to add additional weight of meat, confirming the American dictum of Professor Stewart that " every additional pound put upon an animal costs more than the previous pound of growth."

Lawes on High-pressure and Profitable Feeding. — Sir John Bennett Lawes was an able and persistent advocate for early maturity. He often pointed out, however, that from an economical point of view the high-pressure system of feeding might easily be overdone. He said : " Every day of an animal's life, a certain amount of food is required for sustenance purposes alone. An animal which does not increase in weight is kept at a loss, as it merely turns food into manure. On the other hand, if you require to produce as much

weight of beef in one year as is produced under ordinary feeding in three years, it can only be done by a large expenditure in costly foods ; and, except for show purposes, this very rapid fattening is not necessarily the most profitable. As the rate of increase is limited, however highly an animal is fed, much waste of food takes place under a high-pressure system of feeding ; while on the other hand, an animal is unprofitable if it does not increase in weight every day. Between these two extremes there ought to be some point which marks the minimum cost at which a pound of beef can be produced. I have once or twice tried to construct a table for my own satisfaction, but without much success."

Methods of Fattening Cattle in Winter.

Winter is the season in which cattle-feeding is carried on to the largest extent in this country, the animals being housed for the purpose either loose in courts, boxes, or hammels, or tied up in stalls. The construction and equipment of house accommodation for cattle are dealt with in vol. i. pp. 151-167, and at this stage it would be well to consult what is said there. It is especially important that fattening cattle should be kept in a thoroughly healthy, comfortable condition, for unless this is attended to the progress of the animals will not be satisfactory.

What Food is to be Used ?—It has been seen that in the methods of feeding other classes of cattle, cows, calves, and store cattle, in winter, there is almost endless variety. In the winter fattening of cattle the variation of practice is quite as great. The system of cropping and the supply of home-grown food are leading factors in determining the method of feeding pursued. The farmer should, of course, consider carefully the market price of the various recognised articles of food and of his own produce, and after due deliberation decide whether his own home-grown or purchased foods will be cheapest and most profitable. Other things being equal, he will give the preference to his home-grown food, for, as already pointed out, there is economy in making the farm produce "walk itself to market."

An important point at this time is to estimate the supply of fodder and roots, and so apportion the daily use of these as to extend them evenly over the season.

Feeding Rations. — The fattening cattle will most likely be at various stages in their advance towards maturity. Some, already in high condition, may be intended for the Christmas markets, when winter-fed beef usually brings the maximum price. Others, most probably younger animals, will be leaner, and may require from 4 to 6 months' feeding. Both classes will be accustomed to the house-feeding before winter sets in (for all fattening animals should be housed as soon as the cold nights of September begin to be felt), and both should now be liberally fed. They should not be gorged, but have as much as they can eat, given to them at fixed intervals in as palatable and tempting a form as possible, and in such quantities as will ensure that, without any food being left or wasted, the animals will be well satisfied. As to the gross bulk, there may be little difference in the food given to the cattle, but the riper animals will get the richer food. As the cattle approach maturity, the more concentrated foods, such as cake and grain or meal, are increased, and the bulkier commodities, such as roots and straw, may be slightly lessened.

As to this variation of food, no hard and fast lines can be laid down. It would be worse than useless to attempt to do so,—it would be positively unsafe. The immediate wants, the condition, progress, and appetite, of each individual animal must be carefully considered, and in accordance with these and these alone is it safe to arrange or modify the daily meals. Thus, again, it is seen that the office of cattle-man is a responsible one. The success or failure of the feeding operations is largely dependent upon him. By careful and constant attention to the adapting of the meals to the wants and capacities of the animals, he may greatly facilitate the fattening, as well as economise valuable food.

Balancing Food properly.—As to the importance of having the foods properly balanced, a good deal has already been said. And at this critical time, when the feeder is arranging or modifying the food to suit his fattening cattle at the different stages of their pro-

gress, he may be urged to consider carefully the question of mixing foods, so that the various ingredients shall be present in the proportions most perfectly adapted to the requirements of the animal. See in particular the information given under the heading of "Animal Nutrition," p. 291 of this vol.

Scottish Feeding Customs.

On nearly all Scottish farms turnips still form a dominant or important element in the rations of feeding cattle. Yellow turnips are used at the outset, and these may last for one month, two months, or longer, as the case may be. The more advanced animals, especially those intended for the Christmas market, will receive swedes as soon as practicable, perhaps about the beginning of November. All changes in the food should be introduced gradually. In putting cattle on roots in winter, small quantities should be given at the outset, full meals being allowed only after the animals have become accustomed to the new mode of treatment.

Daily Allowance of Turnips.—What quantity of turnips should a feeding bullock receive daily? This is a vexed question, as to which opinions of practical men have undergone, and are still undergoing, considerable change. Not a few still give the animals all they can comfortably consume in two meals daily. That would perhaps mount up to, or even exceed, 120 lb., according to the size of the animal. That assuredly is improvident feeding, a more prudent and more profitable system being to give much smaller quantities of roots and larger proportions of other foods. The general tendency now is in the latter direction. The majority of the more successful feeders nowadays limit the allowance of roots to from 60 to 90 lb. per day, still less being allowed in many cases.

Feeding Hours.—The general plan is to give the turnips in two meals, about 8 or 9 A.M., and from 1.30 to 3 P.M. Some give the roots as the first meal in the morning, following with oat-straw or hay, cake or meal, or both, about 11 A.M., turnips again early in the afternoon, followed by straw or hay, and cake or meal, as in the forenoon. Others think it better to give about half the daily allowance of cake and meal, say at 6 A.M., to be followed about two or three hours later by turnips and oat-straw or hay; the afternoon meals coming in the same order, beginning with cake and meal at 1 P.M., and ending with straw and hay at 8 P.M. Others, again, give a very small feed of straw or hay as the first mouthful in the morning, say from 6 to 7 A.M.

Turnips or Cake for Breakfast?— Some experienced feeders contend that it is unsafe to give cattle a feed of cold watery turnips upon an empty stomach in the morning, yet many successful feeders have all their lives pursued the system of giving roots as the first meal, and say they have never discovered any evil effects from it. Upon the whole, the weight of experience is in favour of giving a small allowance of cake and meal as the first feed in the morning.

Daily Allowance of Cake and Grain. —Where the allowance of turnips is restricted to from 60 to 90 lb. per day for cattle weighing from 8 to 10 cwt. live-weight, the quantity of cake and meal may vary from 4 to 8 or 10 lb. per day, beginning the winter with the smallest, and finishing off the fattening period with the largest quantity. The concentrated food at the outset often consists of a mixture of decorticated cotton-cake and linseed - cake, or these two and bruised oats, peas, beans, or perhaps Indian corn. Some lessen the proportion of cotton - cake and increase the quantities of linseed-cake and meal as the finishing-time approaches, the maximum allowance of concentrated food being given for a period of about six weeks at the end.

Where a still smaller quanity of turnips is allowed, perhaps 50 lb. or under per day, it is usual to give the roots either in two pulped mixtures, one in the morning and the other in the afternoon, or by themselves in one feed early in the forenoon. In either case, with this small allowance of roots, the quantities of the more concentrated foods must be increased. The necessary bulk will be made up by straw or hay; the essential nutriment mainly in cake or bruised grain.

The Pulping System for Feeding. —When the minimum quantity of turnips is allowed, the pulping system will be found specially serviceable. As already shown, it permits of greater economy of roots than can be secured by any other method. Comparatively speaking, it is perhaps more useful in rearing store stock than in fattening. The laying on of flesh and fat cannot be accomplished without the employment of a certain amount of rich food, which, of course, is as costly in a pulped mixture as by itself. But the pulping method turns the small allowance of roots to better account with fattening as well as with store cattle, and it is easy to add the required cake or grain. With mixed foods used as in the pulping system, it is easier to ensure that the ration shall be properly balanced, with all the essential constituents present in due proportion, than when turnips, cake, grain, and fodder are each given separately. It is possible, also, by careful preparation, and perhaps by a sprinkling of a little condiment or dissolved treacle, to present the pulped mixture in an exceptionally palatable and inviting condition. In the fattening of stock both these points are of much importance.

Cattle-feeding in Aberdeenshire.— The fame of Aberdeenshire beef is worldwide. In the attainment of this the people, the land, and the cattle have each played a creditable part. To reverse the order, the stock of cattle are of the very best class of beef-producing animals, chiefly crosses between the native Black Polls and the Shorthorn breed. Then the land is peculiarly adapted for the raising of turnips of the highest feeding value. It is well known that there are turnips *and turnips*, some considerably richer than others in feeding properties. The roots grown on the well-farmed granite soils of Aberdeenshire are of exceptionally rich quality. And as to the people, the knack of how to make a bullock hard-fat would seem somehow to have become the special birthright of the Aberdeenshire farmer.

Mr M'Combie's System of Feeding.—Aberdeenshire owes not a little of its reputation for cattle-feeding to the late Mr William M'Combie of Tillyfour (1805-1880), who was far in advance of his time as a feeder of cattle. His little volume, *Cattle and Cattle-breeders*,[1] is full of useful hints to breeders and feeders of cattle. He says :—

"The practice of tying up cattle early in Aberdeenshire is now almost universal; the success of the feeder depends upon it, for a few weeks may make a difference of several pounds. I sow annually from 12 to 16 acres of tares, and about the middle of June save a portion of the new grass full of red clover, and from the 1st to the 20th of August both tares and clover are fit for the cattle. I have for many years fed from 300 to 400 cattle; and if I was not to take them up in time, I could pay no rent at all. A week's house-feeding in August, September, and October, is as good as three weeks in the dead of winter. I begin to put the cattle into the yards from the 1st to the middle of August, drafting first the largest cattle intended for the great Christmas market. This drafting gives a great relief to the grass-parks, and leaves abundance to the cattle in the fields. During the months of August, September, and October, cattle do best in the yards, the byres being too hot; but when the cold weather sets in, there is no way, where many cattle are kept, in which they will do so well as at the stall.

Tares and Clover for Fattening Cattle.—"I never give feeding cattle unripe tares; they must be three-parts ripe before being cut. I mix the tares when they are sown with a third of white peas and a third of oats. When three-parts ripe, especially the white peas, they are very good feeding. Fresh clover, given along with tares, peas, &c., forms a capital mixture. I sow a proportion of yellow Aberdeen turnips early, to succeed the tares and clover. It is indispensable for the improvement of the cattle that they receive their turnips clean, dry, and fresh.

Allowance of Cake, Corn, &c.— "I change the feeding cattle from tares and clover on to Aberdeen yellow turnips, and afterwards to swedes, if possible by the middle of October. I do not like soft turnips for feeding cattle. The cattle that I intend for

[1] William Blackwood & Sons, Edinburgh and London.

the great Christmas market have at first from 2 to 4 lb. of cake a-day by the 1st of November. In a week or two I increase the cake to at least 4 lb. a-day, and give a feed of bruised oats or barley, which I continue up to the 12th or 14th of December, when they leave for the Christmas market. The cake is apportioned to the condition of the different animals, and some of the leanest cattle get the double of others which are riper."

Cattle-feeding in Easter Ross.— The district of Easter Ross has long been famous for the large number of "prime beeves" it sends to the London Christmas market. The system of feeding pursued is very liberal and carefully thought out. The majority of the cattle there fattened for the London market are put up for finishing at the end of the grazing season, when they are approaching three years old. They are well-grown cattle of first-class quality, mostly crosses between the Shorthorn and Aberdeen-Angus breeds. They are well grazed, and are in good condition when housed for hard feeding.

Mr John Gordon, Balmuchy, Fearn, one of the largest feeders in Easter Ross, states that when his feeding cattle are housed he starts them with 2 lb. decorticated cotton-cake and 2 lb. linseed-cake, gradually increasing to 3 lb. each, and then by degrees withdrawing 1 lb. of the cotton-cake and substituting a like quantity of linseed-cake. About six weeks before the animals are sent away to the London Christmas market, they get in addition to the cake 2 lb. each of bruised oats or finely ground peas or beans, very slightly moistened with water. Half the daily allowance of cake is given at 6 A.M., and a feed of cut turnips follows at 9 A.M. While the animals are eating their turnips the byres are cleaned out and the cattle groomed, and as soon as the turnips are eaten, a moderate supply of sweet oat-straw or hay is given. The cattle are then allowed perfect rest till 1 P.M., and in the afternoon they receive cake, roots, and straw or hay as in the forenoon, with a "bite" of oat-straw or hay at 8 P.M. Mr Gordon considers it of great importance to have the feeding, grooming, and cleaning done with the regularity of clock-work, and remarks that a cattleman will never be a successful feeder unless he knows how to give a beast as much as it can eat and yet not a "pick" more. He must also watch the bowels of the animals carefully, as if an animal is purging or costive it cannot be doing well.

Anthrax and Imported Food. — In later years, with the object of lessening the risk of anthrax being conveyed to stock by the use of foreign foods, Mr Gordon has confined his choice of feeding materials to home-made linseed-cake and home-grown grain, the mixture being made up of equal portions of the following — viz., pure Aberdeen made linseed-cake, bean meal, dried distillery grains, ground wheat, ground barley, and ground oats.

A Popular Scotch "Blend." — The following mixture of foods is largely used in the Lothians and other parts of Scotland both for sheep and cattle — viz., Decorticated cotton-cake, linseed-cake, bran, maize, ground locust-beans, and peas in equal proportions, and all mixed together. When oats are cheap and maize dear, the former may take the place of the latter.

Cattle-feeding in England.

In many cases English methods of cattle-feeding differ considerably from the prevailing practice in Scotland. The warmer climate and longer period of growth provide the farmer in the south of England with greater variety of winter food than can be grown to advantage upon average Scottish farms. Comparatively fewer turnips are grown in England than in Scotland, and, as a rule, southern farmers place less reliance than northern farmers upon turnips as food for cattle. Mangels are largely grown in England, and in spring they are given freely to cattle being fattened.

In the south, cattle may, of course, in average seasons remain longer out on the pasture-fields in autumn than in the colder regions north of the Tweed, but in too many cases English farmers sustain losses by being too long in housing their feeding cattle towards the end of the grazing season. Feeding cattle should be housed overnight as soon as the chilly evenings set in; though they may have a run out daily for some time after.

Roots and Green Food for Feed- ing Cattle. — As indicated, a greater quantity of green food, other than roots, is grown in England than in Scotland for cattle. This is extensively used in autumn and early winter before the turnips or mangels are available. Many of the best feeders in England feed extensively upon grass-land during summer, giving large quantities of cake and meal on the fields. Any of the cattle not quite fattened on the fields are housed at the end of the grazing season, and finished upon hay, hay-chaff, a small allowance of roots, and about 6 lb. of cake, with about 2 to 4 lb. of meal per head per day.

Hereford Examples. — Farmers in the county of Hereford have been exceptionally successful in the feeding of young Hereford steers, which they turn out in admirable condition for slaughter at from 18 to 20 months old. The animals are fed liberally from their birth onwards, and in the autumn of their second year the steers get on the grass an allowance, beginning with 4 lb. daily, of cotton-cake and ground corn, wheat, barley, or oats. About the end of September they are housed, and receive the best quality of hay and pulped roots, and from 8 to 9 lb. per day of linseed-cake, cotton-cake, and bruised corn. By Christmas they are in prime condition for slaughter, and their average dead-weight would then, at from 18 to 20 months old, be about 640 lb. — i.e., 8 score per quarter. The cake and corn is given in two feeds, the first thing in the morning and about 4 P.M.

Norfolk Systems. — In Norfolk, with the four-course system of cropping, there is little scope for grazing, but an abundance of turnips and straw. Here, therefore, roots are extensively employed in the feeding of cattle. And, as in Aberdeenshire, the turnips grown in Norfolk are credited with exceptionally high feeding qualities. Cattle, for most part animals rising two years old, are purchased in autumn, and fattened during winter in courts and yards, upon turnips, straw, hay, cake, and grain. Some farmers expend up to £5 for artificial food for each animal, this artificial food consisting chiefly of cake, with varying quantities of home-grown corn, lentils, and maize, all ground and mixed.

A Berkshire System.—Mr Chas. H. Eady, who manages the extensive home-farm of Lady Wantage at Lockinge, in Berks, says that the usual system of cattle-fattening in the stalls for ordinary market is as follows : The men begin their duties at 5.30 A.M., giving each animal—

 ½ bushel chaff (hay and straw).
 3 lb. linseed-cake.
 2 lb. barley-meal.

About 7 o'clock each animal gets about 8 lb. hay, and at midday they get ½ bushel roots (swede or mangel). The afternoon feed, commencing at 3.30 P.M., is—

 ½ bushel chaff.
 3 lb. linseed-cake.
 2 lb. barley-meal.
 8 lb. hay.

Water is always before the animals.

Frequent Feeding.—In Mr R. W. Hudson's feeding-courts at Danesfield, Great Marlow, where hundreds of prime Devons are fattened annually, the principle followed is to feed little and often, the belief being that by this method better flesh is obtained and the beasts ripen quicker than by the old method of giving four meals with a munching of hay always at hand. Here is the time-table at the home farm at Danesfield as supplied by Mr Colin Campbell, the agent :—

 6.30 A.M. Cake.
 7 ,, Roots and chaff.
 8 ,, Hay.
 10 ,, Roots.
 12 ,, Meals.
 2 P.M. Cake.
 4 ,, Roots and chaff.
 6 ,, Hay.

The quantities vary according to the cattle being fed. All mangers are cleaned out before each meal, and every beast has water laid on before it. "Under this system," says Mr Campbell, "our bullocks put on from 1½ to 2½ lb. per day.

Mr M'Calmont's System.—At Mr M'Calmont's home farm at Crockfords, near Newmarket, a number of fine Galloway crosses are matured annually. Mr Fred C. Paine, the farm manager, states that he always feeds the roots by themselves. They mix overnight linseed-cake, bean-meal, lentil-meal, and a little cotton-cake, with sainfoin and straw-chaff in equal proportions, together with a popular sugar meal, the quantity allowed being about 3 lb. of the last-named per head daily, and 10 lb. of the mixture. This is given to the cattle early in the morning, and while eating this they are freshly littered up. About 9.30 A.M. they are allowed ¾ bushel of swedes per head (from November to the middle of February and thereafter, mangels). In the afternoon the feeding cattle receive a similar quantity of roots, the rule being to let them have as many as they will eat. The concentrated food ration is increased, as the cattle get on, to 14 lb. each, say 7 lb. linseed-cake and 7 lb. bean-meal, and 4 lb. sugar-meal extra, mixed with chaff.

Potatoes for Cattle.—In some parts of England potatoes are made use of in feeding cattle in most years, although the practice does not find universal favour. It is usual to begin with only a few pounds of potatoes, and increase gradually. They do not require to be steamed, and so long as there is no dirt adhering to them there is little danger either of choking or colic. One very successful feeder regularly turns out 50 ripe bullocks about Christmas time which are finished by the aid of the potato-crop. They will consume up to 56 lb. a-day of potatoes, but that quantity, it need hardly be said, is not recommended unless given by very skilled hands. Tubers that are slightly tainted with disease may be fed in this way. The rest of the ration consists of the usual allowances of cake, grain, and fodder.

Oatmeal Balls.—To finish a bullock well and give it that firm touch which butchers value so highly, one very successful English feeder and exhibitor pins his faith to oatmeal balls. The oatmeal is damped with water and the balls are rolled in the hand and placed before the bullock. All that is necessary is to sufficiently wet the meal to enable it to stick together.

Feeding without Roots.—For feeding cattle without roots the following plan is recommended by an experienced feeder : "One pailful of cut hay or straw three times a-day, mixed with bean-meal, Indian corn meal, linseed-cake meal, and cotton-cake meal in equal proportions. Four to ten lb. of the meals to each beast according to size, &c. Mix the whole day's feed, chop and meal together, in a large box. Then take 1 lb. of treacle for each animal and dissolve in sufficient boiling-water; after which pour the sweetened liquor over the mixture of chop and meals in the box, and turn the whole over to let it mix thoroughly. Next cover up the feed in the box and let it stand twenty-four hours. Give a pailful three times a-day with a little salt. If the cattle have to be pushed very fast, they may get each 2 lb. daily of cotton- and linseed-cake mixed, in addition to the above feed."

Winter Feeding on Fields. — Although the system must necessarily involve a heavier consumption of feeding material to maintain the animal heat, some English farmers nevertheless derive satisfactory results by fattening cattle in dry well-sheltered fields during winter. Mr Richard Stratton, The Duffryn, Newport, Monmouth, one of the most experienced cattle-feeders in the country, says : "I give feeding cattle caké and meal on grass up to 14 lb. per head per day in winter, when they do well on dry pasture, with shelter under banks and hedges. I prefer feeding in this way to either tying up or in open yards. Straw is scarce and dear here, and the system saves litter, and prevents all waste of manure. I begin in October with about 6 lb. of cake and meal, and finish off with 12 or 14 lb. in December or January, given at 7 A.M. and 5 P.M.; the animals going away fat when from 2 years and 6 months to 2 years and 9 months old. But my practice in feeding varies according to the prices of the different commodities. Sometimes I use cake, sometimes corn ; also hay or straw, according to the market prices of these. Again, as to roots, if scarce and dear, I sell them and use artificial foods; if plentiful and cheap, I consume them." Mr Stratton's farm, it should be mentioned, is in a warm locality and well sheltered.

Cattle Feeding in Ireland.

Irish farmers devote their attention to the rearing and selling of store cattle rather than to fattening. The mild open climate of their country favours this system, which is found to be more profitable and better adapted for men with limited means than finishing the cattle for the butcher. There are, however, a good many Irish farmers who fatten cattle, and most of them do it successfully.

Feeding on Pastures.

The extent to which cattle are fattened on pastures has been slowly but steadily increasing. It is now carried on to a large extent both in England and Scotland, and to a much smaller extent in Ireland. In Hereford, in particular, farmers make a special feature of the grass feeding of their famous beef-producing cattle.

Concentrated Food on Pastures. — Only in few cases, where the pasture is exceptionally rich in quality, are cattle fattened on the grass without extra food. The mixtures and quantities of extra food given to feeding cattle on pastures vary greatly according to the size, age, and condition of the cattle, the character of the pasture, the prices of the feeding-stuffs, the supply of home-grown food, and the time available for the fattening process. Cotton-cake and linseed-cake are used to a large extent, along with ground oats or barley and sometimes wheat, or some of the other foods in the market. The extra food is given in boxes on the fields twice or thrice daily. The quantities range from about 4 to 10 lb. per day.

Rock-salt and water are always within reach of the animals. Feeding cattle are put on to fresh pastures at intervals of a few weeks, the more frequently the better both for the animals and the pasture.

"Soiling."

The system of "soiling" might be humorously described as grazing cattle in the house! It consists of retaining the animals in the house, — the byre, hammel, or cattle-court, — and cutting and carting the green food to them, instead of allowing the animals to browse over the pastures and pick up the grasses for themselves.

Advantages of "Soiling." — Several advantages are claimed for this system over the older and more simple and natural method of grazing. The chief of these are—(1) that a given extent of land will carry a heavier stocking of cattle; (2) that more actual food will be produced during the season; (3) that the quantity of food grown is more fully utilised; (4) that the animals thrive better, because they are protected from extremes of temperature, from the attentions of insects, and from undue exercise; and (5) that a greater quantity of manure is made upon the farm.

More Food Better Used.—It is unquestionable that by the frequent and systematic cutting of the grasses as they grow up, a greater weight of food will be grown during the season than when the pasture is cropped irregularly by stock in the ordinary method of grazing. Then with careful cutting and carting, every particle of the food is placed before the stock in a palatable condition, so that the material grown is more fully utilised than when it is trodden upon and unevenly eaten by cattle.

Animals Thriving Better. — Provided the animals are kept in comfortable, well-ventilated compartments, with plenty of fresh air, they will most likely give a better return for the food, in yield of milk or in accumulation of fat, than they would on the pastures exposed to sun and wind and to the torturing of insects. That young animals would develop bone and muscle more rapidly is very doubtful; but it has been abundantly proved that adult animals will accumulate fat more quickly in this confinement than upon pasture fields.

Disadvantages of "Soiling."—"Soiling" is altogether a more artificial system than ordinary grazing. It necessitates the employment of more money per acre, not only in a larger head of stock, but also in providing the necessary house accommodation, and the considerably larger force of labour. The heavy labour bill is indeed the greatest disadvantage of the system as opposed to grazing.

Then, again, there is this further consideration, that substantial outlay may be incurred in providing food to the animals in the house before the grass is sufficiently grown to admit of being cut. Successional forage crops are grown for this purpose, as well as to supplement the grass at other times. All this involves additional outlay, employing more capital per acre.

Utility of the System.—Still there are many circumstances under which the system may — especially with fattening cattle and dairy cows—be pursued with excellent results. It is specially suitable for warm climates, where forage crops may be easily grown, and where cattle would be disturbed by the excessive heat in the open fields. Then, where the supply of water for fields is insufficient, house-feeding may be followed in preference to grazing.

It is not likely, however, that in the best grazing districts, or in the colder parts, "soiling" will ever displace the long-established system of summering stock on the open fields. Indeed, it has to be noted that with all the advantages claimed for it the system of "soiling" cattle is not gaining ground in this country.

Review of Feeding Experiments.

In the *Transactions of the Highland and Agricultural Society of Scotland* for 1909 there appears an exhaustive review of the results of over two hundred experiments in the feeding of cattle conducted in this country in the seventy-six years between 1832 and 1909. The review, which was compiled for the Society by Mr Herbert Ingle, B.Sc., F.I.C., from reports appearing in various publications, is unique in its scope and character. The results are given in tabular form, showing amongst other details—

(1) The average daily ration.

(2) The rate in increase in live-weight.

(3) The quantity of digestible matter in the ration.

(4) The starch equivalent of the digestible fat, carbohydrates, amides, and fibre in the ration.

(5) The albuminoid ratio of the ration.

(6) The amount of digestible matter consumed per 1 lb. of increase in live-weight.

A striking feature in the review is the fact that the results of such a large number of experiments expressed in such definite terms as are here adopted should be found to be so fully in accord with orthodox expectations, with what modern experience and scientific teaching would lead one to look for. Generally speaking, a survey of this exhaustive review supports the conclusion that in the feeding of their cattle the most up-to-date farmers of the present day are pursuing lines that are sound and economic both in a scientific and practical sense.

Increase in Live-weight. — The daily increase in live-weight per head averaged 1.803 lb. for 199 lots. With these 199 lots arranged in order of daily gain in steps of a quarter of a pound, the following table shows the distribution :—

Average daily gain per head.	Number of Lots.
Between 0.25 and 0.5 lb.	2
,, 0.50 ,, 0.75 ,,	5
,, 0.75 ,, 1.0 ,,	5
,, 1.0 ,, 1.25 ,,	4
,, 1.25 ,, 1.50 ,,	29
,, 1.50 ,, 1.75 ,,	39
,, 1.75 ,, 2.0 ,,	56
,, 2.0 ,, 2.25 ,,	27
,, 2.25 ,, 2.50 ,,	17
,, 2.50 ,, 2.75 ,,	8
,, 2.75 ,, 3.0 ,,	6
,, 3.0 ,, 3.25 ,,	1
	199

Digestible Matter per lb. of Increase. — The amount of digestible matter consumed for each pound of live-weight increase obtained is no doubt one of the most important measures of a system of feeding. In 199 trials the total average weight of digestible matter consumed per day per 1000 lb. of live-weight was 13.92 lb., the highest being 22.7 lb. and the lowest 7.4 lb. The weight of digestible matter consumed for each 1 lb. of live-weight increase averaged 9.00 lb. for the 199 lots. In the majority of the cases the amount was between 9 and 10 lb., more than 78 per cent being between 6 and 11 lb., and more than 50 per cent between 7 and 10 lb.

Digestible Albuminoids. — The amount of digestible albuminoids supplied per day for 1000 lb. live-weight averaged 1.675 lb. for 199 lots, the lowest being 0.11 lb., and the highest 3.68 lb. The following table shows (1) the amount of digestible albuminoids consumed per day per 1000 lb. weight by the lots in various grades, (2) the average gain per day in live-weight, and (3) the amount of digestible matter consumed for 1 lb. of live-weight increase :—

				Average gain per day in lb.	Digest. matter for 1 lb. increase.
Receiving less than		0.25 lb. per day	3 lots	0.97	13.53
,,	between 0.25 and 0.5	,,	6 ,,	1.70	7.95
,,	,, 0.50 and 0.75	,,	9 ,,	1.69	8.92
,,	,, 0.75 and 1.0	,,	12 ,,	1.95	8.62
,,	,, 1.0 and 1.25	,,	20 ,,	1.80	8.72
,,	,, 1.25 and 1.50	,,	25 ,,	1.82	8.65
,,	,, 1.50 and 1.75	,,	28 ,,	1.72	9.60
,,	,, 1.75 and 2.0	,,	39 ,,	1.85	9.36
,,	,, 2.0 and 2.25	,,	19 ,,	1.86	8.29
,,	,, 2.25 and 2.50	,,	18 ,,	1.85	8.27
,,	,, 2.50 and 2.75	,,	12 ,,	1.93	8.38
,,	,, 2.75 and 3.0	,,	4 ,,	2.00	8.93
,,	,, 3.0 and 3.25	,,	3 ,,	1.57	8.90
,,	above 3.25	,,	1 lot	1.19	11.30
			199		

It will be observed that the amount of digestible albuminoids given per day was between 1.0 and 2.0 lb. in 55 per cent of the trials.

Albuminoid Ratio. — As would be expected, a wide range is exhibited in the albuminoid ratios of the rations. The average for the whole of 199 lots was 1 : 7.65, the widest being 1 : 69.5, and the narrowest 1 : 2.7.

Influence of Age on Feeding. — The particular ages at which cattle can be fattened most profitably is a matter of much interest. Some light is thrown upon this point by the following table showing the quantity of albuminoids *plus* the starch equivalent of the other food constituents consumed per 1 lb. of live-weight increase in 142 lots :—

6 months old	.	.	.	8 lots consuming 8.17 lb. for 1 lb. increase.
12 ,, ,,	.	.	14 ,,	8.15 ,, ,,
18 ,, ,,	.	.	11 ,,	8.74 ,, ,,
2 years ,,	.	.	51 ,,	9.99 ,, ,,
2½ ,, ,,	.	.	18 ,,	8.97 ,, ,,
3 ,, ,,	.	.	38 ,,	8.55 ,, ,,
4 ,, ,,	.	.	2 ,,	9.80 ,, ,,

142

As would be expected, the younger animals made increase with less consumption of food, but the three-year-old cattle appear to have utilised their food better than those two years old.

Leguminous Fodders. — The particulars relating to the merits of the different classes of fodders indicate that a leguminous fodder such as clover-hay is exceptionally valuable in the fattening of cattle. In 17 cases where clover-hay formed a constituent of the daily ration, the average amount of digestible matter consumed per 1 lb. of increase of live-weight was only 7.47 lb., and if two of these cases be excluded, the value becomes only 7.01 lb., while the average daily gain per head of the animals is 2.13 lb. These figures compare very favourably with the means of the whole, which are 9 lb. and 1.803 lb. respectively. This very significant result is worthy of note by cattle-feeders. The good effects of clover-hay as a constituent of a feeding ration is doubtless partly due to its high content of albuminoids, but, in the opinion of Mr Ingle, is probably also connected with the nature and amount of its ash constituents.

PREPARING CATTLE FOR SHOWS.

The following notes on the selection and preparing of cattle for showing are from the pen of Mr Robert Bruce, himself for many years an exceptionally successful breeder and exhibitor of cattle of different varieties :—

Before referring to the preparation of animals for exhibition at the principal shows, it may be well to speak of the selection of the subjects, upon which much trouble and expense have to be expended before the owner can expect to put creditable exhibits in the judging ring.

It has to be realised that in these days the competition is keener than at any former time in the history of our show system, and that there is a yearly increasing number of thoroughly capable men in charge of showyard stock, who turn out the animals under their care in the "pink of condition."

The great demand for high-class specimens of all our different breeds of cattle, which has been experienced for a considerable length of time, and the remunerative prices obtained for winning animals, have led to an increasing number of owners of pedigree herds and showyard exhibitors.

Improvement in Show Stock. — Those who can look back upon the cattle that appeared at our National shows prior to about 1870 cannot fail to realise the great changes that have taken place in the preparation and management of showyard animals, and that the average merit of showyard specimens is much higher to-day than at any former period. The very fact that such is the case has led casual critics to assert that we do not now see in our showyards the same splendid specimens which we did in former times. Opinions by the majority of such critics are formed upon comparison, and there can be no doubt but that there is now a much narrower margin of excellence between the different exhibits in a class than was the case on former times, and in consequence the winners certainly do not appear to be the same outstanding specimens they were when the average quality of the showyard animals was of a much lower standard.

Selecting Show Stock.—Bearing in mind the competition that has now to be faced, much care ought to be devoted to the selection of the animals intended for the showyard. At the early age at which the selection has to be made, there must be in the whole matter a considerable amount of chance, and even those who may be considered experts prefer to put several of their youngsters upon the probation list before making the final selection.

Good Breeding Essential.—In making a selection, the all-important matter of breeding must receive attention and only well-bred specimens put on such a list. In using the term "well-bred" there is no intention to confine the meaning of the expression to any particular strain of blood that for the time being may be fashionable with owners of the particular breed to which the animal or animals may belong. An animal to be "well-bred" must be the produce of two good parents, and in forming an estimate of the qualifications of the parents, form, constitution, and temperament must be carefully considered.

Form and Constitution.—Without the principal points which go to make up the true form, as recognised in the different breeds, it must only be a waste of food to attempt showyard preparation; and as the life to which a subject is subjected while under training for the showyard is by no means a natural one, strength of constitution is essential.

Temperament.—Granted that form is apparent, and strength of constitution may be reckoned upon through having been inherited from the parents, the possession of a docile temperament is of much importance. No doubt the last-named qualification may be acquired, and much depends upon the man in charge if it has to be developed, but every practical breeder knows that a quiet disposition is hereditary, and is a most important factor in the selection of animals with a view to training for showyard contests.

If, therefore, three or four of the best-bred and most promising-looking calves of a breeder's lot are selected under the above conditions, we may at once proceed to consider the next step towards preparation for showyard honours.

Proper Age.—Seeing that at most of the important breeding shows the ages of animals date from the 1st January, —in some cases from 1st December—it will be at once evident that it is important the selected calves should have been born as early in the season as possible, so that when they come to be shown they may not be handicapped in the matter of age.

Calf-rearing.—The general practice pursued by the majority of breeders in the management of their pure-bred calves is to allow them to run with and suck their dams until they are eight or nine months old, having taught them in the meantime to eat cake or other concentrated foods before weaning them from the milk. In these days of keen competition a more artificial system of calf-rearing is generally pursued by successful exhibitors.

The system of calf-rearing pursued by many is to allow the calf to suck its dam for a week or ten days, when it is taken off and fed from the bucket or pail, and when it has once learned to drink it is supplied with milk long after the usual weaning age. The importance of continuing the use of milk after the animal has passed the calf stage is well understood and largely practised by showyard exhibitors. The success of such a system depends much upon the care with which the quantity given is regulated, more especially during the first two months after birth. It must be recognised that drinking the milk is unnatural, and every care must be exercised to avoid overloading the stomach and upsetting the digestive organs.

Such a system entails considerable trouble and labour, which to a great extent may be avoided by allowing the calves to suck, and in the earlier stages of their lives accustoming them to take to any nurse by frequent changes from one to the other, so that they are ready to take to any cow, and continue to suck so long as it may be considered necessary they should have milk.

A few weeks after birth calves will begin to nibble at food, and ought to receive tit-bits in the form of a handful of sweet meadow-hay, and after a time small quantities of pulped or finely cut roots or cabbages, with a little meal or finely ground linseed-cake.

Mixed Feeding-Cakes.—This brings us to observe that there are in these days a large number of cakes on the market, many of which are prepared in a way to relieve owners of stock of much of the trouble of mixing and regulating the quantities of meals and cakes each animal ought to receive. The time was when feeders had only linseed and other seed-cakes, and had to supplement them with bran and meals given either dry, damp, or scalded. The prepared cakes referred to are sold as corn-cakes, feeding-cakes or composite cakes, and if purchased with a satisfactorily guaranteed analysis and relatively cheap, their use will be found to be labour-saving and economical in comparison with meals. As a rule, the cakes referred to have some sweetening substance in them, such as locust-beans, and are readily eaten by young animals.

Housing Calves. — The calves, whether drinking or sucking the milk, ought to be kept during the spring months in properly ventilated, well-sheltered boxes, facing south if possible, with an opportunity to spend the greater portion of the day in the open air in fine weather.

Salt and Chalk for Calves.—It is a good plan to have a lump of rock-salt and one of chalk placed in the boxes for the calves to lick when they feel inclined to do so. The benefit of salt is quite generally understood, and if chalk is also available it will be seen that it is applied to counteract acidity in the stomach, which may occur now and again, especially when milk is drunk from the pail.

Quantities of Food and Peculiarities of Animals.—No hard and fast rules can possibly be laid down regarding the amounts of the different kinds of food which ought to be given to animals at any stage of their showyard preparation, and especially during their calfhood. Nor would it be well to prescribe a ration, seeing that the most important element in the matter is the extreme necessity of attention to the individuality of the animals in training. The most successful cattle-feeders are those who thoroughly realise that every animal under their care is possessed of a distinct individuality.

Some consume much more of certain kinds of food than others, and all have their fancies, which must be attended to before a full measure of success in the attainment of early development can be expected.

Value of Showyard Honours.—In connection with this matter it is well to realise that the expense connected with showyard preparation must exceed the immediate return of profit as calculated by increase of weight for value of food consumed. Showyard honours are looked upon as advertisements for herds, and, as a rule, are indirectly profitable, although there may be a debit balance standing against the winning animals.

During the first summer the best and most promising of the youngsters on the probationary list should be kept in their boxes, while the others may join the herd and be treated in the ordinary way, being stronger and more forward than those they now join, owing to the few months' showyard preparation they have received.

During the summer months the nurse cows should be brought into the yard morning and evening to suckle the calves that are not fostered on the pail, and month by month the youngsters should get a gradually increasing quantity of linseed and other cakes, with pulped or finely cut mangolds and what fresh-cut grass they will eat. At no time at this or any period of their showyard preparation should more of any kind of food be given than will be at once cleaned up, and every capable cattleman, by careful attention to the state of the bowels, will very soon gauge the amount of the different foods that can be profitably assimilated by each animal under his care.

Exercising and Handling.—It need hardly be said that early in life the calves must be taught to lead in the hand, and during the summer months beyond the exercise they get in the yard they ought to be led out a distance of not less than half a mile each second day at least. Indeed some very successful trainers have their animals led out as regularly every week-day as they are fed. Many prizes are lost in the show-rings through the inability of the animals to walk out with that freedom of action judges look for in high-class specimens of showyard cattle.

Attention to Animals' Feet.—In connection with this subject no amount of exercise can be of any use unless the animals' feet are attended to, and without special appliances for either slinging or throwing the animals, it is a most difficult matter to turn up the feet so that the soles can be properly dressed. If, however, while animals are quite young, they are accustomed to have their legs lifted and their feet attended to, there need be no difficulty in keeping their feet right at any period of their lives.

Grooming.—During the whole time of preparation the animals' skins ought to be kept clean and free from vermin, so that the growth of the hair may be encouraged and a healthy tone preserved. An occasional washing with one or other of the non-poisonous sheep-dips, to be followed in a few days with a thorough washing with soap and water and a cold douche, will destroy the vermin and tend to maintain the skin and hair in a healthy condition.

Increasing Food with Advancing Age.—With increasing age and greater appetite care must be exercised that the increased diet is composed of the best quality of such foods as the animals eat with relish, the quantities of each being based upon evidence deduced by observing closely the effect of any increase or change of diet on the health and digestion of the animals.

Use of Condiments. — Many ani-mals intended for breeding purposes make their appearance in the judging rings at an early age, the large proportion of the males being exhibited at shows and sales from twelve to fifteen months old. Such being the case, early development is essential if prizes or good prices can be looked for, and however much some breeders may believe in spices and other condiments, not a few of the most successful trainers who have led many winners in the keenest of competitions have had no occasion to use such expensive materials. No doubt these condiments may be useful in the case of what are termed "shy feeders," but such animals are, as a rule, disappointing thrivers even after being pampered, and in practice it will be found that close attention to the requirements and tastes of individual animals must be looked upon as being of infinitely more importance than any dependence upon condiments.

Importance of Practical Experience.—Cattle-feeders of the present day enjoy privileges that were quite unknown in days gone by, when the values and effects of the different kinds of foods had to be found out by feeders themselves. In later times scientists have done much for the feeder, yet no amount of scientific knowledge can avail unless it be combined with a knowledge obtained by practical experience.

In much that is written upon the subject, the fact often seems to be overlooked that cattle-feeders have to do with living subjects having their own individual peculiarities, so that however scientifically a ration may be prepared, it may completely fail in producing the desired effect.

FLOCK MANAGEMENT.

The flocks of the United Kingdom form a substantial asset in its agriculture. Their numbers are large, and the enterprising and skilful manner in which they are managed reflects credit upon their owners. In the breeding, rearing, and feeding of sheep in this country there have been just as marked advances in recent times as in the management of our herds of cattle.

In another part of this volume (pp. 138-206) the many valuable breeds of sheep kept in the United Kingdom are fully described, and so also are the methods of management pursued in pure-bred flocks. In view of the fulness of that information, the details to be given here regarding the rearing and feeding of ordinary sheep stocks need not be extended to great length.

LAMBING SEASON.

With the owners of breeding flocks the lambing season is a busy and anxious time. The results of the year's operations depend largely upon how the flock fares at this season. It is therefore of the utmost importance that the most careful attention should be given to the treatment of the ewes and their offspring in the tender days of the latter. These matters, as already indicated, are dealt with so fully in the section relating to flocks of the pure breeds that little need be added here. In their main features, the systems of management suitable for the lambing season in pure-bred flocks are equally well adapted for that period in ordinary mixed-bred flocks. The attentions of the shepherd should be just as thorough and careful in the one case as in the other.

Lambing - Pens. — On many farms there are elaborate and costly lambing sheds and pens built of stone and lime. On others the lambing-pens are merely temporary erections, formed, perhaps, of hurdles and straw; while in many cases no lambing-pens of any kind are provided. Costly erections are not necessary, but lambing-sheds or lambing-pens of one kind or other should be provided upon all farms carrying breeding-sheep, and for all kinds of sheep, whether the hardy mountain breeds or the more tender southern varieties. Little roofed space may suffice, but there should be a dry bed and shelter from the prevailing winds to make it unnecessary to put any of the ewes and lambs under roof, yet the means of doing so should exist. The sudden occurrence of a storm without proper shelter being at hand for ewes with very young or tender lambs might result in serious losses.

Hardiness of Hill Sheep.—Hill sheep are not as a rule brought into lambing-pens as is done with lowland breeds. They produce their young on the hillsides, and in average seasons the death-rate amongst hill lambs is wonderfully small. The vitality of these creatures when newly dropped is quite marvellous. Still, it is desirable that, even for the hardy hill sheep, some provision should be made whereby the more weakly lambs may have shelter in excessively wet cold weather. For this purpose, it will be found useful to have some artificial shelter provided at suitable points throughout the farms. Little huts constructed perhaps of turf, hurdles, and bundles of straw or rushes, will entail little outlay or trouble in formation, and during inclement weather will be found of great benefit to the ewes and lambs. Ewes with weakly lambs can be accommodated comfortably in these scattered huts for a few days and nights, the shepherd carrying or having conveyed to them some hay, corn, and roots.

Lambing Hospital.—A few pens in a corner of the lambing-fold by themselves should always be set apart for hospital purposes. They may be formed of hurdles and straw at very little trouble and expense, and would be of great benefit wherever a breeding flock is kept.

Supplementary Shelter. — In addition to the regular lambing-fold it would be well to provide additional shelter in the form of small covered pens or huts at convenient well-sheltered parts of the farm, for weakly ewes and lambs during a storm.

Shepherd's Hut.—It is advisable to have a sleeping-place or shelter for the shepherd beside the lambing - fold. It may be a fixed structure or may rest on wheels and be made of iron or wood.

In many cases shepherds are provided with medicine - chests furnished with a considerable variety of medicines and stimulants, comprising laudanum, linseed-oil, castor-oil, spirits of nitre, Epsom salts, powdered ginger, powdered chalk, tincture of aconite, carbolic acid, Gallipoli ·oil, and whisky or brandy, &c.

Assistance in Lambing.—As a rule, experienced shepherds are very expert and successful in assisting ewes in lambing. Young shepherds do not acquire the skill and deftness required for this service from books, but from practice in association with older men, and it is the duty of all shepherds to equip themselves thoroughly for the work as early as practicable. Before giving assistance to a ewe while lambing, the shepherd should smear his hands as well as the vagina of the ewe with "carbolic oil"—that is, a mixture of 1 part of carbolic acid to 10 parts of pure olive-oil; and a little of this germ-killer should also be smeared on the broken umbilical cord at the navel, especially if the weather is wet and the land slushy.

The exact moment for rendering assistance can be known only by experience. It is necessary to watch and wait, for a hasty parturition often superinduces inflammation, if not of the womb, of the external parts of the ewe.

Inflammation after Lambing.—Unless the utmost care is exercised there is great risk of losing the ewe after a case of hard labour, by "bearing" or "straining"—after-pains—and inflammation. Formerly the rate of mortality from inflammation after lambing was often high, but it has been abundantly proved that by timely treatment the danger may be effectually averted. It has already been pointed out that in all cases the shepherd, before assisting a ewe, should smear his hand in a mixture of carbolic acid and olive or Gallipoli oil—about 1 part of the former to 10 parts of the latter. Then, after the removal of the lamb, about two tablespoonfuls of the carbolic acid and oil should be injected into the womb, while any of the external parts which seem inflamed should be smeared with the same mixture. This treatment should be repeated every three or four hours, as may be found necessary. The strength of the carbolic mixture should be regulated — from 10 to 20 parts of Gallipoli oil to 1 of carbolic acid — according to the symptoms of the case. The handiest instrument for this purpose, and one which has proved itself invaluable in the lambing-fold, is made by fixing a 6- or 7-inch injection-tube suitable for a female into an indiarubber enema-tube bulb. It is portable and convenient, forcing the germ-killing fluid into all the recesses of the inflamed womb.

Where the symptoms of inflammation are serious, a strong mixture should be applied promptly and frequently.

It should be mentioned that the credit of discovering this invaluable preventive belongs to Mr Charles Scott, author of *The Practice of Sheep Farming.*

Assisting Lambs in Feeding.—When lambs do not succeed at once in finding the teat, the shepherd should give assistance, and if the supply of milk should not be sufficient the shepherd may have to partly feed the lamb on cow's milk. For this purpose he should have with him a supply of fresh cow's milk every day.

Cow's Milk for Lambs.—Caution is required in beginning a young lamb upon cow's milk. At the outset it should be given in small allowances and often. It is best when given immediately it is drawn from the cow, but if it has been allowed to cool it may be raised to its natural heat by being placed in a cup upon the kitchen-range for a moment, or by a clean hot iron being inserted in the milk.

Removing Ewes and Lambs.—Ewes are kept on the lambing-ground until they have recovered from the effects of lambing, the lambs have become strong, and the ewes and lambs are well acquainted with each other. The time required for all this depends on the nature of the lambing and the state of the weather. When quite recovered, the ewes, with their lambs, are put into a field of new grass, where the milk will flush upon the ewes, much to the advantage of the lambs.

Mothering Lambs.—When ewes and lambs are turned out to pasture, or out of the lambing-fold, the shepherd ought for the first ten days to see, at least twice a-day, that every lamb is with its own mother, and especially in the case of twins, to see that they are both having regular access to the right ewe. Distinctive marks with paint on ewes and lambs are helpful in this work of *mothering.*

Much trouble is imposed upon shepherds when ewes will not take their own lambs; but this does not often happen. Another duty which requires tactful conduct on the part of the shepherd is the introducing of a strange lamb to a ewe that may have lost her own lamb. But by patience and kindness difficulties are usually got over.

Stimulants for Weak Lambs.— When a lamb has become so prostrate as to necessitate removal from the mother, it should not only be placed upon a woollen cloth near a moderate fire, but have a little stimulant administered as well. Some experienced shepherds recommend from a half to a whole teaspoonful of gin or whisky in a little warm water, sweetened with moist sugar; a very little of its mother's milk—or the milk of another newly-lambed ewe, if its own mother is not alive—should also be given without delay. The ewe should be milked into a small jug or cup, and the milk at once conveyed to the lamb, which may be fed by a teaspoon. If the milk gets cold before being given to the lamb, it should be heated to the normal temperature by the addition of a few drops of hot water, or, better still, by a clean hot piece of iron inserted into it.

Carrying Lambs. — Young lambs should be handled as little as possible. When they have to be carried, this should be done by the two fore-legs. Never seize or carry a lamb by the body.

Cleaning Ewes' Udders.—Any loose wool should always be removed from the udders of ewes at lambing, so as to prevent the lamb from swallowing pieces of wool, and forming hair-balls in the stomach. These balls often prove fatal to lambs. They are sometimes formed by lambs on bare and dirty pasture where pieces of wool are lying about.

The Lambing Period. — It may at first thought seem curious that within the narrow limits of the British Isles there should be such a length of time as there is between the dates of lambing in the earliest and the latest districts. The lambing period in this country actually extends over six months, beginning with Dorset sheep in the extreme south of England in November, and ending with mountain sheep in the north of Scotland in the month of May.

After Lambing.

Lambing in a flock is usually completed in four or five weeks. The after-treatment of the flock varies in accordance with the class of sheep, and the objects in view.

Castration.—The male lambs not to be kept as rams are castrated when from ten days to five weeks old. In some cases, indeed, castration is performed when the lambs are only two or three days old, but the more general plan is to delay from two to four weeks.

In hill stocks castration is not usually performed until the lambs are fully a month old; in other words, the ewes commence to lamb in the third week in April, and the "marking" takes place about the end of May, varying a little according to circumstances and local custom. Some farmers have a decided objection against too early castration, as it tends to give a feminine appearance to the wedders, stunting the growth of horn, and weakening the neck too much.

Great caution is required in castrating lambs. It should not be done in rainy, cold, or frosty weather; nor should the lambs be heated by being driven before the operation. They should be caught and handled gently. One assistant should catch the lambs, and another hold them while the shepherd operates.

There are different methods of castrating. One method is to make two slight incisions, one for each testicle; another, to cut off the point of the scrotum and pull both testicles through this large opening — the testicles in both cases being pulled out by the shepherd's teeth. The amputated wound takes a considerable time to heal, whereas the two simple incisions heal by the first intention. It is argued, however, by those who prefer the latter plan, that there is an advantage in the larger opening, as all discharges are more readily got rid of.

Docking.—Advantage is taken of the opportunity afforded at castration to dock the tail, which in Scotland is left as long as to reach the meeting of the hams. In docking, the division should be made with a large sharp knife in a joint, when the wound will soon heal. The lamb, after being docked, is let down to the ground by the tail, which has the effect of adjusting the parts in connection with the castration. Ewe lambs are also docked at this time, but they are not held up, being merely caught and held by the shepherd between his legs until the amputation is done.

In England, docking is performed at the third joint, which gives a stumpy appearance to the tail. The object of docking is to keep the sheep clean behind from filth and vermin; but as the tail is a protection against cold in winter, it should not be docked so short in Scotland as is done in England. Tup lambs, in order to strengthen the backbone, are allowed to retain their full tails until one year old.

Risks from Castration and Docking.—The scrotum does not bleed in castration, but the tail often bleeds in docking for some time in two minute and forcible streams, though usually the bleeding soon ceases. Should it continue as long as to sicken the lamb, a small cord should be tied firmly round the end of the tail, but not allowed to remain on above twenty-four hours, as the ligatured point would die by stoppage of the circulation of the blood, and slough off. In some instances inflammation ensues, and the scrotum swells, and even suppurates, when the wound should be carefully examined and the matter discharged.

To avoid irritation to the wounded scrotum, the new-cut lambs should for a few days be put on old grass or new grass, where the stubble is specially short.

A Preventive.—Some farmers use a mixture of pure olive-oil and spirit of turpentine for dropping into the scrotum after extracting the testicles, and the results they claim are satisfactory. Perhaps a still better preventive of inflammation would be a few drops of a solution of carbolic acid and oil poured into the scrotum. The knives used in castrating should be dipped into a disinfecting solution now and again, to keep them clean and free from disease germs.

Rig or Chaser.—Sometimes one of the testicles does not descend into the scrotum, when the lamb ultimately becomes what is called a rig or chaser—one which constantly follows and torments the females of the flock, when near him. It is not, as a rule, safe to rely upon such a ram for breeding.

Look to the Pastures.—The state of the new grass-fields occupied by ewes and lambs requires consideration. Ewes bite very close to the ground, and eat constantly as long as the lambs are with them; and as they are put on the new grass in spring, before vegetation is much advanced, they soon render the pasture bare in the most favourable circumstances, and especially so when the weather is unfavourable to vegetation. In cold weather, in spring, bitten grass soon becomes brown. Whenever the pasture is seen to fail, the ewes should be removed to another field. But in removing ewes and lambs from a short to a full bite of grass, caution is required in choosing the proper time for the removal. It should be done in dry weather, and in the afternoon.

Shepherding on Arable Farms.—On low country or arable farms with the softer breeds of sheep, from 200 to 300 ewes are about as many as one shepherd can superintend during the day; and it may be necessary to have an assistant for him in the night, to gather the ewes into shelter at nightfall, and to take a weakly lamb, or all the lambs that have dropped during the night, into sheds erected on purpose, or into sheltered stells, as a protection against bad weather. To ascertain the state of his flock, he should go through them with a lantern at least every two hours, and oftener if necessary.

Shepherding Hill Sheep.—The hardy breeds of hill sheep need less attention, especially during the night. Indeed, the general plan is to leave the flock undisturbed during the dead of the night. The ewes and lambs are turned out to the dry lair over-night, and there the shepherd looks over them carefully, perhaps as late as eleven o'clock, while he or his substitute returns to them as early as 3 or 4 A.M., when daylight is making its appearance.

Ailments among Lambs. — Young lambs, as long as they are dependent on their mother for food, are subject to few diseases. A change to new luxuriant grass in damp weather may bring on the *skit* or diarrhœa, and exposure to cold may produce the same effect. As long as the lamb feeds and plays, there is little danger; but should it appear dull, its eyes watery and heavy, and its joints somewhat stiff, remedial means should immediately be used. In the first place, it is usual to give a gentle aperient, say, half an ounce of Epsom salts, with half a drachm of ginger, and this may be followed by a tablespoonful of sheep's cordial, consisting of equal parts of brandy and sweet spirits of nitre.

Ailments amongst Ewes. — After recovery from lambing, the complaint the ewe is most subject to is inflammation in the udder, or *udder-clap* or *garget.* The shepherd must give careful attention to this, and apply the usual remedies where required. Directions for the treatment of ailments amongst live stock are given at the end of this volume.

Abortion among Ewes.

Ewes in lamb are liable to abortion, or slipping of the lamb, as it is termed, as well as cows, but not to so great an extent, nor does the complaint so often become epidemic in its character. It is known, however, that there is a form of abortion amongst ewes which is caused by a specific germ (see p. 337 of this volume). Various other causes produce abortion amongst ewes, such as severe weather in winter, having to endure much fatigue in snow, leaping ditches, being frightened by dogs, over-driving, feeding on unripe watery turnips, &c.

Unripe Roots and Abortion.—The clearest evidence as to the evil influence of exclusive feeding of in-lamb ewes upon unripe watery roots was obtained by Professor Axe in the season 1882-1883. The turnip crop in that season was unusually abundant, and, owing to the mild winter of 1882-1883, continued to grow, and remained throughout the season in an unripe and exceptionally watery condition. Of the total number of ewes (about 7800) fed exclusively on roots, no fewer than 19 per cent aborted ; while, where the roots were supplemented by frequent changes to grass, the rate of abortion fell to 3 per cent, and to 1¼ per cent where the roots were supplemented by corn and cake, or some other substantial aliment.

In reference to the high-pressure system of forcing the growth of roots by the free application of artificial manures, and the growing practice of sowing roots late and beginning their consumption early, Professor Axe remarks that these are inconsistent with full maturation and ripening of roots, and that on this account " the desirability of a guarded and judicious employment of this de-

scription of food in the management of breeding stock cannot be too forcibly insisted upon."

He also very strongly objects to the " too common system which condemns pregnant ewes to live exclusively on filth-laden shells " behind other sheep, which get the best of the fresh roots.

Foot-rot and Abortion. — It was shown clearly that foot-rot contributed largely to the cases of abortion. In flocks where it prevailed to any extent the rate of abortion was 4½ per cent greater than in those in which there was no foot-rot.

Twins and Abortion.—The cases of abortion were much more numerous with twin than with single lambs. Indeed, for every abortion with a single lamb there were six abortions with twin-lambs —pointing, as Professor Axe says, " to the existence of some debilitating cause unfitting the ewes with twins to meet the greater demands on their nutritive resources, while influencing in a less degree those with singles."

Preventive Measures.—The following preventive measures are recommended by Professor Axe :—

" 1. That from the time ewes are placed on turnips to the time when they lamb down, they should receive a liberal amount of dry food, to be regulated according to the nature of the season and the condition of the roots.

" 2. The quantity of roots should at all times be limited, and besides shells, a fresh break should be given every day after the hoar-frost has disappeared, and in the early spring the tops should be removed.

" 3. Change from the fold to the open pasture twice or thrice a-week, or for a few hours each day, if convenient, is desirable, and especially when the lair is bad.

" 4. Protection from cold winds and driving rains should be provided in stormy weather.

" 5. Plenty of trough-room should be provided, and ample space allowed for the ewes to fall back.

" 6. All troughs should be shifted daily, and set well apart.

" 7. Dry food should be given at the same time as the fresh break of roots, to prevent crowding at the troughs.

" 8. Rock-salt should be at all times accessible.

" 9. Animals suffering from foot-rot, or other forms of lameness, should be removed from the fold, and placed on dry litter, and receive such other attention as the nature of the case may indicate." [1]

SHEEP IN SUMMER AND AUTUMN.

The summer is the season of least anxiety with flock-owners and their shepherds. Unless abnormally unfavourable weather should be experienced the duties of shepherds in the summer months are not likely to be arduous, yet the really efficient shepherds keep a constant and careful watch over the flocks in their charge throughout the whole year.

Ewes and Lambs.

The treatment of ewes and lambs during summer varies greatly, according to the locality and character of the grazing, the class of sheep, and the ends in view with the lambs and their mothers.

In ordinary unpedigreed flocks, where the ewes are to be kept for further breeding, and the lambs for breeding or for fattening later on, they graze together till weaning time, no extra food being given in ordinary circumstances. Where ewes and lambs are to be fattened for slaughter in the course of the summer or autumn, extra food is allowed all through the season, as is usually the case in pure-bred flocks where the youngsters are intended largely for breeding purposes.

For information regarding methods of feeding ewes and lambs where grazing alone is not relied upon, the reader is referred to the section in this volume dealing with pure-bred sheep, pp. 138-205. The methods of treating ewes and lambs pursued in all parts of the country are so fully stated in those pages that further details here would be mere repetition.

[1] *Jour. Roy. Agric. Soc. of Eng.*, vol. xxi. (1885), p. 199.

Pasturing Sheep on Arable Farms.

The method of pasturing sheep on arable land is regulated according to the class of stock kept and the nature and management of the farm. The stock may be a breeding or "flying" (hogging) one, or a certain modification of either, or both these recognised classes. A ewe stock is generally found where the farm is largely under rotation grasses or permanent pasture. The hogging system, on the other hand, prevails where the farm is worked in rotation, and the soil adapted for turnip culture.

Summer Fattening.—Sheep intended to be fattened on the pastures during summer are usually graded in lots, according to the conveniences on the farm in the way of separate fields. And it is a matter of great importance on grazing farms to have a good many fields of small or moderate size, rather than fewer fields of greater area. Of the sheep to be fattened a draw of the best is made, and these are put into the best piece of pasture. With plenty of good sweet pasture, and perhaps a little cake and grain, they will now fatten rapidly. Bruised oats are much in favour for fattening sheep on pasture.

Store Sheep in Summer.—The sheep to be kept simply in good store condition during summer are of course treated less sumptuously than the fattening sheep. A common plan with a flock of hoggs is to select the leanest and smallest, and assign these to the best of the pasture available for the store sheep, so that upon this (and perhaps a little extra food in the shape of oats) they may so develop as to "match" more evenly with the "tops" at the time of selling.

Shifting Sheep on Pastures.—When sheep are enclosed on fields, it is very desirable that they should be frequently shifted on to fresh pasture. The change will be beneficial both for the sheep and the pasture. It will be all the better for the sheep if the changes can be arranged from poorer to richer food. Where the fields are large they should be divided, perhaps by a temporary fence of wire or iron hurdles.

Water for Sheep.—There is a prevailing idea amongst many farmers that there is little or no necessity to provide water for sheep on pasture. This is a serious mistake, which is responsible for greater losses to flock-owners than would be readily imagined, especially when feeding on cake or other concentrated foods is practised. On succulent pasture with heavy dews sheep may require no further supply of water; but in dry weather and on dry pasture they cannot thrive and maintain good health without access to water.

Salt for Sheep.—Salt is especially necessary for sheep. It gives tone to the system, and should always be within their reach. Common salt may be given to them in partially covered boxes on the fields, or rock-salt may be put within their reach.

Maggot-fly.—During warm weather the shepherd should have his eye upon every sheep on the farm at least twice a-day. At this time they are liable to be attacked by the "maggot-fly." If any animal is seen to be restless, twisting its body, shaking its tail, and running forwards with its head bent down, the shepherd should catch it, and most likely on close examination he will find a colony of maggots located about the hind parts. In hot weather the shepherd should never go to the fields without having in his pocket a bottle of dip-mixture or fly-oil. With this he anoints the part attacked, and shakes out the maggots from the wool. This simple treatment will be quite sufficient.

Unclipped Sheep Falling.—Long-woolled sheep, hoggs especially, before being clipped, are so loaded with wool that, when annoyed by the ked, they are apt to roll upon their backs; and when that happens they are sometimes unable to get up again. They lie *awkward* or *awald*, and would soon die. Shepherds have to watch carefully to guard against deaths from this mishap.

Many collie dogs are quick in observing sheep in this state, and some will run and take hold of the wool, and pull the sheep over on its feet. Shepherds cannot be too alert in visiting sheep on pasture at this season.

Pasturing Sheep on Hill-farms.

The system of management pursued on hill-farms in carrying flocks from spring until weaning-time is usually very simple.

Stocking on Hill-farms.—The classes of sheep kept on hill-farms are arranged to suit the character of the land, the nature of the pasture, the altitude and exposure of the farm. A common plan is to maintain a stock of ewes on the low ground attached to hill-farms, or where the heath is well mixed with green ground, or interspersed by streamlets with green banks. Young sheep are placed on ground similar in character, but with a less admixture of green pasture. Older sheep generally occupy the higher grounds.

Pasture Plants on Hilly Ground.—The intelligent shepherd observes carefully the different kinds and succession of pasture plants suitable for the feeding of sheep, and as these attain sufficient growth he gives his flock a turn upon them. For instance, in most parts during January and February, "mossing" is usually plentiful; in April and May, "deerhair" becomes a standard plant; in June, July, and August, green banks, "haughs," and old pasture land are at their best; in September and October, "prie" and "stool bent" come up; and in November and December, "moss leek" and coarse bent and heath come in for use.

There is thus upon hill-farms, embracing high and low ground, a wonderfully complete succession of pasture plants. It is the object of the careful shepherd to take advantage of these as they come up in turn; and the flock-owner's balance-sheet may be largely influenced by the manner in which these successional growths are observed and utilised.

Heather-burning.

As heath constitutes a large ingredient in the food of mountain sheep, it is important that heath-burning should be carried out systematically, so as to have at all times a succession of young and old heath. Sheep-farmers have long been in the habit of burning a portion of the heath on their farms every year, with the view of allowing it to grow again, that its young shoots may support sheep in those parts of the grazing where there is little grass. Burning causes an abundant growth of young shoots; it is therefore the interest of both landlord and tenant that the heath should be so burned as to produce the greatest growth of young shoots.

Methods of Burning. — Various methods of heather-burning are pursued. The best plan is to burn in regular rotation, so that every piece of heather on the farm be burned at intervals of about eight years or less.

The burning of heather is controlled by the regulations of the property, and is usually carried out at the sight of and with the assistance of the gamekeeper and his gillies, the shepherd helping and pointing out the most suitable parts. Heather takes about three years before it sprouts after burning, but often on the burned ground other plants come up soon which are useful to sheep.

Sheep-washing.

There has from time to time been much discussion as to the utility of washing sheep before clipping them.

Objects in Washing. — There is a two-fold object in washing sheep—to free the wool from earthy material and improve its lustre, and cleanse the skin of the sheep from incrusted matter.

Opposition to Washing. — It is maintained by many flockmasters that any depreciation in the price per pound for unwashed wool is fully compensated by the greater weight of the fleece, and that the advantage to be derived from having the skin of the sheep cleaned by washing may be more than counterbalanced by the risk and trouble of the after-washing. It is better, they think, that the cleaning of the wool should be left to the manufacturer.

Washing is pursued to a large extent in some districts, chiefly where the sheep are kept on arable land, and in others hardly any washing takes place. Perhaps about a third of the sheep stock may be washed.

Study the Market. — The best guide as to the expediency of washing sheep will be the tendency of the wool trade —whether washed or unwashed wool finds the greater favour, or brings relatively the higher price. The advantages from washing are, as a rule, relatively greater when prices of wool are high than when they are low. The loss of weight by washing will most likely be from 1 to 2 lb. per fleece, and washed

wool will usually bring from 1½d. to 3d. per lb. more than unwashed wool. The cost of washing would be from 1d. to 1½d. per head.

Methods of Washing. — There are different methods of washing sheep. It is most frequently done in a pool about 3 feet deep, formed in a small stream; but where a stream does not exist it may be done in a natural pond or at the side of a lake. A pool with a muddy bottom is not suitable. It is important to have grass-land on both sides of the pool.

The sheep to be washed are enclosed on one side of the pool, the animals being one by one pushed or drawn into the water and made to go out at the other side. For a day or more after washing the sheep should be kept on the cleanest grass-land available, where there are no bare earthy banks.

In small flocks washing is sometimes carried out in large tin baths.

Time of Washing. — Washing takes place about eight or ten days before clipping.

Lambs are very rarely washed.

Shearing of Sheep.

This is an interesting event on sheep-farms. In most parts the sheep-shearing is regarded as a joyous occasion—a sort of harvest—in which a liberal allowance of beef and broth and ale is dispensed to the clippers engaged in the laborious work. It is a point of great importance to have dry settled weather for this operation; and as the time approaches, flock-owners watch the weather indications with some anxiety.

Time of Shearing. — The exact time of shearing varies with the locality, the class of sheep, and the season. The clipping season may be said to extend from the middle of May till the end of July. The new growth of wool should be well started before the clipping begins.

If the sheep have been washed, they may be clipped about eight or ten days thereafter.

The tups are first shorn, then the hoggs and wethers, and lastly the ewes.

On Lowland and mixed husbandry farms a covered place is generally selected for clipping. Upon large sheep-farms facilities are provided for clipping at the sorting-pens, where there is often shed accommodation.

In case of dew or rain in the morning, as many dry sheep may be brought into the barn on the previous evening as the number of clippers will shear on the ensuing day.

Force at Clipping.—It is customary for neighbouring sheep-farmers to assist

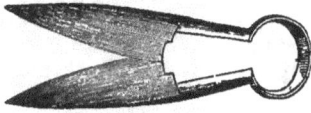

Fig. 747.—*Wool-shears.*

each other in clipping. The emulation amongst a number of men clipping together not only expedites the shearing of the individual flock, but makes the work cheerful, and calls forth the best and quickest specimens of workmanship from each clipper. Many additional hands have to be hired or transferred from other farm-work for the occasion, the number required varying with the size of the flock.

Wool - shears. — The tool with which the wool is clipped off sheep is made of steel, in the form of *shears*, whose broad blades are connected by an elastic ring (fig. 747).

Avoiding Injury to the Sheep.— Shearers who are expert and careful scarcely ever injure sheep in clipping, but when the skin does get cut with the shears the wound should be at once dressed with tar. It is important in clipping to keep the *points of the shears clear of the skin*, which may be done by gently pressing the blades upon the body of the sheep.

Methods of Clipping. — There are various methods of clipping sheep. The process is intricate, and can be learned only by practice. Many clippers, women as well as men, become very expert at the work, and will clip from 25 to 30 sheep per day, some of them even more.

Shearing Lambs.—In the extreme south of England, the practice of clipping lambs has long been pursued. It is by degrees spreading northwards, and is considered by many flock - owners to be decidedly beneficial to the progress of the lambs. In the case of lambs which are to be fattened off in the course of their first winter or following spring, it is specially advantageous to clip them as lambs. Lambs' wool is usually in request at a comparatively high price. It is generally past midsummer before lambs are shorn.

Sheep - shearing Machines. — The shearing of sheep by mechanical appliances is now carried out successfully, and to a large extent, especially on the great sheep - ranges of the colonies. There are several excellent shearing machines in use, all of them working on the principle of the horse - clipper. The first of them was the "Wolseley," brought out in Australia in the closing quarter of the nineteenth century. In many cases hand-power is sufficient, but steam, oil, and other engines are used where the flocks are very large. A

Fig. 748.—*Hand-power sheep-shearer.*

hand - power single clipper (Stewart's), fixed to a post, is shown in fig. 748.

Storing Wool.—As they are taken from the sheep the fleeces are carefully assorted, freed from lumps of dung, straws, thorns, or other rubbish, and rolled up for storing. In some cases the fleeces are immediately put into large canvas sacks or pack - sheets, but, as a rule, this is not done till the time of delivery to the buyer. On large farms a wool - room is provided, but in many cases the wool is stored in a granary or outhouse. The wool should be kept dry and cool, and out of the reach of dust light, and moths.

Weaning Lambs.

The time of the year for the weaning of lambs, like that of the lambing itself, is subject to great variation throughout the country. June, July, and August are the weaning months, southern arable farms coming first, and northern hill-farms last. In some cases in the south weaning takes place as early as May.

In many cases hill lambs are not now weaned. Those to be sold are sent to the marts directly they are taken from the ewes, and the lambs to be retained longer in the flock are allowed to remain with their mothers. This system is harder upon the ewes, but the gain to the young stock is substantial. It is believed that lambs allowed thus to remain with their mothers are less liable to "braxy" in the autumn months than lambs weaned in the ordinary way.

Treatment of Ewes and Lambs.—As to the treatment of ewes and lambs at the weaning time, information is given in the portions of this volume dealing with the management of pure-bred flocks. Nothing further need be said on the matter here beyond urging the importance of the shepherd watching carefully lest any ewe should suffer from a persistent supply of milk. If ewes after weaning are removed to close-eaten dry pasture, there will, as a rule, be little danger; but in extreme cases it may be advisable to relieve the udder by drawing away a little milk by hand, taking care not to empty, but merely to slacken, the udder.

After-treatment of Lambs.—The treatment in the way of feeding given to the lambs after weaning depends mainly upon the purpose for which the youngsters are designed. If they are to be fattened off early on the farm, or sold to others for this purpose, they are fed highly all along. The lambs to be kept for breeding purposes or for fattening at a later time are treated more moderately. The systems pursued in the different parts of the country in the rearing of lambs after weaning are indicated in the description of the management of pure-bred flocks in an earlier part of this volume.

Fattening Lambs.—The rate at which the lambs are forced will, of course, be regulated to suit the time at which it is desired to have them ready for slaughter. In Hampshire and other parts in the south of England, where the fattening of lambs for slaughter at nine to eleven months old is extensively pursued, the system of feeding is most liberal and highly forcing. Until early turnips are ready, the youngsters have frequent changes—perhaps weekly—upon rich pasture, lucerne, and clover aftermath, with all they can well consume of cake and grain. Then on turnips they have artificial food and hay.

The raising of fat lambs for early slaughter is pursued extensively, especially in the south of England, and in these cases both ewes and lambs are fed highly. Lambs being fattened after weaning get ample supplies of highly forcing food, as is shown in the feeding of Hampshire lambs at p. 173 of this volume.

Drafting Lambs.—After weaning the lambs are drafted, so that the various classes may be assigned to the intended purposes. Most probably the stronger of the wether lambs and the greater number (the best) of the ewe lambs will be retained to run on the farm along with the old sheep until later in the season. The others may be sent to arable farms to be wintered on grass and turnips. Those kept behind are drafted to the low country, as the pasture becomes scarce on the high ground, and as the winter approaches.

Marking Sheep.

Sheep are marked for the purposes of identification and classification, in various ways and at different times. There are the farm or flock mark, the age mark, and the pedigree or breeding mark. To provide these, five distinct systems of marking are in use—ear-mark, tar-mark, keel-mark, horn-brand, and tatooing letters and figures in the ears. (A convenient system of ear-marking is shown in fig. 704, p. 167, in this volume.)

Tar should be used sparingly in marking the fleece, so as to avoid as far as possible injuring the selling value of the wool.

Registering Marks. — To facilitate the recovery of strayed sheep, the flock-masters in several counties and districts register their respective marks, and publish these in book or pamphlet form. This is an excellent plan, especially useful in large pastoral districts where there is little fencing.

Dipping Sheep.

In order to protect them from insect attacks, and to generally promote their health and comfort, sheep are dipped, or dressed in some other way, once or twice a-year. With the view of getting rid of "scab," stringent Dipping Orders have been introduced by the Board of Agriculture. These vary from time to time, and it is of the utmost importance that flock-owners and their shepherds should make themselves familiar with all changes in the Orders as they appear.

Former Customs. — Formerly it was the custom to "bathe" the sheep on lowland and arable farms, while the sheep on hill-farms were "smeared." The latter method was preferred for high-lying farms, because "smearing" tends to keep sheep warmer in exposed parts, and to render them less liable to be affected by changes in the weather.

Bathing and smearing have both to a very large extent given place to "dipping," yet it will be useful to indicate briefly how these older methods were carried out.

Bathing. — For bathing, or "pouring" as it was sometimes called, the utensils

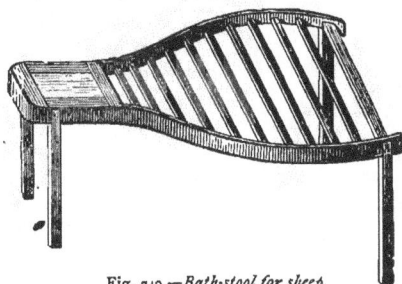

Fig. 749.—*Bath-stool for sheep.*

required are,—a bathing-stool, such as is shown in fig. 749; a bath-jug or a tin bottle with a pipe passed through the cork, and a tub or other vessel to hold the bathing mixture.

The sheep is placed on its belly on the stool, with its legs passed through the rungs, the head being towards the shepherd, who sits on the end of the stool. The shepherd with his thumbs and forefingers sheds the wool along the centre of the back from the head to the tail, and opens the shed with the palms of his hands. A boy then pours the liquid from the tin or jug along the shed, following the shepherd's hands, from the tail to the head of the sheep. Other sheds are made, about 3 inches apart, until the whole animal is covered, and from these sheds the liquid bathes the entire skin of the sheep.

Smearing. — Smearing is done in a manner similar to bathing, although the materials used are different. The smearing mixture consists of tar and butter, made up in such proportion as to be sufficiently consistent to be readily lifted on the finger of the operator. It is applied in the sheds of the wool by the shepherd himself, who takes from the kit or tub beside him a portion of the mixture with his forefinger, and rubs it into the shed. The sheds are made closer than for bathing, perhaps an inch or an inch and a quarter apart. The entire body is thus gone over, so that the sheep becomes enveloped in a close matted covering of wool, tar, and butter.

Dipping.

This is the most expeditious and now almost the universal method of dressing sheep.

Process of Dipping. — The operation of dipping is simple in the extreme. The sheep are either plunged or made to swim through a specially prepared tub, bath, or tank, containing the dipping liquid, after which they are kept on a drainer until the liquid ceases dripping from their fleeces.

The chief recommendations of dipping, therefore, are cheapness, efficiency, and remarkable despatch.

Construction of Dipping-bath.

Dipping-baths of many different patterns are in use throughout the country. Some are small and movable, others large and permanently fixed.

Swimming - bath. — For large flocks

the modern swimming-bath is the most convenient. Directions as to the construction and working of a bath of this kind are given in an admirable treatise on *Sheep-Dipping* by the late Mr David

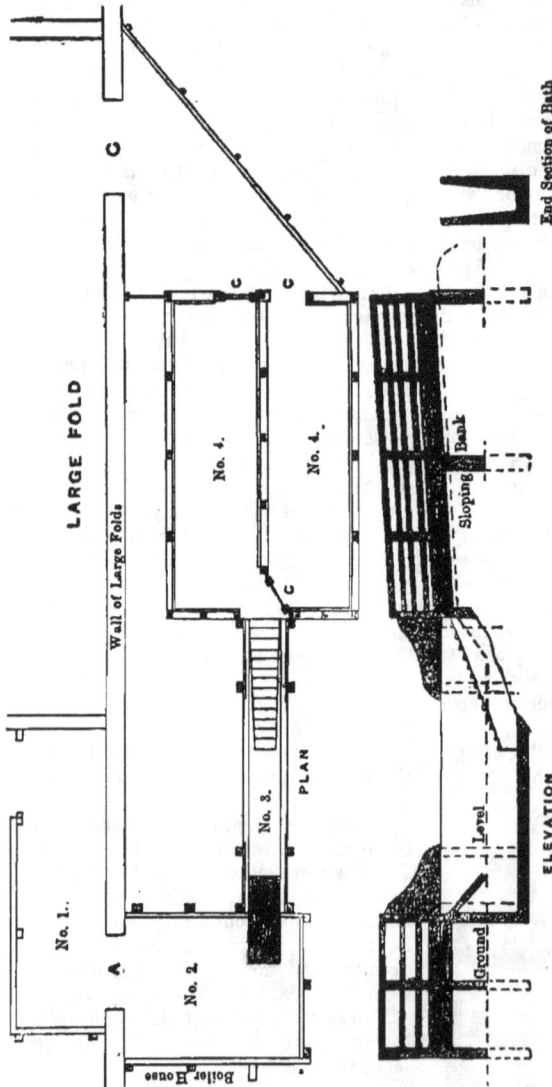

End Section of Bath

LARGE FOLD

Wall of Large Folds

No. 4.

No. 4.

No. 3.

PLAN

No. 1.

No. 2

Boiler House

C

C

C

C

C

Sloping Bank

Level

Ground

ELEVATION

Fig. 750.—Dipping-bath.

Wood. Flock-owners would find it useful to refer to this pamphlet.[1]

Process of Dipping.—The process of dipping in this bath is thus described by

[1] W. Blackwood & Sons.　Price 1s.

Mr Wood: " All being ready for starting, we will suppose a good number has to be dipped : two persons will be needed to bring the sheep forward ; two, or, better still, three should stand at the side of the bath, to guide the sheep through. Let the one nearest the catching or entrance pen take hold of each sheep with one hand as it comes forward and as it walks down the sloping board, and with the other hand press down the hinder part of the sheep, keeping the head above the mixture. It will be found when the sheep has a good coat of wool upon it, that considerable pressure is needed to get it down, but it is of great advantage to do so. Let the sheep then be passed on to the next assistant, and so on until it gets foothold up the sloping gangway."

Plans of Bath.—The bath described by Mr Wood is represented in fig. 750, which shows a bath erected at Bailliemore Farm, Strachur, Argyllshire. The sheep enter the catching or gathering pens at No. 1, which is formed inside one large division of fold ; through gateway A pass into No. 2; through gateway B, thence into the bath, No. 3, passing up into the dripper, No. 4. When drained, they pass out of the upper end of dripper back into a second division of large fold through gateway c. Pens Nos. 1 and 2 will hold about as many sheep as both divisions of dripper 4, 4. The boiler-house is built so as to take advantage of wall of large fold, one side of it forming a side of pen No. 2.

Cost of Bath.—The cost of erecting this bath, exclusive of the boiler-house, and allowing nothing for the carting or the timber, which was grown on the estate, amounted to only about £10.

Stone and Wood Baths.—The main plan of the bath and dipper described by Mr Wood is well suited for swimming-baths of all sizes; but later experience has shown that it is easier for both sheep and shepherd to have the bath deeper set in the ground, so that the top is level with the surface, and a space for standing in made about 3½ feet deep at each side of the dipper. Different materials are used in the construction of dipping-baths. Wood is largely employed; but the best kind of material is the Caithness flagstone—that is, where it or any similar flagstones can be obtained conveniently, and at reasonable cost.

Tossing Sheep into Bath.—The construction of the passage leading into the bath, so as to facilitate the driving of the sheep into the latter, requires consideration. The sheep are of course reluctant to walk into the liquid. It is a good plan to let the floor of the passage terminate in a trap-board, which capsizing forwards, tosses the sheep into the bath in true bathing attitude.

Some farmers consider that the catching or "gripping" pen may be advantageously dispensed with—a short passage or "shedder" being formed between the gathering-pen and the bath. The best method of regulating the passage of the sheep is by hanging a small gate just inside the trap-board, and keeping a lad in charge of it. By adopting this method the services of the "grippers" are unnecessary, and the rough handling the sheep might otherwise experience is avoided.

In some of the modern patent dippers there are ingenious trap-door arrangements, by which, one at a time, the sheep are sunk gently into the bath, being thus dipped without any shock.

Plunge-bath.—For small flocks the small plunge-bath is still most largely used. It is generally constructed of wood or flagstone, and the sheep have to be lifted both into and out of it.

Dipping Mixtures.

The flock-owner has almost unlimited choice as to the material to be used in bathing or dipping his sheep. Prepared sheep-dips are in the market by the score. To say that they are all good would be saying too much. There are at least a dozen, however, which are extensively employed, and each of which is cordially commended by different flock-owners. A certificate is given by the Board of Agriculture for those dips which are found efficient for the cure of scab, when used according to directions.

Non-poisonous Dips. — These dips are roughly classified into poisonous and non-poisonous dips, those which contain poisonous ingredients and those which do not. It is believed, however, that some of the so-called non-poisonous dips are such only in name. Indeed it is affirmed by many farmers that perfectly non-poisonous dips would be ineffectual in destroying keds and other insects unless used at greater strength than directed by their makers. Non-poisonous dips will kill the insects, but not the embryo or eggs. These develop later; and for this reason, those who use non-poisonous dips have to dip twice in order to thoroughly cleanse their sheep. The interval between the two dippings usually extends to ten days or a fortnight.

Composition of Dips.—Non-poisonous dips are, as a rule, made up of carbolic acid in one form or other; an alkali soft soap, with sometimes a slight addition of sulphur. The poisonous dips are in most cases supplied in the form of powder, and are usually made up of arsenic and alkali, soda, or potash, occasionally with the addition of sulphur. Some farmers prepare their own dips, but it is generally safer to use a well-tried manufactured dip.

Time for Dipping.—The most general time for dipping is towards the end of autumn and beginning of winter. It is a common practice to dip lambs when they are weaned, and some repeat the operation about November. In some cases the summer dipping is deferred, and the ewes and lambs dipped together about two weeks before tupping begins. A few dip immediately after clipping. In other cases the dipping of adult sheep is deferred until the New Year, or even until spring, the practice varying with the locality, the liability of the sheep to be struck by the fly, and the prevalence of other parasites. In arranging the times of dipping, farmers must conform to the Orders of the Board of Agriculture.

Weather for Dipping.—It is very essential that dry weather be chosen for the operation, otherwise little benefit will be derived from it. If the sheep are wet the wool will not absorb the dip properly; and if after dipping they are exposed to heavy rain, before the fleece has become perfectly dry, the solution will in all probability be washed out of it.

Dressing for Scab.—When scab appears in a flock the matter must be reported to the Local Authority, who will see that certain dipping and isolating operations are duly carried out.

Tupping Season.

The autumn and early winter is the mating season on sheep-farms, the precise time for introducing the rams to the ewes varying considerably throughout the country.

Flushing Ewes.—It has been found a good plan to "flush" the ewes just before tupping—that is, to give them an exceptionally abundant supply of succulent food for about two weeks before tupping, so as to have them in an improving condition when mated. This treatment hastens tupping, tends to increase the number of twin-lambs and to lessen the number of barren ewes.

If possible, a portion of rich pasture should be preserved for this purpose, or the ewes may have a run of the new grass and stubbles after harvest. On some farms where pasture is not available, a small breadth of rape is grown for the ewes, and in other cases a moderate feed of bruised oats is allowed.

On hill-farms farmers are not so anxious for twin-lambs, for on these lands one good lamb is usually sufficient for a ewe to rear satisfactorily. Hill-farmers, therefore, give less attention than lowland farmers to "flushing" the ewes. Still, many save low pasture upon which to feed the ewes two or three weeks before tupping.

Some flock-owners, however, question the propriety of flushing stock ewes, as they believe that when a big crop of lambs has been got one season by "flushing," the crop of lambs in the following season may be smaller, no matter how much the ewes may be flushed—a view, however, that is not universally held. "Flushing" no doubt can be overdone. Ill effects of severe flushing with such succulent food as mustard may be modified by letting the ewes have mainly dry food between tupping and lambing.

Fertility in Sheep.—An important inquiry into the effects of "flushing" and other factors supposed to influence fertility in sheep was conducted for the Highland and Agricultural Society of Scotland by Dr F. H. A. Marshall in the years 1905, 1906, and 1907. This inquiry confirmed the view that extra feeding at about tupping time results in a larger crop of lambs at the subsequent lambing. In Dr Marshall's report on this inquiry,[1] reference is made to other special causes believed to affect fertility in sheep. Inclement weather during tupping time may lessen the number of twins. It is believed that fertility may be developed by tupping early instead of late in the tupping season, the generative system being most active at the beginning of the season. It seems well established that fertility is a property that can be inherited, and thus it is believed that systematic breeding from twins will tend to increase fertility.

Management in Tupping Season.—The various important matters requiring attention in connection with the tupping season—such as the mating of ewes and tups adapted to each other, the treatment of tups, adjusting the number of ewes to each tup, and observing and recording service—are dealt with fully in the details of management in pure-bred flocks given in an earlier portion of this volume.

SHEEP IN WINTER.

The management of sheep in the winter months demands the utmost care. The system of winter treatment varies greatly,

[1] *Trans. High. and Agric. Soc. of Scotland,* fifth ser., vol. xxii., 1908.

perhaps even more than the treatment in the other seasons. Naturally the anxiety amongst sheep-farmers and shepherds as to the wellbeing of their flocks is greatest in the coldest and stormiest parts, where vast expense and trouble are often involved in carrying flocks safely through severe snowstorms.

SHEEP ON TURNIPS IN WINTER.

The practice of keeping sheep on turnips in winter is pursued largely throughout the country. For the most part the sheep are folded on the roots on the fields where grown, though in some cases the roots are pulled and given to the sheep on pasture or in sheds.

Preparing Turnips for Sheep.—As to methods of preparing unpulled turnips for consumption on the ground by sheep, information is given in vol. ii. pp. 357 and 358. It is important that this work should be carried out in good time and with care, so that the most economical results may be obtained.

Enclosing Sheep on Turnips. — There are two ways of enclosing sheep upon turnips — with hurdles made of iron or wood, and with nets made of twine or wire. Since the introduction of nets, the older method of enclosing with wooden hurdles has become exceptional, and is now seldom adopted unless where the enclosure is to stand for a considerable time, or for temporary enclosures for sorting sheep. Iron hurdles used for enclosing sheep are referred to in vol. i., figs. 119, 120, and 121. The wooden hurdles in use are of various patterns, a specially good light hurdle being that shown in fig. 751. It is formed of any sort of willow or hardwood, as oak-copse, ash-saplings, or hazel. The erecting of hurdles is a simple process which need not be described.

Nets for Enclosing Sheep.—Nets, made of twine of the requisite strength, form a superior enclosure for sheep when supported on stakes driven into the ground. The stakes are best formed of thinnings of trees, and they should be seasoned with the bark on before being cut into stakes. The stakes are usually about 3 inches in diameter and 4 feet 9 inches

long—allowing 9 inches of a hold in the ground, 3 inches between the ground and the bottom of the net, and 3 inches from the top of the net to the top of the stake. They are pointed at one end with the axe, and that end should be the lower one when growing as a tree, as the bark is then in the most natural position for repelling rain.

Setting Sheep-nets.—If the ground is in a soft state, the stakes may simply be driven into the ground with a mallet, the stakes being placed from 2½ to 3 paces asunder. Should the soil be thin and the subsoil hard, a hole sufficiently large for a stake may be made in the subsoil with the tramp-pick used in draining or an iron piercer made for the purpose. The stakes are driven in until their tops may not be less than 4 feet high, along as many sides of the en-

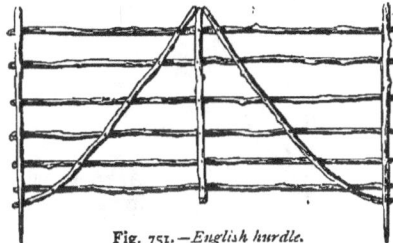

Fig. 751.—English hurdle.

closure as are required at the place to form a complete fence.

The net is set in this manner: Being in a bundle, having been rolled up when not required, the spare ends of the top and bottom ropes, after the stake is run through the outer mesh of the net, are tied to the top and bottom of a stake driven close to the fence, and the net is run out loose in hand towards the right as far as it will extend on the side of the stakes next the turnips. On coming back to the second stake from the fence, with your face to the turnips, the bottom rope first gets a turn to the left round the stake, then the top rope a similar turn round the same stake, so as to keep the meshes of the net straight. The bottom rope is then fastened with the shepherd's knot to this stake, 3 inches from the ground, and the top rope with a similar knot near the top of the stake, adjusting the net along and upwards; and so on, with one stake after another, until the whole net is *set up*, care being taken to have the top of the net parallel with the surface of the ground throughout its entire length.

Shepherd's Knot. — The shepherd's knot is made in this way : Let *a*, fig. 752, be the continuation of the rope fastened to the first stake ; then, standing on the opposite side of the stake from the net, press the second stake with the left hand towards *a*, and at the same time tighten the turn of the rope round the stake with the right hand by taking a hold of the loose end of the rope *d*, and putting it between *a* and the stake at *c*, twist it tight round the stake till it comes to *b*, where it is pulled up under *a*, as seen at *b*, and there its elastic force will secure it tight when the stake is let go. The bottom rope is fastened first, to keep the net at the proper distance from the ground, and then the top rope is fastened to the same stake in the same manner, at the width the net admits, at stake after stake. If both the cord and stake are dry, the knot may slip as soon as made ; but the part of the

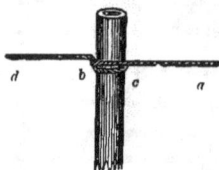

Fig. 752.—*Shepherd's knot, in fastening a net to a stake.*

stake at *b* where the knot is fastened on being wetted, the rope will keep its hold until the cord has acquired the set of the knot. It is difficult to make a new greasy rope retain its hold on a smooth stake even with the assistance of water, but a double turn round the stake will ensure its staying secure.

The shepherd should be provided with net-twine to mend any holes that may break out in the nets.

Wire Nets. — In certain situations, where rabbits and hares are apt to destroy string nets, or where it is not necessary to step over the nets with cut turnips or other food, wire has largely taken the place of twine nets. Wire nets are made with meshes of any size, but 4 inch is the size generally in use, and 3 feet is the most common height. Twine nets are made to set about 40 yards, but wire nets set 10 yards farther. The cheapest are made by machinery, with the wire running prac-

tically horizontally, but the best are made only by hand, with the wire worked from top to bottom and *vice versâ* backwards and forwards. The top and bottom strands are extra strong, and one or two strong strands are worked along the centre. Iron or steel wire is used, and galvanised after manufacture, giving a strong, enduring, and convenient fence at a minimum cost. In setting up, the stobs are first erected as for twine nets, and the end of the wire net unrolled and fixed to the first stob, then the whole roll of netting is unrolled alongside the stobs, pulled tight, and the far end fixed to a stob. After this it is an easy and rapid process to fix to the stobs by twine, or preferably by bellhangers' staples, from which the net is unhooked and rehooked as required when taking down and re-erecting. Sometimes the stobs are driven through the meshes of the net and tied firmly with twine, but this plan is severe on the net.

Extent of Roots given at a time. —Care has to be exercised as to the quantity of turnips made available for sheep in an enclosure at one time. After a week or so, breaks which will serve a couple of days, or three at most, may be given, but this will altogether depend on the weather.

In frosty weather or snow, turnips sufficient for the day only should be given, otherwise the shells will become hard frozen in a very short time, and the sheep are unable to eat them, so that when a thaw sets in these rot. A good plan is to allow the sheep to work on the ground given during the forenoon, and set pickers on in the afternoon, to pick up all the shells for the sheep, no more ground being given than will serve the sheep for the day.

Carting Turnips to Lea Land in Wet Weather.—When the weather becomes excessively wet, and the sheep cannot comfortably consume the roots upon the black earth, the turnips, after being tailed, may be carted from the field and spread on pasture, and the sheep taken from the turnip-breaks until better weather sets in.

Another plan, sometimes adopted in wet weather, is to leave the sheep on the turnip-field only from early morning till about 3 P.M., the rest of the time being spent on pasture, where extra food may be given in boxes. In other cases the turnips are pulped, and given to the sheep on pasture.

Begin Turnip-feeding Early.—The turnip-break should be made ready for the sheep before the grass fails, so that the feeding sheep may not lose any of the condition they have acquired on grass; for it should be borne in mind that it is easier for animals to progress in fattening than to regain lost condition. Much rather leave pastures in a rough state than lose condition in sheep for want of turnips. Rough pasture will never be wasted, but will be serviceable in winter to ewes in lamb and to aged tups. Feeding sheep, therefore, should be put on turnips as early as will maintain the condition they have acquired on grass.

Begin cautiously with Turnips.— It is considered advisable to avoid putting sheep on turnips for the first time in the early part of the day when they are hungry. Danger may be apprehended from luxuriant tops at all times, but when they are wetted by rain, snow, or half-melted rime, they are sure to do harm. The afternoon, when the sheep are full of grass, should be chosen to put them first on turnips; and although they will immediately commence eating the tops, they will not be likely to hurt themselves. But it is a still safer plan to begin by carting cabbage or turnips, a few at a time, to the grass-field, than to put the sheep straight from grass to turnips.

Turnips risky for Ewes.—Sheep for turnips are selected for the purpose. Ewes being at this season with young, are not often put on turnips in the early part of the winter, but continue to occupy the pastures, part of which should be left on purpose for them in a good state, to support them as long as the ground is free of snow. As the lambing-time approaches, and the pastures begin to get bare, a few turnips are often given daily to in-lamb ewes, generally on a pasture-field, and along with a little hay and cake. But care should be taken never to give frozen roots to in-lamb ewes, as this has often been blamed for causing abortion. Many farmers also altogether avoid giving turnips to in-lamb

ewes, in the belief that they are liable to cause inflammation at lambing.

Draft Ewes on Turnips. — Every year a certain number of old ewes, unfit for further breeding, from want of teeth or a supply of milk, are drafted out of the flock to make room for young females, and are fattened upon turnips, with the addition of a little corn or cake and hay.

Young Sheep on Turnips.—It sometimes happens that the hoggs—the castrated male lambs of last year and the ewe lambs not required for breeding,— instead of being sold, have been grazed during the summer, and are fattened on turnips. In many parts of the country lambs are now freely fed on turnips.

Turnip-tops for Sheep.—Care should be taken not to shift the sheep or give them a fresh break when the turnip-tops are covered with white or hoar frost, as numbers of deaths happen from this cause. In fact, farmers put too much value on turnip-tops: if hoggs, fat sheep, or other feeding animals were never to taste them, they would fatten faster. If the tops are cut off a day or two before the fold is shifted, and scattered over the ground, they wither before the hoggs get at them, and loss is avoided. A supply of stored turnips should always be at hand to give to the sheep in case of hard frost.

Dry Food with Turnips.

When sheep are on turnips, they should always be supplied with dry fodder, hay or straw,—that is, where they cannot have a daily run of some rough dry pasture. Clover-hay is the best and most nutritious, but fresh oat-straw answers the purpose very well. The best way of supplying dry food is to chaff the hay or straw and place it in the boxes which are required for the cut turnips later in the season. About ¼ lb. oats per sheep per day, mixed with the chaff, gives excellent results; many of the sheep will become ready for the butcher without further feeding.

South of Scotland Methods.—In the midland and south-eastern counties of Scotland, the fattening of sheep is carried on to a large extent, the moderately dry climate in these parts being favourable for this industry. The sheep are begun on the soft varieties, and are passed on to yellows and swedes in turn. Great numbers of hoggs are fattened in this way. Many are given ample allowances of turnips, just about as much as they can eat without waste. In addition, they get mixtures of oats, decorticated cotton-cake, and other materials, varying from ¼ lb. to 1 lb. or more per head per day, with hay and straw. Linseed-cake, beans, peas, maize, bran, brewers' and distillers' grains, and condimental foods, are all used to a lesser or greater extent.

Sheep-Fodder Racks. — Fodder for sheep is largely given in racks, which are of various forms. A strong and useful fodder-rack for sheep, fit for grass or tares in summer, or turnips in winter, is shown in fig. 753. It was invented by

Fig. 753.—*Kirkwood's wire sheep-fodder rack.*

Rack of wirework 6 feet long, 2 feet 9 inches wide at top, 8 inches wide at bottom, and 2 feet 3½ inches deep.
a Curved cover of sheet-iron with a hatch.
b b Sheet-iron troughs to contain corn, &c.

Mr Kirkwood of Tranent. The troughs are provided with a hole at each end to allow the rain to drain off, and might be used in dry weather for holding salt or oilcake for the day.

Another very useful rack, made by Mr W. Elder, Berwick-on-Tweed, is shown in fig. 754. It is made chiefly of wood and wire, and is useful also as affording shelter.

Substitutes for Feeding-Racks. — Another plan often adopted by farmers is to hang a net on a double row of stakes, the middle of the net forming a receptacle for the hay. Wire-netting with mesh of about 4 inch, set double along a row of stobs, has also been found a cheap and durable means of giving hay to sheep.

Supplying Fodder. — Two racks or

more are required, according to the number of sheep. It is the shepherd's duty to fill them with fodder, which is easily done by carrying a small bundle of fodder every time he visits the sheep. When carts are removing turnips from the field, they carry out the bundles.

Fig. 754.—*Elder's sheep-fodder rack.*

If only as a means of providing shelter, irrespective of fodder, the racks should be kept full. Fodder is consumed more at one time than another; in keen sharp weather the sheep eat it greedily, and when turnips are frozen they have recourse to it. In rainy or soft muggy weather it is eaten with little relish; but it has been observed that sheep eat it steadily and late, and seek shelter near the racks, prior to a storm; while in fine weather they select a lair in the open part of the break.

Fig. 755 is a simple and convenient form of trough for oats or other feeding-stuffs. A convenient length is 9 feet, its form acute at the bottom. An excellent sheep feeding-box is shown in fig. 706, vol. iii. p. 169.

Picking out Turnip-shells. — Until of late years, sheep helped themselves to turnips, and when the bulbs were scooped out to the level of the ground,

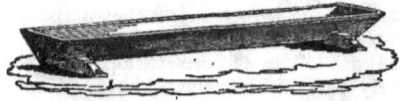

Fig. 755.—*Trough for turnip sheep-feeding.*

their *shells* were raised with a *picker*, the mode of using which is seen in fig. 756. By this mode of action the tap-root of the turnip is cut through and the shell separated from the ground at one stroke. Only half the ground occupied by shells should be picked up at once, so that the sheep may take up a larger space of ground while consuming them. When the ground is dry, the shells should, on the score of economy, be nearly eaten up before a new break of turnips is given; and if any shells are left, the sheep will come over the ground again and eat them.

Cutting Turnips for Sheep.—The feeding of sheep on uncut turnips can be satisfactorily carried out until their teeth become defective: this occurs from the constant eating of hard roots, often in a semi-frozen state, which loosens the front teeth. The farmer can readily judge when other measures become necessary by the appearance of the bulbs, which have their outer skin peeled off by the sheep, and so left.

To meet this difficulty the turnip-cutter comes into requisition. Many thoroughly efficient machines are now available for this purpose, such as that

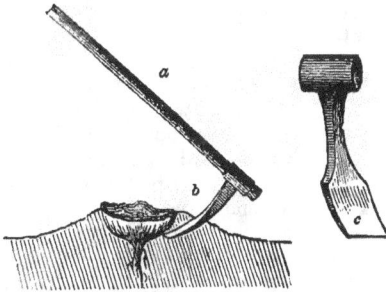

Fig. 756.—*Turnip-picker.*

a Handle 4 feet long.
b Blade 10 inches long, including eye for handle.
c Breadth of blade 2 inches.

shown in fig. 757, which cuts the turnips into finger-pieces. In this form they are readily eaten by the sheep. The plan adopted, if the turnips are to be eaten on the land where grown, is to cast them into heaps alongside the net, a sufficient quantity for one or two days in each heap. The cut turnips are given to the sheep in the troughs or boxes, 7 to 10 boxes being sufficient for 100 sheep.

The heaps being laid down at intervals allows the troughs or boxes to be changed to fresh ground daily, so that the land is equally manured all over the field. One worker can in this manner feed 300 sheep.

The Cutter Cart.—The old-fashioned method of cutting turnips by means of the lever slicer has been largely superseded by the cylinder cutter, fig. 757, or the cutter cart, fig. 758. The cutter

cart is an exceedingly useful invention. It consists of an ordinary farm box-cart with a root cutter of the barrel type placed underneath, driven from the wheels of the cart by tooth-gearing and

Fig. 757.—*Gardner's cylindrical turnip-cutter.*

clutch. By a lever the cutter is easily thrown out of gear. The cart is loaded with roots and set agoing, and the finger-pieces fall regularly as the cart proceeds. To adapt them for use where the cut roots are given to the sheep in boxes, some cutting carts are fitted with a large receptacle or framed box, also made to fix below the cutter barrel, which can be set to catch and carry all the turnips as they fall from the cutter barrel. The feeding boxes are set along

Fig. 758.—*Elder's turnip-cutting cart.*

the field in a row about 30 yards apart. The cart being loaded with turnips, it is pulled along the field, cutting as it goes. As it reaches each box the cut turnips are shovelled from the large receptacle underneath the cart into the feed boxes.

By this plan the feeding can be done all over the field instead of on one spot as with a stationary cutter.

Cake-breaker. — For sheep oilcake must be well broken. This is done by a strong machine such as that shown in fig. 759, made by Barford & Perkins, Peterborough. The oilcake is put into the hopper, the mouth of which is open upwards. The two rollers bruise it to any degree of smallness, by means of pinching-screws. The bruised cake falls down the spout into any vessel below.

Oats and Hay for Hoggs. — Some farmers keep hoggs on turnips all through the season. Others think it better to give them not more than two or three hours daily on the turnips, giving them during the remainder of the time the

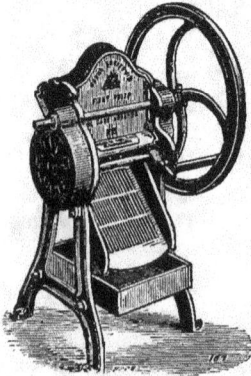

Fig. 759.—*Oilcake-breaker.*

run of a dry pasture-field, where they get ½ lb. of oats per head daily, and a handful of hay when the weather is hard. After the New Year the turnips must be cut for them.

Salt for Sheep. — Salt is frequently given to sheep on turnips, sometimes in the form of rock-salt, and in other cases as common salt. Sheep should have access to water when using salt.

Sheep on Turnips during Snow. — A fall of snow, driven by the wind, may cover the sheltered part of the field, and leave the turnips bare only in the most exposed places. In this case the sheep may have to be fed on the exposed parts, and if so the racks should be so placed there as to afford shelter. If the fall of snow should be very heavy the shepherd may have to get help to clear away

enough of the snow to enable him to get the sheep fed.

Occasionally in stormy districts the sheep may have for a week or more to be fed without roots, say on cake and bruised corn and hay, but it is well to bear in mind that sudden changes in food are undesirable for all kinds of stock, and have therefore to be avoided as much as possible.

Unripe Turnips dangerous. — The danger of giving sheep access to unripe roots is referred to at p. 393. Information is given there as to measures for keeping ewes in good health on roots.

Blackface Sheep in Winter. — "It is always safe policy in stormy weather to supplement the natural food with hay. Blackfaces being naturally very hardy, they require less artificial feeding in winter than almost any other breed of mountain-sheep; yet in excessively severe winters the prudent manager does not leave his sheep to forage for themselves until it is too late to help them. So long as the snow does not get too deep, or is not frozen hard, they take little harm. Blackface sheep are excellent workers in the snow, and will toil bravely for a sustenance under the most trying circumstances. Hand-feeding is only resorted to when it cannot be longer avoided; and in that case the sheep are either removed to a lower district or fed on hay at home." [1]

Sheep in a Wood in a Snowstorm. — During severe snowstorms some farmers put sheep into woods, and supply them there with hay upon the snow round the roots of the trees. A precaution is requisite when the trees are Scots fir; their evergreen branches intercepting the snow are apt to be broken by its weight, and fall upon the sheep and kill them. Heavily loaded branches should therefore be cleared partly of their snow where the sheep are to lodge.

Rape for Sheep. — In the south of Scotland, and more generally in England, rape is grown for sheep. The consumption of rape by sheep is conducted by breaks in exactly the same manner as that of turnips; but rape is never stripped or pulled, the entire crop being consumed on the ground. In England, the rape

[1] *Blackface Sheep*, by J. and C. Scott, 109.

intended for sheep is sown broadcast and very thick. In Scotland, it is often raised in drills like turnips; and although not so convenient for sheep as when sown broadcast, yet the drills permit the land being well cleaned in summer, which renders the rape an ameliorating crop for the land. Rape is extensively used as a catch crop after early potatoes, and often gives an excellent return in fattening hoggs before Christmas.

Shelter for Sheep on Turnips.— Sheep on turnips have little shelter but what is afforded by the fences of the field or plantations. In some cases this is quite sufficient, but in others it is inadequate. Various devices are in use to provide shelter not merely against sudden outbreaks of stormy weather, but with the view of gradually improving the condition of sheep, both in carcase and wool.

An excellent temporary shelter for sheep on turnips may be made by the erection of a double line of hurdles or nets, the space between the lines being filled up with straw. A curve or angle can be introduced, and thus shelter can be provided for every quarter from which storms may come.

Experiments with Foods for Sheep on Roots.—A series of interesting and instructive experiments were carried out during the years 1903-1905 in East Lothian by the staff of the Edinburgh East of Scotland College for the purpose of ascertaining (1) the most profitable feeding-stuff to use along with cut swedish turnips, supplied *ad libitum*, and a daily allowance of from ¼ to ½ lb. of hay in the winter feeding of sheep; and (2) whether the use of feeding-stuffs effects any saving in the *daily* consumption of turnips when the sheep are allowed to take as many as they please. The prices of the foods were taken at— turnips 10s. per ton, hay £3, 10s., Bombay cotton-cake £5, 2s. 6d., dried grains £5, 12s. 6d., decorticated cotton-cake £7, 10s., linseed-cake £8, 5s., maize £5, 7s. 6d., crushed oats £6, 9s. 2d.

At these prices Bombay cotton-cake, linseed-cake, and a mixture of these two, were equally satisfactory feeding-stuffs. Dried grains also fed well, but the carcase-weight was not so good. It resolves the business into a question of the relative prices of the several stuffs. A rise of 10s. per ton will put any one above the profitable line.

THE TOTAL LIVE-WEIGHT INCREASE AND ITS COST.

Lot.	Description of characteristic food.	Total increase in 85 days.	Average increase per head per week.	Gross cost of food per cwt. of live-weight increase.	Net cost of food per cwt. of live-weight increase.
		lb.	lb.	£ s. d.	£ s. d.
I.	Bombay cotton-cake	754	2.07	2 1 3	1 11 7
II.	Bombay cotton-cake and linseed-cake	859	2.35	2 0 9	1 11 1
III.	Linseed-cake	926	2.54	2 1 4	1 11 7
IV.	Bombay cotton-cake and oats . .	727	1.99	2 3 4	1 14 1
V.	Dried distillery grains . . .	796	2.18	2 0 2	1 11 1
VI.	Decorticated cotton-cake and maize .	787	2.16	2 1 11	1 12 4

The average daily consumption of turnips in 1905 was 13.42 lb. per head— fully a pound more than was taken in 1904, and 3 lb. less than in 1903,—and none of the lots varied more than about ½ lb. from this quantity except Lot

VI., which consumed only 11.93 lb. per head per day. Lot VI., however, stands highest for consumption of hay, taking 7.16 oz. per head per day, while the general average is 5.42 oz. The linseed-cake lot also is prominent as a consumer of hay, thus corroborating former results; while the lot fed on Bombay cotton-cake runs to the other extreme, and has to be ranked along with those fed on dried grains and a mixture of Bombay cotton-cake and oats. That Bombay cotton-cake should have the same effect as a bulky ration of dried grains in reducing the consumption of hay seems remarkable. Nevertheless, it is upheld by all three experiments. The quantities of concentrated food taken by the respective lots is in close agreement, all being within $\frac{1}{2}$ of an oz. of the general average of 13.31 oz. per head per day. No appreciable diminution of the amount of swedes consumed was observable between the lots which got concentrated feeding-stuffs and the lot which only got hay with its swedes in the earlier years. Of course, though no *daily* reduction of swedes was caused by cake-feeding, the cake-fed lots were sooner ready for the butcher than the sheep that did not get cake.

Cost of Turnip-feeding for Sheep. —The cost of turnip-feeding varies with the season and the crop as well as in different districts in the same season; but usually the price of turnips for hoggs ranges from 3d. to 5d. a-week, and for ewes and fattening sheep from 4d. to 8d. each sheep. These prices are sometimes exceeded when turnips are scarce in a backward spring. When it comes to extreme prices, however, the flockmaster in many cases can fall back on hay and corn or cake.

House-Feeding of Sheep.

Feeding Sheep in Sheds.—In former times the feeding of sheep in sheds was strongly commended by a few who had experimented upon it with satisfactory results. Others, however, were less successful, and while it was useful for small flocks, it has not come into extensive practice where large flocks are kept. Still, by several enterprising farmers who have carried it out with exceptional care, the practice is pursued with success.

A Ross-shire Example.

For many years the house-feeding of sheep has been carried on with marked success by Mr John Ross, Millcraig, Alness, Ross-shire. He states that by this method he can feed a larger number of sheep, at least a third more, than by outside feeding. Where sheep are fed largely in the house, and littered with peat moss-litter or straw, the fertility of a farm may be so increased that little artificial manure may be required. Sheep can be fattened in a shorter time in the house than outside, and home-grown food will go further. The sheep make steady and often very rapid progress. Mr Ross thinks the saving in death-rate alone would almost pay the interest on the cost of the shed. The sheep are protected from birds and maggots in summer, and from injurious extremes of weather in autumn, winter, and spring, and they fatten all the more rapidly because they undergo so little exercise.

The saving in food is undoubtedly substantial. With care, not a particle of any kind of food need be wasted. All green food and hay are passed through the chaff-cutter, and given in boxes, so that no food can be trampled under foot. The long feeding-trough is not suitable for sheep in houses, and in its place Mr Ross uses five-sided boxes, each side being large enough for two sheep—in all, ten sheep at each box. Much labour, of course, is involved in cutting, carting, and preparing food, as well as in littering and cleaning the shed, yet there is a certain saving, in that the shepherd has no wandering over fields, and no stakes and nets to erect.

The littering has to be carefully attended to. Whether straw or moss-litter is used, it should be raked over daily, and fresh small quantities spread almost every day. Sheep should never be allowed to stand in damp bedding, and if their feet are sound when put into the shed they rarely go wrong.

In the feeding of sheep in houses, distillery "draff," mixed with decorticated cotton-meal and cut hay, and allowed to ferment slightly, gives good results. Where "draff" is not available, a little treacle diluted with water may be used to make a mass of hay and meals palatable to the sheep.

Sheep-feeding Shed.—The shed used by Mr Ross (fig. 760) is 110 feet long by 60 feet wide under one roof. It is divided across the middle into two equal areas by a concrete passage ten feet broad, and raised 3 feet above the floor-level of the shed. This passage affords facilities for storing foods and also for the mixing of them. These two main areas are again divided in the middle, thus providing four compartments of 30 feet by 50 feet, each sufficient to hold from 70 to 100 sheep. It is believed that compartments about this size, and square in shape or nearly so, are better suited than longer and larger enclosures, as in the latter the sheep are apt to run about too much when they are disturbed.

The sides and centre division of the shed are formed of concrete walls 3 feet high, with wooden framing 9 feet high above, to carry the roof. The lower half of the framing is lined with boards, while the upper half is composed of swing-doors, which may be opened or closed at will, thus providing admirable ventilation for the sheep without exposing them to draughts.

The roof of the shed is in one span, covered with corrugated iron, and supported by the centre division and side walls. The south end is half-sparred above the wall, and in the north end there are large doors. There are cart outlets for the convenient cleaning of

Fig. 760.—*Sheep-feeding shed—Exterior and interior.*

the shed. At one end of the centre passage there are stores for straw, hay, and roots, with accommodation for chaff-cutters and turnip-cutters, which are driven by a 1-horse-power petrol engine.

The manure is allowed to accumulate under the sheep until it can be conveniently removed. With the low concrete walls all round, the manure can rise to 3 feet in height without touching the wooden framework of the shed.

A shed such as this, to accommodate from 300 to 400 sheep, will cost about £300. At 10 per cent interest, this represents about 1s. 6d. to 2s. per sheep, but with three sets of sheep turned out each year the cost of the shed is only about 6d. for each sheep.

WINTERING SHEEP ON PASTORAL FARMS.

It is far more difficult to bring hill sheep well through the winter than it is to handle a lowland flock, especially in a winter of severe snowstorms. So long as the snow lies dry, even though it drifts badly, sheep manage, with careful guiding, to find a living; but they are sorely tried when a thaw and frost follow each other closely. The flockmaster who has not a plentiful supply of hay on hand is then in a bad plight.

"Home-Wintering" or "Sending Away."—On semi-pastoral farms, as on arable lands, this question does not arise; on purely pastoral holdings it is different.

On many of the higher and more exposed grazings the sheep have to be brought down to lower ground in winter, even if they are to be wintered on the farm; and, provided that an abundance of natural hay exists, it is better to winter the hoggs as well as the ewes at home. Wethers which are not ready for the butcher when they come off the hill in autumn, it may be necessary to send away for wintering on turnips, if they are to be fattened on grass the following summer; but if store sheep are likely to be cheap in the spring, it will pay best to sell the wethers direct from the hill in autumn to be winter-fattened on arable farms. Sending the hoggs away to be wintered costs 6s. or 7s. a-head, which is more than a second sheep-rent; and sheep that have to go back to hill pasture in the spring are altogether better wintered on hay at home if this should be practicable.

Wintering Sheep in Romney Marsh.—The same difficulty of wintering the hoggs at home has to be met by the Romney Marsh graziers; but whereas the hill-sheep farmers have to contend against winter storms and the failure of the frozen pastures, the Kentish sheep-breeders have to move their young sheep to higher grounds in winter owing to the flooding of their pastures, and not so much in search of better food as of sounder grazing. Many thousands of these Kent hoggs or tegs are sent out to winter in the adjoining counties at the end of September and brought back at the end of March every year, the wintering having cost 8s. or 9s. a-head, and sometimes more.

Saving Hay for Hill-farms.—Care has to be taken during summer to provide sufficient hay for the requirements of the flock in snowstorms. A general practice is to save or hain the enclosed parks which had been used early in spring for weak ewes and lambs. There are usually enclosures of this kind, extending in all to perhaps 6 to 10 acres for every "hirsel" of ewes, and sufficient hay should be obtained here for a flock of 500 ewes during an average winter. It is the duty of the shepherds to cut and secure this hay, and it is important

that the work should be properly and seasonably attended to. For the supply of natural hay specially fertile "haughs" and other patches of green pastures throughout the farm are also saved.

Arable Land on Sheep-farms.—Where at all practicable there should be a certain area of arable land on sheep-farms, so that the supply of natural hay may be augmented by rotation hay, and that a moderate quantity of turnips may be grown. The advantages of this in stormy winters are very great.

Irrigation on Hill-farms.—Since *hay* is the principal food for mountain sheep in snow or black frost, it is of importance to procure this valuable provender in the best state, and of the best description. It has long been known that irrigation promotes, in an extraordinary degree, the growth of natural grasses; and perhaps there are few localities which possess greater facilities for irrigation, though on a limited scale, than the Highland glens of Scotland. Rivulets meander down those glens through haughs of richest alluvium, which bear the finest description of natural pasture plants. Were those rivulets subdivided into irrigating rills, the herbage of the haughs might be multiplied many fold, and hill-farmers are earnestly urged to convert them into irrigated meadows. Although each meadow may be of limited extent, the grass they afford is greatly increased in quantity and value when converted into hay.

One obstruction alone existing to the formation of meadows is, the fencing required to keep stock off while the grass is growing for hay. But the fencing should be made for the sake of the crop protected by it. Hurdles make an excellent fence. This difficulty is now greatly lessened by the introduction of cheap wire-fencing. Besides places for regular irrigation, there are rough patches of pasture, probably stimulated by latent water performing a sort of under-irrigation to the roots of the plants, which should be mown for hay; and to save further trouble, *this* hay should be ricked on the spot, fenced with hurdles, around which the sheep would assemble at times to feed through them in frosty weather from the rick, and wander again over the pasture for the remainder of the day; and when snow came, the stells would be the places of refuge and support. As the hay in the stack is eaten, the hurdles are drawn closer to the stack, to allow the sheep again to reach the hay.

The practice now generally adopted, however, is to lay out the hay in handfuls on the snow, keeping plenty of room between the lines of hay.

On sheep-farms arable land might not itself be capable of yielding rent or profit, but it would most likely add greatly to the value of the adjoining pasture-land. Let it be always kept in view that the more food and shelter provided in winter for stock the less will be the loss incurred during the most inclement season.

Shelter on Sheep - farms. — There is still a marked deficiency of shelter on most pastoral farms in this country— that is, where it is not provided by the configuration and lie of the ground. More tree - planting for the providing of shelter for stock is urgently required, and much may also be done by the planting of suitable bushes such as broom, whin, and juniper. It used to be the custom with some sheep-farmers to fill their pockets on spring mornings with the seeds of the whin and broom, and in their walks over the sheep-farm, scatter these seeds on any likely spot. These eventually provide food for sheep in a stormy winter, besides growing into strong bushes capable of affording excellent shelter.

Stells for Sheep.

To admit of food being supplied with some degree of comfort to sheep during severe snowstorms on high grazings the existence of *stells* is desirable. There are still many store-farmers sceptical of the utility of stells, but on exposed farms their advantages are undeniable. A stell may be formed of a plantation or a high stone wall — either will afford shelter; but a plantation requires to be fenced by a stone wall.

Outside Stell.—Fig. 761 is a good *outside stell*, formed of plantation. The circumscribing stone wall is 6 feet high, the ground within it is planted with trees. Its 4 rounded projections shelter

a corresponding number of recesses embraced between them; so let the wind blow from whatever quarter, two of the recesses will always afford shelter. The size of the stell is regulated by the number of sheep kept.

Sheep Cots or Sheds.—Much diversity of opinion exists regarding the utility of *sheep-cots* on a store-farm. These are rudely formed houses, in which sheep are put under cover in stormy weather, especially at lambing-time. Many object to sheep - cots on high farms, because, when inhabited in winter, even for one night, by as many sheep as would fill them, an unnatural height of temperature is thereby generated. Cots may be serviceable at night when a ewe or two

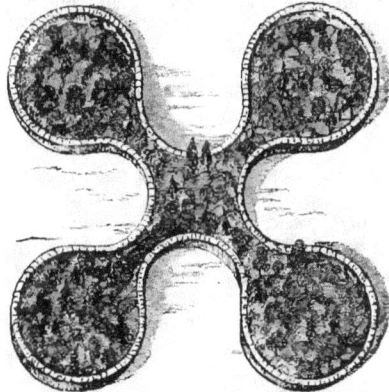

Fig. 761.—*Outside stell sheltered by plantation on every quarter.*

become sick at lambing, or when a lamb has to be mothered upon a ewe that has lost her own lamb; and such cases being few at a time, the cot never becomes overheated.

Paddocks for Sheep.—On an unsheltered breeding-farm it is desirable to have two paddocks, which are sufficient to contain invalid sheep, tups, and twin lambs, until strong enough to join the hirsel.

Forming Plantation Stells.—In making stells of plantations, it is desirable to plant the outside row of trees as far in as their branches shall not drop water upon sheep in their lair, such dropping never failing to chill them with cold, or entangle their wool with icicles. The spruce, by its pyramidal form, has no projecting branches at top, and affords excellent shelter by its evergreen leaves and closeness of sprays, descending to the very ground. The Scots pine would fill up the space behind the spruce; but every soil does not suit the spruce, so in some cases it may be inexpedient to plant it. Larches being deciduous, their branches are bare in winter. Larches grow best amongst the *débris* of rocks and on the sides of ravines ; Scots fir on thin dry soils, however near the rock ; and the spruce in deep moist soils.

Size of Stells.—Stells should be as large as to contain 200 or perhaps as many as 300 sheep on an emergency ; and even in the bustle necessarily occasioned by the dread of a coming storm, so large a number as 200 could be separated from the rest, and accommodated in a sheltered recess accessible from all quarters. Thus 5 such stells as fig. 761 would accommodate a whole hirsel of 1000 sheep.

Suppose, then, that 5 such stells were erected at convenient places—not near any natural shelter, such as a crag, ravine, or deep hollow, but on an open rising plain, over which drift sweeps unobstructed, and remains in less quantity than on any other place—with a stack of hay inside and a store of turnips outside, food would be provided for an emergency. On a sudden blast arriving, the whole hirsel might be safely lodged for the night in the two leeward recesses of one or two of these stells, and, should prognostics threaten a storm, next day all the stells could be inhabited in a short time.

Concave Stells.—Instead of the small circular stell, some recommend a form

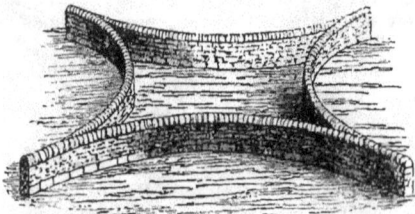

Fig. 762.—*Outside stell without plantation.*

without plantation, having 4 concave sides, and a wall running out from each projecting angle, as in fig. 762—each stell to enclose ½ an acre of ground, to be fenced with a stone wall 6 feet high, if

done by the landlord ; and if by the tenant, 3 feet of stone and 3 feet of turf—which last construction, if done by contract, would not cost more than 2s. per rood of 6 yards. In this form of stell, without a plantation, the wind would strike against a perpendicular face of the wall in either recess, and being directed upwards, would throw the snow down immediately beyond the wall into the inside of the stell. It is for this reason that objections are taken to inside stells.

Inside Stells.—Opinion is not agreed as to the best form of stell for high

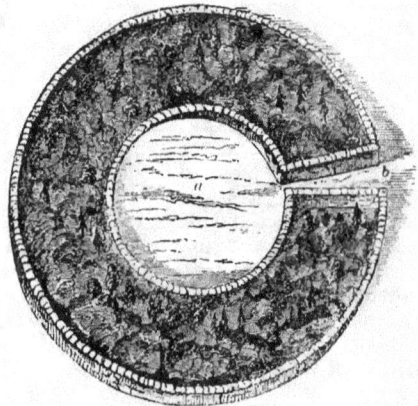

Fig. 763.—*Inside stell sheltered by plantation.*

pastures, where wood is seldom found. At such a height the spruce will not thrive ; and the larch, being deciduous, affords but little shelter with its spear-pointed top. There is nothing left but the evergreen Scots fir for the purpose, and when surrounding a circular stell *a*, fig. 763, it would afford acceptable shelter to a large number of sheep. This stell consists of 2 parallel circles of wall, enclosing a plantation of Scots pine, having a circular space, *a*, in the centre for sheep, as large as to contain any number. For obvious reasons the entrance to the stell should be the same width all through, not wider at the outer end than the inner, as shown in the figure, which has the twofold disadvantage of increasing the velocity of the wind into the circle, and of squeezing the sheep the more the nearer they reach the inner end of the passage.

Circular Stells.—But where trees cannot be planted with a prospect of success, stells may be formed without them, and indeed usually are; and of all forms that have been tried, the *circular* has obtained the preference on hill-farms, as shown in fig. 764. Opinions differ as to size. Some think 8 to 10 yards inside measurement best; others prefer a larger size, perhaps 18 yards.

Giving Hay at Stells. — Circular stells should be fitted up with *hay-racks* round the inside, not in the expensive form of circular woodwork, but of a many-sided regular polygon. It is a bad plan to make sheep eat hay by rotation, as some recommend, because the timid and weak will be kept constantly back, and suffer much privation for days at a time. Let all have room and liberty to eat at one time, and as often as they choose. The hay-stack should be built in the centre of the stell, on a basement of stone, raised 6 inches above the ground to keep the hay dry. The circumference of the stell measures 160 feet round the hay-racks; and were 8 or 9 six-feet hurdles put round the stack, at once to protect the hay and serve as additional hay-racks, they would

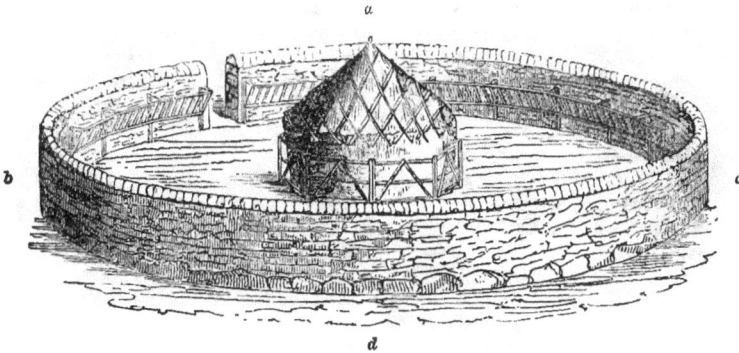

Fig. 764.—*Circular stell, with hay-racks and hay-stack.*

afford 47 feet more, which would give 1 foot of standing-room at the racks to each of 200 sheep at one time.

It is well to have some turnips stored beside the stells for use in a protracted snowstorm.

General Notes.

Bridging Rivulets for Sheep.—Where a rivulet passes through an important part of a farm, it will be advisable to throw *bridges* for sheep across it at convenient places. Bridges are best constructed of stone, and though rough, if put together on correct principles, will be strong; but if stone cannot be found fit for arches, they may do for buttresses, and trees laid close together across the stream, held firmly by transverse pieces, and then covered with tough turf, form a safe roadway.

Young Sheep best for Hill-farms. —The state of hill-pastures modifies the management on hill-farms. The hill-pasture does not rise quickly in spring, nor until early summer; and when it does begin to vegetate it grows rapidly, affording a full bite. It is found that this young and succulent herbage is not congenial to the ewe—it is apt in the autumn to superinduce in her the liver-rot; but it is well adapted for forwarding the condition and increasing the size and bone of young sheep. It is therefore safer for many hill-farmers to purchase lambs from south-country pastoral farmers, who breed Blackface sheep largely, as well as Cheviot, than to keep standing flocks of ewes of their own.

Nomenclature of Sheep.

The various classes of sheep are spoken of by different names throughout the country. A new-born sheep is a *lamb*, and retains the name until weaned from its mother. The generic name is altered according to the sex and state of the animal: when a female, it is a *ewe-lamb ;*

when a male, a *tup-lamb*; and this last is changed to *hogg-lamb* or *wether-lamb* after it has undergone castration.

In Scotland, after a lamb has been weaned, until the first fleece is shorn, it is a *hogg*, a female being a *ewe-hogg*, a male a *tup-hogg*, and a castrated male a *wether-hogg*.

After the first fleece has been shorn, a ewe-hogg becomes a *gimmer* or *shearling-ewe*, a tup-hogg a *shearling-tup*, and the wether-hogg a *dinmont*. After the second shearing, a gimmer is a *ewe*, if *in lamb*; if not in lamb, a *barren gimmer* or *yeld ewe*, and if never put to the ram, a *yeld gimmer*. A shearling-tup is then a 2-*shear tup*, and a dinmont a *wether*, but more correctly a 2-*shear wether*.

A ewe three times shorn is a *twinter ewe* (two-winter ewe); a tup a 3-*shear tup*; and a wether still a *wether*, or more correctly a 3-*shear wether*.

A ewe four times shorn is a *three-winter ewe* or *aged ewe*; a tup, an *aged tup*, a name he retains ever after.

Tup and *ram* are synonymous terms, applied to entire males.

A ewe that has borne a lamb and fails to be with lamb again is a *yeld* or *barren ewe*. After a ewe has ceased to give milk she is a *yeld ewe*.

A ewe when removed from the breeding flock is a *draft ewe* or *broken-mouthed ewe*; gimmers unfit for breeding from are *draft gimmers*; and lambs, dinmonts, or wethers, when drafted, are *sheddings, tails, shots*, or *drafts*.

In many parts of England a somewhat different nomenclature prevails. Sheep bear the name of *lamb* until 8 months old, after which they are *ewe tegs* or *she hoggs* and *wether tegs* until once clipped. Gimmers are *theaves* or "two tooths" until they bear the first lamb, when they are *ewes of 4-teeth*, next year *ewes of 6-teeth*, and the year after *full-mouthed ewes*. Dinmonts are *shear hoggets* until shorn of the fleece, when they are 2-*shear wethers*, and thereafter are *wethers*.

Rig and *chaser* are terms applied to a lamb when one of its testicles does not come into the scrotum.

Chilver is a name sometimes applied in Hampshire to ewe lambs from weaning time till Christmas, when they become tegs.

BRITISH WOOL.

The following notes on the origin, characteristics, and uses of British wool are contributed by Mr S. B. Hollings, Calverley, near Leeds:—

The United Kingdom is a place of variety, no matter from whatever standpoint judgment is given. And in those conditions which are responsible for the production of different types of sheep and wool—such as climatical and physical conditions—this variety is no less distinct than in other respects. Moreover, it is fairly safe to say that there are few countries more free from the disadvantages of unsatisfactory extremes of various kinds.

Sheep-farming is an industry which is by no means disregarded by those who seek their means of livelihood from the land. Still, in the United Kingdom it is not what it once was, on account of competition with colonial and foreign wool and mutton. Judged from the wool standpoint, this is perhaps truest in connection with the longer and most lustrous types, for it may be safely stated that many medium and short breeds — notably those of the white and crispy nature — cannot be seriously competed against, for the reason that they cannot be matched. But in regard to long wool, it should be encouragement to the British farmer to observe that he has the clear lead in the ideal conditions for wool production which are available for him. Pure lustre wool soon turns cross-bred-like in the warm colonies, this meaning loss to a greater or less extent of lustre, length, and uniformity, which are vital characteristics.

Again, in many localities cross-breds only are suitable, and in producing these experience has proved breeding difficulties to exist which necessitate the employment of the shorter-wool types of sheep; and the types these produce, as previously suggested, do not advantageously compete with ours, for the reason that they differ from them so much as to make their use as substitutes impossible save in comparatively few cases. With respect to mutton, in spite of the enormous imports of chilled and frozen carcases, a strong demand still prevails for the home-grown article, and doubtless will be maintained to an extent which, along with the returns for high-class wool, will at least justify the continuance of this industry as much as any other in these days of small profits in all agricultural as well as in other callings.

That this idea is just now becoming prevalent is shown by the growing returns relating to sheep. With the development of the Colonial wool trade —most marked from about 1860—came a decrease in the numbers of sheep reared and quantity of wool grown in Great Britain; and this continued up to quite recent times, say 1905. Then, largely owing to the high prices prevailing, the turn in the right direction set in. During 1907 some 29 million sheep and lambs were depastured in the British Isles, and these yielded a return in wool of 130½ million lb. weight—a quantity of which England contributed, roughly, 57 per cent, Scotland 21½ per cent, Ireland 14½ per cent, and Wales 7 per cent.

Classification of Wools.

Coming to a study of the various breeds of sheep and the types of wool produced by these, the initial difficulty presenting itself is that of a suitable classification. As might be expected, sheep grown under such diverse conditions as obtain in this country, and subject to all the modifications cross-breeding can make, differ both in type of animal and in wool to an extent which makes a perfect classification almost impossible. The following system of classification has been adopted, not so much on account of its accuracy but because of its convenience for our present purpose:—

(1) Long-wool breeds — Lincoln, Leicester, Cotswold, Border Leicester, Wensleydale, Devon, and Romney Marsh.

(2) Short-wool breeds — Southdown, Shropshire Down, Suffolk Down, Hampshire Down, Oxford Down, Ryeland, and Dorset Horn.

(3) Mountain breeds — Blackface, Cheviot, Lonk, Herdwick, Dartmoor, and Exmoor.

The first class consists of types of very large and valuable sheep, chiefly inhabiting the heavier and richer agricultural lands of the western and midland counties of England. They yield wool of a long, strong, and lustrous type, most suitable for the lustrous and demi-lustrous kinds of dress fabrics and linings. Class 2, usually termed "Down-wool breeds," includes sheep of a smaller type, distributed over the more southern portions of England, and these produce wool of a white and crisp type, which is extremely useful for hosieries, flannels, serges, blankets, shawls, &c. From the types in both these classes growers in all wool-producing countries have drawn sheep for the building up and improvement of their flocks to an extent which has rightly earned for the United Kingdom the name of "The World's Stud Farm."

The mountain breeds in class 3, as might naturally be expected, are generally of a somewhat poorer order. Still, these breeds have their great value; without them much land would be sheepless, and as a consequence the range of wool qualities and the variety in price of fabrics—both so necessary for the varying requirements of the trade—would be disadvantageously less. The uses of these are in cheap serges, hosieries, blankets, flannels, and carpets.

A fourth class might very properly be made, consisting of "half-breds," or more correctly "cross-breds," produced by crossing the afore-mentioned types together for purposes of improving both mutton and wool, though chiefly the former. This class is somewhat large, with representatives scattered throughout almost all the sheep-growing areas, the wool yielded being of medium length and quality and suitable for medium-class dress fabrics, serges, hosieries, and woollens. As these are the cross-bred progeny of the breeds referred to, and which will be detailed shortly, there is no necessity for a separate classification.

Long Wools.

Lincoln Wool.—Reverting to class 1 —the Long-wool breeds—the Lincoln must be placed at the head. The Lincoln is the longest and strongest woolled of all British breeds—the wool being 10 inches and upwards in length; it is of excellent lustre for its type, of a fair degree of fineness (being 36's to 40's quality), soft to the handle, and very elastic. The fleece varies from 8 to 12 lb. in weight, though at times it is double this, and it will generally yield three-quarters of its greasy weight in scoured wool. It finds employment in the best dress fabrics, and, because of its exceptional length, strength, and elasticity, it forms the chief material in the "hog top" wrapping for the squeegee rollers of wool-scouring bowls, this material only being really serviceable under practical conditions.

Leicester Wool. — The Leicester is often placed along with the Lincoln breed, especially when the wool is considered. The Leicester wool is of an excellent type, and very lustrous in staple. Its uses are similar to those of the Lincoln wool, though it might also be noted that the addition of Leicester wool to Lincoln gives to that product the quality of softness to a remarkable degree.

Cotswold Wool.—The Cotswold wool is of the demi-lustre type, a shade finer than the Leicester (44's) and a little shorter, with a weight of fleece of about 8 lb., and is of much value in the making of dress goods and linings.

Border Leicester Wool.—The wool produced by this breed is excellent. It is of a demi-class, of good length and fineness (occasionally 46's quality), and is eagerly sought for dress fabrics, linings, &c.

Wensleydale Wool. — This breed, originally containing much Leicester blood, yields a fleece about 8 lb. in weight, of a somewhat curly but very lustrous character, of 40's quality and fair length, which is used for purposes similar to the lustre wools already noted.

Fleece of Devon Long-wools.—Of the two types of these sheep, one, designated South Ham, grows a somewhat fine and silky wool, generally used along with the wool of the ordinary Devon.

Romney Marsh Wool.—While not being of highest excellence in regard to wool, this is one of the most valuable of English breeds. The wool is demi-lustrous, of 46's quality, of good length, strength, and oftentimes with a fleece weight of about 7 lb.

Short-wool Breeds.

Typical of these is the Southdown, a breed which, because of its fineness, whiteness, and softness of wool, might even be called the English Merino. No breed has been more perfected, both as regards mutton and wool, than this. The wool is extremely crimpy, about 3 inches in length, of thick and massive staple, 50's to 56's in quality (this only being a point lower than strong Merino wool), with a fleece weight of 5 lb. In handle it is somewhat harsh and dry, this being due to chalk, which robs the fibre of its nature, and leaves it also somewhat impaired in strength as compared with the lustres. For hosieries, flannels, dress fabrics, serges, &c., it is in great demand.

Shropshire Wool.—For mutton and wool this breed gains the highest praise; in fact, with regard to the latter no breed is in greater demand. The wool is about 5 inches long, of 50's quality, open in fibre, and of excellent spinning quality.

Other Down Wool.—The wool of the Oxford, Hampshire, and Suffolk is fairly similar, being of the Down type just described. The uses are much the same.

Ryeland Wool.—The Ryeland breed yields exceeding fine and open wool of the Down type, but of a small fleece weight. This wool has suffered much in competition with strong colonial wool of the Merino type which forms a satisfactory substitute, and this has interfered with its development to a considerable extent.

Dorset Horn Wool.—This breed is of greater value for mutton purposes than for wool. The fleece is light—3 to 4 lb. in weight, with wool fairly long, fine and bright in appearance, of use as the ordinary Down types.

Mountain Breeds.

Blackface Wool.—The wool of this distinctive breed is not of good quality. It is long, thick 28's to 32's quality, harsh and kempy, of little lustre, and comparatively small weight of fleece, say 4 to 5 lb. It is used mostly in the production of carpets, rugs, &c., of medium class character. In some parts of the extreme north there is a variety of sheep whose fleece is described as wool and hair, the woolly part being shed, plucked, or shorn each year. This wool is of medium length and softness, of fair spinning quality, and is suitable for use as coarse serge and tweed-like fabrics.

Cheviot Wool.— The wool of the Cheviot breed is dense but fairly fine, 46's quality, and long, with a fleece of about 4 lb., being of greatest use in the making of tweeds, and for hosieries and flannels of medium types. Crossed on the Border Leicester, this gives the North or Leicester-Cheviot wool of Yorkshire for which much demand exists, large quantities frequently going to America.

Lonk Wool.—The wool of the Lonk sheep is less characteristic than that of the Cheviot breed, but it is easily disposed of for use in low tweed and serge making.

Herdwick Wool.—This breed yields a fleece of only 3 to 4 lb., the wool being coarse and open. It is of medium length, and fulfils requirements similar to the Blackface and Lonk wools.

Dartmoor and Exmoor Wools.— The wool yield of these breeds is small in weight, short in staple, but is soft. It is used for hosiery, blankets, and flannels.

Welsh Wools.—Generally two kinds exist which are natural to Wales. The first occupies the highest mountains, and yields wool often coloured black, greyish, white, and brown, but of a coarse nature and only medium in length. The second class, which also inhabits the mountains and hills, yields white wool from which the celebrated Welsh flannels are made. The wool is not uniform either in length or fineness, and it contains many kemps. The fleece weight is about 2 to 3 lb.

Irish Sheep and Wools.

As is the case with Welsh sheep, two distinct varieties exist in Ireland—those of the mountain and those of the vale. The mountain sheep somewhat resemble the Welsh and Scotch types, and yield similar wool, though these when carefully tended and grown in less elevated positions show marked improvement in character. Wool from the vale sheep (of which the Roscommon is the only pure breed) is of medium length (6 inches), and of the Down type, but longer of course, and is extremely serviceable in the making of hosieries and flannels, these often being of an excellent quality.

MARKETING OF LIVE STOCK AND DEAD MEAT.

Historical.

The origin of Markets and Fairs (says Mr Loudon M. Douglas, to whom the Editor is indebted for these notes) is wrapt in obscurity, but their history, so far as known, is of a most interesting character. The word market means traffic or trade, and although associated at one time with other pursuits, that meaning has been attached to it from the earliest historical times.

A fair has come to be looked upon as synonymous with a market, although at one period there was a marked difference, the market being liable to be held on any day of the week, whereas a fair was looked upon as a much larger function, and was only held on specified dates, which, in some cases, were proclaimed some time before.

There are many references to the customs of trade in Biblical times, showing that then the principles of marketing were well known. It was to Greece, however, that the principle of marketing owed one of its greatest advantages, as it was the Greeks who invented the idea of a gold and silver coinage whose value should be unquestioned in any country of the world—that is to say, a universal coinage.

In early as in later times, fairs and markets were associated with religious festivals, and, indeed, as far back as Pythagoras (550 B.C.) it was said, on the authority of Cicero, that large numbers of people attended the religious festivals on those days merely with a view to trade. In later times this became a custom, and for many centuries fairs and markets were indissolubly associated with religious festivals, and the practice was carried to so great an extent that the principal fairs and markets during the middle ages were held on Sundays in the churchyards, there being thus a curious blend of business and devotion. The incongruity of the combination, however, was recognised in England in the reign of King Henry VI., during which period it was practically suppressed, on the ground that it was reminiscent of the buyers and sellers in the Temple.

Fairs were at one time common to all countries, but with the introduction of railways and quick transport, together with rapid postal and other means of communication, they have fallen in esteem, and now occupy only a secondary place in market transactions in all civilised countries.

Several of the great fairs of Europe are still carried on; and while some of them are specially devoted to the buying and selling of specific kinds of goods—such, for example, as the Leipzig Book Fair or the Nottingham Michaelmas Goose Fair—there are others which are devoted to the handling of general merchandise. We have also such fairs in the United Kingdom as the Glasgow Fair, Donnybrook Fair, and the Fair of St Bartholomew, the last-mentioned being the greatest fair that has ever existed. The original intention, however, in connection with these fairs — namely, that they should be for the marketing of goods— has been modified to suit the progress of civilisation, and they have dwindled down to mere occasions for a holiday.

It is of interest to recall that fairs and markets have always had some privileges, such as the right of exclusive dealing within a certain area of the fair while it lasted; the administration of justice in connection with the transactions, or anything that has happened during the fair, in a Summary Court, described as the Court of Piepowder (from the French, *pied poudré,* meaning dusty feet), and so described, it is supposed, because of the dusty feet of the suitors. In later days the Clerk of the Markets became the judge of this court, and exercised the jurisdiction previously held by representatives of the community. This privilege has, however, been altogether done away with, and the Clerk of the Markets is now merely a recorder of the transactions which take place within the modern market.

Modern Fairs.

Modern fairs, in so far as the United Kingdom is concerned, are associated almost entirely with agriculture, and the tendency is to dispense with them altogether. In Ireland, where they are very numerous, and where they are largely live-stock markets for the sale of cattle, sheep, and pigs, they serve a useful purpose in remoter districts, where they enable farmers to bring their produce to one centre, and where they may be sure of getting the price which rules in the fair for any particular class of animals. This advantage, however, is also being supplanted by the institution of live-stock scales at various railway stations, where live stock may be sold to agents of purchasers, by weight, in place of by guesswork as obtains in a fair.

"Market overt" was a term which was applied to transactions carried out in open market. In England such markets were held in specified places and on particular days, but in modern life this has been replaced by the institution of shops, which constitute "market overt" in the same way. In Scotland "market overt" does not exist, and the difference will be understood when it is stated that "the owner of goods sold (in 'market overt') by one who has stolen them, or to whom they have been lent, may reclaim them from the purchaser." In England the owner would have no such privilege.

These rights pertaining to fairs, with many others of a like character which need not be recalled, only serve to show

how much they are out of sympathy with modern institutions. They served a useful purpose in their day in enabling produce of various kinds to be conveniently distributed. But wherever railway communication is efficient—and that is fast becoming universal—the need for the fair entirely disappears. The time, therefore, appears to be not far distant when they will cease to exist altogether, and give place to methods more in keeping with modern ideas.

A market at the present day, in so far as agriculture is concerned, means a suitable place—which may either be covered or open—in which the produce of the farm may be bought and sold.

Every town and considerable centre of population has its market, and, in many places, corn markets are held on distinct days of the week from live-stock markets, so that the selling of the cereal produce of the farm need not interfere with the disposal of the live stock.

The Marketing of Dairy Produce.

In connection with dairy farming there are two systems of disposing of the milk. The first is by converting it into butter, and this applies to districts which are remote from populous towns, where milk is valued at considerably less than what it would fetch in the neighbourhood of a large population. It does not pay at any time to make butter in the neighbourhood of a large city, as the value of milk for household purposes is, on the average, at least one-third greater than what it is for butter-making purposes. The farmer, therefore, who is remote from the large town, and who practises dairying, either converts the milk which he sells into butter and feeds the calves and pigs on the separated milk, or he sells his whole milk to a creamery, which may be jointly supported by a large number of farmers; or it may be operated on co-operative lines, in which case it is usual for the farmer to take back again about an equal quantity of separated milk to the whole milk he has supplied. The average price of this separated milk is reckoned at a penny per gallon, and he is therefore able to feed his pigs and calves in the same way as he would if making butter on his farm, but with much less

trouble than if he had a private butter-making establishment. In cheese-making, which is carried on very largely on such farms, there is not the same inducement towards co-operation, as it is entirely questionable whether co-operation in cheese-making is superior to what is accomplished by private enterprise. In either case, however, the residual whey is utilised as feeding, more especially for pigs.

Butter which is made on the farm is, in many cases, delivered to merchants or to consumers direct, but in many districts, more especially in Ireland, small farmers at the present day make what is termed " lump " butter, and which is simply butter produced in a crude way and without any regard to its proper grading. This lump butter is taken to butter fairs, which are held week by week, and is purchased by merchants, who mix all the lump butter together so as to make different grades of a uniform texture and appearance. These merchants term the produce, which they turn out in kiels, firkins, boxes, and packages — " factory " butter, so as to distinguish it from " creamery " butter, or such as is produced from the mixture of the milk. In the one case, the factory butter is the result of mechanical mixture of the various lumps; whereas, in the case of creamery butter, the production is the result of the mixture of the various supplies of milk. It is quite obvious that the creamery is very much more advantageous than the factory system, in so far as the production of a trustworthy and uniform article is concerned. The custom, however, among the small farmers of making their own butter is dying out but slowly, and is not likely to be extinguished until a greater number of creameries have been established, either by private enterprise or by co-operation.

Marketing of Live Stock.

The Markets and Fairs Weighing of Cattle Acts (1887 and 1891) require that all market authorities shall " provide and maintain sufficient and proper buildings or places for weighing cattle brought for sale within the market or fair, and shall keep therein or near thereto a weighing-machine and weights for the purpose of weighing cattle, and shall appoint proper persons to have charge of such machines and weights and to afford the use of such machine and weights to the public for weighing cattle, as may from time to time be required."

By these provisions the business of buying and selling live stock has been placed upon a much better footing, especially for the farmer. The live-weight prices of the various animals are shown clearly, and it is therefore not difficult for the farmer to get fair value for his produce.

The fee which market authorities may charge for these facilities is twopence for each head of cattle and a penny for every five or smaller number of sheep or swine.

There are various live-weight scales made; their construction, however, is practically identical, there being only a slight variation in design. The weighing-machine (fig. 765) consists of an ordinary platform weighing apparatus, and to the platform a cage is attached, the weight of which is allowed for in the counter-balancing. Cattle are simply driven one by one into the cage and are weighed entire, and from such weights it is com-

Fig. 765.

Live stock weighing-machine.

The use of the weighbridge for determining the live-weight of cattle has greatly increased. By its aid and the use of average tables for calculation, the approximate dead-weight of any animal can be determined. By this means both the seller and the buyer know, pretty nearly, the price per lb. which any price for the live animal will give. In this way a more satisfactory manner of marketing is arrived at than by buying or selling by "hand." The live-weight scale also facilitates sales by weight on the basis of either live- or dead-weight.

paratively easy to compute what the nett weight will be. For this purpose several simple rules have been devised by Mr John D. M'Jannet. These rules are as follows :—

For ascertaining the approximate Carcase-weight of Fat Cattle from their Live-weight.

1. For wastefully fat Smithfield Club show cattle, multiply the live-weight by 7 and divide by 10.
2. For extra prime Smithfield Club show cattle, multiply the live-weight by 2 and divide by 3.
3. For prime butchers' bullocks found in ordinary markets, multiply the live-weight by 3 and divide by 5.

4. For fair killing beasts found in ordinary fat-stock markets, multiply the live-weight by 4 and divide by 7.
5. For old fat cows, just take one-half of live-weight.

Live- and Dead-weight of Fat Pigs.

Young fat pigs weighing alive at farm from 83 lb. up to 158 lb. will dress from 62½ to 70 per cent. Fat pigs of prime quality, weighing alive at farm from 160 lb. up to 410 lb., will dress from 67½ to 77½ per cent.

The following table shows approximately the available produce from fat sheep :—

Description.	Unfasted live-weight.		Per cent of mutton.		Average.
	lb.	lb.	Lowest.	Highest.	
Three-part lambs . .	60 to 90		49	52	50 per cent.
Three-part hoggets .	90 to 120		50	54	51 ,,
Cross-bred hoggs . .	120 to 135		52	56	53 ,,
Half-bred ewes . .	140 to 180		50	53	51 ,,
Blackface ewes . .	116 to 136		50	52.7	51 ,,

A specially interesting and useful table is that which follows, and which gives the price per live cwt. of cattle between 80 and 83 stones, when the price "bid" and the weight are known. This table can be extended indefinitely on the same lines, and is a fair model of what such a table should be :—

TABLE SHOWING THE PRICE PER LIVE CWT. OF CATTLE WHERE THE WEIGHT AND THE PRICE BID IS KNOWN.

THE PRICE BID.	LIVE-WEIGHT OF ANIMAL.			
	80 stones or 10 cwt.	81 stones or 10 cwt. 14 lb.	82 stones or 10 cwt. 1 qr.	83 stones or 10 cwt. 1 qr. 14 lb.
	Price per cwt.	Price per cwt.	Price per cwt.	Price per cwt.
£ s. d.	s. d.	s. d.	s. d.	s. d.
13 0 0	26 0	25 8	25 4	25 0
13 2 6	26 3	25 11	25 7	25 3
13 5 0	26 6	26 2	25 10	25 6
13 7 6	26 9	26 5	26 1	25 9
13 10 0	27 0	26 8	26 4	26 0
13 12 6	27 3	26 11	26 7	26 3
13 15 0	27 6	27 2	26 10	26 6
13 17 6	27 9	27 5	27 1	26 9
14 0 0	28 0	27 8	27 4	27 0
14 2 6	28 3	27 11	27 7	27 3
14 5 0	28 6	28 2	27 10	27 6
14 7 6	28 9	28 5	28 1	27 9
14 10 0	29 0	28 8	28 3	28 0
14 12 6	29 3	28 11	28 6	28 2
14 15 0	29 6	29 2	28 9	28 5
14 17 6	29 9	29 5	29 0	28 8
15 0 0	30 0	29 8	29 3	28 11
15 2 6	30 3	29 11	29 6	29 2
15 5 0	30 6	30 1	29 9	29 4
15 7 6	30 9	30 4	30 0	29 7
15 10 0	31 0	30 7	30 3	29 10
15 12 6	31 3	30 10	30 6	30 1
15 15 0	31 6	31 1	30 9	30 4
15 17 6	31 9	31 4	31 0	30 7
16 0 0	32 0	31 7	31 3	30 10
16 2 6	32 3	31 10	31 6	31 1
16 5 0	32 6	32 1	31. 9	31 4
16 7 6	32 9	32 4	31 11	31 7
16 10 0	33 0	32 7	32 2	31 10
16 12 6	33 3	32 10	32 5	32 1
16 15 0	33 6	33 1	32 8	32 4
16 17 6	33 9	33 4	32 11	32 7
17 0 0	34 0	33 7	33 2	32 9
17 2 6	34 3	33 10	33 5	33 0
17 5 0	34 6	34 1	33 8	33 3
17 7 6	34 9	34 4	33 11	33 6
17 10 0	35 0	34 7	34 2	33 9
17 12 6	35 3	34 10	34 5	34 0
17 15 0	35 6	35 1	34 8	34 3
17 17 6	35 9	35 4	34 11	34 6

THE PRICE BID.	LIVE-WEIGHT OF ANIMAL.			
	80 stones or 10 cwt.	81 stones or 10 cwt. 14 lb.	82 stones or 10 cwt. 1 qr.	83 stones or 10 cwt. 1 qr. 14 lb.
	Price per cwt.	Price per cwt.	Price per cwt.	Price per cwt.
£ s. d.	s. d.	s. d.	s. d.	s. d.
18 0 0	36 0	35 7	35 2	34 8
18 2 6	36 3	35 10	35 4	34 11
18 5 0	36 6	36 1	35 7	35 2
18 7 6	36 9	36 4	35 10	35 5
18 10 0	37 0	36 7	36 1	35 8
18 12 6	37 3	36 10	36 4	35 11
18 15 0	37 6	37 1	36 7	36 2
18 17 6	37 9	37 4	36 10	36 5
19 0 0	38 0	37 6	37 1	36 7
19 2 6	38 3	37 9	37 4	36 10
19 5 0	38 6	38 0	37 7	37 1
19 7 6	38 9	38 3	37 9	37 4
19 10 0	39 0	38 6	38 1	37 7
19 12 6	39 3	38 9	38 3	37 10
19 15 0	39 6	39 0	38 6	38 1
19 17 6	39 9	39 3	38 9	38 4
20 0 0	40 0	39 6	39 0	38 7
20 2 6	40 3	39 9	39 3	38 10
20 5 0	40 6	40 0	39 6	39 1
20 7 6	40 9	40 3	39 9	39 4
20 10 0	41 0	40 6	40 0	39 7
20 12 6	41 3	40 9	40 3	39 10
20 15 0	41 6	41 0	40 6	40 0
20 17 6	41 9	41 3	40 9	40 3
21 0 0	42 0	41 6	41 0	40 6
21 2 6	42 3	41 9	41 3	40 9
21 5 0	42 6	42 0	41 6	41 0
21 7 6	42 9	42 3	41 8	41 2
21 10 0	43 0	42 6	41 11	41 5
21 12 6	43 3	42 9	42 2	41 8
21 15 0	43 6	43 0	42 5	41 11
21 17 6	43 9	43 3	42 8	42 2
22 0 0	44 0	43 6	42 11	42 5
22 2 6	44 3	43 9	43 2	42 8
22 5 0	44 6	44 0	43 5	42 10
22 7 6	44 9	44 2	43 8	43 1
22 10 0	45 0	44 5	43 11	43 4
22 12 6	45 3	44 8	44 1	43 7
22 15 0	45 6	44 11	44 4	43 10
22 17 6	45 9	45 2	44 7	44 1

Carcase Competitions.

At the Smithfield Club Show, held in December 1908, the carcase classes were very much extended, and some interesting results were obtained. It was observed that the animals submitted in competition, first as live animals and latterly in the carcase, did not in every case gain the same awards in the two classes, the judgment of the live animal not being borne out when the carcases were examined. It is felt strongly by meat purveyors that more attention should be given to carcase competitions, which they regard as the only satisfactory test of an animal which is destined for food.

It may be of interest to record here the various entries in the classes of cattle, sheep, and pigs which obtained the highest awards at that show; and it may be observed that a comparison of the actual weights realised with the weights stated in the foregoing table, indicates that the rules laid down are fairly accurate.

TABLES OF THREE CARCASE CLASSES ENTERED AT SMITHFIELD CLUB SHOW, LONDON, 1908.

STEERS NOT EXCEEDING TWO YEARS OLD.

Catalogue Number and Name of Exhibitor.	Breed.	Live-weight.	Carcase Weight in 8-lb. stones and lb.	Percentage of Meat.	Placed by Judges Alive.	Carcase Awards.	Realised per 8 lb.
		cwt. qr. lb.	st. lb.				s. d.
584 Coed Côch Trustees .	Welsh	11 3 12	101 6	61.30	...	....	4 0
585 J. J. Cridlan .	Aberd.-Angus	12 1 13	115 4	66.71	4th	H.C.	4 4
586 Sir Walter Gilbey, Bt.	Cross-bred	11 1 12	101 7	64.00	3rd	...	4 6
587 G. Young .	Cross-bred	10 1 2	92 2	64.17	H.C.	...	6 0
588 J. & G. Young .	Cross-bred	10 3 1	94 0	62.41	H.C.	2nd	5 0
589 W. A. Sandeman .	Cross-bred	11 2 20	108 5	66.43	1st	C.	4 8
590 James McWilliam .	Aberd.-Angus	11 1 26	108 4	67.50	...	4th	4 8
591 Joseph Godman .	Cross-bred	11 2 7	107 4	66.41	...	...	4 4
592 R. M. Greaves . .	Welsh	10 1 0	.90 3	63.00	...	...	4 8
593 Finlay Munro . .	Cross-bred	10 1 16	97 4	67.00	...	3rd	5 8
594 R. G. Nash .	Cross-bred	10 1 18	94 1	64.58	2nd	1st & Ch.	7 0
595 Lionel Phillips .	Cross-bred	11 1 8	103 6	65.46	...	...	4 4
596 J. Douglas Fletcher .	Aberd.-Angus	11 2 18	108 4	66.46	...	R. & H.C.	4 4
597 Univ. Coll., N. Wales	Welsh	11 3 10	104 5	63.12	H.C.	...	4 6
598 Sir J. Colman, Bart..	Cross-bred	12 0 18	110 1	64.68	C.	H.C.	4 0
599 Duchess of Newcastle	Cross-bred	9 0 26	84 0	65.00	...	H.C.	5 0
600 Viscount Tredegar .	Shorthorn	13 0 9	124 6	68.10	...	...	3 6
	Averages 1908	11 1 3	102 6	65.11	...	...	4 8⅞
	" 1907	11 1 8	103 2	65.14	...	...	4 5

ONE PURE LONG-WOOLLED WETHER SHEEP ABOVE 12 AND NOT EXCEEDING 24 MONTHS.

Catalogue Number and Name of Exhibitor.	Breed.	Live-weight.	Carcase Weight	Percentage of Meat.	Placed by Judges Alive.	Carcase Awards.	Realised per 8 lb.
		lb.	lb.				
633 J. G. Young . .	Cheviot	165	100	.60.60	2nd	...	4 0
634 Henry Simpson .	Wensleydale	171	111	64.97	...	...	3 8
635 Sir J. Gilmour, Bart.	Blackfaced	147	99	67.35	4th	...	3 8
636 J. D. Fletcher . .	Cheviot	133	88	66.16	C.	2nd	5 6
637 William Kennedy .	Cheviot	125	76	60.80	...	...	4 6
638 Univ. Coll., N. Wales	Welsh M'n	102	63	61.76	3rd	4th.	6 0
639 Robert Graham .	Cheviot	116	72	62.06	...	R. & C.	5 0
640 Sir R. W. Jardine, Bt.	Cheviot	119	76	63.86	...	3rd	5 6
641 W. Vivers & Son .	Cheviot	116	75	64.65	1st	1st	7 2
642 Henry Dudding .	Lincoln	304	205	67.43	...	...	2 6
	Averages 1908	149¾	96½	63.96	...	...	4 9
	" 1907	130	86	65.98	...	...	5 4

ONE PIG, NOT EXCEEDING TWELVE MONTHS OLD, ABOVE 220 LB. AND NOT EXCEEDING 300 LB. LIVE-WEIGHT.

Catalogue Number and Name of Exhibitor.	Breed.	Live-weight.	Carcase Weight.	Percentage of Meat.	Placed by Judges Alive.	Carcase Awards.	Realised per 8 lb.
		lb.	lb.				s. d.
716 H. Peacock . .	Berkshire	304	187	61.51	2nd	2nd	3 10
717 Briant Brothers .	Large White	282	201	71.27	H.C.	3rd	3 6
718 Lionel Phillips . .	Tamworth	285	256	89.78	3rd	...	3 2
719 J. Douglas Fletcher .	M. White	281	233	82.91	4th	4th	3 4
720 Vis. Com. M. C. A. .	Berkshire	288	210	72.91	1st	1st & Ch.	4 8
721 Kenneth M. Clark .	Large Black	254	163	64.17	...	...	3 8
722 Thomas Goodchild .	Large Black	239	183	76.56	...	R. & H.C.	3 6
723 D. E. Higham . .	Berkshire	283	243	85.86	...	...	3 2
724 John Neaverson .	Large White	274	228	83.21	...	...	3 4
725 Harold Sessions .	Large Black	293	237	80.88	...	...	3 4
	Averages 1908	278	217	76.90	...	...	3 6½
	" 1907	276	229	82.87	...	...	3 3
	" 1906	256	213	83.10	...	...	3 9¼
	" 1905	258	213	82.66	...	...	3 10

The Meat-Supply.

Since the nineteenth century entered on its last quarter an entirely new set of conditions has sprung up in connection with the meat-supply of the United Kingdom. Frozen and chilled meats of all descriptions have been imported from various foreign countries, notably the United States, Canada, Argentina, Australia, New Zealand, Holland, and Scandinavia; and, so far as can be seen, the supplies from these various sources are likely to go on increasing. That these supplies are already substantial is clearly seen from a glance at the supplies of dead meats which pass through the Central Markets of London.

According to the returns issued in 1908, four tons out of every five which pass through the London Central Meat Markets, in order to supply the 6,000,000 consumers in London, are of foreign origin. That is to say that the meat-produce derived from the United Kingdom, and which passes through Smithfield Market, amounts to only 20½ per cent of the total. The actual figures for five years are given in the following table:—

TABLE SHOWING THE QUANTITIES OF HOME AND FOREIGN MEAT PASSING THROUGH SMITHFIELD MARKET, LONDON.

Year.	Weight of Market Supplies.	Increase on previous year.		Origin or Sources of Supplies in terms per cent.			
				"English killed" and United Kingdom productions.	Imported productions, chilled or frozen.		
		Weight.	Rate per cent.		North and South American.	Australasian.	Continental.
	tons.						
1869	127,981	...	...	97.7	nil.	nil.	2.3
1877	197,631	69,650	54.4	89.0	7.4	nil.	3.6
1887	259,383	61,752	31.2	77.5	9.5	5.8	7.2
1897	391,707	132,324	51.0	47.9	18.8	20.3	13.0
1907	417,057	25,350	6.4	36.6	24.6	25.7	13.1

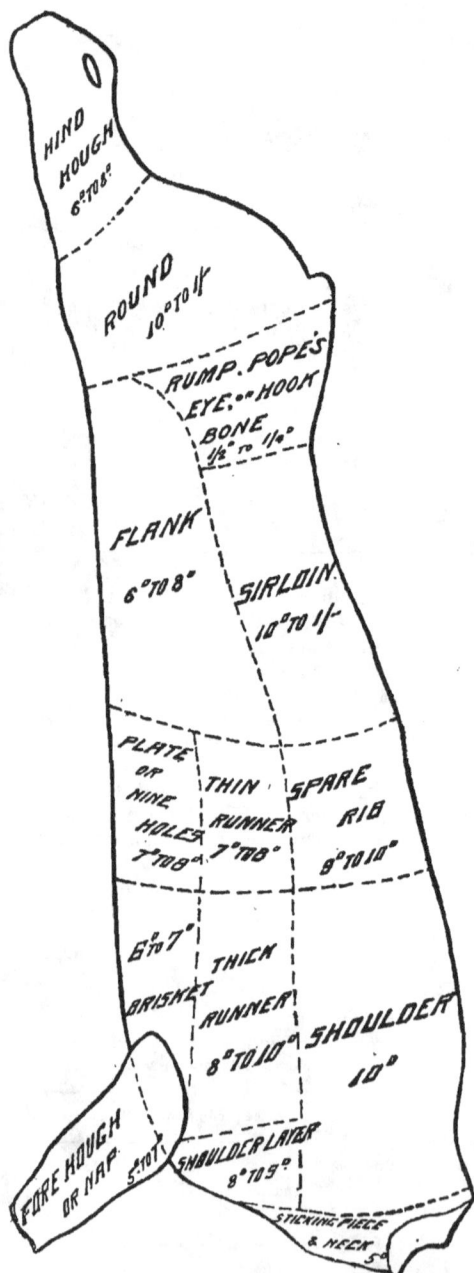

HIND HOUGH
6°TO 8°

ROUND
10°TO 1/-

RUMP. POPE'S
EYE. ·· HOOK
BONE
1/2° TO 1/0°

FLANK
6°TO 8°

SIRLOIN
10°TO 1/-

PLATE OR NINE HOLES
7°TO 8°

THIN RUNNER
7°TO 8°

SPARE RIB
8°TO 10°

6°TO 7°

THICK RUNNER
8°TO 10°

BRISKET

SHOULDER
10°

FORE HOUGH OR NAP
5°TO 7°

SHOULDER LAYER
8°TO 9°

STICKING PIECE & NECK
5°

[Copyright of Loudon M. Douglas.

Fig. 766.—*Side of beef.*

The various cuts of beef which are shown on the diagram are those which obtain throughout the United Kingdom, there being, however, slight modifications in various districts. It will be observed that the "Round," "Rump," and "Sirloin" are the highest-priced parts.

The importation of frozen dead meat to this country began about 1876. The earlier consignments came chiefly from the United States of America, but in 1879 two Scotsmen, named Bell and Coleman, began to bring cargoes of frozen meat from Australia. Since then the supplies have steadily increased, until during 1907 they reached the gigantic total of 18¾ million cwt. of all kinds of meats, valued at about £42,000,000 sterling. Hitherto much of the meat so imported has been frozen to about 18° F., but it has been found that the frozen product commands a much smaller price than "chilled" meat, which is carried at a much higher temperature—namely, at 28° F. Consequently, arrangements have now been made to bring in, from Argentina and the United States, most of the meat in the "chilled" condition. It has been found that chilled meat can be sold so as to compete with the home-grown article, and frequently fetches higher prices. As this means an enormous increase in the revenue derivable from imported meat, it is not to be wondered at that the great shipping companies are strenuously endeavouring to bring all their meats under such conditions.

Marketing of Meats.

So far there has been very little attempt to develop the handling of meats at the farm. A good many farmers slaughter their own pigs in order to make "farm-cured" bacon, but they generally rely upon selling their cattle and sheep on the hoof. In order, therefore, that they may the better understand what the meat-purveyors require, it may be of interest to refer in some little detail to the processes in use in the handling of meats.

When live stock intended for human food passes through markets, it is destined, as a rule, for either a private slaughter-house or a public abattoir.

A private slaughter-house has many objections, the greatest being that it is difficult to control the meat which may be handled there. The conditions also which necessarily exist in a private

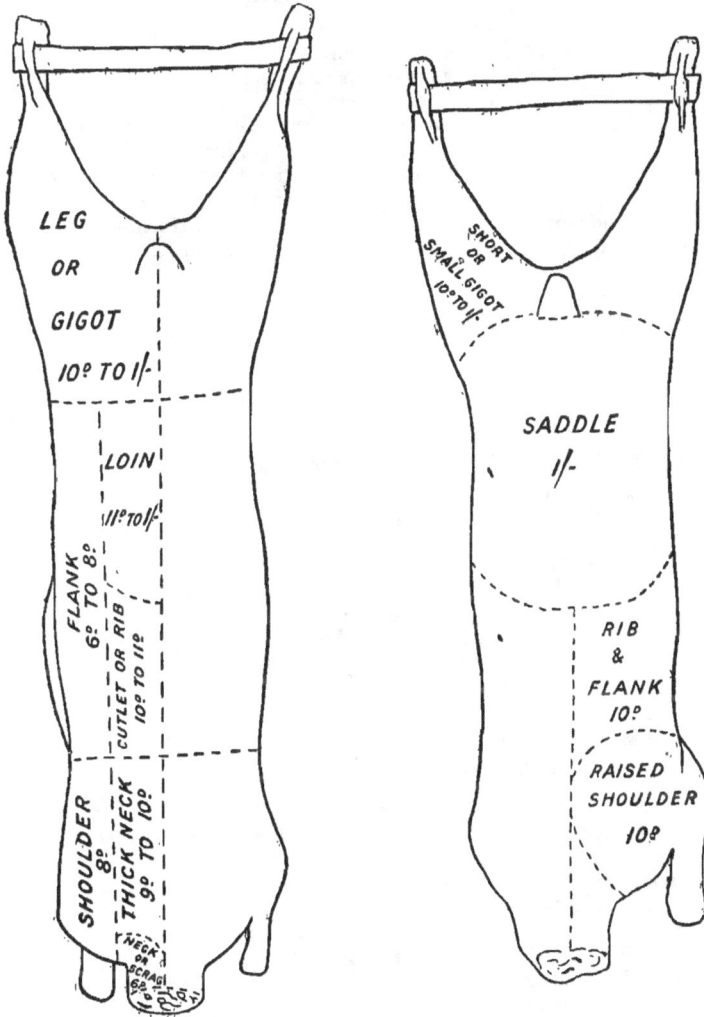

[Copyright of Loudon M. Douglas.

Fig. 767.—*Cutting up of a carcase of mutton.*

The methods in use for dividing up the carcases of mutton vary slightly in different parts of the United Kingdom, but the general custom is shown on the two illustrations, where the names of each portion and the average prices are given. It will be observed that the dearest portions are the legs and saddle, and it is to the increase of these that breeding should be directed. The V space between the legs should be as small as possible, and the greatest development should take place right across the "saddle."

establishment as compared with a public one must be inferior, owing to the fact that the cost of installing proper equip- ment would be too great for a private individual. There are, of course, many exceptions to this, but only in exceptional cases should private slaughter-houses be allowed to exist.

Public slaughter-houses are very frequently erected in conjunction with cattle-markets, so that when live stock is disposed of the animals may be conveniently conveyed to the abattoir adjoining. This relation of the market to the abattoir is also convenient in another way, as it enables complete inspection on the hoof to take place, so that any animal which is suspected of being diseased can be intercepted before it enters the abattoir.

The principal advantages in the handling of animals in an abattoir are, that expert slaughtermen are employed, and complete control and veterinary inspection is possible. Meat can also be matured in a properly constructed abattoir much better than in small premises, as usually plenty of ventilation is provided, and chilling rooms form part of the equipment also. That part of technical detail, however, is not one which very greatly interests the farmer,—what he is concerned with is the product itself. It is his business to supply meat which will conform to the requirements of the meat-purveyor, and what these requirements are can be best understood from a diagram showing

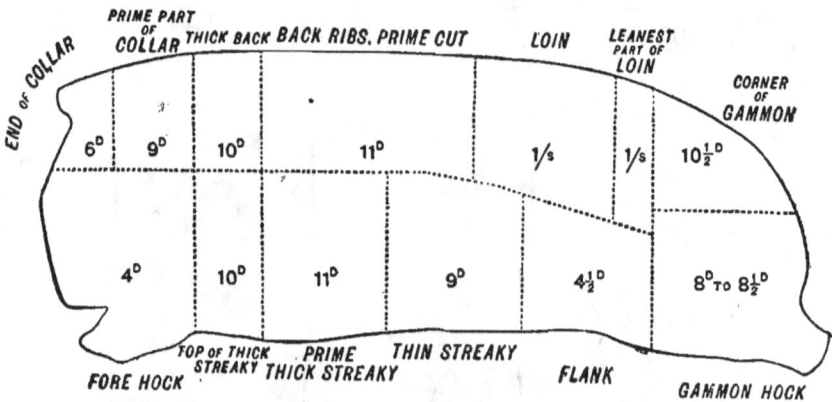

Fig. 768.—*Side of bacon.*

[Copyright of Loudon M. Douglas.

In the cutting up of a side of bacon cured in the "Wiltshire" method, as shown in the illustration, it will be observed that the loin brings the highest price. Breeding should therefore be directed towards increasing the development of the live animals so that the fleshy part of the back from which the loin is derived should form the leading feature.

what parts of the animal bring the highest prices. In the diagram in fig. 766 the British method of cutting up a carcase of beef is illustrated.

The cutting up of the carcases of sheep (fig. 767) is not so detailed an operation, and it does not involve quite so much skill, as the cutting up of beef. The breeding of sheep, however, for the meat-purveyor should be strictly on the lines of producing the largest quantity of meat to the smallest proportion of bone, and it is particularly desired that the meat of the hind-quarters should be extremely full. The transverse sections through the middle should show a large richly coloured "eye." When these two points are accentuated in any sheep, it answers then to the meat-purveyor's requirements.

In so far as pigs are concerned, they may be required for fresh pork or bacon, but in either case the points are pretty much the same, so that in breeding for bacon purposes farmers will answer the requirements of the meat trade all round, and the points to be studied cannot better be illustrated than by reference to the diagram showing the section into which a side of bacon is usually cut, and the prices of each (figs. 768 and 769).

Farmers as Retailers.

At various times attempts have been made to organise societies of farmers in order to retail the products of the farm—not so much on the lines of co-operation as of private enterprise. Such concerns have been instituted in various towns with considerable success, notably in dairying and also in the purveying of meat. Whether this is an advantageous line for farmers to follow or not it is hardly possible to say at the present time, for as yet the experience of it has been limited. If, however, a company of farmers would combine together to open several meat-purveyors' establishments, and in that way save all the cost of marketing and intermediate profits, there seems no reason why such a scheme should not succeed. It is altogether a question of capable management and a proper understanding of how to utilise the by-products of the business.'

Much profit is lost in the Meat Trade of the United Kingdom owing to the fact that a very large number of meat-purveyors slaughter their own animals, and are thus unable to utilise offals to the fullest extent. It would be far better for each one of these to hand over all the offals from his animals to a central depot, where they would be treated in bulk at very much less cost and more efficiently than could possibly be done on the small scale. The residual products are also entirely lost in small slaughter-houses, instead of being converted into fertilisers, which should be the final destiny of the waste material in connection with the handling of all animals used for food.

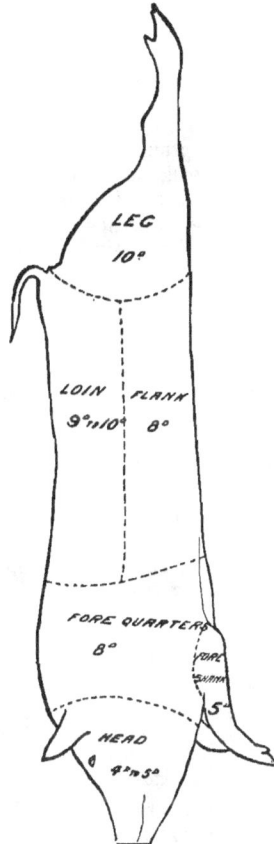

[Copyright of Loudon M. Douglas.

Fig. 769.—*Carcase of pork.*

The sections into which a side of pork is divided are not numerous, the principal being the leg and the loin, which, both in market pork and in connection with bacon-curing, bring the highest prices.

INSECT ENEMIES OF LIVE STOCK.

Dr R. Stewart Macdougall contributes the following notes on the Insect, Mite, and Tick enemies of stock.

Of the four chief classes of jointed-footed animals (Arthropoda) only two, the Insecta and the Arachnoidea (ticks and mites and spiders), contain enemies of stock. These insect and mite and tick enemies affect stock in different ways: e.g., they may be complete parasites, passing their whole life on the affected animal—e.g., lice and mange mites; or they may be parasitic for part of their life—e.g., the bots of cattle and horses and the sheep maggots; or they may visit the animal for a meal of blood and then leave—e.g., cleg and stable-fly.

INSECTS.

A description of the characteristics of insects and of the various orders of insects will be found in vol. ii. p. 442, &c. Of the various groups of insects only three contain stock enemies — viz., the Diptera or two-winged insects, the Mallophaga or biting lice, and the Parasitica section of Hemiptera—viz., the Sucking Lice.

THE TWO-WINGED FLIES.

The insects of this order have a complete metamorphosis (the keds are marked exceptions). The mouth - parts of the adult are fitted for a liquid diet, the mouth-parts of some being modified to form lancets by which wounds can be made previous to the sucking up of blood. The larva of a Dipterous insect is a legless maggot. The families of Diptera containing stock enemies are—

Tabanidæ or true gad-flies.
Œstridæ or bot-flies.
Muscidæ—e.g., stable and sheep maggot-flies.
Hippoboscidæ—e.g., ked and forest-fly.
Pulicidæ (a degraded family), or fleas.

True Gad-Flies.

These insects are harmful only in the adult condition when by their mouth lancets they wound for blood. The wound is made, never by a sting at the end of the body but always by the modified mouth - parts. The eggs are laid elsewhere than on stock, and the maggots of this family are not parasitic on stock.

The Ox Gad-fly (*Tabanus bovinus*) measures up to an inch in length and has a stout body; the thorax is brown-black and hairy and has dark stripes; the abdomen is red-brown, and along the middle line of the back is a row of whitish triangles. The flies are found from about the end of May onwards. Both maggot and pupa live in the soil. Another large Tabanus is *Tabanus sudeticus*, a somewhat darker fly than *bovinus*. Smaller species are *T. autumnalis*, *T. bromius*, *T. maculicornis*, and *T. cordiger*.

T. bovinus and *T. sudeticus* approach stock with a marked humming note; the smaller Tabanidæ land quietly on the beast visited for a meal.

The Cleg (*Hæmatopoda pluvialis*) measures half an inch in length and has a narrow body; there are pale stripes down the thorax, and a pale grey band and greyish spots on the upper surface of the abdomen. The grey wings have light-coloured markings. The maggot and pupa live in the soil. The flies are about in late summer. There are two other British species, *H. crassicornis* and *H. italica*.

The Blinding Breeze Flies (*Chrysops*) may be distinguished from the Cleg by the presence of three small eyes on the top of the head (in addition to the two large compound eyes). The compound eyes are golden-green with purple lines and spots. The flies have a somewhat square-shaped abdomen, and when at rest the wings are held somewhat apart. There are four species—*C. cæcutiens*, *C. relicta*, *C. quadrata*, and *C. sepulcralis*. The two first are the commonest.

The family Chironomidæ or Midges contains many British species. Most are harmless, but several species of the genus Ceratopogon cause great annoyance to man by their "bites." Every one knows the viciousness of these midges in summer and autumn. The larvæ of the midges live, some of them in the soil, others in water. For keeping off the "biting" midges, Theobald gives the following as successful: a mixture of ½ oz. pure carbolic acid, 1½ oz. spirits of lavender, 1 drachm of eucalyptus-oil.

Œstridæ or Bot-Flies.

The adults are hairy flies with abortive or rudimentary mouth-parts. The adult flies cannot wound stock,—it is the larva or maggot which is harmful in this family,—yet stock seem instinctively to fear the flies, and stampede as these approach to lay their eggs. There are three sections of Œstridæ—viz.:

Gastricolæ, in which the larvæ live in the alimentary canal — e.g., horse-bot;
Cuticolæ, in which the larvæ live below the skin—e.g., ox-marble flies;
Cavicolæ, in which the larvæ crawl up the nostrils to the frontal sinuses—e.g., sheep nasal-fly.

Horse Bot-Flies.

Gastrophilus equi.—This yellow-brown fly (fig. 770) measures from one-half to two-thirds of an inch long. There are red hairs on the thorax and yellow-brown hairs on the abdomen. The female fly can be told from the presence of a well-marked ovipositor directed downwards and forwards.

The eggs—pointed at the attached end and blunt at the free end—are glued to the hairs of the horse; they measure $\frac{1}{12}$ inch. The eggs are fixed to the hairs on fore-leg and shoulder, and sometimes to the mane. When the egg is ripe the maggot protrudes itself, and is conveyed by means of the horse's tongue to the horse's mouth, ultimately reaching and fixing itself to the stomach. The first part of the stomach is chiefly infested; but, in case of bad infestation, the stomach generally may be more or less covered. The maggots hold on by means of their mouth-hooks, not leaving till they are full fed. The full-grown larva is rounded, and measures $\frac{3}{4}$ inch; in addition to its mouth-

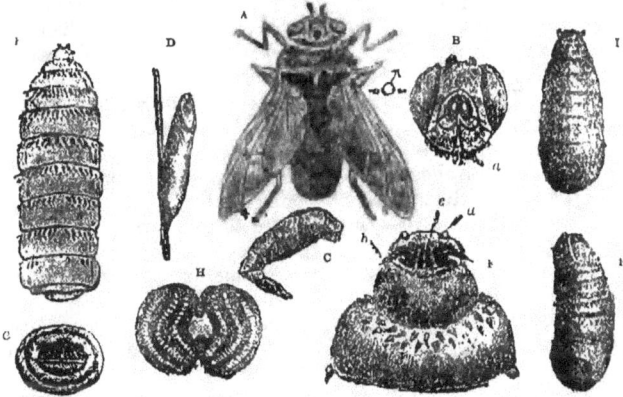

Fig. 770.—*Gastrophilus equi.*

A, Male, twice natural size.
B, Head of same, with *a*, the channel on the face.
C, Abdomen of female, showing ovipositor.
D, Egg on a hair, magnified.
E, Grown larva, magnified.
F, Front end of larva. *a*, antennæ; *b* and *c*, mouth-hooks, much magnified.
G, Last segment of larva seen from behind.
H, Stigmatic plates at hind end of larva, greatly magnified.
I, Pupa.
K, Pupa seen from the side.

(A to D after Brauer; E to K after Nitzsche.)

hooks it has a double row of prickles on each segment. When full grown the larva leaves go, and, passing along the alimentary canal of the host, drops to the ground, where pupation takes place. The flies are commonest in July and August.

Gastrophilus hæmorrhoidalis is a smaller and darker species. The eggs of this fly are darker, and are sometimes laid on the long hairs about the lips. The maggots may fix themselves in the thorax and back of the throat or in the rectum.

The presence of the *Gastrophilus* maggots may set up inflammation and ulcers, and may cause interference with the free passage of food or passage of waste matter: loss of appetite and condition are symptoms of attack.

Horses out at grass should be examined and any eggs removed by thorough grooming. The *hæmorrhoidalis* maggots can be removed by hand from the anal region, and those in the laryngeal region by pushing into the throat a stick covered with an oil-saturated cloth. Where emaciation seems due to the presence of the maggots the following has been recommended as a draught once a fortnight: 2 oz. turpentine and 20 oz. raw linseed-oil mixed.

THE OX-WARBLE FLIES.

There are two species of these—*Hypoderma lineata* and *Hypoderma bovis*. The larvæ of these flies are the cause of immense loss in Britain every year.

Hypoderma lineata.—The Striped Ox-Warble Fly is so named from the longitudinal bands on the front part of the thorax, light stripes alternating with dark. It measures half an inch in length, and is black, with a hairy covering of whitish, red-brown, and black hairs. The flies are found from May till September.

The females lay their eggs in rows, attaching them, as a favourite place, to the hairs just above the hoofs. The cattle, in licking themselves, convey the maggots to their mouth, and the young maggots—spiny at this stage—fix themselves to the gullet. The larva moults, and in doing so loses its spines, and proceeds to wander from the gullet through the tissues of the host, ultimately reaching the back. Here another moult takes place, which leaves the maggot spiny. Lying below the skin, the maggots give rise to great irritation. A swelling, with a hole leading to the outside, marks the position of the larva. The larva lies in the warble with its tail end pointing to the opening; at this tail end the spiracles for respiration are situated. When full grown the maggot presses itself out of the warble and falls to the ground for pupation. Pupation takes place under cover of the last moulted skin, and this puparium or pupa-case hardens and becomes black in colour. In due course the fly, when ready, issues through a cap-like opening at one end of the puparium.

Hypoderma bovis measures over half an inch in length; it is dark coloured, banded with yellow hairs; there are yellow hairs on the face and yellow and black hairs on the thorax; the hairs on the abdomen are yellow-white in front, black in the middle, and yellow-red behind.

There are conflicting views as to the life-history of *H. bovis*. Miss Ormerod's view was that the eggs were laid on the hairs of the back, and that the maggots, on hatching, bored directly through the hide. Others hold that the life-history resembles that of *H. lineata*—viz., that

the eggs or maggots are licked into the mouth, and that after a wandering they reach the tissues of the back. Recently Carpenter, as the result of careful experiment with calves kept under observation, has stated that the eggs are laid chiefly on the legs, not on the back, both fore and hind limbs being struck near the hock. Carpenter also inclines to the view that the larvæ, on hatching, bore through the skin, and after a more or less prolonged wandering reach the back. While maggots may reach the back by way of the mouth, Carpenter's careful experiments seem to prove that this is not necessary. He kept six calves muzzled in such fashion that they neither could lick themselves or their neighbours, and yet five out of the six calves showed warbles.

Once having reached the back, the rest of the life-history is the same as for *H. lineata*.

H. bovis is commonest from mid-summer till the end of July, but not limited to this period. According to the trade reports, February till September is the time for warbled hides, but chiefly April and May.

The Ox-Warble Flies cause loss in various ways. The presence of the flies bent on egg-laying alarms the cattle, which gallop about. The irritation caused by the spring maggots under the skin prevents the cattle grazing at peace, and they fail to put on flesh. After the cattle have been slaughtered there is loss in dressing the carcases from what is called "licked beef" or "butchers' jelly," this inflamed tissue having to be cut and scraped away. Then there is the loss from hides showing the warble holes.

Treatment.—As a fly deterrent various dressings are recommended, the dressing to be applied along the spine at regular intervals. It must be admitted that there is considerable testimony in favour of the value of this treatment. Yet careful experiment indicates that the hairs of the back are not a favourite place for egg-laying, and if so the dressing of the back cannot be a measure for recommendation. As against the maggots, once they are present in the back, it is a sometimes practised measure to treat the warbles separately in order to kill the maggot. The maggot lies in the warble with the tail end—where the spiracles are—uppermost, and advantage is taken of this to apply to each some sticky or greasy or poisonous substance which will suffocate and kill the pest. This mode of treatment, however, is not to be recommended.

The best treatment is to squeeze out the maggots and kill them. This is not a difficult measure when the maggots are approaching the end of their growth, and ought to be the most practised treatment. Regular and methodical attention to this on the part of our stock-owners, if continued over a few seasons, would go far to annihilate this enemy.

Another species of warble-fly — viz., *Hypoderma diana* — is the cause of warbles in deer.

The Sheep Nostril-Fly (Œstrus ovis).

This fly (fig. 771) measures about half an inch in length; the upper surface of the head is light-brown; the upper surface of

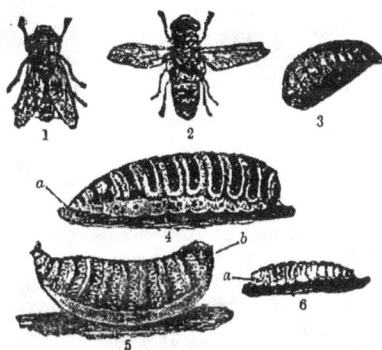

Fig. 771.—Œstrus ovis.

1 and 2, Adult fly.
3, Pupa.
4 and 5, Full-grown larva.
6, Young larva.
a, Head-end.
b, Tail-end.

(After Riley.)

the thorax is light-brown or yellow to grey, and has dark tubercles; the ringed abdomen is brown-yellow, with dark spots; the legs are brown. The wings are glassy, and extend, when the insect is at rest, beyond the body. The newly-hatched maggots are white and worm-like; they have two mouth-hooks and transverse rows of little spines on the under surface of the abdomen; there are also spines at the hind end. When full

grown the maggot measures between ¾ inch and 1 inch.

Life-History.—In warm sunshiny weather the females fly towards the sheep, laying their eggs, just ready to hatch, or newly-hatched maggots, at the sheep's nostril. The maggots, by their mouth-hooks and spines and anal processes, draw and push themselves up the nostril. The maggots feed on the secretions resulting from the irritation caused by their presence and their prickings; they become mature in the frontal and maxillary sinuses of the sheep. When full grown the maggots are sneezed out on to the pasture, where, a little below the surface, or in a sheltering tuft, they become pupæ, the fly maturing under cover of the last moulted skin of the maggot, which becomes dark in colour. The number of maggots in a head varies, a small number being the commoner thing.

Symptoms attending infestation are: discharge from the nostrils of the infected sheep; sneezing and snorting in the endeavour to get rid of the larvæ; tossing of the head; rubbing noses on the ground; a staggering gait; and difficulty in breathing.

Treatment.—Dressings on the nostrils of the sheep to prevent the flies laying their eggs or maggots. A contrivance for this purpose is to have V-shaped salting-troughs in the field, the sides of these being smeared with tar, which reaches the nose of the sheep as they lick the salt.

Infested sheep should be isolated before the maggots are sneezed on to the pasture. In bad cases the sheep should be sent for slaughter. Remedial measures are not of much avail, and fumigation, or an operation to remove the maggots, would be practised only with valuable animals.

FAMILY MUSCIDÆ.

This important family includes the Tsetse-Flies, one of which (*Glossina morsitans*) carries the parasite of tsetse-fly disease, so fatal to the domesticated animals in some parts of Africa; while another species (*Glossina palpalis*) carries the parasite that causes Sleeping Sickness.

Thero are three British blood-sucking species; the others do not draw blood, but arc harmful in other ways. The blood-sucking species are *Stomoxys calcitrans*, *Hæmatobia stimulans*, and *Lyperosia irritans*.

The Stable-Fly (Stomoxys calcitrans).

This fly resembles the common house-fly, but is shorter and stouter; the wings when Stomoxys is at rest are held wider apart, and the head is more erect and carries the cruel proboscis. In stables where these flies settle on the horses' legs the pricks of the proboscis cause the beasts to stamp, and in sensitive fine-skinned animals a swelling may follow the wound. Cows sometimes suffer severely.

The life-history of this fly has recently been worked out by Newstead.[1] Farm-yards and stables are, according to Newstead, favourite haunts of the fly, which also is found in fields, parks, and open woods. The fly is also common in towns. At night the flies may be found resting on beams and rafters in open sheds in farmyards. Fresh dung was offered to flies kept in captivity, and eggs were laid on it. With larvæ fed on moist sheep's dung (the eggs were obtained from captive females), at an average day temperature of 72 per cent, and night temperature of 65° F., the whole life-cycle was completed in 25 to 37 days. With drier material and light admitted the cycle took 42 to 78 days. After prolonged observation, Newstead succeeded in finding the females laying their eggs in the open, in a heap of grass mowings, in September.

HÆMATOBIA STIMULANS.

This fly, smaller than a house-fly, has the head much smaller than that of *S. calcitrans*, but the palpi are much longer. The flics live in the open and suck blood, the pain of their " bite " being severe.

LYPEROSIA IRRITANS (Hæmatobia serrata).

This is the smallest of the blood-sucking Muscids, and measures 4¼ to 5

millimetres in length. It is a pest of cattle. This fly has been introduced to the United States and Canada, where it has been the cause of considerable loss, interfering with the feeding of the cattle and with their digestion, as well as causing a loss of blood. From the habit which these flies have of resting, when in numbers, on the horns of the animals, the name Horn-Fly has been given in America.

The Common House-Fly (Musca domestica).

Although several species of fly are found in our houses, *Musca domestica* is the commonest from midsummer onwards to the autumn. The fly measures 6.5 to 9 millimetres; it is dark ash-grey in colour; there are four longitudinal black streaks down the back of the thorax; the abdomen is checkered with black, and a dark streak runs down the middle of its upper surface. The egg is white and oval; the larva is a white or grey-white maggot, with a pointed head end and a blunt posterior end. In the course of its development it moults twice, and three stages can thus be distinguished. Full grown it measures ⅜ inch. The puparium is oval and brown.

Life-History.—The female lays on an average 120 eggs. The favourite place for egg-laying is horse manure, but the eggs may be laid on the manure of other animals, where this is fresh and not too dry, and in decaying comestibles. The number of eggs laid by a single female may reach 400 to 500. In warm countries the whole life-cycle can be passed through in a fortnight. Gordon Hewitt,[2] in his experiments at Manchester, found the life-cycle to vary from 20 to 30 days: taking the shortest times for the different stages in the various experiments, then, 15 days is the result.

It is clear, then, that enormous numbers of house-flies can be reared in a season, and there is no doubt that this means considerable risk to the community. There is the fouling of food and drink with the dung-stained feet of

---

[1] *Journal of Economic Biology*, 1907, vol. i.

[2] *Memoirs and Proceedings of the Manchester Literary and Philosophical Society*, 1906-07.

the flies, and a soiling with their excrement. As feeders on garbage there is evidence that harmful bacteria can be conveyed to foods and milk; the bacteria of typhoid and tuberculosis can be carried, and probably the prevalence of summer-diarrhœa in children, so baneful in its results, is due to carriage of bacteria by the fly.

Excrement in the open should be covered. Middens are great places for the breeding of *Musca domestica*—the maggots sometimes swarming in such. Howard recommends the application to such of chloride of lime, a shovelful to be thrown over every day or two's addition to the heap. A wise measure would be to protect exposed eatables with gauze.

The Blue-Bottles or Meat-Flies.

Two species of Calliphora—viz., *C. erythrocephala* and *C. vomitoria,*—well known by their loud buzzing noise and their blue colour, lay their eggs on meat and fish or on decaying organic matter, to which they are attracted by their sense of smell.

C. erythrocephala has a red face and a black beard; *C. vomitoria* has a black face and a red beard. The eggs are laid in little heaps, and hatch in twenty-four hours. The voracious maggots are well known, as also are the brown oval pupacases. The length of the life-cycle varies with the food and the temperature. In experiments made by me in August in the open air the whole life-cycle was passed through in a month.

Calliphora erythrocephala also "strikes" sheep: the maggots pass their life on the sheep, not boring deeply into the flesh, however, as the green-bottle (Lucilia) maggots do, and fall to the ground for pupation when they are full grown.

The Green-Bottles.

These belong to the genus Lucilia, and there are two closely resembling species—viz., *Lucilia sericata* and *Lucilia cæsar*. *Lucilia cæsar* is green with a whitish face, and the upper border of the buccal cavity is reddish. *L. sericata* is, if anything, smaller; both face and buccal cavity are white. The colour is bright green, but associated with this is a bluish bloom absent in *cæsar*.

Lucilia sericata.

This fly is, *par excellence*, the sheep maggot-fly. It can be observed flying about the sheep during hot weather. The fly measures about $\frac{1}{3}$ inch long and about $\frac{7}{8}$ inch in spread of wing. The eggs are yellowish-white, and measure about $\frac{1}{16}$ inch. The larva is a legless maggot, but is capable of an active crawling movement; the head end is provided with two mouth-hooks; the hind end is blunt, with tubercles round the margin, and two plates with the spiracles. The pupa cases are brown, and rounded or barrel-shaped.

Life-History.—The female fly may lay as many as 500 eggs, these being fixed to the wool in clusters of 20 or more. The maggots from the eggs feed at first externally, but later bore into the flesh. When full grown they drop to the ground for pupation.

Attack is worse on lambs than on old sheep. The flies are found at work from May onwards. Moist, warm, muggy weather, or warm sunshine after showers, favours the fly.

Symptoms of attack are: Matting together of the wool fibres, a continual wagging of the tail, rubbing and biting of the sheep in their efforts to allay the irritation caused by the maggots, much inflammation, oozing from the sores of an evil-smelling sticky fluid, discoloration of the wool which falls out and in bad cases does not grow again, rapid loss of condition.

Treatment.—Keep the hind-quarters of the sheep clean: a good measure is to clip the wool of the tail and between the hind-legs. Carcases of dead animals should be burned or buried so that they may not serve as breeding-places. Dipping with sulphur as an ingredient. The neighbourhood of wounds should be dressed with an ointment of butter and flowers of sulphur or with spirits of tar. Infested sheep should be isolated. The maggots are not difficult to kill; they should be picked or rubbed off, or where they have got to work the wool may be shorn a little, the affected parts being dressed with fly-oil, or with a mixture of turpentine and rape-oil in equal parts, or with dilute paraffin-oil, finishing off with a dusting of sulphur. Very much depends on the care of the shepherd, and there should be repeated inspection.

*The Flesh-Flies, or Family Sarco-
phaginæ.*

Our best known species is *Sarcophaga
carnaria,* a greyish-looking hairy fly.
The female measures half an inch in
length and the male less. The upper
surface of the thorax is whitish-grey
with longitudinal black stripes. The
dark-coloured abdomen has a number
of whitish-grey markings resembling a
check.- The female fly deposits live
maggots on decomposing animal or vegetable matter, and perhaps in wounds.
The maggots are white-coloured, and
have their upper surface granulated.
Pupation takes place in some convenient
shelter-place; the pupa case is
black-brown. *Sarcophaga carnaria*
maggots are chiefly scavengers, but
on the Continent there is a dangerous species, *S. magnifica,* whose
maggots are found on live animals.

FAMILY HIPPOBOSCIDÆ.

This is a family of flies with
flattened horny body, and parasitic
on various animals, chiefly birds.
Some have well-marked wings; in
others the wings are rudimentary
or may be absent. An interesting
feature in their biology is their
mode of reproduction, the eggs
hatching and the larvæ developing in the
body of the mother right up to the period
when the larvæ are ready for pupation.

The Ked or Kade or Sheep Louse-Fly
(Melophagus ovinus).

This insect (fig. 772) is wingless, and
measures about a quarter of an inch
in size. The colour is brownish; the
body is bristly. There is a tubular
proboscis. The square thorax bears
three pairs of bristly legs, each ending in two strong two-toothed claws
and a plumed bristle. The keds live
among the wool of the sheep, coming
towards the surface on a sunny day,
hiding nearer the skin among the fibres
of the wool in colder weather. They are
not able to live long away from their
hosts, from which they derive shelter,
warmth, and food. After shearing, the
keds are not found so spread over the
body, but collect more on the neck,

shoulders, and ears. The larva is nourished within the body of the mother on
a secretion prepared in uterine glands;
small shining red-brown puparia are fixed
to the wool, and the adult, when ready,
issues by a crack at one end.

Keds—especially in lambs, which remain thin in consequence—give rise to
much irritation by their puncturings.
Infested sheep bite, scratch, and rub
themselves. Where the skin has been
punctured by the proboscis dark spots
show, surrounded by a red area.

Treatment.—Dipping, the bath containing some substance which will poison
or suffocate the pests. After dipping, a
few days should elapse before the sheep

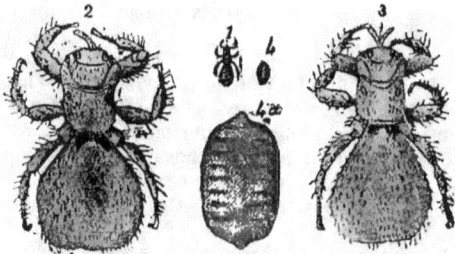

Fig. 772.—*Melophagus ovinus.*

1, Female, natural size.
2, Male, magnified; view of upper surface.
3, Male, magnified; view of lower surface.
4, Pupa case, natural size.
4a, Pupa case, magnified.

are allowed to return to their old feeding-
grounds, so that any keds that may have
previously tumbled to the ground may
have perished. A second dip should
follow the first, say, after a fortnight.

The Forest-Fly (Hippobosca equina).

The New Forest is the chief locality
for this fly, but it has been recorded
from other parts of the south of England and from Wales. The fly is winged
and retains its wings; it is a quarter of
an inch long. The head is yellow, and
there is a dark stripe in the middle of
the face; the thorax is brown, and has
three yellow patches, two in front and
one in the middle. The toothed claws
give the fly a very secure grip of the
horse. The puparia laid by the female
are white at first, but soon darken. The
flies are found fixed to parts of the horse's
body where the skin is soft and the hair
not plentiful. The flies, as they crawl

over the horse or donkey, cause great annoyance and irritation, and strange animals especially plunge and rear or roll themselves about.

The Deer Forest-Fly (Lipoptera cervi).

This insect measures $1/5$ inch; it is yellow-brown in colour and is tough and bristly; the legs are short and hairy. Both males and females can be found all through the winter on the deer, the insects being by this time wingless. To begin with, however, both sexes have wings, but when a host has been reached the wings are shed or torn off and only stumps remain.

The females lay small shining puparia among the hairs of the deer. The flies emerge in summer, and from autumn onwards both males and females may be found running over or clinging to the deer.

Ornithomyia avicularia.

This species is a bird parasite; it is found, for example, on fowls and pigeons, and is generally distributed throughout Britain. The fly measures less than a quarter of an inch; it is greenish-yellow in colour, with the upper surface of the thorax darker; the wings are smoky. The fly retains its wings, and can at pleasure leave one host and fly to another.

PULICIDÆ OR FLEAS.

This family is made up of insects whose wings are reduced to mere scales. Three species of the genus Pulex may be distinguished thus:—

Pulex irritans, the flea of man, without comb-like spines on head and prothorax.

Pulex avium, the flea of fowl and pigeon; 24 to 26 comb-like spines on the prothorax.

Pulex serraticeps, the dog-flea; 7 to 9 comb-like spines on the lower edge of each side of the head and on the sides of the posterior edge of the prothorax.

Fleas are laterally compressed insects with bristles on thorax and abdomen; the six legs are also bristly—the hind pair the longest—and end in claws. The males are smaller than the females and have the end of the abdomen tipped up. Eggs are laid in dusty corners, cracks in

the floor, on mats, and among the hairs of the dog. Legless larvæ hatch out, which can wriggle actively, assisted by the hairs on the segments and by hooks at the hind end. When the maggot is full grown pupation takes place under cover of a silky cocoon that may be covered with dust particles.

The dog-flea also passes to man and the cat. *Pulex avium* attacks the pigeon and fowl and other birds; it can prick man. Sitting hens are annoyed, and where the fleas are plentiful growth of the young birds is interfered with.

Treatment. — Careful and regular sweeping and cleansing of dog-kennel and hen-house with lime-wash. Infested dogs, having first been bathed with soapy water, should be sprinkled with fresh pyrethrum powder. Creolinated water —a 10 per cent solution—is a good wash for flea-infested animals. A little saw-dust soaked in naphthaline and placed in the nests of sitting birds will give them peace.

THE BLOOD-SUCKING LICE OR PEDICULIDÆ.

These insects belong to the Parasitica section of the Order Hemiptera. They have a sucking proboscis capable of expansion and contraction. When not in use this proboscis is invisible, having been withdrawn into its sheath; the sheath carries a number of hooklets which bury themselves in the skin and so hold the sucking-tube steady. The legs have a two-jointed tarsus, and the tarsus ends in a stout claw which bends up and forms, with a projection from the lower extremity of the tibia, an apparatus by which the parasites cling and creep.

The females lay eggs or nits, which are glued to the hairs of the host. Development is rapid; the young forms that issue from the eggs feed and grow and moult themselves to the adult form without any resting pupal stage.

Three genera may be noticed:—

The genus Phthirius has the thorax as broad as the abdomen, the two passing into one another without constriction— *e.g., Phthirius inguinalis,* the crab-louse, against which the best remedy is staves-acre ointment. The genus Pediculus has the thorax narrower than the abdomen; the abdomen, broadest at the middle, narrows at the anterior end, so that the thorax and abdomen do not seem sharply marked off from one another; the eyes are prominent. To this genus belong the head-louse (*Pediculus capitis*) and body-louse (*Pediculus vestimenti*).

The genus Hæmatopinus has the thorax narrower than the abdomen; the abdomen and thorax, owing to the difference in breadth, are sharply marked off from one another. Hæmatopinus species infest the dog, horse, ox, pig, goat, and rodents.

THE BITING LICE OR MALLOPHAGA.

The name Mallophaga means wool-eaters, and indicates that these lice live not on blood, but on epidermal scales, feathers, hairs, scurf, &c.

When present in numbers on the host they are the cause of itching, unrest, and irritation.

The Mallophaga are flattened forms, with their mouth-parts fitted for biting and cutting. The head is large and broad; the first segment of the thorax is distinct, but the other two segments are not marked off from the abdomen; the legs are short and have one or two claws; in one section the legs are fitted more for clinging, in the other more for locomotion. Wings are absent.

There is an incomplete metamorphosis, there being no resting pupal stage in the life-history. Pear-shaped eggs are laid on the hair or feathers, and the young, which hatch, differ externally from the adult only in size; they become sexually mature after some moultings. Mammals may be the hosts, but birds more commonly, hence the Mallophaga are often called the Bird-lice.

The two sections are the Philopteridæ, characterised by their comparative slug-gishness, their feet being more adapted for clinging; and the Liotheidæ, which are more active.

PHILOPTERIDÆ.

Trichodectes. — This genus infests dog, horse, ox, sheep, goat. The species have wide flat heads, beset on the upper surface with hairs. The first segment of the thorax is well marked; the second and third segments are fused to form one piece; the abdomen is nine-jointed, and has scattered hairs, these being most marked at the edges; the abdomen of the female is cleft at the end.

Lipeurus.—This genus has an elon-gated narrow body. Species infest fowls, ducks, geese, turkeys, pigeons, pheasants. *Lipeurus variabilis* of the domestic fowl may be taken as an example. This insect may be found in numbers amongst the primary and secondary feathers, and can move about actively. The head is round; the hind part of the thorax is longer and broader than the fore part; the under surface of the thorax shows a brownish spot; the abdomen is spotted and banded, and has projecting hairs at the edges. The prevailing colour is pale-yellow, with the spots fawn coloured and the bands dark. The male measures 1.9 mm. and the female 2.2 mm.

Goniodes.—The species of the genus Goniodes have flatter, wider bodies than the last.

Goniodes dissimilis is the chicken Goniodes. The head is wider than long; the abdomen is broad and oval, and has on the middle of each segment two bristles; there are curved spots at the edges. The general colour is whitish, with darker spots and fawn-coloured bands. The male measures 2 mm. and the female 2½ mm.

The genus Goniocotes is also made up of flattened wide forms. *Goniocotes holo-gaster*, the chicken-louse, is much smaller than the last, the male measuring about .9 mm. and the female 1.3 mm. The head is as wide as long and broadest just behind the antennæ. Colour yellowish; there are brown-black bands.

LIOTHEIDÆ.

Of this section, fitted for running, *Menopon pallidum* may be taken as an example. This is the commonest and most troublesome of all lice infesting the fowl. It runs with great nimbleness among the feathers. The head is angular and crescent-shaped; the temples bear four bristles; the thorax is the same length as the head in the male, but longer than the head in the female. The abdomen is oval and elongated, and each segment carries a series of bristles. The abdomen of the male is longer and narrower, and has four long bristles at the end. The colour is pale-yellow, with bright fawn spots on the abdomen.

Treatment for Lice, both Sucking and Biting.

There should be scrupulous cleanliness with periodical disinfection of stable, kennel, and hen-house. Dust-baths should be provided for birds. Of ointments which, rubbed into the hair, will suffocate the pests, may be mentioned: (1) one part sulphur to four of lard; (2) stavesacre ointment made of oil of stavesacre one part, lard seven parts; or (3) a decoction of one of stavesacre seeds to four of lard.

For rubbing in or for washing there is a large choice of preparations: (1) a decoction of one ounce of stavesacre seeds to a quart of liquid, half water half vinegar; (2) an infusion of two parts stavesacre to one hundred of vinegar; (3) one part petroleum to ten of rape-oil; (4) a 5 per cent mixture of boiled tobacco or tobacco juice from manufactured tobacco, diluted in the proportion of 1 to 100; (5) creolinated water—i.e., a 5 per cent solution of creolin. When animals are combed the comb should be dipped in a strong solution of soda, or comb and brush should be dipped in paraffin emulsion, or in one quart of water to which has been added 2 oz. of carbonate of soda and ½ oz. of powdered stavesacre.

A repetition of the treatment should follow in some days, so that any of the pests which have hatched from eggs that escaped the first treatment may be killed before they become mature and proceed to egg-laying.

ACARINA OR MITES.

Acarina is an Order of the class Arachnoidea; it denotes the mites as distinct from the spiders and the scorpions.

Mites are small animals with head and thorax soldered together to form a cephalothorax which is united throughout its width to the abdomen: the two parts are so joined that no sign of the union may be visible. The mouth apparatus is fitted for biting or piercing or sucking, the various united pieces forming the rostrum. This rostrum is made up of a pair of mandibles and a pair of pedipalps, the latter consisting of a basal part and a several-jointed palp capable of free movement.

The adult mite has four pairs of legs, which vary in shape according to the habit of life of the mite; the legs end in hooks or hairs or suckers. When the mite hatches from the egg only six legs are present; the fourth pair appears later. Respiration may be by tubules opening on the outside of the body, but in many parasitic forms the breathing is directly through the skin. The sexes are separate, and reproduction is typically oviparous; exceptionally live young may be produced. Five mite families are of importance here.

Family Demodecidæ.

This is a family of very small worm-like mites with the cephalothorax and the transversely striated abdomen distinguishable from one another. The mandibles are little stylets. The adults have four pairs of very short legs; those legs are three-jointed. Eyes are absent, and respiration is through the skin.

Out of the egg there comes a larva, legless in some varieties, in others having three pairs of rudimentary legs; after two moults there is a well-developed mouth and eight legs. These Demodex mites live in the sebaceous glands and the hair follicles of mammals.

Demodex folliculorum (fig. 773) is common in the sebaceous glands of man's face; a favourite place is the skin of the nose, but the presence of this mite in man is of no importance. There are varieties on the dog, cat, horse, ox, sheep, pig, goat, and mouse. The Demodex mites are very small, the largest —that of man—measuring only $1/80$ of an inch or a little over.

The worst Demodex attack is that on the dog, where it causes follicular mange, an affection of the skin characterised by pustules and falling out of the hair. All the stages from larva to adult may swarm in the sebaceous glands and the hair follicles, particularly the latter; the mites are fixed by their rostrum. It is young dogs chiefly that are attacked, and short-haired dogs more than long-haired ones. The disease generally begins about the head, particularly in the neighbourhood of the eyes, and extends gradually to the fore-legs, feet, and sides.

The symptoms vary according to the stage of the disease. To begin with, there is only a slight itching and small papules and a redness marking the places where there has been a slight loss of hair; as the disease spreads the skin wrinkles, larger pimples appear, filled with a purulent material, and when these burst red crusts mark the place; the itching may also increase. If some of the purulent matter be squeezed out and examined under the microscope the parasites are revealed. A characteristic odour, suggestive of mice, comes away from the patient.

Follicular mange does not yield readily to treatment; and where it has persisted and spread, death almost certainly follows. A good dressing is Peruvian balsam dissolved in alcohol—one part balsam to four of alcohol—the solution to be rubbed daily into the skin after the con-

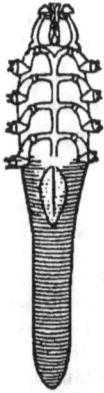

Fig. 773.—*Demodex folliculorum.*

Greatly magnified. (After Lohmann, in 'Das Tierreich.')

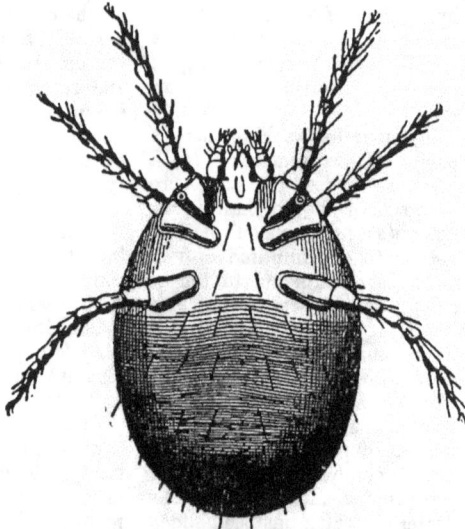

Fig. 774.—*Larva of* Trombidium holosericeum, *ventral surface.* Magnified one hundred times. (After Railliet.)

tents of the pustules have been squeezed out. Another ointment in use is made of five parts creolin and a hundred parts lanoline.

Family Trombidiidæ.

To the carnivorous section of this family belongs *Leptus autumnalis* (fig. 774), the Harvest Bug. This harvest bug is not an adult, but possibly is the larva of the mite known as *Trombidium holosericeum*. *Leptus autumnalis* is six-legged and brick-red. The legs have six joints and are hairy, and each ends in three slender prongs.

This mite swarms in late summer and autumn on grass and undergrowth, and on such plants as gooseberry, currants, raspberry, beans, from which the pest passes to man, horse, ox, dog, cat, hare, rabbit, mole, and fowls.

Dogs, especially hunting dogs, often harbour the parasites fixed about the head and nose and belly and feet, where the mites cause eruptions. On the cat they give rise to little wounds at the root of the tail and the feet between the claws. Dr Johnston, in the *History of the Berwickshire Naturalists' Club*, quotes a correspondent as to this mite on the horse thus: "In the worst case I have ever seen, that on a horse, the skin seemed exactly as if it had been rubbed with a liquid blister."

Fowls and late-hatched chickens are sometimes much annoyed by the Leptus mites which bury their rostrum at the base of the feathers.

Treatment.—A two per cent solution of carbolic acid is a preventive as well as a cure. Chloroform water—one of chloroform to six of water thoroughly mixed—is also of service. Rubbing in sulphur ointment or benzine or phenic acid will get rid of the mites. In the case of fowls the feathers should be dusted with flowers of sulphur.

Family Sarcoptidœ.

In this family we have the mange or itch mites.

Mange, Itch, or Scab Mites.

These tiny mites have rounded or oval bodies and a conical rostrum. There is a metamorphosis in the life-history.

There are three distinct genera of mange mites—viz., Sarcoptes, Psoroptes,[1] Symbiotes,[2] and under the microscope they are distinguished thus :—

SARCOPTES.	PSOROPTES.	SYMBIOTES.
Body rounded.	Body more oval.	Body more oval.
A short rostrum and two little expansions called cheeks.	Rostrum more pointed ; no cheeks.	Rostrum about as wide as long, and blunt ; no cheeks.
Legs short, not reaching far from the body. The four front legs spring from the edge of the body ; the four hind legs are attached to the under surface of the body and almost concealed beneath it.	Legs longer, and all four pairs can be seen projecting from the body.	Legs long, and all four pairs visible.
The tarsus (last joint of the leg) may bear a long unjointed stalk, terminated in a small sucker or suctorial disc.	The tarsus bears a three-jointed stalk terminated by a sucker.	The tarsus bears a short unjointed stalk terminated by a wide sucker.
The mandibles are nipper-like.	The mandibles are more lance-like.	The mandibles are nipper-like.
The mites, which are not found in colonies, mine into and make galleries below the skin.	The mites, found many together, do not burrow into the skin, but live in parts sheltered by hair and wool, and under crusts.	The mites, which are social, live exposed on the outside of the host.

On the same animal all three kinds of mange mite may be found, but each animal has one species of mite which, of the three, is the most harmful for it. In the following table is indicated various hosts with the genus of mite found on these.

HOST.		MAY BE INFESTED BY	
Horse . .	Sarcoptes	Psoroptes on the inner side of legs, tail, mane, genital organs.	Symbiotes on fetlock and limbs.
Ox . .	Sarcoptes, perhaps not a species peculiar to it, but derived from some other animal.	Psoroptes on sides of neck and root of tail, extending over the body except the limbs.	Symbiotes at root of tail.
Sheep . .	Sarcoptes about the head, and in bad cases to fore-limbs.	Psoroptes (Sheep Scab).	Symbiotes on the feet and limbs.
Dog . .	Sarcoptes.		Symbiotes affects the ears.
Cat . .	Sarcoptes.		Symbiotes affects the ears.
Pig . .	Sarcoptes.		
Goat . .	Sarcoptes on head and body generally.		Symbiotes on sides of neck, back, withers, and loins.
Birds . .	Sarcoptes.		

[1] Psoroptes has, as synonyms, the names Dermatodectes (skin-biters) and Dermatocoptes (skin-wounders).

[2] Symbiotes has, as synonyms, the names Chorioptes (hiders) and Dermatophagus (skin-eaters).

Except in the case of sheep, where it is the Psoroptic form of mange which is the most serious and most troublesome form, in all•the other animals noted the most noxious scab is the Sarcoptic.

Transmissibility of Scab.

Generally speaking, the Sarcopt in each species of host is very contagious for the same species—e.g., Sarcopt of the horse is very readily transmissible to other horses.

Man.—The Sarcopt of man placed experimentally on the horse and dog produced on these an eruption which soon passed off. The same gave no result in the cat.

Horse.—The Sarcopt of the horse passes readily to ass and mule, and is transmissible to man; but the disease set up in man is not severe and yields readily to treatment. It is believed that the Sarcopt of the horse can pass to the ox, but it has not been proved to infect more domesticated animals than those already named.

Psoroptic and symbiotic mange of the horse are not communicable to other animals.

Ox.—Neither the psoroptic nor the symbiotic mange of the ox seems communicable as a permanent disease to the other domesticated animals.

Sheep.—The sarcoptic mange of the sheep passes readily to the goat. Placed experimentally on the horse, ox, and dog, an ephemeral but no permanent disease was produced. The psoroptic mange of the sheep is not communicable to the other domesticated animals.

Dog.—The Sarcopt of the dog is communicable to man. Placed experimentally on the various domesticated animals the results were insignificant or negative.

Cat.—The sarcoptic mange of the cat can infect man. It can pass also to the horse, ox, and dog.

Pig.—The Sarcopt of the pig has been shown to be contagious for man. Placed experimentally on the sheep, cat, and dog there was no result, or only a slight infection which soon passed off.

Goat.—The sarcoptic mange of the goat is transmissible to the horse, ox, sheep, and pig. Experiment has shown that the Sarcopt of the pig can be very troublesome to man, whether man receives it directly from the goat or, secondarily, from one of the just mentioned animals.

Life-History of Mange Mites.

The following round of life of Sarcoptes scabiei (fig. 775) may stand as

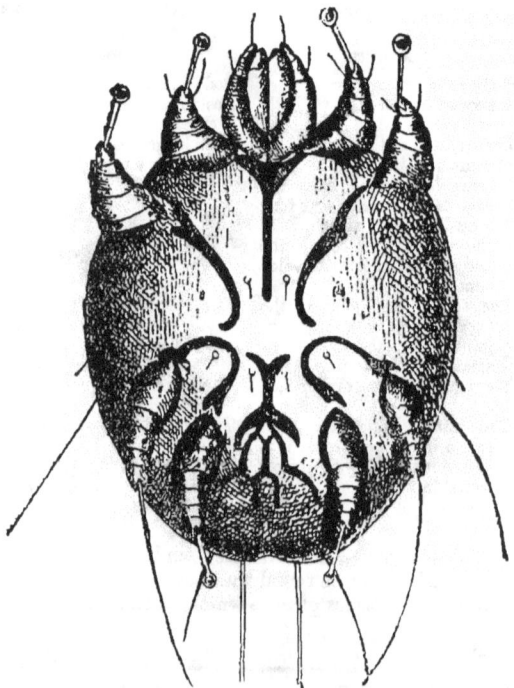

Fig. 775.—Sarcoptes scabiei.
Greatly magnified. (After Lohmann, in 'Das Tierreich.')

typical of mange mites in general (any exceptions will be noted later).

The fertilised female about to lay her eggs burrows (it has already been pointed out that this is the burrowing genus) into the skin, and makes a gallery along which the eggs are laid one by one. The eggs hatch in a few days, and the larvæ, on hatching, pierce their way to the surface of the skin, where they live for a short time. These newly hatched larvæ are 6-legged and not sexually mature. The larva moults several times and grows. The next stage is the nymph stage, in which the fourth pair of legs has appeared. In the next stage pairing takes place. Development from the larval to the adult state is rapid, and the numbers of the pests soon increase.

Sarcoptes scabiei var. equi.

This mange mite, in its attack on the horse, generally starts about the withers, the place of attack being marked by a few hard pimples on the skin. As the disease spreads, neck, shoulders, back, and sides may be invaded, long-haired parts (which would, on the other hand, be chosen by the Psoroptes of the horse) being avoided. Characteristic of attack is an intense itching, the itching being worst at night and in warm conditions; greater in the stable than when the horse is exposed, and greater when the horse is covered with a cloth than when naked. Pimples form, and can be felt as little elevations if the hand be pulled across the skin: these burst when the horse rubs itself, and the secretion from them dries into a crust. To begin with, these crusts are isolated, but as the disease spreads different patches run together and a large crust is formed, under cover of which the young mites may be found. The hair also drops out, and the skin wrinkles and thickens. Care must be exercised against the spread of the disease. The pest spreads easily from horse to horse. The chief agents in the transmission are the larvæ, the nymphs, the newly fertilised females, and the males,—all these being found more towards the external surface. An attacked horse should be isolated. Common means of infection are the brushes, curry-combs, and instruments used in dressing the horse; while the stall where the patient is housed is a source of danger until disinfected.

Scaly Leg of the Fowl (Sarcoptes nutans).

This mite lives underneath the epidermal scales of the legs of the fowl; it also affects and is contagious for turkeys, pheasants, partridges, parroquets, and small cage-birds.

There are slight differences in the structure and mode of life of this Sarcoptes as compared with the preceding general account. First of all, the species seems to be viviparous; then while the male has the legs provided with the usual stalks and suckers, the legs of the female are usually short, and quite lack the bristles and suckers. Again, the female simply burrows into the skin without proceeding to form the little tunnel or gallery described as characteristic for the genus Sarcoptes.

The ripe female is very sluggish, scarcely moving, so that the disease is spread by the males, larvæ, and nymphs, which move about more on the outside of the skin. The progress of the disease is slow; there is a comparatively slight itching; also an elevation of the epidermal scales, chiefly those in front of the ankle and above the toes. Below these scales is a powdery mass glued into a crust by a serous exudate, the whole ultimately forming irregular thick crusts which, if broken off, leave the skin below exposed and bleeding. On the under surface of these crusts is a number of little pits; each of such pits has been the abode of an egg-laying female.

The diseased birds are lame; they have a difficulty in perching, and there is a great falling off in their condition.

Treatment.—Separate mite-infested birds. Thoroughly cleanse and disinfect the places where the birds have been kept with boiling water and whitewash. As to the affected bird itself, the general plan is to soften the crusts by bathing the leg in hot water and then carefully to remove them; then apply a mixture of creasote 1 part and lard 20 parts. Or use a mixture of equal parts of flowers of sulphur and vaseline. The removal of the crusts is often attended with a bleeding, and therefore some prefer to remove only the crusts that are already somewhat loose and are easily removed. After a day or two the dressed limb should be cleansed with soap and water.

Feather-Eating or Depluming Scabies (Sarcoptes lævis).

This is a very contagious and a quickly spreading disease on fowls and pigeons, and is often due to a tiny Sarcopt at the roots of the feathers. The disease is most prevalent in spring and summer. Beginning at the rump, it spreads to other parts of the body, the neck and head being often badly infected. The feathers break off and fall away at the attacked places, and the birds, irritated by the mites which live at the base of the feathers, pull out their feathers. The suffering birds become thin and fall away in their egglaying.

Affected birds—the cock especially—should be isolated. The creosote and lard ointment mentioned above is excellent, as also is oil of cloves rubbed well in.

COMMON SHEEP SCAB (*Psoroptes communis var. ovis*).

This mite is large enough to be visible to the naked eye, the full-grown male $1/50$ inch and the female $1/40$ inch. The egg measures $1/120$ inch. The larvæ from

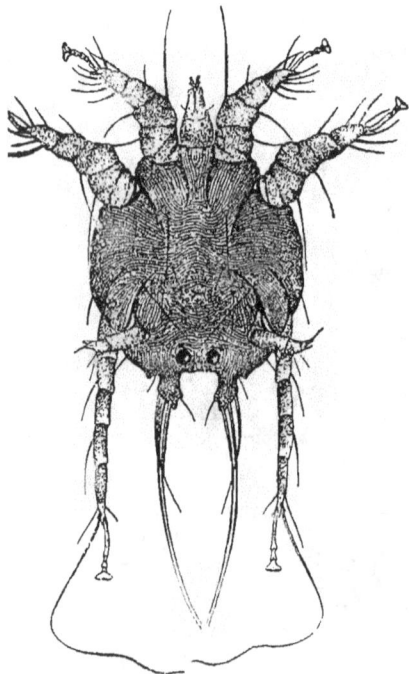

Fig. 776.—*Adult male of* Psoroptes communis *from under surface.*

(After Salmon and Stiles.)

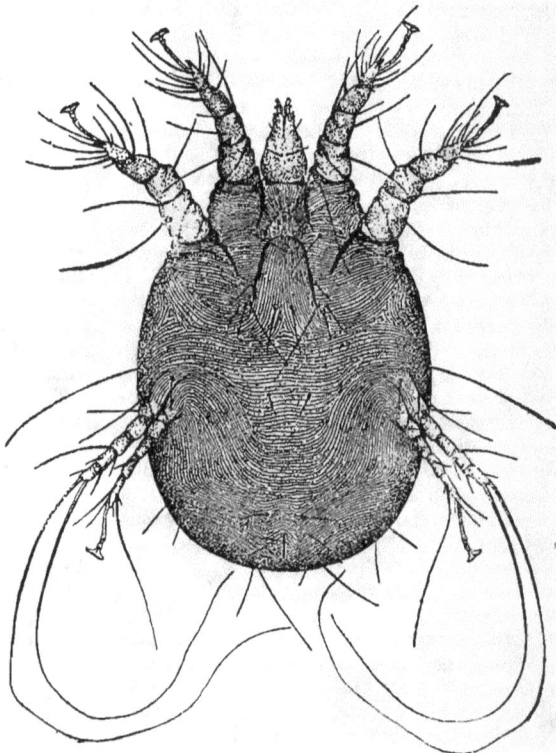

Fig. 777.—*Adult female of* Psoroptes communis *from under surface.*

(After Salmon and Stiles.)

the eggs have only six legs. The adult male (fig. 776) has two abdominal projections which end in long hairs. Each of the six front legs of the male has a stalk ending in a sucker; the two hind legs are small and have neither stalk nor sucker. The legs of the female (fig. 777) differ in appearance before the last moult and after the last moult. Before the last moult only the front four legs have stalk and sucker; the hind two pairs end in hairs. After the last moult the stalk and sucker are present on the two front pairs of legs and on the fourth pair, but the third pair of legs remains without stalk or sucker.

The Psoropt is found, not burrowing into the skin and laying its eggs in a gallery like the Sarcopt, but living externally on the skin, and laying eggs on it or glueing them to the wool near the skin. The parts chosen for infestation are those where the wool is thick—the back, flank, rump, and neck. The six-legged larva hatches from the egg, and the further changes resemble those described above for *Sarcoptes scabiei*.

Symptoms and Result of Attack.— The sheep are restless, and bite and rub themselves against posts, fences, &c., in order to relieve the intense itching that is occasioned by the mites pricking the skin. Little pimples appear as the result of the woundings, and from them there is an exudation of matter; the exudate dries into a crust. The sheep in scraping themselves rub off little pieces of crust and tufts of wool. The area of infection goes on increasing. If material be wanted for microscopic examination it is best procured from the edges of the crusts.

The best method to adopt in order to demonstrate the pest is to make a scraping of a newly formed crust or near the edge of a crust; this should then be placed in a solution of potash and allowed to lie for some time, the material being afterwards placed for examination under the microscope.

From the life-history of this pest care must be taken to avoid the spread of the disease by preventing infection to hitherto clean and healthy sheep. Therefore, yards and sheds that have contained scabby sheep should be thoroughly cleaned and disinfected and allowed to stand empty for a month before being used again for clean sheep. Any posts on the pasture-grounds used for rubbing should be whitewashed, as tags of wool or bits of crust sticking to them may harbour some of the parasites. Hence the danger also to clean sheep passing along a highway were scabby sheep allowed to make use of the highway.

Treatment. — The great method of fighting this Psoropt is by dipping. There is a large number of effective dips on the market, some of them arsenic dips. The Board of Agriculture, while recognising the efficacy of others, mentions the three following:—

Lime and Sulphur Dip.

Mix 25 lb. of flowers of sulphur with 12½ lb. of good quicklime. Triturate the mixture with water until a smooth cream without lumps is obtained.

Transfer this to a boiler capable of boiling 20 gallons, bring the volume of the cream to 20 gallons by the addition of water, boil and stir during half an hour. The liquid should now be of a dark-red colour; if yellowish, continue the boiling until the dark-red colour is obtained, keeping the volume at 20 gallons. After the liquid has cooled, decant it from any small quantity of insoluble residue, and make up the volume to 100 gallons with water.

Carbolic Acid and Soft-Soap Dip.

Dissolve 5 lb. of good soft-soap, with gentle warming, in 3 quarts of liquid carbolic acid (containing not less than 97 per cent of real tar acid). Mix the liquid with enough water to make 100 gallons.

Tobacco and Sulphur Dip.

Steep 35 lb. of finely-ground tobacco (offal tobacco) in 21 gallons of water for four days. Strain off this liquid and remove the last portions of the extract by pressing the residual tobacco. Stir the mixture well to secure an even admixture, and make up the total bulk to 100 gallons with water.

The period of immersion in these dips should not be less than half a minute.

In Leaflet No. 61 of the Board of Agriculture it is stated that of the two forms of baths—hand and swimming—the latter is greatly to be preferred. Its advantages are: "(1) The sheep being in a natural position may be completely immersed, even in a poisonous solution, with comparatively little danger; (2) sheep in lamb may be dipped with much less risk; (3) the motion of swimming allows no portion of the fleece to escape contact with the solution; (4) the work is most easily and therefore most effectively performed; (5) a larger number of sheep can be dipped in a given time and with fewer operators."

TICKS (*Ixodoidea*).

The Ixodidæ (Warburton), Ixodinæ (Neumann), are mites with a terminal rostrum made of mandibles and maxillæ, as in the previous families, but modified in a different way. Parts of the two maxillæ are soldered together to form a so-called dart furnished with backwardlydirected hooks. The other parts of the maxillæ—viz., the maxillary palps—are 4-jointed. The two mandibles complete the rostrum; each has a basal stem and a branched hooked upper part. It is by means of this rostrum that ticks fix themselves so firmly to their hosts; the recurved hooks of the rostrum make it difficult or impossible for one to pull, by main force, a tick from its attachment. In so pulling the body may come away and the mouth-parts be left in the wound.

The legs end in two claws and a little sucker. The skin is leathery but extensible, and a protective dorsal shield is present. Breathing is by tracheæ, which open in spiracles at the bases of the hind pair of legs. The two sexes differ in size, the male being smaller. They also differ in the character of the dorsal shield, which in the male may cover the greater part of the dorsal surface, whereas in the female it is limited to a small region at the front part of the cephalothorax.

Life-History.

In the life-history there is a metamorphosis, there being four stages—adult, egg, larva, nymph. The fertilised female, gorged with blood, drops away from the animal to which it has been fixed and proceeds to lay its eggs. The number of eggs is great, and the egg-laying may be spread over a considerable time, varying with the weather conditions.

From the eggs hatch tiny forms with six legs. These young forms ascend blades of grass or collect at the tip of a twig or branch, and here with marvellous patience they wait until a host passes. The host is clutched at with outstretched fore-legs, and having successfully obtained a lodgment, the larvæ fix themselves and remain for a time, feeding at the expense of the host. After feeding for some time, the larva withdraws its rostrum and drops voluntarily to the ground. Here it remains passive, until at last it undergoes the first moult; the skin ruptures and the nymph appears.

The nymph has a stronger rostrum and longer and stronger legs than the larva. The legs, too, are now eight in number. Spiracles are also present for the first time. Sexual organs are not completely developed. The nymph, like the larva, seeks a host, and attachment is followed by a period of feeding. There is then a similar falling away of the nymph from the host, followed by another moult. As a result of this moult we have the mature ticks, male and female. When these have fixed themselves to a host pairing takes place. The male does not die after one act of copulation, but is able to proceed to other efficient acts of pairing. The mature male on the host feeds, but does not swell much; the female, on the other hand, rapidly increases in size, and at last falls away and prepares for her egg-laying.

There are some species of tick in which, once the larva has reached a host, there is no leaving of the host until the female falls off for her egg-laying: in such cases the moults take place while the larva and nymph remain attached to the animal.

BRITISH TICKS.

The following species are given by Wheler [1] as having been found on one or other of the domesticated animals :—

Ixodes ricinus (Latreille): male 2.35 mm. to 2.80 mm. long, and the female from about 3 mm. when fasting to 10 mm. long when replete. Found on numerous hosts, the favourite seeming to be sheep, goats, cattle, and deer; found also on hedgehogs, moles, bats, and even on birds and lizards.

Ixodes hexagonus (Leach) var. *longispinosus:* male, 2.50 to 3 mm. long; female, 3.00 mm. fasting to 11 mm. when replete. Found especially on stoat, ferret, hedgehog, but also on sheep and cattle.

Ixodes hexagonus (Leach) var. *inchoatus:* male, 2.52 mm. long; female, 2.86 mm. fasting to 6.56 mm. replete. Found abundant on shepherds' dogs on the Border; was never found on the sheep.

Hæmaphysalis punctata (Canestrini and Fanzago): male, 3.10 mm. long; female, 3.44 mm. fasting to 12 mm. when replete. Found on sheep, especially behind the ears,

[1] "British Ticks," by Edward Galton Wheler, in *The Journal of Agricultural Science*, March 1906.

and on goats, cattle, horses; also on the hedgehog.

Dermacentor reticulatus (Fabricius): male, 4.20 mm. long; female, 3.86 mm. fasting to 16 mm. when replete. On sheep occasionally, but also attacks cattle, deer, goats, and even man.

Very much work still remains to be done on ticks ;[1] the actual workers at the family are not numerous, but interest has been greatly stimulated owing to the recognition of the very important part which ticks may play in the spread of grievous diseases. There are numerous pathogenic ticks, and the following diseases are known to be carried by them, the tick being the intermediate host by which the parasite causing the disease is introduced to the suffering animal :—

Red Water or Texas Fever in cattle.
Heart Water in sheep and goats.
Canine Piroplasmosis or Malignant Jaundice.
Rhodesia or Coast Fever in cattle.
Carceag or Piroplasmosis in sheep.

FAMILY GAMASIDÆ.

To this family belongs the Hen Mite (*Dermanyssus gallinœ*). This is an oval, pear-shaped mite which varies in colour

[1] At present an excellent Monograph of the Ixodoidea is being published. The authors are George H. F. Nuttall, M.A., M.D., Ph.D., D.Sc., F.R.S.; Cecil Warburton, M.A., F.L.S.; W. F. Cooper, B.A., F.L.S. ; and L. E. Robinson, A.R.C.Sc. (London).

from white to red, according as it is fasting or replete with blood. The legs are strong and bristly; the body is somewhat flattened, the hind part being widest ; the abdomen is surrounded with bristles. The mites live in colonies in hen-houses and pigeon-lofts, the colony containing all stages at the same time.—larvæ, nymphs, males, females. These hen mites are temporary parasites, hiding in the day-time in crevices and such shelter-places, whence they issue at night and swarm on to the birds, making rest impossible for them. The birds are irritated by the itching which is an accompaniment of the wounding made by the mites' mouth-parts. The pests suck the blood, and what makes them more formidable is their ability to subsist for a long time in absence of a live host. Brood hens are worried, and young birds become anæmic and may die. The mite may also invade the nostrils and external auditory meatus. In addition to attacking the hen and the pigeon, this *Dermanyssus* (skin-pricker) pricks man, the horse, dog, cat, and goat. Horses, like fowls, are attacked at night, and the irritation causes an eruption of small vesicles. From continual rubbing these get broken, the skin becomes raw, and little scabs result.

Treatment.—Do not allow fowls into the stables at night. Bathe with a 5 per cent solution of chlorhydrate of ammonia. To keep fowls free from the mite let there be a regular cleansing and disinfection of their houses and perches.

AILMENTS OF FARM LIVE STOCK.

It is not presumed that the farmer should become so familiar with veterinary science as to be able to dispense with the services of the professional Veterinary Surgeon. The farmer, however, should unquestionably know enough of the ailments which afflict his live stock to enable him to recognise the symptoms of each when he sees them, and also to successfully treat those of the more simple kind, as well as to decide when the veterinary surgeon should be sent

for, and what had best be done until he arrives. The information necessary for these purposes will be found, carefully classified, in the following treatise. This treatise, compiled originally by Mr G. H. C. Wright, LL.B., and edited by Mr F. Tonar, M.R.C.V.S., has been carefully revised for this edition of *The Book of the Farm* by a Fellow of the Royal College of Veterinary Surgeons, who is one of the most experienced and trusted members of the veterinary profession.

SUBJECTS.

HORSES.

DISEASES AFFECTING THE HEAD, EYES, MOUTH, AND NERVOUS SYSTEM.

I.—THE HEAD AND NERVOUS SYSTEM.

Apoplexy or Staggers.

This almost hopeless disease results from an effusion of blood producing pressure on the brain.

Symptoms.—Head carried low. The horse staggers till he falls. Eyes fixed and pupils dilated. Muzzle cold. Hearing and sight affected. Teeth clenched. When convulsions come on, the horse soon dies.

Cause.—Apparently the breaking of a small blood-vessel on or near the brain, occasioned by the derangement of the system from overfeeding, or by violence.

Prevention.—There is practically no means of prevention beyond the ordinary proper management of the animal.

Treatment.—Apply iced or cold water in a douche or stream along the spine and on the back of the head. Give a physic ball (No. 17). Remove dung from lower intestines with the hand, and the urine may require to be removed by the catheter. Administer a warm clyster (No. 13) two or three times a-day, and if it can be done without disturbing the animal too much, give the following every four hours: Bromide of potassium, 4 drs.; brandy, 6 oz.; water, 8 oz. But perfect quietness is the best medicine. Nothing more can be done: this almost invariably fatal disease must be left to take its course.

Tumours in or on the Brain

can seldom, if ever, be recognised and localised during life, and their treatment is practically hopeless.

Brain-fever, Mad Staggers, or Inflammation of the Brain.

Symptoms.—Sleepy and daft condition. Nostrils distended. Flanks heaving. Eye wild. When delirium comes on, the horse becomes violent and dangerous: his struggles will continue till he is exhausted, and the stupor returns.

This disease is sometimes thought to be colic: the difference is, however, very apparent. In colic, the horse is conscious, and only plunges and rolls from pain, often turning his head round to his flanks: in mad staggers the struggles are more violent, and consciousness is lost.

Cause. — Fulness of blood to the head through being over-heated in hot weather.

Treatment.—This disease is so often fatal, and so little can be done to check it by medicine, that bleeding seems the only course to pursue.

Put the horse in a cool stable. Open the jugular vein, and bleed till he is weak.

Give purgative—croton-nut, ½ dr., or croton-oil, 20 drops, in warm gruel, and repeat in 10 gr. doses, or croton-oil 10-drop doses every eight hours till the bowels are open.

Bathe head with iced or cold water. Inject warm water and soap. After purgative has been administered, give every few hours, in gruel, a draught containing chloral hydrate, 4 drs.; bromide of potassium, 4 drs.; Fleming's tincture of aconite, 5 drops; spirit of chloroform, 1 oz.

When recovering, feed moderately for a few days on bran-mashes and a little hay.

Epilepsy or Fits.

A disease of the brain not very common among horses.

Symptoms.—Attack is sudden. Horse stares round, trembles, and falls to the ground. Convulsions more or less severe follow.

When consciousness returns, the horse will feed as if nothing happened.

Cause.—A derangement of the brain; but very little is known about it, and there are no known means of prevention or cure.

A horse subject to these fits is dangerous for either riding or driving.

Treatment.—Give a purgative (No. 17), and if the horse is restless or excitable give a draught composed of chloral hydrate, 4 drs.; tincture of belladonna, 1 oz.; water, 8 oz., morning and evening. Little can be done to guard against a return of this disease.

Epizootic Lymphangitis.

A contagious and eruptive disease affecting horses and mules, and, although less liable to it, the ox sometimes becomes affected. Its first appearance in Britain was in horses brought from South Africa after the war. It is now (1909) believed to be stamped out.

Cause.—It is due to a fairly large, somewhat ovoid, micro-organism, the *Cryptococcus farcinimosus*, possessed of considerable vitality, and the disease is easily communicated from affected to healthy animals.

Symptoms.—Small nodules form on the head, neck, legs, or on any part of the body. They may grow as large as hazel-nuts, burst in succession, and discharge matter. The sores formed have little tendency to heal, and proud flesh may grow from the wounds, forming large ulcers. The lymphatic vessels in their neighbourhood become inflamed, swollen, and stand out like cords. Ulcers may form in the nostrils, but this is not common, although there is often a discharge from one or both nostrils, and enlargement of the glands under the jaw. Unless energetically treated at an early stage, gradual emaciation leads to exhaustion and death.

Treatment.—The best treatment is the complete excision of the diseased part — the ulcers — opening up of the inflamed lymphatic vessels, and the destruction of the diseased tissues with caustics or the hot iron. It should be left to the veterinary surgeon.

Prevention. — Isolation of affected animals, the free use of antiseptics, careful burial or cremation of diseased carcases, and the thorough cleansing and disinfection of the boxes or stables.

Fracture of the Skull.

The bones of the skull are so thick that a fracture can only arise from a fall when a horse rears, or else from brutal violence. This is a common occurrence with pit ponies, and is caused by their knocking their heads against the roof timbers. In most pits they wear a leather skull-cap to prevent this. It is generally fatal.

Treatment. — The parts should be elevated and fastened with adhesive plasters, to prevent their moving. To reduce any inflammation, give purgative (No. 18) and a spare diet.

Fracture of the Nose

is caused by a fall, or a blow across the bones of the nose.

Treatment.—Place the finger up the nose, and gently push the bones back into their place, and retain them there by an adhesive plaster.

Fracture of the Jaw.

Generally caused by a kick, fall, or accidental violence.

Treatment.—If the animal is of any value, get the jaw set by a veterinary surgeon, who will place it in a cradle made for that purpose.

Lockjaw or Tetanus.

A disease affecting the nervous system, and one of the most fatal which attacks the horse.

Symptoms.—A difficulty in chewing its food and some stiffness about the jaws is often the first symptom observable.

Water is gulped down, the jaw becomes rigid, and saliva runs from the mouth. Afterwards the muscles of the head, neck, and shoulders become fixed, till the whole muscular system of the body seems cramped. After some days, if the disease is not checked, the horse will die in agony from sheer exhaustion.

Cause.—It is now believed to be contagious, and due to a micro-organism— the bacillus of Nicolaïer, or drumstick bacillus; but the apparent cause is generally some wound, kick, or blow. Docking has been known to cause it. In some cases the only apparent cause has been the existence of bot-worms in the stomach. Exposure to cold and general neglect have also brought it on.

Prevention.—There is no particular means of prevention beyond the proper care of the animal, and maintaining thorough cleanliness of any wound, accidental or otherwise; but if the beginning of this disease is suspected, give at once linseed-oil, 1 pint; aloes, 2 drs.; Fleming's tincture of aconite, 8 drops.

Treatment.—In this disease there is considerable difficulty in giving remedies, owing to the contraction of the muscles of the jaws and the general stiffness. If the cause of the disease is some wound, it had better be poulticed, and dressed with carbolic oil or carbolic acid in 20 parts of water. Open the bowels by giving ½ dr. of powdered croton (or 20 drops croton-oil) in warm water, repeating the powdered croton in doses of 10 grs. (or croton-oil 10 drops) every eight hours till purging commences. If costiveness still continues, administer a clyster of olive-oil, 8 oz.; opium, ¼ oz.; warm gruel, 2 quarts.

Put the horse in a cool rather dark stable where there is plenty of air, keep him very quiet, and let no one but the attendant have anything to do with him. Mix an ounce each of extract of belladonna and bromide of potassium together, and place a piece of it about the size of a bean between the molar teeth every five hours by the aid of the forefinger.

Put a pail of gruel or sloppy mash in the manger, so that the animal can, if inclined, partake of it.

Never try to force food down the

animal's throat: it only aggravates the disease.

If the animal is a valuable one a veterinary surgeon should be called in, who will probably administer antitetanic serum, which is an almost unfailing preventive, and often seems useful in promoting a cure when the disease is taken in time.

Tetanus does not extend to the internal organs; the horse will suffer from hunger. When the horse is recovering, he should be fed moderately with nourishing food (bran-mashes, linseed, and oatmeal in preference), and he will be all the better for a turn or walk out of doors if the weather is suitable.

Megrims.

This disease is said to be caused by an undue pressure of blood in the head.

Symptoms.—The horse will suddenly stop, shake his head, then proceed on his journey at considerable speed. Sometimes he will turn round twice or more, often perspiring profusely, fall down, and either struggle on the ground or lie quietly. The attack may last five minutes, and when it is over the horse will resume work as though nothing had happened : he will, however, be considerably weakened. A horse subject to these attacks is particularly dangerous to ride or drive, and after one attack will always be liable to others in the future. Sometimes horses will die suddenly when seized with an attack.

Cause. — Violent exertion in hot weather; too small a collar or tight bearing-rein ; a high system of feeding.

Prevention.—A proper-fitting collar and not too violent exercise ; a judicious system of feeding, and an occasional dose of purgative medicine, such as No. 20.

Treatment.—Bleeding is of little use, though recommended by many authorities. Let the animal stand a few minutes, dash cold water on his head, push the collar forward, and proceed home as quietly as possible. Then give a physic ball (No. 17), and every six hours in water bromide potassium, ½ oz., for two or three days; afterwards give tonic (No. 21). Mashes and green meat should be given in preference to dry food, and a run out to grass for two months.

Palsy (Paralysis).

A deficiency in nervous power, which affects usually the hind quarters.

Symptoms.—Stiffness in their action, difficulty in turning, disinclination to lie down from the difficulty in rising again, and sometimes a total inability to rise.

Cause.—Pressure on the spinal cord from effusion of blood or serum, or from tumours within the spinal canal. Falls, injury to the spine from blows or from turning in too narrow a stable, old age, and heavy loads.

Prevention.—Humane treatment and ordinary care.

Treatment.—Give mild doses of purgative medicine, such as linseed-oil, 1 pint, which will not only open the bowels but also support the system. Rub stimulating embrocation, as mustard liniment (No. 15), on the part affected, and in cases of doubt, especially along the spine. Give morning and evening nux vomica tincture, 2 drs.; spirit of ammonia, 1 oz. in 10 oz. of water.

Stomach Staggers.

Disturbance of the brain resulting from a deranged and distended stomach.

Symptoms.—The first symptom may be dull, colicky pains, sleepy look, pulse very slow, profuse perspiration. In many cases blindness. Rests his head against the manger or wall, and sometimes moves his legs in a peculiar manner. Staggering gait till the horse falls down and dies in a state of stupor.

Cause.—Unsuitable food or over-feeding. Food in an overloaded stomach will swell and distend it, affecting the nervous system in such a way as to cause staggers.

Prevention. — Proper, regular, and systematic feeding with food of good quality will ensure immunity from this disease.

Treatment.—Give a purgative, such as 1 pint linseed-oil and 1 oz. of tincture of ginger. An hour after the dose of oil give in gruel draught (No. 9), and repeat the dose of oil if action of the bowels is not obtained. Clysters of warm water and soap should also be given every four hours.

Grass Staggers.

A disease manifesting nervous symptoms but arising from the stomach.

Symptoms.—They come on slowly; the horse is dull and listless at first, but gradually passes into a somnolent condition. In time the animal gets weak, reels or staggers about, and if sharply turned, will most likely fall down. It seldom lies down when suffering from this disease.

Cause.—It is mostly seen during the months of July and August, but varies according to the season. It arises from eating rye-grass at a certain stage of its growth, as if eaten in this state it causes the disease. In hot and dry seasons it is most frequent.

Treatment.—Remove the animal into a loose-box, give ball (17) and draught (9), and repeat the draught every four hours. Do not give any more rye-grass, but steamed oats, bran-mashes, and a little hay. Fresh, succulent, natural grasses may be given in very small quantities.

II.—THE EYE.

Cataract

is an opacity of the crystalline lens, and often follows an attack of ophthalmia.

Symptoms.—A speck in the eye, not on the surface, which varies in different eyes in colour, shape, position, and size. They often become large enough to cause blindness.

Cause.—From a blow, after an attack of ophthalmia, or inflammation of the eye.

Treatment.—In the lower animals very little can be done for it.

Amaurosis or Glass Eye.

Paralysis or loss of special sensation in the optic nerve.

Symptoms.—The eye looks larger, pupil dilated, animal stares—in fact, the eye is blind and motionless, and looks more like a glass eye than a natural one.

Cause.—It is seen as a temporary condition in some cases of poisoning, but when permanent it is the result of either partial or total loss of function in the optic nerve.

Treatment is of no use unless it is the effect of a poison; then give ball (17), a pint of linseed-oil, and every two hours give in pint of cold water 4 oz. of brandy and 2 oz. of spirits of ammonia aromat.

Inflammation of the Eye or Simple Ophthalmia.

Symptoms.—Eyelids swollen, watering, and nearly closed. Eye bloodshot, and inside of the eyelids very red. Cornea cloudy. Health not affected.

Cause.—Foreign matter, such as a hay-seed or chaff in the eye; a blow with a whip; or exposure in facing a cold wind. It is sometimes produced in a young horse by over-exercise.

Treatment.—First remove any foreign substance. Give mild purgative (No. 20) and a mash diet; bathe the eye with poppy-heads and warm water every two hours, and if that is not handy, with weak brandy-and-water; if no improvement, bathe with a solution composed of *liquor opii sedativus* 1 oz., in 1 pint of cold water. A useful lotion for inflammation of the eye is sulphate atropine, 4 grs., in 1 oz. of water. Keep the animal in a dark box until better. The inflammation should be cured in a few days; if not, treat as for Ophthalmia, *infra*.

Itching or Tumour of the Eyelids.

Treatment.—Rub the eyelids with mercurial ointment and lard in equal parts, and give sulphur, ½ oz., and nitre, 1 dr., in the food once a-day till the animal is cured.

Removal of the Eyeball.

It is necessary sometimes, when the eye has been severely damaged, or has a cancerous growth in it, to remove it. This can be done only by a veterinary surgeon, who will not only remove the eye, but, if you wish, place a glass one in its place. In using glass eyes always take them out at night, for if kept in very long they cause pain.

Ophthalmia.

Violent inflammation of the eyelids, extending to the cornea and internal structures of the eye.

Symptoms.—Light pains the eye,

which is kept shut; a profuse flow of tears. Pupil is contracted, and iris changes colour. The opacity usually extends from the circumference towards the centre, and the inflammation diminishes one day to increase twofold the next, till in a few weeks, if not checked, the eye becomes opaque and blindness comes on. After an apparent cure the disease will sometimes come on again, either in the same eye or in the other which had not previously been affected.

Cause.—A foul-smelling, ill-ventilated stable, reeking with ammonia and decomposing manure, is a frequent cause of this disease. Confinement in a dark stable and a sudden transition into the glaring sunshine often accounts for it. The tendency to inherit this disease from sires with defective sight is too well known from sad experience to need any comment. The management of horses being now better understood, this disease is becoming rarer every day.

Prevention. — A well-drained and well-lighted stable and cleanliness are the best preventives.

Treatment. — Foment the eye with warm water, and bathe with a lotion composed of sulphate of atropine, 4 grs., in 1 oz. of water. Feed on spare diet; put the horse in a cool, airy, but dark stable, where there is perfect cleanliness. Give purgative (No. 18 or 19). The use of the lancet may in extreme cases be useful; the inside of the eyelid should be exposed, and the lancet drawn lightly along for the purpose of relieving the parts affected by pressure of blood. Cloudiness of the eye, or complete opacity, is a frequent consequence of this disease, which may be treated by bathing with solution of corrosive sublimate, 1 gr., in 2 oz. of water.

Thickening of the Haw.

The haw of the eye is situated in the inner corner of the eye filling the lid. A horse can bring it forward over the eye, and with it wipe away any foreign matter that may have got into it. This haw sometimes enlarges and protrudes, so that it cannot retract.

Treatment.—Give purgative (No. 19), and bathe the eye with poppy-heads and warm water. Should the ulceration continue, bathe with white vitriol, ½ dr.; water, 6 oz.; or paint with a weak solution of silver nitrate. If further treatment is necessary, it must be left to a veterinary surgeon.

Warts on the Eyelids.

Treatment.—Cut off with a pair of scissors and touch with lunar caustic, taking care not to touch the eye, and not to put on more than is necessary. Rubbing the roots with blue vitriol will sometimes effect a cure. Take care also that any bleeding, when cutting, does not touch any other part, as blood from a wart may spread the disease.

Wounds in the Eye or Eyelids.

Generally caused by brutality or carelessness.

Treatment.—Very little can be done except to reduce the inflammation by purgative medicine (No. 20), and bathe with warm water, and apply a lotion composed of atropine, 4 grs., in 1 oz. of water. This lotion is best applied by the aid of a feather, which, when soaked in the lotion, should be drawn gently across the eye. When the eyelids are torn, never cut any of the skin off, but retain it in its proper position by the aid of pins or silver wire. In these cases the horse should be kept in a dark box.

III.—MOUTH, NOSE, TEETH, TONGUE, PALATE.

Glanders and Farcy.

These names have been long applied to what was believed to be two distinct diseases, but is now known to be only different manifestations of one and the same disease. Glanders has been recognised as affecting horses, asses, and mules from remote ages, and is now included under the Diseases of Animals Acts.

Cause. — Contagion. It is due to the *Bacillus mallei;* but overcrowding, insufficient food, want of fresh air, and insanitary conditions may predispose to it.

Symptoms.—Generally a discharge from one nostril, but may be from both; sometimes a cough, enlargement of the glands inside the lower jaw on the same side as the discharge. It is often chronic, and the animal may work for months, taking its food fairly well, and little to cause suspicion but the nasal discharge. But sometimes there is a slight rise of temperature, and the animal seems sensitive to cold; the hair may get erect on coming out of a warm stable, or after a drink of cold water. One of the most characteristic appearances is the presence of ulcers inside the nostrils. These are very rare, apart from glanders. In the acute form there is high fever, the breathing is distressed, and the animal looks very ill. In the form known as Farcy, one or more of the limbs may become swollen, and the lymphatic vessels inflamed, hard, and cord-like; nodules, which may become as large as hazel-nuts, form here and there on the course of the vessels, generally burst, and discharge a yellowish oily matter. Although most common about the limbs, nodules often form about the head, neck, and other parts of the body. The discharge from the nose and from these nodules is the main source of the contagion, and is very dangerous to other animals, and also to man. It is not very rare for an attendant on a glandered horse to become attacked by the disease.

Glanders is scheduled under the Diseases of Animals Acts: intimation of its existence must at once be given to the police, and it is the duty of the Local Authority to cause every glandered animal to be slaughtered as speedily as practicable. They must also cause the detention of each horse, ass, or mule which, in their opinion, has been exposed to the risk of contagion, until such time as they can have the "mallein test" applied to it. Should the animal not react to the mallein test, the "detention notice" ceases to affect it after forty-eight hours. But when the animal reacts, it has to be slaughtered by the Local Authority.

For all such diseased animals so slaughtered the Local Authority must pay to the owner, as compensation, half the value of each animal before it was tested—the sum paid not to exceed £25 in the case of a horse, or £6 for any ass or mule. Where, after slaughter, the animal is found not to have been glandered, full value must be paid, but

not over £50. And when an animal is slaughtered after being clinically affected — manifestly glandered, — the Local Authority shall pay whatever sum they think expedient, but not more than one-fourth the value of the animal, and not less than £2 in the case of a horse, and 10s. for any ass or mule.

Treatment is not permitted, although it is recognised that animals occasionally recover.

Prevention is comprised in careful isolation and disinfection, and the adoption of rigorous police measures.

Lampas.

A fulness of the lower bars of the palate.

Cause. — It generally occurs with young horses, and is a natural result from the congestion caused by the shedding of their milk-teeth and the growth of the permanent ones.

Treatment. — Cut the bars lightly with a penknife several times across, avoiding the artery. Never burn them. Give bicarbonate of potash, 6 drs., morning and evening in drinking-water, and warm bran-mashes. Use lotion (No. 16) for washing the sore places.

Nasal Gleet.

A profuse and unnatural discharge of mucus from one or both nostrils.

Symptoms.—The nasal discharge continues after every other sign of cold has left. Mucus in large quantities, mingled with matter, constantly flows or is blown from the nose, till the horse becomes much weakened. The mallein test may have to be applied to make sure it is not glanders.

Treatment. — Should cough remain, treat as for Cough (p. 447). If the discharge is fœtid, give daily a dose containing sulphate of copper, 1 dr.; ginger, 2 drs.; gentian, 2 drs. If the discharge is not offensive, but only an excessive discharge of the fluid which moistens the nose, give daily, sulphate of copper, 1 dr., made into a ball with flour and treacle. Horses affected by this complaint should always have a lump of rock-salt in their racks, and a little salt mixed with the most nourishing food possible. Tonic (No. 21) may be useful in treating this disease. Nasal gleet of long standing may be due to a diseased tooth or bone in the head, and the opinion of a veterinary surgeon should be obtained.

Polypus.

An excrescence may grow in the nostril or further back and impede the breathing. It must be removed by a veterinary surgeon, and no treatment by an unprofessional man can be of any use.

Bleeding from Nose.

The result of irritation of the nose, glanders, bursting of a blood-vessel in the head or lungs, and sometimes a blow on the head.

Treatment,—Keep animal quiet, head elevated, and pour cold water over it. Give every two hours, in a pint of gruel, tincture of perchloride of iron, 1 oz.; spirits of sweet nitre, 2 oz.

Rabies or Hydrophobia.

See Dogs (p. 493).

Strangles.

A disease more common among colts and horses under four years old than among older ones.

Symptoms. — A cold, cough, sore throat, and profuse discharge of yellow mucus from the nostrils, swelling under the throat, which increases and renders swallowing painful. The tumour is situated in the centre of the throat under the jaw, and feels like one solid mass. Owing to its solidity this disease can readily be distinguished from Glanders (see p. 443) when the tumour is composed of separate parts, which can be easily identified. The centre of the tumour is soft, and when it suppurates and bursts it discharges an immense quantity of pus, quickly healing after the discharge. When the cough subsides, the horse begins to recover from the extreme weakness attending the disease.

Cause. — Probably cold or climatic changes. I have strong reasons for believing that this disease is contagious.

Prevention.—Isolate affected animals.

Treatment.—Blister the tumour with ordinary blister (No. 1 or 2) to hasten its

progress and prevent the inflammation spreading. When the tumour is soft on the top, lance it and suffer the pus to drain out without any pressure. After the discharge, keep the place clean by bathing it well with warm water; rub with vaseline, which will soften the wound and promote its healing. Give twice daily, in a pint of gruel, No. 9, and keep the bowels open with carrots and bran-mashes. Feed on bran-mashes and green food, and keep the animal in a cool and comfortable stable. When recovery is established, give morning and night tonic (No. 21), and keep the horse well. The discharge from the nose will continue some time, but will gradually cease. If this disease is neglected, death will probably follow.

Bastard Strangles.

A low form of strangles, in which abscesses appear on different parts of the body. The treatment should be the same as for ordinary strangles. In this disease there is much more danger of blood-poisoning.

Teeth (Diseases of).

The irregular growth and rough edges of the teeth frequently produce wounds in the mouth. A horse out of condition should be examined, and if his teeth are irregular or have rough edges, they should be rasped down with a file that is made for the purpose. Sometimes it is necessary to cut off part of a tooth which projects far above the level of the others.

Extracting Teeth.

To extract the corner teeth of a three- or four-year-old horse, so as to try and alter his age, is cruel, and any one with experience of horses can easily see on looking into the animal's mouth if such a thing has been done.

Wolf - teeth.—These little teeth are situated in front of the molars, and are believed by some to interfere with the animal's feeding. They can be easily extracted by the aid of a pair of forceps, or else punched out. But unless it is distinctly evident that they are causing trouble they should be left alone.

Molar Teeth.—They sometimes become diseased. The animal quids his food, and frequently when feeding pauses for a few seconds. The breath is very offensive. Their treatment should certainly be left to the veterinary surgeon.

Wounds in the Mouth.

From a cruel bit, &c.

Treatment.—Wash it with a solution of alum, 1 oz., dissolved in twenty-eight times its weight of water; or use lotion (No. 16).

Teething Cough.

A persistent and violent cough.

Symptoms. — Usually seen between the age of three and four. Food refused, head poked out, gums red and swollen, frequent coughing, and sometimes a tooth is found in the manger.

Cause.—Teething, which causes irritation of throat.

Treatment.—Extract any temporary teeth showing signs of getting loose, and blister throat with mustard liniment (No. 15), and give every night and morning, in a pint of gruel, draught (No. 10).

Wounds of Tongue.

Treat as for wounds of the mouth.

Tongue Bladders (Ranula).

Sometimes occur underneath the tongue.

Cause. — Produced by a slight derangement of the system.

Treatment.—Give a physic ball (Nos. 17 or 18), which will reduce any fever. The bladders may be readily removed by opening with a lancet.

Paralysis of the Tongue.

Palsy of the tongue.

Symptoms.—The tongue hangs in a loose manner from the mouth, and becomes swollen and inflamed.

Cause.—A severe injury to tongue, or by dragging on the tongue when giving a ball.

Treatment.—Suspend the tongue in a net-bag tied to the head-stall; give purgative (No. 19) and a drachm of nux vomica night and morning in a half-pint of water.

Amputation of Tongue.

This is sometimes done by veterinary surgeons when the tongue has been extensively lacerated. A horse that has lost part of his tongue must be fed from a deep manger, and in drinking these animals force their heads deeply into a pail of water.

Paralysis of the Lower Lip.

A pendulous condition of the lower lip.

Symptoms.—The animal's health is not interfered with, and he feeds fairly well, but lets a little food drop, his lip hangs down, and a little saliva flows from it.

Cause.—Paralysis of the nerve of the lip, which is usually brought about by the curb-chain being too tight, or a badly fitted bridle, or accidental injuries.

Treatment.—Give a physic ball, containing 5 drs. of aloes, and rub into the lip and sides of the face a little of embrocation (No. 12). Feed on sloppy mashes.

DISEASES AFFECTING THE THROAT, CHEST, RESPIRATORY ORGANS, AND BLOOD.

Bronchitis,

or inflammation of the bronchial tubes.

Symptoms. — Coughing, wheezing, hard breathing, and weakness. The horse may die in a severe attack from suffocation.

Cause.—In cases of neglected cold or catarrh, bronchitis often follows. Exposure to cold or wet. Common in young animals that are starved and neglected.

Treatment.—Give plenty of fresh air, but keep warm. Apply embrocation (No. 12) to the chest; give nitre, 3 drs., and Fleming's tincture of aconite, 10 drops, three times a-day, and increase the dose if necessary. Feed on bran-mashes containing linseed - meal. For drinking - water, give weak infusion of linseed. In acute cases, give in gruel draught No. 10 three times a-day.

Broken Wind.

Symptoms.—In this disease the expiration of the breath takes two efforts, and the inspiration only one; the breathing, therefore, is not regular, as in thick wind.

Cause.—It is due to the rupture of air-cells, and is generally attended by a dry cough. Dusty food, gross feeding, previous inflammation, and violent exercise after heavy feeding.

Treatment.—There is no *cure*. Keep for slow work, and feed on soft nourishing food which occupies a small space.

Crib-biting

is more a vicious habit than a disease.

Symptoms.—The animal seizes the manger or any fixed object, and makes a gulping noise as if trying to swallow air.

Cause. — Indigestion or habit; one horse will learn it from another.

Treatment.—It takes a lot of curing. Anchovy paste on the manger will sometimes effect a cure. Any saddler will make a strap to go round the horse's neck to prevent crib-biting. An invention has been recently brought out to cure it by the aid of electricity. The battery is placed in such a way that whenever the animal seizes and squeezes the top part of the manger he at once receives a severe shock.

Choking.

Substances which have lodged in the gullet can generally be forced down by the use of a flexible tube, similar to that used for cattle; but it should only be done by a veterinary surgeon.

Sore Throat.

A common complaint, and associated with such diseases as strangles, influenza, and scarlet fever, &c.

Symptoms. — Animal has a nasty cough, quids his food, and pokes out his nose.

Treatment.—Blister the throat with embrocation (No. 12), feed on sloppy food, and give in gruel twice a-day (No. 10). Be careful in drenching, as there is a risk of choking the animal.

Rheumatic Fever.

A specific fever due to a constitutional condition of the system.

Symptoms.—Animal restless, breathing hurried, slight cough, shows signs of pain, goes stiff, and joints swell.

Cause. — Hereditary tendency, bad stables, and insufficient food.

Treatment.—Give physic ball (No. 20), put half an ounce of nitre frequently in drinking-water, and give twice a-day the following ball: iodide of potassium, 1 dr.; powdered colchicum, 20 grs.; liquorice-powder, 2 drs., made up with linseed-meal and treacle. Rub the swollen joints every night with embrocation (No. 12), and apply woollen bandages.

Chronic Cough.

A most annoying disease to the rider. This cough frequently follows an attack of inflammation of the lungs.

Symptoms.—If the horse coughs after drinking, the cough will arise from the windpipe. It may not affect the general health.

Cause. — Previous inflammation, neglected cold, and sometimes worms.

Treatment. — If the coat is staring, the cause of the cough will generally be worms, in which case give turpentine, ½ oz., daily, in 4 oz. of linseed-oil; or santonine, 20 grs., and aloes, 3 drs., made into a ball with linseed-meal and treacle, in the morning on an empty stomach, and repeat after two days; or give draught (No. 11). If the cough proceeds from the throat, feed on green food and mashes, and give ball (No. 8). Apply blister (No. 2) to the throat if other remedies fail. Water, in which a little linseed or treacle has been boiled, is useful instead of plain water, for drinking purposes.

Common Cold.

Symptoms. — Slight discharge from the nose, and weeping of the eyes; fever and cough.

Cause.—Changes of temperature and chills.

Treatment. — Clothe warmly, and place in a cool stable. Feed on warm bran-mashes with a little linseed-meal in them, and give in gruel night and morning till fever is reduced—acetate of ammonium, 3 oz.; potassium bi-carbonate, ½ oz.; chloroform, ½ oz.; and apply liniment (No. 14) to the throat, or embrocation (No. 12).

Distemper, Catarrhal Fever, or Influenza.

Most prevalent in spring and autumn, especially when the weather is cold and wet.

• Symptoms.—At first dulness, loss of appetite, and there may be shivering, cough, weakness, inflamed eyes, nose a pale red, watery discharge from nostrils. Later the discharge from the nostrils becomes thick, but seldom offensive, glands of throat and under jaw swell, which make swallowing difficult. Generally there is intense weakness.

There is a violent form of influenza which has lately come into notice called "pink eye." It is attended with high fever, extreme weakness, depression, and loss of appetite, and has been the cause of serious loss in many parts of the country.

Cause.—Contagion, influences of climate producing cold, amounting almost to an epizootic.

Treatment.—Remove into a cool box, clothe warmly, feed on warm bran-mashes and green food, a little hay, or a carrot or two, and give in weak infusion of linseed 1 oz. nitre, instead of pure water for drinking. Sponge the nostrils with vinegar and water. Give draught twice a-day containing spirits of nitrous ether, 1 oz., *liquor ammonii acetatis*, 3 oz., in a pint of water, and rub the throat with embrocation (No. 12). Half fill a nose-bag with hay, and pour boiling water upon it, and keep the horse's head in it till the vapour ceases to rise, but be careful not to burn the horse's nose. In cases of extreme depression, as in pink eye, give every three hours spirits of nitrous ether, 1 oz.; whisky, 4 oz.; water, 6 oz. When recovering, give tonic (No. 21) in a pint of beer twice a-day. Great care should be taken to prevent these attacks producing roaring and other diseases.

Broken Ribs.

The ribs of horses are frequently broken through accidents and kicks.

Treatment.—If the ribs are only broken and not the skin, put a good pitch-plaster over that side of the chest; but if the skin is broken and there is a hole in the chest, it is beyond the power of any one but a veterinary surgeon to effect a cure.

Dropsy of the Chest.

The result usually of pleurisy.

Symptoms can be detected only by placing the ear against the chest, and by percussing the chest wall.

Treatment. — Call in a veterinary surgeon, who may tap the chest and let the fluid out.

·There is a disease amongst colts running on low marshy land of a dropsical nature, but in this disease the swelling is seen on the outside of the chest and along the abdomen.

Treatment.—Take colt in from the grass, give good food, and every night and morning, in a pint of gruel, give tonic (No. 21).

Simple Fever.

Symptoms.—Staring coat, cold legs and feet, dulness, alternate shivering and hot fits, constipation. There is no cough or turning round to the flanks.

Cause.—Sudden change from heat to cold, often produced by the improper ventilation of a stable; checked perspiration.

Treatment.—Place in a cool stable where there is good air without draught, warm clothing, and give soft food while the fever is at its height, and then a more generous diet. Give mild opening medicine, such as linseed-oil, ½ pint. On no account give active purgatives. Clysters of warm water and soap will aid the action of the bowels, and give every four hours a draught containing solution acetate of ammonium, 3 oz.; Fleming's tincture of aconite, 5 drops; spirits of nitrous ether, 1½ oz., in pint of water. The disease is not dangerous, unless complications ensue.

Bleeding

is gradually becoming a thing of the past, but it is sometimes beneficial, especially where there is great blood-pressure, such as brain-fever, mad staggers, and acute founder.

How to Bleed an Animal.—Put a

driving bridle on the horse, bring his head round to the light, turn it to the left side, raise the jugular vein on the right side by pressing on it with the fingers, hold the fleam in the left hand parallel with the vein, and give it a smart blow with the blood-stick; keep the bucket pressed against the neck below the wound, and if the blood does not flow freely, insert the fingers into the mouth to keep the jaw moving. Take from 1 to 3 quarts of blood, afterwards place a pin through the lips of the wound, and wind tow around it. Do not use too large a fleam.

Inflammation of the Jugular Vein after Bleeding.

The wound caused by bleeding is generally held together by a pin and piece of twisted tow; it will usually heal in a couple of days. If the fleam has been carelessly used, or has been dirty, the wound is apt to become inflamed, swell, and discharge matter. Abscesses will then form, and if not checked will prove dangerous.

Treatment. — Wash the wound at once with a solution of carbolic acid, 1 part in about 20 parts of water; but it is a dangerous condition, and as soon as inflammation of the vein is suspected a veterinary surgeon should be called in.

Purpura or Purpura-hæmorrhagica.

A blood disease of a very low type.

Symptoms. — Is seen frequently after severe illnesses, as strangles and influenza. The legs, nose, and lips swell, pink spots are seen inside nose and eyelids; animal refuses food, and looks a pitiful object.

Cause. — Sequel to other diseases, or from bad hygienic conditions.

Prevention. — See that your stable-ventilation, drainage, and food are good.

Treatment. — Is best left to a veterinary surgeon.

Inflammation of the Lungs or Pneumonia.

Symptoms. — Fever and quickening pulse, cold ears and legs, breathing thick, nostrils dilated, restlessness, unwillingness to lie down, and staring coat. Sometimes the attack comes on suddenly and sometimes gradually.

Cause. — Cold, over-driving when out of condition, and contagion.

Treatment. — Remove to a cool airy loose-box, and clothe warmly; rub the legs well, using white oil liniment (No. 14); feed on green food and bran-mashes only; apply embrocation (No. 12) to each side of the chest; give every four hours a draught containing acetate of ammonium, 3 oz.; bicarbonate potash, ½ oz.; Fleming's tincture of aconite, 3 drops; water, 8 oz., till the fever is subdued. When convalescent, give tonic (No. 21), and two months' run at grass if the season permit.

This is a very dangerous disease, and the aid of a veterinary surgeon should be obtained.

While suffering from fever the diet should be sparing, and entirely composed of green food, carrots, and cold bran-mashes. The open air is preferable to a close warm stable: it is of the first importance that the horse should have cool fresh air to breathe. If this disease is neglected, the after-consequences, even should the horse recover, will be most serious, and his constitution will be ruined.

Scarlet Fever.

A feverish disease of the horse, characterised by pink spots in the nose and mouth, and usually associated with a sore throat.

Symptoms. — The animal dull and off its feed, eyes swollen, pink spots inside the nose and eyelids, and frequently a sore throat.

Treatment. — Place the animal into a comfortable loose-box, give thrice a-day, in gruel, draught (No. 9), and when recovering, give tonic (No. 21) in a pint of ale twice a-day. This disease is now generally considered as a mild type of purpura-hæmorrhagica.

Pleurisy.

A disease affecting the membrane covering the lungs and lining the chest.

Symptoms. — Very similar to those of inflammation of the lungs, except that the pulse is hard and small, the breathing shorter and painful, and performed mostly by the abdominal muscles, showing a line at each expiration from the lower border of the ribs to the flank.

Cause.—Chills.

Treatment.—Remove into a cool airy stable, and feed on cold bran-mashes and green food. Rub the chest and sides with embrocation (No. 12), and give twice a-day oil of turpentine, 1 oz.; iodide of potassium, 2 drs.; linseed-oil, 4 oz.; lime-water, 6 oz. Call in a veterinary surgeon, who may resort to the use of the trocar to tap the chest. Complete rest at grass, if possible, and tonic (No. 21) should follow when the animal is recovering.

Heart Disease,

as a rule, causes interference with blood circulation.

Symptoms.—There is really but one true symptom, and that is the irregularity of the pulse, but often associated with this there is weakness, cough, hurried breathing, and sometimes the animal staggers as if in want of breath.

Cause. — Rheumatic usually in its origin.

Treatment.—Rest, but often there is no improvement.

There are several inflammatory diseases of the heart, but it would only be wasting time to enumerate them here, for they are of a complicated nature and not common.

Poll Evil.

A painful swelling on the upper part of the neck behind the ears, generally terminating in an abscess.

Symptoms.—Inflammation and swelling of the ligaments over the atlas bone.

Cause.—Tight reining, blows on the neck and head from striking the manger, or lintel of the door, or given by a savage attendant.

Treatment. — Apply cooling lotion, such as goulard water, to the swelling, and keep the bowls open with purgative (No. 18). If the tumour increases, apply common blister (No. 3) to hasten its discharge, and when it is soft in the middle it should have a seton drawn through the tumour from the top, through the bottom, out at the side below the tumour; this will completely drain the abscess. Then foment and clean with warm water till cured. The aid of a veterinary surgeon should be obtained to ensure the successful treatment of the tumour.

Roaring.

A rough, disagreeable noise made by some animals during respiration, especially if forced to exert themselves.

Symptoms.—A roaring sound when sharp exercise is taken, caused by the difficulty of the air passing through the contracted opening of the larynx.

Cause.—Frequently results from an attack of strangles. Tight reining tends to produce it.

Treatment.—There is no cure in the case of a confirmed roarer. In early stages rub blister (No. 1) on the throat, and give a ball morning and night, composed of nux vomica, ½ dr.; arseniate of iron, 3 grs.; quinetum, 1 dr. Nothing further can be done.

Saddle Galls.

Cause.—A badly fitting saddle, or heavy bad rider.

Treatment. — Apply lotion (No. 7), alter the saddle, and do not work until cured.

Sore Shoulders.

The shoulders of horses sometimes become very sore and painful, and when in this condition, if neglected, large wounds and abscesses soon follow.

Cause. — Badly fitting collar, heavy loads, the draught badly adjusted, using one trace longer than the other, and working horses too young.

Treatment. — Bathe the shoulders night and morning for an hour with warm water, then apply lotion (No. 16) to the parts where the skin is broken. Do not work the animals until they are properly healed, for you can be summoned for working a horse with sore shoulders.

Sitfasts.

These are small hard tumours which form in the substance of the skin where the harness comes in contact with it.

Cause.—Pressure of the saddle. Small pimples or pustules from an unhealthy condition of the skin, and are often due to necrosis—death—of a small patch of skin.

Treatment.—Give rest, foment, and apply cooling lotion. Should they suppurate, wash with tincture of myrrh,

1 oz. ; carbolic acid, ½ oz. ; glycerine, 2 oz. ; and water, 10 oz. If they make no progress towards healing, apply a little blister (No. 1) to the ulcers, and dress the wounds with friars' balsam. But it is often advisable, and brings about a far more speedy recovery, to have the hard core in the centre carefully removed with the knife. Alter the saddle and make it fit.

Stricture of Gullet.

Symptoms. — A contraction of the gullet which prevents the passage of food.

Can be cured only by a veterinary surgeon.

Thick Wind.

Symptoms.—Difficulty of breathing when driven. Short hurried respirations. This complaint is most usual in horses with contracted chests, often resulting from an attack of inflammation of the lungs.

Treatment.—This annoying disease can be mitigated only by careful management, avoiding sharp exercise after feeding, and by never giving a very full meal. The food should be of a very nutritious nature in small bulk. A thick-winded horse may be able to go a good pace without inconvenience, if he is not hurried when he first leaves the stable.

Whistling and Wheezing

are forms of broken wind, which can be mitigated only by using the animal for slow work. A drink made of linseed-meal, one pint, boiled in six pints of water, with a little treacle, may do good, but there is no cure.

Withers (Fistulous).

Symptoms.—This troublesome disease first appears as a swelling on the withers, develops into a tumour, suppuration takes place, and a deep ulcer forms, which may extend down to the bone.

Cause.—Pressure on the withers from an ill-fitting saddle or collar.

Treatment.—Give the horse complete rest till cured; do not work him till then under any pretence. Upon the first appearance of the swelling, foment, and apply lotion (No. 7). If the tumour appears, apply blister (No. 1 or 3). The veterinary surgeon should be called in if this does not stop the inflammation. Keep the bowels open by feeding on green food and bran-mashes.

DISEASES AFFECTING THE STOMACH, LIVER, BOWELS, KIDNEYS, AND OTHER INTERNAL ORGANS, AND PARTURITION.

Bots.

The larvæ of the gad-fly. Most common in spring and early summer. The eggs of the gad-fly are deposited among the hair, and are introduced into the stomach through the horse licking himself. They attach themselves to the lining of the stomach during the winter, injuring and weakening it, till finally they are seen escaping in the spring out of the anus, causing great itching.

Treatment.—No medicine will totally destroy these bots. The use of salt among the food may serve to mitigate the evil, and a draught containing oil of turpentine, 1 oz., linseed-oil, 10 oz., may remove many of them, but very little can be done, and nature must be left to take its course. Green food assists in bringing them away.

Colic or Gripes.

1. Flatulent Colic.

Symptoms.—Stomach and intestines distended with gas; pain and depression.

Cause.—Overloading of the stomach with green food; cold and over-exertion.

Treatment.—Give a purgative (No. 17), and clyster (No. 13), and every two hours give a draught containing opium tincture, 1 oz. ; spirit of ammonia, 1 oz. ; carbolic acid, 15 drops; chloroform, 1 oz., in 12 oz. of water.

2. Spasmodic Colic.

Symptoms.—Acute pain, rolling on the ground, suddenness of attack, excited countenance, and the intermittent nature of the pain. This last characteristic distinguishes the disease from inflammation of the bowels.

Cause. — Chills from drinking cold water when hot, and errors in feeding and watering, are the most common causes.

Treatment.—If taken in time, this disease can usually be cured by giving linseed-oil, 1 pint; oil of turpentine, 1 oz.; tincture of opium, 1 oz.; chloroform, 1 oz. Walk the horse about after giving the dose. If the attack continues, apply hot fomentations to the belly till the aid of the veterinary surgeon can be obtained.

Diarrhœa.

Frequent passing of fluid dung.

Symptoms. — Animal dull, refuses food, slight colicky pains, and frequent dunging, which, if not checked, will terminate in inflammation of the mucous membrane of the bowels.

Cause.—Bad feeding, or feeding on raw potatoes, too succulent green food, cold and irritation of the bowels from worms or innutritious food.

Treatment.—Place animal in a warm box, if cold put a rug on and bandage his legs, keep short of water, and give in half-pint of gruel twice a-day the following: Tincture of catechu, 1 oz.; powdered chalk, ½ oz.; tincture of cardamoms, 1 oz.; opium powder, 1½ dr. To be continued until the diarrhœa ceases.

Constipation.

Generally arising from the nature of the food or torpidity of the liver or intestines.

Prevention.—All dust from chop or chaff should be sifted out of horse's food, and too much mealy or dry food should not be given without access to water.

Treatment. — Give purgative medicine—linseed-oil, 1 pint, and plenty of watery food, gruel, &c., and warm clysters of soap-and-water, repeating the dose of oil when required. For chronic constipation give daily a ball composed of aloes, 1 dr.; nux vomica, ½ dr.; carbonate ammonium, 1 dr.; ginger, 1 dr.; gentian, 1 dr.

Dysentery.

A continual passing of semi-solid dung, tinged with blood.

Symptoms.—It first starts with diarrhœa, which passes into dysentery; the animal becomes restless, occasionally lies down; in the course of a few hours it trembles; clots of blood are passed with the dung, which has a bad smell if not soon checked; a cold sweat breaks out, the legs become cold, the eye glassy, and death closes the scene.

Cause.—Too large a dose of physic, worms and improper feeding, associated with a bad sanitary condition of the stable.

Treatment.—Put animal in a warm box; if cold put a rug and bandages on. Give every six hours until the purging ceases the following drench in half a pint of gruel: Chlorodyne, ¼ oz.; powdered opium, 1 dr.; prepared chalk, ½ oz.; tincture of cardamoms, 1 oz.; old port wine, ½ pint.

Diabetes or Polyuria.

Symptoms.—Excessive discharge of urine, weakness, and unthrifty appearance.

Cause.—Irritation of the kidneys by a too frequent use of diuretics or bad, musty, or mouldy food.

Treatment.—Feed on green food and mashes, and give morning and night in gruel a draught containing dilute hydrochloric acid, 2 drs.; quinetum, 1 dr.; tincture of opium, ½ oz. The part of the loins over the bladder should be covered with a hot cloth. Attend to the quality of the food, and in severe cases call in a veterinary surgeon.

Inflammation of the Bowels.

Very fatal, often resulting in death in a few hours.

Symptoms.—At first uneasiness and dulness; fever, and in some cases shivering fits; nostrils red and mouth hot; breathing and pulse quick; ears and legs cold; and the passing of small quantities of dung at short intervals. The horse will show great pain by kicking at his belly and whisking his tail.

Cause. — A chill when overheated, often from drinking cold water when hot, over-exertion, a too full meal when the animal is tired and worn out.

Prevention.—These inflammatory diseases of the internal organs are too common among draught-horses. There is no more pernicious habit than that of working horses during hot weather, without allowing them for hours together to have any drinking-water till they get into a probably cold stable, where they are allowed to drink their fill and stand for an hour during the dinner-hour till they are chilled inside and out. It seems extraordinary that so many horses stand this treatment. Allow farm-horses frequent moderate drinks of water while at their work, when that work is heating or the weather hot. The exercise after drinking will prevent any chills, and on their return to the stable they will eat their corn without requiring water to an injurious extent. The custom of giving horses large quantities of coarse boiled food was often to blame for causing this disease.

Treatment. — Place in a cool stable and clothe warmly; give warm clysters of thin gruel and Epsom salts, ½ lb. Foment the belly with hot water and rub it with embrocation (No. 12), and every three hours give in gruel tincture of ginger, 1 oz.; tincture of opium, 1 oz.; chloroform, 4 drs. Rub and bandage the legs. Give plenty of warm linseed-gruel. If costiveness continue, give with great caution in gruel small doses of aloes, 2 drs. dissolved, and ½ pint linseed-oil, and send for a veterinary surgeon.

Gastritis or Inflammation of Stomach.

A disease of rare occurrence.

Symptoms.—Animal shows signs of pain, breathes hard, sweats about the shoulders, thirst, flow of saliva, great prostration, legs and ears become cold, the animal staggers, and soon dies.

Cause. — Too much food rapidly swallowed, foreign body in stomach, or from a vegetable or mineral poison.

Treatment.—Give linseed-oil, 1 pint; tincture of opium, 2 oz., and give after every two hours two eggs beaten up in a pint of linseed-gruel, and add to it an ounce of tincture of nux vomica, and one of tincture of belladonna.

Twist of the Bowels.

A twist in a portion of the bowels, which may cause strangulation, mortification, and death.

Symptoms.—Excruciating pain, the animal is up and down, blowing heavy and sweating, nothing seems to give ease, and death comes as a happy release.

Cause.—Mostly rolling when in pain.

Treatment.— Nothing can do any good except opium, which will ease pain until death.

Inflammation of the Bladder.

See Inflammation of the Kidneys for symptoms and treatment, *infra*.

Cause.—Irritant matter in urine, or stone in the bladder.

Inflammation of the Neck of the Bladder.

Symptoms.—Distended bladder and partial to total suppression of urine.

Cause.—Overstraining or cold.

Treatment.—Give purgative (No. 17), and three times daily in gruel a draught containing Fleming's tincture of aconite, 5 drops; tincture of opium, 1½ oz.; bicarbonate of potash, ½ oz. Apply hot fomentations to the loins, and call in a veterinary surgeon, who will, if necessary, draw off the urine with a catheter.

Parturition.

The act on the part of a mare to bring forth her young. The period of pregnancy in the mare is usually eleven months, though it sometimes varies between ten and twelve months. This animal seldom brings forth more than one at a time, nevertheless twins sometimes do occur, but they rarely live long after birth.

Signs of Foaling.—The mare is dull, abdomen sprung, back bent, vulva swollen, and a little mucus discharged. The udder becomes enlarged, wax appears at the ends of teats. As the time draws near the mare becomes restless, paws, keeps on lying down, an anxious expression in the eyes, and frequent passing of dung and urine. The water-bag soon makes its appearance, which ultimately bursts, after which the foal appears.

Treatment. — The mare should be placed in a nice clean loose-box with plenty of straw, and do not disturb her by keeping open the door and looking in frequently.

Inflammation of the Kidneys.

Symptoms.—Fever and peculiar position, standing with legs wide apart; hot loins, and tenderness in that part; suppressed urine, which is dark in colour and may be tinged with blood; straining to void urine. Put the hand up the rectum, and the bladder under the rectum will be empty without undue heat. In cases of inflammation of the neck of the bladder, it will feel hard and full. In cases of inflammation of the bladder, it will feel empty, but there will be great heat.

Cause.—Unwholesome food, particularly musty oats, or a violent overstraining or cold.

Treatment.—Remove into a comfortable box, clothe warmly, give plenty of water, feed on linseed and bran-mashes, foment the loins with hot water. Apply embrocation (No. 12) to the loins over the kidneys, but leave the turpentine out of the embrocation, and give purgative (No. 17); give also warm clysters of soap-and-water. When the purgative has acted give white hellebore, 5 grs.; tartar emetic, 1 dr., mixed into a ball, three times a-day till cured. If possible, find out and remove the cause of the disease, if it arises from improper food.

Inflammation of the Womb.

An inflammatory disease of the womb shortly after foaling.

Symptoms.—Animal becomes dull and stiff, appetite lost, secretion of milk diminished, breathing hurried; the animal grinds her teeth, suffers from colicky pains, frequently lies down, stamps, kicks at her belly, the vulva is swollen and a discharge comes from it, which is at first yellow, but afterwards becomes a chocolate colour, and foetid.

Cause.—Exposure to cold, retention of the after-birth, and injuries received during foaling.

Treatment.—Put hot cloths across the loins, and give every three hours the following draught in a pint of linseed-gruel: tincture of belladona, 1 oz.; spirits, ether (nitrous), 2 oz.; and soda sulphite, ½ oz.

Inflammation of the Liver.

An uncommon disease.

Symptoms.—Dull pain, but no great uneasiness, yellowness of the mouth and nostrils.

Cause.—Overfeeding and insufficient exercise.

Treatment.—Should the attack be severe, call in a veterinary surgeon. Give for a purgative—sulphate of soda, 5 oz.; virgin scammony, 30 grs.; and feed on bran-mashes with a light diet. A useful draught, to be given three times a-day in gruel, is composed of chloride ammonium, 2 drs.; bicarbonate potassium, ½ oz.; Fleming's tincture of aconite, 5 drops; chloroform, ½ oz.

Jaundice or Yellows.

Symptoms.—A yellow tinge in the eyes, skin, and mouth; urine quite yellow; loss of appetite, and constipation.

Cause.—Obstruction of the flow of bile from the liver, disease of the liver or congestion arising from cold or other cause.

Treatment.—Feed on mashes, thin warm gruel, and green food; clothe well if weather is cold; give every morning calomel, ½ dr. If inflammation sets in, give every morning in gruel a draught containing solution acetate ammonium, 4 oz.; Fleming's tincture of aconite, 5 drops; spirits of nitrous ether, 1½ oz. When recovering give tonic (No. 21).

Peritonitis.

Inflammation of the lining membrane of the abdomen.

Symptoms.—Small hard pulse, colicky pains, dulness, constipation, and tenderness on pressure over the abdomen, which feels hard and rounded.

Cause.—It may arise from cold and neglect, but generally from worms or wounds, as after castration.

Treatment.—Hot fomentations persistently applied; give opium or laudanum, with small doses of spirits of ammonia every four or five hours, gruel and linseed-tea to drink, and soft sloppy food.

Poisons.

The only vegetable poison that need be mentioned is yew. The eating of this tree accounts for the death of many horses every year. If the poison is suspected, give at once linseed-oil, 20 oz., and drench with spirits of ammonia, 3 oz.; brandy, 5 oz.; gruel, 1 pint. Repeat dose of oil if it does not operate in twelve hours.

Lead-poisoning.

A disease due to the introduction of lead into the system. Although comparatively common in cattle is rare in the horse.

Symptoms.—The horse has a care-worn expression, staring coat, back arched, legs cramped, colicky pains, and flow of saliva from the mouth.

Cause.—Grazing near rifle-butts or lead-smelting works, drinking water impregnated with lead, licking lead paints, and the barbarous practice of giving shot for broken wind.

Treatment.—Give sulphate of magnesia, 8 oz., in a pint of water, with tincture of belladonna, 1 oz.; tincture of capsicum, ½ oz. Afterwards, give every four hours until the animal is purged, sulphate of magnesia, 1 oz., tincture of belladonna, 1 oz., in half a pint of water.

Arsenic-poisoning.

Due to arsenic either given accidentally or intentionally.

Symptoms.—Colic, staggering gait, quick breathing, cold ears, diarrhœa, and death.

Cause.—It is sometimes caused by grooms giving it to improve the condition of their horses, or by allowing animals to graze where recently dipped sheep have been lying.

Treatment.—Give every two hours the following in half a pint of water: iron sesquioxide, ½ oz.; brandy, ½ pint.

Umbilical Hernia.

A round swelling under the belly of young horses.

Symptoms.—A soft swelling in the centre of the abdomen, ranging in size from a fowl's egg to a cocoa-nut.

Cause.—Due to non-closure of the navel.

Treatment.—Trusses, bandages, or plasters frequently fail, and it may have to be operated on by a veterinary surgeon.

Scrotal Hernia.

Descent of the small intestines into the scrotum.

Cause.—By galloping, or a severe strain, but very often there at birth.

Symptoms.—The scrotum looks large and feels soft, but is not always the same size.

Treatment.—Castrate by the covered operation (p. 497).

Staling of Blood,

or mixture of blood with the urine.

Cause.—Inflammation or injuries of the kidneys.

Treatment.—Feed on green food and mashes; clothe warmly; give Fleming's tincture of aconite, 8 drops, every night. Purgative (No. 17) should be given, and three times a-day a draught in gruel, composed of extracts of ergot, ½ oz.; tannin, ¼ oz.; dilute sulphuric acid, 2 drs. When the appearance of blood in the urine has ceased, give daily Peruvian bark, 1 oz.; sulphate of iron, 1 dr.

Stone in the Bladder (Calculus).

Symptoms.—Irregular voidance of urine, sometimes total suppression, great pain, suddenness of attack, great uneasiness, a sediment from the urine on the floor of the stable, and profuse perspiration during attack.

Cause.—Formation of solids in the bladder, often brought on by weakness or disease of the kidneys.

Treatment.—Give morning and evening, in gruel, a draught containing bicarbonate of potassium, 1 oz.; benzoate ammonium, 1 oz. If the gravel or small stones are not passed, place the case in the hands of a veterinary surgeon, who will treat it for calculus, the removal of which requires an operation, the stone being too large for the horse to pass.

Worms.

Symptoms.—Rough coat and half-starved appearance, at other times an enormous appetite, but no improvement in condition; appearance of a yellow powder about the anus, with irritation and switching of the tail.

Treatment.—When fasting give in gruel draught (No. 11), and repeat in three days.

Prolapse of the Rectum.

Cause.—A drastic purge, injuries, straining during foaling or in a violent fit of colic.

Treatment.—Wash the gut with equal

parts of olive-oil and *liquor opii sedativus*, and gently work it back to its proper place; afterwards depress the tail.

DISEASES AFFECTING THE LIMBS, FEET, AND SKIN.

I.—THE LIMBS.

Broken Knees.

Cause.—A fall. Horses first brought from a stable are liable, from no fault in their build, to stumble and fall through excitement. They are also apt to tread on a rolling stone and fall. A horse that stands over—*i.e.*, whose fore legs are too far under him—and those that shuffle along without lifting their feet, owing to the formation of the shoulder, are very liable to fall forward.

Treatment.—Wash with warm water and remove the dirt. Apply a linseed-meal poultice to allay inflammation; after twelve hours remove the poultice. If a yellow kind of oil exudes from the wound, it shows that the joint has been cut into, and a veterinary surgeon alone can deal with the case, which, to say the least, is a desperate one. If, however, there is no yellow joint-oil to be seen, wash the wound with a weak solution of carbolic acid, or boric acid, 1 part; water, 30 parts; adjust the injured pieces of skin, apply a piece of carbolised tow, bandage with carbolised gauze, and so dress twice a-day. Keep animal tied up until the knees are healed. If fever runs high, give every four hours in gruel a draught containing salicylate sodium, 3 drs. Purgative (No. 19) may be useful if the health of the horse is affected.

Capped Hocks,

or a swelling on the point of the hock, which does not often cause lameness, but is shown by the swelling and tenderness on the point of the hock.

Cause.—Often caused by striking a closing door or gate, but may be due to kicking.

Treatment.—Foment with hot water and bathe with cooling lotion (No. 7); give complete rest till cured. Apply blister (No. 4) if the swelling has a tendency to harden. If this swelling is neglected it may prove incurable.

Capped Elbow.

A hard swelling at the elbow, varying in size.

Symptoms. — Rarely lameness; the swelling is hard, and about the size of a large hen's egg.

Cause.—It is caused by the heel of the shoe in lying, which either irritates or squeezes the skin at the elbow, and sets up inflammation.

Treatment.—If observed when only commencing, treat as for capped hock; but if left until it gets confirmed and callous, even blisters and setons are of little use. Then it may have to be removed by operation by a veterinary surgeon.

Curb.

A swelling on the posterior aspect of the leg below the hock, seen plainly when the horse is viewed sideways.

Cause. — A sprain of the ligament under the hock.

Treatment.—Foment with hot water and apply cooling lotion (No. 7) and a high-heeled shoe. If the swelling does not go down, apply blister (No. 4), and give complete rest. Curby hocks are natural to some horses, but once the horse reaches maturity they seldom cause lameness but are always a blemish.

Cutting or Brushing.

The names given when a horse strikes the inside of the fetlock with the shoe of the other foot. Horses with feet turned out are most liable to this defect. It is often brought on by fatigue or by working a young horse too soon.

Treatment. — Make the shoe fit the hoof of the cutting foot, which should be rasped on the inside to reduce it. Foment the swelling caused by the bruises, and apply lotion (No. 7). See remarks on "Speedy Cut" (p. 459).

Enlargement of the Hock.

Arising from inflammation.

Cause.—A sprain or a blow, such as a kick by another horse: it produces great lameness.

Treatment.—Foment with hot water, apply lotion (No. 7), and give perfect rest. Purgative (No. 19) will help to relieve the inflammation, or a draught in gruel, containing salicylate sodium, 3 drs., every four hours.

If any enlargement remains when the inflammation is reduced, apply blister (No. 4). The object in view must be to prevent a permanent enlargement of the hock.

Fractures

are divided into simple, compound, and compound comminuted fractures.

A simple fracture is when the bone is broken into two pieces, compound when broken and associated with a wound, and a compound comminuted when broken into several pieces and associated with a wound.

In the horse simple fractures are the only ones worth trying to treat. In the case of the other two kinds, the sooner the animal is destroyed the better.

Before trying to set a fractured limb, it is wise to consider whether the animal is worth it, and if placed in slings will he be quiet. Having decided to set the limb, place the animal in slings; take some gutta-percha, place it in hot water, and mould it to the limb, or use some sheet-tin, and after moulding it to the part, cover with some flannel to prevent its cutting at edges. Take the splints thus made, place them on the part to be set, and pack where they do not exactly fit with tow, then take a nice long bandage, wind it tightly around, and do not touch it for a couple of months.

If the animal is a restless one, it will be only wasting time to try and set the limb. It is a false but popular idea that horses' bones will not unite; nothing will unite quicker, if the animal will only nurse its limb.

Grease.

A disease of the skin of the heel, generally of the hind feet.

Symptoms.—Inflammation, with pain and lameness at first; discharge of matter; at first limpid, soon gets thick, fœtid, and irritating; swelling; often going on to ulceration and the formation of fungus-like growths called "grapes."

Cause.—Too little exercise and too much corn; bad or innutritious food; too much coarse boiled food; washing with cold water without afterwards drying the legs, and chills caused by work in wet, muddy ground, after keeping in too warm a stable.

Prevention.—The legs of horses subject to this disease should not be washed unless they are afterwards dried. Let the mud dry on the legs, and then brush it off; it is more than probable if you do this, you will have no more trouble, provided other conditions are favourable.

Treatment. — Wash the heel with warm water and soap, or if very bad, poultice at first with boiled turnips and bran, sprinkling the sores all over with soot before applying the poultice, and rub in ointment composed of oleate of zinc, 1 part, and vaseline, 2 parts; or lard, 1 oz., sugar of lead, 1 dr.; or wash with lotion containing chromic acid, 1 part, water, 8 parts. If the case proves obdurate, use ointment containing white precipitate of mercury, 1 dr.; liquor carbonis detergens, 1 dr.; vaseline, 1 oz. Give a mild alterative, Barbadoes aloes, 4 drs.; Castile soap, 1 dr.; oil of caraways, 10 drops, or condition powders (No. 6). Sulphate of soda, 4 oz., in the food every night may prove a useful aperient. Iodide of arsenic, 4 grs.; liquorice-powder, 2 drs.; gentian, 3 drs., made into a ball with treacle and linseed-meal, is a very good thing for this disease, and a ball should be given every night.

Open Joints.

The following joints are sometimes opened: hock, stifle, knee, and fetlock joint.

Symptoms. — Great pain and lameness; a small wound is seen, and from it flows a yellowish fluid the consistency of glycerine.

Cause. — Kick from another horse, accidents of various kinds, and by a groom pricking the horse with his fork when bedding the animal up.

Treatment is unsatisfactory. Give a dose of physic; place a cold-water bandage around the joint for twenty-four hours; but it is better to call in a veterinary surgeon as soon as an open joint is deemed possible.

Knee-Tied.

A natural defect, for which there is no cure. It is a want of depth under the knee, owing to the hinder knee-bone not being large enough.

Mallenders and Sallenders.

Dry scurfy humours, which, when affecting the front of the hock, are called *Sallenders*, and when under the back of the knee *Mallenders*.

Cause.—Neglect in the stable.

Treatment.—Rest, and apply ointment containing tar, 1 oz.; sugar of lead, ½ oz.; lard, 4 oz.; and give draught morning and evening containing bicarbonate potassium, 6 drs.; spirits of nitrous ether, 1 oz.; tincture gentian, 1 oz.; water, 8 oz. Feed on green food and improve stable management. If the above treatment is not successful, apply a little of blister (No. 4), mixed with three times its weight of lard, and well rubbed in.

Tumours.

There are many kinds of tumours, and they may be either internal or external. The former are usually situated in the brain, womb, abdomen, and liver, and nothing within the power of man can do any good. The external tumours are the ones we are often asked to cure, and they usually appear on the shoulders, neck, under the tail, and at the end of the cord after castration.

Treatment.—There are various ways of removing them, and the best is by the knife. If the tumour is narrow at its base, an easy and safe way to remove it is by winding a piece of green silk tightly around its base and allowing it to drop off. In cutting large tumours out, veterinary surgeons sometimes come in contact with large arteries, and these must be caught up and tied. When a tumour appears after castration, use the hot iron and clams to remove it.

Rheumatism.

Change of temperature and cold often produce stiffness of the joints, varying in intensity.

Treatment.—Keep the animal warm, and rub the part affected with liniment of belladonna, and morning and evening give in ½ pint of water iodide of potassium, 1 dr. It might be necessary in extreme cases to apply blister (No. 1).

Rupture of the Suspensory Ligament.

Lameness from this cause is generally incurable. The suspensory ligament sustains the foot, and the rupture of it allows the fetlock to drop down almost to the ground. If the horse cannot bend his foot, it is not the suspensory ligament that is ruptured.

Cause.—Over-exertion or strain.

Treatment.—Perfect rest, and put on a high-heeled shoe. Bandage the legs,[1] foment, and apply lotion (No. 7); if this does not reduce the swelling, apply blister (No. 4) and give a mild purgative (No. 19).

In most cases the lameness will be permanent.

Hip Knocked Down.

Symptoms.—At first great swelling, the animal goes lame, but when the swelling is reduced the hip that is knocked down looks less than the other when looking at it from behind.

Cause.—Through falling, in knocking against a wall, in passing through a doorway.

Treatment.—Little can be done except placing the animal in slings, and bathing the part with hot water; if an abscess forms, the piece of bone that is knocked off must be cut down upon and removed.

Spavin.

There are two kinds:—

1. Bone-Spavin.

Symptoms. — Bony enlargement on the inside of the hock-joint towards its antero-inferior aspect, producing lameness when first formed, till the parts accommodate themselves to the enlargement. Afterwards, the lameness may be apparent only when the horse is first taken out of the stable, unless it interferes with the movement of the joint, when a small spavin may permanently lame a horse.

---

[1] The frequent bandaging of the legs is apt to produce an unsightly curliness of the hair. The application of alum, 1 oz.; salt, 2 oz.; in 1 quart of water, will do much to remedy it.

Cause. — Hereditary, local injury, sprains of the ligaments and concussion, overwork when young, peculiar formation of hock, and improper shoeing.

Treatment.—Perfect rest and repeated application of blister (No. 4). Should blistering not remove the lameness, firing may have to be resorted to. I have found ossoline effect a cure when other remedies fail. Spavins always constitute unsoundness.

2. Bog-Spavin.

Symptoms. — A tumour, resembling a wind-gall on the hock, formed on the inside of the front of the hock. The swelling is due to distension of the bursa of the hock with joint-oil, and is usually permanent, but does not much interfere with slow work.

Cause. — Sprain and over-exertion. Hereditary conformation.

Treatment.—If it is not considered advisable to keep the horse for slow work without treating the spavin, which, in my opinion, is the wisest course to pursue, apply blister (No. 3) and allow perfect rest, in the hope of effecting a cure, but it is not likely to be permanent.

Speedy Cut.

Horses are apt to strike the inside of the fore leg at the lower part of the knee with the other foot when trotting fast, or lifting their feet high. Horses liable to this are dangerous to ride or drive, the force of the blow being sufficient in some cases to bring them down. Great pain and inflammation and swelling result from the blow.

Prevention.—Cut the hoof away on the inside, and put on a shoe of equal thickness at toe and heel, having only one nail on the inside, and not projecting beyond the part of the hoof which has been rasped. Keep a speedy cutting boot on the injured leg to protect it.

Treatment. — Foment the bruise, apply lotion (No. 7), and allow complete rest till cured. If the bruises have a tendency to harden, apply blister (No. 4).

Wounds.

Wounds are divided into abrasions, incised, punctured, contused, and lacerated wounds.

An Abrasion.

Caused by falls, kicks, barb-wire, and short nails, &c.

Symptoms.—The skin is torn, but the wound is not of any depth.

Treatment.—Wash well with warm water, dress with tincture of myrrh, and dust fuller's earth over it.

Incised Wounds.

Caused by a knife, scythe, or any sharp instrument.

Symptoms.—The wound usually is deep, and the edges cleanly cut.

Treatment. — Having thoroughly cleansed the wound, sew it up with carbolised gut, and dress with liq. carbonis detergens, 1 oz. to 30 oz. of water.

Punctured Wounds.

Caused by long nails, horns of cattle, forks, parts of agricultural implements, and broken shafts.

Symptoms.—A wound of some depth, and though it may not be large to look at, it is the most fatal of all wounds.

Treatment.—If it is bleeding freely, plug it up for some hours with carbolised tow; after the tow is removed, inject into the wound, by the aid of a wound-syringe, the following lotion: glycerine, 3 oz.; carbolic acid, 1 oz.; water, 30 oz.; and keep in the wound a piece of tow soaked in the lotion until it heals. Remember, wounds of this kind must heal from the bottom.

Contused Wounds.

Caused by a severe blow, fall, or kick.

Symptoms.—This is more of a bruise than a wound.

Treatment. — Bathe for two hours twice a-day, and afterwards dress it with lotion (No. 7).

Torn or Lacerated Wounds.

Caused by a bite from a dog or horse, by being entangled in a fence and struggling, and in coming against the latch of a door in passing through it.

Symptoms.—A wound usually of some size with its edges ragged.

Treatment.—Cleanse the wound well with hot water, sew up any part you think necessary, and dress with liq. carbonis detergens, 1 part; water, 30 parts.

Splint.

Symptoms.—A bony enlargement on the inside of the fore leg below the knee. It often produces lameness until fully grown, when the lameness usually disappears, unless the splint interferes with a tendon or joint. Splints are not thought much of unless near a tendon or joint.

Cause.—Young horses are very subject to splints: they arise from injuries to, and a sudden weight thrown upon, the bones of the legs, and usually found on the inside of the canon-bones of the fore legs.

Treatment.—Blister (No. 4) applied once or twice will generally effect a cure. As a horse gets older splints will generally disappear.

Sprain of the Back Tendons.

Symptoms.—Great pain, thickening and inflammation in the leg above the fetlock, preventing the horse bringing his foot flat to the ground. The leg will appear to be round instead of flat.

Cause.—Inflammation of the sheath which encloses the back tendons, the result of hard work or excessive strain.

Treatment. — Perfect rest ; foment with hot water and then poultice with linseed-meal and bathe with lotion (No. 7); keep the bowels open with purgative (No. 19). When the heat subsides, and the horse can put his foot flat to the ground, bandage the leg with bandages steeped in vinegar. Should the inflammation continue, apply embrocation (No. 12), or blister (No. 3), and give two or three months' complete rest.

Sprain of the Coffin-joint.

Symptoms. — Sudden lameness, and heat and tenderness round the coronet.

Treatment. — This kind of sprain should be treated at once, before the inflammation spreads. Apply blister (No. 3), and give occasionally purgative (No. 19). Bandage the leg and give perfect rest.

Sore Shins.

Only common in young horses that have been put too suddenly to work.

Symptoms.—Lameness; if both legs are affected, the animal rests first on one then on the other leg, and the legs have a doughy feel.

Cause.—By a young horse galloping before the bones are properly developed.

Treatment.—Put cold-water bandages on for a few days, and then blister with (No. 1).

Sprain of the Fetlock.

Symptoms.—Lameness, attended with swelling, heat, and tenderness of the fetlock, is probably a sprain of the fetlock.

Treatment.—Apply repeatedly blister (No. 3) till the heat subsides, then bandage lightly to strengthen the fetlock; give perfect rest.

Sprain of the Round Bone or Hip.

Symptoms.—A sprain in connection with the rounded bone of the thigh, by which the horse loses all power of moving that quarter, and drags his leg, resting it on the toe alone.

Cause.—Sudden strain, slip, or fall.

Treatment.—Foment and apply immediately blister (No. 3), and call in a veterinary surgeon.

Sprain of the Shoulder.

Symptoms. — Great pain, especially when going down-hill, and a dragging of the foot forward on the toe. If the foot is drawn forward, the horse shows pain. No outward swelling or heat.

Cause. — Accident from slipping or going over rough ground. Young horses are very liable to this.

Treatment.—Perfect rest; apply hot fomentations to the shoulder and bathe with lotion (No. 7), and, if necessary, blister (No. 3); keep down inflammation by giving purgative (No. 19). A long rest, combined with this treatment, will generally effect a cure.

Shoulder-Slip.

A peculiar outward movement of the shoulder when the animal walks, sometimes, but not always, accompanied by lameness.

Symptoms.—The shoulder-joint looks enlarged, but the muscles of the shoulder are wasted.

Cause.—By horse being put to plough too young; by the one foot being in the furrow and the other out, and by pulling awkwardly and using the shoulders un-equally before getting accustomed to the draught. Injury to the supra-scapular nerve sufficient to cause more or less paralysis.

Treatment.—Blister the shoulder with (No. 1), and turn the animal out to grass for three or four months.

Sprain of the Stifle-joint.

Symptoms. — Dropping of the hind quarters and dragging of the leg; great heat, swelling, and tenderness of the stifle.

Cause.—A blow, slip, sprain, or over-work.

Treatment.—If the stifle has been dislocated from a kick or blow, send for a veterinary surgeon, who alone can judge as to the treatment. In case of sprain, apply warm fomentations and lotion (No. 7) till the inflammation is somewhat reduced, and then apply blister (No. 3); give perfect rest and purgative (No. 19).

Dislocation or Luxation of the Patella.

This disease is usually seen in young horses, and is due to the slipping out-wards of the patella or bone which corresponds to the lid of the human knee.

Symptoms.—One or both stifles may be wrong, the joint looks swollen, and when the animal moves it slips out and in with a peculiar noise.

Cause.—Hard galloping, feeding on hilly pasture, and often a disease of the joints occurring in foals.

Treatment.—In young horses, seems to be of little use; but blisters may be tried. If the swelling is accompanied by heat and pain, apply hot fomenta-tions and cooling lotion first. In older horses it can be reduced by flexing (working backwards and forwards) the leg; push the patella back into its proper place, and apply a blister.

String-halt.

Symptoms.—A sudden snatching up of the hind leg or legs, but usually only one leg, which makes the horse's action peculiar. Probably a nervous disease, and practically incurable. It produces no lameness, but is liable to get worse, and is always considered unsoundness.

Cause.—Often produced by rheuma-tism or by leaving a horse standing in a stable without sufficient exercise, and is hereditary.

Treatment.—Doses containing citrate iron, 2 drs.; and ammonium, 2 drs.; tincture nux vomica, 2 drs.; tincture capsicum, 2 drs.; carbonate of ammon-ium, 2 drs., given in water night and morning, may relieve and strengthen the system with satisfactory results.

Mud-fever.

Symptoms.—Heat and swelling of the legs, and the animal moves stiffly; there is a certain amount of fever, hence the name.

Cause. — The chilling and irritant action of mud, which in cold weather produces inflammation in the legs of horses, especially when the legs are rendered tender by clipping, repeated washing, and imperfect drying.

Prevention.—Do not clip the horse's legs; let the mud dry, and then brush it off; never wash them in frosty weather.

Treatment.—Dress the legs with a mixture of glycerine, 8 oz.; carbolic acid, 1 dr.; and *liq. plumbi acet.*, 1 oz.

Swelled Legs.

Horses of a coarse nature are very subject to swollen legs, especially the hind ones.

Symptoms.—With or without great heat; lameness accompanied by quick pulse and fever, but there may be neither fever nor lameness.

Cause—Overfeeding, too little exer-cise, and change of food.

Treatment.—If there is much fever, foment the legs, bathe with lotion (No. 7), and give a ball containing turpentine, 1 oz.; ginger, ½ dr.; lin-seed-meal, ½ oz.; and two hours after give purgative (No. 19). If there is not much fever but swelling, stiffness, and pain in the legs, foment them and rub lightly with embrocation (No. 12). Give gentle exercise and purgative (No. 19).

Thoroughpin.

Symptoms.—Very similar to wind-gall (see below). An enlargement at the upper and back part of the hock between the tendon and the bone. It usually projects on both sides of the hock, but rarely causes lameness, if taken in time before the swelling becomes callous.

Cause.—Overwork or strain.

Treatment.—Rest, and apply blister (No. 3 or 4) till the swelling is reduced.

Wind-galls.

Symptoms.—Puffy elastic swellings situated just above the fetlock, which may become large and hard, causing lameness.

Cause.—Strain of the tendons, and overwork in young horses.

Treatment. — Bandage with flannel steeped in vinegar till the swelling subsides. If this does not effect a cure, blister (No. 3) should be applied. Wind-galls do not, as a rule, cause unsoundness.

II.—THE FOOT.

Canker.

A disease of the hoof, generally commencing about the frog or heels, and often spreading over the sole.

Symptoms. — This disease is sometimes the result of neglected thrush, and differs from it in its tendency to spread, and in the swelling or enlargement of the affected parts. The diseased frog assumes a soft, fungatory appearance; is liable to bleed on being touched; emits a very fœtid, offensive, although nearly colourless discharge; and unless energetically treated the disease is apt to spread over the whole sole.

Cause.—Hereditary; but often neglect and want of cleanliness.

Treatment. — Cut away the sole where the canker is situated, removing all fungus, and apply acid solution of nitrate of mercury and bandage up the foot, or dust on iodoform night and morning; morning and night bathe with lotion—carbolic acid, 1 part; glycerine, 1 part; and in four days repeat the application of acid solution. If the fungus still grows, call in the aid of a veterinary surgeon.

Contraction of the Foot.

Symptoms.—A natural hoof is nearly circular, but sometimes through neglect or bad shoeing the hoof is made concave,

and the heels contract, producing permanent lameness if not attended to.

Cause.—Neglect in stable management or shoeing. Too much paring away of the frog, bars, and sole. Extreme dryness, or allowing the shoes to remain on too long, will cause the hoof to shrink.

Prevention.—Stopping the feet with cow-dung or moist clay, and removing the shoes.

Treatment.—A contracted foot can hardly ever be cured, but if it is decided to attempt a cure, a veterinary surgeon should be called in.

Corns.

Symptoms.—The horn of the heel—most frequently the inner heel of a fore-foot—becomes reddish, soft, and tender. The horse will flinch when this part is pressed, and occasional or permanent lameness results.

Cause. — Careless shoeing or tight shoes, producing undue pressure at a particular point.

Treatment.—Old corns are difficult to cure; fresh ones may be prevented increasing by proper shoeing, and by paring the corn as far as possible without wounding the sole. A bar-shoe may be put on in serious cases with advantage, and the horse shod with leather.

False Quarter.

Symptoms.—It is due to a division of or a want of secretion by part of the coronary band, which extends as the horn grows downwards, making a fissure or wide groove in the hoof. It is a serious defect, often resulting in inflammation and lameness, and from the thinness of the horn it is very liable to injury during work.

Cause.—Injury to the coronary band, and sometimes the consequence of neglected sand-crack.

Treatment.—Apply blister (No. 5) to the coronet, and treat the fissure as for Sand-crack (see p. 464). Should the secreting coronary band be permanently injured, no remedy will cure the disease.

Laminitis—Founder (acute),

or inflammation of the feet.

Symptoms.—Great restlessness and continual shifting of the animal's weight from one foot to the other; pain, fever, heaving flanks, hot feet. After a time the horse will lie down and will then rest quietly.

Cause. — Violent exertion on hard roads, or cold causing inflammation; feeding on wheat; unusual or inordinate feeding of any kind; from inflammation of the lungs, or bowels; or from drinking largely of cold water when overheated; putting a horse that has been idle suddenly to work, and sometimes occurs after foaling.

Treatment. — Remove the shoes, foment the feet, and poultice with linseed-meal or bran. Give a draught in gruel every six hours, containing bicarbonate of potassium, 1 oz.; Fleming's tincture of aconite, 5 drops; nitrous ether, 1 oz. Feed on mashes and green food, and keep the poultices on for three days. Bathe the feet with lotion containing ammonium chloride, 2 oz.; potassium nitrate, 2 oz., in 16 oz. water. If the inflammation continues after three days of such treatment, apply blister (No. 2) to the pasterns. In most cases the aid of a veterinary surgeon is advisable.

Laminitis—Founder (chronic).

The result of acute founder or inflammation of the foot, and nothing can cure it; shoeing may do good.

Navicular Joint Disease.

Symptoms.—A sprain of the joint made by the shuttle-bone at the back of the coffin-joint in the foot of the horse will, if the cartilage of the bone is inflamed, produce lameness. When first brought out of the stable, the horse will tread on his toes and avoid bringing his heel to the ground; consequently he will go lame down-hill; when resting he will point his feet. This lameness is very deceptive, and has often been judged to be in the shoulder.

Cause. — Hereditary; over-exercise after undue rest.

Treatment.—Foment and apply hot linseed-meal or bran poultices; and give purgative (No. 19). The early advice of a veterinary surgeon should be obtained, and he will best determine how ulceration and ossification of the cartilage can be prevented.

Over-reaching,

or wounding of the heels or coronet with the other foot.

Symptoms.—Often a clicking noise due to the hind shoe striking the fore one when the animal is moving. Often inflammation and pain; and the wound, however slight, should not be neglected.

Treatment.—Wash all dirt from the wound, apply a piece of tow dipped in friars' balsam, and tie it up. In severe cases poultice with linseed-meal or bran. If the wound does not heal, call in a veterinary surgeon.

Pricks or Wounds in the Sole.

Symptoms. — Lameness, which can probably be located by pressing all round the sole with a pair of pincers, the tender part being of course shown by the horse flinching.

Cause.—Commonly a fault in shoeing, or a wound caused by a stone, flint, piece of glass, or a nail picked up on the road.

Treatment.—Having found the tender place, pare that part of the sole down to the quick, and fill up the wound with a little tow dipped in friars' balsam. If the horse is very lame, or if the wound is festering, apply a poultice of linseed-meal or bran. If it does not heal, touch the place with chloride of antimony, which should induce the crust to form. A picked-up nail is often very dangerous, and if there is much lameness a veterinary surgeon should be called at once.

Pumiced Feet.

A result of inflammation of the feet. The exudate thrown out between the wall of the hoof and the coffin-bone during inflammation of the feet forces the latter to press downwards on the sole of the foot, flattening it and causing what is called a " pumiced " foot.

Symptoms. — Hollowness of the middle of the front part of the foot. Fulness or convexity of the sole.

Cause.—Inflammation of the foot, or very hard work, especially on hard roads or streets.

Treatment.—No cure. Blisters or stimulating dressings to the coronets may be tried, to increase the growth of healthy horn. Care—in shoeing—that nothing presses on the pumiced part of the foot, or a bar-shoe, is the only thing that can be done.

Quittor.

A suppurating wound of the coronet, often arising from a neglected prick, a tread, or accidental injury. Wounds of this nature are very serious, and should be left to the veterinary surgeon.

Ringbone.

A most prevalent disease situated in the pastern. In the hind feet, unless the disease is found at the front of the foot, the horse will walk on his toes; in the fore feet, owing to the greater concussion, it is generally at the front and sides, and the animal will walk on his heel.

Symptoms.—Pain and inflammation, with enlargement of the bone above the coronet, generally on both sides of the pastern, which, if not checked, will spread rapidly.

Cause.—Horses having straight upright pasterns are very liable to this disease, owing to their peculiar formation. It may also be hereditary.

Treatment.—Apply hot poultices to the leg and give purgative (No. 19), repeating the dose if necessary. If there is no improvement, blister once or twice with No. 4. Firing is often resorted to with success. Complete rest for some months will be necessary.

Sand-crack.

Symptoms.—Cracks in the fore feet will generally be found on the inner side, and in the hind feet in the front of the hoof.

Cause.—Brittle nature of the hoof, previous disease, heavy work or neglect.

Treatment.—Wash the crack to clean it from gravel and dirt. If the pain and lameness are severe, it may have to be poulticed. Pare and rasp it, and apply ointment composed of oil of tar, 2 oz.; fish-oil, 4 oz.; and stop the foot with cow-dung and moist clay. By passing a red-hot iron above and below the crack, healthy sound horn may be got to grow from the top. If any growth of proud flesh appears in the crack, apply nitric acid, and blister the coronet with No. 2.

Give rest and cover the crack with a plaster made of pitch, and bind the whole up firmly for five days. If the coronet has been divided, the aid of a veterinary surgeon had better be obtained.

Seedy Toe.

A disease of the foot in which an unhealthy horn is secreted that fails to maintain the connection between the horny laminæ and the wall of the hoof.

Symptoms.—There may or may not be a swelling of the wall of the foot, generally situated towards the toe; sometimes attended with lameness. If the part affected is tapped with a hammer it will sound hollow, and by paring the crack or hollow inside the wall-part of the foot the friable unhealthy horn can be found.

Cause.—Previous disease or injury, naturally weak feet, pressure of a part of the shoe, generally the clip.

Treatment.—Remove the cause, if practicable; pare away the diseased portion of the hoof, and apply blister (No. 2) to the coronet. Rest till cured. Afterwards shoe with side-clips.

Side-bones.

Symptoms. — Somewhat similar to Ringbone (p. 464), except that the disease is located above the heel; it is an ossification of the lateral cartilages of the foot. Usually found in heavy draught-horses, and in the fore feet rather than in the hind feet. If the horse has good, well-developed feet, they do not generally cause lameness.

Cause.—Concussion and hereditary predisposition; bad shoeing.

Treatment.—Apply blister (No. 5); if this does not cure the lameness, have Professor Smith's operation performed on the foot by a veterinary surgeon; some of the well-known patent specifics may be tried with a chance of success.

Thrush.

A disease of the frog, which secretes a semi-fluid fœtid matter.

Symptoms.—A discharge of matter from the cleft of the frog. There is not often lameness, and the disease can be detected only by the matter exuding from the frog. If thrush is neglected, it will increase, the frog will become soft, ragged, and split up, the horn will disappear, and canker of the sole may supervene.

Cause.—Generally excessive moisture in the bedding, bad stable management, and constitutional predisposition.

Treatment.—Give purgative (No. 19), clean the frog thoroughly, and pare away all loose horn, apply a lotion to the frog, composed of carbolic acid, 1 part; glycerine, 6 parts, and place tow moistened with this lotion in the cleft of the frog every night. If possible, remove the cause of the disease. It is not necessary or expedient to turn the horse out to grass.

Weakness of the Foot.

Generally a fault in the make of the horse. Sometimes the result of disease. A well-formed foot should be at an angle of 45° from the coronet to the toe; a weak foot will be perhaps 36° to 40°, which is not sufficient to bear the pressure required. No cure for this defect, but careful shoeing may have a palliative effect. Rasping the wall of the hoof and paring too much off the heels is often the cause of this complaint.

Firing

is a painful operation often unnecessarily performed, for many of the horses that are fired are as lame after the operation as they were before. Firing was at one time greatly in vogue, but, like bleeding, it is getting out of fashion, and by-and-by horses with fired legs will be rare. It is thought by some that the lines in firing act as a permanent bandage to the weakened part: such is not the case, but firing does act as a counter-irritant of a severe kind.

Before you resort to firing, blister your horse once or twice, and give it a three months' run at grass; then if it comes up lame, think about firing; but remember there are some cases of lameness that nothing will ever cure.

There are two kinds of firing—lines and dots: line firing is the best for curbs, ringbones, and the back tendons; the dots are preferable for splints and spavins.

Having clipped the hair off the part to be fired, secure your animal, take the iron and make the lines at first superficial, then with a fresh iron deepen them, but never go through the skin in line firing; afterwards rub blister in, and tie the animal's head up for forty-eight hours.

Blistering.

Clip the hair off the chosen part, and rub the blister in for at least ten minutes, then tie the animal up for twenty-four hours.

III.—THE SKIN AND ITS DISEASES.

Hide-bound.

A want of oily matter, which produces hardness of the skin, giving the coat a rough look. It shows that the digestive organs are out of order, and is not so much a disease of the skin itself.

Treatment.—Give purgative (No. 19), and afterwards daily in the food condition powder (No. 6). Powerful tonics should be avoided.

Lice

may be destroyed by applying a lotion composed of tobacco, 4 drs., in a pint of hot water, or by using an ointment composed of white precipitate of mercury, 1 part, lard, 12 parts, well rubbed in. It is best to clip the horse before applying these dressings. Cleanliness and nourishing food will prevent their reappearance.

Mange or Itch.

Symptoms.—Loss of hair, itching, tenderness, and scurfy eruption, from which matter issues. When the scab falls off, a larger blotch will appear. It generally begins at the root of the mane or on the neck.

Cause.—Stable neglect, dirt, and contagion, it being due to the presence of animal parasites—small insects called Acari.

Treatment.—Give purgative (No. 20), and rub the places with ointment composed of sulphur, 1 oz.; lard, 1 oz. If this does not effect a cure, add to the ointment 30 grs. of white precipitate of mercury. This disease is often very obstinate, and patience must be exercised. A little salt should be given with nourishing food, and the skin kept clean, using warm soap-and-water for the purpose. Complete isolation is necessary. Wash the stable, harness, brushes, &c., with solution of chloride of lime, 1 pint in 3 gallons of water, before they are used again.

Ringworm.

Symptoms.—A parasitic fungus, which affects the skin in circular patches; the hair comes off, leaving a dry and scaly eruption.

Cause.—Contagion, neglect, or dirt.

Treatment.—If the animal is in high condition, or in a disordered state, give purgative (No. 20), but if not, give nourishing food, and keep him clean and isolated from other animals. Rub the fungus with ointment composed of oleate of mercury, 1 part; lard, 2 parts, till cured. Clean the stable, harness, brushes, &c., with water containing 1 pint of chloride of lime to 3 gallons of water before they are used again.

Nettle-rash or Surfeit.

Symptoms. — Large pimples, disappearing as quickly as they come, which spread from the neck to different parts of the body.

Cause.—Exposure to chills, or drinking cold water when hot.

Treatment.—Give in a pint of water 2 oz. of spirits of ether and 1 oz. of tincture of ginger, and then treat the same as for Hide-bound (p. 465).

CATTLE.

DISEASES AFFECTING THE HEAD, EYES, MOUTH, AND NERVOUS SYSTEM.

Inflammation of the Brain.

May arise from violence, disease, or as an effect of poisons.

Symptoms.—Great pain and moaning, slow respiration, eyes red, loss of consciousness. Attacks of delirium, and the beast becomes ungovernable till stupefaction results, accompanied by extreme weakness; at length death ensues.

Cause.—Violence, exposure to great heat, want of water, overdriving; sudden change into a rich pasture.

Treatment.—If the beast is in fair condition you should slaughter it at once. If not, give linseed-oil, 2 pints; croton-oil, ½ dr., and three times a-day, in gruel, hydrate of chloral, 1 oz.; bromide of potassium, 1 oz.; and apply ice or cold water to the head. If the animal survives the first stage, blister the crown of the head and sides of the neck with No. 22. Most probably the animal will never recover.

Paralysis.

There may be palsy of the half, or any part, or of the whole of the body.

Symptoms.—The animal may lie, eat, and chew its cud as if nothing were wrong; but when you try to rouse it you will see it make several attempts to get on its legs but fails. When parturient paralysis appears before calving, it is not so serious as the form of paralysis that comes after calving or an attack of milk-fever.

Cause. — By derangement of the stomach, and is called reflex paralysis; by injury to spine, and before or after calving, and is called parturient paralysis.

Treatment. — Give purge (No. 27), and apply liniment (No. 26) to the whole length of the spine, and every night and morning give in a pint of ale the following drench: tincture of nux vomica, 2 oz.; iodide of potassium, 2 drs.; sulphate of iron, 1 dr.; turn the animal twice a-day. When paralysis appears before calving, the cow usually calves before she rises; but if it appears after calving, treatment often does little good, and if fat it may be best to kill the animal. But if a cure is to be tried, pursue the same line of treatment as before calving. Galvanism may be tried.

Thrush in the Mouth.

This usually appears as an epizootic amongst cattle in cold and wet weather.

Symptoms.—Small pimples and vesicles appear on the tongue, lips, and about the mouth; they break and form ulcers, but these ulcers soon heal. There is not much danger in this disease, though a little fever often exists.

Treatment. — Give purge (No. 28); wash the mouth out with alum-water, 1 part of alum to 30 of water, and give night and morning, in a pint of water, 1 oz. of salicylate of soda.

Lockjaw or Tetanus.

A disease which seriously affects the nervous system, producing contraction or spasm of the muscles.

Symptoms.—Sluggishness, and for some days increasing difficulty in mastication and swallowing, till the jaws become almost closed. The contraction of the muscles will then extend to the head, neck, and shoulders, and appear to cramp the whole body. Constipation. Recovery is very doubtful.

Cause.—Generally some wound or blow affecting a muscle, or exposure to cold. Contagion and the access to a wound of the specific organism of the disease, the bacillus of Nicoläier or drumstick bacillus.

Prevention.—If this disease is suspected, give in gruel Epsom salts, 1 lb., and Fleming's tincture of aconite, 10 drops.

Treatment.—Any treatment must be prompt to be efficacious. A veterinary surgeon should be called at once.

Cancer of the Tongue.

This disease, though not often suspected, frequently exists, and the teeth are usually blamed for it; but the disease now known as actino-mycosis is often mistaken for cancer.

Symptoms.—The animal is unthrifty, off its food, frothy saliva flows from the mouth, and it quids its food. On examining the mouth you find the tongue hard in places, and slightly swollen.

Treatment.—As soon as it is detected, kill the animal, or else it will gradually starve to death. The enlargements on the tongue, caused by the presence amongst the tissues of the parasitic fungus known as the actino-myces, are sometimes successfully treated, if not too far advanced, but they should be left to the veterinary surgeon.

Dishorning and Broken Horns.

On the subject of the dishorning of cattle the following conclusions were adopted by Tennessee Agricultural Experimental Station: "(1) For removing the horns, an ordinary meat-saw is perfectly satisfactory. (2) The horns should be removed as close to the head as possible, without cutting the skull proper.

The sawing should be done rapidly, and with long sweeps of the arm if possible. (3) Animals one and two years of age appear to suffer considerably in dishorning. The painful effects decrease with increase of age, so that an animal of ten years old may suffer but very little. This is owing to the layer of flesh surrounding the base of the horn, which is much thicker in young than in old animals. Dishorning causes an abnormal increase of pulsation and temperature, which extends over several days. The appetite is also affected during the twenty-four hours succeeding the operation. (4) Dishorning is more especially to be recommended for those animals that are of vicious temperament, that are what are termed 'masters'; to be applied to bulls and to beef animals that are kept quiet and closely stabled or shipped. (5) From evidence quoted from other sources, it appears that dishorning is not necessarily a cruel practice, but may be conducted to promote ends that are both humane and desirable in live-stock breeding. Mr Sadler, British Consul at Chicago, reports that in his very extensive district the system of dishorning or dehorning cattle is rapidly increasing. Some farmers have dehorned their whole herd. It has been calculated that 200,000 cattle and horses die each year in the United States from horn-thrusts. The advocates of the system of dehorning claim that, besides lessening this loss and that of human life, much shed-room is saved, less hay is consumed, there is less turmoil from restive animals, and that cows, being more quiet and docile, give more milk.

"In the case of calves, the horn is extracted by a gouge or punch when two or three months old, and with full-grown animals the horn is sawn off at the point where the matrix joins the bone horn, and should be done early in the spring. If sawn higher up, the horn grows again; if below, the process of granulation would not take place." [1]

Destroying Horns in Calves.

In young calves, when the horns are felt causing a projection under the skin,

[1] *Veterinary Journal*, November 1888.

they can be prevented growing by the application of a caustic solution which can be obtained from most agricultural chemists.

Broken Horns.

If the horn is severely crushed, it is best to amputate it; but in cases where it is only torn or broken off, wash clean, smear some Archangel tar over it, wrap some tow around, and take a long linen bandage and wind around the horns in the figure-of-8 style.

Ophthalmia. Inflammation of the Eyes.

The symptoms and treatment of the diseases affecting the eyes are practically the same as those given in the section on Horses (see p. 442).

Growth on Eyeball.

There is sometimes seen in cattle a growth on the eyeball. Should this be causing trouble, so that its removal is deemed necessary, a veterinary surgeon should be applied to.

DISEASES AFFECTING THE THROAT, CHEST, AND RESPIRATORY ORGANS.

Abscesses.

Symptoms.—Frequently large lumps appear on the side of the jaw or on other parts of the body. In time they burst and discharge a large amount of matter, often affecting the health of the animal.

Cause.—Generally a blow, prick, or other injury.

Treatment.—A mild purgative (No. 28) should be given in gruel, and the abscess should be fomented with hot water, and opened as soon as it is ready. If making little or no progress, it should be rubbed with blister (No. 22). Tonic (No. 29) may be given in a pint of warm ale morning and evening when recovering.

Anthrax or Splenic Apoplexy.

A contagious and very dangerous disease which affects all animals, and is also inoculable to man.

It is most common in cattle, but is also met with in horses, pigs, and sheep.

Cause.—It is due to the presence of a micro-organism, the *Bacillus anthracis*, in the blood and tissues. The origin of an outbreak is generally obscure, and foreign feeding-stuffs and imported bone manure have been blamed for introducing it. Although contagious, it is not infectious, and seldom spreads from the farm or herd in which an outbreak occurs.

Symptoms.—Very often the first thing observed is that an animal is found dead, and frequently there is some bloody discharge about its nostrils and anus. If seen alive, there is great dulness and depression, high temperature, the head often low and the back a little raised; the abdomen appears full, as a rule, and there may be some shivering about the flanks or shoulders.

In the horse, there is generally swelling about the throat, which may extend down the neck towards the breast; and in the pig often a great swelling from ear to ear.

Serious outbreaks have frequently been caused by the thoughtless slaughtering of affected animals, and allowing the blood to be scattered about—the blood being the chief means of spreading the disease. When a case of anthrax is suspected,—according to the law,—the owner must at once give notice to the police, and take means to isolate the animal, so as to prevent other animals coming in contact with it. A veterinary inspector is sent to inquire into the case, and if anthrax is found to exist the carcase has to be cremated, or buried without the skin being cut, six feet deep, and covered with a thick layer of lime. The inspector will see that the place and everything connected with the case is thoroughly cleaned and disinfected.

Black-quarter or Quarter-ill.

A disease which has been known as affecting young cattle for a very long time, and was scarcely considered contagious until it was proved experimentally. It is generally confined to animals from three months to two years old, and very often the best thriving one in a lot is the victim. All ruminants are thought to be liable to it, but it is only common in cattle.

Cause.—It is due to a micro-organism, the *Bacillus Chauvœi*, rather smaller than the anthrax bacillus, and, unlike the latter, is never found in the blood during life, but only in the tumours and effusions.

Symptoms.—Very like those of anthrax, but there is usually either lameness or the appearance of a swelling on some part where it does not cause lameness. The swelling, at first hot and painful, rapidly enlarges, and begins to crepitate on pressure—*i.e.*, it contains gas. There is generally constipation, and often the animal goes down and refuses to rise.

Treatment.—This is not of much use. It is almost invariably fatal. A strong dose of Epsom salts and common salt may be given when first seen. Some recommend that the tumour should be fomented with very hot water, freely incised, and turpentine and other antiseptics smartly rubbed into it. Unlike anthrax, the flesh seems quite harmless to dogs, pigs, &c.

Prevention.—Some farms seem liable to this disease, and the calves used to be bled, physicked, and setons put in their dewlaps every season. Now a process of inoculation is used, and can be applied by your veterinary surgeon. It is generally well spoken of, but has sometimes given unfortunate results. Plenty of rock-salt within reach of the young stock, and an occasional dose of salts and nitre, will do them good.

Cattle Plague or Rinderpest.

A contagious, infectious, eruptive fever, and the most serious epizootic disease to which the ox is liable.

It seems to find its home in Central Asia, and is always present in India.

The last serious outbreak of it in Britain was in 1865-66, when it spread over most of Britain, and caused the loss of cattle to the value of several millions sterling.

Cause.—It is presumably due to an ultra-microscopic organism possessed of great virulence, as the disease spreads rapidly from animal to animal.

Symptoms. — High fever, dulness, staring coat, maybe shivering, discharge from eyes and nose, appetite lost, and milk arrested. There is generally constipation at first, followed by a fœtid diarrhœa. The most distinctive symptom is redness of the mouth and nostrils, an eruption appears in small spots over which the mucous membrane becomes shed in bran-like scales.

Treatment. — No treatment is permitted.

Prevention. — Owing greatly to its extension southwards from Egypt after 1890, until it spread practically over all the African continent, wild ruminants dying from it as well as domesticated, many attempts were made to find some satisfactory means of prevention. Now an immunising serum for inoculation has been obtained, serum institutes established both in Egypt and India, and the serum is prepared and distributed under Government supervision. Although the disease cannot be "stamped out" by means of the serum, yet its propagation can be wonderfully controlled and restrained.

Choking.

Very often cattle get pieces of turnips or linseed-cake into their throat or gullet, especially if the turnips are cut in large pieces.

Symptoms.—Animal ceases to feed, nose poked out, saliva flows from the mouth; the animal in time becomes hoven, and frequently dungs. If it is choked by a piece of turnip, you can smell it in the breath, and if the piece is near the larynx there may be coughing.

Treatment.—Give a little linseed-oil very slowly. If this does not pass it on, you must use the probang. In using this instrument you must first place the gag in the animal's mouth, and have it held there by two men, who cross their hands, holding the gag in one and grasping a horn with the other,—and be sure that the animal is held steady, its head and neck as straight as possible and in line with the body; then take the probang, oil it well, and gently pass it down the throat, until you reach the offending body. Do not use much force in passing it into the stomach, for it is an easy thing to rupture the gullet. If it cannot be moved by the probang, the veterinary surgeon should be called in, although if the hoven is extreme it may be necessary to tap the rumen at once with a trocar to permit the gas to escape.

Cold or Common Catarrh.

Symptoms. — Dulness; running discharge from the nose; cough; watering eyes; loss of appetite.

Cause.—This common complaint is most frequently met with in spring and autumn; it arises from exposure to draughts and from chills caught in wet weather. If neglected, it will lay the foundation of serious coughs, inflammation of the lungs, and other formidable diseases.

Treatment.—Epsom salts, 1 lb., and ginger, ½ oz., may be given at first in warm gruel; afterwards, morning and evening, in gruel, solution of acetate ammonia, 4 oz.; bicarbonate of potassium, 1 oz.; spirit of chloroform, ½ oz. All drinking-water to be given with the chill off; and feed on bran-mashes and green food.

Sore Throat or Quinsy.

Inflammation of the mucous membrane of the throat, or of the tonsils.

Symptoms. — The animal refuses to feed, pokes its nose out, breathes hard, and makes a peculiar noise in the throat.

Cause.—Cold and an insanitary condition of the byre, associated with bad feeding.

Treatment.—Blister with No. 22, and place a piece about the size of a bean of the following every three hours between the molar teeth: extract of belladonna and chlorate of potash of each an ounce, and made into a paste.

Cough.

Cause. — Neglected colds are apt to develop into coughs which are difficult to cure. Cold caught when the system is relaxed, as is the case with cows after calving, may take the form of a cough which will settle on the lungs and produce serious consequences.

Treatment.—Keep the animal warm; give water in which linseed has been boiled for drinking purposes, and morning and evening, in gruel, draught (No. 25).

Foot-and-mouth Disease.

Aphthous fever. A contagious and infectious eruptive fever, and the most typical epizootic disease affecting the domestic animals.

All ruminating animals are subject to it, and also the pig, whilst the horse and other animals are sometimes affected.

Cause.—It is due to the presence of an ultra-microscopic organism, which has never been demonstrated, but which has practically been proved to exist, and there is no doubt it is only communicated from animal to animal by contagion and infection.

Symptoms.—Fever, often high fever, and sometimes shivering. Soon there is a profuse discharge of saliva which hangs in strings from the mouth, and the animal smacks its lips. There is also a shaking of the feet as if it wished to get rid of something irritating them. If the mouth is examined small bladders or vesicles will be seen on the lips, in the mouth, and on the gums and tongue. Small ones may also be seen on the teats and udder, and on the scrotum in the male. It is the same thing which causes the sores at the front and back of the cleft of the hoof, but seldom observed until they burst and look raw and sore. When the vesicles in the mouth burst the smacking of the lips stops, and red, raw-looking spots are seen. The sores on the teats of milch cows almost prevent milking, and there is danger of inflammation of the udder. After the animal is recovering the hoofs are shed gradually in all the worst cases.

In the sheep it is not generally so bad; the mouth is seldom so sore, but the sores at the feet may form all round the top of the hoof, and do not appear only at the front and back as in the ox.

In the pig the feet lesions are worst; and pigs moved about or sent per rail at the height of the disease may lose their hoofs altogether.

Calves sucking their dams, while suffering from the disease, often die suddenly. Adult animals seldom die from it in this country, but it causes immense losses through destroying the udders of dairy cows, causing abortions and great loss of condition.

Treatment.—A dose of physic at the start, see that their feet do not get wet or dirty, cleanliness and astringent antiseptic washes for the sores, and the provision of suitable food. But the Board of Agriculture does not permit treatment now.

Prevention. — Isolation and strict police measures.

Hoose or Husk

is a peculiar disease produced by the thread-worm or lung parasite of cattle— the *Strongylus micrurus*.

Symptoms. — Peculiar husky cough, wheezing, loss of condition, and death, if means are not taken to destroy the thread-worms in the bronchial tubes. In the later stages of the disease there is much slimy mucus about the mouth and nose. It is very prevalent among calves and growing stock during the late summer when in the fields, and especially so in the autumn when the weather is wet. The cough is particularly noticeable if the animals are chased.

Cause — Prevention. — The losses caused by these parasites render it most important that every care should be taken to keep animals likely to be affected free from the influences calculated to invite an attack. These parasites frequent low, marshy, and undrained ground. During a wet season many kinds of grass-land will be found to contain them. Calves and young stock should be kept in good condition, and, if possible, during the autumn should be housed at night, and not turned out till the dew is off the grass. When animals are allowed plenty of food there is not much risk of the disease, so that young calves should get plenty of milk or other suitable food, and older animals trough food, as crushed grains and cake with some common salt in it.

Treatment. — Stock in the autumn should be daily examined, and upon the first sound of husk the affected beast should be attended to. Give daily to a calf turpentine, ½ oz., beaten up with milk and an egg; keep him well, giving linseed-porridge; and if the calf is young, new milk. In the case of older cattle, give morning and night turpentine, 1 oz., in six times the quantity of sweet oil. I have tried applications of tar to the animals' noses, but it does no good. A new method of treating this disease is to inject remedies directly into the windpipe, but this should be left to a competent veterinary surgeon.

Bronchitis.

Inflammation of the small air-tubes of the lungs.

Symptoms. — Animal dull, breathes quick and short, fits of coughing, and, on placing the ear against the chest, a peculiar wheezing noise is heard.

Cause.—Cold, exposure to wet, and allowing medicine to get into the windpipe in the act of drenching an animal.

Treatment.—Remove to a warm box, throw a couple of sacks over its back, rub each side of the chest with embrocation (No. 12), and give in a pint of gruel twice a-day (No. 25).

Inflammation of the Lungs or Pneumonia.

Symptoms.—Dulness, loss of appetite, cough dry and hard, rapid breathing, hot mouth, very cold ears, horns, and feet, slimy discharge from mouth.

Cause.—Exposure to cold and sudden chills.

Treatment. — Remove into a cool loose-box, and give every six hours, in a pint of gruel, Fleming's tincture of aconite, 20 drops; solution acetate of ammonia, 4 oz.; spirits of nitrous ether, ½ oz. Feed on mashes and green food; rub the chest with blister (No. 23). If the bowels are costive, give in gruel Epsom salts in 1-lb. dozes. Clothe with rugs or sacks about the shoulders and chest.

Influenza or Epizootic Catarrh.

Symptoms. — All the symptoms attending common colds are intensified in the more serious form of influenza. Profuse discharge from eyes and nose, painful cough, obstinate constipation, fever, followed by equally obstinate diarrhœa; swellings about the head, accompanied by great weakness. Usually the disease runs through a herd, and is attended with considerable loss among the cattle.

Cause.—Influences of climate, which seem to make the disease an epizootic.

Treatment. — Isolate the affected beasts. Give linseed-oil, 1½ pint, followed by gruel to drink; if constipation continues, give enemas and every six hours a draught containing acetate of ammonium, 4 oz.; bicarbonate of potash, 1 oz.; spirit of chloroform, ½ oz. Keep the body warm, and for drinking purposes give water in which a little linseed has been boiled. Feed on mashes and green food.

Murrain or Malignant Catarrh.

Symptoms.—This disease, which is one of the most fatal to which cattle are subject, usually begins with a cough, followed by heaving flanks, shivering, tenderness over the loins, horns cold, dung hard, black, and fœtid, bloody matter running from the nose. As the disease advances, blood is mixed with the dung, and the breath becomes offensive. Great weakness sets in, the mouth becomes ulcerated, till finally the beast dies, a mass of corruption.

Cause.—Not well known.

Treatment.—If this dreadful disease is suspected, completely isolate the beast; give every four hours, in warm gruel, salicylic acid, 3 drs.; tincture of cinchona, 2 oz.; brandy, 4 oz., till the opinion of a veterinary surgeon can be obtained, who will decide whether to slaughter the beast or not.

Contagious Pleuro-pneumonia of the Ox.

A contagious and infectious disease affecting cattle only. It has now been stamped out in Great Britain after proving a perfect pest to the farmer and stock-breeder for over fifty years.

Cause.—A very minute micrococcus which can only just be made out under the highest powers of the microscope.

Symptoms.—A short husky cough is often the first symptom, but if the temperature is taken it will be found that fever is present. As the disease progresses the cough becomes more marked, especially when the animal is hurried or excited, the breathing is more frequent, and there is a distinct lift at the flanks. When punched over the ribs the animal may grunt, and it may seem pained on movement.

But it may require a post-mortem examination to distinguish the disease. When the chest is opened there may be a considerable amount of fluid in it, and the lungs and pleura covered by a yellowish white membrane; but the disease may all be on one side. In old cases the lung may be adherent to the ribs. Part of the lung or lungs will feel solid, and when cut into presents a characteristic marbled appearance, the sections varying in colour through pink, greyish, different shades of red, to almost black, and separated by yellowish veins up to about half an inch broad.

Treatment is not now required in Britain, and

Prevention is obtained by keeping it out of the country.

Tuberculosis—Consumption.

This almost ubiquitous disease is more commonly known as consumption when affecting the chest in man than when met with in the lower animals. In some of its various forms it is also known as "struma" and "scrofula," and animals affected by it are often called "piners" and "wasters."

This is the most widely spread and destructive disease to which animals are liable, and nearly all animals are subject to it.

Cause.—It is due to a very fine bacillus, the Bacillus tuberculosis, or bacillus of Koch, as it was first discovered by Professor Koch of Berlin in 1882. Prior to that time the disease was believed to be strongly hereditary, and that over-crowding, bad hygiene, and privation led to its development. These are now considered predisposing causes, but many still think that a hereditary tendency to it may exist in some individuals. Although tuberculosis cannot exist without the presence of the bacillus, it is now recognised by scientists that there are different varieties of the Bacillus tuberculosis. There is the "human type," which is commonly the cause of tuberculosis in man; the "bovine type," which is the cause of tuberculosis in cattle; and the "avian type," which is the cause of tuberculosis in fowls. Although these differ from each other in several respects, yet the differences are not sufficient to cause them to be considered as distinct species, but only as different varieties of the bacillus of Koch.

At the London Conference of 1901 Koch gave it as his opinion that human and bovine tuberculosis differed so much that it was scarcely possible to communicate the latter disease to man, and that owing to this the presence of tubercle bacilli in the milk and flesh of bovine animals might be disregarded.

Owing to the eminence of Koch as a scientific pathologist such a declaration could not be ignored, and a Royal Commission was soon after appointed to investigate the matter.

The second interim Report of the Commission was published in 1907, and states: "There can be no doubt but that in a certain number of cases the tuberculosis occurring in the human subject, especially in children, is the direct result of the introduction into the human body of the bacillus of bovine tuberculosis; and there also can be no doubt that in the majority of these cases the bacillus is introduced through cows' milk. Cows' milk containing bovine tubercle bacilli is clearly a cause of tuberculosis, and of fatal tuberculosis in man."

Although the bacilli of bovine tuberculosis seem to be more virulent when inoculated to other animals — experimental animals — than the bacilli of human tuberculosis, it does not follow that this is the case when inoculated or communicated to man. It seems rather the opposite, and many scientists seem to doubt whether tubercle bacilli of the bovine type ever cause acute tuberculosis of the lungs — often called phthisis or consumption—in the human subject.

Symptoms. — These are often very indefinite: an animal if well cared for and kept in good condition may be full of tubercles without manifesting any symptoms of illness. Sometimes an animal—often a young one—will begin to make a rough noise in breathing, especially when eating with the head down. This is often due to disease with enlargement of the glands about the throat. Sometimes they will bulge externally below the ears or about the lower jaw, burst and discharge matter. Often a cow, especially if a heavy milker, is inclined to become lean, then a dry short cough is heard, she soon looks unthrifty, gradually emaciates, the cough becomes worse, diarrhœa may set in, the skin seems to adhere to the bones, the appetite is impaired, and she becomes a confirmed piner. Sometimes a young animal will become lame, and it is thought to have been injured, a joint may be observed swollen, it continues to enlarge, and turns out tubercular. A quarter of a cow's udder may feel a little

hard, but it continues to give milk, and little is thought of it. It, however, still grows harder, but is not very painful, and milk is secreted in fair amount. This is generally in a hind quarter, and is at length found to be due to tuberculosis. There is scarcely an organ or tissue but may become affected, sometimes tumours — diseased glands—will appear near the point of the shoulder or about the flanks, and in a bull a testicle may become enlarged,—all due to tubercle. In the horse it is not very common, and the symptoms are often vague. There is not very often a cough, but he seems weak, breathless on exertion, and unfit for his work; he loses appetite, but drinks plenty, and often urinates much more than usual. He becomes dry and open in his coat, rough and scaly on his skin, and acquires an unthrifty appearance.

It is thought that horses often contract the disease from mixing with cattle, or from being reared on cow's milk.

Pigs are frequently affected. The disease may spread from pig to pig, but is often due to diseased offal about slaughter-houses and the refuse from creameries. A growing pig may become lame, and one or more joints enlarge. Or it may begin to cough, to be less keen for its food, diarrhœa may result, it loses condition, and tumours may appear in the region of the throat.

Treatment. — It is scarcely worth while treating an animal with tuberculosis. Still, if an animal in fair condition is suspected, it should get every attention and the most nourishing food to enable it to be sent to the butcher as early as possible, in the hope that the carcase may be free of the disease and fit for human food.

Prevention.—There is no subject connected with the health of animals (or of man either) receiving so much attention at the present time (1909) throughout the civilised world as the suppression of tuberculosis, and in no country, as far as animals are concerned, is less being done in that direction than in Britain. Except in the form of carcases or meat intended for human food, and in the case of cows, the milk of which is offered for public sale, no restriction or control whatsoever is exercised by Government in connection with the disease. It is a big and difficult question, but the time seems approaching when the Board of Agriculture will be obliged to take action regarding it.

Meanwhile the breeder and stockowner must rely on his own knowledge and initiative. A very considerable aid to its suppression is the fact that it is now very generally recognised as contagious. As soon as any breeder has reason to suspect that an animal may be affected with the disease it should be rigorously isolated, and its stall or box cleaned and disinfected. And should a veterinary surgeon pronounce the illness to be due to tuberculosis, unless the animal is in a condition to be rapidly fattened, it should be destroyed. On ho account should it be again returned to the herd. There is no doubt but breeders can clear the disease out of their herds by the use of tuberculin, and keep it out, at less expense than it will cost them in loss and illness if no means of any kind are used to prevent it.

Many scientists in different parts of the world have for years been trying to obtain some reliable means of prevention. Von Behring, a German, has prepared several kinds of serum, one at least of which, he affirmed, when inoculated into animals rendered them immune or insusceptible to the disease, but it has been extensively tried in Argentina, under Government auspices, with very questionable benefit. At the present time the most successful method of conferring some degree of immunity on animals is by the intravenous injection of cultures of human tubercle bacilli, and that will require some time yet before it is applicable on a commercial scale. Meantime the breeder should foster the health, the vigour, and robustness of his herd. Never allow an animal to get into low condition; see that there is sufficient airspace, light, and ventilation in the byres, and that young and breeding animals are turned out for a short time every day unless the weather is very bad. The weaklings should never be retained in the herd. If there is any appearance of delicacy or lack of robustness let them go. It does not matter how fine a pedigree an animal may have, if there is neither vigour nor stamina it should not be in a breeding herd.

DISEASES AFFECTING THE STOMACH, LIVER, BOWELS, KIDNEYS, AND INTERNAL ORGANS.

Bloody Flux, see *Dysentery* (p. 475).

Colic or Gripes.

is of two kinds.

1. *Flatulent Colic.*

Arising from retention of food in the third stomach and bowels.

Symptoms. — Fever, moaning and pain; discharge of gas from anus, distention of the abdomen, restlessness.

Cause.—Errors in dieting, green food, being turned out to grass too suddenly in the early summer, especially if a cold day.

Treatment.—Give purgative (No. 27) in gruel, and every four hours, in gruel, solution of ammonia, 1 oz.; spirit of chloroform, 1 oz. Give gentle exercise, and rub the belly with liniment (No. 26). Clysters of warm water may be necessary. Feed on mashes and gruel.

2. *Simple Colic.*

Symptoms. — Spasmodic attacks of pain, increasing in violence. Irritability, and constant striking of the belly with the hind legs or horns; continual restlessness.

Cause. — Chills from drinking cold water when heated; improper food.

Treatment.—Give linseed-oil, 1 pint, repeating the doze if there is costiveness; and every four hours give in gruel: oil of turpentine, ½ oz.; tincture of opium, 1½ oz.; spirits of nitrous ether, 2 oz. Walk the animal about. In obstinate cases send for a veterinary surgeon, and in the meantime rub the belly with liniment (No. 26).

Costiveness or Fardel-bound.

Symptoms. — Excessive costiveness; dung hard, but at intervals loose and slimy. Frequently the abdomen will become distended, and inflammation follows.

Cause.—Excess of dryness in the food, or the peculiar properties of some kind of underwood often eaten by cattle. Often also a symptom of some other disease.

Treatment.—Give linseed-oil, 1 pint; and warm oatmeal-gruel, in which ½ oz. salt has been mixed. If this does not act, give, in gruel, purgative.(No. 27), and, if necessary, a warm clyster of gruel and ½ oz. salt.

Foreign Bodies in the Rumen.

Some cows at times suffer from depraved appetites, and pick up almost anything that comes in their way. Leather, wire, cutlery, rags, &c., have been found in the paunch of an animal.

Symptoms.—They are not very noticeable, and an animal might have a foreign body in its paunch for months without feeling any inconvenience from it, but if the foreign body passes into any vital organ, symptoms such as loss of appetite and colicky pains are soon noticed.

Treatment.—Nothing can do any good in the shape of medicines; if you suspect there is something in the stomach that should not be, consult your veterinary surgeon on the case.

Diarrhœa.

Symptoms.—A frequent discharge of fluid dung mixed with mucus, which soon causes great weakness.

Cause. — Change of food, especially when moved from a poor into a luxuriant pasture. Bad water or atmospheric influence, amounting almost to an epizootic.

Treatment.—Give linseed-oil, 1 pint; tincture of opium, 1½ oz.; oil of turpentine, ½ oz., and repeat the doze, if necessary, which will remove any cause of irritation in the intestines: till this is done, no astringent should be given. When the oil has cleared the system, give morning and evening, in cold gruel: powdered opium, 2 drs.; catechu, 4 drs.; galls, powdered, 4 drs.; prepared chalk, 1 oz. Looseness of the bowels, unattended with pain and weakness, should not be regarded as serious, provided it can be accounted for by change of food; it should be carefully watched, and steps taken to prevent its assuming too violent a character.

Dysentery or Bloody Flux.

Symptoms.—Continual and obstinate purging, the animal is hide-bound, eyes pale, pulse weak, extreme weakness. In time the dung appears like undigested food, and water with clots of blood in it.

Cause. — Internal inflammation from neglected diarrhœa or the eating of poisonous plants.

Treatment.—Clothe warmly, foment and rub the belly with liniment (No. 26). Feed on gruel made of oatmeal and linseed, with 4 oz. of starch and 1 oz. nitre in it. Give three times a-day in gruel: ipecacuanha, 1 dr.; chlorodyne, 40 drops; opium, 2 drs.; chalk, 1 oz.; galls, 2 oz. Give also cold clysters of oatmeal-gruel; and laudanum, 2 drs. This disease is most dangerous and almost hopeless.

Hoove, Hove, or Hoven,

or distention of the rumen by gas, owing to the food being retained in the stomach so long that it begins to ferment.

Symptoms.—Swelling of the belly; heavy breathing; moaning and unwillingness to move. As the gas is evolved, the stomach becomes further distended, —there is even danger of the paunch bursting; the circulation of the blood is impeded; gradually suffocation sets in, till at length the beast falls and dies.

Cause.—Overloading of the stomach so that it is unable to react on its contents, greedy feeding on green food, feeding on clover before the dew is off it, hence it is often termed "dew-blown."

Treatment.—In desperate cases the only cure is to relieve the stomach by means of a stomach-pump, which will be almost beyond an ordinary breeder of stock. In cases of sudden emergency, an incision into the paunch behind the short ribs with a penknife will give relief. A trocar and canula should be used if it can be got. In ordinary cases give at once in a pint of water hyposulphate of soda, 4 oz., repeating the dose till relief is afforded. When recovering, Epsom salts, 1 lb., and ginger, ½ oz., may be given, and but little food allowed till the digestive organs have recovered their strength.

Impaction of the Paunch or Grain-sick.

This disease is seen when animals are allowed to gorge themselves with such foods as succulent grass, chaff, potatoes, turnips, and grains.

Symptoms. — Animal dull, refuses food; disinclined to move and generally lying down; greatly swollen on the left side, but, unlike hoove, it has a doughy feel.

Treatment. — Give purge (No. 27), and with it 1 pint of linseed-oil and 2 oz. of tincture of nux vomica; if this fails, you must get a veterinary surgeon, who may require to perform an operation to remove the contents.

Inflammation of the Bowels.

Symptoms.—Restlessness, pain, perspiration, hard breathing, quick pulse.

Cause.—Sudden chills in hot weather, as from drinking a great quantity of cold water when overheated, most common in working oxen.

Treatment.—Give, morning and evening, linseed-oil, ½ pint; spirits of nitrous ether, 1 oz.; tincture of opium, 1 oz.; and repeat the dose of spirits of nitrous ether and tincture of opium in a little gruel every four hours; very careful feeding on sloppy foods and gruels.

Dropsy of the Abdomen or Ascites.

An accumulation of fluid in the abdominal cavity.

Symptoms. — The beast increases slowly in size; the swelling is on both sides and on the lower part of the abdomen; as the fluid increases the breathing becomes hurried, belly hangs low, the animal looks thin, and if you force your fist against the side of the belly, you feel the impulse of the returning water against it.

Cause.—Debility and organic disease of the liver or spleen.

Treatment.—The chance of success in treatment is not great, for, unless the cause can be removed, the only thing to be done is to tap the abdomen with a trocar and canula to let the fluid out, and if it again accumulates the case is hopeless.

Inflammation of the Fourth Stomach.

Symptoms. — Uneasiness, pawing of the ground, striking at the belly with the feet, showing where the pain is located; dung thin and offensive; pulse hard and quick; breathing accelerated; alternately hot and cold shivering fits.

Cause. — Unwholesome or poisonous food; change from a poor to a rich pasture; prolonged indigestion.

Treatment.—Feed on bran-mashes, but no green food; give linseed-oil, 1 pint; and every six hours, in gruel, tincture of opium, 2 oz.; Fleming's tincture of aconite, 12 drops; spirit of chloroform, 1 oz. The belly may be frequently rubbed with liniment (No. 26).

Gut Tie.

It is only seen in castrated animals, and generally terminates fatally.

Symptoms.—It is usually seen at the age of two or three, rarely before. The animal at first appears dull and loses its cud; after a time colicky pains appear, it strikes its belly with hind legs, goes stiff, breathing becomes hurried, the animal wears an anxious expression, no medicine seems to do any good, and in a few days it dies in great agony.

Cause.—The cord of the testicle encircling a portion of the small intestines and strangulating it.

Treatment.—There is only one thing to be done, and that is an operation by a veterinary surgeon, opening the abdomen in the right flank, and liberating the constricted gut.

Inflammation of the Kidneys.

Symptoms.—Straining to void urine, which is forcibly ejected in small quantities; loins tender and hot. After a time blood and pus may be mixed with the urine and the straining increases; muzzle becomes dry, horns cold, breathing quick. Diarrhœa follows, dung becomes fœtid; pain increases, total suppression of urine takes place, and the animal will die in about three days.

Cause.—Unwholesome food or a chill which has produced inflammation in this particular part.

Treatment.—Foment the loins with hot water, and rub in mustard mixed with water; give clyster of warm gruel with 2 oz. salt in it, adding tincture of opium, 1½ oz., if straining continues. Give at once, in gruel, purgative (No. 27), and three times a-day give, in gruel, a draught containing Fleming's tincture of aconite, 12 drops; solution acetate of ammonium, 3 oz.; and tincture of opium, 1 oz.

Inflammation of the Liver (*Yellows or Jaundice*).

Symptoms.—Yellowness of the eyes and skin; pulse quick; ears and horns hot; muzzle dry; shivering of the right side; stiffness, fulness of the belly; pain when the right side is pressed; urine and dung light brown in colour.

Cause. — Over-fattening; driving in hot weather; injury to the body near the liver, impeding circulation and inducing inflammation.

Treatment. — Give in warm water purgative (No. 27), and feed on bran-mashes. Morning and evening give in warm water chloride of ammonium, 4 drs.; bicarbonate of potassium, 1 oz.; ginger, 4 drs. Keep free from draughts. The animal should be sold when occasion offers; it is never likely to do well after the attack.

Flukes in Liver.

Cattle, like sheep, suffer from flukes in their livers, but not so severely, and it is rarely discovered until their death.

The reasons for cattle not suffering so severely as sheep are—firstly, they do not feed so close to the ground, and thus pick up fewer fluke-eggs; and, secondly, their livers are larger, and can stand the ravages of the fluke better.

Loss of Cud.

Symptoms.—Very often cattle do not chew their cud properly, and a great quantity of saliva dribbles from their mouth.

Cause.—Indigestion.

Treatment. — Change the food, and give a dose of linseed-oil, 1 pint, and, in the case of a calf, give oatmeal-porridge with bicarbonate of soda, 1 dr., night and morning. A little salt given with the food will help to remedy the evil.

Poisons.

The poisons that cattle principally suffer from are yew, rhododendron, arsenic, mercury, and lead.

Yew-poisoning

is perhaps most frequently met with.

Symptoms are those of a virulent poison, and is rapidly fatal, often shivering, cold extremities, staggering; the animal may fall and die rapidly in convulsions, usually in a few hours.

Cause.—By animals being allowed to graze in the vicinity of yew-trees, when they will often crop the tops of the growing twigs, or by the trimmings of these trees being thrown within their reach.

Rhododendron-poisoning.

Symptoms. — This poison is not so quick in its action; the animal staggers, becomes partially paralysed, colicky pains; animal lies and moans and frequently vomits, the vomit being greenish in colour.

Cause.—Same as yew.

Treatment.—The treatment of these two vegetable poisons is identical. Open the rumen and remove the poisonous stems and leaves, then give purge (No. 28) and half a pint of brandy in some water every three hours.

Arsenic-poisoning.

Symptoms.—Great prostration, shivering, colicky pains, diarrhoea, and death.

Cause.—In being given by accident, and by grazing on land where recently dipped sheep have been lying.

Treatment.—Give the following in a pint of water every hour: the hydrated peroxide of iron. Calcined magnesia is also a chemical antidote. The white of eggs given raw, and powdered charcoal, are also useful.

Mercury-poisoning.

Symptoms.—Flow of saliva from the mouth, breath foetid, gums red and tender, colicky pains, and appetite lost.

Cause.—By dressing cattle with mercurial preparations to cure mange, ringworm, and warbles.

Treatment. — Give purge (No. 28) with half a dozen eggs, and follow every two hours with iodide of potassium, 2 drs.; opium powder, 2 drs., in gruel. Sulphur and sulphate of iron are believed to be useful.

Lead-poisoning.

Symptoms. — Animal dull, abdomen tucked up, eyes staring, unsteady gait, bowels constipated, swelling under jaw, and emaciation. In acute cases blindness and delirium.

Cause. — Grazing near smelting furnaces or rifle-butts, and by eating lead-paint or sheet-lead.

Treatment. — Use the stomach-pump, afterwards give oils, flour-gruel, skimmed milk, and in a pint of cold water sulphuric acid dil., 3 drs. Give every three hours the following: iodide of potassium, 2 drs.; sulphuric acid dil., 3 drs.; 3 eggs; and half a pint of water.

Red Water.

When in an acute form, it is often called Black Water.

Symptoms. — The first thing that draws attention to the animal is usually the red colour of the urine, which froths when it falls to the ground; this is generally accompanied by diarrhœa or scouring, which soon gives way to constipation. The urine gets darker, the appetite fails, the animal gets weaker, and the heart can frequently be heard beating while standing behind the animal. Death often ensues within three or four days.

Cause. — It is not very well known. It is most common in milk cows, occurring generally from ten to fifteen days after calving, and is most common on moorland soils, and where there is a wet retentive subsoil. But in some districts when it is very common it attacks bulls, oxen, and heifers as well, and at all seasons. In this form there is now every reason to believe that it is due to a micro organism which is met with in the blood of affected animals, mostly in the red corpuscles. It was first described by Messrs Smith and Kilborne of the Bureau of Animal Industry of the United States as being the cause of Texas fever, a very fatal disease occurring in cattle in the Southern States. They called the organism the *Pyrosoma bigeminum*, and proved that the disease was not directly contagious as had previously been thought, but that it was communicated to animals by the bites of ticks.

Prevention. — Careful feeding after calving, a limited supply of turnips, some linseed - cake, and other foods allowed. Thorough draining and manuring of the land, the destruction of ticks, and the cutting down of all rank, coarse grass and ragweed which would give shelter to the ticks.

Treatment. — If observed before the appetite and rumination are diminished, give in gruel: Epsom salts, 16 oz., and ginger, ½ oz., but not otherwise; and morning and evening give tincture of perchloride of iron, 1½ oz.; spirit of chloroform, ½ oz., in gruel, and give milk, raw eggs, and stimulants if appetite lost.

Bleeding.

Cattle are bled from the following veins: jugular, the vein below the eye, and the milk-vein. The jugular is usually opened in cases of milk-fever, apoplexy, &c., and is easily got at on either side of the neck. First raise the vein by placing a cord tightly around the neck close to the shoulders, turn the neck a little to the opposite side, and a sharp blow will send the fleam through the skin into the vein. The fleam should be a size larger than that used for a horse. Afterwards, close the wound with a pin, and twist tow or a clean worsted thread around it. The vein below the eye is opened with a lancet in cases of inflammation of the eye, and the milk-vein in cases of inflammation of the udder. Two quarts of blood is a fair quantity to take from an animal.

DISEASES AFFECTING THE GENERATIVE ORGANS.

Abortion.

Symptoms. — When abortion takes place in the early stages of gestation, as it often does in the second month, the symptoms are very slight, and may be unnoticed, especially in the summer when the cattle are at grass. In the later stages of gestation the symptoms are easily recognised. There is restlessness and derangement of health, the udder becomes enlarged, accompanied by calving pains, and discharge from the vagina. But frequently the first symptom is the appearance of the calf.

Cause. — There is so much uncertainty connected with this disease, that it is sufficient to remark here that blows, injuries, exposure to cold, improper food, foul smells, and overdriving are the most immediate causes. But there are some forms of abortion that, once started in a herd of cows, are to all appearance communicable by contagion.

Prevention.—Careful attention, pure clean water, and the removal of any existing injurious influence will do much to make the occurrence of the disease rare. A goat allowed to run amongst the cows is said to be a good preventive. Have the cow isolated as soon as observed, before abortion if possible, and attend to the thorough cleansing and disinfection of everything with which the calf or the discharges could have come in contact.

Treatment. — Should any symptoms of abortion appear, give Epsom salts, 12 oz. ; Fleming's tincture of aconite, 10 drops; chloral hydrate, 1 oz., in a pint of warm water, and repeat the dose of aconite in a half-pint of water three times a-day if there is no improvement. Bury the foetus at once, and if it takes place in a field, remove any cattle in it to another pasture.

Calving, see *Parturition* (p. 481).

Cow-pox.

Symptoms.—Small vesicles followed by pustules on the teats, which, when numerous, may produce inflammation and affect the health.

Cause.—Constitutional, and contagion from other cows, carried by the milker's hand.

Treatment.—Give purgative (No. 28); keep the teats clean, and bathe them with goulard water, or chloride of lime, ½ oz., dissolved in half-gallon of water. The sores will soon heal.

Gonorrhoea or Bull-burnt.

This is a contagious disease of the genital organs, and is propagated through copulation.

Symptoms.—In the cow a glairy discharge is seen coming from the vulva a few days after being bulled; kicking and restlessness on urinating. In the bull this discharge is seen issuing from the penis.

Treatment. — Inject into the vagina twice a-day a little of the following, after syringing with lukewarm water : *liquor opii sedativus,* 1 oz. ; sulphate of zinc, ½ oz. ; water, 1 quart. In the case of the bull, it must be injected into the sheath. Give the animal purgative (No. 28) now and then to keep its bowels open. Sexual connection must not be permitted until all risk of contagion is gone.

Falling Down of the Calf-bed.

Symptoms.—After calving, the womb sometimes follows the calf, and hangs down like a large red bag.

Treatment. — Remove the cleansing carefully if it is still attached, clean the womb with lukewarm water and return it as soon as possible. Give a draught in warm gruel containing tincture of opium, 2 oz. ; chloral hydrate, 1 oz. ; spirit of chloroform, 1 oz. Raise the animal higher behind than in front. Afterwards place a truss on the animal to keep it in.

Flooding after Calving.

A flow of blood from the womb.

Cause.—Rupture of some of the vessels of the womb through using force in extracting a calf.

Treatment. — Keep the cow higher behind than in front; place ice or cold-water cloths across the loins ; give every three hours in a pint of cold water the following : tincture of perchloride of iron, ½ oz. ; tincture of opium, 1 oz. ; and tincture of ergot, 2 oz.

Garget,

or inflammation of part of the udder.

Symptoms.—This is a very serious disease, and usually affects one quarter of the udder, sometimes two, and if the inflammation is not reduced, the milk will become discoloured, or matter may collect in the udder instead of milk, the health become affected, and the cow may be lost.

Cause.—Careless milking; too hasty drying of the cow; injury to the udder; lying on cold wet land in the autumn.

Prevention. — Should there be any appearance of the disease, the calf should, if possible, be put to the mother, and it may, by its sucking and bumping, relieve her of the pressure of milk and disperse the hardness.

Treatment.—Should the disease become established, draw off gently all the contents at frequent intervals, and apply light poultices to the bag, containing belladonna, 3 drs. Give four times a-day, in gruel, nitre, 2 drs.; bicarbonate of potassium, 1 oz.; Fleming's tincture of aconite, 10 drops. Should ulcers form and break, they should be dressed with lotion containing carbolic acid, 1 part; water, 20 parts.

Overstocking or Hefting.

This is not a disease, but the consequence of the cruel practice of placing an elastic band around the teats, or plugging them up with grains of barley, and not milking the animal for twenty-four to thirty-six hours, with the result that the animal arrives in the market with a beautiful udder, and the owner tries to get more for the cow than she is worth. The results of overstocking produced in this way are intense suffering of the animal, inflammation of the udder, and a permanent interference with the secretion of milk, and it undoubtedly comes under the heading, cruelty to animals.

Hard Udder.

Cows' udders frequently become hard, especially with heifers after their first calf.

. Symptoms.—Swelling and inflammation.

Treatment.—Rub a little goose-grease on the udder after each milking, with a good deal of gentle rubbing, and if there is much tenderness give purgative (No. 28) in gruel.

Bloody Milk.

Symptoms.—Generally the first and only symptom is the presence of blood in the milk, and it is very often confined to one teat. The udder may neither be hard nor painful. This disease is especially prevalent among young cows after the first calf.

Cause.—Injuries to the udder; congestion of the gland structure and rupture of some small vessel; sudden change to a rich milk-producing diet; chills; too hasty drying of the cow; careless milking, &c.

Treatment.—Give Epsom salts, 1 lb.; nitre, 1 oz.; and ginger, 1 oz. Follow with tonic (No. 21). Milk the affected teat or teats into separate vessels.

Warts on Teats.

These little but troublesome things can easily be removed by winding green silk around them and allowing them to drop off; or by cutting them off with a pair of scissors, afterwards touching the parts with nitrate of silver. They should be attended to when the animal is dry.

Inflammation of the Womb.

Symptoms.—After calving, inflammation of the womb sometimes sets in, causing fever and loss of milk, and usually accompanied by a fœtid discharge from the uterus; but sometimes the discharge is suppressed. There is generally pain, fever, stiffness, straddling gait, and straining.

Cause.—Generally injury done during parturition, either from violence used in the assistance given, dirty hands or dirty instruments, or otherwise. High condition and improper rich food induce a tendency to this complaint.

Prevention.—A fortnight before calving, a cow's diet should be reduced to the simplest character. If the condition of the beast is very high, Epsom salts, 1 lb., and ½ oz. ginger in gruel; or a pint of linseed-oil, given a few days before calving, will do much to ensure safe recovery.

Treatment—If there is difficulty in passing urine, the aid of a veterinary surgeon should be at once obtained. In the first stages of the disease give a warm clyster containing tincture of opium, 2 oz. If constipation, give a bottle of linseed-oil with a gill of whisky, and if necessary follow with purgative (No. 27) in half-doses till the bowels are opened; then give every six hours in gruel, salicylate of sodium, 4 drs.; tincture of opium, 1½ oz.; solution acetate of ammonium, 4 oz. Feed on mashes, and be careful not to allow the animal to get a chill.

Leucorrhœa or the Whites.

Called so from the colour of the discharges.

Symptoms. — The cow is unthrifty, and a white discharge runs from the vagina, especially when she coughs or lies down.

Cause. — From injury to the womb,

usually after difficult calving or retention of the cleansing.

Treatment.—Give tonic (No. 29) night and morning in a pint of ale; inject into the vagina, by the aid of a syringe, the following twice a-day: sulpho-carbolas of zinc, ½ oz.; water, 1 quart; and feed the animal well.

Parturition or Calving.

The natural presentation of a calf is with the muzzle resting above the fore legs, with the back of the animal upwards. In cases of unnatural presentation, assistance will always be required. Every endeavour should be made to get the calf into a proper position. Experience and skill in extracting the calf are more needed than mere force. Every care must be taken not to wound the cow.

No description within the scope of the present treatise could give a proper idea of the methods used in all cases of unnatural presentation. The aid of an experienced surgeon must therefore always be obtained if the case is beyond the knowledge of the man in charge. Two or three hours after calving it may be prudent to give, in warm gruel, purgative (No. 28). Shortly after calving the cleansing or after-birth should come away. If retained twenty-four hours, with no appearance of coming away, it should be carefully removed before decomposition is too far advanced, as it very often sets up a septic or putrefactive inflammation; but if not removed, a draught in gruel containing Epsom salts, 8 oz.; powdered ergot, 1 oz.; carbonate of ammonia, 4 drs., should be given every day, unless diarrhœa supervene, until it appears. Should decomposition actually commence, the hand must be introduced, and the placenta removed as gently as possible. But it is wise to call in the veterinary surgeon.

Dropsy of the Womb.

An accumulation of fluid in the womb, and is often mistaken for pregnancy.

Symptoms.—The cow looks as though she were pregnant; but when her time is up—that is to say, if she has been to the bull—she shows no sign of calving, and if you place your hand up the rectum nothing but a huge water-bag can be felt.

Treatment consists of tapping the womb and allowing the fluid to escape, and should be left to the veterinary surgeon.

Milk-Fever (Dropping after Calving).

Symptoms. — After calving the cow will appear restless, muzzle hot and dry, udder tender and hot, constipation. Increasing weakness, ending in death, if the treatment is not successful. Sometimes cows have been known to suddenly drop down a few hours after calving without the herdsman previously knowing that anything was wrong.

Cause.—The origin of the disease is as yet not satisfactorily settled; there are many opinions, the enumeration of which would occupy too much space to be profitable for our present purpose.

Prevention. — A fortnight before calving keep the cow on a spare diet, composed in winter of bran-mashes and other opening food; a little linseed meal or cake will help to keep the bowels open. After calving, it has always been my practice to give a drink of thin gruel with 12 oz. Epsom salts in it; and should any signs of derangement appear, add Fleming's tincture of aconite, 10 drops; repeating the dose of aconite every six hours should signs of restlessness continue.

Treatment. — Try and not let the animal injure herself dashing about. When down and unconscious, keep her propped on to her breast with her legs under her in as natural a position as possible.

The treatment of this disease has been quite revolutionised since Schmidt of Kolding, Denmark, published his method of treatment by injecting the udder, in 1897. He used a solution of iodide of potass, ½ dr., in ½ pint of boiled water, into each quarter of the udder. But since that time many medicines have been used, and it is found that the injection of pure aseptic air is very satisfactory. The udder is distended to its fullest and massaged by the hand. Little other treatment is required, and the recoveries by this method, when the cases are taken in time, and the treatment carefully and satisfactorily applied, are about 90 per cent. But great care is required to have everything aseptic, as it is very easy setting up inflammation of the udder, and the cow may recover from milk-fever to die of mammitis. It is therefore advisable to obtain the services of a veterinary surgeon when possible.

Sore Teats.

Symptoms.—After calving, cows are liable to have sores or small cracks or chaps on the teats, making them very tender and painful.

Treatment.—Apply boracic acid ointment or lotion to the teats, having previously bathed them with warm water to remove all scabbiness and dirt. Dry dressings sometimes do better, as oxide of zinc and powdered starch.

Suppression of Urine.

Cows in calf are very subject to this complaint, in consequence of the pressure caused by the calf. It is also a symptom of several other diseases.

Treatment.—Keep the bowels open, by giving in warm gruel purgative (No. 28), assisted by a clyster of warm gruel, and give till relieved, morning and evening, in gruel, tincture of perchloride of iron, 1 oz.; spirit of chloroform, ½ oz.

DISEASES AFFECTING THE LIMBS, FEET, AND SKIN.

Foot-and-mouth Disease.

(See p. 470.)

Foul or Fouls in the Feet.

Symptoms.—Cattle are very liable to this disease of the foot, which produces great lameness. There is a good deal of foetid discharge from the cleft of the foot, also swelling of the pastern.

Cause.—Driving over rough roads or for long distances; injury from a prick, nail, or splinter; standing on moist and dirty bedding.

Treatment. — Put the beast into a dry, clean place. Dress down the diseased hoof with a knife, and wash with hot water and soda. If there is pain and fever, and the lameness excessive, poultice for some days to reduce the inflammation. Then dress the foot with a mixture of tar and powdered sulphate of copper. Should much swelling of the pastern

with some lameness remain, apply blister (No. 23). If there is a wound in the foot caused by a splinter, remove the splinter, apply a hot poultice of linseed-meal, and bind up the foot.

Enlarged Knees.

Cows in byres frequently suffer from an enlargement on the front of the knee through lying on the stony floor.

Treatment.—If there is much pain and swelling, apply warm fomentations assiduously, and cooling lotion (No. 7). See that the knees are protected from the hard floor and from the manger.

Lice.

Symptoms.—Cattle in poor condition often lose their hair, especially on the neck and back, owing to their being infested with lice.

Cause.—Want of cleanliness and poor condition.

Treatment.—Wash the part affected with lotion made from tobacco, 4 drs., dissolved in 1 pint of hot water. Ointment made of lard, 6 oz., white precipitate of mercury, ½ oz., is a certain remedy, but requires careful handling. Improve the quality of the food and keep the animal clean; give tonic (No. 29).

Mange.

Symptoms. — Itching, loss of hair, scurf, scab, or sores, especially on the back.

Cause.—It is caused by a small insect (an *Acarus*); it is favoured by dirt, poverty of the blood, and neglect, and is very contagious when animals get into contact.

Treatment.—Improve the food given, and keep the animal clean. Rub the places affected with ointment composed of sulphur, 1 oz.; lard, 4 oz.; give in gruel mild purgative (No. 28), adding sulphur, 1 oz. If this does not effect a cure, wash the places with corrosive sublimate, ½ oz.; muriatic acid, 1 oz.; soft water, 2 quarts; or by the treatment recommended for lice.

Rheumatism, Lumbago.

Symptoms.—Swelling of the joints; stiffness; listlessness; unwillingness to move, which the beast does with pain.

Cause.—Cold, especially after calving or when weakened by illness. .

Treatment.—Give a draught in gruel or warm water, morning and evening, containing carbonate of ammonia, ½ oz.; bicarbonate of potassium, 1 oz.; gentian, 1 oz.; ginger, 1 oz. Rub the parts affected with liniment of belladonna, 1 part, compound liniment of ammonia, 1 part. Give nourishing food and a little linseed meal or cake, and keep free from chills till quite cured.

Ringworm

is caused by a parasitic fungus growing in the skin.

Symptoms.—Loss of the hair, which comes off in circular patches, leaving a dry and scaly eruption. The face, head, neck, back, and root of the tail are the parts most generally affected.

Cause.—Contagion, neglect, and dirt.

Treatment.—If the animal is in high condition or out of health, give purgative (No. 28), and keep on nourishing food. Rub the parts affected with ointment composed of oleate of mercury, 1 part; lard, 2 parts, or use a lotion of perchloride mercury, 1 part; water, 500 parts. I have used lotions composed of sulphurous acid, but have found the mercurial ointment the most efficacious; sometimes a second application is not even necessary. Wash the cribs, rubbing-posts, &c., which have been used by a beast affected with ringworm with 1 pound chloride of lime dissolved in 2 gallons of water, to avoid spreading the disease among other stock.

Warbles.

Symptoms.—Early in the year and in the spring, from January till May, large lumps about the size of half-a-crown may often be found along the backs of cattle; these increase in size till the contents, the maggots of the bot-fly (*Œstrus bovis*), escape. There is always a small air-hole to be seen in the middle of the lump, and the head of the maggot is often visible.

Cause.—The bot-fly pierces the skin of the beast while out at grass during the hot weather in July and August, and leaves an egg at the bottom of the wound, which hatches, develops, and grows, till at length it emerges a large maggot about May or June in the following year.

Prevention.—None, except the extermination of the flies by diligent destruction of the maggots.

Treatment. — In April or May all cattle should be examined, and the maggots squeezed out between the fingers, which may easily be done, and in my opinion is the best way of ensuring their destruction. If the holes are smeared with M'Dougall's cattle-smear, the maggots are no doubt killed, but they remain in the ulcer, and certainly the most healthy way is simply to crush them out and relieve the beast of them at once.

The damage done by this fly to cattle and hides may be estimated at millions of pounds; every means should therefore be taken to remove this pest from the country.

Wounds.

In severe cases, unless the animal has a fancy value, it would be better to slaughter at once. Simple cases may be cured by bringing the edges of the skin together, and fastening them with carbolised gut and a bandage of carbolised gauze. The bowels must be kept open by doses of purgative (No. 28), and the wound kept clean by bathing with warm water if necessary.

SPECIAL DISEASES AFFECTING CALVES.

Constipation.

Cause.—Frequently the first milk or biesting from the cow has not been given to the calf, and constipation ensues. Also, when milk has been taken to excess it is apt to produce constipation.

Prevention. — In a young calf the natural first milk of the mother is most suitable, and afterwards care should be taken that only as much milk is given as the digestive organs can dispose of.

Treatment. — Give castor-oil, 1 oz., beaten up in the yolk of an egg, with ginger, 1 scr.; repeating the dose if necessary. Clysters may be required, but not so often as in young foals.

Diarrhœa.

Cause.—Injudicious feeding, and at too long intervals; bad smells, · cold, acidity in the stomach, produced by any sudden change of food, or by anything which deranges digestion.

Prevention.—Care should be taken that the milk given to calves should be sweet, and that the air is kept pure.

Treatment.—If the calf refuses its food, and blood is mingled with the dung, accompanied by great pain and weakness, immediate steps must be taken to remove the irritating matter. If the diarrhœa is recent give castor-oil, 2 oz., to relieve the intestines, and after this has had time to act, give morning and evening 1 oz. of calf-cordial (No. 24). If this does not produce any effect, give four times daily, tincture of catechu, 2 drs.; spirit of chloroform, 30 drops; and dilute sulphuric acid, 30 drops, in thin gruel.

Diarrhœa, indigestion, and death are also caused by *hair balls*, which form in the stomach. There is no preventive. Caused by the calves licking each other.

Navel-ill.

Calves sometimes suck one another's navels, which causes swelling and inflammation of it; or it may be caused by the cord breaking off too short, by neglect, exposure to cold, wet, and dirt.

Treatment.—Poultice, if no tendency to bleeding, or apply hot fomentations persistently, and carbolic oil to the raw surface. A very fatal disease.

White Skit.

Whitish diarrhœa seen in calves that live on a milk diet.

Treatment.—Give castor-oil, 1 to 2 oz., according to the size of the animal, and follow up with calf-cordial (No. 24). Keep the animal for a few days on flour or oatmeal gruel.

SHEEP.

DISEASES AFFECTING THE HEAD, EYES, MOUTH, AND NERVOUS SYSTEM.

Apoplexy.

This disease cannot be treated or guarded against; it attacks the fattest sheep on the richest pastures, especially in the spring of the year. The animal seized will drop down suddenly, and in extreme cases die at once. Any animal affected should be killed immediately.

Louping-ill or Trembles.

A disease manifesting nervous symptoms.

Symptoms.—The animal trembles, breathes in a jerky manner, moves its legs in an automatic style, occasional spasms of the muscles of the neck, which usually terminate in paralysis.

Cause.—It is seen only in certain districts, and is supposed to be due to a peculiar formation of the soil or the condition of the grass. The Committee mentioned in connection with braxy, p. 486, state that the cause is a large, feebly motile bacillus, with a great tendency to form spores; that it is a regular inhabitant of the alimentary canal—the intestines; and that it is due to some change in the blood resulting in a diminution of the resisting power of the animal that the bacilli are able to pass the walls of the intestines, invade the tissues, and set up the train of symptoms—usually ending in death—known as louping-ill.

Prevention.—Dip the sheep, remove them to fresh pasture, and give them corn and salt; to every pound of salt add one ounce of sulphate of iron.

Treatment.—Of little use; look to prevention. As a preventive the Committee recommend drenching with cultures of the organism as for braxy.

Blindness.

Sheep are sometimes attacked by temporary blindness, often lasting only about ten days.

Cause.—Changes of temperature; the reflection of the sun on snow; dusty roads on a long journey; and confinement in badly ventilated ships' holds.

Prevention.—Remove the cause.

Treatment.—If left to nature, the blindness will probably pass away. Lambs thus affected require extra care, and should be put to the ewes so that they should not suffer from loss of milk.

Lockjaw or Tetanus.

Symptoms.—Practically the same as those in the case of the horse (p. 440).

Cause.—Cold, especially during lambing-time; also produced by careless castration, wounds, &c.

Prevention.—Shelter and careful attention will do much to avert this disease. (See pp. 440, 441.)

Treatment.—Give castor-oil, 2 oz., repeating the dose every six hours till it takes effect. Give, in gruel, tincture of opium, 1 dr., morning and evening. Fleming's tincture of aconite, 5 drops, may be added to the gruel if there is no relief.

Staggers, Sturdy, Goggles, Fern-sick, Dunt, or Turn-sick.

Symptoms.—Dulness; unsteady walk, generally in a circle; separation from the rest of the flock; blindness. The animal affected will often fall into a ditch and perish, or die gradually. This disease generally attacks young sheep in good condition.

Cause. — A species of parasite—the *Cœnurus cerebralis*—in cysts or bladders containing fluid, which lodge in the brain.

Prevention.—Young sheep in damp situations are very liable to this disease, and care should be taken to avoid putting them into such pastures. The use of lump or rock salt, which they can lick as often as they like, helps to ward off this and similar diseases. Keep your sheep-dog free from tape-worms, for it is the egg of this worm that gets into the brain.

Treatment.—Slaughter is the most profitable course to follow. There is a method of treating this disease by puncturing the soft place in the skull, and removing the bag or cyst; but unless this is done in good time, and performed skilfully, it is rarely successful.

Water on the Brain.

This disease often affects very young lambs.

Symptoms.—Dulness and stupidity; staggering gait; rapid loss of flesh. Death may ensue in about a month.

Cause.—Often congenital, commencing before birth; constitutional weakness.

Treatment.—No cure, so far as the farmer is concerned.

DISEASES AFFECTING THE THROAT, CHEST, RESPIRATORY ORGANS, AND BLOOD.

Catarrh.

Symptoms.—Dulness; loss of appetite; difficulty in swallowing; water running from the eyes and nose; hot mouth and muzzle; constipation; cough, and discharge of yellow mucus from the nose.

Cause.—Chills and exposure.

Treatment. — Give daily, in linseed gruel, nitre, 1 dr.; digitalis, 1 scr., and keep the animal warm.

Hoose or Husk.

Symptoms.—A husky cough, which increases if the animal is hurried. Gradual loss of condition, till the health is undermined. In time the diseased lungs will no longer be able to purify the blood, and the animal will die.

Cause.—The presence of parasites—the *Strongylus filaria*, and sometimes the *Strongylus rufescens* also—in the bronchi and lungs.

Prevention. — Lambs should never be depastured on land fed previously the same year with sheep. If this advice is acted on, and proper care taken of the flock, cases of hoose will be less frequent on most farms.

Treatment.—Give daily: turpentine, 1 dr., for a lamb; 3 drs. for a sheep, in respectively 6 and 12 drs. of sweet-oil, and feed liberally, giving some good linseed-cake with the food. Veterinary surgeons now use injections of parasiticides into the windpipe.

Pneumonia—Inflammation of the Lungs.

Symptoms.—Hard breathing; loss of appetite; fever; cough, which becomes more and more distressing; discharge from the nose; thirst. Afterwards intense weakness sets in, too often followed by unconsciousness and death.

Cause.—Cold, particularly from shearing in cold weather.

Treatment. — Give in linseed-tea, Fleming's tincture of aconite, 5 drops. three times a-day, and with it once a-day tartar of antimony, ½ dr. Keep warm, and in cases of complete prostration,

give, as a stimulant, in the gruel, gin, 1
or 2 oz.

Sheep-pox or Variola ovina.

A contagious and infectious eruptive
fever only affecting sheep.

It is scheduled under the Diseases of
Animals Acts, but it has not been seen
in Britain since 1862.

Cause.—Contagion and infection. In
all probability due to an ultra-microscopic
organism.

Symptoms.—High fever, loss of appetite, and depression, distinct evidence
of serious illness. An eruption of
reddish spots appears about the mouth,
nose, and eyes, inside the arms and
thighs, and about the udder or scrotum.
The spots go on to form vesicles and
pustules; these usually burst and discharge a yellowish matter, which mats
and agglutinates the wool. The animal
acquires a sickly disagreeable odour, and
becomes a loathsome-like object.

Pregnant ewes often abort.

Prevention. — Sheep should not be
admitted into Britain from countries
where sheep-pox exists.

DISEASES AFFECTING THE STOMACH,
LIVER, BOWELS, KIDNEYS, AND
INTERNAL ORGANS, PARTURI-
TION AND MILKING ORGANS.

I.—STOMACH, LIVER, BOWELS, KIDNEYS,
AND INTERNAL ORGANS.

Braxy or Sickness.

Symptoms.—Restlessness; hanging of
the head; aching of the back; grinding
of the teeth; cold extremities; kicking
of the belly with the hind feet; distension of the abdomen; separating from
the rest of the flock.

Cause.—It is due to an anærobic motile bacillus, very similar to the bacillus of
black-quarter. Prevalent in the autumn,
especially among the lambs; often runs
through a flock like an epizootic.

Prevention.—If possible, keep the
sound pastures for the lambs, and avoid
letting them have too succulent pasture
for grazing, and always let rock-salt be
within reach. Do not allow an affected
animal to be bled or slaughtered on
ground that the rest of the flock have

access to, as blood diseases, as well as
inflammation of the bowels, enteritis,
and acute indigestion, are sometimes all
included under the general name of
braxy.

Treatment.—Give purgative (No. 34).
Fleming's tincture of aconite, 5 drops,
may be given in gruel every morning,
and the food should be sparing, with a
little linseed. The treatment of the disease is generally unsatisfactory. Change
their pasture, and if on good arable grass,
put them on the heather, if possible, for
a day or two. A Departmental Committee, of which Professor Hamilton of
Aberdeen was chairman, was appointed
by the Board of Agriculture, in December 1901, to investigate braxy and
louping-ill. This Committee, in 1906,
recommended as a preventive, a drench
prepared from cultures of the organism,
the bacillus, on glucose beef-tea. A small
quantity of this culture to be mixed with
water and given by the mouth, the dose
to be repeated in from 8 to 14 days.

Calculus or Gall-stones, and Kindred Diseases of the Bladder.

Symptoms. — Dulness; loss of appetite; separation from the flock, generally
lying down; quick breathing; when
roused, painful efforts to void urine, only
a few drops of which come away. Saline
deposits will sometimes be found in the
sheath.

Cause.—High and stimulating system
of feeding, especially on saccharine roots,
such as mangel-wurzel; want of exercise;
absence of water for drinking.

Prevention. — Avoid an exclusively
saccharine and starchy diet, and allow
free access to water.

Treatment.—In the latter case warm
fomentations and syringing with tepid
water may get rid of the deposit; afterwards wash out the sheath with an
astringent lotion. But if the seat of the
malady cannot be reached, the sheep
should be at once slaughtered when the
disease is suspected.

Constipation.

Symptoms.—Dulness and costiveness.

Cause.—Especially prevalent among
young lambs, caused by the quality or
too great quantity of the ewes' milk,
which has coagulated in the stomach.

Prevention.—Avoid any irregularity in the ewes' food, and especially too luxuriant pasture.

Treatment.—Put the ewes on shorter pasture, and give every morning as much warm water, with Epsom salts, 1 oz., dissolved in it, as the lamb can take.

Lambs also suffer from *wool balls*, which form in the stomach. There is no cure, but care should be taken that the bags of ewes should be kept as clear of wool as possible, in order to prevent lambs getting it into their stomachs.

Diarrhœa.

Symptoms.—Simple looseness of the bowels without much pain.

Cause. — Fresh, succulent herbage, especially when it has been touched with frost.

Treatment.—Change the pasture or source of irritation, and if weakness comes on, give rhubarb, 1 dr., and afterwards, in warm gruel, powdered opium, 20 grs. ; catechu, ½ dr. ; prepared chalk, ¼ oz.

Dysentery.

Symptoms.—Frequent evacuations of hard lumps of fœtid dung, mixed with slime and blood ; loss of appetite ; pain, fever, and great weakness. Affects sheep of any age, generally in the summer.

Cause.—Aggravated and unchecked diarrhœa ; chills after being clipped. Many believe it to be contagious.

Prevention.—Sheep that are scouring should be watched, the food altered, and any aggravation of the attack checked.

Treatment. — Give in warm water, three times a-day, ipecacuanha, ½ dr. ; powdered opium, 20 grs.; chlorodyne, 10 drops ; chalk, ¼ oz. Gruel, or if animal will eat, flour-porridge, sprinkled with salt, should be given.

Hoove, Hove, or Hoven.

A distension of the stomach by gas, owing to the fermentation of food which has been too long retained in it.

Symptoms. — Enlargement of the belly, especially on the left side, which sounds hollow when tapped. Stupor and death follow, unless the animal is relieved.

Cause.—When sheep are incautiously fed on green clover or turnips, they are apt to eat to excess, and fermentation of food in the stomach sets in before the organs are able to dispose of the accumulation of food.

Prevention.—Green clover and turnips should be given in small quantities at first, and sheep should only be turned into a very succulent pasture for an hour or two till they get accustomed to it, when there will be no danger of hoove.

Treatment.—The insertion of a trocar into the flank will relieve the pressure of gas, and a dose of purgative (No. 33) will open the bowels. A drachm of chloride of lime dissolved in a quarter of a pint of water, and horned into the sheep, will often reduce the pressure of gas. The subsequent food should be rather scanty.

Inflammation of the Liver.

Symptoms.—Loss of appetite ; skin hard and itchy; tongue foul; dung white and fœtid ; weakness.

Cause.—Sudden change from poor to nutritious food is generally the cause.

Prevention.—Extra care when any change of food takes place.

Treatment.—Give daily, in gruel, purgative (No. 32). Foment the body over the liver with hot water, and inject warm water if the bowels are costive. In chronic cases of this disease, salt (4 oz. per head per week) should be given in addition to the purgative above mentioned.

Inflammation of the Stomach.

Symptoms. — Loss of appetite and separation from the flock ; alternate hot and cold shivering fits; restlessness ; straining to empty the bladder.

Cause.—Too nutritious food, or irritating or poisonous plants.

Prevention. — Avoid exposing the lambs to chills and cold east winds after being cut, and exercise care when forcing on sheep for the market.

Treatment.—Foment the belly ; give Fleming's tincture of aconite, 5 drops, twice a-day, and purgative (No. 34), halving the dose in the case of young lambs. Keep warm, and let the food consist of warm gruel with a little boiled linseed in it.

Jaundice.

Symptoms.—Yellowness of the skin and eye; constipation; urine brown; loss of appetite.

Cause. — Richness of the pasture, especially in damp sultry weather.

Prevention.—The use of salt among the artificial food, combined with care as to the nature of the pasture, will make the appearance of this disease rare.

Treatment.—Give purgative (No. 32) and frequent doses of salt, 4 oz. per head per week.

Rot.

Symptoms.—At first the animal will lay on fat very quickly, but afterwards the wool begins to fall off; the eyes become hollow, the belly enlarged; swellings appear on the body; scouring sets in, and finally death ensues.

Cause.—The presence of the *Distoma hepaticum*, or flukes in the liver, especially prevalent in wet seasons. Sheep fed on low-lying, wet, and undrained land are very subject to this disease.

Prevention.—The use of salt in the food and judicious grazing during wet seasons will do much as preventives.

Treatment. — Remove the flock at once to dry uplands or salt-marshes if available, and give salt in the food, 4 oz. per head per week, and as much in the troughs as the sheep will lick up.

II.—PARTURITION AND MILKING ORGANS.

Garget or Inflammation of the Udder.

Symptoms. — Pain when the lambs are sucking, lameness, restlessness, fever, inflammation, and swelling of the udder.

Cause.—Stoppage of the secretion of milk; exposure to cold; injuries.

Prevention.—Care should be taken to prevent the udder being surcharged with milk, especially when the lambs are weaned or dead.

Treatment.—Foment the udder with warm water, and clear it from all wool. If not very painful, let the lamb suck it and knock it about as much as possible. Give purgative (No. 33), and remove all the milk by frequent milkings; rub the udder with belladonna liniment.

Parturient Fever—"Heaving," "Straining."

Symptoms.—Fever, loss of milk, listlessness, frothy saliva, stiffness of the hind quarters, discharge of dark fœtid fluid from the vagina, swelling of the vulva, straining, and pain. The whole constitution will be affected, diarrhœa sets in, followed by death. This disease is nearly always fatal.

Cause.—Probably the result of blood-poisoning, owing to deleterious matter entering the system through wounds of the parturient organs.

Prevention.—Should any wounds be made during parturition, they should be washed with warm water and syringed with lotion (No. 30) daily for some days, and afterwards anointed with glycerine, 8 parts, carbolic acid, 1 part. Great cleanliness should be observed in the lambing-yard, and a free use made of carbolic acid, and the hands washed and nails cut before manipulating the womb.

Treatment.—Besides the injection of carbolic lotion, give every four hours carbolic acid, 20 drops, in water; and if constipated, purgative (No. 33); give plenty of gruel and linseed-tea. If straining and diarrhœa come on, give whisky, 4 oz., and tincture of opium, 1 oz., in gruel. The treatment is, however, generally unsatisfactory.

Abortion,

or the premature expulsion of the fœtus.

Symptoms.—There are no particular symptoms in abortion among ewes.

Cause. — Overdriving; cold; improper food; injuries to, or disease of, the abdomen. A very frequent cause is the disturbance of sheep by dogs.

Prevention and Treatment.—When cases of abortion occur, the cause of the disease must, if possible, be ascertained and removed. It may generally be traced to one of the causes mentioned above. When one sees a flock of ewes occupying a turnip-fold, only vacated by the fat hoggs when it was too bad for them, up to their bellies in mud, one can hardly be surprised that cases of abortion are only too frequent. Given a proper rational system for the management of a breeding flock, and cases of abortion will be rare, and only the result of circumstances which cannot be altogether avoided. Should a ewe appear sickly after abortion, inject lotion composed of carbolic acid, 1 part; water, 40 parts; and give purgative (No. 33).

DISEASES AFFECTING THE LIMBS, FEET, AND SKIN.

Foot-rot.

Symptoms.—Lameness, which may be traced to disease between the claws of the feet, indicated by the discharge of matter and swelling, which, if not checked, will gradually extend to the whole of the foot. Or the horn of the hoof may be broken or fissured, and often a fœtid discharge. In old cases the horn is rough, enlarged, and deformed, and fungoid granulations may project from any openings in the horn, and from sores about the coronet.

Cause.—Often contagion, grazing on low, rich pastures, encouraging overgrowth of the hoofs, which are apt to split and crack and collect dirt.

Prevention.—The maxim, "A stitch in time," &c., if put into practice, will prevent the disease spreading, and will soon cure those affected. As a means of prevention it is useful to pass the whole flock twice a-year through a solution of arsenic, put into a trough, through which the sheep are driven slowly. The solution is thus prepared: Boil 2 lb. of arsenic with 2 lb. of potash (pearl-ash) in 1 gallon of water over a slow fire for half an hour; keep stirring, and when like to boil over, pour in a little cold water; then add 5 gallons of cold water. Put this solution into the trough to the depth of 1¼ inch. The solution is poisonous, so the trough should be kept locked when not in use. A bath of copper sulphate is perhaps as useful and less dangerous. Dissolve 1 lb. of sulphate of copper in 2 gallons of water, and walk the sheep slowly through it as often as required.

Treatment. — Pare away all loose ragged horn, to allow the matter to discharge, cut away any proud flesh with sharp scissors, and have recourse to a stronger bath of copper sulphate— 1 lb. to 1 gallon of water; and in bad cases it may have to be used every four to seven days. A narrow trough about 7 inches wide, sloping a little outwards, with rails on each side 16 to 18 inches wide, or sufficiently wide to admit the bodies of the sheep, and from 15 to 20 feet long.

As copper is poisonous, although not nearly so bad as arsenic, the sheep should be put in a bare court, or on to a hard road until the feet become dry.

Note.—Sheep, when they have travelled far on hard stony roads, get very footsore, and, if possible, should be put on soft cool pasture for a few days, when the feet will soon recover.

Rickets—Weak Backs.

Lameness of hind quarters, resulting from weakness of bones, which, from their constitution, are liable to injury under trifling exertion. This disease only affects the lambs, and when once it appears may run right through the flock. The outbreak may occur at any time while the lambs are young.

Symptoms.—Difficulty in rising.; the fore feet are not affected.

Cause.—Peculiar condition of the soil, pointing to an insufficiency of particular elements necessary to produce a perfect offspring. Food grown on light moor tillage land, dressed with caustic lime, is believed by some to produce rickets.

Prevention.—Avoid in-and-in breeding, and also food grown on land which, as above described, is apt to produce the affection.

Treatment.—Direct treatment is useless.

Sheep-Scab, Shab, or Mange.

An eruption of the skin, produced by parasites—the *Dematodectes ovis*—minute acari which burrow in the skin.

Symptoms.—Constant rubbing against gates, &c.; loss of wool; skin red, rough, and afterwards covered with hard scabs; loss of health and condition.

Cause.—Contagion.

This affection is scheduled under the Diseases of Animals Acts, and the treatment is prescribed by the Board of Agriculture. Affected sheep must be dipped in a "dip" approved by the Board, and the owner cannot be compelled to dip them a second time under 10 days.

At the present time (1909) the Board demands that within certain districts of country, which it terms "compulsory dipping areas," all sheep shall be dipped, in an approved dip, twice a-year, within certain specified dates.

Vermin.

Sheep ticks and lice may be destroyed by the use of one of the many dips which are sold for the purpose.

Maggots may be destroyed by applying spirit of tar, 1 part, olive-oil, 4 parts, to the places affected. It will also keep off the flies.

SWINE.

Anthrax.—(See p. 468.)

Symptoms. — Dulness; urine, and sometimes the dung, mixed with blood, external swellings. When caused by eating the flesh or blood of animals dying of the disease, there is nearly always great swelling about the throat.

Cause.—Generally contagion.

Prevention.—Keep in good condition, and avoid close buildings, putrid food, and bad water. Completely isolate all diseased animals.

Treatment.—Owing to the rapid and fatal nature of this disease, all treatment is unsatisfactory. But it is not so fatal as in horses and cattle, although young pigs are more easily affected by it than grown ones. Give Epsom salts, 3 oz., and oil of turpentine, 2 drs., in a little linseed-gruel, and rub the limbs with oil of turpentine. Call in a veterinary surgeon, who will inject diluted carbolic acid under the skin.

Convulsions.

Symptoms.—Young pigs are subject to convulsions, which take the form of sudden spasms with complete insensibility, frothing from the mouth, and redness of the eyeballs.

Cause.—Disorders of the brain; indigestion; sometimes intestinal worms.

Prevention.—Good water and nourishing food.

Treatment. — Give a purgative — Epsom salts — regulating the dose according to size. Remove as far as pos-

sible the cause of the attack—*i.e.*, expel the worms if they exist (see p. 492), or alter the food if it has produced indigestion. Give sulphate of iron, 1 dr., in the food.

Diarrhœa.

Symptoms.—Looseness of the bowels, which affects the health. If unaccompanied by loss of appetite, fever, or prostration, no treatment is required beyond removing the cause.

Cause.—Often a symptom of some other disease. Often caused by indigestion, putrid food or water.

Prevention.—Proper food and attention.

Treatment. — Give castor-oil, 3 oz., and peppermint-oil, 5 drops, in gruel.

Erysipelas.

Symptoms.—Heat; itching; redness; tenderness and swelling of the skin, generally on the head and neck; loss of appetite. When the swelling goes down at the end of a few days, a dark-red patch will be left.

Cause. — Want of ventilation; dirt; heating food; wounds.

Treatment.—Give, in gruel, jalap, 1 dr.; sulphate of magnesia, 3 oz., mixed in a little water, and as soon as the purgative has acted, give muriate of iron, 10 drops, night and morning, in food. Foment the swellings on the skin with water, 1 quart, in which 1 oz. of sulphate of zinc has been previously dissolved. Good nourishing food should be given, and the animal should be completely isolated.

Hoose or Husk (see p. 471).

Symptoms.—Short, dry cough; frothy discharge from the nose; loss of appetite; thirst; loss of flesh, till finally death results.

Cause. — The presence of worms in the air-passages, amounting almost to an epizootic in certain districts.

Prevention.—Keep in good condition, and isolate from infected animals; pure water, and clean or boiled food.

Treatment. — Give, in milk, salt, 1 teaspoonful (1 dr.); oil of turpentine, 1 teaspoonful, daily. Fumigate in a close building by burning flowers of sulphur on a hot shovel till the animals can bear no more without coughing violently, and repeat the fumigation every week. Give linseed - porridge, nourishing food, and plenty of skim - milk.

Pneumonia—Inflammation of the Lungs.

Symptoms. — Shivering; hot skin; laboured breathing; red eyes, nose, and mouth; cough deep and dry; yellow discharge from the nose.

Cause.—Cold, aggravated by neglect and predisposition to the disease; may result from hoose.

Prevention. — Nourishing food, and warm dry bedding.

Treatment.—Cover with a warm rug, but allow plenty of cool fresh air; rub the chest with mustard, 1 part, and turpentine, 4 parts, and repeat the rubbing. Give spirits of nitrous ether, 2 drs.; tincture of opium, 2 drs., in a little milk twice a-day, and put 2 drs. of nitrate of potash in its food every time the animal is fed.

Measles.

Symptoms. — Fever; cough; loss of appetite; red patches on the skin; pustules under the tongue.

Treatment. — Give, fasting, 1 oz. of sulphur in the food, and repeat the dose till the animal is quite recovered. Keep the animals warm.

Prolapse of the Rectum.

After parturition, and even in young pigs of both sexes, the rectum sometimes protrudes and swells.

Treatment. — The gut should be emptied and washed. It may be returned by inserting the finger into the opening and pressing it into the anus. A nourishing diet should be given. Sometimes a truss will be necessary to keep the gut in its place after it has been returned. It may be necessary to amputate the protruding part. Advice should be obtained if this is found to be necessary.

Rheumatism, Cramp, Lameness.

Symptoms. — Dulness; lameness, especially of the hind quarters; tenderness of the joints; constipation.

Cause. — Damp bedding; lying on cold brick floors; chills. Especially prevalent where pigs lie on fermenting horse-manure.

Prevention. — Allow young pigs plenty of exercise and dry bedding in which they may bury themselves.

Treatment.—Give Epsom salts, 1 oz., and sulphur, 1 oz., in thin gruel, salicylate of soda, 1 dr., and bicarbonate of soda, 2 to 4 drs., twice a-day. Keep warm, and feed on nutritious food of good quality.

Surfeit.

Symptoms. — Fever; swellings in patches on the lips, eyelids, and nostrils, which quickly appear and disappear.

Cause.—Change of food or weather. Generally occurs in the autumn.

Treatment.—Give in gruel, jalap, 1 dr.; Epsom salts, 1 oz., and avoid improper feeding.

Swine-fever or Hog-cholera.

A contagious and infectious disease of pigs which has been known in Britain and America since about the middle of last century, and has often caused very severe losses. Notwithstanding, it was not dealt with in the Contagious Diseases (Animals) Act of 1878, but was included by "The Animals Order," dated 15th December 1879, as was also glanders and farcy. And although the Privy Council and, since its institution, the Board of Agriculture have passed innumerable Orders and struggled at the suppression of the disease, it is still only too prevalent in the country.

Cause.—Two organisms seem to be always present, probably associated, in swine-fever—a small ovoid bacillus, the *Bacillus choleræ suis*, and an invisible infective organism.

Symptoms. — It is sometimes very acute, especially in young pigs. They become suddenly ill, there is high fever, quickened breathing, a rash appears over the thinnest parts of the skin—on the back of the ears and about the belly— they stagger about, and often die in convulsions. Generally it is less rapid, the animal is dull, does not care to come out of its pen, but lies buried under its litter. There is loss of appetite, constipation at first, generally followed by diarrhœa, fever, the eyes look red and watery, the tail hangs limp, red blotches appear about the back of the ears, inside the arms and thighs, and about the belly. These gradually become darker until they may be dark-purple or almost black. The lungs are frequently affected, causing rapid breathing and a short painful cough. Animals often die after two or three days' illness, but they may die after a fortnight, and one will sometimes recover after being very ill, but takes a long time to make a complete recovery. It is doubtful also how long an animal that has been very ill with swine-fever may prove a centre of infection—several months at least.

Treatment.—As it is the digestive tract that is principally affected, only the blandest and least irritating foods should be allowed — skimmed milk, alone or with lime-water, fine well-boiled gruels, beef-tea, which if well boiled can be made from meat which would otherwise be destroyed, and anything which will support the strength without causing irritation. Some also recommend mild antiseptics given internally.

Prevention.—A protective serum has been introduced and can be applied by any veterinary surgeon; but few would care to use it unless the disease has broken out in their immediate neighbourhood or in a large valuable breeding herd.

Isolation and Police Measures.—Any person having a diseased or suspected pig in his possession must at once give notice to the police, and they telegraph the information to the Board of Agriculture, who then deal with the case. An "infected place" is declared, and all movement of pigs out of or into it is stopped except with the licence of the Board. The movement of pigs on any premises in the vicinity is also generally stopped as long as it is thought there is any risk.

The policy adopted in dealing with swine-fever has undergone many changes during the past thirty years, and in 1908 a reversion was made to a system more nearly approaching the stamping out method than has been in use for some time. While the Board retains perfect liberty to deal with separate outbreaks as it considers best, as a rule it now takes over the young and immature pigs and the breeding swine and has them

destroyed, paying full value for healthy pigs and half value for diseased ones, on the understanding that the owner will, as rapidly as possible, have the others killed for the market, and the place cleaned and thoroughly disinfected before any fresh pigs are brought on to it.

As the disease is exceedingly contagious, very strict regulations are laid down by the Board regarding cleansing and disinfection.

Apart from the existence of swine-fever in any place, the Board has divided the whole country into limited districts, which it calls "scheduled areas," presumably to give it some control over the movement of pigs, and pigs cannot be moved from one to another without a licence.

Other Contagious Diseases of Pigs.

There are other two diseases of pigs known to be contagious, one of which—swine erysipelas—is not uncommon in Britain. It sometimes causes considerable losses, and in these cases is generally believed by pig-owners to be swine-fever, but is not nearly so fatal nor so contagious in this country as the latter, and has not been scheduled by the Board.

The other contagious disease of pigs is swine plague, sometimes very destructive on the Continent. It is doubtful if it has ever appeared in Britain. At least, if it has it has not done much harm.

Worms (Intestinal).

Symptoms.— Ill-health; scurfy, dry skin; irregular appetite; itchiness of the anus, and the passing of worms; loss of flesh; cough; scour.

Prevention.—Sound food and water.

Treatment.—Give santonine, 3 grs., on an empty stomach, and four days later repeat the dose; the next day give 3 oz. Epsom salts. Allow access to plenty of coal, slack, or cinders, so that the pigs may eat as much grit as they like.

Trichinosis.

A parasitic disease of the pig, but rarely seen in this country.

Symptoms.—The animal is dull, loss of appetite, goes stiffly, vomits frequently. As the disease advances, the animal persistently stands, and when it lies down it tries to bury itself under the bedding.

Cause.—It is due to a minute worm called the *Trichina spiralis*, which infests the whole body; far more common in Germany and America than in this country.

Treatment.—Slaughter and bury the animals at once, for the diseased pork is poisonous to human beings.

Lice.

These may be easily removed by washing with water saturated with petroleum.

DOGS.

Rabies or Hydrophobia.

The only disease with which dogs are specially liable to be affected that need be noticed here is that dreaded disease, rabies, hydrophobia, or madness, as it is variously called. No disease of animals is more dreaded by man than canine madness, and the cry "mad dog" runs through a district like the sound of an alarm-gun. Scheduled under the Diseases of Animals Acts, it has been stamped out in Britain by the vigorous action of the Board of Agriculture, but there is always a risk of its reintroduction.

Cause.—It is not contagious in the ordinary sense, but is an inoculable disease, doubtless due to some living contagious agent which has not yet been clearly demonstrated. It is inoculable to all animals, including man, its natural mode of communication being by the bites of rabid dogs, and it is never seen in our larger animals but from this cause.

Symptoms.—There is some change in the habits of the dog. He may hide away in a dark corner, or creep under a bed or couch. Will often pick up and swallow pieces of string, rags, leather, straw, feathers, or pieces of wood. He becomes restless, generally wanders from home, and hurries along at a slouching trot, saliva flowing from his mouth, and there may be froth. He will snap at animals or persons who may come in his way, has no fear, and will bite at any

object held out to him. The voice is altered, and becomes something between a bark and a howl. Later, paralysis sets in, the jaw droops, and death soon follows. In the larger animals there is generally excitement, often excitement of the genital organs. The horse may get perfectly delirious,—would bite, kick, and smash everything within his reach. The ox will even try to bite, and will butt at anything, and everything.

Treatment.—Immediate slaughter as soon as definite symptoms are present.

Prevention. — Keep it out of the country; rigorous police measures.

Persons bitten by rabid animals are subjected to a system of inoculation.

RECIPES.

The scientific names are given as found in the *British Pharmacopœia* and *Squire's Companion to the British Pharmacopœia*. Directions for preparing the mixtures are appended to each recipe.

ABBREVIATIONS.

Grains, grs.; scruple, scr.; drachms, drs.; Pounds, lb.; ounces, oz.; quart, qt.; pint, pt.

HORSES.

BLISTERS.

1. Powdered cantharides (*P. cantharis*), 1 oz.
   Olive-oil (*Oleum olivæ*), 8 oz.
Use the ordinary "salad-oil" obtainable from grocers. Mix together in an earthenware pot, and infuse in a water-bath for four hours, and strain. Clip hair off the part before application.

2. Powdered cantharides (*Cantharis*), ¼ lb.
   Lard (*Adeps præparatus*), 1 lb.
   Resin (*Resina*), ¼ lb.
Melt the resin and lard together at a low temperature, then sprinkle in the cantharides, and stir till cold.

3. Perchloride of mercury (corrosive sublimate)
   (*Hydrargyri perchloridum*), 40 grs.
   Methylated spirit (*Spent methyll*), 1 oz.
To be applied with a small brush. Shake together in a bottle until dissolved.

4. Red iodide of mercury (*Hydrargyri iodidum rubrum*), ½ lb.
   Lard (*Adeps præparatus*), 4 lb.
   Mix together. Poison.

5. Iodine (*Iodum*), 2 oz.
   Iodide of potassium (*Potassii iodidum*), 1 oz.
   Camphor (*Camphora*), ½ oz.
   Methylated spirit, 1 pt.

Should be made up by a qualified party. Put the iodine and iodide of potassium in a bottle with 15 oz. of the spirit, shake till dissolved. Dissolve the camphor in 5 oz., then mix together.

CONDITION POWDERS.

6. Fenugrek, 2 parts.
 Carbonate of iron, 1 part.
 Nitrate of potassium (*Potassii nitras*), 2 parts.
 Gentian powder (*Gentianæ radix*), 1 part.
 Sulphur (*Sulphur sublimatum*), 2 parts.
Mix all together and sift. Give 1 oz. daily in the food.

COOLING LOTION.

7. Solution of subacetate of lead (*Liquor plumbi subacetatii*), 1 part.
 Tincture of arnica (*Tinctura arnicæ*), 3 parts.
 Water (*Aqua*), 8 parts.
 Mix.

COUGH BALL.

8. Digitalis (*Digitales folia*), ½ dr.
 Powdered opium (*Opium*), 1 dr.
 Aloes (*Aloe barbadensis*), 1 dr.
 Soft-soap (*Sapo mollis*)   } enough to
 Linseed-meal (*Lini farina*) } make a ball.
Make into a stiff mass. Give one ball every day.

DRAUGHTS.

9. Spirits of ammonia (*Spiritus ammoniæ aromaticus*), 1½ oz.
 Chloroform (*Chloroformum methyll*), 1 oz.
 Bicarbonate of potash (*Potassii bicarbonas*), ½ oz.
 Water (*Aqua*), 10 oz.
Mix. Shake up well before giving in gruel or other bland liquid. Every two hours till improvement, then twice a-day.

FOR BRONCHITIS.

10. Tincture of digitalis (*Tinctura digitalis*), 3 drs.
 Bromide of potassium (*Potassii bromidum*), 2 drs.
 Nitrous ether SPIRIT (*Ætheris nitrosi*), 1 oz.
 Water (*Aqua*), 10 oz.
Dissolve bromide of potassium in water, add the other ingredients, and make up with water to 10 oz. To be given three times a-day.

FOR WORMS.

11. Extract male fern (*Extractum filicis liquidum*), 2 drs.
 Oil of turpentine (*Oleum terebinthinæ*), 1½ oz.
 Linseed-oil (*Oleum lini*), 1 pt.
Mix and shake well together.

EMBROCATION.

12. Hartshorn (*Liq. ammon. dil.*), 1 oz.
 Turpentine (*Oleum terebinthinæ*), 2 oz.
 Spirit of camphor (*Spiritus camphoræ*), 2 oz.
 Laudanum (*Tinctura opii*), ½ oz.
 Olive-oil (*Oleum olivæ*), 6 oz.
Mix the hartshorn with the olive-oil, and shake, then the turpentine, spirit of camphor, and laudanum, shaking after each addition. Shake well before using.

CLYSTER.

13. Oatmeal, 3 qts.
 Salt, 3 oz.
 Olive-oil, ½ pint.
Give warm, and repeat till relief is given.

IRRITANT.

14. Ammonia solution (*Liquor ammoniæ*, F.), 3 oz.
 Soft-soap (*Sapo mollis*), 4 oz.
 Oil of turpentine (*Oleum terebinthinæ*), 8 oz.
 Olive-oil (*Oleum olivæ*), 4 oz.
Rub the soap with the olive-oil to smoothness, then add turpentine and ammonia solution. Bottle, and shake well.

LINIMENT.

15. Mustard (*Sinapis*), 4 oz.
 Oil of turpentine (*Oleum terebinthinæ*), 5 oz.
 Linseed-oil (*Oleum lini*), 1 pt.
Mix together, and shake thoroughly.

LOTION.

16. Tincture of myrrh (*Tinctura myrrhæ*), 1 oz.
 Alum (*Alumcu*), 2 drs.
 Water (*Aqua*), 6 oz.
Mix together.

PURGATIVES.

17. Aloes powdered (*Aloe barbadensis*), 6 drs.
 Ginger (*Zingiber*) 2 drs.
Made into a ball with soap or treacle.

18. Calomel (*Hydrargyri subchloridum*), 1 dr.
 Opium, powdered (*Opium*), 20 grs.
To be made into a ball with linseed-meal and treacle.

19. Aloes (*Aloe barbadensis*), 1½ dr.
 Tartar emetic (*Antimonium tartaratum*), 1 dr.
 Nitre (*Potassii nitras*), 2 drs.
 Digitalis (*Digitales folia*), ½ dr.
To be made into a ball with meal and treacle.

ORDINARY PURGATIVE.

20. Barbadoes aloes (*Aloe barbadensis*), 1½ dr.
 Calomel (*Hydrargyri subchloridum*), 1 dr.
To be made into a ball with meal and treacle.

TONIO.

21. Sulphate of iron (*Ferri sulph.*), 1½ dr.
Sulphate of quinine (*Quininæ sulph.*), 20 grs.
Sulphuric acid, diluted (*Acidum sulphuricum dilutum*), 2 drs.
Water (*Aqua*), 10 oz.
Dissolve the sulphate of iron in water, diffuse quinine in the solution, then add diluted sulphuric acid, and make up to 10 oz.
Give morning and night.

CATTLE.

BLISTER

22. Powdered cantharides (*Cantharis*), 1 oz.
Olive-oil (*Oleum olivæ*), 8 oz.
Use the ordinary "salad-oil" obtainable from grocers. Mix together in an earthenware pot, and infuse in a water-bath for four hours, and strain. Clip hair off the part before application.

23. Powdered cantharides, 1 part.
Venice turpentine, 1 part.
Resin, 1 part.
Lard, 4 parts.
Melt resin and lard together, then stir in the cantharides and Venice turpentine.

CALF-CORDIAL.

24. Prepared chalk (*Creta præparata*), 2 oz.
Powdered catechu (*Catechu*), 1 oz.
Ginger (*Zingiber*), ½ oz.
Opium (*Opium*), 2 drs.
Peppermint-water (*Aquæ menthæ peperitæ*), 1 pt.
Dose for calf; two tablespoonfuls morning and evening; dose for sheep, one tablespoonful morning and evening.
Mix all together.

DRAUGHT FOR COUGHS, &c.

25. Powdered digitalis (*Digitales foliæ*), 1 dr.
Liquor ammonia acetatis (*Liquor ammonii acetatis*), 3 oz.
Spirits of nitrous ether (*Spiritus ætheris nitrosi*), 2 oz.
Extract belladonna (*Extractum belladonnæ*), 2 drs.
To be given in a pint of water.
Melt extract of belladonna in a little warm water; when cold, add the other ingredients. Shake, and make up to a pint with cold water.

LINIMENT.

26. Oil of turpentine (*Oleum terebinthinæ*), 8 oz.
Solution of ammonia (*Liquor ammoniæ*, F.), 3 oz.
Soft-soap (*Sapo mollis*), 4 oz.
Rub down the soft-soap in the turpentine, then add the ammonia, and shake.

PURGATIVES.

27. Epsom salts (*Magnesii sulphas*), 16 oz.
Powdered aloes (*Aloe barbadensis*), 8 drs.
Ginger (ground) (*Zingiber*), 1 oz.
To be given in a quart of warm water or gruel.
Epsom salts for cattle costs 1s. a stone.
Use Barbadoes aloes and ordinary domestic ginger.

MILD PURGATIVE.

28. Epsom salts (*Magnesii sulphas*), 12 oz.
Powdered ginger (*Zingiber*), ½ oz.
To be given in a quart of warm water or gruel.
Salts for cattle, and ordinary ginger.

TONIC.

29. Gentian (*Gentianæ radix*), 1 oz.
Ginger (*Zingiber*), ½ oz.
Carbonate of ammonia (*Ammonii carbonas*), ½ oz.
Carbonate of iron, 2 drs.
To be given in a pint of gruel or water

SHEEP.

LOTION.

30. Carbolic acid (*Acidum carbolicum*), 1 part.
Water (*Aqua*), 50 parts.
Shake.

DRESSING FOR FOOT-ROT.

31. Red nitrate of mercury (*Hydrargyri oxidum rubrum*), 1 oz.
Nitrous acid, 2 oz.
To be mixed with two tablespoonfuls of water; dissolve the red nitrate of mercury in the acid, and then add the water.

PURGATIVES.

32. Calomel (*Hydrargyri subchloridum*), 5 grs.
Powdered opium (*Opium*), 4 grs.
Epsom salts (*Magnesii sulphas*), 1 oz.
To be given in 3 oz. of gruel or water.
All obtainable from a druggist. Mix and give in gruel. Ask for Epsom salts for cattle.

33. Epsom salts (*Magnesii sulphas*), 3 oz.
Ginger (ground) (*Zingiber*), 1 dr.
In thin gruel.
Take ginger used for domestic purposes, mix with the salts, and give in thin gruel.

34. Castor-oil (*Oleum ricini*), 2 oz.
Tincture of opium (laudanum) (*Tinctura opii*), 2 drs.
Use ordinary castor-oil and laudanum; mix, and give.

Note.—The doses given, except where otherwise stated, are intended for fair-grown animals of medium size. Allowance must therefore be made should the age or size of the animal to be treated exceed or otherwise the average.

TABLE OF DOSES.

Horse.	Cattle.	Sheep.	Swine.	Doses.
4 years and over .	2 years .	1 year .	1 year .	1 part.
2 to 3 years . . .	1 " .	6 months	6 months	$\frac{2}{3}$ "
1 year	6 months	3 " .	3 " .	$\frac{1}{4}$ to $\frac{1}{8}$ part.
2 to 10 months . .	1 " .	1 " .	1 " .	$\frac{1}{18}$ to $\frac{1}{10}$ part.

APPENDIX.

ADMINISTERING MEDICINE.

Some notes will be useful as to the methods of administering medicine to the various animals.

The Horse.

Medicine is usually given by the mouth, but sometimes injected under the skin into the blood, by the rectum, and by inhalation.

A drench should never exceed a quart, and before giving it, make sure that it is neither too hot nor too strong, for choking will follow.

A tin bottle is the best for drenching with; if this cannot be had, use a champagne-bottle.

How to fix the animal. — Put on a head-stall or halter; take a piece of rope or plough-line, make a loop at one end, pass the loop first through the nose-band of the halter or head-stall as the case may be, then into the mouth, throw the other end over one of the rafters above, and pull the horse's head up; the medicine should be slowly poured into his mouth, for horses are slow swallowers. Never be guilty of pouring it down the horse's nose, as I have seen some men do, and kill the animal. If the animal makes an attempt to cough whilst you are drenching him, let his head down instantly.

The Ball. —They should never exceed 1½ oz. in weight, and never be given when they have become hard. The best way to give a ball is by the hand, and with a little practice it can be soon learned. Take the tongue gently in the left hand, and draw it to the side of the mouth, place the ball between the fingers of the right hand, quickly run the hand along the roof of the mouth, and leave the ball at the back of the tongue; withdraw the hand, and let go the tongue. The animal will soon swallow, and you will see the ball pass down the left side of the neck. If you are not clever enough to give it in the manner described, use a balling gun or iron. Do not attempt to give a ball on the end of a pointed stick, for you are sure to run the stick into some vital part of the throat, and perhaps ruin the animal.

Cattle.

The cow is best drenched with a bottle or horn, and the quantity should not exceed 2 qts. In giving the medicine, stand on the right side of the cow, seize the nose with the thumb and finger of the left hand, and get some one to hold the horns on the left side. A cow swallows much more quickly than the horse, so it takes but a minute or two to give a drench.

Sheep.

A long-necked sauce-bottle is best to use for sheep. The quantity to be given should not exceed 4 oz. Stand on the right side, span the nose with your finger and thumb, place the finger in the mouth, and slowly run the medicine in at the right side of the mouth.

Swine.

The quantity to be given should not exceed 5 oz. In giving physic to a pig, take a child's old boot, cut a hole in the toe of it about the size of a shilling, place the toe of the boot into the pig's mouth, pour the medicine into the leg portion of the boot, and the pig will bite savagely at the boot and swallow the medicine at the same time.

FOMENTATION.

Fomentation is of great value in all cases of pain and inflammation. Never start to foment a part, however, without having plenty of hot water and time, for it does little or no good unless continued for an hour or two.

In cases of external injuries or inflammation—if it is on the knee or below it—place the leg in a tub full of hot water, if elsewhere soak a piece of flannel or sponge in hot water, and hold on the part.

For internal inflammation, such as in the bowels and the chest, double a blanket, soak it in hot water, and have it held against the chest or belly as the case may be, by a man on each side of the animal, and place over it a waterproof carriage-rug to keep in the heat. The blanket must be dipped into the hot water every three or four minutes. If the blanket is too hot for your hand it is too hot for the horse's skin, so be careful not to scald the animal.

ENEMAS.

Enemas or injections are of various kinds, and are given in cases of constipation to hasten the action of the bowels; in dysentery and diarrhœa to check the action of the bowels; in debility to support the animal, and when in pain to relieve it.

An enema for constipation should consist of linseed-oil, 1 pint; salt, 4 oz.; and warm water, 1 gallon, to be repeated, if required, every four hours.

For diarrhœa and dysentery use *liquor opii sedativus*, 2 oz.; starch, 4 oz.; and warm water, 3 pints. For weakness and debility use half a gallon of warm milk with two eggs, or the same quantity of beef-tea to be given every four or six hours.

To relieve pain use warm water, 1 quart; extract of belladonna, 1 drachm; or *liquor opii sedativus*, 1 oz.; to be given every three hours.

An enema is given by the enema-syringe, and the tail should be depressed for a few minutes after it is given.

BACK-RACKING.

This is occasionally done to remove the hard dung from the bowels, but it is not necessary if an enema has been given. The person who performs this operation should have a small hand, cut the nails short, and oil the hand before introducing it.

POULTICES.

Poultices are applied to certain parts to relieve pain, soften, and draw out any matter that may exist. The poultice should be made of boiled turnips or bran, the softer and warmer the better. A poultice to do any good must be of considerable size, kept on from twelve to twenty-four hours, with hot water continually poured over it, taking care not to scald the animal.

For the foot the poultice should be placed in a stout bag, and fixed around the fetlock by a strap.

CASTRATION.

The horse is usually operated upon at the age of one or two years; but he is sometimes allowed to go uncut until three years old to see if he is worth keeping for an entire horse, or to allow his neck to get developed. The spring or autumn is the best time to perform this operation, as we then avoid the cold winds of winter, and the sultry weather and the troublesome flies of summer.

There are various ways of performing this operation, but the best and most successful way is either by torsion or the hot iron. Some precautions should be taken before operating. Handle the colt for several weeks before, so that when he comes to be cast he will not fight, struggle, and break out into a sweat; feed him sparingly the day before the operation; make sure that both testicles are down and no rupture exists; always see that the ground is soft and free from stones where you intend to cast the animal.

Having haltered the colt, take him to the chosen spot, pass his head through the loop in the rope, pass the two ends between his fore and hind legs, bringing them back, pass them through the loop at the shoulders, and draw tight until the animal is on his side; then tighten up, wind the rope round the fetlock, include the fore legs, and get a man on each side to hold the end of the rope so as to keep the animal on his back.

To remove the stones by torsion, make a bold cut through the bag, release the stone, place the clams around the cord, put the torsion forceps on the cord about half an inch from the clams, and twist the forceps slowly around until you sever the cord; the other stone to be treated in the same way.

To operate with the hot iron: Having placed the stone in the clams, take a red-hot iron and saw the cord slowly through close to the clams.

Horses are now often operated on standing, the testicles being removed by the ecraseur or clams.

As to after-treatment, house the animal for a few days, and then let it run out during the day, housing it again at night.

From castration, lockjaw, bleeding, inflammation of the bowels, or broken back sometimes arise.

If the animal has only one stone down, postpone the operation, for it is almost certain to come down in a few months; if it never appears, the animal is most likely a " rig," and must be operated on as such.

Calves.

When a few weeks old they can be cut standing, by twisting the tail around one hind leg. Stand behind the calf, cut through the bag, twist the stone several times and scrape the cord closely through with your finger-nails or a blunt knife. When they are several months old they require to be cast. This is done by tying the hind legs together with a rope, place a halter round the neck, take the shank end of the halter and run it through the rope that joins the hind legs, tying it back, pass it through the portion that is around the neck, and draw the legs tight, and fasten. The fore legs can be held by a man. Take the stones off with the hot iron as in the case of the horse.

The bull is best castrated standing with the hot iron.

Pigs.

Let a man seize the pig by its hind legs and hold it between his legs. Cut through the bag, twist the stone several times, and scrape through the cord with a blunt knife or your finger-nails.

Lambs.

Let some one hold them on a bench for you; cut the tip of the bag off, and use the hot iron and clams, or do them the same way as the pig. In many parts, one person takes the lamb in his arms, holding its four legs tightly, two in each hand, while the shepherd cuts the top off the bag with a sharp knife, presses out the stones with his fingers, and draws them away with his teeth, then using the hot iron.

SPAYING.

Heifers and sows are sometimes spayed in order that they may fatten more quickly, but a description of this operation would not enable one to do it, and it can be learned only by watching those skilled in it.

DOCKING.

It is best performed when the animal is but a few months old, and at that age the tail can be easily cut off with a stout pocket-knife, and the end seared with the poker. In adults the operation is as simple, but often followed by excess of bleeding, lockjaw, or an abscess at the end of tail. Having parted the hair at the spot where the tail is to be cut off, tie the top hair back, get some one to hold the tail out, and with a sharp stroke of the docking-machine it is divided. Afterwards, hold the tail up, slightly sear it with the searing-iron, then place a piece of tow saturated in perchloride of iron on the end, bring the hair over it, and tightly tie below.

SETONING.

Setons are tapes passed through certain parts of the body, with the object of either draining an abscess, acting as a counter-irritant, or for the purpose of inoculation.

In using a seton for draining an abscess, such as pole-evil or fistulous withers, always bring it out at the lowest part of the abscess, so as to secure drainage.

In using setons as counter-irritants in cases of lameness, diseases of the eye or brain, pass them simply underneath the skin, and be careful not to wound any internal structure.

For inoculation, in cases of black-quarter or pleuro-pneumonia, the seton must be soaked with some irritant, such as embrocation (No. 12) in the case of black-leg, but in pleuro-pneumonia with the serum of a diseased lung.

NURSING THE SICK.

All the doctoring in the world is of no avail unless associated with good nursing.

Sick horses should be placed in a comfortable loose-box, free from draughts, and with plenty of straw in it. In cold weather a rug should be placed on the animal, and its legs bandaged. Animals, like human beings, soon lose their appetite when sick, so that every means should be tried to induce them to feed. The diet must be soft, nourishing, and given frequently in small quantities.

The following foods are recommended: bran-mashes, with bruised oats, sweet hay with a little treacle-water sprinkled over it, scalded oats, a little linseed-cake,

and, when in season, grass, tares, carrots, and parsnips can be given sparingly if the horse is not suffering from any bowel affection. A pail of oaten or linseed gruel should be placed within the reach of the animal, and if it does not drink this, give it treacle-and-water, or water with a tablespoonful of nitre dissolved in it. Take the chill off the water if the weather is cold.

Never allow one kind of food to remain too long in front of the animal; take it out and try something fresh. The animal should, if strong enough, and the weather permits, be taken out every day, and led up and down for half an hour with a rug on. Exercise of this kind strengthens the animal and increases the appetite. See that the manger and bucket from which the horse is fed are clean, for horses are naturally very sensitive as to what they eat, and more so when they are sick. Sick horses should every morning get a thorough wisping down.

Do not work the animal before it has properly recovered, and then gradually.

A USEFUL TABLE.

It is useful for stock-owners to have before them the following table, indicating a normal condition of the pulse, respiration, and temperature of their various animals; also the period of gestation.

	Pulse.		Respiration— Beats per Minute.	Temperature— Degrees Fahrenheit.	Average Duration of Pregnancy.
	Beats per Minute.	Where felt.			
Horse	40	Jaw	10	100	48 weeks.
Cow	45	Jaw	12	101.5	40 "
Sheep	75	At the heart	18	102.5	21 "
Pig	60	At the heart	15	102.5	16 "
Dog	100	Thigh	20	102	9 "

PLATE 9

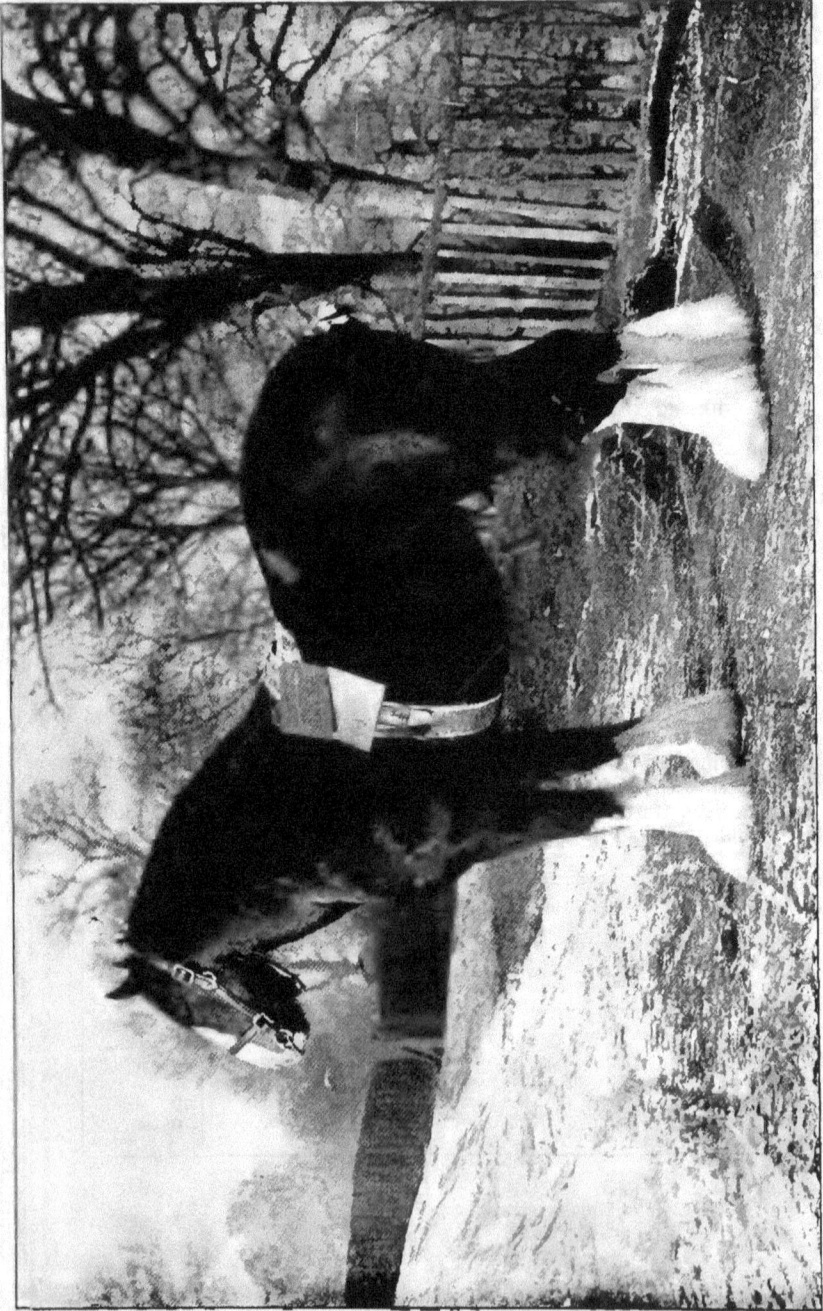

CLYDESDALE STALLION, "HIAWATHA," 10,067.

THE PROPERTY OF JOHN POLLOCK, ESQ., PAPER MILL, LANGSIDE, RENFREW.

PLATE 10

CLYDESDALE STALLION, "BARON'S PRIDE," 9122.

THE PROPERTY OF MESSRS A. AND W. MONTGOMERY, NETHERHALL, AND BANKS, KIRKCUDBRIGHT.

PLATE II

CLYDESDALE MARE, "PYRENE."

THE PROPERTY OF J. E. KERR, ESQ., HARVIESTOUN CASTLE, DOLLAR.

PLATE 12

SHIRE STALLION, "BIRDSALL MENESTREL," 19,357.

THE PROPERTY OF LORD ROTHSCHILD, TRING PARK, HERTS.

PLATE 15

SHIRE MARE, "FASHION SOARIS."

PLATE 14

SUFFOLK MARE, "BOULGE MAID," 4840.

THE PROPERTY OF R. E. WHITE, ESQ., BOULGE HALL, WOODBRIDGE.

PLATE 15

SUFFOLK STALLION, "RENDLESHAM SORCERER," 3077.

THE PROPERTY OF A. J. SMITH, ESQ., RENDLESHAM, WOODRIDGE.

PLATE 16

THOROUGHBRED STALLION, "DIAMOND JUBILEE."

WON THE DERBY IN 1900 FOR HIS MAJESTY THE KING (THEN PRINCE OF WALES).

PLATE 17

HUNTER GELDING, "WHISKY."

THE PROPERTY OF J. H. STOKES, ESQ., GREAT BOWDEN, MARKET HARBORO'.

PLATE 18

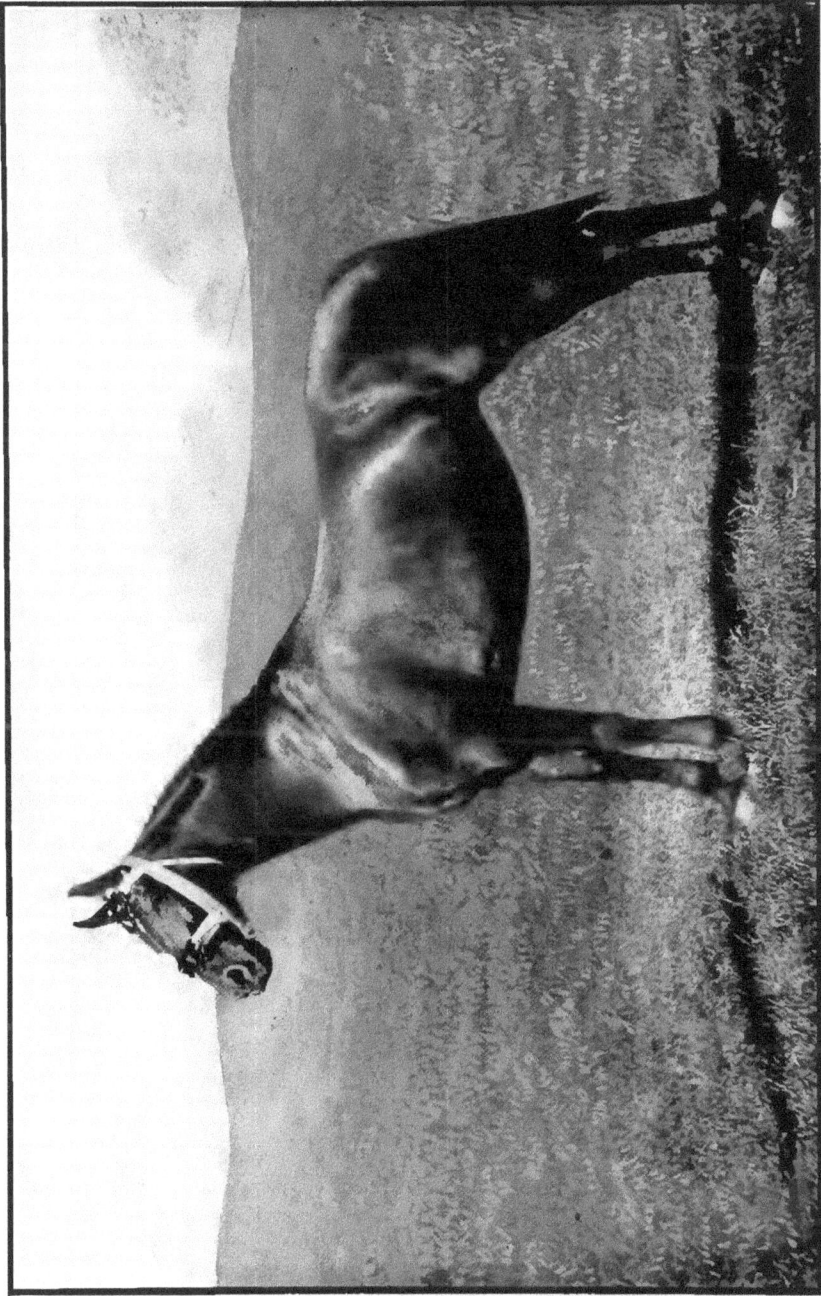

CLEVELAND BAY MARE, "WOODLAND BRIAR," 1318.

THE PROPERTY OF GEORGE GRANDAGE, ESQ., MOOR CROFT, YEADON, LEEDS.

PLATE 19

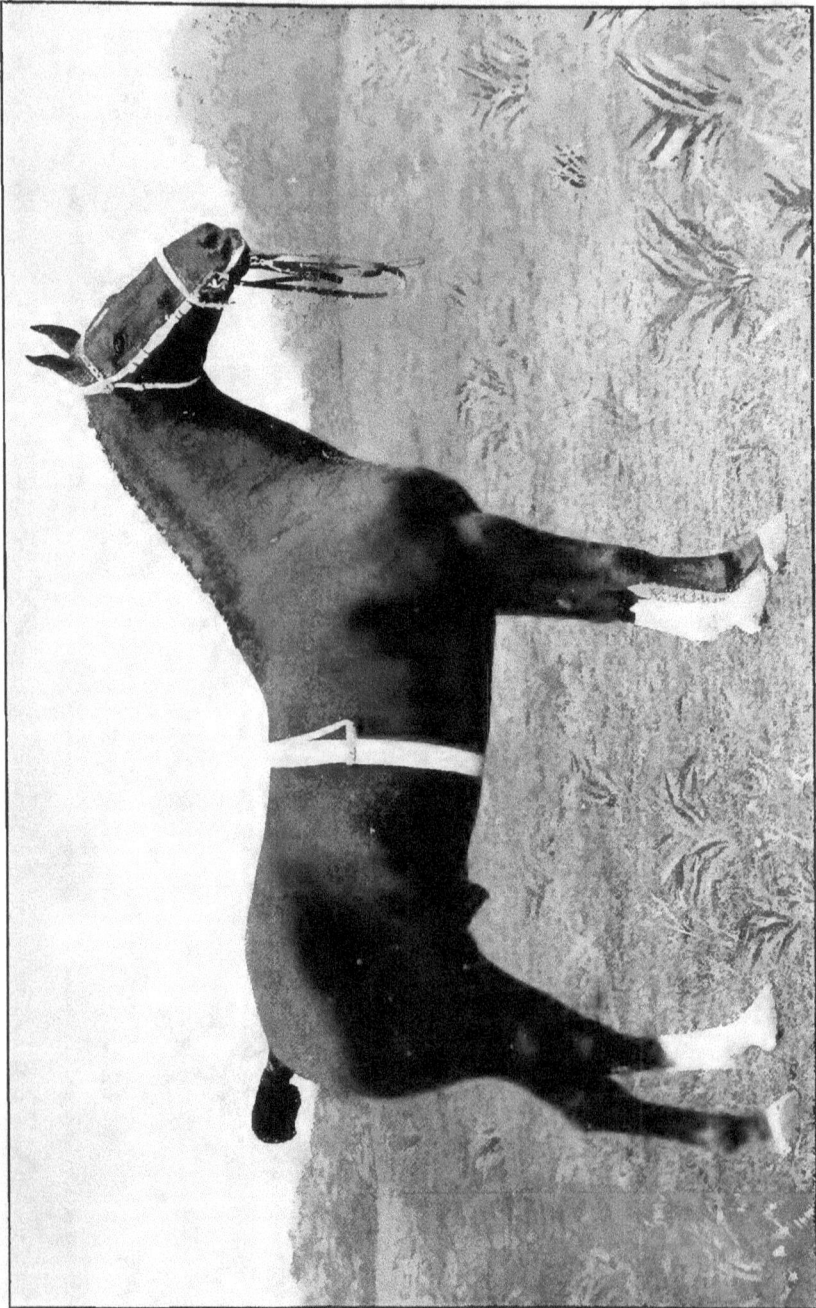

HACKNEY STALLION, "COPMANTHORPE PERFORMER," 9379.

THE PROPERTY OF ARTHUR HALL, ESQ., COPMANTHORPE, YORK

PLATE 20

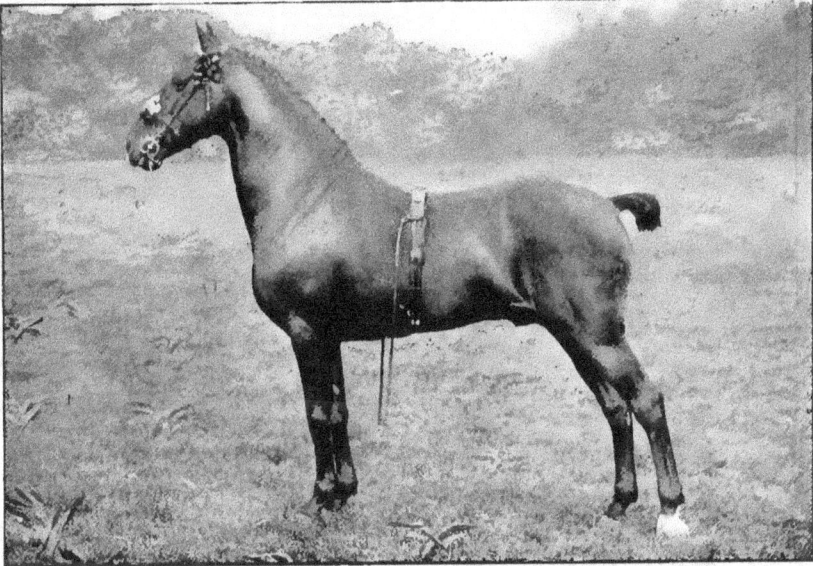

PONY STALLION, "BANTAM KING," 9106.

THE PROPERTY OF W. S. MILLER, ESQ., GLENDERMOT, CRAIGMORE, BUTE.

POLO PONY MARE, "RUBY."

THE PROPERTY OF JOHN BARKER, ESQ., M.P., THE GRANGE, BISHOP-STORTFORD.

PLATE 21

SHETLAND PONY STALLION, "MARQUIS."

THE PROPERTY OF R. W. R. MACKENZIE, ESQ., EARLSHALL, LEUCHARS, FIFE.

WELSH PONY STALLION, "GREYLIGHT."

THE PROPERTY OF E. JONES, ESQ., MANORAVON, LLANDILO, SOUTH WALES.

PLATE 22

HIGHLAND PONY STALLION, "BONNIE LADDIE," 329.

THE PROPERTY OF HIS GRACE THE DUKE OF ATHOLL, K.T.

PLATE 25

CONNEMARA PONY.

THE PROPERTY OF THE IRISH DEPARTMENT OF AGRICULTURE.

FELL PONY MARE, "GREY LADY."

THE PROPERTY OF W. LITTLE, ESQ., GLADENHOLM, AMISFIELD, DUMFRIES.

PLATE 24

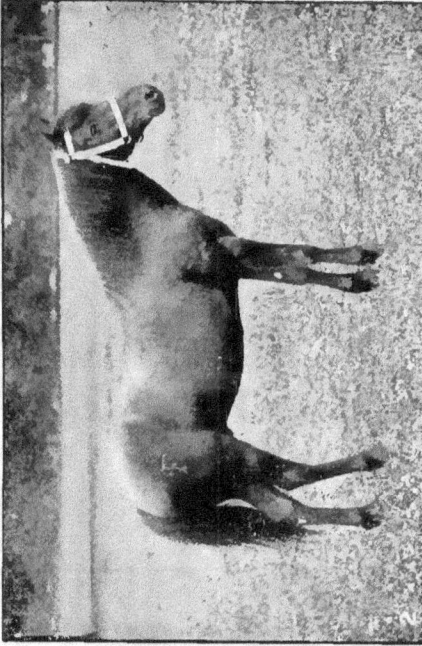

1. EXMOOR PONY STALLION, "TWILIGHT.   2. DARTMOOR PONY.   3. NEW FOREST PONIES.

PLATE 25

GROUP OF PONIES.

ICELAND AND NORWEGIAN.   RUM STALLION.

DONKEY STALLION.

THE PROPERTY OF HAROLD SESSIONS, ESQ., WOOTON MANOR, HENLEY-ON-THAMES.

PLATE 26

SHORTHORN BULL, "RONALD," 79,773.

THE PROPERTY OF HIS MAJESTY KING EDWARD VII.

PLATE 27

SHORTHORN COW, "SWEETHEART."

THE PROPERTY OF LORD CALTHORPE, ELVETHAM PARK, WINCHFIELD, HAMPSHIRE.

PLATE 28

LINCOLN RED SHORTHORN COW, "KEDDINGTON SKIPWORTH 5TH."

THE PROPERTY OF JOSEPH GROUT WILLIAMS, ESQ., PENDLEY MANOR, TRING.

PLATE 29

HEREFORD BULL, "CAMERONIAN," 23,934.

THE PROPERTY OF CAPTAIN E. L. A. HEYGATE, BUCKLAND, LEOMINSTER.

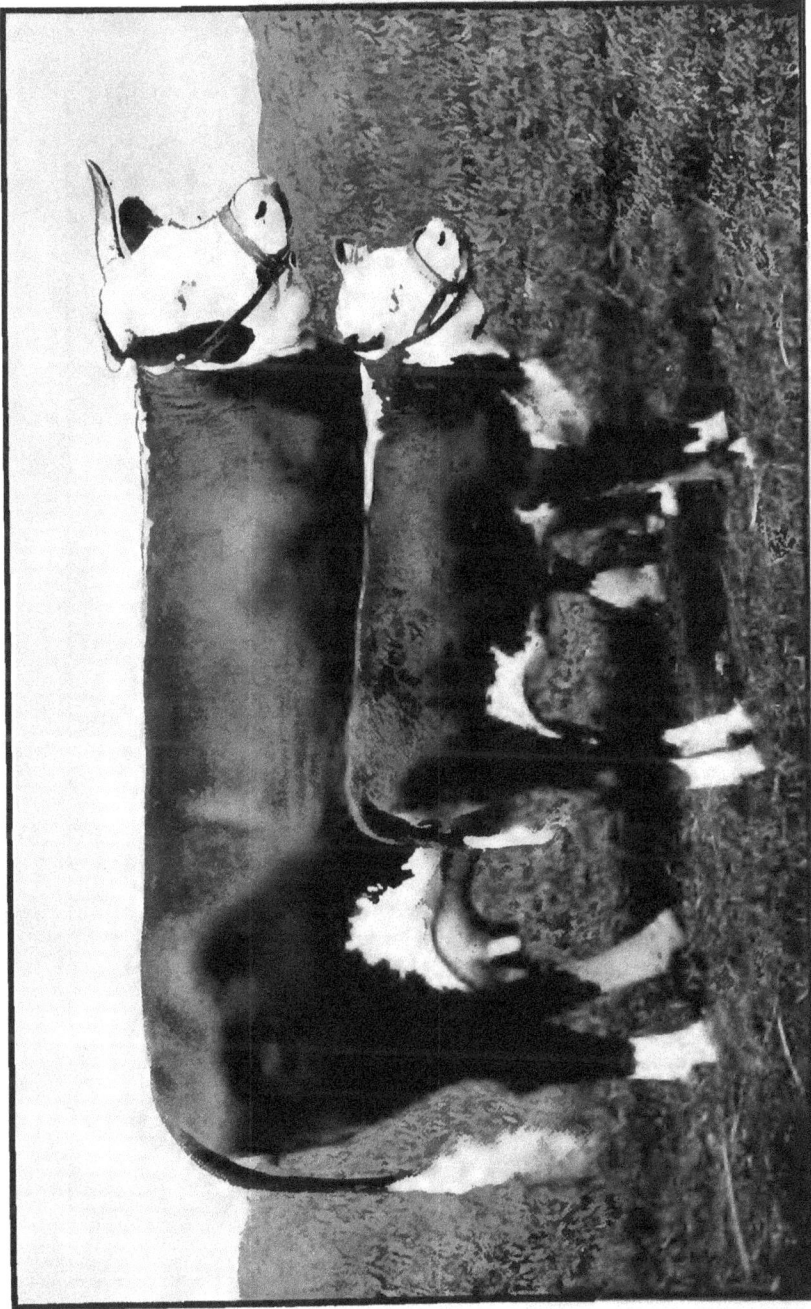

PLATE 30

HEREFORD COW, "MERRIMENT."

THE PROPERTY OF THE EARL OF COVENTRY, CROOME COURT, SEVERN STOKE, WORCESTER.

PLATE 51

DEVON BULL, "POUND PINK 'UN," 5350.

THE PROPERTY OF THE HON. E. W. B. PORTMAN, HESTERCOMBE, TAUNTON.

PLATE 52

SOUTH DEVON BULL, "MACBETH," 1924.

THE PROPERTY OF J. SPARROW WROTH, ESQ., COOMBE, AVETON-GIFFORD.

PLATE 55

SUSSEX COW, "SUNLIGHT 5TH."

THE PROPERTY OF EARL WINTERTON, SHILLINGLEE PARK, PETWORTH, SUSSEX.

PLATE 54

RED POLLED COW, "WAXLIGHT 2ND," 18,965.

THE PROPERTY OF SIR WALTER CORBET, BART., ACTON REYNOLD, SHREWSBURY.

PLATE 35

LONGHORN BULL "MELCOMBE EMPEROR," 416c.

THE PROPERTY OF LORD GERARD, EASTWELL PARK, ASHFORD, KENT.

PLATE 36

ABERDEEN-ANGUS BULL, "JESHURUN," 19,257.

THE PROPERTY OF SIR JOHN MACPHERSON GRANT, BART., BALLINDALLOCH CASTLE, BALLINDALLOCH.

PLATE 37

ABERDEEN-ANGUS COW, "JUANA ERICA," 36,235.

THE PROPERTY OF J. E. KERR, ESQ., OF HARVIESTOUN CASTLE, DOLLAR.

PLATE 58

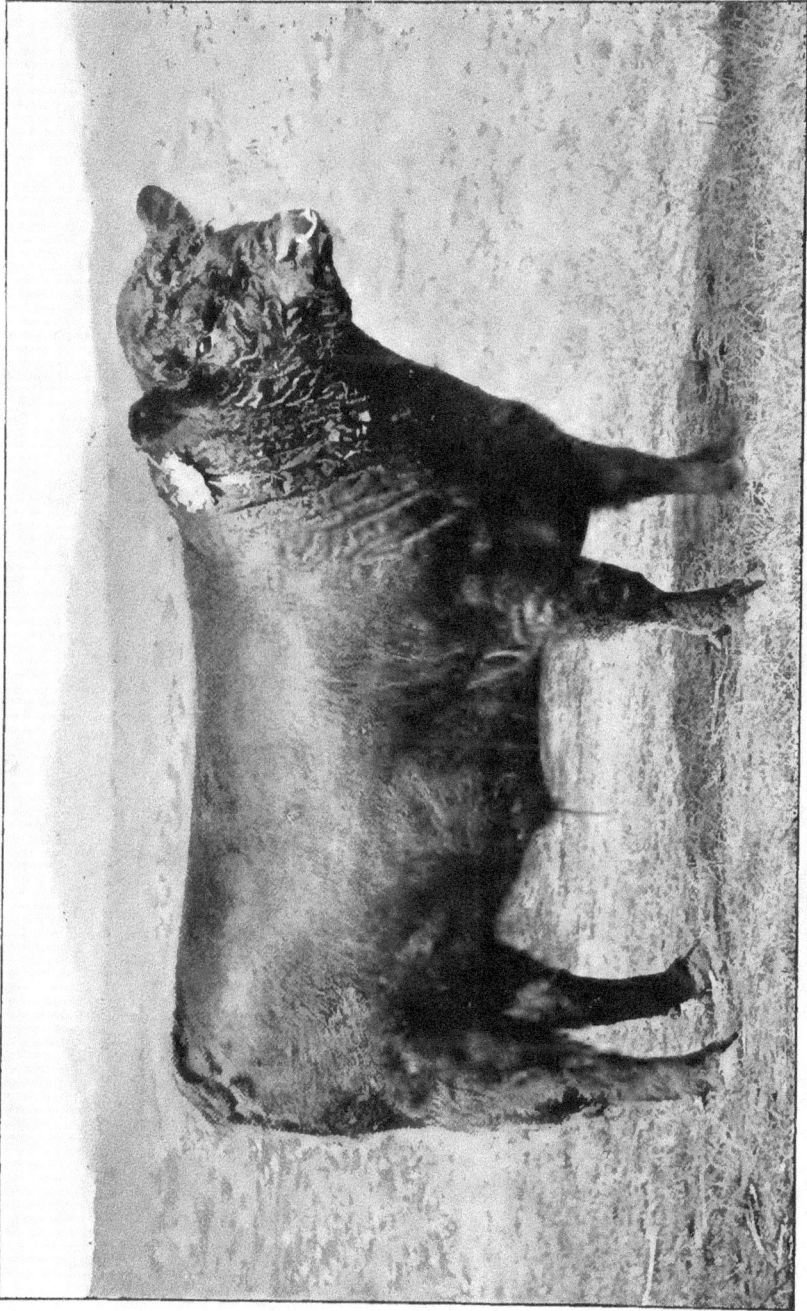

GALLOWAY BULL "KEYSTONE," 9639.

THE PROPERTY OF F. N. M. GOURLAY, ESQ., BROOMFIELD, MONIAIVE, DUMFRIESSHIRE.

PLATE 59

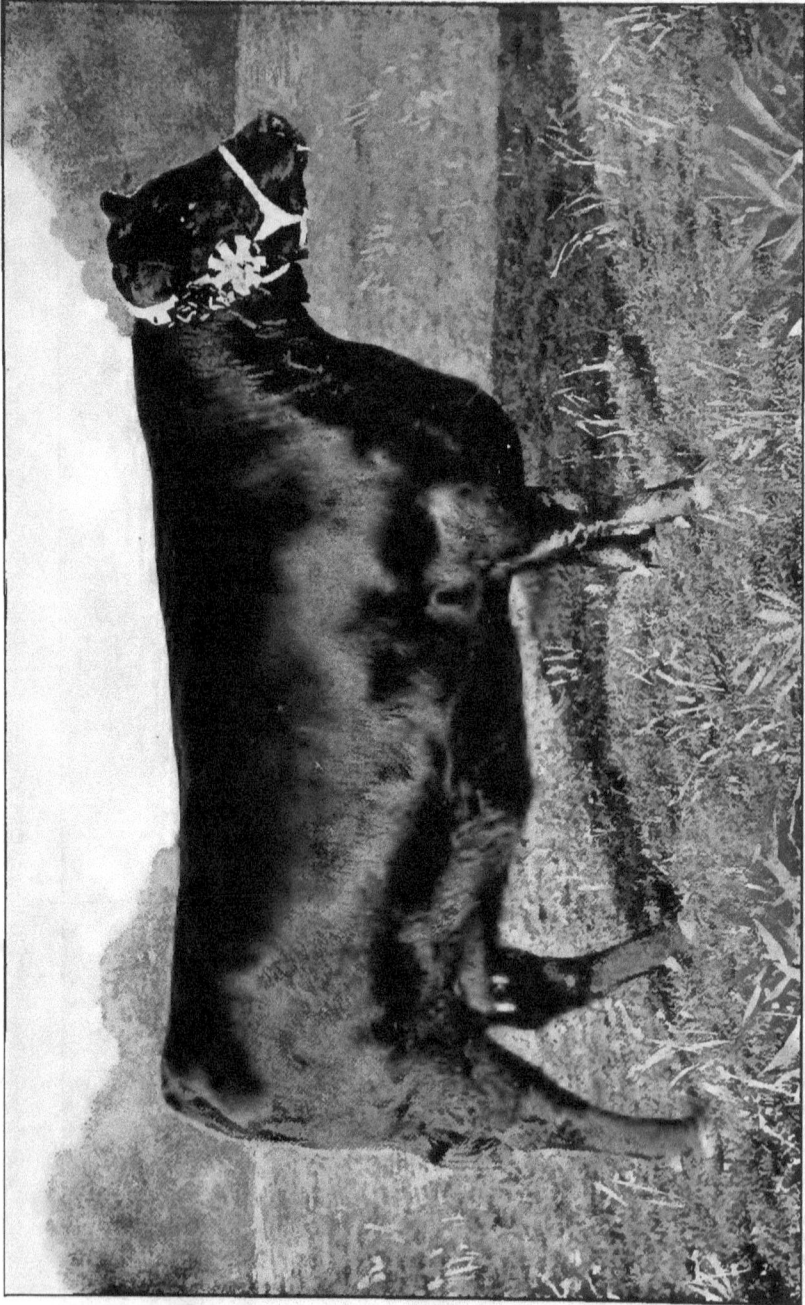

GALLOWAY COW, "ALICE III. OF CASTLEMILK," 16,867.

THE PROPERTY OF SIR WILLIAM BUCHANAN-JARDINE, BART. OF CASTLEMILK, LOCKERBIE.

PLATE 40

HIGHLAND BULL, "LORD CLYDE," 2054.

THE PROPERTY OF THE COUNTESS DOWAGER OF SEAFIELD.

PLATE 41

HIGHLAND COW.

THE PROPERTY OF DONALD A. STEWART, ESQ., INSAW, PORTREE.

PLATE 42

AYRSHIRE COW, "AUCHENTORLIE BLOOMER," 16644.

THE PROPERTY OF LIEUT.-COLONEL FERGUSSON-BUCHANAN OF AUCHENTORLIE, BOWLING, DUMBARTONSHIRE.

PLATE 43

WELSH COW, "MADRYN KATE," 593.

THE PROPERTY OF THE UNIVERSITY COLLEGE OF NORTH WALES, BANGOR.

PLATE 44

JERSEY COW, "LADY VIOLA."

THE PROPERTY OF ALEXANDER MILLER-HALLETT, ESQ., GODDINGTON, CHELSFIELD, KENT.

PLATE 45

GUERNSEY COW, "FI FI," 5438.

THE PROPERTY OF E. A. HAMBRO, ESQ., HAYES PLACE, HAYES, KENT.

PLATE 46

KERRY COW, "WALTON BASHFUL," 371.

PLATE 47

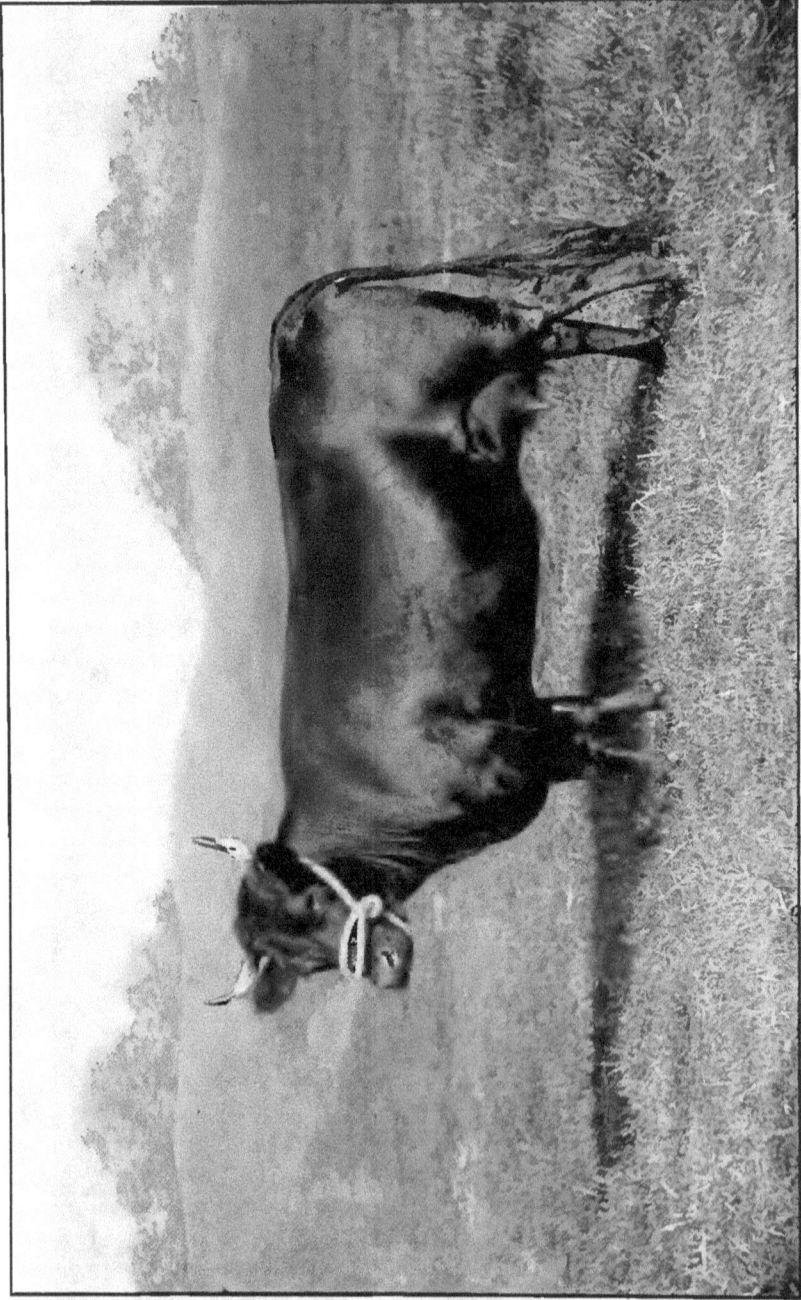

DEXTER COW, "GOXETON FOLLY VARDEN."

THE PROPERTY OF HIS MAJESTY THE KING, SANDRINGHAM, NORFOLK.

PLATE 48

DEXTER-SHORTHORN CATTLE.

DORA.

TIDY BELL, 3RD.

DAISY.

PLATE 49

WILD WHITE CATTLE.

IN CADZOW FOREST.   IN VAYNOL PARK.

CHILLINGHAM WILD BULL.

TAKEN TO CADZOW.

PLATE 50

LEICESTER RAM.

THE PROPERTY OF GEORGE HARRISON, ESQ., GAINFORD HALL, DARLINGTON.

BORDER LEICESTER RAM.

THE PROPERTY OF THE REPRS. OF THE LATE DAVID HUME, ESQ., BARRELWELL, BRECHIN.

PLATE 51

LINCOLN RAM.

THE PROPERTY OF F. MILLER, ESQ., CLIFTON ROAD, BIRKENHEAD.

COTSWOLD RAM.

THE PROPERTY OF MESSRS W. T. GARNE AND SON, ALDSWORTH, NORTHLEACH.

PLATE 52

TWO-SHEAR SOUTHDOWN RAM,

THE PROPERTY OF W. M. CAZALET, ESQ., FAIRLAWN, TONBRIDGE.

SHEARLING SHROPSHIRE RAM.

THE PROPERTY OF THOMAS A. BUTTAR, ESQ., CORSTON, COUPAR-ANGUS, FORFARSHIRE.

PLATE 55

SHEARLING OXFORD DOWN RAM.

THE PROPERTY OF JAMES T. HOBBS, ESQ., MAISEY HAMPTON, FAIRFORD.

HAMPSHIRE DOWN RAM.

THE PROPERTY OF JAMES FLOWER, ESQ., CHILMARK, SALISBURY.

PLATE 54

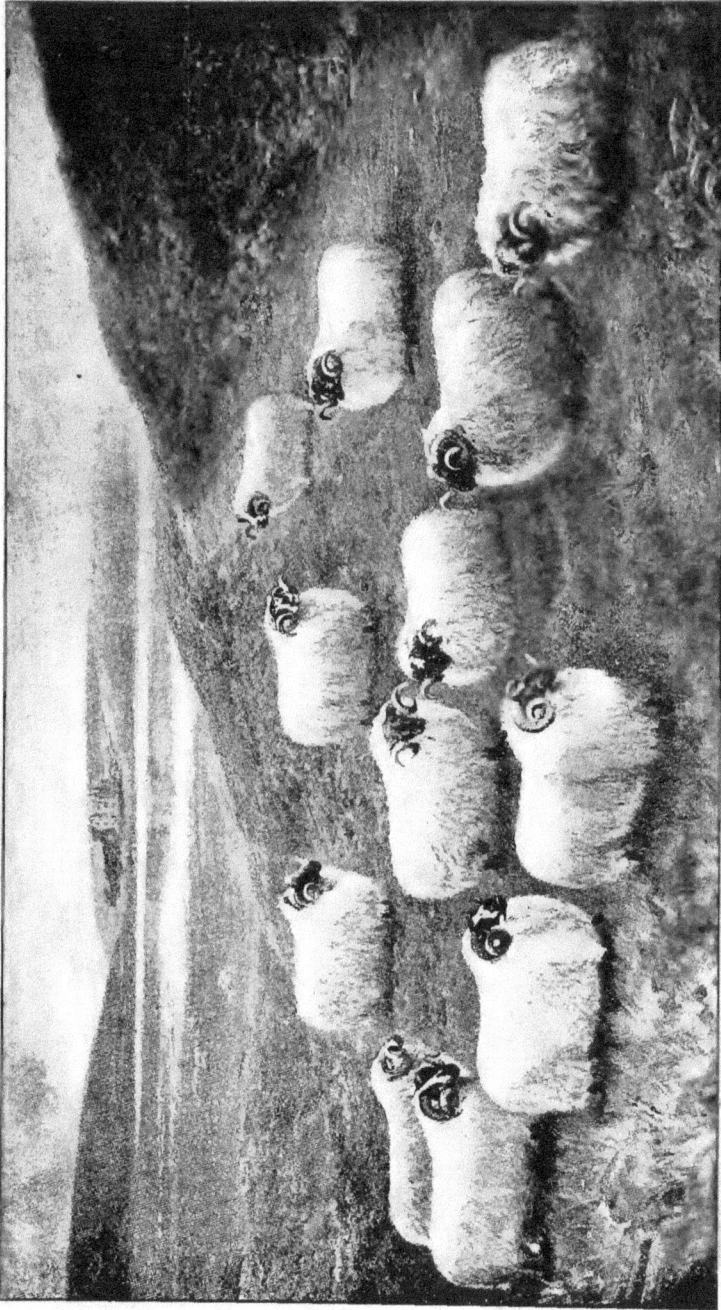

GROUP OF BLACKFACE RAMS.

BRED BY CHARLES HOWATSON, ESQ. OF GLENBUCK, AYRSHIRE.

PLATE 55

TWO-SHEAR BLACKFACE RAM.

THE PROPERTY OF DONALD M'DOUGALL, ESQ., CLAGGAN, KILLIN.

SHEARLING BLACKFACE EWE.

THE PROPERTY OF JOHN MILLER, ESQ., LAMBHILL, STRATHAVEN.

PLATE 56

CHEVIOT RAM.

THE PROPERTY OF JOHN ELLIOT, ESQ., HINDHOPE, JEDBURGH.

HALF-BRED RAM.

THE PROPERTY OF JAMES A. W. MEIN, ESQ. OF HUNTHILL, JEDBURGH.

PLATE 57

RYELAND RAM.

THE PROPERTY OF F. ELLIOTT GOUGH, ESQ., THE MOOR, BODENHAM, LEOMINSTER.

SUFFOLK RAM.

THE PROPERTY OF H. E. SMITH, ESQ., THE GRANGE, WALTON, SUFFOLK.

PLATE 58

DEVON LONG-WOOL SHEARLING RAM.

THE PROPERTY OF FREDERICK WHITE, ESQ., TORWESTON, WILLITON, SOMERSET.

SOUTH DEVON RAM.

THE PROPERTY OF HENRY FAIRWEATHER, ESQ., MALSTON, SHERFORD, KINGSBRIDGE.

PLATE 59

DORSET DOWN RAM.

THE PROPERTY OF G. WOOD HOMER, ESQ., BARDOLF MANOR, DORCHESTER.

DORSET HORN RAM.

THE PROPERTY OF E. A. HAMBRO, ESQ., DELCOMBE FARM, BLANDFORD, DELCOMBE, DORSET.

PLATE 60

DARTMOOR SHEARLING RAM.

THE PROPERTY OF R. SERCOMBE LUSCOMBE, ESQ., WISDORNE, CORNWOOD, DEVON.

EXMOOR RAM.

THE PROPERTY OF HEBER MARDON, ESQ., ASHWICK, DULVERTON, SOMERSET.

PLATE 61

SHEARLING LONK RAM.

THE PROPERTY OF D. HAGUE, ESQ., COPY NOOK, CLITHERO, LANCASHIRE.

WENSLEYDALE RAM.

THE PROPERTY OF THE EXECUTORS OF THE LATE THOMAS WILLIS, ESQ.,
MANOR HOUSE, CARPERBY, YORKS.

PLATE 62

HERDWICK RAM.

THE PROPERTY OF W. J. CROSSLEY, ESQ., M.P., PULLWOODS, AMBLESIDE, WESTMORELAND.

KENT OR ROMNEY MARSH RAM.

THE PROPERTY OF CHARLES FILE, ESQ., ELHAM, CANTERBURY.

PLATE 63

WELSH SHEARLING EWES.

THE PROPERTY OF THE UNIVERSITY COLLEGE OF NORTH WALES, BANGOR.

KERRY HILL (WALES) RAM.

THE PROPERTY OF LAWTON MOORE, ESQ., BRAMPTON BRIAN, HEREFORDSHIRE.

PLATE 64

ROSCOMMON RAM.

THE PROPERTY OF MATHEW FLANAGAN, ESQ., TOMONA, TULSK, CO. ROSCOMMON.

DERBYSHIRE GRITSTONE RAM.

THE PROPERTY OF WILLIAM TRUEMAN, ESQ., GOYTS BRIDGE, BUXTON.

PLATE 65

LARGE WHITE SOW.

THE PROPERTY OF SANDERS SPENCER, ESQ., ST IVES, HUNTS.

LARGE WHITE BOAR, "BROOMHOUSE MERCULES," 9031.

THE PROPERTY OF W. S. WALLACE, ESQ., BROOMHOUSE, CORSTORPHINE, EDINBURGH.

PLATE 66

LARGE WHITE ULSTER BOAR.

THE PROPERTY OF J. CUNNINGHAM, ESQ., BELMOUNT, ANTRIM.

MIDDLE WHITE SOW.

THE PROPERTY OF SIR GILBERT GREENALL, BART., WALTON HALL, WARRINGTON.

PLATE 67

LARGE BLACK SOW, "BRENT DAME," 2355.

THE PROPERTY OF HENRY J. KINGWELL, ESQ., GREAT AISH, SOUTH BRENT, DEVON.

BERKSHIRE SOW, "PEEL EDIE."

THE PROPERTY OF J. JEFFERSTON, ESQ., PEEL HALL, CHESTER.

PLATE 68

TAMWORTH SOW, "CHOLDERTON FAVOURITE."

THE PROPERTY OF H. C. STEPHENS, ESQ., OF CHOLDERTON, SALISBURY.

LINCOLN CURLY-COATED SOW.

THE PROPERTY OF HENRY CAUDWELL, ESQ., MIDVILLE, BOSTON, LINCOLNSHIRE.

INDEX TO VOLUMES I., II., III.

(DIVISIONS I. TO VI.)

www.ingramcontent.com/pod-product-compliance
Lightning Source LLC
Chambersburg PA
CBHW071129270326
41929CB00012B/1688